A LANGE medical

Harper's Illustrated Biochemistry

TWENTY-NINTH EDITION

Robert K. Murray, MD, PhD
Emeritus Professor of Biochemistry
University of Toronto
Toronto, Ontario

David A. Bender, PhD
Professor (Emeritus) of Nutritional Biochemistry
University College London
London, United Kingdom

Kathleen M. Botham, PhD, DSc
Professor of Biochemistry
Department of Veterinary Basic Sciences
Royal Veterinary College
University of London
London, United Kingdom

Peter J. Kennelly, PhD
Professor and Head
Department of Biochemistry
Virginia Polytechnic Institute and State University
Blacksburg, Virginia

Victor W. Rodwell, PhD
Professor (Emeritus) of Biochemistry
Purdue University
West Lafayette, Indiana

P. Anthony Weil, PhD
Professor of Molecular Physiology and Biophysics
Vanderbilt University School of Medicine
Nashville, Tennessee

New York   Chicago   San Francisco   Lisbon   London   Madrid   Mexico City
Milan   New Delhi   San Juan   Seoul   Singapore   Sydney   Toronto

Harper's Illustrated Biochemistry, Twenty-Ninth Edition

1 2 3 4 5 6 7 8 9 0 DOW/DOW 16 15 14 13 12

ISBN 978-0-07-176576-3
MHID 0-07-176576-X
ISSN 1043-9811

Notice

Medicine is an ever-changing science. As new research and clinical experience broaden our knowledge, changes in treatment and drug therapy are required. The authors and the publisher of this work have checked with sources believed to be reliable in their efforts to provide information that is complete and generally in accord with the standards accepted at the time of publication. However, in view of the possibility of human error or changes in medical sciences, neither the authors nor the publisher nor any other party who has been involved in the preparation or publication of this work warrants that the information contained herein is in every respect accurate or complete, and they disclaim all responsibility for any errors or omissions or for the results obtained from use of the information contained in this work. Readers are encouraged to confirm the information contained herein with other sources. For example and in particular, readers are advised to check the product information sheet included in the package of each drug they plan to administer to be certain that the information contained in this work is accurate and that changes have not been made in the recommended dose or in the contraindications for administration. This recommendation is of particular importance in connection with new or infrequently used drugs.

This book was set in Minion Pro by Thomson Digital.
The editors were Michael Weitz and Brian Kearns.
The production supervisor was Sherri Souffrance.
Project management was provided by Aakriti Kathuria, Thomson Digital.
The designer was Elise Lansdon.
Cover image: Adapted from Protein Data Bank entry 2KKA. A detailed explanation
    may be found on page ix.
RR Donnelley was printer and binder.

International Edition ISBN 978-0-07-179277-6; MHID 0-07-179277-5. Copyright © 2012. Exclusive rights by The McGraw-Hill Companies, Inc., for manufacture and export. This book cannot be re-exported from the country to which it is consigned by McGraw-Hill. The International Edition is not available in North America.

Co-Authors

Daryl K. Granner, MD
Professor (Emeritus) of Molecular Physiology
and Biophysics and Medicine
Vanderbilt University
Nashville, Tennessee

Peter L. Gross, MD, MSc, FRCP(C)
Associate Professor
Department of Medicine
McMaster University
Hamilton, Ontario

Molly Jacob, MB BS, MD, PhD
Professor and Chair
Department of Biochemistry
Christian Medical College
Vellore, Tamil Nadu

Frederick W. Keeley, PhD
Associate Director and Senior Scientist
Research Institute, Hospital for Sick Children,
Toronto, and Professor of Biochemistry
Department of Biochemistry
University of Toronto
Toronto, Ontario

Peter A. Mayes, PhD, DSc
Emeritus Professor of Veterinary Biochemistry
Royal Veterinary College
University of London
London, United Kingdom

Margaret L. Rand, PhD
Associate Senior Scientist
Hospital for Sick Children
Toronto and Professor
Departments of Laboratory
Medicine & Pathobiology and Biochemistry
University of Toronto
Toronto, Ontario

Joe Varghese, MB BS, MD
Assistant Professor
Department of Biochemistry
Christian Medical College
Vellore, Tamil Nadu

Key Features of *Harper's Illustrated Biochemistry, 29th Edition*

No other text clarifies the link between biochemistry and the molecular basis of disease like Harper's Illustrated Biochemistry, 29th Edition

Key Features

- Every chapter has been updated to reflect the latest advances in knowledge and technology

- Sixteen case histories add clinical relevance to the material

- A careful balance of detail and brevity not found in any other book on the subject

- NEW chapters on aging, cancer, and clinical chemistry

- NEW multiple-choice questions to test knowledge and comprehension

- NEW chapter-opening statement of objectives, followed by brief discussion of the biomedical importance of topics discussed within the chapter

- Increased number of tables that encapsulate important information, such as vitamin and mineral requirements

- Expanded discussion of non-coding RNAs, DNA damage repair and human diseases, activities of miRNAs, and powerful new assays to monitor and characterize transcription genome-wide

- Enhanced discussion of iron metabolism in health and disease

- High-quality full-color illustrations with integrated coverage of biochemical diseases and clinical information

Chapter-opening statements of objective

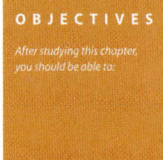

CHAPTER

50

Plasma Proteins & Immunoglobulins

Robert K. Murray, MD, PhD, Molly Jacob, MB BS, MD, PhD, & Joe Varghese, MB BS, MD

OBJECTIVES

After studying this chapter, you should be able to:

- List the major functions of blood.
- Explain the functions of the major plasma proteins, including albumin, haptoglobin, transferrin, ceruloplasmin, α_1-antitrypsin, and α_2-macroglobulin.
- Describe how iron homeostasis is maintained and how it is affected in certain disorders.
- Describe the general structures and functions of the five classes of immunoglobulins and the uses of monoclonal antibodies.
- Appreciate that the complement system is involved in a number of important biological processes.
- Indicate the causes of Wilson disease, Menkes disease, the lung and liver diseases associated with α_1-antitrypsin deficiency, amyloidosis, multiple myeloma, and agammaglobulinemia.

BIOMEDICAL IMPORTANCE

The fundamental role of blood in the maintenance of homeostasis (see Chapter 51) and the ease with which blood can be obtained have meant that the study of its constituents has been of central importance in the development of biochemistry and clinical biochemistry. The basic properties of a number of **plasma proteins**, including the **immunoglobulins** (antibodies), are described in this chapter. Changes in the amounts of various plasma proteins and immunoglobulins occur in many diseases and can be monitored by electrophoresis or other suitable procedures. As indicated in an earlier chapter, alterations of the activities of certain **enzymes** found in plasma are of diagnostic use in a number of pathologic conditions. Plasma proteins involved in blood coagulation are discussed in Chapter 51.

THE BLOOD HAS MANY FUNCTIONS

The functions of blood—except for specific cellular ones such as oxygen transport and cell-mediated immunologic defense—are carried out by plasma and its constituents (**Table 50–1**).

Plasma consists of water, electrolytes, metabolites, nutrients, proteins, and hormones. The water and electrolyte composition of plasma is practically the same as that of

all extracellular fluids. Laboratory determinations of levels of Na^+, K^+, Ca^{2+}, Mg^{2+}, Cl^-, HCO_3^-, $PaCO_2$, and of blood pH are important in the management of many patients.

PLASMA CONTAINS A COMPLEX MIXTURE OF PROTEINS

The concentration of total protein in human plasma is approximately 7.0–7.5 g/dL and comprises the major part of the solids of the plasma. The proteins of the plasma are actually a complex mixture that includes not only simple proteins but also conjugated proteins such as **glycoproteins** and various types of **lipoproteins**. Use of proteomic techniques is allowing the isolation and characterization of previously unknown plasma proteins, some present in very small amounts (eg, detected in hemodialysis fluid and in the plasma of patients with cancer), thus expanding **the plasma proteome**. Thousands of **antibodies** are present in human plasma, although the amount of any one antibody is usually quite low under normal circumstances. The relative dimensions and molecular masses of some of the most important plasma proteins are shown in **Figure 50–1**.

The **separation** of individual proteins from a complex mixture is frequently accomplished by the use of solvents or electrolytes (or both) to remove different protein fractions in accordance with their solubility characteristics. This is the basis

629

TABLE 48–1   Types of Collagen and Their Genes[1]

Type	Genes	Tissue
I	COL1A1, COL1A2	Most connective tissues, including bone
II	COL2A1	Cartilage, vitreous humor
III	COL3A1	Extensible connective tissues such as skin, lung, and the vascular system
IV	COL4A1–COL4A6	Basement membranes
V	COL5A1–COL5A3	Minor component in tissues containing collagen I
VI	COL6A1–COL6A3	Most connective tissues
VII	COL7A1	Anchoring fibrils
VIII	COL8A1–COL8A2	Endothelium, other tissues
IX	COL9A1–COL9A3	Tissues containing collagen II
X	COL10A1	Hypertrophic cartilage
XI	COL11A1, COL11A2, COL2A1	Tissues containing collagen II
XII	COL12A1	Tissues containing collagen I
XIII	COL13A1	Many tissues
XIV	COL14A1	Tissues containing collagen I
XV	COL15A1	Many tissues
XVI	COL16A1	Many tissues
XVII	COL17A1	Skin hemidesmosomes
XVIII	COL18A1	Many tissues (eg, liver, kidney)
XIX	COL19A1	Rhabdomyosarcoma cells

Source: Adapted slightly from Prockop DJ, Kivirikko KI: Collagens: molecular biology, diseases, and potentials for therapy. Annu Rev Biochem 1995;64:403. Copyright © 1995 by Annual Reviews, www.annualreviews.org. Reprinted with permission.

[1]The types of collagen are designated by Roman numerals. Constituent procollagen chains, called proα chains, are numbered using Arabic numerals, followed by the collagen type in parentheses. For instance, type I procollagen is assembled from two proα1(I) and one proα2(I) chains. It is thus a heterotrimer, whereas type 2 procollagen is assembled from three proα1(II) chains and is thus a homotrimer. The collagen genes are named according to the collagen type, written in Arabic numerals for the gene symbol, followed by an A and the number of the proα chain that they encode. Thus, the COL1A1 and COL1A2 genes encode the α1 and α2 chains of type I collagen, respectively. At least 28 types of collagen have now been recognized.

concerned with the fibril-forming **collagens I and II,** the major collagens of skin and bone and of cartilage, respectively. However, mention will be made of some of the other collagens.

COLLAGEN IS THE MOST ABUNDANT PROTEIN IN THE ANIMAL WORLD

All collagen types have a **triple helical structure.** In some collagens, the entire molecule is triple helical, whereas in others the triple helix may involve only a fraction of the structure. Mature collagen type I, containing approximately 1000 amino acids, belongs to the former type; in it, each polypeptide sub-unit or alpha chain is twisted into a left-handed polyproline helix of three residues per turn (**Figure 48–1**). Three of these alpha chains are then wound into a **right-handed superhelix,**

TABLE 48–2   Classification of Collagens, Based Primarily on the Structures That They Form

Class	Type
Fibril-forming	I, II, III, V, and XI
Network-like	IV, VIII, X
FACITs[1]	IX, XII, XIV, XVI, XIX
Beaded filaments	VI
Anchoring fibrils	VII
Transmembrane domain	XIII, XVII
Others	XV, XVIII

Source: Based on Prockop DJ, Kivirikko KI: Collagens: molecular biology, diseases, and potentials for therapy. Annu Rev Biochem 1995;64:403. Copyright © 1995 by Annual Reviews. Reprinted with permission.

[1]FACITs = fibril-associated collagens with interrupted triple helices. Additional collagens to these listed above have been recognized.

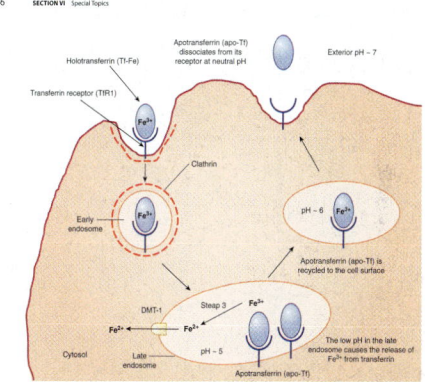

FIGURE 50–6   The transferrin cycle. Holotransferrin (Tf-Fe) binds to transferrin receptor 1 (TfR1) present in clathrin-coated pits on the cell surface. The TfR1–Tf-Fe complex is endocytosed and endocytic vesicles fuse to form early endosomes. The early endosomes mature to late endosomes, which have an acidic pH inside. The low pH causes release of iron from its binding sites on transferrin. Apotransferrin (apo-Tf) remains bound to TfR1. Ferric iron is converted to its ferrous form by the ferrireductase, Steap 3. Ferrous iron is then transported into the cytosol via DMT1. The TfR1-apo-Tf complex is recycled back to the cell surface. At the cell surface, apo-Tf is released from TfR1. TfR1 then binds to new Tf-Fe. This completes the transferrin cycle. (Based on Figure 17–48 in Lodish H et al: *Molecular Cell Biology,* 4th ed. WH Freeman, 2000).

levels are low, ferritin is not synthesized while TfR is, in order to promote uptake of iron from transferrin in blood.

The mechanisms involved in the regulation of syntheses of ferritin and TfR1 have been elucidated (**Figure 50–8**). This is brought about through regulation of the stability of the mRNAs for ferritin and TfR1. The ferritin and TfR1 mRNAs contain **iron response elements (IREs),** which form hairpin loops at their 5′ and 3′ untranslated regions (UTRs), respectively. IREs are bound by **iron regulatory proteins (IRPs).** The IRPs are sensitive to intracellular iron levels and are induced by low levels. They bind IREs only when intracellular iron levels are low. Binding of IRP to the IRE at the 3′ UTR of TfR1 mRNA stabilizes the TfR1 mRNA, thus increasing TfR1 synthesis and expression on the cell surface. On the other hand, binding of IRP to the IRE at the 5′ UTR of ferritin mRNA blocks translation of ferritin. Similarly, in the absence of IRP binding to IRE (which happens in the presence of high levels of iron), translation of ferritin mRNA is facilitated and TfR1 mRNA

is rapidly degraded. The net result is that, when intracellular iron levels are high, ferritin is synthesized but TfR1 is not, and when intracellular iron levels are low, TfR1 is synthesized and ferritin is not. This is a classical example of control of expression of proteins at the translational level.

Hepcidin Is the Chief Regulator of Systemic Iron Homeostasis

Hepcidin is a protein that is known to play a central role in iron homeostasis in the body. It is synthesized by the liver as an 84-amino-acid precursor protein (prohepcidin). Prohepcidin is cleaved to generate bioactive hepcidin, which is a 25-amino-acid peptide. **Hepcidin binds to the cellular iron exporter, ferroportin, and triggers its internalization and degradation.** Thus, as shown in **Figure 50–9,** hepcidin decreases iron absorption in the intestine (producing a "mucosal block") and also prevents recycling of iron from macrophages. These effects

---

Exam Questions

Section V

1. Regarding membrane lipids, select the one FALSE answer.
   A. The major phospholipid by mass in human membranes is generally phosphatidylcholine.
   B. Glycolipids are located on the inner and outer leaflets of the plasma membrane.
   C. Phosphatidic acid is a precursor of phosphatidylserine, but not of sphingomyelin.
   D. Phosphatidylcholine and phosphatidylethanolamine are located primarily on the outer leaflet of the plasma membrane.
   E. The flip-flop of phospholipids in membranes is very slow.

2. Regarding membrane proteins, select the one FALSE answer.
   A. Because of steric considerations, alpha-helices cannot exist in membranes.
   B. A hydropathy plot helps one to estimate whether a segment of a protein is predominantly hydrophobic or hydrophilic.
   C. Certain proteins are anchored to the outer leaflet of plasma membranes via glycophosphatidylinositol (GPI) structures.
   D. Adenylyl cyclase is a marker enzyme for the plasma membrane.
   E. Myelin has a very high content of lipid compared with protein.

3. Regarding membrane transport, select the one FALSE statement.
   A. Potassium has a lower charge density than sodium and tends to move more quickly through membranes than does sodium.
   B. The flow of ions through ion channels is an example of passive transport.
   C. Facilitated diffusion requires a protein transporter.
   D. Inhibition of the Na⁺-K⁺-ATPase will inhibit sodium-dependent uptake of glucose in intestinal cells.
   E. Insulin, by recruiting glucose transporters to the plasma membrane, increases uptake of glucose in fat cells but not in muscle.

4. Regarding the Na⁺K⁺-ATPase, select the one FALSE statement.
   A. Its action maintains the high intracellular concentration of sodium compared with potassium.
   B. It can use as much as 30% of the total ATP expenditure of a cell.
   C. It is inhibited by digitalis, a drug that is useful in certain cardiac conditions.
   D. It is located in the plasma membrane of cells.
   E. Phosphorylation is involved in its mechanism of action, leading to its classification as a P-type ATP-driven active transporter.

5. What molecules enable cells to respond to a specific extracellular signaling molecule?
   A. Specific receptor carbohydrates localized to the inner plasma membrane surface.
   B. Plasma lipid bilayer.
   C. Ion channels.

6. Indicate the term generally applied to the extracellular messenger molecules that bind to transmembrane receptor proteins:
   A. Competitive inhibitor
   B. Ligand
   C. Scatchard curve
   D. Substrate
   E. Key

7. In autocrine signaling
   A. Messenger molecules reach their target cells via passage through bloodstream.
   B. Messenger molecules travel only short distances through the extracellular space to cells that are in close proximity to the cell that is generating the message.
   C. The cell producing the messenger expresses receptors on its surface that can respond to that messenger.
   D. The messenger molecules are usually rapidly degraded and hence can only work over short distances.

8. Regardless of how a signal is initiated, the ligand-binding event is propagated via second messengers or protein recruitment. What is the ultimate outcome of these binding events?
   A. A protein in the middle of an intracellular signaling pathway is activated.
   B. A protein at the top of an intracellular signaling pathway is activated.
   C. A protein at the top of an extracellular signaling pathway is activated.
   D. A protein at the top of an intracellular signaling pathway is deactivated.
   E. A protein at the bottom of an intracellular signaling pathway is activated.

9. What features of the nuclear receptor superfamily suggest that these proteins have evolved from a common ancestor?
   A. They all bind the same ligand with high affinity.
   B. They all function within the nucleus.
   C. They are all subject to regulatory phosphorylation.
   D. They all contain regions of high amino acid sequence similarity/identity.
   E. They all bind DNA.

10. What effect does degradation of receptor–ligand complexes after internalization have upon the ability of a cell to respond if immediately re-exposed to the same hormone?
   A. The cellular response is attenuated due to a decrease in cellular receptor number.
   B. Cellular response is enhanced due to reduced receptor–ligand competition.
   C. The cellular response is unchanged to subsequent stimuli.
   D. Cell hormone response is now bimodal; enhanced for a short time and then blunted.

Increased number of tables

Hundreds of full-color illustrations

More than 250 multiple-choice questions

Contents

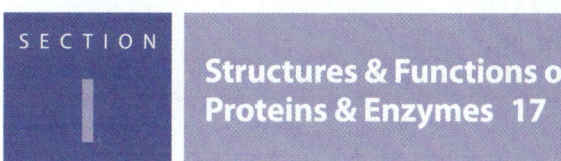

SECTION III

**Metabolism of Proteins
& Amino Acids 265**

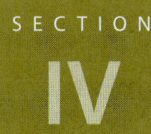

SECTION IV

**Structure, Function, &
Replication of Informational
Macromolecules 323**

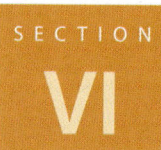

Preface

The authors and publishers are pleased to present the twenty-ninth edition of *Harper's Illustrated Biochemistry*. The first edition of this text, entitled *Harper's Biochemistry*, was published in 1939 under the sole authorship of Dr Harold Harper, University of California, San Francisco. Subsequently, various authors have contributed to the text.

Cover Illustration for the Twenty-Ninth Edition

The cover illustration for the 29th edition commemorates Elizabeth H. Blackburn, Carol W. Greider, and Jack W. Szostak, who shared the 2009 Nobel Prize in Physiology or Medicine for their seminal work on telomeres and the enzyme telomerase. Telomeres comprise up to 200 copies of a repeating DNA sequence called a G-quadruplex, a structure named for the unique cyclic arrangement of four sets of four guanine bases hydrogen-bonded in head-to-tail fashion that stabilize this structure. In the illustration, the phosphodiester backbone of the DNA is represented by a ribbon and the guanine bases by filled hexagons fused to filled pentagons. The spectral color gradation from purple to red facilitates tracing the progression of the polynucleotide chain. The four sets of cyclic tetra-guanine units can be seen in center stacked from top to bottom and tilted roughly 45° from left to right (Adapted from Protein Data Bank ID no. 2KKA).

As a consequence of the unidirectional nature of DNA replication, each time a chromosome is replicated, the number of G-quadruplex units is reduced. When the supply of telomere units is completely exhausted, replication ceases and the cell transitions to a senescent state. Scientists speculate that the telomere serves as a countdown clock that limits the number of times a somatic cell can divide, and hence its lifespan.

Changes in the Twenty-Ninth Edition

Consistent with our goal of providing students with a text that describes and illustrates biochemistry in a medically relevant, up-to-date, comprehensive, and yet relatively concise manner, in addition to updating every chapter, significant new material appears in this edition.

Each chapter now begins with a brief statement of its objectives followed by a brief account of its biomedical importance. A major addition is the inclusion of over 250 multiple-choice exam questions with answers given in an answer bank.

Major Additional Changes Include Three Entirely New Chapters:

"Biochemistry of Aging"
"Biochemistry of Cancer"
"Clinical Chemistry"

Additional Significant Changes Include:

- Inclusion of aspects of epidemiology in the chapter on "Bioinformatics and Computational Biology".
- New figures that illustrate key approaches for identifying possible active sites, ligand-binding sites, and other interaction sites (Section I), and various aspects of metabolism (Section II).
- New tables that summarize aspects of metabolic diseases, including those of purine, pyrimidine, and amino acid metabolism (Section III).
- Expanded discussion of non-coding RNAs, DNA damage repair and human diseases, epigenetic factors that control eukaryotic gene expression, the activities of miRNAs, and powerful new assays to monitor and characterize transcription genome-wide (Section IV).
- New tables that address vitamin and mineral requirements and a greatly expanded discussion of iron metabolism in health and disease (Section VI).

Organization of the Book

Following two introductory chapters, the text is divided into six main sections. All sections and chapters emphasize the medical relevance of biochemistry.

Section I addresses the structures and functions of proteins and enzymes. This section also contains a chapter on Bioinformatics and Computational Biology, reflecting the increasing importance of these topics in modern biochemistry, biology, and medicine.

Section II explains how various cellular reactions either utilize or release energy, and traces the pathways by which carbohydrates and lipids are synthesized and degraded. Also described are the many functions of these molecules.

Section III deals with the amino acids, their metabolic fates, certain features of protein catabolism, and the biochemistry of the porphyrins and bile pigments.

Section IV describes the structure and function of nucleotides and nucleic acids, DNA replication and repair, RNA synthesis and modification, protein synthesis, the principles of recombinant DNA technology, and new understanding of how gene expression is regulated.

Section V deals with aspects of extracellular and intracellular communication. Topics include membrane structure and function, the molecular bases of the actions of hormones, and the field of signal transduction.

Section VI includes fifteen special topics: nutrition, digestion, and absorption; vitamins and minerals; free radicals and antioxidants; intracellular trafficking and sorting of proteins; glycoproteins; the extracellular matrix; muscle and the cytoskeleton; plasma proteins and immunoglobulins; hemostasis and thrombosis; red and white blood cells; the metabolism of xenobiotics; the biochemistry of aging; the biochemistry of cancer; clinical chemistry; and sixteen biochemically oriented case histories. The latter chapter concludes with a brief epilog indicating some major challenges for medicine for which biochemistry and related disciplines will play important roles in finding solutions.

Appendix lists useful web sites and biochemical journals and others with significant biochemical content.

Acknowledgments

The authors thank Michael Weitz for his role in the planning of this edition, and Brian Kearns for his key role in getting this edition ready for publication. We also thank Mala Arora and her colleagues at Thomson Digital for their efforts in editing, typesetting, and artwork, and Calvin "Nic" Steussy of Purdue University for his assistance in generating the cover illustration.

Suggestions from students and colleagues around the world have been most helpful in the formulation of this edition. We look forward to receiving similar input in the future.

Rob Murray acknowledges with thanks Joe Varghese and Molly Jacob as co-authors of Chapters 50, 55, and 56, Fred Keeley for his many contributions to Chapter 48, Peter Gross for co-authorship of Chapters 51 and 57, and Margaret Rand for co-authorship of Chapter 51. Special thanks are extended to Reinhart Reithmeier, Alan Volchuk, and David Williams for reviewing and making invaluable suggestions for the revision of Chapters 40 and 46.

Robert K. Murray
David A. Bender
Kathleen M. Botham
Peter J. Kennelly
Victor W. Rodwell
P. Anthony Weil

Biochemistry & Medicine

Robert K. Murray, MD, PhD

OBJECTIVES

After studying this chapter, you should be able to:

- Explain what biochemistry is about and appreciate its central role in the life sciences.
- Understand the relationship of biochemistry to health and disease and to medicine.
- Appreciate how the Human Genome Project has given rise to, or stimulated interest in numerous disciplines that are already illuminating many aspects of biology and medicine.

INTRODUCTION

Biochemistry can be defined as *the science of the chemical basis of life* (Gk *bios* "life"). The **cell** is the structural unit of living systems. Thus, biochemistry can also be described as *the study of the chemical constituents of living cells and of the reactions and processes they undergo*. By this definition, biochemistry encompasses large areas of **cell biology, molecular biology,** and **molecular genetics**.

The Aim of Biochemistry Is to Describe and Explain, in Molecular Terms, All Chemical Processes of Living Cells

The **major objective** of biochemistry is **the complete understanding, at the molecular level, of all of the chemical processes associated with living cells.** To achieve this objective, biochemists have sought to isolate the numerous molecules found in cells, determine their structures, and analyze how they function. Many techniques have been used for these purposes; some of them are summarized in **Table 1–1**.

Other objectives of biochemistry include helping to **understand the origins of life on Earth** and to integrate biochemical knowledge into efforts to **maintain health** and to **understand diseases and treat them effectively.**

A Knowledge of Biochemistry is Essential to All Life Sciences

The biochemistry of the nucleic acids lies at the heart of **genetics;** in turn, the use of genetic approaches has been critical for elucidating many areas of biochemistry. **Cell biology** is very closely allied to biochemistry. **Physiology,** the study of body function, overlaps with biochemistry almost completely. **Immunology** employs numerous biochemical techniques, and many immunologic approaches have found wide use by biochemists. **Pharmacology** and **pharmacy** rest on a sound knowledge of biochemistry and physiology; in particular, most drugs are metabolized by enzyme-catalyzed reactions. Poisons act on biochemical reactions or processes; this is the subject matter of **toxicology.** Biochemical approaches are being used increasingly to study basic aspects of **pathology** (the study of disease), such as inflammation, cell injury, and cancer. Many workers in **microbiology, zoology,** and **botany** employ biochemical approaches almost exclusively. These relationships are not surprising, because life as we know it depends on biochemical reactions and processes. In fact, the old barriers among the life sciences are breaking down, and biochemistry is increasingly becoming their **common language.**

A Reciprocal Relationship Between Biochemistry & Medicine Has Stimulated Mutual Advances

The two major concerns for workers in the health sciences—and particularly physicians—are the understanding and maintenance of **health** and the understanding and effective

TABLE 1–1 The Principal Methods and Preparations Used in Biochemical Laboratories

Methods for Separating and Purifying Biomolecules[1]
Salt fractionation (eg, precipitation of proteins with ammonium sulfate)
Chromatography: Paper, ion exchange, affinity, thin-layer, gas–liquid, high-pressure liquid, gel filtration
Electrophoresis: Paper, high-voltage, agarose, cellulose acetate, starch gel, polyacrylamide gel, SDS-polyacrylamide gel
Ultracentrifugation

Methods for Determining Biomolecular Structures
Elemental analysis
UV, visible, infrared, and NMR spectroscopy
Use of acid or alkaline hydrolysis to degrade the biomolecule under study into its basic constituents
Use of a battery of enzymes of known specificity to degrade the biomolecule under study (eg, proteases, nucleases, glycosidases)
Mass spectrometry
Specific sequencing methods (eg, for proteins and nucleic acids)
X-ray crystallography

Preparations for Studying Biochemical Processes
Whole animal (includes transgenic animals and animals with gene knockouts)
Isolated perfused organ
Tissue slice
Whole cells
Homogenate
Isolated cell organelles
Subfractionation of organelles
Purified metabolites and enzymes
Isolated genes (including polymerase chain reaction and site-directed mutagenesis)

[1]Most of these methods are suitable for analyzing the components present in cell homogenates and other biochemical preparations. The sequential use of several techniques will generally permit purification of most biomolecules. The reader is referred to texts on methods of biochemical research for details.

treatment of **diseases.** Biochemistry impacts enormously on both of these fundamental concerns of medicine. In fact, the interrelationship of biochemistry and medicine is a wide, two-way street. Biochemical studies have illuminated many aspects of health and disease, and conversely, the study of various aspects of health and disease has opened up new areas of biochemistry. Some examples of this two-way street are shown in **Figure 1–1.** For instance, knowledge of protein structure and function was necessary to elucidate the single biochemical difference between **normal hemoglobin** and **sickle cell hemoglobin.** On the other hand, analysis of sickle cell hemoglobin has contributed significantly to our understanding of the structure and function of both normal hemoglobin and other proteins. Analogous examples of reciprocal benefit between biochemistry and medicine could be cited for the other paired items shown in Figure 1–1. Another example is the pioneering work of Archibald Garrod, a physician in England during the early 1900s. He studied patients with a number of relatively rare disorders (alkaptonuria, albinism, cystinuria, and pentosuria; these are described in later chapters) and established that these conditions were genetically determined. Garrod designated these conditions as **inborn errors of metabolism.** His insights provided a major foundation for the development of the field of human biochemical genetics. More recent efforts to understand the basis of the genetic disease known as **familial hypercholesterolemia,** which results in severe atherosclerosis at an early age, have led to dramatic progress in understanding of cell receptors and of mechanisms of uptake of cholesterol into cells. Studies of **oncogenes** and **tumor suppressor genes** in cancer cells have directed attention to the molecular mechanisms involved in the control of normal cell growth. These and many other examples emphasize how the study of disease can open up areas of cell function for basic biochemical research.

The **relationship between medicine and biochemistry** has important implications for the former. As long as medical treatment is firmly grounded in the knowledge of biochemistry and other basic sciences, the practice of medicine will have **a rational basis** that can be adapted to accommodate new knowledge. This contrasts with unorthodox health cults and at least some "alternative medicine" practices that are often founded on little more than myth and wishful thinking and generally lack any intellectual basis.

Biochemistry is one important area of science. The many ways in which **science is important for physicians** (and equally so for other workers in health care or biology, whether concerned with humans or animals) have been well stated in an article by Cooke (2010). They include (i) offering a foundational understanding on which one's practice should be built, (ii) stimulating curiosity and creating the scientific habits that are essential for continual learning throughout one's career, (iii) showing how our present knowledge has been acquired, and (iv) emphasizing the immensity of what is as yet unknown. Of course, it is vital that the application of science to helping a patient must be practised with humanity and the highest ethical standards.

NORMAL BIOCHEMICAL PROCESSES ARE THE BASIS OF HEALTH

The World Health Organization (WHO) defines **health** as a state of "complete physical, mental, and social well-being and not merely the absence of disease and infirmity." From a strictly biochemical viewpoint, health may be considered that

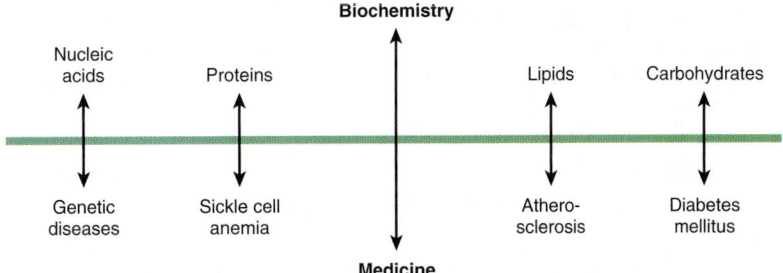

FIGURE 1–1 **Examples of the two-way street connecting biochemistry and medicine.** Knowledge of the biochemical molecules shown in the top part of the diagram has clarified our understanding of the diseases shown on the bottom half—and conversely, analyses of the diseases shown below have cast light on many areas of biochemistry. Note that sickle cell anemia is a genetic disease and that both atherosclerosis and diabetes mellitus have genetic components.

situation in which all of the many thousands of intra- and extracellular reactions that occur in the body are proceeding at rates commensurate with the organism's maximal survival in the physiologic state. However, this is an extremely reductionist view, and it should be apparent that caring for the health of patients requires not only a wide knowledge of **biologic principles** but also of **psychologic** and **social** principles.

Biochemical Research Has Impact on Nutrition & Preventive Medicine

One major prerequisite for the maintenance of health is that there be optimal dietary intake of a number of chemicals; the chief of these are **vitamins,** certain **amino acids,** certain **fatty acids,** various **minerals,** and **water.** Because much of the subject matter of both **biochemistry and nutrition** is concerned with the study of various aspects of these chemicals, there is a close relationship between these two sciences. Moreover, more emphasis is being placed on systematic attempts to maintain health and forestall disease, that is, on **preventive medicine.** Thus, nutritional approaches to—for example—the prevention of atherosclerosis and cancer are receiving increased emphasis. Understanding nutrition depends to a great extent on knowledge of biochemistry.

Most & Perhaps All Diseases Have a Biochemical Basis

We believe that most if not all diseases are manifestations of abnormalities of molecules, chemical reactions, or biochemical processes. The **major factors responsible for causing diseases** in animals and humans are listed in **Table 1–2.** All of them affect one or more critical chemical reactions or molecules in the body. Numerous examples of the biochemical bases of diseases will be encountered in this text. In most of these conditions, biochemical studies contribute to both the diagnosis and treatment. Some **major uses of biochemical investigations and of laboratory tests in relation to diseases** are summarized in Table 56–1. Chapter 56 describes many

aspects of the field of **clinical biochemistry,** which is mainly concerned with the use of biochemical tests to assist in the diagnosis of disease and also in the overall management of patients with various disorders. Chapter 57 further helps to illustrate the relationship of biochemistry to disease by discussing in some detail biochemical aspects of 16 different medical cases.

Some of the **major challenges that medicine and related health sciences** face are also outlined very briefly at the end of Chapter 57. In addressing these challenges, biochemical studies are already and will continue to be interwoven with studies in various other disciplines, such as genetics, cell biology, immunology, nutrition, pathology, and pharmacology. Many biochemists are vitally interested in contributing to solutions to key issues such as how can the survival of mankind be assured, and also in educating the public to support the use of the scientific method in solving major problems (eg, environmental and others) that confront us.

TABLE 1–2 The Major Causes of Diseases[1]

1. **Physical agents:** Mechanical trauma, extremes of temperature, sudden changes in atmospheric pressure, radiation, electric shock.

2. **Chemical agents, including drugs:** Certain toxic compounds, therapeutic drugs, etc.

3. **Biologic agents:** Viruses, bacteria, fungi, higher forms of parasites.

4. **Oxygen lack:** Loss of blood supply, depletion of the oxygen-carrying capacity of the blood, poisoning of the oxidative enzymes.

5. **Genetic disorders:** Congenital, molecular.

6. **Immunologic reactions:** Anaphylaxis, autoimmune disease.

7. **Nutritional imbalances:** Deficiencies, excesses.

8. **Endocrine imbalances:** Hormonal deficiencies, excesses.

[1]**Note:** All of the causes listed act by influencing the various biochemical mechanisms in the cell or in the body.
Source: Adapted, with permission, from Robbins SL, Cotram RS, Kumar V: *The Pathologic Basis of Disease,* 3rd ed. Saunders, 1984. Copyright © 1984 Elsevier Inc. with permission from Elsevier.

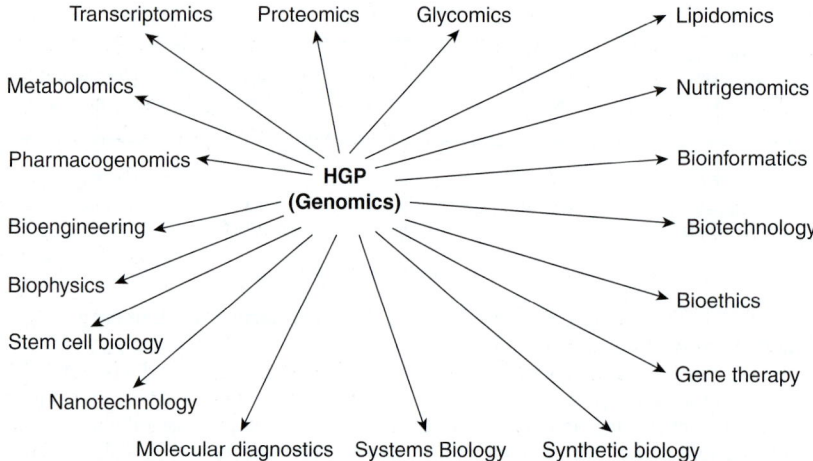

FIGURE 1–2 The Human Genome Project (HGP) has influenced many disciplines and areas of research. Biochemistry itself is not shown in this figure, because it was underway long before the HGP commenced. However, a number of the disciplines shown (eg, bioinformatics, genomics, glycomics, lipidomics, metabolomics, molecular diagnostics, proteomics, and transcriptomics) are very active areas of research by biochemists.

Impact of the Human Genome Project (HGP) on Biochemistry, Biology, & Medicine

Remarkable progress was made in the late 1990s in sequencing the human genome by the HGP. This culminated in July 2000, when leaders of the two groups involved in this effort (the International Human Genome Sequencing Consortium and Celera Genomics, a private company) announced that over 90% of the genome had been sequenced. Draft versions of the sequence were published in early 2001. With the exception of a few gaps, the sequence of the entire human genome was completed in 2003, 50 years after the description of the double-helical nature of DNA by Watson and Crick.

The **implications** of the HGP for **biochemistry,** all of **biology,** and for **medicine** and **related health sciences** are tremendous, and only a few points are mentioned here. It is **now possible to isolate any gene and usually determine its structure and function** (eg, by sequencing and knockout experiments). Many **previously unknown genes** have been revealed; their products have already been established, or are under study. New light has been thrown on **human evolution,** and procedures for **tracking disease genes** have been greatly refined. Reference to the HGP will be made in various chapters of this text.

As the ramifications of the HGP increase, it is vital for readers to understand **the major contributions to understanding human health and disease** that have been made, and are being made, by **studies of the genomes of model organisms,** particularly *Drosophila melanogaster* (the fruit fly) and **Caenorhabditis elegans** (the round worm). This has been clearly stated by Bruce Alberts (2010) in reflecting on the recent impressive progress made in deciphering the genomes of these two organisms. Because these organisms can be experimentally manipulated and have short generation times,

relatively rapid progress can be made in understanding the normal functions of their genes and also how abnormalities of their genes can cause disease. Hopefully these advances can be translated into approaches that help humans. According to Alberts, "As incredible as it seems, future research on flies and worms will quite often provide the shortest and most efficient path to curing human diseases." This applies to disorders as different as cancer and Alzheimer disease.

Figure 1–2 shows **areas of great current interest** that have developed either directly as a result of the progress made in the HGP, or have been spurred on by it. As an outgrowth of the HGP, many so-called **-omics** fields have sprung up, involving comprehensive studies of the structures and functions of the molecules with which each is concerned. Definitions of the fields listed below are given in the Glossary of this chapter. The products of genes (RNA molecules and proteins) are being studied using the technics of **transcriptomics** and **proteomics.** One spectacular example of the speed of progress in transcriptomics is the explosion of knowledge about small RNA molecules as regulators of gene activity. Other -omics fields include **glycomics, lipidomics, metabolomics, nutrigenomics,** and **pharmacogenomics.** To keep pace with the amount of information being generated, **bioinformatics** has received much attention. Other related fields to which the impetus from the HGP has carried over are **biotechnology, bioengineering, biophysics,** and **bioethics. Nanotechnology** is an active area, which, for example, may provide novel methods of diagnosis and treatment for cancer and other disorders. **Stem cell biology** is at the center of much current research. **Gene therapy** has yet to deliver the promise that it offers, but it seems probable that will occur sooner or later. Many new **molecular diagnostic tests** have developed in areas such as genetic, microbiologic, and immunologic testing and diagnosis. **Systems biology** is also burgeoning. **Synthetic biology** is perhaps the most intriguing of all. This has the potential for creating living organisms (eg, initially small bacteria) from genetic material in vitro. These could perhaps be designed to carry out specific tasks (eg, to mop up petroleum spills). As in the case of stem cells, this area will attract much attention from bioethicists and others. Many of the above topics are referred to later in this text.

All of the above have made the present time a very exciting one for studying or to be directly involved in biology and medicine. The outcomes of research in the various areas mentioned above will impact tremendously on the future of biology, medicine, and the health sciences.

SUMMARY

- Biochemistry is the science concerned with studying the various molecules that occur in living cells and organisms and with their chemical reactions. Because life depends on biochemical reactions, biochemistry has become the basic language of all biologic sciences.

- Biochemistry is concerned with the entire spectrum of life forms, from relatively simple viruses and bacteria to complex human beings.

- Biochemistry and medicine and other health care disciplines are intimately related. Health in all species depends on a harmonious balance of biochemical reactions occurring in the body, and disease reflects abnormalities in biomolecules, biochemical reactions, or biochemical processes.

- Advances in biochemical knowledge have illuminated many areas of medicine. Conversely, the study of diseases has often revealed previously unsuspected aspects of biochemistry. Biochemical approaches are often fundamental in illuminating the causes of diseases and in designing appropriate therapies.

- The judicious use of various biochemical laboratory tests is an integral component of diagnosis and monitoring of treatment.

- A sound knowledge of biochemistry and of other related basic disciplines is essential for the rational practice of medicine and related health sciences.

- Results of the HGP and of research in related areas will have a profound influence on the future of biology, medicine, and other health sciences. The importance of genomic research on model organisms such as *D melanogaster* and *C elegans* for understanding human diseases is emphasized.

REFERENCES

Alberts B: Model organisms and human health. Science 2010;330:1724.

Alberts B: Lessons from genomics. Science 2011;331:511. (In this issue of Science and succeeding issues in February 2011 various scientists comment on the significance of the tenth anniversary of the publications of the sequencing of the human genome).

Cammack R, Attwood T, Campbell P, et al (editors): Oxford Dictionary of Biochemistry and Molecular Biology. 2nd ed. Oxford University Press. 2006.

Cooke M. Science for physicians. Science 2010;329;1573.

Feero WG, Guttmacher AE, Collins FS: Genomic medicine—an updated primer. N Eng J Med 2010;362:2001.

Fruton JS: *Proteins, Enzymes, Genes: The Interplay of Chemistry and Biology*. Yale University Press, 1999. (Provides the historical background for much of today's biochemical research.)

Garrod AE: Inborn errors of metabolism. (Croonian Lectures.) Lancet 1908;2:1:73,142,214.

Gibson DG, Glass JI, Lartigue C, et al: Creation of a bacterial cell controlled by a chemically synthesized genome. Science 2010;329:52.

Kornberg A: Basic research: The lifeline of medicine. FASEB J 1992;6:3143.

Kornberg A: Centenary of the birth of modern biochemistry. FASEB J 1997;11:1209.

Online Mendelian Inheritance in Man (OMIM): Center for Medical Genetics, Johns Hopkins University and National Center for Biotechnology Information, National Library of Medicine, 1997. http://www.ncbi.nlm.nih.gov/omim/ (The numbers assigned to the entries in OMIM will be cited in selected chapters of this work. Consulting this extensive collection of diseases and other relevant entries—specific proteins, enzymes, etc—will greatly expand the reader's knowledge and understanding of various topics referred to and discussed in this text. The online version is updated almost daily.)

Scriver CR, Beaudet AL, Valle D, et al (editors): *The Metabolic and Molecular Bases of Inherited Disease*, 8th ed. McGraw-Hill, 2001 (This text is now available online and updated as *The Online Metabolic & Molecular Bases of Inherited Disease* at www.ommbid.com. Subscription is required, although access may be available via university and hospital libraries and other sources).

Scherer S: *A Short Guide to the Human Genome*. CSHL Press, 2008.

Weatherall DJ: Systems biology and red cells. N Engl J Med 2011;364;376.

GLOSSARY

Bioengineering: The application of engineering to biology and medicine.

Bioethics: The area of ethics that is concerned with the application of moral and ethical principles to biology and medicine.

Bioinformatics: The discipline concerned with the collection, storage, and analysis of biologic data, mainly DNA and protein sequences (see Chapter 10).

Biophysics: The application of physics and its technics to biology and medicine.

Biotechnology: The field in which biochemical, engineering, and other approaches are combined to develop biological products of use in medicine and industry.

Gene Therapy: Applies to the use of genetically engineered genes to treat various diseases (see Chapter 39).

Genomics: The genome is the complete set of genes of an organism (eg, the human genome) and genomics is the in-depth study of the structures and functions of genomes (see Chapter 10 and other chapters).

Glycomics: The glycome is the total complement of simple and complex carbohydrates in an organism. Glycomics is the systematic study of the structures and functions of glycomes (eg, the human glycome; see Chapter 47).

Lipidomics: The lipidome is the complete complement of lipids found in an organism. Lipidomics is the in-depth study of the structures and functions of all members of the lipidome and of their interactions, in both health and disease.

Metabolomics: The metabolome is the complete complement of metabolites (small molecules involved in metabolism) found in an organism. Metabolomics is the in-depth study of their structures, functions, and changes in various metabolic states.

Molecular Diagnostics: The use of molecular approaches (eg, DNA probes) to assist in the diagnosis of various biochemical, genetic, immunologic, microbiologic, and other medical conditions.

Nanotechnology: The development and application to medicine and to other areas of devices (such as nanoshells, see Glossary of Chapter 55) which are only a few nanometers in size. (10^{-9} m = 1 nm).

Nutrigenomics: The systematic study of the effects of nutrients on genetic expression and also of the effects of genetic variations on the handling of nutrients.

Pharmacogenomics: The use of genomic information and technologies to optimize the discovery and development of drug targets and drugs (see Chapter 54).

Proteomics: The proteome is the complete complement of proteins of an organism. Proteomics is the systematic study of the structures and functions of proteomes, including variations in health and disease (see Chapter 4).

Stem Cell Biology: A stem cell is an undifferentiated cell that has the potential to renew itself and to differentiate into any of the adult cells found in the organism. Stem cell biology is concerned with the biology of stem cells and their uses in various diseases.

Synthetic Biology: The field that combines biomolecular technics with engineering approaches to build new biological functions and systems.

Systems Biology: The field of science in which complex biologic systems are studied as integrated wholes (as opposed to the reductionist approach of, eg, classic biochemistry).

Transcriptomics: The transcriptome is the complete set of RNA transcripts produced by the genome at a fixed period in time. Transcriptomics is the comprehensive study of gene expression at the RNA level (see Chapter 36 and other chapters).

2

Water & pH

Peter J. Kennelly, PhD & Victor W. Rodwell, PhD

OBJECTIVES

After studying this chapter, you should be able to:

- Describe the properties of water that account for its surface tension, viscosity, liquid state at ambient temperature, and solvent power.
- Use structural formulas to represent several organic compounds that can serve as hydrogen bond donors or acceptors.
- Explain the role played by entropy in the orientation, in an aqueous environment, of the polar and nonpolar regions of macromolecules.
- Indicate the quantitative contributions of salt bridges, hydrophobic interactions, and van der Waals forces to the stability of macromolecules.
- Explain the relationship of pH to acidity, alkalinity, and the quantitative determinants that characterize weak and strong acids.
- Calculate the shift in pH that accompanies the addition of a given quantity of acid or base to the pH of a buffered solution.
- Describe what buffers do, how they do it, and the conditions under which a buffer is most effective under physiologic or other conditions.
- Illustrate how the Henderson–Hasselbalch equation can be used to calculate the net charge on a polyelectrolyte at a given pH.

BIOMEDICAL IMPORTANCE

Water is the predominant chemical component of living organisms. Its unique physical properties, which include the ability to solvate a wide range of organic and inorganic molecules, derive from water's dipolar structure and exceptional capacity for forming hydrogen bonds. The manner in which water interacts with a solvated biomolecule influences the structure both of the biomolecule and of water itself. An excellent nucleophile, water is a reactant or product in many metabolic reactions. Regulation of water balance depends upon hypothalamic mechanisms that control thirst, on antidiuretic hormone (ADH), on retention or excretion of water by the kidneys, and on evaporative loss. Nephrogenic diabetes insipidus, which involves the inability to concentrate urine or adjust to subtle changes in extracellular fluid osmolarity, results from the unresponsiveness of renal tubular osmoreceptors to ADH.

Water has a slight propensity to dissociate into hydroxide ions and protons. The concentration of protons, or **acidity,** of aqueous solutions is generally reported using the logarithmic pH scale. Bicarbonate and other buffers normally maintain the pH of extracellular fluid between 7.35 and 7.45. Suspected disturbances of acid–base balance are verified by measuring the pH of arterial blood and the CO_2 content of venous blood. Causes of acidosis (blood pH <7.35) include diabetic ketosis and lactic acidosis. Alkalosis (pH >7.45) may follow vomiting of acidic gastric contents.

WATER IS AN IDEAL BIOLOGIC SOLVENT

Water Molecules Form Dipoles

A water molecule is an irregular, slightly skewed tetrahedron with oxygen at its center (**Figure 2–1**). The two hydrogens and the unshared electrons of the remaining two sp^3-hybridized orbitals occupy the corners of the tetrahedron. The 105° angle between the hydrogen differs slightly from the ideal tetrahedral angle, 109.5°. Ammonia is also tetrahedral, with a 107° angle between its hydrogens. The strongly electronegative oxygen atoms in water attract electrons away from the hydrogen nuclei, leaving them with a partial positive charge, while its

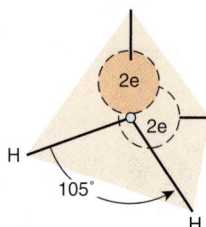

FIGURE 2–1 The water molecule has tetrahedral geometry.

two unshared electron pairs constitute a region of local negative charge.

A molecule with electrical charge distributed asymmetrically about its structure is referred to as a **dipole.** Water's strong dipole is responsible for its high **dielectric constant.** As described quantitatively by Coulomb's law, the strength of interaction F between oppositely charged particles is inversely proportional to the dielectric constant ε of the surrounding medium. The dielectric constant for a vacuum is unity; for hexane it is 1.9; for ethanol, 24.3; and for water, 78.5. Water therefore greatly decreases the force of attraction between charged and polar species relative to water-free environments with lower dielectric constants. Its strong dipole and high dielectric constant enable water to dissolve large quantities of charged compounds such as salts.

Water Molecules Form Hydrogen Bonds

A partially unshielded hydrogen nucleus covalently bound to an electron-withdrawing oxygen or nitrogen atom can interact with an unshared electron pair on another oxygen or nitrogen atom to form a **hydrogen bond.** Since water molecules contain both of these features, hydrogen bonding favors the self-association of water molecules into ordered arrays (**Figure 2–2**). Hydrogen bonding profoundly influences the physical properties of water and accounts for its exceptionally high viscosity, surface tension, and boiling point. On average, each molecule in liquid water associates through hydrogen bonds with 3.5 others. These bonds are both relatively weak and transient, with a half-life of a few nanoseconds or less. Rupture of a hydrogen bond in liquid water requires only about 4.5 kcal/mol, less than 5% of the energy required to rupture a covalent O—H bond.

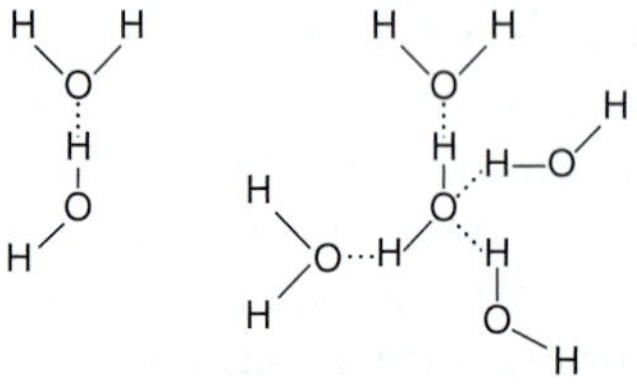

FIGURE 2–2 **Left:** Association of two dipolar water molecules by a hydrogen bond (dotted line). **Right:** Hydrogen-bonded cluster of four water molecules. Note that water can serve simultaneously both as a hydrogen donor and as a hydrogen acceptor.

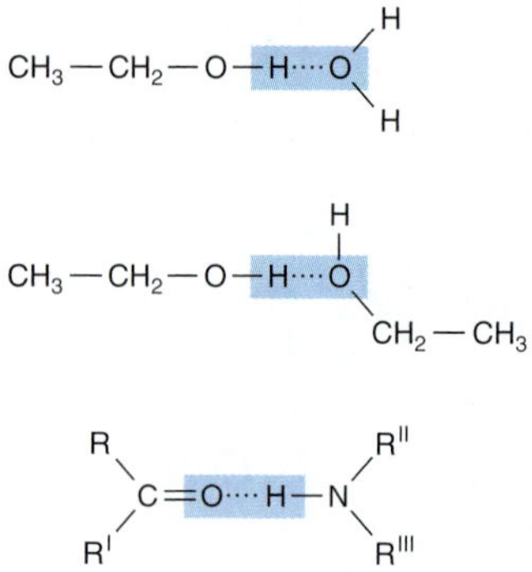

FIGURE 2–3 **Additional polar groups participate in hydrogen bonding.** Shown are hydrogen bonds formed between alcohol and water, between two molecules of ethanol, and between the peptide carbonyl oxygen and the peptide nitrogen hydrogen of an adjacent amino acid.

Hydrogen bonding enables water to dissolve many organic biomolecules that contain functional groups which can participate in hydrogen bonding. The oxygen atoms of aldehydes, ketones, and amides, for example, provide lone pairs of electrons that can serve as hydrogen acceptors. Alcohols, carboxylic acids, and amines can serve both as hydrogen acceptors and as donors of unshielded hydrogen atoms for formation of hydrogen bonds (**Figure 2–3**).

INTERACTION WITH WATER INFLUENCES THE STRUCTURE OF BIOMOLECULES

Covalent and Noncovalent Bonds Stabilize Biologic Molecules

The covalent bond is the strongest force that holds molecules together (**Table 2–1**). Noncovalent forces, while of lesser magnitude, make significant contributions to the structure, stability, and functional competence of macromolecules in living

TABLE 2–1 Bond Energies for Atoms of Biologic Significance

Bond Type	Energy (kcal/mol)	Bond Type	Energy (kcal/mol)
O—O	34	O=O	96
S—S	51	C—H	99
C—N	70	C=S	108
S—H	81	O—H	110
C—C	82	C=C	147
C—O	84	C=N	147
N—H	94	C=O	164

cells. These forces, which can be either attractive or repulsive, involve interactions both within the biomolecule and between it and the water that forms the principal component of the surrounding environment.

Biomolecules Fold to Position Polar & Charged Groups on Their Surfaces

Most biomolecules are **amphipathic;** that is, they possess regions rich in charged or polar functional groups as well as regions with hydrophobic character. Proteins tend to fold with the R-groups of amino acids with hydrophobic side chains in the interior. Amino acids with charged or polar amino acid side chains (eg, arginine, glutamate, serine) generally are present on the surface in contact with water. A similar pattern prevails in a phospholipid bilayer, where the charged "head groups" of phosphatidyl serine or phosphatidyl ethanolamine contact water while their hydrophobic fatty acyl side chains cluster together, excluding water. This pattern maximizes the opportunities for the formation of energetically favorable charge–dipole, dipole–dipole, and hydrogen bonding interactions between polar groups on the biomolecule and water. It also minimizes energetically unfavorable contacts between water and hydrophobic groups.

Hydrophobic Interactions

Hydrophobic interaction refers to the tendency of nonpolar compounds to self-associate in an aqueous environment. This self-association is driven neither by mutual attraction nor by what are sometimes incorrectly referred to as "hydrophobic bonds." Self-association minimizes the disruption of energetically favorable interactions between the surrounding water molecules.

While the hydrogens of nonpolar groups such as the methylene groups of hydrocarbons do not form hydrogen bonds, they do affect the structure of the water that surrounds them. Water molecules adjacent to a hydrophobic group are restricted in the number of orientations (degrees of freedom) that permit them to participate in the maximum number of energetically favorable hydrogen bonds. Maximal formation of multiple hydrogen bonds, which maximizes enthalpy, can be maintained only by increasing the order of the adjacent water molecules, with an accompanying decrease in entropy.

It follows from the second law of thermodynamics that the optimal free energy of a hydrocarbon–water mixture is a function of both maximal enthalpy (from hydrogen bonding) and minimum entropy (maximum degrees of freedom). Thus, nonpolar molecules tend to form droplets that minimize exposed surface area and reduce the number of water molecules whose motional freedom becomes restricted. Similarly, in the aqueous environment of the living cell the hydrophobic portions of biopolymers tend to be buried inside the structure of the molecule, or within a lipid bilayer, minimizing contact with water.

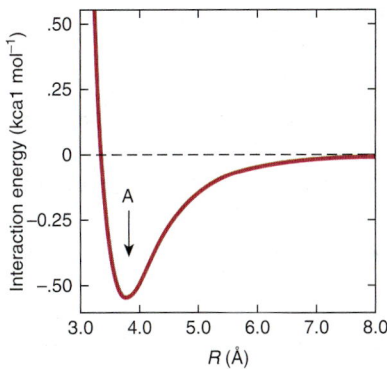

FIGURE 2–4 **The strength of van der Waals interactions varies with the distance, *R*, between interacting species.** The force of interaction between interacting species increases with decreasing distance until they are separated by the van der Waals contact distance (see arrow marked A). Repulsion due to interaction between the electrons of each atom or molecule then supervenes. While individual van der Waals interactions are extremely weak, the cumulative effect is nevertheless substantial for macromolecules such as DNA and proteins with many atoms in close contact.

Electrostatic Interactions

Interactions between charged groups help shape biomolecular structure. Electrostatic interactions between oppositely charged groups within or between biomolecules are termed **salt bridges.** Salt bridges are comparable in strength to hydrogen bonds but act over larger distances. They therefore often facilitate the binding of charged molecules and ions to proteins and nucleic acids.

van der Waals Forces

van der Waals forces arise from attractions between transient dipoles generated by the rapid movement of electrons of all neutral atoms. Significantly weaker than hydrogen bonds but potentially extremely numerous, van der Waals forces decrease as the sixth power of the distance separating atoms (**Figure 2–4**). Thus, they act over very short distances, typically 2–4 Å.

Multiple Forces Stabilize Biomolecules

The DNA double helix illustrates the contribution of multiple forces to the structure of biomolecules. While each individual DNA strand is held together by covalent bonds, the two strands of the helix are held together exclusively by noncovalent interactions such as hydrogen bonds between nucleotide bases (Watson–Crick base pairing) and van der Waals interactions between the stacked purine and pyrimidine bases. The double helix presents the charged phosphate groups and polar hydroxyl groups from the ribose sugars of the DNA backbone to water while burying the relatively hydrophobic nucleotide bases inside. The extended backbone maximizes the distance between negatively charged phosphates, minimizing unfavorable electrostatic interactions.

WATER IS AN EXCELLENT NUCLEOPHILE

Metabolic reactions often involve the attack by lone pairs of electrons residing on electron-rich molecules termed **nucleophiles** upon electron-poor atoms called **electrophiles.** Nucleophiles and electrophiles do not necessarily possess a formal negative or positive charge. Water, whose two lone pairs of sp^3 electrons bear a partial negative charge (Figure 2–1), is an excellent nucleophile. Other nucleophiles of biologic importance include the oxygen atoms of phosphates, alcohols, and carboxylic acids; the sulfur of thiols; and the nitrogen of amines and the imidazole ring of histidine. Common electrophiles include the carbonyl carbons in amides, esters, aldehydes, and ketones and the phosphorus atoms of phosphoesters.

Nucleophilic attack by water typically results in the cleavage of the amide, glycoside, or ester bonds that hold biopolymers together. This process is termed **hydrolysis.** Conversely, when monomer units are joined together to form biopolymers such as proteins or glycogen, water is a product, for example, during the formation of a peptide bond between two amino acids:

While hydrolysis is a thermodynamically favored reaction, the amide and phosphoester bonds of polypeptides and oligonucleotides are stable in the aqueous environment of the cell. This seemingly paradoxic behavior reflects the fact that the thermodynamics governing the equilibrium of a reaction do not determine the rate at which it will proceed. In the cell, protein catalysts called **enzymes** accelerate the rate of hydrolytic reactions when needed. **Proteases** catalyze the hydrolysis of proteins into their component amino acids, while **nucleases** catalyze the hydrolysis of the phosphoester bonds in DNA and RNA. Careful control of the activities of these enzymes is required to ensure that they act only on appropriate target molecules at appropriate times.

Many Metabolic Reactions Involve Group Transfer

Many of the enzymic reactions responsible for synthesis and breakdown of biomolecules involve the transfer of a chemical group G from a donor D to an acceptor A to form an acceptor group complex, A—G:

$$D—G + A \rightleftarrows A—G + D$$

The hydrolysis and phosphorolysis of glycogen, for example, involve the transfer of glucosyl groups to water or to orthophosphate. The equilibrium constant for the hydrolysis of covalent bonds strongly favors the formation of split products. Conversely, in many cases the group transfer reactions responsible for the biosynthesis of macromolecules involve the thermodynamically unfavored formation of covalent bonds. Enzyme catalysts play a critical role in surmounting these barriers by virtue of their capacity to directly link two normally separate reactions together. By marrying an energetically unfavorable group transfer reaction with a thermodynamically favorable reaction, such as the hydrolysis of ATP, a new coupled reaction can be generated whose net *overall* change in free energy favors biopolymer synthesis.

Given the nucleophilic character of water and its high concentration in cells, why are biopolymers such as proteins and DNA relatively stable? And how can synthesis of biopolymers occur in an aqueous, seemingly prohydrolytic, environment? Central to both questions are the properties of enzymes. In the absence of enzymic catalysis, even reactions that are highly favored thermodynamically do not necessarily take place rapidly. Precise and differential control of enzyme activity and the sequestration of enzymes in specific organelles determine under what physiologic conditions a given biopolymer will be synthesized or degraded. Newly synthesized biopolymers are not immediately hydrolyzed because the active sites of biosynthetic enzymes sequester substrates in an environment from which water can be excluded.

Water Molecules Exhibit a Slight but Important Tendency to Dissociate

The ability of water to ionize, while slight, is of central importance for life. Since water can act both as an acid and as a base, its ionization may be represented as an intermolecular proton transfer that forms a hydronium ion (H_3O^+) and a hydroxide ion (OH^-):

$$H_2O + H_2O \rightleftarrows H_3O^+ + OH^-$$

The transferred proton is actually associated with a cluster of water molecules. Protons exist in solution not only as H_3O^+, but also as multimers such as $H_5O_2^+$ and $H_7O_3^+$. The proton is nevertheless routinely represented as H^+, even though it is in fact highly hydrated.

Since hydronium and hydroxide ions continuously recombine to form water molecules, an *individual* hydrogen or oxygen cannot be stated to be present as an ion or as part of a water molecule. At one instant it is an ion; an instant later it is part of a water molecule. Individual ions or molecules are therefore not considered. We refer instead to the *probability* that at any instant in time a given hydrogen will be present as an ion or as part of a water molecule. Since 1 g of water contains 3.46×10^{22} molecules, the ionization of water can be described statistically. To state that the probability that a hydrogen exists as an ion is 0.01 means that at any given moment in time, a hydrogen atom has 1 chance in 100 of being an ion and 99 chances out of 100 of being part of a water molecule. The actual probability of a hydrogen atom in pure water existing as a hydrogen ion is approximately 1.8×10^{-9}. The probability of its being part of a water molecule thus is almost unity. Stated another way, for every hydrogen ion or hydroxide ion in pure water, there are 1.8 billion or 1.8×10^{9} water molecules. Hydrogen ions and hydroxide ions nevertheless contribute significantly to the properties of water.

For dissociation of water,

$$K = \frac{\left[H^{+}\right]\left[OH^{-}\right]}{\left[H_{2}O\right]}$$

where the brackets represent molar concentrations (strictly speaking, molar activities) and K is the **dissociation constant.** Since 1 mole (mol) of water weighs 18 g, 1 liter (L) (1000 g) of water contains $1000 \div 18 = 55.56$ mol. Pure water thus is 55.56 molar. Since the probability that a hydrogen in pure water will exist as a hydrogen ion is 1.8×10^{-9}, the molar concentration of H^{+} ions (or of OH^{-} ions) in pure water is the product of the probability, 1.8×10^{-9}, times the molar concentration of water, 55.56 mol/L. The result is 1.0×10^{-7} mol/L.

We can now calculate K for pure water:

$$K = \frac{\left[H^{+}\right]\left[OH^{-}\right]}{\left[H_{2}O\right]} = \frac{\left[10^{-7}\right]\left[10^{-7}\right]}{\left[55.56\right]}$$

$$= 0.018 \times 10^{-14} = 1.8 \times 10^{-16}\,\text{mol/L}$$

The molar concentration of water, 55.56 mol/L, is too great to be significantly affected by dissociation. It is therefore considered to be essentially constant. This constant may therefore be incorporated into the dissociation constant K to provide a useful new constant K_{w} termed the **ion product** for water. The relationship between K_{w} and K is shown below:

$$K = \frac{\left[H^{+}\right]\left[OH^{-}\right]}{\left[H_{2}O\right]} = 1.8 \times 10^{-16}\,\text{mol/L}$$

$$K_{w} = (K)[H_{2}O] = \left[H^{+}\right]\left[OH^{-}\right]$$

$$= \left(1.8 \times 10^{-16}\,\text{mol/L}\right)(55.56\,\text{mol/L})$$

$$= 1.00 \times 10^{-14}\,(\text{mol/L})^{2}$$

Note that the dimensions of K are moles per liter and those of K_{w} are moles^{2} per liter^{2}. As its name suggests, the ion product K_{w} is numerically equal to the product of the molar concentrations of H^{+} and OH^{-}:

$$K_{w} = \left[H^{+}\right]\left[OH^{-}\right]$$

At 25°C, $K_{w} = (10^{-7})^{2}$, or 10^{-14} (mol/L)^{2}. At temperatures below 25°C, K_{w} is somewhat less than 10^{-14}, and at temperatures above 25°C it is somewhat greater than 10^{-14}. Within the stated limitations of the effect of temperature, K_{w} equals 10^{-14} (mol/L)^{2}

for all aqueous solutions, even solutions of acids or bases. We use K_{w} to calculate the pH of acidic and basic solutions.

pH IS THE NEGATIVE LOG OF THE HYDROGEN ION CONCENTRATION

The term **pH** was introduced in 1909 by Sörensen, who defined pH as the negative log of the hydrogen ion concentration:

$$pH = -\log\left[H^{+}\right]$$

This definition, while not rigorous, suffices for many biochemical purposes. To calculate the pH of a solution:

1. Calculate the hydrogen ion concentration $[H^{+}]$.
2. Calculate the base 10 logarithm of $[H^{+}]$.
3. pH is the negative of the value found in step 2.

For example, for pure water at 25°C,

$$pH = -\log\left[H^{+}\right] = -\log 10^{-7} = -(-7) = 7.0$$

This value is also known as the *power* (English), *puissant* (French), or *potennz* (German) of the exponent, hence the use of the term "p."

Low pH values correspond to high concentrations of H^{+} and high pH values correspond to low concentrations of H^{+}.

Acids are **proton donors** and bases are **proton acceptors. Strong acids** (eg, HCl, $H_{2}SO_{4}$) completely dissociate into anions and protons even in strongly acidic solutions (low pH). **Weak acids** dissociate only partially in acidic solutions. Similarly, **strong bases** (eg, KOH, NaOH)—but not **weak bases,** eg: $Ca(OH)_{2}$, are completely dissociated even at high pH. Many biochemicals are weak acids. Exceptions include phosphorylated intermediates, whose phosphoryl group contains two dissociable protons, the first of which is strongly acidic.

The following examples illustrate how to calculate the pH of acidic and basic solutions.

Example 1: What is the pH of a solution whose hydrogen ion concentration is 3.2×10^{-4} mol/L?

$$pH = -\log\left[H^{+}\right]$$

$$= -\log\left(3.2 \times 10^{-4}\right)$$

$$= -\log(3.2) - \log\left(10^{-4}\right)$$

$$= -0.5 + 4.0$$

$$= 3.5$$

Example 2: What is the pH of a solution whose hydroxide ion concentration is 4.0×10^{-4} mol/L? We first define a quantity **pOH** that is equal to $-\log [OH^{-}]$ and that may be derived from the definition of K_{w}:

$$K_{w} = \left[H^{+}\right]\left[OH^{-}\right] = 10^{-14}$$

Therefore,

$$\log\left[H^{+}\right] + \log\left[OH^{-}\right] = \log 10^{-14}$$

or

$$pH + pOH = 14$$

To solve the problem by this approach:

$$[OH^-] = 4.0 \times 10^{-4}$$

$$pOH = -\log [OH^-]$$

$$= -\log (4.0 \times 10^{-4})$$

$$= -\log (4.0) - \log (10^{-4})$$

$$= -0.60 + 4.0$$

$$= 3.4$$

Now

$$pH = 14 - pOH = 14 - 3.4$$

$$= 10.6$$

The examples above illustrate how the logarithmic pH scale facilitates recording and comparing hydrogen ion concentrations that differ by orders of magnitude from one another, ie, 0.00032 M (pH 3.5) and 0.000000000025 M (pH 10.6).

Example 3: What are the pH values of (a) 2.0×10^{-2} mol/L KOH and of (b) 2.0×10^{-6} mol/L KOH? The OH$^-$ arises from two sources, KOH and water. Since pH is determined by the total [H$^+$] (and pOH by the total [OH$^-$]), both sources must be considered. In the first case (a), the contribution of water to the total [OH$^-$] is negligible. The same cannot be said for the second case (b):

	Concentration (mol/L)	
	(a)	**(b)**
Molarity of KOH	2.0×10^{-2}	2.0×10^{-6}
[OH$^-$] from KOH	2.0×10^{-2}	2.0×10^{-6}
[OH$^-$] from water	1.0×10^{-7}	1.0×10^{-7}
Total [OH$^-$]	2.00001×10^{-2}	2.1×10^{-6}

Once a decision has been reached about the significance of the contribution by water, pH may be calculated as above.

The above examples assume that the strong base KOH is completely dissociated in solution and that the concentration of OH$^-$ ions was thus equal to that due to the KOH plus that present initially in the water. This assumption is valid for dilute solutions of strong bases or acids, but not for weak bases or acids. Since weak electrolytes dissociate only slightly in solution, we must use the **dissociation constant** to calculate the concentration of [H$^+$] (or [OH$^-$]) produced by a given molarity of a weak acid (or base) before calculating total [H$^+$] (or total [OH$^-$]) and subsequently pH.

Functional Groups That Are Weak Acids Have Great Physiologic Significance

Many biochemicals possess functional groups that are weak acids or bases. Carboxyl groups, amino groups, and phosphate esters, whose second dissociation falls within the physiologic range, are present in proteins and nucleic acids, most coenzymes, and most intermediary metabolites. Knowledge of the dissociation of weak acids and bases thus is basic to understanding the influence of intracellular pH on structure and biologic activity. Charge-based separations such as electrophoresis and ion exchange chromatography are also best understood in terms of the dissociation behavior of functional groups.

We term the protonated species (eg, HA or R—NH$_3^+$) the **acid** and the unprotonated species (eg, A$^-$ or R—NH$_2$) its **conjugate base.** Similarly, we may refer to a **base** (eg, A$^-$ or R—NH$_2$) and its **conjugate acid** (eg, HA or R—NH$_3^+$). Representative weak acids (left), their conjugate bases (center), and pK_a values (right) include the following:

R—CH$_2$—COOH	R—CH$_2$—COO$^-$	pK_a = 4 – 5
R—CH$_2$—NH$_3^+$	R—CH$_2$—NH$_2$	pK_a = 9 – 10
H$_2$CO$_3$	HCO$_3^-$	pK_a = 6.4
H$_2$PO$_4^-$	HPO$_4^{-2}$	pK_a = 7.2

We express the relative strengths of weak acids and bases in terms of their dissociation constants. Shown below are the expressions for the dissociation constant (K_a) for two representative weak acids, R—COOH and R—NH$_3^+$.

$$R—COOH \rightleftarrows R—COO^- + H^+$$

$$K_a = \frac{[R—COO^-][H^+]}{[R—COOH]}$$

$$R—NH_3^+ \rightleftarrows R—NH_2 + H^+$$

$$K_a = \frac{[R—NH_2][H^+]}{[R—NH_3^+]}$$

Since the numeric values of K_a for weak acids are negative exponential numbers, we express K_a as pK_a, where

$$pK_a = -\log K_a$$

Note that pK_a is related to K_a as pH is to [H$^+$]. The stronger the acid, the lower is its pK_a value.

pK_a is used to express the relative strengths of both acids and bases. For any weak acid, its conjugate is a strong base. Similarly, the conjugate of a strong base is a weak acid. **The relative strengths of bases are expressed in terms of the pK_a of their conjugate acids.** For polyprotic compounds containing more than one dissociable proton, a numerical subscript is assigned to each dissociation, numbered starting from unity in decreasing order of relative acidity. For a dissociation of the type

$$R—NH_3^+ \rightarrow R—NH_2 + H^+$$

the pK_a is the pH at which the concentration of the acid R–NH$_3^+$ equals that of the base R—NH$_2$.

From the above equations that relate K_a to [H$^+$] and to the concentrations of undissociated acid and its conjugate base, when

$$[R—COO^-] = [R—COOH]$$

or when

$$\left[R{-}NH_2\right] = \left[R{-}NH_3^{+}\right]$$

then

$$K_a = \left[H^{+}\right]$$

Thus, when the associated (protonated) and dissociated (conjugate base) species are present at equal concentrations, the prevailing hydrogen ion concentration $[H^{+}]$ is numerically equal to the dissociation constant, K_a. If the logarithms of both sides of the above equation are taken and both sides are multiplied by –1, the expressions would be as follows:

$$K_a = \left[H^{+}\right]$$
$$-\log K_a = -\log\left[H^{+}\right]$$

Since $-\log K_a$ is defined as pK_a, and $-\log [H^{+}]$ defines pH, the equation may be rewritten as

$$pK_a = pH$$

ie, **the pK_a of an acid group is the pH at which the protonated and unprotonated species are present at equal concentrations.** The pK_a for an acid may be determined by adding 0.5 equivalent of alkali per equivalent of acid. The resulting pH will equal the pK_a of the acid.

The Henderson–Hasselbalch Equation Describes the Behavior of Weak Acids & Buffers

The Henderson–Hasselbalch equation is derived below.
A weak acid, HA, ionizes as follows:

$$HA \rightleftarrows H^{+} + A^{-}$$

The equilibrium constant for this dissociation is

$$K_a = \frac{\left[H^{+}\right]\left[A^{-}\right]}{[HA]}$$

Cross-multiplication gives

$$\left[H^{+}\right]\left[A^{-}\right] = K_a [HA]$$

Divide both sides by $[A^{-}]$:

$$\left[H^{+}\right] = K_a \frac{[HA]}{\left[A^{-}\right]}$$

Take the log of both sides:

$$\log\left[H^{+}\right] = \log\left(K_a \frac{[HA]}{\left[A^{-}\right]}\right)$$
$$= \log K_a + \log \frac{[HA]}{\left[A^{-}\right]}$$

Multiply through by –1:

$$-\log\left[H^{+}\right] = -\log K_a - \log \frac{[HA]}{\left[A^{-}\right]}$$

Substitute pH and pK_a for $-\log [H^{+}]$ and $-\log K_a$, respectively; then

$$pH = pK_a - \log \frac{[HA]}{\left[A^{-}\right]}$$

Inversion of the last term removes the minus sign and gives the **Henderson–Hasselbalch equation**

$$\mathbf{pH = pK_a + \log \frac{\left[A^{-}\right]}{[HA]}}$$

The Henderson–Hasselbalch equation has great predictive value in protonic equilibria. For example,

1. When an acid is exactly half-neutralized, $[A^{-}] = [HA]$. Under these conditions,

$$pH = pK_a + \log \frac{\left[A^{-}\right]}{[HA]} = pK_a + \log\left(\frac{1}{1}\right) = pK_a + 0$$

Therefore, at half-neutralization, $pH = pK_a$.

2. When the ratio $[A^{-}]/[HA] = 100:1$,

$$pH = pK_a + \log \frac{\left[A^{-}\right]}{[HA]}$$
$$pH = pK_a + \log\left(100/1\right) = pK_a + 2$$

3. When the ratio $[A^{-}]/[HA] = 1:10$,

$$pH = pK_a + \log\left(1/10\right) = pK_a + (-1)$$

If the equation is evaluated at ratios of $[A^{-}]/[HA]$ ranging from 10^{3} to 10^{-3} and the calculated pH values are plotted, the resulting graph describes the titration curve for a weak acid (**Figure 2–5**).

Solutions of Weak Acids & Their Salts Buffer Changes in pH

Solutions of weak acids or bases and their conjugates exhibit **buffering,** the ability to resist a change in pH following addition of strong acid or base. Since many metabolic reactions are accompanied by the release or uptake of protons, most intracellular reactions are buffered. Oxidative metabolism produces CO_2, the anhydride of carbonic acid, which if not buffered would produce severe acidosis. Maintenance of a constant pH involves buffering by phosphate, bicarbonate, and proteins, which accept or release protons to resist a change in pH. For experiments using tissue extracts or enzymes, constant pH is maintained by the addition of buffers such as MES ([2-N-morpholino]-ethanesulfonic acid, pK_a 6.1), inorganic orthophosphate (pK_{a2} 7.2), HEPES (N-hydroxyethylpiperazine-N'-2-ethanesulfonic acid, pK_a 6.8), or Tris (tris[hydroxymethyl]aminomethane, pK_a 8.3). The value of pK_a relative to the desired pH is the major determinant of which buffer is selected.

Buffering can be observed by using a pH meter while titrating a weak acid or base (**Figure 2–5**). We can also calculate the pH shift that accompanies addition of acid or base to a buffered solution. In the example, the buffered solution

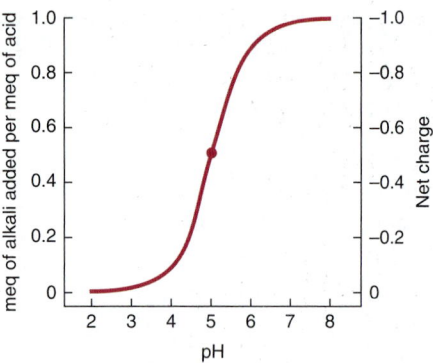

FIGURE 2–5 **Titration curve for an acid of the type HA.** The heavy dot in the center of the curve indicates the pK_a 5.0.

(a weak acid, pK_a = 5.0, and its conjugate base) is initially at one of four pH values. We will calculate the pH shift that results when 0.1 meq of KOH is added to 1 meq of each solution:

Initial pH	5.00	5.37	5.60	5.86
$[A^-]_{initial}$	0.50	0.70	0.80	0.88
$[HA]_{initial}$	0.50	0.30	0.20	0.12
$([A^-]/[HA])_{initial}$	1.00	2.33	4.00	7.33
Addition of 0.1 meq of KOH produces				
$[A^-]_{final}$	0.60	0.80	0.90	0.98
$[HA]_{final}$	0.40	0.20	0.10	0.02
$([A^-]/[HA])_{final}$	1.50	4.00	9.00	49.0
log $([A^-]/[HA])_{final}$	0.18	0.60	0.95	1.69
Final pH	5.18	5.60	5.95	6.69
ΔpH	**0.18**	**0.60**	**0.95**	**1.69**

Notice that the change in pH per milliequivalent of OH^- added depends on the initial pH. The solution resists changes in pH most effectively at pH values close to the pK_a. **A solution of a weak acid and its conjugate base buffers most effectively in the pH range pK_a ± 1.0 pH unit.**

Figure 2–5 also illustrates the net charge on one molecule of the acid as a function of pH. A fractional charge of −0.5 does not mean that an individual molecule bears a fractional charge but that the *probability* is 0.5 that a given molecule has a unit negative charge at any given moment in time. Consideration of the net charge on macromolecules as a function of pH provides the basis for separatory techniques such as ion exchange chromatography and electrophoresis.

Acid Strength Depends on Molecular Structure

Many acids of biologic interest possess more than one dissociating group. The presence of adjacent negative charge

TABLE 2–2 **Relative Strengths of Selected Acids of Biologic Significance[1]**

Monoprotic Acids		
Formic	pK	3.75
Lactic	pK	3.86
Acetic	pK	4.76
Ammonium ion	pK	9.25
Diprotic Acids		
Carbonic	pK_1	6.37
	pK_2	10.25
Succinic	pK_1	4.21
	pK_2	5.64
Glutaric	pK_1	4.34
	pK_2	5.41
Triprotic Acids		
Phosphoric	pK_1	2.15
	pK_2	6.82
	pK_3	12.38
Citric	pK_1	3.08
	pK_2	4.74
	pK_3	5.40

[1]**Note:** Tabulated values are the pK_a values (−log of the dissociation constant) of selected monoprotic, diprotic, and triprotic acids.

hinders the release of a proton from a nearby group, raising its pK_a. This is apparent from the pK_a values for the three dissociating groups of phosphoric acid and citric acid (**Table 2–2**). The effect of adjacent charge decreases with distance. The second pK_a for succinic acid, which has two methylene groups between its carboxyl groups, is 5.6, whereas the second pK_a for glutaric acid, which has one additional methylene group, is 5.4.

pK_a Values Depend on the Properties of the Medium

The pK_a of a functional group is also profoundly influenced by the surrounding medium. The medium may either raise or lower the pK_a depending on whether the undissociated acid or its conjugate base is the charged species. The effect of dielectric constant on pK_a may be observed by adding ethanol to water. The pK_a of a carboxylic acid *increases*, whereas that of an amine *decreases* because ethanol decreases the ability of water to solvate a charged species. The pK_a values of dissociating groups in the interiors of proteins thus are profoundly affected by their local environment, including the presence or absence of water.

SUMMARY

- Water forms hydrogen-bonded clusters with itself and with other proton donors or acceptors. Hydrogen bonds account for the surface tension, viscosity, liquid state at room temperature, and solvent power of water.

- Compounds that contain O or N can serve as hydrogen bond donors and/or acceptors.

- Macromolecules exchange internal surface hydrogen bonds for hydrogen bonds to water. Entropic forces dictate that macromolecules expose polar regions to an aqueous interface and bury nonpolar regions.

- Salt bridges, hydrophobic interactions, and van der Waals forces participate in maintaining molecular structure.

- pH is the negative log of $[H^+]$. A low pH characterizes an acidic solution, and a high pH denotes a basic solution.

- The strength of weak acids is expressed by pK_a, the negative log of the acid dissociation constant. Strong acids have low pK_a values and weak acids have high pK_a values.

- Buffers resist a change in pH when protons are produced or consumed. Maximum buffering capacity occurs $\pm$ 1 pH unit on either side of pK_a. Physiologic buffers include bicarbonate, orthophosphate, and proteins.

REFERENCES

Reese KM: Whence came the symbol pH. Chem & Eng News 2004;82:64.

Segel IM: *Biochemical Calculations.* Wiley, 1968.

Skinner JL: Following the motions of water molecules in aqueous solutions. Science 2010;328:985.

Stillinger FH: Water revisited. Science 1980;209:451.

Suresh SJ, Naik VM: Hydrogen bond thermodynamic properties of water from dielectric constant data. J Chem Phys 2000;113:9727.

Wiggins PM: Role of water in some biological processes. Microbiol Rev 1990;54:432.

C H A P T E R

3

Amino Acids & Peptides

Peter J. Kennelly, PhD & Victor W. Rodwell, PhD

OBJECTIVES

After studying this chapter, you should be able to:

- Name, and draw the structures of, the 20 amino acids present in proteins.
- Write the three- and one-letter designations for each of the common amino acids.
- List the ionizable groups of the common amino acids and their pK_a values.
- Calculate the pH of an unbuffered aqueous solution of a polyfunctional amino acid and the change in pH that occurs following the addition of a given quantity of strong acid or alkali.
- Define pI and indicate its relationship to the net charge on a polyfunctional electrolyte.
- Explain how pH, pK_a and pI can be used to predict the mobility of a polyelectrolyte, such as an amino acid, in a direct-current electrical field.
- Describe the contribution of each type of R group of the common amino acids to their chemical properties.
- Describe the directionality, nomenclature, and primary structure of peptides.
- Identify the bond in a peptide that exhibits partial double-bond character and its conformational consequences in a peptide.
- Identify those bonds in the peptide backbone that are capable of free rotation and the Greek letters used to designate them.

BIOMEDICAL IMPORTANCE

In addition to providing the monomer units from which the long polypeptide chains of proteins are synthesized, the L-α-amino acids and their derivatives participate in cellular functions as diverse as nerve transmission and the biosynthesis of porphyrins, purines, pyrimidines, and urea. Short polymers of amino acids called *peptides* perform prominent roles in the neuroendocrine system as hormones, hormone-releasing factors, neuromodulators, or neurotransmitters. Humans and other higher animals lack the capability to synthesize 10 of the 20 common L-α-amino acids in amounts adequate to support infant growth or to maintain health in adults. Consequently, the human diet must contain adequate quantities of these nutritionally essential amino acids. While human proteins contain only L-α-amino acids, microorganisms make extensive use of D-α-amino acids. *Bacillus subtilis*, for example, secretes a mixture of D-methionine, D-tyrosine, D-leucine, and D-tryptophan to trigger biofilm disassembly, and *Vibrio cholerae* incorporates D-leucine and D-methionine into the peptide component of their peptidoglycan layer. Many bacteria elaborate peptides that contain both D- and L-α-amino acids, several of which possess therapeutic value, including the antibiotics bacitracin and gramicidin A and the antitumor agent bleomycin. Certain other microbial peptides are toxic. The cyanobacterial peptides microcystin and nodularin are lethal in large doses, while small quantities promote the formation of hepatic tumors.

PROPERTIES OF AMINO ACIDS

The Genetic Code Specifies 20 L-α-Amino Acids

Of the over 300 naturally occurring amino acids, 20 constitute the predominant monomer units of proteins. While a three-letter genetic code could potentially accommodate more than 20 amino acids, several amino acids are specified by multiple codons (see Table 37–1). Redundant usage limits the available codons to the 20 L-α-amino acids listed in **Table 3–1**. Both one- and three-letter abbreviations for each amino acid can

be used to represent the amino acids in peptides and proteins (Table 3–1). Some proteins contain additional amino acids that arise by modification of an amino acid already present in a peptide. Examples include conversion of peptidyl proline and lysine to 4-hydroxyproline and 5-hydroxylysine; the conversion of peptidyl glutamate to γ-carboxyglutamate; and the methylation, formylation, acetylation, prenylation, and phosphorylation of certain aminoacyl residues. These modifications extend the biologic diversity of proteins by altering their solubility, stability, and interaction with other proteins.

Selenocysteine, the 21st L-α-Amino Acid

Selenocysteine is an L-α-amino acid found in proteins from every domain of life. Humans contain approximately two dozen selenoproteins that include certain peroxidases and reductases, selenoprotein P which circulates in the plasma, and the iodothyronine deiodinases responsible for converting the prohormone thyroxine (T4) to the thyroid hormone 3,3′5-triiodothyronine (T3) (Chapter 41). As the name implies,

TABLE 3–1 L-α-Amino Acids Present in Proteins

Name	Symbol	Structural Formula	pK_1	pK_2	pK_3
With Aliphatic Side Chains			α-COOH	α-NH$_3^+$	R Group
Glycine	Gly [G]	H—CH—COO$^-$ | NH$_3^+$	2.4	9.8	
Alanine	Ala [A]	CH$_3$—CH—COO$^-$ | NH$_3^+$	2.4	9.9	
Valine	Val [V]	H$_3$C \ CH—CH—COO$^-$ / | H$_3$C NH$_3^+$	2.2	9.7	
Leucine	Leu [L]	H$_3$C \ CH—CH$_2$—CH—COO$^-$ / | H$_3$C NH$_3^+$	2.3	9.7	
Isoleucine	Ile [I]	CH$_3$ \ CH$_2$ \ CH—CH—COO$^-$ / | CH$_3$ NH$_3^+$	2.3	9.8	
With Side Chains Containing Hydroxylic (OH) Groups					
Serine	Ser [S]	CH$_2$—CH—COO$^-$ | | OH NH$_3^+$	2.2	9.2	about 13
Threonine	Thr [T]	CH$_3$—CH—CH—COO$^-$ | | OH NH$_3^+$	2.1	9.1	about 13
Tyrosine	Tyr [Y]	See below.			

(continued)

TABLE 3–1 L-α-Amino Acids Present in Proteins (*continued*)

Name	Symbol	Structural Formula	pK_1	pK_2	pK_3
With Side Chains Containing Sulfur Atoms			α-COOH	α-NH_3^+	R Group
Cysteine	Cys [C]	CH_2—CH—COO$^-$; SH; NH_3^+	1.9	10.8	8.3
Methionine	Met [M]	CH_2—CH_2—CH—COO$^-$; S—CH_3; NH_3^+	2.1	9.3	
With Side Chains Containing Acidic Groups or Their Amides					
Aspartic acid	Asp [D]	$^-$OOC—CH_2—CH—COO$^-$; NH_3^+	2.1	9.9	3.9
Asparagine	Asn [N]	H_2N—C(=O)—CH_2—CH—COO$^-$; NH_3^+	2.1	8.8	
Glutamic acid	Glu [E]	$^-$OOC—CH_2—CH_2—CH—COO$^-$; NH_3^+	2.1	9.5	4.1
Glutamine	Gln [Q]	H_2N—C(=O)—CH_2—CH_2—CH—COO$^-$; NH_3^+	2.2	9.1	
With Side Chains Containing Basic Groups					
Arginine	Arg [R]	H—N—CH_2—CH_2—CH_2—CH—COO$^-$; C=NH_2^+; NH_2; NH_3^+	1.8	9.0	12.5
Lysine	Lys [K]	CH_2—CH_2—CH_2—CH_2—CH—COO$^-$; NH_3^+; NH_3^+	2.2	9.2	10.8
Histidine	His [H]	HN imidazole N—CH_2—CH—COO$^-$; NH_3^+	1.8	9.3	6.0
Containing Aromatic Rings					
Histidine	His [H]	See above.			
Phenylalanine	Phe [F]	phenyl—CH_2—CH—COO$^-$; NH_3^+	2.2	9.2	
Tyrosine	Tyr [Y]	HO—phenyl—CH_2—CH—COO$^-$; NH_3^+	2.2	9.1	10.1
Tryptophan	Trp [W]	indole—CH_2—CH—COO$^-$; NH_3^+	2.4	9.4	
Imino Acid					
Proline	Pro [P]	pyrrolidine ring N^+H_2—COO$^-$	2.0	10.6	

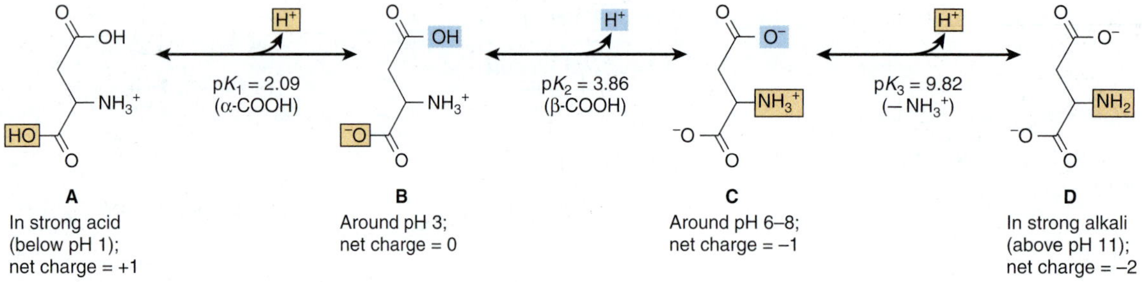

FIGURE 3–1 Protonic equilibria of aspartic acid.

a selenium atom replaces the sulfur of its structural analog, cysteine. The pK_3 of selenocysteine, 5.2, is three units lower than that of cysteine. Unlike other unusual amino acids, selenocysteine is not the product of a posttranslational modification. Rather, it is inserted directly into a growing polypeptide during translation. Selenocysteine thus is commonly referred to as the "21st amino acid." However, unlike other 20 genetically encoded amino acids, selenocysteine is specified by a much larger and more complex genetic element than the basic three-letter codon (see Chapter 27).

Only L-α-Amino Acids Occur in Proteins

With the sole exception of glycine, the α-carbon of every amino acid is chiral. Although some protein amino acids are dextrorotatory and some levorotatory, all share the absolute configuration of L-glyceraldehyde and thus are defined as L-α-amino acids. Several free L-α-amino acids fulfill important roles in metabolic processes. Examples include ornithine, citrulline, and argininosuccinate that participate in urea synthesis, tyrosine in formation of thyroid hormones, and glutamate in neurotransmitter biosynthesis. D-amino acids that occur naturally include free D-serine and D-aspartate in brain tissue, D-alanine and D-glutamate in the cell walls of gram-positive bacteria, and D-amino acids in certain peptides and antibiotics produced by bacteria, fungi, reptiles, and other nonmammalian species.

Amino Acids May Have Positive, Negative, or Zero Net Charge

Charged and uncharged forms of the ionizable —COOH and —NH_3^+ weak acid groups that exist in solution in protonic equilibrium:

$$R-COOH \rightleftharpoons R-COO^- + H^+$$

$$R-NH_3^+ \rightleftharpoons R-NH_2 + H^+$$

While both R—COOH and R—NH_3^+ are weak acids, R—COOH is a far stronger acid than R—NH_3^+. Thus, at physiologic pH (pH 7.4), carboxyl groups exist almost entirely as R—COO⁻ and amino groups predominantly as R—NH_3^+.

Figure 3–1 illustrates the effect of pH on the charged state of aspartic acid.

Molecules that contain an equal number of ionizable groups of opposite charge and that therefore bear no *net* charge are termed **zwitterions**. Amino acids in blood and most tissues thus should be represented as in **A,** below.

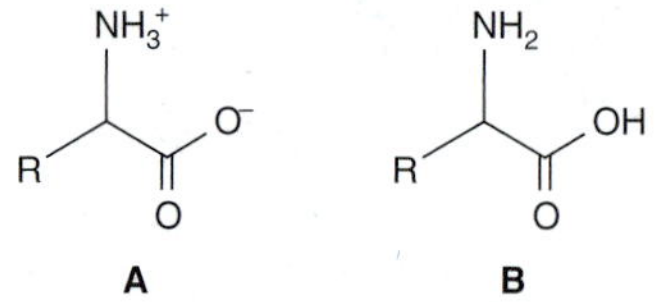

Structure **B** cannot exist in an aqueous solution because at any pH low enough to protonate the carboxyl group, the amino group would also be protonated. Similarly, at any pH sufficiently high for an uncharged amino group to predominate, a carboxyl group will be present as R—COO⁻. The uncharged representation **B** is, however, often used for reactions that do not involve protonic equilibria.

pK_a Values Express the Strengths of Weak Acids

The acid strengths of weak acids are expressed as their pK_a. For molecules with multiple dissociable protons, the pK_a for each acidic group is designated by replacing the subscript "a" with a number (Table 3–1). The imidazole group of histidine and the guanidino group of arginine exist as resonance hybrids with positive charge distributed between both nitrogens (histidine) or all three nitrogens (arginine) (**Figure 3–2**). The net charge on an amino acid—the algebraic sum of all the positively and negatively charged groups present—depends upon the pK_a values of its functional groups and on the pH of the surrounding medium. Altering the charge on amino acids and their derivatives by varying the pH facilitates the physical separation of amino acids, peptides, and proteins (see Chapter 4).

At its Isoelectric pH (pI), an Amino Acid Bears No Net Charge

Zwitterions are one example of an **isoelectric** species—the form of a molecule that has an equal number of positive and

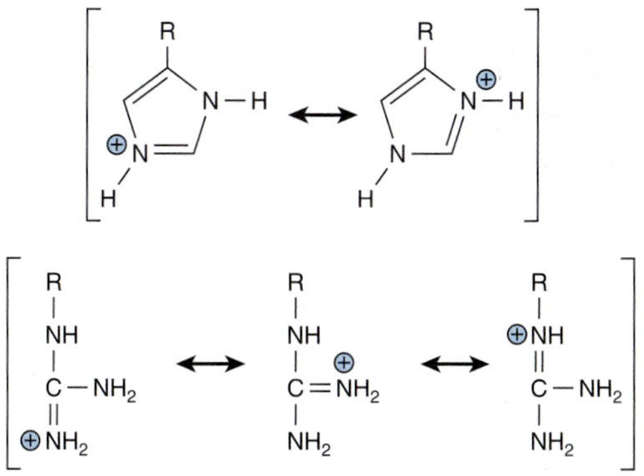

FIGURE 3–2 **Resonance hybrids of the protonated forms of the R groups of histidine and arginine.**

TABLE 3–2 **Typical Range of pK_a Typical Range of pK_a Values for Ionizable Groups in Proteins**

Dissociating Group	pK_a Range
α-Carboxyl	3.5–4.0
Non-α COOH of Asp or Glu	4.0–4.8
Imidazole of His	6.5–7.4
SH of Cys	8.5–9.0
OH of Tyr	9.5–10.5
α-Amino	8.0–9.0
ε-Amino of Lys	9.8–10.4
Guanidinium of Arg	~12.0

negative charges and thus is electrically neutral. **The isoelectric pH, also called the pI, is the pH midway between pK_a values for the ionizations on either side of the isoelectric species.** For an amino acid such as alanine that has only two dissociating groups, there is no ambiguity. The first pK_a (R—COOH) is 2.35 and the second pK_a (R—NH$_3^+$) is 9.69. The isoelectric pH (pI) of alanine thus is

$$pI = \frac{pK_1 + pK_2}{2} = \frac{2.35 + 9.69}{2} = 6.02$$

For polyprotic acids, pI is also the pH midway between the pK_a values on either side of the isoionic species. For example, the pI for aspartic acid is

$$pI = \frac{pK_1 + pK_2}{2} = \frac{2.09 + 3.96}{2} = 3.02$$

For lysine, pI is calculated from

$$pI = \frac{pK_2 + pK_3}{2}$$

Similar considerations apply to all polyprotic acids (eg, proteins), regardless of the number of dissociating groups present. In the clinical laboratory, knowledge of the pI guides selection of conditions for electrophoretic separations. For example, electrophoresis at pH 7.0 will separate two molecules with pI values of 6.0 and 8.0, because at pH 7.0 the molecule with a pI of 6.0 will have a net positive charge, and that with a pI of 8.0 a net negative charge. Similar considerations apply to understanding chromatographic separations on ionic supports such as diethylaminoethyl (DEAE) cellulose (see Chapter 4).

pK_a Values Vary with the Environment

The environment of a dissociable group affects its pK_a. The pK_a values of the R groups of free amino acids in an aqueous solution (Table 3–1) thus provide only an approximate guide to the pK_a values of the same amino acids when present in proteins. A polar environment favors the charged form (R—COO⁻ or R—NH$_3^+$), and a nonpolar environment favors the uncharged form (R—COOH or R—NH$_2$). A nonpolar environment thus *raises* the pK_a of a carboxyl group (making it a weaker acid) but *lowers* that of an amino group (making it a stronger acid). The presence of adjacent charged groups can reinforce or counteract solvent effects. The pK_a of a functional group thus will depend upon its location within a given protein. Variations in pK_a can encompass whole pH units (**Table 3–2**). pK_a values that diverge from those listed by as much as 3 pH units are common at the active sites of enzymes. An extreme example, a buried aspartic acid of thioredoxin, has a pK_a above 9—a shift of more than 6 pH units.

The Solubility of Amino Acids Reflects their Ionic Character

The charges conferred by the dissociable functional groups of amino acids ensure that they are readily solvated by—and thus soluble in—polar solvents such as water and ethanol but insoluble in nonpolar solvents such as benzene, hexane, or ether.

Amino acids do not absorb visible light and thus are colorless. However, tyrosine, phenylalanine, and especially tryptophan absorb high-wavelength (250–290 nm) ultraviolet light. Because it absorbs ultraviolet light about ten times more efficiently than phenylalanine or tyrosine, tryptophan makes the major contribution to the ability of most proteins to absorb light in the region of 280 nm.

THE α-R GROUPS DETERMINE THE PROPERTIES OF AMINO ACIDS

Since glycine, the smallest amino acid, can be accommodated in places inaccessible to other amino acids, it often occurs where peptides bend sharply. The hydrophobic R groups of alanine, valine, leucine, and isoleucine and the aromatic R groups of phenylalanine, tyrosine, and tryptophan typically occur primarily in the interior of cytosolic proteins. The charged

R groups of basic and acidic amino acids stabilize specific protein conformations via ionic interactions or salt bridges. These interactions also function in "charge relay" systems during enzymatic catalysis and electron transport in respiring mitochondria. Histidine plays unique roles in enzymatic catalysis. The pK_a of its imidazole proton permits histidine to function at neutral pH as either a base or an acid catalyst without the need for any environmentally induced shift. The primary alcohol group of serine and the primary thioalcohol (—SH) group of cysteine are excellent nucleophiles, and can function as such during enzymatic catalysis. However, the secondary alcohol group of threonine, while a good nucleophile, is not known to fulfill an analogous role in catalysis. The —OH groups of serine, tyrosine, and threonine also participate in regulation of the activity of enzymes whose catalytic activity depends on the phosphorylation state of these residues.

FUNCTIONAL GROUPS DICTATE THE CHEMICAL REACTIONS OF AMINO ACIDS

Each functional group of an amino acid exhibits all of its characteristic chemical reactions. For carboxylic acid groups, these reactions include the formation of esters, amides, and acid anhydrides; for amino groups, acylation, amidation, and esterification; and for —OH and —SH groups, oxidation, and esterification. The most important reaction of amino acids is the formation of a peptide bond (shaded).

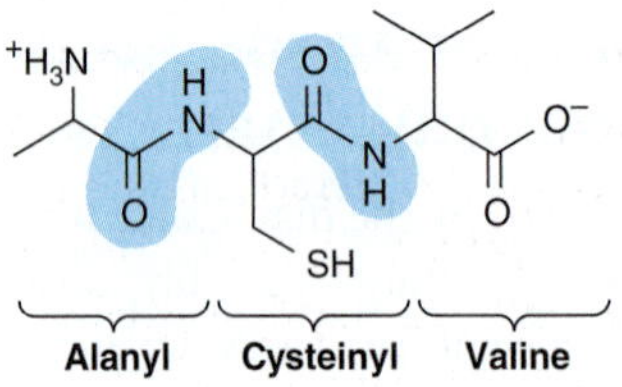

| Alanyl | Cysteinyl | Valine |

Amino Acid Sequence Determines Primary Structure

The number and order of all the amino acid residues in a polypeptide constitute its primary structure. Amino acids present in peptides are called aminoacyl residues and are named by replacing the *-ate* or *-ine* suffixes of free amino acids with *-yl* (eg, alan*yl*, aspart*yl*, tyros*yl*). Peptides are then named as derivatives of the *carboxy* terminal aminoacyl residue. For example, Lys-Leu-Tyr-Gln is called lys*yl*-leuc*yl*-tyros*yl*-glutam*ine*. The *-ine* ending on glutamine indicates that its α-carboxyl group is *not* involved in peptide bond formation.

Peptide Structures Are Easy to Draw

Prefixes such as *tri-* or *octa-* denote peptides with three or eight *residues*, respectively. By convention, peptides are written with the residue that bears the free α-amino group on the left. To draw a peptide, use a zigzag to represent the main chain or backbone. Add the main chain atoms, which occur in the repeating order: α-nitrogen, α-carbon, and carbonyl carbon. Now add a hydrogen atom to each α-carbon and to each peptide nitrogen, and an oxygen atom to the carbonyl carbon. Finally, add the appropriate R groups (shaded) to each α-carbon atom.

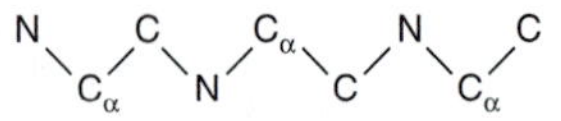

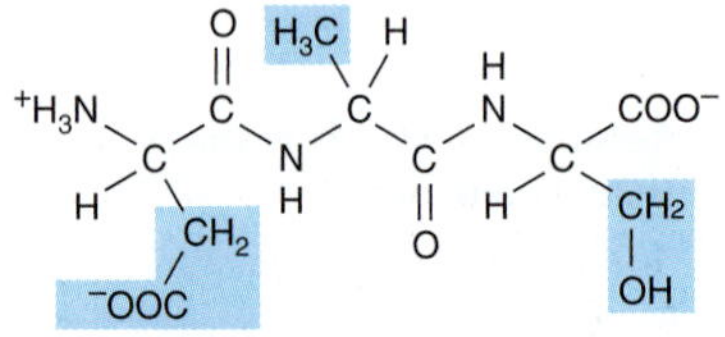

Three-letter abbreviations linked by straight lines represent an unambiguous primary structure. Lines are omitted for single-letter abbreviations.

Glu - Ala - Lys - Gly - Tyr - Ala
  E   A   K   G   Y   A

Some Peptides Contain Unusual Amino Acids

In mammals, peptide hormones typically contain only the 20 genetically encoded α-amino acids linked by standard peptide bonds. Other peptides may, however, contain nonprotein amino acids, derivatives of the protein amino acids, or amino acids linked by an atypical peptide bond. For example, the amino terminal glutamate of glutathione, a tripeptide that participates in protein folding and in the metabolism of xenobiotics (Chapter 53), is linked to cysteine by a non-α peptide bond (**Figure 3–3**). The amino terminal glutamate of thyrotropin-releasing hormone (TRH) is cyclized to pyroglutamic acid, and the carboxyl group of the carboxyl terminal prolyl residue is amidated. The nonprotein amino acids D-phenylalanine and ornithine are present in the cyclic peptide antibiotics tyrocidin and gramicidin S, while the heptapeptide opioids dermorphin and deltophorin in the skin of South American tree frogs contain D-tyrosine and D-alanine.

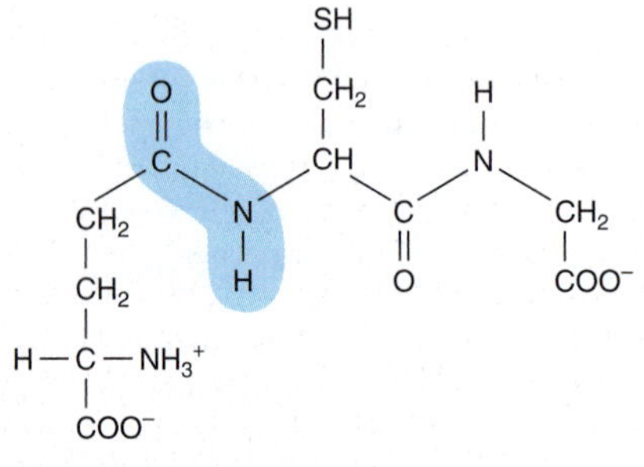

FIGURE 3–3 Glutathione (γ-glutamyl-cysteinyl-glycine). Note the non-α peptide bond that links Glu to Cys.

Peptides are Polyelectrolytes

The peptide bond is uncharged at any pH of physiologic interest. Formation of peptides from amino acids is therefore accompanied by a net loss of one positive and one negative charge per peptide bond formed. Peptides nevertheless are charged at physiologic pH owing to their terminal carboxyl and amino groups and, where present, their acidic or basic R groups. As for amino acids, the net charge on a peptide depends on the pH of its environment and on the pK_a values of its dissociating groups.

The Peptide Bond Has a Partial Double-Bond Character

Although peptides are written as if a single bond linked the α-carboxyl and α-nitrogen atoms, this bond in fact exhibits a partial double-bond character:

There thus is no freedom of rotation about the bond that connects the carbonyl carbon and the nitrogen of a peptide bond. Consequently, the O, C, N, and H atoms of a peptide bond are coplanar. The imposed semirigidity of the peptide bond has important consequences for the manner in which peptides and proteins fold to generate higher order of structure. Encircling brown arrows (**Figure 3–4**) indicate free rotation about the remaining bonds of the polypeptide backbone.

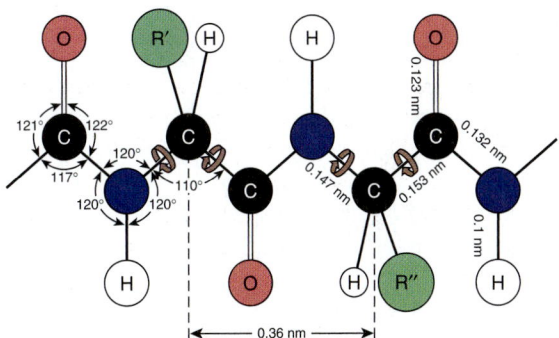

FIGURE 3–4 Dimensions of a fully extended polypeptide chain. The four atoms of the peptide bond are coplanar. Free rotation can occur about the bonds that connect the α-carbon with the α-nitrogen and with the α-carbonyl carbon (brown arrows). The extended polypeptide chain is thus a semirigid structure with two-thirds of the atoms of the backbone held in a fixed planar relationship one to another. The distance between adjacent α-carbon atoms is 0.36 nm (3.6 Å). The interatomic distances and bond angles, which are not equivalent, are also shown. (Redrawn and reproduced, with permission, from Pauling L, Corey LP, Branson HR: The structure of proteins: Two hydrogen-bonded helical configurations of the polypeptide chain. Proc Natl Acad Sci USA 1951;37:205.)

Noncovalent Forces Constrain Peptide Conformations

Folding of a peptide probably occurs coincident with its biosynthesis (see Chapter 37). The physiologically active conformation reflects the collective contributions of the amino acid sequence, steric hindrance, and noncovalent interactions (eg, hydrogen bonding, hydrophobic interactions) between residues. Common conformations include α-helices and β-pleated sheets (see Chapter 5).

ANALYSIS OF THE AMINO ACID CONTENT OF BIOLOGIC MATERIALS

To determine the identity of each amino acid present in a protein, it is first treated with hot hydrochloric acid to hydrolyze the peptide bonds. There are several methods for separation and identification of amino acids derived from a protein hydrolysate or from urine or other biologic fluids. One approach is to react the amino acids with 6-amino-N-hydroxysuccinimidyl carbamate to form fluorescent derivatives that can be separated by high-pressure liquid chromatography (see Chapter 4). An alternative approach, which requires only minimal equipment, employs partition chromatography on a solid support, typically a sheet of filter paper (paper chromatography) or a thin layer of powdered cellulose or silica gel on an inert support (thin-layer chromatography, or TLC). The amino acids present are resolved by a mobile phase that contains a mixture of miscible polar and nonpolar components (eg, n-butanol, formic acid, and water). As the mobile phase moves up the sheet, its polar components associate with the polar groups of the support. The solvent therefore becomes progressively less polar as it migrates up the sheet. The amino acids therefore partition between a polar stationary phase and a less polar mobile phase ("partition chromatography"). Nonpolar amino acids (eg, Leu, Ile) migrate the farthest as they spend the greatest proportion of their time in the mobile phase. Polar amino acids (eg, Glu, Lys) travel the least distance from the origin as they spend a high proportion of their time in the stationary phase consisting of a layer of polar solvent molecules immobilized by their association with the cellulose or silica support. Following removal of the solvent by air drying, amino acids are visualized using ninhydrin, which forms purple products with α-amino acids, but a yellow adduct with proline and hydroxyproline.

SUMMARY

- Both D-amino acids and non-α-amino acids occur in nature, but only L-α-amino acids are present in proteins.
- All amino acids possess at least two weakly acidic functional groups, R—NH$_3^+$ and R—COOH. Many also possess additional

weakly acidic functional groups such as —OH, —SH, guanidino, or imidazole moieties.

- The pK_a values of all functional groups of an amino acid dictate its net charge at a given pH. pI is the pH at which an amino acid bears no net charge and thus does not move in a direct current electrical field.

- Of the biochemical reactions of amino acids, the most important is the formation of peptide bonds.

- The R groups of amino acids determine their unique biochemical functions. Amino acids are classified as basic, acidic, aromatic, aliphatic, or sulfur containing based on the properties of their R groups.

- Peptides are named for the number of amino acid residues present, and as derivatives of the carboxyl terminal residue. The primary structure of a peptide is its amino acid sequence, starting from the amino-terminal residue.

- The partial double-bond character of the bond that links the carbonyl carbon and the nitrogen of a peptide renders four atoms of the peptide bond coplanar and restricts the number of possible peptide conformations.

REFERENCES

Doolittle RF: Reconstructing history with amino acid sequences. Protein Sci 1992;1:191.

Gladyschev VN, Hatfield DL: Selenocysteine-containing proteins in mammals. J Biomed Sci 1999;6:151.

Kolodkin-Gal I: D-Amino acids trigger biofilm disassembly. Science 2010;328:627.

Kreil G: D-Amino acids in animal peptides. Annu Rev Biochem 1997;66:337.

Nokihara K, Gerhardt J: Development of an improved automated gas-chromatographic chiral analysis system: application to nonnatural amino acids and natural protein hydrolysates. Chirality 2001;13:431.

Papp LV: From selenium to selenoproteins: synthesis, identity, and their role in human health. Antioxidants Redox Signal 2007;9:775.

Sanger F: Sequences, sequences, and sequences. Annu Rev Biochem 1988;57:1.

Stadtman TC: Selenocysteine. Annu Rev Biochem 1996;65:83.

Wilson NA, Barbar E, Fuchs JA, et al: Aspartic acid 26 in reduced *Escherichia coli* thioredoxin has a pK_a greater than 9. Biochemistry 1995;34:8931.

Proteins: Determination of Primary Structure

Peter J. Kennelly, PhD & Victor W. Rodwell, PhD

- Describe multiple chromatographic methods commonly employed for the isolation of proteins from biologic materials.
- Explain how scientists analyze the sequence or structure of a protein to extract insights into its possible physiologic function.
- List several of the posttranslational alterations that proteins undergo during their lifetime and the influence of such modifications upon a protein's function and fate.
- Describe the chemical basis of the Edman method for determining primary structure.
- Give three reasons why mass spectrometry (MS) has largely supplanted chemical methods for the determination of the primary structure of proteins and the detection of posttranslational modifications.
- Explain why MS can detect posttranslational modifications that are not detected by Edman sequencing or DNA sequencing.
- Describe how DNA cloning and molecular biology made the determination of the primary structures of proteins much more rapid and efficient.
- Explain what is meant by "the proteome" and cite examples of its ultimate potential significance.
- Comment on the contributions of genomics, computer algorithms, and databases to the identification of the open reading frames (ORFs) that encode a given protein.

BIOMEDICAL IMPORTANCE

Proteins are physically and functionally complex macromolecules that perform multiple critically important roles. For example, an internal protein network, the cytoskeleton (Chapter 49) maintains cellular shape and physical integrity. Actin and myosin filaments form the contractile machinery of muscle (Chapter 49). Hemoglobin transports oxygen (Chapter 6), while circulating antibodies defend against foreign invaders (Chapter 50). Enzymes catalyze reactions that generate energy, synthesize and degrade biomolecules, replicate and transcribe genes, process mRNAs, etc (Chapter 7). Receptors enable cells to sense and respond to hormones and other environmental cues (Chapters 41 and 42). Proteins are subject to physical and functional changes that mirror the life cycle of the organisms in which they reside. A typical protein is "born" at translation (Chapter 37), matures through posttranslational processing events such as selective proteolysis (Chapters 9 and 37), alternates between working and resting states through the intervention of regulatory factors (Chapter 9), ages through oxidation, deamidation, etc (Chapter 52), and "dies" when degraded to its component amino acids (Chapter 29). An important goal of molecular medicine is to identify biomarkers such as proteins and/or modifications to proteins whose presence, absence, or deficiency is associated with specific physiologic states or diseases (**Figure 4–1**).

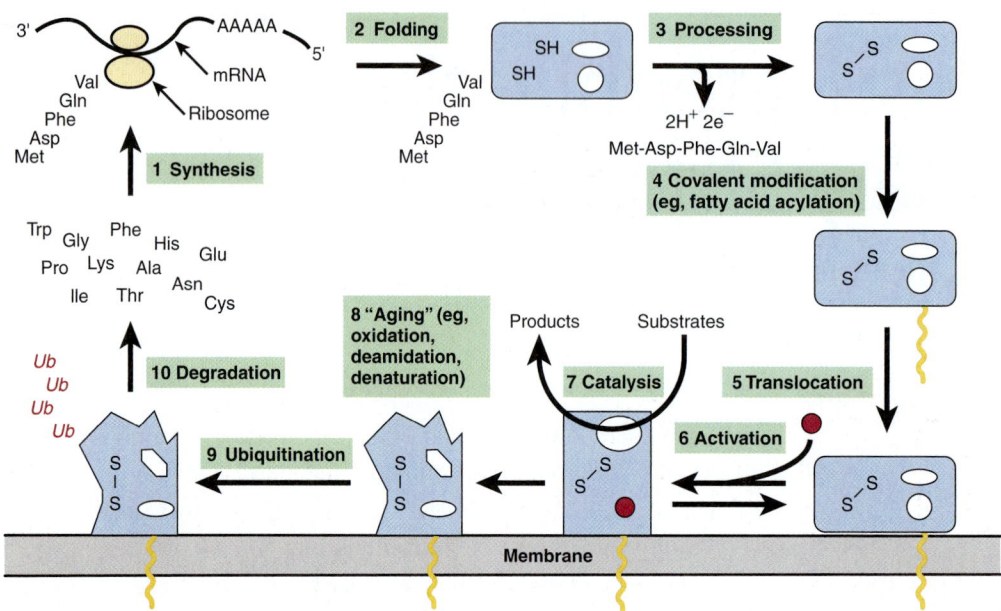

FIGURE 4–1 **Diagrammatic representation of the life cycle of a hypothetical protein.** (1) The life cycle begins with the synthesis on a ribosome of a polypeptide chain, whose primary structure is dictated by an mRNA. (2) As synthesis proceeds, the polypeptide begins to fold into its native conformation (blue). (3) Folding may be accompanied by processing events such as proteolytic cleavage of an *N*-terminal leader sequence (Met-Asp-Phe-Gln-Val) or the formation of disulfide bonds (S—S). (4) Subsequent covalent modifications may, for example, attach a fatty acid molecule (yellow) for (5) translocation of the modified protein to a membrane. (6) Binding an allosteric effector (red) may trigger the adoption of a catalytically active conformation. (7) Over time, proteins get damaged by chemical attack, deamidation, or denaturation, and (8) may be "labeled" by the covalent attachment of several ubiquitin molecules (*Ub*). (9) The ubiquitinated protein is subsequently degraded to its component amino acids, which become available for the synthesis of new proteins.

PROTEINS & PEPTIDES MUST BE PURIFIED PRIOR TO ANALYSIS

Highly purified protein is essential for the detailed examination of its physical and functional properties. Cells contain thousands of different proteins, each in widely varying amounts. The isolation of a specific protein in quantities sufficient for analysis of its properties thus presents a formidable challenge that may require successive application of multiple purification techniques. Selective precipitation exploits differences in relative solubility of individual proteins as a function of pH (isoelectric precipitation), polarity (precipitation with ethanol or acetone), or salt concentration (salting out with ammonium sulfate). Chromatographic techniques separate one protein from another based upon difference in their size (size exclusion chromatography), charge (ion-exchange chromatography), hydrophobicity (hydrophobic interaction chromatography), or ability to bind a specific ligand (affinity chromatography).

Column Chromatography

In column chromatography, the stationary phase matrix consists of small beads loaded into a cylindrical container of glass, plastic, or steel called a column. Liquid-permeable frits confine the beads within this space while allowing the mobile-phase liquid to flow or percolate through the column. The stationary phase beads can be chemically derivatized to coat their surface with the acidic, basic, hydrophobic, or ligand-like groups required for ion exchange, hydrophobic interaction, or affinity chromatography. As the mobile-phase liquid emerges from the column, it is automatically collected in a series of small portions called fractions. **Figure 4–2** depicts the basic arrangement of a simple bench-top chromatography system.

HPLC—High-Pressure Liquid Chromatography

First-generation column chromatography matrices consisted of long, intertwined oligosaccharide polymers shaped into spherical beads roughly a tenth of a millimeter in diameter. Unfortunately, their relatively large size perturbed mobile-phase flow and limited the available surface area. Reducing particle size offered the potential to greatly increase resolution. However, the resistance created by the more tightly packed matrix required the use of very high pressures that would crush the soft and spongy polysaccharide beads and similar materials, eg, acrylamide. Eventually, methods were developed to manufacture silicon particles of the necessary size and shape,

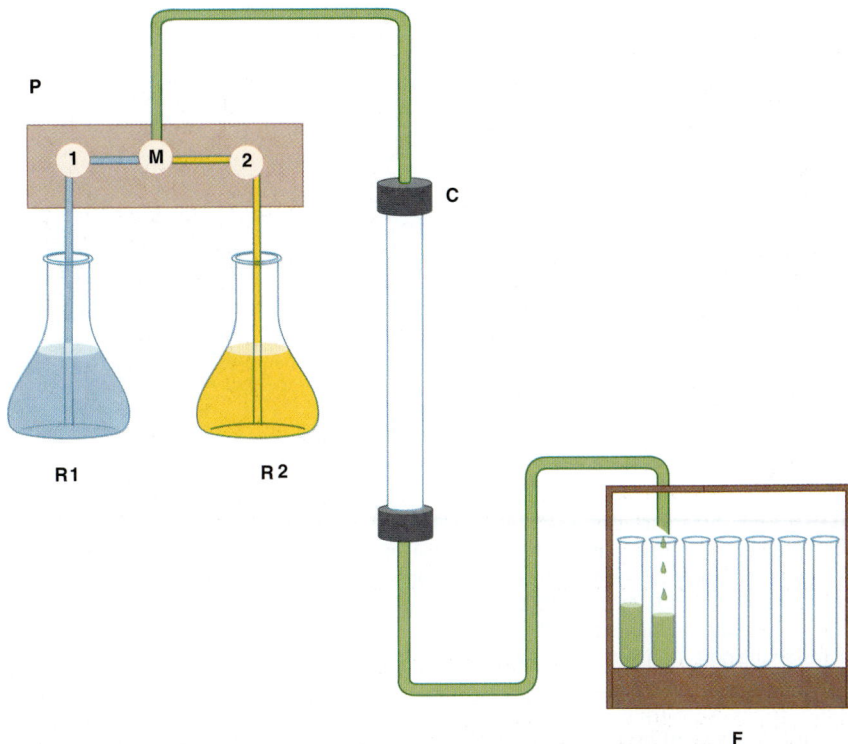

FIGURE 4–2 Components of a typical liquid chromatography apparatus. R1 and R2: Reservoirs of mobile-phase liquid. P: Programable pumping system containing two pumps, 1 and 2, and a mixing chamber, M. The system can be set to pump liquid from only one reservoir, to switch reservoirs at some predetermined point to generate a step gradient, or to mix liquids from the two reservoirs in proportions that vary over time to create a continuous gradient. C: Glass, metal, or plastic column containing stationary phase. F: Fraction collector for collecting portions, called *fractions,* of the eluant liquid in separate test tubes.

to derivatize their surface with various functional groups, and to pack them into stainless steel columns capable of withstanding pressures of several thousand psi. Because of their greater resolving power, high-pressure liquid chromatography systems have largely displaced the once familiar glass columns in the protein purification laboratory.

Size-exclusion Chromatography

Size exclusion—or gel filtration—chromatography separates proteins based on their **Stokes radius;** the radius of the sphere they occupy as they tumble in solution. The Stokes radius is a function of molecular mass and shape. A tumbling elongated protein occupies a larger volume than a spherical protein of the same mass. Size-exclusion chromatography employs porous beads (**Figure 4–3**). The pores are analogous to indentations in a riverbank. As objects move downstream, those that enter an indentation are retarded until they drift back into the main current. Similarly, proteins with Stokes radii too large to enter the pores (excluded proteins), remain in the flowing mobile phase, and emerge *before* proteins that can enter the pores (included proteins). Proteins thus emerge from a gel filtration column in descending order of their Stokes radii.

Ion-Exchange Chromatography

In ion-exchange chromatography, proteins interact with the stationary phase by charge–charge interactions. Proteins with a net positive charge at a given pH will tightly adhere to beads with negatively charged functional groups such as carboxylates or sulfates (cation exchangers). Similarly, proteins with a net negative charge adhere to beads with positively charged functional groups, typically tertiary, or quaternary amines (anion exchangers). Nonadherent proteins flow through the matrix and are washed away. Bound proteins are then selectively displaced by gradually raising the ionic strength of the mobile phase, thereby weakening charge–charge interactions. Proteins elute in inverse order of the strength of their interactions with the stationary phase.

Hydrophobic Interaction Chromatography

Hydrophobic interaction chromatography separates proteins based on their tendency to associate with a stationary phase matrix coated with hydrophobic groups (eg, phenyl Sepharose, octyl Sephadex). Proteins with exposed hydrophobic surfaces adhere to the matrix via hydrophobic interactions that are

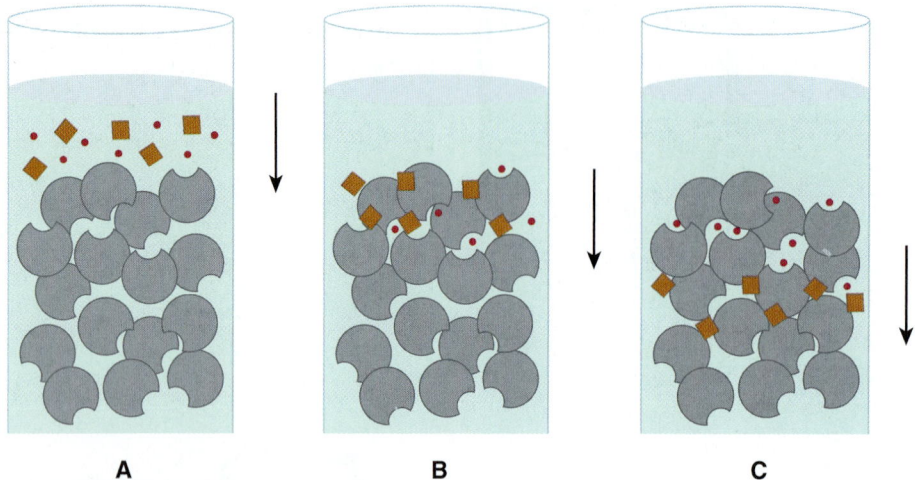

FIGURE 4–3 **Size-exclusion chromatography.** A: A mixture of large molecules (brown) and small molecules (red) are applied to the top of a gel filtration column. B: Upon entering the column, the small molecules enter pores in the stationary phase matrix (gray) from which the large molecules are excluded. C: As the mobile phase (blue) flows down the column, the large, excluded molecules flow with it, while the small molecules, which are temporarily sheltered from the flow when inside the pores, lag farther and farther behind.

enhanced by employing a mobile phase of high ionic strength. After nonadherent proteins are washed away, the polarity of the mobile phase is decreased by gradually lowering the salt concentration of the flowing mobile phase. If the interaction between protein and stationary phase is particularly strong, ethanol or glycerol may be added to the mobile phase to decrease its polarity and further weaken hydrophobic interactions.

Affinity Chromatography

Affinity chromatography exploits the high selectivity of most proteins for their ligands. Enzymes may be purified by affinity chromatography using immobilized substrates, products, coenzymes, or inhibitors. In theory, only proteins that interact with the immobilized ligand adhere. Bound proteins are then eluted either by competition with free, soluble ligand or, less selectively, by disrupting protein–ligand interactions using urea, guanidine hydrochloride, mildly acidic pH, or high salt concentrations. Commercially available stationary phase matrices contain ligands such as NAD^+ or ATP analogs. Purification of recombinantly expressed proteins is frequently facilitated by modifying the cloned gene to add a new fusion domain designed to interact with a specific matrix-bound ligand (Chapter 7).

Protein Purity is Assessed by Polyacrylamide Gel Electrophoresis (PAGE)

The most widely used method for determining the purity of a protein is SDS-PAGE—polyacrylamide gel electrophoresis (PAGE) in the presence of the anionic detergent sodium dodecyl sulfate (SDS). Electrophoresis separates charged biomolecules based on the rates at which they migrate in an applied electrical field. For SDS-PAGE, acrylamide is polymerized and cross-linked to form a porous matrix. SDS binds to proteins at a ratio of one molecule of SDS per two peptide bonds, causing the polypeptide to unfold or denature. When used in conjunction with 2-mercaptoethanol or dithiothreitol to reduce and break disulfide bonds (**Figure 4–4**), SDS-PAGE separates the component polypeptides of multimeric proteins. The large number of anionic SDS molecules, each bearing a charge of –1, overwhelms the charge contributions of the amino acid functional

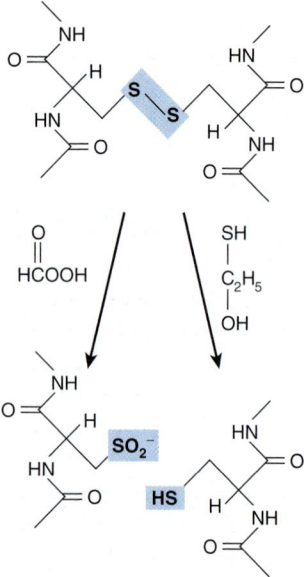

FIGURE 4–4 **Oxidative cleavage of adjacent polypeptide chains linked by disulfide bonds (highlighted in blue) by performic acid (*left*) or reductive cleavage by β-mercaptoethanol (*right*) forms two peptides that contain cysteic acid residues or cysteinyl residues, respectively.**

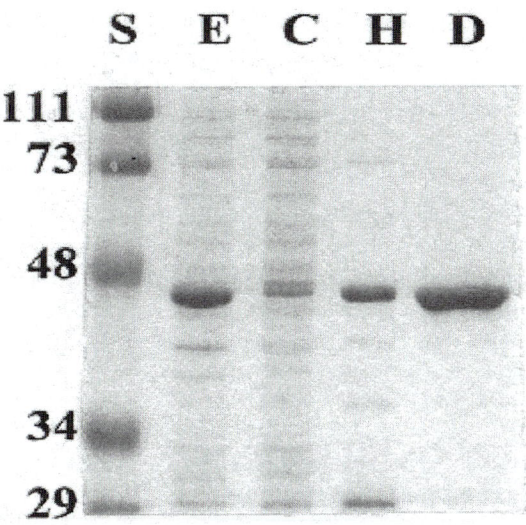

FIGURE 4–5 **Use of SDS-PAGE to observe successive purification of a recombinant protein.** The gel was stained with Coomassie blue. Shown are protein standards (lane S) of the indicated M_r, in kDa, crude cell extract (E), cytosol (C), high-speed supernatant liquid (H), and the DEAE-Sepharose fraction (D). The recombinant protein has a mass of about 45 kDa.

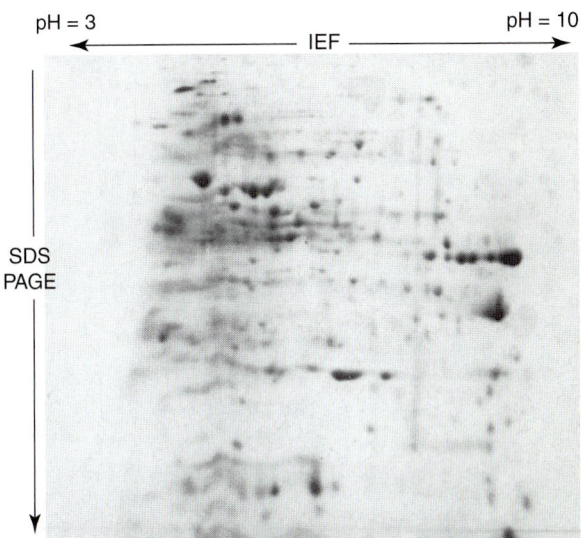

FIGURE 4–6 **Two-dimensional IEF-SDS-PAGE.** The gel was stained with Coomassie blue. A crude bacterial extract was first subjected to isoelectric focusing (IEF) in a pH 3–10 gradient. The IEF gel was then placed horizontally on the top of an SDS-PAGE gel, and the proteins then further resolved by SDS-PAGE. Notice the greatly improved resolution of distinct polypeptides relative to ordinary SDS-PAGE gel (Figure 4–5).

groups endogenous to the polypeptides. Since the charge-to-mass ratio of each SDS-polypeptide complex is approximately equal, the physical resistance each peptide encounters as it moves through the acrylamide matrix determines the rate of migration. Since large complexes encounter greater resistance, polypeptides separate based on their relative molecular mass (M_r). Individual polypeptides trapped in the acrylamide gel after removal of the electrical field are visualized by staining with dyes such as Coomassie blue (**Figure 4–5**).

Isoelectric Focusing (IEF)

Ionic buffers called ampholytes and an applied electric field are used to generate a pH gradient within a polyacrylamide matrix. Applied proteins migrate until they reach the region of the matrix where the pH matches their isoelectric point (pI), the pH at which a molecule's net charge is 0. IEF is used in conjunction with SDS-PAGE for two-dimensional electrophoresis, which separates polypeptides based on pI in one dimension and on M_r in the second (**Figure 4–6**). Two-dimensional electrophoresis is particularly well suited for separating the components of complex mixtures of proteins.

SANGER WAS THE FIRST TO DETERMINE THE SEQUENCE OF A POLYPEPTIDE

Mature insulin consists of the 21-residue A chain and the 30-residue B chain linked by disulfide bonds. Frederick Sanger reduced the disulfide bonds (Figure 4–4), separated the A and

B chains, and cleaved each chain into smaller peptides using trypsin, chymotrypsin, and pepsin. The resulting peptides were then isolated and treated with acid to hydrolyze a portion of the peptide bonds and generate peptides with as few as two or three amino acids. Each peptide was reacted with 1-fluoro-2,4-dinitrobenzene (Sanger's reagent), which derivatizes the exposed α-amino groups of the amino-terminal residues. The amino acid content of each peptide was then determined and the amino-terminal amino acid identified. The ε-amino group of lysine also reacts with Sanger's reagent; but since an amino-terminal lysine reacts with 2 mol of Sanger's reagent, it is readily distinguished from a lysine in the interior of a peptide. Working from di- and tripeptides up through progressively larger fragments, Sanger was able to reconstruct the complete sequence of insulin, an accomplishment for which he received a Nobel Prize, in 1958.

THE EDMAN REACTION ENABLES PEPTIDES & PROTEINS TO BE SEQUENCED

Pehr Edman introduced phenylisothiocyanate (Edman's reagent) to selectively label the amino-terminal residue of a peptide. In contrast to Sanger's reagent, the phenylthiohydantoin (PTH) derivative can be removed under mild conditions to generate a new amino-terminal residue (**Figure 4–7**). Successive rounds of derivatization with Edman's reagent can therefore be used to sequence many residues of a single sample of peptide. Even so, the determination of the complete

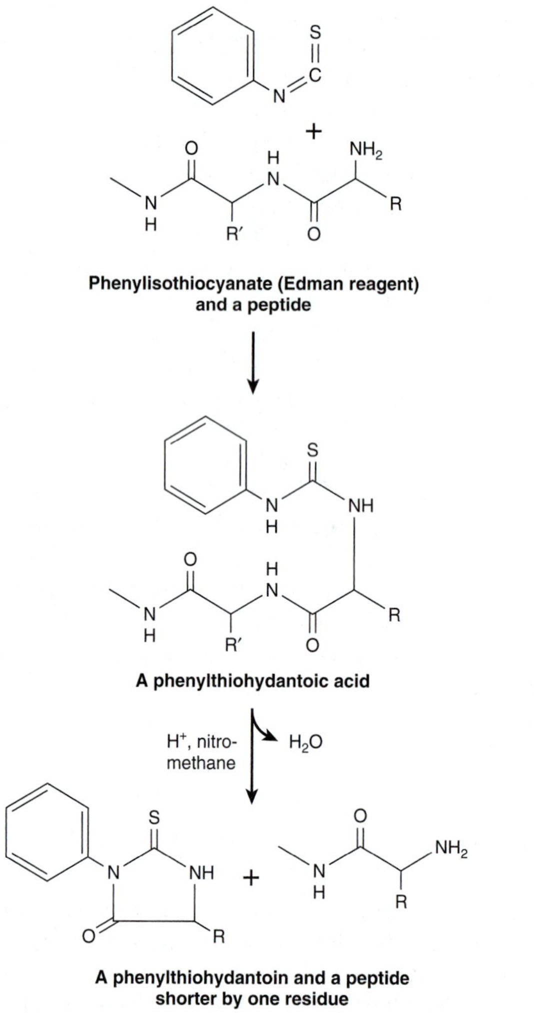

**Phenylisothiocyanate (Edman reagent)
and a peptide**

A phenylthiohydantoic acid

H+, nitro-
methane   H₂O

**A phenylthiohydantoin and a peptide
shorter by one residue**

FIGURE 4–7 **The Edman reaction.** Phenylisothiocyanate derivatizes the amino-terminal residue of a peptide as a phenylthiohydantoic acid. Treatment with acid in a nonhydroxylic solvent releases a phenylthiohydantoin, which is subsequently identified by its chromatographic mobility, and a peptide one residue shorter. The process is then repeated.

sequence of a protein by chemical methods remains a time- and labor-intensive process to this day.

The heterogeneous chemical properties of the amino acids meant that every step in the procedure represented a compromise between efficiency for any particular amino acid or set of amino acids and the flexibility needed to accommodate all 20. Consequently, each step in the process operates at less than 100% efficiency, which leads to the accumulation of polypeptide fragments with varying N-termini. Eventually, it becomes impossible to distinguish the correct PTH amino acid for that position in the peptide from the contaminants. As a result, the read length for Edman sequencing varies from 5 to 30 amino acid residues depending upon the quantity and purity of the peptide.

In order to determine the complete sequence of a polypeptide several hundred residues in length, a protein must first be cleaved into smaller peptides, using either a protease or a reagent such as cyanogen bromide. Following purification by reversed phase high-pressure liquid chromatography (HPLC), these peptides are then analyzed by Edman sequencing. In order to assemble these short peptide sequences to solve the complete sequence of the intact polypeptide, it is necessary to analyze peptides whose sequences overlap one another. This is accomplished by generating multiple sets of peptides using more than one method of cleavage. The large quantities of purified protein required to test multiple protein fragmentation and peptide purification conditions constitutes the second major drawback of direct chemical protein sequencing techniques.

MOLECULAR BIOLOGY REVOLUTIONIZED THE DETERMINATION OF PRIMARY STRUCTURE

The reactions that sequentially derivatize and cleave PTH amino acids from the amino-terminal end of a peptide typically are conducted in an automated sequenator. DNA sequencing, by contrast, is both far more rapid and more economical. Recombinant techniques permit researchers to manufacture a virtually infinite supply of DNA using the original sample as template (Chapter 39). DNA sequencing methods, whose chemistry was also developed by Sanger, routinely enable polydeoxyribonucleotide sequences a few hundred residues in length to be determined in a single analysis, while automated sequencers can "read" sequences several thousand nucleotides in length. Knowledge of the genetic code enables the sequence of the encoded polypeptide to be determined by simply translating the oligonucleotide sequence of its gene. Conversely, early molecular biologists designed complementary oligonucleotide probes to identify the DNA clone containing the gene of interest by reversing this process and using a segment of chemically determined amino acid sequence as template. The advent of DNA cloning thus ushered in the widespread use of a hybrid approach in which Edman chemistry was employed to sequence a small portion of the protein, then exploiting this information to determine the remaining sequence by DNA cloning and sequencing.

GENOMICS ENABLES PROTEINS TO BE IDENTIFIED FROM SMALL AMOUNTS OF SEQUENCE DATA

Today the number of organisms for which the complete DNA sequence of their genomes has been determined and made available to the scientific community numbers in the hundreds (see Chapter 10). These sequences encompass nearly all of the "model organisms" commonly employed in biomedical research laboratories: *Homo sapiens*, mouse, rat, *Escherichia coli*, *Drosophila melanogaster*, *Caenorhabditis elegans*, yeast, etc, as well as numerous pathogens. Meanwhile, across the globe, arrays of automated DNA sequenators continue to generate genome sequence data ever more rapidly and economically. Thus, for most research scientists the sequence of the protein(s) with which they are working has already been determined and lies waiting to be accessed in a database such as GenBank (Chapter 10). All that the scientist needs is to acquire sufficient amino acid sequence information from the protein, sometimes as little as five or six consecutive residues, to make an unambiguous identification. While the requisite amino acid sequence information can be obtained using the Edman technique, today mass spectrometry (MS) has emerged as the method of choice for protein identification.

MASS SPECTROMETRY CAN DETECT COVALENT MODIFICATIONS

The superior sensitivity, speed, and versatility of MS have replaced the Edman technique as the principal method for determining the sequences of peptides and proteins. MS is significantly more sensitive and tolerant of variations in sample quality. Moreover, since mass and charge are common properties of a wide range of biomolecules, MS can be used to analyze metabolites, carbohydrates, and posttranslational modifications such as phosphorylation or hydroxylation that add readily identified increments of mass to a protein (**Table 4–1**). These modifications are difficult to detect using the Edman technique and undetectable in the DNA-derived amino acid sequence.

MASS SPECTROMETERS COME IN VARIOUS CONFIGURATIONS

In a simple, single quadrupole mass spectrometer a sample is placed under vacuum and allowed to vaporize in the presence of a proton donor to impart a positive charge. An electrical field then propels the cations toward a curved flight tube

TABLE 4–1 **Mass Increases Resulting from Common Post-Translational Modifications**

Modification	Mass Increase (Da)
Phosphorylation	80
Hydroxylation	16
Methylation	14
Acetylation	42
Myristylation	210
Palmitoylation	238
Glycosylation	162

where they encounter a magnetic field, which deflects them at a right angle to their original direction of flight (**Figure 4–8**). The current powering the electromagnet is gradually increased until the path of each ion is bent sufficiently to strike a detector mounted at the end of the flight tube. For ions of identical net charge, the force required to bend their path to the same extent is proportionate to their mass.

The flight tube for a time-of-flight (TOF) mass spectrometer is linear. Following vaporization of the sample in the presence of a proton donor, an electric field is briefly applied to accelerate the ions toward the detector at the end of the flight tube. For molecules of identical charge, the velocity to which they are accelerated—and hence the time required to reach the detector—is inversely proportional to their mass.

Quadrupole mass spectrometers generally are used to determine the masses of molecules of 4000 Da or less, whereas time-of-flight mass spectrometers are used to determine the large masses of complete proteins. Various combinations of multiple quadrupoles, or reflection of ions back down the linear flight tube of a TOF mass spectrometer, are used to create more sophisticated instruments.

Peptides Can Be Volatilized for Analysis by Electrospray Ionization or Matrix-Assisted Laser Desorption

The analysis of peptides and proteins by mass spectrometry initially was hindered by difficulties in volatilizing large organic molecules. While small organic molecules could be readily vaporized by heating in a vacuum (**Figure 4–9**), proteins, oligonucleotides, etc, were destroyed under these conditions. Only when reliable techniques were devised for dispersing peptides, proteins, and other large biomolecules into the vapor phase was it possible to apply MS for their structural analysis and sequence determination. Dispersion into the vapor phase is accomplished by **electrospray ionization** and **matrix-assisted laser desorption and ionization,** aka **MALDI.** In electrospray ionization, the molecules to be analyzed are dissolved in a volatile solvent and introduced into the sample chamber in a minute stream through a capillary (Figure 4–9). As the droplet of liquid emerges into the sample chamber, the solvent rapidly disperses leaving the macromolecule suspended in the gaseous phase. The charged probe serves to ionize the sample. Electrospray ionization is frequently used to analyze peptides and proteins as they elute from an HPLC or other chromatography column already dissolved in a volatile solvent. In MALDI, the sample is mixed with a liquid matrix containing a light-absorbing dye and a source of protons. In the sample chamber, the mixture is excited using a laser, causing the surrounding matrix to disperse into the vapor phase so rapidly as to avoid heating embedded peptides or proteins (Figure 4–9).

Peptides inside the mass spectrometer can be broken down into smaller units by collisions with neutral helium or argon atoms (collision-induced dissociation) and the masses

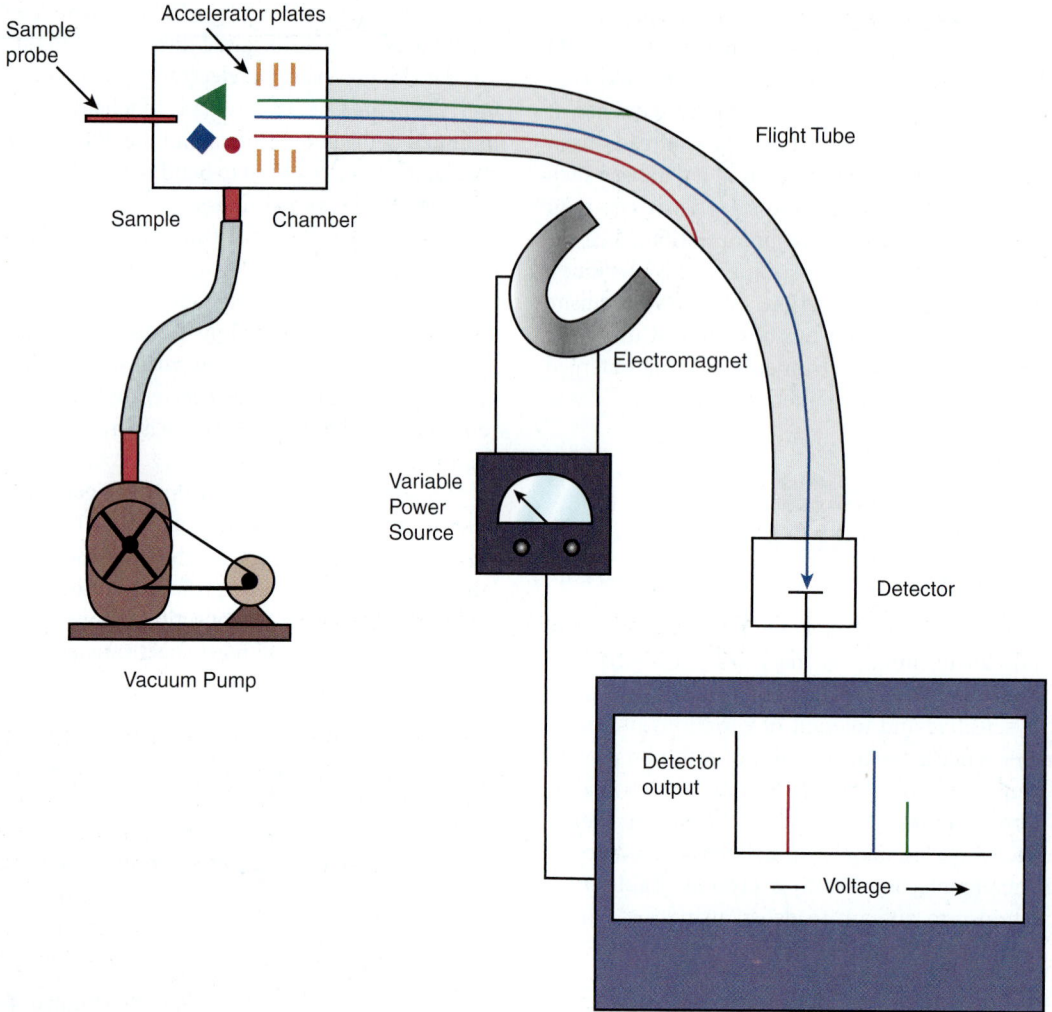

FIGURE 4–8 **Basic components of a simple mass spectrometer.** A mixture of molecules, represented by a red circle, green triangle, and blue diamond, is vaporized in an ionized state in the sample chamber. These molecules are then accelerated down the flight tube by an electrical potential applied to the accelerator grid (yellow). An adjustable field strength electromagnet applies a magnetic field that deflects the flight of the individual ions until they strike the detector. The greater the mass of the ion, the higher the magnetic field required to focus it onto the detector.

of the individual fragments determined. Since peptide bonds are much more labile than carbon–carbon bonds, the most abundant fragments will differ from one another by units equivalent to one or two amino acids. Since—with the exceptions of (1) leucine and isoleucine and (2) glutamine and lysine—the molecular mass of each amino acid is unique, the sequence of the peptide can be reconstructed from the masses of its fragments.

Tandem Mass Spectrometry

Complex peptide mixtures can now be analyzed, without prior purification, by tandem MS, which employs the equivalent of two mass spectrometers linked in series. For this reason, such tandem instruments are often referred to as **MS–MS.** The first mass spectrometer separates individual peptides based upon their differences in mass. By adjusting the field strength of the

first magnet, a single peptide can be directed into the second mass spectrometer, where fragments are generated and their masses determined. Alternatively, they can be held in an **ion trap** placed between the two quadrupoles and selectively passed to the second quadrupoles instead of being lost when the first quadrupoles is set to select ions of a different mass.

Tandem Mass Spectrometry Can Detect Metabolic Abnormalities

Tandem MS can be used to screen blood samples from newborns for the presence and concentrations of amino acids, fatty acids, and other metabolites. Abnormalities in metabolite levels can serve as diagnostic indicators for a variety of genetic disorders, such as phenylketonuria, ethylmalonic encephalopathy, and glutaric acidemia type 1.

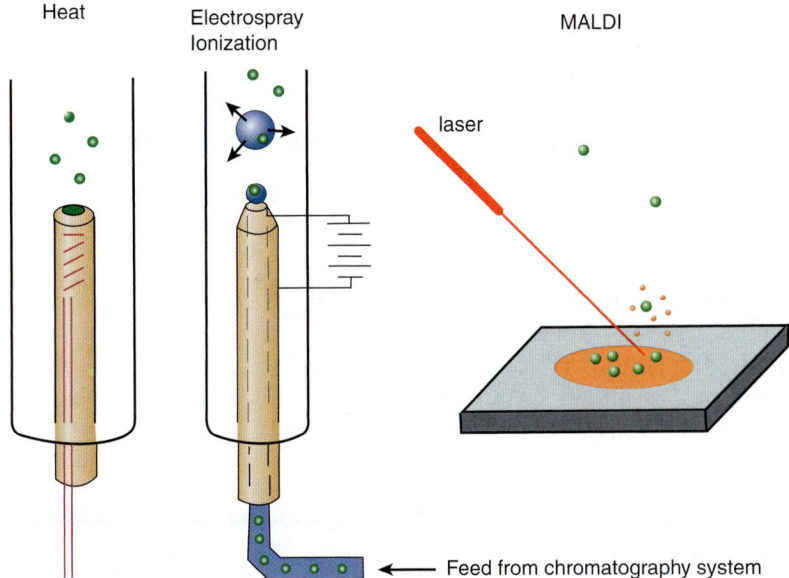

FIGURE 4–9 **Three common methods for vaporizing molecules in the sample chamber of a mass spectrometer.**

PROTEOMICS & THE PROTEOME

The Goal of Proteomics Is to Identify the Entire Complement of Proteins Elaborated by a Cell Under Diverse Conditions

While the sequence of the human genome is known, the picture provided by genomics alone is both static and incomplete. Proteomics aims to identify the entire complement of proteins elaborated by a cell under diverse conditions. As genes are switched on and off, proteins are synthesized in particular cell types at specific times of growth or differentiation and in response to external stimuli. Muscle cells express proteins not expressed by neural cells, and the type of subunits present in the hemoglobin tetramer undergo change pre- and postpartum. Many proteins undergo post-translational modifications during maturation into functionally competent forms or as a means of regulating their properties. Knowledge of the human genome therefore represents only the beginning of the task of describing living organisms in molecular detail and understanding the dynamics of processes such as growth, aging, and disease. As the human body contains thousands of cell types, each containing thousands of proteins, the **proteome**—the set of all the proteins expressed by an individual cell at a particular time—represents a moving target of formidable dimensions.

Two-Dimensional Electrophoresis & Gene Array Chips Are Used to Survey Protein Expression

One goal of proteomics is the identification of proteins whose levels of expression correlate with medically significant events.

The presumption is that proteins whose appearance or disappearance is associated with a specific physiologic condition or disease are linked, either directly or indirectly, to their root causes and mechanisms. Determination of the proteomes characteristic of each cell type requires the utmost efficiency in the isolation and identification of individual proteins. The contemporary approach utilizes robotic automation to speed sample preparation and large two-dimensional gels to resolve cellular proteins. Individual polypeptides are then extracted and analyzed by Edman sequencing or mass spectroscopy. While only about 1000 proteins can be resolved on a single gel, two-dimensional electrophoresis has a major advantage, in that it examines the proteins themselves.

An alternative approach, called multidimensional protein identification technology (MudPIT) employs successive rounds of chromatography to resolve the peptides produced from the digestion of a complex biologic sample into several simpler fractions that can be analyzed separately by MS. **Gene arrays, sometimes called DNA chips,** in which the expression of the mRNAs that encode proteins is detected, offer a complementary approach to proteomics. While changes in the expression of the mRNA encoding a protein do not necessarily reflect comparable changes in the level of the corresponding protein, gene arrays are more sensitive than two-dimensional gels, particularly with respect to low abundance proteins, and thus can examine a wider range of gene products.

Bioinformatics Assists Identification of Protein Functions

The functions of a large proportion of the proteins encoded by the human genome are presently unknown. The development of protein arrays or chips for directly testing the potential functions of proteins on a mass scale remains in its infancy. However, recent advances in bioinformatics permit researchers to compare amino acid sequences to discover clues to potential properties, physiologic roles, and mechanisms of action of proteins. Algorithms exploit the tendency of nature to employ variations of a structural theme to perform similar functions in several proteins [eg, the Rossmann nucleotide binding fold to bind NAD(P)H, nuclear targeting sequences, and EF hands to bind Ca^{2+}]. These domains generally are detected in the primary structure by conservation of particular amino acids at key positions. Insights into the properties and physiologic role of a newly discovered protein thus may be inferred by comparing its primary structure with that of known proteins.

SUMMARY

- Long amino acid polymers or polypeptides constitute the basic structural unit of proteins, and the structure of a protein provides insight into how it fulfills its functions.

- Proteins undergo post-transitional alterations during their lifetime that influence their function and determine their fate.

- The Edman method has largely been replaced by MS, a sensitive and versatile tool for determining primary structure, for identifying post-translational modifications, and for detecting metabolic abnormalities.

- DNA cloning and molecular biology coupled with protein chemistry provide a hybrid approach that greatly increases the speed and efficiency for determination of primary structures of proteins.

- Genomics—the analysis of the entire oligonucleotide sequence of an organism's complete genetic material—has provided further enhancements.

- Computer algorithms facilitate identification of the ORFs that encode a given protein by using partial sequences and peptide mass profiling to search sequence databases.

- Scientists are now trying to determine the primary sequence and functional role of every protein expressed in a living cell, known as its proteome.

- A major goal is the identification of proteins and of their post-translational modifications whose appearance or disappearance correlates with physiologic phenomena, aging, or specific diseases.

REFERENCES

Arnaud CH: Mass spec tackles proteins. Chem Eng News 2006;84:17.

Austin CP: The impact of the completed human genome sequence on the development of novel therapeutics for human disease. Annu Rev Med 2004;55:1.

Deutscher MP (editor): *Guide to Protein Purification*. Methods Enzymol, vol. 182, Academic Press, 1990 (Entire volume).

Gwynne P, Heebner G: Mass spectrometry in drug discovery and development: from physics to pharma. Science 2006;313:1315.

Kislinger T, Gramolini AO, MacLennan DH, et al: Multidimensional protein identification technology (MudPIT): technical overview of a profiling method optimized for the comprehensive proteomic investigation of normal and diseased heart tissue. J Am Soc Mass Spectrom 2005;16:1207.

Kolialexi A, Anagnostopoulos AK, Mavrou A, et al: Application of proteomics for diagnosis of fetal aneuploidies and pregnancy complications. J Proteomics 2009;72:731.

Levy PA: An overview of newborn screening. J Dev Behav Pediatr 2010;31:622.

Schena M, Shalon D, Davis RW, et al: Quantitative monitoring of gene expression patterns with a complementary DNA microarray. Science 1995;270:467.

Scopes RK: *Protein Purification. Principles and Practice*, 3rd ed. Springer, 1994.

Semsarian C, Seidman CE: Molecular medicine in the 21st century. Intern Med J 2001;31:53.

Sharon M, Robinson CV: The role of mass spectrometry in structure elucidation of dynamic protein complexes. Annu Rev Biochem 2007;76:167.

Shendure J, Ji H: Next-generation DNA sequencing. Nature Biotechnol 2008; 26:1135.

Sikaroodi M, Galachiantz Y, Baranova Al: Tumor markers: the potential of "omics" approach. Curr Mol Med 2010;10:249.

Woodage T, Broder S: The human genome and comparative genomics: understanding human evolution, biology, and medicine. J Gastroenterol 2003;15:68.

5

Proteins: Higher Orders of Structure

Peter J. Kennelly, PhD & Victor W. Rodwell, PhD

OBJECTIVES

After studying this chapter, you should be able to:

- Indicate the advantages and drawbacks of several approaches to classifying proteins.
- Explain and illustrate the primary, secondary, tertiary, and quaternary structure of proteins.
- Identify the major recognized types of secondary structure and explain supersecondary motifs.
- Describe the kind and relative strengths of the forces that stabilize each order of protein structure.
- Describe the information summarized by a Ramachandran plot.
- Indicate the present state of knowledge concerning the stepwise process by which proteins are thought to attain their native conformation.
- Identify the physiologic roles in protein maturation of chaperones, protein disulfide isomerase, and peptidylproline *cis–trans*-isomerase.
- Describe the principal biophysical techniques used to study tertiary and quaternary structure of proteins.
- Explain how genetic and nutritional disorders of collagen maturation illustrate the close linkage between protein structure and function.
- For the prion diseases, outline the overall events in their molecular pathology and name the life forms each affects.

BIOMEDICAL IMPORTANCE

In nature, form follows function. In order for a newly synthesized polypeptide to mature into a biologically functional protein capable of catalyzing a metabolic reaction, powering cellular motion, or forming the macromolecular rods and cables that provide structural integrity to hair, bones, tendons, and teeth, it must fold into a specific three-dimensional arrangement, or **conformation.** In addition, during maturation **post-translational modifications** may add new chemical groups or remove transiently-needed peptide segments. Genetic or nutritional deficiencies that impede protein maturation are deleterious to health. Examples of the former include Creutzfeldt–Jakob disease, scrapie, Alzheimer's disease, and bovine spongiform encephalopathy ("mad cow disease"). Scurvy represents a nutritional deficiency that impairs protein maturation.

CONFORMATION VERSUS CONFIGURATION

The terms configuration and conformation are often confused. **Configuration** refers to the geometric relationship between a given set of atoms, for example, those that distinguish L- from D-amino acids. Interconversion of *configurational* alternatives requires breaking (and reforming) covalent bonds. **Conformation** refers to the spatial relationship of every atom in a molecule. Interconversion between conformers occurs without covalent bond rupture, with retention of configuration, and typically via rotation about single bonds.

PROTEINS WERE INITIALLY CLASSIFIED BY THEIR GROSS CHARACTERISTICS

Scientists initially approached structure–function relationships in proteins by separating them into classes based upon properties such as solubility, shape, or the presence of nonprotein groups. For example, the proteins that can be extracted from cells using aqueous solutions of physiologic pH and ionic strength are classified as **soluble.** Extraction of **integral membrane proteins** requires dissolution of the membrane with detergents. **Globular proteins** are compact, roughly spherical molecules that have **axial ratios** (the ratio of their shortest to longest dimensions) of not over 3. Most enzymes are globular proteins. By contrast, many structural proteins adopt highly extended conformations. These **fibrous proteins** possess axial ratios of 10 or more.

Lipoproteins and **glycoproteins** contain covalently bound lipid and carbohydrate, respectively. Myoglobin, hemoglobin, cytochromes, and many other **metalloproteins** contain tightly associated metal ions. While more precise classification schemes have emerged based upon similarity, or **homology,** in amino acid sequence and three-dimensional structure, many early classification terms remain in use.

PROTEINS ARE CONSTRUCTED USING MODULAR PRINCIPLES

Proteins perform complex physical and catalytic functions by positioning specific chemical groups in a precise three-dimensional arrangement. The polypeptide scaffold containing these groups must adopt a conformation that is both functionally efficient and physically strong. At first glance, the biosynthesis of polypeptides comprised of tens of thousands of individual atoms would appear to be extremely challenging. When one considers that a typical polypeptide can adopt $\geq 10^{50}$ distinct conformations, folding into the conformation appropriate to their biologic function would appear to be even more difficult. As described in Chapters 3 and 4, synthesis of the polypeptide backbones of proteins employs a small set of common building blocks or modules, the amino acids, joined by a common linkage, the peptide bond. Similarly, a stepwise modular pathway simplifies the folding and processing of newly synthesized polypeptides into mature proteins.

FOUR ORDERS OF THE PROTEIN STRUCTURE

The modular nature of protein synthesis and folding are embodied in the concept of orders of the protein structure: **primary structure**—the sequence of the amino acids in a polypeptide chain; **secondary structure**—the folding of short (3- to 30- residue), contiguous segments of polypeptide into geometrically ordered units; **tertiary structure**—the assembly of secondary structural units into larger functional units such as the mature polypeptide and its component domains; and **quaternary structure**—the number and types of polypeptide units of oligomeric proteins and their spatial arrangement.

SECONDARY STRUCTURE

Peptide Bonds Restrict Possible Secondary Conformations

Free rotation is possible about only two of the three covalent bonds of the polypeptide backbone: the α-carbon (Cα) to the carbonyl carbon (Co) bond, and the Cα to nitrogen bond (Figure 3–4). The partial double-bond character of the peptide bond that links Co to the α-nitrogen requires that the carbonyl carbon, carbonyl oxygen, and α-nitrogen remain coplanar, thus preventing rotation. The angle about the Cα—N bond is termed the phi (Φ) angle, and that about the Co—Cα bond the psi (Ψ) angle. For amino acids other than glycine, most combinations of phi and psi angles are disallowed because of steric hindrance (**Figure 5–1**). The conformations of proline are even more restricted due to the absence of free rotation of the N—Cα bond.

Regions of ordered secondary structure arise when a series of aminoacyl residues adopt similar phi and psi angles. Extended segments of polypeptide (eg, loops) can possess a variety of such angles. The angles that define the two most

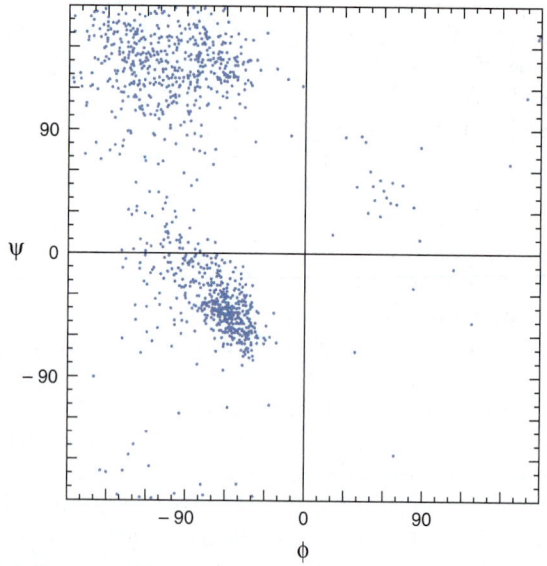

FIGURE 5–1 Ramachandran plot of the main chain phi (Φ) and psi (Ψ) angles for approximately 1000 nonglycine residues in eight proteins whose structures were solved at high resolution. The dots represent allowable combinations, and the spaces prohibited combinations, of phi and psi angles. (Reproduced, with permission, from Richardson JS: The anatomy and taxonomy of protein structures. Adv Protein Chem 1981;34:167. Copyright © 1981. Reprinted with permission from Elsevier.)

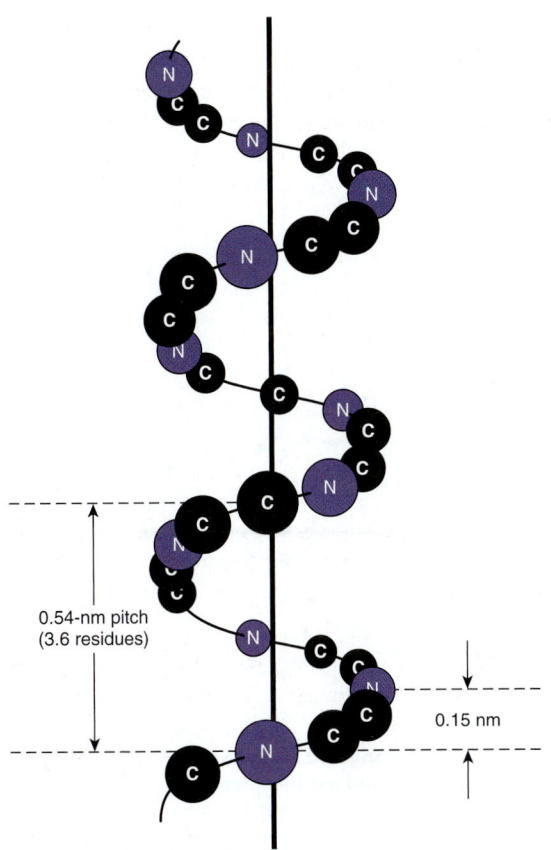

0.54-nm pitch
(3.6 residues)

0.15 nm

FIGURE 5–2 Orientation of the main chain atoms of a peptide about the axis of an α helix.

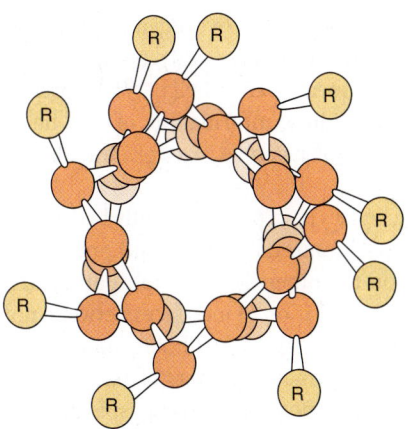

FIGURE 5–3 **View down the axis of an α helix.** The side chains (R) are on the outside of the helix. The van der Waals radii of the atoms are larger than shown here; hence, there is almost no free space inside the helix. (Slightly modified and reproduced, with permission, from Stryer L: *Biochemistry,* 3rd ed. Freeman, 1988. Copyright © 1988 W.H. Freeman and Company.)

common types of secondary structure, the **α helix** and the **β sheet,** fall within the lower and upper left-hand quadrants of a Ramachandran plot, respectively (Figure 5–1).

Alpha Helix

The polypeptide backbone of an α helix is twisted by an equal amount about each α-carbon with a phi angle of approximately −57° and a psi angle of approximately −47°. A complete turn of the helix contains an average of 3.6 amino-acyl residues, and the distance it rises per turn (its *pitch*) is 0.54 nm (**Figure 5–2**). The R groups of each aminoacyl residue in an α helix face outward (**Figure 5–3**). Proteins contain only L-amino acids, for which a right-handed α helix is by far the more stable, and only right-handed α helices are present in proteins. Schematic diagrams of proteins represent α helices as coils or cylinders.

The stability of an α helix arises primarily from hydrogen bonds formed between the oxygen of the peptide bond carbonyl and the hydrogen atom of the peptide bond nitrogen of the fourth residue down the polypeptide chain (**Figure 5–4**). The ability to form the maximum number of hydrogen bonds, supplemented by van der Waals interactions in the core of this tightly packed structure, provides the thermodynamic driving force for the formation of an α helix. Since the peptide bond

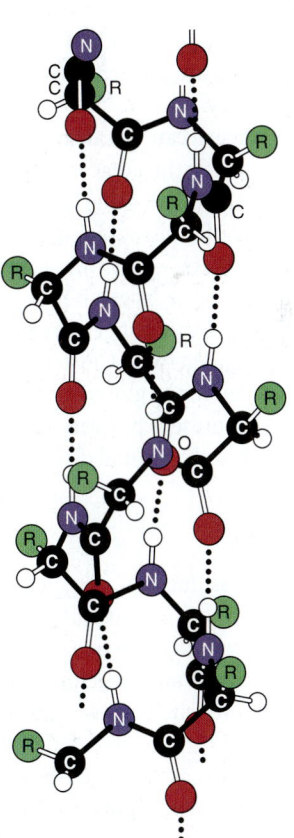

FIGURE 5–4 **Hydrogen bonds (dotted lines) formed between H and O atoms stabilize a polypeptide in an α-helical conformation.** (Reprinted, with permission, from Haggis GH, et al, (1964), *"Introduction to Molecular Biology"*. Science 146:1455–1456. Reprinted with permission from AAAS.)

nitrogen of proline lacks a hydrogen atom to contribute to a hydrogen bond, proline can only be stably accommodated within the first turn of an α helix. When present elsewhere, proline disrupts the conformation of the helix, producing a bend. Because of its small size, glycine also often induces bends in α helices.

Many α helices have predominantly hydrophobic R groups on one side of the axis of the helix and predominantly hydrophilic ones on the other. These **amphipathic helices** are well adapted to the formation of interfaces between polar and nonpolar regions such as the hydrophobic interior of a protein and its aqueous environment. Clusters of amphipathic helices can create a channel, or pore, that permits specific polar molecules to pass through hydrophobic cell membranes.

Beta Sheet

The second (hence "beta") recognizable regular secondary structure in proteins is the β sheet. The amino acid residues of a β sheet, when viewed edge-on, form a zigzag or pleated pattern in which the R groups of adjacent residues point in opposite directions. Unlike the compact backbone of the α helix, the peptide backbone of the β sheet is highly extended. But like the α helix, β sheets derive much of their stability from hydrogen bonds between the carbonyl oxygens and amide hydrogens of peptide bonds. However, in contrast to the α helix, these bonds are formed with adjacent segments of the β sheet (**Figure 5–5**).

Interacting β sheets can be arranged either to form a **parallel** β sheet, in which the adjacent segments of the polypeptide chain proceed in the same direction amino to carboxyl, or an **antiparallel** sheet, in which they proceed in opposite directions (Figure 5–5). Either configuration permits the maximum number of hydrogen bonds between segments, or strands, of the sheet. Most β sheets are not perfectly flat but tend to have a right-handed twist. Clusters of twisted strands of β sheet form the core of many globular proteins (**Figure 5–6**). Schematic diagrams represent β sheets as arrows that point in the amino to the carboxyl terminal direction.

Loops & Bends

Roughly half of the residues in a "typical" globular protein reside in α helices or β sheets, and half in loops, turns, bends, and other extended conformational features. Turns and bends refer to short segments of amino acids that join two units of the secondary structure, such as two adjacent strands of an antiparallel β sheet. A β turn involves four aminoacyl residues, in which the first residue is hydrogen-bonded to the fourth, resulting in a tight 180° turn (**Figure 5–7**). Proline and glycine often are present in β turns.

Loops are regions that contain residues beyond the minimum number necessary to connect adjacent regions of secondary structure. Irregular in conformation, loops nevertheless serve key biologic roles. For many enzymes, the loops that bridge domains responsible for binding substrates

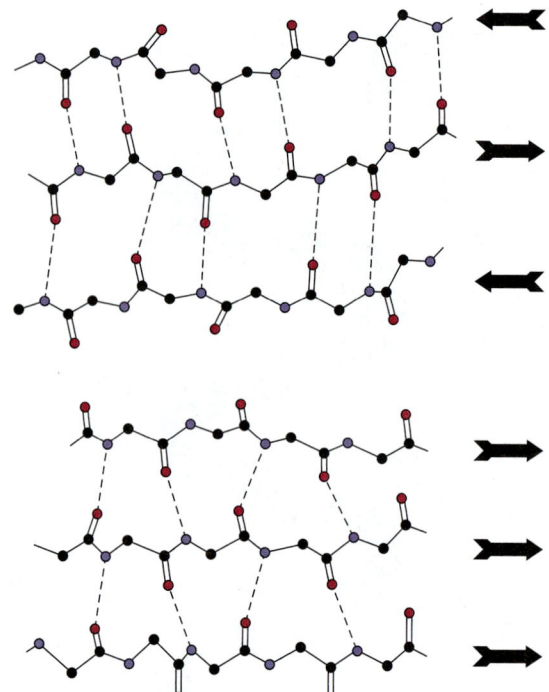

FIGURE 5–5 **Spacing and bond angles of the hydrogen bonds of antiparallel and parallel pleated β sheets.** Arrows indicate the direction of each strand. Hydrogen bonds are indicated by dotted lines with the participating α-nitrogen atoms (hydrogen donors) and oxygen atoms (hydrogen acceptors) shown in blue and red, respectively. Backbone carbon atoms are shown in black. For clarity in presentation, R groups and hydrogen atoms are omitted. *Top:* Antiparallel β sheet. Pairs of hydrogen bonds alternate between being close together and wide apart and are oriented approximately perpendicular to the polypeptide backbone. *Bottom:* Parallel β sheet. The hydrogen bonds are evenly spaced but slant in alternate directions.

often contain aminoacyl residues that participate in catalysis. **Helix-loop-helix motifs** provide the oligonucleotide-binding portion of many DNA-binding proteins such as repressors and transcription factors. Structural motifs such as the helix-loop-helix motif that are intermediate in scale between secondary and tertiary structures are often termed **supersecondary structures.** Since many loops and bends reside on the surface of proteins and are thus exposed to solvent, they constitute readily accessible sites, or **epitopes,** for recognition and binding of antibodies.

While loops lack apparent structural regularity, many adopt a specific conformation stabilized through hydrogen bonding, salt bridges, and hydrophobic interactions with other portions of the protein. However, not all portions of proteins are necessarily ordered. Proteins may contain "disordered" regions, often at the extreme amino or carboxyl terminal, characterized by high conformational flexibility. In many instances, these disordered regions assume an ordered conformation upon binding of a ligand. This structural flexibility enables such regions to act as ligand-controlled switches that affect protein structure and function.

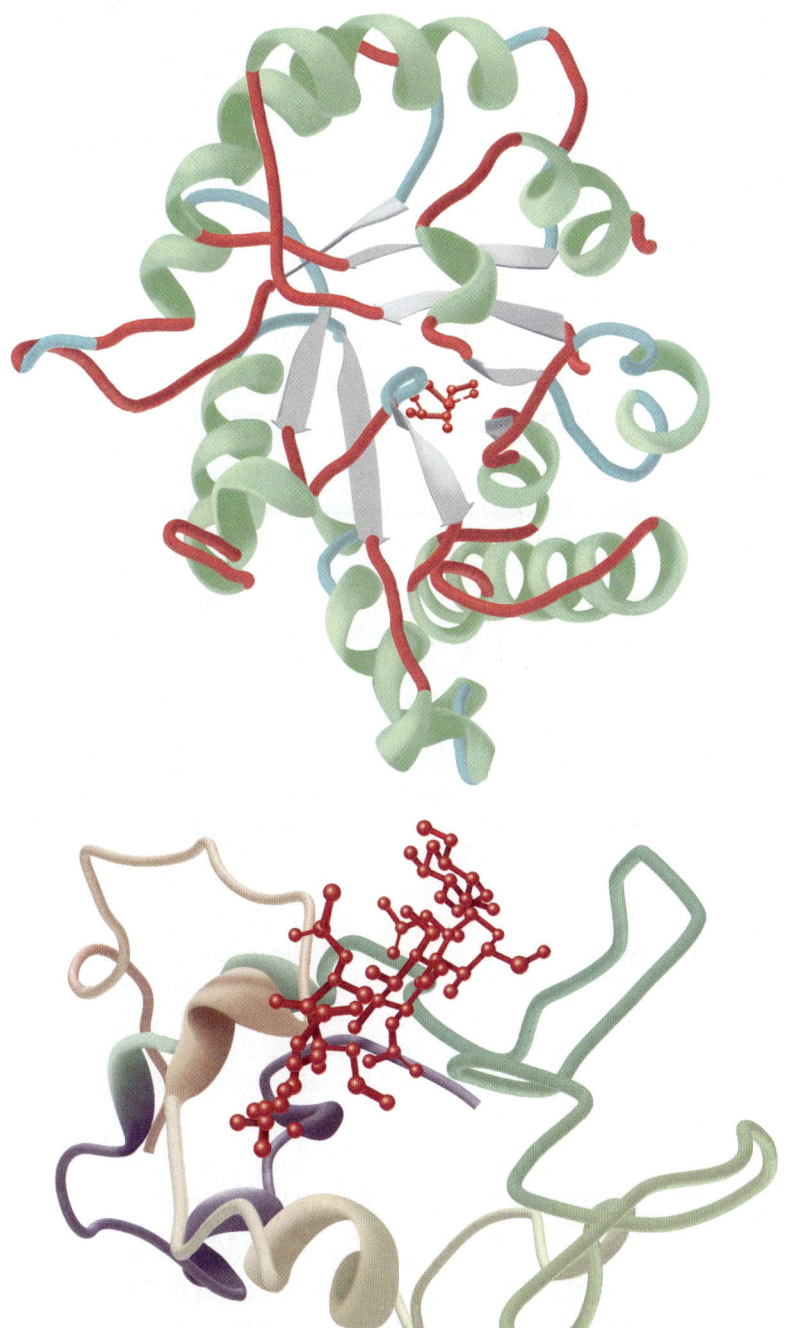

FIGURE 5–6 **Examples of the tertiary structure of proteins.** Top: The enzyme triose phosphate isomerase complexed with the substrate analog 2-phosphoglycerate (red). Note the elegant and symmetrical arrangement of alternating β sheets (light blue) and α helices (green), with the β sheets forming a β-barrel core surrounded by the helices. (Adapted from Protein Data Bank ID no. 1o5x.) Bottom: Lysozyme complexed with the substrate analog penta-N-acetyl chitopentaose (red). The color of the polypeptide chain is graded along the visible spectrum from purple (N-terminal) to tan (C-terminal). Notice how the concave shape of the domain forms a binding pocket for the pentasaccharide, the lack of β sheet, and the high proportion of loops and bends. (Adapted from Protein Data Bank ID no. 1sfb.)

Tertiary & Quaternary Structure

The term "tertiary structure" refers to the entire three-dimensional conformation of a polypeptide. It indicates, in three-dimensional space, how secondary structural features—helices, sheets, bends, turns, and loops—assemble to form domains and how these domains relate spatially to one another. A **domain** is a section of the protein structure sufficient to perform a particular chemical or physical task such as binding of a substrate or other ligand. Most domains are modular in nature, and contiguous in both primary sequence and three-dimensional space (**Figure 5–8**). Simple proteins, particularly those that interact with a single substrate, such as lysozyme or triose phosphate isomerase (Figure 5–6) and the oxygen storage protein myoglobin (Chapter 6), often consist of a single domain. By contrast, lactate dehydrogenase is comprised of two domains, an N-terminal NAD⁺-binding domain and a C-terminal binding domain for the second substrate, pyruvate (Figure 5–8). Lactate dehydrogenase is one of the family of oxidoreductases that share a common N-terminal NAD(P)⁺-binding domain known as the **Rossmann fold.** By fusing the Rossmann fold domain to a variety of C-terminal

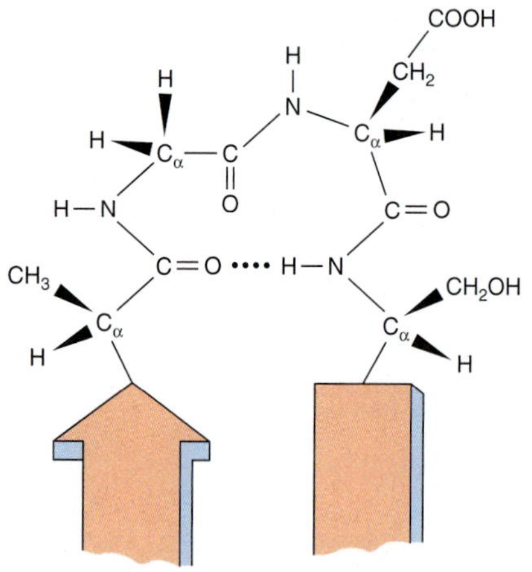

FIGURE 5–7 **A β turn that links two segments of anti-parallel β sheet.** The dotted line indicates the hydrogen bond between the first and fourth amino acids of the four-residue segment Ala-Gly-Asp-Ser.

domains, a large family of oxidoreductases have evolved that utilize NAD(P)$^+$/NAD(P)H for the oxidation and reduction of a wide range of metabolites. Examples include alcohol dehydrogenase, glyceraldehyde-3-phosphate dehydrogenase, malate dehydrogenase, quinone oxidoreductase, 6-phosphogluconate dehydrogenase, D-glycerate dehydrogenase, formate dehydrogenase, and 3α, 20β-hydroxysteroid dehydrogenase.

Not all domains bind substrates. Hydrophobic membrane domains anchor proteins to membranes or enable them to span membranes. Localization sequences target proteins to specific subcellular or extracellular locations such as the nucleus, mitochondria, secretory vesicles, etc. Regulatory domains trigger changes in protein function in response to the binding of allosteric effectors or covalent modifications (Chapter 9). Combining domain modules provides a facile route for generating proteins of great structural complexity and functional sophistication (**Figure 5–9**).

Proteins containing multiple domains can also be assembled through the association of multiple polypeptides, or protomers. Quaternary structure defines the polypeptide composition of a protein and, for an oligomeric protein, the spatial relationships between its protomers or subunits. **Monomeric** proteins consist of a single polypeptide chain. **Dimeric** proteins contain two polypeptide chains. **Homodimers** contain two copies of the same polypeptide chain, while in a **heterodimer** the polypeptides differ. Greek letters (α, β, γ, etc) are used to distinguish different subunits of a hetero-oligomeric protein, and subscripts indicate the number of each subunit type. For example, α$_4$ designates a homotetrameric protein, and α$_2$β$_2$γ a protein with five subunits of three different types.

Since even small proteins contain many thousands of atoms, depictions of protein structure that indicate the position of every atom are generally too complex to be readily interpreted. Simplified schematic diagrams thus are used to depict the key features of a protein's tertiary and quaternary

structure. Ribbon diagrams (Figures 5–6 and 5–8) trace the conformation of the polypeptide backbone, with cylinders and arrows indicating regions of α helix and β sheet, respectively. In an even simpler representation, line segments that link the α carbons indicate the path of the polypeptide backbone. These schematic diagrams often include the side chains of selected amino acids that emphasize specific structure–function relationships.

MULTIPLE FACTORS STABILIZE THE TERTIARY & QUATERNARY STRUCTURE

Higher orders of protein structure are stabilized primarily—and often exclusively—by noncovalent interactions. Principal among these are hydrophobic interactions that drive most hydrophobic amino acid side chains into the interior of the protein, shielding them from water. Other significant contributors include hydrogen bonds and salt bridges between the carboxylates of aspartic and glutamic acid and the oppositely charged side chains of protonated lysyl, argininyl, and histidyl residues. While individually weak relative to a typical covalent bond of 80–120 kcal/mol, collectively these numerous interactions confer a high degree of stability to the biologically functional conformation of a protein, just as a Velcro fastener harnesses the cumulative strength of a multitude of tiny plastic loops and hooks.

Some proteins contain covalent disulfide (S—S) bonds that link the sulfhydryl groups of cysteinyl residues. Formation of disulfide bonds involves oxidation of the cysteinyl sulfhydryl groups and requires oxygen. Intrapolypeptide disulfide bonds further enhance the stability of the folded conformation of a peptide, while interpolypeptide disulfide bonds stabilize the quaternary structure of certain oligomeric proteins.

THREE-DIMENSIONAL STRUCTURE IS DETERMINED BY X-RAY CRYSTALLOGRAPHY OR BY NMR SPECTROSCOPY

X-Ray Crystallography

Following the solution in 1960 by John Kendrew of the three-dimensional structure of myoglobin, x-ray crystallography revealed the structures of thousands of biological macromolecules ranging from proteins to many oligonucleotides and a few viruses. For the solution of its structure by x-ray crystallography, a protein is first precipitated under conditions that form large, well-ordered crystals. To establish appropriate conditions, crystallization trials use a few microliters of protein solution and a matrix of variables (temperature, pH, presence of salts or organic solutes such as polyethylene glycol) to establish optimal conditions for crystal formation.

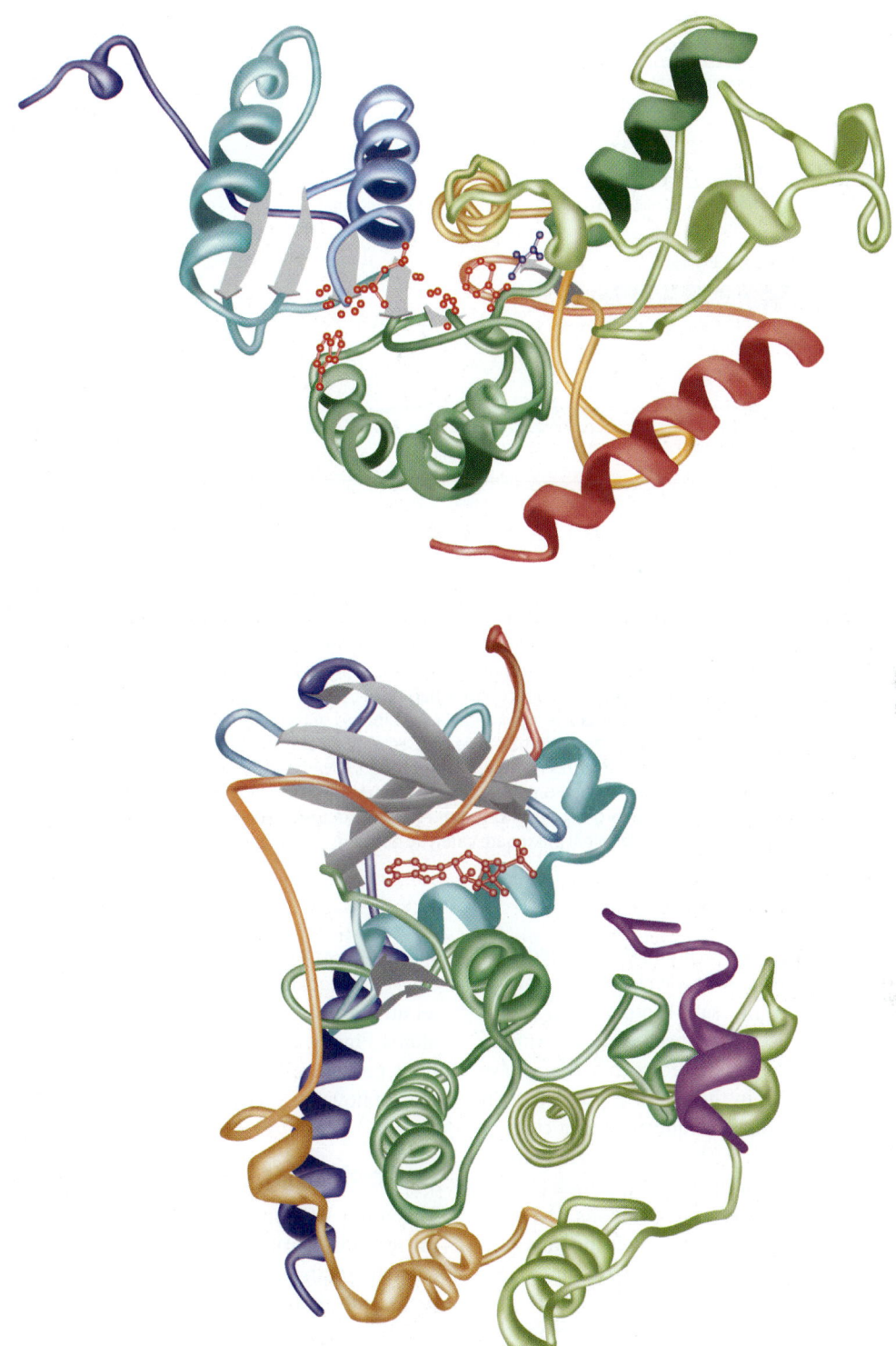

FIGURE 5–8 Polypeptides containing two domains. Top: Shown is the three-dimensional structure of a monomer unit of the tetrameric enzyme lactate dehydrogenase with the substrates NADH (red) and pyruvate (blue) bound. Not all bonds in NADH are shown. The color of the polypeptide chain is graded along the visible spectrum from blue (N-terminal) to orange (C-terminal). Note how the N-terminal portion of the polypeptide forms a contiguous domain, encompassing the left portion of the enzyme, responsible for binding NADH. Similarly, the C-terminal portion forms a contiguous domain responsible for binding pyruvate. (Adapted from Protein Data Bank ID no. 3ldh.) Bottom: Shown is the three-dimensional structure of the catalytic subunit of the cAMP-dependent protein kinase (Chapter 42) with the substrate analogs ADP (red) and peptide (purple) bound. The color of the polypeptide chain is graded along the visible spectrum from blue (N-terminal) to orange (C-terminal). Protein kinases transfer the γ-phosphate group of ATP to protein and peptide substrates (Chapter 9). Note how the N-terminal portion of the polypeptide forms a contiguous domain rich in β sheet that binds ADP. Similarly, the C-terminal portion forms a contiguous, α helix-rich domain responsible for binding the peptide substrate. (Adapted from Protein Data Bank ID no. 1jbp.)

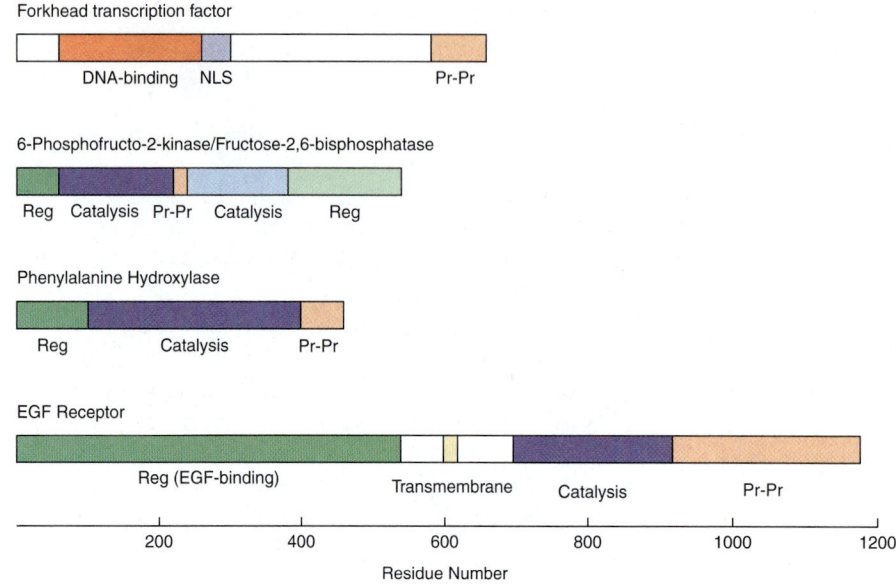

FIGURE 5–9 **Some multidomain proteins.** The rectangles represent the polypeptide sequences of a forkhead transcription factor; 6-phosphofructo-2-kinase/fructose-2,6-bisphosphatase, a bifunctional enzyme whose activities are controlled in a reciprocal fashion by allosteric effectors and covalent modification (Chapter 20); phenylalanine hydroxylase (Chapters 27 and 29), whose activity is stimulated by phosphorylation of its regulatory domain; and the epidermal growth factor (EGF) receptor (Chapter 41), a transmembrane protein whose intracellular protein kinase domain is regulated via the binding of the peptide hormone EGF to its extracellular domain. Regulatory domains are colored green, catalytic domains dark blue and light blue, protein–protein interaction domains light orange, DNA binding domains dark orange, nuclear localization sequences medium blue, and transmembrane domains yellow. The kinase and bisphosphatase activities of 6-phosphofructo-2-kinase/fructose-2,6-bisphosphatase are catalyzed by the N- and C-terminal proximate catalytic domains, respectively.

Crystals mounted in quartz capillaries are first irradiated with monochromatic x-rays of approximate wavelength 0.15 nm to confirm that they are protein, not salt. Protein crystals may then be frozen in liquid nitrogen for subsequent collection of a high-resolution data set. The patterns formed by the x-rays that are diffracted by the atoms in their path are recorded on a photographic plate or its computer equivalent as a circular pattern of spots of varying intensity. The data inherent in these spots are then analyzed using a mathematical approach termed a *Fourier synthesis,* which summates wave functions. The wave amplitudes are related to spot intensity, but since the waves are not in phase, the relationship between their phases must next be determined.

The traditional approach to solution of the "phase problem" employs **isomorphous displacement.** Prior to irradiation, an atom with a distinctive x-ray "signature" is introduced into a crystal at known positions in the primary structure of the protein. Heavy atom isomorphous displacement generally uses mercury or uranium, which bind to cysteine residues. An alternative approach uses the expression of plasmid-encoded recombinant proteins in which selenium replaces the sulfur of methionine. Expression uses a bacterial host auxotrophic for methionine biosynthesis and a defined medium in which selenomethionine replaces methionine. Alternatively, if the

unknown structure is similar to one that has already been solved, **molecular replacement** on an existing model provides an attractive way to phase the data without the use of heavy atoms. Finally, the results from the phasing and Fourier summations provide an electron density profile or three-dimensional map of how the atoms are connected or related to one another.

Laue X-Ray Crystallography

The ability of some crystallized enzymes to catalyze chemical reactions strongly suggests that structures determined by crystallography are indeed representative of the structures present in the free solution. Classic crystallography provides, however, an essentially static picture of a protein that may undergo significant structural changes in vivo, such as those that accompany enzymic catalysis. The Laue approach uses diffraction of *polychromatic* x-rays, and many crystals. The time-consuming process of rotating the crystal in the x-ray beam is avoided, which permits the use of extremely short exposure times. Detection of the motions of residues or domains of an enzyme during catalysis uses crystals that contain an inactive or "caged" substrate analog. An intense flash of visible light cleaves the caged precursor to release free substrate and initiate catalysis in a precisely controlled manner. Using this approach, data can be collected over time periods as short as a few nanoseconds.

Nuclear Magnetic Resonance Spectroscopy

Nuclear magnetic resonance (NMR) spectroscopy, a powerful complement to x-ray crystallography, measures the absorbance of radio frequency electromagnetic energy by certain atomic nuclei. "NMR-active" isotopes of biologically relevant elements include ^1H, ^{13}C, ^{15}N, and ^{31}P. The frequency, or chemical shift, at which a particular nucleus absorbs energy is a function of both the functional group within which it resides and the proximity of other NMR-active nuclei. Once limited to metabolites and relatively small macromolecules, ≤ 30 kDa, today proteins and protein complexes of >100 kDa can be analyzed by NMR. Two-dimensional NMR spectroscopy permits a three-dimensional representation of a protein to be constructed by determining the proximity of these nuclei to one another. NMR spectroscopy analyzes proteins in aqueous solution. Not only does this obviate the need to form crystals (a particular advantage when dealing with difficult to crystallize membrane proteins), it renders possible real-time observation of the changes in conformation that accompany ligand binding or catalysis. It also offers the possibility of perhaps one day being able to observe the structure and dynamics of proteins (and metabolites) within living cells.

Molecular Modeling

A valuable adjunct to the empirical determination of the three-dimensional structure of proteins is the use of computer technology for molecular modeling. When the three-dimensional structure is known, **molecular dynamics** programs can be used to simulate the conformational dynamics of a protein and the manner in which factors such as temperature, pH, ionic strength, or amino acid substitutions influence these motions. **Molecular docking** programs simulate the interactions that take place when a protein encounters a substrate, inhibitor, or other ligand. Virtual screening for molecules likely to interact with key sites on a protein of biomedical interest is extensively used to facilitate the discovery of new drugs.

Molecular modeling is also employed to infer the structure of proteins for which x-ray crystallographic or NMR structures are not yet available. Secondary structure algorithms weigh the propensity of specific residues to become incorporated into α helices or β sheets in previously studied proteins to predict the secondary structure of other polypeptides. In **homology modeling,** the known three-dimensional structure of a protein is used as a template upon which to erect a model of the *probable* structure of a related protein. Scientists are working to devise computer programs that will reliably predict the three-dimensional conformation of a protein directly from its primary sequence, thereby permitting determination of the structures of the many unknown proteins for which templates currently are lacking.

PROTEIN FOLDING

Proteins are conformationally dynamic molecules that can fold into their functionally competent conformation in a time frame of milliseconds, and often can refold if their conformation becomes disrupted, or denatured. How is this remarkable process achieved? Folding into the native state does not involve a haphazard search of all possible structures. Denatured proteins are not just random coils. Native contacts are favored, and regions of the native structure persist even in the denatured state. Discussed below are factors that facilitate folding and refolding, and current concepts and proposed mechanisms based on more than 40 years of largely in vitro experimentation.

Native Conformation of a Protein is Thermodynamically Favored

The number of distinct combinations of phi and psi angles specifying potential conformations of even a relatively small—15 kDa—polypeptide is unbelievably vast. Proteins are guided through this vast labyrinth of possibilities by thermodynamics. Since the biologically relevant—or native—conformation of a protein generally is the one that is most energetically favored, knowledge of the native conformation is specified in the primary sequence. However, if one were to wait for a polypeptide to find its native conformation by random exploration of all possible conformations, the process would require billions of years to complete. Clearly, in nature, protein folding takes place in a more orderly and guided fashion.

Folding Is Modular

Protein folding generally occurs via a stepwise process. In the first stage, as the newly synthesized polypeptide emerges from the ribosome, short segments fold into secondary structural units that provide local regions of organized structure. Folding is now reduced to the selection of an appropriate arrangement of this relatively small number of secondary structural elements. In the second stage, the hydrophobic regions segregate into the interior of the protein away from solvent, forming a "molten globule," a partially folded polypeptide in which the modules of the secondary structure rearrange until the mature conformation of the protein is attained. This process is orderly, but not rigid. Considerable flexibility exists in the ways and in the order in which elements of secondary structure can be rearranged. In general, each element of the secondary or super-secondary structure facilitates proper folding by directing the folding process toward the native conformation and away from unproductive alternatives. For oligomeric proteins, individual protomers tend to fold before they associate with other subunits.

Auxiliary Proteins Assist Folding

Under appropriate laboratory conditions, many proteins will spontaneously refold after being **denatured** (ie, unfolded) by treatment with acid or base, chaotropic agents, or detergents. However, refolding under these conditions is slow—minutes to hours. Moreover, many proteins fail to spontaneously refold in vitro. Instead they form insoluble **aggregates,** disordered complexes of unfolded or partially folded polypeptides held together predominantly by hydrophobic interactions. Aggregates represent unproductive dead ends in the folding process. Cells employ auxiliary proteins to speed the process of folding and to guide it toward a productive conclusion.

Chaperones

Chaperone proteins participate in the folding of over half of all mammalian proteins. The hsp70 (70 kDa heat shock protein) family of chaperones binds short sequences of hydrophobic amino acids that emerge while a new polypeptide is being synthesized, shielding them from solvent. Chaperones prevent aggregation, thus providing an opportunity for the formation of appropriate secondary structural elements and their subsequent coalescence into a molten globule. The hsp60 family of chaperones, sometimes called **chaperonins,** differ in sequence and structure from hsp70 and its homologs. Hsp60 acts later in the folding process, often together with an hsp70 chaperone. The central cavity of the donut-shaped hsp60 chaperone provides a sheltered environment in which a polypeptide can fold until all hydrophobic regions are buried in its interior, thus preempting any tendency toward aggregation.

Protein Disulfide Isomerase

Disulfide bonds between and within polypeptides stabilize tertiary and quaternary structures. However, disulfide bond formation is nonspecific. Under oxidizing conditions, a given cysteine can form a disulfide bond with the —SH of any accessible cysteinyl residue. By catalyzing disulfide exchange, the rupture of an S—S bond and its reformation with a different partner cysteine, protein disulfide isomerase facilitates the formation of disulfide bonds that stabilize a protein's native conformation.

Proline-*cis, trans*-Isomerase

All X-Pro peptide bonds—where X represents any residue—are synthesized in the *trans* configuration. However, of the X-Pro bonds of mature proteins, approximately 6% are *cis*. The *cis* configuration is particularly common in β turns. Isomerization from *trans* to *cis* is catalyzed by the enzyme proline-*cis, trans*-isomerase (**Figure 5–10**).

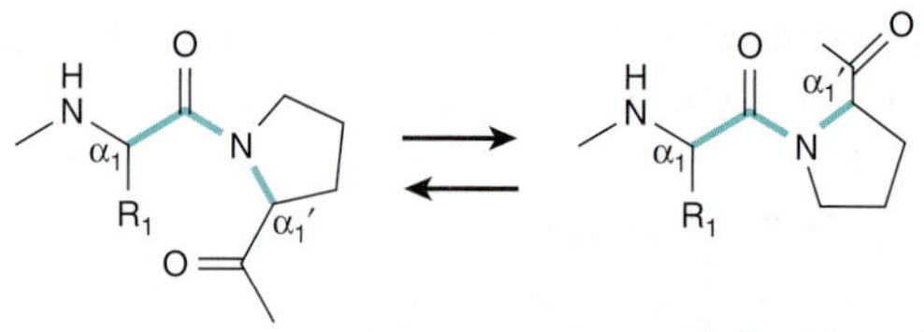

Folding Is a Dynamic Process

Proteins are conformationally dynamic molecules that can fold and unfold hundreds or thousands of times in their lifetime. How do proteins, once unfolded, refold and restore their

FIGURE 5–10 Isomerization of the *N*-α$_1$ prolyl peptide bond from a *cis* to a *trans* configuration relative to the backbone of the polypeptide.

functional conformation? First, unfolding rarely leads to the complete randomization of the polypeptide chain inside the cell. Unfolded proteins generally retain a number of contacts and regions of the secondary structure that facilitate the refolding process. Second, chaperone proteins can "rescue" unfolded proteins that have become thermodynamically trapped in a misfolded dead end by unfolding hydrophobic regions and providing a second chance to fold productively. Glutathione can reduce inappropriate disulfide bonds that may be formed upon exposure to oxidizing agents such as O_2, hydrogen peroxide, or superoxide (Chapter 52).

PERTURBATION OF PROTEIN CONFORMATION MAY HAVE PATHOLOGIC CONSEQUENCES

Prions

The transmissible spongiform encephalopathies, or **prion diseases,** are fatal neurodegenerative diseases characterized by spongiform changes, astrocytic gliomas, and neuronal loss resulting from the deposition of insoluble protein aggregates in neural cells. They include Creutzfeldt–Jakob disease in humans, scrapie in sheep, and bovine spongiform encephalopathy (mad cow disease) in cattle. A variant form of Creutzfeldt-Jacob disease (vCJD) that afflicts younger patients is associated with early-onset psychiatric and behavioral disorders. Prion diseases may manifest themselves as infectious, genetic, or sporadic disorders. Because no viral or bacterial gene encoding the pathologic prion protein could be identified, the source and mechanism of transmission of prion disease long remained elusive.

Today it is recognized that prion diseases are protein conformation diseases transmitted by altering the conformation, and hence the physical properties, of proteins endogenous to the host. Human prion-related protein (PrP), a glycoprotein encoded on the short arm of chromosome 20, normally is monomeric and rich in α helix. Pathologic prion proteins serve as the templates for the conformational transformation of normal PrP, known as PrPc, into PrPsc. PrPsc is rich in β sheet with many hydrophobic aminoacyl side chains exposed to solvent. As each new PrPsc molecule is formed, it triggers the production of yet more pathologic variants in a conformational chain reaction. Because PrPsc molecules associate strongly with one other through their exposed hydrophobic regions, the accumulating PrPsc units coalesce to form insoluble protease-resistant aggregates. Since one pathologic prion or prion-related protein can serve as template for the conformational transformation of many times its number of PrPc molecules, prion diseases can be transmitted by the protein alone without involvement of DNA or RNA.

Alzheimer's Disease

Refolding or misfolding of another protein endogenous to human brain tissue, β-amyloid, is a prominent feature of Alzheimer's disease. While the main cause of Alzheimer's disease remains elusive, the characteristic senile plaques and neurofibrillary bundles contain aggregates of the protein β-amyloid, a 4.3 kDa polypeptide produced by proteolytic cleavage of a larger protein known as amyloid precursor protein. In Alzheimer's disease patients, levels of β-amyloid become elevated, and this protein undergoes a conformational transformation from a soluble α helix-rich state to a state rich in β sheet and prone to self-aggregation. Apolipoprotein E has been implicated as a potential mediator of this conformational transformation.

Beta-Thalassemias

Thalassemias are caused by genetic defects that impair the synthesis of one of the polypeptide subunits of hemoglobin (Chapter 6). During the burst of hemoglobin synthesis that occurs during erythrocyte development, a specific chaperone called α-hemoglobin-stabilizing protein (AHSP) binds to free hemoglobin α-subunits awaiting incorporation into the hemoglobin multimer. In the absence of this chaperone, free α-hemoglobin subunits aggregate, and the resulting precipitate has cytotoxic effects on the developing erythrocyte. Investigations using genetically modified mice suggest a role for AHSP in modulating the severity of β-thalassemia in human subjects.

COLLAGEN ILLUSTRATES THE ROLE OF POSTTRANSLATIONAL PROCESSING IN PROTEIN MATURATION

Protein Maturation Often Involves Making & Breaking of Covalent Bonds

The maturation of proteins into their final structural state often involves the cleavage or formation (or both) of covalent bonds, a process of **post-translational modification.** Many polypeptides are initially synthesized as larger precursors called **proproteins.** The "extra" polypeptide segments in these proproteins often serve as leader sequences that target a polypeptide to a particular organelle or facilitate its passage through a membrane. Other segments ensure that the potentially harmful activity of a protein such as the proteases trypsin and chymotrypsin

remains inhibited until these proteins reach their final destination. However, once these transient requirements are fulfilled and the now superfluous peptide regions are removed by selective proteolysis. Other covalent modifications may take place that add new chemical functionalities to a protein. The maturation of collagen illustrates both of these processes.

Collagen Is a Fibrous Protein

Collagen is the most abundant of the fibrous proteins that constitute more than 25% of the protein mass in the human body. Other prominent fibrous proteins include keratin and myosin. These fibrous proteins represent a primary source of structural strength for cells (ie, the cytoskeleton) and tissues. Skin derives its strength and flexibility from an intertwined mesh of collagen and keratin fibers, while bones and teeth are buttressed by an underlying network of collagen fibers analogous to steel strands in reinforced concrete. Collagen also is present in connective tissues such as ligaments and tendons. The high degree of tensile strength required to fulfill these structural roles requires elongated proteins characterized by repetitive amino acid sequences and a regular secondary structure.

Collagen Forms a Unique Triple Helix

Tropocollagen, the repeating unit of a mature collagen fiber, consists of three collagen polypeptides, each containing about 1000 amino acids, bundled together in a unique conformation, the collagen triple helix (**Figure 5–11**). A mature collagen fiber forms an elongated rod with an axial ratio of about 200. Three intertwined polypeptide strands, which twist to the left, wrap around one another in a right-handed fashion to form the collagen triple helix. The opposing handedness of this superhelix and its component polypeptides makes the collagen triple helix highly resistant to unwinding—a principle also applied to the steel cables of suspension bridges. A collagen triple helix has 3.3 residues per turn and a rise per residue nearly twice that of an α helix. The R groups of each polypeptide strand of the triple helix pack so closely that, in order to fit, one of the three must be H. Thus, every third amino acid residue in collagen is a glycine residue. Staggering of the three strands provides appropriate positioning of the requisite glycines throughout the helix. Collagen is also rich in proline and hydroxyproline, yielding a repetitive Gly-X-Y pattern (Figure 5–11) in which Y generally is proline or hydroxyproline.

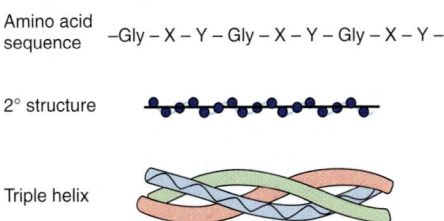

FIGURE 5–11 Primary, secondary, and tertiary structures of collagen.

Collagen triple helices are stabilized by hydrogen bonds between residues in *different* polypeptide chains, a process helped by the hydroxyl groups of hydroxyprolyl residues. Additional stability is provided by covalent cross links formed between modified lysyl residues both within and between polypeptide chains.

Collagen Is Synthesized as a Larger Precursor

Collagen is initially synthesized as a larger precursor polypeptide, procollagen. Numerous prolyl and lysyl residues of procollagen are hydroxylated by prolyl hydroxylase and lysyl hydroxylase, enzymes that require ascorbic acid (vitamin C; see Chapters 27 and 44). Hydroxyprolyl and hydroxylysyl residues provide additional hydrogen bonding capability that stabilizes the mature protein. In addition, glucosyl and galactosyl transferases attach glucosyl or galactosyl residues to the hydroxyl groups of specific hydroxylysyl residues.

The central portion of the precursor polypeptide then associates with other molecules to form the characteristic triple helix. This process is accompanied by the removal of the globular amino terminal and carboxyl terminal extensions of the precursor polypeptide by selective proteolysis. Certain lysyl residues are modified by lysyl oxidase, a copper-containing protein that converts ε-amino groups to aldehydes. The aldehydes can either undergo an aldol condensation to form a C=C double bond or to form a Schiff base (eneimine) with the ε-amino group of an unmodified lysyl residue, which is subsequently reduced to form a C—N single bond. These covalent bonds cross-link the individual polypeptides and imbue the fiber with exceptional strength and rigidity.

Nutritional & Genetic Disorders Can Impair Collagen Maturation

The complex series of events in collagen maturation provide a model that illustrates the biologic consequences of incomplete polypeptide maturation. The best-known defect in collagen biosynthesis is scurvy, a result of a dietary deficiency of vitamin C required by prolyl and lysyl hydroxylases. The resulting deficit in the number of hydroxyproline and hydroxylysine residues undermines the conformational stability of collagen fibers, leading to bleeding gums, swelling joints, poor wound healing, and ultimately death. Menkes' syndrome, characterized by kinky hair and growth retardation, reflects a dietary deficiency of the copper required by lysyl oxidase, which catalyzes a key step in the formation of the covalent cross-links that strengthen collagen fibers.

Genetic disorders of collagen biosynthesis include several forms of osteogenesis imperfecta, characterized by fragile bones. In the Ehlers–Danlos syndrome, a group of connective tissue disorders that involve impaired integrity of supporting structures, defects in the genes that encode α collagen-1, procollagen *N*-peptidase, or lysyl hydroxylase result in mobile joints and skin abnormalities (see also Chapter 48).

SUMMARY

- Proteins may be classified based on their solubility, shape, or function or on the presence of a prosthetic group, such as heme.

- The gene-encoded primary structure of a polypeptide is the sequence of its amino acids. Its secondary structure results from folding of polypeptides into hydrogen-bonded motifs such as the α helix, the β pleated sheet, β bends, and loops. Combinations of these motifs can form supersecondary motifs.

- Tertiary structure concerns the relationships between secondary structural domains. Quaternary structure of proteins with two or more polypeptides (oligomeric proteins) concerns the spatial relationships between various types of polypeptides.

- Primary structures are stabilized by covalent peptide bonds. Higher orders of structure are stabilized by weak forces—multiple hydrogen bonds, salt (electrostatic) bonds, and association of hydrophobic R groups.

- The phi (Φ) angle of a polypeptide is the angle about the C_α—N bond; the psi (Ψ) angle is that about the C_α—C_o bond. Most combinations of phi-psi angles are disallowed due to steric hindrance. The phi-psi angles that form the α helix and the β sheet fall within the lower and upper left-hand quadrants of a Ramachandran plot, respectively.

- Protein folding is a poorly understood process. Broadly speaking, short segments of newly synthesized polypeptide fold into secondary structural units. Forces that bury hydrophobic regions from solvent then drive the partially folded polypeptide into a "molten globule" in which the modules of the secondary structure are rearranged to give the native conformation of the protein.

- Proteins that assist folding include protein disulfide isomerase, proline-*cis*, *trans*-isomerase, and the chaperones that participate in the folding of over half of mammalian proteins. Chaperones shield newly synthesized polypeptides from solvent and provide an environment for elements of secondary structure to emerge and coalesce into molten globules.

- X-Ray crystallography and NMR are key techniques used to study higher orders of protein structure.

- Prions—protein particles that lack nucleic acid—cause fatal transmissible spongiform encephalopathies such as Creutzfeldt–Jakob disease, scrapie, and bovine spongiform encephalopathy. Prion diseases involve an altered secondary-tertiary structure of a naturally occurring protein, PrPc. When PrPc interacts with its pathologic isoform PrPSc, its conformation is transformed from a predominantly α-helical structure to the β-sheet structure characteristic of PrPSc.

- Collagen illustrates the close linkage between protein structure and biologic function. Diseases of collagen maturation include Ehlers–Danlos syndrome and the vitamin C deficiency disease scurvy.

REFERENCES

Caughey B, Baron GS, Chesebro B, et al: Getting a grip on prions: oligomers, amyloids, and pathological membrane interactions. Annu Rev Biochem 2009;78:177.

Chiti F, Dobson CM: Protein misfolding, functional amyloid, and human disease. Annu Rev Biochem 2006;75:519.

Foster MP, McElroy CA, Amero CD: Solution NMR of large molecules and assemblies. Biochemistry 2007;46:331.

Gothel SF, Marahiel MA: Peptidyl-prolyl *cis–trans* isomerases, a superfamily of ubiquitous folding catalysts. Cell Mol Life Sci 1999;55:423.

Hardy J: Toward Alzheimer therapies based on genetic knowledge. Annu Rev Med 2004;55:15.

Hartl FU, Hayer-Hartl M: Converging concepts of protein folding in vitro and in vivo. Nat Struct Biol 2009;16:574.

Ho BK, Thomas A, Brasseur R: Revisiting the Ramachandran plot: hard-sphere repulsion, electrostatics, and H-bonding in the α-helix. Protein Sci 2003;12:2508.

Hristova K, Wimley WC, Mishra VK, et al: An amphipathic alpha-helix at a membrane interface: a structural study using a novel X-ray diffraction method. J Mol Biol 1999;290:99.

Irani DN, Johnson RT: Diagnosis and prevention of bovine spongiform encephalopathy and variant Creutzfeldt–Jakob disease. Annu Rev Med 2003;54:305.

Jorgensen WL: The many roles of computation in drug discovery. Science 2004;303:1813.

Khare SD, Dokholyan NV: Molecular mechanisms of polypeptide aggregation in human disease. Curr Protein Pept Sci 2007;8:573.

Kim J, Holtzman DM: Prion-like behavior of amyloid-ß. Science 2010;330:918.

Kong Y, Zhou S, Kihm AJ, et al: Loss of alpha-hemoglobin-stabilizing protein impairs erythropoiesis and exacerbates beta-thalassemia. J Clin Invest 2004;114:1457.

Myllyharju J: Prolyl 4-hydroxylases, the key enzymes of collagen biosynthesis. Matrix Biol 2003;22:15.

Rider MH, Bertrand L, Vertommen D, et al: 6-Phosphofructo-2-kinase/fructose-2,6-bisphosphatase: head-to-head with a bifunctional enzyme that controls glycolysis. Biochem J 2004;381:561.

Shoulders MD, Raines RT: Collagen structure and stability. Annu Rev Biochem 2009; 78:929.

Stoddard BL, Cohen BE, Brubaker M, et al: Millisecond Laue structures of an enzyme-product complex using photocaged substrate analogs. Nat Struct Biol 1998;5:891.

Wegrzyn RD, Deuerling E: Molecular guardians for newborn proteins: ribosome-associated chaperones and their role in protein folding. Cellular Mol Life Sci 2005;62:2727.

Young JC, Moarefi I, Hartl FU: *Hsp*90: a specialized but essential protein-folding tool. J Cell Biol 2001;154:267.

Proteins: Myoglobin & Hemoglobin

Peter J. Kennelly, PhD & Victor W. Rodwell, PhD

OBJECTIVES

After studying this chapter, you should be able to:

- Describe the most important structural similarities and differences between myoglobin and hemoglobin.
- Sketch binding curves for the oxygenation of myoglobin and hemoglobin.
- Identify the covalent linkages and other close associations between heme and globin in oxymyoglobin and oxyhemoglobin.
- Explain why the physiologic function of hemoglobin requires that its O_2-binding curve be sigmoidal rather than hyperbolic.
- Explain the role of a hindered environment on the ability of hemoglobin to bind carbon monoxide.
- Define P_{50} and indicate its significance in oxygen transport and delivery.
- Describe the structural and conformational changes in hemoglobin that accompany its oxygenation and subsequent deoxygenation.
- Explain the role of 2,3-bisphosphoglycerate (BPG) in oxygen binding and delivery.
- Outline the role of hemoglobin in CO_2 and proton transport, and describe accompanying changes in the pK_a of the relevant imidazolium group.
- Describe the structural consequences to HbS of lowering pO_2.
- Identify the metabolic defect that occurs as a consequence of α- and -β thalassemias.

BIOMEDICAL IMPORTANCE

The heme proteins myoglobin and hemoglobin maintain a supply of oxygen essential for oxidative metabolism. Myoglobin, a monomeric protein of red muscle, stores oxygen as a reserve against oxygen deprivation. Hemoglobin, a tetrameric protein of erythrocytes, transports O_2 to the tissues and returns CO_2 and protons to the lungs. Cyanide and carbon monoxide kill because they disrupt the physiologic function of the heme proteins cytochrome oxidase and hemoglobin, respectively. The secondary-tertiary structure of the subunits of hemoglobin resembles myoglobin. However, the tetrameric structure of hemoglobin permits cooperative interactions that are central to its function. For example, 2,3-BPG promotes the efficient release of O_2 by stabilizing the quaternary structure of deoxyhemoglobin. Hemoglobin and myoglobin illustrate both protein structure–function relationships and the molecular basis of genetic diseases such as sickle cell disease and the thalassemias.

HEME & FERROUS IRON CONFER THE ABILITY TO STORE & TO TRANSPORT OXYGEN

Myoglobin and hemoglobin contain **heme,** a cyclic tetrapyrrole consisting of four molecules of pyrrole linked by methyne bridges. This planar network of conjugated double bonds absorbs visible light and colors heme deep red. The substituents at the β-positions of heme are methyl (M), vinyl (V), and propionate (Pr) groups arranged in the order M, V, M, V, M, Pr, Pr, M (**Figure 6–1**). The atom of ferrous iron (Fe^{2+}) resides at the center of the planar tetrapyrrole. Other proteins with metal-containing tetrapyrrole prosthetic groups include the cytochromes (Fe and Cu) and chlorophyll (Mg) (see Chapter 31). Oxidation and reduction of the Fe and Cu atoms of cytochromes are essential to their biologic function as carriers of electrons. By contrast, oxidation of the Fe^{2+} of myoglobin or hemoglobin to Fe^{3+} destroys their biologic activity.

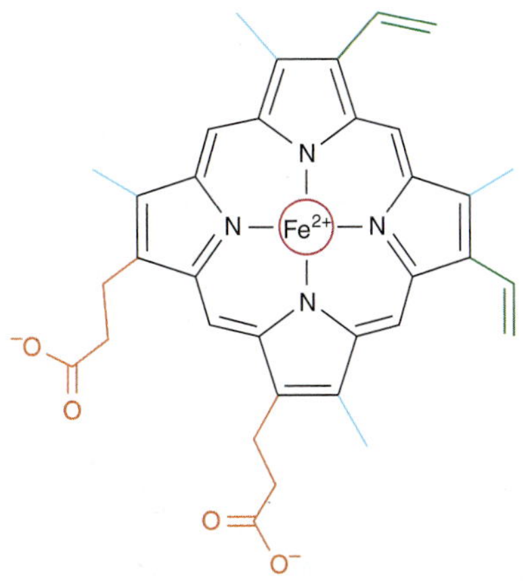

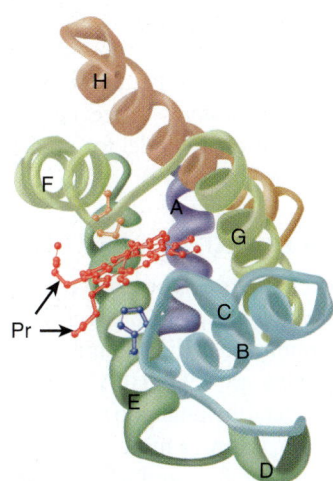

FIGURE 6–1   **Heme.** The pyrrole rings and methyne bridge carbons are coplanar, and the iron atom (Fe^{2+}) resides in almost the same plane. The fifth and sixth coordination positions of Fe^{2+} are directly perpendicular to—and directly above and below—the plane of the heme ring. Observe the nature of the methyl (blue), vinyl (green), and propionate (orange) substituent groups on the β carbons of the pyrrole rings, the central iron atom (red), and the location of the polar side of the heme ring (at about 7 o'clock) that faces the surface of the myoglobin molecule.

Myoglobin Is Rich in α Helix

Oxygen stored in red muscle myoglobin is released during O_2 deprivation (eg, severe exercise) for use in muscle mitochondria for aerobic synthesis of ATP (see Chapter 13). A 153-aminoacyl residue polypeptide (MW 17,000), myoglobin folds into a compact shape that measures $4.5 \times 3.5 \times 2.5$ nm (**Figure 6–2**). An unusually high proportion, about 75%, of the

residues are present in eight right-handed 7–20 residue α helices. Starting at the amino terminal, these are termed helices A–H. Typical of globular proteins, the surface of myoglobin is rich in amino acids bearing polar and potentially charged side chains, while—with only two exceptions—the interior contains only residues such as Leu, Val, Phe, and Met that possess nonpolar R groups. The exceptions are His E7 and His F8, the seventh and eighth residues in helices E and F, which lie close to the heme iron, where they function in O_2 binding.

Histidines F8 & E7 Perform Unique Roles in Oxygen Binding

The heme of myoglobin lies in a crevice between helices E and F oriented with its polar propionate groups facing the surface of the globin (Figure 6–2). The remainder resides in the nonpolar interior. The fifth coordination position of the iron is occupied by a nitrogen from the imidazole ring of the **proximal histidine,** His F8. The **distal histidine,** His E7, lies on the side of the heme ring opposite to His F8.

The Iron Moves Toward the Plane of the Heme when Oxygen Is Bound

The iron of unoxygenated myoglobin lies 0.03 nm (0.3 Å) outside the plane of the heme ring, toward His F8. The heme therefore "puckers" slightly. When O_2 occupies the sixth coordination position, the iron moves to within 0.01 nm (0.1 Å) of the plane of the heme ring. Oxygenation of myoglobin thus is accompanied by motion of the iron, of His F8, and of residues linked to His F8.

Apomyoglobin Provides a Hindered Environment for the Heme Iron

When O_2 binds to myoglobin, the bond between the first oxygen atom and the Fe^{2+} is perpendicular to the plane of the heme ring. The bond that links the first and second oxygen atoms lies at an angle of 121° to the plane of the heme, orienting the second oxygen away from the distal histidine (**Figure 6–3**, left).

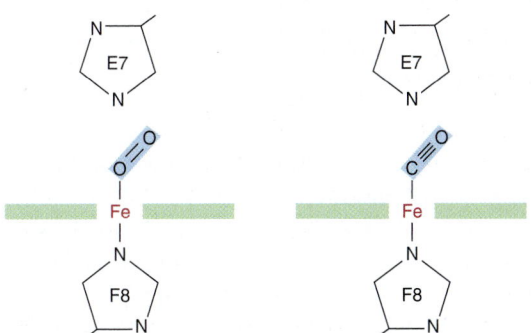

FIGURE 6–2   **Three-dimensional structure of myoglobin.** Shown is a ribbon diagram tracing the polypeptide backbone of myoglobin. The color of the polypeptide chain is graded along the visible spectrum from blue (N-terminal) to tan (C-terminal). The heme prosthetic group is red. The α-helical regions are designated A through H. The distal (E7) and proximal (F8) histidine residues are highlighted in blue and orange, respectively. Note how the polar propionate substituents (Pr) project out of the heme toward solvent. (Adapted from Protein Data Bank ID no. 1a6n.)

FIGURE 6–3   **Angles for bonding of oxygen and carbon monoxide (CO) to the heme iron of myoglobin.** The distal E7 histidine hinders bonding of CO at the preferred (90°) angle to the plane of the heme ring.

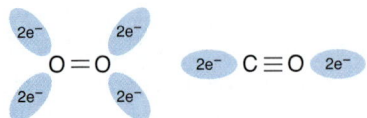

FIGURE 6–4 **Orientation of the lone pairs of electrons relative to the O = O and C ≡ O bonds of oxygen and carbon monoxide.** In molecular oxygen, formation of the double bond between the two oxygen atoms is facilitated by the adoption of an sp^2 hybridization state by the valence electron of each oxygen atom. As a consequence, the two atoms of the oxygen molecule and each lone pair of electrons are coplanar and separated by an angle of roughly 120° (left). By contrast, the two atoms of carbon monoxide are joined by a triple bond, which requires that the carbon and oxygen atoms adopt an sp hybridization state. In this state the lone pairs of electrons and triple bonds are arranged in a linear fashion, where they are separated by an angle of 180° (right).

This permits maximum overlap between the iron and one of the lone pairs of electrons on the sp^2 hybridized oxygen atoms, which lie at an angle of roughly 120° to the axis of the O=O double bond (**Figure 6–4**, left). Isolated heme binds carbon monoxide (CO) 25,000 times more strongly than oxygen. Since CO is present in small quantities in the atmosphere and arises in cells from the catabolism of heme, why is it that CO does not completely displace O_2 from heme iron? The accepted explanation is that the apoproteins of myoglobin and hemoglobin create a **hindered environment.** When CO binds to isolated heme, all the three atoms (Fe, C, and O) lie perpendicular to the plane of the heme. This geometry maximizes the overlap between the lone pair of electrons on the sp hybridized oxygen of the CO molecule and the Fe^{2+} iron (Figure 6–4, right). However, in myoglobin and hemoglobin the distal histidine sterically precludes this preferred, high-affinity orientation of CO, but not that of O_2. Binding at a less favored angle reduces the strength of the heme-CO bond to about 200 times that of the heme-O_2 bond (Figure 6–3, right) at which level the great excess of O_2 over CO normally present dominates. Nevertheless, about 1% of myoglobin typically is present combined with CO.

THE OXYGEN DISSOCIATION CURVES FOR MYOGLOBIN & HEMOGLOBIN SUIT THEIR PHYSIOLOGIC ROLES

Why is myoglobin unsuitable as an O_2 transport protein but well suited for O_2 storage? The relationship between the concentration, or partial pressure, of O_2 (Po_2) and the quantity of O_2 bound is expressed as an O_2 saturation isotherm (**Figure 6–5**). The oxygen-binding curve for myoglobin is hyperbolic. Myoglobin therefore loads O_2 readily at the Po_2 of the lung capillary bed (100 mm Hg). However, since myoglobin releases only a small fraction of its bound O_2 at the Po_2 values typically encountered in active muscle (20 mm Hg) or other tissues (40 mm Hg), it represents an ineffective vehicle for delivery of O_2. When strenuous exercise lowers the Po_2 of muscle tissue to

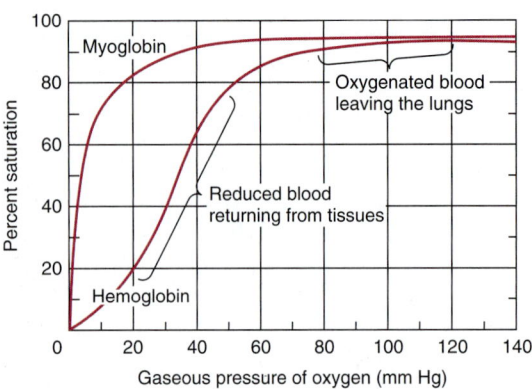

FIGURE 6–5 **Oxygen-binding curves of both hemoglobin and myoglobin.** Arterial oxygen tension is about 100 mm Hg; mixed venous oxygen tension is about 40 mm Hg; capillary (active muscle) oxygen tension is about 20 mm Hg; and the minimum oxygen tension required for cytochrome oxidase is about 5 mm Hg. Association of chains into a tetrameric structure (hemoglobin) results in much greater oxygen delivery than would be possible with single chains. (Modified, with permission, from Scriver CR et al (editors): *The Molecular and Metabolic Bases of Inherited Disease,* 7th ed. McGraw-Hill, 1995.)

about 5 mm Hg, myoglobin releases O_2 for mitochondrial synthesis of ATP, permitting continued muscular activity.

THE ALLOSTERIC PROPERTIES OF HEMOGLOBINS RESULT FROM THEIR QUATERNARY STRUCTURES

The properties of individual hemoglobins are consequences of their quaternary as well as of their secondary and tertiary structures. The quaternary structure of hemoglobin confers striking additional properties, absent from monomeric myoglobin, which adapts it to its unique biologic roles. The **allosteric** (Gk *allos* "other," *steros* "space") properties of hemoglobin provide, in addition, a model for understanding other allosteric proteins (see Chapter 18).

Hemoglobin Is Tetrameric

Hemoglobins are tetramers composed of pairs of two different polypeptide subunits (**Figure 6–6**). Greek letters are used to designate each subunit type. The subunit composition of the principal hemoglobins are $\alpha_2\beta_2$ (HbA; normal adult hemoglobin), $\alpha_2\gamma_2$ (HbF; fetal hemoglobin), $\alpha_2\beta_2^S$ (HbS; sickle cell hemoglobin), and $\alpha_2\delta_2$ (HbA$_2$; a minor adult hemoglobin). The primary structures of the β, γ, and δ chains of human hemoglobin are highly conserved.

Myoglobin & the β Subunits of Hemoglobin Share Almost Identical Secondary and Tertiary Structures

Despite differences in the kind and number of amino acids present, myoglobin and the β polypeptide of hemoglobin A

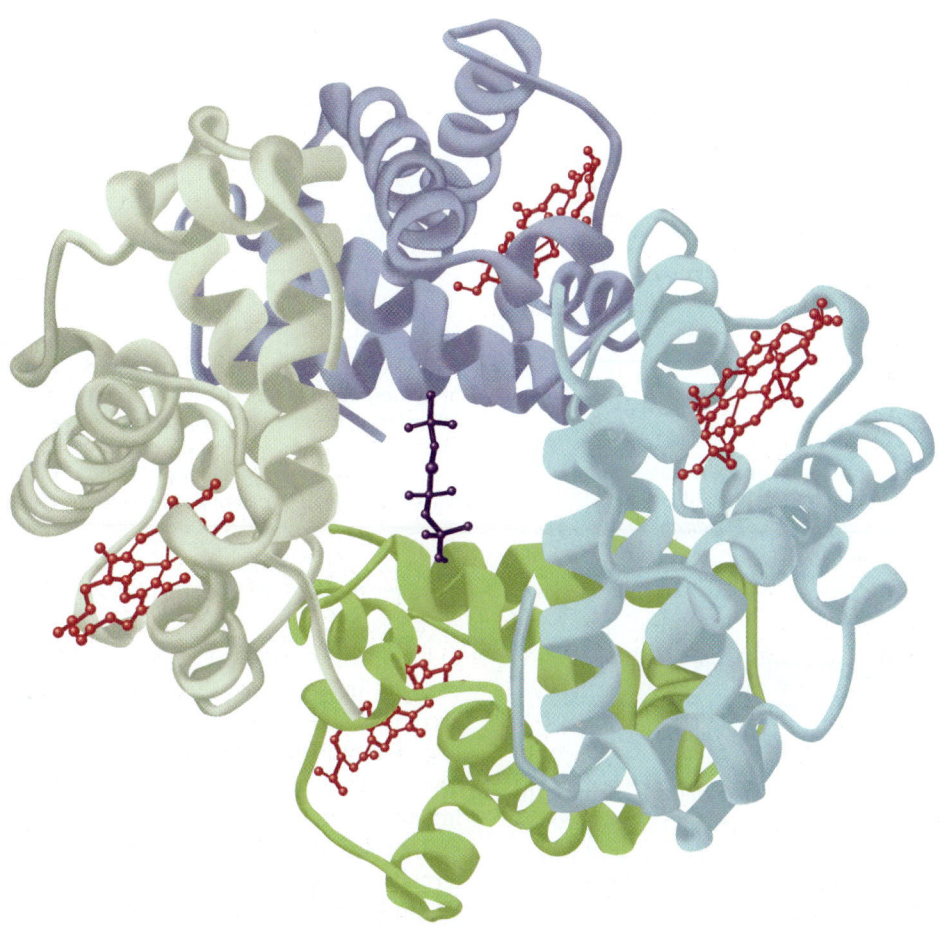

FIGURE 6–6 Hemoglobin. Shown is the three-dimensional structure of deoxyhemoglobin with a molecule of 2,3-bisphosphoglycerate (dark blue) bound. The two α subunits are colored in the darker shades of green and blue, the two β subunits in the lighter shades of green and blue, and the heme prosthetic groups in red. (Adapted from Protein Data Bank ID no. 1b86.)

have almost identical secondary and tertiary structures. Similarities include the location of the heme and the helical regions, and the presence of amino acids with similar properties at comparable locations. Although it possesses seven rather than eight helical regions, the α polypeptide of hemoglobin also closely resembles myoglobin.

Oxygenation of Hemoglobin Triggers Conformational Changes in the Apoprotein

Hemoglobins bind four molecules of O_2 per tetramer, one per heme. A molecule of O_2 binds to a hemoglobin tetramer more readily if other O_2 molecules are already bound (Figure 6–5). Termed **cooperative binding,** this phenomenon permits hemoglobin to maximize both the quantity of O_2 loaded at the Po_2 of the lungs and the quantity of O_2 released at the Po_2 of the peripheral tissues. Cooperative interactions, an exclusive property of multimeric proteins, are critically important to aerobic life.

P_{50} Expresses the Relative Affinities of Different Hemoglobins for Oxygen

The quantity P_{50}, a measure of O_2 concentration, is the partial pressure of O_2 that half-saturates a given hemoglobin. Depending on the organism, P_{50} can vary widely, but in all instances, it will exceed the Po_2 of the peripheral tissues. For example, the values of P_{50} for HbA and HbF are 26 and 20 mm Hg, respectively. In the placenta, this difference enables HbF to extract oxygen from the HbA in the mother's blood. However, HbF is suboptimal postpartum since its high affinity for O_2 limits the quantity of O_2 delivered to the tissues.

The subunit composition of hemoglobin tetramers undergoes complex changes during development. The human fetus initially synthesizes a $\xi_2\varepsilon_2$ tetramer. By the end of the first trimester, ξ and ε subunits have been replaced by α and γ subunits, forming HbF ($\alpha_2\gamma_2$), the hemoglobin of late fetal life. While synthesis of β subunits begins in the third trimester, β subunits do not completely replace γ subunits to yield adult HbA ($\alpha_2\beta_2$) until some weeks postpartum (**Figure 6–7**).

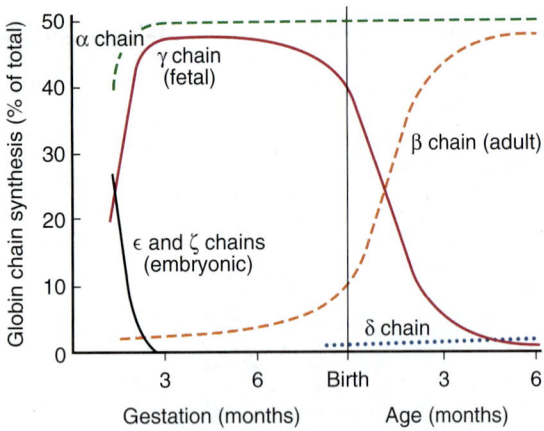

FIGURE 6–7 **Developmental pattern of the quaternary structure of fetal and newborn hemoglobins.** (Reproduced, with permission, from Ganong WF: *Review of Medical Physiology*, 20th ed. McGraw-Hill, 2001.)

Oxygenation of Hemoglobin Is Accompanied by Large Conformational Changes

The binding of the first O_2 molecule to deoxyHb shifts the heme iron toward the plane of the heme ring from a position about 0.04 nm beyond it (**Figure 6–8**). This motion is transmitted to the proximal (F8) histidine and to the residues attached thereto, which in turn causes the rupture of salt bridges

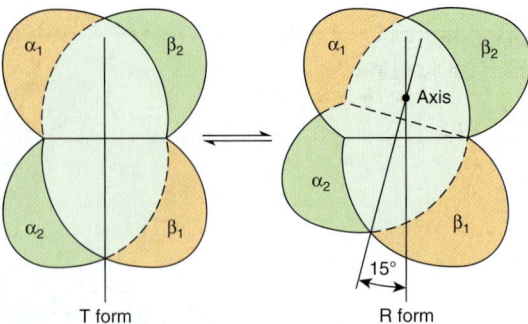

FIGURE 6–9 **During transition of the T form to the R form of hemoglobin, the $\alpha_2\beta_2$ pair of subunits (green) rotates through 15° relative to the pair of $\alpha_1\beta_1$ subunits (yellow).** The axis of rotation is eccentric, and the $\alpha_2\beta_2$ pair also shifts toward the axis somewhat. In the representation, the tan $\alpha_1\beta_1$ pair is shown fixed while the green $\alpha_2\beta_2$ pair of subunits both shifts and rotates.

between the carboxyl terminal residues of all four subunits. As a result, one pair of α/β subunits rotates 15° with respect to the other, compacting the tetramer (**Figure 6–9**). Profound changes in secondary, tertiary, and quaternary structures accompany the O_2-induced transition of hemoglobin from the low-affinity **T (taut)** state to the high-affinity **R (relaxed) state.** These changes significantly increase the affinity of the remaining unoxygenated hemes for O_2, as subsequent binding events require the rupture of fewer salt bridges (**Figure 6–10**). The terms T and R also are used to refer to the low-affinity and high-affinity conformations of allosteric enzymes, respectively.

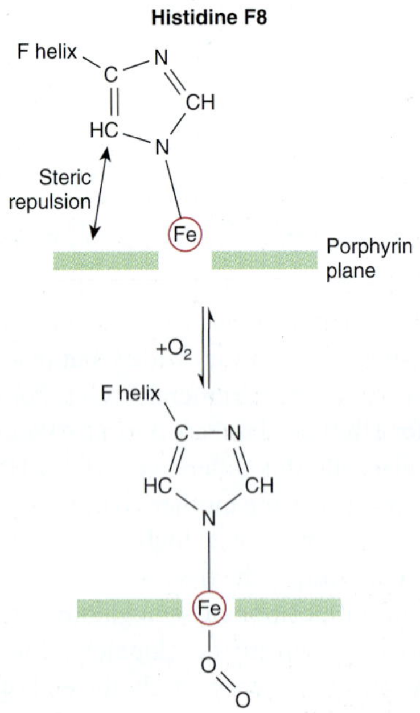

After Releasing O_2 at the Tissues, Hemoglobin Transports CO_2 & Protons to the Lungs

In addition to transporting O_2 from the lungs to peripheral tissues, hemoglobin transports CO_2, the byproduct of respiration, and protons from peripheral tissues to the lungs. Hemoglobin carries CO_2 as carbamates formed with the amino terminal nitrogens of the polypeptide chains:

$$CO_2 + Hb-NH_3^+ \rightleftharpoons 2H^+ + Hb-\overset{H}{\underset{}{N}}-\overset{O}{\overset{\|}{C}}-O^-$$

Carbamates change the charge on amino terminals from positive to negative, favoring salt bridge formation between α and β chains.

Hemoglobin carbamates account for about 15% of the CO_2 in venous blood. Much of the remaining CO_2 is carried as bicarbonate, which is formed in erythrocytes by the hydration of CO_2 to carbonic acid (H_2CO_3), a process catalyzed by carbonic anhydrase. At the pH of venous blood, H_2CO_3 dissociates into bicarbonate and a proton.

$$CO_2 + H_2O \underset{\text{CARBONIC ANHYDRASE}}{\rightleftharpoons} H_2CO_3 \underset{\text{(Spontaneous)}}{\rightleftharpoons} HCO_3^- + H^+$$

Carbonic acid

Histidine F8

F helix

Steric repulsion

Fe

Porphyrin plane

$+O_2$

F helix

Fe

FIGURE 6–8 **The iron atom moves into the plane of the heme on oxygenation.** Histidine F8 and its associated residues are pulled along with the iron atom. (Slightly modified and reproduced, with permission, from Stryer L: *Biochemistry*, 3rd ed. Freeman, 1988. Copyright © 1988 W. H. Freeman and Company.)

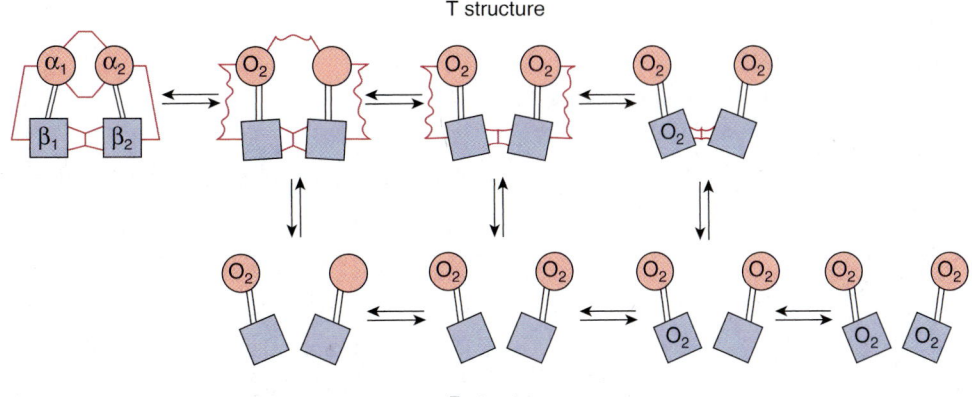

T structure

R structure

FIGURE 6–10   Transition from the T structure to the R structure. In this model, salt bridges (red lines) linking the subunits in the T structure break progressively as oxygen is added, and even those salt bridges that have not yet ruptured are progressively weakened (wavy red lines). The transition from T to R does not take place after a fixed number of oxygen molecules have been bound but becomes more probable as each successive oxygen binds. The transition between the two structures is influenced by protons, carbon dioxide, chloride, and BPG; the higher their concentration, the more oxygen must be bound to trigger the transition. Fully oxygenated molecules in the T structure and fully deoxygenated molecules in the R structure are not shown because they are unstable. (Modified and redrawn, with permission, from Perutz MF: Hemoglobin structure and respiratory transport. Sci Am [Dec] 1978; 239:92.)

Deoxyhemoglobin binds one proton for every two O_2 molecules released, contributing significantly to the buffering capacity of blood. The somewhat lower pH of peripheral tissues, aided by carbamation, stabilizes the T state and thus enhances the delivery of O_2. In lungs, the process reverses. As O_2 binds to deoxyhemoglobin, protons are released and combine with bicarbonate to form carbonic acid. Dehydration of H_2CO_3, catalyzed by carbonic anhydrase, forms CO_2, which is exhaled. Binding of oxygen thus drives the exhalation of CO_2 (**Figure 6–11**). This reciprocal coupling of proton and O_2 binding is termed the **Bohr effect**. The Bohr effect is dependent upon **cooperative interactions between the hemes of the hemoglobin tetramer.** Myoglobin, a monomer, exhibits no Bohr effect.

Protons Arise from Rupture of Salt Bridges When O_2 Binds

Protons responsible for the Bohr effect arise from rupture of salt bridges during the binding of O_2 to T-state hemoglobin. Conversion to the oxygenated R state breaks salt bridges involving β chain residue His 146. The subsequent dissociation of protons from His 146 drives the conversion of bicarbonate to carbonic acid (Figure 6–11). Upon the release of O_2, the T structure and its salt bridges re-form. This conformational change increases the pK_a of the β chain His 146 residues, which bind protons. By facilitating the re-formation of salt bridges, an increase in proton concentration enhances the release of O_2 from oxygenated (R-state) hemoglobin. Conversely, an increase in Po_2 promotes proton release.

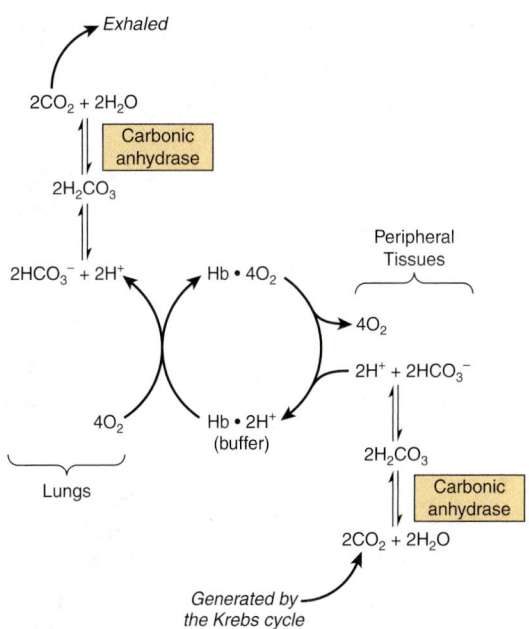

FIGURE 6–11   The Bohr effect. Carbon dioxide generated in peripheral tissues combines with water to form carbonic acid, which dissociates into protons and bicarbonate ions. Deoxyhemoglobin acts as a buffer by binding protons and delivering them to the lungs. In the lungs, the uptake of oxygen by hemoglobin releases protons that combine with bicarbonate ion, forming carbonic acid, which when dehydrated by carbonic anhydrase becomes carbon dioxide, which then is exhaled.

2,3- BPG Stabilizes the T Structure of Hemoglobin

A low Po_2 in peripheral tissues promotes the synthesis of 2,3-BPG in erythrocytes from the glycolytic intermediate 1,3-BPG.

The hemoglobin tetramer binds one molecule of BPG in the central cavity formed by its four subunits (Figure 6–6). However, the space between the H helices of the β chains lining the cavity is sufficiently wide to accommodate BPG only when hemoglobin is in the T state. BPG forms salt bridges with the terminal amino groups of both β chains via Val NA1 and with Lys EF6 and His H21 (**Figure 6–12**). BPG therefore stabilizes deoxygenated (T-state) hemoglobin by forming additional salt bridges that must be broken prior to conversion to the R state.

Residue H21 of the γ subunit of HbF is Ser rather than His. Since Ser cannot form a salt bridge, BPG binds more weakly to HbF than to HbA. The lower stabilization afforded to the T state by BPG accounts for HbF having a higher affinity for O_2 than HbA.

Adaptation to High Altitude

Physiologic changes that accompany prolonged exposure to high altitude include an increase in the number of erythrocytes and in their concentrations of hemoglobin and of BPG.

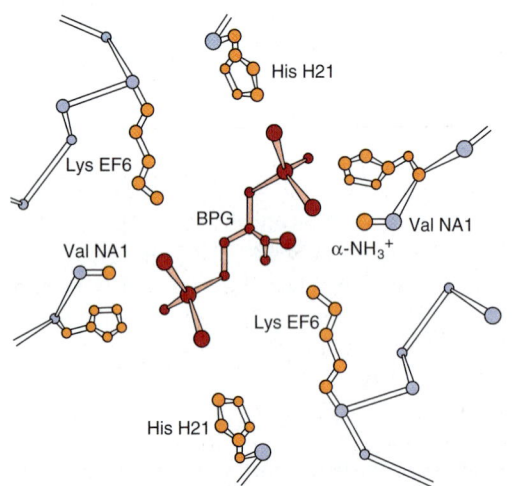

FIGURE 6–12 **Mode of binding of 2,3-bisphosphoglycerate (BPG) to human deoxyhemoglobin.** BPG interacts with three positively charged groups on each β chain. (Based on Arnone A: X-ray diffraction study of binding of 2,3-diphosphoglycerate to human deoxyhemoglobin. Nature 1972; 237:146. Copyright © 1972. Adapted by permission from Macmillan Publishers Ltd.)

Elevated BPG lowers the affinity of HbA for O_2 (increases P_{50}), which enhances the release of O_2 at peripheral tissues.

NUMEROUS MUTATIONS AFFECTING HUMAN HEMOGLOBINS HAVE BEEN IDENTIFIED

Mutations in the genes that encode the α or β subunits of hemoglobin potentially can affect its biologic function. However, almost all of the over 1,100 known genetic mutations affecting human hemoglobins are both extremely rare and benign, presenting no clinical abnormalities. When a mutation does compromise biologic function, the condition is termed a **hemoglobinopathy.** It is estimated that more than 7% of the globe's population are carriers for hemoglobin disorders. The URL http://globin.cse.psu.edu/ (Globin Gene Server) provides information about—and links for—normal and mutant hemoglobins. Selected examples are described below.

Methemoglobin & Hemoglobin M

In methemoglobinemia, the heme iron is ferric rather than ferrous. Methemoglobin thus can neither bind nor transport O_2. Normally, the enzyme methemoglobin reductase reduces the Fe^{3+} of methemoglobin to Fe^{2+}. Methemoglobin can arise by oxidation of Fe^{2+} to Fe^{3+} as a side effect of agents such as sulfonamides, from hereditary hemoglobin M, or consequent to reduced activity of the enzyme methemoglobin reductase.

In hemoglobin M, histidine F8 (His F8) has been replaced by tyrosine. The iron of HbM forms a tight ionic complex with the phenolate anion of tyrosine that stabilizes the Fe^{3+} form. In α-chain hemoglobin M variants, the R-T equilibrium favors the T state. Oxygen affinity is reduced, and the Bohr effect is absent. β-chain hemoglobin M variants exhibit R-T switching, and the Bohr effect is therefore present.

Mutations that favor the R state (eg, hemoglobin Chesapeake) increase O_2 affinity. These hemoglobins therefore fail to deliver adequate O_2 to peripheral tissues. The resulting tissue hypoxia leads to **polycythemia,** an increased concentration of erythrocytes.

Hemoglobin S

In HbS, the nonpolar amino acid valine has replaced the polar surface residue Glu6 of the β subunit, generating a hydrophobic **"sticky patch"** on the surface of the β subunit of both oxyHbS and deoxyHbS. Both HbA and HbS contain a complementary sticky patch on their surfaces that is exposed only in the deoxygenated T state. Thus, at low Po_2, deoxyHbS can polymerize to form long, insoluble fibers. Binding of deoxyHbA terminates fiber polymerization, since HbA lacks the second sticky patch necessary to bind another Hb molecule (**Figure 6–13**). These twisted helical fibers distort the erythrocyte into a characteristic sickle shape, rendering it vulnerable

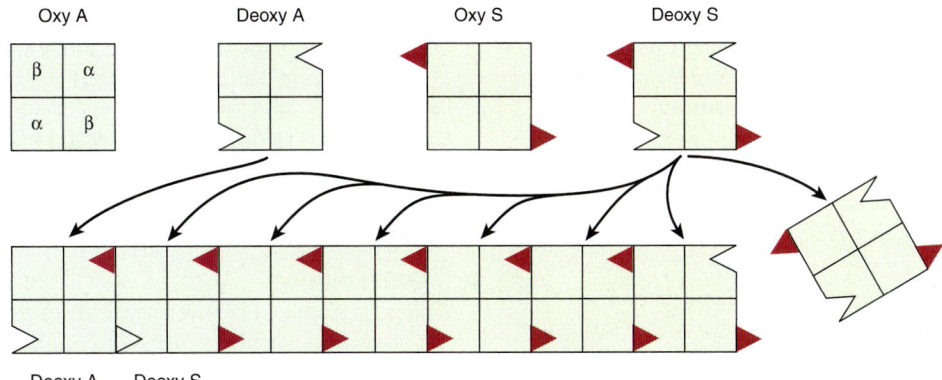

FIGURE 6–13 **Representation of the sticky patch (▲) on hemoglobin S and its "receptor" (△) on deoxyhemoglobin A and deoxyhemoglobin S.** The complementary surfaces allow deoxyhemoglobin S to polymerize into a fibrous structure, but the presence of deoxyhemoglobin A will terminate the polymerization by failing to provide sticky patches. (Modified and reproduced, with permission, from Stryer L: *Biochemistry*, 3rd ed. Freeman, 1988. Copyright © 1988 W. H. Freeman and Company.)

to lysis in the interstices of the splenic sinusoids. They also cause multiple secondary clinical effects. A low Po_2, such as that at high altitudes, exacerbates the tendency to polymerize. Emerging treatments for sickle cell disease include inducing HbF expression to inhibit the polymerization of HbS, stem cell transplantation, and, in the future, gene therapy.

BIOMEDICAL IMPLICATIONS

Myoglobinuria

Following massive crush injury, myoglobin released from damaged muscle fibers colors the urine dark red. Myoglobin can be detected in plasma following a myocardial infarction, but assay of serum enzymes (see Chapter 7) provides a more sensitive index of myocardial injury.

Anemias

Anemias, reductions in the number of red blood cells or of hemoglobin in the blood, can reflect impaired synthesis of hemoglobin (eg, in iron deficiency; Chapter 50) or impaired production of erythrocytes (eg, in folic acid or vitamin B_{12} deficiency; Chapter 44). Diagnosis of anemias begins with spectroscopic measurement of blood hemoglobin levels.

Thalassemias

The genetic defects known as thalassemias result from the partial or total absence of one or more α or β chains of hemoglobin. Over 750 different mutations have been identified, but only three are common. Either the α chain (alpha thalassemias) or β chain (beta thalassemias) can be affected. A superscript indicates whether a subunit is completely absent (α^0 or β^0) or whether its synthesis is reduced (α^- or β^-). Apart from marrow transplantation, treatment is symptomatic.

Certain mutant hemoglobins are common in many populations, and a patient may inherit more than one type. Hemoglobin disorders thus present a complex pattern of clinical phenotypes. The use of DNA probes for their diagnosis is considered in Chapter 39.

Glycated Hemoglobin (HbA$_{1c}$)

When blood glucose enters the erythrocytes, it glycates the ε-amino group of lysyl residues and the amino terminals of hemoglobin. The fraction of hemoglobin glycated, normally about 5%, is proportionate to blood glucose concentration. Since the half-life of an erythrocyte is typically 60 days, the level of glycated hemoglobin (HbA$_{1c}$) reflects the mean blood glucose concentration over the preceding 6–8 weeks. Measurement of HbA$_{1c}$ therefore provides valuable information for management of diabetes mellitus.

SUMMARY

- Myoglobin is monomeric; hemoglobin is a tetramer of two subunit types ($\alpha_2\beta_2$ in HbA). Despite having different primary structures, myoglobin and the subunits of hemoglobin have nearly identical secondary and tertiary structures.

- Heme, an essentially planar, slightly puckered, cyclic tetrapyrrole has a central Fe^{2+} linked to all four nitrogen atoms of the heme, to histidine F8, and, in oxyMb and oxyHb, also to O_2.

- The O_2-binding curve for myoglobin is hyperbolic, but for hemoglobin it is sigmoidal, a consequence of cooperative interactions in the tetramer. Cooperativity maximizes the ability of hemoglobin both to load O_2 at the Po_2 of the lungs and to deliver O_2 at the Po_2 of the tissues.

- Relative affinities of different hemoglobins for oxygen are expressed as P_{50}, the Po_2 that half-saturates them with O_2. Hemoglobins saturate at the partial pressures of their respective respiratory organ, eg, the lung or placenta.

- On oxygenation of hemoglobin, the iron, histidine F8, and linked residues move toward the heme ring. Conformational changes that accompany oxygenation include rupture of salt bonds and loosening of the quaternary structure, facilitating binding of additional O_2.

- 2,3- BPG in the central cavity of deoxyHb forms salt bonds with the β subunits that stabilize deoxyHb. On oxygenation, the central cavity contracts, BPG is extruded, and the quaternary structure loosens.

- Hemoglobin also functions in CO_2 and proton transport from tissues to lungs. Release of O_2 from oxyHb at the tissues is accompanied by uptake of protons due to lowering of the pK_a of histidine residues.

- In sickle cell hemoglobin (HbS), Val replaces the $\beta6$ Glu of HbA, creating a "sticky patch" that has a complement on deoxyHb (but not on oxyHb). DeoxyHbS polymerizes at low O_2 concentrations, forming fibers that distort erythrocytes into sickle shapes.

- Alpha and beta thalassemias are anemias that result from reduced production of α and β subunits of HbA, respectively.

REFERENCES

Frauenfelder H, McMahon BH, Fenimore PW: Myoglobin: The hydrogen atom of biology and a paradigm of complexity. Proc Natl Acad Sci USA 2003;100:8615.

Hardison RC, Chui DH, Riemer C, et al: Databases of human hemoglobin variants and other resources at the globin gene server. Hemoglobin 2001;25:183.

Lukin JA, Ho C: The structure–function relationship of hemoglobin in solution at atomic resolution. Chem Rev 2004;104:1219.

Ordway GA, Garry DJ: Myoglobin: An essential hemoprotein in striated muscle. J Exp Biol 2004;207:3441.

Papanikolaou E, Anagnou NP: Major challenges for gene therapy of thalassemia and sickle cell dsease. Curr Gene Ther 2010;10:404.

Schrier SL, Angelucci E: New strategies in the treatment of the thalassemias. Annu Rev Med 2005;56:157.

Steinberg MH, Brugnara C: Pathophysiological-based approaches to treatment of sickle-cell disease. Annu Rev Med 2003;54:89.

Umbreit J: Methemoglobin—it's not just blue: A concise review. Am J Hematol 207;82:134.

Weatherall DJ, Akinyanju O, Fucharoen S, et al: Inherited disorders of hemoglobin. In: *Disease Control Priorities in Developing Countries,* Jamison DT, Breman JG, Measham AR (editors). Oxford University Press and the World Bank, 2006;663–680.

Weatherall DJ, Clegg JB, Higgs DR, et al: The hemoglobinopathies. In: *The Metabolic Basis of Inherited Disease,* 8th ed. Scriver CR, Sly WS, Childs B, et al (editors). McGraw-Hill, 2000;4571.

Weatherall DJ, Clegg JD: *The Thalassemia Syndromes.* Blackwell Science, 2001.

Yonetani T, Laberge M: Protein dynamics explain the allosteric behaviors of hemoglobin. Biochim Biophys Acta 2008;1784:1146.

Enzymes: Mechanism of Action

Peter J. Kennelly, PhD & Victor W. Rodwell, PhD

OBJECTIVES

After studying this chapter, you should be able to:

- Illustrate the structural relationships between B vitamins and coenzymes.
- Outline the four principal mechanisms by which enzymes achieve catalysis.
- Describe how an "induced fit" facilitates substrate recognition and catalysis.
- Outline the underlying principles of enzyme-linked immunoassays.
- Explain how coupling an enzyme to an NAD(P)$^+$-dependent dehydrogenase can simplify assay of its activity.
- Identify enzymes and proteins whose plasma levels are used for the diagnosis and prognosis of a myocardial infarction.
- Describe the application of restriction endonucleases and of restriction fragment length polymorphisms in the detection of genetic diseases.
- Explain the utility of site-directed mutagenesis for the identification of residues involved in catalysis, in the recognition of substrates or allosteric effectors, or in the mechanism of enzyme action.
- Describe how the addition of fused affinity "tags" via recombinant DNA technology can facilitate purification of a protein expressed from its cloned gene.
- Indicate the function of specific proteases in the purification of affinity-tagged enzymes.
- Discuss the events that led to the discovery that RNAs can act as enzymes.

BIOMEDICAL IMPORTANCE

Enzymes are biologic polymers that catalyze the chemical reactions that make life, as we know it, possible. The presence and maintenance of a complete and balanced set of enzymes is essential for the breakdown of nutrients to supply energy and chemical building blocks; the assembly of those building blocks into proteins, DNA, membranes, cells, and tissues; and the harnessing of energy to power cell motility, neural function, and muscle contraction. The vast majority of enzymes are proteins. Notable exceptions include **ribosomal RNAs** and a handful of self-cleaving or self-splicing RNA molecules known collectively as **ribozymes.** The ability to assay the activity of specific enzymes in blood, other tissue fluids, or cell extracts aids in the diagnosis and prognosis of disease. Deficiencies in the quantity or catalytic activity of key enzymes can result from genetic defects, nutritional deficits, or toxins.

Defective enzymes can result from genetic mutations or infection by viral or bacterial pathogens (eg, *Vibrio cholerae*). Medical scientists address imbalances in enzyme activity by using pharmacologic agents to inhibit specific enzymes and are investigating gene therapy as a means to remedy deficits in enzyme level or function.

In addition to serving as the catalysts for all metabolic processes, their impressive catalytic activity, substrate specificity, and stereospecificity enable enzymes to fulfill key roles in other processes related to human health and well-being. The absolute stereospecificity of enzymes is of a particular value for use as soluble or immobilized catalysts for specific reactions in the synthesis of a drug or antibiotic. Proteases and amylases augment the capacity of detergents to remove dirt and stains. Enzymes play an important role in producing or enhancing the nutrient value of food products for both humans and animals. The protease rennin, for example, is utilized in the production of cheeses

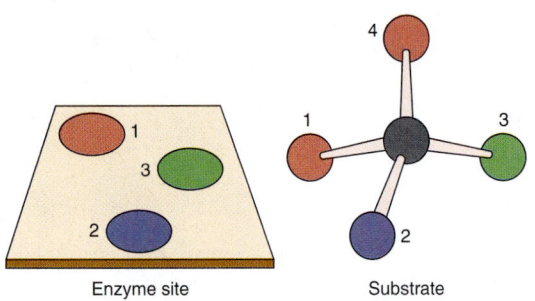

FIGURE 7–1 **Planar representation of the "three-point attachment" of a substrate to the active site of an enzyme.** Although atoms 1 and 4 are identical, once atoms 2 and 3 are bound to their complementary sites on the enzyme, only atom 1 can bind. Once bound to an enzyme, apparently identical atoms thus may be distinguishable, permitting a stereospecific chemical change.

while lactase is employed to remove lactose from milk for the benefit of lactose-intolerant persons deficient in this hydrolytic enzyme (Chapter 43).

ENZYMES ARE EFFECTIVE & HIGHLY SPECIFIC CATALYSTS

The enzymes that catalyze the conversion of one or more compounds (**substrates**) into one or more different compounds (**products**) enhance the rates of the corresponding noncatalyzed reaction by factors of at least 10^6. Like all catalysts, enzymes are neither consumed nor permanently altered as a consequence of their participation in a reaction.

In addition to being highly efficient, enzymes are also extremely selective catalysts. Unlike most catalysts used in synthetic chemistry, enzymes are specific both for the type of reaction catalyzed and for a single substrate or a small set of closely related substrates. Enzymes are also stereospecific catalysts and typically catalyze reactions of only one stereoisomer of a given compound—for example, D- but not L-sugars, L- but not D-amino acids. Since they bind substrates through at least "three points of attachment," enzymes can even convert nonchiral substrates to chiral products. **Figure 7–1** illustrates why the enzyme-catalyzed reduction of the nonchiral substrate pyruvate produces only L-lactate, not a racemic mixture of D- and L-lactate. The exquisite specificity of enzyme catalysts imbues living cells with the ability to simultaneously conduct and independently control a broad spectrum of chemical processes.

ENZYMES ARE CLASSIFIED BY REACTION TYPE

The commonly used names for most enzymes describe the type of reaction catalyzed, followed by the suffix -ase. For example, dehydrogenases remove hydrogen atoms, proteases hydrolyze proteins, and isomerases catalyze rearrangements in configuration. Modifiers may precede the name to indicate

the substrate (*xanthine* oxidase), the source of the enzyme (*pancreatic* ribonuclease), its regulation (*hormone-sensitive* lipase), or a feature of its mechanism of action (*cysteine* protease). Where needed, alphanumeric designators are added to identify multiple forms of an enzyme (eg, RNA polymerase *III*; protein kinase *C*β).

To address ambiguities, the International Union of Biochemists (IUB) developed an unambiguous system of enzyme nomenclature in which each enzyme has a unique name and code number that identify the type of reaction catalyzed and the substrates involved. Enzymes are grouped into the following six classes.

1. **Oxidoreductases**—enzymes that catalyze oxidations and reductions.
2. **Transferases**—enzymes that catalyze transfer of moieties such as glycosyl, methyl, or phosphoryl groups.
3. **Hydrolases**—enzymes that catalyze *hydrolytic* cleavage of C—C, C—O, C—N and other covalent bonds.
4. **Lyases**—enzymes that catalyze cleavage of C—C, C—O, C—N and other covalent bonds by *atom elimination*, generating double bonds.
5. **Isomerases**—enzymes that catalyze geometric or structural changes *within* a molecule.
6. **Ligases**—enzymes that catalyze the joining together (ligation) of two molecules in reactions coupled to the hydrolysis of ATP.

Despite the clarity of the IUB system, the names are lengthy and relatively cumbersome, so we generally continue to refer to enzymes by their traditional, albeit sometimes ambiguous names. The IUB name for hexokinase illustrates both the clarity of the IUB system and its complexities. The IUB name of hexokinase is ATP:D-hexose 6-phosphotransferase E.C. 2.7.1.1. This name identifies hexokinase as a member of class 2 (transferases), subclass 7 (transfer of a phosphoryl group), subsubclass 1 (alcohol is the phosphoryl acceptor), and "hexose-6" indicates that the alcohol phosphorylated is on carbon six of a hexose. However, we continue to call it hexokinase.

PROSTHETIC GROUPS, COFACTORS, & COENZYMES PLAY IMPORTANT ROLES IN CATALYSIS

Many enzymes contain small nonprotein molecules and metal ions that participate directly in substrate binding or in catalysis. Termed **prosthetic groups, cofactors,** and **coenzymes,** these extend the repertoire of catalytic capabilities beyond those afforded by the limited number of functional groups present on the aminoacyl side chains of peptides.

Prosthetic Groups Are Tightly Integrated Into an Enzyme's Structure

Prosthetic groups are tightly and stably incorporated into a protein's structure by covalent or noncovalent forces. Examples include pyridoxal phosphate, flavin mononucleotide (FMN), flavin adenine dinucleotide (FAD), thiamin pyrophosphate, biotin, and the metal ions of Co, Cu, Mg, Mn, and Zn. Metals are the most common prosthetic groups. The roughly one-third of all enzymes that contain tightly bound metal ions are termed **metalloenzymes.** Metal ions that participate in redox reactions generally are complexed to prosthetic groups such as heme (Chapter 6) or iron–sulfur clusters (Chapter 12). Metals also may facilitate the binding and orientation of substrates, the formation of covalent bonds with reaction intermediates (Co^{2+} in coenzyme B_{12}), or by acting as Lewis acids or bases to render substrates more **electrophilic** (electron-poor) or **nucleophilic** (electron-rich), and hence more reactive.

Cofactors Associate Reversibly With Enzymes or Substrates

Cofactors serve functions similar to those of prosthetic groups, but bind in a transient, dissociable manner either to the enzyme or to a substrate such as ATP. Unlike the stably associated prosthetic groups, cofactors must be present in the medium surrounding the enzyme for catalysis to occur. The most common cofactors also are metal ions. Enzymes that require a metal ion cofactor are termed **metal-activated enzymes** to distinguish them from the **metalloenzymes** for which metal ions serve as prosthetic groups.

Coenzymes Serve as Substrate Shuttles

Coenzymes serve as recyclable shuttles—or group transfer agents—that transport many substrates from one point within the cell to another. The function of these shuttles is twofold. First, they stabilize species such as hydrogen atoms (FADH) or hydride ions (NADH) that are too reactive to persist for any significant time period in the presence of the water or organic molecules that permeate the cell interior. They also serve as an adaptor or handle that facilitates the recognition and binding of small chemical groups, such as acetate (coenzyme A), by their target enzymes. Other chemical moieties transported by coenzymes include methyl groups (folates) and oligosaccharides (dolichol).

Many Coenzymes, Cofactors & Prosthetic Groups Are Derivatives of B Vitamins

The water-soluble B vitamins supply important components of numerous coenzymes. Several coenzymes contain, in addition, the adenine, ribose, and phosphoryl moieties of AMP or ADP (**Figure 7–2**). **Nicotinamide** is a component of the redox coenzymes NAD and NADP, whereas **riboflavin** is a component of the redox coenzymes FMN and FAD. **Pantothenic acid** is a component of the acyl group carrier **coenzyme A.** As its pyrophosphate, **thiamin** participates in decarboxylation of α-keto acids, and the **folic acid** and **cobamide** coenzymes function in one-carbon metabolism.

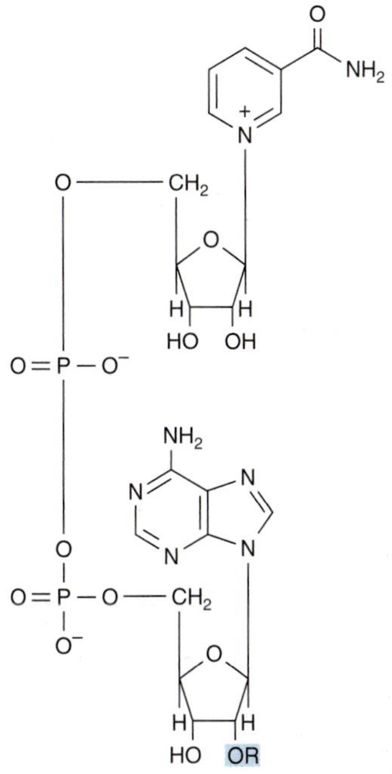

FIGURE 7–2 Structure of NAD+ and NADP+. For NAD+, R=H. For NADP+, R=PO_3^{2-}.

CATALYSIS OCCURS AT THE ACTIVE SITE

An important early 20th-century insight into enzymic catalysis sprang from the observation that the presence of substrates renders enzymes more resistant to the denaturing effects of elevated temperatures. This observation led Emil Fischer to propose that enzymes and their substrates interact to form an enzyme–substrate (ES) complex whose thermal stability was greater than that of the enzyme itself. This insight profoundly shaped our understanding of both the chemical nature and kinetic behavior (Chapter 8) of enzymic catalysis.

Fischer reasoned that the exquisitely high specificity with which enzymes discriminate their substrates when forming an ES complex was analogous to the manner in which a mechanical lock distinguishes the proper key. In most enzymes, the "lock" is formed by a cleft or pocket on the protein's surface that forms part of a region called the **active site** (Figures 5–6 and 5–8). As implied by the adjective "active," the active site is much more than simply a recognition site for binding substrates. Within the active site, substrates are brought into close proximity to one another in optimal alignment with the cofactors, prosthetic groups, and amino acid side chains responsible for catalyzing their chemical transformation into products (**Figure 7–3**). Catalysis is further enhanced by the capacity of the active site to shield substrates from water and generate an environment whose polarity, hydrophobicity, acidity, or alkalinity can differ markedly from that of the surrounding cytoplasm.

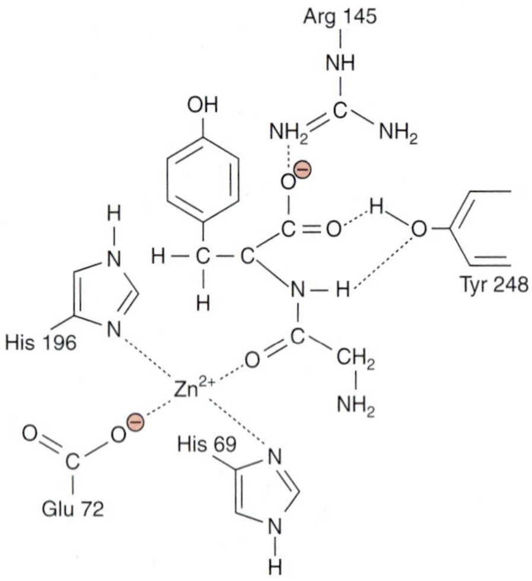

FIGURE 7–3 Two-dimensional representation of a dipeptide substrate, glycyl-tyrosine, bound within the active site of carboxypeptidase A.

ENZYMES EMPLOY MULTIPLE MECHANISMS TO FACILITATE CATALYSIS

Enzymes use various combinations of four general mechanisms to achieve dramatic catalytic enhancement of the rates of chemical reactions.

Catalysis by Proximity

For molecules to react, they must come within bond-forming distance of one another. The higher their concentration, the more frequently they will encounter one another, and the greater will be the rate of their reaction. When an enzyme binds substrate molecules at its active site, it creates a region of high local substrate concentration. This environment also orients the substrate molecules spatially in a position ideal for them to interact, resulting in rate enhancements of at least a thousandfold.

Acid–Base Catalysis

The ionizable functional groups of aminoacyl side chains and (where present) of prosthetic groups contribute to catalysis by acting as acids or bases. Acid-base catalysis can be either specific or general. By "specific" we mean only protons (H_3O^+) or OH^- ions. In **specific acid catalysis** or **specific base catalysis,** the rate of reaction is sensitive to changes in the concentration of protons of but *independent* of the concentrations of other acids (proton donors) or bases (proton acceptors) present in the solution or at the active site. Reactions whose rates are responsive to *all* the acids or bases present are said to be subject to **general acid catalysis** or **general base catalysis.**

Catalysis by Strain

Enzymes that catalyze *lytic* reactions that involve breaking a covalent bond typically bind their substrates in a conformation that is somewhat unfavorable for the bond that will undergo cleavage. This conformation mimics that of the **transition state intermediate,** a transient species that represents the transition state, or half-way point, in the transformation of substrates to products. The resulting strain stretches or distorts the targeted bond, weakening it and making it more vulnerable to cleavage. Nobel Laureate Linus Pauling was the first to suggest a role for transition state stabilization as a general mechanism by which enzymes accelerate the rates of chemical reactions. Knowledge of the transition state of an enzyme-catalyzed reaction is frequently exploited by chemists to design and create more effective enzyme inhibitors, called **transition state analogs,** as potential pharmacophores.

Covalent Catalysis

The process of **covalent catalysis** involves the formation of a covalent bond between the enzyme and one or more substrates. The modified enzyme then becomes a reactant. Covalent catalysis introduces a new reaction pathway whose activation energy is lower—and the reaction therefore is faster—than the reaction pathway in homogeneous solution. Chemical modification of the enzyme is, however, transient. Completion of the reaction returns the enzyme to its original unmodified state. Its role thus remains catalytic. Covalent catalysis is particularly common among enzymes that catalyze group transfer reactions. Residues on the enzyme that participate in covalent catalysis generally are cysteine or serine and occasionally histidine. Covalent catalysis often follows a "ping-pong" mechanism—one in which the first substrate is bound and its product released prior to the binding of the second substrate (**Figure 7–4**).

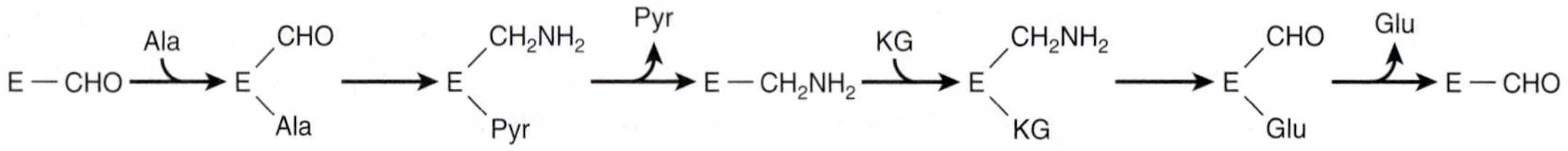

FIGURE 7–4 "Ping-pong" mechanism for transamination. E—CHO and E—CH$_2$NH$_2$ represent the enzyme-pyridoxal phosphate and enzyme-pyridoxamine complexes, respectively. (Ala, alanine; Glu, glutamate; KG, α-ketoglutarate; Pyr, pyruvate.)

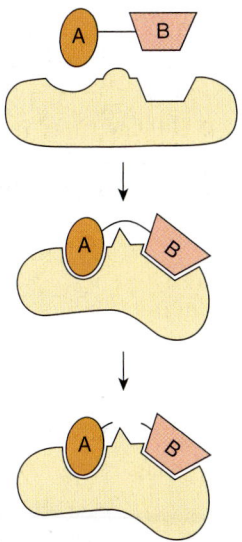

FIGURE 7–5 **Two-dimensional representation of Koshland's induced fit model of the active site of a lyase.** Binding of the substrate A—B induces conformational changes in the enzyme that align catalytic residues which participate in catalysis and strains the bond between A and B, facilitating its cleavage.

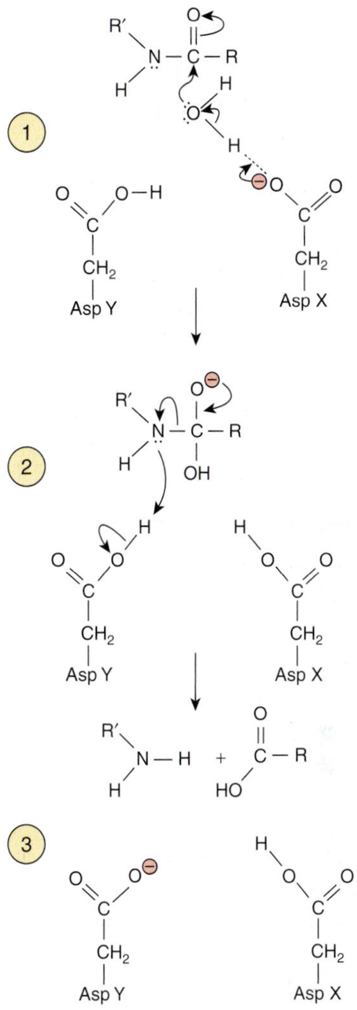

SUBSTRATES INDUCE CONFORMATIONAL CHANGES IN ENZYMES

While Fischer's "lock and key model" accounted for the exquisite specificity of enzyme-substrate interactions, the implied rigidity of the enzyme's active site failed to account for the dynamic changes that we now know accompany catalysis. This drawback was addressed by Daniel Koshland's **induced fit** model, which states that when substrates approach and bind to an enzyme they induce a conformational change analogous to placing a hand (substrate) into a glove (enzyme) (**Figure 7–5**). The enzyme in turn induces reciprocal changes in its substrates, harnessing the energy of binding to facilitate the transformation of substrates into products. The induced fit model has been amply confirmed by biophysical studies of enzyme motion during substrate binding.

HIV PROTEASE ILLUSTRATES ACID–BASE CATALYSIS

Enzymes of the **aspartic protease family,** which includes the digestive enzyme pepsin, the lysosomal cathepsins, and the protease produced by the human immunodeficiency virus (HIV) share a common catalytic mechanism. Catalysis involves two conserved aspartyl residues, which act as acid–base catalysts. In the first stage of the reaction, an aspartate

FIGURE 7–6 **Mechanism for catalysis by an aspartic protease such as HIV protease.** Curved arrows indicate directions of electron movement. ① Aspartate X acts as a base to activate a water molecule by abstracting a proton. ② The activated water molecule attacks the peptide bond, forming a transient tetrahedral intermediate. ③ Aspartate Y acts as an acid to facilitate breakdown of the tetrahedral intermediate and release of the split products by donating a proton to the newly formed amino group. Subsequent shuttling of the proton on Asp X to Asp Y restores the protease to its initial state.

functioning as a general base (Asp X, **Figure 7–6**) extracts a proton from a water molecule, making it more nucleophilic. The resulting nucleophile then attacks the electrophilic carbonyl carbon of the peptide bond targeted for hydrolysis, forming a **tetrahedral transition state intermediate.** A second aspartate (Asp Y, Figure 7–6) then facilitates the decomposition of this tetrahedral intermediate by donating a proton to the amino group produced by rupture of the peptide bond. The two different active site aspartates can act simultaneously as a general base or as a general acid because their immediate environment favors ionization of one, but not the other.

CHYMOTRYPSIN & FRUCTOSE-2, 6-BISPHOSPHATASE ILLUSTRATE COVALENT CATALYSIS

Chymotrypsin

While catalysis by aspartic proteases involves the direct hydrolytic attack of water on a peptide bond, catalysis by the **serine protease** chymotrypsin involves prior formation of a covalent acyl-enzyme intermediate. A highly reactive seryl residue, serine 195, participates in a charge-relay network with histidine 57 and aspartate 102. While these three residues are far apart in primary structure, in the active site of the mature, folded protein they are within bond-forming distance of one another. Aligned in the order Asp 102-His 57-Ser 195, they constitute a **"charge-relay network"** that functions as a **"proton shuttle."**

Binding of substrate initiates proton shifts that in effect transfer the hydroxyl proton of Ser 195 to Asp 102 (**Figure 7–7**). The enhanced nucleophilicity of the seryl oxygen facilitates its attack on the carbonyl carbon of the peptide bond of the substrate, forming a covalent **acyl-enzyme intermediate.** The proton on Asp 102 then shuttles via His 57 to the amino group liberated when the peptide bond is cleaved. The portion of the original peptide with a free amino group then leaves the active site and is replaced by a water molecule. The charge-relay network now activates the water molecule by withdrawing a proton through His 57 to Asp 102. The resulting hydroxide ion attacks the acyl-enzyme intermediate and a reverse proton shuttle returns a proton to Ser 195, restoring its original state. While modified during the process of catalysis, chymotrypsin emerges unchanged on completion of the reaction. The proteases trypsin and elastase employ a similar catalytic mechanism, but the numbers of the residues in their Ser-His-Asp proton shuttles differ.

Fructose-2,6-Bisphosphatase

Fructose-2,6-bisphosphatase, a regulatory enzyme of gluconeogenesis (Chapter 20), catalyzes the hydrolytic release of the phosphate on carbon 2 of fructose 2,6-bisphosphate. **Figure 7–8** illustrates the roles of seven active site residues. Catalysis involves a "catalytic triad" of one Glu and two His residues and a covalent phosphohistidyl intermediate.

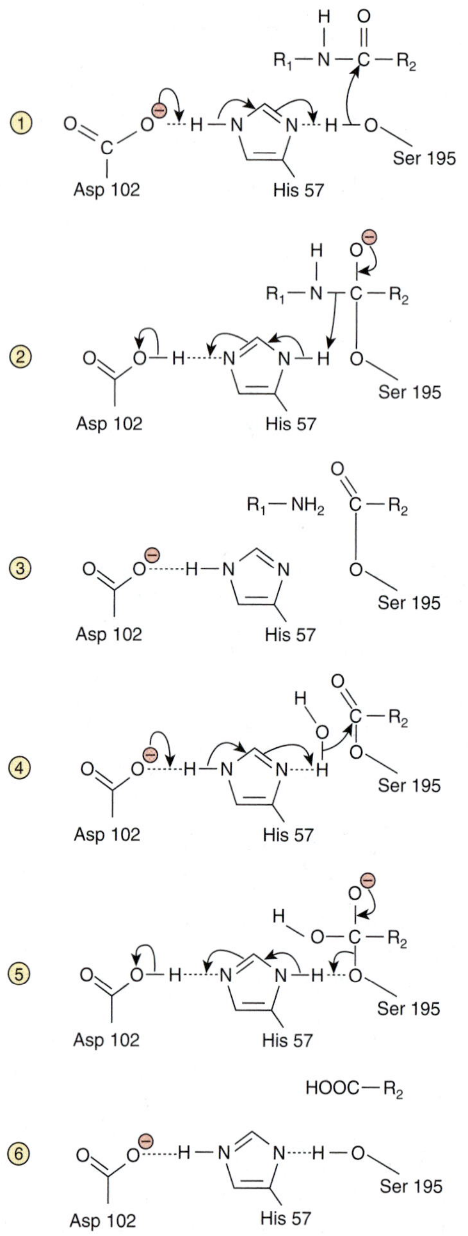

CATALYTIC RESIDUES ARE HIGHLY CONSERVED

Members of an enzyme family such as the aspartic or serine proteases employ a similar mechanism to catalyze a common reaction type, but act on different substrates. Most enzyme families arose through gene duplication events that created a second copy of the gene that encodes a particular enzyme. The proteins encoded by the two genes can then evolve independently to recognize different substrates—resulting, for

FIGURE 7–7 Catalysis by chymotrypsin. ① The charge-relay system removes a proton from Ser 195, making it a stronger nucleophile. ② Activated Ser 195 attacks the peptide bond, forming a transient tetrahedral intermediate. ③ Release of the amino terminal peptide is facilitated by donation of a proton to the newly formed amino group by His 57 of the charge-relay system, yielding an acyl-Ser 195 intermediate. ④ His 57 and Asp 102 collaborate to activate a water molecule, which attacks the acyl-Ser 195, forming a second tetrahedral intermediate. ⑤ The charge-relay system donates a proton to Ser 195, facilitating breakdown of tetrahedral intermediate to release the carboxyl terminal peptide⑥.

example, in chymotrypsin, which cleaves peptide bonds on the carboxyl terminal side of large hydrophobic amino acids, and trypsin, which cleaves peptide bonds on the carboxyl terminal side of basic amino acids. Proteins that diverged from a common ancestor are said to be **homologous** to one

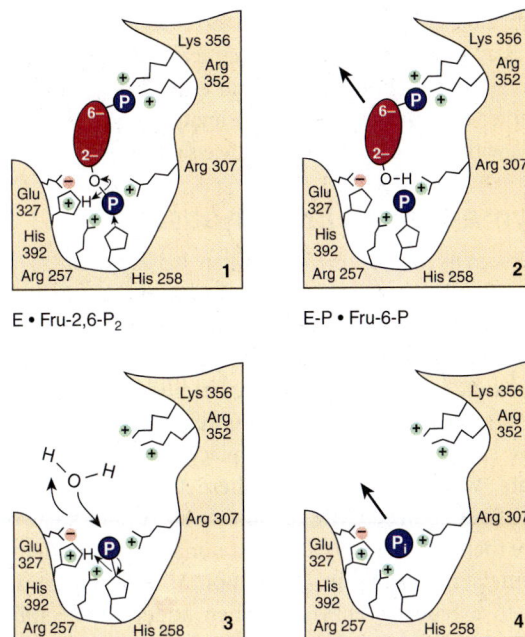

E • Fru-2,6-P₂

E-P • Fru-6-P

E-P • H₂O

E • Pᵢ

FIGURE 7–8 Catalysis by fructose-2,6-bisphosphatase.
(1) Lys 356 and Arg 257, 307, and 352 stabilize the quadruple
negative charge of the substrate by charge-charge interactions.
Glu 327 stabilizes the positive charge on His 392. (2) The
nucleophile His 392 attacks the C-2 phosphoryl group and
transfers it to His 258, forming a phosphoryl-enzyme intermediate.
Fructose 6-phosphate now leaves the enzyme. (3) Nucleophilic
attack by a water molecule, possibly assisted by Glu 327 acting as a
base, forms inorganic phosphate. (4) Inorganic orthophosphate is
released from Arg 257 and Arg 307. (Reproduced, with permission,
from Pilkis SJ, et al: 6-Phosphofructo-2-kinase/fructose-2,6-
bisphosphatase: A metabolic signaling enzyme. Annu Rev Biochem
1995;64:799. © 1995 by Annual Reviews, www.annualreviews.org.)

another. The common ancestry of enzymes can be inferred
from the presence of specific amino acids in the same position in each family member. These residues are said to be
conserved residues. Table 7–1 illustrates the primary structural conservation of two components of the charge-relay
network for several serine proteases. Among the most highly
conserved residues are those that participate directly in
catalysis.

ISOZYMES ARE DISTINCT ENZYME FORMS THAT CATALYZE THE SAME REACTION

Higher organisms often elaborate several physically distinct
versions of a given enzyme, each of which catalyzes the same
reaction. Like the members of other protein families, these
protein catalysts or **isozymes** arise through gene duplication.
Isozymes may exhibit subtle differences in properties such
as sensitivity to particular regulatory factors (Chapter 9) or
substrate affinity (eg, hexokinase and glucokinase) that adapt
them to specific tissues or circumstances. Some isozymes may
also enhance survival by providing a "backup" copy of an
essential enzyme.

THE CATALYTIC ACTIVITY OF ENZYMES FACILITATES THEIR DETECTION

The relatively small quantities of enzymes present in cells
complicate determination of their presence and concentration. However, the amplification conferred by their ability to
rapidly transform thousands of molecules of a specific substrate into products imbues each enzyme with the ability to
reveal its presence. Assays of the catalytic activity of enzymes
are frequently used in research and clinical laboratories.
Under appropriate conditions (see Chapter 8), the rate of the
catalytic reaction being monitored is proportionate to the
amount of enzyme present, which allows its concentration to
be inferred.

Single-Molecule Enzymology

The limited sensitivity of traditional enzyme assays neces-
sitates the use of a large group, or ensemble, of enzyme molecules in order to produce measurable quantities of product.
The data obtained thus reflect the average catalytic capability
of individual molecules. Recent advances in **nanotechnology**
have made it possible to observe, usually by fluorescence
microscopy, catalytic events involving individual enzyme
and substrate molecules. Consequently, scientists can now
measure the rate of single catalytic events and sometimes

TABLE 7–1 Amino Acid Sequences in the Neighborhood of the Catalytic Sites of Several Bovine Proteases

Enzyme	Sequence Around Serine Ⓢ														Sequence Around Histidine Ⓗ									
Trypsin	D	S	C	Q	D	G	Ⓢ	G	G	P	V	V	C	S	G	K	V	V	S	A	A	Ⓗ	C	Y	K	S	G
Chymotrypsin A	S	S	C	M	G	D	Ⓢ	G	G	P	L	V	C	K	K	N	V	V	T	A	A	Ⓗ	G	G	V	T	T
Chymotrypsin B	S	S	C	M	G	D	Ⓢ	G	G	P	L	V	C	Q	K	N	V	V	T	A	A	Ⓗ	C	G	V	T	T
Thrombin	D	A	C	E	G	D	Ⓢ	G	G	P	F	V	M	K	S	P	V	L	T	A	A	Ⓗ	C	L	L	Y	P

Note: Regions shown are those on either side of the catalytic site seryl Ⓢ and histidyl Ⓗ residues.

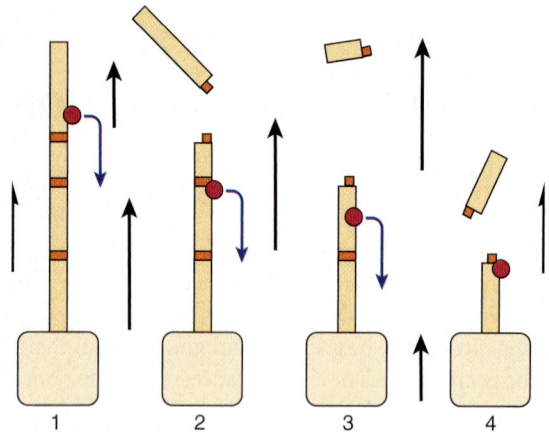

FIGURE 7–9 **Direct observation of single DNA cleavage events catalyzed by a restriction endonuclease.** DNA molecules immobilized to beads (pale yellow) are placed in a flowing stream of buffer (black arrows), which causes them to assume an extended conformation. Cleavage at one of the restriction sites (orange) by an endonuclease leads to a shortening of the DNA molecule, which can be observed directly in a microscope since the nucleotide bases in DNA are fluorescent. Although the endonuclease (red) does not fluoresce, and hence is invisible, the progressive manner in which the DNA molecule is shortened (1→ 4) reveals that the endonuclease binds to the free end of the DNA molecule and moves along it from site to site.

the individual steps in catalysis by a process called **single-molecule enzymology** (**Figure 7–9**).

Drug Discovery Requires Enzyme Assays Suitable for High-Throughput Screening

Enzymes constitute one of the primary classes of biomolecules targeted for the development of drugs and other therapeutic agents. Many antibiotics, for example, inhibit enzymes that are unique to microbial pathogens. The discovery of new drugs is greatly facilitated when a large number of potential pharmacophores can be assayed in a rapid, automated fashion—a process referred to as **high-throughput screening.** High-throughput screening (HTS) takes advantage of recent advances in robotics, optics, data processing, and microfluidics to conduct and analyze many thousands of simultaneous assays of the activity of a given enzyme. The most commonly used high-throughput screening devices employ 4–100 µL volumes in 96, 384, or 1536 well plastic plates and fully automated equipment capable of dispensing substrates, coenzymes, enzymes, and potential inhibitors in a multiplicity of combinations and concentrations. High-throughput screening is ideal for surveying the numerous products of **combinatorial chemistry,** the simultaneous synthesis of large libraries of chemical compounds that contain all possible combinations of a set of chemical precursors. Enzyme assays that produce a chromagenic or fluorescent product are ideal, since optical detectors are readily engineered to permit the rapid analysis of multiple samples. At present, the

sophisticated equipment required for truly large numbers of assays is available only in pharmaceutical houses, government-sponsored laboratories, and research universities. As described in Chapter 8, its principal use is the analysis of inhibitory compounds with ultimate potential for use as drugs.

Enzyme-Linked Immunoassays

The sensitivity of enzyme assays can be exploited to detect proteins that lack catalytic activity. **Enzyme-linked immunosorbent assays** (ELISAs) use antibodies covalently linked to a "reporter enzyme" such as alkaline phosphatase or horseradish peroxidase whose products are readily detected, generally by the absorbance of light or by fluorescence. Serum or other biologic samples to be tested are placed in a plastic microtiter plate, where the proteins adhere to the plastic surface and are immobilized. Any remaining absorbing areas of the well are then "blocked" by adding a nonantigenic protein such as bovine serum albumin. A solution of antibody covalently linked to a reporter enzyme is then added. The antibodies adhere to the immobilized antigen and are themselves immobilized. Excess free antibody molecules are then removed by washing. The presence and quantity of bound antibody is then determined by adding the substrate for the reporter enzyme.

NAD(P)-Dependent Dehydrogenases Are Assayed Spectrophotometrically

The physicochemical properties of the reactants in an enzyme-catalyzed reaction dictate the options for the assay of enzyme activity. Spectrophotometric assays exploit the ability of a substrate or product to absorb light. The reduced coenzymes NADH and NADPH, written as NAD(P)H, absorb light at a wavelength of 340 nm, whereas their oxidized forms $NAD(P)^+$ do not (**Figure 7–10**). When $NAD(P)^+$ is reduced, the absorbance at 340 nm therefore increases in proportion to—and at a rate determined by—the quantity of NAD(P)H produced. Conversely, for a dehydrogenase that catalyzes the oxidation of NAD(P)H, a decrease in absorbance at 340 nm will be observed. In each case, the rate of change in optical density at 340 nm will be proportionate to the quantity of the enzyme present.

Many Enzymes Are Assayed by Coupling to a Dehydrogenase

The assay of enzymes whose reactions are not accompanied by a change in absorbance or fluorescence is generally more difficult. In some instances, the product or remaining substrate can be transformed into a more readily detected compound. In other instances, the reaction product may have to be separated from unreacted substrate prior to measurement. An alternative strategy is to devise a synthetic substrate whose product absorbs light or fluoresces. For example, *p*-nitrophenyl phosphate is an artificial substrate for certain phosphatases and for chymotrypsin that does not absorb visible light. However, following hydrolysis, the resulting *p*-nitrophenylate anion absorbs light at 419 nm.

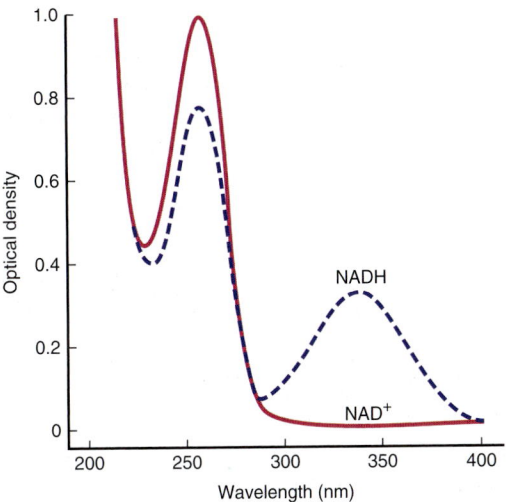

FIGURE 7–10 Absorption spectra of NAD⁺ and NADH.
Densities are for a 44 mg/L solution in a cell with a 1 cm
light path. NADP⁺ and NADPH have spectra analogous
to NAD⁺ and NADH, respectively.

Another quite general approach is to employ a "coupled" assay (**Figure 7–11**). Typically, a dehydrogenase whose substrate is the product of the enzyme of interest is added in catalytic excess. The rate of appearance or disappearance of NAD(P)H then depends on the rate of the enzyme reaction to which the dehydrogenase has been coupled.

THE ANALYSIS OF CERTAIN ENZYMES AIDS DIAGNOSIS

The analysis of enzymes in blood plasma has played a central role in the diagnosis of several disease processes. Many enzymes are functional constituents of blood. Examples include

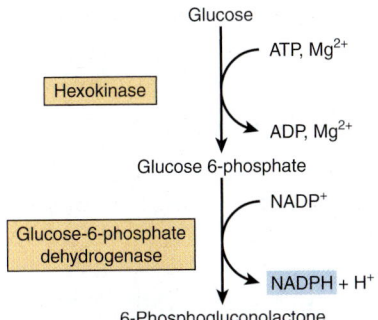

**FIGURE 7–11 Coupled enzyme assay for hexokinase
activity.** The production of glucose 6-phosphate by hexokinase
is coupled to the oxidation of this product by glucose-6-phosphate
dehydrogenase in the presence of added enzyme and NADP⁺.
When an excess of glucose-6-phosphate dehydrogenase is present,
the rate of formation of NADPH, which can be measured at 340 nm,
is governed by the rate of formation of glucose-6-phosphate
by hexokinase.

pseudocholinesterase, lipoprotein lipase, and components of the cascade that trigger blood clotting and clot dissolution. Other enzymes are released into plasma following cell death or injury. While these latter enzymes perform no physiologic function in plasma, their appearance or levels can assist in the diagnosis and prognosis of diseases and injuries affecting specific tissues. Following injury, the plasma concentration of a released enzyme may rise early or late, and may decline rapidly or slowly. Proteins from the cytoplasm tend to appear more rapidly than those from subcellular organelles. The speed with which enzymes and other proteins are removed from plasma varies with their susceptibility to proteolysis and permeability through renal glomeruli.

Quantitative analysis of the activity of released enzymes or other proteins, typically in plasma or serum but also in urine or various cells, provides information concerning diagnosis, prognosis, and response to treatment. Assays of enzyme *activity* typically employ standard kinetic assays of initial reaction rates. **Table 7–2** lists several enzymes of value in clinical diagnosis. These enzymes are, however, not absolutely specific for the indicated disease. For example, elevated blood levels of prostatic acid phosphatase are associated typically with prostate cancer, but also with certain other cancers and noncancerous conditions. Consequently, enzyme assay data must be considered together with other factors elicited through a comprehensive clinical examination. Factors to be considered in interpreting enzyme data include patient age, sex, prior history, possible drug use, and the sensitivity and the diagnostic specificity of the enzyme test.

TABLE 7–2 Principal Serum Enzymes Used in Clinical Diagnosis

Serum Enzyme	Major Diagnostic Use
Aminotransferases	
Aspartate aminotransferase (AST, or SGOT)	Myocardial infarction
Alanine aminotransferase (ALT, or SGPT)	Viral hepatitis
Amylase	Acute pancreatitis
Ceruloplasmin	Hepatolenticular degeneration (Wilson's disease)
Creatine kinase	Muscle disorders and myocardial infarction
γ-Glutamyl transferase	Various liver diseases
Lactate dehydrogenase isozyme 5	Liver diseases
Lipase	Acute pancreatitis
Phosphatase, acid	Metastatic carcinoma of the prostate
Phosphatase, alkaline (isozymes)	Various bone disorders, obstructive liver diseases

Note: Many of the above enzymes are not specific to the disease listed.

Enzymes Assist Diagnosis of Myocardial Infarction

An enzyme useful for diagnostic enzymology should be relatively specific for the tissue or organ under study, should appear in the plasma or other fluid at a time useful for diagnosis (the "diagnostic window"), and should be amenable to automated assay. The enzymes used to confirm a myocardial infarction (MI) illustrate the concept of a "diagnostic window," and provide a historical perspective on the use of different enzymes for this purpose.

Detection of an enzyme must be possible within a few hours of an MI to confirm a preliminary diagnosis and permit initiation of appropriate therapy. Enzymes that only appear in the plasma 12 h or more following injury are thus of limited utility. The first enzymes used to diagnose MI were aspartate aminotransferase (AST), alanine aminotransferase (ALT), and lactate dehydrogenase. AST and ALT proved less than ideal, however, as they appear in plasma relatively slowly and are not specific to heart muscle. While LDH also is released relatively slowly into plasma, it offered the advantage of tissue specificity as a consequence of its quaternary structure.

Lactate dehydrogenase (LDH) is a tetrameric enzyme consisting of two monomer types: H (for heart) and M (for muscle) that combine to yield five LDH isozymes: HHHH (I_1), HHHM (I_2), HHMM (I_3), HMMM (I_4), and MMMM (I_5). Tissue-specific expression of the H and M genes determines the relative proportions of each subunit in different tissues. Isozyme I_1 predominates in heart tissue, and isozyme I_5 in the liver. Thus, tissue injury releases a characteristic pattern of LDH isozymes that can be separated by electrophoresis and detected using a coupled assay (**Figure 7–12**). Today, LDH has been superseded as a marker for MI by other proteins that appear more rapidly in plasma.

Creatine kinase (CK) has three isozymes: CK-MM (skeletal muscle), CK-BB (brain), and CK-MB (heart and skeletal muscle). CK-MB has a useful diagnostic window. It appears within 4–6 h of an MI, peaks at 24 h, and returns to baseline by 48–72 h. As for LDH, individual CK isozymes are separable by electrophoresis, thus facilitating detection. Assay of plasma CK levels continues in use to assess skeletal muscle disorders such as Duchene muscular dystrophy. Today, however, in most clinical laboratories the measurement of plasma troponin levels has replaced CK as the preferred diagnostic marker for MI.

Troponins

Troponin is a complex of three proteins involved in muscle contraction in *skeletal* and *cardiac muscle* but not in *smooth muscle* (see Chapter 49). Immunological measurement of plasma levels of cardiac troponins I and T provide sensitive and specific indicators of damage to heart muscle. Troponin levels rise for 2–6 h after an MI and remain elevated for 4–10 days. In addition to MI, other heart muscle damage also elevates serum troponin levels. Cardiac troponins thus serve as a marker of all heart muscle damage. The search for additional markers for heart disease, such as ischemia-modified albumin, and the simultaneous assessment of a spectrum of diagnostic markers via proteomics, continues to be an active area of clinical research.

Enzymes also can be employed in the clinical laboratory as tools for determining the concentration of critical metabolites. For example, glucose oxidase is frequently utilized to measure plasma glucose concentration. Enzymes are employed with increasing frequency as tools for the treatment of injury and disease. Tissue plasminogen activator (tPA) or streptokinase is used in the treatment of acute MI,

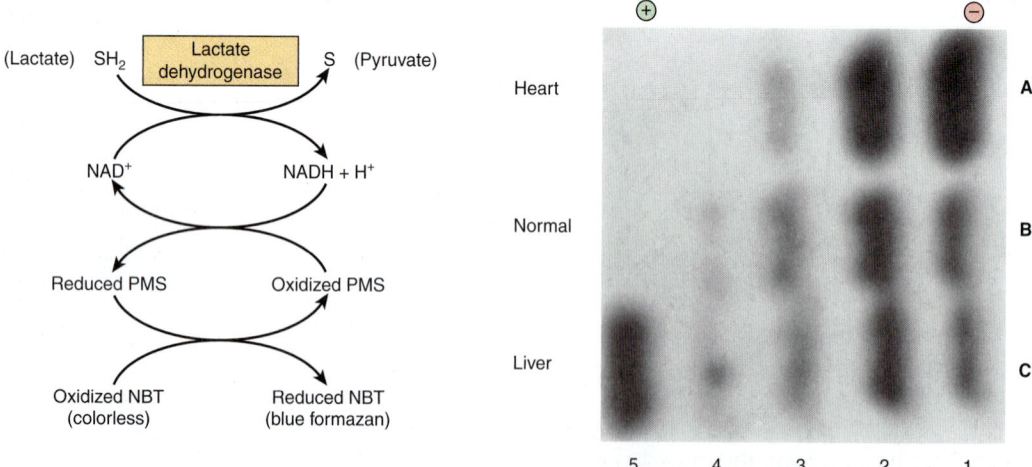

FIGURE 7–12 **Normal and pathologic patterns of lactate dehydrogenase (LDH) isozymes in human serum.** LDH isozymes of serum were separated by electrophoresis and visualized using the coupled reaction scheme shown on the left. (NBT, nitroblue tetrazolium; PMS, phenazine methylsulfate.) At right is shown the stained electropherogram. Pattern A is serum from a patient with a myocardial infarct; B is normal serum; and C is serum from a patient with liver disease. Arabic numerals denote specific LDH isozymes.

while trypsin has been used in the treatment of cystic fibrosis (see Chapter 54).

ENZYMES FACILITATE DIAGNOSIS OF GENETIC AND INFECTIOUS DISEASES

Many diagnostic techniques take advantage of the specificity and efficiency of the enzymes that act on oligonucleotides such as DNA (Chapter 39). Enzymes known as **restriction endonucleases,** for example, cleave double-stranded DNA at sites specified by a sequence of four, six, or more base pairs called **restriction sites.** Cleavage of a sample of DNA with a restriction enzyme produces a characteristic set of smaller DNA fragments (see Chapter 39). Deviations in the normal product pattern, called **restriction fragment length polymorphisms (RFLPs),** occur if a mutation renders a restriction site unrecognizable to its cognate restriction endonuclease or, alternatively, generates a new recognition site. RFLPs are currently utilized to facilitate prenatal detection of a number of hereditary disorders, including sickle cell trait, beta-thalassemia, infant phenylketonuria, and Huntington's disease.

The **polymerase chain reaction (PCR)** employs a thermostable DNA polymerase and appropriate oligonucleotide primers to produce thousands of copies of a defined segment of DNA from a minute quantity of starting material (see Chapter 39). PCR enables medical, biological, and forensic scientists to detect and characterize DNA present initially at levels too low for direct detection. In addition to screening for genetic mutations, PCR can be used to detect and identify pathogens and parasites such as *Trypanosoma cruzi,* the causative agent of Chagas' disease, and *Neisseria meningitides,* the causative agent of bacterial meningitis, through the selective amplification of their DNA.

RECOMBINANT DNA PROVIDES AN IMPORTANT TOOL FOR STUDYING ENZYMES

Recombinant DNA technology has emerged as an important asset in the study of enzymes. Highly purified samples of enzymes are necessary for the study of their structure and function. The isolation of an individual enzyme, particularly one present in low concentration, from among the thousands of proteins present in a cell can be extremely difficult. If the gene for the enzyme of interest has been cloned, it generally is possible to produce large quantities of its encoded protein in *Escherichia coli* or yeast. However, not all animal proteins can be expressed in an active form in microbial cells, nor do microbes perform certain post-translational processing tasks. For these reasons, a gene may be expressed in cultured animal

cell systems employing the baculovirus expression vector to transform cultured insect cells. For more details concerning recombinant DNA techniques, see Chapter 39.

Recombinant Fusion Proteins Are Purified by Affinity Chromatography

Recombinant DNA technology can also be used to create modified proteins that are readily purified by affinity chromatography. The gene of interest is linked to an oligonucleotide sequence that encodes a carboxyl or amino terminal extension to the encoded protein. The resulting modified protein, termed as a **fusion protein,** contains a domain tailored to interact with a specific affinity support. One popular approach is to attach an oligonucleotide that encodes six consecutive histidine residues. The expressed "His tag" protein binds to chromatographic supports that contain an immobilized divalent metal ion such as Ni^{2+} or Cd^{2+}. Alternatively, the substrate-binding domain of glutathione S-transferase (GST) can serve as a "GST tag." **Figure 7–13** illustrates the purification of a GST-fusion protein using an affinity support containing bound glutathione. Fusion proteins also often encode a cleavage site for a highly specific protease such as thrombin in the region that links the two portions of the protein. This permits removal of the added fusion domain following affinity purification.

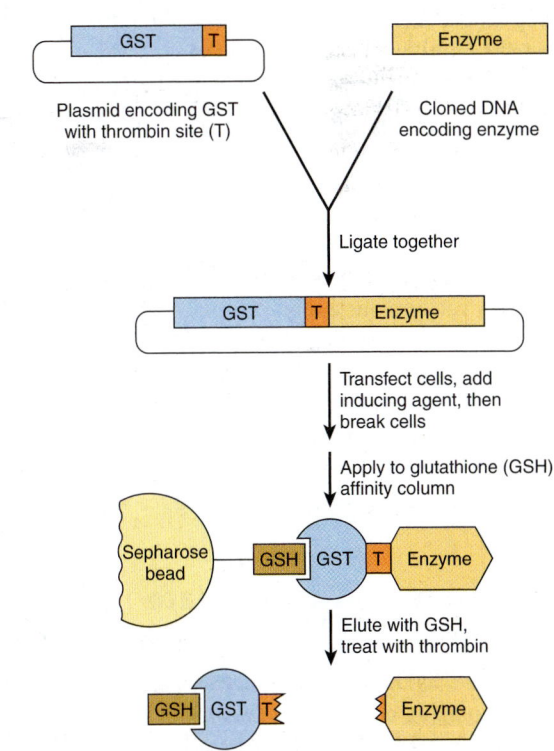

FIGURE 7–13 Use of glutathione S-transferase (GST) fusion proteins to purify recombinant proteins. (GSH, glutathione.)

Site-Directed Mutagenesis Provides Mechanistic Insights

Once the ability to express a protein from its cloned gene has been established, it is possible to employ **site-directed mutagenesis** to change specific aminoacyl residues by altering their codons. Used in combination with kinetic analyses and x-ray crystallography, this approach facilitates identification of the specific roles of given aminoacyl residues in substrate binding and catalysis. For example, the inference that a particular aminoacyl residue functions as a general acid can be tested by replacing it with an aminoacyl residue incapable of donating a proton.

RIBOZYMES: ARTIFACTS FROM THE RNA WORLD

Cech Discovered the First Catalytic RNA Molecule

The participation of enzyme catalysts in the posttranslational maturation of certain proteins has analogies in the RNA world. Many RNA molecules undergo processing that both removes segments of oligonucleotide and re-ligates the remaining segments to form the mature product (Chapter 36). Not all of these catalysts are proteins, however. While examining the processing of ribosomal RNA (rRNA) molecules in the ciliated protozoan *Tetrahymena*, Thomas Cech and his coworkers observed, in the early 1980s, that processing of the 26S rRNA proceeded smoothly in vitro even in the total *absence* of protein. The source of this splicing activity was traced to a 413 bp catalytic segment that retained its catalytic activity even when replicated in *E. coli* (Chapter 39). Prior to that time, polynucleotides had been thought to serve solely as information storage and transmission entities, and that catalysis was restricted solely to proteins.

Several other ribozymes have since been discovered. The vast majority catalyze nucleophilic displacement reactions that target the phosphodiester bonds of the RNA backbone. In small self-cleaving RNAs, such as hammerhead or hepatitis delta virus RNA, the attacking nucleophile is water and the result is hydrolysis. For the large group I intron ribozymes, the attacking nucleophile is the 3′-hydroxyl of the terminal ribose of another segment of RNA and the result is a splicing reaction.

The Ribosome—The Ultimate Ribozyme

The ribozyme was the first recognized "molecular machine." A massive complex comprised of scores of protein subunits and several large ribosomal RNA molecules, the ribosome performs the vitally important and highly complex process of synthesizing long polypeptide chains following the instructions encoded in messenger RNA molecules (Chapter 37).

For many years, it was assumed that ribosomal RNAs played a passive, structural role, or perhaps helped in the recognition of cognate mRNAs through a base pairing mechanism.

The RNA World Hypothesis

The discovery of ribozymes had a profound influence on evolutionary theory. For many years, scientists had hypothesized that the first biologic catalysts were formed when amino acids contained in the primordial soup coalesced to form the first simple proteins. With the realization that RNA could both carry information and catalyze simple chemical reactions, a new "RNA World" hypothesis emerged in which RNA constituted the first biological macromolecule. Eventually, DNA emerged as a more chemically stable oligonucleotide for long-term information storage while proteins, by virtue of their much greater variety of chemical functional groups, dominated catalysis. If one assumes that some sort of RNA-protein hybrid was formed as an intermediate in the transition from ribonucleotide to polypeptide catalysts, one need look no further than the ribosome to find the presumed missing link.

Why did not proteins take over all catalytic functions? Presumably, in the case of the ribosome the process was both too complex and too essential to permit much opportunity for possible competitors to gain a foothold. In the case of the small self-cleaving RNAs and self-splicing introns, they may represent one of the few cases in which RNA autocatalysis is more efficient than development of a new protein catalyst.

SUMMARY

- Enzymes are efficient catalysts whose stringent specificity extends to the kind of reaction catalyzed, and typically to a single substrate.

- Organic and inorganic prosthetic groups, cofactors, and coenzymes play important roles in catalysis. Coenzymes, many of which are derivatives of B vitamins, serve as "shuttles" for commonly-used groups such as amines, electrons, and acetyl groups.

- During catalysis, enzymes frequently redirect the conformational changes induced by substrate binding to effect complementary changes in the substrate that facilitate its transformation into product.

- Catalytic mechanisms employed by enzymes include the introduction of strain, approximation of reactants, acid-base catalysis, and covalent catalysis. HIV protease illustrates acid-base catalysis; chymotrypsin and fructose-2,6-bisphosphatase illustrate covalent catalysis.

- Aminoacyl residues that participate in catalysis are highly conserved among all classes of a given enzyme. Site-directed mutagenesis, used to change residues suspected of being important in catalysis or substrate binding, provides insights into mechanisms of enzyme action.

- The catalytic activity of enzymes reveals their presence, facilitates their detection, and provides the basis for enzyme-linked immunoassays. Many enzymes can be assayed spectrophotometrically by coupling them to an NAD(P)$^+$-dependent dehydrogenase.

- Combinatorial chemistry generates extensive libraries of potential enzyme activators and inhibitors that can be tested by high-throughput screening.

- Assay of plasma enzymes aids diagnosis and prognosis of myocardial infarction, acute pancreatitis, and various bone and liver disorders.

- Restriction endonucleases facilitate diagnosis of genetic diseases by revealing restriction fragment length polymorphisms, and the polymerase chain reaction (PCR) amplifies DNA initially present in quantities too small for analysis.

- Attachment of a polyhistidyl, glutathione S-transferase (GST), or other "tag" to the N- or C-terminus of a recombinant protein facilitates its purification by affinity chromatography on a solid support that contains an immobilized ligand such as a divalent cation (eg, Ni^{2+}) or GST. Specific proteases can then remove affinity "tags" and generate the native enzyme.

- Not all enzymes are proteins. Several ribozymes are known that can cut and re-splice the phosphodiester bonds of RNA. In the ribosome, it is the rRNA and not the polypeptide components that are primarily responsible for catalysis.

REFERENCES

Brik A, Wong C-H: HIV-1 protease: mechanism and drug discovery. Org Biomol Chem 2003;1:5.

Burtis CA, Ashwood ER, Bruns DE: *Tietz Textbook of Clinical Chemistry and Molecular Diagnostics.* 4th ed. Elsevier, 2006.

Cornish PV, Ha T: A survey of single-molecule techniques in chemical biology. ACS Chem Biol 2007;2:53.

Doudna JA, Lorsch JR: Ribozyme catalysis: not different, just worse. Nature Struct Biol 2005;12:395.

Frey PA, Hegeman AD: *Enzyme Reaction Mechanisms.* Oxford University Press. 2006.

Geysen HM, Schoenen F, Wagner D, Wagner R: Combinatorial compound libraries for drug discovery: an ongoing challenge. Nature Rev Drug Disc 2003;2:222.

Goddard J-P, Reymond J-L: Enzyme assays for high-throughput screening. Curr Opin Biotech 2004;15:314.

Gupta S, de Lemos JA: Use and misuse of cardiac troponins in clinical practice. Prog Cardiovasc Dis 2007;50:151.

Hedstrom L: Serine protease mechanism and specificity. Chem Rev 2002;102:4501.

Melanson SF, Tanasijevic MJ: Laboratory diagnosis of acute myocardial injury. Cardiovascular Pathol 2005;14:156.

Pereira DA, Williams JA: Origin and evolution of high throughput screening. Br J Pharmacol 2007;152:53.

René AWF, Titman CM, Pratap CV, et al: A molecular switch and proton wire synchronize the active sites in thiamine enzymes. Science 2004;306:872.

Schmeing TM, Ramakrishnan V: What recent ribosome structures have revealed about the mechanism of translation. Nature 2009;461:1234.

Schafer B, Gemeinhardt H, Greulich KO: Direct microscopic observation of the time course of single-molecule DNA restriction reactions. Agnew Chem Int Ed 2001;40:4663.

Silverman RB: *The Organic Chemistry of Enzyme-Catalyzed Reactions.* Academic Press, 2002.

Sundaresan V, Abrol R: Towards a general model for protein-substrate stereoselectivity. Protein Sci 2002;11:1330.

Todd AE, Orengo CA, Thornton JM: Plasticity of enzyme active sites. Trends Biochem Sci 2002;27:419.

Urich T, Gomes CM, Kletzin A, et al: X-ray structure of a self-compartmentalizing sulfur cycle metalloenzyme. Science 2006;311:996.

Walsh CT: *Enzymatic Reaction Mechanisms.* Freeman, 1979.

Enzymes: Kinetics

Peter J. Kennelly, PhD & Victor W. Rodwell, PhD

OBJECTIVES

After studying this chapter, you should be able to:

- Describe the scope and overall purposes of the study of enzyme kinetics.
- Indicate whether ΔG, the overall change in free energy for a reaction, is dependent on reaction mechanism.
- Indicate whether ΔG is a function of the *rates* of reactions.
- Explain the relationship between K_{eq}, concentrations of substrates and products at equilibrium, and the ratio of the rate constants k_1/k_{-1}.
- Outline how temperature and the concentration of hydrogen ions, enzyme, and substrate affect the rate of an enzyme-catalyzed reaction.
- Indicate why laboratory measurement of the rate of an enzyme-catalyzed reaction typically employs initial rate conditions.
- Describe the application of linear forms of the Michaelis–Menten equation to the determination of K_m and V_{max}.
- Give one reason why a linear form of the Hill equation is used to evaluate the substrate-binding kinetics exhibited by some multimeric enzymes.
- Contrast the effects of an increasing concentration of substrate on the kinetics of simple competitive and noncompetitive inhibition.
- Describe the ways in which substrates add to, and products depart from, an enzyme that follows a ping–pong mechanism and do the same for an enzyme that follows a rapid-equilibrium mechanism.
- Illustrate the utility of enzyme kinetics in ascertaining the mode of action of drugs.

BIOMEDICAL IMPORTANCE

Enzyme kinetics is the field of biochemistry concerned with the quantitative measurement of the rates of enzyme-catalyzed reactions and the systematic study of factors that affect these rates. Kinetic analysis can reveal the number and order of the individual steps by which enzymes transform substrates into products. Together with site-directed mutagenesis and other techniques that probe the protein structure, kinetic analyses can reveal details of the catalytic mechanism of a given enzyme.

A complete, balanced set of enzyme activities is of fundamental importance for maintaining homeostasis. An understanding of enzyme kinetics thus is important to understanding how physiologic stresses such as anoxia, metabolic acidosis or alkalosis, toxins, and pharmacologic agents affect that balance. The involvement of enzymes in virtually all physiologic processes makes them the targets of choice for drugs

that cure or ameliorate human disease. Applied enzyme kinetics represents the principal tool by which scientists identify and characterize therapeutic agents that selectively inhibit the rates of specific enzyme-catalyzed processes. Enzyme kinetics thus plays a central and critical role in drug discovery and comparative pharmacodynamics, as well as in elucidating the mode of action of drugs.

CHEMICAL REACTIONS ARE DESCRIBED USING BALANCED EQUATIONS

A **balanced chemical equation** lists the initial chemical species (substrates) present and the new chemical species (products) formed for a particular chemical reaction, all in their correct proportions or **stoichiometry.** For example, balanced equation

(1) describes the reaction of one molecule each of substrates A and B to form one molecule each of products P and Q:

$$A + B \rightleftarrows P + Q \qquad (1)$$

The double arrows indicate reversibility, an intrinsic property of all chemical reactions. Thus, for reaction (1), if A and B can form P and Q, then P and Q can also form A and B. Designation of a particular reactant as a "substrate" or "product" is therefore somewhat arbitrary since the products for a reaction written in one direction are the substrates for the reverse reaction. The term "products" is, however, often used to designate the reactants whose formation is thermodynamically favored. Reactions for which thermodynamic factors strongly favor formation of the products to which the arrow points often are represented with a single arrow as if they were "irreversible":

$$A + B \rightarrow P + Q \qquad (2)$$

Unidirectional arrows are also used to describe reactions in living cells where the products of reaction (2) are immediately consumed by a subsequent enzyme-catalyzed reaction. The rapid removal of product P or Q therefore effectively precludes occurrence of the reverse reaction, rendering equation (2) **functionally irreversible under physiologic conditions.**

CHANGES IN FREE ENERGY DETERMINE THE DIRECTION & EQUILIBRIUM STATE OF CHEMICAL REACTIONS

The Gibbs free energy change ΔG (also called either free energy or Gibbs energy) describes both the *direction* in which a chemical reaction will tend to proceed and the concentrations of reactants and products that will be present at equilibrium. ΔG for a chemical reaction equals the sum of the free energies of formation of the reaction products ΔG_p minus the sum of the free energies of formation of the substrates ΔG_s. ΔG^0 denotes the change in free energy that accompanies transition from the standard state, one-molar concentrations of substrates and products, to equilibrium. A more useful biochemical term is $\Delta G^{0'}$, which defines ΔG^0 at a standard state of 10^{-7} M protons, pH 7.0 (Chapter 11). If the free energy of formation of the products is lower than that of the substrates, the signs of ΔG^0 and $\Delta G^{0'}$ will be negative, indicating that the reaction as written is favored in the direction left to right. Such reactions are referred to as **spontaneous**. The **sign** and the **magnitude** of the free energy change determine how far the reaction will proceed. Equation (3) illustrates the relationship between the equilibrium constant K_{eq} and ΔG^0:

$$\Delta G^0 = -RT \ln K_{eq} \qquad (3)$$

where R is the gas constant (1.98 cal/mol°K or 8.31 J/mol°K) and T is the absolute temperature in degrees Kelvin. K_{eq} is equal to the product of the concentrations of the reaction products, each raised to the power of their stoichiometry, divided by the product of the substrates, each raised to the power of their stoichiometry:

For the reaction $A + B \rightleftarrows P + Q$

$$K_{eq} = \frac{[P][Q]}{[A][B]} \qquad (4)$$

and for reaction (5)

$$A + A \rightleftarrows P \qquad (5)$$

$$K_{eq} = \frac{[P]}{[A]^2} \qquad (6)$$

ΔG^0 may be calculated from equation (3) if the molar concentrations of substrates and products present at equilibrium are known. If ΔG^0 is a negative number, K_{eq} will be greater than unity, and the concentration of products at equilibrium will exceed that of the substrates. If ΔG^0 is positive, K_{eq} will be less than unity, and the formation of substrates will be favored.

Note that, since ΔG^0 is a function exclusively of the initial and final states of the reacting species, it can provide information only about the *direction* and *equilibrium state* of the reaction. ΔG^0 is independent of the **mechanism** of the reaction and therefore provides no information concerning **rates** of reactions. Consequently—and as explained below—although a reaction may have a large negative ΔG^0 or $\Delta G^{0'}$, it may nevertheless take place at a negligible rate.

THE RATES OF REACTIONS ARE DETERMINED BY THEIR ACTIVATION ENERGY

Reactions Proceed via Transition States

The concept of the **transition state** is fundamental to understanding the chemical and thermodynamic basis of catalysis. Equation (7) depicts a group transfer reaction in which an entering group E displaces a leaving group L, attached initially to R:

$$E + R - L \rightleftarrows E - R + L \qquad (7)$$

The net result of this process is to transfer group R from L to E. Midway through the displacement, the bond between R and L has weakened but has not yet been completely severed, and the new bond between E and R is yet incompletely formed. This transient intermediate—in which neither free substrate nor product exists—is termed the **transition state, E···R···L.** Dotted lines represent the "partial" bonds that are undergoing formation and rupture. **Figure 8–1** provides a more detailed illustration of the transition state intermediate formed during the transfer of a phosphoryl group.

Reaction (7) can be thought of as consisting of two "partial reactions," the first corresponding to the formation (F) and the second to the subsequent decay (D) of the transition state intermediate. As for all reactions, characteristic changes

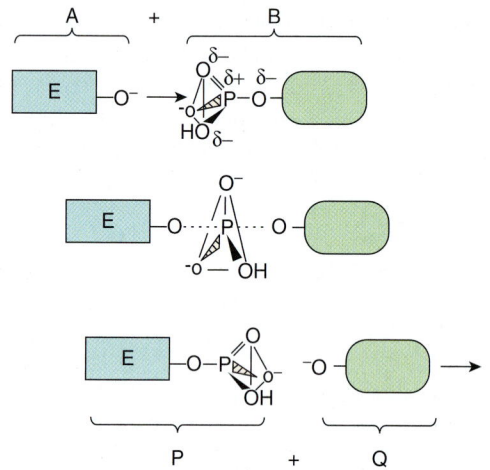

FIGURE 8–1 Formation of a transition state intermediate during a simple chemical reaction, A + B → P + Q. Shown are three stages of a chemical reaction in which a phosphoryl group is transferred from leaving group L to entering group E. Top: entering group E (A) approaches the other reactant, L-phosphate (B). Notice how the three oxygen atoms linked by the triangular lines and the phosphorus atom of the phosphoryl group form a pyramid. Center: as E approaches L-phosphate, the new bond between E and the phosphate group begins to form (dotted line) as that linking L to the phosphate group weakens. These partially formed bonds are indicated by dotted lines. Bottom: formation of the new product, E-phosphate (P), is now complete as the leaving group L (Q) exits. Notice how the geometry of the phosphoryl group differs between the transition state and the substrate or product. Notice how the phosphorus and three oxygen atoms that occupy the four corners of a pyramid in the substrate and product become coplanar, as emphasized by the triangle, in the transition state.

in free energy, ΔG_F and ΔG_D are associated with each partial reaction:

$$E + R - L \rightleftarrows E \cdots R \cdots L \quad \Delta G_F \tag{8}$$

$$E \cdots R \cdots L \rightleftarrows E - R + L \quad \Delta G_D \tag{9}$$

$$E + R - L \rightleftarrows E - R + L \quad \Delta G = \Delta G_F + \Delta G_D \tag{10}$$

For the overall reaction (10), ΔG is the sum of ΔG_F and ΔG_D. As for any equation of two terms, it is not possible to infer from ΔG either the sign or the magnitude of ΔG_F or ΔG_D.

Many reactions involve multiple transition states, each with an associated change in free energy. For these reactions, the overall ΔG represents the sum of *all* of the free energy changes associated with the formation and decay of *all* of the transition states. **Therefore, it is not possible to infer from the overall ΔG the number or type of transition states through which the reaction proceeds.** Stated another way, overall thermodynamics tells us nothing about kinetics.

ΔG_F Defines the Activation Energy

Regardless of the sign or magnitude of ΔG, ΔG_F for the overwhelming majority of chemical reactions has a positive sign. The formation of transition state intermediates therefore requires surmounting energy barriers. For this reason, ΔG_F

for reaching a transition state is often termed the **activation energy**, E_{act}. The ease—and hence the frequency—with which this barrier is overcome is inversely related to E_{act}. The thermodynamic parameters that determine how *fast* a reaction proceeds thus are the ΔG_F values for formation of the transition states through which the reaction proceeds. For a simple reaction, where ∝ means "proportionate to,"

$$\text{Rate} \propto e^{-E_{act}/RT} \tag{11}$$

The activation energy for the reaction proceeding in the opposite direction to that drawn is equal to $-\Delta G_D$.

NUMEROUS FACTORS AFFECT THE REACTION RATE

The **kinetic theory**—also called the **collision theory**—of chemical kinetics states that for two molecules to react they (1) must approach within bond-forming distance of one another, or "collide," and (2) must possess sufficient kinetic energy to overcome the energy barrier for reaching the transition state. It therefore follows that anything that increases the *frequency* or *energy* of collision between substrates will increase the rate of the reaction in which they participate.

Temperature

Raising the temperature increases the kinetic energy of molecules. As illustrated in **Figure 8–2**, the total number of molecules whose kinetic energy exceeds the energy barrier E_{act} (vertical bar) for formation of products increases from low (A) through intermediate (B) to high (C) temperatures. Increasing the kinetic energy of molecules also increases their rapidity of motion and therefore the frequency with which they collide. This combination of more frequent and more highly energetic, and hence productive, collisions increases the reaction rate.

Reactant Concentration

The frequency with which molecules collide is directly proportionate to their concentrations. For two different molecules A and B, the frequency with which they collide will double if the concentration of either A or B is doubled. If the concentrations of both A and B are doubled, the probability of collision will increase fourfold.

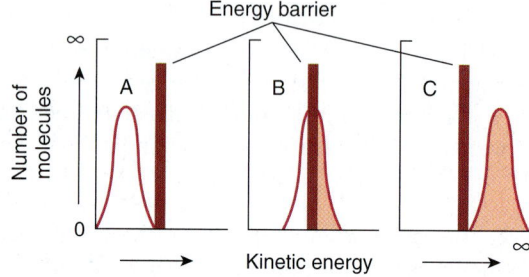

FIGURE 8–2 The energy barrier for chemical reactions. (See text for discussion.)

For a chemical reaction proceeding at constant temperature that involves one molecule each of A and B,

$$A + B \rightarrow P \qquad (12)$$

the number of the molecules that possess kinetic energy sufficient to overcome the activation energy barrier will be a constant. The number of collisions with sufficient energy to produce product P therefore will be directly proportionate to the number of collisions between A and B, and thus to their molar concentrations, denoted by the square brackets:

$$\text{Rate} \propto [A][B] \qquad (13)$$

Similarly, for the reaction represented by

$$A + 2B \rightarrow P \qquad (14)$$

which can also be written as

$$A + B + B \rightarrow P \qquad (15)$$

The corresponding rate expression is

$$\text{Rate} \propto [A][B][B] \qquad (16)$$

or

$$\text{Rate} \propto [A][B]^2 \qquad (17)$$

For the general case, when n molecules of A react with m molecules of B,

$$nA + mB \rightarrow P \qquad (18)$$

the rate expression is

$$\text{Rate} \propto [A]^n [B]^m \qquad (19)$$

Replacing the proportionality sign with an equals sign by introducing a **rate constant k** characteristic of the reaction under study gives equations (20) and (21), in which the subscripts 1 and −1 refer to the forward and reverse reactions, respectively:

$$\text{Rate}_1 = k_1 [A]^n [B]^m \qquad (20)$$

$$\text{Rate}_{-1} = k_{-1} [P] \qquad (21)$$

The sum of the molar ratios of the reactants defines the **kinetic order** of the reaction. Consider reaction (5). The stoichiometric coefficient for the sole reactant, A, is 2. Therefore, the rate of production of P is proportional to the square of [A] and the reaction is said to be *second order* with respect to reactant A. In this instance, the overall reaction is also *second order*. Therefore, k_1 is referred to as a *second-order rate constant*.

Reaction (12) describes a simple second-order reaction between two different reactants, A and B. The stoichiometric coefficient for each reactant is 1. Therefore, while the overall order of the reaction is 2, it is said to be *first order* with respect to A and *first order* with respect to B. In the laboratory, the kinetic order of a reaction with respect to a particular reactant, referred to as the variable reactant or substrate, can be determined by maintaining the concentration of the other reactants at a constant, or fixed, concentration in large excess over the variable reactant. Under these *pseudo-first-order conditions*, the concentration of the fixed reactant(s) remains virtually

constant. Thus, the rate of reaction will depend exclusively on the concentration of the variable reactant, sometimes also called the limiting reactant. The concepts of reaction order and pseudo-first-order conditions apply not only to simple chemical reactions but also to enzyme-catalyzed reactions.

K_{eq} Is a Ratio of Rate Constants

While all chemical reactions are to some extent reversible, at equilibrium the *overall* concentrations of reactants and products remain constant. At equilibrium, the rate of conversion of substrates to products therefore equals the rate at which products are converted to substrates:

$$\text{Rate}_1 = \text{Rate}_{-1} \qquad (22)$$

Therefore,

$$k_1 [A]^n [B]^m = k_{-1} [P] \qquad (23)$$

and

$$\frac{k_1}{k_{-1}} = \frac{[P]}{[A]^n [B]^m} \qquad (24)$$

The ratio of k_1 to k_{-1} is termed the equilibrium constant, K_{eq}. The following important properties of a system at equilibrium must be kept in mind.

1. The equilibrium constant is a ratio of the reaction *rate constants* (not the reaction *rates*).
2. At equilibrium, the reaction *rates* (not the *rate constants*) of the forward and back reactions are equal.
3. Equilibrium is a *dynamic* state. Although there is no *net* change in the concentration of substrates or products, individual substrate and product molecules are continually being interconverted.
4. The numeric value of the equilibrium constant K_{eq} can be calculated either from the concentrations of substrates and products at equilibrium or from the ratio k_1/k_{-1}.

THE KINETICS OF ENZYMATIC CATALYSIS

Enzymes Lower the Activation Energy Barrier for a Reaction

All enzymes accelerate reaction rates by lowering ΔG_F for the formation of transition states. However, they may differ in the way this is achieved. Where the mechanism or the sequence of chemical steps at the active site is essentially equivalent to those for the same reaction proceeding in the absence of a catalyst, **the environment of the active site lowers ΔG_F** by stabilizing the transition state intermediates. To put it another way, the enzyme can be envisioned as binding to the transition state intermediate (Figure 8–1) more tightly than it does to either substrates or products. As discussed in Chapter 7, stabilization can involve (1) acid–base groups suitably positioned to transfer protons to or from the developing transition state intermediate, (2) suitably positioned charged groups or metal ions that stabilize developing charges, or (3) the imposition of steric strain on substrates so that their geometry approaches that of the transition state. HIV protease (see Figure 7–6) illustrates catalysis by an enzyme that lowers the activation barrier by stabilizing a transition state intermediate.

Catalysis by enzymes that proceeds via a *unique* reaction mechanism typically occurs when the transition state intermediate forms a covalent bond with the enzyme (**covalent catalysis**). The catalytic mechanism of the serine protease chymotrypsin (see Figure 7–7) illustrates how an enzyme utilizes covalent catalysis to provide a unique reaction pathway.

ENZYMES DO NOT AFFECT K_{eq}

While enzymes undergo transient modifications during the process of catalysis, they always emerge unchanged at the completion of the reaction. **The presence of an enzyme therefore has no effect on ΔG^0 for the *overall* reaction,** which is a function solely of the **initial and final states** of the reactants. Equation (25) shows the relationship between the equilibrium constant for a reaction and the standard free energy change for that reaction:

$$\Delta G^0 = -RT \ln K_{eq} \qquad (25)$$

This principle is perhaps most readily illustrated by including the presence of the enzyme (Enz) in the calculation of the equilibrium constant for an enzyme-catalyzed reaction:

$$A + B + Enz \rightleftarrows P + Q + Enz \qquad (26)$$

Since the enzyme on both sides of the double arrows is present in equal quantity and identical form, the expression for the equilibrium constant,

$$K_{eq} = \frac{[P][Q][Enz]}{[A][B][Enz]} \qquad (27)$$

reduces to one identical to that for the reaction in the absence of the enzyme:

$$K_{eq} = \frac{[P][Q]}{[A][B]} \qquad (28)$$

Enzymes therefore have no effect on K_{eq}.

MULTIPLE FACTORS AFFECT THE RATES OF ENZYME-CATALYZED REACTIONS

Temperature

Raising the temperature increases the rate of both uncatalyzed and enzyme-catalyzed reactions by increasing the kinetic energy and the collision frequency of the reacting molecules. However, heat energy can also increase the kinetic energy of the enzyme to a point that exceeds the energy barrier for disrupting the noncovalent interactions that maintain its three-dimensional structure. The polypeptide chain then begins to unfold, or **denature,** with an accompanying loss of the catalytic activity. The temperature range over which an enzyme maintains a stable, catalytically competent conformation depends upon—and typically moderately exceeds—the normal temperature of the cells in which it resides. Enzymes from humans generally exhibit stability at temperatures up to 45–55°C. By contrast, enzymes from the thermophilic microorganisms that reside in volcanic hot springs or undersea hydrothermal vents may be stable at temperatures up to or even above 100°C.

The **temperature coefficient (Q_{10})** is the factor by which the rate of a biologic process increases for a 10°C increase in temperature. For the temperatures over which enzymes are stable, the rates of most biological processes typically double for a 10°C rise in temperature ($Q_{10} = 2$). Changes in the rates of enzyme-catalyzed reactions that accompany a rise or fall in body temperature constitute a prominent survival feature for "cold-blooded" life forms such as lizards or fish, whose body temperatures are dictated by the external environment. However, for mammals and other homeothermic organisms, changes in enzyme reaction rates with temperature assume physiologic importance only in circumstances such as fever or hypothermia.

Hydrogen Ion Concentration

The rate of almost all enzyme-catalyzed reactions exhibits a significant dependence on hydrogen ion concentration. Most intracellular enzymes exhibit optimal activity at pH values between 5 and 9. The relationship of activity to hydrogen ion concentration (**Figure 8–3**) reflects the balance between enzyme denaturation at high or low pH and effects on the charged state of the enzyme, the substrates, or both. For enzymes whose mechanism involves acid–base catalysis, the residues involved must be in the appropriate state of protonation for the reaction to proceed. The binding and recognition

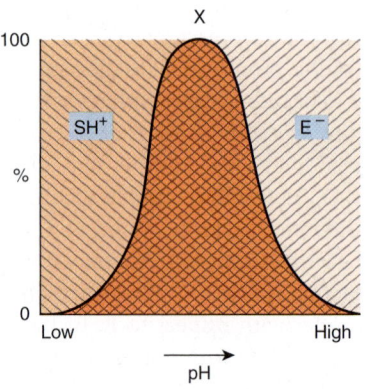

FIGURE 8–3   **Effect of pH on enzyme activity.** Consider, for example, a negatively charged enzyme (E⁻) that binds a positively charged substrate (SH⁺). Shown is the proportion (%) of SH⁺ [\\\] and of E⁻ [///] as a function of pH. Only in the cross-hatched area do both the enzyme and the substrate bear an appropriate charge.

of substrate molecules with dissociable groups also typically involves the formation of salt bridges with the enzyme. The most common charged groups are carboxylate groups (negative) and protonated amines (positive). Gain or loss of critical charged groups adversely affects substrate binding and thus will retard or abolish catalysis.

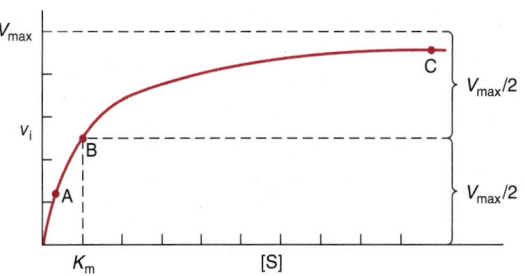

FIGURE 8–4 Effect of substrate concentration on the initial velocity of an enzyme-catalyzed reaction.

ASSAYS OF ENZYME-CATALYZED REACTIONS TYPICALLY MEASURE THE INITIAL VELOCITY

Most measurements of the rates of enzyme-catalyzed reactions employ relatively short time periods, conditions that approximate **initial rate conditions.** Under these conditions, only traces of product accumulate, rendering the rate of the reverse reaction negligible. The **initial velocity** (v_i) of the reaction thus is essentially that of the rate of the forward reaction. Assays of enzyme activity almost always employ a large (10^3–10^7) molar excess of substrate over enzyme. Under these conditions, v_i is proportionate to the concentration of enzyme. Measuring the initial velocity therefore permits one to estimate the quantity of enzyme present in a biologic sample.

SUBSTRATE CONCENTRATION AFFECTS THE REACTION RATE

In what follows, enzyme reactions are treated as if they had only a single substrate and a single product. For enzymes with multiple substrates, the principles discussed below apply with equal validity. Moreover, by employing pseudo-first-order conditions (see above), scientists can study the dependence of reaction rate upon an individual reactant through the appropriate choice of fixed and variable substrates. In other words, under pseudo-first-order conditions the behavior of a multisubstrate enzyme will imitate one having a single substrate. In this instance, however, the observed rate constant will be a

function of the rate constant k_1 for the reaction as well as the concentration of the fixed substrate(s).

For a typical enzyme, as substrate concentration is increased, v_i increases until it reaches a maximum value V_{max} (**Figure 8–4**). When further increases in substrate concentration do not further increase v_i, the enzyme is said to be "saturated" with the substrate. Note that the shape of the curve that relates activity to substrate concentration (Figure 8–4) is hyperbolic. At any given instant, only substrate molecules that are combined with the enzyme as an enzyme-substrate (ES) complex can be transformed into a product. Since the equilibrium constant for the formation of the enzyme-substrate complex is not infinitely large, only a fraction of the enzyme may be present as an ES complex even when the substrate is present in excess (points A and B of **Figure 8–5**). At points A or B, increasing or decreasing [S] therefore will increase or decrease the number of ES complexes with a corresponding change in v_i. At point C (Figure 8–5), however, essentially all the enzyme is present as the ES complex. Since no free enzyme remains available for forming ES, further increases in [S] cannot increase the rate of the reaction. Under these saturating conditions, v_i depends solely on—and thus is limited by—the rapidity with which product dissociates from the enzyme so that it may combine with more substrate.

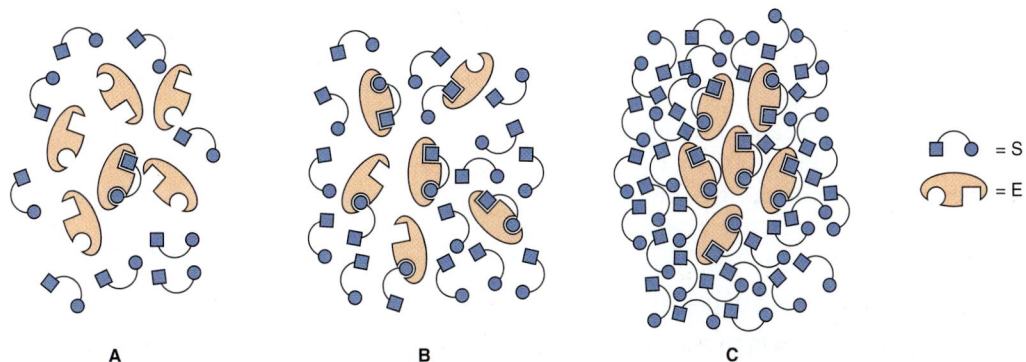

FIGURE 8–5 Representation of an enzyme in the presence of a concentration of substrate that is below K_m (**A**), at a concentration equal to K_m (**B**), and at a concentration well above K_m (**C**). Points A, B, and C correspond to those points in Figure 8–4.

THE MICHAELIS–MENTEN & HILL EQUATIONS MODEL THE EFFECTS OF SUBSTRATE CONCENTRATION

The Michaelis–Menten Equation

The Michaelis–Menten equation (29) illustrates in mathematical terms the relationship between initial reaction velocity v_i and substrate concentration [S], shown graphically in Figure 8–4:

$$v_i = \frac{V_{max}[S]}{K_m + [S]} \qquad (29)$$

The Michaelis constant K_m **is the substrate concentration at which v_i is half the maximal velocity ($V_{max}/2$) attainable at a particular concentration of the enzyme.** K_m thus has the dimensions of substrate concentration. The dependence of initial reaction velocity on [S] and K_m may be illustrated by evaluating the Michaelis–Menten equation under three conditions.

1. When [S] is much less than K_m (point A in Figures 8–4 and 8–5), the term $K_m + [S]$ is essentially equal to K_m. Replacing $K_m + [S]$ with K_m reduces equation (29) to

$$v_i = \frac{V_{max}[S]}{K_m + [S]} \quad v_i \approx \frac{V_{max}[S]}{K_m} \approx \left(\frac{V_{max}}{K_m}\right)[S] \qquad (30)$$

where ≈ means "approximately equal to." Since V_{max} and K_m are both constants, their ratio is a constant. In other words, when [S] is considerably below K_m, v_i is proportionate to $k[S]$. The initial reaction velocity therefore is directly proportional to [S].

2. When [S] is much greater than K_m (point C in Figures 8–4 and 8–5), the term $K_m + [S]$ is essentially equal to [S]. Replacing $K_m + [S]$ with [S] reduces equation (29) to

$$v_i = \frac{V_{max}[S]}{K_m + [S]} \quad v_i \approx \frac{V_{max}[S]}{[S]} \approx V_{max} \qquad (31)$$

Thus, when [S] greatly exceeds K_m, the reaction velocity is maximal (V_{max}) and unaffected by further increases in the substrate concentration.

3. When $[S] = K_m$ (point B in Figures 8–4 and 8–5):

$$v_i = \frac{V_{max}[S]}{K_m + [S]} = \frac{V_{max}[S]}{2[S]} = \frac{V_{max}}{2} \qquad (32)$$

Equation (32) states that when [S] equals K_m, the initial velocity is half-maximal. Equation (32) also reveals that K_m is—and may be determined experimentally from—the substrate concentration at which the initial velocity is half-maximal.

A Linear Form of the Michaelis–Menten Equation Is Used to Determine K_m & V_{max}

The direct measurement of the numeric value of V_{max}, and therefore the calculation of K_m, often requires impractically high concentrations of substrate to achieve saturating conditions.

A linear form of the Michaelis–Menten equation circumvents this difficulty and permits V_{max} and K_m to be extrapolated from initial velocity data obtained at less than saturating concentrations of the substrate. Start with equation (29),

$$v_i = \frac{V_{max}[S]}{K_m + [S]} \qquad (29)$$

invert

$$\frac{1}{v_i} = \frac{K_m + [S]}{V_{max}[S]} \qquad (33)$$

factor

$$\frac{1}{v_i} = \frac{K_m}{V_{max}[S]} + \frac{[S]}{V_{max}[S]} \qquad (34)$$

and simplify

$$\frac{1}{v_i} = \left(\frac{K_m}{V_{max}}\right)\frac{1}{[S]} + \frac{1}{V_{max}} \qquad (35)$$

Equation (35) is the equation for a straight line, $y = ax + b$, where $y = 1/v_i$ and $x = 1/[S]$. A plot of $1/v_i$ as y as a function of $1/[S]$ as x therefore gives a straight line whose y intercept is $1/V_{max}$ and whose slope is K_m/V_{max}. Such a plot is called a **double reciprocal** or **Lineweaver–Burk plot** (**Figure 8–6**). Setting the y term of equation (36) equal to zero and solving for x reveals that the x intercept is $-1/K_m$:

$$0 = ax + b; \text{ therefore, } x = \frac{-b}{a} = \frac{-1}{K_m} \qquad (36)$$

K_m is thus most readily calculated from the negative x intercept.

The greatest virtue of the Lineweaver–Burk plot resides in the facility with which it can be used to determine the kinetic mechanisms of enzyme inhibitors (see below). However, in using a double-reciprocal plot to determine kinetic constants it is important to avoid the introduction of bias through the clustering of data at low values of $1/[S]$. To avoid this bias, prepare a solution of substrate whose dilution into an assay will produce the maximum desired concentration of the substrate. Now use the same volume of solutions prepared by diluting the stock solution by factors of 1:2, 1:3, 1:4, 1:5, etc. The data will then fall on the $1/[S]$ axis at intervals of 1, 2, 3, 4, 5, etc. Alternatively, a

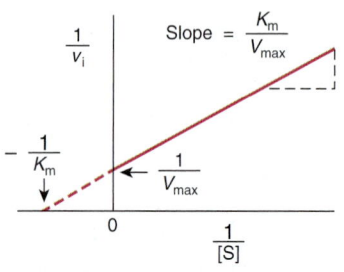

FIGURE 8–6 Double-reciprocal or Lineweaver–Burk plot of $1/v_i$ versus $1/[S]$ used to evaluate K_m and V_{max}.

single-reciprocal plot such as the Eadie–Hofstee (v_i versus v_i/[S]) or Hanes–Woolf ([S]/v_i versus [S]) plot can be used to minimize clustering.

The Catalytic Constant, k_{cat}

Several parameters may be used to compare the relative activity of different enzymes or of different preparations of the same enzyme. The activity of impure enzyme preparations typically is expressed as a *specific activity* (V_{max} divided by the protein concentration). For a homogeneous enzyme, one may calculate its *turnover number* (V_{max} divided by the moles of enzyme present). But if the number of active sites present is known, the catalytic activity of a homogeneous enzyme is best expressed as its *catalytic constant, k_{cat}* (V_{max} divided by the number of active sites, S_t):

$$k_{cat} = \frac{V_{max}}{S_t} \qquad (37)$$

Since the units of concentration cancel out, the units of k_{cat} are reciprocal time.

Catalytic Efficiency, k_{cat}/K_m

By what measure should the efficiency of different enzymes, different substrates for a given enzyme, and the efficiency with which an enzyme catalyzes a reaction in the forward and reverse directions be quantified and compared? While the maximum capacity of a given enzyme to convert substrate to product is important, the benefits of a high k_{cat} can only be realized if K_m is sufficiently low. Thus, *catalytic efficiency* of enzymes is best expressed in terms of the ratio of these two kinetic constants, k_{cat}/K_m.

For certain enzymes, once substrate binds to the active site, it is converted to product and released so rapidly as to render these events effectively instantaneous. For these exceptionally efficient catalysts, the rate-limiting step in catalysis is the formation of the ES complex. Such enzymes are said to be *diffusion-limited,* or catalytically perfect, since the fastest possible rate of catalysis is determined by the rate at which molecules move or diffuse through the solution. Examples of enzymes for which k_{cat}/K_m approaches the diffusion limit of 10^8–10^9 M^{-1}s^{-1} include triosephosphate isomerase, carbonic anhydrase, acetylcholinesterase, and adenosine deaminase.

In living cells, the assembly of enzymes that catalyze successive reactions into multimeric complexes can circumvent the limitations imposed by diffusion. The geometric relationships of the enzymes in these complexes are such that the substrates and products do not diffuse into the bulk solution until the last step in the sequence of catalytic steps is complete. Fatty acid synthetase extends this concept one step further by covalently attaching the growing substrate fatty acid chain to a biotin tether that rotates from active site to active site within the complex until synthesis of a palmitic acid molecule is complete (Chapter 23).

K_m May Approximate a Binding Constant

The affinity of an enzyme for its substrate is the inverse of the dissociation constant K_d for dissociation of the enzyme-substrate complex ES:

$$E + S \underset{k_{-1}}{\overset{k_1}{\rightleftharpoons}} ES \qquad (38)$$

$$K_d = \frac{k_{-1}}{k_1} \qquad (39)$$

Stated another way, the *smaller* the tendency of the enzyme and its substrate to *dissociate,* the *greater* the affinity of the enzyme for its substrate. While the Michaelis constant K_m often approximates the dissociation constant K_d, this is by no means always the case. For a typical enzyme-catalyzed reaction:

$$E + S \underset{k_{-1}}{\overset{k_1}{\rightleftharpoons}} ES \overset{k_2}{\longrightarrow} E + P \qquad (40)$$

The value of [S] that gives $v_i = V_{max}/2$ is

$$[S] = \frac{k_{-1} + k_2}{k_1} = K_m \qquad (41)$$

When $k_{-1} \gg k_2$, then

$$k_{-1} + k_2 \approx k_{-1} \qquad (42)$$

and

$$[S] \approx \frac{k_1}{k_{-1}} = K_d \qquad (43)$$

Hence, $1/K_m$ only approximates $1/K_d$ under conditions where the association and dissociation of the ES complex are rapid relative to catalysis. For the many enzyme-catalyzed reactions for which $k_{-1} + k_2$ is **not** approximately equal to k_{-1}, $1/K_m$ will underestimate $1/K_d$.

The Hill Equation Describes the Behavior of Enzymes That Exhibit Cooperative Binding of Substrate

While most enzymes display the simple **saturation kinetics** depicted in Figure 8–4 and are adequately described by the Michaelis–Menten expression, some enzymes bind their substrates in a **cooperative** fashion analogous to the binding of oxygen by hemoglobin (Chapter 6). Cooperative behavior is an exclusive property of multimeric enzymes that bind substrate at multiple sites.

For enzymes that display positive cooperativity in binding the substrate, the shape of the curve that relates changes in v_i to changes in [S] is sigmoidal (**Figure 8–7**). Neither the Michaelis–Menten expression nor its derived plots can be used to evaluate cooperative kinetics. Enzymologists therefore employ a graphic representation of the **Hill equation** originally derived to describe the cooperative binding of O$_2$ by hemoglobin. Equation (44) represents the Hill equation

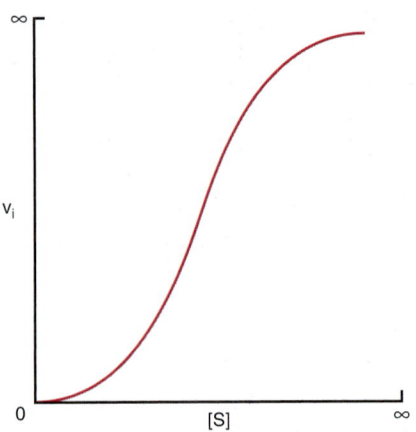

FIGURE 8–7 Representation of sigmoid substrate saturation kinetics.

arranged in a form that predicts a straight line, where k' is a complex constant:

$$\frac{\log v_i}{V_{max} - v_i} = n \log [S] - \log k' \qquad (44)$$

Equation (44) states that when [S] is low relative to k', the initial reaction velocity increases as the nth power of [S].

A graph of $\log v_i/(V_{max} - v_i)$ versus $\log[S]$ gives a straight line (**Figure 8–8**), where the slope of the line n is the **Hill coefficient,** an empirical parameter whose value is a function of the number, kind, and strength of the interactions of the multiple substrate-binding sites on the enzyme. When $n = 1$, all binding sites behave independently and simple Michaelis–Menten kinetic behavior is observed. If n is greater than 1, the enzyme is said to exhibit positive cooperativity. Binding of substrate to one site then enhances the affinity of the remaining sites to bind additional substrate. The greater the value for n, the higher the degree of cooperativity and the more markedly sigmoidal will be the plot of v_i versus [S]. A perpendicular dropped from the point where the y term $\log v_i/(V_{max} - v_i)$ is

zero intersects the x-axis at a substrate concentration termed S_{50}, the substrate concentration that results in half-maximal velocity. S_{50} thus is analogous to the P_{50} for oxygen binding to hemoglobin (Chapter 6).

KINETIC ANALYSIS DISTINGUISHES COMPETITIVE FROM NONCOMPETITIVE INHIBITION

Inhibitors of the catalytic activities of enzymes provide both pharmacologic agents and research tools for the study of the mechanism of enzyme action. The strength of the interaction between an inhibitor and an enzyme depends on forces important in protein structure and ligand binding (hydrogen bonds, electrostatic interactions, hydrophobic interactions, and van der Waals forces; see Chapter 5). Inhibitors can be classified on the basis of their site of action on the enzyme, on whether they chemically modify the enzyme, or on the kinetic parameters they influence. Compounds that mimic the transition state of an enzyme-catalyzed reaction (transition state analogs) or that take advantage of the catalytic machinery of an enzyme (mechanism-based inhibitors) can be particularly potent inhibitors. Kinetically, we distinguish two classes of inhibitors based upon whether raising the substrate concentration does or does not overcome the inhibition.

Competitive Inhibitors Typically Resemble Substrates

The effects of competitive inhibitors can be overcome by raising the concentration of substrate. Most frequently, in competitive inhibition the inhibitor (**I**) binds to the substrate-binding portion of the active site thereby blocking access by the substrate. The structures of most classic competitive inhibitors therefore tend to resemble the structures of a substrate, and thus are termed **substrate analogs.** Inhibition of the enzyme succinate dehydrogenase by malonate illustrates competitive inhibition by a substrate analog. Succinate dehydrogenase catalyzes the removal of one hydrogen atom from each of the two-methylene carbons of succinate (**Figure 8–9**). Both succinate and its structural analog malonate ($^-OOC—CH_2—COO^-$) can bind to the active site of succinate dehydrogenase, forming an ES or an EI complex, respectively. However, since malonate contains

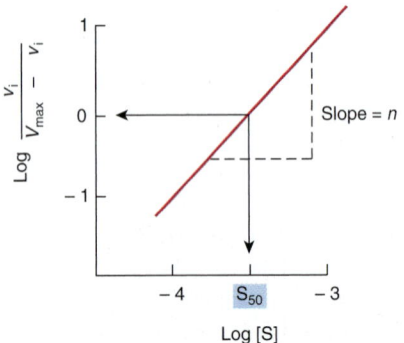

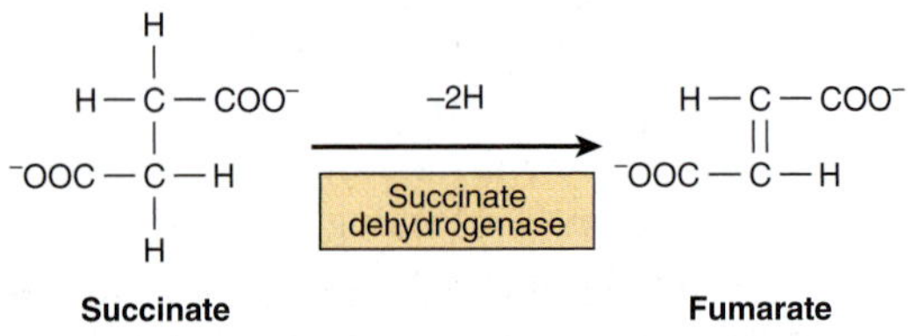

FIGURE 8–8 A graphical representation of a linear form of the Hill equation is used to evaluate S_{50}, the substrate concentration that produces half-maximal velocity, and the degree of cooperativity n.

FIGURE 8–9 The succinate dehydrogenase reaction.

only one methylene carbon, it cannot undergo dehydrogenation.

The formation and dissociation of the EI complex is a dynamic process described by

$$E - I \underset{k_{-1}}{\overset{k_1}{\rightleftharpoons}} E + I \qquad (45)$$

for which the equilibrium constant K_i is

$$K_i = \frac{[E][I]}{[E-I]} = \frac{k_i}{k_{-i}} \qquad (46)$$

In effect, **a competitive inhibitor acts by decreasing the number of free enzyme molecules available to bind substrate, ie, to form ES, and thus eventually to form product,** as described below.

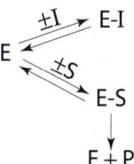

A competitive inhibitor and substrate exert reciprocal effects on the concentration of the EI and ES complexes. Since the formation of ES complexes removes free enzyme available to combine with the inhibitor, increasing [S] decreases the concentration of the EI complex and raises the reaction velocity. The extent to which [S] must be increased to completely overcome the inhibition depends upon the concentration of the inhibitor present, its affinity for the enzyme, K_i, and the affinity, K_m, of the enzyme for its substrate.

Double-Reciprocal Plots Facilitate the Evaluation of Inhibitors

Double-reciprocal plots distinguish between competitive and noncompetitive inhibitors and simplify evaluation of inhibition constants. v_i is determined at several substrate concentrations both in the presence and in the absence of the inhibitor. For classic competitive inhibition, the lines that connect the experimental data points converge at the y-axis (**Figure 8–10**).

Since the y intercept is equal to $1/V_{max}$, this pattern indicates that **when 1/[S] approaches 0, v_i is independent of the presence of inhibitor.** Note, however, that the intercept on the x-axis does vary with *inhibitor* concentration—and that since $-1/K'_m$ is smaller than $1/K_m$, K'_m (the "apparent K_m") becomes larger in the presence of increasing concentrations of the inhibitor. Thus, **a competitive inhibitor has no effect on V_{max} but raises K'_m, the apparent K_m for the substrate.** For a simple competitive inhibition, the intercept on the x-axis is

$$x = \frac{-1}{K_m}\left(1 + \frac{[I]}{K_i}\right) \qquad (47)$$

Once K_m has been determined in the absence of inhibitor, K_i can be calculated from equation (47). K_i values are used to compare different inhibitors of the same enzyme. The lower the value for K_i, the more effective the inhibitor. For example, the statin drugs that act as competitive inhibitors of HMG-CoA reductase (Chapter 26) have K_i values several orders of magnitude lower than the K_m for the substrate HMG-CoA.

Simple Noncompetitive Inhibitors Lower V_{max} But Do Not Affect K_m

In strict noncompetitive inhibition, binding of the inhibitor does not affect binding of the substrate. Formation of both EI and EIS complexes is therefore possible. However, while the enzyme-inhibitor complex can still bind the substrate, its efficiency at transforming substrate to product, reflected by V_{max}, is decreased. Noncompetitive inhibitors bind enzymes at sites distinct from the substrate-binding site and generally bear little or no structural resemblance to the substrate.

For simple noncompetitive inhibition, E and EI possess identical affinity for the substrate, and the EIS complex generates product at a negligible rate (**Figure 8–11**). More complex noncompetitive inhibition occurs when binding of the inhibitor *does* affect the apparent affinity of the enzyme for the substrate, causing the lines to intercept in either the third or fourth quadrants of a double-reciprocal plot (not shown). While certain inhibitors exhibit characteristics of a mixture of competitive and noncompetitive inhibition, the evaluation of these inhibitors exceeds the scope of this chapter.

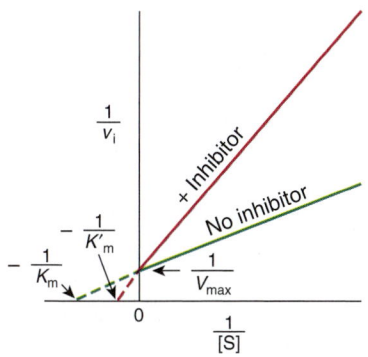

FIGURE 8–10 **Lineweaver–Burk plot of simple competitive inhibition.** Note the complete relief of inhibition at high [S] (ie, low 1/[S]).

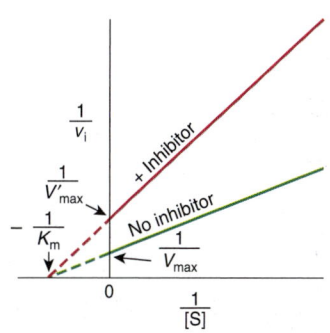

FIGURE 8–11 **Lineweaver–Burk plot for simple noncompetitive inhibition.**

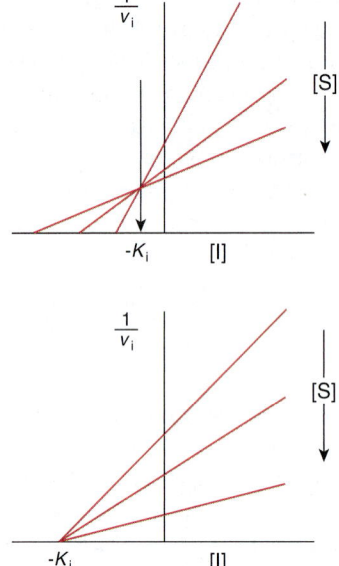

FIGURE 8–12 **Applications of Dixon plots.** Top: competitive inhibition, estimation of K_i. Bottom: noncompetitive inhibition, estimation of K_i.

Dixon Plot

A Dixon plot is sometimes employed as an alternative to the Lineweaver–Burk plot for determining inhibition constants. The initial velocity (v_i) is measured at several concentrations of inhibitor, but at a fixed concentration of the substrate (S). For a simple competitive or noncompetitive inhibitor, a plot of $1/v_i$ versus inhibitor concentration [I] yields a straight line. The experiment is repeated at different fixed concentrations of the substrate. The resulting set of lines intersects to the left of the y-axis. For *competitive* inhibition, a perpendicular dropped to the x-axis from the point of intersection of the lines gives $-K_i$ (**Figure 8–12**, top). For *noncompetitive* inhibition the intercept on the x-axis is $-K_i$ (**Figure 8–12**, bottom). Pharmaceutical publications frequently employ Dixon plots to illustrate the comparative potency of competitive inhibitors.

IC$_{50}$

A less rigorous alternative to K_i as a measure of inhibitory potency is the concentration of inhibitor that produces 50% inhibition, **IC$_{50}$**. Unlike the equilibrium dissociation constant K_i, the numeric value of IC$_{50}$ varies as a function of the specific circumstances of substrate concentration, etc, under which it is determined.

Tightly Bound Inhibitors

Some inhibitors bind to enzymes with such high affinity, $K_i \leq 10^{-9}$ M, that the concentration of inhibitor required to measure K_i falls below the concentration of enzyme typically present in an assay. Under these circumstances, a significant fraction of

the total inhibitor may be present as an EI complex. If so, this violates the assumption, implicit in classical steady-state kinetics, that the concentration of free inhibitor is independent of the concentration of enzyme. The kinetic analysis of these tightly bound inhibitors requires specialized kinetic equations that incorporate the concentration of enzyme to estimate K_i or IC$_{50}$ and to distinguish competitive from noncompetitive tightly bound inhibitors.

Irreversible Inhibitors "Poison" Enzymes

In the above examples, the inhibitors form a dissociable, dynamic complex with the enzyme. Fully active enzyme can therefore be recovered simply by removing the inhibitor from the surrounding medium. However, a variety of other inhibitors act irreversibly by chemically modifying the enzyme. These modifications generally involve making or breaking covalent bonds with aminoacyl residues essential for substrate binding, catalysis, or maintenance of the enzyme's functional conformation. Since these covalent changes are relatively stable, an enzyme that has been "poisoned" by an irreversible inhibitor such as a heavy metal atom or an acylating reagent remains inhibited even after the removal of the remaining inhibitor from the surrounding medium.

Mechanism-Based Inhibition

"Mechanism-based" or "suicide" inhibitors are specialized substrate analogs that contain a chemical group that can be transformed by the catalytic machinery of the target enzyme. After binding to the active site, catalysis by the enzyme generates a highly reactive group that forms a covalent bond to, and **blocks the function of a catalytically essential residue.** The specificity and persistence of suicide inhibitors, which are both enzyme-specific and unreactive outside the confines of the enzyme's active site, render them promising leads for the development of enzyme-specific drugs. The kinetic analysis of suicide inhibitors lies beyond the scope of this chapter. Neither the Lineweaver–Burk nor the Dixon approach is applicable since suicide inhibitors violate a key boundary condition common to both approaches, namely that the activity of the enzyme does not decrease during the course of the assay.

MOST ENZYME-CATALYZED REACTIONS INVOLVE TWO OR MORE SUBSTRATES

While several enzymes have a single substrate, many others have two—and sometimes more—substrates and products. The fundamental principles discussed above, while illustrated for single-substrate enzymes, apply also to multisubstrate enzymes. The mathematical expressions used to evaluate multisubstrate reactions are, however, complex. While a detailed analysis of the full range of multisubstrate reactions exceeds the scope of this chapter, some common types of kinetic

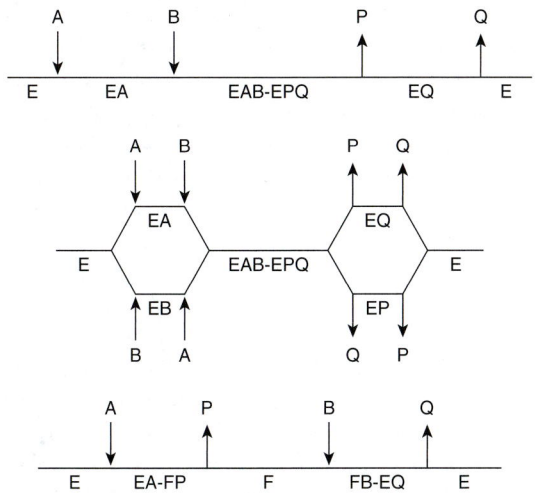

FIGURE 8–13 **Representations of three classes of Bi–Bi reaction mechanisms.** Horizontal lines represent the enzyme. Arrows indicate the addition of substrates and departure of products. Top: an ordered Bi–Bi reaction, characteristic of many NAD(P)H-dependent oxidoreductases. Center: a random Bi–Bi reaction, characteristic of many kinases and some dehydrogenases. Bottom: a ping–pong reaction, characteristic of aminotransferases and serine proteases.

behavior for two-substrate, two-product reactions (termed "BiBi" reactions) are considered below.

Sequential or Single-Displacement Reactions

In **sequential reactions,** both substrates must combine with the enzyme to form a ternary complex before catalysis can proceed (**Figure 8–13,** top). Sequential reactions are sometimes referred to as single-displacement reactions because the group undergoing transfer is usually passed directly, in a single step, from one substrate to the other. Sequential Bi–Bi reactions can be further distinguished on the basis of whether the two substrates add in a **random** or in a **compulsory** order. For random-order reactions, either substrate A or substrate B may combine first with the enzyme to form an EA or an EB complex (**Figure 8–13,** center). For compulsory-order reactions, A must first combine with E before B can combine with the EA complex. One explanation for why some enzymes employ compulsory-order mechanisms can be found in Koshland's induced fit hypothesis: the addition of A induces a conformational change in the enzyme that aligns residues that recognize and bind B.

Ping–Pong Reactions

The term **"ping–pong"** applies to mechanisms in which one or more products are released from the enzyme before all the substrates have been added. Ping-pong reactions involve covalent catalysis and a transient, modified form of the enzyme (see Figure 7–4). Ping-pong Bi–Bi reactions are often referred

to as **double displacement reactions.** The group undergoing transfer is first displaced from substrate A by the enzyme to form product P and a modified form of the enzyme (F). The subsequent group transfer from F to the second substrate B, forming product Q and regenerating E, constitutes the second displacement (**Figure 8–13,** bottom).

Most Bi–Bi Reactions Conform to Michaelis–Menten Kinetics

Most Bi–Bi reactions conform to a somewhat more complex form of Michaelis–Menten kinetics in which V_{max} refers to the reaction rate attained when both substrates are present at saturating levels. Each substrate has its own characteristic K_m value, which corresponds to the concentration that yields half-maximal velocity when the second substrate is present at saturating levels. As for single-substrate reactions, double-reciprocal plots can be used to determine V_{max} and K_m. v_i is measured as a function of the concentration of one substrate (the variable substrate) while the concentration of the other substrate (the fixed substrate) is maintained constant. If the lines obtained for several fixed-substrate concentrations are plotted on the same graph, it is possible to distinguish a ping–pong mechanism, which yields parallel lines (**Figure 8–14**), from a sequential mechanism, which yields a pattern of intersecting lines (not shown).

Product inhibition studies are used to complement kinetic analyses and to distinguish between ordered and random Bi–Bi reactions. For example, in a random-order Bi–Bi reaction, each product will act as a competitive inhibitor in the absence of its coproducts regardless of which substrate is designated the variable substrate. However, for a sequential mechanism (Figure 8–13, top), only product Q will give the pattern indicative of competitive inhibition when A is the

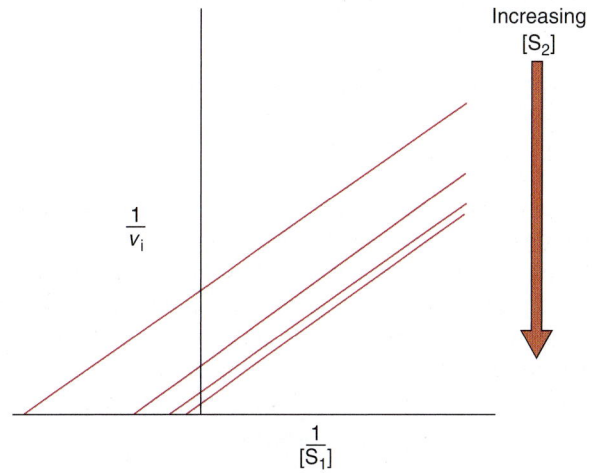

FIGURE 8–14 **Lineweaver–Burk plot for a two-substrate ping–pong reaction.** An increase in concentration of one substrate (S_1) while that of the other substrate (S_2) is maintained constant for changes both the x and y intercepts, but not the slope.

variable substrate, while only product P will produce this pattern with B as the variable substrate. The other combinations of product inhibitor and variable substrate will produce forms of complex noncompetitive inhibition.

KNOWLEDGE OF ENZYME KINETICS, MECHANISM, AND INHIBITION AIDS DRUG DEVELOPMENT

Many Drugs Act as Enzyme Inhibitors

The goal of pharmacology is to identify agents that can

1. Destroy or impair the growth, invasiveness, or development of invading pathogens.
2. Stimulate endogenous defense mechanisms.
3. Halt or impede aberrant molecular processes triggered by genetic, environmental, or biologic stimuli with minimal perturbation of the host's normal cellular functions.

By virtue of their diverse physiologic roles and high degree of substrate selectivity, enzymes constitute natural targets for the development of pharmacologic agents that are both potent and specific. Statin drugs, for example, lower cholesterol production by inhibiting 3-hydroxy-3-methylglutaryl coenzyme A reductase (Chapter 26), while emtricitabine and tenofovir disoproxil fumarate block replication of the human immunodeficiency virus by inhibiting the viral reverse transcriptase (Chapter 34). Pharmacologic treatment of hypertension often includes the administration of an inhibitor of angiotensin-converting enzyme, thus lowering the level of angiotensin II, a vasoconstrictor (Chapter 42).

Enzyme Kinetics Defines Appropriate Screening Conditions

Enzyme kinetics plays a crucial role in drug discovery. Knowledge of the kinetic behavior of the enzyme of interest is necessary, first and foremost, to select appropriate assay conditions for detecting the presence of an inhibitor. The concentration of substrate, for example, must be adjusted such that sufficient product is generated to permit facile detection of the enzyme's activity without being so high that it masks the presence of an inhibitor. Second, enzyme kinetics provides the means for quantifying and comparing the potency of different inhibitors and defining their mode of action. Noncompetitive inhibitors are particularly desirable, because—by contrast to competitive inhibitors—their effects can never be completely overcome by increases in substrate concentration.

Most Drugs Are Metabolized In Vivo

Drug development often involves more than the kinetic evaluation of the interaction of inhibitors with the target enzyme.

Drugs may be acted upon by enzymes present in the patient or pathogen, a process termed **drug metabolism.** For example, penicillin and other β-lactam antibiotics block cell wall synthesis in bacteria by irreversibly poisoning the enzyme alanyl alanine carboxypeptidase-transpeptidase. Many bacteria, however, produce β-lactamases that hydrolyze the critical β-lactam function in penicillin and related drugs. One strategy for overcoming the resulting antibiotic resistance is to simultaneously administer a β-lactamase inhibitor with a β-lactam antibiotic.

Metabolic transformation is sometimes required to convert an inactive drug precursor, or **prodrug,** into its biologically active form (Chapter 53). 2′-Deoxy-5-fluorouridylic acid, a potent inhibitor of thymidylate synthase, a common target of cancer chemotherapy, is produced from 5-fluorouracil via a series of enzymatic transformations catalyzed by a phosphoribosyl transferase and the enzymes of the deoxyribonucleoside salvage pathway (Chapter 33). The effective design and administration of prodrugs requires knowledge of the kinetics and mechanisms of the enzymes responsible for transforming them into their biologically active forms.

SUMMARY

- The study of enzyme kinetics—the factors that affect the rates of enzyme-catalyzed reactions—reveals the individual steps by which enzymes transform substrates into products.

- ΔG, the overall change in free energy for a reaction, is independent of reaction mechanism and provides no information concerning *rates* of reactions.

- K_{eq}, a ratio of reaction *rate constants*, may be calculated from the concentrations of substrates and products at equilibrium or from the ratio k_1/k_{-1}. Enzymes do not affect K_{eq}.

- Reactions proceed via transition states for which ΔG_F is the activation energy. Temperature, hydrogen ion concentration, enzyme concentration, substrate concentration, and inhibitors all affect the rates of enzyme-catalyzed reactions.

- Measurement of the rate of an enzyme-catalyzed reaction generally employs initial rate conditions, for which the virtual absence of product precludes the reverse reaction.

- Linear forms of the Michaelis–Menten equation simplify determination of K_m and V_{max}.

- A linear form of the Hill equation is used to evaluate the cooperative substrate-binding kinetics exhibited by some multimeric enzymes. The slope n, the Hill coefficient, reflects the number, nature, and strength of the interactions of the substrate-binding sites. A value of n greater than 1 indicates positive cooperativity.

- The effects of simple competitive inhibitors, which typically resemble substrates, are overcome by raising the concentration of the substrate. Simple noncompetitive inhibitors lower V_{max} but do not affect K_m.

- For simple competitive and noncompetitive inhibitors, the inhibitory constant K_i is equal to the equilibrium dissociation constant for the relevant enzyme-inhibitor complex. A simpler and less rigorous term for evaluating the effectiveness of an inhibitor is IC_{50}, the concentration of inhibitor that produces

50% inhibition under the particular circumstances of the experiment.

■ Substrates may add in a random order (either substrate may combine first with the enzyme) or in a compulsory order (substrate A must bind before substrate B).

■ In ping–pong reactions, one or more products are released from the enzyme before all the substrates have been added.

■ Applied enzyme kinetics facilitates the identification and characterization of drugs that selectively inhibit specific enzymes. Enzyme kinetics thus plays a central and critical role in drug discovery, in comparative pharmacodynamics, and in determining the mode of action of drugs.

REFERENCES

Cook PF, Cleland WW: *Enzyme Kinetics and Mechanism.* Garland Science, 2007.

Copeland RA: *Evaluation of Enzyme Inhibitors in Drug Discovery.* John Wiley & Sons, 2005.

Cornish-Bowden A: *Fundamentals of Enzyme Kinetics.* Portland Press Ltd, 2004.

Dixon M: The determination of enzyme inhibitor constants. Biochem J 1953;55:170.

Dixon M: The graphical determination of K_m and K_i. Biochem J 1972;129:197.

Fersht A: *Structure and Mechanism in Protein Science: A Guide to Enzyme Catalysis and Protein Folding.* Freeman, 1999.

Fraser CM, Rappuoli R: Application of microbial genomic science to advanced therapeutics. Annu Rev Med 2005;56:459.

Henderson PJF: A linear equation that describes the steady-state kinetics of enzymes and subcellular particles interacting with tightly bound inhibitors. Biochem J 1972;127:321.

Schramm, VL: Enzymatic transition-state theory and transition-state analogue design. J Biol Chem 2007;282:28297.

Schultz AR: *Enzyme Kinetics: From Diastase to Multi-enzyme Systems.* Cambridge University Press, 1994.

Segel IH: *Enzyme Kinetics.* Wiley Interscience, 1975.

Wlodawer A: Rational approach to AIDS drug design through structural biology. Annu Rev Med 2002;53:595.

Enzymes: Regulation of Activities

9

Peter J. Kennelly, PhD & Victor W. Rodwell, PhD

OBJECTIVES

*After studying this chapter,
you should be able to:*

- Explain the concept of whole-body homeostasis and its response to fluctuations in the external environment.
- Discuss why the cellular concentrations of substrates for most enzymes tend to be close to K_m.
- List multiple mechanisms by which active control of metabolite flux is achieved.
- Describe the advantages of certain enzymes being elaborated as proenzymes.
- Illustrate the physiologic events that trigger the conversion of a proenzyme to the corresponding active enzyme.
- Describe typical structural changes that accompany conversion of a proenzyme to the active enzyme.
- Describe the basic features of a typical binding site for metabolites and second messengers that regulate catalytic activity of certain enzymes.
- Indicate two general ways in which an allosteric effector can modify catalytic activity.
- Outline the roles of protein kinases, protein phosphatases, and of regulatory and hormonal and second messengers in initiating a metabolic process.

BIOMEDICAL IMPORTANCE

The nineteenth-century physiologist Claude Bernard enunciated the conceptual basis for metabolic regulation. He observed that living organisms respond in ways that are both quantitatively and temporally appropriate to permit them to survive the multiple challenges posed by changes in their external and internal environments. Walter Cannon subsequently coined the term "homeostasis" to describe the ability of animals to maintain a constant intracellular environment despite changes in their external environment. We now know that organisms respond to changes in their external and internal environment by balanced, coordinated adjustments in the rates of specific metabolic reactions. Perturbations of the sensor-response machinery responsible for maintaining homeostatic balance can be deleterious to human health. Cancer, diabetes, cystic fibrosis, and Alzheimer's disease, for example, are all characterized by regulatory dysfunctions triggered by pathogenic agents or genetic mutations. Many oncogenic viruses elaborate protein-tyrosine kinases that modify the regulatory events that control patterns of gene expression, contributing to the initiation

and progression of cancer. The toxin from *Vibrio cholerae*, the causative agent of cholera, disables sensor-response pathways in intestinal epithelial cells by ADP-ribosylating the GTP-binding proteins (G-proteins) that link cell surface receptors to adenylyl cyclase. The consequent activation of the cyclase leads to the unrestricted flow of water into the intestines, resulting in massive diarrhea and dehydration. *Yersinia pestis,* the causative agent of plague, elaborates a protein-tyrosine phosphatase that hydrolyzes phosphoryl groups on key cytoskeletal proteins. Dysfunctions in the proteolytic systems responsible for the degradation of defective or abnormal proteins are believed to play a role in neurodegenerative diseases such as Alzheimer and Parkinson's. In addition to their immediate function as regulators of enzyme activity, protein degradation, etc, covalent modifications such as phosphorylation, acetylation, and ubiquitination provide a protein-based code for the storage and hereditary transmission of information (Chapter 35). Such DNA-independent information systems are referred to as **epigenetic.** Knowledge of factors that control the rates of enzyme-catalyzed reactions thus is essential to an understanding of the molecular basis of disease and its transmission. This

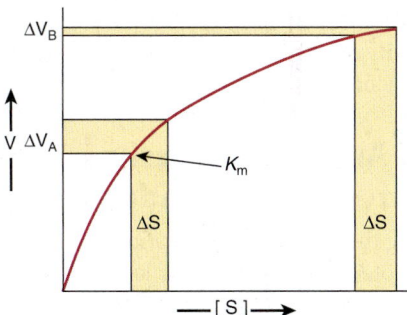

FIGURE 9–1 Differential response of the rate of an enzyme-catalyzed reaction, Δ*V*, to the same incremental change in substrate concentration at a substrate concentration close to K_m (Δ*V*$_A$) or far above K_m (Δ*V*$_B$).

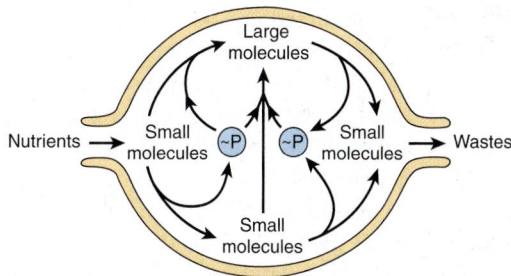

FIGURE 9–2 An idealized cell in steady state. Note that metabolite flow is unidirectional.

chapter outlines the patterns by which metabolic processes are controlled, and provides illustrative examples. Subsequent chapters provide additional examples.

REGULATION OF METABOLITE FLOW CAN BE ACTIVE OR PASSIVE

Enzymes that operate at their maximal rate cannot respond to increases in substrate concentration, and can respond only to precipitous decreases in substrate concentration. The K_m values for most enzymes, therefore, tend to be close to the average intracellular concentration of their substrates, so that changes in substrate concentration generate corresponding changes in the metabolite flux (**Figure 9–1**). Responses to changes in substrate level represent an important but *passive* means for coordinating metabolite flow and maintaining homeostasis in quiescent cells. However, they offer a limited scope for responding to changes in environmental variables. The mechanisms that regulate enzyme efficiency in an *active* manner in response to internal and external signals are discussed below.

Metabolite Flow Tends to Be Unidirectional

Despite the existence of short-term oscillations in metabolite concentrations and enzyme levels, living cells exist in a dynamic steady state in which the mean concentrations of metabolic intermediates remain relatively constant over time. While all chemical reactions are to some extent reversible, in living cells the reaction products serve as substrates for—and are removed by—other enzyme-catalyzed reactions (**Figure 9–2**). Many nominally reversible reactions thus occur unidirectionally. This succession of coupled metabolic reactions is accompanied by an overall change in free energy that favors unidirectional metabolite flow (Chapter 11). The unidirectional flow of metabolites through a pathway with a large overall negative change in free energy is analogous to the flow of water through a pipe in which one end is lower than the

other. Bends or kinks in the pipe simulate individual enzyme-catalyzed steps with a small negative or positive change in free energy. Flow of water through the pipe nevertheless remains unidirectional due to the overall change in height, which corresponds to the overall change in free energy in a pathway (**Figure 9–3**).

COMPARTMENTATION ENSURES METABOLIC EFFICIENCY & SIMPLIFIES REGULATION

In eukaryotes, anabolic and catabolic pathways that interconvert common products may take place in specific subcellular compartments. For example, many of the enzymes that degrade proteins and polysaccharides reside inside organelles called lysosomes. Similarly, fatty acid biosynthesis occurs in the cytosol, whereas fatty acid oxidation takes place within mitochondria (Chapters 22 and 23). Segregation of certain metabolic pathways within specialized cell types provides a further means for physical compartmentation.

Fortunately, many apparently antagonistic pathways can coexist in the absence of physical barriers, provided that thermodynamics dictates that each proceeds with the formation of one or more *unique intermediates*. For any reaction or series of reactions, the change in free energy that takes place when metabolite flow proceeds in the "forward" direction is equal in magnitude *but opposite in sign* from that required to

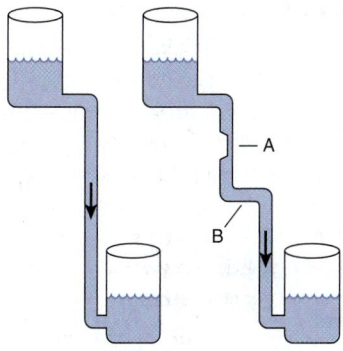

FIGURE 9–3 Hydrostatic analogy for a pathway with a rate-limiting step (A) and a step with a Δ*G* value near 0 (B).

proceed in the reverse direction. Some enzymes within these pathways catalyze reactions, such as isomerizations, that can act as bidirectional catalysts in vivo because the difference in free energy between substrates and products is close to zero. However, they represent the exception rather than the rule. Virtually all metabolic pathways proceed via one or more steps for which ΔG is significant. For example glycolysis, the breakdown of glucose to form two molecules of pyruvate, has a favorable overall ΔG of −96 kJ/mol, a value much too large to simply operate in "reverse" when wishing to convert excess pyruvate to glucose. Consequently, gluconeogenesis proceeds via a pathway in which the three most energetically disfavored steps from glycolysis are replaced by new reactions catalyzed by distinct enzymes (Chapter 20).

The ability of enzymes to discriminate between the structurally similar coenzymes NAD⁺ and NADP⁺ also results in a form of compartmentation. The reduced forms of both coenzymes are not readily distinguishable. However, the reactions that generate and later consume electrons that are destined for ATP generation are segregated in NADH, away from those used in the reductive steps of many biosynthetic pathways, which are carried by NADPH.

Controlling an Enzyme That Catalyzes a Rate-Limiting Reaction Regulates an Entire Metabolic Pathway

While the flux of metabolites through metabolic pathways involves catalysis by numerous enzymes, active control of homeostasis is achieved by the regulation of only a select subset of these enzymes. The ideal enzyme for regulatory intervention is one whose quantity or catalytic efficiency dictates that the reaction it catalyzes is slow relative to all others in the pathway. Decreasing the catalytic efficiency or the quantity of the catalyst responsible for the "bottleneck" or **rate-limiting reaction** immediately reduces metabolite flux through the entire pathway. Conversely, an increase in either its quantity or catalytic efficiency enhances flux through the pathway as a whole. For example, acetyl-CoA carboxylase catalyzes the synthesis of malonyl-CoA, the first committed reaction of fatty acid biosynthesis (Chapter 23). When synthesis of malonyl-CoA is inhibited, subsequent reactions of fatty acid synthesis cease for lack of substrates. As natural "governors" of metabolic flux, the enzymes that catalyze rate-limiting steps also constitute efficient targets for regulatory intervention by drugs. For example, "statin" drugs curtail synthesis of cholesterol by inhibiting HMG-CoA reductase, which catalyzes the rate-limiting reaction of cholesterogenesis.

REGULATION OF ENZYME QUANTITY

The catalytic capacity of the rate-limiting reaction in a metabolic pathway is the product of the concentration of enzyme molecules and their intrinsic catalytic efficiency. It therefore follows that catalytic capacity can be influenced both by changing the quantity of enzyme present and by altering its intrinsic catalytic efficiency.

Proteins Are Continuously Synthesized and Degraded

By measuring the rates of incorporation of ¹⁵N-labeled amino acids into protein and the rates of loss of ¹⁵N from protein, Schoenheimer deduced that body proteins are in a state of "dynamic equilibrium" in which they are continuously synthesized and degraded—a process referred to as **protein turnover.** This holds even for those proteins that are present at an essentially constant, or **constitutive,** steady-state level over time. On the other hand, the concentrations of many enzymes are influenced by a wide range of physiologic, hormonal, or dietary factors.

The absolute quantity of an enzyme reflects the net balance between its rate of synthesis and its rate of degradation. In human subjects, alterations in the levels of specific enzymes can be effected by a change in the rate constant for the overall processes of synthesis (k_s), degradation (k_{deg}), or both.

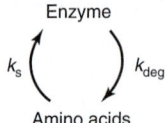

Control of Enzyme Synthesis

The synthesis of certain enzymes depends upon the presence of **inducers,** typically substrates or structurally related compounds that stimulate the transcription of the gene that encodes them (Chapters 36 and 37). *Escherichia coli* grown on glucose will, for example, only catabolize lactose after addition of a β-galactoside, an inducer that triggers synthesis of a β-galactosidase and a galactoside permease (Figure 38–3). Inducible enzymes of humans include tryptophan pyrrolase, threonine dehydratase, tyrosine-α-ketoglutarate aminotransferase, enzymes of the urea cycle, HMG-CoA reductase, and cytochrome P450. Conversely, an excess of a metabolite may curtail synthesis of its cognate enzyme via **repression.** Both induction and repression involve *cis* elements, specific DNA sequences located upstream of regulated genes, and *trans*-acting regulatory proteins. The molecular mechanisms of induction and repression are discussed in Chapter 38. The synthesis of other enzymes can be stimulated by the interaction of hormones and other extracellular signals with specific cell-surface receptors. Detailed information on the control of protein synthesis in response to hormonal stimuli can be found in Chapter 42.

Control of Enzyme Degradation

In animals many proteins are degraded by the ubiquitin-proteasome pathway, the discovery of which earned Aaron Ciechanover, Avram Hershko, and Irwin Rose a Nobel Prize. Degradation takes place in the 26S proteasome, a large macromolecular complex made up of more than 30 polypeptide subunits arranged in the form of a hollow cylinder. The active sites of its proteolytic subunits face the interior of the cylinder, thus preventing indiscriminate degradation of cellular proteins. Proteins are targeted to the interior of the proteasome by "ubiquitination," the covalent attachment of one or more ubiquitin molecules. Ubiquitin is a small, approximately 75 residue, protein that is highly conserved among eukaryotes. Ubiquitination is catalyzed by a large family of enzymes called E3 ligases, which attach ubiquitin to the side-chain amino group of lysyl residues.

The ubiquitin-proteasome pathway is responsible both for the regulated degradation of selected cellular proteins (for example, cyclins—Chapter 35) and for the removal of defective or aberrant protein species. The key to the versatility and selectivity of the ubiquitin-proteasome system resides in both the variety of intracellular E3 ligases and their ability to discriminate between the different physical or conformational states of target proteins. Thus, the ubiquitin-proteasome pathway can selectively degrade proteins whose physical integrity and functional competency have been compromised by the loss of or damage to a prosthetic group, oxidation of cysteine or histidine residues, or deamidation of asparagine or glutamine residues. Recognition by proteolytic enzymes also can be regulated by covalent modifications such as phosphorylation; binding of substrates or allosteric effectors; or association with membranes, oligonucleotides, or other proteins. A growing body of evidence suggests that dysfunctions of the ubiquitin-proteasome pathway contribute to the accumulation of aberrantly folded protein species characteristic of several neurodegenerative diseases.

MULTIPLE OPTIONS ARE AVAILABLE FOR REGULATING CATALYTIC ACTIVITY

In humans the induction of protein synthesis is a complex multistep process that typically requires hours to produce significant changes in overall enzyme level. By contrast, changes in intrinsic catalytic efficiency effected by binding of dissociable ligands (**allosteric regulation**) or by **covalent modification** achieve regulation of enzymic activity within seconds. Consequently, changes in protein level generally dominate when meeting long-term adaptive requirements, whereas changes in catalytic efficiency are best suited for rapid and transient alterations in metabolite flux.

ALLOSTERIC EFFECTORS REGULATE CERTAIN ENZYMES

Feedback inhibition refers to the process by which the end product of a multistep biosynthetic pathway binds to and inhibits an enzyme catalyzing one of the early steps in that

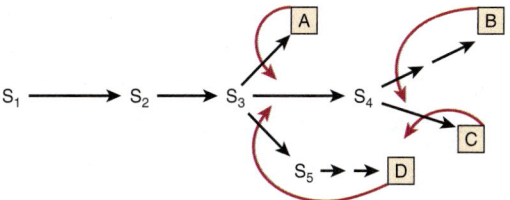

FIGURE 9–4 Sites of feedback inhibition in a branched biosynthetic pathway. S_1–S_5 are intermediates in the biosynthesis of end products A–D. Straight arrows represent enzymes catalyzing the indicated conversions. Curved red arrows represent feedback loops and indicate sites of feedback inhibition by specific end products.

pathway. In the following example, for the biosynthesis of D from A catalyzed by enzymes Enz_1 through Enz_3:

$$\begin{array}{ccccc} & Enz_1 & Enz_2 & Enz_3 & \\ A & \rightarrow & B & \rightarrow & C & \rightarrow & D \end{array}$$

high concentrations of D inhibit the conversion of A to B. In this example, the feedback inhibitor D acts as a **negative allosteric effector** of Enz_1. Inhibition results, not from the "backing up" of intermediates, but from the ability of D to bind to and inhibit Enz_1. Generally, D binds at an **allosteric site,** one spatially distinct from the catalytic site of the target enzyme. Feedback inhibitors thus typically bear little or no structural similarity to the substrates of the enzymes they inhibit. For example, NAD^+ and 3-phosphogylcerate, the substrates for 3-phosphgylcerate dehydrogenase, which catalyzes the first committed step in serine biosynthesis, bear no resemblance to the feedback inhibitor serine. In branched biosynthetic pathways, such as those responsible for nucleotide biosynthesis (Chapter 33), the initial reactions supply intermediates required for the synthesis of multiple end products. **Figure 9–4** shows a hypothetical branched biosynthetic pathway in which curved arrows lead from feedback inhibitors to the enzymes whose activity they inhibit. The sequences $S_3 \rightarrow A$, $S_4 \rightarrow B$, $S_4 \rightarrow C$, and $S_3 \rightarrow \rightarrow D$ each represent linear reaction sequences that are feedback-inhibited by their end products. Branch point enzymes thus can be targeted to route metabolite flow.

Feedback inhibitors typically inhibit the first committed step in a particular biosynthetic sequence. The kinetics of feedback inhibition may be competitive, noncompetitive, partially competitive, or mixed. Layering multiple feedback loops can provide additional fine control. For example, as shown in **Figure 9–5**, the presence of excess product B decreases the requirement for substrate S_2. However, S_2 is also required for synthesis of A, C, and D. Therefore, for this pathway, excess B curtails synthesis of all four end products, regardless of the need for the other three. To circumvent this potential difficulty, each end product may only partially inhibit catalytic activity. The effect of an excess of two or more end products may be strictly additive or, alternatively, greater than their individual effect (cooperative feedback inhibition).

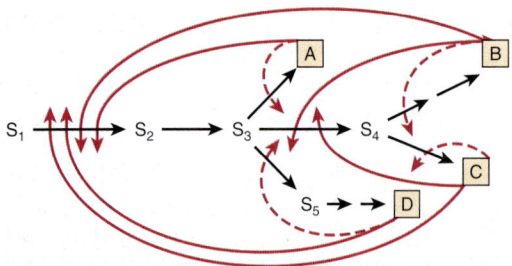

FIGURE 9–5 **Multiple feedback inhibition in a branched biosynthetic pathway.** Superimposed on simple feedback loops (dashed red arrows) are multiple feedback loops (solid red arrows) that regulate enzymes common to biosynthesis of several end products.

Aspartate Transcarbamoylase Is a Model Allosteric Enzyme

Aspartate transcarbamoylase (ATCase), the catalyst for the first reaction unique to pyrimidine biosynthesis (Figure 33–9), is a target of feedback regulation by two nucleotide triphosphates: cytidine triphosphate (CTP) and adenosine triphosphate. CTP, an end product of the pyrimidine biosynthetic pathway, inhibits ATCase, whereas the purine nucleotide ATP activates it. Moreover, high levels of ATP can overcome inhibition by CTP, enabling synthesis of *pyrimidine* nucleotides to proceed when *purine* nucleotide levels are elevated.

Allosteric & Catalytic Sites Are Spatially Distinct

Jacques Monod proposed the existence of allosteric sites that are physically distinct from the catalytic site. He reasoned that the lack of structural similarity between a feedback inhibitor and the substrate(s) for the enzyme whose activity it regulates indicated that these effectors are not **isosteric** with a substrate but **allosteric** ("occupy another space"). **Allosteric enzymes** thus are those for which catalysis at the active site may be modulated by the presence of effectors at an allosteric site. The existence of spatially distinct active and allosteric sites has since been verified in several enzymes using many lines of evidence. For example, x-ray crystallography revealed that the ATCase of *E coli* consists of six catalytic subunits and six regulatory subunits, the latter of which bind the nucleotide triphosphates that modulate activity. In general, binding of an allosteric regulator induces a conformational change in the enzyme that encompasses the active site.

Allosteric Effects May Be on K_m or on V_{max}

To refer to the kinetics of allosteric inhibition as "competitive" or "noncompetitive" with substrate carries misleading mechanistic implications. We refer instead to two classes of allosterically regulated enzymes: K-series and V-series enzymes. For

K-series allosteric enzymes, the substrate saturation kinetics is competitive in the sense that K_m is raised without an effect on V_{max}. For V-series allosteric enzymes, the allosteric inhibitor lowers V_{max} without affecting the K_m. Alterations in K_m or V_{max} often are the product of conformational changes at the catalytic site induced by binding of the allosteric effector at its site. For a K-series allosteric enzyme, this conformational change may weaken the bonds between substrate and substrate-binding residues. For a V-series allosteric enzyme, the primary effect may be to alter the orientation or charge of catalytic residues, lowering V_{max}. Intermediate effects on K_m and V_{max}, however, may be observed consequent to these conformational changes.

FEEDBACK REGULATION IS NOT SYNONYMOUS WITH FEEDBACK INHIBITION

In both mammalian and bacterial cells, some end products "feed back" to control their own synthesis, in many instances by feedback inhibition of an early biosynthetic enzyme. We must, however, distinguish between **feedback regulation,** a phenomenologic term devoid of mechanistic implications, and **feedback inhibition,** a mechanism for regulation of enzyme activity. For example, while dietary cholesterol decreases hepatic synthesis of cholesterol, this feedback **regulation** does not involve feedback **inhibition.** HMG-CoA reductase, the rate-limiting enzyme of cholesterogenesis, is affected, but cholesterol does not inhibit its activity. Rather, regulation in response to dietary cholesterol involves curtailment by cholesterol or a cholesterol metabolite of the expression of the gene that encodes HMG-CoA reductase (enzyme repression) (Chapter 26).

MANY HORMONES ACT THROUGH ALLOSTERIC SECOND MESSENGERS

Nerve impulses and the binding of many hormones to cell surface receptors elicit changes in the rate of enzyme-catalyzed reactions within target cells by inducing the release or synthesis of specialized allosteric effectors called **second messengers.** The primary, or "first," messenger is the hormone molecule or nerve impulse. Second messengers include 3′, 5′-cAMP, synthesized from ATP by the enzyme adenylyl cyclase in response to the hormone epinephrine, and Ca^{2+}, which is stored inside the endoplasmic reticulum of most cells. Membrane depolarization resulting from a nerve impulse opens a membrane channel that releases calcium ions into the cytoplasm, where they bind to and activate enzymes involved in the regulation of muscle contraction and the mobilization of stored glucose from glycogen. Glucose then supplies the increased energy demands of muscle contraction. Other second messengers include 3′,5′-cGMP, nitric oxide, and the polyphosphoinositols produced by the hydrolysis of inositol phospholipids by hormone-regulated phospholipases. Specific examples of the participation of second messengers in the regulation of cellular processes can be found in Chapters 19, 42, and 48.

REGULATORY COVALENT MODIFICATIONS CAN BE REVERSIBLE OR IRREVERSIBLE

In mammalian cells, a wide range of regulatory covalent modifications occur. **Partial proteolysis** and **phosphorylation,** for example, are frequently employed to regulate the catalytic activity of enzymes. On the other hand, histones and other DNA binding proteins in chromatin are subject to extensive modification by **acetylation, methylation, ADP-ribosylation,** as well as phosphorylation. The latter modifications, which modulate the manner in which the proteins within chromatin interact with each other as well as the DNA itself, constitute the basis for the "histone code." The resulting changes in chromatin structure within the region affected can render genes more accessible to the protein responsible for their transcription, thereby enhancing gene expression or, on a larger scale, facilitating replication of the entire genome (Chapter 38). On the other hand, changes in chromatin structure that restrict the accessibility of genes to transcription factors, DNA-dependent RNA polymerases, etc, thereby inhibiting transcription, are said to **silence** gene expression.

The histone code represents a classic example of **epigenetics,** the hereditary transmission of information by a means other than the sequence of nucleotides that comprise the genome. In this instance, the pattern of gene expression within a newly formed "daughter" cell will be determined, in part, by the particular set of histone covalent modifications embodied in the chromatin proteins inherited from the "parental" cell.

Acetylation, ADP-ribosylation, methylation, and phosphorylation are all examples of "reversible" covalent modifications. In this instance, reversible refers to the fact that the modified protein can be restored to its original, modification-free state. It does not, however, refer to the mechanisms by which such restoration takes place. Thermodynamics dictates that if the enzyme-catalyzed reaction by which the modification was introduced is thermodynamically favorable, the free energy change involved in simply trying to run the reaction in reverse will be unfavorable. The phosphorylation of proteins on seryl, threonyl, or tyrosyl residues, catalyzed by protein kinases, is thermodynamically favored as a consequence of utilizing the high-energy gamma phosphoryl group of ATP. Phosphate groups are removed, not by recombining the phosphate with ADP to form ATP, but by a hydrolytic reaction catalyzed by enzymes called protein phosphatases. Similarly, acetyltransferases employ a high-energy donor substrate, NAD^+, while deacetylases catalyze a direct hydrolysis that generates free acetate.

Because the high entropic barrier prevents the reunification of the two portions of a protein produced by hydrolysis of

a peptide bond, proteolysis constitutes a physiologically irreversible modification. Once a proprotein is activated, it will continue to carry out its catalytic or other functions until it is removed by degradation or some other means. Zymogen activation thus represents a simple and economical, albeit one way, mechanism for restraining the latent activity of a protein until the appropriate circumstances are encountered. It is therefore not surprising that partial proteolysis is employed frequently to regulate proteins that work in the gastrointestinal tract or bloodstream rather than in the interior of cells.

PROTEASES MAY BE SECRETED AS CATALYTICALLY INACTIVE PROENZYMES

Certain proteins are synthesized and secreted as inactive precursor proteins known as **proproteins.** Selective, or "partial," proteolysis converts a proprotein by one or more successive proteolytic "clips" to a form that exhibits the characteristic activity of the mature protein, for example, its catalytic activity. The proprotein forms of enzymes are termed **proenzymes** or **zymogens.** Proteins synthesized as proproteins include the hormone insulin (proprotein = proinsulin), the digestive enzymes pepsin, trypsin, and chymotrypsin (proproteins = pepsinogen, trypsinogen, and chymotrypsinogen, respectively), several factors of the blood clotting and blood clot dissolution cascades (see Chapter 51), and the connective tissue protein collagen (proprotein = procollagen).

Proenzymes Facilitate Rapid Mobilization of an Activity in Response to Physiologic Demand

The synthesis and secretion of proteases as catalytically inactive proenzymes protect the tissue of origin (eg, the pancreas) from autodigestion, such as can occur in pancreatitis. Certain physiologic processes such as digestion are intermittent but fairly regular and predictable in frequency. Others such as blood clot formation, clot dissolution, and tissue repair are brought "on line" only in response to pressing physiologic or pathophysiologic need. The processes of blood clot formation and dissolution clearly must be temporally coordinated to achieve homeostasis. Enzymes needed intermittently but rapidly often are secreted in an initially inactive form since new synthesis and secretion of the required proteins might be insufficiently rapid to respond to a pressing pathophysiologic demand such as the loss of blood (see Chapter 51).

Activation of Prochymotrypsin Requires Selective Proteolysis

Selective proteolysis involves one or more highly specific proteolytic clips that may or may not be accompanied by separation of the resulting peptides. Most importantly, selective

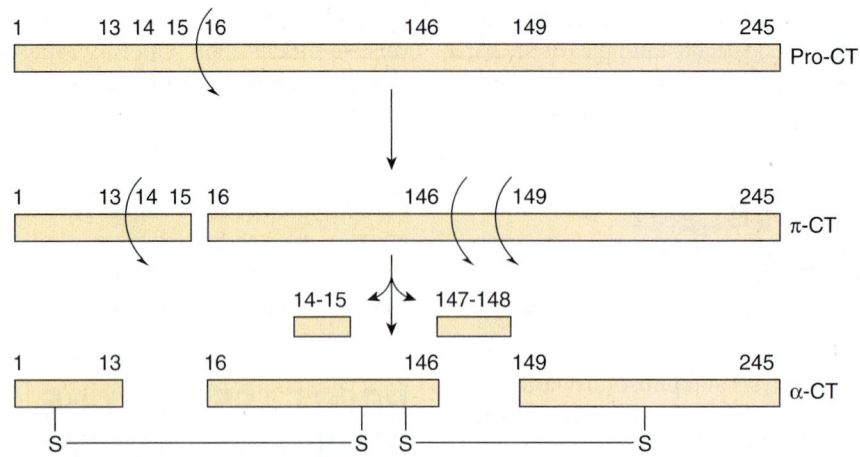

FIGURE 9–6   Two-dimensional representation of the sequence of proteolytic events that ultimately result in formation of the catalytic site of chymotrypsin, which includes the Asp 102-His57-Ser195 catalytic triad (see Figure 7–7). Successive proteolysis forms prochymotrypsin (pro-CT), π-chymotrypsin (π-Ct), and ultimately α-chymotrypsin (α-CT), an active protease whose three peptides (A, B, C) remain associated by covalent inter-chain disulfide bonds.

proteolysis often results in conformational changes that "create" the catalytic site of an enzyme. Note that while the catalytically essential residues His 57 and Asp 102 reside on the B peptide of α-chymotrypsin, Ser 195 resides on the C peptide (**Figure 9–6**). The conformational changes that accompany selective proteolysis of prochymotrypsin (chymotrypsinogen) align the three residues of the charge-relay network (see Figure 7–7), forming the catalytic site. Note also that contact and catalytic residues can be located on different peptide chains but still be within bond-forming distance of bound substrate.

REVERSIBLE COVALENT MODIFICATION REGULATES KEY MAMMALIAN PROTEINS

Mammalian proteins are the targets of a wide range of covalent modification processes. Modifications such as prenylation, glycosylation, hydroxylation, and fatty acid acylation introduce unique structural features into newly synthesized proteins that tend to persist for the lifetime of the protein. Among the covalent modifications that regulate protein function (eg, methylation, acetylation), the most common by far is phosphorylation–dephosphorylation. **Protein kinases** phosphorylate proteins by catalyzing transfer of the terminal phosphoryl group of ATP to the hydroxyl groups of seryl, threonyl, or tyrosyl residues, forming O-phosphoseryl, O-phosphothreonyl, or O-phosphotyrosyl residues, respectively (**Figure 9–7**). Some protein kinases target the side chains of histidyl, lysyl, arginyl, and aspartyl residues. The unmodified form of the protein can be regenerated by hydrolytic removal of phosphoryl groups, catalyzed by **protein phosphatases.**

A typical mammalian cell possesses thousands of phosphorylated proteins and several hundred protein kinases and protein phosphatases that catalyze their interconversion. The ease of interconversion of enzymes between their phospho- and dephospho- forms accounts, in part, for the frequency with which phosphorylation–dephosphorylation is utilized as a mechanism for regulatory control. Phosphorylation–dephosphorylation permits the functional properties of the affected enzyme to be altered only for as long as it serves a specific need. Once the need has passed, the enzyme can be converted back to its original form, poised to respond to the next stimulatory event. A second factor underlying the widespread use of protein phosphorylation–dephosphorylation lies in the chemical properties of the phosphoryl group itself. In order to alter an enzyme's functional properties, any modification of its chemical structure must influence the protein's three-dimensional configuration. The high charge density of protein-bound phosphoryl groups—generally –2 at physiologic pH—and their propensity to form strong salt bridges with arginyl and lysyl residues renders them potent agents for modifying protein structure and function. Phosphorylation

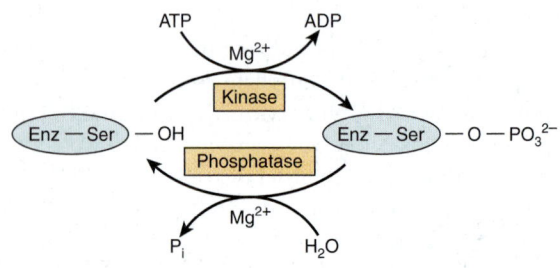

FIGURE 9–7   Covalent modification of a regulated enzyme by phosphorylation–dephosphorylation of a seryl residue.

generally influences an enzyme's intrinsic catalytic efficiency or other properties by inducing conformational changes. Consequently, the amino acids targeted by phosphorylation can be and typically are relatively distant from the catalytic site itself.

Covalent Modifications Regulate Metabolite Flow

In many respects, sites of protein phosphorylation and other covalent modifications can be considered another form of allosteric site. However, in this case, the "allosteric ligand" binds covalently to the protein. Both phosphorylation–dephosphorylation and feedback inhibition provide short-term, readily reversible regulation of metabolite flow in response to specific physiologic signals. Both act without altering gene expression. Both act on early enzymes of a protracted biosynthetic metabolic pathway, and both act at allosteric rather than catalytic sites. Feedback inhibition, however, involves a single protein and lacks hormonal and neural features. By contrast, regulation of mammalian enzymes by phosphorylation–dephosphorylation involves several proteins and ATP, and is under direct neural and hormonal control.

PROTEIN PHOSPHORYLATION IS EXTREMELY VERSATILE

Protein phosphorylation–dephosphorylation is a highly versatile and selective process. Not all proteins are subject to phosphorylation, and of the many hydroxyl groups on a protein's surface, only one or a small subset are targeted. While the most common enzyme function affected is the protein's catalytic efficiency, phosphorylation can also alter its location within the cell, susceptibility to proteolytic degradation, or responsiveness to regulation by allosteric ligands. Phosphorylation can increase an enzyme's catalytic efficiency, converting it to its active form in one protein, while phosphorylation of another protein converts it to an intrinsically inefficient, or inactive, form (**Table 9–1**).

Many proteins can be phosphorylated at multiple sites. Others are subject to regulation both by phosphorylation–dephosphorylation and by the binding of allosteric ligands, or by phosphorylation–dephosphorylation and another covalent modification. Phosphorylation–dephosphorylation at any one site can be catalyzed by multiple protein kinases or protein phosphatases. Many protein kinases and most protein phosphatases act on more than one protein and are themselves interconverted between active and inactive forms by the binding of second messengers or by covalent modification by phosphorylation–dephosphorylation.

The interplay between protein kinases and protein phosphatases, between the functional consequences of phosphorylation at different sites, between phosphorylation sites and allosteric sites, or between phosphorylation sites and other sites of covalent modification provides the basis for regulatory networks that integrate multiple environmental input signals

TABLE 9–1 Examples of Mammalian Enzymes Whose Catalytic Activity Is Altered by Covalent Phosphorylation-Dephosphorylation

Enzyme	Activity State	
	Low	High
Acetyl-CoA carboxylase	EP	E
Glycogen synthase	EP	E
Pyruvate dehydrogenase	EP	E
HMG-CoA reductase	EP	E
Glycogen phosphorylase	E	EP
Citrate lyase	E	EP
Phosphorylase b kinase	E	EP
HMG-CoA reductase kinase	E	EP

Abbreviations: E, dephosphoenzyme; EP, phosphoenzyme.

to evoke an appropriate coordinated cellular response. In these sophisticated regulatory networks, individual enzymes respond to different environmental signals. For example, if an enzyme can be phosphorylated at a single site by more than one protein kinase, it can be converted from a catalytically efficient to an inefficient (inactive) form, or vice versa, in response to any one of several signals. If the protein kinase is activated in response to a signal different from the signal that activates the protein phosphatase, the phosphoprotein becomes a decision node. The functional output, generally catalytic activity, reflects the phosphorylation state. This state or degree of phosphorylation is determined by the relative activities of the protein kinase and protein phosphatase, a reflection of the presence and relative strength of the environmental signals that act through each.

The ability of many protein kinases and protein phosphatases to target more than one protein provides a means for an environmental signal to coordinately regulate multiple metabolic processes. For example, the enzymes 3-hydroxy-3-methylglutaryl-CoA reductase and acetyl-CoA carboxylase—the rate-controlling enzymes for cholesterol and fatty acid biosynthesis, respectively—are phosphorylated and inactivated by the AMP-activated protein kinase. When this protein kinase is activated either through phosphorylation by yet another protein kinase or in response to the binding of its allosteric activator 5′-AMP, the two major pathways responsible for the synthesis of lipids from acetyl-CoA are both inhibited.

INDIVIDUAL REGULATORY EVENTS COMBINE TO FORM SOPHISTICATED CONTROL NETWORKS

Cells carry out a complex array of metabolic processes that must be regulated in response to a broad spectrum of environmental factors. Hence, interconvertible enzymes and the

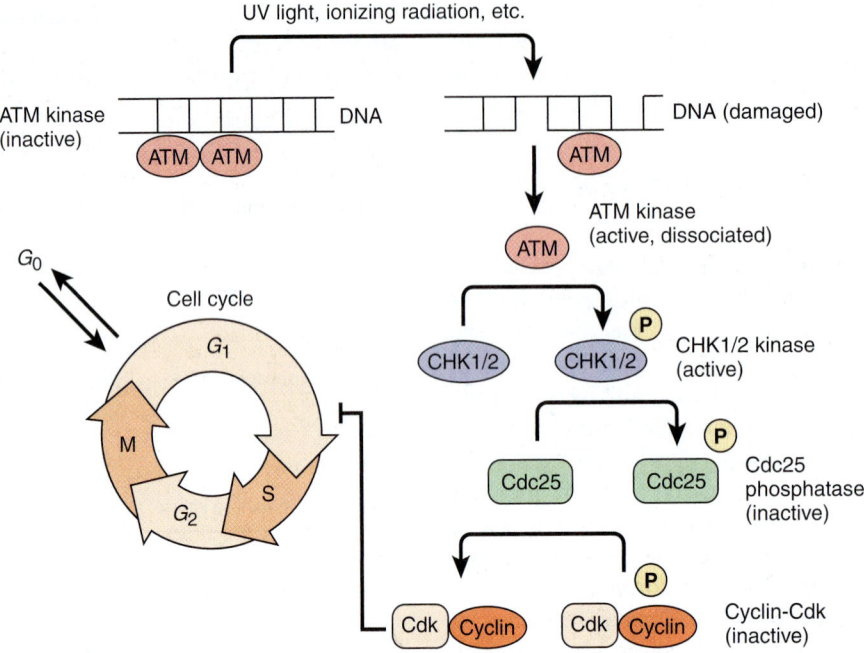

FIGURE 9–8 **A simplified representation of the G_1 to S checkpoint of the eukaryotic cell cycle.** The circle shows the various stages in the eukaryotic cell cycle. The genome is replicated during S phase, while the two copies of the genome are segregated and cell division occurs during M phase. Each of these phases is separated by a G, or growth, phase characterized by an increase in cell size and the accumulation of the precursors required for the assembly of the large macromolecular complexes formed during S and M phases.

enzymes responsible for their interconvesion do not act as isolated "on" and "off" switches. In order to meet the demands of maintaining homeostasis, these building blocks are linked to form integrated regulatory networks.

One well-studied example of such a network is the eukaryotic cell cycle that controls cell division. Upon emergence from the G_0 or quiescent state, the extremely complex process of cell division proceeds through a series of specific phases designated G_1, S, G_2, and M (**Figure 9–8**). Elaborate monitoring systems, called checkpoints, assess key indicators of progress to ensure that no phase of the cycle is initiated until the prior phase is complete. Figure 9–8 outlines, in simplified form, part of the checkpoint that controls the initiation of DNA replication, called the S phase. A protein kinase called ATM is associated with the genome. If the DNA contains a double-stranded break, the resulting change in the conformation of the chromatin activates ATM. Upon activation, one subunit of the activated ATM dimer dissociates and initiates a series, or cascade, of protein phosphorylation–dephosphorylation events mediated by the CHK1 and CHK2 protein kinases, the Cdc25 protein phosphatase, and finally a complex between a cyclin and a cyclin-dependent protein kinase, or Cdk. Activation of the Cdk-cyclin complex blocks the G_1 to S transition, thus preventing the replication of damaged DNA. Failure at this checkpoint can lead to mutations in DNA that may lead to cancer or other diseases. Each step in the cascade provides a conduit for monitoring additional indicators of cell status prior to entering S phase.

SUMMARY

- Homeostasis involves maintaining a relatively constant intracellular and intra-organ environment despite wide fluctuations in the external environment. This is achieved via appropriate changes in the rates of biochemical reactions in response to physiologic need.

- The substrates for most enzymes are usually present at a concentration close to their K_m. This facilitates passive control of the rates of product formation in response to changes in levels of metabolic intermediates.

- Active control of metabolite flux involves changes in the concentration, catalytic activity, or both of an enzyme that catalyzes a committed, rate-limiting reaction.

- Selective proteolysis of catalytically inactive proenzymes initiates conformational changes that form the active site. Secretion as an inactive proenzyme facilitates rapid mobilization of activity in response to injury or physiologic need and may protect the tissue of origin (eg, autodigestion by proteases).

- Binding of metabolites and second messengers to sites distinct from the catalytic site of enzymes triggers conformational changes that alter V_{max} or K_m.

- Phosphorylation by protein kinases of specific seryl, threonyl, or tyrosyl residues—and subsequent dephosphorylation by protein phosphatases—regulates the activity of many human enzymes. The protein kinases and phosphatases that participate in regulatory cascades that respond to hormonal or second messenger signals constitute regulatory networks that can process and integrate complex environmental information to produce an appropriate and comprehensive cellular response.

REFERENCES

Ciechanover A, Schwartz AL: The ubiquitin system: pathogenesis of human diseases and drug targeting. Biochim Biophys Acta 2004;1695:3.

Elgin SC, Reuter G. In Allis CD, Jenuwein T, Reinberg D, et al (editors): Epigenetics, Cold Spring Harbor Laboratory Press, 2007.

Johnson LN, Lewis RJ: Structural basis for control by phosphorylation. Chem Rev 2001;101:2209.

Muoio DM, Newgard CB: Obseity-related derangements in metabolic regulation. Anu Rev Biochem 2006;75:403.

Scriver CR, Sly WS, Childs B, et al (editors): *The Metabolic and Molecular Bases of Inherited Disease,* 8th ed. McGraw-Hill, 2000.

Stieglitz K, Stec B, Baker DP, et al: Monitoring the transition from the T to the R state in *E. coli* aspartate transcarbamoylase by x-ray crystallography: crystal structures of the E50A mutant enzyme in four distinct allosteric states. J Mol Biol 2004; 341:853.

Tu BP, Kudlicki A, Rowicka M, et al: Logic of the yeast metabolic cycle: temporal compartmentalization of cellular processes. Science 2005;310:1152.

Walsh CT: *Posttranslational Modification of Proteins. Expanding Nature's Inventory,* Roberts and Company Publishers, 2006.

Bioinformatics & Computational Biology

Peter J. Kennelly, PhD & Victor W. Rodwell, PhD

OBJECTIVES

After studying this chapter, you should be able to:

- Describe the major features of genomics, proteomics, and bioinformatics.
- Summarize the principal features and medical relevance of the Encode project.
- Describe the functions served by HapMap, *Entrez* Gene, BLAST, and the dbGAP databases.
- Describe the major features of computer-aided drug design and discovery.
- Describe possible future applications of computational models of individual pathways and pathway networks.
- Outline the possible medical utility of "virtual cells."

BIOMEDICAL IMPORTANCE

The first scientific models of pathogenesis, such as Louis Pasteur's seminal germ theory of disease, were binary in nature: each disease possessed a single, definable causal agent. Malaria was caused by the amoeba *Plasmodium falciparum,* tuberculosis by the bacterium *Mycobacterium tuberculosis,* sickle cell disease by a mutation in a gene encoding one of the subunits of hemoglobin, poliomyelitis by poliovirus, and scurvy by a deficiency in ascorbic acid. The strategy for treating or preventing disease thus could be reduced to a straightforward process of tracing the causal agent, and then devising some means of eliminating it, neutralizing its effects, or blocking its route of transmission. This approach has been successfully employed to understand and treat a wide range of infectious and genetic diseases. However, it has become clear that the determinants of many pathologies—including cancer, coronary heart disease, type II diabetes, and Alzheimer's disease—are **multifactorial** in nature. Rather than having a specific causal agent or agents whose presence is both necessary and sufficient, the appearance and progression of the aforementioned diseases reflect the complex interplay between each individual's genetic makeup, other inherited or **epigenetic** factors, and environmental factors such as diet, lifestyle, toxins, viruses, or bacteria.

The challenge posed by **multifactorial diseases** demands a quantum increase in the breadth and depth of our knowledge of living organisms capable of matching their sophistication and complexity. We must identify the many as yet unknown proteins encoded within the genomes of humans and the organisms with which they interact, their cellular functions and interactions. We must be able to trace the factors, both external and internal, that compromise human health and wellbeing by analyzing the impact of dietary, genetic, and environmental factors across entire communities or populations. The sheer mass of information that must be processed lies well beyond the ability of the human mind to review and analyze unaided. To understand, as completely and comprehensively as possible, the molecular mechanisms that underlie the behavior of living organisms, the manner in which perturbations can lead to disease or dysfunction, and how such perturbing factors spread throughout a population, biomedical scientists have turned to sophisticated computational tools to collect and evaluate biologic information on a mass scale.

GENOMICS: AN INFORMATION AVALANCHE

Physicians and scientists have long understood that the genome, the complete complement of genetic information of a living organism, represented a rich source of information concerning topics ranging from basic metabolism to evolution to aging. However, the massive size of the human genome, 3×10^9 nucleotide base pairs, required a paradigm shift in the manner in which scientists approached the determination of DNA sequences. Recent advances in bioinformatics and computational biology have, in turn, been fueled by the need to develop new approaches to "mine" the mass of sequence data generated by the application of increasingly efficient and economical technology to the genomes of hundreds of new organisms and, most recently, of several individual human beings.

The Human Genome Project

The successful completion of the Human Genome Project (HGP) represents the culmination of more than six decades of achievements in molecular biology, genetics, and biochemistry. The chronology below lists several of the milestone events that led to the determination of the entire sequence of the human genome.

1944—DNA is shown to be the hereditary material
1953—Concept of the double helix is posited
1966—The genetic code is solved
1972—Recombinant DNA technology is developed
1977—Practical DNA sequencing technology emerges
1983—The gene for Huntington's disease is mapped
1985—The Polymerase Chain Reaction (PCR) is invented
1986—DNA sequencing becomes automated
1986—The gene for Duchenne muscular dystrophy is identified
1989—The gene for cystic fibrosis is identified
1990—The Human Genome Project is launched in the United States
1994—Human genetic mapping is completed
1996—The first human gene map is established
1999—The Single Nucleotide Polymorphism Initiative is started
1999—The first sequence of a human chromosome, number 22, is completed
2000—"First draft" of the human genome is completed
2003—Sequencing of the first human genome is completed
2010—Scientists embark on the sequencing of 1000 individual genomes to determine degree of genetic diversity in humans

By 2011 >180 eukaryotic, prokaryotic, and archaeal genomes had been sequenced. Examples include the genomes of chicken, cat, dog, elephant, rat, rabbit, orangutan, woolly mammoth, and duck-billed platypus, and the genomes of several individuals including Craig Venter and James Watson. The year 2010 saw completion of the Neanderthal genome, whose initial analysis suggests that up to 2% of the DNA in the genome of present-day humans outside of Africa originated in Neanderthals or in Neanderthal ancestors.

As of this writing, the genome sequences of >5,000 biological entities, ranging from viruses and bacteria to plants and animals have been reported. Ready access to genome sequences from organisms spanning all three phylogenetic domains and to the powerful algorithms requisite for manipulating and transforming the data derived from these sequences has transformed basic research in biology, microbiology, pharmacology, and biochemistry.

Genomes and Medicine

By comparing the genomes of pathogenic and nonpathogenic strains of a particular microorganism, genes likely to encode determinants of virulence can be highlighted by virtue of their presence in only the virulent strain. Similarly, comparison of the genomes of a pathogen with its host can identify genes unique to the former. Drugs targeting the protein products of the pathogen-specific genes should, in theory, produce little or no side effects for the infected host. The coming decade will witness the expansion of the "Genomics Revolution" into the day-to-day practice of medicine and agriculture as physicians and scientists exploit new knowledge of the human genome and of the genomes of the organisms that colonize, feed, and infect *Homo sapiens*. Whereas the first human genome project required several years, hundreds of people, and many millions of dollars to complete, quantum leaps in efficiency and economy have led one company to project that up to one million persons will have their individual genome sequences determined by the year 2014. The ability to diagnose and treat patients guided by knowledge of their genetic makeup, an approach popularly referred to as "designer medicine," will render medicine safer and more effective.

Exome Sequencing

A harbinger of this new era has been provided by the application of "exome sequencing" to the diagnosis of rare or cryptic genetic diseases. The exome consists of those segments of DNA, called exons, that code for the amino acid sequences of proteins (Chapter 36). Since exons comprise only about 1% of the human genome, the exome represents a much smaller and more tractable target than the complete genome. Comparison of exome sequences has identified the genes responsible for a growing list of diseases that includes retinitis pigmentosa, Freeman–Sheldon syndrome, and Kabuki syndrome.

The Potential Challenges of Designer Medicine

While genome-based "designer medicine" promises to be very effective, it also confronts humanity with profound challenges in the areas of ethics, law, and public policy. Who owns and controls access to this information? Can a life or health insurance company, for instance, deny coverage to an individual based upon the risk factors apparent from their genome sequence? Does a prospective employer have the right to know a potential employee's genetic makeup? Do prospective spouses have the right to know their fiancées' genetic risk factors? Ironically, the resolution of these issues may prove a more lengthy and laborious process than did the determination of the first human genome sequence.

BIOINFORMATICS

Bioinformatics exploits the formidable information storage and processing capabilities of the computer to develop tools for the collection, collation, retrieval, and analysis of biologic data on a mass scale. That many bioinformatic resources (see below) can be accessed via the Internet provides them with global reach and impact. The central objective of a typical bioinformatics project is to assemble all of the available information relevant to a particular topic in a single location, often referred to as a **library** or **database,** in a uniform format that renders the data amenable to manipulation and analysis by computer algorithms.

Bioinformatic Databases

The size and capabilities of bioinformatic databases can vary widely depending upon the scope and nature of their objectives. The PubMed database (http://www.ncbi.nlm.nih.gov/sites/entrez?db=pubmed) compiles citations for all articles published in thousands of journals devoted to biomedical and biological research. Currently, PubMed contains >20 million citations. By contrast, the RNA Helicase Database (http://www.rnahelicase.org/) confines itself to the sequence, structure, and biochemical and cellular function of a single family of enzymes, the RNA helicases.

Challenges of Database Construction

The construction of a comprehensive and user-friendly database presents many challenges. First, biomedical information comes in a wide variety of forms. For example, the coding information in a genome, although voluminous, is composed of simple linear sequences of four nucleotide bases. While the number of amino acid residues that define a protein's primary structure is minute relative to the number of base pairs in a genome, a description of a protein's x-ray structure requires that the location of each atom be specified in three-dimensional space.

Second, the designer must correctly anticipate the manner in which users may wish to search or analyze the information within a database, and devise algorithms for coping with these variables. Even the seemingly simple task of searching a gene database commonly employs, alone or in various combinations, criteria as diverse as the name of the gene, the name of the protein that it encodes, the biologic function of the gene product, a nucleotide sequence within the gene, a sequence of amino acids within the protein it encodes, the organism in which it is present, or the name of the investigators who determined the sequence of that gene.

EPIDEMIOLOGY ESTABLISHED THE MEDICAL POTENTIAL OF INFORMATION PROCESSING

The power of basic biomedical research resides in the laboratory scientist's ability to manipulate homogenous, well-defined research targets under carefully controlled circumstances. The ability to independently vary the qualitative and quantitative characteristics of both target and input variables permits

cause–effect relationships to be determined in a direct and reliable manner. These advantages are obtained, however, by employing "model" organisms such as mice or cultured human cell lines as stand-ins for the human patients that represent the ultimate targets for, and beneficiaries of, this research. Laboratory animals do not always react as do *Homo sapiens,* nor can a dish of cultured fibroblast, kidney, or other cells replicate the incredible complexity of a human being.

Although unable to conduct rigorously controlled experiments on human subjects, careful observation of real world behavior has long proven to be a source of important biomedical insights. Hippocrates, for example, noted that while certain **epidemic** diseases appeared in a sporadic fashion, **endemic** diseases such as malaria exhibited clear association with particular locations, age group, etc. **Epidemiology** refers to the branch of the biomedical sciences that employs bioinformatic approaches to extend our ability and increase the accuracy with which we can identify factors that contribute to or detract from human health through the study of real world populations.

Early Epidemiology of Cholera

One of the first recorded epidemiological studies, conducted by Dr. John Snow, employed simple geospatial analysis to track the source of a cholera outbreak. Epidemics of cholera, typhus, and other infectious diseases were relatively common in the crowded, unsanitary conditions of nineteenth century London. By mapping the locations of the victims' residences, Snow was able to trace the source of the contagion to the contamination of one of the public pumps that supplied citizens with their drinking water (**Figure 10–1**). Unfortunately, the limited capacity of hand calculations or graphing rendered the success of analyses such as Snow's critically dependent upon the choice of the working hypothesis used to select the variables to be measured and processed. Thus, while nineteenth century Londoners also widely recognized that haberdashers were particularly prone to display erratic and irrational behavior (eg, "as Mad as a Hatter"), nearly a century would pass before the cause was traced to the mercury compounds used to prepare the felt from which the hats were constructed.

Impact of Bioinformatics on Epidemiological Analysis

As the process of data analysis has become automated, the sophistication and success rate of epidemiological analyses has risen accordingly. Today, complex computer algorithms enable researchers to assess the influence of a broad range of health-related parameters when tracking the identity and source or reconstructing the transmission of a disease or condition: height; weight; age; gender; body mass index; diet; ethnicity; medical history; profession; drug, alcohol, or tobacco use; exercise; blood pressure; habitat; marital status; blood type; serum cholesterol level; etc. Equally important,

FIGURE 10–1 **This version of the map drawn by Dr. John Snow compares the location of the residences of victims of an 1854 London cholera epidemic (Dots) with the locations of the pumps that supplied their drinking water (X's).** Contaminated water from the pump on Broad Street, lying roughly in the center of the cluster of victims, proved to be the source of the epidemic in this neighborhood.

modern bioinformatics may soon enable epidemiologists to dissect the identities and interactions of the multiple factors underlying complex diseases such as cancer or Alzheimer's disease.

The accumulation of individual genome sequences will shortly introduce a powerful new dimension to the host of biological, environmental, and behavioral factors to be compared and contrasted with each person's medical history. One of the first fruits of these studies has been the identification of genes responsible for a few of the >3000 known or suspected Mendelian disorders whose causal genetic abnormalities have yet to be traced. The ability to evaluate contributions of and the interactions among an individual's genetic makeup, behavior, environment, diet, and lifestyle holds the promise of eventually revealing the answers to the age-old question of why some persons exhibit greater vitality, stamina, longevity, and resistance to disease than others—in other words, the root sources of health and wellness.

BIOINFORMATIC AND GENOMIC RESOURCES

The large collection of databases that have been developed for the assembly, annotation, analysis and distribution of biological and biomedical data reflects the breadth and variety of contemporary molecular, biochemical, epidemiological, and clinical research. Below are listed examples of the following prominent bioinformatics resources: UniProt, GenBank, and the Protein Database (PDB) represent three of the oldest and most widely used bioinformatics databases. Each complements the other by focusing on a different aspect of macromolecular structure.

Uniprot

The UniProt Knowledgebase (http://www.pir.uniprot.org/) can trace its origins to the *Atlas of Protein Sequence and Structure,* a printed encyclopedia of protein sequences first published in

1968 by Margaret Dayhoff and the National Biomedical Research Foundation at Georgetown University. The aim of the Atlas was to facilitate studies of protein evolution using the amino acid sequences being generated consequent to the development of the Edman method for protein sequencing (Chapter 4). In partnership with the Martinsreid Institute for Protein Sequences and the International Protein Information Database of Japan, the Atlas made the transition to electronic form as the Protein Information Resource (PIR) Protein Sequence Database in 1984. In 2002, PIR integrated its database of protein sequence and function with the Swiss-Prot Protein Database established by Amos Bairoch in 1986 under the auspices of the Swiss Institute of Bioinformatics and the European Bioinformatics Institute, to form the world's most comprehensive resource on protein structure and function, the UniProt Knowledgebase.

GenBank

The goal of GenBank (http://www.ncbi.nlm.nih.gov/Genbank/), the National Institutes of Health's (NIH) genetic sequence database, is to collect and store all known biological nucleotide sequences and their translations in a searchable form. Established in 1979 by Walter Goad of Los Alamos National Laboratory, GenBank currently is maintained by the National Center for Biotechnology Information at the NIH. GenBank constitutes one of the cornerstones of the International Sequence Database Collaboration, a consortium that includes the DNA Database of Japan and the European Molecular Biology Laboratory.

PDB

The RCSB Protein Data Base (PDB) (http://www.rcsb.org/pdb/home/home.do), a repository of the three-dimensional structures of proteins, polynucleotides, and other biological macromolecules, was established in 1971 by Edgar Meyer and Walter Hamilton of Brookhaven National Laboratories. In 1998, responsibility for the PDB was transferred to the Research Collaboration for Structural Bioinformatics formed by Rutgers University, the University of California at San Diego, and the University of Wisconsin. The PDB contains well over 50,000 three-dimensional structures for proteins, as well as proteins bound with substrates, substrate analogs, inhibitors, or other proteins. The user can rotate these structures freely in three-dimensional space, highlight specific amino acids, and select from a variety of formats such as space filling, ribbon, backbone, etc. (Chapters 5, 6, and 10).

SNPs & Tagged SNPs

While the genome sequence of any two individuals is 99.9% identical, human DNA contains ~10 million sites where individuals differ by a single-nucleotide base. These sites are called **Single-Nucleotide Polymorphisms** or **SNPs.** When sets of SNPs localized to the same chromosome are inherited together in blocks, the pattern of SNPs in each block is termed a **haplotype.** By comparing the haplotype distributions between groups of individuals that differ in some physiological characteristic, such as susceptibility to a disease, biomedical scientists can identify SNPs that are associated with specific phenotypic traits. This process can be facilitated by focusing on **Tag SNPs,** a subset of the SNPs in a given block sufficient to provide a unique marker for a given haplotype. Detailed study of each region should reveal variants in genes that contribute to a specific disease or response.

HapMap

In 2002, scientists from the United States, Canada, China, Japan, Nigeria, and the United Kingdom launched the International **Haplotype Map (HapMap) Project** (http://hapmap.ncbi.nlm.nih.gov/), a comprehensive effort to identify SNPs associated with common human diseases and differential responses to pharmaceuticals. The resulting **HapMap Database** should lead to earlier and more accurate diagnosis, improved prevention, and patient management. Knowledge of an individual's genetic profile will also be used to guide the selection of safer and more effective drugs or vaccines, a process termed **pharmacogenomics.** These genetic markers will also provide labels with which to identify and track specific genes as scientists seek to learn more about the critical processes of genetic inheritance and selection.

ENCODE

Identification of all the *functional elements* of the genome will vastly expand our understanding of the molecular events that underlie human development, health, and disease. To address this goal, in late 2003, the National Human Genome Research Institute (NHGRI) initiated the **ENCODE (Encyclopedia of DNA Elements) Project.** Based at the University of California at Santa Cruz, ENCODE (http://www.genome.gov/10005107) is a collaborative effort that combines laboratory and computational approaches to identify every functional element in the human genome. Consortium investigators with diverse backgrounds and expertise collaborate in the development and evaluation of new high-throughput techniques, technologies, and strategies to address current deficiencies in our ability to identify functional elements.

The pilot phase of ENCODE targeted ~1% (30 Mb) of the human genome for rigorous computational and experimental analysis. A variety of methods were employed to identify, or **annotate,** the function of each portion of the DNA in 500 base pair steps. These pilot studies revealed that the human genome contains a large number and variety of functionally active components interwoven to form complex networks. The successful prosecution of this pilot study has resulted in the funding of a series of Scale-Up Projects aimed at tackling the remaining 99% of the genome.

Entrez Gene

Entrez Gene (http://www.ncbi.nlm.nih.gov/sites/entrez?db=gene), a database maintained by the National Center for

Biotechnology Information (NCBI), provides a variety of information about individual human genes. The information includes the sequence of the genome in and around the gene, exon-intron boundaries, the sequence of the mRNA(s) produced from the gene, and any known phenotypes associated with a given mutation of the gene in question. *Entrez* Gene also lists, where known, the function of the encoded protein and the impact of known single-nucleotide polymorphisms within its coding region.

dbGAP

dbGAP, the **Database of Genotype and Phenotype,** is an NCBI database that complements *Entrez* Gene (http://www.ncbi.nlm.nih.gov/gap). dbGAP compiles the results of research into the links between specific genotypes and phenotypes. To protect the confidentiality of sensitive clinical data, the information contained in dbGAP is organized into open- and controlled-access sections. Access to sensitive data requires that the user apply for authorization to a data access committee.

Additional Databases

Other databases dealing with human genetics and health include **OMIM,** Online Mendelian Inheritance in Man (http://www.ncbi.nlm.nih.gov/sites/entrez?db=omim), **HGMD,** the Human Gene Mutation Database (http://www.hgmd.cf.ac.uk/ac/index.php), the **Cancer Genome Atlas** (http://cancergenome.nih.gov/), and **GeneCards** (http://www.genecards.org/), which tries to collect all relevant information on a given gene from databases worldwide to create a single, comprehensive "card" for each.

COMPUTATIONAL BIOLOGY

The primary objective of **computational biology** is to develop computer models that apply physical, chemical, and biological principles to reproduce the behavior of biologic molecules and processes. Unlike bioinformatics, whose major focus is the collection and evaluation of existing data, computational biology is experimental and exploratory in nature. By performing virtual experiments and analyses "*in silico,*" meaning performed on a computer or through a computer simulation, computational biology offers the promise of greatly accelerating the pace and efficiency of scientific discovery.

Computational biologists are attempting to develop predictive models that will (1) permit the three-dimensional structure of a protein to be determined directly from its primary sequence, (2) determine the function of unknown proteins from their sequence and structure, (3) screen for potential inhibitors of a protein *in silico,* and (4) construct virtual cells that reproduce the behavior and predict the responses of their living counterparts to pathogens, toxins, diet, and drugs. The creation of computer algorithms that accurately imitate the behavior of proteins, enzymes, cells, etc., promises to enhance the speed, efficiency, and the safety of biomedical research. Computational biology will also enable scientists to perform

Language	Word	Alignment
English	PHYSIOLOGICAL	PHYSIOLOGI - CAL
French	PHYSIOLOGIQUE	PHYSIOLOGI - QUE
German	PHYSIOLOGISCH	PHYSIOLOGISCH
Dutch	FYSIOLOGISCH	F - YSIOLOGISCH
Spanish	FYSIOLOGICO	F - YSIOLOGI - CO
Polish	FIZJOLOGICZNY	F - IZJOLOGI - CZNY

FIGURE 10–2 Representation of a multiple sequence alignment. Languages evolve in a fashion that mimics that of genes and proteins. Shown is the English word "physiological" in several languages. The alignment demonstrates their conserved features. Identities with the English word are shown in dark red; linguistic similarities in dark blue. Multiple sequence alignment algorithms identify conserved nucleotide and amino acid letters in DNA, RNA, and polypeptides in an analogous fashion.

experiments *in silico* whose scope, hazard, or nature renders them inaccessible to, or inappropriate for, conventional laboratory or clinical approaches.

IDENTIFICATION OF PROTEINS BY HOMOLOGY

One important method for the identification, also called **annotation,** of novel proteins and gene products compares the sequences of novel proteins with those of proteins whose functions or structures had been determined. Simply put, homology searches and multiple sequence comparisons operate on the principle that proteins that perform similar functions will share conserved domains or other sequence features or **motifs,** and vice versa (**Figure 10–2**). Of the many algorithms developed for this purpose, the most widely used is **BLAST.**

BLAST

BLAST (Basic Local Alignment Search Tool) and other sequence comparison/alignment algorithms trace their origins to the efforts of early molecular biologists to determine whether observed similarities in sequence among proteins that perform parallel metabolic functions were indicative of progressive changes in a common ancestral protein. The major evolutionary question addressed was whether the similarities reflected (1) descent from a common ancestral protein (**divergent evolution**) or (2) the independent selection of a common mechanism for meeting some specific cellular need (**convergent evolution**), as would be anticipated if one particular solution was overwhelmingly superior to the alternatives. Calculation of the minimum number of nucleotide changes required to interconvert putative protein isoforms allows inferences to be drawn concerning whether or not the similarities and differences exhibit a pattern indicative of progressive change from a shared origin.

BLAST has evolved into a family of programs optimized to address specific needs and data sets. Thus, **blastp** compares an *amino acid* query sequence against a *protein* sequence database, **blastn** compares a *nucleotide* query sequence against a *nucleotide* sequence database, **blastx** compares a *nucleotide* query sequence translated in all reading frames against a *protein* sequence database to reveal potential translation products, **tblastn** compares a *protein* query sequence against a *nucleotide* sequence database dynamically translated in all six reading frames, and **tblastx** compares the six-frame translations of a *nucleotide* query sequence against the six-frame translations of a *nucleotide* sequence database. Unlike multiple sequence alignment programs that rely on *global* alignments, the **BLAST** algorithms emphasize regions of *local* alignment to detect relationships among sequences with only isolated regions of similarity. This approach provides speed and increased sensitivity for distant sequence relationships. Input or "query" sequences are broken into "words" (default size 11 for nucleotides, 3 for amino acids). Word hits to databases are then extended in both directions.

IDENTIFICATION OF "UNKNOWN" PROTEINS

A substantial portion of the genes discovered by genome sequencing projects code for "unknown" or hypothetical polypeptides for which homologs of known function are lacking. Bioinformaticists are developing tools to enable scientists to deduce the three-dimensional structure and function of cryptic proteins directly from their amino acid sequences. Currently, the list of unknown proteins uncovered by genomics contains thousands of entries, with new entries being added as more genome sequences are solved. The ability to generate structures and infer function *in silico* promises to significantly accelerate protein identification and provide insight into the mechanism by which proteins fold. This knowledge will aid in understanding the underlying mechanisms of various protein folding diseases, and will assist molecular engineers to design new proteins to perform novel functions.

The Folding Code

Comparison of proteins whose three-dimensional structures have been determined by x-ray crystallography or NMR spectroscopy can reveal patterns that link specific primary sequence features to specific primary, secondary, and tertiary structures—sometimes called the folding code. The first algorithms used the frequency with which individual amino acids occurred in α-helices, β-sheets, turns, and loops to predict the number and location of these elements within the sequence of a polypeptide, known as secondary structure topography. By extending this process, for example, by weighing the impact of hydrophobic interactions in the formation of the protein core, algorithms of remarkable predictive reliability are being developed. However, while current programs perform well in generating the conformations of proteins comprised of a single domain, projecting the likely structure of membrane proteins and those composed of multiple domains remains problematic.

Relating Three-Dimensional Structure to Function

Scientists are also attempting to discern patterns of three-dimensional structure that correlate to specific physiologic functions. The space-filling representation of the enzyme HMG-CoA reductase and its complex with the drug lovastatin (**Figure 10–3**) provides some perspective on the challenges inherent in identifying ligand-binding sites from scratch.

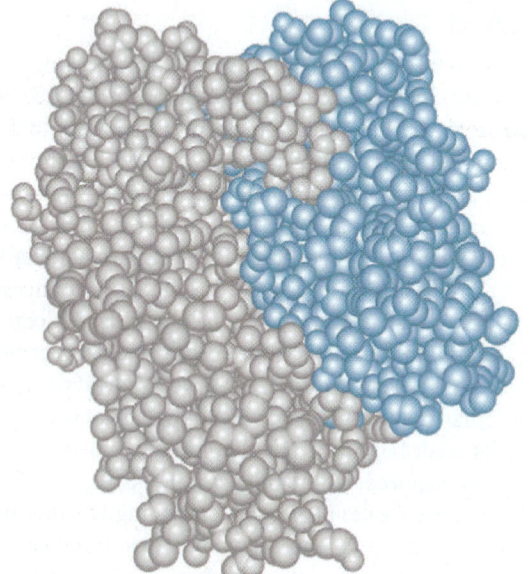 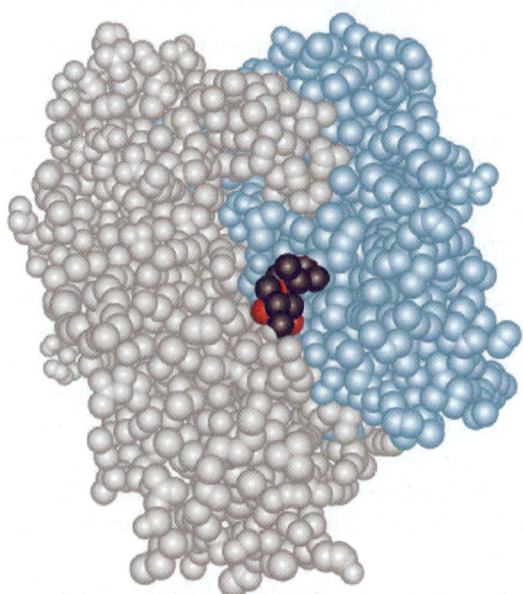

FIGURE 10–3 **Shown are space-filling representations of the homodimeric HMG-CoA reductase from *Pseudomonas mevalonii* with (right) and without (left) the statin drug lovastatin bound.** Each atom is represented by a sphere the size of its van der Waals' radius. The two polypeptide chains are colored gray and blue. The carbon atoms of lovastatin are colored black and the oxygen atoms red. Compare this model with the backbone representations of proteins shown in Chapters 5 and 6. (Adapted from Protein Data Bank ID no. 1t02.)

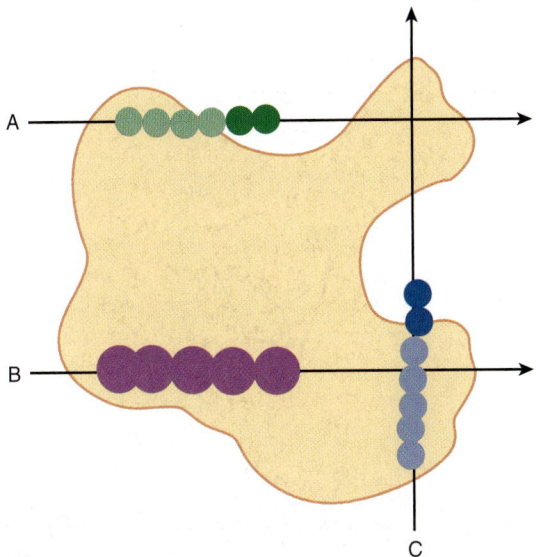

FIGURE 10–4 **A simplified representation of a ligand site prediction program.** Ligand site prediction programs such as POCKET, LIGSITE, or Pocket-Finder convert the three-dimensional structure of a protein into a set of coordinates for its component atoms. A two-dimensional slice of the space filled by these coordinates is presented as an irregularly shaped outline (yellow). A round probe is then passed repeatedly through these coordinates along a series of lines paralleling each of the three coordinate axes (A, B, C). Lightly shaded circles represent positions of the probe where its radius overlaps with one or more atoms in the Cartesian coordinate set. Darkly shaded circles represent positions where no protein atom coordinates fall within the probe's radius. In order to qualify as a pocket or crevice within the protein, and not just open space outside of it, the probe must eventually encounter protein atoms lying on the other side of the opening (C).

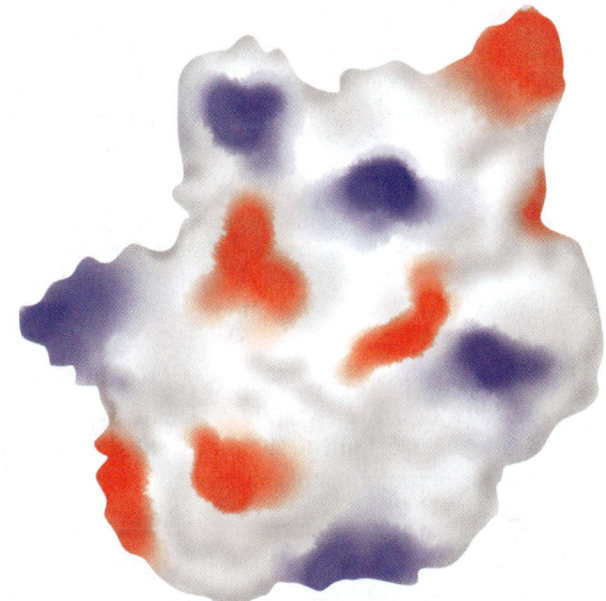

FIGURE 10–5 **Representation of a GRASP diagram indicating the electrostatic topography of a protein.** Shown is a space-filling representation of a hypothetical protein. Areas shaded in red indicate the presence of amino acid side chains or other moieties on the protein surface predicted to bear a negative charge at neutral pH. Blue indicates the presence of predicted positively charged groups. White denotes areas predicted to be electrostatically neutral.

Where a complete three-dimensional structure can be determined or predicted, the protein's surface can be scanned for the types of pockets and crevices indicative of likely binding sites for substrates, allosteric effectors, etc., by any one of a variety of methods such as tracing its surface with balls of a particular dimension (**Figure 10–4**). Surface maps generated with the program Graphical Representation and Analysis of Surface Properties, commonly referred to as **GRASP diagrams,** highlight the locations of neutral, negatively charged, and positively charged functional groups on a protein's surface (**Figure 10–5**) to infer a more detailed picture of the biomolecule that binds to or "docks" at that site. The predicted structure of the ligands that bind to an unknown protein, along with other structural characteristics and sequence motifs can then provide scientists with the clues needed to make an "educated guess" regarding its biological function(s).

COMPUTER-AIDED DRUG DESIGN

Computer-Aided Drug Design (CADD) employs the same type of molecular-docking algorithms used to identify ligands for unknown proteins. However, in this case the set of potential ligands to be considered is not confined to those occurring in nature and is aided by empirical knowledge of the structure or functional characteristics of the target protein.

Molecular Docking Algorithms

For proteins of known three-dimensional structure, molecular-docking approaches employ programs that attempt to fit a series of potential ligand "pegs" into a designated binding site or "hole" on a protein template. To identify optimum ligands, docking programs must account for matching shapes as well as the presence and position of complementary hydrophobic, hydrophilic, and charge attributes. The binding affinities of the inhibitors selected on the basis of early docking studies were disappointing, as the rigid models for proteins and ligands employed were incapable of replicating the conformational changes that occur in both ligand and protein as a consequence of binding, a phenomenon referred to as "induced fit" (Chapter 7).

Imbuing proteins and ligands with conformational flexibility requires massive computing power, however. Hybrid approaches have thus evolved that employ a set, or ensemble, of templates representing slightly different conformations of the protein (**Figure 10–6**) and either ensembles of ligand conformers (**Figure 10–7**) or ligands in which only a few select bonds are permitted to rotate freely. Once the set

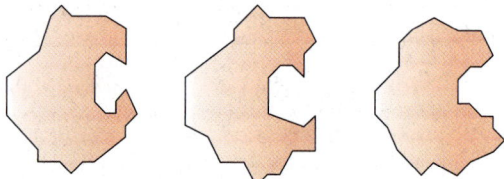

FIGURE 10–6 **Two-dimensional representation of a set of conformers of a protein.** Notice how the shape of the binding site changes.

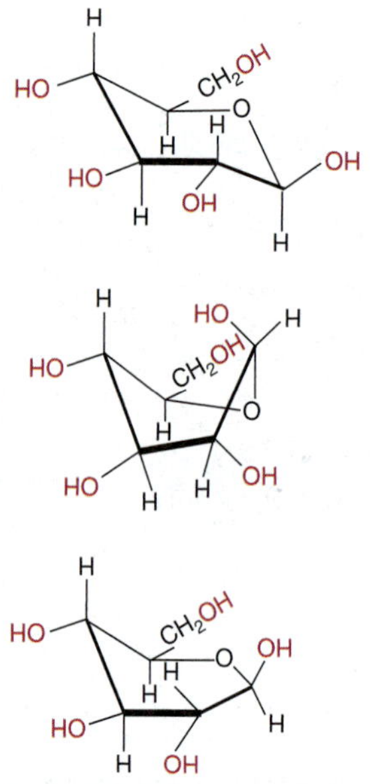

of potential ligands has been narrowed, more sophisticated docking analyses can be undertaken to identify high-affinity ligands able to interact with their protein target across the latter's spectrum of conformational states.

Structure–Activity Relationships

If no structural template is available for the protein of interest, computers can be used to assist the search for high-affinity inhibitors by calculating and projecting **Structure–Activity Relationships (SARs)**. In this process, the measured binding affinities for several known inhibitors are compared and contrasted to determine whether specific chemical features make positive or negative thermodynamic contributions to ligand binding. This information can then be used to search databases of chemical compounds to identify those which possess the most promising combination of positive versus negative features.

SYSTEMS BIOLOGY & VIRTUAL CELLS

The Goal of Systems Biology Is to Construct Molecular Circuit Diagrams

What if a scientist could detect, in a few moments, the effect of inhibiting a particular enzyme, of replacing a particular gene, the response of a muscle cell to insulin, the proliferation of a cancer cell, or the production of beta amyloid by entering the appropriate query into a computer? The goal of **systems biology** is to construct the molecular equivalent of circuit diagrams that faithfully depict the components of a particular functional unit and the interactions between them in logical or mathematical terms. These functional units can range in size and complexity from the enzymes and metabolites within a biosynthetic pathway to the network of proteins that controls the cell division cycle to, ultimately, entire cells, organs, and organisms. These models can then be used to perform "virtual" experiments that can enhance the speed and efficiency of empirical investigations by identifying the most promising lines of investigation and assisting in the evaluation of results. Perhaps more significantly, the ability to conduct virtual

FIGURE 10–7 **Conformers of a simple ligand.** Shown are three of the many different conformations of glucose, commonly referred to as chair (Top), twist boat (Middle), and half chair (Bottom). Note the differences not only in shape and compactness but in the position of the hydroxyl groups, potential participants in hydrogen bonds, as highlighted in red.

experiments extends the reach of the investigator, within the limits of the accuracy of the model, beyond the reach of current empirical technology.

Already, significant progress is being made. By constructing virtual molecular networks, scientists have been able to determine how cyanobacteria assemble a reliable circadian clock using only four proteins. Models of the T cell receptor signaling pathway have revealed how its molecular circuitry has been arranged to produce switch-like responses upon stimulation by agonist peptide-major histocompatability complexes (MHC) on an antigen-presenting cell. Scientists can use the gaps encountered in modeling molecular and cellular systems to guide the identification and annotation of the remaining protein pieces, in the same way that someone who solves a jigsaw puzzle surveys the remaining pieces for matches to the gaps in the puzzle. This reverse engineering approach has been successfully used to define the function of type II glycerate 2-kinases in bacteria and to identify "cryptic" folate synthesis and transport genes in plants.

Virtual Cells

Recently, scientists have been able to successfully create a sustainable metabolic network, composed of nearly two hundred proteins—an important step toward the creation of a **virtual cell.** The "holy grail" of systems biologists is to replicate the behavior of living human cells *in silico*. The potential benefits of such virtual cells are enormous. Not only will they permit optimum sites for therapeutic intervention to be identified in a rapid and unbiased manner, but unintended side effects may be revealed prior to the decision to invest time and resources in the synthesis, analysis, and trials of a potential pharmacophore. The ability to conduct fast, economic toxicological screening of materials ranging from herbicides to cosmetics will benefit human health. Virtual cells can also aid in diagnosis. By manipulating a virtual cell to reproduce the metabolic profile of a patient, underlying genetic abnormalities may be revealed. The interplay of the various environmental, dietary, and genetic factors that contribute to multifactorial diseases such as cancer can be systematically analyzed. Preliminary trials of potential gene therapies can be assessed safely and rapidly *in silico*.

The duplication of a living cell *in silico* represents an extremely formidable undertaking. Not only must the virtual cell possess all of the proteins and metabolites for the type of cell to be modeled (eg, from brain, liver, nerve, muscle, or adipose), but these must be present in the appropriate concentration and subcellular location. The model must also account for the functional dynamics of its components, binding affinities, catalytic efficiency, covalent modifications, etc. To render a virtual cell capable of dividing or differentiating will entail a further quantum leap in complexity and sophistication.

Molecular Interaction Maps Employ Symbolic Logic

The models constructed by systems biologists can take a variety of forms depending upon the uses for which they are intended and the data available to guide their construction. If one wishes to model the flux of metabolites through an anabolic or catabolic pathway, it is not enough to know the identities and the reactants involved in each enzyme-catalyzed reaction. To obtain mathematically precise values, it is necessary to know the concentrations of the metabolites in question, the quantity of each of each enzyme present, and their catalytic parameters.

For most users, it is sufficient that a model describe and predict the qualitative nature of the interactions between components. Does an allosteric ligand activate or inhibit the enzyme? Does dissociation of a protein complex lead to the degradation of one or more of its components? For this purpose, a set of symbols depicting the symbolic logic of these interactions was needed. Early representations frequently used the symbols previously developed for constructing flow charts (computer programming) or electronic circuits (**Figure 10–8**, top). Ultimately, however, systems biologists designed dedicated symbols (Figure 10–8, bottom) to depict these molecular circuit diagrams, more commonly referred to as **Molecular Interaction Maps (MIM),** an example of which is shown in **Figure 10–9**. Unfortunately, as is the case with enzyme nomenclature (Chapter 7) a consistent, universal set of symbols has yet to emerge.

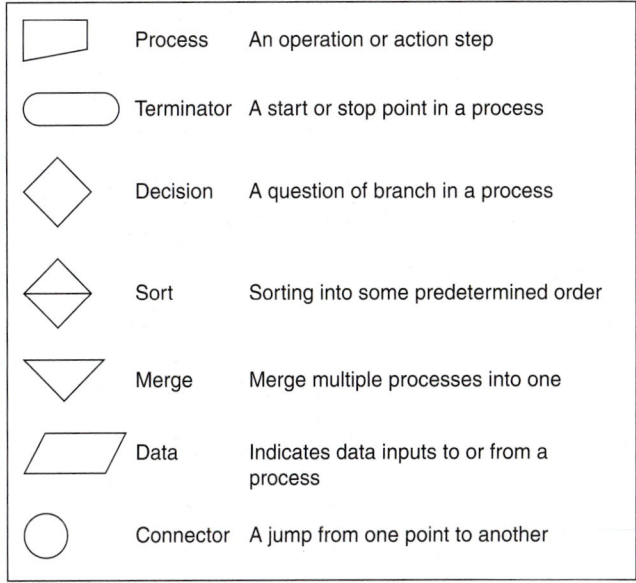

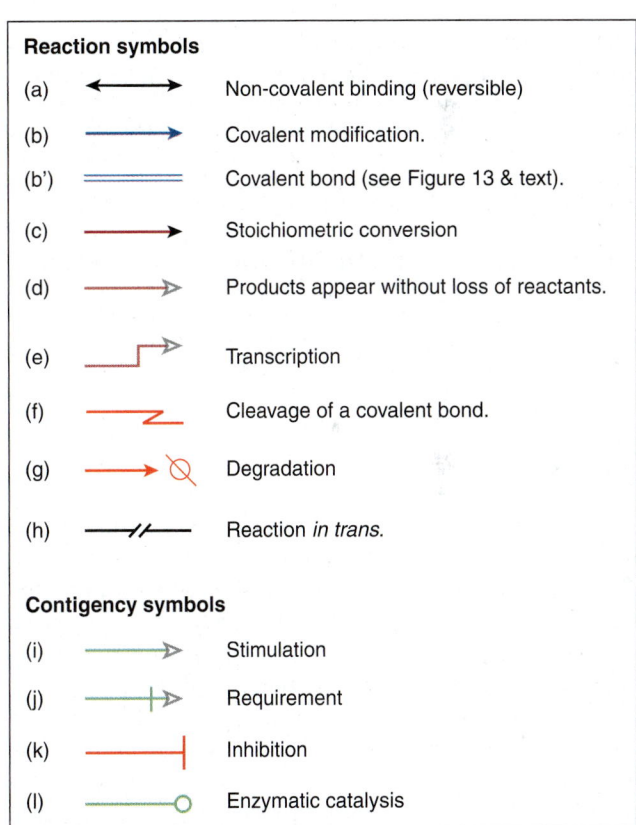

FIGURE 10–8 Symbols used to construct molecular circuit diagrams in systems biology. (Top) Sample flowchart symbols. (Bottom) Graphical symbols for molecular interaction maps (Adapted from Kohn KW et al: Molecular interaction maps of bioregulatory networks: a general rubric for systems biology. Mol Biol Cell 2006;17:1).

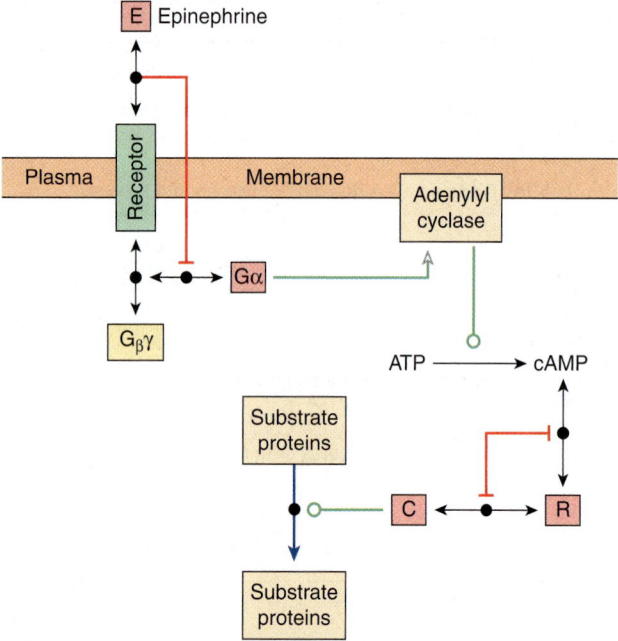

FIGURE 10–9 **Representation of a molecular interaction network (MIN) depicting a signal transduction cascade leading to the phosphorylation of substrate proteins by the catalytic subunit, C, of the cyclic AMP-dependent protein kinase in response to epinephrine.** Proteins are depicted as rectangles or squares. Double headed arrows indicate formation of noncovalent complex represented by dot at the midpoint of the arrow. Red lines with T-shaped heads indicate inhibitory interaction. Green arrow with hollow head indicates a stimulatory interaction. Green line with open circle at end indicates catalysis. Blue arrow with P indicates covalent modification by phosphorylation. (Symbols adapted from Kohn KW et al: Molecular interaction maps of bioregulatory networks: a general rubric for systems biology. Mol Biol Cell 2006; 17:1.)

- The objective of epidemiology is to extract medical insights from the behavior of heterogeneous human populations by the application of sophisticated statistical tools.
- Major challenges in the construction of user-friendly databases include devising means for storing and organizing complex data that accommodate a wide range of potential search criteria.
- The goal of the Encode Project is to identify all the functional elements within the human genome.
- The HapMap, *Entrez* Gene, and dbGAP databases contain data concerning the relation of genetic mutations to pathological conditions.
- Computational biology uses computer algorithms to identify unknown proteins and conduct virtual experiments.
- BLAST is used to identify unknown proteins and genes by searching for sequence homologs of known function.
- Computational biologists are developing programs that will predict the three-dimensional structure of proteins directly from their primary sequence.
- Computer-aided drug design speeds drug discovery by trying to dock potential inhibitors to selected protein targets *in silico*.
- A major goal of systems biologists is to create faithful models of individual pathways and networks in order to elucidate functional principles and perform virtual experiments.
- The ultimate goal of systems biologists is to create virtual cells that can be used to more safely and efficiently diagnose and treat diseases, particularly those of a multifactorial nature.
- Systems biologists commonly construct schematic representations known as molecular interaction maps in which symbolic logic is employed to illustrate the relationships between the components making up a pathway or some other functional unit

CONCLUSION

The rapidly evolving fields of bioinformatics and computational biology hold unparalleled promise for the future of both medicine and basic biology. Some promises are at present perceived clearly, others dimly, while yet others remain unimagined. A major objective of computational biologists is to develop computational tools that will enhance the efficiency, effectiveness, and speed of drug development. Epidemiologists employ computers to extract patterns within a population indicative of specific causes of and contributors to both disease and wellness. There seems little doubt that their impact on medical practice in the 21st century will equal or surpass that of the discovery of bacterial pathogenesis in the 19th century.

SUMMARY

- Genomics has yielded a massive quantity of information of great potential value to scientists and physicians.
- Bioinformatics involves the design of computer algorithms and construction of databases that enable biomedical scientists to access and analyze the growing avalanche of biomedical data.

REFERENCES

Altschul SF, Gish W, Miller W, et al: Basic local alignment search tool. J Mol Biol 1990;215:403.

Collins FS, Barker AD: Mapping the human cancer genome. Sci Am 2007;296:50.

Collins FS, Green ED, Guttmacher AE, et al: A vision for the future of genomics research. A blueprint for the genomic era. Nature 2003;422:835.

Couzin J: The HapMap gold rush: researchers mine a rich deposit. Science 2006;312:1131.

Cravatt BF, Wright AT, Kozarich JW: Activity-based protein profiling: from enzyme chemistry to proteomic chemistry. Annu Rev Biochem 2008;77:383.

Debes JD, Urrutia R: Bioinformatics tools to understand human diseases. Surgery 2004;135:579.

Dunning Hotopp JC, Grifantini R, Kumar N, et al: Comparative genomics of *Neisseria meningitides*: core genome, islands of horizontal transfer and pathogen-specific genes. Microbiology 2006;152:3691.

Ekins S, Mestres J, Testa B: *In silico* pharmacology for drug discovery: applications to targets and beyond. Br J Pharmacol 2007;152:21.

Ekins S, Mestres J, Testa B: *In silico* pharmacology for drug discovery: methods for virtual ligand screening and profiling. Br J Pharmacol 2007;152:9.

Kaiser J: Affordable 'exomes' fill gaps in a catalog of rare diseases. Science 2010;330:903.

Kim JH: Bioinformatics and genomic medicine. Genet Med 2002;4:62S.

Kohn KW, Aladjem MI, Weinstein JN, et al: Molecular interaction maps of bioregulatory networks: a general rubric for systems biology. Mol Biol Cell 2006; 17:1.

Koonin EV, Galperin MY: *Sequence—Evolution—Function. Computational Approaches to Comparative Genomics.* Kluwer Academic Publishers, 2003.

Laurie ATR, Jackson RM: Methods for prediction of protein–ligand binding sites for structure-based drug design and virtual ligand screening. Curr Prot Peptide Sci 2006; 7:395–406.

McInnes C: Virtual screening strategies in drug discovery. Curr Opin Cell Biol 2007;11:494.

Nebert DW, Zhang G, Vesell ES: From human genetics and genomics to pharmacogenetics and pharmacogenomics: past lessons, future directions. Drug Metab Rev 2008; 40:187.

Sansom C: Genes and disease. The Scientist 2008;30:34.

Slepchenko BM, Schaff JC, Macara I, et al: Quantitative cell biology with the Virtual Cell. Trends Cell Biol 2003; 13:570.

Sudmant PH, Kitzman JO, Antonacci F, et al: Diversity of human gene copy number variation and multicopy genes. Science 2010; 330:641.

Villoutreix BO, Renault N, Lagorce D, et al: Free resources to assist structure-based virtual ligand screening experiments. Curr Protein Pept Sci 2007;8:381.

Exam Questions

Section I

1. The propensity of water atoms to form hydrogen bonds with one another is the primary factor responsible for all of the following properties of water EXCEPT:
   A. Its atypically high boiling point.
   B. Its high heat of vaporization.
   C. Its high surface tension.
   D. Its ability to dissolve hydrocarbons.
   E. Its expansion upon freezing.

2. Select the one of the following statements that is NOT CORRECT:
   A. The side-chains of the amino acids cysteine and methionine absorb visible light.
   B. Glycine is often present in regions where a polypeptide forms a sharp bend, reversing the direction of a polypeptide.
   C. Polypeptides are named as derivatives of the C-terminal aminoacyl residue.
   D. The C, N, O, and H atoms of a peptide bond are coplanar.
   E. A linear pentapeptide contains four peptide bonds.

3. Select the one of the following statements that is NOT CORRECT:
   A. Buffers of human tissue include bicarbonate, proteins, and orthophosphate.
   B. A weak acid or a weak base exhibits its greatest buffering capacity when the pH is equal to its pK_a plus or minus one pH unit.
   C. The isoelectric pH (pI) of lysine can be calculated using the formula $(pK_2+pK_3)/2$.
   D. The mobility of a monofunctional weak acid in a direct current electrical field reaches its maximum when the pH of its surrounding environment is equal to its pK_a.
   E. For simplicity, the strengths of weak bases are generally expressed as the pK_a of their conjugate acids.

4. Select the one of the following statements that is NOT CORRECT:
   A. If the pK_a of a weak acid is 4.0, 50% of the molecules will be in the dissociated state when the pH of the surrounding environment is 4.0.
   B. A weak acid with a pK_a of 4.0 will be a more effective buffer at pH 3.8 than at pH 5.7.
   C. At a pH equal to its pI a polypeptide carries no charged groups.
   D. Strong acids and bases are so named because they undergo complete dissociation when dissolved in water.
   E. The pK_a of an ionizable group can be influenced by the physical and chemical properties of its surrounding environment.

5. Select the one of the following statements that is NOT CORRECT:
   A. To calculate K_{eq}, the equilibrium constant for a reaction, divide the initial rate of the forward reaction rate ($rate_1$) by the initial velocity of the reverse reaction ($rate_{-1}$).
   B. The presence of an enzyme has no effect on K_{eq}.

   C. For a reaction conducted at constant temperature the fraction of the potential reactant molecules possessing sufficient kinetic energy to exceed the activation energy of the reaction is a constant.
   D. Enzymes and other catalysts lower the activation energy of reactions.
   E. The algebraic sign of ΔG, the Gibbs free energy change for a reaction, indicates the direction in which a reaction will proceed.

6. Select the one of the following statements that is NOT CORRECT:
   A. As used in biochemistry, the standard state concentration for products and reactants other than protons is 1 molar.
   B. ΔG is a function of the logarithm of K_{eq}.
   C. As used in reaction kinetics, the term "spontaneity" refers to whether the reaction as written is favored to proceed from left to right.
   D. $\Delta G°$ denotes the change in free energy that accompanies transition from the standard state to equilibrium.
   E. Upon reaching equilibrium, the rates of the forward and reverse reaction both drop to zero.

7. Select the one of the following statements that is NOT CORRECT:
   A. Enzymes lower the activation energy for a reaction.
   B. One means by which enzymes lower the activation energy is by destabilizing transition state intermediates.
   C. Active site histidyl residues frequently aid catalysis by acting as proton donors or acceptors.
   D. Covalent catalysis is employed by some enzymes to provide a unique reaction pathway.
   E. The presence of an enzyme has no effect on $\Delta G°$.

8. Select the one of the following statements that is NOT CORRECT:
   A. For most enzymes, the relationship of [S] and the initial reaction rate, v_i, yields a hyperbolic curve.
   B. When [S] is much lower than K_m, the term $K_m + [S]$ in the Michaelis–Menten equation closely approaches K_m. Under these conditions, the rate of catalysis is a linear function of [S].
   C. The molar concentrations of substrate and products are equal when the rate of an enzyme-catalyzed reaction reaches half of its potential maximum ($V_{max}/2$).
   D. An enzyme is said to have become saturated with substrate when successively raising [S] fails to produce a significant increase in v_i.
   E. When making steady-state rate measurements, the concentration of substrates should greatly exceed that of the enzyme catalyst.

9. Select the one of the following statements that is NOT CORRECT:
   A. Certain monomeric enzymes exhibit sigmoidal initial rate kinetics.
   B. The Hill equation is used to perform quantitative analysis of cooperative behavior of enzymes or carrier proteins such as hemoglobin or calmodulin.
   C. For an enzyme that exhibits cooperative binding of substrate, a value of n (the Hill coefficient) greater than unity is said to exhibit positive cooperativity.

D. An enzyme that catalyzes a reaction between two or more substrates is said to operate by a sequential mechanism if the substrates must bind in a fixed order.

E. Prosthetic groups enable enzymes to add capabilities beyond those provided by amino acid side-chains.

10. Select the one of the following statements that is NOT CORRECT:

A. IC_{50} is a simple operational term for expressing the potency of an inhibitor.

B. Lineweaver-Burk and Dixon plots employ rearranged versions of the Michaelis–Menten equation to generate linear representations of kinetic behavior and inhibition.

C. A plot of $1/v_i$ versus $1/[S]$ can be used to evaluate the type and magnitude of an inhibitor.

D. Simple non-competitive inhibitors lower the apparent K_m for a substrate.

E. Non-competitive inhibitors typically bear little or no structural resemblance to the substrate(s) of an enzyme-catalyzed reaction.

11. Select the one of the following statements that is NOT CORRECT:

A. For a given enzyme, the intracellular concentrations of its substrates tend to be close to their K_m values.

B. The sequestration of certain pathways within intracellular organelles facilitates the task of metabolic regulation.

C. The earliest step in a biochemical pathway where regulatory control can be efficiently exerted is the first committed step.

D. Feedback regulation refers to the allosteric control of an early step in a biochemical pathway by the end product(s) of that pathway.

E. Metabolic control is most effective when one of the rapid steps in a pathway is targeted for regulation.

12. Select the one of the following statements that is NOT CORRECT:

A. A major objective of proteomics is to identify all of the proteins present in a cell under different conditions as well as their states of modification.

B. Mass spectrometry has largely replaced the Edman method for sequencing of peptides and proteins.

C. Sanger's reagent was an improvement on Edman's because the former generates a new amino terminus, allowing several consecutive cycles of sequencing to take place.

D. Since mass is a universal property of all atoms and molecules, mass spectrometry is ideally suited to the detection of post-translational modifications in proteins.

E. Time-of-flight mass spectrometers take advantage of the relationship F = ma.

13. Select the one of the following statements that is NOT CORRECT:

A. Ion-exchange chromatography separates proteins based upon the sign and magnitude of their charge at a given pH.

B. Two-dimensional gel electrophoresis separates proteins first on the basis of their pI values and second on their charge to mass ratio using SDS.

C. Affinity chromatography exploits the selectivity of protein–ligand interactions to isolate a specific protein from a complex mixture.

D. Many recombinant proteins are expressed with an additional domain fused to their N- or C-terminus. One common component of these fusion domains is a ligand binding site designed expressly to facilitate purification by affinity chromatography.

E. Tandem mass spectrometry can analyze peptides derived from complex protein mixtures without their prior separation.

14. Select the one of the following statements that is NOT CORRECT:

A. Protein folding is assisted by intervention of specialized auxiliary proteins called chaperones.

B. Protein folding tends to be modular, with areas of local secondary structure forming first, then coalescing into a molten globule.

C. Protein folding is driven first and foremost by the thermodynamics of the water molecules surrounding the nascent polypeptide.

D. The formation of S-S bonds in a mature protein is facilitated by the enzyme protein disulfide isomerase.

E. Only a few unusual proteins, such as collagen, require posttranslational processing by partial proteolysis to attain their mature conformation.

15. Select the one of the following statements that is NOT CORRECT:

A. Posttranslational modifications of proteins can affect both their function and their metabolic fate.

B. The native conformational state generally is that which is thermodynamically favored.

C. The complex three-dimensional structures of most proteins are formed and stabilized by the cumulative effects of a large number of weak interactions.

D. Research scientists employ gene arrays for the high-throughput detection of the presence and expression level of proteins.

E. Examples of weak interactions that stabilize protein folding include hydrogen bonds, salt bridges, and van der Waals forces.

16. Select the one of the following statements that is NOT CORRECT:

A. Changes in configuration involve the rupture of covalent bonds.

B. Changes in conformation involve the rotation of one or more single bonds.

C. The Ramachandran plot illustrates the degree to which steric hindrance limits the permissible angles of the single bonds in the backbone of a peptide or protein.

D. Formation of an α-helix is stabilized by the hydrogen bonds between each peptide bond carboxyl oxygen and the N-H group of the next peptide bond.

E. In a β-sheet the R-groups of adjacent residues point in opposite directions relative to the plane of the sheet.

17. Select the one of the following statements that is NOT CORRECT:

A. The descriptor $\alpha_2\beta_2\gamma_3$ denotes a protein with seven subunits of three different types.

B. Loops are extended regions that connect adjacent regions of secondary structure.

C. More than half of the residues in a typical protein reside in either α-helices or β-sheets.

D. Most β-sheets have a right-handed twist.

E. Prions are viruses that cause protein-folding diseases that attack the brain.

18. Select the one of the following statements that is NOT CORRECT:

    A. The Bohr effect refers to the release of protons that occurs when oxygen binds to deoxyhemoglobin.
    B. Shortly after birth of the human infant synthesis of the α-chain undergoes rapid induction until it comprises 50% of the hemoglobin tetramer.
    C. The β-chain of fetal hemoglobin is present throughout gestation.
    D. Thalassemias are genetic defects due to partial or total absence of the α- or β-chains of hemoglobin.
    E. The taut conformation of hemoglobin is stabilized by several salt bridges that form between the subunits.

19. Select the one of the following statements that is NOT CORRECT:

    A. Steric hindrance by histidine E7 plays a critical role in weakening the affinity of hemoglobin for carbon monoxide (CO).
    B. Carbonic anhydrase plays a critical role in respiration by virtue of its capacity to break down 2,3-bisphosphoglycerate in the lungs.
    C. Hemoglobin S is distinguished by a genetic mutation that substitutes Glu6 on the ß subunit with Val, creating a sticky patch on the surface.
    D. Oxidation of the heme iron from the +2 to the +3 state abolishes the ability of hemoglobin to bind oxygen.
    E. The functional differences between hemoglobin and myoglobin reflect to a large degree differences in their quaternary structure.

20. Select the one of the following statements that is NOT CORRECT:

    A. The charge-relay network of trypsin makes the active site serine a strong nucleophile.
    B. The Michaelis constant is the substrate concentration at which the rate of the reaction is half-maximal.
    C. During transamination reactions, both substrates are bound to the enzyme before either product is released.
    D. Histidine residues act both as acids and as bases during catalysis by an aspartate protease.
    E. Many coenzymes and cofactors are derived from vitamins.

21. Select the one of the following statements that is NOT CORRECT:

    A. Interconvertible enzymes fulfill key roles in integrated regulatory networks.
    B. Phosphorylation of an enzyme often alters its catalytic efficiency.
    C. "Second messengers" act as intracellular extensions or surrogates for hormones and nerve impulses impinging on cell surface receptors.
    D. The ability of protein kinases to catalyze the reverse reaction that removes the phosphate group is key to the versatility of this molecular regulatory mechanism.
    E. Zymogen activation by partial proteolysis is irreversible under physiological conditions.

22. Select the one of the following statements that is NOT CORRECT:

    A. The HapMap Database focuses on the location and identity of single nucleotide polymorphisms in humans.
    B. Genbank is a repository of data on the phenotypic results of gene knockouts in humans.
    C. The Protein Database or PDB stores the three-dimensional structures of proteins as determined by X-ray crystallography or nuclear magnetic resonance spectroscopy (NMR).
    D. The objective of the ENCODE project is to identify all of the functional elements of the genome.
    E. BLAST compares protein and nucleotide sequences in order to identify areas of similarity.

23. Select the one of the following statements that is NOT CORRECT:

    A. A major obstacle to computer-aided drug design is the extraordinary demands in computing capacity required to permit proteins and ligands a realistic degree of conformational flexibility.
    B. Conformational flexibility is needed to permit ligand and protein to influence one another as described by Fischer's lock and key model for protein-ligand binding.
    C. Construction of a virtual cell could provide a means to rapidly and efficiently detect many undesirable effects of potential drugs without the need for expensive laboratory testing.
    D. Systems biology highlights the manner in which the connections between enzymatic or other units in a cell affect their performance.
    E. Systems biologists frequently employ the symbolic logic of computer programs and electronic circuits to describe the interactions between proteins, genes, and metabolites.

24. Select the one of the following statements that is NOT CORRECT:

    A. GRASP representations highlight areas of a protein's surface possessing local positive or negative character.
    B. Molecular dynamics simulations seek to reconstruct the types and range of movement that conformationally-flexible proteins undergo.
    C. Researchers use rolling ball programs to locate indentations and crevices on the surface of a protein because these represent likely sites for attack by proteases.
    D. Molecular docking simulations often restrict free rotation to only a small set of bonds in a ligand to match the computing power available.
    E. Discerning the evolutionary relationships between proteins constitutes one of the most effective means of predicting the likely functions of a newly discovered polypeptide.

CHAPTER

11

Bioenergetics: The Role of ATP

Kathleen M. Botham, PhD, DSc & Peter A. Mayes, PhD, DSc

OBJECTIVES

After studying this chapter, you should be able to:

- State the first and second laws of thermodynamics and understand how they apply to biologic systems.
- Explain what is meant by the terms free energy, entropy, enthalpy, exergonic, and endergonic.
- Appreciate how reactions that are endergonic may be driven by coupling to those that are exergonic in biologic systems.
- Understand the role of high-energy phosphates, ATP, and other nucleotide triphosphates in the transfer of free energy from exergonic to endergonic processes, enabling them to act as the "energy currency" of cells.

BIOMEDICAL IMPORTANCE

Bioenergetics, or biochemical thermodynamics, is the study of the energy changes accompanying biochemical reactions. Biologic systems are essentially **isothermic** and use chemical energy to power living processes. How an animal obtains suitable fuel from its food to provide this energy is basic to the understanding of normal nutrition and metabolism. Death from **starvation** occurs when available energy reserves are depleted, and certain forms of malnutrition are associated with energy imbalance (**marasmus).** Thyroid hormones control the rate of energy release (metabolic rate), and disease results when they malfunction. Excess storage of surplus energy causes **obesity,** an increasingly common disease of Western society, which predisposes to many diseases, including cardiovascular disease and diabetes mellitus type 2, and lowers life expectancy.

FREE ENERGY IS THE USEFUL ENERGY IN A SYSTEM

Gibbs change in free energy (ΔG) is that portion of the total energy change in a system that is available for doing work—ie, the useful energy, also known as the chemical potential.

Biologic Systems Conform to the General Laws of Thermodynamics

The first law of thermodynamics states that **the total energy of a system, including its surroundings, remains constant.** It implies that within the total system, energy is neither lost nor gained during any change. However, energy may be transferred from one part of the system to another or may be transformed into another form of energy. In living systems, chemical energy may be transformed into heat or into electrical, radiant, or mechanical energy.

The second law of thermodynamics states that **the total entropy of a system must increase if a process is to occur spontaneously. Entropy** is the extent of disorder or randomness of the system and becomes maximum as equilibrium is approached. Under conditions of constant temperature and pressure, the relationship between the free-energy change (ΔG) of a reacting system and the change in entropy (ΔS) is expressed by the following equation, which combines the two laws of thermodynamics:

$$\Delta G = \Delta H - T\Delta S$$

where ΔH is the change in **enthalpy** (heat) and T is the absolute temperature.

In biochemical reactions, since ΔH is approximately equal to ΔE, the total change in internal energy of the reaction, the above relationship may be expressed in the following way:

$$\Delta G = \Delta E - T\Delta S$$

If ΔG is negative, the reaction proceeds spontaneously with loss of free energy; ie, it is **exergonic.** If, in addition, ΔG is of great magnitude, the reaction goes virtually to completion and is essentially irreversible. On the other hand, if ΔG is positive, the reaction proceeds only if free energy can be gained; ie, it is **endergonic.** If, in addition, the magnitude of ΔG is great, the system is stable, with little or no tendency for a reaction to occur. If ΔG is zero, the system is at equilibrium and no net change takes place.

When the reactants are present in concentrations of 1.0 mol/L, ΔG^0 is the standard free-energy change. For biochemical reactions, a standard state is defined as having a pH of 7.0. The standard free-energy change at this standard state is denoted by $\Delta G^{0\prime}$.

The standard free-energy change can be calculated from the equilibrium constant K_{eq}.

$$\Delta G^{0\prime} = -RT \ln K'_{eq}$$

where R is the gas constant and T is the absolute temperature (Chapter 8). It is important to note that the actual ΔG may be larger or smaller than $\Delta G^{0\prime}$ depending on the concentrations of the various reactants, including the solvent, various ions, and proteins.

In a biochemical system, an enzyme only speeds up the attainment of equilibrium; it never alters the final concentrations of the reactants at equilibrium.

ENDERGONIC PROCESSES PROCEED BY COUPLING TO EXERGONIC PROCESSES

The vital processes—eg, synthetic reactions, muscular contraction, nerve impulse conduction, and active transport—obtain energy by chemical linkage, or **coupling,** to oxidative reactions. In its simplest form, this type of coupling may

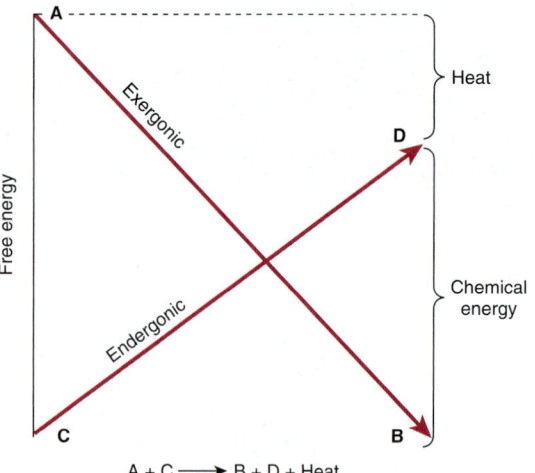

FIGURE 11–1 **Coupling of an exergonic to an endergonic reaction.**

be represented as shown in **Figure 11–1.** The conversion of metabolite A to metabolite B occurs with release of free energy and is coupled to another reaction in which free energy is required to convert metabolite C to metabolite D. The terms **exergonic** and **endergonic,** rather than the normal chemical terms "exothermic" and "endothermic," are used to indicate that a process is accompanied by loss or gain, respectively, of free energy in any form, not necessarily as heat. In practice, an endergonic process cannot exist independently, but must be a component of a coupled exergonic–endergonic system where the overall net change is exergonic. The exergonic reactions are termed **catabolism** (generally, the breakdown or oxidation of fuel molecules), whereas the synthetic reactions that build up substances are termed **anabolism.** The combined catabolic and anabolic processes constitute **metabolism.**

If the reaction shown in Figure 11–1 is to go from left to right, then the overall process must be accompanied by loss of free energy as heat. One possible mechanism of coupling could be envisaged if a common obligatory intermediate (I) took part in both reactions, ie,

$$A + C \rightarrow I \rightarrow B + D$$

Some exergonic and endergonic reactions in biologic systems are coupled in this way. This type of system has a built-in mechanism for biologic control of the rate of oxidative processes since the common obligatory intermediate allows the rate of utilization of the product of the synthetic path (D) to determine by mass action the rate at which A is oxidized. Indeed, these relationships supply a basis for the concept of **respiratory control,** the process that prevents an organism from burning out of control. An extension of the coupling concept is provided by dehydrogenation reactions, which are coupled to hydrogenations by an intermediate carrier (**Figure 11–2**).

An alternative method of coupling an exergonic to an endergonic process is to synthesize a compound of high-energy potential in the exergonic reaction and to incorporate this new

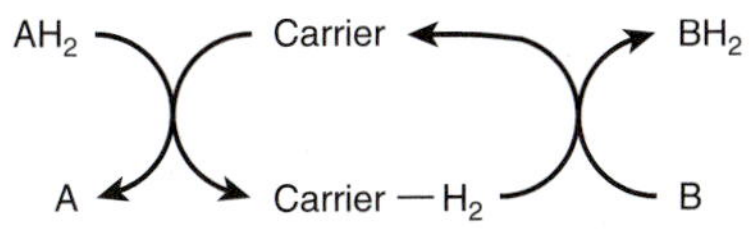

FIGURE 11–2 Coupling of dehydrogenation and hydrogenation reactions by an intermediate carrier.

compound into the endergonic reaction, thus effecting a transference of free energy from the exergonic to the endergonic pathway (**Figure 11–3**). The biologic advantage of this mechanism is that the compound of high potential energy, ~Ⓔ, unlike I in the previous system, need not be structurally related to A, B, C, or D, allowing Ⓔ to serve as a transducer of energy from a wide range of exergonic reactions to an equally wide range of endergonic reactions or processes, such as biosyntheses, muscular contraction, nervous excitation, and active transport. In the living cell, the principal high-energy intermediate or carrier compound (designated ~Ⓔ in Figure 11–3) is **adenosine triphosphate (ATP)** (**Figure 11-4**).

HIGH-ENERGY PHOSPHATES PLAY A CENTRAL ROLE IN ENERGY CAPTURE AND TRANSFER

In order to maintain living processes, all organisms must obtain supplies of free energy from their environment. **Autotrophic** organisms utilize simple exergonic processes; eg, the energy of sunlight (green plants), the reaction $Fe^{2+} \rightarrow Fe^{3+}$ (some bacteria). On the other hand, **heterotrophic** organisms obtain free energy by coupling their metabolism to the breakdown of complex organic molecules in their environment. In all these organisms, ATP plays a central role in the transference of free energy from the exergonic to the endergonic processes (Figure 11–3). ATP is a nucleoside

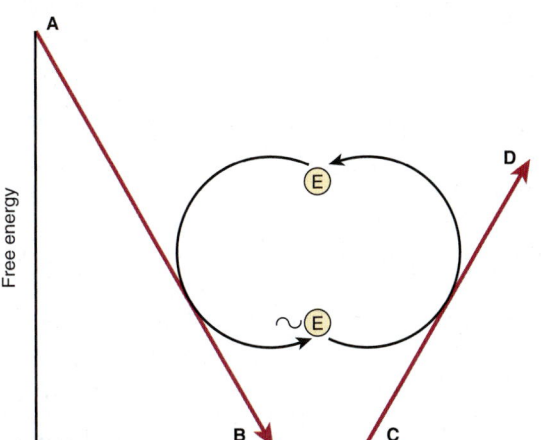

FIGURE 11–3 Transfer of free energy from an exergonic to an endergonic reaction via a high-energy intermediate compound (~Ⓔ).

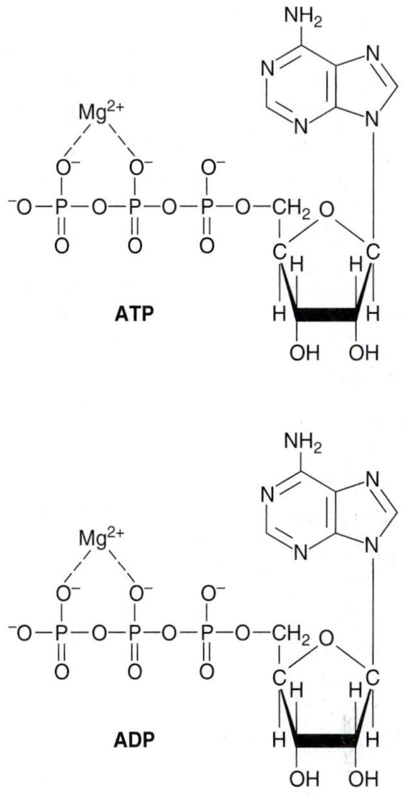

FIGURE 11–4 Adenosine triphosphate (ATP) and adenosine diphosphate shown as the magnesium complexes.

triphosphate containing adenine, ribose, and three phosphate groups. In its reactions in the cell, it functions as the Mg^{2+} complex (Figure 11–4).

The importance of phosphates in intermediary metabolism became evident with the discovery of the role of ATP, adenosine diphosphate (ADP) (Figure 11–4), and inorganic phosphate (P_i) in glycolysis (Chapter 18).

The Intermediate Value for the Free Energy of Hydrolysis of ATP Has Important Bioenergetic Significance

The standard free energy of hydrolysis of a number of biochemically important phosphates is shown in **Table 11–1**. An estimate of the comparative tendency of each of the phosphate groups to transfer to a suitable acceptor may be obtained from the $\Delta G^{0\prime}$ of hydrolysis at 37°C. The value for the hydrolysis of the terminal phosphate of ATP divides the list into two groups. **Low-energy phosphates,** exemplified by the ester phosphates found in the intermediates of glycolysis, have $G^{0\prime}$ values smaller than that of ATP, while in **high-energy phosphates** the value is higher than that of ATP. The components of this latter group, including ATP, are usually anhydrides (eg, the 1-phosphate of 1,3-bisphosphoglycerate), enolphosphates (eg, phosphoenolpyruvate), and phosphoguanidines (eg, creatine phosphate, arginine phosphate).

TABLE 11–1 Standard Free Energy of Hydrolysis of Some Organophosphates of Biochemical Importance

Compound	$\Delta G^{0\prime}$ kJ/mol	$\Delta G^{0\prime}$ kcal/mol
Phosphoenolpyruvate	−61.9	−14.8
Carbamoyl phosphate	−51.4	−12.3
1,3-Bisphosphoglycerate (to 3-phosphoglycerate)	−49.3	−11.8
Creatine phosphate	−43.1	−10.3
ATP → AMP + PP$_i$	−32.2	−7.7
ATP → ADP + P$_i$	−30.5	−7.3
Glucose 1-phosphate	−20.9	−5.0
PP$_i$	−19.2	−4.6
Fructose 6-phosphate	−15.9	−3.8
Glucose 6-phosphate	−13.8	−3.3
Glycerol 3-phosphate	−9.2	−2.2

Abbreviations: PP$_i$, pyrophosphate; P$_i$, inorganic orthophosphate.
Note: All values taken from Jencks (1976), except that for PP$_i$ which is from Frey and Arabshahi (1995). Values differ between investigators, depending on the precise conditions under which the measurements were made.

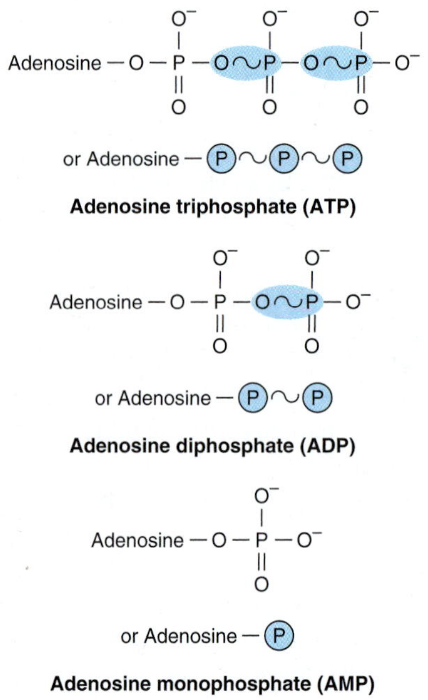

The symbol ~ⓅΡ indicates that the group attached to the bond, on transfer to an appropriate acceptor, results in transfer of the larger quantity of free energy. For this reason, the term **group transfer potential**, rather than "high-energy bond," is preferred by some. Thus, ATP contains two high-energy phosphate groups and ADP contains one, whereas the phosphate in AMP (adenosine monophosphate) is of the low-energy type since it is a normal ester link (**Figure 11–5**).

The intermediate position of ATP allows it to play an important role in energy transfer. The high free-energy change on hydrolysis of ATP is due to relief of charge repulsion of adjacent negatively charged oxygen atoms and to stabilization of the reaction products, especially phosphate, as resonance hybrids (**Figure 11–6**). Other "high-energy compounds" are thiol esters involving coenzyme A (eg, acetyl-CoA), acyl carrier protein, amino acid esters involved in protein synthesis, *S*-adenosylmethionine (active methionine), UDPGlc (uridine diphosphate glucose), and PRPP (5-phosphoribosyl-1-pyrophosphate).

HIGH-ENERGY PHOSPHATES ACT AS THE "ENERGY CURRENCY" OF THE CELL

ATP is able to act as a donor of high-energy phosphate to form those compounds below it in Table 11–1. Likewise, with the necessary enzymes, ADP can accept high-energy phosphate

to form ATP from those compounds above ATP in the table. In effect, an **ATP/ADP cycle** connects those processes that generate ~ⓅΡ to those processes that utilize ~ⓅΡ (**Figure 11–7**), continuously consuming and regenerating ATP. This occurs at a very rapid rate since the total ATP/ADP pool is extremely

FIGURE 11–5 Structure of ATP, ADP, and AMP showing the position and the number of high-energy phosphates (~ⓅΡ).

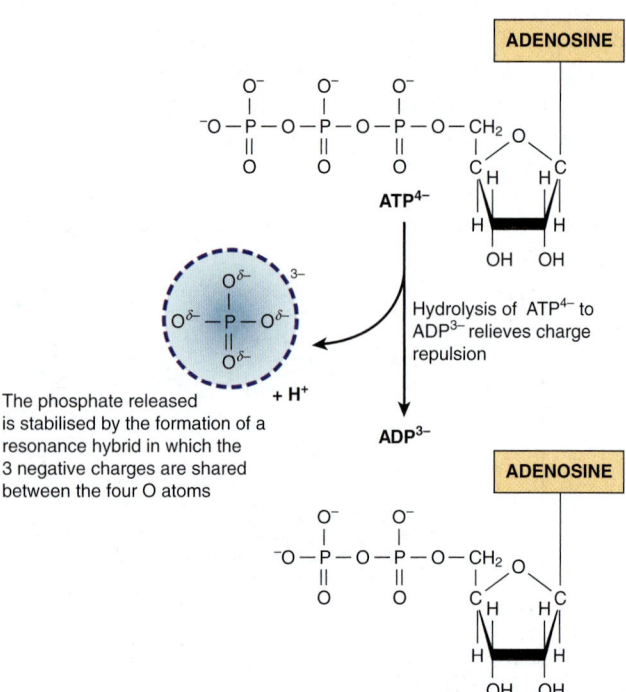

FIGURE 11–6 The free-energy change on hydrolysis of ATP to ADP.

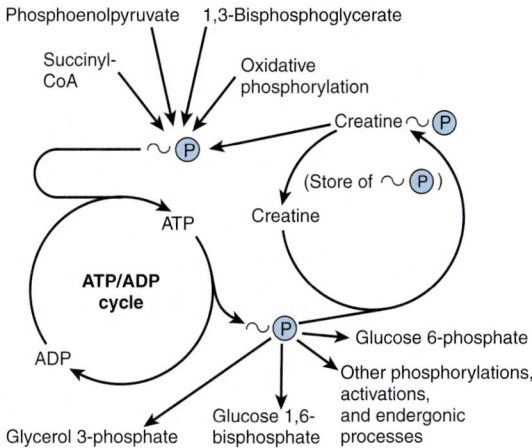

FIGURE 11–7 Role of ATP/ADP cycle in transfer of high-energy phosphate.

small and sufficient to maintain an active tissue for only a few seconds.

There are three major sources of ~Ⓟ taking part in **energy conservation** or **energy capture:**

1. **Oxidative phosphorylation.** The greatest quantitative source of ~Ⓟ in aerobic organisms. Free energy comes from respiratory chain oxidation using molecular O_2 within mitochondria (Chapter 12).
2. **Glycolysis.** A net formation of two ~Ⓟ results from the formation of lactate from one molecule of glucose, generated in two reactions catalyzed by phosphoglycerate kinase and pyruvate kinase, respectively (Figure 18–2).
3. **The citric acid cycle.** One ~Ⓟ is generated directly in the cycle at the succinate thiokinase step (Figure 17–3).

Phosphagens act as storage forms of high-energy phosphate and include creatine phosphate, which occurs in vertebrate skeletal muscle, heart, spermatozoa, and brain, and arginine phosphate, which occurs in invertebrate muscle. When ATP is rapidly being utilized as a source of energy for muscular contraction, phosphagens permit its concentrations to be maintained, but when the ATP/ADP ratio is high, their concentration can increase to act as a store of high-energy phosphate (**Figure 11–8**).

When ATP acts as a phosphate donor to form those compounds of lower free energy of hydrolysis (Table 11–1),

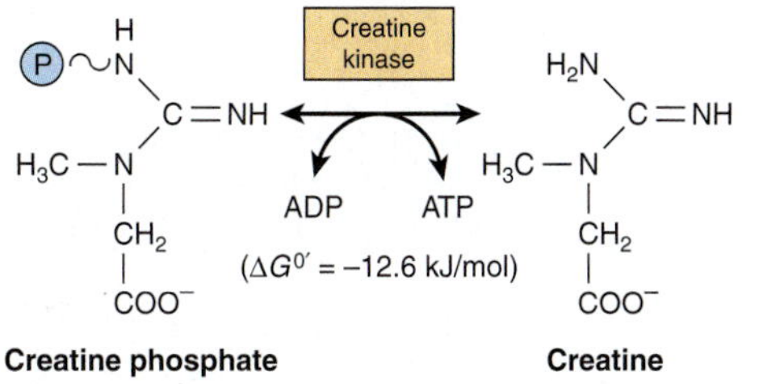

the phosphate group is invariably converted to one of low energy, eg

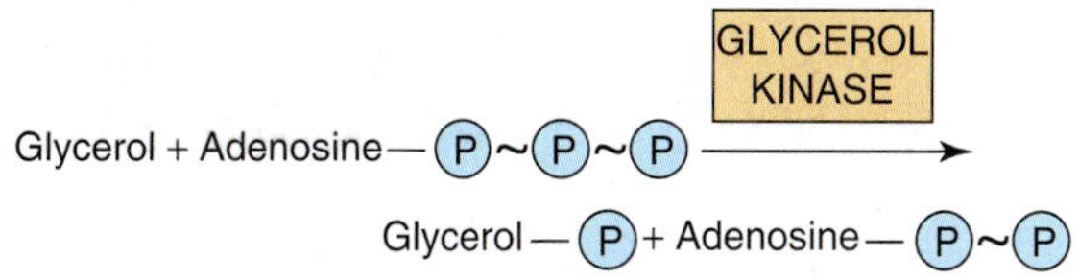

Glycerol + Adenosine—Ⓟ~Ⓟ~Ⓟ $\xrightarrow{\text{GLYCEROL KINASE}}$

Glycerol—Ⓟ + Adenosine—Ⓟ~Ⓟ

ATP Allows the Coupling of Thermodynamically Unfavorable Reactions to Favorable Ones

The phosphorylation of glucose to glucose 6-phosphate, the first reaction of glycolysis (Figure 18–2), is highly endergonic and cannot proceed under physiologic conditions:

$$\text{Glucose} + P_i \rightarrow \text{Glucose 6-phosphate} + H_2O \qquad (1)$$

$$(\Delta G^{0'} = +13.8 \text{ kJ/mol})$$

To take place, the reaction must be coupled with another—more exergonic—reaction such as the hydrolysis of the terminal phosphate of ATP.

$$\text{ATP} \rightarrow \text{ADP} + P_i \ (\Delta G^{0'} = -30.5 \text{ kJ/mol}) \qquad (2)$$

When (1) and (2) are coupled in a reaction catalyzed by hexokinase, phosphorylation of glucose readily proceeds in a highly exergonic reaction that under physiologic conditions is irreversible. Many "activation" reactions follow this pattern.

Adenylyl Kinase (Myokinase) Interconverts Adenine Nucleotides

This enzyme is present in most cells. It catalyzes the following reaction:

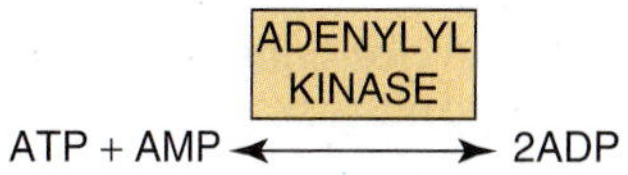

$$\text{ATP} + \text{AMP} \xleftrightarrow{\text{ADENYLYL KINASE}} 2\text{ADP}$$

This allows:

1. High-energy phosphate in ADP to be used in the synthesis of ATP.
2. AMP, formed as a consequence of several activating reactions involving ATP, to be recovered by rephosphorylation to ADP.
3. AMP to increase in concentration when ATP becomes depleted and act as a metabolic (allosteric) signal to increase the rate of catabolic reactions, which in turn lead to the generation of more ATP (Chapter 20).

When ATP Forms AMP, Inorganic Pyrophosphate (PP$_i$) Is Produced

ATP can also be hydrolyzed directly to AMP, with the release of PP$_i$ (Table 11–1). This occurs, for example, in the activation of long-chain fatty acids (Chapter 22).

FIGURE 11–8 Transfer of high-energy phosphate between ATP and creatine.

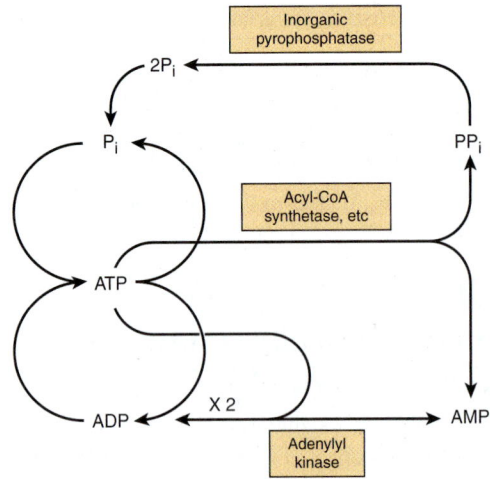

FIGURE 11–9 **Phosphate cycles and interchange of adenine nucleotides.**

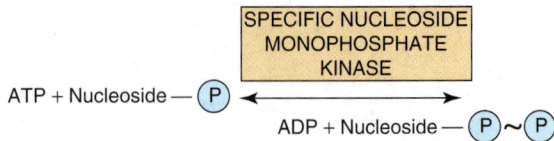

All of these triphosphates take part in phosphorylations in the cell. Similarly, specific nucleoside monophosphate kinases catalyze the formation of nucleoside diphosphates from the corresponding monophosphates.

$$\boxed{\text{SPECIFIC NUCLEOSIDE MONOPHOSPHATE KINASE}}$$

ATP + Nucleoside — P ⇌ ADP + Nucleoside — P ~ P

Thus, adenylyl kinase is a specialized monophosphate kinase.

SUMMARY

- Biologic systems use chemical energy to power living processes.

- Exergonic reactions take place spontaneously with loss of free energy (ΔG is negative). Endergonic reactions require the gain of free energy (ΔG is positive) and occur only when coupled to exergonic reactions.

- ATP acts as the "energy currency" of the cell, transferring free energy derived from substances of higher energy potential to those of lower energy potential.

REFERENCES

de Meis L: The concept of energy-rich phosphate compounds: water, transport ATPases, and entropy energy. Arch Biochem Biophys 1993;306:287.

Frey PA, Arabshahi A: Standard free-energy change for the hydrolysis of the alpha, beta-phosphoanhydride bridge in ATP. Biochemistry 1995;34:11307.

Harold FM: *The Vital Force: A Study of Bioenergetics.* Freeman, 1986.

Harris DA: *Bioenergetics at a Glance: An Illustrated Introduction.* Blackwell Publishing, 1995.

Haynie D: *Biological Thermodynamics.* Cambridge University Press, 2008.

Jencks WP: Free energies of hydrolysis and decarboxylation. In: *Handbook of Biochemistry and Molecular Biology,* vol 1. *Physical and Chemical Data.* Fasman GD (editor). CRC Press, 1976:296–304.

Klotz IM: *Introduction to Biomolecular Energetics.* Academic Press, 1986.

Nicholls D, Ferguson F: *Bioenergetics.* Elsevier, 2003.

$$\boxed{\text{ACYL-CoA SYNTHETASE}}$$

ATP + CoA • SH + R • COOH $\longrightarrow$ AMP + PP$_i$ + R • CO — SCoA

This reaction is accompanied by loss of free energy as heat, which ensures that the activation reaction will go to the right and is further aided by the hydrolytic splitting of PP$_i$, catalyzed by **inorganic pyrophosphatase,** a reaction that itself has a large $\Delta G^{0'}$ of −19.2 kJ/mol. Note that activations via the pyrophosphate pathway result in the loss of two ~Ⓟ rather than one, as occurs when ADP and P$_i$ are formed.

$$\boxed{\text{INORGANIC PYROPHOSPHATASE}}$$

PP$_i$ + H$_2$O $\longrightarrow$ 2P$_i$

A combination of the above reactions makes it possible for phosphate to be recycled and the adenine nucleotides to interchange (**Figure 11–9**).

Other Nucleoside Triphosphates Participate in the Transfer of High-Energy Phosphate

By means of the enzyme **nucleoside diphosphate kinase,** UTP, GTP, and CTP can be synthesized from their diphosphates, eg, UDP reacts with ATP to form UTP.

Biologic Oxidation

Kathleen M. Botham, PhD, DSc & Peter A. Mayes, PhD, DSc

OBJECTIVES

After studying this chapter, you should be able to:

- Understand the meaning of redox potential and explain how it can be used to predict the direction of flow of electrons in biologic systems.
- Identify the four classes of enzymes (oxidoreductases) involved in oxidation and reduction reactions.
- Describe the action of oxidases and provide examples of where they play an important role in metabolism.
- Indicate the two main functions of dehydrogenases and explain the importance of NAD- and riboflavin-linked dehydrogenases in metabolic pathways such as glycolysis, the citric acid cycle, and the respiratory chain.
- Identify the two types of enzymes classified as hydroperoxidases; indicate the reactions they catalyze and explain why they are important.
- Give the two steps of reactions catalyzed by oxygenases and identify the two subgroups of this class of enzymes.
- Appreciate the role of cytochrome P450 in drug detoxification and steroid synthesis.
- Describe the reaction catalyzed by superoxide dismutase and explain how it protects tissues from oxygen toxicity.

BIOMEDICAL IMPORTANCE

Chemically, **oxidation** is defined as the removal of electrons and **reduction** as the gain of electrons. Thus, oxidation is always accompanied by reduction of an electron acceptor. This principle of oxidation–reduction applies equally to biochemical systems and is an important concept underlying understanding of the nature of biologic oxidation. Note that many biologic oxidations can take place without the participation of molecular oxygen, eg, dehydrogenations. The life of higher animals is absolutely dependent upon a supply of oxygen for **respiration,** the process by which cells derive energy in the form of ATP from the controlled reaction of hydrogen with oxygen to form water. In addition, molecular oxygen is incorporated into a variety of substrates by enzymes designated as **oxygenases;** many drugs, pollutants, and chemical carcinogens (xenobiotics) are metabolized by enzymes of this class, known as the **cytochrome P450 system.** Administration of oxygen can be lifesaving in the treatment of patients with respiratory or circulatory failure.

FREE ENERGY CHANGES CAN BE EXPRESSED IN TERMS OF REDOX POTENTIAL

In reactions involving oxidation and reduction, the free energy change is proportionate to the tendency of reactants to donate or accept electrons. Thus, in addition to expressing free energy change in terms of $\Delta G^{0\prime}$ (Chapter 11), it is possible, in an analogous manner, to express it numerically as an **oxidation–reduction** or **redox potential** (E'_0). The redox potential of a system (E_0) is usually compared with the potential of the hydrogen electrode (0.0 V at pH 0.0). However, for biologic systems, the redox potential (E'_0) is normally expressed at pH 7.0, at which pH the electrode potential of the hydrogen electrode is −0.42 V. The redox potentials of some redox systems of special interest in mammalian biochemistry are shown in **Table 12–1**. The relative positions of redox systems in the table allow prediction of the direction of flow of electrons from one redox couple to another.

TABLE 12–1 Some Redox Potentials of Special Interest in Mammalian Oxidation Systems

System	E'_0 Volts
H^+/H_2	−0.42
$NAD^+/NADH$	−0.32
Lipoate; ox/red	−0.29
Acetoacetate/3-hydroxybutyrate	−0.27
Pyruvate/lactate	−0.19
Oxaloacetate/malate	−0.17
Fumarate/succinate	+0.03
Cytochrome b; Fe^{3+}/Fe^{2+}	+0.08
Ubiquinone; ox/red	+0.10
Cytochrome c_1; Fe^{3+}/Fe^{2+}	+0.22
Cytochrome a; Fe^{3+}/Fe^{2+}	+0.29
Oxygen/water	+0.82

Enzymes involved in oxidation and reduction are called **oxidoreductases** and are classified into four groups: **oxidases, dehydrogenases, hydroperoxidases,** and **oxygenases.**

OXIDASES USE OXYGEN AS A HYDROGEN ACCEPTOR

Oxidases catalyze the removal of hydrogen from a substrate using oxygen as a hydrogen acceptor.* They form water or hydrogen peroxide as a reaction product (**Figure 12–1**).

Some Oxidases Contain Copper

Cytochrome oxidase is a hemoprotein widely distributed in many tissues, having the typical heme prosthetic group present in myoglobin, hemoglobin, and other cytochromes (Chapter 6). It is the terminal component of the chain of respiratory carriers found in mitochondria (Chapter 13) and transfers electrons resulting from the oxidation of substrate molecules by dehydrogenases to their final acceptor, oxygen. The action of the enzyme is blocked by carbon monoxide, cyanide, and hydrogen sulfide, and this causes poisoning by preventing cellular respiration. It has also been termed "cytochrome a_3." However, it is now known that the heme a_3 is combined with another heme, heme a, in a single protein to form the cytochrome oxidase enzyme complex, and so it is more correctly termed **cytochrome aa_3.** It contains two molecules of heme, each having one Fe atom that oscillates between Fe^{3+}

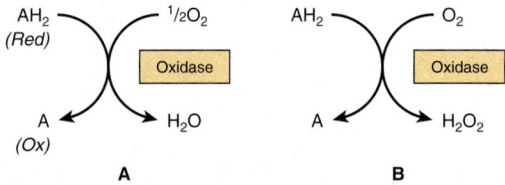

FIGURE 12–1 Oxidation of a metabolite catalyzed by an oxidase (A) forming H_2O and (B) forming H_2O_2.

and Fe^{2+} during oxidation and reduction. Furthermore, two atoms of Cu are present, each associated with a heme unit.

Other Oxidases Are Flavoproteins

Flavoprotein enzymes contain **flavin mononucleotide (FMN)** or **flavin adenine dinucleotide (FAD)** as prosthetic groups. FMN and FAD are formed in the body from the vitamin **riboflavin** (Chapter 44). FMN and FAD are usually tightly—but not covalently—bound to their respective apoenzyme proteins. Metalloflavoproteins contain one or more metals as essential cofactors. Examples of flavoprotein enzymes include **L-amino acid oxidase,** an FMN-linked enzyme found in kidney with general specificity for the oxidative deamination of the naturally occurring L-amino acids; **xanthine oxidase,** which contains molybdenum and plays an important role in the conversion of purine bases to uric acid (Chapter 33), and is of particular significance in uricotelic animals (Chapter 28); and **aldehyde dehydrogenase,** an FAD-linked enzyme present in mammalian livers, which contains molybdenum and nonheme iron and acts upon aldehydes and N-heterocyclic substrates. The mechanisms of oxidation and reduction of these enzymes are complex. Evidence suggests a two-step reaction as shown in **Figure 12–2.**

DEHYDROGENASES CANNOT USE OXYGEN AS A HYDROGEN ACCEPTOR

There are a large number of enzymes in the dehydrogenase class. They perform the following two main functions:

1. Transfer of hydrogen from one substrate to another in a coupled oxidation–reduction reaction (**Figure 12–3**). These dehydrogenases are specific for their substrates but often utilize common coenzymes or hydrogen carriers, eg, NAD^+. Since the reactions are reversible, these properties enable reducing equivalents to be freely transferred within the cell. This type of reaction, which enables one substrate to be oxidized at the expense of another, is particularly useful in enabling oxidative processes to occur in the absence of oxygen, such as during the anaerobic phase of glycolysis (Figure 18–2).

2. Transfer of electrons in the **respiratory chain** of electron transport from substrate to oxygen (Figure 13–3).

---

*The term "oxidase" is sometimes used collectively to denote all enzymes that catalyze reactions involving molecular oxygen.

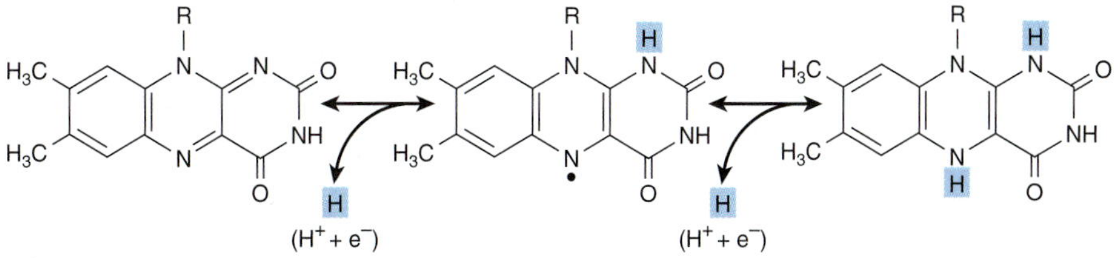

FIGURE 12–2 Oxidoreduction of isoalloxazine ring in flavin nucleotides via a semiquinone (free radical) intermediate (center).

Many Dehydrogenases Depend on Nicotinamide Coenzymes

These dehydrogenases use **nicotinamide adenine dinucleotide (NAD⁺)** or **nicotinamide adenine dinucleotide phosphate (NADP⁺)**—or both—which are formed in the body from the vitamin **niacin** (Chapter 44). The coenzymes are reduced by the specific substrate of the dehydrogenase and reoxidized by a suitable electron acceptor (**Figure 12–4**). They are able to freely and reversibly dissociate from their respective apoenzymes.

Generally, **NAD-linked dehydrogenases** catalyze oxidoreduction reactions in the oxidative pathways of metabolism, particularly in glycolysis (Chapter 18), in the citric acid cycle (Chapter 17), and in the respiratory chain of mitochondria (Chapter 13). NADP-linked dehydrogenases are found characteristically in reductive syntheses, as in the extramitochondrial pathway of fatty acid synthesis (Chapter 23) and steroid synthesis Chapter 26)—and also in the pentose phosphate pathway (Chapter 21).

Other Dehydrogenases Depend on Riboflavin

The **flavin groups** associated with these dehydrogenases are similar to FMN and FAD occurring in oxidases. They are generally more tightly bound to their apoenzymes than are the nicotinamide coenzymes. Most of the **riboflavin-linked dehydrogenases** are concerned with electron transport in (or to) the respiratory chain (Chapter 13). **NADH dehydrogenase** acts as a carrier of electrons between NADH and the components of higher redox potential (Figure 13–3). Other dehydrogenases such as **succinate dehydrogenase, acyl-CoA dehydrogenase,** and **mitochondrial glycerol-3-phosphate dehydrogenase** transfer reducing equivalents directly from the substrate to the respiratory chain (Figure 13–5). Another role of the flavin-dependent dehydrogenases is in the dehydrogenation (by **dihydrolipoyl dehydrogenase**) of reduced lipoate, an intermediate in the oxidative decarboxylation of pyruvate and α-ketoglutarate (Figures 13–5 & 18–5). The **electron-transferring flavoprotein (ETF)** is an intermediary carrier of electrons between acyl-CoA dehydrogenase and the respiratory chain (Figure 13–5).

Cytochromes May Also Be Regarded as Dehydrogenases

The **cytochromes** are iron-containing hemoproteins in which the iron atom oscillates between Fe^{3+} and Fe^{2+} during oxidation and reduction. Except for cytochrome oxidase (previously described), they are classified as dehydrogenases. In the respiratory chain, they are involved as carriers of electrons from flavoproteins on the one hand to cytochrome oxidase on the other (Figure 13–5). Several identifiable cytochromes occur in the respiratory chain, ie, cytochromes b, c_1, c, and cytochrome oxidase. Cytochromes are also found in other locations, eg, the endoplasmic reticulum (cytochromes P450 and b_5), and in plant cells, bacteria, and yeasts.

HYDROPEROXIDASES USE HYDROGEN PEROXIDE OR AN ORGANIC PEROXIDE AS SUBSTRATE

Two type of enzymes found both in animals and plants fall into this category: **peroxidases** and **catalase.**

Hydroperoxidases protect the body against harmful peroxides. Accumulation of peroxides can lead to generation of free radicals, which in turn can disrupt membranes and perhaps cause diseases including cancer and atherosclerosis (see Chapters 15 and 44).

Peroxidases Reduce Peroxides Using Various Electron Acceptors

Peroxidases are found in milk and in leukocytes, platelets, and other tissues involved in eicosanoid metabolism (Chapter 23).

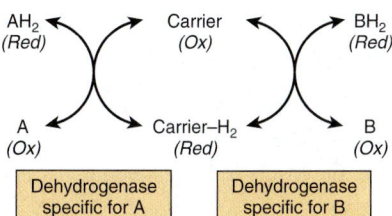

FIGURE 12–3 Oxidation of a metabolite catalyzed by coupled dehydrogenases.

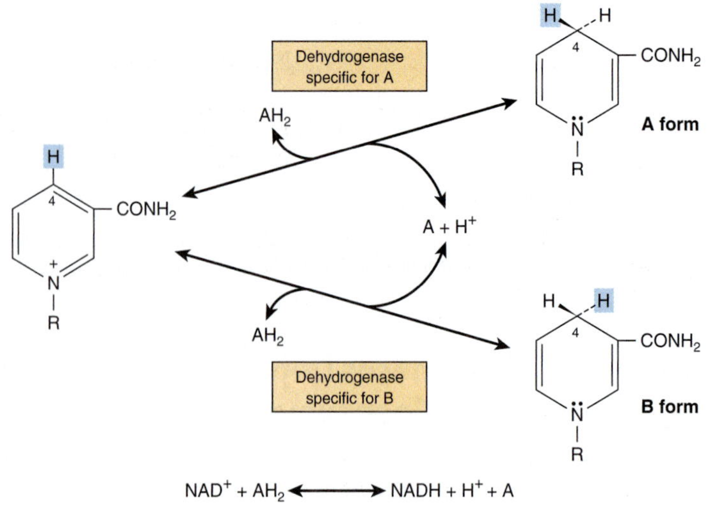

FIGURE 12–4 **Mechanism of oxidation and reduction of nicotinamide coenzymes.** There is stereospecificity about position 4 of nicotinamide when it is reduced by a substrate AH_2. One of the hydrogen atoms is removed from the substrate as a hydrogen nucleus with two electrons (hydride ion, H^-) and is transferred to the 4 position, where it may be attached in either the A or the B form according to the specificity determined by the particular dehydrogenase catalyzing the reaction. The remaining hydrogen of the hydrogen pair removed from the substrate remains free as a hydrogen ion.

The prosthetic group is protoheme. In the reaction catalyzed by peroxidase, hydrogen peroxide is reduced at the expense of several substances that will act as electron acceptors, such as ascorbate, quinones, and cytochrome *c*. The reaction catalyzed by peroxidase is complex, but the overall reaction is as follows:

$$H_2O_2 + AH_2 \xrightarrow{\text{PEROXIDASE}} 2H_2O + A$$

In erythrocytes and other tissues, the enzyme **glutathione peroxidase,** containing **selenium** as a prosthetic group, catalyzes the destruction of H_2O_2 and lipid hydroperoxides through the conversion of reduced glutathione to its oxidized form, protecting membrane lipids and hemoglobin against oxidation by peroxides (Chapter 21).

Catalase Uses Hydrogen Peroxide as Electron Donor & Electron Acceptor

Catalase is a hemoprotein containing four heme groups. In addition to possessing peroxidase activity, it is able to use one molecule of H_2O_2 as a substrate electron donor and another molecule of H_2O_2 as an oxidant or electron acceptor.

$$2H_2O_2 \xrightarrow{\text{CATALASE}} 2H_2O_2 + O_2$$

Under most conditions in vivo, the peroxidase activity of catalase seems to be favored. Catalase is found in blood, bone marrow, mucous membranes, kidney, and liver. It functions to destroy hydrogen peroxide formed by the action of oxidases. **Peroxisomes** are found in many tissues, including liver. They are rich in oxidases and in catalase. Thus, the enzymes that produce H_2O_2 are grouped with the enzyme that breaks it down. However, mitochondrial and microsomal electron transport systems as well as xanthine oxidase must be considered as additional sources of H_2O_2.

OXYGENASES CATALYZE THE DIRECT TRANSFER & INCORPORATION OF OXYGEN INTO A SUBSTRATE MOLECULE

Oxygenases are concerned with the synthesis or degradation of many different types of metabolites. They catalyze the incorporation of oxygen into a substrate molecule in two steps: (1) oxygen is bound to the enzyme at the active site and (2) the bound oxygen is reduced or transferred to the substrate. Oxygenases may be divided into two subgroups, dioxygenases and monooxygenases.

Dioxygenases Incorporate Both Atoms of Molecular Oxygen into the Substrate

The basic reaction catalyzed by dioxygenases is shown below:

$$A + O_2 \rightarrow AO_2$$

Examples include the liver enzymes, **homogentisate dioxygenase** (oxidase) and **3-hydroxyanthranilate dioxygenase** (oxidase), which contain iron; and **L-tryptophan dioxygenase** (tryptophan pyrolase) (Chapter 29), which utilizes heme.

Monooxygenases (Mixed-Function Oxidases, Hydroxylases) Incorporate Only One Atom of Molecular Oxygen into the Substrate

The other oxygen atom is reduced to water, an additional electron donor or cosubstrate (Z) being necessary for this purpose:

$$A{-}H + O_2 + ZH_2 \rightarrow A{-}OH + H_2O + Z$$

$$NAD^+ + AH_2 \longleftrightarrow NADH + H^+ + A$$

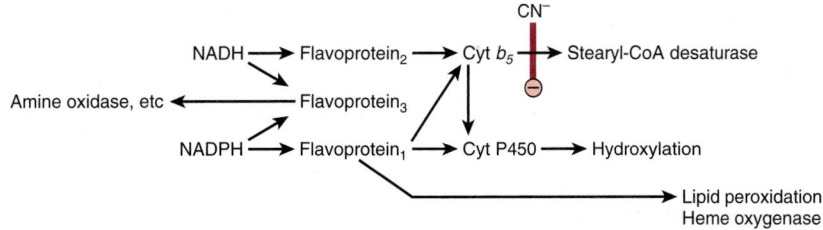

FIGURE 12–5 **Electron transport chain in the endoplasmic reticulum.** Cyanide (CN⁻) inhibits the indicated step.

Cytochromes P450 Are Monooxgenases Important for the Detoxification of Many Drugs & for the Hydroxylation of Steroids

Cytochromes P450 are an important superfamily of heme-containing monooxygenases, and >50 such enzymes have been found in the human genome. These cytochromes are located mainly in the endoplasmic reticulum in the liver and intestine, but are also found in the mitochondria in some tissues. Both NADH and NADPH donate reducing equivalents for the reduction of these cytochromes (**Figure 12–5**), which in turn are oxidized by substrates in a series of enzymatic reactions collectively known as the **hydroxylase cycle** (**Figure 12–6**). In the endoplasmic reticulum of the liver, cytochromes P450 are found together with **cytochrome b_5** and have a major role in drug metabolism and detoxification; they are responsible for

about 75% of the modification and degradation of drugs which occurs in the body. The rate of detoxification of many medicinal drugs by cytochromes P450 determines the duration of their action. Benzpyrene, aminopyrine, aniline, morphine, and benzphetamine are hydroxylated, increasing their solubility and aiding their excretion. Many drugs such as phenobarbital have the ability to induce the synthesis of cytochromes P450.

Mitochondrial cytochrome P450 systems are found in steroidogenic tissues such as adrenal cortex, testis, ovary, and placenta and are concerned with the biosynthesis of steroid hormones from cholesterol (hydroxylation at C_{22} and C_{20} in side-chain cleavage and at the 11β and 18 positions). In addition, renal systems catalyzing 1α- and 24-hydroxylations of 25-hydroxycholecalciferol in vitamin D metabolism—and cholesterol 7α-hydroxylase and sterol 27-hydroxylase involved in bile acid biosynthesis from cholesterol in the liver (Chapter 26)—are P450 enzymes.

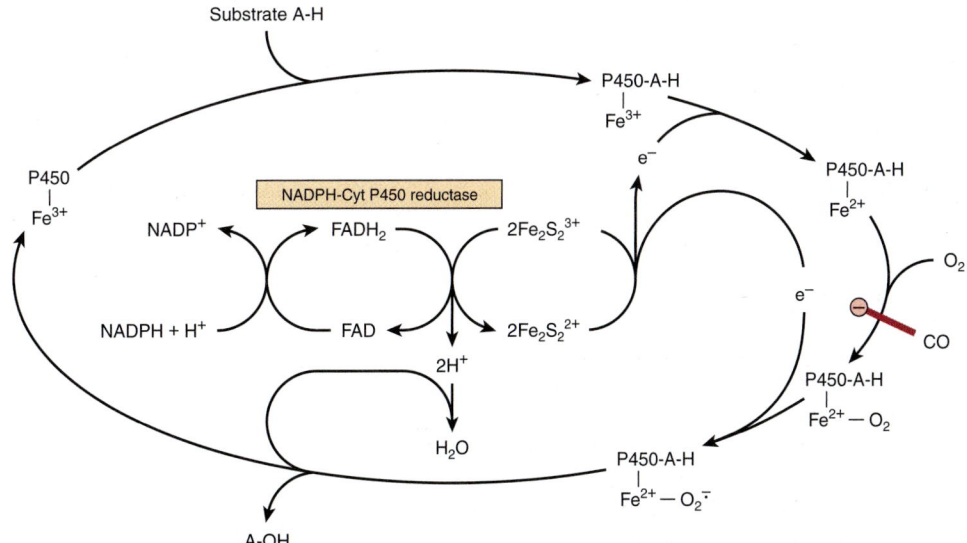

FIGURE 12–6 **Cytochrome P450 hydroxylase cycle.** The system shown is typical of steroid hydroxylases of the adrenal cortex. Liver microsomal cytochrome P450 hydroxylase does not require the iron-sulfur protein Fe_2S_2. Carbon monoxide (CO) inhibits the indicated step.

SUPEROXIDE DISMUTASE PROTECTS AEROBIC ORGANISMS AGAINST OXYGEN TOXICITY

Transfer of a single electron to O_2 generates the potentially damaging **superoxide anion free radical ($O_2^{\cdot-}$),** which gives rise to free-radical chain reactions (Chapter 15), amplifying its destructive effects. The ease with which superoxide can be formed from oxygen in tissues and the occurrence of **superoxide dismutase,** the enzyme responsible for its removal in all aerobic organisms (although not in obligate anaerobes), indicate that the potential toxicity of oxygen is due to its conversion to superoxide.

Superoxide is formed when reduced flavins—present, for example, in xanthine oxidase—are reoxidized univalently by molecular oxygen:

$$Enz - Flavin - H_2 + O_2 \rightarrow Enz - Flavin - H + O_2^{\cdot-} + H^+$$

Superoxide can reduce oxidized cytochrome c

$$O_2^{\cdot-} + Cyt\, c\, (Fe^{3+}) \rightarrow O_2 + Cyt\, c\, (Fe^{2+})$$

or be removed by superoxide dismutase.

In this reaction, superoxide acts as both oxidant and reductant. Thus, superoxide dismutase protects aerobic organisms against the potential deleterious effects of superoxide. The enzyme occurs in all major aerobic tissues in the mitochondria and the cytosol. Although exposure of animals to an atmosphere of 100% oxygen causes an adaptive increase in superoxide dismutase, particularly in the lungs, prolonged exposure leads to lung damage and death. Antioxidants, eg, α-tocopherol (vitamin E), act as scavengers of free radicals and reduce the toxicity of oxygen (Chapter 44).

SUMMARY

- In biologic systems, as in chemical systems, oxidation (loss of electrons) is always accompanied by reduction of an electron acceptor.

- Oxidoreductases have a variety of functions in metabolism; oxidases and dehydrogenases play major roles in respiration; hydroperoxidases protect the body against damage by free radicals; and oxygenases mediate the hydroxylation of drugs and steroids.

- Tissues are protected from oxygen toxicity caused by the superoxide free radical by the specific enzyme superoxide dismutase.

REFERENCES

Babcock GT, Wikstrom M: Oxygen activation and the conservation of energy in cell respiration. Nature 1992;356:301.

Coon MJ: Cytochrome P450: Nature's most versatile biological catalyst. Annu Rev Pharmacol Toxicol 2005;4:1.

Harris DA: *Bioenergetics at a Glance: An Illustrated Introduction.* Blackwell Publishing, 1995.

Johnson F, Giulivi C: Superoxide dismutases and their impact upon human health. Mol Aspects Med 2005;26.

Nicholls DG, Ferguson SJ: *Bioenergetics3.* Academic Press, London 2002.

Raha S, Robinson BH: Mitochondria, oxygen free radicals, disease and aging. Trends Biochem Sci 2000;25:502.

The Respiratory Chain & Oxidative Phosphorylation

Kathleen M. Botham, PhD, DSc & Peter A. Mayes, PhD, DSc

OBJECTIVES

After studying this chapter, you should be able to:

- Describe the double membrane structure of mitochondria and indicate the location of various enzymes.

- Appreciate that energy from the oxidation of fuel substrates (fats, carbohydrates, amino acids) is almost all liberated in mitochondria as reducing equivalents, which are passed by a process termed electron transport through a series of redox carriers or complexes embedded in the inner mitochondrial membrane known as the respiratory chain until they are finally reacted with oxygen to form water.

- Describe the four protein complexes involved in the transfer of electrons through the respiratory chain and explain the roles of flavoproteins, iron sulfur proteins, and coenzyme Q.

- Understand how coenzyme Q accepts electrons from NADH via Complex I and from $FADH_2$ via Complex II.

- Indicate how electrons are passed from reduced coenzyme Q to cytochrome *c* via Complex III in the Q cycle.

- Explain the process by which reduced cytochrome *c* is oxidized and oxygen is reduced to water via Complex IV.

- Understand how electron transport through the respiratory chain generates a proton gradient across the inner mitochondrial membrane, leading to the buildup of a proton motive force that generates ATP by the process of oxidative phosphorylation.

- Describe the structure of the ATP synthase enzyme and explain how it works as a rotary motor to produce ATP from ADP and Pi.

- Identify the five conditions controlling the rate of respiration in mitochondria and understand that oxidation of reducing equivalents via the respiratory chain and oxidative phosphorylation are tightly coupled in most circumstances, so that one cannot proceed unless the other is functioning.

- Indicate examples of common poisons that block respiration or oxidative phosphorylation and identify their site of action.

- Explain, with examples, how uncouplers may act as poisons by dissociating oxidation via the respiratory chain from oxidative phosphorylation, but may also have a physiological role in generating body heat.

- Explain the role of exchange transporters present in the inner mitochondrial membrane in allowing ions and metabolites to pass through while preserving electrochemical and osmotic equilibrium.

BIOMEDICAL IMPORTANCE

Aerobic organisms are able to capture a far greater proportion of the available free energy of respiratory substrates than anaerobic organisms. Most of this takes place inside mitochondria, which have been termed the "powerhouses" of the cell. Respiration is coupled to the generation of the high-energy intermediate, ATP, by **oxidative phosphorylation.** A number of drugs (eg, **amobarbital**) and poisons (eg, **cyanide, carbon monoxide**) inhibit oxidative phosphorylation, usually with fatal consequences. Several inherited defects of mitochondria involving components of the respiratory chain and oxidative phosphorylation have been reported. Patients present with **myopathy** and **encephalopathy** and often have **lactic acidosis.**

SPECIFIC ENZYMES ACT AS MARKERS OF COMPARTMENTS SEPARATED BY THE MITOCHONDRIAL MEMBRANES

Mitochondria have an **outer membrane** that is permeable to most metabolites and an **inner membrane** that is selectively permeable, enclosing a **matrix** within (**Figure 13–1**). The outer membrane is characterized by the presence of various enzymes, including **acyl-CoA synthetase** and **glycerolphosphate acyltransferase. Adenylyl kinase** and **creatine kinase** are found in **the intermembrane space.** The phospholipid **cardiolipin** is concentrated in the inner membrane together with the enzymes of **the respiratory chain, ATP synthase,** and various **membrane transporters.**

THE RESPIRATORY CHAIN OXIDIZES REDUCING EQUIVALENTS & ACTS AS A PROTON PUMP

Most of the energy liberated during the oxidation of carbohydrate, fatty acids, and amino acids is made available within mitochondria as reducing equivalents (—H or electrons) (**Figure 13–2**). Note that the enzymes of the citric acid cycle and β-oxidation (Chapters 22 & 17) are contained in mitochondria, together with the respiratory chain, which collects and transports reducing equivalents, directing them to their final reaction with oxygen to form water, and the machinery for oxidative phosphorylation, the process by which the liberated free energy is trapped as **high-energy phosphate.**

Components of the Respiratory Chain Are Contained in Four Large Protein Complexes Embedded in the Inner Mitochondrial Membrane

Electrons flow through the respiratory chain through a redox span of 1.1 V from NAD⁺/NADH to $O_2/2H_2O$ (Table 12–1), passing through three large protein complexes: **NADH-Q oxidoreductase (Complex I),** where electrons are transferred from NADH to **coenzyme Q (Q)** (also called **ubiquinone); Q-cytochrome *c* oxidoreductase (Complex III),** which passes the electrons on to **cytochrome *c*;** and **cytochrome *c* oxidase (Complex IV),** which completes the chain, passing the electrons to O_2 and causing it to be reduced to H_2O (**Figure 13–3**). Some substrates with more positive redox potentials than NAD⁺/NADH (eg, succinate) pass electrons to Q via a fourth complex, **succinate-Q reductase (Complex II),** rather than Complex I. The four complexes are embedded in the inner mitochondrial membrane, but Q and cytochrome *c* are mobile. Q diffuses rapidly within the membrane, while cytochrome *c* is a soluble protein. The flow of electrons through Complexes I, III, and IV results in the pumping of protons from the matrix across the inner mitochondrial membrane into the intermembrane space (Figure 13–7).

Flavoproteins and Iron–Sulfur Proteins (Fe–S) Are Components of the Respiratory Chain Complexes

Flavoproteins (Chapter 12) are important components of Complexes I and II. The oxidized flavin nucleotide (FMN or FAD) can be reduced in reactions involving the transfer of two electrons (to form FMNH₂ or FADH₂), but they can also

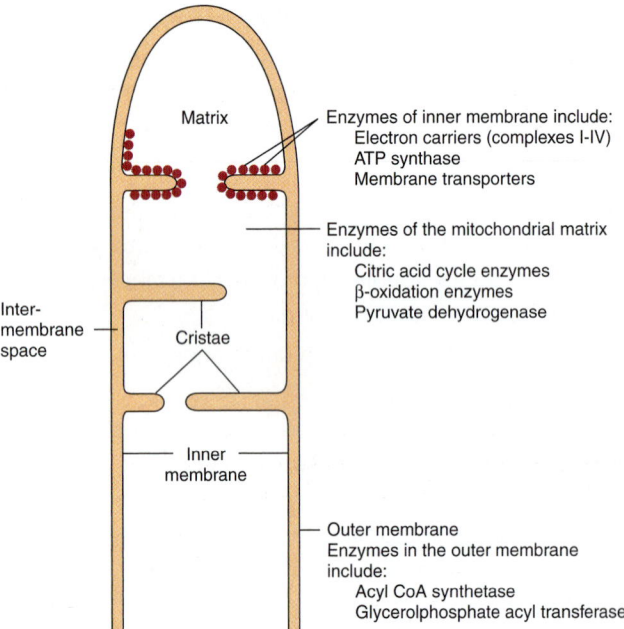

Matrix

Enzymes of inner membrane include:
Electron carriers (complexes I-IV)
ATP synthase
Membrane transporters

Enzymes of the mitochondrial matrix include:
Citric acid cycle enzymes
β-oxidation enzymes
Pyruvate dehydrogenase

Inter-
membrane
space

Cristae

Inner
membrane

Outer membrane
Enzymes in the outer membrane include:
Acyl CoA synthetase
Glycerolphosphate acyl transferase

FIGURE 13–1   Structure of the mitochondrial membranes. Note that the inner membrane contains many folds or cristae.

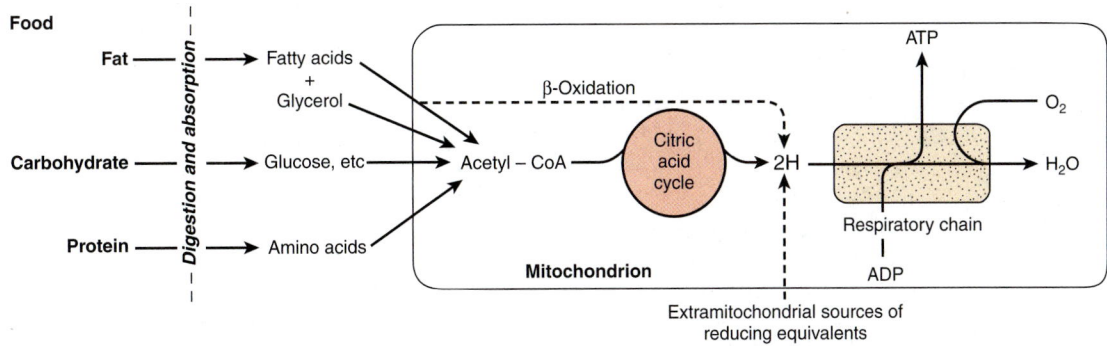

FIGURE 13–2 **Role of the respiratory chain of mitochondria in the conversion of food energy to ATP.** Oxidation of the major foodstuffs leads to the generation of reducing equivalents (2H) that are collected by the respiratory chain for oxidation and coupled generation of ATP.

accept one electron to form the semiquinone (Figure 12–2). **Iron–sulfur proteins (nonheme iron proteins, Fe–S)** are found in Complexes I, II, and III. These may contain one, two, or four Fe atoms linked to inorganic sulfur atoms and/or via cysteine-SH groups to the protein (**Figure 13–4**). The Fe–S take part in single electron transfer reactions in which one Fe atom undergoes oxidoreduction between Fe^{2+} and Fe^{3+}.

Q Accepts Electrons via Complexes I and II

NADH-Q oxidoreductase or Complex I is a large L-shaped multisubunit protein that catalyzes electron transfer from NADH to Q, coupled with the transfer of four H^+ across the membrane:

$$NADH + Q + 5H^+_{matrix} \rightarrow$$
$$NAD + QH_2 + 4H^+_{intermembrane\,space}$$

Electrons are transferred from NADH to FMN initially, then to a series of Fe-S centers, and finally to Q (**Figure 13–5**). In Complex II (succinate-Q reductase), $FADH_2$ is formed during the conversion of succinate to fumarate in the citric acid cycle (Figure 17–3) and electrons are then passed via several Fe–S centers to Q (Figure 13–5). Glycerol-3-phosphate (generated in the breakdown of triacylglycerols or from glycolysis, Figure 18–2) and acyl-CoA also pass electrons to Q via different pathways involving flavoproteins (Figure 13–5).

The Q Cycle Couples Electron Transfer to Proton Transport in Complex III

Electrons are passed from QH_2 to cytochrome c via Complex III (Q-cytochrome **c** oxidoreductase):

$$QH_2 + 2Cyt\,c_{oxidized} + 2H^+_{matrix} \rightarrow$$
$$Q + 2\,Cyt\,c_{reduced} + 4H^+_{intermembrane\,space}$$

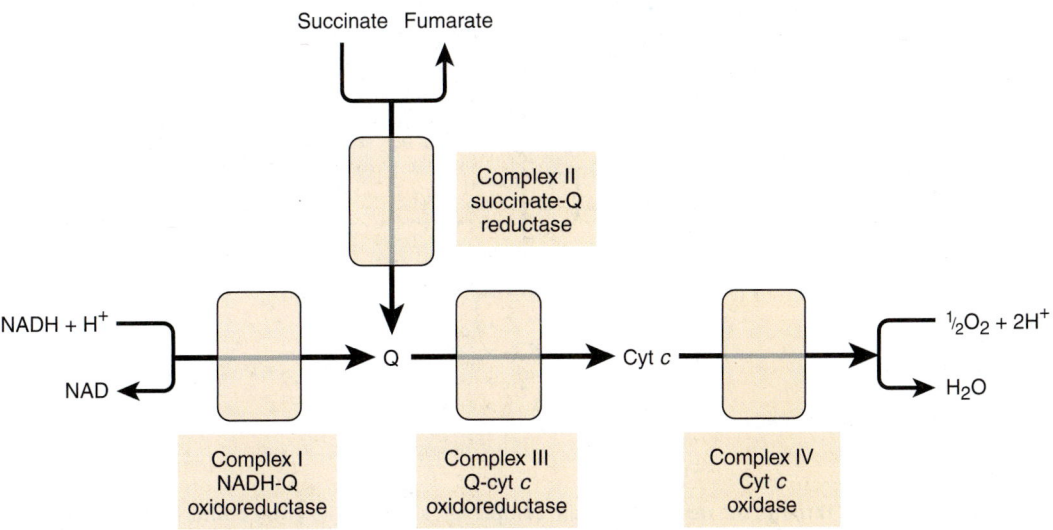

FIGURE 13–3 **Overview of electron flow through the respiratory chain.** (cyt, cytochrome; Q, coenzyme Q or ubiquinone.)

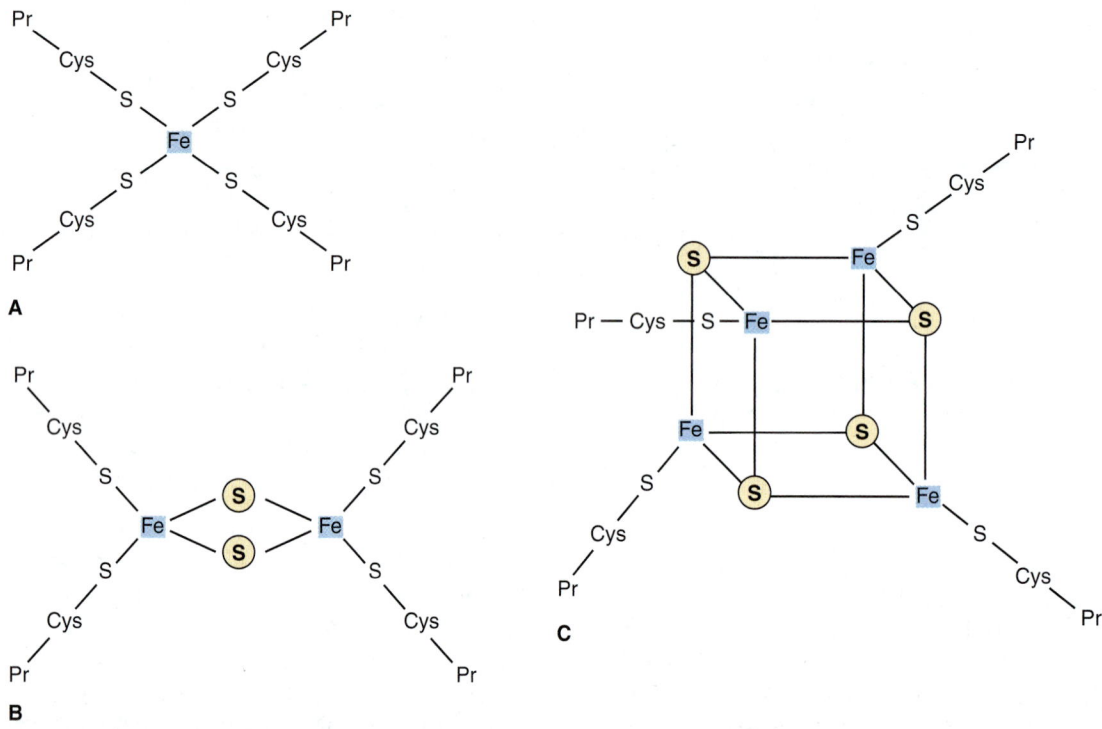

FIGURE 13–4 **Iron–sulfur proteins (Fe–S). (A)** The simplest Fe–S with one Fe bound by four cysteines. **(B)** 2Fe–2S center. (C) 4Fe–4S center. (Cys, cysteine; Pr, apoprotein; Ⓢ, Inorganic sulfur.)

The process is believed to involve **cytochromes c_1, b_L and b_H and a Rieske Fe–S** (an unusual Fe–S in which one of the Fe atoms is linked to two histidine residues rather than two cysteine residues) (Figure 13–5) and is known as the **Q cycle** (**Figure 13–6**). Q may exist in three forms: the oxidized quinone, the reduced quinol, or the semiquinone (Figure 13–6).

The semiquinone is formed transiently during the cycle, one turn of which results in the oxidation of $2QH_2$ to Q, releasing $4H^+$ into the intermembrane space, and the reduction of one Q to QH_2, causing $2H^+$ to be taken up from the matrix (Figure 13–6). Note that while Q carries two electrons, the cytochromes carry only one, thus the oxidation of one QH_2

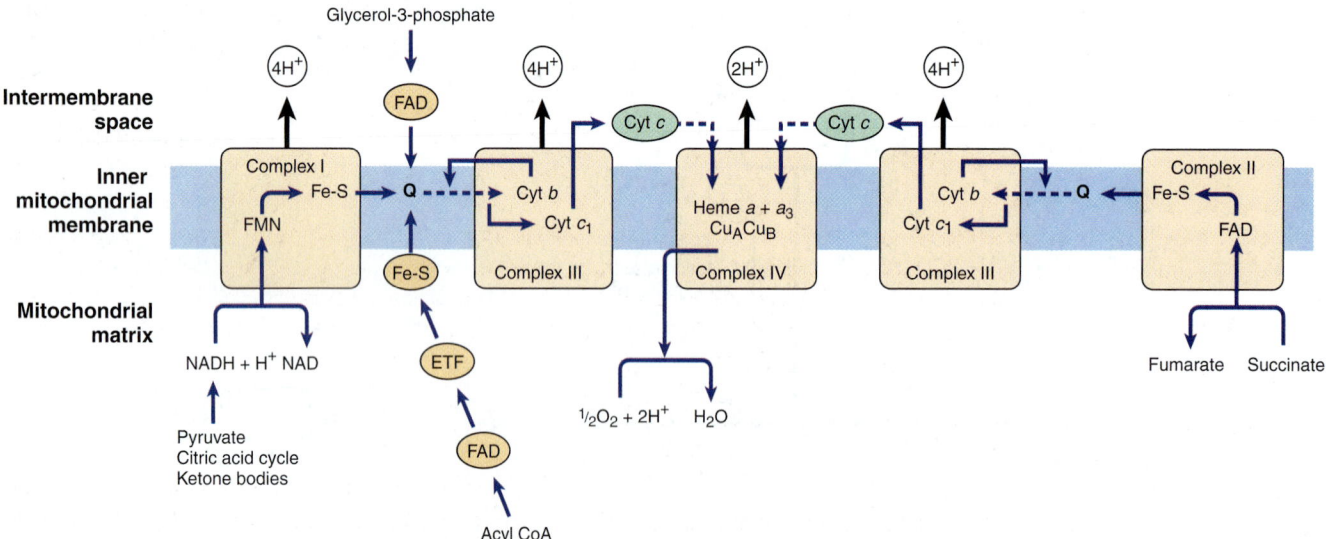

FIGURE 13–5 **Flow of electrons through the respiratory chain complexes, showing the entry points for reducing equivalents from important substrates.** Q and cyt c are mobile components of the system as indicated by the dotted arrows. The flow through Complex III (the Q cycle) is shown in more detail in Figure 13–6. (ETF, electron transferring flavoprotein; Fe–S, iron–sulfur protein; cyt, cytochrome; Q, coenzyme Q or ubiquinone.)

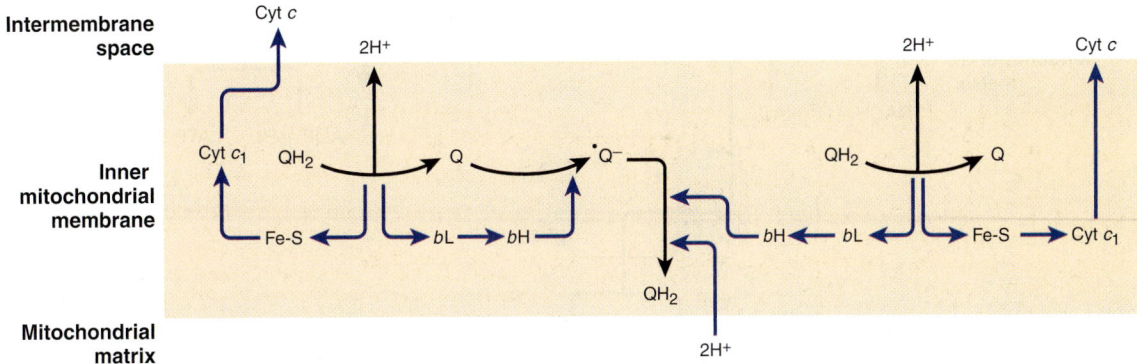

QH₂: Reduced (quinol) form (QH₂)   **Q:** Fully oxidized (quinone) form   **˙Q⁻:** Semiquinone (free radical) form

FIGURE 13–6 **The Q cycle.** During the oxidation of QH_2 to Q, one electron is donated to cyt c via a Rieske Fe–S and cyt c_1 and the second to a Q to form the semiquinone via cyt b_L and cyt b_{H}, with $2H^+$ being released into the intermembrane space. A similar process then occurs with a second QH_2, but in this case the second electron is donated to the semiquinone, reducing it to QH_2, and $2H^+$ are taken up from the matrix. (cyt, cytochrome; Fe–S, iron-sulfur protein; Q, coenzyme Q or ubiquinone.)

is coupled to the reduction of two molecules of cytochrome c via the Q cycle.

Molecular Oxygen Is Reduced to Water via Complex IV

Reduced cytochrome c is oxidized by Complex IV (cytochrome c oxidase), with the concomitant reduction of O_2 to two molecules of water:

$$4Cyt\ c_{reduced} + O_2 + 8H^+_{matrix} \rightarrow$$
$$4Cyt\ c_{oxidized} + 2H_2O + 4H^+_{intermembrane\ space}$$

This transfer of four electrons from cytochrome c to O_2 involves **two heme groups, a and a_3**, and **Cu** (Figure 13–5). Electrons are passed initially to a Cu center (Cu_A), which contains 2Cu atoms linked to two protein cysteine-SH groups (resembling an Fe–S), then in sequence to heme a, heme a_3, a second Cu center, Cu_B, which is linked to heme a_3, and finally to O_2. Of the eight H^+ removed from the matrix, four are used to form two water molecules and four are pumped into the intermembrane space. Thus, for every pair of electrons passing down the chain from NADH or $FADH_2$, $2H^+$ are pumped across the membrane by Complex IV. The O_2 remains tightly bound to Complex IV until it is fully reduced, and this minimizes the release of potentially damaging intermediates such as superoxide anions or peroxide which are formed when O_2 accepts one or two electrons, respectively (Chapter 12).

ELECTRON TRANSPORT VIA THE RESPIRATORY CHAIN CREATES A PROTON GRADIENT WHICH DRIVES THE SYNTHESIS OF ATP

The flow of electrons through the respiratory chain generates ATP by the process of **oxidative phosphorylation. The chemiosmotic theory,** proposed by Peter Mitchell in 1961, postulates that the two processes are coupled by a proton gradient across the inner mitochondrial membrane so that **the proton motive force** caused by the electrochemical potential difference (negative on the matrix side) drives the mechanism of ATP synthesis. As we have seen, Complexes I, III, and IV act as **proton pumps.** Since the inner mitochondrial membrane is impermeable to ions in general and particularly to protons, these accumulate in the intermembrane space, creating the proton motive force predicted by the chemiosmotic theory.

A Membrane-Located ATP Synthase Functions as a Rotary Motor to Form ATP

The proton motive force drives a membrane-located **ATP synthase** that forms ATP in the presence of P_i + ADP. ATP synthase is embedded in the inner membrane, together with the respiratory chain complexes (**Figure 13–7**). Several subunits of the protein form a ball-like shape arranged around an axis known as F_1, which projects into the matrix and contains

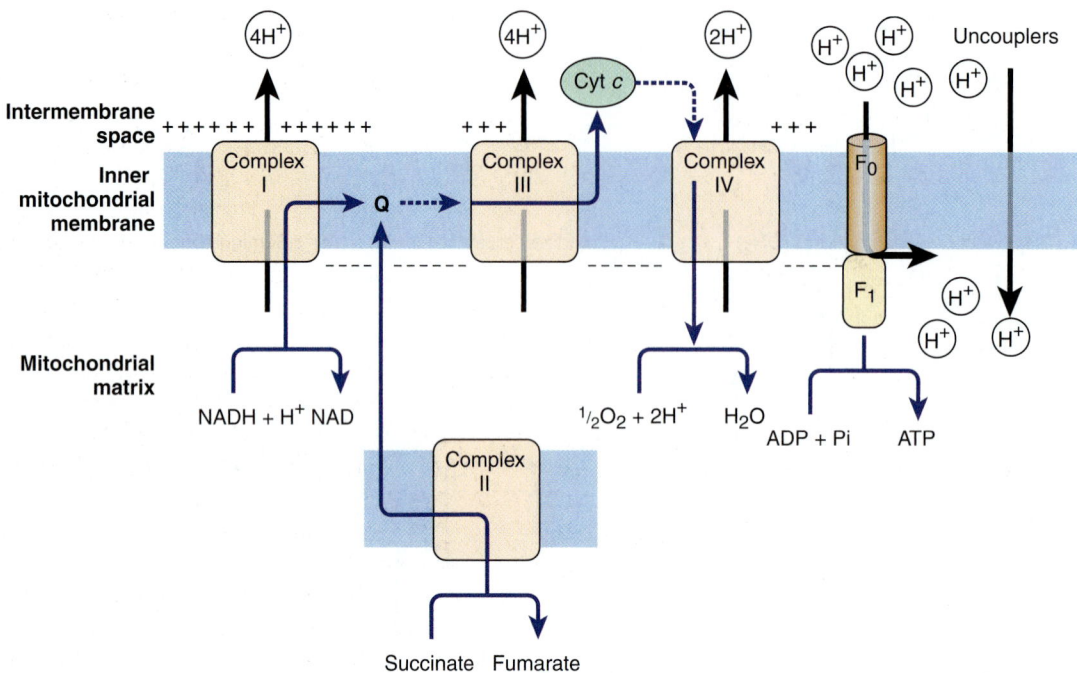

FIGURE 13–7   **The chemiosmotic theory of oxidative phosphorylation.** Complexes I, III, and IV act as proton pumps creating a proton gradient across the membrane, which is negative on the matrix side. The proton motive force generated drives the synthesis of ATP as the protons flow back into the matrix through the ATP synthase enzyme (see Figure 13–8). Uncouplers increase the permeability of the membrane to ions, collapsing the proton gradient by allowing the H^+ to pass across without going through the ATP synthase, and thus uncouple electron flow through the respiratory complexes from ATP synthesis. (cyt, cytochrome; Q, coenzyme Q or ubiquinone.)

the phosphorylation mechanism (**Figure 13–8**). F_1 is attached to a membrane protein complex known as F_0, which also consists of several protein subunits. F_0 spans the membrane and forms a proton channel. The flow of protons through F_0

causes it to rotate, driving the production of ATP in the F_1 complex (Figures 13–7 and 13–8). This is thought to occur by a **binding change mechanism** in which the conformation of the β-subunits in F_1 is changed as the axis rotates from one that

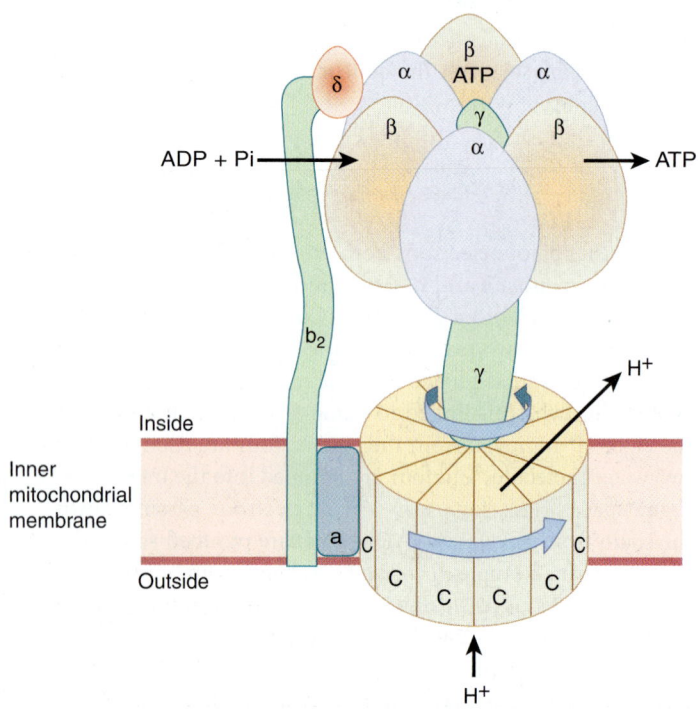

FIGURE 13–8   **Mechanism of ATP production by ATP synthase.** The enzyme complex consists of an F_0 subcomplex which is a disk of "C" protein subunits. Attached is a γ subunit in the form of a "bent axle." Protons passing through the disk of "C" units cause it and the attached γ subunit to rotate. The γ subunit fits inside the F_1 subcomplex of three α and three β subunits, which are fixed to the membrane and do not rotate. ADP and P_i are taken up sequentially by the β subunits to form ATP, which is expelled as the rotating γ subunit squeezes each β subunit in turn and changes its conformation. Thus, three ATP molecules are generated per revolution. For clarity, not all the subunits that have been identified are shown—eg, the "axle" also contains an ε subunit.

binds ATP tightly to one that releases ATP and binds ADP and P_i so that the next ATP can be formed. Estimates suggest that for each NADH oxidized, Complexes I and III translocate four protons each and Complex IV translocates two.

THE RESPIRATORY CHAIN PROVIDES MOST OF THE ENERGY CAPTURED DURING CATABOLISM

ADP captures, in the form of high-energy phosphate, a significant proportion of the free energy released by catabolic processes. The resulting ATP has been called the **energy "currency"** of the cell because it passes on this free energy to drive those processes requiring energy (Figure 11–6).

There is a net direct capture of two high-energy phosphate groups in the glycolytic reactions (Table 18–1). Two more high-energy phosphates per mole of glucose are captured in the citric acid cycle during the conversion of succinyl CoA to succinate. All of these phosphorylations occur at the **substrate level**. For each mol of substrate oxidized via Complexes I, III, and IV in the respiratory chain (ie, via NADH), 2.5 mol of ATP are formed per 0.5 mol of O_2 consumed; ie, the P:O ratio = 2.5 (Figure 13–7). On the other hand, when 1 mol of substrate (eg, succinate or 3-phophoglycerate) is oxidized via Complexes II, III, and IV, only 1.5 mol of ATP are formed; ie, P:O = 1.5. These reactions are known as **oxidative phosphorylation at the respiratory chain level**. Taking these values into account, it can be estimated that nearly 90% of the high-energy phosphates produced from the complete oxidation of 1 mol glucose is obtained via oxidative phosphorylation coupled to the respiratory chain (Table 18–1).

Respiratory Control Ensures a Constant Supply of ATP

The rate of respiration of mitochondria can be controlled by the availability of ADP. This is because oxidation and phosphorylation are **tightly coupled;** ie, oxidation cannot proceed via the respiratory chain without concomitant phosphorylation of ADP. **Table 13–1** shows the five conditions controlling the rate

TABLE 13–1 States of Respiratory Control

	Conditions Limiting the Rate of Respiration
State 1	Availability of ADP and substrate
State 2	Availability of substrate only
State 3	The capacity of the respiratory chain itself, when all substrates and components are present in saturating amounts
State 4	Availability of ADP only
State 5	Availability of oxygen only

of respiration in mitochondria. Most cells in the resting state are in **state 4,** and respiration is controlled by the availability of ADP. When work is performed, ATP is converted to ADP, allowing more respiration to occur, which in turn replenishes the store of ATP. Under certain conditions, the concentration of inorganic phosphate can also affect the rate of functioning of the respiratory chain. As respiration increases (as in exercise), the cell approaches **state 3 or 5** when either the capacity of the respiratory chain becomes saturated or the PO_2 decreases below the K_m for heme a_3. There is also the possibility that the ADP/ATP transporter, which facilitates entry of cytosolic ADP into and ATP out of the mitochondrion, becomes rate limiting.

Thus, the manner in which biologic oxidative processes allow the free energy resulting from the oxidation of foodstuffs to become available and to be captured is stepwise, efficient, and controlled—rather than explosive, inefficient, and uncontrolled, as in many nonbiologic processes. The remaining free energy that is not captured as high-energy phosphate is liberated as **heat**. This need not to be considered "wasted" since it ensures that the respiratory system as a whole is sufficiently exergonic to be removed from equilibrium, allowing continuous unidirectional flow and constant provision of ATP. It also contributes to maintenance of body temperature.

MANY POISONS INHIBIT THE RESPIRATORY CHAIN

Much information about the respiratory chain has been obtained by the use of inhibitors, and, conversely, this has provided knowledge about the mechanism of action of several poisons (**Figure 13–9**). They may be classified as inhibitors of the respiratory chain, inhibitors of oxidative phosphorylation, and uncouplers of oxidative phosphorylation.

Barbiturates such as amobarbital inhibit electron transport via Complex I by blocking the transfer from Fe–S to Q. At sufficient dosage, they are fatal in vivo. **Antimycin A** and **dimercaprol** inhibit the respiratory chain at Complex III. The classic poisons H_2S, **carbon monoxide,** and **cyanide** inhibit Complex IV and can therefore totally arrest respiration. **Malonate** is a competitive inhibitor of Complex II.

Atractyloside inhibits oxidative phosphorylation by inhibiting the transporter of ADP into and ATP out of the mitochondrion (**Figure 13–10**). The antibiotic **oligomycin** completely blocks oxidation and phosphorylation by blocking the flow of protons through ATP synthase (Figure 13–9).

Uncouplers dissociate oxidation in the respiratory chain from phosphorylation (Figure 13–7). These compounds are toxic in vivo, causing respiration to become uncontrolled, since the rate is no longer limited by the concentration of ADP or P_i. The uncoupler that has been used most frequently is **2,4-dinitrophenol,** but other compounds act in a similar manner. **Thermogenin (or the uncoupling protein)** is a physiological uncoupler found in brown adipose tissue that functions to generate body heat, particularly for the newborn and during hibernation in animals (Chapter 25).

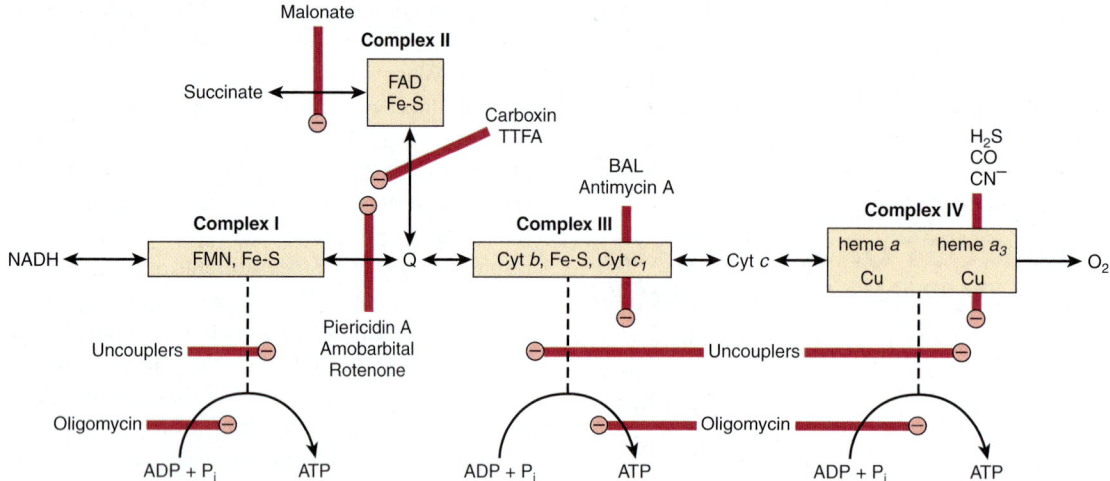

FIGURE 13–9   Sites of inhibition (⊖) of the respiratory chain by specific drugs, chemicals, and antibiotics. (BAL, dimercaprol; TTFA, an Fe-chelating agent. Other abbreviations as in Figure 13–5.)

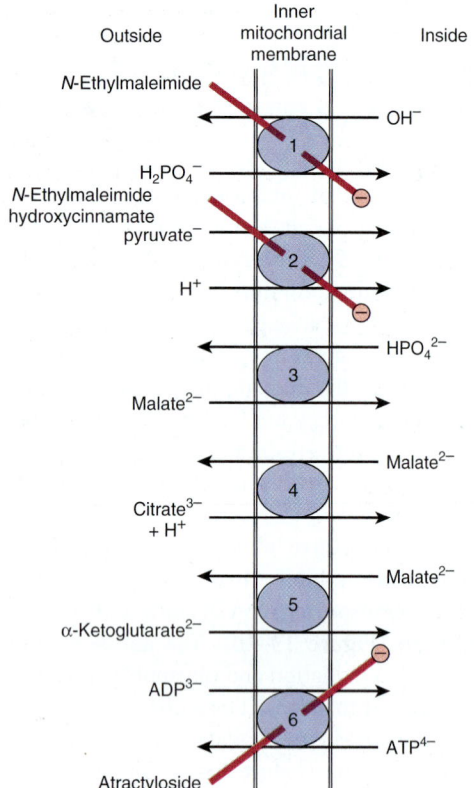

FIGURE 13–10   **Transporter systems in the inner mitochondrial membrane.** ① Phosphate transporter, ② pyruvate symport, ③ dicarboxylate transporter, ④ tricarboxylate transporter, ⑤ α-ketoglutarate transporter, ⑥ adenine nucleotide transporter. N-Ethylmaleimide, hydroxycinnamate, and atractyloside inhibit (⊖) the indicated systems. Also present (but not shown) are transporter systems for glutamate/aspartate (Figure 13–13), glutamine, ornithine, neutral amino acids, and carnitine (Figure 22–1).

THE CHEMIOSMOTIC THEORY CAN ACCOUNT FOR RESPIRATORY CONTROL AND THE ACTION OF UNCOUPLERS

The electrochemical potential difference across the membrane, once established as a result of proton translocation, inhibits further transport of reducing equivalents through the respiratory chain unless discharged by back-translocation of protons across the membrane through the ATP synthase. This in turn depends on availability of ADP and P_i.

Uncouplers (eg, dinitrophenol) are amphipathic (Chapter 15) and increase the permeability of the lipoid inner mitochondrial membrane to protons, thus reducing the electrochemical potential and short-circuiting the ATP synthase (Figure 13–7). In this way, oxidation can proceed without phosphorylation.

THE RELATIVE IMPERMEABILITY OF THE INNER MITOCHONDRIAL MEMBRANE NECESSITATES EXCHANGE TRANSPORTERS

Exchange diffusion systems involving transporter proteins that span the membrane are present in the membrane for exchange of anions against OH^- ions and cations against H^+ ions. Such systems are necessary for uptake and output of ionized metabolites while preserving electrical and osmotic equilibrium. The inner mitochondrial membrane is freely permeable to uncharged small molecules, such as oxygen, water, CO_2, NH_3, and to monocarboxylic acids, such as 3-hydroxybutyric, acetoacetic, and acetic. Long-chain fatty acids are transported into mitochondria via the carnitine system (Figure 22–1), and there is also a special carrier for pyruvate involving a symport that utilizes the H⁺ gradient from outside to inside the mitochondrion (Figure 13–10). However, dicarboxylate and tricarboxylate anions and amino acids require specific transporter or carrier systems to facilitate their passage across the membrane. Monocarboxylic acids penetrate more readily in their undissociated, more lipid-soluble form.

The transport of di- and tricarboxylate anions is closely linked to that of inorganic phosphate, which penetrates readily as the $H_2PO_4^-$ ion in exchange for OH⁻. The net uptake of malate by the dicarboxylate transporter requires inorganic phosphate for exchange in the opposite direction. The net uptake of citrate, isocitrate, or *cis*-aconitate by the tricarboxylate transporter requires malate in exchange. α-Ketoglutarate transport also requires an exchange with malate. The adenine nucleotide transporter allows the exchange of ATP and ADP but not AMP. It is vital in allowing ATP exit from mitochondria to the sites of extramitochondrial utilization and in allowing the return of ADP for ATP production within the mitochondrion (**Figure 13–11**). Since in this translocation four negative charges are removed from the matrix for every three taken in, the electrochemical gradient across the membrane (the proton motive force) favors the export of ATP. Na⁺ can be exchanged for H⁺, driven by the proton gradient. It is believed that active uptake of Ca²⁺ by mitochondria occurs with a net charge transfer of 1 (Ca⁺ uniport), possibly through a Ca^{2+}/H^+ antiport. Calcium release from mitochondria is facilitated by exchange with Na⁺.

Ionophores Permit Specific Cations to Penetrate Membranes

Ionophores are lipophilic molecules that complex specific cations and facilitate their transport through biologic membranes,

eg, **valinomycin** (K⁺). The classic uncouplers such as dinitrophenol are, in fact, proton ionophores.

A Proton-Translocating Transhydrogenase Is a Source of Intramitochondrial NADPH

Energy-linked transhydrogenase, a protein in the inner mitochondrial membrane, couples the passage of protons down the electrochemical gradient from outside to inside the mitochondrion with the transfer of H from intramitochondrial NADH to NADPH for intramitochondrial enzymes such as glutamate dehydrogenase and hydroxylases involved in steroid synthesis.

Oxidation of Extramitochondrial NADH Is Mediated by Substrate Shuttles

NADH cannot penetrate the mitochondrial membrane, but it is produced continuously in the cytosol by 3-phosphoglyceraldehyde dehydrogenase, an enzyme in the glycolysis sequence (Figure 18–2). However, under aerobic conditions, extramitochondrial NADH does not accumulate and is presumed to be oxidized by the respiratory chain in mitochondria. The transfer of reducing equivalents through the mitochondrial membrane requires substrate pairs, linked by suitable dehydrogenases on each side of the mitochondrial membrane. The mechanism of transfer using the **glycerophosphate shuttle** is shown in **Figure 13–12**. Since the mitochondrial enzyme is linked to the respiratory chain via a flavoprotein rather than NAD, only 1.5 mol rather than 2.5 mol of ATP are formed per atom of oxygen consumed. Although this shuttle is present in some tissues (eg, brain, white muscle), in others (eg, heart muscle) it is deficient. It is therefore believed that the **malate shuttle** system (**Figure 13–13**) is of more universal utility. The complexity of this system is due to the impermeability of the mitochondrial membrane to oxaloacetate, which must react with glutamate to form aspartate and α-ketoglutarate by transamination before transport through the mitochondrial membrane and reconstitution to oxaloacetate in the cytosol.

Ion Transport in Mitochondria Is Energy Linked

Mitochondria maintain or accumulate cations such as K⁺, Na⁺, Ca²⁺, and Mg²⁺, and P_i. It is assumed that a primary proton pump drives cation exchange.

The Creatine Phosphate Shuttle Facilitates Transport of High-Energy Phosphate from Mitochondria

This shuttle (**Figure 13–14**) augments the functions of **creatine phosphate** as an energy buffer by acting as a dynamic system for transfer of high-energy phosphate from mitochondria in

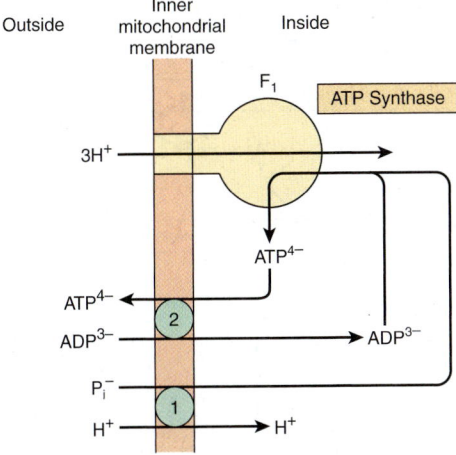

FIGURE 13–11 Combination of phosphate transporter ① with the adenine nucleotide transporter ② in ATP synthesis. The H⁺/P_i symport shown is equivalent to the P_i/OH⁻ antiport shown in Figure 13–10.

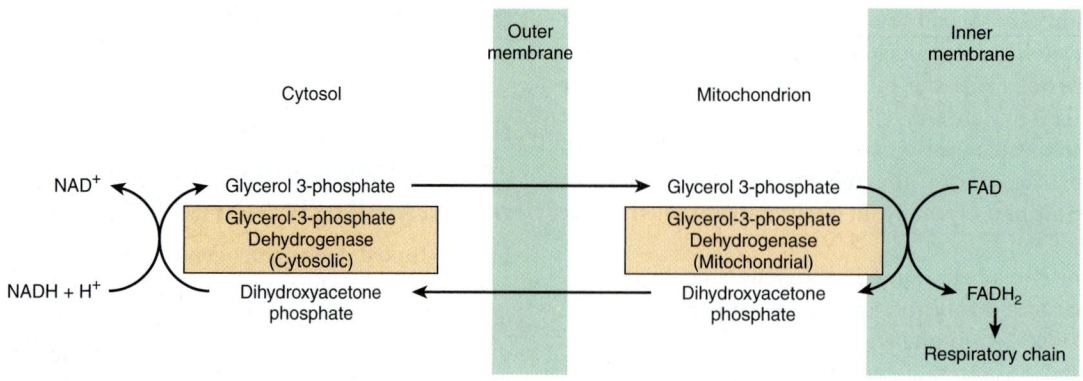

FIGURE 13–12 **Glycerophosphate shuttle for transfer of reducing equivalents from the cytosol into the mitochondrion.**

active tissues such as heart and skeletal muscle. An isoenzyme of **creatine kinase** (CK_m) is found in the mitochondrial intermembrane space, catalyzing the transfer of high-energy phosphate to creatine from ATP emerging from the adenine nucleotide transporter. In turn, the creatine phosphate is transported into the cytosol via protein pores in the outer mitochondrial membrane, becoming available for generation of extramitochondrial ATP.

CLINICAL ASPECTS

The condition known as **fatal infantile mitochondrial myopathy and renal dysfunction** involves severe diminution or absence of most oxidoreductases of the respiratory chain. **MELAS** (mitochondrial encephalopathy, lactic acidosis, and stroke) is an inherited condition due to NADH-Q oxidoreductase (Complex I) or cytochrome oxidase (Complex IV) deficiency. It is caused by a mutation in mitochondrial DNA and may be involved in

Alzheimer's disease and **diabetes mellitus.** A number of drugs and poisons act by inhibition of oxidative phosphorylation.

SUMMARY

- Virtually all energy released from the oxidation of carbohydrate, fat, and protein is made available in mitochondria as reducing equivalents (—H or e^-). These are funneled into the respiratory chain, where they are passed down a redox gradient of carriers to their final reaction with oxygen to form water.

- The redox carriers are grouped into four respiratory chain complexes in the inner mitochondrial membrane. Three of the four complexes are able to use the energy released in the redox gradient to pump protons to the outside of the membrane, creating an electrochemical potential between the matrix and the inner membrane space.

- ATP synthase spans the membrane and acts like a rotary motor using the potential energy of the proton gradient or

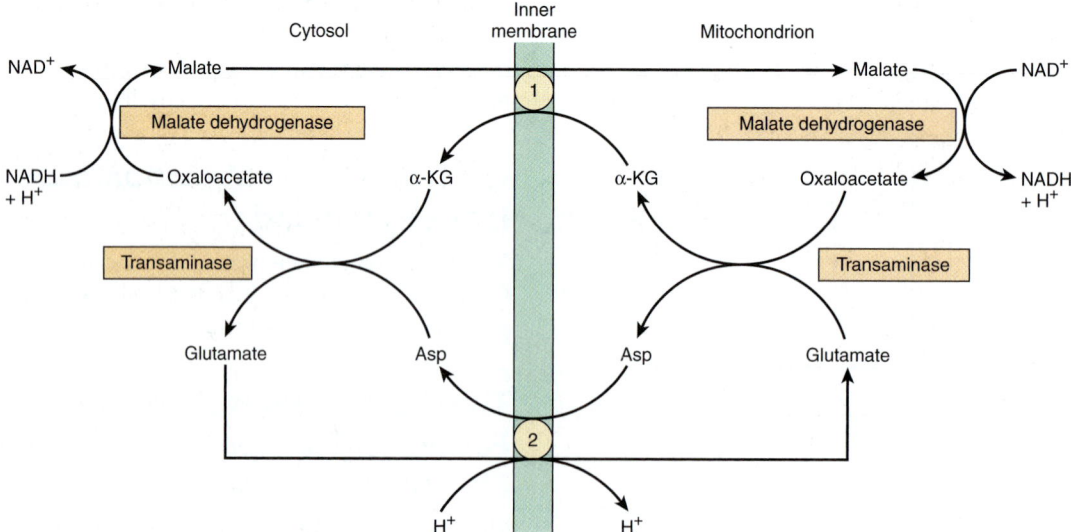

FIGURE 13–13 **Malate shuttle for transfer of reducing equivalents from the cytosol into the mitochondrion.**
① α-Ketoglutarate transporter and ② glutamate/aspartate transporter (note the proton symport with glutamate).

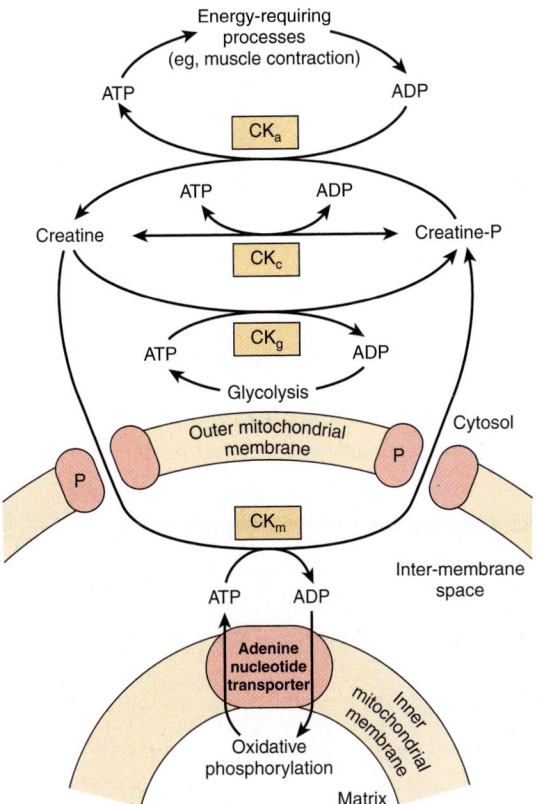

FIGURE 13–14 **The creatine phosphate shuttle of heart and skeletal muscle.** The shuttle allows rapid transport of high-energy phosphate from the mitochondrial matrix into the cytosol. (CK_a, creatine kinase concerned with large requirements for ATP, eg, muscular contraction; CK_c, creatine kinase for maintaining equilibrium between creatine and creatine phosphate and ATP/ADP; CK_g, creatine kinase coupling glycolysis to creatine phosphate synthesis; CK_m, mitochondrial creatine kinase mediating creatine phosphate production from ATP formed in oxidative phosphorylation; P, pore protein in outer mitochondrial membrane.)

proton motive force to synthesize ATP from ADP and P_i. In this way, oxidation is closely coupled to phosphorylation to meet the energy needs of the cell.

■ Since the inner mitochondrial membrane is impermeable to protons and other ions, special exchange transporters span the membrane to allow ions such as OH^-, ATP^{4-}, ADP^{3-}, and metabolites to pass through without discharging the electrochemical gradient across the membrane.

■ Many well-known poisons such as cyanide arrest respiration by inhibition of the respiratory chain.

REFERENCES

Hinkle PC: P/O ratios of mitochondrial oxidative phosphorylation. Biochem Biophys Acta 2005;1706:1.

Kocherginsky N: Acidic lipids, H(+)-ATPases, and mechanism of oxidative phosphorylation. Physico-chemical ideas 30 years after P. Mitchell's Nobel Prize award. Prog Biophys Mol Biol 2009;99:20.

Mitchell P: Keilin's respiratory chain concept and its chemiosmotic consequences. Science 1979;206:1148.

Nakamoto RK, Baylis Scanlon JA, Al-Shawi MK: The rotary mechanism of the ATP synthase. Arch Biochem Biophys 2008;476:43.

Smeitink J, van den Heuvel L, DiMauro S: The genetics and pathology of oxidative phosphorylation. Nat Rev Genet 2001;2:342.

Tyler DD: *The Mitochondrion in Health and Disease.* VCH Publishers, 1992.

Wallace DC: Mitochondrial DNA in aging and disease. Sci Am 1997;277:22.

Yoshida M, Muneyuki E, Hisabori T: ATP synthase—a marvelous rotary engine of the cell. Nat Rev Mol Cell Biol 2001;2:669.

Carbohydrates of Physiologic Significance

David A. Bender, PhD & Peter A. Mayes, PhD, DSc

O B J E C T I V E S

After studying this chapter, you should be able to:

- Explain what is meant by the terms monosaccharide, disaccharide, oligosaccharide and polysaccharide.
- Explain the different ways in which the structures of glucose and other monosaccharides can be represented, and describe the various types of isomerism of sugars and the pyranose and furanose ring structures.
- Describe the formation of glycosides and the structures of the important disaccharides and polysaccharides.
- Explain what is meant by the glycemic index of a carbohydrate.
- Describe the roles of carbohydrates in cell membranes and lipoproteins.

BIOMEDICAL IMPORTANCE

Carbohydrates are widely distributed in plants and animals; they have important structural and metabolic roles. In plants, glucose is synthesized from carbon dioxide and water by photosynthesis and stored as starch or used to synthesize the cellulose of the plant cell walls. Animals can synthesize carbohydrates from amino acids, but most are derived ultimately from plants. **Glucose** is the most important carbohydrate; most dietary carbohydrate is absorbed into the bloodstream as glucose formed by hydrolysis of dietary starch and disaccharides, and other sugars are converted to glucose in the liver. Glucose is the major metabolic fuel of mammals (except ruminants) and a universal fuel of the fetus. It is the precursor for synthesis of all the other carbohydrates in the body, including **glycogen** for storage; **ribose** and **deoxyribose** in nucleic acids; **galactose** for synthesis of lactose in milk, in glycolipids, and in combination with protein in glycoproteins and proteoglycans. Diseases associated with carbohydrate metabolism include **diabetes mellitus, galactosemia, glycogen storage diseases,** and **lactose intolerance.**

CARBOHYDRATES ARE ALDEHYDE OR KETONE DERIVATIVES OF POLYHYDRIC ALCOHOLS

Carbohydrates are classified as follows:

1. Monosaccharides are those sugars that cannot be hydrolyzed into simpler carbohydrates. They may be classified as **trioses, tetroses, pentoses, hexoses,** or **heptoses,** depending upon the number of carbon atoms (3–7), and as **aldoses** or **ketoses,** depending upon whether they have an aldehyde or ketone group. Examples are listed in **Table 14–1.** In addition to aldehydes and ketones, the polyhydric alcohols (sugar alcohols or **polyols**), in which the aldehyde or ketone group has been reduced to an alcohol group, also occur naturally in foods. They are synthesized by reduction of monosaccharides for use in the manufacture of foods for weight reduction and for diabetics. They are poorly absorbed, and have about half the energy yield of sugars.

2. Disaccharides are condensation products of two monosaccharide units, for example, lactose, maltose, sucrose, and trehalose.

3. Oligosaccharides are condensation products of three to ten monosaccharides. Most are not digested by human enzymes.

4. Polysaccharides are condensation products of more than ten monosaccharide units; examples are the starches and dextrins, which may be linear or branched polymers. Polysaccharides are sometimes classified as hexosans or pentosans, depending on the identity of the constituent monosaccharides (hexoses and pentoses, respectively). In addition to starches and dextrins, foods contain a wide variety of other polysaccharides that are collectively known as nonstarch polysaccharides; they are not digested by human enzymes, and are the major component of dietary fiber. Examples are cellulose from plant cell walls (a glucose polymer) and inulin, the storage carbohydrate in some plants (a fructose polymer).

TABLE 14–1 **Classification of Important Sugars**

	Aldoses	Ketoses
Trioses ($C_3H_6O_3$)	Glycerose (glyceraldehyde)	Dihydroxyacetone
Tetroses ($C_4H_8O_4$)	Erythrose	Erythrulose
Pentoses ($C_5H_{10}O_5$)	Ribose	Ribulose
Hexoses ($C_6H_{12}O_6$)	Glucose	Fructose
Heptoses ($C_7H_{14}O_7$)	—	Sedoheptulose

BIOMEDICALLY, GLUCOSE IS THE MOST IMPORTANT MONOSACCHARIDE

The Structure of Glucose Can Be Represented in Three Ways

The straight-chain structural formula (aldohexose; **Figure 14–1A**) can account for some of the properties of glucose, but a cyclic structure (a **hemiacetal** formed by reaction between the aldehyde group and a hydroxyl group) is thermodynamically favored and accounts for other properties. The cyclic structure is normally drawn as shown in **Figure 14–1B**, the Haworth projection, in which the molecule is viewed from the side and above the plane of the ring; the bonds nearest to the viewer are bold and thickened, and the hydroxyl groups are above or below the plane of the ring. The six-membered ring containing one oxygen atom is actually in the form of a chair (**Figure 14–1C**).

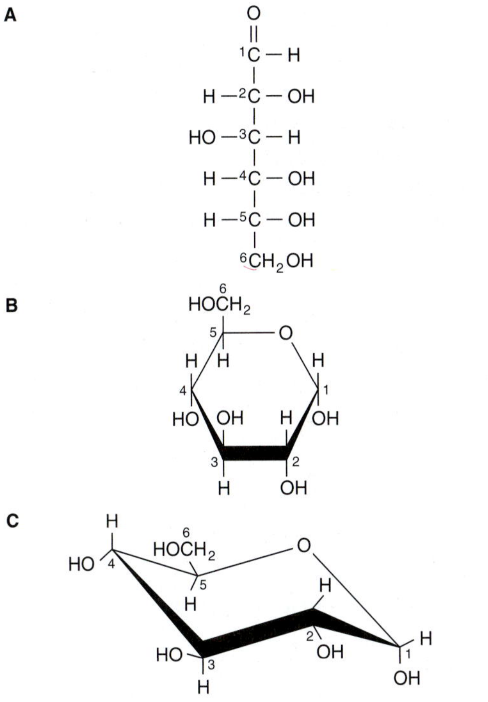

Sugars Exhibit Various Forms of Isomerism

Glucose, with four asymmetric carbon atoms, can form 16 isomers. The more important types of isomerism found with glucose are as follows.

1. D and L isomerism: The designation of a sugar isomer as the D form or of its mirror image as the L form is determined by its spatial relationship to the parent compound of the carbohydrates, the three-carbon sugar glycerose (glyceraldehyde). The L and D forms of this sugar, and of glucose, are shown in **Figure 14–2**. The orientation of the —H and —OH groups around the carbon atom adjacent to the terminal primary alcohol carbon (carbon 5 in glucose) determines whether the sugar belongs to the D or L series. When the —OH group on this carbon is on the right (as seen in Figure 14–2), the sugar is the D isomer; when it is on the left, it is the L isomer. Most of the naturally occurring monosaccharides are D sugars, and the enzymes responsible for their metabolism are specific for this configuration.

The presence of asymmetric carbon atoms also confers **optical activity** on the compound. When a beam of plane-polarized light is passed through a solution of an **optical isomer,** it rotates either to the right, dextrorotatory (+), or to the left, levorotatory (−). The direction of rotation of polarized light is independent of the stereochemistry of the sugar, so it may be designated D(−), D(+), L(−), or L(+). For example, the naturally occurring form of fructose is the D(−) isomer. In solution, glucose is dextrorotatory, and glucose solutions are sometimes known as **dextrose.**

2. Pyranose and furanose ring structures: The ring structures of monosaccharides are similar to the ring structures of

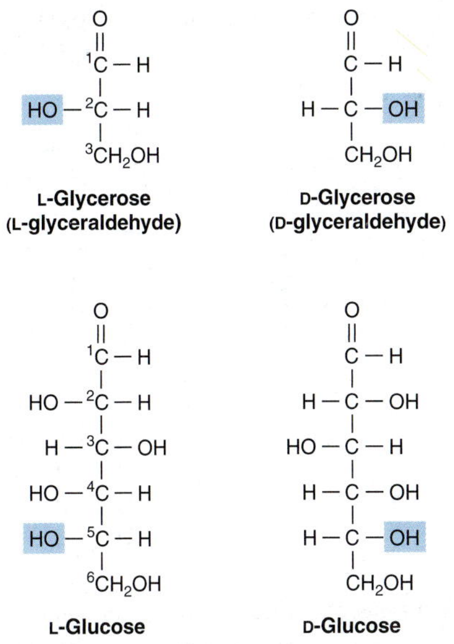

FIGURE 14–1 D-Glucose. **(A)** Straight-chain form. **(B)** α-D-glucose; Haworth projection. **(C)** α-D-glucose; chair form.

FIGURE 14–2 D- and L-isomerism of glycerose and glucose.

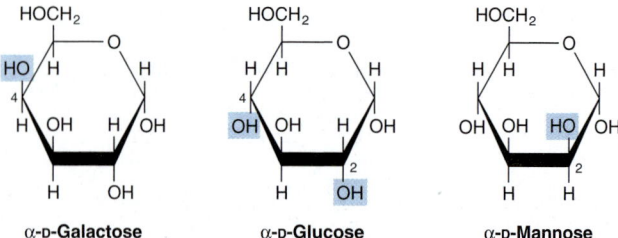

FIGURE 14–5 Epimers of glucose.

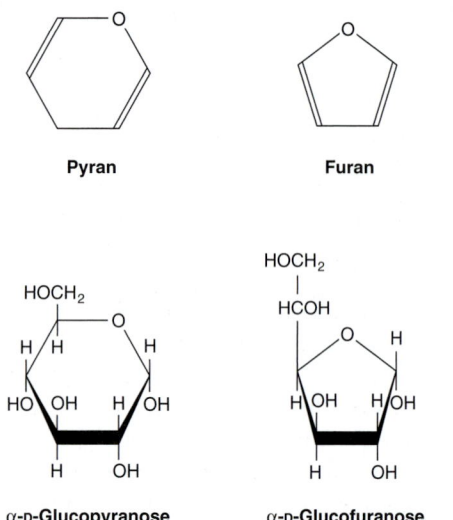

FIGURE 14–3 Pyranose and furanose forms of glucose.

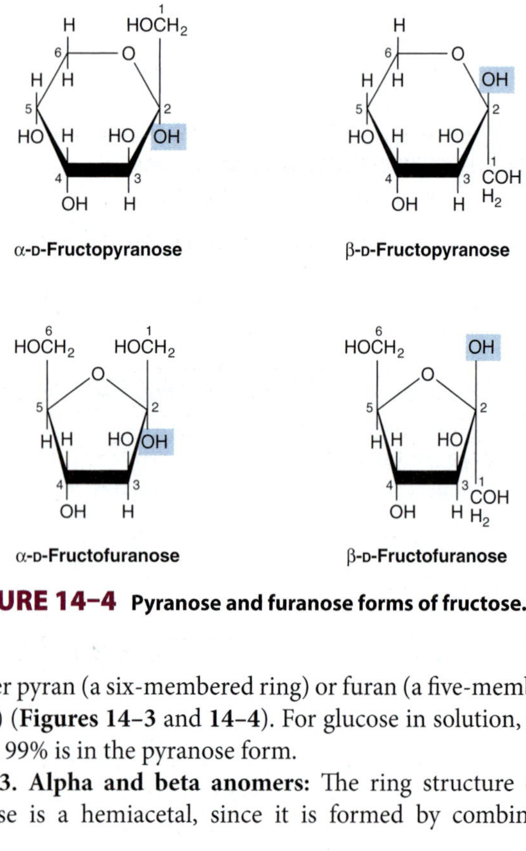

α-D-Glucopyranose   α-D-Glucofuranose

FIGURE 14–4 Pyranose and furanose forms of fructose.

α-D-Fructopyranose   β-D-Fructopyranose

α-D-Fructofuranose   β-D-Fructofuranose

of an aldehyde and an alcohol group. Similarly, the ring structure of a ketose is a hemiketal. Crystalline glucose is α-D-glucopyranose. The cyclic structure is retained in the solution, but isomerism occurs about position 1, the carbonyl or **anomeric carbon atom,** to give a mixture of α-glucopyranose (38%) and β-glucopyranose (62%). Less than 0.3% is represented by α and β anomers of glucofuranose.

4. Epimers: Isomers differing as a result of variations in configuration of the —OH and —H on carbon atoms 2, 3, and 4 of glucose are known as epimers. Biologically, the most important epimers of glucose are mannose (epimerized at carbon 2) and galactose (epimerized at carbon 4) (**Figure 14–5**).

5. Aldose-ketose isomerism: Fructose has the same molecular formula as glucose but differs in its structure, since there is a potential keto group in position 2, the anomeric carbon of fructose (**Figures 14–4 and 14–6**), whereas in glucose there is a potential aldehyde group in position 1, the anomeric carbon (**Figures 14–2 and 14–7**).

Many Monosaccharides Are Physiologically Important

Derivatives of trioses, tetroses, and pentoses and of a seven-carbon sugar (sedoheptulose) are formed as metabolic intermediates in glycolysis (Chapter 18) and the pentose phosphate pathway (Chapter 21). Pentoses are important in nucleotides, nucleic acids, and several coenzymes (**Table 14–2**). Glucose, galactose, fructose, and mannose are physiologically the most important hexoses (**Table 14–3**). The biochemically important ketoses are shown in Figure 14–6, and aldoses in Figure 14–7.

In addition, carboxylic acid derivatives of glucose are important, including D-glucuronate (for glucuronide formation and in glycosaminoglycans) and its metabolic derivative,

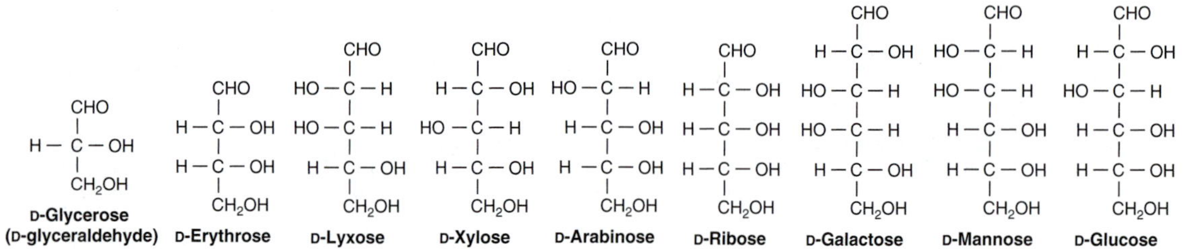

either pyran (a six-membered ring) or furan (a five-membered ring) (**Figures 14–3 and 14–4**). For glucose in solution, more than 99% is in the pyranose form.

3. Alpha and beta anomers: The ring structure of an aldose is a hemiacetal, since it is formed by combination

FIGURE 14–6 Examples of aldoses of physiologic significance.

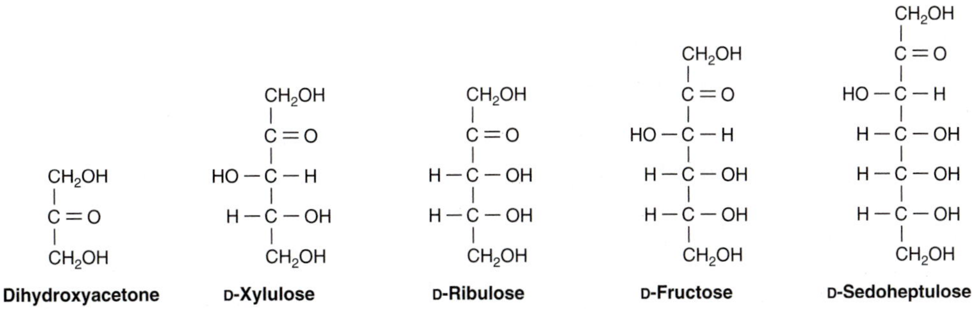

FIGURE 14–7 Examples of ketoses of physiologic significance.

L-iduronate (in glycosaminoglycans) (**Figure 14–8**) and L-gulonate (an intermediate in the uronic acid pathway; see Figure 21–4).

Sugars Form Glycosides with Other Compounds & with Each Other

Glycosides are formed by condensation between the hydroxyl group of the anomeric carbon of a monosaccharide, and a second compound that may or may not (in the case of an **aglycone**) be another monosaccharide. If the second group is a hydroxyl, the O-glycosidic bond is an **acetal** link because it results from a reaction between a hemiacetal group (formed from an aldehyde and an —OH group) and another —OH group. If the hemiacetal portion is glucose, the resulting compound is a **glucoside;** if galactose, a **galactoside;** and so on. If the second group is an amine, an N-glycosidic bond is formed, for example, between adenine and ribose in nucleotides such as ATP (Figure 11–4).

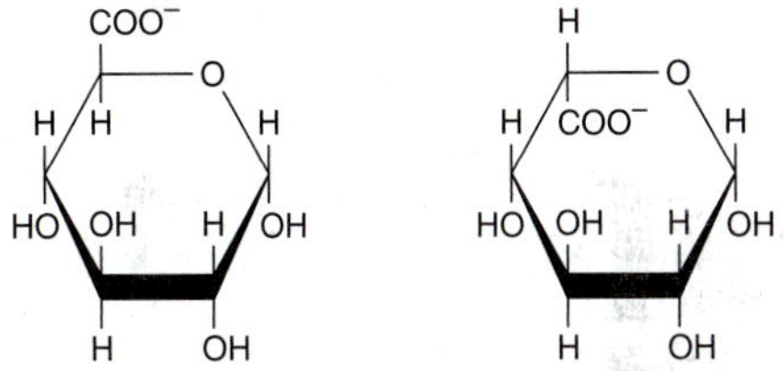

FIGURE 14–8 α-D-Glucuronate (*left*) and β-L-iduronate (*right*).

TABLE 14–2 Pentoses of Physiologic Importance

Sugar	Source	Biochemical and Clinical Importance
D-Ribose	Nucleic acids and metabolic intermediate	Structural component of nucleic acids and coenzymes, including ATP, NAD(P), and flavin coenzymes
D-Ribulose	Metabolic intermediate	Intermediate in the pentose phosphate pathway
D-Arabinose	Plant gums	Constituent of glycoproteins
D-Xylose	Plant gums, proteoglycans, glycosaminoglycans	Constituent of glycoproteins
L-Xylulose	Metabolic intermediate	Excreted in the urine in essential pentosuria

TABLE 14–3 Hexoses of Physiologic Importance

Sugar	Source	Biochemical Importance	Clinical Significance
D-Glucose	Fruit juices, hydrolysis of starch, cane or beet sugar, maltose and lactose	The main metabolic fuel for tissues; "blood sugar"	Excreted in the urine (glucosuria) in poorly controlled diabetes mellitus as a result of hyperglycemia
D-Fructose	Fruit juices, honey, hydrolysis of cane or beet sugar and inulin, enzymic isomerization of glucose syrups for food manufacture	Readily metabolized either via glucose or directly	Hereditary fructose intolerance leads to fructose accumulation and hypoglycemia
D-Galactose	Hydrolysis of lactose	Readily metabolized to glucose; synthesized in the mammary gland for synthesis of lactose in milk. A constituent of glycolipids and glycoproteins	Hereditary galactosemia as a result of failure to metabolize galactose leads to cataracts
D-Mannose	Hydrolysis of plant mannan gums	Constituent of glycoproteins	

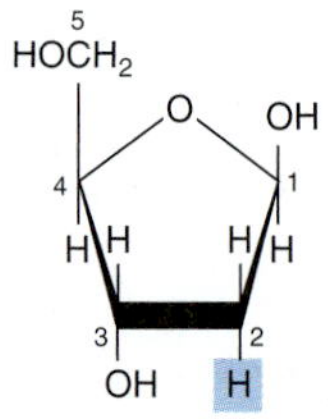

FIGURE 14–9 2-Deoxy-D-ribofuranose (β form).

Glycosides are widely distributed in nature; the aglycone may be methanol, glycerol, a sterol, a phenol, or a base such as adenine. The glycosides that are important in medicine because of their action on the heart (**cardiac glycosides**) all contain steroids as the aglycone. These include derivatives of digitalis and strophanthus such as **ouabain,** an inhibitor of the Na⁺–K⁺-ATPase of cell membranes. Other glycosides include antibiotics such as **streptomycin.**

Deoxy Sugars Lack an Oxygen Atom

Deoxy sugars are those in which one hydroxyl group has been replaced by hydrogen. An example is **deoxyribose** (**Figure 14–9**) in DNA. The deoxy sugar L-fucose (Figure 14–13) occurs in glycoproteins; 2-deoxyglucose is used experimentally as an inhibitor of glucose metabolism.

Amino Sugars (Hexosamines) Are Components of Glycoproteins, Gangliosides, & Glycosaminoglycans

The amino sugars include D-glucosamine, a constituent of hyaluronic acid (**Figure 14–10**), D-galactosamine (also known as chondrosamine), a constituent of chondroitin, and D-mannosamine. Several **antibiotics** (eg, **erythromycin**) contain amino sugars, which are important for their antibiotic activity.

Maltose, Sucrose, & Lactose Are Important Disaccharides

The disaccharides are sugars composed of two monosaccharide residues linked by a glycoside bond (**Figure 14–11**). The physiologically important disaccharides are maltose, sucrose,

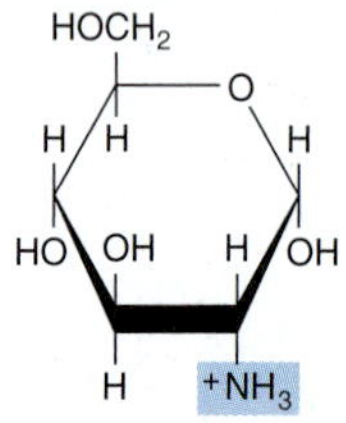

FIGURE 14–10 Glucosamine (2-amino-D-glucopyranose) (α form). Galactosamine is 2-amino-D-galactopyranose. Both glucosamine and galactosamine occur as N-acetyl derivatives in more complex carbohydrates, for example, glycoproteins.

Maltose

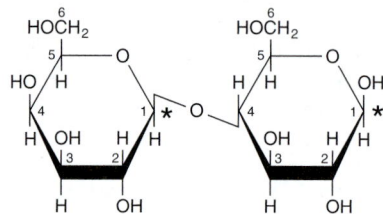

O-α-D-glucopyranosyl-(1 → 4)-α-D-glucopyranose

Lactose

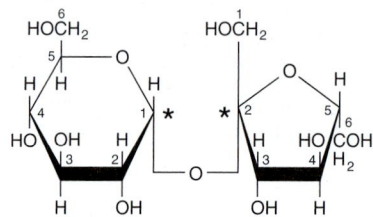

O-β-D-Galactopyranosyl-(1 → 4)-β-D-glucopyranose

Sucrose

O-α-D-Glucopyranosyl-(1 → 2)-β-D-fructofuranoside

FIGURE 14–11 **Structures of important disaccharides.** α and β refer to the configuration at the anomeric carbon atom (*). When the anomeric carbon of the second residue takes part in the formation of the glycosidic bond, as in sucrose, the residue becomes a glycoside known as a furanoside or a pyranoside. As the disaccharide no longer has an anomeric carbon with a free potential aldehyde or ketone group, it no longer exhibits reducing properties. The configuration of the β-fructofuranose residue in sucrose results from turning the β-fructofuranose molecule depicted in Figure 14–4 through 180° and inverting it.

and lactose (**Table 14–4**). Hydrolysis of sucrose yields a mixture of glucose and fructose called "invert sugar" because fructose is strongly levorotatory and changes (inverts) the weaker dextrorotatory action of sucrose.

POLYSACCHARIDES SERVE STORAGE & STRUCTURAL FUNCTIONS

Polysaccharides include the following physiologically important carbohydrates:

Starch is a homopolymer of glucose forming an α-glucosidic chain, called a **glucosan** or **glucan.** It is the most important

TABLE 14–4 Disaccharides of Physiologic Importance

Sugar	Composition	Source	Clinical Significance
Sucrose	O-α-D-glucopyranosyl-(1→2)-β-D-fructofuranoside	Cane and beet sugar, sorghum and some fruits and vegetables	Rare genetic lack of sucrase leads to sucrose intolerance—diarrhea and flatulence
Lactose	O-α-D-galactopyranosyl-(1→4)-β-D-glucopyranose	Milk (and many pharmaceutical preparations as a filler)	Lack of lactase (alactasia) leads to lactose intolerance—diarrhea and flatulence; may be excreted in the urine in pregnancy
Maltose	O-α-D-glucopyranosyl-(1→4)-α-D-glucopyranose	Enzymic hydrolysis of starch (amylase); germinating cereals and malt	
Isomaltose	O-α-D-glucopyranosyl-(1→6)-α-D-glucopyranose	Enzymic hydrolysis of starch (the branch points in amylopectin)	
Lactulose	O-α-D-galactopyranosyl-(1→4)-β-D-fructofuranose	Heated milk (small amounts), mainly synthetic	Not hydrolyzed by intestinal enzymes, but fermented by intestinal bacteria; used as a mild osmotic laxative
Trehalose	O-α-D-glucopyranosyl-(1→1)-α-D-glucopyranoside	Yeasts and fungi; the main sugar of insect hemolymph	

dietary carbohydrate in cereals, potatoes, legumes, and other vegetables. The two main constituents are **amylose** (13–20%), which has a nonbranching helical structure, and **amylopectin** (80–87%), which consists of branched chains composed of 24–30 glucose residues with α1 → 4 linkages in the chains and by α1 → 6 linkages at the branch points (**Figure 14–12**).

The extent to which starch in foods is hydrolyzed by amylase is determined by its structure, the degree of crystallization or hydration (the result of cooking), and whether it is enclosed in intact (and indigestible) plant cells walls. The **glycemic index** of a starchy food is a measure of its digestibility, based on the extent to which it raises the blood concentration of glucose compared with an equivalent amount of glucose or a reference food such as white bread or boiled rice. Glycemic index ranges from 1 (or 100%) for starches that are readily hydrolyzed in the small intestine to 0 for those that are not hydrolysed at all.

Glycogen (**Figure 14–13**) is the storage polysaccharide in animals and is sometimes called animal starch. It is a more highly branched structure than amylopectin with chains of 12–14 α-D-glucopyranose residues (in α1 → 4 glucosidic linkage) with branching by means of α1 → 6 glucosidic bonds. Muscle glycogen granules (β-particles) are spherical and contain up to 60,000 glucose residues; in liver there are similar granules and also rosettes of glycogen granules that appear to be aggregated β-particles.

Inulin is a polysaccharide of fructose (and hence a fructosan) found in tubers and roots of dahlias, artichokes, and dandelions. It is readily soluble in water and is used to determine the glomerular filtration rate, but it is not hydrolyzed by intestinal enzymes. **Dextrins** are intermediates in the hydrolysis of starch. **Cellulose** is the chief constituent of plant cell walls. It is insoluble and consists of β-D-glucopyranose units linked by β1 → 4 bonds to form long, straight chains strengthened by cross-linking hydrogen bonds. Mammals lack any enzyme that hydrolyzes the β1 → 4 bonds, and so cannot digest cellulose. It is an important source of "bulk" in the diet, and the major component of dietary fiber. Microorganisms in the gut

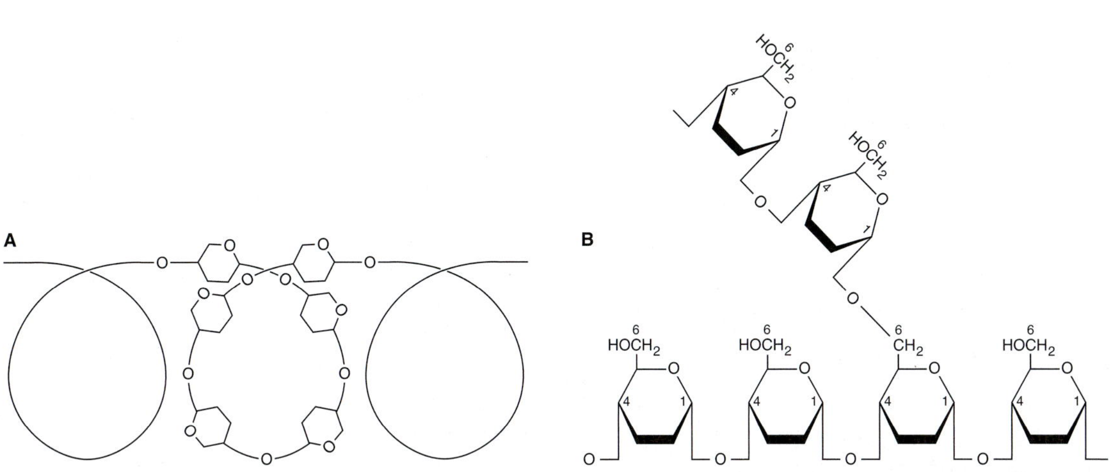

FIGURE 14–12 **Structure of starch. (A)** Amylose, showing helical coil structure. **(B)** Amylopectin, showing 1 → 6 branch point.

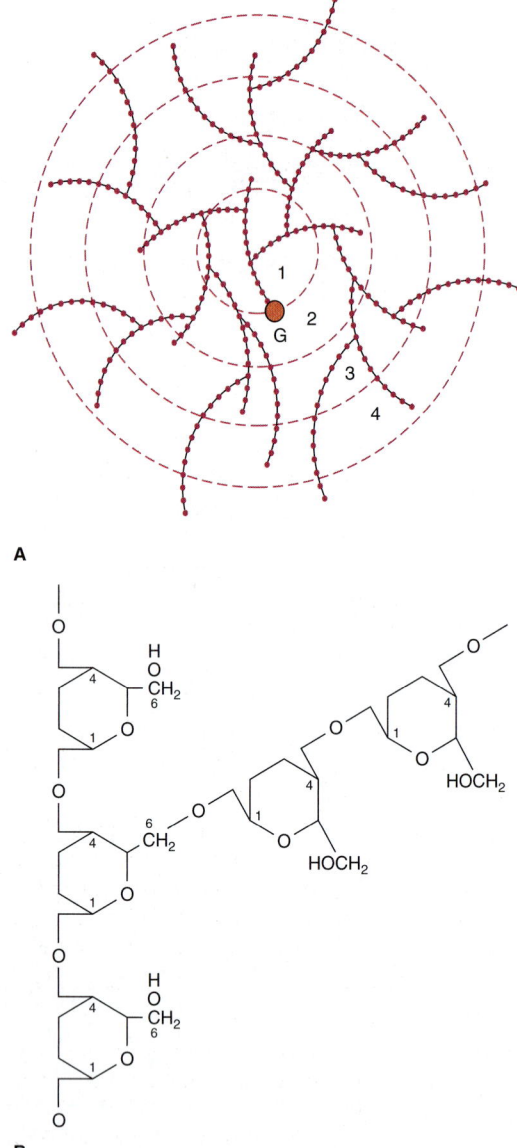

A

B

FIGURE 14–13 **The glycogen molecule. (A)** General structure. **(B)** Enlargement of structure at a branch point. The molecule is a sphere ~21 nm in diameter that can be seen in electron micrographs. It has a molecular mass of ~10^7 Da and consists of polysaccharide chains, each containing about 13 glucose residues. The chains are either branched or unbranched and are arranged in 12 concentric layers (only four are shown in the figure). The branched chains (each has two branches) are found in the inner layers and the unbranched chains in the outer layer. (G, glycogenin, the primer molecule for glycogen synthesis.)

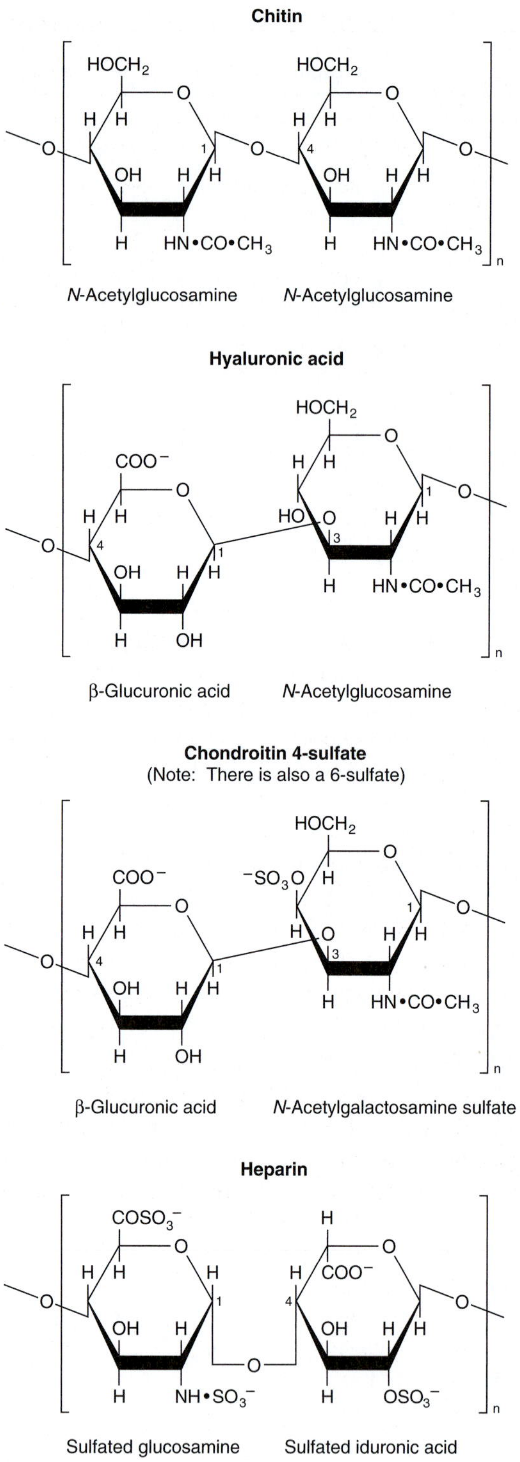

Chitin

N-Acetylglucosamine   *N*-Acetylglucosamine

Hyaluronic acid

β-Glucuronic acid   *N*-Acetylglucosamine

Chondroitin 4-sulfate
(Note: There is also a 6-sulfate)

β-Glucuronic acid   *N*-Acetylgalactosamine sulfate

Heparin

Sulfated glucosamine   Sulfated iduronic acid

FIGURE 14–14 **Structure of some complex polysaccharides and glycosaminoglycans.**

of ruminants and other herbivores can hydrolyze the linkage and ferment the products to short-chain fatty acids as a major energy source. There is some bacterial metabolism of cellulose in the human colon. **Chitin** is a structural polysaccharide in the exoskeleton of crustaceans and insects, and also in mushrooms. It consists of *N*-acetyl-ᴅ-glucosamine units joined by β1 → 4 glycosidic bonds (**Figure 14–14**).

Glycosaminoglycans (mucopolysaccharides) are complex carbohydrates containing **amino sugars** and **uronic acids.** They may be attached to a protein molecule to form a **proteoglycan.** Proteoglycans provide the ground or packing substance of connective tissue. They hold large quantities of water and occupy space, thus cushioning or lubricating other structures, because of the large number of —OH groups and negative charges on the molecule which, by repulsion, keep

TABLE 14–5 Carbohydrates Found in Glycoproteins

Hexoses	Mannose (Man), Galactose (Gal)
Acetyl hexosamines	N-Acetylglucosamine (GlcNAc), N-acetylgalactosamine (GalNAc)
Pentoses	Arabinose (Ara), Xylose (Xyl)
Methyl pentose	L-Fucose (Fuc, see Figure 14–15)
Sialic acids	N-Acyl derivatives of neuraminic acid; the predominant sialic acid is N-acetylneuraminic acid (NeuAc, see Figure 14–16)

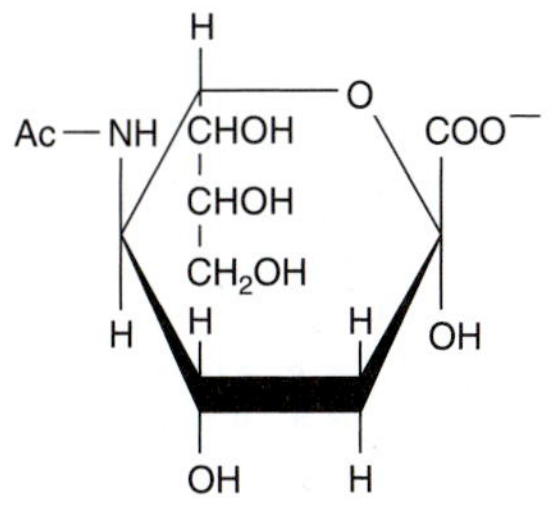

FIGURE 14–16 Structure of *N*-acetylneuraminic acid, a sialic acid (Ac = CH_3—CO—).

the carbohydrate chains apart. Examples are **hyaluronic acid, chondroitin sulfate,** and **heparin** (Figure 14–14).

Glycoproteins (also known as mucoproteins) are proteins containing branched or unbranched oligosaccharide chains (**Table 14–5, Figure 14–15**); they occur in cell membranes (Chapters 40 and 47) and many other situations; serum albumin is a glycoprotein. The **sialic acids** are *N*- or *O*-acyl derivatives of neuraminic acid (**Figure 14–16**). **Neuraminic acid** is a nine-carbon sugar derived from mannosamine (an epimer of glucosamine) and pyruvate. Sialic acids are constituents of both **glycoproteins** and **gangliosides.**

CARBOHYDRATES OCCUR IN CELL MEMBRANES & IN LIPOPROTEINS

Approximately 5% of the weight of cell membranes is carbohydrate in glycoproteins and glycolipids. Their presence on the outer surface of the plasma membrane (the **glycocalyx**) has been shown with the use of plant **lectins,** protein agglutinins that bind specific glycosyl residues. For example, **concanavalin A** binds α-glucosyl and α-mannosyl residues. **Glycophorin** is a major integral membrane glycoprotein of human erythrocytes. It has 130 amino acid residues and spans the lipid membrane, with polypeptide regions outside both the external and internal (cytoplasmic) surfaces. Carbohydrate chains are attached to the amino terminal portion outside the external surface. Carbohydrates are also present in apo-protein B of plasma lipoproteins.

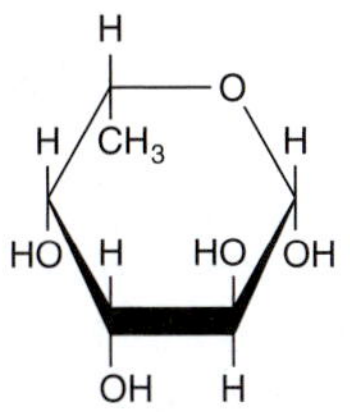

FIGURE 14–15 β-L-Fucose (6-deoxy-β-L-galactose).

SUMMARY

- Carbohydrates are major constituents of animal food and animal tissues. They are characterized by the type and number of monosaccharide residues in their molecules.

- Glucose is the most important carbohydrate in mammalian biochemistry because nearly all carbohydrate in food is converted to glucose for metabolism.

- Sugars have large numbers of stereoisomers because they contain several asymmetric carbon atoms.

- The physiologically important monosaccharides include glucose, the "blood sugar," and ribose, an important constituent of nucleotides and nucleic acids.

- The important disaccharides include maltose (glucosyl glucose), an intermediate in the digestion of starch; sucrose (glucosyl fructose), important as a dietary constituent containing fructose; and lactose (galactosyl glucose), in milk.

- Starch and glycogen are storage polymers of glucose in plants and animals, respectively. Starch is the major source of energy in the diet.

- Complex carbohydrates contain other sugar derivatives such as amino sugars, uronic acids, and sialic acids. They include proteoglycans and glycosaminoglycans, which are associated with structural elements of the tissues, and glycoproteins, which are proteins containing oligosaccharide chains; they are found in many situations including the cell membrane.

REFERENCES

Champ M, Langkilde A-M, Brouns F, et al: Advances in dietary fibre characterisation. Nutrition Res Rev 2003;16:(1)71–82.

Davis BG, Fairbanks AJ: *Carbohydrate Chemistry.* Oxford University Press, 2002.

Kiessling LL, Splain RA: Chemical approaches to glycobiology. Ann Rev Biochem 2010;79:619–53.

Lindhorst TK, Thisbe K: *Essentials of Carbohydrate Chemistry and Biochemistry,* 3rd ed. Wiley-VCH, 2007.

Sinnott M: *Carbohydrate Chemistry and Biochemistry: Structure and Mechanisms,* Royal Society of Chemistry, 2007.

Lipids of Physiologic Significance

Kathleen M. Botham, PhD, DSc & Peter A. Mayes, PhD, DSc

O B J E C T I V E S

After studying this chapter, you should be able to:

- Define simple and complex lipids and identify the lipid classes in each group.
- Indicate the structure of saturated and unsaturated fatty acids, explain how the chain length and degree of unsaturation influence their melting point, give examples, and explain the nomenclature.
- Understand the difference between *cis* and *trans* carbon-carbon double bonds.
- Describe how eicosanoids are formed by modification of the structure of unsaturated fatty acids; identify the various eicosanoid classes and indicate their functions.
- Outline the general structure of triacylglycerols and indicate their function.
- Outline the general structure of phospholipids and glycosphingolipids and indicate the functions of the different classes.
- Appreciate the importance of cholesterol as the precursor of many biologically important steroids, including steroid hormones, bile acids, and vitamins D.
- Recognize the cyclic nucleus common to all steroids and explain the difference between the "chair" and "boat" forms of the six-carbon rings and that the rings may be either *cis* or *trans* in relation to each other, making many stereoisomers possible.
- Explain why free radicals are damaging to tissues and identify the three stages in the chain reaction of lipid peroxidation that produces them continuously.
- Understand how antioxidants protect lipids from peroxidation by either inhibiting chain initiation or breaking the chain and give physiological and nonphysiological examples.
- Understand that many lipid molecules are amphipathic, having both hydrophobic and hydrophilic groups in their structure, and explain how this influences their behavior in an aqueous environment and enables certain classes, including phospholipids, sphingolipids, and cholesterol, to form the basic structure of biologic membranes.

BIOMEDICAL IMPORTANCE

The lipids are a heterogeneous group of compounds, including fats, oils, steroids, waxes, and related compounds, that are related more by their physical than by their chemical properties. They have the common property of being (1) relatively **insoluble in water** and (2) **soluble in nonpolar solvents** such as ether and chloroform. They are important dietary constituents not only because of their high energy value, but also because **fat-soluble vitamins** and **essential fatty acids** are contained in the fat of natural foods. Fat is stored in **adipose tissue,** where it also serves as a thermal insulator in the subcutaneous tissues and around certain organs. Nonpolar lipids act as **electrical insulators,** allowing rapid propagation of depolarization waves along **myelinated nerves.** Combinations of lipid and protein (lipoproteins) serve as the means of **transporting lipids** in the blood. Lipids have essential roles in nutrition and health and knowledge of lipid biochemistry is necessary for the understanding of many important biomedical conditions, including **obesity, diabetes mellitus, and atherosclerosis.**

LIPIDS ARE CLASSIFIED AS SIMPLE OR COMPLEX

1. **Simple lipids:** Esters of fatty acids with various alcohols.
   a. **Fats:** Esters of fatty acids with glycerol. **Oils** are fats in the liquid state.
   b. **Waxes:** Esters of fatty acids with higher molecular weight monohydric alcohols.
2. **Complex lipids:** Esters of fatty acids containing groups in addition to an alcohol and a fatty acid.
   a. **Phospholipids:** Lipids containing, in addition to fatty acids and an alcohol, a phosphoric acid residue. They frequently have nitrogen-containing bases and other substituents, for example, in **glycerophospholipids** the alcohol is glycerol and in **sphingophospholipids** the alcohol is sphingosine.
   b. **Glycolipids (glycosphingolipids):** Lipids containing a fatty acid, sphingosine, and carbohydrate.
   c. **Other complex lipids:** Lipids such as sulfolipids and amino lipids. Lipoproteins may also be placed in this category.
3. **Precursor and derived lipids:** These include fatty acids, glycerol, steroids, other alcohols, fatty aldehydes, ketone bodies (Chapter 22), hydrocarbons, lipid-soluble vitamins, and hormones.

Because they are uncharged, acylglycerols (glycerides), cholesterol, and cholesteryl esters are termed **neutral lipids.**

FATTY ACIDS ARE ALIPHATIC CARBOXYLIC ACIDS

Fatty acids occur in the body mainly as esters in natural fats and oils, but are found in the unesterified form as **free fatty acids,** a transport form in the plasma. Fatty acids that occur in natural fats usually contain an even number of carbon atoms. The chain may be **saturated** (containing no double bonds) or **unsaturated** (containing one or more double bonds) (**Figure 15–1**).

Fatty Acids Are Named after Corresponding Hydrocarbons

The most frequently used systematic nomenclature names the fatty acid after the hydrocarbon with the same number and arrangement of carbon atoms, with **-oic** being substituted for the final **-e** (Genevan system). Thus, saturated acids end in **-anoic,** for example, octanoic acid, and unsaturated acids with double bonds end in **-enoic,** for example, octadecenoic acid (oleic acid).

Carbon atoms are numbered from the carboxyl carbon (carbon no. 1). The carbon atoms adjacent to the carboxyl

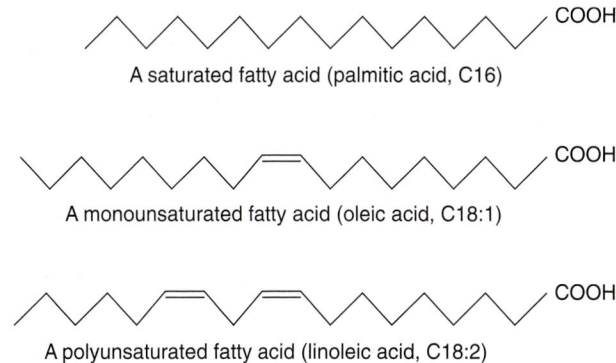

A saturated fatty acid (palmitic acid, C16)

A monounsaturated fatty acid (oleic acid, C18:1)

A polyunsaturated fatty acid (linoleic acid, C18:2)

FIGURE 15–1 Fatty acids.

carbon (nos. 2, 3, and 4) are also known as the α, β, and γ carbons, respectively, and the terminal methyl carbon is known as the ω- or n-carbon.

Various conventions use Δ for indicating the number and position of the double bonds (**Figure 15–2**); for example, Δ^9 indicates a double bond between carbons 9 and 10 of the fatty acid; ω9 indicates a double bond on the ninth carbon counting from the ω-carbon. In animals, additional double bonds are introduced only between the existing double bond (eg, ω9, ω6, or ω3) and the carboxyl carbon, leading to three series of fatty acids known as the ω9, ω6, and ω3 families, respectively.

Saturated Fatty Acids Contain No Double Bonds

Saturated fatty acids may be envisaged as based on acetic acid (CH_3—COOH) as the first member of the series in which— CH_2— is progressively added between the terminal CH_3—and —COOH groups. Examples are shown in **Table 15–1**. Other higher members of the series are known to occur, particularly in waxes. A few branched-chain fatty acids have also been isolated from both plant and animal sources.

Unsaturated Fatty Acids Contain One or More Double Bonds

Unsaturated fatty acids (Figure 15–1, **Table 15–2**) may be further subdivided as follows:

1. **Monounsaturated** (monoethenoid, monoenoic) acids, containing one double bond.

18:1;9 or Δ^9 18:1

$$\overset{18}{C}H_3(CH_2)_7\overset{10}{C}H = \overset{9}{C}H(CH_2)_7\overset{1}{C}OOH$$

or

ω9,C18:1 or n–9, 18:1

$$\overset{\omega}{C}H_3\overset{2}{C}H_2\overset{3}{C}H_2\overset{4}{C}H_2\overset{5}{C}H_2\overset{6}{C}H_2\overset{7}{C}H_2\overset{8}{C}H_2\overset{9}{C}H = \overset{10}{C}H(CH_2)_7\overset{18}{C}OOH$$

FIGURE 15–2 Oleic acid. n – 9 is equivalent to ω9.

TABLE 15–1   Saturated Fatty Acids

Common Name	Number of C Atoms	
Acetic	2	Major end product of carbohydrate fermentation by rumen organisms
Butyric	4	In certain fats in small amounts (especially butter). An end product of carbohydrate fermentation by rumen organisms[1]
Valeric	5	
Caproic	6	
Lauric	12	Spermaceti, cinnamon, palm kernel, coconut oils, laurels, butter
Myristic	14	Nutmeg, palm kernel, coconut oils, myrtles, butter
Palmitic	16	Common in all animal and plant fats
Stearic	18	

[1]Also formed in the cecum of herbivores and to a lesser extent in the colon of humans.

2. **Polyunsaturated** (polyethenoid, polyenoic) acids, containing two or more double bonds.
3. **Eicosanoids:** These compounds, derived from eicosa (20-carbon) polyenoic fatty acids, comprise the **prostanoids, leukotrienes (LTs),** and **lipoxins (LXs).**

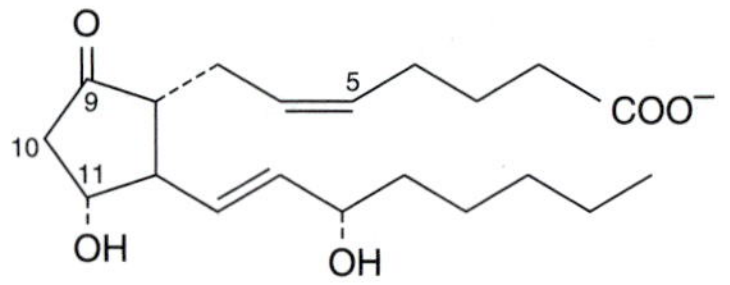

FIGURE 15–3 Prostaglandin E$_2$ (PGE$_2$).

Prostanoids include **prostaglandins (PGs), prostacyclins (PGIs),** and **thromboxanes (TXs).**

Prostaglandins exist in virtually every mammalian tissue, acting as local hormones; they have important physiologic and pharmacologic activities. They are synthesized in vivo by cyclization of the center of the carbon chain of 20-carbon (eicosanoic) polyunsaturated fatty acids (eg, arachidonic acid) to form a cyclopentane ring (**Figure 15–3**). A related series of compounds, the **thromboxanes,** have the cyclopentane ring interrupted with an oxygen atom (oxane ring) (**Figure 15–4**). Three different eicosanoic fatty acids give rise to three groups of eicosanoids characterized by the number of double bonds in the side chains, for example, PG$_1$, PG$_2$, and PG$_3$. Different substituent groups attached to the rings give rise to series of prostaglandins and thromboxanes, labeled A, B, etc—for example, the "E" type of prostaglandin (as in PGE$_2$) has a keto group in

TABLE 15–2   Unsaturated Fatty Acids of Physiologic and Nutritional Significance

Number of C Atoms and Number and Position of Common Double Bonds	Family	Common Name	Systematic Name	Occurrence
Monoenoic acids (one double bond)				
16:1;9	ω7	Palmitoleic	cis-9-Hexadecenoic	In nearly all fats.
18:1;9	ω9	Oleic	cis-9-Octadecenoic	Possibly the most common fatty acid in natural fats; particularly high in olive oil.
18:1;9	ω9	Elaidic	trans-9-Octadecenoic	Hydrogenated and ruminant fats.
Dienoic acids (two double bonds)				
18:2;9,12	ω6	Linoleic	all-cis-9,12-Octadecadienoic	Corn, peanut, cottonseed, soy bean, and many plant oils.
Trienoic acids (three double bonds)				
18:3;6,9,12	ω6	γ-Linolenic	all-cis-6,9,12-Octadecatrienoic	Some plants, eg, oil of evening primrose, borage oil; minor fatty acid in animals.
18:3;9,12,15	ω3	α-Linolenic	all-cis-9,12,15-Octadecatrienoic	Frequently found with linoleic acid but particularly in linseed oil.
Tetraenoic acids (four double bonds)				
20:4;5,8,11,14	ω6	Arachidonic	all-cis-5,8,11,14-Eicosatetraenoic	Found in animal fats; important component of phospholipids in animals.
Pentaenoic acids (five double bonds)				
20:5;5,8,11,14,17	ω3	Timnodonic	all-cis-5,8,11,14,17-Eicosapentaenoic	Important component of fish oils, eg, cod liver, mackerel, menhaden, salmon oils.
Hexaenoic acids (six double bonds)				
22:6;4,7,10,13,16,19	ω3	Cervonic	all-cis-4,7,10,13,16,19-Docosahexaenoic	Fish oils, phospholipids in brain.

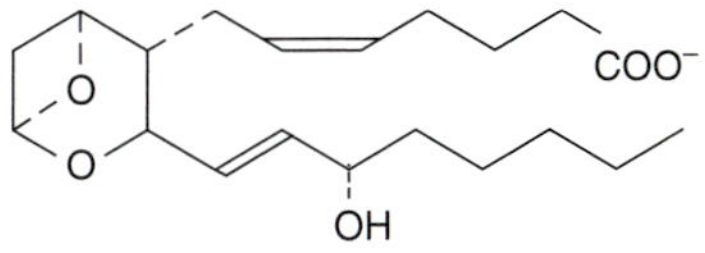

FIGURE 15–4 Thromboxane A$_2$ (TXA$_2$).

position 9, whereas the "F" type has a hydroxyl group in this position. The **leukotrienes** and **lipoxins** (**Figure 15–5**) are a third group of eicosanoid derivatives formed via the lipoxygenase pathway (Figure 23–11). They are characterized by the presence of three or four conjugated double bonds, respectively. Leukotrienes cause bronchoconstriction as well as being potent proinflammatory agents, and play a part in **asthma.**

Most Naturally Occurring Unsaturated Fatty Acids Have *cis* Double Bonds

The carbon chains of saturated fatty acids form a zigzag pattern when extended at low temperatures (Figure 15–1). At higher temperatures, some bonds rotate, causing chain shortening, which explains why biomembranes become thinner with increases in temperature. A type of **geometric isomerism** occurs in unsaturated fatty acids, depending on the orientation of atoms or groups around the axes of double bonds, which do not allow rotation. If the acyl chains are on the same side of the bond, it is *cis-*, as in oleic acid; if on opposite sides, it is *trans-*, as in elaidic acid, the *trans* isomer of oleic acid (**Figure 15–6**). Double bonds in naturally occurring unsaturated long-chain fatty acids are nearly all in the *cis* configuration, the molecules being "bent" 120° at the double bond. Thus, oleic acid has an L shape, whereas elaidic acid remains "straight." Increase in the number of *cis* double bonds in a fatty acid leads to a variety of possible spatial configurations of the molecule—for example, arachidonic acid, with four *cis* double bonds, is bent into a U shape (**Figure 15–7**). This has profound significance for molecular packing in cell membranes and on the positions occupied by fatty acids in more complex molecules such as phospholipids. *Trans* double bonds alter these spatial relationships. **Trans fatty acids** are present in certain foods, arising as a by-product of the saturation of fatty acids during hydrogenation, or "hardening," of natural oils in the manufacture of margarine. An additional small contribution comes from the ingestion of ruminant fat that contains *trans* fatty acids arising from the action of microorganisms in the rumen. Consumption of *trans*

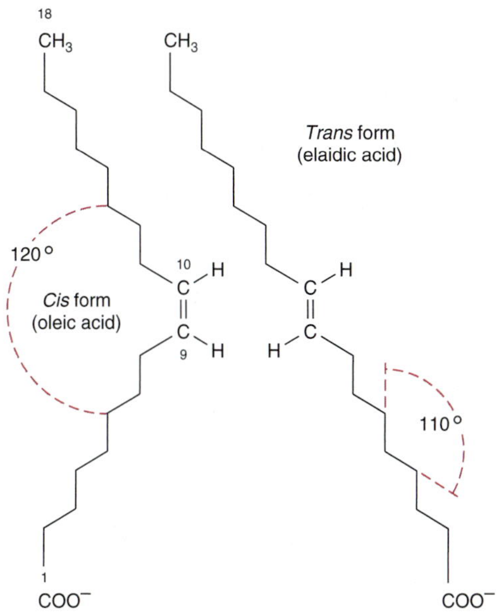

FIGURE 15–6 Geometric isomerism of Δ^9, 18:1 fatty acids (oleic and elaidic acids).

fatty acids is now known to be detrimental to health and is associated with increased risk of diseases including cardiovascular disease and diabetes mellitus. This has led to improved technology to produce soft margarine low in *trans* fatty acids or containing none at all.

Physical and Physiologic Properties of Fatty Acids Reflect Chain Length and Degree of Unsaturation

The melting points of even-numbered carbon fatty acids increase with chain length and decrease according to unsaturation. A triacylglycerol containing three saturated fatty acids of 12 carbons or more is solid at body temperature, whereas if the fatty acid residues are 18:2, it is liquid to below 0°C. In practice, natural acylglycerols contain a mixture of fatty acids tailored to suit their functional roles. The membrane lipids, which must be fluid at all environmental temperatures, are more unsaturated than storage lipids. Lipids in tissues that are subject to cooling, for example, in hibernators or in the extremities of animals, are more unsaturated.

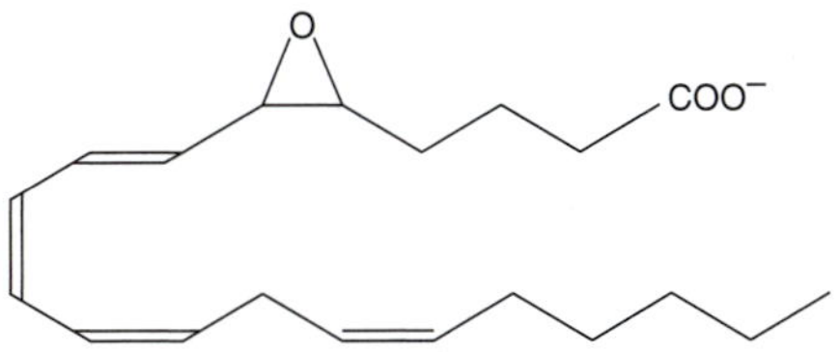

FIGURE 15–5 Leukotriene A$_4$ (LTA$_4$).

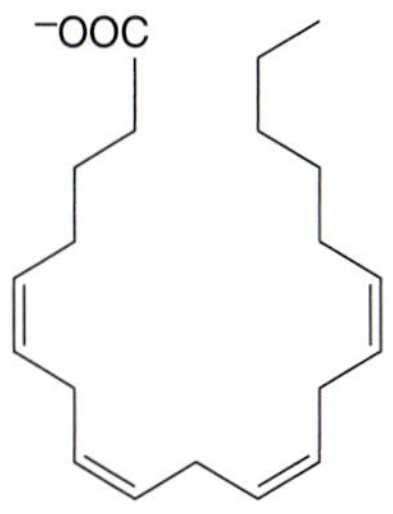

FIGURE 15–7 Arachidonic acid.

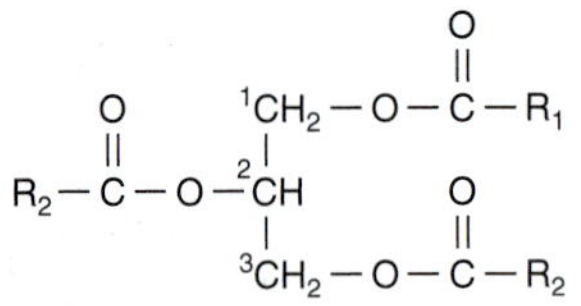

$$R_2-C-O-^2CH$$

FIGURE 15–8 Triacylglycerol.

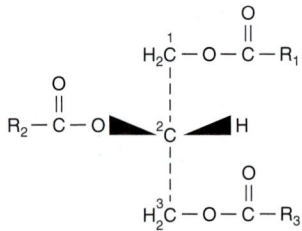

FIGURE 15–9 Triacyl-*sn*-glycerol.

TRIACYLGLYCEROLS (TRIGLYCERIDES)* ARE THE MAIN STORAGE FORMS OF FATTY ACIDS

The triacylglycerols (**Figure 15–8**) are esters of the trihydric alcohol glycerol and fatty acids. Mono- and diacylglycerols, wherein one or two fatty acids are esterified with glycerol, are also found in the tissues. These are of particular significance in the synthesis and hydrolysis of triacylglycerols.

Carbons 1 & 3 of Glycerol Are Not Identical

To number the carbon atoms of glycerol unambiguously, the -*sn* (stereochemical numbering) system is used. It is important to realize that carbons 1 and 3 of glycerol are not identical when viewed in three dimensions (shown as a projection formula in **Figure 15–9**). Enzymes readily distinguish between them and are nearly always specific for one or the other carbon; for example, glycerol is always phosphorylated on *sn*-3 by glycerol kinase to give glycerol 3-phosphate and not glycerol 1-phosphate.

PHOSPHOLIPIDS ARE THE MAIN LIPID CONSTITUENTS OF MEMBRANES

Phospholipids may be regarded as derivatives of **phosphatidic acid** (**Figure 15–10**), in which the phosphate is esterified with the —OH of a suitable alcohol. Phosphatidic acid is important as an intermediate in the synthesis of triacylglycerols as well as phosphoglycerols but is not found in any great quantity in tissues.

Phosphatidylcholines (Lecithins) Occur in Cell Membranes

Phosphoacylglycerols containing choline (Figure 15–10) are the most abundant phospholipids of the cell membrane and represent a large proportion of the body's store of choline. Choline is important in nervous transmission, as acetylcholine,

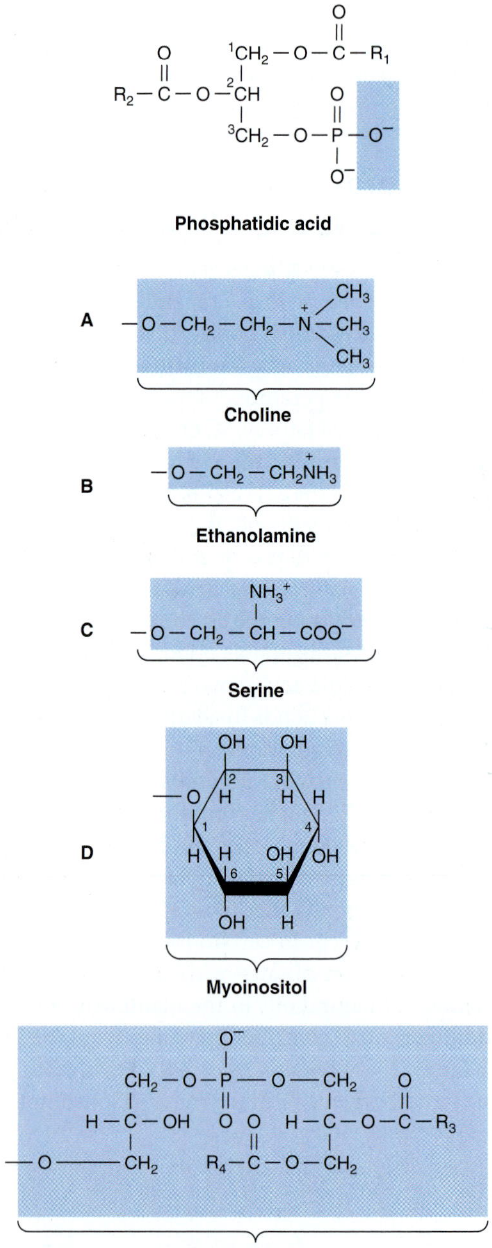

Phosphatidic acid

A Choline

B Ethanolamine

C Serine

D Myoinositol

E Phosphatidylglycerol

FIGURE 15–10 **Phosphatidic acid and its derivatives.** The O− shown shaded in phosphatidic acid is substituted by the substituents shown to form in **(A)** 3-phosphatidylcholine, **(B)** 3-phosphatidylethanolamine, **(C)** 3-phosphatidylserine, **(D)** 3-phosphatidylinositol, and **(E)** cardiolipin (diphosphatidylglycerol).

---

*According to the standardized terminology of the International Union of Pure and Applied Chemistry and the International Union of Biochemistry, the monoglycerides, diglycerides, and triglycerides should be designated monoacylglycerols, diacylglycerols, and triacylglycerols, respectively. However, the older terminology is still widely used, particularly in clinical medicine.

and as a store of labile methyl groups. **Dipalmitoyl lecithin** is a very effective surface-active agent and a major constituent of the **surfactant** preventing adherence, due to surface tension, of the inner surfaces of the lungs. Its absence from the lungs of premature infants causes **respiratory distress syndrome.** Most phospholipids have a saturated acyl radical in the *sn*-1 position but an unsaturated radical in the *sn*-2 position of glycerol.

Phosphatidylethanolamine (cephalin) and phosphatidylserine (found in most tissues) are also found in cell membranes and differ from phosphatidylcholine only in that ethanolamine or serine, respectively, replaces choline (Figure 15–10). Phosphatidylserine also plays a role in **apoptosis** (programmed cell death).

Phosphatidylinositol Is a Precursor of Second Messengers

The inositol is present in **phosphatidylinositol** as the stereoisomer, myoinositol (Figure 15–10). **Phosphatidylinositol 4,5-bisphosphate** is an important constituent of cell membrane phospholipids; upon stimulation by a suitable hormone agonist, it is cleaved into **diacylglycerol** and **inositol trisphosphate,** both of which act as internal signals or second messengers.

Cardiolipin Is a Major Lipid of Mitochondrial Membranes

Phosphatidic acid is a precursor of **phosphatidylglycerol,** which in turn gives rise to **cardiolipin** (Figure 15–10). This phospholipid is found only in mitochondria and is essential for the mitochondrial function. Decreased cardiolipin levels or alterations in its structure or metabolism cause mitochondrial dysfunction in aging and in pathological conditions including heart failure, hypothyroidism, and Barth syndrome (cardioskeletal myopathy).

Lysophospholipids Are Intermediates in the Metabolism of Phosphoglycerols

These are phosphoacylglycerols containing only one acyl radical, for example, **lysophosphatidylcholine (lysolecithin)** (**Figure 15–11**), important in the metabolism and interconversion of phospholipids. It is also found in oxidized lipoproteins

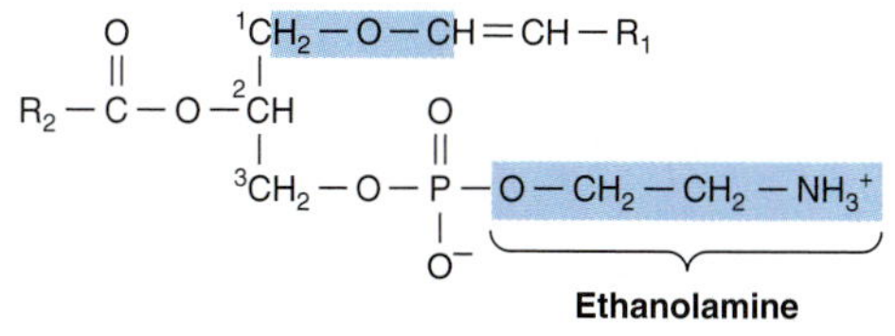

FIGURE 15–12 Plasmalogen.

and has been implicated in some of their effects in promoting **atherosclerosis.**

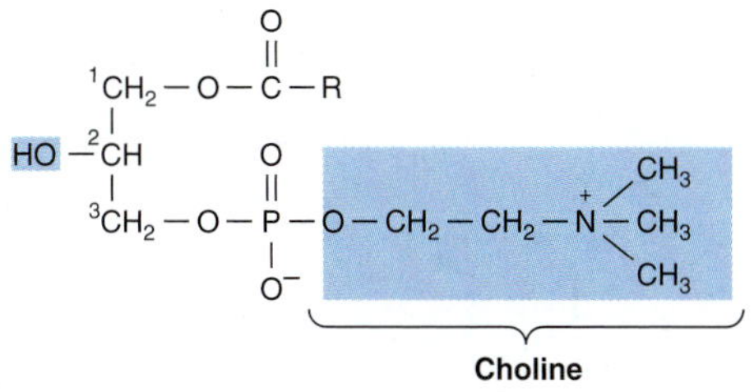

Plasmalogens Occur in Brain & Muscle

These compounds constitute as much as 10% of the phospholipids of brain and muscle. Structurally, the plasmalogens resemble phosphatidylethanolamine but possess an ether link on the *sn*-1 carbon instead of the ester link found in acylglycerols. Typically, the alkyl radical is an unsaturated alcohol (**Figure 15–12**). In some instances, choline, serine, or inositol may be substituted for ethanolamine.

Sphingomyelins Are Found in the Nervous System

Sphingomyelins are found in large quantities in brain and nerve tissue. On hydrolysis, the sphingomyelins yield a fatty acid, phosphoric acid, choline, and a complex amino alcohol, **sphingosine** (**Figure 15–13**). No glycerol is present. The combination of sphingosine plus fatty acid is known as **ceramide,** a structure also found in the glycosphingolipids (see next Section below).

GLYCOLIPIDS (GLYCOSPHINGOLIPIDS) ARE IMPORTANT IN NERVE TISSUES & IN THE CELL MEMBRANE

Glycolipids are widely distributed in every tissue of the body, particularly in nervous tissue such as brain. They occur particularly in the outer leaflet of the plasma membrane, where they contribute to **cell surface carbohydrates.**

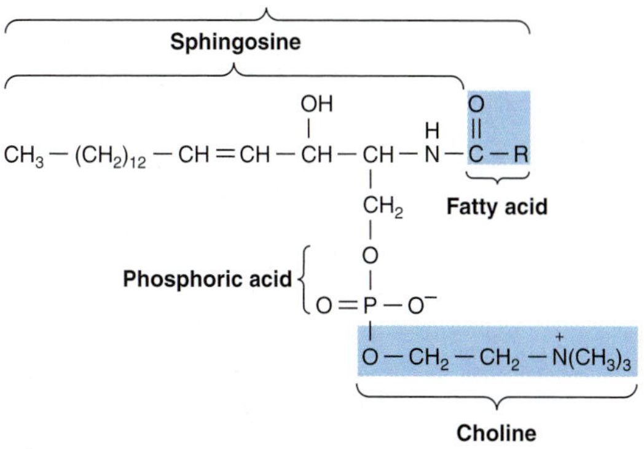

FIGURE 15–11 Lysophosphatidylcholine (lysolecithin).

FIGURE 15–13 A sphingomyelin.

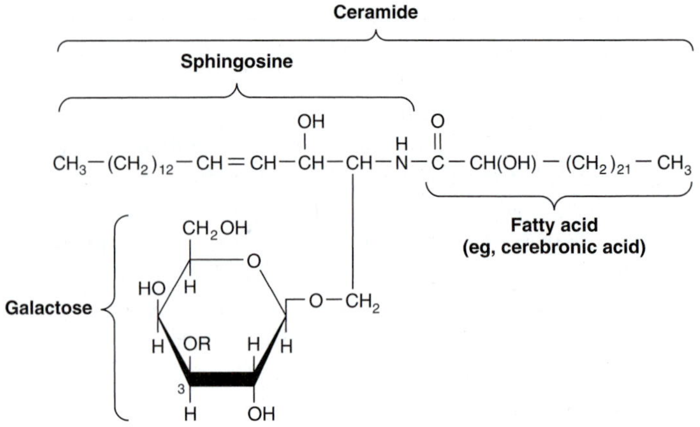

FIGURE 15–14   Structure of galactosylceramide (galactocerebroside, R = H), and sulfogalactosylceramide (a sulfatide, R = SO_4^{2-}).

The major glycolipids found in animal tissues are glycosphingolipids. They contain ceramide and one or more sugars. **Galactosylceramide** is a major glycosphingolipid of brain and other nervous tissue, found in relatively low amounts elsewhere. It contains a number of characteristic C24 fatty acids, for example, cerebronic acid.

Galactosylceramide (**Figure 15–14**) can be converted to sulfogalactosylceramide (**sulfatide**), present in high amounts in **myelin**. Glucosylceramide is the predominant simple glycosphingolipid of extraneural tissues, also occurring in the brain in small amounts. **Gangliosides** are complex glycosphingolipids derived from glucosylceramide that contain in addition one or more molecules of a **sialic acid**. Neuraminic acid (NeuAc; see Chapter 14) is the principal sialic acid found in human tissues. Gangliosides are also present in nervous tissues in high concentration. They appear to have receptor and other functions. The simplest ganglioside found in tissues is GM_3, which contains ceramide, one molecule of glucose, one molecule of galactose, and one molecule of NeuAc. In the shorthand nomenclature used, G represents ganglioside; M is a monosialo-containing species; and the subscript 3 is a number assigned on the basis of chromatographic migration. GM1 (**Figure 15–15**), a more complex ganglioside derived from GM_3, is of considerable biologic interest, as it is known to be the receptor in human intestine for cholera toxin. Other gangliosides can contain anywhere from one to five molecules of sialic acid, giving rise to di-, trisialogangliosides, etc.

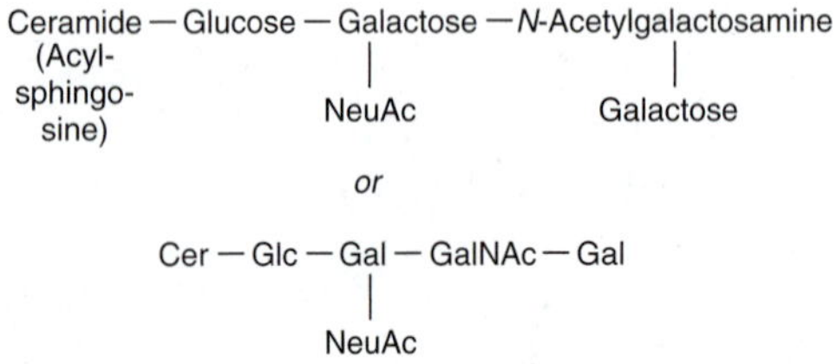

STEROIDS PLAY MANY PHYSIOLOGICALLY IMPORTANT ROLES

Cholesterol is probably the best-known steroid because of its association with **atherosclerosis** and heart disease. However, biochemically it is also of significance because it is the precursor of a large number of equally important **steroids** that include the bile acids, adrenocortical hormones, sex hormones, D vitamins, cardiac glycosides, sitosterols of the plant kingdom, and some alkaloids.

All steroids have a similar cyclic nucleus resembling phenanthrene (rings A, B, and C) to which a cyclopentane ring (D) is attached. The carbon positions on the steroid nucleus are numbered as shown in **Figure 15–16**. It is important to realize that in structural formulas of steroids, a simple hexagonal ring denotes a completely saturated six-carbon ring with all valences satisfied by hydrogen bonds unless shown otherwise; that is, it is not a benzene ring. All double bonds are shown as such. Methyl side chains are shown as single bonds unattached at the farther (methyl) end. These occur typically at positions 10 and 13 (constituting C atoms 19 and 18). A side chain at position 17 is usual (as in cholesterol). If the compound has one or more hydroxyl groups and no carbonyl or carboxyl groups, it is a **sterol**, and the name terminates in -ol.

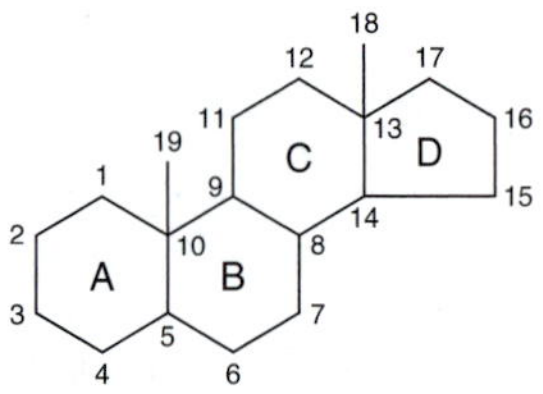

FIGURE 15–15   GM1 ganglioside, a monosialoganglioside, the receptor in human intestine for cholera toxin.

FIGURE 15–16   The steroid nucleus.

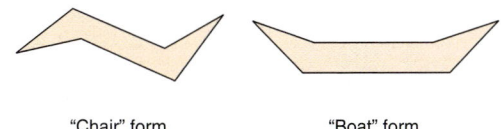

"Chair" form   "Boat" form

FIGURE 15–17 Conformations of stereoisomers of the steroid nucleus.

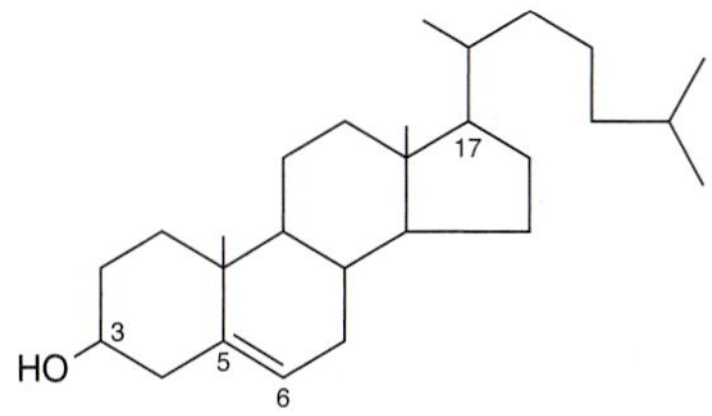

FIGURE 15–19 Cholesterol, 3-hydroxy-5,6-cholestene.

Because of Asymmetry in the Steroid Molecule, Many Stereoisomers Are Possible

Each of the six-carbon rings of the steroid nucleus is capable of existing in the three-dimensional conformation either of a "chair" or a "boat" (**Figure 15–17**). In naturally occurring steroids, virtually all the rings are in the "chair" form, which is the more stable conformation. With respect to each other, the rings can be either *cis* or *trans* (**Figure 15–18**). The junction between the A and B rings can be *cis* or *trans* in naturally occurring steroids. That between B and C is *trans*, as is usually the C/D junction. Bonds attaching substituent groups above the plane of the rings (β bonds) are shown with bold solid lines, whereas those bonds attaching groups below (α bonds) are indicated with broken lines. The A ring of a 5α steroid is always *trans* to the B ring, whereas it is *cis* in a 5β steroid. The methyl groups attached to C10 and C13 are invariably in the β configuration.

Cholesterol Is a Significant Constituent of Many Tissues

Cholesterol (**Figure 15–19**) is widely distributed in all cells of the body but particularly in nervous tissue. It is a major constituent of the plasma membrane and of plasma lipoproteins. It is often found as **cholesteryl ester,** where the hydroxyl group on position 3 is esterified with a long-chain fatty acid. It occurs in animals but not in plants or bacteria.

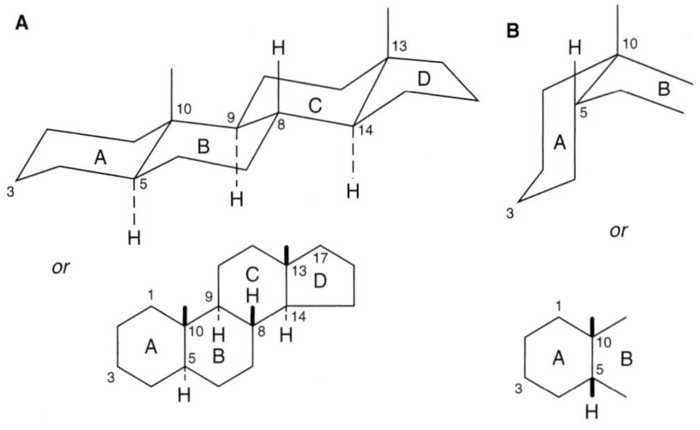

Ergosterol Is a Precursor of Vitamin D

Ergosterol occurs in plants and yeast and is important as a precursor of vitamin D (**Figure 15–20**). When irradiated with ultraviolet light, ring B is opened to form vitamin D_2 in a process similar to the one that forms vitamin D_3 from 7-dehydrocholesterol in the skin (Figure 44–3).

Polyprenoids Share the Same Parent Compound as Cholesterol

Although not steroids, polyprenoids are related because they are synthesized, like cholesterol (Figure 26–2), from five-carbon isoprene units (**Figure 15–21**). They include **ubiquinone** (Chapter 13), which participates in the respiratory chain in mitochondria, and the long-chain alcohol **dolichol** (**Figure 15–22**), which takes part in glycoprotein synthesis by transferring carbohydrate residues to asparagine residues of the polypeptide (Chapter 47). Plant-derived isoprenoid compounds include rubber, camphor, the fat-soluble vitamins A, D, E, and K, and β-carotene (provitamin A).

LIPID PEROXIDATION IS A SOURCE OF FREE RADICALS

Peroxidation (**auto-oxidation**) of lipids exposed to oxygen is responsible not only for deterioration of foods (**rancidity**), but also for damage to tissues in vivo, where it may be a cause of cancer, inflammatory diseases, atherosclerosis, and aging. The deleterious effects are considered to be caused by **free radicals** (ROO•, RO•, OH•) produced during peroxide formation from fatty acids containing methylene-interrupted double bonds, that is, those found in the naturally occurring polyunsaturated fatty acids (**Figure 15–23**). Lipid peroxidation is a chain reaction providing a continuous supply of free radicals that initiate

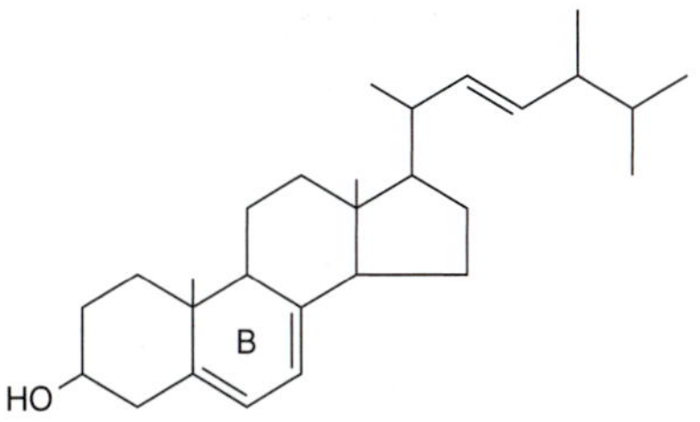

FIGURE 15–18 Generalized steroid nucleus, showing (A) an all-*trans* configuration between adjacent rings and (B) a *cis* configuration between rings A and B.

FIGURE 15–20 Ergosterol.

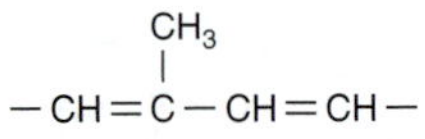

FIGURE 15–21 Isoprene unit.

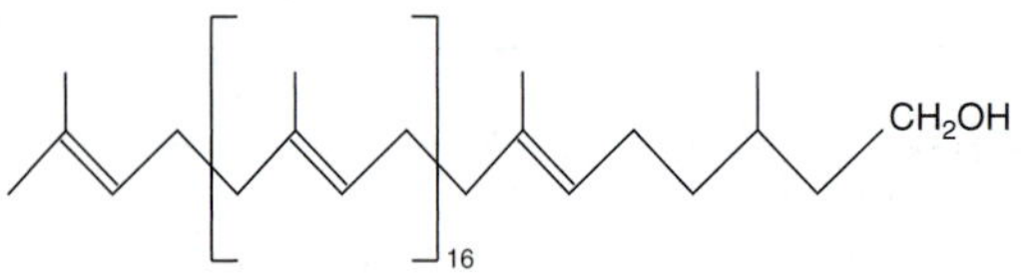

FIGURE 15–22 Dolichol—a C95 alcohol.

further peroxidation and thus has potentially devastating effects. The whole process can be depicted as follows:

1. Initiation:

$$ROOH + Metal^{(n)+} \rightarrow ROO^\bullet + Metal^{(n-1)+} + H^+$$

$$X^\bullet + RH \rightarrow R^\bullet + XH$$

2. Propagation:

$$R^\bullet + O_2 \rightarrow ROO^\bullet$$

$$ROO^\bullet + RH \rightarrow ROOH + R^\bullet, etc$$

3. Termination:

$$ROO^\bullet + ROO^\bullet \rightarrow ROOR + O_2$$

$$ROO^\bullet + R^\bullet \rightarrow ROOR$$

$$R^\bullet + R^\bullet \rightarrow RR$$

To control and reduce lipid peroxidation, both humans in their activities and nature invoke the use of **antioxidants.** Propyl gallate, butylated hydroxyanisole (BHA), and butylated hydroxytoluene (BHT) are antioxidants used as food additives. Naturally occurring antioxidants include vitamin E (tocopherol), which is lipid soluble, and urate and vitamin C, which are water soluble. Beta-carotene is an antioxidant at low PO_2. Antioxidants fall into two classes: (1) preventive antioxidants, which reduce the rate of chain initiation and (2) chain-breaking antioxidants, which interfere with chain propagation. Preventive antioxidants include catalase and other peroxidases such as glutathione peroxidase (Figure 21–3) that react with ROOH; selenium, which is an essential component of glutathione peroxidase and regulates its activity, and chelators of metal ions such as EDTA (ethylenediaminetetraacetate) and DTPA (diethylenetriaminepentaacetate). In vivo, the principal chain-breaking antioxidants are superoxide dismutase, which acts in the aqueous phase to trap superoxide free radicals

$(O_2^{\bar{\;}})$ urate, and vitamin E, which acts in the lipid phase to trap $ROO^\bullet$ radicals (Figure 44–6).

Peroxidation is also catalyzed in vivo by heme compounds and by **lipoxygenases** found in platelets and leukocytes. Other products of auto-oxidation or enzymic oxidation of physiologic significance include **oxysterols** (formed from cholesterol) and **isoprostanes** (formed from the peroxidation of polyunsaturated fatty acids such as arachidonic acid).

AMPHIPATHIC LIPIDS SELF-ORIENT AT OIL: WATER INTERFACES

They Form Membranes, Micelles, Liposomes, & Emulsions

In general, lipids are insoluble in water since they contain a predominance of nonpolar (hydrocarbon) groups. However, fatty acids, phospholipids, sphingolipids, bile salts, and, to a lesser extent, cholesterol contain polar groups. Therefore, a part of the molecule is **hydrophobic,** or water insoluble; and a part is **hydrophilic,** or water soluble. Such molecules are described as **amphipathic (Figure 15–24)**. They become oriented at oil–water interfaces with the polar group in the water phase and the nonpolar group in the oil phase. A bilayer of such amphipathic lipids is the basic structure in biologic **membranes** (Chapter 40). When a critical concentration of these lipids is present in an aqueous medium, they form **micelles. Liposomes** may be formed by sonicating an amphipathic lipid in an aqueous medium. They consist of spheres of lipid bilayers that enclose part of the aqueous medium. Aggregations of bile

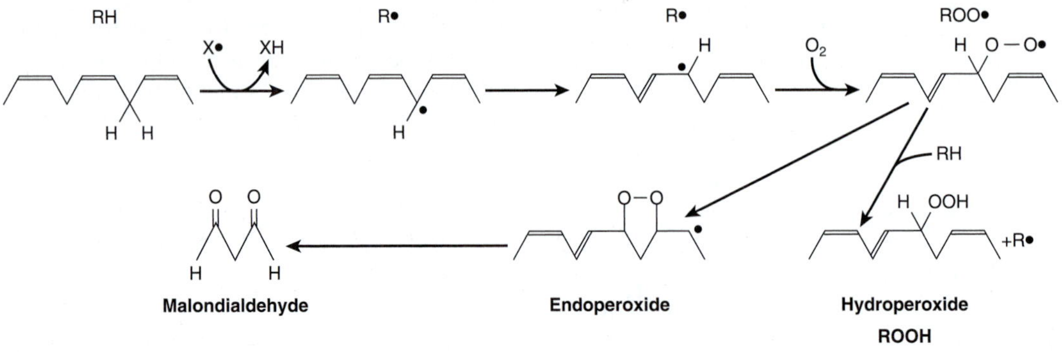

FIGURE 15–23 **Lipid peroxidation.** The reaction is initiated by an existing free radical (X•), by light, or by metal ions. Malondialdehyde is only formed by fatty acids with three or more double bonds and is used as a measure of lipid peroxidation together with ethane from the terminal two carbons of ω3 fatty acids and pentane from the terminal five carbons of ω6 fatty acids.

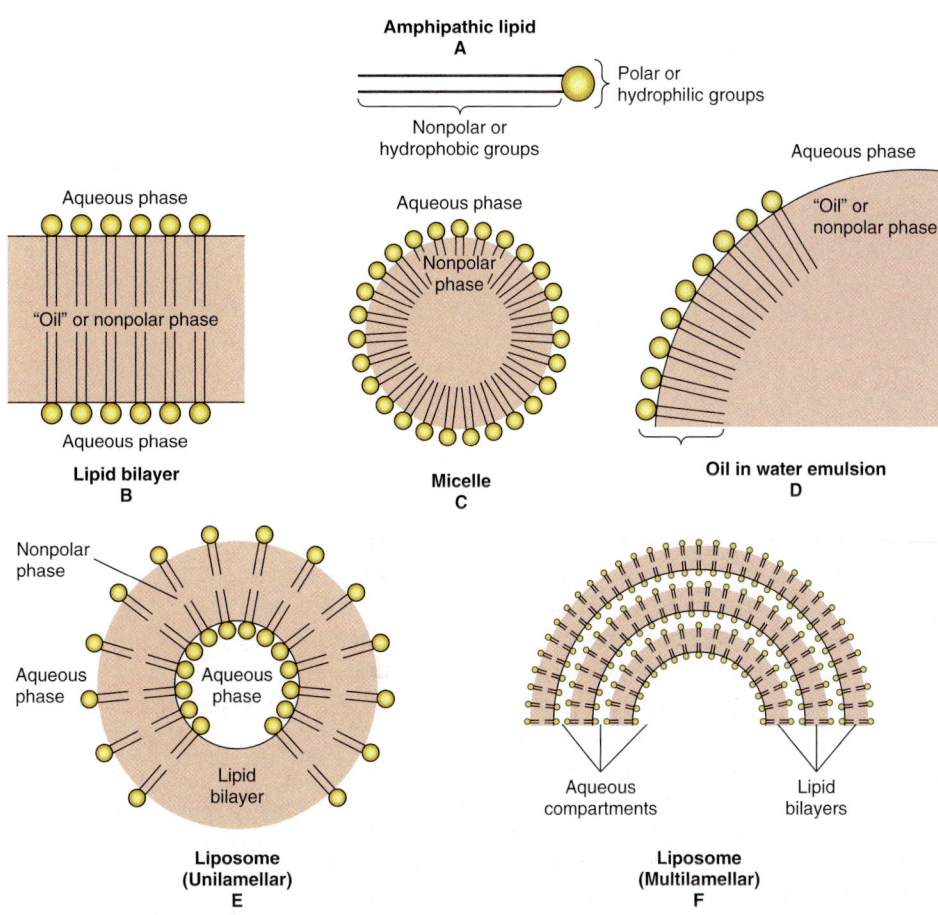

FIGURE 15–24 Formation of lipid membranes, micelles, emulsions, and liposomes from amphipathic lipids, for example, phospholipids.

salts into micelles and liposomes and the formation of mixed micelles with the products of fat digestion are important in facilitating absorption of lipids from the intestine. Liposomes are of potential clinical use—particularly when combined with tissue-specific antibodies—as carriers of drugs in the circulation, targeted to specific organs, for example, in cancer therapy. In addition, they are used for gene transfer into vascular cells and as carriers for topical and transdermal delivery of drugs and cosmetics. **Emulsions** are much larger particles, formed usually by nonpolar lipids in an aqueous medium. These are stabilized by emulsifying agents such as amphipathic lipids (eg, lecithin), which form a surface layer separating the main bulk of the nonpolar material from the aqueous phase (Figure 15–24).

SUMMARY

- Lipids have the common property of being relatively insoluble in water (hydrophobic) but soluble in nonpolar solvents. Amphipathic lipids also contain one or more polar groups, making them suitable as constituents of membranes at lipid–water interfaces.

- The lipids of major physiologic significance are fatty acids and their esters, together with cholesterol and other steroids.

- Long-chain fatty acids may be saturated, monounsaturated, or polyunsaturated, according to the number of double bonds present. Their fluidity decreases with chain length and increases according to degree of unsaturation.

- Eicosanoids are formed from 20-carbon polyunsaturated fatty acids and make up an important group of physiologically and pharmacologically active compounds known as prostaglandins, thromboxanes, leukotrienes, and lipoxins.

- The esters of glycerol are quantitatively the most significant lipids, represented by triacylglycerol ("fat"), a major constituent of some lipoprotein classes and the storage form of lipid in adipose tissue. Phosphoacylglycerols are amphipathic lipids and have important roles—as major constituents of membranes and the outer layer of lipoproteins, as surfactant in the lung, as precursors of second messengers, and as constituents of nervous tissue.

- Glycolipids are also important constituents of nervous tissue such as brain and the outer leaflet of the cell membrane, where they contribute to the carbohydrates on the cell surface.

- Cholesterol, an amphipathic lipid, is an important component of membranes. It is the parent molecule from which all other steroids in the body, including major hormones such as the adrenocortical and sex hormones, D vitamins, and bile acids, are synthesized.

■ Peroxidation of lipids containing polyunsaturated fatty acids leads to generation of free radicals that damage tissues and cause disease.

REFERENCES

Benzie IFF: Lipid peroxidation: a review of causes, consequences, measurement and dietary influences. Int J Food Sci Nutr 1996;47:233.

Christie WW: *Lipid Analysis,* 3rd ed. The Oily Press, 2003.

Dowhan W, Bodanov H, Mileykovskaya E: Functional roles of lipids in membranes. In: *Biochemistry of Lipids, Lipoproteins and Membranes,* 5th ed. Vance DE, Vance JE (editors). Elsevier, 2008:1–37.

Gunstone FD, Harwood JL, Dijkstra AJ: *The Lipid Handbook with CD-Rom.* CRC Press, 2007.

Gurr MI, Harwood JL, Frayn K: *Lipid Biochemistry.* Blackwell Publishing, 2002.

Overview of Metabolism & the Provision of Metabolic Fuels

16

David A. Bender, PhD & Peter A. Mayes, PhD, DSc

OBJECTIVES

After studying this chapter, you should be able to:

- Explain what is meant by anabolic, catabolic and amphibolic metabolic pathways.
- Describe in outline the metabolism of carbohydrates, lipids and amino acids at the level of tissues and organs, and at the subcellular level, and how metabolic fuels are interconvertible.
- Describe the ways in which flux of metabolites through metabolic pathways is regulated.
- Describe how a supple of metabolic fuels is provided in both the fed and fasting states; the formation of metabolic fuels reserves in the fed state and their mobilization in fasting.

BIOMEDICAL IMPORTANCE

Metabolism is the term used to describe the interconversion of chemical compounds in the body, the pathways taken by individual molecules, their interrelationships, and the mechanisms that regulate the flow of metabolites through the pathways. Metabolic pathways fall into three categories. (1) **Anabolic pathways,** which are those involved in the synthesis of larger and more complex compounds from smaller precursors—for example, the synthesis of protein from amino acids and the synthesis of reserves of triacylglycerol and glycogen. Anabolic pathways are endothermic. (2) **Catabolic pathways,** which are involved in the breakdown of larger molecules, commonly involving oxidative reactions; they are exothermic, producing reducing equivalents, and, mainly via the respiratory chain, ATP. (3) **Amphibolic pathways,** which occur at the "crossroads" of metabolism, acting as links between the anabolic and catabolic pathways, for example, the citric acid cycle.

Knowledge of normal metabolism is essential for an understanding of abnormalities underlying disease. Normal metabolism includes adaptation to periods of starvation, exercise, pregnancy, and lactation. Abnormal metabolism may result from nutritional deficiency, enzyme deficiency, abnormal secretion of hormones, or the actions of drugs and toxins.

A 70-kg adult human being requires about 8–12 MJ (1920–2900 kcal) from metabolic fuels each day, depending on the physical activity. Larger animals require less, and smaller animals more, per kg body weight, and growing children and animals have a proportionally higher requirement to allow for the energy cost of growth. For human beings, this requirement is met from carbohydrates (40–60%), lipids (mainly triacylglycerol, 30–40%), and protein (10–15%), as well as alcohol. The mix of carbohydrate, lipid, and protein being oxidized varies, depending on whether the subject is in the fed or fasting state, and on the duration and intensity of physical work.

The requirement for metabolic fuels is relatively constant throughout the day, since the average physical activity increases metabolic rate only by about 40–50% over the basal or resting metabolic rate. However, most people consume their daily intake of metabolic fuels in two or three meals, so there is a need to form reserves of carbohydrate (glycogen in liver and muscle) and lipid (triacylglycerol in adipose tissue) in the period following a meal, for use during the intervening time when there is no intake of food.

If the intake of metabolic fuels is consistently greater than energy expenditure, the surplus is stored, largely as triacylglycerol in adipose tissue, leading to the development of **obesity** and its associated health hazards. By contrast, if the intake of metabolic fuels is consistently lower than energy expenditure, there are negligible reserves of fat and carbohydrate, and amino acids arising from protein turnover are used for energy-yielding metabolism rather than replacement protein synthesis, leading to **emaciation,** wasting, and, eventually, death (see Chapter 43).

In the fed state, after a meal, there is an ample supply of carbohydrate, and the metabolic fuel for most tissues is glucose. In the fasting state glucose must be spared for use by the central nervous system (which is largely dependent on glucose) and the red blood cells (which are wholly reliant on glucose). Therefore, tissues that can use fuels other than glucose do so; muscle and liver oxidize fatty acids and the liver synthesizes ketone bodies from fatty acids to export to muscle and other tissues. As glycogen reserves become depleted, amino acids arising from protein turnover are used for **gluconeogenesis.**

The formation and utilization of reserves of triacylglycerol and glycogen, and the extent to which tissues take up and oxidize glucose, are largely controlled by the hormones **insulin** and **glucagon.** In **diabetes mellitus,** there is either impaired synthesis and secretion of insulin (type I diabetes, sometimes called juvenile onset, or insulin-dependent diabetes) or impaired sensitivity of tissues to insulin action (type II diabetes, sometimes called adult onset or noninsulin-dependent diabetes), leading to severe metabolic derangement. In cattle, the demands of heavy lactation can lead to ketosis, as can the demands of twin pregnancy in sheep.

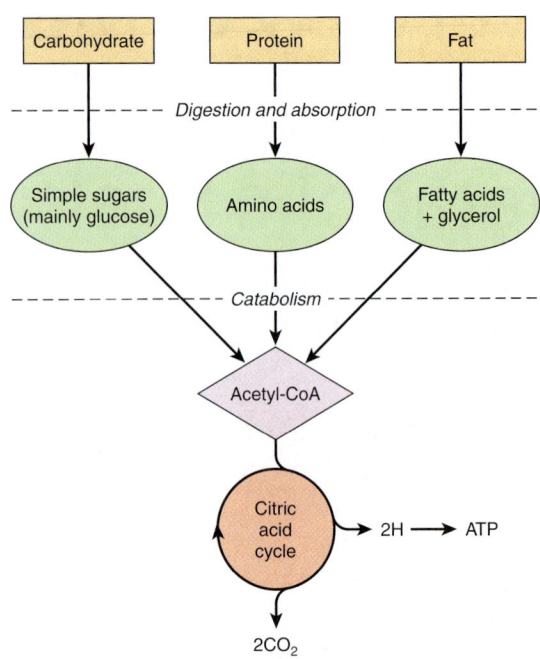

FIGURE 16–1 **Outline of the pathways for the catabolism of dietary carbohydrate, protein, and fat.** All the pathways lead to the production of acetyl-CoA, which is oxidized in the citric acid cycle, ultimately yielding ATP by the process of oxidative phosphorylation.

PATHWAYS THAT PROCESS THE MAJOR PRODUCTS OF DIGESTION

The nature of the diet sets the basic pattern of metabolism. There is a need to process the products of digestion of dietary carbohydrate, lipid, and protein. These are mainly glucose, fatty acids and glycerol, and amino acids, respectively. In ruminants (and, to a lesser extent, other herbivores), dietary cellulose is fermented by symbiotic microorganisms to short-chain fatty acids (acetic, propionic, butyric), and metabolism in these animals is adapted to use these fatty acids as major substrates. All the products of digestion are metabolized to a **common product, acetyl-CoA,** which is then oxidized by the **citric acid cycle (Figure 16–1).**

Carbohydrate Metabolism Is Centered on the Provision & Fate of Glucose

Glucose is the major fuel of most tissues (**Figure 16–2**). It is metabolized to pyruvate by the pathway of **glycolysis.** Aerobic tissues metabolize pyruvate to **acetyl-CoA,** which can enter the citric acid cycle for complete oxidation to CO_2 and H_2O, linked to the formation of ATP in the process of **oxidative phosphorylation** (Figure 13–2). Glycolysis can also occur anaerobically (in the absence of oxygen) when the end product is lactate.

Glucose and its metabolites also take part in other processes, eg. (1) Synthesis of the storage polymer **glycogen** in skeletal muscle and liver. (2) The **pentose phosphate pathway,** an alternative to part of the pathway of glycolysis. It is a source of reducing equivalents (NADPH) for fatty acid synthesis and the source of **ribose** for nucleotide and nucleic

acid synthesis. (3) Triose phosphates give rise to the **glycerol moiety** of triacylglycerols. (4) Pyruvate and intermediates of the citric acid cycle provide the carbon skeletons for the synthesis of nonessential or dispensable **amino acids,** and acetyl-CoA is the precursor of **fatty acids** and **cholesterol** (and hence of all steroids synthesized in the body). **Gluconeogenesis** is the process of forming glucose from noncarbohydrate precursors, for example, lactate, amino acids, and glycerol.

Lipid Metabolism Is Concerned Mainly with Fatty Acids & Cholesterol

The source of long-chain fatty acids is either dietary lipid or de novo synthesis from acetyl-CoA derived from carbohydrate or amino acids. Fatty acids may be oxidized to **acetyl-CoA** (β-oxidation) or esterified with glycerol, forming **triacylglycerol** (fat) as the body's main fuel reserve.

Acetyl-CoA formed by β-oxidation may undergo three fates (**Figure 16–3**).

1. As with acetyl-CoA arising from glycolysis, it is **oxidized** to $CO_2 + H_2O$ via the citric acid cycle.
2. It is the precursor for synthesis of **cholesterol** and other **steroids.**
3. In the liver, it is used to form **ketone bodies** (acetoacetate and 3-hydroxybutyrate), which are important fuels in prolonged fasting and starvation.

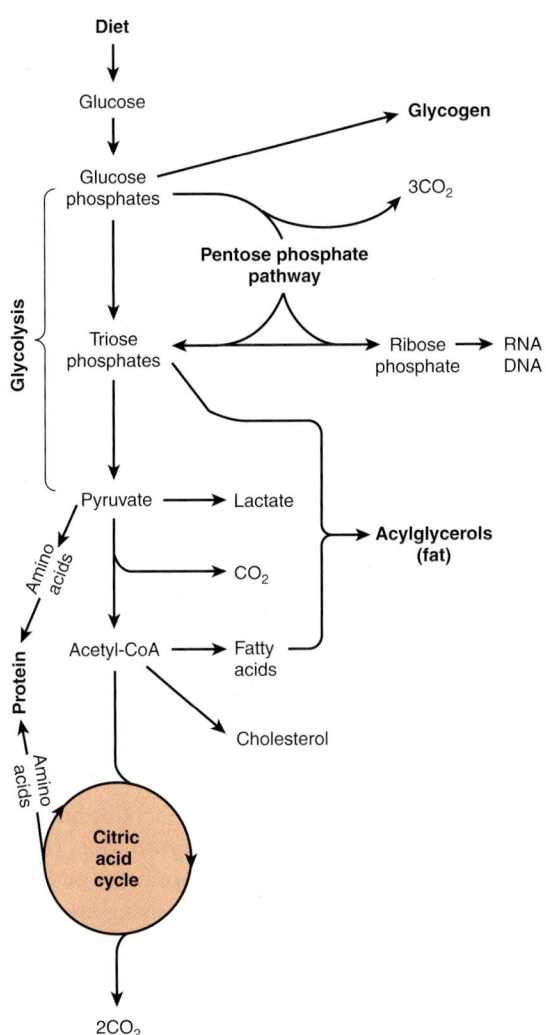

FIGURE 16–2 **Overview of carbohydrate metabolism showing the major pathways and end products.** Gluconeogenesis is not shown.

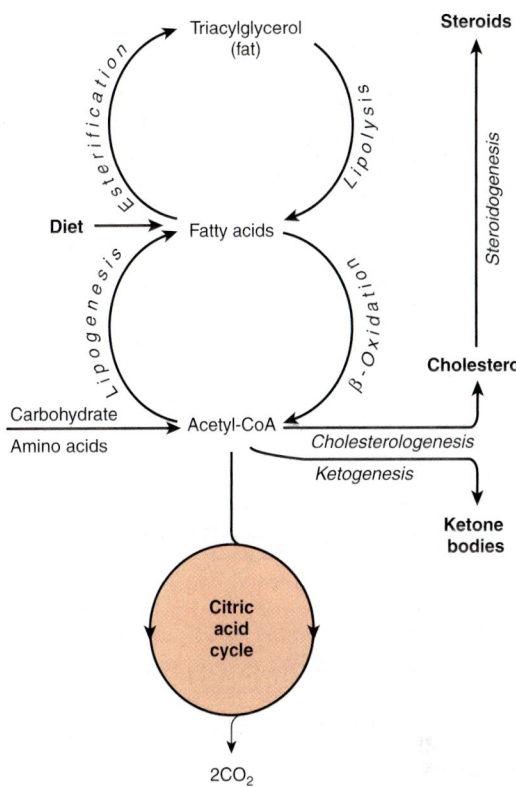

FIGURE 16–3 **Overview of fatty acid metabolism showing the major pathways and end products.** The ketone bodies are acetoacetate, 3-hydroxybutyrate, and acetone.

Much of Amino Acid Metabolism Involves Transamination

The amino acids are required for protein synthesis (**Figure 16–4**). Some must be supplied in the diet (the **essential or indispensable amino acids**), since they cannot be synthesized in the body. The remainder are **nonessential or dispensable amino acids,** which are supplied in the diet, but can also be formed from metabolic intermediates by **transamination** using the amino group from other amino acids. After **deamination,** amino nitrogen is excreted as **urea,** and the carbon skeletons that remain after transamination may (1) be oxidized to CO_2 via the citric acid cycle, (2) be used to synthesize glucose (gluconeogenesis), or (3) form ketone bodies or acetyl CoA, which may be oxidized or be used for synthesis of fatty acids.

Several amino acids are also the precursors of other compounds, for example, purines, pyrimidines, hormones such as epinephrine and thyroxine, and neurotransmitters.

METABOLIC PATHWAYS MAY BE STUDIED AT DIFFERENT LEVELS OF ORGANIZATION

In addition to studies in the whole organism, the location and integration of metabolic pathways is revealed by studies at several levels of organization. (1) At the **tissue and organ level** the nature of the substrates entering and metabolites leaving tissues and organs is defined. (2) At the **subcellular level** each cell organelle (eg, the mitochondrion) or compartment (eg, the cytosol) has specific roles that form part of a subcellular pattern of metabolic pathways.

At the Tissue & Organ Level, the Blood Circulation Integrates Metabolism

Amino acids resulting from the digestion of dietary protein and **glucose** resulting from the digestion of carbohydrates are absorbed via the hepatic portal vein. The liver has the role of regulating the blood concentration of these water-soluble metabolites (**Figure 16–5**). In the case of glucose, this is achieved by taking up glucose in excess of immediate requirements and using it to synthesize glycogen (**glycogenesis,** Chapter 19) or to fatty acids (**lipogenesis,** Chapter 23). Between meals, the liver acts to maintain the blood glucose

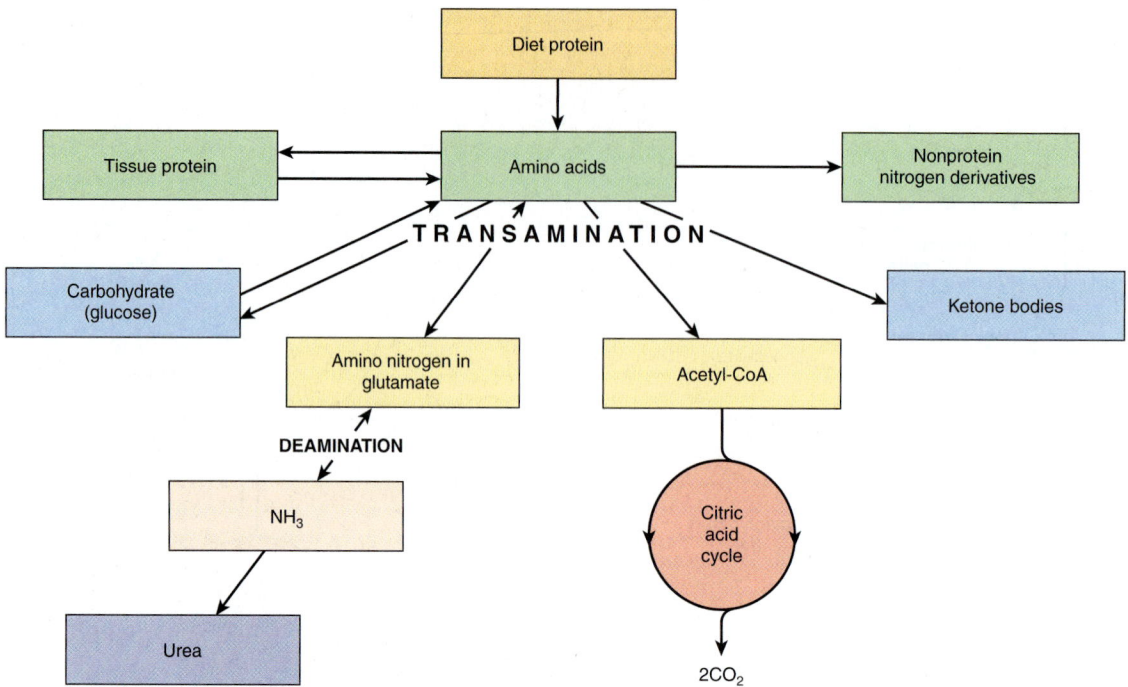

FIGURE 16–4 Overview of amino acid metabolism showing the major pathways and end products.

concentration by breaking down glycogen (**glycogenolysis,** Chapter 19) and, together with the kidney, by converting non-carbohydrate metabolites such as lactate, glycerol, and amino acids to glucose (**gluconeogenesis,** Chapter 20). The maintenance of an adequate concentration of blood glucose is vital

for those tissues for which it is the major fuel (the brain) or the only fuel (erythrocytes). The liver also **synthesizes the major plasma proteins** (eg, albumin) and **deaminates amino acids** that are in excess of requirements, synthesizing urea, which is transported to the kidney and excreted (Chapter 28).

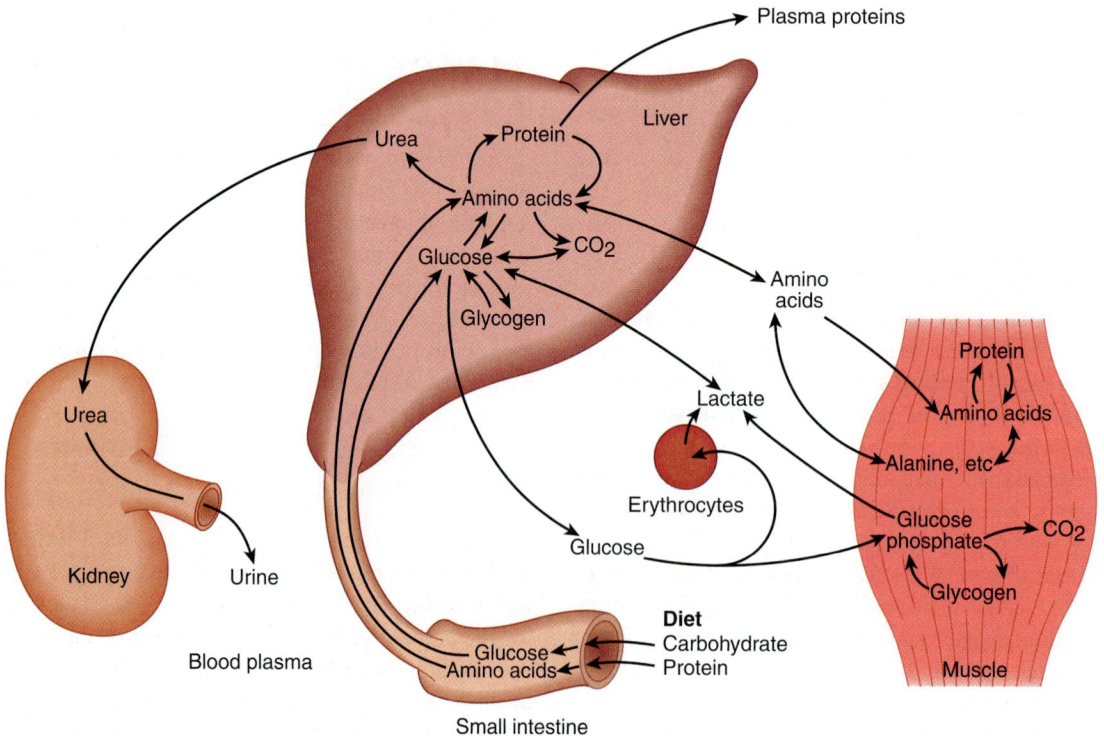

FIGURE 16–5 **Transport and fate of major carbohydrate and amino acid substrates and metabolites.** Note that there is little free glucose in muscle, since it is rapidly phosphorylated upon entry.

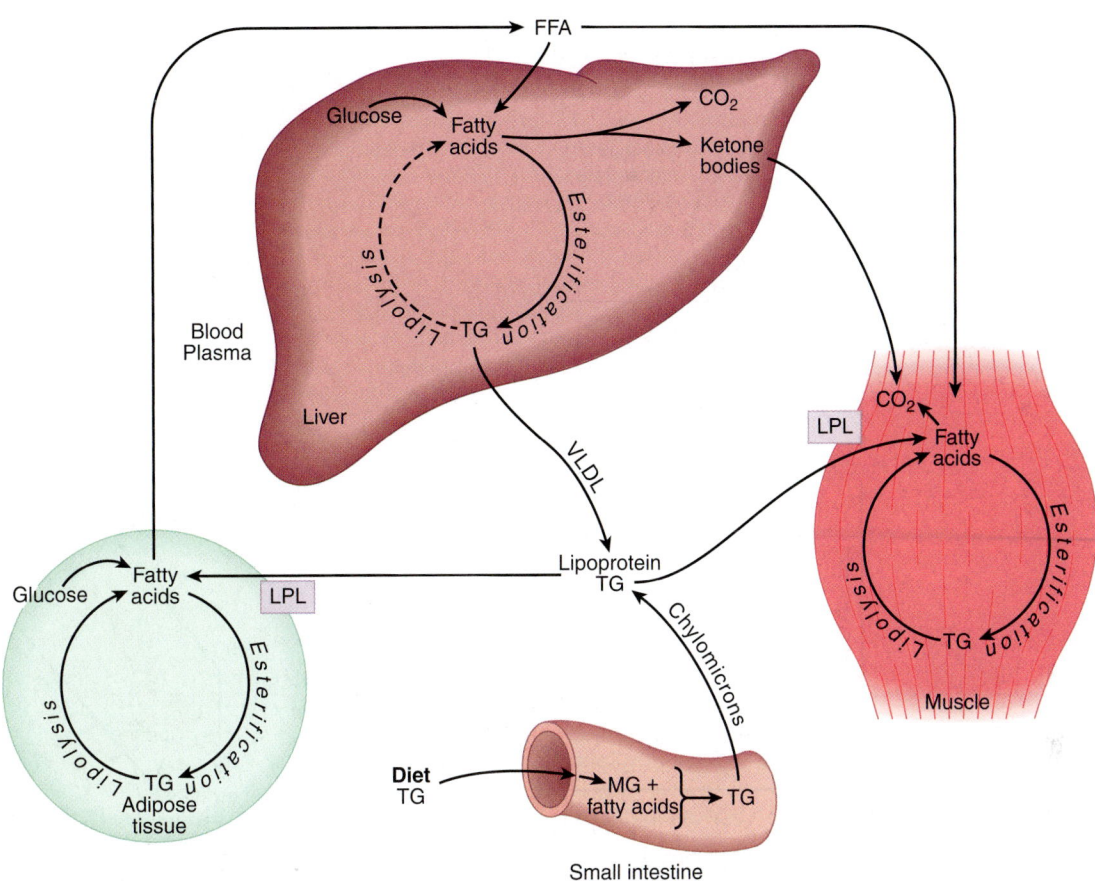

FIGURE 16–6 **Transport and fate of major lipid substrates and metabolites.** (FFA, free fatty acids; LPL, lipoprotein lipase; MG, monoacylglycerol; TG, triacylglycerol; VLDL, very low density lipoprotein.)

Skeletal muscle utilizes glucose as a fuel, both aerobically, forming CO_2, and anaerobically, forming lactate. It stores glycogen as a fuel for use in muscle contraction and synthesizes muscle protein from plasma amino acids. Muscle accounts for approximately 50% of body mass and consequently represents a considerable store of protein that can be drawn upon to supply amino acids for gluconeogenesis in starvation (Chapter 20).

Lipids in the diet (**Figure 16–6**) are mainly triacylglycerol, and are hydrolyzed to monoacylglycerols and fatty acids in the gut, then re-esterified in the intestinal mucosa. Here they are packaged with protein and secreted into the lymphatic system and thence into the bloodstream as **chylomicrons,** the largest of the plasma **lipoproteins.** Chylomicrons also contain other lipid-soluble nutrients. Unlike glucose and amino acids, chylomicron triacylglycerol is not taken up directly by the liver. It is first metabolized by tissues that have **lipoprotein lipase,** which hydrolyzes the triacylglycerol, releasing fatty acids that are incorporated into tissue lipids or oxidized as fuel. The chylomicron remnants are cleared by the liver. The other major source of long-chain fatty acids is synthesis (**lipogenesis**) from carbohydrate, in adipose tissue and the liver.

Adipose tissue triacylglycerol is the main fuel reserve of the body. It is hydrolyzed (**lipolysis**) and glycerol and free fatty acids are released into the circulation. Glycerol is a substrate for gluconeogenesis. The fatty acids are transported bound to serum albumin; they are taken up by most tissues (but not brain or erythrocytes) and either esterified to triacylglycerols for storage or oxidized as a fuel. In the liver, newly synthesized triacylglycerol and triacylglycerol from chylomicron remnants (see Figure 25–3) is secreted into the circulation in **very low density lipoprotein (VLDL).** This triacylglycerol undergoes a fate similar to that of chylomicrons. Partial oxidation of fatty acids in the liver leads to **ketone body** production (**ketogenesis,** Chapter 22). Ketone bodies are exported to extrahepatic tissues, where they act as a fuel in prolonged fasting and starvation.

At the Subcellular Level, Glycolysis Occurs in the Cytosol & the Citric Acid Cycle in the Mitochondria

Compartmentation of pathways in separate subcellular compartments or organelles permits integration and regulation of metabolism. Not all pathways are of equal importance in all cells. **Figure 16–7** depicts the subcellular compartmentation of metabolic pathways in a liver parenchymal cell.

The central role of the **mitochondrion** is immediately apparent, since it acts as the focus of carbohydrate, lipid, and

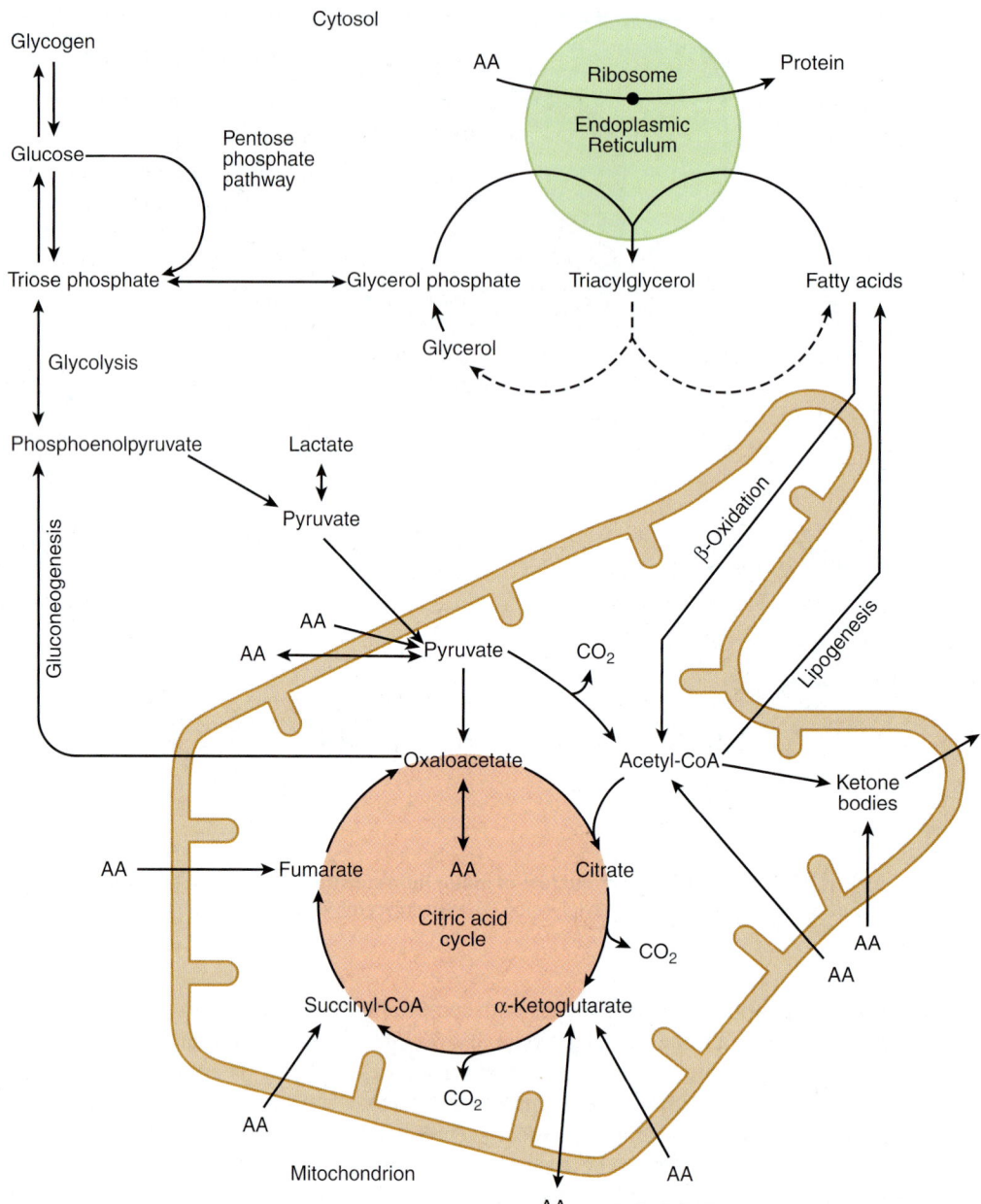

FIGURE 16–7 Intracellular location and overview of major metabolic pathways in a liver parenchymal cell. (AA →, metabolism of one or more essential amino acids; AA ↔, metabolism of one or more nonessential amino acids.)

amino acid metabolism. It contains the enzymes of the citric acid cycle (Chapter 17), β-oxidation of fatty acids and ketogenesis (Chapter 22), as well as the respiratory chain and ATP synthase (Chapter 13).

Glycolysis (Chapter 18), the pentose phosphate pathway (Chapter 21), and fatty acid synthesis (Chapter 23) all occur in the cytosol. In gluconeogenesis (Chapter 20), substrates such as lactate and pyruvate, which are formed in the cytosol, enter the mitochondrion to yield **oxaloacetate** as a precursor for the synthesis of glucose in the cytosol.

The membranes of the **endoplasmic reticulum** contain the enzyme system for **triacylglycerol synthesis** (Chapter 24), and the **ribosomes** are responsible for **protein synthesis** (Chapter 37).

THE FLUX OF METABOLITES THROUGH METABOLIC PATHWAYS MUST BE REGULATED IN A CONCERTED MANNER

Regulation of the overall flux through a pathway is important to ensure an appropriate supply of the products of that pathway. It is achieved by control of one or more key reactions in the pathway, catalyzed by **regulatory enzymes**. The physicochemical factors that control the rate of an enzyme-catalyzed reaction, such as substrate concentration, are of primary importance in the control of the overall rate of a metabolic pathway (Chapter 9).

Nonequilibrium Reactions Are Potential Control Points

In a reaction at equilibrium, the forward and reverse reactions occur at equal rates, and there is therefore no net flux in either direction.

$$A \leftrightarrow B \leftrightarrow C \leftrightarrow D$$

In vivo, under "steady-state" conditions, there is a net flux from left to right because there is a continuous supply of A and removal of D. In practice, there are normally one or more **nonequilibrium** reactions in a metabolic pathway, where the reactants are present in concentrations that are far from equilibrium. In attempting to reach equilibrium, large losses of free energy occur, making this type of reaction essentially irreversible.

$$A \leftrightarrow B \overset{\text{Heat}}{\leftrightarrow} C \leftrightarrow D$$

Such a pathway has both flow and direction. The enzymes catalyzing nonequilibrium reactions are usually present in low concentration and are subject to a variety of regulatory mechanisms. However, most reactions in metabolic pathways cannot be classified as equilibrium or nonequilibrium, but fall somewhere between the two extremes.

The Flux-Generating Reaction Is the First Reaction in a Pathway That Is Saturated with the Substrate

It may be identified as a nonequilibrium reaction in which the K_m of the enzyme is considerably lower than the normal substrate concentration. The first reaction in glycolysis, catalyzed by hexokinase (Figure 18–2), is such a flux-generating step because its K_m for glucose of 0.05 mmol/L is well below the normal blood glucose concentration of 5 mmol/L. Later reactions then control the rate of flux through the pathway.

ALLOSTERIC & HORMONAL MECHANISMS ARE IMPORTANT IN THE METABOLIC CONTROL OF ENZYME-CATALYZED REACTIONS

A hypothetical metabolic pathway is shown in **Figure 16–8**, in which reactions $A \leftrightarrow B$ and $C \leftrightarrow D$ are equilibrium reactions and $B \rightarrow C$ is a nonequilibrium reaction. The flux through such a pathway can be regulated by the availability of substrate A. This depends on its supply from the blood,

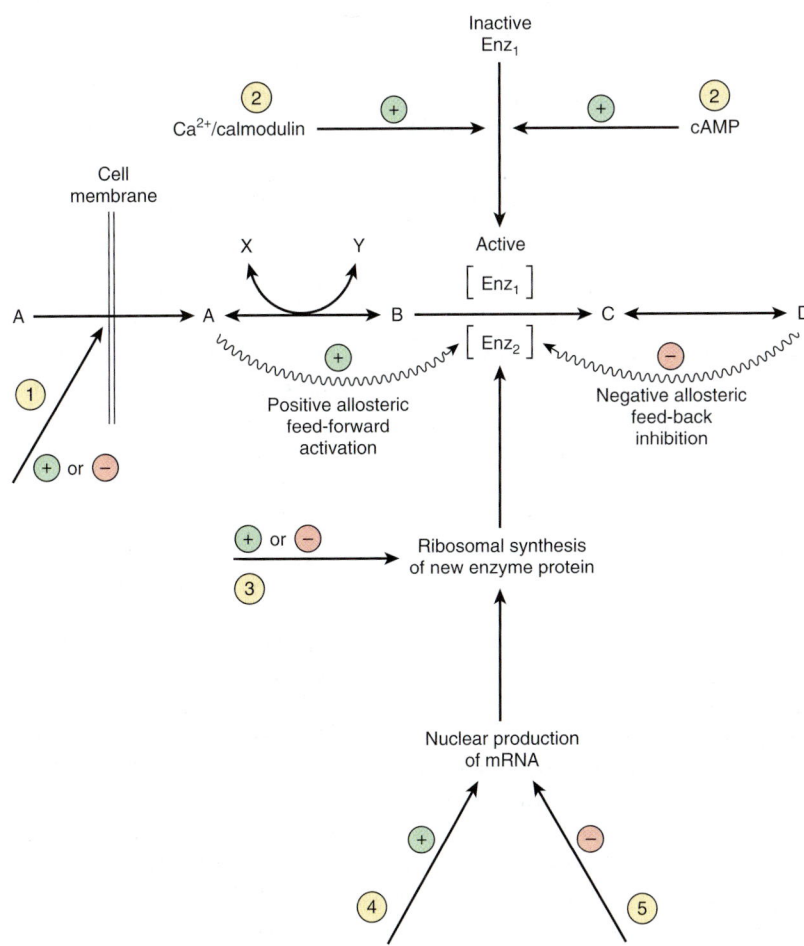

FIGURE 16–8 **Mechanisms of control of an enzyme-catalyzed reaction.** Circled numbers indicate possible sites of action of hormones: ① alteration of membrane permeability; ② conversion of an inactive to an active enzyme, usually involving phosphorylation/ dephosphorylation reactions; ③ alteration of the rate translation of mRNA at the ribosomal level; ④ induction of new mRNA formation; and ⑤ repression of mRNA formation. ① and ② are rapid, whereas ③ through ⑤ are slower ways of regulating enzyme activity.

which in turn depends on either food intake or key reactions that release substrates from tissue reserves into the bloodstream, for example, glycogen phosphorylase in liver (Figure 19–1) and hormone-sensitive lipase in adipose tissue (Figure 25–8). It also depends on the transport of substrate A into the cell. The flux is also determined by removal of the end product D and the availability of cosubstrates or cofactors represented by X and Y. Enzymes catalyzing non-equilibrium reactions are often allosteric proteins subject to the rapid actions of "feed-back" or "feed-forward" control by **allosteric modifiers,** in immediate response to the needs of the cell (Chapter 9). Frequently, the end product of a biosynthetic pathway inhibits the enzyme catalyzing the first reaction in the pathway. Other control mechanisms depend on the action of **hormones** responding to the needs of the body as a whole; they may act rapidly by altering the activity of existing enzyme molecules, or slowly by altering the rate of enzyme synthesis (see Chapter 42).

MANY METABOLIC FUELS ARE INTERCONVERTIBLE

Carbohydrate in excess of requirements for immediate energy-yielding metabolism and formation of glycogen reserves in muscle and liver can readily be used for synthesis of fatty acids, and hence triacylglycerol in both adipose tissue and liver (whence it is exported in very low-density lipoprotein). The importance of lipogenesis in humans is unclear; in Western countries dietary fat provides 35–45% of energy intake, while in less-developed countries, where carbohydrate may provide 60–75% of energy intake, the total intake of food is so low that there is little surplus for lipogenesis anyway. A high intake of fat inhibits lipogenesis in the adipose tissue and liver.

Fatty acids (and ketone bodies formed from them) cannot be used for the synthesis of glucose. The reaction of pyruvate dehydrogenase, forming acetyl-CoA, is irreversible, and for every two-carbon unit from acetyl-CoA that enters the citric acid cycle, there is a loss of two carbon atoms as carbon dioxide before oxaloacetate is reformed. This means that acetyl-CoA (and hence any substrates that yield acetyl-CoA) can never be used for gluconeogenesis. The (relatively rare) fatty acids with an odd number of carbon atoms yield propionyl CoA as the product of the final cycle of β oxidation, and this can be a substrate for gluconeogenesis, as can the glycerol released by lipolysis of adipose tissue triacylglycerol reserves.

Most of the amino acids in excess of requirements for protein synthesis (arising from the diet or from tissue protein turnover) yield pyruvate, or four- and five-carbon intermediates of the citric acid cycle (Chapter 29). Pyruvate can be carboxylated to oxaloacetate, which is the primary substrate for gluconeogenesis, and the other intermediates of the cycle also result in a net increase in the formation of oxaloacetate, which is then available for gluconeogenesis. These amino acids are classified as **glucogenic.** Two amino acids (lysine and leucine)

yield only acetyl-CoA on oxidation, and hence cannot be used for gluconeogenesis, and four others (ie, phenylalanine, tyrosine, tryptophan, and isoleucine) give rise to both acetyl-CoA and intermediates that can be used for gluconeogenesis. Those amino acids that give rise to acetyl-CoA are referred to as **ketogenic,** because in prolonged fasting and starvation much of the acetyl-CoA is used for synthesis of ketone bodies in the liver.

A SUPPLY OF METABOLIC FUELS IS PROVIDED IN BOTH THE FED & FASTING STATES

Glucose Is Always Required by the Central Nervous System and Erythrocytes

Erythrocytes lack mitochondria and hence are wholly reliant on (anaerobic) glycolysis and the pentose phosphate pathway at all times. The brain can metabolize ketone bodies to meet about 20% of its energy requirements; the remainder must be supplied by glucose. The metabolic changes that occur in the fasting state and starvation are the consequences of the need to preserve glucose and the limited reserves of glycogen in liver and muscle for use by the brain and red blood cells, and to ensure the provision of alternative metabolic fuels for other tissues. In pregnancy, the fetus requires a significant amount of glucose, as does the synthesis of lactose in lactation (**Figure 16–9**).

In the Fed State, Metabolic Fuel Reserves Are Laid Down

For several hours after a meal, while the products of digestion are being absorbed, there is an abundant supply of metabolic fuels. Under these conditions, glucose is the major fuel for oxidation in most tissues; this is observed as an increase in the respiratory quotient (the ratio of carbon dioxide produced/oxygen consumed) from about 0.8 in the fasting state to near 1 (**Table 16–1**).

Glucose uptake into muscle and adipose tissue is controlled by **insulin,** which is secreted by the β-islet cells of the pancreas in response to an increased concentration of glucose in the portal blood. In the fasting state, the glucose transporter of muscle and adipose tissue (GLUT-4) is in intracellular vesicles. An early response to insulin is the migration of these vesicles to the cell surface, where they fuse with the plasma membrane, exposing active glucose transporters. These insulin sensitive tissues only take up glucose from the bloodstream to any significant extent in the presence of the hormone. As insulin secretion falls in the fasting state, so the receptors are internalized again, reducing glucose uptake. However, in skeletal muscle, the increase in cytoplasmic calcium ion concentration in response to nerve stimulation stimulates the migration of the vesicles to the cell surface and exposure of

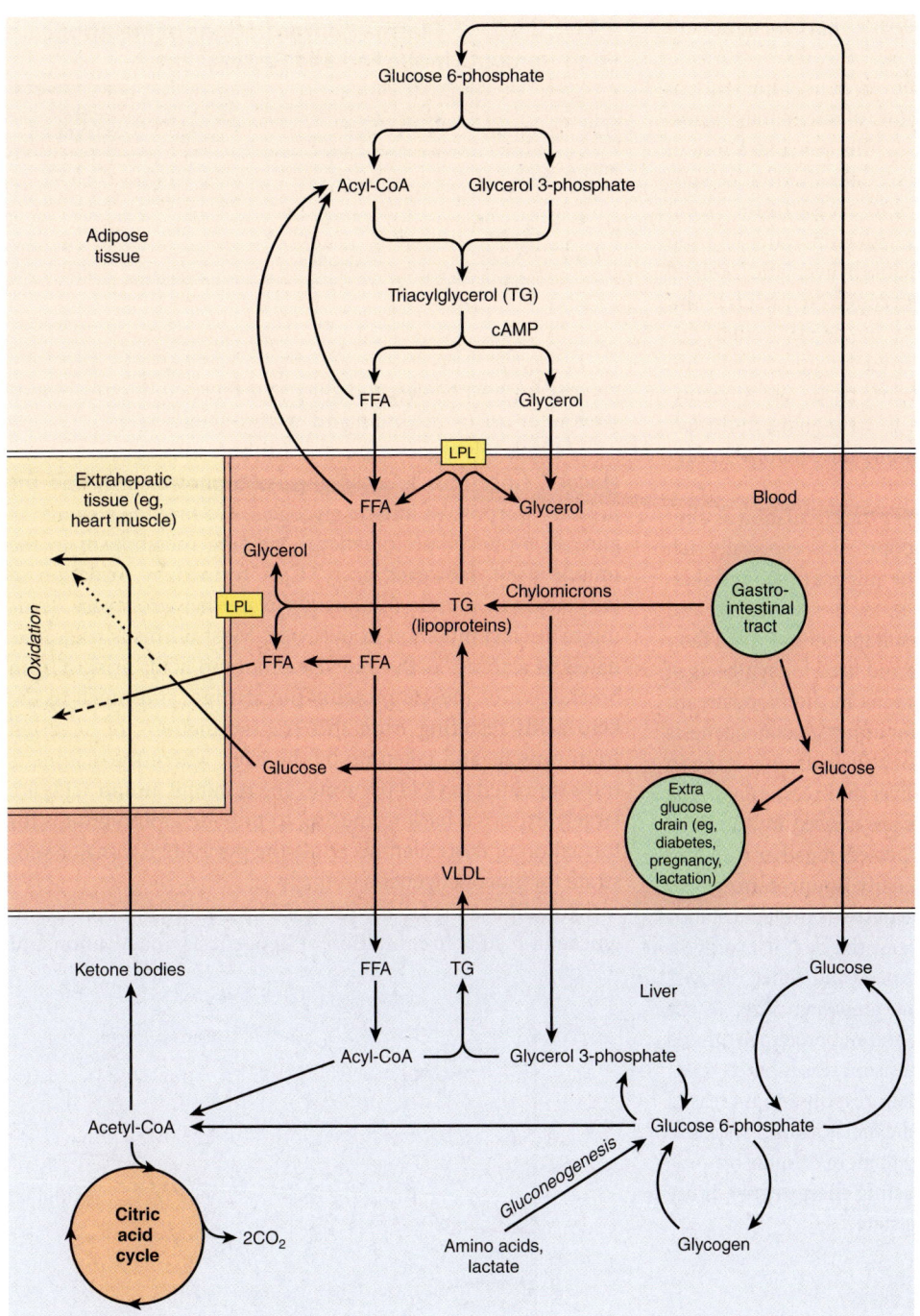

FIGURE 16–9 **Metabolic interrelationships among adipose tissue, the liver, and extrahepatic tissues.** In tissues such as heart, metabolic fuels are oxidized in the following order of preference: ketone bodies > fatty acids > glucose. (FFA, free fatty acids; LPL, lipoprotein lipase; VLDL, very low density lipoproteins.)

TABLE 16–1 Energy Yields, Oxygen Consumption, and Carbon Dioxide Production in the Oxidation of Metabolic Fuels

	Energy Yield (kJ/g)	O_2 Consumed (L/g)	CO_2 Produced (L/g)	RQ (CO_2 Produced/ O_2 Consumed)	Energy (kJ)/L O_2
Carbohydrate	16	0.829	0.829	1.00	20
Protein	17	0.966	0.782	0.81	20
Fat	37	2.016	1.427	0.71	20
Alcohol	29	1.429	0.966	0.66	20

active glucose transporters whether or not there is significant insulin stimulation.

The uptake of glucose into the liver is independent of insulin, but liver has an isoenzyme of hexokinase (glucokinase) with a high K_m, so that as the concentration of glucose entering the liver increases, so does the rate of synthesis of glucose 6-phosphate. This is in excess of the liver's requirement for energy-yielding metabolism, and is used mainly for synthesis of **glycogen.** In both liver and skeletal muscle, insulin acts to stimulate glycogen synthetase and inhibit glycogen phosphorylase. Some of the additional glucose entering the liver may also be used for lipogenesis and hence triacylglycerol synthesis. In adipose tissue, insulin stimulates glucose uptake, its conversion to fatty acids, and their esterification to triacylglycerol. It inhibits intracellular lipolysis and the release of free fatty acids.

The products of lipid digestion enter the circulation as **chylomicrons,** the largest of the plasma lipoproteins, especially rich in triacylglycerol (see Chapter 25). In the adipose tissue and skeletal muscle, extracellular lipoprotein lipase is synthesized and activated in response to insulin; the resultant nonesterified fatty acids are largely taken up by the tissue and used for synthesis of triacylglycerol, while the glycerol remains in the bloodstream and is taken up by the liver and used for either gluconeogenesis and glycogen synthesis or lipogenesis. Fatty acids remaining in the bloodstream are taken up by the liver and reesterified. The lipid-depleted chylomicron remnants are cleared by the liver, and the remaining triacylglycerol is exported, together with that synthesized in the liver, in **very low density lipoprotein.**

Under normal conditions, the rate of tissue protein catabolism is more or less constant throughout the day; it is only in **cachexia** associated with advanced cancer and other diseases that there is an increased rate of protein catabolism. There is net protein catabolism in the fasting state, and net protein synthesis in the fed state, when the rate of synthesis increases by 20–25%. The increased rate of protein synthesis in response to increased availability of amino acids and metabolic fuel is again a response to insulin action. Protein synthesis is an energy expensive process; it may account for up to 20% of resting energy expenditure after a meal, but only 9% in the fasting state.

Metabolic Fuel Reserves Are Mobilized in the Fasting State

There is a small fall in plasma glucose in the fasting state, and then little change as fasting is prolonged into starvation. Plasma free fatty acids increase in fasting, but then rise little more in starvation; as fasting is prolonged, the plasma concentration of ketone bodies (acetoacetate and 3-hydroxybutyrate) increases markedly (**Table 16–2, Figure 16–10**).

In the fasting state, as the concentration of glucose in the portal blood falls, insulin secretion decreases, and skeletal muscle and adipose tissue take up less glucose. The increase in secretion of **glucagon** by α cells of the pancreas inhibits glycogen synthetase, and activates glycogen phosphorylase in the liver. The resulting glucose 6-phosphate is hydrolyzed by

TABLE 16–2 Plasma Concentrations of Metabolic Fuels (mmol/L) in the Fed and Fasting States

	Fed	40 h Fasting	7 Days Starvation
Glucose	5.5	3.6	3.5
Free fatty acids	0.30	1.15	1.19
Ketone bodies	Negligible	2.9	4.5

glucose 6-phosphatase, and glucose is released into the bloodstream for use by the brain and erythrocytes.

Muscle glycogen cannot contribute directly to plasma glucose, since muscle lacks glucose 6-phosphatase, and the primary purpose of muscle glycogen is to provide a source of glucose 6-phosphate for energy-yielding metabolism in the muscle itself. However, acetyl-CoA formed by oxidation of fatty acids in muscle inhibits pyruvate dehydrogenase, leading to an accumulation of pyruvate. Most of this is transaminated to alanine, at the expense of amino acids arising from breakdown of muscle protein. The alanine, and much of the keto acids resulting from this transamination are exported from muscle, and taken up by the liver, where the alanine is transaminated to yield pyruvate. The resultant amino acids are largely exported back to muscle, to provide amino groups for formation of more alanine, while the pyruvate is a major substrate for gluconeogenesis in the liver.

In adipose tissue, the decrease in insulin and increase in glucagon results in inhibition of lipogenesis, inactivation and

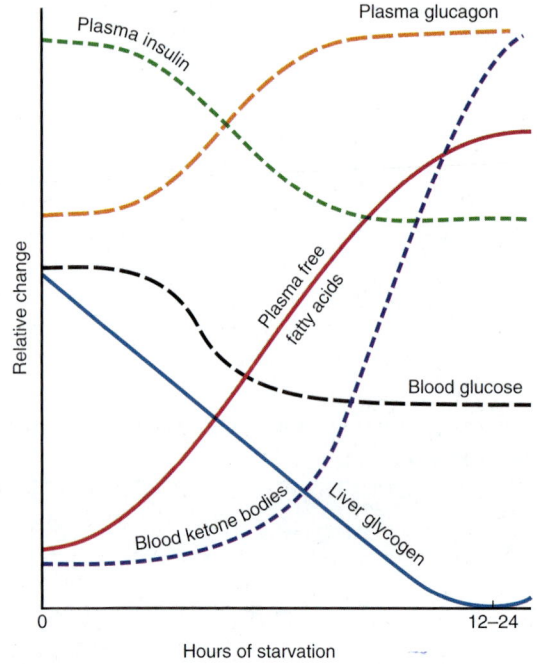

FIGURE 16–10 Relative changes in metabolic parameters during the onset of starvation.

TABLE 16–3 Summary of the Major Metabolic Features of the Principal Organs

Organ	Major Pathways	Main Substrates	Major Products Exported	Specialist Enzymes
Liver	Glycolysis, gluconeogenesis, lipogenesis, β-oxidation, citric acid cycle, ketogenesis, lipoprotein metabolism, drug metabolism, synthesis of bile salts, urea, uric acid, cholesterol, plasma proteins	Free fatty acids, glucose (in fed state), lactate, glycerol, fructose, amino acids, alcohol	Glucose, triacylglycerol in VLDL[1], ketone bodies, urea, uric acid, bile salts, cholesterol, plasma proteins	Glucokinase, glucose 6-phosphatase, glycerol kinase, phosphoenolpyruvate carboxykinase, fructokinase, arginase, HMG CoA synthase, HMG CoA lyase, alcohol dehydrogenase
Brain	Glycolysis, citric acid cycle, amino acid metabolism, neurotransmitter synthesis	Glucose, amino acids, ketone bodies in prolonged starvation	Lactate, end products of neurotransmitter metabolism	Those for synthesis and catabolism of neurotransmitters
Heart	β-Oxidation and citric acid cycle	Ketone bodies, free fatty acids, lactate, chylomicron and VLDL triacylglycerol, some glucose	—	Lipoprotein lipase, very active electron transport chain
Adipose tissue	Lipogenesis, esterification of fatty acids, lipolysis (in fasting)	Glucose, chylomicron and VLDL triacylglycerol	Free fatty acids, glycerol	Lipoprotein lipase, hormone-sensitive lipase, enzymes of pentose phosphate pathway
Fast twitch muscle	Glycolysis	Glucose, glycogen	Lactate, (alanine and ketoacids in fasting)	—
Slow twitch muscle	β-Oxidation and citric acid cycle	Ketone bodies, chylomicron and VLDL triacylglycerol	—	Lipoprotein lipase, very active electron transport chain
Kidney	Gluconeogenesis	Free fatty acids, lactate, glycerol, glucose	Glucose	Glycerol kinase, phosphoenolpyruvate carboxykinase
Erythrocytes	Anaerobic glycolysis, pentose phosphate pathway	Glucose	Lactate	Hemoglobin, enzymes of pentose phosphate pathway

[1]VLDL very low density lipoprotein.

internalization of lipoprotein lipase, and activation of intracellular hormone-sensitive lipase (Chapter 25). This leads to release from adipose tissue of increased amounts of glycerol (which is a substrate for gluconeogenesis in the liver) and free fatty acids, which are used by liver, heart, and skeletal muscle as their preferred metabolic fuel, therefore sparing glucose.

Although muscle preferentially takes up and metabolizes free fatty acids in the fasting state, it cannot meet all of its energy requirements by β-oxidation. By contrast, the liver has a greater capacity for β-oxidation than it requires to meet its own energy needs, and as fasting becomes more prolonged, it forms more acetyl-CoA than can be oxidized. This acetyl-CoA is used to synthesize the **ketone bodies** (Chapter 22), which are major metabolic fuels for skeletal and heart muscle and can meet up to 20% of the brain's energy needs. In prolonged starvation, glucose may represent less than 10% of whole body energy-yielding metabolism.

Were there no other source of glucose, liver and muscle glycogen would be exhausted after about 18 h fasting. As fasting becomes more prolonged, so an increasing amount of the amino acids released as a result of protein catabolism is utilized in the liver and kidneys for gluconeogenesis (**Table 16–3**).

CLINICAL ASPECTS

In prolonged starvation, as adipose tissue reserves are depleted, there is a very considerable increase in the net rate of protein catabolism to provide amino acids, not only as substrates for gluconeogenesis, but also as the main metabolic fuel of all tissues. Death results when essential tissue proteins are catabolized and not replaced. In patients with **cachexia** as a result of release of **cytokines** in response to tumors and a number of other pathologic conditions, there is an increase in the rate of tissue protein catabolism, as well as a considerably increased metabolic rate, so they are in a state of advanced starvation. Again, death results when essential tissue proteins are catabolized and not replaced.

The high demand for glucose by the fetus, and for lactose synthesis in lactation, can lead to ketosis. This may be seen as mild ketosis with hypoglycemia in human beings; in lactating cattle and in ewes carrying a twin pregnancy, there may be very pronounced ketoacidosis and profound hypoglycemia.

In poorly controlled type 1 **diabetes mellitus,** patients may become hyperglycemic, partly as a result of lack of insulin to stimulate uptake and utilization of glucose, and partly because in the absence of insulin there is increased gluconeogenesis from amino acids in the liver. At the same time, the lack of insulin results in increased lipolysis in adipose tissue, and the resultant free fatty acids are substrates for ketogenesis in the liver.

Utilization of these ketone bodies in muscle (and other tissues) may be impaired because of the lack of oxaloacetate (all tissues have a requirement for some glucose metabolism to maintain an adequate amount of oxaloacetate for citric acid cycle activity). In uncontrolled diabetes, the ketosis may be severe enough to result in pronounced acidosis (**ketoacidosis**) since acetoacetate and 3-hydroxybutyrate are relatively strong acids. Coma results from both the acidosis and also the considerably increased osmolality of extracellular fluid (mainly as a result of the hyperglycemia, and diuresis resulting from the excretion of glucose and ketone bodies in the urine).

SUMMARY

- The products of digestion provide the tissues with the building blocks for the biosynthesis of complex molecules and also with the fuel for metabolic processes.

- Nearly all products of digestion of carbohydrate, fat, and protein are metabolized to a common metabolite, acetyl-CoA, before oxidation to CO_2 in the citric acid cycle.

- Acetyl-CoA is also the precursor for synthesis of long-chain fatty acids and steroids (including cholesterol) and ketone bodies.

- Glucose provides carbon skeletons for the glycerol of triacylglycerols and nonessential amino acids.

- Water-soluble products of digestion are transported directly to the liver via the hepatic portal vein. The liver regulates the blood concentrations of glucose and amino acids. Lipids and lipid-soluble products of digestion enter the bloodstream from the lymphatic system, and the liver clears the remnants after extra-hepatic tissues have taken up fatty acids.

- Pathways are compartmentalized within the cell. Glycolysis, glycogenesis, glycogenolysis, the pentose phosphate pathway, and lipogenesis occur in the cytosol. The mitochondria contain the enzymes of the citric acid cycle, β-oxidation of fatty acids,

and the respiratory chain and ATP synthase. The membranes of the endoplasmic reticulum contain the enzymes for a number of other processes, including triacylglycerol synthesis and drug metabolism.

- Metabolic pathways are regulated by rapid mechanisms affecting the activity of existing enzymes, that is, allosteric and covalent modification (often in response to hormone action) and slow mechanisms affecting the synthesis of enzymes.

- Dietary carbohydrate and amino acids in excess of requirements can be used for fatty acid and hence triacylglycerol synthesis.

- In fasting and starvation, glucose must be provided for the brain and red blood cells; in the early fasting state, this is supplied from glycogen reserves. In order to spare glucose, muscle and other tissues do not take up glucose when insulin secretion is low; they utilize fatty acids (and later ketone bodies) as their preferred fuel.

- Adipose tissue releases free fatty acids in the fasting state. In prolonged fasting and starvation these are used by the liver for synthesis of ketone bodies, which are exported to provide the major fuel for muscle.

- Most amino acids, arising from the diet or from tissue protein turnover, can be used for gluconeogenesis, as can the glycerol from triacylglycerol.

- Neither fatty acids, arising from the diet or from lipolysis of adipose tissue triacylglycerol, nor ketone bodies, formed from fatty acids in the fasting state, can provide substrates for gluconeogenesis.

REFERENCES

Bender DA: *Introduction to Nutrition and Metabolism,* 4th ed. CRC Press, 2007.

Brosnan JT: Comments on the metabolic needs for glucose and the role of gluconeogenesis. Eur J Clin Nutr 1999;53:S107–S111.

Frayn KN: Integration of substrate flow in vivo: some insights into metabolic control. Clin Nutr 1997;16:277–282.

Frayn KN: *Metabolic Regulation: A Human Perspective,* 3rd ed. Wiley-Blackwell, 2010.

Zierler K: Whole body metabolism of glucose. Am J Physiol 1999;276:E409–E426.

The Citric Acid Cycle: The Catabolism of Acetyl-CoA

David A. Bender, PhD & Peter A. Mayes, PhD, DSc

OBJECTIVES

After studying this chapter, you should be able to:

- Describe the reactions of the citric acid cycle and the reactions that lead to the production of reducing equivalents that are oxidized in the mitochondrial electron transport chain to yield ATP.
- Explain the importance of vitamins in the citric acid cycle.
- Explain how the citric acid cycle provides both a route for catabolism of amino acids and also a route for their synthesis.
- Describe the main anaplerotic pathways that permit replenishment of citric acid cycle intermediates, and how the withdrawal of oxaloacetate for gluconeogenesis is controlled.
- Describe the role of the citric acid cycle in fatty acid synthesis.
- Explain how the activity of the citric acid cycle is controlled by the availability of oxidized cofactors.
- Explain how hyperammonemia can lead to loss of consciousness.

BIOMEDICAL IMPORTANCE

The citric acid cycle (Krebs cycle, tricarboxylic acid cycle) is a sequence of reactions in mitochondria that oxidizes the acetyl moiety of acetyl-CoA and reduces coenzymes that are reoxidized through the electron transport chain, linked to the formation of ATP.

The citric acid cycle is the final common pathway for the oxidation of carbohydrate, lipid, and protein because glucose, fatty acids, and most amino acids are metabolized to acetyl-CoA or intermediates of the cycle. It also has a central role in gluconeogenesis, lipogenesis, and interconversion of amino acids. Many of these processes occur in most tissues, but liver is the only tissue in which all occur to a significant extent. The repercussions are therefore profound when, for example, large numbers of hepatic cells are damaged as in acute **hepatitis** or replaced by connective tissue (as in **cirrhosis**). The few genetic defects of citric acid cycle enzymes that have been reported are associated with severe neurological damage as a result of very considerably impaired ATP formation in the central nervous system.

Hyperammonemia, as occurs in advanced liver disease, leads to loss of consciousness, coma, and convulsions as a result of impaired activity of the citric acid cycle, leading to reduced formation of ATP. Ammonia both depletes citric acid cycle intermediates (by withdrawing α-ketoglutarate for the formation of glutamate and glutamine) and also inhibits the oxidative decarboxylation of α-ketoglutarate.

THE CITRIC ACID CYCLE PROVIDES SUBSTRATE FOR THE RESPIRATORY CHAIN

The cycle starts with reaction between the acetyl moiety of acetyl-CoA and the four-carbon dicarboxylic acid oxaloacetate, forming a six-carbon tricarboxylic acid, citrate. In the subsequent reactions, two molecules of CO_2 are released and oxaloacetate is regenerated (**Figure 17–1**). Only a small quantity of oxaloacetate is needed for the oxidation of a large quantity of acetyl-CoA; it can be considered as playing a **catalytic role,** since it is regenerated at the end of the cycle.

The citric acid cycle is an integral part of the process by which much of the free energy liberated during the oxidation of fuels is made available. During the oxidation of acetyl-CoA, coenzymes are reduced and subsequently reoxidized in the respiratory chain, linked to the formation of ATP (oxidative phosphorylation, **Figure 17–2;** see also Chapter 13). This process is **aerobic,** requiring oxygen as the final oxidant of the

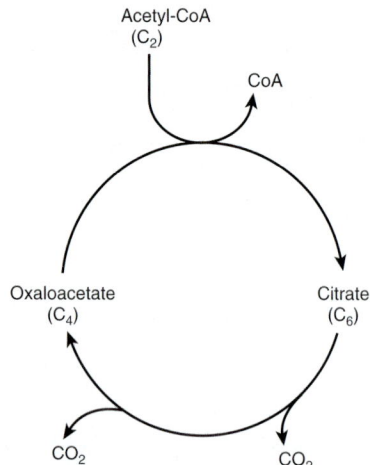

FIGURE 17–1 The citric acid cycle, illustrating the catalytic role of oxaloacetate.

reduced coenzymes. The enzymes of the citric acid cycle are located in the **mitochondrial matrix,** either free or attached to the inner mitochondrial membrane and the crista membrane, where the enzymes and coenzymes of the respiratory chain are also found (Chapter 13).

REACTIONS OF THE CITRIC ACID CYCLE LIBERATE REDUCING EQUIVALENTS & CO_2

The initial reaction between acetyl-CoA and oxaloacetate to form citrate is catalyzed by **citrate synthase,** which forms a carbon–carbon bond between the methyl carbon of acetyl-CoA and the carbonyl carbon of oxaloacetate (**Figure 17–3**). The thioester bond of the resultant citryl-CoA is hydrolyzed, releasing citrate and CoASH—an exothermic reaction.

Citrate is isomerized to isocitrate by the enzyme **aconitase** (aconitate hydratase); the reaction occurs in two steps: dehydration to *cis*-aconitate and rehydration to isocitrate. Although citrate is a symmetric molecule, aconitase reacts with citrate asymmetrically, so that the two carbon atoms that are lost in subsequent reactions of the cycle are not those that were added from acetyl-CoA. This asymmetric behavior is the result of **channeling**—transfer of the product of citrate synthase directly onto the active site of aconitase, without entering free solution. This provides integration of citric acid cycle activity and the provision of citrate in the cytosol as a source of acetyl-CoA for fatty acid synthesis. Citrate is only available in free solution to be transported from the mitochondria to the cytosol for fatty acid synthesis when aconitase is inhibited by accumulation of its product, isocitrate.

The poison **fluoracetate** is found in some of plants, and their consumption can be fatal to grazing animals. Some fluo-

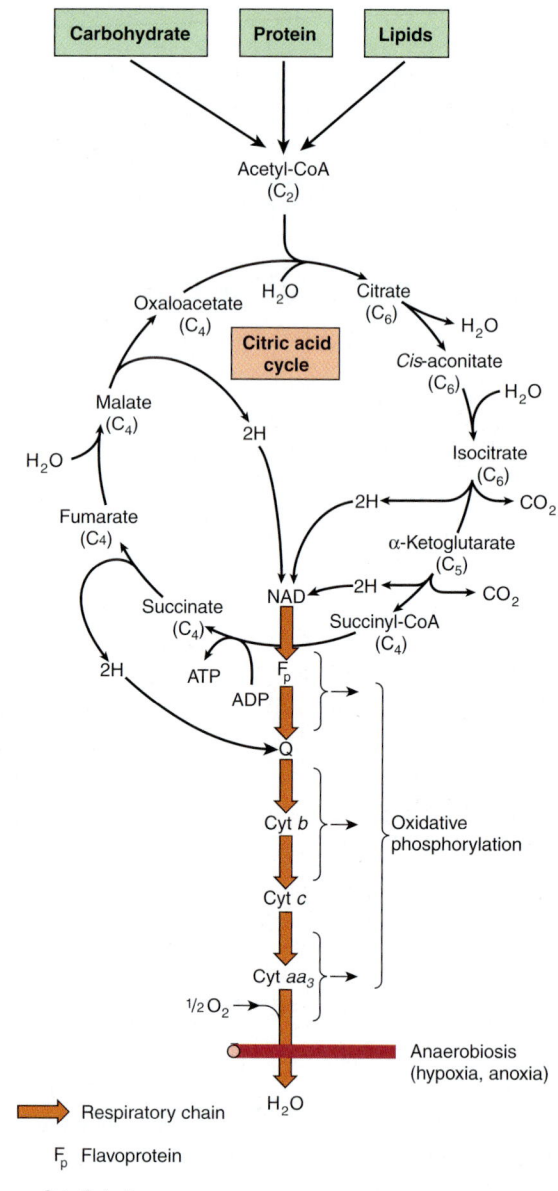

FIGURE 17–2 **The citric acid cycle: the major catabolic pathway for acetyl-CoA in aerobic organisms.** Acetyl-CoA, the product of carbohydrate, protein, and lipid catabolism, is taken into the cycle and oxidized to CO_2 with the release of reducing equivalents (2H). Subsequent oxidation of 2H in the respiratory chain leads to phosphorylation of ADP to ATP. For one turn of the cycle, nine ATP are generated via oxidative phosphorylation and one ATP (or GTP) arises at substrate level from the conversion of succinyl-CoA to succinate.

rinated compounds used as anticancer agents and industrial chemicals (including pesticides) are metabolized to fluoroacetate. It is toxic because fluoroacetyl-CoA condenses with oxaloacetate to form fluorocitrate, which inhibits aconitase, causing citrate to accumulate.

Isocitrate undergoes dehydrogenation catalyzed by **isocitrate dehydrogenase** to form, initially, oxalosuccinate,

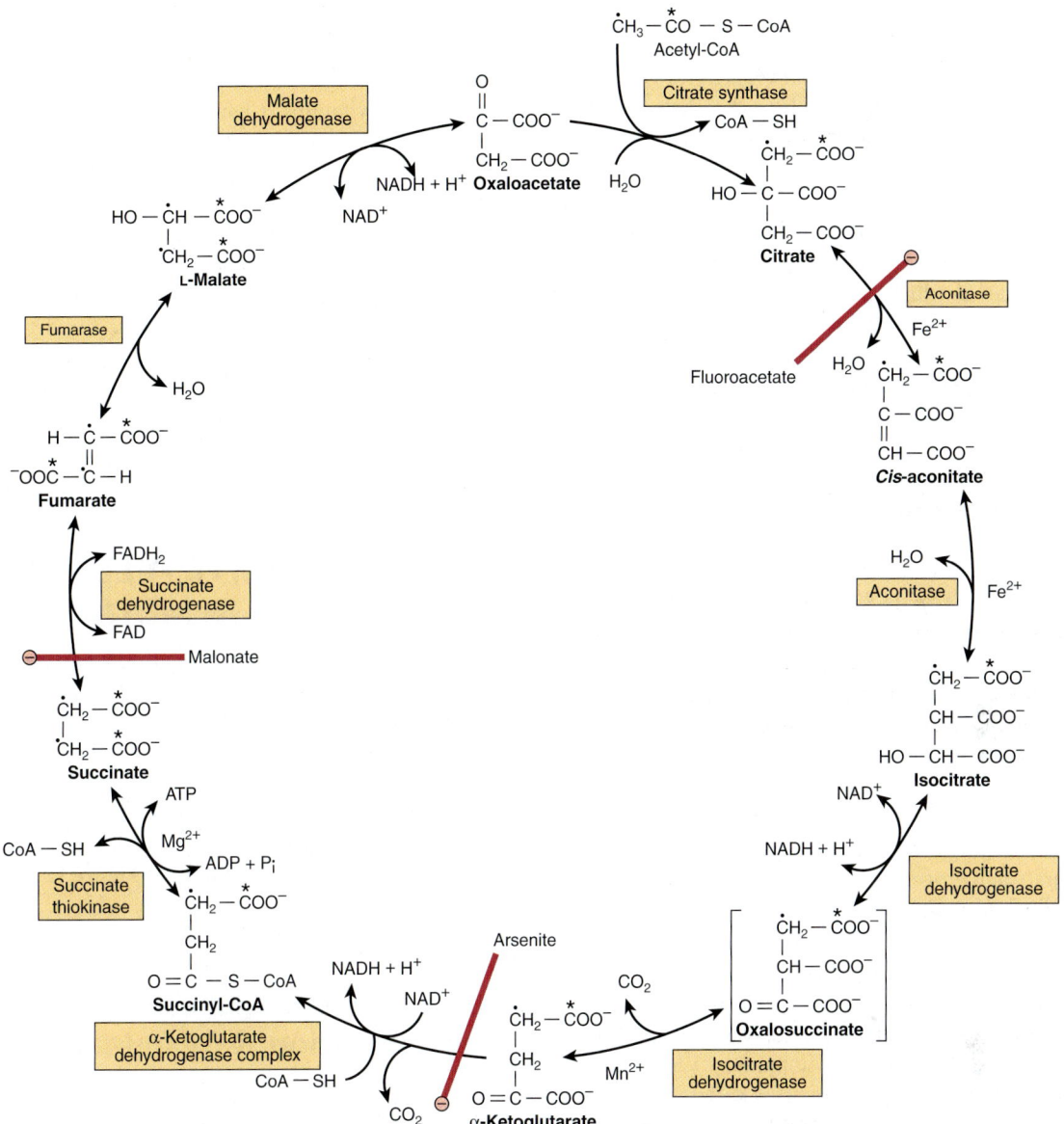

FIGURE 17–3 **The citric acid (Krebs) cycle.** Oxidation of NADH and $FADH_2$ in the respiratory chain leads to the formation of ATP via oxidative phosphorylation. In order to follow the passage of acetyl-CoA through the cycle, the two carbon atoms of the acetyl radical are shown labeled on the carboxyl carbon (*) and on the methyl carbon (·). Although two carbon atoms are lost as CO_2 in one turn of the cycle, these atoms are not derived from the acetyl-CoA that has immediately entered the cycle, but from that portion of the citrate molecule that was derived from oxaloacetate. However, on completion of a single turn of the cycle, the oxaloacetate that is regenerated is now labeled, which leads to labeled CO_2 being evolved during the second turn of the cycle. Because succinate is a symmetric compound, "randomization" of label occurs at this step so that all four carbon atoms of oxaloacetate appear to be labeled after one turn of the cycle. During gluconeogenesis, some of the label in oxaloacetate is incorporated into glucose and glycogen (Figure 20–1). The sites of inhibition (⊖) by fluoroacetate, malonate, and arsenite are indicated.

which remains enzyme bound and undergoes decarboxylation to α-ketoglutarate. The decarboxylation requires Mg^{2+} or Mn^{2+} ions. There are three isoenzymes of isocitrate dehydrogenase. One, which uses NAD^+, is found only in mitochondria. The other two use $NADP^+$ and are found in mitochondria and the cytosol. Respiratory-chain-linked oxidation of isocitrate occurs through the NAD^+-dependent enzyme.

α-Ketoglutarate undergoes **oxidative decarboxylation** in a reaction catalyzed by a multienzyme complex similar to that involved in the oxidative decarboxylation of pyruvate (Figure 18–5). The **α-ketoglutarate dehydrogenase complex** requires the same cofactors as the pyruvate dehydrogenase complex—thiamin diphosphate, lipoate, NAD^+, FAD, and CoA—and results in the formation of succinyl-CoA. The equilibrium of this reaction is so much in favor of succinyl-CoA formation that it must be considered to be physiologically unidirectional. As in the case of pyruvate oxidation (Chapter 18), arsenite inhibits the reaction, causing the substrate, **α-ketoglutarate,** to accumulate. High concentrations of ammonia inhibit α-ketoglutarate dehydrogenase.

Succinyl-CoA is converted to succinate by the enzyme **succinate thiokinase (succinyl-CoA synthetase).** This is the only example of substrate level phosphorylation in the citric acid cycle. Tissues in which gluconeogenesis occurs (the liver and kidney) contain two isoenzymes of succinate thiokinase, one specific for GDP and the other for ADP. The GTP formed is used for the decarboxylation of oxaloacetate to phosphoenolpyruvate in gluconeogenesis, and provides a regulatory link between citric acid cycle activity and the withdrawal of oxaloacetate for gluconeogenesis. Nongluconeogenic tissues have only the isoenzyme that uses ADP.

When ketone bodies are being metabolized in extrahepatic tissues, there is an alternative reaction catalyzed by **succinyl-CoA–acetoacetate-CoA transferase (thiophorase),** involving transfer of CoA from succinyl-CoA to acetoacetate, forming acetoacetyl-CoA and succinate (Chapter 22).

The onward metabolism of succinate, leading to the regeneration of oxaloacetate, is the same sequence of chemical reactions as occurs in the β-oxidation of fatty acids: dehydrogenation to form a carbon–carbon double bond, addition of water to form a hydroxyl group, and a further dehydrogenation to yield the oxo-group of oxaloacetate.

The first dehydrogenation reaction, forming fumarate, is catalyzed by **succinate dehydrogenase,** which is bound to the inner surface of the inner mitochondrial membrane. The enzyme contains FAD and iron–sulfur (Fe:S) protein, and directly reduces ubiquinone in the electron transport chain. **Fumarase (fumarate hydratase)** catalyzes the addition of water across the double bond of fumarate, yielding malate. Malate is converted to oxaloacetate by **malate dehydrogenase,** a reaction requiring NAD$^+$. Although the equilibrium of this reaction strongly favors malate, the net flux is to oxaloacetate because of the continual removal of oxaloacetate (to form citrate, as a substrate for gluconeogenesis, or to undergo transamination to aspartate) and also the continual reoxidation of NADH.

TEN ATP ARE FORMED PER TURN OF THE CITRIC ACID CYCLE

As a result of oxidations catalyzed by the dehydrogenases of the citric acid cycle, three molecules of NADH and one of FADH$_2$ are produced for each molecule of acetyl-CoA catabolized in one turn of the cycle. These reducing equivalents are transferred to the respiratory chain (see Figure 13–3), where reoxidation of each NADH results in formation of ~2.5 ATP, and of FADH$_2$, ~1.5 ATP. In addition, 1 ATP (or GTP) is formed by substrate-level phosphorylation catalyzed by succinate thiokinase.

VITAMINS PLAY KEY ROLES IN THE CITRIC ACID CYCLE

Four of the B vitamins (Chapter 44) are essential in the citric acid cycle and hence energy-yielding metabolism: (1) **riboflavin,** in the form of flavin adenine dinucleotide (FAD), a cofactor for succinate dehydrogenase; (2) **niacin,** in the form of nicotinamide adenine dinucleotide (NAD), the electron acceptor for isocitrate dehydrogenase, α-ketoglutarate dehydrogenase, and malate dehydrogenase; (3) thiamin (**vitamin B$_1$**), as thiamin diphosphate, the coenzyme for decarboxylation in the α-ketoglutarate dehydrogenase reaction; and (4) **pantothenic acid,** as part of coenzyme A, the cofactor attached to "active" carboxylic acid residues such as acetyl-CoA and succinyl-CoA.

THE CITRIC ACID CYCLE PLAYS A PIVOTAL ROLE IN METABOLISM

The citric acid cycle is not only a pathway for oxidation of two carbon units, but is also a major pathway for interconversion of metabolites arising from **transamination** and **deamination** of amino acids (Chapters 28 & 29), and providing the substrates for **amino acid synthesis** by transamination (Chapter 27), as well as for **gluconeogenesis** (Chapter 20) and **fatty acid synthesis** (Chapter 23). Because it functions in both oxidative and synthetic processes, it is **amphibolic** (**Figure 17–4**).

The Citric Acid Cycle Takes Part in Gluconeogenesis, Transamination, & Deamination

All the intermediates of the cycle are potentially **glucogenic,** since they can give rise to oxaloacetate, and hence net production of glucose (in the liver and kidney, the organs that carry out gluconeogenesis; see Chapter 20). The key enzyme that catalyzes net transfer out of the cycle into gluconeogenesis is **phosphoenolpyruvate carboxykinase,** which catalyzes the decarboxylation of oxaloacetate to phosphoenolpyruvate, with GTP acting as the phosphate donor (see Figure 20–1). The GTP required for this reaction is provided (in liver and kidney) by the GDP-dependent isoenzyme of succinate thiokinase. This ensures that oxaloacetate will not be withdrawn from the cycle for gluconeogenesis if this would lead to depletion of citric acid cycle intermediates, and hence reduced generation of ATP.

Net transfer into the cycle occurs as a result of several reactions. Among the most important of such **anaplerotic** reactions is the formation of oxaloacetate by the carboxylation of pyruvate, catalyzed by **pyruvate carboxylase.** This reaction is important in maintaining an adequate concentration of oxaloacetate for the condensation reaction with acetyl-CoA. If acetyl-CoA accumulates, it acts as both an allosteric activator

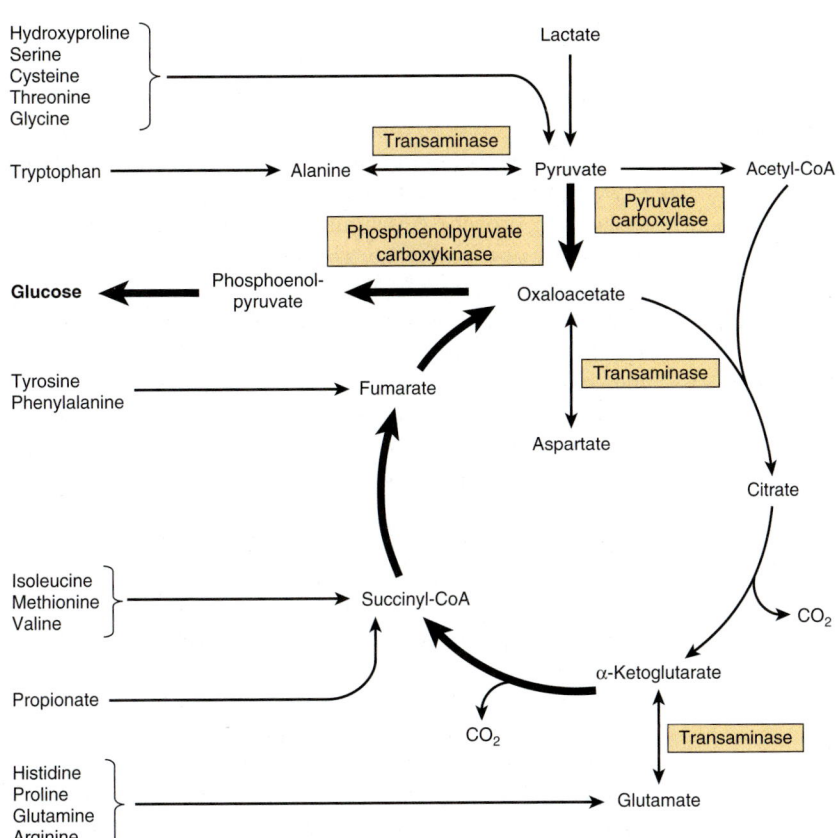

FIGURE 17–4 Involvement of the citric acid cycle in transamination and gluconeogenesis. The bold arrows indicate the main pathway of gluconeogenesis.

of pyruvate carboxylase and an inhibitor of pyruvate dehydrogenase, thereby ensuring a supply of oxaloacetate. Lactate, an important substrate for gluconeogenesis, enters the cycle via oxidation to pyruvate and then carboxylation to oxaloacetate. **Glutamate** and **glutamine** are important anaplerotic substrates because they yield α-ketoglutarate as a result of the reactions catalyzed by glutaminase and glutamate dehydrogenase. Transamination of **aspartate** leads directly to the formation of oxaloacetate, and a variety of compounds that are metabolized to yield **propionyl CoA,** which can be carboxylated and isomerized to succinyl CoA are also important anaplerotic substrates.

Aminotransferase (transaminase) reactions form pyruvate from alanine, oxaloacetate from aspartate, and α-ketoglutarate from glutamate. Because these reactions are reversible, the cycle also serves as a source of carbon skeletons for the synthesis of these amino acids. Other amino acids contribute to gluconeogenesis because their carbon skeletons give rise to citric acid cycle intermediates. Alanine, cysteine, glycine, hydroxyproline, serine, threonine, and tryptophan yield pyruvate; arginine, histidine, glutamine, and proline yield α-ketoglutarate; isoleucine, methionine, and valine yield succinyl-CoA; tyrosine and phenylalanine yield fumarate (see Figure 17–4).

The citric acid cycle itself does not provide a pathway for the complete oxidation of the carbon skeletons of amino acids

that give rise to intermediates such as α-ketoglutarate, succinyl CoA, fumarate and oxaloacetate, because this results in an increase in the amount of oxaloacetate. For complete oxidation to occur, oxaloacetate must undergo phosphorylation and carboxylation to phosphoenolpyruvate (at the expense of GTP) then dephosphorylation to pyruvate (catalyzed by pyruvate kinase) and oxidative decarboxylation to acetyl Co (catalyzed by pyruvate dehydrogenase).

In ruminants, whose main metabolic fuel is short-chain fatty acids formed by bacterial fermentation, the conversion of propionate, the major glucogenic product of rumen fermentation, to succinyl-CoA via the methylmalonyl-CoA pathway (Figure 20–2) is especially important.

The Citric Acid Cycle Takes Part in Fatty Acid Synthesis

Acetyl-CoA, formed from pyruvate by the action of pyruvate dehydrogenase, is the major substrate for long-chain fatty acid synthesis in nonruminants (**Figure 17–5**). (In ruminants, acetyl-CoA is derived directly from acetate.) Pyruvate dehydrogenase is a mitochondrial enzyme, and fatty acid synthesis is a cytosolic pathway; the mitochondrial membrane is impermeable to acetyl-CoA. Acetyl-CoA is made available in the cytosol from citrate synthesized in the mitochondrion, transported into the cytosol, and cleaved in a reaction catalyzed by

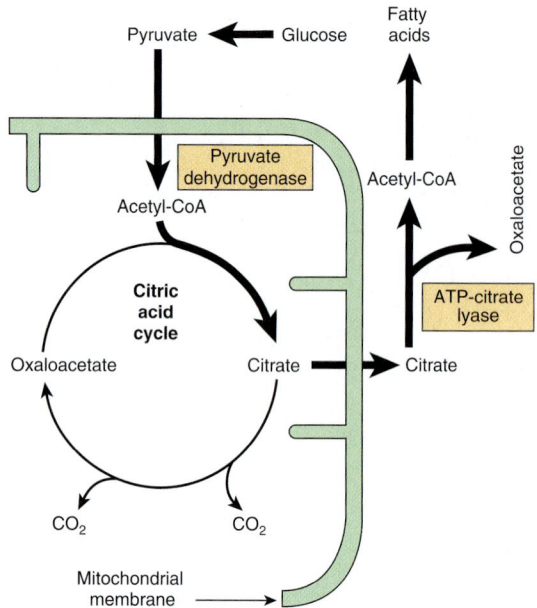

FIGURE 17–5 **Participation of the citric acid cycle in fatty acid synthesis from glucose.** See also Figure 23–5.

ATP-citrate lyase (Figure 17–5). Citrate is only available for transport out of the mitochondrion when aconitase is inhibited by its product and therefore saturated with its substrate, so that citrate cannot be channeled directly from citrate synthase onto aconitase. This ensures that citrate is used for fatty acid synthesis only when there is an adequate amount to ensure continued activity of the cycle.

Regulation of the Citric Acid Cycle Depends Primarily on a Supply of Oxidized Cofactors

In most tissues, where the primary role of the citric acid cycle is in energy-yielding metabolism, **respiratory control** via the respiratory chain and oxidative phosphorylation regulates citric acid cycle activity (Chapter 13). Thus, activity is immediately dependent on the supply of NAD⁺, which in turn, because of the tight coupling between oxidation and phosphorylation, is dependent on the availability of ADP and hence, ultimately on the rate of utilization of ATP in chemical and physical work. In addition, individual enzymes of the cycle are regulated. The most likely sites for regulation are the nonequilibrium reactions catalyzed by pyruvate dehydrogenase, citrate synthase, isocitrate dehydrogenase, and α-ketoglutarate dehydrogenase. The dehydrogenases are activated by Ca^{2+}, which increases in concentration during contraction of muscle and secretion by other tissues, when there is increased energy demand. In a tissue such as brain, which is largely dependent on carbohydrate to supply acetyl-CoA, control of the citric acid cycle may occur at pyruvate dehydrogenase. Several enzymes are responsive to the energy status as shown by the [ATP]/[ADP] and [NADH]/

[NAD⁺] ratios. Thus, there is allosteric inhibition of citrate synthase by ATP and long-chain fatty acyl-CoA. Allosteric activation of mitochondrial NAD-dependent isocitrate dehydrogenase by ADP is counteracted by ATP and NADH. The α-ketoglutarate dehydrogenase complex is regulated in the same way as is pyruvate dehydrogenase (Figure 18–6). Succinate dehydrogenase is inhibited by oxaloacetate, and the availability of oxaloacetate, as controlled by malate dehydrogenase, depends on the [NADH]/[NAD⁺] ratio. Since the K_m for oxaloacetate of citrate synthase is of the same order of magnitude as the intramitochondrial concentration, it is likely that the concentration of oxaloacetate controls the rate of citrate formation. Which of these mechanisms are important in vivo is still to be resolved.

Hyperammonemia, as occurs in advanced liver disease and a number of (rare) genetic diseases of amino acid metabolism, leads to loss of consciousness, coma and convulsions, and may be fatal. This is because of the withdrawal of α-ketoglutarate to form glutamate (catalyzed by glutamate dehydrogenase) and then glutamine (catalyzed by glutamine synthetase), leading to reduced concentrations of all citric acid cycle intermediates, and hence reduced generation of ATP. The equilibrium of glutamate dehydrogenase is finely poised, and the direction of reaction depends on the ratio of NAD⁺ : NADH and the concentration of ammonium ions. In addition, ammonia inhibits α-ketoglutarate dehydrogenase, and possibly also pyruvate dehydrogenase.

SUMMARY

- The citric acid cycle is the final pathway for the oxidation of carbohydrate, lipid, and protein. Their common end-metabolite, acetyl-CoA, reacts with oxaloacetate to form citrate. By a series of dehydrogenations and decarboxylations, citrate is degraded, reducing coenzymes, releasing 2 CO_2, and regenerating oxaloacetate.

- The reduced coenzymes are oxidized by the respiratory chain linked to formation of ATP. Thus, the cycle is the major pathway for the formation of ATP and is located in the matrix of mitochondria adjacent to the enzymes of the respiratory chain and oxidative phosphorylation.

- The citric acid cycle is amphibolic, since in addition to oxidation it is important in the provision of carbon skeletons for gluconeogenesis, fatty acid synthesis, and interconversion of amino acids.

REFERENCES

Baldwin JE, Krebs HA: The evolution of metabolic cycles. Nature 1981;291:381.

Bowtell JL, Bruce M: Glutamine: an anaplerotic precursor. Nutrition 2002;18:222.

Briere JJ, Favier J, Giminez-Roqueplo A-P, et al: Tricarboxylic acid cycle dysfunction as a cause of human diseases and tumor formation. Am J Physiol Cell Physiol 2006;291:C1114.

Brunengraber H, Roe CR: Anaplerotic molecules: current and future. J Inherit Metab Dis 2006;29:327.

De Meirleir L: Defects of pyruvate metabolism and the Krebs cycle. J Child Neurol 2002;Suppl 3:3S26.

Gibala MJ, Young ME: Anaplerosis of the citric acid cycle: role in energy metabolism of heart and skeletal muscle. Acta Physiol Scand 2000;168:657.

Hertz L, Kala G: Energy metabolism in brain cells: effects of elevated ammonia concentrations. Metab Brain Dis 2007; 22: 199–218.

Jitrapakdee S, Vidal-Puig A, Wallace JC: Anaplerotic roles of pyruvate carboxylase in mammalian tissues. Cell Mol Life Sci 2006; 63:843.

Jitrapakdee S, St Maurice M, Rayment I, et al: Structure, mechanism and regulation of pyruvate carboxylase. Biochem J 2008;413:369

Kay J, Weitzman PDJ (editors): *Krebs' Citric Acid Cycle—Half a Century and Still Turning.* Biochemical Society, 1987.

Kornberg H: Krebs and his trinity of cycles. Nat Rev Mol Cell Biol 2000;1:225.

Ott P, Clemmesen O, Larsen FS: Cerebral metabolic disturbances in the brain during acute liver failure: from hyperammonemia to energy failure and proteolysis. Neurochem Int 2005;47:13.

Owen OE, Kalhan SC: The key role of anaplerosis and cataplerosis for citric acid cycle function. J Biol Chem 2002;277:30409.

Pithukpakorn, M: Disorders of pyruvate metabolism and the tricarboxylic acid cycle. Mol Genet Metab 2005;85:243.

Sumegi B, Sherry AD: Is there tight channelling in the tricarboxylic acid cycle metabolon? Biochem Soc Trans 1991;19:1002.

Glycolysis & the Oxidation of Pyruvate

David A. Bender, PhD & Peter A. Mayes, PhD, DSc

OBJECTIVES

After studying this chapter, you should be able to:

- Describe the pathway of glycolysis and its control and explain how glycolysis can operate under anaerobic conditions.
- Describe the reaction of pyruvate dehydrogenase and its regulation.
- Explain how inhibition of pyruvate metabolism leads to lactic acidosis

BIOMEDICAL IMPORTANCE

Most tissues have at least some requirement for glucose. In the brain, the requirement is substantial, and even in prolonged fasting the brain can meet no more than about 20% of its energy needs from ketone bodies. Glycolysis, the major pathway for glucose metabolism, occurs in the cytosol of all cells. It is unique, in that it can function either aerobically or anaerobically, depending on the availability of oxygen and the electron transport chain. Erythrocytes, which lack mitochondria, are completely reliant on glucose as their metabolic fuel, and metabolize it by anaerobic glycolysis. However, to oxidize glucose beyond pyruvate (the end product of glycolysis) requires both oxygen and mitochondrial enzyme systems: the pyruvate dehydrogenase complex, the citric acid cycle (Chapter 17), and the respiratory chain (Chapter 13).

Glycolysis is both the principal route for glucose metabolism and also the main pathway for the metabolism of fructose, galactose, and other dietary carbohydrates. The ability of glycolysis to provide ATP in the absence of oxygen is especially important, because this allows skeletal muscle to perform at very high levels of work output when oxygen supply is insufficient, and it allows tissues to survive anoxic episodes. However, heart muscle, which is adapted for aerobic performance, has relatively low glycolytic activity and poor survival under conditions of **ischemia.** Diseases in which enzymes of glycolysis (eg, pyruvate kinase) are deficient are mainly seen as **hemolytic anemias** or, if the defect affects skeletal muscle (eg, phosphofructokinase), as **fatigue.** In fast-growing cancer cells, glycolysis proceeds at a high rate, forming large amounts of pyruvate, which is reduced to lactate and exported. This produces a relatively acidic local environment in the tumor, which may have implications for cancer therapy. The lactate is used for gluconeogenesis in the liver (Chapter 20), an energy-expensive process, which is responsible for much of the **hypermetabolism** seen in **cancer cachexia. Lactic acidosis** results from various causes, including impaired activity of pyruvate dehydrogenase, especially in thiamin (vitamin B_1) deficiency.

GLYCOLYSIS CAN FUNCTION UNDER ANAEROBIC CONDITIONS

Early in the investigations of glycolysis it was realized that fermentation in yeast was similar to the breakdown of glycogen in muscle. It was noted that when a muscle contracts in an anaerobic medium, that is, one from which oxygen is excluded, **glycogen disappears** and **lactate appears.** When oxygen is admitted, aerobic recovery takes place and lactate is no longer produced. However, if contraction occurs under aerobic conditions, lactate does not accumulate and pyruvate is the major end product of glycolysis. Pyruvate is oxidized further to CO_2 and water (**Figure 18–1**). When oxygen is in short supply, mitochondrial reoxidation of NADH formed during glycolysis is impaired, and NADH is reoxidized by reducing pyruvate to lactate, so permitting glycolysis to proceed (Figure 18–1). While glycolysis can occur under anaerobic conditions, this has a price, for it limits the amount of ATP formed per mole of glucose oxidized, so that much more glucose must be metabolized under anaerobic than aerobic conditions. In yeast and some other microorganisms, pyruvate formed in anaerobic glycolysis is not reduced to lactate, but is decarboxylated and reduced to ethanol.

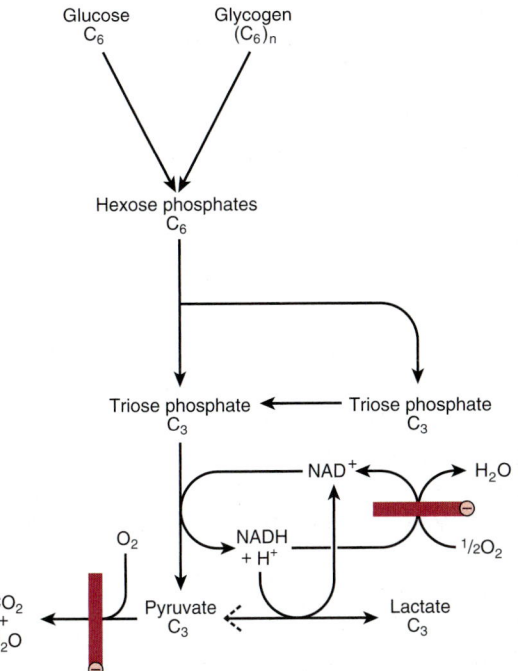

FIGURE 18–1 **Summary of glycolysis.** ⊖, blocked by anaerobic conditions or by absence of mitochondria containing key respiratory enzymes, as in erythrocytes.

THE REACTIONS OF GLYCOLYSIS CONSTITUTE THE MAIN PATHWAY OF GLUCOSE UTILIZATION

The overall equation for glycolysis from glucose to lactate is as follows:

$$Glucose + 2\ ADP + 2\ P_i \rightarrow 2\ Lactate + 2\ ATP + 2\ H_2O$$

All of the enzymes of glycolysis (**Figure 18–2**) are found in the cytosol. Glucose enters glycolysis by phosphorylation to glucose 6-phosphate, catalyzed by **hexokinase,** using ATP as the phosphate donor. Under physiologic conditions, the phosphorylation of glucose to glucose 6-phosphate can be regarded as irreversible. Hexokinase is inhibited allosterically by its product, glucose 6-phosphate.

In tissues other than the liver (and pancreatic β-islet cells), the availability of glucose for glycolysis (or glycogen synthesis in muscle, Chapter 19, and lipogenesis in adipose tissue, Chapter 23) is controlled by transport into the cell, which in turn is regulated by **insulin.** Hexokinase has a high affinity (low K_m) for glucose, and in the liver it is saturated under normal conditions, and so acts at a constant rate to provide glucose 6-phosphate to meet the liver's needs. Liver cells also contain an isoenzyme of hexokinase, **glucokinase,** which has a K_m very much higher than the normal intracellular concentration of glucose. The function of glucokinase in the liver is to remove glucose from the blood following a meal, provid-

ing glucose 6-phosphate in excess of requirements for glycolysis, which is used for glycogen synthesis and lipogenesis. Glucokinase is also found in pancreatic β-islet cells, where it functions to detect high concentrations of glucose; onward metabolites of the glucose 6-phosphate formed stimulate the secretion of insulin.

Glucose 6-phosphate is an important compound at the junction of several metabolic pathways: glycolysis, gluconeogenesis, the pentose phosphate pathway, glycogenesis, and glycogenolysis. In glycolysis, it is converted to fructose 6-phosphate by **phosphohexose isomerase,** which involves an aldose-ketose isomerization. This reaction is followed by another phosphorylation catalyzed by the enzyme **phosphofructokinase** (phosphofructokinase-1) forming fructose 1,6-bisphosphate. The phosphofructokinase reaction may be considered to be functionally irreversible under physiologic conditions; it is both inducible and subject to allosteric regulation, and has a major role in regulating the rate of glycolysis. Fructose 1,6-bisphosphate is cleaved by **aldolase** (fructose 1,6-bisphosphate aldolase) into two triose phosphates, glyceraldehyde 3-phosphate and dihydroxyacetone phosphate, which are interconverted by the enzyme **phosphotriose isomerase.**

Glycolysis continues with the oxidation of glyceraldehyde 3-phosphate to 1,3-bisphosphoglycerate. The enzyme catalyzing this oxidation, **glyceraldehyde 3-phosphate dehydrogenase,** is NAD dependent. Structurally, it consists of four identical polypeptides (monomers) forming a tetramer. Four—SH groups are present on each polypeptide, derived from cysteine residues within the polypeptide chain. One of the—SH groups is found at the active site of the enzyme (**Figure 18–3**). The substrate initially combines with this—SH group, forming a thiohemiacetal that is oxidized to a thiol ester; the hydrogens removed in this oxidation are transferred to NAD⁺. The thiol ester then undergoes phosphorolysis; inorganic phosphate (P_i) is added, forming 1,3-bisphosphoglycerate, and the—SH group is reconstituted.

In the next reaction, catalyzed by **phosphoglycerate kinase,** phosphate is transferred from 1,3-bisphosphoglycerate onto ADP, forming ATP (substrate-level phosphorylation) and 3-phosphoglycerate. Since two molecules of triose phosphate are formed per molecule of glucose undergoing glycolysis, two molecules of ATP are formed in this reaction per molecule of glucose undergoing glycolysis. The toxicity of arsenic is the result of competition of arsenate with inorganic phosphate (P_i) in this reaction to give 1-arseno-3-phosphoglycerate, which undergoes spontaneous hydrolysis to 3-phosphoglycerate without forming ATP. 3-Phosphoglycerate is isomerized to 2-phosphoglycerate by **phosphoglycerate mutase.** It is likely that 2,3-bisphosphoglycerate (diphosphoglycerate, DPG) is an intermediate in this reaction.

The subsequent step is catalyzed by **enolase** and involves a dehydration, forming phosphoenolpyruvate. Enolase is inhibited by **fluoride,** and when blood samples are taken for measurement of glucose, it is collected in tubes containing

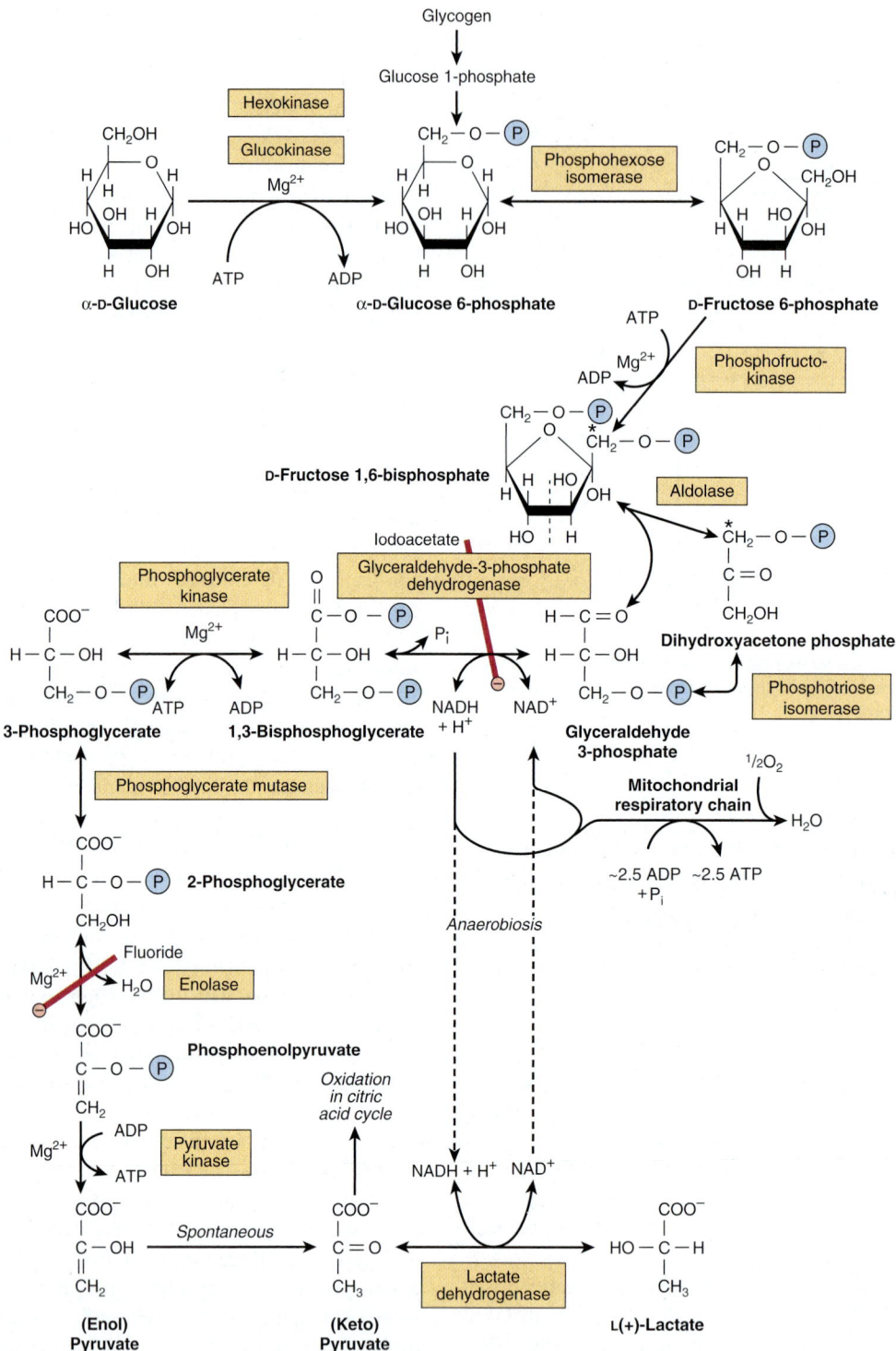

FIGURE 18–2 **The pathway of glycolysis.** (Ⓟ, —PO$_3^{2-}$; P$_i$, HOPO$_3^{2-}$; ⊖, inhibition.) *Carbons 1–3 of fructose bisphosphate form dihydroxyacetone phosphate, and carbons 4–6 form glyceraldehyde 3-phosphate. The term "bis-," as in bisphosphate, indicates that the phosphate groups are separated, whereas the term "di-," as in adenosine diphosphate, indicates that they are joined.

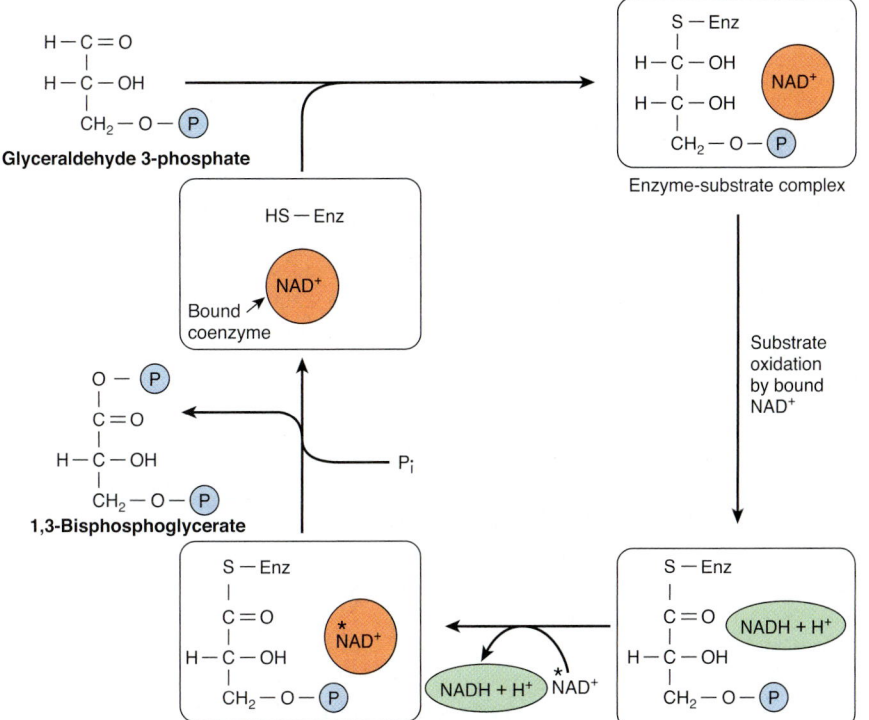

Glyceraldehyde 3-phosphate

1,3-Bisphosphoglycerate

Enzyme-substrate complex

Substrate oxidation by bound NAD$^+$

FIGURE 18–3 Mechanism of oxidation of glyceraldehyde 3-phosphate. (Enz, glyceraldehyde 3-phosphate dehydrogenase.) The enzyme is inhibited by the—SH poison iodoacetate, which is thus able to inhibit glycolysis. The NADH produced on the enzyme is not so firmly bound to the enzyme as is NAD$^+$. Consequently, NADH is easily displaced by another molecule of NAD$^+$.

fluoride to inhibit glycolysis. The enzyme is also dependent on the presence of either Mg^{2+} or Mn^{2+}. The phosphate of phosphoenolpyruvate is transferred to ADP by **pyruvate kinase** to form two molecules of ATP per molecule of glucose oxidized. The reaction of pyruvate kinase is essentially irreversible under physiological conditions, partly because of the large free energy change involved and partly because the immediate product of the enzyme-catalyzed reaction is enol-pyruvate, which undergoes spontaneous isomerization to pyruvate, so that the product of the reaction is not available to undergo the reverse reaction.

The redox state of the tissue now determines which of the two pathways is followed. Under **anaerobic** conditions, the NADH cannot be reoxidized through the respiratory chain to oxygen. Pyruvate is reduced by the NADH to lactate, catalyzed by **lactate dehydrogenase.** There are different tissue-specific isoenzymes lactate dehydrogenases that have clinical significance (Chapter 7). The reoxidation of NADH via lactate formation allows glycolysis to proceed in the absence of oxygen by regenerating sufficient NAD$^+$ for another cycle of the reaction catalyzed by glyceraldehyde 3-phosphate dehydrogenase. Under **aerobic conditions,** pyruvate is taken up into mitochondria, and after oxidative decarboxylation to acetyl-CoA is oxidized to CO$_2$ by the citric acid cycle (Chapter 17). The reducing equivalents from the NADH formed in glycolysis are taken up into mitochondria for oxidation via one of the two shuttles described in Chapter 13.

Tissues That Function Under Hypoxic Conditions Produce Lactate

This is true of skeletal muscle, particularly the white fibers, where the rate of work output, and hence the need for ATP formation, may exceed the rate at which oxygen can be taken up and utilized. Glycolysis in erythrocytes always terminates in lactate, because the subsequent reactions of pyruvate oxidation are mitochondrial, and erythrocytes lack mitochondria. Other tissues that normally derive much of their energy from glycolysis and produce lactate include brain, gastrointestinal tract, renal medulla, retina, and skin. Lactate production is also increased in septic shock, and many cancers also produce lactate. The liver, kidneys, and heart usually take up lactate and oxidize it, but produce it under hypoxic conditions.

When lactate production is high, as in vigorous exercise, septic shock, and cancer cachexia, much is used in the liver for gluconeogenesis (Chapter 20), leading to an increase in metabolic rate to provide the ATP and GTP needed. The increase in oxygen consumption as a result of increased oxidation of metabolic fuels to provide the ATP and GTP needed for gluconeogenesis is seen as **oxygen debt** after vigorous exercise.

Under some conditions, lactate may be formed in the cytosol, but then enter the mitochondrion to be oxidized to pyruvate for onward metabolism. This provides a pathway for the transfer of reducing equivalents from the cytotol into the mitochondrion

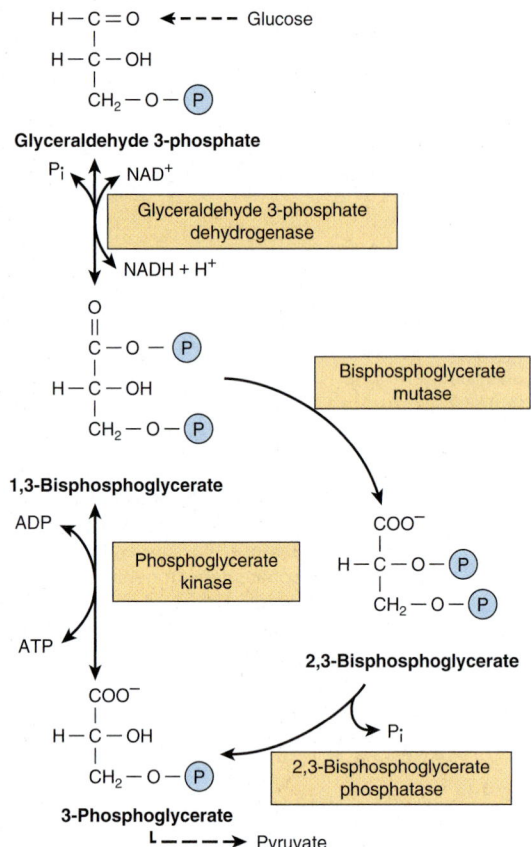

FIGURE 18–4 **2,3-Bisphosphoglycerate pathway in erythrocytes.**

for the electron transport chain in addition to the glycerophosphate (Figure 13–12) and malate (Figure 13–13) shuttles.

GLYCOLYSIS IS REGULATED AT THREE STEPS INVOLVING NONEQUILIBRIUM REACTIONS

Although most of the reactions of glycolysis are reversible, three are markedly exergonic and must therefore be considered physiologically irreversible. These reactions, catalyzed by **hexokinase** (and glucokinase), **phosphofructokinase,** and **pyruvate kinase,** are the major sites of regulation of glycolysis. Phosphofructokinase is significantly inhibited at normal intracellular concentrations of ATP; as discussed in Chapter 20, this inhibition can be rapidly relieved by 5′ AMP that is formed as ADP begins to accumulate, signaling the need for an increased rate of glycolysis. Cells that are capable of **gluconeogenesis** (reversing the glycolytic pathway, Chapter 20) have different enzymes that catalyze reactions to reverse these irreversible steps; glucose 6-phosphatase, fructose 1,6-bisphosphatase and, to reverse the reaction of pyruvate kinase, pyruvate carboxylase, and phosphoenolpyruvate carboxykinase. **Fructose** enters glycolysis by phosphorylation to fructose 1-phosphate, and bypasses the main regulatory steps, so resulting in forma-

tion of more pyruvate (and acetyl-CoA) than is required for ATP formation (Chapter 21). In the liver and adipose tissue, this leads to increased lipogenesis, and a high intake of fructose may be a factor in the development of obesity.

In Erythrocytes, the First Site of ATP Formation in Glycolysis May Be Bypassed

In erythrocytes, the reaction catalyzed by **phosphoglycerate kinase** may be bypassed to some extent by the reaction of **bisphosphoglycerate mutase,** which catalyzes the conversion of 1,3-bisphosphoglycerate to 2,3-bisphosphoglycerate, followed by hydrolysis to 3-phosphoglycerate and P_i, catalyzed by **2,3-bisphosphoglycerate phosphatase (Figure 18–4)**. This alternative pathway involves no net yield of ATP from glycolysis. However, it does serve to provide 2,3-bisphosphoglycerate, which binds to hemoglobin, decreasing its affinity for oxygen, and so making oxygen more readily available to tissues (see Chapter 6).

THE OXIDATION OF PYRUVATE TO ACETYL-CoA IS THE IRREVERSIBLE ROUTE FROM GLYCOLYSIS TO THE CITRIC ACID CYCLE

Pyruvate, formed in the cytosol, is transported into the mitochondrion by a proton symporter (Figure 13–10). Inside the mitochondrion, it is oxidatively decarboxylated to acetyl-CoA by a multienzyme complex that is associated with the inner mitochondrial membrane. This **pyruvate dehydrogenase complex** is analogous to the α-ketoglutarate dehydrogenase complex of the citric acid cycle (Figure 17–3). Pyruvate is decarboxylated by the **pyruvate dehydrogenase** component of the enzyme complex to a hydroxyethyl derivative of the thiazole ring of enzyme-bound **thiamin diphosphate,** which in turn reacts with oxidized lipoamide, the prosthetic group of **dihydrolipoyl transacetylase,** to form acetyl lipoamide (**Figure 18–5**). Thiamin is vitamin B_1 (Chapter 44) and in deficiency, glucose metabolism is impaired, and there is significant (and potentially life-threatening) lactic and pyruvic acidosis. Acetyl lipoamide reacts with coenzyme A to form acetyl-CoA and reduced lipoamide. The reaction is completed when the reduced lipoamide is reoxidized by a flavoprotein, **dihydrolipoyl dehydrogenase,** containing FAD. Finally, the reduced flavoprotein is oxidized by NAD$^+$, which in turn transfers reducing equivalents to the respiratory chain. The overall reaction is:

$$\text{Pyruvate} + \text{NAD}^+ + \text{CoA} \rightarrow \text{Acetyl-CoA} + \text{NADH} + \text{H}^+ + \text{CO}_2$$

The pyruvate dehydrogenase complex consists of a number of polypeptide chains of each of the three component enzymes, and the intermediates do not dissociate, but remain bound to the enzymes. Such a complex of enzymes, in which

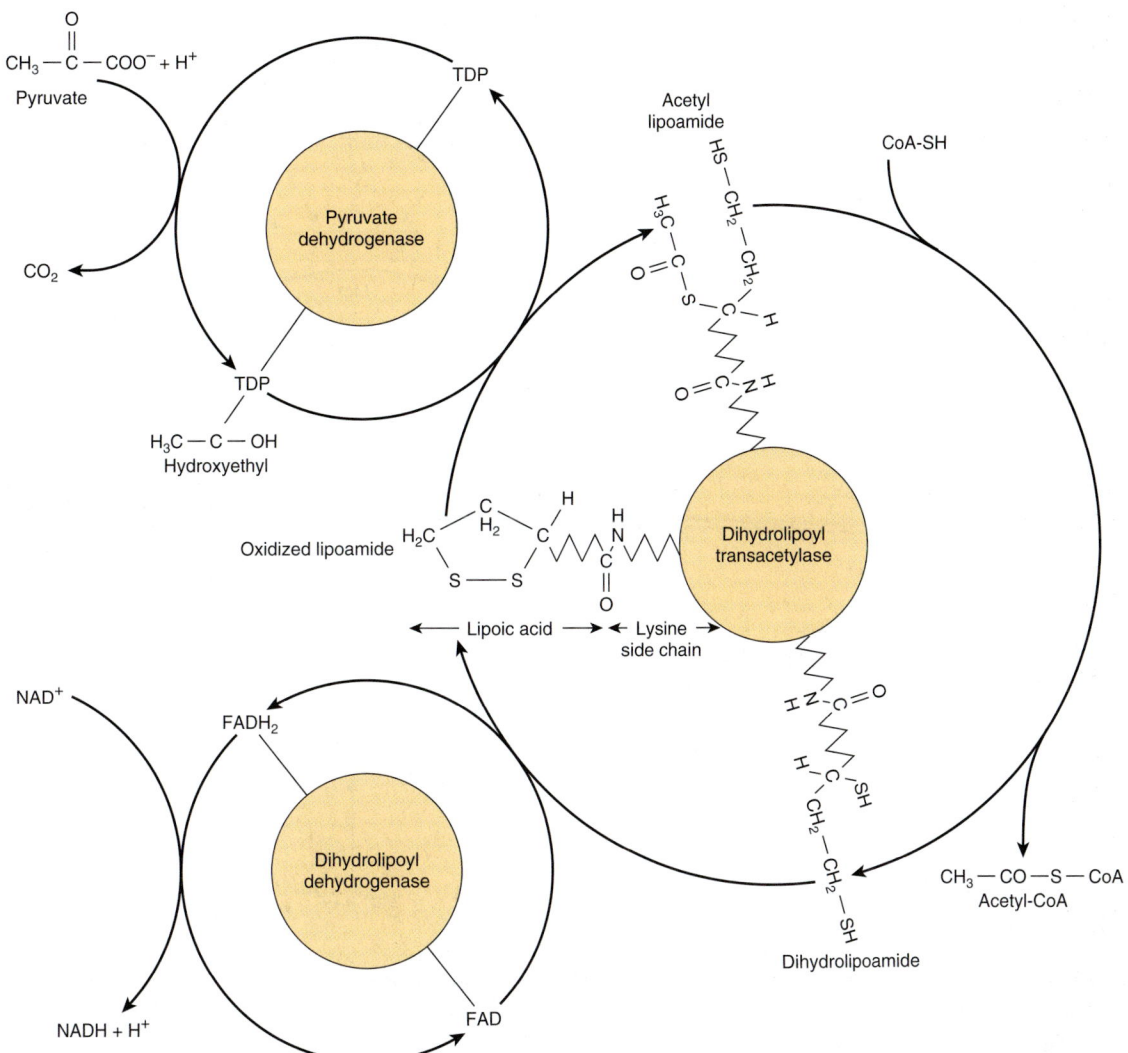

FIGURE 18–5 **Oxidative decarboxylation of pyruvate by the pyruvate dehydrogenase complex.** Lipoic acid is joined by an amide link to a lysine residue of the transacetylase component of the enzyme complex. It forms a long flexible arm, allowing the lipoic acid prosthetic group to rotate sequentially between the active sites of each of the enzymes of the complex. (FAD, flavin adenine dinucleotide; NAD⁺, nicotinamide adenine dinucleotide; TDP, thiamin diphosphate.)

the substrates are channeled from one enzyme to the next, increases the rate of reaction and prevents side reactions, increasing overall efficiency.

Pyruvate Dehydrogenase Is Regulated by End-Product Inhibition & Covalent Modification

Pyruvate dehydrogenase is inhibited by its products, acetyl-CoA, and NADH (**Figure 18–6**). It is also regulated by phosphorylation (catalyzed by a kinase) of three serine residues on the pyruvate dehydrogenase component of the multienzyme complex, resulting in decreased activity and by dephosphorylation (catalyzed by a phosphatase) that causes an increase in activity. The kinase is activated by increases in the [ATP]/[ADP], [acetyl-CoA]/[CoA], and [NADH]/[NAD⁺] ratios.

Thus, pyruvate dehydrogenase, and therefore glycolysis, is inhibited both when there is adequate ATP (and reduced coenzymes for ATP formation) available, and also when fatty acids are being oxidized. In fasting, when free fatty acid concentrations increase, there is a decrease in the proportion of the enzyme in the active form, leading to a sparing of carbohydrate. In adipose tissue, where glucose provides acetyl-CoA for lipogenesis, the enzyme is activated in response to insulin.

Oxidation of Glucose Yields Up to 32 Mol of ATP Under Aerobic Conditions, But Only 2 Mol When O₂ Is Absent

When 1 mol of glucose is combusted in a calorimeter to CO_2 and water, approximately 2870 kJ are liberated as heat. When

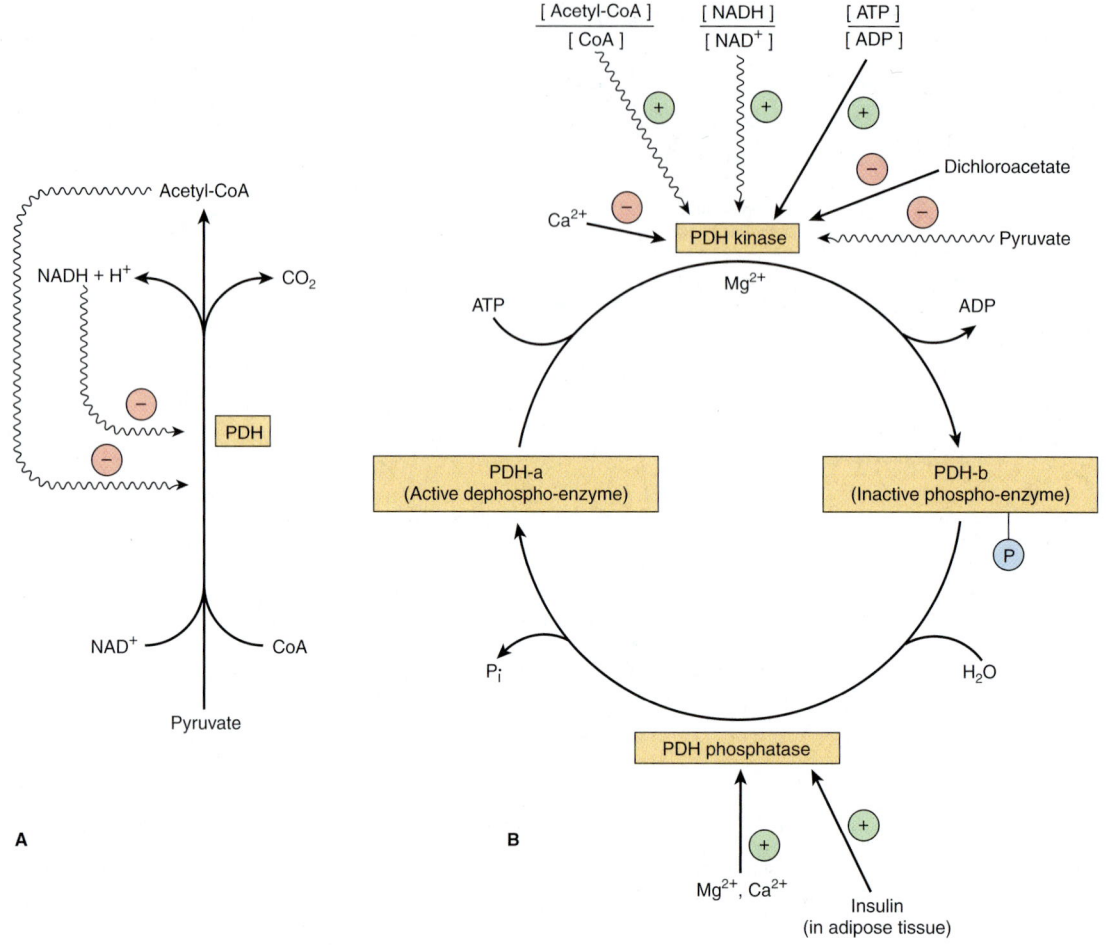

FIGURE 18–6 **Regulation of pyruvate dehydrogenase (PDH).** Arrows with wavy shafts indicate allosteric effects. **(A)** Regulation by end-product inhibition. **(B)** Regulation by interconversion of active and inactive forms.

oxidation occurs in the tissues, approximately 32 mol of ATP are generated per molecule of glucose oxidized to CO_2 and water. In vivo, ΔG for the ATP synthase reaction has been calculated as approximately 51.6 kJ. It follows that the total energy captured in ATP per mole of glucose oxidized is 1651 kJ, or approximately 58% of the energy of combustion. Most of the ATP is formed by oxidative phosphorylation resulting from the reoxidation of reduced coenzymes by the respiratory chain. The remainder is formed by substrate level phosphorylation (**Table 18–1**).

CLINICAL ASPECTS

Inhibition of Pyruvate Metabolism Leads to Lactic Acidosis

Arsenite and mercuric ions react with the—SH groups of lipoic acid and inhibit pyruvate dehydrogenase, as does a **dietary deficiency of thiamin** (Chapter 44), allowing pyruvate to accumulate. Many alcoholics are thiamin deficient (both because of a poor diet and also because alcohol inhibits thiamin absorption), and may develop potentially fatal pyruvic and lactic acidosis. Patients with **inherited pyruvate dehydro-**

genase deficiency, which can be the result of defects in one or more of the components of the enzyme complex, also present with lactic acidosis, particularly after a glucose load. Because of the dependence of the brain on glucose as a fuel, these metabolic defects commonly cause neurologic disturbances.

Inherited aldolase A deficiency and pyruvate kinase deficiency in erythrocytes cause **hemolytic anemia.** The exercise capacity of patients with **muscle phosphofructokinase deficiency** is low, particularly if they are on high-carbohydrate diets. By providing lipid as an alternative fuel, for example, during starvation, when blood-free fatty acid and ketone bodies are increased, work capacity is improved.

SUMMARY

- Glycolysis is the cytosolic pathway of all mammalian cells for the metabolism of glucose (or glycogen) to pyruvate and lactate.

- It can function anaerobically by regenerating oxidized NAD^+ (required in the glyceraldehyde-3-phosphate dehydrogenase reaction), by reducing pyruvate to lactate.

- Lactate is the end product of glycolysis under anaerobic conditions (eg, in exercising muscle) or, in erythrocytes, when

TABLE 18–1 ATP Formation in the Catabolism of Glucose

Pathway	Reaction Catalyzed by	Method of ATP Formation	ATP per mol of Glucose
Glycolysis	Glyceraldehyde 3-phosphate dehydrogenase	Respiratory chain oxidation of 2 NADH	5[1]
	Phosphoglycerate kinase	Substrate-level phosphorylation	2
	Pyruvate kinase	Substrate-level phosphorylation	2
			9
	Consumption of ATP for reactions of hexokinase and phosphofructokinase		−2
			Net 7
Citric acid cycle	Pyruvate dehydrogenase	Respiratory chain oxidation of 2 NADH	5
	Isocitrate dehydrogenase	Respiratory chain oxidation of 2 NADH	5
	α-Ketoglutarate dehydrogenase	Respiratory chain oxidation of 2 NADH	5
	Succinate thiokinase	Substrate level phosphorylation	2
	Succinate dehydrogenase	Respiratory chain oxidation of 2 $FADH_2$	3
	Malate dehydrogenase	Respiratory chain oxidation of 2 NADH	5
			Net 25
	Total per mol of glucose under aerobic conditions		32
	Total per mol of glucose under anaerobic conditions		2

[1]This assumes that NADH formed in glycolysis is transported into mitochondria by the malate shuttle (Figure 13–13). If the glycerophosphate shuttle is used, then only 1.5 ATP will be formed per mol of NADH. Note that there is a considerable advantage in using glycogen rather than glucose for anaerobic glycolysis in muscle, since the product of glycogen phosphorylase is glucose 1-phosphate (Figure 19–1), which is interconvertible with glucose 6-phosphate. This saves the ATP that would otherwise be used by hexokinase, increasing the net yield of ATP from 2 to 3 per glucose.

there are no mitochondria to permit the further oxidation of pyruvate.

■ Glycolysis is regulated by three enzymes catalyzing nonequilibrium reactions: hexokinase, phosphofructokinase, and pyruvate kinase.

■ In erythrocytes, the first site in glycolysis for generation of ATP may be bypassed, leading to the formation of 2,3-bisphosphoglycerate, which is important in decreasing the affinity of hemoglobin for O_2.

■ Pyruvate is oxidized to acetyl-CoA by a multienzyme complex, pyruvate dehydrogenase, which is dependent on the vitamin-derived cofactor thiamin diphosphate.

■ Conditions that involve an impairment of pyruvate metabolism frequently lead to lactic acidosis.

REFERENCES

Behal RH, Buxton DB, Robertson JG, Olson MS: Regulation of the pyruvate dehydrogenase multienzyme complex. Annu Rev Nutr 1993;13:497.

Boiteux A, Hess B: Design of glycolysis. Philos Trans R Soc Lond B Biol Sci 1981;293:5.

Fothergill-Gilmore LA: The evolution of the glycolytic pathway. Trends Biochem Sci 1986;11:47.

Gladden LB: Lactate metabolism: a new paradigm for the third millennium. J Physiol 2004;558:5.

Kim J-W, Dang CV: Multifaceted roles of glycolytic enzymes. Trends Biochem Sci 2005;30:142.

Levy B: Lactate and shock state: the metabolic view. Curr Opin Crit Care 2006;1:315.

Maj MC, Cameron JM, Robinson BH: Pyruvate dehydrogenase phosphatase deficiency: orphan disease or an under-diagnosed condition? Mol Cell Endocrinol 2006;249:1.

Martin E, Rosenthal RE, Fiskum G: Pyruvate dehydrogenase complex: metabolic link to ischemic brain injury and target of oxidative stress. J Neurosci Res 2005;79:240.

Patel MS, Korotchkina LG: Regulation of the pyruvate dehydrogenase complex. Biochem Soc Trans 2006;34:217.

Philp A, Macdonald AL, Watt PW: Lactate—a signal coordinating cell and systemic function. J Exp Biol 2005;208:4561.

Pumain R, Laschet J: A key glycolytic enzyme plays a dual role in GABAergic neurotransmission and in human epilepsy. Crit Rev Neurobiol 2006;18:197.

Rider MH, Bertrand L, Vertommen D, et al: 6-Phosphofructo-2-kinase/fructose-2,6-bisphosphatase: head-to-head with a bifunctional enzyme that controls glycolysis. Biochem J 2004;381:561.

Robergs RA, Ghiasvand F, Parker D: Biochemistry of exercise-induced metabolic acidosis. Am J Physiol 2004;287:R502.

Sugden MC, Holness MJ: Mechanisms underlying regulation of the expression and activities of the mammalian pyruvate dehydrogenase kinases. Arch Physiol Biochem 2006;112:139.

Wasserman DH: Regulation of glucose fluxes during exercise in the postabsorptive state. Annu Rev Physiol 1995;57:191.

Metabolism of Glycogen

David A. Bender, PhD & Peter A. Mayes, PhD, DSc

O B J E C T I V E S

After studying this chapter, you should be able to:

- Describe the structure of glycogen and its importance as a carbohydrate reserve.
- Describe the synthesis and breakdown of glycogen and how the processes are regulated in response to hormone action.
- Describe the various types of glycogen storage diseases.

BIOMEDICAL IMPORTANCE

Glycogen is the major storage carbohydrate in animals, corresponding to starch in plants; it is a branched polymer of α-D-glucose (Figure 14–13). It occurs mainly in liver and muscle, with modest amounts in the brain. Although the liver content of glycogen is greater than that of muscle, because the muscle mass of the body is considerably greater than that of the liver, about three-quarters of total body glycogen is in muscle (**Table 19–1**).

Muscle glycogen provides a readily available source of glucose 1-phosphate for glycolysis within the muscle itself. Liver glycogen functions to store and export glucose to maintain the **blood glucose** concentration in the fasting state. The liver concentration of glycogen is about 450 mM after a meal, falling to about 200 mM after an overnight fast; after 12–18 h of fasting, liver glycogen is almost totally depleted. Although muscle glycogen does not directly yield free glucose (because muscle lacks glucose 6-phosphatase), pyruvate formed by glycolysis in muscle can undergo transamination to alanine, which is exported from muscle and used for gluconeogenesis in the liver (Figure 20–4). **Glycogen storage diseases** are a group of inherited disorders characterized by deficient mobilization of glycogen or deposition of abnormal forms of glycogen, leading to liver damage and muscle weakness; some glycogen storage diseases result in early death.

The highly branched structure of glycogen (Figure 14–13) provides a large number of sites for glycogenolysis, permitting rapid release of glucose 1-phosphate for muscle activity. Endurance athletes require a slower, more sustained release of glucose 1-phosphate. The formation of branch points in glycogen is slower than the addition of glucose units to a linear chain, and some endurance athletes practice **carbohydrate loading**—exercise to exhaustion (when muscle glycogen in largely depleted) followed by a high-carbohydrate meal, which results in rapid glycogen synthesis, with fewer branch points than normal.

GLYCOGENESIS OCCURS MAINLY IN MUSCLE & LIVER

The Pathway of Glycogen Biosynthesis Involves a Special Nucleotide of Glucose

As in glycolysis, glucose is phosphorylated to glucose 6-phosphate, catalyzed by **hexokinase** in muscle and **glucokinase** in liver (**Figure 19–1**). Glucose 6-phosphate is isomerized to glucose 1-phosphate by **phosphoglucomutase.** The enzyme itself is phosphorylated, and the phosphate group takes part in a reversible reaction in which glucose 1,6-bisphosphate is an intermediate. Next, glucose 1-phosphate reacts with uridine triphosphate (UTP) to form the active nucleotide **uridine diphosphate glucose (UDPGlc)** and pyrophosphate (**Figure 19–2**), catalyzed by **UDPGlc pyrophosphorylase.** The reaction proceeds in the direction of UDPGlc formation because **pyrophosphatase** catalyzes hydrolysis of pyrophosphate to 2 × phosphate, so removing one of the reaction products. UDPGlc pyrophosphorylase has a low K_m for glucose 1-phosphate and is present in relatively large amounts, so that it is not a regulatory step in glycogen synthesis.

TABLE 19–1 Storage of Carbohydrate in a 70 kg Human Being

	Percentage of Tissue Weight	Tissue Weight	Body Content (g)
Liver glycogen	5.0	1.8 kg	90
Muscle glycogen	0.7	35 kg	245
Extracellular glucose	0.1	10 L	10

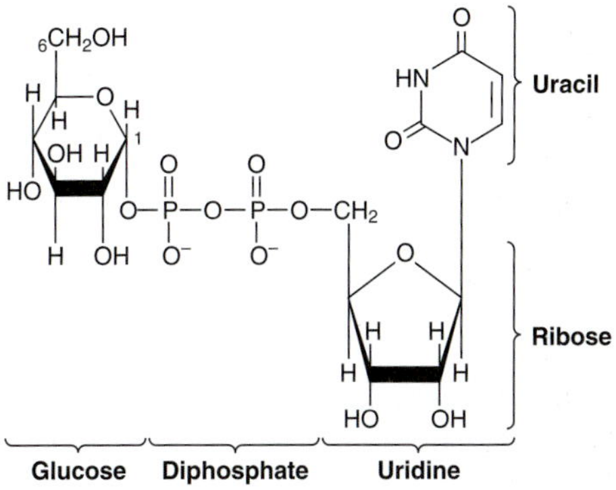

The initial steps in glycogen synthesis involve the protein **glycogenin,** a 37 kDa protein that is glucosylated on a specific tyrosine residue by UDPGlc. Glycogenin catalyzes the transfer of a further seven glucose residues from UDPGlc, in $1 \rightarrow 4$ linkage, to form a **glycogen primer** that is the substrate for glycogen synthase. In muscle the glycogenin remains at the core of the glycogen granule (Figure 14–13), but in liver many glycogen

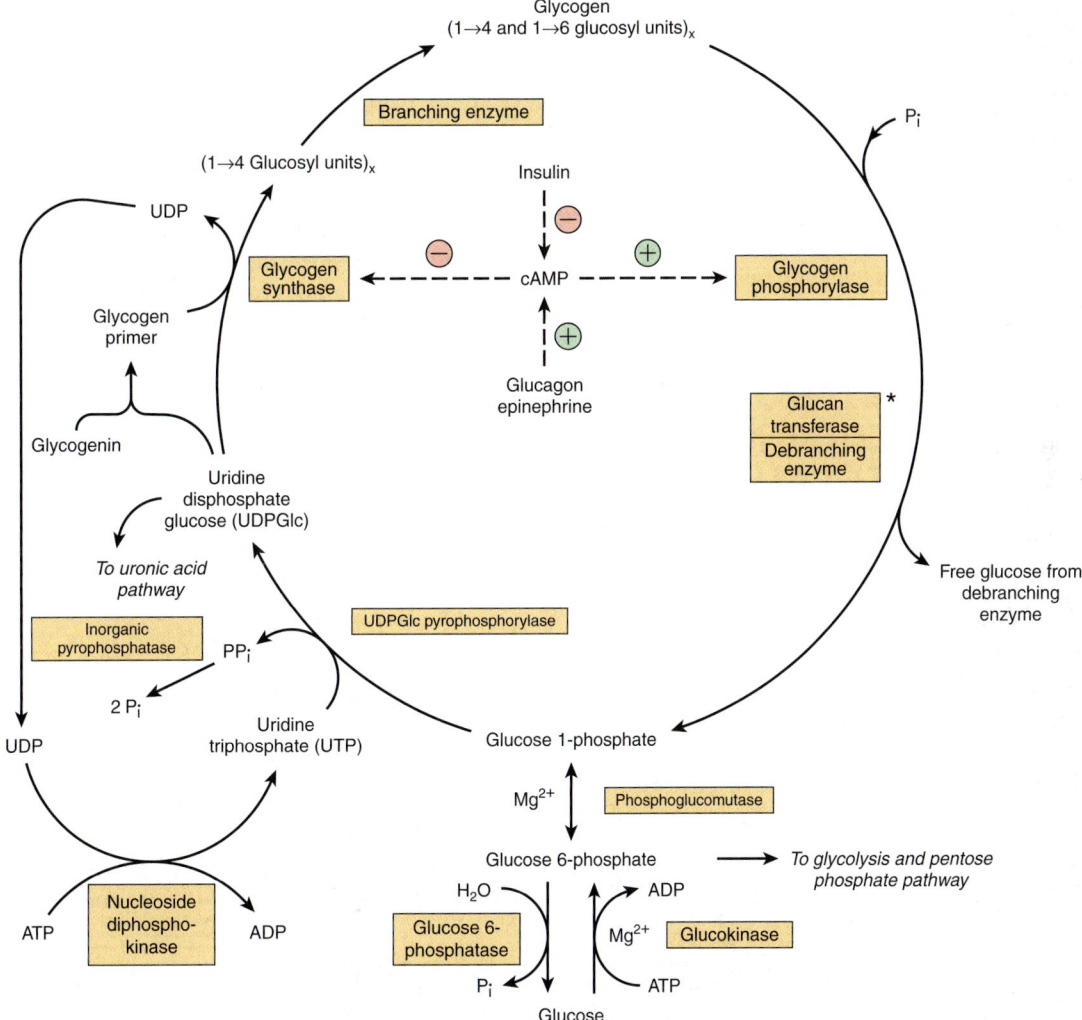

FIGURE 19–2 Uridine diphosphate glucose (UDPGlc).

granules do not have a central glycogenin molecule. **Glycogen synthase** catalyzes the formation of a glycoside bond between C-1 of the glucose of UDPGlc and C-4 of a terminal glucose residue of glycogen, liberating uridine diphosphate (UDP). The

FIGURE 19–1 **Pathways of glycogenesis and of glycogenolysis in the liver.** ($\oplus$, Stimulation; $\ominus$, inhibition.) Insulin decreases the level of cAMP only after it has been raised by glucagon or epinephrine; that is, it antagonizes their action. Glucagon is active in heart muscle but not in skeletal muscle. *Glucan transferase and debranching enzyme appear to be two separate activities of the same enzyme.

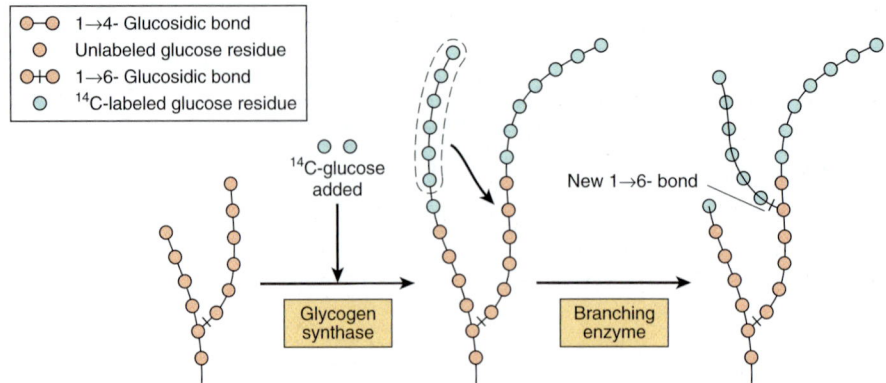

FIGURE 19–3 **The biosynthesis of glycogen.** The mechanism of branching as revealed by feeding ^{14}C-labeled glucose and examining liver glycogen at intervals.

addition of a glucose residue to a preexisting glycogen chain, or "primer," occurs at the nonreducing, outer end of the molecule, so that the branches of the glycogen molecule become elongated as successive $1 \rightarrow 4$ linkages are formed (**Figure 19–3**).

Branching Involves Detachment of Existing Glycogen Chains

When a growing chain is at least 11 glucose residues long, **branching enzyme** transfers a part of the $1 \rightarrow 4$-chain (at least six glucose residues) to a neighboring chain to form a $1 \rightarrow 6$ linkage, establishing a **branch point.** The branches grow by further additions of $1 \rightarrow 4$-glucosyl units and further branching.

GLYCOGENOLYSIS IS NOT THE REVERSE OF GLYCOGENESIS, BUT IS A SEPARATE PATHWAY

Glycogen phosphorylase catalyzes the rate-limiting step in glycogenolysis— the phosphorolytic cleavage (phosphorolysis; cf hydrolysis) of the $1 \rightarrow 4$ linkages of glycogen to yield glucose 1-phosphate (**Figure 19–4**). There are distinct isoenzymes of glycogen phosphorylase in liver, muscle and brain, encoded by distinct genes. Glycogen phosphorylase requires pyridoxal phosphate (see Chapter 44) as its coenzyme. Unlike the reactions of amino acid metabolism (Chapter 29), in which the aldehyde is the reactive group, in phosphorylase it is the phosphate group that it catalytically active.

The terminal glucosyl residues from the outermost chains of the glycogen molecule are removed sequentially until approximately four glucose residues remain on either side of a $1 \rightarrow 6$ branch (Figure 19–4). The **debranching enzyme** has two distinct catalytic sites in a single polypeptide chain. One is a glucan transferase that transfers a trisaccharide unit from one branch to the other, exposing the $1 \rightarrow 6$ branch point. The other is a 1,6-glycosidase that catalyzes hydrolysis of the $1 \rightarrow 6$

glycoside bond to liberate free glucose. Further phosphorylase action can then proceed. The combined action of phosphorylase and these other enzymes leads to the complete breakdown of glycogen.

The reaction catalyzed by phosphoglucomutase is reversible, so that glucose 6-phosphate can be formed from glucose 1-phosphate. In **liver,** but not muscle, **glucose 6-phosphatase** catalyzes hydrolysis of glucose 6-phosphate, yielding glucose that is exported, leading to an increase in the blood glucose concentration. Glucose 6-phosphatase is in the lumen of the smooth endoplasmic reticulum, and genetic defects of the glucose 6-phosphate transporter can cause a variant of type I glycogen storage disease (see **Table 19–2**).

Glycogen granules can also be engulphed by **lysosomes,** where acid maltase catalyzes the hydrolysis of glycogen to glucose. This may be especially important in glucose homeostasis in neonates, but genetic lack of lysosomal acid maltase leads to type II glycogen storage disease (Pompe's disease,

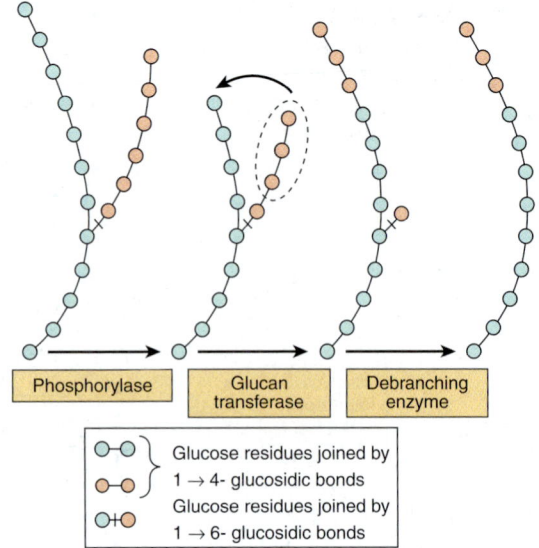

FIGURE 19–4 **Steps in glycogenolysis.**

TABLE 19–2 Glycogen Storage Diseases

Type	Name	Enzyme Deficiency	Clinical Features
0	—	Glycogen synthase	Hypoglycemia; hyperketonemia; early death
Ia	Von Gierke's disease	Glucose 6-phosphatase	Glycogen accumulation in liver and renal tubule cells; hypoglycemia; lactic acidemia; ketosis; hyperlipemia
Ib	—	Endoplasmic reticulum glucose 6-phosphate transporter	As type Ia; neutropenia and impaired neutrophil function leading to recurrent infections
II	Pompe's disease	Lysosomal $\alpha_1 \rightarrow 4$ and $\alpha_1 \rightarrow 6$ glucosidase (acid maltase)	Accumulation of glycogen in lysosomes: juvenile onset variant, muscle hypotonia, death from heart failure by age 2; adult onset variant, muscle dystrophy
IIIa	Limit dextrinosis, Forbe's or Cori's disease	Liver and muscle debranching enzyme	Fasting hypoglycemia; hepatomegaly in infancy; accumulation of characteristic branched polysaccharide (limit dextrin); muscle weakness
IIIb	Limit dextrinosis	Liver debranching enzyme	As type IIIa, but no muscle weakness
IV	Amylopectinosis, Andersen's disease	Branching enzyme	Hepatosplenomegaly; accumulation of polysaccharide with few branch points; death from heart or liver failure before age 5
V	Myophosphorylase deficiency, McArdle's syndrome	Muscle phosphorylase	Poor exercise tolerance; muscle glycogen abnormally high (2.5%–4%); blood lactate very low after exercise
VI	Hers' disease	Liver phosphorylase	Hepatomegaly; accumulation of glycogen in liver; mild hypoglycemia; generally good prognosis
VII	Tarui's disease	Muscle and erythrocyte phosphofructokinase 1	Poor exercise tolerance; muscle glycogen abnormally high (2.5%–4%); blood lactate very low after exercise; also hemolytic anemia
VIII		Liver phosphorylase kinase	Hepatomegaly; accumulation of glycogen in liver; mild hypoglycemia; generally good prognosis
IX		Liver and muscle phosphorylase kinase	Hepatomegaly; accumulation of glycogen in liver and muscle; mild hypoglycemia; generally good prognosis
X		cAMP-dependent protein kinase A	Hepatomegaly; accumulation of glycogen in liver

Table 19–2). The lysosomal catabolism of glycogen is under hormonal control.

CYCLIC AMP INTEGRATES THE REGULATION OF GLYCOGENOLYSIS & GLYCOGENESIS

The principal enzymes controlling glycogen metabolism—glycogen phosphorylase and glycogen synthase—are regulated in opposite directions by allosteric mechanisms and covalent modification by reversible phosphorylation and dephosphorylation of enzyme protein in response to hormone action (Chapter 9). Phosphorylation of glycogen phosphorylase increases its activity; phosphorylation of glycogen synthase reduces its activity.

Phosphorylation is increased in response to cyclic AMP (cAMP) (**Figure 19–5**) formed from ATP by **adenylyl cyclase** at the inner surface of cell membranes in response to hormones such as **epinephrine, norepinephrine,** and **glucagon.** cAMP is hydrolyzed by **phosphodiesterase,** so terminating hormone action; in liver insulin increases the activity of phosphodiesterase.

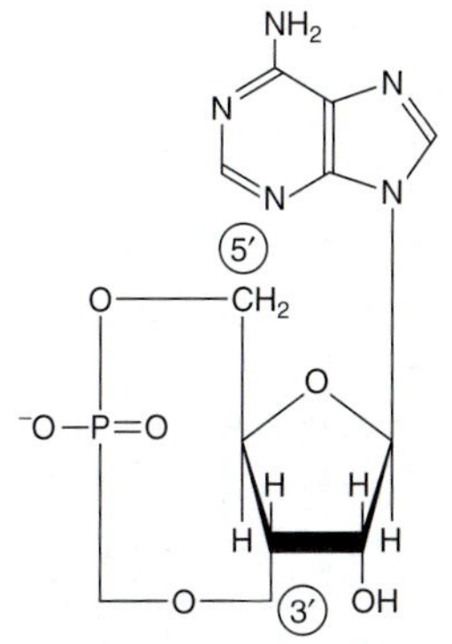

FIGURE 19–5 3′,5′-Adenylic acid (cyclic AMP; cAMP).

The Control of Glycogen Phosphorylase Differs Between Liver & Muscle

In the liver, the role of glycogen is to provide free glucose for export to maintain the blood concentration of glucose; in muscle the role of glycogen is to provide a source of glucose 6-phosphate for glycolysis in response to the need for ATP for muscle contraction. In both tissues, the enzyme is activated by phosphorylation catalyzed by phosphorylase kinase (to yield phosphorylase a) and inactivated by dephosphorylation catalyzed by phosphoprotein phosphatase (to yield phosphorylase b), in response to hormonal and other signals.

There is instantaneous overriding of this hormonal control. Active phosphorylase a in both tissues is allosterically inhibited by ATP and glucose 6-phosphate; in liver, but not muscle, free glucose is also an inhibitor. Muscle phosphorylase differs from the liver isoenzyme in having a binding site for 5′ AMP, which acts as an allosteric activator of the (inactive) dephosphorylated b-form of the enzyme. 5′ AMP acts as a potent signal of the energy state of the muscle cell; it is formed as the concentration of ADP begins to increase (indicating the need for increased substrate metabolism to permit ATP formation), as a result of the reaction of adenylate kinase: $2 \times ADP \leftrightarrow ATP + 5′ AMP$.

cAMP Activates Glycogen Phosphorylase

Phosphorylase kinase is activated in response to cAMP (**Figure 19–6**). Increasing the concentration of cAMP activates **cAMP-dependent protein kinase,** which catalyzes the phosphorylation by ATP of inactive **phosphorylase kinase b** to active **phosphorylase kinase a,** which in turn, phosphorylates phosphorylase b to phosphorylase a. In the liver, cAMP is formed in response to glucagon, which is secreted in response to falling blood glucose. Muscle is insensitive to glucagon; in muscle, the signal for increased cAMP formation is the action of norepinephrine, which is secreted in response to fear or fright, when there is a need for increased glycogenolysis to permit rapid muscle activity.

Ca²⁺ Synchronizes the Activation of Glycogen Phosphorylase With Muscle Contraction

Glycogenolysis in muscle increases several 100-fold at the onset of contraction; the same signal (increased cytosolic Ca^{2+} ion concentration) is responsible for initiation of both contraction and glycogenolysis. Muscle phosphorylase kinase, which activates glycogen phosphorylase, is a tetramer of four different subunits, α, β, γ, and δ. The α and β subunits contain serine residues that are phosphorylated by cAMP-dependent protein kinase. The δ subunit is identical to the Ca^{2+}-binding protein **calmodulin** (Chapter 42), and binds four Ca^{2+}. The binding of

Ca^{2+} activates the catalytic site of the γ subunit even while the enzyme is in the dephosphorylated b state; the phosphorylated a form is only fully activated in the presence of high concentrations of Ca^{2+}.

Glycogenolysis in Liver Can Be cAMP-Independent

In the liver, there is cAMP-independent activation of glycogenolysis in response to stimulation of α_1 **adrenergic** receptors by epinephrine and norepinephrine. This involves mobilization of Ca^{2+} into the cytosol, followed by the stimulation of a **Ca²⁺/calmodulin-sensitive phosphorylase kinase.** cAMP-independent glycogenolysis is also activated by vasopressin, oxytocin, and angiotensin II acting either through calcium or the phosphatidylinositol bisphosphate pathway (Figure 42–10).

Protein Phosphatase-1 Inactivates Glycogen Phosphorylase

Both phosphorylase a and phosphorylase kinase a are dephosphorylated and inactivated by **protein phosphatase-1.** Protein phosphatase-1 is inhibited by a protein, **inhibitor-1,** which is active only after it has been phosphorylated by cAMP-dependent protein kinase. Thus, cAMP controls both the activation and inactivation of phosphorylase (Figure 19–6). **Insulin** reinforces this effect by inhibiting the activation of phosphorylase b. It does this indirectly by increasing uptake of glucose, leading to increased formation of glucose 6-phosphate, which is an inhibitor of phosphorylase kinase.

Glycogen Synthase & Phosphorylase Activities Are Reciprocally Regulated

There are different isoenzymes of glycogen synthase in liver, muscle, and brain. Like phosphorylase, glycogen synthase exists in both phosphorylated and nonphosphorylated states, and the effect of phosphorylation is the reverse of that seen in phosphorylase (**Figure 19–7**). Active **glycogen synthase a** is dephosphorylated and inactive **glycogen synthase b** is phosphorylated.

Six different protein kinases act on glycogen synthase, and there are at least 9 different serine residues in the enzyme that can be phosphorylated. Two of the protein kinases are Ca^{2+}/calmodulin dependent (one of these is phosphorylase kinase). Another kinase is cAMP-dependent protein kinase, which allows cAMP-mediated hormonal action to inhibit glycogen synthesis synchronously with the activation of glycogenolysis. Insulin also promotes glycogenesis in muscle at the same time as inhibiting glycogenolysis by raising glucose 6-phosphate concentrations, which stimulates the dephosphorylation and activation of glycogen synthase.

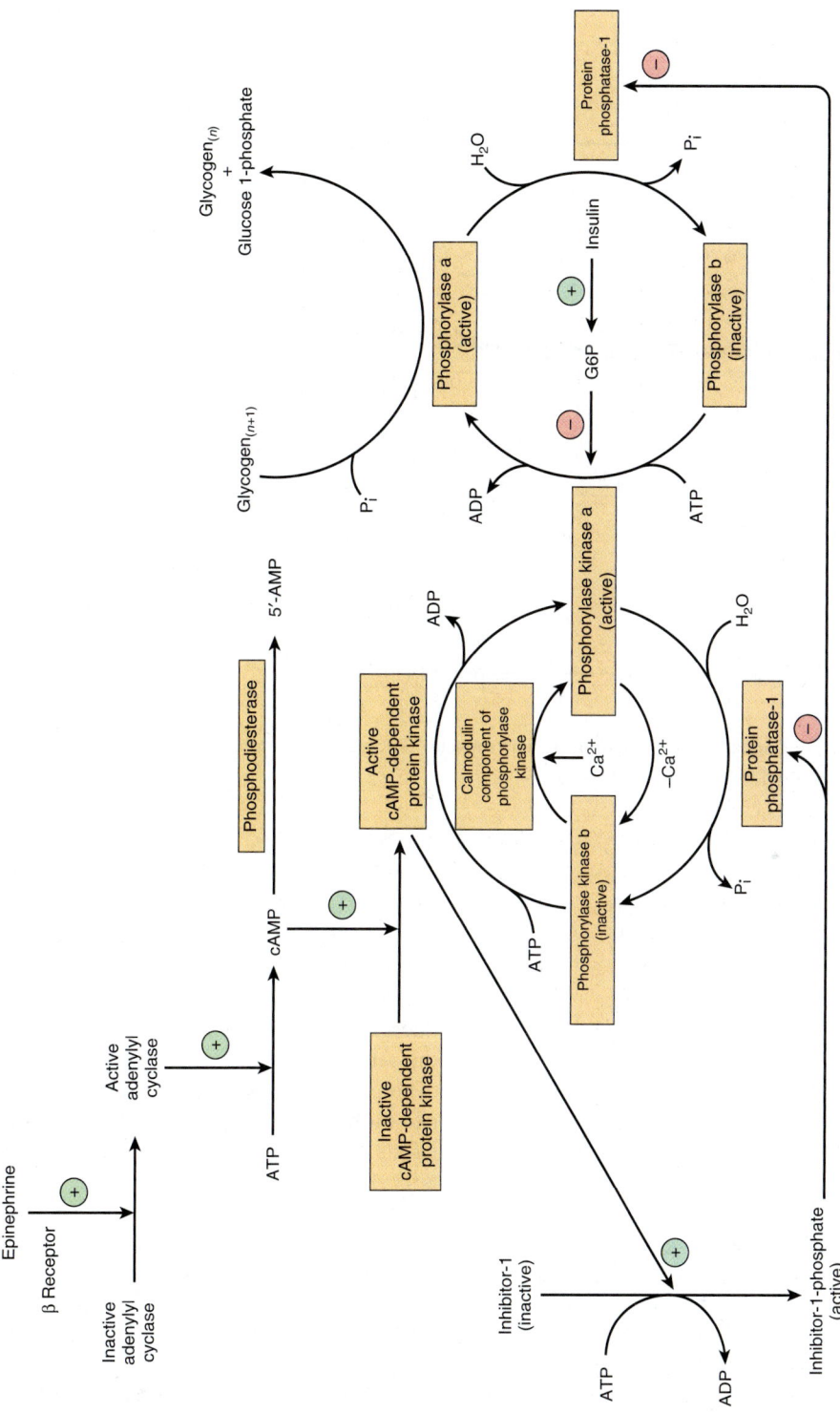

FIGURE 19–6 **Control of phosphorylase in muscle.** The sequence of reactions arranged as a cascade allows amplification of the hormonal signal at each step. (G6P, glucose 6-phosphate; *n*, number of glucose residues.)

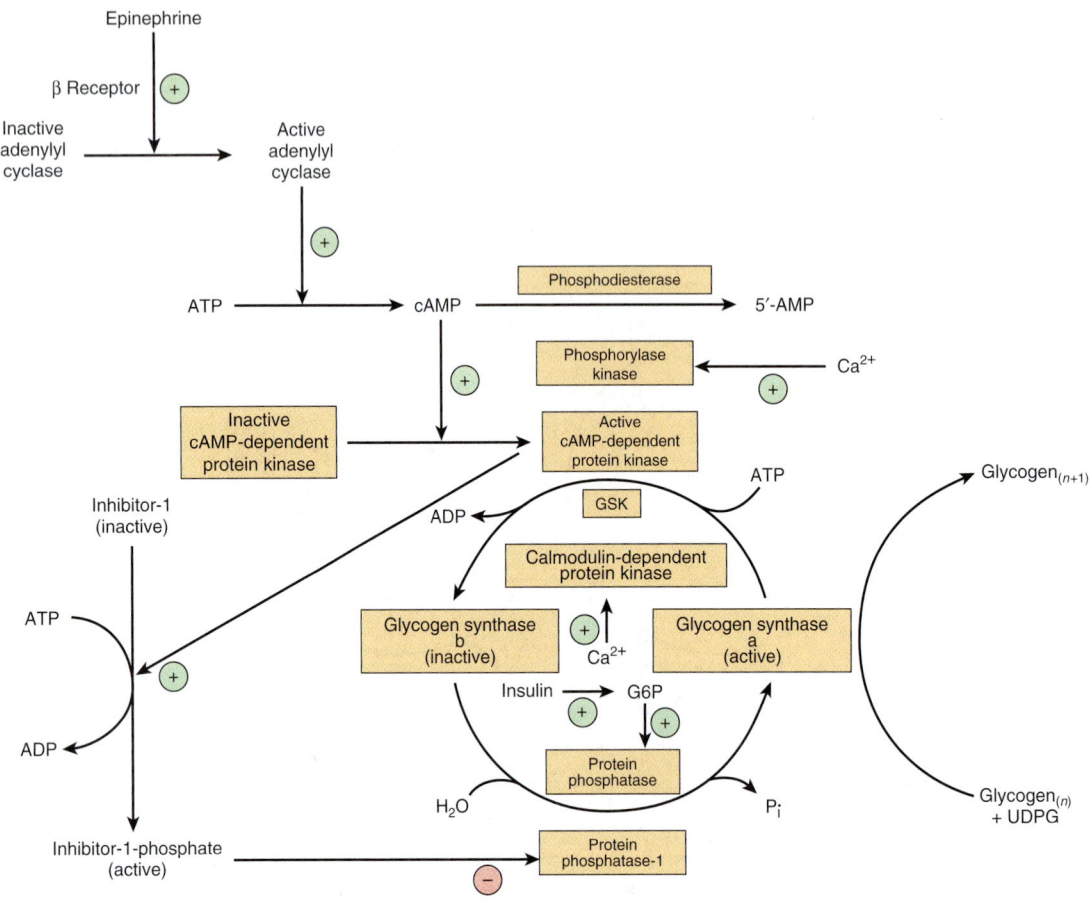

FIGURE 19–7 **Control of glycogen synthase in muscle.** (GSK, glycogen synthase kinase; G6P, glucose 6-phosphate; *n*, number of glucose residues.)

Dephosphorylation of glycogen synthase b is carried out by protein phosphatase-1, which is under the control of cAMP-dependent protein kinase.

REGULATION OF GLYCOGEN METABOLISM IS EFFECTED BY A BALANCE IN ACTIVITIES BETWEEN GLYCOGEN SYNTHASE & PHOSPHORYLASE

At the same time as phosphorylase is activated by a rise in concentration of cAMP (via phosphorylase kinase), glycogen synthase is converted to the inactive form; both effects are mediated via **cAMP-dependent protein kinase (Figure 19–8)**. Thus, inhibition of glycogenolysis enhances net glycogenesis, and inhibition of glycogenesis enhances net glycogenolysis. Also, the dephosphorylation of phosphorylase a, phosphorylase kinase, and glycogen synthase b is catalyzed by a single enzyme with broad specificity—**protein phosphatase-1.** In turn, protein phosphatase-1 is inhibited by cAMP-dependent protein kinase via inhibitor-1. Thus, glycogenolysis can be terminated and glycogenesis can be stimulated, or vice versa, synchronously, because both processes are dependent on the activity of cAMP-dependent protein kinase. Both phosphorylase kinase and glycogen synthase may be reversibly phosphorylated at more than one site by separate kinases and phosphatases. These secondary phosphorylations modify the sensitivity of the primary sites to phosphorylation and dephosphorylation (**multisite phosphorylation**). Also, they allow insulin, by way of increased glucose 6-phosphate, to have effects that act reciprocally to those of cAMP (see Figures 19–6 & 19–7).

CLINICAL ASPECTS

Glycogen Storage Diseases Are Inherited

"Glycogen storage disease" is a generic term to describe a group of inherited disorders characterized by deposition of an abnormal type or quantity of glycogen in tissues, or failure to mobilize glycogen. The principal diseases are summarized in Table 19–2.

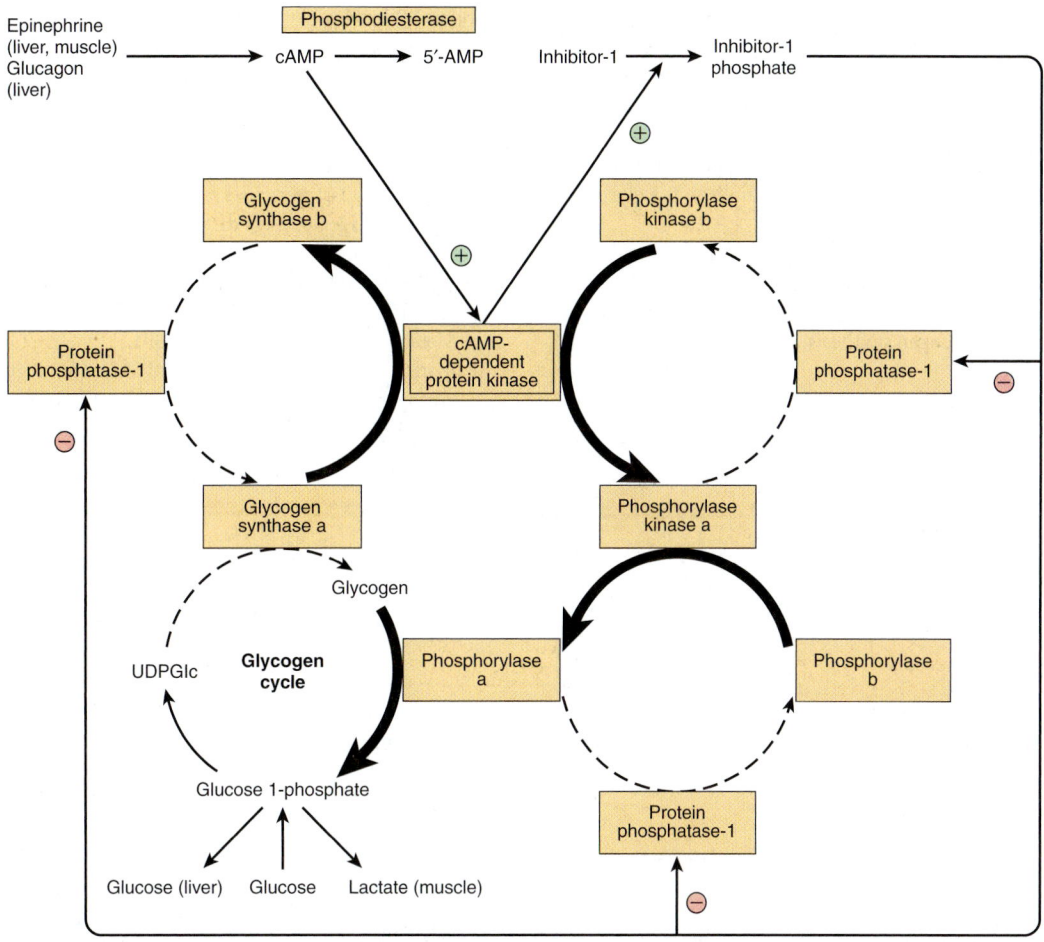

FIGURE 19–8 **Coordinated control of glycogenolysis and glycogenesis by cAMP-dependent protein kinase.** The reactions that lead to glycogenolysis as a result of an increase in cAMP concentrations are shown with bold arrows, and those that are inhibited by activation of protein phosphatase-1 are shown with dashed arrows. The reverse occurs when cAMP concentrations decrease as a result of phosphodiesterase activity, leading to glycogenesis.

SUMMARY

- Glycogen represents the principal storage carbohydrate in the body, mainly in the liver and muscle.

- In the liver, its major function is to provide glucose for extrahepatic tissues. In muscle, it serves mainly as a ready source of metabolic fuel for use in muscle. Muscle lacks glucose 6-phosphatase and cannot release free glucose from glycogen.

- Glycogen is synthesized from glucose by the pathway of glycogenesis. It is broken down by a separate pathway, glycogenolysis.

- Cyclic AMP integrates the regulation of glycogenolysis and glycogenesis by promoting the simultaneous activation of phosphorylase and inhibition of glycogen synthase. Insulin acts reciprocally by inhibiting glycogenolysis and stimulating glycogenesis.

- Inherited deficiencies of enzymes of glycogen metabolism in both liver and muscle cause glycogen storage diseases.

REFERENCES

Alanso MD, Lomako J, Lomako WM, et al: A new look at the biogenesis of glycogen. FASEB J. 1995;9:1126.

Bollen M, Keppens S, Stalmans W: Specific features of glycogen metabolism in the liver. Biochem J 1998;336:19.

Ferrer JC, Favre C, Gomis RR, et al: Control of glycogen deposition. FEBS Lett 2003;546:127–132.

Forde JE, Dale TC: Glycogen synthase kinase 3: a key regulator of cellular fate. Cell Mol Life Sci 2007;64:1930.

Graham TE, Yuan Z, Hill AK, et al: The regulation of muscle glycogen: the granule and its proteins. Acta Physiol (Oxf) 2010;199:489.

Greenberg CC, Jurczak MJ, Danos AM, et al: Glycogen branches out: new perspectives on the role of glycogen metabolism in the integration of metabolic pathways. Am J Physiol Endocrinol Metab 2006;291:E1.

Jensen J, Lai YC: Regulation of muscle glycogen synthase phosphorylation and kinetic properties by insulin, exercise, adrenaline and role in insulin resistance. Arch Physiol Biochem 2009;115:13.

Jentjens R, Jeukendrup A: Determinants of post-exercise glycogen synthesis during short-term recovery. Sports Med 2003;33:117.

McGarry JD, Kuwajima M, Newgard CB, et al: From dietary glucose to liver glycogen: the full circle round. Annu Rev Nutr 1987;7:51.

Meléndez-Hevia E, Waddell TG, Shelton ED: Optimization of molecular design in the evolution of metabolism: the glycogen molecule. Biochem J 1993;295:477.

Ozen H: Glycogen storage diseases: new perspectives. World J Gastroenterol 2007;13:2541.

Radziuk J, Pye S: Hepatic glucose uptake, gluconeogenesis and the regulation of glycogen synthesis. Diabetes Metab Res Rev 2001;17(4):250.

Roach PJ: Glycogen and its metabolism. Curr Mol Med 2002;2(2):101.

Roden M, Bernroider E: Hepatic glucose metabolism in humans—its role in health and disease. Best Pract Res Clin Endocrinol Metab 2003;17:365.

Rybicka KK: Glycosomes—the organelles of glycogen metabolism. Tissue Cell 1996; 28:254.

Shearer J, Graham TE: New perspectives on the storage and organization of muscle glycogen. Can J Appl Physiol 2002;27:179.

Shin YS: Glycogen storage disease: clinical, biochemical, and molecular heterogeneity. Semin Pediatr Neurol 2006;13:115.

Wolfsdorf JI, Holm IA: Glycogen storage diseases. Phenotypic, genetic, and biochemical characteristics, and therapy. Endocrinol Metab Clin North Am 1999;28:801.

Yeaman SJ, Armstrong JL, Bonavaud SM, et al: Regulation of glycogen synthesis in human muscle cells. Biochem Soc Trans 2001;29:537.

Gluconeogenesis & the Control of Blood Glucose

20

David A. Bender, PhD & Peter A. Mayes, PhD, DSc

OBJECTIVES

After studying this chapter, you should be able to:

- Explain the importance of gluconeogenesis in glucose homeostasis.
- Describe the pathway of gluconeogenesis, how irreversible enzymes of glycolysis are bypassed, and how glycolysis and gluconeogenesis are regulated reciprocally.
- Explain how plasma glucose concentration is maintained within narrow limits in the fed and fasting states.

BIOMEDICAL IMPORTANCE

Gluconeogenesis is the process of synthesizing glucose or glycogen from noncarbohydrate precursors. The major substrates are the glucogenic amino acids (Chapter 29), lactate, glycerol, and propionate. Liver and kidney are the major gluconeogenic tissues; the kidney may contribute up to 40% of total glucose synthesis in the fasting state and more in starvation. The key gluconeogenic enzymes are expressed in the small intestine, but it is unclear whether or not there is significant glucose production by the intestine in the fasting state.

A supply of glucose is necessary especially for the nervous system and erythrocytes. After an overnight fast, glycogenolysis (Chapter 19) and gluconeogenesis make approximately equal contributions to blood glucose; as glycogen reserves are depleted, so gluconeogenesis becomes progressively more important.

Failure of gluconeogenesis is usually fatal. **Hypoglycemia** causes brain dysfunction, which can lead to coma and death. Glucose is also important in maintaining the level of intermediates of the citric acid cycle even when fatty acids are the main source of acetyl-CoA in the tissues. In addition, gluconeogenesis clears lactate produced by muscle and erythrocytes, and glycerol produced by adipose tissue. In ruminants, propionate is a product of rumen metabolism of carbohydrates, and is a major substrate for gluconeogenesis.

Excessive gluconeogenesis occurs in **critically ill patients** in response to injury and infection, contributing to **hyperglycemia** which is associated with a poor outcome. Hyperglycemia leads to changes in osmolality of body fluids, impaired blood flow, intracellular acidosis and increased superoxide radical production (Chapter 45), resulting in deranged endothelial

and immune system function and impaired blood coagulation. Excessive gluconeogenesis is also a contributory factor to hyperglycemia in **type 2 diabetes** because of impaired sensitivity of gluconeogenesis to downregulation in response to insulin.

GLUCONEOGENESIS INVOLVES GLYCOLYSIS, THE CITRIC ACID CYCLE, PLUS SOME SPECIAL REACTIONS

Thermodynamic Barriers Prevent a Simple Reversal of Glycolysis

Three nonequilibrium reactions in glycolysis (Chapter 18), catalyzed by hexokinase, phosphofructokinase and pyruvate kinase, prevent simple reversal of glycolysis for glucose synthesis (**Figure 20–1**). They are circumvented as follows.

Pyruvate & Phosphoenolpyruvate

Reversal of the reaction catalyzed by pyruvate kinase in glycolysis involves two endothermic reactions. Mitochondrial **pyruvate carboxylase** catalyzes the carboxylation of pyruvate to oxaloacetate, an ATP-requiring reaction in which the vitamin biotin is the coenzyme. Biotin binds CO_2 from bicarbonate as carboxybiotin prior to the addition of the CO_2 to pyruvate (Figure 44–17). The resultant oxaloacetate is reduced to malate, exported from the mitochondrion into the cytosol and there oxidized back to oxaloacetate. A second enzyme,

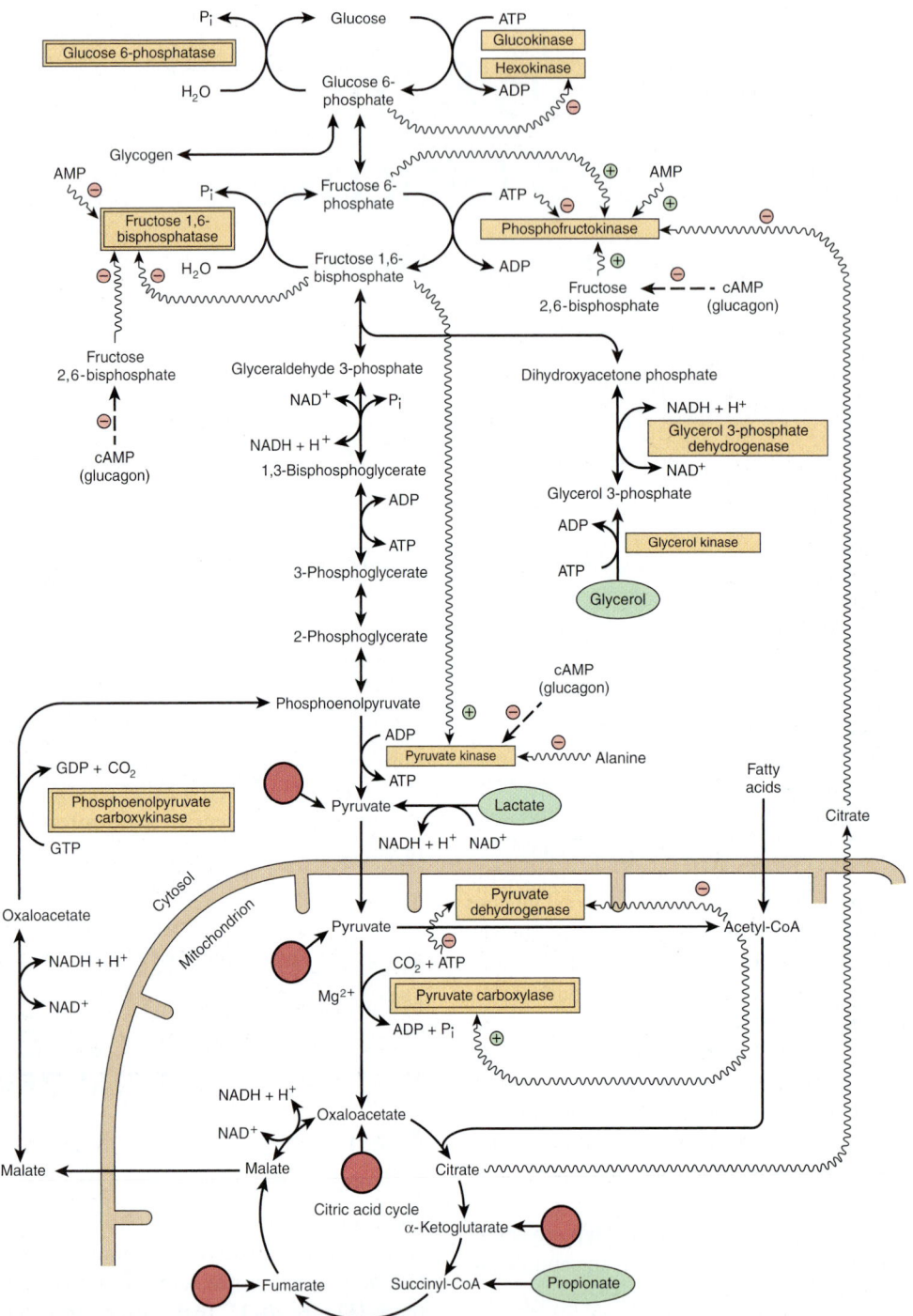

FIGURE 20–1   **Major pathways and regulation of gluconeogenesis and glycolysis in the liver.**
Entry points of glucogenic amino acids after transamination are indicated by arrows extended from
circles (see also Figure 17–4). The key gluconeogenic enzymes are enclosed in double-bordered boxes.
The ATP required for gluconeogenesis is supplied by the oxidation of fatty acids. Propionate is of
quantitative importance only in ruminants. Arrows with wavy shafts signify allosteric effects; dash-shafted
arrows, covalent modification by reversible phosphorylation. High concentrations of alanine act as a
"gluconeogenic signal" by inhibiting glycolysis at the pyruvate kinase step.

phosphoenolpyruvate carboxykinase, catalyzes the decarboxylation and phosphorylation of oxaloacetate to phosphoenolpyruvate using GTP as the phosphate donor. In liver and
kidney, the reaction of succinate thiokinase in the citric acid

cycle (Chapter 17) produces GTP (rather than ATP as in other
tissues), and this GTP is used for the reaction of phosphoenolpyruvate carboxykinase, thus providing a link between
citric acid cycle activity and gluconeogenesis, to prevent

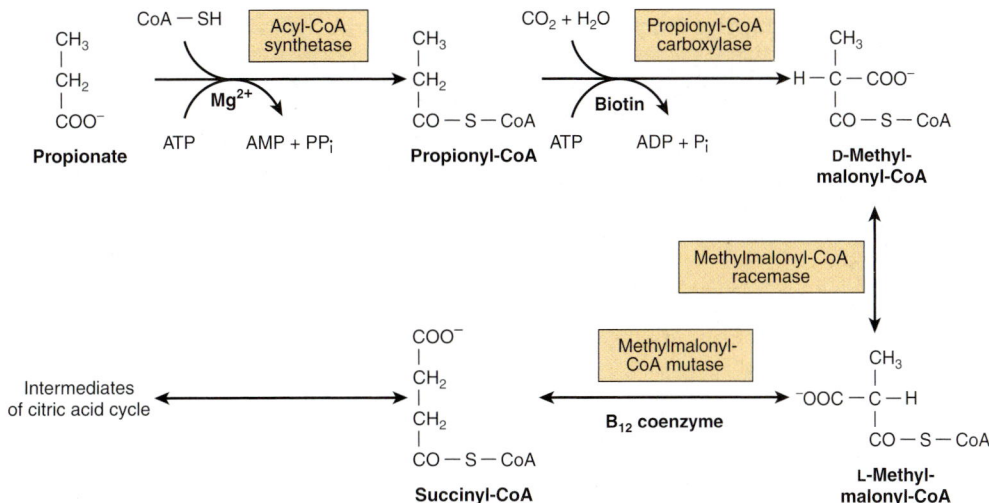

FIGURE 20–2 Metabolism of propionate.

excessive removal of oxaloacetate for gluconeogenesis, which would impair citric acid cycle activity.

Fructose 1,6-Bisphosphate & Fructose 6-Phosphate

The conversion of fructose 1,6-bisphosphate to fructose 6-phosphate, for the reversal of glycolysis, is catalyzed by **fructose 1,6-bisphosphatase.** Its presence determines whether a tissue is capable of synthesizing glucose (or glycogen) not only from pyruvate, but also from triose phosphates. It is present in liver, kidney, and skeletal muscle, but is probably absent from heart and smooth muscle.

Glucose 6-Phosphate & Glucose

The conversion of glucose 6-phosphate to glucose is catalyzed by **glucose 6-phosphatase.** It is present in liver and kidney, but absent from muscle and adipose tissue, which, therefore, cannot export glucose into the bloodstream.

Glucose 1-Phosphate & Glycogen

The breakdown of glycogen to glucose 1-phosphate is catalyzed by phosphorylase. Glycogen synthesis involves a different pathway via uridine diphosphate glucose and **glycogen synthase** (Figure 19–1).

The relationships between gluconeogenesis and the glycolytic pathway are shown in Figure 20–1. After transamination or deamination, glucogenic amino acids yield either pyruvate or intermediates of the citric acid cycle. Therefore, the reactions described above can account for the conversion of both lactate and glucogenic amino acids to glucose or glycogen.

Propionate is a major precursor of glucose in ruminants; it enters gluconeogenesis via the citric acid cycle. After esterification with CoA, propionyl-CoA is carboxylated

to D-methylmalonyl-CoA, catalyzed by **propionyl-CoA carboxylase,** a biotin-dependent enzyme (**Figure 20–2**). **Methylmalonyl-CoA racemase** catalyzes the conversion of D-methylmalonyl-CoA to L-methylmalonyl-CoA, which then undergoes isomerization to succinyl-CoA catalyzed by **methylmalonyl-CoA mutase.** In nonruminants, including human beings, propionate arises from the β-oxidation of odd-chain fatty acids that occur in ruminant lipids (Chapter 22), as well as the oxidation of isoleucine and the side chain of cholesterol, and is a (relatively minor) substrate for gluconeogenesis. Methylmalonyl-CoA mutase is a vitamin B_{12}-dependent enzyme, and in deficiency methylmalonic acid is excreted in the urine (**methylmalonicaciduria**).

Glycerol is released from adipose tissue as a result of lipolysis of lipoprotein triacylglycerol in the fed state; it may be used for reesterification of free fatty acids to triacylglycerol in adipose tissue or liver, or may be a substrate for gluconeogenesis in the liver. In the fasting state, glycerol released from lipolysis of adipose tissue triacylglycerol is used solely as a substrate for gluconeogenesis in the liver and kidneys.

SINCE GLYCOLYSIS & GLUCONEOGENESIS SHARE THE SAME PATHWAY BUT IN OPPOSITE DIRECTIONS, THEY MUST BE REGULATED RECIPROCALLY

Changes in the availability of substrates are responsible for most changes in metabolism either directly or indirectly acting via changes in hormone secretion. Three mechanisms are responsible for regulating the activity of enzymes concerned in carbohydrate metabolism: (1) changes in the rate of enzyme

TABLE 20–1 Regulatory and Adaptive Enzymes Associated with Carbohydrate Metabolism

| | Activity in | | | | | |
	Carbohydrate Feeding	Fasting and Diabetes	Inducer	Repressor	Activator	Inhibitor
Glycogenolysis, glycolysis, and pyruvate oxidation						
Glycogen synthase	↑	↓			Insulin, glucose 6-phosphate	Glucagon
Hexokinase						Glucose 6-phosphate
Glucokinase	↑	↓	Insulin	Glucagon		
Phosphofructokinase-1	↑	↓	Insulin	Glucagon	5′ AMP, fructose 6-phosphate, fructose 2,6-bisphosphate, P_i	Citrate, ATP, glucagon
Pyruvate kinase	↑	↓	Insulin, fructose	Glucagon	Fructose 1,6-bisphosphate, insulin	ATP, alanine, glucagon, norepinephrine
Pyruvate dehydrogenase	↑	↓			CoA, NAD^+, insulin, ADP, pyruvate	Acetyl CoA, NADH, ATP (fatty acids, ketone bodies)
Gluconeogenesis						
Pyruvate carboxylase	↓	↑	Glucocorticoids, glucagon, epinephrine	Insulin	Acetyl CoA	ADP
Phosphoenolpyruvate carboxykinase	↓	↑	Glucocorticoids, glucagon, epinephrine	Insulin	Glucagon?	
Glucose 6-phosphatase	↓	↑	Glucocorticoids, glucagon, epinephrine	Insulin		

synthesis, (2) covalent modification by reversible phosphorylation, and (3) allosteric effects.

Induction & Repression of Key Enzymes Requires Several Hours

The changes in enzyme activity in the liver that occur under various metabolic conditions are listed in **Table 20–1**. The enzymes involved catalyze nonequilibrium (physiologically irreversible) reactions. The effects are generally reinforced because the activity of the enzymes catalyzing the reactions in the opposite direction varies reciprocally (see Figure 20–1). The enzymes involved in the utilization of glucose (ie, those of glycolysis and lipogenesis) become more active when there is a superfluity of glucose, and under these conditions the enzymes of gluconeogenesis have low activity. Insulin, secreted in response to increased blood glucose, enhances the synthesis of the key enzymes in glycolysis. It also antagonizes the effect of the glucocorticoids and glucagon-stimulated cAMP, which induce synthesis of the key enzymes of gluconeogenesis.

Covalent Modification by Reversible Phosphorylation Is Rapid

Glucagon and **epinephrine,** hormones that are responsive to a decrease in blood glucose, inhibit glycolysis and stimulate gluconeogenesis in the liver by increasing the concentration of cAMP. This in turn activates cAMP-dependent protein kinase, leading to the phosphorylation and inactivation of **pyruvate kinase.** They also affect the concentration of fructose 2,6-bisphosphate and therefore glycolysis and gluconeogenesis, as described below.

Allosteric Modification Is Instantaneous

In gluconeogenesis, pyruvate carboxylase, which catalyzes the synthesis of oxaloacetate from pyruvate, requires acetyl-CoA as an **allosteric activator.** The addition of acetyl-CoA results in a change in the tertiary structure of the protein, lowering the K_m for bicarbonate. This means that as acetyl-CoA is formed from pyruvate, it automatically ensures the provision of oxaloacetate and, therefore, its further oxidation in the citric acid cycle, by activating pyruvate carboxylase. The activation of pyruvate carboxylase and the reciprocal inhibition of pyruvate dehydrogenase by acetyl-CoA derived from the oxidation of fatty acids explain the action of fatty acid oxidation in sparing the oxidation of pyruvate and in stimulating gluconeogenesis. The reciprocal relationship between these two enzymes alters the metabolic fate of pyruvate as the tissue changes from carbohydrate oxidation (glycolysis) to gluconeogenesis during the transition from the fed to fasting state (see Figure 20–1). A major role of fatty acid oxidation in promoting gluconeogenesis is to supply the ATP required.

Phosphofructokinase (phosphofructokinase-1) occupies a key position in regulating glycolysis and is also subject to feedback control. It is inhibited by citrate and by normal intracellular concentrations of ATP and is activated by 5′ AMP. 5′ AMP acts as an indicator of the energy status of the cell. The presence of **adenylyl kinase** in liver and many other tissues allows rapid equilibration of the reaction

$$2ADP \leftrightarrow ATP + 5'\,AMP$$

Thus, when ATP is used in energy-requiring processes resulting in formation of ADP, [AMP] increases. A relatively small decrease in [ATP] causes a several-fold increase in [AMP], so that [AMP] acts as a metabolic amplifier of a small change in [ATP], and hence a sensitive signal of the energy state of the cell. The activity of phosphofructokinase-1 is thus regulated in response to the energy status of the cell to control the quantity of carbohydrate undergoing glycolysis prior to its entry into the citric acid cycle. Simultaneously, AMP activates phosphorylase, increasing glycogenolysis. A consequence of the inhibition of phosphofructokinase-1 is an accumulation of glucose 6-phosphate, which in turn inhibits further uptake of glucose in extrahepatic tissues by inhibition of hexokinase.

Fructose 2,6-Bisphosphate Plays a Unique Role in the Regulation of Glycolysis & Gluconeogenesis in Liver

The most potent positive allosteric activator of phosphofructokinase-1 and inhibitor of fructose 1,6-bisphosphatase in liver is **fructose 2,6-bisphosphate.** It relieves inhibition of phosphofructokinase-1 by ATP and increases the affinity for fructose 6-phosphate. It inhibits fructose 1,6-bisphosphatase by increasing the K_m for fructose 1,6-bisphosphate. Its concentration is under both substrate (allosteric) and hormonal control (covalent modification) (**Figure 20–3**).

Fructose 2,6-bisphosphate is formed by phosphorylation of fructose 6-phosphate by **phosphofructokinase-2.** The same enzyme protein is also responsible for its breakdown, since it has **fructose 2,6-bisphosphatase** activity. This **bifunctional enzyme** is under the allosteric control of fructose 6-phosphate, which stimulates the kinase and inhibits the phosphatase. Hence, when there is an abundant supply of glucose, the concentration of fructose 2,6-bisphosphate increases, stimulating glycolysis by activating phosphofructokinase-1 and inhibiting fructose 1,6-bisphosphatase. In the fasting state, glucagon stimulates the production of cAMP, activating cAMP-dependent protein kinase,

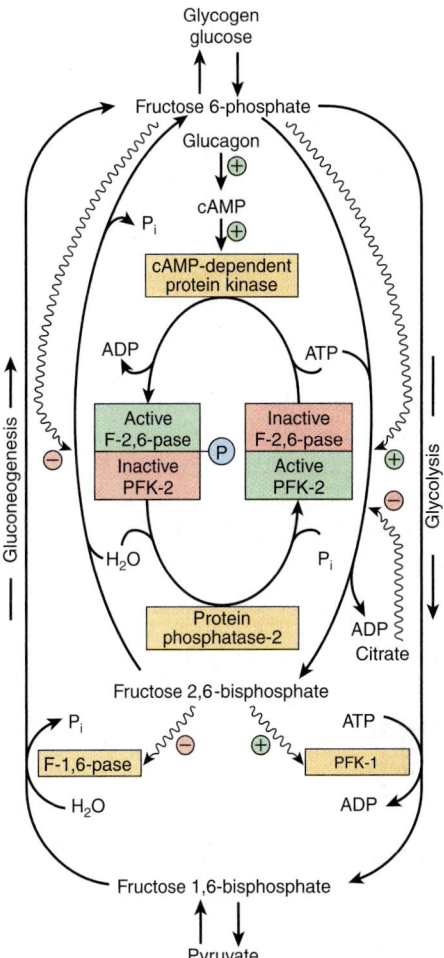

FIGURE 20–3 Control of glycolysis and gluconeogenesis in the liver by fructose 2,6-bisphosphate and the bifunctional enzyme PFK-2/F-2,6-Pase (6-phosphofructo-2-kinase/fructose 2,6-bisphosphatase). (F-1,6-Pase, fructose 1,6-bisphosphatase; PFK-1, phosphofructokinase-1 [6-phosphofructo-1-kinase].) Arrows with wavy shafts indicate allosteric effects.

which in turn inactivates phosphofructokinase-2 and activates fructose 2,6-bisphosphatase by phosphorylation. Hence, gluconeogenesis is stimulated by a decrease in the concentration of fructose 2,6-bisphosphate, which inactivates phosphofructokinase-1 and relieves the inhibition of fructose 1,6-bisphosphatase. Xylulose 5-phosphate, an intermediate of the pentose phosphate pathway (Chapter 21) activates the protein phosphatase that dephosphorylates the bifunctional enzyme, so increasing the formation of fructose 2,6-bisphosphate and increasing the rate of glycolysis. This leads to increased flux through glycolysis and the pentose phosphate pathway and increased fatty acid synthesis (Chapter 23).

Substrate (Futile) Cycles Allow Fine Tuning & Rapid Response

The control points in glycolysis and glycogen metabolism involve a cycle of phosphorylation and dephosphorylation catalyzed by glucokinase and glucose 6-phosphatase; phosphofructokinase-1 and fructose 1,6-bisphosphatase; pyruvate kinase, pyruvate carboxylase, and phosphoenolpyruvate carboxykinase; and glycogen synthase and phosphorylase. It would seem obvious that these opposing enzymes are regulated in such a way that when those involved in glycolysis are active, those involved in gluconeogenesis are inactive, since otherwise there would be cycling between phosphorylated and nonphosphorylated intermediates, with net hydrolysis of ATP. While this is so, in muscle both phosphofructokinase and fructose 1,6-bisphosphatase have some activity at all times, so that there is indeed some measure of (wasteful) substrate cycling. This permits the very rapid increase in the rate of glycolysis necessary for muscle contraction. At rest the rate of phosphofructokinase activity is some 10-fold higher than that of fructose 1,6-bisphosphatase; in anticipation of muscle contraction, the activity of both enzymes increases, fructose 1,6-bisphosphatase 10 times more than phosphofructokinase, maintaining the same net rate of glycolysis. At the start of muscle contraction, the activity of phosphofructokinase increases further, and that of fructose 1,6-bisphosphatase falls, so increasing the net rate of glycolysis (and hence ATP formation) as much as a 1000-fold.

THE BLOOD CONCENTRATION OF GLUCOSE IS REGULATED WITHIN NARROW LIMITS

In the postabsorptive state, the concentration of blood glucose in most mammals is maintained between 4.5–5.5 mmol/L. After the ingestion of a carbohydrate meal, it may rise to 6.5–7.2 mmol/L, and in starvation, it may fall to 3.3–3.9 mmol/L. A sudden decrease in blood glucose (eg, in response to insulin

overdose) causes convulsions, because of the dependence of the brain on a supply of glucose. However, much lower concentrations can be tolerated if hypoglycemia develops slowly enough for adaptation to occur. The blood glucose level in birds is considerably higher (14.0 mmol/L) and in ruminants considerably lower (approximately 2.2 mmol/L in sheep and 3.3 mmol/L in cattle). These lower normal levels appear to be associated with the fact that ruminants ferment virtually all dietary carbohydrate to short-chain fatty acids, and these largely replace glucose as the main metabolic fuel of the tissues in the fed state.

BLOOD GLUCOSE IS DERIVED FROM THE DIET, GLUCONEOGENESIS, & GLYCOGENOLYSIS

The digestible dietary carbohydrates yield glucose, galactose, and fructose that are transported to the liver via the **hepatic portal vein.** Galactose and fructose are readily converted to glucose in the liver (Chapter 21).

Glucose is formed from two groups of compounds that undergo gluconeogenesis (see Figures 17–4 and 20–1): (1) those which involve a direct net conversion to glucose, including most **amino acids** and **propionate;** and (2) those which are the products of the metabolism of glucose in tissues. Thus **lactate,** formed by glycolysis in skeletal muscle and erythrocytes, is transported to the liver and kidney where it reforms glucose, which again becomes available via the circulation for oxidation in the tissues. This process is known as the **Cori cycle,** or the **lactic acid cycle (Figure 20–4).**

In the fasting state, there is a considerable output of alanine from skeletal muscle, far in excess of its concentration in

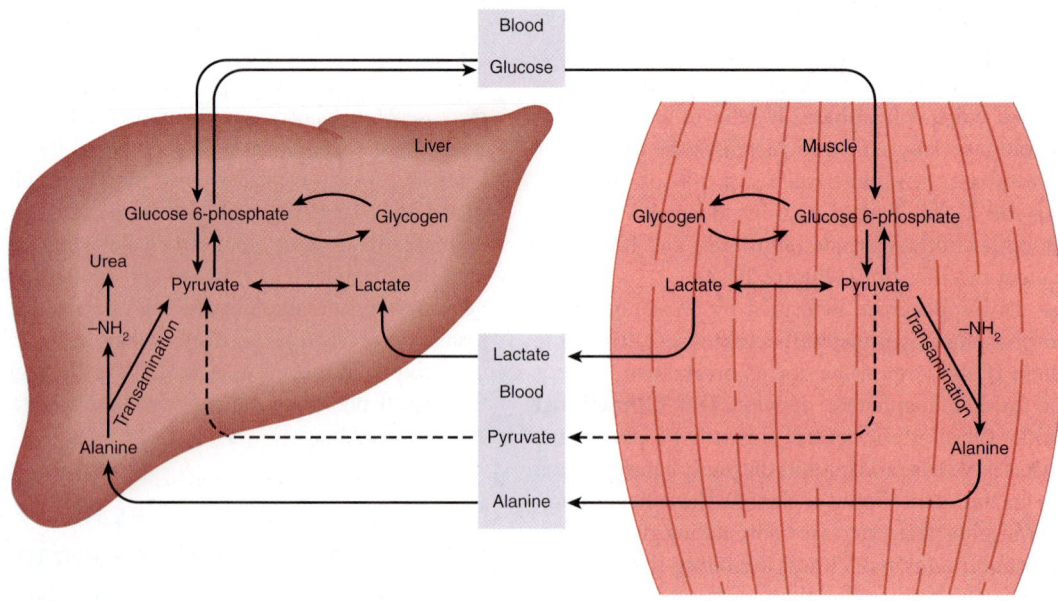

FIGURE 20–4 **The lactic acid (Cori cycle) and glucose-alanine cycles.**

the muscle proteins that are being catabolized. It is formed by transamination of pyruvate produced by glycolysis of muscle glycogen, and is exported to the liver, where, after transamination back to pyruvate, it is a substrate for gluconeogenesis. This **glucose–alanine cycle** (see Figure 20–4) thus provides an indirect way of utilizing muscle glycogen to maintain blood glucose in the fasting state. The ATP required for the hepatic synthesis of glucose from pyruvate is derived from the oxidation of fatty acids.

Glucose is also formed from liver glycogen by glycogenolysis (Chapter 19).

Metabolic & Hormonal Mechanisms Regulate the Concentration of Blood Glucose

The maintenance of stable levels of glucose in the blood is one of the most finely regulated of all homeostatic mechanisms, involving the liver, extrahepatic tissues, and several hormones. Liver cells are freely permeable to glucose (via the GLUT 2 transporter), whereas cells of extrahepatic tissues (apart from pancreatic β-islets) are relatively impermeable, and their glucose transporters are regulated by insulin. As a result, uptake from the bloodstream is the rate-limiting step in the utilization of glucose in extrahepatic tissues. The role of various glucose transporter proteins found in cell membranes is shown in **Table 20–2**.

Glucokinase Is Important in Regulating Blood Glucose After a Meal

Hexokinase has a low K_m for glucose, and in the liver is saturated and acting at a constant rate under all normal conditions. Glucokinase has a considerably higher K_m (lower affinity) for glucose, so that its activity increases with increases in the concentration of glucose in the hepatic portal vein (**Figure 20–5**). It promotes hepatic uptake of large amounts of glucose after a carbohydrate meal. It is absent from the liver of ruminants, which have little glucose entering the portal circulation from the intestines.

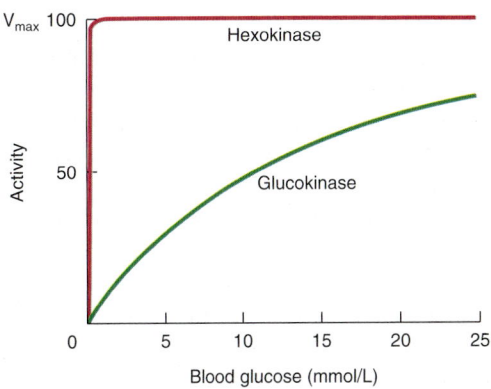

FIGURE 20–5 **Variation in glucose phosphorylating activity of hexokinase and glucokinase with increasing blood glucose concentration.** The K_m for glucose of hexokinase is 0.05 mmol/L and of glucokinase is 10 mmol/L.

At normal systemic-blood glucose concentrations (4.5–5.5 mmol/L), the liver is a net producer of glucose. However, as the glucose level rises, the output of glucose ceases, and there is a net uptake.

Insulin Plays a Central Role in Regulating Blood Glucose

In addition to the direct effects of hyperglycemia in enhancing the uptake of glucose into the liver, the hormone insulin plays a central role in regulating blood glucose. It is produced by the β cells of the islets of Langerhans in the pancreas in response to hyperglycemia. The β-islet cells are freely permeable to glucose via the GLUT 2 transporter, and the glucose is phosphorylated by glucokinase. Therefore, increasing blood glucose increases metabolic flux through glycolysis, the citric acid cycle, and the generation of ATP. The increase in [ATP] inhibits ATP-sensitive K^+ channels, causing depolarization of the cell membrane, which increases Ca^{2+} influx via voltage-sensitive Ca^{2+} channels, stimulating exocytosis of insulin. Thus, the concentration of insulin in the blood parallels that of

TABLE 20–2 Major Glucose Transporters

	Tissue Location	Functions
Facilitative bidirectional transporters		
GLUT 1	Brain, kidney, colon, placenta, erythrocytes	Glucose uptake
GLUT 2	Liver, pancreatic β cell, small intestine, kidney	Rapid uptake or release of glucose
GLUT 3	Brain, kidney, placenta	Glucose uptake
GLUT 4	Heart and skeletal muscle, adipose tissue	Insulin-stimulated glucose uptake
GLUT 5	Small intestine	Absorption of glucose
Sodium-dependent unidirectional transporter		
SGLT 1	Small intestine and kidney	Active uptake of glucose against a concentration gradient

the blood glucose. Other substances causing release of insulin from the pancreas include amino acids, free fatty acids, ketone bodies, glucagon, secretin, and the sulfonylurea drugs tolbutamide and glyburide. These drugs are used to stimulate insulin secretion in type 2 diabetes mellitus (NIDDM, noninsulin-dependent diabetes mellitus) via the ATP-sensitive K^+ channels. Epinephrine and norepinephrine block the release of insulin. Insulin lowers blood glucose immediately by enhancing glucose transport into adipose tissue and muscle by recruitment of glucose transporters (GLUT 4) from the interior of the cell to the plasma membrane. Although it does not affect glucose uptake into the liver directly, insulin does enhance long-term uptake as a result of its actions on the enzymes controlling glycolysis, glycogenesis, and gluconeogenesis (Chapter 19 and Table 20–1).

Glucagon Opposes the Actions of Insulin

Glucagon is the hormone produced by the α cells of the pancreatic islets. Its secretion is stimulated by hypoglycemia. In the liver, it stimulates glycogenolysis by activating phosphorylase. Unlike epinephrine, glucagon does not have an effect on muscle phosphorylase. Glucagon also enhances gluconeogenesis from amino acids and lactate. In all these actions, glucagon acts via generation of cAMP (see Table 20–1). Both hepatic glycogenolysis and gluconeogenesis contribute to the **hyperglycemic effect** of glucagon, whose actions oppose those of insulin. Most of the endogenous glucagon (and insulin) is cleared from the circulation by the liver (**Table 20–3**).

Other Hormones Affect Blood Glucose

The **anterior pituitary gland** secretes hormones that tend to elevate blood glucose and therefore antagonize the action of insulin. These are growth hormone, ACTH (corticotropin), and possibly other "diabetogenic" hormones. Growth hormone secretion is stimulated by hypoglycemia; it decreases glucose uptake in muscle. Some of this effect may be indirect, since it stimulates mobilization of free fatty acids from adipose tissue, which themselves inhibit glucose utilization. The **glucocorticoids** (11-oxysteroids) are secreted by the adrenal cortex, and are also synthesized in an unregulated manner in adipose tissue. They act to increase gluconeogenesis as a result of enhanced hepatic catabolism of amino acids, due to induction of aminotransferases (and other enzymes such as tryptophan dioxygenase) and key enzymes of gluconeogenesis. In addition, glucocorticoids inhibit the utilization of glucose in extrahepatic tissues. In all these actions, glucocorticoids act in a manner antagonistic to insulin. A number of **cytokines** secreted by macrophages infiltrating adipose tissue also have insulin antagonistic actions; together with glucocorticoids secreted by adipose tissue, this explains the insulin resistance that commonly occurs in obese people.

Epinephrine is secreted by the adrenal medulla as a result of stressful stimuli (fear, excitement, hemorrhage, hypoxia, hypoglycemia, etc) and leads to glycogenolysis in liver and muscle owing to stimulation of phosphorylase via generation of cAMP. In muscle, glycogenolysis results in increased glycolysis, whereas in liver it results in the release of glucose into the bloodstream.

FURTHER CLINICAL ASPECTS

Glucosuria Occurs When the Renal Threshold for Glucose Is Exceeded

When the blood glucose rises to relatively high levels, the kidney also exerts a regulatory effect. Glucose is continuously filtered by the glomeruli, but is normally completely reabsorbed in the renal tubules by active transport. The capacity of the tubular system to reabsorb glucose is limited to a rate of about 2 mmol/min, and in hyperglycemia (as occurs in poorly controlled diabetes mellitus), the glomerular filtrate may contain more glucose than can be reabsorbed, resulting in **glucosuria.** Glucosuria occurs when the venous blood glucose concentration exceeds about 10 mmol/L; this is termed the **renal threshold** for glucose.

Hypoglycemia May Occur During Pregnancy & in the Neonate

During pregnancy, fetal glucose consumption increases and there is a risk of maternal and possibly fetal hypoglycemia, particularly if there are long intervals between meals or at night. Furthermore, premature and low-birth-weight babies are more susceptible to hypoglycemia, since they have little adipose tissue to provide free fatty acids. The enzymes of gluconeogenesis may not be completely functional at this time, and gluconeogenesis is anyway dependent on a supply of free fatty acids for energy. Little glycerol, which would normally be released from adipose tissue, is available for gluconeogenesis.

TABLE 20–3 **Tissue Responses to Insulin and Glucagon**

	Liver	Adipose Tissue	Muscle
Increased by insulin	Fatty acid synthesis Glycogen synthesis Protein synthesis	Glucose uptake Fatty acid synthesis	Glucose uptake Glycogen synthesis Protein synthesis
Decreased by insulin	Ketogenesis Gluconeogenesis	Lipolysis	
Increased by glucagon	Glycogenolysis Gluconeogenesis Ketogenesis	Lipolysis	

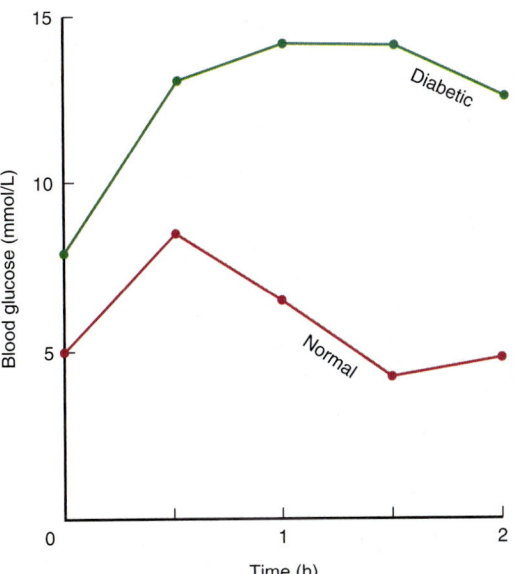

FIGURE 20–6 **Glucose tolerance test.** Blood glucose curves of a normal and a diabetic person after oral administration of 1 g of glucose/kg body weight. Note the initial raised concentration in the fasting diabetic. A criterion of normality is the return of the curve to the initial value within 2 h.

The Body's Ability to Utilize Glucose May Be Ascertained by Measuring Glucose Tolerance

Glucose tolerance is the ability to regulate the blood glucose concentration after the administration of a test dose of glucose (normally 1 g/kg body weight) (**Figure 20–6**). **Diabetes mellitus** (type 1, or insulin-dependent diabetes mellitus; IDDM) is characterized by decreased glucose tolerance as a result of decreased secretion of insulin as a result of progressive destruction of pancreatic β-islet cells. Glucose tolerance is also impaired in type 2 diabetes mellitus (NIDDM) as a result of impaired sensitivity of tissues to insulin action. Insulin resistance associated with obesity (and especially abdominal obesity) leading to the development of hyperlipidemia, then atherosclerosis and coronary heart disease, as well as overt diabetes, is known as the **metabolic syndrome.** Impaired glucose tolerance also occurs in conditions where the liver is damaged, in some infections, and in response to some drugs, as well as in conditions that lead to hyperactivity of the pituitary or adrenal cortex because of the antagonism of the hormones secreted by these glands to the action of insulin.

Administration of insulin (as in the treatment of diabetes mellitus) lowers the blood glucose concentration and increases its utilization and storage in the liver and muscle as glycogen. An excess of insulin may cause **hypoglycemia,** resulting in convulsions and even in death unless glucose is administered promptly. Increased tolerance to glucose is observed in pituitary or adrenocortical insufficiency, attributable to a decrease in the antagonism to insulin by the hormones normally secreted by these glands.

The Energy Cost of Gluconeogenesis Explains Why Very Low Carbohydrate Diets Promote Weight Loss

Very low carbohydrate diets, providing only 20 g per day of carbohydrate or less (compared with a desirable intake of 100–120 g/day), but permitting unlimited consumption of fat and protein, have been promoted as an effective regime for weight loss, although such diets are counter to all advice on a prudent diet for health. Since there is a continual demand for glucose, there will be a considerable amount of gluconeogenesis from amino acids; the associated high ATP cost must then be met by oxidation of fatty acids.

SUMMARY

- Gluconeogenesis is the process of synthesizing glucose or glycogen from noncarbohydrate precursors. It is of particular importance when carbohydrate is not available from the diet. Significant substrates are amino acids, lactate, glycerol, and propionate.

- The pathway of gluconeogenesis in the liver and kidney utilizes those reactions in glycolysis that are reversible plus four additional reactions that circumvent the irreversible nonequilibrium reactions.

- Since glycolysis and gluconeogenesis share the same pathway but operate in opposite directions, their activities must be regulated reciprocally.

- The liver regulates the blood glucose concentration after a meal because it contains the high-K_m glucokinase that promotes increased hepatic utilization of glucose.

- Insulin is secreted as a direct response to hyperglycemia; it stimulates the liver to store glucose as glycogen and facilitates uptake of glucose into extrahepatic tissues.

- Glucagon is secreted as a response to hypoglycemia and activates both glycogenolysis and gluconeogenesis in the liver, causing release of glucose into the blood.

REFERENCES

Barthel A, Schmoll D: Novel concepts in insulin regulation of hepatic gluconeogenesis. Am J Physiol Endocrinol Metab 2003;285:E685.

Boden G: Gluconeogenesis and glycogenolysis in health and diabetes. J Investig Med 2004;52:375.

Dzugaj, A: Localization and regulation of muscle fructose 1,6-bisphosphatase, the key enzyme of glyconeogenesis. Adv Enzyme Regul 2006;46:51.

Jiang G, Zhang BB: Glucagon and regulation of glucose metabolism. Am J Physiol Endocrinol Metab 2003;284:E671.

Jitrapakdee S, Vidal-Puig A, Wallace JC: Anaplerotic roles of pyruvate carboxylase in mammalian tissues. Cell Mol Life Sci 2006;63:843.

Jitrapakdee S, St Maurice M, Rayment, et al: Structure, mechanism and regulation of pyruvate carboxylase. Biochem J 2008;413:369.

Klover PJ, Mooney RA: Hepatocytes: critical for glucose homeostasis. Int J Biochem Cell Biol 2004;36:753.

McGuinness OP: Defective glucose homeostasis during infection. Ann Rev Nutr 2005;25:9.

Mithiuex G, Andreelli F, Magnan C: Intestinal gluconeogenesis: key signal of central control of energy and glucose homeostasis. Curr Opin Clin Nutr Metab Care 2009;12:419.

Mlinar B, Marc J, Janez A, et al: Molecular mechanisms of insulin resistance and associated diseases. Clin Chim Acta 2007;375:20.

Nordlie RC, Foster JD, Lange AJ: Regulation of glucose production by the liver. Ann Rev Nutr 1999;19:379.

Pilkis SJ, Claus TH: Hepatic gluconeogenesis/glycolysis: regulation and structure/function relationships of substrate cycle enzymes. Ann Rev Nutr 1991;11:465.

Pilkis SJ, Granner DK: Molecular physiology of the regulation of hepatic gluconeogenesis and glycolysis. Ann Rev Physiol 1992;54:885.

Postic C, Shiota M, Magnuson MA: Cell-specific roles of glucokinase in glucose homeostasis. Rec Prog Horm Res 2001;56:195.

Previs SF, Brunengraber DZ, Brunengraber H: Is there glucose production outside of the liver and kidney? Ann Rev Nutr 2009;29:43.

Quinn PG, Yeagley D: Insulin regulation of PEPCK gene expression: a model for rapid and reversible modulation. Curr Drug Targets Immune Endocr Metabol Disord 2005;5:423.

Reaven GM: The insulin resistance syndrome: definition and dietary approaches to treatment. Ann Rev Nutr 2005;25:391.

Roden M, Bernroider E: Hepatic glucose metabolism in humans—its role in health and disease. Best Pract Res Clin Endocrinol Metab 2003;17:365.

Saggerson D: Malonyl-CoA, a key signaling molecule in mammalian cells. Ann Rev Nutr 2008;28:253.

Schuit FC, Huypens P, Heimberg H, Pipeleers DG: Glucose sensing in pancreatic beta-cells: a model for the study of other glucose-regulated cells in gut, pancreas, and hypothal amus. Diabetes 2001;50:1.

Suh SH, Paik IY, Jacobs K: Regulation of blood glucose homeostasis during prolonged exercise. Mol Cells 2007;23:272.

Wahren J, Ekberg K: Splanchnic regulation of glucose production. Ann Rev Nutr 2007;27:329.

Young, A: Inhibition of glucagon secretion. Adv Pharmacol 2005;52:151.

The Pentose Phosphate Pathway & Other Pathways of Hexose Metabolism

David A. Bender, PhD & Peter A. Mayes, PhD, DSc

OBJECTIVES

After studying this chapter, you should be able to:

- Describe the pentose phosphate pathway and its roles as a source of NADPH and in the synthesis of ribose for nucleotide synthesis.
- Describe the uronic acid pathway and its importance for synthesis of glucuronic acid for conjugation reactions and (in animals for which it is not a vitamin) vitamin C.
- Describe and explain the consequences of large intakes of fructose.
- Describe the synthesis and physiological importance of galactose.
- Explain the consequences of genetic defects of glucose 6-phosphate dehydrogenase deficiency (favism), the uronic acid pathway (essential pentosuria), and fructose and galactose metabolism.

BIOMEDICAL IMPORTANCE

The pentose phosphate pathway is an alternative route for the metabolism of glucose. It does not lead to formation of ATP but has two major functions: (1) the formation of **NADPH** for synthesis of fatty acids and steroids, and maintaining reduced glutathione for antioxidant activity, and (2) the synthesis of **ribose** for nucleotide and nucleic acid formation. Glucose, fructose, and galactose are the main hexoses absorbed from the gastrointestinal tract, derived from dietary starch, sucrose, and lactose, respectively. Fructose and galactose can be converted to glucose, mainly in the liver.

Genetic deficiency of **glucose 6-phosphate dehydrogenase,** the first enzyme of the pentose phosphate pathway, is a major cause of acute hemolysis of red blood cells, resulting in **hemolytic anemia.** Glucuronic acid is synthesized from glucose via the **uronic acid pathway,** of minor quantitative importance, but of major significance for the conjugation and excretion of metabolites and foreign chemicals (xenobiotics) as **glucuronides.** A deficiency in the pathway leads to the condition of **essential pentosuria.** The lack of one enzyme of the pathway (gulonolactone oxidase) in primates and some other animals explains why **ascorbic acid** (vitamin C) is a dietary requirement for humans but not most other mammals. Deficiencies in the enzymes of fructose and galactose metabolism lead to metabolic diseases such as **essential fructosuria, hereditary fructose intolerance, and galactosemia.**

THE PENTOSE PHOSPHATE PATHWAY FORMS NADPH & RIBOSE PHOSPHATE

The pentose phosphate pathway (hexose monophosphate shunt) is a more complex pathway than glycolysis (**Figure 21–1**). Three molecules of glucose 6-phosphate give rise to three molecules of CO_2 and three 5-carbon sugars. These are rearranged to regenerate two molecules of glucose 6-phosphate and one molecule of the glycolytic intermediate, glyceraldehyde 3-phosphate. Since two molecules of glyceraldehyde 3-phosphate can regenerate glucose 6-phosphate, the pathway can account for the complete oxidation of glucose.

REACTIONS OF THE PENTOSE PHOSPHATE PATHWAY OCCUR IN THE CYTOSOL

Like glycolysis, the enzymes of the pentose phosphate pathway are cytosolic. Unlike glycolysis, oxidation is achieved by dehydrogenation using **NADP+**, not **NAD+,** as the hydrogen acceptor. The sequence of reactions of the pathway may be divided into two phases: an **irreversible oxidative phase** and a **reversible nonoxidative phase.** In the first phase, glucose 6-phosphate undergoes dehydrogenation and

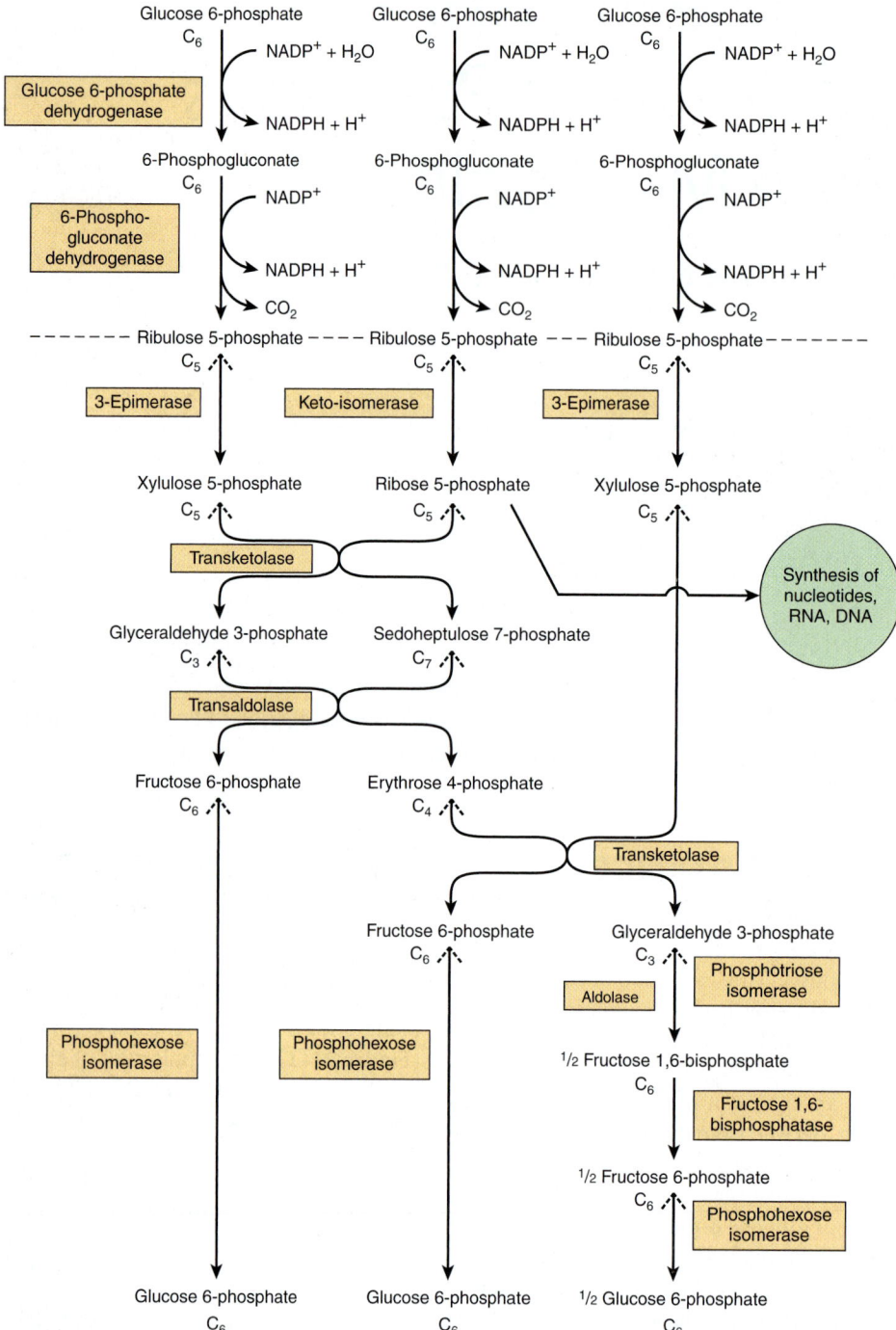

FIGURE 21–1 **Flow chart of pentose phosphate pathway and its connections with the pathway of glycolysis.** The full pathway, as indicated, consists of three interconnected cycles in which glucose 6-phosphate is both substrate and end-product. The reactions above the broken line are nonreversible, whereas all reactions under that line are freely reversible apart from that catalyzed by fructose 1,6-bisphosphatase.

decarboxylation to yield a pentose, ribulose 5-phosphate. In the second phase, ribulose 5-phosphate is converted back to glucose 6-phosphate by a series of reactions involving mainly two enzymes: **transketolase** and **transaldolase** (see Figure 21–1).

The Oxidative Phase Generates NADPH

Dehydrogenation of glucose 6-phosphate to 6-phosphogluconate occurs via the formation of 6-phosphogluconolactone catalyzed by **glucose 6-phosphate dehydrogenase,** an NADP-dependent enzyme (**Figures 21–1** and **21–2**). The hydrolysis of

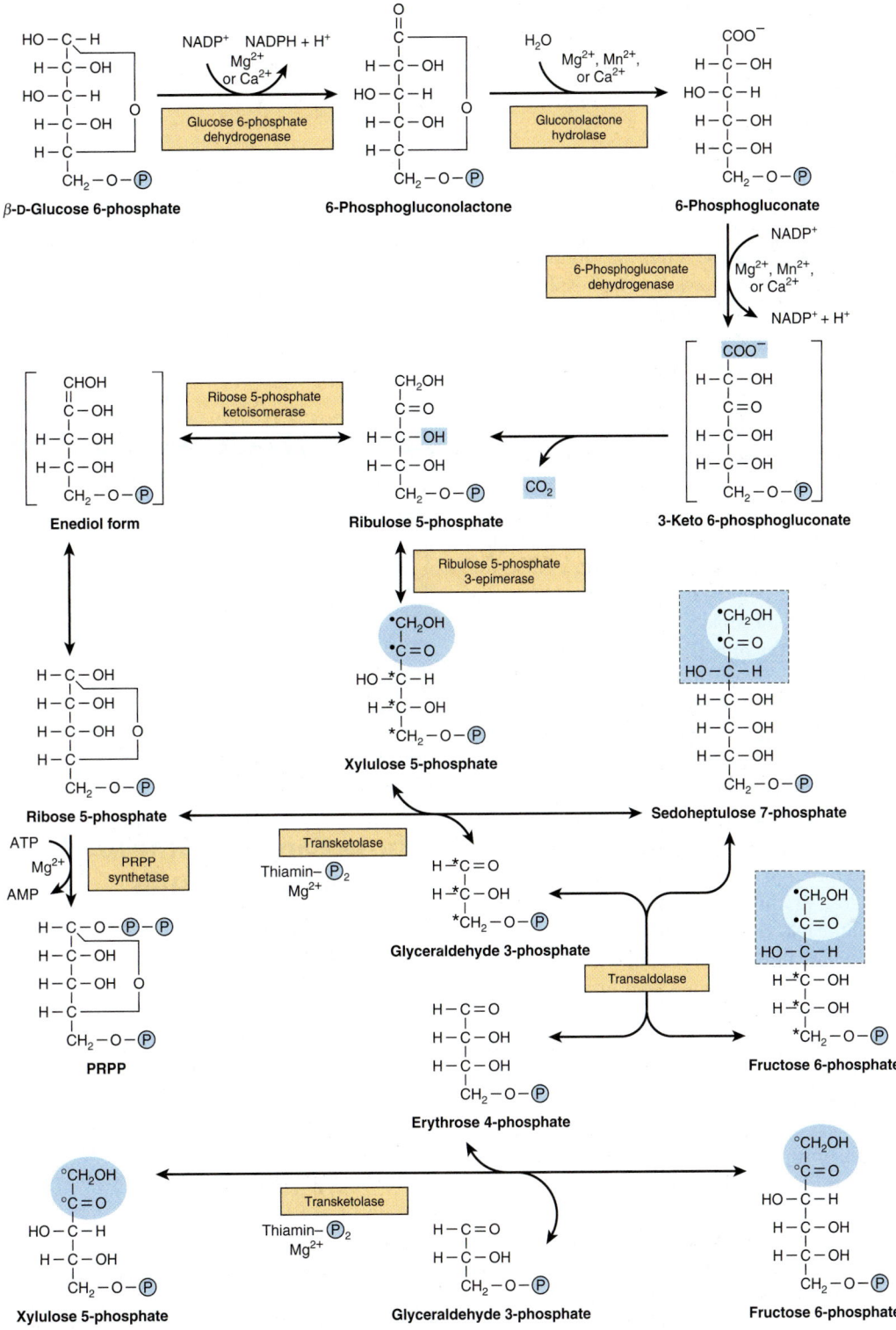

FIGURE 21–2 **The pentose phosphate pathway.** (P, $-PO_3^{2-}$; PRPP, 5-phosphoribosyl 1-pyrophosphate.)

6-phosphogluconolactone is accomplished by the enzyme **gluconolactone hydrolase.** A second oxidative step is catalyzed by **6-phosphogluconate dehydrogenase,** which also requires NADP⁺ as hydrogen acceptor. Decarboxylation follows with the formation of the ketopentose ribulose 5-phosphate.

In the endoplasmic reticulum, an isoenzyme of glucose 6-phosphate dehydrogenase, hexose 6-phosphate dehydrogenase, provides NADPH for hydroxylation (mixed function oxidase) reactions, and also for 11-β-hydroxysteroid dehydrogenase-1. This enzyme catalyzes the reduction of (inactive) cortisone to (active) cortisol in liver, the nervous system, and adipose tissue. It is the major source of intracellular cortisol in these tissues and may be important in obesity and the metabolic syndrome.

The Nonoxidative Phase Generates Ribose Precursors

Ribulose 5-phosphate is the substrate for two enzymes. **Ribulose 5-phosphate 3-epimerase** alters the configuration about carbon 3, forming the epimer xylulose 5-phosphate, also a ketopentose. **Ribose 5-phosphate ketoisomerase** converts ribulose 5-phosphate to the corresponding aldopentose, ribose 5-phosphate, which is used for nucleotide and nucleic acid synthesis. **Transketolase** transfers the two-carbon unit comprising carbons 1 and 2 of a ketose onto the aldehyde carbon of an aldose sugar. It therefore affects the conversion of a ketose sugar into an aldose with two carbons less and an aldose sugar into a ketose with two carbons more. The reaction requires Mg^{2+} and **thiamin diphosphate** (vitamin B_1) as coenzyme. Measurement of erythrocyte transketolase and its activation by thiamin diphosphate provides an index of vitamin B_1 nutritional status (Chapter 44). The two-carbon moiety transferred is probably glycolaldehyde bound to thiamin diphosphate. Thus, transketolase catalyzes the transfer of the two-carbon unit from xylulose 5-phosphate to ribose 5-phosphate, producing the seven-carbon ketose sedoheptulose 7-phosphate and the aldose glyceraldehyde 3-phosphate. These two products then undergo transaldolation. **Transaldolase** catalyzes the transfer of a three-carbon dihydroxyacetone moiety (carbons 1–3) from the ketose sedoheptulose 7-phosphate onto the aldose glyceraldehyde 3-phosphate to form the ketose fructose 6-phosphate and the four-carbon aldose erythrose 4-phosphate. Transaldolase has no cofactor, and the reaction proceeds via the intermediate formation of a Schiff base of dihydroxyacetone to the ε-amino group of a lysine residue in the enzyme. In a further reaction catalyzed by **transketolase,** xylulose 5-phosphate serves as a donor of glycolaldehyde. In this case, erythrose 4-phosphate is the acceptor, and the products of the reaction are fructose 6-phosphate and glyceraldehyde 3-phosphate.

In order to oxidize glucose completely to CO_2 via the pentose phosphate pathway, there must be enzymes present in the tissue to convert glyceraldehyde 3-phosphate to glucose 6-phosphate. This involves reversal of glycolysis and the gluconeogenic enzyme **fructose 1,6-bisphosphatase.** In tissues that lack this enzyme, glyceraldehyde 3-phosphate follows the normal pathway of glycolysis to pyruvate.

The Two Major Pathways for the Catabolism of Glucose Have Little in Common

Although glucose 6-phosphate is common to both pathways, the pentose phosphate pathway is markedly different from glycolysis. Oxidation utilizes NADP rather than NAD, and CO_2, which is not produced in glycolysis, is a characteristic product. No ATP is generated in the pentose phosphate pathway, whereas it is a major product of glycolysis.

The two pathways are, however, connected. Xylulose 5-phosphate activates the protein phosphatase that dephosphorylates the 6-phosphofructo-2-kinase/fructose 2,6-bisphosphatase bifunctional enzyme (Chapter 18). This activates the kinase and inactivates the phosphatase, leading to increased formation of fructose 2,6-bisphosphate, increased activity of phosphofructokinase-1, and hence increased glycolytic flux. Xylulose 5-phosphate also activates the protein phosphatase that initiates the nuclear translocation and DNA binding of the carbohydrate response element binding protein, leading to increased synthesis of fatty acids (Chapter 23) in response to a high carbohydrate diet.

Reducing Equivalents Are Generated in Those Tissues Specializing in Reductive Syntheses

The pentose phosphate pathway is active in liver, adipose tissue, adrenal cortex, thyroid, erythrocytes, testis, and lactating mammary gland. Its activity is low in nonlactating mammary gland and skeletal muscle. Those tissues in which the pathway is active use NADPH in reductive syntheses, for example, of fatty acids, steroids, amino acids via glutamate dehydrogenase, and reduced glutathione. The synthesis of glucose 6-phosphate dehydrogenase and 6-phosphogluconate dehydrogenase may also be induced by insulin in the fed state, when lipogenesis increases.

Ribose Can Be Synthesized in Virtually All Tissues

Little or no ribose circulates in the bloodstream, so tissues have to synthesize the ribose they require for nucleotide and nucleic acid synthesis using the pentose phosphate pathway (see Figure 21–2). It is not necessary to have a completely functioning pentose phosphate pathway for a tissue to synthesize ribose 5-phosphate. Muscle has only low activity of glucose 6-phosphate dehydrogenase and 6-phosphogluconate dehydrogenase, but, like most other tissues, it is capable of synthesizing ribose 5-phosphate by reversal of the nonoxidative phase of the pentose phosphate pathway utilizing fructose 6-phosphate.

THE PENTOSE PHOSPHATE PATHWAY & GLUTATHIONE PEROXIDASE PROTECT ERYTHROCYTES AGAINST HEMOLYSIS

In red blood cells, the pentose phosphate pathway is the sole source of NADPH for the reduction of oxidized glutathione catalyzed by **glutathione reductase,** a flavoprotein containing FAD. Reduced glutathione removes H_2O_2 in a reaction

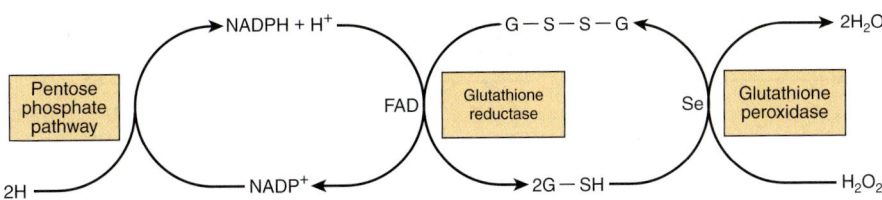

FIGURE 21–3 Role of the pentose phosphate pathway in the glutathione peroxidase reaction of erythrocytes. (G-SH, reduced glutathione; G-S-S-G, oxidized glutathione; Se, selenium-containing enzyme.)

catalyzed by **glutathione peroxidase,** an enzyme that contains the **selenium** analog of cysteine (selenocysteine) at the active site (**Figure 21–3**). The reaction is important since accumulation of H_2O_2 may decrease the life span of the erythrocyte by causing oxidative damage to the cell membrane, leading to hemolysis. In other tissues, NADPH can also be generated by the reaction catalyzed by the malic enzyme.

GLUCURONATE, A PRECURSOR OF PROTEOGLYCANS & CONJUGATED GLUCURONIDES, IS A PRODUCT OF THE URONIC ACID PATHWAY

In liver, the **uronic acid pathway** catalyzes the conversion of glucose to glucuronic acid, ascorbic acid (except in human beings and other species for which ascorbate is a vitamin), and pentoses (**Figure 21–4**). It is also an alternative oxidative pathway for glucose that, like the pentose phosphate pathway, does not lead to the formation of ATP. Glucose 6-phosphate is isomerized to glucose 1-phosphate, which then reacts with uridine triphosphate (UTP) to form uridine diphosphate glucose (UDPGlc) in a reaction catalyzed by **UDPGlc pyrophosphorylase,** as occurs in glycogen synthesis (Chapter 19). UDPGlc is oxidized at carbon 6 by NAD-dependent **UDPGlc dehydrogenase** in a two-step reaction to yield UDP-glucuronate.

UDP-glucuronate is the source of glucuronate for reactions involving its incorporation into proteoglycans or for reaction with substrates such as steroid hormones, bilirubin, and a number of drugs that are excreted in urine or bile as glucuronide conjugates (Figure 31–13).

Glucuronate is reduced to L-gulonate, the direct precursor of **ascorbate** in those animals capable of synthesizing this vitamin, in an NADPH-dependent reaction. In human beings and other primates, as well as guinea pigs, bats, and some birds and fishes, ascorbic acid cannot be synthesized because of the absence of **L-gulonolactone oxidase.** L-Gulonate is oxidized to 3-keto-L-gulonate, which is then decarboxylated to L-xylulose. L-Xylulose is converted to the D isomer by an NADPH-dependent reduction to xylitol, followed by oxidation in an NAD-dependent reaction to D-xylulose. After conversion to D-xylulose 5-phosphate, it is metabolized via the pentose phosphate pathway.

INGESTION OF LARGE QUANTITIES OF FRUCTOSE HAS PROFOUND METABOLIC CONSEQUENCES

Diets high in sucrose or in high-fructose syrups (HFS) used in manufactured foods and beverages lead to large amounts of fructose (and glucose) entering the hepatic portal vein.

Fructose undergoes more rapid glycolysis in the liver than does glucose because it bypasses the regulatory step catalyzed by phosphofructokinase (**Figure 21–5**). This allows fructose to flood the pathways in the liver, leading to increased fatty acid synthesis, esterification of fatty acids, and secretion of VLDL, which may raise serum triacylglycerols and ultimately raise LDL cholesterol concentrations. A specific kinase, **fructokinase,** in liver, kidney, and intestine, catalyzes the phosphorylation of fructose to fructose 1-phosphate. This enzyme does not act on glucose, and, unlike glucokinase, its activity is not affected by fasting or by insulin, which may explain why fructose is cleared from the blood of diabetic patients at a normal rate. Fructose 1-phosphate is cleaved to D-glyceraldehyde and dihydroxyacetone phosphate by **aldolase B,** an enzyme found in the liver, which also functions in glycolysis in the liver by cleaving fructose 1,6-bisphosphate. D-Glyceraldehyde enters glycolysis via phosphorylation to glyceraldehyde 3-phosphate catalyzed by **triokinase.** The two triose phosphates, dihydroxyacetone phosphate, and glyceraldehyde 3-phosphate, may either be degraded by glycolysis or may be substrates for aldolase and hence gluconeogenesis, which is the fate of much of the fructose metabolized in the liver.

In extrahepatic tissues, hexokinase catalyzes the phosphorylation of most hexose sugars, including fructose, but glucose inhibits the phosphorylation of fructose since it is a better substrate for hexokinase. Nevertheless, some fructose can be metabolized in adipose tissue and muscle. Fructose is found in seminal plasma and in the fetal circulation of ungulates and whales. Aldose reductase is found in the placenta of the ewe and is responsible for the secretion of sorbitol into the fetal blood. The presence of sorbitol dehydrogenase in the liver, including the fetal liver, is responsible for the conversion of sorbitol into fructose. This pathway is also responsible for the occurrence of fructose in seminal fluid.

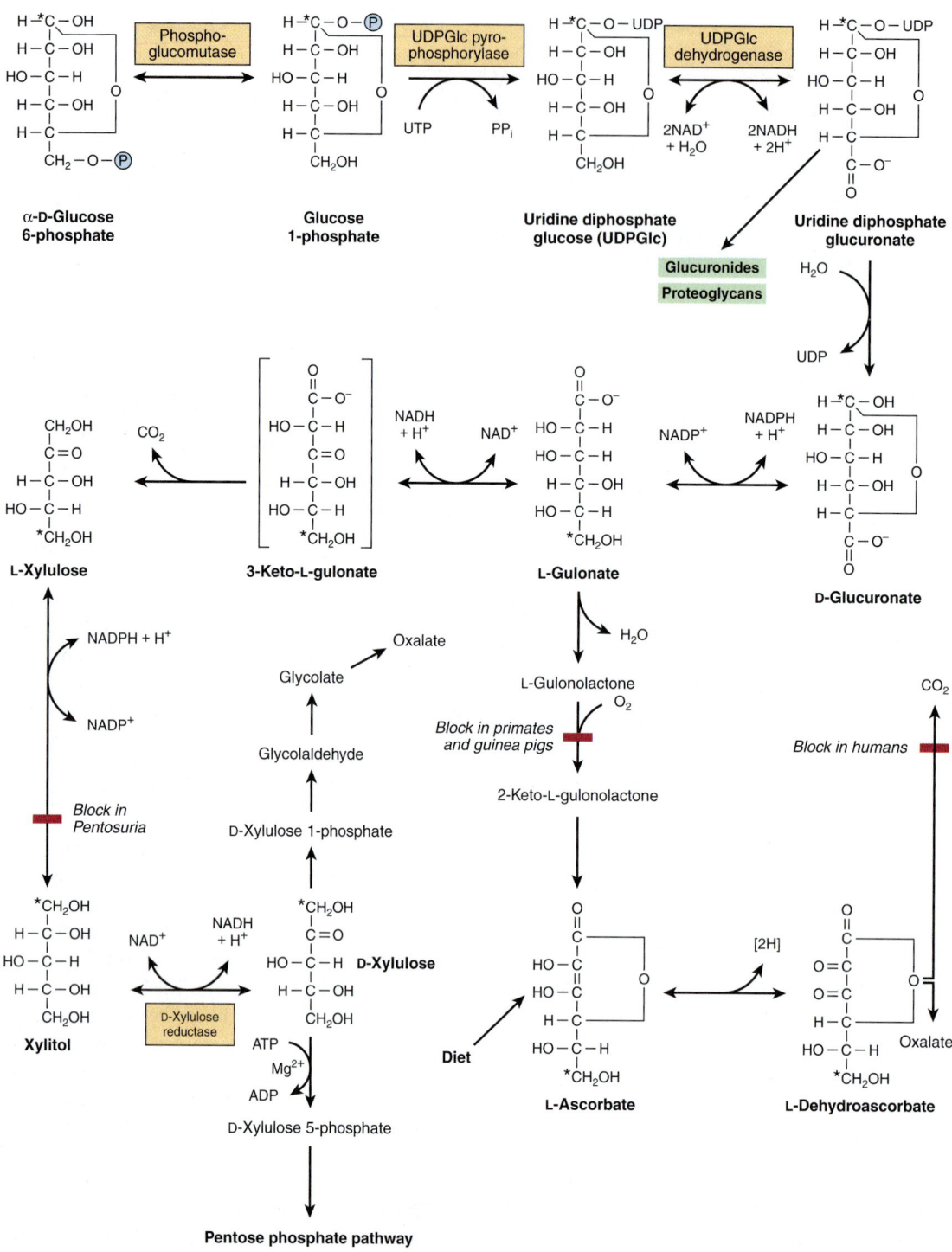

FIGURE 21–4   Uronic acid pathway. (*Indicates the fate of carbon 1 of glucose; Ⓟ, —PO₃²⁻.)

GALACTOSE IS NEEDED FOR THE SYNTHESIS OF LACTOSE, GLYCOLIPIDS, PROTEOGLYCANS, & GLYCOPROTEINS

Galactose is derived from intestinal hydrolysis of the disaccharide **lactose,** the sugar of milk. It is readily converted in the liver to glucose. **Galactokinase** catalyzes the phos-

phorylation of galactose, using ATP as phosphate donor (**Figure 21–6**). Galactose 1-phosphate reacts with UDPGlc to form uridine diphosphate galactose (UDPGal) and glucose 1-phosphate, in a reaction catalyzed by **galactose 1-phosphate uridyl transferase.** The conversion of UDP-Gal to UDPGlc is catalyzed by **UDPGal 4-epimerase.** The reaction involves oxidation, and then reduction, at carbon 4, with NAD⁺ as a coenzyme. The UDPGlc is then incorporated into glycogen (Chapter 19).

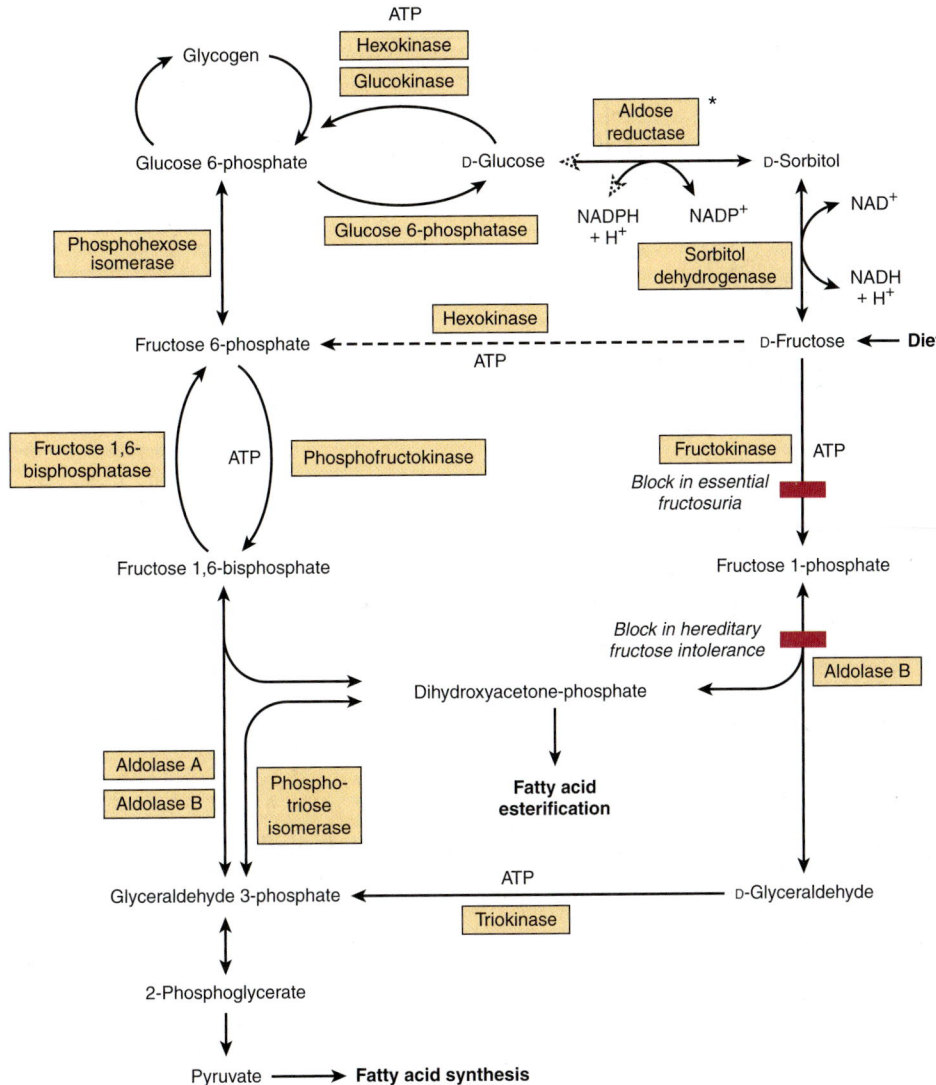

FIGURE 21–5 **Metabolism of fructose.** Aldolase A is found in all tissues, whereas aldolase B is the predominant form in liver. (*Not found in liver.)

The epimerase reaction is freely reversible, so glucose can be converted to galactose, and galactose is not a dietary essential. Galactose is required in the body not only for the formation of lactose in lactation, but also as a constituent of glycolipids (cerebrosides), proteoglycans, and glycoproteins. In the synthesis of lactose in the mammary gland, UDPGal condenses with glucose to yield lactose, catalyzed by **lactose synthase** (see Figure 21–6).

Glucose Is the Precursor of Amino Sugars (Hexosamines)

Amino sugars are important components of **glycoproteins** (Chapter 47), of certain **glycosphingolipids** (eg, gangliosides; Chapter 15), and of glycosaminoglycans (Chapter 48). The major amino sugars are the hexosamines **glucosamine, galactosamine,** and **mannosamine,** and the nine-carbon compound

sialic acid. The principal sialic acid found in human tissues is *N*-acetylneuraminic acid (NeuAc). A summary of the metabolic interrelationships among the amino sugars is shown in **Figure 21–7**.

CLINICAL ASPECTS

Impairment of the Pentose Phosphate Pathway Leads to Erythrocyte Hemolysis

Genetic defects of glucose 6-phosphate dehydrogenase, with consequent impairment of the generation of NADPH, are common in populations of Mediterranean and Afro-Caribbean origin. The gene is on the X chromosome, so it is mainly males who are affected. Some 400 million people carry a mutated

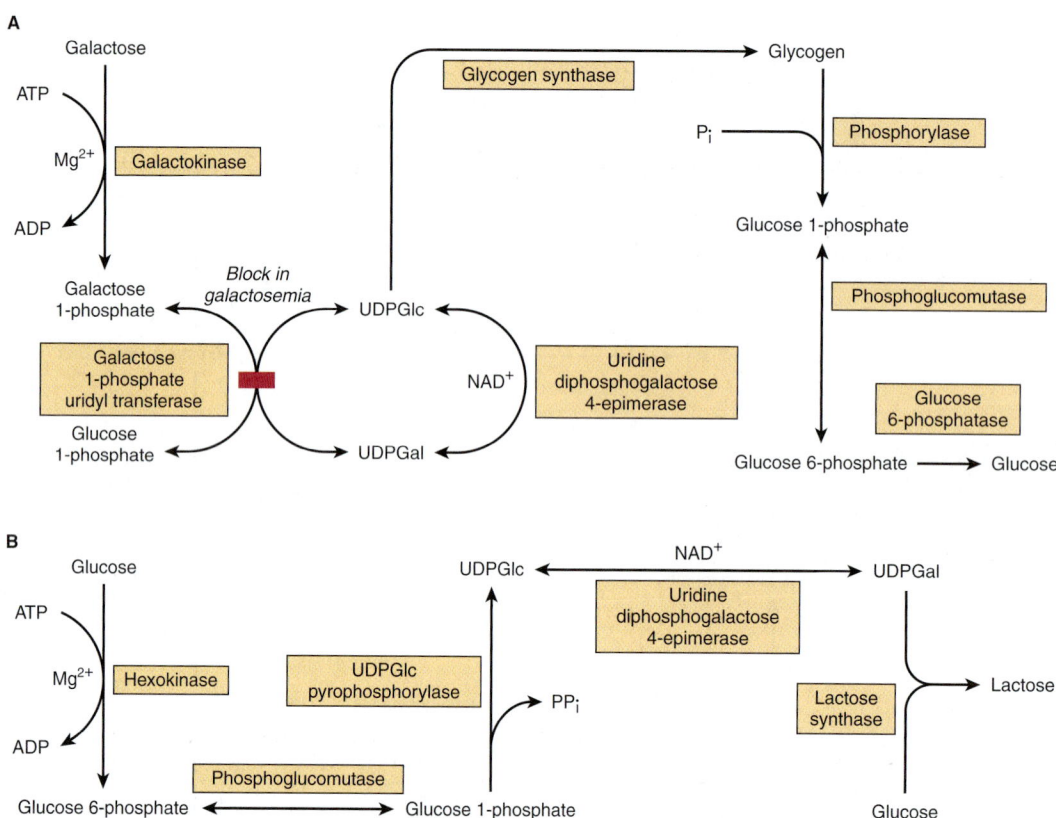

FIGURE 21–6 Pathway of conversion of (A) galactose to glucose in the liver and (B) glucose to lactose in the lactating mammary gland.

gene for glucose 6-phosphate dehydrogenase, making it the most common genetic defect, but most are asymptomatic. In some populations, glucose 6-phosphatase deficiency is common enough for it to be regarded as a genetic polymorphism. The distribution of mutant genes parallels that of malaria, suggesting that being heterozygous confers resistance against malaria. The defect is manifested as red cell hemolysis (**hemolytic anemia**) when susceptible individuals are subjected to oxidative stress (Chapter 52) from infection, drugs such as the antimalarial primaquine, and sulfonamides, or when they have eaten fava beans (*Vicia faba*—hence the name of the disease, **favism**).

There are two main variants of favism. In the Afro-Caribbean variant the enzyme is unstable, so that while average red-cell activities are low, it is only the older erythrocytes that are affected by oxidative stress, and the hemolytic crises tend to be self-limiting. By contrast, in the Mediterranean variant the enzyme is stable, but has low activity in all erythrocytes. Hemolytic crises in these people are more severe and can be fatal. Glutathione peroxidase is dependent upon a supply of NADPH, which in erythrocytes can be formed only via the pentose phosphate pathway. It reduces organic peroxides and H_2O_2, as part of the body's defense against lipid peroxidation (Figure 15–21). Measurement of erythrocyte **glutathione reductase,** and its activation by FAD is used to assess **vitamin B$_2$** nutritional status (Chapter 44).

Disruption of the Uronic Acid Pathway Is Caused by Enzyme Defects & Some Drugs

In the rare benign hereditary condition **essential pentosuria,** considerable quantities of **L-xylulose** appear in the urine, because of the absence of the enzyme necessary to reduce L-xylulose to xylitol. Although pentosuria is benign, with no clinical consequences, xylulose is a reducing sugar and can give false positive results when urinary glucose is measured using alkaline copper reagents. Various drugs increase the rate at which glucose enters the uronic acid pathway. For example, administration of barbital or chlorobutanol to rats results in a significant increase in the conversion of glucose to glucuronate, L-gulonate, and ascorbate. Aminopyrine and antipyrine increase the excretion of L-xylulose in pentosuric subjects. Pentosuria also occurs after consumption of relatively large amounts of fruits such as pears that are rich sources of pentoses (**alimentary pentosuria**).

Loading of the Liver with Fructose May Potentiate Hypertriacylglycerolemia, Hypercholesterolemia, & Hyperuricemia

In the liver, fructose increases fatty acid and triacylglycerol synthesis and VLDL secretion, leading to

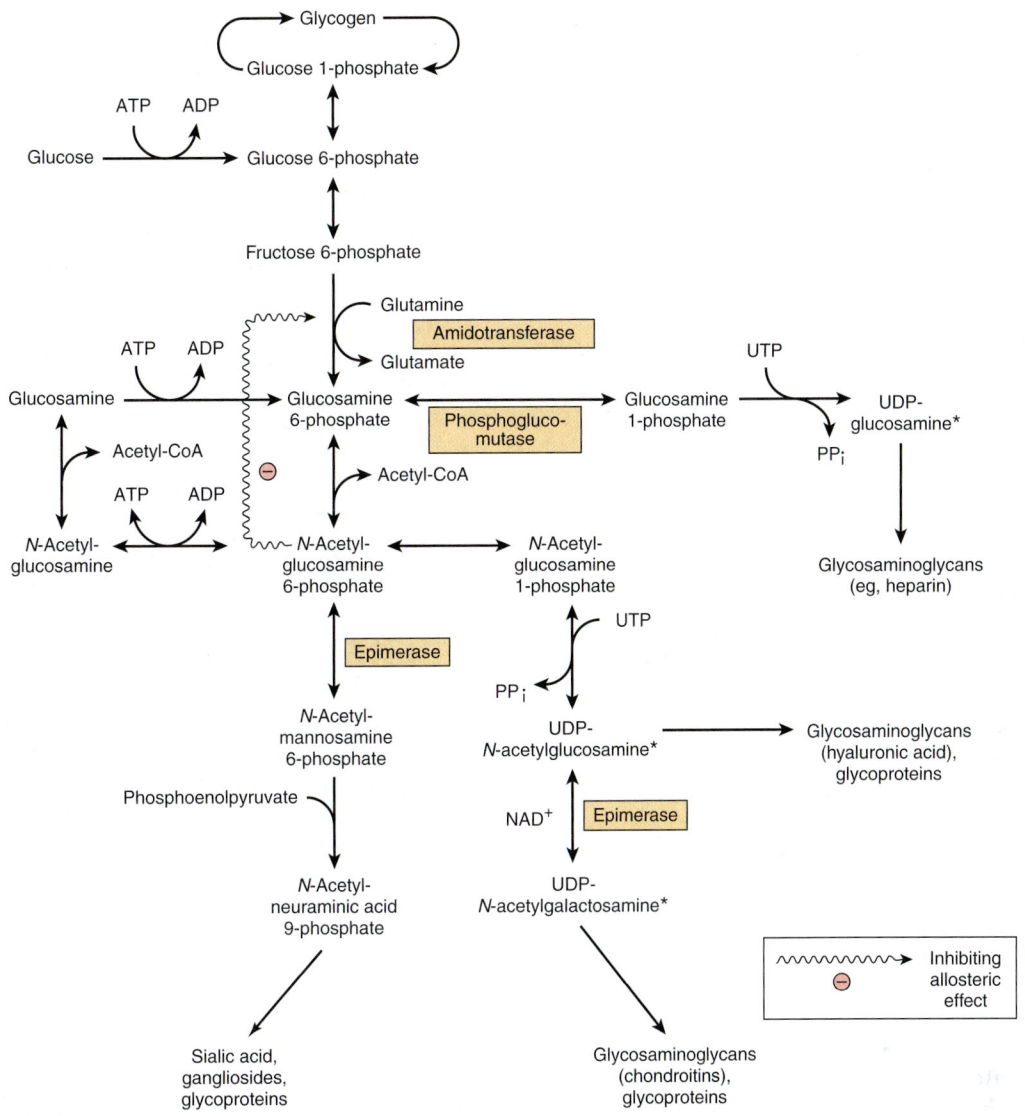

FIGURE 21–7 **Summary of the interrelationships in metabolism of amino sugars.** (*Analogous to UDPGlc.) Other purine or pyrimidine nucleotides may be similarly linked to sugars or amino sugars. Examples are thymidine diphosphate (TDP)-glucosamine and TDP-*N*-acetylglucosamine.

hypertriacylglycerolemia—and increased LDL cholesterol—which can be regarded as potentially atherogenic (Chapter 26). This is because fructose enters glycolysis via fructokinase, and the resulting fructose 1-phosphate bypasses the regulatory step catalyzed by phosphofructokinase (Chapter 18). In addition, acute loading of the liver with fructose, as can occur with intravenous infusion or following very high fructose intakes, causes sequestration of inorganic phosphate in fructose 1-phosphate and diminished ATP synthesis. As a result, there is less inhibition of de novo purine synthesis by ATP, and uric acid formation is increased, causing hyperuricemia, which is the cause of gout (Chapter 33). Since fructose is absorbed from the small intestine by (passive) carrier-mediated diffusion, high oral doses may lead to osmotic diarrhea.

Defects in Fructose Metabolism Cause Disease

A lack of hepatic fructokinase causes **essential fructosuria,** which is a benign and asymptomatic condition. The absence of aldolase B, which cleaves fructose 1-phosphate, leads to **hereditary fructose intolerance,** which is characterized by profound hypoglycemia and vomiting after consumption of fructose (or sucrose, which yields fructose on digestion) (Figure 21–5). Diets low in fructose, sorbitol, and sucrose are beneficial for both conditions. One consequence of hereditary fructose intolerance and of a related condition as a result of **fructose 1,6-bisphosphatase deficiency** is fructose-induced **hypoglycemia** despite the presence of high glycogen reserves, because fructose 1-phosphate and 1,6-bisphosphate allosterically inhibit liver glycogen phosphorylase. The sequestration of inorganic phosphate also leads to depletion of ATP and hyperuricemia.

Fructose & Sorbitol in the Lens Are Associated with Diabetic Cataract

Both fructose and sorbitol are found in the lens of the eye in increased concentrations in diabetes mellitus and may be involved in the pathogenesis of **diabetic cataract.** The **sorbitol (polyol) pathway** (not found in liver) is responsible for fructose formation from glucose (see Figure 21–5) and increases in activity as the glucose concentration rises in those tissues that are not insulin-sensitive, ie, the lens, peripheral nerves, and renal glomeruli. Glucose is reduced to sorbitol by **aldose reductase,** followed by oxidation of sorbitol to fructose in the presence of NAD^+ and sorbitol dehydrogenase (polyol dehydrogenase). Sorbitol does not diffuse through cell membranes, but accumulates, causing osmotic damage. Simultaneously, myoinositol levels fall. In experimental animals, sorbitol accumulation and myoinositol depletion, as well as diabetic cataract, can be prevented by aldose reductase inhibitors. One inhibitor has been licensed in Japan for treatment of diabetic neuropathy, although there is little or no evidence that inhibitors are effective in preventing cataract or slowing the progression of diabetic neuropathy in human beings.

Enzyme Deficiencies in the Galactose Pathway Cause Galactosemia

Inability to metabolize galactose occurs in the **galactosemias,** which may be caused by inherited defects of galactokinase, uridyl transferase, or 4-epimerase (Figure 21–6A), though deficiency of **uridyl transferase** is best known. Galactose is a substrate for aldose reductase, forming galactitol, which accumulates in the lens of the eye, causing cataract. The condition is more severe if it is the result of a defect in the uridyl transferase since galactose 1-phosphate accumulates and depletes the liver of inorganic phosphate. Ultimately, liver failure and mental deterioration result. In uridyl transferase deficiency, the epimerase is present in adequate amounts, so that the galactosemic individual can still form UDPGal from glucose. This explains how it is possible for normal growth and development of affected children to occur despite the galactose-free diets used to control the symptoms of the disease.

SUMMARY

- The pentose phosphate pathway, present in the cytosol, can account for the complete oxidation of glucose, producing NADPH and CO_2 but not ATP.

- The pathway has an oxidative phase, which is irreversible and generates NADPH, and a nonoxidative phase, which is reversible and provides ribose precursors for nucleotide synthesis. The complete pathway is present mainly in those tissues having a requirement for NADPH for reductive syntheses, eg, lipogenesis or steroidogenesis, whereas the nonoxidative phase is present in all cells requiring ribose.

- In erythrocytes, the pathway has a major function in preventing hemolysis by providing NADPH to maintain glutathione in the reduced state as the substrate for glutathione peroxidase.

- The uronic acid pathway is the source of glucuronic acid for conjugation of many endogenous and exogenous substances before excretion as glucuronides in urine and bile.

- Fructose bypasses the main regulatory step in glycolysis, catalyzed by phosphofructokinase, and stimulates fatty acid synthesis and hepatic triacylglycerol secretion.

- Galactose is synthesized from glucose in the lactating mammary gland and in other tissues where it is required for the synthesis of glycolipids, proteoglycans, and glycoproteins.

REFERENCES

Ali M, Rellos P, Cox TM: Hereditary fructose intolerance. J Med Gen 1998;35:353.

Cappellini MD, Fiorelli G: Glucose 6-phosphate dehydrogenase deficiency. Lancet 2008;371:64.

Dunlop M: Aldose reductase and the role of the polyol pathway in diabetic nephropathy. Kidney Int 2000;77:S3.

Grant CM: Metabolic reconfiguration is a regulated response to oxidative stress. J Biol 2008;7:1.

Hers HG, Hue L: Gluconeogenesis and related aspects of glycolysis. Annu Rev Biochem 1983;52:617.

Horecker BL: The pentose phosphate pathway. J Biol Chem 2002;277:47965.

Le KA, Tappy L: Metabolic effects of fructose. Curr Opin Clin Nutr Metab Care 2006;9:469.

Leslie ND: Insights into the pathogenesis of galactosemia. Ann Rev Nutr 2003;23:59.

Manganelli G, Fico A, Martini G, et al: (2010). Discussion on pharmacogenetic interaction in G6PD deficiency and methods to identify potential hemolytic drugs. Cardiovasc Hematol Disord Drug Targets 2010;10:143.

Mayes PA: Intermediary metabolism of fructose. Amer J Clin Nutr 1993;58:754.

Van den Berghe G: Inborn errors of fructose metabolism. Ann Rev Nutr 1994;14:41.

Veech RL: A humble hexose monophosphate pathway metabolite regulates short- and long-term control of lipogenesis. Proc Natl Acad Sci USA 2003;100:5578.

Wamelink MM, Struys EA, Jakobs C: (2008). The biochemistry, metabolism and inherited defects of the pentose phosphate pathway: a review. J Inherit Metab Dis 2008;31:703.

Wong D: Hereditary fructose intolerance. Mol Genet Metab 2005;85:165.

Oxidation of Fatty Acids: Ketogenesis

Kathleen M. Botham, PhD, DSc & Peter A. Mayes, PhD, DSc

OBJECTIVES

After studying this chapter, you should be able to:

- Describe the processes by which fatty acids are transported in the blood and activated and transported into the matrix of the mitochondria for breakdown to obtain energy.
- Outline the β-oxidation pathway by which fatty acids are metabolized to acetyl-CoA and explain how this leads to the production of large quantities of ATP from the reducing equivalents produced during β-oxidation and further metabolism of the acetyl-CoA via the citric acid cycle.
- Identify the three compounds termed "ketone bodies" and describe the reactions by which they are formed in liver mitochondria.
- Appreciate that ketone bodies are important fuels for extrahepatic tissues and indicate the conditions in which their synthesis and use are favored.
- Indicate the three stages in the metabolism of fatty acids where ketogenesis is regulated.
- Understand that overproduction of ketone bodies leads to ketosis and, if prolonged, ketoacidosis, and identify pathological conditions when this occurs.
- Give examples of diseases associated with impaired fatty acid oxidation.

BIOMEDICAL IMPORTANCE

Although fatty acids are broken down by oxidation to acetyl-CoA and also synthesized from acetyl-CoA, fatty acid oxidation is not the simple reverse of fatty acid biosynthesis but an entirely different process taking place in a separate compartment of the cell. The separation of fatty acid oxidation in mitochondria from biosynthesis in the cytosol allows each process to be individually controlled and integrated with tissue requirements. Each step in fatty acid oxidation involves acyl-CoA derivatives, is catalyzed by separate enzymes, utilizes NAD⁺ and FAD as coenzymes, and generates ATP. It is an aerobic process, requiring the presence of oxygen.

Increased fatty acid oxidation is a characteristic of starvation and of diabetes mellitus, and leads to **ketone body** production by the liver **(ketosis).** Ketone bodies are acidic and when produced in excess over long periods, as in diabetes, cause **ketoacidosis,** which is ultimately fatal. Because gluconeogenesis is dependent upon fatty acid oxidation, any impairment in fatty acid oxidation leads to **hypoglycemia.** This occurs in various states of **carnitine deficiency** or deficiency of essential enzymes in fatty acid oxidation, for example, **carnitine palmitoyltransferase,** or inhibition of fatty acid oxidation by poisons, for example, **hypoglycin.**

OXIDATION OF FATTY ACIDS OCCURS IN MITOCHONDRIA

Fatty Acids Are Transported in the Blood as Free Fatty Acids

Free fatty acids (FFA)—also called unesterified (UFA) or nonesterified (NEFA) fatty acids—are fatty acids that are in the **unesterified state.** In plasma, longer chain FFA are combined with **albumin,** and in the cell they are attached to a **fatty acid binding protein,** so that in fact they are never really "free." Shorter chain fatty acids are more water-soluble and exist as the unionized acid or as a fatty acid anion.

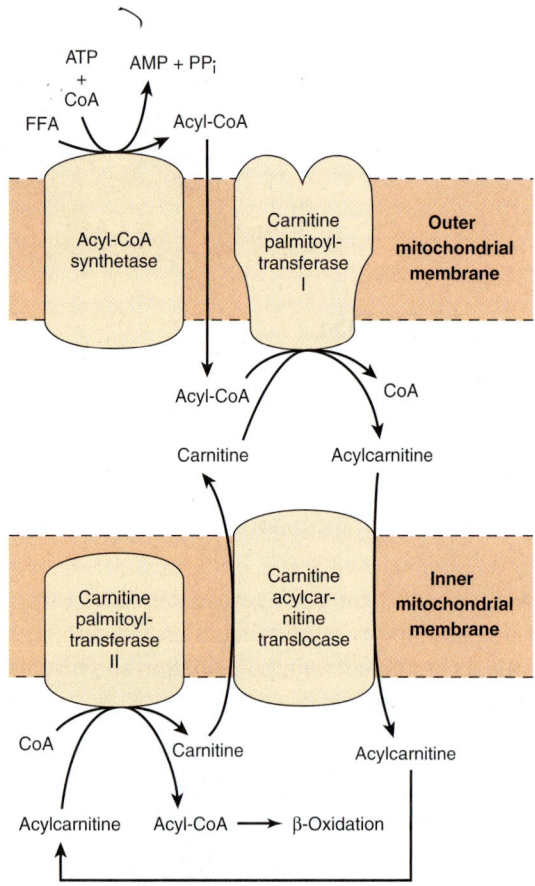

FIGURE 22–1   **Role of carnitine in the transport of long-chain fatty acids through the inner mitochondrial membrane.** Long-chain acyl-CoA cannot pass through the inner mitochondrial membrane, but its metabolic product, acylcarnitine, can.

Fatty Acids Are Activated Before Being Catabolized

Fatty acids must first be converted to an active intermediate before they can be catabolized. This is the only step in the complete degradation of a fatty acid that requires energy from ATP. In the presence of ATP and coenzyme A, the enzyme **acyl-CoA synthetase (thiokinase)** catalyzes the conversion of a fatty acid (or FFA) to an "active fatty acid" or acyl-CoA, which uses one high-energy phosphate with the formation of AMP and PP_i (**Figure 22–1**). The PP_i is hydrolyzed by **inorganic pyrophosphatase** with the loss of a further high-energy phosphate, ensuring that the overall reaction goes to completion. Acyl-CoA synthetases are found in the endoplasmic reticulum, peroxisomes, and inside and on the outer membrane of mitochondria.

Long-Chain Fatty Acids Penetrate the Inner Mitochondrial Membrane as Carnitine Derivatives

Carnitine (β-hydroxy-γ-trimethylammonium butyrate), $(CH_3)_3 N^+$—CH_2—$CH(OH)$—CH_2—COO^-, is widely distributed and is particularly abundant in muscle. Long-chain acyl-CoA

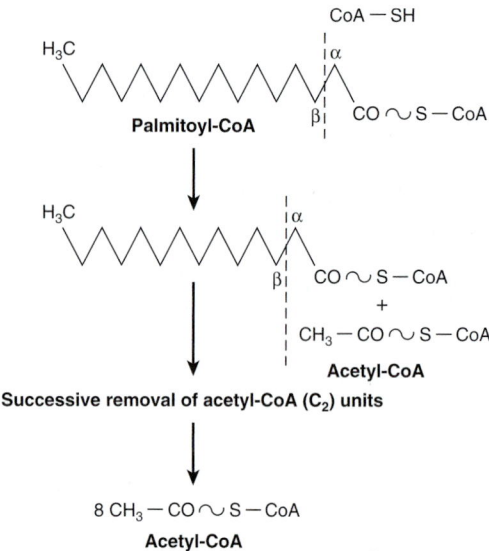

FIGURE 22–2   Overview of β-oxidation of fatty acids.

(or FFA) cannot penetrate the inner membrane of mitochondria. In the presence of carnitine, however, **carnitine palmitoyltransferase-I,** located in the outer mitochondrial membrane, converts long-chain acyl-CoA to **acylcarnitine,** which is able to penetrate the inner membrane and gain access to the β-oxidation system of enzymes (Figure 22–1). **Carnitine-acylcarnitine translocase** acts as an inner membrane exchange transporter. Acylcarnitine is transported in, coupled with the transport out of one molecule of carnitine. The acylcarnitine then reacts with CoA, catalyzed by **carnitine palmitoyltransferase-II,** located on the inside of the inner membrane, reforming acyl-CoA in the mitochondrial matrix, and carnitine is liberated.

β-OXIDATION OF FATTY ACIDS INVOLVES SUCCESSIVE CLEAVAGE WITH RELEASE OF ACETYL-CoA

In **β-oxidation** (**Figure 22–2**), two carbons at a time are cleaved from acyl-CoA molecules, starting at the carboxyl end. The chain is broken between the α(2)- and β(3)-carbon atoms—hence the name β-oxidation. The two-carbon units formed are acetyl-CoA; thus, palmitoyl-CoA forms eight acetyl-CoA molecules.

The Cyclic Reaction Sequence Generates FADH₂ & NADH

Several enzymes, known collectively as "fatty acid oxidase," are found in the mitochondrial matrix or inner membrane adjacent to the respiratory chain. These catalyze the oxidation of acyl-CoA to acetyl-CoA, the system being coupled with the phosphorylation of ADP to ATP (**Figure 22–3**).

The first step is the removal of two hydrogen atoms from the 2(α)- and 3(β)-carbon atoms, catalyzed by **acyl-CoA**

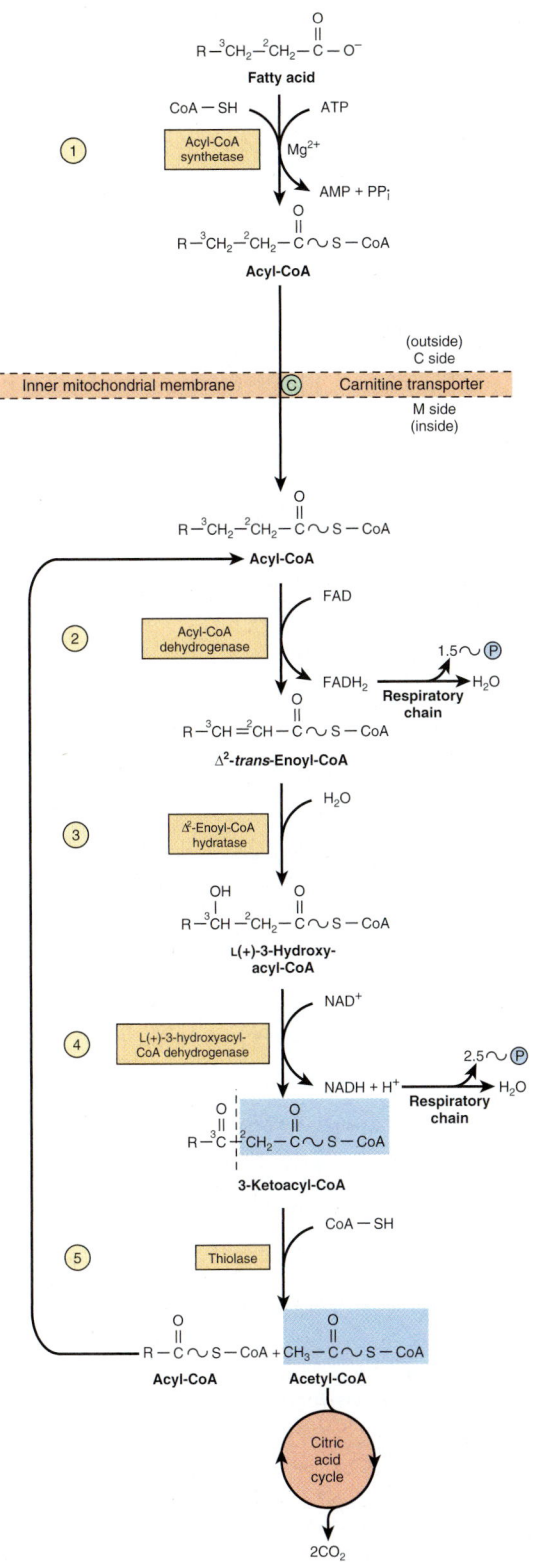

FIGURE 22–3 β-Oxidation of fatty acids. Long-chain acyl-CoA is cycled through reactions ②–⑤, acetyl-CoA being split off, each cycle, by thiolase (reaction ⑤). When the acyl radical is only four carbon atoms in length, two acetyl-CoA molecules are formed in reaction ⑤.

dehydrogenase and requiring FAD. This results in the formation of Δ^2-*trans*-enoyl-CoA and FADH$_2$. The reoxidation of FADH$_2$ by the respiratory chain requires the mediation of another flavoprotein, termed **electron-transferring flavoprotein** (Chapter 12). Water is added to saturate the double bond and form 3-hydroxyacyl-CoA, catalyzed by Δ^2-**enoyl-CoA hydratase**. The 3-hydroxy derivative undergoes further dehydrogenation on the 3-carbon catalyzed by **L(+)-3-hydroxyacyl-CoA dehydrogenase** to form the corresponding 3-ketoacyl-CoA compound. In this case, NAD$^+$ is the coenzyme involved. Finally, 3-ketoacyl-CoA is split at the 2,3-position by **thiolase** (3-ketoacyl-CoA-thiolase), forming acetyl-CoA and a new acyl-CoA two carbons shorter than the original acyl-CoA molecule. The acyl-CoA formed in the cleavage reaction reenters the oxidative pathway at reaction 2 (Figure 22–3). In this way, a long-chain fatty acid may be degraded completely to acetyl-CoA (C$_2$ units). Since acetyl-CoA can be oxidized to CO$_2$ and water via the citric acid cycle (which is also found within the mitochondria), the complete oxidation of fatty acids is achieved.

Oxidation of a Fatty Acid with an Odd Number of Carbon Atoms Yields Acetyl-CoA Plus a Molecule of Propionyl-CoA

Fatty acids with an odd number of carbon atoms are oxidized by the pathway of β-oxidation, producing acetyl-CoA, until a three-carbon (propionyl-CoA) residue remains. This compound is converted to succinyl-CoA, a constituent of the citric acid cycle (Figure 20–2). Hence, **the propionyl residue from an odd-chain fatty acid is the only part of a fatty acid that is glucogenic.**

Oxidation of Fatty Acids Produces a Large Quantity of ATP

Transport of electrons from FADH$_2$ and NADH via the respiratory chain leads to the synthesis of four high-energy phosphates (Chapter 13) for each of the seven cycles needed for the breakdown of the C16 fatty acid, palmitate, to acetyl-CoA ($7 \times 4 = 28$). A total of 8 mol of acetyl-CoA is formed, and each gives rise to 10 mol of ATP on oxidation in the citric acid cycle, making $8 \times 10 = 80$ mol. Two must be subtracted for the initial activation of the fatty acid, yielding a net gain of 106 mol of ATP per mole of palmitate, or $106 \times 51.6^* = 5470$ kJ. This represents 68% of the free energy of combustion of palmitic acid.

Peroxisomes Oxidize Very Long Chain Fatty Acids

A modified form of β-oxidation is found in **peroxisomes** and leads to the formation of acetyl-CoA and H$_2$O$_2$ (from the flavoprotein-linked dehydrogenase step), which is broken

---

$^*\Delta G$ for the ATP reaction, as explained in Chapter 18.

down by catalase (Chapter 12). Thus, this dehydrogenation in peroxisomes is not linked directly to phosphorylation and the generation of ATP. The system facilitates the oxidation of **very long chain fatty acids** (eg, C_{20}, C_{22}). These enzymes are induced by high-fat diets and in some species by hypolipidemic drugs such as clofibrate.

The enzymes in peroxisomes do not attack shorter chain fatty acids; the β-oxidation sequence ends at octanoyl-CoA. Octanoyl and acetyl groups are both further oxidized in mitochondria. Another role of peroxisomal β-oxidation is to shorten the side chain of cholesterol in bile acid formation (Chapter 26). Peroxisomes also take part in the synthesis of ether glycerolipids (Chapter 24), cholesterol, and dolichol (Figure 26–2).

OXIDATION OF UNSATURATED FATTY ACIDS OCCURS BY A MODIFIED β-OXIDATION PATHWAY

The CoA esters of unsaturated fatty acids are degraded by the enzymes normally responsible for β-oxidation until either a Δ^3-*cis*-acyl-CoA compound or a Δ^4-*cis*-acyl-CoA compound is formed, depending upon the position of the double bonds (**Figure 22–4**). The former compound is isomerized (Δ^3**cis** → Δ^2-*trans*-**enoyl-CoA isomerase**) to the corresponding Δ^2-*trans*-CoA stage of β-oxidation for subsequent hydration and oxidation. Any Δ^4-*cis*-acyl-CoA either remaining, as in the case of linoleic acid, or entering the pathway at this point after conversion by acyl-CoA dehydrogenase to Δ^2-*trans*-Δ^4-*cis*-dienoyl-CoA, is then metabolized as indicated in Figure 22–4.

KETOGENESIS OCCURS WHEN THERE IS A HIGH RATE OF FATTY-ACID OXIDATION IN THE LIVER

Under metabolic conditions associated with a high rate of fatty acid oxidation, the liver produces considerable quantities of **acetoacetate** and **D(–)-3-hydroxybutyrate** (β-hydroxybutyrate). Acetoacetate continually undergoes spontaneous decarboxylation to yield **acetone**. These three substances are collectively known as the **ketone bodies** (also called acetone bodies or [incorrectly*] "ketones") (**Figure 22–5**). Acetoacetate and 3-hydroxybutyrate are interconverted by the mitochondrial enzyme **D(–)-3-hydroxybutyrate dehydrogenase;** the equilibrium is controlled by the mitochondrial [NAD⁺]/[NADH] ratio, ie, the **redox state**. The concentration of total ketone

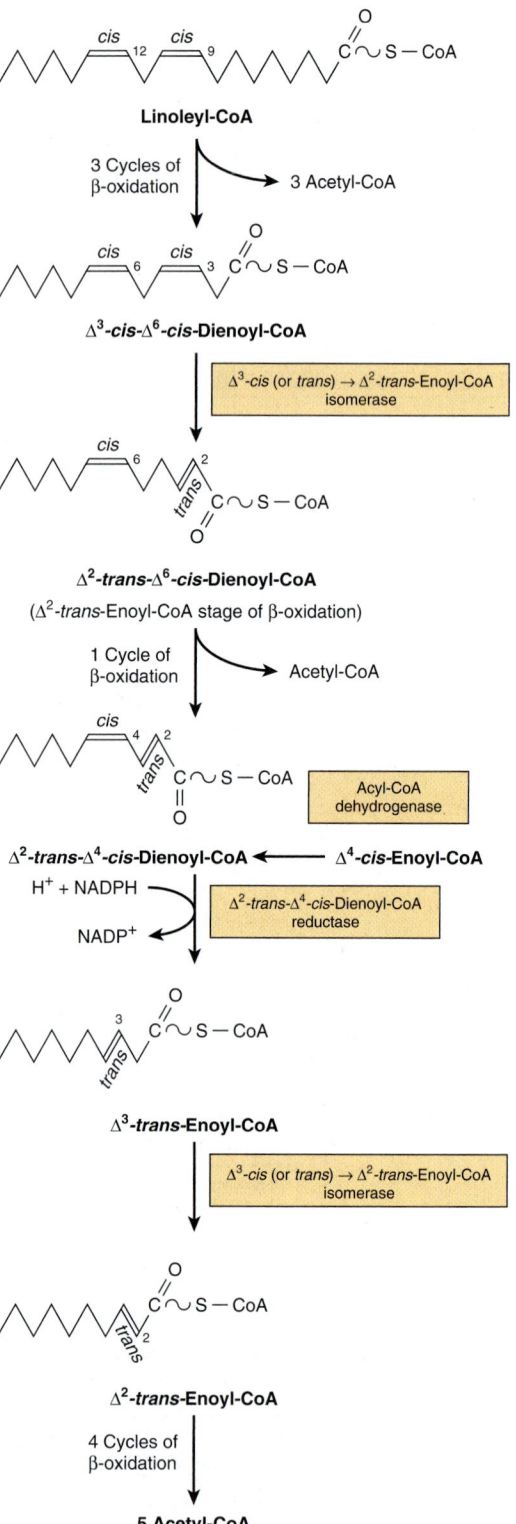

FIGURE 22–4 Sequence of reactions in the oxidation of unsaturated fatty acids, for example, linoleic acid. Δ^4-*cis*-fatty acids or fatty acids forming Δ^4-*cis*-enoyl-CoA enter the pathway at the position shown. NADPH for the dienoyl-CoA reductase step is supplied by intramitochondrial sources such as glutamate dehydrogenase, isocitrate dehydrogenase, and NAD(P)H transhydrogenase.

---

*The term "ketones" should not be used because 3-hydroxybutyrate is not a ketone and there are ketones in blood that are not ketone bodies, for example, pyruvate and fructose.

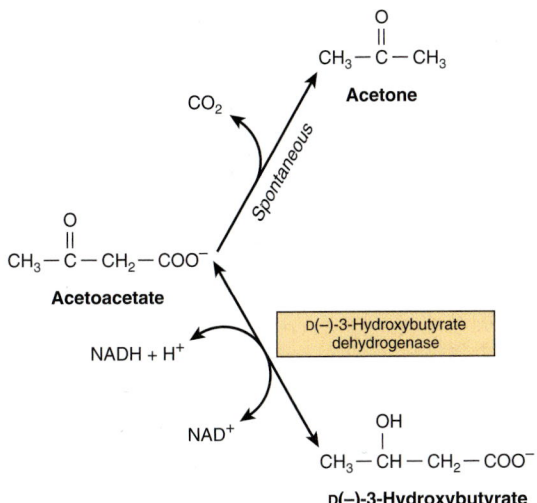

FIGURE 22–5 Interrelationships of the ketone bodies. D(−)-3-hydroxybutyrate dehydrogenase is a mitochondrial enzyme.

3-Hydroxy-3-Methylglutaryl-CoA (HMG-CoA) Is an Intermediate in the Pathway of Ketogenesis

Enzymes responsible for ketone body formation are associated mainly with the mitochondria. Two acetyl-CoA molecules formed in β-oxidation condense to form acetoacetyl-CoA by a reversal of the **thiolase** reaction. Acetoacetyl-CoA, which is the starting material for ketogenesis, also arises directly from the terminal four carbons of a fatty acid during β-oxidation (**Figure 22–7**). Condensation of acetoacetyl-CoA with another molecule of acetyl-CoA by **3-hydroxy-3-methylglutaryl-CoA synthase** forms 3-hydroxy-3-methylglutaryl-CoA (**HMG-CoA**). **3-Hydroxy-3-methylglutaryl-CoA lyase** then causes acetyl-CoA to split off from the HMG-CoA, leaving free acetoacetate. The carbon atoms split off in the acetyl-CoA molecule are derived from the original acetoacetyl-CoA molecule. **Both enzymes must be present in mitochondria for ketogenesis to take place.** This occurs solely in liver and rumen epithelium. D(−)-3-Hydroxybutyrate is quantitatively the predominant ketone body present in the blood and urine in ketosis.

Ketone Bodies Serve as a Fuel for Extrahepatic Tissues

While an active enzymatic mechanism produces acetoacetate from acetoacetyl-CoA in the liver, acetoacetate once formed cannot be reactivated directly except in the cytosol, where it is used in a much less active pathway as a precursor in cholesterol synthesis. This accounts for the net production of ketone bodies by the liver.

bodies in the blood of well-fed mammals does not normally exceed 0.2 mmol/L except in ruminants, where 3-hydroxybutyrate is formed continuously from butyric acid (a product of ruminal fermentation) in the rumen wall. In vivo, the liver appears to be the only organ in nonruminants to add significant quantities of ketone bodies to the blood. Extrahepatic tissues utilize them as respiratory substrates. The net flow of ketone bodies from the liver to the extrahepatic tissues results from active hepatic synthesis coupled with very low utilization. The reverse situation occurs in extrahepatic tissues (**Figure 22–6**).

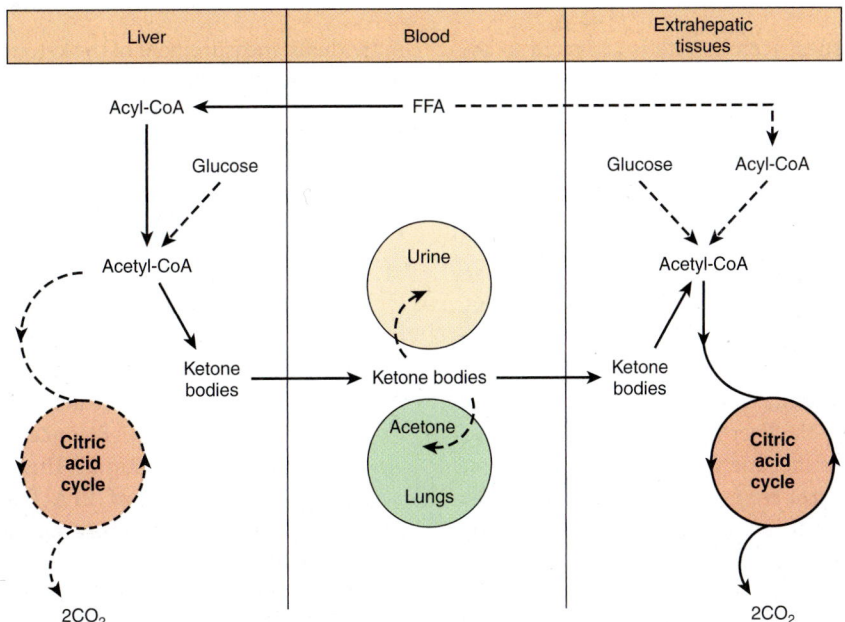

FIGURE 22–6 Formation, utilization, and excretion of ketone bodies. (The main pathway is indicated by the solid arrows.)

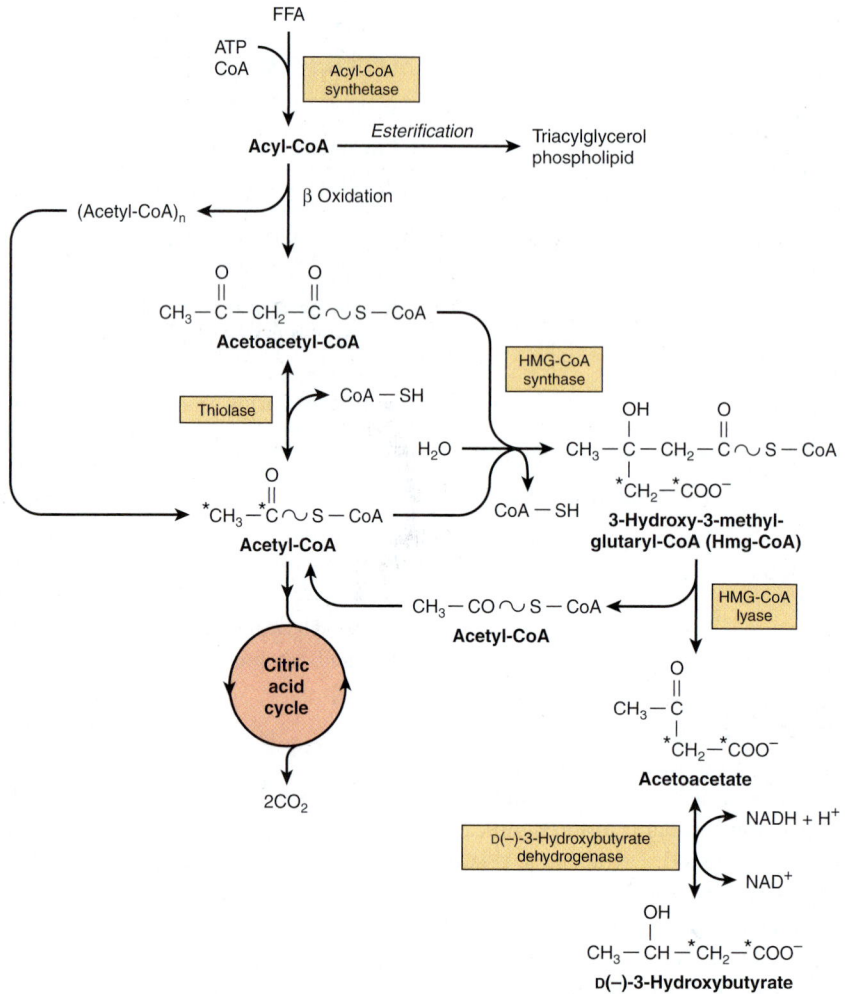

FIGURE 22–7   **Pathways of ketogenesis in the liver.** (FFA, free fatty acids.)

In extrahepatic tissues, acetoacetate is activated to ace-toacetyl-CoA by **succinyl-CoA-acetoacetate CoA trans-ferase.** CoA is transferred from succinyl-CoA to form acetoacetyl-CoA (**Figure 22–8**). With the addition of a CoA, the acetoacetyl-CoA is split into two acetyl-CoAs by thiolase and oxidized in the citric acid cycle. If the blood level is raised, oxidation of ketone bodies increases until, at a concentration of ~12 mmol/L, the oxidative machinery is saturated. When this occurs, a large proportion of oxy-gen consumption may be accounted for by the oxidation of ketone bodies.

In most cases, **ketonemia is due to increased produc-tion of ketone bodies** by the liver rather than to a deficiency in their utilization by extrahepatic tissues. While acetoacetate and D(–)-3-hydroxybutyrate are readily oxidized by extrahe-patic tissues, acetone is difficult to oxidize in vivo and to a large extent is volatilized in the lungs.

In moderate ketonemia, the loss of ketone bodies via the urine is only a few percent of the total ketone body produc-tion and utilization. Since there are renal threshold-like effects (there is not a true threshold) that vary between species and

individuals, measurement of the ketonemia, not the ketonuria, is the preferred method of assessing the severity of ketosis.

KETOGENESIS IS REGULATED AT THREE CRUCIAL STEPS

1. Ketosis does not occur in vivo unless there is an increase in the level of circulating FFA that arise from lipolysis of triacylglycerol in adipose tissue. **FFA are the precursors of ketone bodies in the liver.** The liver, both in fed and in fasting conditions, extracts ~30% of the FFA passing through it, so that at high concentrations the flux passing into the liver is substantial. **Therefore, the factors regu-lating mobilization of FFA from adipose tissue are important in controlling ketogenesis (Figures 22–9 and 25–8).**

2. After uptake by the liver, FFA are either **β-oxidized** to CO_2 or ketone bodies or **esterified** to triacylglycerol and phospholipid. There is regulation of entry of fatty acids into the oxidative pathway by **carnitine palmitoyl-**

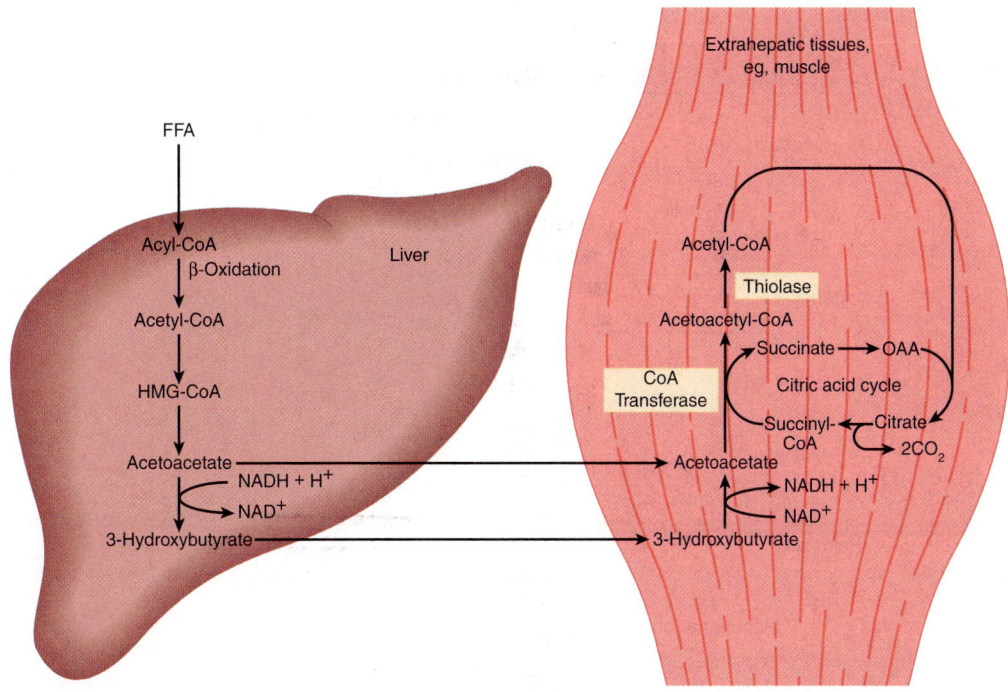

FIGURE 22–8 Transport of ketone bodies from the liver and pathways of utilization and oxidation in extrahepatic tissues.

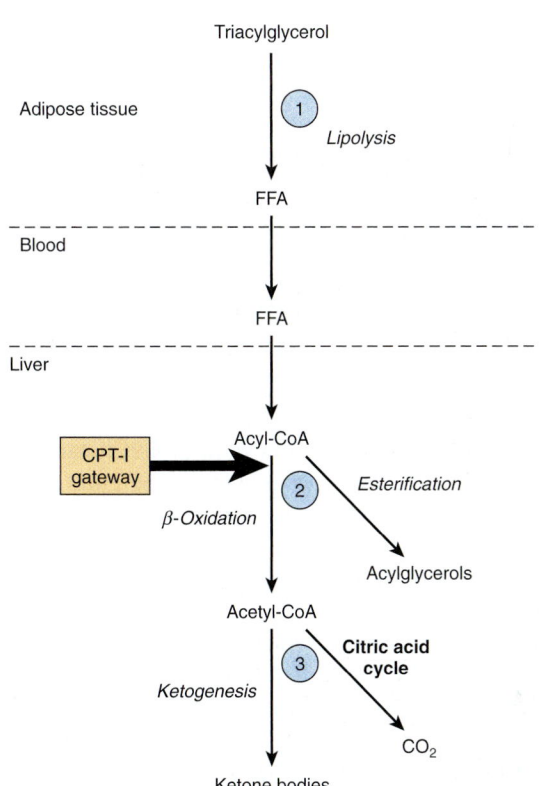

FIGURE 22–9 Regulation of ketogenesis. ①–③ show three crucial steps in the pathway of metabolism of free fatty acids (FFA) that determine the magnitude of ketogenesis. (CPT-I, carnitine palmitoyltransferase-I.)

transferase-I (CPT-I), and the remainder of the fatty acid taken up is esterified. CPT-I activity is low in the fed state, leading to depression of fatty acid oxidation, and high in starvation, allowing fatty acid oxidation to increase. **Malonyl-CoA,** the initial intermediate in fatty acid biosynthesis (Figure 23–1) formed by acetyl-CoA carboxylase in the fed state, is a potent inhibitor of CPT-I (**Figure 22–10**). Under these conditions, FFA enter the liver cell in low concentrations and are nearly all esterified to acylglycerols and transported out of the liver in **very low density lipoproteins** (VLDL). However, as the concentration of FFA increases with the onset of starvation, acetyl-CoA carboxylase is inhibited directly by acyl-CoA, and (malonyl-CoA) decreases, releasing the inhibition of CPT-I and allowing more acyl-CoA to be β-oxidized. These events are reinforced in starvation by a decrease in the **(insulin)/(glucagon) ratio.** Thus, β-oxidation from FFA is controlled by the CPT-I gateway into the mitochondria, and the balance of the FFA uptake not oxidized is esterified.

3. In turn, the acetyl-CoA formed in β-oxidation is oxidized in the citric acid cycle, or it enters the pathway of ketogenesis to form ketone bodies. As the level of serum FFA is raised, proportionately more FFA is converted to ketone bodies and less is oxidized via the citric acid cycle to CO_2. The partition of acetyl-CoA between the ketogenic pathway and the pathway of oxidation to CO_2 is regulated so that the total free energy captured in ATP which results from the oxidation of FFA remains constant as their concentration in the serum changes. This may be appreciated when it is realized

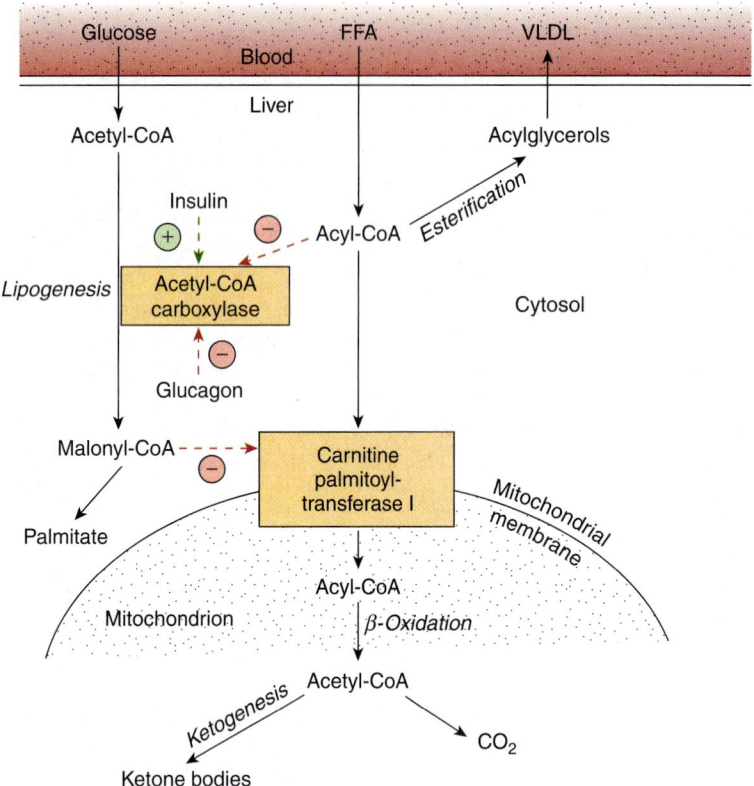

FIGURE 22–10 **Regulation of long-chain fatty acid oxidation in the liver.** (FFA, free fatty acids; VLDL, very low density lipoprotein.) Positive (⊕) and negative (⊖) regulatory effects are represented by broken arrows and substrate flow by solid arrows.

that complete oxidation of 1 mol of palmitate involves a net production of 106 mol of ATP via β-oxidation and CO_2 production in the citric acid cycle (see above), whereas only 26 mol of ATP are produced when acetoacetate is the end product and only 21 mol when 3-hydroxybutyrate is the end product. Thus, ketogenesis may be regarded as a mechanism that allows the liver to oxidize increasing quantities of fatty acids within the constraints of a tightly coupled system of oxidative phosphorylation.

A fall in the concentration of oxaloacetate, particularly within the mitochondria, can impair the ability of the citric acid cycle to metabolize acetyl-CoA and divert fatty acid oxidation toward ketogenesis. Such a fall may occur because of an increase in the (NADH)/(NAD⁺) ratio caused by increased β-oxidation of fatty acids affecting the equilibrium between oxaloacetate and malate, leading to a decrease in the concentration of oxaloacetate, and when gluconeogenesis is elevated, which occurs when blood glucose levels are low. The activation of pyruvate carboxylase, which catalyzes the conversion of pyruvate to oxaloacetate, by acetyl-CoA partially alleviates this problem, but in conditions such as starvation and untreated diabetes mellitus, ketone bodies are overproduced causing ketosis.

CLINICAL ASPECTS

Impaired Oxidation of Fatty Acids Gives Rise to Diseases Often Associated with Hypoglycemia

Carnitine deficiency can occur particularly in the newborn—and especially in preterm infants—owing to inadequate biosynthesis or renal leakage. Losses can also occur in hemodialysis. This suggests a vitamin-like dietary requirement for carnitine in some individuals. Symptoms of deficiency include hypoglycemia, which is a consequence of impaired fatty acid oxidation and lipid accumulation with muscular weakness. Treatment is by oral supplementation with carnitine.

Inherited **CPT-I deficiency** affects only the liver, resulting in reduced fatty acid oxidation and ketogenesis, with hypoglycemia. **CPT-II deficiency** affects primarily skeletal muscle and, when severe, the liver. The sulfonylurea drugs (**glyburide [glibenclamide]** and **tolbutamide**), used in the treatment of type 2 diabetes mellitus, reduce fatty acid oxidation and, therefore, hyperglycemia by inhibiting CPT-I.

Inherited defects in the enzymes of β-oxidation and ketogenesis also lead to nonketotic hypoglycemia, coma, and fatty liver. Defects are known in long- and short-chain 3-hydroxyacyl-CoA dehydrogenase (deficiency of the long-chain enzyme may be a cause of **acute fatty liver of pregnancy**). **3-Ketoacyl-CoA thiolase** and **HMG-CoA lyase deficiency** also affect the degradation of leucine, a ketogenic amino acid (Chapter 29).

Jamaican vomiting sickness is caused by eating the unripe fruit of the akee tree, which contains the toxin **hypoglycin**. This inactivates medium- and short-chain acyl-CoA dehydrogenase, inhibiting β-oxidation and causing hypoglycemia. **Dicarboxylic aciduria** is characterized by the excretion of C_6–C_{10} ω-dicarboxylic acids and by nonketotic hypoglycemia, and is caused by a lack of mitochondrial **medium-chain acyl-CoA dehydrogenase**. **Refsum's disease** is a rare neurologic disorder due to a metabolic defect that results in the accumulation of phytanic acid, which is found in dairy products and ruminant fat and meat. Phytanic acid is thought to have pathological effects on membrane function, protein prenylation, and gene expression. **Zellweger's (cerebrohepatorenal) syndrome** occurs in individuals with a rare inherited absence of peroxisomes in all tissues. They accumulate C_{26}–C_{38} polyenoic acids in brain tissue and also exhibit a generalized loss of peroxisomal functions. The disease causes severe neurological symptoms, and most patients die in the first year of life.

Ketoacidosis Results from Prolonged Ketosis

Higher than normal quantities of ketone bodies present in the blood or urine constitute **ketonemia** (hyperketonemia) or **ketonuria,** respectively. The overall condition is called **ketosis.** The basic form of ketosis occurs in **starvation** and involves depletion of available carbohydrate coupled with mobilization of FFA. This general pattern of metabolism is exaggerated to produce the pathologic states found in **diabetes mellitus, the type 2 form of which is increasingly common in Western countries; twin lamb disease;** and **ketosis in lactating cattle.** Nonpathologic forms of ketosis are found under conditions of high-fat feeding and after severe exercise in the postabsorptive state.

Acetoacetic and 3-hydroxybutyric acids are both moderately strong acids and are buffered when present in blood or other tissues. However, their continual excretion in quantity progressively depletes the alkali reserve, causing **ketoacidosis.** This may be fatal in uncontrolled **diabetes mellitus.**

SUMMARY

- Fatty acid oxidation in mitochondria leads to the generation of large quantities of ATP by a process called β-oxidation that cleaves acetyl-CoA units sequentially from fatty acyl chains. The acetyl-CoA is oxidized in the citric acid cycle, generating further ATP.

- The ketone bodies (acetoacetate, 3-hydroxybutyrate, and acetone) are formed in hepatic mitochondria when there is a high rate of fatty acid oxidation. The pathway of ketogenesis involves synthesis and breakdown of 3-hydroxy-3-methylglutaryl-CoA (HMG-CoA) by two key enzymes, HMG-CoA synthase, and HMG-CoA lyase.

- Ketone bodies are important fuels in extrahepatic tissues.

- Ketogenesis is regulated at three crucial steps: (1) control of FFA mobilization from adipose tissue; (2) the activity of carnitine palmitoyltransferase-I in liver, which determines the proportion of the fatty acid flux that is oxidized rather than esterified; and (3) partition of acetyl-CoA between the pathway of ketogenesis and the citric acid cycle.

- Diseases associated with impairment of fatty acid oxidation lead to hypoglycemia, fatty infiltration of organs, and hypoketonemia.

- Ketosis is mild in starvation but severe in diabetes mellitus and ruminant ketosis.

REFERENCES

Eaton S, Bartlett K, Pourfarzam M: Mammalian mitochondrial β-oxidation. Biochem J 1996;320:345.

Fukao T, Lopaschuk GD, Mitchell GA: Pathways and control of ketone body metabolism: on the fringe of lipid metabolism. Prostaglandins Leukot Essent Fatty Acids 2004;70:243.

Gurr MI, Harwood JL, Frayn K: *Lipid Biochemistry.* Blackwell Publishing, 2002.

Reddy JK, Mannaerts GP: Peroxisomal lipid metabolism. Annu Rev Nutr 1994;14:343.

Scriver CR, Beaudet AL, Sly WS, et al (editors): *The Metabolic and Molecular Bases of Inherited Disease,* 8th ed. McGraw-Hill, 2001.

Wood PA: Defects in mitochondrial beta-oxidation of fatty acids. Curr Opin Lipidol 1999;10:107.

23

Biosynthesis of Fatty Acids & Eicosanoids

Kathleen M. Botham, PhD, DSc & Peter A. Mayes, PhD, DSc

OBJECTIVES

After studying this chapter, you should be able to:

- Describe the reaction catalyzed by acetyl-CoA carboxylase and understand the mechanisms by which its activity is regulated to control the rate of fatty acid synthesis.
- Outline the structure of the fatty acid synthase multienzyme complex, indicating the sequence of enzymes in the two peptide chains of the homodimer.
- Explain how long-chain fatty acids are synthesized by the repeated condensation of two carbon units, with formation of the 16-carbon palmitate being favored in most tissues, and identify the co-factors required.
- Indicate the sources of reducing equivalents (NADPH) for fatty acid synthesis.
- Understand how fatty acid synthesis is regulated by nutritional status and identify other control mechanisms that operate in addition to modulation of the activity of acetyl-CoA carboxylase.
- Identify the nutritionally essential fatty acids and explain why they cannot be formed in the body.
- Explain how polyunsaturated fatty acids are synthesized by desaturase and elongation enzymes.
- Outline the cyclooxygenase and lipoxygenase pathways responsible for the formation of the various classes of eicosanoids.

BIOMEDICAL IMPORTANCE

Fatty acids are synthesized by an **extramitochondrial system,** which is responsible for the complete synthesis of palmitate from acetyl-CoA in the cytosol. In most mammals, glucose is the primary substrate for lipogenesis, but in ruminants it is acetate, the main fuel molecule produced by the diet. Critical diseases of the pathway have not been reported in humans. However, inhibition of lipogenesis occurs in type 1 (insulin-dependent) **diabetes mellitus,** and variations in the activity of the process affect the nature and extent of **obesity.**

Unsaturated fatty acids in phospholipids of the cell membrane are important in maintaining membrane fluidity. A high ratio of polyunsaturated fatty acids to saturated fatty acids (P:S ratio) in the diet is considered to be beneficial in preventing coronary heart disease. Animal tissues have limited capacity for desaturating fatty acids, and require certain dietary polyunsaturated fatty acids derived from plants. These **essential fatty acids** are used to form eicosanoic (C_{20}) fatty acids, which give rise to the **eicosanoids** prostaglandins, thromboxanes, leukotrienes, and lipoxins. Prostaglandins mediate **inflammation, pain,** and induce **sleep** and also regulate **blood coagulation** and **reproduction. Nonsteroidal anti-inflammatory drugs (NSAIDs)** such as **aspirin and ibuprofen** act by inhibiting prostaglandin synthesis. Leukotrienes have muscle contractant and chemotactic properties and are important in allergic reactions and inflammation.

THE MAIN PATHWAY FOR DE NOVO SYNTHESIS OF FATTY ACIDS (LIPOGENESIS) OCCURS IN THE CYTOSOL

This system is present in many tissues, including liver, kidney, brain, lung, mammary gland, and adipose tissue. Its cofactor requirements include NADPH, ATP, Mn^{2+}, biotin, and HCO_3^-

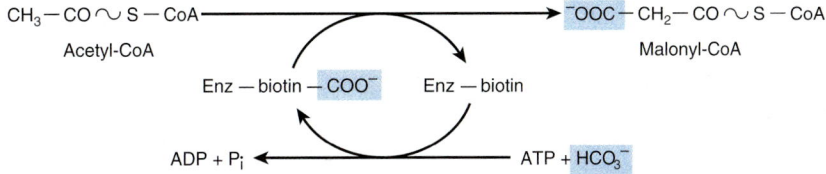

FIGURE 23–1 **Biosynthesis of malonyl-CoA.** (Enz, acetyl-CoA carboxylase.)

Production of Malonyl-CoA Is the Initial & Controlling Step in Fatty Acid Synthesis

(as a source of CO_2). **Acetyl-CoA** is the immediate substrate, and **free palmitate** is the end product.

Bicarbonate as a source of CO_2 is required in the initial reaction for the carboxylation of acetyl-CoA to **malonyl-CoA** in the presence of ATP and **acetyl-CoA carboxylase.** Acetyl-CoA carboxylase has a requirement for the B vitamin **biotin** (**Figure 23–1**). The enzyme is a **multienzyme protein** containing a variable number of identical subunits, each containing biotin, biotin carboxylase, biotin carboxyl carrier protein, and transcarboxylase, as well as a regulatory allosteric site. The reaction takes place in two steps: (1) carboxylation of biotin involving ATP and (2) transfer of the carboxyl group to acetyl-CoA to form malonyl-CoA.

The Fatty Acid Synthase Complex Is a Homodimer of Two Polypeptide Chains Containing Six Enzyme Activities

In mammals, the individual enzymes of the fatty acid synthase system are linked in a multienzyme polypeptide complex that incorporates **the acyl carrier protein (ACP),** which takes over the role of CoA. It contains the vitamin **pantothenic acid** in the form of 4'-phosphopantetheine (Figure 44–18). In the primary structure of the protein, the enzyme domains are linked in the sequence as shown in **Figure 23–2**. X-ray crystallography of the three-dimensional structure, however, has shown that the complex is a homodimer, with two identical subunits, each containing 6 enzymes and an ACP, arranged in an X shape (Figure 23–2). The position of the ACP and thioesterase domains cannot be resolved as yet by X-ray crystallography, possibly because they are too flexible, but they are thought to lie close to the 3-ketoacylreductase enzyme. The use of one multienzyme

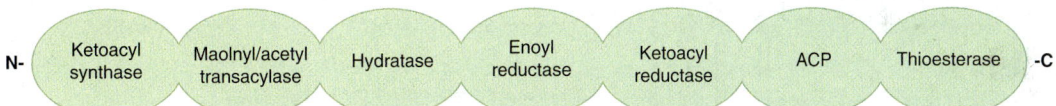

Sequence of enzyme domains in primary structure of fatty acid synthase monomer

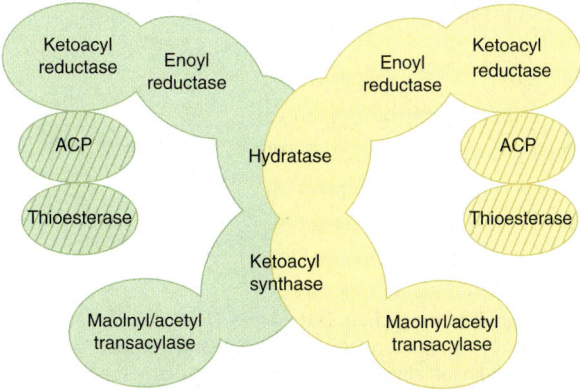

Fatty acid synthase homodimer

FIGURE 23–2 **Fatty acid synthase multienzyme complex.** The complex is a dimer of two identical polypeptide monomers in which six enzymes and the acyl carrier protein (ACP) are linked in the primary structure in the sequence shown. X-ray crystallography of the three-dimensional structure has demonstrated that the two monomers in the complex are arranged in an X-shape. The position of the ACP and thioesterase is not yet resolved, but they are thought to be close to the 3 ketoacylreductase enzyme domain.

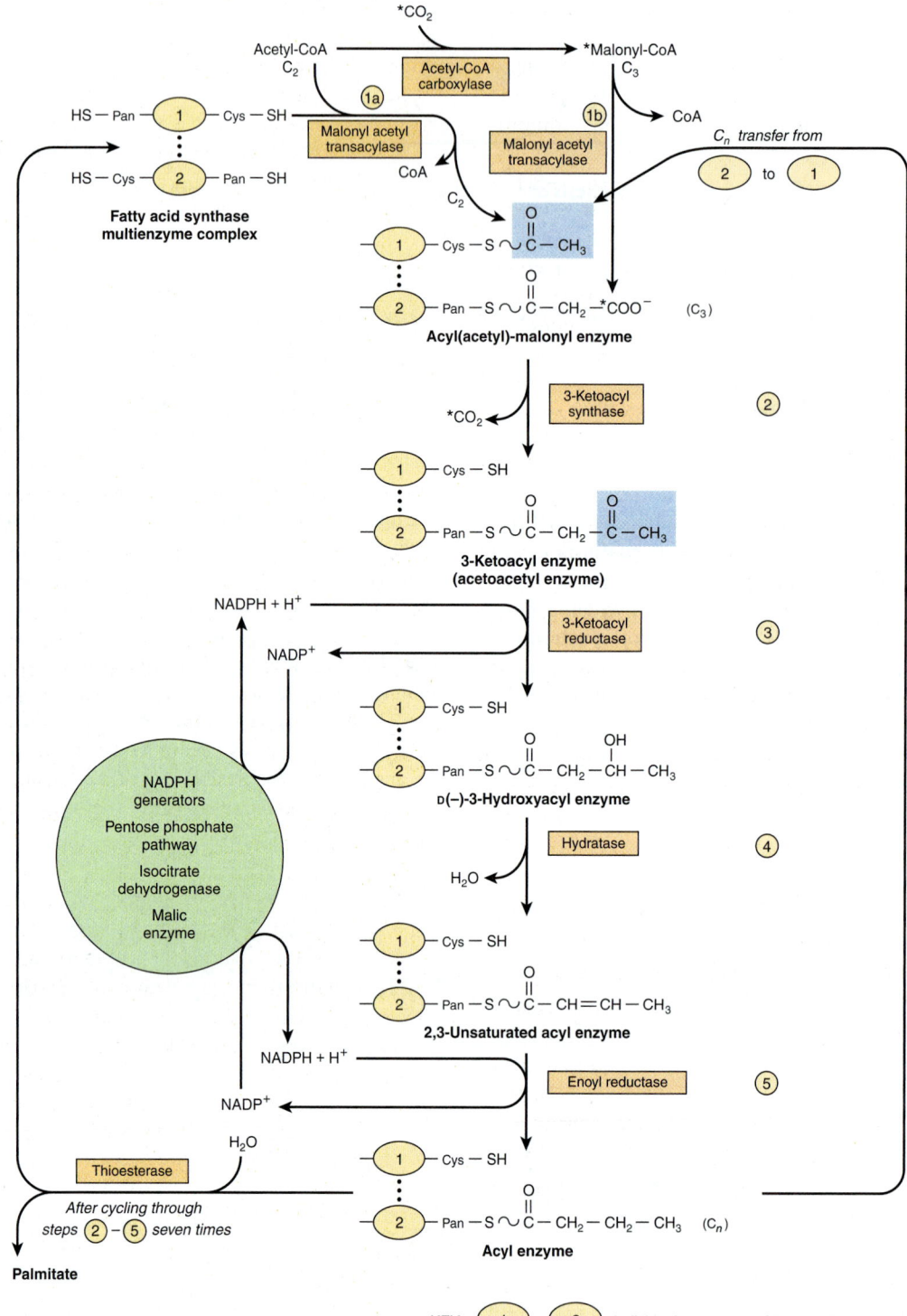

FIGURE 23–3 **Biosynthesis of long-chain fatty acids.** Details of how addition of a malonyl residue causes the acyl chain to grow by two carbon atoms. (Cys, cysteine residue; Pan, 4′-phosphopantetheine.) The blocks highlighted in blue contain initially a C_2 unit derived from acetyl-CoA (as illustrated) and subsequently the C_n unit formed in reaction 5.

functional unit has the advantages of achieving the effect of compartmentalization of the process within the cell without the erection of permeability barriers, and synthesis of all enzymes in the complex is coordinated since it is encoded by a single gene.

Initially, a priming molecule of acetyl-CoA combines with a cysteine —SH group (**Figure 23–3**, reaction 1a), while malonyl-CoA combines with the adjacent —SH on the 4′-phosphopantetheine of ACP of the other monomer (reaction 1b). These reactions are catalyzed by **malonyl acetyl transacylase,** to form **acetyl (acyl)-malonyl enzyme.** The acetyl group attacks the methylene group of the malonyl residue, catalyzed by **3-ketoacyl synthase,** and liberates CO_2, forming 3-ketoacyl enzyme (acetoacetyl enzyme) (reaction 2), freeing the cysteine —SH group. Decarboxylation allows the reaction to go to completion, pulling the whole sequence of reactions in the forward direction. The 3-ketoacyl group is reduced, dehydrated, and reduced again (reactions 3–5) to form the corresponding saturated acyl-S-enzyme. A new malonyl-CoA molecule combines with the —SH of 4′-phosphopantetheine, displacing the saturated acyl residue onto the free cysteine —SH group. The sequence of reactions is repeated six more times until a saturated 16-carbon acyl radical (palmitoyl) has been assembled. It is liberated from the enzyme complex by the activity of the sixth enzyme in the complex, **thioesterase** (deacylase). The free palmitate must be activated to acyl-CoA before it can proceed via any other metabolic pathway. Its possible fates are esterification into acylglycerols, chain elongation or de-saturation, or esterification to cholesteryl ester. In mammary gland, there is a separate thioesterase specific for acyl residues of C_8, C_{10}, or C_{12}, which are subsequently found in milk lipids.

The equation for the overall synthesis of palmitate from acetyl-CoA and malonyl-CoA is

$$CH_3CO—S—CoA + 7HOOCCHCO—S—CoA + 14NADPH + 14H^+$$
$$\rightarrow CH_3(CH_2)_{14}COOH + 7CO_2 + 6H_2O + 8CoA—SH + 14NADP^+$$

The acetyl-CoA used as a primer forms carbon atoms 15 and 16 of palmitate. The addition of all the subsequent C_2 units is via malonyl-CoA. Propionyl CoA acts as primer for the synthesis of long-chain fatty acids having an odd number of carbon atoms, found particularly in ruminant fat and milk.

The Main Source of NADPH for Lipogenesis Is the Pentose Phosphate Pathway

NADPH is involved as donor of reducing equivalents in both the reduction of the 3-ketoacyl and of the 2,3-unsaturated acyl derivatives (Figure 23–3, reactions 3 and 5). The oxidative reactions of the pentose phosphate pathway (see Chapter 21) are the chief source of the hydrogen required for the reductive synthesis of fatty acids. Significantly, tissues specializing in active lipogenesis—ie, liver, adipose tissue, and the lactating mammary gland—also possess an active pentose phosphate pathway. Moreover, both metabolic pathways are found in the cytosol of the cell; so, there are no membranes or permeability barriers against the transfer of NADPH. Other sources of NADPH include the reaction that converts malate to pyruvate catalyzed by the "**malic enzyme**" (NADP malate dehydrogenase) (**Figure 23–4**) and the extramitochondrial **isocitrate dehydrogenase** reaction (probably not a substantial source, except in ruminants).

Acetyl-CoA Is the Principal Building Block of Fatty Acids

Acetyl-CoA is formed from glucose via the oxidation of pyruvate within the mitochondria. However, it does not diffuse readily into the extramitochondrial cytosol, the principal site of fatty acid synthesis. Citrate, formed after condensation of acetyl-CoA with oxaloacetate in the citric acid cycle within mitochondria, is translocated into the extramitochondrial compartment via the tricarboxylate transporter, where in the presence of CoA and ATP, it undergoes cleavage to acetyl-CoA and oxaloacetate catalyzed by **ATP-citrate lyase,** which increases in activity in the well-fed state. The acetyl-CoA is then available for malonyl-CoA formation and synthesis to palmitate (Figure 23–4). The resulting oxaloacetate can form malate via NADH-linked malate dehydrogenase, followed by the generation of NADPH via the malic enzyme. The NADPH becomes available for lipogenesis, and the pyruvate can be used to regenerate acetyl-CoA after transport into the mitochondrion. This pathway is a means of transferring reducing equivalents from extramitochondrial NADH to NADP. Alternatively, malate itself can be transported into the mitochondrion, where it is able to re-form oxaloacetate. Note that the citrate (tricarboxylate) transporter in the mitochondrial membrane requires malate to exchange with citrate (see Figure 13–10). There is little ATP-citrate lyase or malic enzyme in ruminants, probably because in these species acetate (derived from carbohydrate digestion in the rumen and activated to acetyl-CoA extramitochondrially) is the main source of acetyl-CoA.

Elongation of Fatty Acid Chains Occurs in the Endoplasmic Reticulum

This pathway (**the "microsomal system"**) elongates saturated and unsaturated fatty acyl-CoAs (from C_{10} upward) by two carbons, using malonyl-CoA as the acetyl donor and NADPH as the reductant, and is catalyzed by the microsomal **fatty acid elongase** system of enzymes (**Figure 23–5**). Elongation of stearyl-CoA in brain increases rapidly during myelination in order to provide C_{22} and C_{24} fatty acids for sphingolipids.

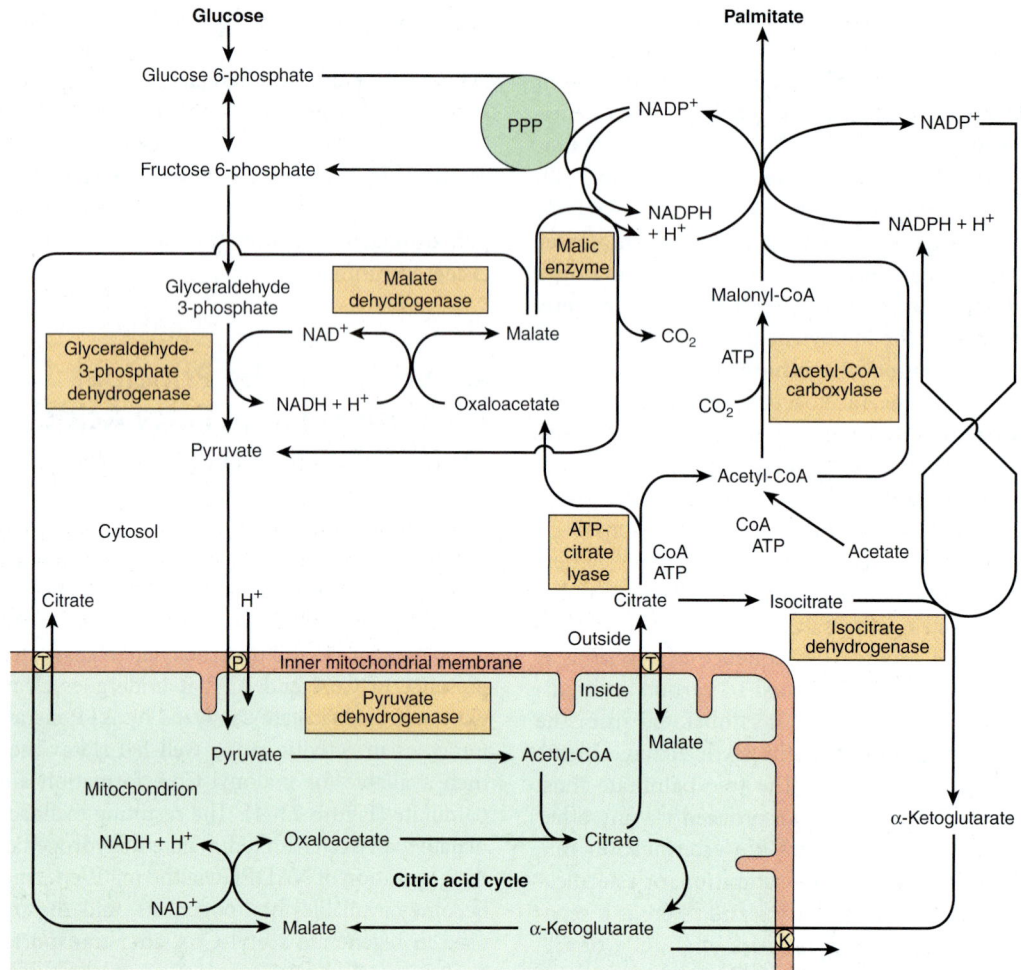

FIGURE 23–4 **The provision of acetyl-CoA and NADPH for lipogenesis.** (K, α-ketoglutarate transporter; P, pyruvate transporter; PPP, pentose phosphate pathway; T, tricarboxylate transporter.)

THE NUTRITIONAL STATE REGULATES LIPOGENESIS

Excess carbohydrate is stored as fat in many animals in anticipation of periods of caloric deficiency such as starvation, hibernation, etc, and to provide energy for use between meals in animals, including humans, that take their food at spaced intervals. Lipogenesis converts surplus glucose and intermediates such as pyruvate, lactate, and acetyl-CoA to fat, assisting the anabolic phase of this feeding cycle. The nutritional state of the organism is the main factor regulating the rate of lipogenesis. Thus, the rate is high in the well-fed animal whose diet contains a high proportion of carbohydrate. It is depressed by restricted caloric intake, high-fat diet, or a deficiency of insulin, as in diabetes mellitus. These latter conditions are associated with increased concentrations of plasma-free fatty acids, and an inverse relationship has been demonstrated between hepatic lipogenesis and the concentration of serum-free fatty acids. Lipogenesis is increased when sucrose is fed instead of glucose because fructose bypasses the phosphofructokinase control point in glycolysis and floods the lipogenic pathway (Figure 21–5).

SHORT- & LONG-TERM MECHANISMS REGULATE LIPOGENESIS

Long-chain fatty acid synthesis is controlled in the short term by allosteric and covalent modification of enzymes and in the long term by changes in gene expression governing rates of synthesis of enzymes.

Acetyl-CoA Carboxylase Is the Most Important Enzyme in the Regulation of Lipogenesis

Acetyl-CoA carboxylase is an allosteric enzyme and is activated by **citrate,** which increases in concentration in the well-fed state and is an indicator of a plentiful supply of acetyl-CoA. Citrate promotes the conversion of the enzyme from an inactive dimer to an active polymeric form, with a molecular mass of several million. Inactivation is promoted by phosphorylation of the enzyme and by long-chain acyl-CoA

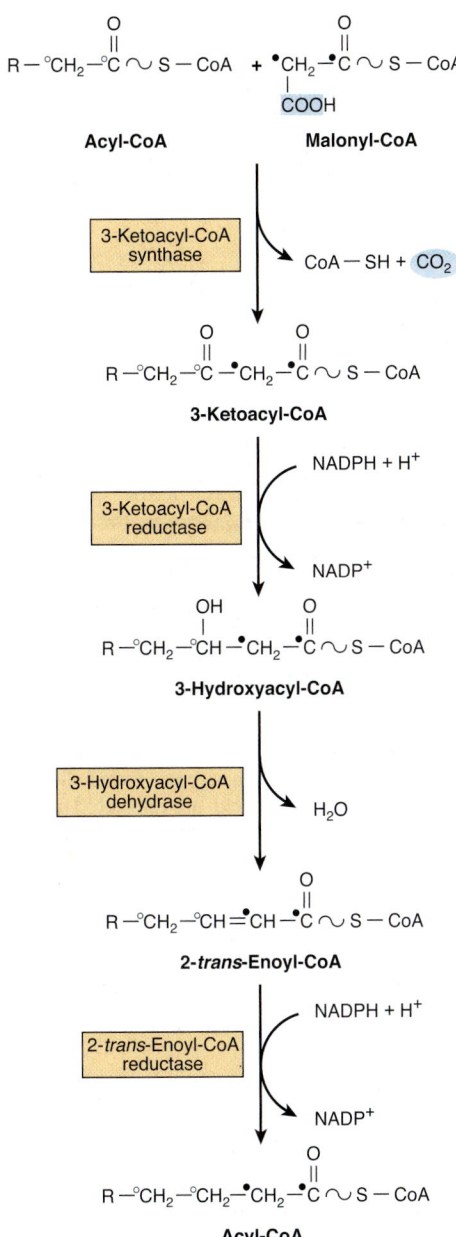

FIGURE 23–5 **Microsomal elongase system for fatty acid chain elongation.** NADH is also used by the reductases, but NADPH is preferred.

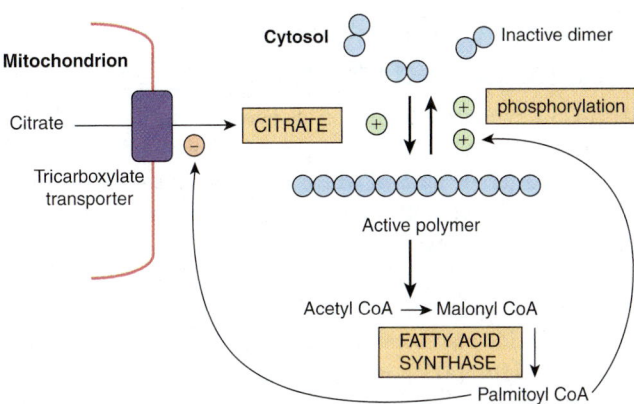

FIGURE 23–6 **Regulation of acetyl CoA carboxylase.** Acetyl-CoA carboxylase is activated by citrate, which promotes the conversion of the enzyme from an inactive dimer to an active polymeric form. Inactivation is promoted by phosphorylation of the enzyme and by long-chain acyl-CoA molecules such as palmitoyl CoA. In addition, acyl-CoA inhibits the tricarboxylate transporter, which transports citrate out of mitochondria into the cytosol, thus decreasing the citrate concentration in the cytosol and favoring inactivation of the enzyme.

molecules, an example of negative feedback inhibition by a product of a reaction (**Figure 23–6**). Thus, if acyl-CoA accumulates because it is not esterified quickly enough or because of increased lipolysis or an influx of free fatty acids into the tissue, it will automatically reduce the synthesis of new fatty acid. Acyl-CoA also inhibits the mitochondrial **tricarboxylate transporter,** thus preventing activation of the enzyme by egress of citrate from the mitochondria into the cytosol (Figure 23–6).

Acetyl-CoA carboxylase is also regulated by hormones such as **glucagon, epinephrine,** and **insulin** via changes in its phosphorylation state (details in **Figure 23–7**).

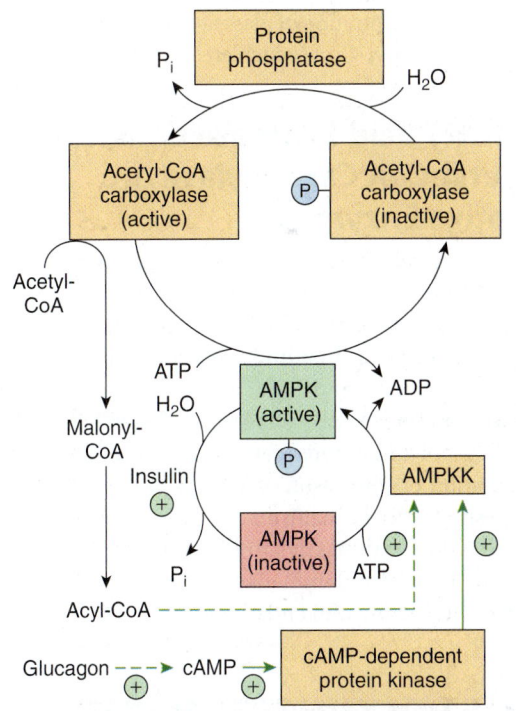

FIGURE 23–7 **Regulation of acetyl-CoA carboxylase by phosphorylation/dephosphorylation.** The enzyme is inactivated by phosphorylation by AMP-activated protein kinase (AMPK), which in turn is phosphorylated and activated by AMP-activated protein kinase kinase (AMPKK). Glucagon (and epinephrine) increase cAMP, and thus activate this latter enzyme via cAMP-dependent protein kinase. The kinase kinase enzyme is also believed to be activated by acyl-CoA. Insulin activates acetyl-CoA carboxylase via dephosphorylation of AMPK.

Pyruvate Dehydrogenase Is Also Regulated by Acyl-CoA

Acyl-CoA causes an inhibition of pyruvate dehydrogenase by inhibiting the ATP–ADP exchange transporter of the inner mitochondrial membrane, which leads to increased intramitochondrial (ATP)/(ADP) ratios and therefore to conversion of active to inactive pyruvate dehydrogenase (see Figure 18–6), thus regulating the availability of acetyl-CoA for lipogenesis. Furthermore, oxidation of acyl-CoA due to increased levels of free fatty acids may increase the ratios of (acetyl-CoA)/(CoA) and (NADH)/(NAD$^+$) in mitochondria, inhibiting pyruvate dehydrogenase.

Insulin Also Regulates Lipogenesis by Other Mechanisms

Insulin stimulates lipogenesis by several other mechanisms as well as by increasing acetyl-CoA carboxylase activity. It increases the transport of glucose into the cell (eg, in adipose tissue), increasing the availability of both pyruvate for fatty acid synthesis and glycerol 3-phosphate for esterification of the newly formed fatty acids, and also converts the inactive form of pyruvate dehydrogenase to the active form in adipose tissue, but not in liver. Insulin also—by its ability to depress the level of intracellular cAMP—**inhibits lipolysis** in adipose tissue and reducing the concentration of plasma-free fatty acids and, therefore, long-chain acyl-CoA, which are inhibitors of lipogenesis.

The Fatty Acid Synthase Complex & Acetyl-CoA Carboxylase Are Adaptive Enzymes

These enzymes adapt to the body's physiologic needs by increasing in total amount in the fed state and by decreasing in during intake of a high-fat diet and in conditions such as starvation, and diabetes mellitus. **Insulin** is an important hormone causing gene expression and induction of enzyme biosynthesis, and **glucagon** (via cAMP) antagonizes this effect. Feeding fats containing polyunsaturated fatty acids coordinately regulates the inhibition of expression of key enzymes of glycolysis and lipogenesis. These mechanisms for longer term regulation of lipogenesis take several days to become fully manifested and augment the direct and immediate effect of free fatty acids and hormones such as insulin and glucagon.

SOME POLYUNSATURATED FATTY ACIDS CANNOT BE SYNTHESIZED BY MAMMALS & ARE NUTRITIONALLY ESSENTIAL

Certain long-chain unsaturated fatty acids of metabolic significance in mammals are shown in **Figure 23–8**. Other C_{20}, C_{22}, and C_{24} polyenoic fatty acids may be derived from oleic, linoleic,

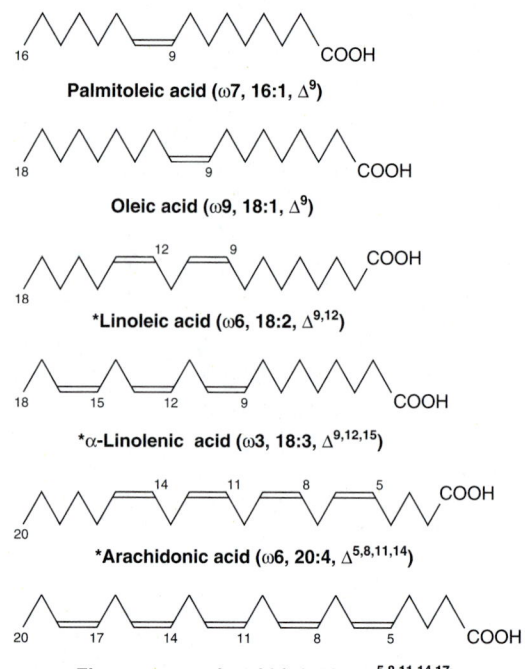

FIGURE 23–8 **Structure of some unsaturated fatty acids.** Although the carbon atoms in the molecules are conventionally numbered—ie, numbered from the carboxyl terminal—the ω numbers (eg, ω7 in palmitoleic acid) are calculated from the reverse end (the methyl terminal) of the molecules. The information in parentheses shows, for instance, that α-linolenic acid contains double bonds starting at the third carbon from the methyl terminal, has 18 carbons and 3 double bonds, and has these double bonds at the 9th, 12th, and 15th carbons from the carboxyl terminal. (*Classified as "essential fatty acids.")

and α-linolenic acids by chain elongation. Palmitoleic and oleic acids are not essential in the diet because the tissues can introduce a double bond at the Δ^9 position of a saturated fatty acid. **Linoleic and α-linolenic acids** are the only fatty acids known to be essential for the complete nutrition of many species of animals, including humans, and are termed the **nutritionally essential fatty acids**. In most mammals, **arachidonic acid** can be formed from linoleic acid (Figure 23–11). Double bonds can be introduced at the Δ^4, Δ^5, Δ^6, and Δ^9 positions (see Chapter 15) in most animals, but never beyond the Δ^9 position. In contrast, plants are able to synthesize the nutritionally essential fatty acids by introducing double bonds at the Δ^{12} and Δ^{15} positions.

MONOUNSATURATED FATTY ACIDS ARE SYNTHESIZED BY A Δ^9 DESATURASE SYSTEM

Several tissues including the liver are considered to be responsible for the formation of nonessential monounsaturated fatty acids from saturated fatty acids. The first double bond introduced into a saturated fatty acid is nearly always in the Δ^9 position. An enzyme system—Δ^9 **desaturase** (**Figure 23–9**)—in the endo-

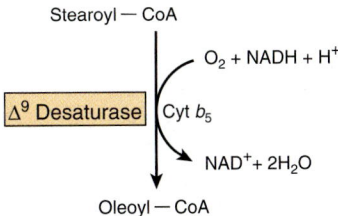

FIGURE 23–9 Microsomal Δ^9 desaturase.

plasmic reticulum catalyzes the conversion of palmitoyl-CoA or stearoyl-CoA to palmitoleoyl-CoA or oleoyl-CoA, respectively. Oxygen and either NADH or NADPH are necessary for the reaction. The enzymes appear to be similar to a mono-oxygenase system involving cytochrome b_5 (Chapter 12).

SYNTHESIS OF POLYUNSATURATED FATTY ACIDS INVOLVES DESATURASE & ELONGASE ENZYME SYSTEMS

Additional double bonds introduced into existing monounsaturated fatty acids are always separated from each other by a methylene group (methylene interrupted) except in bacteria. Since animals have a Δ^9 desaturase, they are able to synthesize the ω9 (oleic acid) family of unsaturated fatty acids completely by a combination of chain elongation and desaturation (**Figure 23–10**). However, as indicated above, linoleic (ω6) or α-linolenic (ω3) acids required for the synthesis of the other members of the ω6 or ω3 families must be supplied in the diet. Linoleate may be converted to arachidonate via **γ-linolenate** by the pathway shown in **Figure 23–11**. The nutritional requirement for arachidonate may thus be dispensed with if there is adequate linoleate in the diet. The desaturation and chain elongation system is greatly diminished in the starving state, in response to glucagon and epinephrine administration, and in the absence of insulin as in type 1 diabetes mellitus.

DEFICIENCY SYMPTOMS ARE PRODUCED WHEN THE ESSENTIAL FATTY ACIDS (EFA) ARE ABSENT FROM THE DIET

Rats fed a purified nonlipid diet containing vitamins A and D exhibit a reduced growth rate and reproductive deficiency which may be cured by the addition of **linoleic, α-linolenic, and arachidonic acids** to the diet. These fatty acids are found in high concentrations in vegetable oils (Table 15–2) and in small amounts in animal carcasses. Essential fatty acids are required for prostaglandin, thromboxane, leukotriene, and lipoxin formation (see below), and they also have various other functions that are less well defined. They are found in the structural lipids of the cell, often in the 2 position of phospholipids, and are concerned with the structural integrity of the mitochondrial membrane.

Arachidonic acid is present in membranes and accounts for 5–15% of the fatty acids in phospholipids. Docosahexaenoic acid (DHA; ω3, 22:6), which is synthesized to a limited extent from α-linolenic acid or obtained directly from fish oils, is present in high concentrations in retina, cerebral cortex, testis, and sperm. DHA is particularly needed for development of the brain and retina and is supplied via the placenta and milk. Patients with **retinitis pigmentosa** are reported to have low blood levels of DHA. In **essential fatty acid deficiency,** nonessential polyenoic acids of the ω9 family, particularly $\Delta^{5,8,11}$-eicosatrienoic acid (ω9 20:3) (Figure 23–10), replace the essential fatty acids in phospholipids, other complex lipids, and membranes. The triene:tetraene ratio in plasma lipids can be used to diagnose the extent of essential fatty acid deficiency.

Trans Fatty Acids Are Implicated in Various Disorders

Small amounts of trans-unsaturated fatty acids are found in ruminant fat (eg, butter fat has 2–7%), where they arise from the action of micro-organisms in the rumen, but the main source

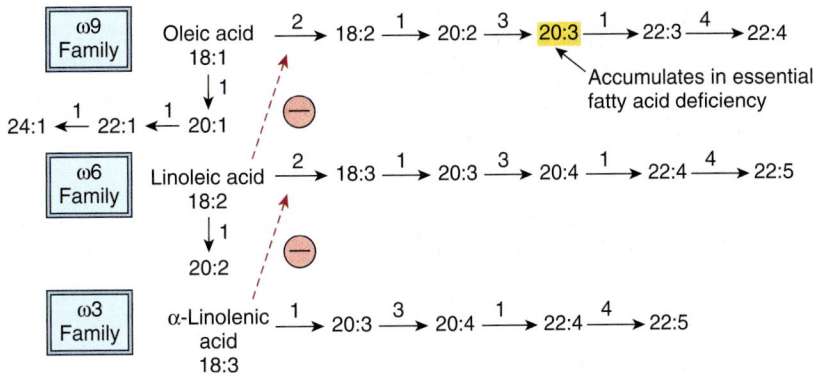

FIGURE 23–10 Biosynthesis of the ω9, ω6, and ω3 families of polyunsaturated fatty acids. Each step is catalyzed by the microsomal chain elongation or desaturase system: 1, elongase; 2, Δ^6 desaturase; 3, Δ^5 desaturase; 4, Δ^4 desaturase. (⊖, Inhibition.)

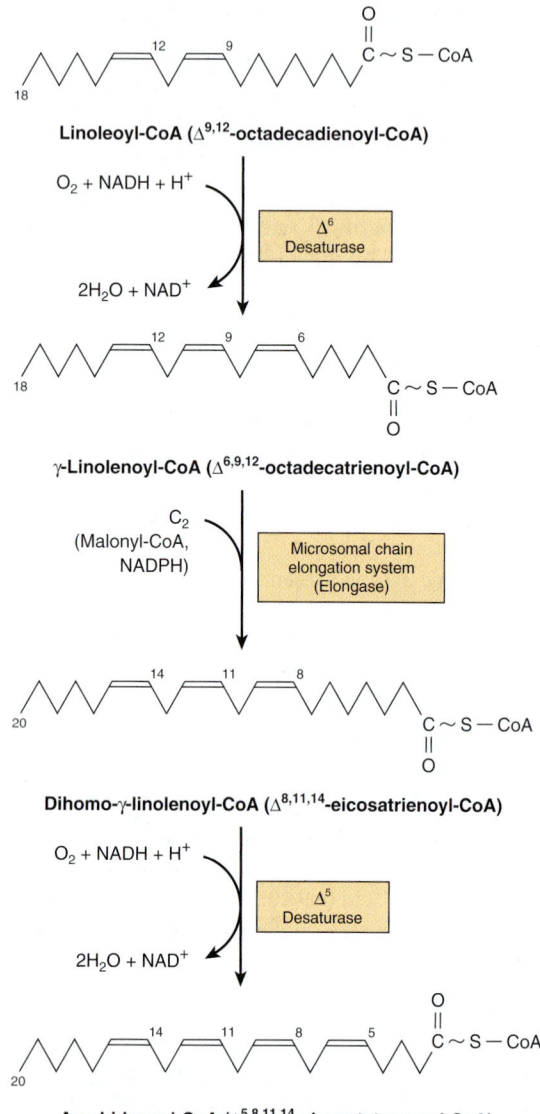

FIGURE 23–11 **Conversion of linoleate to arachidonate.** Cats cannot carry out this conversion owing to the absence of Δ^6 desaturase and must obtain arachidonate in their diet.

in the human diet is from partially hydrogenated vegetable oils (eg, margarine). Trans fatty acids compete with essential fatty acids and may exacerbate essential fatty acid deficiency. Moreover, they are structurally similar to saturated fatty acids (Chapter 15) and have comparable effects in the promotion of hypercholesterolemia and atherosclerosis (Chapter 26).

EICOSANOIDS ARE FORMED FROM C₂₀ POLYUNSATURATED FATTY ACIDS

Arachidonate and some other C_{20} polyunsaturated fatty acids give rise to **eicosanoids,** physiologically and pharmacologically active compounds known as **prostaglandins (PG),**

thromboxanes (TX), leukotrienes (LT), and lipoxins (LX) (Chapter 15). Physiologically, they are considered to act as local hormones functioning through G-protein-linked receptors to elicit their biochemical effects.

There are three groups of eicosanoids that are synthesized from C_{20} eicosanoic acids derived from the essential fatty acids **linoleate** and **α-linolenate,** or directly from dietary arachidonate and eicosapentaenoate (**Figure 23–12**). Arachidonate, which may be obtained from the diet, but is usually derived from the 2 position of phospholipids in the plasma membrane by the action of phospholipase A₂ (Figure 24–6), is the substrate for the synthesis of the PG₂, TX₂ series (**prostanoids**) by the **cyclooxygenase pathway,** or the LT₄ and LX₄ series by the **lipoxygenase pathway,** with the two pathways competing for the arachidonate substrate (Figure 23–11).

THE CYCLOOXYGENASE PATHWAY IS RESPONSIBLE FOR PROSTANOID SYNTHESIS

Prostanoid synthesis (**Figure 23–13**) involves the consumption of two molecules of O_2 catalyzed by **cyclooxygenase (COX)** (also called **prostaglandin H synthase**), an enzyme that has two activities, a **cyclooxygenase** and **peroxidase.** COX is present as two isoenzymes, **COX-1 and COX-2.** The product, an endoperoxide (PGH), is converted to prostaglandins D and E as well as to a thromboxane (TXA₂) and prostacyclin (PGI₂). Each cell type produces only one type of prostanoid. The NSAID **aspirin** inhibits COX-1 and COX-2. Other NSAIDs include **indomethacin** and **ibuprofen,** and usually inhibit cyclooxygenases by competing with arachidonate. Since inhibition of COX-1 causes the stomach irritation often associated with taking NSAIDs, attempts have been made to develop drugs which selectively inhibit COX-2 (**coxibs**). Unfortunately, however, the success of this approach has been limited and some coxibs have been withdrawn or suspended from the market due to undesirable side effects and safety issues. Transcription of COX-2—but not of COX-1—is completely inhibited by **anti-inflammatory corticosteroids.**

Essential Fatty Acids Do Not Exert All Their Physiologic Effects Via Prostaglandin Synthesis

The role of essential fatty acids in membrane formation is unrelated to prostaglandin formation. Prostaglandins do not relieve symptoms of essential fatty acid deficiency, and an essential fatty acid deficiency is not caused by inhibition of prostaglandin synthesis.

Cyclooxygenase Is a "Suicide Enzyme"

"Switching off" of prostaglandin activity is partly achieved by a remarkable property of cyclooxygenase—that of self-catalyzed

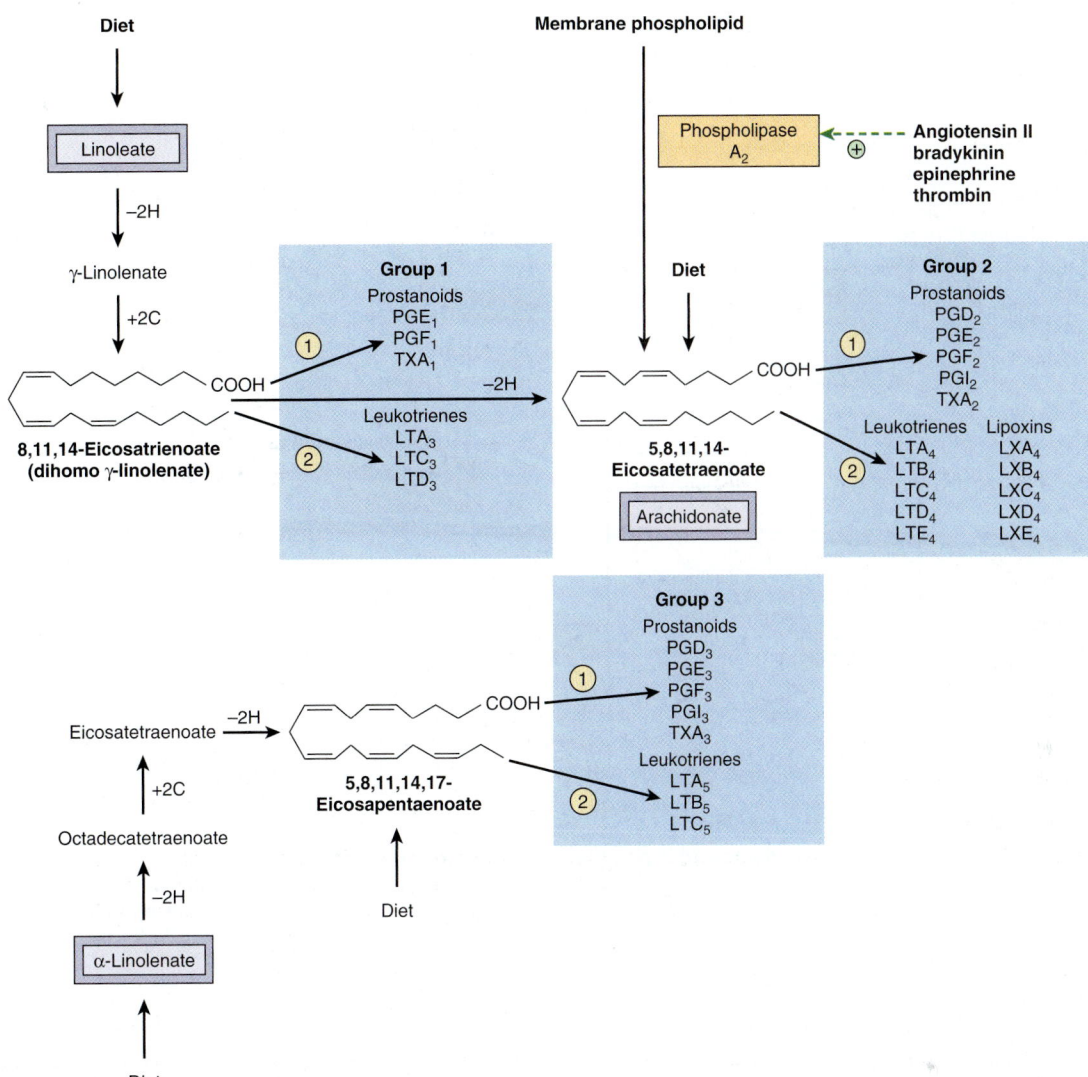

FIGURE 23–12 **The three groups of eicosanoids and their biosynthetic origins.** (①, cyclooxygenase pathway; ②, lipoxygenase pathway; LT, leukotriene; LX, lipoxin; PG, prostaglandin; PGI, prostacyclin; TX, thromboxane.) The subscript denotes the total number of double bonds in the molecule and the series to which the compound belongs.

destruction; ie, it is a **"suicide enzyme."** Furthermore, the inactivation of prostaglandins by **15-hydroxyprostaglandin dehydrogenase** is rapid. Blocking the action of this enzyme with sulfasalazine or indomethacin can prolong the half-life of prostaglandins in the body.

LEUKOTRIENES & LIPOXINS ARE FORMED BY THE LIPOXYGENASE PATHWAY

The **leukotrienes** are a family of conjugated trienes formed from eicosanoic acids in leukocytes, mastocytoma cells, platelets, and macrophages by the **lipoxygenase pathway** in response to both immunologic and nonimmunologic stimuli. Three different lipoxygenases (dioxygenases) insert oxygen into the 5, 12, and 15 positions of arachidonic acid, giving

rise to hydroperoxides (HPETE). Only **5-lipoxygenase** forms leukotrienes (details in **Figure 23–14**). **Lipoxins** are a family of conjugated tetraenes also arising in leukocytes. They are formed by the combined action of more than one lipoxygenase (Figure 23–14).

CLINICAL ASPECTS

Symptoms of Essential Fatty Acid Deficiency in Humans Include Skin Lesions & Impairment of Lipid Transport

In adults subsisting on ordinary diets, no signs of essential fatty acid deficiencies have been reported. However, infants receiving formula diets low in fat and patients maintained for long

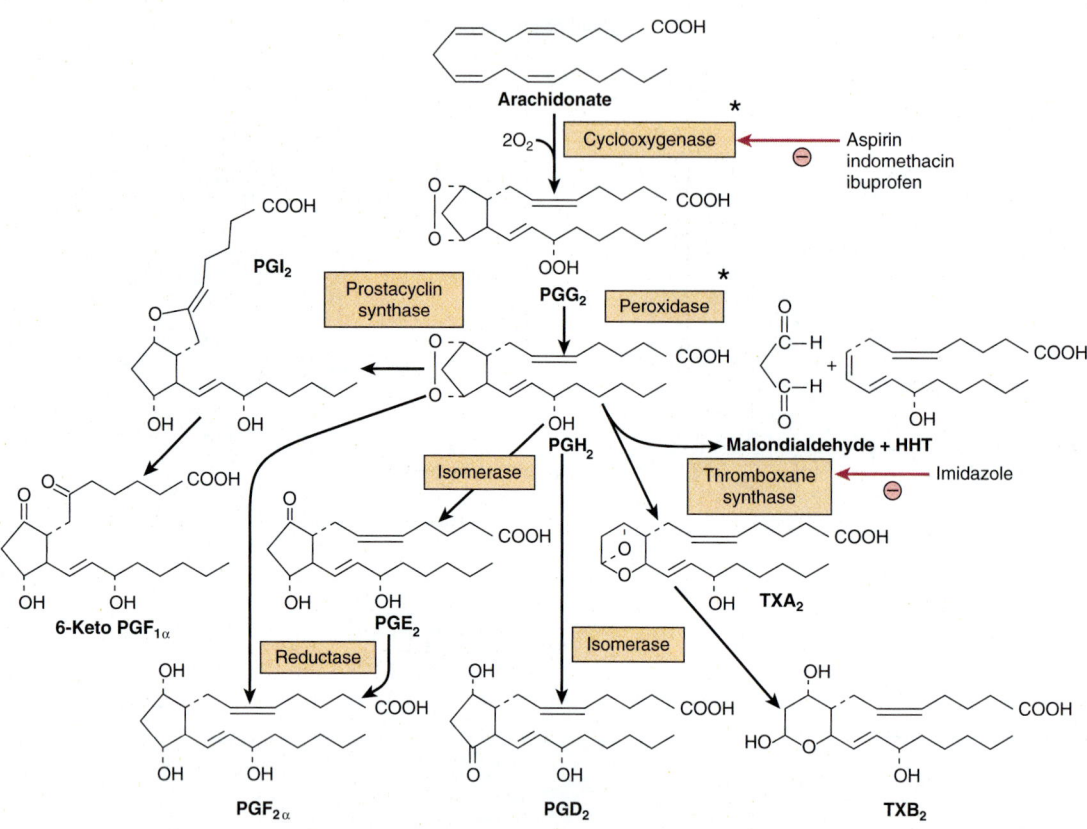

FIGURE 23–13 **Conversion of arachidonic acid to prostaglandins and thromboxanes of series 2.** (HHT, hydroxyheptadecatrienoate; PG, prostaglandin; PGI, prostacyclin; TX, thromboxane.) (*Both of these starred activities are attributed to the cyclooxygenase enzyme [prostaglandin H synthase]. Similar conversions occur in prostaglandins and thromboxanes of series 1 and 3.)

periods exclusively by intravenous nutrition low in essential fatty acids show deficiency symptoms that can be prevented by an essential fatty acid intake of 1–2% of the total caloric requirement.

Abnormal Metabolism of Essential Fatty Acids Occurs in Several Diseases

Abnormal metabolism of essential fatty acids, which may be connected with dietary insufficiency, has been noted in cystic fibrosis, acrodermatitis enteropathica, hepatorenal syndrome, Sjögren–Larsson syndrome, multisystem neuronal degeneration, Crohn's disease, cirrhosis and alcoholism, and Reye's syndrome. Elevated levels of very long chain polyenoic acids have been found in the brains of patients with Zellweger's syndrome (Chapter 22). Diets with a high P:S (polyunsaturated:saturated fatty acid) ratio reduce serum cholesterol levels and are considered to be beneficial in terms of the risk of development of coronary heart disease.

Prostanoids Are Potent, Biologically Active Substances

Thromboxanes are synthesized in platelets and upon release cause vasoconstriction and platelet aggregation. Their

synthesis is specifically inhibited by low-dose aspirin. **Prostacyclins (PGI$_2$)** are produced by blood vessel walls and are potent inhibitors of platelet aggregation. Thus, thromboxanes and prostacyclins are antagonistic. PG$_3$ and TX$_3$, formed from eicosapentaenoic acid (EPA), inhibit the release of arachidonate from phospholipids and the formation of PG$_2$ and TX$_2$. PGI$_3$ is as potent an antiaggregator of platelets as PGI$_2$, but TXA$_3$ is a weaker aggregator than TXA$_2$, changing the balance of activity and favoring longer clotting times. As little as 1 ng/mL of plasma prostaglandins causes contraction of smooth muscle in animals. Potential therapeutic uses include prevention of conception, induction of labor at term, termination of pregnancy, prevention or alleviation of gastric ulcers, control of inflammation and of blood pressure, and relief of asthma and nasal congestion. In addition, PGD$_2$ is a potent sleep-promoting substance. Prostaglandins increase cAMP in platelets, thyroid, corpus luteum, fetal bone, adenohypophysis, and lung but reduce cAMP in renal tubule cells and adipose tissue (Chapter 25).

Leukotrienes & Lipoxins Are Potent Regulators of Many Disease Processes

Slow-reacting substance of anaphylaxis (**SRS-A**) is a mixture of leukotrienes C$_4$, D$_4$, and E$_4$. This mixture of leukotrienes

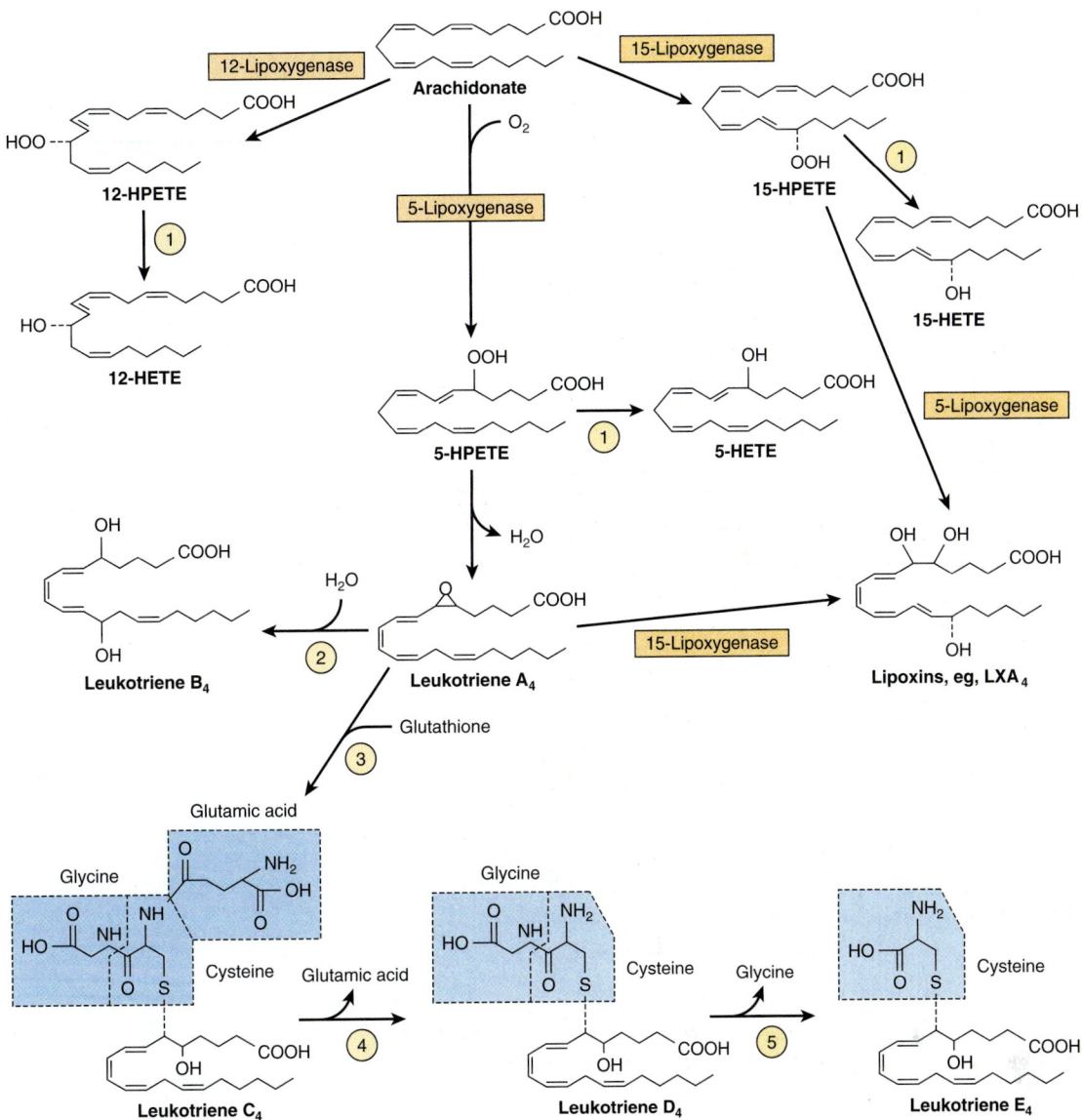

FIGURE 23–14 **Conversion of arachidonic acid to leukotrienes and lipoxins of series 4 via the lipoxygenase pathway.** Some similar conversions occur in series 3 and 5 leukotrienes. (①, peroxidase; ②, leukotriene A₄ epoxide hydrolase; ③, glutathione *S*-transferase; ④, γ-glutamyltranspeptidase; ⑤, cysteinyl-glycine dipeptidase; HETE, hydroxyeicosatetraenoate; HPETE, hydroperoxyeicosatetraenoate.)

is a potent constrictor of the bronchial airway musculature. These leukotrienes together with **leukotriene B₄** also cause vascular permeability and attraction and activation of leukocytes and are important regulators in many diseases involving inflammatory or immediate hypersensitivity reactions, such as asthma. Leukotrienes are vasoactive, and 5-lipoxygenase has been found in arterial walls. Evidence supports an anti-inflammatory role for lipoxins in vasoactive and immunoregulatory function, eg, as counter-regulatory compounds (**chalones**) of the immune response.

SUMMARY

■ The synthesis of long-chain fatty acids (lipogenesis) is carried out by two enzyme systems: acetyl-CoA carboxylase and fatty acid synthase.

■ The pathway converts acetyl-CoA to palmitate and requires NADPH, ATP, Mn²⁺, biotin, and pantothenic acid as cofactors.

■ Acetyl-CoA carboxylase converts acetyl-CoA to malonyl-CoA, and then fatty acid synthase, a multienzyme complex consisting of two identical polypeptide chains, each containing six separate enzymatic activities and ACP, catalyzes the formation of palmitate from one acetyl-CoA and seven malonyl-CoA molecules.

■ Lipogenesis is regulated at the acetyl-CoA carboxylase step by allosteric modifiers, phosphorylation/dephosphorylation, and induction and repression of enzyme synthesis. The enzyme is allosterically activated by citrate and deactivated by long-chain acyl-CoA. Dephosphorylation (eg, by insulin) promotes its activity, while phosphorylation (eg, by glucagon or epinephrine) is inhibitory.

- Biosynthesis of unsaturated long-chain fatty acids is achieved by desaturase and elongase enzymes, which introduce double bonds and lengthen existing acyl chains, respectively.

- Higher animals have Δ^4, Δ^5, Δ^6, and Δ^9 desaturases but cannot insert new double bonds beyond the 9 position of fatty acids. Thus, the essential fatty acids linoleic ($\omega 6$) and α-linolenic ($\omega 3$) must be obtained from the diet.

- Eicosanoids are derived from C_{20} (eicosanoic) fatty acids synthesized from the essential fatty acids and make up important groups of physiologically and pharmacologically active compounds, including the prostaglandins, thromboxanes, leukotrienes, and lipoxins.

REFERENCES

Fischer S: Dietary polyunsaturated fatty acids and eicosanoid formation in humans. Adv Lipid Res 1989;23:169.

Fitzpatrick FA: Cyclooxygenase enzymes: regulation and function. Curr Pharm Des 2004;10:577.

McMahon B, Mitchell S, Brady HR, et al: Lipoxins: revelations on resolution. Trends Pharmacol Sci 2001;22:391.

Miyazaki M, Ntambi JM: Fatty acid desaturation and chain elongation in mammals. In: *Biochemistry of Lipids, Lipoproteins and Membranes,* 5th ed. Vance DE, Vance JE (editors). Elsevier, 2008;191–212.

Sith S, Witkowski A, Joshi AK: Structural and functional organisation of the animal fatty acid synthase. Prog Lipid Res 2003;42:289.

Smith WL, Murphy RC: The eicosanoids: cyclooxygenase, lipoxygenase, and epoxygenase pathways. In: *Biochemistry of Lipids, Lipoproteins and Membranes,* 5th ed. Vance DE, Vance JE (editors). Elsevier, 2008;331–362.

Sul HS, Smith S: Fatty acid synthesis in eukaryotes. In: *Biochemistry of Lipids, Lipoproteins and Membranes,* 5th ed. Vance DE, Vance JE (editors). Elsevier, 2008;155–190.

Tong L: Acetyl-coenzyme A carboxylase: crucial metabolic enzyme and an attractive target for drug discovery. Cell Mol Life Sci 2005;62:1784.

Wijendran V, Hayes KC: Dietary n-6 and n-3 fatty acid balance and cardiovascular health. Annu Rev Nutr 2004;24:597.

Metabolism of Acylglycerols & Sphingolipids

Kathleen M. Botham, PhD, DSc & Peter A. Mayes, PhD, DSc

OBJECTIVES

After studying this chapter, you should be able to:

- Appreciate that the catabolism of triacylglycerols involves hydrolysis by a lipase to free fatty acids and glycerol and indicate the fate of these metabolites.
- Understand that glycerol-3-phosphate is the substrate for the formation of both triacylglycerols and phosphoglycerols and that a branch point at phosphatidate leads to the synthesis of inositol phospholipids and cardiolipin via one branch and triacylglycerols and other phospholipids via the second branch.
- Explain that plasmalogens and platelet activating factor (PAF) are formed by a complex pathway starting from dihydroxyacetone phosphate.
- Illustrate the role of various phospholipases in the degradation and remodeling of phospholipids.
- Appreciate that ceramide is produced from the amino acid serine and is the precursor from which all sphingolipids are formed.
- Indicate how sphingomyelin and glycosphingolipids are produced by reacting ceramide with phosphatidylcholine (with the release of diacylglycerol) or sugar residue(s), respectively.
- Identify examples of disease processes caused by defects in phospholipid or sphingolipid synthesis or breakdown.

BIOMEDICAL IMPORTANCE

Acylglycerols constitute the majority of lipids in the body. Triacylglycerols are the major lipids in fat deposits and in food, and their roles in lipid transport and storage and in various diseases such as obesity, diabetes, and hyperlipoproteinemia will be described in subsequent chapters. The amphipathic nature of phospholipids and sphingolipids makes them ideally suitable as the main lipid component of cell membranes. Phospholipids also take part in the metabolism of many other lipids. Some phospholipids have specialized functions; eg, dipalmitoyl lecithin is a major component of **lung surfactant,** which is lacking in **respiratory distress syndrome** of the newborn. Inositol phospholipids in the cell membrane act as precursors of **hormone second messengers,** and **platelet-activating factor** is an alkylphospholipid. Glycosphingolipids, containing sphingosine and sugar residues as well as fatty acid that are found in the outer leaflet of the plasma membrane with their oligosaccharide chains facing outward, form part of the glycocalyx of the cell surface and are important (1) in cell adhesion and cell recognition, (2) as receptors for bacterial toxins (eg, the toxin that causes cholera), and (3) as ABO blood group substances. A dozen or so **glycolipid storage diseases** have been described (eg, Gaucher's disease and Tay-Sachs disease), each due to a genetic defect in the pathway for glycolipid degradation in the lysosomes.

HYDROLYSIS INITIATES CATABOLISM OF TRIACYLGLYCEROLS

Triacylglycerols must be hydrolyzed by a **lipase** to their constituent fatty acids and glycerol before further catabolism can proceed. Much of this hydrolysis (lipolysis) occurs in adipose tissue with release of free fatty acids into the plasma, where they are found combined with serum albumin (Figure 25–7). This is followed by free fatty acid uptake into tissues (including liver, heart, kidney, muscle, lung, testis, and adipose tissue, but not readily by brain), where they are oxidized or re-esterified. The utilization of glycerol depends upon whether such tissues have the enzyme **glycerol kinase,** which is found in significant

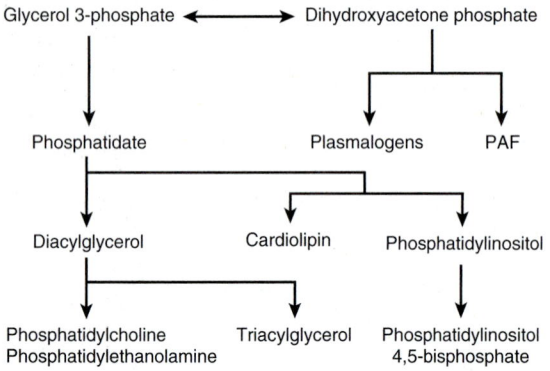

FIGURE 24–1 **Overview of acylglycerol biosynthesis.** (PAF, platelet-activating factor.)

amounts in liver, kidney, intestine, brown adipose tissue, and the lactating mammary gland.

TRIACYLGLYCEROLS & PHOSPHOGLYCEROLS ARE FORMED BY ACYLATION OF TRIOSE PHOSPHATES

The major pathways of triacylglycerol and phosphoglycerol biosynthesis are outlined in **Figure 24–1**. Important substances such as triacylglycerols, phosphatidylcholine, phosphatidylethanolamine, phosphatidylinositol, and cardiolipin, a constituent of mitochondrial membranes, are formed from **glycerol-3-phosphate.** Significant branch points in the pathway occur at the phosphatidate and diacylglycerol steps. Phosphoglycerols containing an ether link (—C—O—C—), the best known of which are plasmalogens and platelet-activating factor (PAF), are derived from dihydroxyacetone phosphate. Glycerol 3-phosphate and dihydroxyacetone phosphate are intermediates in glycolysis, making a very important connection between carbohydrate and lipid metabolism (see Chapter 16).

Phosphatidate Is the Common Precursor in the Biosynthesis of Triacylglycerols, Many Phosphoglycerols, & Cardiolipin

Both glycerol and fatty acids must be activated by ATP before they can be incorporated into acylglycerols. **Glycerol kinase** catalyzes the activation of glycerol to *sn*-glycerol 3-phosphate. If the activity of this enzyme is absent or low, as in muscle or adipose tissue, most of the glycerol 3-phosphate is formed from dihydroxyacetone phosphate by **glycerol-3-phosphate dehydrogenase** (**Figure 24–2**).

Biosynthesis of Triacylglycerols

Two molecules of acyl-CoA, formed by the activation of fatty acids by **acyl-CoA synthetase** (Chapter 22),

combine with glycerol 3-phosphate to form **phosphatidate** (1,2-diacylglycerol phosphate). This takes place in two stages, catalyzed by **glycerol-3-phosphate acyltransferase** and **1-acylglycerol-3-phosphate acyltransferase.** Phosphatidate is converted by **phosphatidate phosphohydrolase** and **diacylglycerol acyltransferase (DGAT)** to 1,2-diacylglycerol and then triacylglycerol. DGAT catalyzes the only step specific for triacylglycerol synthesis and is thought to be rate limiting in most circumstances. In intestinal mucosa, **monoacylglycerol acyltransferase** converts **monoacylglycerol** to 1,2-diacylglycerol in the **monoacylglycerol pathway.** Most of the activity of these enzymes resides in the endoplasmic reticulum, but some is found in mitochondria. Although phosphatidate phosphohydrolase protein is found mainly in the cytosol, the active form of the enzyme is membrane bound.

In the biosynthesis of phosphatidylcholine and phosphatidylethanolamine (Figure 24–2), choline or ethanolamine must first be activated by phosphorylation by ATP followed by linkage to CDP. The resulting CDP-choline or CDP-ethanolamine reacts with 1,2-diacylglycerol to form either phosphatidylcholine or phosphatidylethanolamine, respectively. Phosphatidylserine is formed from phosphatidylethanolamine directly by reaction with serine (Figure 24–2). Phosphatidylserine may re-form phosphatidylethanolamine by decarboxylation. An alternative pathway in liver enables phosphatidylethanolamine to give rise directly to phosphatidylcholine by progressive methylation of the ethanolamine residue. In spite of these sources of choline, it is considered to be an essential nutrient in many mammalian species, although this has not been established in humans.

The regulation of triacylglycerol, phosphatidylcholine, and phosphatidylethanolamine biosynthesis is driven by the availability of free fatty acids. Those that escape oxidation are preferentially converted to phospholipids, and when this requirement is satisfied they are used for triacylglycerol synthesis.

Cardiolipin (diphosphatidylglycerol; Figure 15–10) is a phospholipid present in mitochondria. It is formed from phosphatidylglycerol, which in turn is synthesized from CDP-diacylglycerol (Figure 24–2) and glycerol 3-phosphate according to the scheme shown in **Figure 24–3**. Cardiolipin, found in the inner membrane of mitochondria, has a key role in mitochondrial structure and function, and is also thought to be involved in programmed cell death (**apoptosis**).

Biosynthesis of Glycerol Ether Phospholipids

This pathway is located in peroxisomes. Dihydroxyacetone phosphate is the precursor of the glycerol moiety of glycerol ether phospholipids (**Figure 24–4**). This compound combines with acyl-CoA to give 1-acyldihydroxyacetone phosphate. The ether link is formed in the next reaction, producing 1-alkyldihydroxyacetone phosphate, which is then converted to 1-alkylglycerol 3-phosphate. After further acylation in the 2 position, the resulting 1-alkyl-2-acylglycerol 3-phosphate (analogous to phosphatidate in Figure 24–2) is hydrolyzed to give the free glycerol derivative. **Plasmalogens,** which comprise much of the phospholipid in mitochondria, are formed by

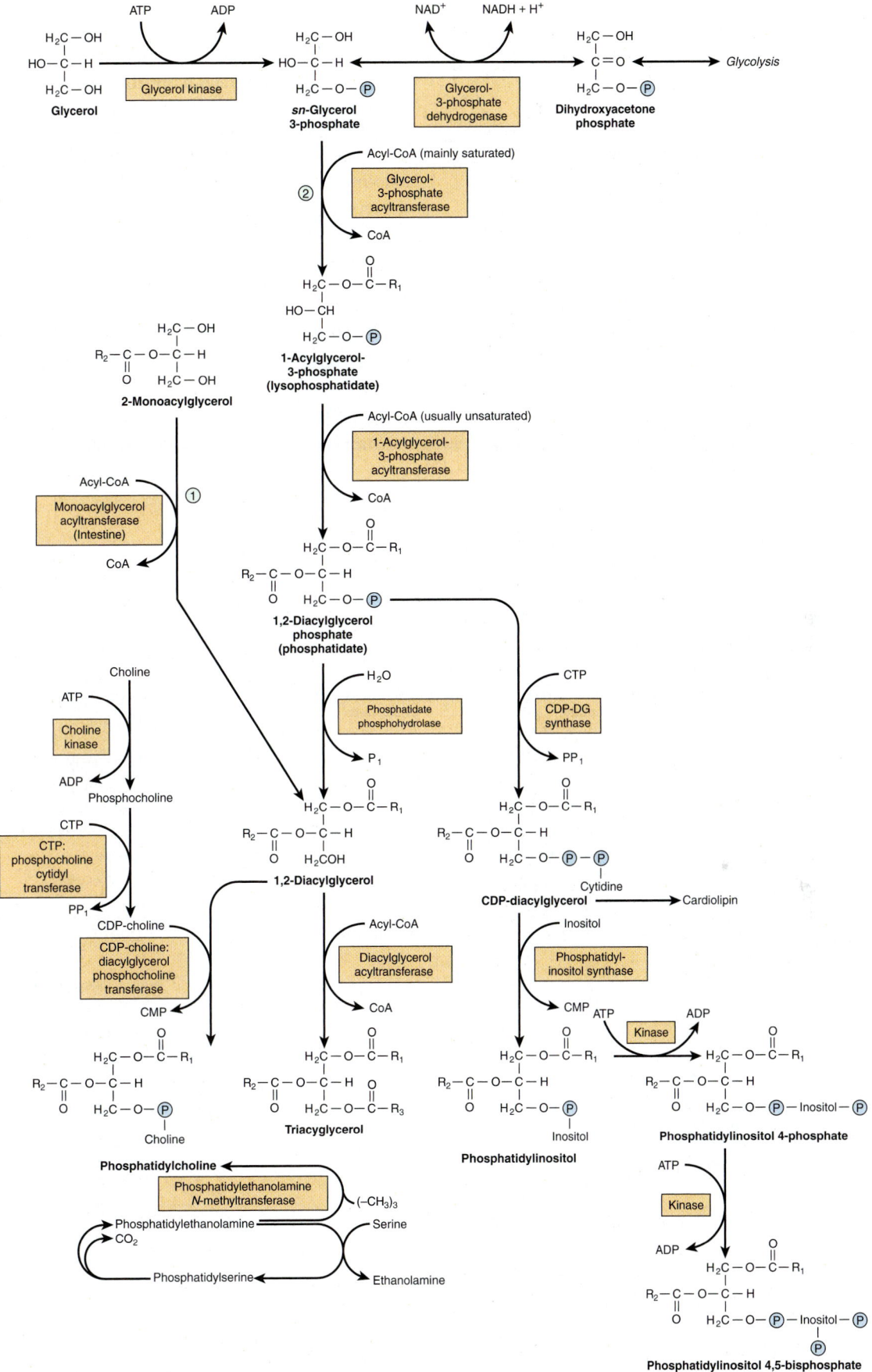

FIGURE 24–2 **Biosynthesis of triacylglycerol and phospholipids.** (①, Monoacylglycerol pathway; ②, glycerol phosphate pathway.) Phosphatidylethanolamine may be formed from ethanolamine by a pathway similar to that shown for the formation of phosphatidylcholine from choline.

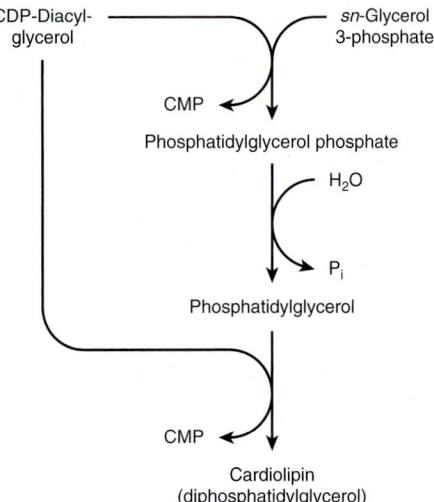

FIGURE 24–3 Biosynthesis of cardiolipin.

desaturation of the analogous 3-phosphoethanolamine derivative (Figure 24–4). **Platelet-activating factor (PAF)** (1-alkyl-2--acetyl-*sn*-glycerol-3-phosphocholine) is synthesized from the corresponding 3-phosphocholine derivative. It is formed by many blood cells and other tissues and aggregates platelets at concentrations as low as 10^{-11} mol/L. It also has hypotensive and ulcerogenic properties and is involved in a variety of biologic responses, including inflammation, chemotaxis, and protein phosphorylation.

Phospholipases Allow Degradation & Remodeling of Phosphoglycerols

Although phospholipids are actively degraded, each portion of the molecule turns over at a different rate—eg, the turnover time of the phosphate group is different from that of the

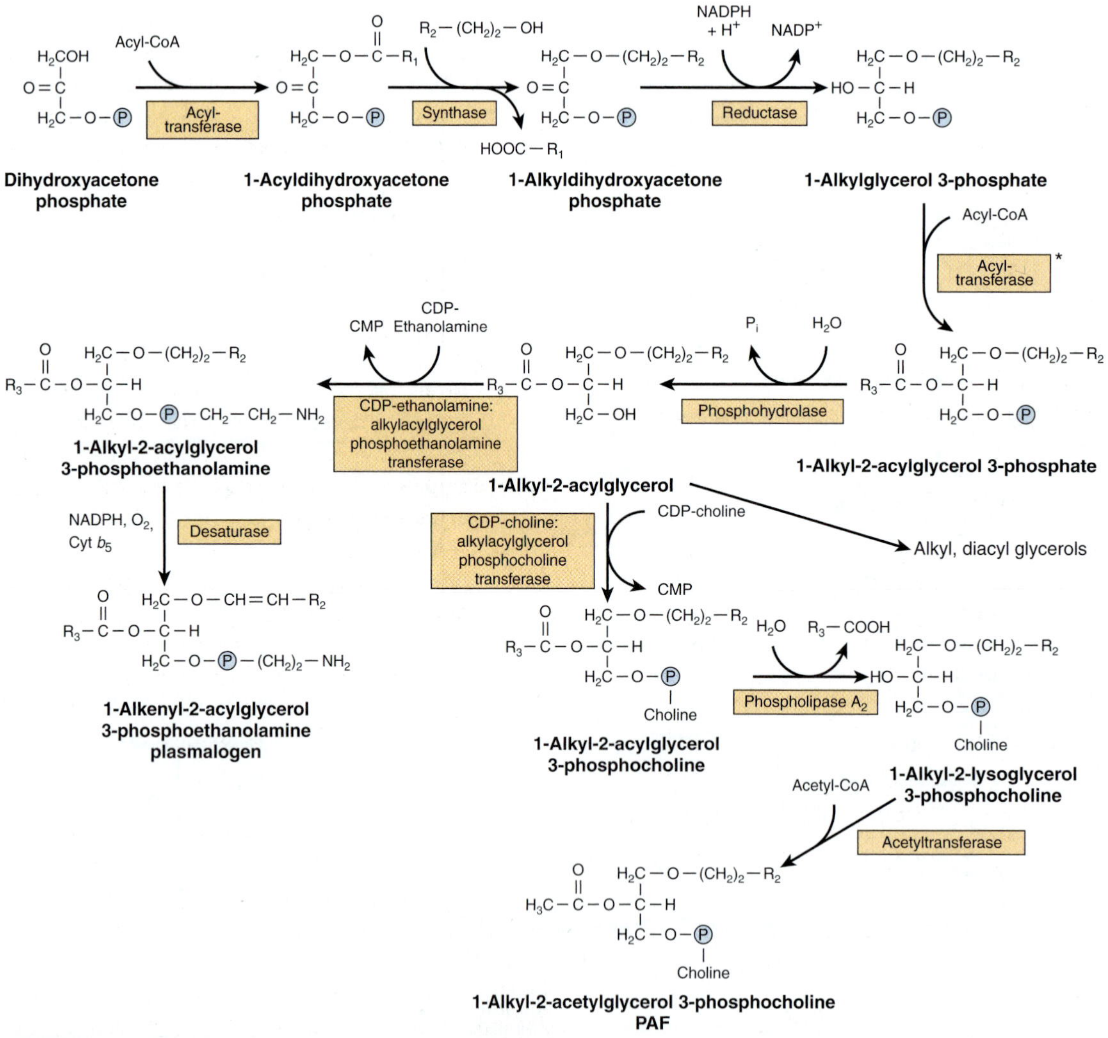

FIGURE 24–4 Biosynthesis of ether lipids, including plasmalogens, and platelet-activating factor **(PAF).** In the de novo pathway for PAF synthesis, acetyl-CoA is incorporated at stage*, avoiding the last two steps in the pathway shown here.

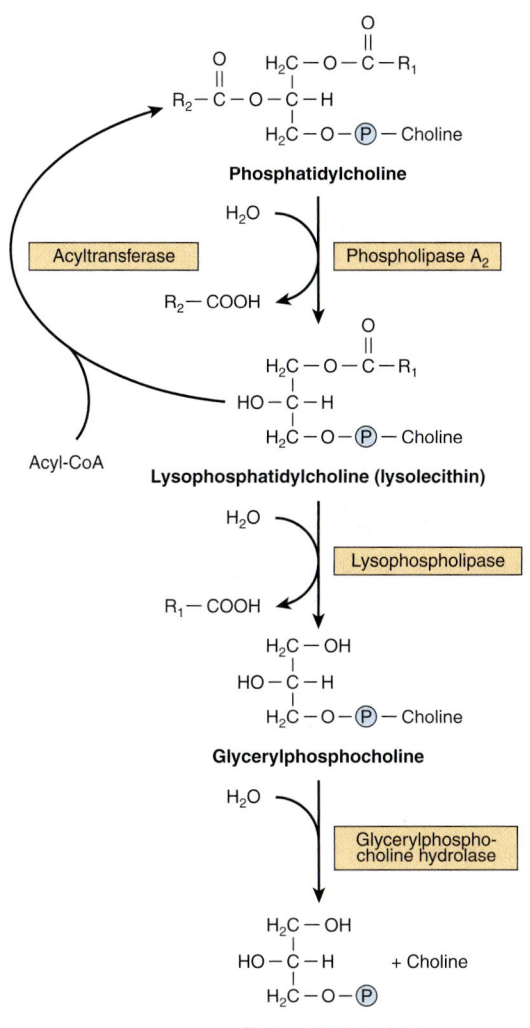

FIGURE 24–5 Metabolism of phosphatidylcholine (lecithin).

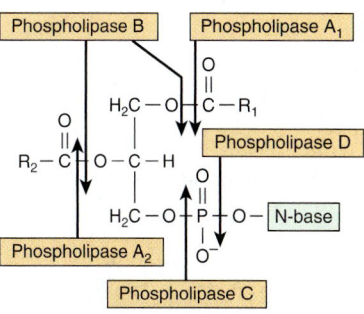

FIGURE 24–6 Sites of the hydrolytic activity of phospholipases on a phospholipid substrate.

to cholesterol to form cholesteryl ester and lysolecithin, and is considered to be responsible for much of the cholesteryl ester in plasma lipoproteins. Long-chain saturated fatty acids are found predominantly in the 1 position of phospholipids, whereas the polyunsaturated fatty acids (eg, the precursors of prostaglandins) are incorporated more frequently into the 2 position. The incorporation of fatty acids into lecithin occurs in three ways; by complete synthesis of the phospholipid; by transacylation between cholesteryl ester and lysolecithin; and by direct acylation of lysolecithin by acyl-CoA. Thus, a continuous exchange of the fatty acids is possible, particularly with regard to introducing essential fatty acids into phospholipid molecules.

ALL SPHINGOLIPIDS ARE FORMED FROM CERAMIDE

Ceramide is synthesized in the endoplasmic reticulum from the amino acid serine as shown in **Figure 24–7**. Ceramide is an important signaling molecule (second messenger) regulating pathways including programmed cell death **(apoptosis)**, the **cell cycle**, and **cell differentiation and senescence.**

Sphingomyelins (Figure 15–13) are phospholipids and are formed when ceramide reacts with phosphatidylcholine to form sphingomyelin plus diacylglycerol (**Figure 24–8A**). This occurs mainly in the Golgi apparatus and to a lesser extent in the plasma membrane.

Glycosphingolipids Are a Combination of Ceramide with One or More Sugar Residues

The simplest glycosphingolipids (**cerebrosides**) are **galactosylceramide (GalCer)** and **glucosylceramide (GlcCer)**. GalCer is a major lipid of **myelin,** whereas GlcCer is the major glycosphingolipid of **extraneural tissues** and a precursor of most of the more complex glycosphingolipids. GalCer (**Figure 24–8B**) is formed in a reaction between ceramide and UDPGal (formed by epimerization from UDPGlc— Figure 21–6).

1-acyl group. This is due to the presence of enzymes that allow partial degradation followed by resynthesis (**Figure 24–5**). **Phospholipase A$_2$** catalyzes the hydrolysis of glycerophospholipids to form a free fatty acid and lysophospholipid, which in turn may be reacylated by acyl-CoA in the presence of an acyltransferase. Alternatively, lysophospholipid (eg, lysolecithin) is attacked by **lysophospholipase,** forming the corresponding glyceryl phosphoryl base, which may then be split by a hydrolase liberating glycerol 3-phosphate plus base. **Phospholipases A$_1$, A$_2$, B, C, and D** attack the bonds indicated in **Figure 24–6**. **Phospholipase A$_2$** is found in pancreatic fluid and snake venom as well as in many types of cells; **phospholipase C** is one of the major toxins secreted by bacteria; and **phospholipase D** is known to be involved in mammalian signal transduction.

Lysolecithin (lysophosphatidylcholine) may be formed by an alternative route that involves **lecithin: cholesterol acyltransferase (LCAT).** This enzyme, found in plasma, catalyzes the transfer of a fatty acid residue from the 2 position of lecithin

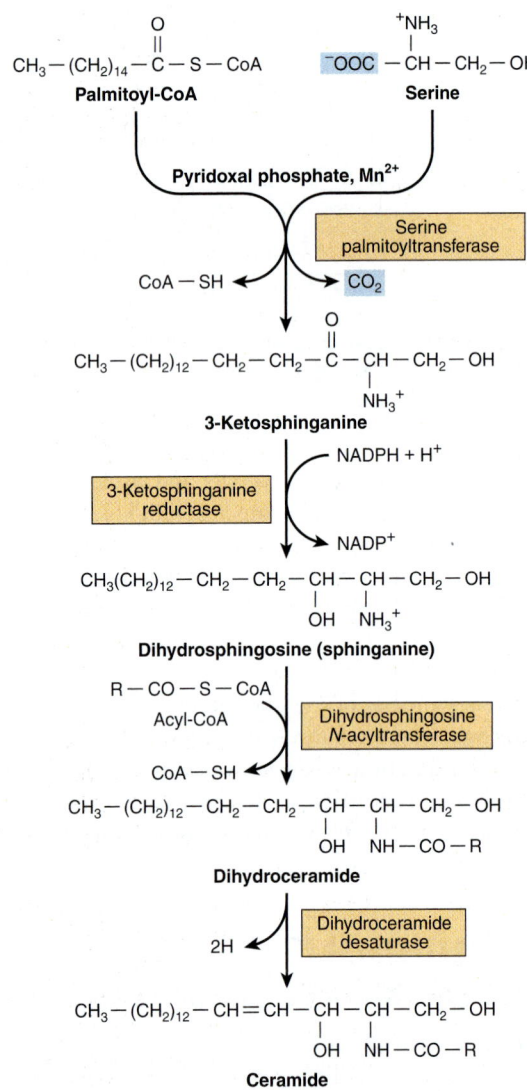

FIGURE 24–7 Biosynthesis of ceramide.

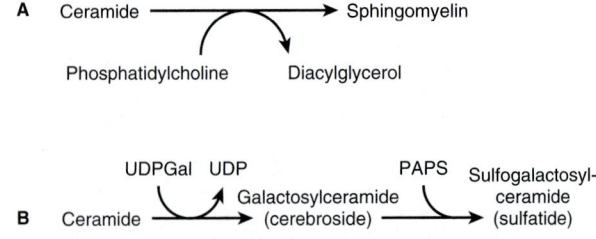

FIGURE 24–8 Biosynthesis of (A) sphingomyelin, (B) galactosylceramide and its sulfo derivative. (PAPS, "active sulfate," adenosine 3′-phosphate-5′-phosphosulfate.)

Sulfogalactosylceramide and other sulfolipids such as the **sulfo(galacto)-glycerolipids** and the **steroid sulfates** are formed after further reactions involving 3′-phosphoadenosine-5′-phosphosulfate (PAPS; "active sulfate"). **Gangliosides** are synthesized from ceramide by the stepwise addition of activated sugars (eg, UDPGlc and UDPGal) and a **sialic acid,** usually N-acetylneuraminic acid (**Figure 24–9**). A large number of gangliosides of increasing molecular weight may be formed. Most of the enzymes transferring sugars from nucleotide sugars (glycosyl transferases) are found in the Golgi apparatus.

Glycosphingolipids are constituents of the outer leaflet of plasma membranes and are important in **cell adhesion** and **cell recognition.** Some are antigens, eg, ABO blood group substances. Certain gangliosides function as receptors for bacterial toxins (eg, for **cholera toxin,** which subsequently activates adenylyl cyclase).

CLINICAL ASPECTS

Deficiency of Lung Surfactant Causes Respiratory Distress Syndrome

Lung surfactant is composed mainly of lipid with some proteins and carbohydrate and prevents the alveoli from

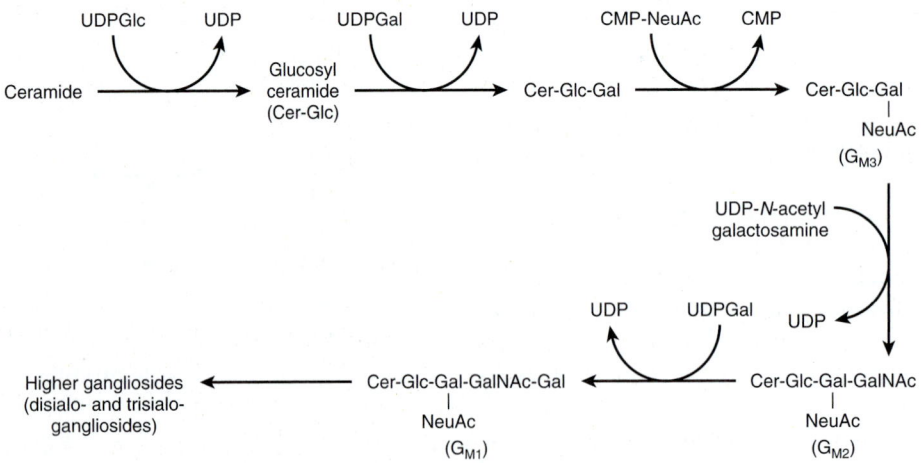

FIGURE 24–9 Biosynthesis of gangliosides. (NeuAc, N-acetylneuraminic acid.)

TABLE 24–1 Examples of Sphingolipidoses

Disease	Enzyme Deficiency	Lipid Accumulating	Clinical Symptoms
Tay-Sachs disease	Hexosaminidase A	Cer—Glc—Gal(NeuAc)÷GalNAc G_{M2} Ganglioside	Mental retardation, blindness, muscular weakness
Fabry's disease	α-Galactosidase	Cer—Glc—Gal—÷Gal Globotriaosylceramide	Skin rash, kidney failure (full symptoms only in males; X-linked recessive)
Metachromatic leukodystrophy	Arylsulfatase A	Cer—Gal—÷OSO₃ 3-Sulfogalactosylceramide	Mental retardation and psychologic disturbances in adults; demyelination
Krabbe's disease	β-Galactosidase	Cer—÷Gal Galactosylceramide	Mental retardation; myelin almost absent
Gaucher's disease	β-Glucosidase	Cer—÷Glc Glucosylceramide	Enlarged liver and spleen, erosion of long bones, mental retardation in infants
Niemann-Pick disease	Sphingomyelinase	Cer—÷P—choline Sphingomyelin	Enlarged liver and spleen, mental retardation; fatal in early life
Farber's disease	Ceramidase	Acyl—÷Sphingosine Ceramide	Hoarseness, dermatitis, skeletal deformation, mental retardation; fatal in early life

Abbreviations: Cer, ceramide; Gal, galactose; Glc, glucose; NeuAc, *N*-acetylneuraminic acid; ÷, site of deficient enzyme reaction.

collapsing. The phospholipid **dipalmitoyl-phosphatidylcholine** decreases surface tension at the air–liquid interface and thus greatly reduces the work of breathing, but other surfactant lipid and protein components are also important in surfactant function. Deficiency of lung surfactant in the lungs of many preterm newborns gives rise to **infant respiratory distress syndrome (IRDS).** Administration of either natural or artificial surfactant is of therapeutic benefit.

Phospholipids & Sphingolipids Are Involved in Multiple Sclerosis and Lipidoses

Certain diseases are characterized by abnormal quantities of these lipids in the tissues, often in the nervous system. They may be classified into two groups: (1) true demyelinating diseases and (2) sphingolipidoses.

In **multiple sclerosis,** which is a demyelinating disease, there is loss of both phospholipids (particularly ethanolamine plasmalogen) and of sphingolipids from white matter. Thus, the lipid composition of white matter resembles that of gray matter. The cerebrospinal fluid shows raised phospholipid levels.

The **sphingolipidoses (lipid storage diseases)** are a group of inherited diseases that are caused by a genetic defect in the catabolism of lipids containing sphingosine. They are part of a larger group of lysosomal disorders and exhibit several constant features: (1) complex lipids containing ceramide accumulate in cells, particularly neurons, causing neurodegeneration and shortening the lifespan. (2) The rate of **synthesis** of the stored lipid is normal. (3) The enzymatic defect is in the **lysosomal degradation pathway** of sphingolipids. (4) The extent to which the activity of the affected enzyme is decreased

is similar in all tissues. There is no effective treatment for many of the diseases, although some success has been achieved with **enzyme replacement therapy** and **bone marrow transplantation** in the treatment of Gaucher's and Fabry's diseases. Other promising approaches are **substrate deprivation therapy** to inhibit the synthesis of sphingolipids and **chemical chaperone therapy. Gene therapy** for lysosomal disorders is also currently under investigation. Some examples of the more important lipid storage diseases are shown in **Table 24–1.**

Multiple sulfatase deficiency results in accumulation of sulfogalactosylceramide, steroid sulfates, and proteoglycans owing to a combined deficiency of arylsulfatases A, B, and C and steroid sulfatase.

SUMMARY

- Triacylglycerols are the major energy-storing lipids, whereas phosphoglycerols, sphingomyelin, and glycosphingolipids are amphipathic and have structural functions in cell membranes as well as other specialized roles.

- Triacylglycerols and some phosphoglycerols are synthesized by progressive acylation of glycerol 3-phosphate. The pathway bifurcates at phosphatidate, forming inositol phospholipids and cardiolipin on the one hand and triacylglycerol and choline and ethanolamine phospholipids on the other.

- Plasmalogens and platelet-activating factor (PAF) are ether phospholipids formed from dihydroxyacetone phosphate.

- Sphingolipids are formed from ceramide (*N*-acylsphingosine). Sphingomyelin is present in membranes of organelles involved in secretory processes (eg, Golgi apparatus). The simplest glycosphingolipids are a combination of ceramide plus a sugar residue (eg, GalCer in myelin). Gangliosides are more complex glycosphingolipids containing more sugar residues plus sialic acid. They are present in the outer layer of the plasma membrane, where they contribute to the glycocalyx and are important as antigens and cell receptors.

■ Phospholipids and sphingolipids are involved in several disease processes, including infant respiratory distress syndrome (lack of lung surfactant), multiple sclerosis (demyelination), and sphingolipidoses (inability to break down sphingolipids in lysosomes due to inherited defects in hydrolase enzymes).

REFERENCES

McPhail LC: Glycerolipid in signal transduction. *Biochemistry of Lipids, Lipoproteins and Membranes,* 4th ed. Vance DE, Vance JE (editors). Elsevier, 2002:315–340.

Merrill AH: Sphingolipids. *Biochemistry of Lipids, Lipoproteins and Membranes,* 5th ed. Vance DE, Vance JE (editors). Elsevier, 2008:363–398.

Meyer KC, Zimmerman JJ: Inflammation and surfactant. Paediatr Respir Rev 2002;3:308.

Prescott SM, Zimmerman GA, Stafforini DM, et al: Platelet-activating factor and related lipid mediators. Annu Rev Biochem 2000;69:419.

Ruvolo PP: Intracellular signal transduction pathways activated by ceramide and its metabolites. Pharmacol Res 2003;47:383.

Scriver CR, Beaudet AL, Sly WS, et al (editors): *The Metabolic and Molecular Bases of Inherited Disease,* 8th ed. McGraw-Hill, 2001.

Vance DE, Vance JE (editors): Phospholipid biosynthesis in eukaryotes. In: *Biochemistry of Lipids, Lipoproteins and Membranes,* 5th ed. Elsevier, 2008: 213–244.

van Echten G, Sandhoff K: Ganglioside metabolism. Enzymology, topology, and regulation. J Biol Chem 1993;268:5341.

Lipid Transport & Storage

Kathleen M. Botham, PhD, DSc & Peter A. Mayes, PhD, DSc

OBJECTIVES

After studying this chapter, you should be able to:

- Identify the four major groups of plasma lipoproteins and the four major lipid classes they carry.
- Illustrate the structure of a lipoprotein particle.
- Indicate the major types of apolipoprotein found in the different lipoprotein classes.
- Explain that triacylglycerol is carried from the intestine (after intake from the diet) to the liver in chylomicrons and from the liver to extrahepatic tissues in very low density lipoprotein (VLDL), and these particles are synthesized in intestinal and liver cells, respectively, by similar processes.
- Illustrate the processes by which chylomicrons are metabolized by lipases to form chylomicron remnants, which are then removed from the circulation by the liver.
- Explain how VLDL is metabolized by lipases to VLDL remnants (also called intermediate-density lipoprotein (IDL)) which may be cleared by the liver or converted to low-density lipoprotein (LDL), which functions to deliver cholesterol from the liver to extrahepatic tissues and is taken up via the LDL (apoB100,E) receptor.
- Explain how high-density lipoprotein (HDL), which returns cholesterol from extrahepatic tissues to the liver in reverse cholesterol transport, is synthesized, indicate the mechanisms by which it accepts cholesterol from tissues, and show how it is metabolized in the HDL cycle.
- Understand how the liver plays a central role in lipid transport and metabolism and how hepatic VLDL secretion is regulated by the diet and hormones.
- Be aware of the roles of LDL and HDL in promoting and retarding, respectively, the development of atherosclerosis.
- Indicate the causes of alcoholic and nonalcoholic fatty liver disease.
- Appreciate that adipose tissue is the main store of triacylglycerol in the body and explain the processes by which fatty acids are released and how they are regulated.
- Understand the role of brown adipose tissue in the generation of body heat.

BIOMEDICAL IMPORTANCE

Fat absorbed from the diet and lipids synthesized by the liver and adipose tissue must be transported between the various tissues and organs for utilization and storage. Since lipids are insoluble in water, the problem of how to transport them in the aqueous blood plasma is solved by associating nonpolar lipids (triacylglycerol and cholesteryl esters) with amphipathic lipids (phospholipids and cholesterol) and proteins to make water-miscible lipoproteins.

TABLE 25–1   Composition of the Lipoproteins in Plasma of Humans

Lipoprotein	Source	Diameter (nm)	Density (g/mL)	Composition		Main Lipid Components	Apolipoproteins
				Protein (%)	Lipid (%)		
Chylomicrons	Intestine	90–1000	< 0.95	1–2	98–99	Triacylglycerol	A-I, A-II, A-IV,[1] B-48, C-I, C-II, C-III, E
Chylomicron remnants	Chylomicrons	45–150	< 1.006	6–8	92–94	Triacylglycerol, phospholipids, cholesterol	B-48, E
VLDL	Liver (intestine)	30–90	0.95–1.006	7–10	90–93	Triacylglycerol	B-100, C-I, C-II, C-III
IDL	VLDL	25–35	1.006–1.019	11	89	Triacylglycerol, cholesterol	B-100, E
LDL	VLDL	20–25	1.019–1.063	21	79	Cholesterol	B-100
HDL	Liver, intestine, VLDL, chylomicrons					Phospholipids, cholesterol	A-I, A-II, A-IV, C-I, C-II, C-III, D,[2] E
HDL₁		20–25	1.019–1.063	32	68		
HDL₂		10–20	1.063–1.125	33	67		
HDL₃		5–10	1.125–1.210	57	43		
Preβ-HDL[3]		< 5	> 1.210				A-I
Albumin/free fatty acids	Adipose tissue		> 1.281	99	1	Free fatty acids	

[1]Secreted with chylomicrons but transfers to HDL.
[2]Associated with HDL₂ and HDL₃ subfractions.
[3]Part of a minor fraction known as very high density lipoproteins (VHDL).
Abbreviations: HDL, high-density lipoproteins; IDL, intermediate-density lipoproteins; LDL, low-density lipoproteins; VLDL, very low density lipoproteins.

In a meal-eating omnivore such as the human, excess calories are ingested in the anabolic phase of the feeding cycle, followed by a period of negative caloric balance when the organism draws upon its carbohydrate and fat stores. Lipoproteins mediate this cycle by transporting lipids from the intestines as chylomicrons—and from the liver as very low density lipoproteins (VLDL)—to most tissues for oxidation and to adipose tissue for storage. Lipid is mobilized from adipose tissue as free fatty acids (FFA) bound to serum albumin. Abnormalities of lipoprotein metabolism cause various **hypo-** or **hyperlipoproteinemias.** The most common of these is in **diabetes mellitus,** where insulin deficiency causes excessive mobilization of FFA and underutilization of chylomicrons and VLDL, leading to **hypertriacylglycerolemia.** Most other pathologic conditions affecting lipid transport are due primarily to inherited defects, some of which cause **hypercholesterolemia** and premature **atherosclerosis** (Table 26–1). **Obesity**—particularly abdominal obesity—is a risk factor for increased mortality, hypertension, type 2 diabetes mellitus, hyperlipidemia, hyperglycemia, and various endocrine dysfunctions.

LIPIDS ARE TRANSPORTED IN THE PLASMA AS LIPOPROTEINS

Four Major Lipid Classes Are Present in Lipoproteins

Plasma lipids consist of **triacylglycerols** (16%), **phospholipids** (30%), **cholesterol** (14%), and **cholesteryl esters** (36%) and a much smaller fraction of unesterified long-chain fatty acids (free fatty acids or FFA) (4%). This latter fraction, the **FFA,** is metabolically the most active of the plasma lipids.

Four Major Groups of Plasma Lipoproteins Have Been Identified

Since fat is less dense than water, the density of a lipoprotein decreases as the proportion of lipid to protein increases (**Table 25–1**). Four major groups of lipoproteins have been identified that are important physiologically and in clinical diagnosis. These are (1) **chylomicrons,** derived from intestinal absorption of triacylglycerol and other lipids; (2) **very low** density lipoproteins (VLDL, or pre-β-lipoproteins), derived from the liver for the export of triacylglycerol; (3) **low-density lipoproteins** (LDL, or β-lipoproteins), representing a final stage in the catabolism of VLDL; and (4) **high-density lipoproteins** (HDL, or α-lipoproteins), involved in cholesterol transport and also in VLDL and chylomicron metabolism. Triacylglycerol is the predominant lipid in chylomicrons and VLDL, whereas cholesterol and phospholipid are the predominant lipids in LDL and HDL, respectively (Table 25–1). Lipoproteins may be separated according to their electrophoretic properties into **α-, β-, and pre-β-lipoproteins.**

Lipoproteins Consist of a Nonpolar Core & a Single Surface Layer of Amphipathic Lipids

The **nonpolar lipid core** consists of mainly **triacylglycerol** and **cholesteryl ester** and is surrounded by a **single surface layer** of **amphipathic phospholipid** and **cholesterol** molecules (**Figure 25–1**). These are oriented so that their polar groups face outward to the aqueous medium, as in the cell membrane (Chapter 15). The protein moiety of a lipoprotein is known as an **apolipoprotein** or **apoprotein,** constituting nearly 70% of some HDL and as little as 1% of chylomicrons. Some apolipoproteins are integral and cannot be removed, whereas others are free to transfer to other lipoproteins.

The Distribution of Apolipoproteins Characterizes the Lipoprotein

One or more apolipoproteins (proteins or polypeptides) are present in each lipoprotein. The major apolipoproteins of HDL (α-lipoprotein) are designated A (Table 25–1). The main apolipoprotein of LDL (β-lipoprotein) is apolipoprotein B (B-100), which is found also in VLDL. Chylomicrons contain a truncated form of apo B (B-48) that is synthesized in the intestine, while B-100 is synthesized in the liver. Apo B-100 is one of the longest single polypeptide chains known, having 4536 amino acids and a molecular mass of 550,000 Da. Apo B-48 (48% of B-100) is formed after transcription of the apo B-100 gene by the introduction of a stop signal into the mRNA transcript by an RNA editing enzyme. Apo C-I, C-II, and C-III are smaller polypeptides (molecular mass 7000–9000 Da) freely transferable between several different lipoproteins. Apo E, found in VLDL, HDL, chylomicrons, and chylomicron remnants, is also freely transferable; it accounts for 5–10% of total VLDL apolipoproteins in normal subjects.

Apolipoproteins carry out several roles: (1) they can form part of the structure of the lipoprotein, eg, apo B; (2) they are enzyme cofactors, eg, C-II for lipoprotein lipase, A-I for lecithin:cholesterol acyltransferase, or enzyme inhibitors, eg, apo A-II and apo C-III for lipoprotein lipase, apo C-I for cholesteryl ester transfer protein; and (3) they act as ligands for interaction with lipoprotein receptors in tissues, eg, apo B-100 and apo E for the LDL receptor, apo E for the LDL-receptor-related protein (LRP), which has been identified as the remnant receptor, and apo A-I for the HDL receptor. The functions of apo A-IV and apo D, however, are not yet clearly defined, although apo D is believed to be an important factor in human neurodegenerative disorders.

FREE FATTY ACIDS ARE RAPIDLY METABOLIZED

The FFA (also termed nonesterified fatty acids or unesterified fatty acids) arise in the plasma from the breakdown of triacylglycerol in adipose tissue or as a result of the action of lipoprotein lipase on the plasma triacylglycerols. They are found **in combination with albumin,** a very effective solubilizer, in concentrations varying between 0.1 and 2.0 µeq/mL of plasma. Levels are low in the fully fed condition and rise to 0.7–0.8 µeq/mL in the starved state. In uncontrolled **diabetes mellitus,** the level may rise to as much as 2 µeq/mL.

FFA are removed from the blood extremely rapidly and oxidized (fulfilling 25–50% of energy requirements in starvation) or esterified to form triacylglycerol in the tissues. In starvation, esterified lipids from the circulation or in the tissues are oxidized as well, particularly in heart and skeletal muscle cells, where considerable stores of lipid are to be found.

The FFA uptake by tissues is related directly to the plasma-FFA concentration, which in turn is determined by the rate of lipolysis in adipose tissue. After dissociation of the fatty acid-albumin complex at the plasma membrane, fatty acids bind to a **membrane fatty acid transport protein** that acts

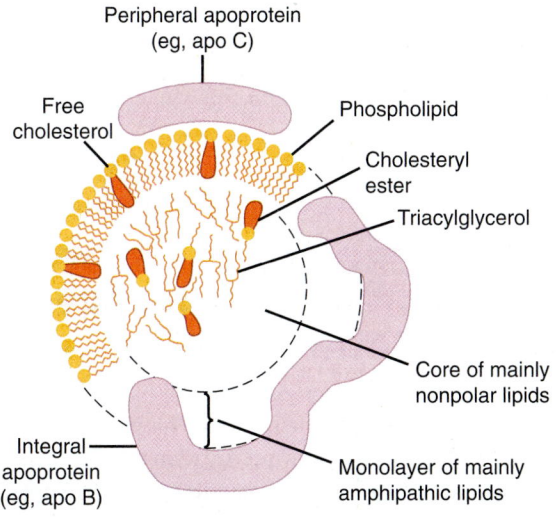

FIGURE 25–1 **Generalized structure of a plasma lipoprotein.** The similarities with the structure of the plasma membrane are to be noted. Small amounts of cholesteryl ester and triacylglycerol are found in the surface layer and a little free cholesterol in the core.

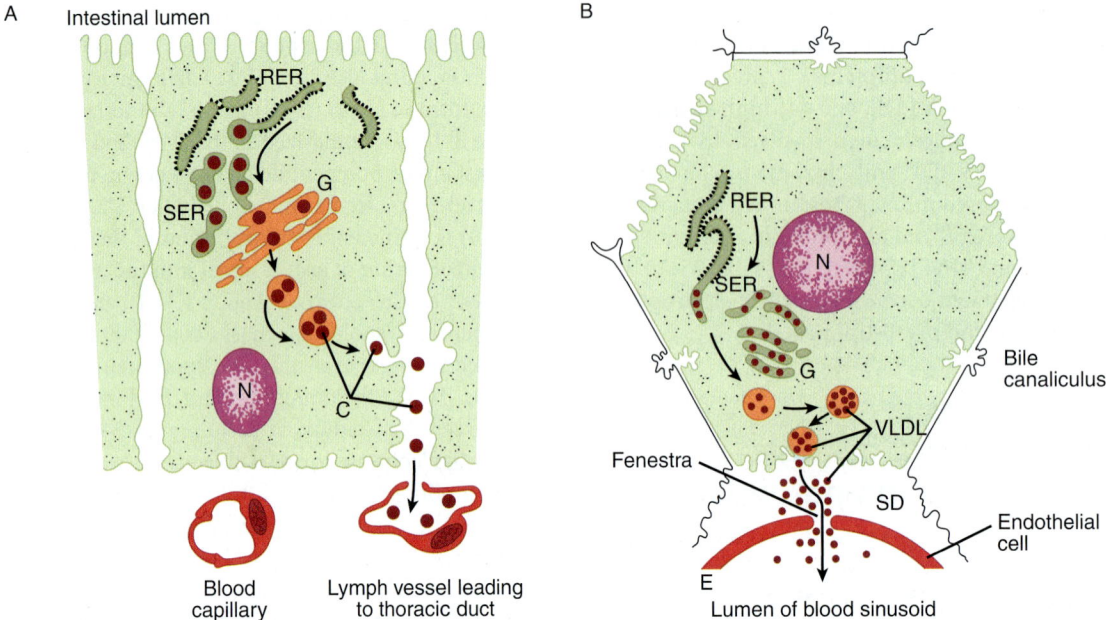

FIGURE 25–2 **The formation and secretion of (A) chylomicrons by an intestinal cell and (B) very low density lipoproteins by a hepatic cell.** (C, chylomicrons; E, endothelium; G, Golgi apparatus; N, nucleus; RER, rough endoplasmic reticulum; SD, space of Disse, containing blood plasma; SER, smooth endoplasmic reticulum; VLDL, very low density lipoproteins.) Apolipoprotein B, synthesized in the RER, is incorporated into particles with triacylglycerol, cholesterol, and phospholipids in the SER. After the addition of carbohydrate residues in G, they are released from the cell by reverse pinocytosis. Chylomicrons pass into the lymphatic system. VLDL are secreted into the space of Disse and then into the hepatic sinusoids through fenestrae in the endothelial lining.

as a transmembrane cotransporter with Na^+. On entering the cytosol, FFA are bound by intracellular **fatty-acid-binding proteins.** The role of these proteins in intracellular transport is thought to be similar to that of serum albumin in extracellular transport of long-chain fatty acids.

TRIACYLGLYCEROL IS TRANSPORTED FROM THE INTESTINES IN CHYLOMICRONS & FROM THE LIVER IN VERY LOW DENSITY LIPOPROTEINS

By definition, **chylomicrons** are found in **chyle** formed only by the lymphatic system **draining the intestine.** They are responsible for the transport of all dietary lipids into the circulation. Small quantities of VLDL are also to be found in chyle; however, most **VLDL in the plasma** are of hepatic origin. **They are the vehicles of transport of triacylglycerol from the liver to the extrahepatic tissues.**

There are striking similarities in the mechanisms of formation of chylomicrons by intestinal cells and of VLDL by hepatic parenchymal cells (**Figure 25–2**), perhaps because—apart from the mammary gland—the intestine and liver are the only tissues from which particulate lipid is secreted. Newly

secreted or "nascent" chylomicrons and VLDL contain only a small amount of apolipoproteins C and E, and the full complement is acquired from HDL in the circulation (**Figures 25–3** and **25–4**). Apo B, however, is an integral part of the lipoprotein particles, it is incorporated inside the cells and is essential for chylomicron and VLDL formation. In **abetalipoproteinemia** (a rare disease), lipoproteins containing apo B are not formed and lipid droplets accumulate in the intestine and liver.

A more detailed account of the factors controlling hepatic VLDL secretion is given below.

CHYLOMICRONS & VERY LOW DENSITY LIPOPROTEINS ARE RAPIDLY CATABOLIZED

The clearance of chylomicrons from the blood is rapid, the half-time of disappearance being under 1 h in humans. Larger particles are catabolized more quickly than smaller ones. Fatty acids originating from chylomicron triacylglycerol are delivered mainly to adipose tissue, heart, and muscle (80%), while ~20% goes to the liver. However, **the liver does not metabolize native chylomicrons or VLDL significantly;** thus, the fatty acids in the liver must be secondary to their metabolism in extrahepatic tissues.

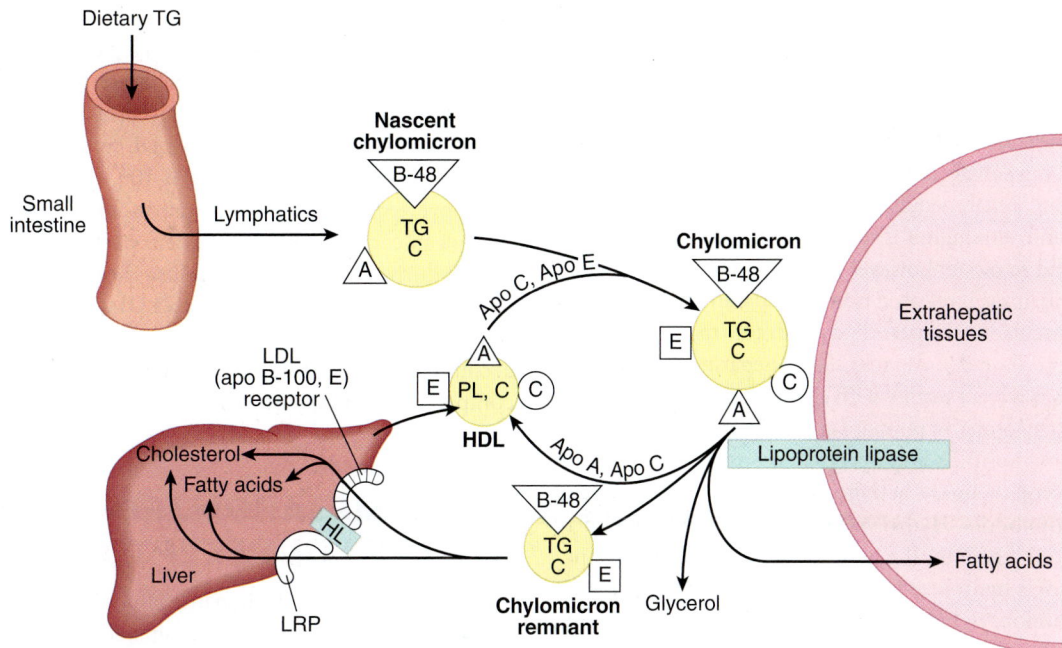

FIGURE 25–3 **Metabolic fate of chylomicrons.** (A, apolipoprotein A; B-48, apolipoprotein B-48; C, apolipoprotein C; C, cholesterol and cholesteryl ester; E, apolipoprotein E; HDL, high-density lipoprotein; HL, hepatic lipase; LRP, LDL-receptor-related protein; PL, phospholipid; TG, triacylglycerol.) Only the predominant lipids are shown.

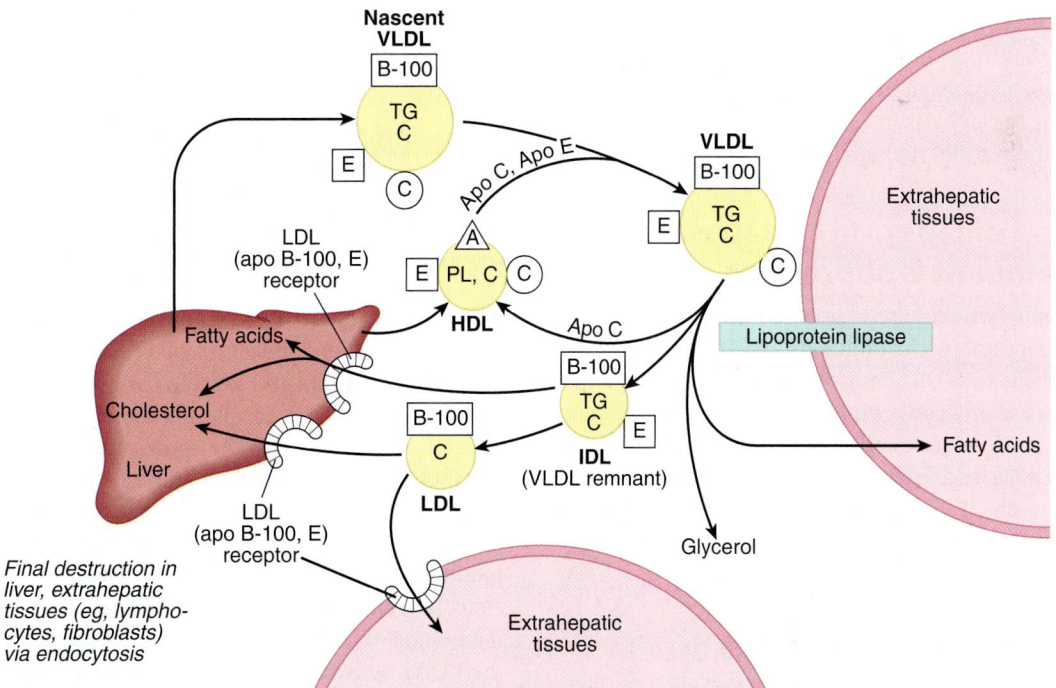

FIGURE 25–4 **Metabolic fate of very low density lipoproteins (VLDL) and production of low-density lipoproteins (LDL).** (A, apolipoprotein A; B-100, apolipoprotein B-100; C, apolipoprotein C; C, cholesterol and cholesteryl ester; E, apolipoprotein E; HDL, high-density lipoprotein; IDL, intermediate-density lipoprotein; PL, phospholipid; TG, triacylglycerol.) Only the predominant lipids are shown. It is possible that some IDL is also metabolized via the low density lipoproten receptor-related protein (LRP.)

Triacylglycerols of Chylomicrons & VLDL Are Hydrolyzed by Lipoprotein Lipase

Lipoprotein lipase is located on the walls of blood capillaries, anchored to the endothelium by negatively charged proteoglycan chains of heparan sulfate. It has been found in heart, adipose tissue, spleen, lung, renal medulla, aorta, diaphragm, and lactating mammary gland, although it is not active in adult liver. It is not normally found in blood; however, following injection of **heparin**, lipoprotein lipase is released from its heparan sulfate binding sites into the circulation. **Hepatic lipase** is bound to the sinusoidal surface of liver cells and is also released by heparin. This enzyme, however, does not react readily with chylomicrons or VLDL but is involved in chylomicron remnant and HDL metabolism.

Both **phospholipids** and **apo C-II** are required as cofactors for lipoprotein lipase activity, while **apo A-II and apo C-III** act as inhibitors. Hydrolysis takes place while the lipoproteins are attached to the enzyme on the endothelium. Triacylglycerol is hydrolyzed progressively through a diacylglycerol to a monoacylglycerol and finally to FFA plus glycerol. Some of the released FFA return to the circulation, attached to albumin, but the bulk is transported into the tissue (Figures 25–3 and 25–4). Heart lipoprotein lipase has a low K_m for triacylglycerol, about one-tenth of that for the enzyme in adipose tissue. This enables the delivery of fatty acids from triacylglycerol to be **redirected from adipose tissue to the heart in the starved state** when the plasma triacylglycerol decreases. A similar redirection to the mammary gland occurs during lactation, allowing uptake of lipoprotein triacylglycerol fatty acid for **milk fat** synthesis. The **VLDL receptor** plays an important part in the delivery of fatty acids from VLDL triacylglycerol to adipocytes by binding VLDL and bringing it into close contact with lipoprotein lipase. In adipose tissue, **insulin** enhances lipoprotein lipase synthesis in adipocytes and its translocation to the luminal surface of the capillary endothelium.

The Action of Lipoprotein Lipase Forms Remnant Lipoproteins

Reaction with lipoprotein lipase results in the loss of 70–90% of the triacylglycerol of chylomicrons and in the loss of apo C (which returns to HDL) but not apo E, which is retained. The resulting **chylomicron remnant** is about half the diameter of the parent chylomicron and is relatively enriched in cholesterol and cholesteryl esters because of the loss of triacylglycerol (Figure 25–3). Similar changes occur to VLDL, with the formation of **VLDL remnants** (also called **intermediate-density lipoprotein (IDL)** (Figure 25–4).

The Liver Is Responsible for the Uptake of Remnant Lipoproteins

Chylomicron remnants are taken up by the liver by receptor-mediated endocytosis, and the cholesteryl esters and triacylglycerols are hydrolyzed and metabolized. Uptake is mediated by **apo E** (Figure 25–3), via two apo E-dependent receptors, the **LDL (apo B-100, E) receptor** and the **LRP (LDL receptor-related protein)**. Hepatic lipase has a dual role: (1) it acts as a ligand to facilitate remnant uptake and (2) it hydrolyzes remnant triacylglycerol and phospholipid.

After metabolism to IDL, VLDL may be taken up by the liver directly via the LDL (apo B-100, E) receptor, or it may be converted to LDL. Only one molecule of apo B-100 is present in each of these lipoprotein particles, and this is conserved during the transformations. Thus, each LDL particle is derived from a single precursor VLDL particle (Figure 25–4). In humans, a relatively large proportion of IDL forms LDL, accounting for the increased concentrations of LDL in humans compared with many other mammals.

LDL IS METABOLIZED VIA THE LDL RECEPTOR

The liver and many extrahepatic tissues express the **LDL (apo B-100, E) receptor.** It is so designated because it is specific for apo B-100 but not B-48, which lacks the carboxyl terminal domain of B-100 containing the LDL receptor ligand, and it also takes up lipoproteins rich in apo E. Approximately 30% of LDL is degraded in extrahepatic tissues and 70% in the liver. A positive correlation exists between the incidence of **atherosclerosis** and the plasma concentration of LDL cholesterol. The LDL (apoB-100, E) receptor is defective in **familial hypercholesterolemia,** a genetic condition which blood LDL cholesterol levels are increased, causing premature atherosclerosis (Table 26-1). For further discussion of the regulation of the LDL receptor, see Chapter 26.

HDL TAKES PART IN BOTH LIPOPROTEIN TRIACYLGLYCEROL & CHOLESTEROL METABOLISM

HDL is synthesized and secreted from both liver and intestine (**Figure 25–5**). However, apo C and apo E are synthesized in the liver and transferred from liver HDL to intestinal HDL when the latter enters the plasma. A major function of HDL is to act as a repository for the apo C and apo E required in the metabolism of chylomicrons and VLDL. Nascent HDL consists of discoid phospholipid bilayers containing apo A and free cholesterol. These lipoproteins are similar to the particles found in the plasma of patients with a deficiency of the plasma enzyme **lecithin:cholesterol acyltransferase (LCAT)** and in the plasma of patients with **obstructive jaundice.** LCAT—and the LCAT activator apo A-I—bind to the discoidal particles, and the surface phospholipid and free cholesterol are converted into cholesteryl esters and lysolecithin (Chapter 24). The nonpolar cholesteryl esters move into the hydrophobic interior of the bilayer, whereas lysolecithin is transferred to plasma albumin. Thus, a nonpolar core is generated, forming a spherical,

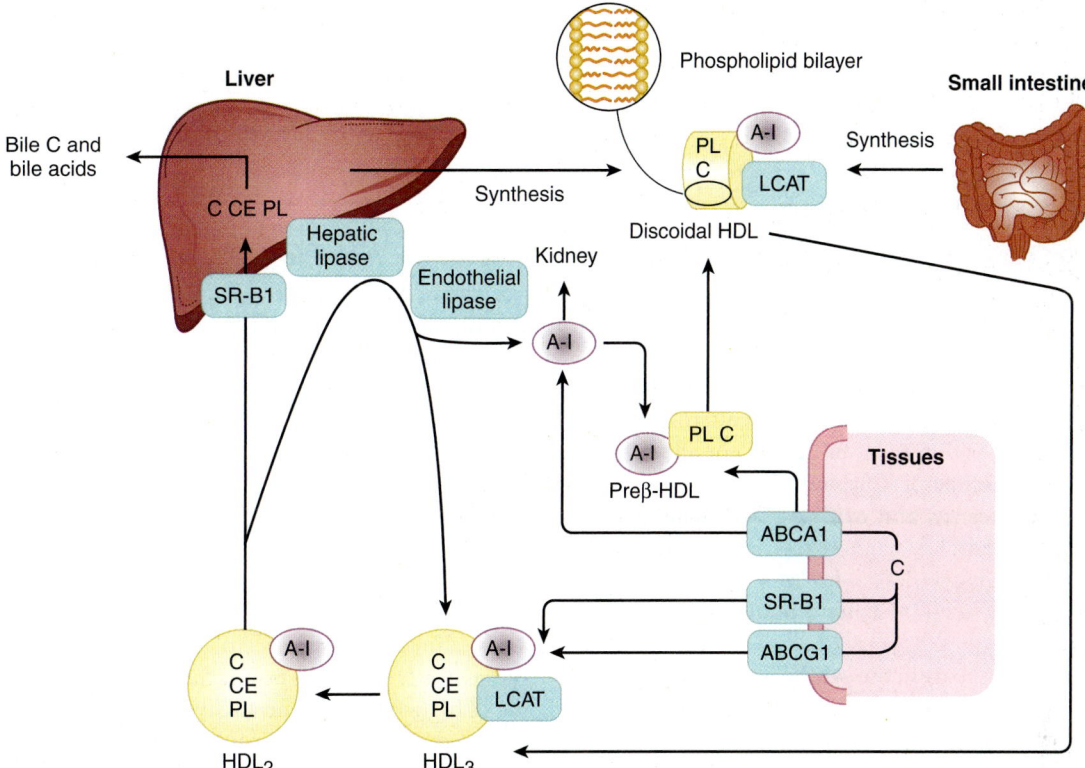

FIGURE 25–5 **Metabolism of high-density lipoprotein (HDL) in reverse cholesterol transport.**
(A-I, apolipoprotein A-I; ABCA 1, ATP-binding cassette transporter A1; ABCG1, ATP-binding cassette transporter G1; C, cholesterol; CE, cholesteryl ester; LCAT, lecithin:cholesterol acyltransferase; PL, phospholipid; SR-B1, scavenger receptor B1.) Preβ-HDL, HDL$_2$, HDL$_3$—see Table 25–1. Surplus surface constituents from the action of lipoprotein lipase on chylomicrons and VLDL are another source of preβ-HDL. Hepatic lipase activity is increased by androgens and decreased by estrogens, which may account for higher concentrations of plasma HDL$_2$ in women.

pseudomicellar HDL covered by a surface film of polar lipids and apolipoproteins. This aids the removal of excess unesterified cholesterol from lipoproteins and tissues as described below. The **class B scavenger receptor B1 (SR-B1)** has been identified as an **HDL receptor with a dual role in HDL metabolism.** In the liver and in steroidogenic tissues, it binds HDL via apo A-I, and cholesteryl ester is selectively delivered to the cells, although the particle itself, including apo A-I, is not taken up. In the tissues, on the other hand, SR-B1 mediates the acceptance of cholesterol effluxed from the cells by HDL, which then transports it to the liver for excretion via the bile (either as cholesterol or after conversion to bile acids) in the process known as **reverse cholesterol transport** (Figure 25–5). HDL$_3$, generated from discoidal HDL by the action of LCAT, accepts cholesterol from the tissues via the **SR-B1** and the cholesterol is then esterified by LCAT, increasing the size of the particles to form the less dense HDL$_2$. HDL$_3$ is then reformed, either after selective delivery of cholesteryl ester to the liver via the SR-B1 or by hydrolysis of HDL$_2$ phospholipid and triacylglycerol by hepatic lipase and endothelial lipase. This interchange of HDL$_2$ and HDL$_3$ is called the **HDL cycle** (Figure 25–5). Free apo A-I is released by these processes and forms **preβ-HDL** after associating with a minimum amount of phospholipid

and cholesterol. Surplus apo A-I is destroyed in the kidney. A second important mechanism for reverse cholesterol transport involves the **ATP-binding cassette transporters A1 (ABCA1) and G1 (ABCG1).** These transporters are members of a family of transporter proteins that couple the hydrolysis of ATP to the binding of a substrate, enabling it to be transported across the membrane. ABCG1 mediates the transport of cholesterol from cells to HDL, while ABCA1 preferentially promotes efflux to poorly lipidated particles such as preβ-HDL or apo A-1, which are then converted to HDL$_3$ via discoidal HDL (Figure 25–5). Preβ-HDL is the most potent form of HDL inducing cholesterol efflux from the tissues.

HDL concentrations vary reciprocally with plasma triacylglycerol concentrations and directly with the activity of lipoprotein lipase. This may be due to surplus surface constituents, eg, phospholipid and apo A-I, being released during hydrolysis of chylomicrons and VLDL and contributing toward the formation of preβ-HDL and discoidal HDL. HDL$_2$ concentrations are **inversely related to the incidence of atherosclerosis,** possibly because they reflect the efficiency of reverse cholesterol transport. HDL$_c$ (HDL$_1$) is found in the blood of diet-induced hypercholesterolemic animals. It is rich in cholesterol, and its sole apolipoprotein is apo E. It appears that all plasma lipoproteins are interrelated components of one or more metabolic cycles that together are responsible for the complex process of plasma lipid transport.

THE LIVER PLAYS A CENTRAL ROLE IN LIPID TRANSPORT & METABOLISM

The liver carries out the following major functions in lipid metabolism:

1. It facilitates the digestion and absorption of lipids by the production of **bile**, which contains cholesterol and bile salts synthesized within the liver de novo or after uptake of lipoprotein cholesterol (Chapter 26).
2. It actively **synthesizes and oxidizes fatty acids** (Chapters 22 & 23) and also synthesizes triacylglycerols and phospholipids (Chapter 24).
3. It **converts fatty acids to ketone bodies (ketogenesis)** (Chapter 22).
4. It plays an integral part in the **synthesis and metabolism of plasma lipoproteins** (this chapter).

Hepatic VLDL Secretion Is Related to Dietary & Hormonal Status

The cellular events involved in VLDL formation and secretion have been described above (Figure 25–2) and are shown in **Figure 25–6**. Hepatic triacylglycerol synthesis provides the immediate stimulus for the formation and secretion of VLDL. The fatty acids used are derived from two possible sources: (1) synthesis within the liver from **acetyl-CoA** derived mainly from carbohydrate (perhaps not so important in humans) and (2) uptake of **FFA** from the circulation. The first source is predominant in the well-fed condition, when fatty acid synthesis is high and the level of circulating FFA is low. As triacylglycerol does not normally accumulate in the liver in these conditions, it must be inferred that it is transported from the liver in VLDL as rapidly as it is synthesized. FFA from the circulation are the main source during starvation, the feeding of high-fat diets, or in diabetes mellitus, when hepatic lipogenesis is inhibited. Synthesis of VLDL takes place in the endoplasmic reticulum (ER) and requires the **microsomal triacylglycerol transfer proten (MTP)**, which transfers triacylglycerol from the cytosol into the ER lumen where it is incorporated into particles with cholesterol, phospholipids and apoB-100 (Figure 25–6). Factors that enhance both the synthesis of triacylglycerol and the secretion of VLDL by the liver include (1) the fed state rather than the starved state; (2) the feeding of diets high in carbohydrate (particularly if they contain sucrose or fructose), leading to high rates of lipogenesis and esterification of fatty acids; (3) high levels of circulating FFA; (4) ingestion of ethanol; and (5) the presence of high concentrations of insulin and low concentrations of glucagon, which enhance fatty acid synthesis and esterification and inhibit their oxidation (Figure 25–6).

CLINICAL ASPECTS

Imbalance in the Rate of Triacylglycerol Formation & Export Causes Fatty Liver

For a variety of reasons, lipid—mainly as triacylglycerol—can accumulate in the liver (Figure 25–6). Extensive accumulation is regarded as a pathologic condition. **Nonalcoholic fatty liver disease (NAFLD)** is the most common liver disorder worldwide. When accumulation of lipid in the liver becomes chronic, inflammatory and fibrotic changes may develop leading to **nonalcoholic steatohepatitis (NASH),** which can progress to liver diseases including **cirrhosis, hepatocarcinoma, and liver failure.**

Fatty livers fall into two main categories. The first type is associated with **raised levels of plasma free fatty acids** resulting from mobilization of fat from adipose tissue or from the hydrolysis of lipoprotein triacylglycerol by lipoprotein lipase in extrahepatic tissues. The production of VLDL does not keep pace with the increasing influx and esterification of free fatty acids, allowing triacylglycerol to accumulate, which in turn causes a fatty liver. This occurs during **starvation** and the feeding of **high-fat diets.** The ability to secrete VLDL may also be impaired (eg, in starvation). In uncontrolled **diabetes mellitus, twin lamb disease,** and **ketosis in cattle,** fatty infiltration is sufficiently severe to cause visible pallor (fatty appearance) and enlargement of the liver with possible liver dysfunction.

The second type of fatty liver is usually due to a **metabolic block in the production of plasma lipoproteins,** thus allowing triacylglycerol to accumulate. Theoretically, the lesion may be due to (1) a block in apolipoprotein synthesis (or an increase in its degradation before it can be incorporated into VLDL), (2) a block in the synthesis of the lipoprotein from lipid and apolipoprotein, (3) a failure in provision of phospholipids that are found in lipoproteins, or (4) a failure in the secretory mechanism itself.

One type of fatty liver that has been studied extensively in rats is caused by a deficiency of **choline,** which has therefore been called a **lipotropic factor.** The antibiotic puromycin, ethionine (α-amino-γ-mercaptobutyric acid), carbon tetrachloride, chloroform, phosphorus, lead, and arsenic all cause fatty liver and a marked reduction in concentration of VLDL in rat blood. Choline will not protect the organism against these agents, but appears to aid in recovery. The action of carbon tetrachloride probably involves formation of free radicals causing lipid peroxidation. Some protection against this is provided by the antioxidant action of **vitamin E**-supplemented diets. The action of ethionine is thought to be caused by a reduction in availability of ATP due to its replacing methionine in S-adenosylmethionine, trapping available adenine and preventing synthesis of ATP. **Orotic acid** also causes fatty liver; it is believed to interfere with glycosylation of the lipoprotein, thus inhibiting release, and may also impair the recruitment of triacylglycerol to the particles. A deficiency of vitamin E enhances the hepatic necrosis of the choline deficiency type of fatty liver. Added vitamin E or a source of **selenium** has a protective effect by combating lipid

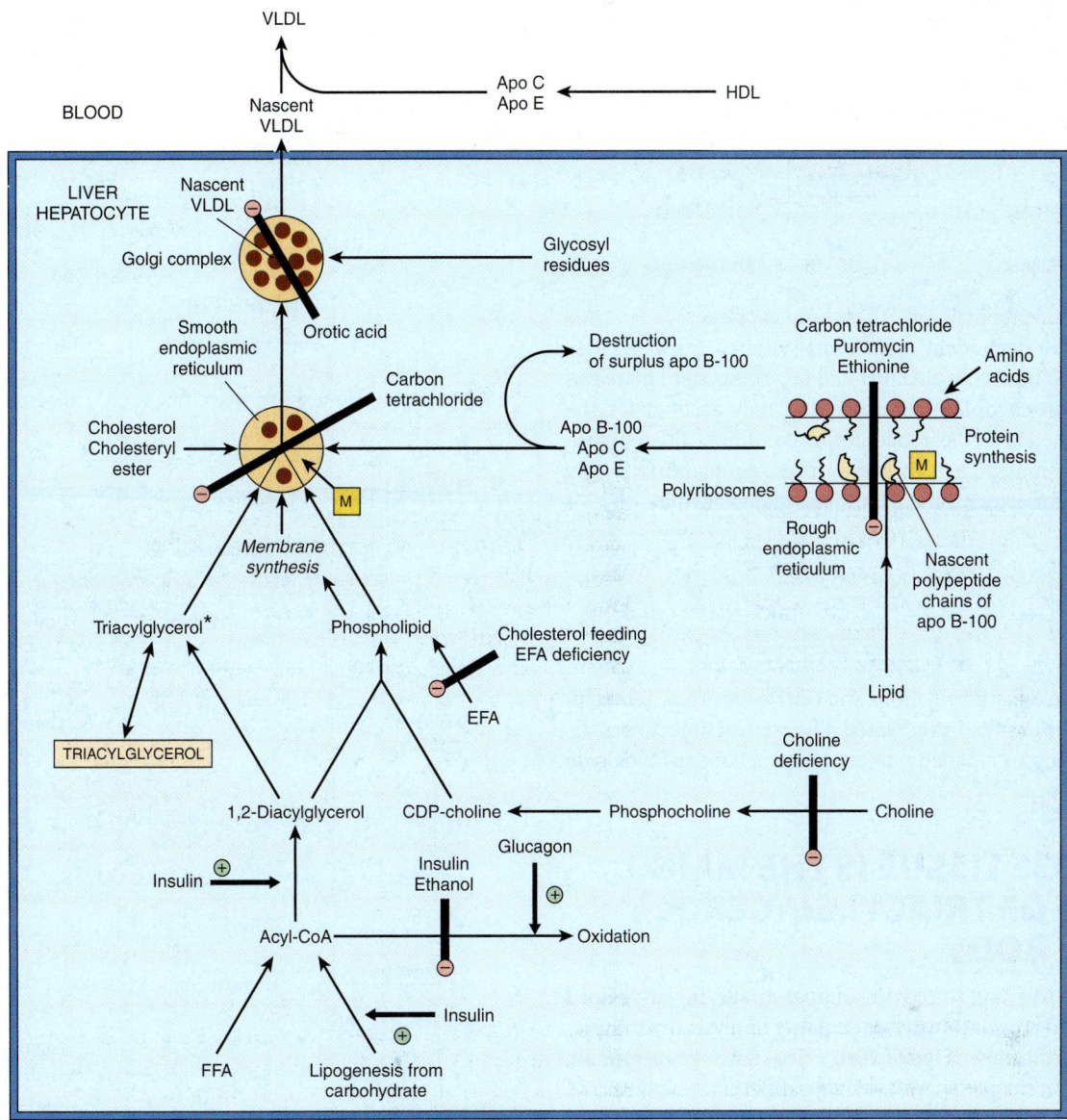

FIGURE 25–6 **The synthesis of very low density lipoprotein (VLDL) in the liver and the possible loci of action of factors causing accumulation of triacylglycerol and fatty liver.** (Apo, apolipoprotein; EFA, essential fatty acids; FFA, free fatty acids; HDL, high-density lipoproteins; M, microsomal triacylglycerol transfer protein.) The pathways indicated form a basis for events depicted in Figure 25–2. The main triacylglycerol pool in liver is not on the direct pathway of VLDL synthesis from acyl-CoA. Thus, FFA, insulin, and glucagon have immediate effects on VLDL secretion as their effects impinge directly on the small triacylglycerol precursor pool*. In the fully fed state, apo B-100 is synthesized in excess of requirements for VLDL secretion and the surplus is destroyed in the liver. During translation of apo B-100 in the rough endoplasmic reticulum, microsomal transfer protein-mediated lipid transport enables lipid to become associated with the nascent polypeptide chain. After release from the ribosomes, these particles fuse with more lipids from the smooth endoplasmic reticulum, producing nascent VLDL.

peroxidation. In addition to protein deficiency, essential fatty acid and vitamin deficiencies (eg, linoleic acid, pyridoxine, and pantothenic acid) can cause fatty infiltration of the liver.

Ethanol Also Causes Fatty Liver

Alcoholic fatty liver is the first stage in **alcoholic liver disease (ALD)** which is caused by **alcoholism** and ultimately leads to **cirrhosis.** The fat accumulation in the liver is caused by a combination of impaired fatty acid oxidation and increased lipo-

genesis, which is thought to be due to changes in the [NADH]/[NAD⁺] redox potential in the liver, and also to interference with the action of transcription factors regulating the expression of the enzymes involved in the pathways. Oxidation of ethanol by **alcohol dehydrogenase** leads to excess production of NADH, which competes with reducing equivalents from other substrates, including fatty acids, for the respiratory chain. This inhibits their oxidation and causes increased esterification of fatty acids to form triacylglycerol, resulting in the fatty liver. Oxidation of ethanol leads to the formation of acetaldehyde, which is oxidized by **aldehyde dehydrogenase,** producing acetate. The increased (NADH)/(NAD⁺) ratio also causes increased (lactate)/(pyruvate), resulting in **hyperlacticacidemia,** which decreases excretion of uric acid, aggravating **gout.**

$$CH_3 - CH_2 - OH \xrightarrow[\text{NAD}^+ \quad \text{NADH} + \text{H}^+]{\text{ALCOHOL DEHYDROGENASE}} CH_3 - CHO$$

Ethanol   **Acetaldehyde**

Some metabolism of ethanol takes place via a cytochrome P450-dependent microsomal ethanol oxidizing system (MEOS) involving NADPH and O_2. This system increases in activity in **chronic alcoholism** and may account for the increased metabolic clearance in this condition. Ethanol also inhibits the metabolism of some drugs, eg, barbiturates, by competing for cytochrome P450-dependent enzymes.

$$CH_3 - CH_2 - OH + NADPH + H^+ + O_2 \xrightarrow{\text{MEOS}} CH_3 - CHO + NADP^+ + 2H_2O$$

Ethanol   **Acetaldehyde**

In some Asian populations and Native Americans, alcohol consumption results in increased adverse reactions to acetaldehyde owing to a genetic defect of mitochondrial aldehyde dehydrogenase.

ADIPOSE TISSUE IS THE MAIN STORE OF TRIACYLGLYCEROL IN THE BODY

Triacylglycerols are stored in adipose tissue in large **lipid droplets** and are continually undergoing lipolysis (hydrolysis) and re-esterification (**Figure 25–7**). These two processes are entirely different pathways involving different reactants and enzymes. This allows the processes of esterification or lipolysis to be regulated separately by many nutritional, metabolic, and hormonal factors. The balance between these two processes determines the magnitude of the FFA pool in adipose tissue, which in turn determines the level of FFA circulating in the plasma. Since the latter has most profound effects upon the metabolism of other tissues, particularly liver and muscle, the factors operating in adipose tissue that regulate the outflow of FFA exert an influence far beyond the tissue itself.

The Provision of Glycerol 3-Phosphate Regulates Esterification: Lipolysis Is Controlled by Hormone-Sensitive Lipase

Triacylglycerol is synthesized from acyl-CoA and glycerol 3-phosphate (Figure 24–2). Since the enzyme **glycerol kinase** is not expressed in adipose tissue, glycerol cannot be utilized for the provision of glycerol 3-phosphate, which must be supplied from glucose via glycolysis.

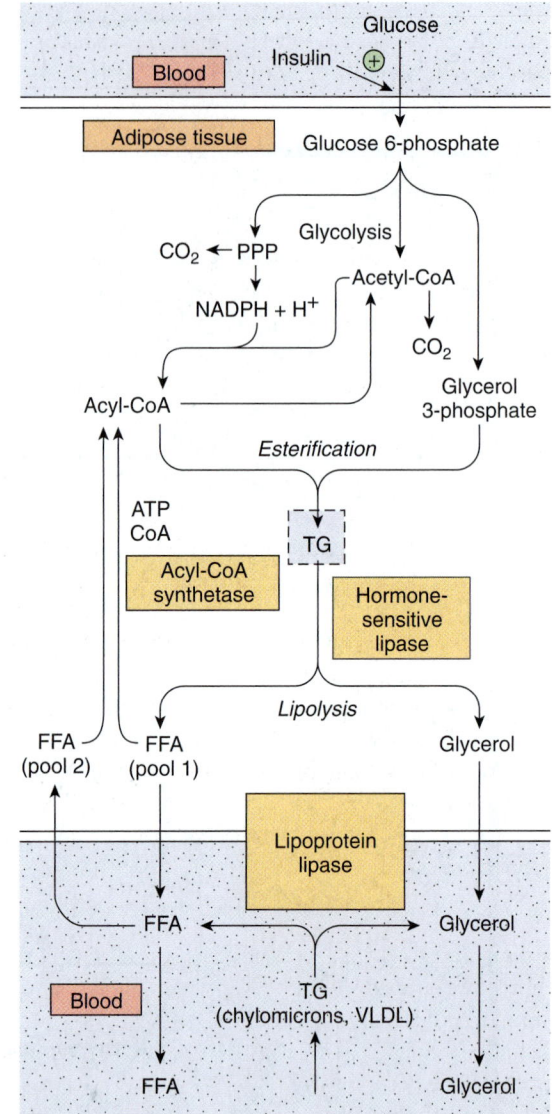

FIGURE 25–7 **Triacylglycerol metabolism in adipose tissue.** Hormone-sensitive lipase is activated by ACTH, TSH, glucagon, epinephrine, norepinephrine, and vasopressin and inhibited by insulin, prostaglandin E₁, and nicotinic acid. Details of the formation of glycerol 3-phosphate from intermediates of glycolysis are shown in Figure 24–2. (FFA, free fatty acids; PPP, pentose phosphate pathway; TG, triacylglycerol; VLDL, very low density lipoprotein.)

Triacylglycerol undergoes hydrolysis by a **hormone-sensitive lipase** to form FFA and glycerol. This lipase is distinct from lipoprotein lipase, which catalyzes lipoprotein triacylglycerol hydrolysis before its uptake into extrahepatic tissues (see above). Since the glycerol cannot be utilized, it enters the blood and is taken up and transported to tissues such as the liver and kidney, which possess an active glycerol kinase. The FFA formed by lipolysis can be reconverted in adipose tissue to acyl-CoA by **acyl-CoA synthetase** and reesterified with glycerol 3-phosphate to form triacylglycerol. Thus, **there is a continuous cycle of lipolysis and re-esterification within the tissue.** However, when the rate of re-esterification is not sufficient to match the rate of lipolysis, FFA accumulate and diffuse into the plasma, where they bind to albumin and raise the concentration of plasma-free fatty acids.

Increased Glucose Metabolism Reduces the Output of FFA

When the utilization of glucose by adipose tissue is increased, the FFA outflow decreases. However, the release of glycerol continues, demonstrating that the effect of glucose is not mediated by reducing the rate of lipolysis. The effect is due to the provision of glycerol 3-phosphate, which enhances esterification of FFA. Glucose can take several pathways in adipose tissue, including oxidation to CO_2 via the citric acid cycle, oxidation in the pentose phosphate pathway, conversion to long-chain fatty acids, and formation of acylglycerol via glycerol 3-phosphate (Figure 25–7). When glucose utilization is high, a larger proportion of the uptake is oxidized to CO_2 and converted to fatty acids. However, as total glucose utilization decreases, the greater proportion of the glucose is directed to the formation of glycerol 3-phosphate for the esterification of acylCoA, which helps to minimize the efflux of FFA.

HORMONES REGULATE FAT MOBILIZATION

Adipose Tissue Lipolysis Is Inhibited by Insulin

The rate of release of FFA from adipose tissue is affected by many hormones that influence either the rate of esterification or the rate of lipolysis. **Insulin** inhibits the release of FFA from adipose tissue, which is followed by a fall in circulating plasma free fatty acids. Insulin also enhances lipogenesis and the synthesis of acylglycerol and increases the oxidation of glucose to CO_2 via the pentose phosphate pathway. All of these effects are dependent on the presence of glucose and can be explained, to a large extent, on the basis of the ability of insulin to enhance the uptake of glucose into adipose cells via the **GLUT 4 transporter**. In addition, insulin increases the activity of the enzymes pyruvate dehydrogenase, acetyl-CoA carboxylase, and glycerol phosphate acyltransferase, reinforcing the effects of increased glucose uptake on the enhancement of fatty acid and acylglycerol synthesis. These three enzymes are regulated in a coordinate manner by phosphorylation–dephosphorylation mechanisms.

Another principal action of insulin in adipose tissue is to inhibit the activity of **hormone-sensitive lipase,** reducing the release not only of FFA but also of glycerol. Adipose tissue is much more sensitive to insulin than many other tissues, which points to adipose tissue as a major site of insulin action in vivo.

Several Hormones Promote Lipolysis

Other hormones accelerate the release of FFA from adipose tissue and raise the plasma-free fatty acid concentration by increasing the rate of lipolysis of the triacylglycerol stores (**Figure 25–8**). These include **epinephrine, norepinephrine, glucagon, adrenocorticotropic hormone (ACTH), α- and β-melanocyte-stimulating hormones (MSH), thyroid-stimulating hormone (TSH), growth hormone (GH), and vasopressin.** Many of these activate hormone-sensitive lipase. For an optimal effect, most of these lipolytic processes require the presence of **glucocorticoids** and **thyroid hormones.** These hormones act in a **facilitatory** or **permissive** capacity with respect to other lipolytic endocrine factors.

The hormones that act rapidly in promoting lipolysis, ie, catecholamines (epinephrine and nor-epinephrine), do so by stimulating the activity of **adenylyl cyclase,** the enzyme that converts ATP to cAMP. The mechanism is analogous to that responsible for hormonal stimulation of glycogenolysis (Chapter 19). cAMP, by stimulating **cAMP-dependent protein kinase,** activates hormone-sensitive lipase. Thus, processes which destroy or preserve cAMP influence lipolysis. cAMP is degraded to 5′-AMP by the enzyme **cyclic 3′,5′-nucleotide phosphodiesterase.** This enzyme is inhibited by methylxanthines such as **caffeine** and **theophylline. Insulin** antagonizes the effect of the lipolytic hormones. Lipolysis appears to be more sensitive to changes in concentration of insulin than are glucose utilization and esterification. The antilipolytic effects of insulin, nicotinic acid, and prostaglandin E_1 are accounted for by inhibition of the synthesis of cAMP at the adenylyl cyclase site, acting through a G_i protein. Insulin also stimulates phosphodiesterase and the lipase phosphatase that inactivates hormone-sensitive lipase. The effect of growth hormone in promoting lipolysis is dependent on synthesis of proteins involved in the formation of cAMP. Glucocorticoids promote lipolysis via synthesis of new lipase protein by a cAMP-independent pathway, which may be inhibited by insulin, and also by promoting transcription of genes involved in the cAMP signal cascade. These findings help to explain the role of the pituitary gland and the adrenal cortex in enhancing fat mobilization. Adipose tissue secretes hormones such as **adiponectin,** which modulates glucose and lipid metabolism in muscle and liver, and **leptin,** which regulates energy homeostasis. Current evidence suggests that the main role of leptin in humans is to suppress appetite when food intake is sufficient. If it is lacking, food intake may be uncontrolled, causing obesity.

The sympathetic nervous system, through liberation of norepinephrine in adipose tissue, plays a central role in the mobilization of FFA. Thus, the increased lipolysis caused by many of the factors described above can be reduced or abolished by denervation of adipose tissue or by ganglionic blockade.

Perilipin Regulates the Balance Between Triacylglycerol Storage and Lipolysis in Adipocytes

Perilipin, a protein involved in the formation of lipid droplets in adipocytes, inhibits lipolysis in basal conditions by preventing access of the lipase enzymes to the stored triacylglycerols.

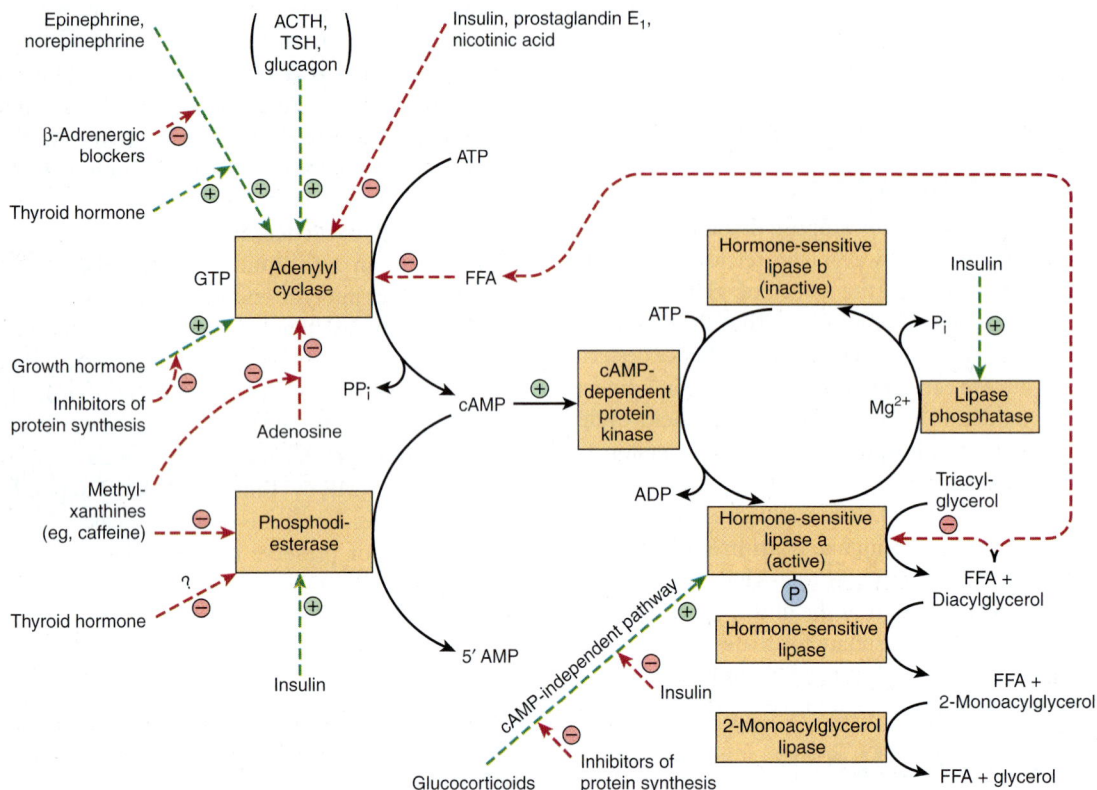

FIGURE 25–8 **Control of adipose tissue lipolysis.** (FFA, free fatty acids; TSH, thyroid-stimulating hormone.) Note the cascade sequence of reactions affording amplification at each step. The lipolytic stimulus is "switched off" by removal of the stimulating hormone; the action of lipase phosphatase; the inhibition of the lipase and adenylyl cyclase by high concentrations of FFA; the inhibition of adenylyl cyclase by adenosine; and the removal of cAMP by the action of phosphodiesterase. ACTH, TSH, and glucagon may not activate adenylyl cyclase in vivo since the concentration of each hormone required in vitro is much higher than is found in the circulation. Positive (⊕) and negative (⊖) regulatory effects are represented by broken lines and substrate flow by solid lines.

On stimulation with hormones which promote triacylglycerol degradation, however, the protein targets hormone-sensitive lipase to the lipid droplet surface and thus promotes lipolysis. Perilipin, therefore, enables the storage and breakdown of triacylglycerol to be coordinated according to the metabolic needs of the body.

Human Adipose Tissue May Not Be an Important Site of Lipogenesis

In adipose tissue, there is no significant incorporation of glucose or pyruvate into long-chain fatty acids, ATP-citrate lyase, a key enzyme in lipogenesis, does not appear to be present, and other lipogenic enzymes—eg, glucose-6-phosphate dehydrogenase and the malic enzyme—do not undergo adaptive changes. Indeed, it has been suggested that in humans there is a **"carbohydrate excess syndrome"** due to a unique limitation in ability to dispose of excess carbohydrate by lipogenesis. In birds, lipogenesis is confined to the liver, where it is particularly important in providing lipids for egg formation, stimulated by estrogens.

BROWN ADIPOSE TISSUE PROMOTES THERMOGENESIS

Brown adipose tissue is involved in metabolism, particularly at times when heat generation is necessary. Thus, the tissue is extremely active in some species, for example, during arousal from hibernation, in animals exposed to cold (nonshivering thermogenesis), and in heat production in the newborn. Though not a prominent tissue in humans, it is present in normal individuals, where it could be responsible for **"diet-induced thermogenesis."** It is noteworthy that brown adipose tissue is reduced or absent in obese persons. The tissue is characterized by a well-developed blood supply and a high content of mitochondria and cytochromes, but low activity of ATP synthase. Metabolic emphasis is placed on oxidation of both glucose and fatty acids. **Norepinephrine** liberated from sympathetic nerve endings is important in increasing lipolysis in the tissue and increasing synthesis of lipoprotein lipase to enhance utilization of triacylglycerol-rich lipoproteins from the circulation. Oxidation and phosphorylation are not coupled in mitochondria of this tissue, and the phosphorylation

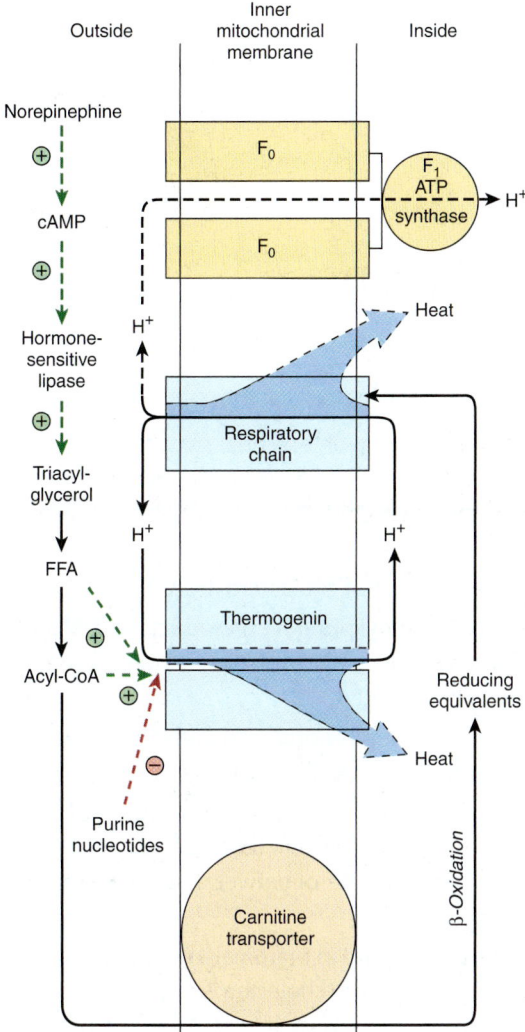

FIGURE 25–9 **Thermogenesis in brown adipose tissue.**
Activity of the respiratory chain produces heat in addition to
translocating protons (Chapter 13). These protons dissipate more
heat when returned to the inner mitochondrial compartment via
thermogenin instead of via the F_1 ATP synthase, the route that
generates ATP (Figure 13–7). The passage of H^+ via thermogenin
is inhibited by purine nucleotides when brown adipose tissue is
unstimulated. Under the influence of norepinephrine, the inhibition
is removed by the production of free fatty acids (FFA) and acyl-CoA.
Note the dual role of acyl-CoA in both facilitating the action of
thermogenin and supplying reducing equivalents for the respiratory
chain. ⊕ and ⊖ signify positive or negative regulatory effects.

that does occur is at the substrate level, eg, at the succinate
thiokinase step and in glycolysis. Thus, **oxidation produces
much heat, and little free energy is trapped in ATP.** A ther-
mogenic uncoupling protein, **thermogenin,** acts as a proton
conductance pathway dissipating the electrochemical potential across the mitochondrial membrane (**Figure 25–9**).

SUMMARY

- Since nonpolar lipids are insoluble in water, for transport
  between the tissues in the aqueous blood plasma they are

combined with amphipathic lipids and proteins to make
water-miscible lipoproteins.

- Four major groups of lipoproteins are recognized. Chylomicrons
  transport lipids resulting from digestion and absorption. Very
  low density lipoproteins (VLDL) transport triacylglycerol from
  the liver. Low-density lipoproteins (LDL) deliver cholesterol
  to the tissues, and high-density lipoproteins (HDL) remove
  cholesterol from the tissues and return it to the liver for excretion
  in the process known as reverse cholesterol transport.

- Chylomicrons and VLDL are metabolized by hydrolysis of
  their triacylglycerol, and lipoprotein remnants are left in
  the circulation. These are taken up by liver, but some of the
  remnants (IDL), resulting from VLDL form LDL, which is
  taken up by the liver and other tissues via the LDL receptor.

- Apolipoproteins constitute the protein moiety of lipoproteins.
  They act as enzyme activators (eg, apo C-II and apo A-I) or as
  ligands for cell receptors (eg, apo A-I, apo E, and apo B-100).

- Triacylglycerol is the main storage lipid in adipose tissue.
  Upon mobilization, FFA and glycerol are released. FFA are an
  important fuel source.

- Brown adipose tissue is the site of "nonshivering
  thermogenesis." It is found in hibernating and newborn
  animals and is present in small quantity in humans.
  Thermogenesis results from the presence of an uncoupling
  protein, thermogenin, in the inner mitochondrial membrane.

REFERENCES

Arner P: Human fat cell lipolysis: biochemistry, regulation and
  clinical role. Best Pract Res Clin Endocrinol Metab 2005;19:471.

Brasaemle DL: Thematic review series: adipocyte biology.
  The perilipin family of structural lipid droplet proteins:
  stabilization of lipid droplets and control of lipolysis.
  J Lipid Res 2007;48:2547.

Fielding CJ, Fielding PE: Dynamics of lipoprotein transport in the
  circulatory system. In *Biochemistry of Lipids, Lipoproteins and
  Membranes,* 5th ed. Vance DE, Vance JE (editors). Elsevier,
  2008;533–554.

Goldberg IJ, Merkel M: Lipoprotein lipase: physiology, biochemistry
  and molecular biology. Front Biosci 2001;6:D388.

Kershaw EE, Flier JS: Adipose tissue as an endocrine organ. J Clin
  Endocrinol Metab 2004;89:2548.

Lass A, Zimmermann R, Oberer M, et al: Lipolysis—a highly
  regulated multi-enzyme complex mediates the catabolism of
  cellular fat stores. Prog Lipid Res 2011;50:14.

Lenz A, Diamond FB: Obesity: the hormonal milieu. Curr Opin
  Endocrinol Diabetes Obes 2008;15:9.

Redgrave TG: Chylomicron metabolism. Biochem Soc Trans
  2004;32:79.

Schreuder TC, Verwer BJ, van Nieuwkerk CM, et al: Nonalcoholic
  fatty liver disease: an overview of current insights in pathogenesis,
  diagnosis and treatment. World J Gastroenterol 2008;14:2474.

Sell H, Deshaies Y, Richard D: The brown adipocyte: update on its
  metabolic role. Int J Biochem Cell Biol 2004;36:2098.

Vance JE, Adeli K: Assembly and secretion of triacylglycerol-
  rich lipoproteins. In *Biochemistry of Lipids, Lipoproteins and
  Membranes,* 5th ed. Vance DE, Vance JE (editors). Elsevier,
  2008;507–532.

Cholesterol Synthesis, Transport, & Excretion

Kathleen M. Botham, PhD, DSc & Peter A. Mayes, PhD, DSc

OBJECTIVES

After studying this chapter, you should be able to:

- Appreciate the importance of cholesterol as an essential structural component of cell membranes and as a precursor of all other steroids in the body, and indicate its pathological role in cholesterol gallstone disease and atherosclerosis development.

- Identify the five stages in the biosynthesis of cholesterol from acetyl-CoA.

- Understand the role of 3-hydroxy-3-methylglutaryl CoA reductase (HMG-CoA reductase) in controlling the rate of cholesterol synthesis and explain the mechanisms by which its activity is regulated.

- Appreciate that cholesterol balance in cells is tightly regulated and indicate the factors involved in maintaining the correct balance.

- Explain the role of plasma lipoproteins, including chylomicrons, very low density lipoprotein (VLDL), low-density lipoprotein (LDL), and high-density lipoprotein (HDL), in the transport of cholesterol between tissues in the plasma.

- Name the two main primary bile acids found in mammals, outline the pathways by which they are synthesized from cholesterol in the liver, and understand the role of cholesterol 7α-hydroxylase in regulating the process.

- Appreciate the importance of bile acid synthesis not only in the digestion and absorption of fats but also as a major excretory route for cholesterol.

- Indicate how secondary bile acids are produced from primary bile acids by intestinal bacteria.

- Explain what is meant by the "enterohepatic circulation" and why it is important.

- Identify the lifestyle factors that influence plasma cholesterol concentrations and thus affect the risk of coronary heart disease.

- Understand that the class of lipoprotein in which cholesterol is carried is important in determining the effects of plasma cholesterol on atherosclerosis development, with high levels of VLDL or LDL being deleterious and high levels of HDL being beneficial.

- Give examples of inherited and noninherited conditions affecting lipoprotein metabolism that cause hypo- or hyperlipoproteinemia.

BIOMEDICAL IMPORTANCE

Cholesterol is present in tissues and in plasma either as free cholesterol or combined with a long-chain fatty acid as cholesteryl ester, the storage form. In plasma, both forms are transported in lipoproteins (Chapter 25). Cholesterol is an amphipathic lipid and as such is an essential structural component of membranes, where it is important for the maintenance of the correct permeability and fluidity, and of the outer layer of plasma lipoproteins. It is synthesized in many tissues from acetyl-CoA and is the precursor of all other steroids in the body, including corticosteroids, sex hormones, bile acids, and vitamin D. As a typical product of animal metabolism, cholesterol occurs in foods of animal origin such as egg yolk, meat, liver, and brain. Plasma **low-density lipoprotein (LDL)** is the vehicle that supplies cholesterol and cholesteryl ester to many tissues. Free cholesterol is removed from tissues by plasma **high-density lipoprotein (HDL)** and transported to the liver, where it is eliminated from the body either unchanged or after conversion to bile acids in the process known as **reverse cholesterol transport** (Chapter 25). Cholesterol is a major constituent of **gallstones.** However, its chief role in pathologic processes is as a factor in the genesis of **atherosclerosis** of vital arteries, causing cerebrovascular, coronary, and peripheral vascular disease.

CHOLESTEROL IS DERIVED ABOUT EQUALLY FROM THE DIET & FROM BIOSYNTHESIS

A little more than half the cholesterol of the body arises by synthesis (about 700 mg/d), and the remainder is provided by the average diet. The liver and intestine account for approximately 10% each of total synthesis in humans. Virtually all tissues containing nucleated cells are capable of cholesterol synthesis, which occurs in the endoplasmic reticulum and the cytosolic compartments.

Acetyl-CoA Is the Source of All Carbon Atoms in Cholesterol

The biosynthesis of cholesterol may be divided into five steps: (1) synthesis of **mevalonate** from acetyl-CoA (**Figure 26–1**); (2) formation of **isoprenoid units** from mevalonate by loss of CO_2 (**Figure 26–2**); (3) condensation of six isoprenoid units form **squalene** (Figure 26–2); (4) cyclization of squalene give rise to the parent steroid, **lanosterol.** (5) formation of cholesterol from lanosterol (**Figure 26–3**).

Step 1—Biosynthesis of Mevalonate: HMG-CoA (3-hydroxy-3-methylglutaryl-CoA) is formed by the reactions used in mitochondria to synthesize ketone bodies (Figure 22–7). However, since cholesterol synthesis is extramitochondrial, the two pathways are distinct. Initially, two molecules of acetyl-CoA condense to form acetoacetyl-CoA catalyzed by cytosolic **thiolase.** Acetoacetyl-CoA

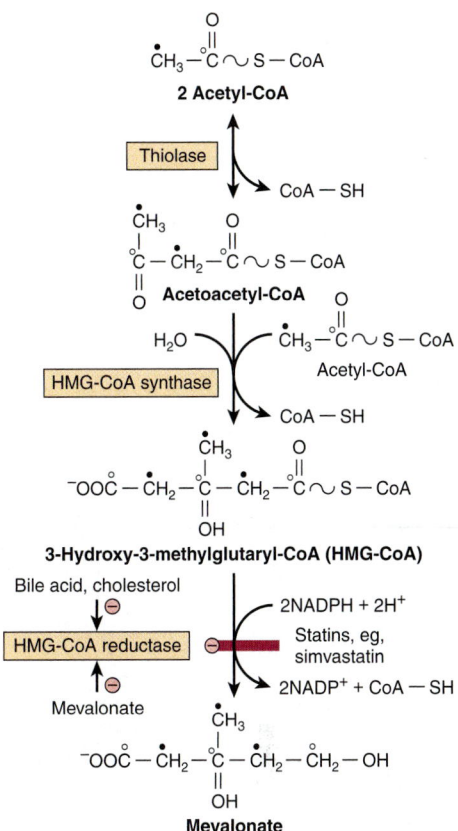

FIGURE 26–1 **Biosynthesis of mevalonate.** HMG-CoA reductase is inhibited by statins. The open and solid circles indicate the fate of each of the carbons in the acetyl moiety of acetyl-CoA.

condenses with a further molecule of acetyl-CoA catalyzed by **HMG-CoA synthase** to form HMG-CoA, which is reduced to **mevalonate** by NADPH in a reaction catalyzed by **HMG-CoA reductase.** This last step is the principal regulatory step in the pathway of cholesterol synthesis and is the site of action of the most effective class of cholesterol-lowering drugs, the statins, which are HMG-CoA reductase inhibitors (Figure 26–1).

Step 2—Formation of Isoprenoid Units: Mevalonate is phosphorylated sequentially using ATP by three kinases, and after decarboxylation (Figure 26–2) the active isoprenoid unit, **isopentenyl diphosphate,** is formed.

Step 3—Six Isoprenoid Units Form Squalene: Isopentenyl diphosphate is isomerized by a shift of the double bond to form **dimethylallyl diphosphate,** and then condensed with another molecule of isopentenyl diphosphate to form the ten-carbon intermediate **geranyl diphosphate** (Figure 26–2). A further condensation with isopentenyl diphosphate forms **farnesyl diphosphate.** Two molecules of farnesyl diphosphate condense at the diphosphate end to form **squalene.** Initially, inorganic pyrophosphate is eliminated, forming presqualene diphosphate, which is then reduced by NADPH with elimination of a further inorganic pyrophosphate molecule.

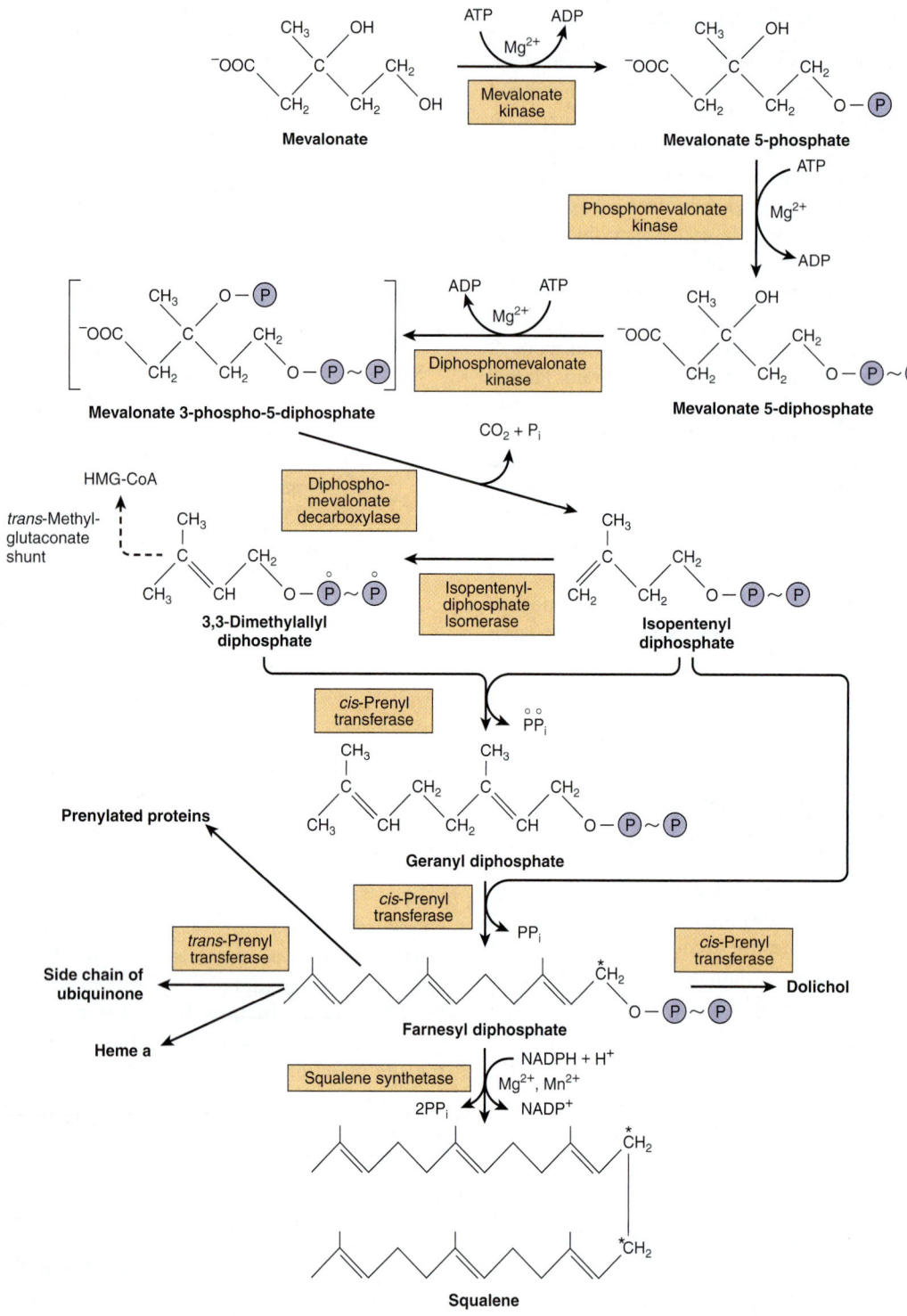

FIGURE 26–2 **Biosynthesis of squalene, ubiquinone, dolichol, and other polyisoprene derivatives.** (HMG-CoA, 3-hydroxy-3-methylglutaryl-CoA.) A farnesyl residue is present in heme a of cytochrome oxidase. The carbon marked with an asterisk becomes C_{11} or C_{12} in squalene. Squalene synthetase is a microsomal enzyme; all other enzymes indicated are soluble cytosolic proteins, and some are found in peroxisomes.

Step 4—Formation of Lanosterol: Squalene can fold into a structure that closely resembles the steroid nucleus (Figure 26–3). Before ring closure occurs, squalene is converted to squalene 2,3-epoxide by a mixed-function oxidase in the endoplasmic reticulum, **squalene epoxidase.** The methyl group on C_{14} is transferred to C_{13} and that on C_8 to C_{14} as cyclization occurs, catalyzed by **oxidosqualene-lanosterol cyclase.**

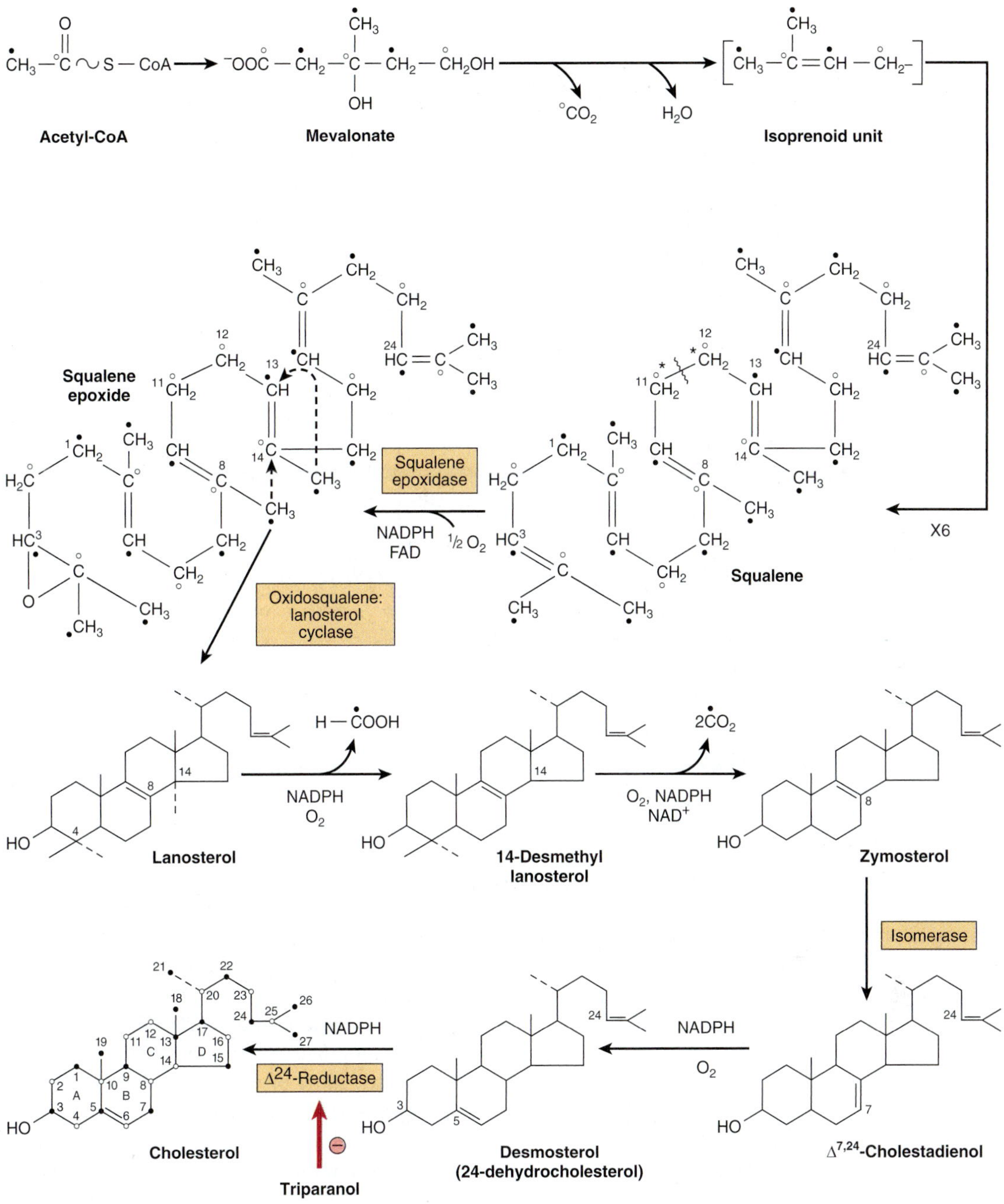

FIGURE 26–3 **Biosynthesis of cholesterol.** The numbered positions are those of the steroid nucleus and the open and solid circles indicate the fate of each of the carbons in the acetyl moiety of acetyl-CoA. (*Refer to labeling of squalene in Figure 26–2.)

Step 5—Formation of Cholesterol: The formation of cholesterol from **lanosterol** takes place in the membranes of the endoplasmic reticulum and involves changes in the steroid nucleus and the side chain (Figure 26–3). The methyl groups on C_{14} and C_4 are removed to form 14-desmethyl lanosterol and then zymosterol. The double bond at C_8–C_9 is subsequently moved to C_5–C_6 in two steps, forming **desmosterol.** Finally, the double bond of the side chain is reduced, producing cholesterol.

Farnesyl Diphosphate Gives Rise to Dolichol & Ubiquinone

The polyisoprenoids **dolichol** (Figure 15–20 and Chapter 47) and **ubiquinone** (Figure 13–5) are formed from farnesyl diphosphate by the further addition of up to 16 (dolichol) or 3–7 (ubiquinone) isopentenyl diphosphate residues (Figure 26–2). Some **GTP-binding proteins** in the cell membrane are prenylated with farnesyl or geranylgeranyl

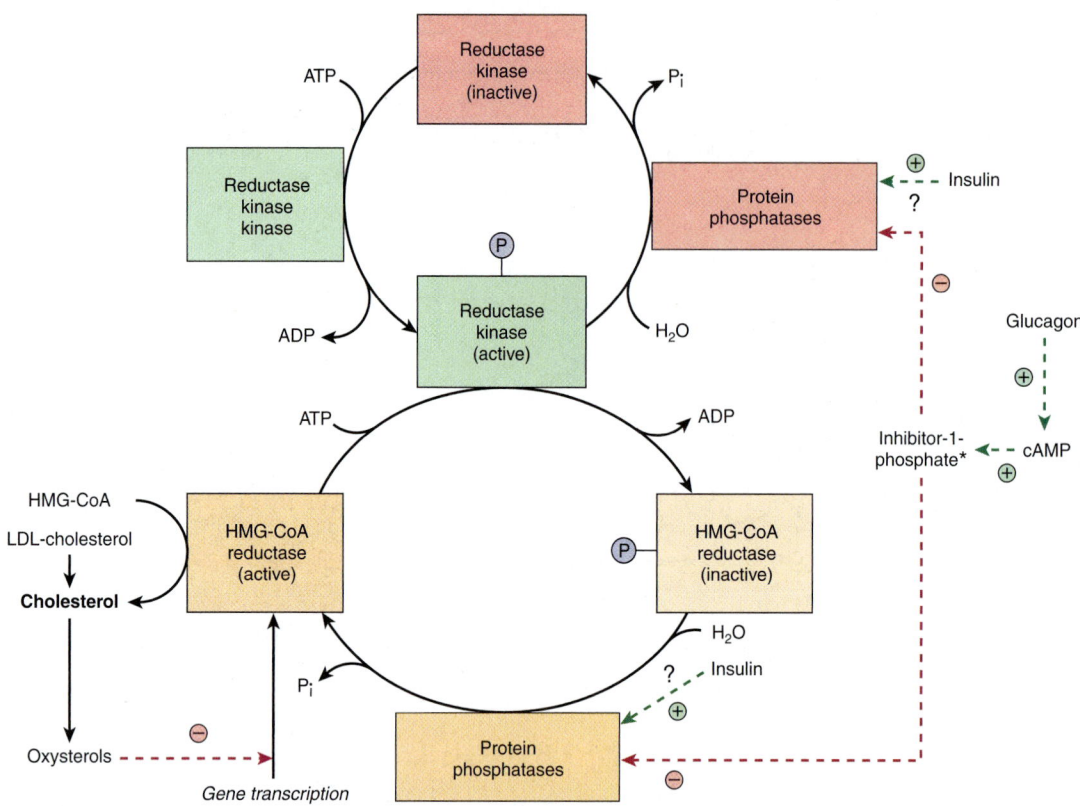

FIGURE 26–4 **Possible mechanisms in the regulation of cholesterol synthesis by HMG-CoA reductase.** Insulin has a dominant role compared with glucagon. (*See Figure 19–6.)

(20 carbon) residues. **Protein prenylation** is believed to facilitate the anchoring of proteins into lipoid membranes and may also be involved in protein–protein interactions and membrane-associated protein trafficking.

CHOLESTEROL SYNTHESIS IS CONTROLLED BY REGULATION OF HMG-COA REDUCTASE

Regulation of cholesterol synthesis is exerted near the beginning of the pathway, at the HMG-CoA reductase step. The decreased synthesis of cholesterol in starving animals is accompanied by reduced activity of the enzyme. However, it is only hepatic synthesis that is inhibited by dietary cholesterol. HMG-CoA reductase in liver is inhibited by mevalonate, the immediate product of the reaction, and by cholesterol, the main product of the pathway. Cholesterol and metabolites repress transcription of the HMG-CoA reductase via activation of a **sterol regulatory element-binding protein (SREBP)** transcription factor. SREBPs are a family of proteins that regulate the transcription of a range of genes involved in the cellular uptake and metabolism of cholesterol and other lipids. A **diurnal variation** occurs both in cholesterol synthesis and reductase activity. In addition

to these mechanisms regulating the rate of protein synthesis, the enzyme activity is also modulated more rapidly by post-translational modification (**Figure 26–4**). **Insulin or thyroid hormone** increases HMG-CoA reductase activity, whereas **glucagon or glucocorticoids** decrease it. Activity is reversibly modified by phosphorylation–dephosphorylation mechanisms, some of which may be cAMP-dependent and therefore immediately responsive to glucagon. Attempts to lower plasma cholesterol in humans by reducing the amount of cholesterol in the diet produce variable results. Generally, a decrease of 100 mg in dietary cholesterol causes a decrease of approximately 0.13 mmol/L of serum.

MANY FACTORS INFLUENCE THE CHOLESTEROL BALANCE IN TISSUES

In tissues, cholesterol balance is regulated as follows (**Figure 26–5**). An increase in cell cholesterol is caused by uptake of cholesterol-containing lipoproteins by receptors, for example, the LDL receptor or the scavenger receptor, uptake of free cholesterol from cholesterol-rich lipoproteins to the cell membrane, cholesterol synthesis, and hydrolysis of cholesteryl esters by the enzyme **cholesteryl ester**

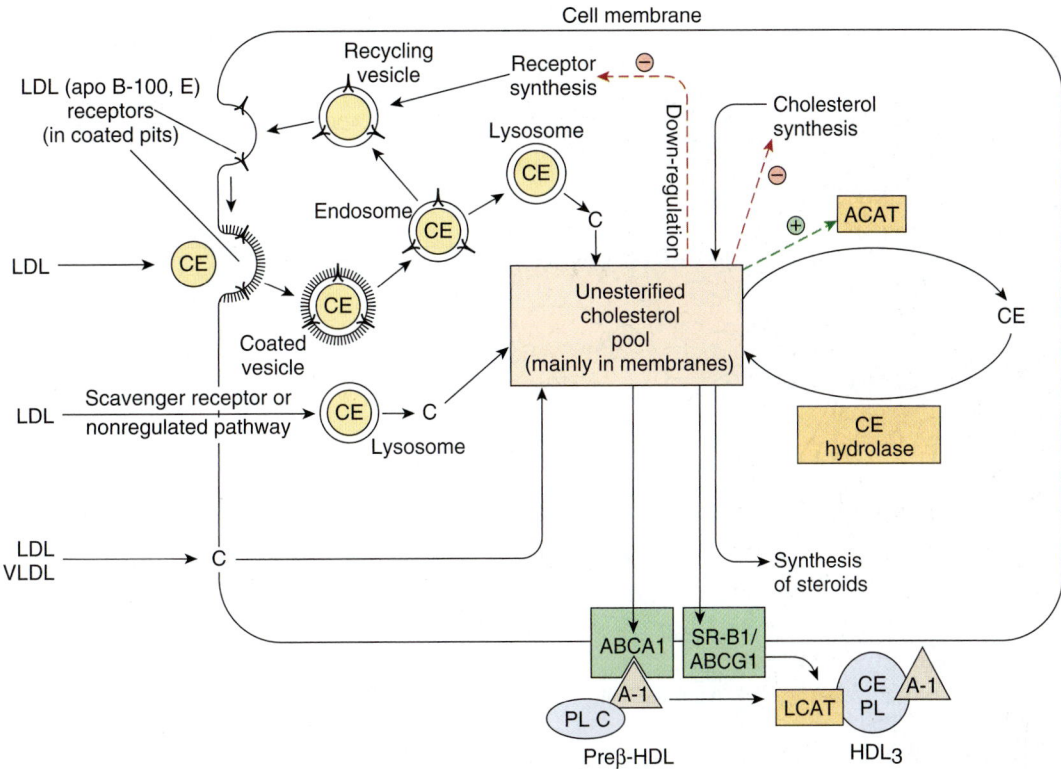

FIGURE 26–5 **Factors affecting cholesterol balance at the cellular level.** Reverse cholesterol transport may be mediated via the ABCA-1 transporter protein (with preβ-HDL as the exogenous acceptor) or the SR-B1 or ABCG-1 (with HDL₃ as the exogenous acceptor). (A-I, apolipoprotein A-I; ACAT, acyl-CoA:cholesterol acyltransferase; C, cholesterol; CE, cholesteryl ester; LCAT, lecithin:cholesterol acyltransferase; LDL, low-density lipoprotein; PL, phospholipid; VLDL, very low density lipoprotein.) LDL and HDL are not shown to scale.

hydrolase. A decrease is due to efflux of cholesterol from the membrane to HDL via the ABCA1, ABCG1, or SR-B1 (Figure 25–5); esterification of cholesterol by **ACAT** (acyl-CoA:cholesterol acyltransferase); and utilization of cholesterol for synthesis of other steroids, such as hormones, or bile acids in the liver.

The LDL Receptor Is Highly Regulated

LDL (apo B-100, E) receptors occur on the cell surface in pits that are coated on the cytosolic side of the cell membrane with a protein called **clathrin.** The glycoprotein receptor spans the membrane, the B-100 binding region being at the exposed amino terminal end. After binding, LDL is taken up intact by endocytosis. The apoprotein and cholesteryl ester are then hydrolyzed in the lysosomes, and cholesterol is translocated into the cell. The receptors are recycled to the cell surface. This influx of cholesterol inhibits the transcription of the genes encoding HMG-CoA synthase, HMG-CoA reductase, and other enzymes involved in cholesterol synthesis, as well as the LDL receptor itself, via the SREBP pathway, and thus coordinately suppresses cholesterol synthesis and uptake. In addition, ACAT activity is stimulated, promoting cholesterol esterification. In this way, the LDL receptor activity on the cell surface is regulated by the cholesterol requirement for membranes, steroid hormones, or bile acid synthesis (Figure 26–5).

CHOLESTEROL IS TRANSPORTED BETWEEN TISSUES IN PLASMA LIPOPROTEINS

Cholesterol is transported in plasma in lipoproteins, with the greater part in the form of cholesteryl ester (**Figure 26–6**), and in humans the highest proportion is found in LDL. Dietary cholesterol equilibrates with plasma cholesterol in days and with tissue cholesterol in weeks. Cholesteryl ester in the diet is hydrolyzed to cholesterol, which is then absorbed by the intestine together with dietary unesterified cholesterol and other lipids. With cholesterol synthesized in the intestines, it is then incorporated into chylomicrons (Chapter 25). Of the cholesterol absorbed, 80–90% is esterified with long-chain fatty acids in the intestinal mucosa. Ninety-five percent of the chylomicron cholesterol is delivered to the liver in chylomicron remnants, and most of the cholesterol secreted by the liver in very low density lipoprotein (VLDL) is retained during the formation of intermediate-density lipoprotein (IDL) and ultimately LDL, which is taken up by the LDL receptor in liver and extrahepatic tissues (Chapter 25).

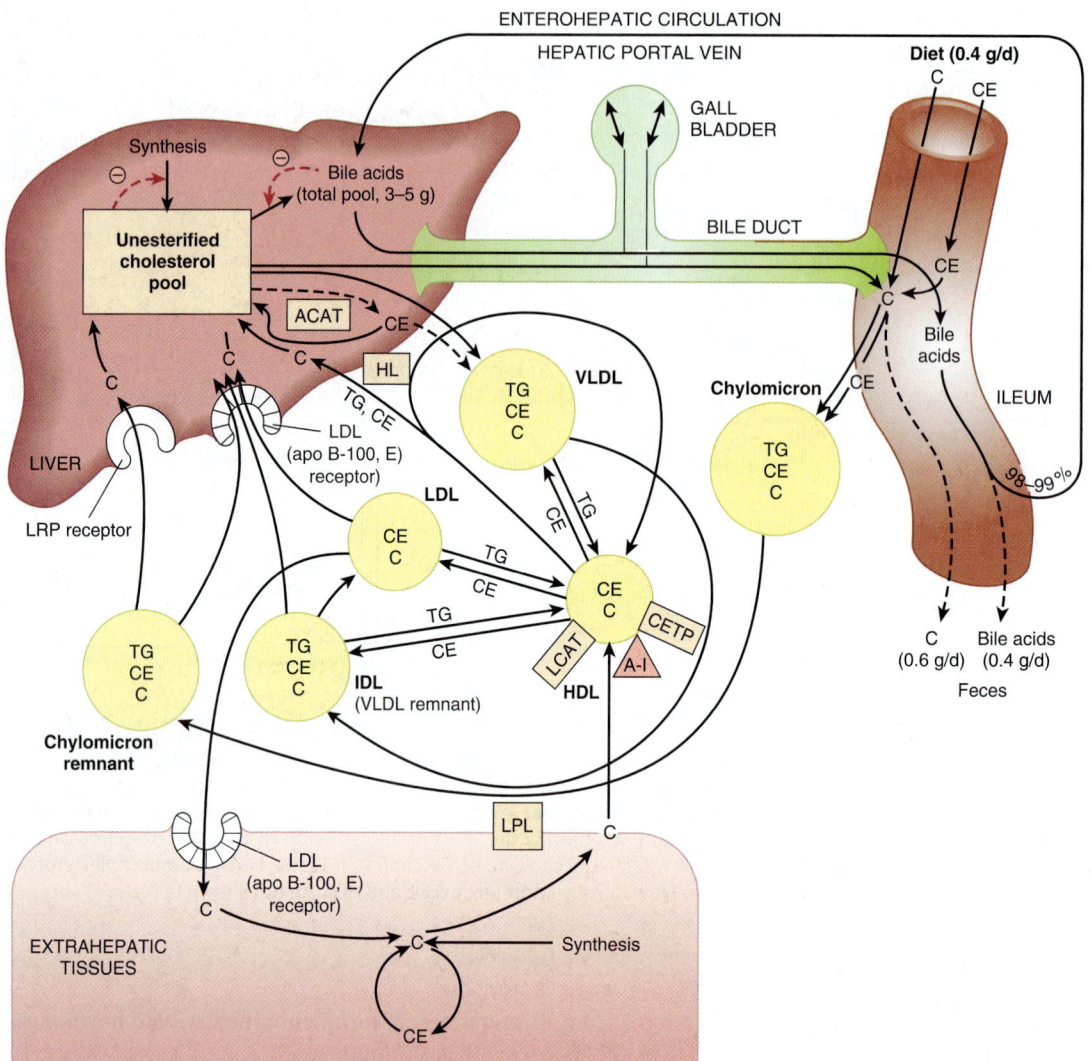

FIGURE 26–6 **Transport of cholesterol between the tissues in humans.** (A-I, apolipoprotein A-I; ACAT, acyl-CoA:cholesterol acyltransferase; C, unesterified cholesterol; CE, cholesteryl ester; CETP, cholesteryl ester transfer protein; HDL, high-density lipoprotein; HL, hepatic lipase; IDL, intermediate density lipoprotein; LCAT, lecithin:cholesterol acyltransferase; LDL, low-density lipoprotein; LPL, lipoprotein lipase; LRP, LDL receptor-related protein; TG, triacylglycerol; VLDL, very low density lipoprotein.)

Plasma LCAT Is Responsible for Virtually All Plasma Cholesteryl Ester in Humans

Lecithin: cholesterol acyltransferase (LCAT) activity is associated with HDL containing apo A-I. As cholesterol in HDL becomes esterified, it creates a concentration gradient and draws in cholesterol from tissues and from other lipoproteins (Figures 26–5 and 26–6), thus enabling HDL to function in **reverse cholesterol transport** (Figure 25–5).

Cholesteryl Ester Transfer Protein Facilitates Transfer of Cholesteryl Ester from HDL to Other Lipoproteins

Cholesteryl ester transfer protein, associated with HDL, is found in plasma of humans and many other species. It

facilitates transfer of cholesteryl ester from HDL to VLDL, IDL, and LDL in exchange for triacylglycerol, relieving product inhibition of the LCAT activity in HDL. Thus, in humans, much of the cholesteryl ester formed by LCAT finds its way to the liver via VLDL remnants (IDL) or LDL (Figure 26–6). The triacylglycerol-enriched HDL_2 delivers its cholesterol to the liver in the HDL cycle (Figure 25–5).

CHOLESTEROL IS EXCRETED FROM THE BODY IN THE BILE AS CHOLESTEROL OR BILE ACIDS (SALTS)

Cholesterol is excreted from the body via the bile either in the unesterified form or after conversion into bile acids in the liver.

Coprostanol is the principal sterol in the feces; it is formed from cholesterol by the bacteria in the lower intestine.

Bile Acids Are Formed from Cholesterol

The **primary bile acids** are synthesized in the liver from cholesterol. These are **cholic acid** (found in the largest amount) and **chenodeoxycholic acid** (**Figure 26–7**). The 7α-hydroxylation of cholesterol is the first and principal regulatory step in the biosynthesis of bile acids and is catalyzed by **cholesterol 7α-hydroxylase**, a microsomal enzyme. A typical monooxygenase, it requires oxygen, NADPH, and cytochrome P450. Subsequent hydroxylation steps are also catalyzed by monooxygenases. The pathway of bile acid biosynthesis divides early into one subpathway leading to **cholyl-CoA**, characterized

by an extra α-OH group on position 12, and another pathway leading to **chenodeoxycholyl-CoA** (Figure 26–7). A second pathway in mitochondria involving the 27-hydroxylation of cholesterol by **sterol 27-hydroxylase** as the first step is responsible for a significant proportion of the primary bile acids synthesized. The primary bile acids (Figure 26–7) enter the bile as glycine or taurine conjugates. Conjugation takes place in liver peroxisomes. In humans, the ratio of the glycine to the taurine conjugates is normally 3:1. In the alkaline bile (pH 7.6–8.4), the bile acids and their conjugates are assumed to be in a salt form—hence the term "bile salts."

Primary bile acids are further metabolized in the intestine by the activity of the intestinal bacteria. Thus, deconjugation and 7α-dehydroxylation occur, producing the **secondary bile acids, deoxycholic acid,** and **lithocholic acid.**

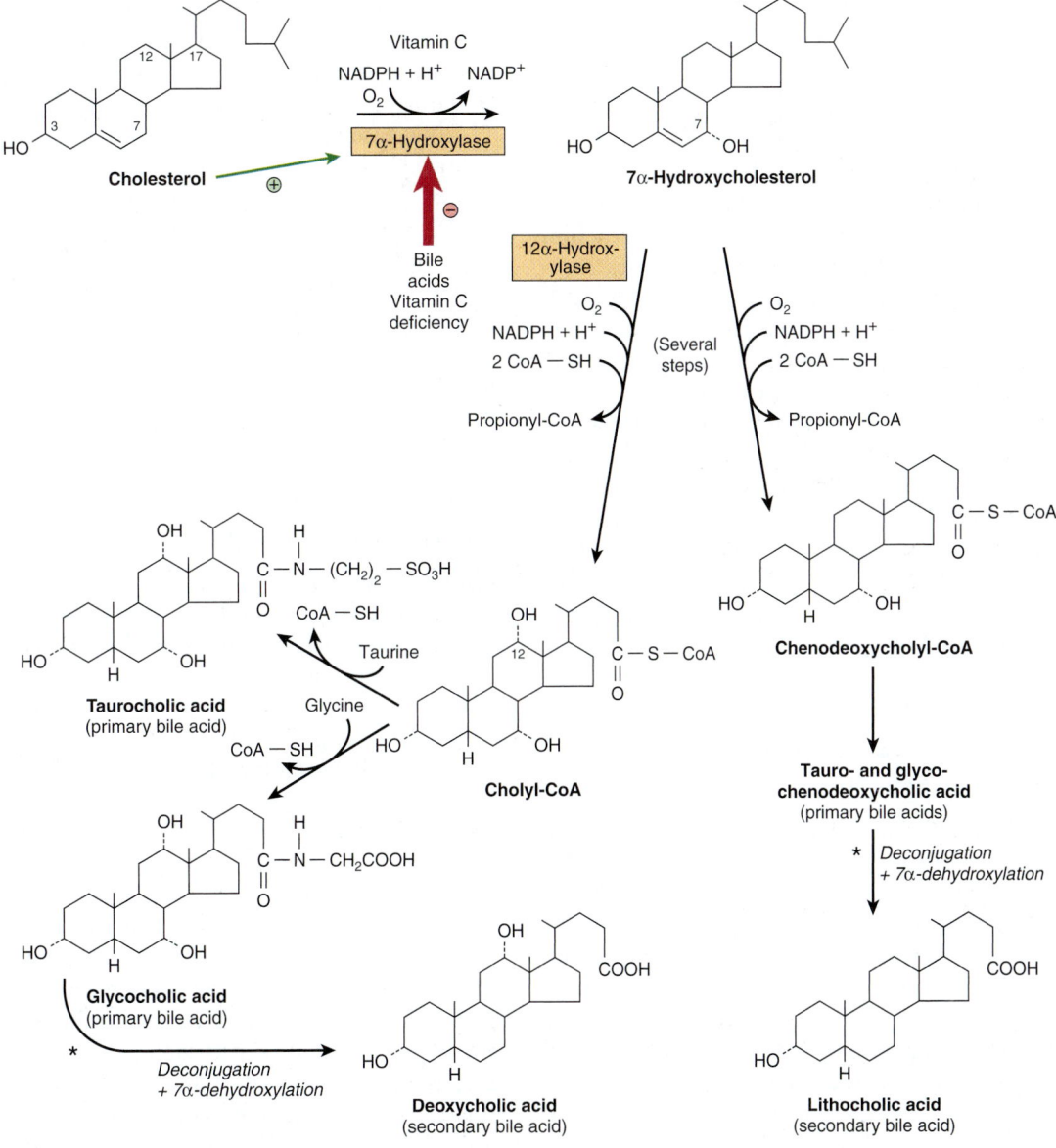

FIGURE 26–7 **Biosynthesis and degradation of bile acids.** A second pathway in mitochondria involves hydroxylation of cholesterol by sterol 27-hydroxylase. (*Catalyzed by microbial enzymes.)

Most Bile Acids Return to the Liver in the Enterohepatic Circulation

Although products of fat digestion, including cholesterol, are absorbed in the first 100 cm of small intestine, the primary and secondary bile acids are absorbed almost exclusively in the ileum, and 98–99% is returned to the liver via the portal circulation. This is known as the **enterohepatic circulation** (Figure 26–6). However, lithocholic acid, because of its insolubility, is not reabsorbed to any significant extent. Only a small fraction of the bile salts escapes absorption and is therefore eliminated in the feces. Nonetheless, this represents a major pathway for the elimination of cholesterol. Each day the pool of bile acids (about 3–5 g) is cycled through the intestine six to ten times and an amount of bile acid equivalent to that lost in the feces is synthesized from cholesterol, so that a pool of bile acids of constant size is maintained. This is accomplished by a system of feedback controls.

Bile Acid Synthesis Is Regulated at the 7α-Hydroxylase Step

The principal rate-limiting step in the biosynthesis of bile acids is at the **cholesterol 7α-hydroxylase reaction** (Figure 26–7). The activity of the enzyme is feedback regulated via the nuclear bile acid-binding receptor, **farnesoid X receptor (FXR)**. When the size of the bile acid pool in the enterohepatic circulation increases, FXR is activated, and transcription of the cholesterol 7α-hydroxylase gene is suppressed. Chenodeoxycholic acid is particularly important in activating FXR. Cholesterol 7α-hydroxylase activity is also enhanced by cholesterol of endogenous and dietary origin and regulated by insulin, glucagon, glucocorticoids, and thyroid hormone.

CLINICAL ASPECTS

Serum Cholesterol Is Correlated with the Incidence of Atherosclerosis & Coronary Heart Disease

While elevated plasma cholesterol levels (>5.2 mmol/L) are believed to be a major factor in promoting atherosclerosis, it is now recognized that triacylglycerols are independent risk factors. Atherosclerosis is characterized by the deposition of cholesterol and cholesteryl ester from the plasma lipoproteins into the artery wall. Diseases in which there is a prolonged elevation of levels of VLDL, IDL, chylomicron remnants, or LDL in the blood (eg, diabetes mellitus, lipid nephrosis, hypothyroidism, and other conditions of hyperlipidemia) are often accompanied by premature or more severe atherosclerosis. There is also an inverse relationship between HDL (HDL_2) concentrations and coronary heart disease, making the **LDL:HDL cholesterol ratio**

a good predictive parameter. This is consistent with the function of HDL in reverse cholesterol transport. Susceptibility to atherosclerosis varies widely among species, and humans are one of the few in which the disease can be induced by diets high in cholesterol.

Diet Can Play an Important Role in Reducing Serum Cholesterol

Hereditary factors play the most important role in determining individual serum cholesterol concentrations; however, dietary and environmental factors also play a part, and the most beneficial of these is the substitution in the diet of **polyunsaturated and monounsaturated fatty acids** for saturated fatty acids. Plant oils such as corn oil and sunflower seed oil contain a high proportion of polyunsaturated fatty acids, while olive oil contains a high concentration of monounsaturated fatty acids. On the other hand, butter fat, beef fat, and palm oil contain a high proportion of saturated fatty acids. Sucrose and fructose have a greater effect in raising blood lipids, particularly triacylglycerols, than do other carbohydrates.

The reason for the cholesterol-lowering effect of polyunsaturated fatty acids is still not fully understood. It is clear, however, that one of the mechanisms involved is the upregulation of LDL receptors by poly- and monounsaturated as compared with saturated fatty acids, causing an increase in the catabolic rate of LDL, the main atherogenic lipoprotein. In addition, saturated fatty acids cause the formation of smaller VLDL particles that contain relatively more cholesterol, and they are utilized by extrahepatic tissues at a slower rate than are larger particles—tendencies that may be regarded as atherogenic.

Lifestyle Affects the Serum Cholesterol Level

Additional factors considered to play a part in coronary heart disease include **high blood pressure, smoking, male gender, obesity (particularly abdominal obesity), lack of exercise, and drinking soft as opposed to hard water.** Factors associated with elevation of plasma FFA followed by increased output of triacylglycerol and cholesterol into the circulation in VLDL include **emotional stress and coffee drinking.** Premenopausal women appear to be protected against many of these deleterious factors, and this is thought to be related to the beneficial effects of **estrogen.** There is an association between moderate alcohol consumption and a lower incidence of coronary heart disease. This may be due to elevation of HDL concentrations resulting from increased synthesis of apo A-I and changes in activity of cholesteryl ester transfer protein. It has been claimed that red wine is particularly beneficial, perhaps because of its content of antioxidants. Regular exercise lowers plasma LDL but raises HDL. Triacylglycerol concentrations are also reduced, due most likely to increased insulin sensitivity, which enhances the expression of lipoprotein lipase.

TABLE 26–1 Primary Disorders of Plasma Lipoproteins (Dyslipoproteinemias)

Name	Defect	Remarks
Hypolipoproteinemias Abetalipoproteinemia	No chylomicrons, VLDL, or LDL are formed because of defect in the loading of apo B with lipid.	Rare; blood acylglycerols low; intestine and liver accumulate acylglycerols. Intestinal malabsorption. Early death avoidable by administration of large doses of fat-soluble vitamins, particularly vitamin E.

Familial alpha-lipoprotein deficiency
  Tangier disease
  Fish-eye disease
  Apo-A-I deficiencies
	All have low or near absence of HDL.	Tendency toward hypertriacylglycerolemia as a result of absence of apo C-II, causing inactive LPL. Low LDL levels. Atherosclerosis in the elderly.

Hyperlipoproteinemias Familial lipoprotein lipase deficiency (type I)	Hypertriacylglycerolemia due to deficiency of LPL, abnormal LPL, or apo C-II deficiency causing inactive LPL.	Slow clearance of chylomicrons and VLDL. Low levels of LDL and HDL. No increased risk of coronary disease.
Familial hypercholesterolemia (type IIa)	Defective LDL receptors or mutation in ligand region of apo B-100.	Elevated LDL levels and hypercholesterolemia, resulting in atherosclerosis and coronary disease.
Familial type III hyperlipoproteinemia (broad beta disease, remnant removal disease, familial dysbetalipoproteinemia	Deficiency in remnant clearance by the liver is due to abnormality in apo E. Patients lack isoforms E3 and E4 and have only E2, which does not react with the E receptor.[1]	Increase in chylomicron and VLDL remnants of density < 1.019 (β-VLDL). Causes hypercholesterolemia, xanthomas, and atherosclerosis.
Familial hypertriacylglycerolemia (type IV)	Overproduction of VLDL often associated with glucose intolerance and hyperinsulinemia.	Cholesterol levels rise with the VLDL concentration. LDL and HDL tend to be subnormal. This type of pattern is commonly associated with coronary heart disease, type II diabetes mellitus, obesity, alcoholism, and administration of progestational hormones.
Familial hyperalphalipoproteinemia	Increased concentrations of HDL.	A rare condition apparently beneficial to health and longevity.
Hepatic lipase deficiency	Deficiency of the enzyme leads to accumulation of large triacylglycerolrich HDL and VLDL remnants.	Patients have xanthomas and coronary heart disease.
Familial lecithin:cholesterol acyltransferase (LCAT) deficiency	Absence of LCAT leads to block in reverse cholesterol transport. HDL remains as nascent disks incapable of taking up and esterifying cholesterol.	Plasma concentrations of cholesteryl esters and lysolecithin are low. Present is an abnormal LDL fraction, lipoprotein X, found also in patients with cholestasis. VLDL is abnormal (β-VLDL).
Familial lipoprotein(a) excess	Lp(a) consists of 1 mol of LDL attached to 1 mol of apo(a). Apo(a) shows structural homologies to plasminogen.	Premature coronary heart disease due to atherosclerosis, plus thrombosis due to inhibition of fibrinolysis.

[1]There is an association between patients possessing the apo E4 allele and the incidence of Alzheimer's disease. Apparently, apo E4 binds more avidly to β-amyloid found in neuritic plaques.

When Diet Changes Fail, Hypolipidemic Drugs Will Reduce Serum Cholesterol & Triacylglycerol

A family of drugs known as statins have proved highly efficacious in lowering plasma cholesterol and preventing heart disease. Statins act by inhibiting HMG-CoA reductase and up-regulating the LDL receptor activity. Examples currently in use include **atorvastatin, simvastatin, fluvastatin,** and **pravastatin. Ezetimibe** reduces blood cholesterol levels by inhibiting the absorption of cholesterol by the intestine by blocking uptake via the **Neimann-Pick C-like 1 protein.** Other drugs used include fibrates such as **clofibrate, gemfibrozil and nico-tinic acid,** which act mainly to lower plasma triacylglycerols by decreasing the secretion of triacylglycerol and cholesterol-containing VLDL by the liver.

Primary Disorders of the Plasma Lipoproteins (Dyslipoproteinemias) Are Inherited

Inherited defects in lipoprotein metabolism lead to the primary condition of either **hypo-** or **hyperlipoproteinemia** (**Table 26–1**). In addition, diseases such as diabetes mellitus, hypothyroidism, kidney disease (nephrotic syndrome), and atherosclerosis are associated with secondary abnormal lipoprotein patterns that are very similar to one or another of the primary inherited conditions. Virtually all of the primary conditions are due to a defect at a stage in lipoprotein formation, transport, or destruction (see Figures 25–4, 26–5, and 26–6). Not all of the abnormalities are harmful.

SUMMARY

- Cholesterol is the precursor of all other steroids in the body, for example, corticosteroids, sex hormones, bile acids, and vitamin D. It also plays an important structural role in membranes and in the outer layer of lipoproteins.

- Cholesterol is synthesized in the body entirely from acetyl-CoA. Three molecules of acetyl-CoA form mevalonate via the important regulatory reaction for the pathway, catalyzed by HMG-CoA reductase. Next, a five-carbon isoprenoid unit is formed, and six of these condense to form squalene. Squalene undergoes cyclization to form the parent steroid lanosterol, which, after the loss of three methyl groups and other changes, forms cholesterol.

- Cholesterol synthesis in the liver is regulated partly by cholesterol in the diet. In tissues, cholesterol balance is maintained between the factors causing gain of cholesterol (eg, synthesis, uptake via the LDL or scavenger receptors) and the factors causing loss of cholesterol (eg, steroid synthesis, cholesteryl ester formation, excretion). The activity of the LDL receptor is modulated by cellular cholesterol levels to achieve this balance. In reverse cholesterol transport, HDL takes up cholesterol from the tissues and LCAT esterifies it and deposits it in the core of the particles. The cholesteryl ester in HDL is taken up by the liver, either directly or after transfer to VLDL, IDL, or LDL via the cholesteryl ester transfer protein.

- Excess cholesterol is excreted from the liver in the bile as cholesterol or bile salts. A large proportion of bile salts is absorbed into the portal circulation and returned to the liver as part of the enterohepatic circulation.

- Elevated levels of cholesterol present in VLDL, IDL, or LDL are associated with atherosclerosis, whereas high levels of HDL have a protective effect.

- Inherited defects in lipoprotein metabolism lead to a primary condition of hypo- or hyperlipoproteinemia. Conditions such as diabetes mellitus, hypothyroidism, kidney disease, and atherosclerosis exhibit secondary abnormal lipoprotein patterns that resemble certain primary conditions.

REFERENCES

Agellon LB: Metabolism and function of bile acids. In *Biochemistry of Lipids, Lipoproteins and Membranes,* 5th ed. Vance DE, Vance JE (editors). Elsevier, 2008:423–440.

Chiang JL: Regulation of bile acid synthesis: pathways, nuclear receptors and mechanisms. J Hepatol 2004;40:539.

Denke MA: Dietary fats, fatty acids and their effects on lipoproteins. Curr Atheroscler Rep 2006;8:466.

Djoussé L, Gaziano JM: Dietary cholesterol and coronary disease risk: a systematic review. Curr Atheroscler Rep 2009;11:418.

Fernandez ML, West KL: Mechanisms by which dietary fatty acids modulate plasma lipids. J Nutr 2005;135:2075.

Jiang XC, Zhou HW: Plasma lipid transfer proteins. Curr Opin Lipidol 2006;17:302.

Liscum L: Cholesterol biosynthesis. In *Biochemistry of Lipids, Lipoproteins and Membranes,* 5th ed. Vance DE, Vance JE (editors). Elsevier, 2008:399–422.

Ness GC, Chambers CM: Feedback and hormonal regulation of hepatic 3-hydroxy-3-methylglutaryl coenzyme A reductase: the concept of cholesterol buffering capacity. Proc Soc Exp Biol Med 2000;224:8.

Parks DJ, Blanchard SG, Bledsoe RK, et al: Bile acids: natural ligands for a nuclear orphan receptor. Science 1999;284:1365.

Perez-Sala D: Protein isoprenylation in biology and disease: general overview and perspectives from studies with genetically engineered animals. Front Biosci 2007;12:4456.

Exam Questions

Section II

1. A number of compounds inhibit oxidative phosphorylation—the synthesis of ATP from ADP and inorganic phosphate linked to oxidation of substrates in mitochondria. Which of the following describes the action of oligomycin?
   A. It discharges the proton gradient across the mitochondrial inner membrane.
   B. It discharges the proton gradient across the mitochondrial outer membrane.
   C. It inhibits the electron transport chain directly by binding to one of the electron carriers in the mitochondrial inner membrane.
   D. It inhibits the transport of ADP into, and ATP out of, the mitochondrial matrix.
   E. It inhibits the transport of protons back into the mitochondrial matrix through the stalk of the primary particle.

2. A number of compounds inhibit oxidative phosphorylation—the synthesis of ATP from ADP and inorganic phosphate linked to oxidation of substrates in mitochondria. Which of the following describes the action of an uncoupler?
   A. It discharges the proton gradient across the mitochondrial inner membrane.
   B. It discharges the proton gradient across the mitochondrial outer membrane.
   C. It inhibits the electron transport chain directly by binding to one of the electron carriers in the mitochondrial inner membrane.
   D. It inhibits the transport of ADP into, and ATP out of, the mitochondrial matrix.
   E. It inhibits the transport of protons back into the mitochondrial matrix through the stalk of the primary particle.

3. A student takes some tablets she is offered at a disco, and without asking what they are she swallows them. A short time later she starts to hyperventilate, and becomes very hot. What is the most likely action of the tablets she has taken?
   A. An inhibitor of mitochondrial ATP synthesis.
   B. An inhibitor of mitochondrial electron transport.
   C. An inhibitor of the transport of ADP into mitochondria to be phosphorylated.
   D. An inhibitor of the transport of ATP out of mitochondria into the cytosol.
   E. An uncoupler of mitochondrial electron transport and oxidative phosphorylation.

4. AA eats a very poor diet and consumes two bottles of vodka a day. He is admitted to a hospital in coma. He is hyperventilating and has low blood pressure and high cardiac output. A chest x-ray shows that his heart is enlarged. Laboratory tests show that he is suffering from thiamin (vitamin B_1) deficiency. Which of the following enzymes is most likely to be affected?
   A. Lactate dehydrogenase
   B. Pyruvate carboxylase
   C. Pyruvate decarboxylase
   D. Pyruvate dehydrogenase
   E. Pyruvate kinase

5. In monitoring glycemic control in diabetic patients, urine or blood glucose can be measured in two ways: chemically, using an alkaline copper reagent that detects reducing compounds; and biochemically, using the enzyme glucose oxidase. In a series of experiments both methods were used on the same urine sample and the result was positive for the alkaline copper reagent and negative using glucose oxidase. Which of the following is the most likely diagnosis for the person being tested?
   A. A diabetic with good glycemic control.
   B. A diabetic with poor glycemic control.
   C. A normal healthy person who has fasted overnight.
   D. A normal healthy person who has just eaten a meal.
   E. A healthy person who has essential pentosuria.

6. A 25-year-old man undertakes a prolonged fast for religious reasons. Which of the following metabolites will be elevated in his blood plasma after 24 hours?
   A. Glucose
   B. Glycogen
   C. Ketone bodies
   D. Non-esterified fatty acids
   E. Triacylglycerol

7. A 25-year-old man undertakes a prolonged fast for religious reasons. Which of the following metabolites will be most elevated in his blood plasma after 3 days?
   A. Glucose
   B. Glycogen
   C. Ketone bodies
   D. Non-esterified fatty acids
   E. Triacylglycerol

8. A 25-year-old man visits his GP complaining of abdominal cramps and diarrhea after drinking milk. What is the most likely cause of his problem?
   A. Bacterial and yeast overgrowth in the large intestine.
   B. Infection with the intestinal parasite *Giardia lamblia*.
   C. Lack of pancreatic amylase.
   D. Lack of small intestinal lactase.
   E. Lack of small intestinal sucrose-isomaltase.

9. Which one of following statements about favism (lack of glucose 6-phosphate dehydrogenase) and the pentose phosphate pathway is CORRECT?
   A. In favism red blood cells are more susceptible to oxidative stress because of a lack of NADPH for fatty acid synthesis.
   B. People who lack glucose 6-phosphate dehydrogenase cannot synthesize fatty acids because of a lack of NADPH in liver and adipose tissue.
   C. The pentose phosphate pathway is especially important in tissues that are synthesizing fatty acids.
   D. The pentose phosphate pathway is the only source of NADPH for fatty acid synthesis.
   E. The pentose phosphate pathway provides an alternative to glycolysis only in the fasting state.

10. Which one of following statements about glycogen synthesis and utilization is CORRECT?

    A. Glycogen is synthesized in the liver in the fed state, then exported to other tissues in low-density lipoproteins.
    B. Glycogen reserves in liver and muscle will meet energy requirements for several days in prolonged fasting.
    C. Liver synthesizes more glycogen when the hepatic portal blood concentration of glucose is high because of the activity of glucokinase in the liver.
    D. Muscle synthesises glycogen in the fed state because glycogen phosphorylase is activated in response to insulin.
    E. The plasma concentration of glycogen increases in the fed state.

11. In glycolysis, the conversion of 1 mol of fructose 1,6-bisphosphate to 2 mol of pyruvate results in the formation of

    A. 1 mol NAD^+ and 2 mol of ATP.
    B. 1 mol NADH and 1 mol of ATP.
    C. 2 mol NAD^+ and 4 mol of ATP.
    D. 2 mol NADH and 2 mol of ATP.
    E. 2 mol NADH and 4 mol of ATP.

12. Which one of the following statements about fatty acid metabolism is CORRECT?

    A. Acylcarnitine is formed from acyl-CoA and carnitine at the inner face of the inner mitochondrial membrane.
    B. Acyl-CoA can only cross the inner mitochondrial membrane in exchange for free CoA leaving the mitochondrial matrix.
    C. Creatinine is essential for transport of fatty acids into the mitochondrial matrix.
    D. In the fed state, the main source of fatty acids for tissues is triacylglycerol in chylomicrons and very low-density lipoproteins (VLDL).
    E. β-Oxidation of fatty acids occurs in the cytosol.

13. Which one of the following statements about synthesis of fats is CORRECT?

    A. Synthesis of triacylglycerol in adipose tissue can only occur when gluconeogenesis is occurring.
    B. Synthesis of triacylglycerol in adipose tissue is stimulated when the insulin/glucagon ratio is low.
    C. Synthesis of triacylglycerol in the liver can only occur when glycolysis is active.
    D. Triacylglycerol is synthesized from glycerol 3-phosphate and acyl CoA in adipose tissue.
    E. Triacylglycerol is synthesized from monoacylglycerol and acyl CoA in adipose tissue.

14. Which of the plasma lipoproteins is best described as follows: synthesized in the intestinal mucosa, containing a high concentration of triacylglycerol and mainly cleared from the circulation by adipose tissue and muscle?

    A. Chylomicrons
    B. High-density lipoprotein
    C. Intermediate density lipoprotein
    D. Low-density lipoprotein
    E. Very low-density lipoprotein

15. Which of the plasma lipoproteins is best described as follows: synthesized in the liver, containing a high concentration of triacylglycerol and mainly cleared from the circulation by adipose tissue and muscle?

    A. Chylomicrons
    B. High-density lipoprotein
    C. Intermediate density lipoprotein
    D. Low-density lipoprotein
    E. Very low-density lipoprotein

16. Which of the plasma lipoproteins is best described as follows: formed in the circulation by removal of triacylglycerol from very low-density lipoprotein, and containing cholesterol taken up from high-density lipoprotein, cleared by the liver?

    A. Chylomicrons
    B. High-density lipoprotein
    C. Intermediate density lipoprotein
    D. Low-density lipoprotein
    E. Very low-density lipoprotein

17. Which of the following will be elevated in the bloodstream about 2 hours after eating a high-fat meal?

    A. Chylomicrons
    B. High-density lipoprotein
    C. Ketone bodies
    D. Non-esterified fatty acids
    E. Very low-density lipoprotein

18. Which of the following will be elevated in the bloodstream about 4 hours after eating a high-fat meal?

    A. Chylomicrons
    B. High-density lipoprotein
    C. Ketone bodies
    D. Non-esterified fatty acids
    E. Very low-density lipoprotein

19. After they are produced from acetyl-CoA in the liver, ketone bodies are mainly used for which one of the following processes?

    A. Excretion as waste products.
    B. Energy generation in the liver.
    C. Conversion to fatty acids for storage of energy.
    D. Generation of energy in the tissues.
    E. Generation of energy in red blood cells.

20. Which one of the following statements concerning the biosynthesis of cholesterol is CORRECT?

    A. The rate-limiting step is the formation of 3-hydroxy 3-methylglutaryl-CoA (HMG-CoA) by the enzyme HMG-CoA synthase.
    B. Synthesis occurs in the cytosol of the cell.
    C. All the carbon atoms in the cholesterol synthesized originate from acetyl-CoA.
    D. Squalene is the first cyclic intermediate in the pathway.
    E. The initial substrate is mevalonate.

21. The class of drugs called statins have proved very effective against hypercholesterolemia, a major cause of atherosclerosis and associated cardiovascular disease. These drugs reduce plasma cholesterol levels by:

    A. Preventing absorption of cholesterol from the intestine.
    B. Increasing the excretion of cholesterol from the body via conversion to bile acids.
    C. Inhibiting the conversion of 3-hydroxy-3-methylglutaryl-CoA to mevalonate in the pathway for cholesterol biosynthesis.

D. Increasing the rate of degradation of 3-hydroxy-3-methylglutaryl CoA reductase.

E. Stimulating the activity of the LDL receptor in the liver.

22. Which one of the following statements about the free energy change (ΔG) in a biochemical reaction is CORRECT?

A. If ΔG is negative, the reaction proceeds spontaneously with a loss of free energy.

B. In an exergonic reaction, ΔG is positive.

C. The standard free energy change when reactants are present in concentrations of 1.0 mol/L and the pH is 7.0 is represented as ΔG^0

D. In an endergonic reaction, ΔG is negative.

E. If ΔG is 0, the reaction is essentially irreversible.

23. Which one of the following statements about the citric acid cycle is CORRECT?

A. It produces most of the ATP in anaerobic organisms.

B. It oxidizes acetyl-CoA derived from fatty acid oxidation.

C. It provides acetyl-CoA for the synthesis fatty acids.

D. It slows down when energy levels are low.

E. It provides ATP mainly by substrate-linked phosphorylation.

24. Which one of the following is the major product of fatty acid synthase?

A. Acetyl-CoA

B. Oleate

C. Palmitoyl-CoA

D. Acetoacetate

E. Palmitate

25. Which one of the following statements concerning chylomicrons is CORRECT?

A. Chylomicrons are made inside intestinal cells and secreted into lymph, where they acquire apolipoproteins B and C.

B. The core of chylomicrons contains triacylglycerol and phospholipids.

C. The enzyme hormone sensitive lipase acts on chylomicrons to release fatty acids from triacylglycerol when they are bound to the surface of endothelial cells in blood capillaries.

D. Chylomicron remnants differ from chylomicrons in that they are smaller and contain a lower proportion of triacylglycerol.

E. Chylomicrons are taken up by the liver.

26. For each turn of the citric acid cycle, 3NADH and 1 FADH$_2$ molecules are formed and oxidized via the respiratory chain producing:

A. 10 ATP molecules

B. 4 ATP molecules

C. 9 ATP molecules

D. 7 ATP molecules

E. 12 ATP molecules

27. The subcellular site of the breakdown of long chain fatty acids to acetyl-CoA via β-oxidation is:

A. The cytosol

B. The matrix of the mitochondria

C. The endoplasmic reticulum

D. The mitochondrial intermembrane space

E. The Golgi apparatus

28. Which one of the following statements concerning fatty acid molecules is CORRECT?

A. They consist of a carboxylic acid head group attached to a carbohydrate chain.

B. They are called polyunsaturated when they contain one or more carbon–carbon double bonds.

C. Their melting points increase with increasing unsaturation.

D. They almost always have their double bonds in the cis configuration when they occur naturally.

E. They occur in the body mainly in the form of free (non-esterified) fatty acids.

29. The flow of electrons through the respiratory chain and the production of ATP are normally tightly coupled. The processes are uncoupled by which of the following:

A. Cyanide

B. Oligomycin

C. Thermogenin

D. Carbon monoxide

E. Hydrogen sulphide

Biosynthesis of the Nutritionally Nonessential Amino Acids

Victor W. Rodwell, PhD

- Explain why the absence from the diet of some amino acids is not deleterious to human health.
- Appreciate the distinction between "essential" and "nutritionally essential" amino acids, and identify the amino acids that are nutritionally nonessential.
- Name the citric acid cycle and the glycolytic intermediates that are precursors of aspartate, asparagine, glutamate, glutamine, glycine, and serine.
- Appreciate the key role of transaminases in amino acid metabolism.
- Explain the process by which the hydroxyproline and hydroxylysine of proteins are formed.
- Provide a biochemical explanation for why a severe deprivation of vitamin C (ascorbic acid) results in the nutritional disease scurvy, and describe the clinical consequences of this nutritional disorder.
- Appreciate that, despite the toxicity of selenium, selenocysteine is an essential component of several mammalian proteins.
- Outline the reaction catalyzed by a mixed-function oxidase.
- Identify the role of tetrahydrobiopterin in tyrosine biosynthesis.
- Indicate the role of a modified tRNA in the cotranslational insertion of selenocysteine into proteins.

BIOMEDICAL IMPORTANCE

Medical implications of the material in this chapter relate to the amino acid deficiency states that can result if nutritionally essential amino acids are absent from the diet, or are present in inadequate amounts. Amino acid deficiency states endemic in certain regions of West Africa include **kwashiorkor,** which results when a child is weaned onto a starchy diet poor in protein, and **marasmus,** in which both caloric intake and specific amino acids are deficient. Patients with short bowel syndrome unable to absorb sufficient quantities of calories and nutrients suffer from significant nutritional and metabolic abnormalities. Both the nutritional disorder **scurvy,** a dietary deficiency of vitamin C, and specific genetic disorders are associated with an impaired ability of connective tissue to form hydroxyproline and hydroxylysine. The resulting conformational instability of collagen results in bleeding gums, swelling joints, poor wound healing, and ultimately in death. **Menkes' syndrome,** characterized by kinky hair and growth retardation, results from a dietary deficiency of copper, which is an essential cofactor for lysyl oxidase, an enzyme that functions in formation of the covalent cross-links that strengthen collagen fibers. Genetic disorders of collagen biosynthesis include several forms of **osteogenesis imperfecta,** characterized by fragile bones, and **Ehlers–Danlos syndrome,** a group of connective tissue disorders that result in mobile joints and skin abnormalities due to defects in the genes that encode enzymes including lysyl hydroxylase.

NUTRITIONALLY ESSENTIAL & NUTRITIONALLY NONESSENTIAL AMINO ACIDS

As applied to amino acids, the terms "essential" and "nonessential" are misleading since all 20 common amino acids are essential to ensure health. Of these 20 amino acids, 8 *must* be present in the human diet, and thus are best termed "*nutritionally essential.*" The other 12 amino acids are "*nutritionally nonessential*" since they need not be present in the diet (**Table 27–1**). The distinction between these two classes of amino acids was established in the 1930s by feeding human subjects purified amino acids in place of protein. Subsequent biochemical investigations revealed the reactions and intermediates involved in the biosynthesis of all 20 amino acids. Amino acid deficiency disorders are endemic in certain regions of West Africa where diets rely heavily on grains that are poor sources of tryptophan and lysine. These nutritional disorders include kwashiorkor, which results when a child is weaned onto a starchy diet poor in protein, and marasmus, in which both caloric intake and specific amino acids are deficient.

Lengthy Metabolic Pathways Form the Nutritionally Essential Amino Acids

The existence of nutritional requirements suggests that dependence on an external supply of a given nutrient can be of greater survival value than the ability to biosynthesize it. Why? If a specific nutrient is present in the food, an organism that can synthesize it will transfer to its progeny genetic information of *negative* survival value. The survival value is negative rather than nil because ATP and nutrients are required to synthesize "unnecessary" DNA—even if specific encoded genes are no longer expressed. The number of enzymes required by prokaryotic cells to synthesize the nutritionally essential amino acids is large relative to the number of enzymes required to synthesize the nutritionally nonessential amino acids (**Table 27–2**). This suggests a survival advantage in retaining the ability to manufacture "easy" amino acids while losing the ability to make "difficult" amino acids. The metabolic pathways that form the nutritionally essential amino acids occur in plants and bacteria, but not in humans, and thus are not discussed. This chapter addresses the reactions and intermediates involved in the biosynthesis by human tissues of the 12 nutritionally *nonessential* amino acids and selected nutritional and metabolic disorders associated with their metabolism.

TABLE 27–1   Amino Acid Requirements of Humans

Nutritionally Essential	Nutritionally Nonessential
Arginine[1]	Alanine
Histidine	Asparagine
Isoleucine	Aspartate
Leucine	Cysteine
Lysine	Glutamate
Methionine	Glutamine
Phenylalanine	Glycine
Threonine	Hydroxyproline[2]
Tryptophan	Hydroxylysine[2]
Valine	Proline
	Serine
	Tyrosine

[1]Nutritionally "semiessential." Synthesized at rates inadequate to support growth of children.
[2]Not necessary for protein synthesis, but is formed during post-translational processing of collagen.

TABLE 27–2   Enzymes Required for the Synthesis of Amino Acids from Amphibolic Intermediates

Number of Enzymes Required to Synthesize			
Nutritionally Essential		Nutritionally Nonessential	
Arg[1]	7	Ala	1
His	6	Asp	1
Thr	6	Asn[2]	1
Met	5 (4 shared)	Glu	1
Lys	8	Gln[1]	1
Ile	8 (6 shared)	Hyl[3]	1
Val	1 (7 shared)	Hyp[4]	1
Leu	3 (7 shared)	Pro[1]	3
Phe	10	Ser	3
Trp	5 (8 shared)	Gly[5]	1
	59	Cys[6]	2
		Tyr[7]	1
			17

[1]From Glu, [2]From Asp, [3]From Lys, [4]From Pro, [5]From Ser, [6]From Ser plus S[2-], [7]From Phe.

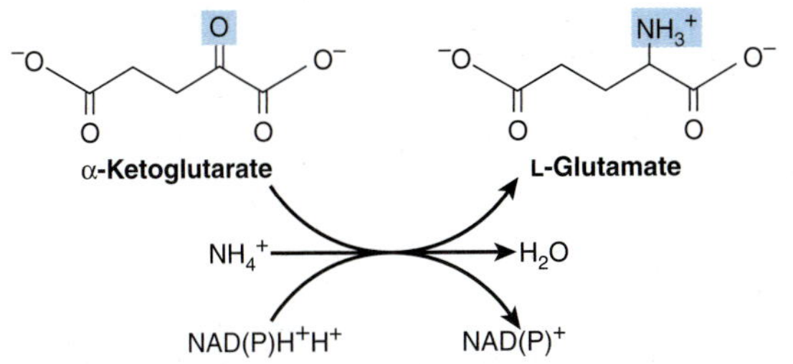

FIGURE 27–1 The glutamate dehydrogenase reaction.

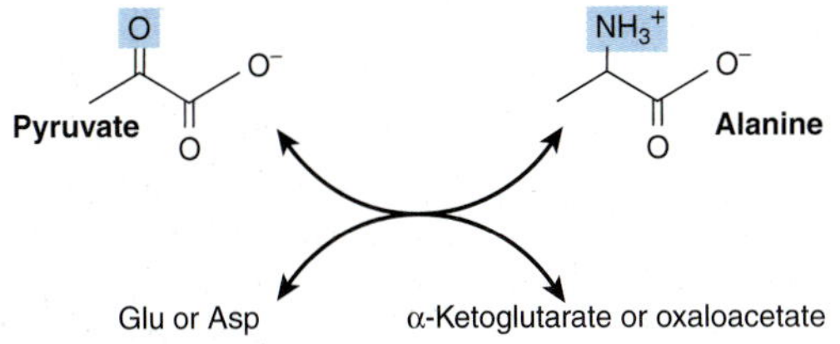

FIGURE 27–3 Formation of alanine by transamination of pyruvate. The amino donor may be glutamate or aspartate. The other product thus is α-ketoglutarate or oxaloacetate.

BIOSYNTHESIS OF THE NUTRITIONALLY NONESSENTIAL AMINO ACIDS

Glutamate

The first reaction in biosynthesis of the "glutamate family" of amino acids is the reductive amidation of α-ketoglutarate catalyzed by glutamate dehydrogenase (**Figure 27–1**). The reaction is shown as unidirectional in the direction of glutamate synthesis because the reaction strongly favors glutamate. This is physiologically important because high concentrations of ammonium ion are cytotoxic.

Glutamine

The amidation of glutamate to glutamine catalyzed by glutamine synthetase involves the intermediate formation of γ-glutamyl phosphate (**Figure 27–2**). Following the ordered binding of glutamate and ATP, glutamate attacks the γ-phosphorus of ATP, forming γ-glutamyl phosphate and ADP. NH_4^+ then binds, and as NH_3 attacks γ-glutamyl phosphate to form a tetrahedral intermediate. Release of P_i and of a proton from the γ-amino group of the tetrahedral intermediate then allows release of the product, glutamine.

Alanine & Aspartate

Transamination of pyruvate forms alanine (**Figure 27–3**). Similarly, transamination of oxaloacetate forms aspartate.

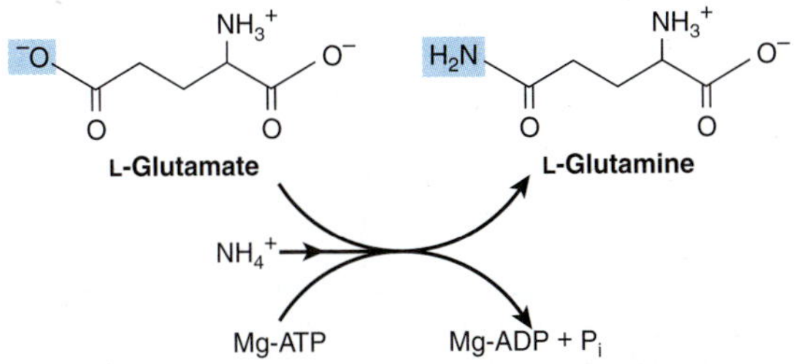

Glutamate Dehydrogenase, Glutamine Synthetase & Aminotransferases Play Central Roles in Amino Acid Biosynthesis

The combined action of the enzymes glutamate dehydrogenase, glutamine synthetase, and the aminotransferases (Figures 27–1, 27–2 and 27–3) converts inorganic ammonium ion into the α-amino nitrogen of amino acids.

Asparagine

The conversion of aspartate to asparagine, catalyzed by asparagine synthetase (**Figure 27–4**), resembles the glutamine synthetase reaction (Figure 27–2), but glutamine, rather than ammonium ion, provides the nitrogen. Bacterial asparagine synthetases can, however, also use ammonium ion. The reaction involves the intermediate formation of aspartyl phosphate. The coupled hydrolysis of PP_i to P_i by pyrophosphatase ensures that the reaction is strongly favored.

Serine

Oxidation of the α-hydroxyl group of the glycolytic intermediate 3-phosphoglycerate by 3-phosphoglycerate dehydrogenase converts it to 3-phosphohydroxypyruvate. Transamination and subsequent dephosphorylation then form serine (**Figure 27–5**).

Glycine

Glycine aminotransferases can catalyze the synthesis of glycine from glyoxylate and glutamate or alanine. Unlike most

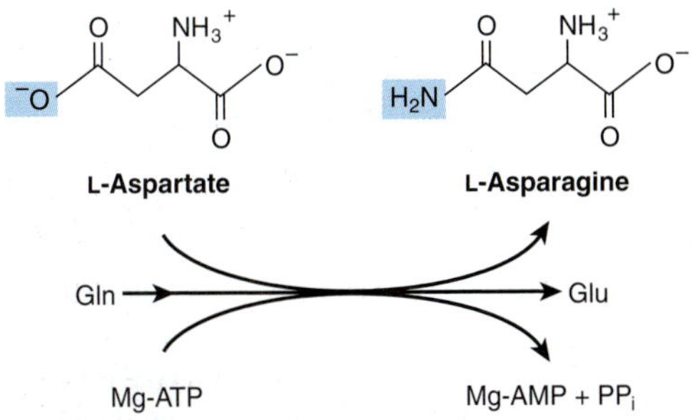

FIGURE 27–2 The glutamine synthetase reaction.

FIGURE 27–4 The asparagine synthetase reaction. Note similarities to and differences from the glutamine synthetase reaction (Figure 27–2).

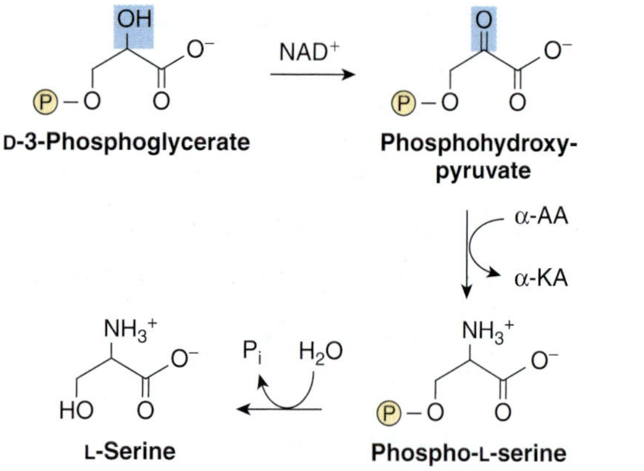

FIGURE 27–5 Serine biosynthesis. (α-AA, α-amino acids; α-KA, α-keto acids.)

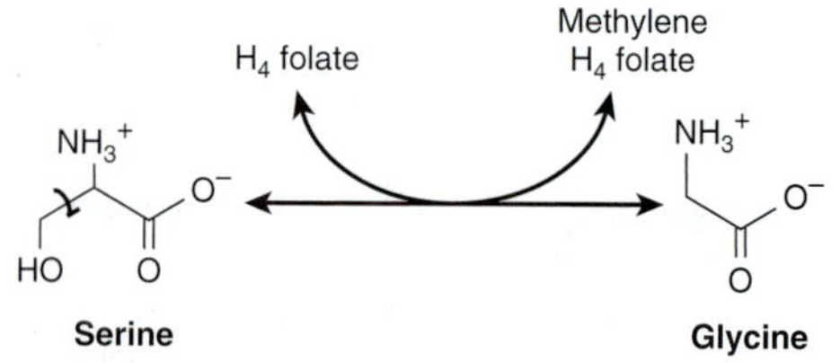

FIGURE 27–7 The serine hydroxymethyltransferase reaction. The reaction is freely reversible. (H_4 folate, tetrahydrofolate.)

aminotransferase reactions, these strongly favor glycine synthesis. Additional important mammalian routes for glycine formation are from choline (**Figure 27–6**) and from serine (**Figure 27–7**).

Proline

The initial reaction of proline biosynthesis converts the γ-carboxyl group of glutamate to the mixed acid anhydride of glutamate γ-phosphate (**Figure 27–8**). Subsequent reduction forms glutamate γ-semialdehyde, which following spontaneously cyclization is reduced to L-proline.

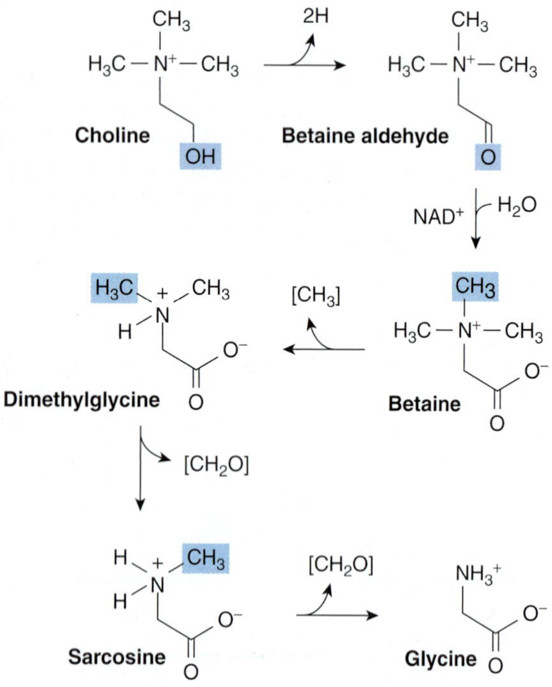

FIGURE 27–6 Formation of glycine from choline. The enzymes that catalyze the reactions shown are choline dehydrogenase, betaine dehydrogenase, betaine-homocysteine N-methyltransferase, sarcosine demethylase, and sarcosine oxidase, respectively.

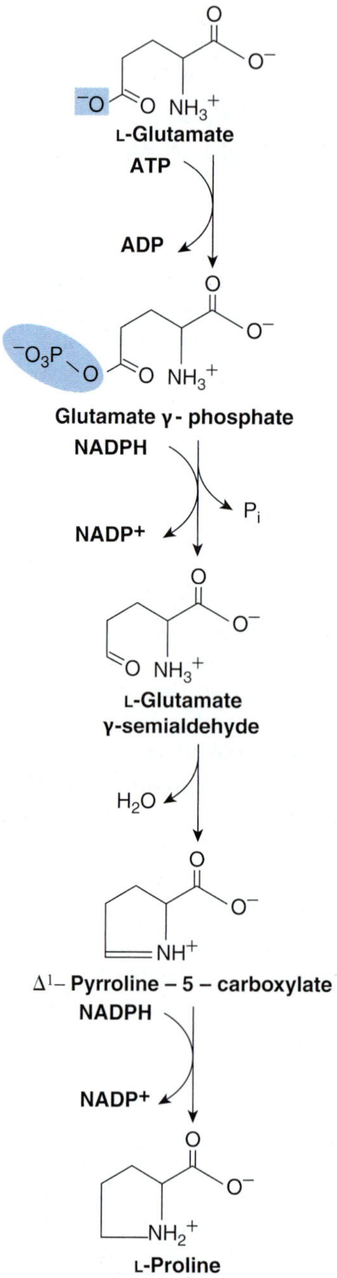

FIGURE 27–8 Biosynthesis of proline from glutamate. The catalysts for these reactions are glutamate 5-kinase, glutamate semialdehyde dehydrogenase, noncatalyzed ring closure, and pyrroline 5-carboxylate reductase.

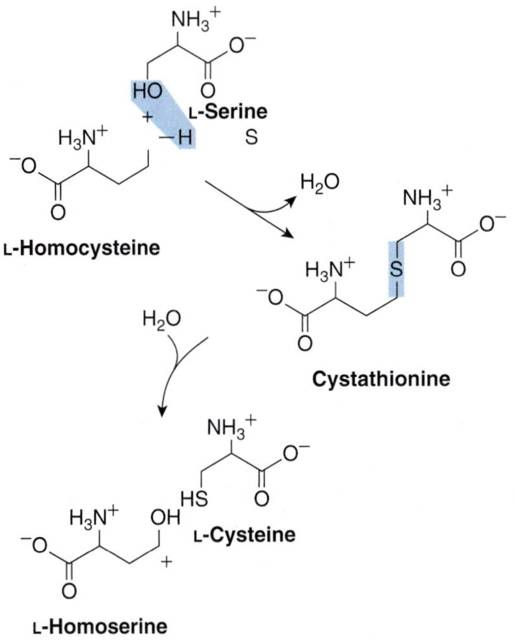

FIGURE 27–9 **Conversion of homocysteine and serine to homoserine and cysteine.** The sulfur of cysteine derives from methionine and the carbon skeleton from serine.

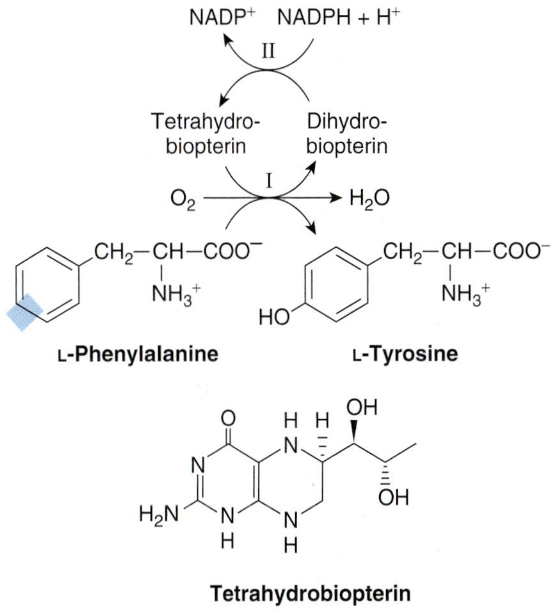

FIGURE 27–10 **The phenylalanine hydroxylase reaction.** Two distinct enzymatic activities are involved. Activity II catalyzes reduction of dihydrobiopterin by NADPH, and activity I the reduction of O_2 to H_2O and of phenylalanine to tyrosine. This reaction is associated with several defects of phenylalanine metabolism discussed in Chapter 29.

Cysteine

While not nutritionally essential, cysteine is formed from methionine, which is nutritionally essential. Following conversion of methionine to homocysteine (see Figure 29–19), homocysteine and serine form cystathionine, whose hydrolysis forms cysteine and homoserine (**Figure 27–9**).

Tyrosine

Phenylalanine hydroxylase converts phenylalanine to tyrosine (**Figure 27–10**). If the diet contains adequate quantities of the nutritionally essential amino acid phenylalanine, tyrosine is nutritionally nonessential. However, since the phenylalanine hydroxylase reaction is irreversible, dietary tyrosine cannot replace phenylalanine. Catalysis by this mixed-function oxygenase incorporates one atom of O_2 into the *para* position of phenylalanine and reduces the other atom to water. Reducing power, provided as tetrahydrobiopterin derives ultimately from NADPH (Figure 27–10).

Hydroxyproline & Hydroxylysine

Hydroxyproline and hydroxylysine occur principally in collagen. Since there is no tRNA for either hydroxylated amino acid, neither dietary hydroxyproline nor dietary hydroxylysine is incorporated during protein synthesis. Peptidyl hydroxyproline and hydroxylysine arise from proline and lysine, but only after these amino acids have been incorporated into peptides. Hydroxylation of peptidyl prolyl and peptidyl lysyl residues, catalyzed by **prolyl hydroxylase** and **lysyl hydroxylase** of skin,

skeletal muscle, and granulating wounds requires, in addition to the substrate, molecular O_2, ascorbate, Fe^{2+}, and α-ketoglutarate (**Figure 27–11**). For every mole of proline or lysine hydroxylated, one mole of α-ketoglutarate is decarboxylated to succinate. The hydroxylases are mixed-function oxygenases. One atom of O_2 is incorporated into proline or lysine, the other into succinate (Figure 27–11). A deficiency of the vitamin C required for these two hydroxylases results in **scurvy**, in which bleeding gums, swelling joints, and impaired wound healing result from the impaired stability of collagen (see Chapters 5 and 48).

Valine, Leucine & Isoleucine

While leucine, valine, and isoleucine are all nutritionally essential amino acids, tissue aminotransferases reversibly interconvert all three amino acids and their corresponding α-keto acids. These α-keto acids thus can replace their amino acids in the diet.

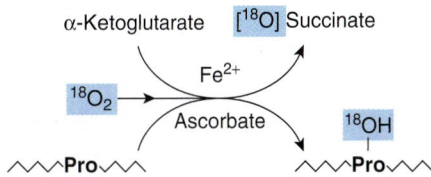

FIGURE 27–11 **The prolyl hydroxylase reaction.** The substrate is a proline-rich peptide. During the course of the reaction, molecular oxygen is incorporated into both succinate and proline. Lysyl hydroxylase catalyzes an analogous reaction.

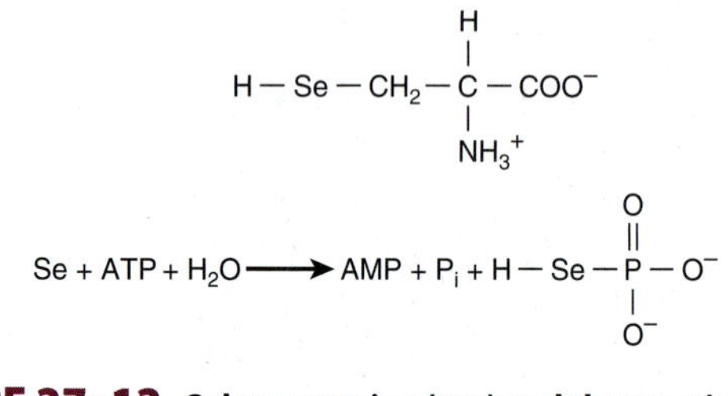

$$H-Se-CH_2-\overset{\overset{\displaystyle H}{|}}{\underset{\underset{\displaystyle NH_3^+}{|}}{C}}-COO^-$$

$$Se + ATP + H_2O \longrightarrow AMP + P_i + H-Se-\overset{\overset{\displaystyle O}{\|}}{\underset{\underset{\displaystyle O^-}{|}}{P}}-O^-$$

FIGURE 27–12 Selenocysteine (*top*) and the reaction catalyzed by selenophosphate synthetase (*bottom*).

Selenocysteine, the 21st Amino Acid

While the occurrence of selenocysteine (**Figure 27–12**) in proteins is uncommon, at least 25 human selenoproteins are known. Selenocysteine is present at the active site of several human enzymes that catalyze redox reactions. Examples include thioredoxin reductase, glutathione peroxidase, and the deiodinase that converts thyroxine to triiodothyronine. Where present, selenocysteine participates in the catalytic mechanism of these enzymes. Significantly, the replacement of selenocysteine by cysteine can actually impair catalytic activity. Impairments in human selenoproteins have been implicated in tumorigenesis and atherosclerosis, and are associated with selenium deficiency cardiomyopathy (Keshan disease).

Biosynthesis of selenocysteine requires cysteine, selenate (SeO_4^{2-}), ATP, a specific tRNA, and several enzymes. Serine provides the carbon skeleton of selenocysteine. Selenophosphate, formed from ATP and selenate (Figure 27–12), serves as the selenium donor. Unlike hydroxyproline or hydroxylysine, selenocysteine arises cotranslationally during its incorporation into peptides. The UGA anticodon of the unusual tRNA called tRNA^{Sec} normally signals STOP. The ability of the protein synthetic apparatus to identify a selenocysteine-specific UGA codon involves the selenocysteine insertion element, a stem-loop structure in the untranslated region of the mRNA. tRNA^{Sec} is first charged with serine by the ligase that charges tRNA^{Ser}. Subsequent replacement of the serine oxygen by selenium involves selenophosphate formed by selenophosphate synthetase (Figure 27–12). Successive enzyme-catalyzed reactions convert cysteyl-tRNA^{Sec} to aminoacrylyl-tRNA^{Sec} and then to selenocysteyl-tRNA^{Sec}. In the presence of a specific elongation factor that recognizes selenocysteyl-tRNA^{Sec}, selenocysteine can then be incorporated into proteins.

SUMMARY

- All vertebrates can form certain amino acids from amphibolic intermediates or from other dietary amino acids. The intermediates and the amino acids to which they give rise are α-ketoglutarate (Glu, Gln, Pro, Hyp), oxaloacetate (Asp, Asn), and 3-phosphoglycerate (Ser, Gly).

- Cysteine, tyrosine, and hydroxylysine are formed from nutritionally essential amino acids. Serine provides the carbon skeleton and homocysteine the sulfur for cysteine biosynthesis.

- In Scurvy, a nutritional disease that results from a deficiency of vitamin C, impaired hydroxylation of peptidyl proline and peptidyl lysine results in a failure to provide the substrates for cross-linking of maturing collagens.

- Phenylalanine hydroxylase converts phenylalanine to tyrosine. The reaction catalyzed by this mixed function oxidase is irreversible.

- Neither dietary hydroxyproline nor hydroxylysine is incorporated into proteins because no codon or tRNA dictates their insertion into peptides.

- Peptidyl hydroxyproline and hydroxylysine are formed by hydroxylation of peptidyl proline or lysine in reactions catalyzed by mixed-function oxidases that require vitamin C as cofactor.

- Selenocysteine, an essential active site residue in several mammalian enzymes, arises by cotranslational insertion from a previously modified tRNA.

REFERENCES

Beckett GJ, Arthur JR: Selenium and endocrine systems. J Endocrinol 2005;184:455.

Donovan J, Copeland PR: The efficiency of selenocysteine incorporation is regulated by translation initiation factors. J Mol Biol 2010;400:659.

Kilberg MS: Asparagine synthetase chemotherapy. Annu Rev Biochem 2006;75:629.

Lobanov AV, Hatfield DL, Gladyshev VN: Reduced reliance on the trace element selenium during evolution of mammals. Genome Biol 2008;9:R62.

Scriver CR, Sly WS, Childs B, et al (editors): *The Metabolic and Molecular Bases of Inherited Disease,* 8th ed. McGraw-Hill, 2001.

Stickel F, Inderbitzin D, Candinas D: Role of nutrition in liver transplantation for end-stage chronic liver disease. Nutr Rev 2008;66:47.

Catabolism of Proteins & of Amino Acid Nitrogen

Victor W. Rodwell, PhD

- Describe protein turnover, indicate the mean rate of protein turnover in healthy individuals, and provide examples of human proteins that are degraded at rates greater than the mean rate.
- Describe the events in protein turnover by both ATP-dependent and ATP-independent pathways, and the roles in protein degradation of ubiquitin, cell surface receptors, circulating asialoglycoproteins, and lysosomes.
- Indicate how the ultimate end products of nitrogen catabolism in mammals differ from those in birds and in fish.
- Illustrate the central roles of transaminases (aminotransferases), of glutamate dehydrogenase, and of glutaminase in human nitrogen metabolism.
- Use structural formulas to represent the reactions that convert NH_3, CO_2, and the amide nitrogen of aspartate into urea.
- Indicate the subcellular locations of the enzymes that catalyze urea biosynthesis, and the roles of allosteric regulation and of acetylglutamate in the regulation of this process.
- Explain why metabolic defects in different enzymes of urea biosynthesis, although distinct at the molecular level, present similar clinical signs and symptoms.
- Describe both the classical approaches and the role of tandem mass spectrometry in screening neonates for inherited metabolic diseases.

BIOMEDICAL IMPORTANCE

This chapter describes how the nitrogen of amino acids is converted to urea and the rare metabolic disorders that accompany defects in urea biosynthesis. In normal adults, nitrogen intake matches nitrogen excreted. Positive nitrogen balance, an excess of ingested over excreted nitrogen, accompanies growth and pregnancy. Negative nitrogen balance, where output exceeds intake, may follow surgery, advanced cancer, and the nutritional disorders kwashiorkor, and marasmus. Ammonia, which is highly toxic, arises in humans primarily from the α-amino nitrogen of amino acids. Tissues therefore convert ammonia to the amide nitrogen of the nontoxic amino acid glutamine. Subsequent deamination of glutamine in the liver releases ammonia, which is then converted to urea, which is not toxic. If liver function is compromised, as in cirrhosis or hepatitis, elevated blood ammonia levels generate clinical

signs and symptoms. Each enzyme of the urea cycle provides examples of metabolic defects and their physiologic consequences, and the cycle as a whole serves as a molecular model for the study of human metabolic defects.

PROTEIN TURNOVER OCCURS IN ALL FORMS OF LIFE

The continuous degradation and synthesis (turnover) of cellular proteins occur in all forms of life. Each day, humans turn over 1–2% of their total body protein, principally muscle protein. High rates of protein degradation occur in tissues that are undergoing structural rearrangement, for example, uterine tissue during pregnancy, skeletal muscle in starvation, and tadpole tail tissue during metamorphosis. Approximately 75% of the amino acids liberated by protein degradation are reutilized. Excess free amino acids are, however, not stored. Those not immediately incorporated into new protein are rapidly degraded. The major portion of the carbon skeletons of the amino acids is converted to amphibolic intermediates, while in humans the amino nitrogen is converted to urea and excreted in the urine.

PROTEASES & PEPTIDASES DEGRADE PROTEINS TO AMINO ACIDS

The relative susceptibility of a protein to degradation is expressed as its half-life ($t_{1/2}$), the time required to lower its concentration to half the initial value. Half-lives of liver proteins range from under 30 min to over 150 h. Typical "housekeeping" enzymes have $t_{1/2}$ values of over 100 h. By contrast, key regulatory enzymes may have $t_{1/2}$ values as low as 0.5–2 h. PEST sequences, regions rich in proline (P), glutamate (E), serine (S), and threonine (T), target some proteins for rapid degradation. Intracellular proteases hydrolyze internal peptide bonds. The resulting peptides are then degraded to amino acids by endopeptidases that cleave internal peptide bonds, and by aminopeptidases and carboxypeptidases that remove amino acids sequentially from the amino- and carboxyl-termini, respectively.

ATP-Independent Degradation

Degradation of blood glycoproteins (see Chapter 47) follows loss of a sialic acid moiety from the nonreducing ends of their oligosaccharide chains. Asialoglycoproteins are then internalized by liver-cell asialoglycoprotein receptors and degraded by lysosomal proteases. Extracellular, membrane-associated, and long-lived intracellular proteins are degraded in lysosomes by ATP-independent processes.

ATP and Ubiquitin-Dependent Degradation

Degradation of regulatory proteins with short half-lives and of abnormal or misfolded proteins occurs in the cytosol, and requires ATP and ubiquitin. **Ubiquitin,** so named because it is present in all eukaryotic cells, is a small (8.5 kDa, 76 residues) polypeptide that targets many intracellular proteins for degradation. The primary structure of ubiquitin is highly conserved. Only three of 76 residues differ between yeast and human ubiquitin. Ubiquitin molecules are attached by **non-α-peptide bonds** formed between the carboxyl terminal of ubiquitin and the ε-amino groups of lysyl residues in the target protein (**Figure 28–1**). The residue present at its amino terminal affects whether a protein is ubiquitinated. Amino terminal Met or Ser retards, whereas Asp or Arg accelerates ubiquitination. Attachment of a single ubiquitin molecule to transmembrane proteins alters their subcellular localization and targets them for degradation. Soluble proteins undergo **polyubiquitination,** the ligase-catalyzed attachment of four

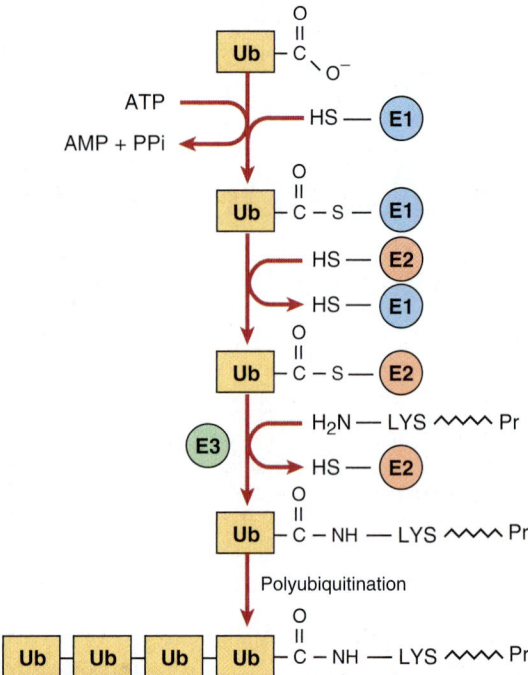

FIGURE 28–1 **Reactions involved in the attachment of ubiquitin (Ub) to proteins.** Three enzymes are involved. E1 is an activating enzyme, E2 a ligase, and E3 a transferase. While depicted as single entities, there are several types of E1, and over 500 types of E2. The terminal COOH of ubiquitin first forms a thioester. The coupled hydrolysis of PP$_i$ by pyrophosphatase ensures that the reaction will proceed readily. A thioester exchange reaction now transfers activated ubiquitin to E2. E3 then catalyzes the transfer of ubiquitin to the ε-amino group of a lysyl residue of the target protein. Additional rounds of ubiquitination result in subsequent polyubiquitination.

or more additional ubiquitin molecules. Subsequent degradation of ubiquitin-tagged proteins takes place in the **proteasome,** a macromolecule with multiple different subunits that also is ubiquitous in eukaryotic cells (see Chapter 46). For the discovery of ubiquitin-mediated protein degradation, Aaron Ciechanover and Avram Hershko of Israel and Irwin Rose of the United States were awarded the 2004 Nobel Prize in Chemistry. Metabolic diseases associated with defects of ubiquitination include the Angelman syndrome and the von Hippel–Lindau syndrome in which there is a defect in the ubiquitin E3 ligase. For additional aspects of protein degradation and of ubiquitination, including its role in the cell cycle, see Chapters 4 and 46.

INTERORGAN EXCHANGE MAINTAINS CIRCULATING LEVELS OF AMINO ACIDS

The maintenance of steady-state concentrations of circulating plasma amino acids between meals depends on the net balance between release from endogenous protein stores and utilization by various tissues. Muscle generates over half of the

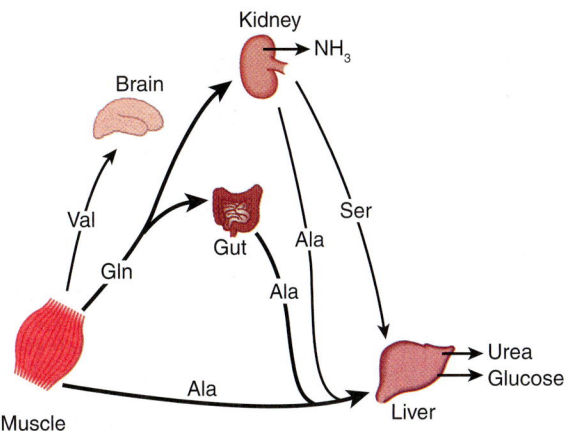

FIGURE 28–2 **Interorgan amino acid exchange in normal postabsorptive humans.** The key role of alanine in amino acid output from muscle and gut and uptake by the liver is shown.

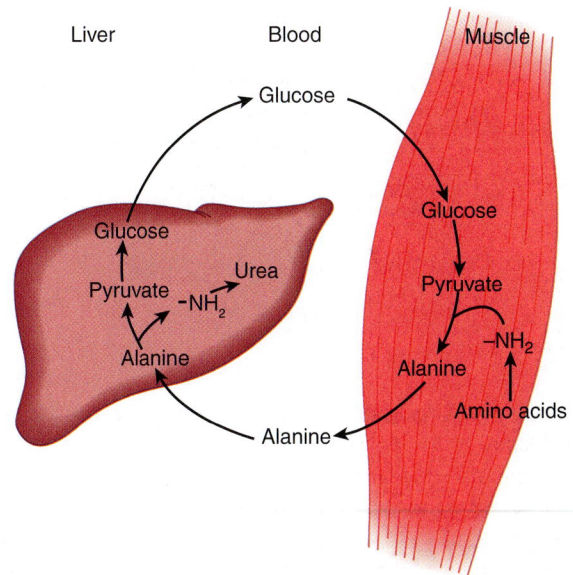

FIGURE 28–3 **The glucose-alanine cycle.** Alanine is synthesized in muscle by transamination of glucose-derived pyruvate, released into the bloodstream, and taken up by the liver. In the liver, the carbon skeleton of alanine is reconverted to glucose and released into the bloodstream, where it is available for uptake by muscle and resynthesis of alanine.

total body pool of free amino acids, and liver is the site of the urea cycle enzymes necessary for disposal of excess nitrogen. Muscle and liver thus play major roles in maintaining circulating amino acid levels.

Figure 28–2 summarizes the postabsorptive state. Free amino acids, particularly alanine and glutamine, are released from muscle into the circulation. Alanine, which appears to be the vehicle of nitrogen transport in the plasma, is extracted primarily by the liver. Glutamine is extracted by the gut and the kidney, both of which convert a significant portion to alanine. Glutamine also serves as a source of ammonia for excretion by the kidney. The kidney provides a major source of serine for uptake by peripheral tissues, including liver and muscle. Branched-chain amino acids, particularly valine, are released by muscle and taken up predominantly by the brain.

Alanine is a key **gluconeogenic amino acid** (**Figure 28–3**). The rate of hepatic gluconeogenesis from alanine is far higher than from all other amino acids. The capacity of the liver for gluconeogenesis from alanine does not reach saturation until the alanine concentration reaches 20–30 times its normal physiologic level. Following a protein-rich meal, the splanchnic tissues release amino acids (**Figure 28–4**) while the peripheral muscles extract amino acids, in both instances predominantly branched-chain amino acids. Branched-chain amino acids thus serve a special role in nitrogen metabolism: in the fasting state, when they provide the brain with an energy source, and after feeding, when they are extracted predominantly by muscle, having been spared by the liver.

ANIMALS CONVERT α-AMINO NITROGEN TO VARIED END PRODUCTS

Depending on their ecological niche and physiology, different animals excrete excess nitrogen as ammonia, as uric acid, or as urea. The aqueous environment of teleostean fish, which

are **ammonotelic** (excrete ammonia), permits them to excrete water continuously to facilitate excretion of ammonia, which is highly toxic. While this approach is appropriate for an aquatic animal, birds must both conserve water and maintain low weight. Birds, which are **uricotelic,** address both problems by excreting nitrogen-rich uric acid (see Figure 33–11) as semisolid guano. Many land animals, including humans, are **ureotelic** and excrete nontoxic, highly water-soluble urea. Since urea is nontoxic to humans, high blood levels in renal disease are a consequence, not a cause, of impaired renal function.

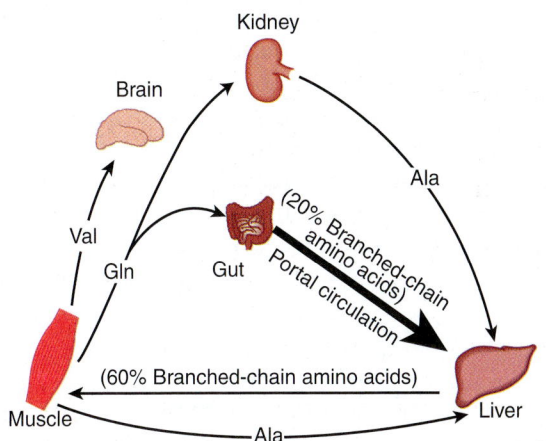

FIGURE 28–4 **Summary of amino acid exchange between organs immediately after feeding.**

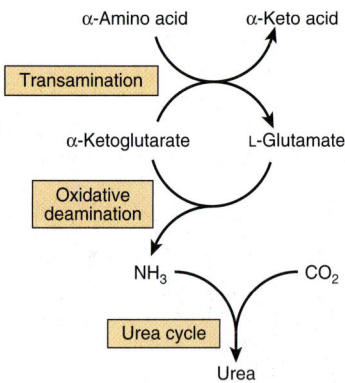

FIGURE 28–5 Overall flow of nitrogen in amino acid catabolism.

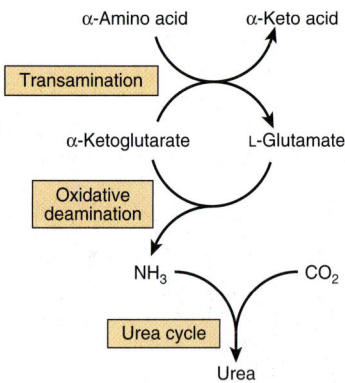

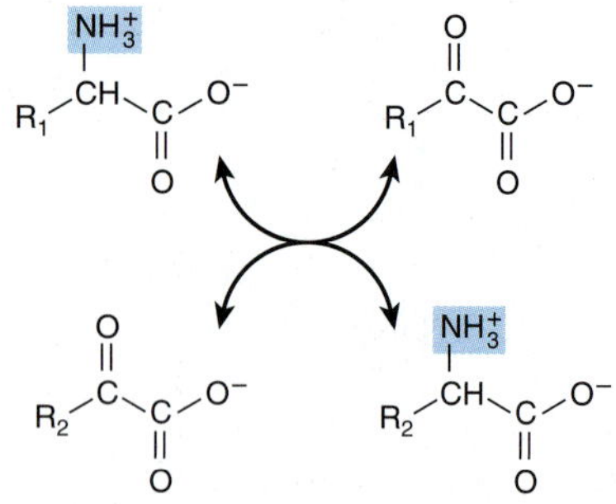

FIGURE 28–6 **Transamination.** The reaction is freely reversible with an equilibrium constant close to unity.

BIOSYNTHESIS OF UREA

Urea biosynthesis occurs in four stages: (1) transamination, (2) oxidative deamination of glutamate, (3) ammonia transport, and (4) reactions of the urea cycle (**Figure 28–5**). The use of complementary DNA probes has shown that the expression in liver of the RNAs for all the enzymes of the urea cycle increases several fold in starvation.

Transamination Transfers α-Amino Nitrogen to α-Ketoglutarate, Forming Glutamate

Transamination reactions interconvert pairs of α-amino acids and α-keto acids (**Figure 28–6**). Transamination reactions, which are freely reversible, also function in amino acid biosynthesis (see Figure 27–3). All of the common amino acids except lysine, threonine, proline, and hydroxyproline participate in transamination. Transamination is not restricted to α-amino groups. The δ-amino group of ornithine (but not the ε-amino group of lysine) readily undergoes transamination.

Alanine-pyruvate aminotransferase (alanine aminotransferase) and glutamate-α-ketoglutarate aminotransferase (glutamate aminotransferase) catalyze the transfer of amino groups to pyruvate (forming alanine) or to α-ketoglutarate (forming glutamate) (**Figure 28–7**). Each aminotransferase is specific for one pair of substrates, but nonspecific for the other pair. Since alanine is also a substrate for glutamate aminotransferase, the α-amino nitrogen from all amino acids that undergo transamination can be concentrated in glutamate. This is important because L-glutamate is the only amino acid that undergoes oxidative deamination at an appreciable rate in mammalian

tissues. The formation of ammonia from α-amino groups thus occurs mainly via the α-amino nitrogen of L-glutamate.

Transamination occurs via a "ping-pong" mechanism characterized by the alternate addition of a substrate and release of a product (Figure 28–7). Following removal of its α-amino nitrogen by transamination, the remaining carbon "skeleton" of an amino acid is degraded by pathways discussed in Chapter 29. As noted earlier, certain diseases are associated with elevated serum levels of aminotransferases (see Table 7–2).

Pyridoxal phosphate (PLP), a derivative of vitamin B_6 (see Figure 44–12) is present at the catalytic site of all aminotransferases, and plays a key role in catalysis. During transamination, PLP serves as a "carrier" of amino groups. An enzyme-bound Schiff base (**Figure 28–8**) is formed between the oxo group of enzyme-bound PLP and the α-amino group of an α-amino acid. The Schiff base can rearrange in various ways. In transamination, rearrangement forms an α-keto acid and an enzyme-bound pyridoxamine phosphate. As noted earlier, certain diseases are associated with elevated serum levels of aminotransferases (see Table 7–2).

L-GLUTAMATE DEHYDROGENASE OCCUPIES A CENTRAL POSITION IN NITROGEN METABOLISM

Transfer of amino nitrogen to α-ketoglutarate forms L-glutamate. Hepatic **L-glutamate dehydrogenase (GDH),** which can use either NAD$^+$ or NADP$^+$, releases this nitrogen as ammonia (**Figure 28–9**). Conversion of α-amino nitrogen to ammonia by the concerted action of glutamate aminotransferase and GDH is often termed "transdeamination." Liver GDH activity is allosterically inhibited by ATP, GTP, and NADH, and is activated by ADP. The GDH reaction is freely reversible, and also functions in amino acid biosynthesis (see Figure 27–1).

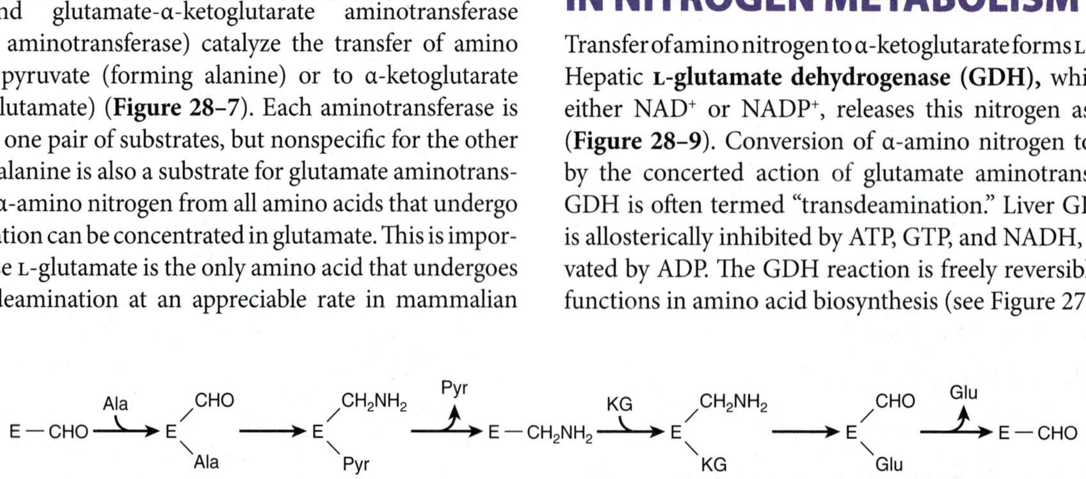

FIGURE 28–7 **"Ping-pong" mechanism for transamination.** E—CHO and E—CH$_2$NH$_2$ represent enzyme-bound pyridoxal phosphate and pyridoxamine phosphate, respectively. (Ala, alanine; Glu, glutamate; KG, α-ketoglutarate; Pyr, pyruvate.)

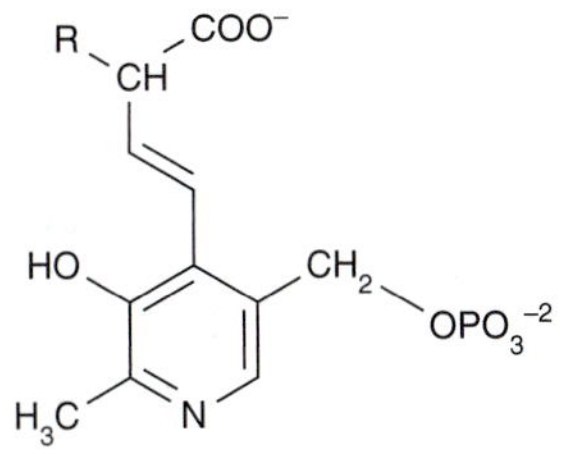

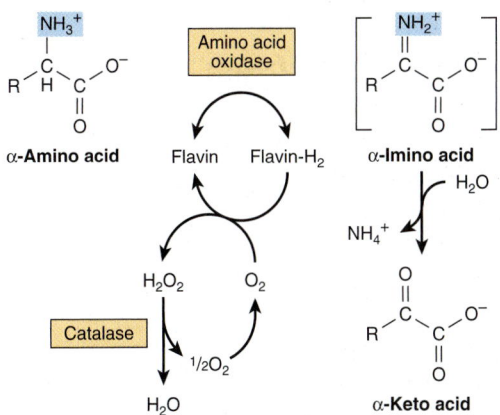

FIGURE 28–8 Structure of a Schiff base formed between pyridoxal phosphate and an amino acid.

Amino Acid Oxidases Remove Nitrogen as Ammonia

While their physiologic importance is uncertain, L-amino acid oxidases of liver and kidney convert an amino acid to an α-imino acid that decomposes to an α-keto acid with release of ammonium ion (**Figure 28–10**). The reduced flavin is reoxidized by molecular oxygen, forming hydrogen peroxide (H_2O_2), which then is split to O_2 and H_2O by **catalase**.

Ammonia Intoxication Is Life Threatening

The ammonia produced by enteric bacteria and absorbed into portal venous blood and the ammonia produced by tissues are rapidly removed from circulation by the liver and converted to urea. Thus, only traces (10–20 μg/dL) normally are present in peripheral blood. This is essential, since ammonia is toxic to the central nervous system. Should portal blood bypass the liver, systemic blood ammonia levels may attain toxic levels. This occurs in severely impaired hepatic function or the development of collateral links between the portal and systemic veins in cirrhosis. Symptoms of **ammonia intoxication** include tremor, slurred speech, blurred vision, coma, and ultimately death. Ammonia may be toxic to the brain in part because it reacts with α-ketoglutarate to form glutamate. The resulting depletion of levels of α-ketoglutarate then impairs function of the tricarboxylic acid (TCA) cycle in neurons.

Glutamine Synthase Fixes Ammonia as Glutamine

Formation of glutamine is catalyzed by mitochondrial **glutamine synthase** (**Figure 28–11**). Since amide bond synthesis

FIGURE 28–10 Oxidative deamination catalyzed by L-amino acid oxidase (L-α-amino acid: O_2 oxidoreductase). The α-imino acid, shown in brackets, is not a stable intermediate.

is coupled to the hydrolysis of ATP to ADP and P_i, the reaction strongly favors glutamine synthesis. During catalysis, glutamate attacks the γ-phosphoryl group of ATP, forming γ-glutamyl phosphate and ADP. Following deprotonation of NH_4^+, NH_3 attacks γ-glutamyl phosphate, and glutamine and P_i are released. In addition to providing glutamine to serve as a carrier of nitrogen, carbon and energy between organs (Figure 28–2), glutamine synthase plays a major role in ammonia detoxification and acid-base homeostasis. A rare deficiency in neonate glutamine synthase results in severe brain damage, multiorgan failure, and death.

Glutaminase & Asparaginase Deamidate Glutamine & Asparagine

There are two human isoforms of mitochondrial glutaminase, termed liver-type and renal type glutaminase. Products of different genes, the glutaminases differ with respect to their structure, kinetics, and regulation. Hepatic glutaminase levels

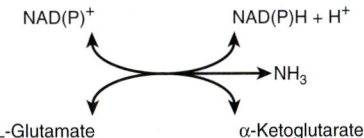

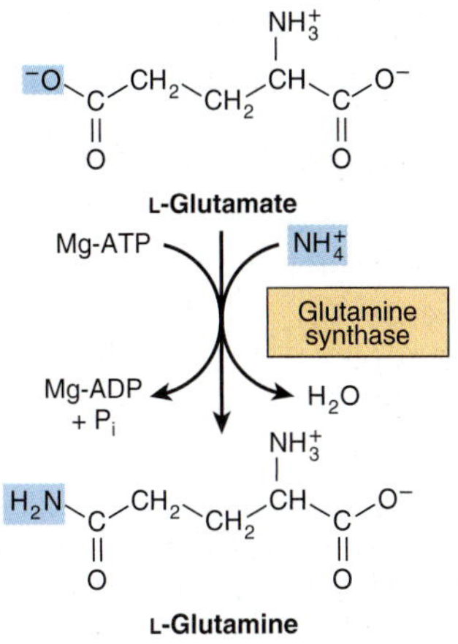

FIGURE 28–9 The L-glutamate dehydrogenase reaction. NAD(P)$^+$ means that either NAD$^+$ or NADP$^+$ can serve as the oxidoreductant. The reaction is reversible, but favors glutamate formation.

FIGURE 28–11 The glutamine synthase reaction strongly favors glutamine synthesis.

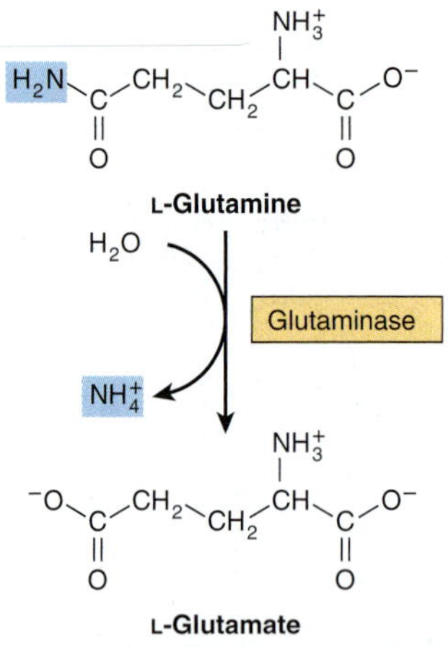

FIGURE 28–12 The glutaminase reaction proceeds essentially irreversibly in the direction of glutamate and NH_4^+ formation. Note that the *amide* nitrogen, not the α-amino nitrogen, is removed.

rise in response to high protein intake while renal kidney-type glutaminase increases in metabolic acidosis. Hydrolytic release of the amide nitrogen of glutamine as ammonia, catalyzed by **glutaminase** (**Figure 28–12**), strongly favors glutamate formation. An analogous reaction is catalyzed by L-asparaginase. The concerted action of glutamine synthase and glutaminase thus catalyzes the interconversion of free ammonium ion and glutamine.

Formation & Secretion of Ammonia Maintains Acid-Base Balance

Excretion into urine of ammonia produced by renal tubular cells facilitates cation conservation and regulation of acid-base balance. Ammonia production from intracellular renal amino acids, especially glutamine, increases in **metabolic acidosis** and decreases in **metabolic alkalosis.**

UREA IS THE MAJOR END PRODUCT OF NITROGEN CATABOLISM IN HUMANS

Synthesis of 1 mol of urea requires 3 mol of ATP, 1 mol each of ammonium ion and of aspartate, and employs five enzymes (**Figure 28–13**). Of the six participating amino acids, *N*-acetylglutamate functions solely as an enzyme activator. The others serve as carriers of the atoms that ultimately become urea. The major metabolic role of **ornithine, citrulline,** and **argininosuccinate** in mammals is urea synthesis. Urea synthesis is a cyclic process. While ammonium ion, CO_2, ATP, and aspartate are consumed, the ornithine consumed in reaction 2 is regenerated in reaction 5. There thus is no net loss or gain of ornithine, citrulline, argininosuccinate, or arginine. Some reactions of urea synthesis occur in the matrix

of the mitochondrion, and other reactions in the cytosol (Figure 28–13).

Carbamoyl Phosphate Synthase I Initiates Urea Biosynthesis

Condensation of CO_2, ammonia, and ATP to form **carbamoyl phosphate** is catalyzed by mitochondrial **carbamoyl phosphate synthase I.** A cytosolic form of this enzyme, carbamoyl phosphate synthase II, uses glutamine rather than ammonia as the nitrogen donor and functions in pyrimidine biosynthesis (see Figure 33–9). The concerted action of glutamate dehydrogenase and carbamoyl phosphate synthase 1 thus shuttles amino nitrogen into carbamoyl phosphate, a compound with high group transfer potential.

Carbamoyl phosphate synthase I, the rate-limiting enzyme of the urea cycle, is active only in the presence of *N*-**acetylglutamate,** an allosteric activator that enhances the affinity of the synthase for ATP. Synthesis of 1 mol of carbamoyl phosphate requires 2 mol of ATP. One ATP serves as the phosphoryl donor for formation of the mixed acid anhydride bond of carbamoyl phosphate. The second ATP provides the driving force for synthesis of the amide bond of carbamoyl phosphate. The other products are 2 mol of ADP and 1 mol of P_i (reaction 1, Figure 28–13). The reaction proceeds stepwise. Reaction of bicarbonate with ATP forms carbonyl phosphate and ADP. Ammonia then displaces ADP, forming carbamate and orthophosphate. Phosphorylation of carbamate by the second ATP then forms carbamoyl phosphate.

Carbamoyl Phosphate Plus Ornithine Forms Citrulline

L-**Ornithine transcarbamoylase** catalyzes transfer of the carbamoyl group of carbamoyl phosphate to ornithine, forming citrulline and orthophosphate (reaction 2, Figure 28–13). While the reaction occurs in the mitochondrial matrix, both the formation of ornithine and the subsequent metabolism of citrulline take place in the cytosol. Entry of ornithine into mitochondria and exodus of citrulline from mitochondria therefore involve mitochondrial inner membrane permeases (Figure 28–13).

Citrulline Plus Aspartate Forms Argininosuccinate

Argininosuccinate synthase links aspartate and citrulline via the amino group of aspartate (reaction 3, Figure 28–13) and provides the second nitrogen of urea. The reaction requires ATP and involves intermediate formation of citrullyl-AMP. Subsequent displacement of AMP by aspartate then forms argininosuccinate.

Cleavage of Argininosuccinate Forms Arginine & Fumarate

Cleavage of argininosuccinate is catalyzed by **argininosuccinate lyase.** The reaction proceeds with retention of all three

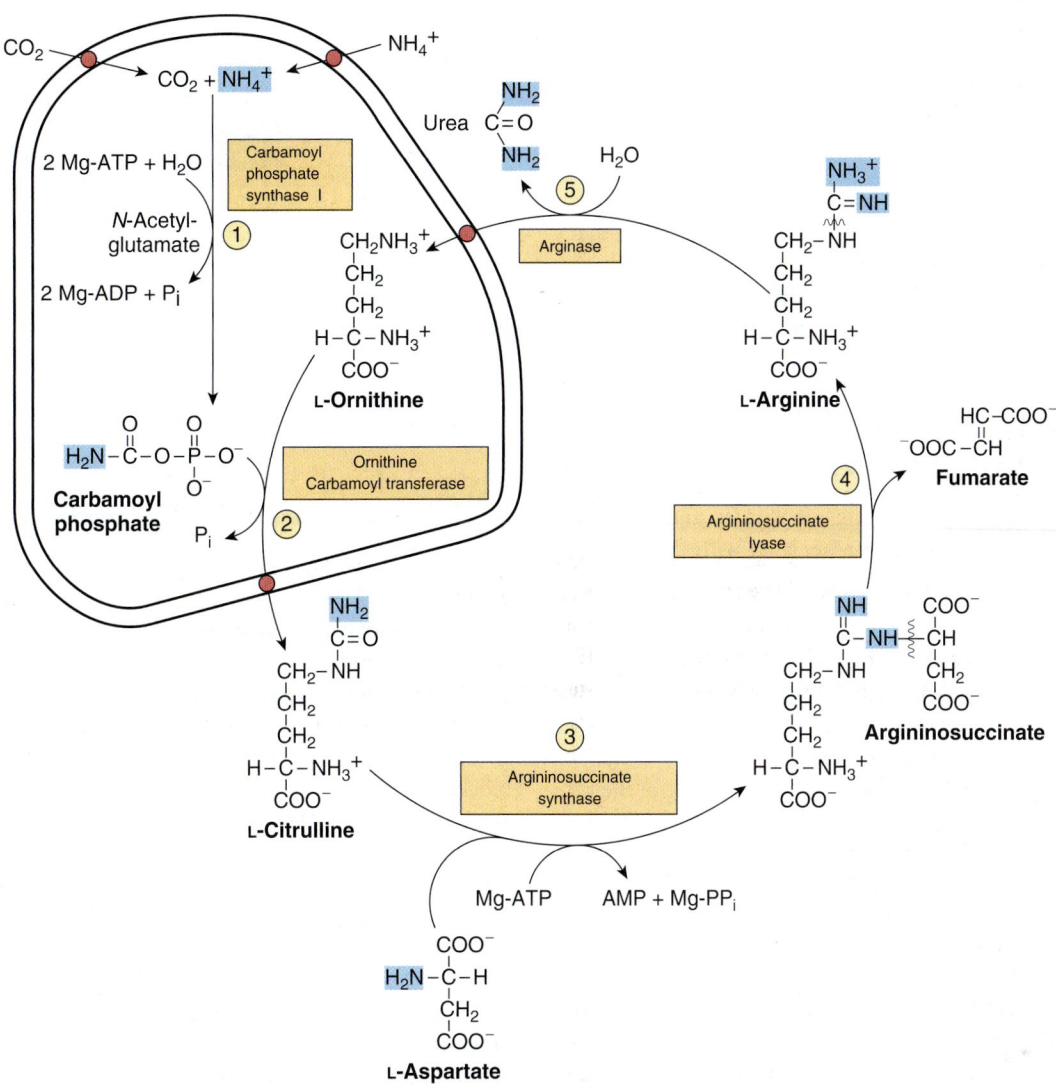

FIGURE 28–13 **Reactions and intermediates of urea biosynthesis.** The nitrogen-containing groups that contribute to the formation of urea are shaded. Reactions ① and ② occur in the matrix of liver mitochondria and reactions ③, ④, and ⑤ in liver cytosol. CO_2 (as bicarbonate), ammonium ion, ornithine, and citrulline enter the mitochondrial matrix via specific carriers (see red dots) present in the inner membrane of liver mitochondria.

nitrogens in arginine and release of the aspartate skeleton as fumarate (reaction 4, Figure 28–13). Subsequent addition of water to fumarate forms L-malate, whose subsequent NAD^+-dependent oxidation forms oxaloacetate. These two reactions are analogous to reactions of the citric acid cycle (see Figure 17–3), but are catalyzed by **cytosolic fumarase and malate dehydrogenase.** Transamination of oxaloacetate by glutamate aminotransferase then re-forms aspartate. The carbon skeleton of aspartate-fumarate thus acts as a carrier of the nitrogen of glutamate into a precursor of urea.

Cleavage of Arginine Releases Urea & Re-Forms Ornithine

Hydrolytic cleavage of the guanidino group of arginine, catalyzed by liver **arginase,** releases urea (reaction 5,

Figure 28–13). The other product, ornithine, reenters liver mitochondria and participates in additional rounds of urea synthesis. Ornithine and lysine are potent inhibitors of arginase, and compete with arginine. Arginine also serves as the precursor of the potent muscle relaxant nitric oxide (NO) in a Ca^{2+}-dependent reaction catalyzed by NO synthase (see Figure 49–15).

Carbamoyl Phosphate Synthase I Is the Pacemaker Enzyme of the Urea Cycle

The activity of carbamoyl phosphate synthase I is determined by N-acetylglutamate, whose steady-state level is dictated by the balance between its rate of synthesis from acetyl-CoA and glutamate and its rate of hydrolysis to acetate and glutamate, reactions catalyzed by *N*-acetylglutamate synthase (NAGS) and *N*-acetylglutamate hydrolase, respectively.

$$\text{Acetyl-CoA} + \text{L-glutamate} \rightarrow N\text{-acetyl-L-glutamate} + \text{CoASH}$$

$$N\text{-acetyl-L-glutamate} + H_2O \rightarrow \text{L-glutamate} + \text{acetate}$$

Major changes in diet can increase the concentrations of individual urea cycle enzymes 10- to 20-fold. For example, starvation elevates enzyme levels, presumably to cope with the increased production of ammonia that accompanies enhanced starvation-induced degradation of protein.

GENERAL FEATURES OF METABOLIC DISORDERS

The comparatively rare, but well-characterized and medically devastating metabolic disorders associated with the enzymes of urea biosynthesis illustrate the following general principles of inherited metabolic diseases.

1. Similar or identical clinical signs and symptoms can characterize various genetic mutations in a gene that encodes a given enzyme or in enzymes that catalyze successive reactions in a metabolic pathway.
2. Rational therapy is based on an understanding of the relevant biochemical enzyme-catalyzed reactions in both normal and impaired individuals.
3. The identification of intermediates and of ancillary products that accumulate prior to a metabolic block provides the basis for metabolic screening tests that can implicate the reaction that is impaired.
4. Definitive diagnosis involves quantitative assay of the activity of the enzyme suspected to be defective.
5. The DNA sequence of the gene that encodes a given mutant enzyme is compared to that of the wild-type gene to identify the specific mutation(s) that cause the disease.
6. The exponential increase in DNA sequencing of human genes has identified dozens of mutations of an affected gene that are benign or are associated with symptoms of varying severity of a given metabolic disorder.

METABOLIC DISORDERS ARE ASSOCIATED WITH EACH REACTION OF THE UREA CYCLE

Defects in each enzyme of the urea cycle have been described. Many of the causative mutations have been mapped, and specific defects in the encoded enzymes have been identified. Five well-documented diseases represent defects in the biosynthesis of enzymes of the urea cycle. Molecular genetic analysis has pinpointed the loci of mutations associated with each deficiency, each of which exhibits considerable genetic and phenotypic variability (**Table 28–1**).

Urea cycle disorders are characterized by hyperammonemia, encephalopathy, and respiratory alkalosis. Four of the five

TABLE 28–1 Enzymes of Inherited Metabolic Disorders of the Urea Cycle

Enzyme	Enzyme Catalog Number	OMIM[1] Reference	Figure and Reaction
Carbamoyl-phosphate synthase	6.3.4.16	237300	28–11 ①
Ornithine carbamoyl transferase	2.1.3.3	311250	28–11 ②
Argininosuccinate synthase	6.3.4.5	215700	28–11 ③
Argininosuccinate lyase	4.3.2.1	608310	28–11 ④
Arginase	3.5.3.1	608313	28–11 ⑤

[1]Online Mendelian inheritance in man database: ncbi.nlm.nih.gov/omim/

metabolic diseases, deficiencies of carbamoyl phosphate synthase, ornithine carbamoyl transferase, argininosuccinate synthase, and argininosuccinate lyase, result in the accumulation of precursors of urea, principally ammonia and glutamine. Ammonia intoxication is most severe when the metabolic block occurs at reactions 1 or 2 (Figure 28–13), for if citrulline can be synthesized, some ammonia has already been removed by being covalently linked to an organic metabolite.

Clinical symptoms common to all urea cycle disorders include vomiting, avoidance of high-protein foods, intermittent ataxia, irritability, lethargy, and severe mental retardation. The most dramatic clinical presentation occurs in full-term infants who initially appear normal, then exhibit progressive lethargy, hypothermia, and apnea due to high plasma ammonia levels. The clinical features and treatment of all five disorders are similar. Significant improvement and minimization of brain damage can accompany a low-protein diet ingested as frequent small meals to avoid sudden increases in blood ammonia levels. The goal of dietary therapy is to provide sufficient protein, arginine, and energy to promote growth and development while simultaneously minimizing the metabolic perturbations.

Carbamoyl Phosphate Synthase I

N-Acetylglutamate is essential for the activity of carbamoyl phosphate synthase I (reaction 1, Figure 28–13). Defects in carbamoyl phosphate synthase I are responsible for the relatively rare (estimated frequency 1:62,000) metabolic disease termed "hyperammonemia type 1."

N-Acetylglutamate Synthase

N-Acetylglutamate synthase (NAGS) catalyzes the formation from acetyl-CoA and glutamate of the *N*-acetylglutamate essential for carbamoyl phosphate synthase I activity.

$$\text{L-glutamate} + \text{acetyl-CoA} \rightarrow N\text{-acetyl-L-glutamate} + \text{CoASH}$$

While the clinical and biochemical features of NAGS deficiency are indistinguishable from those arising from a defect in carbamoyl phosphate synthase I, a deficiency in NAGS may respond to administered *N*-acetylglutamate.

Ornithine Permease

The hyperornithinemia, hyperammonemia, and homocitrullinuria syndrome (**HHH syndrome**) results from mutation of the ORNT1 gene that encodes the mitochondrial membrane ornithine permease. The failure to import cytosolic ornithine into the mitochondrial matrix renders the urea cycle inoperable, with consequent hyperammonemia, and hyperornithinemia due to the accompanying accumulation of cytosolic ornithine. In the absence of its normal acceptor (ornithine), mitochondrial carbamoyl phosphate carbamoylates lysine to homocitrulline, resulting in homocitrullinuria.

Ornithine Transcarbamoylase

The X-chromosome linked deficiency termed "hyperammonemia type 2" reflects a defect in ornithine transcarbamoylase (reaction 2, Figure 28–13). The mothers also exhibit hyperammonemia and an aversion to high-protein foods. Levels of glutamine are elevated in blood, cerebrospinal fluid, and urine, probably as a result of enhanced glutamine synthesis in response to elevated levels of tissue ammonia.

Argininosuccinate Synthase

In addition to patients who lack detectable argininosuccinate synthase activity (reaction 3, Figure 28–13), a 25-fold elevated K_m for citrulline has been reported. In the resulting citrullinemia, plasma and cerebrospinal fluid citrulline levels are elevated, and 1–2 g of citrulline are excreted daily.

Argininosuccinate Lyase

Argininosuccinicaciduria, accompanied by elevated levels of argininosuccinate in blood, cerebrospinal fluid, and urine, is associated with friable, tufted hair (trichorrhexis nodosa). Both early- and late-onset types are known. The metabolic defect is in argininosuccinate lyase (reaction 4, Figure 28–13). Diagnosis by the measurement of erythrocyte argininosuccinate lyase activity can be performed on umbilical cord blood or amniotic fluid cells.

Arginase

Hyperargininemia is an autosomal recessive defect in the gene for arginase (reaction 5, Figure 28–13). Unlike other urea cycle disorders, the first symptoms of hyperargininemia typically do not appear until age 2 to 4 years. Blood and cerebrospinal fluid levels of arginine are elevated. The urinary amino acid pattern, which resembles that of lysine-cystinuria (see Chapter 29), may reflect competition by arginine with lysine and cysteine for reabsorption in the renal tubule.

Analysis of Neonate Blood by Tandem Mass Spectrometry Can Detect Metabolic Diseases

Metabolic diseases caused by the absence or functional impairment of metabolic enzymes can be devastating. Early dietary intervention, however, can in many instances ameliorate the otherwise inevitable dire effects. The early detection of such metabolic diseases is thus is of primary importance. Since the initiation in the United States of newborn screening programs in the 1960s, all states now conduct metabolic screening of newborns, although the scope of screen employed varies among states. The powerful and sensitive technique of **tandem mass spectrometry** (see Chapter 4) can in a few minutes detect over 40 analytes of significance in the detection of metabolic disorders. Most states employ tandem MS to screen newborns to detect metabolic disorders such as organic acidemias, aminoacidemias, disorders of fatty acid oxidation, and defects in the enzymes of the urea cycle. However, at present there remain significant differences in analyte coverage between states. An article in *Clinical Chemistry* 2006 39:315 reviews the theory of tandem MS, its application to the detection of metabolic disorders, and situations that can yield false positives, and includes a lengthy table of detectable analytes and the relevant metabolic diseases.

Can Gene Therapy Offer Promise for Correcting Defects in Urea Biosynthesis?

Gene therapy of defects in the enzymes of the urea cycle is an area of active investigation. Despite encouraging results in animal models using an adenoviral vector to treat citrullinemia, at present gene therapy provides no effective solution for human subjects.

SUMMARY

- Human subjects degrade 1–2% of their body protein daily at rates that vary widely between proteins and with physiologic state. Key regulatory enzymes often have short half-lives.

- Proteins are degraded by both ATP-dependent and ATP-independent pathways. Ubiquitin targets many intracellular proteins for degradation. Liver cell surface receptors bind and internalize circulating asialoglycoproteins destined for lysosomal degradation.

- Fish excrete highly toxic NH_3 directly. Birds convert NH_3 to uric acid. Higher vertebrates convert NH_3 to urea.

- Transamination channels amino acid nitrogen into glutamate. GDH occupies a central position in nitrogen metabolism.

- Glutamine synthase converts NH_3 to nontoxic glutamine. Glutaminase releases NH_3 for use in urea synthesis.

- NH$_3$, CO$_2$, and the amide nitrogen of aspartate provide the atoms of urea.

- Hepatic urea synthesis takes place in part in the mitochondrial matrix and in part in the cytosol.

- Changes in enzyme levels and allosteric regulation of carbamoyl phosphate synthase I by N-acetylglutamate regulate urea biosynthesis.

- Metabolic diseases are associated with defects in each enzyme of the urea cycle, of the membrane-associated ornithine permease, and of NAGS.

- Tandem mass spectrometry is the technique of choice for screening neonates for inherited metabolic diseases.

REFERENCES

Brooks P, Fuertes G, Murray RZ, et al: Subcellular localization of proteasomes and their regulatory complexes in mammalian cells. Biochem J 2000;346:155.

Caldovic L, Morizono H, Tuchman M: Mutations and polymorphisms in the human N-acetylglutamate synthase (NAGS) gene. Hum Mutat 2007;28:754.

Crombez EA, Cederbaum SD: Hyperargininemia due to liver arginase deficiency. Mol Genet Metab 2005;84:243.

Elpeleg O, Shaag A, Ben-Shalom E, et al: N-acetylglutamate synthase deficiency and the treatment of hyperammonemic encephalopathy. Ann Neurol 2002;52:845.

Garg U, Dasouki M: Expanded newborn screening of inherited metabolic disorders by tandem mass spectrometry. Clinical and laboratory aspects. Clin Biochem 2006;39:315.

Gyato K, Wray J, Huang ZJ, et al: Metabolic and neuropsychological phenotype in women heterozygous for ornithine transcarbamylase deficiency. Ann Neurol 2004;55:80.

Häberle J, Denecke J, Schmidt E, et al: Diagnosis of N-acetylglutamate synthase deficiency by use of cultured fibroblasts and avoidance of nonsense-mediated mRNA decay. J Inherit Metab Dis 2003;26:601.

Häberle J, Görg B, Rutsch F, et al: Congenital glutamine deficiency with glutamine synthetase mutations. N Engl J Med 2005;353:1926.

Häberle J, Pauli S, Schmidt E, et al: Mild citrullinemia in caucasians is an allelic variant of argininosuccinate synthetase deficiency (citrullinemia type 1). Mol Genet Metab 2003;80:302.

Iyer R, Jenkinson CP, Vockley JG, et al: The human arginases and arginase deficiency. J Inherit Metab Dis 1998;21:86.

Pickart CM: Mechanisms underlying ubiquitination. Annu Rev Biochem 2001;70:503.

Scriver CR: Garrod's foresight; our hindsight. J Inherit Metab Dis 2001;24:93.

Scriver CR, Sly WS, Childs B, et al (editors): *The Metabolic and Molecular Bases of Inherited Disease,* 8th ed. McGraw-Hill, 2001.

Yi JJ, Ehlers MD: Emerging roles for ubiquitin and protein degradation in neuronal function. Pharmacol Rev 2007;59:206.

Catabolism of the Carbon Skeletons of Amino Acids

Victor W. Rodwell, PhD

OBJECTIVES

After studying this chapter, you should be able to:

- Name the principal catabolites of the carbon skeletons of the common amino acids and the major metabolic fates of these catabolites.

- Write an equation for an aminotransferase (transaminase) reaction and illustrate the role played by the coenzyme.

- Outline the metabolic pathways for each of the common amino acids, and identify reactions associated with clinically significant metabolic disorders.

- Provide examples of aminoacidurias that arise from defects in glomerular tubular reabsorption, and the consequences of impaired intestinal absorption of tryptophan.

- Explain why metabolic defects in different enzymes of the catabolism of a specific amino acid can be associated with similar clinical signs and symptoms.

- Describe the implications of a metabolic defect in glutamate-γ-semialdehyde dehydrogenase for the catabolism of proline and of 4-hydroxyproline.

- Explain how the α-amino nitrogen of proline and of lysine is removed by processes other than transamination.

- Draw analogies between the reactions that participate in the catabolism of fatty acids and of the branched-chain amino acids.

- Identify the specific metabolic defects in hypervalinemia, maple syrup urine disease, intermittent branched-chain ketonuria, isovaleric acidemia, and methylmalonic aciduria.

BIOMEDICAL IMPORTANCE

The prior chapter described the removal and metabolic fate of the nitrogen atoms of the common L-α-amino acids. This chapter will address the metabolic fates of the resulting hydrocarbon skeletons of these amino acids. Discussed are the enzymes and intermediates formed during the conversion of the carbon skeletons to amphibolic intermediates, and several metabolic diseases or "inborn errors of metabolism" associated with these processes. While most disorders of amino acid catabolism are rare, if left untreated they can result in irreversible brain damage and early mortality. Prenatal or early postnatal detection of metabolic disorders and timely initiation of treatment thus are essential. The ability to detect the activities of enzymes in cultured amniotic fluid cells facilitates prenatal diagnosis by amniocentesis. All states now conduct screening tests of newborns for as many as 30 metabolic diseases. These tests include, but are not limited to, disorders associated with defects in the catabolism of amino acids. The most reliable screening tests use tandem mass spectrometry to detect, in a few drops of neonate blood, catabolites suggestive of a given metabolic defect. The metabolites detected pinpoint the metabolic defect as the lowered or absent activity of a given enzyme. Treatment consists primarily of feeding diets low in the amino acid whose catabolism is impaired.

Mutations in the exons or in the regulatory regions of a gene that encodes an enzyme of amino acid metabolism can result in the failure to synthesize that enzyme or in the synthesis of a partially or completely nonfunctional enzyme. Mutations may have no significant effect of the activity of the encoded enzyme. By contrast, mutations that compromise the overall three-dimensional structure or the structure of catalytic or regulatory sites may be associated with adverse metabolic consequences. Low catalytic efficiency of a mutant enzyme can result from impaired positioning of residues involved in catalysis, or in binding a substrate, coenzyme, or metal ion. Mutations may also impair the ability of certain enzymes to respond appropriately to the signals that modulate their activity by altering an enzyme's affinity for an allosteric regulator of activity. Since different mutations can have similar effects on any of the above factors, various mutations may give rise to the same clinical signs and symptoms. At a molecular level, these therefore are distinct molecular diseases. To supplement the disorders of amino acid metabolism discussed in this chapter, readers should consult major reference works on this topic such as Scriver et al 2001.

AMINO ACIDS ARE CATABOLIZED TO INTERMEDIATES FOR CARBOHYDRATE AND LIPID BIOSYNTHESIS

Nutritional studies in the period 1920–1940, reinforced and confirmed by studies using isotopically labeled amino acids conducted from 1940 to 1950, established the interconvertibility of the carbon atoms of fat, carbohydrate, and protein. These studies also revealed that all or a portion of the carbon skeleton of every amino acid is convertible either to carbohydrate (13 amino acids), fat (one amino acid), or both fat and carbohydrate (five amino acids) (**Table 29–1**). **Figure 29–1** outlines overall aspects of these interconversions.

TABLE 29–1   Fate of the Carbon Skeletons of the Common L-α-Amino Acids

Converted to Amphibolic Intermediates That Form		
Carbohydrate (Glycogenic)	Fat (Ketogenic)	Glycogen and Fat (Glycogenic and Ketogenic)
Ala	Hyp	Leu	Ile
Arg	Met		Lys
Asp	Pro		Phe
Cys	Ser		Trp
Glu	Thr		Tyr
Gly	Val		
His			

TRANSAMINATION TYPICALLY INITIATES AMINO ACID CATABOLISM

Removal of α-amino nitrogen by transamination, a reaction catalyzed by an aminotransferase or transaminase (see Figure 28–6), is the first catabolic reaction of all the common amino acids except proline, hydroxyproline, threonine, or lysine. The hydrocarbon skeleton that remains is then degraded to amphibolic intermediates as outlined in Figure 29–1.

Asparagine and Aspartate Form Oxaloacetate

All four carbons of asparagine and of aspartate form **oxaloacetate** via reactions catalyzed by **asparaginase** and

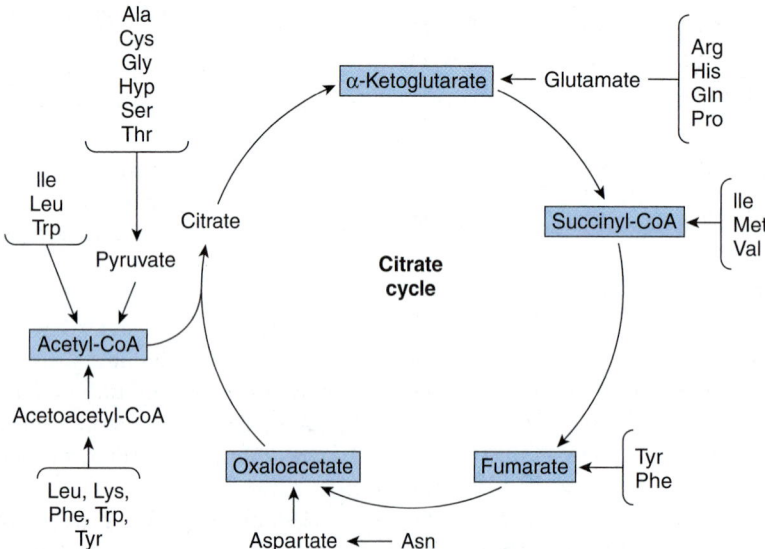

FIGURE 29–1   Overview of the amphibolic intermediates that result from catabolism of the common amino acids.

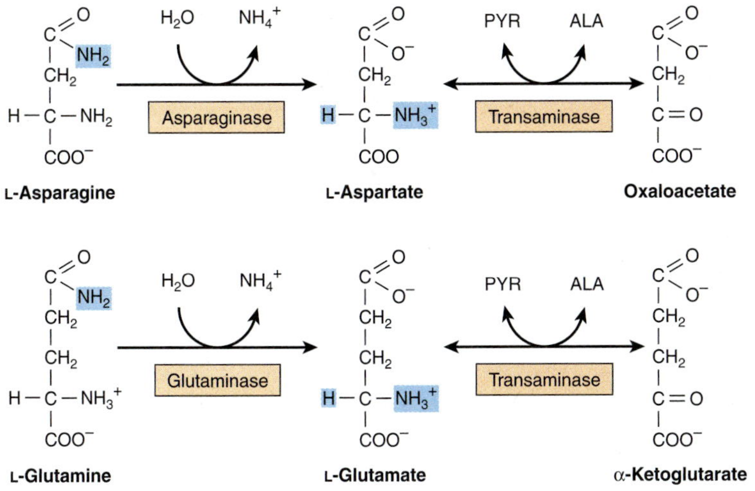

FIGURE 29–2 **Catabolism to amphibolic intermediates of L-asparagine (*top*) and of L-glutamine (*bottom*).** (PYR, pyruvate; ALA, L-alanine.) In this and subsequent figures, blue highlights emphasize the portions of the molecules that are undergoing chemical change.

a **transaminase** (**Figure 29–2**, top). Metabolic defects in transaminases, which fulfill central amphibolic functions, may be incompatible with life. Consequently, no known metabolic defect is associated with this short catabolic pathway.

Glutamine and Glutamate Form α-Ketoglutarate

The catabolism of glutamine and of glutamate parallels that of asparagine and aspartate in reactions catalyzed by **glutaminase** and **transaminase** that forms **α-ketoglutarate** (**Figure 29–2**, bottom). While both glutamate and aspartate are substrates for the same transaminase, deamidation of their corresponding amides is catalyzed by different enzymes: asparaginase and glutaminase. Possibly for the reason stated earlier, there are no known metabolic defects of the glutamine-glutamate catabolic pathway.

Significant metabolic disorders are, however, associated with the catabolism of many other amino acids. Discussed below under the catabolism of each amino acid, these disorders are summarized in **Table 29–2**. This table lists the impaired enzyme, its IUB enzyme catalog (EC) number, a cross-reference to a specific figure and numbered reaction, and a numerical link to the Online Mendelian Inheritance in Man database (**OMIM**).

Proline

The catabolism of proline takes place in mitochondria. Since proline does not participate in transamination, the nitrogen of this imino acid is retained throughout its oxidation to Δ¹-pyrroline-5-carboxylate, ring opening to glutamate-γ-semialdehyde, and oxidation to glutamate, and is only removed during transamination of glutamate to α-ketoglutarate (**Figure 29–3**). There are two metabolic disorders of proline catabolism. Both are inherited as autosomal recessive traits and are consistent with a normal adult life. The metabolic block in

type I hyperprolinemia is at **proline dehydrogenase.** There is no associated impairment of hydroxyproline catabolism. The metabolic block in **type II hyperprolinemia** is at **glutamate-γ-semialdehyde dehydrogenase,** a mitochondrial matrix enzyme that also participates in the catabolism of arginine, ornithine, and hydroxyproline (see below). Since proline and hydroxyproline catabolism are affected, both Δ¹-pyrroline-5-carboxylate and Δ¹-pyrroline-3-hydroxy-5-carboxylate (see Figure 29–12) are excreted.

Arginine and Ornithine

The initial reactions in arginine catabolism are conversion to ornithine followed by transamination of ornithine to glutamate-γ-semialdehyde (**Figure 29–4**). Subsequent catabolism of glutamate-γ-semialdehyde to **α-ketoglutarate** occurs as described for proline (see Figure 29–3). Mutations in **ornithine δ-aminotransferase** (ornithine transaminase) elevate plasma and urinary ornithine and are associated with **gyrate atrophy of the choroid and retina.** Treatment involves restricting dietary arginine. In the **hyperornithinemia–hyperammonemia syndrome,** a defective mitochondrial **ornithine-citrulline antiporter** (see Figure 28–13) impairs transport of ornithine into mitochondria for use in urea synthesis.

Histidine

Catabolism of histidine proceeds via urocanate, 4-imidazolone-5-propionate, and *N*-formiminoglutamate (Figlu). Formimino group transfer to tetrahydrofolate forms glutamate, then **α-ketoglutarate** (**Figure 29–5**). In **folic acid deficiency,** transfer of the formimino group is impaired, and Figlu is excreted. Excretion of Figlu following a dose of histidine thus can be used to detect folic acid deficiency. Benign disorders of histidine catabolism include **histidinemia** and **urocanic aciduria** associated with impaired **histidase.**

TABLE 29–2 **Metabolic Diseases of Amino Acid Metabolism**

Defective Enzyme	Enzyme Catalog Number	OMIM[1] Reference	Major Signs and Symptoms	Figure and Reaction
S-Adenosylhomocysteine hydrolase	3.3.1.1	180960	Hypermethioninemia	29–19 ③
Arginase	3.5.3.1	207800	Argininemia	29–4 ①
Cystathionine-β-synthase	4.2.1.22	236200	Homocystinuria	29–19 ④
Fumarylacetoacetate hydrolase	3.7.1.12	276700	Type-I tyrosinemia (Tyrosinosis)	29–13 ④
Glycine N-methyl-transferase	2.1.1.20	606664	Hypermethioninemia	29–13 ②
Histidine ammonia lyase (Histidase)	4.3.1.3	609457	Histidinemia & urocanic acidemia	29–5 ①
Homogentisate oxidase	1.13.11.5	607474	Alkaptonuria. Homogentisate excreted.	29–13 ③
p-Hydroxyphenylpyruvate hydroxylase	1.13.11.27	276710	Neonatal tyrosinemia	29–13 ②
Isovaleryl-CoA dehydrogenase	1.3.99.10	607036	Isovaleric acidemia	29–20 ③
Branched chain α -ketoacid decarboxylase complex		248600	Branched-chain ketonuria (MSUD)	29–20 ①
Methionine adenosyltransferase	2.5.1.6	250850	Hypermethioninemia	29–18 ①
Ornithine-δ-aminotransferase	2.6.1.13	258870	Ornithemia, gyrate atrophy	29–4 ②
Phenylalanine hydroxylase	1.14.16.1	261600	Type I, Classic phenylketonuria	27–10 ①
Proline dehydrogenase	1.5.99.8	606810	Type I hyperprolinemia	29–3 ①
Δ¹-Pyrroline-5-carboxylate dehydrogenase	1.5.1.12	606811	Type II hyperprolinemia & hyper 4-hydroxyprolinemia	29–3 ②
Saccharopine dehydrogenase	1.5.1.7	268700	Saccharopinuria	29–15 ②
Tyrosine aminotransferase	2.6.1.15	613018	Type-II tyrosinemia	29–13 ①

[1]Online Mendelian Inheritance in Man database: ncbi.nlm.nih.gov/omim/

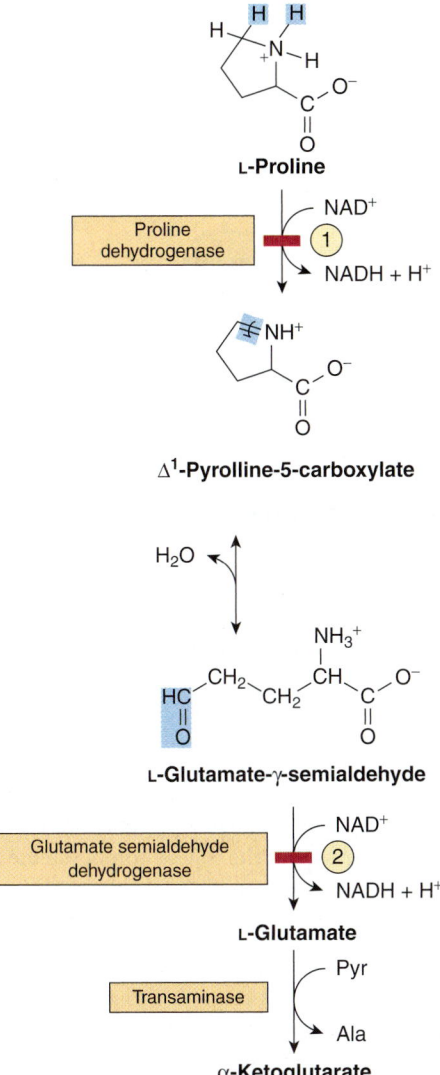

FIGURE 29–3 **Catabolism of proline.** Red bars and circled numerals indicate the locus of the inherited metabolic defects in ① type-I hyperprolinemia and ② type-II hyperprolinemia.

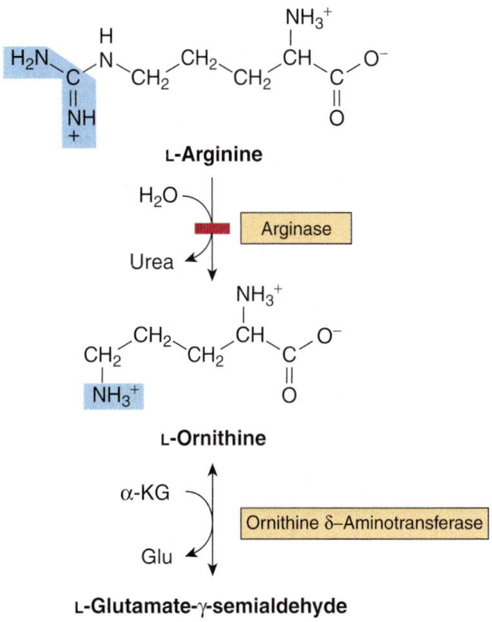

FIGURE 29–4 **Catabolism of arginine.** Arginase-catalyzed cleavage of L-arginine forms urea and L-ornithine. This reaction (red bar) represents the site of the inherited metabolic defect in hyperargininemia. Subsequent transamination of L-ornithine to glutamate-γ-semialdehyde is followed by conversion to α-ketoglutarate (see Figure 29–3).

CATABOLISM OF GLYCINE, SERINE, ALANINE, CYSTEINE, THREONINE, AND 4-HYDROXYPROLINE

Glycine

The **glycine cleavage complex** of liver mitochondria splits glycine to CO_2 and NH_4^+ and forms N^5, N^{10}-methylene tetrahydrofolate.

$$Glycine + H_4folate + NAD^+ \rightarrow CO_2 + NH_3$$
$$+ 5,10\text{-}CH_2\text{-}H_4folate + NADH + H^+$$

The glycine cleavage system (**Figure 29–6**) consists of three enzymes and an "H-protein" that has a covalently attached dihydrolipoyl moiety. Figure 29–6 also illustrates the individual reactions and intermediates in glycine cleavage. In **nonketotic hyperglycinemia,** a rare inborn error of glycine degradation presently known only in Finland, glycine accumulates in all body tissues including the central nervous system. The defect in **primary hyperoxaluria** is the failure to catabolize glyoxylate formed by the deamination of glycine. Subsequent oxidation of glyoxylate to oxalate results in urolithiasis, nephrocalcinosis, and early mortality from renal failure or hypertension. **Glycinuria** results from a defect in renal tubular reabsorption.

Serine

Following conversion to glycine, catalyzed by **serine hydroxymethyltransferase,** serine catabolism merges with that of glycine (**Figure 29–7**).

Alanine

Transamination of α-alanine forms pyruvate. Probably on account of its central role in metabolism there is no known metabolic defect of α-alanine catabolism.

Cystine and Cysteine

Cystine is first reduced to cysteine by **cystine reductase** (**Figure 29–8**). Two different pathways then convert cysteine to pyruvate (**Figure 29–9**). There are numerous abnormalities of cysteine metabolism. Cystine, lysine, arginine, and ornithine are excreted in **cystine-lysinuria (cystinuria),** a defect in renal reabsorption of these amino acids. Apart from cystine calculi, cystinuria is benign. The mixed disulfide of L-cysteine

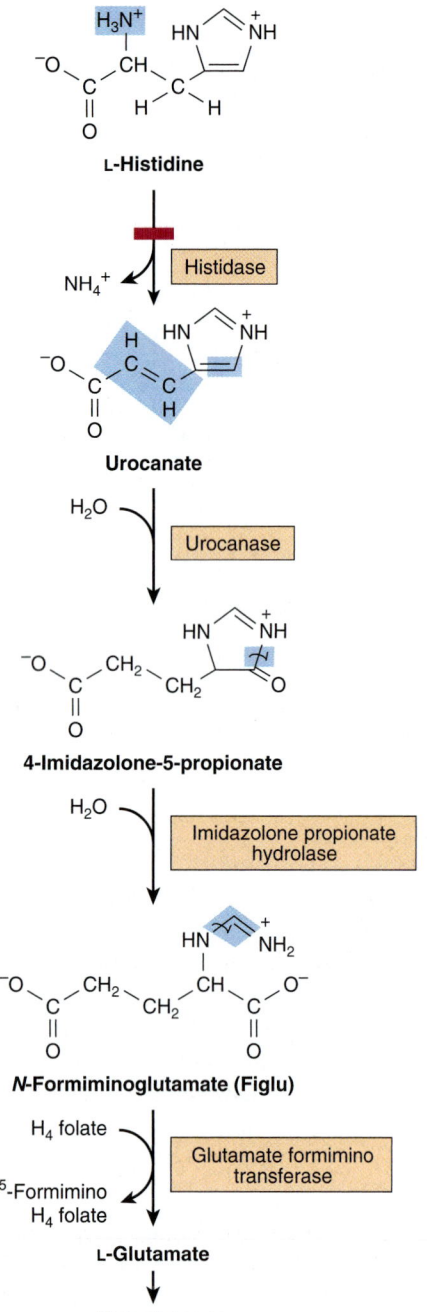

FIGURE 29–5 Catabolism of L-histidine to α-ketoglutarate. (H₄ folate, tetrahydrofolate.) The red bar indicates the site of an inherited metabolic defect.

and L-homocysteine (**Figure 29–10**) excreted by cystinuric patients is more soluble than cystine and reduces formation of cystine calculi.

Several metabolic defects result in vitamin B_6-responsive or vitamin B_6-unresponsive **homocystinurias**. These include a deficiency in the reaction catalyzed by cystathionine β-synthase:

$$\text{Serine} + \text{homocysteine} \rightarrow \text{cystathionine} + H_2O$$

Consequences include osteoporosis and mental retardation. Defective carrier-mediated transport of cystine results in **cystinosis (cystine storage disease)** with deposition of cystine

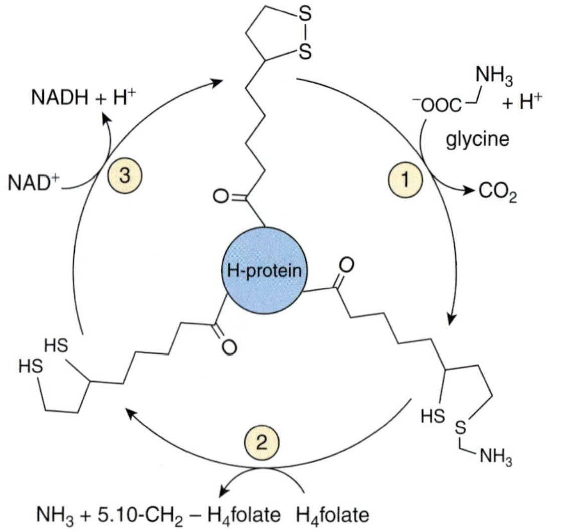

FIGURE 29–6 **The glycine cleavage system of liver mitochondria.** The glycine cleavage complex consists of three enzymes and an "H-protein" that has covalently attached dihyrolipoate. Catalysts for the numbered reactions are ① glycine dehydrogenase (decarboxylating), ② an ammonia-forming aminomethyltransferase, and ③ dihydrolipoamide dehydrogenase. (H₄folate, tetrahydrofolate).

crystals in tissues and early mortality from acute renal failure. Epidemiologic and other data link plasma homocysteine levels to cardiovascular risk, but the role of homocysteine as a causal cardiovascular risk factor remains controversial.

Threonine

Threonine aldolase cleaves threonine to acetaldehyde and glycine. Catabolism of glycine is discussed above. Oxidation of acetaldehyde to acetate is followed by formation of acetyl-CoA (**Figure 29–11**).

4-Hydroxyproline

Catabolism of 4-hydroxy-L-proline forms, successively, L-Δ¹-pyrroline-3-hydroxy-5-carboxylate, γ-hydroxy-L-glutamate-γ-semialdehyde, erythro-γ-hydroxy-L-glutamate, and α-keto-γ-hydroxyglutarate. An aldol-type cleavage then forms glyoxylate plus pyruvate (**Figure 29–12**). A defect in **4-hydroxyproline dehydrogenase** results in **hyperhydroxyprolinemia**, which is benign. There is no associated impairment of proline catabolism. As noted above under

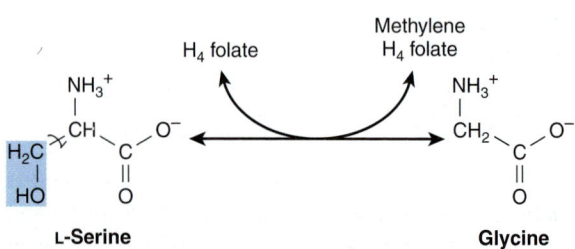

FIGURE 29–7 Interconversion of serine and glycine by serine hydroxymethyltransferase. (H₄folate, tetrahydrofolate).

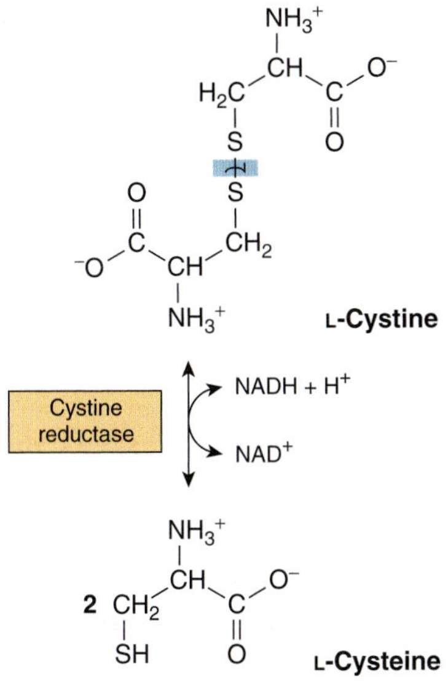

FIGURE 29–8 Reduction of cystine to cysteine in the cystine reductase reaction.

proline, a defect in **glutamate-γ-semialdehyde dehydrogenase** is accompanied by excretion of Δ^1-pyrroline-3-hydroxy-5-carboxylate.

ADDITIONAL AMINO ACIDS THAT FORM ACETYL-CoA

Tyrosine

Figure 29–13 illustrates the intermediates and enzymes that participate in the catabolism of tyrosine to amphibolic intermediates. Following transamination of tyrosine to *p*-hydroxyphenylpyruvate, successive reactions form maleylacetoacetate, fumarylacetoacetate, fumarate, acetoacetate, and ultimately acetyl-CoA and acetate.

Several metabolic disorders are associated with the tyrosine catabolic pathway. The probable metabolic defect in **type I tyrosinemia (tyrosinosis)** is at **fumarylacetoacetate hydrolase** (reaction 4, Figure 29–1). Therapy employs a diet low in tyrosine and phenylalanine. Untreated acute and chronic tyrosinosis leads to death from liver failure. Alternate metabolites of tyrosine are also excreted in **type II tyrosinemia (Richner–Hanhart syndrome),** a defect in **tyrosine aminotransferase** (reaction 1, Figure 29–13), and in **neonatal tyrosinemia,** due to the lowered *p*-hydroxyphenylpyruvate hydroxylase activity (reaction 2, Figure 29–13). Therapy employs a diet low in protein.

The metabolic defect in **alkaptonuria** is a defective **homogentisate oxidase,** the enzyme that catalyzes reaction 3 of Figure 29–13. The urine darkens on exposure to air due to oxidation of excreted homogentisate. Late in the disease, there is arthritis and connective tissue pigmentation (ochronosis) due to oxidation of homogentisate to benzoquinone acetate, which polymerizes and binds to connective tissue. First described in the sixteenth century based on the observation that the urine

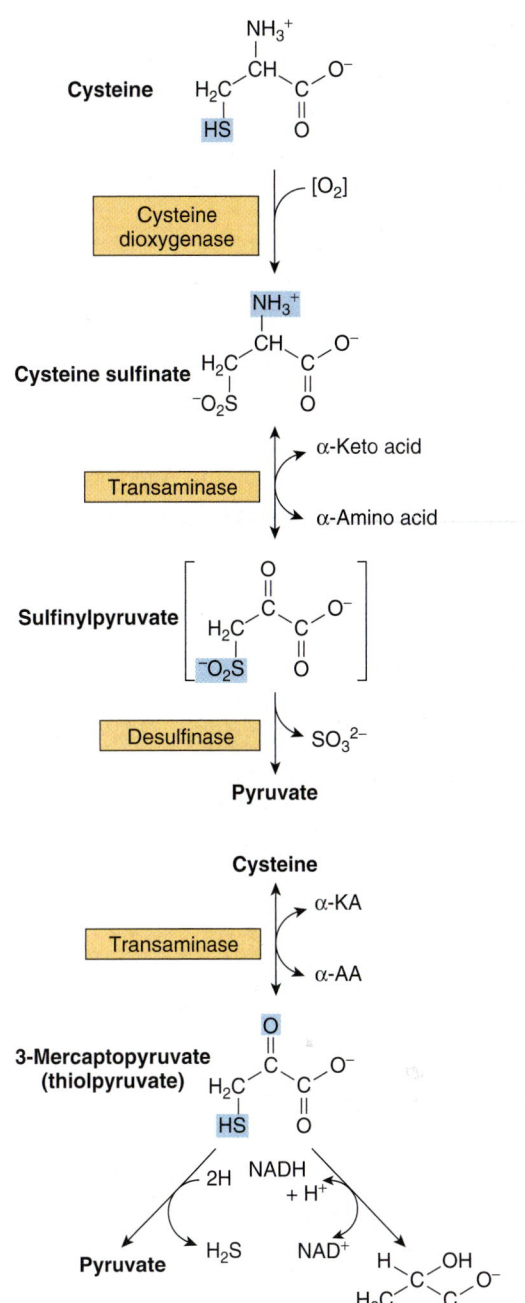

FIGURE 29–9 Two pathways catabolize L-cysteine: the cysteine sulfinate pathway (*top*) and the 3-mercaptopyruvate pathway (*bottom*).

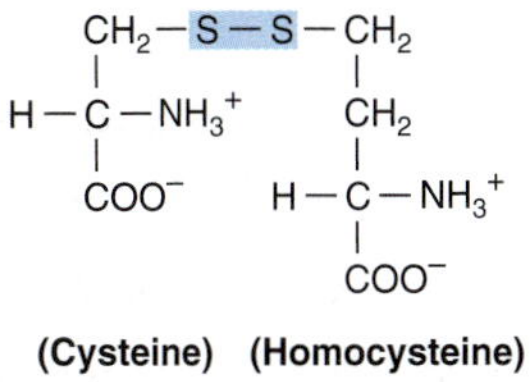

FIGURE 29–10 Structure of the mixed disulfide of cysteine and homocysteine.

darkened on exposure to air, alkaptonuria provided the basis for Sir Archibald Garrod's early twentieth century classic ideas concerning heritable metabolic disorders. Based on the presence of ochronosis and on chemical evidence, the earliest known case of alkaptonuria is, however, its 1977 detection in an Egyptian mummy dating from 1500 B.C.

Phenylalanine

Phenylalanine is first converted to tyrosine (see Figure 27–10). Subsequent reactions are those of tyrosine (Figure 29–13). **Hyperphenylalaninemias** arise from defects in phenylalanine hydroxylase (**type I, classic phenylketonuria (PKU),** frequency 1 in 10,000 births), in dihydrobiopterin reductase (**types II and III),** or in dihydrobiopterin biosynthesis (**types IV and V)** (see Figure 27–10). Alternative catabolites are excreted (**Figure 29–14**). A diet low in phenylalanine can prevent the mental retardation of PKU.

DNA probes facilitate prenatal diagnosis of defects in phenylalanine hydroxylase or dihydrobiopterin reductase. Elevated blood phenylalanine may not be detectable until 3–4 days postpartum. False-positives in premature infants

L-Threonine

Threonine aldolase → Glycine

Acetaldehyde

Aldehyde dehydrogenase

H_2O, NAD^+ → $NADH + H^+$

Acetate

Acetate thiokinase

CoASH, Mg-ATP → H_2O, Mg-ADP

Acetyl-CoA

FIGURE 29–11 Intermediates in the conversion of threonine to glycine and acetyl-CoA.

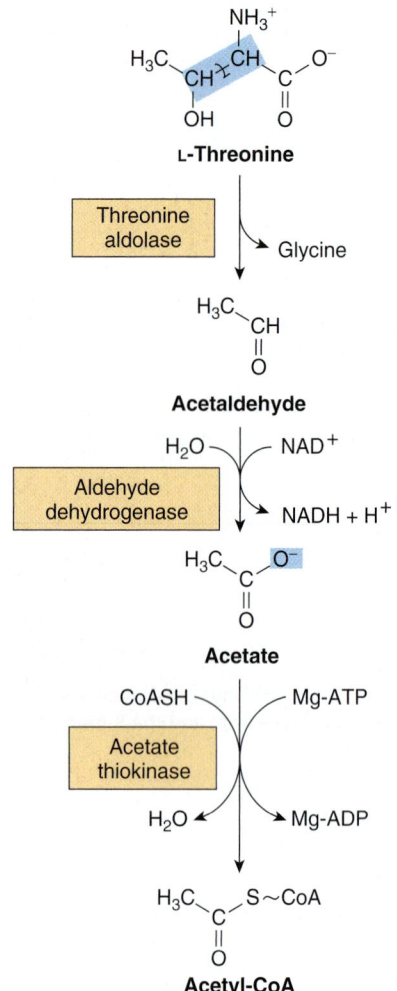

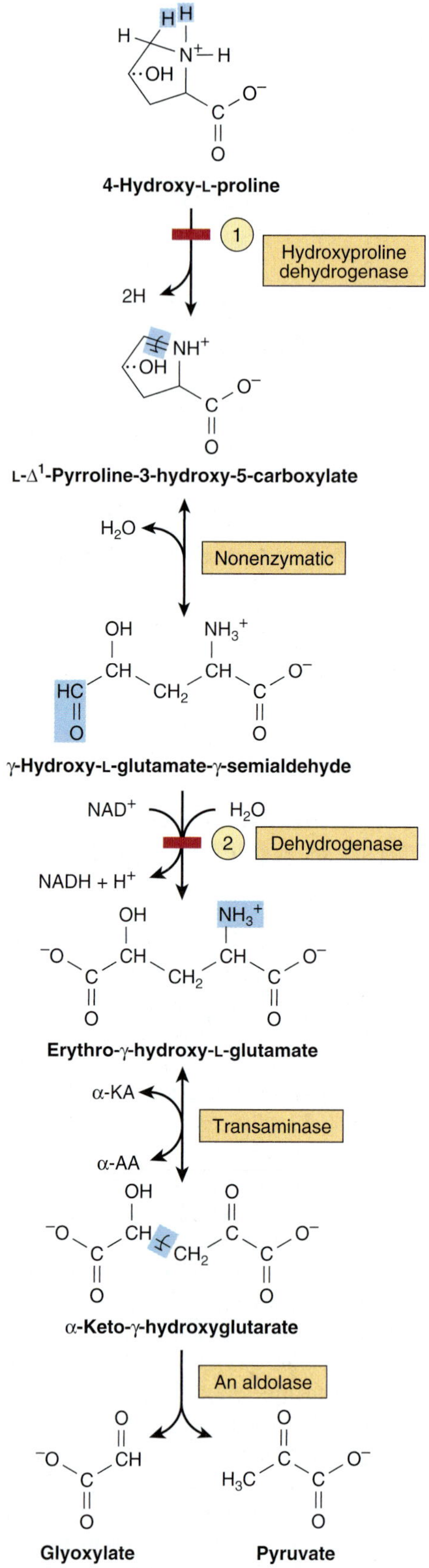

4-Hydroxy-L-proline

① Hydroxyproline dehydrogenase

2H

L-Δ¹-Pyrroline-3-hydroxy-5-carboxylate

H_2O ← Nonenzymatic

γ-Hydroxy-L-glutamate-γ-semialdehyde

NAD^+, H_2O → $NADH + H^+$

② Dehydrogenase

Erythro-γ-hydroxy-L-glutamate

α-KA ← Transaminase

α-AA

α-Keto-γ-hydroxyglutarate

An aldolase

Glyoxylate   **Pyruvate**

FIGURE 29–12 Intermediates in L-hydroxyproline catabolism. (α-AA, α-amino acid; α-KA, α-keto acid.) Red bars indicate the sites of the inherited metabolic defects in ① hyperhydroxyprolinemia and ② type II hyperprolinemia.

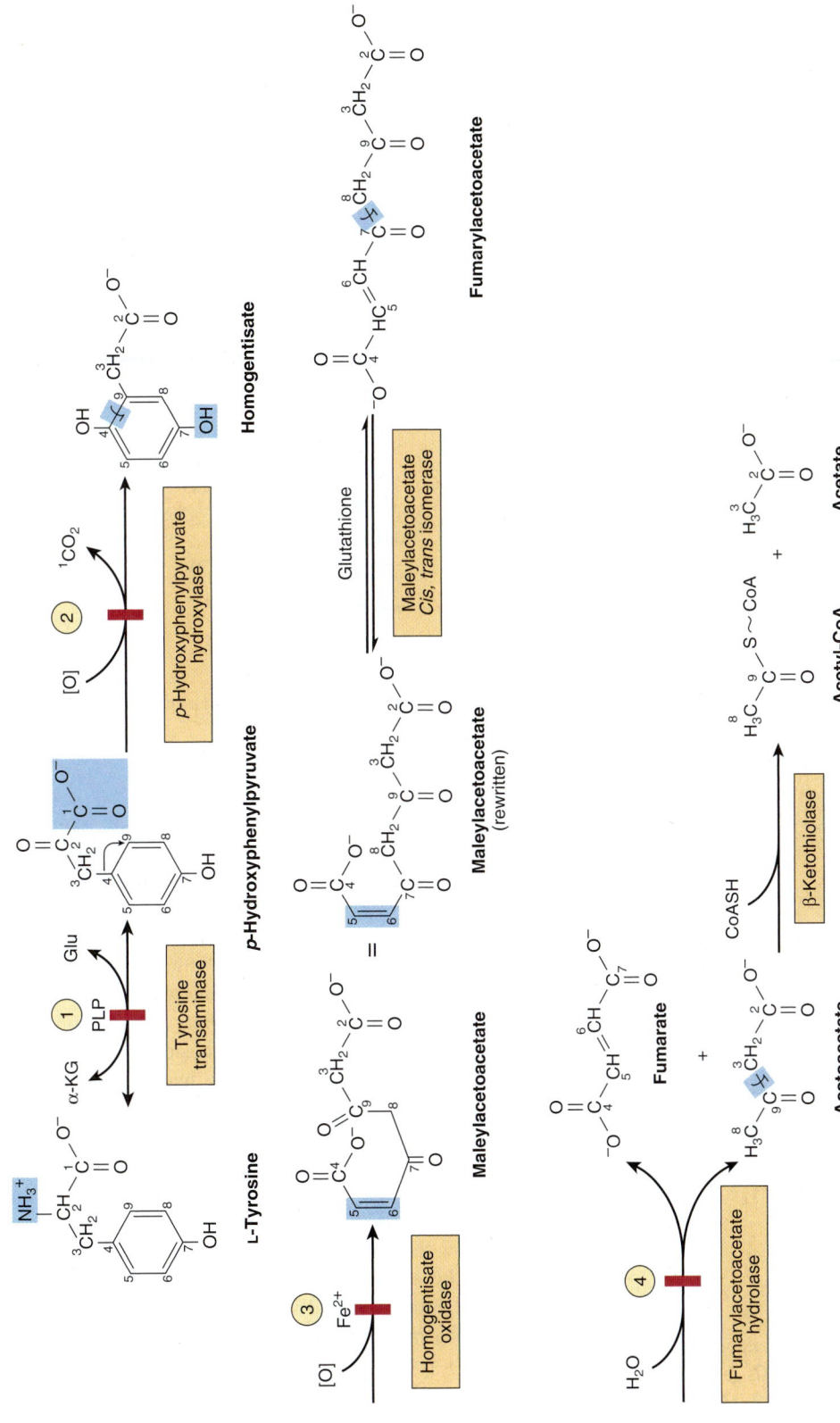

FIGURE 29–13 Intermediates in tyrosine catabolism. Carbons are numbered to emphasize their ultimate fate. (α-KG, α-ketoglutarate; Glu, glutamate; PLP, pyridoxal phosphate.) Red bars indicate the probable sites of the inherited metabolic defects in ① type II tyrosinemia; ② neonatal tyrosinemia; ③ alkaptonuria; and ④ type I tyrosinemia, or tyrosinosis.

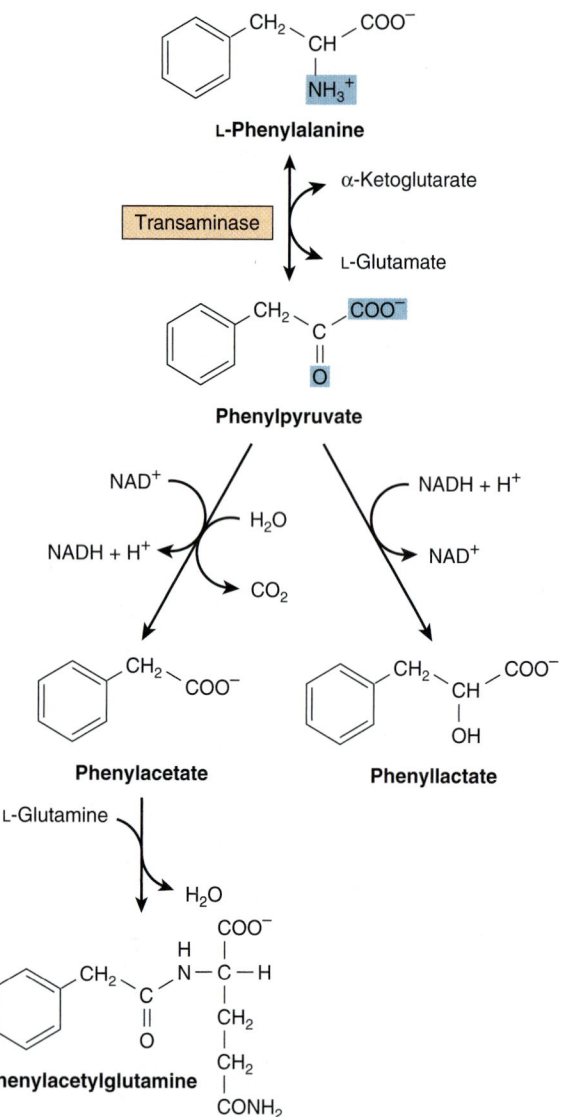

FIGURE 29–14 **Alternative pathways of phenylalanine catabolism in phenylketonuria.** The reactions also occur in normal liver tissue but are of minor significance.

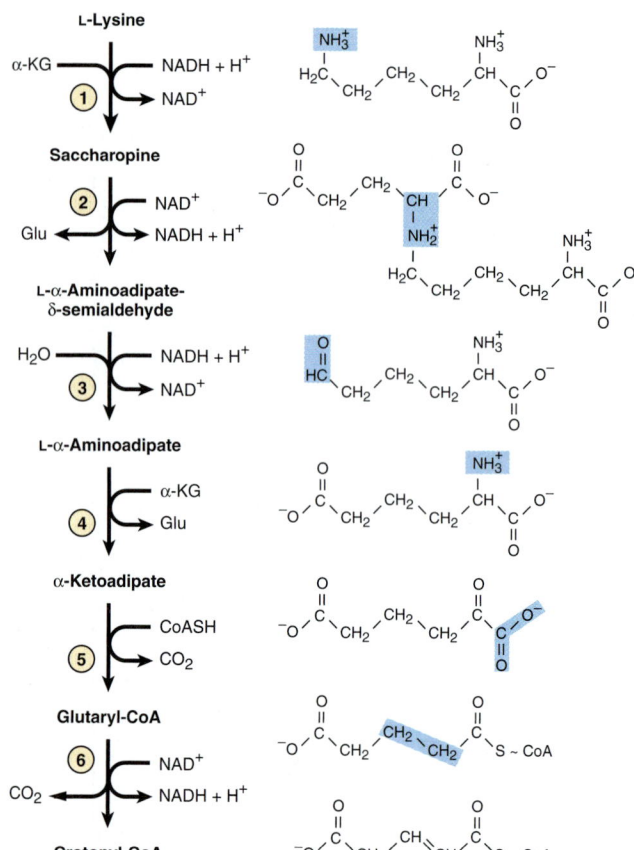

FIGURE 29–15 **Reactions and intermediates in the catabolism of L-lysine.**

may reflect delayed maturation of enzymes of phenylalanine catabolism. An older and less reliable screening test employs $FeCl_3$ to detect urinary phenylpyruvate. $FeCl_3$ screening for PKU of the urine of newborn infants is compulsory in many countries, but in the United States has been largely supplanted by tandem mass spectrometry.

Lysine

The first six reactions of L-lysine catabolism in human liver form crotonyl-CoA, which is then degraded to acetyl-CoA by the reactions of fatty acid catabolism (see Figure 22–3). In what follows, circled numerals refer to the corresponding numbered reactions of **Figure 29–15**. Reactions 1 and 2 convert the Schiff base formed between α-ketoglutarate and the ε-amino group of lysine to L-α-aminoadipate-δ-semialdehyde. Reactions 1 and 2 both are catalyzed by a single bifunctional

enzyme, aminoadipate semialdehyde synthase (also called lysine 2-oxoglutarate reductase-saccharopine dehydrogenase). Reduction of L-α-aminoadipate-δ-semialdehyde to L-α-aminoadipate (reaction 3) is followed by transamination to α-ketoadipate (reaction 4). Conversion to the thioester glutaryl-CoA (reaction 5) is followed by the decarboxylation of glutaryl-CoA to crotonyl-CoA (reaction 6). Subsequent reactions are those of the fatty acid catabolism.

Metabolic defects associated with reactions of the lysine catabolic pathway include hyperlysinemias. Hyperlysinemia can result from a defect in activity 1 or 2 of the bifunctional enzyme aminoadipate semialdehyde synthase. Hyperlysinemia is accompanied by elevated levels of blood saccharopine only if the defect involves activity 2. A metabolic defect at reaction 6 results in an inherited metabolic disease that is associated with striatal and cortical degeneration, and is characterized by elevated concentrations of glutarate and its metabolites glutaconate and 3-hydroxyglutarate. The challenge in management of these metabolic defects is to restrict dietary intake of L-lysine without producing malnutrition.

Tryptophan

Tryptophan is degraded to amphibolic intermediates via the kynurenine-anthranilate pathway (**Figure 29–16**) Tryptophan

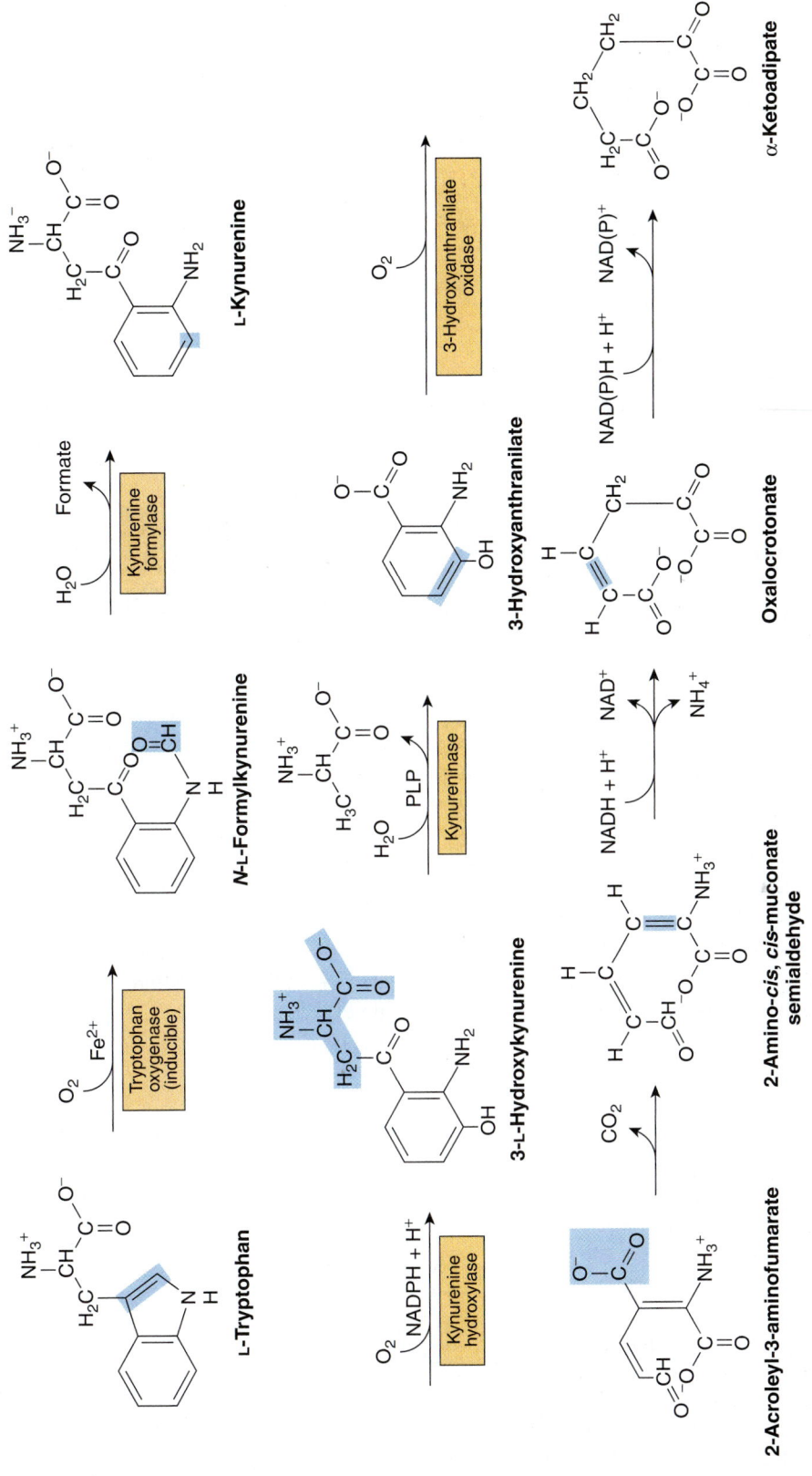

FIGURE 29–16 Reactions and intermediates in the catabolism of L-tryptophan. (PLP, pyridoxal phosphate.)

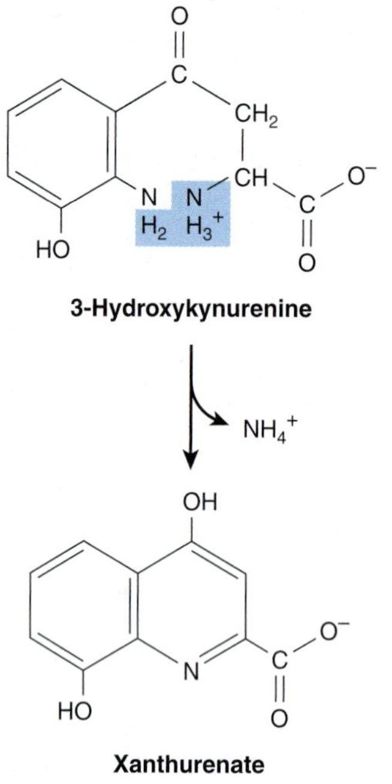

FIGURE 29–17 Formation of xanthurenate in vitamin B$_6$ deficiency. Conversion of the tryptophan metabolite 3-hydroxykynurenine to 3-hydroxyanthranilate is impaired (see Figure 29–16). A large portion is therefore converted to xanthurenate.

oxygenase (tryptophan pyrrolase) opens the indole ring, incorporates molecular oxygen, and forms N-formylkynurenine. Tryptophan oxygenase, an iron porphyrin metalloprotein that is inducible in liver by adrenal corticosteroids and by tryptophan, is feedback inhibited by nicotinic acid derivatives, including NADPH. Hydrolytic removal of the formyl group of N-formylkynurenine, catalyzed by **kynurenine formylase,** produces kynurenine. Since **kynureninase** requires pyridoxal phosphate, excretion of xanthurenate (**Figure 29–17**)

in response to a tryptophan load is diagnostic of vitamin B$_6$ deficiency. **Hartnup disease** reflects impaired intestinal and renal transport of tryptophan and other neutral amino acids. Indole derivatives of unabsorbed tryptophan formed by intestinal bacteria are excreted. The defect limits tryptophan availability for niacin biosynthesis and accounts for the pellagra-like signs and symptoms.

Methionine

Methionine reacts with ATP forming S-adenosylmethionine, "active methionine" (**Figure 29–18**). Subsequent reactions form propionyl-CoA (**Figure 29–19**), which three subsequent reactions convert to succinyl-CoA (see Figure 20–2).

THE INITIAL REACTIONS ARE COMMON TO ALL THREE BRANCHED-CHAIN AMINO ACIDS

The first three reactions of the catabolism of isoleucine, leucine, and valine (**Figure 29–20**) are analogous to reactions of fatty acid catabolism (see Figure 22–3). Following transamination (Figure 29–20, reaction 1), the carbon skeletons of the resulting α-keto acids undergo oxidative decarboxylation and conversion to coenzyme A thioesters. This multistep process is catalyzed by the mitochondrial **branched-chain α-keto acid dehydrogenase complex,** whose components are functionally identical to those of the pyruvate dehydrogenase complex (PDH) (see Figure 18–5). Like PDH, the branched-chain α-ketoacid dehydrogenase complex consists of five components.

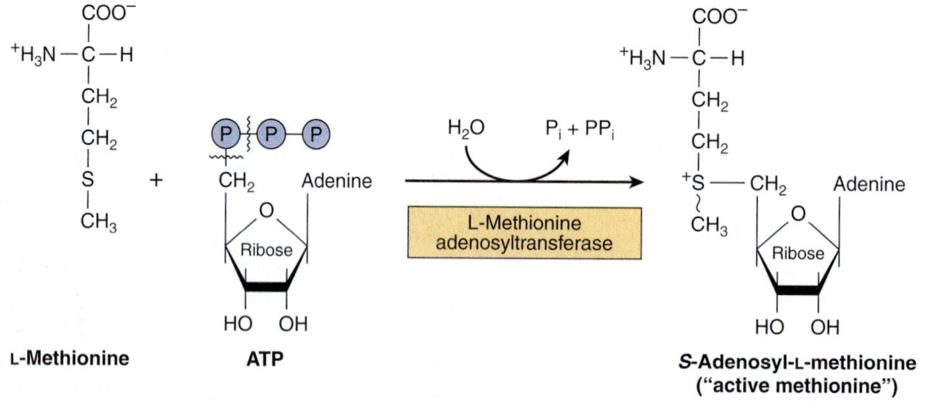

FIGURE 29–18 Formation of S-adenosylmethionine. ~ CH$_3$ represents the high group transfer potential of "active methionine."

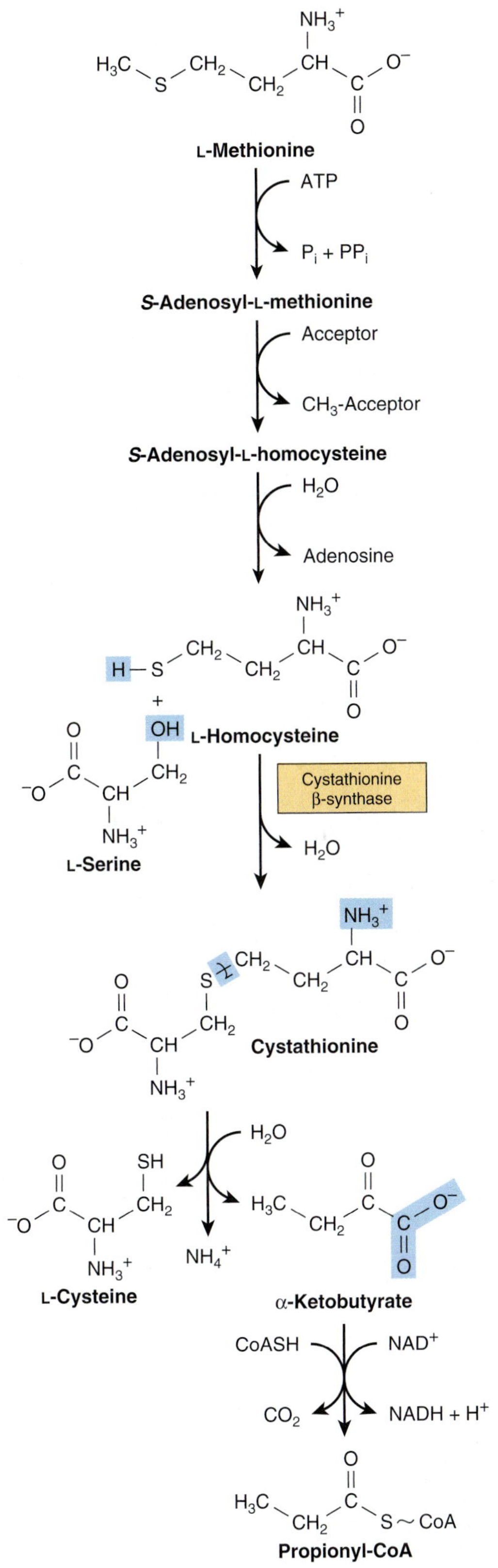

FIGURE 29–19 Conversion of methionine to propionyl-CoA.

E1: thiamin pyrophosphate (TPP)-dependent branched-chain α-ketoacid decarboxylase.

E2: dihydrolipoyl transacylase (contains lipoamide).

E3: dihydrolipoamide dehydrogenase (contains FAD).

Protein kinase.

Protein phosphatase.

As for PDH [see Figure 18–6(B)], the protein kinase and protein phosphatase regulate activity of the branched-chain α-keto acid dehydrogenase complex via phosphorylation (inactivation) and dephosphorylation (activation).

Dehydrogenation of the resulting coenzyme A thioesters (reaction 3, Figure 29–20) proceeds like the dehydrogenation of lipid-derived fatty acyl-CoA thioesters (see Figure 22–3). Subsequent reactions that are unique for each amino acid skeleton are given in **Figures 29–21, 29–22,** and **29–23.**

METABOLIC DISORDERS OF BRANCHED-CHAIN AMINO ACID CATABOLISM

As the name implies, the odor of urine in **maple syrup urine disease (branched-chain ketonuria,** or **MSUD)** suggests maple syrup, or burnt sugar. The biochemical defect in MSUD involves the **α-keto acid decarboxylase complex** (reaction 2, Figure 29–20). Plasma and urinary levels of leucine, isoleucine, valine, and their α-keto acids and α-hydroxy acids (reduced α-keto acids) are elevated, but the urinary ketoacids derive principally from leucine. Signs and symptoms of MSUD include often fatal ketoacidosis, neurological derangements, mental retardation, and a maple syrup odor of urine. The mechanism of toxicity is unknown. Early diagnosis by enzymatic analysis is essential to avoid brain damage and early mortality by replacing dietary protein by an amino acid mixture that lacks leucine, isoleucine, and valine.

The molecular genetics of MSUD are heterogeneous. MSUD can result from mutations in the genes that encode E1α, E1β, E2, and E3. Based on the locus affected, genetic subtypes of MSUD are recognized. Type IA MSUD arises from mutations in the E1α gene, type IB in the E1β gene, type II in the E2 gene, and type III in the E3 gene (**Table 29–3**). In **intermittent branched-chain ketonuria,** the α-keto acid decarboxylase retains some activity, and symptoms occur later in life. In **isovaleric acidemia,** ingestion of protein-rich foods elevates isovalerate, the deacylation product of isovaleryl-CoA. The impaired enzyme in **isovaleric acidemia** is **isovaleryl-CoA dehydrogenase** (reaction 3, Figure 29–20). Vomiting, acidosis, and coma follow ingestion of excess protein. Accumulated isovaleryl-CoA is hydrolyzed to isovalerate and excreted.

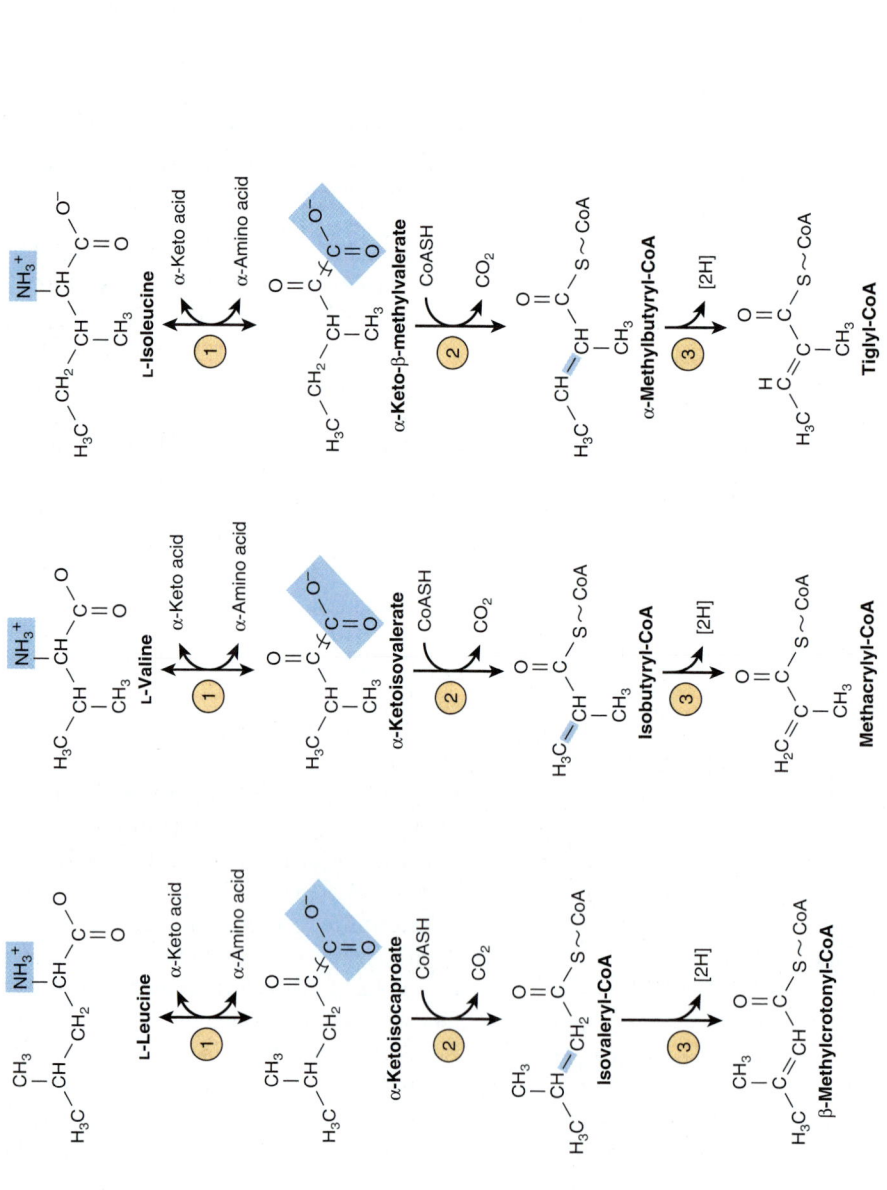

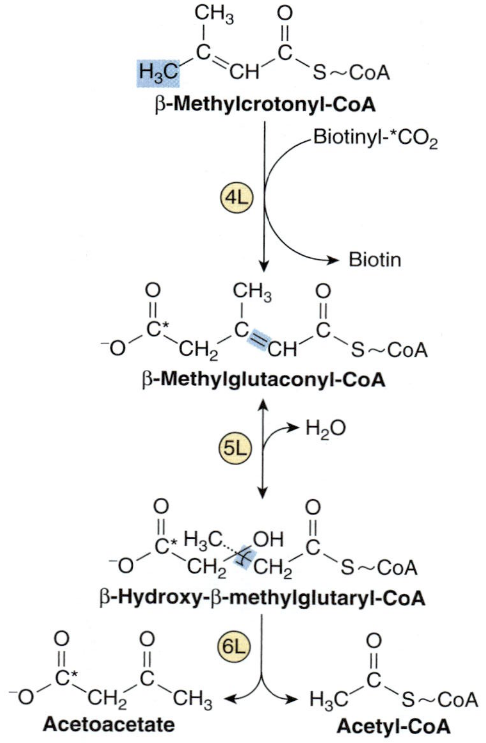

FIGURE 29–21   Catabolism of the β-methylcrotonyl-CoA formed from L-leucine. Asterisks indicate carbon atoms derived from CO_2.

FIGURE 29–20   The first three reactions in the catabolism of leucine, valine, and isoleucine.

Note the analogy of reactions 2 and 3 to reactions of the catabolism of fatty acids (see Figure 22–3). The analogy to fatty acid catabolism continues, as shown in subsequent figures.

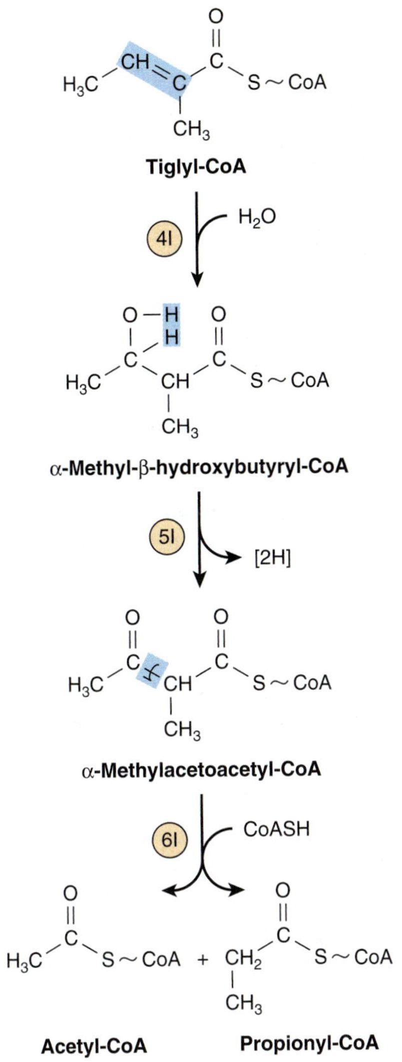

FIGURE 29–22 Subsequent catabolism of the tiglyl-CoA formed from L-isoleucine.

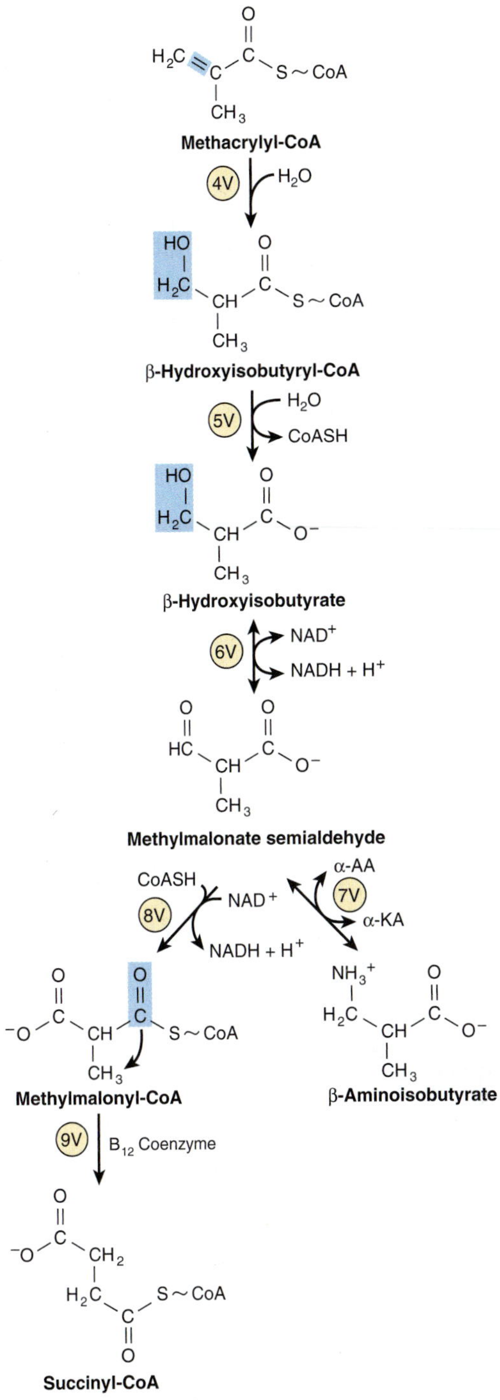

TABLE 29–3 Maple Syrup Urine Disease Can Reflect Impaired Function of Various Components of the α-Ketoacid Decarboxylase Complex

Branched-chain α-Ketoacid Decarboxylase Component		OMIM[1] Reference	Maple Syrup Urine Disease
E1α	α-Ketoacid decarboxylase	608348	Type 1A
E1β	α-Ketoacid decarboxylase	248611	Type 1B
E2	Dihydrolipoyl transacylase	608770	Type II
E3	Dihydrolipoamide dehydrogenase	238331	Type III

[1]Online Mendelian Inheritance in Man database: ncbi.nlm.nih.gov/omim/

FIGURE 29–23 Subsequent catabolism of the methacrylyl-CoA formed from L-valine (see Figure 29–20). (α-AA, α-amino acid; α-KA, α-keto acid.)

SUMMARY

- Excess amino acids are catabolized to amphibolic intermediates that serve as sources of energy or for the biosynthesis of carbohydrates and lipids.

- Transamination is the most common initial reaction of amino acid catabolism. Subsequent reactions remove any additional nitrogen and restructure hydrocarbon skeletons for conversion to oxaloacetate, α-ketoglutarate, pyruvate, and acetyl-CoA.

- Metabolic diseases associated with glycine catabolism include glycinuria and primary hyperoxaluria.

- Two distinct pathways convert cysteine to pyruvate. Metabolic disorders of cysteine catabolism include cystine-lysinuria, cystine storage disease, and the homocystinurias.

- Threonine catabolism merges with that of glycine after threonine aldolase cleaves threonine to glycine and acetaldehyde.

- Following transamination, the carbon skeleton of tyrosine is degraded to fumarate and acetoacetate. Metabolic diseases of tyrosine catabolism include tyrosinosis, Richner–Hanhart syndrome, neonatal tyrosinemia, and alkaptonuria.

- Metabolic disorders of phenylalanine catabolism include PKU and several hyperphenylalaninemias.

- Neither nitrogen of lysine undergoes direct transamination. The same effect is, however, achieved by the intermediate formation of saccharopine. Metabolic diseases of lysine catabolism include periodic and persistent forms of hyperlysinemia-ammonemia.

- The catabolism of leucine, valine, and isoleucine presents many analogies to fatty acid catabolism. Metabolic disorders of branched-chain amino acid catabolism include hypervalinemia, maple syrup urine disease, intermittent branched-chain ketonuria, isovaleric acidemia, and methylmalonic aciduria.

REFERENCES

Blacher J, Safar ME: Homocysteine, folic acid, B vitamins and cardiovascular risk. J Nutr Health Aging 2001;5:196.

Bliksrud YT, Brodtkorb E, Andresen PA, et al: Tyrosinemia type I, de novo mutation in liver tissue suppressing an inborn splicing defect. J Mol Med 2005;83:406.

Dobrowolski, Pey AL, Koch R, et al: Biochemical characterization of mutant phenylalanine hydroxylase enzymes and correlation with clinical presentation in hyperphenylalaninaemic patients. J Inherit Metab Dis 2009:32;10.

Flusser H, Korman SH, Sato K, et al: Mild glycine encephalopathy (NKH) in a large kindred due to a silent exonic GLDC splice mutation. Neurology 2005;64:1426.

Garg U, Dasouki M: Expanded newborn screening of inherited metabolic disorders by tandem mass spectrometry. Clinical and laboratory aspects. Clin Biochem 2006;39:315.

Gerstner B, Gratopp A, Marcinkowski M, et al: Glutaric acid and its metabolites cause apoptosis in immature oligodendrocytes: a novel mechanism of white matter degeneration in glutaryl-CoA dehydrogenase deficiency. Pediatr Res 2005;57;771.

Häussinger D, Schliess F: Glutamine metabolism and signaling in the liver. Front Biosci 2007;12:371.

Heldt K, Schwahn B, Marquardt I, et al: Diagnosis of maple syrup urine disease by newborn screening allows early intervention without extraneous detoxification. Mol Genet Metab 2005;84:313.

Moshal K, Camel CK, Kartha GK, et al: Cardiac dys-synchronization and arrhythmia in hyperhomocysteinemia. Curr Neurovasc Res 2007;4:289.

Muller E, Kolker S: Reduction of lysine intake while avoiding malnutrition: major goals and major problems in dietary treatment of glutaryl-CoA dehydrogenase deficiency. J Inherit Metab Dis 2004;27:903.

Sacksteder KA, Biery BJ, Morrell JC, et al: Identification of the alpha-aminoadipic semialdehyde synthase gene which is defective in familial hyperlysinemia. Am J Hum Genet 2000;66:1736.

Scriver CR, Sly WS, Childs B, et al (editors): The Metabolic and Molecular Bases of Inherited Disease, 8th ed. McGraw-Hill, 2001.

Stenn FF, Milgram JW, Lee SL, et al: Biochemical identification of homogentisic acid pigment in an ochronotic Egyptian mummy. Science 1977;197:566.

Waters PJ, Scriver CR, Parniak MA: Homomeric and heteromeric interactions between wild-type and mutant phenylalanine hydroxylase subunits: evaluation of two-hybrid approaches for functional analysis of mutations causing hyperphenylalaninemia. Mol Genet Metab 2001;73:230.

Conversion of Amino Acids to Specialized Products

Victor W. Rodwell, PhD

- Describe how amino acids participate in a variety of biosynthetic processes other than protein synthesis.
- Outline how arginine participates in the biosynthesis of creatine, nitric oxide (NO), putrescine, spermine, and spermidine.
- Indicate the contribution of cysteine and of β-alanine to the structure of coenzyme A and of cysteine to the structure of taurocholic acid.
- Discuss the role of glycine in drug catabolism.
- Document the role of glycine in the biosynthesis of heme, purines, creatine, and sarcosine.
- Identify the reaction that converts an amino acid to the neurotransmitter histamine.
- Document the role of *S*-adenosylmethionine as a source of methyl groups in metabolism.
- Recognize the tryptophan metabolites serotonin and melatonin, and the conversion of tryptophan to tryptamine and subsequently to indole 3-acetate.
- Indicate the role of tyrosine in the formation of norepinephrine, epinephrine, triiodothyronine, and thyroxine.
- Illustrate the key roles of peptidyl serine, threonine and tyrosine in metabolic regulation and signal transduction pathways.
- Outline the roles of glycine, arginine, and *S*-adenosylmethionine in the biosynthesis of creatine.
- Describe the role of creatine phosphate in energy homeostasis.
- Describe the formation of γ-aminobutyrate (GABA) and the rare metabolic disorders associated with defects in GABA catabolism.

BIOMEDICAL IMPORTANCE

Certain proteins contain amino acids that have been post-translationally modified to permit them to perform specific functions. Examples include the carboxylation of glutamate to form γ-carboxyglutamate, which functions in Ca^{2+} binding, the hydroxylation of proline for incorporation into the collagen triple helix, and the hydroxylation of lysine to hydroxylysine, whose subsequent modification and cross-linking stabilizes maturing collagen fibers. In addition to serving as the building blocks for protein synthesis, amino acids serve as precursors of biologic materials such as heme, purines, pyrimidines, hormones, neurotransmitters, and biologically active peptides. Histamine plays a central role in many allergic reactions. Neurotransmitters derived from amino acids include γ-aminobutyrate, 5-hydroxytryptamine (serotonin), dopamine, norepinephrine, and epinephrine. Many drugs used to treat neurologic and psychiatric conditions act by altering the metabolism of these neurotransmitters. Discussed below are the metabolism and metabolic roles of selected α- and non-α amino acids.

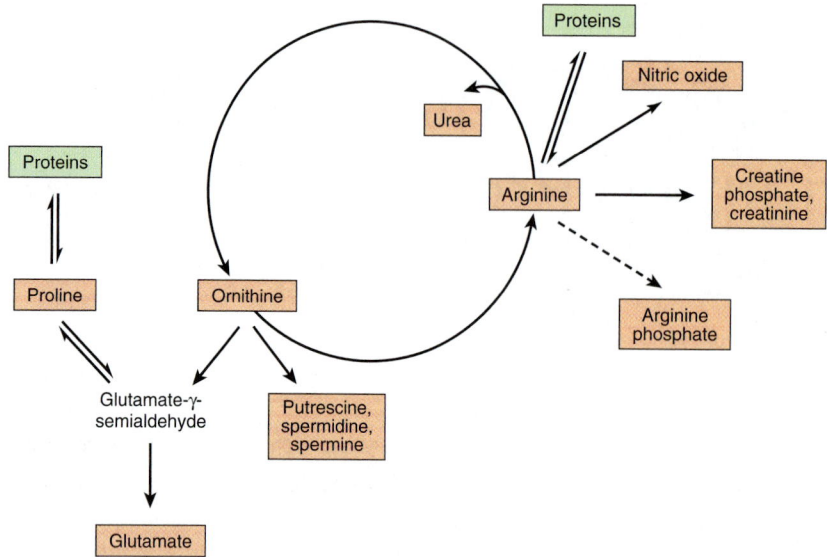

FIGURE 30–1   **Arginine, ornithine, and proline metabolism.** Reactions with solid arrows all occur in mammalian tissues. Putrescine and spermine synthesis occurs in both mammals and bacteria. Arginine phosphate of invertebrate muscle functions as a phosphagen analogous to creatine phosphate of mammalian muscle.

L-α-AMINO ACIDS

Alanine

Alanine serves as a carrier of ammonia and of the carbons of pyruvate from skeletal muscle to liver via the Cori cycle (see Figure 20–4), and together with glycine constitutes a major fraction of the free amino acids in plasma.

Arginine

Figure 30–1 summarizes the metabolic fates of arginine. In addition to serving as a carrier of nitrogen atoms in urea biosynthesis (see Figure 28–13), the guanidino group of arginine is incorporated into creatine, and following conversion to ornithine, its carbon skeleton becomes that of the polyamines putrescine and spermine.

The reaction catalyzed by NO synthase (**Figure 30–2**), a five-electron oxidoreductase with multiple cofactors, converts one nitrogen of the guanidine group of arginine to L-ornithine and NO, an intercellular signaling molecule that serves as a neurotransmitter, smooth muscle relaxant, and vasodilator (see Chapter 49).

Cysteine

Cysteine participates in the biosynthesis of coenzyme A (see Figure 44–18) by reacting with pantothenate to form 4-phosphopantothenoyl-cysteine (**Figure 30–3**). Three enzyme-catalyzed reactions convert cysteine to taurine, which can displace the coenzyme A moiety of cholyl-CoA to form the bile acid taurocholic acid (see Figure 26–7). The conversion of cysteine to taurine is initiated by its oxidation to cysteine sulfinate, catalyzed by the nonheme Fe^{2+} enzyme cysteine dioxygenase. Decarboxylation of cysteine sulfinate

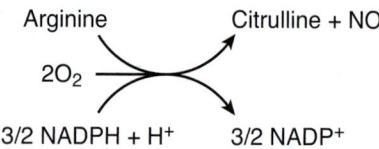

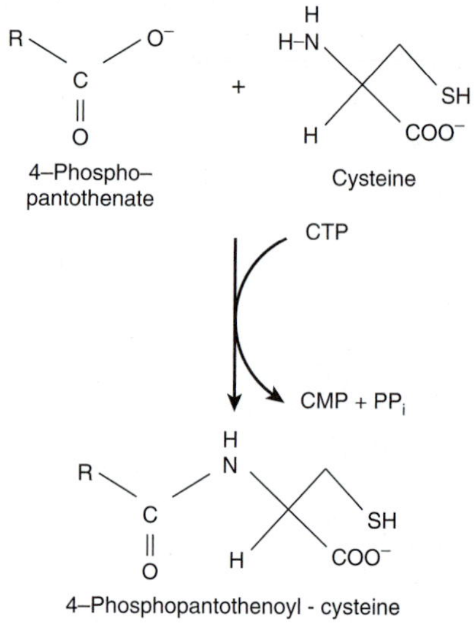

FIGURE 30–2   **The reaction catalyzed by nitric oxide synthase.**

FIGURE 30–3   The reaction catalyzed by phosphopantothenate cysteine ligase.

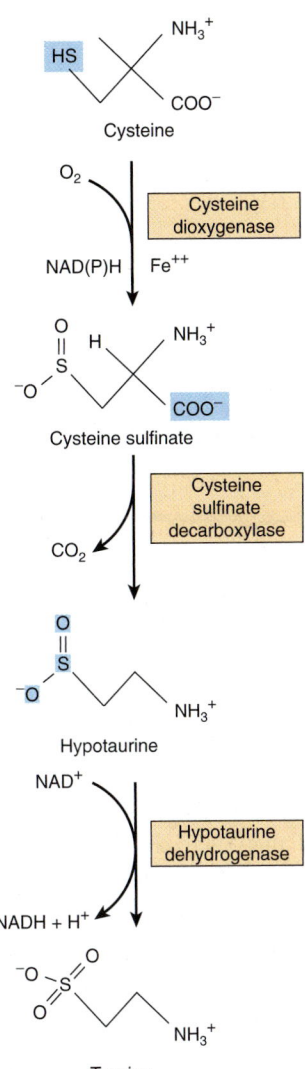

FIGURE 30–4 **Conversion of cysteine to taurine.** The reactions are catalyzed by cysteine dioxygenase, cysteine sulfinate decarboxylase, and hypotaurine decarboxylase, respectively.

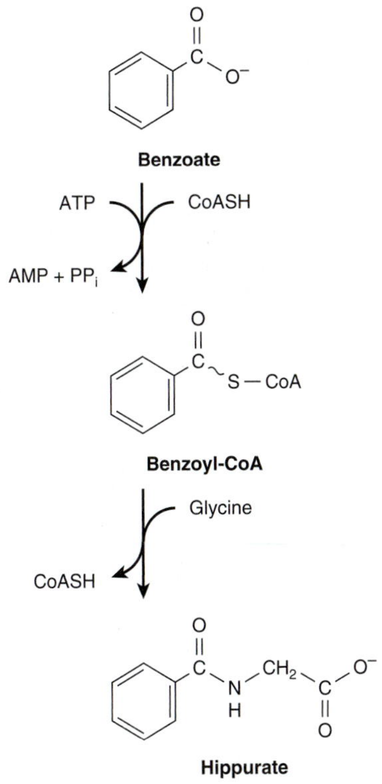

FIGURE 30–5 **Biosynthesis of hippurate.** Analogous reactions occur with many acidic drugs and catabolites.

Histidine

Decarboxylation of histidine by the pyridoxal 5′-phosphate-dependent enzyme histidine decarboxylase forms histamine (**Figure 30–6**). A biogenic amine that functions in allergic reactions and gastric secretion, histamine is present in all tissues. Its concentration in the brain hypothalamus varies in accordance with a circadian rhythm. Histidine compounds present in the human body include ergothioneine, carnosine, and dietary anserine (**Figure 30–7**). While their physiological functions are unknown, carnosine (β-alanyl-histidine) and homocarnosine (γ-aminobutyryl-histidine) are major constituents of excitable tissues, brain, and skeletal muscle. Urinary levels of 3-methylhistidine are unusually low in patients with **Wilson's disease.**

Methionine

The major nonprotein fate of methionine is conversion to *S*-adenosylmethionine, the principal source of methyl groups

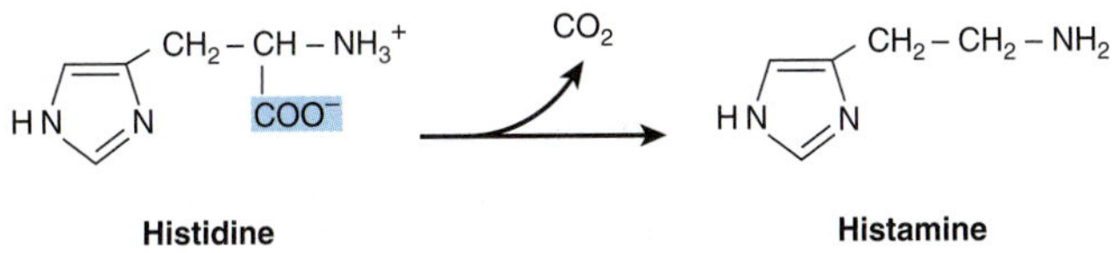

Glycine

Metabolites and pharmaceuticals excreted as water-soluble glycine conjugates include glycocholic acid (see Chapter 26) and hippuric acid formed from the food additive benzoate (**Figure 30–5**). Many drugs, drug metabolites, and other compounds with carboxyl groups are excreted in the urine as glycine conjugates. Glycine is incorporated into creatine, and the nitrogen and α-carbon of glycine are incorporated into the pyrrole rings and the methylene bridge carbons of heme (see Chapter 31), and the entire glycine molecule becomes atoms 4, 5, and 7 of purines (see Figure 33–1).

by cysteine sulfinate decarboxylase forms hypotaurine, whose oxidation by hypotaurine dehydrogenase forms taurine (**Figure 30–4**).

FIGURE 30–6 **The reaction catalyzed by histidine decarboxylase.**

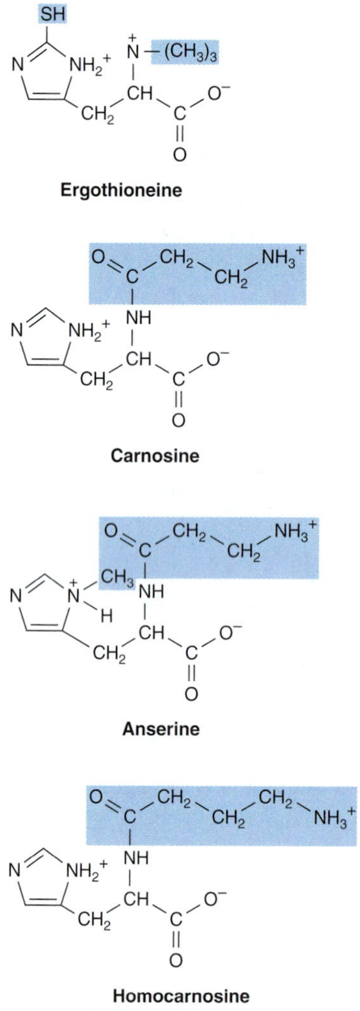

FIGURE 30–7 **Derivatives of histidine.** Colored boxes surround the components not derived from histidine. The SH group of ergothioneine derives from cysteine.

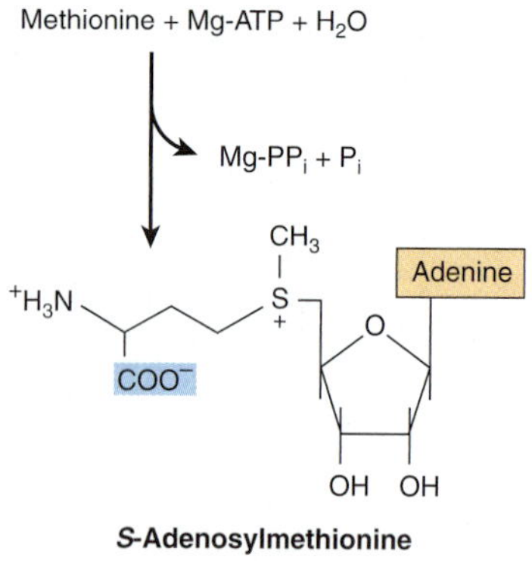

FIGURE 30–8 **Biosynthesis of S-adenosylmethionine, catalyzed by methionine adenosyltransferase.**

in the body. S-adenosylmethionine is synthesized from methionine and ATP, a reaction catalyzed by methionine adenosyltransferase (MAT) (**Figure 30–8**). Human tissues contain three MAT isozymes (MAT-1 and MAT-3 of liver and MAT-2 of nonhepatic tissues). Although **hypermethioninemia** can result from severely decreased hepatic MAT-1 and MAT-3 activity, if there is residual MAT-1 or MAT-3 activity and MAT-2 activity is normal, a high tissue concentration of methionine will assure synthesis of adequate amounts of S-adenosylmethionine.

Following decarboxylation of S-adenosylmethionine by methionine decarboxylase, three carbons and the α-amino group of methionine contribute to the biosynthesis of the polyamines spermine and spermidine (**Figure 30–9**). These polyamines function in cell proliferation and growth, are growth factors for cultured mammalian cells, and stabilize intact cells, subcellular organelles, and membranes. Pharmacologic doses of polyamines are hypothermic and hypotensive. Since they bear multiple positive charges, polyamines associate readily with

DNA and RNA. Figure 30–9 summarizes the biosynthesis of polyamines from methionine and ornithine, and **Figure 30–10** the catabolism of polyamines.

Serine

Serine participates in the biosynthesis of sphingosine (see Chapter 24), and of purines and pyrimidines, where it provides carbons 2 and 8 of purines and the methyl group of thymine (see Chapter 33). Genetic defects in cystathionine β-synthase, a heme protein that catalyzes the pyridoxal 5′-phosphate-dependent condensation of serine with homocysteine to form cystathionine, result in **homocystinuria**.

$$\text{Serine} + \text{Homocysteine} \rightarrow \text{Cystathionine} + \text{H}_2\text{O}$$

Tryptophan

Following hydroxylation of tryptophan to 5-hydroxytryptophan by liver tyrosine hydroxylase, subsequent decarboxylation forms serotonin (5-hydroxytryptamine), a potent vasoconstrictor and stimulator of smooth muscle contraction. Catabolism of serotonin is initiated by monoamine oxidase-catalyzed oxidative deamination to 5-hydroxyindole-3-acetate (**Figure 30–11**). The psychic stimulation that follows administration of iproniazid results from its ability to prolong the action of serotonin by inhibiting monoamine oxidase. In carcinoid (argentaffinoma), tumor cells overproduce serotonin. Urinary metabolites of serotonin in patients with carcinoid include N-acetylserotonin glucuronide and the glycine conjugate of 5-hydroxyindoleacetate. Serotonin and 5-methoxytryptamine are metabolized to the corresponding acids by monoamine oxidase. N-Acetylation of serotonin, followed by its O-methylation in the pineal body, forms melatonin. Circulating melatonin is taken up by all tissues, including brain, but is rapidly metabolized by hydroxylation followed by conjugation with sulfate or with glucuronic acid. Kidney tissue, liver tissue, and fecal bacteria all convert tryptophan to tryptamine, then to indole 3-acetate. The principal normal urinary catabolites of tryptophan are 5-hydroxyindoleacetate and indole 3-acetate.

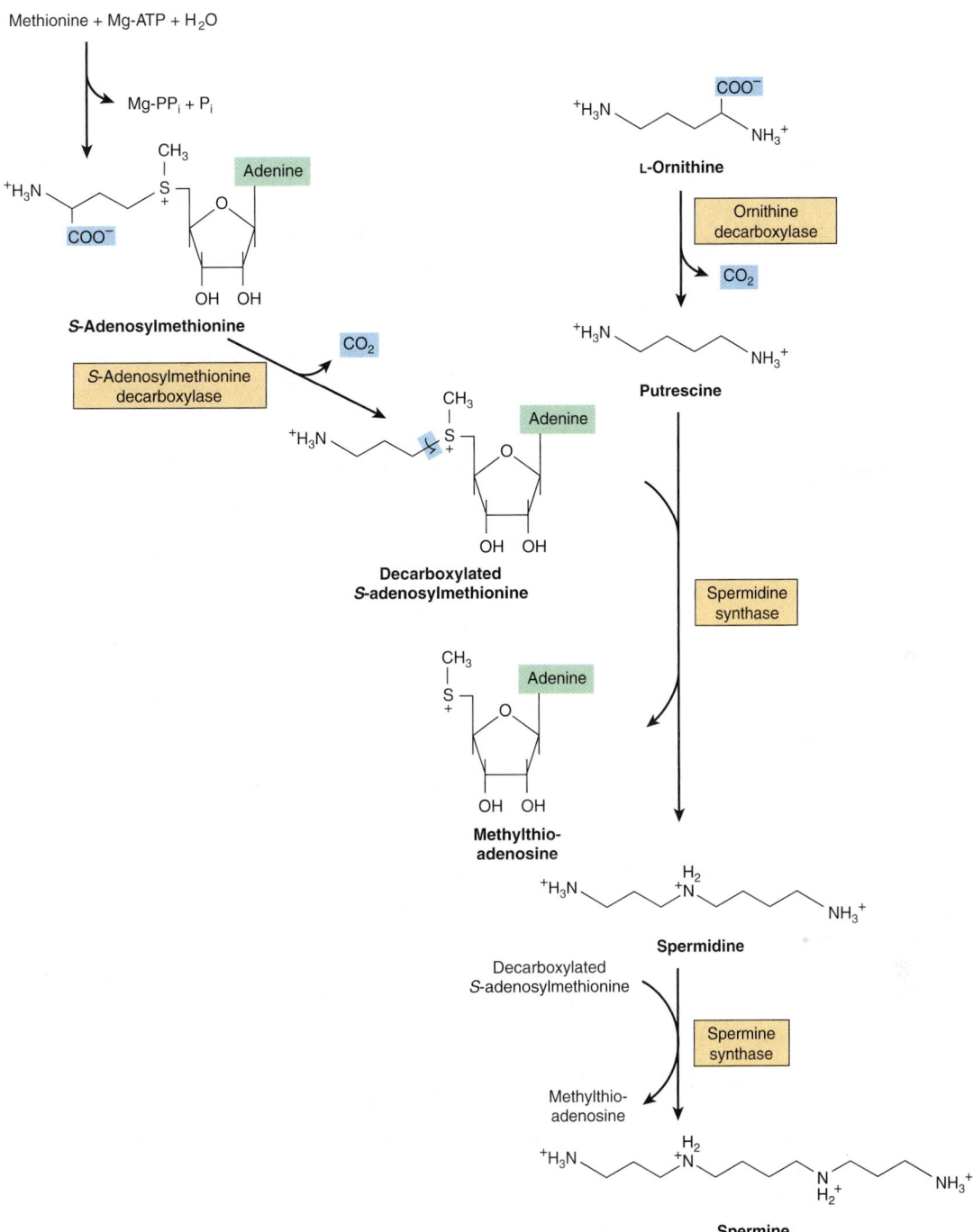

FIGURE 30–9 Intermediates and enzymes that participate in the biosynthesis of spermidine and spermine.

Tyrosine

Neural cells convert tyrosine to epinephrine and norepinephrine (**Figure 30–12**). While dopa is also an intermediate in the formation of melanin, different enzymes hydroxylate tyrosine in melanocytes. Dopa decarboxylase, a pyridoxal phosphate-dependent enzyme, forms dopamine. Subsequent hydroxylation by dopamine β-oxidase then forms norepinephrine. In the adrenal medulla, phenylethanolamine-*N*-methyltransferase utilizes *S*-adenosylmethionine to methylate the primary amine

of norepinephrine, forming epinephrine (Figure 30–12). Tyrosine is also a precursor of triiodothyronine and thyroxine (see Chapter 41).

Phosphoserine, Phosphothreonine, & Phosphotyrosine

The phosphorylation and dephosphorylation of specific seryl, threonyl, or tyrosyl residues of proteins regulate the activity of certain enzymes of lipid and carbohydrate metabolism (see

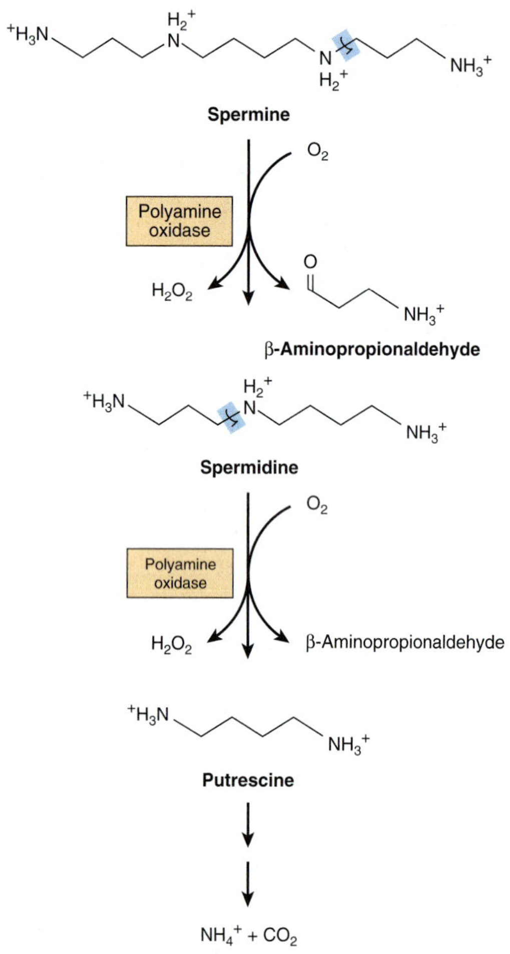

FIGURE 30–10   **Catabolism of polyamines.**

Chapters 9 and 19–26) and of proteins that participate in signal transduction cascades (see Chapter 42).

Sarcosine (*N*-Methylglycine)

The biosynthesis and catabolism of sarcosine (*N*-methylglycine) occur in mitochondria. Formation of sarcosine from dimethylglycine is catalyzed by the flavoprotein dimethylglycine dehydrogenase, which requires reduced pteroylpentaglutamate (TPG).

$$\text{Dimethylglycine} + \text{FADH}_2 + \text{H}_4\text{TPG} + \text{H}_2\text{O}$$
$$\rightarrow \text{Sarcosine} + N\text{-formyl-TPG}$$

Traces of sarcosine can also arise by methylation of glycine, a reaction catalyzed by *S*-adenosylmethionine glycine methyltransferase.

$$\text{Glycine} + S\text{-Adenosylmethionine} \rightarrow \text{Sarcosine}$$
$$+ S\text{-Adenosylhomocysteine}$$

Catabolism of sarcosine to glycine, catalyzed by the flavoprotein sarcosine dehydrogenase, also requires reduced TPG.

$$\text{Sarcosine} + \text{FAD} + \text{H}_4\text{TPG} + \text{H}_2\text{O}$$
$$\rightarrow \text{Glycine} + \text{FADH}_2 + N\text{-formyl-TPG}$$

The demethylation reactions that form and degrade sarcosine represent important sources of one-carbon units. FADH_2 is reoxidized via the electron transport chain (see Chapter 13).

Creatine & Creatinine

Creatinine is formed in muscle from creatine phosphate by irreversible, nonenzymatic dehydration, and loss of phosphate (**Figure 30–13**). Since the 24-h urinary excretion of creatinine is proportionate to muscle mass, it provides a measure of whether a complete 24-h urine specimen has been collected. Glycine, arginine, and methionine all participate in creatine biosynthesis. Synthesis of creatine is completed by methylation of guanidoacetate by *S*-adenosylmethionine (Figure 30–13).

NON-α-AMINO ACIDS

Non-α-amino acids present in tissues in a free form include β-alanine, β-aminoisobutyrate, and γ-aminobutyrate (GABA). β-Alanine is also present in combined form in coenzyme A (see Figure 44–18) and in the β-alanyl dipeptides carnosine, anserine and homocarnosine (see below).

β-Alanine & β-Aminoisobutyrate

β-Alanine and β-aminoisobutyrate are formed during catabolism of the pyrimidines uracil and thymine, respectively (see Figure 33–9). Traces of β-alanine also result from the hydrolysis of β-alanyl dipeptides by the enzyme carnosinase. β-Aminoisobutyrate also arises by transamination of methylmalonate semialdehyde, a catabolite of L-valine (see Figure 29–24).

The initial reaction of β-alanine catabolism is transamination to malonate semialdehyde. Subsequent transfer of coenzyme A from succinyl-CoA forms malonyl-CoA semialdehyde, which is then oxidized to malonyl-CoA and decarboxylated to the amphibolic intermediate acetyl-CoA. Analogous reactions characterize the catabolism of β-aminoisobutyrate. Transamination forms methylmalonate semialdehyde, which is converted to the amphibolic intermediate succinyl-CoA by reactions 8V and 9V of Figure 29–24. Disorders of β-alanine and β-aminoisobutyrate metabolism arise from defects in enzymes of the pyrimidine catabolic pathway. Principal among these are disorders that result from a total or partial deficiency of dihydropyrimidine dehydrogenase (see Figure 33–9).

β-Alanyl Dipeptides

The β-alanyl dipeptides carnosine and anserine (*N*-methylcarnosine) (Figure 30–7) activate myosin ATPase, chelate copper, and enhance copper uptake. β-Alanyl-imidazole buffers the pH of anaerobically contracting skeletal muscle. Biosynthesis of carnosine is catalyzed by carnosine synthetase

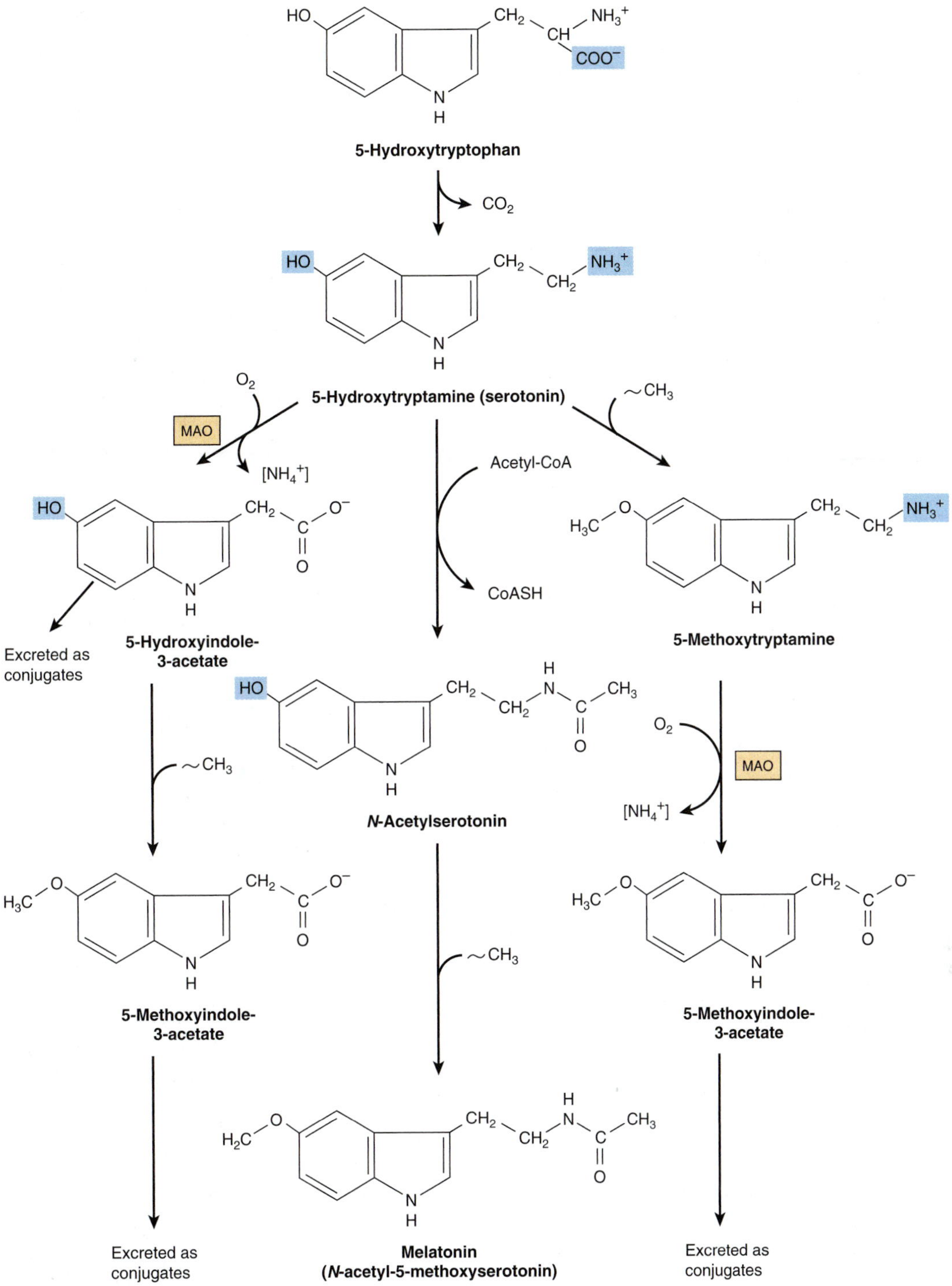

FIGURE 30–11 **Biosynthesis and metabolism of serotonin and melatonin.** ([NH$_4^+$], by transamination; MAO, monoamine oxidase; ~CH$_3$, from S-adenosylmethionine.)

in a two-stage reaction that involves initial formation of an enzyme-bound acyl-adenylate of β-alanine and subsequent transfer of the β-alanyl moiety to L-histidine.

$$ATP + β\text{-Alanine} \rightarrow β\text{-Alanyl-AMP} + PP_i$$

$$β\text{-Alanyl-AMP} + L\text{-Histidine} \rightarrow Carnosine + AMP$$

Hydrolysis of carnosine to β-alanine and L-histidine is catalyzed by carnosinase. The heritable disorder carnosinase deficiency is characterized by **carnosinuria.**

Homocarnosine (Figure 30–7), present in human brain at higher levels than carnosine, is synthesized in brain tissue by carnosine synthetase. Serum carnosinase does not hydrolyze

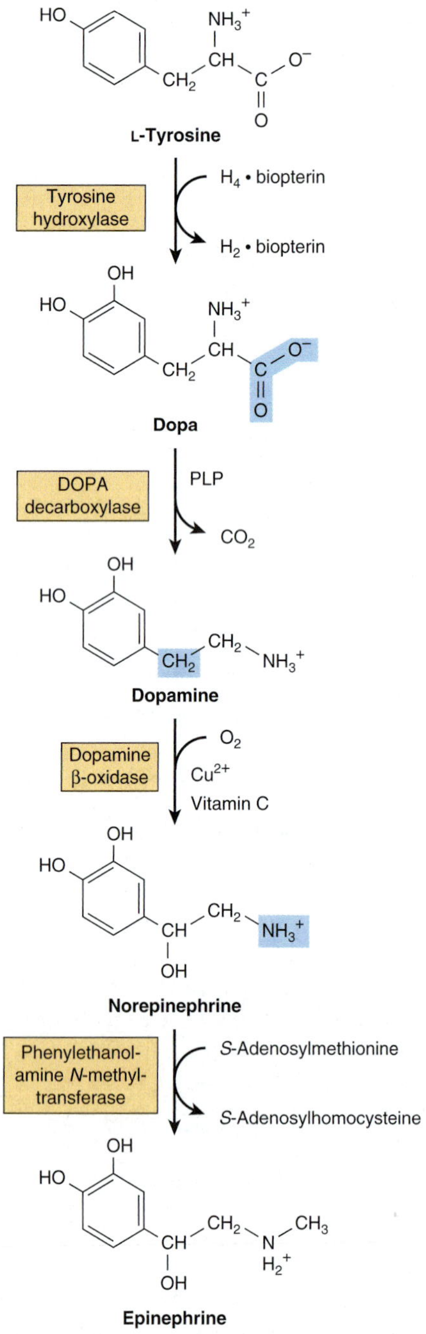

FIGURE 30–12 Conversion of tyrosine to epinephrine and norepinephrine in neuronal and adrenal cells. (PLP, pyridoxal phosphate.)

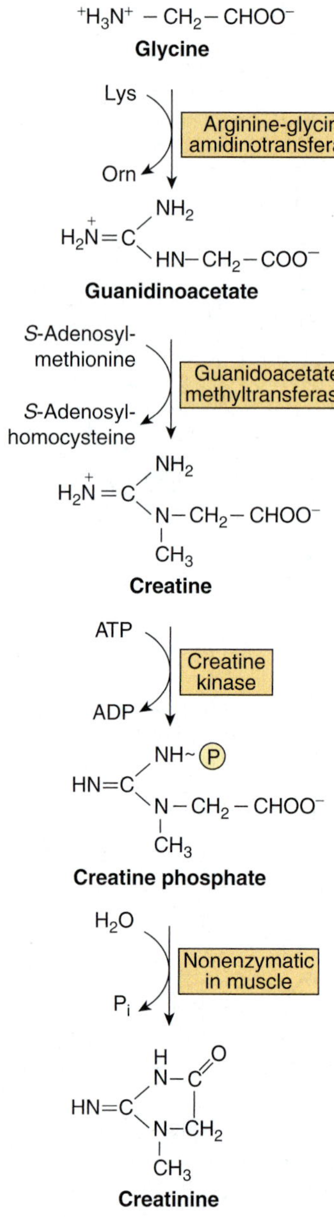

FIGURE 30–13 Biosynthesis of creatine and creatinine. Conversion of glycine and the guanidine group of arginine to creatine and creatine phosphate. Also shown is the nonenzymic hydrolysis of creatine phosphate to creatinine.

homocarnosine. **Homocarnosinosis,** a rare genetic disorder, is associated with progressive spastic paraplegia and mental retardation.

γ-Aminobutyrate

γ-Aminobutyrate (GABA) functions in brain tissue as an inhibitory neurotransmitter by altering transmembrane potential differences. GABA is formed by decarboxylation

of glutamate by L-glutamate decarboxylase (**Figure 30–14**). Transamination of γ-aminobutyrate forms succinate semialdehyde, which can be reduced to γ-hydroxybutyrate by L-lactate dehydrogenase, or be oxidized to succinate and thence via the citric acid cycle to CO_2 and H_2O (Figure 30–14). A rare genetic disorder of GABA metabolism involves a defective GABA aminotransferase, an enzyme that participates in the catabolism of GABA subsequent to its postsynaptic release in brain tissue. Defects in succinic semialdehyde dehydrogenase (Figure 30–14) are responsible for another rare metabolic disorder of γ-aminobutyrate catabolism characterized by **4-hydroxybutyric aciduria.**

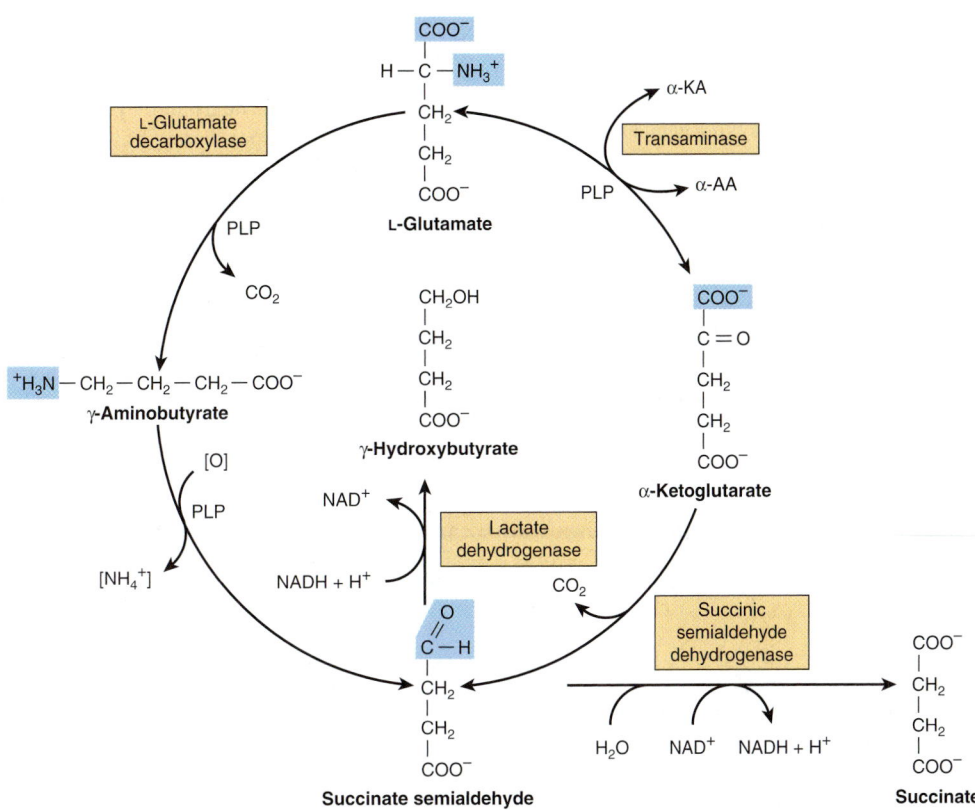

FIGURE 30–14 Metabolism of γ-aminobutyrate. (α-KA, α-keto acids; α-AA, α-amino acids; PLP, pyridoxal phosphate.)

SUMMARY

- In addition to serving structural and functional roles in proteins, α-amino acids participate in a wide variety of other biosynthetic processes.

- Arginine provides the formamidine group of creatine and the nitrogen of NO. Via ornithine, arginine provides the skeleton of the polyamines putrescine, spermine, and spermidine.

- Cysteine provides the thioethanolamine portion of coenzyme A, and following its conversion to taurine, part of the bile acid taurocholic acid.

- Glycine participates in the biosynthesis of heme, purines, creatine, and N-methylglycine (sarcosine). Many drugs and drug metabolites are excreted as glycine conjugates, which increases water solubility for urinary excretion.

- Decarboxylation of histidine forms the neurotransmitter histamine. Histidine compounds present in the human body include ergothioneine, carnosine, and dietary anserine.

- S-Adenosylmethionine, the principal source of methyl groups in metabolism, contributes its carbon skeleton to the biosynthesis of the polyamines spermine and spermidine.

- In addition to its roles in phospholipid and sphingosine biosynthesis, serine provides carbons 2 and 8 of purines and the methyl group of thymine.

- Key tryptophan metabolites include serotonin and melatonin. Kidney and liver tissue, and also fecal bacteria, convert tryptophan to tryptamine and thence to indole 3-acetate, The principal tryptophan catabolites in urine are indole 3-acetate and 5-hydroxyindoleacetate.

- Tyrosine forms norepinephrine and epinephrine, and following iodination the thyroid hormones triiodothyronine and thyroxine.

- The enzyme-catalyzed interconversion of the phospho- and dephospho-forms of peptide bound serine, threonine, and tyrosine plays key roles in metabolic regulation, including signal transduction.

- Glycine, arginine, and S-adenosylmethionine all participate in the biosynthesis of creatine, which as creatine phosphate serves as a major energy reserve in muscle and brain tissue. Excretion in the urine of its catabolite creatine is proportionate to muscle mass.

- β-Alanine and β-aminoisobutyrate both are present in tissues as free amino acids. β-Alanine also occurs in bound form in coenzyme A, carnosine, anserine, and homocarnosine. Catabolism of β-alanine involves stepwise conversion to acetyl-CoA. Analogous reactions catabolize β-aminoisobutyrate to succinyl-CoA. Disorders of β-alanine and β-aminoisobutyrate metabolism arise from defects in enzymes of pyrimidine catabolism.

- Decarboxylation of glutamate forms the inhibitory neurotransmitter γ-aminobutyrate (GABA). Two rare metabolic disorders are associated with defects in GABA catabolism.

REFERENCES

Conti M, Beavo J: Biochemistry and physiology of cyclic nucleotide phosphodiesterases: essential components in cyclic nucleotide signaling. Annu Rev Biochem 2007; 76:481.

Dominy JE Jr, Hwang J, Guo S, et al: Synthesis of cysteine dioxygenase's amino acid cofactor is regulated by substrate and represents a novel post-translational regulation of activity. J Biol Chem 2008;283:12188.

Joseph CA, Maroney MJ: Cysteine dioxygenase: structure and mechanism. Chem Commun (Camb) 2007;28:3338.

Lindemose S, Nielsen PE, Mollegaard NE: Polyamines preferentially interact with bent adenine tracts in double-stranded DNA. Nucleic Acids Res 2005;33:1790.

Manegold C, Hoffmann GF, Degen I, et al: Aromatic L-amino acid decarboxylase deficiency: clinical features, drug therapy and followup. J Inherit Metab Dis 2009;32:371.

Moinard C, Cynober L, de Bandt JP: Polyamines: metabolism and implications in human diseases. Clin Nutr 2005;24:184.

Pearl PL, Gibson KM, Cortez MA, et al: Succinic semialdehyde dehydrogenase deficiency: Lessons from mice and men. J Inherit Metab Dis 2009;32:343.

Pearl PL, Taylor JL, Trzcinski S, et al: The pediatric neurotransmitter disorders. J Child Neurol 2007;22:606.

Scriver CR, Sly WS, Childs B, et al (editors): *The Metabolic and Molecular Bases of Inherited Disease*, 8th ed. McGraw-Hill, 2001.

Wu F, Yu J, Gehring H. Inhibitory and structural studies of novel coenzyme-substrate analogs of human histidine decarboxylase. FASEB J. 2008;3:890.

Porphyrins & Bile Pigments

Robert K. Murray, MD, PhD

O B J E C T I V E S

After studying this chapter, you should be able to:

- Know the relationship between porphyrins and heme
- Be familiar with how heme is synthesized
- Understand the causes and general clinical pictures of the various porphyrias
- Know how bilirubin is derived from heme and how it is handled in the body
- Understand the nature of jaundice and appreciate how to approach determining its cause in a patient.

BIOMEDICAL IMPORTANCE

The biochemistry of the porphyrins and of the bile pigments is presented in this chapter. These topics are closely related, because heme is synthesized from porphyrins and iron, and the products of degradation of heme are the bile pigments and iron.

Knowledge of the biochemistry of the porphyrins and of heme is basic to understanding the varied functions of **hemoproteins** (see below) in the body. The **porphyrias** are a group of diseases caused by abnormalities in the pathway of biosynthesis of the various porphyrins. Although porphyrias are not very prevalent, physicians must be aware of them. A much more prevalent clinical condition is **jaundice**, due to elevation of bilirubin in the plasma. This elevation is due to overproduction of bilirubin or to failure of its excretion and is seen in numerous diseases ranging from hemolytic anemias to viral hepatitis and to cancer of the pancreas.

METALLOPORPHYRINS & HEMOPROTEINS ARE IMPORTANT IN NATURE

Porphyrins are cyclic compounds formed by the linkage of four pyrrole rings through methyne ($=HC—$) bridges (**Figure 31–1**). A characteristic property of porphyrins is the formation of complexes with metal ions bound to the nitrogen atom of the pyrrole rings. Examples are the **iron porphyrins** such as **heme** of hemoglobin and the **magnesium-containing** porphyrin **chlorophyll,** the photosynthetic pigment of plants.

Proteins that contain heme (**hemoproteins**) are widely distributed in nature. Examples of their importance in humans and animals are listed in **Table 31–1**.

Natural Porphyrins Have Substituent Side Chains on the Porphyrin Nucleus

The **porphyrins** found in nature are compounds in which various **side chains** are substituted for the eight hydrogen atoms numbered in the porphyrin nucleus shown in Figure 31–1. As a simple means of showing these substitutions, Fischer proposed a shorthand formula in which the methyne bridges are omitted and each pyrrole ring is shown as in **Figure 31–2**, with the eight substituent positions numbered as indicated. Various porphyrins are represented in **Figures 31–2, 31–3,** and **31–4**.

The arrangement of the acetate (A) and propionate (P) substituents in the uroporphyrin shown in Figure 31–2 is asymmetric (in ring IV, the expected order of the A and P substituents is reversed). A porphyrin with this type of **asymmetric substitution** is classified as a **type III** porphyrin. A porphyrin with a completely symmetric arrangement of the substituents is classified as a **type I** porphyrin. Only types I and III are found in nature, and the type III series is far more abundant (Figure 31–3) and more important because it includes heme.

Heme and its immediate precursor, **protoporphyrin IX** (Figure 31–4), are both **type III** porphyrins (ie, the methyl groups are asymmetrically distributed, as in type III coproporphyrin). However, they are sometimes identified as belonging to series IX, because they were designated ninth in a series of isomers postulated by Hans Fischer, the pioneer worker in the field of porphyrin chemistry.

Pyrrole

Porphyrin
(C$_{20}$H$_{14}$N$_4$)

FIGURE 31–1   The porphyrin molecule. Rings are labeled I, II, III, and IV. Substituent positions on the rings are labeled 1, 2, 3, 4, 5, 6, 7, and 8. The methyne bridges (=HC—) are labeled α, β, γ, and δ. The numbering system used is that of Hans Fischer.

HEME IS SYNTHESIZED FROM SUCCINYL-CoA & GLYCINE

Heme is synthesized in living cells by a pathway that has been much studied. The two starting materials are **succinyl-CoA,** derived from the citric acid cycle in mitochondria, and the amino acid **glycine.** Pyridoxal phosphate is also necessary in this reaction to "activate" glycine. The product of the condensation reaction between succinyl-CoA and glycine is **α-amino-β-ketoadipic acid,** which is rapidly decarboxylated to form **δ-aminolevulinate** (ALA) (**Figure 31–5**). This reaction sequence is catalyzed by **ALA synthase,** the rate-controlling enzyme in porphyrin biosynthesis in the mammalian liver. Synthesis of ALA occurs in **mitochondria.** In the cytosol, two molecules of ALA are condensed by the enzyme

TABLE 31–1   Examples of Some Important Human and Animal Hemoproteins[1]

Protein	Function
Hemoglobin	Transport of oxygen in blood
Myoglobin	Storage of oxygen in muscle
Cytochrome *c*	Involvement in electron transport chain
Cytochrome P450	Hydroxylation of xenobiotics
Catalase	Degradation of hydrogen peroxide
Tryptophan pyrrolase	Oxidation of tryptophan

[1]The functions of the above proteins are described in various chapters of this text.

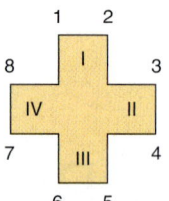 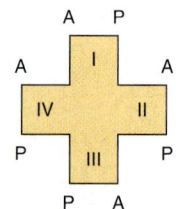

FIGURE 31–2   Uroporphyrin III. (A [acetate] = —CH$_2$COOH; P [propionate] = —CH$_2$CH$_2$COOH.) Note the asymmetry of substituents in ring IV (see the text).

ALA dehydratase to form two molecules of water and one of **porphobilinogen** (PBG) (Figure 31–5). ALA dehydratase is a zinc-containing enzyme and is sensitive to inhibition by **lead,** as can occur in lead poisoning.

The formation of a **cyclic tetrapyrrole**—ie, a porphyrin—occurs by condensation of four molecules of PBG (**Figure 31–6**). These four molecules condense in a head-to-tail manner to form a linear tetrapyrrole, **hydroxymethylbilane** (HMB). The reaction is catalyzed by **uroporphyrinogen I synthase,** also named PBG deaminase or HMB synthase. HMB cyclizes spontaneously to form **uroporphyrinogen I** (left-hand side of Figure 31–6) or is converted to **uroporphyrinogen III** by the action of **uroporphyrinogen III synthase** (right-hand side of Figure 31–6). Under normal conditions, the uroporphyrinogen formed is almost exclusively the III isomer, but in certain of the porphyrias (discussed below), the type-I isomers of porphyrinogens are formed in excess.

Note that both of these uroporphyrinogens have the pyrrole rings connected by **methylene bridges** (—CH$_2$—), which do not form a conjugated ring system. Thus, these compounds are **colorless** (as are all porphyrinogens). However, the porphyrinogens are readily auto-oxidized to their respective colored porphyrins. These oxidations are catalyzed by light and by the porphyrins that are formed.

Uroporphyrinogen III is converted to **coproporphyrinogen III** by decarboxylation of all of the acetate (A) groups, which changes them to methyl (M) substituents. The reaction is catalyzed by **uroporphyrinogen decarboxylase,** which is also capable of converting uroporphyrinogen I to coproporphyrinogen I (**Figure 31–7**). Coproporphyrinogen III then enters the mitochondria, where it is converted to **protoporphyrinogen III** and then to **protoporphyrin III.** Several steps are involved in this conversion. The mitochondrial enzyme **coproporphyrinogen oxidase** catalyzes the decarboxylation and oxidation of two propionic side chains to form protoporphyrinogen. This enzyme is able to act only on type-III coproporphyrinogen, which would explain why type I protoporphyrins do not generally occur in nature. The oxidation of protoporphyrinogen to **protoporphyrin** is catalyzed by another mitochondrial enzyme, **protoporphyrinogen oxidase.** In the mammalian liver, the conversion of coproporphyrinogen to protoporphyrin requires molecular oxygen.

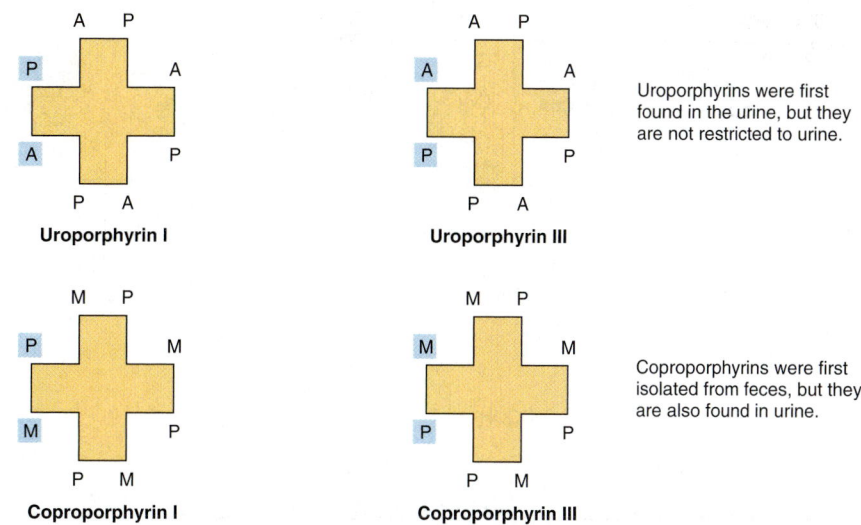

FIGURE 31–3 **Uroporphyrins and coproporphyrins.** (A, acetate; M, methyl; P, propionate.)

Formation of Heme Involves Incorporation of Iron into Protoporphyrin

The final step in heme synthesis involves the incorporation of ferrous iron into protoporphyrin in a reaction catalyzed by **ferrochelatase (heme synthase),** another mitochondrial enzyme (Figure 31–4).

A summary of the steps in the biosynthesis of the porphyrin derivatives from PBG is given in **Figure 31–8.** The last three enzymes in the pathway and ALA synthase are located in the **mitochondrion,** whereas the other enzymes are **cytosolic.** Both **erythroid** and **nonerythroid** ("housekeeping") forms of **ALA synthase** are found. Heme biosynthesis occurs in most mammalian cells with the exception of mature erythrocytes, which do not contain mitochondria. However, approximately 85% of heme synthesis occurs in erythroid precursor cells in the **bone marrow** and the majority of the remainder in **hepatocytes.**

The **porphyrinogens** described earlier are **colorless,** containing six extra hydrogen atoms as compared to the corresponding colored porphyrins. These **reduced porphyrins** (the porphyrinogens) and not the corresponding porphyrins are

the actual intermediates in the biosynthesis of protoporphyrin and of heme.

ALA Synthase Is the Key Regulatory Enzyme in Hepatic Biosynthesis of Heme

ALA synthase occurs in both **hepatic** (ALAS1) and **erythroid** (ALAS2) forms. The rate-limiting reaction in the synthesis of heme in liver is that catalyzed by ALAS1 (Figure 31–5), a regulatory enzyme. It appears that **heme,** probably acting through an aporepressor molecule, acts as **a negative regulator** of the synthesis of ALAS1. This repression-derepression mechanism is depicted diagrammatically in **Figure 31–9.** Thus, the rate of synthesis of ALAS1 increases greatly in the absence of heme and is diminished in its presence. The turnover rate of ALAS1 in the rat liver is normally rapid (half-life about 1 h), a common feature of an enzyme catalyzing a rate-limiting reaction. Heme also affects translation of the enzyme and its transfer from the cytosol to the mitochondrion.

Many **drugs** when administered to humans can result in a marked increase in ALAS1. Most of these drugs are

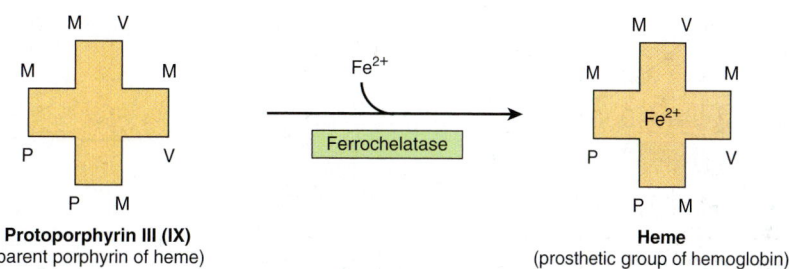

FIGURE 31–4 **Addition of iron to protoporphyrin to form heme.** (V [vinyl] = —CH=CH$_2$.)

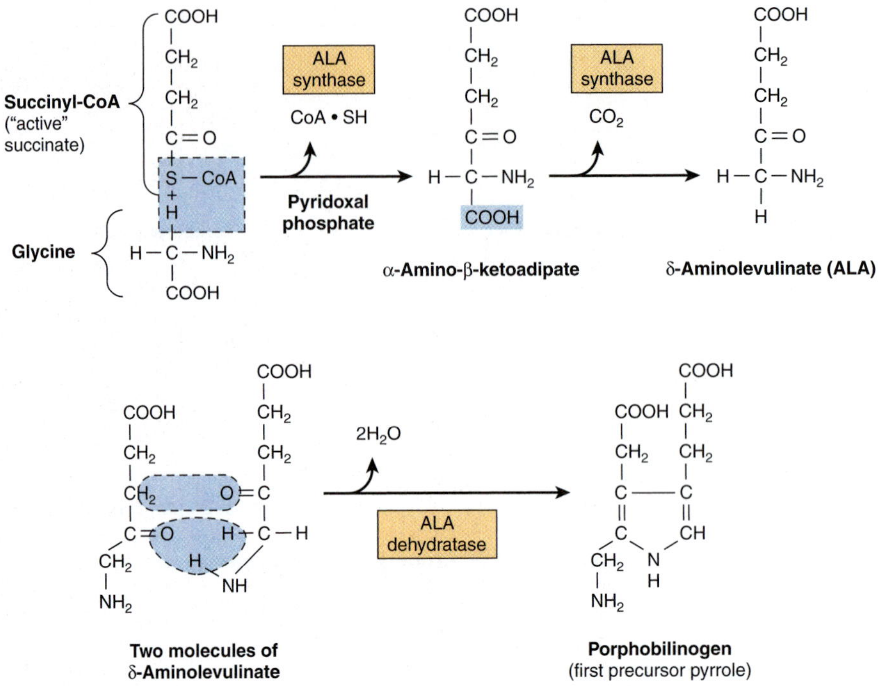

FIGURE 31–5 visual (top of page): Biosynthesis of porphobilinogen showing Succinyl-CoA ("active" succinate), Glycine, ALA synthase, α-Amino-β-ketoadipate, δ-Aminolevulinate (ALA), and two molecules of δ-Aminolevulinate converting to Porphobilinogen via ALA dehydratase.

FIGURE 31–5 **Biosynthesis of porphobilinogen.** ALA synthase occurs in the mitochondria, whereas ALA dehydratase is present in the cytosol.

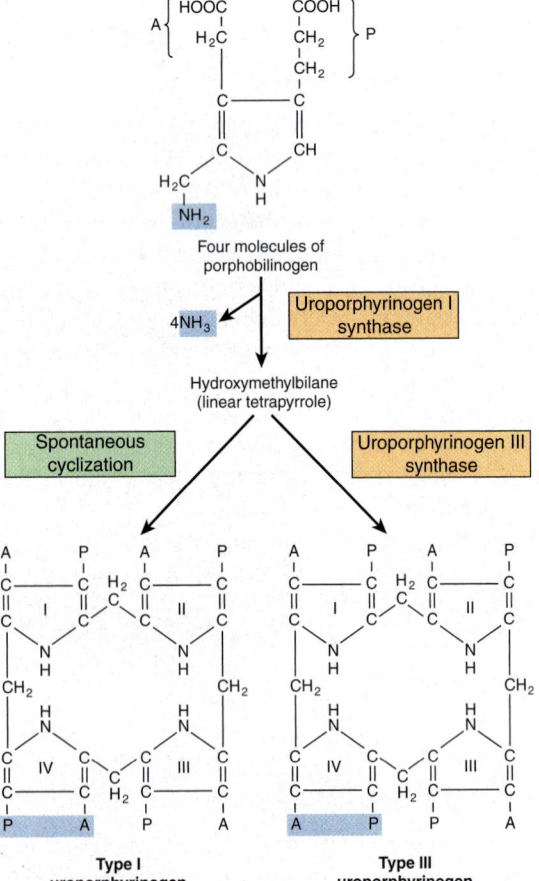

FIGURE 31–6 **Conversion of porphobilinogen to uroporphyrinogens.** Uroporphyrinogen synthase I is also called porphobilinogen (PBG) deaminase or hydroxymethylbilane (HMB) synthase.

metabolized by a system in the liver that utilizes a specific hemo-protein, **cytochrome P450** (see Chapter 53). During their metabolism, the utilization of heme by cytochrome P450 is greatly increased, which in turn diminishes the intracellular heme concentration. This latter event effects a derepression of ALAS1 with a corresponding increased rate of heme synthesis to meet the needs of the cells.

Several factors affect drug-mediated derepression of ALAS1 in the liver—eg, the administration of **glucose** can prevent it, as can the administration of **hematin** (an oxidized form of heme).

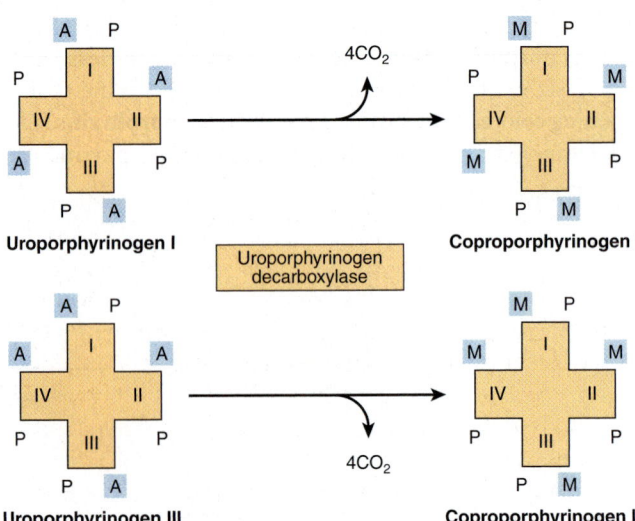

FIGURE 31–7 **Decarboxylation of uroporphyrinogens to coproporphyrinogens in cytosol.** (A, acetyl; M, methyl; P, propionyl.)

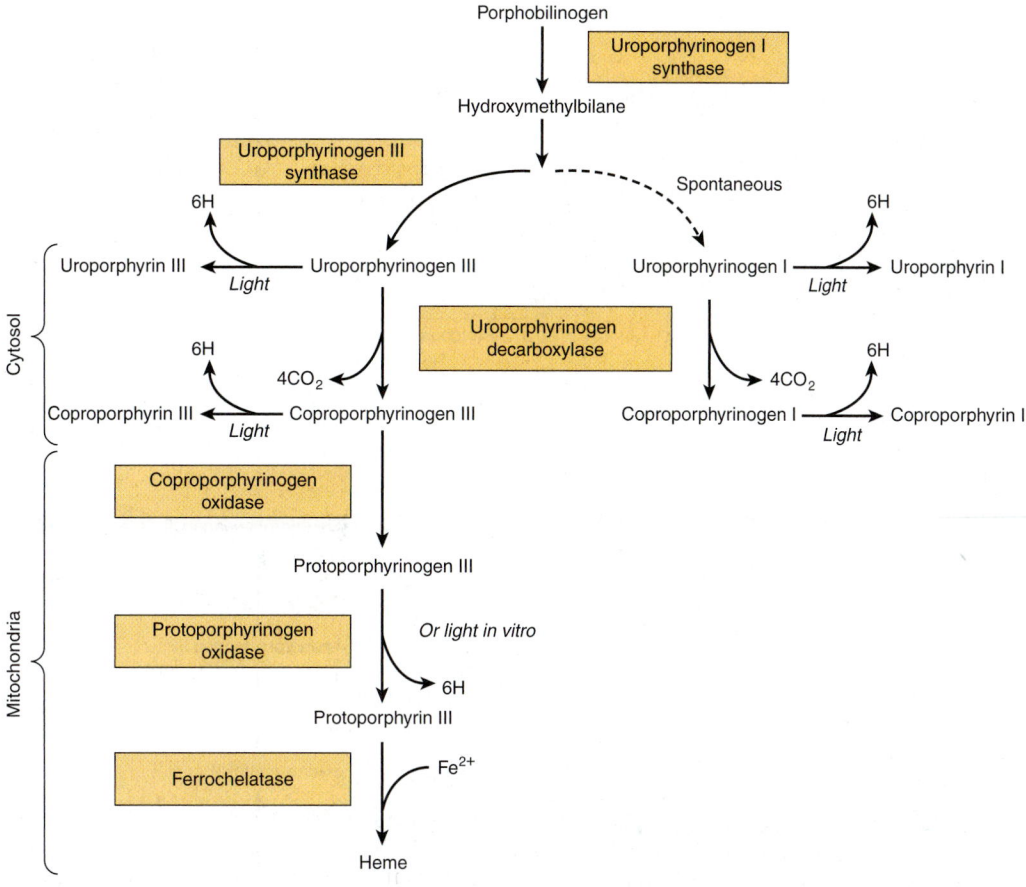

FIGURE 31–8 **Steps in the biosynthesis of the porphyrin derivatives from porphobilinogen.** Uroporphyrinogen I synthase is also called porphobilinogen deaminase or hydroxymethylbilane synthase.

The importance of some of these regulatory mechanisms is further discussed below when the porphyrias are described.

Regulation of the **erythroid** form of ALAS (ALAS2) differs from that of ALAS1. For instance, it is not induced by the drugs that affect ALAS1, and it does not undergo feedback regulation by heme.

PORPHYRINS ARE COLORED & FLUORESCE

The various **porphyrinogens** are **colorless,** whereas the various **porphyrins** are all **colored.** In the study of porphyrins or porphyrin derivatives, the **characteristic absorption spectrum** that each exhibits—in both the visible and the ultraviolet regions of the spectrum—is of great value. An example is the absorption curve for a solution of porphyrin in 5% hydrochloric acid (**Figure 31–10**). Note particularly the sharp absorption band **near 400 nm.** This is a distinguishing feature of the porphyrin ring and is characteristic of all porphyrins regardless of the side chains present. This band is termed the **Soret band** after its discoverer, the French physicist Charles Soret.

When porphyrins dissolved in strong mineral acids or in organic solvents are illuminated by ultraviolet light, they emit a strong red **fluorescence.** This fluorescence is so characteristic that it is often used to detect small amounts of free porphyrins. The **double bonds** joining the pyrrole rings in the porphyrins are responsible for the characteristic absorption and fluorescence of these compounds; these double bonds are absent in the porphyrinogens.

An interesting application of the photodynamic properties of porphyrins is their possible use in the treatment of certain types of cancer, a procedure called **cancer phototherapy.** Tumors often take up more porphyrins than do normal tissues. Thus, **hematoporphyrin** or other related compounds are administered to a patient with an appropriate tumor. The tumor is then exposed to an **argon laser,** which excites the porphyrins, producing cytotoxic effects.

Spectrophotometry Is Used to Test for Porphyrins & Their Precursors

Coproporphyrins and **uroporphyrins** are of clinical interest because they are excreted in increased amounts in the porphyrias. These compounds, when present in urine or feces, can be separated from each other by extraction with appropriate solvent mixtures. They can then be identified and quantified using spectrophotometric methods.

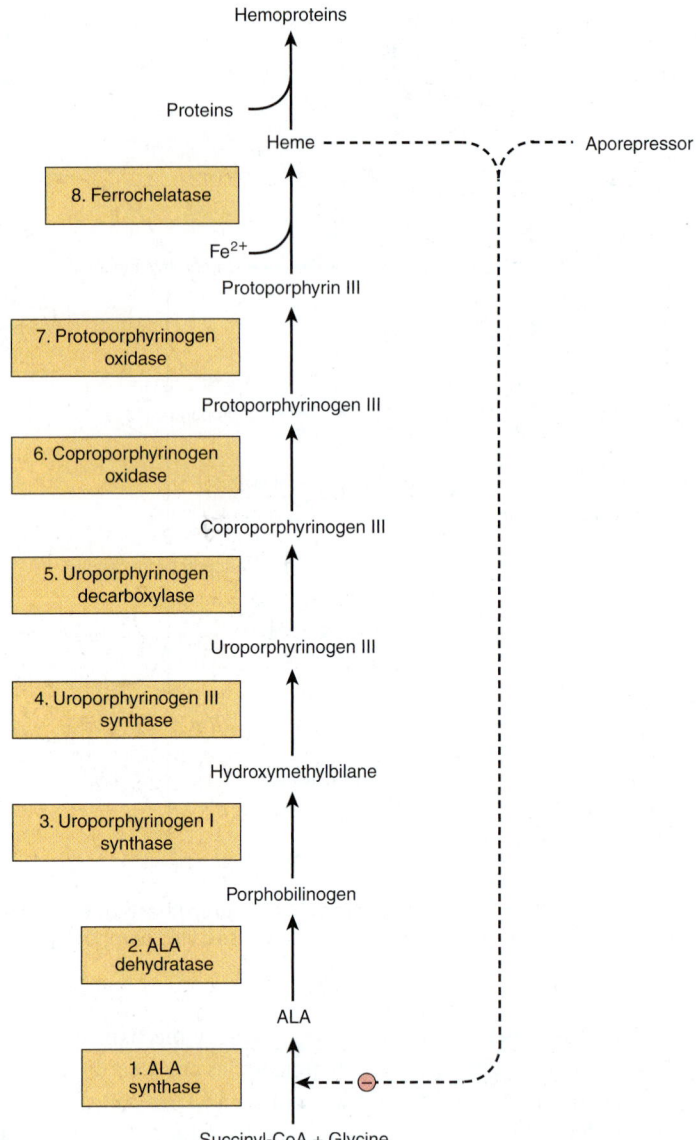

FIGURE 31–9 **Intermediates, enzymes, and regulation of heme synthesis.** The enzyme numbers are those referred to in column 1 of Table 31–2. Enzymes 1, 6, 7, and 8 are located in mitochondria, the others in the cytosol. Mutations in the gene encoding enzyme 1 cause X-linked sideroblastic anemia. Mutations in the genes encoding enzymes 2–8 cause the porphyrias, though only a few cases due to deficiency of enzyme 2 have been reported. Regulation of hepatic heme synthesis occurs at ALA synthase (ALAS1) by a repression-derepression mechanism mediated by heme and its hypothetical aporepressor. The dotted lines indicate the negative (−) regulation by repression. Enzyme 3 is also called porphobilinogen deaminase or hydroxymethylbilane synthase.

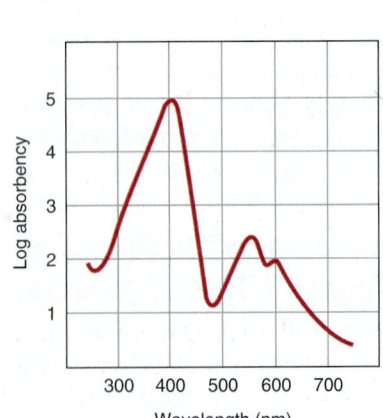

FIGURE 31–10 **Absorption spectrum of hematoporphyrin (0.01% solution in 5% HCl).**

ALA and **PBG** can also be measured in urine by appropriate colorimetric tests.

THE PORPHYRIAS ARE GENETIC DISORDERS OF HEME METABOLISM

The **porphyrias** are a group of disorders due to abnormalities in the pathway of biosynthesis of heme; they can be **genetic** or **acquired.** They are not prevalent, but it is important to consider them in certain circumstances (eg, in the differential diagnosis of abdominal pain and of a variety of neuropsychiatric findings); otherwise, patients will be subjected to inappropriate treatments. It has been speculated that King George III had a type of porphyria, which may account for his periodic confinements in Windsor Castle and perhaps for some of his

TABLE 31–2 Summary of Major Findings in the Porphyrias[1]

Enzyme Involved[2]	Type, Class, and OMIM Number	Major Signs and Symptoms	Results of Laboratory Tests
1. ALA synthase (erythroid form)	X-linked sideroblastic anemia[3] (erythropoietic) (OMIM 301300)	Anemia	Red cell counts and hemoglobin decreased
2. ALA dehydratase	ALA dehydratase deficiency (hepatic) (OMIM 125270)	Abdominal pain, neuropsychiatric symptoms	Urinary ALA and coproporphyrin III increased
3. Uroporphyrinogen I synthase[4]	Acute intermittent porphyria (hepatic) (OMIM 176000)	Abdominal pain, neuropsychiatric symptoms	Urinary ALA and PBG increased
4. Uroporphyrinogen III synthase	Congenital erythropoietic (erythropoietic) (OMIM 263700)	Photosensitivity	Urinary, fecal, and red cell uroporphyrin I increased
5. Uroporphyrinogen decarboxylase	Porphyria cutanea tarda (hepatic) (OMIM 176100)	Photosensitivity	Urinary uroporphyrin I increased
6. Coproporphyrinogen oxidase	Hereditary coproporphyria (hepatic) (OMIM 121300)	Photosensitivity, abdominal pain, neuropsychiatric symptoms	Urinary ALA, PBG, and coproporphyrin III and fecal coproporphyrin III increased
7. Protoporphyrinogen oxidase	Variegate porphyria (hepatic) (OMIM 176200)	Photosensitivity, abdominal pain, neuropsychiatric symptoms	Urinary ALA, PBG, and coproporphyrin III and fecal protoporphyrin IX increased
8. Ferrochelatase	Protoporphyria (erythropoietic) (OMIM 177000)	Photosensitivity	Fecal and red cell protoporphyrin IX increased

[1]Only the biochemical findings in the active stages of these diseases are listed. Certain biochemical abnormalities are detectable in the latent stages of some of the above conditions. Conditions 3, 5, and 8 are generally the most prevalent porphyrias. Condition 2 is rare.
[2]The numbering of the enzymes in this Table corresponds to that used in Figure 31–9.
[3]X-linked sideroblastic anemia is not a porphyria but is included here because ALA synthase is involved.
[4]This enzyme is also called PBG deaminase or hydroxymethylbilane synthase.
Abbreviations: ALA, δ-aminolevulinic acid; PBG, porphobilinogen.

views regarding American colonists. Also, the **photosensitivity** (favoring nocturnal activities) and severe **disfigurement** exhibited by some victims of congenital erythropoietic porphyria have led to the suggestion that these individuals may have been the prototypes of so-called **were-wolves.** No evidence to support this notion has been adduced.

Biochemistry Underlies the Causes, Diagnoses, & Treatments of the Porphyrias

Six major types of **porphyria** have been described, resulting from depressions in the activities of enzymes 3 through 8 shown in Figure 31–9 (see also **Table 31–2**). Assay of the activity of one or more of these enzymes using an appropriate source (eg, red blood cells) is thus important in making a definitive diagnosis in a suspected case of porphyria. Individuals with low activities of enzyme 1 (ALAS2) develop anemia, not porphyria (see Table 31–2). Patients with low activities of enzyme 2 (ALA dehydratase) have been reported, but very rarely; the resulting condition is called ALA dehydratase-deficient porphyria.

In general, the porphyrias described are **inherited** in an autosomal dominant manner, with the exception of congenital erythropoietic porphyria, which is inherited in a recessive mode. The precise abnormalities in the genes directing synthesis of the enzymes involved in heme biosynthesis have been determined in some instances. Thus, the use of appropri-

ate gene probes has made possible the **prenatal diagnosis** of some of the porphyrias.

As is true of most **inborn errors,** the signs and symptoms of porphyria result from either a **deficiency** of metabolic products beyond the enzymatic block or from an **accumulation** of metabolites behind the block.

If the enzyme lesion occurs **early** in the pathway prior to the formation of porphyrinogens (eg, enzyme 3 of Figure 31–9, which is affected in acute intermittent porphyria), **ALA** and **PBG** will accumulate in body tissues and fluids (**Figure 31–11**). Clinically, patients complain of **abdominal pain** and **neuropsychiatric symptoms.** The precise biochemical cause of these symptoms has not been determined but may relate to elevated levels of ALA or PBG or to a deficiency of heme.

On the other hand, enzyme blocks **later** in the pathway result in the **accumulation of the porphyrinogens** indicated in Figures 31–9 and 31–11. Their oxidation products, the corresponding porphyrin derivatives, cause **photosensitivity,** a reaction to visible light of about 400 nm. The porphyrins, when exposed to light of this wavelength, are thought to become "excited" and then react with molecular oxygen to form oxygen radicals. These latter species **injure lysosomes** and other organelles. Damaged lysosomes release their degradative enzymes, causing variable degrees of skin damage, including scarring.

The porphyrias can be **classified** on the basis of the **organs or cells** that are most affected. These are generally organs or

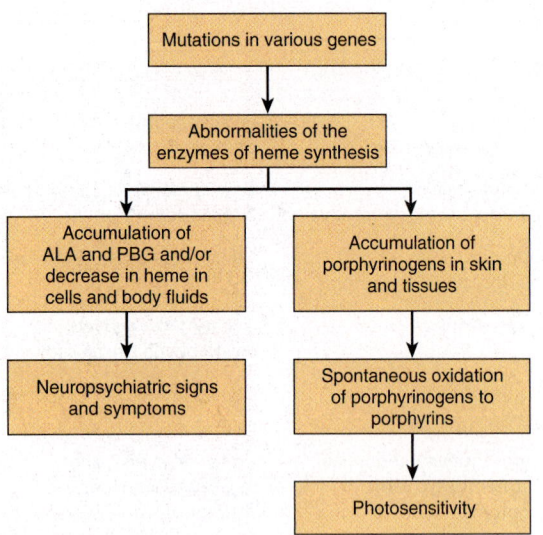

FIGURE 31–11 Biochemical causes of the major signs and symptoms of the porphyrias.

cells in which synthesis of heme is particularly active. The **bone marrow** synthesizes considerable hemoglobin, and the **liver** is active in the synthesis of another hemoprotein, cytochrome P450. Thus, one classification of the porphyrias is to designate them as predominantly either **erythropoietic** or **hepatic;** the types of porphyrias that fall into these two classes are so characterized in Table 31–2. Porphyrias can also be classified as **acute** or **cutaneous** on the basis of their clinical features. Why do specific types of porphyria affect certain organs more markedly than others? A partial answer is that the levels of metabolites that cause damage (eg, ALA, PBG, specific porphyrins, or lack of heme) can vary markedly in different organs or cells depending upon the differing activities of their heme-forming enzymes.

As described above, **ALAS1** is the key regulatory enzyme of the heme biosynthetic pathway in liver. A large number of **drugs** (eg, barbiturates, griseofulvin) induce the enzyme. Most of these drugs do so by inducing cytochrome P450 (see Chapter 53), which uses up heme and thus derepresses (induces) ALAS1. In patients with porphyria, increased activities of ALAS1 result in increased levels of potentially harmful heme precursors prior to the metabolic block. Thus, taking **drugs that cause induction of cytochrome P450** (so-called microsomal inducers) can precipitate attacks of porphyria.

The **diagnosis** of a specific type of porphyria can generally be established by consideration of the **clinical** and **family history,** the **physical examination,** and appropriate **laboratory tests.** The major findings in the six principal types of porphyria are listed in Table 31–2.

High levels of **lead** can affect heme metabolism by combining with SH groups in enzymes such as ferrochelatase and ALA dehydratase. This affects porphyrin metabolism. Elevated levels of protoporphyrin are found in red blood cells, and elevated levels of ALA and of coproporphyrin are found in urine.

It is hoped that **treatment** of the porphyrias at the gene level will become possible. In the meantime, treatment is essentially symptomatic. It is important for patients to **avoid drugs** that cause induction of cytochrome P450. Ingestion of large amounts of **carbohydrates** (glucose loading) or administration of **hematin** (a hydroxide of heme) may repress ALAS1, resulting in diminished production of harmful heme precursors. Patients exhibiting photosensitivity may benefit from administration of **β-carotene;** this compound appears to lessen production of free radicals, thus diminishing photosensitivity. **Sunscreens** that filter out visible light can also be helpful to such patients.

CATABOLISM OF HEME PRODUCES BILIRUBIN

Under normal conditions in human adults, some 200 billion erythrocytes are destroyed per day. Thus, a 70-kg human turns over approximately **6 g of hemoglobin** daily. When hemoglobin is destroyed in the body, **globin** is degraded to its constituent amino acids, which are reused, and the **iron** of heme enters the iron pool, also for reuse. The iron-free **porphyrin** portion of heme is also degraded, mainly in the reticuloendothelial cells of the liver, spleen, and bone marrow.

The **catabolism of heme** from all of the heme proteins appears to be carried out in the microsomal fractions of cells by a complex enzyme system called **heme oxygenase.** By the time the heme derived from heme proteins reaches the oxygenase system, the iron has usually been oxidized to the **ferric** form, constituting **hemin.** The heme oxygenase system is substrate-inducible. As depicted in **Figure 31–12,** the hemin is **reduced** to heme with NADPH, and, with the help of more NADPH, oxygen is added to the α-methyne bridge between pyrroles I and II of the porphyrin. The ferrous iron is **again oxidized** to the ferric form. With the further addition of oxygen, **ferric ion** is released, **carbon monoxide** is produced, and an equimolar quantity of **biliverdin** results from the splitting of the tetrapyrrole ring.

In birds and amphibia, the green **biliverdin IX** is excreted; in mammals, a soluble enzyme called **biliverdin reductase** reduces the methyne bridge between pyrrole III and pyrrole IV to a methylene group to produce **bilirubin,** a yellow pigment (Figure 31–12).

It is estimated that **1 g of hemoglobin** yields **35 mg of bilirubin.** The daily bilirubin formation in human adults is approximately 250–350 mg, deriving mainly not only from hemoglobin, but from ineffective erythropoiesis and from various other heme proteins such as cytochrome P450.

The chemical conversion of heme to bilirubin by reticuloendothelial cells can be observed in vivo as the purple color of the heme in a **hematoma** is slowly converted to the yellow pigment of bilirubin.

Bilirubin formed in peripheral tissues is **transported to the liver** by plasma albumin. The **further metabolism** of bilirubin occurs primarily in the liver. It can be divided into three processes. (1) uptake of bilirubin by liver parenchymal cells,

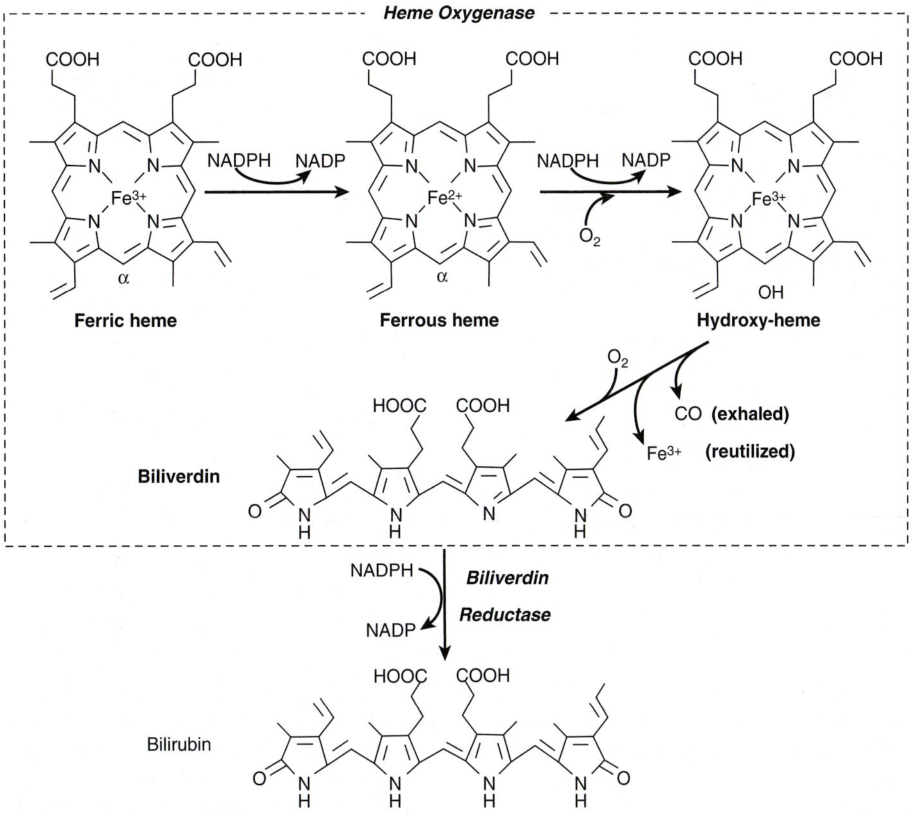

FIGURE 31–12 **Schematic representation of the microsomal heme oxygenase system.** The order of side-chain groups in bilirubin (from left to right) is M, V, M, P, P, M, M, V (M, methyl; V, vinyl; P, propionyl). (Redrawn, with permission, by Antony McDonagh from Schmid R, McDonough AF in: *The Porphyrins*. Dolphin D [ed.]. Academic Press, 1978. Reprinted with permission from Elsevier.)

(2) conjugation of bilirubin with glucuronate in the endoplasmic reticulum, and (3) secretion of conjugated bilirubin into the bile. Each of these processes will be considered separately.

THE LIVER TAKES UP BILIRUBIN

Bilirubin is only **sparingly soluble** in water, but its solubility in plasma is increased by noncovalent binding to albumin. Each molecule of albumin appears to have one high-affinity site and one low-affinity site for bilirubin. In 100 mL of plasma, approximately 25 mg of bilirubin can be tightly bound to albumin at its high-affinity site. Bilirubin in excess of this quantity can be bound only loosely and thus can easily be detached and diffuse into tissues. A number of compounds such as **antibiotics and other drugs** compete with bilirubin for the high-affinity binding site on albumin. Thus, these compounds can displace bilirubin from albumin and have significant clinical effects.

In the liver, the bilirubin is removed from albumin and taken up at the sinusoidal surface of the hepatocytes by a carrier-mediated saturable system. This **facilitated transport system** has a very large capacity, so that even under pathologic conditions the system does not appear to be rate limiting in the metabolism of bilirubin.

Since this facilitated transport system allows the equilibrium of bilirubin across the sinusoidal membrane of the hepatocyte, the net uptake of bilirubin will be dependent upon the **removal of bilirubin** via subsequent metabolic pathways.

Once bilirubin enters the hepatocytes, it can **bind to certain cytosolic proteins**, which help to keep it solubilized prior to conjugation. **Ligandin** (a member of the family of glutathione S-transferases) and **protein Y** are the involved proteins. They may also help to prevent efflux of bilirubin back into the blood stream.

Conjugation of Bilirubin with Glucuronic Acid Occurs in the Liver

Bilirubin is **nonpolar** and would persist in cells (eg, bound to lipids) if not rendered water-soluble. Hepatocytes convert bilirubin to a **polar** form, which is readily excreted in the bile, by adding glucuronic acid molecules to it. This process is called **conjugation** and can employ polar molecules other than glucuronic acid (eg, sulfate). Many steroid hormones and drugs are also converted to water-soluble derivatives by conjugation in preparation for excretion (see **Chapter 53**).

The conjugation of bilirubin is catalyzed by a specific **glucuronosyltransferase.** The enzyme is mainly located in

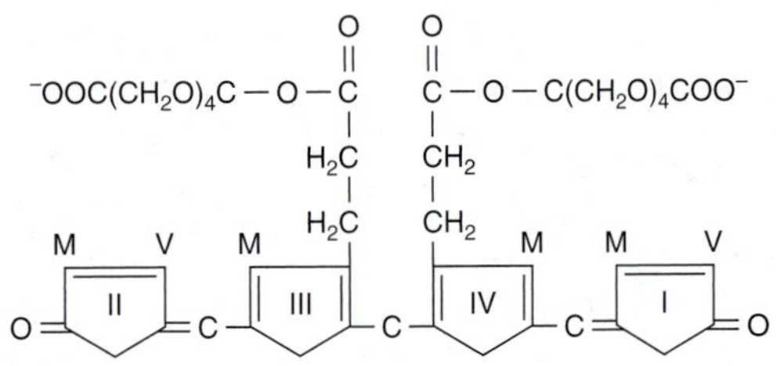

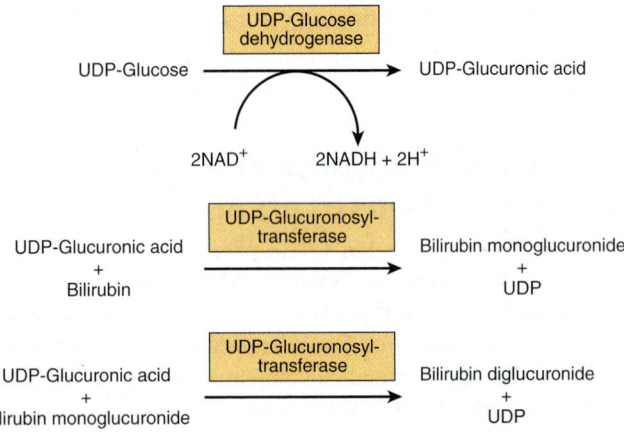

FIGURE 31–13 **Structure of bilirubin diglucuronide (conjugated, "direct-reacting" bilirubin).** Glucuronic acid is attached via ester linkage to the two propionic acid groups of bilirubin to form an acylglucuronide.

the endoplasmic reticulum, uses UDP-glucuronic acid as the glucuronosyl donor, and is referred to as bilirubin-UGT. Bilirubin monoglucuronide is an intermediate and is subsequently converted to the **diglucuronide** (**Figures 31–13** and **31–14**). Most of the bilirubin excreted in the bile of mammals is in the form of bilirubin diglucuronide. However, when bilirubin conjugates exist abnormally in **human plasma** (eg, in obstructive jaundice), they are predominantly **monoglucuronides.** Bilirubin-UGT activity can be **induced** by a number of clinically useful drugs, including phenobarbital. More information about glucuronosylation is presented below in the discussion of inherited disorders of bilirubin conjugation.

Bilirubin Is Secreted into Bile

Secretion of conjugated bilirubin into the bile occurs by an **active transport** mechanism, which is probably rate-limiting for the entire process of hepatic bilirubin metabolism. The protein involved is **MRP-2** (multidrug-resistance-like protein 2), also called multispecific organic anion transporter (MOAT). It is located in the **plasma membrane** of the bile canalicular membrane and handles a number of organic

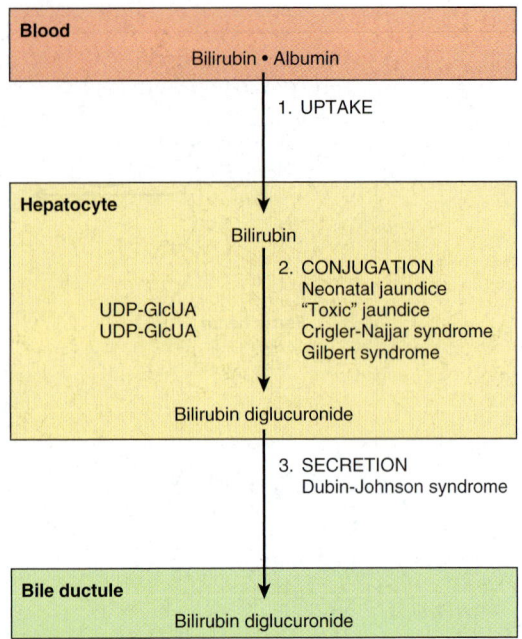

FIGURE 31–15 **Diagrammatic representation of the three major processes (uptake, conjugation, and secretion) involved in the transfer of bilirubin from blood to bile.** Certain proteins of hepatocytes, such as ligandin (a member of the glutathione S-transferase family of enzymes) and Y protein, bind intracellular bilirubin and may prevent its efflux into the blood stream. The process affected in a number of conditions causing jaundice is also shown.

anions. It is a member of the family of ATP-binding cassette transporters. The hepatic transport of conjugated bilirubin into the bile is **inducible** by those same drugs that are capable of inducing the conjugation of bilirubin. Thus, the conjugation and excretion systems for bilirubin behave as a coordinated functional unit.

Figure 31–15 summarizes the three major processes involved in the **transfer of bilirubin** from blood to bile. Sites that are affected in a number of conditions causing jaundice (see below) are also indicated.

FIGURE 31–14 **Conjugation of bilirubin with glucuronic acid.** The glucuronate donor, UDP-glucuronic acid, is formed from UDP-glucose as depicted. The UDP-glucuronosyltransferase is also called bilirubin-UGT.

Conjugated Bilirubin Is Reduced to Urobilinogen by Intestinal Bacteria

As the conjugated bilirubin reaches the terminal ileum and the large intestine, the glucuronides are removed by specific bacterial enzymes (**β-glucuronidases**), and the pigment is subsequently **reduced** by the fecal flora to a group of colorless tetrapyrrolic compounds called **urobilinogens.** In the terminal ileum and large intestine, a small fraction of the urobilinogens is reabsorbed and reexcreted through the liver to constitute the **enterohepatic urobilinogen cycle.** Under abnormal conditions, particularly when excessive bile pigment is formed or liver disease interferes with this intrahepatic cycle, urobilinogen may also be excreted in **the urine.**

Normally, most of the colorless urobilinogens formed in the colon by the fecal flora are **oxidized** there to **urobilins** (colored compounds) and are excreted in the feces. Darkening of feces upon standing in air is due to the oxidation of residual urobilinogens to urobilins.

HYPERBILIRUBINEMIA CAUSES JAUNDICE

When bilirubin in the blood exceeds 1 mg/dL (17.1 μmol/L), **hyperbilirubinemia** exists. Hyperbilirubinemia may be due to the **production** of more bilirubin than the normal liver can excrete, or it may result from the **failure** of a damaged liver to **excrete** bilirubin produced in normal amounts. In the absence of hepatic damage, **obstruction** of the excretory ducts of the liver—by preventing the excretion of bilirubin—will also cause hyperbilirubinemia. In all these situations, bilirubin accumulates in the blood, and when it reaches a certain concentration (approximately 2–2.5 mg/dL), it diffuses into the tissues, which then become yellow. That condition is called **jaundice** or **icterus.**

In clinical studies of jaundice, **measurement of bilirubin in the serum** is of a great value. A method for quantitatively assaying the bilirubin content of the serum was first devised by **van den Bergh** by application of **Ehrlich's test** for bilirubin in urine. The Ehrlich reaction is based on the coupling of diazotized sulfanilic acid (Ehrlich's diazo reagent) and bilirubin to produce a reddish-purple azo compound. In the original procedure as described by Ehrlich, **methanol** was used to provide a solution in which both bilirubin and the diazo regent were soluble. van den Bergh inadvertently omitted the methanol on an occasion when assay of bile pigment in human bile was being attempted. To his surprise, normal development of the color occurred "directly." This form of bilirubin that would react without the addition of methanol was thus termed **"direct-reacting."** It was then found that this same direct reaction would also occur in serum from cases of jaundice due to biliary obstruction. However, it was still necessary to add methanol to detect bilirubin in normal serum or the one that was present in excess in serum from cases of hemolytic jaundice where no evidence of obstruction was to be found. To that form of bilirubin, which could be measured only after the addition of methanol, the term **"indirect-reacting"** was applied.

It was subsequently discovered that the **indirect bilirubin is "free"** (unconjugated) bilirubin en route to the liver from the reticuloendothelial tissues, where the bilirubin was originally produced by the breakdown of heme porphyrins. Since this bilirubin is not water-soluble, it requires methanol to initiate coupling with the diazo reagent. In the liver, the free bilirubin becomes **conjugated** with glucuronic acid, and the conjugate, predominantly bilirubin diglucuronide, can then be excreted into the bile. Furthermore, conjugated bilirubin, being water-soluble, can react directly with the diazo reagent, so that the "direct bilirubin" of van den Bergh is actually a bilirubin conjugate (bilirubin glucuronide).

TABLE 31–3 Some Causes of Unconjugated and Conjugated Hyperbilirubinemia

Unconjugated	Conjugated
Hemolytic anemias	Obstruction of the biliary tree
Neonatal "physiological jaundice"	Dubin–Johnson syndrome
Crigler–Najjar syndromes types I and II	Rotor syndrome
Gilbert syndrome	Liver diseases such as the various types of hepatitis
Toxic hyperbilirubinemia	

These causes are discussed briefly in the text. Common causes of obstruction of the biliary tree are a stone in the common bile duct and cancer of the head of the pancreas. Various liver diseases (eg, the various types of hepatitis) are frequent causes of predominantly conjugated hyperbilirubinemia.

Depending on the type of bilirubin present in plasma—ie, unconjugated or conjugated—hyperbilirubinemia may be classified as **retention hyperbilirubinemia,** due to overproduction, or **regurgitation hyperbilirubinemia,** due to reflux into the bloodstream because of biliary obstruction.

Separation and quantitation of unconjugated bilirubin and the conjugated species can be performed using high-pressure **liquid chromatography.**

Because of its **hydrophobicity,** only unconjugated bilirubin can cross the blood-brain barrier into the central nervous system; thus, encephalopathy due to hyperbilirubinemia (**kernicterus**) can occur only in connection with unconjugated bilirubin, as found in retention hyperbilirubinemia. On the other hand, because of its water-solubility, only conjugated bilirubin can appear in urine. Accordingly, **choluric jaundice** (choluria is the presence of bile pigments in the urine) occurs only in regurgitation hyperbilirubinemia, and **acholuric jaundice** occurs only in the presence of an excess of unconjugated bilirubin.

Table 31–3 lists some causes of unconjugated and conjugated hyperbilirubinemia. These conditions are briefly described in the following sections.

Elevated Amounts of Unconjugated Bilirubin in Blood Occur in a Number of Conditions

Hemolytic Anemias

Hemolytic anemias are important causes of unconjugated hyperbilirubinemia, though unconjugated hyperbilirubinemia is **usually only slight** (<4 mg/dL; <68.4 μmol/L) even in the event of extensive hemolysis because of the healthy liver's large capacity for handling bilirubin.

Neonatal "Physiologic Jaundice"

This **transient** condition is the most common cause of unconjugated hyperbilirubinemia. It results from an accelerated hemolysis around the time of **birth** and an immature hepatic system for the uptake, conjugation, and secretion of bilirubin. Not only is the bilirubin-UGT activity reduced, but there probably is reduced synthesis of the substrate for that enzyme, UDP-glucuronic acid. Since the increased amount of bilirubin is unconjugated, it is capable of penetrating the blood–brain barrier when its concentration in plasma exceeds that which can be tightly bound by albumin (20–25 mg/dL). This can result in a hyperbilirubinemic toxic encephalopathy, or **kernicterus,** which can cause mental retardation. Because of the recognized inducibility of this bilirubin-metabolizing system, **phenobarbital** has been administered to jaundiced neonates and is effective in this disorder. In addition, exposure to **blue light** (phototherapy) promotes the hepatic excretion of unconjugated bilirubin by converting some of the bilirubin to other derivatives such as maleimide fragments and geometric isomers that are excreted in the bile.

Crigler–Najjar Syndrome, Type I; Congenital Nonhemolytic Jaundice

The type-I **Crigler–Najjar syndrome** is a **rare** autosomal recessive disorder. It is characterized by severe congenital jaundice (serum bilirubin usually exceeds 20 mg/dL) due to mutations in the gene encoding **bilirubin-UGT** activity in hepatic tissues. The disease is often fatal within the first 15 months of life. Children with this condition have been treated with **phototherapy,** resulting in some reduction in plasma bilirubin levels. Phenobarbital has no effect on the formation of bilirubin glucuronides in patients with the type I Crigler–Najjar syndrome. A **liver transplant** may be curative.

It should be noted that the gene encoding human bilirubin-UGT is part of a large UGT gene complex situated on chromosome 2. Many different substrates are subjected to glucuronosylation; therefore, **many glucuronosyltransferases** are required. The complex contains some 13 substrate-specific first exons, each with its own promoter. Four are pseudogenes, so nine different isoforms with differing glucuronosyltransferase activities are encoded. Exon A1 is that involved with conjugation of bilirubin. In the case of bilirubin, exon A1 is spliced to DNA containing exons 2–5, producing bilirubin-UGT. Other transferases are produced by splicing other first exons (members of A 2–13) to exons 2–5.

Crigler–Najjar Syndrome, Type II

This **rare** inherited disorder also results from mutations in the gene encoding bilirubin-UGT, but some activity of the enzyme is retained and the condition has a **more benign** course than type I. Serum bilirubin concentrations usually do not exceed 20 mg/dL. Patients with this condition can respond to treatment with large doses of **phenobarbital.**

Gilbert Syndrome

Again, this relatively prevalent condition is caused by mutations in the gene encoding bilirubin-UGT. It is more common among **males.** Approximately 30% of the enzyme's activity is preserved and the condition is entirely **harmless.**

Toxic Hyperbilirubinemia

Unconjugated hyperbilirubinemia can result from **toxin-induced** liver dysfunction such as that caused by chloroform, arsphenamines, carbon tetrachloride, acetaminophen, hepatitis virus, cirrhosis, and *Amanita* mushroom poisoning. These acquired disorders are due to hepatic parenchymal cell damage, which impairs conjugation.

Obstruction in the Biliary Tree Is the Most Common Cause of Conjugated Hyperbilirubinemia
Obstruction of the Biliary Tree

Conjugated hyperbilirubinemia commonly results from blockage of the hepatic or common bile ducts, most often due to a **gallstone** or to **cancer** of the head of the pancreas (**Figure 31–16**). Because of the obstruction, bilirubin diglucuronide cannot be excreted. It thus regurgitates into the hepatic veins and lymphatics, and conjugated bilirubin appears in the blood and urine (**choluric jaundice**). Also, the **stools** are usually pale in color, and should be examined routinely in any case of jaundice.

The term **cholestatic jaundice** is used to include all cases of extrahepatic obstructive jaundice. It also covers those

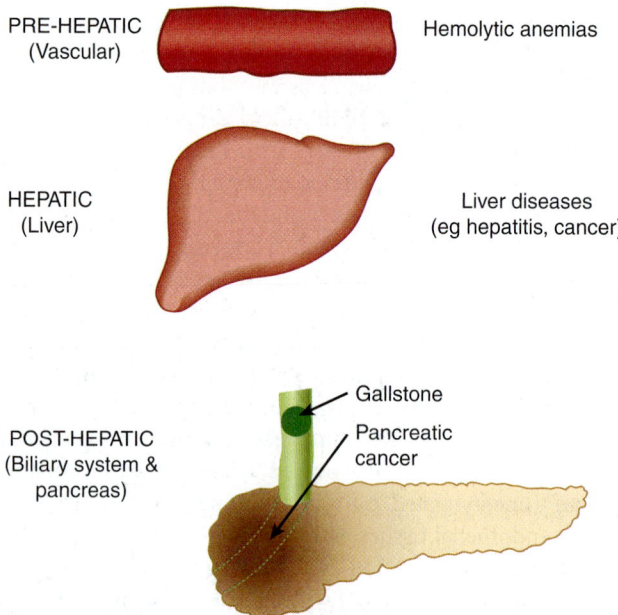

PRE-HEPATIC (Vascular)   Hemolytic anemias

HEPATIC (Liver)   Liver diseases (eg hepatitis, cancer)

POST-HEPATIC (Biliary system & pancreas)   Gallstone   Pancreatic cancer

FIGURE 31–16 **Diagrammatic representation of some major causes of jaundice. Prehepatic** indicates events in the blood stream; the major cause would be various forms of hemolytic anemia (see Chapter 52). **Hepatic** signifies events in the liver, such as the various types of hepatitis or other forms of liver disease (eg, cancer). **Posthepatic** refers to events in the biliary tree; the major causes of posthepatic jaundice are obstruction of the common bile duct by a gallstone (biliary calculus) or by cancer of the head of the pancreas.

TABLE 31–4 Laboratory Results in Normal Patients and Patients with Three Different Causes of Jaundice

Condition	Serum Bilirubin	Urine Urobilinogen	Urine Bilirubin	Fecal Urobilinogen
Normal	Direct: 0.1–0.4 mg/dL Indirect: 0.2–0.7 mg/dL	0–4 mg/24 h	Absent	40–280 mg/24 h
Hemolytic anemia	↑Indirect	Increased	Absent	Increased
Hepatitis	↑Direct and indirect	Decreased if micro-obstruction is present	Present if micro-obstruction occurs	Decreased
Obstructive jaundice[1]	↑Direct	Absent	Present	Trace to absent

[1]The most common causes of obstructive (posthepatic) jaundice are cancer of the head of the pancreas and a gallstone lodged in the common bile duct. The presence of bilirubin in the urine is sometimes referred to as choluria—therefore, hepatitis and obstruction of the common bile duct cause choluric jaundice, whereas the jaundice of hemolytic anemia is referred to as acholuric. The laboratory results in patients with hepatitis are variable, depending on the extent of damage to parenchymal cells and the extent of micro-obstruction to bile ductules. Serum levels of **alanine aminotransferase** (ALT) and **aspartate aminotransferase** (AST) are usually markedly elevated in hepatitis, whereas serum levels of **alkaline phosphatase** are elevated in obstructive liver disease.

cases of jaundice that exhibit conjugated hyperbilirubinemia due to micro-obstruction of intrahepatic biliary ductules by swollen, damaged hepatocytes (eg, as may occur in infectious hepatitis).

Dubin–Johnson Syndrome

This **benign** autosomal recessive disorder consists of **conjugated hyperbilirubinemia** in childhood or during adult life. The hyperbilirubinemia is caused by mutations in the gene encoding **MRP-2** (see above), the protein involved in the **secretion** of conjugated bilirubin into bile. The centrilobular hepatocytes contain an abnormal black pigment that may be derived from epinephrine.

Rotor Syndrome

This is a rare **benign** condition characterized by chronic conjugated hyperbilirubinemia and normal liver histology. Its precise cause has not been identified.

Some Conjugated Bilirubin Can Bind Covalently to Albumin

When levels of conjugated bilirubin remain high in plasma, a fraction can **bind covalently to albumin** (δ [delta] bilirubin). Because it is bound covalently to albumin, this fraction has a **longer half-life** in plasma than does conventional conjugated bilirubin. Thus, it remains elevated during the recovery phase of obstructive jaundice after the remainder of the conjugated bilirubin has declined to normal levels; this explains why some patients continue to appear jaundiced after conjugated bilirubin levels have returned to normal.

Urobilinogen & Bilirubin in Urine Are Clinical Indicators

Normally, there are mere traces of **urobilinogen** in the urine. In **complete obstruction of the bile duct,** no urobilinogen is found in the urine, since bilirubin has no access to the intestine, where it can be converted to urobilinogen. In this case, the presence of bilirubin (conjugated) in the urine without urobilinogen suggests obstructive jaundice, either intrahepatic or posthepatic.

In **jaundice secondary to hemolysis,** the increased production of bilirubin leads to increased production of **urobilinogen,** which appears in the urine in large amounts. Bilirubin is not usually found in the urine in hemolytic jaundice (because unconjugated bilirubin does not pass into the urine), so that the **combination of increased urobilinogen and absence of bilirubin** is suggestive of hemolytic jaundice. Increased blood destruction from any cause brings about an increase in urine urobilinogen.

Table 31–4 summarizes **laboratory results** obtained on patients with three different causes of jaundice—**hemolytic anemia** (a prehepatic cause), **hepatitis** (a hepatic cause), and **obstruction of the common bile duct** (a posthepatic cause) (see **Figure 31–16**). Laboratory tests on **blood** (evaluation of the possibility of a hemolytic anemia and measurement of prothrombin time) and on **serum** (eg, electrophoresis of proteins; activities of the enzymes ALT, AST, and alkaline phosphatase) are also important in helping to distinguish between prehepatic, hepatic, and posthepatic causes of jaundice.

SUMMARY

■ Hemoproteins, such as hemoglobin and the cytochromes, contain heme. Heme is an iron-porphyrin compound (Fe^{2+}-protoporphyrin IX) in which four pyrrole rings are joined by methyne bridges. The eight side groups (methyl, vinyl, and propionyl substituents) on the four pyrrole rings of heme are arranged in a specific sequence.

■ Biosynthesis of the heme ring occurs in mitochondria and cytosol via eight enzymatic steps. It commences with formation of δ-aminolevulinate from succinyl-CoA and glycine in a reaction catalyzed by the ALA synthase, the regulatory enzyme of the pathway.

■ Genetically determined abnormalities of seven of the eight enzymes involved in heme biosynthesis result in the inherited porphyrias. Red blood cells and liver are the major sites of metabolic expression of the porphyrias. Photosensitivity and neurologic problems are common complaints. Intake of certain compounds (such as lead) can cause acquired porphyrias. Increased amounts of porphyrins or their precursors can be detected in blood and urine, facilitating diagnosis.

■ Catabolism of the heme ring is initiated by the enzyme heme oxygenase, producing a linear tetrapyrrole.

■ Biliverdin is an early product of catabolism and on reduction yields bilirubin. The latter is transported by albumin from peripheral tissues to the liver, where it is taken up by hepatocytes. The iron of heme and the amino acids of globin are conserved and reutilized.

■ In the liver, bilirubin is made water-soluble by conjugation with two molecules of glucuronic acid and is secreted into the bile. The action of bacterial enzymes in the gut produces urobilinogen and urobilin, which are excreted in the feces and urine.

■ Jaundice is due to elevation of the level of bilirubin in the blood. The causes of jaundice can be classified as prehepatic (eg, hemolytic anemias), hepatic (eg, hepatitis), and posthepatic (eg, obstruction of the common bile duct). Measurements of plasma total and nonconjugated bilirubin, of urinary urobilinogen and bilirubin, and of certain serum enzymes as well as inspection and analysis of stool samples help distinguish between these causes.

REFERENCES

Beckett G, Walker S, Rae T, Ashby P: Liver disease (Chapter 13). In: *Lecture Notes: Clinical Biochemistry.* 8th ed. Wiley-Blackwell, 2010.

Deacon AC, Whatley SD, Elder GH: Porphyrins and disorders of porphyrin metabolism. In: *Tietz Textbook of Clinical Chemistry and Molecular Diagnostics,* 4th ed. Ch. 32. Burtis CA, Ashwood ER, Bruns DE (editors). Elsevier Saunders, 2006.

Desnick RJ, Astrin KH: The porphyrias. In: *Harrison's Principles of Internal Medicine,* 17th ed. Fauci AS et al (editors). Ch. 352. McGraw-Hill, 2008.

Dufour DR: Liver disease. In: *Tietz Textbook of Clinical Chemistry and Molecular Diagnostics,* 4th ed. Burtis CA, Ashwood ER, Bruns DE (editors). Ch. 47. Elsevier Saunders, 2006.

Higgins T, Beutler E, Doumas BT: Hemoglobin, iron and bilirubin. In: *Tietz Textbook of Clinical Chemistry and Molecular Diagnostics,* 4th ed. Burtis CA, Ashwood ER, Bruns DE (editors). Ch. 31. Elsevier Saunders, 2006.

Pratt DS, Kaplan MM: Evaluation of liver function. In: *Harrison's Principles of Internal Medicine,* 17th ed. Fauci AS et al (editors). Ch. 296. McGraw-Hill, 2008.

Pratt DS, Kaplan MM: Jaundice. In: *Harrison's Principles of Internal Medicine,* 17th ed. Fauci AS et al (editors). Ch. 43. McGraw-Hill, 2008.

Wolkoff AW: The hyperbilirubinemias. In: *Harrison's Principles of Internal Medicine,* 17th ed. Fauci AS et al (editors). Ch. 297. McGraw-Hill, 2008.

Exam Questions

Section III

1. Select the one of the following statements that is NOT CORRECT?

   A. Selenocysteine is present at the active sites of certain human enzymes.

   B. Selenocysteine is inserted into proteins by a post-translational process.

   C. Transamination of dietary α-keto acids can replace the dietary essential amino acids leucine, isoleucine, and valine.

   D. Conversion of peptidyl proline to peptidyl 4-hydroxyproline is accompanied by the incorporation of oxygen into succinate.

   E. Serine and glycine are interconverted in a single reaction in which tetrahydrofolate derivatives participate.

2. Select the one of the following statements that is NOT CORRECT?

   A. Δ^1-Pyrroline-5-carboxylate is an intermediate in both the biosynthesis and catabolism of L-proline.

   B. Human tissues can form dietarily nonessential amino acids from amphibolic intermediates or from dietarily essential amino acids.

   C. Human liver tissue can form serine from the glycolytic intermediate 3-phosphoglycerate.

   D. The reaction catalyzed by phenylalanine hydroxylase interconverts phenylalanine and tyrosine.

   E. The reducing power of tetrahydrobiopterin derives ultimately from NADPH.

3. Identify the metabolite that does NOT serve as a precursor of a dietarily essential amino acid:

   A. α-Ketoglutarate

   B. 3-Phosphoglycerate

   C. Glutamate

   D. Aspartate

   E. Histamine

4. Select the CORRECT answer. The first reaction in the degradation of the majority of the common amino acids involves the participation of:

   A. NAD^+

   B. Pyridoxal phosphate

   C. Thiamine pyrophosphate (TPP)

   D. FAD

   E. NAD^+ and TPP

5. Identify the amino acid that is the major contributor to hepatic gluconeogenesis.

   A. Alanine

   B. Glutamine

   C. Glycine

   D. Lysine

   E. Ornithine

6. Select the one of the following statements that is NOT CORRECT?

   A. The rate of hepatic gluconeogenesis from glutamine is far higher than that of any other amino acid.

   B. Angelman syndrome is associated with a defective ubiquitin E3 ligase.

   C. Following a protein-rich meal, the splanchnic tissues release predominantly branched-chain amino acids, which are taken up by peripheral muscle tissue.

   D. The L-α-amino oxidase-catalyzed conversion of an α-amino acid to its corresponding α-keto acid is accompanied by the release of NH_4^+.

   E. Similar or even identical signs and symptoms can be associated with different mutations of the gene that encodes a given enzyme.

7. Select the one of the following statements that is NOT CORRECT?

   A. PEST sequences target some proteins for rapid degradation.

   B. ATP and ubiquitin typically participate in the degradation of membrane-associated proteins and proteins with long half-lives.

   C. Ubiquitin molecules are attached to target proteins via non-α peptide bonds.

   D. The discoverers of ubiquitin-mediated protein degradation received a Nobel Prize.

   E. Degradation of ubiquitin-tagged proteins takes place in the proteasome, a multi-subunit macromolecule present in all eukaryotes.

8. For metabolic disorders of the urea cycle, which statement is NOT CORRECT?

   A. Ammonia intoxication is most severe when the metabolic block occurs prior to reaction 3 of the urea cycle, catalyzed by argininosuccinate synthase.

   B. Clinical symptoms include mental retardation and the avoidance of protein-rich foods.

   C. Clinical signs include hyperammonemia and respiratory acidosis.

   D. Aspartate provides the second nitrogen of argininosuccinate.

   E. Dietary management focuses on a low-protein diet ingested as frequent small meals.

9. Select the one of the following statements that is NOT CORRECT?

   A. One metabolic function of glutamine is to sequester nitrogen in a non-toxic form.

   B. Liver glutamate dehydrogenase is allosterically inhibited by ATP and activated by ADP.

   C. Urea is formed both from absorbed ammonia produced by enteric bacteria and from ammonia generated by tissue metabolic activity.

   D. The concerted action of glutamate dehydrogenase and glutamate aminotransferase may be termed transdeamination.

   E. Fumarate generated during biosynthesis of argininosuccinate ultimately forms oxaloacetate in reactions catalyzed by mitochondrial fumarase and malate dehydrogenase.

10. Select the one of the following statements that is NOT CORRECT?

    A. Histamine arises by decarboxylation of histidine.
    B. Threonine provides the thioethanol moiety for biosynthesis of coenzyme A.
    C. Ornithine serves as a precursor of both spermine and spermidine.
    D. Serotonin and melatonin are metabolites of tryptophan.
    E. Glycine, arginine, and methionine each contribute atoms for biosynthesis of creatine.

11. Select the one of the following statements that is NOT CORRECT?

    A. Excreted creatinine is a function of muscle mass, and can be used to determine whether a patient has provided a complete 24-hour urine specimen.
    B. Many drugs and drug catabolites are excreted in urine as glycine conjugates.
    C. Decarboxylation of glutamine forms the inhibitory neurotransmitter GABA (γ-aminobutyrate).
    D. The concentration of histamine in brain hypothalamus exhibits a circadian rhythm.
    E. The major nonprotein metabolic fate of methionine is conversion to *S*-adenosylmethionine.

12. Which of the following is NOT a hemoprotein?

    A. Myoglobin
    B. Cytochrome c
    C. Catalase
    D. Cytochrome P450
    E. Albumin

13. A 30-year-old male presented at clinic with a history of intermittent abdominal pain and episodes of confusion and psychiatric problems. Laboratory tests revealed increases of urinary δ-aminolevulinate and porphobilinogen. Mutational analysis revealed a mutation in the gene for uroporphyrinogen I synthase (porphobilinogen deaminase). The probable diagnosis was:

    A. X-linked sideroblastic anemia.
    B. Acute intermittent porphyria.
    C. Congenital erythropoietic porphyria.
    D. Porphyria cutanea tarda.
    E. Variegate porphyria.

14. Select the one of the following statements that is NOT CORRECT?

    A. Bilirubin is a cyclic tetrapyrrole.
    B. Bilirubin is carried in the plasma to the liver bound to albumin.
    C. High levels of bilirubin can cause damage to the brains of newborn infants.
    D. Bilirubin contains methyl and vinyl groups.
    E. Bilirubin does not contain iron.

15. A 62-year-old female presented at clinic with intense jaundice, steadily increasing over the preceding 3 months. She gave a history of severe upper abdominal pain, radiating to the back, and had lost considerable weight. She had noted that her stools had been very pale for some time. Lab tests revealed a very high level of direct bilirubin, and also elevated urinary bilirubin. The plasma level of alanine aminotransferase (ALT) was only slightly elevated, whereas the level of alkaline phosphatase was markedly elevated. Abdominal ultrasonography revealed no evidence of gallstones. Of the following, which is the most likely diagnosis?

    A. Gilbert syndrome.
    B. Hemolytic anemia.
    C. Type 1 Criggler–Najjar syndrome.
    D. Carcinoma of the pancreas.
    E. Infectious hepatitis.

SECTION IV

Structure, Function, & Replication of Informational Macromolecules

CHAPTER 32

Nucleotides

Victor W. Rodwell, PhD

OBJECTIVES

After studying this chapter, you should be able to:

- Write structural formulas to represent the amino- and oxo-tautomers of a purine and of a pyrimidine and state which tautomer predominates under physiologic conditions.
- Reproduce the structural formulas for the principal nucleotides present in DNA and in RNA and the less common nucleotides 5-methylcytosine, 5-hydroxymethylcytosine, and pseudouridine (Ψ).
- Represent D-ribose or 2-deoxy-D-ribose linked as either a *syn* or an *anti* conformer of a purine, name the bond between the sugar and the base, and indicate which conformer predominates under most physiologic conditions.
- Number the C and N atoms of a pyrimidine nucleotide and of a purine nucleoside, including using a primed numeral for C atoms of the sugars.
- Compare the phosphoryl group transfer potential of each phosphoryl group of a nucleoside triphosphate.
- Outline the physiologic roles of the cyclic phosphodiesters cAMP and cGMP.
- Appreciate that polynucleotides are directional macromolecules composed of mononucleotides linked by 3′ → 5′-phosphodiester bonds.
- Understand that in the abbreviated representations of polynucleotide structures such as pTpGpT or TGCATCA, the 5′-end is always shown at the left and all phosphodiester bonds are 3′ → 5′.
- For specific synthetic analogs of purine and pyrimidine bases and their derivatives that have served as anticancer drugs, indicate in what ways these compounds inhibit metabolism.

BIOMEDICAL IMPORTANCE

In addition to serving as precursors of nucleic acids, purine and pyrimidine nucleotides participate in metabolic functions as diverse as energy metabolism, protein synthesis, regulation of enzyme activity, and signal transduction. When linked to vitamins or vitamin derivatives, nucleotides form a portion of many coenzymes. As the principal donors and acceptors of phosphoryl groups in metabolism, nucleoside tri- and diphosphates such as ATP and ADP are the principal players in the energy transductions that accompany metabolic interconversions and oxidative phosphorylation. Linked to sugars or lipids, nucleosides constitute key biosynthetic intermediates. The sugar derivatives UDP-glucose and UDP-galactose participate in sugar interconversions and in the biosynthesis of starch and glycogen. Similarly,

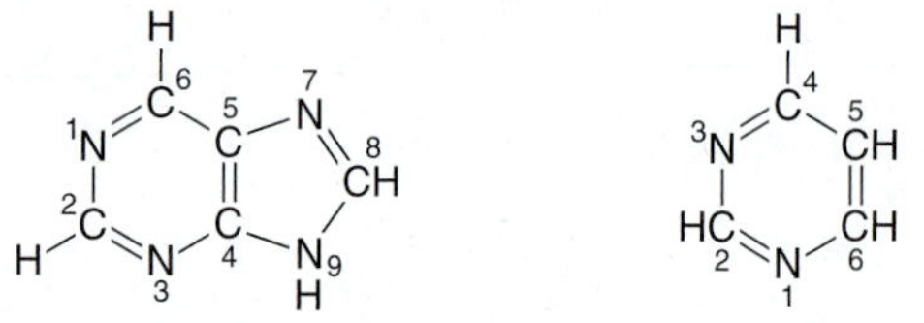

FIGURE 32–1 **Purine and pyrimidine.** The atoms are numbered according to the international system.

nucleoside-lipid derivatives such as CDP-acylglycerol are intermediates in lipid biosynthesis. Roles that nucleotides perform in metabolic regulation include ATP-dependent phosphorylation of key metabolic enzymes, allosteric regulation of enzymes by ATP, ADP, AMP, and CTP, and control by ADP of the rate of oxidative phosphorylation. The cyclic nucleotides cAMP and cGMP serve as the second messengers in hormonally regulated events, and GTP and GDP play key roles in the cascade of events that characterize signal transduction pathways. Medical applications include the use of synthetic purine and pyrimidine analogs that contain halogens, thiols, or additional nitrogen atoms in the chemotherapy of cancer and AIDS, and as suppressors of the immune response during organ transplantation.

CHEMISTRY OF PURINES, PYRIMIDINES, NUCLEOSIDES & NUCLEOTIDES

Purines & Pyrimidines Are Heterocyclic Compounds

Purines and pyrimidines are nitrogen-containing **heterocycles,** cyclic structures that contain, in addition to carbon, other (hetero) atoms such as nitrogen. Note that the smaller pyrimidine molecule has the *longer* name and the larger purine molecule the *shorter* name, and that their six-atom rings are numbered in opposite directions (**Figure 32–1**). Purines or pyrimidines with an –NH$_2$ group are weak bases (pK_a values 3–4), although the proton present at low pH is associated, not as one might expect with the exocyclic amino group, but with a ring nitrogen, typically N1 of adenine, N7 of guanine, and

FIGURE 32–2 **Tautomerism of the oxo and amino functional groups of purines and pyrimidines.**

N3 of cytosine. The planar character of purines and pyrimidines facilitates their close association, or "stacking," that stabilizes double-stranded DNA (see Chapter 34). The oxo and amino groups of purines and pyrimidines exhibit keto-enol and amine-imine **tautomerism** (**Figure 32–2**), although physiologic conditions strongly favor the amino and oxo forms.

Nucleosides Are *N*-Glycosides

Nucleosides are derivatives of purines and pyrimidines that have a sugar linked to a ring nitrogen of a purine or pyrimidine. Numerals with a prime (eg, 2′ or 3′) distinguish atoms of the sugar from those of the heterocycle. The sugar in **ribonucleosides** is D-ribose, and in **deoxyribonucleosides** is 2-deoxy-D-ribose. Both sugars are linked to the heterocycle by a **β-*N*-glycosidic bond,** almost always to the *N*-1 of a pyrimidine or to *N*-9 of a purine (**Figure 32–3**).

Nucleotides Are Phosphorylated Nucleosides

Mononucleotides are nucleosides with a phosphoryl group esterified to a hydroxyl group of the sugar. The 3′- and 5′-nucleotides are nucleosides with a phosphoryl group on the 3′- or 5′-hydroxyl group of the sugar, respectively. Since most nucleotides are 5′-, the prefix "5′-" usually is omitted when naming them. UMP and dAMP thus represent nucleotides with a phosphoryl group on C-5 of the pentose. Additional phosphoryl groups, ligated by **acid anhydride bonds** to the phosphoryl group of a mononucleotide, form **nucleoside diphosphates** and **triphosphates** (**Figure 32–4**).

Heterocylic *N*-Glycosides Exist as *Syn* and *Anti* Conformers

Steric hindrance by the heterocycle dictates that there is no freedom of rotation about the β-*N*-glycosidic bond of nucleosides

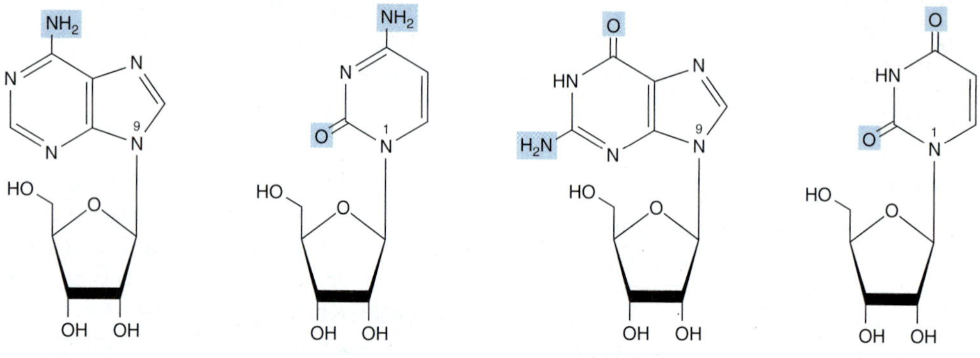

FIGURE 32–3 Ribonucleosides, drawn as the *syn* conformers.

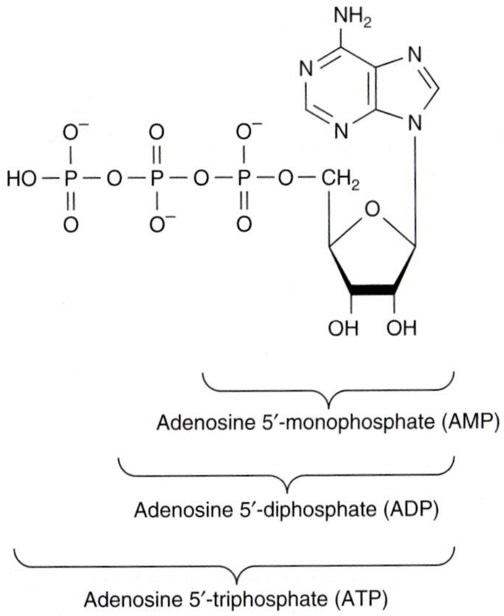

FIGURE 32–4 **ATP, its diphosphate, and its monophosphate.**

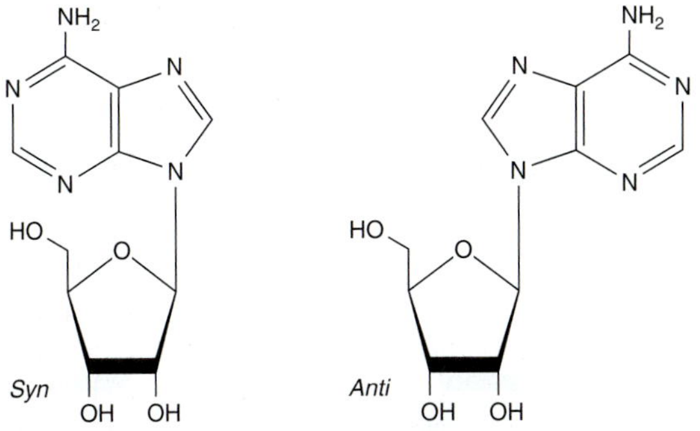

FIGURE 32–5 The *syn* and *anti* conformers of adenosine differ with respect to orientation about the *N*-glycosidic bond.

or nucleotides. Both therefore exist as noninterconvertible *syn* or *anti* conformers (**Figure 32–5**). Unlike tautomers, *syn* and *anti* conformers can only be interconverted by cleavage and reformation of the glycosidic bond. Both *syn* and *anti* conformers occur in nature, but the *anti* conformers predominate.

Table 32–1 lists the major purines and pyrimidines and their nucleoside and nucleotide derivatives. Single-letter

TABLE 32–1 Purine Bases, Ribonucleosides, and Ribonucleotides

Purine or Pyrimidine	X = H	X = Ribose	X = Ribose Phosphate
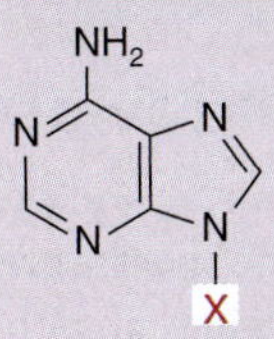	Adenine	Adenosine	Adenosine monophosphate (AMP)
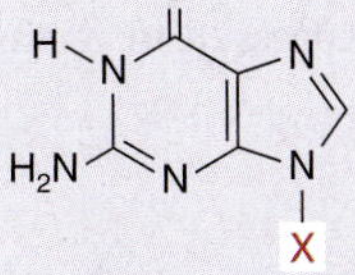	Guanine	Guanosine	Guanosine monophosphate (GMP)
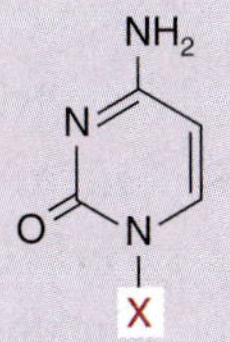	Cytosine	Cytidine	Cytidine monophosphate (CMP)
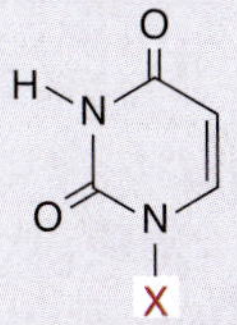	Uracil	Uridine	Uridine monophosphate (UMP)
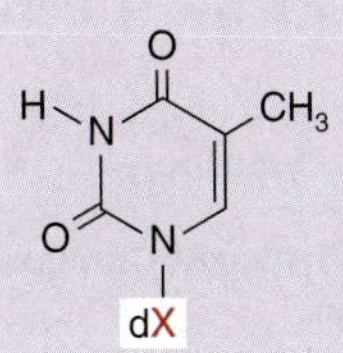	Thymine	Thymidine	Thymidine monophosphate (TMP)

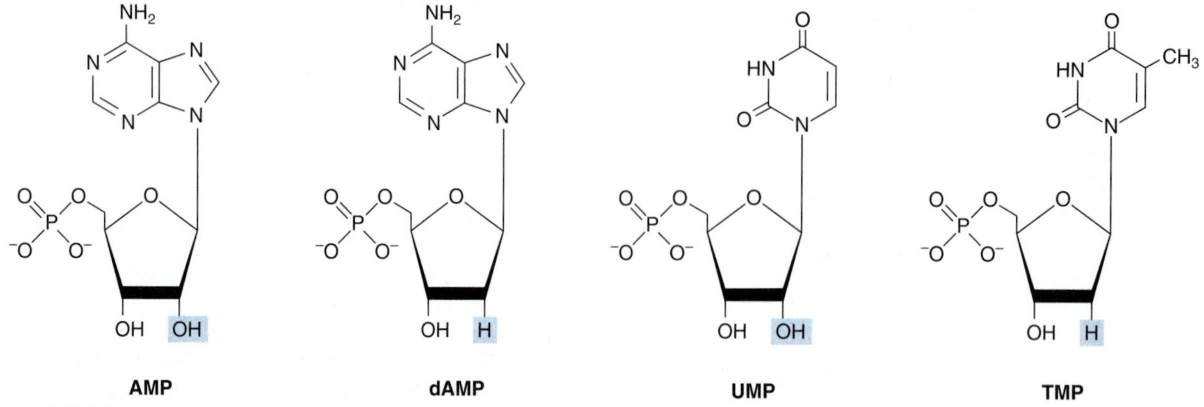

FIGURE 32–6   Structures of AMP, dAMP, UMP, and TMP.

abbreviations are used to identify adenine (A), guanine (G), cytosine (C), thymine (T), and uracil (U), whether free or present in nucleosides or nucleotides. The prefix "d" (deoxy) indicates that the sugar is 2′-deoxy-D-ribose (for example, in dATP) (**Figure 32–6**).

Modification of Polynucleotides Can Generate Additional Structures

Small quantities of additional purines and pyrimidines occur in DNA and RNAs. Examples include 5-methylcytosine of bacterial and human DNA, 5-hydroxymethylcytosine of bacterial and viral nucleic acids, and mono- and the di-*N*-methylated adenine and guanine of mammalian messenger RNAs (**Figure 32–7**) that function in oligonucleotide recognition and in regulating the half-lives of RNAs. Free nucleotides include hypoxanthine, xanthine, and uric acid (**Figure 32–8**) that are intermediates in the catabolism of adenine and guanine (see Chapter 33). Methylated heterocycles of plants

include the xanthine derivatives caffeine of coffee, theophylline of tea, and theobromine of cocoa (**Figure 32–9**).

Nucleotides Are Polyfunctional Acids

The primary and secondary phosphoryl groups of nucleosides have pK_a values of about 1.0 and 6.2, respectively. Nucleotides therefore bear significant negative charge at physiologic pH. The pK_a values of the secondary phosphoryl groups are such that they can serve both as proton donors and as proton acceptors at pH values approximately two or more units above or below neutrality.

Nucleotides Absorb Ultraviolet Light

The conjugated double bonds of purine and pyrimidine derivatives absorb ultraviolet light. While spectra are pH-dependent, at pH 7.0 all the common nucleotides absorb light at a wavelength close to 260 nm. The concentration of nucleotides and nucleic acids thus often is expressed in terms of "absorbance

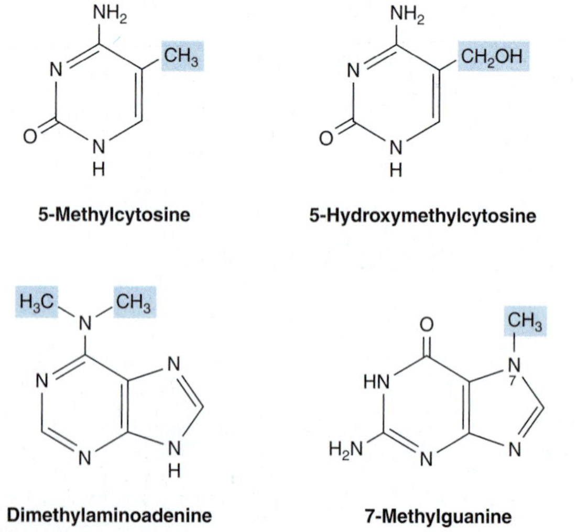

5-Methylcytosine

5-Hydroxymethylcytosine

Dimethylaminoadenine

7-Methylguanine

FIGURE 32–7   Four uncommon naturally occurring pyrimidines and purines.

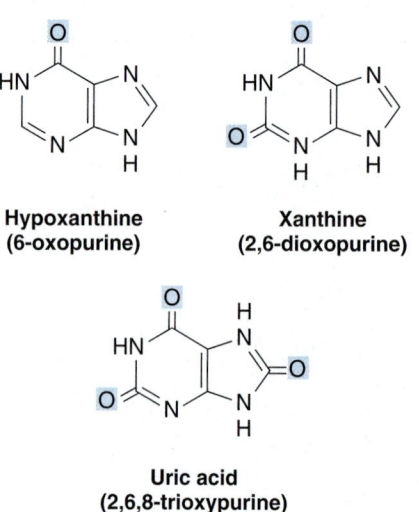

Hypoxanthine (6-oxopurine)

Xanthine (2,6-dioxopurine)

Uric acid (2,6,8-trioxypurine)

FIGURE 32–8   Structures of hypoxanthine, xanthine, and uric acid, drawn as the oxo tautomers.

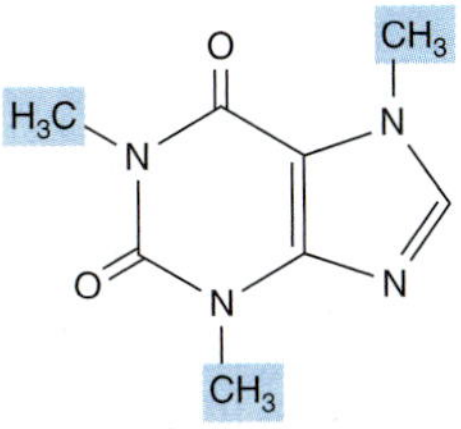

FIGURE 32–9 **Caffeine, a trimethylxanthine.** The dimethylxanthines theobromine and theophylline are similar but lack the methyl group at *N*-1 and at *N*-7, respectively.

at 260 nm." The mutagenic effect of ultraviolet light is due to its absorption by nucleotides in DNA that results in chemical modifications (see Chapter 35).

Nucleotides Serve Diverse Physiologic Functions

In addition to their roles as precursors of nucleic acids, ATP, GTP, UTP, CTP, and their derivatives each serve unique physiologic functions discussed in other chapters. Selected examples include the role of ATP as the principal biologic transducer of free energy, and the second messenger cAMP (**Figure 32–10**). The mean intracellular concentration of ATP, the most abundant free nucleotide in mammalian cells, is about 1 mmol/L. Since little cAMP is required, the intracellular cAMP concentration (about 1 nmol/L) is six orders of magnitude below that of ATP. Other examples include adenosine 3′-phosphate-5′-phosphosulfate (**Figure 32–11**), the sulfate donor for sulfated proteoglycans (see Chapter 48) and for sulfate conjugates of drugs; and the methyl group donor *S*-adenosylmethionine (**Figure 32–12**). GTP serves as an allosteric regulator and as an energy source for protein synthesis, and cGMP (Figure 32–10) serves as a second messenger in response to nitric oxide (NO) during relaxation of smooth muscle (see Chapter 49).

UDP-sugar derivatives participate in sugar epimerizations and in biosynthesis of glycogen (see Chapter 19), glucosyl disaccharides, and the oligosaccharides of glycoproteins and proteoglycans (see Chapters 47 and 48). UDP-glucuronic acid forms the urinary glucuronide conjugates of bilirubin

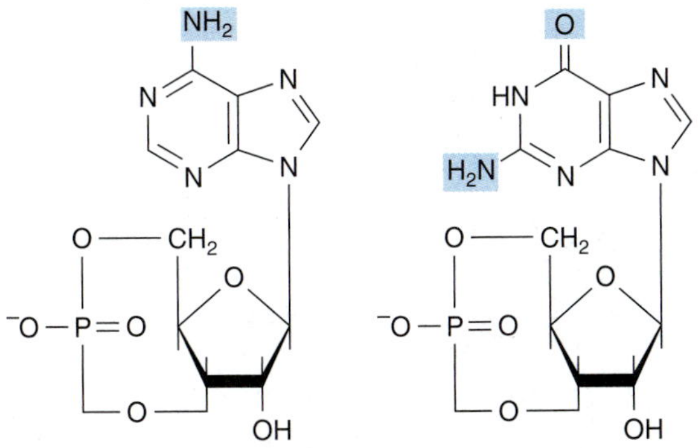

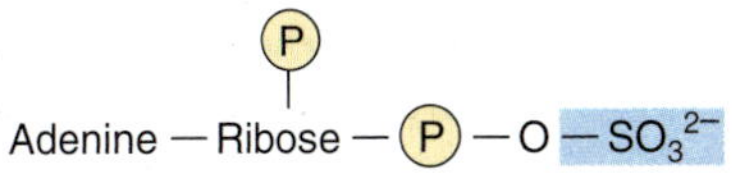

FIGURE 32–11 **Adenosine 3′-phosphate-5′-phosphosulfate.**

(see Chapter 31) and of many drugs, including aspirin. CTP participates in biosynthesis of phosphoglycerides, sphingomyelin, and other substituted sphingosines (see Chapter 24). Finally, many coenzymes incorporate nucleotides as well as structures similar to purine and pyrimidine nucleotides (see **Table 32–2**).

Nucleoside Triphosphates Have High-Group Transfer Potential

Nucleotide triphosphates have two acid anhydride bonds and one ester bond. Unlike esters, acid anhydrides have a high-group transfer potential. $\Delta G^{0\prime}$ for the hydrolysis of each of the two terminal (β and γ) phosphoryl groups of a nucleoside triphosphate is about −7 kcal/mol (−30 kJ/mol). This high group transfer potential not only permits purine and pyrimidine nucleoside triphosphates to function as group transfer reagents, most commonly of the γ-phosphoryl group, but also on occasion transfer of a nucleotide monophosphate moiety with an accompanying release of PP$_i$. Cleavage of an acid anhydride bond typically is coupled with a highly endergonic process such as covalent bond synthesis, for example, the polymerization of nucleoside triphosphates to form a nucleic acid (see Chapter 34).

SYNTHETIC NUCLEOTIDE ANALOGS ARE USED IN CHEMOTHERAPY

Synthetic analogs of purines, pyrimidines, nucleosides, and nucleotides modified in the heterocyclic ring or in the sugar moiety have numerous applications in clinical medicine. Their

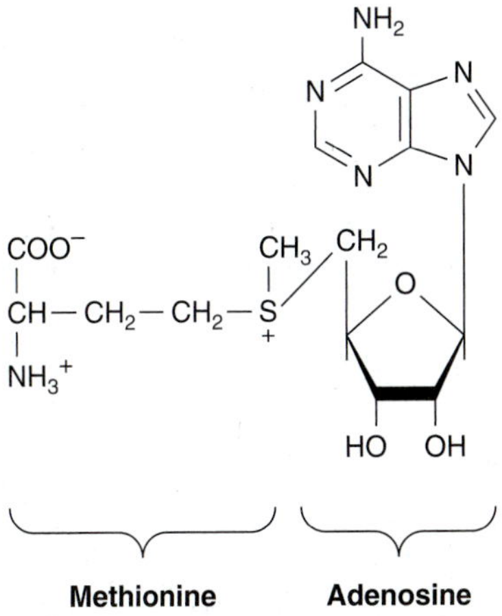

FIGURE 32–10 cAMP, 3′,5′-cyclic AMP, and cGMP, 3′, 5′-cyclic GMP.

FIGURE 32–12 *S*-Adenosylmethionine.

TABLE 32–2 **Many Coenzymes and Related Compounds Are Derivatives of Adenosine Monophosphate**

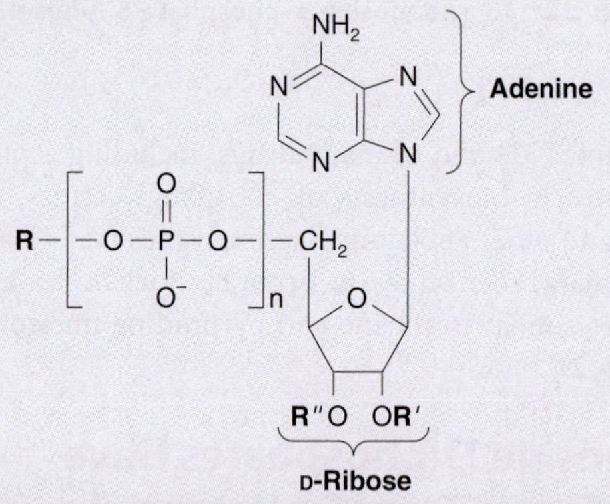

Coenzyme	R	R′	R″	n
Active methionine	Methionine[1]	H	H	0
Amino acid adenylates	Amino acid	H	H	1
Active sulfate	SO_3^{2-}	H	PO_3^{2-}	1
3′,5′-Cyclic AMP		H	PO_3^{2-}	1
NAD[2]	Nicotinamide	H	H	2
NADP[2]	Nicotinamide	PO_3^{2-}	H	2
FAD	Riboflavin	H	H	2
Coenzyme A	Pantothenate	H	PO_3^{2-}	2

[1]Replaces phosphoryl group.
[2]R is a B vitamin derivative.

toxic effects reflect either inhibition of enzymes essential for nucleic acid synthesis or their incorporation into nucleic acids with resulting disruption of base-pairing. Oncologists employ 5-fluoro- or 5-iodouracil, 3-deoxyuridine, 6-thioguanine and 6-mercaptopurine, 5- or 6-azauridine, 5- or 6-azacytidine, and 8-azaguanine (**Figure 32–13**), which are incorporated into DNA prior to cell division. The purine analog allopurinol, used in treatment of hyperuricemia and gout, inhibits purine biosynthesis and xanthine oxidase activity. Cytarabine is used in chemotherapy of cancer, and azathioprine, which is catabolized to 6-mercaptopurine, is employed during organ transplantation to suppress immunologic rejection (**Figure 32–14**).

Nonhydrolyzable Nucleoside Triphosphate Analogs Serve as Research Tools

Synthetic, nonhydrolyzable analogs of nucleoside triphosphates (**Figure 32–15**) allow investigators to distinguish the effects of nucleotides due to phosphoryl transfer from effects mediated by occupancy of allosteric nucleotide-binding sites on regulated enzymes.

DNA & RNA ARE POLYNUCLEOTIDES

The 5′-phosphoryl group of a mononucleotide can esterify a second hydroxyl group, forming a **phosphodiester.** Most commonly, this second hydroxyl group is the 3′-OH of the pentose of a second nucleotide. This forms a **dinucleotide**

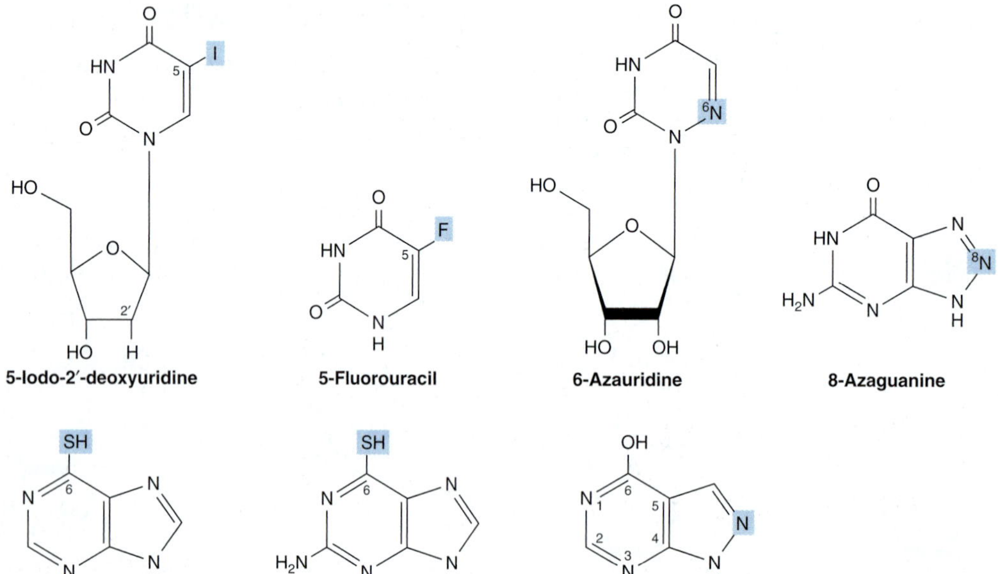

FIGURE 32–13 Selected synthetic pyrimidine and purine analogs.

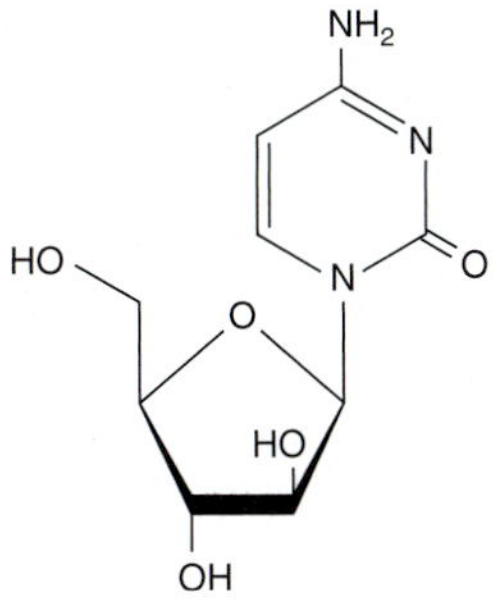

Cytarabine

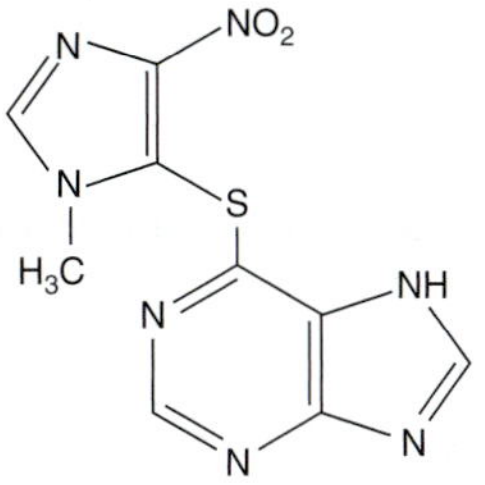

Azathioprine

FIGURE 32–14 Arabinosylcytosine (cytarabine) and azathioprine.

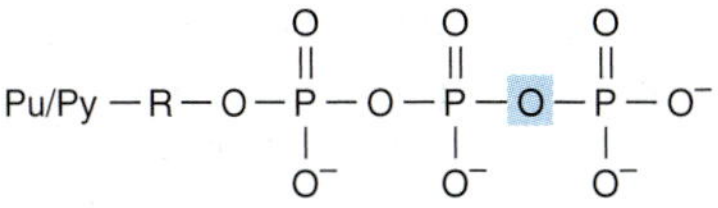

Parent nucleoside triphosphate

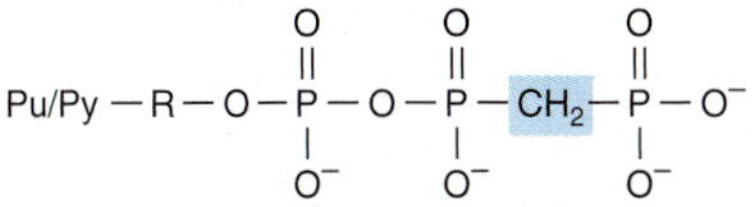

β,γ-Methylene derivative

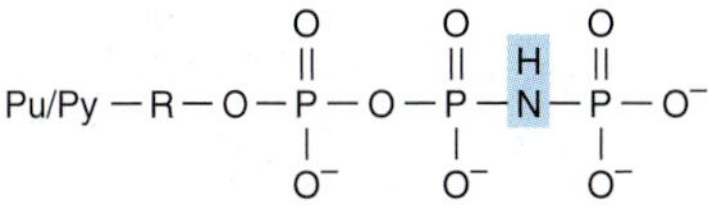

β,γ-Imino derivative

FIGURE 32–15 Synthetic derivatives of nucleoside triphosphates incapable of undergoing hydrolytic release of the terminal phosphoryl group. (Pu/Py, a purine or pyrimidine base; R, ribose or deoxyribose.) Shown are the parent (hydrolyzable) nucleoside triphosphate (*top*) and the unhydrolyzable β-methylene (*center*) and γ-imino derivatives (*bottom*).

in which the pentose moieties are linked by a 3′,5′-phosphodiester bond to form the "backbone" of RNA and DNA. The formation of a dinucleotide may be represented as the elimination of water between two mononucleotides. Biologic formation of dinucleotides does not, however, occur in this way because the reverse reaction, hydrolysis of the phosphodiester bond, is strongly favored on thermodynamic grounds. However, despite an extremely favorable ΔG, in the absence of catalysis by **phosphodiesterases** hydrolysis of the phosphodiester bonds of DNA occurs only over long periods of time. DNA therefore persists for considerable periods, and has been detected even in fossils. RNAs are far less stable than DNA since the 2′-hydroxyl group of RNA (absent from DNA) functions as a nucleophile during hydrolysis of the 3′,5′-phosphodiester bond.

Posttranslational modification of preformed **polynucleotides** can generate additional structures such as **pseudouridine,** a nucleoside in which D-ribose is linked to C-5 of uracil by a carbon-to-carbon bond rather than by the usual β-*N*-glycosidic bond. The nucleotide pseudouridylic acid (ψ) arises by rearrangement of a UMP of a preformed tRNA. Similarly, methylation by *S*-adenosylmethionine of a UMP of preformed tRNA forms TMP (thymidine monophosphate), which contains ribose rather than deoxyribose.

Polynucleotides Are Directional Macromolecules

Phosphodiester bonds link the 3′- and 5′-carbons of adjacent monomers. Each end of a nucleotide polymer thus is distinct.

We therefore refer to the "5′-end" or the "3′-end" of a polynucleotide, the 5′-end being the one with a free or phosphorylated 5′-hydroxyl.

The base sequence or **primary structure** of a polynucleotide can be represented as shown below. The phosphodiester bond is represented by P or p, bases by a single letter, and pentoses by a vertical line.

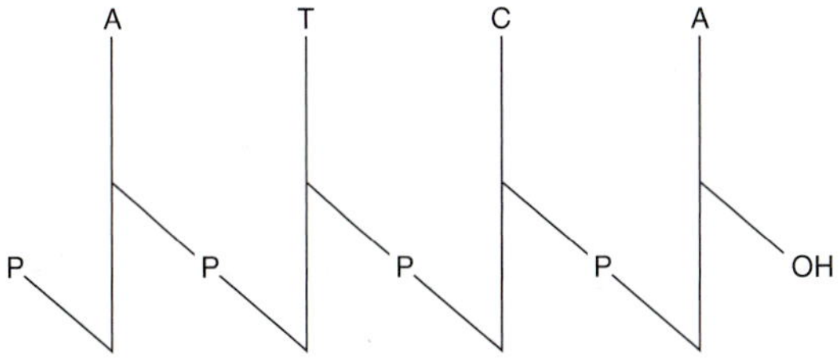

Where all the phosphodiester bonds are 3′ → 5′, a more compact notation is possible:

pGpGpApTpCpA

This representation indicates that the 5′-hydroxyl—but not the 3′-hydroxyl—is phosphorylated. The most compact representation, for example GGATC, shows only the base sequence with the 5′-end on the left and the 3′-end on the right. The phosphoryl groups are assumed to be present, but not shown.

SUMMARY

- Under physiologic conditions, the amino and oxo tautomers of purines, pyrimidines, and their derivatives predominate.
- Nucleic acids contain, in addition to A, G, C, T, and U, traces of 5-methylcytosine, 5-hydroxymethylcytosine, pseudouridine (ψ), and *N*-methylated heterocycles.

- Most nucleosides contain D-ribose or 2-deoxy-D-ribose linked to *N*-1 of a pyrimidine or to *N*-9 of a purine by a β-glycosidic bond whose *syn* conformers predominate.

- A primed numeral locates the position of the phosphate on the sugars of mononucleotides (eg, 3′-GMP, 5′-dCMP). Additional phosphoryl groups linked to the first by acid anhydride bonds form nucleoside diphosphates and triphosphates.

- Nucleoside triphosphates have high group transfer potential and participate in covalent bond syntheses. The cyclic phosphodiesters cAMP and cGMP function as intracellular second messengers.

- Mononucleotides linked by 3′ → 5′-phosphodiester bonds form polynucleotides, directional macromolecules with distinct 3′- and 5′-ends. For pTpGpT or TGCATCA, the 5′-end is at the left, and all phosphodiester bonds are 3′ → 5′.

- Synthetic analogs of purine and pyrimidine bases and their derivatives serve as anticancer drugs either by inhibiting an enzyme of nucleotide biosynthesis or by being incorporated into DNA or RNA.

REFERENCES

Adams RLP, Knowler JT, Leader DP: *The Biochemistry of the Nucleic Acids,* 11th ed. Chapman & Hall, 1992.

Blackburn GM, Gait MJ: *Nucleic Acids in Chemistry & Biology.* IRL Press, 1990.

Pacher P, Nivorozhkin A, Szabo C: Therapeutic effects of xanthine oxidase inhibitors: renaissance half a century after the discovery of allopurinol. Pharmacol Rev 2006;58:87.

33

Metabolism of Purine & Pyrimidine Nucleotides

Victor W. Rodwell, PhD

- Compare and contrast the roles of dietary nucleic acids and of de novo biosynthesis in the production of purines and pyrimidines destined for polynucleotide biosynthesis.
- Explain why antifolate drugs and analogs of the amino acid glutamine inhibit purine biosynthesis.
- Outline the sequence of reactions that convert IMP, first to AMP and GMP, and subsequently to their corresponding nucleoside triphosphates.
- Describe the formation from ribonucleotides of deoxyribonucleotides (dNTPs).
- Indicate the regulatory role of PRPP in hepatic purine biosynthesis and the specific reaction of hepatic purine biosynthesis that is feedback inhibited by AMP and by GMP.
- State the relevance of coordinated control of purine and pyrimidine nucleotide biosynthesis.
- Identify reactions that are inhibited by anticancer drugs.
- Write the structure of the end product of purine catabolism. Comment on its solubility and indicate its role in gout, Lesch–Nyhan syndrome, and von Gierke disease.
- Identify reactions whose impairment leads to modified pathologic signs and symptoms.
- Indicate why there are few clinically significant disorders of pyrimidine catabolism.

BIOMEDICAL IMPORTANCE

Even when humans consume a diet rich in nucleoproteins, dietary purines and pyrimidines are not incorporated directly into tissue nucleic acids. Humans synthesize the nucleic acids, ATP, NAD$^+$, coenzyme A, etc, from amphibolic intermediates. However, *injected* purine or pyrimidine analogs, including potential anticancer drugs, may be incorporated into DNA. The biosyntheses of purine and pyrimidine ribonucleotide triphosphates (NTPs) and dNTPs are precisely regulated events. Coordinated feedback mechanisms ensure their production in appropriate quantities and at times that match varying physiologic demand (eg, cell division). Human diseases that involve abnormalities in purine metabolism include gout,

Lesch–Nyhan syndrome, adenosine deaminase deficiency, and purine nucleoside phosphorylase deficiency. Diseases of pyrimidine biosynthesis are rarer, but include orotic acidurias. Unlike the low solubility of uric acid formed by catabolism of purines, the end-products of pyrimidine catabolism (carbon dioxide, ammonia, β-alanine, and γ-aminoisobutyrate) are highly water soluble. One genetic disorder of pyrimidine catabolism is β-hydroxybutyric aciduria, due to total or partial deficiency of the enzyme dihydropyrimidine dehydrogenase. This disorder of pyrimidine catabolism, also known as combined uraciluria-thyminuria, is also a disorder of β-amino acid biosynthesis, since the formation of β-alanine and of β-aminoisobutyrate is impaired. A nongenetic form can be triggered by administration of 5-fluorouracil to patients with low levels of dihydropyrimidine dehydrogenase.

PURINES & PYRIMIDINES ARE DIETARILY NONESSENTIAL

Normal human tissues can synthesize purines and pyrimidines from amphibolic intermediates in quantities and at times appropriate to meet variable physiologic demand. Ingested nucleic acids and nucleotides therefore are dietarily nonessential. Following their degradation in the intestinal tract, the resulting mononucleotides may be absorbed or converted to purine and pyrimidine bases. The purine bases are then oxidized to uric acid, which may be absorbed and excreted in the urine. While little or no dietary purine or pyrimidine is incorporated into tissue nucleic acids, injected compounds are incorporated. The incorporation of injected [³H]thymidine into newly synthesized DNA thus can be used to measure the rate of DNA synthesis.

BIOSYNTHESIS OF PURINE NUCLEOTIDES

With the exception of parasitic protozoa, all forms of life synthesize purine and pyrimidine nucleotides. Synthesis from amphibolic intermediates proceeds at controlled rates appropriate for all cellular functions. To achieve homeostasis, intracellular mechanisms sense and regulate the pool sizes of NTPs, which rise during growth or tissue regeneration when cells are rapidly dividing.

Purine and pyrimidine nucleotides are synthesized in vivo at rates consistent with physiologic need. Early investigations of nucleotide biosynthesis first employed birds, and later *Escherichia coli*. Isotopic precursors of uric acid fed to pigeons established the source of each atom of a purine (**Figure 33–1**) and initiated study of the intermediates of purine biosynthesis. Avian

tissues also served as a source of cloned genes that encode enzymes of purine biosynthesis and the regulatory proteins that control the rate of purine biosynthesis.

The three processes that contribute to purine nucleotide biosynthesis are, in order of decreasing importance.

1. Synthesis from amphibolic intermediates (synthesis de novo).
2. Phosphoribosylation of purines.
3. Phosphorylation of purine nucleosides.

INOSINE MONOPHOSPHATE (IMP) IS SYNTHESIZED FROM AMPHIBOLIC INTERMEDIATES

Figure 33–2 illustrates the intermediates and the 11 enzyme-catalyzed reactions that convert α-D-ribose 5-phosphate to inosine monophosphate (IMP). The first intermediate formed in the de novo pathway for purine biosynthesis is 5-phosphoribosyl 5-pyrophosphate (PRPP; structure II, Figure 33–2). PRPP is also an intermediate in the biosynthesis of pyrimidine nucleotides, NAD^+, and $NADP^+$. Separate branches then lead from IMP to AMP and GMP (**Figure 33–3**). Subsequent phosphoryl transfer from ATP converts AMP and GMP to ADP and GDP. Conversion of GDP to GTP involves a second phosphoryl transfer from ATP, whereas conversion of ADP to ATP is achieved primarily by oxidative phosphorylation (see Chapter 13).

Multifunctional Catalysts Participate in Purine Nucleotide Biosynthesis

In prokaryotes, each reaction of Figure 33–2 is catalyzed by a different polypeptide. By contrast, the enzymes of eukaryotes are polypeptides that possess multiple catalytic activities whose adjacent catalytic sites facilitate channeling of intermediates between sites. Three distinct multifunctional enzymes catalyze reactions ③, ④, and ⑥; reactions ⑦ and ⑧; and reactions ⑩ and ⑪ of Figure 33–2.

Antifolate Drugs and Glutamine Analogs Block Purine Nucleotide Biosynthesis

The carbons added in reactions ④ and ⑩ of Figure 33–2 are contributed by derivatives of tetrahydrofolate. Purine deficiency states, while rare in humans, generally reflect a deficiency of folic acid. Compounds that inhibit formation of tetrahydrofolates and therefore block purine synthesis have been used in cancer chemotherapy. Inhibitory compounds and the reactions they inhibit include **azaserine** (reaction ⑤, Figure 33–2), **diazanorleucine** (reaction ②, Figure 33–2),

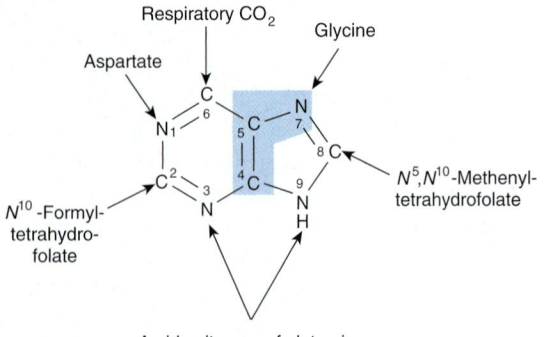

FIGURE 33–1   **Sources of the nitrogen and carbon atoms of the purine ring.** Atoms 4, 5, and 7 (**blue highlight**) derive from glycine.

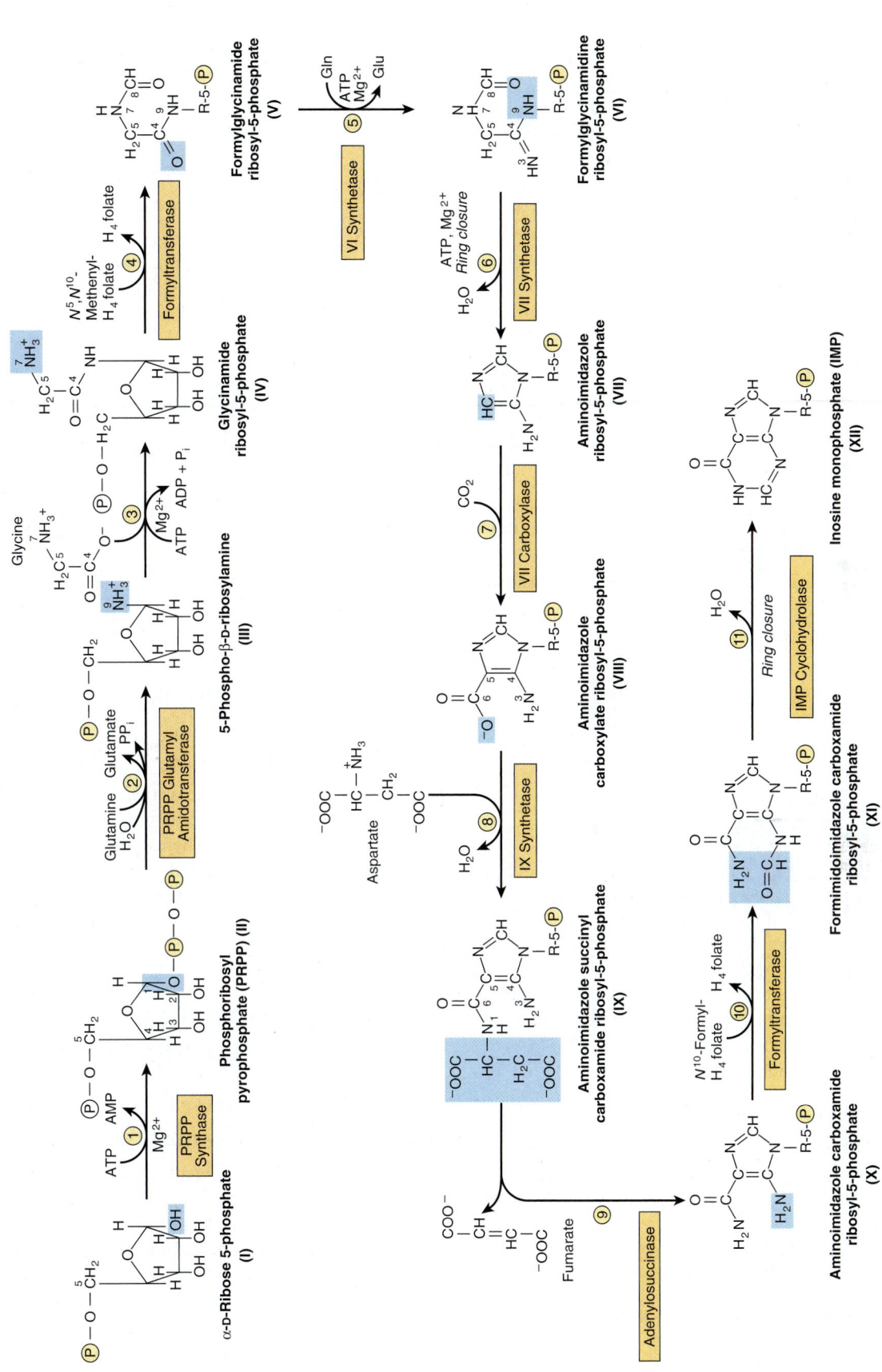

FIGURE 33-2 Purine biosynthesis from ribose 5-phosphate and ATP. See the text for explanations. (Ⓟ, PO_3^{2-} or PO_2^{-}.)

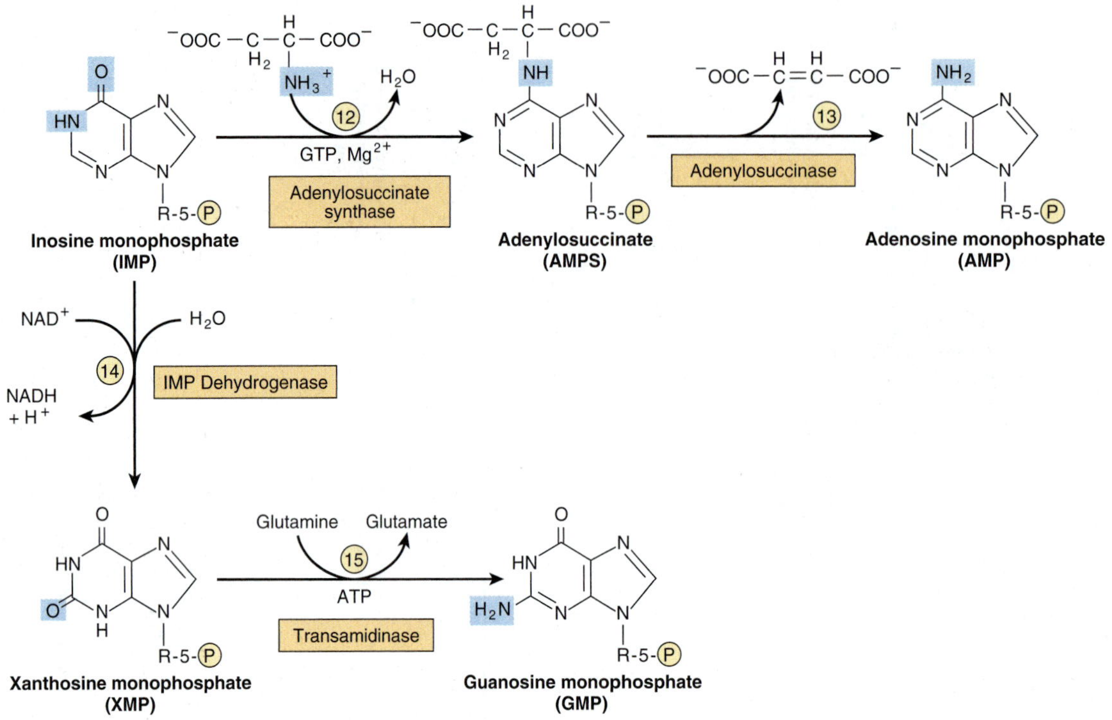

FIGURE 33–3   Conversion of IMP to AMP and GMP.

6-mercaptopurine (reactions ⑬ and ⑭, Figure 33–3), and **mycophenolic acid** (reaction ⑭, Figure 33–3).

"SALVAGE REACTIONS" CONVERT PURINES & THEIR NUCLEOSIDES TO MONONUCLEOTIDES

Conversion of purines, their ribonucleo**sides,** and their deoxyribonucleo**sides** to mononucleo**tides** involves "salvage reactions" that require far less energy than de novo synthesis. The more important mechanism involves phosphoribosylation by PRPP (structure II, Figure 33–2) of a free purine (Pu) to form a purine 5'-mononucleotide (Pu-RP).

$$Pu + PR\text{-}PP \rightarrow Pu\text{-}RP + PP_i$$

Phosphoryl transfer from ATP, catalyzed by adenosine- and hypoxanthine-phosphoribosyl transferases, converts adenine, hypoxanthine, and guanine to their mononucleotides (**Figure 33–4**).

A second salvage mechanism involves phosphoryl transfer from ATP to a purine ribonucleo**side** (Pu-R):

$$Pu\text{-}R + ATP \rightarrow PuR\text{-}P + ADP$$

Phosphorylation of the purine nucleotides, catalyzed by adenosine kinase, converts adenosine and deoxyadenosine to AMP and dAMP. Similarly, deoxycytidine kinase

phosphorylates deoxycytidine and 2'-deoxyguanosine, forming dCMP and dGMP.

Liver, the major site of purine nucleotide biosynthesis, provides purines and purine nucleosides for salvage and for utilization by tissues incapable of their biosynthesis. Human brain tissue has a low level of PRPP glutamyl amidotransferase (reaction ②, Figure 33–2) and hence depends in part on exogenous purines. Erythrocytes and polymorphonuclear leukocytes cannot synthesize 5-phosphoribosylamine (structure III, Figure 33–2) and therefore utilize exogenous purines to form nucleotides.

HEPATIC PURINE BIOSYNTHESIS IS STRINGENTLY REGULATED

AMP & GMP Feedback Regulate PRPP Glutamyl Amidotransferase

Biosynthesis of IMP is energetically expensive. In addition to ATP, glycine, glutamine, aspartate, and reduced tetrahydrofolate derivatives all are consumed. It thus is of survival advantage to closely regulate purine biosynthesis in response to varying physiological need. The overall determinant of the rate of de novo purine nucleotide biosynthesis is the concentration of PRPP. This depends on the rate of PRPP synthesis, utilization, degradation, and regulation. The rate of PRPP synthesis depends on the availability of ribose 5-phosphate and on the activity of PRPP synthase, (reaction ② **Figure 33–5**),

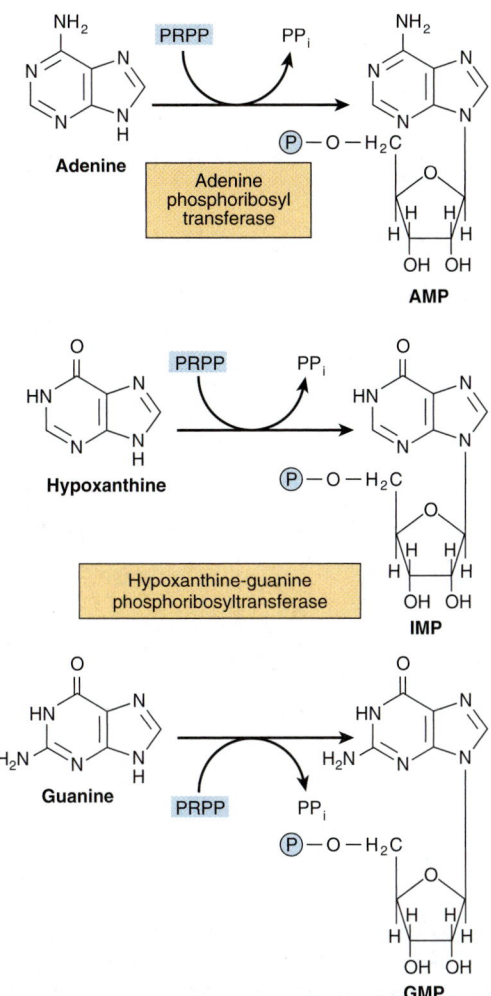

FIGURE 33–4 **Phosphoribosylation of adenine, hypoxanthine, and guanine to form AMP, IMP, and GMP, respectively.**

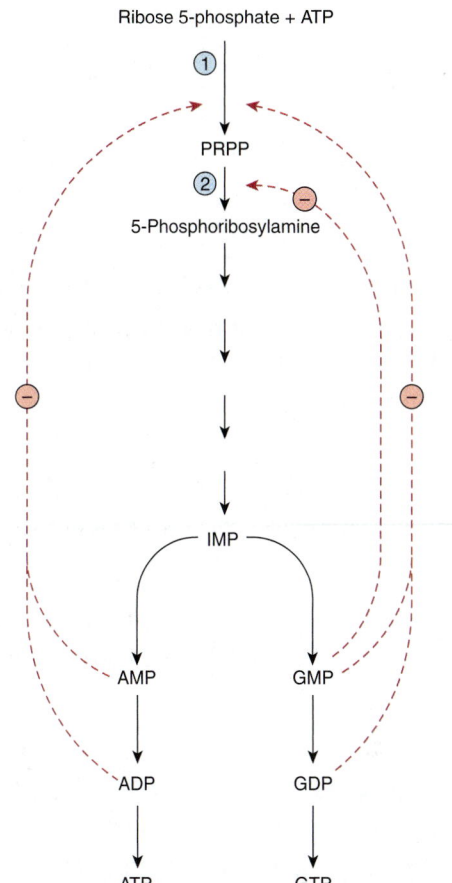

FIGURE 33–5 **Control of the rate of de novo purine nucleotide biosynthesis.** Reactions ① and ② are catalyzed by PRPP synthase and by PRPP glutamyl amidotransferase, respectively. Solid lines represent chemical flow. Broken red lines represent feedback inhibition by intermediates of the pathway.

an enzyme whose activity is feedback inhibited by AMP, ADP, GMP, and GDP. Elevated levels of these nucleoside phosphates thus signal a physiologically appropriate overall decrease in their biosynthesis.

AMP & GMP Feedback Regulate Their Formation from IMP

In addition to regulation at the level of PRPP biosynthesis, additional mechanisms regulate conversion of IMP to ATP and GTP. These are summarized in **Figure 33–6**. AMP feedback inhibits adenylosuccinate synthase (reaction ⑫, Figure 33–3), and GMP inhibits IMP dehydrogenase (reaction ⑭, Figure 33–3). Furthermore, conversion of IMP to adenylosuccinate en route to AMP (reaction ⑫, Figure 33–3) requires GTP, and conversion of xanthinylate (XMP) to GMP requires ATP. This cross-regulation between the pathways of IMP metabolism thus serves to balance the biosynthesis of

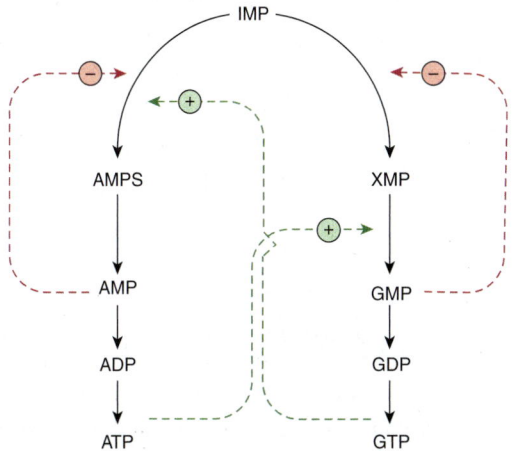

FIGURE 33–6 **Regulation of the conversion of IMP to adenosine nucleotides and guanosine nucleotides.** Solid lines represent chemical flow. Broken green lines represent positive feedback loops ⊕, and broken red lines represent negative feedback loops ⊖. Abbreviations include AMPS (adenylosuccinate) and XMP (xanthosine monophosphate), whose structures are given in Figure 33–3.

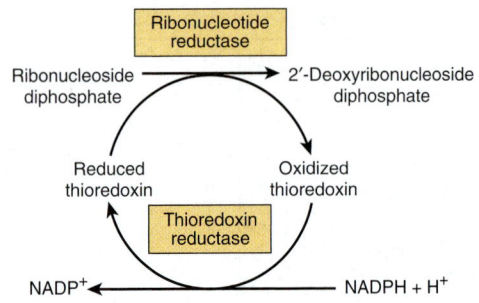

FIGURE 33–7 Reduction of ribonucleoside diphosphates to 2′-deoxyribonucleoside diphosphates.

purine nucleoside triphosphates by decreasing the synthesis of one purine nucleotide when there is a deficiency of the other nucleotide. AMP and GMP also inhibit hypoxanthine-guanine phosphoribosyltransferase, which converts hypoxanthine and guanine to IMP and GMP (Figure 33–4), and GMP feedback inhibits PRPP glutamyl amidotransferase (reaction ②, Figure 33–2).

REDUCTION OF RIBONUCLEOSIDE DIPHOSPHATES FORMS DEOXYRIBONUCLEOSIDE DIPHOSPHATES

Reduction of the 2′-hydroxyl of purine and pyrimidine ribonucleotides, catalyzed by the **ribonucleotide reductase complex** (**Figure 33–7**), provides the deoxyribonucleoside diphosphates (dNDPs) needed for both the synthesis and repair of DNA (see Chapter 35). The enzyme complex is functional only when cells are actively synthesizing DNA. Reduction requires thioredoxin, thioredoxin reductase, and NADPH. The immediate reductant, reduced thioredoxin, is produced by NADPH:thioredoxin reductase (Figure 33–7). The reduction of ribonucleoside diphosphates (NDPs) to dNDPs is subject to complex regulatory controls that achieve balanced production of dNTPs for synthesis of DNA (**Figure 33–8**).

BIOSYNTHESIS OF PYRIMIDINE NUCLEOTIDES

Figure 33–9 illustrates the intermediates and enzymes of pyrimidine nucleotide biosynthesis. The catalyst for the initial reaction is *cytosolic* carbamoyl phosphate synthase II, a different enzyme from the *mitochondrial* carbamoyl phosphate synthase I of urea synthesis (see Figure 28–13). Compartmentation thus provides an independent pool of carbamoyl phosphate for each process. PRPP, an early participant in purine nucleotide synthesis (Figure 33–2), is a much later participant

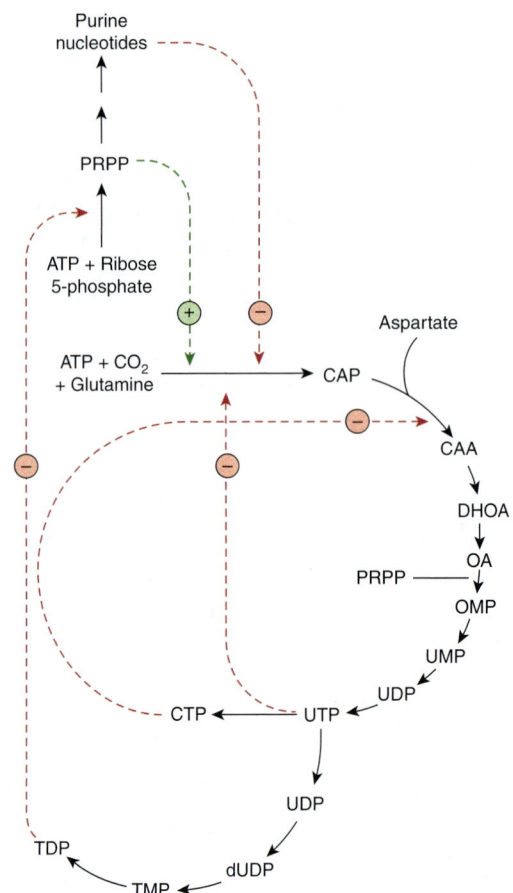

FIGURE 33–8 Regulatory aspects of the biosynthesis of purine and pyrimidine ribonucleotides and reduction to their respective 2′-deoxyribonucleotides. The broken green line represents a positive feedback loop. Broken red lines represent negative feedback loops. Abbreviations for the intermediates in the biosynthesis of pyrimidine nucleotides whose structures are given in Figure 33–9 are: (CAA, carbamoyl aspartate; DHOA, dihydroorotate; OA, orotic acid; OMP, orotidine monophosphate; and PRPP phosphoribosyl pyrophosphate).

in pyrimidine biosynthesis. Inspection of the reaction components in Figure 33–9 will reveal that, like the biosynthesis of pyrimidines, the biosynthesis of the purine nucleosides is energetically costly.

Multifunctional Proteins Catalyze the Early Reactions of Pyrimidine Biosynthesis

Five of the first six enzyme activities of pyrimidine biosynthesis reside on **multifunctional polypeptides.** One polypeptide catalyzes the first three reactions of Figure 33–9. A second bifunctional enzyme catalyzes reactions ⑤ and ⑥ of Figure 33–9. The close proximity of multiple active sites on a multifunctional polypeptide facilitates efficient channeling of the intermediates of pyrimidine biosynthesis.

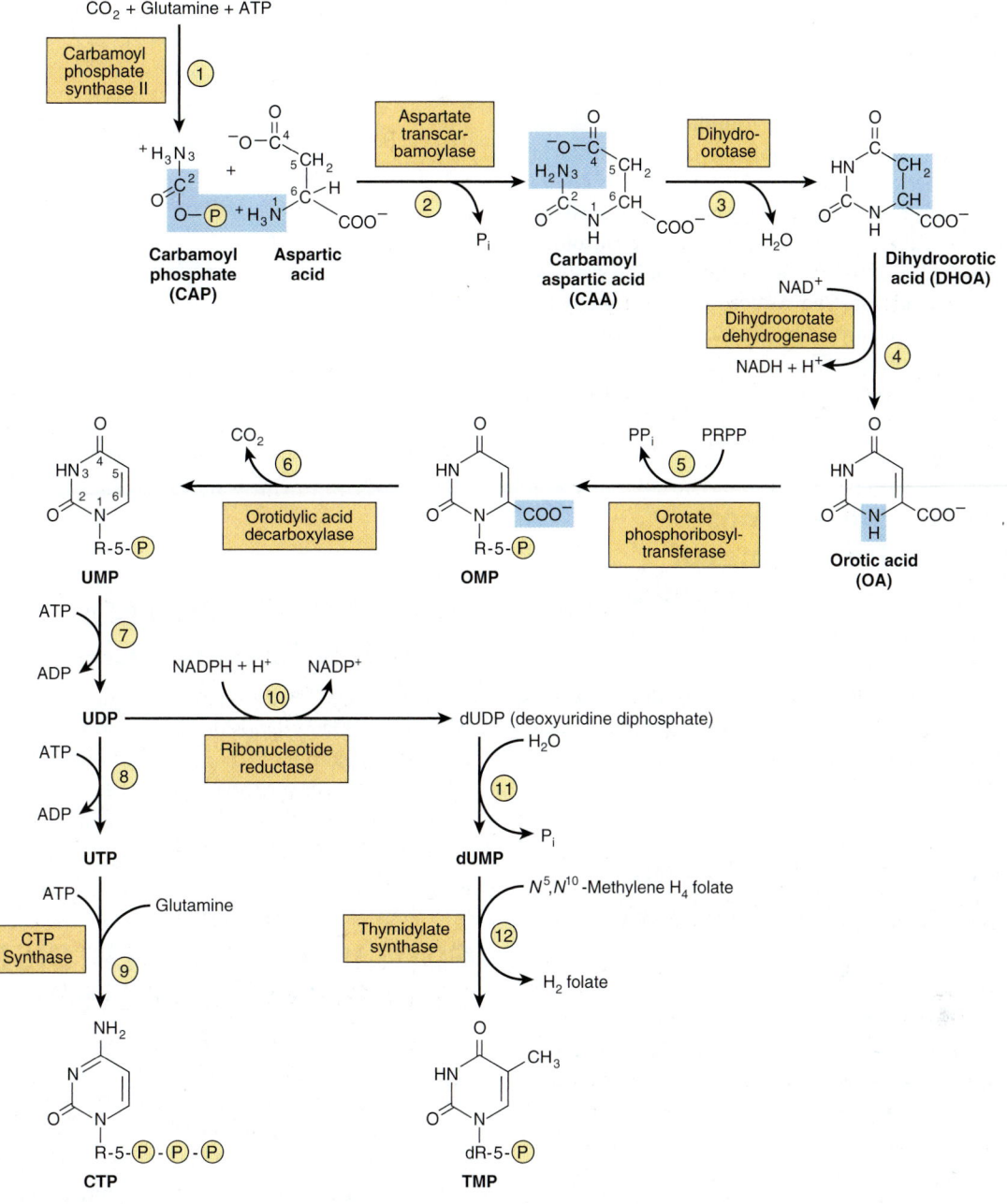

FIGURE 33–9 The biosynthetic pathway for pyrimidine nucleotides.

THE DEOXYRIBONUCLEOSIDES OF URACIL & CYTOSINE ARE SALVAGED

Adenine, guanine and hypoxanthine released during the turnover of nucleic acids, especially messenger RNAs, are reconverted to nucleoside triphosphates via so-called salvage pathways. While mammalian cells reutilize few *free* pyrimidines, "salvage reactions" convert the pyrimidine ribonucleosides uridine and cytidine and the pyrimidine deoxyribonucleosides thymidine and deoxycytidine to their respective nucleotides.

$$Guanine + PRPP \rightarrow GMP + PP_i$$

$$Hypoxanthine + PRPP \rightarrow IMP + PP_i$$

Phosphoryltransferases (kinases) catalyze transfer of the γ-phosphoryl group of ATP to the diphosphates of 2′-deoxycytidine, 2′-deoxyguanosine, and 2′-deoxyadenosine, converting them to the corresponding nucleoside triphosphates.

$$NDP + ATP \rightarrow NTP + ADP$$

$$dNDP + ATP \rightarrow dNTP + ADP$$

Methotrexate Blocks Reduction of Dihydrofolate

The reaction catalyzed by thymidylate synthase (reaction ⑫ of Figure 33–9) is the only reaction of pyrimidine nucleotide biosynthesis that requires a tetrahydrofolate derivative. During this reaction the methylene group of N^5,N^{10}-methylene-tetrahydrofolate is reduced to the methyl group that is transferred to the 5-position of the pyrimidine ring, and tetrahydrofolate is oxidized to dihydrofolate. For further pyrimidine synthesis to occur, dihydrofolate must be reduced back to tetrahydrofolate. This reduction, catalyzed by dihydrofolate reductase, is inhibited by **methotrexate.** Dividing cells, which must generate TMP and dihydrofolate, thus are especially sensitive to inhibitors of dihydrofolate reductase such as the anticancer drug methotrexate.

Certain Pyrimidine Analogs Are Substrates for Enzymes of Pyrimidine Nucleotide Biosynthesis

Allopurinol and the anticancer drug **5-fluorouracil** (see Figure 32–13) are alternate substrates for orotate phosphoribosyltransferase (reaction ⑤, Figure 33–9). Both drugs are phosphoribosylated, and allopurinol is converted to a nucleotide in which the ribosyl phosphate is attached to N–1 of the pyrimidine ring.

REGULATION OF PYRIMIDINE NUCLEOTIDE BIOSYNTHESIS

Gene Expression & Enzyme Activity Both Are Regulated

The activities of the first and second enzymes of pyrimidine nucleotide biosynthesis are controlled by allosteric regulation. Carbamoyl phosphate synthase II (reaction ①, Figure 33–9) is inhibited by UTP and purine nucleotides but activated by PRPP. Aspartate transcarbamoylase (reaction ②, Figure 33–9) is inhibited by CTP but activated by ATP (**Figure 33–10**). In addition, the first three and the last two enzymes of the pathway are regulated by coordinate repression and derepression.

Purine & Pyrimidine Nucleotide Biosynthesis Are Coordinately Regulated

Purine and pyrimidine biosynthesis parallel one another quantitatively, that is, mole for mole, suggesting coordinated control of their biosynthesis. Several sites of *cross-regulation* characterize the pathways that lead to the biosynthesis of purine and pyrimidine nucleotides. PRPP synthase (reaction ①, Figure 33–2), which forms a precursor essential for both processes, is feedback inhibited by both purine and pyrimidine nucleotides.

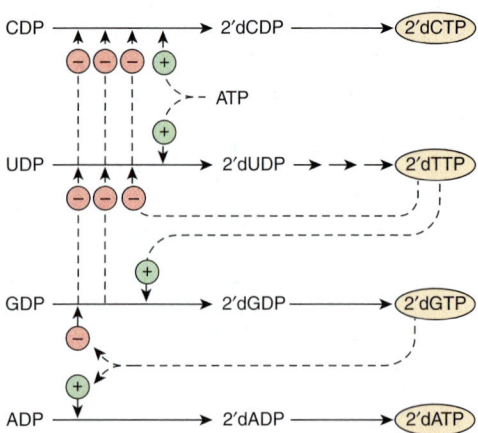

FIGURE 33–10   Control of pyrimidine nucleotide biosynthesis. Solid lines represent chemical flow. Broken green lines represent positive ⊕, and broken red lines negative ⊖ feedback regulation.

HUMANS CATABOLIZE PURINES TO URIC ACID

Humans convert adenosine and guanosine to uric acid (**Figure 33–11**). Adenosine is first converted to inosine by adenosine deaminase. In mammals other than higher primates, uricase converts uric acid to the water-soluble product allantoin. However, since humans lack uricase, the end product of purine catabolism in humans is uric acid.

GOUT IS A METABOLIC DISORDER OF PURINE CATABOLISM

Various genetic defects in PRPP synthase (reaction ①, Figure 33–2) present clinically as gout. Each defect—for example, an elevated V_{max}, increased affinity for ribose 5-phosphate, or resistance to feedback inhibition—results in overproduction and overexcretion of purine catabolites. When serum urate levels exceed the solubility limit, sodium urate crystalizes in soft tissues and joints and causes an inflammatory reaction, **gouty arthritis.** However, most cases of gout reflect abnormalities in renal handling of uric acid.

OTHER DISORDERS OF PURINE CATABOLISM

While purine deficiency states are rare in human subjects, there are numerous genetic disorders of purine catabolism. **Hyperuricemias** may be differentiated based on whether patients excrete normal or excessive quantities of total urates. Some hyperuricemias reflect specific enzyme defects. Others are secondary to diseases such as cancer or psoriasis that enhance tissue turnover.

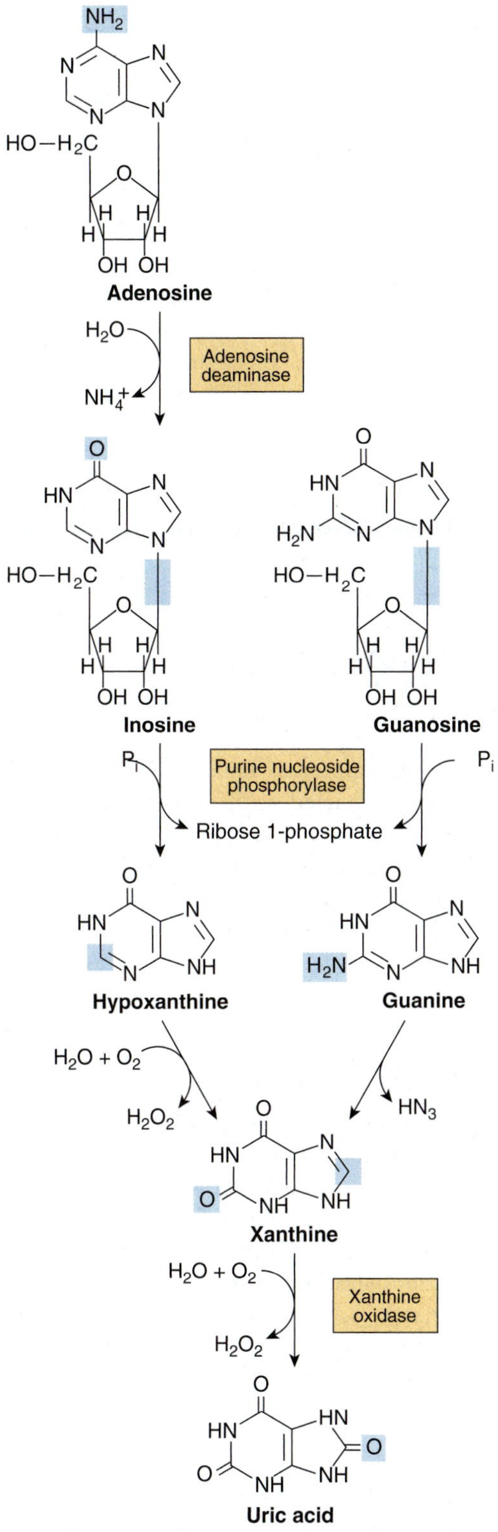

FIGURE 33–11 **Formation of uric acid from purine nucleosides by way of the purine bases hypoxanthine, xanthine, and guanine.** Purine deoxyribonucleosides are degraded by the same catabolic pathway and enzymes, all of which exist in the mucosa of the mammalian gastrointestinal tract.

Lesch–Nyhan Syndrome

The Lesch–Nyhan syndrome, an overproduction hyperuricemia characterized by frequent episodes of uric acid lithiasis and a bizarre syndrome of self-mutilation, reflects a defect in **hypoxanthine-guanine phosphoribosyl transferase,** an enzyme of purine salvage (Figure 33–4). The accompanying rise in intracellular PRPP results in purine overproduction. Mutations that decrease or abolish hypoxanthine-guanine phosphoribosyltransferase activity include deletions, frame-shift mutations, base substitutions, and aberrant mRNA splicing.

von Gierke Disease

Purine overproduction and hyperuricemia in von Gierke disease (**glucose-6-phosphatase deficiency**) occurs secondary to enhanced generation of the PRPP precursor ribose 5-phosphate. An associated lactic acidosis elevates the renal threshold for urate, elevating total body urates.

Hypouricemia

Hypouricemia and increased excretion of hypoxanthine and xanthine are associated with **xanthine oxidase deficiency** (Figure 33–11) due to a genetic defect or to severe liver damage. Patients with a severe enzyme deficiency may exhibit xanthinuria and xanthine lithiasis.

Adenosine Deaminase & Purine Nucleoside Phosphorylase Deficiency

Adenosine deaminase deficiency (Figure 33–11) is associated with an immunodeficiency disease in which both thymus-derived lymphocytes (T cells) and bone marrow-derived lymphocytes (B cells) are sparse and dysfunctional. Patients suffer from severe immunodeficiency. In the absence of enzyme replacement or bone marrow transplantation, infants often succumb to fatal infections. **Purine nucleoside phosphorylase deficiency** is associated with a severe deficiency of T cells but apparently normal B cell function. Immune dysfunctions appear to result from accumulation of dGTP and dATP, which inhibit ribonucleotide reductase and thereby deplete cells of DNA precursors. **Table 33–1** summarizes known disorders of purine metabolism.

CATABOLISM OF PYRIMIDINES PRODUCES WATER-SOLUBLE METABOLITES

Unlike the low solubility products of purine catabolism, catabolism of the pyrimidines forms highly water-soluble products—CO_2, NH_3, β-alanine, and β-aminoisobutyrate (**Figure 33–12**). Humans transaminate β-aminoisobutyrate to methylmalonate semialdehyde, which then forms succinyl-CoA (see Figure 20–2). Excretion of β-aminoisobutyrate increases in leukemia and severe x-ray radiation exposure due to increased destruction of DNA. However, many persons of Chinese or

TABLE 33–1   Metabolic Disorders of Purine and Pyrimidine Metabolism

Defective Enzyme	Enzyme Catalog Number	OMIM Reference	Major Signs and Symptoms	Figure and Reaction
Purine Metabolism				
Hypoxanthine-guanine phosphoribosyl transferase	2.4.2.8	308000	Lesch–Nyhan syndrome. Uricemia, self-mutilation	33–4 ②
PRPP synthase	2.7.6.1	311860	Gout; gouty arthritis	33–2 ①
Adenosine deaminase	3.5.4.6	102700	Severely compromised immune system	33–1 ①
Purine nucleoside phosphorylase	2.4.2.1	164050	Autoimmune disorders; benign and opportunistic infections	33–11 ②
Pyrimidine Metabolism				
Dihydropyrimidine dehydrogenase	1.3.1.2	274270	Can develop toxicity to 5-fluorouracil, also a substrate for this dehydrogenase	33–12 ②
Orotate phosphoribosyl transferase and orotidylic acid decarboxylase	2.4.2.10 and 4.1.1.23	258900	Orotic acid aciduria type 1; megaloblastic anemia	33–9 ⑤ and ⑥
Orotidylic acid decarboxylase	4.1.1.23	258920	Orotic aciduria Type 2	33–9 ⑥

Japanese ancestry routinely excrete β-aminoisobutyrate. Disorders of β-alanine and β-aminoisobutyrate metabolism arise from defects in enzymes of pyrimidine catabolism. These include **β-hydroxybutyric aciduria,** a disorder due to total or partial deficiency of the enzyme **dihydropyrimidine dehydrogenase** (Figure 33–12). The genetic disease reflects an absence of the enzyme. A disorder of pyrimidine catabolism, known also as combined uraciluria-thyminuria, it is also a disorder of β-amino acid metabolism, since the *formation* of β-alanine and of β-aminoisobutyrate is impaired. When due to an inborn error, there are serious neurological complications. A nongenetic form is triggered by the administration of the anticancer drug 5-fluorouracil (see Figure 32–13) to patients with low levels of dihydropyrimidine dehydrogenase.

Pseudouridine Is Excreted Unchanged

No human enzyme catalyzes hydrolysis or phosphorolysis of the pseudouridine (ψ) derived from the degradation of RNA molecules. This unusual nucleotide therefore is excreted unchanged in the urine of normal subjects. Pseudouridine was indeed first isolated from human urine (**Figure 33–13**).

OVERPRODUCTION OF PYRIMIDINE CATABOLITES IS ONLY RARELY ASSOCIATED WITH CLINICALLY SIGNIFICANT ABNORMALITIES

Since the end products of pyrimidine catabolism are highly water soluble, pyrimidine overproduction results in few

clinical signs or symptoms. Table 33–1 lists exceptions. In hyperuricemia associated with severe overproduction of PRPP, there is overproduction of pyrimidine nucleotides and increased excretion of β-alanine. Since N^5,N^{10}-methylene-tetrahydrofolate is required for thymidylate synthesis, disorders of folate and vitamin B_{12} metabolism result in deficiencies of TMP.

Orotic Aciduria

The orotic aciduria that accompanies the **Reye syndrome** probably is a consequence of the inability of severely damaged mitochondria to utilize carbamoyl phosphate, which then becomes available for cytosolic overproduction of orotic acid. **Type-I orotic aciduria** reflects a deficiency of both orotate phosphoribosyltransferase and orotidylate decarboxylase (reactions ⑤ and ⑥, Figure 33–9); the rarer **Type-II orotic aciduria** is due to a deficiency only of orotidylate decarboxylase (reaction ⑥, Figure 33–9).

Deficiency of a Urea Cycle Enzyme Results in Excretion of Pyrimidine Precursors

Increased excretion of orotic acid, uracil, and uridine accompanies a deficiency in liver mitochondrial ornithine transcarbamoylase (see reaction ②, Figure 28–13). Excess carbamoyl phosphate exits to the cytosol, where it stimulates pyrimidine nucleotide biosynthesis. The resulting mild **orotic aciduria** is increased by high-nitrogen foods.

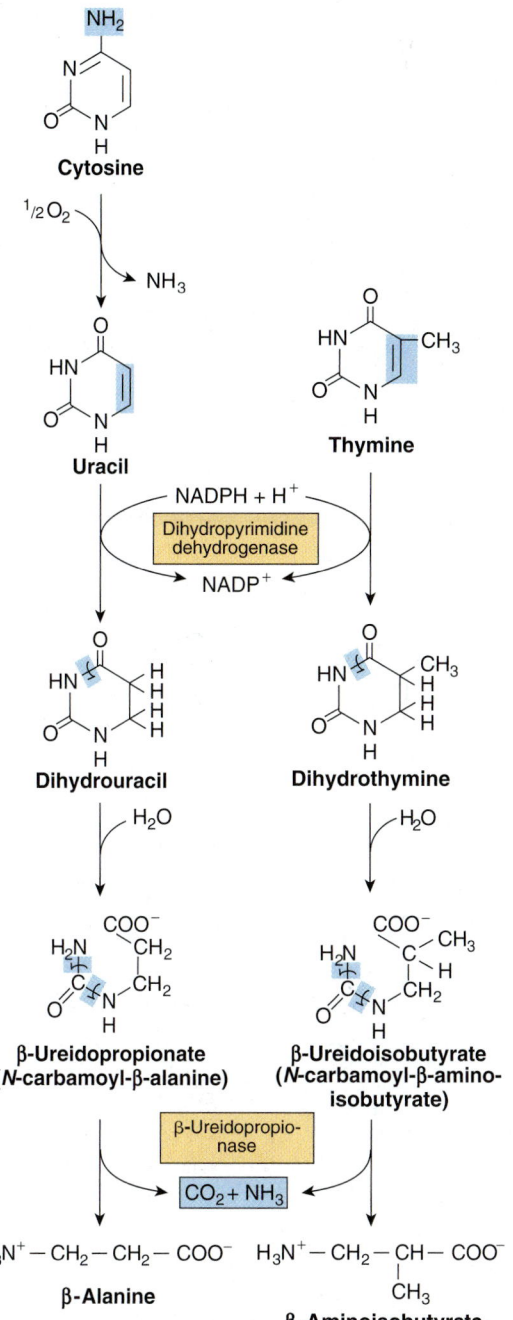

FIGURE 33–12 Catabolism of pyrimidines. Hepatic β-ureidopropionase catalyzes the formation of both β-alanine and β-aminoisobutyrate from their pyrimidine precursors.

Drugs May Precipitate Orotic Aciduria

Allopurinol (see Figure 32–13), an alternative substrate for orotate phosphoribosyltransferase (reaction ⑤, Figure 33–9), competes with orotic acid. The resulting nucleotide product also inhibits orotidylate decarboxylase (reaction ⑥, Figure 33–9), resulting in **orotic aciduria** and **orotidinuria**. 6-Azauridine, following conversion to 6-azauridylate, also

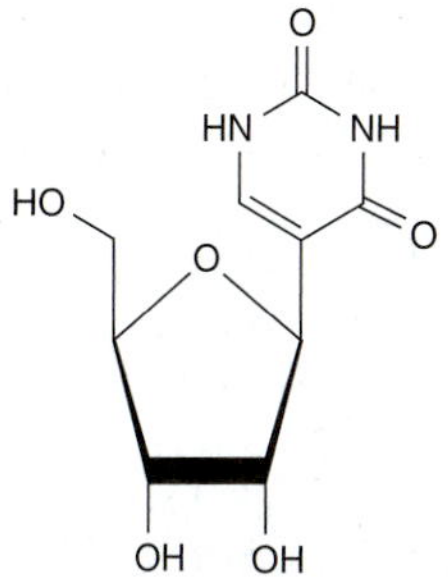

FIGURE 33–13 Pseudouridine, in which ribose is linked to C5 of uridine.

competitively inhibits orotidylate decarboxylase (reaction ⑥, Figure 33–9), enhancing excretion of orotic acid and orotidine. Four genes that encode urate transporters have been identified. Two of the encoded proteins are localized to the apical membrane of proximal tubular cells.

SUMMARY

■ Ingested nucleic acids are degraded to purines and pyrimidines. New purines and pyrimidines are formed from amphibolic intermediates and thus are dietarily nonessential.

■ Several reactions of IMP biosynthesis require folate derivatives and glutamine. Consequently, antifolate drugs and glutamine analogs inhibit purine biosynthesis.

■ Oxidation and amination of IMP forms AMP and GMP, and subsequent phosphoryl transfer from ATP forms ADP and GDP. Further phosphoryl transfer from ATP to GDP forms GTP. ADP is converted to ATP by oxidative phosphorylation. Reduction of NDPs forms dNDPs.

■ Hepatic purine nucleotide biosynthesis is stringently regulated by the pool size of PRPP and by feedback inhibition of PRPP glutamyl amidotransferase by AMP and GMP.

■ Coordinated regulation of purine and pyrimidine nucleotide biosynthesis ensures their presence in proportions appropriate for nucleic acid biosynthesis and other metabolic needs.

■ Humans catabolize purines to uric acid (pK_a 5.8), present as the relatively insoluble acid at acidic pH or as its more soluble sodium urate salt at a pH near neutrality. Urate crystals are diagnostic of gout. Other disorders of purine catabolism include Lesch–Nyhan syndrome, von Gierke disease, and hypouricemias.

■ Since pyrimidine catabolites are water-soluble, their overproduction does not result in clinical abnormalities. Excretion of pyrimidine precursors can, however, result from a deficiency of ornithine transcarbamoylase because excess carbamoyl phosphate is available for pyrimidine biosynthesis.

REFERENCES

Chow EL, Cherry JD, Harrison R, et al: Reassessing Reye syndrome. Arch Pediatr Adolesc Med 2003;157:1241.

Christopherson RI, Lyons SD, Wilson PK: Inhibitors of de novo nucleotide biosynthesis as drugs. Acc Chem Res 2002;35:961.

Kamal MA, Christopherson RI: Accumulation of 5-phosphoribosyl-1-pyrophosphate in human CCRF-CEM leukemia cells treated with antifolates. Int J Biochem Cell Biol 2004;36:957.

Lipkowitz MS, Leal-Pinto E, Rappoport JZ, et al: Functional reconstitution, membrane targeting, genomic structure, and chromosomal localization of a human urate transporter. J Clin Invest 2001;107:1103.

Martinez J, Dugaiczyk LJ, Zielinski R, et al: Human genetic disorders, a phylogenetic perspective. J Mol Biol 2001; 308:587.

Moyer RA, John DS: Acute gout precipitated by total parenteral nutrition. J Rheumatol 2003;30:849.

Nofech-Mozes Y, Blaser SI, Kobayashi J, et al: Neurologic abnormalities in patients with adenosine deaminase deficiency. Pediatr Neurol 2007;37:218.

Rafey MA, Lipkowitz MS, Leal-Pinto E, et al: Uric acid transport. Curr Opin Nephrol Hypertens 2003;12:511.

Scriver CR, Sly WS, Childs B, et al (editors): *The Metabolic and Molecular Bases of Inherited Disease,* 8th ed. McGraw-Hill, 2001.

Torres RJ, Puig JG: Hypoxanthine-guanine phosophoribosyltransferase (HPRT) deficiency: Lesch–Nyhan syndrome. Orphanet J Rare Dis 2007;2:48.

Wu VC, Huang JW, Hsueh PR, et al: Renal hypouricemia is an ominous sign in patients with severe acute respiratory syndrome. Am J Kidney Dis 2005;45:88.

Nucleic Acid Structure & Function

P. Anthony Weil, PhD

OBJECTIVES

After studying this chapter, you should be able to:

- Understand the chemical monomeric and polymeric structure of the genetic material, deoxyribonucleic acid, or DNA, which is found within the nucleus of eukaryotic cells.

- Explain why genomic nuclear eukaryotic DNA is double stranded and highly negatively charged.

- Understand the outline of how the genetic information of DNA can be faithfully duplicated.

- Understand how the genetic information of DNA is transcribed, or copied into myriad, distinct forms of ribonucleic acid (RNA).

- Appreciate that one form of information-rich RNA, the so-called messenger RNA (mRNA), can be subsequently translated into proteins, the molecules that form the structures, shapes, and ultimately functions of individual cells, tissues, and organs.

BIOMEDICAL IMPORTANCE

The discovery that genetic information is coded along the length of a polymeric molecule composed of only four types of monomeric units was one of the major scientific achievements of the 20th century. This polymeric molecule, **deoxyribonucleic acid (DNA),** is the chemical basis of heredity and is organized into genes, the fundamental units of genetic information. The basic information pathway—that is, DNA, which directs the synthesis of RNA, which in turn both directs and regulates protein synthesis—has been elucidated. Genes do not function autonomously; their replication and function are controlled by various gene products, often in collaboration with components of various signal transduction pathways. Knowledge of the structure and function of nucleic acids is essential in understanding genetics and many aspects of pathophysiology as well as the genetic basis of disease.

DNA CONTAINS THE GENETIC INFORMATION

The demonstration that DNA contained the genetic information was first made in 1944 in a series of experiments by Avery, MacLeod, and McCarty. They showed that the genetic determination of the character (type) of the capsule of a specific pneumococcus could be transmitted to another of a different capsular type by introducing purified DNA from the former coccus into the latter. These authors referred to the agent (later shown to be DNA) accomplishing the change as "transforming factor." Subsequently, this type of genetic manipulation has become commonplace. Similar experiments have recently been performed utilizing yeast, cultured plant and mammalian cells, and insect and mammalian embryos as recipients and molecularly cloned DNA as the donor of genetic information.

DNA Contains Four Deoxynucleotides

The chemical nature of the monomeric deoxynucleotide units of DNA—**deoxyadenylate, deoxyguanylate, deoxycytidylate, and thymidylate**—is described in Chapter 32. These monomeric units of DNA are held in polymeric form by 3',5'-phosphodiester bonds constituting a single strand, as depicted in **Figure 34–1**. The informational content of DNA (the genetic code) resides in the sequence in which these monomers—purine and pyrimidine deoxyribonucleotides—are ordered. The polymer as depicted possesses a polarity; one end has a 5'-hydroxyl or phosphate terminal while the other has a 3'-phosphate or hydroxyl terminal. The importance of this polarity will become evident. Since the genetic information resides in the

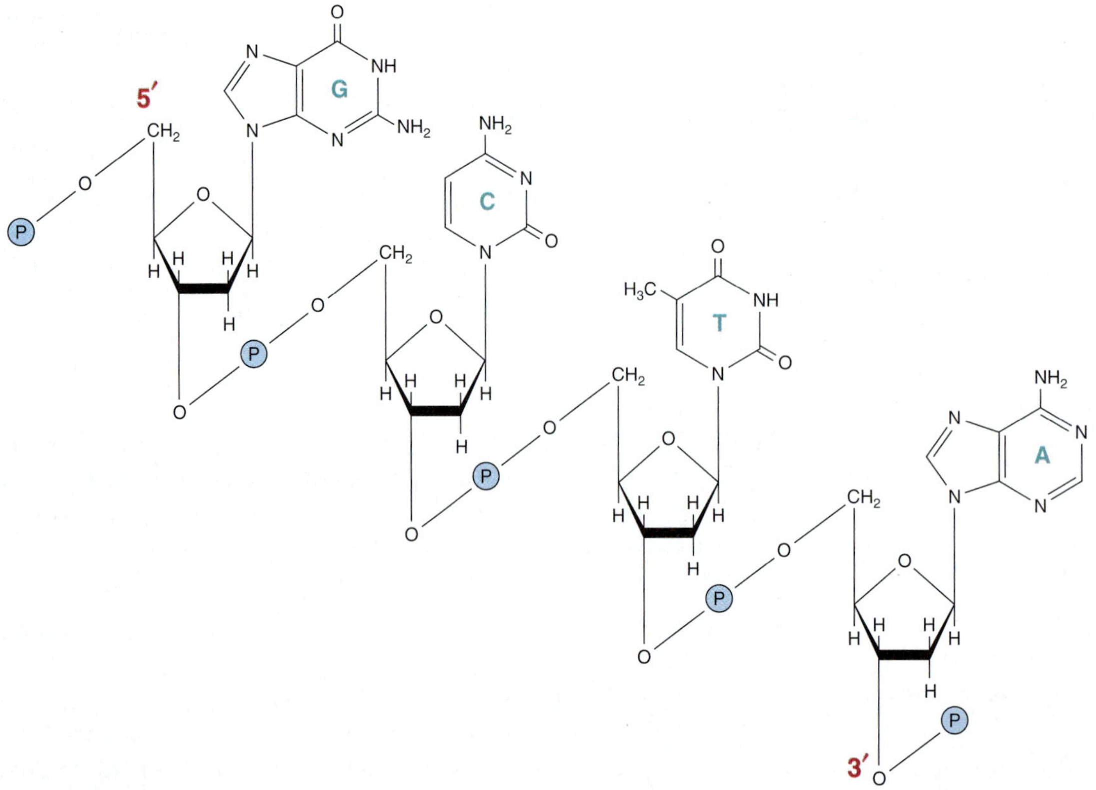

FIGURE 34–1 A segment of one strand of a DNA molecule in which the purine and pyrimidine bases guanine (G), cytosine (C), thymine (T), and adenine (A) are held together by a phosphodiester backbone between 2′-deoxyribosyl moieties attached to the nucleobases by an *N*-glycosidic bond. Note that the backbone has a polarity (ie, a direction). Convention dictates that a single-stranded DNA sequence is written in the 5′ to 3′ direction (ie, pGpCpTpA, where G, C, T, and A represent the four bases and p represents the interconnecting phosphates).

order of the monomeric units within the polymers, there must exist a mechanism of reproducing or replicating this specific information with a high degree of fidelity. That requirement, together with X-ray diffraction data from the DNA molecule and the observation of Chargaff that in DNA molecules the concentration of deoxyadenosine (A) nucleotides equals that of thymidine (T) nucleotides (A = T), while the concentration of deoxyguanosine (G) nucleotides equals that of deoxycytidine (C) nucleotides (G = C), led Watson, Crick, and Wilkins to propose in the early 1950s a model of a double-stranded DNA molecule. The model they proposed is depicted in **Figure 34–2**. The two strands of this double-stranded helix are held in register by both **hydrogen bonds** between the purine and pyrimidine bases of the respective linear molecules and by **van der Waals** and **hydrophobic interactions** between the stacked adjacent base pairs. The pairings between the purine and pyrimidine nucleotides on the opposite strands are very specific and are dependent upon hydrogen bonding of **A with T** and **G with C** (Figure 34–2).

This common form of DNA is said to be right-handed because as one looks down the double helix, the base residues form a spiral in a clockwise direction. In the double-stranded molecule, restrictions imposed by the rotation about the phosphodiester bond, the favored anticonfiguration of

the glycosidic bond (Figure 32–5), and the predominant tautomers (see Figure 32–2) of the four bases (A, G, T, and C) allow A to pair only with T and G only with C, as depicted in **Figure 34–3**. This base-pairing restriction explains the earlier observation that in a double-stranded DNA molecule the content of A equals that of T and the content of G equals that of C. The two strands of the double-helical molecule, each of which possesses a polarity, are **antiparallel;** that is, one strand runs in the 5′ to 3′ direction and the other in the 3′ to 5′ direction. In the double-stranded DNA molecules, the genetic information resides in the sequence of nucleotides on one strand, the **template strand.** This is the strand of DNA that is copied during **ribonucleic acid (RNA)** synthesis. It is sometimes referred to as the **noncoding strand.** The opposite strand is considered the **coding strand** because it matches the sequence of the RNA transcript (but containing uracil in place of thymine; see Figure 34–8) that encodes the protein.

The two strands, in which opposing bases are held together by interstrand hydrogen bonds, wind around a central axis in the form of a **double helix.** In the test tube, double-stranded DNA can exist in at least six forms (A–E and Z). The B form is usually found under physiologic conditions (low salt, high degree of hydration). A single turn of B-DNA about the long axis of the molecule contains 10 bp.

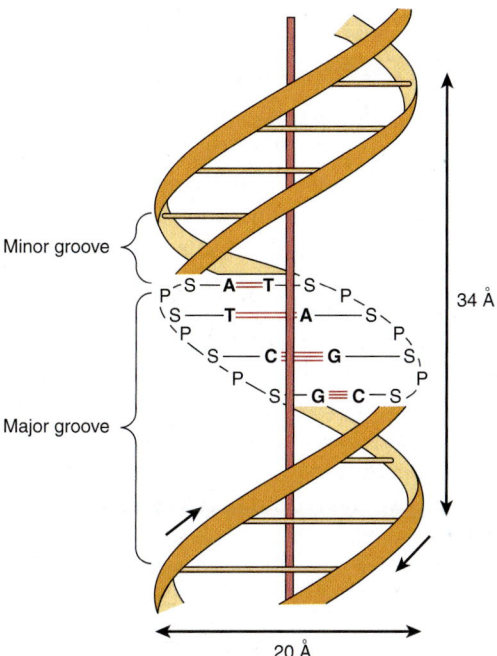

FIGURE 34–2 **A diagrammatic representation of the Watson and Crick model of the double-helical structure of the B form of DNA.** The horizontal arrow indicates the width of the double helix (20 Å), and the vertical arrow indicates the distance spanned by one complete turn of the double helix (34 Å). One turn of B-DNA includes 10 base pairs (bp), so the rise is 3.4 Å per bp. The central axis of the double helix is indicated by the vertical rod. The short arrows designate the polarity of the antiparallel strands. The major and minor grooves are depicted. (A, adenine; C, cytosine; G, guanine; P, phosphate; S, sugar [deoxyribose]; T, thymine.) Hydrogen bonds between A/T and G/C bases indicated by short, red, horizontal lines.

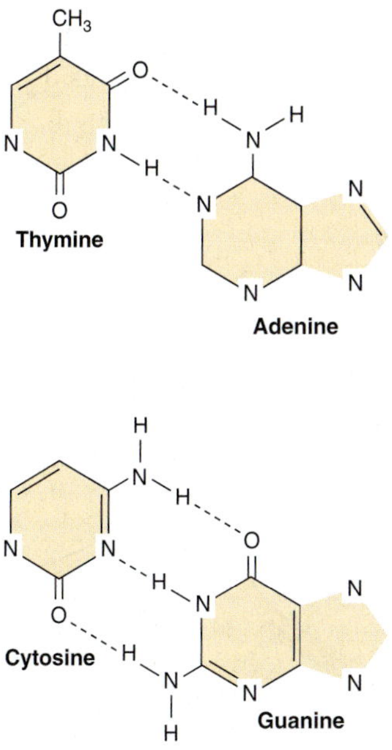

FIGURE 34–3 **DNA base pairing between adenine and thymine involves the formation of two hydrogen bonds.** Three such bonds form between cytidine and guanine. The broken lines represent hydrogen bonds.

The distance spanned by one turn of B-DNA is 3.4 nm (34 Å). The width (helical diameter) of the double helix in B-DNA is 2 nm (20 Å).

As depicted in Figure 34–3, three hydrogen bonds, formed by hydrogen bonded to electronegative N or O atoms, hold the deoxyguanosine nucleotide to the deoxycytidine nucleotide, whereas the other pair, the A–T pair, is held together by two hydrogen bonds. Thus, the G–C bonds are more resistant to denaturation, or strand separation, termed "melting," than A–T-rich regions of DNA.

The Denaturation of DNA Is Used to Analyze Its Structure

The double-stranded structure of DNA can be separated into two component strands in solution by increasing the temperature or decreasing the salt concentration. Not only do the two stacks of bases pull apart, but the bases themselves unstack while still connected in the polymer by the phosphodiester backbone. Concomitant with this denaturation of the DNA molecule is an increase in the optical absorbance of the purine and pyrimidine bases—a phenomenon referred to as **hyperchromicity** of denaturation. Because of the stacking of the bases and the hydrogen bonding between the stacks, the double-stranded DNA molecule exhibits properties of a rigid rod and in solution is a viscous material that loses its viscosity upon denaturation.

The strands of a given molecule of DNA separate over a temperature range. The midpoint is called the **melting temperature, or T_m**. The T_m is influenced by the base composition of the DNA and by the salt concentration of the solution. DNA rich in G–C pairs, which have three hydrogen bonds, melts at a higher temperature than that rich in A–T pairs, which have two hydrogen bonds. A 10-fold increase of monovalent cation concentration increases the T_m by 16.6°C. The organic solvent formamide, which is commonly used in recombinant DNA experiments, destabilizes hydrogen bonding between bases, thereby lowering the T_m. Formamide addition allows the strands of DNA or DNA–RNA hybrids to be separated at much lower temperatures and minimizes the phosphodiester bond breakage that can occur at higher temperatures.

Renaturation of DNA Requires Base Pair Matching

Importantly, separated strands of DNA will renature or reassociate when appropriate physiologic temperature and salt conditions are achieved; this reannealing process is often referred to as **hybridization**. The rate of reassociation depends upon the concentration of the complementary strands. Reassociation of the two complementary DNA strands of a chromosome after transcription is a physiologic example of renaturation (see below). At a given temperature and salt concentration, a particular nucleic acid strand will associate tightly only with a complementary strand. Hybrid molecules will also form under appropriate conditions. For example, DNA will form a hybrid with a complementary DNA (cDNA) or with a cognate messenger RNA (mRNA; see below). When hybridization is combined with gel electrophoresis techniques that separate nucleic acids by size coupled with radioactive or fluorescent probe labeling to provide a detectable signal, the resulting analytic techniques are called **Southern (DNA/DNA)** and **Northern (RNA–DNA) blotting,** respectively. These procedures allow for very distinct, high-sensitivity identification of specific nucleic acid species from complex mixtures of DNA or RNA (see Chapter 39).

There Are Grooves in the DNA Molecule

Examination of the model depicted in Figure 34–2 reveals a **major groove** and a **minor groove** winding along the molecule parallel to the phosphodiester backbones. In these grooves, proteins can interact specifically with exposed atoms of the nucleotides (via specific hydrophobic and ionic interactions) thereby recognizing and binding to specific nucleotide sequences as well as the unique shapes formed therefrom. Binding usually occurs without disrupting the base pairing of the double-helical DNA molecule. As discussed in Chapters 36 and 38, regulatory proteins control the expression of specific genes via such interactions.

DNA Exists in Relaxed & Supercoiled Forms

In some organisms such as bacteria, bacteriophages, many DNA-containing animal viruses, as well as organelles such as mitochondria (see Figure 35–8), the ends of the DNA molecules are joined to create a closed circle with no covalently free ends. This of course does not destroy the polarity of the molecules, but it eliminates all free 3′ and 5′ hydroxyl and phosphoryl groups. Closed circles exist in relaxed or supercoiled forms. Supercoils are introduced when a closed circle is twisted around its own axis or when a linear piece of duplex DNA, whose ends are fixed, is twisted. This energy-requiring process puts the molecule under torsional stress, and the greater the number of supercoils, the greater the stress or torsion (test this by twisting a rubber band). **Negative supercoils** are formed when the molecule is twisted in the direction opposite from the clockwise turns of the right-handed double helix found in B-DNA. Such DNA is said to be underwound. The energy required to achieve this state is, in a sense, stored in the supercoils. The transition to another form that requires energy is thereby facilitated by the underwinding (see Figure 35–19). One such transition is strand separation, which is a prerequisite for DNA replication and transcription. Supercoiled DNA is therefore a preferred form in biologic systems. Enzymes that catalyze topologic changes of DNA are called **topoisomerases.** Topoisomerases can relax or insert supercoils, using ATP as an energy source. Homologs of this enzyme exist in all organisms and are important targets for cancer chemotherapy.

DNA PROVIDES A TEMPLATE FOR REPLICATION & TRANSCRIPTION

The genetic information stored in the nucleotide sequence of DNA serves two purposes. It is the source of information for the synthesis of all protein molecules of the cell and organism, and it provides the information inherited by daughter cells or offspring. Both of these functions require that the DNA molecule serve as a template—in the first case for the transcription of the information into RNA and in the second case for the replication of the information into daughter DNA molecules.

When each strand of the double-stranded parental DNA molecule separates from its complement during replication, each independently serves as a template on which a new complementary strand is synthesized (**Figure 34–4**). The two newly formed double-stranded daughter DNA molecules, each containing one strand (but complementary rather than

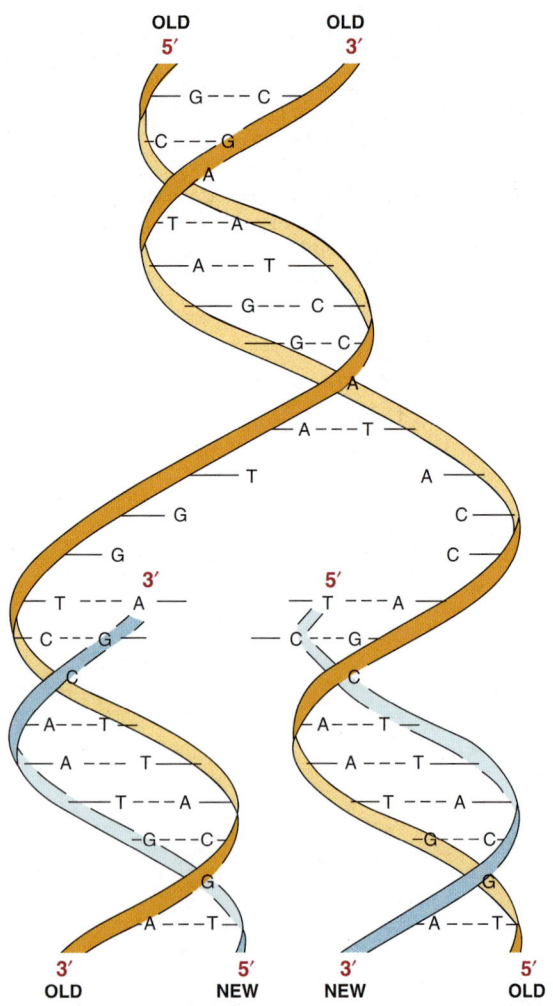

FIGURE 34–4   The double-stranded structure of DNA and the template function of each old strand (orange) on which a new complementary strand (blue) is synthesized.

identical) from the parent double-stranded DNA molecule, are then sorted between the two daughter cells (**Figure 34–5**). Each daughter cell contains DNA molecules with information identical to that which the parent possessed; yet in each daughter cell, the DNA molecule of the parent cell has been only semiconserved.

THE CHEMICAL NATURE OF RNA DIFFERS FROM THAT OF DNA

RNA is a polymer of purine and pyrimidine ribonucleotides linked together by 3′,5′-phosphodiester bonds analogous to those in DNA (**Figure 34–6**). Although sharing many features with DNA, RNA possesses several specific differences:

1. In RNA, the sugar moiety to which the phosphates and purine and pyrimidine bases are attached is ribose rather than the 2′-deoxyribose of DNA.
2. The pyrimidine components of RNA can differ from those of DNA. Although RNA contains the ribonucleotides of adenine, guanine, and cytosine, it does not possess thymine except in the rare case mentioned below. Instead of thymine, RNA contains the ribonucleotide of uracil.
3. RNA typically exists as a single strand, whereas DNA exists as a double-stranded helical molecule. However, given the proper complementary base sequence with opposite polarity, the single strand of RNA—as demonstrated in

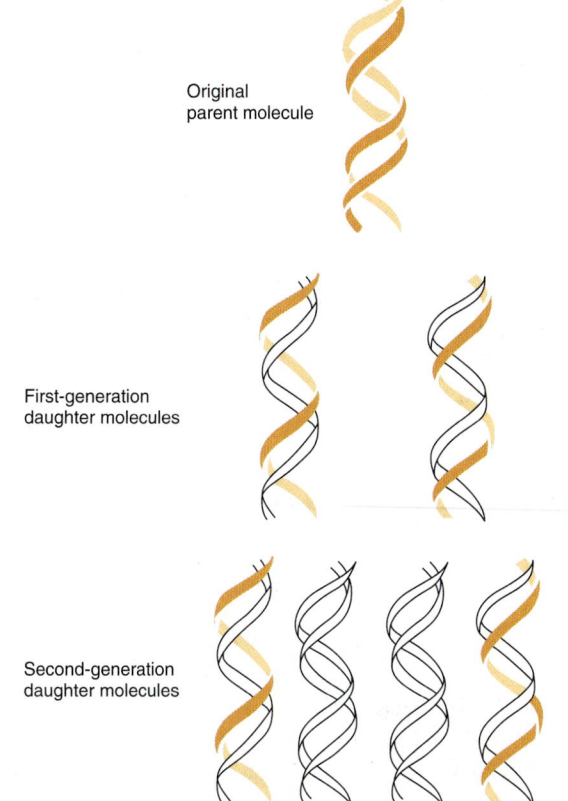

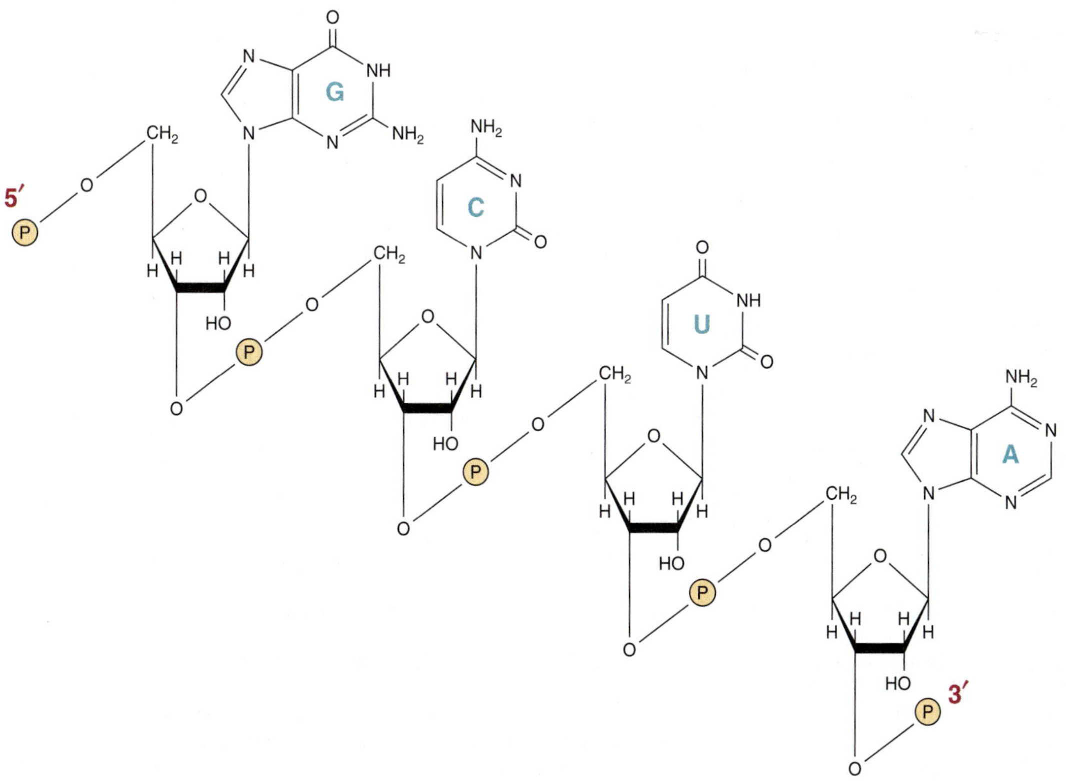

Original parent molecule

First-generation daughter molecules

Second-generation daughter molecules

FIGURE 34–5 **DNA replication is semiconservative.** During a round of replication, each of the two strands of DNA is used as a template for synthesis of a new, complementary strand.

FIGURE 34–6
A segment of a ribonucleic acid (RNA) molecule in which the purine and pyrimidine bases—guanine (G), cytosine (C), uracil (U), and adenine (A)—are held together by phosphodiester bonds between ribosyl moieties attached to the nucleobases by *N*-glycosidic bonds. Note that the polymer has a polarity as indicated by the labeled 3′- and 5′-attached phosphates.

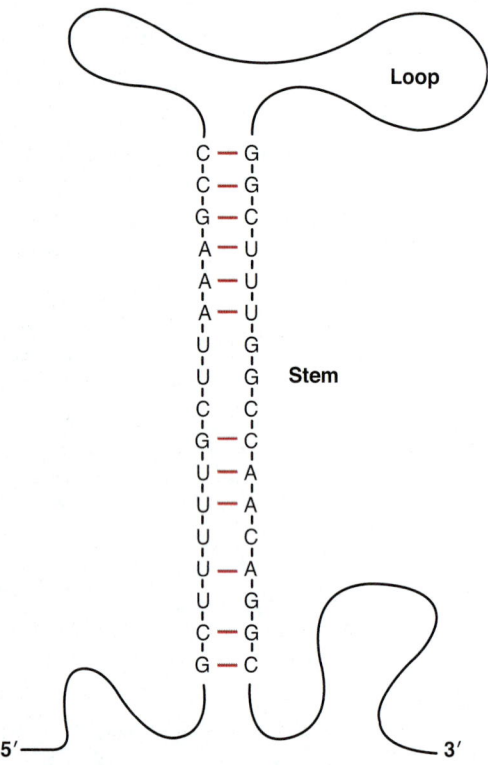

FIGURE 34–7 **Diagrammatic representation of the secondary structure of a single-stranded RNA molecule in which a stem loop, or "hairpin," has been formed.** Formation of this structure is dependent upon the indicated intramolecular base pairing (colored horizontal lines between bases). Note that A forms hydrogen bonds with U in RNA.

Figure 34–7 and Figure 34–11—is capable of folding back on itself like a hairpin and thus acquiring double-stranded characteristics: G pairing with C, and A with U.

4. Since the RNA molecule is a single strand complementary to only one of the two strands of a gene, its guanine content does not necessarily equal its cytosine content, nor does its adenine content necessarily equal its uracil content.

5. RNA can be hydrolyzed by alkali to 2′,3′ cyclic diesters of the mononucleotides, compounds that cannot be formed from alkali-treated DNA because of the absence of a 2′-hydroxyl group. The alkali lability of RNA is useful both diagnostically and analytically.

Information within the single strand of RNA is contained in its sequence ("primary structure") of purine and pyrimidine nucleotides within the polymer. The sequence is complementary to the template strand of the gene from which it was transcribed. Because of this complementarity, an RNA molecule can bind specifically via the base-pairing rules to its template DNA strand (A–T, **G**–C, **C**–G, **U**–A; RNA base bolded); it will not bind ("hybridize") with the other (coding) strand of its gene. The sequence of the RNA molecule (except for U replacing T) is the same as that of the coding strand of the gene (**Figure 34–8**).

Nearly All the Several Species of Stable, Abundant RNAs Are Involved in Some Aspect of Protein Synthesis

Those cytoplasmic RNA molecules that serve as templates for protein synthesis (ie, that transfer genetic information from DNA to the protein-synthesizing machinery) are designated **mRNAs.** Many other very abundant cytoplasmic RNA molecules (**ribosomal RNAs; rRNAs**) have structural roles wherein they contribute to the formation and function of ribosomes (the organellar machinery for protein synthesis) or serve as adapter molecules (**transfer RNAs; tRNAs**) for the translation of RNA information into specific sequences of polymerized amino acids.

Interestingly, some RNA molecules have intrinsic catalytic activity. The activity of these **ribozymes** often involves the cleavage of a nucleic acid. Two well-studied RNA enzymes, or ribozymes, are the peptidyl transferase that catalyzes peptide bond formation on the ribosome, and ribozymes involved in the RNA splicing.

In all eukaryotic cells, there are **small nuclear RNA (snRNA)** species that are not directly involved in protein synthesis but play pivotal roles in RNA processing. These relatively small molecules vary in size from 90 to about 300 nucleotides (**Table 34–1**). The properties of the several classes of cellular RNAs are detailed below.

The genetic material for some animal and plant viruses is RNA rather than DNA. Although some RNA viruses never have their information transcribed into a DNA molecule, many animal RNA viruses—specifically, the retroviruses (the HIV virus, for example)—are transcribed by **viral RNA-dependent DNA polymerase, the so-called reverse transcriptase,** to produce a double-stranded DNA copy of their RNA genome.

DNA strands:

Coding → 5′ —T G G A A T T G T G A G C G G A T A A C A A T T T C A C A C A G G A A A C A G C T A T G A C C A T G—3′
Template → 3′ —A C C T T A A C A C T C G C C T A T T G T T A A A G T G T G T C C T T T G T C G A T A C T G G T A C—5′

RNA transcript: 5′ —— ppp A U U G U G A G C G G A U A A C A A U U U C A C A C A G G A A A C A G C U A U G A C C A U G   3′

FIGURE 34–8 **The relationship between the sequences of an RNA transcript and its gene, in which the coding and template strands are shown with their polarities.** The RNA transcript with a 5′ to 3′ polarity is complementary to the template strand with its 3′ to 5′ polarity. Note that the sequence in the RNA transcript and its polarity is the same as that in the coding strand, except that the U of the transcript replaces the T of the gene; the initiating nucleotide of RNAs contain a terminal 5′-triphosphate (ie. pppA-above).

TABLE 34–1 Some of the Species of Small-Stable RNAs Found in Mammalian Cells

Name	Length (nucleotides)	Molecules per Cell	Localization
U1	165	1×10^6	Nucleoplasm
U2	188	5×10^5	Nucleoplasm
U3	216	3×10^5	Nucleolus
U4	139	1×10^5	Nucleoplasm
U5	118	2×10^5	Nucleoplasm
U6	106	3×10^5	Perichromatin granules
4.5S	95	3×10^5	Nucleus and cytoplasm
7SK	280	5×10^5	Nucleus and cytoplasm

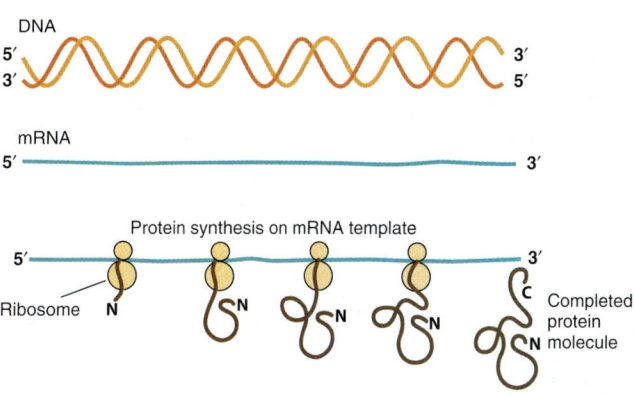

FIGURE 34–9 The expression of genetic information in DNA into the form of an mRNA transcript with 5′ to 3′ polarity shown. The mRNA is subsequently translated by ribosomes into a specific protein molecule that also exhibits polarity N-terminal (**N**) to C-terminal (**C**).

In many cases, the resulting double-stranded DNA transcript is integrated into the host genome and subsequently serves as a template for gene expression and from which new viral RNA genomes and viral mRNAs can be transcribed. Genomic insertion of such integrating "proviral" DNA molecules can, depending on the site involved, be mutagenic, inactivating a gene or disregulating its expression (see Figure 35–11).

There Exist Several Distinct Classes of RNA

In all prokaryotic and eukaryotic organisms, four main classes of RNA molecules exist: mRNA, tRNA, rRNA, and small RNAs. Each differs from the others by abundance, size, function, and general stability.

Messenger RNA

This class is the most heterogeneous in abundance, size and stability; for example, in brewer's yeast, specific mRNAs are present in 100s/cell to, on average, ≤0.1/mRNA/cell in a genetically homogeneous population. As detailed in Chapters 36 and 38, both specific transcriptional and posttranscription mechanisms contribute to this large dynamic range in mRNA content. In mammalian cells, mRNA abundance likely varies over a 10^4-fold range. All members of the class function as messengers conveying the information in a gene to the protein-synthesizing machinery, where each mRNA serves as a template on which a specific sequence of amino acids is polymerized to form a specific protein molecule, the ultimate gene product (**Figure 34–9**).

Eukaryotic mRNAs have unique chemical characteristics. The 5′ terminal of mRNA is "capped" by a 7-methylguanosine triphosphate that is linked to an adjacent 2′-O-methyl ribonucleoside at its 5′-hydroxyl through the three phosphates (**Figure 34–10**). The mRNA molecules frequently contain internal 6-methyladenylates and other 2′-O-ribose-methylated nucleotides. The cap is involved in the recognition of mRNA by

the translation machinery, and also helps stabilize the mRNA by preventing the attack of 5′-exonucleases. The protein-synthesizing machinery begins translating the mRNA into proteins beginning downstream of the 5′ or capped terminal. The other end of mRNA molecules, the 3′-hydroxyl terminal, has an attached, nongenetically-encoded polymer of adenylate residues 20–250 nucleotides in length. The poly(A) "tail" at the 3′-hydroxyl terminal of mRNAs maintains the intracellular stability of the specific mRNA by preventing the attack of 3′-exonucleases and also facilitates translation (Figure 37–7). A few mRNAs, including those for some histones, do not contain a poly(A) tail. Both the mRNA "cap" and "poly(A) tail" are added posttranscriptionally by nontemplate-directed enzymes to mRNA precursor molecules (pre-mRNA). mRNA represents 2%–5% of total eukaryotic cellular RNA.

In mammalian cells, including cells of humans, the mRNA molecules present in the cytoplasm are not the RNA products immediately synthesized from the DNA template but must be formed by processing from the pre-mRNA before entering the cytoplasm. Thus, in mammalian nuclei, the immediate products of gene transcription (primary transcripts) are very heterogeneous and can be greater than 10- to 50-fold longer than mature mRNA molecules. As discussed in Chapter 36, pre-mRNA molecules are processed to generate the mRNA molecules, which then enter the cytoplasm to serve as templates for protein synthesis.

Transfer RNA

tRNA molecules vary in length from 74 to 95 nucleotides. They are also generated by nuclear processing of a precursor molecule (Chapter 36). The tRNA molecules serve as adapters for the translation of the information in the sequence of nucleotides of the mRNA into specific amino acids. There are at least 20 species of tRNA molecules in every cell, at least one (and often several) corresponding to each of the 20 amino acids required for protein synthesis. Although each specific tRNA

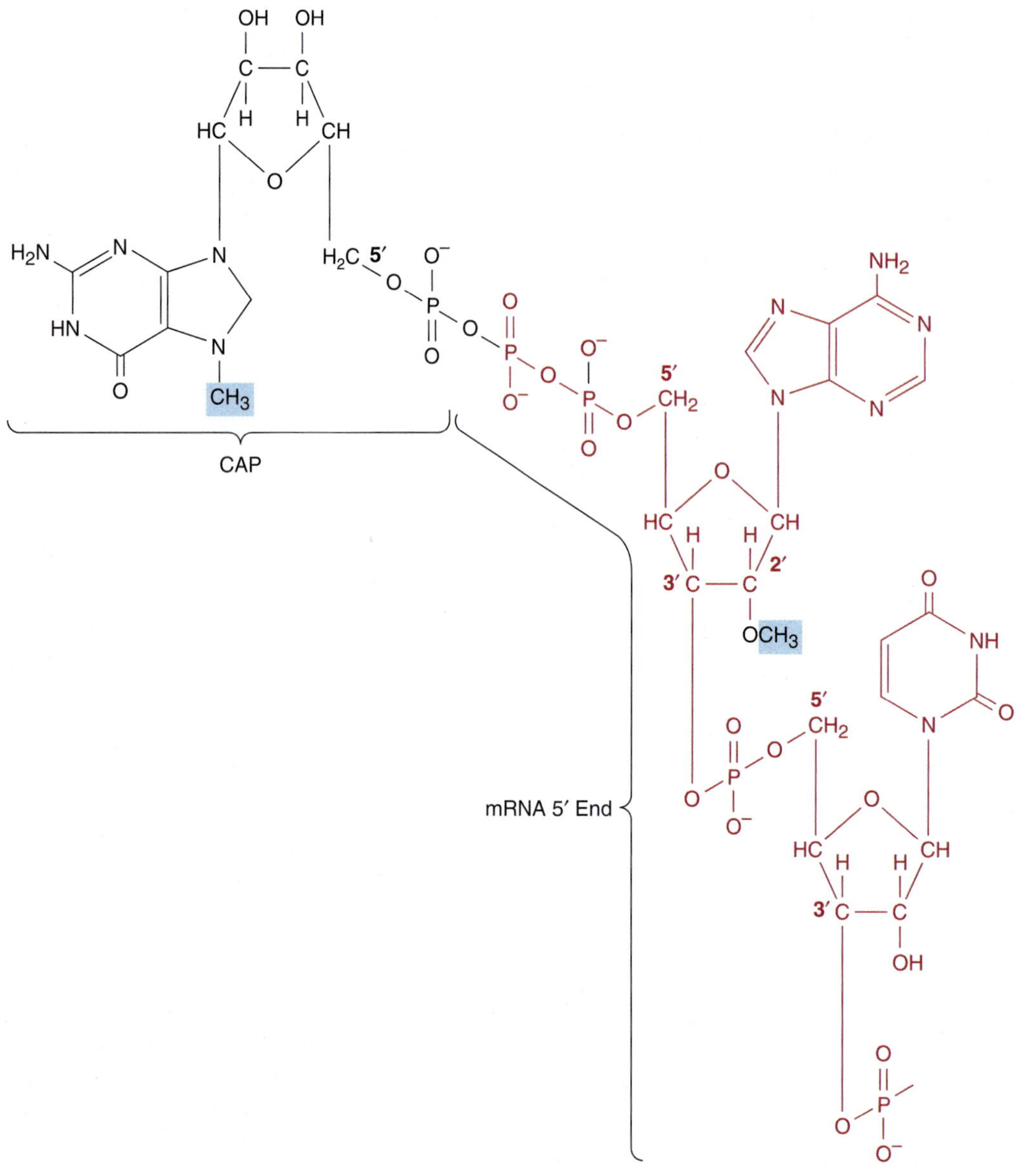

FIGURE 34–10 **The cap structure attached to the 5′ terminal of most eukaryotic messenger RNA molecules.** A 7-methylguanosine triphosphate (black) is attached at the 5′ terminal of the mRNA (shown in color), which usually also contains a 2′-O-methylpurine nucleotide. These modifications (the cap and methyl group) are added after the mRNA is transcribed from DNA.

differs from the others in its sequence of nucleotides, the tRNA molecules as a class have many features in common. The primary structure—that is, the nucleotide sequence—of all tRNA molecules allows extensive folding and intrastrand complementarity to generate a secondary structure that appears in two dimensions like a cloverleaf (**Figure 34–11**).

All tRNA molecules contain four main arms. The **acceptor arm** terminates in the nucleotides CpCpAOH. These three nucleotides are added posttranscriptionally by a specific nucleotidyl transferase enzyme. The tRNA-appropriate amino acid is attached, or "charged" onto, the 3′-OH group of the A moiety of the acceptor arm (see Figure 37–1). The **D, TψC,**

and **extra arms** help define a specific tRNA. tRNAs compose roughly 20% of total cellular RNA.

Ribosomal RNA

A ribosome is a cytoplasmic nucleoprotein structure that acts as the machinery for the synthesis of proteins from the mRNA templates. On the ribosomes, the mRNA and tRNA molecules interact to translate into a specific protein molecule information transcribed from the gene. During periods of active protein synthesis, many ribosomes can be associated with any mRNA molecule to form an assembly called the **polysome** (Figure 37–7).

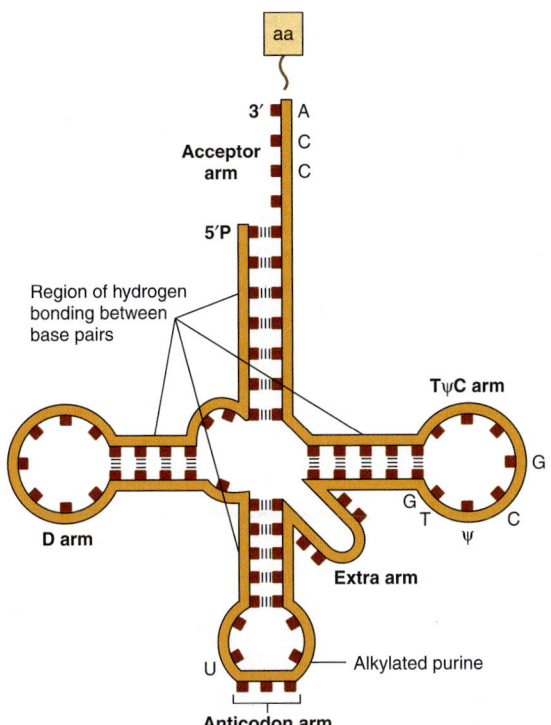

FIGURE 34–11 **Typical aminoacyl tRNA in which the amino acid (aa) is attached to the 3′ CCA terminal.** The anticodon, TψC, and dihydrouracil (D) arms are indicated, as are the positions of the intramolecular hydrogen bonding between these base pairs. Ψ is pseudouridine, an isomer of uridine formed posttranscriptionally. (Watson JD, et al, *Molecular Biology of the Gene,* 6th Edition, © 2008, p. 243. Adapted by permission of Pearson Education, Inc., Upper Saddle River, NJ.)

The components of the mammalian ribosome, which has a molecular weight of about 4.2×10^6 and a sedimentation velocity coefficient of 80S (S = Svedberg units, a parameter sensitive to molecular size and shape) are shown in **Table 34–2**. The mammalian ribosome contains two major nucleoprotein subunits—a larger one with a molecular weight of 2.8×10^6 (60S) and a smaller subunit with a molecular weight of 1.4×10^6 (40S). The 60S subunit contains a 5S rRNA, a 5.8S rRNA, and a 28S rRNA; there are also more than 50 specific polypeptides. The 40S subunit is smaller and contains a single 18S rRNA and approximately 30 distinct polypeptide chains. All of the rRNA molecules except the 5S rRNA, which is independently transcribed, are processed from a single 45S precursor RNA molecule in the nucleolus (Chapter 36). The highly methylated rRNA molecules are packaged in the nucleolus with the specific ribosomal proteins. In the cytoplasm, the ribosomes remain quite stable and capable of many translation cycles. The exact functions of the rRNA molecules in the ribosomal particle are not fully understood, but they are necessary for ribosomal assembly and also play key roles in the binding of mRNA to ribosomes and its translation. Recent studies indicate that the large rRNA component performs the peptidyl transferase activity and thus is a ribozyme. The rRNAs (28S + 18S) represent roughly 70% of total cellular RNA.

Small RNA

A large number of discrete, highly conserved, and small RNA species are found in eukaryotic cells; some are quite stable. Most of these molecules are complexed with proteins to form ribonucleoproteins and are distributed in the nucleus, the cytoplasm, or both. They range in size from 20 to 1000 nucleotides and are present in 100,000–1,000,000 copies per cell, collectively representing ≤5% of cellular RNA.

Small Nuclear RNAs

snRNAs, a subset of the small RNAs (Table 34–1), are significantly involved in rRNA and mRNA processing and gene regulation. Of the several snRNAs, U1, U2, U4, U5, and U6 are involved in intron removal and the processing of mRNA precursors into mRNA (Chapter 36). The U7 snRNA is involved in production of the correct 3′ ends of histone mRNA—which lacks a poly(A) tail. 7SK RNA associates with several proteins to form a ribonucleoprotein complex, termed P-TEFb, that modulates mRNA gene transcription elongation by RNA polymerase II (see Chapter 36).

Micro-RNAs, miRNAs, and Small Interfering RNAs, siRNAs, and Noncoding RNAs

One of the most exciting and unanticipated discoveries in the last decade of eukaryotic regulatory biology was the

TABLE 34–2 Components of Mammalian Ribosomes

| Component | Mass (MW) | Protein | | RNA | | |
		Number	Mass	Size	Mass	Bases
40S subunit	1.4×10^6	33	7×10^5	18S	7×10^5	1900
60S subunit	2.8×10^6	50	1×10^6	5S	3.5×10^4	120
				5.8S	4.5×10^4	160
				28S	1.6×10^6	4700

Note: The ribosomal subunits are defined according to their sedimentation velocity in Svedberg (**S**) units (40S or 60S). The number of unique proteins and their total mass (MW) and the RNA components of each subunit in size (Svedberg units), mass, and number of bases are listed.

identification and characterization of miRNAs, a class of small RNAs found in most eukaryotes (Chapter 38). Nearly all known **miRNAs and siRNAs cause inhibition of gene expression** by decreasing specific protein production, albeit via distinct mechanisms. miRNAs are typically 21–25 nucleotides in length and are generated by nucleolytic processing of the products of distinct genes/transcription units (see Figure 36–17). miRNA precursors are single stranded but have extensive intramolecular secondary structure. These precursors range in size from about 500 to 1000 nucleotides; the small processed mature **miRNAs typically hybridize, via the formation of imperfect RNA–RNA duplexes within the 3'-untranslated regions** (3'UTRs; see Figure 38–19) of specific target mRNAs, **leading, via poorly understood mechanisms, to translation arrest.** To date, hundreds of distinct miRNAs have been described in humans; estimates suggest that there are ~1000 human miRNA-encoding genes. As with miRNAs, siRNAs are also derived by the specific nucleolytic cleavage of larger, RNAs to again form small 21–25 nucleotide-long products. These short **siRNAs usually form perfect RNA–RNA hybrids** with their distinct targets potentially anywhere within the length of the mRNA where the complementary sequence exists. Formation of such RNA–RNA duplexes between siRNA and mRNA results in reduced specific protein production because the **siRNA–mRNA complexes are degraded** by dedicated nucleolytic machinery; some or all of this mRNA degradation occurs in specific cytoplasmic organelles termed **P bodies** (Figure 37–11). Given their exquisite genetic specificity both miRNAs and siRNAs represent exciting new **potential agents for therapeutic drug development.** In addition, siRNAs are frequently used to decrease or "knock-down" specific protein levels (via siRNA homology–directed mRNA degradation) in experimental contexts in the laboratory, an extremely useful and powerful alternative to gene-knockout technology (Chapter 39).

Another exciting recent discovery in the RNA realm is the identification and characterization of **long noncoding RNAs,** or **ncRNAs.** The long ncRNAs, which as their name implies, do not code for protein range in size from ~300 to 1000s of nucleotides in length. These RNAs are typically transcribed from the large regions of eukaryotic genomes that do not code for protein. In fact transcriptome analyses via the next generation sequencing technology (see Chapter 39) indicate that **>90% of all eukaryotic genomic DNA is transcribed.** ncRNAs make up a significant portion of this transcription. ncRNAs play many roles ranging from contributing to structural aspects of chromatin to regulation of mRNA gene transcription by RNA polymerase II. Future work will further characterize this important new class of RNA molecules.

Interestingly, bacteria also contain small, heterogeneous regulatory RNAs termed sRNAs. Bacterial sRNAs range in size from 50 to 500 nucleotides, and like eukaryotic mi/siRNAs, also control a large array of genes. sRNAs often repress, but sometimes activate protein synthesis by binding to specific mRNA.

SPECIFIC NUCLEASES DIGEST NUCLEIC ACIDS

Enzymes capable of degrading nucleic acids have been recognized for many years. These nucleases can be classified in several ways. Those which exhibit specificity for DNA are referred to as **deoxyribonucleases.** Those which specifically hydrolyze RNA are **ribonucleases.** Some nucleases degrade both DNA and RNA. Within both of these classes are enzymes capable of cleaving internal phosphodiester bonds to produce either 3'-hydroxyl and 5'-phosphoryl terminals or 5'-hydroxyl and 3'-phosphoryl terminals. These are referred to as **endonucleases.** Some are capable of hydrolyzing both strands of a **double-stranded** molecule, whereas others can only cleave **single strands** of nucleic acids. Some nucleases can hydrolyze only unpaired single strands, while others are capable of hydrolyzing single strands participating in the formation of a double-stranded molecule. There exist classes of endonucleases that recognize specific sequences in DNA; the majority of these are the **restriction endonucleases,** which have in recent years become important tools in molecular genetics and medical sciences. A list of some currently recognized restriction endonucleases is presented in Table 39–2.

Some nucleases are capable of hydrolyzing a nucleotide only when it is present at a terminal of a molecule; these are referred to as **exonucleases.** Exonucleases act in one direction ($3' \rightarrow 5'$ or $5' \rightarrow 3'$) only. In bacteria, a $3' \rightarrow 5'$ exonuclease is an integral part of the DNA replication machinery and there serves to edit—or proofread—the most recently added deoxynucleotide for base-pairing errors.

SUMMARY

- DNA consists of four bases—A, G, C, and T—that are held in linear array by phosphodiester bonds through the 3' and 5' positions of adjacent deoxyribose moieties.
- DNA is organized into two strands by the pairing of bases A to T and G to C on complementary strands. These strands form a double helix around a central axis.
- The 3×10^9 bp of DNA in humans are organized into the haploid complement of 23 chromosomes. The exact sequence of these 3 billion nucleotides defines the uniqueness of each individual.
- DNA provides a template for its own replication and thus maintenance of the genotype and for the transcription of the roughly 25,000 protein coding human genes as well as a large array on nonprotein coding regulatory RNAs.
- RNA exists in several different single-stranded structures, most of which are directly or indirectly involved in protein synthesis or its regulation. The linear array of nucleotides in RNA consists of A, G, C, and U, and the sugar moiety is ribose.
- The major forms of RNA include mRNA, rRNA, tRNA, and snRNAs (miRNAs). Certain RNA molecules act as catalysts (ribozymes).

REFERENCES

Chapman EJ, Carrington JC: Specialization and evolution of endogenous small RNA pathways. Nature Rev Genetics 2007;8:884.

Costa FF: Non-coding RNAs: meet thy masters. Bioessays 2010;32:599–608.

Dunkle JA, Cate JH: Ribosome structure and dynamics during translation. Annu Rev Biophys 2010;39:227–244.

Green R, Noller HF: Ribosomes and translation. Annu Rev Biochem 1997;66:689.

Guthrie C, Patterson B: Spliceosomal snRNAs. Ann Rev Genet 1988;22:387.

Keene JD: Minireview: global regulation and dynamics of ribonucleic acid. Endocrinology 2010;151: 1391–1397.

Loewer S, Cabili MN, Guttman M, et al: Large intergenic non-coding RNA–RoR modulates reprogramming of human induced pluripotent stem cells. Nat Genet 2010;42:113–1137.

Moore M: From birth to death: the complex lives of eukaryotic mRNAs. Science 2005;309:1514.

Narla A, Ebert BL: Ribosomopathies: human disorders of ribosome dysfunction. Blood 2010;115:3196–3205.

Phizicky EM, Hopper AK: tRNA biology charges to the front. Genes Devlop 2010;24:1832–1860.

Wang G-S, Cooper TA: Splicing in disease: disruption of the splicing code and the decoding machinery. Nature Rev Genetics 2007;8:749.

Watson JD, Crick FHC: Molecular structure of nucleic acids. Nature 1953;171:737.

Watson JD: *The Double Helix*. Atheneum, 1968.

Watson JD, Baker TA, Bell SP, et al: *Molecular Biology of the Gene,* 6th ed. Benjamin-Cummings, 2007.

DNA Organization, Replication, & Repair

P. Anthony Weil, PhD

- Appreciate that roughly 3×10^9 base pairs of DNA that compose the haploid genome of humans are divided uniquely between 23 linear DNA units, the chromosomes. Humans, being diploid, have 23 pairs of chromosomes: 22 autosomes and 2 sex chromosomes.

- Understand that human genomic DNA, if extended end-to-end, would be meters in length, yet still fits within the nucleus of the cell, an organelle that is only microns (μ; 10^{-6} meters) in diameter. Such condensation in DNA length is induced following its association with the highly positively charged histone proteins resulting in the formation of a unique DNA-histone complex termed the nucleosome. Nucleosomes have DNA wrapped around the surface of an octamer of histones.

- Explain that strings of nucleosomes form along the linear sequence of genomic DNA to form chromatin, which itself can be more tightly packaged and condensed, which ultimately leads to the formation of the chromosomes.

- Appreciate that while the chromosomes are the macroscopic functional units for DNA recombination, gene assortment, and cellular division, it is DNA function at the level of the individual nucleotides that composes regulatory sequences linked to specific genes that are essential for life.

- Explain the steps, phase of the cell cycle, and the molecules responsible for the replication, repair, and recombination of DNA, and understand the negative effects of errors in any of these processes upon cellular and organismal integrity and health.

BIOMEDICAL IMPORTANCE*

The genetic information in the DNA of a chromosome can be transmitted by exact replication or it can be exchanged by a number of processes, including crossing over, recombination, transposition, and conversion. These provide a means of ensuring adaptability and diversity for the organism but, when these processes go awry, can also result in disease. A number of enzyme systems are involved in DNA replication, alteration, and repair.

Mutations are due to a change in the base sequence of DNA and may result from the faulty replication, movement, or repair of DNA and occur with a frequency of about one in every 10^6 cell divisions. Abnormalities in gene products (either in RNA, protein function, or amount) can be the result of mutations that occur in coding or regulatory-region DNA. A mutation in a germ cell is transmitted to offspring (so-called vertical transmission of hereditary disease). A number of factors, including viruses, chemicals, ultraviolet light, and ionizing radiation, increase the rate of mutation. Mutations often affect somatic cells and so are passed on to successive generations of cells, but only within an organism (ie, horizontally). It is becoming apparent that a number of diseases—and perhaps most cancers—are due to the combined effects of vertical transmission of mutations as well as horizontal transmission of induced mutations.

---

*So far as is possible, the discussion in this chapter and in Chapters 36, 37, and 38 will pertain to mammalian organisms, which are, of course, among the higher eukaryotes. At times it will be necessary to refer to observations in prokaryotic organisms such as bacteria and viruses, or lower eukaryotic model systems such as Drosophila, C. elegans or yeast. However, in such cases the information will be of a kind that can be extrapolated to mammalian organisms.

CHROMATIN IS THE CHROMOSOMAL MATERIAL IN THE NUCLEI OF CELLS OF EUKARYOTIC ORGANISMS

Chromatin consists of very long double-stranded **DNA (dsDNA) molecules** and a nearly equal mass of rather small basic proteins termed **histones** as well as a smaller amount of **nonhistone proteins** (most of which are acidic and larger than histones) and a small quantity of **RNA**. The nonhistone proteins include enzymes involved in DNA replication and repair, and the proteins involved in RNA synthesis, processing, and transport to the cytoplasm. The dsDNA helix in each chromosome has a length that is thousands of times the diameter of the cell nucleus. One purpose of the molecules that comprise chromatin, particularly the histones, is to condense the DNA; however, it is important to note that the histones also integrally participate in gene regulation (Chapters 36, 38, and 42); indeed histones contribute importantly to all DNA-directed molecular transactions. Electron microscopic studies of chromatin have demonstrated dense spherical particles called **nucleosomes,** which are approximately 10 nm in diameter and connected by DNA filaments (**Figure 35-1**). Nucleosomes are composed of DNA wound around a collection of histone molecules.

Histones Are the Most Abundant Chromatin Proteins

Histones are a small family of closely related basic proteins. **H1 histones** are the ones least tightly bound to chromatin (**Figures 35-1, 35-2, and 35-3**) and are, therefore, easily removed with a salt solution, after which chromatin becomes more soluble. The organizational unit of this soluble chromatin is the nucleosome. **Nucleosomes contain four major types of histones: H2A, H2B, H3, and H4.** The structures of all four histones—H2A, H2B, H3, and H4, the so-called core histones that form the nucleosome—have been highly conserved between species, although variants of the histones exist and are used for specialized purposes. This extreme conservation implies that the function of histones is identical in all eukaryotes and that the entire molecule is involved quite specifically in carrying out this function. The carboxyl terminal two-thirds of the histone molecules are hydrophobic, while their amino terminal thirds are particularly rich in basic amino acids. **These four core histones are subject to at least six types of covalent modification or posttranslational modifications (PTMs):** acetylation, methylation, phosphorylation, ADP-ribosylation, monoubiquitylation, and sumoylation. These histone modifications play an important role in chromatin structure and function, as illustrated in **Table 35-1**.

The histones interact with each other in very specific ways. **H3 and H4 form a tetramer** containing two molecules of each (H3–H4)$_2$, while **H2A and H2B form dimers** (H2A–H2B). Under physiologic conditions, these histone oligomers associate to form the **histone octamer** of the composition (H3–H4)$_2$–(H2A–H2B)$_2$.

The Nucleosome Contains Histone & DNA

When the histone octamer is mixed with purified dsDNA under appropriate ionic conditions, the same X-ray diffraction pattern is formed as that observed in freshly isolated chromatin. Electron microscopic studies confirm the existence of reconstituted nucleosomes. Furthermore, the reconstitution of nucleosomes from DNA and histones H2A, H2B, H3, and H4 is independent of the organismal or cellular origin of the various components. Neither the histone H1 nor the nonhistone proteins are necessary for the reconstitution of the nucleosome core.

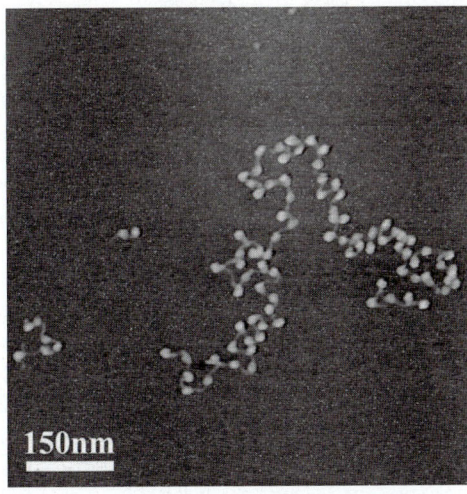

FIGURE 35-1 Electron micrograph of nucleosomes (white, ball-shaped) attached to strands of DNA (thin, gray line); see also **Figure 35-2.** (Reproduced, with permission, from Shao Z: Probing nanometer structures with atomic force microscopy. News Physiol Sci, 1999;14:142–149. Courtesy of Professor Zhifeng Shao, University of Virginia.)

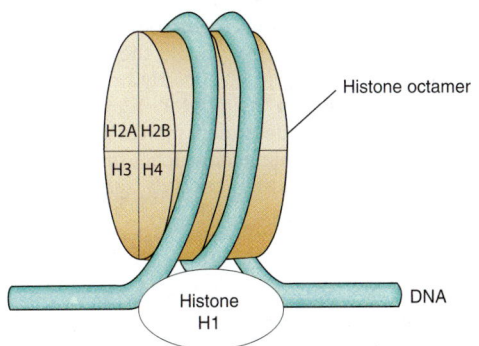

FIGURE 35-2 Model for the structure of the nucleosome, in which DNA is wrapped around the surface of a flat protein cylinder consisting of two each of histones H2A, H2B, H3, and H4 that form the histone octamer. The ~145 bp of DNA, consisting of 1.75 superhelical turns, are in contact with the histone octamer. The position of histone H1, when it is present, is indicated by the dashed outline at the bottom of the figure. Histone H1 interacts with DNA as it enters and exits the nucleosome.

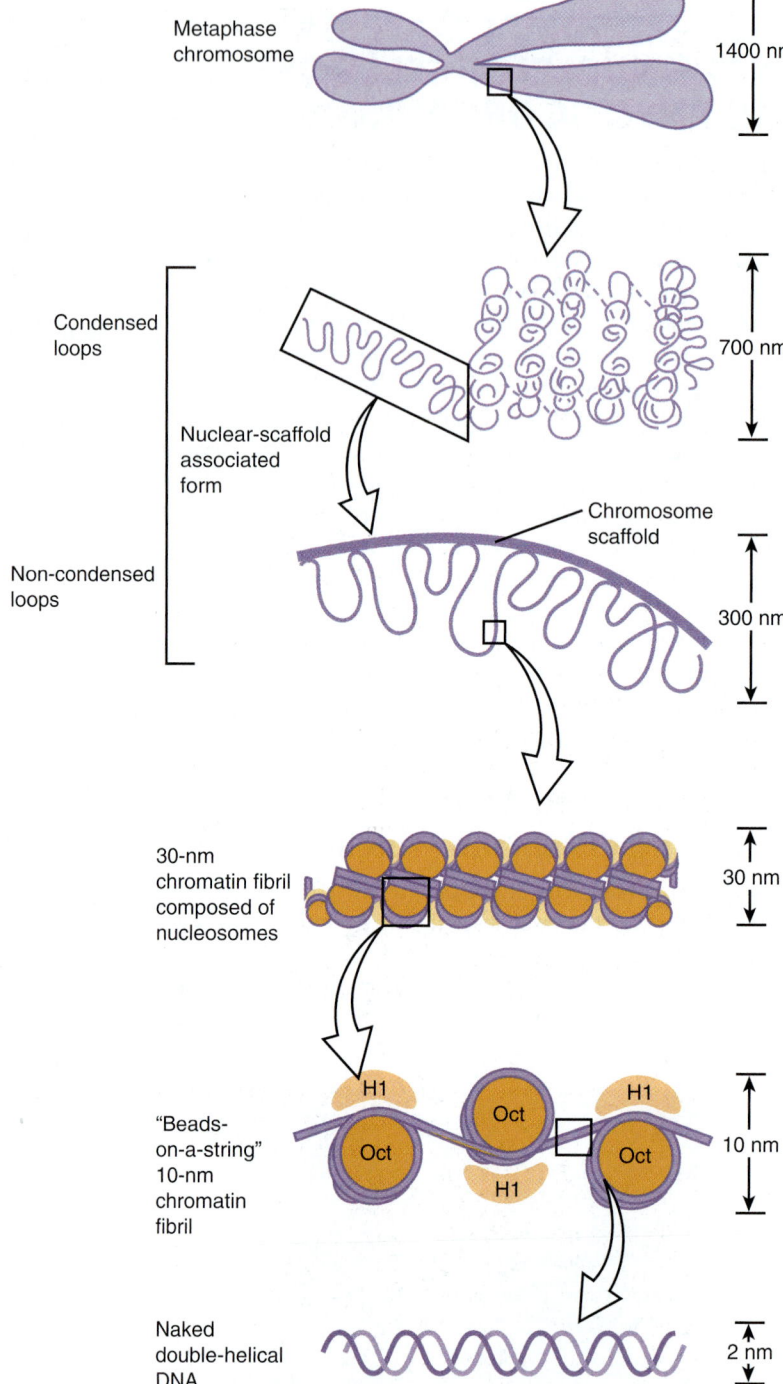

FIGURE 35–3 **Shown is the extent of DNA packaging in metaphase chromosomes (top) to noted duplex DNA (bottom).** Chromosomal DNA is packaged and organized at several levels as shown (see Table 35–2). Each phase of condensation or compaction and organization (bottom to top) decreases overall DNA accessibility to an extent that the DNA sequences in metaphase chromosomes are almost totally transcriptionally inert. In toto, these five levels of DNA compaction result in nearly a 10^4-fold linear decrease in end-to-end DNA length. Complete condensation and decondensation of the linear DNA in chromosomes occur in the space of hours during the normal replicative cell cycle (see Figure 35–20).

In the nucleosome, the DNA is supercoiled in a left-handed helix over the surface of the disk-shaped histone octamer (Figure 35–2). The majority of core histone proteins interact with the DNA on the inside of the supercoil without protruding, although the amino terminal tails of all the histones are thought to extend outside of this structure and are available for regulatory PTMs (see Table 35–1).

The $(H3–H4)_2$ tetramer itself can confer nucleosome-like properties on DNA and thus has a central role in the formation of the nucleosome. The addition of two H2A–H2B dimers stabilizes the primary particle and firmly binds two additional half-turns of DNA previously bound only loosely to the $(H3–H4)_2$. Thus, 1.75 superhelical turns of DNA are wrapped around the surface of the histone octamer, protecting 145–150 bp of DNA and forming the nucleosome core particle (Figure 35–2). In chromatin, core particles are separated by an about 30-bp region of DNA termed **"linker."** Most of the DNA is in a repeating series of these structures, giving the so-called

TABLE 35–1 Possible Roles of Modified Histones

1. Acetylation of histones H3 and H4 is associated with the activation or inactivation of gene transcription.

2. Acetylation of core histones is associated with chromosomal assembly during DNA replication.

3. Phosphorylation of histone H1 is associated with the condensation of chromosomes during the replication cycle.

4. ADP-ribosylation of histones is associated with DNA repair.

5. Methylation of histones is correlated with activation and repression of gene transcription.

6. Monoubiquitylation is associated with gene activation, repression, and heterochromatic gene silencing.

7. Sumoylation of histones (SUMO; small ubiquitin-related modifier) is associated with transcription repression.

beads-on-a-string appearance when examined by electron microscopy (see Figure 35–1).

The assembly of nucleosomes is mediated by one of several nuclear chromatin assembly factors facilitated by histone chaperones, a group of proteins that exhibit high-affinity histone binding. As the nucleosome is assembled, histones are released from the histone chaperones. Nucleosomes appear to exhibit preference for certain regions on specific DNA molecules, but the basis for this nonrandom distribution, termed **phasing**, is not yet completely understood. Phasing is likely related both to the relative physical flexibility of particular nucleotide sequences to accommodate the regions of kinking within the supercoil, as well as the presence of other DNA-bound factors that limit the sites of nucleosome deposition.

HIGHER ORDER STRUCTURES PROVIDE FOR THE COMPACTION OF CHROMATIN

Electron microscopy of chromatin reveals two higher orders of structure—the 10-nm fibril and the 30-nm chromatin fiber—beyond that of the nucleosome itself. The disk-like nucleosome structure has a 10-nm diameter and a height of 5 nm. The **10-nm fibril** consists of nucleosomes arranged with their edges separated by a small distance (30 bp of DNA) with their flat faces parallel to the fibril axis (Figure 35–3). The 10-nm fibril is probably further supercoiled with six or seven nucleosomes per turn to form the **30-nm chromatin fiber** (Figure 35–3). Each turn of the supercoil is relatively flat, and the faces of the nucleosomes of successive turns would be nearly parallel to each other. H1 histones appear to stabilize the 30-nm fiber, but their position and that of the variable length spacer DNA are not clear. It is probable that nucleosomes can form a variety of packed structures. In order to form a mitotic chromosome, the 30-nm fiber must be compacted in length another 100-fold (see below).

In **interphase chromosomes,** chromatin fibers appear to be organized into 30,000–100,000 bp **loops or domains** anchored in a scaffolding (or supporting matrix) within the nucleus, the so-called **nuclear matrix.** Within these domains, some DNA sequences may be located nonrandomly. It has been suggested that each looped domain of chromatin corresponds to one or more separate genetic functions, containing both coding and noncoding regions of the cognate gene or genes. This nuclear architecture is likely dynamic, having important regulatory effects upon gene regulation. Recent data suggest that certain genes or gene regions are mobile within the nucleus, moving obligatorily to discrete loci within the nucleus upon activation. Further work will determine both if this is a general phenomenon, and what molecular mechanisms are responsible.

SOME REGIONS OF CHROMATIN ARE "ACTIVE" & OTHERS ARE "INACTIVE"

Generally, every cell of an individual metazoan organism contains the same genetic information. Thus, the differences between cell types within an organism must be explained by differential expression of the common genetic information. Chromatin containing active genes (ie, transcriptionally or potentially transcriptionally active chromatin) has been shown to differ in several ways from that of inactive regions. The nucleosome structure of active chromatin appears to be altered, sometimes quite extensively, in highly active regions. DNA in active chromatin contains large regions (about 100,000 bases long) that are relatively more **sensitive to digestion by a nuclease** such as DNase I. DNase I makes single-strand cuts in nearly any segment of DNA (ie, low-sequence specificity). It will digest DNA that is not protected, or bound by protein, into its component deoxynucleotides. The sensitivity to DNase I of active chromatin regions reflects only a potential for transcription rather than transcription itself and in several systems can be correlated with a relative lack of 5-methyldeoxycytidine (meC) in the DNA and particular histone variants and/or PTMs (phosphorylation, acetylation, etc; see Table 35–1).

Within the large regions of active chromatin there exist shorter stretches of 100–300 nucleotides that exhibit an even greater (another 10-fold) sensitivity to DNase I. These **hypersensitive sites** probably result from a structural conformation that favors access of the nuclease to the DNA. These regions are often located immediately upstream from the active gene and are the location of interrupted nucleosomal structure caused by the binding of nonhistone regulatory transcription factor proteins (see Chapters 36 and 38). In many cases, it seems that if a gene is capable of being transcribed, it very often has a DNase-hypersensitive site(s) in the chromatin immediately upstream. As noted above, nonhistone regulatory proteins involved in transcription control and those involved in maintaining access to the template strand lead to the formation of hypersensitive sites. Such sites often provide the first clue about the presence and location of a transcription control element.

By contrast, transcriptionally inactive chromatin is densely packed during interphase as observed by electron microscopic studies and is referred to as **heterochromatin;** transcriptionally active chromatin stains less densely and is referred to as **euchromatin.** Generally, euchromatin is replicated earlier than heterochromatin in the mammalian cell cycle (see below). The chromatin in these regions of inactivity is often high in meC content, and histones therein contain relatively lower levels of covalent modifications.

There are two types of heterochromatin: constitutive and facultative. **Constitutive heterochromatin** is always condensed and thus essentially inactive. It is found in the regions near the chromosomal centromere and at chromosomal ends (telomeres). **Facultative heterochromatin** is at times condensed, but at other times it is actively transcribed and, thus, uncondensed and appears as euchromatin. Of the two members of the X chromosome pair in mammalian females, one X chromosome is almost completely inactive transcriptionally and is heterochromatic. However, the heterochromatic X chromosome decondenses during gametogenesis and becomes transcriptionally active during early embryogenesis—thus, it is facultative heterochromatin.

Certain cells of insects, for example, *Chironomus* and *Drosophila,* contain giant chromosomes that have been replicated for multiple cycles without separation of daughter chromatids. These copies of DNA line up side by side in precise register and produce a banded chromosome containing regions of condensed chromatin and lighter bands of more extended chromatin. Transcriptionally active regions of these **polytene chromosomes** are especially decondensed into **"puffs"** that can be shown to contain the enzymes responsible for transcription and to be the sites of RNA synthesis (**Figure 35–4**). Using highly sensitive fluorescently labeled hybridization probes, specific gene sequences can be mapped, or "painted," within the nuclei of human cells, even without polytene chromosome formation, using FISH (fluorescence in situ hybridization; Chapter 39) techniques.

DNA IS ORGANIZED INTO CHROMOSOMES

At metaphase, mammalian **chromosomes** possess a twofold symmetry, with the identical duplicated **sister chromatids** connected at a **centromere,** the relative position of which is characteristic for a given chromosome (**Figure 35–5**). The centromere is an adenine-thymine (A–T)-rich region containing repeated DNA sequences that range in size from 10^2 (brewers' yeast) to 10^6 (mammals) base pairs (bp). Metazoan centromeres are bound by nucelosomes containing the histone H3 variant protein CENP-A and other specific centromere-binding proteins. This complex, called the **kinetochore,** provides the anchor for the mitotic spindle. It thus

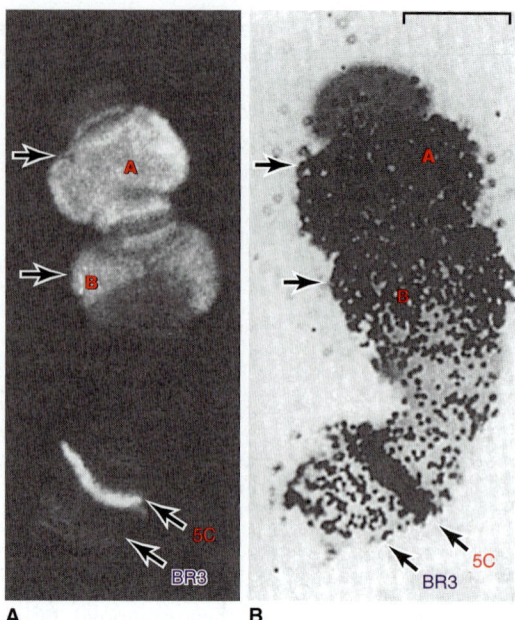

A   **B**

FIGURE 35–4 **Illustration of the tight correlation between the presence of RNA polymerase II (Table 36–2) and messenger RNA synthesis.** A number of genes, labeled A, B (top), and 5C, but not genes at locus (band) BR3 (5C, BR3, bottom) are activated when *Chironomus tentans* larvae are subjected to heat shock (39°C for 30 min). **(A)** Distribution of RNA polymerase II in isolated chromosome IV from the salivary gland (**at arrows**). The enzyme was detected by immunofluorescence using an antibody directed against the polymerase. The 5C and BR3 are specific bands of chromosome IV, and the arrows indicate puffs. **(B)** Autoradiogram of a chromosome IV that was incubated in ^{3}H-uridine to label the RNA. Note the correspondence of the immunofluorescence and presence of the radioactive RNA (black dots). Bar = 7 μm. (Reproduced, with permission, from Sass H: RNA polymerase B in polytene chromosomes. Cell 1982;28:274. Copyright © 1982. Reprinted with permission from Elsevier.)

is an essential structure for chromosomal segregation during mitosis.

The ends of each chromosome contain structures called **telomeres.** Telomeres consist of short TG-rich repeats. Human telomeres have a variable number of repeats of the sequence 5′-TTAGGG-3′, which can extend for several kilobases. **Telomerase,** a multisubunit RNA template-containing complex related to viral RNA-dependent DNA polymerases (reverse transcriptases), is the enzyme responsible for telomere synthesis and thus for maintaining the length of the telomere. Since telomere shortening has been associated with both malignant transformation and aging, this enzyme has become an attractive target for cancer chemotherapy and drug development. Each sister chromatid contains one dsDNA molecule. During interphase, the packing of the DNA molecule is less dense than it is in the condensed chromosome during metaphase. Metaphase chromosomes are nearly completely transcriptionally inactive.

The human haploid genome consists of about 3×10^9 bp and about 1.7×10^7 nucleosomes. Thus, each of the 23 chromatids

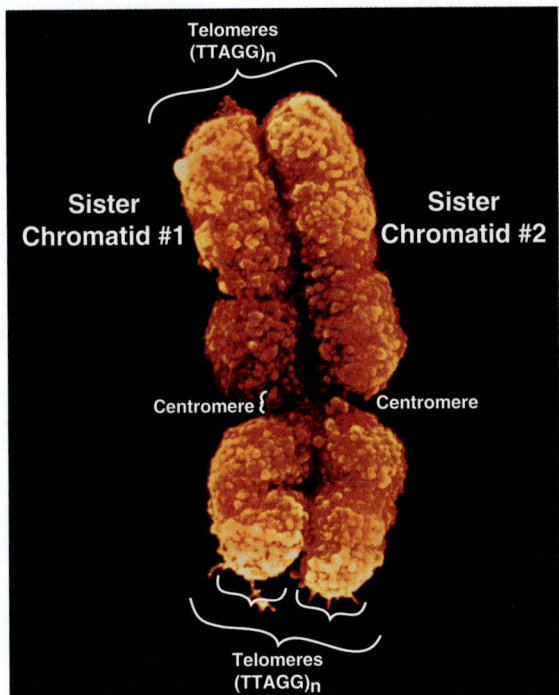

FIGURE 35–5 **The two sister chromatids of mitotic human chromosome 12.** The location of the A+T-rich centromeric region connecting sister chromatids is indicated, as are two of the four telomeres residing at the very ends of the chromatids that are attached one to the other at the centromere. (Courtesy of Biophoto Associates/Photo Researchers, Inc.)

TABLE 35–2 The Packing or Compaction Ratios of Each of the Orders of DNA Structure

Chromatin Form	Packing Ratio
Naked double-helical DNA	~1.0
10-nm fibril of nucleosomes	7–10
30-nm chromatin fiber of superhelical nucleosomes	40–60
Condensed metaphase chromosome of loops	8000

chromosome. In metaphase chromosomes, the 30-nm chromatin fibers are also folded into a series of **looped domains,** the proximal portions of which are anchored to a nonhistone proteinaceous nuclear matrix scaffolding within the nucleus (Figure 35–3). The packing ratios of each of the orders of DNA structure are summarized in **Table 35–2.** The packaging of nucleoproteins within chromatids is not random, as evidenced by the characteristic patterns observed when chromosomes are stained with specific dyes such as quinacrine or Giemsa stain (**Figure 35–6**).

From individual to individual within a single species, the pattern of staining (banding) of the entire chromosome complement is highly reproducible; nonetheless, it differs significantly between species, even those closely related. Thus, the packaging of the nucleoproteins in chromosomes of higher eukaryotes must in some way be dependent upon species-specific characteristics of the DNA molecules.

A combination of specialized staining techniques and high-resolution microscopy has allowed cytogeneticists to quite precisely map many genes to specific regions of mouse and human chromosomes. With the recent elucidation of the

in the human haploid genome would contain on the average 1.3×10^8 nucleotides in one dsDNA molecule. Therefore, the length of each DNA molecule must be compressed about 8000-fold to generate the structure of a condensed metaphase

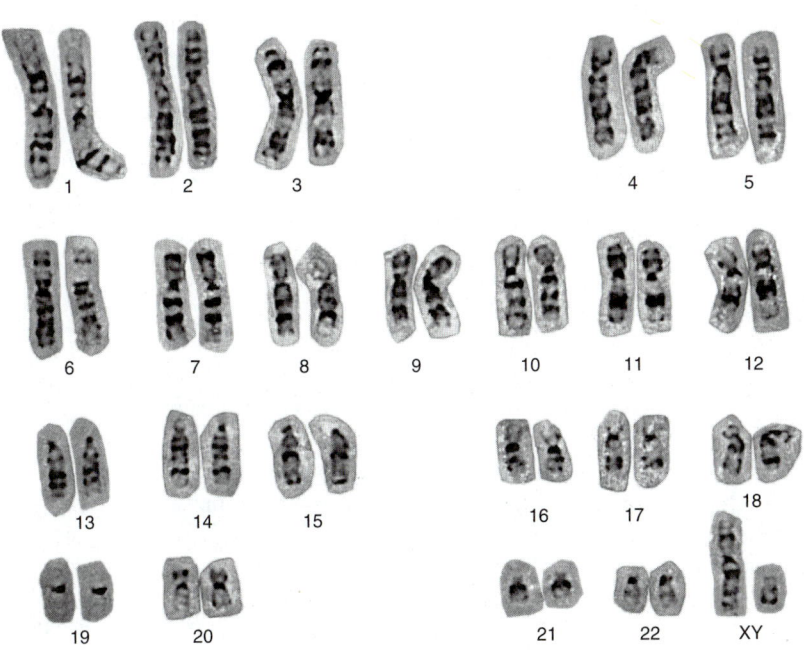

FIGURE 35–6 **A human karyotype (of a man with a normal 46,XY constitution), in which the metaphase chromosomes have been stained by the Giemsa method and aligned according to the Paris Convention.** (Courtesy of H Lawce and F Conte.)

human and mouse genome sequences (among others), it has become clear that many of these visual mapping methods were remarkably accurate.

Coding Regions Are Often Interrupted by Intervening Sequences

The **protein coding regions of DNA,** the transcripts of which ultimately appear in the cytoplasm as single mRNA molecules, are usually **interrupted in the eukaryotic genome by large intervening sequences of nonprotein-coding DNA.** Accordingly, the **primary transcripts of DNA, mRNA precursors,** (originally termed **hnRNA** because this species of RNA was quite heterogeneous in size [length] and mostly restricted to the nucleus), contain noncoding intervening sequences of RNA that must be removed in a process which also joins together the appropriate coding segments to form the mature mRNA. Most coding sequences for a single mRNA are interrupted in the genome (and thus in the primary transcript) by at least one—and in some cases as many as 50—noncoding intervening sequences (**introns**). In most cases, the introns are much longer than the coding regions (**exons**). The processing of the primary transcript, which involves precise removal of introns and splicing of adjacent exons, is described in Chapter 36.

The function of the intervening sequences, or introns, is not totally clear. Introns may serve to separate functional domains (exons) of coding information in a form that permits genetic rearrangement by recombination to occur more rapidly than if all coding regions for a given genetic function were contiguous. Such an enhanced rate of genetic rearrangement of functional domains might allow more rapid evolution of biologic function. In some instances other protein or noncoding RNAs are localized within the intronic DNA of certain genes (Chapter 34). The relationships among chromosomal DNA, gene clusters on the chromosome, the exon–intron structure of genes, and the final mRNA product are illustrated in **Figure 35–7.**

MUCH OF THE MAMMALIAN GENOME APPEARS REDUNDANT & MUCH IS NOT HIGHLY TRANSCRIBED

The haploid genome of each human cell consists of 3×10^9 bp of DNA subdivided into 23 chromosomes. The entire haploid genome contains sufficient DNA to code for nearly 1.5 million average-sized genes. However, studies of mutation rates and of the complexities of the genomes of higher organisms strongly suggest that humans have significantly fewer than 100,000 proteins encoded by the ~1% of the human genome that is composed of exonic DNA. Indeed current estimates suggest there are 25,000 or less protein-coding genes in humans. This implies that most of the DNA is nonprotein-coding—that is, its information is never translated into an amino acid sequence of a protein molecule. Certainly, some of the excess DNA sequences serve to regulate the expression of genes during development, differentiation, and adaptation to the environment, either by serving as binding sites for regulatory proteins or by encoding regulatory RNAs (ie, miRNAs and ncRNAs). Some excess clearly makes up the intervening sequences or introns (24% of the total human genome) that split the coding regions of genes, and another portion of the excess appears to be composed of many families of repeated sequences for which clear functions have not yet been defined, though some small RNAs transcribed from these repeats can modulate transcription, either directly by interacting with the transcription machinery or indirectly by affecting the activity of the chromatin template. A summary of the salient features of the human genome is presented in Chapter 39. Interestingly, the ENCODE Project Consortium (Chapter 39) has shown that for the 1% of the genome studied most of the genomic sequence was indeed transcribed at a low rate. Further research will elucidate the role(s) played by such transcripts.

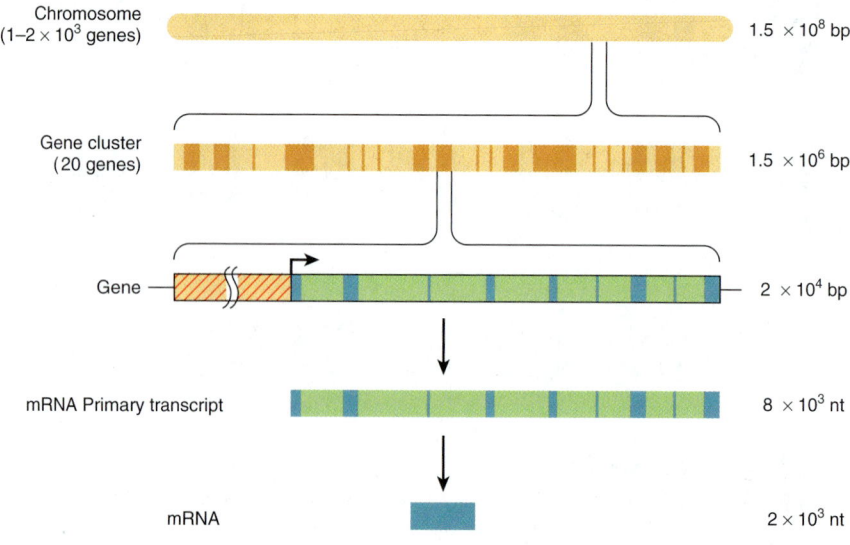

FIGURE 35–7 The relationship between chromosomal DNA and mRNA. The human haploid DNA complement of 3×10^9 bp is distributed between 23 chromosomes. Genes are often clustered on these chromosomes. An average gene is 2×10^4 bp in length, including the regulatory region (red-hatched area), which is usually located at the 5' end of the gene. The regulatory region is shown here as being adjacent to the transcription initiation site (arrow). Most eukaryotic genes have alternating exons and introns. In this example, there are nine exons (blue colored areas) and eight introns (green colored areas). The introns are removed from the primary transcript by the processing reactions, and the exons are ligated together in sequence to form the mature mRNA. (nt, nucleotides.)

Chromosome (1–2 × 10³ genes)   1.5 × 10⁸ bp

Gene cluster (20 genes)   1.5 × 10⁶ bp

Gene   2 × 10⁴ bp

mRNA Primary transcript   8 × 10³ nt

mRNA   2 × 10³ nt

The DNA in a eukaryotic genome can be divided into different "sequence classes." These are unique-sequence DNA, or nonrepetitive DNA and repetitive-sequence DNA. In the haploid genome, unique-sequence DNA generally includes the single copy genes that code for proteins. The repetitive DNA in the haploid genome includes sequences that vary in copy number from 2 to as many as 10^7 copies per cell.

More Than Half the DNA in Eukaryotic Organisms Is in Unique or Nonrepetitive Sequences

This estimation (and the distribution of repetitive-sequence DNA) is based on a variety of DNA–RNA hybridization techniques and, more recently, on direct DNA sequencing. Similar techniques are used to estimate the number of active genes in a population of unique-sequence DNA. In brewers' yeast (*Saccharomyces cerevisiae*, a lower eukaryote), about two-thirds of its 6200 genes are expressed, but only ~1/5 are required for viability under laboratory growth conditions. In typical, tissues in a higher eukaryote (eg, mammalian liver and kidney), between 10,000 and 15,000 genes are actively expressed. Different combinations of genes are expressed in each tissue, of course, and how this is accomplished is one of the major unanswered questions in biology.

In Human DNA, at Least 30% of the Genome Consists of Repetitive Sequences

Repetitive-sequence DNA can be broadly classified as moderately repetitive or as highly repetitive. The highly repetitive sequences consist of 5–500 base pair lengths repeated many times in tandem. These sequences are often clustered in centromeres and telomeres of the chromosome and some are present in about 1–10 million copies per haploid genome. The majority of these sequences are transcriptionally inactive and some of these sequences play a structural role in the chromosome (Figure 35–5; see Chapter 39).

The moderately repetitive sequences, which are defined as being present in numbers of less than 10^6 copies per haploid genome, are not clustered but are interspersed with unique sequences. In many cases, these long interspersed repeats are transcribed by RNA polymerase II and contain caps indistinguishable from those on mRNA.

Depending on their length, moderately repetitive sequences are classified as **long interspersed repeat sequences (LINEs)** or **short interspersed repeat sequences (SINEs)**. Both types appear to be **retroposons;** that is, they arose from movement from one location to another **(transposition)** through an RNA intermediate by the action of reverse transcriptase that transcribes an RNA template into DNA. Mammalian genomes contain 20,000–50,000 copies of the 6–7 kbp LINEs. These represent species-specific families of repeat elements. SINEs are shorter (70–300 bp), and there may be more than 100,000 copies per genome. Of the SINEs in the human genome, one family, the **Alu family,** is present in about 500,000 copies per haploid genome and accounts for ~10% of the human genome. Members of the human Alu family and their closely related analogs in other animals are transcribed as integral components of mRNA precursors or as discrete RNA molecules, including the well-studied 4.5S RNA and 7S RNA. These particular family members are highly conserved within a species as well as between mammalian species. Components of the short interspersed repeats, including the members of the Alu family, may be mobile elements, capable of jumping into and out of various sites within the genome (see below). These transposition events can have disastrous results, as exemplified by the insertion of Alu sequences into a gene, which, when so mutated, causes neurofibromatosis. Additionally, Alu B1 and B2 SINE RNAs have been shown to regulate mRNA production at the levels of transcription and mRNA splicing.

Microsatellite Repeat Sequences

One category of repeat sequences exists as both dispersed and grouped tandem arrays. The sequences consist of 2–6 bp repeated up to 50 times. These **microsatellite sequences** most commonly are found as dinucleotide repeats of AC on one strand and TG on the opposite strand, but several other forms occur, including CG, AT, and CA. The AC repeat sequences occur at 50,000–100,000 locations in the genome. At any locus, the number of these repeats may vary on the two chromosomes, thus providing heterozygosity of the number of copies of a particular microsatellite number in an individual. This is a heritable trait, and because of their number and the ease of detecting them using the **polymerase chain reaction (PCR)** (Chapter 39), such repeats are useful in constructing genetic linkage maps. Most genes are associated with one or more microsatellite markers, so the relative position of genes on chromosomes can be assessed, as can the association of a gene with a disease. Using PCR, a large number of family members can be rapidly screened for a certain **microsatellite polymorphism.** The association of a specific polymorphism with a gene in affected family members—and the lack of this association in unaffected members—may be the first clue about the genetic basis of a disease.

Trinucleotide sequences that increase in number (microsatellite instability) can cause disease. The unstable p(CGG)$_n$ repeat sequence is associated with the fragile X syndrome. Other trinucleotide repeats that undergo dynamic mutation (usually an increase) are associated with Huntington's chorea (CAG), myotonic dystrophy (CTG), spinobulbar muscular atrophy (CAG), and Kennedy disease (CAG).

ONE PERCENT OF CELLULAR DNA IS IN MITOCHONDRIA

The majority of the polypeptides in mitochondria (about 54 out of 67) are coded by nuclear genes, while the rest are coded by genes found in mitochondrial (mt) DNA. Human

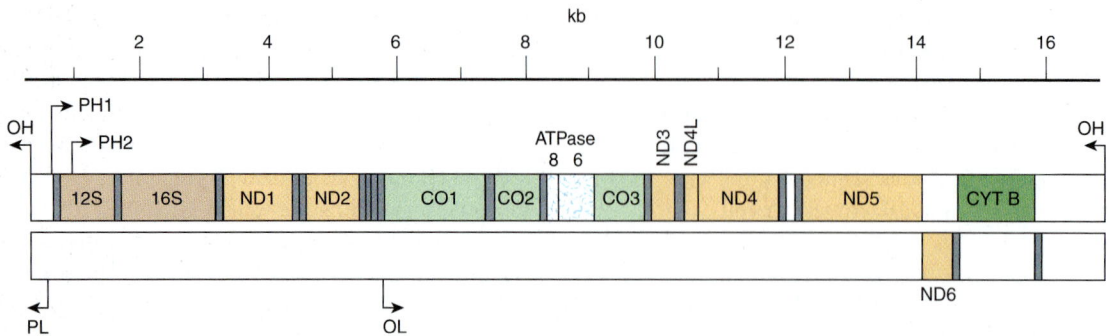

FIGURE 35–8   Maps of human mitochondrial genes. The maps represent the so-called heavy (upper strand) and light (lower map) strands of linearized mitochondrial (mt) DNA, showing the genes for the subunits of NADH-coenzyme Q oxidoreductase (ND1 through ND6), cytochrome *c* oxidase (CO1 through CO3), cytochrome *b* (CYT B), and ATP synthase (ATPase 8 and 6) and for the 12S and 16S ribosomal mt rRNAs. The transfer RNAs are denoted by small blue boxes. The origin of heavy-strand (OH) and light-strand (OL) replication and the promoters for the initiation of heavy-strand (PH1 and PH2) and light-strand (PL) transcription are indicated by arrows. (Reproduced, with permission, from Moraes CT et al: Mitochondrial DNA deletions in progressive external ophthalmoplegia and Kearns-Sayre syndrome. N Engl J Med 1989;320:1293. Copyright ©1989. Massachusetts Medical Society. All rights reserved.)

mitochondria contain 2–10 copies of a small circular dsDNA molecule that makes up approximately 1% of total cellular DNA. This mtDNA codes for mt-specific ribosomal and transfer RNAs and for 13 proteins that play key roles in the respiratory chain (Chapter 13). The linearized structural map of the human mitochondrial genes is shown in **Figure 35–8**. Some of the features of mtDNA are shown in **Table 35–3**.

An important feature of human mitochondrial mtDNA is that—because all mitochondria are contributed by the ovum during zygote formation—it is transmitted by maternal nonmendelian inheritance. Thus, in diseases resulting from mutations of mtDNA, an affected mother would in theory pass the disease to all of her children but only her daughters would transmit the trait. However, in some cases, deletions in mtDNA occur during oogenesis and thus are not inherited from the mother. A number of diseases have now been shown to be due to mutations of mtDNA. These include a variety of myopathies, neurologic disorders, and some cases of diabetes mellitus.

GENETIC MATERIAL CAN BE ALTERED & REARRANGED

An alteration in the sequence of purine and pyrimidine bases in a gene due to a change—a removal or an insertion—of one or more bases may result in an altered gene product. Such alteration in the genetic material results in a **mutation** whose consequences are discussed in detail in Chapter 37.

Chromosomal Recombination Is One Way of Rearranging Genetic Material

Genetic information can be exchanged between similar or homologous chromosomes. The exchange, or **recombination** event, occurs primarily during meiosis in mammalian cells and requires alignment of homologous metaphase chromosomes, an alignment that almost always occurs with great exactness. A process of crossing over occurs as shown in **Figure 35–9**. This usually results in an equal and reciprocal exchange of genetic information between homologous chromosomes. If the homologous chromosomes possess different alleles of the same genes, the crossover may produce noticeable and heritable genetic linkage differences. In the rare case where the alignment of homologous chromosomes is not exact, the crossing over or recombination event may result in an unequal exchange of information. One chromosome may receive less genetic material and thus a deletion, while the other partner of the chromosome pair receives more genetic material and thus an

TABLE 35–3   Major Features of Human Mitochondrial DNA

- Is circular, double-stranded, and composed of heavy (H) and light (L) chains or strands
- Contains 16,569 bp
- Encodes 13 protein subunits of the respiratory chain (of a total of about 67)
    Seven subunits of NADH dehydrogenase (complex I)
    Cytochrome *b* of complex III
    Three subunits of cytochrome oxidase (complex IV)
    Two subunits of ATP synthase
- Encodes large (16S) and small (12S) mt ribosomal RNAs
- Encodes 22 mt tRNA molecules
- Genetic code differs slightly from the standard code
    UGA (standard stop codon) is read as Trp
    AGA and AGG (standard codons for Arg) are read as stop codons
- Contains very few untranslated sequences
- High mutation rate (5 to 10 times that of nuclear DNA)
- Comparisons of mtDNA sequences provide evidence about evolutionary origins of primates and other species

Source: Adapted from Harding AE: Neurological disease and mitochondrial genes. Trends Neurol Sci 1991;14:132. Copyright © 1991. Reprinted with permission from Elsevier.

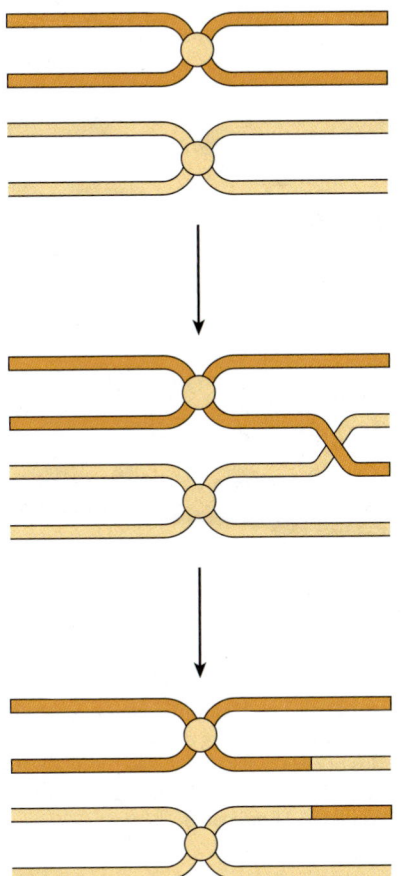

FIGURE 35–9 **The process of crossing over between homologous metaphase chromosomes to generate recombinant chromosomes.** See also Figure 35–12.

insertion or duplication (Figure 35–9). Unequal crossing over does occur in humans, as evidenced by the existence of hemoglobins designated Lepore and anti-Lepore (**Figure 35–10**). The farther apart two sequences are on an individual chromosome, the greater the likelihood of a crossover recombination event. This is the basis for genetic mapping methods. **Unequal crossover** affects tandem arrays of repeated DNAs whether they are related globin genes, as in Figure 35–10, or more abundant repetitive DNA. Unequal crossover through slippage in the

pairing can result in expansion or contraction in the copy number of the repeat family and may contribute to the expansion and fixation of variant members throughout the repeat array.

Chromosomal Integration Occurs with Some Viruses

Some bacterial viruses (bacteriophages) are capable of recombining with the DNA of a bacterial host in such a way that the genetic information of the bacteriophage is incorporated in a linear fashion into the genetic information of the host. This integration, which is a form of recombination, occurs by the mechanism illustrated in **Figure 35–11**. The backbone of the circular bacteriophage genome is broken, as is that of the DNA molecule of the host; the appropriate ends are resealed with the proper polarity. The bacteriophage DNA is figuratively straightened out ("linearized") as it is integrated into the bacterial DNA molecule—frequently a closed circle as well. The site at which the bacteriophage genome integrates or recombines with the bacterial genome is chosen by one of two mechanisms. If the bacteriophage contains a DNA sequence **homologous** to a sequence in the host DNA molecule, then a recombination event analogous to that occurring between homologous chromosomes can occur. However, some bacteriophages synthesize proteins that bind specific sites on bacterial chromosomes to a **nonhomologous** site characteristic of the bacteriophage DNA molecule. Integration occurs at the site and is said to be "**site specific.**"

Many animal viruses, particularly the oncogenic viruses—either directly or, in the case of RNA viruses such as HIV that causes AIDS, their DNA transcripts generated by the action of the viral **RNA-dependent DNA polymerase,** or **reverse transcriptase**—can be integrated into chromosomes of the mammalian cell. The integration of the animal virus DNA into the animal genome generally is not "site specific" but does display site preferences.

Transposition Can Produce Processed Genes

In eukaryotic cells, small DNA elements that clearly are not viruses are capable of transposing themselves in and out of

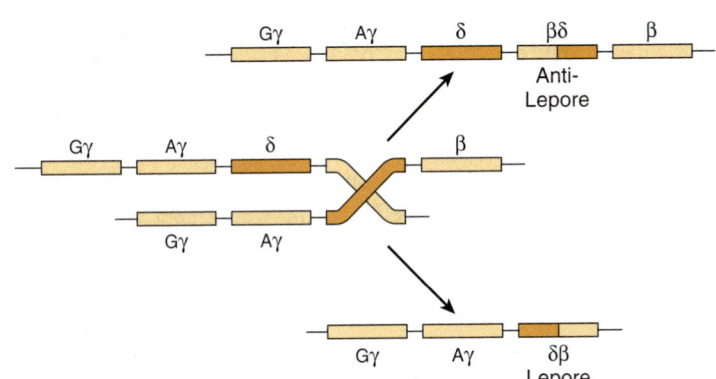

FIGURE 35–10 **The process of unequal crossover in the region of the mammalian genome that harbors the structural genes encoding hemoglobins and the generation of the unequal recombinant products hemoglobin delta-beta Lepore and beta-delta anti-Lepore.** The examples given show the locations of the crossover regions within amino acid coding regions of the indicated genes (ie, β and δ globin genes). (Redrawn and reproduced, with permission, from Clegg JB, Weatherall DJ: β⁰ Thalassemia: time for a reappraisal? Lancet 1974;2:133. Copyright © 1974. Reprinted with permission from Elsevier.)

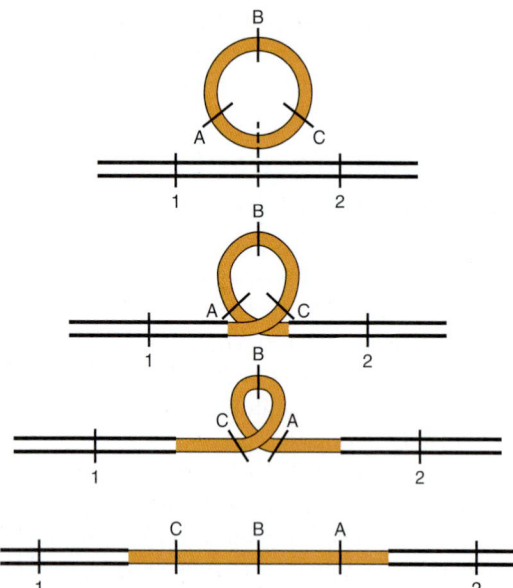

FIGURE 35–11 The integration of a circular genome from a virus (with genes A, B, and C) into the DNA molecule of a host (with genes 1 and 2) and the consequent ordering of the genes.

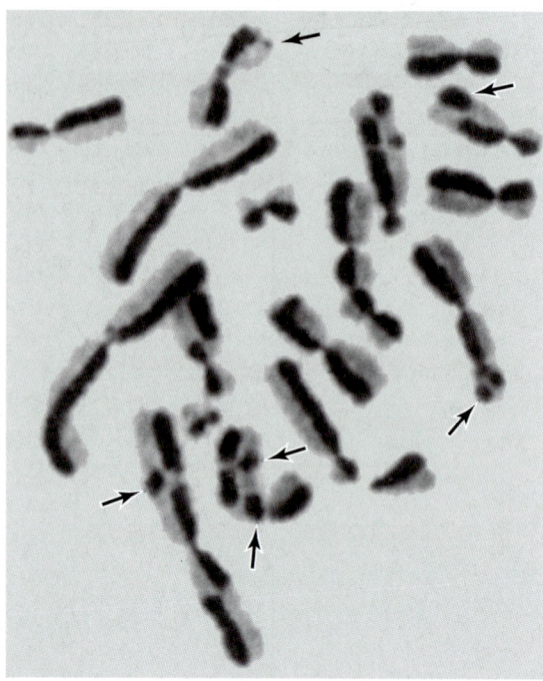

FIGURE 35–12 Sister chromatid exchanges between human chromosomes. The exchanges are detectable by Giemsa staining of the chromosomes of cells replicated for two cycles in the presence of bromodeoxyuridine. The arrows indicate some regions of exchange. (Courtesy of S Wolff and J Bodycote.)

the host genome in ways that affect the function of neighboring DNA sequences. These mobile elements, sometimes called **"jumping DNA,"** or jumping genes, can carry flanking regions of DNA and, therefore, profoundly affect evolution. As mentioned above, the Alu family of moderately repeated DNA sequences has structural characteristics similar to the termini of retroviruses, which would account for the ability of the latter to move into and out of the mammalian genome.

Direct evidence for the transposition of other small DNA elements into the human genome has been provided by the discovery of **"processed genes"** for immunoglobulin molecules, α-globin molecules, and several others. These processed genes consist of DNA sequences identical or nearly identical to those of the messenger RNA for the appropriate gene product. That is, the 5′-nontranslated region, the coding region without intron representation, and the 3′ poly(A) tail are all present contiguously. This particular DNA sequence arrangement must have resulted from the reverse transcription of an appropriately processed messenger RNA molecule from which the intron regions had been removed and the poly(A) tail added. The only recognized mechanism this reverse transcript could have used to integrate into the genome would have been a transposition event. In fact, these "processed genes" have short terminal repeats at each end, as do known transposed sequences in lower organisms. In the absence of their transcription and thus genetic selection for function, many of the processed genes have been randomly altered through evolution so that they now contain nonsense codons that preclude their ability to encode a functional, intact protein (see Chapter 37). Thus, they are referred to as **"pseudogenes."**

Gene Conversion Produces Rearrangements

Besides unequal crossover and transposition, a third mechanism can effect rapid changes in the genetic material. Similar sequences on homologous or nonhomologous chromosomes may occasionally pair up and eliminate any mismatched sequences between them. This may lead to the accidental fixation of one variant or another throughout a family of repeated sequences and thereby homogenize the sequences of the members of repetitive DNA families. This latter process is referred to as **gene conversion.**

Sister Chromatids Exchange

In diploid eukaryotic organisms such as humans, after cells progress through the S phase they contain a tetraploid content of DNA. This is in the form of sister chromatids of chromosome pairs (Figure 35–6). Each of these sister chromatids contains identical genetic information since each is a product of the semiconservative replication of the original parent DNA molecule of that chromosome. Crossing over can occur between these genetically identical sister chromatids. Of course, these **sister chromatid exchanges** (**Figure 35–12**) have no genetic consequence as long as the exchange is the result of an equal crossover.

Immunoglobulin Genes Rearrange

In mammalian cells, some interesting gene rearrangements occur normally during development and differentiation. For example, in mice the V_L and C_L genes for a single immunoglobulin molecule (see Chapter 38) are widely separated in the germ line DNA. In the DNA of a differentiated immunoglobulin-producing (plasma) cell, the same V_L and C_L genes have been moved physically closer together in the genome and into the same transcription unit. However, even then, this rearrangement of DNA during differentiation does not bring the V_L and C_L genes into contiguity in the DNA. Instead, the DNA contains an interspersed or interruption sequence of about 1200 bp at or near the junction of the V and C regions. The interspersed sequence is transcribed into RNA along with the V_L and C_L genes, and the interspersed information is removed from the RNA during its nuclear processing (Chapters 36 and 38).

DNA SYNTHESIS & REPLICATION ARE RIGIDLY CONTROLLED

The primary function of DNA replication is understood to be the provision of progeny with the genetic information possessed by the parent. Thus, the replication of DNA must be complete and carried out in such a way as to maintain genetic stability within the organism and the species. The process of DNA replication is complex and involves many cellular functions and several verification procedures to ensure fidelity in replication. About 30 proteins are involved in the replication of the *Escherichia coli* chromosome, and this process is more complex in eukaryotic organisms. The first enzymologic observations on DNA replication were made by Arthur Kornberg, who described in *E coli* the existence of an enzyme now called DNA polymerase I. This enzyme has multiple catalytic activities, a complex structure, and a requirement for the triphosphates of the four deoxyribonucleosides of adenine, guanine, cytosine, and thymine. The polymerization reaction catalyzed by DNA polymerase I of *E coli* has served as a prototype for all DNA polymerases of both prokaryotes and eukaryotes, even though it is now recognized that the major role of this polymerase is proofreading and repair.

In all cells, replication can occur only from a single-stranded DNA (ssDNA) template. Therefore, mechanisms must exist to target the site of initiation of replication and to unwind the dsDNA in that region. The replication complex must then form. After replication is complete in an area, the parent and daughter strands must re-form dsDNA. In eukaryotic cells, an additional step must occur. The dsDNA must re-form the chromatin structure, including nucleosomes, that existed prior to the onset of replication. Although this entire process is not completely understood in eukaryotic cells, replication has been quite precisely described in prokaryotic cells, and the general principles are the same in both. The major steps are listed in **Table 35–4**, illustrated in **Figure 35–13**, and discussed, in sequence, below. A number of proteins, most with specific enzymatic action, are involved in this process (**Table 35–5**).

The Origin of Replication

At the **origin of replication (ori),** there is an association of sequence-specific dsDNA-binding proteins with a series of

direct repeat DNA sequences. In bacteriophage λ, the oriλ is bound by the λ-encoded O protein to four adjacent sites. In *E coli*, the oriC is bound by the protein dnaA. In both cases, a complex is formed consisting of 150–250 bp of DNA and multimers of the DNA-binding protein. This leads to the local denaturation and unwinding of an adjacent A+T-rich region of DNA. Functionally similar **autonomously replicating sequences (ARS) or replicators** have been identified in yeast cells. The ARS contains a somewhat degenerate 11-bp sequence called the **origin replication element (ORE).** The ORE binds a set of proteins, analogous to the dnaA protein of *E coli*, the group of proteins is collectively called the **origin recognition complex (ORC).** ORC homologs have been found in all eukaryotes examined. The ORE is located adjacent to an approximately 80-bp A+T-rich sequence that is easy to unwind. This is called the **DNA unwinding element (DUE).** The DUE is the origin of replication in yeast and is bound by the MCM protein complex.

Consensus sequences similar to ori or ARS in structure have not been precisely defined in mammalian cells, though several of the proteins that participate in ori recognition and function have been identified and appear quite similar to their yeast counterparts in both amino acid sequence and function.

Unwinding of DNA

The interaction of proteins with ori defines the start site of replication and provides a short region of ssDNA essential for initiation of synthesis of the nascent DNA strand. This process requires the formation of a number of protein–protein and protein–DNA interactions. A critical step is provided by a DNA helicase that allows for processive unwinding of DNA. In uninfected *E coli*, this function is provided by a complex of dnaB helicase and the dnaC protein. Single-stranded DNA-binding proteins (SSBs) stabilize this complex. In λ phage-infected *E coli*, the phage protein P binds to dnaB and the P/dnaB complex binds to oriλ by interacting with the O protein. dnaB is not an active helicase when in the P/dnaB/O complex. Three *E coli* heat shock proteins (dnaK, dnaJ, and GrpE) cooperate to remove the P protein and activate the dnaB helicase. In cooperation with SSB, this leads to DNA unwinding and active replication. In this way, the replication

TABLE 35–4 Steps Involved in DNA Replication in Eukaryotes

1. Identification of the origins of replication
2. Unwinding (denaturation) of dsDNA to provide an ssDNA template
3. Formation of the replication fork; synthesis of RNA primer
4. Initiation of DNA synthesis and elongation
5. Formation of replication bubbles with ligation of the newly synthesized DNA segments
6. Reconstitution of chromatin structure

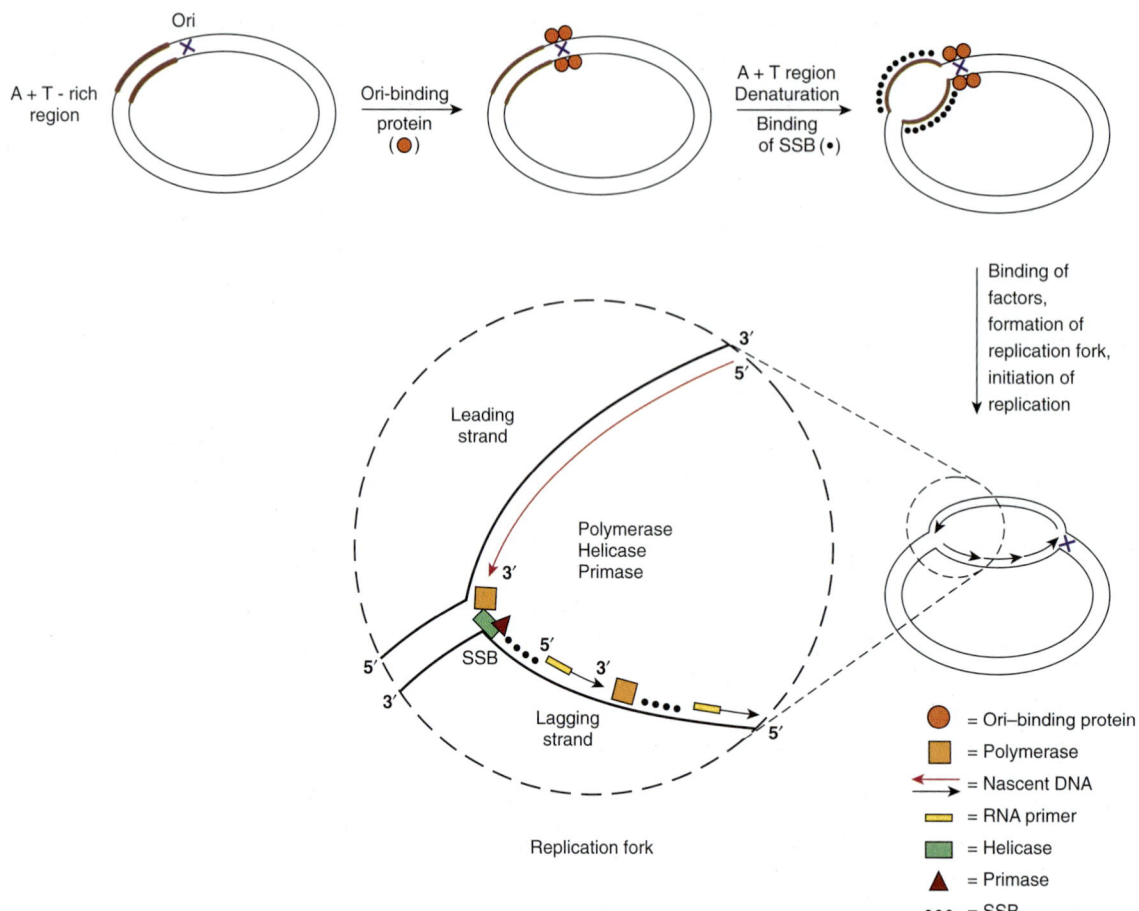

FIGURE 35–13 **Steps involved in DNA replication.** This figure describes DNA replication in an *E coli* cell, but the general steps are similar in eukaryotes. A specific interaction of a protein (the dnaA protein) to the origin of replication (oriC) results in local unwinding of DNA at an adjacent A+T-rich region. The DNA in this area is maintained in the single-strand conformation (ssDNA) by single-strand-binding proteins (SSBs). This allows a variety of proteins, including helicase, primase, and DNA polymerase, to bind and to initiate DNA synthesis. The replication fork proceeds as DNA synthesis occurs continuously (long red arrow) on the leading strand and discontinuously (short black arrows) on the lagging strand. The nascent DNA is always synthesized in the 5′ to 3′ direction, as DNA polymerases can add a nucleotide only to the 3′ end of a DNA strand.

of the λ phage is accomplished at the expense of replication of the host *E coli* cell.

Formation of the Replication Fork

A replication fork consists of four components that form in the following sequence: (1) the DNA helicase unwinds a short segment of the parental duplex DNA; (2) a primase initiates synthesis of an RNA molecule that is essential for priming DNA synthesis; (3) the DNA polymerase initiates nascent, daughter-strand synthesis; and (4) SSBs bind to ssDNA and prevent premature reannealing of ssDNA to dsDNA. These reactions are illustrated in Figure 35–13.

The DNA polymerase III enzyme (the *dnaE* gene product in *E coli*) binds to template DNA as part of a multiprotein complex that consists of several polymerase accessory factors (β, γ, δ, δ′, and τ). DNA polymerases only synthesize DNA in the 5′–3′ direction, and only one of the several different types of polymerases is involved at the replication fork. Because the

DNA strands are antiparallel (Chapter 34), the polymerase functions asymmetrically. On the **leading (forward) strand,** the DNA is synthesized continuously. On the **lagging (retrograde) strand,** the DNA is synthesized in short (1–5 kb; see Figure 35–16) fragments, the so-called **Okazaki fragments.** Several Okazaki fragments (up to a thousand) must be sequentially synthesized for each replication fork. To ensure that this happens, the helicase acts on the lagging strand to unwind dsDNA in a 5′–3′ direction. The helicase associates with the primase to afford the latter proper access to the template. This allows the RNA primer to be made and, in turn, the polymerase to begin replicating the DNA. This is an important reaction sequence since DNA polymerases cannot initiate DNA synthesis de novo. The mobile complex between helicase and primase has been called a **primosome.** As the synthesis of an Okazaki fragment is completed and the polymerase is released, a new primer has been synthesized. The same polymerase molecule remains associated with the replication fork and proceeds to synthesize the next Okazaki fragment.

TABLE 35–5 Classes of Proteins Involved in Replication

Protein	Function
DNA polymerases	Deoxynucleotide polymerization
Helicases	Processive unwinding of DNA
Topoisomerases	Relieve torsional strain that results from helicase-induced unwinding
DNA primase	Initiates synthesis of RNA primers
Single-strand binding proteins	Prevent premature reannealing of dsDNA
DNA ligase	Seals the single strand nick between the nascent chain and Okazaki fragments on lagging strand

TABLE 35–6 A Comparison of Prokaryotic and Eukaryotic DNA Polymerases

E coli	Eukaryotic	Function
I		Gap filling following DNA replication, repair, and recombination
II		DNA proofreading and repair
	β	DNA repair
	γ	Mitochondrial DNA synthesis
III	ε	Processive, leading strand synthesis
DnaG	α	Primase
	δ	Processive, lagging strand synthesis

The DNA Polymerase Complex

A number of different DNA polymerase molecules engage in DNA replication. These share three important properties: (1) **chain elongation,** (2) **processivity,** and (3) **proofreading.** Chain elongation accounts for the rate (in nucleotides per second; nt/s) at which polymerization occurs. Processivity is an expression of the number of nucleotides added to the nascent chain before the polymerase disengages from the template. The proofreading function identifies copying errors and corrects them. In *E coli,* DNA polymerase III (pol III) functions at the replication fork. Of all polymerases, it catalyzes the highest rate of chain elongation and is the most processive. It is capable of polymerizing 0.5 Mb of DNA during one cycle on the leading strand. Pol III is a large (>1 MDa), multisubunit protein complex in *E coli.* DNA pol III associates with the two identical β subunits of the DNA sliding "clamp"; this association dramatically increases pol III-DNA complex stability, processivity (100 to >50,000 nucleotides) and rate of chain elongation (20–50 nt/s) generating the high degree of processivity the enzyme exhibits.

Polymerase I (pol I) and II (pol II) are mostly involved in proofreading and DNA repair. Eukaryotic cells have counterparts for each of these enzymes plus a large number of additional DNA polymerases primarily involved in DNA repair. A comparison is shown in **Table 35–6.**

In mammalian cells, the polymerase is capable of polymerizing at a rate that is somewhat slower than the rate of polymerization of deoxynucleotides by the bacterial DNA polymerase complex. This reduced rate may result from interference by nucleosomes.

Initiation & Elongation of DNA Synthesis

The initiation of DNA synthesis (**Figure 35–14**) requires **priming by a short length of RNA,** about 10–200 nucleotides long. In *E coli* this is catalyzed by dnaG (primase), in eukaryotes DNA Pol α synthesizes these RNA primers. The priming process involves nucleophilic attack by the 3′-hydroxyl group of the RNA primer on the phosphate of the first entering deoxynucleoside triphosphate (*N* in Figure 35–14) with the splitting off of pyrophosphate; this transition to DNA synthesis is catalyzed by the appropriate DNA polymerases (DNA pol III in *E coli*; DNA pol δ and ε in eukaryotes). The 3′-hydroxyl group of the recently attached deoxyribonucleoside monophosphate is then free to carry out a **nucleophilic attack** on the next entering deoxyribonucleoside triphosphate (N + 1 in Figure 35–14), again at its α phosphate moiety, with the splitting off of pyrophosphate. Of course, selection of the proper deoxyribonucleotide whose terminal 3′-hydroxyl group is to be attacked is dependent upon **proper base pairing with the other strand** of the DNA molecule according to the rules proposed originally by Watson and Crick (**Figure 35–15**). When an adenine deoxyribonucleoside monophosphoryl moiety is in the template position, a thymidine triphosphate will enter and its α phosphate will be attacked by the 3′-hydroxyl group of the deoxyribonucleoside monophosphoryl most recently added to the polymer. By this stepwise process, the template dictates which deoxyribonucleoside triphosphate is complementary and by hydrogen bonding holds it in place while the 3′-hydroxyl group of the growing strand attacks and incorporates the new nucleotide into the polymer. These segments of DNA attached to an RNA initiator component are the **Okazaki fragments** (**Figure 35–16**). In mammals, after many Okazaki fragments are generated, the replication complex begins to remove the RNA primers, to fill in the gaps left by their removal with the proper base-paired deoxynucleotide, and then to seal the fragments of newly synthesized DNA by enzymes referred to as **DNA ligases.**

Replication Exhibits Polarity

As has already been noted, DNA molecules are double stranded and the two strands are antiparallel. The replication of DNA in prokaryotes and eukaryotes occurs on both strands simultaneously. However, an enzyme capable of polymerizing DNA in the 3′ to 5′ direction does not exist in any organism, so that both of the newly replicated DNA strands cannot grow in the same direction simultaneously. Nevertheless, the same enzyme does

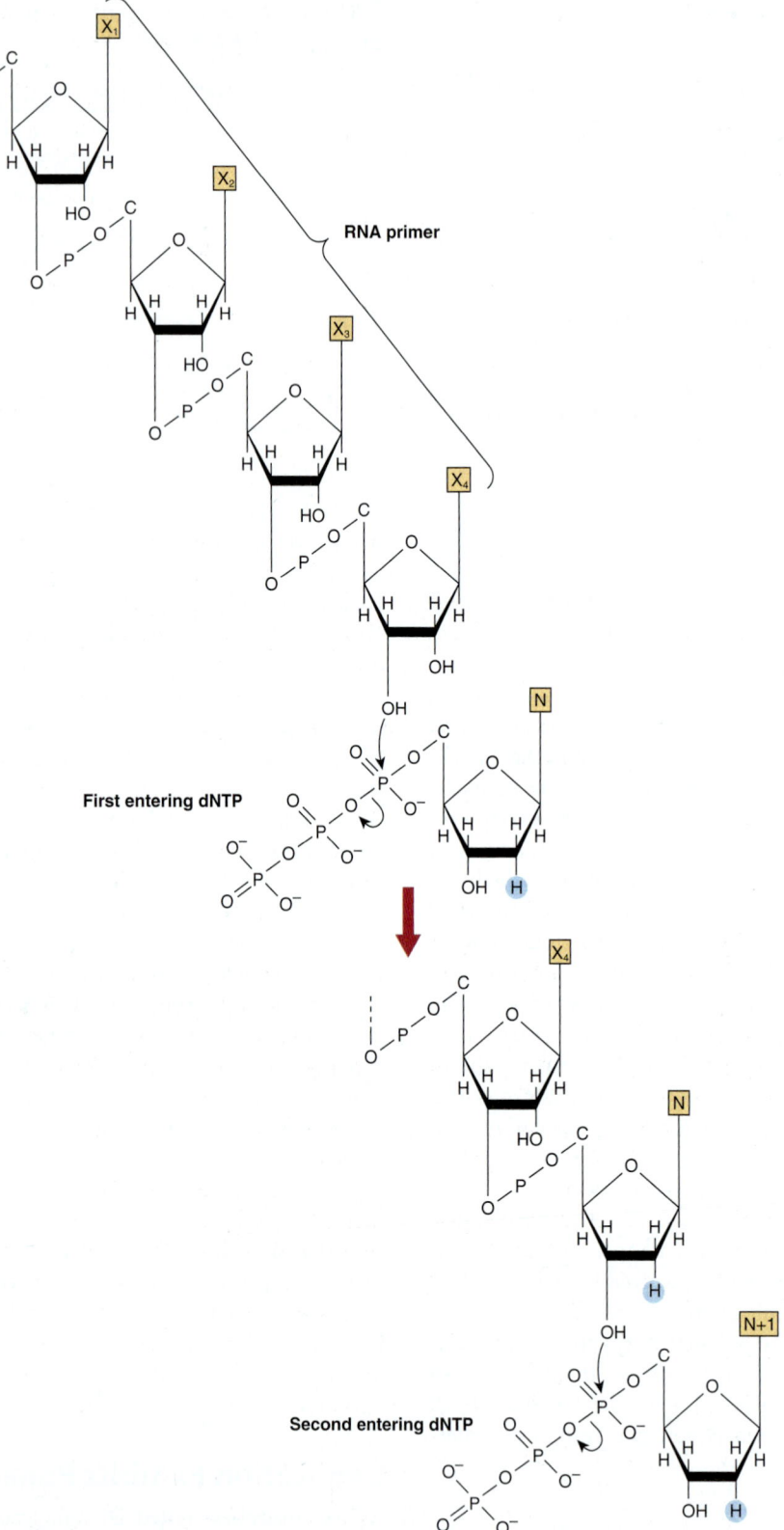

FIGURE 35–14 The initiation of DNA synthesis upon a primer of RNA and the subsequent attachment of the second deoxyribonucleoside triphosphate.

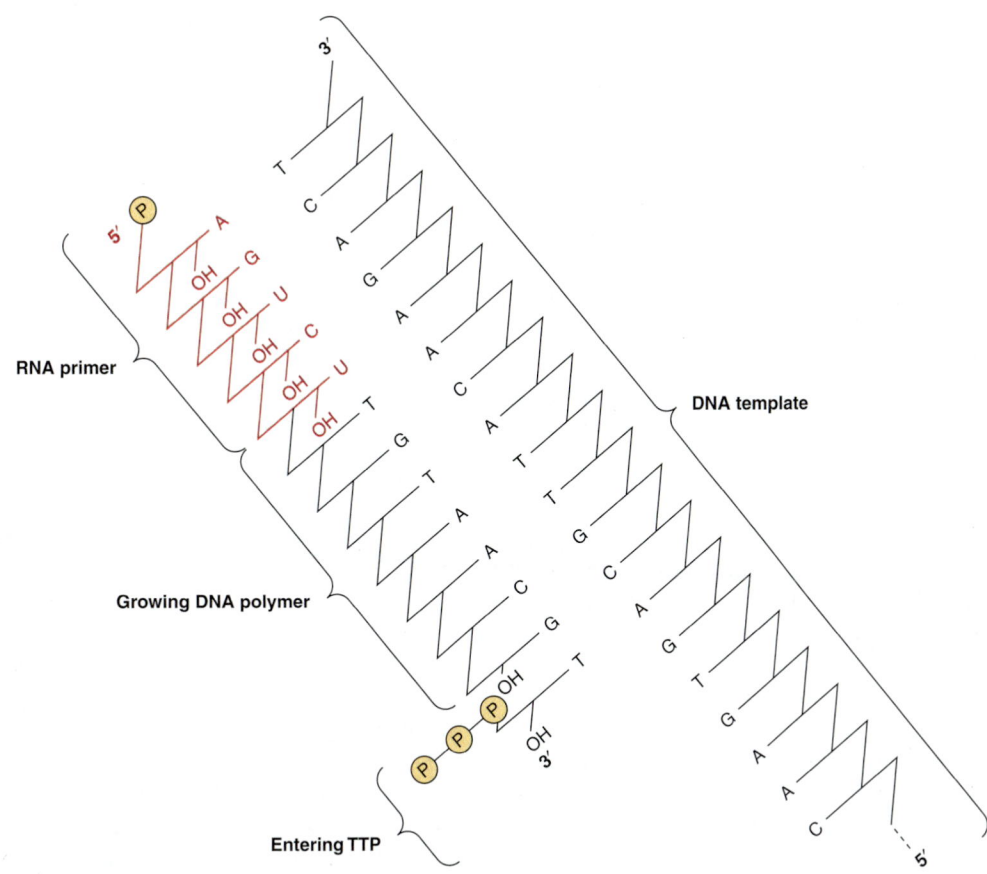

FIGURE 35–15 The RNA-primed synthesis of DNA demonstrating the template function of the complementary strand of parental DNA.

replicate both strands at the same time. The single enzyme replicates one strand ("leading strand") in a continuous manner in the 5′ to 3′ direction, with the same overall forward direction. It replicates the other strand ("lagging strand") discontinuously while polymerizing the nucleotides in short spurts of 150–250 nucleotides, again in the 5′ to 3′ direction, but at the same time it faces toward the back end of the preceding RNA primer rather than toward the unreplicated portion. This process of **semidiscontinuous DNA synthesis** is shown diagrammatically in Figures 35–13 and 35–16.

Formation of Replication Bubbles

Replication proceeds from a single ori in the circular bacterial chromosome, composed of roughly 5×10^6 bp of DNA.

This process is completed in about 30 min, a replication rate of 3×10^5 bp/min. The entire mammalian genome replicates in approximately 9 h, the average period required for formation of a tetraploid genome from a diploid genome in a replicating cell. If a mammalian genome (3×10^9 bp) replicated at the same rate as bacteria (ie, 3×10^5 bp/min) from but a single ori, replication would take over 150 h! Metazoan organisms get around this problem using two strategies. First, replication is bidirectional. Second, replication proceeds from multiple origins in each chromosome (a total of as many as 100 in humans). Thus, replication occurs in both directions along all of the chromosomes, and both strands are replicated simultaneously. This replication process generates "**replication bubbles**" (**Figure 35–17**).

The multiple ori sites that serve as origins for DNA replication in eukaryotes are poorly defined except in a few animal

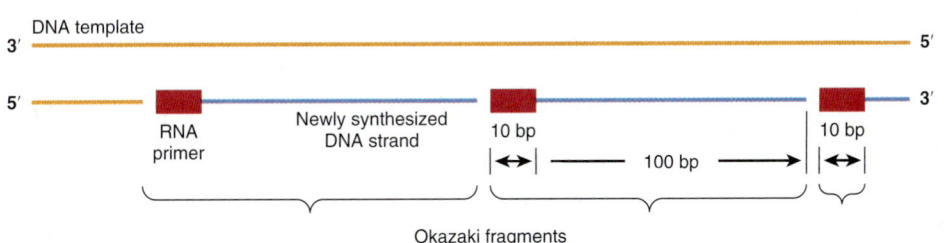

FIGURE 35–16 The discontinuous polymerization of deoxyribonucleotides on the lagging strand; formation of Okazaki fragments during lagging strand DNA synthesis is illustrated. Okazaki fragments are 100–250 nucleotides long in eukaryotes, 1000–2000 nucleotides in prokaryotes.

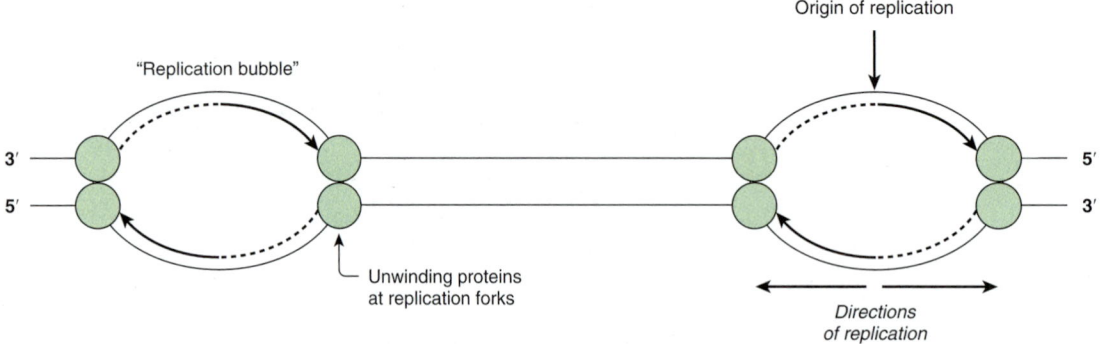

FIGURE 35–17 **The generation of "replication bubbles" during the process of DNA synthesis.** The bidirectional replication and the proposed positions of unwinding proteins at the replication forks are depicted.

viruses and in yeast. However, it is clear that initiation is regulated both spatially and temporally, since clusters of adjacent sites initiate replication synchronously. Replication firing, or DNA replication initiation at a replicator/ori, is influenced by a number of distinct properties of chromatin structure that are just beginning to be understood. It is clear, however, that there are more replicators and excess ORC than needed to replicate the mammalian genome within the time of a typical S-phase. Therefore, mechanisms for controlling the excess ORC-bound replicators must exist. Understanding the control of the formation and firing of replication complexes is one of the major challenges in this field.

During the replication of DNA, there must be a separation of the two strands to allow each to serve as a template by hydrogen bonding its nucleotide bases to the incoming deoxynucleoside triphosphate. The separation of the DNA double helix is promoted by SSBs in *E coli,* a protein termed replication protein A (RPA) in eukaryotes. These molecules stabilize the single-stranded structure as the replication fork progresses. The stabilizing proteins bind cooperatively and stoichiometrically to the single strands without interfering with the abilities of the nucleotides to serve as templates (Figure 35–13). In addition to separating the two strands of the double helix, there must be an unwinding of the molecule (once every 10 nucleotide pairs) to allow strand separation. The hexameric DNA β protein complex unwinds DNA in *E coli,* whereas the hexameric MCM complex unwinds eukaryotic DNA. This unwinding happens in segments adjacent to the replication bubble. To counteract this unwinding, there are multiple "swivels" interspersed in the DNA molecules of all organisms. The swivel function is provided by specific enzymes that introduce **"nicks" in one strand of the unwinding double helix,** thereby allowing the unwinding process to proceed. The nicks are quickly resealed without requiring energy input, because of the formation of a high-energy covalent bond between the nicked phosphodiester backbone and the nicking-sealing enzyme. The nicking-resealing enzymes are called **DNA topoisomerases.** This process is depicted diagrammatically in **Figure 35–18** and there compared with the ATP-dependent resealing carried out by the DNA ligases.

Topoisomerases are also capable of unwinding supercoiled DNA. Supercoiled DNA is a higher-ordered structure occurring in circular DNA molecules wrapped around a core, as depicted in **Figures 35–2** and **35–19.**

There exists in one species of animal viruses (retroviruses) a class of enzymes capable of synthesizing a single-stranded and then a dsDNA molecule from a single-stranded RNA template. This polymerase, RNA-dependent DNA polymerase, or **"reverse transcriptase,"** first synthesizes a DNA–RNA hybrid molecule utilizing the RNA genome as a template. A specific virus-encoded nuclease, **RNase H,** degrades the hybridized template RNA strand, and the remaining DNA strand in turn serves as a template to form a dsDNA molecule containing the information originally present in the RNA genome of the animal virus.

Reconstitution of Chromatin Structure

There is evidence that nuclear organization and chromatin structure are involved in determining the regulation and initiation of DNA synthesis. As noted above, the rate of polymerization in eukaryotic cells, which have chromatin and nucleosomes, is slower than that in prokaryotic cells, which lack canonical nucleosomes. It is also clear that chromatin structure must be re-formed after replication. Newly replicated DNA is rapidly assembled into nucleosomes, and the preexisting and newly assembled histone octamers are randomly distributed to each arm of the replication fork. These reactions are facilitated through the actions of histone chaperone proteins working in concert with chromatin remodeling complexes.

DNA Synthesis Occurs During the S Phase of the Cell Cycle

In animal cells, including human cells, the replication of the DNA genome occurs only at a specified time during the life span of the cell. This period is referred to as the **synthetic or S phase.** This is usually temporally separated from the **mitotic, or M phase,** by nonsynthetic periods referred to as **gap 1 (G₁)**

Step 1 DNA topoisomerase I = **E**   DNA ligase = **E**

$$E + ATP \rightleftharpoons E-P-R-A$$
(Enzyme-AMP)

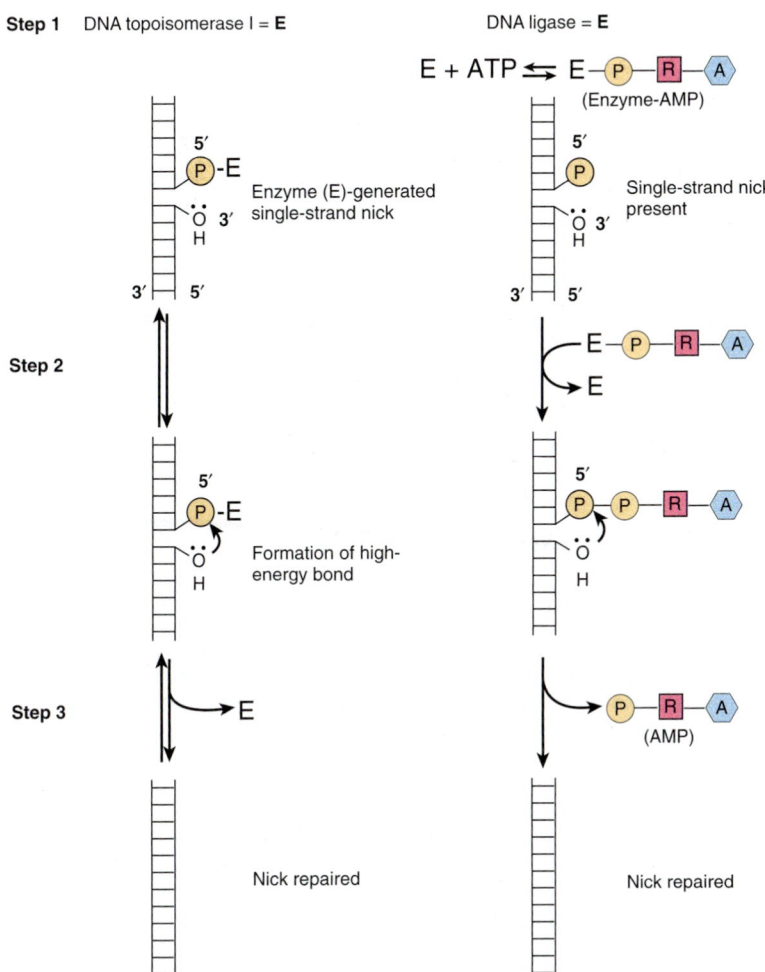

Enzyme (E)-generated single-strand nick

Single-strand nick present

Step 2

Formation of high-energy bond

Step 3

(AMP)

Nick repaired

Nick repaired

FIGURE 35–18 **Comparison of two types of nick-sealing reactions on DNA.** The series of reactions at left is catalyzed by DNA topoisomerase I, that at right by DNA ligase; P, phosphate; R, ribose; A, adenine. (Slightly modified and reproduced, with permission, from Lehninger AL: *Biochemistry*, 2nd ed. Worth, 1975. Copyright © 1975 by Worth Publishers. Used, with permission, from W. H. Freeman and Company.)

and **gap 2 (G$_2$) phases,** occurring before and after the S phase, respectively (**Figure 35–20**). Among other things, the cell prepares for DNA synthesis in G$_1$ and for mitosis in G$_2$. The cell regulates the DNA synthesis process by allowing it to occur only once per cell cycle and only at specific times in cells preparing to divide by a mitotic process.

All eukaryotic cells have gene products that govern the transition from one phase of the cell cycle to another. The **cyclins** are a family of proteins whose concentration increases and decreases at specific times, that is, "cycle" during the cell cycle—thus their name. The cyclins turn on, at the appropriate time, different **cyclin-dependent protein kinases (CDKs)** that phosphorylate substrates essential for progression through the cell cycle (**Figure 35–21**). For example, cyclin D levels rise in late G$_1$ phase and allow progression beyond the **start (yeast)** or **restriction point (mammals),** the point beyond which cells irrevocably proceed into the S or DNA synthesis phase.

The D cyclins activate CDK4 and CDK6. These two kinases are also synthesized during G$_1$ in cells undergoing active division. The D cyclins and CDK4 and CDK6 are nuclear proteins that assemble as a complex in late G$_1$ phase. The complex is an active serine-threonine protein kinase. One substrate for this kinase is the retinoblastoma (Rb) protein. Rb is a cell-cycle regulator because it binds to and inactivates a transcription factor (E2F) necessary for the transcription of certain genes (histone genes, DNA replication proteins, etc) needed for progression from G$_1$ to S phase. The phosphorylation of Rb by CDK4 or CDK6 results in the release of E2F from Rb-mediated transcription repression—thus, gene activation ensues and cell-cycle progression takes place.

Other cyclins and CDKs are involved in different aspects of cell-cycle progression (**Table 35–7**). Cyclin E and CDK2 form a complex in late G$_1$. Cyclin E is rapidly degraded, and the released CDK2 then forms a complex with cyclin A. This sequence is necessary for the initiation of DNA synthesis in S phase. A complex between cyclin B and CDK1 is rate-limiting for the G2/M transition in eukaryotic cells.

Many of the cancer-causing viruses (oncoviruses) and cancer-inducing genes (oncogenes) are capable of alleviating or disrupting the apparent restriction that normally controls the entry of mammalian cells from G$_1$ into the S phase. From the foregoing, one might have surmised that excessive production of a cyclin, loss of a specific CDK inhibitor, or

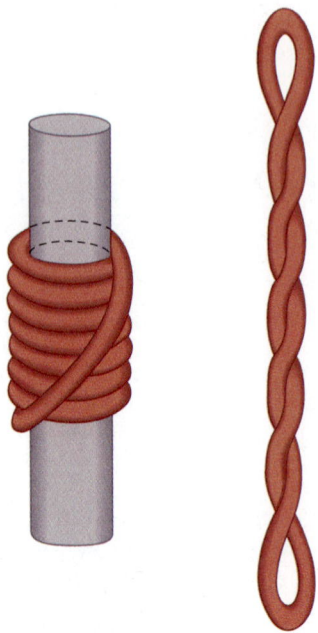

FIGURE 35–19 **Supercoiling of DNA.** A left-handed toroidal (solenoidal) supercoil, at left, will convert to a right-handed interwound supercoil, at right, when the cylindric core is removed. Such a transition is analogous to that which occurs when nucleosomes are disrupted by the high salt extraction of histones from chromatin.

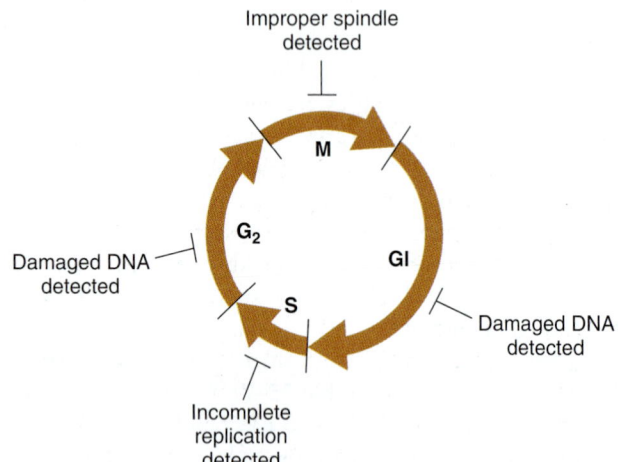

FIGURE 35–20 **Progress through the mammalian cell cycle is continuously monitored via multiple cell-cycle checkpoints.** DNA, chromosome, and chromosome segregation integrity is continuously monitored throughout the cell cycle. If DNA damage is detected in either the G$_1$ or the G$_2$ phase of the cell cycle, if the genome is incompletely replicated, or if normal chromosome segregation machinery is incomplete (ie, a defective spindle), cells will not progress through the phase of the cycle in which defects are detected. In some cases, if the damage cannot be repaired, such cells undergo programmed cell death (apoptosis).

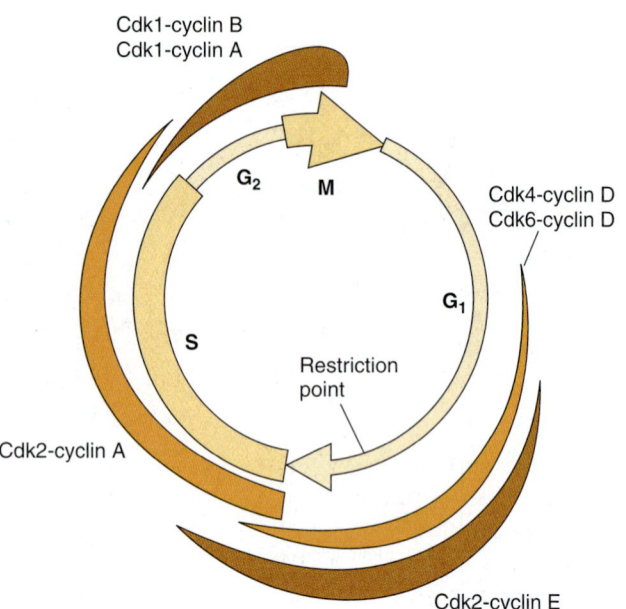

FIGURE 35–21 **Schematic illustration of the points during the mammalian cell cycle during which the indicated cyclins and cyclin-dependent kinases are activated.** The thickness of the various colored lines is indicative of the extent of activity.

production or activation of a cyclin/CDK at an inappropriate time might result in abnormal or unrestrained cell division. In this context, it is noteworthy that the *bcl* oncogene associated with B-cell lymphoma appears to be the cyclin D1 gene. Similarly, the oncoproteins (or transforming proteins) produced by several DNA viruses target the Rb transcription repressor for inactivation, inducing cell division inappropriately, while inactivation of Rb, itself a tumor suppressor gene, leads to uncontrolled cell growth and tumor formation.

During the S phase, mammalian cells contain greater quantities of DNA polymerase than during the nonsynthetic phases of the cell cycle. Furthermore, those enzymes responsible for formation of the substrates for DNA synthesis—that is, deoxyribonucleoside triphosphates—are also increased in activity, and their activity will diminish following the synthetic phase until the reappearance of the signal for renewed DNA synthesis. During the S phase, the **nuclear DNA is completely replicated once and only once.** It seems that once chromatin has been replicated, it is marked so as to prevent its further

TABLE 35–7 Cyclins and Cyclin-Dependent Kinases Involved in Cell-Cycle Progression

Cyclin	Kinase	Function
D	CDK4, CDK6	Progression past restriction point at G$_1$/S boundary
E, A	CDK2	Initiation of DNA synthesis in early S phase
B	CDK1	Transition from G$_2$ to M

TABLE 35–8 Types of Damage to DNA

I. Single-base alteration
  A. Depurination
  B. Deamination of cytosine to uracil
  C. Deamination of adenine to hypoxanthine
  D. Alkylation of base
  E. Insertion or deletion of nucleotide
  F. Base-analog incorporation

II. Two-base alteration
  A. UV light-induced thymine–thymine (pyrimidine) dimer
  B. Bifunctional alkylating agent cross-linkage

III. Chain breaks
  A. Ionizing radiation
  B. Radioactive disintegration of backbone element
  C. Oxidative free radical formation

IV. Cross-linkage
  A. Between bases in same or opposite strands
  B. Between DNA and protein molecules (eg, histones)

replication until it again passes through mitosis. This process is termed replication licensing. The molecular mechanisms for this phenomenon appear to involve dissociation and/or cyclin-CDK phosphorylation and subsequent degradation of several origin binding proteins that play critical roles in replication complex formation. Consequently origins fire only once per cell cycle.

In general, a given pair of chromosomes will replicate simultaneously and within a fixed portion of the S phase upon every replication. On a chromosome, clusters of replication units replicate coordinately. The nature of the signals that regulate DNA synthesis at these levels is unknown, but the regulation does appear to be an intrinsic property of each individual chromosome that is mediated by the several replication origins contained therein.

All Organisms Contain Elaborate Evolutionarily Conserved Mechanisms to Repair Damaged DNA

Repair of damaged DNA is critical for maintaining genomic integrity and thereby preventing the propagation of mutations, either horizontally, that is DNA sequence changes in somatic cells, or vertically, where nonrepaired lesions are present in sperm or oocyte DNA and hence can be transmitted to progeny. DNA is subjected to a huge array of chemical, physical, and biological assaults on a daily basis (**Table 35-8**), hence recognition and repair of DNA lesions is essential. Consequently, eukaryotic cells contain five major DNA repair pathways, each of which contain multiple, sometimes shared proteins; these DNA repair proteins typically have orthologues in prokaryotes. The mechanisms of DNA repair include **Nucleotide Excision Repair, NER; Mismatch Repair, MMR; Base Excision Repair, BER; Homologous Recombination, HR;** and **Nonhomologous End-Joining, NHEJ** repair pathways (**Figure 35–22**). The experiment of testing the importance of many of these DNA repair proteins to human biology has been performed by nature—mutations in a large number of these genes lead to human disease (**Table 35–9**). Moreover, systematic gene-directed experiments with laboratory mice have clearly ascribed critical gene integrity maintenance functions to these genes as well. In the mouse genetic studies, it was observed that indeed targeted mutations within these genes induce defects in DNA

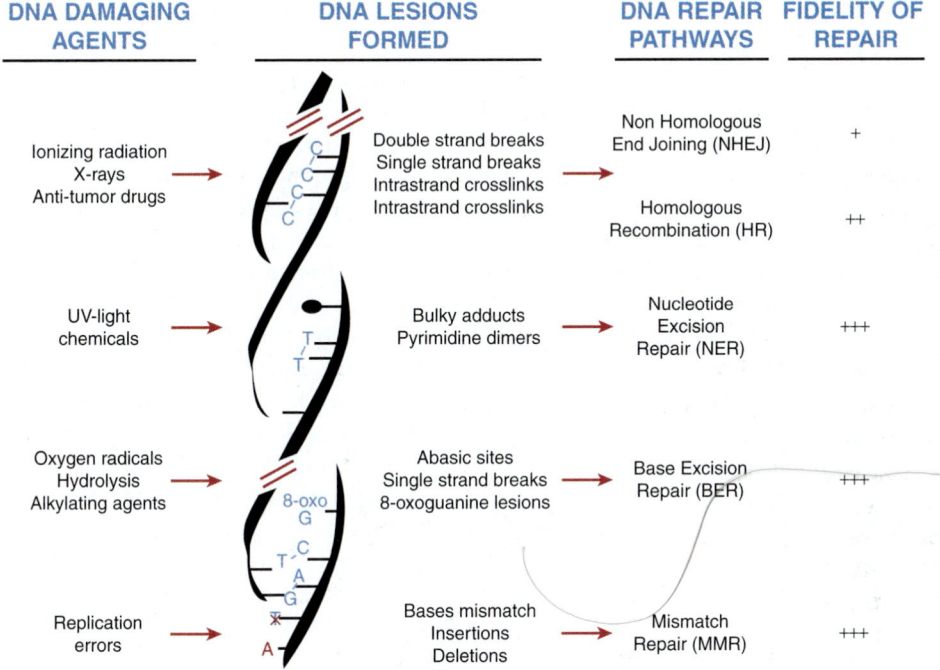

| DNA DAMAGING AGENTS | DNA LESIONS FORMED | DNA REPAIR PATHWAYS | FIDELITY OF REPAIR |

Ionizing radiation X-rays Anti-tumor drugs → Double strand breaks, Single strand breaks, Intrastrand crosslinks, Intrastrand crosslinks → Non Homologous End Joining (NHEJ) +, Homologous Recombination (HR) ++

UV-light chemicals → Bulky adducts, Pyrimidine dimers → Nucleotide Excision Repair (NER) +++

Oxygen radicals Hydrolysis Alkylating agents → Abasic sites, Single strand breaks, 8-oxoguanine lesions → Base Excision Repair (BER) +++

Replication errors → Bases mismatch, Insertions, Deletions → Mismatch Repair (MMR) +++

FIGURE 35–22 Mammals use multiple DNA repair pathways of variable accuracy to repair the myriad forms of DNA damage genomic DNA is subjected to. Listed are the major types of DNA damaging agents, the DNA lesions so formed (schematized and listed), the DNA repair pathway responsible for repairing the different lesions, and the relative fidelity of these pathways. (Modified, with permission, from: "DNA-Damage Response in Tissue-Specific and Cancer Stem Cells" *Cell Stem Cell* 8:16–29 (2011) copyright © 2011 Elsevier Inc.

TABLE 35–9 Human Diseases of DNA Damage Repair

Defective Non Homologous End Joining Repair (NHEJ)
Severe combined immunodeficiency disease (SCID)
Radiation sensitive severe combined immunodeficiency disease (RS-SCID)

Defective Homologous Repair (HR)
AT-like disorder (ATLD)
Nijmegen breakage syndrome (NBS)
Bloom syndrome (BS)
Werner syndrome (WS)
Rothmund thomson syndrome (RTS)
Breast cancer suspectility 1 and 2 (BRCA1, BRCA2)

Defective DNA Nucleotide Exicision Repair (NER)
Xeromderma pigmentosum (XP)
Cockayne syndrome (CS)
Trichothiodystrophy (TTD)

Defective DNA Base Excision Repair (BER)
MUTYH-associated polyposis (MAP)

Defective DNA Mismatch Repair (MMR)
Hereditary non-polyposis colorectal cancer (HNPCC)

repair while often also dramatically increasing susceptibility to cancer.

One of the most intensively studied mechanisms of DNA repair is the mechanism used to repair DNA **double-strand breaks,** or **DSBs;** these will be discussed in some detail here. There are two pathways, **HR** and **NHEJ,** that eukaryotic cells utilize to remove DSBs. The choice between the two depends upon the phase of the cell cycle (Figures 35–20 and 35–21) and the exact type of DSB breaks to be repaired (Table 35–8). During the G_0/G_1 phases of the cell cycle, DSBs are corrected by the NHEJ pathway, whereas during cell cycle phases S, and G_2/M, HR is utilized. All steps of DNA damage repair are catalyzed by evolutionarily conserved molecules, which include DNA damage **Sensors, Transducers,** and damage repair **Mediators.** Collectively, these cascades of proteins participate in the cellular response to DNA damage. Importantly, the ultimate cellular outcomes of DNA damage and cellular attempts to repair DNA damage range from **Cell-Cycle Delay** to allow for DNA repair, to **Cell-Cycle Arrest,** to **Apoptosis** or **Senescence** (see **Figure 35–23;** and further detail below). The molecules involved in these complex and highly integrated processes range from damage-specific histone modifications (ie, dimethylated lysine 20 Histone H4; H4K20me2) and histone isotype variants such as histone **H2AX** (*cf.* Table 35–1), poly ADP ribose polymerase, **PARP,** the MRN protein complex (Mre11-Rad50-NBS1 subunits); to DNA damage-activated kinase recognition/signaling proteins [**ATM** (Ataxia Telangiectasia, Mutated) and ATM-related kinase, **ATR,** the multisubunit DNA-dependent protein kinase (**DNA-PK and Ku70/80**), and Checkpoint kinases 1 and 2 (**CHK1, CHK2**)]. These multiple kinases phosphorylate, and consequently modulate the activities of dozens of proteins, such as numerous DNA repair, checkpoint control, and cell-cycle control proteins like CDC25A, B, C, Wee1, p21, p16, and p19 [all Cyclin-CDK regulators (see

Figure 9–8; and below); various exo- and endonucleases; DNA single-strand-specific DNA-binding proteins (RPA); PCNA and specific DNA polymerases (DNA pol delta,δ; and eta,η)]. Several of these (types) of proteins/enzymes have been discussed above in the context of DNA replication. DNA repair and its relationship to cell-cycle control are very active areas of research given their central roles in cell biology and potential for generating and preventing cancer.

DNA & Chromosome Integrity Is Monitored Throughout the Cell Cycle

Given the importance of normal DNA and chromosome function to survival, it is not surprising that eukaryotic cells have developed elaborate mechanisms to monitor the integrity of the genetic material. As detailed above, a number of complex multisubunit enzyme systems have evolved to repair damaged DNA at the nucleotide sequence level. Similarly, DNA mishaps at the chromosome level are also monitored and repaired. As shown in Figure 35–20, both DNA and chromosomal integrity are continuously monitored throughout the cell cycle. The four specific steps at which this monitoring occurs have been termed **checkpoint controls.** If problems are detected at any of these checkpoints, progression through the cycle is interrupted and transit through the cell cycle is halted until the damage is repaired. The molecular mechanisms underlying detection of DNA damage during the G_1 and G_2 phases of the cycle are understood better than those operative during S and M phases.

The **tumor suppressor p53,** a protein of apparent MW 53 kDa on SDS-PAGE, plays a key role in both G_1 and G_2 checkpoint control. Normally a very unstable protein, p53 is a DNA-binding transcription factor, **one of a family of related proteins (ie, p53, p63, and p73),** that is somehow stabilized in response to DNA damage, perhaps by direct p53-DNA interactions. Like the histones discussed above, p53 is subject to a panoply of regulatory PTMs, all of which likely modify its multiple biological activities. Increased levels of p53 activate transcription of an ensemble of genes that collectively serve to delay transit through the cycle. One of these induced proteins, p21^{CIP}, is a potent CDK–cyclin inhibitor (CKI) that is capable of efficiently inhibiting the action of all CDKs. Clearly, inhibition of CDKs will halt progression through the cell cycle (see Figures 35–19 and 35–20). If DNA damage is too extensive to repair, the affected cells undergo **apoptosis** (programmed cell death) in a p53-dependent fashion. In this case, p53 induces the activation of a collection of genes that induce apoptosis. Cells lacking functional p53 fail to undergo apoptosis in response to high levels of radiation or DNA-active chemotherapeutic agents. It may come as no surprise, then, that *p53* is one of the most frequently mutated genes in human cancers. Indeed recent genomic sequencing studies of multiple tumor DNA samples suggest that over 80% of human cancers carry p53 loss of function mutations. Additional research into the mechanisms of checkpoint control will prove invaluable for the development of effective anticancer therapeutic options.

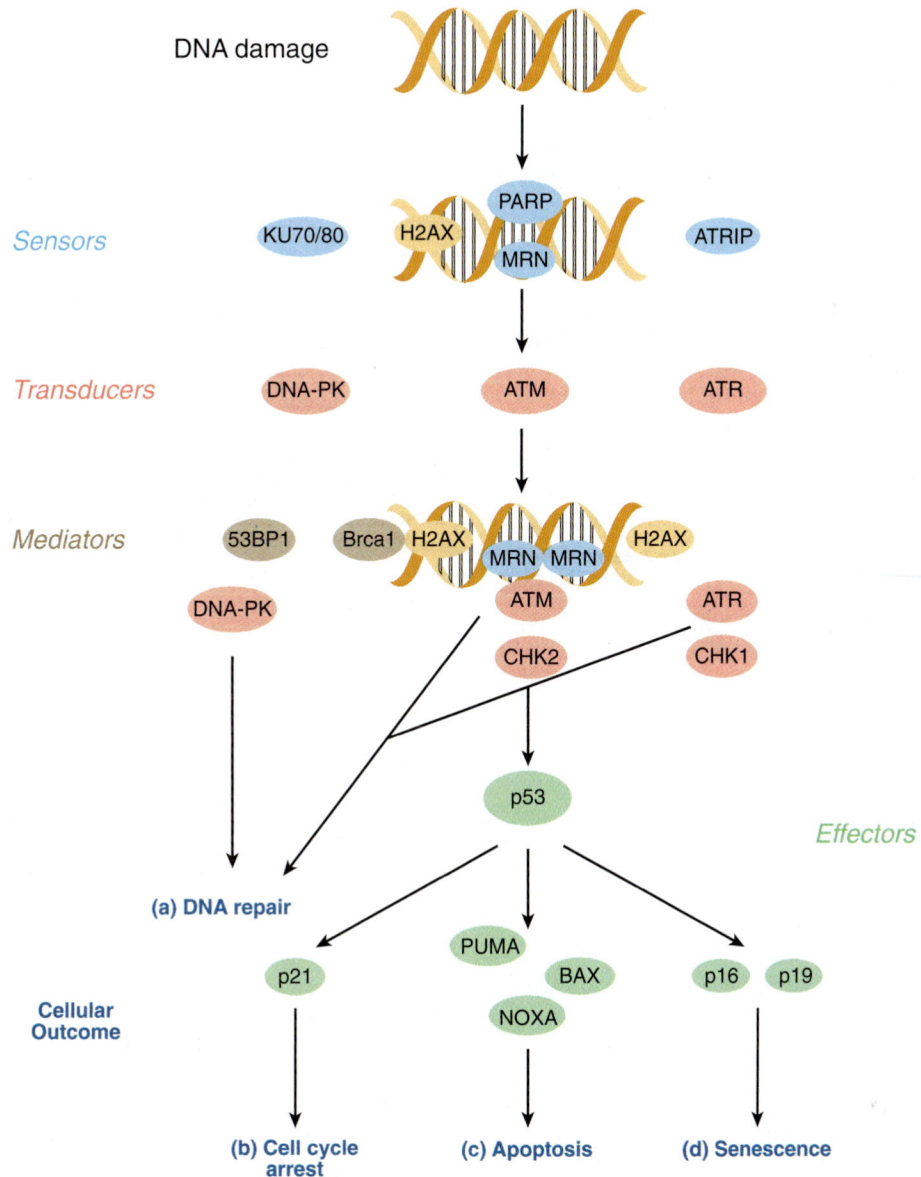

FIGURE 35–23 **The multistep mechanism of DNA double-strand break repair.** Shown top to bottom are the proteins (protein complexes) that: identify DSBs in genomic DNA (Sensors), transduce and amplify the recognized DNA damage (Transducers and Mediators), as well as the molecules that dictate the ultimate outcomes of the DNA damage response (Effectors). Damaged DNA can be: (a) repaired directly (DNA repair), or, via p53-mediated pathways and depending upon the severity of DNA damage and p53-activated genes induced, (b), cells can be arrested in the cell cycle by p21/WAF1 the potent CDK–cyclin complex inhibitor to allow time for extensively damaged DNA to be repaired, or (c), and (d) if the extent of DNA damage is too great to repair, cells can either apotose or senesce; both of these processes prevent the cell containing such damaged DNA from ever dividing and hence inducing cancer or other deleterious biological outcomes. (Based on: "DNA-Damage Response in Tissue-Specific and Cancer Stem Cells" *Cell Stem Cell* 8:16–29 (2011) copyright © 2011 Elsevier Inc.)

SUMMARY

- DNA in eukaryotic cells is associated with a variety of proteins, resulting in a structure called chromatin.

- Much of the DNA is associated with histone proteins to form a structure called the nucleosome. Nucleosomes are composed of an octamer of histones around which about 150 bp of DNA is wrapped.

- Histones are subject to an extensive array of dynamic covalent modifications that have important regulatory consequences.

- Nucleosomes and higher-order structures formed from them serve to compact the DNA.

- DNA in transcriptionally active regions is relatively more sensitive to nuclease attack in vitro; some regions, so-called hypersensitive sites are exceptionally sensitive and are often found to contain transcription control sites.

- Highly transcriptionally active DNA (genes) is often clustered in regions of each chromosome. Within these regions, genes may be separated by inactive DNA in nucleosomal structures. In eukaryotes the transcription unit—that portion of a gene that is copied by RNA polymerase—often consists of coding regions of DNA (exons) interrupted by intervening sequences of noncoding DNA (introns).

- After transcription, during RNA processing, introns are removed and the exons are ligated together to form the mature mRNA that appears in the cytoplasm; this process is termed RNA splicing.

- DNA in each chromosome is exactly replicated according to the rules of base pairing during the S phase of the cell cycle.

- Each strand of the double helix is replicated simultaneously but by somewhat different mechanisms. A complex of proteins, including DNA polymerase, replicates the leading strand continuously in the 5′ to 3′ direction. The lagging strand is replicated discontinuously, in short pieces of 150–250 nucleotides, in the 3′ to 5′ direction.

- DNA replication is initiated at special sites termed origins, or ori's and generate replication bubbles. Each chromosome contains multiple origins. The entire process takes about 9 h in a typical human cell and only occurs during the S phase of the cell cycle.

- A variety of mechanisms that employ different enzyme systems repair damaged cellular DNA after exposure of cells to chemical and physical mutagens.

REFERENCES

Blanpain C, Mohrin M, Sotiropoulou PA, et al: DNA-damage response in tissue-specific and cancer stem cells. Cell Stem Cell 2011;8:16–29.

Bohgaki T, Bohgaki M, Hakem R: DNA double-strand break signaling and human disorders. Genome Integr 2010;1:15–29.

Campos EL, Fillingham J, Li G, et al: The program for processing newly synthesized histones H3.1 and H4. Nat Struct Mol Biol 2010;17:1343–1351.

Campos EL, Reinberg D: Histones: annotating chromatin. Annu Rev Genet 2009;43:559–599.

Dalal Y, Furuyama T, Vermaak D, et al: Structure, dynamics, and evolution of centromeric nucleosomes. Proc Nat Academy of Sciences 2007;104:41.

Deng W, Blobel GA: Do chromatin loops provide epigenetic gene expression states? Curr Opin Genet Develop 2010;20:548–554.

Encode Consortium: Identification and analysis of functional elements in 1% of the human genome by the ENCODE pilot project. Nature 2007;447:799.

Gilbert DM: In search of the holy replicator. Nature Rev Mol Cell Biol 2004;5:848.

Hakem R: DNA-damage repair; the good, the bad, and the ugly. EMBO J 2008;27:589–605.

Johnson A, O'Donnell M: Cellular DNA replicases: components and dynamics at the replication fork. Ann Rev Biochemistry 2005;74:283.

Krishnan KJ, Reeve AK, Samuels DC, et al: What causes mitochondrial DNA deletions in human cells? Nat Genet 2008;40:275.

Lander ES, Linton LM, Birren B, et al: Initial sequencing and analysis of the human genome. Nature 2001;409:860.

Luger K, Mäder AW, Richmond RK, et al: Crystal structure of the nucleosome core particle at 2.8 Å resolution. Nature 1997;389:251.

Margueron R, Reinberg D: Chromatin structure and the inheritance of epigenetic information. Nat Rev Genet 2010;11:285–296.

Misteli T: Beyond the sequence: cellular organization of genome function. Cell 2007;128:787.

Misteli T, Soutoglou E: The emerging role of nuclear architecture in DNA repair and genome maintenance. Nat Rev Mol Cell Biol 2010;10:243–254.

Orr HT, Zoghbi, HY: Trinucleotide repeat disorders. Annu Rev Neurosci 2007;30:375.

Ponicsan SL, Kugel JF, Goodrich JA: Genomic gems: SINE RNAs regulate mRNA production. Curr Opin Genet Develop 2010; 20:149–155.

Sullivan, Blower MD, Karpen GH: Determining centromere identity: cyclical stories and forking paths. Nat Rev Genet 2001; 2:584.

Takizawa T, Meaburn KJ, Misteli T: The meaning of gene positioning. Cell 2008;135:9–13.

Talbert PB, Henikoff S: Histone variants–ancient wrap artists of the epigenome. Nat Rev Mol Cell Biol 2010;11:264–275.

Venter JC, Adams MD, Myers EW, et al: The sequence of the human genome. Science 2002;291:1304.

Zaidi SK, Young DW, Montecino MA, et al: Mitotic bookmarking of genes: a novel dimension to epigenetic control. Nat Rev Genet 2010;11:583–589.

Zilberman D, Henikoff S: Genome wide analysis of DNA methylation patterns. Development 2007;134:3959.

RNA Synthesis, Processing, & Modification

P. Anthony Weil, PhD

- Describe the molecules involved and the mechanism of RNA synthesis.

- Explain how eukaryotic DNA-dependent RNA polymerases, in collaboration with an array of specific accessory factors, can differentially transcribe genomic DNA to produce specific mRNA precursor molecules.

- Describe the structure of eukaryotic mRNA precursors, which are highly modified at both termini.

- Appreciate the fact that the majority of mammalian mRNA-encoding genes are interrupted by multiple non-protein coding sequences termed introns, which are interspersed between protein coding regions termed exons.

- Explain that since intron RNA does not encode protein, the intronic RNA must be specifically and accurately removed in order to generate functional mRNAs from the mRNA precursor molecules in a series of precise molecular events termed RNA splicing.

- Explain the steps and molecules that catalyze mRNA splicing, a process that converts the end-modified mRNA precursor molecules into mRNAs that are functional for translation.

BIOMEDICAL IMPORTANCE

The synthesis of an RNA molecule from DNA is a complex process involving one of the group of RNA polymerase enzymes and a number of associated proteins. The general steps required to synthesize the primary transcript are initiation, elongation, and termination. Most is known about initiation. A number of DNA regions (generally located upstream from the initiation site) and protein factors that bind to these sequences to regulate the initiation of transcription have been identified. Certain RNAs—mRNAs in particular—have very different life spans in a cell. The RNA molecules synthesized in mammalian cells are made as precursor molecules that have to be processed into mature, active RNA. It is important to understand the basic principles of messenger RNA (mRNA) synthesis and metabolism, for modulation of this process results in altered rates of protein synthesis and thus a variety of both metabolic and phenotypic changes. This is how all organisms adapt to changes of environment. It is also how differentiated cell structures and functions are established and maintained. Errors or changes in synthesis, processing, splicing, stability, or function of mRNA transcripts are a cause of disease.

RNA EXISTS IN FOUR MAJOR CLASSES

All eukaryotic cells have four major classes of RNA (**Table 36–1**): ribosomal RNA (rRNA), mRNA, transfer RNA (tRNA), and small RNAs, the small nuclear RNAs and microRNAs (snRNA and miRNA). The first three are involved in protein synthesis, while the small RNAs are involved in mRNA splicing and modulation of gene expression by altering mRNA function. The various classes of RNA are different in their diversity, stability, and abundance in cells.

RNA IS SYNTHESIZED FROM A DNA TEMPLATE BY AN RNA POLYMERASE

The processes of DNA and RNA synthesis are similar in that they involve (1) the general steps of initiation, elongation, and termination with 5′–3′ polarity; (2) large, multicomponent initiation complexes; and (3) adherence to

TABLE 36–1 Classes of Eukaryotic RNA

RNA	Types	Abundance	Stability
Ribosomal (rRNA)	28S, 18S, 5.8S, 5S	80% of total	Very stable
Messenger (mRNA)	~10^5 Different species	2–5% of total	Unstable to very stable
Transfer (tRNA)	~60 Different species	~15% of total	Very stable
Small RNAs			
Small nuclear (snRNA)	~30 Different species	≤1% of total	Very stable
Micro (miRNA)	100s–1000	<1% of total	Stable

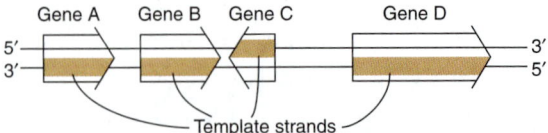

FIGURE 36–1 **Genes can be transcribed off both strands of DNA.** The arrowheads indicate the direction of transcription (polarity). Note that the template strand is always read in the 3′–5′ direction. The opposite strand is called the coding strand because it is identical (except for T for U changes) to the mRNA transcript (the primary transcript in eukaryotic cells) that encodes the protein product of the gene.

Watson–Crick base-pairing rules. However, DNA and RNA synthesis do differ in several important ways, including the following: (1) ribonucleotides are used in RNA synthesis rather than deoxyribonucleotides; (2) U replaces T as the complementary base for A in RNA; (3) a primer is not involved in RNA synthesis as RNA polymerases have the ability to initiate synthesis de novo; (4) only portions of the genome are vigorously transcribed or copied into RNA, whereas the entire genome must be copied, once and only once during DNA replication; and (5) there is no highly active, efficient proofreading function during RNA transcription.

The process of synthesizing RNA from a DNA template has been characterized best in prokaryotes. Although in mammalian cells, the regulation of RNA synthesis and the processing of the RNA transcripts are different from those in prokaryotes, the process of RNA synthesis per se is quite similar in these two classes of organisms. Therefore, the description of RNA synthesis in prokaryotes, where it is best understood, is applicable to eukaryotes even though the enzymes involved and the regulatory signals, though related, are different.

The Template Strand of DNA Is Transcribed

The sequence of ribonucleotides in an RNA molecule is complementary to the sequence of deoxyribonucleotides in one strand of the double-stranded DNA molecule (Figure 34–8). The strand that is transcribed or copied into an RNA molecule is referred to as the **template strand** of the DNA. The other DNA strand, the **nontemplate strand,** is frequently referred to as the **coding strand** of that gene. It is called this because, with the exception of T for U changes, it corresponds exactly to the sequence of the messenger RNA primary transcript, which encodes the (protein) product of the gene. In the case of a double-stranded DNA molecule containing many genes, the template strand for each gene will not necessarily be the same strand of the DNA double helix (**Figure 36–1**). Thus, a given strand of a double-stranded DNA molecule will serve as the template strand for some genes and the coding strand of other

genes. Note that the nucleotide sequence of an RNA transcript will be the same (except for U replacing T) as that of the coding strand. The information in the template strand is read out in the 3′–5′ direction. Though not shown in Figure 36–1 there are instances of genes embedded within other genes.

DNA-Dependent RNA Polymerase Initiates Transcription at a Distinct Site, the Promoter

DNA-dependent RNA polymerase is the enzyme responsible for the polymerization of ribonucleotides into a sequence complementary to the template strand of the gene (see **Figures 36–2** and **36–3**). The enzyme attaches at a specific site—the promoter—on the template strand. This is followed by initiation of RNA synthesis at the starting point, and the process continues until a termination sequence is reached (Figure 36–3). A **transcription unit** is defined as that region of DNA that includes the signals for transcription initiation, elongation, and termination. The RNA product, which is synthesized in the 5′–3′ direction, is the **primary transcript.** Transcription rates vary from gene to gene but can be quite

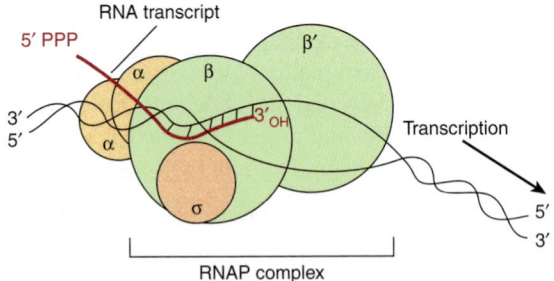

FIGURE 36–2 RNA polymerase (RNAP) catalyzes the polymerization of ribonucleotides into an RNA sequence that is complementary to the template strand of the gene. The RNA transcript has the same polarity (5′–3′) as the coding strand but contains U rather than T. *E coli* RNAP consists of a core complex of two α subunits and two β subunits (β and β′). The holoenzyme contains the σ subunit bound to the $\alpha_2 \beta\beta'$ core assembly. The ω subunit is not shown. The transcription "bubble" is an approximately 20-bp area of melted DNA, and the entire complex covers 30–75 bp, depending on the conformation of RNAP.

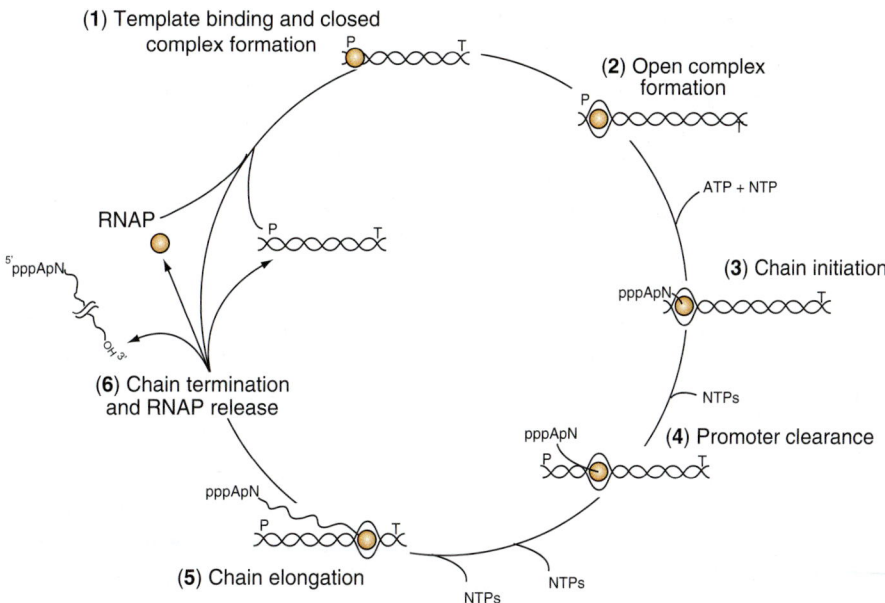

FIGURE 36–3 **The transcription cycle.** Transcription can be described in six steps: **(1) Template binding and closed RNA polymerase-promoter complex formation:** RNA polymerase (RNAP) binds to DNA and then locates a promoter (**P**), **(2) Open promoter complex formation:** once bound to the promoter, RNAP melts the two DNA strands to form an open promoter complex; this complex is also referred to as the preinitiation complex or PIC. Strand separation allows the polymerase to access the coding information in the template strand of DNA **(3) Chain initiation:** using the coding information of the template RNAP catalyzes the coupling of the first base (often a purine) to the second, template-directed ribonucleoside triphosphate to form a dinucleotide (in this example forming the dinucleotide 5′ pppApN$_{OH}$ 3′). **(4) Promoter clearance:** after RNA chain length reaches ~10–20 nt, the polymerase undergoes a conformational change and then is able to move away from the promoter, transcribing down the transcription unit. **(5) Chain elongation:** Successive residues are added to the 3′-OH terminus of the nascent RNA molecule until a transcription termination signal (**T**) is encountered. **(6) Chain termination and RNAP release:** Upon encountering the transcription termination site RNAP undergoes an additional conformational change that leads to release of the completed RNA chain, the DNA template and RNAP. RNAP can rebind to DNA beginning the promoter search process and the cycle is repeated. Note that all of the steps in the transcription cycle are facilitated by additional proteins, and indeed are often subjected to regulation by positive and/or negative-acting factors.

high. An electron micrograph of transcription in action is presented in **Figure 36–4**. In prokaryotes, this can represent the product of several contiguous genes; in mammalian cells, it usually represents the product of a single gene. If a transcription unit contains only a single gene, then the 5′ termini of the primary RNA transcript and the mature cytoplasmic RNA are identical. **Thus, the starting point of transcription corresponds to the 5′ nucleotide of the mRNA.** This is designated position +1, as is the corresponding nucleotide in the DNA. The numbers increase as the sequence proceeds *downstream* from the start site. This convention makes it easy to locate particular regions, such as intron and exon boundaries. The nucleotide in the promoter adjacent to the transcription initiation site in the upstream direction is designated –1, and these negative numbers increase as the sequence proceeds *upstream*, away from the initiation site. This +/– numbering system provides a conventional way of defining the location of regulatory elements in the promoter.

The primary transcripts generated by RNA polymerase II—one of the three distinct nuclear DNA-dependent RNA polymerases in eukaryotes—are promptly capped by 7-methyl-guanosine triphosphate caps (Figure 34–10) that persist and eventually appear on the 5′ end of mature cytoplasmic mRNA. These caps are necessary for the subsequent

processing of the primary transcript to mRNA, for the translation of the mRNA, and for protection of the mRNA against exonucleolytic attack.

Bacterial DNA-Dependent RNA Polymerase Is a Multisubunit Enzyme

The DNA-dependent RNA polymerase (RNAP) of the bacterium *Escherichia coli* exists as an approximately 400 kDa core complex consisting of two identical α subunits, similar but not identical β and β′ subunits, and an ω subunit. The β subunit binds Mg^{2+} ions and composes the catalytic subunit (Figure 36–2). The core RNA polymerase, ββ′α$_2$ω, often termed E, associates with a specific protein factor (the sigma [σ] factor) to form holoenzyme, ββ′α$_2$ωσ, or Eσ. The σ subunit helps the core enzyme recognize and bind to the specific deoxynucleotide sequence of the promoter region (**Figure 36–5**) to form the **preinitiation complex (PIC)**. There are multiple, distinct σ-factor encoding genes in all bacterial species. Sigma factors have a dual role in the process of promoter recognition; σ association with core RNA polymerase decreases its affinity for nonpromoter DNA while simultaneously increasing holoenzyme affinity for promoter DNA. The multiple σ-factors compete for interaction with limiting core RNA polymerase

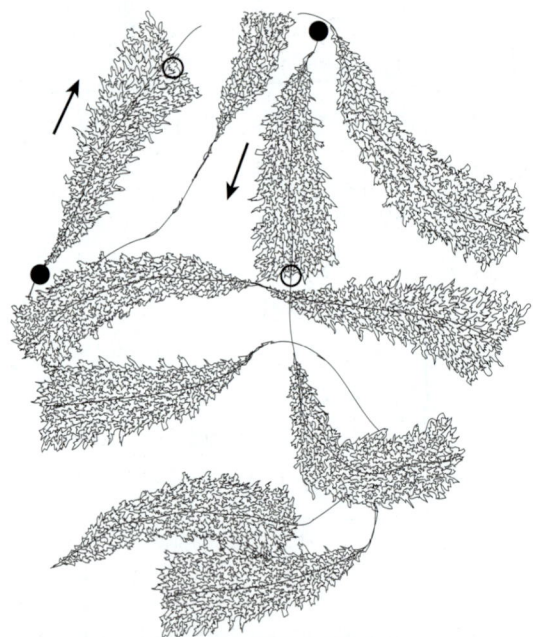

FIGURE 36–4 **Schematic representation of an electron photomicrograph of multiple copies of amphibian rRNA-encoding genes in the process of being transcribed.** The magnification is about 6000×. Note that the length of the transcripts increases as the RNA polymerase molecules progress along the individual rRNA genes from transcription start sites (filled circles) to transcription termination sites (open circles). RNA polymerase I (not visualized here) is at the base of the nascent rRNA transcripts. Thus, the proximal end of the transcribed gene has short transcripts attached to it, while much longer transcripts are attached to the distal end of the gene. The arrows indicate the direction (5′–3′) of transcription.

(ie, **E**). Each of these unique σ-factors act as a regulatory protein that modifies the **promoter recognition specificity** of the resulting unique RNA polymerase holoenzyme (ie, $E\sigma_1$, $E\sigma_2$,...). The appearance of different σ-factors and their association with core RNA polymerase forming novel holoenzyme forms can be correlated temporally with various programs of gene expression in prokaryotic systems such as sporulation, growth in various poor nutrient sources, and the response to heat shock.

Mammalian Cells Possess Three Distinct Nuclear DNA-Dependent RNA Polymerases

The properties of mammalian nuclear polymerases are described in **Table 36–2**. Each of these DNA-dependent RNA polymerases is responsible for transcription of different sets of genes. The sizes of the RNA polymerases range from MW 500,000 to MW 600,000. These enzymes exhibit more complex subunit profiles than prokaryotic RNA polymerases. They all have two large subunits and a number of smaller subunits—as many as 14 in the case of RNA pol III. However, the eukaryotic RNA polymerase subunits do exhibit extensive amino acid sequence homologies with prokaryotic RNA polymerases. This homology has been shown recently to extend to the level of three-dimensional structures. The functions of each of the subunits are not yet fully understood.

A peptide toxin from the mushroom *Amanita phalloides*, α-amanitin, is a specific differential inhibitor of the eukaryotic nuclear DNA-dependent RNA polymerases and as such has proved to be a powerful research tool (Table 36–2). α-Amanitin blocks the translocation of RNA polymerase during phosphodiester bond formation.

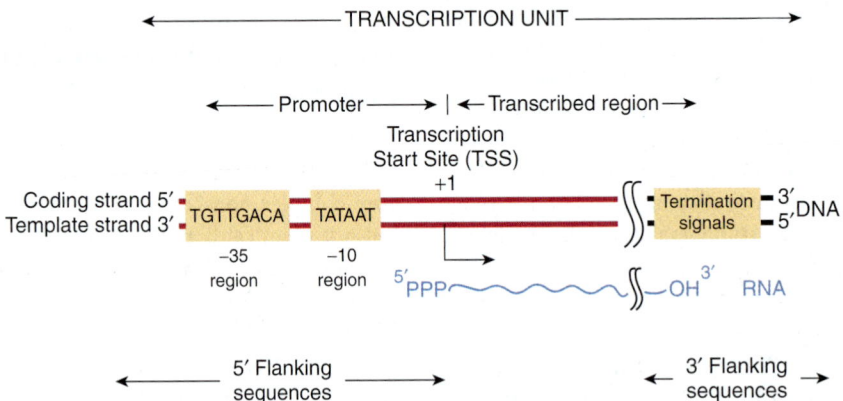

FIGURE 36–5 **Bacterial promoters, such as that from *E coli* shown here, share two regions of highly conserved nucleotide sequence.** These regions are located 35 and 10 bp upstream (in the 5′ direction of the coding strand) from the transcription start site (**TSS**), which is indicated as +1. By convention, all nucleotides upstream of the transcription initiation site (at +1) are numbered in a negative sense and are referred to as 5′-flanking sequences, while sequences downstream are numbered in a positive sense with the TSS as +1. Also by convention, the promoter DNA regulatory sequence elements such as the -35 and TATA box elements are described in the 5′–3′ direction and as being on the coding strand. These elements function only in double-stranded DNA, however. Other transcriptional regulatory elements, however, can often act in a direction independent fashion, and such *cis*-elements are drawn accordingly in any schematic (see also Figure 36–8). Note that the transcript produced from this transcription unit has the same polarity or "sense" (ie, 5′–3′ orientation) as the coding strand. Termination *cis*-elements reside at the end of the transcription unit (see Figure 36–6 for more detail). By convention, the sequences downstream of the site at which transcription termination occurs are termed 3′-flanking sequences.

TABLE 36-2 Nomenclature and Properties of Mammalian Nuclear DNA-Dependent RNA Polymerases

Form of RNA Polymerase	Sensitivity to α-Amanitin	Major Products
I	Insensitive	rRNA
II	High sensitivity	mRNA, miRNA, SnRNA
III	Intermediate sensitivity	tRNA, 5s rRNA

RNA SYNTHESIS IS A CYCLICAL PROCESS & INVOLVES RNA CHAIN INITIATION, ELONGATION, & TERMINATION

The process of RNA synthesis in bacteria—depicted in Figure 36–3—is cyclical and involves multiple steps. First RNA polymerase holoenzyme (E-σ) must bind DNA and locate a promoter (P; Figure 36–3). Once the promoter is located, the Eσ–promoter DNA complex undergoes a temperature-dependent conformational change and unwinds, or melts the DNA in and around the transcription start site (at +1). This complex is termed the **preinitiation complex, or PIC.** This unwinding allows the active site of the Eσ to access the template strand, which of course dictates the sequence of ribonucleotides to be polymerized into RNA. The first nucleotide (typically, though not always a purine) then associates with the nucleotide-binding site on the β subunit of the enzyme, and in the presence of the next appropriate nucleotide bound to the polymerase, RNAP catalyzes the formation of the first phosphodiester bond, and the nascent chain is now attached to the polymerization site on the β subunit of RNAP. This reaction is termed **initiation.** The analogy to the A and P sites on the ribosome should be noted; see Figure 37–9, below. The nascent dinucleotide retains the 5′-triphosphate of the initiating nucleotide (Figure 36–3, ATP).

RNA polymerase continues to incorporate nucleotides 3 to ~10, at which point the polymerase undergoes another conformational change and moves away from the promoter; this reaction is termed **promoter clearance.** The **elongation phase** then commences, here the nascent RNA molecule grows 5′–3′ as consecutive NTP incorporation steps continue cyclically, antiparallel to its template. The enzyme polymerizes the ribonucleotides in the specific sequence dictated by the template strand and interpreted by Watson–Crick base-pairing rules. Pyrophosphate is released following each cycle of polymerization. As for DNA synthesis, this pyrophosphate (PP$_i$) is rapidly degraded to 2 mol of inorganic phosphate (P$_i$) by ubiquitous pyrophosphatases, thereby providing irreversibility on the overall synthetic reaction. The decision, to stay at the promoter in a poised or stalled state, or transition to elongation appears to be an important regulatory step in both prokaryotic and eukaryotic mRNA gene transcription.

As the **elongation** complex containing RNA polymerase progresses along the DNA molecule, **DNA unwinding** must occur in order to provide access for the appropriate base pairing to the nucleotides of the coding strand. The extent of this transcription bubble (ie, DNA unwinding) is constant throughout transcription and has been estimated to be about 20 base pairs per polymerase molecule. Thus, it appears that the size of the unwound DNA region is dictated by the polymerase and is independent of the DNA sequence in the complex. RNA polymerase has an intrinsic "unwindase" activity that opens the DNA helix (ie, see PIC formation above). The fact that the DNA double helix must unwind, and the strands part at least transiently for transcription implies some disruption of the nucleosome structure of eukaryotic cells. Topoisomerase both precedes and follows the progressing RNA polymerase to prevent the formation of superhelical tensions that would serve to increase the energy required to unwind the template DNA ahead of RNAP.

Termination of the synthesis of the RNA molecule in bacteria is signaled by a sequence in the template strand of the DNA molecule—a signal that is recognized by a **termination protein, the rho (ρ) factor.** Rho is an ATP-dependent RNA-stimulated helicase that disrupts the ternary transcription elongation complex composed of RNA polymerase-nascent RNA and DNA. In some cases, bacterial RNAP can directly recognize DNA-encoded termination signals (Figure 36–3; T) without assistance by the rho factor. After termination of synthesis of the RNA, the enzyme separates from the DNA template and probably dissociates to free core enzyme and free factor. With the assistance of another σ-factor, the core enzyme then recognizes a promoter at which the synthesis of a new RNA molecule commences. In eukaryotic cells, termination is less well understood but the proteins catalyzing RNA processing, termination, and polyadenylation proteins all appear to load onto RNA polymerase II soon after initiation (see below). More than one RNA polymerase molecule may transcribe the same template strand of a gene simultaneously, but the process is phased and spaced in such a way that at any one moment each is transcribing a different portion of the DNA sequence (Figures 36–1 and 36–4).

THE FIDELITY & FREQUENCY OF TRANSCRIPTION IS CONTROLLED BY PROTEINS BOUND TO CERTAIN DNA SEQUENCES

Analysis of the DNA sequence of specific genes has allowed the recognition of a number of sequences important in gene transcription. From the large number of bacterial genes studied, it is possible to construct consensus models of transcription initiation and termination signals.

The question, "How does RNAP find the correct site to initiate transcription?" is not trivial when the complexity of the genome is considered. E coli has 4×10^3 transcription initiation sites (ie, gene promoters) in 4.2×10^6 base pairs (bp) of DNA. The situation is even more complex in humans, where as many as 10⁵ transcription initiation sites are distributed throughout 3×10^9 bp of DNA. RNAP can bind, with low affinity, to many regions of DNA, but it scans the DNA sequence—at a rate of $\geq 10^3$ bp/s—until it recognizes certain specific regions of DNA to which it binds with higher affinity. These regions are termed promoters, and it is the association of RNAP with promoters that ensures accurate initiation of transcription. The promoter recognition-utilization process is the target for regulation in both bacteria and humans.

Bacterial Promoters Are Relatively Simple

Bacterial promoters are approximately 40 nucleotides (40 bp or four turns of the DNA double helix) in length, a region small enough to be covered by an *E coli* RNA holopolymerase molecule. In a consensus promoter, there are two short, conserved sequence elements. Approximately 35-bp upstream of the transcription start site there is a consensus sequence of eight nucleotide pairs (consensus: 5′-TGTTGACA-3′) to which the RNAP binds to form the so-called **closed complex.** More proximal to the transcription start site—about 10 nucleotides upstream—is a six-nucleotide-pair A+T-rich sequence (consensus: 5′-TATAAT-3′). These conserved sequence elements together comprise the promoter, and are shown schematically in Figure 36–5. The latter sequence has a low melting temperature because of its lack of GC nucleotide pairs. Thus, the so-called **TATA "box"** is thought to ease the dissociation of the two DNA strands so that RNA polymerase bound to the promoter region can have access to the nucleotide sequence of its immediately downstream template strand. Once this process occurs, the combination of RNA polymerase plus promoter is called the **open complex.** Other bacteria have slightly different consensus sequences in their promoters, but all generally have two components to the promoter; these tend to be in the same position relative to the transcription start site, and in all cases the sequences between the two promoter elements have no similarity but still **provide critical spacing functions** that facilitate recognition of –35 and –10 sequences by RNA polymerase holoenzyme. Within a bacterial cell, different sets of genes are often coordinately regulated. One important way that this is accomplished is through the fact that these co-regulated genes share particular –35 and –10 promoter sequences. These unique promoters are recognized by different σ-factors bound to core RNA polymerase (ie, $E\sigma_1$, $E\sigma_2$,…).

Rho-dependent transcription **termination signals** in *E coli* also appear to have a distinct consensus sequence, as shown in **Figure 36–6.** The conserved consensus sequence, which is about 40 nucleotide pairs in length, can be seen to contain a hyphenated or interrupted inverted repeat followed by a series of AT base pairs. As transcription proceeds through the hyphenated, inverted repeat, the generated transcript can form the intramolecular hairpin structure, also depicted in Figure 36–6.

Transcription continues into the AT region, and with the aid of the ρ termination protein the RNA polymerase stops, dissociates from the DNA template, and releases the nascent transcript.

As discussed in detail in Chapter 38 bacterial gene transcription is controlled through the action of repressor and activator proteins. These proteins typically bind to unique and specific DNA sequences that lie adjacent to promoters. These repressors and activators affect the ability of the RNA polymerase to bind promoter DNA and/or form open complexes. The net effect is to stimulate or inhibit

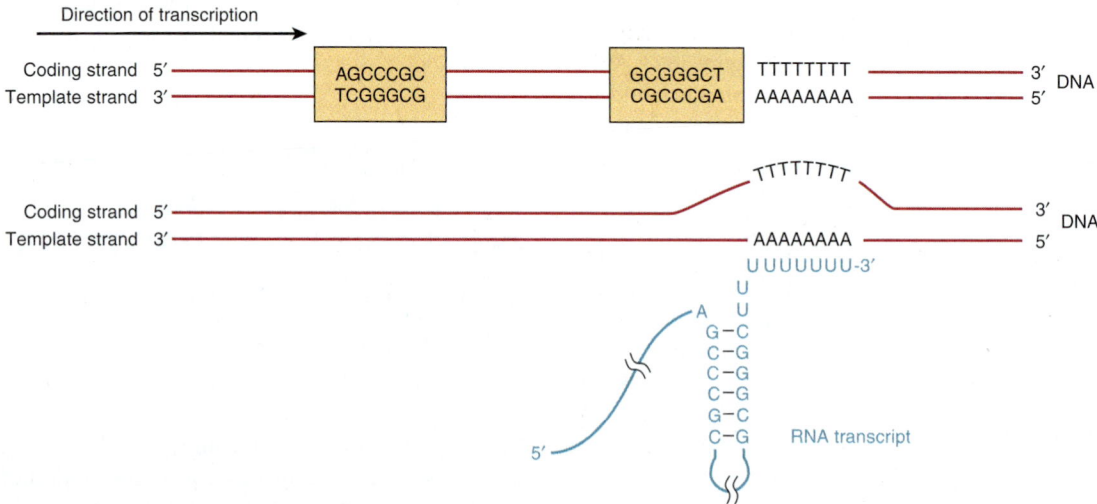

FIGURE 36–6 **The predominant bacterial transcription termination signal contains an inverted, hyphenated repeat (the two boxed areas) followed by a stretch of AT base pairs (top).** The inverted repeat, when transcribed into RNA, can generate the secondary structure in the RNA transcript (**bottom**). Formation of this RNA hairpin causes RNA polymerase to pause and subsequently the ρ (rho) termination factor interacts with the paused polymerase and induces chain termination through mechanisms not yet fully understood.

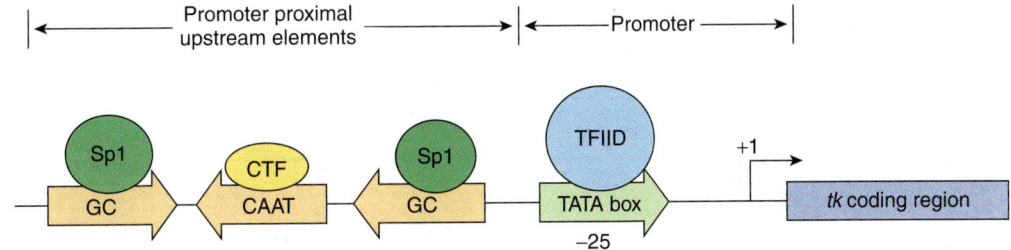

FIGURE 36–7 **Transcription elements and binding factors in the herpes simplex virus thymidine kinase *(tk)* gene.** DNA-dependent RNA polymerase II (not shown) binds to the region encompassing the TATA box (which is bound by transcription factor TFIID) and TSS at +1 (see also Figure 36–9) to form a multicomponent preinitiation complex capable of initiating transcription at a single nucleotide (+1). The frequency of this event is increased by the presence of upstream *cis*-acting elements (the GC and CAAT boxes) located either near to the promoter (promoter proximal) or distant from the promoter (distal elements; Figure 36–8). Proximal and distal DNA *cis*-elements are bound by *trans*-acting transcription factors, in this example Sp1 and CTF (also called C/EBP, NF1, NFY). These *cis*-elements can function independently of orientation (**arrows**).

PIC formation and transcription initiation—consequently blocking or enhancing specific RNA synthesis.

Eukaryotic Promoters Are More Complex

It is clear that the signals in DNA that control transcription in eukaryotic cells are of several types. Two types of sequence elements are promoter-proximal. One of these defines **where transcription is to commence** along the DNA, and the other contributes to the mechanisms that control **how frequently** this event is to occur. For example, in the thymidine kinase gene of the herpes simplex virus, which utilizes transcription factors of its mammalian host for its early gene expression program, there is a single unique transcription start site, and accurate transcription from this start site depends upon a nucleotide sequence located 32 nucleotides upstream from the start site (ie, at –32) (**Figure 36–7**). This region has the sequence of **TATAAAG** and bears remarkable similarity to the functionally related **TATA box** that is located about 10 bp upstream from the prokaryotic mRNA start site (Figure 36–5). Mutation or inactivation of the TATA box markedly reduces transcription of this and many other genes that contain this consensus *cis*-active element (see **Figures 36–7** and **36–8**). The TATA box is usually located 25–30 bp upstream from the transcription start site in mammalian genes that contain it. The consensus sequence for a TATA box is TATAAA, though numerous variations have been characterized. The human TATA box is bound by the 34 kDa **TATA-binding protein (TBP)**, which is a subunit

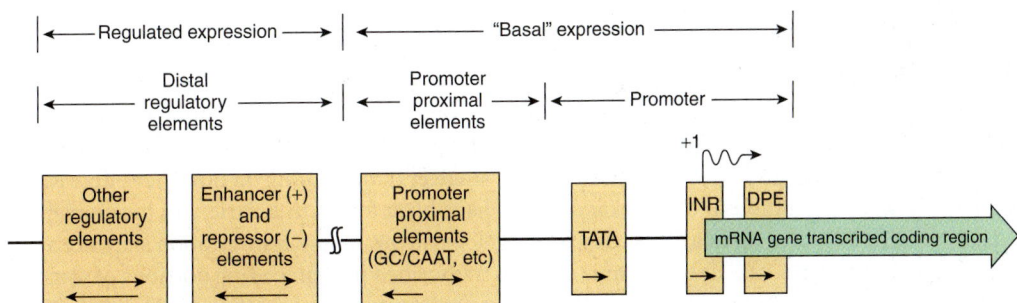

FIGURE 36–8 **Schematic diagram showing the transcription control regions in a hypothetical mRNA-producing, eukaryotic gene transcribed by RNA polymerase II.** Such a gene can be divided into its coding and regulatory regions, as defined by the transcription start site (**arrow**; +1). The coding region contains the DNA sequence that is transcribed into mRNA, which is ultimately translated into protein. The regulatory region consists of two classes of elements. One class is responsible for ensuring basal expression. The "promoter," is often composed of the TATA box and/or Inr and/or DPE elements, directs RNA polymerase II to the correct site (fidelity). However, in certain genes that lack TATA, the so-called TATA-less promoters, an initiator (Inr) and/or DPE elements may direct the polymerase to this site. Another component, the upstream elements, specifies the frequency of initiation; such elements can either be proximal (50–200 bp) or distal (1000–10^5 bp) to the promoter as shown. Among the best studied of the proximal elements is the CAAT box, but several other elements (bound by the transactivator proteins Sp1, NF1, AP1, etc) may be used in various genes. The distal elements enhance or repress expression, several of which mediate the response to various signals, including hormones, heat shock, heavy metals, and chemicals. Tissue-specific expression also involves specific sequences of this sort. The orientation dependence of all the elements is indicated by the arrows within the boxes. For example, the proximal promoter elements (TATA box, INR, DPE) must be in the 5′–3′ orientation, while the proximal upstream elements often work best in the 5′–3′ orientation, but some can be reversed. The locations of some elements are not fixed with respect to the transcription start site. Indeed, some elements responsible for regulated expression can be located interspersed with the upstream elements or can be located downstream from the start site.

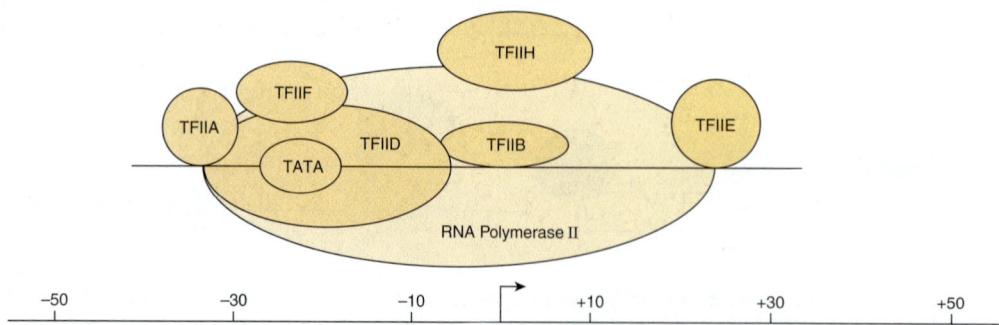

FIGURE 36–9   **The eukaryotic basal transcription complex.** Formation of the basal transcription complex begins when TFIID binds to the TATA box. It directs the assembly of several other components by protein-DNA and protein-protein interactions; TFIIA, B, E, F, H, and polymerase II (pol II). The entire complex spans DNA from position −30 to +30 relative to the transcription start site (TSS; +1, marked by bent arrow). The atomic level, X-ray-derived structures of RNA polymerase II alone and of the TBP subunit of TFIID bound to TATA promoter DNA in the presence of either TFIIB or TFIIA have all been solved at 3 Å resolution. The structures of TFIID and TFIIH complexes have been determined by electron microscopy at 30 Å resolution. Thus, the molecular structures of the transcription machinery are beginning to be elucidated. Much of this structural information is consistent with the models presented here.

in at least two multisubunit complexes, TFIID and SAGA/P-CAF. The non-TBP subunits of TFIID are proteins called **TBP-associated factors (TAFs)**. This complex of TBP and TAFs is referred to as TFIID. Binding of the TBP-TAF TFIID complex to the TATA box sequence is thought to represent a first step in the formation of the transcription complex on the promoter.

Some number of eukaryotic mRNA-encoding genes lack a consensus TATA box. In such instances, additional *cis*-elements, an **initiator sequence (Inr)** and/or the **downstream promoter element (DPE)**, direct the RNA polymerase II transcription machinery to the promoter and in so doing provide basal transcription starting from the correct site. The Inr element spans the start site (from −3 to +5) and consists of the general consensus sequence TCA_{+1} G/T T T/C (A_{+1} indicates the first nucleotide transcribed). The proteins that bind to Inr in order to direct pol II binding include TFIID. Promoters that have both a TATA box and an Inr may be stronger or more vigorously transcribed than those that have just one of these elements. The DPE has the consensus sequence A/GGA/T CGTG and is localized about 25-bp downstream of the +1 start site. Like the Inr, DPE sequences are also bound by the TAF subunits of TFIID. In a survey of thousands of eukaryotic protein coding genes, roughly 30% contained a TATA box and Inr, 25% contained Inr and DPE, 15% contained all three elements, whereas ~30% contained just the Inr.

Sequences generally, though not always, just upstream from the start site determine how frequently a transcription event occurs. Mutations in these regions reduce the frequency of transcriptional starts 10-fold to 20-fold. Typical of these DNA elements are the GC and CAAT boxes, so named because of the DNA sequences involved. As illustrated in Figure 36–7, each of these boxes binds a specific protein, Sp1 in the case of the GC box and CTF by the CAAT box; both bind through their distinct **DNA-binding domains (DBDs)**. The frequency of transcription initiation is a consequence of these protein-DNA interactions and complex interactions between particular domains of the transcription factors (distinct from the DBD domains—so-called

activation domains; ADs) of these proteins and the rest of the transcription machinery (RNA polymerase II, the **basal, or general factors, GTFs, TFIIA, B, D, E, F, H** and other coregulatory factors such as Mediator, chromatin remodellers and chromatin modifying factors). (See below and **Figures 36–9** and **36–10**.) The protein-DNA interaction at the TATA box involving RNA polymerase II and other components of the basal transcription machinery ensures the fidelity of initiation.

Together, the promoter and promoter-proximal *cis*-active upstream elements confer fidelity and frequency of initiation upon a gene. The TATA box has a particularly rigid requirement for both position and orientation. As with bacterial promoters, single-base changes in any of these *cis*-elements can have dramatic effects on function by reducing the binding affinity of the cognate *trans*-factors (either TFIID/TBP or Sp1, CTF, and similar factors). The spacing of the TATA box, Inr, and DPE is also critical.

A third class of sequence elements can either increase or decrease the rate of transcription initiation of eukaryotic genes. These elements are called either **enhancers** or **repressors (or silencers),** depending on how they effect RNA synthesis. They have been found in a variety of locations both upstream and downstream of the transcription start site and even within the transcribed protein coding portions of some genes. Enhancers and silencers can exert their effects when located thousands or even tens of thousands of bases away from transcription units located on the same chromosome. Surprisingly, enhancers and silencers can function in an orientation-independent fashion. Literally, hundreds of these elements have been described. In some cases, the sequence requirements for binding are rigidly constrained; in others, considerable sequence variation is allowed. Some sequences bind only a single protein, but the majority bind several different proteins. Together, these many transfactors binding to promoter distal and proximal *cis*-elements regulate transcription in response to a vast array of biological signals. Such transcriptional regulatory events contribute importantly to control of gene expression.

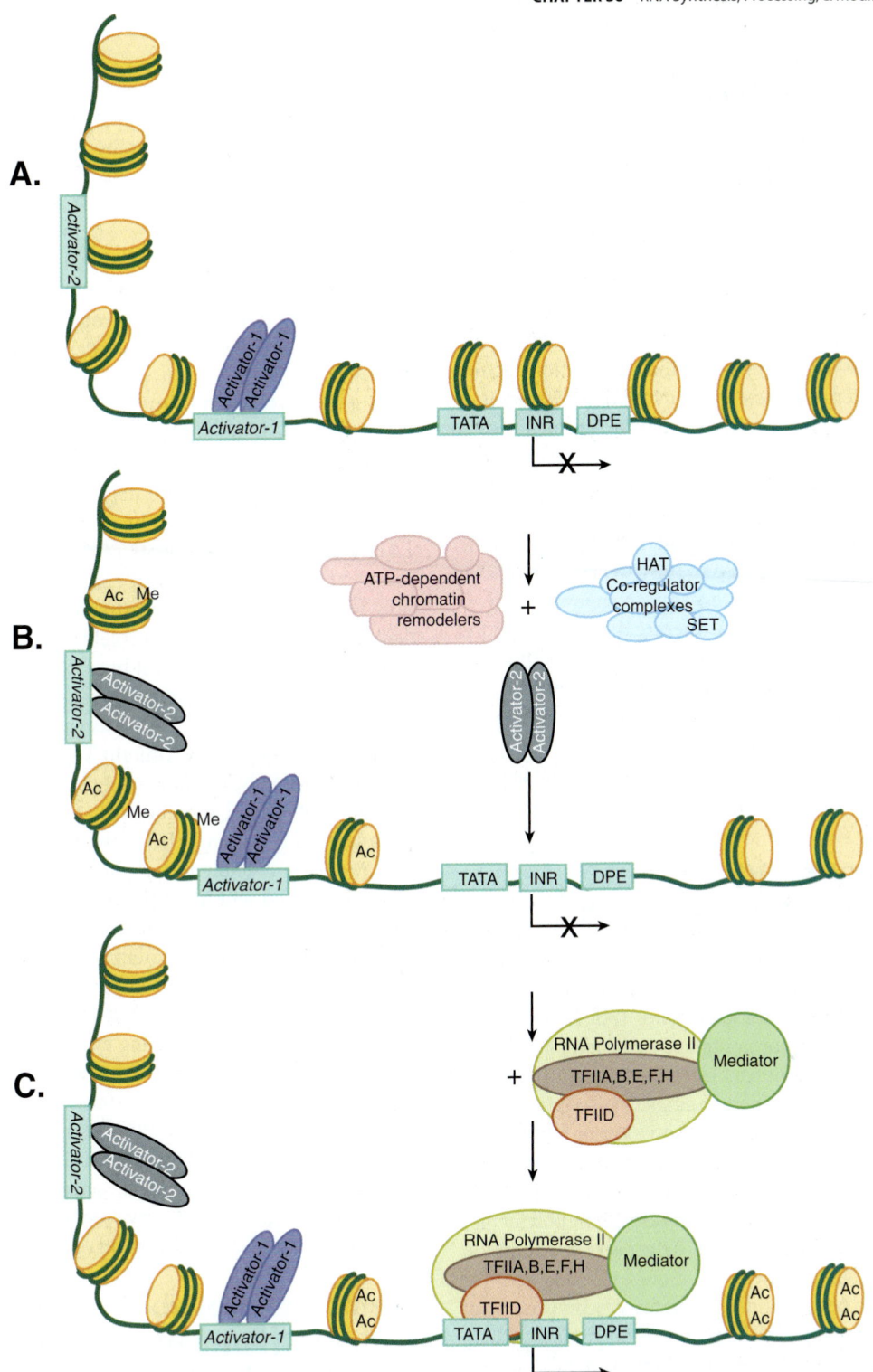

FIGURE 36–10 **Nucleosome eviction by chromatin-active coregulators facilitates PIC formation and transcription.** Shown in **A**, is an inactive mRNA encoding gene (see X over TSS) with a single dimeric transcription factor (Activator-1; violet ovals) bound to its cognate enhancer binding site (*Activator-1*). This particular enhancer element was nucleosome-free and hence available for interaction with its cognate activator binding protein. However, this gene is still inactive (X over TSS) due to the fact that a portion of its enhancer (in this illustration the enhancer is bipartitite and composed of *Activator-1* and *Activator-2*, DNA-binding sites) *and* the entirety of the promoter are covered by nucleosomes. **(B)** Enhancer DNA-bound Activator-1 interacts with any of a number of distinct ATP-dependent chromatin remodelers and chromatin-modifying Co-regulator complexes. These coregulators together have the ability to both move and/or remove nucleosomes (ATP-dependent remodelers) as well as to covalently modify nucleosomal histones using intrinsic acetylases (HAT; resulting in acetylation [Ac]) and methylases (SET; resulting in methylation [Me], among other PTMs; Table 35–1), carried by subunits of these complexes. **(C)** The resulting changes in nucleosome position and nucleosome occupancy thus allow for the binding of the second Activator-2 dimer to *Activator-2* DNA sequences, which leads to the binding of the transcription machinery (TFIIA,B,D,E,F,H; Polymerase II and Mediator) to the promoter (TATA-INR-DPE) and the formation of an active PIC, which leads to activated transcription.

Specific Signals Regulate Transcription Termination

The **signals for the termination of transcription** by eukaryotic RNA polymerase II are only poorly understood. It appears that the termination signals exist far downstream of the coding sequence of eukaryotic genes. For example, the transcription termination signal for mouse β-globin occurs at several positions 1000–2000 bases beyond the site at which the mRNA poly(A) tail will eventually be added. Less is known about the termination process or whether specific termination factors similar to the bacterial ρ factor are involved. However, it is known that formation of the mRNA 3′ terminal, which is generated posttranscriptionally, is somehow coupled to events or structures formed at the time and site of initiation. Moreover, mRNA formation, and in this case mRNA 3′-end formation depends on a special structure present on the C-terminus of the largest subunit of RNA polymerase II, the **C-terminal domain,** or **CTD** (see below), and this process appears to involve at least two steps as follows. After RNA polymerase II has traversed the region of the transcription unit encoding the 3′ end of the transcript, RNA endonucleases cleave the primary transcript at a position about 15 bases 3′ of the consensus sequence **AAUAAA** that serves in eukaryotic transcripts as a cleavage and polyadenylation signal. Finally, this newly formed 3′ terminal is polyadenylated in the nucleoplasm, as described below.

THE EUKARYOTIC TRANSCRIPTION COMPLEX

A complex apparatus consisting of as many as 50 unique proteins provides accurate and regulatable transcription of eukaryotic genes. The RNA polymerase enzymes (pol I, pol II, and pol III) transcribe information contained in the template strand of DNA into RNA. These polymerases must recognize a specific site in the promoter in order to initiate transcription at the proper nucleotide. In contrast to the situation in prokaryotes though, in vitro eukaryotic RNA polymerases alone are not able to discriminate between promoter sequences and other, non-promoter regions of DNA. All eukaryotic RNA polymerase forms require other proteins known as general transcription factors or GTFs. RNA polymerase II requires TFIIA, B, D (or TBP), E, F, and H to both facilitate promoter-specific binding of the enzyme and formation of the preinitiation complex (PIC). RNA polymerases I and III require their own polymerase-specific GTFs. Moreover, RNA polymerase II and GTFs do not respond to activator proteins and can only catalyze basal or (non)-unregulated transcription in vitro. Another set of proteins—the **coactivators,** or **coregulators**—work in conjunction with DNA-binding transactivator proteins to communicate with Pol II/GTFs to regulate the rate of transcription (see below).

Formation of the Pol II Transcription Complex

In bacteria, a σ-factor–polymerase holoenzyme complex, Eσ, selectively binds to promoter DNA to form the PIC. The situation is much more complex in eukaryotic genes. mRNA-encoding genes, which are transcribed by pol II, are described as an example. In the case of pol II-transcribed genes, the function of σ-factors is assumed by a number of proteins. PIC formation requires pol II and the six GTFs (TFIIA, TFIIB, TFIID, TFIIE, TFIIF, TFIIH). These GTFs serve to promote RNA polymerase II transcription on essentially all genes. Some of these GTFs are composed of multiple subunits. **TFIID, which binds to the TATA box promoter element through its TBP subunit,** is the only one of these factors that is independently capable of specific, high affinity binding to promoter DNA. TFIID consists of 15 subunits, TBP and 14 TBP Associated Factors, or TAFs.

TBP binds to the TATA box in the minor groove of DNA (most transcription factors bind in the major groove) and causes an approximately 100-degree bend or kink of the DNA helix. This bending is thought to facilitate the interaction of TAFs with other components of the transcription initiation complex, the multicomponent eukaryotic promoter and possibly with factors bound to upstream elements. Although initially defined as a component solely required for transcription of pol II gene promoters, TBP, by virtue of its association with distinct, polymerase-specific sets of TAFs, is also an important component of pol I and pol III transcription initiation complexes even if they do not contain TATA boxes.

The binding of TFIID marks a specific promoter for transcription. Of several subsequent in vitro steps, the first is the binding of TFIIA, then TFIIB to the TFIID-promoter complex. This results in a stable ternary complex, which is then more precisely located and more tightly bound at the transcription initiation site. This complex then attracts and tethers the pol II–TFIIF complex to the promoter. Addition of TFIIE and TFIIH are the final steps in the assembly of the PIC. TFIIE appears to join the complex with pol II–TFIIF, and TFIIH is then recruited. Each of these binding events extends the size of the complex so that finally about 60 bp (from −30 to +30 relative to +1, the nucleotide from which transcription commences) are covered (Figure 36–9). The PIC is now complete and capable of basal transcription initiated from the correct nucleotide. In genes that lack a TATA box, the same factors are required. In such cases, the Inr or DPE serve to (see Figure 36–8) position the complex for accurate initiation of transcription.

Promoter Accessibility and Hence PIC Formation Is Often Modulated by Nucleosomes

On certain eukaryotic genes, the transcription machinery (pol II, etc) cannot access the promoter sequences (ie, TATA–INR–DPE) because these essential promoter elements are wrapped up in nucleosomes (Figures 35–2 and 35–3 and 36–10). Only after transcription factors bind to enhancer DNA upstream of the promoter and recruit chromatin remodeling and modifying coregulatory factors such as the Swi/Snf, SRC-1, p300/CBP (see Chapter 42,) or P/CAF factors, are the repressing nucleosomes removed (Figure 36–10). Once the promoter is "open" following nucleosome eviction, GTFs and RNA polymerase II can bind and initiate mRNA gene transcription. Note that the binding of transactivators and coregulators can be sensitive to, and/or directly control the covalent modification status of the histones within the nucleosomes in and around the promoter and enhancer, and thereby increase or decrease the ability of all the other components required for PIC formation to interact with a particular gene. This so-called **epigenetic code of histone and protein modifications** can contribute importantly to gene transcription control. Indeed, mutations in proteins that catalyze (code writers), remove (code erasers), or differentially bind (code readers) modified histones can lead to human disease.

Phosphorylation Activates Pol II

Eukaryotic pol II consists of 12 subunits. The two largest subunits (MW 150 and 190 kDa) are homologous to the bacterial β and β′ subunits. In addition to the increased number of subunits, eukaryotic pol II differs from its prokaryotic counterpart in that it has a series of heptad repeats with consensus sequence Tyr-Ser-Pro-Thr-Ser-Pro-Ser at the carboxyl terminus of the largest pol II subunit. This **carboxyl terminal repeat domain (CTD)** has 26 repeated units in brewers' yeast and 52 units in mammalian cells. The CTD is a substrate for several enzymes (kinases, phosphatases, prolyl isomerases, glycosylases), Phosphorylation of the CTD was the first CTD PTM discovered. The kinase component of TFIIH can modify the CTD. Covalently modified CTD is binding site for a wide array of proteins, and it has been shown to interact with many mRNA modifying and processing enzymes and nuclear transport proteins. The association of these factors with the CTD of RNA polymerase II (and other components of the basal machinery) thus serves to couple transcription initiation with mRNA capping splicing, 3′ end formation and transport to the cytoplasm (see below). Pol II polymerization is activated when phosphorylated on the Ser and Thr residues and displays reduced activity when the CTD is dephosphorylated. CTD phosphorylation/dephosphorylation is critical for promoter clearance, elongation, termination, and even appropriate mRNA processing. Pol II lacking the CTD tail is incapable of activating transcription, and cells expressing pol II lacking the CTD are inviable. These results underscore the importance of this domain.

Pol II can associate with other proteins termed **Mediator** or **Med** proteins to form a complex sometimes referred to as the pol II holoenzyme; this complex can form on the promoter or in solution prior to PIC formation (see below). The Med proteins are essential for appropriate regulation of pol II transcription by serving myriad roles, both activating and repressing transcription. Thus Mediator, like TFIID is a transcriptional coregulator (**Figure 36–11**). Complex forms of RNA polymerase II

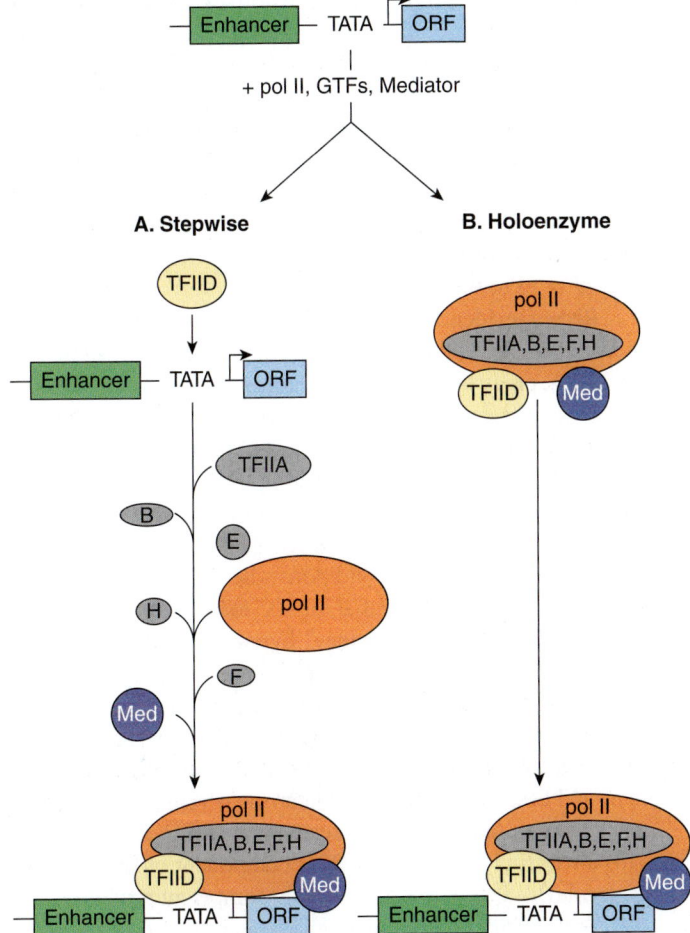

FIGURE 36–11 **Models for the formation of an RNA polymerase II preinitiation complex.** Shown at top is a typical mRNA encoding transcription unit: enhancer-promoter (TATA)-initiation site (bent arrow) and transcribed region (ORF; open reading frame). PICs have been shown to form by at least two distinct mechanisms: **(A)** the stepwise binding of GTFs, pol II, and Mediator, or **(B)** by the binding of a single multiprotein complex composed of pol II, Med, and the six GTFs. DNA-binding transactivator proteins specifically bind enhancers and in part facilitate PIC formation (or PIC function) by binding directly to the TFIID-TAF subunits or Med subunits of Mediator (not shown, see Figure 36–10); the molecular mechanism(s) by which such protein–protein interactions stimulate transcription remain a subject of intense investigation.

holoenzyme (pol II plus Med) have been described in human cells that contain over 30 Med proteins (Med1–Med31).

The Role of Transcription Activators & Coregulators

TFIID was originally considered to be a single protein, TBP. However, several pieces of evidence led to the important discovery that TFIID is actually a complex consisting of TBP and the 14 TAFs. The first evidence that TFIID was more complex than just the TBP molecules came from the observation that TBP binds to a 10-bp segment of DNA, immediately over the TATA box of the gene, whereas native holo-TFIID covers a 35 bp or larger region (Figure 36–9). Second, TBP has a molecular mass of 20–40 kDa (depending on the species), whereas the TFIID complex has a mass of about 1000 kDa. Finally, and perhaps most importantly, TBP supports basal transcription but not the augmented transcription provided by certain activators, for example, Sp1 bound to the GC box. TFIID, on the other hand, supports both basal and enhanced transcription by Sp1, Oct1, AP1, CTF, ATF, etc. (**Table 36–3**). The TAFs are essential for this activator-enhanced transcription. There are likely several forms of TFIID that differ slightly in their complement of TAFs. Thus different combinations of TAFs with TBP—or one of several recently discovered TBP-like factors (TLFs)—bind to different promoters, and recent reports suggest that this may account for the tissue or cell-selective gene activation noted in various promoters and for the different strengths of certain promoters. TAFs, since they are required for the action of activators, are often called coactivators or coregulators. There are thus three classes of transcription factors involved in the regulation of pol II genes: pol II and GTFs, coregulators, and DNA-binding activator-repressors (**Table 36–4**). How these classes of proteins interact to govern both the site and frequency of transcription is a question of central importance and active

TABLE 36–3 Some of the Mammalian RNA Polymerase II Transcription Control Elements, Their Consensus Sequences, and the Factors That Bind to Them

Element	Consensus Sequence	Factor
TATA box	TATAAA	TBP/TFIID
CAAT box	CCAATC	C/EBP*, NF-Y*
GC box	GGGCGG	Sp1*
	CAACTGAC	Myo D
	T/CGGA/CN$_5$GCCAA	NF1*
Ig octamer	ATGCAAAT	Oct1, 2, 4, 6*
AP1	TGAG/CTC/AA	Jun, Fos, ATF*
Serum response	GATGCCCATA	SRF
Heat shock	(NGAAN)$_3$	HSF

Note: A complete list would include hundreds of examples. The asterisks mean that there are several members of this family.

TABLE 36–4 Three Classes of Transcription Factors Involved in mRNA Gene Transcription

General Mechanisms	Specific Components
Basal components	RNA Polymerase II, TBP, TFIIA, B, D, E, F, and H
Coregulators	TAFs (TBP + TAFs) = TFIID; certain genes Mediator, Meds Chromatin modifiers Chromatin remodelers
Activators	SP1, ATF, CTF, AP1, etc

investigation. It is currently thought that coregulators both act as a bridge between the DNA-binding transactivators and pol II/GTFs and modify chromatin.

Two Models Can Explain the Assembly of the Preinitiation Complex

The formation of the PIC described above is based on the sequential addition of purified components as observed through in vitro experiments. An essential feature of this model is that PIC assembly takes place on a DNA template where the transcription proteins all have ready access to DNA. Accordingly, transcription activators, which have autonomous DNA binding and activation domains (see Chapter 38), are thought to function by stimulating PIC formation. Here the TAF or mediator complexes are viewed as bridging factors that communicate between the upstream-bound activators, and the GTFs and pol II. This view assumes that there is **stepwise assembly** of the PIC—promoted by various interactions between activators, coactivators, and PIC components, and is illustrated in panel A of Figure 36–11. This model was supported by observations that many of these proteins can indeed bind to one another in vitro.

Recent evidence suggests that there is another possible mechanism of PIC formation and thus transcription regulation. First, large preassembled complexes of GTFs and pol II are found in cell extracts, and these complexes can associate with the promoter in a single step. Second, the rate of transcription achieved when activators are added to limiting concentrations of pol II holoenzyme can be matched by increasing the concentration of the pol II holoenzyme in the absence of activators. Thus, at least in vitro, one can establish conditions where activators are not in themselves absolutely essential for PIC formation. These observations led to the **"recruitment" hypothesis,** which has now been tested experimentally. Simply stated, the role of activators and some coactivators may be solely to recruit a preformed holoenzyme-GTF complex to the promoter. The requirement for an activation domain is circumvented when either a component of TFIID or the pol II holoenzyme is artificially tethered, using recombinant DNA techniques, to the DBD of an activator. This anchoring, through the DBD component of the activator molecule, leads to a transcriptionally competent structure, and there is no further requirement for the activation domain of the activator. In this view, the role of activation domains is to direct preformed holoenzyme–GTF complexes to the promoter; they do not assist in PIC assembly (see panel B, Figure 36–11). In this model, the efficiency of the recruitment process directly determines the rate of transcription at a given promoter.

RNA MOLECULES ARE PROCESSED BEFORE THEY BECOME FUNCTIONAL

In prokaryotic organisms, the primary transcripts of mRNA-encoding genes begin to serve as translation templates even before their transcription has been completed. This can occur because the site of transcription is not compartmentalized into a nucleus as it is in eukaryotic organisms. Thus, transcription and translation are coupled in prokaryotic cells. Consequently, prokaryotic mRNAs are subjected to little processing prior to carrying out their intended function in protein synthesis. Indeed, appropriate regulation of some genes (eg, the *Trp* operon) relies upon this coupling of transcription and translation. Prokaryotic rRNA and tRNA molecules are transcribed in units considerably longer than the ultimate molecule. In fact, many of the tRNA transcription units encode more than one tRNA molecule. Thus, in prokaryotes, the processing of these rRNA and tRNA precursor molecules is required for the generation of the mature functional molecules.

Nearly all eukaryotic RNA primary transcripts undergo extensive processing between the time they are synthesized and the time at which they serve their ultimate function, whether it be as mRNA, miRNAs, or as a component of the translation machinery such as rRNA or tRNA. Processing occurs primarily within the nucleus. The processes of **transcription, RNA processing, and even RNA transport from the nucleus, are highly coordinated.** Indeed, a transcriptional coactivator termed SAGA in yeasts and P/CAF in human cells is thought to link transcription activation to RNA processing by recruiting a second complex termed TREX to transcription elongation, splicing, and nuclear export. TREX (<u>tr</u>anscription-<u>ex</u>port) represents a likely molecular link between transcription elongation complexes, the RNA splicing machinery, and nuclear export (see **Figure 36–12**). This coupling presumably dramatically increases both the fidelity and rate of processing and movement of mRNA to the cytoplasm for translation.

The Coding Portions (Exons) of Most Eukaryotic Genes Are Interrupted by Introns

The RNA sequences that appear in mature RNAs are termed **exons.** In mRNA encoding genes, exons are often interrupted by long sequences of DNA that neither appear in mature mRNA, nor contribute to the genetic information ultimately translated into the amino acid sequence of a protein molecule

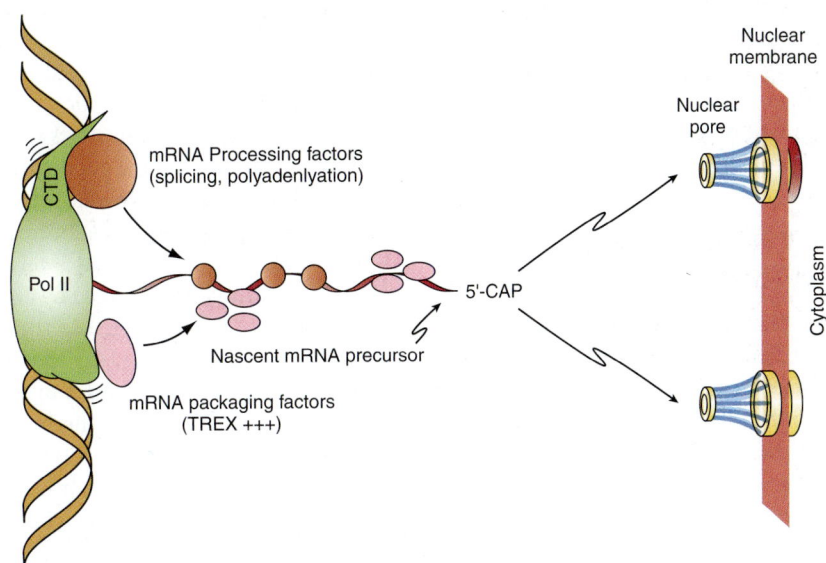

FIGURE 36–12 **RNA polymerase II-mediated mRNA gene transcription is cotranscriptionally coupled to RNA processing and transport.** Shown is RNA pol II actively transcribing an mRNA encoding gene (elongation top to bottom of figure). RNA processing factors (ie, SR/RNP-motif-containing splicing factors as well as polyadenylation and termination factors) interact with the C-terminal domain (CTD) of pol II, while mRNA packaging factors such as THO/TREX complex are recruited to the nascent mRNA primary transcript either through direct pol II interactions as shown or through interactions with SR/splicing factors resident on the nascent mRNA. Note that the CTD is not drawn to scale. The evolutionarily conserved CTD of the Rpb1 subunit of pol II is in reality 5–10 times the length of the polymerase due to its many prolines and consequent unstructured nature, and thus a significant docking site for RNA processing and transport proteins. In both cases, nascent mRNA chains are thought to be more rapidly and accurately processed due to the rapid recruitment of these many factors to the growing mRNA (precursor) chain. Following appropriate mRNA processing, the mature mRNA is delivered to the nuclear pores dotting the nuclear membrane, where, upon transport through the pores, the mRNAs can be engaged by ribosomes and translated into protein. (Adapted from Jensen et al: *Mol Cell.* 2005;11:1129–1138.)

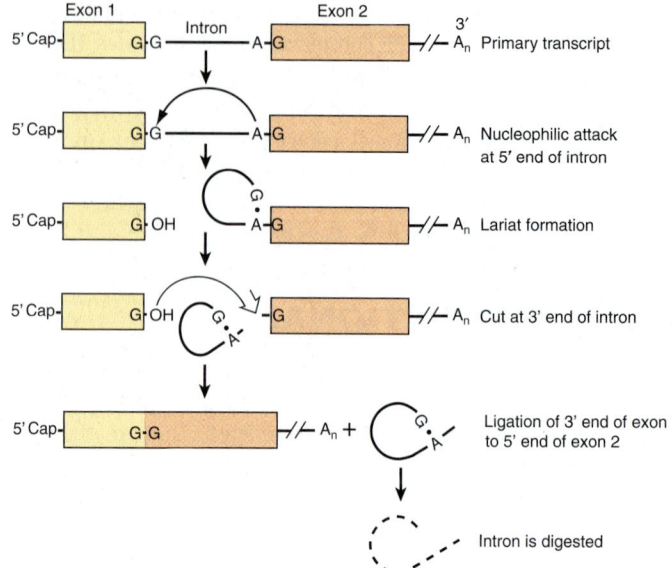

FIGURE 36–13 **The processing of the primary transcript to mRNA.** In this hypothetical transcript, the 5′ (**left**) end of the intron is cut (↓) and a lariat forms between the G at the 5′ end of the intron and an A near the 3′ end, in the consensus sequence UACUA**A**C. This sequence is called the branch site, and it is the 3′ most **A** that forms the 5′–2′ bond with the G. The 3′ (**right**) end of the intron is then cut (⇓). This releases the lariat, which is digested, and exon 1 is joined to exon 2 at G residues.

(see Chapter 35). In fact, these sequences often interrupt the coding region of structural genes. These **intervening sequences,** or **introns,** exist within most but not all mRNA encoding genes of higher eukaryotes. In human mRNA encoding genes, **exons average ~150 nt,** while **introns are much more heterogenous, ranging from 10–100 nt to 30,000 nucleotides in length.** The intron RNA sequences are cleaved out of the transcript, and the exons of the transcript are appropriately spliced together in the nucleus before the resulting mRNA molecule appears in the cytoplasm for translation (**Figures 36–13** and **36–14**). One speculation for this exon–intron gene organization is that exons, which often encode an activity domain, or functional module of a protein, represent a convenient means of shuffling genetic information, permitting organisms to quickly test the results of combining novel protein functional domains.

Introns Are Removed & Exons Are Spliced Together

Several different splicing reaction mechanisms for intron removal have been described. The one most frequently used in eukaryotic cells is described below. Although the sequences of nucleotides in the introns of the various eukaryotic transcripts—and even those within a single transcript—are quite heterogeneous, there are reasonably conserved sequences at each of the two exon–intron (splice) junctions

and at the branch site, which is located 20–40 nucleotides upstream from the 3′ splice site (see consensus sequences in Figure 36–14). A special multicomponent complex, the **spliceosome,** is involved in converting the primary transcript into mRNA. Spliceosomes consist of the primary transcript, five snRNAs (U1, U2, U4, U5, and U6) and more than 60 proteins, many of which contain conserved "RNP" and "SR" protein motifs. Collectively, the five snRNAs and RNP/SR–containing proteins form a **small nuclear ribonucleoprotein termed an snRNA complex.** It is likely that this penta-snRNP spliceosome forms prior to interaction with mRNA precursors. snRNPs are thought to position the exon and intron RNA segments for the necessary splicing reactions. The splicing reaction starts with a cut at the junction of the 5′-exon (donor or left) and intron (Figure 36–13). This is accomplished by a nucleophilic attack by an adenylyl residue in the branch point sequence located just upstream from the 3′ end of this intron. The free 5′ terminal then forms a loop or lariat structure that is linked by an unusual 5′–2′ phosphodiester bond to the reactive A in the PyNPyPyPuAPy branch site sequence (Figure 36–14). This adenylyl residue is typically located 20–30 nucleotides upstream from the 3′ end of the intron being removed. The branch site identifies the 3′ splice site. A second cut is made at the junction of the intron with the 3′ exon (donor on right). In this second transesterification reaction, the 3′ hydroxyl of the upstream exon attacks

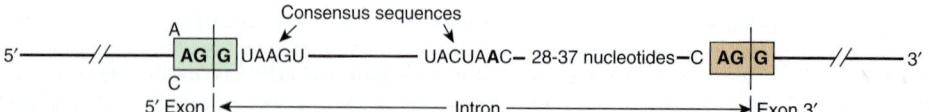

FIGURE 36–14 **Consensus sequences at splice junctions.** The 5′ (donor; **left**) and 3′ (acceptor; **right**) sequences are shown. Also shown is the yeast consensus sequence (UACUAAC) for the branch site. In mammalian cells, this consensus sequence is PyNPyPyPuAPy, where Py is a pyrimidine, Pu is a purine, and N is any nucleotide. The branch site is located 20–40 nucleotides upstream from the 3′ splice site.

the 5′ phosphate at the downstream exon–intron boundary, and the lariat structure containing the intron is released and hydrolyzed. The 5′ and 3′ exons are ligated to form a continuous sequence.

The snRNAs and associated proteins are required for formation of the various structures and intermediates. U1 within the snRNP complex binds first by base pairing to the 5′ exon–intron boundary. U2 within the snRNP complex then binds by base pairing to the branch site, and this exposes the nucleophilic A residue. U4/U5/U6 within the snRNP complex mediates an ATP-dependent protein-mediated unwinding that results in disruption of the base-paired U4–U6 complex with the release of U4. U6 is then able to interact first with U2, then with U1. These interactions serve to approximate the 5′ splice site, the branch point with its reactive A, and the 3′ splice site. This alignment is enhanced by U5. This process also results in the formation of the loop or lariat structure. The two ends are cleaved, probably by the U2–U6 within the snRNP complex. U6 is certainly essential, since yeasts deficient in this snRNA are not viable. It is important to note that RNA serves as the catalytic agent. This sequence of events is then repeated in genes containing multiple introns. In such cases, a definite pattern is followed for each gene, though the introns are not necessarily removed in sequence—1, then 2, then 3, etc.

Alternative Splicing Provides for Different mRNAs

The processing of mRNA molecules is a site for regulation of gene expression. Alternative patterns of mRNA splicing result from tissue-specific adaptive and developmental control mechanisms. Interestingly, recent studies suggest that alternative splicing is controlled, at least in part, through chromatin epigenetic marks (ie, Table 35–1). This form of coupling of transcription and mRNA processing may either be kinetic and/or mediated through interactions between specific histone PTMs and alternative splicing factors that can load onto nascent mRNA gene transcripts during the process of transcription (Figure 36–12).

As mentioned above, the sequence of exon–intron splicing events generally follows a hierarchical order for a given gene. The fact that very complex RNA structures are formed during splicing—and that a number of snRNAs and proteins are involved—affords numerous possibilities for a change of this order and for the generation of different mRNAs. Similarly, the use of alternative termination-cleavage polyadenylation sites also results in mRNA variability. Some schematic examples of these processes, all of which occur in nature, are shown in **Figure 36–15**.

Faulty splicing can cause disease. At least one form of β-thalassemia, a disease in which the β-globin gene of hemoglobin is severely underexpressed, appears to result from a nucleotide change at an exon–intron junction, precluding removal of the intron and therefore leading to diminished or absent synthesis of the β-chain protein. This is a consequence of the fact that the normal translation reading frame of the

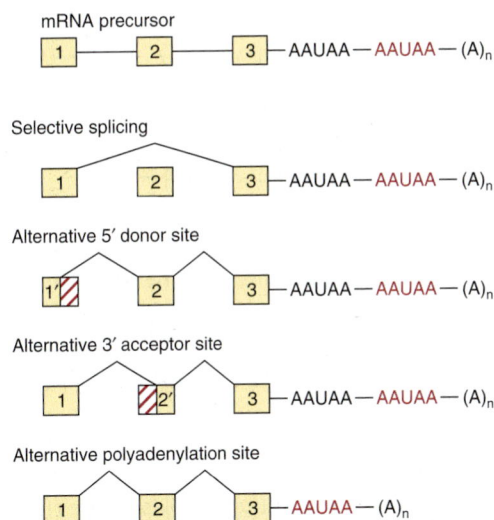

FIGURE 36–15 **Mechanisms of alternative processing of mRNA precursors.** This form of mRNA processing involves the selective inclusion or exclusion of exons, the use of alternative 5′ donor or 3′ acceptor sites, and the use of different polyadenylation sites.

mRNA is disrupted by a defect in the fundamental process of RNA splicing, underscoring the accuracy that the process of RNA–RNA splicing must maintain.

Alternative Promoter Utilization Provides a Form of Regulation

Tissue-specific regulation of gene expression can be provided by alternative splicing, as noted above, by control elements in the promoter or by the use of alternative promoters. The glucokinase (*GK*) gene consists of 10 exons interrupted by 9 introns. The sequence of exons 2–10 is identical in liver and pancreatic β cells, the primary tissues in which *GK* protein is expressed. Expression of the *GK* gene is regulated very differently—by two different promoters—in these two tissues. The liver promoter and exon 1L are located near exons 2–10; exon 1L is ligated directly to exon 2. By contrast, the pancreatic β-cell promoter is located about 30 kbp upstream. In this case, the 3′ boundary of exon 1B is ligated to the 5′ boundary of exon 2. The liver promoter and exon 1L are excluded and removed during the splicing reaction (see **Figure 36–16**). The existence of multiple distinct promoters allows for cell- and tissue-specific expression patterns of a particular gene (mRNA). In the case of *GK*, insulin and cAMP (Chapter 42) control *GK* transcription in liver, while glucose controls *GK* expression in β cells.

Both Ribosomal RNAs & Most Transfer RNAs Are Processed from Larger Precursors

In mammalian cells, the three rRNA molecules (28S, 18S, 5.8S) are transcribed as part of a single large 45S precursor molecule. **The precursor is subsequently processed in the**

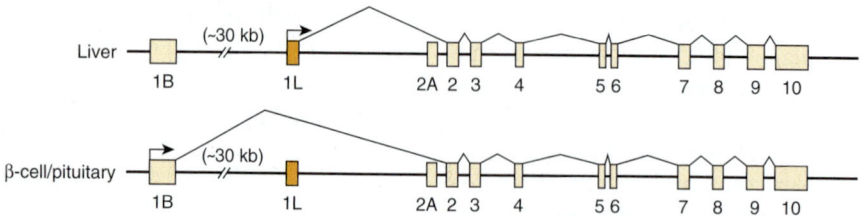

FIGURE 36–16 **Alternative promoter use in the liver and pancreatic β-cell glucokinase (*GK*) genes.** Differential regulation of the glucokinase gene is accomplished by the use of tissue-specific promoters. The β-cell *GK* gene promoter and exon 1B are located about 30 kbp upstream from the liver promoter and exon 1L. Each promoter has a unique structure and is regulated differently. Exons 2–10 are identical in the two genes, and the GK proteins encoded by the liver and β-cell mRNAs have identical kinetic properties.

nucleolus to provide these three RNA components for the ribosome subunits found in the cytoplasm. The rRNA genes are located in the nucleoli of mammalian cells. Hundreds of copies of these genes are present in every cell. This large number of genes is required to synthesize sufficient copies of each type of rRNA to form the 10^7 ribosomes required for each cell replication. Whereas a single mRNA molecule may be copied into 10^5 protein molecules, providing a large amplification, the rRNAs are end products. This lack of amplification requires both a large number of genes and a high transcription rate, typically synchronized with cell growth rate. Similarly, tRNAs are often synthesized as precursors, with extra sequences both 5′ and 3′ of the sequences comprising the mature tRNA. A small fraction of tRNAs contain introns.

RNAs CAN BE EXTENSIVELY MODIFIED

Essentially all RNAs are covalently modified after transcription. It is clear that at least some of these modifications are regulatory.

Messenger RNA Is Modified at the 5′ & 3′ Ends

As mentioned above, mammalian mRNA molecules contain a 7-methylguanosine cap structure at their 5′ terminal (Figure 34–10), and most have a poly(A) tail at the 3′ terminal. The cap structure is added to the 5′ end of the newly transcribed mRNA precursor in the nucleus prior to transport of the mRNA molecule to the cytoplasm. The **5′ cap** of the RNA transcript is required both for efficient translation initiation and protection of the 5′ end of mRNA from attack by 5′ → 3′ exonucleases. The secondary methylations of mRNA molecules, those on the 2′-hydroxy and the N^7 of adenylyl residues, occur after the mRNA molecule has appeared in the cytoplasm.

Poly(A) tails are added to the 3′ end of mRNA molecules in a posttranscriptional processing step. The mRNA is first cleaved about 20 nucleotides downstream from an AAUAA

recognition sequence. Another enzyme, **poly(A) polymerase,** adds a poly(A) tail which is subsequently extended to as many as 200 A residues. The **poly(A) tail** appears to protect the 3′ end of mRNA from 3′ → 5′ exonuclease attack and facilitate translation. The presence or absence of the poly(A) tail does not determine whether a precursor molecule in the nucleus appears in the cytoplasm, because all poly(A)-tailed nuclear mRNA molecules do not contribute to cytoplasmic mRNA, nor do all cytoplasmic mRNA molecules contain poly(A) tails (histone mRNAs are most notable in this regard). Following nuclear transport cytoplasmic enzymes in mammalian cells can both add and remove adenylyl residues from the poly(A) tails; this process has been associated with an alteration of mRNA stability and translatability.

The size of some cytoplasmic mRNA molecules, even after the poly(A) tail is removed, is still considerably greater than the size required to code for the specific protein for which it is a template, often by a factor of 2 or 3. **The extra nucleotides occur in untranslated (nonprotein coding) regions** both 5′ and 3′ of the coding region; the longest untranslated sequences are usually at the 3′ end. The exact function of **5′ UTR and 3′ UTR** sequences is unknown, but they have been implicated in RNA processing, transport, storage, degradation, and translation; each of these reactions potentially contributes additional levels of control of gene expression. **The micro-RNAs typically target sequences within the 3′ UTR.** Many of these posttranscriptional events involving mRNAs occur in cytoplasmic organelles termed P bodies (Chapter 37).

Micro-RNAs Are Derived from Large Primary Transcripts Through Specific Nucleolytic Processing

The majority of miRNAs are transcribed by RNA pol II into **primary transcripts** termed **pri-miRNAs.** pri-miRNAs are 5′-capped and 3′-polyadenylated (**Figure 36–17**). pri-miRNAs are synthesized from transcription units encoding one or several distinct miRNAs; these transcription units are either located independently in the genome or within the intronic DNA of other genes. miRNA-encoding genes must therefore minimally possess a distinct promoter, coding region and

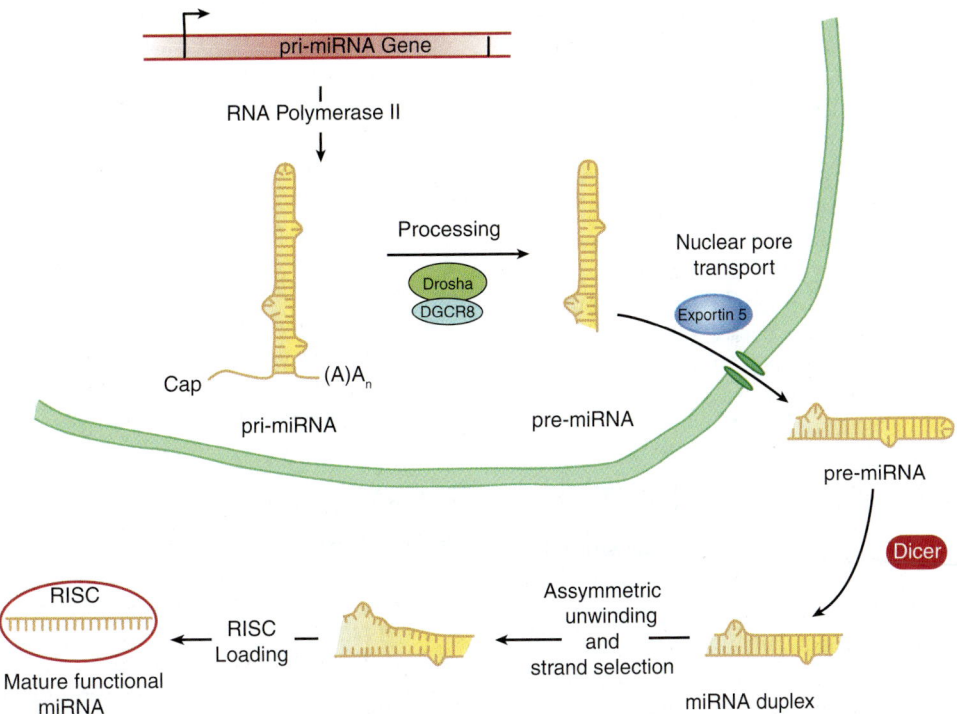

FIGURE 36–17 **Biogenesis of miRNAs.** miRNA encoding genes are transcribed by RNA pol II into a primary miRNA transcript (pri-miRNA), which is 5′-capped and polyadenylated as is typical of mRNA coding primary transcripts. This pri-miRNA is subjected to processing within the nucleus by the action of the Drosha-DGCR8 nuclease, which trims sequences from both 5′ and 3′ ends generating the pre-miRNA. This partially processed double-stranded RNA is transported through the nuclear pore by exportin-5. The cytoplasmic pre-miRNA is then trimmed further by the action of the multisubunit nuclease termed Dicer, to form the miRNA duplex. One of the two resulting 21–22 nucleotide-long RNA strands is selected, the duplex unwound, and the selected strand loaded into the RISC complex, thereby generating the mature, functional miRNA.

polyadenylation/termination signals. pri-miRNAs have extensive 2° structure, and this intramolecular structure is maintained following processing by the **Drosha-DGCR8 nuclease;** the portion containing the RNA hairpin is preserved, transported through the nuclear pore and once in the cytoplasm, further processed to a **21 or 22-mer** by the **dicer nuclease.** Ultimately, one of the two strands is selected for loading into the **RISC,** or **RNA-induced silencing complex** to form a mature, functional miRNA. siRNAs are produced similarly. Once in the RISC complex, miRNAs can modulate mRNA function (see Chapter 39). Recent data suggest that regulatory miRNA-encoding genes may be linked, and hence co-evolve with their target genes.

RNA Editing Changes mRNA After Transcription

The central dogma states that for a given gene and gene product there is a linear relationship between the coding sequence in DNA, the mRNA sequence, and the protein sequence (Figure 35–7). Changes in the DNA sequence should be reflected in a change in the mRNA sequence and, depending on codon usage, in protein sequence. However, exceptions to this dogma have been recently documented. Coding information can be changed at the mRNA level by **RNA editing.** In such cases, the coding sequence of the mRNA differs from that in the cognate DNA. An example is the apolipoprotein B

(apoB) gene and mRNA. In liver, the single *apoB* gene is transcribed into an mRNA that directs the synthesis of a 100-kDa protein, apoB100. In the intestine, the same gene directs the synthesis of the primary transcript; however, a cytidine deaminase converts a CAA codon in the mRNA to UAA at a single specific site. Rather than encoding glutamine, this codon becomes a termination signal, and a 48-kDa protein (apoB48) is the result. ApoB100 and apoB48 have different functions in the two organs. A growing number of other examples include a glutamine to arginine change in the glutamate receptor and several changes in trypanosome mitochondrial mRNAs, generally involving the addition or deletion of uridine. The exact extent of RNA editing is unknown, but current estimates suggest that <0.01% of mRNAs are edited in this fashion. Recently, editing of miRNAs has been described suggesting that these two forms of posttranscriptional control mechanisms could cooperatively contribute to gene regulation.

Transfer RNA Is Extensively Processed & Modified

As described in Chapters 34 & 37, the tRNA molecules serve as adapter molecules for the translation of mRNA into protein sequences. The tRNAs contain many modifications of the standard bases A, U, G, and C, including methylation, reduction, deamination, and rearranged glycosidic bonds.

Further posttranscriptional modification of the tRNA molecules includes nucleotide alkylations and the attachment of the characteristic CpCpA$_{OH}$ terminal at the 3′ end of the molecule by the enzyme nucleotidyl transferase. The 3′ OH of the A ribose is the point of attachment for the specific amino acid that is to enter into the polymerization reaction of protein synthesis. The methylation of mammalian tRNA precursors probably occurs in the nucleus, whereas the cleavage and attachment of CpCpA$_{OH}$ are cytoplasmic functions, since the terminals turn over more rapidly than do the tRNA molecules themselves. Enzymes within the cytoplasm of mammalian cells are required for the attachment of amino acids to the CpCpA$_{OH}$ residues (see Chapter 37).

RNA CAN ACT AS A CATALYST

In addition to the catalytic action served by the snRNAs in the formation of mRNA, several other enzymatic functions have been attributed to RNA. **Ribozymes** are RNA molecules with catalytic activity. These generally involve transesterification reactions, and most are concerned with RNA metabolism (splicing and endoribonuclease). Recently, a rRNA component has been implicated in hydrolyzing an aminoacyl ester and thus to play a central role in peptide bond function (peptidyl transferases; see Chapter 37). These observations, made using RNA molecules derived from the organelles from plants, yeast, viruses, and higher eukaryotic cells, show that RNA can act as an enzyme, and have revolutionized thinking about enzyme action and the origin of life itself.

SUMMARY

- RNA is synthesized from a DNA template by the enzyme RNA polymerase.

- While bacteria contain but a single RNA polymerase (β,β' α_2) there are three distinct nuclear DNA-dependent RNA polymerases in mammals: RNA polymerases I, II, and III. These enzymes catalyze the transcription of rRNA(Pol I), mRNA/miRNAs (Pol II), and tRNA and 5S rRNA (Pol III) encoding genes.

- RNA polymerases interact with unique *cis*-active regions of genes, termed promoters, in order to form preinitiation complexes (PICs) capable of initiation. In eukaryotes, the process of pol II PIC formation requires, in addition to polymerase, multiple general transcription factors (GTFs), TFIIA, B, D, E, F, and H.

- Eukaryotic PIC formation can occur on accessible promoters either step-wise—by the sequential, ordered interactions of GTFs and RNA polymerase with DNA promoters—or in one step by the recognition of the promoter by a pre-formed GTF-RNA polymerase holoenzyme complex.

- Transcription exhibits three phases: initiation, elongation, and termination. All are dependent upon distinct DNA *cis*-elements and can be modulated by distinct *trans*-acting protein factors.

- The presence of nucleosomes can occlude the binding of both transfactors and the transcription machinery to their cognate DNA *cis*-elements, thereby inhibiting transcription.

- Most eukaryotic RNAs are synthesized as precursors that contain excess sequences which are removed prior to the generation of mature, functional RNA. These processing reactions provide additional potential steps for regulation of gene expression.

- Eukaryotic mRNA synthesis results in a pre-mRNA precursor that contains extensive amounts of excess RNA (introns) that must be precisely removed by RNA splicing to generate functional, translatable mRNA composed of exonic coding and 5′ and 3′ noncoding sequences.

- All steps—from changes in DNA template, sequence, and accessibility in chromatin to RNA stability and translatability—are subject to modulation and hence are potential control sites for eukaryotic gene regulation.

REFERENCES

Bourbon H-M, Aguilera A, Ansari AZ, et al: A unified nomenclature for protein subunits of mediator complexes linking transcriptional regulators to RNA polymerase II. Mol Cell 2004;14:553.

Busby S, Ebright RH: Promoter structure, promoter recognition, and transcription activation in prokaryotes. Cell 1994;79:743.

Cramer P, Bushnell DA, Kornberg RD: Structural basis of transcription: RNA polymerase II at 2.8 angstrom resolution. Science 2001;292:1863.

Fedor MJ: Ribozymes. Curr Biol 1998;8:R441.

Gott JM, Emeson RB: Functions and mechanisms of RNA editing. Ann Rev Genet 2000;34:499.

Harel-Sharvit L, Eldad N, Haimovich G, et al: RNA polymerase II subunits link transcription and mRNA decay to translation. Cell 2010;143:552–563.

Kawauchi J, Mischo H, Braglia P, et al: Budding yeast RNA polymerases I and II employ parallel mechanisms of transcriptional termination. Genes Dev 2008;22:1082.

Keaveney M, Struhl K: Activator-mediated recruitment of the RNA polymerase machinery is the predominant mechanism for transcriptional activation in yeast. Mol Cell 1998;1:917.

Maniatis T, Reed R: An extensive network of coupling among gene expression machines. Nature 2002;416:499.

Mapendano CK, Lykke-Andersen S, Kjems J, et al: Crosstalk between mRNA 3′ end processing and transcription initiation Mol Cell 2010;40:410–422.

Nechaev S, Adelman K: Pol II waiting in the starting gates: regulating the transition from transcription initiation into productive elongation. Biochim Biophys Acta 2011;1809:34–45.

Orphanides G, Reinberg D: A unified theory of gene expression. Cell 2002;108:439.

Price DH: Poised polymerases: on your mark ... get set ... go!. Mol Cell 2008;30:7.

Reed R, Cheng H: TREX, SR proteins and export of mRNA. Curr Opin Cell Biol 2005;17:269.

Trcek T, Singer RH: The cytoplasmic fate of an mRNP is determined cotranscriptionally: exception or rule? Genes Dev 2010; 24:1827–1831.

Tucker M, Parker R: Mechanisms and control of mRNA decapping in *Saccharomyces cerevisiae*. Ann Rev Biochem 2000;69:571.

West S, Proudfoot NJ, Dye MJ, et al: Molecular dissection of mammalian RNA polymerase II transcriptional termination. Mol Cell 2008;29:600.

Protein Synthesis & the Genetic Code

P. Anthony Weil, PhD

OBJECTIVES

After studying this chapter, you should be able to:

- Understand that the genetic code is a three-letter nucleotide code, which is encoded in the linear array of the exon DNA (composed of triplets of A, G, C, and T) of protein coding genes, and that this three-letter code is translated into mRNA (composed of triplets of A, G, C, and U) to specify the linear order of amino acid addition during protein synthesis via the process of translation.

- Appreciate that the universal genetic code is degenerate, unambiguous, non-overlapping, and punctuation free.

- Explain that the genetic code is composed of 64 codons, 61 of which encode amino acids while 3 induce the termination of protein synthesis.

- Explain how the transfer RNAs (tRNAs) serve as the ultimate informational agents that decode the genetic code of mRNAs.

- Understand the mechanism of the energy-intensive process of protein synthesis that occurs on RNA-protein complexes termed ribosomes.

- Appreciate that protein synthesis, like DNA replication and transcription, is precisely controlled through the action of multiple accessory factors that are responsive to multiple extra- and intracellular regulatory signaling inputs.

BIOMEDICAL IMPORTANCE

The letters A, G, T, and C correspond to the nucleotides found in DNA. Within the protein-coding genes, these nucleotides are organized into three-letter code words called **codons,** and the collection of these codons makes up the **genetic code.** It was impossible to understand protein synthesis—or to explain mutations—before the genetic code was elucidated. The code provides a foundation for explaining the way in which protein defects may cause genetic disease and for the diagnosis and perhaps the treatment of these disorders. In addition, the pathophysiology of many viral infections is related to the ability of these infectious agents to disrupt host cell protein synthesis. Many antibacterial drugs are effective because they selectively disrupt protein synthesis in the invading bacterial cell but do not affect protein synthesis in eukaryotic cells.

GENETIC INFORMATION FLOWS FROM DNA TO RNA TO PROTEIN

The genetic information within the nucleotide sequence of DNA is transcribed in the nucleus into the specific nucleotide sequence of an RNA molecule. The sequence of nucleotides in the RNA transcript is complementary to the nucleotide sequence of the template strand of its gene in accordance with the base-pairing rules. Several different classes of RNA combine to direct the synthesis of proteins.

In prokaryotes there is a linear correspondence between the gene, the **messenger RNA (mRNA)** transcribed from the gene, and the polypeptide product. The situation is more complicated in higher eukaryotic cells, in which the primary transcript is much larger than the mature mRNA. The large mRNA precursors contain coding regions (**exons**) that will form the mature mRNA and long intervening sequences (introns) that separate the exons. The mRNA is processed within the nucleus, and the introns, which make up much more of this RNA than the exons, are removed. Exons are spliced together to form mature mRNA, which is transported to the cytoplasm, where it is translated into protein.

The cell must possess the machinery necessary to translate information accurately and efficiently from the nucleotide sequence of an mRNA into the sequence of amino acids of the corresponding specific protein. Clarification of our understanding of this process, which is termed **translation,** awaited deciphering of the genetic code. It was realized early that mRNA molecules themselves have no affinity for amino acids and, therefore, that the translation of the information in the mRNA nucleotide sequence into the amino acid sequence of a protein requires an intermediate adapter molecule. This adapter molecule must recognize a specific nucleotide sequence on the one hand as well as a specific amino acid on the other. With such an adapter molecule, the cell can direct a specific amino acid into the proper sequential position of a protein during its synthesis as dictated by the nucleotide sequence of the specific mRNA. In fact, the functional groups of the amino acids do not themselves actually come into contact with the mRNA template.

THE NUCLEOTIDE SEQUENCE OF AN mRNA MOLECULE CONTAINS A SERIES OF CODONS THAT SPECIFY THE AMINO ACID SEQUENCE OF THE ENCODED PROTEIN

Twenty different amino acids are required for the synthesis of the cellular complement of proteins; thus, there must be at least 20 distinct codons that make up the genetic code. Since there are only four different nucleotides in mRNA, each codon must consist of more than a single purine or pyrimidine nucleotide. Codons consisting of two nucleotides each could provide for only 16 (4^2)-specific codons, whereas codons of three nucleotides could provide 64 (4^3)-specific codons.

It is now known that each codon consists of a sequence of three nucleotides; that is, **it is a triplet code** (see **Table 37–1**). The initial deciphering of the **genetic code** depended heavily on in vitro synthesis of nucleotide polymers, particularly triplets in repeated sequence. These synthetic triplet ribonucleotides were used as mRNAs to program protein synthesis in the test tube, and allowed investigators to deduce the genetic code.

THE GENETIC CODE IS DEGENERATE, UNAMBIGUOUS, NONOVERLAPPING, WITHOUT PUNCTUATION, & UNIVERSAL

Three of the 64 possible codons do not code for specific amino acids; these have been termed **nonsense codons.** These nonsense codons are utilized in the cell as **termination signals;**

TABLE 37–1 The Genetic Code[1] (Codon Assignments in Mammalian Messenger RNAs)

First Nucleotide	Second Nucleotide				Third Nucleotide
	U	C	A	G	
U	Phe	Ser	Tyr	Cys	U
	Phe	Ser	Tyr	Cys	C
	Leu	Ser	Term	Term[2]	A
	Leu	Ser	Term	Trp	G
C	Leu	Pro	His	Arg	U
	Leu	Pro	His	Arg	C
	Leu	Pro	Gln	Arg	A
	Leu	Pro	Gln	Arg	G
A	Ile	Thr	Asn	Ser	U
	Ile	Thr	Asn	Ser	C
	Ile[2]	Thr	Lys	Arg[2]	A
	Met	Thr	Lys	Arg[2]	G
G	Val	Ala	Asp	Gly	U
	Val	Ala	Asp	Gly	C
	Val	Ala	Glu	Gly	A
	Val	Ala	Glu	Gly	G

[1]The terms first, second, and third nucleotide refer to the individual nucleotides of a triplet codon read 5'–3', left to right. A, adenine nucleotide; C, cytosine nucleotide; G, guanine nucleotide; Term, chain terminator codon; U, uridine nucleotide. AUG, which codes for Met, serves as the initiator codon in mammalian cells and also encodes for internal methionines in a protein. (Abbreviations of amino acids are explained in Chapter 3.)
[2]In mammalian mitochondria, AUA codes for Met and UGA for Trp, and AGA and AGG serve as chain terminators.

they specify where the polymerization of amino acids into a protein molecule is to stop. **The remaining 61 codons code for the 20 naturally occurring amino acids** (Table 37–1). Thus, there is **"degeneracy"** in the genetic code—that is, multiple codons decode the same amino acid. Some amino acids are encoded by several codons; for example six different codons, UCU, UCC, UCA, UCG, AGU, and AGC all specify serine. Other amino acids, such as methionine and tryptophan, have a single codon. In general, the third nucleotide in a codon is less important than the first two in determining the specific amino acid to be incorporated, and this accounts for most of the degeneracy of the code. However, for any specific codon, only a single amino acid is indicated; with rare exceptions, the genetic code is **unambiguous**—that is, given a specific codon, only a single amino acid is indicated. **The distinction between ambiguity and degeneracy is an important concept.**

The unambiguous but degenerate code can be explained in molecular terms. The recognition of specific codons in the mRNA by the tRNA adapter molecules is dependent upon their **anticodon region** and specific base-pairing rules. Each tRNA molecule contains a specific sequence, complementary to a codon, which is termed its anticodon. For a given codon in the mRNA, only a single species of tRNA molecule possesses the proper anticodon. Since each tRNA molecule can be charged with only one specific amino acid, each codon therefore specifies only one amino acid. However, some tRNA molecules can utilize the anticodon to recognize more than one codon. **With few exceptions, given a specific codon, only a specific amino acid will be incorporated—although, given a specific amino acid, more than one codon may be used.**

As discussed below, the reading of the genetic code during the process of protein synthesis does not involve any overlap of codons. **Thus, the genetic code is nonoverlapping.** Furthermore, once the reading is commenced at a specific codon, there is **no punctuation** between codons, and the message is read in a continuing sequence of nucleotide triplets until a translation stop codon is reached.

Until recently, the genetic code was thought to be universal. It has now been shown that the set of tRNA molecules in mitochondria (which contain their own separate and distinct set of translation machinery) from lower and higher eukaryotes, including humans, reads four codons differently from the tRNA molecules in the cytoplasm of even the same cells. As noted in Table 37–1, the codon AUA is read as Met, and UGA codes for Trp in mammalian mitochondria. In addition, in mitochondria, the codons AGA and AGG are read as stop or chain terminator codons rather than as Arg. As a result of these organelle-specific changes in genetic code, mitochondria require only 22 tRNA molecules to read their genetic code, whereas the cytoplasmic translation system possesses a full complement of 31 tRNA species. These exceptions noted, **the genetic code is universal.** The frequency of use of each amino acid codon varies considerably between species and among different tissues within a species. The specific tRNA levels generally mirror these codon usage biases. Thus, a particular abundantly used codon is decoded by a similarly abundant-specific tRNA which recognizes that particular codon. Tables of **codon usage** are becoming more accurate as more genes and genomes are sequenced; such information can prove vital for large-scale production of proteins for therapeutic purposes (ie, insulin, erythropoietin). Such proteins are often produced in nonhuman cells using recombinant DNA technology (Chapter 39). The main features of the genetic code are listed in **Table 37–2.**

TABLE 37–2   Features of the Genetic Code

- Degenerate
- Unambiguous
- Nonoverlapping
- Not punctuated
- Universal

AT LEAST ONE SPECIES OF TRANSFER RNA (tRNA) EXISTS FOR EACH OF THE 20 AMINO ACIDS

tRNA molecules have extraordinarily similar functions and three-dimensional structures. The adapter function of the tRNA molecules requires the charging of each specific tRNA with its specific amino acid. Since there is no affinity of nucleic acids for specific functional groups of amino acids, this recognition must be carried out by a protein molecule capable of recognizing both a specific tRNA molecule and a specific amino acid. At least 20-specific enzymes are required for these specific recognition functions and for the proper attachment of the 20 amino acids to specific tRNA molecules. The energy requiring process of **recognition and attachment (charging)** proceeds in two steps and is catalyzed by one enzyme for each of the 20 amino acids. These enzymes are termed **aminoacyl-tRNA synthetases.** They form an activated intermediate of aminoacyl-AMP–enzyme complex (**Figure 37–1**). The specific aminoacyl-AMP–enzyme complex then recognizes a specific tRNA to which it attaches the aminoacyl moiety at the 3′-hydroxyl adenosyl terminal. The charging reactions have an error rate of less than 10^{-4} and so are quite accurate. The amino acid remains attached to its specific tRNA in an ester linkage until it is polymerized at a specific position in the fabrication of a polypeptide precursor of a protein molecule.

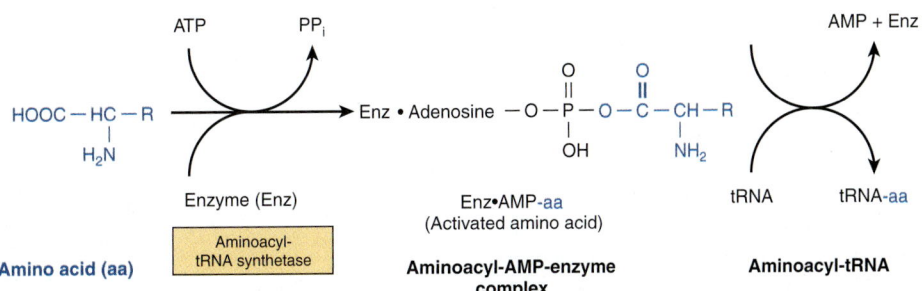

FIGURE 37–1   Formation of aminoacyl-tRNA. A two-step reaction, involving the enzyme amino-acyl-tRNA synthetase, results in the formation of aminoacyl-tRNA. The first reaction involves the formation of an AMP-amino acid–enzyme complex. This activated amino acid is next transferred to the corresponding tRNA molecule. The AMP and enzyme are released, and the latter can be reutilized. The charging reactions have an error rate (ie, esterifying the incorrect amino acid on tRNA$_x$) of less than 10^{-4}.

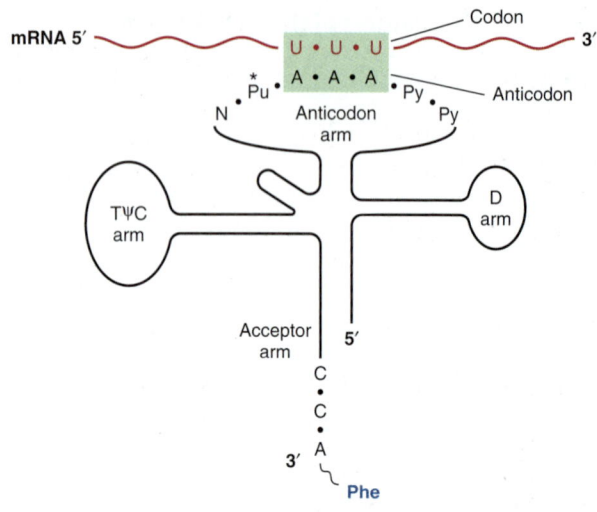

FIGURE 37–2   Recognition of the codon by the anticodon.
One of the codons for phenylalanine is UUU. tRNA charged with phenylalanine (Phe) has the complementary sequence AAA; hence, it forms a base-pair complex with the codon. The anticodon region typically consists of a sequence of seven nucleotides: variable (N), modified purine (Pu*), X, Y, Z (here, A A A), and two pyrimidines (Py) in the 3′ –5′ direction.

The regions of the tRNA molecule referred to in Chapter 34 (and illustrated in Figure 34–11) now become important. The ribothymidine pseudouridine cytidine (TψC) arm is involved in binding of the aminoacyl-tRNA to the ribosomal surface at the site of protein synthesis. The D arm is one of the sites important for the proper recognition of a given tRNA species by its proper aminoacyl-tRNA synthetase. The acceptor arm, located at the 3′-hydroxyl adenosyl terminal, is the site of attachment of the specific amino acid.

The anticodon region consists of seven nucleotides, and it recognizes the three-letter codon in mRNA (**Figure 37–2**). The sequence read from the 3′–5′ direction in that anticodon loop consists of a variable base–modified purine–XYZ–pyrimidine–pyrimidine-5′. Note that this direction of reading the anticodon is 3′–5′, whereas the genetic code in Table 37–1 is read 5′–3′, since the codon and the anticodon loop of the mRNA and tRNA molecules, respectively, are **antiparallel** in their complementarity just like all other intermolecular interactions between nucleic acid strands.

The degeneracy of the genetic code resides mostly in the last nucleotide of the codon triplet, suggesting that the base pairing between this last nucleotide and the corresponding nucleotide of the anticodon is not strictly by the Watson–Crick rule. This is called **wobble;** the pairing of the codon and anti-codon can "wobble" at this specific nucleotide-to-nucleotide pairing site. For example, the two codons for arginine, AGA and AGG, can bind to the same anticodon having a uracil at its 5′ end (UCU). Similarly, three codons for glycine—GGU, GGC, and GGA—can form a base pair from one anticodon, 3′ CCI 5′ (ie, I can base pair with U, C and A).

I is a purine inosine nucleotide generated by deamination of adenine (see Figure 33–2 for structure), another of the peculiar bases often appearing in tRNA molecules.

MUTATIONS RESULT WHEN CHANGES OCCUR IN THE NUCLEOTIDE SEQUENCE

Although the initial change may not occur in the template strand of the double-stranded DNA molecule for that gene, after replication, daughter DNA molecules with mutations in the template strand will segregate and appear in the population of organisms.

Some Mutations Occur by Base Substitution

Single-base changes (**point mutations**) may be **transitions** or **transversions.** In the former, a given pyrimidine is changed to the other pyrimidine or a given purine is changed to the other purine. Transversions are changes from a purine to either of the two pyrimidines or the change of a pyrimidine into either of the two purines, as shown in **Figure 37–3**.

If the nucleotide sequence of the gene containing the mutation is transcribed into an RNA molecule, then the RNA molecule will of course possess the base change at the corresponding location.

Single-base changes in the mRNA molecules may have one of several effects when translated into protein:

1. There may be no detectable effect because of the degeneracy of the code; such mutations are often referred to as **silent mutations.** This would be more likely if the changed base in the mRNA molecule were to be at the third nucleotide of a codon. Because of wobble, the translation of a codon is least sensitive to a change at the third position.

2. A **missense effect** will occur when a different amino acid is incorporated at the corresponding site in the protein molecule. This mistaken amino acid—or missense, depending upon its location in the specific protein—might be acceptable, partially acceptable, or unacceptable to the function of that protein molecule. From a careful examination of the genetic code, one can conclude that most single-base changes would result in the

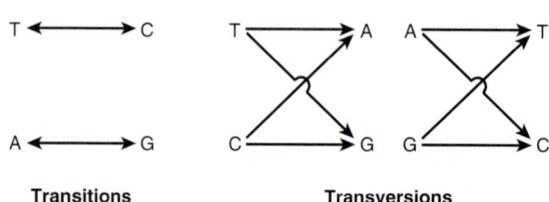

Transitions   Transversions

FIGURE 37–3   Diagrammatic representation of transition mutations and transversion mutations.

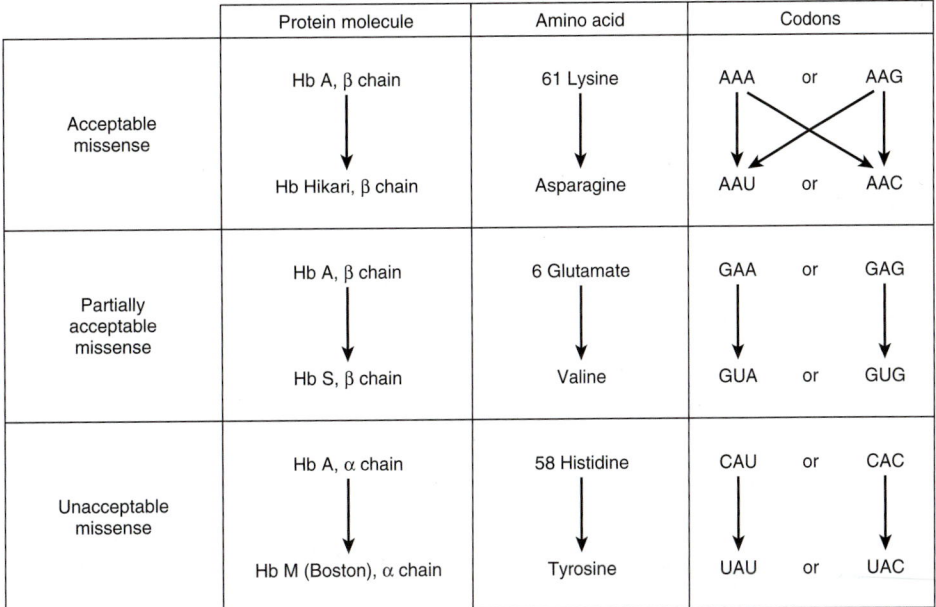

FIGURE 37–4 **Examples of three types of missense mutations resulting in abnormal hemoglobin chains.** The amino acid alterations and possible alterations in the respective codons are indicated. The hemoglobin Hikari β-chain mutation has apparently normal physiologic properties but is electrophoretically altered. Hemoglobin S has a β-chain mutation and partial function; hemoglobin S binds oxygen but precipitates when deoxygenated; this causes red blood cells to sickle, and represents the cellular and molecular basis of sickle cell disease (see Figure 6–12). Hemoglobin M Boston, an α-chain mutation, permits the oxidation of the heme ferrous iron to the ferric state and so will not bind oxygen at all.

replacement of one amino acid by another with rather similar functional groups. This is an effective mechanism to avoid drastic change in the physical properties of a protein molecule. If an acceptable missense effect occurs, the resulting protein molecule may not be distinguishable from the normal one. A partially acceptable missense will result in a protein molecule with partial but abnormal function. If an unacceptable missense effect occurs, then the protein molecule will not be capable of functioning normally.

3. A **nonsense** codon may appear that would then result in the **premature termination** of amino acid incorporation into a peptide chain and the production of only a fragment of the intended protein molecule. The probability is high that a prematurely terminated protein molecule or peptide fragment will not function in its assigned role. Examples of the different types of mutations, and their effects on the coding potential of mRNA are shown in **Figures 37–4** and **37–5**.

Frameshift Mutations Result from Deletion or Insertion of Nucleotides in DNA That Generates Altered mRNAs

The deletion of a single nucleotide from the coding strand of a gene results in an altered reading frame in the mRNA. The machinery translating the mRNA does not recognize that a base was missing, since there is no punctuation in the reading of codons. Thus, a major alteration in the sequence of polymerized amino acids, as depicted in example 1, Figure 37–5, results. Altering the reading frame results in a garbled translation of the mRNA distal to the single nucleotide deletion. Not only is the sequence of amino acids distal to this deletion garbled, but reading of the message can also result in the appearance of a nonsense codon and thus the production of a polypeptide both garbled and prematurely terminated (example 3, Figure 37–5).

If three nucleotides or a multiple of three are deleted from a coding region, the corresponding mRNA when translated will provide a protein from which is missing the corresponding number of amino acids (example 2, Figure 37–5). Because the reading frame is a triplet, the reading phase will not be disturbed for those codons distal to the deletion. If, however, deletion of one or two nucleotides occurs just prior to or within the normal termination codon (nonsense codon), the reading of the normal termination signal is disturbed. Such a deletion might result in reading through the now "mutated" termination signal until another nonsense codon is encountered (example 1, Figure 37–5).

Insertions of one or two or nonmultiples of three nucleotides into a gene result in an mRNA in which the reading frame is distorted upon translation, and the same effects

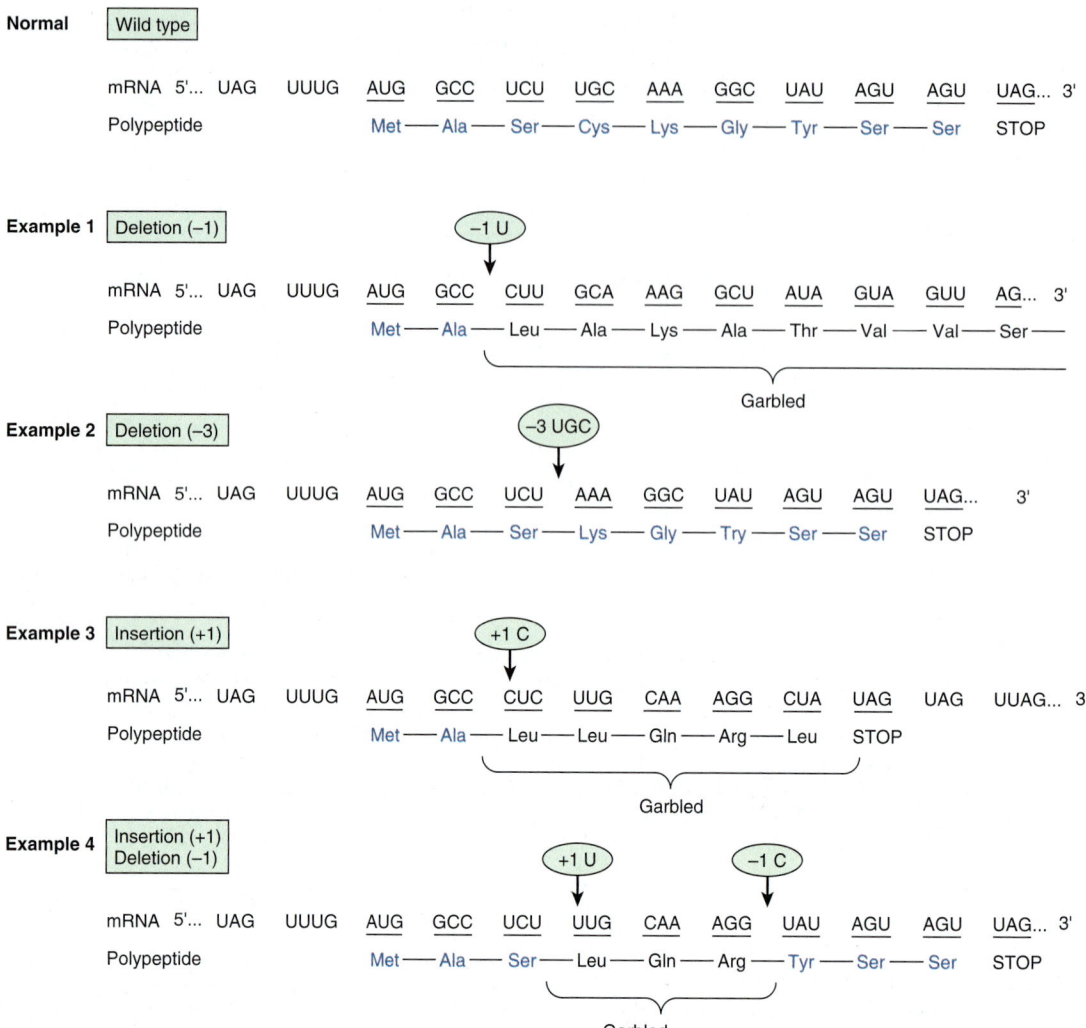

FIGURE 37–5 **Examples of the effects of deletions and insertions in a gene on the sequence of the mRNA transcript and of the polypeptide chain translated therefrom.** The arrows indicate the sites of deletions or insertions, and the numbers in the ovals indicate the number of nucleotide residues deleted or inserted. Colored type indicates amino acids in correct order.

that occur with deletions are reflected in the mRNA translation. This may result in garbled amino acid sequences distal to the insertion and the generation of a **nonsense codon** at or distal to the insertion, or perhaps reading through the normal termination codon. Following a deletion in a gene, an insertion (or vice versa) can reestablish the proper reading frame (example 4, Figure 37–5). The corresponding mRNA, when translated, would contain a garbled amino acid sequence between the insertion and deletion. Beyond the reestablishment of the reading frame, the amino acid sequence would be correct. One can imagine that different combinations of deletions, of insertions, or of both would result in formation of a protein wherein a portion is abnormal, but this portion is surrounded by the normal amino acid sequences. Such phenomena have been demonstrated convincingly in a number of human diseases.

Suppressor Mutations Can Counteract Some of the Effects of Missense, Nonsense, & Frameshift Mutations

The above discussion of the altered protein products of gene mutations is based on the presence of normally functioning tRNA molecules. However, in prokaryotic and lower eukaryotic organisms, abnormally functioning tRNA molecules have been discovered that are themselves the results of mutations. Some of these abnormal tRNA molecules are capable of binding to and decoding altered codons, thereby suppressing the effects of mutations in distinct-mutated mRNA-encoding structural genes. These **suppressor tRNA molecules,** usually formed as a result of alterations in their anticodon regions, are capable of suppressing certain missense mutations, nonsense mutations, and frameshift mutations. However, since the suppressor tRNA molecules are not capable of distinguishing between a normal codon and one resulting from a gene mutation, their presence in the microbial cell usually results in decreased viability. For instance, the nonsense suppressor tRNA molecules can suppress the normal termination signals to allow a read-through when it is not desirable. Frameshift suppressor tRNA molecules may read a normal codon plus a component of a juxtaposed codon to provide a frameshift, also when it is not desirable. Suppressor tRNA molecules may exist in mammalian cells, since read-through of translation has on occasion been observed. In the laboratory context such suppressor tRNAs, coupled with mutated variants of aminoacyl tRNA synthetases, can be utilized to incorporate unnatural amino acids into defined locations within altered genes that carry engineered nonsense mutations. The resulting labeled proteins can be used for in vivo and in vitro cross-linking and biophysical studies. This new tool adds significantly to biologists interested in studying the mechanisms of a wide range of biological processes.

LIKE TRANSCRIPTION, PROTEIN SYNTHESIS CAN BE DESCRIBED IN THREE PHASES: INITIATION, ELONGATION, & TERMINATION

The general structural characteristics of ribosomes and their self-assembly process are discussed in Chapter 34. These particulate entities serve as the machinery on which the mRNA nucleotide sequence is translated into the sequence of amino acids of the specified protein. The translation of the mRNA commences near its 5′ terminal with the formation of the corresponding amino terminal of the protein molecule. The message is read from 5′–3′, concluding with the formation of the carboxyl terminal of the protein. Again, the concept of **polarity** is apparent. As described in Chapter 36, the transcription of a gene into the corresponding mRNA or its precursor first forms the 5′ terminal of the RNA molecule. In prokaryotes, this allows for the beginning of mRNA translation before the transcription of the gene is completed. In eukaryotic organisms, the process of transcription is a nuclear one; mRNA translation occurs in the cytoplasm. This precludes simultaneous transcription and translation in eukaryotic organisms and makes possible the processing necessary to generate mature mRNA from the primary transcript.

Initiation Involves Several Protein–RNA Complexes

Initiation of protein synthesis requires that an mRNA molecule be selected for translation by a ribosome (**Figure 37–6**). Once the mRNA binds to the ribosome, the latter finds the correct reading frame on the mRNA, and translation begins. This process involves tRNA, rRNA, mRNA, and at least 10 eukaryotic initiation factors (eIFs), some of which have

multiple (three to eight) subunits. Also involved are GTP, ATP, and amino acids. Initiation can be divided into four steps: (1) dissociation of the ribosome into its 40S and 60S subunits; (2) binding of a ternary complex consisting of the initiator methionyl-tRNA, (met-tRNA^i), GTP, and eIF-2 to the 40S ribosome to form the 43S preinitiation complex; (3) binding of mRNA to the 40S preinitiation complex to form the 48S initiation complex; and (4) combination of the 48S initiation complex with the 60S ribosomal subunit to form the 80S initiation complex.

Ribosomal Dissociation

Two initiation factors, eIF-3 and eIF-1A, bind to the newly dissociated 40S ribosomal subunit. This delays its reassociation with the 60S subunit and allows other translation initiation factors to associate with the 40S subunit.

Formation of the 43S Preinitiation Complex

The first step in this process involves the binding of GTP by eIF-2. This binary complex then binds to met tRNA^i, a tRNA specifically involved in binding to the initiation codon AUG. (There are two tRNAs for methionine. One specifies methionine for the initiator codon, the other for internal methionines. Each has a unique nucleotide sequence; both are aminoacylated by the same methionyl-tRNA synthetase.) This ternary complex binds to the 40S ribosomal subunit to form the 43S preinitiation complex, which is stabilized by association with eIF-3 and eIF-1A.

eIF-2 is one of two control points for protein synthesis initiation in eukaryotic cells. eIF-2 consists of α, β, and γ subunits. eIF-2α is phosphorylated (on serine 51) by at least four different protein kinases (HCR, PKR, PERK, and GCN2) that are activated when a cell is under stress and when the energy expenditure required for protein synthesis would be deleterious. Such conditions include amino acid and glucose starvation, virus infection, intracellular presence of large quantities of misfolded proteins, serum deprivation, hyperosmolality, and heat shock. PKR is particularly interesting in this regard. This kinase is activated by viruses and provides a host defense mechanism that decreases protein synthesis, including viral protein synthesis, thereby inhibiting viral replication. Phosphorylated eIF-2α binds tightly to and inactivates the GTP–GDP recycling protein eIF-2B. Thus preventing formation of the 43S preinitiation complex and blocking protein synthesis.

Formation of the 48S Initiation Complex

The 5′ terminals of most mRNA molecules in eukaryotic cells are "capped," as described in Chapter 36. This methyl-guanosyl triphosphate cap facilitates the binding of mRNA to the 43S preinitiation complex. A cap-binding protein complex, eIF–4F (4F), which consists of eIF–4E (4E) and the eIF–4G (4G) –eIF4A (4A) complex, binds to the cap through the 4E protein. Then eIF–4B (4B) binds and reduces the complex secondary structure of the 5′ end of the mRNA through ATPase and ATP-dependent

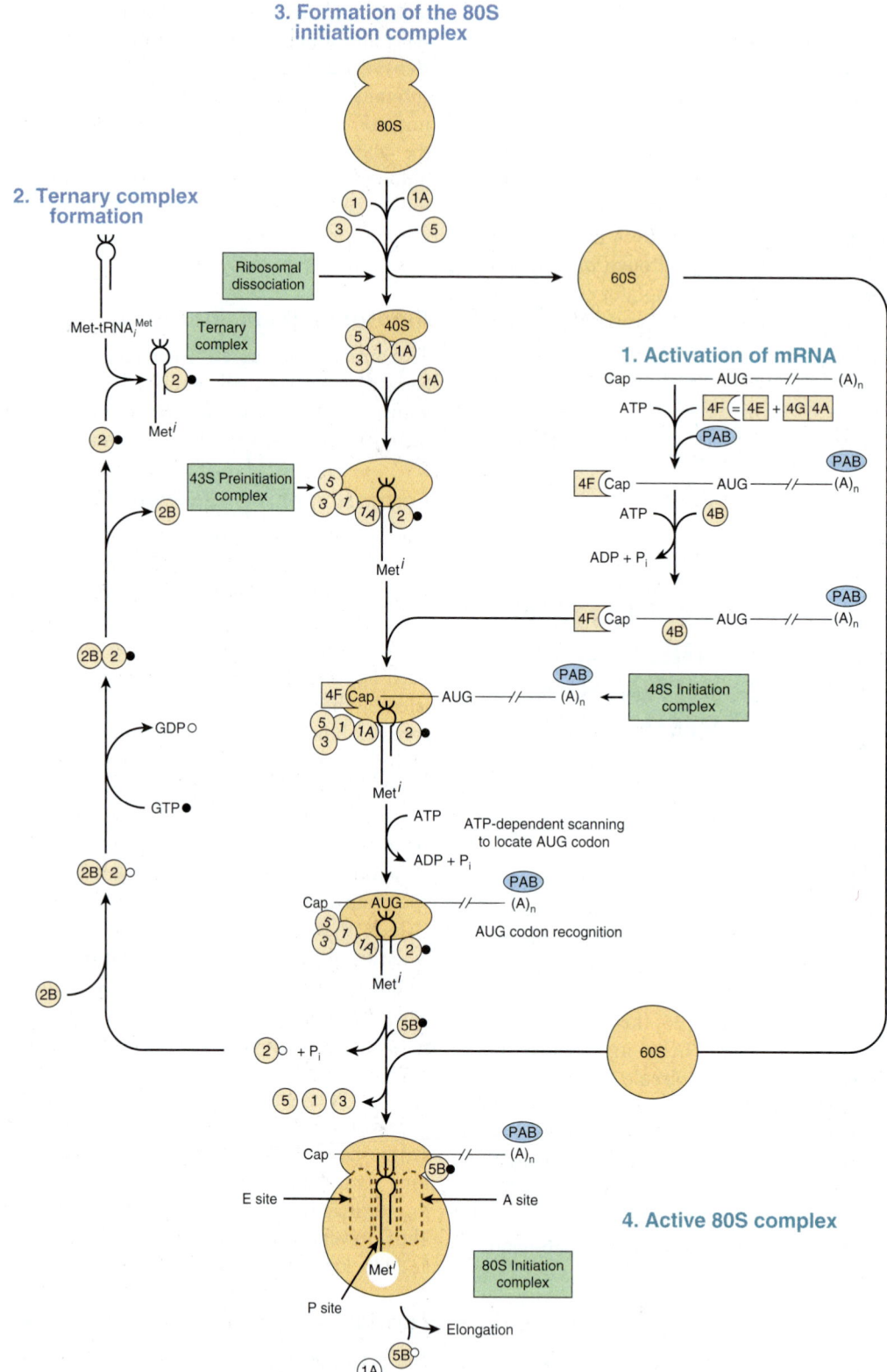

FIGURE 37–6 **Diagrammatic representation of the initiation phase of protein synthesis on an eukaryotic mRNA template containing a 5′ cap (Cap) and 3′ poly(A) terminal [(A)ₙ].** This process proceeds in several steps: (1) activation of mRNA (**right**); (2) formation of the ternary complex consisting of met-tRNAⁱ, initiation factor eIF-2, and GTP (**left**); (3) scanning in the 43S complex to locate the AUG initiator coding, forming the 48S initiation complex (**center**); and (4) formation of the active 80S initiation complex (**bottom, center**). (See text for details.) (GTP, •; GDP,∘.) The various initiation factors appear in abbreviated form as circles or squares, for example, eIF-3, (③), eIF–4F, (4F), ([4F]). 4-F is a complex consisting of 4E and 4A bound to 4G (see Figure 37–7). The poly A binding protein, which interacts with the mRNA 3′-poly A tail, is abbreviated PAB. The constellation of protein factors and the 40S ribosomal subunit comprise the 43S preinitiation complex. When bound to mRNA, this forms the 48S preinitiation complex.

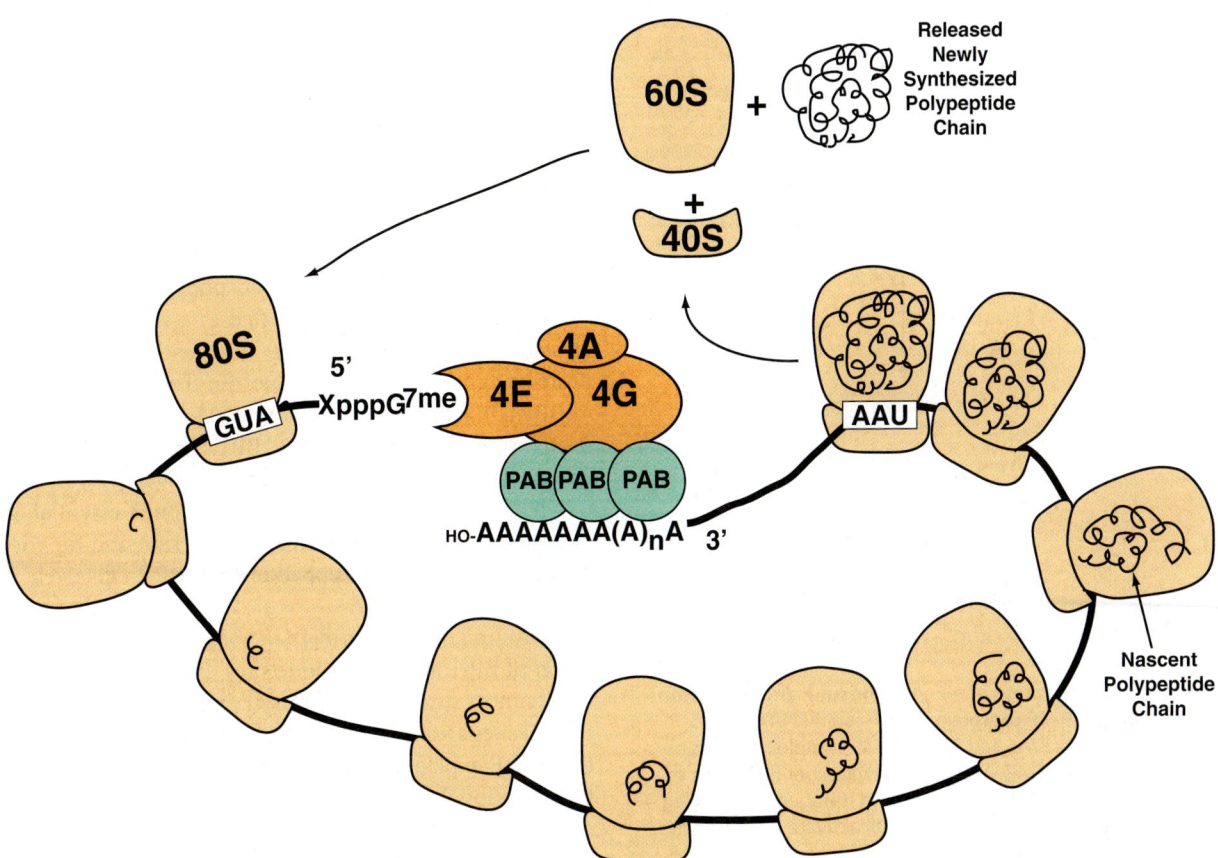

FIGURE 37–7 Schematic illustrating the circularization of mRNA through protein–protein interactions between m⁷G-bound eIF4F and poly A tail-bound PolyA binding protein. eIF4F, composed of eIF4A, 4E, and 4G subunits binds the mRNA 5′-m7G "Cap" (-XpppG⁷ᵐᵉ) upstream of the translation initiation codon (AUG) with high affinity. The eIF4G subunit of the complex also binds poly A binding protein (PAB) with high affinity. Since PAB is bound tightly to the mRNA 3′-poly A tail (OH-AAAAAAA(A)ₙA), circularization results. Shown are multiple 80S ribosomes that are in the process of translating the circularized mRNA into protein (black curlicues), forming a polysome. Upon encountering a termination codon (UAA), translation termination occurs leading to release and dissociation of the 80S ribosome into 60S, 40S subunits and newly translated protein. Dissociated ribosomal subunits can recycle through another round of translation (see Figure 37–6).

helicase activities. The association of mRNA with the 43S pre-initiation complex to form the 48S initiation complex requires ATP hydrolysis. eIF-3 is a key protein because it binds with high affinity to the 4G component of 4F, and it links this complex to the 40S ribosomal subunit. Following association of the 43S preinitiation complex with the mRNA cap, and reduction ("melting") of the secondary structure near the 5′ end of the mRNA through the action of the 4B helicase and ATP, the complex translocates 5′ → 3′ and scans the mRNA for a suitable initiation codon. Generally this is the 5′-most AUG, but the precise initiation codon is determined by so-called **Kozak consensus sequences** that surround the AUG:

$$
\begin{array}{ccc}
-3 & -1 & +4 \\
| & | & \\
\end{array}
$$
GCCA/ GCC**AUG**G

Most preferred is the presence of a purine at positions −3 and +4 relative to the AUG.

Role of the Poly(A) Tail in Initiation

Biochemical and genetic experiments in yeast have revealed that the 3′ poly(A) tail and its binding protein, PAB1, are required for efficient initiation of protein synthesis. Further studies showed that the poly(A) tail stimulates recruitment of the 40S ribosomal subunit to the mRNA through a complex set of interactions. PAB1 (**Figure 37–7**), bound to the poly(A) tail, interacts with eIF–4G, and 4E subunit of eIF–4F that is bound to the cap. A circular structure is formed that helps direct the 40S ribosomal subunit to the 5′ end of the mRNA and also likely stabilizes mRNAs from exonucleotytic degradation. This helps explain how the cap and poly(A) tail structures have a synergistic effect on protein synthesis. Indeed, differential protein–protein interactions between general and specific mRNA translational repressors and eIF-4E result in m⁷GCap-dependent translation control (**Figure 37–8**).

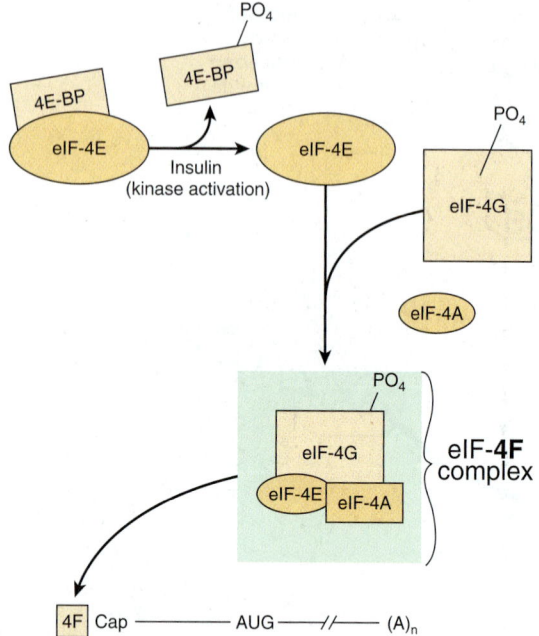

FIGURE 37–8 **Activation of eIF–4E by insulin and formation of the cap binding eIF–4F complex.** The 4F-cap mRNA complex is depicted as in Figures 37–6 and 37–7. The 4F complex consists of eIF-4E (4E), eIF-4A, and eIF-4G. 4E is inactive when bound by one of a family of binding proteins (4EBPs). Insulin and mitogenic factors (eg, IGF-1, PDGF, interleukin-2, and angiotensin II) activate the PI3 kinase/AKT kinase pathways, which activate the mTOR kinase, and results in the phosphorylation of 4E-BP (see Figure 42–8). Phosphorylated 4E-BP dissociates from 4E, and the latter is then able to form the 4F complex and bind to the mRNA cap. These growth polypeptides also induce phosphorylation of 4G itself by the mTOR and MAP kinase pathways. Phosphorylated 4F binds much more avidly to the cap than does nonphosphorylated 4F.

Formation of the 80S Initiation Complex

The binding of the 60S ribosomal subunit to the 48S initiation complex involves hydrolysis of the GTP bound to eIF-2 by eIF-5. This reaction results in release of the initiation factors bound to the 48S initiation complex (these factors then are recycled) and the rapid association of the 40S and 60S subunits to form the 80S ribosome. At this point, the met-tRNA^i is on the P site of the ribosome, ready for the elongation cycle to commence.

The Regulation of eIF–4E Controls the Rate of Initiation

The 4F complex is particularly important in controlling the rate of protein translation. As described above, 4F is a complex consisting of 4E, which binds to the m^{7}G cap structure at the 5′ end of the mRNA, and 4G, which serves as a scaffolding protein. In addition to binding 4E, 4G binds to eIF-3, which links the complex to the 40S ribosomal subunit. It also binds 4A and 4B, the ATPase–helicase complex that helps unwind the RNA (Figure 37–8).

4E is responsible for recognition of the mRNA cap structure, a rate-limiting step in translation. This process is further regulated by phosphorylation. Insulin and mitogenic growth factors result in the phosphorylation of 4E on ser 209 (or thr 210). Phosphorylated 4E binds to the cap much more avidly than does the nonphosphorylated form, thus enhancing the rate of initiation. Components of the MAP kinase, PI3K, mTOR, RAS, and S6 kinases pathways (see Figure 42–8) appear to be involved in these phosphorylation reactions.

The activity of 4E is regulated in a second way, and this also involves phosphorylation. A recently discovered set of proteins bind to and inactivate 4E. These proteins include 4EBP1 (BP1, also known as PHAS-1) and the closely related proteins 4E-BP2 and 4E-BP3. BP1 binds with high affinity to 4E. The [4E]•[BP1] association prevents 4E from binding to 4G (to form 4F). Since this interaction is essential for the binding of 4F to the ribosomal 40S subunit and for correctly positioning this on the capped mRNA, BP-1 effectively inhibits translation initiation.

Insulin and other growth factors result in the phosphorylation of BP-1 at seven unique sites. Phosphorylation of BP-1 results in its dissociation from 4E, and it cannot rebind until critical sites are dephosphorylated. These effects on the activation of 4E explain in part how insulin causes a marked posttranscriptional increase of protein synthesis in liver, adipose, and muscle tissue.

Elongation Is Also a Multistep, Accessory Factor-Facilitated Process

Elongation is a cyclic process on the ribosome in which one amino acid at a time is added to the nascent peptide chain (**Figure 37–9**). The peptide sequence is determined by the order of the codons in the mRNA. Elongation involves several steps catalyzed by proteins called elongation factors (EFs). These steps are (1) binding of aminoacyl-tRNA to the A site, (2) peptide bond formation, (3) translocation of the ribosome on the mRNA, and (4) expulsion of the deacylated tRNA from the P- and E-sites.

Binding of Aminoacyl-tRNA to the A Site

In the complete 80S ribosome formed during the process of initiation, both the A site (aminoacyl or acceptor site) and E site (deacylated tRNA exit site) are free. The binding of the appropriate aminoacyl-tRNA in the A site requires proper codon recognition. **Elongation factor 1A (EF1A)** forms a ternary complex with GTP and the entering aminoacyl-tRNA (Figure 37–9). This complex then allows the correct aminoacyl-tRNA to enter the A site with the release of EF1A•GDP and phosphate. GTP hydrolysis is catalyzed by an active site on the ribosome; hydrolysis induces a conformational change in the ribosome concomitantly increasing affinity for the tRNA. As shown in Figure 37–9, EF1A-GDP then recycles to EF1A-GTP with the aid of other soluble protein factors and GTP.

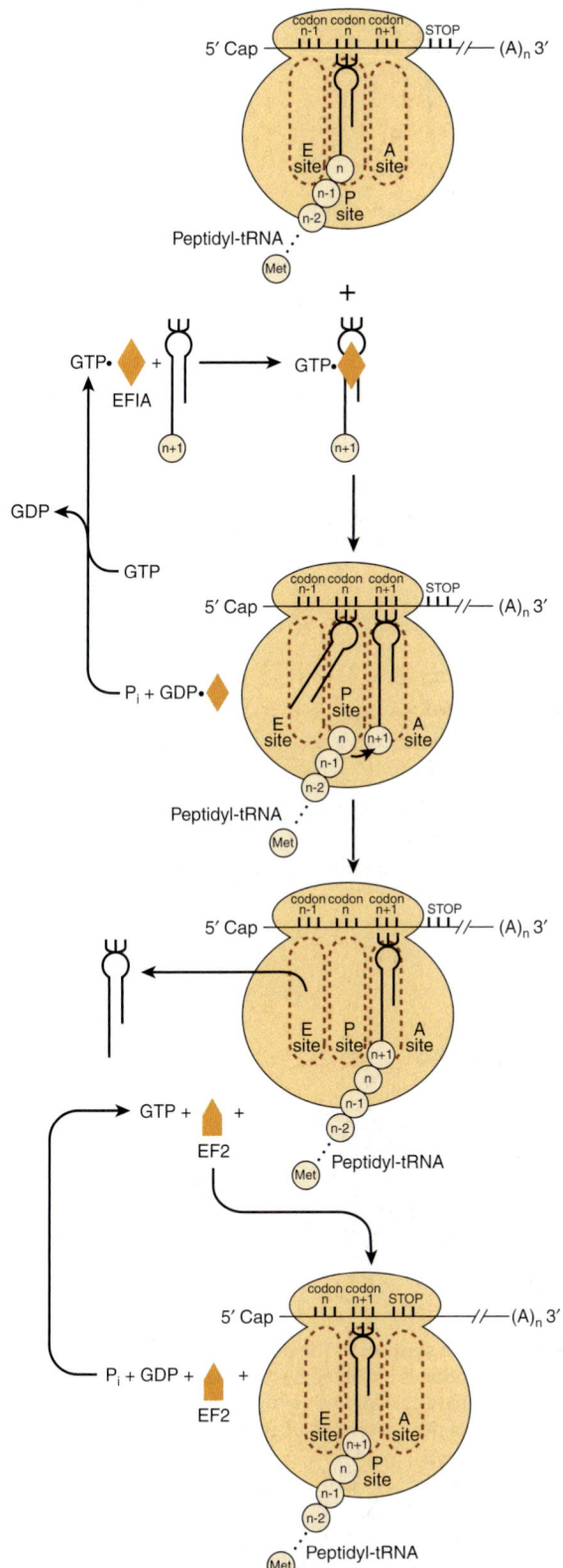

FIGURE 37–9 **Diagrammatic representation of the peptide elongation process of protein synthesis.** The small circles labeled n − 1, n, n + 1, etc., represent the amino acid residues of the newly formed protein molecule and corresponding codons in the mRNA. EFIA and EF2 represent elongation factors 1 and 2, respectively. The peptidyl-tRNA, aminoacyl-tRNA, and Exit sites on the ribosome are represented by P site, A site, and E site, respectively.

TABLE 37–3 Evidence That rRNA Is A Peptidyltransferase

- Ribosomes can make peptide bonds even when proteins are removed or inactivated
- Certain parts of the rRNA sequence are highly conserved in all species
- These conserved regions are on the surface of the RNA molecule.
- RNA can be catalytic
- Mutations that result in antibiotic resistance at the level of protein synthesis are more often found in rRNA than in the protein components of the ribosome
- X-ray crystal structure of large subunit bound to tRNAs suggest detailed mechanism

Peptide Bond Formation

The α-amino group of the new aminoacyl-tRNA in the A site carries out a nucleophilic attack on the esterified carboxyl group of the peptidyl-tRNA occupying the P site (peptidyl or polypeptide site). At initiation, this site is occupied by the initiator met-tRNA[i]. This reaction is catalyzed by a **peptidyltransferase,** a component of the 28S RNA of the 60S ribosomal subunit. This is another example of ribozyme activity and indicates an important—and previously unsuspected—direct role for RNA in protein synthesis (**Table 37–3**). Because the amino acid on the aminoacyl-tRNA is already "activated," no further energy source is required for this reaction. The reaction results in attachment of the growing peptide chain to the tRNA in the A site.

Translocation

The now deacylated tRNA is attached by its anticodon to the P site at one end and by the open CCA tail to an **exit (E) site** on the large ribosomal subunit (middle portion of Figure 37–9). At this point, **elongation factor 2 (EF2)** binds to and displaces the peptidyl tRNA from the A site to the P site. In turn, the deacylated tRNA is on the E site, from which it leaves the ribosome. The EF2–GTP complex is hydrolyzed to EF2–GDP, effectively moving the mRNA forward by one codon and leaving the A site open for occupancy by another ternary complex of amino acid tRNA–EF1AGTP and another cycle of elongation.

The charging of the tRNA molecule with the aminoacyl moiety requires the hydrolysis of an ATP to an AMP, equivalent to the hydrolysis of two ATPs to two ADPs and phosphates. The entry of the aminoacyl-tRNA into the A site results in the hydrolysis of one GTP to GDP. Translocation of the newly formed peptidyl-tRNA in the A site into the P site by EF2 similarly results in hydrolysis of GTP to GDP and phosphate. Thus, the energy requirements for the formation of one peptide bond include the equivalent of the hydrolysis of two ATP molecules to ADP and of two GTP molecules to GDP, or the hydrolysis of four high-energy phosphate bonds. A eukaryotic ribosome can incorporate as many as six amino acids per second; prokaryotic ribosomes incorporate as many as 18 per second. Thus, the energy requiring process of peptide

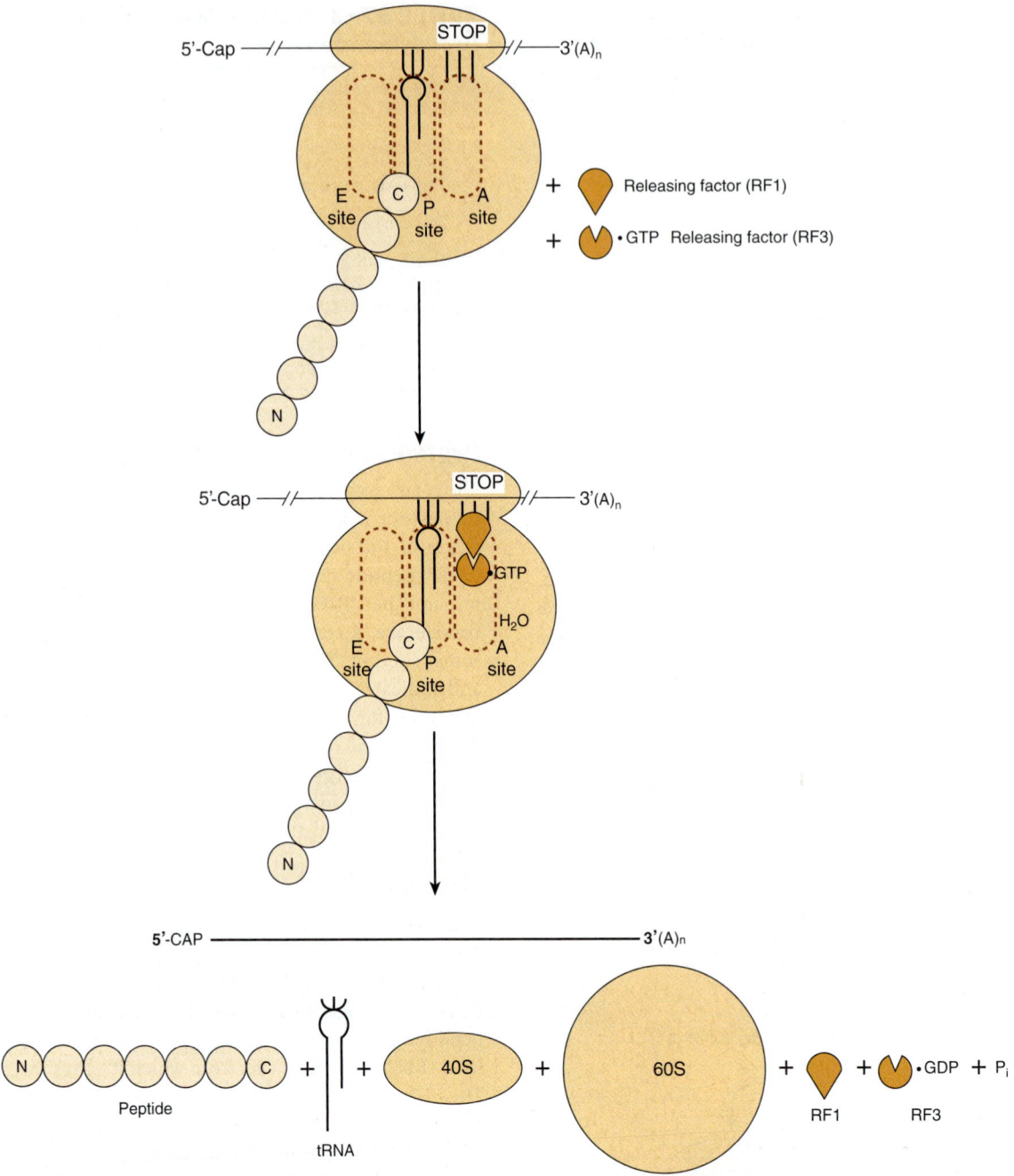

FIGURE 37–10 **Diagrammatic representation of the termination process of protein synthesis.** The peptidyl-tRNA, aminoacyl-tRNA and exit sites are indicated as P site, A site, and E site, respectively. The termination (stop) codon is indicated by the three vertical bars and stop. Releasing factor RF1 binds to the stop codon. Releasing factor RF3, with bound GTP, binds to RF1. Hydrolysis of the peptidyl–tRNA complex is shown by the entry of H_2O. N and C indicate the amino and carboxyl terminal amino acids of the nascent polypeptide chain, respectively, and illustrate the polarity of protein synthesis.

synthesis occurs with great speed and accuracy until a termination codon is reached.

Termination Occurs When a Stop Codon Is Recognized

In comparison to initiation and elongation, termination is a relatively simple process (**Figure 37–10**). After multiple cycles of elongation culminating in polymerization of the specific amino acids into a protein molecule, the stop or terminating codon of mRNA (UAA, UAG, UGA) appears in the A site. Normally, there is no tRNA with an anticodon capable of recognizing such a termination signal. **Releasing factor RF1** recognizes that a stop codon resides in the A site (Figure 37–10). RF1 is bound by a complex consisting of **releasing factor RF3** with bound GTP. This complex, with the peptidyl transferase, promotes hydrolysis of the bond between the peptide and the tRNA occupying the P site. Thus, a water molecule rather than an amino acid is added. This hydrolysis releases the protein and the tRNA from the P site. Upon hydrolysis and release, the **80S ribosome dissociates** into its 40S and 60S subunits, which are then recycled (Figure 37–7). Therefore, the releasing factors are proteins that hydrolyze the peptidyl-tRNA bond when a stop codon occupies the A site. The mRNA is then released from the ribosome, which dissociates into its component 40S and 60S subunits, and another cycle can be repeated.

Polysomes Are Assemblies of Ribosomes

Many ribosomes can translate the same mRNA molecule simultaneously. Because of their relatively large size, the ribosome particles cannot attach to an mRNA any closer than 35 nucleotides apart. Multiple ribosomes on the same mRNA molecule form a **polyribosome,** or "polysome" (Figure 37–7). In an unrestricted system, the number of ribosomes attached to an mRNA (and thus the size of polyribosomes) correlates positively with the length of the mRNA molecule.

Polyribosomes actively synthesizing proteins can exist as free particles in the cellular cytoplasm or may be attached to sheets of membranous cytoplasmic material referred to as **endoplasmic reticulum.** Attachment of the particulate polyribosomes to the endoplasmic reticulum is responsible for its "rough" appearance as seen by electron microscopy. The proteins synthesized by the attached polyribosomes are extruded into the cisternal space between the sheets of rough endoplasmic reticulum and are exported from there. Some of the protein products of the rough endoplasmic reticulum are packaged by the Golgi apparatus for eventual export (see Chapter 46). The polyribosomal particles free in the cytosol are responsible for the synthesis of proteins required for intracellular functions.

Nontranslating mRNAs Can Form Ribonucleoprotein Particles That Accumulate in Cytoplasmic Organelles Termed P Bodies

mRNAs, bound by specific packaging proteins and exported from the nucleus as ribonucleoproteins particles (RNPs) sometimes do not immediately associate with ribosomes to be translated. Instead, specific mRNAs can associate with the protein constituents that form P bodies, small dense compartments that incorporate mRNAs as mRNPs (**Figure 37–11**).

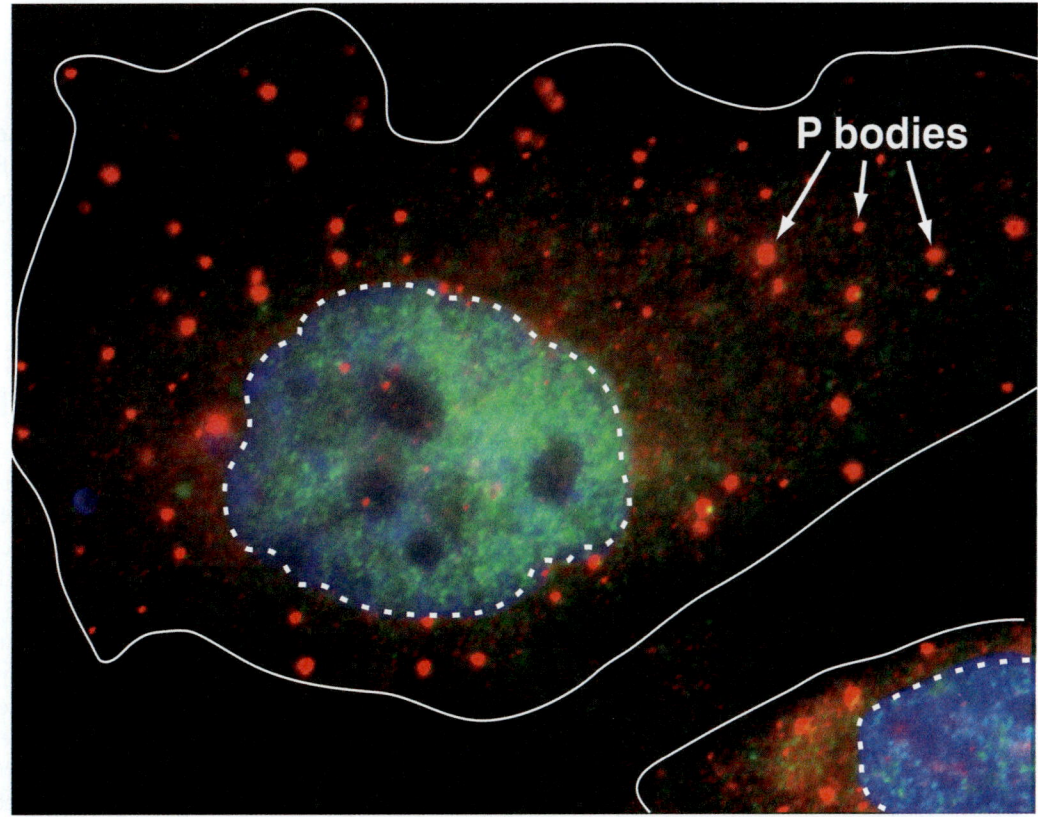

FIGURE 37–11 **The P body is a cytoplasmic organelle that modulates mRNA metabolism.** Shown is a photomicrograph of two mammalian cells in which a single distinct protein constituent of the P body has been visualized using the cognate specific fluorescently labeled antibody. P bodies appear as light circles of varying size throughout the cytoplasm. Cell membranes indicated by a solid white line, nuclei by a dashed line. Nuclei were counterstained using a fluorescent dye with different fluorescence excitation/emission spectra from the labeled antibody used to identify P bodies; the nuclear stain intercalates between the DNA base pairs. Modified from http://www.mcb.arizona.edu/parker/WHAT/what.htm. (Used with permission of Dr Roy Parker.)

These cytoplasmic organelles are related to similar small mRNA-containing granules found in neurons and certain maternal cells. P bodies are sites of translation repression and mRNA decay. Over 35 distinct proteins have been suggested to reside exclusively or extensively within P bodies. These proteins range from mRNA decapping enzymes, RNA helicases and RNA exonucleases (5′–3′ and 3′–5′), to components involved in miRNA function and mRNA quality control. However, incorporation of an mRNP is not an unequivocal mRNA "death sentence." Indeed, though the mechanisms are not yet fully understood, certain mRNAs appear to be temporarily stored in P bodies and then retrieved and utilized for protein translation. This suggests that an equilibrium exists where the cytoplasmic functions of mRNA (translation and degradation) are controlled by the dynamic interaction of mRNA with polysomes and P bodies.

The Machinery of Protein Synthesis Can Respond to Environmental Threats

Ferritin, an iron-binding protein, prevents ionized iron (Fe^{2+}) from reaching toxic levels within cells. Elemental iron stimulates ferritin synthesis by causing the release of a cytoplasmic protein that binds to a specific region in the 5′ nontranslated region of ferritin mRNA. Disruption of this protein–mRNA interaction activates ferritin mRNA and results in its translation. This mechanism provides for rapid control of the synthesis of a protein that sequesters Fe^{2+}, a potentially toxic molecule. Similarly environmental stress and starvation inhibit the positive roles of mTOR (Figure 37–8; Figure 42–8) on promoting activation of eIF4F and 48S complex formation.

Many Viruses Co-Opt the Host Cell Protein Synthesis Machinery

The protein synthesis machinery can also be modified in deleterious ways. **Viruses replicate by using host cell processes,** including those involved in protein synthesis. Some viral mRNAs are translated much more efficiently than those of the host cell (eg, encephalomyocarditis virus). Others, such as reovirus and vesicular stomatitis virus, replicate efficiently, and thus their very abundant mRNAs have a competitive advantage over host cell mRNAs for limited translation factors. Other viruses inhibit host cell protein synthesis by preventing the association of mRNA with the 40S ribosome.

Poliovirus and other picornaviruses gain a selective advantage by disrupting the function of the 4F complex. The mRNAs of these viruses do not have a cap structure to direct the binding of the 40S ribosomal subunit (see above). Instead, the 40S ribosomal subunit contacts an **internal ribosomal entry site (IRES)** in a reaction that requires 4G but not 4E. The virus gains a selective advantage by having a protease that attacks 4G and removes the amino terminal 4E binding site. Now the 4E–4G complex (4F) cannot form, so the 40S ribosomal subunit cannot be directed to capped mRNAs. Host cell translation is thus

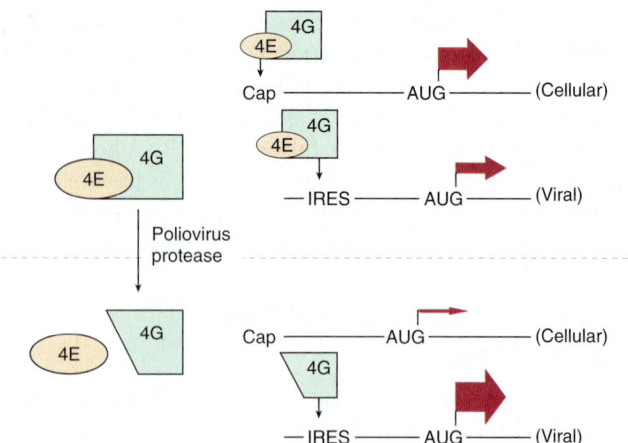

FIGURE 37–12 **Picornaviruses disrupt the 4F complex.** The 4E–4G complex (4F) directs the 40S ribosomal subunit to the typical capped mRNA (see text). 4G alone is sufficient for targeting the 40S subunit to the internal ribosomal entry site (IRES) of viral mRNAs. To gain selective advantage, certain viruses (eg, poliovirus) express a protease that cleaves the 4E binding site from the amino terminal end of 4G. This truncated 4G can direct the 40S ribosomal subunit to mRNAs that have an IRES but not to those that have a cap. The widths of the arrows indicate the rate of translation initiation from the AUG codon in each example. Other viruses utilize distinct processes to effect selective initiation of translation on their cognate viral mRNAs via IRES elements.

abolished. The 4G fragment can direct binding of the 40S ribosomal subunit to IRES-containing mRNAs, so viral mRNA translation is very efficient (**Figure 37–12**). These viruses also promote the dephosphorylation of BP1 (PHAS-1), thereby decreasing cap (4E)-dependent translation (Figure 37–8).

POSTTRANSLATIONAL PROCESSING AFFECTS THE ACTIVITY OF MANY PROTEINS

Some animal viruses, notably HIV, poliovirus, and hepatitis A virus, synthesize long polycistronic proteins from one long mRNA molecule. The protein molecules translated from these long mRNAs are subsequently cleaved at specific sites to provide the several specific proteins required for viral function. In animal cells, many cellular proteins are synthesized from the mRNA template as a precursor molecule, which then must be modified to achieve the active protein. The prototype is insulin, which is a small protein having two polypeptide chains with interchain and intrachain disulfide bridges. The molecule is synthesized as a single chain precursor, or **prohormone,** which folds to allow the disulfide bridges to form. A specific protease then clips out the segment that connects the two chains which form the functional insulin molecule (see Figure 41–12).

Many other peptides are synthesized as proproteins that require modifications before attaining biologic activity. Many of the posttranslational modifications involve the removal of amino terminal amino acid residues by specific aminopeptidases. Collagen, an abundant protein in the extracellular spaces of higher eukaryotes, is synthesized as procollagen. Three procollagen polypeptide molecules, frequently not identical in sequence, align themselves in a particular way that is dependent upon the existence of specific amino terminal peptides (Figure 5–11). Specific enzymes then carry out hydroxylations and oxidations of specific amino acid residues within the procollagen molecules to provide cross-links for greater stability. Amino terminal peptides are cleaved off the molecule to form the final product—a strong, insoluble collagen molecule. Many other posttranslational modifications of proteins occur. Covalent modification by acetylation, phosphorylation, methylation, ubiquitylation, and glycosylation is common, for example (ie, Chapter 5; Table 35–1).

MANY ANTIBIOTICS WORK BY SELECTIVELY INHIBITING PROTEIN SYNTHESIS IN BACTERIA

Ribosomes in bacteria and in the mitochondria of higher eukaryotic cells differ from the mammalian ribosome described in Chapter 34. The bacterial ribosome is smaller (70S rather than 80S) and has a different, somewhat simpler complement of RNA and protein molecules. This difference can be exploited for clinical purposes because many effective antibiotics interact specifically with the proteins and RNAs of prokaryotic ribosomes and thus only inhibit bacterial protein synthesis. This results in growth arrest or death of the bacterium. The most useful members of this class of antibiotics (eg, tetracyclines, lincomycin, erythromycin, and chloramphenicol) do not interact with components of eukaryotic ribosomes and thus are not toxic to eukaryotes. Tetracycline prevents the binding of aminoacyl-tRNAs to the bacterial ribosome A site. Chloramphenicol and the macrolide class of antibiotics work by binding to 23S rRNA, which is interesting in view of the newly appreciated role of rRNA in peptide bond formation through its peptidyltransferase activity. It should be mentioned that the close similarity between prokaryotic and mitochondrial ribosomes can lead to complications in the use of some antibiotics.

Other antibiotics inhibit protein synthesis on all ribosomes (**puromycin**) or only on those of eukaryotic cells (**cycloheximide**). Puromycin (**Figure 37–13**) is a structural analog of tyrosinyl-tRNA. Puromycin is incorporated via the A site on the ribosome into the carboxyl terminal position of a peptide but causes the premature release of the polypeptide. Puromycin, as a tyrosinyl-tRNA analog, effectively inhibits protein synthesis in both prokaryotes and eukaryotes. Cycloheximide inhibits peptidyltransferase in the 60S ribosomal subunit in eukaryotes, presumably by binding to an rRNA component.

Diphtheria toxin, an exotoxin of *Corynebacterium diphtheriae* infected with a specific lysogenic phage, catalyzes the ADP-ribosylation of EF-2 on the unique amino acid diphthamide in mammalian cells. This modification inactivates EF-2 and thereby specifically inhibits mammalian protein

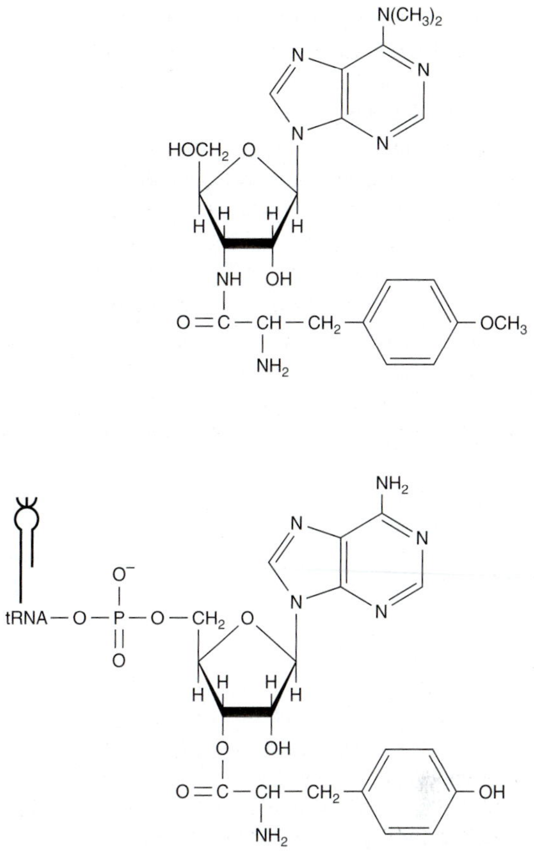

FIGURE 37–13 The comparative structures of the antibiotic puromycin (top) and the 3′ terminal portion of tyrosinyl-tRNA (bottom).

synthesis. Many animals (eg, mice) are resistant to diphtheria toxin. This resistance is due to inability of diphtheria toxin to cross the cell membrane rather than to insensitivity of mouse EF-2 to diphtheria toxin-catalyzed ADP-ribosylation by NAD.

Ricin, an extremely toxic molecule isolated from the castor bean, inactivates eukaryotic 28S ribosomal RNA by providing the N-glycolytic cleavage or removal of a single adenine.

Many of these compounds—puromycin and cycloheximide in particular—are not clinically useful but have been important in elucidating the role of protein synthesis in the regulation of metabolic processes, particularly enzyme induction by hormones.

SUMMARY

- The flow of genetic information follows the sequence DNA → RNA → protein.

- The genetic information in the structural region of a gene is transcribed into an RNA molecule such that the sequence of the latter is complementary to that in one strand of the DNA.

- Ribosomal RNA (rRNA), transfer RNA (tRNA), and messenger RNA (mRNA), are directly involved in protein synthesis.

- miRNAs regulate mRNA function at the level of translation and/or stability.

- The information in mRNA is in a tandem array of codons, each of which is three nucleotides long.

- The mRNA is read continuously from a start codon (AUG) to a termination codon (UAA, UAG, UGA).

- The open reading frame, or ORF, of the mRNA is the series of codons, each specifying a certain amino acid, that determines the precise amino acid sequence of the protein.

- Protein synthesis, like DNA and RNA synthesis, follows the 5′–3′ polarity of mRNA and can be divided into three processes: initiation, elongation, and termination.

- Mutant proteins arise when single-base substitutions result in codons that specify a different amino acid at a given position, when a stop codon results in a truncated protein, or when base additions or deletions alter the reading frame, so different codons are read.

- A variety of compounds, including several antibiotics, inhibit protein synthesis by affecting one or more of the steps involved in protein synthesis.

REFERENCES

Altmann M, Linder P: Power of yeast for analysis of eukaryotic translation initiation. J Biol Chem 2010;285:31907–13192.

Beckham CJ, Parker R: P bodies, stress granules, and viral life cycles. Cell Host Microbe 2008;3:206.

Buchan JR, Parker R: Eukaryotic stress granules: the ins and outs of translation. Mol Cell 2009;36:932–941.

Crick FH, Barnett L, Brenner S, et al: The genetic code. Nature 1961;192:1227.

Hinnebusch AG: Molecular Mechanism of Scanning and Start Codon Selection in Eukaryotes. Microbiology & Molecular Biology Reviews 2011;75:434–467.

Kimball SR, Jefferson, LS: Control of translation initiation through integration of signals generated by hormones, nutrients, and exercise. J Biol Chem 2009;285:29027–29032.

Kozak M: Structural features in eukaryotic mRNAs that modulate the initiation of translation. J Biol Chem 1991;266:1986.

Liu CC, Schultz PG: Adding new chemistries to the genetic code. Annu Rev Biochem 2010;79:413–444.

Maquat LE, Tarn WY, Isken O: The pioneer round of translation: features and functions. Cell 2010;142:368–374.

Silvera D, Formenti SC, Schneider RJ: Translational control in cancer. Nat Rev Cancer 2010;10:254–266.

Sonenberg N, Hinnebusch AG: Regulation of translation initiation in eukaryotes: mechanisms and biological targets. Cell 2010;136:731–745.

Spriggs KA, Bushell M, Willis AE: Translational regulation of gene expression during conditions of cell stress. Mol Cell 2010;40:228–237.

Steitz TA, Moore PB: RNA, the first macromolecular catalyst: the ribosome is a ribozyme. Trends Biochem Sci 2003;28:411.

Wang Q, Parrish AR, Wang L: Expanding the genetic code for biological studies. Chem Biol 2009;16:323–336.

Weatherall DJ: Phenotype-genotype relationships in mogenic disease: Lessons from the thalassaemias. Nature Reviews Genetics 2001;2:245.

Regulation of Gene Expression

P. Anthony Weil, PhD

OBJECTIVES

After studying this chapter,
you should be able to:

- Explain that the many steps involved in the vectorial processes of gene expression, which range from targeted modulation of gene copy number, to gene rearrangement, to transcription, to mRNA processing and transport from the nucleus, to translation, to protein post-translational modification and degradation, are all subject to regulatory control, both positive and negative. Changes in any, or multiple of these processes, can increase or decrease the amount and/or activity of the cognate gene product.

- Appreciate that DNA binding transcription factors, proteins that bind to specific DNA sequences that are often located near to transcriptional promoter elements, can either activate or repress gene transcription.

- Recognize that DNA binding transcription factors are often modular proteins that are composed of structurally and functionally distinct domains, which can directly or indirectly control mRNA gene transcription, either through contacts with RNA polymerase and its cofactors, or through interactions with coregulators that modulate nucleosome structure via covalent modifications and/or displacement.

- Understand that nucleosome-directed regulatory events typically increase or decrease the accessibility of the underlying DNA such as enhancer or promoter sequences, although nucleosome modification can also create new binding sites for other coregulators.

- Understand that the processes of transcription, RNA processing, and nuclear export of RNA are all coupled.

BIOMEDICAL IMPORTANCE

Organisms adapt to environmental changes by altering gene expression. The mechanisms controlling gene expression have been studied in detail and often involve modulation of gene transcription. Control of transcription ultimately results from changes in the mode of interaction of specific regulatory molecules, usually proteins, with various regions of DNA in the controlled gene. Such interactions can either have a positive or negative effect on transcription. Transcription control can result in tissue-specific gene expression, and gene regulation is influenced by hormones, heavy metals, and chemicals.

In addition to transcription level controls, gene expression can also be modulated by gene amplification, gene rearrangement, posttranscriptional modifications, RNA stabilization, translational control, protein modification, and protein stabilization. Many of the mechanisms that control gene expression are used to respond to developmental cues, growth factors, hormones, environmental agents, and therapeutic drugs. Dysregulation of gene expression can lead to human disease. Thus, a molecular understanding of these processes will lead to development of agents that alter pathophysiologic mechanisms or inhibit the function or arrest the growth of pathogenic organisms.

REGULATED EXPRESSION OF GENES IS REQUIRED FOR DEVELOPMENT, DIFFERENTIATION, & ADAPTATION

The genetic information present in each normal somatic cell of a metazoan organism is practically identical. The exceptions are found in those few cells that have amplified or rearranged genes in order to perform specialized cellular functions or cells that have undergone oncogenic transformation. Expression of the genetic information must be regulated during ontogeny and differentiation of the organism and its cellular components. Furthermore, in order for the organism to adapt to its environment and to conserve energy and nutrients, the expression of genetic information must be cued to extrinsic signals and respond only when necessary. As organisms have evolved, more sophisticated regulatory mechanisms have appeared which provide the organism and its cells with the responsiveness necessary for survival in a complex environment. Mammalian cells possess about 1000 times more genetic information than does the bacterium *Escherichia coli.* Much of this additional genetic information is probably involved in regulation of gene expression during the differentiation of tissues and biologic processes in the multicellular organism and in ensuring that the organism can respond to complex environmental challenges.

In simple terms, there are only two types of gene regulation: **positive regulation** and **negative regulation** (**Table 38–1**). When the expression of genetic information is quantitatively increased by the presence of a specific regulatory element, regulation is said to be positive; when the expression of genetic information is diminished by the presence of a specific regulatory element, regulation is said to be negative. The element or molecule mediating negative regulation is said to be a **negative regulator,** a **silencer** or **repressor;** that mediating positive regulation is a **positive regulator,** an **enhancer** or **activator.** However, a **double negative** has the effect of acting as a positive. Thus, an effector that inhibits the function of a negative regulator will appear to bring about a positive regulation. Many regulated systems that appear to be induced are in fact **derepressed** at the molecular level. (See Chapter 9 for explanation of these terms.)

TABLE 38–1   Effects of Positive and Negative Regulation on Gene Expression

	Rate of Gene Expression	
	Negative Regulation	Positive Regulation
Regulator present	Decreased	Increased
Regulator absent	Increased	Decreased

BIOLOGIC SYSTEMS EXHIBIT THREE TYPES OF TEMPORAL RESPONSES TO A REGULATORY SIGNAL

Figure 38–1 depicts the extent or amount of gene expression in three types of temporal response to an inducing signal. A **type A response** is characterized by an increased extent of gene expression that is dependent upon the continued presence of the inducing signal. When the inducing signal is removed, the amount of gene expression diminishes to its basal level, but the amount repeatedly increases in response to the reappearance of the specific signal. This type of response is commonly observed in prokaryotes in response to sudden changes of the intracellular concentration of a nutrient. It is also observed in many higher organisms after exposure to inducers such as hormones, nutrients, or growth factors (Chapter 42).

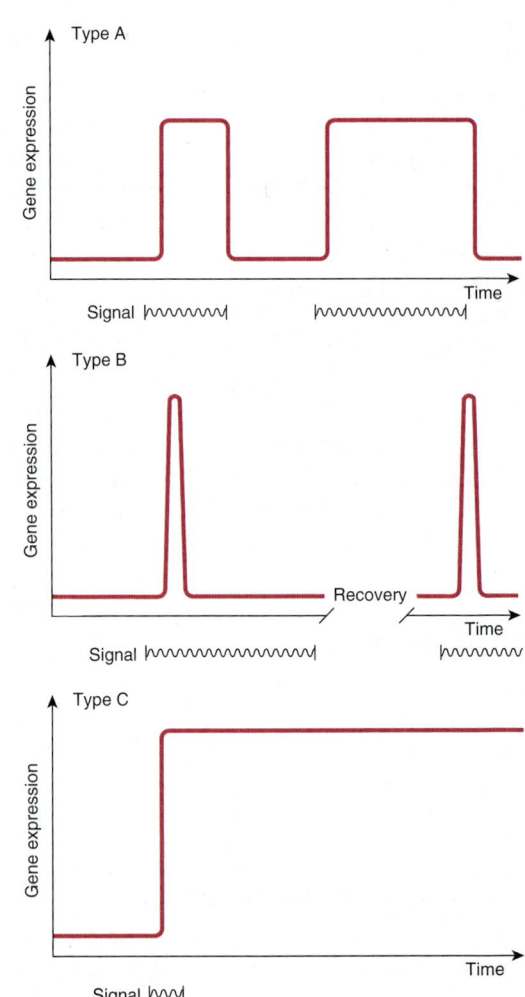

FIGURE 38–1 Diagrammatic representations of the responses of the extent of expression of a gene to specific regulatory signals (such as a hormone as a function of time).

A **type B response** exhibits an increased amount of gene expression that is transient even in the continued presence of the regulatory signal. After the regulatory signal has terminated and the cell has been allowed to recover, a second transient response to a subsequent regulatory signal may be observed. This phenomenon of response–desensitization recovery characterizes the action of many pharmacologic agents, but it is also a feature of many naturally occurring processes. This type of response commonly occurs during development of an organism, when only the transient appearance of a specific gene product is required although the signal persists.

The **type C response** pattern exhibits, in response to the regulatory signal, an increased extent of gene expression that persists indefinitely even after termination of the signal. The signal acts as a trigger in this pattern. Once expression of the gene is initiated in the cell, it cannot be terminated even in the daughter cells; it is therefore an irreversible and inherited alteration. This type of response typically occurs during the development of differentiated function in a tissue or organ.

Simple Unicellular and Multicellular Organisms Serve as Valuable Models for the Study of Gene Expression in Mammalian Cells

Analysis of the regulation of gene expression in prokaryotic cells helped establish the principle that information flows from the gene to a messenger RNA to a specific protein molecule. These studies were aided by the advanced genetic analyses that could be performed in prokaryotic and lower eukaryotic organisms such as baker's yeast, *Saccharomyces cerevisiae*, and the fruit fly, *Drosophila melanogaster*, among others. In recent years, the principles established in these studies, coupled with a variety of molecular biology techniques, have led to remarkable progress in the analysis of gene regulation in higher eukaryotic organisms, including mammals. In this chapter, the initial discussion will center on prokaryotic systems. The impressive genetic studies will not be described, but the physiology of gene expression will be discussed. However, nearly all of the conclusions about this physiology have been derived from genetic studies and confirmed by molecular genetic and biochemical experiments.

Some Features of Prokaryotic Gene Expression Are Unique

Before the physiology of gene expression can be explained, a few specialized genetic and regulatory terms must be defined for prokaryotic systems. In prokaryotes, the genes involved in a metabolic pathway are often present in a linear array called an **operon**, for example, the *lac* operon. An operon can be regulated by a single promoter or regulatory region. The **cistron** is the smallest unit of genetic expression. As described in Chapter 9, some enzymes and other protein molecules are composed of two or more nonidentical subunits. Thus, the "one gene, one enzyme" concept is not necessarily valid. The cistron is the

genetic unit coding for the structure of the subunit of a protein molecule, acting as it does as the smallest unit of genetic expression. Thus, the one gene, one enzyme idea might more accurately be regarded as a **one cistron, one subunit** concept. A single mRNA that encodes more than one separately translated protein is referred to as a **polycistronic mRNA.** For example, the polycistronic *lac* operon mRNA is translated into three separate proteins (see below). Operons and polycistronic mRNAs are common in bacteria but not in eukaryotes.

An **inducible gene** is one whose expression increases in response to an **inducer** or **activator**, a specific positive regulatory signal. In general, inducible genes have relatively low basal rates of transcription. By contrast, genes with high basal rates of transcription are often subject to downregulation by repressors.

The expression of some genes is **constitutive,** meaning that they are expressed at a reasonably constant rate and not known to be subject to regulation. These are often referred to as **housekeeping genes.** As a result of mutation, some inducible gene products become constitutively expressed. A mutation resulting in constitutive expression of what was formerly a regulated gene is called a **constitutive mutation.**

Analysis of Lactose Metabolism in *E coli* Led to the Operon Hypothesis

Jacob and Monod in 1961 described their **operon model** in a classic paper. Their hypothesis was to a large extent based on observations on the regulation of lactose metabolism by the intestinal bacterium *E coli*. The molecular mechanisms responsible for the regulation of the genes involved in the metabolism of lactose are now among the best-understood in any organism. β-Galactosidase hydrolyzes the β-galactoside lactose to galactose and glucose. The structural gene for β-galactosidase (*lacZ*) is clustered with the genes responsible for the permeation of lactose into the cell (*lacY*) and for thiogalactoside transacetylase (*lacA*). The structural genes for these three enzymes, along with the *lac* promoter and *lac* operator (a regulatory region), are physically associated to constitute the **lac operon** as depicted in **Figure 38–2.** This genetic arrangement of the structural genes and their regulatory genes allows for **coordinate expression** of the three enzymes concerned with lactose metabolism. Each of these linked genes is transcribed into one large polycistronic mRNA molecule that contains multiple independent

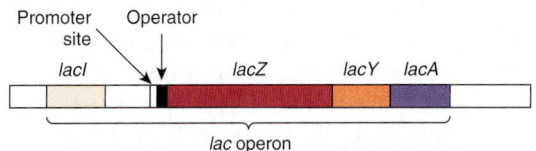

FIGURE 38–2 **The positional relationships of the structural and regulatory genes of the *lac* operon.** *lacZ* encodes β-galactosidase, *lacY* encodes a permease, and *lacA* encodes a thiogalactoside transacetylase. *lacI* encodes the *lac* operon repressor protein.

translation start (AUG) and stop (UAA) codons for each of the three cistrons. Thus, each protein is translated separately, and they are not processed from a single large precursor protein.

It is now conventional to consider that a gene includes regulatory sequences as well as the region that encodes the primary transcript. Although there are many historical exceptions, a gene is generally italicized in lower case and the encoded protein, when abbreviated, is expressed in roman type with the first letter capitalized. For example, the gene *lacI* encodes the repressor protein LacI. When *E coli* is presented with lactose or some specific lactose analogs under appropriate nonrepressing conditions (eg, high concentrations of lactose, no or very low glucose in media; see below), the expression of the activities of β-galactosidase, galactoside permease, and thiogalactoside transacetylase is increased 100-fold to 1000-fold. This is a type A response, as depicted in Figure 38–1. The kinetics of induction can be quite rapid; *lac*-specific mRNAs are fully induced within 5–6 min after addition of lactose to a culture; β-galactosidase protein is maximal within 10 min. Under fully induced conditions, there can be up to 5000 β-galactosidase molecules per cell, an amount about 1000 times greater than the basal, uninduced level. Upon removal of the signal, that is, the inducer, the synthesis of these three enzymes declines.

When *E coli* is exposed to both lactose and glucose as sources of carbon, the organisms first metabolize the glucose and then temporarily stop growing until the genes of the *lac* operon become induced to provide the ability to metabolize lactose as a usable energy source. Although lactose is present from the beginning of the bacterial growth phase, the cell does not induce those enzymes necessary for catabolism of lactose until the glucose has been exhausted. This phenomenon was first thought to be attributable to repression of the *lac* operon by some catabolite of glucose; hence, it was termed catabolite repression. It is now known that catabolite repression is in fact mediated by a **catabolite gene activator protein (CAP)** in conjunction with **cAMP** (Figure 17–5). This protein is also referred to as the cAMP regulatory protein (CRP). The expression of many inducible enzyme systems or operons in *E coli* and other prokaryotes is sensitive to catabolite repression, as discussed below.

The physiology of induction of the *lac* operon is well understood at the molecular level (**Figure 38–3**). Expression of the normal *lacI* gene of the *lac* operon is constitutive; it is expressed at a constant rate, resulting in formation of the subunits of the **lac repressor.** Four identical subunits with molecular weights of 38,000 assemble into a tetrameric Lac repressor molecule. The LacI repressor protein molecule, the product of *lacI*, has a high affinity (dissociation constant, K_d about 10^{-13} mol/L) for the operator locus. The **operator locus** is a region of double-stranded DNA that exhibits a twofold rotational symmetry and an inverted palindrome (indicated by arrows about the dotted axis) in a region that is 21 bp long, as shown below:

$$\overrightarrow{}$$
$$:$$
$$5' - \textbf{AATTGT}\text{GAGC G GATAACAATT}$$
$$3' - \text{TTA ACACTCG C CTAT}\textbf{TGTTAA}$$
$$:$$
$$\overleftarrow{}$$

At any one time, only two of the four subunits of the repressor appear to bind to the operator, and within the 21-base-pair region nearly every base of each base pair is involved in LacI recognition and binding. The binding occurs mostly in the **major groove** without interrupting the base-paired, double-helical nature of the operator DNA. The **operator locus** is between the **promoter site,** at which the DNA-dependent RNA polymerase attaches to commence transcription, and the transcription initiation site of the **lacZ gene,** the structural gene for β-galactosidase (Figure 38–2). When attached to the operator locus, the LacI repressor molecule prevents transcription of the distal structural genes, *lacZ, lacY,* and *lacA* by interfering with the binding of RNA polymerase to the promoter; RNA polymerase and LacI repressor cannot be effectively bound to the *lac* operon at the same time. Thus, the LacI repressor molecule is a **negative regulator;** in its presence (and in the absence of inducer; see below), expression from the *lacZ, lacY,* and *lacA* genes is very, very low. There are normally 20–40 repressor tetramer molecules in the cell, a concentration of tetramer sufficient to effect, at any given time, >95% occupancy of the one *lac* operator element in a bacterium, thus ensuring low (but not zero) basal *lac* operon gene transcription in the absence of inducing signals.

A lactose analog that is capable of inducing the *lac* operon while not itself serving as a substrate for β-galactosidase is an example of a **gratuitous inducer.** An example is isopropylthiogalactoside (IPTG). The addition of lactose or of a gratuitous inducer such as IPTG to bacteria growing on a poorly utilized carbon source (such as succinate) results in prompt induction of the *lac* operon enzymes. Small amounts of the gratuitous inducer or of lactose are able to enter the cell even in the absence of permease. The LacI repressor molecules—both those attached to the operator loci and those free in the cytosol—have a high affinity for the inducer. Binding of the inducer to repressor molecule induces a conformational change in the structure of the repressor and causes it to dissociate from operator DNA because its affinity for the operator is now 10^4 times lower (K_d about 10^{-9} mol/L) than that of LacI in the absence of IPTG. DNA-dependent RNA polymerase can now bind to the promoter (ie, Figures 36–3 and 36–8), and transcription will begin, although this process is relatively inefficient (see below). In such a manner, **an inducer derepresses the lac operon** and allows transcription of the structural genes for β-galactosidase, galactoside permease, and thiogalactoside transacetylase. Translation of the polycistronic mRNA can occur even before transcription is completed. Derepression of the *lac* operon allows the cell to synthesize the enzymes necessary to catabolize lactose as an energy source. Based on the physiology just described, IPTG-induced expression of transfected plasmids bearing the *lac* operator–promoter ligated to appropriate bioengineered constructs is commonly used to express mammalian recombinant proteins in *E coli*.

In order for the RNA polymerase to form a PIC at the promoter site most efficiently, there must also be present the **CAP** to which cAMP is bound. By an independent mechanism, the bacterium accumulates cAMP only when it is starved for

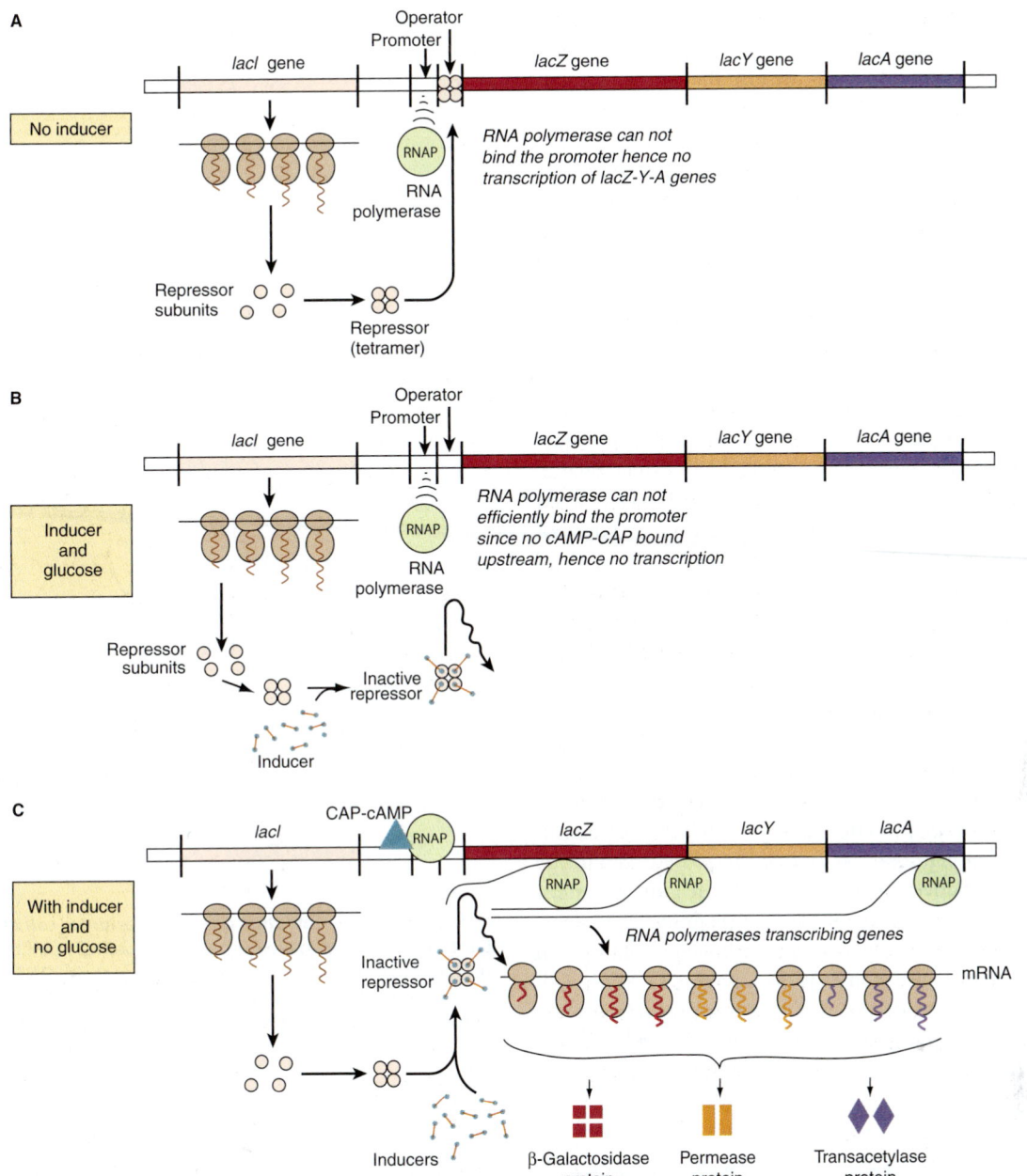

FIGURE 38–3 **The mechanism of repression and derepression of the *lac* operon.** When no inducer is present **(A)** the constitutively synthesized *lacI* gene products forms a repressor tetramer molecule that binds at the operator locus. Repressor-operator binding prevents the binding of RNA polymerase and consequently prevents transcription of the *lacZ*, *lacY*, and *lacA* structural genes into a polycistonic mRNA. When inducer is present, but glucose is also present in the culture medium **(B),** the tetrameric repressor molecules are conformationally altered by inducer, and cannot efficiently bind to the operator locus (affinity of binding reduced >1000-fold). However, RNA polymerase will not efficiently bind the promoter and initiate transcription, therefore the operon is not transcribed. However, when inducer is present *and* glucose is depleted from the medium **(C)** adenyl cylase is activated and cAMP is produced. This cAMP binds with high affinity to its binding protein the Cyclic AMP Activator Protein, or CRP. The camp-CAP complex binds to its recognition sequence (CRE, the cAMP Response Element) located ~15 bp upstream of the promoter. Direct protein–protein contacts between the CRE-bound CAP and the RNA polymerase increases promoter binding >20-fold; hence RNAP will efficiently transcribe the structural genes *lacZ*, *lacY*, and *lacA*, and the polycistronic mRNA molecule formed can be translated into the corresponding protein molecules β-galactosidase, permease, and transacetylase as shown, whick allows for the catabolism of lactose as the sole carbon source for growth.

a source of carbon. In the presence of glucose—or of glycerol in concentrations sufficient for growth—the bacteria will lack sufficient cAMP to bind to CAP because the glucose inhibits adenylyl cyclase, the enzyme that converts ATP to cAMP (see Chapter 41). Thus, in the presence of glucose or glycerol, cAMP-saturated CAP is lacking, so that the DNA-dependent RNA polymerase cannot initiate transcription of the *lac* operon at the maximal rate. However, in the presence of the CAP–cAMP complex, which binds to DNA just upstream of the promoter site, transcription occurs at maximal levels (Figure 38–3). Studies indicate that a region of CAP directly contacts the RNA polymerase α-subunit, and these protein–protein interactions facilitate the binding of RNAP to the promoter. Thus, the CAP–cAMP regulator is acting as a **positive regulator** because its presence is required for optimal gene expression. The *lac* operon is therefore controlled by two distinct, ligand-modulated DNA binding *trans*-factors; one that acts positively (cAMP–CRP complex) to facilitate productive binding of RNA polymerase to the promoter and one that acts negatively (LacI repressor) that antagonizes RNA polymerase promoter binding. Maximal activity of the *lac* operon occurs when glucose levels are low (high cAMP with CAP activation) and lactose is present (LacI is prevented from binding to the operator).

When the *lacI* gene has been mutated so that its product, LacI, is not capable of binding to operator DNA, the organism will exhibit **constitutive expression** of the *lac* operon. In a contrary manner, an organism with a *lacI* gene mutation that produces a LacI protein which prevents the binding of an inducer to the repressor will remain repressed even in the presence of the inducer molecule, because the inducer cannot bind to the repressor on the operator locus in order to derepress the operon. Similarly, bacteria harboring mutations in their *lac* operator locus such that the operator sequence will not bind a normal repressor molecule constitutively express the *lac* operon genes. Mechanisms of positive and negative regulation comparable to those described here for the *lac* system have been observed in eukaryotic cells (see below).

The Genetic Switch of Bacteriophage Lambda (λ) Provides Another Paradigm for Protein–DNA Interactions and Transcriptional Regulation in Eukaryotic Cells

Like some eukaryotic viruses (eg, herpes simplex virus and HIV), some bacterial viruses can either reside in a dormant state within the host chromosomes or can replicate within the bacterium and eventually lead to lysis and killing of the bacterial host. Some *E. coli* harbor such a "temperate" virus, bacteriophage lambda (λ). When lambda infects an organism of that species, it injects its 45,000-bp, double-stranded, linear DNA genome into the cell (**Figure 38–4**). Depending upon the nutritional state of the cell, the lambda DNA will either **integrate** into the host genome (**lysogenic pathway**) and remain dormant until activated (see below), or it will commence **replicating** until it has

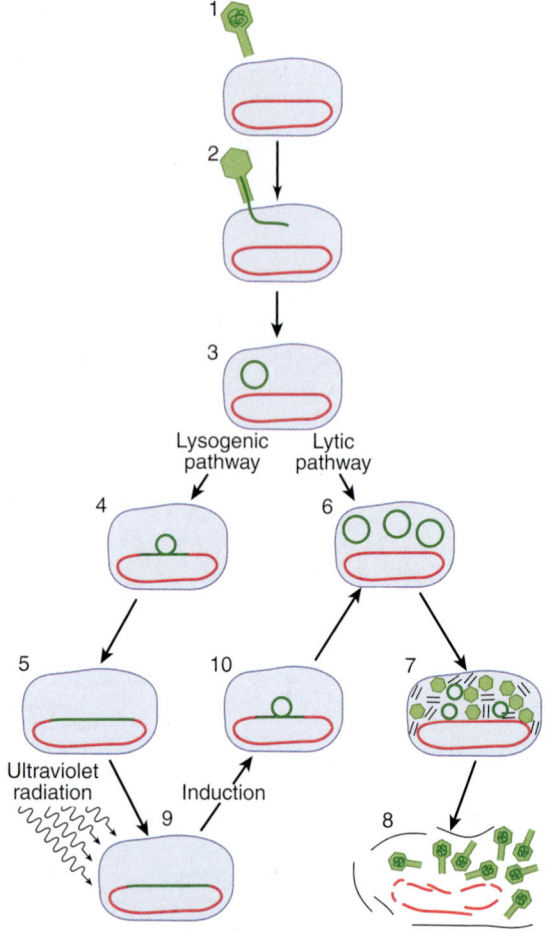

FIGURE 38–4 Infection of the bacterium *E. coli* by phage lambda begins when a virus particle attaches itself to specific receptors on the bacterial cell (**1**) and injects its DNA (dark green line) into the cell (**2, 3**). Infection can take either of two courses depending on which of two sets of viral genes is turned on. In the lysogenic pathway, the viral DNA becomes integrated into the bacterial chromosome (**red**) (**4, 5**), where it replicates passively as the bacterial DNA and cell divides. This dormant genomically integrated virus is called a prophage, and the cell that harbors it is called a lysogen. In the alternative lytic mode of infection, the viral DNA replicates itself (**6**) and directs the synthesis of viral proteins (**7**). About 100 new virus particles are formed. The proliferating viruses induce lysis of the cell (**8**). A prophage can be "induced" by a DNA damaging agent such as ultraviolet radiation (**9**). The inducing agent throws a switch, so that a different set of genes is turned on. Viral DNA loops out of the chromosome (**10**) and replicates; the virus proceeds along the lytic pathway. (Reproduced, with permission, from Ptashne M, Johnson AD, Pabo CO: A genetic switch in a bacterial virus. Sci Am [Nov] 1982;247:128.)

made about 100 copies of complete, protein-packaged virus, at which point it causes lysis of its host (**lytic pathway**). The newly generated virus particles can then infect other susceptible hosts. Poor growth conditions favor lysogeny while good growth conditions promote the lytic pathway of lambda growth.

When integrated into the host genome in its dormant state, lambda will remain in that state until activated by exposure of its bacterial host to DNA-damaging agents. In response to

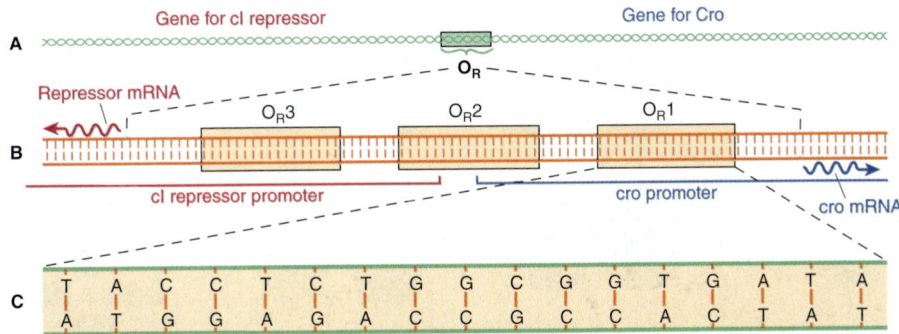

FIGURE 38–5 Right operator (O$_R$) is shown in increasing detail in this series of drawings. The operator is a region of the viral DNA some 80 bp long **(A).** To its left lies the gene encoding lambda repressor (cI), to its right the gene *(cro)* encoding the regulator protein Cro. When the operator region is enlarged **(B),** it is seen to include three subregions, O$_R$1, O$_R$2, and O$_R$3, each 17 bp long. They are recognition sites to which both repressor and Cro can bind. The recognition sites overlap two promoters—sequences of bases to which RNA polymerase binds in order to transcribe these genes into mRNA (wavy lines), that are translated into protein. Site O$_R$1 is enlarged **(C)** to show its base sequence. Note that in the O$_R$ region of the lambda chromosome, both strands of DNA act as a template for transcription. (Reproduced, with permission, from Ptashne M, Johnson AD, Pabo CO: A genetic switch in a bacterial virus. Sci Am [Nov] 1982;247:128.)

such a noxious stimulus, the dormant bacteriophage becomes "induced" and begins to transcribe and subsequently translate those genes of its own genome that are necessary for its excision from the host chromosome, its DNA replication, and the synthesis of its protein coat and lysis enzymes. This event acts like a trigger or type C (Figure 38–1) response; that is, once dormant lambda has committed itself to induction, there is no turning back until the cell is lysed and the replicated bacteriophage released. This switch from a dormant or **prophage state** to a **lytic infection** is well understood at the genetic and molecular levels and will be described in detail here; though less well understood at the molecular level, HIV and herpes viruses can behave similarly.

The lytic/lysogenic genetic switching event in lambda is centered around an 80-bp region in its double-stranded DNA genome referred to as the "right operator" (O$_R$) **(Figure 38–5A).** The **right operator** is flanked on its left side by the structural gene for the lambda repressor protein, *cI*, and on its right side by the structural gene encoding another regulatory protein called *cro*. When lambda is in its prophage state—that is, integrated into the host genome—the *cI* repressor gene is the *only* lambda gene that is expressed. When the bacteriophage is undergoing lytic growth, the *cI* repressor gene is not expressed, but the *cro* gene—as well as many other lambda genes—is expressed. That is, **when the repressor gene is on, the *cro* gene is off, and when the *cro* gene is on, the *cI* repressor gene is off.** As we shall see, these two genes regulate each other's expression and thus, ultimately, the decision between lytic and lysogenic growth of lambda. **This decision between repressor gene transcription and *cro* gene transcription is a paradigmatic example of a molecular transcriptional switch.**

The 80-bp lambda right operator, O$_R$, can be subdivided into three discrete, evenly spaced, 17-bp *cis*-active DNA elements

that represent the binding sites for either of two bacteriophage lambda regulatory proteins. Importantly, the nucleotide sequences of these three tandemly arranged sites are similar but not identical **(Figure 38–5B).** The three related *cis*-elements, termed operators O$_R$1, O$_R$2, and O$_R$3, can be bound by either cI or Cro proteins. However, the relative affinities of cI and Cro for each of the sites vary, and this differential binding affinity is central to the appropriate operation of the lambda phage lytic or lysogenic "molecular switch." The DNA region between the *cro* and repressor genes also contains two promoter sequences that direct the binding of RNA polymerase in a specified orientation, where it commences transcribing adjacent genes. One promoter directs RNA polymerase to transcribe in the **rightward direction** and, thus, to transcribe *cro* and other distal genes, while the other promoter directs the transcription of the *cI* repressor gene in the **leftward direction** (Figure 38–5B).

The product of the repressor gene, the 236-amino-acid, 27 kDa *cI* **repressor protein,** exists as a **two-domain** molecule in which the **amino terminal domain binds to operator DNA** and the **carboxyl terminal domain promotes the association** of one repressor protein with another to form a dimer. A **dimer** of repressor molecules binds to **operator DNA** much more tightly than does the monomeric form **(Figure 38–6A to 38–6C).**

The product of the *cro* gene, the 66-amino-acid, 9-kDa **Cro protein,** has a single domain but also binds the operator DNA more tightly as a **dimer (Figure 38–6D).** The Cro protein's single domain mediates both operator binding and dimerization.

In a lysogenic bacterium—that is, a bacterium containing an integrated dormant lambda prophage—the lambda repressor dimer binds **preferentially to O$_R$1** but in so doing, by a cooperative interaction, enhances the binding (by a factor

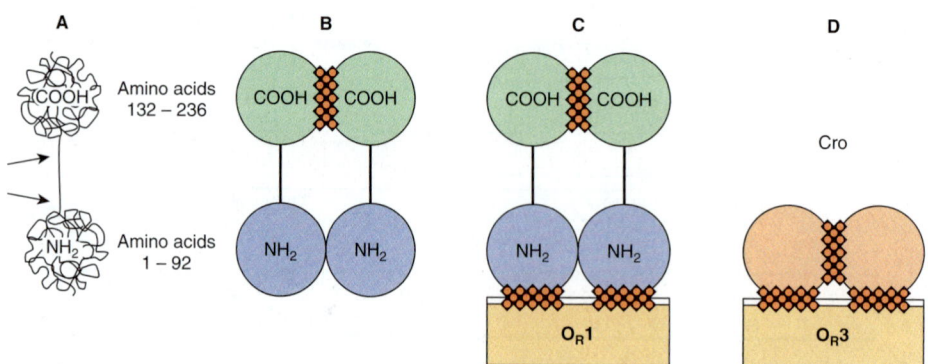

FIGURE 38–6 **Schematic molecular structures of cI (lambda repressor, shown in A, B, and C) and Cro (D).** The lambda repressor protein is a polypeptide chain 236 amino acids long. The chain folds itself into a dumbbell shape with two substructures: an amino terminal (NH$_2$) domain and a carboxyl terminal (COOH) domain. The two domains are linked by a region of the chain that is less structured and susceptible to cleavage by proteases (indicated by the two arrows in **A**). Single repressor molecules (monomers) tend to reversibly associate to form dimers. **(B)** A dimer is held together mainly by contact between the carboxyl terminal domains (hatching). Repressor dimers bind to (and can dissociate from) the recognition sites in the operator region; they display differential affinites for the three operator sites, O$_R$1 > O$_R$2 > O$_R$3 **(C)**. It is the DBD of the repressor molecule that makes contact with the DNA (hatching). Cro **(D)** has a single domain with sites that promote dimerization and other sites that promote binding of dimers to operator, cro exhibits the highest affinity for O$_R$3, opposite the sequence binding preference of the cI protein. (Reproduced, with permission, from Ptashne M, Johnson AD, Pabo CO: A genetic switch in a bacterial virus. Sci Am [Nov] 1982;247:128.)

of 10) of another repressor dimer to O$_R$2 (**Figure 38–7**). The affinity of repressor for O$_R$3 is the least of the three operator subregions. The binding of repressor to O$_R$1 has two major effects. The occupation of O$_R$1 by repressor **blocks the binding of RNA polymerase to the rightward promoter** and in that way prevents expression of *cro*. Second, as mentioned above, repressor dimer bound to O$_R$1 enhances the binding of repressor dimer to O$_R$2. The binding of repressor to O$_R$2 has the important added effect of **enhancing the binding of RNA polymerase to the leftward promoter** that overlaps O$_R$3 and thereby enhances transcription and subsequent expression of the repressor gene. This enhancement of transcription is mediated through direct protein–protein interactions between promoter-bound RNA polymerase and O$_R$2-bound repressor, much as described above for CAP protein and RNA polymerase on the *lac* operon. Thus, the lambda repressor is both a **negative regulator,** by preventing transcription of *cro,* and a **positive regulator,** by enhancing transcription of its own gene, *cI.* This dual effect of repressor is responsible for the stable state of the dormant lambda bacteriophage; not only does the repressor prevent expression of the genes necessary for lysis, but it also promotes expression of itself to stabilize this state of differentiation. In the event that intracellular repressor protein concentration becomes very high, this excess repressor will bind to O$_R$3 and by so doing diminish transcription of the repressor gene from the leftward promoter, by blocking RNAP binding to the cI promoter, until the repressor concentration drops and repressor dissociates itself from O$_R$3. Interestingly, similar examples of repressor proteins also having the ability to activate transcription have been observed in eukaryotes.

With such a stable, repressive, cI-mediated, lysogenic state, one might wonder how the lytic cycle could ever be entered. However, this process does occur quite efficiently. When a DNA-damaging signal, such as ultraviolet light, strikes the lysogenic host bacterium, fragments of single-stranded DNA are generated that activate a specific **co-protease** coded by a bacterial gene and referred to as *recA* (Figure 38–7). The activated recA protease hydrolyzes the portion of the repressor protein that connects the amino terminal and carboxyl terminal domains of that molecule (see Figure 38–6A). Such cleavage of the repressor domains causes the **repressor dimers to dissociate,** which in turn causes **dissociation of the repressor molecules from O$_R$2** and eventually from O$_R$1. The effects of removal of repressor from O$_R$1 and O$_R$2 are predictable. RNA polymerase immediately has access to the rightward promoter and commences transcribing the *cro* gene, and the enhancement effect of the repressor at O$_R$2 on leftward transcription is lost (Figure 38–7).

The resulting newly synthesized Cro protein also binds to the operator region as a dimer, but its order of preference is opposite to that of repressor (Figure 38–7). That is, **Cro binds most tightly to O$_R$3,** but there is no cooperative effect of Cro at O$_R$3 on the binding of Cro to O$_R$2. At increasingly higher concentrations of Cro, the protein will bind to O$_R$2 and eventually to O$_R$1.

Occupancy of O$_R$3 by Cro immediately turns off transcription from the leftward *cI* promoter and in that way **prevents any further expression of the repressor gene.** The molecular switch is thus completely "thrown" in the lytic direction. The *cro* gene is now expressed, and the repressor gene is fully turned off. This event is irreversible, and the expression of

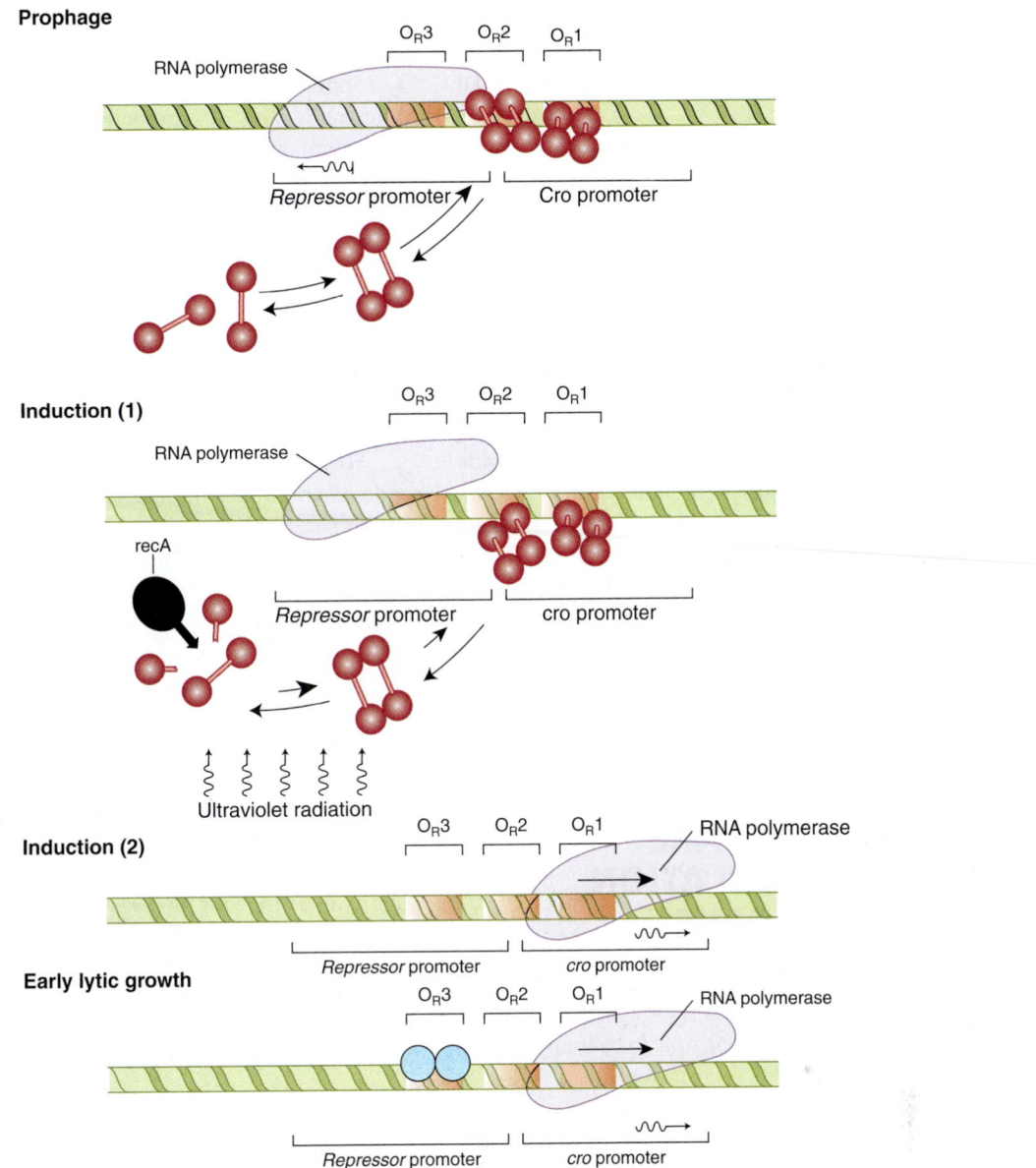

FIGURE 38–7 **Configuration of the lytic/lysogenic switch is shown at four stages of the lambda life cycle.** The lysogenic pathway (in which the virus remains dormant as a prophage) is selected when a repressor dimer binds to O_R1, thereby making it likely that O_R2 will be filled immediately by another dimer. In the prophage (**top**), the repressor dimers bound at O_R1 and O_R2 prevent RNA polymerase from binding to the rightward promoter and so block the synthesis of Cro (negative control). The repressors also enhance the binding of polymerase to the leftward promoter (positive control), with the result that the repressor gene is transcribed into RNA (wavy line) and more repressor is synthesized, maintaining the lysogenic state. The prophage is induced (**middle**) when ultraviolet radiation activates the protease recA, which cleaves repressor monomers. The equilibrium of free monomers, free dimers, and bound dimers is thereby shifted, and dimers leave the operator sites. RNA polymerase is no longer encouraged to bind to the leftward promoter, so that repressor is no longer synthesized. As induction proceeds, all the operator sites become vacant, thus polymerase can bind to the rightward promoter and Cro is synthesized. During early lytic growth, a single Cro dimer binds to O_R3 (light blue shaded circles), the site for which it has the highest affinity. Consequently, RNA polymerase cannot bind to the leftward promoter, but the rightward promoter remains accessible. Polymerase continues to bind there, transcribing *cro* and other early lytic genes. Lytic growth ensues (**bottom**). (Reproduced, with permission, from Ptashne M, Johnson AD, Pabo CO: A genetic switch in a bacterial virus. Sci Am [Nov] 1982;247:128.)

other lambda genes begins as part of the lytic cycle. When Cro repressor concentration becomes quite high, it will eventually occupy O_R1 and in so doing reduce the expression of its own gene, a process that is necessary in order to effect the final stages of the lytic cycle.

The three-dimensional structures of Cro and of the lambda repressor protein have been determined by X-ray crystallography, and models for their binding and effecting the above-described molecular and genetic events have been proposed and tested. Both bind to DNA using helix-turn-helix DNA-binding domain (DBD) motifs (see below). To date, this system provides arguably the best understanding of the molecular events involved in gene activation and repression.

Detailed analysis of the lambda repressor led to the important concept that transcription regulatory proteins have several functional domains. For example, lambda repressor binds to DNA with high affinity. Repressor monomers form dimers, cooperatively interact with each other, and repressor interacts with RNA polymerase, to enhance or block promoter binding or RNAP open complex formation (see Figure 36–3). The protein–DNA interface and the three protein–protein interfaces all involve separate and distinct domains of the repressor molecule. As will be noted below (see Figure 38–19), this is a characteristic shared by most (perhaps all) molecules that regulate transcription.

SPECIAL FEATURES ARE INVOLVED IN REGULATION OF EUKARYOTIC GENE TRANSCRIPTION

Most of the DNA in prokaryotic cells is organized into genes, and the templates always have the potential to be transcribed if appropriate positive and negative *trans*-factors are activated. A very different situation exists in mammalian cells, in which relatively little of the total DNA is organized into mRNA encoding genes and their associated regulatory regions. The function of the extra DNA is being actively investigated (ie, Chapter 39; the ENCODE Projects). More importantly, as described in Chapter 35, the DNA in eukaryotic cells is extensively folded and packed into the protein–DNA complex called chromatin. Histones are an important part of this complex since they both form the structures known as nucleosomes (see Chapter 35) and also factor significantly into gene regulatory mechanisms as outlined below.

The Chromatin Template Contributes Importantly to Eukaryotic Gene Transcription Control

Chromatin structure provides an additional level of control of gene transcription. As discussed in Chapter 35, large regions of chromatin are transcriptionally inactive while others are either active or potentially active. With few exceptions, each cell contains the same complement of genes. The development of specialized organs, tissues, and cells and their function in the intact organism depend upon the differential expression of genes.

Some of this differential expression is achieved by having different regions of chromatin available for transcription in cells from various tissues. For example, the DNA containing the β-globin gene cluster is in **"active" chromatin** in the reticulocyte but in **"inactive" chromatin** in muscle cells. All the factors involved in the determination of active chromatin have not been elucidated. The presence of nucleosomes and of complexes of histones and DNA (see Chapter 35) certainly provides a barrier against the ready association of transcription factors with specific DNA regions. The dynamics of the formation and disruption of nucleosome structure are therefore an important part of eukaryotic gene regulation.

Histone covalent modification, also dubbed **the histone code,** is an important determinant of gene activity. Histones are subjected to a wide range of specific posttranslational modifications (Table 35–1). These modifications are dynamic and reversible. Histone acetylation and deacetylation are best understood. The surprising discovery that histone acetylase and other enzymatic activities are associated with the coregulators involved in regulation of gene transcription (see Chapter 42) has provided a new concept of gene regulation. Acetylation is known to occur on lysine residues in the amino terminal tails of histone molecules, and has been consistently correlated with transcription, or alternatively transcriptional potential. Histone acetylation reduces the positive charge of these tails and likely contributes to a decrease in the binding affinity of histone for the negatively charged DNA. Such covalent modification of the histones creates new binding sites for additional proteins such as ATP-dependent chromatin remodeling complexes, which contain subunits that carry structural domains that specifically bind to histones that have been subjected to coregulator-deposited PTMs. These complexes can increase accessibility of adjacent DNA sequences by removing nucleosomal histones. Together then coregulators (chromatin modifiers and chromatin remodellers), working in conjunction, can open up gene promoters and regulatory regions, facilitating binding of other *trans*-factors and RNA polymerase II and GTFs (see Figures 36–10 and 36–11). Histone deacetylation catalyzed by transcriptional corepressors would have the opposite effect. Different proteins with specific acetylase and deacetylase activities are associated with various components of the transcription apparatus. The proteins that catalyze the histone PTMs are sometimes referred to as **"code writers"** while the proteins that recognize, bind and interpret these histone PTMs are termed **"code readers"** and the enzymes that remove histone PTMs are called **"code erasers"** Collectively then, there histone PTMs represent a very dynamic, potentially information-rich source of regulatory information. The exact rules and mechanisms defining the specificity of these various processes are under investigation. Some specific examples are illustrated in Chapter 42. A variety of commercial enterprises are working to develop drugs that specifically alter the ability of the proteins that modulate the histone code.

There is evidence that the **methylation of deoxycytidine residues, 5MeC,** (in the sequence 5′-^{me}CpG-3′) in DNA may effect changes in chromatin so as to preclude its active transcription, as described in Chapter 35. For example, in mouse liver, only the unmethylated ribosomal genes can be expressed, and there is evidence that many animal viruses are not transcribed when their DNA is methylated. Acute demethylation of 5MeC residues in specific regions of steroid hormone inducible genes has been associated with an increased rate of transcription of the gene. However, it is not yet possible to generalize that methylated DNA is transcriptionally inactive, that all inactive chromatin is methylated, or that active DNA is not methylated.

Finally, the binding of specific transcription factors to cognate DNA elements may result in disruption of nucleosomal structure. Many eukaryotic genes have multiple protein-binding DNA elements. The serial binding of transcription factors to these elements—in a combinatorial fashion—may either directly disrupt the structure of the nucleosome, prevent its re-formation, or recruit, via protein–protein interactions, multiprotein coregulator complexes that have the ability to covalently modify and/or remodel nucleosomes. These reactions result in chromatin-level structural changes that in the end increase DNA accessibility to other factors and the transcription machinery (cf. above).

Eukaryotic DNA that is in an "active" region of chromatin can be transcribed. As in prokaryotic cells, a **promoter** dictates where the RNA polymerase will initiate transcription, but the promoter in mammalian cells (Chapter 36) is more complex. In addition, the *trans*-acting factors generally come from other chromosomes (and so act in *trans*), whereas this consideration is moot in the case of the single chromosome-containing prokaryotic cells. Additional complexity is added by elements or factors that enhance or repress transcription, define tissue-specific expression, and modulate the actions of many effector molecules. Finally, recent results suggest that gene activation and repression might occur when particular genes move into or out of different subnuclear compartments or locations.

Epigenetic Mechanisms Contribute Importantly to the Control of Gene Transcription

The molecules and regulatory biology described above contributes importantly to transcriptional regulation. Indeed, in recent years the role of covalent modification of DNA and histone and nonhistone proteins and the newly discovered ncRNAs has received tremendous attention in the field of gene regulation research, particularly through investigation into how such chemical modifications and/or molecules stably alter gene expression patterns without altering the underlying DNA gene sequence. This field of study has been termed **epigenetics.** As mentioned in Chapter 35, one aspect of these mechanisms, PTMs of histones has been dubbed the **histone code** or histone epigenetic code. The term "epigenetics" means "above

genetics" and refers to the fact that these regulatory mechanisms do not change the underlying, regulated DNA sequence, but rather simply the expression patterns of this DNA. Epigenetic mechanisms play key roles in the establishment, maintenance, and reversibility of transcriptional states. A key feature of epigenetic mechanisms is that the controlled transcriptional on/off states can be maintained through multiple rounds of cell division. This observation indicates that there must be robust mechanisms to maintain and stably propagate these epigenetic states.

Two forms of epigenetic signals, *cis*- and *trans*-epigenetic signals, can be described; these are schematically illustrated in **Figure 38–8**. A simple *trans*-signaling event composed of positive transcriptional feedback mediated by an abundant, diffusible transactivator that partitions between mother and daughter cell at each division is depicted in Figure 38–8A. As long as the indicated, transcription factor is expressed at a sufficient level to allow all subsequent daughter cells to inherit the *trans*-epigenetic signal (transcription factor), such cells will have the cellular or molecular phenotype dictated by the other target genes of this transcriptional activator. Shown in Figure 38–8 panel B is an example of how a *cis*-epigenetic signal (such as a specific 5MeCpG methylation mark) can be stably propagated to the two daughter cells following cell division. The hemi-methylated (ie, only one of the two DNA strands is 5MeC modified) DNA mark generated during DNA replication directs the methylation of the newly replicated strand through the action of ubiquitous maintenance DNA methylases. This 5MeC methylation results in both DNA daughter strands having the complete *cis*-epigenetic mark.

Both *cis*- and *trans*-epigenetic signals result in stable and hereditable expression states, and therefore represent type C gene expression responses (ie, Figure 38–1). However, it is important to note that both states can be reversed if either the *trans*- or *cis*-epigenetic signals are removed by, for example, extinguishing the expression of the enforcing transcription factor (*trans*-signal) or by removing a DNA *cis*-epigenetic signal (via DNA demethylation). Enzymes have been described that, at least in vitro, can remove both protein PTMs and 5MeC modifications.

Stable transmission of epigenetic on/off states can be effected by multiple molecular mechanisms. Shown in **Figure 38–9** are three ways by which *cis*-epigenetic marks can be propagated through a round of DNA replication. The first example of epigenetic mark transmission involves the propagation of DNA 5MeC marks, and occurs as described above in Figure 38–8. The second example of epigenetic state transmission illustrates how a nucleosomal histone PTM (in this example, Lysine K-27 trimethylated histone H3; H3K27me3) can be propagated. In this example immediately following DNA replication, both H3K27me3-marked and H3-unmarked nucleosomes randomly reform on both daughter DNA strands. The polycomb repressive complex 2 (PRC2), composed of EED-SUZ12-EZH2 and RbAP subunits, binds to the nucleosome containing the preexisting H3K27me3 mark via the EED subunit. Binding of PRC2 to this Histone mark stimulates the

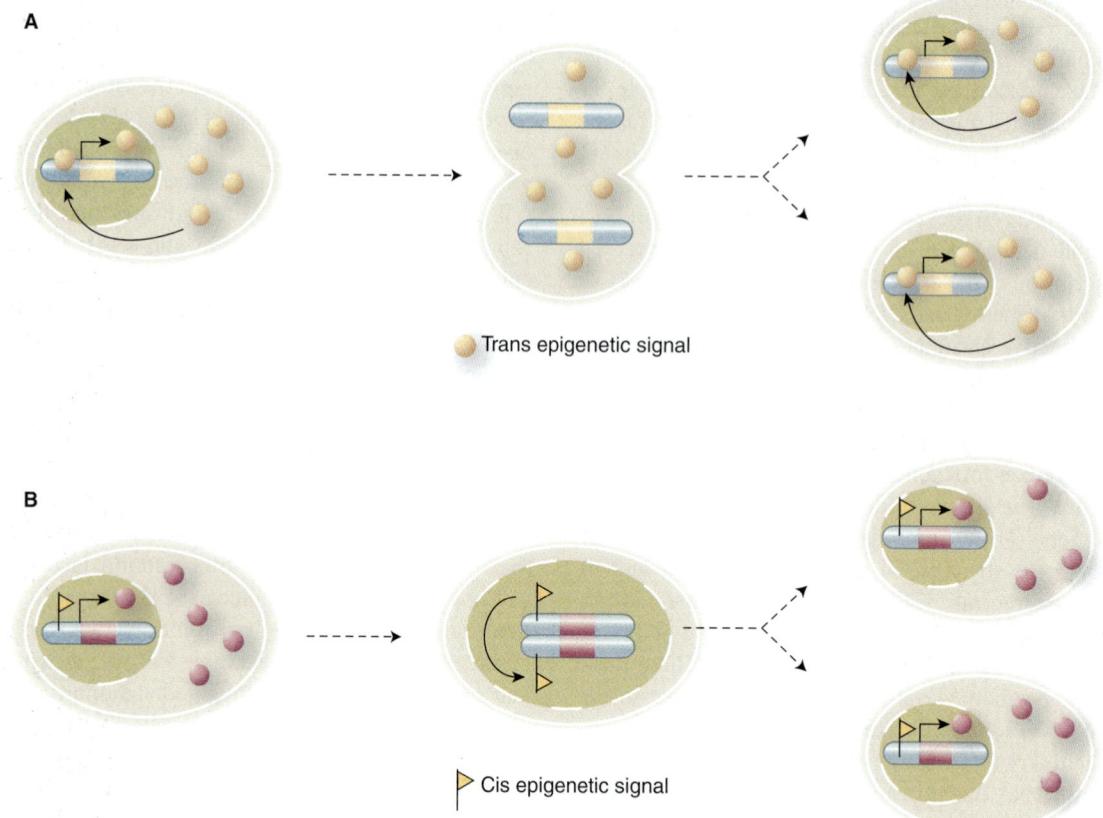

FIGURE 38–8   *cis-* and *trans*-epigenetic signals. (**A**) An example of an epigenetic signal that acts in trans. A DNA binding transactivator protein (yellow circle) is transcribed from its cognate gene (yellow bar) located on a particular chromosome (blue). The expressed protein is freely diffusible between nuclear and cytoplasmic compartments. Note that excess transactivator re-enters the nucleus following cell division, binds to its own gene and activates transcription in both daughter cells. This cycle re-establishes the positive feedback loop in effect prior to cell division, and thereby enforces stable expression of this transcriptional activator protein in both cells. (**B**) A *cis*-epigenetic signal; a gene (pink) located on a particular chromosome (blue) carries a *cis*-epigenetic signal (small yellow flag) within the regulatory region upstream of the pink gene transcription unit. In this case, the epigenetic signal is associated with active gene transcription and subsequent gene product production (pink circles). During DNA replication, the newly replicated chromatid serves as a template that elicits and templates the introduction of the same epigenetic signal, or mark, on the newly synthesized, unmarked chromatid. Consequently, both daughter cells contain the pink gene in a similarly *cis*-epigenetically marked state, which ensures expression in an identical fashion in both cells. See text for more detail. (Image Taken from: Roberto Bonasio, R, Tu, S, Reinberg D (2010), "Molecular Signals of Epigenetic States". *Science* 330:612–616. Reprinted with permission from AAAS.)

methylase activity of the PRC2 subunit EZH2, which results in the local methylation of nucleosomal H3. Histone H3 methylation thus causes the full, stable transmission of the H3K27me3 epigenetic mark to both chromatids. Finally, locus/sequence-specific targeting of nucleosomal histone epigenetic *cis*-signals can be attained through the action of ncRNAs as depicted in Figure 38–9, panel C. Here a specific ncRNA interacts with target DNA sequences and the resulting RNA–DNA complex is recognized by RBP, an RNA-binding protein. Then, likely through a specific adaptor protein (A), the RNA–DNA–RBP complex recruits a chromatin modifying complex (CMC) that locally modifies nucleosomal histones. Again, this mechanism leads to the transmission of a stable epigenetic mark.

Additional work will be required to establish the complete molecular details of these epigenetic processes, determine how ubiquitously these mechanisms operate, identify the full complement of molecules involved, and genes controlled.

Epigenetic signals are critically important to gene regulation as evidenced by the fact that mutations and/or overexpression of many of the molecules that contribute to epigenetic control lead to human disease.

Certain DNA Elements Enhance or Repress Transcription of Eukaryotic Genes

In addition to gross changes in chromatin affecting transcriptional activity, certain DNA elements facilitate or enhance initiation at the promoter and hence are termed **enhancers. Enhancer elements, which typically contain multiple binding sites for transactivator proteins,** differ from the promoter in notable ways. They can exert their positive influence on transcription even when separated by tens of thousands of base pairs from a promoter; they work when oriented in

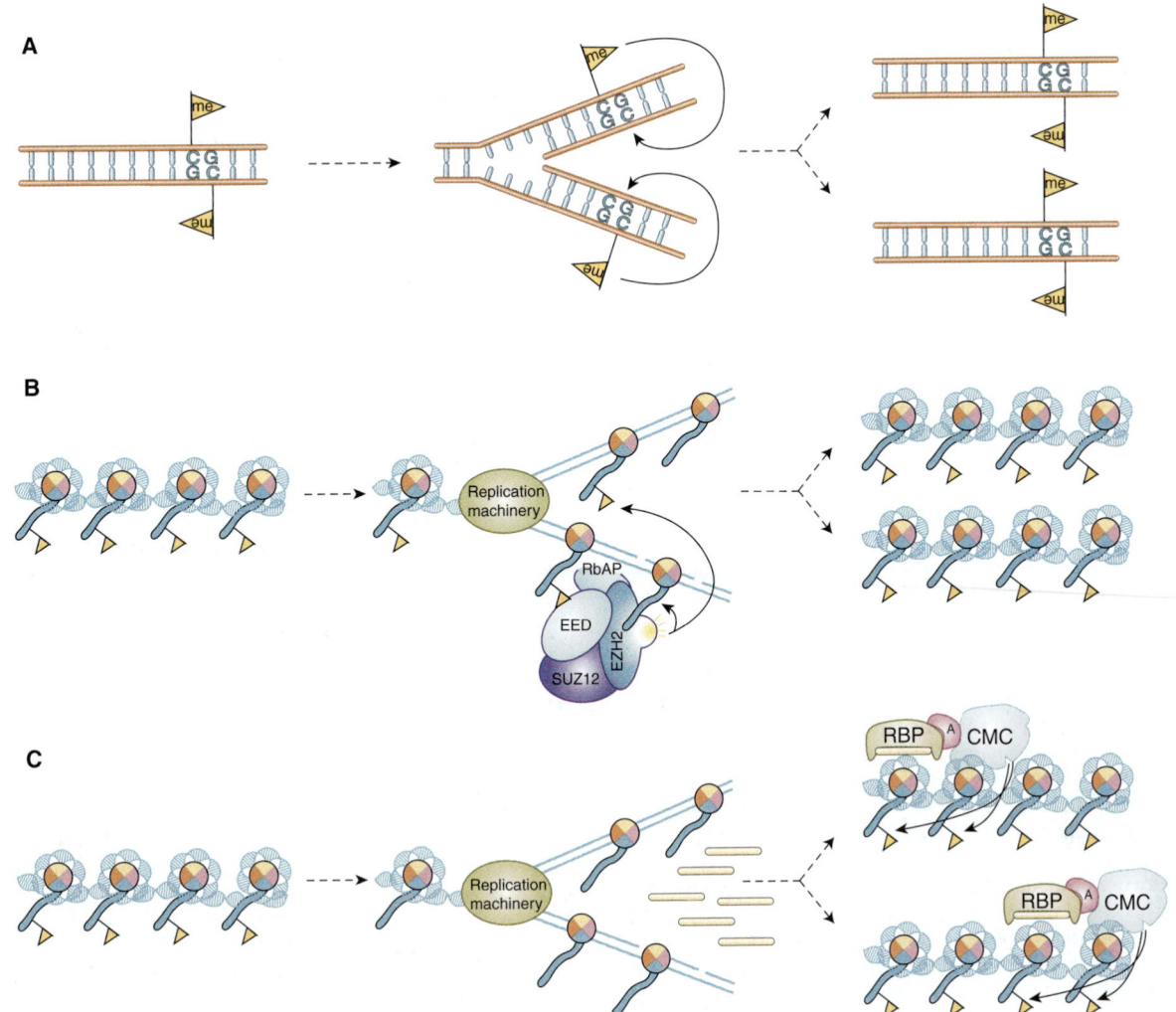

FIGURE 38–9 **Mechanisms for the transmission and propagation of epigenetic signals following a round of DNA replication.**
(**A**) Propagation of a 5MeC signal (yellow flag; see Figure 38-8B). (**B**) Propagation of a histone PTM mark epigenetic signal (H3K27me) that is mediated through the action of the PRC2 CMC, a four subunit protein composed of EED, EZH2 histone methylase, RbAP and SUZ12. Note that in this context PRC2 is a both a histone code reader (via the methylated histone binding domain in EED) and histone code writer (via the SET domain histone methylase within EZH2). Location-specific deposition of the histone PTM *cis*-epigenetic signal is targeted by the recognition of the H3K27me marks in preexisting nucleosomal histones (yellow flag). (**C**) Another example of the transmission of a histone epigenetic signal (yellow flag) except here signal-targeting is mediated through the action of small ncRNAs, which work in concert with an RNA-binding protein (RBP), an Adaptor (A) protein, and a CMC. See text for more detail. (Image Taken from: Roberto Bonasio, R, Tu, S, Reinberg D (2010), "Molecular Signals of Epigenetic States". *Science* 330:612–616. Reprinted with permission from AAAS.)

either direction; and they can work upstream (5′) or downstream (3′) from the promoter. Enhancers are promiscuous; they can stimulate any promoter in the vicinity and may act on more than one promoter. The viral SV40 enhancer can exert an influence on, for example, the transcription of β-globin by increasing its transcription 200-fold in cells containing both the SV40 enhancer and the β-globin gene on the same plasmid (see below and **Figure 38–10**); in this case the SV40 enhancer β-globin gene was constructed using recombinant DNA technology—see Chapter 39. The enhancer element does not produce a product that in turn acts on the promoter, since it is active only when it exists within the same DNA molecule as (ie, *cis* to) the promoter. Enhancer-binding proteins are responsible for this effect. The exact mechanisms by which

these transcription activators work are subject to intensive investigation. Certainly, enhancer-binding *trans*-factors have been shown to interact with a plethora of other transcription proteins. These interactions include chromatin-modifying co-activators, mediator, as well as the individual components of the basal RNA polymerase II transcription machinery. Ultimately, transfactor-enhancer DNA-binding events result in an increase in the binding of the basal transcription machinery to the promoter. Enhancer elements and associated binding proteins often convey nuclease hypersensitivity to those regions where they reside (Chapter 35). A summary of the properties of enhancers is presented in **Table 38–2**.

One of the best-understood mammalian enhancer systems is that of the β-interferon gene. This gene is induced upon viral

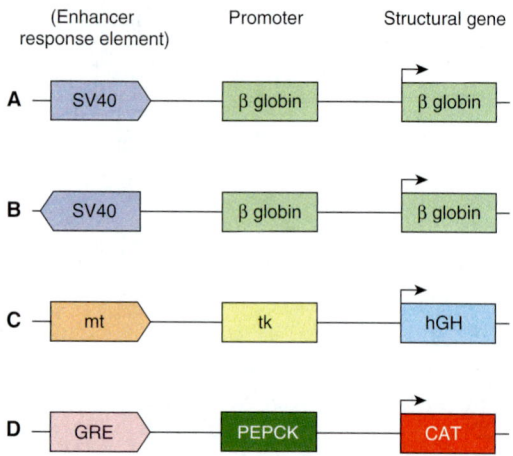

FIGURE 38–10 **A schematic illustrating the action of enhancers and other *cis*-acting regulatory elements.** These model chimeric genes, all constructed by recombinant DNA techniques (Chapter 39) in vitro, consist of a reporter (structural) gene that encodes a protein that can be readily assayed, and that is not normally produced in the cells to be studied, a promoter that ensures accurate initiation of transcription, and the indicated regulatory elements. In all cases, high-level transcription from the indicated chimeras depends upon the presence of enhancers, which stimulate transcription ≥ 100-fold over basal transcriptional levels (ie, transcription of the same chimeric genes containing just promoters fused to the structural genes). Examples **(A)** and **(B)** illustrate the fact that enhancers (eg, SV40) work in either orientation and upon a heterologous promoter. Example **(C)** illustrates that the metallothionein (mt) regulatory element (which under the influence of cadmium or zinc induces transcription of the endogenous mt gene and hence the metal-binding mt protein) will work through the thymidine kinase (tk) promoter to enhance transcription of the human growth hormone (hGH) gene. The engineered genetic constructions were introduced into the male pronuclei of single-cell mouse embryos and the embryos placed into the uterus of a surrogate mother to develop as transgenic animals. Offspring have been generated under these conditions, and in some the addition of zinc ions to their drinking water effects an increase in growth hormone expression in liver. In this case, these transgenic animals have responded to the high levels of growth hormone by becoming twice as large as their normal litter mates. Example **(D)** illustrates that a glucocorticoid response element (GRE) will work through homologous (PEPCK gene) or heterologous promoters (not shown; ie, tk) promoter, SV40 promoter, β-globin promoter, etc) to drive expression of the chloramphenicol acetyl transferase (CAT) reporter gene.

infection of mammalian cells. One goal of the cell, once virally infected, is to attempt to mount an antiviral response—if not to save the infected cell, then to help to save the entire organism from viral infection. Interferon production is one mechanism by which this is accomplished. This family of proteins is secreted by virally infected cells. Secreted interferon interacts with neighboring cells to cause an inhibition of viral replication by a variety of mechanisms, thereby limiting the extent of viral infection. The enhancer element controlling induction of the β-interferon gene, which is located between nucleotides −110 and −45 relative to the transcription start site (+1), is well characterized. This enhancer is composed of four distinct clustered *cis*-elements, each of which is bound by unique *trans*-factors. One *cis*-element is bound by the transacting factor NF-κB, one

TABLE 38–2 Summary of the Properties of Enhancers

- Work when located long distances from the promoter
- Work when upstream or downstream from the promoter
- Work when oriented in either direction
- Can work with homologous or heterologous promoters
- Work by binding one or more proteins
- Work by facilitating binding of the basal transcription complex to the *cis*-linked promoter
- Work by recruiting chromatin-modifying coregulatory complexes

by a member of the IRF (interferon regulatory factor) family of *trans*-factors, and a third by the heterodimeric leucine zipper factor ATF-2/c-Jun (see below). The fourth factor is the ubiquitous, abundant architectural transcription factor known as HMG I(Y). Upon binding to its A+T-rich binding sites, HMG I(Y) induces a significant bend in the DNA. There are four such HMG I(Y) binding sites interspersed throughout the enhancer. These sites play a critical role in forming a particular 3D structure, along with the aforementioned three *trans*-factors, by inducing a series of critically spaced DNA bends. Consequently, HMG I(Y) induces the cooperative formation of a unique, stereospecific, 3D structure within which all four factors are active when viral infection signals are sensed by the cell. The structure formed by the cooperative assembly of these four factors is termed the β-interferon enhanceosome (see **Figure 38–11**), so named because of its obvious structural similarity to the nucleosome, also a unique three-dimensional protein–DNA structure that wraps DNA about an assembly of proteins (see Figures 35–1 and 35–2). The enhanceosome, once formed, induces a large increase in β-interferon gene transcription upon virus infection. It is not simply the protein occupancy of the linearly apposed *cis*-element sites that induces β-interferon gene transcription—rather, it is the formation of the enhanceosome proper that provides appropriate surfaces for the recruitment of coactivators that results in the enhanced formation of the PIC on the *cis*-linked promoter and thus transcription activation.

The *cis*-acting elements that decrease or **repress** the expression of specific genes have also been identified. Because fewer of these elements have been studied, it is not possible to formulate generalizations about their mechanism of action—though again, as for gene activation, chromatin level covalent modifications of histones and other proteins by (repressor)-recruited multisubunit corepressors have been implicated.

Tissue-Specific Expression May Result From Either the Action of Enhancers or Repressors or a Combination of Both *Cis*-Acting Regulatory Elements

Many genes are now recognized to harbor enhancer or activator elements in various locations relative to their coding

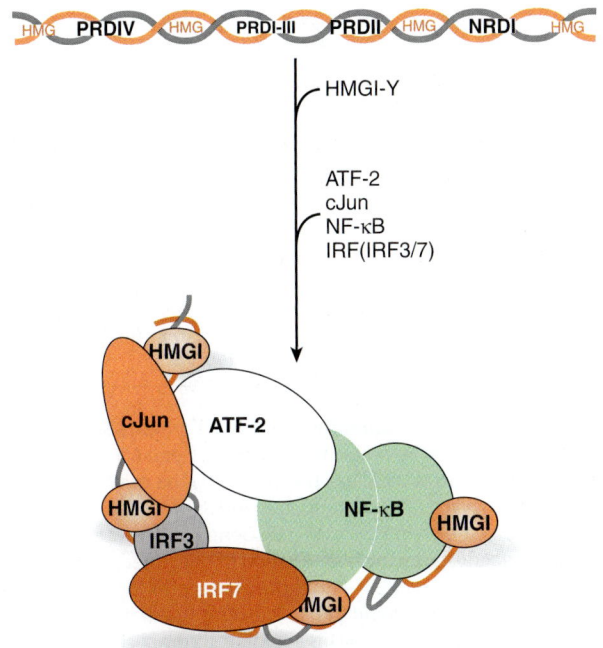

FIGURE 38–11   **Formation and putative structure of the enhanceosome formed on the human β-interferon gene enhancer.** Diagrammatically represented at the top is the distribution of the multiple *cis*-elements (HMG, PRDIV, PRDI-III, PRDII, NRDI) composing the β-interferon gene enhancer. The intact enhancer mediates transcriptional induction of the β-interferon gene (over 100-fold) upon virus infection of human cells. The *cis*-elements of this modular enhancer represent the binding sites for the *trans*-factors HMG I(Y), cJun-ATF-2, IRF3-IRF7, and NF-κB, respectively. The factors interact with these DNA elements in an obligatory, ordered, and highly cooperative fashion as indicated by the arrow. Initial binding of four HMG I(Y) proteins induces sharp DNA bends in the enhancer, causing the entire 70–80 bp region to assume a high level of curvature. This curvature is integral to the subsequent highly cooperative binding of the other *trans*-factors since this enables the DNA-bound factors to make important, direct protein–protein interactions that both contribute to the formation and stability of the enhanceosome and generate a unique 3D surface that serves to recruit chromatin-modifying coregulators that carry enzymatic activities (eg, Swi/Snf: ATPase, chromatin remodeler and P/CAF: histone acetyltransferase) as well as the general transcription machinery (RNA polymerase II and GTFs). Although four of the five *cis*-elements (PRDIV, PRDI-III, PRDII, NRDI) independently can modestly stimulate (~10-fold) transcription of a reporter gene in transfected cells (see Figures 38–10 and 38–12), all five *cis*-elements, in appropriate order, are required to form an enhancer that can appropriately stimulate mRNA gene transcription (ie, ≥100-fold) in response to viral infection of a human cell. This distinction indicates the strict requirement for appropriate enhanceosome architecture for efficient *trans*-activation. Similar enhanceosomes, involving distinct *cis*- and *trans*-factors and coregulators, are proposed to form on many other mammalian genes.

regions. In addition to being able to enhance gene transcription, some of these enhancer elements clearly possess the ability to do so in a tissue-specific manner. Thus, the enhancer element associated with the immunoglobulin genes between the J and C regions enhances the expression of those genes preferentially in lymphoid cells. Similarly by fusing known

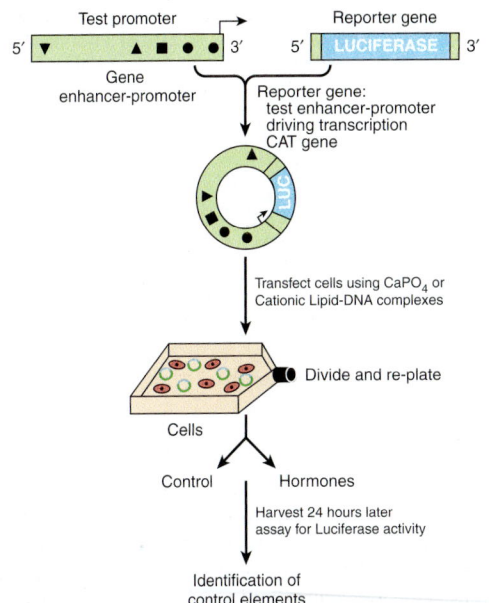

FIGURE 38–12   **The use of reporter genes to define DNA regulatory elements.** A DNA fragment bearing regulatory *cis*-elements (triangles, square, circles in diagram) from the gene in question—in this example, approximately 2 kb of 5′-flanking DNA and cognate promoter—is ligated into a plasmid vector that contains a suitable reporter gene—in this case, the enzyme firefly luciferase, abbreviated LUC. Whatever reporter gene is utilized in these experiments, the reporter can not be present in the transfected cells. Consequently, any detection of these activities in a cell extract means that the cell was successfully transfected by the plasmid. Not shown here, but typically one cotransfects an additional reporter such as Renilla luciferase to serve as a transfection efficiency control. Assay conditions for the firefly and Renilla luciferases are different, hence the two activities can be sequentially assayed using the same cell extract. An increase of firefly luciferase activity over the basal level, for example, after addition of one or more hormones, means that the region of DNA inserted into the reporter gene plasmid contains functional hormone response elements (HRE). Progressively shorter pieces of DNA, regions with internal deletions, or regions with point mutations can be constructed and inserted to pinpoint the response element (see Figure 38–13 for deletion mapping of the relevant HREs).

or suspected tissue-specific enhancers to reporter genes (see below) and introducing these chimeric enhancer-reporter constructs microsurgically into single-cell embryo, one can create a transgenic animal (see Chapter 39), and rigorously test whether a given test enhancer truly drives expression in a cell- or tissue-specific fashion. This **transgenic animal** approach has proved useful in studying tissue-specific gene expression.

Reporter Genes Are Used to Define Enhancers & Other Regulatory Elements

By ligating regions of DNA suspected of harboring regulatory sequences to various reporter genes (the **reporter** or **chimeric gene approach**) (**Figures 38–10, 38–12, & 38–13**), one can determine which regions in the vicinity of structural genes have an influence on their expression. Pieces of DNA thought to harbor regulatory elements are ligated to a suitable reporter

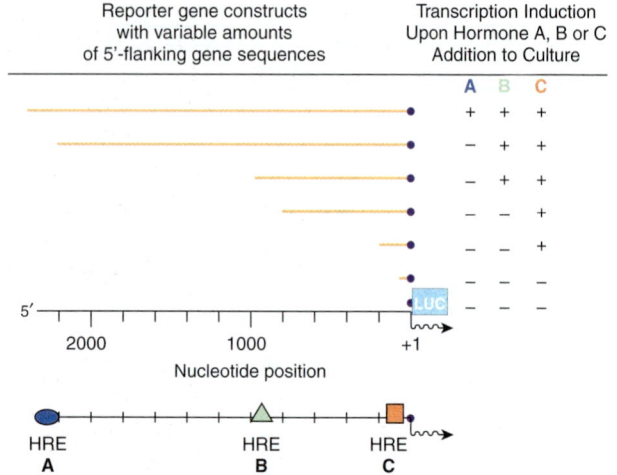

FIGURE 38–13 **Mapping hormone response elements (HREs) (A), (B), and (C) using the reporter gene–transfection approach.** A family of reporter genes, constructed as described in Figure 38–10, can be transfected individually into a recipient cell. By analyzing when certain hormone responses are lost in comparison to the 5' deletion end point, specific hormone-responsive elements can be located.

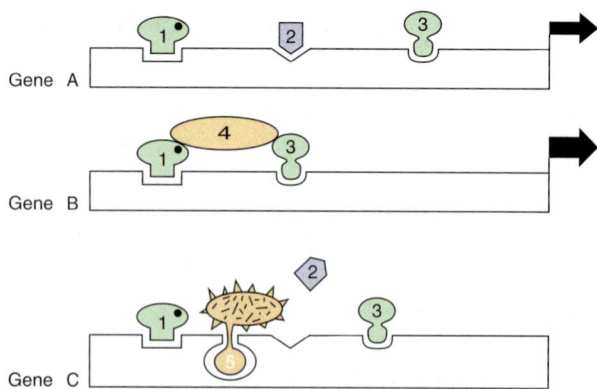

FIGURE 38–14 **Combinations of DNA elements and proteins provide diversity in the response of a gene.** Gene A is activated (the width of the arrow indicates the extent) by the combination of transcripitonal activator proteins 1, 2, and 3 (probably with coactivators, as shown in Figure 36–10). Gene B is activated, in this case more effectively, by the combination of one, three, and four; note that transcription factor 4 does not contact DNA directly in this example. The activators could form a linear bridge that links the basal machinery to the promoter, or this could be accomplished by looping out of the DNA. In either case, the purpose is to direct the basal transcription machinery to the promoter. Gene C is inactivated by the combination of transcription factors 1, 5, and 3; in this case, factor 5 is shown to preclude the essential binding of factor 2 to DNA, as occurs in example A. If activator 1 helps repressor 5 bind and if activator 1 binding requires a ligand (solid dot), it can be seen how the ligand could activate one gene in a cell (gene A) and repress another (gene C) in the same cell.

gene and introduced into a host cell (Figure 38–12). Basal expression of the reporter gene will be increased if the DNA contains an enhancer. Addition of a hormone or heavy metal to the culture medium will increase expression of the reporter gene if the DNA contains a hormone or metal response element (Figure 38–13). The location of the element can be pinpointed by using progressively shorter pieces of DNA, deletions, or point mutations (Figure 38–13).

This strategy, typically **using transfected cells in culture** (ie, cells induced to take up exogenous DNAs), has led to the identification of hundreds of enhancers, repressors, tissue-specific elements, and hormone, heavy metal, and drug–response elements. The activity of a gene at any moment reflects the interaction of these numerous *cis*-acting DNA elements with their respective *trans*-acting factors. Overall, transcriptional output is determined by the balance of positive and negative signaling to the transcription machinery. The challenge now is to figure out how this occurs at the molecular level.

Combinations of DNA Elements & Associated Proteins Provide Diversity in Responses

Prokaryotic genes are often regulated in an on–off manner in response to simple environmental cues. Some eukaryotic genes are regulated in the simple on–off manner, but the process in most genes, especially in mammals, is much more complicated. Signals representing a number of complex environmental stimuli may converge on a single gene. The response of the gene to these signals can have several physiologic characteristics. First, the response may extend over a considerable range.

This is accomplished by having additive and synergistic positive responses counterbalanced by negative or repressing effects. In some cases, either the positive or the negative response can be dominant. Also required is a mechanism whereby an effector such as a hormone can activate some genes in a cell while repressing others and leaving still others unaffected. When all of these processes are coupled with tissue-specific element factors, considerable flexibility is afforded. These physiologic variables obviously require an arrangement much more complicated than an on–off switch. The array of DNA elements in a promoter specifies—with associated factors—how a given gene will respond and how long a particular response is maintained. Some simple examples are illustrated in **Figure 38–14**.

Transcription Domains Can Be Defined by Locus Control Regions & Insulators

The large number of genes in eukaryotic cells and the complex arrays of transcription regulatory factors present an organizational problem. Why are some genes available for transcription in a given cell whereas others are not? If enhancers can regulate several genes from tens of kilobase distances and are not position- and orientation-dependent, how are they prevented from triggering transcription of all *cis*-linked genes in the vicinity? Part of the solution to these problems is arrived at by having the chromatin arranged in functional units that restrict patterns of gene expression. This may be achieved by having the chromatin form a structure with the nuclear matrix or other physical entity, or compartment within the nucleus. Alternatively, some regions are controlled by complex DNA elements called **locus control regions (LCRs).** An LCR—with associated bound proteins—controls the expression of a cluster of genes. The best-defined LCR regulates expression of the globin gene family over a large region of DNA. Another mechanism is provided by **insulators.** These DNA elements, also in association with one or more proteins, prevent an enhancer from acting on a promoter on the other side of an insulator in another transcription domain. Insulators thus serve as transcriptional **boundary elements.**

SEVERAL MOTIFS COMPOSE THE DNA BINDING DOMAINS OF REGULATORY TRANSCRIPTION FACTOR PROTEINS

The specificity involved in the control of transcription requires that regulatory proteins bind with high affinity and specificity to the correct region of DNA. Three unique motifs—the **helix-turn-helix,** the **zinc finger,** and the **leucine zipper**—account for many of these specific protein–DNA interactions. Examples of proteins containing these motifs are given in **Table 38–3.**

Comparison of the binding activities of the proteins that contain these motifs leads to several important generalizations.

1. Binding must be of high affinity to the specific site and of low affinity to other DNA.
2. Small regions of the protein make direct contact with DNA; the rest of the protein, in addition to providing

TABLE 38–3 Examples of Transcription Factors That Contain Various DNA Binding Motifs

Binding Motif	Organism	Regulatory Protein
Helix-turn-helix	*E coli*	lac repressor CAP
	Phage	λcI, cro, and 434 repressors
	Mammals	Homeobox proteins Pit-1, Oct1, Oct2
Zinc finger	*E coli a*	Gene 32 protein
	Yeast	Gal4
	Drosophila	Serendipity, Hunchback
	Xenopus	TFIIIA
	Mammals	Steroid receptor family, Sp1
Leucine zipper	Yeast	GCN4
	Mammals	C/EBP, fos, Jun, Fra-1, CRE binding protein, *c-myc*, *n-myc*, *l-myc*

the *trans*-activation domains, may be involved in the dimerization of monomers of the binding protein, may provide a contact surface for the formation of heterodimers, may provide one or more ligand-binding sites, or may provide surfaces for interaction with coactivators or corepressors.

3. The protein–DNA interactions are maintained by hydrogen bonds, ionic interactions and van der Waals forces.
4. The motifs found in these proteins are unique; their presence in a protein of unknown function suggests that the protein may bind to DNA.
5. Proteins with the helix-turn-helix or leucine zipper motifs form dimers, and their respective DNA-binding sites are symmetric palindromes. In proteins with the zinc finger motif, the binding site is repeated two to nine times. These features allow for cooperative interactions between binding sites and enhance the degree and affinity of binding.

The Helix-Turn-Helix Motif

The first motif described was the **helix-turn-helix.** Analysis of the 3D structure of the lambda Cro transcription regulator has revealed that each monomer consists of three antiparallel β sheets and three α helices (**Figure 38–15**). The dimer forms by association of the antiparallel $β_3$ sheets. The $α_3$ helices form the DNA recognition surface, and the rest of the molecule appears to be involved in stabilizing these structures. The average diameter of an α helix is 1.2 nm, which is the approximate width of the major groove in the B form of DNA.

The DNA recognition domain of each Cro monomer interacts with 5 bp and the dimer binding sites span 3.4 nm, allowing fit into successive half turns of the major groove on the same surface (Figure 38–15). X-ray analyses of the λ cI repressor, CAP (the cAMP receptor protein of *E. coli*), tryptophan repressor, and phage 434 repressor, all also display this dimeric helix-turn-helix structure that is present in eukaryotic DNA-binding proteins as well (see Table 38–3).

The Zinc Finger Motif

The **zinc finger** was the second DNA binding motif whose atomic structure was elucidated. It was known that the protein TFIIIA, a positive regulator of 5S RNA gene transcription, required zinc for activity. Structural and biophysical analyses revealed that each TFIIIA molecule contains nine zinc ions in a repeating coordination complex formed by closely spaced cysteine–cysteine residues followed 12–13 amino acids later by a histidine–histidine pair (**Figure 38–16**). In some instances—notably the steroid–thyroid nuclear hormone receptor family—the His–His doublet is replaced by a second Cys–Cys pair. The protein containing zinc fingers appears to lie on one face of the DNA helix, with successive fingers alternatively positioned in one turn in the major groove. As is the case with the recognition domain in the helix-turn-helix protein, each TFIIIA

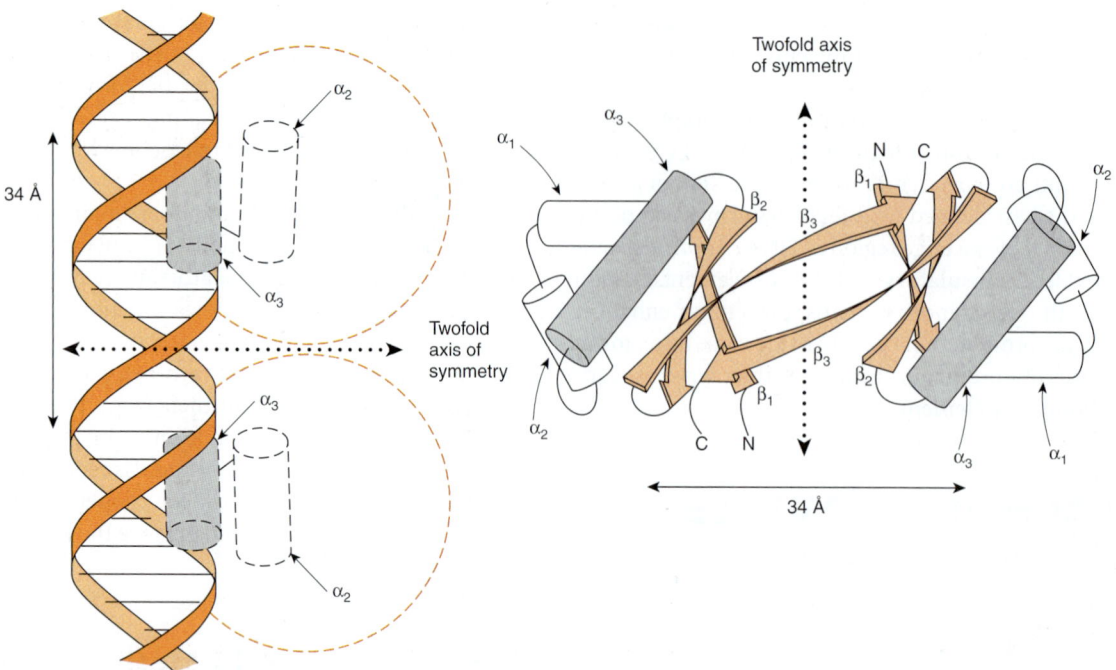

FIGURE 38–15 **A schematic representation of the 3D structure of Cro protein and its binding to DNA by its helix-turn-helix motif (left).** The Cro monomer consists of three antiparallel β sheets (β_1–β_3) and three α-helices (α_1–α_3). The helix-turn-helix motif is formed because the α_3 and α_2 helices are held at about 90 degrees to each other by a turn of four amino acids. The α_3 helix of Cro is the DNA recognition surface (**shaded**). Two monomers associate through the antiparallel β_3 sheets to form a dimer that has a twofold axis of symmetry (**right**). A Cro dimer binds to DNA through its α_3 helices, each of which contacts about 5 bp on the same surface of the major groove (see Figure 38–6). The distance between comparable points on the two DNA α-helices is 34 Å, which is the distance required for one complete turn of the double helix. (Courtesy of B Mathews.)

zinc finger contacts about 5 bp of DNA. The importance of this motif in the action of steroid hormones is underscored by an "experiment of nature." A single amino acid mutation in either of the two zinc fingers of the 1,25(OH)$_2$-D$_3$ receptor protein results in resistance to the action of this hormone and the clinical syndrome of rickets.

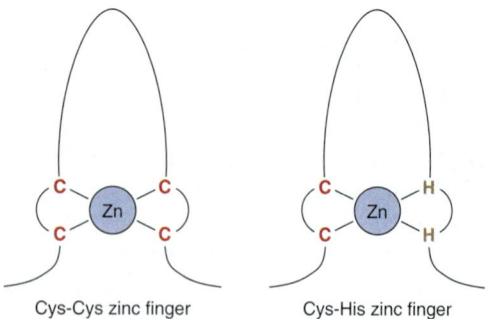

FIGURE 38–16 **Zinc fingers are a series of repeated domains (two to nine) in which each is centered on a tetrahedral coordination with zinc.** In the case of TFIIIA, the coordination is provided by a pair of cysteine residues (C) separated by 12–13 amino acids from a pair of histidine (H) residues. In other zinc finger proteins, the second pair also consists of C residues. Zinc fingers bind in the major groove, with adjacent fingers making contact with 5 bp along the same face of the helix.

The Leucine Zipper Motif

Careful analysis of a 30-amino-acid sequence in the carboxyl terminal region of the enhancer binding protein C/EBP revealed a novel structure, **the leucine zipper motif.** As illustrated in **Figure 38–17,** this region of the protein forms an α helix in which there is a periodic repeat of leucine residues at every seventh position. This occurs for eight helical turns and four leucine repeats. Similar structures have been found in a number of other proteins associated with the regulation of transcription in mammalian and yeast cells. This structure allows two identical or nonidentical monomers (eg, Jun-Jun or Fos-Jun) to "zip together" in a coiled coil and form a tight dimeric complex (Figure 38–17). This protein–protein interaction may serve to enhance the association of the separate DBDs with their target (Figure 38–17).

THE DNA BINDING & TRANSACTIVATION DOMAINS OF MOST REGULATORY PROTEINS ARE SEPARATE

DNA binding could result in a general conformational change that allows the bound protein to activate transcription, or these two functions could be served by separate and independent

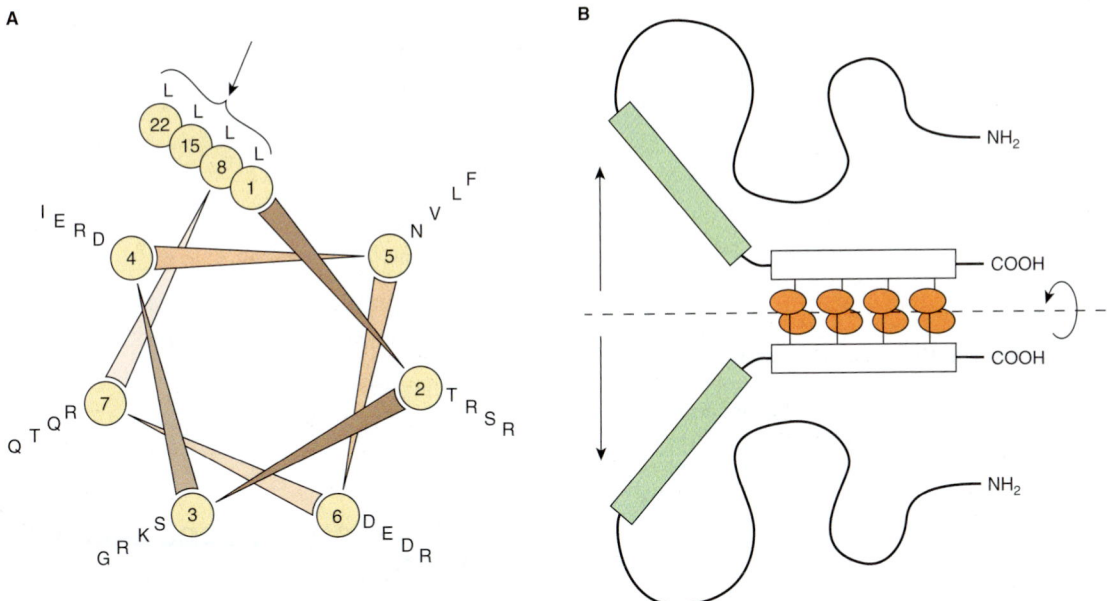

FIGURE 38–17 **The leucine zipper motif. (A)** It shows a helical wheel analysis of a carboxyl terminal portion of the DNA binding protein C/EBP. The amino acid sequence is displayed end-to-end down the axis of a schematic α-helix. The helical wheel consists of seven spokes that correspond to the seven amino acids that comprise every two turns of the α-helix. Note that leucine residues (L) occur at every seventh position (in this schematic C/EBP amino acid residues 1, 8, 15, 22; see arrow). Other proteins with "leucine zippers" have a similar helical wheel pattern. **(B)** It is a schematic model of the DNA-binding domain of C/EBP. Two identical C/EBP polypeptide chains are held in dimer formation by the leucine zipper domain of each polypeptide (denoted by the rectangles and attached ovals). This association is required to hold the DNA binding domains of each polypeptide (the shaded rectangles) in the proper conformation for DNA binding. (Courtesy of S McKnight.)

domains. Domain swap experiments suggest that the latter is typically the case.

The *GAL1* gene product is involved in galactose metabolism in yeast. Transcription of this gene is positively regulated by the GAL4 protein, which binds to an upstream activator sequence (UAS), or enhancer, through an amino terminal domain. The amino terminal 73-amino-acid DBD of GAL4 was removed and replaced with the DBD of LexA, an *E coli* DNA-binding protein. This domain swap resulted in a molecule that did not bind to the *GAL1* UAS and, of course, did not activate the *GAL1* gene (**Figure 38–18**). If, however, the *lexA* operator—the DNA sequence normally bound by the *lexA*DBD—was inserted into the promoter region of the *GAL* gene thereby replacing the normal *GAL1* enhancer, the hybrid protein bound to this promoter (at the *lexA* operator) and it activated transcription of *GAL1*. This experiment, which has been repeated a number of times, affords solid evidence that the carboxyl terminal region of GAL4 causes transcriptional activation. These data also demonstrate that the DBD and transactivation domains (ADs) are independent and noninteractive. The hierarchy involved in assembling gene transcription-activating complexes includes proteins that bind DNA and transactivate; others that form protein–protein complexes which bridge DNA-binding proteins to transactivating proteins; and others that form protein–protein complexes with components of coregulators or the basal transcription apparatus. A given protein may thus have several modular surfaces or domains that serve different functions (see **Figure 38–19**). As

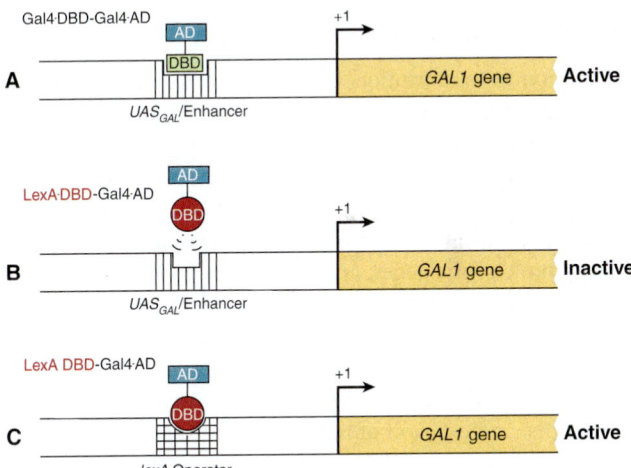

FIGURE 38–18 **Domain-swap experiments demonstrate the independent nature of DNA binding and transcription activation domains.** The *GAL1* gene promoter contains an upstream activating sequence (UAS) or enhancer that is bound by the regulatory transcription factor GAL4 **(A)**. GAL4, like the lambda cI protein is modular, and contains an N-terminal DBD and an C-terminal activation domain, or AD. When the GAL4 transcription factor binds the *GAL1* enhancer/UAS, activation of *GAL1* gene transcription ensues (Active). A chimeric protein, in which the amino terminal DNA-binding domain (DBD) of GAL4 is removed and replaced with the DBD of the *E. coli* protein LexA (LexA DBD-GAL4 AD), fails to stimulate *GAL1* transcription because the LexA DBD cannot bind to the GAL1 enhancer/UAS **(B)**. By contrast, the LexA DBD–GAL4 AD fusion protein does increase *GAL1* transcription when the *lexA* operator (the natural target for the LexA DBD) is inserted into the *GAL1* promoter region **(C)**, replacing the normal *GAL1* UAS.

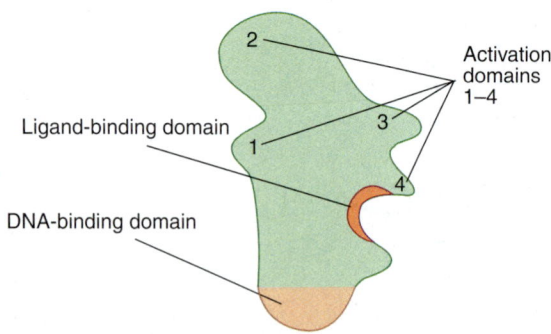

FIGURE 38–19 **Proteins that regulate transcription have several domains.** This hypothetical transcription factor has a DBD that is distinct from a ligand-binding domain (LBD) and several activation domains (ADs) (1–4). Other proteins may lack the DBD or LBD and all may have variable numbers of domains that contact other proteins, including coregulators and those of the basal transcription complex (see also Chapters 41 & 42).

described in Chapter 36, the primary purpose of these complex assemblies is to facilitate the assembly and/or activity of the basal transcription apparatus on the *cis*-linked promoter.

GENE REGULATION IN PROKARYOTES & EUKARYOTES DIFFERS IN IMPORTANT RESPECTS

In addition to transcription, eukaryotic cells employ a variety of mechanisms to regulate gene expression (**Table 38–4**). The nuclear membrane of eukaryotic cells physically segregates gene transcription from translation, since ribosomes exist only in the cytoplasm. Many more steps, especially in RNA processing, are involved in the expression of eukaryotic genes than of prokaryotic genes, and these steps provide additional sites for regulatory influences that cannot exist in prokaryotes. These RNA processing steps in eukaryotes, described in detail in Chapter 36, include capping of the 5′ ends of the primary transcripts, addition of a polyadenylate tail to the 3′ ends of transcripts, and excision of intron regions to generate spliced exons in the mature mRNA molecule. To date, analyses of eukaryotic gene expression provide evidence that regulation occurs at the level of **transcription, nuclear RNA processing, mRNA stability, and translation.** In addition, gene amplification and rearrangement influence gene expression.

Owing to the advent of recombinant DNA technology, much progress has been made in recent years in the understanding of eukaryotic gene expression. However, because most eukaryotic organisms contain so much more genetic information than do prokaryotes and because manipulation of their genes is so much more difficult, molecular aspects of eukaryotic gene regulation are less well understood than the examples discussed earlier in this chapter. This section briefly describes a few different types of eukaryotic gene regulation.

TABLE 38–4 **Gene Expression is Regulated by Transcription and in Numerous Other Ways at the RNA Level in Eukaryotic Cells**

• Gene amplification
• Gene rearrangement
• RNA processing
• Alternate mRNA splicing
• Transport of mRNA from nucleus to cytoplasm
• Regulation of mRNA stability
• Compartmentalization
• miRNA/ncRNA silencing and activation

miRNAs Modulate Gene Expression by Altering mRNA Function

As noted in Chapter 35 the recently discovered class of eukaryotic small RNAs, termed miRNAs, contribute importantly to the control of gene expression. These ~22 nucleotide RNAs regulate the translatability of specific mRNAs by either inhibiting translation or inducing mRNA degradation, though in a few cases miRNAs have been shown to stimulate mRNA function (translation). At least a portion of the miRNA-driven modulation of mRNA activity is thought to occur in the **P body** (Figure 37–11). miRNA action can result in dramatic changes in protein production and hence gene expression. miRNAs have been implicated in numerous human diseases such as heart disease, cancer, muscle wasting, viral infection and diabetes.

miRNAs, like the DNA-binding transcription factors described in detail above, are transactive, and once synthesized and appropriately processed, interact with specific proteins and bind target mRNAs, typically in 3′ untranslated mRNA regions (Figure 36–17). Binding of miRNAs to mRNA targets is directed by normal base-pairing rules. In general, if **miRNA–mRNA** base pairing has one or more mismatches, translation of the cognate "target" mRNA is inhibited, whereas if **miRNA–mRNA base pairing** is perfect over all 22 nucleotides, the corresponding mRNA is degraded.

Given the tremendous and ever growing import of miRNAs, many scientists and biotechnology companies are actively studying miRNA biogenesis, transport, and function in hopes of curing human disease. Time will tell the magnitude and universality of miRNA-mediated gene regulation. It is likely that in the near future scientists will unveil the medical significance of these intriguing small RNAs.

Eukaryotic Genes Can Be Amplified or Rearranged During Development or in Response to Drugs

During early development of metazoans, there is an abrupt increase in the need for specific molecules such as ribosomal RNA and messenger RNA molecules for proteins that make

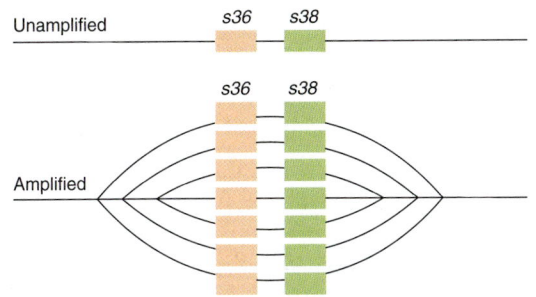

FIGURE 38–20 **Schematic representation of the amplification of chorion protein genes *s36* and *s38*.** (Reproduced, with permission, from Chisholm R: Gene amplification during development. Trends Biochem Sci 1982;7:161. Copyright © 1982. Reprinted, with permission, from Elsevier.)

up such organs as the eggshell. One way to increase the rate at which such molecules can be formed is to increase the number of genes available for transcription of these specific molecules. Among the repetitive DNA sequences within the genome are hundreds of copies of ribosomal RNA genes. These genes preexist repetitively in the DNA of the gametes and thus are transmitted in high copy numbers from generation to generation. In some specific organisms such as the fruit fly (drosophila), there occurs during oogenesis an amplification of a few preexisting genes such as those for the chorion (eggshell) proteins. Subsequently, these amplified genes, presumably generated by a process of repeated initiations during DNA synthesis, provide multiple sites for gene transcription (**Figures 36–4** and **38–20**).

As noted in Chapter 36, the coding sequences responsible for the generation of specific protein molecules are frequently not contiguous in the mammalian genome. In the case of antibody encoding genes, this is particularly true. As described in detail in Chapter 50, immunoglobulins are composed of two polypeptides, the so-called heavy (about 50 kDa) and light (about 25 kDa) chains. The mRNAs encoding these two protein subunits are encoded by gene sequences that are subjected to extensive DNA sequence-coding changes. These DNA coding changes are integral to generating the requisite recognition diversity central to appropriate immune function.

IgG heavy and light chain mRNAs are encoded by several different segments that are tandemly repeated in the germline. Thus, for example, the IgG light chain is composed of variable (V_L), joining (J_L), and constant (C_L) domains or segments. For particular subsets of IgG light chains, there are roughly 300 tandemly repeated V_L gene coding segments, 5 tandemly arranged J_L coding sequences, and roughly 10 C_L gene coding segments. All of these multiple, distinct coding regions are located in the same region of the same chromosome, and each type of coding segment (V_L, J_L, and C_L) is tandemly repeated in head-to-tail fashion within the segment repeat region. By having multiple V_L, J_L, and C_L segments to choose from, an immune cell has a greater repertoire of sequences to work with to develop both immunologic flexibility and specificity. However, a given functional IgG light chain transcription unit—like all other "normal" mammalian transcription units—contains

only the coding sequences for a single protein. Thus, before a particular IgG light chain can be expressed, *single* V_L, J_L, and C_L coding sequences must be recombined to generate a *single*, contiguous transcription unit excluding the multiple nonutilized segments (ie, the other approximately 300 unused V_L segments, the other 4 unused J_L segments, and the other 9 unused C_L segments). This deletion of unused genetic information is accomplished by selective DNA recombination that removes the unwanted coding DNA while retaining the required coding sequences: one V_L, one J_L, and one C_L sequence. (V_L sequences are subjected to additional point mutagenesis to generate even more variability—hence the name.) The newly recombined sequences thus form a single transcription unit that is competent for RNA polymerase II-mediated transcription into a single monocistronic mRNA. Although the IgG genes represent one of the best-studied instances of directed DNA rearrangement modulating gene expression, other cases of gene regulatory DNA rearrangement have been described in the literature. Indeed, as detailed below, drug-induced gene amplification is an important complication of cancer chemotherapy.

In recent years, it has been possible to promote the amplification of specific genetic regions in cultured mammalian cells. In some cases, a several 1000-fold increase in the copy number of specific genes can be achieved over a period of time involving increasing doses of selective drugs. In fact, it has been demonstrated in patients receiving methotrexate for cancer that malignant cells can develop **drug resistance** by increasing the number of genes for dihydrofolate reductase, the target of methotrexate. Gene amplification and deletion events involving 10 to 1,000,000 of bp of DNA such as these occur spontaneously in vivo—ie, in the absence of exogenously supplied selective agents—and these unscheduled extra rounds of replication can become stabilized in the genome under appropriate selective pressures.

Alternative RNA Processing Is Another Control Mechanism

In addition to affecting the efficiency of promoter utilization, eukaryotic cells employ alternative RNA processing to control gene expression. This can result when alternative promoters, intron–exon splice sites, or polyadenylation sites are used. Occasionally, heterogeneity within a cell results, but more commonly the same primary transcript is processed differently in different tissues. A few examples of each of these types of regulation are presented below.

The use of **alternative transcription start sites** results in a different 5′ exon on mRNAs encoding mouse amylase and myosin light chain, rat glucokinase, and drosophila alcohol dehydrogenase and actin. **Alternative polyadenylation sites** in the μ immunoglobulin heavy chain primary transcript result in mRNAs that are either 2700 bases long (μ_m) or 2400 bases long (μ_s). This results in a different carboxyl terminal region of the encoded proteins such that the μ_m protein remains attached to the membrane of the B lymphocyte and the μ_s immunoglobulin is secreted. **Alternative splicing and processing** results in

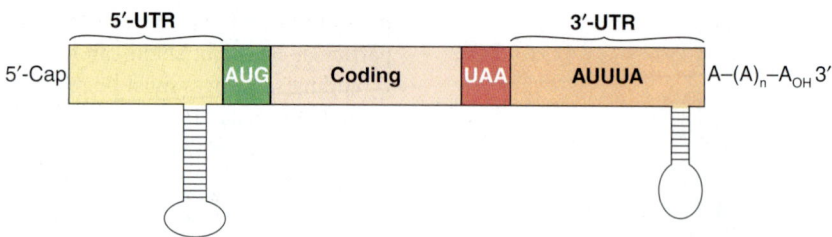

FIGURE 38–21   **Structure of a typical eukaryotic mRNA showing elements that are involved in regulating mRNA stability.** The typical eukaryotic mRNA has a 5′ noncoding sequence, or untranslated region (5′ UTR), a coding region, and a 3′ untranslated region (3′ UTR). Essentially all mRNAs are capped at the 5′ end, and most have a polyadenylate sequence, 100–200 nucleotides long at their 3′ end. The 5′ cap and 3′ poly(A) tail protect the mRNA against exonuclease attack and are bound by specific proteins that interact to facilitate translation (see Figure 37–7). Stem-loop structures in the 5′ and 3′ NCS, and the AU-rich region in the 3′ NCS are thought to represent the binding sites for specific proteins that modulate mRNA stability. miRNAs typically target sequences in the 3′ UTR.

the formation of seven unique α-tropomyosin mRNAs in seven different tissues. It is not clear how these processing-splicing decisions are made or whether these steps can be regulated.

Regulation of Messenger RNA Stability Provides Another Control Mechanism

Although most mRNAs in mammalian cells are very stable (half-lives measured in hours), some turn over very rapidly (half-lives of 10–30 minutes). In certain instances, mRNA stability is subject to regulation. This has important implications since there is usually a direct relationship between mRNA amount and the translation of that mRNA into its cognate protein. Changes in the stability of a specific mRNA can therefore have major effects on biologic processes.

Messenger RNAs exist in the cytoplasm as ribonucleoprotein particles (RNPs). Some of these proteins protect the mRNA from digestion by nucleases, while others may under certain conditions promote nuclease attack. It is thought that mRNAs are stabilized or destabilized by the interaction of proteins with these various structures or sequences. Certain effectors, such as hormones, may regulate mRNA stability by increasing or decreasing the amount of these proteins.

It appears that **the ends of mRNA molecules are involved in mRNA stability (Figure 38–21)**. The 5′ cap structure in eukaryotic mRNA prevents attack by 5′ exonucleases, and the poly(A) tail prohibits the action of 3′ exonucleases. In mRNA molecules with those structures, it is presumed that a single endonucleolytic cut allows exonucleases to attack and digest the entire molecule. Other structures (sequences) in the 5′ untranslated region (5′ UTR), the coding region, and the 3′ UTR are thought to promote or prevent this initial endonucleolytic action (Figure 38–21). A few illustrative examples will be cited.

Deletion of the 5′ UTR results in a threefold to fivefold prolongation of the half-life of c-*myc* mRNA. Shortening the coding region of histone mRNA results in a prolonged half-life. A form of autoregulation of mRNA stability indirectly involves the coding region. Free tubulin binds to the first four amino acids of a nascent chain of tubulin as it emerges from the ribosome. This appears to activate an RNase associated with the ribosome which then digests the tubulin mRNA.

Structures at the 3′ end, including the poly(A) tail, enhance or diminish the stability of specific mRNAs. The absence of a poly(A) tail is associated with rapid degradation of mRNA, and the removal of poly(A) from some RNAs results in their destabilization. Histone mRNAs lack a poly(A) tail but have a sequence near the 3′ terminal that can form a stem-loop structure, and this appears to provide resistance to exonucleolytic attack. Histone H4 mRNA, for example, is degraded in the 3′–5′ direction but only after a single endonucleolytic cut occurs about nine nucleotides from the 3′ end in the region of the putative stem-loop structure. Stem-loop structures in the 3′ noncoding sequence are also critical for the regulation, by iron, of the mRNA encoding the transferrin receptor. Stem-loop structures are also associated with mRNA stability in bacteria, suggesting that this mechanism may be commonly employed.

Other sequences in the 3′ ends of certain eukaryotic mRNAs appear to be involved in the destabilization of these molecules. Some of this is mediated through the action of specific miRNAs as discussed above. In addition, of particular interest are AU-rich regions, many of which contain the sequence AUUUA. This sequence appears in mRNAs that have a very short half-life, including some encoding oncogene proteins and cytokines. The importance of this region is underscored by an experiment in which a sequence corresponding to the 3′ UTR of the short-half-life colony-stimulating factor (CSF) mRNA, which contains the AUUUA motif, was added to the 3′ end of the β-globin mRNA. Instead of becoming very stable, this hybrid β-globin mRNA now had the short-half-life characteristic of CSF mRNA. Much of this mRNA metabolism likely occurs in cytoplasmic P bodies.

From the few examples cited, it is clear that a number of mechanisms are used to regulate mRNA stability and hence function—just as several mechanisms are used to regulate the synthesis of mRNA. Coordinate regulation of these two processes confers on the cell remarkable adaptability.

SUMMARY

- The genetic constitutions of metazoan somatic cells are nearly all identical.

- Phenotype (tissue or cell specificity) is dictated by differences in gene expression of this complement of genes.

- Alterations in gene expression allow a cell to adapt to environmental changes, developmental cues, and physiological signals.

- Gene expression can be controlled at multiple levels by changes in transcription, RNA processing, localization, and stability or utilization. Gene amplification and rearrangements also influence gene expression.

- Transcription controls operate at the level of protein–DNA and protein–protein interactions. These interactions display protein domain modularity and high specificity.

- Several different classes of DNA-binding domains have been identified in transcription factors.

- Chromatin and DNA modifications contribute importantly in eukaryotic transcription control by modulating DNA accessibility and specifying recruitment of specific coactivators and corepressors to target genes.

- Several epigenetic mechanisms for gene control have been described and the molecular mechanisms through which these processes operate are beginning to be elucidated at the molecular level.

- miRNA and siRNAs modulate mRNA translation and stability; these mechanisms complement transcription controls to regulate gene expression.

REFERENCES

Bhaumik SR, Smith E, Shilatifard A, et al: Covalent modifications of histones during development and disease pathogenesis. Nat Struct Mol Biol 2007;14;1008.

Bird AP, Wolffe AP: Methylation-induced repression—belts, braces and chromatin. Cell 1999;99:451.

Bonasio R, Tu S, Reinberg D: Molecular signals of epigenetic states. Science 2010;330:612–616.

Busby S, Ebright RH: Promoter structure, promoter recognition, and transcription activation in prokaryotes. Cell 1994;79:743.

Gerstein MB, Lu ZJ, Van Nostrand EL, et al: Integrative analysis of the *Caenorhabditis elegans* genome by the modENCODE project. Science 2010;330:1775–1787.

Jacob F, Monod J: Genetic regulatory mechanisms in protein synthesis. J Mol Biol 1961;3:318.

Klug A: The discovery of zinc fingers and their applications in gene regulation and genome manipulation. Annu Rev Biochem 2010;79:213–231.

Lemon B, Tjian R: Orchestrated response: a symphony of transcription factors for gene control. Genes Dev 2000; 14:2551.

Letchman DS: Transcription factor mutations and disease. N Engl J Med 1996;334:28.

Margueron R, Reinberg D: The polycomb complex PRC2 and its mark in life. Nature 2011;469:343–349.

Näär AM, Lemon BD, Tjian R: Transcriptional coactivator complexes. Annu Rev Biochem 2001;70:475.

Nabel CS, Kohli RM: Demystifying DNA Demethylation Science 2011; 333:1229–1230.

Oltz EM: Regulation of antigen receptor gene assembly in lymphocytes. Immunol Res 2001;23:121.

Ørom UA, Derrien T, Beringer M, et al: Long noncoding RNAs with enhancer-like function in human cells. Cell 2010;143: 46–58.

Ptashne M: *A Genetic Switch,* 2nd ed. Cell Press and Blackwell Scientific Publications, 1992.

Roeder RG: Transcriptional regulation and the role of diverse coactivators in animal cells. FEBS Lett 2005;579:909.

Ruthenburg AJ, Li H, Patel DJ, et al: Multivalent engagement of chromatin modifications by linked binding molecules. Nature Rev Mol Cell Bio 2007;8:983.

Segal E, Fondufe-Mittendorf Y, Chen L, et al: A genomic code for nucleosome positioning. Nature 2006;442:772.

Small EM, Olson EN: Pervasive roles of microRNAs in cardiovascular biology. Nature 2011;469:336–342.

The modENCODE Consortium, Roy S, Ernst J, Kharchenko PV, et al: Identification of functional elements and regulatory circuits by Drosophila modENCODE. Science 2010;330,1787–1797.

Valencia-Sanchez MA, Liu J, Hannon GJ, et al: Control of translation and mRNA degradation by miRNAs and siRNAs. Genes Dev 2006;20:515.

Yang XJ, Seto E: HATs and HDACs: from structure, function and regulation to novel strategies for therapy and prevention. Oncogene 2007;26:5310.

Weake VM, Workman JL: Inducible gene expression: diverse regulatory mechanisms. Nat Rev Genet 2010;11:426–437.

Wu R, Bahl CP, Narang SA: Lactose operator-repressor interaction. Curr Top Cell Regul 1978;13:137.

Zhang Z, Pugh BF: High-resolution genome-wide mapping of the primary structure of chromatin. Cell 2011;144:175-186.

Molecular Genetics, Recombinant DNA, & Genomic Technology

P. Anthony Weil, PhD

O B J E C T I V E S

After studying this chapter, you should be able to:

- Explain the basic procedures and methods involved in recombinant DNA technology and genetic engineering.
- Appreciate the rationale behind the methods used to synthesize, analyze, and sequence DNA and RNA.
- Explain how to identify and quantify individual proteins, both soluble and insoluble (ie, membrane bound or compartmentalized intracellularly) proteins, as well as proteins bound to specific sequences of genomic DNA and RNA.

BIOMEDICAL IMPORTANCE*

The development of recombinant DNA, high-density DNA microarrays, high-throughput screening, low-cost genome-scale analyses, DNA sequencing and other molecular genetic methodologies has revolutionized biology and is having an increasing impact on clinical medicine. Though much has been learned about human genetic disease from pedigree analysis and study of affected proteins, in many cases where the specific genetic defect is unknown, these approaches cannot be used. The new technologies circumvent these limitations by going directly to the DNA molecule for information. Manipulation of a DNA sequence and the construction of chimeric molecules—so-called genetic engineering—provides a means of studying how a specific segment of DNA works. Novel biochemical and molecular genetic tools and direct DNA sequencing allow investigators to query and manipulate genomic sequences as well as to examine the entire complement of cellular RNA, protein profiles and protein PTM status at the molecular level.

Understanding this technology is important for several reasons: (1) it offers a rational approach to understanding the molecular basis of a number of diseases. For example, familial hypercholesterolemia, sickle-cell disease, the thalassemias, cystic fibrosis, muscular dystrophy as well as more complex multifactorial diseases like vascular and heart disease, cancer, and diabetes. (2) Human proteins can be produced in abundance for therapy (eg, insulin, growth hormone, and tissue plasminogen activator). (3) Proteins for vaccines (eg, hepatitis B) and for diagnostic testing (eg, Ebola and AIDS

tests) can be obtained. (4) This technology is used both to diagnose existing diseases as well as to predict the risk of developing a given disease and individual response to pharmacological therapeutics. (5) Special techniques have led to remarkable advances in forensic medicine. (6) Gene therapy for potentially curing diseases caused by a single-gene deficiency such as sickle-cell disease, the thalassemias, adenosine deaminase deficiency, and others may be devised.

RECOMBINANT DNA TECHNOLOGY INVOLVES ISOLATION & MANIPULATION OF DNA TO MAKE CHIMERIC MOLECULES

Isolation and manipulation of DNA, including end-to-end joining of sequences from very different sources to make chimeric molecules (eg, molecules containing both human and bacterial DNA sequences in a sequence-independent fashion), is the essence of recombinant DNA research. This involves several unique techniques and reagents.

Restriction Enzymes Cleave DNA Chains at Specific Locations

Certain endonucleases—enzymes that cut DNA at specific DNA sequences within the molecule (as opposed to exonucleases, which digest from the ends of DNA molecules)—are a key tool in recombinant DNA research. These enzymes were called **restriction enzymes** because their presence in a given

---

*See glossary of terms at the end of this chapter.

bacterium restricted the growth of certain bacterial viruses called bacteriophages. Restriction enzymes cut DNA of any source into unique, short pieces in a sequence-specific manner—in contrast to most other enzymatic, chemical, or physical methods, which break DNA randomly. These defensive enzymes (hundreds have been discovered) protect the host bacterial DNA from the DNA genome of foreign organisms (primarily infective phages) by specifically inactivating the invading phage DNA by digestion. The viral RNA-inducible interferon system (Chapter 38; Figure 38–11) provides the same sort of molecular defense against RNA viruses in mammalian cells. However, restriction endonucleases are present only in cells that also have a companion enzyme that site-specifically methylates the host DNA, rendering it an unsuitable substrate for digestion by that particular restriction enzyme. Thus, **site-specific DNA methylases** and restriction enzymes that target the exact same sites always exist in pairs in a bacterium.

Restriction enzymes are named after the bacterium from which they are isolated. For example, *EcoRI* is from *Escherichia coli*, and *BamHI* is from *Bacillus amyloliquefaciens* (**Table 39–1**). The first three letters in the restriction enzyme name consist of the first letter of the genus (*E*) and the first two letters of the species (*co*). These may be followed by a strain designation (*R*) and a roman numeral (*I*) to indicate the order of discovery (eg, *EcoRI* and *EcoRII*). Each enzyme recognizes and cleaves a specific double-stranded DNA sequence that is typically 4–7 bp long. These DNA cuts result in **blunt ends** (eg, *HpaI*) or overlapping (**sticky or cohesive) ends** (eg, *BamHI*) (**Figure 39–1**), depending on the mechanism used by the enzyme. Sticky ends are particularly useful in constructing hybrid or chimeric DNA molecules (see below). If the four nucleotides are distributed randomly in a given DNA molecule, one can calculate how frequently a given enzyme will cut a length of DNA. For each position in the DNA molecule, there are four possibilities (A, C, G, and T); therefore, a restriction enzyme that recognizes a 4-bp sequence cuts, on average, once every 256 bp (4^4), whereas another enzyme that recognizes a 6-bp sequence cuts once every 4096 bp (4^6). A given piece of DNA has a characteristic linear array of sites for the various enzymes dictated by the linear sequence of its bases; hence, a **restriction map** can be constructed. When DNA is digested with a particular enzyme, the ends of all the fragments have the same DNA sequence. The fragments produced can be isolated by electrophoresis on agarose or polyacrylamide gels (see the discussion of blot transfer, below); this is an essential step in DNA cloning as well as various DNA analyses, and a major use of these enzymes.

A number of other enzymes that act on DNA and RNA are an important part of recombinant DNA technology. Many of these are referred to in this and subsequent chapters (**Table 39–2**).

Restriction Enzymes & DNA Ligase Are Used to Prepare Chimeric DNA Molecules

Sticky, or comlementary cohesive-end ligation of DNA fragments is technically easy, but some special techniques are often

TABLE 39–1 Selected Restriction Endonucleases and Their Sequence Specificities

Endonuclease	Sequence Recognized Cleavage Sites Shown	Bacterial Source
BamHI	↓ GGATCC CCTACC ↑	Bacillus amyloliquefaciens H
BglII	↓ AGATCT TCTAGA ↑	Bacillus glolbigii
EcoRI	↓ GAATTC CTTAAC ↑	Escherichia coli RY13
EcoRII	↓ CCTGG GGACC ↑	Escherichia coli R245
HindIII	↓ AAGCTT TTCGAA ↑	Haemophilus influenzae R_d
HhaI	↓ GCGC CGCG ↑	Haemophilus haemolyticus
HpaI	↓ GTTAAC CAATTC ↑	Haemophilus parainfluenzae
MstII	↓ CCTnAGG GGAnTCC ↑	Microcoleus strain
PstI	↓ CTGCAG GACGTC ↑	Providencia stuartii 164
TaqI	↓ TCGA AGCT ↓	Thermus aquaticus YTI

Abbreviations: A, adenine; C, cytosine; G, guanine, T, thymine. Arrows show the site of cleavage; depending on the site, the ends of the resulting cleaved double-stranded DNA are termed sticky ends (*BamHI*) or blunt ends (*HpaI*). The length of the recognition sequence can be 4 bp (*TaqI*), 5 bp (*EcoRII*), 6 bp (EcoRI), or 7 bp (*MstII*) or longer. By convention, these are written in the 5′ to 3′ direction for the upper strand of each recognition sequence, and the lower strand is shown with the opposite (ie, 3′ to 5′) polarity. Note that most recognition sequences are palindromes (ie, the sequence reads the same in opposite directions on the two strands). A residue designated n means that any nucleotide is permitted.

required to overcome problems inherent in this approach. Sticky ends of a vector may reconnect with themselves, with no net gain of DNA. Sticky ends of fragments also anneal so that heterogeneous tandem inserts form. Also, sticky-end sites

FIGURE 39–1   **Results of restriction endonuclease digestion.** Digestion with a restriction endonuclease can result in the formation of DNA fragments with sticky, or cohesive, ends **(A)** or blunt ends **(B)**; phosphodiester backbone, black lines; interstrand hydrogen bonds between purine and pyrimidine bases, blue. This is an important consideration in devising cloning strategies.

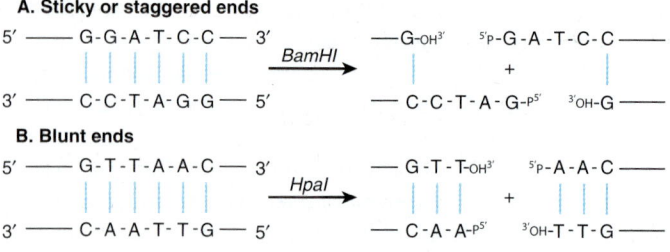

may not be available or in a convenient position. To circumvent these problems, an enzyme that generates blunt ends can be used. Blunt ends can be ligated directly; however, ligation is not directional. Two alternatives thus exist: new ends are added using the enzyme terminal transferase or synthetic sticky ends are added. If poly d(G) is added to the 3′ ends of the vector and poly d(C) is added to the 3′ ends of the foreign DNA using terminal transferase, the two molecules can only anneal to each other, thus circumventing the problems listed above. This procedure is called homopolymer tailing. Alternatively, synthetic blunt-ended duplex oligonucleotide linkers containing the recognition sequence for a convenient restriction enzyme sequence are ligated to the blunt-ended DNA. Direct blunt-end ligation is accomplished using the bacteriophage T4 enzyme DNA ligase. This technique, though less efficient than sticky-end ligation, has the advantage of joining together any pairs of ends. If blunt ends or homopolymer tailing methods are used there is no easy way to retrieve the insert. As an adjunct to

the use of restriction endonucleases scientists have recently begun utilizing specific prokaryotic or eukaryotic recombinases (such as bacterial lox P sites, which are recognized by the CRE recombinase, or yeast FRT sites recognized by the Flp recombinase) to catalyze specific incorporation of two DNA fragments that carry the appropriate recognition sequences. These enzymes catalyze homologous recombination (Figure 35–9) between the relevant recognition sites.

Cloning Amplifies DNA

A **clone** is a large population of identical molecules, bacteria, or cells that arise from a common ancestor. Molecular cloning allows for the production of a large number of identical DNA molecules, which can then be characterized or used for other purposes. This technique is based on the fact that chimeric or hybrid DNA molecules can be constructed in **cloning vectors**—typically bacterial plasmids, phages, or cosmids—which then

TABLE 39–2   **Some of the Enzymes Used in Recombinant DNA Research**

Enzyme	Reaction	Primary Use
Alkaline phosphatase	Dephosphorylates 5′ ends of RNA and DNA	Removal of 5′-PO$_4$ groups prior to kinase labeling; also used to prevent self-ligation
BAL 31 nuclease	Degrades both the 3′ and 5′ ends of DNA	Progressive shortening of DNA molecules
DNA ligase	Catalyzes bonds between DNA molecules	Joining of DNA molecules
DNA polymerase I	Synthesizes double-stranded DNA from single-stranded DNA	Synthesis of double-stranded cDNA; nick translation; generation of blunt ends from sticky ends
Thermostable DNA polymerases	Synthesize DNA at elevated temperatures (60–80° C)	Polymerase chain reaction (DNA synthesis)
DNase I	Under appropriate conditions, produces single-stranded nicks in DNA	Nick translation; mapping of hypersensitive sites; mapping protein-DNA interactions
Exonuclease III	Removes nucleotides from 3′ ends of DNA	DNA sequencing; mapping of DNA-protein interactions
λ Exonuclease	Removes nucleotides from 5′ ends of DNA	DNA sequencing
Polynucleotide kinase	Transfers terminal phosphate (γ position) from ATP to 5′-OH groups of DNA or RNA	^{32}P end-labeling of DNA or RNA
Reverse transcriptase	Synthesizes DNA from RNA template	Synthesis of cDNA from mRNA; RNA (5′ end) mapping studies
S1 nuclease	Degrades single-stranded DNA	Removal of "hairpin" in synthesis of cDNA; RNA mapping studies (both 5′ and 3′ ends)
Terminal transferase	Adds nucleotides to the 3′ ends of DNA	Homopolymer tailing

(Adapted and reproduced, with permission, from Emery AEH: Page 41 in: *An Introduction to Recombinant DNA.* Wiley, 1984. Copyright © 1984 John Wiley & Sons Limited. Reproduced with permission.)

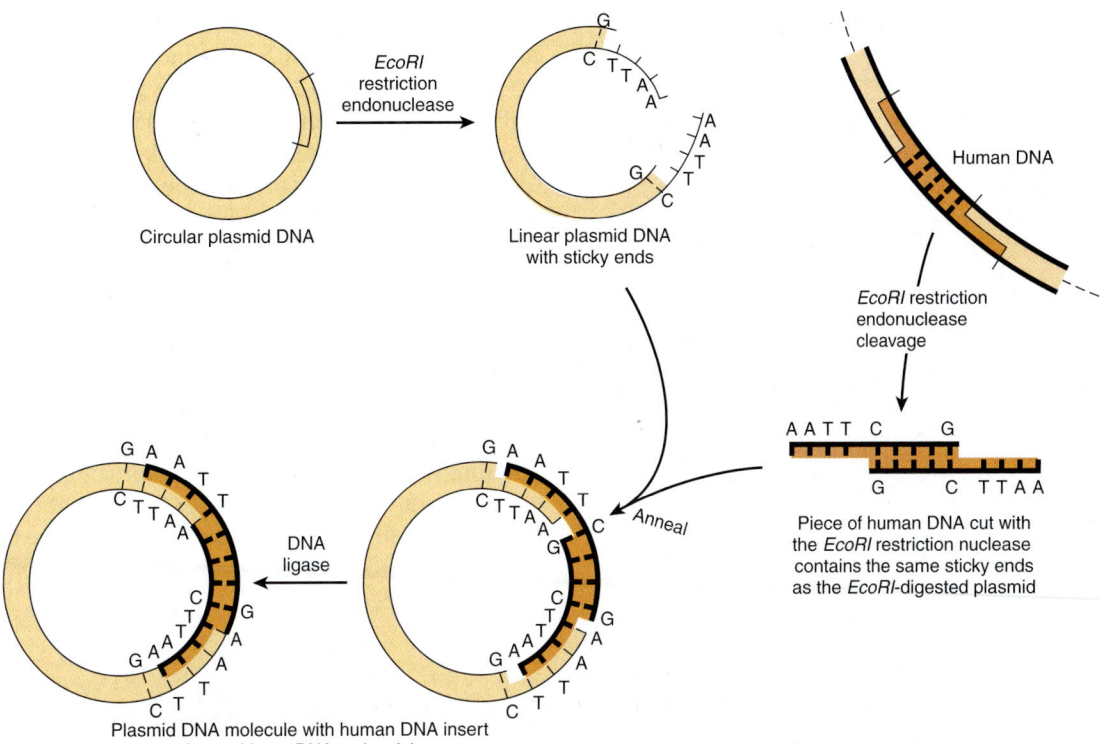

FIGURE 39–2 **Use of restriction nucleases to make new recombinant or chimeric DNA molecules.** When inserted back into a bacterial cell (by the process called DNA-mediated transformation), typically only a single plasmid is taken up by a single cell, and the plasmid DNA replicates not only itself but also the physically linked new DNA insert. Since recombining the sticky ends, as indicated, typically regenerates the same DNA sequence recognized by the original restriction enzyme, the cloned DNA insert can be cleanly cut back out of the recombinant plasmid circle with this endonuclease. If a mixture of all of the DNA pieces created by treatment of total human DNA with a single restriction nuclease is used as the source of human DNA, a million or so different types of recombinant DNA molecules can be obtained, each pure in its own bacterial clone. (Modified and reproduced, with permission, from Cohen SN: The manipulation of genes. Sci Am [July] 1975;233:25. Copyright © The Estate of Bunji Tagawa.)

continue to replicate in a host cell under their own control systems. In this way, the chimeric DNA is amplified. The general procedure is illustrated in **Figure 39–2**.

Bacterial **plasmids** are small, circular, duplex DNA molecules whose natural function is to confer antibiotic resistance to the host cell. Plasmids have several properties that make them extremely useful as cloning vectors. They exist as single or multiple copies within the bacterium and replicate independently from the bacterial DNA while using primarily the host replication machinery. The complete DNA sequence of many plasmids is known; hence, the precise location of restriction enzyme cleavage sites for inserting the foreign DNA is available. Plasmids are smaller than the host chromosome and are therefore easily separated from the latter, and the desired plasmid-inserted DNA can be readily removed by cutting the plasmid with the enzyme specific for the restriction site into which the original piece of DNA was inserted.

Phages (bacterial viruses) often have linear DNA molecules into which foreign DNA can be inserted at several restriction enzyme sites. The chimeric DNA is collected after the phage proceeds through its lytic cycle and produces mature,

infective phage particles. A major advantage of phage vectors is that while plasmids accept DNA pieces about 6–10 kb long, phages can accept DNA fragments 10–20 kb long, a limitation imposed by the amount of DNA that can be packed into the phage head during virus propagation.

Larger fragments of DNA can be cloned in **cosmids,** which combine the best features of plasmids and phages. Cosmids are plasmids that contain the DNA sequences, so-called **cos sites,** required for packaging lambda DNA into the phage particle. These vectors grow in the plasmid form in bacteria, but since much of the unnecessary lambda DNA has been removed, more chimeric DNA can be packaged into the particle head. It is not unusual for cosmids to carry inserts of chimeric DNA that are 35–50 kb long. Even larger pieces of DNA can be incorporated into bacterial artificial chromosome **(BAC),** yeast artificial chromosome **(YAC),** or *E coli* bacteriophage P1-based **(PAC)** vectors. These vectors will accept and propagate DNA inserts of several hundred kilobases or more and have largely replaced the plasmid, phage, and cosmid vectors for some cloning and eukaryotic gene mapping applications. A comparison of these vectors is shown in **Table 39–3**.

TABLE 39–3   **Cloning Capacities of Common Cloning Vectors**

Vector	DNA Insert Size (kb)
Plasmid pUC19	0.01–10
Lambda charon 4A	10–20
Cosmids	35–50
BAC, P1	50–250
YAC	500–3000

Because insertion of DNA into a functional region of the vector will interfere with the action of this region, care must be taken not to interrupt an essential function of the vector. This concept can be exploited, however, to provide a selection technique. For example, a common early plasmid vector **pBR322** has both **tetracycline (tet)** and **ampicillin (amp)** resistance genes. A single *PstI* restriction enzyme site within the amp resistance gene is commonly used as the insertion site for a piece of foreign DNA. In addition to having sticky ends (Table 39–1 and Figure 39–1), the DNA inserted at this site disrupts the amp resistance gene and makes the bacterium carrying this plasmid amp-sensitive (**Figure 39–3**). Thus, cells carrying the parental plasmid, which provides resistance to both antibiotics, can be readily distinguished and separated

from cells carrying the chimeric plasmid, which is resistant only to tetracycline. YACs contain selection, replication, and segregation functions that work in both bacteria and yeast cells and therefore can be propagated in either organism.

In addition to the vectors described in Table 39–3 that are designed primarily for propagation in bacterial cells, vectors for mammalian cell propagation and insert gene (cDNA)/protein expression have also been developed. These vectors are all based upon various eukaryotic viruses that are composed of RNA or DNA genomes. Notable examples of such **viral vectors** are those utilizing **adenoviral (Ad),** or **adenovirus-associated viral (AAV)** (DNA-based) and **retroviral** (RNA-based) genomes. Though somewhat limited in the size of DNA sequences that can be inserted, such **mammalian viral cloning vectors** make up for this shortcoming because they will efficiently infect a wide range of different cell types. For this reason, various mammalian viral vectors are being investigated for use in **gene therapy** and are commonly used for laboratory experiments.

A Library Is a Collection of Recombinant Clones

The combination of restriction enzymes and various cloning vectors allows the entire genome of an organism to be individually packed into a vector. A collection of these different

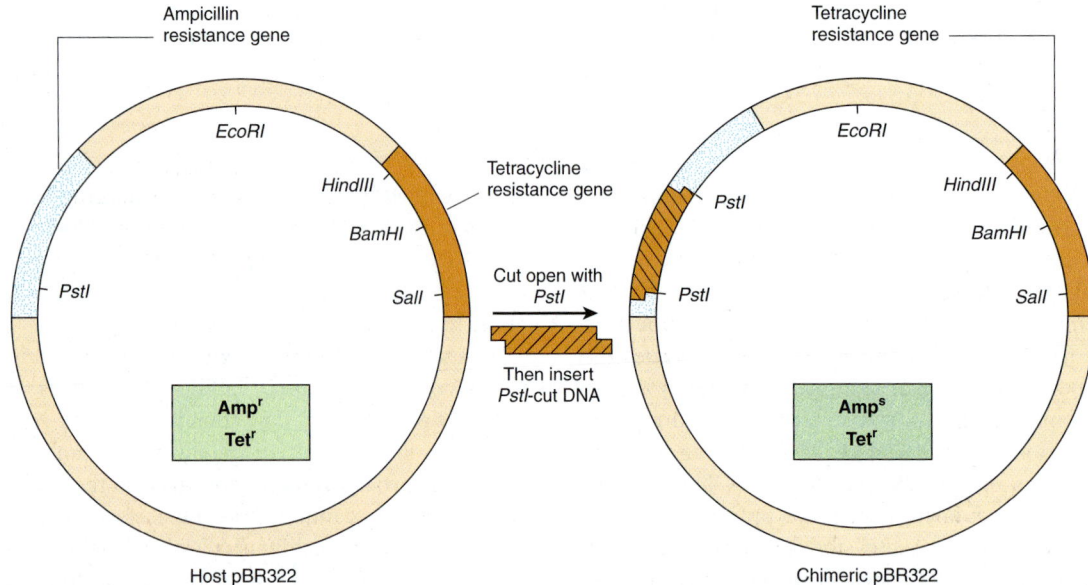

FIGURE 39–3   **A method of screening recombinants for inserted DNA fragments.** Using the plasmid pBR322, a piece of DNA is inserted into the unique *PstI* site. This insertion disrupts the gene coding for a protein that provides ampicillin resistance to the host bacterium. Hence, cells carrying the chimeric plasmid will no longer survive when plated on a substrate medium that contains this antibiotic. The differential sensitivity to tetracycline and ampicillin can therefore be used to distinguish clones of plasmid that contain an insert. A similar scheme relying upon production of an in-frame fusion of a newly inserted DNA producing a peptide fragment capable of complementing an inactive, N-terminally truncated form of the enzyme β-galactosidase, a component of the *lac* operon (Figure 38-2) allows for blue-white colony formation on agar plates containing a dye hydrolyzable by β-galactoside. β-Galactosidase-positive colonies are blue; such colonies contain plasmids in which a DNA was successfully inserted.

recombinant clones is called a library. A **genomic library** is prepared from the total DNA of a cell line or tissue. A **cDNA library** comprises complementary DNA copies of the population of mRNAs in a tissue. Genomic DNA libraries are often prepared by performing **partial digestion of total DNA** with a restriction enzyme that cuts DNA frequently (eg, a four base cutter such as *TaqI*). The idea is to generate rather large fragments so that most genes will be left intact. The BAC, YAC, and P1 vectors are preferred since they can accept very large fragments of DNA and thus offer a better chance of isolating an intact eukaryotic mRNA-encoding gene on a single DNA fragment.

A vector in which the protein coded by the gene introduced by recombinant DNA technology is actually synthesized is known as an **expression vector.** Such vectors are now commonly used to detect specific cDNA molecules in libraries and to produce proteins by genetic engineering techniques. These vectors are specially constructed to contain very active inducible promoters, proper in-phase translation initiation codons, both transcription and translation termination signals, and appropriate protein processing signals, if needed. Some expression vectors even contain genes that code for protease inhibitors, so that the final yield of product is enhanced. Interestingly as the cost of synthetic DNA synthesis has dropped, many investigators often synthesize an entire cDNA (gene) of interest (in 100–150 nt segments) incorporating the codon preferences of the host used for expression in order to maximize protein production. New efficiencies in synthetic DNA synthesis now allow for the de novo synthesis of complete genes and even genomes. These advances usher in new and exciting possibilities in synthetic biology while concomitantly introducing potential ethical conundrums.

Probes Search Libraries or Complex Samples for Specific Genes or cDNA Molecules

A variety of molecules can be used to "probe" libraries in search of a specific gene or cDNA molecule or to define and quantitate DNA or RNA separated by electrophoresis through various gels. Probes are generally pieces of DNA or RNA labeled with a ^{32}P-containing nucleotide—or fluorescently labeled nucleotides (more commonly now). Importantly, neither modification (^{32}P or fluorescent-label) affects the hybridization properties of the resulting labeled nucleic acid probes. The probe must recognize a complementary sequence to be effective. A cDNA synthesized from a specific mRNA can be used to screen either a cDNA library for a longer cDNA or a genomic library for a complementary sequence in the coding region of a gene. A popular technique for finding specific genes entails taking a short amino acid sequence and, employing the codon usage for that species (see Chapter 37), making an oligonucleotide probe (or probe mixture) that will detect the corresponding DNA fragment in a genomic library. If the sequences match exactly, probes 15–20 nucleotides long will hybridize. cDNA probes are used to detect DNA fragments on Southern blot transfers and to detect and quantitate RNA on

Northern blot transfers. Specific antibodies can also be used as probes provided that the vector used synthesizes protein molecules that are recognized by them.

Blotting & Hybridization Techniques Allow Visualization of Specific Fragments

Visualization of a specific DNA or RNA fragment among the many thousands of "contaminating" molecules in a complex sample requires the convergence of a number of techniques, collectively termed **blot transfer. Figure 39–4** illustrates the

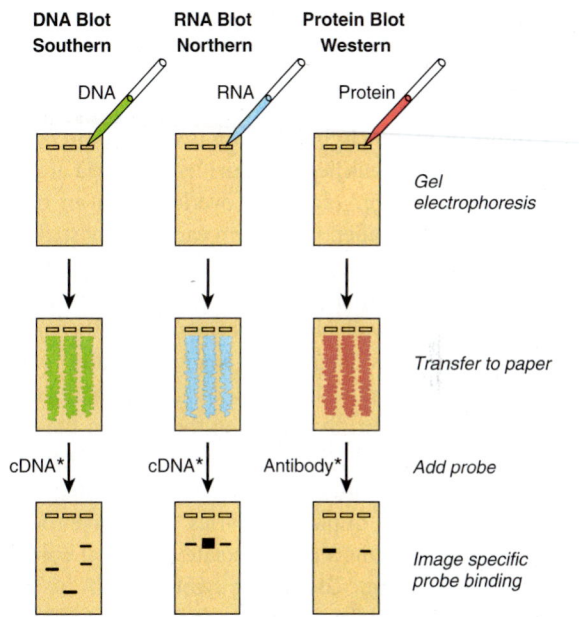

FIGURE 39–4 **The blot transfer procedure.** In a Southern, or DNA blot transfer, DNA isolated from a cell line or tissue is digested with one or more restriction enzymes. This mixture is pipetted into a well in an agarose or polyacrylamide gel and exposed to a direct electrical current. DNA, being negatively charged, migrates toward the anode; the smaller fragments move the most rapidly. After a suitable time, the DNA within the gel is denatured by exposure to mild alkali and transferred to nitrocellulose or nylon paper, resulting in an exact replica of the pattern on the gel, by the blotting technique devised by Southern. The DNA is bound to the paper by exposure to heat or UV, and the paper is then exposed to the labeled cDNA probe, which hybridizes to complementary strands on the filter. After thorough washing, the paper is exposed to X-ray film or an imaging screen, which is developed to reveal several specific bands corresponding to the DNA fragment that recognized the sequences in the cDNA probe. The RNA, or Northern, blot is conceptually similar. RNA is subjected to electrophoresis before blot transfer. This requires some different steps from those of DNA transfer, primarily to ensure that the RNA remains intact, and is generally somewhat more difficult. In the protein, or Western, blot, proteins are electrophoresed and transferred to special paper that avidly binds proteins and then probed with a specific antibody or other probe molecule. (Asterisks signify labeling, either radioactive or fluorescent.) In the case of Southwestern blotting (see the text; not shown), a protein blot similar to that shown above under "Western" is exposed to labeled nucleic acid, and protein-nucleic acid complexes formed are detected by autoradiography or imaging.

Southern (DNA), Northern (RNA), and Western (protein) blot transfer procedures. (The first is named for the person who devised the technique [Edward Southern], and the other names began as laboratory jargon but are now accepted terms.) These procedures are useful in determining how many copies of a gene are in a given tissue or whether there are any alterations in a gene (deletions, insertions, or rearrangements) because the requisite electrophoresis step separates the molecules on the basis of size. Occasionally, if a specific base is changed and a restriction site is altered, these procedures can detect a point mutation. The Northern and Western blot transfer techniques are used to size and quantitate specific RNA and protein molecules, respectively. A fourth hybridization technique, the Southwestern blot, examines protein–DNA interactions (not shown). In this method, proteins are separated by electrophoresis, blotted to a membrane, renatured, and analyzed for an interaction with a particular sequence by incubation with a specific labeled nucleic acid probe.

Colony or plaque hybridization is the method by which specific clones are identified and purified. Bacteria are grown as colonies on an agar plate and overlaid with an oriented nitrocellulose filter paper. Cells from each colony stick to the filter and are permanently fixed thereto by heat or UV, which with NaOH treatment also lyses the cells and denatures the DNA so that it is available to hybridize with the probe. A radioactive probe is added to the filter, and (after washing) the hybrid complex is localized by exposing the filter to x-ray film or imaging screen. By matching the spot on the autoradiograph (exposed and developed x-ray film) to a colony, the latter can be picked from the plate. A similar strategy is used to identify fragments in phage libraries. Successive rounds of this procedure result in a clonal isolate (bacterial colony) or individual phage plaque containing a unique DNA insert.

All of the hybridization procedures discussed in this section depend on the specific base-pairing properties of complementary nucleic acid strands described above. Perfect matches hybridize readily and withstand high temperatures in the hybridization and washing reactions. Specific complexes also form in the presence of low salt concentrations. Less than perfect matches do not tolerate such stringent conditions (ie, elevated temperatures and low salt concentrations); thus, hybridization either never occurs or is disrupted during the washing step. Gene families, in which there is some degree of homology, can be detected by varying the stringency of the hybridization and washing steps. Cross-species comparisons of a given gene can also be made using this approach. Hybridization conditions capable of detecting just a single base-pair (bp) mismatch between probe and target have been devised.

Manual & Automated Techniques Are Available to Determine the Sequence of DNA

The segments of specific DNA molecules obtained by recombinant DNA technology can be analyzed to determine their nucleotide sequence. This method depends upon having a large number of identical DNA molecules. This requirement can be satisfied by cloning the fragment of interest, using the techniques described above, or by using PCR methods (see below). The manual enzymatic method (Sanger) employs specific dideoxynucleotides that terminate DNA strand synthesis at specific nucleotides as the strand is synthesized on purified template nucleic acid. The reactions are adjusted so that a population of DNA fragments representing termination at every nucleotide is obtained. By having a radioactive label incorporated at the termination site, one can separate the fragments according to size using polyacrylamide gel electrophoresis. An autoradiograph is made, and each of the fragments produces an image (band) on an x-ray film or imaging plate. These are read in order to give the DNA sequence (Figure 39–5). Another manual method is that of Maxam and Gilbert, which employs chemical methods to cleave the DNA molecules where they contain the specific nucleotides. Techniques that do not require the use of radioisotopes are employed in automated DNA sequencing. Most commonly employed is an automated procedure in which four different fluorescent labels—one representing each nucleotide—are used. Each emits a specific signal upon excitation by a laser beam of a particular wavelength that is measured by densitive detectors, and this can be recorded by a computer. The newest DNA sequencing machines use fluorescently labeled nucleotides but detect incorporation using microscopic optics. These machines have reduced the cost of DNA sequencing dramatically, over 100X. These reductions in cost have ushered in the era of personalized genome sequencing. Indeed, using this new technology the sequence of the co-discoverer of the double helix, James Watson, was completely determined.

Oligonucleotide Synthesis Is Now Routine

The automated chemical synthesis of moderately long oligonucleotides (~100 nucleotides) of precise sequence is now a routine laboratory procedure. Each synthetic cycle takes but a few minutes, so an entire molecule can be made by synthesizing relatively short segments that can then be ligated to one another. As mentioned above, the process has been miniaturized and can be significantly parallelized to allow the synthesis of 100s to 1000s of defined sequence oligonucleotides simultaneously. Oligonucleotides are now indispensable for DNA sequencing, library screening, protein–DNA binding assays, the polymerase chain reaction (PCR) (see below), site-directed mutagenesis, synthetic gene synthesis, and numerous other applications.

The Polymerase Chain Reaction (PCR) Method Amplifies DNA Sequences

The PCR is a method of amplifying a target sequence of DNA. The development of PCR has revolutionized the ways in which both DNA and RNA can be studied. PCR provides a sensitive, selective, and extremely rapid means of amplifying any

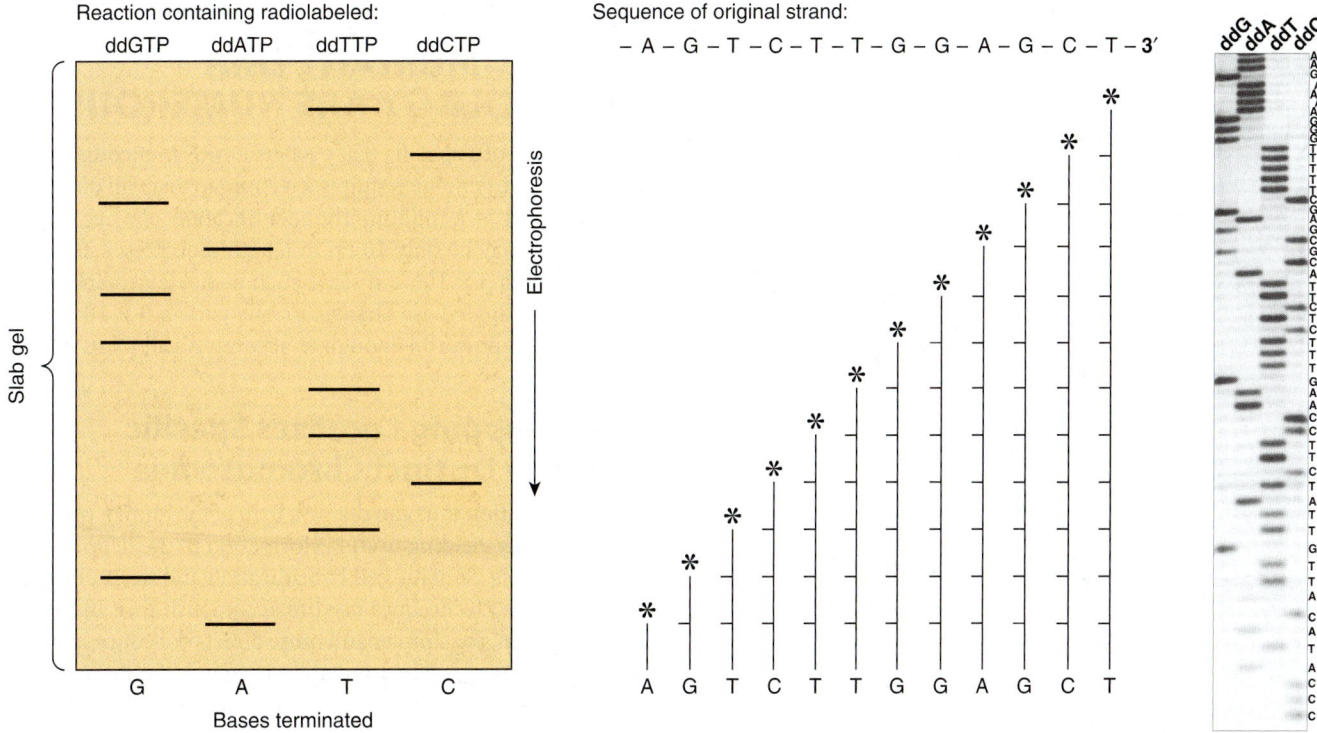

FIGURE 39–5 **Sequencing of DNA by the chain termination method devised by Sanger.** The ladder-like arrays represent from bottom to top all of the successively longer fragments of the original DNA strand. Knowing which specific dideoxynucleotide reaction was conducted to produce each mixture of fragments, one can determine the sequence of nucleotides from the unlabeled end toward the labeled end (*) by reading up the gel. The base-pairing rules of Watson and Crick (A–T, G–C) dictate the sequence of the other (complementary) strand. (Asterisks signify site of radiolabeling.) Shown (**left, middle**) are the terminated synthesis products of a hypothetical fragment of DNA, sequence shown. An autoradiogram (**right**) of an actual set of DNA sequencing reactions that utilized the four ^{32}P-labeled dideoxynucleotides indicated at the top of the scanned autoradiogram (ie, dideoxy(dd)G, ddA, ddT, ddC). Electrophoresis was from top to bottom. The deduced DNA sequence is listed on the right side of the gel. Note the log-linear relationship between distance of migration (ie, top to bottom of gel) and DNA fragment length. Current state-of-the-art DNA sequencers no longer utilize gel electrophoresis for fractionation of labeled synthesis products. Moreover in the NGS sequencing platforms, synthesis is followed by monitoring incorporation of the four fluorescently labeled dXTPs.

desired sequence of DNA. Specificity is based on the use of two oligonucleotide primers that hybridize to complementary sequences on opposite strands of DNA and flank the target sequence (**Figure 39–6**). The DNA sample is first heated to separate the two strands of the template DNA containing the target sequence; the primers, added in vast excess, are allowed to anneal to the DNA; and each strand is copied by a DNA polymerase, starting at the primer sites in the presence of all four dXTPs. The two DNA strands each serve as a template for the synthesis of new DNA from the two primers. Repeated cycles of heat denaturation, annealing of the primers to their complementary sequences, and extension of the annealed primers with DNA polymerase result in the exponential amplification of DNA segments of defined length (a doubling at each cycle). Early PCR reactions used an *E coli* DNA polymerase that was destroyed by each heat denaturation cycle and hence needed to be re-added at the beginning of each cycle. Substitution of a heat-stable DNA polymerase from *Thermus aquaticus* (or the corresponding DNA polymerase from many other thermophilic bacteria), an organism that lives and replicates at 70–80°C, obviates this problem and has made possible

automation of the reaction since the polymerase reactions can be run at 70°C. This has also improved the specificity and the yield of DNA.

DNA sequences as short as 50–100 bp and as long as 10 kb can be amplified. Twenty cycles provide an amplification of 10^6 (ie, 2^{20}) and 30 cycles, 10^9 (2^{30}). Each cycle takes ≤5–10 min so that even large DNA molecules can be amplified rapidly. The PCR allows the DNA in a single cell, hair follicle, or spermatozoon to be amplified and analyzed. Thus, the applications of PCR to forensic medicine are obvious. The PCR is also used (1) to detect infectious agents, especially latent viruses; (2) to make prenatal genetic diagnoses; (3) to detect allelic polymorphisms; (4) to establish precise tissue types for transplants; and (5) to study evolution, using DNA from archeological samples (6) for quantitative RNA analyses after RNA copying and mRNA quantitation by the so-called RTPCR method (cDNA copies of mRNA generated by a retroviral reverse transcriptase) or (7) to score in vivo protein–DNA occupancy using chromatin immunoprecipitation assays to facilitate NGS sequencing (see below). There are an equal number of applications of PCR to problems in basic science, and new uses are developed every year.

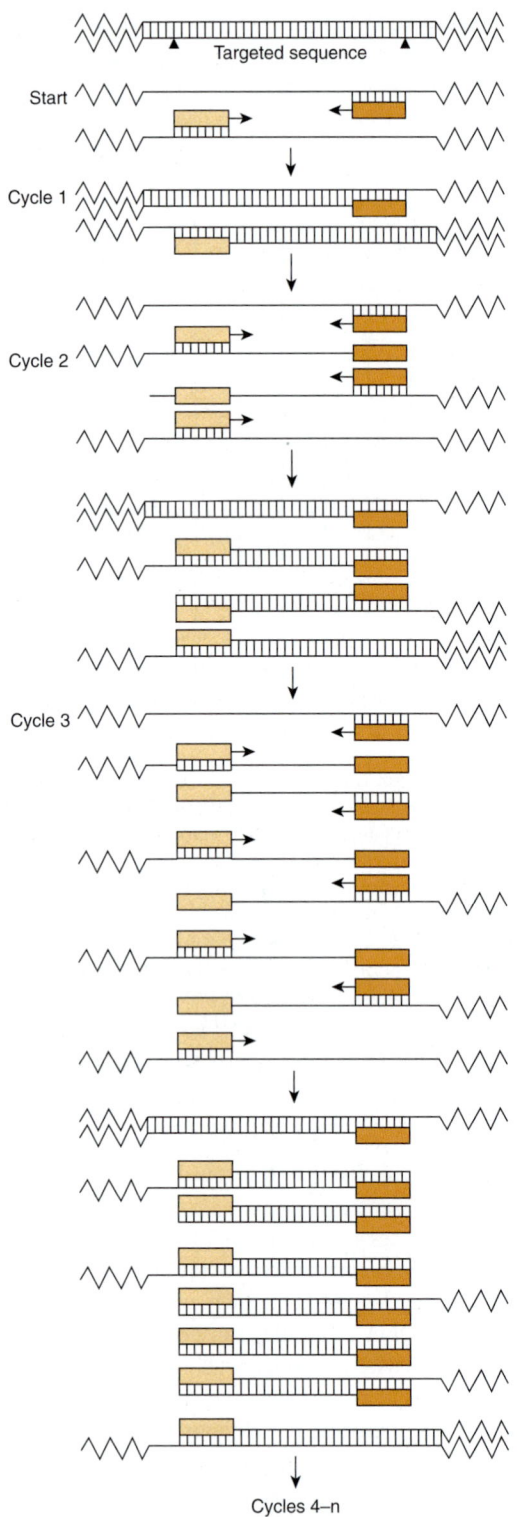

FIGURE 39–6   The polymerase chain reaction is used to amplify specific gene sequences. Double-stranded DNA is heated to separate it into individual strands. These bind two distinct primers that are directed at specific sequences on opposite strands and that define the segment to be amplified. DNA polymerase extends the primers in each direction and synthesizes two strands complementary to the original two. This cycle is repeated several times, giving an amplified product of defined length and sequence. Note that the two primers are present in vast excess.

PRACTICAL APPLICATIONS OF RECOMBINANT DNA TECHNOLOGY ARE NUMEROUS

The isolation of a specific (ca. 1000 bp) mRNA-encoding gene from an entire genome requires a technique that will discriminate one part in a million. The identification of a regulatory region that may be only 10 bp in length requires a sensitivity of one part in 3×10^8; a disease such as sickle-cell anemia is caused by a single base change, or one part in 3×10^9. DNA technology is powerful enough to accomplish all these things.

Gene Mapping Localizes Specific Genes to Distinct Chromosomes

Gene localization thus can define a map of the human genome. This is already yielding useful information in the definition of human disease. Somatic cell hybridization and in situ hybridization are two techniques used to accomplish this. In **in situ hybridization,** the simpler and more direct procedure, a radioactive probe is added to a metaphase spread of chromosomes on a glass slide. The exact area of hybridization is localized by layering photographic emulsion over the slide and, after exposure, lining up the grains with some histologic identification of the chromosome. **Fluorescence in situ hybridization (FISH), which utilizes fluorescent rather than radioactively labeled probes,** is a very sensitive technique that is also used for this purpose. This often places the gene at a location on a given band or region on the chromosome. Some of the human genes localized using these techniques are listed in **Table 39–4**. This table represents only a sampling of mapped genes since tens of thousands of genes have been mapped as a result of the recent sequencing of the human genome. Once the defect is localized to a region of DNA that has the characteristic structure of a gene, a synthetic cDNA copy of the gene can be constructed, which contains only mRNA encoding exons, and expressed in an appropriate vector and its function can be assessed—or the putative peptide, deduced from the open reading frame in the coding region, can be synthesized. Antibodies directed against this peptide can be used to assess whether this peptide is expressed in normal persons and whether it is absent, or altered in those with the genetic syndrome.

Proteins Can Be Produced for Research, Diagnosis & Commerce

A practical goal of recombinant DNA research is the production of materials for biomedical applications. This technology has two distinct merits: (1) it can supply large amounts of material that could not be obtained by conventional purification methods (eg, interferon, tissue plasminogen activating factor, etc). (2) It can provide human material (eg, insulin and growth hormone). The advantages in both cases are obvious. Although the primary aim is to supply products—generally proteins—for treatment (insulin) and diagnosis (AIDS testing) of human

TABLE 39–4 Localization of Human Genes[1]

Gene	Chromosome	Disease
Insulin	11p15	Diabetes
Prolactin	6p23-q12	Sheehan Syndrome
Growth hormone	17q21-qter	Growth hormone deficiency
α-Globin	16p12-pter	α-Thalassemia
β-Globin	11p12	β-Thalassemia, sickle cell
Adenosine deaminase	20q13-qter	Adenosine deaminase deficiency
Phenylalanine hydroxylase	12q24	Phenylketonuria
Hypoxanthine-guanine phosphoribosyltransferase	Xq26-q27	Lesch–Nyhan syndrome
DNA segment G8	4p	Huntington chorea

[1]This table indicates the chromosomal location of several genes and the diseases associated with deficient or abnormal production of the gene products. The chromosome involved is indicated by the first number or letter. The other numbers and letters refer to precise localizations, as defined in McKusick, Victor A., MD, *Mendelian Inheritance in Man: Catalogs of Autosomal Dominant, Autosomal Recessive, and X-Linked Phenotypes.* Copyright © 1983 Johns Hopkins University Press. Reprinted with permission from the Johns Hopkins University Press.

and other animal diseases and for disease prevention (hepatitis B vaccine), there are other potential commercial applications, especially in agriculture. An example of the latter is the attempt to engineer plants that are more resistant to drought or temperature extremes, more efficient at fixing nitrogen, or that produce seeds containing the complete complement of essential amino acids (rice, wheat, corn, etc).

Recombinant DNA Technology Is Used in the Molecular Analysis of Disease

Normal Gene Variations

There is a normal variation of DNA sequence just as is true of more obvious aspects of human structure. Variations of DNA sequence, **polymorphisms,** occur approximately once in every 500–1000 nucleotides. A recent comparison of the nucleotide sequence of the genome of James Watson, the co-discoverer of DNA structure, identified about 3,300,000 single-nucleotide polymorphisms (SNPs) relative to the "standard" initially sequenced human reference genome. Interestingly, >80% of the SNPs found in Watson's DNA had already been identified in other individuals. There are also genomic deletions and insertions of DNA (ie, **copy number variations; CNV**) as well as single-base substitutions. In healthy people, these alterations obviously occur in noncoding regions of DNA or at sites that cause no change in function of the encoded protein. This heritable polymorphism of DNA structure can be associated with certain diseases within a large kindred and can be used to search for the specific gene involved, as is illustrated below. It can also be used in a variety of applications in forensic medicine.

Gene Variations Causing Disease

Classic genetics taught that most genetic diseases were due to point mutations which resulted in an impaired protein. This may still be true, but if on reading previous chapters one predicted that genetic disease could result from derangement of any of the steps leading from replication to transcription to RNA processing/transport and protein synthesis, one would have made a proper assessment. This point is again nicely illustrated by examination of the β-globin gene. This gene is located in a cluster on chromosome 11 (**Figure 39–7**), and an expanded version of the gene is illustrated in **Figure 39–8**.

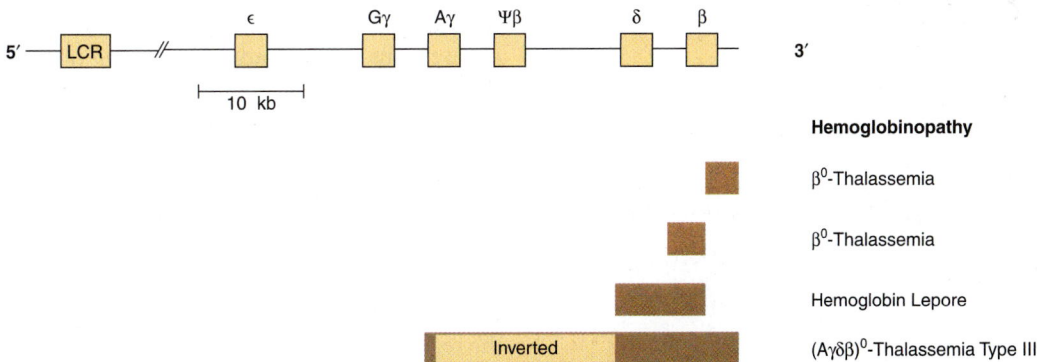

FIGURE 39–7 **Schematic representation of the β-globin gene cluster and of the lesions in some genetic disorders.** The β-globin gene is located on chromosome 11 in close association with the two γ-globin genes and the δ-globin gene. The β-gene family is arranged in the order 5′-ε-Gγ-Aγ-Ψβ-δ-β-3′. The ε locus is expressed in early embryonic life (as $\alpha_2\varepsilon_2$). The γ genes are expressed in fetal life, making fetal hemoglobin (HbF, $\alpha_2\gamma_2$). Adult hemoglobin consists of HbA ($\alpha_2\beta_2$) or HbA$_2$ ($\alpha_2\delta_2$). The Ψβ is a pseudogene that has sequence homology with β but contains mutations that prevent its expression. A locus control region (LCR), a powerful enhancer located upstream (5′) from the gene, controls the rate of transcription of the entire β-globin gene cluster. Deletions (solid bar) of the β locus cause β-thalassemia (deficiency or absence [β^0] of β-globin). A deletion of δ and β causes hemoglobin Lepore (only hemoglobin is present). An inversion (Aγδβ)^0 in this region (largest bar) disrupts gene function and also results in thalassemia (type III). Each type of thalassemia tends to be found in a certain group of people, eg, the (Aγδβ)^0 deletion inversion occurs in persons from India. Many more deletions in this region have been mapped, and each causes some type of thalassemia.

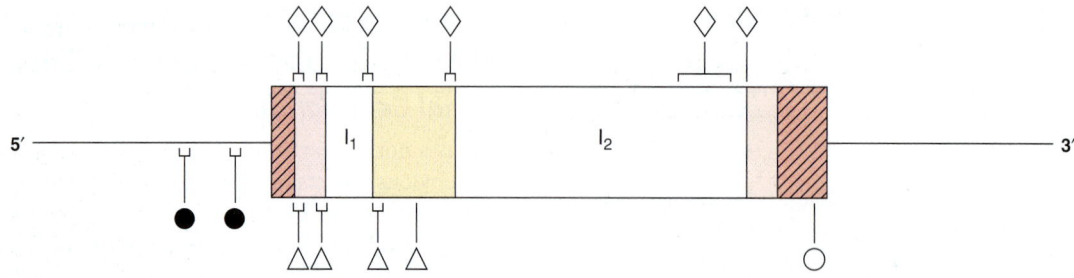

FIGURE 39–8   **Mutations in the β-globin gene causing β-thalassemia.** The β-globin gene is shown in the 5′ to 3′ orientation. The cross-hatched areas indicate the 5′ and 3′ nontranslated regions. Reading from the 5′ to 3′ direction, the shaded areas are exons 1–3 and the clear spaces are introns 1 (I₁) and 2 (I₂). Mutations that affect transcription control (●) are located in the 5′ flanking-region DNA. Examples of nonsense mutations (△), mutations in RNA processing (◇), and RNA cleavage mutations (○) have been identified and are indicated. In some regions, many distinct mutations have been found. These are indicated by the brackets.

Defective production of β-globin results in a variety of diseases and is due to many different lesions in and around the β-globin gene (**Table 39–5**).

Point Mutations

The classic example is **sickle-cell disease,** which is caused by mutation of a single base out of the 3 × 10⁹ in the genome, a T-to-A DNA substitution, which in turn results in an A-to-U change in the mRNA corresponding to the sixth codon of the β-globin gene. The altered codon specifies a different amino acid (valine rather than glutamic acid), and this causes a structural abnormality of the β-globin molecule. Other point mutations in and around the β-globin gene result in decreased or, in some instances, no production of β-globin; β-thalassemia is the result of these mutations. (The thalassemias are characterized by defects in the synthesis of hemoglobin subunits, and so β-thalassemia results when there is insufficient production of β-globin.) Figure 39–8 illustrates that point mutations affecting each of the many processes involved in generating a normal mRNA (and therefore a normal protein) have been implicated as a cause of β-thalassemia.

TABLE 39–5   Structural Alterations of the β-Globin Gene

Alteration	Function Affected	Disease
Point mutations	Protein folding	Sickle cell disease
	Transcriptional control	β-Thalassemia
		β-Thalassemia
	Frameshift and nonsense mutations	β-Thalassemia
	RNA processing	
Deletion	mRNA production	β⁰-Thalassemia
		Hemoglobin Lepore
Rearrangement	mRNA production	β-Thalassemia type III

Deletions, Insertions, & Rearrangements of DNA

Studies of bacteria, viruses, yeasts, fruit flies, and now humans show that pieces of DNA can move from one place to another within a genome. The deletion of a critical piece of DNA, the rearrangement of DNA within a gene, or the insertion or amplification of a piece of DNA within a coding or regulatory region can all cause changes in gene expression resulting in disease. Again, a molecular analysis of thalassemias produces numerous examples of these processes—particularly deletions—as causes of disease (Figure 39–7). The globin gene clusters seem particularly prone to this lesion. Deletions in the α-globin cluster, located on chromosome 16, cause α-thalassemia. There is a strong ethnic association for many of these deletions, so that northern Europeans, Filipinos, blacks, and Mediterranean peoples have different lesions all resulting in the absence of hemoglobin A and α-thalassemia.

A similar analysis could be made for a number of other diseases. Point mutations are usually defined by sequencing the gene in question, though occasionally, if the mutation destroys or creates a restriction enzyme site, the technique of restriction fragment analysis can be used to pinpoint the lesion. Deletions or insertions of DNA larger than 50 bp can often be detected by the Southern blotting procedure while PCR-based assays can detect much smaller changes in DNA structure.

Pedigree Analysis

Sickle-cell disease again provides an excellent example of how recombinant DNA technology can be applied to the study of human disease. The substitution of T for A in the template strand of DNA in the β-globin gene changes the sequence in the region that corresponds to the sixth codon from

$$
\begin{array}{ll}
\text{C C T G A G G} & \text{Coding strand} \\
\text{G G A C } \textcircled{T}\text{ C C} & \text{Template strand}
\end{array}
$$

to

$$
\begin{array}{ll}
\downarrow \\
\text{C C T G T G G} & \text{Coding strand} \\
\text{G G A C } \textcircled{A}\text{ C C} & \text{Template strand} \\
\quad\quad\uparrow
\end{array}
$$

and destroys a recognition site for the restriction enzyme *MstII* (CCTNAGG; denoted by the small vertical arrows; Table 39–1).

A. MstII restriction sites around and in the β-globin gene

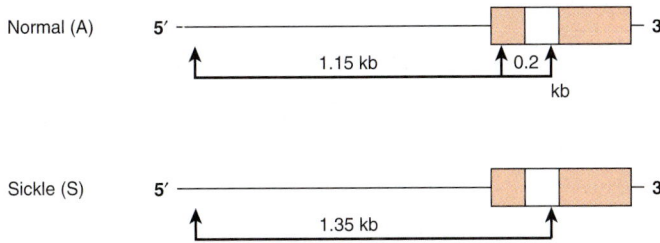

B. Pedigree analysis

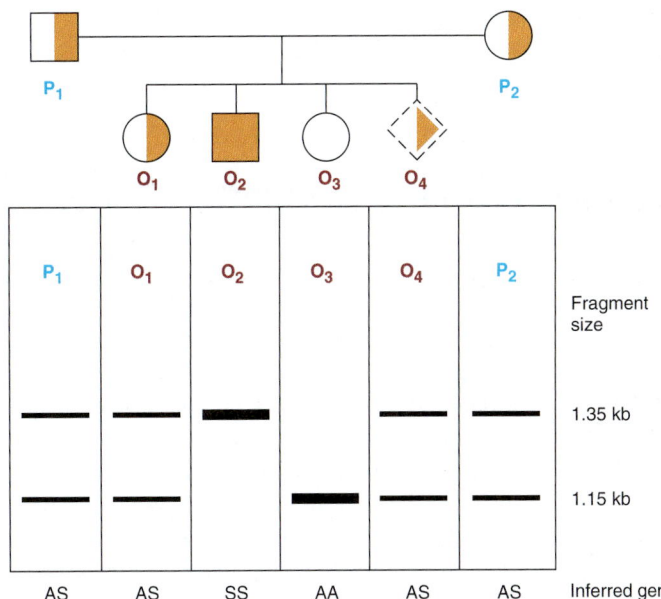

FIGURE 39–9 **Pedigree analysis of sickle-cell disease.** The top part of the figure **(A)** shows the first part of the β-globin gene and the *MstII* restriction enzyme sites in the normal (A) and sickle-cell (S) β-globin genes. Digestion with the restriction enzyme *MstII* results in DNA fragments 1.15 kb and 0.2 kb long in normal individuals. The T-to-A change in individuals with sickle-cell disease abolishes one of the three *MstII* sites around the β-globin gene; hence, a single restriction fragment 1.35 kb in length is generated in response to *MstII*. This size difference is easily detected on a Southern blot. (The 0.2-kb fragment would run off the gel in this illustration.) **(B)** Pedigree analysis shows three possibilities: AA = normal (open circle); AS = heterozygous (half-solid circles, half-solid square); SS = homozygous (solid square). This approach can allow for prenatal diagnosis of sickle-cell disease (dash-sided square). See the text.

Other *MstII* sites 5′ and 3′ from this site (**Figure 39–9**) are not affected and so will be cut. Therefore, incubation of DNA from normal (AA), heterozygous (AS), and homozygous (SS) individuals results in three different patterns on Southern blot transfer (Figure 39–9). This illustrates how a DNA pedigree can be established using the principles discussed in this chapter. Pedigree analysis has been applied to a number of genetic diseases and is most useful in those caused by deletions and insertions or the rarer instances in which a restriction endonuclease cleavage site is affected, as in the example cited here. Such analyses are now facilitated by the PCR reaction, which can amplify and hence provide sufficient DNA for analysis from just a few nucleated cells.

Prenatal Diagnosis

If the genetic lesion is understood and a specific probe is available, prenatal diagnosis is possible. DNA from cells collected from as little as 10 mL of amniotic fluid (or by chorionic villus biopsy) can be analyzed by Southern blot transfer. A fetus with the restriction pattern AA in Figure 39–9 does not have sickle-cell disease, nor is it a carrier. A fetus with the SS pattern will develop the disease. Probes are now available for this type of analysis of many genetic diseases.

Restriction Fragment Length Polymorphism and SNPs

The differences in DNA sequence cited above can result in variations of restriction sites and thus in the length of restriction fragments. Similarly, single nucleotide polymorphisms, or **SNPs,** can be detected by the sensitive PCR method. An inherited difference in the pattern of restriction enzyme digestion (eg, a DNA variation occurring in more than 1% of the general population) is known as a **restriction fragment length polymorphism (RFLP).** Extensive RFLP and SNP maps of the human genome have been constructed. This is proving useful in the Human Genome Analysis Project and is an important component of the effort to understand various single-gene and multigenic diseases. RFLPs result from single-base changes (eg, sickle-cell disease) or from deletions or insertions (CNVs) of DNA into a restriction fragment (eg, the thalassemias) and have proved to be useful diagnostic tools. They have been found at known gene loci and in sequences that have no known function; thus, RFLPs may disrupt the function of the gene or may have no apparent biologic consequences. As mentioned above, 80% of the SNPs in the genome of a single known individual had already been mapped independently through the efforts of the SNP-mapping component of the International HapMap Project.

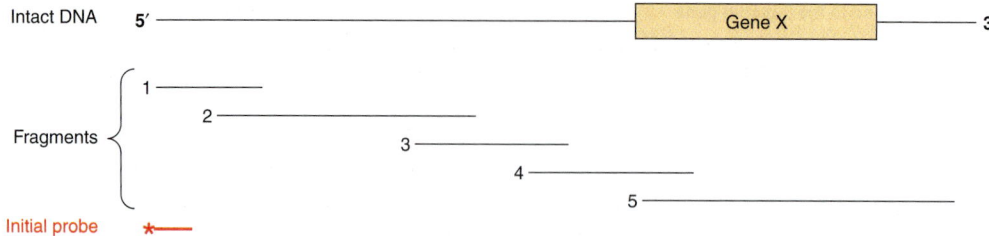

FIGURE 39–10 **The technique of chromosome walking.** Gene X is to be isolated from a large piece of DNA. The exact location of this gene is not known, but a probe (*——) directed against a fragment of DNA (shown at the 5′ end in this representation) is available, as is a library of clones containing a series of overlapping DNA insert fragments. For the sake of simplicity, only five of these are shown. The initial probe will hybridize only with clones containing fragment 1, which can then be isolated and used as a probe to detect fragment 2. This procedure is repeated until fragment 4 hybridizes with fragment 5, which contains the entire sequence of gene X.

RFLPs and SNPs are inherited, and they segregate in a mendelian fashion. A major use of SNPs/RFLPs is in the definition of inherited diseases in which the functional deficit is unknown. SNPs/RFLPs can be used to establish linkage groups, which in turn, by the process of **chromosome walking,** will eventually define the disease locus. In chromosome walking (**Figure 39–10**), a fragment representing one end of a long piece of DNA is used to isolate another that overlaps but extends the first. The direction of extension is determined by restriction mapping, and the procedure is repeated sequentially until the desired sequence is obtained. Collections of mapped, overlapping BAC- or PAC-cloned human genomic DNAs are commercially available. The X chromosome-linked disorders are particularly amenable to the approach of chromosome walking since only a single allele is expressed. Hence, 20% of the defined RFLPs are on the X chromosome and a complete linkage map (and genomic sequence) of this chromosome have been determined. The gene for the X-linked disorder, Duchenne-type muscular dystrophy, was found using RFLPs. Similarly, the defect in Huntington disease was localized to the terminal region of the short arm of chromosome 4, and the defect that causes polycystic kidney disease is linked to the α-globin locus on chromosome 16.

Microsatellite DNA Polymorphisms

Short (2–6 bp), inherited, tandem repeat units of DNA occur about 50,000–100,000 times in the human genome (Chapter 35). Because they occur more frequently—and in view of the routine application of sensitive PCR methods—they are replacing RFLPs as the marker loci for various genome searches.

RFLPs & VNTRs in Forensic Medicine

Variable numbers of tandemly repeated (VNTR) units are one common type of "insertion" that results in an RFLP. The VNTRs can be inherited, in which case they are useful in establishing genetic association with a disease in a family or kindred; or they can be unique to an individual and thus serve as a molecular fingerprint of that person.

Direct Sequencing of Genomic DNA

As noted above, recent advances in DNA sequencing technology, the so-called next generation sequencing (NGS) platforms, have dramatically reduced the per base cost of DNA sequencing. The initial sequence of the human genome cost roughly $350,000,000. The cost of sequencing the same 3×10^9 bp diploid human genome using the new NGS platforms is estimated to be <0.03% of the original. This dramatic reduction in cost has stimulated various international initiatives to sequence the entire genomes of thousands of individuals of various racial and ethnic backgrounds in order to determine the true extent of DNA/genome polymorphisms present within the population. The resulting cornucopia of genetic information, and the ever-decreasing cost of genomic DNA sequencing is dramatically increasing our ability to diagnose and, ultimately treat human disease. Obviously, when personal genome sequencing does become commonplace, dramatic changes in the practice of medicine will result because therapies will ultimately be custom tailored to the exact genetic makeup of each individual.

Gene Therapy and Stem Cell Biology

Diseases caused by deficiency of a single gene product (Table 39–4) are all theoretically amenable to replacement therapy. The strategy is to clone a normal copy of the relevant gene (eg, the gene that codes for adenosine deaminase) into a vector that will readily be taken up and incorporated into the genome of a host cell. Bone marrow precursor cells are being investigated for this purpose because they presumably will resettle in the marrow and replicate there. The introduced gene would begin to direct the expression of its protein product, and this would correct the deficiency in the host cell.

As an alternative to "replacing" defective genes to cure human disease, many scientists are investigating the feasibility of identifying and characterizing pluripotent stem cells that have the ability to differentiate into any cell type in the body. Recent results in this field have shown that adult human somatic cells can readily be converted into apparent **induced pluripotent stem cells (iPSCs)** by transfection with cDNAs encoding a handful of DNA binding transcription factors. These and other new developments in the fields of gene therapy and stem cell biology promise exciting new potential therapies for curing human disease.

Transgenic Animals

The somatic cell gene replacement therapy described above would obviously not be passed on to offspring. Other strategies to alter germ cell lines have been devised but have been tested only in experimental animals. A certain percentage of genes injected into a fertilized mouse ovum will be incorporated into the genome and found in both somatic and germ cells. Hundreds of transgenic animals have been established, and these are useful for analysis of tissue-specific effects on gene expression and effects of overproduction of gene products (eg, those from the growth hormone gene or oncogenes) and in discovering genes involved in development—a process that heretofore has been difficult to study. The transgenic approach has been used to correct a genetic deficiency in mice. Fertilized ova obtained from mice with genetic hypogonadism were injected with DNA containing the coding sequence for the gonadotropin-releasing hormone (GnRH) precursor protein. This gene was expressed and regulated normally in the hypothalamus of a certain number of the resultant mice, and these animals were in all respects normal. Their offspring also showed no evidence of GnRH deficiency. This is, therefore, evidence of somatic cell expression of the transgene and of its maintenance in germ cells.

Targeted Gene Disruption or Knockout

In transgenic animals, one is adding one or more copies of a gene to the genome, and there is no way to control where that gene eventually resides. A complementary—and much more difficult—approach involves the selective removal of a gene from the genome. Gene knockout animals (usually mice) are made by creating a mutation that totally disrupts the function of a gene. This is then used to replace one of the two genes in an embryonic stem cell that can be used to create a heterozygous transgenic animal. The mating of two such animals will, by Mendelian genetics, result in a homozygous mutation in 25% of offspring. Several thousand strains of mice with knockouts of specific genes have been developed. Techniques for disrupting genes in specific cells, tissues, or organs have been developed, so-called conditional, or directed, knockouts. This can be accomplished by taking advantage of particular promoter-enhancer combinations driving expression of DNA recombinases, or alternatively expression of miRNAs, both of which inactivate gene expression. These methods are particularly useful in cases where gene ablation during early development causes embryonic lethality.

RNA and Protein Profiling, and Protein–DNA Interaction Mapping

The "-omic" revolution of the last decade has culminated in the determination of the nucleotide sequences of entire genomes, including those of budding and fission yeasts, numerous bacteria, the fruit fly, the worm *Caenorhabditis elegans,* plants, the mouse, rat, chicken, monkey and, most notably, humans. Additional genomes are being sequenced at an accelerating pace. The availability of all of this DNA sequence information, coupled with engineering advances, has lead to the development of several revolutionary methodologies, most of which are based upon **high-density microarray technology** or **NGS platforms.** In the case of microarrays, it is now possible to deposit thousands of specific, known, definable DNA sequences (more typically now synthetic oligonucleotides) on a glass microscope-style slide in the space of a few square centimeters. By coupling such DNA microarrays with highly sensitive detection of hybridized fluorescently labeled nucleic acid probes derived from mRNA, investigators can rapidly and accurately generate profiles of gene expression (eg, specific cellular mRNA content) from cell and tissue samples as small as 1 g or less. Thus, entire **transcriptome information** (the entire collection of cellular RNAs) for such cell or tissue sources can readily be obtained in only a few days. In the case of NGS sequencing, mRNAs are converted to cDNAs using reverse transcription, and these cDNAs are amplified by PCR and directly sequenced; this method is termed **RNA-Seq.** Methods are being developed to allow direct RNA sequencing obviating the need for the cDNA/PCR step. These methods allow for the description of the entire transcriptome. Recent methodological advances (**GRO-Seq, Global Run-On sequencing,** and **NET-seq, native elongating transcript sequencing**) allow for sequencing of RNA within elongating RNA polymerase-DNA-RNA ternary complexes, thereby allowing nucleotide-level descriptions, genome-wide, of transcription in living cells. Such transcriptome information allows one to quantitatively predict the collection of proteins that might be expressed in a particular cell, tissue, or organ in normal and disease states based upon the mRNAs present in those cells. Complementing this high-throughput, transcript-profiling method is the recent development of methods to map the location, or occupancy of specific proteins bound to discrete sites within living cells. This method, illustrated in **Figure 39–11,** is termed **chromatin immunoprecipitation (ChIP).** Proteins are crosslinked in situ in cells or tissues, chromatin isolated, sheared, and specific protein DNA complexes purified using antibodies recognizing a particular protein, or protein iso-form. DNA bound to this protein is recovered and analyzed using PCR and either gel electrophoresis, direct sequencing (**ChIP-SEQ**) or microarray analysis (**ChIP-chip**). Both ChIP-SEQ and ChIP-chip methods allow investigators to identify the entire genome-wide locations of a single protein throughout all the chromosomes; ChIP-SEQ allows mapping at nucleotide-level resolugion. Finally, methods for high-sensitivity, high-throughput **mass spectrometry of metabolites (metabolomics) and complex protein samples (proteomics)** have been developed. Newer mass spectrometry methods allow one to identify hundreds to thousands of proteins in samples extracted from very small numbers of cells (<1 g). Such analyses can now be used to quantitate the amounts of proteins in two samples as well as

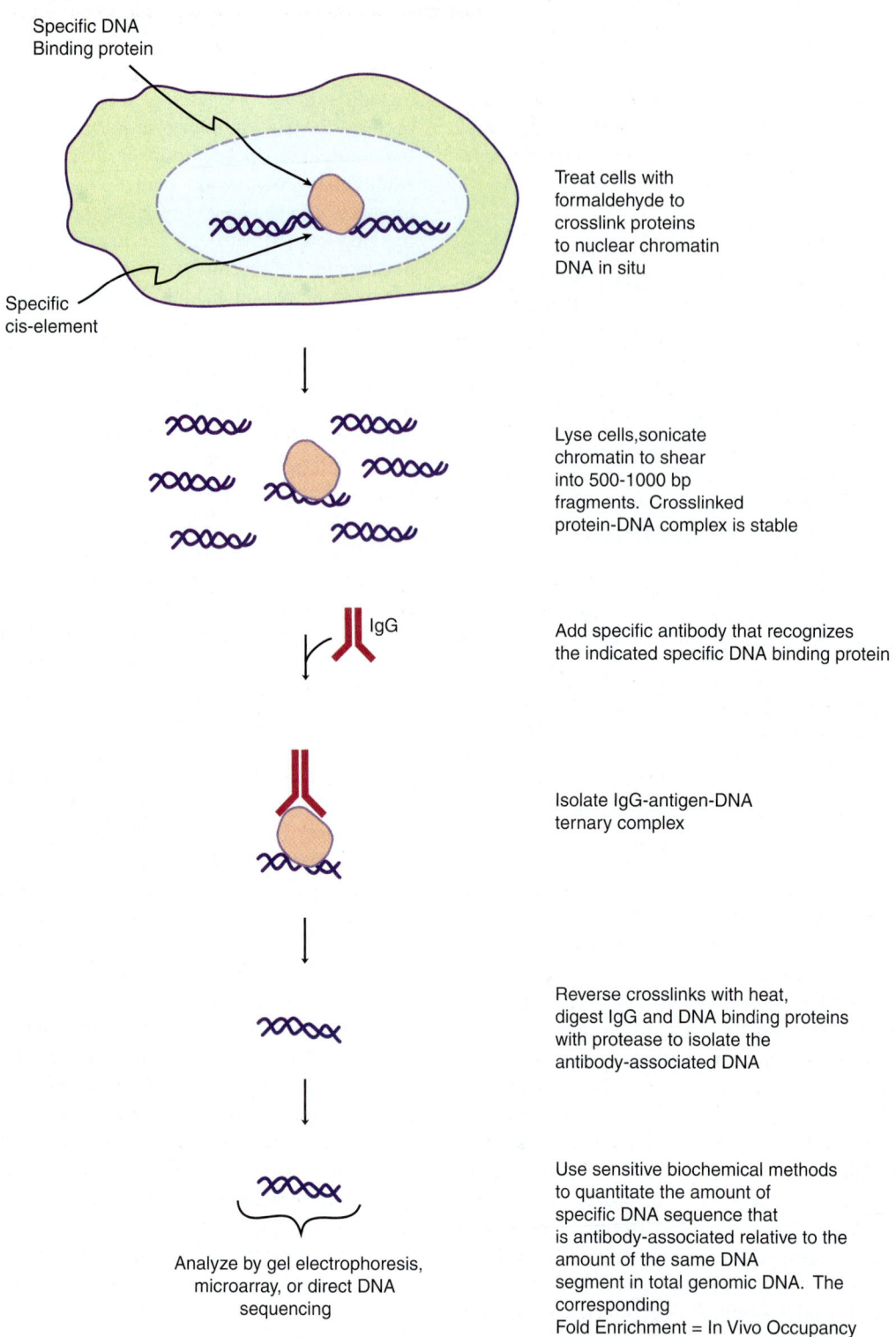

Specific DNA
Binding protein

Specific
cis-element

Treat cells with
formaldehyde to
crosslink proteins
to nuclear chromatin
DNA in situ

Lyse cells,sonicate
chromatin to shear
into 500-1000 bp
fragments. Crosslinked
protein-DNA complex is stable

IgG

Add specific antibody that recognizes
the indicated specific DNA binding protein

Isolate IgG-antigen-DNA
ternary complex

Reverse crosslinks with heat,
digest IgG and DNA binding proteins
with protease to isolate the
antibody-associated DNA

Analyze by gel electrophoresis,
microarray, or direct DNA
sequencing

Use sensitive biochemical methods
to quantitate the amount of
specific DNA sequence that
is antibody-associated relative to the
amount of the same DNA
segment in total genomic DNA. The
corresponding
Fold Enrichment = In Vivo Occupancy

FIGURE 39–11 **Outline of the chromatin immunoprecipitation (ChIP) technique.** This method allows for the precise localization of a particular protein (or modified protein if an appropriate antibody is available; eg, phosphorylated or acetylated histones, transcription factors, etc) on a particular sequence element in living cells. Depending upon the method used to analyze the immunopurified DNA, quantitative or semi-quantitative information, at near nucleotide level resolution, can be obtained. Protein-DNA occupancy can be scored genome-wide in two ways. First, by ChIP-chip, a method that uses a hybridization readout. In ChIP-chip total genomic DNA is labeled with one particular fluorphore and the immunopurified DNA is labeled with a spectrally distinct fluorphore. These differentially labeled DNAs are mixed and hybridized to microarray 'chips' (microscope slides) that contain specific DNA fragments, or more commonly now, synthetic oligonucleotide 50-70 nucleotides long. These gene-specific oligonucleotides are deposited and covalently attached at predetermined, known X,Y coordinates on the slide. The labeled DNAs are hybridized, the slides washed and hybridization to each gene-specific oligonucleotide probe is scored using differential laser scanning and sensitive photodetection at micron resolution. The hybridization signal intensities are quantified and the ratio of IP DNA/genomic DNA signals is used to score occupancy levels. The second method, termed ChIP-seq, directly sequences immunopurified DNAs using NGS/deep sequencing methods. Both approaches rely upon efficient bioinformatic algorithms to deal with the very large datasets that are generated. ChIP-chip and ChIP-seq techniques provide a (semi-)quantitative measure of in vivo protein occupancy.

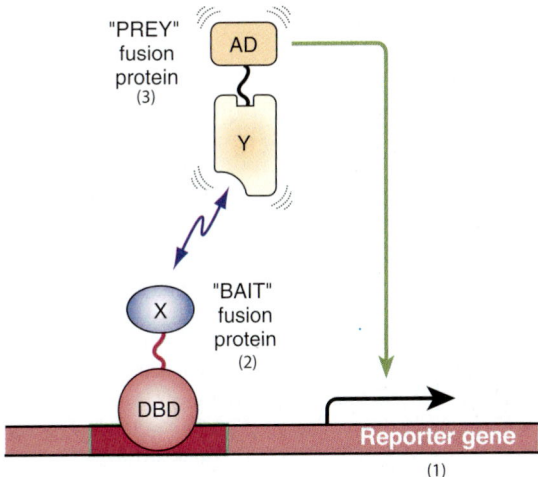

FIGURE 39–12 **Overview of two hybrid system for identifying and characterizing protein–protein interactions.** Shown are the basic components and operation of the two hybrid system, originally devised by Fields and Song (Nature 340:245–246 [1989]) to function in the bakers yeast system. **(1)** A reporter gene, either a selectable marker (ie, a gene conferring prototrophic growth on selective media, or producing an enzyme for which a colony colorimetric assay exists, such as β-galactosidase) that is expressed only when a transcription factor binds upstream to a *cis*-linked enhancer (dark red bar). **(2)** A "bait" fusion protein **(DBD-X)** produced from a chimeric gene expressing a modular DNA binding domain **(DBD**; often derived from the yeast Gal 4 protein or the bacterial Lex A protein, both high-affinity, high-specificity DNA binding proteins) fused in-frame to a protein of interest, here X. In two hybrid experiments, one is testing whether any protein can interact with protein X. Prey protein X may be fused in its entirety or often alternatively just a portion of protein X is expressed in-frame with the DBD. **(3)** A "prey" protein **(Y-AD),** which represents a fusion of a specific protein fused in-frame to a transcriptional activation domain **(AD;** often derived from either the *Herpes simplex* virus VP16 protein or the yeast Gal 4 protein). This system serves as a useful test of protein–protein interactions between proteins X and Y because in the absence of a functional transactivator binding to the indicated enhancer, no transcription of the reporter gene occurs (ie, see Figure 38–16). Thus, one observes transcription only if protein X–protein Y interaction occurs, thereby bringing a functional AD to the *cis*-linked transcription unit, in this case activating transcription of the reporter gene. In this scenario, protein DBD-X alone fails to activate reporter transcription because the X-domain fused to the DBD does not contain an AD. Similarly, protein Y-AD alone fails to activate reporter gene transcription because it lacks a DBD to target the Y-AD protein to the enhancer. Only when both proteins are expressed in a single cell and bind the enhancer and, via DBD-X–Y-AD protein–protein interactions, regenerate a functional transactivator binary "protein," does reporter gene transcription result in activation and mRNA synthesis (line from AD to reporter gene).

the level of certain PTMs, such as phosphorylation. This critical information tells investigators which of the many mRNAs detected in transcriptome mapping studies are actually translated into protein, generally the ultimate dictator of phenotype. New genetic means for identifying protein–protein interactions and protein function have also been devised. Systematic genome-wide gene expression knockdown, using SiRNAs (miRNAs), or synthetic lethal genetic interaction screens, have been applied to assess the contribution of individual genes to a variety of processes in model systems (yeast, worms, and flies) and mammalian cells (human and mouse). Specific network mappings of protein–protein interactions on a genome-wide basis have been identified using high-throughput variants of the **two hybrid interaction** test (**Figure 39–12**). This simple yet powerful method can be performed in bacteria, yeast, or metazoan cells, and allows for detecting specific protein–protein interactions in living cells. Reconstruction experiments indicate that protein–protein interactions with affinities of K_d ~1 μM or tighter can readily be detected with this method. Together, these technologies provide powerful new tools with which to dissect the intricacies of human biology.

Microarray techniques, high-throughput DNA sequencing, two-hybrid, genetic knockdown, and mass spectrometric protein and metabolite identification experiments have led

to the generation of huge amounts of data. Appropriate data management and interpretation of the deluge of information forthcoming from such studies have relied upon statistical methods, and this new technology, coupled with the flood of DNA sequence information, has led to the development of the fields of **bioinformatics** (Chapter 11) **and systems biology,** new disciplines whose goals are to help manage, analyze, and integrate this flood of biologically important information. Future work at the intersection of bioinformatics, transcript-protein/PTM profiling, and systems biology will revolutionize our understanding of physiology and medicine.

SUMMARY

- A variety of very sensitive techniques can now be applied to the isolation and characterization of genes and to the quantitation of gene products.

- In DNA cloning, a particular segment of DNA is removed from its normal environment using PCR or one of many restriction endonucleases. This is then ligated into a vector in which the DNA segment can be amplified and produced in abundance.

- Cloned or in vitro synthesized DNA can be used as a probe in one of several types of hybridization reactions to detect

- other related or adjacent pieces of DNA, or it can be used to quantitate gene products such as mRNA.

- ■ Manipulation of the DNA to change its structure, so-called genetic engineering, is a key element in cloning (eg, the construction of chimeric molecules) and can also be used to study the function of a certain fragment of DNA and to analyze how genes are regulated.

- ■ Chimeric DNA molecules are introduced into cells to make transfected cells or into the fertilized oocyte to make transgenic animals.

- ■ Techniques involving cloned or synthetic DNA are used to locate genes to specific regions of chromosomes, identify the genes responsible for diseases, study how faulty gene regulation causes disease, diagnose genetic diseases, and increasingly to treat genetic diseases.

REFERENCES

Brass AL, Dykxhoorn DM, Benita Y, et al: Identification of host proteins required for HIV infection through a functional genomic screen. Science 2008;319:921.

Core LJ, Waterfall JJ, Lis JT: Nascent RNA sequencing reveals widespread pausing and divergent initiation at human promoters. Science 2008;322:1845–1848.

Churchman LS, Weissman JS: Nascent transcript sequencing visualizes transcription at nucleotide resolution. Nature 2011;469:368–373.

Friedman A, Perrimon N: Genome-wide high-throughput screens in functional genomics. Curr Opin Gen Dev 2004;14:470.

Gandhi TK, Zhong J, Mathivanan S, et al: Analysis of the human protein interactome and comparison with yeast, worm and fly interaction datasets. Nat Genet 2006;38:285.

Gerstein MB, Lu ZJ, Van Nostrand EL, et al: Integrative analysis of the *Caenorhabditis elegans* genome by the modENCODE project. Science 2010;330:1775–1787.

Gibson DG, Glass JI, Lartigue C, et al: Creation of a bacterial cell controlled by a chemically synthesized genome. Science 2010;329:52–56.

Gilchrist DA, Fargo DC, Adelman K: Using ChIP-chip and ChIP-seq to study the regulation of gene expression: genome-wide localization studies reveal widespread regulation of transcription elongation. Methods 2009;48:398–408.

Isaacs FJ, Carr PA, Wang HH, et al: Precise manipulation of chromosomes in vivo enables genome-wide codon replacement. Science. 2011;333:348–353.

Kodzius R, Kojima M, Nishiyori H, et al: CAGE: cap analysis of gene expression. Nat Meth 2006;3:211–222.

Martin JB, Gusella JF: Huntington's disease: pathogenesis and management. N Engl J Med 1986;315:1267.

Myers RM, Stamatoyannopoulos J, Snyder M, et al: A user's guide to the encyclopedia of DNA elements (ENCODE). PLoS Biol. 2011;9:e1001046.

Sambrook J, Fritsch EF, Maniatis T: *Molecular Cloning: A Laboratory Manual.* Cold Spring Harbor Laboratory Press, 1989.

Schlabach MR, Luo J, Solimini NL, et al: Cancer proliferation gene discovery through functional genomics. Science 2008;319:620.

Suter B, Kittanakom S, Stagljar I: Interactive proteomics: what lies ahead? Biotechniques 2008;44:681.

Spector DL, Goldman RD, Leinwand LA: *Cells: A Laboratory Manual.* Cold Spring Harbor Laboratory Press, 1998.

Takahashi K, Tanabe K, Ohnuki M, et al: Induction of pluripotent stem cells from adult human fibroblasts by defined factors. Cell 2007;131:861.

The ENCODE Project Consortium: Identification and analysis of functional in 1% of the human genome by the ENCODE Pilot Project. Nature 2007;447:799.

Watson JD, Gilman M, Witkowski JA, et al: *Recombinant DNA,* 2nd ed. Scientific American Books. Freeman, 1992.

Weatherall DJ: *The New Genetics and Clinical Practice,* 3rd ed. Oxford University Press, 1991.

Wernig M, Meissner A, Foreman R, et al: In vitro reprogramming of fibroblasts into a pluripotent ES-cell-like state. Nature 2007;448:318.

Wheeler DA, Srinivasan M, Egholm M, et al: The complete genome of an individual by massively parallel DNA sequencing. Nature 2008;451:872.

Wold B, Myers RM: Sequence census methods for functional genomics. Nat Meth 2008;5:19.

GLOSSARY

ARS: Autonomously replicating sequence; the origin of replication in yeast.

Autoradiography: The detection of radioactive molecules (eg, DNA, RNA, and protein) by visualization of their effects on photographic or x-ray film.

Bacteriophage: A virus that infects a bacterium.

Blunt-ended DNA: Two strands of a DNA duplex having ends that are flush with each other.

CAGE: Cap analysis of gene expression. A method that allows the selective capture, amplification, cloning and sequencing of mRNAs via the 5′-Cap structure.

cDNA: A single-stranded DNA molecule that is complementary to an mRNA molecule and is synthesized from it by the action of reverse transcriptase.

Chimeric molecule: A molecule (eg, DNA, RNA, and protein) containing sequences derived from two different species.

ChIP, chromatin immunoprecipitation: A technique that the determination of the exact localization of a particular protein, or protein isoform, on any particular genomic location in a living cell. The method is based upon cross-linking of living cells, cell disruption, DNA fragmentation, and immunoprecipitation with specific antibodies that purify the cognate protein crosss-linked to DNA. Cross-links are reversed, associated DNA purified and specific sequences that are purified are measured using any of several different methods.

ChIP-chip, chromatin immunoprecipitation assayed via a microarray chip hybridization read-out: A hybridization-based method that uses chromatin immunoprecipitation (ChIP) techniques to map, genome-wide, the in vivo sites of binding of specific proteins within chromatin in living cells. Sequence-binding is determined by annealing fluorescently labeled DNA samples to microarrays (array).

ChIP-Seq, chromatin immunoprecipitation assayed via a NGS/deep sequencing read-out: Genomic DNA binding location in a ChIP determined by high-throughput deep\sequencing, rather than hybridization to microarrays.

Clone: A large number of organisms, cells or molecules that are identical with a single parental organism cell or molecule.

Copy number variation (CNV): Change in the copy number of specific genomic regions of DNA between two or more individuals. CNVs can be as large as 10^6 bp of DNA and include deletions or insertions.

Cosmid: A plasmid into which the DNA sequences from bacteriophage lambda that are necessary for the packaging of DNA (cos sites) have been inserted; this permits the plasmid DNA to be packaged in vitro.

ENCODE project: Encyclopedia of DNA elements project; an effort of multiple laboratories throughout the world to provide a detailed, biochemically informative representation of the human genome using high-throughput sequencing methods to identify and catalog the functional elements within a single restricted portion (~1%; 30,000,000 bp) of one human chromosome.

Endonuclease: An enzyme that cleaves internal bonds in DNA or RNA.

Epigenetic code: The patterns of modification of chromosomal DNA (ie, cytosine methylation) and nucleosomal histone post-translational modifications. These changes in modification status can lead to dramatic alterations in gene expression. Notably though, the actual underlying DNA sequence involved does not change.

Excinuclease: The excision nuclease involved in nucleotide exchange repair of DNA.

Exome: The nucleotide sequence of the entire complement of mRNA exons expressed in a particular cell, tissue, organ or organism. The exome differs from the transcriptome that represents the entire collection of genome transcripts; the exome represents a subset of the RNA sequences composing the transcriptome.

Exon: The sequence of a gene that is represented (expressed) as mRNA.

Exonuclease: An enzyme that cleaves nucleotides from either the 3′ or 5′ ends of DNA or RNA.

Fingerprinting: The use of RFLPs or repeat sequence DNA to establish a unique pattern of DNA fragments for an individual.

FISH: Fluorescence in situ hybridization, a method used to map the location of specific DNA sequences within fixed nuclei.

Footprinting: DNA with protein bound is resistant to digestion by DNase enzymes. When a sequencing reaction is performed using such DNA, a protected area, representing the "footprint" of the bound protein, will be detected because nucleases are unable to cleave the DNA directly bound by the protein.

GRO-Seq, global run-on sequencing: A method where nascent transcripts are specifically captured and sequenced using NGS/deep sequencing. This methods allows for the mapping of the location of active transcription complexes.

Hairpin: A double-helical stretch formed by base pairing between neighboring complementary sequences of a single strand of DNA or RNA.

Hybridization: The specific reassociation of complementary strands of nucleic acids (DNA with DNA, DNA with RNA, or RNA with RNA).

Insert: An additional length of base pairs in DNA, generally introduced by the techniques of recombinant DNA technology.

Intron: The sequence of an mRNA-encoding gene that is transcribed but excised before translation. tRNA genes can also contain introns.

Library: A collection of cloned fragments that represents, in aggregate, the entire genome. Libraries may be either genomic DNA (in which both introns and exons are represented) or cDNA (in which only exons are represented).

Ligation: The enzyme-catalyzed joining in phosphodiester linkage of two stretches of DNA or RNA into one; the respective enzymes are DNA and RNA ligases.

Lines: Long interspersed repeat sequences.

Microsatellite polymorphism: Heterozygosity of a certain microsatellite repeat in an individual.

Microsatellite repeat sequences: Dispersed or group repeat sequences of 2–5 bp repeated up to 50 times. May occur at 50–100 thousand locations in the genome.

miRNAs: MicroRNAs, 21–22 nucleotide long RNA species derived from RNA polymerase II transcription units, and 500–1500 bp in length via RNA processing. These RNAs, recently discovered, are thought to play crucial roles in gene regulation.

NET-seq, native elongating sequencing: Genome-wide analysis of eukaryotic mRNA nascent chain 3′-ends maped at nucleotide-level resolution. RNA Polymerase II elongation complexes are captured by immunopurification with anti-Pol II IgG and nascent RNAs containing a free 3′OH group are tagged via ligation with an RNA linker and subsequently amplified by PCR and subjected to deep sequencing.

Nick translation: A technique for labeling DNA based on the ability of the DNA polymerase from *E coli* to degrade a strand of DNA that has been nicked and then to resynthesize the strand; if a radioactive nucleoside triphosphate is employed, the rebuilt strand becomes labeled and can be used as a radioactive probe.

Northern blot: A method for transferring RNA from an agarose or polyacrylamide gel to a nitrocellulose filter, on which the RNA can be detected by a suitable probe.

Oligonucleotide: A short, defined sequence of nucleotides joined together in the typical phosphodiester linkage.

Ori: The origin of DNA replication.

PAC: A high-capacity (70–95 kb) cloning vector based upon the lytic *E coli* bacteriophage P1 that replicates in bacteria as an extrachromosomal element.

Palindrome: A sequence of duplex DNA that is the same when the two strands are read in opposite directions.

Plasmid: A small, extrachromosomal, circular molecule of DNA that replicates independently of the host DNA.

Polymerase chain reaction (PCR): An enzymatic method for the repeated copying (and thus amplification) of the two strands of DNA that make up a particular gene sequence.

Primosome: The mobile complex of helicase and primase that is involved in DNA replication.

Probe: A molecule used to detect the presence of a specific fragment of DNA or RNA in, for instance, a bacterial colony that is formed from a genetic library or during analysis by blot transfer techniques; common probes are cDNA molecules, synthetic oligodeoxynucleotides of defined sequence, or antibodies to specific proteins.

Proteome: The entire collection of expressed proteins in an organism.

Pseudogene: An inactive segment of DNA arising by mutation of a parental active gene; typically generated by transposition of a cDNA copy of an mRNA.

Recombinant DNA: The altered DNA that results from the insertion of a sequence of deoxynucleotides not previously present into an existing molecule of DNA by enzymatic or chemical means.

Restriction enzyme: An endodeoxynuclease that causes cleavage of both strands of DNA at highly specific sites dictated by the base sequence.

Reverse transcription: RNA-directed synthesis of DNA, catalyzed by reverse transcriptase.

RNA-Seq: A method where cellular RNA populations are converted, via linker ligation and PCR into cDNAs that are then subjected to deep sequencing to determine the complete sequence of essentially all RNAs in the preparation.

RT-PCR: A method used to quantitate mRNA levels that relies upon a first step of cDNA copying of mRNAs catalyzed by reverse transcriptase prior to PCR amplification and quantitation.

Signal: The end product observed when a specific sequence of DNA or RNA is detected by autoradiography or some other method. Hybridization with a complementary radioactive polynucleotide (eg, by Southern or Northern blotting) is commonly used to generate the signal.

Sines: Short interspersed repeat sequences.

SiRNAs: Silencing RNAs, 21–25 nt in length generated by selective nucleolytic degradation of double-stranded RNAs of cellular or viral origin. SiRNAs anneal to various specific sites within target in RNAs leading to mRNA degradation, hence gene "knockdown."

SNP: Single nucleotide polymorphism. Refers to the fact that single nucleotide genetic variation in genome sequence exists at discrete loci throughout the chromosomes. Measurement of allelic SNP differences is useful for gene mapping studies.

snRNA: Small nuclear RNA. This family of RNAs is best known for its role in mRNA processing.

Southern blot: A method for transferring DNA from an agarose gel to nitrocellulose filter, on which the DNA can be detected by a suitable probe (eg, complementary DNA or RNA).

Southwestern blot: A method for detecting protein–DNA interactions by applying a labeled DNA probe to a transfer membrane that contains a renatured protein.

Spliceosome: The macromolecular complex responsible for precursor mRNA splicing. The spliceosome consists of at least five small nuclear RNAs (snRNA; U1, U2, U4, U5, and U6) and many proteins.

Splicing: The removal of introns from RNA accompanied by the joining of its exons.

Sticky-ended DNA: Complementary single strands of DNA that protrude from opposite ends of a DNA duplex or from the ends of different duplex molecules (see also Blunt-ended DNA, above).

Tandem: Used to describe multiple copies of the same sequence (eg, DNA) that lie adjacent to one another.

Terminal transferase: An enzyme that adds nucleotides of one type (eg, deoxyadenonucleotidyl residues) to the 3' end of DNA strands.

Transcription: Template DNA-directed synthesis of nucleic acids, typically DNA-directed synthesis of RNA.

Transcriptome: The entire collection of expressed RNAs in a cell, tissue, organ, or organism.

Transgenic: Describing the introduction of new DNA into germ cells by its injection into the nucleus of the ovum.

Translation: Synthesis of protein using mRNA as template.

Vector: A plasmid or bacteriophage into which foreign DNA can be introduced for the purposes of cloning.

Western blot: A method for transferring protein to a nitrocellulose filter, on which the protein can be detected by a suitable probe (eg, an antibody).

Exam Questions

Section IV

1. Which of the following statements about β,γ-methylene and β,γ-imino derivatives of purine and pyrimidine triphosphates is CORRECT?
   A. They are potential anticancer drugs.
   B. They are precursors of B-vitamins.
   C. They readily undergo hydrolytic removal of the terminal phosphate.
   D. They can be used to implicate involvement of nucleotide triphosphates by effects other than phosphoryl transfer.
   E. They serve as polynucleotide precursors.

2. Which of the following statements about nucleotide structures is NOT CORRECT?
   A. Nucleotides are polyfunctional acids.
   B. Caffeine and theobromine differ structurally solely with respect to the number of methyl groups attached to their ring nitrogens.
   C. The atoms of the purine ring portion of pyrimidines are numbered in the same direction as those of a pyrimidine.
   D. NAD^+, FMN, "active methionine" and coenzyme A all are derivatives of ribonucleotides.
   E. 3′,5′-Cyclic AMP and GMP (cAMP and cGMP) serve as second messengers in human biochemistry.

3. Which of the following statements about purine nucleotide metabolism is NOT CORRECT?
   A. An early step in purine biosynthesis is the formation of PRPP (phosphoribosyl 1-pyrophosphate).
   B. Inosine monophosphate (IMP) is a precursor of both AMP and GMP.
   C. Orotic acid is an intermediate in pyrimidine nucleotide biosynthesis.
   D. Humans catabolize uridine and pseudouridine by analogous reactions.
   E. Ribonucleotide reductase converts nucleoside diphosphates to the corresponding deoxyribonucleoside diphosphates.

4. Which of the following statements is NOT CORRECT?
   A. Metabolic disorders are only infrequently associated with defects in the catabolism of purines.
   B. Immune dysfunctions are associated both with a defective adenosine deaminase and with a defective purine nucleoside phosphorylase.
   C. The Lesch–Nyhan syndrome reflects a defect in hypoxanthine-guanine phosphoribosyl transferase.
   D. Xanthine lithiasis can be due to a severe defect in xanthine oxidase.
   E. Hyperuricemia can result from conditions such as cancer characterized by enhanced tissue turnover.

5. Which of the following components are found in DNA?
   A. A phosphate group, adenine, and ribose.
   B. A phosphate group, guanine, and deoxyribose.
   C. Cytosine and ribose.
   D. Thymine and deoxyribose.
   E. A phosphate group and adenine.

6. The backbone of a DNA molecule is composed of which of the following?
   A. Alternating sugars and nitrogenous bases.
   B. Nitrogenous bases alone.
   C. Phosphate groups alone.
   D. Alternating phosphate and sugar groups.
   E. Five carbon sugars alone.

7. The interconnecting bonds that connecting the nucleotides of RNA and DNA are termed:
   A. *N*-glycosidic bonds.
   B. 3′-5′-phosphodiester linkages.
   C. Phosphomonoesters.
   D. 3′-2′-phosphodiester linkages.
   E. Peptide nucleic acid bonds.

8. Which component of the DNA duplex causes the molecule to have a net negative charge at physiological pH?
   A. Deoxyribose
   B. Ribose
   C. Phosphate groups
   D. Chlorine ion
   E. Adenine

9. Which molecular feature listed causes duplex DNA to exhibit a near constant width along its long axis?
   A. A purine nitrogenous base always pairs with another purine nitrogenous base.
   B. A pyrimidine nitrogenous base always pairs with another pyrimidine nitrogenous base.
   C. A pyrimidine nitrogenous base always pairs with a purine nitrogenous base.
   D. Repulsion between phosphate groups keeps the strands a uniform distance apart.
   E. Attraction between phosphate groups keeps the strands a uniform distance apart.

10. The model for DNA replication first proposed by Watson and Crick's posited that every newly replicated double stranded daughter duplex DNA molecule
   A. Was composed of the two strands from the parent DNA molecule.
   B. Contained solely the two newly synthesized strands of DNA.
   C. Contained two strands that are random mixtures of new and old DNA within each strand.
   D. Was composed of one strand derived from the original parental DNA duplex and one strand that was newly synthesized.
   E. Was composed of nucleotide sequences completely distinct from either parental DNA strand.

11. Name the mechanism through which RNAs are synthesized from DNA?
   A. Replicational duplication
   B. Translation
   C. Translesion repair
   D. Transesterification
   E. Transcription

12. Which of the forces or interactions listed below play the predominant role in driving RNA secondary and tertiary structure formation?

    A. Hydrophilic repulsion
    B. Formation of complementary base pair regions
    C. Hydrophobic interaction
    D. van der Waals interactions
    E. Salt bridge formation

13. Name the enzyme that synthesizes RNA from a double stranded DNA template.

    A. RNA-dependent RNA polymerase
    B. DNA-dependent RNA convertase
    C. RNA-dependent replicase
    D. DNA-dependent RNA polymerase
    E. Reverse transcriptase

14. Define the most notable characteristic difference with regard to gene expression between eukaryotes and prokaryotes.

    A. Ribosomal RNA nucleotide lengths
    B. Mitochondria
    C. Lysosomes and peroxisomes
    D. Sequestration of the genomic material in the nucleus
    E. Chlorophyll

15. Which entry below correctly describes the ~number of bp of DNA_____, which is separated into _____chromosomes in a typical diploid human cell in a nonreplicating state?

    A. 64 billion, 23
    B. 6.4 trillion, 46
    C. 23 billion, 64
    D. 64 billion, 46
    E. 6.4 billion, 46

16. What is the approximate number of base pairs associated with a single nucleosome?

    A. 146
    B. 292
    C. 73
    D. 1460
    E. 900

17. All but one of the following histones are found located within the superhelix formed between DNA and the histone octamer; this histone is

    A. Histone H2B
    B. Histone H3
    C. Histone H1
    D. Histone H3
    E. Hisone H4

18. Chromatin can be broadly defined as active and repressed; a subclass of chromatin that is specifically inactivated at certain times within an organism's life and/or in particular sets of differentiated cells is termed

    A. Constitutive euchromatin
    B. Facultative heterochromatin
    C. Euchromatin
    D. Constitutive heterochromatin

19. Which of the following hypothesizes that the physical and functional status of a certain region of genomic chromatin is dependent upon the patterns of specific histone post-translational modifications (PTMs), and/or DNA methylation status?

    A. Morse code
    B. PTM hypothesis
    C. Nuclear body hypothesis
    D. Epigenetic code
    E. Genetic code

20. What is the name of the unusual repeated stretch of DNA localized at the tips of all eukaryotic chromosomes?

    A. Kinetochore
    B. Telomere
    C. Centriole
    D. Chromomere
    E. Micromere

21. Given that DNA polymerases are unable to synthesize DNA without a primer, what molecule serves as the primer for these enzymes during DNA replication?

    A. Five carbon sugars
    B. Deoxyribose alone
    C. A short RNA molecule
    D. Proteins with free hydroxyl groups
    E. Phosphomonoester

22. The discontinuous DNA replication that occurs during replication is catalyzed via the production of small DNA segments termed

    A. Okazaki fragments
    B. Toshihiro pieces
    C. Onishi oligonucleotides
    D. Crick strands
    E. Watson fragments

23. What molecule or force supplies the energy that drives the relief of mechanical strain by DNA gyrase?

    A. Pyrimidine to purine conversion
    B. Hydrolysis of GTP
    C. Hydrolysis of ATP
    D. Glycolysis
    E. A proton gradient molecule or force

24. What is the name of the phase of the cell cycle between the conclusion of cell division and the beginning of DNA synthesis?

    A. G_1
    B. S
    C. G_2
    D. M
    E. G_0

25. At what stage of the cell cycle are key protein kinases, like cyclin-dependent kinase, activated?

    A. Right before mitosis
    B. At the beginning of S phase
    C. Near the end of G_1 phase
    D. At the end of the G_2 phase
    E. All of the above

26. What disease is often associated with a breakdown of a cell's ability to regulate/control its own division?

    A. Kidney disease
    B. Cancer

C. Emphysema

D. Diabetes

E. Heart disease

27. What is the molecular mechanism that is responsible for the quick decrease in the Cdk activity that leads to exit from the M phase and the entry into G_1?

A. Drop in mitotic cyclin concentration

B. Decreased G_1 cyclin concentration

C. Rise in G_2 cyclin concentration

D. Rise in mitotic cyclin concentration

E. Rise in G_1 cyclin concentration

28. The site to which RNA polymerase binds on the DNA template prior to the initiation of transcription.

A. Intron/exon junction

B. Open reading frame DNA the terminator

C. Terminator

D. Initiator methionine codon

E. Promoter

29. The large eukaryotic rRNA genes, such as 18S and 28S RNA-encoding genes, are transcribed by which of the following RNA polymerases?

A. RNA polymerase III

B. RNA-dependent RNA polymerase δ

C. RNA polymerase I

D. RNA polymerase II

E. Mitochondrial RNA polymerase

30. Eukaryotic RNA polymerases all have a requirement for a large variety of accessory proteins to enable them to bind promoters and form physiologically relevant transcription complexes; these proteins are termed

A. Basal or general transcription factors

B. Activators

C. Accessory factors

D. Elongation factors

E. Facilitator polypeptides

31. The DNA segment from which the primary transcript is copied or transcribed is called

A. Coding region

B. Initiator methionine domain

C. Translation unit

D. Transcriptome

E. Initial codon

32. What class of DNA are the eukaryotic rDNA cistrons?

A. Single copy DNA

B. Highly repetitive DNA

C. Moderately repetitive DNA

D. Mixed sequence DNA

33. Modifications to the nucleotides of the pre-tRNAs, pre-rRNAs and pre-mRNAs occur

A. Postprandially

B. Postmitotically

C. Pretranscriptionally

D. Posttranscriptionally

E. Prematurely

34. RNA polymerase II promoters are located on which side of the transcription unit?

A. Internal

B. 3′

C. Nearest the C-terminus

D. Nearest the N-terminus

E. 5′

35. With regard to eukaryotic mRNAs, one of the following is not a normal property of mRNAs.

A. Eukaryotic mRNAs have special modifications at their 5′ (cap) and 3′ (polyA tail) termini.

B. Are attached to ribosomes when they are translated.

C. They are found in the cytoplasm within peroxisomes.

D. Most have a significant noncoding segment that does not direct assembly of amino acids.

E. Contain continuous nucleotide sequences that encode a particular polypeptide.

36. The bond connecting the initiation nucleotide of the mRNA with the 5me-G Cap structure is a

A. 3′-5′ phosphodiester bridge

B. 5′-5′ triphosphate bridge

C. 3′-3′ triphosphate bridge

D. 3′-5′ triphosphate bridge

E. 5′-3′ triphosphate bridge

37. What sequence feature of mature mRNAs listed below is thought to protect mRNAs from degradation?

A. Special post-translational modifications

B. 3′ $Poly(C)_n$ tail

C. 5me-G Cap

D. Introns

E. Lariat structures

38. What could the consequences of inaccurate mRNA splicing be for the RNA?

A. A single base error at a splice junction will cause a large deletion.

B. A single base error at a splice junction will cause a large insertion.

C. A single base error at a splice junction will cause a large inversion.

D. C and E

E. A single base error at a splice junction will change the reading frame and result in mRNA mistranslation.

39. What is the macromolecular complex that associates with introns during mRNA splicing?

A. Splicer

B. Dicer

C. Nuclear body

D. Spliceosome

E. Slicer

40. What reaction does reverse transcriptase catalyze?

A. Translation of RNA to DNA.

B. Transcription of DNA to RNA.

C. Conversion of ribonucleotides into deoxyribonucleotides.

D. Transcription of RNA to DNA.

E. Conversion of a ribonucleotide to deoxynucleotides in the DNA double helix.

41. RNAi or dsRNA-mediated RNA interference mediates
    A. RNA ligation
    B. RNA silencing
    C. RNA inversion
    D. RNA restoration
    E. RNA quelling

42. While the genetic code has 64 codons, there are only 20 naturally occurring amino acids. Consequently, some amino acids are encoded by more than one codon. This feature of the genetic code is an illustration of the genetic code being
    A. Degenerate
    B. Duplicitive
    C. Nonoverlapping
    D. Overlapping
    E. Redundant

43. The genetic code contains _____ termination codons?
    A. 3
    B. 21
    C. 61
    D. 64
    E. 20

44. If a tRNA has the sequence 5′-CAU-3′, what codon would it recognize (ignore wobble base pairing).
    A. 3′-UAC-5′
    B. 3′-AUG-5′
    C. 5′-ATG-3′
    D. 5′-AUC-3′
    E. 5′-AUG-3′

45. What is on the 3′ end of all functional, mature tRNAs?
    A. The cloverleaf loop
    B. The anticodon
    C. The sequence CCA
    D. The codon

46. Most aminoacyl-tRNA synthetases possess an activity that is shared with DNA polymerases. This activity is a _____ function.
    A. Proofreading
    B. Hydrolysis
    C. Proteolytic
    D. Helicase
    E. Endonucleolytic

47. The three distinct phases of protein synthesis, in the CORRECT order are
    A. initiation, termination, elongation
    B. termination, initiation, elongation
    C. initiation, elongation, termination
    D. elongation, initiation, termination
    E. elongation, termination, initiation

48. Which amino acid is the initiating amino acid for all proteins?
    A. Cysteine
    B. Threonine
    C. Tryptophan
    D. Methionine
    E. Glutamic acid

49. The initiator tRNA is placed within the active 80S complex at which of the three canonical ribosomal "sites" during protein synthesis
    A. E site
    B. I site
    C. P site
    D. A site
    E. Releasing factor binding site

50. Name the enzyme that forms the peptide bond during protein synthesis and define its chemical composition.
    A. Pepsynthase, protein
    B. Peptidyl transferase, RNA
    C. Peptidase, glycolipid
    D. Peptidyl transferase, protein
    E. GTPase, glycopeptide

51. Mutations in the middle of an open reading frame that create a stop codon are termed:
    A. Frameshift mutation
    B. Missense mutation
    C. No-nonsense mutation
    D. Point mutation
    E. Nonsense mutation

52. What is the directionality of polypeptide synthesis?
    A. C-terminal to N-terminal direction
    B. N-terminal to 3′ direction
    C. N-terminal to C-terminal direction
    D. 3′ to 5′ direction
    E. 5′ to 3′ direction

53. Which of the following cis-acting elements typically resides adjacent to or overlaps with many prokaryotic promoters?
    A. Regulatory gene
    B. Structural gene(s)
    C. Repressor
    D. Operator
    E. Terminator

54. What is the term applied to a segment of a bacterial chromosome where genes for the enzymes of a particular metabolic pathway are clustered and subject to coordinate control?
    A. Operon
    B. Operator
    C. Promoter
    D. Terminal controller
    E. Origin

55. What is the term applied to the complete collection of proteins present in a particular cell type?
    A. Genome
    B. Peptide collection
    C. Transcriptome
    D. Translatome
    E. Proteome

56. How does nucleosome formation on genomic DNA affect the initiation and/or elongation phases of transcription?
    A. Nucleosomes inhibit access of enzymes involved in all phases of transcription.

B. Nucleosomes recruit histone and DNA modifying enzymes, and the actions of these recruited enzymes affect the access of transcription proteins to DNA.

C. Nucleosomes induce DNA degradation where the DNA contacts the histones.

D. Nucleosomes have no significant effect on transcription.

57. Which types of molecules interact with eukaryotic mRNA gene core promoter sites to facilitate the association of RNA polymerase II?

A. Termination factors

B. Sequence-specific transcription factors (transactivators)

C. Elongation factors.

D. GTPases

E. General, or basal transcription factors (ie, the GTFs)

58. Most eukaryotic transcription factors contain at least two domains, each of which mediate different aspects of transcription factor function; these domains are

A. RNA-binding domain and repression domain.

B. Activation domain and repression domain.

C. DNA-binding domain and activation domain.

D. DNA-binding domain and ligand binding domain.

E. RNA-binding domain and the activation domain.

59. Transcription factors bound at enhancers stimulate the initiation of transcription at the cis-linked core promoter through the action of intermediaries termed

A. Coactivators

B. Cotranscription proteins

C. Corepressors

D. Receptors

E. Coordinators

60. What reactions among transcription proteins greatly expand the diversity of regulatory factors that can be generated from a small number of polypeptides?

A. Recombination

B. Homodimerization

C. Heterozygosity

D. Heterodimerization

E. Trimerization

61. The gene region containing the TATA box and extending to the transcription start site (TSS) is often termed the _____.

A. Polymerase home

B. Initiator

C. Start selector

D. Core promoter

E. Operator

62. Which of the following possible mechanisms for how enhancers can stimulate transcription from great distances are currently thought to be CORRECT?

A. Enhancers can reversibly excise the intervening DNA between enhancers and promoters.

B. RNA polymerase II binds avidly to enhancer sequences.

C. Enhancers unwind DNA.

D. Enhancers can search through DNA and bind directly to the associated core promoter.

E. Enhancers and core promoters are brought into close proximity through DNA loop formation mediated by DNA binding proteins.

63. Which of the following histone amino acids are typically acetylated?

A. Lysine

B. Arginine

C. Asparagine

D. Histidine

E. Leucine

64. Place the following steps in order; what are the steps that occur sequentially during a transcription activation event following the binding of a transcriptional activator to its cognate activator binding site on genomic DNA.

1. The chromatin remodeling complex binds to the core histones at the target region.

2. The combined actions of the various molecular complexes increase promoter accessibility to the transcriptional machinery.

3. The activator recruits a coactivator to a region of chromatin targeted for transcription.

4. Transcriptional machinery assembles at the site where transcription will be initiated.

5. The coactivator acetylates the core histones of nearby nucleosomes.

A. 1 – 2 – 3 – 4 – 5

B. 3 – 1 – 5 – 2 – 4

C. 3 – 5 – 1 – 2 – 4

D. 5 – 3 – 1 – 2 – 4

E. 3 – 5 – 1 – 4 – 2

65. What strategy in transcription factor research allows for the simultaneous identification of all of the genomic sites bound by a given transcription factor under a given set of physiological conditions, monitoring and hence allowing for insights into how gene transcription networks are coordinately regulated?

A. Systematic deletion mapping.

B. DNAase I sensitivity.

C. Chromatin immunoprecipitation (ChIP).

D. FISH.

E. Fluorescence lifetime imaging microscopy.

66. Which sequences extend between the 5′methylguanosine cap present on eukaryotic mRNAs to the AUG initiation codon?

A. Stop codon

B. Last exon

C. Last intron

D. 3′ UTR

E. 5′ UTR

67. Which of the following features of eukaryotic mRNA contribute importantly to message half-life?

A. 5′ UTR sequences

B. The promoter

C. The operator

D. 3′ UTR and poly(A) tail

E. The first intron

Biochemistry of Extracellular & Intracellular Communication

C H A P T E R

40

Membranes: Structure & Function

Robert K. Murray, MD, PhD & Daryl K. Granner, MD

OBJECTIVES

After studying this chapter, you should be able to:

- Know that biological membranes are mainly composed of a lipid bilayer and associated proteins and glycoproteins. The major lipids are phospholipids, cholesterol, and glycosphingolipids.
- Appreciate that membranes are asymmetric, dynamic structures containing a mixture of integral and peripheral proteins.
- Know the fluid mosaic model of membrane structure and that it is widely accepted, with lipid rafts, caveolae, and tight junctions being specialized features.
- Understand the concepts of passive diffusion, facilitated diffusion, active transport, endocytosis, and exocytosis.
- Recognize that transporters, ion channels, the Na$^+$–K$^+$-ATPase, receptors, and gap junctions are important participants in membrane function.
- Know that a variety of disorders result from abnormalities of membrane structure and function, including familial hypercholesterolemia, cystic fibrosis, hereditary spherocytosis, and many others.

BIOMEDICAL IMPORTANCE

Membranes are highly fluid, dynamic structures consisting of a lipid bilayer and associated proteins. **Plasma membranes** form closed compartments around the cytoplasm to define cell boundaries. The plasma membrane has **selective permeabilities** and acts as a barrier, thereby maintaining differences in composition between the inside and outside of the cell. The selective permeabilities for substrates and ions are provided mainly by specific proteins named **transporters** and **ion channels.** The plasma membrane also exchanges material with the extracellular environment by **exocytosis** and **endocytosis,** and there are special areas of membrane structure—**gap junctions**—through which adjacent cells

exchange material. In addition, the plasma membrane plays key roles in **cell–cell interactions** and in **transmembrane signaling.**

Membranes also form **specialized compartments** within the cell. Such intracellular membranes help **shape** many of the morphologically distinguishable structures (organelles), eg, mitochondria, ER, Golgi, secretory granules, lysosomes, and the nucleus. Membranes localize **enzymes,** function as integral elements in **excitation-response coupling,** and provide sites of **energy transduction,** such as in photosynthesis and oxidative phosphorylation.

Changes in membrane components can affect water balance and ion flux, and therefore many processes within the cell. Specific deficiencies or alterations of certain membrane components (eg, caused by mutations genes encoding membrane proteins) lead to a variety of **diseases** (see Table 40–7). In short, normal cellular function depends on normal membranes.

MAINTENANCE OF A NORMAL INTRA- & EXTRACELLULAR ENVIRONMENT IS FUNDAMENTAL TO LIFE

Life originated in an aqueous environment; enzyme reactions, cellular and subcellular processes, and so forth have therefore evolved to work in this milieu, encapsulated within a cell.

The Body's Internal Water Is Compartmentalized

Water makes up about **60%** of the lean body mass of the human body and is distributed in two large compartments.

Intracellular Fluid (ICF)

This compartment constitutes **two-thirds** of total body water and provides a specialized environment for the cell (1) to make, store, and utilize energy; (2) to repair itself; (3) to replicate; and (4) to perform cell-specific functions.

Extracellular Fluid (ECF)

This compartment contains about **one-third** of total body water and is distributed between the plasma and interstitial compartments. The extracellular fluid is a **delivery system.** It brings to the cells nutrients (eg, glucose, fatty acids, and amino acids), oxygen, various ions and trace minerals, and a variety of regulatory molecules (hormones) that coordinate the functions of widely separated cells. Extracellular fluid **removes** CO_2, waste products, and toxic or detoxified materials from the immediate cellular environment.

The Ionic Compositions of Intracellular & Extracellular Fluids Differ Greatly

As illustrated in **Table 40–1**, the **internal environment** is rich in K^+ and Mg^{2+}, and phosphate is its major inorganic anion. The cytosol of cells contains a high concentration of protein that acts as a major intracellular buffer. **Extracellular fluid** is characterized by high Na^+ and Ca^{2+} content, and Cl^- is the major anion. Why are there such differences? It is thought that the primordial sea in which life originated was rich in K^+ and Mg^{2+}. It, therefore, follows that enzyme reactions and other biologic processes evolved to function best in that environment—hence, the high concentration of these ions within cells. Vast changes would have been required for evolution of a completely new set of biochemical and physiologic machinery; instead, as it happened, **cells developed barriers**—membranes

TABLE 40–1 Comparison of the Mean Concentrations of Various Molecules Outside and Inside a Mammalian Cell

Substance	Extracellular Fluid	Intracellular Fluid
Na^+	140 mmol/L	10 mmol/L
K^+	4 mmol/L	140 mmol/L
Ca^{2+} (free)	2.5 mmol/L	0.1 µmol/L
Mg^{2+}	1.5 mmol/L	30 mmol/L
Cl^-	100 mmol/L	4 mmol/L
HCO_3^-	27 mmol/L	10 mmol/L
PO_4^{3-}	2 mmol/L	60 mmol/L
Glucose	5.5 mmol/L	0–1 mmol/L
Protein	2 g/dL	16 g/dL

with associated "pumps" such as the Na^+–K^+-ATPase (see below)—to maintain the internal microenvironment.

MEMBRANES ARE COMPLEX STRUCTURES COMPOSED OF LIPIDS, PROTEINS, & CARBOHYDRATE-CONTAINING MOLECULES

We shall mainly discuss the membranes present in eukaryotic cells, although many of the principles described also apply to the membranes of prokaryotes. The various cellular membranes have **different compositions,** as reflected in the **ratio of protein to lipid** (**Figure 40–1**). This is not surprising, given their divergent functions. Membranes are sheet-like enclosed structures consisting of an asymmetric lipid bilayer with distinct inner and outer surfaces. These sheet-like structures are **noncovalent assemblies** that form spontaneously in water due to the amphipathic nature of lipids. Many different proteins are located in membranes, where they carry out specific functions.

The Major Lipids in Mammalian Membranes Are Phospholipids, Glycosphingolipids & Cholesterol

Phospholipids

Of the two major phospholipid classes present in membranes, **phosphoglycerides** are the more common and consist of a glycerol backbone to which are attached two fatty acids in ester linkages and a phosphorylated alcohol (**Figure 40–2**). The **fatty acid** constituents are usually even-numbered carbon molecules, most commonly containing 16 or 18 carbons. They are unbranched and can be saturated or unsaturated with one or more *cis* double bonds. The **simplest** phosphoglyceride is

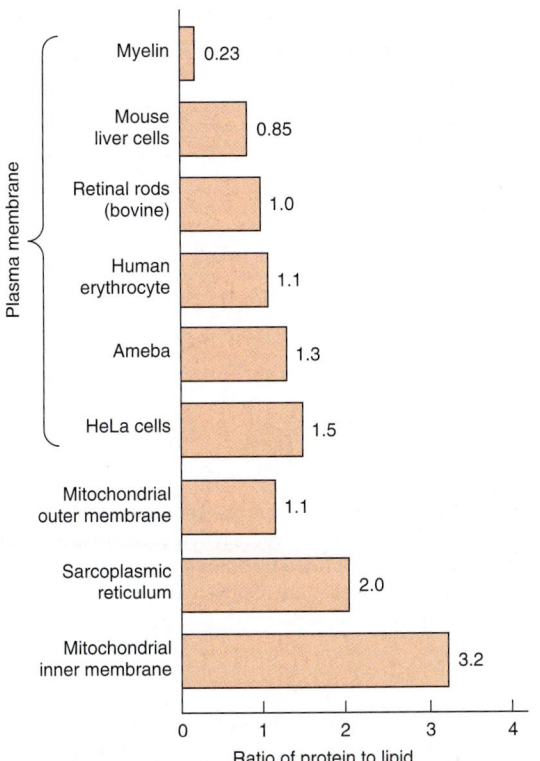

FIGURE 40–1 **Ratio of protein to lipid in different membranes.** Proteins equal or exceed the quantity of lipid in nearly all membranes. The outstanding exception is myelin, an electrical insulator found on many nerve fibers.

phosphatidic acid, which is 1,2-diacylglycerol 3-phosphate, a key intermediate in the formation of other phosphoglycerides (Chapter 24). In most phosphoglycerides present in membranes, the 3-phosphate is esterified to an **alcohol** such as choline, ethanolamine, glycerol, inositol or serine (Chapter 15). Phosphatidylcholine is generally the major phosphoglyceride by mass in the membranes of human cells.

The second major class of phospholipids comprises **sphingomyelin** (Figure 15–13), which contains a sphingosine backbone rather than glycerol. A fatty acid is attached by an

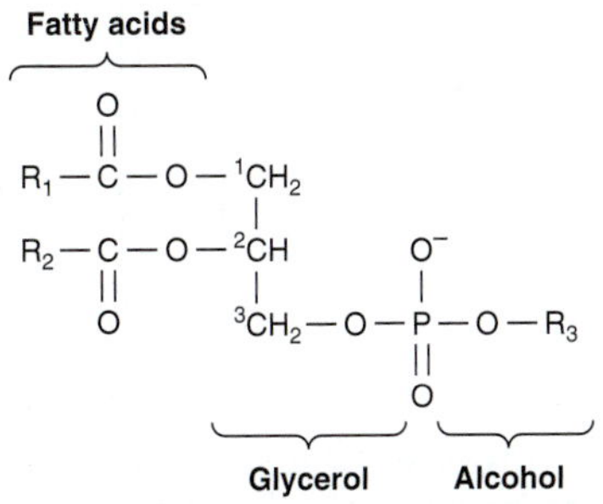

Fatty acids

$$
\begin{array}{c}
\quad\quad\;\; \overset{\displaystyle O}{\underset{\displaystyle \|}{}} \\
R_1 - C - O - {}^{1}CH_2 \\
\quad\quad\;\; \overset{\displaystyle O}{\underset{\displaystyle \|}{}}\quad\quad\quad O^{-} \\
R_2 - C - O - {}^{2}CH \quad\quad\; | \\
\quad\quad\;\; \overset{\displaystyle O}{\underset{\displaystyle \|}{}}\quad\quad {}^{3}CH_2 - O - P - O - R_3 \\
\quad\quad\quad\quad\quad\quad\quad\quad\quad\quad \| \\
\quad\quad\quad\quad\quad\quad\quad\quad\quad\quad O
\end{array}
$$

Glycerol   Alcohol

FIGURE 40–2 **A phosphoglyceride showing the fatty acids (R₁ and R₂), glycerol, and a phosphorylated alcohol component.** Saturated fatty acids are usually attached to carbon 1 of glycerol, and unsaturated fatty acids to carbon 2. In phosphatidic acid, R_3 is hydrogen.

amide linkage to the amino group of sphingosine, forming **ceramide.** When the primary hydroxyl group of sphingosine is esterified to phosphorylcholine, sphingomyelin is formed. As the name implies, sphingomyelin is prominent in myelin sheaths.

Glycosphingolipids

The glycosphingolipids (GSLs) are sugar-containing lipids built on a backbone of **ceramide;** they include **galactosyl-** and **glucosyl- ceramide** (cerebrosides) and the **gangliosides.** Their structures are described in Chapter 15. They are mainly located in the plasma membranes of cells, displaying their sugar components to the exterior of the cell.

Sterols

The most common sterol in the membranes of animal cells is **cholesterol** (Chapter 15), which resides mainly in their **plasma membranes,** but can also be found in lesser quantities in mitochondria, Golgi complexes, and nuclear membranes. Cholesterol intercalates among the phospholipids of the membrane, with its hydroxyl group at the aqueous interface and the remainder of the molecule within the leaflet. Its effect on the fluidity of membranes will be discussed subsequently. From a nutritional viewpoint, it is important to know that cholesterol is not present in plants.

Lipids can be separated from one another and quantitated by techniques such as column, thin-layer, and gas-liquid chromatography and their structures can be established by mass spectrometry and other techniques.

Membrane Lipids Are Amphipathic

All major lipids in membranes contain both hydrophobic and hydrophilic regions and are therefore termed **amphipathic.** If the hydrophobic region were separated from the rest of the molecule, it would be insoluble in water but soluble in oil. Conversely, if the hydrophilic region were separated from the rest of the molecule, it would be insoluble in oil but soluble in water. The amphipathic nature of a phospholipid is represented in **Figure 40–3** and also Figure 15–24. Thus, the **polar head groups** of the phospholipids and the hydroxyl group of cholesterol interface with the aqueous environment; a similar situation applies to the **sugar moieties** of the GSLs (see below).

Saturated fatty acids have straight tails, whereas unsaturated fatty acids, which generally exist in the *cis* form in membranes, make kinked tails (Figure 40–3). As more kinks are inserted in the tails, the lipids become less tightly packed and the membrane more fluid. The problem caused by the presence of **trans fatty acids** in membrane lipids is described in Chapter 15.

Detergents are amphipathic molecules that are important in biochemistry as well as in the household. The molecular structure of a detergent is not unlike that of a phospholipid. Certain detergents are widely used to **solubilize** membrane proteins and in their **purification.** The hydrophobic end of the detergent binds to hydrophobic regions of the proteins,

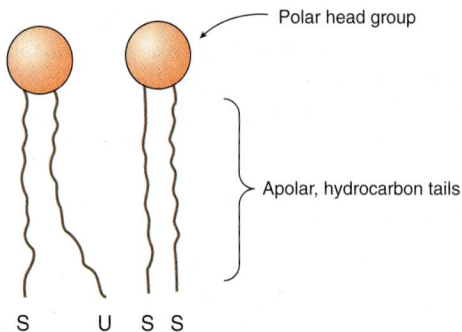

FIGURE 40–3 **Diagrammatic representation of a phospholipid or other membrane lipid.** The polar head group is hydrophilic, and the hydrocarbon tails are hydrophobic or lipophilic. The fatty acids in the tails are saturated (S) or unsaturated (U); the former are usually attached to carbon 1 of glycerol and the latter to carbon 2 (see Figure 40–2). Note the kink in the tail of the unsaturated fatty acid (U), which is important in conferring increased membrane fluidity.

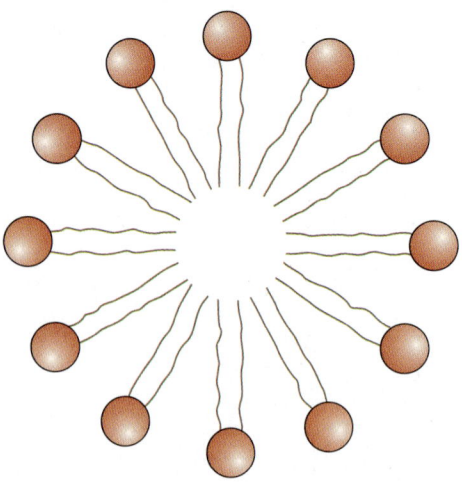

FIGURE 40–4 **Diagrammatic cross-section of a micelle.** The polar head groups are bathed in water, whereas the hydrophobic hydrocarbon tails are surrounded by other hydrocarbons and thereby protected from water. Micelles are relatively small (compared with lipid bilayers) spherical structures.

displacing most of their bound lipids. The polar end of the detergent is free, bringing the proteins into solution as detergent–protein complexes, usually also containing some residual lipids.

Membrane Lipids Form Bilayers

The amphipathic character of phospholipids suggests that the two regions of the molecule have incompatible solubilities; however, in a solvent such as water, phospholipids organize themselves into a form that thermodynamically serves the solubility requirements of both regions. A **micelle** (**Figure 40–4** and Figure 15–24) is such a structure; the hydrophobic regions are shielded from water, while the hydrophilic polar groups are immersed in the aqueous environment. However, micelles are usually **relatively small** in size (eg, ~200 nm) and thus are limited in their potential to form membranes. Detergents commonly form micelles.

As was recognized in 1925 by Gorter and Grendel, a **bimolecular layer,** or **lipid bilayer,** can also satisfy the thermodynamic requirements of amphipathic molecules in an aqueous environment. Bilayers are the **key structures** in biological membranes. A bilayer exists as a sheet in which the hydrophobic regions of the phospholipids are sequestered from the aqueous environment, while the hydrophilic regions are exposed to water (**Figure 40–5** and Figure 15–24). The ends or edges of the bilayer sheet can be eliminated by folding the sheet back upon itself to form an enclosed vesicle with no edges. The closed bilayer provides one of the most essential properties of membranes. It is **impermeable to most water-soluble molecules** since they would be insoluble in the hydrophobic core of the bilayer.

Lipid bilayers are formed by **self-assembly,** driven by the **hydrophobic effect** (Chapter 2). When lipid molecules come together in a bilayer, the entropy of the surrounding solvent molecules increases due to the release of immobilized water.

Two questions arise from consideration of the above. First, how many biologic materials are **lipid-soluble** and can therefore readily enter the cell? Gases such as oxygen, CO_2, and nitrogen—small molecules with little interaction with solvents—readily diffuse through the hydrophobic regions of the membrane. The **permeability coefficients** of several ions and of a number of other molecules in a lipid bilayer are shown in **Figure 40–6**. The three electrolytes shown (Na^+, K^+, and Cl^-) cross the bilayer much more slowly than water. In general, the permeability coefficients of small molecules in a lipid bilayer **correlate with their solubilities in nonpolar solvents.** For instance, **steroids** more readily traverse the lipid bilayer compared with electrolytes. The high permeability coefficient of **water** itself is surprising, but is partly explained by its small size and relative lack of charge. Many **drugs** are hydrophobic and can readily cross membranes and enter cells.

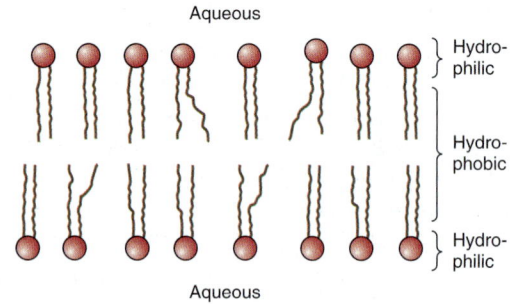

FIGURE 40–5 **Diagram of a section of a bilayer membrane formed from phospholipid molecules.** The unsaturated fatty acid tails are kinked and lead to more spacing between the polar head groups, hence to more room for movement. This in turn results in increased membrane fluidity. (Slightly modified and reproduced, with permission, from Stryer L: *Biochemistry*, 7th ed. Freeman, 2012. Copyright © 2012 by W H. Freeman and Company.)

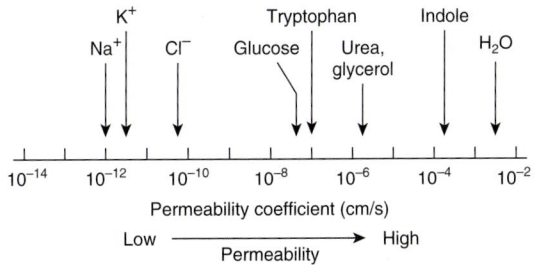

FIGURE 40–6 **Permeability coefficients of water, some ions, and other small molecules in lipid bilayer membranes.** The permeability coefficient is a measure of the ability of a molecule to diffuse across a permeability barrier. Molecules that move rapidly through a given membrane are said to have a high permeability coefficient. (Slightly modified and reproduced, with permission, from Stryer L: *Biochemistry,* 7th ed. Freeman, 2012. Copyright © 2012)

The second question concerns **molecules that are not lipid-soluble:** How are the transmembrane concentration gradients for nonlipid-soluble molecules maintained? The answer is that **membranes contain proteins,** many of which span the lipid bilayer. Such proteins form **channels** for the movement of ions and small molecules or serve as **transporters** for molecules that otherwise could not pass the bilayer. These structures are described below.

Membrane Proteins Are Associated with the Lipid Bilayer

Membrane **phospholipids** act as a solvent for membrane proteins, creating an environment in which the latter can function. As described in Chapter 5, the **α-helical structure of proteins** minimizes the hydrophilic character of the peptide bonds themselves. Thus, proteins can be amphipathic and form an integral part of the membrane by having hydrophilic regions protruding at the inside and outside faces of the membrane but connected by a hydrophobic region traversing the hydrophobic core of the bilayer. In fact, those portions of membrane proteins that traverse membranes do contain substantial numbers of hydrophobic amino acids and almost invariably have a high α-helical content. For many membranes, a stretch of **~20 amino acids** in an α-helix **will span the bilayer.**

It is possible to calculate whether a particular sequence of amino acids present in a protein is consistent with a **transmembrane location.** This can be done by consulting a Table that lists the hydrophobicities of each of the 20 common amino acids and the free energy values for their transfer from the interior of a membrane to water. Hydrophobic amino acids have positive values; polar amino acids have negative values. The total free energy values for transferring successive sequences of 20 amino acids in the protein are plotted, yielding a so-called **hydropathy plot.** Values of over 20 kcal mol^{-1} are consistent with—but do not prove—the interpretation that the hydrophobic sequence is a transmembrane segment.

Another aspect of the interaction of lipids and proteins is that some proteins are **anchored** to one leaflet of the bilayer by

covalent linkages to certain lipids. **Palmitate** and **myristate** are fatty acids involved in such linkages to specific cytosolic proteins. A number of cell surface proteins (see Chapter 47) are linked to the plasma membrane via **glycophosphatidylinositol (GPI) structures.**

Different Membranes Have Different Protein Compositions

The **number of different proteins** in a membrane varies from less than a dozen in the sarcoplasmic reticulum of muscle cells to over 100 in plasma membranes. Membrane proteins can be separated from one another using **sodium dodecyl sulfate polyacrylamide gel electrophoresis** (SDS–PAGE), a technique that separates proteins based on their molecular mass. Using standard proteins of known molecular mass as a comparison, one can estimate the approximate molecular mass of an unknown protein via SDS–PAGE. SDS is a powerful detergent that disrupts protein–lipid interactions and thereby solubilizes membrane proteins. SDS also disrupts protein–protein interactions and unfolds or denatures proteins. In the absence of SDS, few membrane proteins would remain soluble.

Proteins are the **major functional molecules** of membranes and consist of **enzymes, pumps** and **transporters, channels, structural components, antigens** (eg, for histocompatibility), and **receptors** for various molecules. Because every type of membrane possesses a different complement of proteins, there is no such thing as a typical membrane structure. The enzymatic properties of several different membranes are shown in **Table 40–2.**

Membranes Are Dynamic Structures

Membranes and their components are **dynamic structures.** The **lipids and proteins** in membranes undergo **turnover,** just as they do in other compartments of the cell. Different lipids have

TABLE 40–2 Enzymatic Markers of Different Membranes[1]

Membrane	Enzyme
Plasma	5′-Nucleotidase
	Adenylyl cyclase
	Na$^+$–K$^+$-ATPase
Endoplasmic reticulum	Glucose-6-phosphatase
Golgi apparatus	
Cis	GlcNAc transferase I
Medial	Golgi mannosidase II
Trans	Galactosyl transferase
TGN	Sialyl transferase
Inner mitochondrial membrane	ATP synthase

[1]Membranes contain many proteins, some of which have enzymatic activity. Some of these enzymes are located only in certain membranes and can therefore be used as markers to follow the purification of these membranes.
Abbreviation: TGN, *trans* Golgi network.

different turnover rates, and the turnover rates of individual species of membrane proteins may vary widely. The membrane itself can turn over even more rapidly than any of its constituents. This is discussed in more detail in the section on endocytosis.

Another indicator of the dynamic nature of membranes is that a variety of studies have shown that lipids and certain proteins exhibit **lateral diffusion** in the plane of their membranes. Some proteins do not exhibit lateral diffusion because they are anchored to the underlying actin cytoskeleton. In contrast, the **transverse** movement of lipids across the membrane (**flip–flop**) is extremely slow (see below) and does not occur at all in the case of membrane proteins.

Membranes Are Asymmetric Structures

Proteins have unique orientations in membranes, making the **outside surfaces different from the inside surfaces.** An **inside–outside asymmetry** is also provided by the external location of the carbohydrates attached to membrane proteins. In addition, specific proteins are located exclusively on the outsides or insides of membranes.

There are also **regional heterogeneities** in membranes. Some, such as occur at the villous borders of mucosal cells, are almost macroscopically visible. Others, such as those at gap junctions, tight junctions, and synapses, occupy much smaller regions of the membrane and generate correspondingly smaller local asymmetries.

There is also inside–outside **asymmetry of the phospholipids.** The **choline-containing phospholipids** (phosphatidylcholine and sphingomyelin) are located mainly in the **outer leaflet;** the **aminophospholipids** (phosphatidylserine and phosphatidylethanolamine) are preferentially located in the **inner leaflet.** Obviously, if this asymmetry is to exist at all, there must be limited transverse mobility (flip–flop) of the membrane phospholipids. In fact, phospholipids in synthetic bilayers exhibit an **extraordinarily slow rate of flip–flop;** the half-life of the asymmetry can be measured in several weeks.

The mechanisms involved in the **establishment of lipid asymmetry** are not well understood. The enzymes involved in the synthesis of phospholipids are located on the cytoplasmic side of microsomal membrane vesicles. Translocases (**flippases**) exist that transfer certain phospholipids (eg, phosphatidylcholine) from the inner to the outer leaflet. Specific **proteins that preferentially bind** individual phospholipids also appear to be present in the two leaflets, contributing to the asymmetric distribution of these lipid molecules. In addition, **phospholipid exchange proteins** recognize specific phospholipids and transfer them from one membrane (eg, the endoplasmic reticulum [ER]) to others (eg, mitochondrial and peroxisomal). A related issue is **how lipids enter membranes.** This has not been studied as intensively as the topic of how proteins enter membranes (see Chapter 46) and knowledge is still relatively meager. Many membrane lipids are synthesized in the ER. At least three pathways have been recognized. (1) Transport from the ER in vesicles, which then transfer the contained lipids to the recipient membrane. (2) Entry via direct contact of one membrane (eg, the ER) with another, facilitated by specific proteins. (3) Transport via the phospholipid exchange proteins (also known as lipid transfer proteins) mentioned above. This only exchanges lipids, but does not cause net transfer.

There is **further asymmetry** with regard to glycosphingolipids and **glycoproteins;** the **sugar moieties** of these molecules all **protrude outward** from the plasma membrane and are absent from its inner face. Thus, cells are "sugar coated".

Membranes Contain Integral & Peripheral Proteins

It is useful to classify membrane proteins into two types: **integral** and **peripheral** (**Figure 40–7**). Most membrane proteins fall into the **integral class,** meaning that they interact extensively with the phospholipids and **require the use of detergents** for their solubilization. Also, they generally **span the bilayer** as a bundle of α-helical transmembrane segments. Integral proteins are usually **globular** and are themselves **amphipathic.** They consist of two hydrophilic ends separated by an intervening hydrophobic region that traverses the hydrophobic core of the bilayer. As the structures of integral membrane proteins were being elucidated, it became apparent that certain ones (eg, transporter molecules, ion channels, various receptors, and G proteins) **span the bilayer many times** (see Figure 46–7), whereas other simple membrane proteins (eg, glycophorin A) **span the membrane only once.** Integral proteins are **asymmetrically distributed** across the membrane bilayer. This asymmetric orientation is conferred at the time of their insertion in the lipid bilayer during biosynthesis in the ER. The molecular mechanisms involved in insertion of proteins into membranes and the topic of membrane assembly are discussed in Chapter 46.

Peripheral proteins do not interact directly with the hydrophobic cores of the phospholipids in the bilayer and thus **do not require use of detergents** for their release. They are bound to the hydrophilic regions of specific integral proteins and head groups of phospholipids and can be released from them by treatment with **salt solutions of high ionic strength.** For example, **ankyrin,** a peripheral protein, is bound to the inner aspect of the integral protein "band 3" of erythrocyte membrane. **Spectrin,** a cytoskeletal structure within the erythrocyte, is in turn bound to ankyrin and thereby plays an important role in maintenance of the biconcave shape of the erythrocyte.

ARTIFICIAL MEMBRANES MODEL MEMBRANE FUNCTION

Artificial membrane systems can be prepared by appropriate techniques. These systems generally consist of mixtures of one or more **phospholipids** of natural or synthetic origin that can be treated (eg, by using **mild sonication**) to form spherical vesicles in which the lipids form a bilayer. Such vesicles, surrounded by a lipid bilayer with an aqueous interior, are termed **liposomes** (see Figure 15–24).

Some of the advantages and uses of artificial membrane systems in the study of membrane function are as follows:

1. The **lipid content** of the membranes can be varied, allowing systematic examination of the effects of varying lipid composition on certain functions.
2. **Purified membrane proteins or enzymes** can be incorporated into these vesicles in order to assess what factors (eg, specific lipids or ancillary proteins) the proteins require to reconstitute their function.
3. The **environment** of these systems can be rigidly controlled and systematically varied (eg, ion concentrations and ligands).
4. When liposomes are formed, they can be made to **entrap** certain compounds inside themselves, eg, drugs and isolated genes. There is interest in using liposomes to distribute drugs to certain tissues, and if components (eg, antibodies to certain cell surface molecules) could be incorporated into liposomes so that they would be targeted to specific tissues or tumors, the therapeutic impact would be considerable. **DNA** entrapped inside liposomes appears to be less sensitive to attack by nucleases; this approach may prove useful in attempts at **gene therapy.**

THE FLUID MOSAIC MODEL OF MEMBRANE STRUCTURE IS WIDELY ACCEPTED

The **fluid mosaic model** of membrane structure proposed in 1972 by Singer and Nicolson (**Figure 40–7**) is now widely accepted. The model is often likened to **icebergs** (membrane proteins) **floating in a sea** of predominantly fluid phospholipid molecules. Early **evidence** for the model was the finding that certain **integral proteins** (detected by fluorescent labeling techniques) rapidly and randomly redistributed in the plasma membrane of a hybrid cell formed by the artificially induced fusion of two different (mouse and human) parent cells. Biophysical studies of integral proteins showed that they spanned the membrane and had a globular nature. It has subsequently been demonstrated that **phospholipids** undergo even more rapid redistribution in the plane of the membrane. This diffusion within the plane of the membrane, termed **lateral diffusion,** can be quite rapid for a phospholipid; in fact, within the plane of the membrane, one molecule of phospholipid can move several micrometers per second.

The **phase changes**—and thus the **fluidity** of membranes—are largely **dependent upon the lipid composition** of the membrane. In a lipid bilayer, the hydrophobic chains of the fatty acids can be highly aligned or ordered to provide a rather stiff structure. As the temperature increases, the hydrophobic side chains undergo a **transition** from the **ordered state** (more gel-like or crystalline phase) to a **disordered** one, taking on a more liquid-like or fluid arrangement. The temperature at which the structure undergoes the transition from ordered to disordered (ie, melts) is called the **"transition temperature"** (T_m). The **longer** and more **saturated** fatty acid chains interact more strongly with each other via their longer hydrocarbon chains and thus cause higher values of T_m—ie, higher temperatures are required to increase the fluidity of the bilayer. On the other hand, **unsaturated bonds** that exist in the *cis* configuration tend to increase the fluidity of a bilayer by decreasing the compactness of the side chain packing without diminishing hydrophobicity (Figure 40–3). The phospholipids

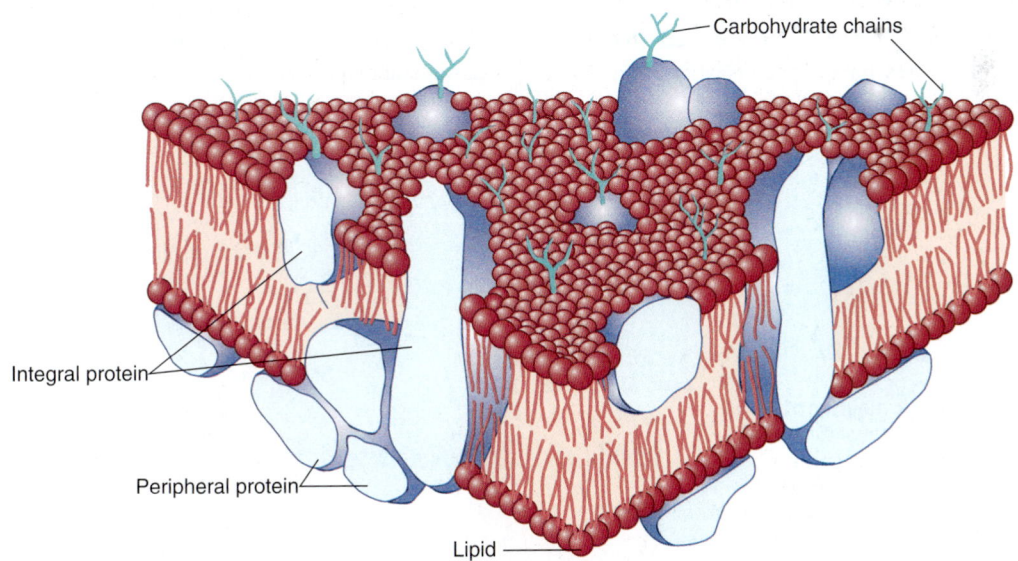

FIGURE 40–7 **The fluid mosaic model of membrane structure.** The membrane consists of a bimolecular lipid layer with proteins inserted in it or bound to either surface. Integral membrane proteins are firmly embedded in the lipid layers. Some of these proteins completely span the bilayer and are called transmembrane proteins, while others are embedded in either the outer or inner leaflet of the lipid bilayer. Loosely bound to the outer or inner surface of the membrane are the peripheral proteins. Many of the proteins and all the glycolipids have externally exposed oligosaccharide chains. (Reproduced, with permission, from Junqueira LC, Carneiro J: *Basic Histology: Text & Atlas,* 10th ed., McGraw-Hill, 2003.)

of cellular membranes generally contain at least one unsaturated fatty acid with at least one *cis* double bond.

Cholesterol modifies the fluidity of membranes. At temperatures below the T_m, it interferes with the interaction of the hydrocarbon tails of fatty acids and thus **increases fluidity.** At temperatures above the T_m, it limits disorder because it is more rigid than the hydrocarbon tails of the fatty acids and cannot move in the membrane to the same extent, thus **limiting fluidity**. At high cholesterol–phospholipid ratios, transition temperatures are altogether indistinguishable.

The **fluidity** of a membrane significantly affects its **functions.** As membrane fluidity increases, so does its **permeability** to water and other small hydrophilic molecules. The lateral mobility of integral proteins increases as the fluidity of the membrane increases. If the active site of an integral protein involved in a given function is exclusively in its hydrophilic regions, changing lipid fluidity will probably have little effect on the activity of the protein; however, if the protein is involved in a transport function in which transport components span the membrane, lipid-phase effects may significantly alter the **transport rate.** The **insulin receptor** is an excellent example of altered function with changes in fluidity. As the concentration of **unsaturated fatty acids** in the membrane is increased (by growing cultured cells in a medium rich in such molecules), **fluidity increases.** This alters the receptor so that it binds more insulin. At normal body temperature (37°C), the lipid bilayer is in a fluid state. Bacteria can modify the composition of their membrane lipids to adapt to changes in temperature.

Lipid Rafts, Caveolae, & Tight Junctions Are Specialized Features of Plasma Membranes

Plasma membranes contain **certain specialized structures** whose biochemical natures have been investigated in some detail.

Lipid rafts are specialized areas of the exoplasmic leaflet of the lipid bilayer enriched in cholesterol, sphingolipids, and certain proteins (see **Figure 40–8**). They are involved in **signal transduction** and **other processes.** It is thought that clustering certain components of signaling systems closely together may increase the efficiency of their function.

Caveolae may derive from lipid rafts. Many, if not all, contain the protein **caveolin-1,** which may be involved in their formation from rafts. Caveolae are observable by electron microscopy as flask-shaped indentations of the cell membrane facing the cytosol (**Figure 40–9**). Proteins detected in caveolae include various components of the signal transduction system (eg, the insulin receptor and some G proteins), the folate receptor, and endothelial nitric oxide synthase (eNOS). Caveolae and lipid rafts are active areas of research, and ideas concerning them and their possible roles in various disorders are rapidly evolving.

Tight junctions are other structures found in surface membranes. They are often located below the apical surfaces of epithelial cells and **prevent the diffusion of macromolecules between cells.** They are composed of **various proteins,** including occludin, various claudins, and junctional adhesion molecules.

Yet **other specialized structures** found in surface membranes include **desmosomes, adherens junctions,** and **microvilli;** their chemical natures and functions are not discussed here. The nature of **gap junctions** is described below.

MEMBRANE SELECTIVITY ALLOWS ADJUSTMENTS OF CELL COMPOSITION & FUNCTION

If the plasma membrane is relatively impermeable, **how do most molecules enter a cell?** How is **selectivity** of this movement established? Answers to such questions are important in understanding how cells adjust to a constantly changing

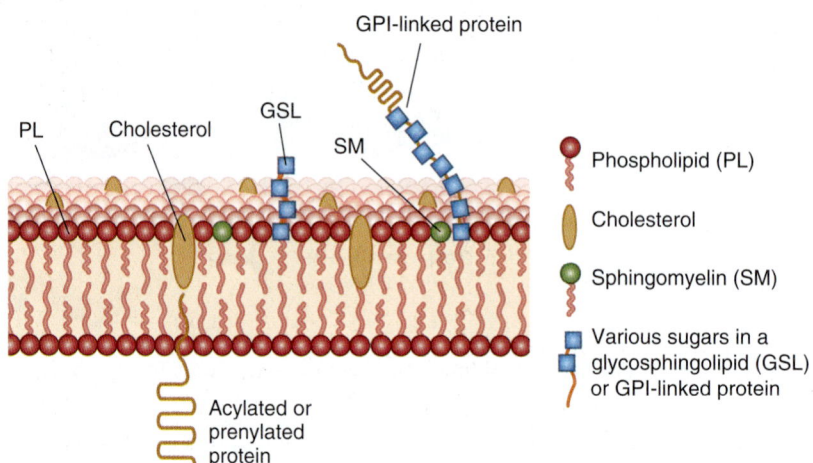

FIGURE 40–8 Schematic diagram of a lipid raft. Lipid rafts are somewhat thicker than the remainder of the bilayer. They are enriched in sphingolipids (eg, sphingomyelin), glycosphingolipids (eg, the ganglioside GM_1), saturated phospholipids, and cholesterol. They also contain certain GPI-linked proteins (outer leaflet) and acylated and prenylated proteins (inner leaflet). GPI-linked proteins are discussed in Chapter 47. Acylation and prenylation are post-translational modifications of certain membrane proteins.

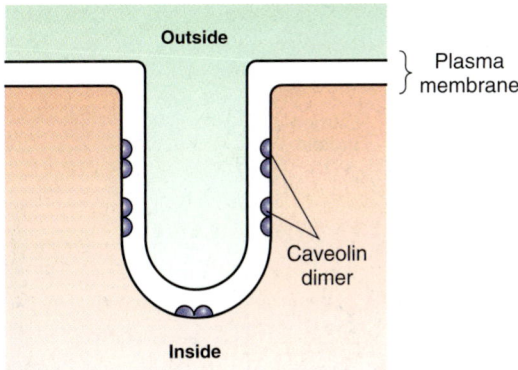

FIGURE 40–9 **Schematic diagram of a caveola.** A caveola is an invagination in the plasma membrane. The protein caveolin appears to play an important role in the formation of caveolae and occurs as a dimer. Each caveolin monomer is anchored to the inner leaflet of the plasma membrane by three palmitoyl molecules (not shown).

TABLE 40–3 Transfer of Material and Information Across Membranes

Cross-membrane movement of small molecules
Diffusion (passive and facilitated)
Active transport
Cross-membrane movement of large molecules
Endocytosis
Exocytosis
Signal transmission across membranes
Cell surface receptors
1. Signal transduction (eg, glucagon → cAMP)
2. Signal internalization (coupled with endocytosis, eg, the LDL receptor)
Movement to intracellular receptors (steroid hormones; a form of diffusion)
Intercellular contact and communication

Passive (simple) diffusion is the flow of solute from a higher to a lower concentration due to random thermal movement

Facilitated diffusion is passive transport of a solute from a higher concentration to a lower concentration, mediated by a specific protein transporter

Active transport is transport of a solute across a membrane in the direction of increasing concentration, and thus requires energy (frequently derived from the hydrolysis of ATP); a specific transporter (pump) is involved

The other terms used in this table are explained later in this chapter or elsewhere in this text

extracellular environment. Metazoan organisms also must have **means of communicating** between adjacent and distant cells, so that complex biologic processes can be coordinated. These **signals** must arrive at and be transmitted by the membrane, or they must be **generated** as a consequence of some interaction with the membrane. Some of the major mechanisms used to accomplish these different objectives are listed in **Table 40–3**.

Passive Diffusion Involving Transporters & Ion Channels Moves Many Small Molecules Across Membranes

Molecules can **passively** traverse the bilayer down electrochemical gradients by **simple diffusion** or by **facilitated diffusion.** This spontaneous movement toward equilibrium contrasts with **active transport,** which **requires energy** because it constitutes movement against an electrochemical gradient. **Figure 40–10** provides a schematic representation of these mechanisms. We shall first describe various aspects of passive transport, and then discuss aspects of active transport.

First, let us define the various terms. **Simple diffusion** is the passive flow of a solute from a higher to a lower concentration due to random thermal movement. **Facilitated diffusion** is passive transport of a solute from a higher to a lower concentration mediated by a specific protein transporter. **Active transport** is transport of a solute across a membrane against a concentration gradient, and thus requires energy (frequently derived from the hydrolysis of ATP); a specific transporter (**pump**) is involved.

As mentioned earlier in this chapter, some solutes such as gases can enter the cell by diffusing down an electrochemical gradient across the membrane and do not require metabolic energy. The **simple diffusion** of a solute across the membrane is limited by the **thermal agitation** of that specific molecule, by the **concentration gradient** across the membrane, and by the **solubility** of that solute (the permeability coefficient, Figure 40–6) in the hydrophobic core of the membrane bilayer. **Solubility** is inversely proportionate to the number of hydrogen bonds that must be broken in order for a solute in the external aqueous phase to become incorporated in the hydrophobic bilayer. Electrolytes, poorly soluble in lipid, do not form hydrogen bonds with water, but they do **acquire a shell of water** from hydration by electrostatic interaction. The size of the shell is directly proportionate to the **charge density** of the electrolyte. Electrolytes with a large charge density have a larger shell of hydration and thus a slower diffusion rate. Na^+, for example, has a higher charge density than K^+. **Hydrated Na^+** is therefore **larger** than **hydrated K^+;** hence, the latter tends to move more easily through the membrane.

The following factors affect **net diffusion** of a substance. (1) Its **concentration gradient** across the membrane: solutes move from high to low concentration. (2) The **electrical potential** across the membrane: solutes move toward the solution that has the opposite charge. The inside of the cell usually has a negative charge. (3) The **permeability coefficient** of the substance for the membrane. (4) The hydrostatic **pressure gradient** across the membrane: increased pressure will increase the rate and force of the collision between the molecules and the membrane. (5) **Temperature:** increased temperature will increase particle motion and thus increase the frequency of collisions between external particles and the membrane.

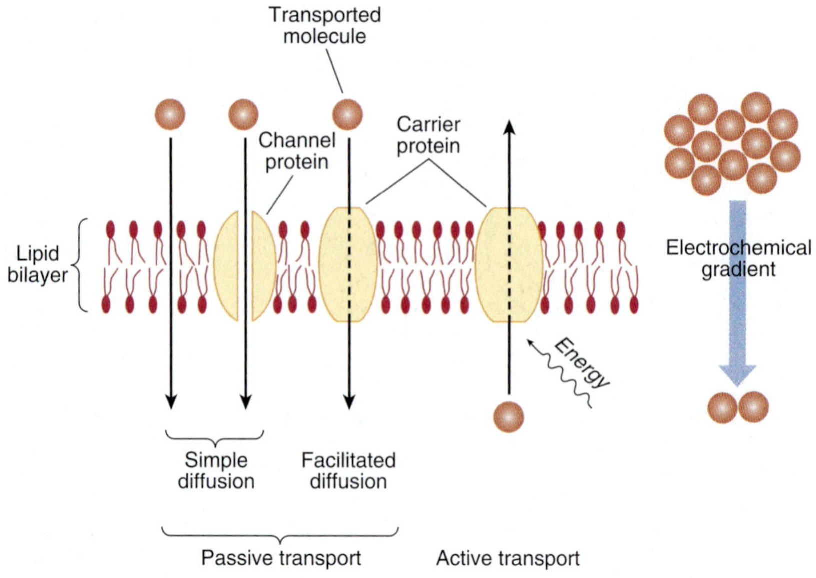

FIGURE 40–10 **Many small, uncharged molecules pass freely through the lipid bilayer by simple diffusion.** Larger uncharged molecules, and some small uncharged molecules, are transferred by specific carrier proteins (transporters) or through channels or pores. Passive transport is always down an electrochemical gradient, toward equilibrium. Active transport is against an electrochemical gradient and requires an input of energy, whereas passive transport does not. (Redrawn and reproduced, with permission, from Alberts B et al: *Molecular Biology of the Cell.* Garland, 1983.)

Transporters Are Specific Proteins Involved in Facilitated Diffusion & Also Active Transport

Facilitated diffusion involves either certain **transporters** or **ion channels** (see **Figure 40–11**). **Other transporters** (mostly ATP-driven) are involved in **active transport.** A multitude of **transporters and channels** exist in biological membranes that route the entry of ions into and out of cells. They are described in the following sections. **Table 40–4** summarizes some important points of difference between transporters and ion channels.

Transport systems can be described in a functional sense according to the number of molecules moved and the direction of movement (**Figure 40–12**) or according to whether movement is toward or away from equilibrium. The following **classification** depends primarily on the former. A **uniport** system moves one type of molecule bidirectionally. In **cotransport** systems, the transfer of one solute depends upon the stoichiometric simultaneous or sequential transfer of another solute. A **symport** moves two solutes in the same direction. Examples are the proton-sugar transporter in bacteria and the Na^+-sugar transporters (for glucose and certain other sugars)

and Na^+-amino acid transporters in mammalian cells. **Antiport** systems move two molecules in opposite directions (eg, Na^+ in and Ca^{2+} out).

Hydrophilic molecules that cannot pass freely through the lipid bilayer membrane do so passively by **facilitated diffusion** or by **active transport.** Passive transport is driven by the transmembrane gradient of substrate. Active transport always occurs against an electrical or chemical gradient, and so it requires energy, usually ATP. Both types of transport involve **specific carrier proteins** (transporters) and both show **specificity** for ions, sugars, and amino acids. Passive and active transports **resemble a substrate–enzyme interaction.** Points of resemblance of both to enzyme action are as follows: (1) There is a specific binding site for the solute. (2) The carrier is saturable, so it has a maximum rate of transport (V_{max}; **Figure 40–13**). (3) There is a binding constant (K_m) for the solute, and so the whole system has a K_m (Figure 40–13). (4) Structurally similar competitive inhibitors block transport. Transporters are thus like enzymes, but generally do not modify their substrates.

Cotransporters use the gradient of one substrate created by active transport to drive the movement of the other substrate. The Na^+ gradient produced by the Na^+–K^+-ATPase is used to drive the transport of a number of important

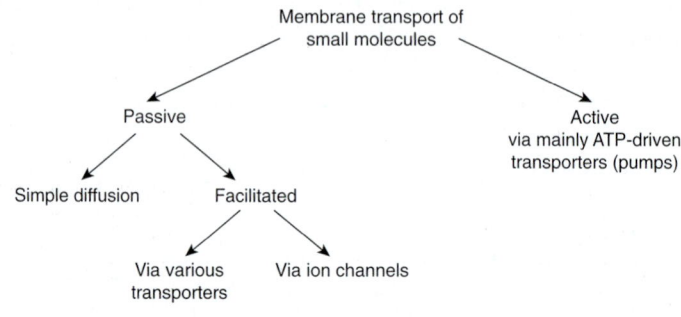

FIGURE 40–11 **A schematic diagram of the two types of membrane transport of small molecules.**

TABLE 40–4 Comparison of Transporters and Ion Channels

Transporters	Ion Channels
Bind solute and undergo conformational changes, transferring the solute across the membrane	Form pores in membranes
Involved in passive (facilitated diffusion) and active transport	Involved only in passive transport
Transport is significantly slower than via ion channels	Transport is significantly faster than via transporters

Note: Transporters are also known as carriers or permeases. Active transporters are often called pumps.

metabolites. The ATPase is a very important example of **primary transport,** while the Na$^+$-dependent systems are examples of **secondary transport** that rely on the gradient produced by another system. Thus, inhibition of the Na$^+$–K$^+$-ATPase in cells also blocks the Na$^+$-dependent uptake of substances like glucose.

Facilitated Diffusion Is Mediated by a Variety of Specific Transporters

Some specific solutes diffuse down electrochemical gradients across membranes **more rapidly** than might be expected from their size, charge, or partition coefficient. This is because specific transporters are involved. This **facilitated diffusion** exhibits properties distinct from those of simple diffusion. The rate of facilitated diffusion, a uniport system, can be **saturated;** ie, the number of sites involved in diffusion of the specific solutes appears finite. Many facilitated diffusion systems are **stereospecific** but, like simple diffusion, are **driven by the transmembrane electrochemical gradient.**

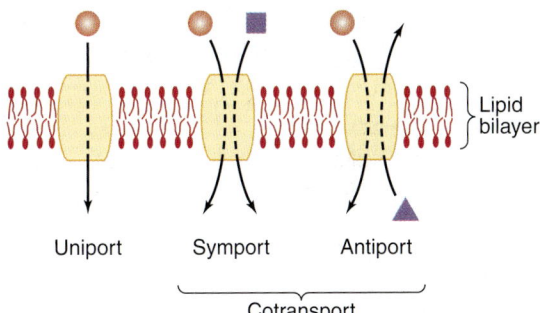

FIGURE 40–12 **Schematic representation of types of transport systems.** Transporters can be classified with regard to the direction of movement and whether one or more unique molecules are moved. A uniport can also allow movement in the opposite direction, depending on the concentrations inside and outside a cell of the molecule transported. (Redrawn and reproduced, with permission, from Alberts B et al: *Molecular Biology of the Cell.* Garland, 1983.)

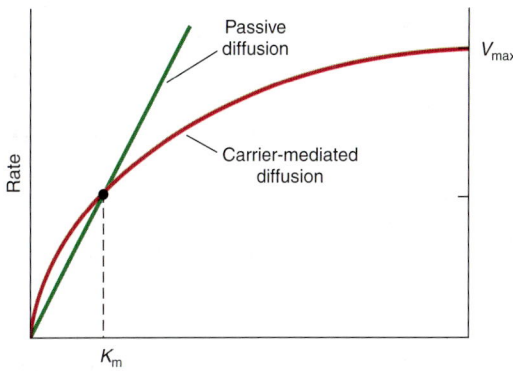

FIGURE 40–13 **A comparison of the kinetics of carrier-mediated (facilitated) diffusion with passive diffusion.** The rate of movement in the latter is directly proportionate to solute concentration, whereas the process is saturable when carriers are involved. The concentration at half-maximal velocity is equal to the binding constant (K_m) of the carrier for the solute. (V_{max}, maximal rate.)

A **"ping–pong"** mechanism (**Figure 40–14**) helps explain facilitated diffusion. In this model, the carrier protein exists in two principal conformations. In the **"ping"** state, it is exposed to high concentrations of solute, and molecules of the solute bind to specific sites on the carrier protein. Binding induces a **conformational change** that exposes the carrier to a lower concentration of solute (**"pong"** state). This process is completely **reversible,** and net flux across the membrane depends upon the concentration gradient. The **rate** at which solutes enter a cell by facilitated diffusion is determined by the following factors: (1) the concentration gradient across the membrane; (2) the amount of carrier available (this is a key control step); (3) the affinity of the solute–carrier interaction; (4) the rapidity of the conformational change for both the loaded and the unloaded carrier.

Hormones can regulate facilitated diffusion by changing the number of transporters available. **Insulin** via a complex signaling pathway increases glucose transport in fat and muscle by recruiting glucose transporters (GLUT) from an intracellular reservoir. Insulin also enhances amino acid transport in liver and other tissues. One of the coordinated actions of **glucocorticoid hormones** is to enhance transport of amino acids into liver, where the amino acids then serve as a substrate for gluconeogenesis. **Growth hormone** increases amino acid transport in all cells, and **estrogens** do this in the uterus. There are at least five different carrier systems for amino acids in animal cells. Each is specific for a group of closely related amino acids, and most operate as Na$^+$-symport systems (**Figure 40–12**).

Ion Channels Are Transmembrane Proteins That Allow the Selective Entry of Various Ions

Natural membranes contain transmembrane channels, pore-like structures composed of proteins that constitute selective

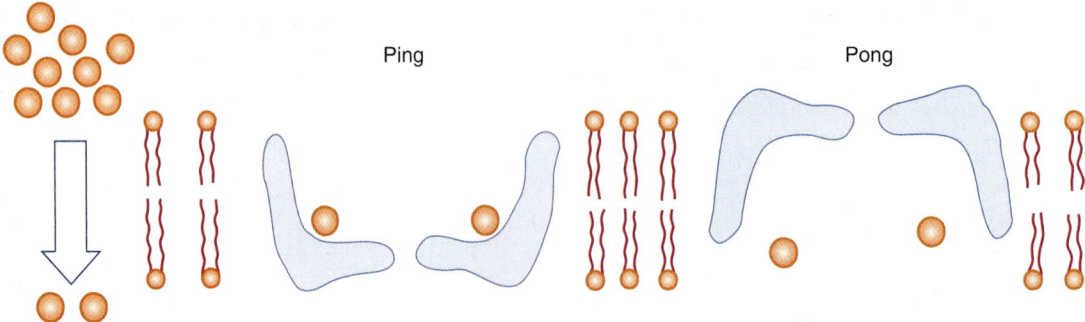

FIGURE 40–14   **The "ping–pong" model of facilitated diffusion.** A protein carrier (blue structure) in the lipid bilayer associates with a solute in high concentration on one side of the membrane. A conformational change ensues ("ping" to "pong"), and the solute is discharged on the side favoring the new equilibrium. The empty carrier then reverts to the original conformation ("pong" to "ping") to complete the cycle.

ion channels. Cation-conductive channels have an average diameter of about 5–8 nm. The **permeability** of a channel depends upon the size, the extent of hydration, and the extent of charge density on the ion. **Specific channels** for Na$^+$, K$^+$, Ca^{2+}, and Cl$^-$ have been identified. One such **Na$^+$ channel** is illustrated in **Figure 40–15**. It is seen to consist of four subunits. Each subunit consists of six α-helical transmembrane domains. The amino and carboxyl terminals are located in the cytoplasm, with both extracellular and intracellular loops being present. The actual pore in the channel through which the ions pass is not shown. A pore constitutes the center (diameter about 5–8 nm) of a structure formed by apposition of the subunits. Ion channels are very **selective,** in most cases permitting the passage of only one type of ion (Na$^+$, Ca^{2+}, etc). The **selectivity filter** of K$^+$ channels is made up of a ring of carbonyl groups donated by the subunits. The carbonyls displace bound water from the ion, and thus restrict its size to appropriate precise dimensions for passage through the channel. Many variations on the above structural theme are found, but all ion channels are basically made up of transmembrane subunits that come together to form a central pore through which ions pass selectively.

The membranes of **nerve cells** contain well-studied ion channels that are responsible for the generation of action potentials. The activity of some of these channels is controlled by neurotransmitters; hence, channel activity can be **regulated.**

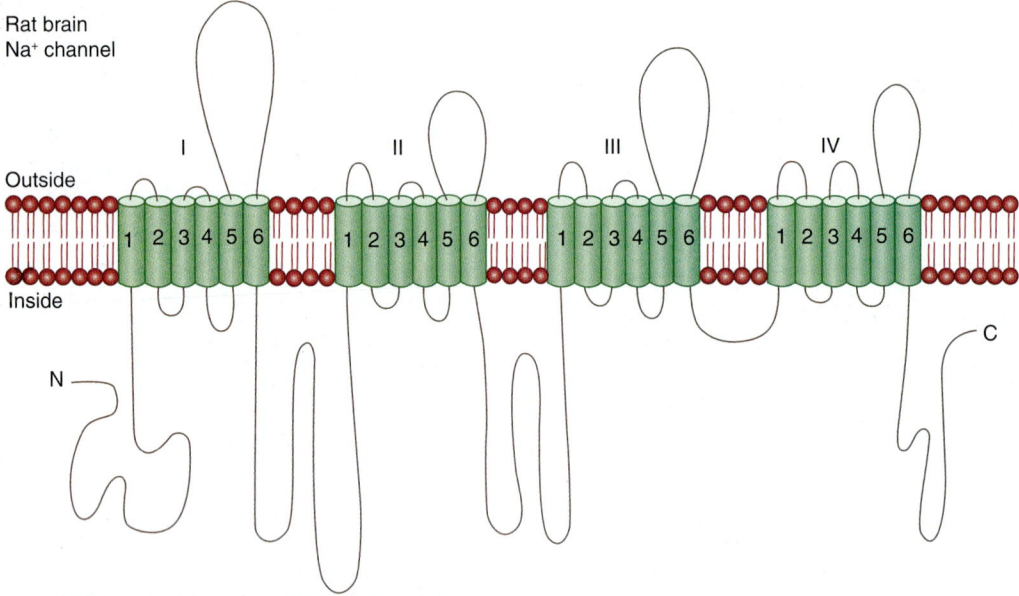

FIGURE 40–15   **Diagrammatic representation of the structures of an ion channel (a Na$^+$ channel of rat brain).** The Roman numerals indicate the four subunits of the channel and the Arabic numerals the α-helical transmembrane domains of each subunit. The actual pore through which the ions (Na$^+$) pass is not shown, but is formed by apposition of the various subunits. The specific areas of the subunits involved in the opening and closing of the channel are also not indicated. (After WK Catterall. Modified and reproduced, with permission, from Hall ZW: *An Introduction to Molecular Neurobiology.* Sinauer, 1992.)

TABLE 40–5 Some Properties of Ion Channels

- They are composed of transmembrane protein subunits.
- Most are highly selective for one ion; a few are nonselective.
- They allow impermeable ions to cross membranes at rates approaching diffusion limits.
- They can permit ion fluxes of 10^6–10^7/s.
- Their activities are regulated.
- The main types are voltage-gated, ligand-gated, and mechanically gated.
- They are usually highly conserved across species.
- Most cells have a variety of Na^+, K^+, Ca^{2+}, and Cl^- channels.
- Mutations in genes encoding them can cause specific diseases[1].
- Their activities are affected by certain drugs.

[1]Some diseases caused by mutations of ion channels are briefly discussed in Chapter 49.

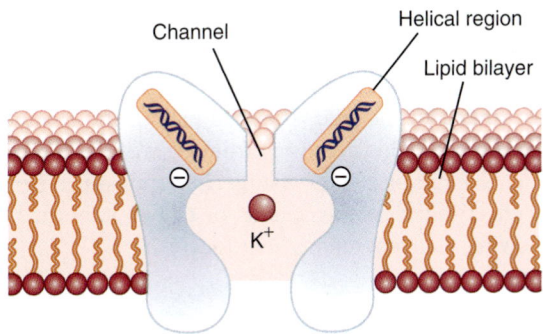

FIGURE 40–16 Schematic diagram of the structure of a K⁺ channel (KvAP) from *Streptomyces lividans.* A single K⁺ is shown in a large aqueous cavity inside the membrane interior. Two helical regions of the channel protein are oriented with their carboxylate ends pointing to where the K⁺ is located. The channel is lined by carboxyl oxygen. (Modified, with permission, from Doyle DA, et al, (1998), "The Structure of the Potassium Channel: Molecular Basis of K⁺ Conduction and Selectivity". Science 280:69. Reprinted with permission from AAAS.)

Ion channels are open transiently and thus are **"gated."** Gates can be controlled by opening or closing. In **ligand-gated channels,** a specific molecule binds to a receptor and opens the channel. **Voltage-gated channels** open (or close) in response to changes in membrane potential. **Mechanically gated** channels respond to mechanical stimuli (pressure and touch).

Some properties of ion channels are listed in **Tables 40–4** and **40–5**; other aspects of ion channels are discussed briefly in Chapter 48.

Detailed Studies of a K⁺ Channel & of a Voltage-Gated Channel Have Yielded Major Insights into Their Actions

There are at least **four features** of ion channels that must be elucidated: (1) their overall structures; (2) how they conduct ions so rapidly; (3) their selectivity; and (4) their gating properties. As described below, considerable progress in tackling these difficult problems has been made.

Especial progress has been made by Roderick MacKinnon, who received the Nobel Prize for elucidating the structure and function of **a K⁺ channel** (KvAP) present in *Streptomyces lividans*. A variety of techniques were used, including site-directed mutagenesis and x-ray crystallography. The channel is an integral membrane protein composed of four identical subunits, each with two transmembrane segments, creating an inverted teepee-like structure (**Figure 40–16**). The part of the channels that confers ion selectivity (the **selectivity filter**) measures 12 Å long (a relatively short length of the membrane, so K⁺ does not have far to travel in the membrane) and is situated at the wide end of the inverted teepee. The large, water-filled cavity and helical dipoles shown in Figure 40–16 help overcome the relatively large electrostatic energy barrier for a cation to cross the membrane. The **selectivity filter** is lined with carbonyl oxygen atoms (contributed by a TVGYG sequence), providing a number of sites with which K⁺ can interact. K⁺ ions,

which dehydrate as they enter the narrow selectivity filter, fit with proper coordination into the filter, but Na⁺ is too small to interact with the carbonyl oxygen atoms in correct alignment and is rejected. Two K⁺ ions, when close to each other in the filter, repel one another. This repulsion overcomes interactions between K⁺ and the surrounding protein molecule and allows very rapid conduction of K⁺ with high selectivity.

Other studies on a **voltage-gated ion channel** (HvAP) in *Aeropyrum pernix* have revealed many features of its voltage-sensing and voltage-gating mechanisms. This channel is made up of four subunits, each with six transmembrane segments. One of the six segments (S4 and part of S3) is the voltage sensor. It behaves like a **charged paddle** (**Figure 40–17**), in that it can move through the interior of the membrane transferring

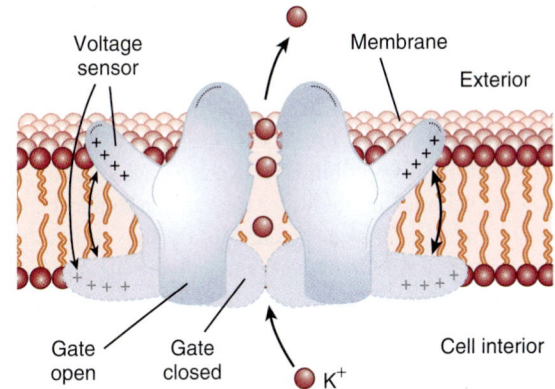

FIGURE 40–17 Schematic diagram of the voltage-gated K⁺ channel of *Aeorpyrum pernix.* The voltage sensors behave like charged paddles that move through the interior of the membrane. Four voltage sensors (only two are shown here) are linked mechanically to the gate of the channel. Each sensor has four positive charges contributed by arginine residues. (Modified, with permission, from Sigworth FJ: Nature 2003;423:21. Copyright © 2003. Macmillan Publishers Ltd.)

four positive charges (due to four Arg residues in each sub-unit) from one membrane surface to the other in response to changes in voltage. There are four voltage sensors in each channel, linked to the gate. The gate part of the channel is constructed from S6 helices (one from each of the subunits). Movements of this part of the channel in response to changing voltage effectively close the channel or reopen it, in the latter case allowing a current of ions to cross.

Ionophores Are Molecules That Act as Membrane Shuttles for Various Ions

Certain microbes synthesize small cyclic organic molecules, **ionophores, such as valinomycin** that function as shuttles for the movement of ions (K^+ in the case of valinomycin) across membranes. These ionophores contain hydrophilic centers that bind specific ions and are surrounded by peripheral hydropho-bic regions; this arrangement allows the molecules to dissolve effectively in the membrane and diffuse transversely therein. Others, like the well-studied polypeptide **gramicidin** (an anti-biotic), fold up to form hollow channels.

Microbial toxins such as **diphtheria toxin** and activated **serum complement components** can produce large pores in cellular membranes and thereby provide macromolecules with direct access to the internal milieu. The toxin **α-hemolysin** (produced by certain species of *Streptococcus*) consists of seven subunits which come together to form a β-barrel that allows metabolites like ATP to leak out of cells, resulting in cell lysis.

Aquaporins Are Proteins That Form Water Channels in Certain Membranes

In certain cells (eg, red cells and cells of the collecting ductules of the kidney), the movement of water by simple diffusion is aug-mented by movement through **water channels.** These channels are composed of tetrameric transmembrane proteins named **aquaporins.** At least 10 distinct aquaporins (AP-1 to AP-10) have been identified. Crystallographic and other studies have revealed how these channels permit passage of water but exclude passage of ions and protons. In essence, the pores are too narrow to permit passage of ions. Protons are excluded by the fact that the oxygen atom of water binds to two asparagine residues lin-ing the channel, making the water unavailable to participate in a H^+ relay, and thus preventing entry of protons. Mutations in the gene encoding AP-2 have been shown to be the cause of one type of **nephrogenic diabetes insipidus,** a condition in which there is an inability to concentrate urine. Peter Agre won a Nobel Prize for his work on the structure and function of aquaporins.

ACTIVE TRANSPORT SYSTEMS REQUIRE A SOURCE OF ENERGY

The process of active transport differs from diffusion in that molecules are transported against concentration gradients; hence, **energy is required.** This energy can come from the

TABLE 40–6 Major Types of ATP-Driven Active Transporters

Type	Example with Subcellular Location
P-type	Ca^{2+} ATPase (SR); Na^+–K^+-ATPase (PM)
F-type	mt ATP synthase of oxidative phosphorylation
V-type	The ATPase that pumps protons into lysosomes and synaptic vesicles
ABC transporter	CFTR protein (PM); MDR-1 protein (PM)

P (in P-type) signifies phosphorylation (these proteins autophosphorylate).
F (in F-type) signifies energy coupling factors.
V (in V-type) signifies vacuolar.
ABC signifies ATP-binding cassette transporter (all have two nucleotide-binding domains and two transmembrane segments).
SR, sarcoplasmic reticulum of muscle; PM, plasma membrane; mt, mitochondrial; CFTR, cystic fibrosis transmembrane regulator protein, a Cl^- transporter, and the protein implicated in the causation of cystic fibrosis (see later in this chapter and also Chapter 57); MDR-1 protein (multidrug resistance-1 protein), a protein that pumps many chemotherapeutic agents out of cancer cells and is thus an important contributor to the resistance of certain cancer cells to treatment.

hydrolysis of ATP, from **electron movement,** or from **light.** The maintenance of electrochemical gradients in biologic sys-tems is so important that it consumes approximately **30% of the total energy expenditure** in a cell.

As shown in **Table 40–6, four major classes** of ATP-driven active transporters (**P, F, V, and ABC transporters**) have been recognized. The nomenclature is explained in the legend to the Table. The first example of the P class, the **Na^+–K^+-ATPase,** is discussed below. The **Ca^{2+} ATPase** of muscle is discussed in Chapter 48. The second class is referred to as F-type. The most important example of this class is the **mt ATP synthase,** described in Chapter 13. V-type active trans-porters pump protons into lysosomes and other structures. ABC transporters include the **CFTR protein,** a chloride channel involved in the causation of cystic fibrosis (described later in this chapter and in Chapter 54). Another important member of this class is the **multidrug resistance-1 protein** (MDR-1 protein). This transporter will pump a variety of drugs, including many anti-cancer agents, out of cells. It is a very important cause of cancer cells exhibiting **resistance to chemotherapy,** although many other mechanisms are also implicated.

The Na^+–K^+-ATPase of the Plasma Membrane Is a Key Enzyme in Regulating Intracellular Concentrations of Na^+ and K^+

In general, cells maintain a **low** intracellular Na^+ concentra-tion and a **high** intracellular K^+ concentration (Table 40–1), along with a net negative electrical potential inside. The pump that maintains these ionic gradients is an **ATPase** that is acti-vated by Na^+ and K^+ (Na^+–K^+-ATPase; see **Figure 40–18**). It pumps three Na^+ out and two K^+ into cells. The ATPase is an integral membrane protein that contains a transmembrane

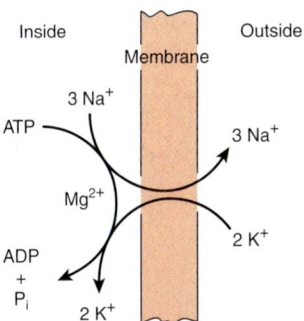

FIGURE 40–18 Stoichiometry of the Na⁺–K⁺-ATPase pump. This pump moves three Na⁺ ions from inside the cell to the outside and brings two K⁺ ions from the outside to the inside for every molecule of ATP hydrolyzed to ADP by the membrane-associated ATPase. Ouabain and other cardiac glycosides inhibit this pump by acting on the extracellular surface of the membrane. (Courtesy of R Post.)

domain allowing the passage of ions, and cytosolic domains that couple ATP hydrolysis to transport. It has catalytic centers for both ATP and Na⁺ on the cytoplasmic (inner) side of the plasma membrane (PM), with K⁺ binding sites located on the extracellular side of the membrane. Phosphorylation by ATP induces **a conformational change** in the protein leading to transfer of three Na⁺ ions from the inner to the outer side of the PM. Two molecules of K⁺ bind to sites on the protein on the external surface of the PM, resulting in dephosphorylation of the protein and transfer of the K⁺ ions across the membrane to the interior. Thus, three Na⁺ ions are transported out for every two K⁺ ions entering. This creates a charge imbalance between the inside and the outside of the cell, making the **inside more negative** (an **electrogenic** effect). **Ouabain** or **digitalis** (two important cardiac drugs) inhibit this ATPase by binding to the extracellular domain. This enzyme can consume ~30% of cellular energy. The Na⁺–K⁺-ATPase can be **coupled** to various other transporters, such as those involved in transport of glucose (see below).

TRANSMISSION OF NERVE IMPULSES INVOLVES ION CHANNELS AND PUMPS

The membrane enclosing **neuronal cells** maintains an asymmetry of inside–outside voltage (electrical potential) and is also **electrically excitable** due to the presence of voltage-gated channels. When appropriately stimulated by a chemical signal mediated by a specific synaptic membrane receptor (see discussion of the transmission of biochemical signals, below), channels in the membrane are opened to allow the rapid influx of Na⁺ or Ca²⁺ (with or without the efflux of K⁺), so that the voltage difference rapidly collapses and that segment of the membrane is depolarized. However, as a result of the action of the ion pumps in the membrane, the gradient is quickly restored.

When large areas of the membrane are **depolarized** in this manner, the electrochemical disturbance propagates in wave-like form down the membrane, generating a **nerve impulse.** **Myelin sheets,** formed by Schwann cells, wrap around nerve fibers and provide an **electrical insulator** that surrounds most of the nerve and greatly speeds up the propagation of the wave (signal) by allowing ions to flow in and out of the membrane only where the membrane is free of the insulation (at the nodes of Ranvier). The myelin membrane has a high content of **lipid,** accounting for its excellent insulating property. Relatively few proteins are found in the myelin membrane; those present appear to hold together multiple membrane bilayers to form the hydrophobic insulating structure that is impermeable to ions and water. **Certain diseases,** eg, multiple sclerosis and the Guillain-Barré syndrome, are characterized by demyelination and impaired nerve conduction.

TRANSPORT OF GLUCOSE INVOLVES SEVERAL MECHANISMS

A discussion of the **transport of glucose** summarizes many of the points made in this chapter. Glucose must enter cells as the first step in energy utilization. A number of different glucose transporters (GLUTs) are involved, varying in different tissues (see Table 20–2). In **adipocytes and skeletal muscle,** glucose enters by a specific transport system (GLUT4) that is enhanced by insulin. Changes in transport are primarily due to alterations of V_{max} (presumably from more or fewer transporters), but changes in K_m may also be involved.

Glucose transport **in the small intestine** involves some different aspects of the principles of transport discussed above. Glucose and Na⁺ bind to different sites on a **Na⁺-glucose symporter** located at the **apical surface.** Na⁺ moves into the cell down its electrochemical gradient and "drags" glucose with it (**Figure 40–19**). Therefore, the greater the Na⁺ gradient, the more glucose enters; and if Na⁺ in extracellular fluid is low, glucose transport stops. To maintain a steep Na⁺ gradient, this Na⁺-glucose symporter is dependent on gradients generated by the **Na⁺-K⁺-ATPase,** which maintains a low intracellular Na⁺ concentration. Similar mechanisms are used to transport **other sugars** as well as **amino acids** across the apical lumen in polarized cells such as are found in the intestine and kidney. The transcellular movement of glucose in this case involves one additional component: a uniport (Figure 40–19) that allows the glucose accumulated within the cell to move across the **basolateral membrane** and involves **a glucose uniporter** (GLUT2).

The treatment of severe cases of **diarrhea** (such as is found in cholera) makes use of the above information. In **cholera** (see Chapter 57), massive amounts of fluid can be passed as watery stools in a very short time, resulting in severe dehydration and possibly death. **Oral rehydration therapy,** consisting primarily of **NaCl and glucose,** has been developed by the World Health Organization (WHO). The transport of glucose and

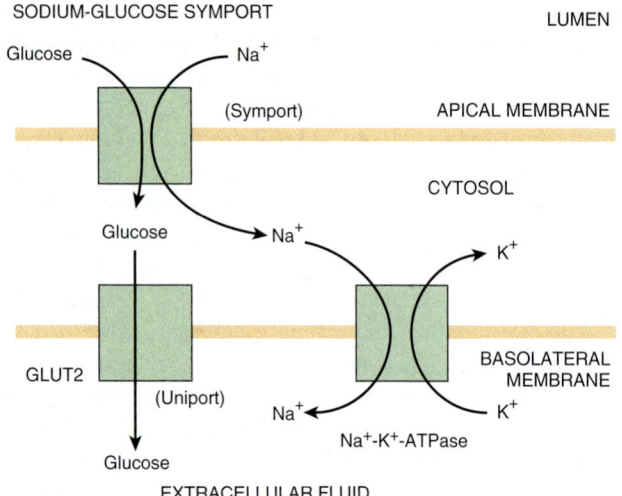

FIGURE 40–19 **The transcellular movement of glucose in an intestinal cell.** Glucose follows Na⁺ across the luminal epithelial membrane. The Na⁺ gradient that drives this symport is established by Na⁺–K⁺ exchange, which occurs at the basal membrane facing the extracellular fluid compartment via the action of the Na⁺–K⁺-ATPase. Glucose at high concentration within the cell moves "downhill" into the extracellular fluid by facilitated diffusion (a uniport mechanism), via GLUT2 (a glucose transporter, see Table 20–2). The sodium-glucose symport actually carries 2 Na⁺ for each glucose.

Na⁺ across the intestinal epithelium forces (via osmosis) movement of water from the lumen of the gut into intestinal cells, resulting in rehydration. Glucose alone or NaCl alone would not be effective.

CELLS TRANSPORT CERTAIN MACROMOLECULES ACROSS THE PLASMA MEMBRANE BY ENDOCYTOSIS AND EXOCYTOSIS

The process by which cells take up large molecules is called **endocytosis.** Some of these molecules (eg, polysaccharides, proteins, and polynucleotides), when hydrolyzed inside the cell, **yield nutrients.** Endocytosis also provides a mechanism for **regulating** the content of certain membrane components, hormone receptors being a case in point. Endocytosis can be used to learn more about how cells function. DNA from one cell type can be used to transfect a different cell and alter the latter's function or phenotype. A specific gene is often employed in these experiments, and this provides a unique way to study and analyze the regulation of that gene. **DNA transfection** depends upon endocytosis, which is responsible for the entry of DNA into the cell. Such experiments commonly use calcium phosphate since Ca²⁺ stimulates endocytosis and precipitates DNA, which makes the DNA a better object for endocytosis. Cells also **release macromolecules** by **exocytosis.** Endocytosis and exocytosis both involve vesicle formation with or from the plasma membrane.

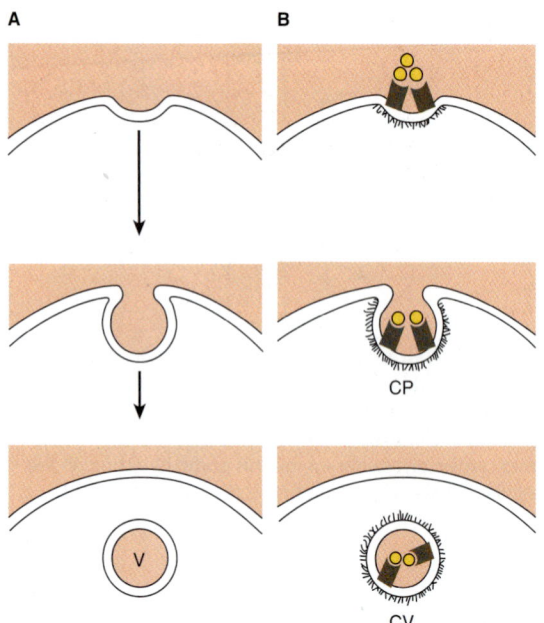

FIGURE 40–20 **Two types of pinocytosis.** An endocytotic vesicle (V) forms as a result of invagination of a portion of the plasma membrane. Fluid-phase pinocytosis **(A)** is random and nondirected. Absorptive (receptor-mediated endocytosis) **(B)** is selective and occurs in coated pits (CP) lined with the protein clathrin (the fuzzy material). Targeting is provided by receptors (brown symbols) specific for a variety of molecules. This results in the formation of an internalized coated vesicle (CV).

Endocytosis Involves Ingestion of Parts of the Plasma Membrane

Almost all eukaryotic cells are continuously re-cycling parts of their plasma membranes. Endocytotic vesicles are generated when segments of the plasma membrane invaginate, enclosing a small volume of extracellular fluid and its contents. The vesicle then pinches off as the fusion of plasma membranes seals the neck of the vesicle at the original site of invagination (**Figure 40–20**). This vesicle fuses with other membrane structures and thus achieves the transport of its contents to other cellular compartments or even back to the cell exterior. Most endocytotic vesicles fuse with **primary lysosomes** to form **secondary lysosomes,** which contain hydrolytic enzymes and are therefore specialized organelles for intracellular disposal. The macromolecular contents are digested to yield amino acids, simple sugars, or nucleotides, and they are transported out of the vesicles to be reused by the cell. Endocytosis requires (1) energy, usually from the hydrolysis of ATP; (2) Ca²⁺; and (3) contractile elements in the cell (probably the microfilament system) (Chapter 48).

There are **two** general types of endocytosis. **Phagocytosis** occurs only in specialized cells such as macrophages and granulocytes. Phagocytosis involves the ingestion of large particles such as viruses, bacteria, cells, or debris. Macrophages are extremely active in this regard and may ingest 25% of their volume per hour. In so doing, a macrophage may internalize

3% of its plasma membrane each minute or the entire membrane every 30 min.

Pinocytosis ("cell drinking") is a property of all cells and leads to the cellular uptake of fluid and fluid contents. There are **two types. Fluid-phase pinocytosis** is a nonselective process in which the uptake of a solute by formation of small vesicles is simply proportionate to its concentration in the surrounding extracellular fluid. The formation of these vesicles is an extremely active process. Fibroblasts, for example, internalize their plasma membrane at about one-third the rate of macrophages. This process occurs more rapidly than membranes are made. The surface area and volume of a cell do not change much, so membranes must be replaced by exocytosis or by being recycled as fast as they are removed by endocytosis.

The other type of pinocytosis, **absorptive pinocytosi** or **receptor-mediated endocytosis,** is primarily responsible for the uptake of specific macromolecules for which there are binding sites on the plasma membrane. These high-affinity receptors permit the selective concentration of ligands from the medium, minimize the uptake of fluid or soluble unbound macromolecules, and markedly increase the rate at which specific molecules enter the cell. The **vesicles** formed during absorptive pinocytosis are derived from invaginations (pits) that are coated on the cytoplasmic side with a filamentous material and are appropriately named **coated pits.** In many systems, the protein **clathrin** is the filamentous material. It has a three-limbed structure (called a **triskelion**), with each limb being made up of one light and one heavy chain of clathrin. The polymerization of clathrin into a vesicle is directed by **assembly particles,** composed of four **adapter proteins.** These interact with certain amino acid sequences in the receptors that become cargo, ensuring selectivity of uptake. The lipid **phophatidylinositol 4.5-bisphosphate (PIP$_2$)** (see Chapter 15) also plays an important role in vesicle assembly. In addition, the protein **dynamin,** which both binds and hydrolyzes GTP, is necessary for the pinching off of clathrincoated vesicles from the cell surface. Coated pits may constitute as much as 2% of the surface of some cells. Other aspects of vesicles are discussed in Chapter 46.

As an example, **the low-density lipoprotein** (LDL) molecule and its **receptor** (Chapter 25) are internalized by means of coated pits containing the LDL receptor. These endocytotic vesicles containing LDL and its receptor fuse to lysosomes in the cell. The receptor is released and recycled back to the cell surface membrane, but the apoprotein of LDL is degraded and the cholesteryl esters metabolized. Synthesis of the LDL receptor is regulated by secondary or tertiary consequences of pinocytosis, eg, by metabolic products—such as cholesterol—released during the degradation of LDL. Disorders of the LDL receptor and its internalization are medically important and are discussed in Chapters 25 & 26.

Absorptive pinocytosis of **extracellular glycoproteins** requires that the glycoproteins carry specific carbohydrate recognition signals. These recognition signals are bound by membrane receptor molecules, which play a role analogous to that of the LDL receptor. A **galactosyl receptor** on the surface of hepatocytes is instrumental in the absorptive pinocytosis of **asialoglycoproteins** from the circulation (Chapter 47). **Acid hydrolases** taken up by absorptive pinocytosis in fibroblasts are recognized by their **mannose 6-phosphate** moieties. Interestingly, the mannose 6-phosphate moiety also seems to play an important role in the intracellular targeting of the acid hydrolases to the lysosomes of the cells in which they are synthesized (Chapter 47).

There is a **dark side** to receptor-mediated endocytosis in that **viruses** which cause such diseases as hepatitis (affecting liver cells), poliomyelitis (affecting motor neurons), and AIDS (affecting T cells) initiate their damage by entering cells by this mechanism. **Iron toxicity** also begins with excessive uptake due to endocytosis.

Exocytosis Releases Certain Macromolecules from Cells

Most cells **release** macromolecules to the exterior by **exocytosis.** This process is also involved in membrane remodeling, when the components synthesized in the ER and Golgi are carried in vesicles that fuse with the plasma membrane. The **signal** for exocytosis is often a hormone which, when it binds to a cell-surface receptor, induces a local and transient change in Ca^{2+} concentration. **Ca^{2+}** triggers exocytosis. **Figure 40–21** provides a comparison of the mechanisms of exocytosis and endocytosis.

Molecules released by exocytosis have at least **three fates.** (1) They are membrane proteins and remain associated with the cell surface. (2) They can become part of the extracellular matrix, eg, collagen and glycosaminoglycans. (3) They can enter extracellular fluid and signal other cells. Insulin, parathyroid hormone, and the catecholamines are all packaged in granules and processed within cells, to be released upon appropriate stimulation.

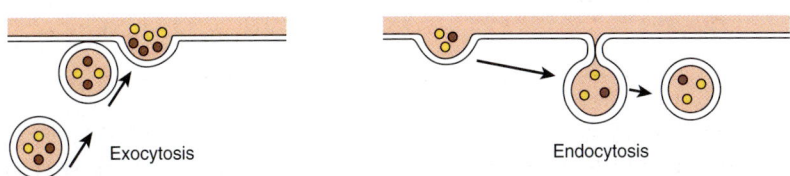

FIGURE 40–21 **A comparison of the mechanisms of endocytosis and exocytosis.** Exocytosis involves the contact of two inside-surface (cytoplasmic side) monolayers, whereas endocytosis results from the contact of two outer-surface monolayers.

VARIOUS SIGNALS ARE TRANSMITTED ACROSS MEMBRANES

Specific biochemical signals such as neurotransmitters, hormones, and immunoglobulins bind to specific **receptors** (integral proteins) exposed to the outside of cellular membranes and transmit information through these membranes to the cytoplasm. This process, called **transmembrane signaling** (see Chapter 42), involves the generation of a number of signaling molecules, including cyclic nucleotides, calcium, phosphoinositides, and diacylglycerol. Many of the steps involve **phosphorylation** of receptors and downstream proteins.

GAP JUNCTIONS ALLOW DIRECT FLOW OF MOLECULES FROM ONE CELL TO ANOTHER

Gap junctions are structures that permit **direct transfer of small molecules** (up to ~1200 Da) from one cell to its neighbor. They are composed of a family of proteins called **connexins** that form a hexagonal structure consisting of 12 such proteins. Six connexins form a connexin hemichannel and join to a similar structure in a neighboring cell to make a **complete connexon channel** (**Figure 40–22**). One gap junction contains several connexons. Different connexins are found in different tissues. **Mutations** in genes encoding connexins have been found to be associated with a number of conditions, including cardiovascular abnormalities, one type of deafness, and the X-linked form of Charcot-Marie-Tooth disease (a demyelinating neurologic disorder).

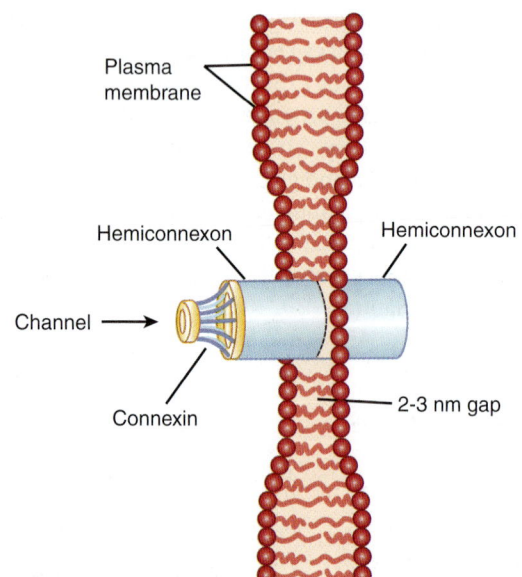

FIGURE 40–22 **Schematic diagram of a gap junction.** One connexon is made from two hemiconnexons. Each hemiconnexon is made from six connexin molecules. Small solutes are able to diffuse through the central channel, providing a direct mechanism of cell–cell communication.

MUTATIONS AFFECTING MEMBRANE PROTEINS CAUSE DISEASES

In view of the fact that membranes are located in so many organelles and are involved in so many processes, it is not surprising that **mutations** affecting their protein constituents should result in many diseases or disorders. While some mutations directly affect the function of membrane proteins, the majority of mutations cause protein misfolding and impair traffic (see Chapter 46) of the membrane proteins from their site of synthesis in the ER to the plasma membrane or other intracellular sites. Examples of diseases or disorders due to abnormalities in membrane proteins are listed in **Table 40–7**. These mainly reflect mutations in proteins of the **plasma membrane,** with one affecting **lysosomal function** (I-cell disease).

Proteins in **plasma membranes** can be classified as **receptors, transporters, ion channels, enzymes,** and **structural components.** Members of all of these classes are often **glycosylated,** so that mutations affecting this process (see Chapter 47) may alter their function. Mutations in **receptors** can cause defects in transmembrane signaling, a common occurrence in **cancer** (see Chapter 55). Many **genetic diseases** or disorders have been ascribed to mutations affecting various proteins involved in the **transport** of amino acids, sugars, lipids, urate, anions, cations, water, and vitamins across the **plasma membrane.**

Mutations in genes encoding proteins in **other membranes** can also have harmful consequences. For example, mutations in genes encoding **mitochondrial membrane proteins** involved in oxidative phosphorylation can cause neurologic and other problems (eg, Leber hereditary optic neuropathy; LHON, **a condition in which some success with gene therapy was reported in 2008**).

Membrane proteins can also be affected by **conditions other than mutations.** Formation of **autoantibodies** to the acetylcholine receptor in skeletal muscle causes myasthenia gravis. **Ischemia** can quickly affect the integrity of various ion channels in membranes. **Overexpression** of P-glycoprotein (MDR-1), a drug pump, results in multidrug resistance (MDR) in cancer cells. Abnormalities of membrane constituents other than proteins can also be harmful. With regard to **lipids,** excess of cholesterol (eg, in familial hypercholesterolemia), of lysophospholipid (eg, after bites by certain snakes, whose venom contains phospholipases), or of glycosphingolipids (eg, in a sphingolipidosis), can all affect membrane function.

Cystic Fibrosis Is Due to Mutations in the Gene Encoding CFTR, a Chloride Transporter

Cystic fibrosis (CF) is a **recessive** genetic disorder prevalent among whites in North America and certain parts of northern Europe. It is characterized by chronic bacterial infections of the airways and sinuses, fat maldigestion due to pancreatic exocrine insufficiency, infertility in males due to abnormal development of the vas deferens, and **elevated levels of**

TABLE 40–7 Some Diseases or Pathologic States Resulting From or Attributed to Abnormalities of Membranes[1]

Disease	Abnormality
Achondroplasia (OMIM 100800)	Mutations in the gene encoding the fibroblast growth factor receptor 3
Familial hypercholesterolemia (OMIM 143890)	Mutations in the gene encoding the LDL receptor
Cystic fibrosis (OMIM 219700)	Mutations in the gene encoding the CFTR protein, a Cl⁻ transporter
Congenital long QT syndrome (OMIM 192500)	Mutations in genes encoding ion channels in the heart
Wilson disease (OMIM 277900)	Mutations in the gene encoding a copper-dependent ATPase
I-cell disease (OMIM 252500)	Mutations in the gene encoding GlcNAc phosphotransferase, resulting in absence of the Man 6-P signal for lysosomal localization of certain hydrolases
Hereditary spherocytosis (OMIM 182900)	Mutations in the genes encoding spectrin or other structural proteins in the red cell membrane
Metastasis of cancer cells	Abnormalities in the oligosaccharide chains of membrane glycoproteins and glycolipids are thought to be of importance
Paroxysmal nocturnal hemoglobinuria (OMIM 311770)	Mutation resulting in deficient attachment of the GPI anchor (see Chapter 47) to certain proteins of the red cell membrane

[1]The disorders listed are discussed further in other chapters. The table lists examples of mutations affecting two receptors, one transporter, several ion channels (ie, congenital long QT syndrome), two enzymes, and one structural protein. Examples of altered or defective glycosylation of glycoproteins are also presented. Most of the conditions listed involve the plasma membrane.

chloride in sweat (>60 mmol/L). In 1989, it was shown that mutations in a gene encoding a protein named **cystic fibrosis transmembrane regulator protein (CFTR)** were responsible for CF. CFTR is a **cyclic AMP-regulated Cl⁻ transporter.** The major clinical features of CF and further information about the gene responsible for CF and about CFTR are presented in Case 5, Chapter 57.

SUMMARY

- Membranes are complex structures composed of lipids, proteins, and carbohydrate-containing molecules.

- The basic structure of all membranes is the lipid bilayer. This bilayer is formed by two sheets of phospholipids in which the hydrophilic polar head groups are directed away from each other and are exposed to the aqueous environment on the outer and inner surfaces of the membrane. The hydrophobic nonpolar tails of these molecules are oriented toward each other, in the direction of the center of the membrane.

- Membranes are dynamic structures. Lipids and certain proteins show rapid lateral diffusion. Flip–flop is very slow for lipids and nonexistent for proteins.

- The fluid mosaic model forms a useful basis for thinking about membrane structure.

- Membrane proteins are classified as integral if they are firmly embedded in the bilayer and as peripheral if they are attached to the outer or inner surface.

- The 20 or so membranes in a mammalian cell have different compositions and functions and they define compartments, or specialized environments, within the cell that have specific functions (eg, lysosomes).

- Certain hydrophobic molecules freely diffuse across membranes, but the movement of others is restricted because of their size or charge.

- Various passive and active (usually ATP-dependent) mechanisms are employed to maintain gradients of such molecules across different membranes.

- Certain solutes, eg, glucose, enter cells by facilitated diffusion along a downhill gradient from high to low concentration using specific carrier proteins (transporters).

- The major ATP-driven pumps are classified as P (phosphorylated), F (energy factors), V (vacuolar), and ABC transporters. Member of these classes include the Na⁺–K⁺-ATPase and the Ca²⁺ ATPase of the sarcoplasmic reticulum; the mt ATP synthase; the ATPase acidifying lysosomes; and the CFTR protein and the MDR-1 protein.

- Ligand- or voltage-gated ion channels are often employed to move charged molecules (Na⁺, K⁺, Ca²⁺, etc.) across membranes down their electrochemical gradients

- Large molecules can enter or leave cells through mechanisms such as endocytosis or exocytosis. These processes often require binding of the molecule to a receptor, which affords specificity to the process.

- Receptors may be integral components of membranes (particularly the plasma membrane). The interaction of a ligand with its receptor may not involve the movement of either into the cell, but the interaction results in the generation of a signal that influences intracellular processes (transmembrane signaling).

- Mutations that affect the structure of membrane proteins (receptors, transporters, ion channels, enzymes, and structural proteins) may cause diseases; examples include cystic fibrosis and familial hypercholesterolemia.

REFERENCES

Alberts B, Johnson A, Lewis J, et al: *Molecular Biology of the Cell,* 5th ed. Garland Science, 2008.

Cooper GM, Hausman RE: *The Cell, A Molecular Approach.* Sinauer Assoc Inc., 2009.

Doherty GJ, McMahon HT: Mechanisms of endocytosis. Annu Rev Biochem 2009;78:857.

Lodish H, Berk A, Kaiser CA, et al: *Molecular Cell Biology,* 6th ed. WH Freeman & Co, 2008.

Longo N: Inherited defects of membrane transport. In: *Harrison's Principles of Internal Medicine,* 17th ed. Fauci AS, et al (editors). Chapter 359. McGraw-Hill, 2008.

Pollard TD, Earnshaw WC: *Cell Biology,* 2nd ed. Saunders Elsevier, 2008.

Singer SJ: Some early history of membrane molecular biology. Annu Rev Physiol 2004;66:1.

Vance DE, Vance J (editors): *Biochemistry of Lipids, Lipoproteins and Membranes,* 5th ed. Elsevier, 2008.

Voelker DR: Genetic and biochemical analysis of non-vesicular lipid traffic. Annu Rev Biochem 2009;78:827.

The Diversity of the Endocrine System

P. Anthony Weil, PhD

O B J E C T I V E S

After studying this chapter, you should be able to:

- Explain the basic principles of endocrine hormone action, including the determinants of hormone target cell response and the determinants of hormone concentration at target cells.
- Understand the broad diversity and mechanisms of action of endocrine hormones.
- Appreciate the complex steps involved in the production, transport, and storage of hormones.

ACTH	Adrenocorticotropic hormone	**IGF-I**	Insulin-like growth factor-I
ANF	Atrial natriuretic factor	**LH**	Luteotropic hormone
cAMP	Cyclic adenosine monophosphate	**LPH**	Lipotropin
CBG	Corticosteroid-binding globulin	**MIT**	Monoiodotyrosine
CG	Chorionic gonadotropin	**MSH**	Melanocyte-stimulating hormone
cGMP	Cyclic guanosine monophosphate	**OHSD**	Hydroxysteroid dehydrogenase
CLIP	Corticotropin-like intermediate lobe peptide	**PNMT**	Phenylethanolamine-*N*-methyltransferase
DBH	Dopamine β-hydroxylase	**POMC**	Pro-opiomelanocortin
DHEA	Dehydroepiandrosterone	**SHBG**	Sex hormone-binding globulin
DHT	Dihydrotestosterone	**StAR**	Steroidogenic acute regulatory (protein)
DIT	Diiodotyrosine	**TBG**	Thyroxine-binding globulin
DOC	Deoxycorticosterone	**TEBG**	Testosterone-estrogen-binding globulin
EGF	Epidermal growth factor	**TRH**	Thyrotropin-releasing hormone
FSH	Follicle-stimulating hormone	**TSH**	Thyrotropin-stimulating hormone
GH	Growth hormone		

BIOMEDICAL IMPORTANCE

The survival of multicellular organisms depends on their ability to adapt to a constantly changing environment. Intercellular communication mechanisms are necessary requirements for this adaptation. The nervous system and the endocrine system provide this intercellular, organism-wide communication. The nervous system was originally viewed as providing a fixed communication system, whereas the endocrine system supplied hormones, which are mobile messages. In fact, there is a remarkable convergence of these regulatory systems. For example, neural regulation of the endocrine system is important in the production and secretion of some hormones; many neurotransmitters resemble hormones in their synthesis, transport, and mechanism of action; and many hormones are synthesized in the nervous system. The word "hormone" is derived from a Greek term that means to arouse to activity. As classically defined, a hormone is a substance that is synthesized in one organ and transported by the circulatory system to act on another tissue. However, this original description is too restrictive because hormones can act on adjacent cells (paracrine action) and on the cell in which they were synthesized (autocrine action) without entering the systemic circulation. A diverse array of hormones—each with distinctive mechanisms of action and properties of biosynthesis, storage, secretion, transport, and metabolism—has evolved to provide homeostatic responses. This biochemical diversity is the topic of this chapter.

THE TARGET CELL CONCEPT

There are about 200 types of differentiated cells in humans. Only a few produce hormones, but virtually all of the 75 trillion cells in a human are targets of one or more of the >50 known

TABLE 41–1 Determinants of the Concentration of a Hormone at the Target Cell

The rate of synthesis and secretion of the hormones.
The proximity of the target cell to the hormone source (dilution effect).
The dissociation constants of the hormone with specific plasma transport proteins (if any).
The conversion of inactive or suboptimally active forms of the hormone into the fully active form.
The rate of clearance from plasma by other tissues or by digestion, metabolism, or excretion.

TABLE 41–2 Determinants of the Target Cell Response

The number, relative activity, and state of occupancy of the specific receptors on the plasma membrane or in the cytoplasm or nucleus.
The metabolism (activation or inactivation) of the hormone in the target cell.
The presence of other factors within the cell that are necessary for the hormone response.
Up- or downregulation of the receptor consequent to the interaction with its ligand.
Postreceptor desensitzation of the cell, including down-regulation of the receptor.

hormones. The concept of the target cell is a useful way of looking at hormone action. It was thought that hormones affected a single cell type—or only a few kinds of cells—and that a hormone elicited a unique biochemical or physiologic action. We now know that a given hormone can affect several different cell types; that more than one hormone can affect a given cell type; and that hormones can exert many different effects in one cell or in different cells. With the discovery of specific cell-surface and intracellular hormone receptors, the definition of a target has been expanded to include any cell in which the hormone (ligand) binds to its receptor, whether or not a biochemical or physiologic response has yet been determined.

Several factors determine the response of a target cell to a hormone. These can be thought of in two general ways: (1) as factors that affect the concentration of the hormone at the target cell (**Table 41–1**) and (2) as factors that affect the actual response of the target cell to the hormone (**Table 41–2**).

HORMONE RECEPTORS ARE OF CENTRAL IMPORTANCE

Receptors Discriminate Precisely

One of the major challenges faced in making the hormone-based communication system work is illustrated in **Figure 41–1**. Hormones are present at very low concentrations in the extracellular fluid, generally in the atto- to nanomolar range (10^{-15} to 10^{-9} mol/L). This concentration is much lower than that of the many structurally similar molecules (sterols, amino acids, peptides, and proteins) and other molecules that circulate at concentrations in the micro- to millimolar (10^{-6} to 10^{-3} mol/L) range. Target cells, therefore, must distinguish not only between different hormones present in small amounts but also between a given hormone and the 10^6- to 10^9-fold excess of other similar molecules. This high degree of discrimination is provided by cell-associated recognition molecules called receptors. Hormones initiate their biologic effects by binding to specific receptors, and since any effective control system also must provide a means of stopping a response, hormone-induced actions generally but not always terminate when the effector dissociates from the receptor (Figure 38–1; Type A response).

A target cell is defined by its ability to selectively bind a given hormone to its cognate receptor. Several biochemical features of this interaction are important in order for hormone-receptor interactions to be physiologically relevant: (1) binding should be specific, ie, displaceable by agonist or antagonist; (2) binding should be saturable; and (3) binding should occur within the concentration range of the expected biologic response.

Both Recognition & Coupling Domains Occur on Receptors

All receptors have at least two functional domains. A recognition domain binds the hormone ligand and a second region generates a signal that couples hormone recognition to some intracellular function. This coupling, or signal transduction, occurs in two general ways. Polypeptide and protein hormones and the catecholamines bind to receptors located in the plasma membrane and thereby generate a signal that regulates various intracellular functions, often by changing the activity of an

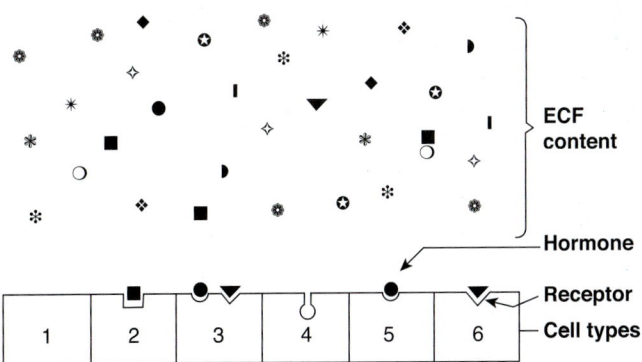

FIGURE 41–1 Specificity and selectivity of hormone receptors. Many different molecules circulate in the extracellular fluid (ECF), but only a few are recognized by hormone receptors. Receptors must select these molecules from among high concentrations of the other molecules. This simplified drawing shows that a cell may have no hormone receptors (1), have one receptor (2+5+6), have receptors for several hormones (3), or have a receptor but no hormone in the vicinity (4).

enzyme. In contrast, steroid, retinoid, and thyroid hormones interact with intracellular receptors, and it is this ligand–receptor complex that directly provides the signal, generally to specific genes whose rate of transcription is thereby affected.

The domains responsible for hormone recognition and signal generation have been identified in the protein polypeptide and catecholamine hormone receptors. Steroid, thyroid, and retinoid hormone receptors have several functional domains: one site binds the hormone; another binds to specific DNA regions; a third is involved in the interaction with other coregulator proteins that result in the activation (or repression) of gene transcription; and a fourth may specify binding to one or more other proteins that influence the intracellular trafficking of the receptor.

The dual functions of binding and coupling ultimately define a receptor, and it is the coupling of hormone binding to signal transduction—so-called **receptor–effector coupling**—that provides the first step in amplification of the hormonal response. This dual purpose also distinguishes the target cell receptor from the plasma carrier proteins that bind hormone but do not generate a signal (see Table 41–6).

Receptors Are Proteins

Several classes of peptide hormone receptors have been defined. For example, the insulin receptor is a heterotetramer composed of two copies of two different protein subunits ($\alpha_2\beta_2$) linked by multiple disulfide bonds in which the extracellular α subunit binds insulin and the membrane-spanning β subunit transduces the signal through the tyrosine protein kinase domain located in the cytoplasmic portion of this polypeptide. The receptors for insulin-like growth factor I (IGF-I) and epidermal growth factor (EGF) are generally similar in structure to the insulin receptor. The growth hormone and prolactin receptors also span the plasma membrane of target cells but do not contain intrinsic protein kinase activity. Ligand binding to these receptors, however, results in the association and activation of a completely different protein kinase signaling pathway, the JakStat pathway. Polypeptide hormone and catecholamine receptors, which transduce signals by altering the rate of production of cAMP through G-proteins, are characterized by the presence of seven domains that span the plasma membrane. Protein kinase activation and the generation of cyclic AMP (cAMP, 3′5′-adenylic acid; see Figure 19–5) is a downstream action of this class of receptor (see Chapter 42 for further details).

A comparison of several different steroid receptors with thyroid hormone receptors revealed a remarkable conservation of the amino acid sequence in certain regions, particularly in the DNA-binding domains. This led to the realization that receptors of the steroid or thyroid type are members of a large superfamily of nuclear receptors. Many related members of this family currently have no known ligand and thus are called orphan receptors. The nuclear receptor superfamily plays a critical role in the regulation of gene transcription by hormones, as described in Chapter 42.

HORMONES CAN BE CLASSIFIED IN SEVERAL WAYS

Hormones can be classified according to chemical composition, solubility properties, location of receptors, and the nature of the signal used to mediate hormonal action within the cell. A classification based on the last two properties is illustrated in **Table 41–3**, and general features of each group are illustrated in **Table 41–4**.

The hormones in group I are lipophilic. After secretion, these hormones associate with plasma transport or carrier proteins, a process that circumvents the problem of solubility while prolonging the plasma half-life of the hormone. The relative percentages of bound and free hormone are determined by the amount, binding affinity, and binding capacity of the transport protein. The free hormone, which is the biologically active form, readily traverses the lipophilic plasma membrane of all cells and encounters receptors in either the cytosol or nucleus of target cells. The ligand–receptor complex is assumed to be the intracellular messenger in this group.

The second major group consists of water-soluble hormones that bind to specific receptors spanning the plasma membrane of the target cell. Hormones that bind to these surface receptors of cells communicate with intracellular metabolic processes through intermediary molecules called **second messengers** (the hormone itself is the first messenger), which are generated as a consequence of the ligand–receptor interaction. The second messenger concept arose from an observation that epinephrine binds to the plasma membrane of certain cells and increases intracellular cAMP. This was followed by a series of experiments in which cAMP was found to mediate the effects of many hormones. Hormones that employ this mechanism are shown in group II.A of Table 41–3. Atrial natriuretic factor (ANF) uses cGMP as its second messenger (group II.B). Several hormones, many of which were previously thought to affect cAMP, appear to use ionic calcium (Ca^{2+}) or metabolites of complex phosphoinositides (or both) as the intracellular second messenger signal. These are shown in group II.C of the table. The intracellular messenger for group II.D is a protein kinase-phosphatase cascades; several have been identified, and a given hormone may use more than one kinase cascade. A few hormones fit into more than one category, and assignments change as new information is discovered.

DIVERSITY OF THE ENDOCRINE SYSTEM

Hormones Are Synthesized in a Variety of Cellular Arrangements

Hormones are synthesized in discrete organs designed solely for this specific purpose, such as the thyroid (triiodothyronine), adrenal (glucocorticoids and mineralocorticoids), and the pituitary (TSH, FSH, LH, growth hormone, prolactin, ACTH). Some organs are designed to perform two distinct

TABLE 41–3 Classification of Hormones by Mechanism of Action

I. *Hormones that bind to intracellular receptors*

Androgens
Calcitriol (1,25[OH]$_2$-D$_3$)
Estrogens
Glucocorticoids
Mineralocorticoids
Progestins
Retinoic acid
Thyroid hormones (T$_3$ and T$_4$)

II. *Hormones that bind to cell surface receptors*

  A. The second messenger is cAMP

    α$_2$-Adrenergic catecholamines
    β-Adrenergic catecholamines
    Adrenocorticotropic hormone (ACTH)
    Antidiuretic hormone
    Calcitonin
    Chorionic gonadotropin, human (CG)
    Corticotropin-releasing hormone
    Follicle-stimulating hormone (FSH)
    Glucagon
    Lipotropin (LPH)
    Luteinizing hormone (LH)
    Melanocyte-stimulating hormone (MSH)
    Parathyroid hormone (PTH)
    Somatostatin
    Thyroid-stimulating hormone (TSH)

  B. The second messenger is cGMP

    Atrial natriuretic factor
    Nitric oxide

  C. The second messenger is calcium or phosphatidylinositols (or both)

    Acetylcholine (muscarinic)
    α$_1$-Adrenergic catecholamines
    Angiotensin II
    Antidiuretic hormone (vasopressin)
    Cholecystokinin
    Gastrin
    Gonadotropin-releasing hormone
    Oxytocin
    Platelet-derived growth factor (PDGF)
    Substance P
    Thyrotropin-releasing hormone (TRH)

  D. The second messenger is a kinase or phosphatase cascade

    Adiponectin
    Chorionic somatomammotropin
    Epidermal growth factor (EGF)
    Erythropoietin (EPO)
    Fibroblast growth factor (FGF)
    Growth hormone (GH)
    Insulin
    Insulin-like growth factors I and II
    Leptin
    Nerve growth factor (NGF)
    Platelet-derived growth factor
    Prolactin

TABLE 41–4 General Features of Hormone Classes

	Group I	Group II
Types	Steroids, iodothyronines, calcitriol, retinoids	Polypeptides, proteins, glycoproteins, catecholamines
Solubility	Lipophilic	Hydrophilic
Transport proteins	Yes	No
Plasma half-life	Long (hours to days)	Short (minutes)
Receptor	Intracellular	Plasma membrane
Mediator	Receptor-hormone complex	cAMP, cGMP, Ca^{2+}, metabolites of complex phosphinositols, kinase cascades

but closely related functions. For example, the ovaries produce mature oocytes and the reproductive hormones estradiol and progesterone. The testes produce mature spermatozoa and testosterone. Hormones are also produced in specialized cells within other organs such as the small intestine (glucagon-like peptide), thyroid (calcitonin), and kidney (angiotensin II). Finally, the synthesis of some hormones requires the parenchymal cells of more than one organ—eg, the skin, liver, and kidney are required for the production of 1,25(OH)$_2$-D$_3$ (calcitriol). Examples of this diversity in the approach to hormone synthesis, each of which has evolved to fulfill a specific purpose, are discussed below.

Hormones Are Chemically Diverse

Hormones are synthesized from a wide variety of chemical building blocks. A large series is derived from cholesterol. These include the glucocorticoids, mineralocorticoids, estrogens, progestins, and 1,25(OH)$_2$-D$_3$ (**Figure 41–2**). In some cases, a steroid hormone is the precursor molecule for another hormone. For example, progesterone is a hormone in its own right but is also a precursor in the formation of glucocorticoids, mineralocorticoids, testosterone, and estrogens. Testosterone is an obligatory intermediate in the biosynthesis of estradiol and in the formation of dihydrotestosterone (DHT). In these examples, described in detail below, the final product is determined by the cell type and the associated set of enzymes in which the precursor exists.

The amino acid tyrosine is the starting point in the synthesis of both the catecholamines and thyroid hormones tetraiodothyronine (thyroxine; T$_4$) and triiodothyronine (T$_3$) (Figure 41–2). T$_3$ and T$_4$ are unique in that they require the addition of iodine (as I$^-$) for bioactivity. Since dietary iodine is very scarce in many parts of the world, an intricate mechanism for accumulating and retaining I$^-$ has evolved.

Many hormones are polypeptides or glycoproteins. These range in size from the small thyrotropin-releasing

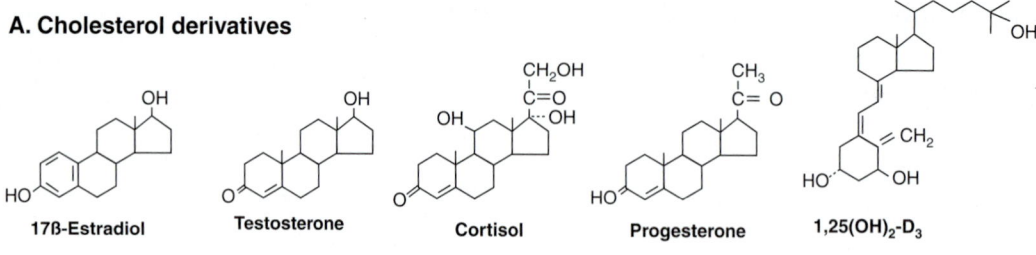

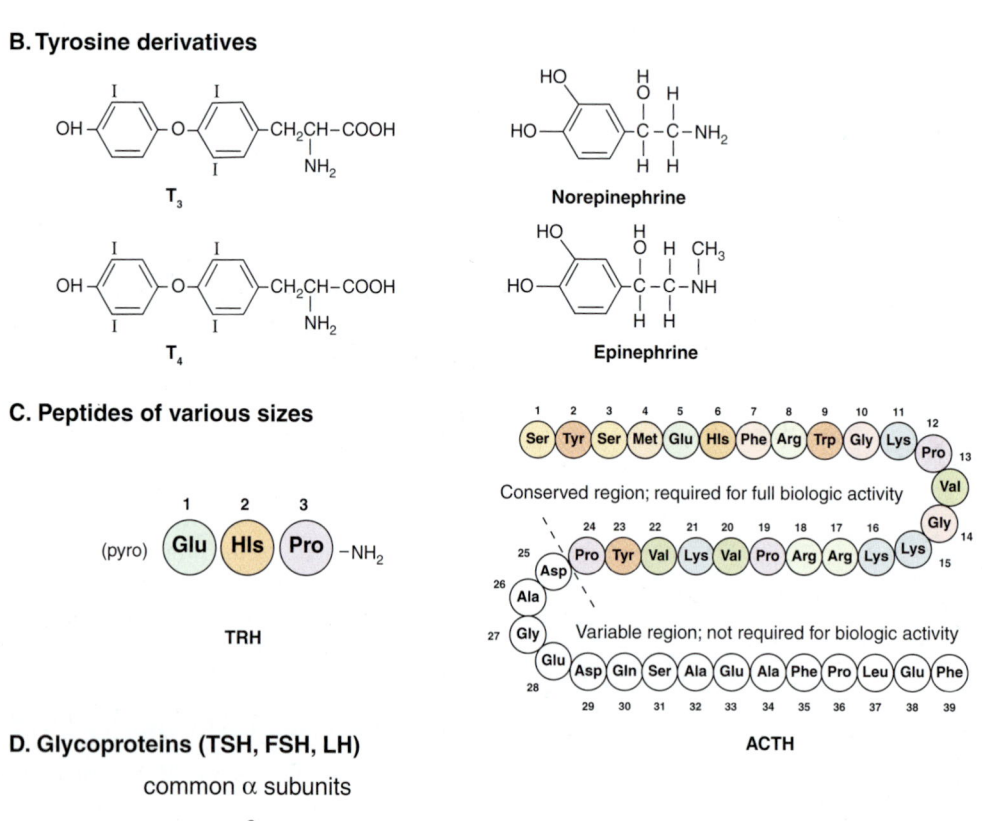

FIGURE 41–2 Chemical diversity of hormones: (A) cholesterol derivatives; (B) tyrosine derivatives; (C) peptides of various sizes; (D) glycoproteins (TSH, FSH, and LH) with common α subunits and unique β subunits.

hormone (TRH), a tripeptide, to single-chain polypeptides like adrenocorticotropic hormone (ACTH; 39 amino acids), parathyroid hormone (PTH; 84 amino acids), and growth hormone (GH; 191 amino acids) (Figure 41–2). Insulin is an AB chain heterodimer of 21 and 30 amino acids, respectively. Follicle-stimulating hormone (FSH), luteinizing hormone (LH), thyroid-stimulating hormone (TSH), and chorionic gonadotropin (CG) are glycoprotein hormones of αβ heterodimeric structure. The α chain is identical in all of these hormones, and distinct β chains impart hormone uniqueness. These hormones have a molecular mass in the range of 25–30 kDa depending on the degree of glycosylation and the length of the β chain.

Hormones Are Synthesized & Modified for Full Activity in a Variety of Ways

Some hormones are synthesized in final form and secreted immediately. Included in this class are hormones derived from cholesterol. Some, such as the catecholamines are synthesized in final form and stored in the producing cells, while others, like insulin, are synthesized from precursor molecules in the producing cell, and then are processed and secreted upon a physiologic cue (plasma glucose concentrations). Finally, still others are converted to active forms from precursor molecules in the periphery (T_3 and DHT). All of these examples are discussed in more detail below.

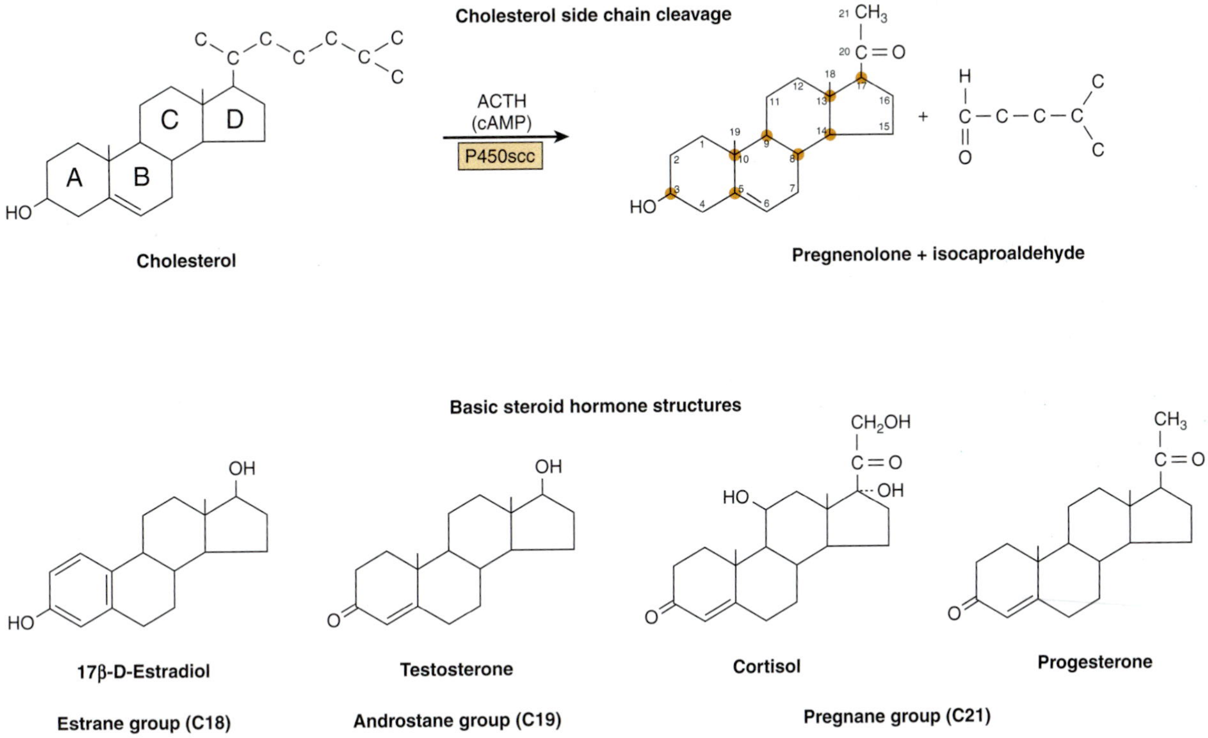

FIGURE 41–3 **Cholesterol side-chain cleavage and basic steroid hormone structures.** The basic sterol rings are identified by the letters A–D. The carbon atoms are numbered 1–21, starting with the A ring.

MANY HORMONES ARE MADE FROM CHOLESTEROL

Adrenal Steroidogenesis

The adrenal steroid hormones are synthesized from cholesterol, which is mostly derived from the plasma, but a small portion is synthesized in situ from acetyl-CoA via mevalonate and squalene. Much of the cholesterol in the adrenal is esterified and stored in cytoplasmic lipid droplets. Upon stimulation of the adrenal by ACTH, an esterase is activated, and the free cholesterol formed is transported into the mitochondrion, where a **cytochrome P450 side chain cleavage enzyme (P450scc)** converts cholesterol to pregnenolone. Cleavage of the side chain involves sequential hydroxylations, first at C_{22} and then at C_{20}, followed by side chain cleavage (removal of the six-carbon fragment isocaproaldehyde) to give the 21-carbon steroid (**Figure 41–3**, top). An ACTH-dependent **steroidogenic acute regulatory (StAR) protein** is essential for the transport of cholesterol to P450scc in the inner mitochondrial membrane.

All mammalian steroid hormones are formed from cholesterol via pregnenolone through a series of reactions that occur in either the mitochondria or endoplasmic reticulum of the producing cell. Hydroxylases that require molecular oxygen and NADPH are essential, and dehydrogenases, an isomerase, and a lyase reaction are also necessary for certain steps. There is cellular specificity in adrenal steroidogenesis. For instance, 18-hydroxylase and 19-hydroxysteroid dehydrogenises, which

are required for aldosterone synthesis, are found only in the zona glomerulosa cells (the outer region of the adrenal cortex), so that the biosynthesis of this mineralocorticoid is confined to this region. A schematic representation of the pathways involved in the synthesis of the three major classes of adrenal steroids is presented in **Figure 41–4**. The enzymes are shown in the rectangular boxes, and the modifications at each step are shaded.

Mineralocorticoid Synthesis

Synthesis of aldosterone follows the mineralocorticoid pathway and occurs in the zona glomerulosa. Pregnenolone is converted to progesterone by the action of two smooth endoplasmic reticulum enzymes, **3β-hydroxysteroid dehydrogenase (3β-OHSD)** and **Δ^{5,4}-isomerase**. Progesterone is hydroxylated at the C_{21} position to form 11-deoxycorticosterone (DOC), which is an active (Na$^+$-retaining) mineralocorticoid. The next hydroxylation, at C_{11}, produces corticosterone, which has glucocorticoid activity and is a weak mineralocorticoid (it has <5% of the potency of aldosterone). In some species (eg, rodents), it is the most potent glucocorticoid. C_{21} hydroxylation is necessary for both mineralocorticoid and glucocorticoid activity, but most steroids with a C_{17} hydroxyl group have more glucocorticoid and less mineralocorticoid action. In the zona glomerulosa, which does not have the smooth endoplasmic reticulum enzyme 17α-hydroxylase, a mitochondrial 18-hydroxylase is present. The **18-hydroxylase (aldosterone synthase)** acts on corticosterone to form

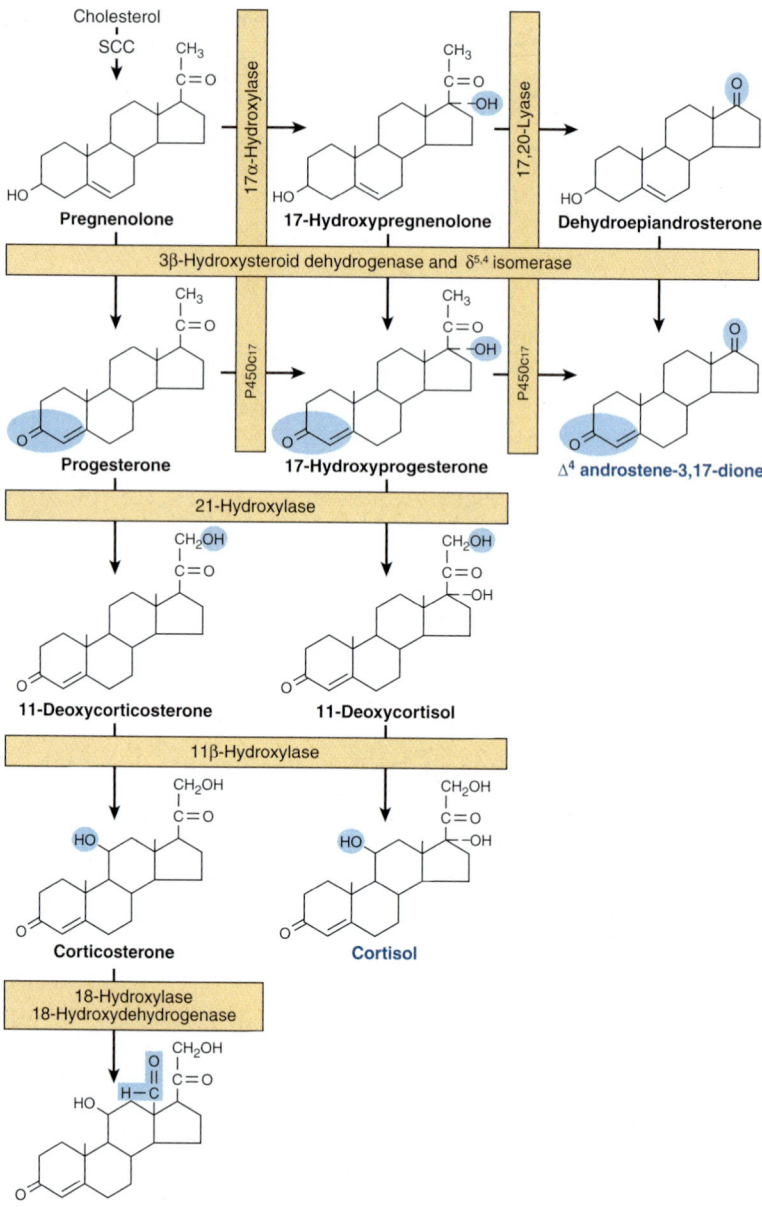

FIGURE 41–4 **Pathways involved in the synthesis of the three major classes of adrenal steroids (mineralocorticoids, glucocorticoids, and androgens).** Enzymes are shown in the rectangular boxes, and the modifications at each step are shaded. Note that the 17α-hydroxylase and 17,20-lyase activities are both part of one enzyme, designated P450c17. (Slightly modified and reproduced, with permission, from Harding BW: In: *Endocrinology*, vol 2. DeGroot LJ (editors). Grune & Stratton, 1979. Copyright © 1979 Elsevier Inc. Reprinted with permission from Elsevier.)

18-hydroxycorticosterone, which is changed to aldosterone by conversion of the 18-alcohol to an aldehyde. This unique distribution of enzymes and the special regulation of the zona glomerulosa by K^+ and angiotensin II have led some investigators to suggest that, in addition to the adrenal being two glands, the adrenal cortex is actually two separate organs.

Glucocorticoid Synthesis

Cortisol synthesis requires three hydroxylases located in the fasciculata and reticularis zones of the adrenal cortex that act sequentially on the C_{17}, C_{21}, and C_{11} positions. The first two reactions are rapid, while C_{11} hydroxylation is relatively slow. If the C_{11} position is hydroxylated first, the action of **17α-hydroxylase** is impeded and the mineralocorticoid pathway is followed

(forming corticosterone or aldosterone, depending on the cell type). 17α-Hydroxylase is a smooth endoplasmic reticulum enzyme that acts upon either progesterone or, more commonly, pregnenolone. 17α-Hydroxyprogesterone is hydroxylated at C_{21} to form 11-deoxycortisol, which is then hydroxylated at C_{11} to form cortisol, the most potent natural glucocorticoid hormone in humans. 21-Hydroxylase is a smooth endoplasmic reticulum enzyme, whereas 11β-hydroxylase is a mitochondrial enzyme. Steroidogenesis thus involves the repeated shuttling of substrates into and out of the mitochondria.

Androgen Synthesis

The major androgen or androgen precursor produced by the adrenal cortex is dehydroepiandrosterone (DHEA). Most

17-hydroxypregnenolone follows the glucocorticoid pathway, but a small fraction is subjected to oxidative fission and removal of the two-carbon side chain through the action of 17,20-lyase. The lyase activity is actually part of the same enzyme (P450c17) that catalyzes 17α-hydroxylation. This is therefore a **dual-function protein.** The lyase activity is important in both the adrenals and the gonads and acts exclusively on 17α-hydroxy-containing molecules. Adrenal androgen production increases markedly if glucocorticoid biosynthesis is impeded by the lack of one of the hydroxylases (**adrenogenital syndrome**). DHEA is really a prohormone since the actions of 3β-OHSD and $\Delta^{5,4}$-isomerase convert the weak androgen DHEA into the more potent **androstenedione.** Small amounts of androstenedione are also formed in the adrenal by the action of the lyase on 17α-hydroxyprogesterone. Reduction of androstenedione at the C_{17} position results in the formation of **testosterone,** the most potent adrenal androgen. Small amounts of testosterone are produced in the adrenal by this mechanism, but most of this conversion occurs in the testes.

Testicular Steroidogenesis

Testicular androgens are synthesized in the interstitial tissue by the Leydig cells. The immediate precursor of the gonadal steroids, as for the adrenal steroids, is cholesterol. The rate-limiting step, as in the adrenal, is delivery of cholesterol to the inner membrane of the mitochondria by the transport protein StAR. Once in the proper location, cholesterol is acted upon by the side chain cleavage enzyme P450scc. The conversion of cholesterol to pregnenolone is identical in adrenal, ovary, and testis. In the latter two tissues, however, the reaction is promoted by LH rather than ACTH.

The conversion of pregnenolone to testosterone requires the action of five enzyme activities contained in three proteins: (1) 3β-hydroxysteroid dehydrogenase (3β-OHSD) and $\Delta^{5,4}$-isomerase; (2) 17α-hydroxylase and 17,20-lyase; and (3) 17β-hydroxysteroid dehydrogenase (17β-OHSD). This sequence, referred to as the **progesterone (or Δ^4) pathway,** is shown on the right side of **Figure 41–5.** Pregnenolone can also be converted to testosterone by the **dehydroepiandrosterone (or Δ^5) pathway,** which is illustrated on the left side of Figure 41–5. The Δ^5 route appears to be most used in human testes.

The five enzyme activities are localized in the microsomal fraction in rat testes, and there is a close functional association between the activities of 3β-OHSD and $\Delta^{5,4}$-isomerase and between those of a 17α-hydroxylase and 17,20-lyase. These enzyme pairs, both contained in a single protein, are shown in the general reaction sequence in Figure 41–5.

DHT Is Formed from Testosterone in Peripheral Tissues

Testosterone is metabolized by two pathways. One involves oxidation at the 17 position, and the other involves reduction of the A ring double bond and the 3-ketone. Metabolism by the first pathway occurs in many tissues, including liver, and produces 17-ketosteroids that are generally inactive or less active than the parent compound. Metabolism by the second pathway, which is less efficient, occurs primarily in target tissues and produces the potent metabolite DHT.

The most significant metabolic product of testosterone is DHT, since in many tissues, including prostate, external genitalia, and some areas of the skin, this is the active form of the hormone. The plasma content of DHT in the adult male is about one-tenth that of testosterone, and ~400 μg of DHT is produced daily as compared with about 5 mg of testosterone. About 50–100 μg of DHT are secreted by the testes. The rest is produced peripherally from testosterone in a reaction catalyzed by the NADPH-dependent **5α-reductase** (**Figure 41–6**). Testosterone can thus be considered a prohormone since it is converted into a much more potent compound (DHT) and since most of this conversion occurs outside the testes. Some estradiol is formed from the peripheral aromatization of testosterone, particularly in males.

Ovarian Steroidogenesis

The estrogens are a family of hormones synthesized in a variety of tissues. 17β-Estradiol is the primary estrogen of ovarian origin. In some species, estrone, synthesized in numerous tissues, is more abundant. In pregnancy, relatively more estriol is produced, and this comes from the placenta. The general pathway and the subcellular localization of the enzymes involved in the early steps of estradiol synthesis are the same as those involved in androgen biosynthesis. Features unique to the ovary are illustrated in **Figure 41–7.**

Estrogens are formed by the aromatization of androgens in a complex process that involves three hydroxylation steps, each of which requires O_2 and NADPH. The **aromatase enzyme complex** is thought to include a P450 monooxygenase. Estradiol is formed if the substrate of this enzyme complex is testosterone, whereas estrone results from the aromatization of androstenedione.

The cellular source of the various ovarian steroids has been difficult to unravel, but a transfer of substrates between two cell types is involved. Theca cells are the source of androstenedione and testosterone. These are converted by the aromatase enzyme in granulosa cells to estrone and estradiol, respectively. Progesterone, a precursor for all steroid hormones, is produced and secreted by the corpus luteum as an end-product hormone because these cells do not contain the enzymes necessary to convert progesterone to other steroid hormones (**Figure 41–8**).

Significant amounts of estrogens are produced by the peripheral aromatization of androgens. In human males, the peripheral aromatization of testosterone to estradiol (E_2) accounts for 80% of the production of the latter. In females, adrenal androgens are important substrates since as much as 50% of the E_2 produced during pregnancy comes from the aromatization of androgens. Finally, conversion of androstenedione to estrone is the major source of estrogens in postmenopausal women. Aromatase activity is present in adipose cells and also in liver, skin, and other tissues. Increased activity

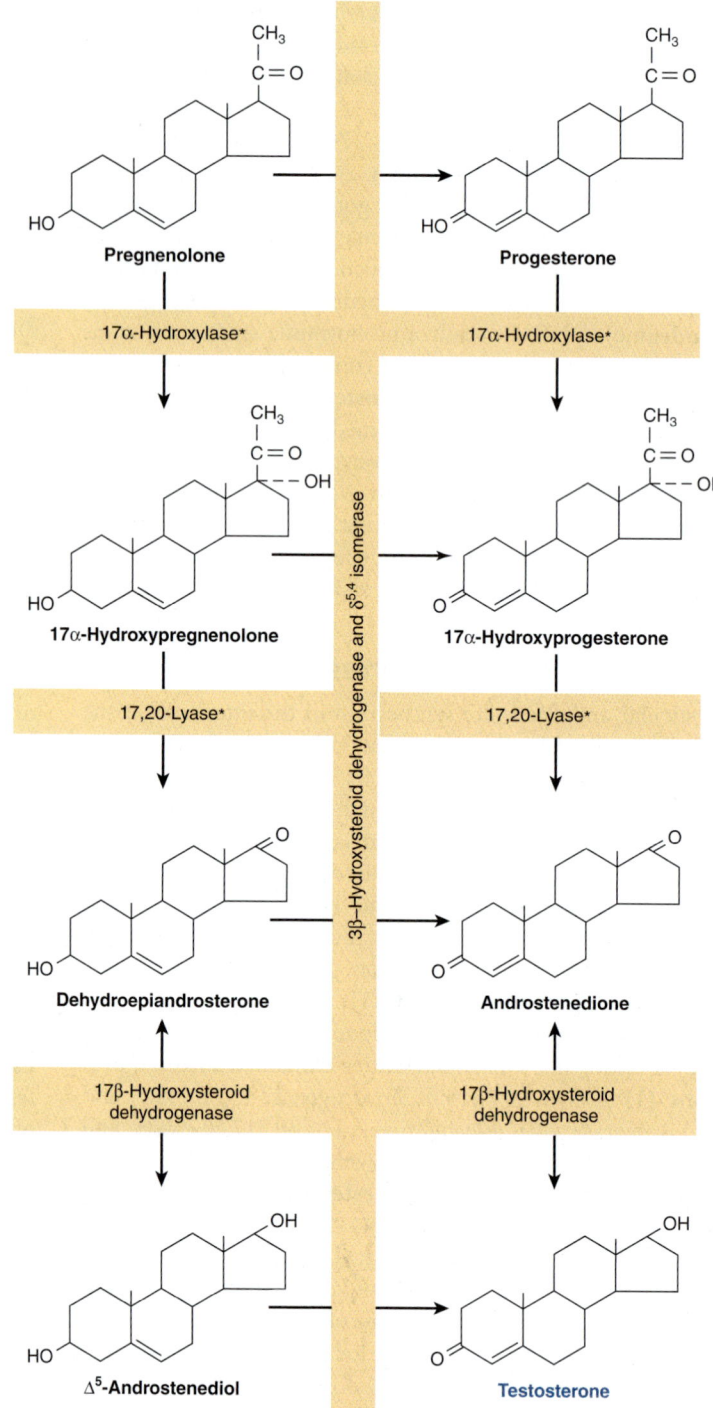

FIGURE 41–5 **Pathways of testosterone biosynthesis.**
The pathway on the left side of the figure is called the Δ^5 or
dehydroepiandrosterone pathway; the pathway on the right
side is called the Δ^4 or progesterone pathway. The asterisk
indicates that the 17α-hydroxylase and 17,20-lyase activities
reside in a single protein, P450c17.

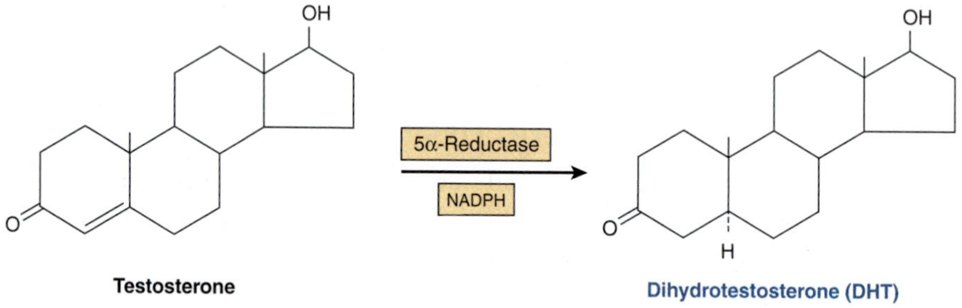

FIGURE 41–6 **Dihydrotestosterone is
formed from testosterone through action
of the enzyme 5α-reductase.**

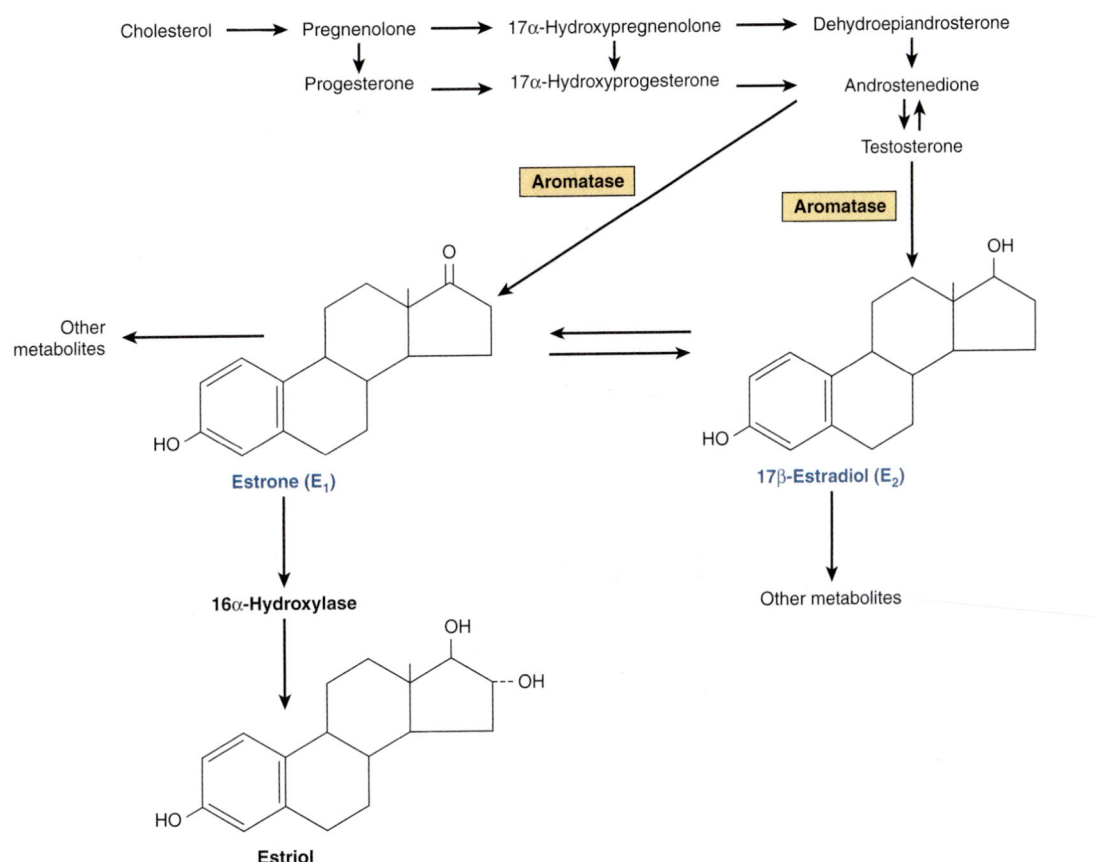

FIGURE 41–7 **Biosynthesis of estrogens.** (Slightly modified and reproduced, with permission, from Ganong WF: *Review of Medical Physiology*, 21st ed. McGraw-Hill, 2005.)

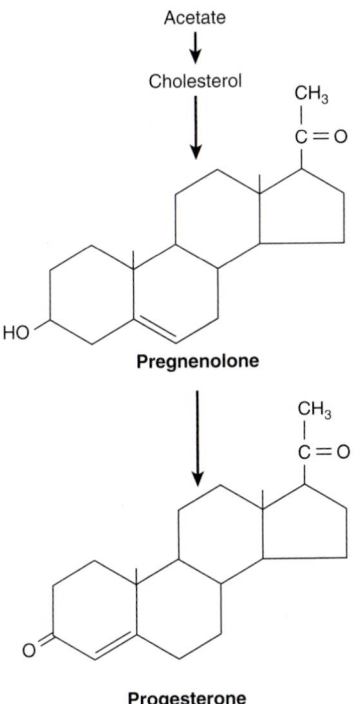

FIGURE 41–8 **Biosynthesis of progesterone in the corpus luteum.**

of this enzyme may contribute to the "estrogenization" that characterizes such diseases as cirrhosis of the liver, hyperthyroidism, aging, and obesity. Aromatase inhibitors show promise as therapeutic agents in breast cancer and possibly in other female reproductive tract malignancies.

1,25(OH)$_2$-D$_3$ (Calcitriol) Is Synthesized from a Cholesterol Derivative

1,25(OH)$_2$-D$_3$ is produced by a complex series of enzymatic reactions that involve the plasma transport of precursor molecules to a number of different tissues (**Figure 41–9**). One of these precursors is vitamin D—really not a vitamin, but this common name persists. The active molecule, 1,25(OH)$_2$-D$_3$, is transported to other organs where it activates biologic processes in a manner similar to that employed by the steroid hormones.

Skin

Small amounts of the precursor for 1,25(OH)$_2$-D$_3$ synthesis are present in food (fish liver oil, and egg yolk), but most of the precursor for 1,25(OH)$_2$-D$_3$ synthesis is produced in the malpighian layer of the epidermis from 7-dehydrocholesterol in an ultraviolet light-mediated, nonenzymatic **photolysis** reaction. The extent of this conversion is related directly to the intensity

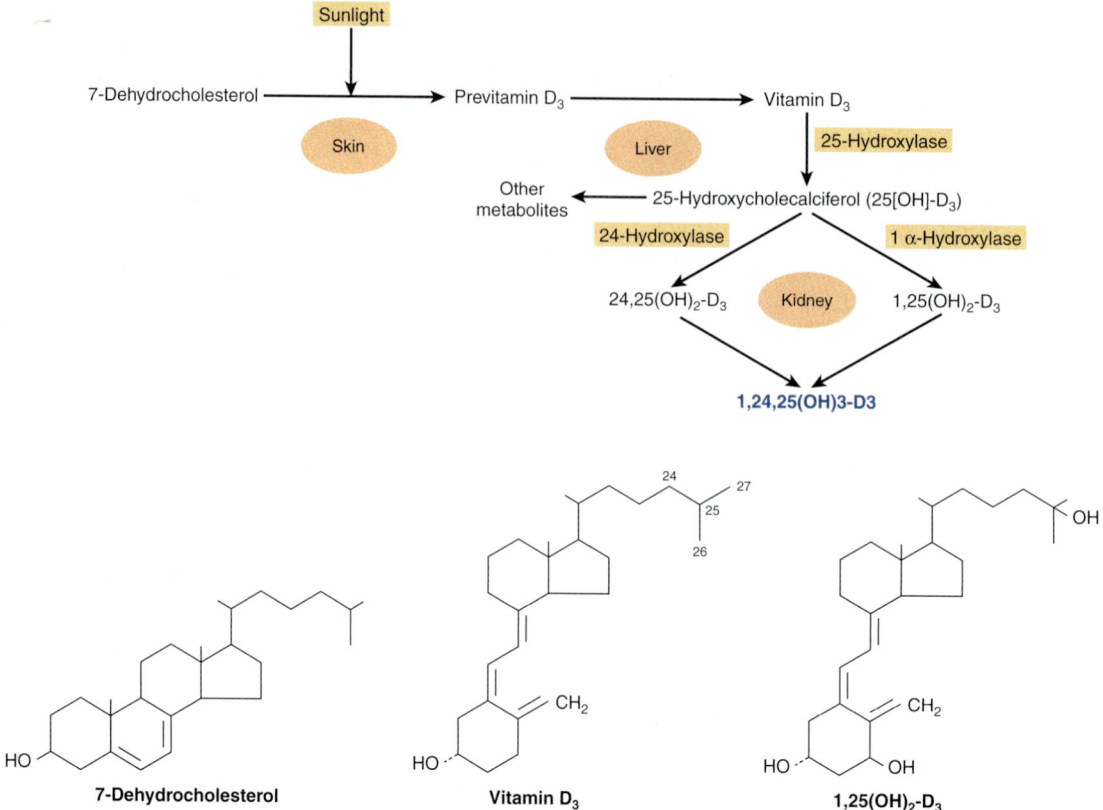

FIGURE 41–9 **Formation and hydroxylation of vitamin D₃.** 25-Hydroxylation takes place in the liver, and the other hydroxylations occur in the kidneys. 25,26(OH)₂-D₃ and 1,25,26(OH)₃-D₃ are probably formed as well. The structures of 7-dehydrocholesterol, vitamin D₃, and 1,25(OH)₂-D₃ are also shown. (Modified and reproduced, with permission, from Ganong WF: *Review of Medical Physiology*, 21st ed. McGraw-Hill, 2005.)

of the exposure and inversely to the extent of pigmentation in the skin. There is an age-related loss of 7-dehydrocholesterol in the epidermis that may be related to the negative calcium balance associated with old age.

Liver

A specific transport protein called the **vitamin D–binding protein** binds vitamin D_3 and its metabolites and moves vitamin D_3 from the skin or intestine to the liver, where it undergoes 25-hydroxylation, the first obligatory reaction in the production of $1,25(OH)_2$-D_3. 25-Hydroxylation occurs in the endoplasmic reticulum in a reaction that requires magnesium, NADPH, molecular oxygen, and an uncharacterized cytoplasmic factor. Two enzymes are involved: an NADPH-dependent cytochrome P450 reductase and a cytochrome P450. This reaction is not regulated, and it also occurs with low efficiency in kidney and intestine. The $25(OH)_2$-D_3 enters the circulation, where it is the major form of vitamin D found in plasma, and is transported to the kidney by the vitamin D–binding protein.

Kidney

$25(OH)_2$-D_3 is a weak agonist and must be modified by hydroxylation at position C_1 for full biologic activity. This is accomplished in mitochondria of the renal proximal convoluted tubule by a three-component monooxygenase reaction that requires NADPH, Mg^{2+}, molecular oxygen, and at least three enzymes: (1) a flavoprotein, renal ferredoxin reductase; (2) an iron sulfur protein, renal ferredoxin; and (3) cytochrome P450. This system produces $1,25(OH)_2$-D_3, which is the most potent naturally occurring metabolite of vitamin D.

CATECHOLAMINES & THYROID HORMONES ARE MADE FROM TYROSINE

Catecholamines Are Synthesized in Final Form & Stored in Secretion Granules

Three amines—dopamine, norepinephrine, and epinephrine—are synthesized from tyrosine in the chromaffin cells of the adrenal medulla. The major product of the adrenal medulla is epinephrine. This compound constitutes about 80% of the catecholamines in the medulla, and it is not made in extramedullary tissue. In contrast, most of the norepinephrine present in organs innervated by sympathetic nerves is made in situ (about 80% of the total), and most of the rest is made in other nerve endings

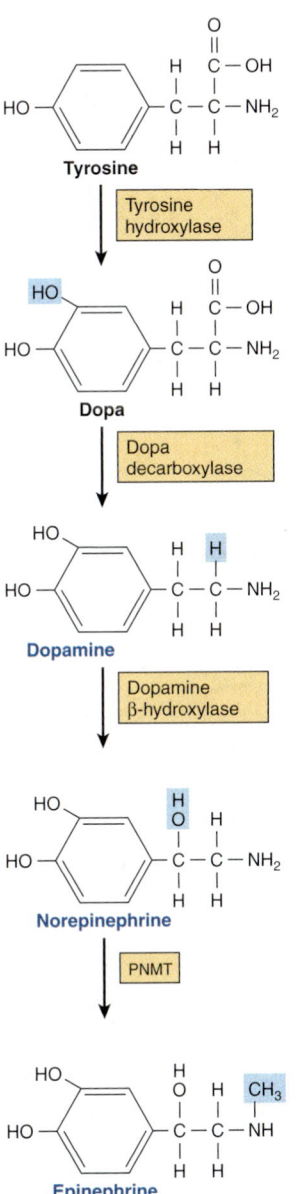

FIGURE 41–10 **Biosynthesis of catecholamines.** (PNMT, phenylethanolamine-*N*-methyltransferase.)

and reaches the target sites via the circulation. Epinephrine and norepinephrine may be produced and stored in different cells in the adrenal medulla and other chromaffin tissues.

The conversion of tyrosine to epinephrine requires four sequential steps: (1) ring hydroxylation; (2) decarboxylation; (3) side-chain hydroxylation to form norepinephrine; and (4) *N*-methylation to form epinephrine. The biosynthetic pathway and the enzymes involved are illustrated in **Figure 41–10.**

Tyrosine Hydroxylase Is Rate-Limiting for Catecholamine Biosynthesis

Tyrosine is the immediate precursor of catecholamines, and **tyrosine hydroxylase** is the rate-limiting enzyme in

catecholamine biosynthesis. Tyrosine hydroxylase is found in both soluble and particle-bound forms only in tissues that synthesize catecholamines; it functions as an oxidoreductase, with tetrahydropteridine as a cofactor, to convert L-tyrosine to L-dihydroxyphenylalanine (**L-dopa**). As the rate-limiting enzyme, tyrosine hydroxylase is regulated in a variety of ways. The most important mechanism involves feedback inhibition by the catecholamines, which compete with the enzyme for the pteridine cofactor. Catecholamines cannot cross the blood–brain barrier; hence, in the brain they must be synthesized locally. In certain central nervous system diseases (eg, Parkinson's disease), there is a local deficiency of dopamine synthesis. L-Dopa, the precursor of dopamine, readily crosses the blood–brain barrier and so is an important agent in the treatment of Parkinson disease.

Dopa Decarboxylase Is Present in All Tissues

This soluble enzyme requires pyridoxal phosphate for the conversion of L-dopa to 3,4-dihydroxyphenylethylamine (**dopamine**). Compounds that resemble L-dopa, such as α-methyldopa, are competitive inhibitors of this reaction. α-Methyldopa is effective in treating some kinds of hypertension.

Dopamine β-Hydroxylase (DBH) Catalyzes the Conversion of Dopamine to Norepinephrine

DBH is a monooxygenase and uses ascorbate as an electron donor, copper at the active site, and fumarate as modulator. DBH is in the particulate fraction of the medullary cells, probably in the secretion granule; thus, the conversion of dopamine to **norepinephrine** occurs in this organelle.

Phenylethanolamine-N-Methyltransferase (PNMT) Catalyzes the Production of Epinephrine

PNMT catalyzes the *N*-methylation of norepinephrine to form **epinephrine** in the epinephrine-forming cells of the adrenal medulla. Since PNMT is soluble, it is assumed that norepinephrine-to-epinephrine conversion occurs in the cytoplasm. The synthesis of PNMT is induced by glucocorticoid hormones that reach the medulla via the intra-adrenal portal system. This special system provides for a 100-fold steroid concentration gradient over systemic arterial blood, and this high intra-adrenal concentration appears to be necessary for the induction of PNMT.

T₃ & T₄ Illustrate the Diversity in Hormone Synthesis

The formation of **triiodothyronine (T₃)** and **tetraiodothyronine (thyroxine; T₄)** (see Figure 41–2) illustrates many of the principles of diversity discussed in this chapter. These hormones require a rare element (iodine) for bioactivity; they are synthesized as part of a very large precursor molecule (thyroglobulin); they are stored in an intracellular reservoir (colloid); and there is peripheral conversion of T_4 to T_3, which is a much more active hormone.

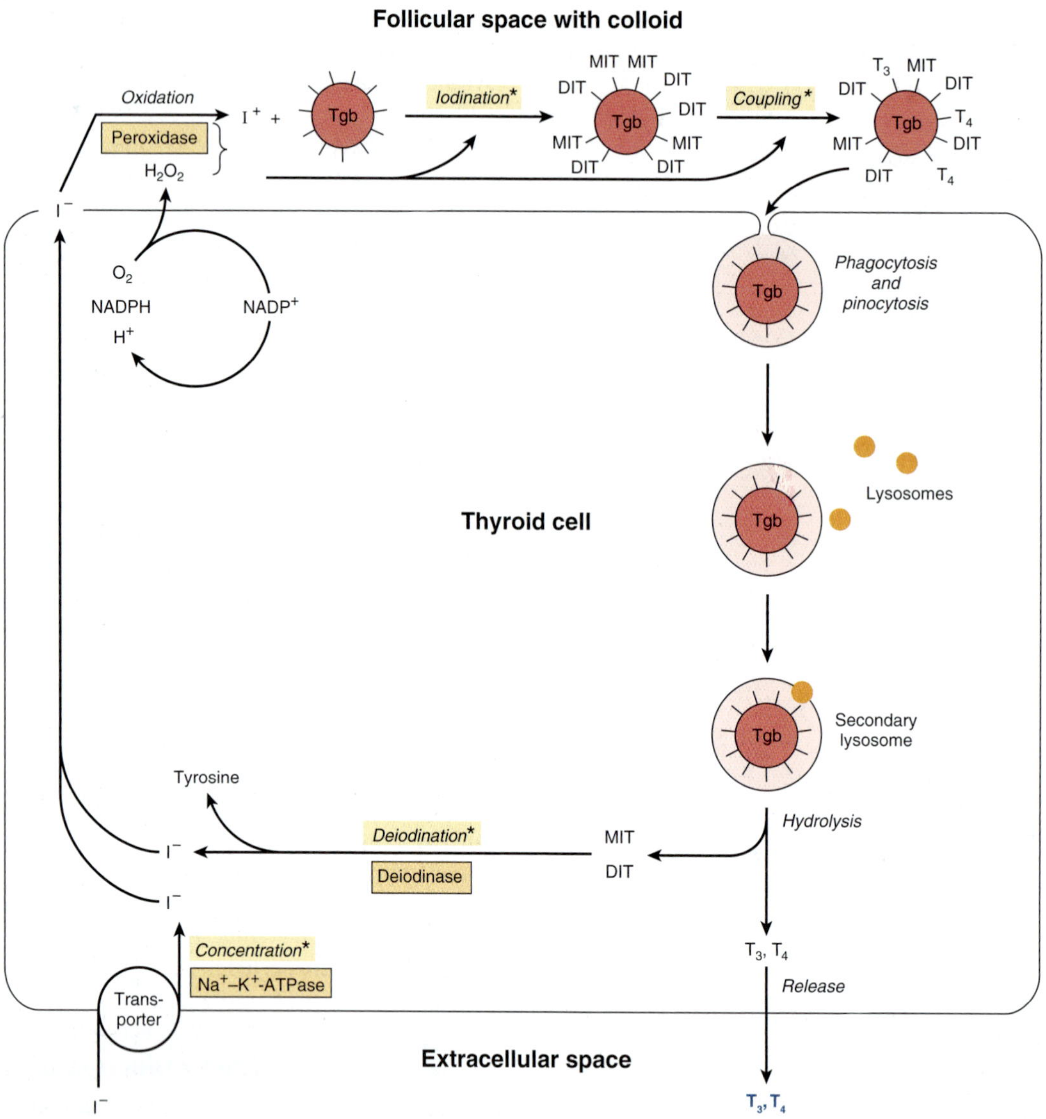

Follicular space with colloid

FIGURE 41–11 **Model of iodide metabolism in the thyroid follicle.** A follicular cell is shown facing the follicular lumen (top) and the extracellular space (bottom). Iodide enters the thyroid primarily through a transporter (bottom left). Thyroid hormone synthesis occurs in the follicular space through a series of reactions, many of which are peroxidase-mediated. Thyroid hormones, stored in the colloid in the follicular space, are released from thyroglobulin by hydrolysis inside the thyroid cell. (DIT, diiodotyrosine; MIT, monoiodotyrosine; Tgb, thyroglobulin; T_3, triiodothyronine; T_4, tetraiodothyronine.) Asterisks indicate steps or processes where inherited enzyme deficiencies cause congenital goiter and often result in hypothyroidism.

The thyroid hormones T_3 and T_4 are unique in that iodine (as iodide) is an essential component of both. In most parts of the world, iodine is a scarce component of soil, and for that reason there is little in food. A complex mechanism has evolved to acquire and retain this crucial element and to convert it into a form suitable for incorporation into organic compounds. At the same time, the thyroid must synthesize thyronine from tyrosine, and this synthesis takes place in thyroglobulin (**Figure 41–11**).

Thyroglobulin is the precursor of T_4 and T_3. It is a large iodinated, glycosylated protein with a molecular mass of 660 kDa. Carbohydrate accounts for 8–10% of the weight of thyroglobulin

and iodide for about 0.2–1%, depending upon the iodine content in the diet. Thyroglobulin is composed of two large subunits. It contains 115 tyrosine residues, each of which is a potential site of iodination. About 70% of the iodide in thyroglobulin exists in the inactive precursors, **monoiodotyrosine (MIT)** and **diiodotyrosine (DIT)**, while 30% is in the **iodothyronyl residues,** T_4 and T_3. When iodine supplies are sufficient, the T_4:T_3 ratio is about 7:1. In **iodine deficiency,** this ratio decreases, as does the DIT:MIT ratio. Thyroglobulin, a large molecule of about 5000 amino acids, provides the conformation required for tyrosyl coupling and iodide organification necessary in the formation of the diaminoacid thyroid hormones. It is synthesized in the basal portion of the cell and moves to the lumen, where it is a storage form of T_3 and T_4 in the colloid; several weeks' supply of these hormones exist in the normal thyroid. Within minutes after stimulation of the thyroid by TSH, colloid reenters the cell and there is a marked increase of phagolysosome activity. Various acid proteases and peptidases hydrolyze the thyroglobulin into its constituent amino acids, including T_4 and T_3, which are discharged into the extracellular space (see Figure 41–11). Thyroglobulin is thus a very large prohormone.

Iodide Metabolism Involves Several Discrete Steps

The thyroid is able to concentrate I^- against a strong electrochemical gradient. This is an energy-dependent process and is linked to the $Na^+–K^+$-ATPase-dependent thyroidal I^- transporter. The ratio of iodide in thyroid to iodide in serum (T:S ratio) is a reflection of the activity of this transporter. This activity is primarily controlled by TSH and ranges from 500:1 in animals chronically stimulated with TSH to 5:1 or less in hypophysectomized animals (no TSH). The T:S ratio in humans on a normal iodine diet is about 25:1.

The thyroid is the only tissue that can oxidize I^- to a higher valence state, an obligatory step in I^- organification and thyroid hormone biosynthesis. This step involves a heme-containing peroxidase and occurs at the luminal surface of the follicular cell. Thyroperoxidase, a tetrameric protein with a molecular mass of 60 kDa, requires hydrogen peroxide as an oxidizing agent. The H_2O_2 is produced by an NADPH-dependent enzyme resembling cytochrome c reductase. A number of compounds inhibit I^- oxidation and therefore its subsequent incorporation into MIT and DIT. The most important of these are the thiourea drugs. They are used as antithyroid drugs because of their ability to inhibit thyroid hormone biosynthesis at this step. Once iodination occurs, the iodine does not readily leave the thyroid. Free tyrosine can be iodinated, but it is not incorporated into proteins since no tRNA recognizes iodinated tyrosine.

The coupling of two DIT molecules to form T_4—or of an MIT and DIT to form T_3—occurs within the thyroglobulin molecule. A separate coupling enzyme has not been found, and since this is an oxidative process it is assumed that the same thyroperoxidase catalyzes this reaction by stimulating free radical formation of iodotyrosine. This hypothesis is supported by the observation that the same drugs which inhibit I^- oxidation also inhibit coupling. The formed thyroid hormones remain as integral parts of thyroglobulin until the latter is degraded, as described above.

A deiodinase removes I^- from the inactive mono and diiodothyronine molecules in the thyroid. This mechanism provides a substantial amount of the I^- used in T_3 and T_4 biosynthesis. A peripheral deiodinase in target tissues such as pituitary, kidney, and liver selectively removes I^- from the 5′ position of T_4 to make T_3 (see Figure 41–2), which is a much more active molecule. In this sense, T_4 can be thought of as a prohormone, though it does have some intrinsic activity.

Several Hormones Are Made from Larger Peptide Precursors

Formation of the critical disulfide bridges in insulin requires that this hormone be first synthesized as part of a larger precursor molecule, proinsulin. This is conceptually similar to the example of the thyroid hormones, which can only be formed in the context of a much larger molecule. Several other hormones are synthesized as parts of large precursor molecules, not because of some special structural requirement but rather as a mechanism for controlling the available amount of the active hormone. PTH and angiotensin II are examples of this type of regulation. Another interesting example is the POMC protein, which can be processed into many different hormones in a tissue-specific manner. These examples are discussed in detail below.

Insulin Is Synthesized as a Preprohormone & Modified Within the β Cell

Insulin has an AB heterodimeric structure with one intrachain (A6–A11) and two interchain disulfide bridges (A7–B7 and A20–B19) (**Figure 41–12**). The A and B chains could be synthesized in the laboratory, but attempts at a biochemical synthesis of the mature insulin molecule yielded very poor results. The reason for this became apparent when it was discovered that insulin is synthesized as a **preprohormone** (molecular weight ~11,500), which is the prototype for peptides that are processed from larger precursor molecules. The hydrophobic 23-amino-acid pre-, or leader, sequence directs the molecule into the cisternae of the endoplasmic reticulum and then is removed. This results in the 9000-MW proinsulin molecule, which provides the conformation necessary for the proper and efficient formation of the disulfide bridges. As shown in Figure 41–12, the sequence of proinsulin, starting from the amino terminal, is B chain—connecting (C) peptide—A chain. The proinsulin molecule undergoes a series of site-specific peptide cleavages that result in the formation of equimolar amounts of mature insulin and C-peptide. These enzymatic cleavages are summarized in Figure 41–12.

PTH Is Secreted as an 84-Amino-Acid Peptide

The immediate precursor of PTH is **proPTH**, which differs from the native 84-amino-acid hormone by having a highly basic hexapeptide amino terminal extension. The primary gene product and the immediate precursor for proPTH is the 115-amino-acid **preproPTH**. This differs from proPTH by having an additional 25-amino-acid amino terminal extension that, in common with the other leader or signal sequences characteristic of secreted proteins, is hydrophobic. The complete structure of preproPTH and the sequences of proPTH and PTH are illustrated in **Figure 41–13**. $PTH_{1–34}$ has full

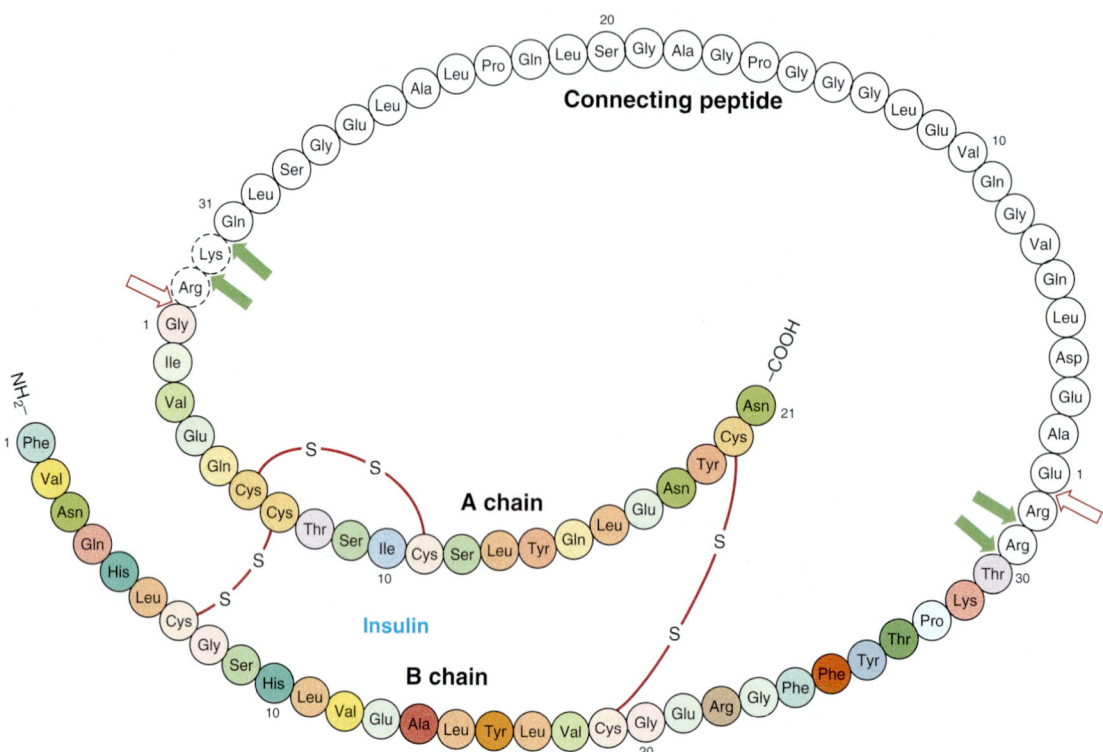

FIGURE 41–12   **Structure of human proinsulin.** Insulin and C-peptide molecules are connected at two sites by dipeptide links. An initial cleavage by a trypsin-like enzyme (open arrows) followed by several cleavages by a carboxypeptidase-like enzyme (solid arrows) results in the production of the heterodimeric (AB) insulin molecule (colored) and the C-peptide (white).

biologic activity, and the region 25–34 is primarily responsible for receptor binding.

The biosynthesis of PTH and its subsequent secretion are regulated by the plasma ionized calcium (Ca^{2+}) concentration through a complex process. An acute decrease of Ca^{2+} results in a marked increase of PTH mRNA, and this is followed by an increased rate of PTH synthesis and secretion. However, about 80–90% of the proPTH synthesized cannot be accounted for as intact PTH in cells or in the incubation medium of experimental systems. This finding led to the conclusion that most of the proPTH synthesized is quickly degraded. It was later discovered that this rate of degradation decreases when Ca^{2+} concentrations are low, and it increases when Ca^{2+} concentrations are high. A Ca^{2+} receptor on the surface of the parathyroid cell mediates these effects. Very specific fragments of PTH are generated during its proteolytic digestion (Figure 41–13). A number of proteolytic enzymes, including cathepsins B and D, have been identified in parathyroid tissue. Cathepsin B cleaves PTH into two fragments: PTH_{1-36} and PTH_{37-84}. PTH_{37-84} is not further degraded; however, PTH_{1-36} is rapidly and progressively cleaved into di- and tripeptides. Most of the proteolysis of PTH occurs within the gland, but a number of studies confirm that PTH, once secreted, is proteolytically degraded in other tissues, especially the liver, by similar mechanisms.

Angiotensin II Is Also Synthesized from a Large Precursor

The renin–angiotensin system is involved in the regulation of blood pressure and electrolyte metabolism (through production of aldosterone). The primary hormone involved in these processes is angiotensin II, an octapeptide made from angiotensinogen (**Figure 41–14**). Angiotensinogen, a large α_2-globulin made in liver, is the substrate for renin, an enzyme produced in the juxtaglomerular cells of the renal afferent arteriole. The position of these cells makes them particularly sensitive to blood pressure changes, and many of the physiologic regulators of renin release act through renal baroreceptors. The juxtaglomerular cells are also sensitive to changes of Na^+ and Cl^- concentration in the renal tubular fluid; therefore, any combination of factors that decreases fluid volume (dehydration, decreased blood pressure, fluid, or blood loss) or decreases NaCl concentration stimulates renin release. Renal sympathetic nerves that terminate in the juxtaglomerular cells mediate the central nervous system and postural effects on renin release independently of the baroreceptor and salt effects, a mechanism that involves the β-adrenergic receptor. Renin acts upon the substrate angiotensinogen to produce the decapeptide angiotensin I.

Angiotensin-converting enzyme, a glycoprotein found in lung, endothelial cells, and plasma, removes two carboxyl

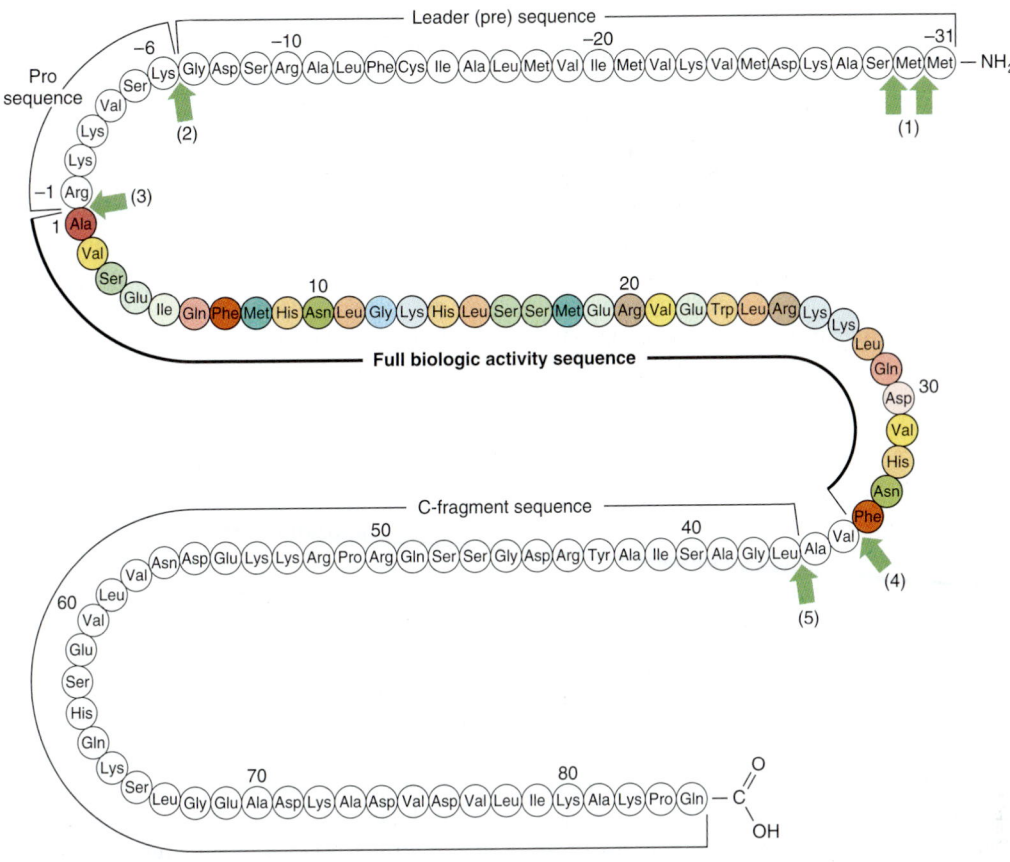

FIGURE 41–13 **Structure of bovine preproparathyroid hormone.** Arrows indicate sites cleaved by processing enzymes in the parathyroid gland and in the liver after secretion of the hormone (1–5). The biologically active region of the molecule (colored) is flanked by sequence not required for activity on target receptors. (Slightly modified and reproduced, with permission, from Habener JF: Recent advances in parathyroid hormone research. Clin Biochem 1981;14:223. Copyright © 1981. Reprinted with permission from Elsevier.)

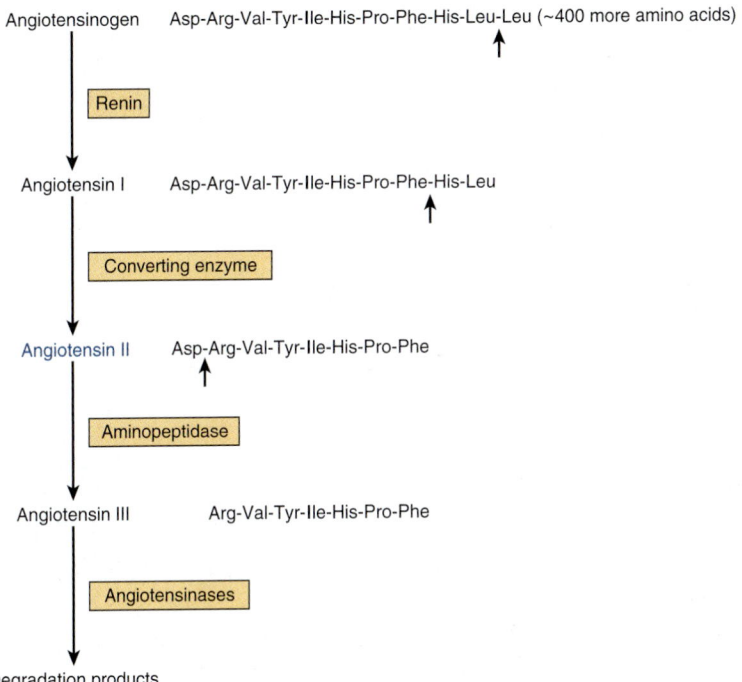

FIGURE 41–14 **Formation and metabolism of angiotensins.** Small arrows indicate cleavage sites.

terminal amino acids from the decapeptide angiotensin I to form angiotensin II in a step that is not thought to be rate limiting. Various nonapeptide analogs of angiotensin I and other compounds act as competitive inhibitors of converting enzyme and are used to treat renin-dependent hypertension. These are referred to as **angiotensin-converting enzyme (ACE) inhibitors.** Angiotensin II increases blood pressure by causing vasoconstriction of the arteriole and is a very potent vasoactive substance. It inhibits renin release from the juxtaglomerular cells and is a potent stimulator of aldosterone production. This results in Na^+ retention, volume expansion, and increased blood pressure.

In some species, angiotensin II is converted to the heptapeptide angiotensin III (Figure 41–14), an equally potent stimulator of aldosterone production. In humans, the plasma level of angiotensin II is four times greater than that of angiotensin III, so most effects are exerted by the octapeptide. Angiotensins II and III are rapidly inactivated by angiotensinases.

Angiotensin II binds to specific adrenal cortex glomerulosa cell receptors. The hormone-receptor interaction does not activate adenylyl cyclase, and cAMP does not appear to mediate the action of this hormone. The actions of angiotensin II, which are to stimulate the conversion of cholesterol to pregnenolone and of corticosterone to 18-hydroxycorticosterone and aldosterone, may involve changes in the concentration of intracellular calcium and of phospholipid metabolites by mechanisms similar to those described in Chapter 42.

Complex Processing Generates the Pro-Opiomelanocortin (POMC) Peptide Family

The POMC family consists of peptides that act as hormones (ACTH, LPH, MSH) and others that may serve as neurotransmitters or neuromodulators (endorphins) (**Figure 41–15**). POMC is synthesized as a precursor molecule of 285 amino acids and is processed differently in various regions of the pituitary.

The POMC gene is expressed in the anterior and intermediate lobes of the pituitary. The most conserved sequences between species are within the amino terminal fragment, the ACTH region, and the β-endorphin region. POMC or related products are found in several other vertebrate tissues, including the brain, placenta, gastrointestinal tract, reproductive tract, lung, and lymphocytes.

The POMC protein is processed differently in the anterior lobe than in the intermediate lobe. The intermediate lobe of the pituitary is rudimentary in adult humans, but it is active in human fetuses and in pregnant women during late gestation and is also active in many animal species. Processing of the POMC protein in the peripheral tissues (gut, placenta, and male reproductive tract) resembles that in the intermediate lobe. There are three basic peptide groups: (1) ACTH, which can give rise to α-MSH and corticotropin-like intermediate lobe peptide (CLIP); (2) β-lipotropin (β-LPH), which can yield γ-LPH, β-MSH, and β-endorphin (and thus α- and γ-endorphins); and (3) a large amino terminal peptide, which generates γ-MSH (not shown). The diversity of these products is due to the many dibasic amino acid clusters that are potential cleavage sites for trypsin-like enzymes. Each of the peptides mentioned is preceded by Lys-Arg, Arg-Lys, Arg-Arg, or Lys-Lys residues. After the prehormone segment is cleaved, the next cleavage, in both anterior and intermediate lobes, is between ACTH and β-LPH, resulting in an amino terminal peptide with ACTH and a β-LPH segment (Figure 41–15). $ACTH_{1-39}$ is subsequently cleaved from the amino terminal peptide, and in the anterior lobe essentially no further cleavages occur. In the intermediate lobe, $ACTH_{1-39}$ is cleaved into α-MSH (residues 1–13) and CLIP (18–39); β-LPH (42–134) is converted to γ-LPH (42–101) and β-endorphin (104–134). β-MSH (84–101) is derived from γ-LPH, while γ-MSH (50–74) is derived from a POMC N-terminal fragment (1–74).

There are extensive additional tissue-specific modifications of these peptides that affect activity. These modifications include phosphorylation, acetylation, glycosylation, and amidation.

Mutations of the α-MSH receptor are linked to a common, early-onset form of obesity. This observation has redirected attention to the POMC peptide hormones.

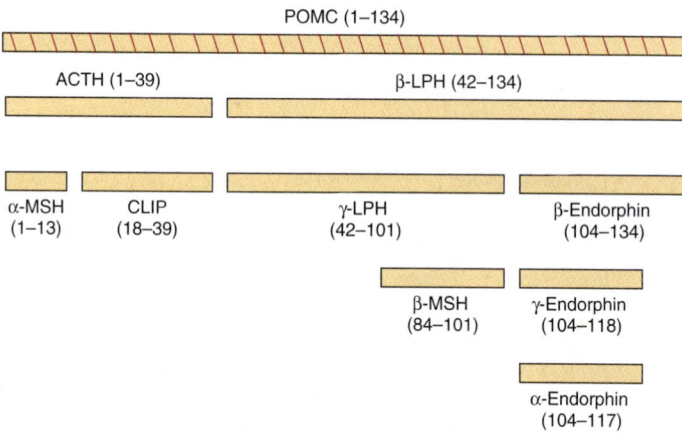

FIGURE 41–15 **Products of pro-opiomelanocortin (POMC) cleavage.** (CLIP, corticotropin-like intermediate lobe peptide; LPH, lipotropin; MSH, melanocyte-stimulating hormone.)

THERE IS VARIATION IN THE STORAGE & SECRETION OF HORMONES

As mentioned above, the steroid hormones and $1,25(OH)_2$-D_3 are synthesized in their final active form. They are also secreted as they are made, and thus there is no intracellular reservoir of these hormones. The catecholamines, also synthesized in active form, are stored in granules in the chromaffin cells in the adrenal medulla. In response to appropriate neural stimulation, these granules are released from the cell through exocytosis, and the catecholamines are released into the circulation. A several-hour reserve supply of catecholamines exists in the chromaffin cells.

PTH also exists in storage vesicles. As much as 80–90% of the pro PTH synthesized is degraded before it enters this final storage compartment, especially when Ca^{2+} levels are high in the parathyroid cell (see above). PTH is secreted when Ca^{2+} is low in the parathyroid cells, which contain a several-hour supply of the hormone.

The human pancreas secretes about 40–50 units of insulin daily; this represents about 15–20% of the hormone stored in the β cells. Insulin and the C-peptide (see Figure 41–12) are normally secreted in equimolar amounts. Stimuli such as glucose, which provokes insulin secretion, therefore trigger the processing of proinsulin to insulin as an essential part of the secretory response.

A several-week supply of T_3 and T_4 exists in the thyroglobulin that is stored in colloid in the lumen of the thyroid follicles. These hormones can be released upon stimulation by TSH. This is the most exaggerated example of a prohormone, as a molecule containing ~5000 amino acids must be first synthesized, then degraded, to supply a few molecules of the active hormones T_4 and T_3.

The diversity in storage and secretion of hormones is illustrated in **Table 41–5**.

SOME HORMONES HAVE PLASMA TRANSPORT PROTEINS

The class I hormones are hydrophobic in chemical nature and thus are not very soluble in plasma. These hormones, principally the steroids and thyroid hormones, have specialized plasma transport proteins that serve several purposes. First,

TABLE 41–5 Diversity in the Storage of Hormones

Hormone	Supply Stored in Cell
Steroids and $1,25(OH)_2$-D_3	None
Catecholamines and PTH	Hours
Insulin	Days
T_3 and T_4	Weeks

TABLE 41–6 Comparison of Receptors with Transport Proteins

Feature	Receptors	Transport Proteins
Concentration	Very low (thousands/cell)	Very high (billions/μL)
Binding affinity	High (pmol/L to nmol/L range)	Low (μmol/L range)
Binding specificity	Very high	Low
Saturability	Yes	No
Reversibility	Yes	Yes
Signal transduction	Yes	No

these proteins circumvent the solubility problem and thereby deliver the hormone to the target cell. They also provide a circulating reservoir of the hormone that can be substantial, as in the case of the thyroid hormones. Hormones, when bound to the transport proteins, cannot be metabolized, thereby prolonging their plasma half-life ($t_{1/2}$). The binding affinity of a given hormone to its transporter determines the bound versus free ratio of the hormone. This is important because only the free form of a hormone is biologically active. In general, the concentration of free hormone in plasma is very low, in the range of 10^{-15} to 10^{-9} mol/L. It is important to distinguish between plasma transport proteins and hormone receptors. Both bind hormones but with very different characteristics (**Table 41–6**).

The hydrophilic hormones—generally class II and of peptide structure—are freely soluble in plasma and do not require transport proteins. Hormones such as insulin, growth hormone, ACTH, and TSH circulate in the free, active form and have very short plasma half-lives. A notable exception is IGF-I, which is transported bound to members of a family of binding proteins.

Thyroid Hormones Are Transported by Thyroid-Binding Globulin

Many of the principles discussed above are illustrated in a discussion of thyroid-binding proteins. One-half to two-thirds of T_4 and T_3 in the body is in an extrathyroidal reservoir. Most of this circulates in bound form, ie, bound to a specific binding protein, **thyroxine-binding globulin (TBG)**. TBG, a glycoprotein with a molecular mass of 50 kDa, binds T_4 and T_3 and has the capacity to bind 20 μg/dL of plasma. Under normal circumstances, TBG binds—noncovalently—nearly all of the T_4 and T_3 in plasma, and it binds T_4 with greater affinity than T_3 (**Table 41–7**). The plasma half-life of T_4 is correspondingly four to five times that of T_3. The small, unbound (free) fraction is responsible for the biologic activity. Thus, in spite of the great difference in total amount, the free fraction of T_3 approximates that of T_4, and given that T_3 is intrinsically more active than T_4, most biologic activity is attributed to T_3. TBG does not bind any other hormones.

TABLE 41–7 Comparison of T_4 and T_3 in Plasma

Total Hormone (µg/dL)	Free Hormone			$t_{1/2}$ in Blood (days)
	Percentage of Total	ng/dL	Molarity	
T_4 8	0.03	~2.24	3.0×10^{-11}	6.5
T_3 0.15	0.3	~0.4	0.6×10^{-11}	1.5

TABLE 41–8 Approximate Affinities of Steroids for Serum-Binding Proteins

	SHBG[1]	CBG[1]
Dihydrotestosterone	1	>100
Testosterone	2	>100
Estradiol	5	>10
Estrone	>10	>100
Progesterone	>100	~2
Cortisol	>100	~3
Corticosterone	>100	~5

[1]Affinity expressed as K_d (nmol/L).

Glucocorticoids Are Transported by Corticosteroid-Binding Globulin

Hydrocortisone (cortisol) also circulates in plasma in protein-bound and free forms. The main plasma binding protein is an α-globulin called **transcortin,** or **corticosteroid-binding globulin (CBG).** CBG is produced in the liver, and its synthesis, like that of TBG, is increased by estrogens. CBG binds most of the hormone when plasma cortisol levels are within the normal range; much smaller amounts of cortisol are bound to albumin. The avidity of binding helps determine the biologic half-lives of various glucocorticoids. Cortisol binds tightly to CBG and has a $t_{1/2}$ of 1.5–2 h, while corticosterone, which binds less tightly, has a $t_{1/2}$ of <1 h (**Table 41–8**). The unbound (free) cortisol constitutes ~8% of the total and represents the biologically active fraction. Binding to CBG is not restricted to glucocorticoids. Deoxycorticosterone and progesterone interact with CBG with sufficient affinity to compete for cortisol binding. Aldosterone, the most potent natural mineralocorticoid, does not have a specific plasma transport protein. Gonadal steroids bind very weakly to CBG (Table 41–8).

Gonadal Steroids Are Transported by Sex-Hormone-Binding Globulin

Most mammals, humans included, have a plasma β-globulin that binds testosterone with specificity, relatively high affinity, and limited capacity (Table 41–8). This protein, usually called **sex-hormone-binding globulin (SHBG)** or testosterone-estrogen-binding globulin (TEBG), is produced in the liver. Its production is increased by estrogens (women have twice the serum concentration of SHBG as men), certain types of liver disease, and hyperthyroidism; it is decreased by androgens, advancing age, and hypothyroidism. Many of these conditions also affect the production of CBG and TBG. Since SHBG and albumin bind 97–99% of circulating testosterone, only a small fraction of the hormone in circulation is in the free (biologically active) form. The primary function of SHBG may be to restrict the free concentration of testosterone in the serum. Testosterone binds to SHBG with higher affinity than does estradiol (Table 41–8). Therefore, a change in the level of SHBG causes a greater change in the free testosterone level than in the free estradiol level.

Estrogens are bound to SHBG and progestins to CBG. SHBG binds estradiol about five times less avidly than it binds testosterone or DHT, while progesterone and cortisol have little affinity for this protein (Table 41–8). In contrast, progesterone and cortisol bind with nearly equal affinity to CBG, which in turn has little avidity for estradiol and even less for testosterone, DHT, or estrone.

These binding proteins also provide a circulating reservoir of hormone, and because of the relatively large binding capacity they probably buffer against sudden changes in the plasma level. Because the metabolic clearance rates of these steroids are inversely related to the affinity of their binding to SHBG, estrone is cleared more rapidly than estradiol, which in turn is cleared more rapidly than testosterone or DHT.

SUMMARY

- The presence of a specific receptor defines the target cells for a given hormone.
- Receptors are proteins that bind specific hormones and generate an intracellular signal (receptor–effector coupling).
- Some hormones have intracellular receptors; others bind to receptors on the plasma membrane.
- Hormones are synthesized from a number of precursor molecules, including cholesterol, tyrosine per se, and all the constituent amino acids of peptides and proteins.
- A number of modification processes alter the activity of hormones. For example, many hormones are synthesized from larger precursor molecules.
- The complement of enzymes in a particular cell type allows for the production of a specific class of steroid hormone.
- Most of the lipid-soluble hormones are bound to rather specific plasma transport proteins.

REFERENCES

Bain DL, Heneghan AF, Connaghan-Jones KD, et al: Nuclear receptor structure: implications for function. Ann Rev Physiol 2007;69:201.

Bartalina L: Thyroid hormone-binding proteins: update 1994. Endocr Rev 1994;13:140.

Beato M, Herrlich P, Schütz G: Steroid hormone receptors: many actors in search of a plot. Cell 1995;83:851.

Cheung E, Kraus WL: Genomic Analyses of Hormone Signaling and Gene Regulation. Annu Rev Physiol 2010;72:191–218.

Cristina Casals-Casas C, Desvergne B: Endocrine Disruptors: From Endocrine to Metabolic Disruption. Annu Rev Physiol 2011;73:23.1–23.28.

Dai G, Carrasco L, Carrasco N: Cloning and characterization of the thyroid iodide transporter. Nature 1996;379:458.

DeLuca HR: The vitamin D story: a collaborative effort of basic science and clinical medicine. FASEB J 1988;2:224.

Douglass J, Civelli O, Herbert E: Polyprotein gene expression: Generation of diversity of neuroendocrine peptides. Annu Rev Biochem 1984;53:665.

Farooqi IS, O'Rahilly S: Monogenic obesity in humans. Ann Rev Med 2005;56:443.

Miller WL: Molecular biology of steroid hormone biosynthesis. Endocr Rev 1988;9:295.

Nagatsu T: Genes for human catecholamine-synthesizing enzymes. Neurosci Res 1991;12:315.

Russell DW, Wilson JD: Steroid 5 alpha-reductase: two genes/two enzymes. Annu Rev Biochem 1994;63:25.

Russell J, Bar A, Sherwood LM, et al: Interaction between calcium and 1,25-dihydroxyvitamin D_3 in the regulation of preproparathyroid hormone and vitamin D receptor mRNA in avian parathyroids. Endocrinology 1993;132:2639.

Steiner DF, Smeekens SP, Ohagi S, et al: The new enzymology of precursor processing endoproteases. J Biol Chem 1992;267:23435.

Taguchi A, White M: Insulin-like signaling, nutrient homeostasis, and life span. Ann Rev Physiol 2008;70:191.

42

Hormone Action & Signal Transduction

P. Anthony Weil, PhD

- Explain the roles of stimulus, hormone release, signal generation, and effector response in a variety of hormone-regulated physiological processes.
- Explain the role of receptors and GTP-binding G-proteins in hormone signal transduction, particularly with regard to the generation of second messengers.
- Appreciate the complex patterns of signal transduction pathway cross-talk in mediating complicated physiological outputs.
- Understand the key roles that protein-ligand, protein-protein, protein post-translational modification (eg., phosphorylation and acetylation), and protein-DNA interactions play in mediating hormone-directed physiological processes.
- Appreciate that hormone-modulated receptors, second messengers, and associated signaling molecules represent a rich source of potential drug target development given their key roles in the regulation of physiology.

BIOMEDICAL IMPORTANCE

The homeostatic adaptations an organism makes to a constantly changing environment are in large part accomplished through alterations of the activity and amount of proteins. Hormones provide a major means of facilitating these changes. A hormone–receptor interaction results in generation of an intracellular signal that can either regulate the activity of a select set of genes, thereby altering the amount of certain proteins in the target cell, or affect the activity of specific proteins, including enzymes and transporter or channel proteins. The signal can influence the location of proteins in the cell and can affect general processes such as protein synthesis, cell growth, and replication, often through effects on gene expression. Other signaling molecules—including cytokines, interleukins, growth factors, and metabolites—use some of the same general mechanisms and signal transduction pathways. Excessive, deficient, or inappropriate production and release of hormones and of these other regulatory molecules are major causes of disease. Many pharmacotherapeutic agents are aimed at correcting or otherwise influencing the pathways discussed in this chapter.

HORMONES TRANSDUCE SIGNALS TO AFFECT HOMEOSTATIC MECHANISMS

The general steps involved in producing a coordinated response to a particular stimulus are illustrated in **Figure 42–1**. The stimulus can be a challenge or a threat to the organism, to an organ, or to the integrity of a single cell within that organism. Recognition of the stimulus is the first step in the adaptive response. At the organismic level, this generally involves the nervous system and the special senses (sight, hearing, pain, smell, and touch). At the organismic or cellular level, recognition involves physicochemical factors such as pH, O_2 tension, temperature, nutrient supply, noxious metabolites, and osmolarity. Appropriate recognition results in the release of one or more hormones that will govern generation of the necessary adaptive response. For purposes of this discussion, the hormones are categorized as described in Chapter 41, ie, based on the location of their specific cellular receptors and the type of signals generated. Group I hormones interact with an intracellular receptor and group II hormones with receptor

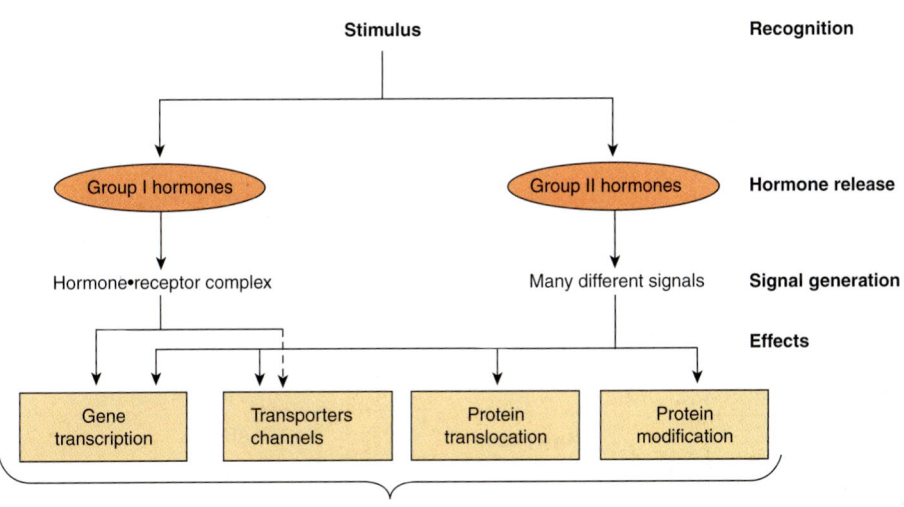

FIGURE 42–1 **Hormonal involvement in responses to a stimulus.** A challenge to the integrity of the organism elicits a response that includes the release of one or more hormones. These hormones generate signals at or within target cells, and these signals regulate a variety of biologic processes that provide for a coordinated response to the stimulus or challenge. See Figure 42–8 for a specific example.

recognition sites located on the extracellular surface of the plasma membrane of target cells. The cytokines, interleukins, and growth factors should also be considered in this latter category. These molecules, of critical importance in homeostatic adaptation, are hormones in the sense that they are produced in specific cells, have the equivalent of autocrine, paracrine, and endocrine actions, bind to cell surface receptors, and activate many of the same signal transduction pathways employed by the more traditional group II hormones.

SIGNAL GENERATION

The Ligand–Receptor Complex Is the Signal for Group I Hormones

The lipophilic group I hormones diffuse through the plasma membrane of all cells but only encounter their specific, high-affinity intracellular receptors in target cells. These receptors can be located in the cytoplasm or in the nucleus of target cells. The hormone–receptor complex first undergoes an **activation reaction.** As shown in **Figure 42–2**, receptor activation occurs by at least two mechanisms. For example, glucocorticoids diffuse across the plasma membrane and encounter their cognate receptor in the cytoplasm of target cells. Ligand–receptor binding results in a conformational change in the receptor leading to the dissociation of heat shock protein 90 (hsp90). This step appears to be necessary for subsequent nuclear localization of the glucocorticoid receptor. This receptor also contains a nuclear localization sequence that is now free to assist in the translocation from cytoplasm to nucleus. The activated receptor moves into the nucleus (Figure 42–2) and binds with high affinity to a specific DNA sequence called the **hormone response element (HRE).** In the case illustrated, this is a glucocorticoid response element, or GRE. Consensus sequences for HREs are shown in **Table 42–1**. The DNA-bound, liganded

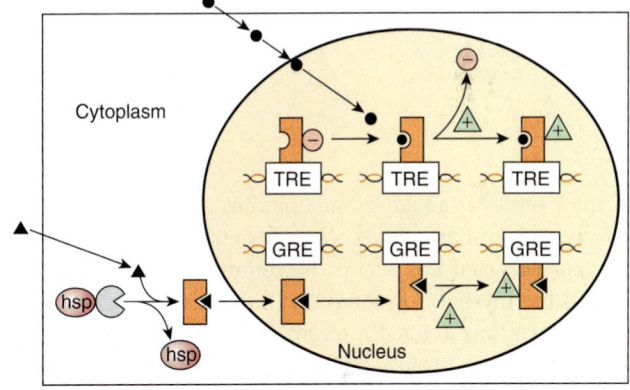

FIGURE 42–2 **Regulation of gene expression by two different class I hormones, thyroid hormone and glucocorticoids.** The hydrophobic steroid hormones readily gain access to the cytoplasmic compartment of target cells by diffusion through the plasma membrane. Glucocorticoid hormones (solid triangles) encounter their cognate receptor (GR) in the cytoplasm, where GR exists in a complex with heat shock protein 90 (hsp). Ligand binding causes dissociation of hsp and a conformational change of the receptor. The receptor–ligand complex then traverses the nuclear membrane and binds to DNA with specificity and high affinity at a glucocorticoid response element (GRE). This event affects the architecture of a number of transcription coregulators (green triangles), and enhanced transcription ensues. By contrast, thyroid hormones and retinoic acid (•) directly enter the nucleus, where their cognate heterodimeric (TR-RXR; see Figure 42–12) receptors are already bound to the appropriate response elements with an associated transcription repressor complex (red circles). Hormone–receptor binding occurs, which again induces conformational changes in receptor leading to a reorganization of receptor (TR)-coregulator interactions (ie, molecules such as N-CoR or SMRT [see Table 42–6]). Ligand binding results in dissociation of the repressor complex from the receptor, allowing an activator complex, consisting of the TR-TRE and coactivator, to assemble. The gene is then actively transcribed.

TABLE 42–1   **The DNA Sequences of Several Hormone Response Elements (HREs)[1]**

Hormone or Effector	HRE	DNA Sequence
Glucocorticoids	GRE	
Progestins	PRE	GGTACA NNN TGTTCT
Mineralocorticoids	MRE	← →
Androgens	ARE	
Estrogens	ERE	AGGTCA --- TGACCT → →
Thyroid hormone	TRE	
Retinoic acid	RARE	AGGTCA N1-5 AGGTCA
Vitamin D	VDRE	→ →
cAMP	CRE	TGACGTCA

[1]Letters indicate nucleotide; N means any one of the four can be used in that position. The arrows pointing in opposite directions illustrate the slightly imperfect inverted palindromes present in many HREs; in some cases these are called "half binding sites," or half-sites, because each binds one monomer of the receptor. The GRE, PRE, MRE, and ARE consist of the same DNA sequence. Specificity may be conferred by the intracellular concentration of the ligand or hormone receptor, by flanking DNA sequences not included in the consensus, or by other accessory elements. A second group of HREs includes those for thyroid hormones, estrogens, retinoic acid, and vitamin D. These HREs are similar except for the orientation and spacing between the half palindromes. Spacing determines the hormone specificity. VDRE ($N = 3$), TRE ($N = 4$), and RARE ($N = 5$) bind to direct repeats rather than to inverted repeats. Another member of the steroid receptor superfamily, the retinoid X receptor (RXR), forms heterodimers with VDR, TR, and RARE, and these constitute the functional forms of these transacting factors. cAMP affects gene transcription through the CRE.

receptor serves as a high-affinity binding site for one or more co-activator proteins and accelerated gene transcription typically ensues when this occurs. By contrast, certain hormones such as the thyroid hormones and retinoids diffuse from the extracellular fluid across the plasma membrane and go directly into the nucleus. In this case, the cognate receptor is already bound to the HRE (the thyroid hormone response element [TRE], in this example). However, this DNA-bound receptor fails to activate transcription because it exists in complex with a corepressor. Indeed, this receptor–corepressor complex serves as an active repressor of gene transcription. The association of ligand with these receptors results in dissociation of the corepressor(s). The liganded receptor is now capable of binding one or more co-activators with high affinity, resulting in the recruitment of RNA polymerase II and the GTFs and activation of gene transcription. The relationship of hormone receptors to other nuclear receptors and to coregulators is discussed in more detail below.

By selectively affecting gene transcription and the consequent production of appropriate target mRNAs, the amounts of specific proteins are changed and metabolic processes are influenced. The influence of each of these hormones is quite specific; generally, a given hormone affects <1% of the genes, mRNA, or proteins in a target cell; sometimes only a few are affected. The nuclear actions of steroid, thyroid, and retinoid hormones are quite well defined. Most evidence suggests that these hormones exert their dominant effect on modulating

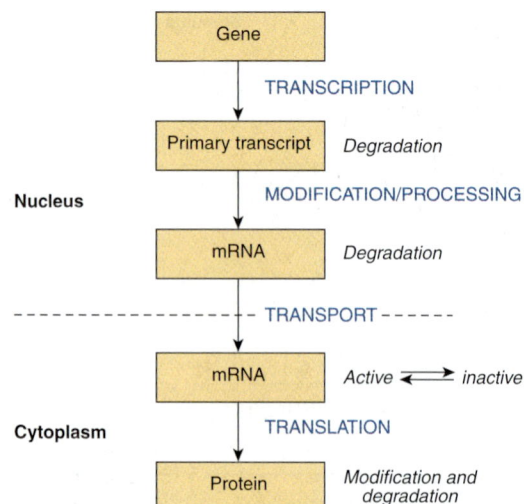

FIGURE 42–3   **The "information pathway."** Information flows from the gene to the primary transcript to mRNA to protein. Hormones can affect any of the steps involved and can affect the rates of processing, degradation, or modification of the various products.

gene transcription, but they—and many of the hormones in the other classes discussed below—can act at any step of the "information pathway," as illustrated in **Figure 42–3**, to control specific gene expression and, ultimately, a biological response. Direct actions of steroids in the cytoplasm and on various organelles and membranes have also been described. Recently, microRNAs have been implicated in mediating some of the diverse actions of the peptide hormone insulin.

GROUP II (PEPTIDE & CATECHOLAMINE) HORMONES HAVE MEMBRANE RECEPTORS & USE INTRACELLULAR MESSENGERS

Many hormones are water soluble, have no transport proteins (and therefore have a short plasma half-life), and initiate a response by binding to a receptor located in the plasma membrane (Tables 41–3 and 41–4). The mechanism of action of this group of hormones can best be discussed in terms of the **intracellular signals** they generate. These signals include **cAMP** (cyclic AMP; 3′,5′-adenylic acid; see Figure 19–5), a nucleotide derived from ATP through the action of adenylyl cyclase; **cGMP,** a nucleotide formed by guanylyl cyclase; **Ca^{2+}**; and **phosphatidylinositides; such small molecules are termed second messengers** as their synthesis is triggered by the presence of the primary hormone (molecule) binding its receptor. Many of these second messengers affect gene transcription, as described in the previous paragraph; but they also influence a variety of other biologic processes, as shown in Figure 42–3.

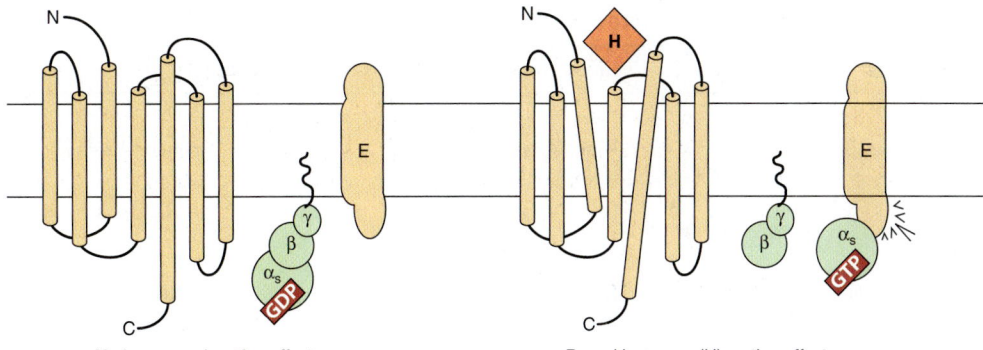

No hormone: inactive effector   Bound hormone (H): active effector

FIGURE 42–4 **Components of the hormone receptor–G protein effector system.** Receptors that couple to effectors through G proteins (GPCR) typically have seven membrane-spanning domains. In the absence of hormone (**left**), the heterotrimeric G-protein complex (α,β,γ) is in an inactive guanosine diphosphate (GDP)-bound form and is probably not associated with the receptor. This complex is anchored to the plasma membrane through prenylated groups on the βγ subunits (**wavy lines**) and perhaps by myristoylated groups on α subunits (not shown). On binding of hormone (H) to the receptor, there is a presumed conformational change of the receptor—as indicated by the tilted membrane spanning domains—and activation of the G–protein complex. This results from the exchange of GDP with guanosine triphosphate (GTP) on the α subunit, after which α and βγ dissociate. The α subunit binds to and activates the effector (E). E can be adenylyl cyclase, Ca^{2+}, Na^+, or Cl^- channels ($α_s$), or it could be a K^+ channel ($α_i$), phospholipase Cβ ($α_q$), or cGMP phosphodiesterase ($α_t$). The βγ subunit can also have direct actions on E. (Modified and reproduced, with permission, from Granner DK in: *Principles and Practice of Endocrinology and Metabolism,* 2nd ed. Becker KL (editor). Lippincott, 1995.)

G Protein–Coupled Receptors

Many of the group II hormones bind to receptors that couple to effectors through a **GTP-binding protein (G-proteins)** intermediary. **These receptors typically have seven hydrophobic plasma membrane-spanning domains.** This is illustrated by the seven interconnected helices extending through the lipid bilayer in **Figure 42–4**. Receptors of this class, which signal through guanine nucleotide-bound protein intermediates, are known as **G protein-coupled receptors (GPCRs).** To date, hundreds of G protein-linked receptor genes have been identified; this represents the largest family of cell surface receptors in humans. A wide variety of responses are mediated by the GPCRs.

cAMP Is the Intracellular Signal for Many Responses

Cyclic AMP was the first intracellular second messenger signal identified in mammalian cells. Several components comprise a system for the generation, degradation, and action of cAMP.

Adenylyl Cyclase

Different peptide hormones can either stimulate (s) or inhibit (i) the production of cAMP from adenylyl cyclase, which is encoded by at least nine different genes (**Table 42–2**). Two parallel systems, a stimulatory (s) one and an inhibitory (i) one, converge upon a catalytic molecule (C). Each consists of a receptor, R_s or R_i, and a regulatory complex, G_s and G_i. G_s and G_i are each **heterotrimeric G-protein composed of α, β, and γ subunits.** Since the α subunit in G_s differs from that in G_i, the proteins, which are distinct gene products, are designated $α_s$ and $α_i$. The α subunits bind guanine nucleotides.

The β and γ subunits are always associated (βγ) and appear to function as a heterodimer. The binding of a hormone to R_s or R_i results in a receptor-mediated activation of G, which entails the exchange of GDP by GTP on α and the concomitant dissociation of βγ from α.

The $α_s$ protein has intrinsic GTPase activity. The active form, $α_s$·GTP, is inactivated upon hydrolysis of the GTP to GDP; the trimeric G_s complex (αβγ) is then re-formed and is ready for another cycle of activation. Cholera and pertussis toxins

TABLE 42–2 Subclassification of Group II.A Hormones

Hormones that Stimulate Adenylyl Cyclase (H_s)	Hormones that Inhibit Adenylyl Cyclase (H_i)
ACTH	Acetylcholine
ADH	$α_2$-Adrenergics
β-Adrenergics	Angiotensin II
Calcitonin	Somatostatin
CRH	
FSH	
Glucagon	
hCG	
LH	
LPH	
MSH	
PTH	
TSH	

TABLE 42–3 **Classes and Functions of Selected G Proteins[1]**

Class or Type	Stimulus	Effector	Effect
G_s			
α_s	Glucagon, β-adrenergics	↑Adenylyl cyclase	Glyconeogenesis, lipolysis, glycogenolysis
		↑Cardiac Ca^{2+}, Cl^-, and Na^+ channels	Olfaction
α_{olf}	Odorant	↑Adenylyl cyclase	
G_i			
$\alpha_{i-1,2,3}$	Acetylcholine, α_2-adrenergics	↓Adenylyl cyclase	Slowed heart rate
		↑Potassium channels	
	M_2 cholinergics	↓Calcium channels	
α_o	Opioids, endorphins	↑Potassium channels	Neuronal electrical activity
α_t	Light	↑cGMP phosphodiesterase	Vision
G_q			
α_q	M_1 cholinergics		
	α_1-Adrenergics	↑Phospholipase C-β1	↓Muscle contraction
			and
α_{11}	α_1-Adrenergics	↑Phospholipase C-β2	↓Blood pressure
G_{12}			
α_{12}	Thrombin	Rho	Cell shape changes

[1]The four major classes or families of mammalian G proteins (G_s, G_i, G_q, and G_{12}) are based on protein sequence homology. Representative members of each are shown, along with known stimuli, effectors, and well-defined biologic effects. Nine isoforms of adenylyl cyclase have been identified (isoforms I—IX). All isoforms are stimulated by α_s; α_i isoforms inhibit types V and VI, and α_o inhibits types I and V. At least 16 different α subunits have been identified.
Source: Modified and reproduced, with permission, from Granner DK in: *Principles and Practice of Endocrinology and Mefabolism,* 2nd ed. Becker KL (editor). Lippincott, 1995.

catalyze the ADPribosylation of a_s and α_{i-2} (see **Table 42–3**), respectively. In the case of α_s, this modification disrupts the intrinsic GTPase activity; thus, α_s cannot reassociate with βγ and is therefore irreversibly activated. ADP ribosylation of α_{i-2} prevents the dissociation of α_{i-2} from βγ, and free α_{i-2} thus cannot be formed. α_s activity in such cells is therefore unopposed.

There is a large family of G proteins, and these are part of the superfamily of GTPases. The G protein family is classified according to sequence homology into four subfamilies, as illustrated in Table 42–3. There are 21 α, 5 β, and 8 γ subunit genes. Various combinations of these subunits provide a large number of possible αβγ and cyclase complexes.

The α subunits and the βγ complex have actions independent of those on adenylyl cyclase (see Figure 42–4 and Table 42–3). Some forms of α_i stimulate K^+ channels and inhibit Ca^{2+} channels, and some α_s molecules have the opposite effects. Members of the G_q family activate the phospholipase C group of enzymes. The βγ complexes have been associated with K^+ channel stimulation and phospholipase C activation. G proteins are involved in many important biologic processes in addition to hormone action. Notable examples include olfaction (α_{OLF}) and vision (α_t). Some examples are listed in Table 42–3. GPCRs are implicated in a number of diseases and are major targets for pharmaceutical agents.

Protein Kinase

As discussed in Chapter 38, in prokaryotic cells, cAMP binds to a specific protein called catabolite regulatory protein (CRP)

that binds directly to DNA and influences gene expression. By contrast, in eukaryotic cells, cAMP binds to a protein kinase called **protein kinase A (PKA),** a heterotetrameric molecule consisting of two regulatory subunits (R) and two catalytic subunits (C). cAMP binding results in the following reaction:

$$4\,cAMP + R_2C_2 \rightleftarrows R_2 \cdot (4\,cAMP) + 2C$$

The R_2C_2 complex has no enzymatic activity, but the binding of cAMP by R induces dissociation of the R–C complex, thereby activating the latter (**Figure 42–5**). The active C subunit catalyzes the transfer of the γ phosphate of ATP to a serine or threonine residue in a variety of proteins. The consensus phosphorylation sites are -ArgArg/Lys-X-Ser/Thr- and -Arg-Lys-X-X-Ser-, where X can be any amino acid.

Protein kinase activities were originally described as being "cAMP-dependent" or "cAMP-independent." This classification has changed, as protein phosphorylation is now recognized as being a major regulatory mechanism. Several hundred protein kinases have now been described. The kinases are related in sequence and structure within the catalytic domain, but each is a unique molecule with considerable variability with respect to subunit composition, molecular weight, auto-phosphorylation, K_m for ATP, and substrate specificity. Both kinase and protein phosphatase activities can be targeted by interaction with specific kinase binding proteins. In the case of PKA, such targeting proteins are termed **AKAPs (A kinase anchoring proteins),** they serve as scaffolds, which localize PKA near to substrates thereby focusing PKA activity

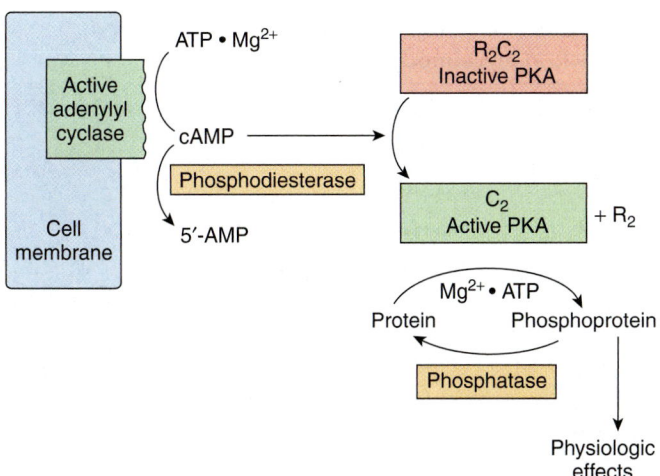

FIGURE 42–5 **Hormonal regulation of cellular processes through cAMP-dependent protein kinase (PKA).** PKA exists in an inactive form as an R$_2$C$_2$ heterotetramer consisting of two regulatory (R) and two catalytic (C) subunits. The cAMP generated by the action of adenylyl cyclase (activated as shown in Figure 42–4) binds to the regulatory subunit of PKA. This results in dissociation of the regulatory and catalytic subunits and activation of the latter. The active catalytic subunits phosphorylate a number of target proteins on serine and threonine residues. Phosphatases remove phosphate from these residues and thus terminate the physiologic response. A phosphodiesterase can also terminate the response by converting cAMP to 5′-AMP.

toward physiological substrates and facilitating spatiotemporal biological regulation while also allowing for common, shared proteins to elicit specific physiological responses. Multiple AKAPs have been described; they can bind PKA and other kinases as well as phosphatases, phosophodiestases (which hydrolyze cAMP), and protein kinase substrates.

Phosphoproteins

The effects of cAMP in eukaryotic cells are all thought to be mediated by protein phosphorylation-dephosphorylation, principally on serine and threonine residues. The control of any of the effects of cAMP, including such diverse processes as steroidogenesis, secretion, ion transport, carbohydrate and fat metabolism, enzyme induction, gene regulation, synaptic transmission, and cell growth and replication, could be conferred by a specific protein kinase, by a specific phosphatase, or by specific substrates for phosphorylation. These substrates help define a target tissue and are involved in defining the extent of a particular response within a given cell. For example, the effects of cAMP on gene transcription are mediated by the protein **cyclic AMP response element binding protein (CREB).** CREB binds to a cAMP responsive element (CRE) (see Table 42–1) in its nonphosphorylated state and is a weak activator of transcription. When phosphorylated by PKA, CREB binds the coactivator **CREB-binding protein CBP/p300** (see below) and as a result is a much more potent transcription activator. CBP and the related p300 contain histone

acetyltransferase activities, and hence serve as chromatin-active transcriptional coregulators (Chapters 36, 38). Interestingly, CBP/p300 can also acetylate certain transcription factors thereby stimulating their ability to bind DNA and modulate transcription.

Phosphodiesterases

Actions caused by hormones that increase cAMP concentration can be terminated in a number of ways, including the hydrolysis of cAMP to 5′-AMP by phosphodiesterases (see Figure 42–5). The presence of these hydrolytic enzymes ensures a rapid turnover of the signal (cAMP) and hence a rapid termination of the biologic process once the hormonal stimulus is removed. There are at least 11 known members of the phosphodiesterase family of enzymes. These are subject to regulation by their substrates, cAMP and cGMP; by hormones; and by intracellular messengers such as calcium, probably acting through calmodulin. Inhibitors of phosphodiesterase, most notably methylated xanthine derivatives such as caffeine, increase intracellular cAMP and mimic or prolong the actions of hormones through this signal.

Phosphoprotein Phosphatases

Given the importance of protein phosphorylation, it is not surprising that regulation of the protein dephosphorylation reaction is another important control mechanism (see Figure 42–5). The phosphoprotein phosphatases are themselves subject to regulation by phosphorylation-dephosphorylation reactions and by a variety of other mechanisms, such as protein–protein interactions. In fact, the substrate specificity of the phosphoserine-phosphothreonine phosphatases may be dictated by distinct regulatory subunits whose binding is regulated hormonally. One of the best-studied roles of regulation by the dephosphorylation of proteins is that of glycogen metabolism in muscle. Two major types of phosphoserine-phosphothreonine phosphatases have been described. Type I preferentially dephosphorylates the β subunit of phosphorylase kinase, whereas type II dephosphorylates the α subunit. Type I phosphatase is implicated in the regulation of glycogen synthase, phosphorylase, and phosphorylase kinase. This phosphatase is itself regulated by phosphorylation of certain of its subunits, and these reactions are reversed by the action of one of the type II phosphatases. In addition, two heat-stable protein inhibitors regulate type I phosphatase activity. Inhibitor-1 is phosphorylated and activated by cAMP-dependent protein kinases, and inhibitor-2, which may be a subunit of the inactive phosphatase, is also phosphorylated, possibly by glycogen synthase kinase-3. Phosphatases that attack phosphotyrosine are also important in signal transduction (see Figure 42–8).

cGMP Is Also an Intracellular Signal

Cyclic GMP is made from GTP by the enzyme guanylyl cyclase, which exists in soluble and membrane-bound forms. Each of these isozymes has unique physiologic properties. The atriopeptins, a family of peptides produced in cardiac atrial tissues, cause natriuresis, diuresis, vasodilation, and inhibition of aldosterone secretion. These peptides (eg, atrial natriuretic factor) bind to and activate the membrane-bound form of guanylyl cyclase. This results in an increase of cGMP by as much as 50-fold in some cases, and this is thought to mediate the effects mentioned above. Other evidence links cGMP to vasodilation. A series of compounds, including nitroprusside, nitroglycerin, nitric oxide, sodium nitrite, and sodium azide, all cause smooth muscle relaxation and are potent vasodilators. These agents increase cGMP by activating the soluble form of guanylyl cyclase, and inhibitors of cGMP phosphodiesterase (the drug sildenafil [Viagra], for example) enhance and prolong these responses. The increased cGMP activates cGMP-dependent protein kinase (PKG), which in turn phosphorylates a number of smooth muscle proteins. Presumably, this is involved in relaxation of smooth muscle and vasodilation.

Several Hormones Act Through Calcium or Phosphatidylinositols

Ionized calcium is an important regulator of a variety of cellular processes, including muscle contraction, stimulus-secretion coupling, blood clotting cascade, enzyme activity, and membrane excitability. It is also an intracellular messenger of hormone action.

Calcium Metabolism

The extracellular calcium (Ca^{2+}) concentration is ~5 mmol/L and is very rigidly controlled. Although substantial amounts of calcium are associated with intracellular organelles such as mitochondria and the endoplasmic reticulum, the intracellular concentration of free or ionized calcium (Ca^{2+}) is very low: 0.05–10 μmol/L. In spite of this large concentration gradient and a favorable trans-membrane electrical gradient, Ca^{2+} is restrained from entering the cell. A considerable amount of energy is expended to ensure that the intracellular Ca^{2+} is controlled, as a prolonged elevation of Ca^{2+} in the cell is very toxic. A Na^+/Ca^{2+} exchange mechanism that has a high-capacity but low-affinity pumps Ca^{2+} out of cells. There also is a Ca^{2+}/proton ATPase-dependent pump that extrudes Ca^{2+} in exchange for H^+. This has a high affinity for Ca^{2+} but a low capacity and is probably responsible for fine-tuning cytosolic Ca^{2+}. Furthermore, Ca^{2+}-ATPases pump Ca^{2+} from the cytosol to the lumen of the endoplasmic reticulum. There are three ways of changing cytosolic Ca^{2+}: (1) Certain hormones (class II.C, Table 41–3) by binding to receptors that are themselves Ca^{2+} channels, enhance membrane permeability to Ca^{2+}, and thereby increase Ca^{2+} influx. (2) Hormones also indirectly promote Ca^{2+} influx by modulating the membrane potential at the plasma membrane. Membrane depolarization opens voltage-gated Ca^{2+} channels and allows for Ca^{2+} influx. (3) Ca^{2+} can be mobilized from the endoplasmic reticulum, and possibly from mitochondrial pools.

An important observation linking Ca^{2+} to hormone action involved the definition of the intracellular targets of

TABLE 42–4 Enzymes and Proteins Regulated by Calcium or Calmodulin

- Adenylyl cyclase
- Ca^{2+}-dependent protein kinases
- Ca^{2+}–Mg^{2+}-ATPase
- Ca^{2+}-phospholipid-dependent protein kinase
- Cyclic nucleotide phosphodiesterase
- Some cytoskeletal proteins
- Some ion channels (eg, L-type calcium channels)
- Nitric oxide synthase
- Phosphorylase kinase
- Phosphoprotein phosphatase 2B
- Some receptors (eg, NMDA-type glutamate receptor)

Ca^{2+} action. The discovery of a Ca^{2+}-dependent regulator of phosphodiesterase activity provided the basis for a broad understanding of how Ca^{2+} and cAMP interact within cells.

Calmodulin

The calcium-dependent regulatory protein is calmodulin, a 17-kDa protein that is homologous to the muscle protein troponin C in structure and function. Calmodulin has four Ca^{2+} binding sites, and full occupancy of these sites leads to a marked conformational change, which allows calmodulin to activate enzymes and ion channels. The interaction of Ca^{2+} with calmodulin (with the resultant change of activity of the latter) is conceptually similar to the binding of cAMP to PKA and the subsequent activation of this molecule. Calmodulin can be one of numerous subunits of complex proteins and is particularly involved in regulating various kinases and enzymes of cyclic nucleotide generation and degradation. A partial list of the enzymes regulated directly or indirectly by Ca^{2+}, probably through calmodulin, is presented in **Table 42–4**.

In addition to its effects on enzymes and ion transport, Ca^{2+}/calmodulin regulates the activity of many structural elements in cells. These include the actin–myosin complex of smooth muscle, which is under β-adrenergic control, and various microfilament-mediated processes in noncontractile cells, including cell motility, cell conformation changes, mitosis, granule release, and endocytosis.

Calcium Is a Mediator of Hormone Action

A role for Ca^{2+} in hormone action is suggested by the observations that the effect of many hormones is (1) blunted by Ca^{2+}-free media or when intracellular calcium is depleted; (2) can be mimicked by agents that increase cytosolic Ca^{2+}, such as the Ca^{2+} ionophore A23187; and (3) influences cellular calcium flux. The regulation of glycogen metabolism in liver by vasopressin and β-adrenergic catecholamines

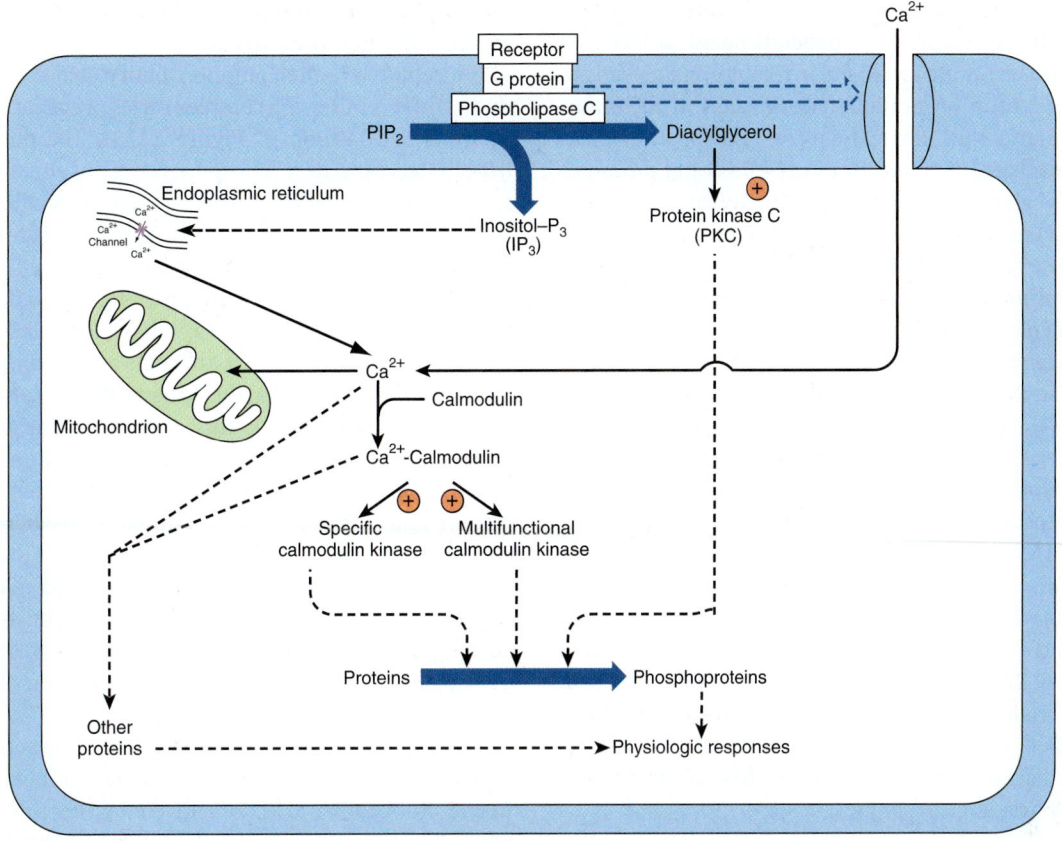

FIGURE 42–6 **Certain hormone–receptor interactions result in the activation of phospholipase C (PLC).** PLC activation appears to involve a specific G protein, which also may activate a calcium channel. Phospholipase C generates inositol trisphosphate (IP_3), which liberates stored intracellular Ca^{2+}, and diacylglycerol (DAG), a potent activator of protein kinase C (PKC). In this scheme, the activated PKC phosphorylates specific substrates, which then alter physiologic processes. Likewise, the Ca^{2+}–calmodulin complex can activate specific kinases, two of which are shown here. These actions result in phosphorylation of substrates, and this leads to altered physiologic responses. This figure also shows that Ca^{2+} can enter cells through voltage- or ligand-gated Ca^{2+} channels. The intracellular Ca^{2+} is also regulated through storage and release by the mitochondria and endoplasmic reticulum. (Courtesy of JH Exton.)

provides a good example. This is shown schematically in Figures 19–6 and 19–7.

A number of critical metabolic enzymes are regulated by Ca^{2+}, phosphorylation, or both, including glycogen synthase, pyruvate kinase, pyruvate carboxylase, glycerol-3-phosphate dehydrogenase, and pyruvate dehydrogenase.

Phosphatidylinositide Metabolism Affects Ca^{2+}-Dependent Hormone Action

Some signal must provide communication between the hormone receptor on the plasma membrane and the intracellular Ca^{2+} reservoirs. This is accomplished by products of phosphatidylinositol metabolism. Cell surface receptors such as those for acetylcholine, antidiuretic hormone, and α_1-type catecholamines are, when occupied by their respective ligands, potent activators of phospholipase C. Receptor binding and activation of phospholipase C are coupled by the G_q isoforms (Table 42–3 & **Figure 42–6**). Phospholipase C catalyzes the hydrolysis of phosphatidylinositol 4,5-bisphosphate to inositol trisphosphate (IP_3) and 1,2-diacylglycerol (**Figure 42–7**).

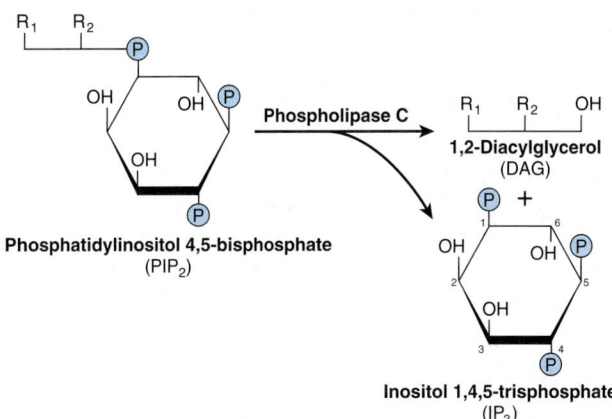

FIGURE 42–7 **Phospholipase C cleaves PIP_2 into diacylglycerol and inositol trisphosphate.** R_1 generally is stearate, and R_2 is usually arachidonate. IP_3 can be dephosphorylated (to the inactive I-1,4-P_2) or phosphorylated (to the potentially active I-1,3,4,5-P_4).

Diacylglycerol is itself capable of activating **protein kinase C (PKC),** the activity of which also depends upon Ca^{2+}. IP_3, by interacting with a specific intracellular receptor, is an effective releaser of Ca^{2+} from intracellular storage sites in the endoplasmic reticulum. Thus, the hydrolysis of phosphatidylinositol 4,5-bisphosphate leads to activation of PKC and promotes an increase of cytoplasmic Ca^{2+}. As shown in Figure 42–4, the activation of G proteins can also have a direct action on Ca^{2+} channels. The resulting elevations of cytosolic Ca^{2+} activate Ca^{2+}–calmodulin-dependent kinases and many other Ca^{2+}–calmodulin-dependent enzymes.

Steroidogenic agents—including ACTH and cAMP in the adrenal cortex; angiotensin II, K^+, serotonin, ACTH, and cAMP in the zona glomerulosa of the adrenal; LH in the ovary; and LH and cAMP in the Leydig cells of the testes—have been associated with increased amounts of phosphatidic acid, phosphatidylinositol, and polyphosphoinositides (see Chapter 15) in the respective target tissues. Several other examples could be cited.

The roles that Ca^{2+} and polyphosphoinositide breakdown products might play in hormone action are presented in Figure 42–6. In this scheme, the activated protein kinase C can phosphorylate specific substrates, which then alter physiologic processes. Likewise, the Ca^{2+}–calmodulin complex can activate specific kinases. These then modify substrates and thereby alter physiologic responses.

Some Hormones Act Through a Protein Kinase Cascade

Single protein kinases such as PKA, PKC, and Ca^{2+}-calmodulin (CaM)-kinases, which result in the phosphorylation of serine and threonine residues in target proteins, play a very important role in hormone action. The discovery that the EGF receptor contains an intrinsic tyrosine kinase activity that is activated by the binding of the ligand EGF was an important breakthrough. The insulin and IGF-I receptors also contain intrinsic ligand-activated tyrosine kinase activity. Several receptors—generally those involved in binding ligands involved in growth control, differentiation, and the inflammatory response—either have intrinsic tyrosine kinase activity or are associated with proteins that are tyrosine kinases. Another distinguishing feature of this class of hormone action is that these kinases preferentially phosphorylate tyrosine residues, and tyrosine phosphorylation is infrequent (<0.03% of total amino acid phosphorylation) in mammalian cells. A third distinguishing feature is that the ligand–receptor interaction that results in a tyrosine phosphorylation event initiates a cascade that may involve several protein kinases, phosphatases, and other regulatory proteins.

Insulin Transmits Signals by Several Kinase Cascades

The insulin, epidermal growth factor (EGF), and IGF-I receptors have intrinsic protein tyrosine kinase activities located in their cytoplasmic domains. These activities are stimulated when their ligands bind to the cognate receptor. The receptors are then autophosphorylated on tyrosine residues, and this initiates a complex series of events (summarized in simplified fashion in **Figure 42–8**). The phosphorylated insulin receptor next phosphorylates insulin receptor substrates (there are at least four of these molecules, called IRS 1–4) on tyrosine residues. Phosphorylated IRS binds to the Src homology 2 (SH2) domains of a variety of proteins that are directly involved in mediating different effects of insulin. One of these proteins, PI-3 kinase, links insulin receptor activation to insulin action through activation of a number of molecules, including the kinase PDK1 (phosphoinositide-dependent kinase-1). This enzyme propagates the signal through several other kinases, including PKB (also known as AKT), SKG, and aPKC (see legend to Figure 42–8 for definitions and expanded abbreviations). An alternative pathway downstream from PDK1 involves p70S6K and perhaps other as yet unidentified kinases. A second major pathway involves mTOR. This enzyme is directly regulated by amino acid levels and insulin and is essential for p70S6K activity. This pathway provides a distinction between the PKB and p70S6K branches downstream from PKD1. These pathways are involved in protein translocation, enzyme activity, and the regulation, by insulin, of genes involved in metabolism (Figure 42–8). Another SH2 domain-containing protein is GRB2, which binds to IRS-1 and links tyrosine phosphorylation to several proteins, the result of which is activation of a cascade of threonine and serine kinases. A pathway showing how this insulin–receptor interaction activates the mitogen-activated protein (MAP) kinase pathway and the anabolic effects of insulin is illustrated in Figure 42–8. The exact roles of many of these docking proteins, kinases, and phosphatases remain to be established.

The Jak/STAT Pathway is Used by Hormones and Cytokines

Tyrosine kinase activation can also initiate a phosphorylation and dephosphorylation cascade that involves the action of several other protein kinases and the counterbalancing actions of phosphatases. Two mechanisms are employed to initiate this cascade. Some hormones, such as growth hormone, prolactin, erythropoietin, and the cytokines, initiate their action by activating a tyrosine kinase, but this activity is not an integral part of the hormone receptor. The hormone–receptor interaction promotes binding and activation of **cytoplasmic protein tyrosine kinases,** such as **Tyk-2, Jak1,** or **Jak2.**

These kinases phosphorylate one or more cytoplasmic proteins, which then associate with other docking proteins through binding to SH2 domains. One such interaction results in the activation of a family of cytosolic proteins called **signal transducers and activators of transcription (STATs).** The phosphorylated STAT protein dimerizes and translocates into the nucleus, binds to a specific DNA element

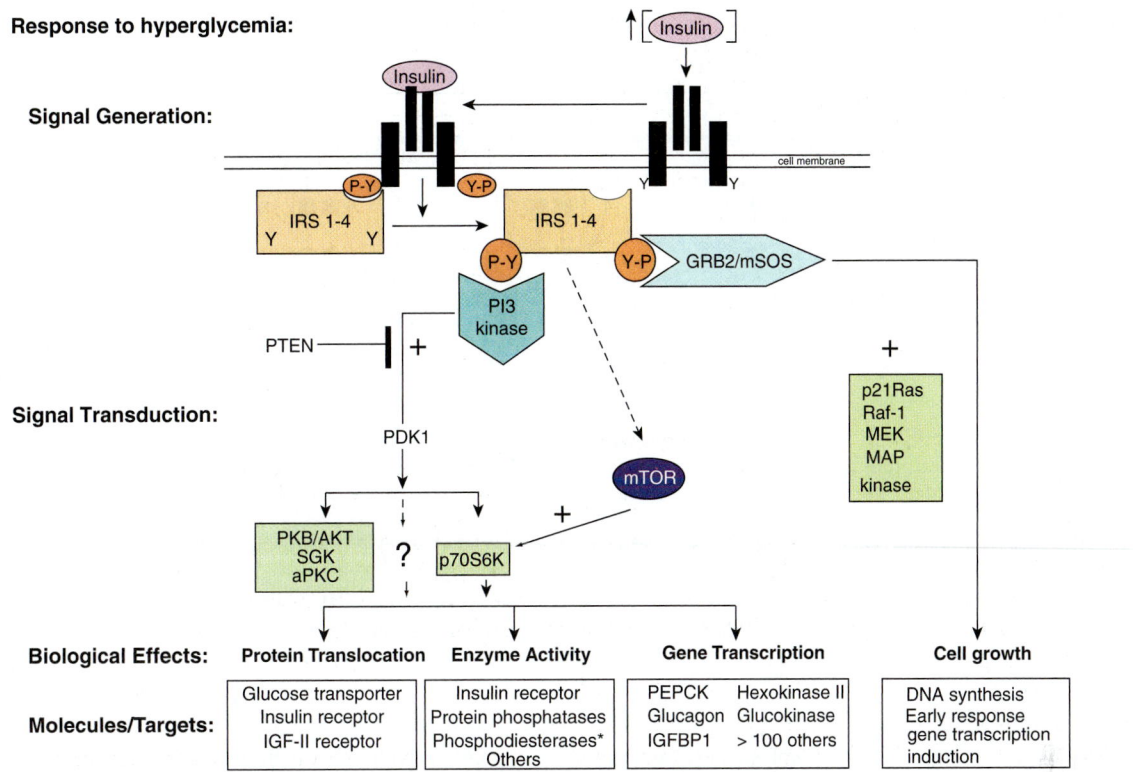

FIGURE 42–8 **Insulin signaling pathways.** The insulin signaling pathways provide an excellent example of the "recognition → hormone release → signal generation → effects" paradigm outlined in Figure 42–1. Insulin is released into the bloodstream from pancreatic β-cells in response to hyperglycemia. Binding of insulin to a target cell-specific plasma membrane heterotetrameric insulin receptor (IR) results in a cascade of intracellular events. First, the intrinsic tyrosine kinase activity of the insulin receptor is activated, and marks the initial event. Receptor activation results in increased tyrosine phosphorylation (conversion of specific Y residues → Y-P) within the receptor. One or more of the insulin receptor substrate (IRS) molecules (IRS 1–4) then bind to the tyrosine-phosphorylated receptor and themselves are specifically tyrosine phosphorylated. IRS proteins interact with the activated IR via N-terminal PH (plectsrin homology) and PTB (phosphotyrosine binding) domains. IR-docked IRS proteins are tyrosine phosphorylated and the resulting P-Y-residues form the docking sites for several additional signaling proteins (ie, PI-3 kinase, GRB2, and mTOR). GRB2 and PI3K bind to IRS P-Y residues via their SH (*Src* Homology) domains, Binding to IRS-Y-P residues leads to activation of the activity of many intracellular signaling molecules such as GTPases, protein kinases, and lipid kinases, all of which play key roles in certain metabolic actions of insulin. The two best-described pathways are shown. In detail, phosphorylation of an IRS molecule (probably IRS-2) results in docking and activation of the lipid kinase, PI-3 kinase; PI-3K generates novel inositol lipids that act as "second messenger" molecules. These, in turn, activate PDK1 and then a variety of downstream signaling molecules, including protein kinase B (PKB/AKT), SGK, and aPKC. An alternative pathway involves the activation of p70S6K and perhaps other as yet unidentified kinases. Next, phosphorylation of IRS (probably IRS-1) results in docking of GRB2/mSOS and activation of the small GTPase, p21Ras, which initiates a protein kinase cascade that activates Raf-1, MEK, and the p42/p44 MAP kinase isoforms. These protein kinases are important in the regulation of proliferation and differentiation of many cell types. The mTOR pathway provides an alternative way of activating p70S6K and appears to be involved in nutrient signaling as well as insulin action. Each of these cascades may influence different physiologic processes, as shown. All of the phosphorylation events are reversible through the action of specific phosphatases. As an example, the lipid phosphatase PTEN dephosphorylates the product of the PI-3 kinase reaction, thereby antagonizing the pathway and terminating the signal. Representative effects of major actions of insulin are shown in each of the boxes. The asterisk after phosphodiesterase indicates that insulin indirectly affects the activity of many enzymes by activating phosphodiesterases and reducing intracellular cAMP levels. (aPKC, atypical protein kinase C; GRB2, growth factor receptor binding protein 2; IGFBP, insulin-like growth factor binding protein; IRS 1–4, insulin receptor substrate isoforms 1–4; MAP kinase, mitogen-activated protein kinase; MEK, MAP kinase kinase and ERK kinase; mSOS, mammalian son of sevenless; mTOR, mammalian target of rapamycin; p70S6K, p70 ribosomal protein S6 kinase; PDK1, phosphoinositide-dependent kinase; PI-3 kinase, phosphatidylinositol 3-kinase; PKB, protein kinase B; PTEN, phosphatase and tensin homolog deleted on chromosome 10; SGK, serum and glucocorticoid-regulated kinase.)

such as the interferon response element (IRE), and activates transcription. This is illustrated in **Figure 42–9**. Other SH2 docking events may result in the activation of PI-3 kinase, the MAP kinase pathway (through SHC or GRB2), or G protein-mediated activation of phospholipase C (PLCγ) with the attendant production of diacylglycerol and activation of protein kinase C. It is apparent that there is a potential for "cross-talk" when different hormones activate these various signal transduction pathways.

The NF-κB Pathway Is Regulated by Glucocorticoids

The transcription factor **NF-κβ is a heterodimeric complex** typically composed of two **subunits termed p50 and p65**

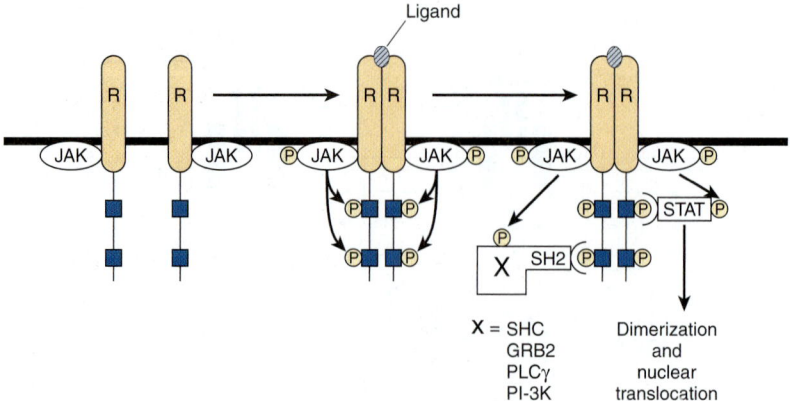

FIGURE 42–9 **Initiation of signal transduction by receptors linked to Jak kinases.** The receptors (R) that bind prolactin, growth hormone, interferons, and cytokines lack endogenous tyrosine kinase. Upon ligand binding, these receptors dimerize and an associated protein (Jak1, Jak2, or TYK) is phosphorylated. Jak-P, an active kinase, phosphorylates the receptor on tyrosine residues. The STAT proteins associate with the phosphorylated receptor and then are themselves phosphorylated by Jak-P. The phosphorylated STAT protein, STAT ⓟ dimerizes, translocates to the nucleus, binds to specific DNA elements, and regulates transcription. The phosphotyrosine residues of the receptor also bind to several SH2 domain-containing proteins (X-SH2). This results in activation of the MAP kinase pathway (through SHC or GRB2), PLCγ, or PI-3 kinase.

(**Figure 42–10**). Normally, NF-κβ is kept sequestered in the cytoplasm in a transcriptionally inactive form by members of the **Inhibitor of NF-κβ (IκB) family.** Extracellular stimuli such as proinflammatory cytokines, reactive oxygen species, and mitogens lead to activation of the IκB kinase complex, IKK, which is a heterohexameric structure consisting of α, β, and γ subunits. IKK phosphorylates IκB on two serine residues, and this targets IκB for ubiquitination and subsequent degradation by the proteasome. Following IκB degradation, free NF-κB translocates to the nucleus, where it binds to a number of gene promoters and activates transcription, particularly of

genes involved in the **inflammatory response.** Transcriptional regulation by NF-κB is mediated by a variety of coactivators such as CREB binding protein (CBP), as described below (see Figure 42–13).

Glucocorticoid hormones are therapeutically useful agents for the treatment of a variety of inflammatory and immune diseases. Their anti-inflammatory and immunomodulatory actions are explained in part by the inhibition of NF-κB and its subsequent actions. Evidence for three mechanisms for the inhibition of NF-κB by glucocorticoids has been presented: (1) glucocorticoids increase IκB mRNA, which leads to

FIGURE 42–10 **Regulation of the NF-κβ pathway.** NF-κβ consists of two subunits, p50 and p65, which when present in the nucleus regulate transcription of the multitude of genes important for the inflammatory response. NF-κB is restricted from entering the nucleus by IκB, an inhibitor of NF-κB. IκB binds to—and masks—the nuclear localization signal of NF-κB. This cytoplasmic protein is phosphorylated by an IKK complex which is activated by cytokines, reactive oxygen species, and mitogens. Phosphorylated IκB can be ubiquitinylated and degraded, thus releasing its hold on NF-κB. Glucocorticoids, potent anti-inflammatory agents, are thought to affect at least three steps in this process (1, 2, 3), as described in the text.

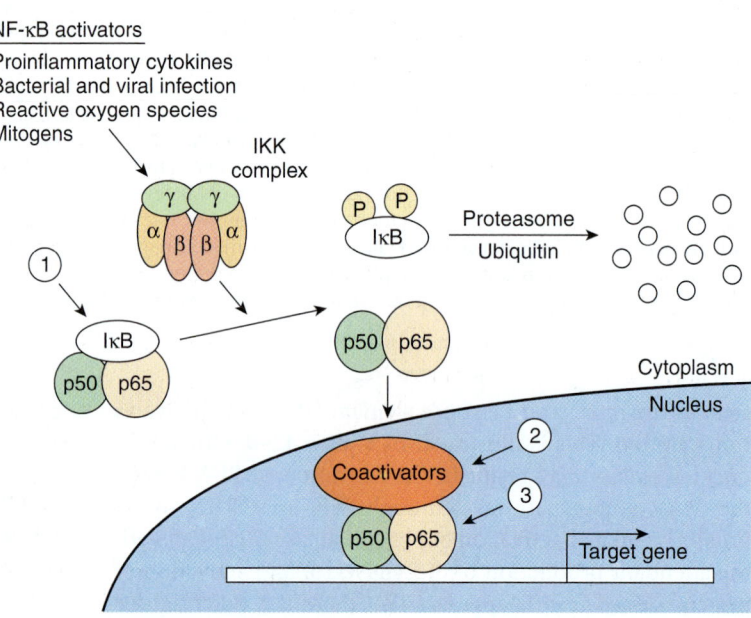

an increase of IκB protein and more efficient sequestration of NF-κB in the cytoplasm. (2) The glucocorticoid receptor competes with NF-κB for binding to coactivators. (3) The glucocorticoid receptor directly binds to the p65 subunit of NF-κB and inhibits its activation (Figure 42–10).

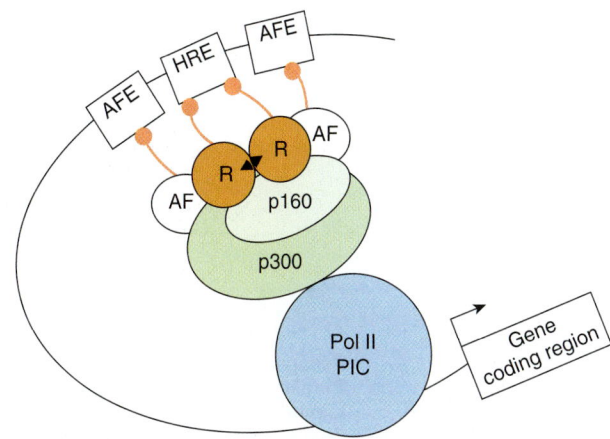

FIGURE 42–11 **The hormone response transcription unit.** The hormone response transcription unit is an assembly of DNA elements and bound proteins that interact, through protein–protein interactions, with a number of coactivator or corepressor molecules. An essential component is the hormone response element that binds the ligand (▲)-bound receptor (R). Also important are the accessory factor elements (AFEs) with bound transcription factors. More than two dozen of these accessory factors (AFs), which are often members of the nuclear receptor superfamily, have been linked to hormone effects on transcription. The AFs can interact with each other, with the liganded nuclear receptors, or with coregulators. These components communicate with the basal transcription machinery, forming the Polymerase II PIC (ie, RNAP II and GTFs) through a coregulator complex that can consist of one or more members of the p160, corepressor, mediator-related, or CBP/p300 families (see Table 42–6). Recall (Chapters 36, 38) that many of the transcription coregulators carry intrinsic enzymatic activities, which covalently modify the DNA, transcription proteins, and the histones present in the nucleosomes (not shown here) in and around the enhancer (HRE, AFE) and promoter. Collectively the hormone, hormone receptor, chromatin, DNA and transcription machinery integrate and process hormone signals to regulate transcription in a physiological fashion.

HORMONES CAN INFLUENCE SPECIFIC BIOLOGIC EFFECTS BY MODULATING TRANSCRIPTION

The signals generated as described above have to be translated into an action that allows the cell to effectively adapt to a challenge (Figure 42–1). Much of this adaptation is accomplished through alterations in the rates of transcription of specific genes. Many different observations have led to the current view of how hormones affect transcription. Some of these are as follows: (1) actively transcribed genes are in regions of "open" chromatin (experimentally defined as relative susceptibility to the enzyme DNase I), which allows for the access of transcription factors to DNA. (2) Genes have regulatory regions, and transcription factors bind to these to modulate the frequency of transcription initiation. (3) The hormone–receptor complex can be one of these transcription factors. The DNA sequence to which this binds is called a HRE (see Table 42–1 for examples). (4) Alternatively, other hormone-generated signals can modify the location, amount, or activity of transcription factors and thereby influence binding to the regulatory or response element. (5) Members of a large superfamily of nuclear receptors act with—or in a manner analogous to—the hormone receptors described above. (6) These nuclear receptors interact with another large group of coregulatory molecules to effect changes in the transcription of specific genes.

Several HREs Have Been Defined

HREs resemble enhancer elements in that they are not strictly dependent on position and location or orientation. They generally are found within a few hundred nucleotides upstream (5′) of the transcription initiation site, but they may be located within the coding region of the gene, in introns. HREs were defined by the strategy illustrated in Figure 38–11. The consensus sequences illustrated in Table 42–1 were arrived at through analysis of many genes regulated by a given hormone using simple, heterologous reporter systems (see Figure 38–10). Although these simple HREs bind the hormone–receptor complex more avidly than surrounding DNA—or DNA from an unrelated source—and confer hormone responsiveness to a reporter gene, it soon became apparent that the regulatory circuitry of natural genes must be much more complicated. Glucocorticoids, progestins, mineralocorticoids, and androgens have vastly different physiologic actions. How could the specificity required for these effects be achieved through regulation of gene expression by the same HRE (Table 42–1)? Questions like this have led to experiments which have allowed for elaboration of a very complex model of transcription

regulation. For example, the HRE must associate with other DNA elements (and associated binding proteins) to function optimally. The extensive sequence similarity noted between steroid hormone receptors, particularly in their DNA-binding domains (DBD), led to discovery of the **nuclear receptor superfamily** of proteins. These—and a large number of **coregulator proteins**—allow for a wide variety of DNA–protein and protein–protein interactions and the specificity necessary for highly regulated physiologic control. A schematic of such an assembly is illustrated in **Figure 42–11**.

There Is a Large Family of Nuclear Receptor Proteins

The nuclear receptor superfamily consists of a diverse set of transcription factors that were discovered because of a sequence similarity in their DBDs. This family, now with >50 members, includes the nuclear hormone receptors discussed above, a number of other receptors whose ligands were discovered after the receptors were identified, and many putative or orphan receptors for which a ligand has yet to be discovered.

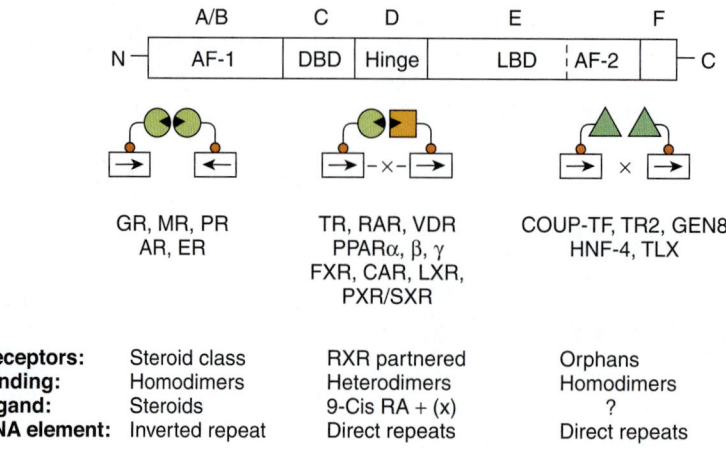

FIGURE 42–12 **The nuclear receptor superfamily.** Members of this family are divided into six structural domains (A–F). Domain A/B is also called AF-1, or the modulator region, because it is involved in activating transcription. The C domain consists of the DNA-binding domain (DBD). The D region contains the hinge, which provides flexibility between the DBD and the ligand-binding domain (LBD, region E). The C terminal part of region E contains AF-2, another domain important for activation of transcription. The F region is poorly defined. The functions of these domains are discussed in more detail in the text. Receptors with known ligands, such as the steroid hormones, bind as homodimers on inverted repeat half-sites. Other receptors form heterodimers with the partner RXR on direct repeat elements. There can be nucleotide spacers of one to five bases between these direct repeats (DR1–5). Another class of receptors for which ligands have not been definitively determined (orphan receptors) bind as homodimers to direct repeats and occasionally as monomers to a single half-site.

These nuclear receptors have several common structural features (**Figure 42–12**). All have a centrally located DBD that allows the receptor to bind with high affinity to a response element. The DBD contains two zinc finger binding motifs (see Figure 38–14) that direct binding either as homodimers, as heterodimers (usually with a retinoid X receptor [RXR] partner), or as monomers. The target response element consists of one or two DNA half-site consensus sequences arranged as an inverted or direct repeat. The spacing between the latter helps determine binding specificity. Thus, in general, a direct repeat with three, four, or five nucleotide spacer regions specifies the binding of the vitamin D, thyroid, and retinoic acid receptors, respectively, to the same consensus response element (Table 42–1). A multifunctional **ligand-binding domain (LBD)** is located in the carboxyl terminal half of the receptor. The LBD binds hormones or metabolites with selectivity and thus specifies a particular biologic response. The LBD also contains domains that mediate the binding of heat shock proteins, dimerization, nuclear localization, and transactivation. The latter function is facilitated by the carboxyl terminal transcription activation function (**AF-2 domain**), which forms a surface required for the interaction with coactivators. A highly variable **hinge region** separates the DBD from the LBD. This region provides flexibility to the receptor, so it can assume different DNA-binding conformations. Finally, there is a highly variable amino terminal region that contains another transactivation domain referred to as **AF-1**. The AF-1 domain likely provides for distinct physiologic functions through the binding of different coregulator proteins. This region of the receptor, through the use of different promoters, alternative splice sites, and multiple translation initiation sites, provides for receptor isoforms that share DBD and LBD identity but

exert different physiologic responses because of the association of various coregulators with this variable amino terminal AF-1 domain.

It is possible to sort this large number of receptors into groups in a variety of ways. Here, they are discussed according to the way they bind to their respective DNA elements (Figure 42–12). Classic hormone receptors for glucocorticoids (GR), mineralocorticoids (MR), estrogens (ER), androgens (AR), and progestins (PR) bind as homodimers to inverted repeat sequences. Other hormone receptors such as thyroid (TR), retinoic acid (RAR), and vitamin D (VDR) and receptors that bind various metabolite ligands such as PPAR α, β, and γ, FXR, LXR, PXR/SXR, and CAR bind as heterodimers, with retinoid X receptor (RXR) as a partner, to direct repeat sequences (see Figure 42–12 and **Table 42–5**). Another group of orphan receptors that as yet have no known ligand bind as homodimers or monomers to direct repeat sequences.

As illustrated in Table 42–5, the discovery of the nuclear receptor superfamily has led to an important understanding of how a variety of metabolites and xenobiotics regulate gene expression and thus the metabolism, detoxification, and elimination of normal body products and exogenous agents such as pharmaceuticals. Not surprisingly, this area is a fertile field for investigation of new therapeutic interventions.

A Large Number of Nuclear Receptor Coregulators Also Participate in Regulating Transcription

Chromatin remodeling (histone modifications, DNA methylation), transcription factor modification by various enzyme

TABLE 42–5 Nuclear Receptors with Special Ligands[1]

Receptor		Partner	Ligand	Process Affected
Peroxisome	PPAR$_\alpha$	RXR (DR1)	Fatty acids	Peroxisome proliferation
Proliferator-activated	PPAR$_\beta$		Fatty acids	Lipid and carbohydrate metabolism
	PPAR$_\gamma$		Fatty acids	
			Eicosanoids, thiazolidinediones	
Farnesoid X	FXR	RXR (DR4)	Farnesol, bile acids	Bile acid metabolism
Liver X	LXR	RXR (DR4)	Oxysterols	Cholesterol metabolism
Xenobiotic X	CAR	RXR (DR5)	Androstanes	Protection against certain drugs, toxic metabolites, and xenobiotics
			Phenobarbital	
			Xenobiotics	
	PXR	RXR (DR3)	Pregnanes	
			Xenobiotics	

[1]Many members of the nuclear receptor superfamily were discovered by cloning, and the corresponding ligands were subsequently identified. These ligands are not hormones in the classic sense, but they do have a similar function in that they activate specific members of the nuclear receptor superfamily. The receptors described here form heterodimers with RXR and have variable nucleotide sequences separating the direct repeat binding elements (DR1-5). These receptors regulate a variety of genes encoding cytochrome p450s (CYP), cytosolic binding proteins, and ATP-binding cassette (ABC) transporters to influence metabolism and protect cells against drugs and noxious agents.

activities, and the communication between the nuclear receptors and the basal transcription apparatus are accomplished by protein–protein interactions with one or more of a class of coregulator molecules. The number of these coregulator molecules now exceeds 100, not counting species variations and splice variants. The first of these to be described was the **CREB-binding protein, CBP.** CBP, through an amino terminal domain, binds to phosphorylated serine 137 of CREB and mediates transactivation in response to cAMP. It thus is described as a coactivator. CBP and its close relative, p300, interact directly or indirectly with a number of signaling molecules, including activator protein-1 (AP-1), STATs, nuclear receptors, and CREB (**Figure 42–13**). **CBP/p300** also binds to the p160 family of coactivators described below and to a number of other proteins, including viral transcription factor Ela, the p90^{rsk} protein kinase, and RNA helicase A. It is important

to note, as mentioned above, that **CBP/p300 also has intrinsic histone acetyltransferase (HAT) activity.** Some of the many actions of CBP/p300, which appear to depend on intrinsic enzyme activities and its ability to serve as a scaffold for the binding of other proteins, are illustrated in Figure 42–11. Other coregulators serve similar functions.

Several other families of coactivator molecules have been described. Members of the **p160 family of coactivators,** all of about 160 kDa, include (1) SRC-1 and NCoA-1; (2) GRIP 1, TIF2, and NCoA-2; and (3) p/CIP, ACTR, AIB1, RAC3, and TRAM-1 (**Table 42–6**). The different names for members within a subfamily often represent species variations or minor splice variants. There is about 35% amino acid identity between members of the different subfamilies. The p160 coactivators share several properties. They (1) bind nuclear receptors in an agonist- and AF-2 transactivation domain-dependent manner,

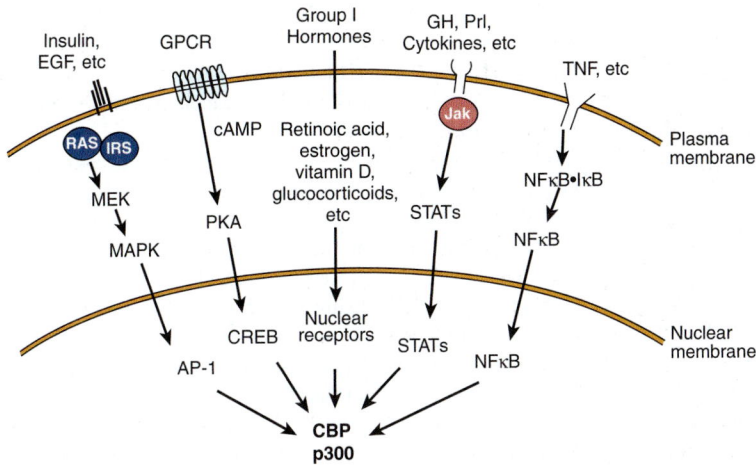

FIGURE 42–13 **Several signal transduction pathways converge on CBP/p300.** Many ligands that associate with membrane or nuclear receptors eventually converge on CBP/p300. Several different signal transduction pathways are employed. (EGF, epidermal growth factor; GH, growth hormone; Prl, prolactin; TNF, tumor necrosis factor; other abbreviations are expanded in the text.)

TABLE 42–6 Some Mammalian Coregulator Proteins

I. 300-kDa family of coactivators	
A. CBP	CREB-binding protein
B. p300	Protein of 300 kDa

II. 160-kDa family of coactivators	
A. SRC-1	Steroid receptor coactivator 1
NCoA-1	Nuclear receptor coactivator 1
B. TIF2	Transcriptional intermediary factor 2
GRIP1	Glucocorticoid receptor-interacting protein
NCoA-2	Nuclear receptor coactivator 2
C. p/CIP	p300/CBP cointegrator-associated protein 1
ACTR	Activator of the thyroid and retinoic acid receptors
AIB	Amplified in breast cancer
RAC3	Receptor-associated coactivator 3
TRAM-1	TR activator molecule 1

III. Corepressors	
A. NCoR	Nuclear receptor corepressor
B. SMRT	Silencing mediator for RXR and TR

IV. Mediator subunits	
A. TRAPs	Thyroid hormone receptor-associated proteins
B. DRIPs	Vitamin D receptor-interacting proteins
C. ARC	Activator-recruited cofactor

(2) have a conserved amino terminal basic helix-loop-helix (bHLH) motif (see Chapter 38), (3) have a weak carboxyl terminal transactivation domain and a stronger amino terminal transactivation domain in a region that is required for the CBP/p160 interaction, (4) contain at least three of the **LXXLL motifs** required for protein–protein interaction with other coactivators, and (5) often have HAT activity. The role of HAT is particularly interesting, as mutations of the HAT domain disable many of these transcription factors. Current thinking holds that these HAT activities acetylate histones, which facilitates the remodeling of chromatin into a transcription-efficient environment. Histone acetylation/deacetylation thus plays a critical role in gene expression. Finally, it is important to note that other protein substrates for HAT-mediated acetylation, such as DNA binding transcription activators and other coregulators have been reported. Such nonhistone PTM events likely also factor importantly into the overall regulatory response.

A small number of proteins, including NCoR and SMRT, comprise the **corepressor family.** They function, at least in part, as described in Figure 42–2. Another family includes the TRAPs, DRIPs, and ARC (Table 42–6). These proteins represent subunits of the Mediator (Chapter 36) and range in size from 80 kDa to 240 kDa and are thought to link the nuclear receptor–coactivator complex to RNA polymerase II and the other components of the basal transcription apparatus.

The exact role of these coactivators is presently under intensive investigation. Many of these proteins have intrinsic enzymatic activities. This is particularly interesting in view of the fact that acetylation, phosphorylation, methylation, sumoylation, and ubiquitination—as well as proteolysis and cellular translocation—have been proposed to alter the activity of some of these coregulators and their targets.

It appears that certain combinations of coregulators—and thus different combinations of activators and inhibitors—are responsible for specific ligand-induced actions through various receptors. Furthermore, these interactions on a given promoter are dynamic. In some cases, complexes consisting of over 45 transcription factors have been observed on a single gene.

SUMMARY

- Hormones, cytokines, interleukins, and growth factors use a variety of signaling mechanisms to facilitate cellular adaptive responses.

- The ligand–receptor complex serves as the initial signal for members of the nuclear receptor family.

- Class II peptide/protein and catecholamine hormones, which bind to cell surface receptors, generate a variety of intracellular signals. These include cAMP, cGMP, Ca^{2+}, phosphatidylinositides, and protein kinase cascades.

- Many hormone responses are accomplished through alterations in the rate of transcription of specific genes.

- The nuclear receptor superfamily of proteins plays a central role in the regulation of gene transcription.

- Nuclear receptors, which may have hormones, metabolites, or drugs as ligands, bind to specific DNA elements as homodimers or as heterodimers with RXR. Some—orphan receptors—have no known ligand but bind DNA and influence transcription.

- Another large family of coregulator proteins remodel chromatin, modify other transcription factors, and bridge the nuclear receptors to the basal transcription apparatus.

REFERENCES

Arvanitakis L, Geras-Raaka E, Gershengorn MC: Constitutively signaling G-protein-coupled receptors and human disease. Trends Endocrinol Metab 1998;9:27

Aylon Y, Oren M: Living with p53, Dying of p53. Cell 2007;130:597.

Beene DL, Scott JD: A-kinase anchoring proteins take shape. Current Opinion in Cell Biol 2007;19:192.

Brummer T, Schmitz-Perffer C, Daly RJ: Docking proteins. FEBS Journal 2010; 277:4356–4369.

Cheung E, Kraus WL: Genomic Analyses of Hormone Signaling and Gene Regulation. Annu Rev Physiol 2010;72:191–218.

Darnell JE Jr, Kerr IM, Stark GR: Jak-STAT pathways and transcriptional activation in response to IFNs and other extracellular signaling proteins. Science 1994;264:1415.

Fantl WJ, Johnson DE, Williams LT: Signalling by receptor tyrosine kinases. Annu Rev Biochem 1993;62:453.

Hanoune J, Defer N: Regulation and role of adenylyl cyclase isoforms. Annu Rev Pharmacol Toxicol 2001;41:145.

Jaken S: Protein kinase C isozymes and substrates. Curr Opin Cell Biol 1996;8:168.

Lee C-H, Olson P, Evans RM: Mini-review: Lipid metabolism, metabolic diseases and peroxisome proliferators-activated receptor. Endocrinology 2003;144:2201.

Métivier R, Gallais R, Tiffoche C, et al: Cyclical DNA methylation of a transcriptionally active promter. Nature 2008;452:45.

Métivier R, Reid G, Gannon F: Transcription in four dimensions: nuclear receptor-directed initiation of gene expression. EMBO Journal 2006;7:161.

Montminy M: Transcriptional regulation by cyclic AMP. Annu Rev Biochem 1997;66:807

Morris AJ, Malbon CC: Physiological regulation of G protein-linked signaling. Physiol Rev 1999;79:1373.

O'Malley B: Coregulators: from whence came these "master genes." Mol Endocrinology 2007;21:1009.

Rosenfeld MG, Lunyak VV, Glass CK: Sensors and signals: a coactivator/corepressor/epigenetic code for integrating signal-dependent programs of transcriptional response. Genes and Dev 2006;20:1405.

Sonoda J, Pei L, Evans RM: Nuclear receptors: decoding metabolic disease. Fed of European Biochem Soc 2007;582:2.

Tang X, Tang G, Ozcan S: Role of microRNAs in diabetes. Biochim Biophys Acta 2008;1779:697.

Walton KM, Dixon JE: Protein tyrosine phosphatases. Annu Rev Biochem 1993;62:101.

Exam Questions

Section V

1. Regarding membrane lipids, select the one FALSE answer.
   A. The major phospholipid by mass in human membranes is generally phosphatidylcholine.
   B. Glycolipids are located on the inner and outer leaflets of the plasma membrane.
   C. Phosphatidic acid is a precursor of phosphatidylserine, but not of sphingomyelin.
   D. Phosphatidylcholine and phosphatidylethanolamine are located primarily on the outer leaflet of the plasma membrane.
   E. The flip–flop of phospholipids in membranes is very slow.

2. Regarding membrane proteins, select the one FALSE answer.
   A. Because of steric considerations, alpha-helices cannot exist in membranes.
   B. A hydropathy plot helps one to estimate whether a segment of a protein is predominantly hydrophobic or hydrophilic.
   C. Certain proteins are anchored to the outer leaflet of plasma membranes via glycophosphatidylinositol (GPI) structures.
   D. Adenylyl cyclase is a marker enzyme for the plasma membrane.
   E. Myelin has a very high content of lipid compared with protein.

3. Regarding membrane transport, select the one FALSE statement.
   A. Potassium has a lower charge density than sodium and tends to move more quickly through membranes than does sodium.
   B. The flow of ions through ion channels is an example of passive transport.
   C. Facilitated diffusion requires a protein transporter.
   D. Inhibition of the Na^+ K^+-ATPase will inhibit sodium-dependent uptake of glucose in intestinal cells.
   E. Insulin, by recruiting glucose transporters to the plasma membrane, increases uptake of glucose in fat cells but not in muscle.

4. Regarding the Na^+K^+-ATPase, select the one FALSE statement.
   A. Its action maintains the high intracellular concentration of sodium compared with potassium.
   B. It can use as much as 30% of the total ATP expenditure of a cell.
   C. It is inhibited by digitalis, a drug that is useful in certain cardiac conditions.
   D. It is located in the plasma membrane of cells.
   E. Phosphorylation is involved in its mechanism of action, leading to its classification as a P-type ATP-driven active transporter.

5. What molecules enable cells to respond to a specific extracellular signaling molecule?
   A. Specific receptor carbohydrates localized to the inner plasma membrane surface.
   B. Plasma lipid bilayer.
   C. Ion channels.

D. Receptors that specifically recognize and bind that particular messenger molecule.
   E. Intact nuclear membranes.

6. Indicate the term generally applied to the extracellular messenger molecules that bind to transmembrane receptor proteins:
   A. Competitive inhibitor
   B. Ligand
   C. Scatchard curve
   D. Substrate
   E. Key

7. In autocrine signaling
   A. Messenger molecules reach their target cells via passage through bloodstream.
   B. Messenger molecules travel only short distances through the extracellular space to cells that are in close proximity to the cell that is generating the message.
   C. The cell producing the messenger expresses receptors on its surface that can respond to that messenger.
   D. The messenger molecules are usually rapidly degraded and hence can only work over short distances.

8. Regardless of how a signal is initiated, the ligand-binding event is propagated via second messengers or protein recruitment. What is the ultimate outcome of these binding events?
   A. A protein in the middle of an intracellular signaling pathway is activated.
   B. A protein at the top of an intracellular signaling pathway is activated.
   C. A protein at the top of an extracellular signaling pathway is activated.
   D. A protein at the top of an intracellular signaling pathway is deactivated.
   E. A protein at the bottom of an intracellular signaling pathway is activated.

9. What features of the nuclear receptor superfamily suggest that these proteins have evolved from a common ancestor?
   A. They all bind the same ligand with high affinity.
   B. They all function within the nucleus.
   C. They are all subject to regulatory phosphorylation.
   D. They all contain regions of high amino acid sequence similarity/identity.
   E. They all bind DNA.

10. What effect does degradation of receptor–ligand complexes after internalization have upon the ability of a cell to respond if immediately re-exposed to the same hormone?
    A. The cellular response is attenuated due to a decrease in cellular receptor number.
    B. Cellular response is enhanced due to reduced receptor–ligand competition.
    C. The cellular response is unchanged to subsequent stimuli.
    D. Cell hormone response is now bimodal; enhanced for a short time and then blunted.

11. Typically, what is the first reaction after most receptor protein-tyrosine kinases (RTKs) bind their ligand?

   A. Receptor trimerization
   B. Receptor degradation
   C. Receptor denaturation
   D. Receptor dissociation
   E. Receptor dimerization

12. Where is the kinase catalytic domain of the receptor protein–tyrosine kinases found?

   A. On the extracellular surface of the receptor, immediately adjacent to the ligand binding domain.
   B. On an independent protein that rapidly binds the receptor upon ligand binding.
   C. On the cytoplasmic domain of the receptor.
   D. Within the transmembrane spanning portion of the receptor.

13. The subunits of the heterotrimeric G proteins are called the __,__and__ subunits.

   A. α, β, and χ
   B. α, β, and δ
   C. α, γ, and δ
   D. α, β, and γ
   E. γ, δ, and η

14. Of the receptors listed below, which can conduct a flow of ions across the plasma membrane when bound to their cognate ligand?

   A. Receptor Tyrosine Kinases (RTKs).
   B. G-protein coupled receptors (GPCRs).
   C. G protein coupled receptors
   D. Steroid hormone receptors
   E. Ligand-gated channels

15. Which of the following is NOT a natural ligand that binds to G-protein coupled receptors?

   A. Hormones
   B. Steroid hormones
   C. Chemoattractants
   D. Opium derivatives
   E. Neurotransmitters

16. Place the events of signaling listed below in the CORRECT order.

   1. G protein binds to activated receptor forming a receptor-G protein complex.
   2. Release of GDP by the G protein.
   3. Change in conformation of the cytoplasmic loops of the receptor.
   4. Binding of GTP by the G protein.
   5. Increase in the affinity of the receptor for a G protein on the cytoplasmic surface of the membrane.
   6. Binding of a hormone or neurotransmitter to a G-protein coupled receptor.
   7. Conformational shift in the α subunit of the G protein.

   A. 6 – 3 – 5 – 1 – 2 – 4 – 7
   B. 6 – 5 – 4 – 1 – 7 – 2 – 3
   C. 6 – 3 – 5 – 1 – 7 – 2 – 4
   D. 6 – 7 – 3 – 5 – 1 – 2 – 4
   E. 6 – 3 – 5 – 1 – 7 – 2 – 4

17. Which heterotrimeric G proteins couple receptors to adenylyl cyclase via the activation of GTP-bound G_α subunits?

   A. G_r family
   B. G_q family
   C. G_i family
   D. $G_{12/13}$ family
   E. G_s family

18. What must happen in order to prevent overstimulation by a hormone?

   A. Hormones must be degraded.
   B. G proteins must be recycled and then degraded.
   C. Receptors must be blocked from continuing to activate G proteins.
   D. Receptors must dimerize.

19. Which of the following hormones termed the "flight-or-fight" hormone is secreted by the adrenal medulla?

   A. Epinephrine
   B. Oxytocin
   C. Insulin
   D. Glucagon
   E. Somatostatin

20. Which hormone is secreted by α-cells in the pancreas in response to low blood glucose levels?

   A. Insulin
   B. Glucagon
   C. Estradiol
   D. Epinephrine
   E. Somatostatin

21. In liver cells the expression of genes encoding gluconeogenic enzymes such as phophoenolpyruvate carboxykinase is induced in response to which of the following molecules?

   A. cGMP
   B. Insulin
   C. ATP
   D. cAMP
   E. Cholesterol

22. What happens to protein kinase A (PKA) following the binding of cAMP?

   A. The regulatory subunits of PKA dissociate, thereby activating the catalytic subunits.
   B. PKA catalytic subunits then bind to two regulatory subunits, thereby activating the catalytic subunits.
   C. The inhibitory regulatory subunits dissociate from the catalytic subunits, completely inactivating the enzyme.
   D. The stimulatory regulatory subunits dissociate from the catalytic subunits, inhibiting the enzyme.
   E. Phosphodiesterase binds to the catalytic subunits, which results in enzyme inactivation.

Nutrition, Digestion, & Absorption

David A. Bender, PhD & Peter A. Mayes, PhD, DSc

O B J E C T I V E S

After studying this chapter, you should be able to:

- Describe the digestion and absorption of carbohydrates, lipids, proteins, vitamins, and minerals.
- Explain how energy requirements can be measured and estimated and how measuring the respiratory quotient permits estimation of the mix of metabolic fuels being oxidized.
- Describe the consequences of undernutrition: marasmus, cachexia, and kwashiorkor.
- Explain how protein requirements are determined and why more of some proteins than others is required to maintain nitrogen balance.

BIOMEDICAL IMPORTANCE

In addition to water, the diet must provide metabolic fuels (mainly carbohydrates and lipids), protein (for growth and turnover of tissue proteins, as well as a source of metabolic fuel), fiber (for bulk in the intestinal lumen), minerals (containing elements with specific metabolic functions), and vitamins and essential fatty acids (organic compounds needed in smaller amounts for other metabolic and physiologic functions). The polysaccharides, triacylglycerols, and proteins that make up the bulk of the diet must be hydrolyzed to their constituent monosaccharides, fatty acids, and amino acids, respectively, before absorption and utilization. Minerals and vitamins must be released from the complex matrix of food before they can be absorbed and utilized.

Globally, **undernutrition** is widespread, leading to impaired growth, defective immune systems, and reduced work capacity. By contrast, in developed countries, there is excessive food consumption (especially of fat), leading to obesity, and the development of diabetes, cardiovascular disease, and some cancers. Deficiencies of vitamin A, iron, and iodine pose major health concerns in many countries, and deficiencies of other vitamins and minerals are a major cause of ill health. In developed countries nutrient deficiency is rare, although there are vulnerable sections of the population at risk. Intakes of minerals and vitamins that are adequate to prevent deficiency may be inadequate to promote optimum health and longevity.

Excessive secretion of gastric acid, associated with *Helicobacter pylori* infection, can result in the development of gastric and duodenal **ulcers;** small changes in the composition of bile can result in crystallization of cholesterol as **gallstones;** failure of exocrine pancreatic secretion (as in **cystic fibrosis**) leads to undernutrition and steatorrhea. **Lactose intolerance** is the result of lactase deficiency, leading to diarrhea and intestinal discomfort when lactose is consumed. Absorption of intact peptides that stimulate antibody responses causes **allergic reactions; celiac disease** is an allergic reaction to wheat gluten.

DIGESTION & ABSORPTION OF CARBOHYDRATES

The digestion of carbohydrates is by hydrolysis to liberate oligosaccharides, then free mono- and disaccharides. The increase in blood glucose after a test dose of a carbohydrate compared with that after an equivalent amount of glucose (as glucose or from a reference starchy food) is known as the **glycemic index.** Glucose and galactose have an index of 1 (or 100%), as do lactose, maltose, isomaltose, and trehalose, which give rise to these monosaccharides on hydrolysis. Fructose and the sugar alcohols are absorbed less rapidly and have a lower glycemic index, as does sucrose. The glycemic index of starch varies between near 1 (or 100%) to near 0 as a result of variable rates of hydrolysis, and that of nonstarch polysaccharides is 0. Foods that have a low glycemic index are considered to be more beneficial since they cause less fluctuation in insulin secretion. Resistant starch and nonstarch polysaccharides provide substrates for bacterial fermentation in the large intestine, and the resultant butyrate and other short chain fatty acids provide a significant source of fuel for intestinal enterocytes. There is some evidence that butyrate also has antiproliferative activity, and so provides protection against colorectal cancer.

Amylases Catalyze the Hydrolysis of Starch

The hydrolysis of starch is catalyzed by salivary and pancreatic amylases, which catalyze random hydrolysis of $\alpha(1\rightarrow4)$ glycoside bonds, yielding dextrins, then a mixture of glucose, maltose, and maltotriose and small branched dextrins (from the branchpoints in amylopectin).

Disaccharidases Are Brush Border Enzymes

The disaccharidases, maltase, sucrase-isomaltase (a bifunctional enzyme catalyzing hydrolysis of sucrose and isomaltose), lactase, and trehalase are located on the brush border of the intestinal mucosal cells, where the resultant monosaccharides and those arising from the diet are absorbed. Congenital deficiency of lactase occurs rarely in infants, leading to lactose intolerance and failure to thrive when fed on breast milk or normal infant formula. Congenital deficiency of sucrase-isomaltase occurs among the Inuit, leading to sucrose intolerance, with persistent diarrhea and failure to thrive when the diet contains sucrose.

In most mammals, and most human beings, lactase activity begins to fall after weaning and is almost completely lost by late adolescence, leading to **lactose intolerance.** Lactose remains in the intestinal lumen, where it is a substrate for bacterial fermentation to lactate, resulting in abdominal discomfort and diarrhea after consumption of relatively large amounts. In two population groups, people of north European origin and nomadic tribes of sub-Saharan Africa and Arabia, lactase persists after weaning and into adult life. Marine mammals secrete a high-fat milk that contains no carbohydrate, and their pups lack lactase.

There Are Two Separate Mechanisms for the Absorption of Monosaccharides in the Small Intestine

Glucose and galactose are absorbed by a sodium-dependent process. They are carried by the same transport protein (SGLT 1) and compete with each other for intestinal absorption (**Figure 43–1**). Other monosaccharides are absorbed by carrier-mediated diffusion. Because they are not actively transported, fructose and sugar alcohols are only absorbed down their concentration gradient, and after a moderately high intake, some may remain in the intestinal lumen, acting as a substrate for bacterial fermentation. Large intakes of fructose and sugar alcohols can lead to osmotic diarrhea.

DIGESTION & ABSORPTION OF LIPIDS

The major lipids in the diet are triacylglycerols and, to a lesser extent, phospholipids. These are hydrophobic molecules and have to be hydrolyzed and emulsified to very small droplets

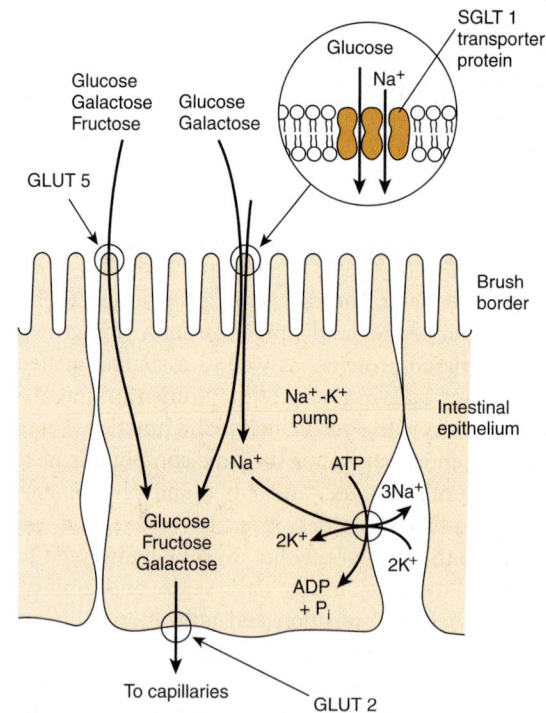

FIGURE 43–1   Transport of glucose, fructose, and galactose across the intestinal epithelium. The SGLT 1 transporter is coupled to the Na⁺–K⁺ pump, allowing glucose and galactose to be transported against their concentration gradients. The GLUT 5 Na⁺-independent facilitative transporter allows fructose, as well as glucose and galactose, to be transported down their concentration gradients. Exit from the cell for all sugars is via the GLUT 2 facilitative transporter.

(micelles, 4–6 nm in diameter) before they can be absorbed. The fat-soluble vitamins, A, D, E, and K, and a variety of other lipids (including cholesterol) are absorbed dissolved in the lipid micelles. Absorption of fat-soluble vitamins is impaired on a very low fat diet.

Hydrolysis of triacylglycerols is initiated by lingual and gastric lipases, which attack the sn-3 ester bond forming 1,2-diacylglycerols and free fatty acids, which act as emulsifying agents. Pancreatic lipase is secreted into the small intestine and requires a further pancreatic protein, colipase, for activity. It is specific for the primary ester links—ie, positions 1 and 3 in triacylglycerols—resulting in 2-monoacylglycerols and free fatty acids as the major end products of luminal triacylglycerol digestion. Pancreatic esterase in the intestinal lumen hydrolyzes monoacylglycerols, but they are poor substrates, and only ~25% of ingested triacylglycerol is completely hydrolyzed to glycerol and fatty acids before absorption (**Figure 43–2**). Bile salts, formed in the liver and secreted in the bile, permit emulsification of the products of lipid digestion into micelles together with dietary phospholipids and cholesterol secreted in the bile (about 2 g/day) as well as dietary cholesterol (about 0.5 g/day). Because the micelles are soluble, they allow the products of digestion, including the fat-soluble vitamins, to be transported through the aqueous environment of the intestinal lumen to come into close contact with the brush border of the mucosal cells, allowing uptake into the epithelium. The bile salts remain in the intestinal lumen, where most are absorbed from the ileum into the **enterohepatic circulation** (Chapter 26). Within the intestinal epithelium, 1-monoacyglycerols are hydrolyzed to fatty acids and glycerol and 2-monoacylglycerols are reacylated to triacylglycerols via the **monoacylglycerol pathway.** Glycerol released in the intestinal lumen is absorbed into the hepatic portal vein; glycerol released within the epithelium is reutilized for triacylglycerol synthesis via the normal phosphatidic acid pathway (Chapter 24). Long-chain fatty acids are esterified to yield to triacylglycerol in the mucosal cells and together with the other products of lipid digestion, secreted as chylomicrons into the lymphatics, entering the bloodstream via the thoracic duct (Chapter 25). Short- and medium-chain fatty acids are mainly absorbed into the hepatic portal vein as free fatty acids.

Cholesterol is absorbed dissolved in lipid micelles and is mainly esterified in the intestinal mucosa before being incorporated into chylomicrons. Plant sterols and stanols (in which the B ring is saturated) compete with cholesterol for esterification, but are poor substrates, so that there is an increased amount of unesterified cholesterol in the mucosal cells. Unesterified cholesterol and other sterols are actively transported out of the mucosal cells into the intestinal lumen. This means that plant sterols and stanols effectively inhibit the absorption of not only dietary cholesterol, but also the larger amount that is secreted in the bile, so lowering the whole body cholesterol content, and hence the plasma cholesterol concentration.

DIGESTION & ABSORPTION OF PROTEINS

Native proteins are resistant to digestion because few peptide bonds are accessible to the proteolytic enzymes without prior denaturation of dietary proteins (by heat in cooking and by the action of gastric acid).

Several Groups of Enzymes Catalyze the Digestion of Proteins

There are two main classes of proteolytic digestive enzymes **(proteases),** with different specificities for the amino acids forming the peptide bond to be hydrolyzed. **Endopeptidases** hydrolyze peptide bonds between specific amino acids throughout the molecule. They are the first enzymes to act, yielding a larger number of smaller fragments. Pepsin in the gastric juice catalyzes hydrolysis of peptide bonds adjacent to amino acids with bulky side-chains (aromatic and branched-chain amino acids and methionine). Trypsin, chymotrypsin, and elastase are secreted into the small intestine by the pancreas. Trypsin catalyzes hydrolysis of lysine and arginine esters, chymotrypsin esters of aromatic amino acids, and elastase esters of small neutral aliphatic amino acids. **Exopeptidases** catalyze the hydrolysis of peptide bonds, one at a time, from the ends of peptides. **Carboxypeptidases,** secreted in the pancreatic juice, release amino acids from the free carboxyl terminal; **aminopeptidases,** secreted by the intestinal mucosal cells, release amino acids from the amino terminal. **Dipeptidases** and **tripeptidases** in the brush border of intestinal mucosal cells catalyze the hydrolysis of di- and tripeptides, which are not substrates for amino- and carboxypeptidases.

The proteases are secreted as inactive **zymogens;** the active site of the enzyme is masked by a small region of the peptide chain that is removed by hydrolysis of a specific peptide bond. Pepsinogen is activated to pepsin by gastric acid and by activated pepsin. In the small intestine, trypsinogen, the precursor of trypsin, is activated by enteropeptidase, which is secreted by the duodenal epithelial cells; trypsin can then activate chymotrypsinogen to chymotrypsin, proelastase to elastase, procarboxypeptidase to carboxypeptidase, and proaminopeptidase to aminopeptidase.

Free Amino Acids & Small Peptides Are Absorbed by Different Mechanisms

The end product of the action of endopeptidases and exopeptidases is a mixture of free amino acids, di- and tripeptides, and oligopeptides, all of which are absorbed. Free amino acids are absorbed across the intestinal mucosa by sodium-dependent active transport. There are several different amino acid transporters, with specificity for the nature of the amino acid side-chain (large or small, neutral, acidic, or basic). The various amino acids carried by any one transporter compete with each other for absorption and tissue uptake. Dipeptides and tripeptides enter

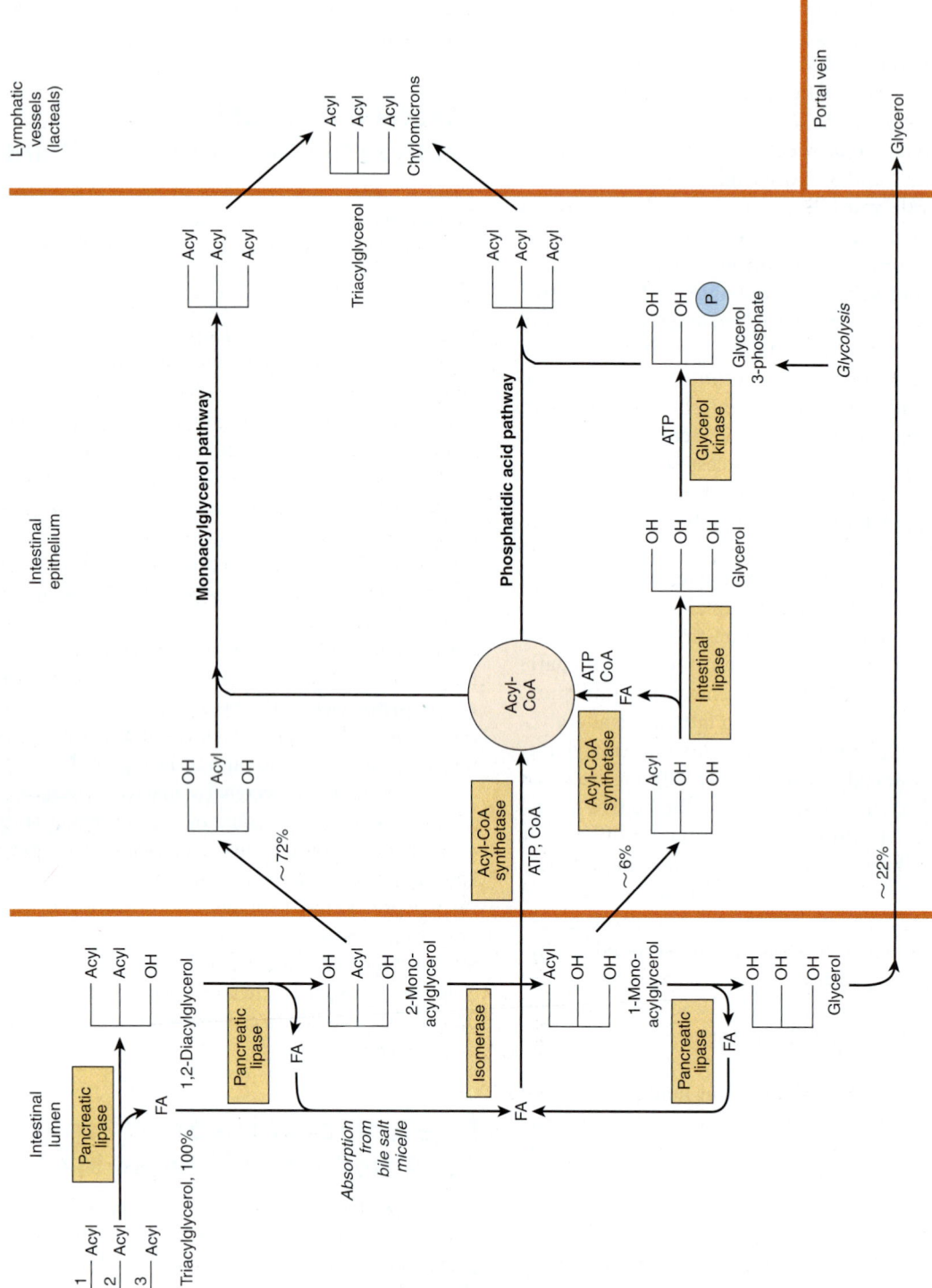

FIGURE 43–2 Digestion and absorption of triacylglycerols. The values given for percentage uptake may vary widely but indicate the relative importance of the three routes shown.

the brush border of the intestinal mucosal cells, where they are hydrolyzed to free amino acids, which are then transported into the hepatic portal vein. Relatively large peptides may be absorbed intact, either by uptake into mucosal epithelial cells (transcellular) or by passing between epithelial cells (paracellular). Many such peptides are large enough to stimulate antibody formation—this is the basis of **allergic reactions** to foods.

DIGESTION & ABSORPTION OF VITAMINS & MINERALS

Vitamins and minerals are released from food during digestion, although this is not complete, and the availability of vitamins and minerals depends on the type of food and, especially for minerals, the presence of chelating compounds. The fat-soluble vitamins are absorbed in the lipid micelles that are the result of fat digestion; water-soluble vitamins and most mineral salts are absorbed from the small intestine either by active transport or by carrier-mediated diffusion followed by binding to intracellular proteins to achieve concentrative uptake. Vitamin B_{12} absorption requires a specific transport protein, intrinsic factor (Chapter 44); calcium absorption is dependent on vitamin D; zinc absorption probably requires a zinc-binding ligand secreted by the exocrine pancreas, and the absorption of iron is limited (see below).

Calcium Absorption Is Dependent on Vitamin D

In addition to its role in regulating calcium homeostasis, vitamin D is required for the intestinal absorption of calcium. Synthesis of the intracellular calcium-binding protein, **calbindin,** required for calcium absorption, is induced by vitamin D. Vitamin D also acts to recruit calcium transporters to the cell surface, so increasing calcium absorption rapidly—a process that is independent of new protein synthesis.

Phytic acid (inositol hexaphosphate) in cereals binds calcium in the intestinal lumen, preventing its absorption. Other minerals, including zinc, are also chelated by phytate. This is mainly a problem among people who consume large amounts of unleavened whole-wheat products; yeast contains an enzyme, **phytase,** that dephosphorylates phytate, so rendering it inactive. High concentrations of fatty acids in the intestinal lumen, as a result of impaired fat absorption, can also reduce calcium absorption by forming insoluble calcium salts; a high intake of oxalate can sometimes cause deficiency since calcium oxalate is insoluble.

Iron Absorption Is Limited and Strictly Controlled, but Enhanced by Vitamin C and Alcohol

Although iron deficiency is a common problem in both developed and developing countries, about 10% of the population are genetically at risk of iron overload (**hemochromatosis),** and in order to reduce the risk of adverse effects of nonenzymic generation of free radicals by iron salts, absorption is strictly regulated. Inorganic iron is transported into the mucosal cell by a proton-linked divalent metal ion transporter, and accumulated intracellularly by binding to **ferritin.** Iron leaves the mucosal cell via a transport protein ferroportin, but only if there is free **transferrin** in plasma to bind to. Once transferrin is saturated with iron, any that has accumulated in the mucosal cells is lost when the cells are shed. Expression of the ferroportin gene (and possibly also that for the divalent metal ion transporter) is downregulated by hepcidin, a peptide secreted by the liver when body iron reserves are adequate. In response to hypoxia, anemia, or hemorrhage, the synthesis of hepcidin is reduced, leading to increased synthesis of ferroportin and increased iron absorption (**Figure 43–3**). As a result of this mucosal barrier, only ~10% of dietary iron is absorbed, and only 1–5% from many plant foods (Chapter 50).

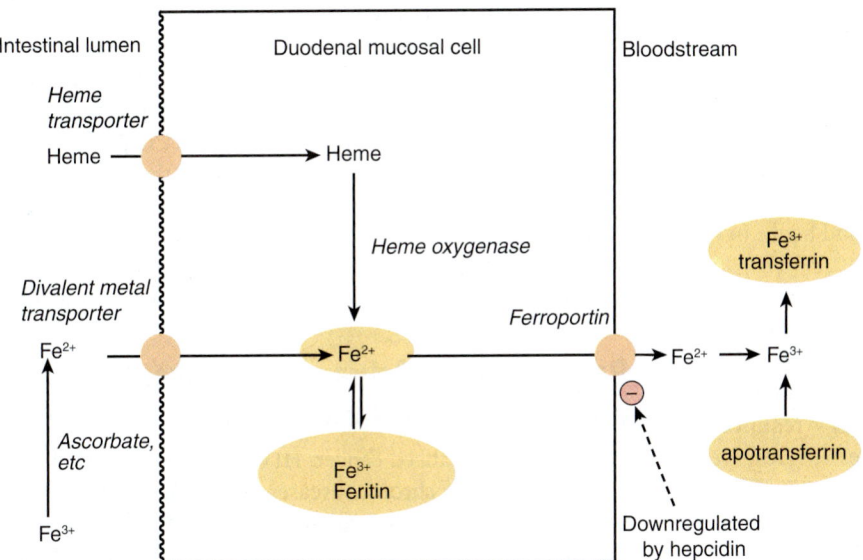

FIGURE 43–3 Absorption of iron. Hepcidin secreted by the liver downregulates synthesis of ferroportin and limits iron absorption.

Inorganic iron is absorbed in the Fe^{2+} (reduced) state, and hence, the presence of reducing agents enhances absorption. The most effective compound is **vitamin C,** and while intakes of 40–80 mg of vitamin C/day are more than adequate to meet requirements, an intake of 25–50 mg per meal enhances iron absorption, especially when iron salts are used to treat iron deficiency anemia. Alcohol and fructose also enhance iron absorption. Heme iron from meat is absorbed separately and is considerably more available than inorganic iron. However, the absorption of both inorganic and heme iron is impaired by calcium—a glass of milk with a meal significantly reduces iron availability.

ENERGY BALANCE: OVER- & UNDERNUTRITION

After the provision of water, the body's first requirement is for metabolic fuels—fats, carbohydrates, and amino acids from proteins (Table 16–1). Food intake in excess of energy expenditure leads to **obesity,** while intake less than expenditure leads to emaciation and wasting, **marasmus,** and **kwashiorkor.** Both obesity and severe undernutrition are associated with increased mortality. The body mass index = weight (in kg)/height^2 (in m) is commonly used as a way of expressing relative obesity; a desirable range is between 20 and 25.

Energy Requirements Are Estimated by Measurement of Energy Expenditure

Energy expenditure can be determined directly by measuring heat output from the body, but is normally estimated indirectly from the consumption of oxygen. There is an energy expenditure of ~20 kJ/L of oxygen consumed, regardless of whether the fuel being metabolized is carbohydrate, fat, or protein (Table 16–1).

Measurement of the ratio of the volume of carbon dioxide produced: volume of oxygen consumed (**respiratory quotient, RQ**) is an indication of the mixture of metabolic fuels being oxidized (Table 16–1).

A more recent technique permits estimation of total energy expenditure over a period of 1–2 weeks, using dual isotopically labeled water, $^2H_2^{18}O$. ^2H is lost from the body only in water, while ^{18}O is lost in both water and carbon dioxide; the difference in the rate of loss of the two labels permits estimation of total carbon dioxide production, and hence oxygen consumption and energy expenditure.

Basal metabolic rate (BMR) is the energy expenditure by the body when at rest, but not asleep, under controlled conditions of thermal neutrality, measured about 12 h after the last meal, and depends on weight, age, and gender. Total energy expenditure depends on the BMR, the energy required for physical activity and the energy cost of synthesizing reserves in the fed state. It is therefore possible to estimate an individual's energy requirement from body weight, age, gender, and level of physical activity. Body weight affects BMR because there is

a greater amount of active tissue in a larger body. The decrease in BMR with increasing age, even when body weight remains constant, is the result of muscle tissue replacement by adipose tissue, which is metabolically less active. Similarly, women have a significantly lower BMR than do men of the same body weight and age because women's bodies contain proportionally more adipose tissue.

Energy Requirements Increase with Activity

The most useful way of expressing the energy cost of physical activities is as a multiple of BMR. Sedentary activities use only about 1.1–1.2 × BMR. By contrast, vigorous exertion, such as climbing stairs, cross-country walking uphill, etc, may use 6–8 × BMR.

Ten Percent of the Energy Yield of a Meal May Be Expended in Forming Reserves

There is a considerable increase in metabolic rate after a meal (**diet-induced thermogenesis**). A small part of this is the energy cost of secreting digestive enzymes and of active transport of the products of digestion; the major part is the result of synthesizing reserves of glycogen, triacylglycerol, and protein.

There Are Two Extreme Forms of Undernutrition

Marasmus can occur in both adults and children and occurs in vulnerable groups of all populations. **Kwashiorkor** affects only children and has been reported only in developing countries. The distinguishing feature of kwashiorkor is that there is fluid retention, leading to edema, and fatty infiltration of the liver. Marasmus is a state of extreme emaciation; it is the outcome of prolonged negative energy balance. Not only have the body's fat reserves been exhausted, but there is wastage of muscle as well, and as the condition progresses there is loss of protein from the heart, liver, and kidneys. The amino acids released by the catabolism of tissue proteins are used as a source of metabolic fuel and as substrates for gluconeogenesis to maintain a supply of glucose for the brain and red blood cells (Chapter 20). As a result of the reduced synthesis of proteins, there is impaired immune response and more risk from infections. Impairment of cell proliferation in the intestinal mucosa occurs, resulting in reduction in the surface area of the intestinal mucosa, and reduction in the absorption of such nutrients as are available.

Patients with Advanced Cancer and AIDS Are Malnourished

Patients with advanced cancer, HIV infection and AIDS, and a number of other chronic diseases are frequently undernourished, a condition called **cachexia.** Physically, they show all the signs of marasmus, but there is considerably more loss of body protein than that occurs in starvation. The secretion of cytokines in response to infection and cancer increases the catabolism of tissue protein by the ATP-dependent ubiquitin-proteasome pathway, so increasing energy expenditure. This differs from marasmus, in which protein synthesis is reduced, but catabolism in unaffected. Patients are **hypermetabolic,** ie, they have a considerably increased BMR. In addition to activation of the ubiquitin-proteasome pathway of protein catabolism, three other factors are involved. Many tumors metabolize glucose anaerobically to release lactate. This is then used for gluconeogenesis in the liver, which is energy consuming with a net cost of six ATP for each mol of glucose cycled (see Figure 20–4). There is increased stimulation of mitochondrial **uncoupling proteins** by **cytokines** leading to thermogenesis and increased oxidation of metabolic fuels. **Futile cycling of lipids** occurs because hormone sensitive lipase is activated by a proteoglycan secreted by tumors resulting in liberation of fatty acids from adipose tissue and ATP-expensive reesterification to triacylglycerols in the liver, which are exported in VLDL.

Kwashiorkor Affects Undernourished Children

In addition to the wasting of muscle tissue, loss of intestinal mucosa and impaired immune responses seen in marasmus, children with **kwashiorkor** show a number of characteristic features. The defining feature is **edema,** associated with a decreased concentration of plasma proteins. In addition, there is enlargement of the liver as a result of accumulation of fat. It was formerly believed that the cause of kwashiorkor was a lack of protein, with a more or less adequate energy intake; however, analysis of the diets of affected children shows that this is not so. Protein deficiency leads to stunting of growth, and children with kwashiorkor are less stunted than those with marasmus. Furthermore, the edema begins to improve early in treatment, when the child is still receiving a low protein diet.

Very commonly, an infection precipitates kwashiorkor. Superimposed on general food deficiency, there is probably a deficiency of antioxidant nutrients such as zinc, copper, carotene, and vitamins C and E. The **respiratory burst** in response to infection leads to the production of oxygen and halogen **free radicals** as part of the cytotoxic action of stimulated macrophages. This added oxidant stress triggers the development of kwashiorkor (see Chapter 54).

PROTEIN & AMINO ACID REQUIREMENTS

Protein Requirements Can Be Determined by Measuring Nitrogen Balance

The state of protein nutrition can be determined by measuring the dietary intake and output of nitrogenous compounds from the body. Although nucleic acids also contain nitrogen, protein is the major dietary source of nitrogen and measurement of total nitrogen intake gives a good estimate of protein intake (mg $N \times 6.25$ = mg protein, as N is 16% of most proteins). The output of N from the body is mainly in urea and smaller quantities of other compounds in urine, undigested protein in feces; significant amounts may also be lost in sweat and shed skin. The difference between intake and output of nitrogenous compounds is known as **nitrogen balance.** Three states can be defined. In a healthy adult, nitrogen balance is in **equilibrium,** when intake equals output, and there is no change in the total body content of protein. In a growing child, a pregnant woman, or a person in recovery from protein loss, the excretion of nitrogenous compounds is less than the dietary intake and there is net retention of nitrogen in the body as protein—**positive nitrogen balance.** In response to trauma or infection, or if the intake of protein is inadequate to meet requirements, there is net loss of protein nitrogen from the body—**negative nitrogen balance.** Except when replacing protein losses, nitrogen equilibrium can be maintained at any level of protein intake above requirements. A high intake of protein does not lead to positive nitrogen balance; although it increases the rate of protein synthesis, it also increases the rate of protein catabolism, so that nitrogen equilibrium is maintained, albeit with a higher rate of protein turnover. Both protein synthesis and catabolism are ATP expensive, and this increased rate of protein turnover explains the increased diet-induced thermogenesis seen in people consuming a high protein diet.

The continual catabolism of tissue proteins creates the requirement for dietary protein, even in an adult who is not growing; although some of the amino acids released can be re-utilized, much is used for gluconeogenesis in the fasting state. Nitrogen balance studies show that the average daily requirement is 0.66 g of protein/kg body weight (0.825 allowing for individual variation), ~55 g/day, or 0.825% of energy intake. Average intakes of protein in developed countries are of the order of 80–100 g/day, ie, 14–15% of energy intake. Because growing children are increasing the protein in the body, they have a proportionally greater requirement than adults and should be in positive nitrogen balance. Even so, the need is relatively small compared with the requirement for protein turnover. In some countries, protein intake is inadequate to meet these requirements, resulting in stunting of growth. There is little or no evidence that athletes and body builders require large amounts of protein; simply consuming more of a normal diet providing about 14% of energy from protein will provide more than enough protein for increased muscle protein synthesis—the main requirement is for an increased energy intake to permit increased protein synthesis.

There Is a Loss of Body Protein in Response to Trauma & Infection

One of the metabolic reactions to a major trauma, such as a burn, a broken limb, or surgery, is an increase in the net catabolism of tissue proteins, both in response to cytokines and glucocorticoid hormones, and as a result of excessive utilization of threonine and cysteine in the synthesis of **acute-phase proteins.** As much as 6–7% of the total body protein may be lost over 10 days. Prolonged bed rest results in considerable loss of protein because of atrophy of muscles. Protein catabolism may be increased in response to cytokines, and without the stimulus of exercise it is not completely replaced. Lost protein is replaced during **convalescence,** when there is positive nitrogen balance. Again, as in the case of athletes, a normal diet is adequate to permit this replacement protein synthesis.

The Requirement Is Not Just for Protein, but for Specific Amino Acids

Not all proteins are nutritionally equivalent. More of some is needed to maintain nitrogen balance than others because different proteins contain different amounts of the various amino acids. The body's requirement is for amino acids in the correct proportions to replace tissue proteins. The amino acids can be divided into two groups: **essential** and **nonessential.** There are nine essential or indispensable amino acids, which cannot be synthesized in the body: histidine, isoleucine, leucine, lysine, methionine, phenylalanine, threonine, tryptophan, and valine. If one of these is lacking or inadequate, then regardless of the total intake of protein, it will not be possible to maintain nitrogen balance since there will not be enough of that amino acid for protein synthesis.

Two amino acids, cysteine and tyrosine, can be synthesized in the body, but only from essential amino acid precursors—cysteine from methionine and tyrosine from phenylalanine. The dietary intakes of cysteine and tyrosine thus affect the requirements for methionine and phenylalanine. The remaining 11 amino acids in proteins are considered to be nonessential or dispensable since they can be synthesized as long as there is enough total protein in the diet. If one of these amino acids is omitted from the diet, nitrogen balance can still be maintained. However, only three amino acids, alanine, aspartate, and glutamate, can be considered to be truly dispensable; they are synthesized from common metabolic intermediates (pyruvate, oxaloacetate, and ketoglutarate, respectively). The remaining amino acids are considered as nonessential, but under some circumstances the requirement may outstrip the capacity for their synthesis.

SUMMARY

- Digestion involves hydrolyzing food molecules into smaller molecules for absorption through the gastrointestinal epithelium. Polysaccharides are absorbed as monosaccharides, triacylglycerols as 2-monoacylglycerols, fatty acids and glycerol, and proteins as amino acids and small peptides.

- Digestive disorders arise as a result of (1) enzyme deficiency, eg, lactase and sucrase; (2) malabsorption, eg, of glucose and galactose as a result of defects in the Na^+-glucose cotransporter (SGLT 1); (3) absorption of unhydrolyzed polypeptides leading to immune responses, eg, as in celiac disease; and (4) precipitation of cholesterol from bile as gallstones.

- In addition to water, the diet must provide metabolic fuels (carbohydrate and fat) for body growth and activity, protein for synthesis of tissue proteins, fiber for bulk in the intestinal contents, minerals for specific metabolic functions (Chapter 44), polyunsaturated fatty acids of the n–3 and n–6 families, and vitamins–organic compounds needed in small amounts for other essential functions (Chapter 44).

- Twenty different amino acids are required for protein synthesis, of which nine are essential in the human diet. The quantity of protein required can be determined by studies of nitrogen balance and is affected by protein quality—the amounts of essential amino acids present in dietary proteins compared with the amounts required for tissue porotein synthesis.

- Undernutrition occurs in two extreme forms: marasmus, in adults and children, and kwashiorkor in children. Chronic illness can also lead to undernutrition (cachexia) as a result of hypermetabolism.

- Overnutrition leads to excess energy intake and is associated with chronic noncommunicable diseases such as obesity, type 2 diabetes, atherosclerosis, cancer, and hypertension.

REFERENCES

Bender DA, Bender AE: *Nutrition: A Reference Handbook.* Oxford University Press, 1997.

Fuller MF, Garllick PJ: Human amino acid requirements: can the controversy be resolved? Ann Rev Nutr 1994;14:217.

Geissler C, Powers HJ (editors): *Human Nutrition,* 12th ed. Elsevier, 2010.

Gibney MJ, Lanham-New S, Cassidy A, et al: *Introduction to Human Nutrition, The Nutrition Society Textbook Series,* 2nd ed. Wiley–Blackwell, 2009.

Institute of Medicine: *Dietary Reference Intakes for Energy, Carbohydrate, Fiber, Fat, Fatty Acids, Cholesterol, Protein, and Amino Acids (Macronutrients).* National Academies Press, 2002.

Pencharz PB, Ball RO: Different approaches to define individual amino acid requirements. Ann Rev Nutr 2003;23:101.

Royal College of Physicians: *Nutrition and Patients—A Doctor's Responsibility.* Royal College of Physicians, 2002.

Swallow DM: Genetic influences on carbohydrate digestion. Nutr Res Rev 2003;16:37.

World Health Organization Technical Report Series 894: *Obesity—Preventing and Managing the Global Epidemic.* WHO, 2000.

World Health Organization Technical Report Series 916: *Diet and the Prevention of Chronic Diseases.* WHO, 2003.

World Health Organization Technical report Series 935: *Protein and Amino Acid Requirements in Human Nutrition.* WHO, 2007.

Micronutrients: Vitamins & Minerals

David A. Bender, PhD

BIOMEDICAL IMPORTANCE

Vitamins are a group of organic nutrients, required in small quantities for a variety of biochemical functions that, generally, cannot be synthesized by the body and must therefore be supplied in the diet.

The lipid-soluble vitamins are hydrophobic compounds that can be absorbed efficiently only when there is normal fat absorption. Like other lipids, they are transported in the blood in lipoproteins or attached to specific binding proteins. They have diverse functions—eg, vitamin A, vision and cell differentiation; vitamin D, calcium and phosphate metabolism, and cell differentiation; vitamin E, antioxidant; and vitamin K, blood clotting. As well as dietary inadequacy, conditions affecting the digestion and absorption of the lipid-soluble vitamins, such as a very low fat diet, steatorrhea and disorders of the biliary system, can all lead to deficiency syndromes, including night blindness and xerophthalmia (vitamin A); rickets in young children and osteomalacia in adults (vitamin D); neurologic disorders and hemolytic anemia of the newborn (vitamin E); and hemorrhagic disease of the newborn (vitamin K). Toxicity can result from excessive intake of vitamins A and D. Vitamin A and the carotenes (many of which are precursors of vitamin A), and vitamin E are antioxidants (Chapter 45) and have possible roles in prevention of atherosclerosis and cancer.

The water-soluble vitamins are vitamins B and C; they function mainly as enzyme cofactors. Folic acid acts as a carrier of one-carbon units. Deficiency of a single vitamin of the B complex is rare since poor diets are most often associated with **multiple deficiency states.** Nevertheless, specific syndromes are characteristic of deficiencies of individual vitamins, eg, beriberi (thiamin); cheilosis, glossitis, seborrhea (riboflavin); pellagra (niacin); megaloblastic anemia, methylmalonic aciduria, and pernicious anemia (vitamin B_{12}); megaloblastic anemia (folic acid); and scurvy (vitamin C).

Inorganic mineral elements that have a function in the body must be provided in the diet. When the intake is insufficient, deficiency signs may arise, eg, anemia (iron), and cretinism and goiter (iodine). Excessive intakes may be toxic.

The Determination of Micronutrient Requirements Depends on the Criteria of Adequacy Chosen

For any nutrient, there is a range of intakes between that which is clearly inadequate, leading to **clinical deficiency disease,** and that which is so much in excess of the body's metabolic capacity that there may be signs of **toxicity.** Between these two extremes is a level of intake that is adequate for normal health and the maintenance of metabolic integrity. Individuals do not all have the same requirement for nutrients, even when calculated on the basis of body size or energy expenditure. There is a range of individual requirements of up to 25% around the mean. Therefore, in order to assess the adequacy of diets, it is necessary to set a reference level of intake high enough to ensure that no one either suffers from deficiency or is at risk of toxicity. If it is assumed that individual requirements are distributed in a statistically normal fashion around the observed mean requirement, then a range of $\pm 2 \times$ the standard deviation (SD) around the mean includes the requirements of 95% of the population. Reference or recommended intakes are therefore set at the average requirement plus $2 \times SD$, and so meet or exceed the requirements of 97.5% of the population.

TABLE 44–1   Reference Nutrient Intakes of Vitamins and Minerals, UK 1991

Age	Vit B₁ (mg)	Vit B₂ (mg)	Niacin (mg)	Vit B₆ (mg)	Vit B₁₂ (μg)	Folate (μg)	Vit C (mg)	Vit A (μg)	Vit D (μg)	Ca (mg)	P (mg)	Mg (mg)	Fe (mg)	Zn (mg)	Cu (mg)	Se (μg)	I (μg)
0–3 m	0.2	0.4	3	0.2	0.3	50	25	350	8.5	525	400	55	1.7	4.0	0.2	10	50
4–6 m	0.2	0.4	3	0.2	0.3	50	25	350	8.5	525	400	60	4.3	4.0	0.3	13	60
7–9 m	0.2	0.4	4	0.3	0.4	50	25	350	7	525	400	75	7.8	5.0	0.3	10	60
10–12 m	0.3	0.4	5	0.4	0.4	50	25	350	7	525	400	80	7.8	5.0	0.3	10	60
1–3 y	0.5	0.6	8	0.7	0.5	70	30	400	7	350	270	85	6.9	5.0	0.4	15	70
4–6 y	0.7	0.8	11	0.9	0.8	100	30	500	–	450	350	120	6.1	6.5	0.6	20	100
7–10 y	0.7	1.0	12	1.0	1.0	150	30	500	–	550	450	200	8.7	7.0	0.7	30	110
Males																	
11–14 y	0.9	1.2	15	1.2	1.2	200	35	600	–	1000	775	280	11.3	9.0	0.8	45	130
15–18 y	1.1	1.3	18	1.5	1.5	200	40	700	–	1000	775	300	11.3	9.5	1.0	70	140
19–50 y	1.0	1.3	17	1.4	1.5	200	40	700	–	700	550	300	8.7	9.5	1.2	75	140
50+ y	0.9	1.3	16	1.4	1.5	200	40	700	10	700	550	300	8.7	9.5	1.2	75	140
Females																	
11–14 y	0.7	1.1	12	1.0	1.2	200	35	600	–	800	625	280	14.8	9.0	0.8	45	130
15–18 y	0.8	1.1	14	1.2	1.5	200	40	600	–	800	6254	300	14.8	7.0	1.0	60	140
19–50 y	0.8	1.1	13	1.2	1.5	200	40	600	–	700	550	270	14.8	7.0	1.2	60	140
50+ y	0.8	1.1	12	1.2	1.5	200	40	600	10	700	550	270	8.7	7.0	1.2	60	140
Pregnant	+0.1	+0.3	–	–	–	+100	+10	+100	10	–	–	–					
Lactating	+0.1	+0.5	+2	–	+0.5	+60	+30	+350	10	+550	+440	+ 50		+6.0	+0.3	+15	

Source: Department of Health. *Dietary Reference Values for Food Energy and Nutrients for the United Kingdom.* HMSO, London, 1991.

Reference and recommended intakes of vitamins and minerals published by different national and international authorities (**Tables 44–1** to **44–4**) differ because of different interpretations of the available data, and the availability of new experimental data in more recent publications.

THE VITAMINS ARE A DISPARATE GROUP OF COMPOUNDS WITH A VARIETY OF METABOLIC FUNCTIONS

A vitamin is defined as an organic compound that is required in the diet in small amounts for the maintenance of normal metabolic integrity. Deficiency causes a specific disease, which is cured or prevented only by restoring the vitamin to the diet (**Table 44–5**). However, **vitamin D,** which is formed in the skin from 7-dehydrocholesterol on exposure to sunlight, and **niacin,** which can be formed from the essential amino acid tryptophan, do not strictly comply with this definition.

LIPID-SOLUBLE VITAMINS

TWO GROUPS OF COMPOUNDS HAVE VITAMIN A ACTIVITY

Retinoids comprise **retinol, retinaldehyde,** and **retinoic acid** (preformed vitamin A, found only in foods of animal origin); carotenoids, found in plants, are composed of carotenes and related compounds; many are precursors of vitamin A, as they can be cleaved to yield retinaldehyde, then retinol and retinoic acid (**Figure 44–1**). The α-, β-, and γ-carotenes and cryptoxanthin are quantitatively the most important provitamin A carotenoids. β-Carotene and other provitamin A carotenoids are cleaved in the intestinal mucosa by carotene dioxygenase, yielding retinaldehyde, which is reduced to retinol, esterified and secreted in chylomicrons together with esters formed from dietary retinol. The intestinal activity of carotene dioxygenase is low, so that a relatively large proportion of ingested β-carotene may appear in the circulation unchanged. While the principal site of carotene dioxygenase attack is the central bond of β-carotene, asymmetric cleavage may also occur, leading to the formation of 8'-, 10'-, and 12'-apo-carotenals, which

TABLE 44–2 Population Reference Intakes of Vitamins and Minerals, European Union, 1993

Age	Vit A (µg)	Vit B₁ (mg)	Vit B₂ (mg)	Niacin (mg)	Vit B₆ (mg)	Folate (µg)	Vit B₁₂ (µg)	Vit C (mg)	Ca (mg)	P (mg)	Fe (mg)	Zn (mg)	Cu (mg)	Se (µg)	I (µg)
6–12 m	350	0.3	0.4	5	0.4	50	0.5	20	400	300	6	4	0.3	8	50
1–3 y	400	0.5	0.8	9	0.7	100	0.7	25	400	300	4	4	0.4	10	70
4–6 y	400	0.7	1.0	11	0.9	130	0.9	25	450	350	4	6	0.6	15	90
7–10 y	500	0.8	1.2	13	1.1	150	1.0	30	550	450	6	7	0.7	25	100
Males															
11–14 y	600	1.0	1.4	15	1.3	180	1.3	35	1000	775	10	9	0.8	35	120
15–17 y	700	1.2	1.6	18	1.5	200	1.4	40	1000	775	13	9	1.0	45	130
18 + y	700	1.1	1.6	18	1.5	200	1.4	45	700	550	9	9.5	1.1	55	130
Females															
11–14 y	600	0.9	1.2	14	1.1	180	1.3	35	800	625	18	9	0.8	35	120
15–17 y	600	0.9	1.3	14	1.1	200	1.4	40	800	625	17	7	1.0	45	130
18 + y	600	0.9	1.3	14	1.1	200	1.4	45	700	550	16[1]	7	1.1	55	130
Pregnant	700	1.0	1.6	14	1.3	400	1.6	55	700	550	[1]	7	1.1	55	130
Lactating	950	1.1	1.7	16	1.4	350	1.9	70	1200	950	16	12	1.4	70	160

Source: Scientific Committee for Food *Nutrient and energy intakes for the European Community,* Commission of the European Communities, Luxembourg, 1993.

are oxidized to retinoic acid, but cannot be used as sources of retinol or retinaldehyde.

Although it would appear that one molecule of β-carotene should yield two of retinol, this is not so in practice; 6 µg of β-carotene is equivalent to 1 µg of preformed retinol. The total amount of vitamin A in foods is therefore expressed as micrograms of retinol equivalents = µg preformed vitamin A + 1/6 × µg β-carotene + 1/12 × µg other provitamin A carotenoids. Before pure vitamin A was available for chemical analysis, the vitamin A content of foods was determined by biological assay and the results expressed as international units (iu). 1 iu = 0.3 µg retinol; 1 µg retinol = 3.33 iu. Although obsolete, iu is sometimes still used in food labeling. In 2001 The USA/Canadian Dietary Reference Values report introduced the term *retinol activity equivalent* to take account of the incomplete absorption and metabolism of carotenoids; 1 RAE = 1 µg all-*trans*-retinol, 12 µg β-carotene, 24 µg α-carotene or β-cryptoxanthin. On this basis, 1 iu of vitamin A activity is equal to 3.6 µg β-carotene or 7.2 µg of other provitamin A carotenoids.

Vitamin A Has a Function in Vision

In the retina, retinaldehyde functions as the prosthetic group of the light-sensitive opsin proteins, forming **rhodopsin** (in rods) and **iodopsin** (in cones). Any one cone cell contains only one type of opsin and is sensitive to only one color. In the pigment epithelium of the retina, all-*trans*-retinol is isomerized to 11-*cis*-retinol and oxidized to 11-*cis*-retinaldehyde. This reacts with a lysine residue in opsin, forming the holoprotein rhodopsin. As shown in **Figure 44–2**, the absorption of light by rhodopsin causes isomerization of the retinaldehyde from 11-*cis* to all-*trans*, and a conformational change in opsin. This results in the release of retinaldehyde from the protein, and the initiation of a nerve impulse. The formation of the initial excited form of rhodopsin, bathorhodopsin, occurs within picoseconds of illumination. There is then a series of conformational changes leading to the formation of metarhodopsin II, which initiates a guanine nucleotide amplification cascade and then a nerve impulse. The final step is hydrolysis to release all-*trans*-retinaldehyde and opsin. The key to initiation of the visual cycle is the availability of 11-*cis*-retinaldehyde, and hence vitamin A. In deficiency, both the time taken to adapt to darkness and the ability to see in poor light are impaired.

Retinoic Acid Has a Role in the Regulation of Gene Expression and Tissue Differentiation

A major role of vitamin A is in the control of cell differentiation and turnover. All-*trans*-retinoic acid and 9-*cis*-retinoic acid (Figure 44–1) regulate growth, development, and tissue differentiation; they have different actions in different tissues. Like the thyroid and steroid hormones and vitamin D, retinoic acid binds to nuclear receptors that bind to response elements of DNA and regulate the transcription of specific genes. There

TABLE 44–3 Recommended Dietary Allowances and Acceptable Intakes for Vitamins and Minerals, USA and Canada, 1997–2001

Age	Vit A (μg)	Vit D (μg)	Vit E (mg)	Vit K (μg)	Vit B$_1$ (mg)	Vit B$_2$ (mg)	Niacin (mg)	Vit B$_6$ (mg)	Folate (μg)	Vit B$_{12}$ (μg)	Vit C (mg)	Ca (mg)	P (mg)	Fe (mg)	Zn (mg)	Cu (mg)	Se (μg)	I (μg)
0–6 m	400	5	4	2.0	0.2	0.3	2	0.1	65	0.4	40	210	100	–	2.0	200	15	110
7–12 m	500	5	5	2.5	0.3	0.4	4	0.3	80	0.5	50	270	275	11	3	220	20	130
1–3 y	300	5	6	30	0.5	0.5	6	0.5	150	0.9	15	500	460	7	3	340	20	90
4–8 y	400	5	7	55	0.5	0.6	8	0.6	200	1.2	25	800	500	10	5	440	30	90
Males																		
9–13 y	600	5	11	60	0.9	0.9	12	1.0	300	1.8	45	1300	1250	8	8	700	40	120
14–18 y	900	5	15	75	1.2	1.3	16	1.3	400	2.4	75	1300	1250	11	11	890	55	150
19–30 y	900	5	15	120	1.2	1.3	16	1.3	400	2.4	90	1000	700	8	11	900	55	150
31–50 y	900	5	15	120	1.2	1.3	16	1.3	400	2.4	90	1000	700	8	11	900	55	150
51–70 y	900	10	15	120	1.2	1.3	16	1.7	400	2.4	90	1200	700	8	11	900	55	150
> 70 y	900	15	15	120	1.2	1.3	16	1.7	400	2.4	90	1200	700	8	11	900	55	150
Females																		
9–13 y	600	5	11	60	0.9	0.9	12	1.0	300	1.8	45	1300	1250	8	8	700	40	120
14–18 y	700	5	15	75	1.0	1.0	14	1.2	400	2.4	65	1300	1250	15	9	890	55	150
19–30 y	700	5	15	90	1.1	1.1	14	1.3	400	2.4	75	1000	700	18	8	900	55	150
31–50 y	700	5	15	90	1.1	1.1	14	1.3	400	2.4	75	1000	700	18	8	900	55	150
51–70 y	700	10	15	90	1.1	1.1	14	1.5	400	2.4	75	1200	700	8	8	900	55	150
> 70 y	700	15	15	90	1.1	1.1	14	1.5	400	2.4	75	1200	700	8	8	900	55	150
Pregnant	770	5	15	90	1.4	1.4	18	1.9	600	2.6	85	1000	700	27	11	1000	60	220
Lactating	900	5	16	90	1.4	1.6	17	2.0	500	2.8	120	1000	700	9	12	1300	70	290

(Figures for infants under 12 months are adequate intakes, based on the observed mean intake of infants fed principally on breast milk; for nutrients other than vitamin K figures are RDA, based on estimated average requirement + 2 sd; figures for vitamin K are adequate intakes, based on observed average intakes.)
Source: Standing Committee on the Scientific Evaluation of Dietary Reference Intakes, Food and Nutrition Board, Institute of Medicine Dietary Reference Intakes for calcium, phosphorus, magnesium, vitamin D and fluoride, 1997; dietary reference intakes for thiamin, riboflavin, niacin, vitamin B$_6$, folate, vitamin B$_{12}$, pantothenic acid, biotin and choline, 1998; dietary reference intakes for vitamin C, vitamin E, selenium and carotenoids, 2000; dietary reference intakes for vitamin A, vitamin K, arsenic, boron, chromium, copper, iodine, iron, manganese, molybdenum, nickel, silicon, vanadium and zinc, 2001, National Academy Press, Washington DC.

are two families of nuclear retinoid receptors: the retinoic acid receptors (RAR) bind all-*trans*-retinoic acid or 9-*cis*-retinoic acid, and the retinoid X receptors (RXR) bind 9-*cis*-retinoic acid. Retinoid X receptors also form dimers with vitamin D, thyroid, and other a nuclear acting hormone receptors. Deficiency of vitamin A impairs vitamin D function because of lack of 9-*cis*-retinoic acid to form receptor dimers, while excessive vitamin A also impairs vitamin D function, because of formation of RXR homodimers, meaning that there are not enough RXR available to form heterodimers with the vitamin D receptor.

Vitamin A Deficiency Is a Major Public Health Problem Worldwide

Vitamin A deficiency is the most important preventable cause of blindness. The earliest sign of deficiency is a loss of sensitivity to green light, followed by impairment to adapt to dim light, followed by night blindness. More prolonged deficiency leads to **xerophthalmia:** keratinization of the cornea and blindness. Vitamin A also has an important role in differentiation of immune system cells, and even mild deficiency leads to increased susceptibility to infectious diseases. Also, the synthesis of retinol binding protein is reduced in response to infection (it is a negative **acute phase protein**), decreasing the circulating concentration of the vitamin, and further impairing immune responses.

Vitamin A Is Toxic in Excess

There is only a limited capacity to metabolize vitamin A, and excessive intakes lead to accumulation beyond the capacity of binding proteins, so that unbound vitamin A causes tissue damage. Symptoms of toxicity affect the central nervous system (headache, nausea, ataxia, and anorexia, all associated with increased cerebrospinal fluid pressure); the liver

TABLE 44–4 Recommended Nutrient Intakes for Vitamins, FAO 2001

Age	Vit A (µg)	Vit D (µg)	Vit K (µg)	Vit B$_1$ (mg)	Vit B$_2$ (mg)	Niacin (mg)	Vit B$_6$ (mg)	Folate (µg)	Vit B$_{12}$ (µg)	Vit C (mg)	Panto (mg)	Biotin (µg)
0–6 m	375	5	5	0.2	0.3	2	0.1	80	0.4	25	1.7	5
7–12 m	400	5	10	0.3	0.4	4	0.3	80	0.5	30	1.8	6
1–3 y	400	5	15	0.5	0.5	6	0.5	160	0.9	30	2.0	8
4–6 y	450	5	20	0.6	0.6	8	0.6	200	1.2	30	3.0	12
7–9 y	500	5	25	0.9	0.9	12	1.0	300	1.8	35	4.0	20
Males												
10–18 y	600	5	35–55	1.2	1.3	16	1.3	400	2.4	40	5.0	30
19–50 y	600	5	65	1.2	1.3	16	1.3	400	2.4	45	5.0	30
50–65 y	600	10	65	1.2	1.3	16	1.7	400	2.4	45	5.0	30
> 65 y	600	15	65	1.2	1.3	16	1.7	400	2.4	45	5.0	30
Female												
10–18 y	600	5	35–55	1.1	1.0	16	1.2	400	2.4	40	5.0	25
19–50 y	600	5	55	1.1	1.1	14	1.3	400	2.4	45	5.0	30
50–65 y	600	10	55	1.1	1.1	14	1.5	400	2.4	45	5.0	30
> 65 y	600	15	55	1.1	1.1	14	1.5	400	2.4	45	5.0	30
Pregnant	800	5	55	1.4	1.4	18	1.9	600	2.6	55	6.0	30
Lactating	850	5	55	1.5	1.6	17	2.0	500	2.8	70	7.0	35

Source: Food and Agriculture Organization of the United Nations and World Health Organization, *Human Vitamin and Mineral Requirements*, FAO, 2001.

(hepatomegaly with histologic changes and hyperlipidemia); calcium homeostasis (thickening of the long bones, hypercalcemia, and calcification of soft tissues); and the skin (excessive dryness, desquamation, and alopecia).

VITAMIN D IS REALLY A HORMONE

Vitamin D is not strictly a vitamin since it can be synthesized in the skin, and under most conditions that is the major source of the vitamin. Only when sunlight exposure is inadequate is a dietary source required. Its main function is in the regulation of calcium absorption and homeostasis; most of its actions are mediated by way of nuclear receptors that regulate gene expression. It also has a role in regulating cell proliferation and differentiation. There is evidence that intakes considerably higher than are required to maintain calcium homeostasis reduce the risk of insulin resistance, obesity and the metabolic syndrome, as well as various cancers. Deficiency, leading to rickets in children and osteomalacia in adults, continues to be a problem in northern latitudes, where sunlight exposure is inadequate.

Vitamin D Is Synthesized in the Skin

7-Dehydrocholesterol (an intermediate in the synthesis of cholesterol that accumulates in the skin) undergoes a nonenzymic reaction on exposure to ultraviolet light, yielding previtamin D (**Figure 44–3**). This undergoes a further reaction over a period of hours to form cholecalciferol, which is absorbed into the bloodstream. In temperate climates, the plasma concentration of vitamin D is highest at the end of summer and lowest at the end of winter. Beyond latitudes about 40° north or south, there is very little ultraviolet radiation of the appropriate wavelength in winter.

Vitamin D Is Metabolized to the Active Metabolite, Calcitriol, in Liver & Kidney

Cholecalciferol, either synthesized in the skin or from food, undergoes two hydroxylations to yield the active metabolite, 1,25-dihydroxyvitamin D or calcitriol (**Figure 44–4**). Ergocalciferol from fortified foods undergoes similar hydroxylation to yield ercalcitriol. In the liver, cholecalciferol is hydroxylated to form the 25-hydroxy-derivative, calcidiol. This is released into the circulation bound to a vitamin D binding globulin, which is the main storage form of the vitamin. In the kidney, calcidiol undergoes either 1-hydroxylation to yield the active metabolite 1,25-dihydroxyvitamin D (calcitriol), or 24-hydroxylation to yield a probably inactive metabolite, 24,25-dihydroxyvitamin D (24-hydroxycalcidiol).

TABLE 44–5 The Vitamins

Vitamin		Functions	Deficiency Disease
Lipid-soluble			
A	Retinol, β-carotene	Visual pigments in the retina; regulation of gene expression and cell differentiation (β-carotene is an antioxidant)	Night blindness, xerophthalmia; keratinization of skin
D	Calciferol	Maintenance of calcium balance; enhances intestinal absorption of Ca^{2+} and mobilizes bone mineral; regulation of gene expression and cell differentiation	Rickets = poor mineralization of bone; osteomalacia = bone demineralization
E	Tocopherols, tocotrienols	Antioxidant, especially in cell membranes; roles in cell signaling	Extremely rare—serious neurologic dysfunction
K	Phylloquinone: menaquinones	Coenzyme in formation of γ-carboxyglutamate in enzymes of blood clotting and bone matrix	Impaired blood clotting, hemorrhagic disease
Water-soluble			
B₁	Thiamin	Coenzyme in pyruvate and α-ketoglutarate dehydrogenases, and transketolase; regulates Cl^- channel in nerve conduction	Peripheral nerve damage (beriberi) or central nervous system lesions (Wernicke-Korsakoff syndrome)
B₂	Riboflavin	Coenzyme in oxidation and reduction reactions (FAD and FMN); prosthetic group of flavoproteins	Lesions of corner of mouth, lips, and tongue, seborrheic dermatitis
Niacin	Nicotinic acid, nicotinamide	Coenzyme in oxidation and reduction reactions, functional part of NAD and NADP; role in intracellular calcium regulation and cell signaling	Pellagra—photosensitive dermatitis, depressive psychosis
B₆	Pyridoxine, pyridoxal, pyridoxamine	Coenzyme in transamination and decarboxylation of amino acids and glycogen phosphorylase; modulation of steroid hormone action	Disorders of amino acid metabolism, convulsions
	Folic acid	Coenzyme in transfer of one-carbon fragments	Megaloblastic anemia
B₁₂	Cobalamin	Coenzyme in transfer of one-carbon fragments and metabolism of folic acid	Pernicious anemia = megaloblastic anemia with degeneration of the spinal cord
	Pantothenic acid	Functional part of CoA and acyl carrier protein: fatty acid synthesis and metabolism	Peripheral nerve damage (nutritional melalgia or "burning foot syndrome")
H	Biotin	Coenzyme in carboxylation reactions in gluconeogenesis and fatty acid synthesis; role in regulation of cell cycle	Impaired fat and carbohydrate metabolism, dermatitis
C	Ascorbic acid	Coenzyme in hydroxylation of proline and lysine in collagen synthesis; antioxidant; enhances absorption of iron	Scurvy—impaired wound healing, loss of dental cement, subcutaneous hemorrhage

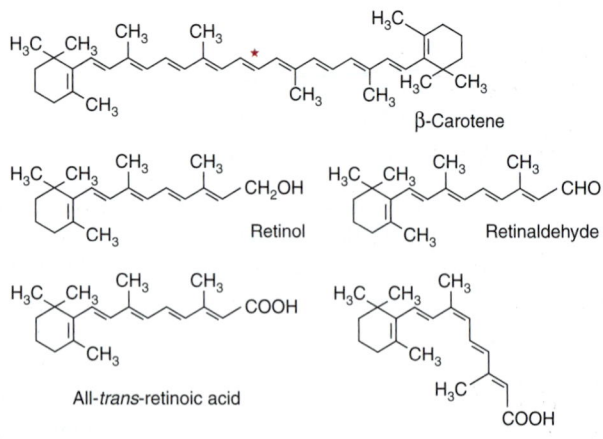

FIGURE 44–1 β-Carotene and the major vitamin A vitamers.
Asterisk shows the site of cleavage of β-carotene by carotene dioxygenase, to yield retinaldehyde.

Vitamin D Metabolism Is Both Regulated by and Regulates Calcium Homeostasis

The main function of vitamin D is in the control of calcium homeostasis, and in turn, vitamin D metabolism is regulated by factors that respond to plasma concentrations of calcium and phosphate. Calcitriol acts to reduce its own synthesis by inducing the 24-hydroxylase and repressing the 1-hydroxylase in the kidney. The principal function of vitamin D is to maintain the plasma calcium concentration. Calcitriol achieves this in three ways: it increases intestinal absorption of calcium; it reduces excretion of calcium (by stimulating resorption in the distal renal tubules); and it mobilizes bone mineral. In addition, calcitriol is involved in insulin secretion, synthesis and secretion of parathyroid and thyroid hormones, inhibition of production of interleukin by activated T-lymphocytes and of immunoglobulin by activated B-lymphocytes, differentiation

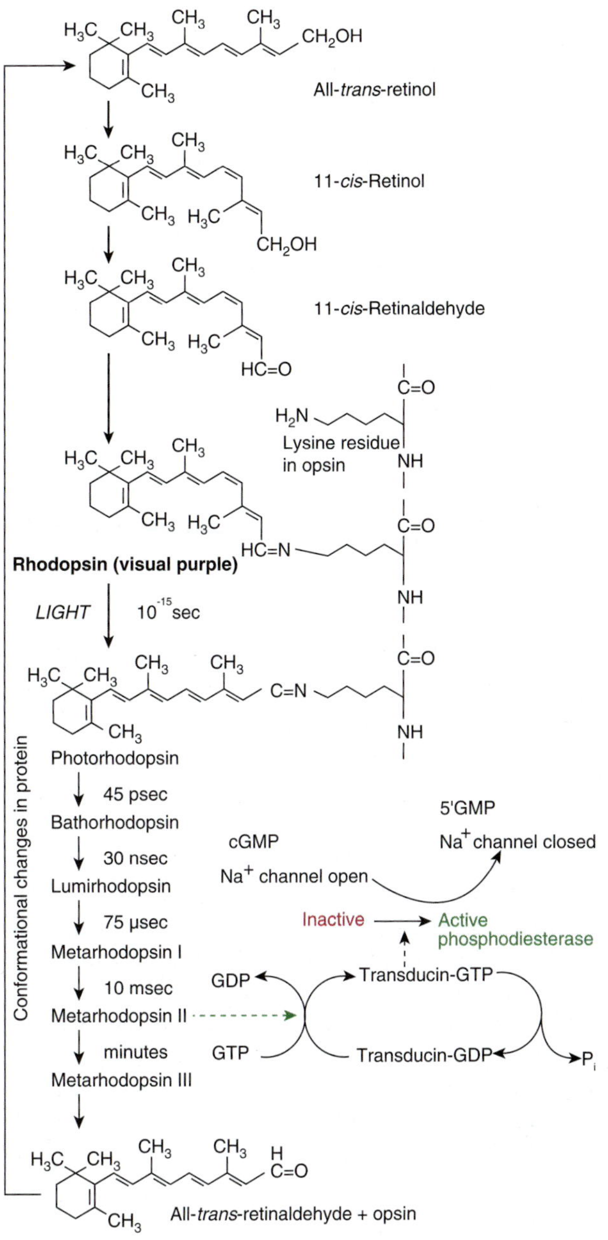

FIGURE 44–2 The role of retinaldehyde in the visual cycle.

of monocyte precursor cells, and modulation of cell proliferation. In most of these actions, it acts like a steroid hormone, binding to nuclear receptors and enhancing gene expression, although it also has rapid effects on calcium transporters in the intestinal mucosa. For further details of the role of calcitriol in calcium homeostasis, see Chapter 47.

Higher Intakes of Vitamin D may be Beneficial

There is growing evidence that higher vitamin D status is protective against various cancers, including prostate and colorectal cancer, and also against prediabetes and the metabolic syndrome. Desirable levels of intake may be considerably higher than current reference intakes, and could certainly not be met from unfortified foods. While increased sunlight exposure would meet the need, it carries the risk of developing skin cancer.

Vitamin D Deficiency Affects Children & Adults

In the vitamin D deficiency disease **rickets,** the bones of children are undermineralized as a result of poor absorption of calcium. Similar problems occur as a result of deficiency during the adolescent growth spurt. **Osteomalacia** in adults results from the demineralization of bone, especially in women who have little exposure to sunlight, especially after several pregnancies. Although vitamin D is essential for prevention and treatment of osteomalacia in the elderly, there is little evidence that it is beneficial in treating **osteoporosis.**

Vitamin D Is Toxic in Excess

Some infants are sensitive to intakes of vitamin D as low as 50 µg/day, resulting in an elevated plasma concentration of calcium. This can lead to contraction of blood vessels, high blood pressure, and **calcinosis**—the calcification of soft tissues. Although excess dietary vitamin D is toxic, excessive exposure to sunlight does not lead to vitamin D poisoning, because there is a limited capacity to form the precursor, 7-dehydrocholesterol, and prolonged exposure of previtamin D to sunlight leads to formation of inactive compounds.

VITAMIN E DOES NOT HAVE A PRECISELY DEFINED METABOLIC FUNCTION

No unequivocal unique function for vitamin E has been defined. It acts as a lipid-soluble **antioxidant** in cell membranes, where many of its functions can be provided by synthetic

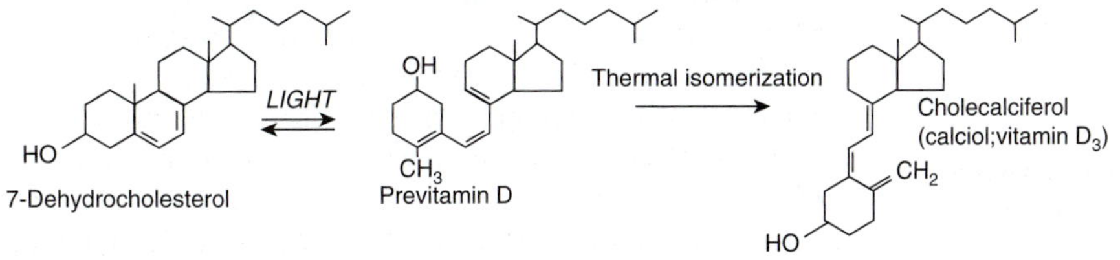

FIGURE 44–3 The synthesis of vitamin D in the skin.

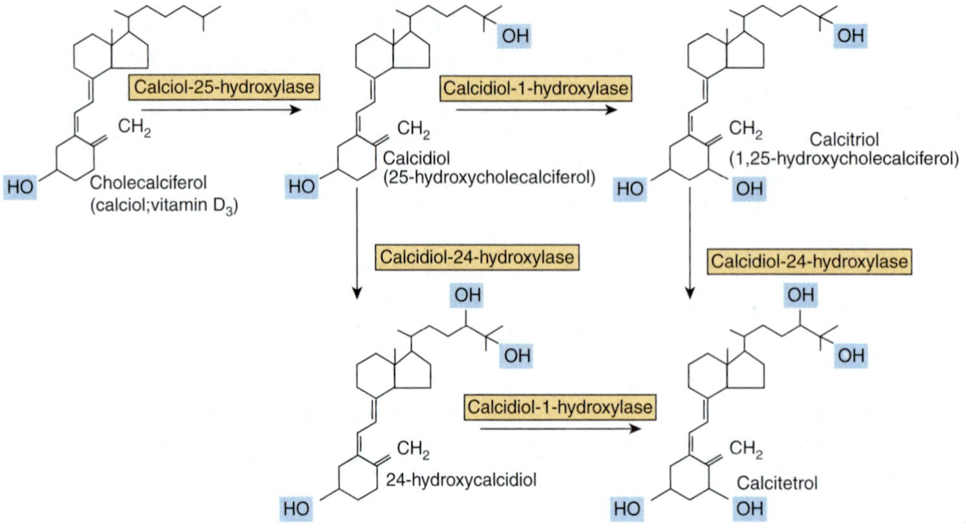

FIGURE 44–4 Metabolism of vitamin D.

antioxidants, and is important in maintaining the fluidity of cell membranes. It also has a (relatively poorly defined) role in cell signaling. Vitamin E is the generic descriptor for two families of compounds, the **tocopherols** and the **tocotrienols** (**Figure 44–5**). The different vitamers have different biologic potency; the most active is D-α-tocopherol, and it is usual to express vitamin E intake in terms of milligrams D-α-tocopherol equivalents. Synthetic DL-α-tocopherol does not have the same biologic potency as the naturally occurring compound.

Vitamin E Is the Major Lipid-Soluble Antioxidant in Cell Membranes and Plasma Lipoproteins

The main function of vitamin E is as a chain-breaking, free-radical-trapping antioxidant in cell membranes and plasma lipoproteins by reacting with the lipid peroxide radicals formed by peroxidation of polyunsaturated fatty acids (Chapter 45). The tocopheroxyl radical product is relatively unreactive, and ultimately forms nonradical compounds. Commonly, the tocopheroxyl radical is reduced back to tocopherol by reaction

with vitamin C from plasma (**Figure 44–6**). The resultant monodehydroascorbate radical then undergoes enzymic or nonenzymic reaction to yield ascorbate and dehydroascorbate, neither of which is a radical.

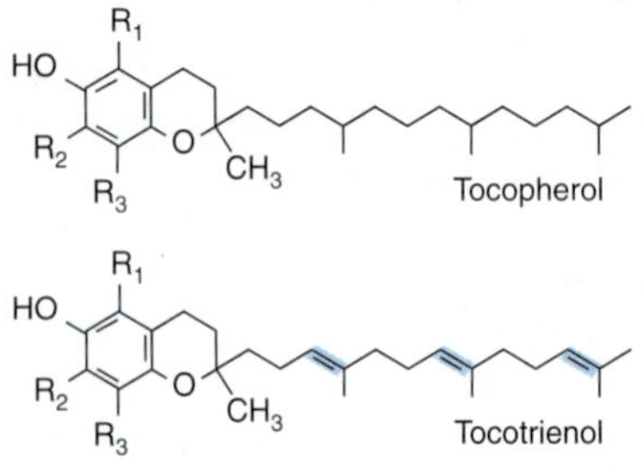

Vitamin E Deficiency

In experimental animals, vitamin E deficiency results in resorption of fetuses and testicular atrophy. Dietary deficiency of vitamin E in humans is unknown, although patients with severe fat malabsorption, cystic fibrosis, and some forms of chronic liver disease suffer deficiency because they are unable to absorb the vitamin or transport it, exhibiting nerve and muscle membrane damage. Premature infants are born with inadequate reserves of the vitamin. The erythrocyte membranes are abnormally fragile as a result of lipid peroxidation, leading to hemolytic anemia.

VITAMIN K IS REQUIRED FOR SYNTHESIS OF BLOOD-CLOTTING PROTEINS

Vitamin K was discovered as a result of investigations into the cause of a bleeding disorder, hemorrhagic (sweet clover) disease of cattle and of chickens fed on a fat-free diet. The missing factor in the diet of the chickens was vitamin K, while the cattle feed contained **dicumarol,** an antagonist of the vitamin. Antagonists of vitamin K are used to reduce blood coagulation in patients at risk of thrombosis; the most widely used is **warfarin.**

Three compounds have the biological activity of vitamin K (**Figure 44–7**): **phylloquinone,** the normal dietary source, found in green vegetables; **menaquinones,** synthesized by intestinal bacteria, with differing lengths of side chain; and **menadione** and menadiol diacetate, synthetic compounds

FIGURE 44–5 **Vitamin E vitamers.** In α-tocopherol and tocotrienol R_1, R_2, and R_3 are all —CH_3 groups. In the β-vitamers R_2 is H, in the γ-vitamers R_1 is H, and in the δ-vitamers R_1 and R_2 are both H.

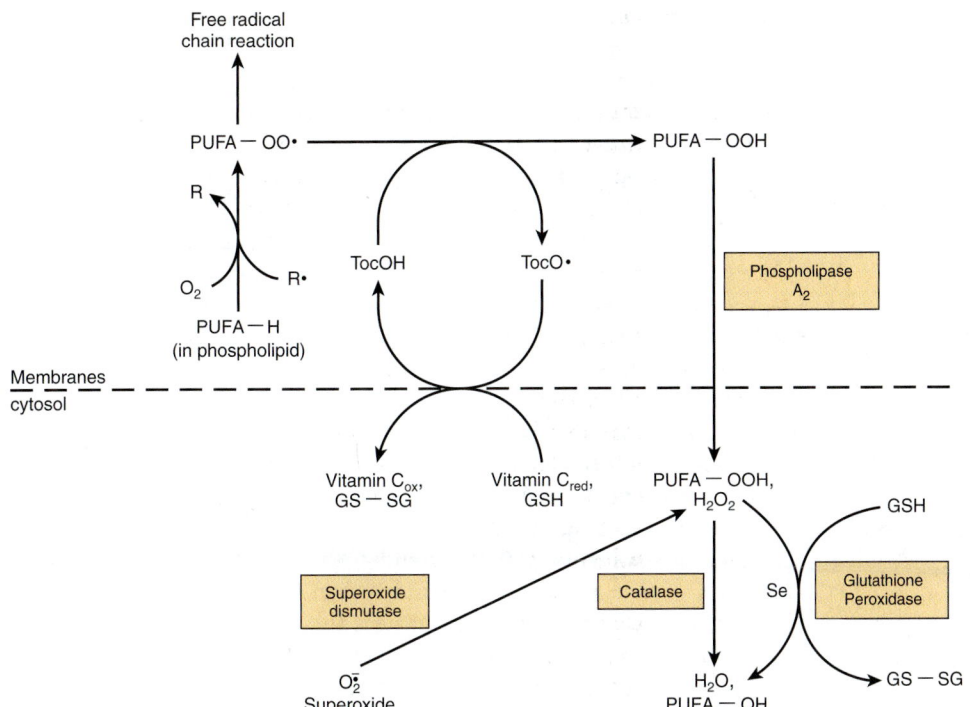

FIGURE 44–6 Interaction between antioxidants in the lipid phase (cell membranes) and the aqueous phase (cytosol). (R•, free radical; PUFA-OO•, peroxyl radical of polyunsaturated fatty acid in membrane phospholipid; PUFA-OOH, hydroxyperoxy polyunsaturated fatty acid in membrane phospholipid, released into the cytosol as hydroxyperoxy polyunsaturated fatty acid by the action of phospholipase A$_2$; PUFA-OH, hydroxy polyunsaturated fatty acid; Toc-OH vitamin E [α-tocopherol]; TocO•, tocopheroxyl radical; Se, selenium; GSH, reduced glutathione; GS-SG, oxidized glutathione, which is reduced to GSH after reaction with NADPH, catalyzed by glutathione reductase; PUFA-H, polyunsaturated fatty acid.)

that can be metabolized to phylloquinone. Menaquinones are absorbed to some extent, but it is not clear to what extent they are biologically active as it is possible to induce signs of vitamin K deficiency simply by feeding a phylloquinone-deficient diet, without inhibiting intestinal bacterial action.

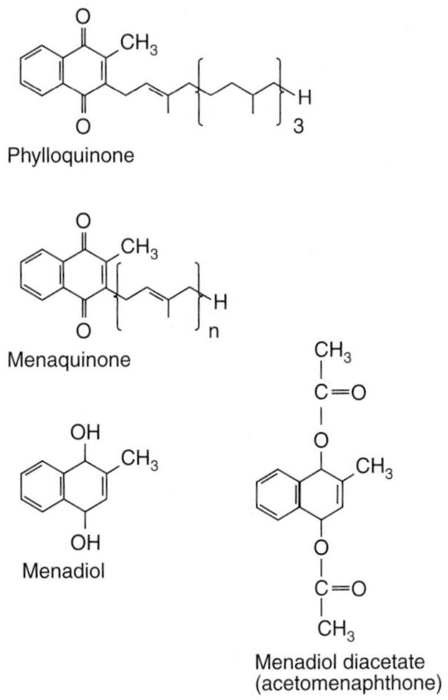

FIGURE 44–7 The vitamin K vitamers. Menadiol (or menadione) and menadiol diacetate are synthetic compounds that are converted to menaquinone in the liver.

Vitamin K Is the Coenzyme for Carboxylation of Glutamate in Postsynthetic Modification of Calcium-Binding Proteins

Vitamin K is the cofactor for the carboxylation of glutamate residues in the postsynthetic modification of proteins to form the unusual amino acid γ-carboxyglutamate (Gla) (**Figure 44–8**). Initially, vitamin K hydroquinone is oxidized to the epoxide, which activates a glutamate residue in the protein substrate to a carbanion, which reacts nonenzymically with carbon dioxide to form γ-carboxyglutamate. Vitamin K epoxide is reduced to the quinone by a warfarin-sensitive reductase, and the quinone is reduced to the active hydroquinone by either the same warfarin-sensitive reductase or a warfarin-insensitive quinone reductase. In the presence of warfarin, vitamin K epoxide cannot be reduced, but accumulates and is excreted. If enough vitamin K (as the quinone) is provided in the diet, it can be reduced to the active hydroquinone by the warfarin-insensitive enzyme, and carboxylation can continue, with stoichiometric utilization of vitamin K and excretion of the epoxide. A high dose of vitamin K is the antidote to an overdose of warfarin.

Prothrombin and several other proteins of the blood-clotting system (Factors VII, IX, and X, and proteins C and S, Chapter 50) each contain 4–6 γ-carboxyglutamate residues. γ-Carboxyglutamate chelates calcium ions, and so permits the binding of the blood-clotting proteins to membranes. In vitamin K deficiency, or in the presence of warfarin,

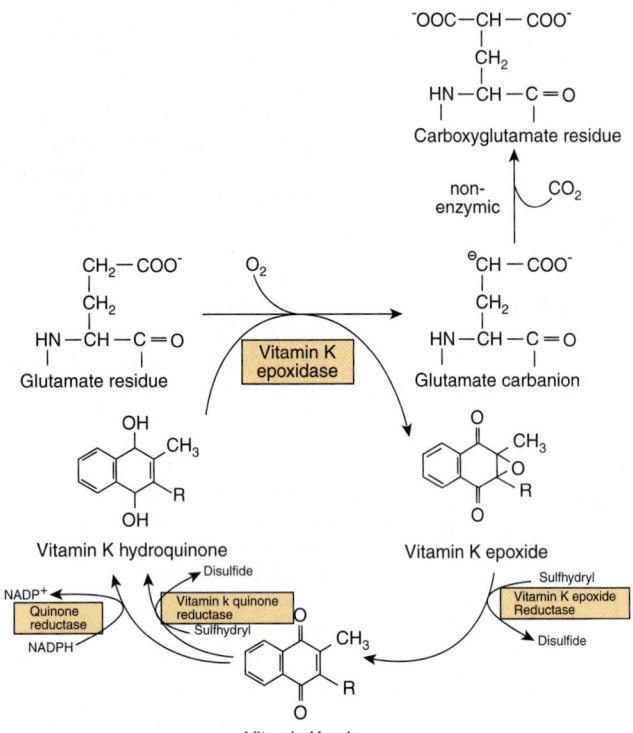

FIGURE 44–8 The role of vitamin K in the synthesis of γ-carboxyglutamate.

an abnormal precursor of prothrombin (preprothrombin) containing little or no γ-carboxyglutamate, and incapable of chelating calcium, is released into the circulation.

Vitamin K Is Also Important in Synthesis of Bone and other Calcium-Binding Proteins

A number of other proteins undergo the same vitamin K-dependent carboxylation of glutamate to γ-carboxyglutamate, including osteocalcin and the matrix Gla protein in bone, nephrocalcin in kidney and the product of the growth arrest specific gene Gas6, which is involved in both the regulation of differentiation and development in the nervous system, and control of apoptosis in other tissues. All of these γ-carboxyglutamate containing proteins bind calcium, which causes a conformational change so that they interact with membrane phospholipids. The release into the circulation of osteocalcin provides an index of vitamin D status.

WATER-SOLUBLE VITAMINS

VITAMIN B₁ (THIAMIN) HAS A KEY ROLE IN CARBOHYDRATE METABOLISM

Thiamin has a central role in energy-yielding metabolism, and especially the metabolism of carbohydrates (**Figure 44–9**). **Thiamin diphosphate** is the coenzyme for three multienzyme complexes that catalyze oxidative decarboxylation reactions: pyruvate dehydrogenase in carbohydrate metabolism (Chapter 17); α-ketoglutarate dehydrogenase in the citric acid cycle (Chapter 17); and the branched-chain keto-acid dehydrogenase involved in the metabolism of leucine, isoleucine, and valine (Chapter 29). In each case, the thiamin diphosphate provides a reactive carbon on the thiazole moiety that forms a carbanion, which then adds to the carbonyl group, eg, pyruvate. The addition compound is then decarboxylated, eliminating CO_2. Thiamin diphosphate is also the coenzyme for transketolase, in the pentose phosphate pathway (Chapter 21).

Thiamin triphosphate has a role in nerve conduction; it phosphorylates, and so activates, a chloride channel in the nerve membrane.

Thiamin Deficiency Affects the Nervous System & the Heart

Thiamin deficiency can result in three distinct syndromes: a chronic peripheral neuritis, **beriberi,** which may or may not be associated with **heart failure** and **edema;** acute pernicious (fulminating) beriberi (shoshin beriberi), in which heart failure and metabolic abnormalities predominate, without peripheral neuritis; and **Wernicke encephalopathy** with **Korsakoff psychosis,** which is associated especially with alcohol and narcotic abuse. The role of thiamin diphosphate in pyruvate dehydrogenase means that in deficiency there is impaired conversion of pyruvate to acetyl CoA. In subjects on a relatively high carbohydrate diet, this results in increased plasma concentrations of lactate and pyruvate, which may cause life-threatening **lactic acidosis.**

Thiamin Nutritional Status Can Be Assessed by Erythrocyte Transketolase Activation

The activation of apo-transketolase (the enzyme protein) in erythrocyte lysate by thiamin diphosphate added in vitro has become the accepted index of thiamin nutritional status.

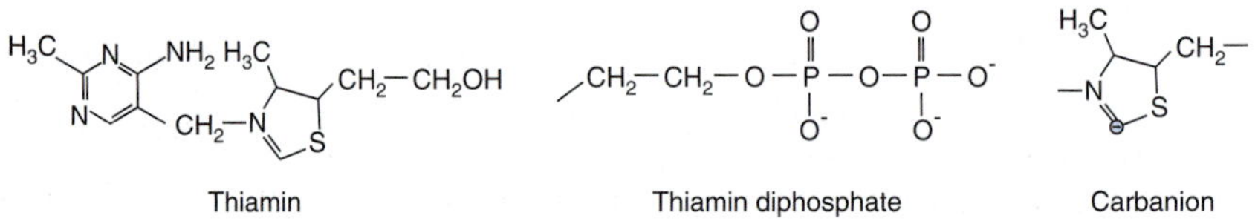

FIGURE 44–9 Thiamin, thiamin diphosphate, and the carbanion form.

Thiamin   Thiamin diphosphate   Carbanion

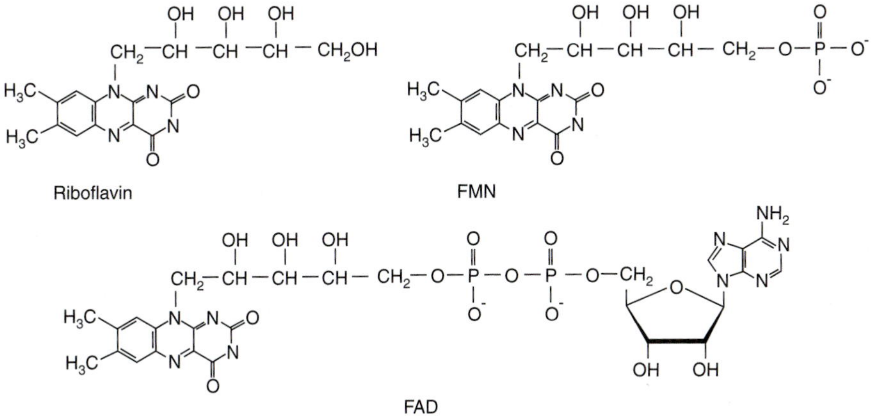

FIGURE 44–10 Riboflavin and the coenzymes flavin mononucleotide (FMN) and flavin adenine dinucleotide (FAD).

VITAMIN B$_2$ (RIBOFLAVIN) HAS A CENTRAL ROLE IN ENERGY-YIELDING METABOLISM

Riboflavin provides the reactive moieties of the coenzymes **flavin mononucleotide (FMN)** and **flavin adenine dinucleotide (FAD)** (**Figure 44–10**). FMN is formed by ATP-dependent phosphorylation of riboflavin, whereas FAD is synthesized by further reaction with ATP in which its AMP moiety is transferred to FMN. The main dietary sources of riboflavin are milk and dairy products. In addition, because of its intense yellow color, riboflavin is widely used as a food additive.

Flavin Coenzymes Are Electron Carriers in Oxidoreduction Reactions

These include the mitochondrial respiratory chain, key enzymes in fatty acid and amino acid oxidation, and the citric acid cycle. Reoxidation of the reduced flavin in oxygenases and mixed-function oxidases proceeds by way of formation of the flavin radical and flavin hydroperoxide, with the intermediate generation of superoxide and perhydroxyl radicals and hydrogen peroxide. Because of this, flavin oxidases make a significant contribution to the total oxidant stress in the body (Chapter 45).

Riboflavin Deficiency Is Widespread but Not Fatal

Although riboflavin is centrally involved in lipid and carbohydrate metabolism, and deficiency occurs in many countries, it is not fatal, because there is very efficient conservation of tissue riboflavin. Riboflavin released by the catabolism of enzymes is rapidly incorporated into newly synthesized enzymes. Deficiency is characterized by cheilosis, desquamation and inflammation of the tongue, and a seborrheic dermatitis. Riboflavin nutritional status is assessed by measurement of the activation of erythrocyte glutathione reductase by FAD added in vitro.

NIACIN IS NOT STRICTLY A VITAMIN

Niacin was discovered as a nutrient during studies of **pellagra**. It is not strictly a vitamin since it can be synthesized in the body from the essential amino acid tryptophan. Two compounds, **nicotinic acid** and **nicotinamide**, have the biologic activity of niacin; its metabolic function is as the nicotinamide ring of the coenzymes **NAD** and **NADP** in oxidation/reduction reactions (**Figure 44–11**). Some 60 mg of tryptophan is equivalent to 1 mg of dietary niacin. The niacin content of foods is expressed as

$$\text{mg niacin equivalents} = \text{mg preformed niacin} + 1/60 \times \text{mg tryptophan}$$

Since most of the niacin in cereals is biologically unavailable, this is discounted.

NAD Is the Source of ADP-Ribose

In addition to its coenzyme role, NAD is the source of ADP-ribose for the **ADP-ribosylation** of proteins and polyADP-ribosylation of nucleoproteins involved in the **DNA repair mechanism.** Cyclic ADP-ribose and nicotinic acid adenine dinucleotide, formed from NAD, act to increase intracellular calcium in response to neurotransmitters and hormones.

Pellagra Is Caused by Deficiency of Tryptophan & Niacin

Pellagra is characterized by a photosensitive dermatitis. As the condition progresses, there is dementia and possibly diarrhea. Untreated pellagra is fatal. Although the nutritional etiology of pellagra is well established, and tryptophan or niacin prevents

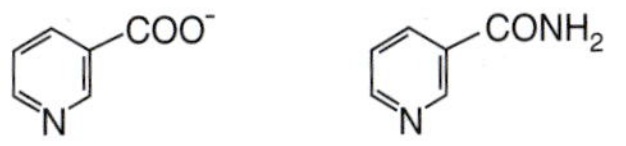

Niacin (nicotinic acid and nicotinamide) See also Figure 7–2

FIGURE 44–11 Niacin (nicotinic acid and nicotinamide).

or cures the disease, additional factors, including deficiency of riboflavin or vitamin B_6, both of which are required for synthesis of nicotinamide from tryptophan, may be important. In most outbreaks of pellagra, twice as many women as men are affected, probably the result of inhibition of tryptophan metabolism by estrogen metabolites.

Pellagra Can Occur as a Result of Disease Despite an Adequate Intake of Tryptophan & Niacin

A number of genetic diseases that result in defects of tryptophan metabolism are associated with the development of pellagra, despite an apparently adequate intake of both tryptophan and niacin. **Hartnup disease** is a rare genetic condition in which there is a defect of the membrane transport mechanism for tryptophan, resulting in large losses as a result of intestinal malabsorption and failure of the renal reabsorption mechanism. In **carcinoid syndrome,** there is metastasis of a primary liver tumor of enterochromaffin cells, which synthesize 5-hydroxytryptamine. Overproduction of 5-hydroxytryptamine may account for as much as 60% of the body's tryptophan metabolism, causing pellagra because of the diversion away from NAD synthesis.

Niacin Is Toxic in Excess

Nicotinic acid has been used to treat hyperlipidemia when of the order of 1–6 g/day are required, causing dilatation of blood vessels and flushing, along with skin irritation. Intakes of both nicotinic acid and nicotinamide in excess of 500 mg/day also cause liver damage.

VITAMIN B_6 IS IMPORTANT IN AMINO ACID & GLYCOGEN METABOLISM & IN STEROID HORMONE ACTION

Six compounds have vitamin B_6 activity (**Figure 44–12**): **pyridoxine, pyridoxal, pyridoxamine,** and their 5′-phosphates. The active coenzyme is pyridoxal 5′-phosphate. Some 80% of the body's total vitamin B_6 is pyridoxal phosphate in muscle, mostly associated with glycogen phosphorylase. This is not available in deficiency, but is released in starvation, when glycogen reserves become depleted, and is then available, especially in liver and kidney, to meet increased requirement for gluconeogenesis from amino acids.

Vitamin B_6 Has Several Roles in Metabolism

Pyridoxal phosphate is a coenzyme for many enzymes involved in amino acid metabolism, especially transamination and decarboxylation. It is also the cofactor of glycogen

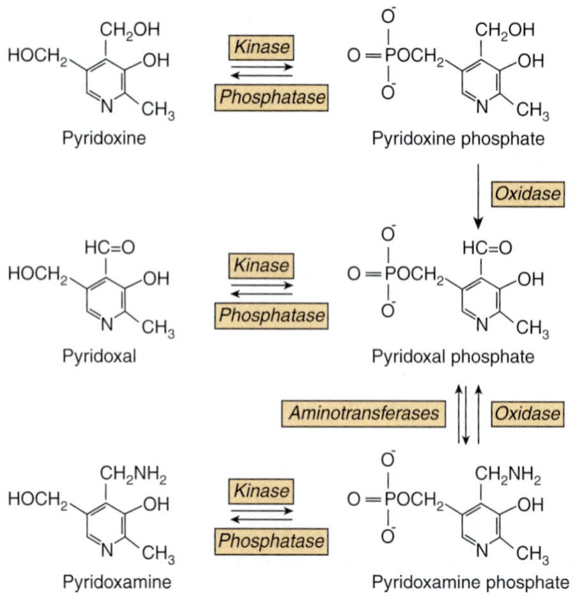

FIGURE 44–12 Interconversion of the vitamin B_6 vitamers.

phosphorylase, where the phosphate group is catalytically important. In addition, B_6 is important in steroid hormone action. Pyridoxal phosphate removes the hormone-receptor complex from DNA binding, terminating the action of the hormones. In vitamin B_6 deficiency, there is increased sensitivity to the actions of low concentrations of estrogens, androgens, cortisol, and vitamin D.

Vitamin B_6 Deficiency Is Rare

Although clinical deficiency disease is rare, there is evidence that a significant proportion of the population have marginal vitamin B_6 status. Moderate deficiency results in abnormalities of tryptophan and methionine metabolism. Increased sensitivity to steroid hormone action may be important in the development of **hormone-dependent cancer** of the breast, uterus, and prostate, and vitamin B_6 status may affect the prognosis.

Vitamin B_6 Status Is Assessed by Assaying Erythrocyte Transaminases

The most widely used method of assessing vitamin B_6 status is by the activation of erythrocyte transaminases by pyridoxal phosphate added in vitro, expressed as the activation coefficient.

In Excess, Vitamin B_6 Causes Sensory Neuropathy

The development of sensory neuropathy has been reported in patients taking 2–7 g of pyridoxine per day for a variety of reasons (there is some slight evidence that it is effective in treating **premenstrual syndrome**). There was some residual damage after withdrawal of these high doses; other reports suggest that intakes in excess of 200 mg/d are associated with neurologic damage.

VITAMIN B$_{12}$ IS FOUND ONLY IN FOODS OF ANIMAL ORIGIN

The term "vitamin B$_{12}$" is used as a generic descriptor for the **cobalamins**—those **corrinoids** (cobalt-containing compounds possessing the corrin ring) having the biologic activity of the vitamin (**Figure 44–13**). Some corrinoids that are growth factors for micro-organisms not only have no vitamin B$_{12}$ activity, but may also be antimetabolites of the vitamin. Although it is synthesized exclusively by microorganisms, for practical purposes vitamin B$_{12}$ is found only in foods of animal origin, there being no plant sources of this vitamin. This means that strict vegetarians (vegans) are at risk of developing B$_{12}$ deficiency. The small amounts of the vitamin formed by bacteria on the surface of fruits may be adequate to meet requirements, but preparations of vitamin B$_{12}$ made by bacterial fermentation are available.

Vitamin B$_{12}$ Absorption Requires Two Binding Proteins

Vitamin B$_{12}$ is absorbed bound to **intrinsic factor**, a small glycoprotein secreted by the parietal cells of the gastric mucosa. Gastric acid and pepsin release the vitamin from protein

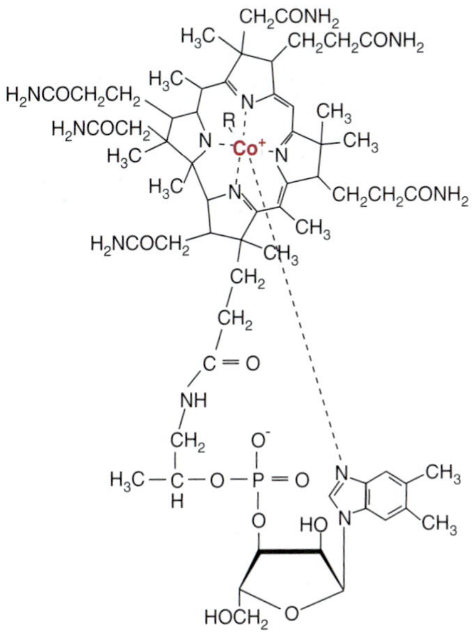

FIGURE 44–13 **Vitamin B$_{12}$.** Four coordination sites on the central cobalt atom are chelated by the nitrogen atoms of the corrin ring, and one by the nitrogen of the dimethylbenzimidazole nucleotide. The sixth coordination site may be occupied by: CN$^-$ (cyanocobalamin), OH$^-$ (hydroxocobalamin), H$_2$O (aquocobalamin, —CH$_3$ (methyl cobalamin), or 5'-deoxyadenosine (adenosylcobalamin).

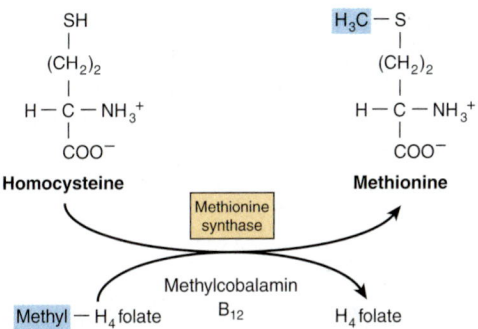

FIGURE 44–14 **Homocysteine and the "folate trap."** Vitamin B$_{12}$ deficiency leads to impairment of methionine synthase, resulting in accumulation of homocysteine and trapping folate as methyltetrahydrofolate.

binding in food and make it available to bind to **cobalophilin,** a binding protein secreted in the saliva. In the duodenum, cobalophilin is hydrolyzed, releasing the vitamin for binding to intrinsic factor. **Pancreatic insufficiency** can therefore be a factor in the development of vitamin B$_{12}$ deficiency, resulting in the excretion of cobalophilin-bound vitamin B$_{12}$. Intrinsic factor binds only the active vitamin B$_{12}$ vitamers and not other corrinoids. Vitamin B$_{12}$ is absorbed from the distal third of the ileum via receptors that bind the intrinsic factor-vitamin B$_{12}$ complex, but not free intrinsic factor or free vitamin.

There Are Three Vitamin B$_{12}$–Dependent Enzymes

Methylmalonyl CoA mutase, leucine aminomutase, and **methionine synthase** (**Figure 44–14**) are vitamin B$_{12}$-dependent enzymes. Methylmalonyl CoA is formed as an intermediate in the catabolism of valine and by the carboxylation of propionyl CoA arising in the catabolism of isoleucine, cholesterol, and, rarely, fatty acids with an odd number of carbon atoms or directly from propionate, a major product of microbial fermentation in the rumen. It undergoes a vitamin B$_{12}$-dependent rearrangement to succinyl CoA, catalyzed by methylmalonyl CoA mutase (Figure 20–2). The activity of this enzyme is greatly reduced in vitamin B$_{12}$ deficiency, leading to an accumulation of methylmalonyl CoA and urinary excretion of methylmalonic acid, which provides a means of assessing vitamin B$_{12}$ nutritional status.

Vitamin B$_{12}$ Deficiency Causes Pernicious Anemia

Pernicious anemia arises when vitamin B$_{12}$ deficiency impairs the metabolism of folic acid, leading to functional folate deficiency that disturbs erythropoiesis, causing immature precursors of erythrocytes to be released into the circulation (megaloblastic anemia). The most common cause of pernicious anemia is failure of the absorption of vitamin B$_{12}$ rather than dietary deficiency. This can be the result of failure

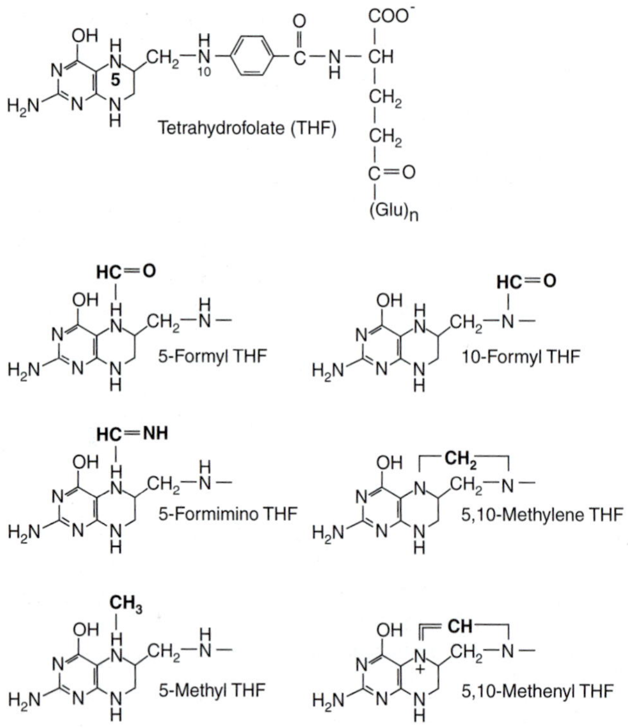

Tetrahydrofolate (THF)

5-Formyl THF

10-Formyl THF

5-Formimino THF

5,10-Methylene THF

5-Methyl THF

5,10-Methenyl THF

FIGURE 44–15 Tetrahydrofolic acid and the one-carbon substituted folates.

of intrinsic factor secretion caused by autoimmune disease affecting parietal cells or from production of anti-intrinsic factor antibodies. There is irreversible degeneration of the spinal cord in pernicious anemia, as a result of failure of methylation of one arginine residue in myelin basic protein. This is the result of methionine deficiency in the central nervous system, rather than secondary folate deficiency.

THERE ARE MULTIPLE FORMS OF FOLATE IN THE DIET

The active form of folic acid (pteroyl glutamate) is tetrahydrofolate (**Figure 44–15**). The folates in foods may have up to seven additional glutamate residues linked by γ-peptide bonds. In addition, all of the one-carbon substituted folates in Figure 44–15 may also be present in foods. The extent to which the different forms of folate can be absorbed varies, and folate intakes are calculated as dietary folate equivalents—the sum of μg food folates + 1.7 × μg of folic acid (used in food enrichment).

Tetrahydrofolate Is a Carrier of One-Carbon Units

Tetrahydrofolate can carry one-carbon fragments attached to N-5 (formyl, formimino, or methyl groups), N-10 (formyl) or bridging N-5–N-10 (methylene or methenyl groups). 5-Formyl-tetrahydrofolate is more stable than folate and is therefore used pharmaceutically (known as **folinic acid**), and the synthetic (racemic) compound (**leucovorin**). The major point of entry for one-carbon fragments into substituted folates is methylene-tetrahydrofolate (**Figure 44–16**), which is formed by the reaction of glycine, serine, and choline with tetrahydrofolate. Serine is the most important source of substituted folates for biosynthetic reactions, and the activity of serine hydroxymethyltransferase is regulated by the state of folate substitution and the availability of folate. The reaction is reversible, and in liver it can form serine from glycine as a substrate for gluconeogenesis. Methylene-, methenyl-, and 10-formyl-tetrahydrofolates are interconvertible. When one-carbon folates are not required, the oxidation of formyl-tetrahydrofolate to yield carbon dioxide provides a means of maintaining a pool of free folate.

Inhibitors of Folate Metabolism Provide Cancer Chemotherapy, Antibacterial, & Antimalarial Drugs

The methylation of deoxyuridine monophosphate (dUMP) to thymidine monophosphate (TMP), catalyzed by thymidylate synthase, is essential for the synthesis of DNA. The one-carbon fragment of methylene-tetrahydrofolate is reduced to a methyl group with release of dihydrofolate, which is then reduced back to tetrahydrofolate by **dihydrofolate reductase**. Thymidylate synthase and dihydrofolate reductase are especially active in tissues with a high rate of cell division. **Methotrexate,**

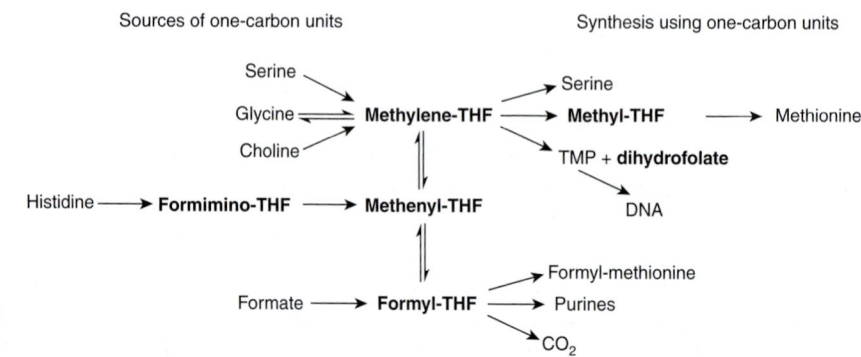

FIGURE 44–16 Sources and utilization of one-carbon substituted folates.

an analog of 10-methyl-tetrahydrofolate, inhibits dihydrofolate reductase and has been exploited as an anticancer drug. The dihydrofolate reductases of some bacteria and parasites differ from the human enzyme; inhibitors of these enzymes can be used as antibacterial drugs (eg, **trimethoprim**) and antimalarial drugs (eg, **pyrimethamine**).

Vitamin B$_{12}$ Deficiency Causes Functional Folate Deficiency—the "Folate Trap"

When acting as a methyl donor, S-adenosyl methionine forms homocysteine, which may be remethylated by methyl-tetrahydrofolate catalyzed by methionine synthase, a vitamin B$_{12}$–dependent enzyme (Figure 44–14). As the reduction of methylene-tetrahydrofolate to methyl-tetrahydrofolate is irreversible and the major source of tetrahydrofolate for tissues is methyltetrahydrofolate, the role of methionine synthase is vital, and provides a link between the functions of folate and vitamin B$_{12}$. Impairment of methionine synthase in vitamin B$_{12}$ deficiency results in the accumulation of methyltetrahydrofolate—the "folate trap." There is therefore functional deficiency of folate, secondary to the deficiency of vitamin B$_{12}$.

Folate Deficiency Causes Megaloblastic Anemia

Deficiency of folic acid itself or deficiency of vitamin B$_{12}$, which leads to functional folic acid deficiency, affects cells that are dividing rapidly because they have a large requirement for thymidine for DNA synthesis. Clinically, this affects the bone marrow, leading to megaloblastic anemia.

Folic Acid Supplements Reduce the Risk of Neural Tube Defects & Hyperhomocysteinemia, & May Reduce the Incidence of Cardiovascular Disease & Some Cancers

Supplements of 400 μg/day of folate begun before conception result in a significant reduction in the incidence of **spina bifida** and other **neural tube defects**. Because of this, there is mandatory enrichment of flour with folic acid in many countries. Elevated blood homocysteine is a significant risk factor for **atherosclerosis, thrombosis,** and **hypertension**. The condition is the result of an impaired ability to form methyltetrahydrofolate by methylene-tetrahydrofolate reductase, causing functional folate deficiency, resulting in failure to remethylate homocysteine to methionine. People with an abnormal variant of methylene-tetrahydrofolate reductase that occurs in 5–10% of the population do not develop hyperhomocysteinemia if they have a relatively high intake of folate. A number of placebo-controlled trials of supplements of folate (commonly together with vitamins B$_6$ and B$_{12}$) have shown the expected lowering of plasma homocysteine, but apart from reduced incidence of stroke there has been no effect on cardiovascular disease.

There is also evidence that low folate status results in impaired methylation of CpG islands in DNA, which is a factor in the development of colorectal and other cancers. A number of studies suggest that folate supplementation or food enrichment may reduce the risk of developing some cancers. However, there is some evidence that folate supplements increase the rate of transformation of preneoplastic colorectal polyps into cancers, so that people with such polyps are at increased risk of developing colorectal cancer if they have a high folate intake.

Folate Enrichment of Foods May Put Some People at Risk

Folate supplements will rectify the megaloblastic anemia of vitamin B$_{12}$ deficiency but may hasten the development of the (irreversible) nerve damage found in vitamin B$_{12}$ deficiency. There is also antagonism between folic acid and the anticonvulsants used in the treatment of epilepsy, and, as noted above, there is some evidence that folate supplements may increase the risk of developing colorectal cancer among people with preneoplastic colorectal polyps.

DIETARY BIOTIN DEFICIENCY IS UNKNOWN

The structures of biotin, biocytin, and carboxybiotin (the active metabolic intermediate) are shown in **Figure 44–17**. Biotin is widely distributed in many foods as biocytin (ε-amino-biotinyllysine), which is released on proteolysis. It is synthesized by intestinal flora in excess of requirements. Deficiency is unknown, except among people maintained for many months on total parenteral nutrition, and a very small number who eat abnormally large amounts of uncooked egg

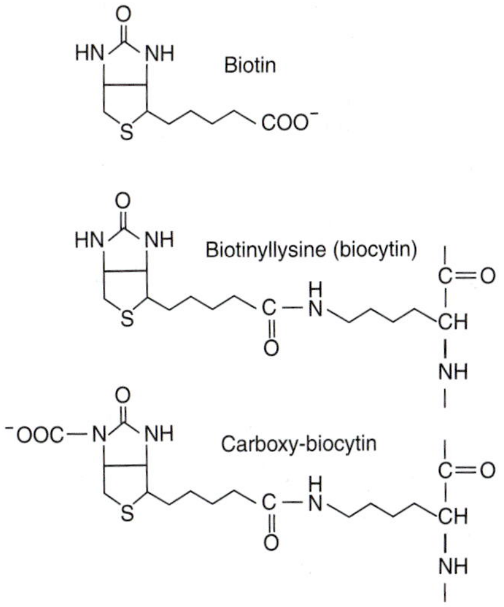

FIGURE 44–17 Biotin, biocytin, and carboxy-biocytin.

white, which contains avidin, a protein that binds biotin and renders it unavailable for absorption.

Biotin Is a Coenzyme of Carboxylase Enzymes

Biotin functions to transfer carbon dioxide in a small number of reactions: acetyl-CoA carboxylase (Figure 23–1), pyruvate carboxylase (Figure 20–1), propionyl-CoA carboxylase (Figure 20–2), and methylcrotonyl-CoA carboxylase. A holocarboxylase synthetase catalyzes the transfer of biotin onto a lysine residue of the apo-enzyme to form the biocytin residue of the holoenzyme. The reactive intermediate is 1-N-carboxy-biocytin, formed from bicarbonate in an ATP-dependent reaction. The carboxy group is then transferred to the substrate for carboxylation.

Biotin also has a role in regulation of the cell cycle, acting to biotinylate key nuclear proteins.

AS PART OF CoA & ACP, PANTOTHENIC ACID ACTS AS A CARRIER OF ACYL GROUPS

Pantothenic acid has a central role in acyl group metabolism when acting as the pantetheine functional moiety of coenzyme A or acyl carrier protein (ACP) (**Figure 44–18**). The pantetheine moiety is formed after combination of pantothenate with cysteine, which provides the –SH prosthetic group of CoA and ACP. CoA takes part in reactions of the citric acid cycle (Chapter 17), fatty acid oxidation (Chapter 22), acetylations, and cholesterol synthesis (Chapter 26). ACP participates in fatty acid synthesis (Chapter 23). The vitamin is widely distributed in all food-stuffs, and deficiency has not been unequivocally reported in humans except in specific depletion studies.

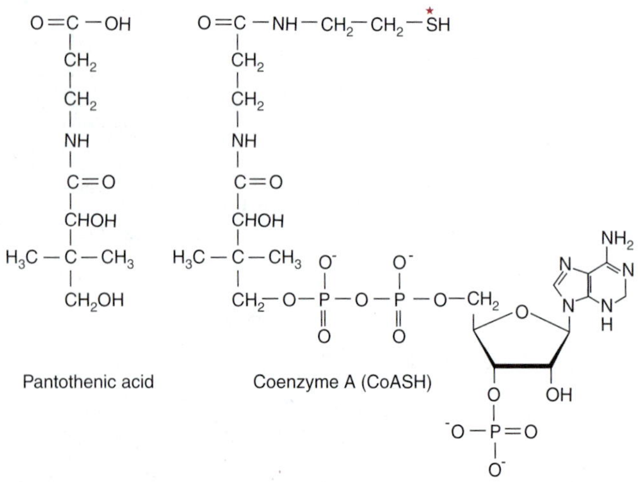

FIGURE 44–18 **Pantothenic acid and coenzyme A.** Asterisk shows site of acylation by fatty acids.

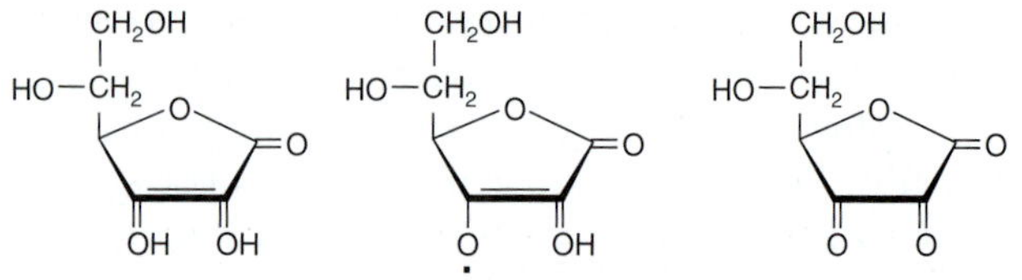

FIGURE 44–19 **Vitamin C.**

ASCORBIC ACID IS A VITAMIN FOR ONLY SOME SPECIES

Vitamin C (**Figure 44–19**) is a vitamin for human beings and other primates, the guinea pig, bats, passeriform birds, and most fishes and invertebrates; other animals synthesize it as an intermediate in the uronic acid pathway of glucose metabolism (Figure 21–4). In those species for which it is a vitamin, there is a block in the pathway as a result of the absence of gulonolactone oxidase. Both ascorbic acid and dehydroascorbic acid have vitamin activity.

Vitamin C Is the Coenzyme for Two Groups of Hydroxylases

Ascorbic acid has specific roles in the copper-containing hydroxylases and the α-ketoglutarate-linked iron-containing hydroxylases. It also increases the activity of a number of other enzymes in vitro, although this is a nonspecific reducing action. In addition, it has a number of nonenzymic effects as a result of its action as a reducing agent and oxygen radical quencher (Chapter 45).

Dopamine β-hydroxylase is a copper-containing enzyme involved in the synthesis of the catecholamines (norepinephrine and epinephrine), from tyrosine in the adrenal medulla and central nervous system. During hydroxylation the Cu^+ is oxidized to Cu^{2+}; reduction back to Cu^+ specifically requires ascorbate, which is oxidized to monodehydroascorbate.

A number of peptide hormones have a carboxy terminal amide that is derived from a terminal glycine residue. This glycine is hydroxylated on the α-carbon by a copper-containing enzyme, **peptidylglycine hydroxylase,** which, again, requires ascorbate for reduction of Cu^{2+}.

A number of iron-containing, ascorbate-requiring hydroxylases share a common reaction mechanism, in which hydroxylation of the substrate is linked to oxidative decarboxylation of α-ketoglutarate. Many of these enzymes are involved in the modification of precursor proteins. **Proline** and **lysine hydroxylases** are required for the postsynthetic modification of **procollagen** to **collagen,** and proline hydroxylase is also required in formation of **osteocalcin** and the C1q component of **complement.** Aspartate β-hydroxylase is required for the postsynthetic modification of the precursor of protein C, the vitamin K-dependent protease that hydrolyzes activated factor V in the blood-clotting cascade (Chapter 50). Trimethyllysine and γ-butyrobetaine hydroxylases are required for the synthesis of carnitine.

Vitamin C Deficiency Causes Scurvy

Signs of vitamin C deficiency include skin changes, fragility of blood capillaries, gum decay, tooth loss, and bone fracture, many of which can be attributed to deficient collagen synthesis.

There May Be Benefits from Higher Intakes of Vitamin C

At intakes above about 100 mg/day, the body's capacity to metabolize vitamin C is saturated, and any further intake is excreted in the urine. However, in addition to its other roles, vitamin C enhances the absorption of inorganic iron, and this depends on the presence of the vitamin in the gut. Therefore, increased intakes may be beneficial. There is very little good evidence that high doses of vitamin C prevent the common cold, although they may reduce the duration and severity of symptoms.

MINERALS ARE REQUIRED FOR BOTH PHYSIOLOGIC & BIOCHEMICAL FUNCTIONS

Many of the essential minerals (**Table 44–6**) are widely distributed in foods, and most people eating a mixed diet are likely to receive adequate intakes. The amounts required vary from grams per day for sodium and calcium, through milligrams per day (eg, iron and zinc), to micrograms per day for the trace elements. In general, mineral deficiencies occur when foods come from one region where the soil may be deficient in some minerals (eg, iodine and selenium, deficiencies of both of which occur in many areas of the world); when foods come from a variety of regions, mineral deficiency is less likely to occur. However, iron deficiency is a general problem, because if iron losses from the body are relatively high (eg, from heavy menstrual blood loss), it is difficult to achieve an adequate intake to replace losses. Foods grown on soil containing high levels of selenium cause toxicity, and excessive intakes of sodium cause hypertension in susceptible people.

SUMMARY

- Vitamins are organic nutrients with essential metabolic functions, generally required in small amounts in the diet because they cannot be synthesized by the body. The lipid-soluble vitamins (A, D, E, and K) are hydrophobic molecules requiring normal fat absorption for their absorption and the avoidance of deficiency.
- Vitamin A (retinol), present in meat, and the provitamin (β-carotene), found in plants, form retinaldehyde, utilized in vision, and retinoic acid, which acts in the control of gene expression.
- Vitamin D is a steroid prohormone yielding the active hormone calcitriol, which regulates calcium and phosphate metabolism; deficiency leads to rickets and osteomalacia. It has a role in controlling cell differentiation and insulin secretion.
- Vitamin E (tocopherol) is the most important lipid-soluble antioxidant in the body, acting in the lipid phase of membranes protecting against the effects of free radicals.
- Vitamin K functions acts as the cofactor of a carboxylase that acts on glutamate residues of precursor proteins of clotting factors and bone and other proteins to enable them to chelate calcium.
- The water-soluble vitamins of the B complex act as enzyme cofactors. Thiamin is a cofactor in oxidative decarboxylation of α-keto acids and of transketolase in the pentose phosphate pathway. Riboflavin and niacin are important cofactors in oxidoreduction reactions, present in flavoprotein enzymes and in NAD and NADP, respectively.
- Pantothenic acid is present in coenzyme A and acyl carrier protein, which act as carriers for acyl groups in metabolic reactions.
- Vitamin B_6 as pyridoxal phosphate is the coenzyme for several enzymes of amino acid metabolism, including the transaminases, and of glycogen phosphorylase. Biotin is the coenzyme for several carboxylase enzymes.
- Vitamin B_{12} and folate provide one-carbon residues for DNA synthesis and other reactions; deficiency results in megaloblastic anemia.
- Vitamin C is a water-soluble antioxidant that maintains vitamin E and many metal cofactors in the reduced state.
- Inorganic mineral elements that have a function in the body must be provided in the diet. When intake is insufficient, deficiency may develop, and excessive intakes may be toxic.

TABLE 44–6 Classification of Minerals According to Their Function

Function	Mineral
Structural function	Calcium, magnesium, phosphate
Involved in membrane function	Sodium, potassium
Function as prosthetic groups in enzymes	Cobalt, copper, iron, molybdenum, selenium, zinc
Regulatory role or role in hormone action	Calcium, chromium, iodine, magnesium, manganese, sodium, potassium
Known to be essential, but function unknown	Silicon, vanadium, nickel, tin
Have effects in the body, but essentiality is not established	Fluoride, lithium
May occur in foods and known to be toxic in excess	Aluminum, arsenic, antimony, boron, bromine, cadmium, cesium, germanium, lead, mercury, silver, strontium

REFERENCES

Bender DA, Bender AE: *Nutrition: A Reference Handbook.* Oxford University Press, 1997.

Bender DA: *Nutritional Biochemistry of the Vitamins,* 2nd ed. Cambridge University Press, 2003.

Department of Health: *Dietary Reference Values for Food Energy and Nutrients for the United Kingdom.* Her Majesty's Stationery Office, 1991.

FAO/WHO: *Human Vitamin and Mineral Requirements: Report of a Joint FAO/WHO Expert Consultation: Bankok, Thailand.* Food and Nutrition Division of the United Nations Food and Agriculture Organization, 2000.

Geissler C, Powers HJ: *Human Nutrition,* 12th ed. Elsevier, 2010.

Gibney MJ, Lanham-New S, Cassidy A, et al: *Introduction to Human Nutrition, The Nutrition Society Textbook Series,* 2nd ed. Wiley–Blackwell, 2009.

Institute of Medicine: *Dietary Reference Intakes for Calcium, Phosphorus, Magnesium, Vitamin D and Fluoride.* National Academy Press, 1997.

Institute of Medicine: *Dietary Reference Values for Thiamin, Riboflavin, Niacin, Vitamin B6, Folate, Vitamin B12, Pantothenic Acid, Biotin and Choline.* National Academy Press, 2000.

Institute of Medicine: *Dietary Reference Values for Vitamin C, Vitamin E, Selenium and Carotenoids.* National Academy Press, 2000.

Institute of Medicine: *Dietary Reference Intakes for Vitamin A, Vitamin K, Arsenic, Boron, Chromium, Copper, Iodine, Iron, Manganese, Molybdenum, Nickel, Silicon, Vanadium and Zinc.* National Academy Press, 2001.

Scientific Advisory Committee on Nutrition of the Food Standards Agency: *Folate and Disease Prevention.* The Stationery Office, 2006.

45

Free Radicals and Antioxidant Nutrients

David A. Bender, PhD

OBJECTIVES

After studying this chapter, you should be able to:

- Describe the damage caused to DNA, lipids, and proteins by free radicals, and the diseases associated with radical damage.
- Describe the main sources of oxygen radicals in the body.
- Describe the mechanisms and dietary factors that protect against radical damage.
- Explain how antioxidants can also act as pro-oxidants, and why intervention trials of antioxidant nutrients have generally yielded disappointing results.

BIOMEDICAL IMPORTANCE

Free radicals are formed in the body under normal conditions. They cause damage to nucleic acids, proteins, and lipids in cell membranes and plasma lipoproteins. This can cause cancer, atherosclerosis and coronary artery disease, and autoimmune diseases. Epidemiological and laboratory studies have identified a number of protective antioxidant nutrients: selenium, vitamins C and E, β-carotene, and other carotenoids, and a variety of polyphenolic compounds derived from plant foods. Many people take supplements of one or more antioxidant nutrients. However, intervention trials show little benefit of antioxidant supplements except among people who were initially deficient, and many trials of β-carotene and vitamin E have shown increased mortality among those taking the supplements.

Free Radical Reactions Are Self-Perpetuating Chain Reactions

Free radicals are highly reactive molecular species with an unpaired electron; they persist for only a very short time (of the order of 10^{-9} to 10^{-12} sec) before they collide with another molecule and either abstract or donate an electron in order to achieve stability. In so doing, they generate a new radical from the molecule with which they collided. The main way in which a free radical can be quenched, so terminating this chain reaction, is if two radicals react together, when the unpaired electrons can become paired in one or other of the parent molecules. This is a rare occurrence, because of the very short half-life of an individual radical and the very low concentrations of radicals in tissues.

The most damaging radicals in biological systems are oxygen radicals (sometimes called reactive oxygen species)—especially superoxide, O_2^-, hydroxyl, $OH^·$, and perhydroxyl, $O_2H^·$. Tissue damage caused by oxygen radicals is often called oxidative damage, and factors that protect against oxygen radical damage are known as antioxidants.

Radicals Can Damage DNA, Lipids, and Proteins

Interaction of radicals with bases in DNA can lead to chemical changes that, if not repaired (Chapter 35), may be inherited in daughter cells. Radical damage to unsaturated fatty acids in cell membranes and plasma lipoproteins leads to the formation of lipid peroxides, then highly reactive dialdehydes that can chemically modify proteins and nucleic acid bases. Proteins are also subject to direct chemical modification by interaction with radicals. Oxidative damage to tyrosine residues in proteins can lead to the formation of dihydroxyphenylalanine that can undergo nonenzymic reactions leading to further formation of oxygen radicals (**Figure 45–1**).

The total body radical burden can be estimated by measuring the products of lipid peroxidation. Lipid peroxides can be measured by the ferrous oxidation in xylenol orange (FOX) assay. Under acidic conditions, they oxidize Fe^{2+} to Fe^{3+}, which forms a chromophore with xylenol orange. The dialdehydes formed from lipid peroxides can be measured by reaction with thiobarbituric acid, when they form a red fluorescent adduct—the results of this assay are generally reported as total thiobarbituric acid reactive substances, TBARS. Peroxidation of *n*-6 polyunsaturated fatty acids leads to the formation of pentane,

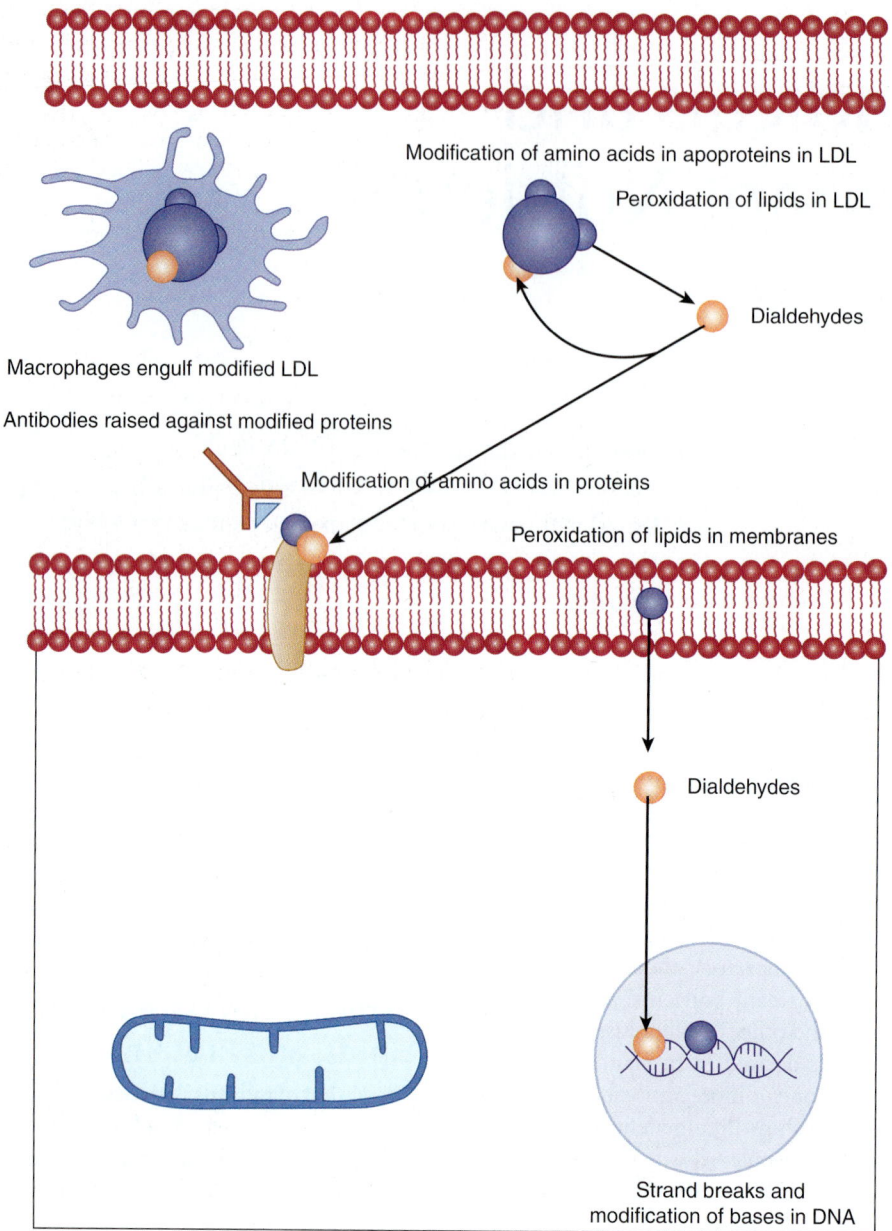

FIGURE 45–1 Tissue damage by radicals.

and of n-3 polyunsaturated fatty acids to ethane, both of which can be measured in exhaled air.

Radical Damage May Cause Mutations, Cancer, Autoimmune Disease, and Atherosclerosis

Radical damage to DNA in germline cells in ovaries and testes can lead to heritable mutations; in somatic cells, the result may be initiation of cancer. The dialdehydes formed as a result of radical-induced lipid peroxidation in cell membranes can also modify bases in DNA.

Chemical modification of amino acids in proteins, either by direct radical action or as a result of reaction with the products of radical-induced lipid peroxidation, leads to proteins that are recognized as nonself by the immune system. The resultant antibodies will also cross-react with normal tissue proteins, so initiating autoimmune disease.

Chemical modification of the proteins or lipids in plasma low-density lipoprotein (LDL) leads to abnormal LDL that is not recognized by the liver LDL receptors, and so is not cleared by the liver. The modified LDL is taken up by macrophage scavenger receptors. Lipid-engorged macrophages infiltrate under blood vessel endothelium (especially when there is already some damage to the endothelium), and are killed by the high content of unesterified cholesterol they have accumulated. This occurs in the development of atherosclerotic plaques, which, in extreme cases, can more or less completely occlude a blood vessel.

There Are Multiple Sources of Oxygen Radicals in the Body

Ionizing radiation (x-rays and UV) can lyse water, leading to the formation of hydroxyl radicals. Transition metal ions, including Cu^+, Co^{2+}, Ni^{2+}, and Fe^{2+} can react nonenzymically with oxygen or hydrogen peroxide, again leading to the formation of hydroxyl radicals. Nitric oxide (the endothelium-derived relaxation factor) is itself a radical, and, more importantly, can react with superoxide to yield peroxynitrite, which decays to form hydroxyl radicals (**Figure 45–2**).

The respiratory burst of activated macrophages (Chapter 52) is increased utilization of glucose via the pentose phosphate pathway (Chapter 21) to reduce $NADP^+$ to NADPH, and increased utilization of oxygen to oxidise NADPH to produce oxygen (and halogen) radicals as cytotoxic agents to kill phagocytosed microorganisms. The respiratory burst oxidase (NADPH oxidase) is a flavoprotein that reduces oxygen to superoxide: $NADPH + 2O_2 \rightarrow NADP^+ + 2O_2^{\cdot-} + 2H^+$. Plasma markers of radical damage to lipids increase considerably in response to even a mild infection.

The oxidation of reduced flavin coenzymes in the mitochondrial (Chapter 13) and microsomal electron transport chains proceeds through a series of steps in which the flavin semiquinone radical is stabilized by the protein to which it is bound, and forms oxygen radicals as transient intermediates. Although the final products are not radicals, because of the unpredictable nature of radicals there is considerable "leakage" of radicals, and some 3%–5% of the daily consumption of 30 mol of oxygen by an adult human being is converted to singlet oxygen, hydrogen peroxide, and superoxide, perhydroxyl, and hydroxyl radicals, rather then undergoing complete reduction to water. This results in daily production of about 1.5 mol of reactive oxygen species.

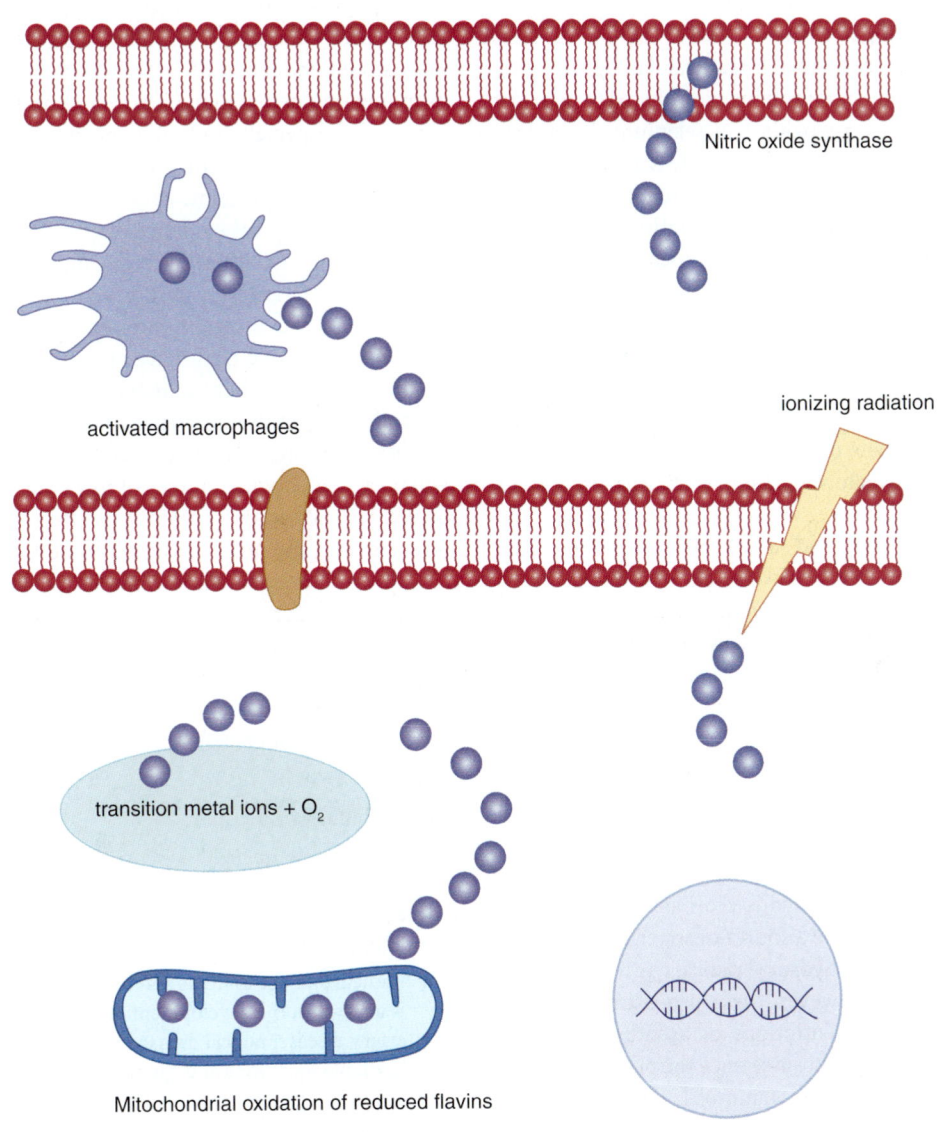

Nitric oxide synthase

activated macrophages

ionizing radiation

transition metal ions + O_2

Mitochondrial oxidation of reduced flavins

FIGURE 45–2 **Sources of radicals.**

There Are Various Mechanisms of Protection Against Radical Damage

The metal ions that undergo nonenzymic reaction to form oxygen radicals are not normally free in solution, but are bound to either the proteins for which they provide the prosthetic group, or to specific transport and storage proteins, so that they are unreactive. Iron is bound to transferrin, ferritin, and hemosiderin, copper to ceruloplasmin, and other metal ions are bound to metallothionein. This binding to transport proteins that are too large to be filtered in the kidneys also prevents loss of metal ions in the urine.

Superoxide is produced both accidentally and also as the reactive oxygen species required for a number of enzyme-catalyzed reactions. A family of superoxide dismutases catalyze the reaction between superoxide and protons to yield oxygen and hydrogen peroxide: $O_2^- + 2H^+ \rightarrow H_2O_2$. The hydrogen peroxide is then removed by catalase and various peroxidases: $2H_2O_2 \rightarrow 2H_2O + O_2$. Most enzymes that produce and require superoxide are in the peroxisomes, together with superoxide dismutase, catalase, and peroxidases.

The peroxides that are formed by radical damage to lipids in membranes and plasma lipoproteins are reduced to hydroxy fatty acids by glutathione peroxidase, a selenium-dependent enzyme (hence the importance of adequate selenium intake to maximize antioxidant activity), and the oxidized glutathione is reduced by NADPH-dependent glutathione reductase (Figure 21–3). Lipid peroxides are also reduced to fatty acids by reaction with vitamin E, forming the relatively stable tocopheroxyl radical, which persist long enough to undergo reduction back to tocopherol by reaction with vitamin C at the surface of the cell or lipoprotein (Figure 44–6). The resultant monodehydroascorbate radical then undergoes enzymic reduction back to ascorbate or a nonenzymic reaction of 2 mol of monodehydroascorbate to yield 1 mol each of ascorbate and dehydroascorbate.

Ascorbate, uric acid and a variety of polyphenols derived from plant foods act as water-soluble radical trapping antioxidants, forming relatively stable radicals that persist long enough to undergo reaction to nonradical products. Ubiquinone and carotenes similarly act as lipid-soluble radical-trapping antioxidants in membranes and plasma lipoproteins.

Antioxidants Can Also Be Pro-Oxidants

Although ascorbate is an antioxidant, reacting with superoxide and hydroxyl to yield monodehydroascorbate and hydrogen peroxide or water, it can also be a source of superoxide radicals by reaction with oxygen, and hydroxyl radicals by reaction with Cu^{2+} ions (**Table 45–1**). However, these pro-oxidant actions require relatively high concentrations of ascorbate that are unlikely to be reached in tissues, since once the plasma concentration of ascorbate reaches about 30 mmol/L, the renal threshold is reached, and at intakes above about 100–120 mg/day the vitamin is excreted in the urine quantitatively with intake.

TABLE 45–1 Antioxidant and Pro-Oxidant Roles of Vitamin C

Antioxidant roles:
Ascorbate + O_2^- → H_2O_2 + monodehydroascorbate; catalase and peroxidases catalyze the reaction: $2H_2O_2 \rightarrow 2H_2O + O_2$
Ascorbate + OH˙ → H_2O + monodehydroascorbate
Pro-oxidant roles:
Ascorbate + O_2 → O_2^- + monodehydroascorbate
Ascorbate + Cu^{2+} → Cu^+ + monodehydroasacorbate
$Cu^+ + H_2O_2$ → $Cu^{2+} + OH^- + OH˙$

A considerable body of epidemiological evidence suggested that carotene is protective against lung and other cancers. However, two major intervention trials in the 1990s showed an increase in death from lung (and other) cancer among people given supplements of β-carotene. The problem is that although β-carotene is indeed a radical-trapping antioxidant under conditions of low partial pressure of oxygen, as in most tissues, at high partial pressures of oxygen (as in the lungs) and especially in high concentrations, β-carotene is an autocatalytic pro-oxidant, and hence can initiate radical damage to lipids and proteins.

Epidemiological evidence also suggests that vitamin E is protective against atherosclerosis and cardiovascular disease. However, meta-analysis of intervention trials with vitamin E shows increased mortality among those taking (high dose) supplements. These trials have all used α-tocopherol, and it is possible that the other vitamers of vitamin E that are present in foods, but not the supplements, may be important. In vitro, plasma lipoproteins form less cholesterol ester hydroperoxide when incubated with sources of low concentrations of perhydroxyl radicals when vitamin E has been removed than when it is present. The problem seems to be that vitamin E acts as an antioxidant by forming a stable radical that persists long enough to undergo metabolism to nonradical products. This means that the radical also persists long enough to penetrate deeper in to the lipoprotein, causing further radical damage, rather than interacting with a water-soluble antioxidant at the surface of the lipoprotein.

SUMMARY

- Free radicals are highly reactive molecular species with an unpaired electron. They can react with, and modify, proteins, nucleic acids and fatty acids in cell membranes and plasma lipoproteins.

- Radical damage to lipids and proteins in plasma lipoproteins is a factor in the development of atherosclerosis and coronary artery disease; radical damage to nucleic acids may induce heritable mutations and cancer; radical damage to proteins may lead to the development of autoimmune diseases.

- Oxygen radicals arise as a result of exposure to ionizing radiation, nonenzymic reactions of transition metal ions, the respiratory burst of activated macrophages, and the normal oxidation of reduced flavin coenzymes.

■ Protection against radical damage is afforded by enzymes that remove superoxide ions and hydrogen peroxide, enzymic reduction of lipid peroxides linked to oxidation of glutathione, nonenzymic reaction of lipid peroxides with vitamin E, and reaction of radicals with compounds such as vitamins C and E, carotene, ubiquinone, uric acid, and dietary polyphenols that form relatively stable radicals that persist long enough to undergo reaction to nonradical products.

■ Except in people who were initially deficient, intervention trials of vitamin E and β-carotene have generally shown increased mortality among those taking the supplements. β-Carotene is only an antioxidant at low concentrations of oxygen; at higher concentrations of oxygen it is an autocatalytic prooxidant. Vitamin E forms a stable radical that is capable of either undergoing reaction with water-soluble antioxidants or penetrating further into lipoproteins and tissues, so increasing radical damage.

REFERENCES

Asplund K: Antioxidant vitamins in the prevention of cardiovascular disease: a systematic review. J Intern Med 2002;251:372.

Bjelakovic G, Nikolova D, Gluud LL, et al: Mortality in randomised trials of antioxidant supplements for primary and secondary prevention. JAMA 2007;297:842.

Burton G, Ingold K: β-Carotene, an unusual type of lipid antioxidant. Science 1984;224:569.

Carr A, Frei B: Does vitamin C act as a pro-oxidant under physiological conditions? FASEB J 1999;13:1007.

Cordero Z, Drogan D, Weikert C, et al: Vitamin E and risk of cardiovascular diseases: a review of epidemiologic and clinical trial studies. Crit Rev Food Sci Nutr 2010;50;420.

Dotan Y, Lichtenberg D, Pinchuk I: No evidence supports vitamin E indiscriminate supplementation. Biofactors 2009;35:469.

Halliwell B, Gutteridge JM, Cross CE: Free radicals, antioxidants and human disease: where are we now? J Lab Clin Med 1992;119:598.

Imlay JA: Pathways of oxidative damage. Ann Rev Microbiol 2003;57:395.

Imlay JA: Cellular Defenses against superoxide and hydrogen peroxide. Ann Rev Biochem 2008;77:755.

Klaunig JE, Kamendulis LM: The role of oxidative stress in carcinogenesis. Ann Rev Pharm Tox 2004;44:239.

Miller ER, Pastor-Barriuso R, Dalal D, et al: Meta-analysis: high-dosage vitamin E supplementation may increase all-cause mortality. Ann Intern Med 2005;142:37.

Omenn GS, Goodman GE, Thornquist MD, et al: Effects of a combination of beta carotene and vitamin A on lung cancer and cardiovascular disease. N Engl J Med 1996;334:1150.

Various authors: Symposium: Antioxidant vitamins and β-carotene in disease prevention. Amer J Clin Nutr 1995;62(suppl 6):12995–15405.

Various authors: Symposium Proceedings: Molecular mechanisms of protective effects of vitamin E in atherosclerosis. J Nutr 2001;131:366–397.

Intracellular Traffic & Sorting of Proteins

Robert K. Murray, MD, PhD

- Know that many proteins are targeted by signal sequences to their correct destinations and that the Golgi apparatus plays an important role in sorting proteins.

- Understand that specialized signals are involved in sorting proteins to mitochondria, the nucleus, and to peroxisomes.

- Appreciate that N-terminal signal peptides play a key role in directing newly synthesized proteins into the lumen of the endoplasmic reticulum.

- Know that chaperones prevent faulty folding of other proteins, that mechanisms exist for disposing of misfolded proteins, and that the endoplasmic reticulum acts as a quality control compartment.

- Comprehend that ubiquitin is a key molecule in protein degradation.

- Recognize the important role of transport vesicles in intracellular transport.

- Appreciate that many diseases result from mutations in genes encoding proteins involved in intracellular transport and be familiar with the terms conformational diseases and diseases of proteostatic deficiency.

BIOMEDICAL IMPORTANCE

Proteins must travel from polyribosomes, where they are synthesized, to many different sites in the cell to perform their particular functions. Some are destined to be components of specific organelles, others for the cytosol or for export, and yet others will be located in the various cellular membranes. Thus, there is considerable **intracellular traffic of proteins.** A major insight was the recognition by Blobel and subsequently others that for proteins to attain their proper locations, they generally contain **information** (a signal or coding sequence) that **targets** them appropriately. Once a number of the signals were defined (see **Table 46–1**), it became apparent that **certain diseases** result from mutations that affect these signals. In this chapter, we discuss the intracellular traffic of proteins and their sorting and briefly consider some of the disorders that result when abnormalities occur.

MANY PROTEINS ARE TARGETED BY SIGNAL SEQUENCES TO THEIR CORRECT DESTINATIONS

The protein biosynthetic pathways in cells can be considered to be **one large sorting system.** Many proteins carry **signals** (usually but not always specific sequences of amino acids) that direct them to their destination, thus ensuring that they will end up in the appropriate membrane or cell compartment; these signals are a fundamental component of the sorting system. Usually, the signal sequences are recognized and interact with complementary areas of other proteins that serve as receptors for those containing the signals.

A **major sorting decision** is made early in protein biosynthesis, when specific proteins are synthesized either on **free** or on **membrane-bound polyribosomes.** It is important to understand that these two types of ribosomes are identical

TABLE 46–1 Some Sequences or Molecules That Direct Proteins to Specific Organelles

Targeting Sequence or Compound	Organelle Targeted
N-terminal signal peptide	ER
Carboxyl-terminal KDEL sequence (Lys-Asp-Glu-Leu) in ER-resident proteins in COPI vesicles	Lumen of ER
Di-acidic sequences (eg, Asp-X-Glu) in membrane proteins in COPII vesicles	Golgi membranes
Amino terminal sequence (20–50 residues)	Mitochondrial matrix
NLS (eg, Pro$_2$-Lys$_3$-Arg-Lys-Val)	Nucleus
PTS (eg, Ser-Lys-Leu)	Peroxisome
Mannose 6-phosphate	Lysosome

Abbreviations: NLS, nuclear localization signal; PTS, peroxisomal-matrix targeting sequence.

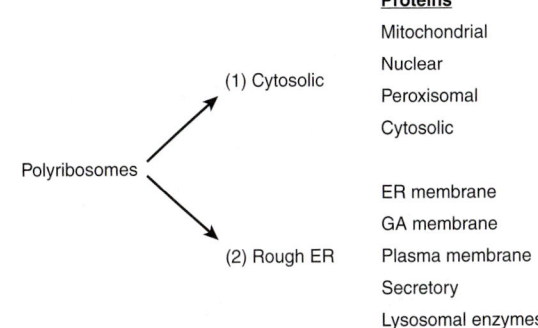

FIGURE 46–1 **Diagrammatic representation of the two branches of protein sorting occurring by synthesis on (1) cytosolic and (2) membrane-bound polyribosomes.** The mitochondrial proteins listed are encoded by nuclear genes; one of the signals used in further sorting of mitochondrial matrix proteins is listed in Table 46–1. (ER, endoplasmic reticulum; GA, Golgi apparatus.)

in structure and potentially interchangeable. However, if a growing polypeptide chain attached to polyribosomes lacks an N-terminal signal peptide (see below), it will not interact with the membrane of the endoplasmic reticulum (ER) and the polyribosomes are described as cytosolic. On the other hand, a growing polypeptide chain that contains an N-terminal signal peptide will interact with the membrane of the ER, and the polyribosomes to which it is attached are described as membrane-bound. This results in two sorting branches, called the **cytosolic branch** and the **rough ER (RER) branch** (**Figure 46–1**).

Proteins synthesized by cytosolic polyribosomes are directed to mitochondria, nuclei, and peroxisomes by specific signals, or remain in the cytosol if they lack a signal. Any protein that contains a targeting sequence that is subsequently removed is designated as a **preprotein.** In some cases, a second peptide is also removed, and in that event the original protein is known as a **preproprotein** (eg, preproalbumin; Chapter 50).

Proteins synthesized and sorted in the **rough ER branch** (Figure 46–1) include many destined for various membranes (eg, of the ER, Golgi apparatus [GA], plasma membrane [PM]) and for secretion. Lysosomal enzymes are also included. These various proteins may thus reside in the membranes or lumen of the ER, or follow the major transport route of intracellular proteins to the GA. The entire pathway of ER → GA→ plasma membrane is often called the **secretory** or **exocytotic pathway.** The secretory pathway was first delineated by the work of George Palade and colleagues, using radioactive amino acids and radioautography to follow the fate of proteins synthesized in the exocrine pancreas. Events along the secretory pathway will be given special attention here. Proteins destined for the GA, the PM, certain other sites, or for secretion are carried in **transport vesicles** (**Figure 46–2**); a brief description of the formation of these important particles will be given subsequently. Certain other proteins destined for secretion are carried in secretory vesicles (Figure 46–2). These are prominent

in the pancreas and certain other glands. Their mobilization and discharge are regulated and often referred to as "**regulated secretion**". In contrast, transport of vesicles occurring continuously through the secretory pathway is referred to as "**constitutive transport**". Passage of enzymes to the lysosomes using the mannose 6-phosphate signal is described in Chapter 47.

The Golgi Apparatus Is Involved in Glycosylation & Sorting of Proteins

The **GA** plays two major roles in protein synthesis. First, it is involved in the **processing of the oligosaccharide chains** of membrane and other N-linked glycoproteins and also contains enzymes involved in O-glycosylation (see Chapter 47). Second, it is involved in the **sorting** of various proteins prior to their delivery to their appropriate intracellular destinations. All parts of the GA participate in the first role, whereas the **trans-Golgi network** (TGN) is particularly involved in the second and is very rich in vesicles.

A Wide Variety of Experimental Techniques Have Been Used to Investigate Trafficking and Sorting

Approaches that have afforded major insights to the processes described in this chapter include (1) electron microscopy; (2) use of yeast mutants; (3) subcellular fractionation; (4) application of recombinant DNA techniques (eg, mutating or eliminating particular sequences in proteins, or fusing new sequences onto them); and (5) development of in vitro systems (eg, to study translocation into the ER and mechanisms of vesicle formation); (6) use of fluorescent tags to follow the movement of proteins; and (7) structural studies on certain proteins, particularly by x-ray crystallography.

The sorting of proteins belonging to **the cytosolic branch** referred to above is described next, starting with mitochondrial proteins.

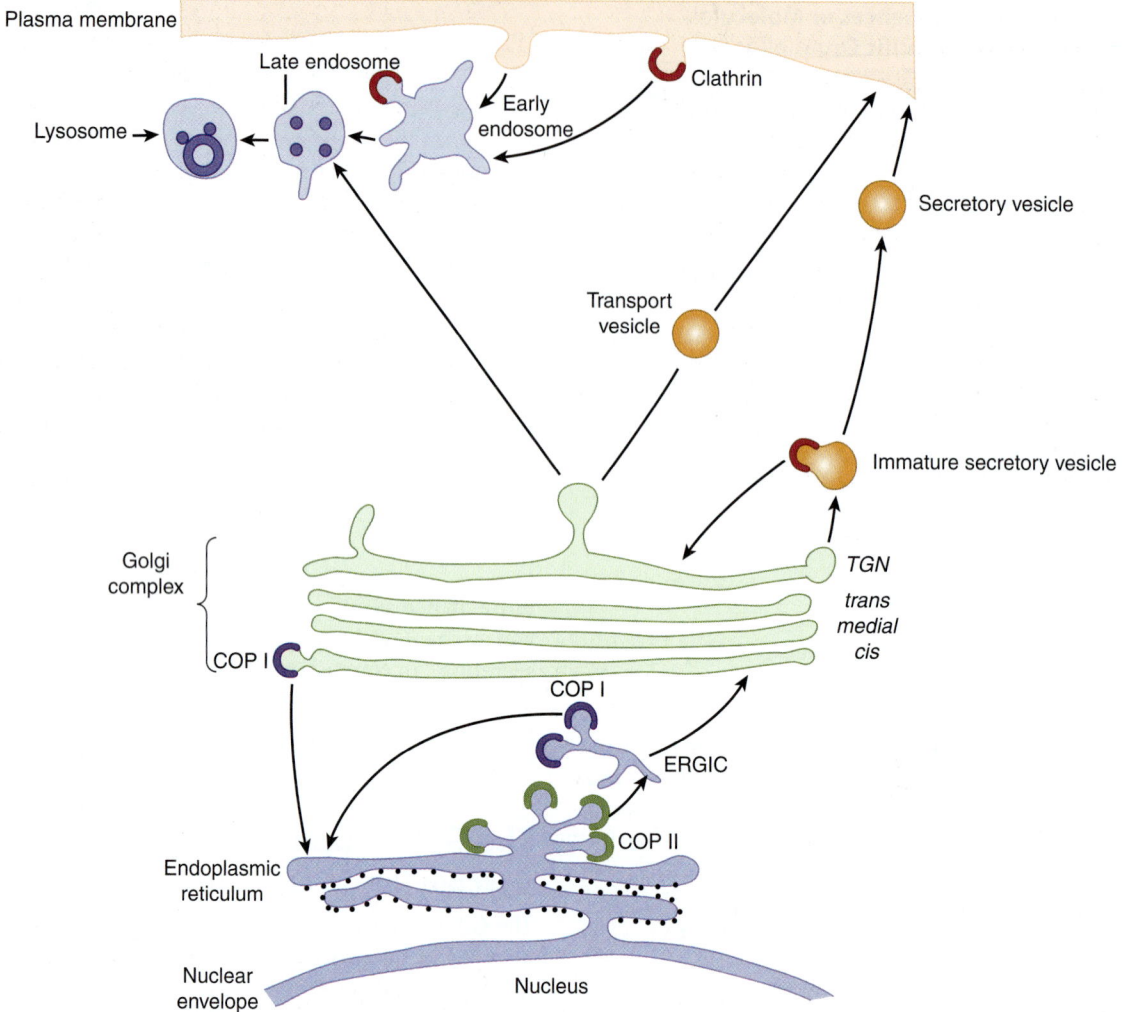

FIGURE 46–2 **Diagrammatic representation of the rough ER branch of protein sorting.** Newly synthesized proteins are inserted into the ER membrane or lumen from membrane-bound polyribosomes (small black circles studding the cytosolic face of the ER). Proteins that are transported out of the ER are carried in COPII vesicles to the cis-Golgi (anterograde transport). Movement of proteins through the Golgi appears to be mainly by cisternal maturation. In the TGN, the exit side of the Golgi, proteins are segregated and sorted. Secretory proteins accumulate in secretory vesicles (regulated secretion), from which they are expelled at the plasma membrane. Proteins destined for the plasma membrane or those that are secreted in a constitutive manner are carried out to the cell surface in transport vesicles (constitutive secretion). Clathrin-coated vesicles are involved in endocytosis, carrying cargo to late endosomes and to lysosomes. Mannose 6-phosphate (not shown; see Chapter 47) acts as a signal for transporting enzymes to lysosomes. COPI vesicles are involved in retrieving proteins from the Golgi to the ER (retrograde transport) and may be involved in some intra-Golgi transport. COP II vesicles are involved in concentrating cargo for export from the ER to the GA. Cargo normally passes through the ERGIC compartment to the GA. (TGN, *trans*-Golgi network; ERGIC, ER-Golgi intermediate complex.) (Courtesy of E Degen.)

THE MITOCHONDRION BOTH IMPORTS & SYNTHESIZES PROTEINS

Mitochondria contain many proteins. Thirteen polypeptides (mostly membrane components of the electron transport chain) are encoded by the **mitochondrial (mt) genome** and synthesized in that organelle using its own protein-synthesizing system. However, the majority (at least several hundred) are encoded by **nuclear genes,** are synthesized outside the mitochondria on cytosolic **polyribosomes,** and must be imported. **Yeast cells** have proved to be a particularly useful system for analyzing the mechanisms of import of mitochondrial proteins, partly because it has proved possible to generate a variety of **mutants** that have illuminated the fundamental processes involved. Most progress has been made in the study of proteins present in **the mitochondrial matrix,** such as the F_1 ATPase subunits. Only the pathway of import of matrix proteins will be discussed in any detail here.

Matrix proteins must pass from cytosolic polyribosomes through the **outer** and **inner mitochondrial membranes** to reach their destination. Passage through the two membranes is called **translocation.** They have an amino terminal leader sequence **(presequence),** about 20–50 amino acids in length (see Table 46–1), which is not highly conserved but is amphipathic and contains many hydrophobic and positively charged amino acids (eg, Lys or Arg). The presequence is equivalent to a signal peptide mediating attachment of polyribosomes to membranes of the ER (see below), but in this instance **targeting proteins to the matrix.** Some general features of the passage of a protein from the cytosol to the mt matrix are shown in **Figure 46–3.**

Translocation occurs **posttranslationally,** after the matrix proteins are released from the cytosolic polyribosomes. Interactions with a number of cytosolic proteins that act as **chaperones** (see below) and as **targeting factors** occur prior to translocation.

Two distinct **translocation complexes** are situated in the outer and inner mitochondrial membranes, referred to (respectively) as TOM (translocase-of-the-outer membrane) and TIM (translocase-of-the-inner membrane). Each complex has been analyzed and found to be composed of a number of proteins, some of which act as **receptors** (eg, Tom20/22) for the incoming proteins and others as **components** (eg, Tom40) **of the transmembrane pores** through which these proteins must pass. Proteins must be in the **unfolded state** to pass through the complexes, and this is made possible by **ATP-dependent binding to several chaperone proteins.** The roles of chaperone proteins in protein folding are discussed later in this chapter. In mitochondria, they are involved in translocation, sorting, folding, assembly, and degradation of imported

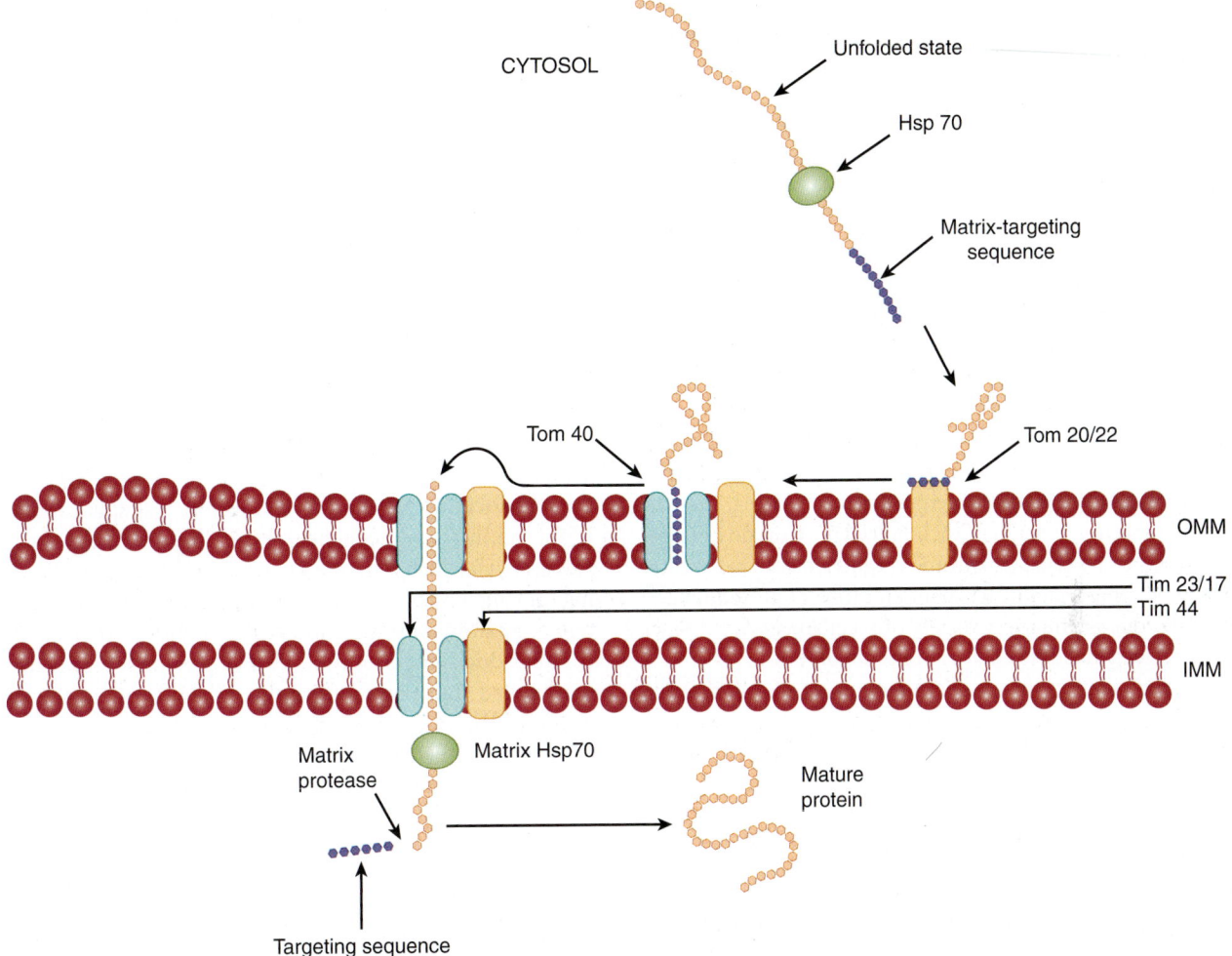

FIGURE 46–3 **Schematic representation of the entry of a protein into the mitochondrial matrix.** The unfolded protein synthesized on cytosolic poyribosomes and containing a matrix-targeting sequence interacts with the cytosolic chaperone Hsp 70. The protein next interacts with the mt outer membrane receptor Tom 20/22, and is transferred to the neighboring import channel Tom 40 (Tom, translocon of the outer membrane). The protein is then translocated across the channel; the channel on the inner mt membrane is largely composed of Tim 23 and Tim 17 proteins (Tim, translocon of the inner membrane). On the inside of the inner mt membrane, it interacts with the matrix chaperone Hsp 70, which in turn interacts with membrane protein Tim 44. The hydrolysis of ATP by mt Hsp70 probably helps drive the translocation, as does the electronegative interior of the matrix. The targeting sequence is subsequently cleaved by the matrix processing enzyme, and the imported protein assumes its final shape, or may interact with an mt chaperonin prior to this. At the site of translocation, the outer and inner mt membranes are in close contact. OMM, outer mitochondrial membrane; IMM, inner mitochondrial membrane. (Modified, with permission, from Lodish H, et al: *Molecular Cell Biology*, 6th ed. W.H. Freeman & Co., 2008.)

proteins. A **proton-motive force** across the inner membrane is required for import; it is made up of the **electric potential** across the membrane (inside negative) and the **pH gradient** (see Chapter 13). The positively charged leader sequence may be helped through the membrane by the negative charge in the matrix. The presequence is split off in the matrix by a **matrix-processing protease (MPP)**. Contact with **other chaperones** present in the matrix is essential to complete the overall process of import. Interaction with mt-Hsp70 (mt = mitochondrial; Hsp = heat shock protein; 70 = ~70 kDa) ensures proper import into the matrix and prevents misfolding or aggregation, while interaction with the mt-Hsp60–Hsp10 system ensures proper folding. The interactions of imported proteins with the above chaperones require **hydrolysis of ATP** to drive them.

The details of how preproteins are translocated have not been fully elucidated. It is possible that the electric potential associated with the inner mitochondrial membrane causes a conformational change in the unfolded preprotein being translocated and that this helps to pull it across. Furthermore, the fact that the matrix is more negative than the intermembrane space may "attract" the positively charged amino terminal of the preprotein to enter the matrix. Close apposition at **contact sites** between the outer and inner membranes is necessary for translocation to occur.

The above describes the major pathway of proteins destined for the mitochondrial matrix. However, certain proteins insert into the **outer mitochondrial membrane** facilitated by the TOM complex. Others stop in the **intermembrane space,** and some insert into the **inner membrane.** Yet others proceed into the matrix and then return to the inner membrane or intermembrane space. A number of proteins contain two signaling sequences—one to enter the mitochondrial matrix and the other to mediate subsequent relocation (eg, into the inner membrane). Certain mitochondrial proteins do not contain presequences (eg, cytochrome *c*, which locates in the inter membrane space), and others contain **internal presequences.** Overall, proteins employ a variety of mechanisms and routes to attain their final destinations in mitochondria.

General features that apply to the import of proteins into organelles, including mitochondria and some of the other organelles to be discussed below, are summarized in **Table 46–2.**

LOCALIZATION SIGNALS, IMPORTINS, & EXPORTINS ARE INVOLVED IN TRANSPORT OF MACROMOLECULES IN & OUT OF THE NUCLEUS

It has been estimated that more than a million macromolecules per minute are transported between the nucleus and the cytoplasm in an active eukaryotic cell. These macromolecules include histones, ribosomal proteins and ribosomal subunits, transcription factors, and mRNA molecules. The transport is bidirectional and occurs through the **nuclear pore complexes**

TABLE 46–2 Some General Features of Protein Import to Organelles

- Import of a protein into an organelle usually occurs in three stages: recognition, translocation, and maturation.
- Targeting sequences on the protein are recognized in the cytoplasm or on the surface of the organelle.
- The protein is generally unfolded for translocation, a state maintained in the cytoplasm by chaperones.
- Threading of the protein through a membrane requires energy and organellar chaperones on the *trans* side of the membrane.
- Cycles of binding and release of the protein to the chaperone result in pulling of its polypeptide chain through the membrane.
- Other proteins within the organelle catalyze folding of the protein, often attaching cofactors or oligosaccharides and assembling them into active monomers or oligomers.

Source: Data from McNew JA, Goodman JM: The targeting and assembly of peroxisomal proteins: some old rules do not apply. Trends Biochem Sci 1998;21:54. Reprinted, with permission, from Elsevier.

(NPCs). These are complex structures with a mass approximately 15 times that of a ribosome and are composed of aggregates of about 30 different proteins. The minimal diameter of an NPC is approximately 9 nm. Molecules smaller than about 40 kDa can pass through the channel of the NPC by **diffusion,** but **special translocation mechanisms** exist for larger molecules. These mechanisms are under intensive investigation, but some important features have already emerged.

Here we shall mainly describe **nuclear import** of certain macromolecules. The general picture that has emerged is that proteins to be imported (cargo molecules) carry a **nuclear localization signal (NLS).** One example of an NLS is the amino acid sequence $(Pro)_2$-$(Lys)_3$-Arg-Lys-Val (see Table 46–1), which is markedly rich in basic residues. Depending on which NLS it contains, a cargo molecule interacts with one of a family of soluble proteins called **importins,** and the complex **docks** transiently at the NPC. Another family of proteins called **Ran** plays a critical regulatory role in the interaction of the complex with the NPC and in its translocation through the NPC. Ran proteins are small monomeric nuclear **GTPases** and, like other GTPases, exist in either GTP-bound or GDP-bound states. They are themselves regulated by **guanine nucleotide exchange factors (GEFs),** which are located in the nucleus, and Ran **GTPase-accelerating proteins (GAPs),** which are predominantly cytoplasmic. The GTP-bound state of Ran is favored in the nucleus and the GDP-bound state in the cytoplasm. The conformations and activities of Ran molecules vary depending on whether GTP or GDP is bound to them (the GTP-bound state is active; see discussion of G proteins in Chapter 42). The **asymmetry** between nucleus and cytoplasm—with respect to which of these two nucleotides is bound to Ran molecules—is thought to be crucial in understanding the roles of Ran in transferring complexes unidirectionally across the NPC. When **cargo molecules** are **released inside the nucleus, the importins recirculate to the**

cytoplasm to be used again. **Figure 46–4** summarizes some of the principal features in the above process.

Proteins similar to importins, referred to as **exportins,** are involved in the export of many macromolecules (various proteins, tRNA molecules, ribosomal subunits and certain mRNA molecules) from the nucleus. Cargo molecules for export carry **nuclear export signals (NESs).** Ran proteins are involved in this process also, and it is now established that the processes of import and export share a number of common features. The family of importins and exportins are referred to as **karyopherins.**

Another system is involved in the translocation of the majority of **mRNA molecules.** These are exported from the nucleus to the cytoplasm as ribonucleoprotein (RNP) complexes attached to a protein named **mRNP exporter.** This is a heterodimeric molecule (ie, composed of two different subunits, TAP and Nxt-1) that carries RNP molecules through the NPC. Ran is not involved. This system appears to use the hydrolysis of **ATP** by an RNA helicase (Dbp5) to drive translocation.

Other **small monomeric GTPases** (eg, ARF, Rab, Ras, and Rho) are important in various cellular processes such as vesicle formation and transport (ARF and Rab; see below), certain growth and differentiation processes (Ras), and formation of the actin cytoskeleton (Rho). A process involving GTP and GDP is also crucial in the transport of proteins across the membrane of the ER (see below).

PROTEINS IMPORTED INTO PEROXISOMES CARRY UNIQUE TARGETING SEQUENCES

The **peroxisome** is an important organelle involved in aspects of the metabolism of many molecules, including fatty acids and other lipids (eg, plasmalogens, cholesterol, bile acids), purines, amino acids, and hydrogen peroxide. The peroxisome is bounded by a single membrane and contains more than 50 enzymes; catalase and urate oxidase are marker enzymes for

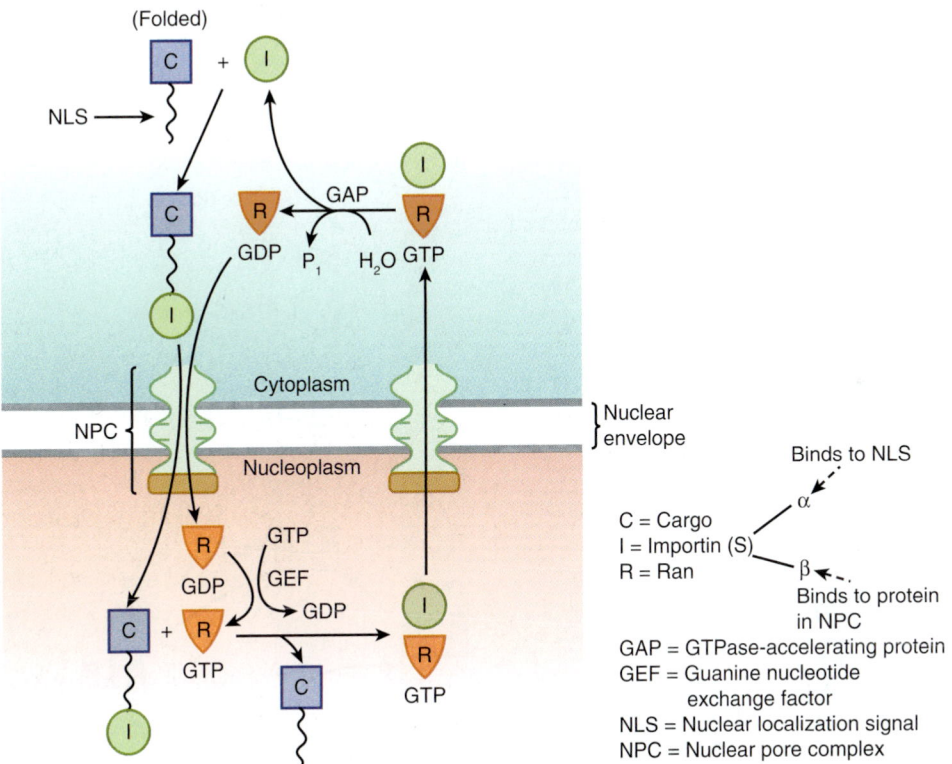

FIGURE 46–4 **Simplified representation of the entry of a protein into the nucleoplasm.** As shown in the top left-hand side of the figure, a cargo molecule in the cytoplasm via its NLS interacts to form a complex with an importin. (This may be either importin α or both importin α and importin β). This complex next interacts with Ran·GDP and traverses the NPC into the nucleoplasm. In the nucleoplasm, Ran·GDP is converted to Ran·GTP by GEF, causing a conformational change in Ran resulting in the cargo molecule being released. The importin-Ran·GTP complex then leaves the nucleoplasm via the NPC to return to the cytoplasm. In the cytoplasm, due to the action of GTPase-accelerating protein (GAP), which converts GTP to GDP, the importin is released to participate in another import cycle. The Ran·GTP is the active form of the complex, with the Ran·GDP form being considered inactive. Directionality is believed to be conferred on the overall process by the dissociation of Ran·GTP in the cytoplasm. (C, cargo molecule; GAP, GTPase-accelerating protein; GEF, guanine nucleotide exchange factor; I, importin; NLS, nuclear localizing signal; NPC, nuclear pore complex.) (Modified, with permission, from Lodish H, et al: *Molecular Cell Biology,* 6th ed. W.H. Freeman & Co., 2008.)

this organelle. Its proteins are **synthesized on cytosolic poly-ribosomes** and fold prior to import. The pathways of import of a number of its proteins and enzymes have been studied, some being **matrix components** (see **Figure 46–5**) and others **membrane components**. At least two **peroxisomal–matrix targeting sequences (PTSs)** have been discovered. One, **PTS1,** is a tripeptide (ie, Ser-Lys-Leu [SKL], but variations of this sequence have been detected) located at the carboxyl terminal of a number of matrix proteins, including catalase. Another, **PTS2,** is at the N-terminus and has been found in at least four matrix proteins (eg, thiolase). Neither of these two sequences is cleaved after entry into the matrix. Proteins containing PTS1 sequences **form complexes** with a cytosolic receptor protein (**Pex5**) and proteins containing PTS2 sequences complex with another receptor protein. The resulting complexes then interact with a membrane receptor complex, **Pex2/10/12,** which translocates them into the matrix. Proteins involved in further transport of proteins into the matrix are also present. Pex5 is recycled to the cytosol. Most peroxisomal **membrane proteins** have been found to contain neither of the above two targeting sequences, but apparently contain others. The import system can handle **intact oligomers** (eg, tetrameric catalase). Import of **matrix proteins** requires **ATP,** whereas import of **membrane proteins** does **not.**

Most Cases of Zellweger Syndrome Are Due to Mutations in Genes Involved in the Biogenesis of Peroxisomes

Interest in import of proteins into peroxisomes has been stimulated by studies on **Zellweger syndrome.** This condition is apparent at birth and is characterized by **profound neurologic impairment,** victims often dying within a year. The number of peroxisomes can vary from being almost normal to being virtually absent in some patients. Biochemical findings include an accumulation of very-long-chain fatty acids, abnormalities of the synthesis of bile acids, and a marked reduction of plasmalogens. The condition is believed to be due to **mutations** in genes encoding certain proteins—so-called **peroxins**—involved in various steps of **peroxisome biogenesis** (such as the import of proteins described above), or in genes encoding certain peroxisomal enzymes themselves. Two closely related conditions are **neonatal adrenoleukodystrophy** and **infantile Refsum disease.** Zellweger syndrome and these two conditions represent a **spectrum** of overlapping features, with Zellweger syndrome being the **most severe** (many proteins affected) and infantile Refsum disease the least severe (only one or a few proteins affected). **Table 46–3** lists these and related conditions.

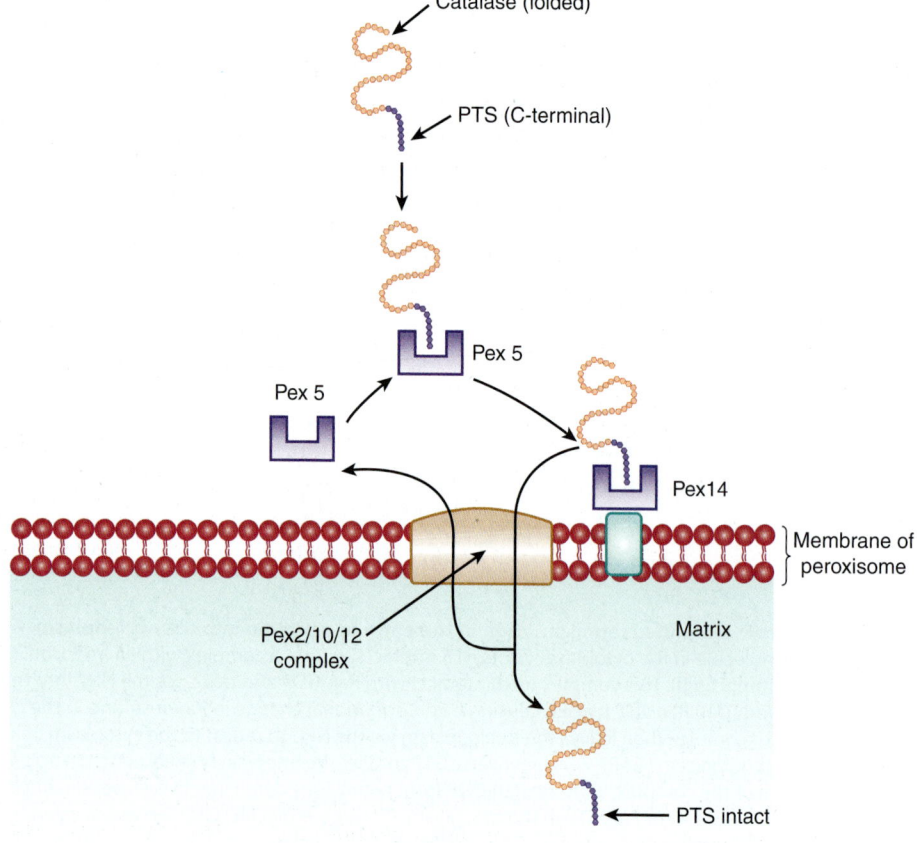

FIGURE 46–5 **Schematic representation of the entry of a protein into the peroxisomal matrix.** The protein to be imported into the matrix is synthesized on cytosolic polyribosomes, assumes its folded shape prior to import, and contains a C-terminal peroxisomal-targeting sequence (PTS). It interacts with cytosolic receptor protein Pex5, and the complex then interacts with a receptor on the peroxisomal membrane, Pex14. In turn, the protein-Pex 14 complex passes to the Pex 2/10/12 complex on the peroxisomal membrane and is translocated. Pex 5 is returned to the cytosol. The protein retains its PTS in the matrix. (Modified, with permission, from Lodish H, et al: *Molecular Cell Biology*, 6th ed. W.H. Freeman & Co., 2008.)

TABLE 46–3 Disorders Due to Peroxisomal Abnormalities

	OMIM Number[1]
Zellweger syndrome	214100
Neonatal adrenoleukodystrophy	202370
Infantile Refsum disease	266510
Hyperpipecolic acidemia	239400
Rhizomelic chondrodysplasia punctata	215100
Adrenoleukodystrophy	300100
Pseudoneonatal adrenoleukodystrophy	264470
Pseudo-Zellweger syndrome	261515
Hyperoxaluria type 1	259900
Acatalasemia	115500
Glutaryl-CoA oxidase deficiency	231690

Source: Reproduced, with permission, from Seashore MR, Wappner RS: *Genetics in Primary Care & Clinical Medicine.* Appleton & Lange, 1996.
[1]OMIM, *Online Mendelian Inheritance in Man.* Each number specifies a reference in which information regarding each of the above conditions can be found.

THE SIGNAL HYPOTHESIS EXPLAINS HOW POLYRIBOSOMES BIND TO THE ENDOPLASMIC RETICULUM

As indicated above, the **rough ER branch** is the second of the two branches involved in the synthesis and sorting of proteins. In this branch, proteins are synthesized on **membrane-bound polyribosomes** and are usually **translocated into the lumen** of the rough ER prior to further sorting (Figure 46–2). Certain membrane proteins, however, are transferred directly into the membrane of the ER without reaching its lumen.

The **signal hypothesis** was proposed by Blobel and Sabatini in 1971 partly to explain the distinction between free and membrane-bound polyribosomes. On the basis of certain experimental findings, they proposed that proteins synthesized on membrane-bound polyribosomes contained an N-terminal peptide extension (**N-terminal signal peptide**) which mediated their attachment to the membranes of the ER, and facilitated transfer into the ER lumen, On the other hand, proteins whose entire synthesis occurs on free polyribosomes would lack this signal peptide. An important aspect of the signal hypothesis was that it suggested—as turns out to be the case—that **all ribosomes have the same structure** and that the distinction between membrane-bound and free ribosomes depends solely on the former carrying proteins that have signal peptides. Because many membrane proteins are synthesized on membrane-bound polyribosomes, the signal hypothesis plays an important role in **concepts of membrane assembly.** Some **characteristics of** N-terminal **signal peptides** are summarized in **Table 46–4.**

TABLE 46–4 Some Properties of Signal Peptides Directing Proteins to the ER

- Usually, but not always, located at the amino terminal
- Contain approximately 12–35 amino acids
- Methionine is usually the amino terminal amino acid
- Contain a central cluster (~6 to 12) of hydrophobic amino acids
- The region near the N-terminus usually carries a net positive charge
- The amino acid residue at the cleavage site is variable, but residues −1 and −3 relative to the cleavage site must be small and neutral

There is much **evidence to support** the signal hypothesis, confirming that the N-terminal signal peptide is involved in the process of translocation across the ER membrane. For example, mutant proteins containing altered signal peptides in which hydrophobic amino acids are replaced by hydrophilic ones, are not inserted into the lumen of the ER. Non-membrane proteins (eg, α-globin) to which signal peptides have been attached by genetic engineering can be inserted into the lumen of the ER, or even secreted.

Many Details of the ER Translocation Process Have Been Revealed

Since the original formulation of the signal hypothesis, many scientists have contributed to revealing details of the overall process of translocation of nascent proteins across the ER membrane into its lumen. **Table 46–5** lists some principal components of the overall process, and major steps in it are summarized in **Figure 46–6.**

Step 1: The signal sequence emerges from the ribosome and binds to the SRP. This temporarily arrests further elongation of the polypeptide chain (elongation arrest) after some 70 amino acids have been polymerized.

Step 2: The SRP–ribosome–nascent protein complex travels to the ER membrane, where it binds to the SRP receptor (SRP-R). The SRP guides the complex to the SR, which prevents premature expulsion of the growing polypeptide into the cytosol.

TABLE 46–5 Principal Components Involved in ER Translocation

- N-terminal signal peptide
- Polyribosomes
- SRP, signal recognition particle
- SR, signal recognition particle receptor
- Sec 61, the translocon
- Signal peptidase
- Associated proteins (eg, TRAM and TRAP)

TRAM, translocating chain associated membrane protein; TRAP, translocon-associated protein complex. TRAM accelerates the translocation of certain proteins. The function of TRAP is not clear.

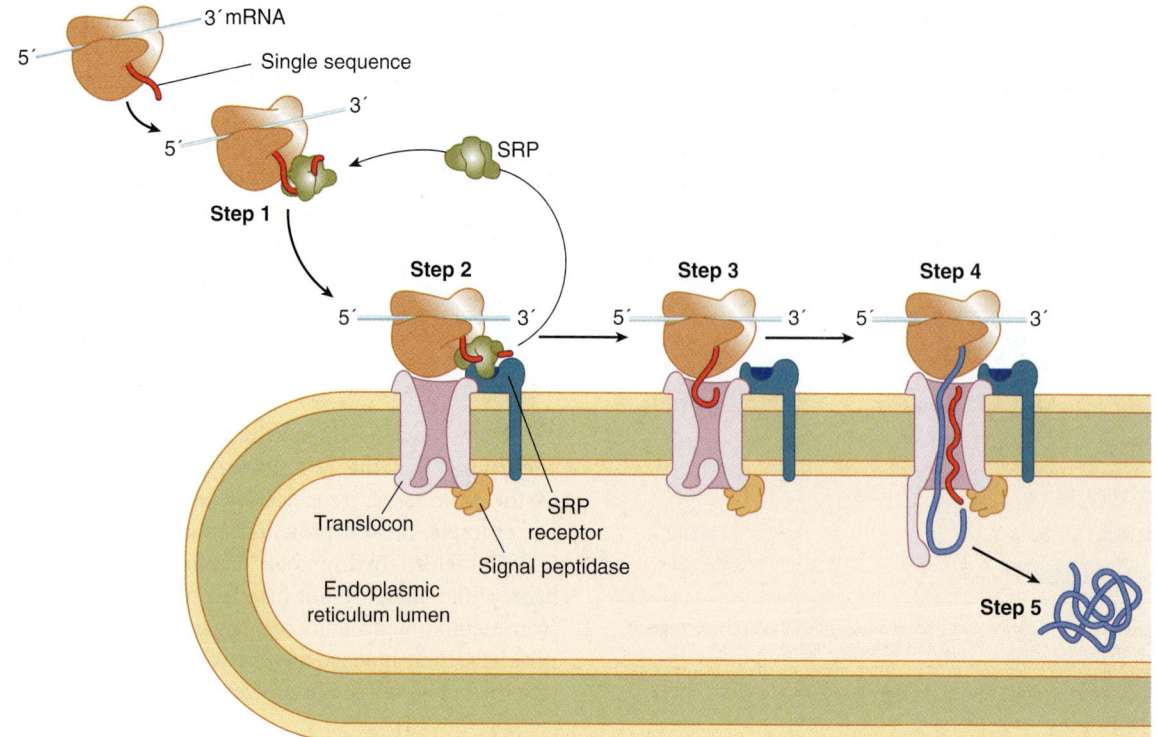

FIGURE 46–6 **Cotranslational targeting of secretory proteins to the ER. Step 1:** As the signal sequence emerges from the ribosome, it is recognized and bound by the signal recognition particle (SRP). **Step 2:** The SRP escorts the complex to the ER membrane where it binds to the SRP receptor (SR). **Step 3:** The SRP is released, the ribosome binds to the translocon, and the signal sequence is inserted into the membrane channel. **Step 4:** The signal sequence opens the translocon. Translation resumes and the growing polypeptide chain is translocated across the membrane. **Step 5:** Cleavage of the signal sequence by signal peptidase releses the polypeptide into the lumen of the ER. Reproduced, with permission, from Cooper GM, Hausman RE: *The Cell: A Molecular Approach.* Sinauer Associates, Inc. 2009.

Step 3: The SRP is released, translation resumes, the ribosome binds to the translocon (Sec 61 complex), and the signal peptide inserts into the channel in the translocon. As translation is still occurring, the entry of the signal peptide into the translocon and its further passage is termed **cotranslational translocation.**

Step 4: The signal peptide induces opening of the channel in the translocon by binding to certain hydrophobic residues in it, thus causing the plug (shown at the bottom on the translocon in Figure 46–6) to move. The growing polypeptide is then fully translocated across the membrane, driven by its ongoing synthesis.

Step 5: Cleavage of the signal peptide by signal peptidase occurs, and the fully translocated polypeptide/protein is released into the lumen of the ER. The signal peptide is presumably degraded by proteases. Ribosomes are released from the ER membrane and dissociate into their two types of subunits.

In yeast, many proteins are targeted to the ER after their translation is completed **(posttranslational translocation)** by a process that does not require the SRP. It does involve cytosolic chaperones (such as Hsp70) to keep the protein unfolded and also the luminal chaperone BiP, which may "pull" the growing polypeptide into the ER lumen. Some mammalian proteins also undergo this process.

Additional Comments on SRP, SRP Receptor, GTP, Sec61, and Glycosylation

The signal recognition particle (SRP) contains **six proteins** and has a **7S RNA** associated with it that is closely related to the Alu family of highly repeated DNA sequences (Chapter 35). Both the RNA molecule and its proteins play various roles (such as binding other molecules) in its function.

The **SRP receptor (SR)** is an ER membrane protein composed of **α** and **β subunits,** the latter spanning the ER membrane.

SRP and both subunits of the SR can bind **GTP.** Both the SRP and SR must be in the GTP form to interact. When they bind, hydrolysis of GTP is stimulated, SRP is released, and the ribosome binds to the translocon allowing the signal peptide to enter it. The SRP and SR act as GTPase-acceleratng proteins (GAPs). When GTP is hydrolyzed to GDP, they **dissociate.** SRP and SR can be regarded as **molecular matchmakers.** SRP picks up the ribosome with its nascent chain and exposed signal peptide and the SR associates with the empty translocon pore, likely via its β subunit. The overall result of their interaction is to bring the ribosome to the translocon.

The **translocon** consists of three membrane proteins (the **Sec61 complex**) that form a **protein-conducting channel** in the ER membrane through which the newly synthesized protein may pass. As mentioned above, the channel appears to be **open only when a signal peptide is present,** preserving conductance across the ER membrane when it closes. The conductance of the channel has been measured experimentally. Closure of the channel when proteins are not being translocated prevents ions such as calcium and other molecules leaking through it, and causing cell dysfunction.

The insertion of the signal peptide into the conducting channel, while the other end of the parent protein is still attached to ribosomes, is termed **cotranslational insertion.** The process of elongation of the remaining portion of the protein being synthesized probably facilitates passage of the nascent protein across the lipid bilayer. It is important that proteins be kept in an **unfolded state** prior to entering the conducting channel—otherwise, they may not be able to gain access to the channel.

Secretory proteins and **soluble proteins destined for organelles distal to the ER** completely traverse the membrane bilayer and are discharged into the lumen of the ER. Many secretory proteins are N-glycosylated. *N*-Glycan chains, if present, are added by the enzyme oligosaccharide:protein transferase (Chapter 47) as these proteins traverse the inner part of the ER membrane—a process called **cotranslational glycosylation.** Subsequently, these glycoproteins are found in the **lumen of the Golgi apparatus,** where further changes in glycan chains occur (Figure 47–9) prior to intracellular distribution or secretion.

In contrast, proteins embedded in **membranes of the ER** as well as in **other membranes** along the secretory pathway only **partially translocate** across the ER membrane (see below). They are able to insert into the ER membrane by lateral transfer through the wall of the translocon.

There is evidence that the ER membrane is involved in **retrograde transport** of various molecules from the ER lumen **to the cytosol.** These molecules include unfolded or misfolded glycoproteins, glycopeptides, and oligosaccharides. At least some of these molecules are **degraded in proteasomes** (see below). The involvement of the translocon in retrotranslocation is not clear; one or more other channels may be involved. Whatever the case, there is **two-way traffic** across the ER membrane.

PROTEINS FOLLOW SEVERAL ROUTES TO BE INSERTED INTO OR ATTACHED TO THE MEMBRANES OF THE ENDOPLASMIC RETICULUM

The routes that proteins follow to be inserted into the membranes of the ER include the following.

Cotranslational Insertion

Figure 46–7 shows a variety of ways in which proteins are distributed in membranes. In particular, the **amino termini** of certain proteins (eg, the LDL receptor) can be seen to be on the extracytoplasmic face, whereas for other proteins (eg,

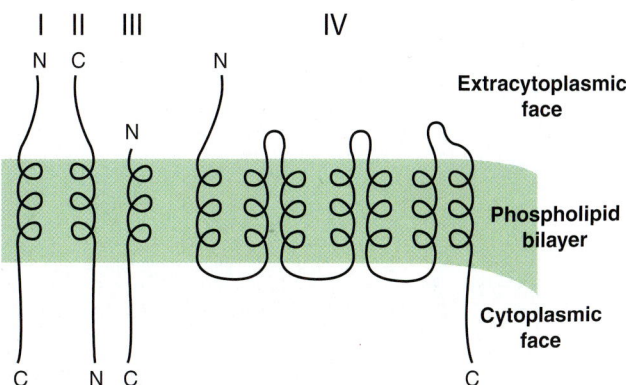

FIGURE 46–7 **Variations in the way in which proteins are inserted into membranes.** This schematic representation, which illustrates a number of possible orientations, shows the segments of the proteins within the membrane as α-helices and the other segments as lines. The orientations form initially in the ER membrane. Type I transmembrane proteins (eg, the LDL receptor and influenza hemagglutinin) cross the membrane once and have their amino termini on the exterior aspect of the membrane. Type II transmembrane proteins (eg, the asialoglycoprotein and transferrin receptors) also cross the membrane once, but have their C-termini on the exterior aspect. Type III transmembrane proteins (eg, cytochrome P450) have a disposition similar to type I proteins, but do not contain a cleavable signal peptide. Type IV transmembrane proteins (eg, G-protein-coupled receptors and glucose transporters) cross the membrane a number of times (7 times for the former and 12 times for glucose transporters); they are also called polytopic membrane proteins. (C, carboxyl terminal; N, amino terminal.) (Adapted, with permission, from Wickner WT, Lodish HF (1985), "Multiple mechanisms of protein insertion into and across membranes". Science 230:400. Reprinted with permission from AAAS.)

the asialoglycoprotein receptor) the **carboxyl termini** are on this face. To explain these dispositions, one must consider the initial biosynthetic events at the ER membrane. The **LDL receptor** enters the ER membrane in a manner analogous to a secretory protein (Figure 46–6); it partly traverses the ER membrane, its signal peptide is cleaved, and its amino terminal protrudes into the lumen (see also Figure 46–13). However, it is retained in the membrane because it contains a highly hydrophobic segment, the halt- or stop-transfer signal (compare **Figure 46–8**). This sequence forms the single transmembrane segment of the protein and is its membrane-anchoring domain. The small patch of ER membrane in which the newly synthesized LDL receptor is located subsequently buds off as a component of a transport vesicle. As described below in the discussion of asymmetry of proteins and lipids in membrane assembly, the disposition of the receptor in the ER membrane is preserved in the vesicle (see Figure 46–13), which eventually fuses with the plasma membrane. In contrast, the **asialoglycoprotein receptor** lacks a cleavable N-terminal signal peptide, but possesses an internal insertion sequence, which inserts into the membrane but is not cleaved. This acts as an anchor, and its carboxyl terminus is extruded through the membrane. The more complex disposition of **a transporter** (eg, for glucose) can be explained

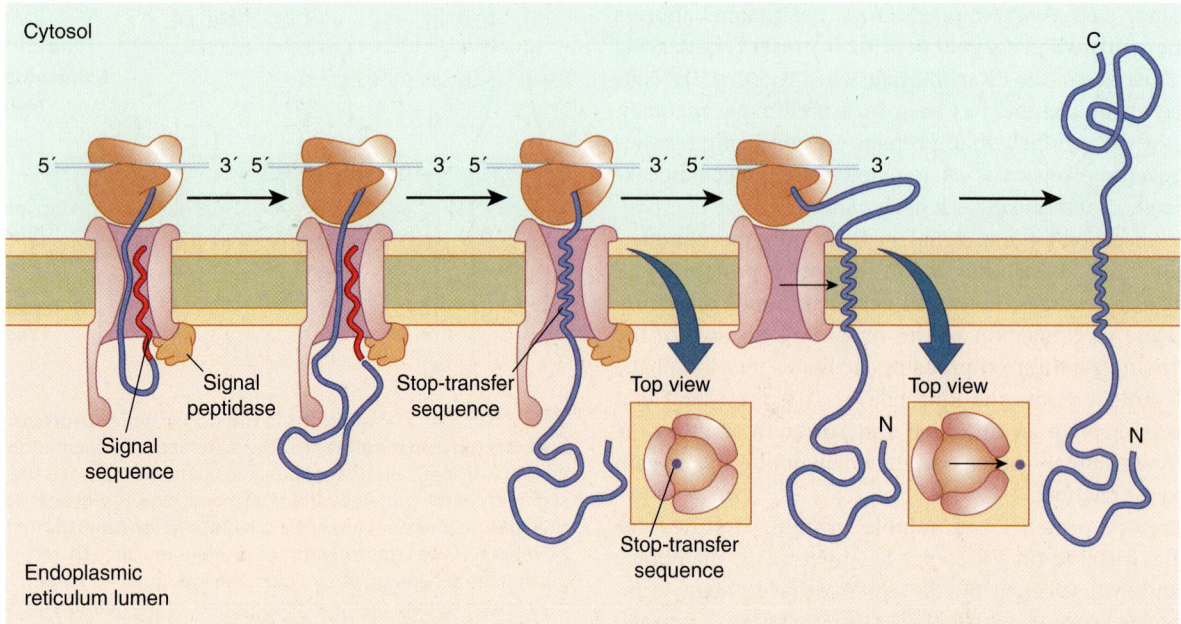

FIGURE 46–8 **Insertion of a membrane protein with a cleavable signal sequence and a single stop-transfer sequence.** The signal sequence is cleaved as the polypeptide chain crosses the membrane, so the amino terminus of the polypeptide chain is exposed in the ER lumen. However, translocation of the polypeptide chain across the membrane is halted when the translocon recognizes a transmembrane stop-transfer sequence. This closes the translocon and allows the protein to exit the channel laterally and become anchored in the ER membrane. Continued translation results in a membrane-spanning protein with its carboxy terminus on the cytosolic side. Reproduced with permission from Cooper GM Hausman RE: *The Cell: A Molecular Approach.* Sinauer Associates, Inc. 2009.

by the fact that alternating transmembrane α-helices act as uncleaved insertion sequences and as halt-transfer signals, respectively. Each pair of helical segments is inserted as a hairpin. Sequences that determine the structure of a protein in a membrane are called **topogenic sequences.** As explained in the legend to Figure 46–7, the above three proteins are examples of type I, type II, and type IV transmembrane proteins, whereas **cytochrome P450** is a member of type III.

Synthesis on Free Polyribosomes & Posttranslational Attachment to the Endoplasmic Reticulum Membrane

An example is **cytochrome b_5,** which appears to directly enter the ER membrane subsequent to translation, assisted by several chaperones.

Other Routes Include Retention in the GA with Retrieval to the ER and also Retrograde Transport from the GA

A number of proteins possess the amino acid sequence **KDEL** (Lys-Asp-Glu-Leu) at their carboxyl terminal (see Table 46–1). KDEL-containing proteins first travel to the **GA** in COPII transport vesicles (see below) and interact there with a specific KDEL receptor protein, which retains them transiently. They then **return in COPI transport vesicles to the ER,** where they dissociate from the receptor, and are thus retrieved. HDEL

sequences (H = histidine) serve a similar purpose. The above processes result in net localization of certain soluble proteins to the ER lumen.

Certain other **non-KDEL-containing proteins** also pass to the Golgi and then return, by **retrograde vesicular transport,** to the ER to be inserted therein These include vesicle components that must be recycled, as well as certain ER membrane proteins. These proteins often possess a C-terminal signal located in the cytosol rich in basic residues.

The foregoing paragraphs demonstrate that a **variety of routes** are involved in assembly of the proteins of the ER membranes. A similar situation probably holds for other membranes (eg, the mitochondrial membranes and the plasma membrane). Precise targeting sequences have been identified in some instances (eg, KDEL sequences).

The topic of membrane biogenesis is discussed further later in this chapter.

CHAPERONES ARE PROTEINS THAT PREVENT FAULTY FOLDING & UNPRODUCTIVE INTERACTIONS OF OTHER PROTEINS

Molecular chaperones have been referred to previously in this Chapter. A number of important properties of these proteins are listed in **Table 46–6,** and the names of some of

TABLE 46–6 Some Properties of Chaperone Proteins

- Present in a wide range of species from bacteria to humans
- Many are so-called heat shock proteins (Hsp)
- Some are inducible by conditions that cause unfolding of newly synthesized proteins (eg, elevated temperature and various chemicals)
- They bind to predominantly hydrophobic regions of unfolded proteins and prevent their aggregation
- They act in part as a quality control or editing mechanism for detecting misfolded or otherwise defective proteins
- Most chaperones show associated ATPase activity, with ATP or ADP being involved in the protein–chaperone interaction
- Found in various cellular compartments such as cytosol, mitochondria, and the lumen of the endoplasmic reticulum

TABLE 46–7 Some Chaperones and Enzymes Involved in Folding That Are Located in the Rough Endoplasmic Reticulum

- BiP (immunoglobulin heavy chain binding protein)
- GRP94 (glucose-regulated protein)
- Calnexin
- Calreticulin
- PDI (protein disulfide isomerase)
- PPI (peptidyl prolyl *cis-trans* isomerase)

particular importance in the ER are listed in **Table 46–7.** Basically, they **stabilize unfolded or partially folded intermediates,** allowing them time to fold properly, and prevent inappropriate interactions, thus combating the formation of nonfunctional structures. Most chaperones exhibit **ATPase activity** and bind ADP and ATP. This activity is important for their effect on protein folding. The ADP–chaperone complex often has a high affinity for the unfolded protein, which, when bound, stimulates release of ADP with replacement by ATP. The ATP–chaperone complex, in turn, releases segments of the protein that have folded properly, and the **cycle** involving ADP and ATP binding is repeated until the protein is released.

Chaperonins are the second major class of chaperones. They form complex **barrel-like structures** in which an unfolded protein is sequestered away from other proteins, giving it time and suitable conditions in which to fold properly. The structure of the bacterial chaperonin GroEL has been studied in detail. It is polymeric, has two ring-like structures, each composed of seven identical subunits, and again ATP is involved in its action.

Several examples of chaperones were introduced above when the sorting of mitochondrial proteins was discussed. The **immunoglobulin heavy chain-binding protein (BiP)** is located in the lumen of the ER. This protein **promotes proper folding by preventing aggregation** and will temporarily bind abnormally folded immunoglobulin heavy chains and many other proteins, preventing them from leaving the ER. Another important chaperone is **calnexin,** a calcium-binding protein located in the ER membrane. This protein binds a wide variety of proteins, including major histocompatibility complex (MHC) antigens and a variety of plasma proteins. As described in Chapter 47, calnexin binds the monoglucosylated species of glycoproteins that occur during processing of glycoproteins, retaining them in the ER until the glycoprotein has folded properly. **Calreticulin,** which is also a calcium-binding protein, has properties similar to those of calnexin;

it is not membrane-bound. Chaperones are not the only proteins in the ER lumen that are concerned with proper folding of proteins. Two **enzymes** are present that play an active role in folding. **Protein disulfide isomerase (PDI)** promotes **rapid formation** and reshuffling of disulfide bonds until the correct set is achieved. **Peptidyl prolyl isomerase (PPI)** accelerates folding of proline-containing proteins by catalyzing the cis–trans isomerization of X-Pro bonds, where X is any amino acid residue.

Thus, the ER functions as a **quality control compartment** of the cell. Newly synthesized proteins attempt to fold with the assistance of chaperones and folding enzymes, and their folding status is monitored by the chaperones. Misfolded or incompletely folded proteins interact with chaperones, which retain them in the ER and prevent them from being exported to their final destinations. If such interactions continue for a prolonged period of time, the misfolded proteins are usually disposed of by endoplasmic reticulum-associated degradation (ERAD, see below). This avoids a harmful build-up of misfolded proteins. In a number of genetic diseases, such as cystic fibrosis (see Chapter 57), retention of misfolded proteins occurs in the ER, In some cases, the retained proteins still exhibit some functional activity. As discussed later in this Chapter, there is much current interest in finding drugs that will interact with such proteins and promote their correct folding and export out of the ER.

ACCUMULATION OF MISFOLDED PROTEINS IN THE ENDOPLASMIC RETICULUM CAN INDUCE THE UNFOLDED PROTEIN RESPONSE (UPR)

Maintenance of **homeostasis in the ER** is important for normal cell function. When the unique environment within the lumen of the ER is perturbed (eg, changes in ER Ca^{2+}, alterations of redox status, exposure to various toxins or some viruses), this can lead to reduced protein folding capacity and the accumulation of misfolded proteins. The accumulation of misfolded proteins in the ER is referred to as **ER stress.** The cell has evolved a mechanism termed the UPR to sense the levels of misfolded proteins and initiate intracellular signaling mechanisms to compensate for the stress conditions and restore ER homeostasis. The UPR is initiated by ER stress sensors, which are transmembrane proteins embedded in the ER membrane. Activation of these stress sensors causes three principal effects: (i) transient inhibition of translation to reduce the amount of newly synthesized proteins, (ii) induction of a transcriptional response that leads to increased expression of ER chaperones and to (iii) increased synthesis of proteins involved in degradation of misfolded ER proteins (discussed below). Therefore, the UPR increases the ER folding capacity and prevents a buildup of unproductive and potentially toxic protein products, in addition to other responses to restore cellular homeostasis. However, if impairment of folding persists, cell death pathways (apoptosis) are activated. A more complete understanding of the UPR is likely to provide new approaches to treating diseases in which ER stress and defective protein folding occur (see **Table 46–8**).

MISFOLDED PROTEINS UNDERGO ENDOPLASMIC RETICULUM-ASSOCIATED DEGRADATION

As shown in Table 46–8, misfolded proteins occur in many genetic diseases. Proteins that misfold in the ER are selectively **transported back across the ER (retrotranslocation or dislocation)** to enter **proteasomes** present in the cytosol. The precise route by which the misfolded proteins pass back across the ER membrane is still under investigation. Two proteins have been implicated: the Sec61 complex described earlier and Derlin 1. The energy for translocation appears to be at least partly supplied by **p97,** an AAA-ATPase (one of a family of **A**TPases **A**ssociated with various cellular **A**ctivities). **Chaperones** present in the lumen of the ER (eg, BiP) and in the cytosol help target misfolded proteins to proteasomes. Prior to entering proteasomes, most proteins are **ubiquitinated** (see the next paragraph) and are escorted to proteasomes by polyubiquitin-binding proteins. Ubiquitin ligases are present in the ER membrane. The above process is referred to as ERAD and is outlined briefly in **Figure 46–9**.

UBIQUITIN IS A KEY MOLECULE IN PROTEIN DEGRADATION

There are two major pathways of protein degradation in eukaryotes. One involves **lysosomal proteases** and does not require ATP. The other pathway involves **ubiquitin** and is ATP-dependent. It plays the major role in the degradation of proteins, and is particularly associated with **disposal of misfolded proteins and regulatory enzymes that have short half-lives.** Research on ubiquitin has expanded rapidly, and it is known to be involved in **cell-cycle regulation**

TABLE 46–8 Some Conformational Diseases That Are Caused by Abnormalities in Intracellular Transport of Specific Proteins and Enzymes Due to Mutations[1]

Disease	Affected Protein
α_1-Antitrypsin deficiency with liver disease (OMIM 107400)	α_1-Antitrypsin
Chediak–Higashi syndrome (OMIM 214500)	Lysosomal trafficking regulator
Combined deficiency of factors V and VIII (OMIM 227300)	ERGIC53, a mannose-binding lectin
Cystic fibrosis (OMIM 219700)	CFTR
Diabetes mellitus [some cases] (OMIM 147670)	Insulin receptor (α-subunit)
Familial hypercholesterolemia, autosomal dominant (OMIM 143890)	LDL receptor
Gaucher disease (OMIM 230800)	β-Glucosidase
Hemophilia A (OMIM 306700) and B (OMIM 306900)	Factors VIII and IX
Hereditary hemochromatosis (OMIM 235200)	HFE
Hermansky–Pudlak syndrome (OMIM 203300)	AP-3 adaptor complex β3A subunit
I-cell disease (OMIM 252500)	*N*-acetylglucosamine 1-phosphotransferase
Lowe oculocerebrorenal syndrome (OMIM 309000)	PIP$_2$ 5-phosphatase
Tay-Sachs disease (OMIM 272800)	β-Hexosaminidase
von Willebrand disease (OMIM 193400)	von Willebrand factor

Abbreviation: PIP$_2$, phosphatidylinositol 4,5-bisphosphate.
Note: Readers should consult textbooks of medicine or pediatrics for information on the clinical manifestations of the conditions listed.
[1]See Schroder M, Kaufman RJ: The mammalian unfolded protein response. Annu Rev Biochem 2005;74:739 and Olkonnen V, Ikonen E: Genetic defects of intracellular membrane transport. N Eng J Med 2000;343:10095.

(degradation of cyclins), **DNA repair, activation of NFκB** (see Chapter 50), **muscle wasting, viral infections,** and **many other** important physiologic and pathologic processes. Ubiquitin is a **small** (76 amino acids), **highly conserved protein** that plays a key role in **marking** various proteins for subsequent **degradation in proteasomes.** The mechanism of attachment of ubiquitin to a target protein (eg, a misfolded form of CFTR, the protein involved in the causation of cystic fibrosis; see Chapters 40 and 57) is shown in **Figure 46–10** and involves **three enzymes: an activating enzyme,** a **conjugating enzyme,** and a **ligase.** There are a number of types of conjugating enzymes, and, surprisingly, some hundreds of different ligases. It is the latter enzyme that confers substrate specificity. Once the molecule of ubiquitin is attached to the protein, a number of others are also attached, resulting in a **polyubiquitinated target protein.**

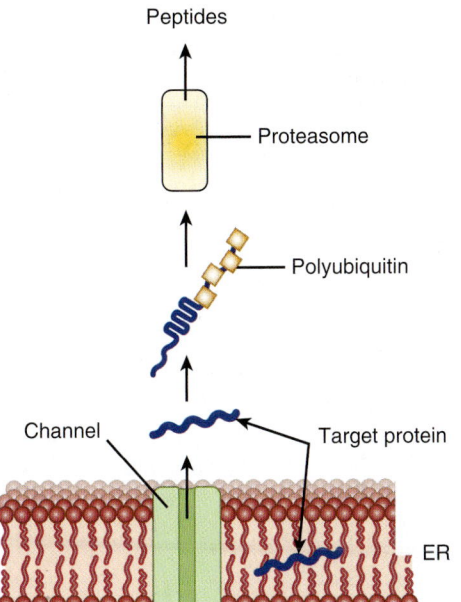

FIGURE 46–9 **Simplified schematic diagram of the events in ERAD.** A target protein which is misfolded undergoes retrograde transport through the ER membrane into the cytosol, where it is subjected to polyubiquitination. Following polyubiquitination, it enters a proteasome, inside which it is degraded to small peptides that exit and may have several fates. Liberated ubiquitin molecules are recycled. Two proteins have been implicated in the retrograde transport through the ER; these are Sec 61 (also involved in transfer of newly synthesized proteins into the lumen of the ER) and another protein named Derlin 1.

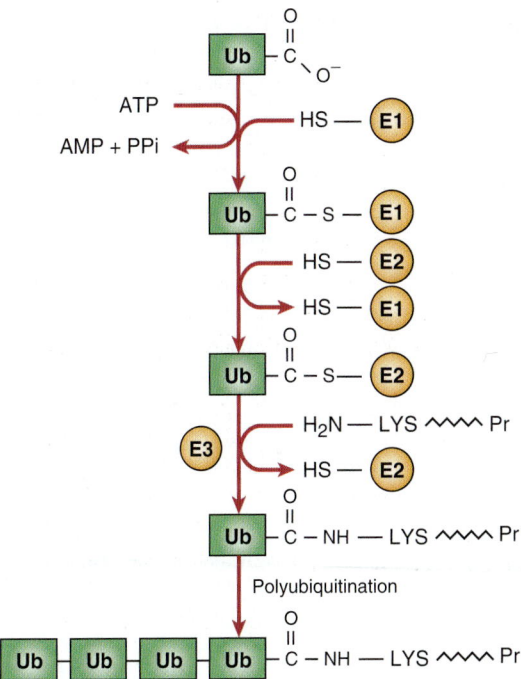

FIGURE 46–10 **Sequence of reactions in addition of ubiquitin to a target protein.** In the reaction catalyzed by E1, the C-terminal COO$^-$ group of ubiquitin is linked in a thioester bond to an SH group of E1. In the reaction catalyzed by E2, the activated ubiquitin is transferred to an SH group of E2. In the reaction catalyzed by E3, ubiquitin is transferred from E2 to an ε-amino group on a lysine of the target protein. Additional rounds of ubiquitination then build up the polyubiquitin chain. (Ub, ubiquitin; E1, activating enzyme; E2, conjugating enzyme; E3, ligase; LYS〰〰Pr, target protein.)

It has been estimated that a **minimum of four ubiquitin molecules** must be attached to commit a target molecule to degradation in a proteasome. Ubiquitin can be **cleaved** from a target protein by deubiquitinating **enzymes** and the liberated ubiquitin can be reused.

Ubiquitinated Proteins Are Degraded in Proteasomes

Polyubiquitinated target proteins enter proteasomes located in the cytosol. The proteasome is a relatively **large cylindrical structure** and is composed of some **50 subunits.** The proteasome has a hollow **core,** and one or two **caps** that play a regulatory role. Target proteins are **unfolded** by ATPases present in the proteasome caps. Proteasomes can hydrolyze a very wide variety of peptide bonds. Target proteins pass into the core to be degraded to small peptides, which then exit the proteasome (see Figure 46–9) to be further degraded by cytosolic peptidases. Both normally and abnormally folded proteins are substrates for the proteasome. Liberated ubiquitin molecules are recycled. The proteasome plays an important role in **presenting small peptides** produced by **degradation of various viruses** and other molecules to **major histocompatibility class I molecules,** a key step in antigen presentation to T lymphocytes.

TRANSPORT VESICLES ARE KEY PLAYERS IN INTRACELLULAR PROTEIN TRAFFIC

Proteins that are synthesized on membrane-bound polyribosomes and are destined for the GA or PM reach these sites inside **transport** vesicles. Those vesicles involved in **anterograde transport** (COPII) from the ER to the GA and in **retrograde transport** (COPI) from the GA to the ER are clathrin-free. Transport and secretory vesicles carrying cargo from the GA to the PM are also clathrin-free. The vesicles involved in endocytosis (see discussions of the LDL receptor in Chapters 25 and 26) are coated with clathrin, as are certain vesicles carrying cargo to lysosomes. For the sake of clarity, the non-clathrin-coated vesicles are referred to in this text as **transport vesicles. Table 46–9** summarizes **the types and functions** of the major vesicles identified to date.

Model of Transport Vesicles Involves SNAREs & Other Factors

Vesicles lie at the heart of intracellular transport of many proteins. Significant progress has been made in understanding

TABLE 46–9 Some Types of Vesicles and Their Functions

Vesicle	Function
COPI	Involved in intra-GA transport and retrograde transport from the GA to the ER
COPII	Involved in export from the ER to either ERGIC or the GA
Clathrin	Involved in transport in post-GA locations including the PM, TGN and endosomes
Secretory vesicles	Involved in regulated secretion from organs such as the pancreas (eg, secretion of insulin)
Vesicles from the TGN to the PM	They carry proteins to the PM and are also involved in constitutive secretion

Abbreviations: GA, Golgi apparatus; ER, endoplasmic reticulum; ERGIC, ER-GA intermediate compartment; PM, plasma membrane; TGN, *trans*-Golgi network.
Note: Each vesicle has its own set of coat proteins. Clathrin is associated with various adaptor proteins such as AP-1, AP-2, and AP-3 (AP, adaptor protein), GGA-1, GGA-2, and GGA-3 (GGA, Golgi-localizing, gamma-adaptin ear homology domain, ARF-binding protein), forming different types of clathrin vesicles. These various clathrin vesicles have different intracellular targets. The proteins of secretory vesicles and vesicles involved in transport from the GA to the PM are not well characterized, nor are the mechanisms involved in their formations and fates.

TABLE 46–10 Some Factors Involved in the Formation of Non-clathrin-coated Vesicles and Their Transport

- **ARF:** ADP-ribosylation factor, a GTPase involved in formation of COPI and also clathrin-coated vesicles.
- **Coat proteins:** A family of proteins found in coated vesicles. Different transport vesicles have different complements of coat proteins.
- **NSF:** NEM-sensitive factor, an ATPase.
- **Sar1:** A GTPase that plays a key role in assembly of COPII vesicles.
- **Sec12:** A guanine nucleotide exchange factor (GEF) that interconverts Sar1·GDP and Sar1·GTP.
- **α-SNAP:** Soluble NSF attachment protein. Along with NSF, this protein is involved in dissociation of SNARE complexes.
- **SNARE:** SNAP receptor. SNAREs are key molecules in the fusion of vesicles with acceptor membranes.
- **t-SNARE:** Target SNARE.
- **v-SNARE:** Vesicle SNARE.
- **Rab proteins:** A family of Ras-related proteins (monomeric GTPases) first observed in rat brain. They are active when GTP is bound. Different Rab molecules dock different vesicles to acceptor membranes.
- **Rab effector proteins:** A family of proteins that interact with Rab molecules; some act to tether vesicles to acceptor membranes.

the events involved in vesicle formation and transport. This has transpired because of the use of a number of approaches. In particular, the use by Schekman and colleagues of **genetic approaches for studying** vesicles in yeast and the development by Rothman and colleagues of **cell-free systems** to study vesicle formation have been crucial. For instance, it is possible to observe, by electron microscopy, budding of vesicles from Golgi preparations incubated with cytosol, ATP and GTP-γ. The overall mechanism is complex, with its own **nomenclature** (**Table 46–10**), and involves a variety of cytosolic and membrane proteins, GTP, ATP, and accessory factors. **Budding, tethering, docking,** and **membrane fusion** are key steps in the life cycles of vesicles with Sar, ARF, and the Rab GTPases (see below) acting as **molecular switches.**

There are common general steps in transport vesicle formation, vesicle targeting and fusion with a target membrane, irrespective of the membrane the vesicle forms from or its intracellular destination. The nature of the coat proteins, GTPases and targeting factors differ depending on where the vesicle forms from and its eventual destination. Transport from the ER to the Golgi is the best studied example and will be used to illustrate these steps. **Anterograde vesicular transport** from the ER to the Golgi involves **COPII vesicles** and the process can be considered to occur in eight steps (**Figure 46–11**). The basic concept is that each transport vesicle is loaded with specific cargo and also one or more **v-SNARE** proteins that direct targeting. Each target membrane bears one or more **complementary t-SNARE proteins** with which the former interact, mediating SNARE protein-dependent vesicle–membrane fusion. In addition, **Rab proteins** also help direct the vesicles to specific membranes and are involved in tethering, prior to vesicle docking at a target membrane.

Step 1: **Budding** is initiated when **Sar1** is activated by binding GTP, which is exchanged for GDP via the action of **Sec12**. This causes a conformational change in Sar1·GTP, embedding it in the ER membrane to form a focal point for vesicle assembly.

Step 2: Various **coat proteins** bind to **Sar1·GTP.** In turn, membrane cargo proteins bind to the coat proteins and soluble cargo proteins inside vesicles bind to receptor regions of the former. Additional coat proteins are assembled to **complete bud formation.** Coat proteins promote budding, contribute to the curvature of buds and also help sort proteins.

Step 3: The **bud pinches off,** completing formation of the coated vesicle. The curvature of the ER membrane and protein–protein and protein–lipid interactions in the bud facilitate pinching off from ER exit sites.

Step 4: **Coat disassembly** (involving **dissociation** of Sar1 and the **shell** of coat proteins) follows **hydrolysis of bound GTP to GDP** by Sar1, promoted by a specific coat protein. Sar1 thus plays key roles in both assembly and dissociation of the coat proteins. Uncoating is necessary for fusion to occur.

Step 5: **Vesicle targeting** is achieved by attachment of **Rab** molecules to vesicles. Rab·GDP molecules in the cytosol are converted to Rab·GTP molecules by a specific GEF and these attach to the vesicles. The Rab·GTP molecules subsequently interact with **Rab effector proteins** on membranes to **tether** the vesicle to the membranes.

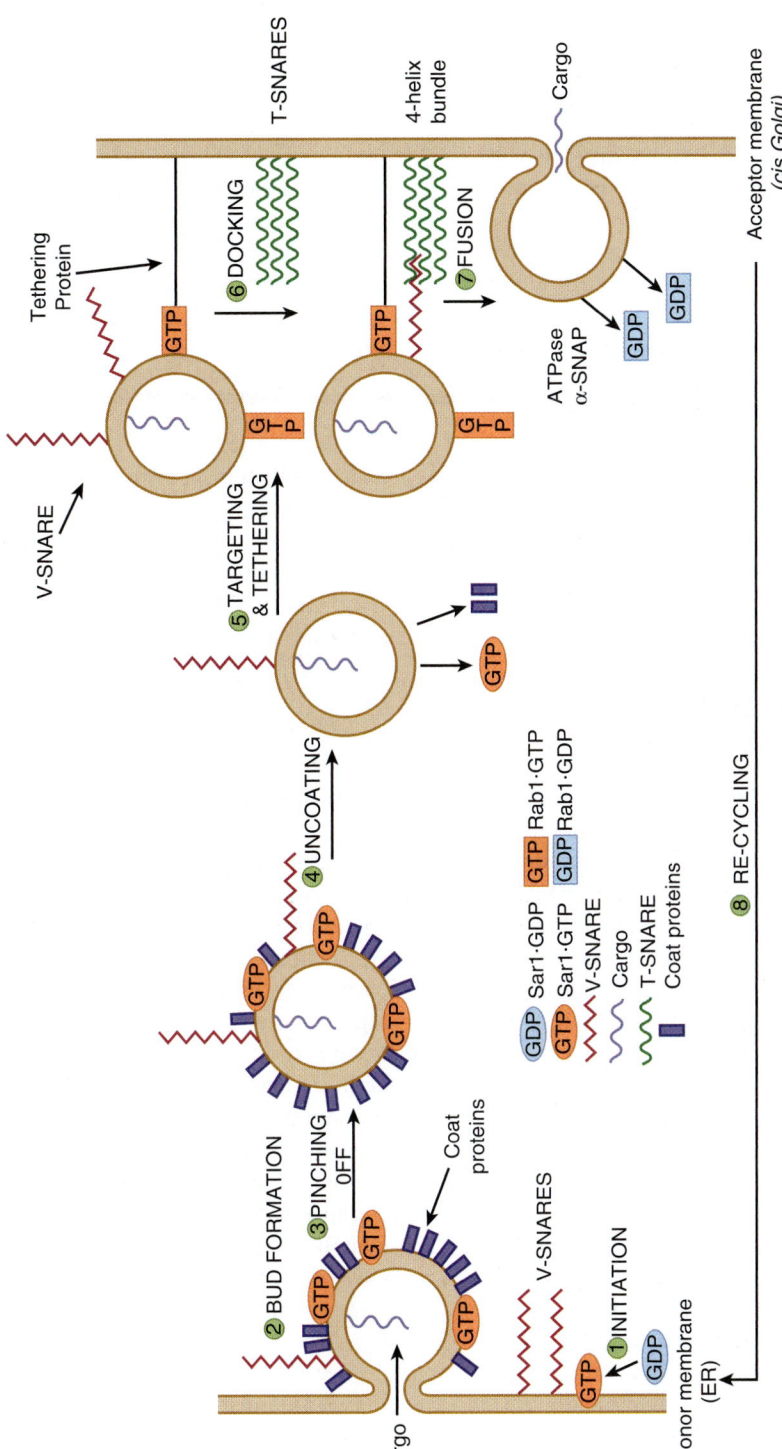

FIGURE 46–11   Model of the steps in a round of anterograde transport involving COPII vesicles. The cycle starts in the bottom left-hand side of the figure, where a molecule of Sar1·GDP is represented as a blue oval labelled GDP. The steps in the cycle are described in the text. The various components are briefly described in Table 46–8. The roles of Rab and Rab effector proteins (see text) in the overall process are not dealt with in this figure. (Adapted, with permission, from Rothman JE: Mechanisms of intracellular protein transport. Nature 1994;372:55.)

Step 6: **v-SNAREs pair with cognate t-SNAREs** in the target membrane to **dock** the vesicles and initiate fusion. Generally one v-SNARE in the vesicle pairs with three t-SNAREs on the acceptor membrane to form a tight **four-helix bundle.**

Step 7: **Fusion** of the vesicle with the acceptor membrane occurs once the v- and t-SNARES are closely aligned. After vesicle fusion and release of contents occurs, GTP is hydrolyzed to GDP, and the Rab·GDP molecules are released into the cytosol. When a SNARE on one membrane interacts with a SNARE on another membrane, linking the two membranes, this is referred to as a trans-SNARE complex or a SNARE pin. Interactions of SNARES on the same membrane form a cis-SNARE complex. In order to **dissociate the four-helix bundle** between the v- and t-SNARES so that they can be reused, two additional proteins are required. These are an **ATPase** (NSF) and **α-SNAP.** NSF hydrolyzes ATP and the energy released dissociates the four-helix bundle making the SNARE proteins available for another round of membrane fusion.

Step 8: Certain components, such as the Rab and SNARE proteins, are **recycled** for subsequent rounds of vesicle fusion.

During the above cycle, SNARES, tethering proteins, Rab and other proteins all **collaborate** to deliver a vesicle and its contents to the appropriate site.

COPI, COPII, and Clathrin-Coated Vesicles Have Been Most Studied

The following points clarify and expand on the previous section.

1. As indicated in Table 46–9, there are a **number of different types of vesicles.** Other types of vesicles may remain to be discovered. Here we focus mainly on COPII, COPI and clathrin-coated vesicles. Each of these types has a different complement of proteins in its coat. The details of assembly for COPI and clathrin-coated vesicles are somewhat different from those described above. For example, **Sar1** is the protein involved in step 1 of formation of COPII vesicles, whereas **ARF** is involved in the formation of COPI and clathrin-coated vesicles. However, the principles concerning assembly of these different types are generally similar.

2. Regarding **selection** of cargo molecules by vesicles, this appears to be primarily **a function of the coat proteins** of vesicles. **Cargo molecules** via their sorting signals may interact with coat proteins either **directly** or via **intermediary proteins** that attach to coat proteins, and they then become enclosed in their appropriate vesicles. A number of **signal sequences** on cargo molecules have been identified (see Table 46–1). For example KDEL sequences direct certain ER-resident proteins in retrograde flow to the ER in COPI vesicles. Di-acidic sequences (eg, Asp-X-Glu, X = any amino acid) and short hydrophobic sequences on membrane proteins are involved in interactions with coat proteins of COPII vesicles.

Proteins in the **apical** or **basolateral** areas of the plasma membranes of polarized epithelial cells can be transported to these sites in **transport vesicles** budding from the TGN. Different Rab proteins likely direct some vesicles to apical regions and others to basolateral regions. In certain cells, proteins are first directed to the basolateral membrane, then endocytosed and transported across the cell by **transcytosis** to the apical region. Yet another mechanism for sorting proteins to the apical region (or in some cases to the basolateral region) involves the **glycosylphosphatidylinositol (GPI) anchor** described in Chapter 47. This structure is also often present in **lipid rafts** (see Chapter 40).

Not all cargo molecules may have a sorting signal. Some highly abundant secretory proteins travel to various cellular destinations in transport vesicles by **bulk flow;** that is, they enter into transport vesicles at the same concentration that they occur in the organelle. The precise extent of bulk flow is not clearly known, although it appears that most proteins are actively sorted (concentrated) into transport vesicles and bulk flow is used by only a select group of cargo proteins.

3. Once proteins in the secretory pathway reach the *cis*-Golgi from the ER in vesicles, they can travel through the GA to the *trans*-Golgi **in vesicles,** or by a process called **cisternal maturation,** or perhaps in some cases **diffusion** via intracisternal connections that have been observed in some cell types. A former view was that the GA is essentially a static organelle, allowing vesicular flow from one static cisterna to the next. There is now, however, evidence to support the view that the cisternae move and transform into one another (ie, cisternal maturation). In this model, vesicular elements from the ER fuse with one another to help form the *cis*-Golgi, which in turn can move forward to become the medial Golgi, etc. COPI vesicles return Golgi enzymes (eg, glycosyltransferases) back from distal cisternae of the GA to more proximal (eg, *cis*) cisternae.

4. Vesicles move through cells along **microtubules** or along **actin filaments.**

5. The fungal metabolite **brefeldin A prevents GTP from binding to ARF,** and thus inhibits formation of COPI vesicles. In its presence, the Golgi apparatus appears to **collapse into the ER.** It may do this by inhibiting the guanine nucleotide exchanger involved in formation of COPI vesicles. Brefeldin A has thus proven to be a useful tool for examining some aspects of Golgi structure and function.

6. **GTP-γ-S** (a nonhydrolyzable analog of GTP often used in investigations of the role of GTP in biochemical processes) **blocks disassembly of the coat** from coated vesicles, leading to a build-up of coated vesicles, facilitating their study.

7. As mentioned above, a family of Ras-like proteins, called the **Rab protein family,** is required in several steps of intracellular protein transport and also in regulated secretion and endocytosis. (Ras proteins are involved in cell

signaling via receptor tyrosine kinases). Like Ras, Rab proteins are **small monomeric GTPases** that attach to the cytosolic faces of membranes (via **geranylgeranyl** lipid anchors). They **attach** in the **GTP-bound state** to the budding vesicle and are also present on acceptor membranes. Rab proteins interact with **Rab effector proteins** that have various roles, such as involvement in tethering and in membrane fusion.

8. The fusion of **synaptic vesicles** with the plasma membrane of **neurons** involves a series of events similar to that described above. For example, one v-SNARE is designated **synaptobrevin** and two t-SNAREs are designated **syntaxin** and **SNAP 25** (synaptosome-associated protein of 25 kDa). **Botulinum B toxin** is one of the most lethal toxins known and the most serious cause of food poisoning. One component of this toxin is a **protease** that appears to **cleave only synaptobrevin,** thus **inhibiting release of acetylcholine** at the neuromuscular junction and possibly proving fatal, depending on the dose taken.

9. Although the above model refers to **non-clathrin-coated vesicles,** it appears likely that many of the events described above apply, at least in principle, to clathrin-coated vesicles.

10. Some proteins are further subjected to **further processing by proteolysis** while inside either transport or secretory vesicles. For example, **albumin** is synthesized by hepatocytes as **preproalbumin** (see Chapter 50). Its signal peptide is removed, converting it to **proalbumin**. In turn, proalbumin, while inside transport vesicles, is converted to **albumin** by action of **furin (Figure 46–12)**. This enzyme cleaves a hexapeptide from proalbumin immediately C-terminal to a dibasic amino acid site (ArgArg). The resulting mature albumin is secreted into the plasma. Hormones such as **insulin** (see Chapter 41) are subjected to similar proteolytic cleavages while inside secretory vesicles.

THE ASSEMBLY OF MEMBRANES IS COMPLEX

There are many cellular membranes, each with its own specific features. No satisfactory scheme describing the assembly of any one of these membranes is available. How various proteins are initially inserted into the membrane of the ER has been discussed above. The transport of proteins, including membrane proteins, to various parts of the cell inside vesicles has also been described. Some general points about membrane assembly remain to be addressed.

Asymmetry of Both Proteins & Lipids Is Maintained During Membrane Assembly

Vesicles formed from membranes of the ER and Golgi apparatus, either naturally or pinched off by homogenization, exhibit **transverse asymmetries** of both lipid and protein. These **asymmetries are maintained** during fusion of transport vesicles with the plasma membrane. The **inside** of the vesicles after fusion becomes the **outside of the plasma membrane,** and the cytoplasmic side of the vesicles remains the cytoplasmic side of the membrane (**Figure 46–13**). Since the transverse asymmetry of the membranes already exists in the vesicles of the ER well before they are fused to the plasma membrane, a major problem of membrane assembly becomes understanding how the integral proteins are inserted into the lipid bilayer of the ER. This problem was addressed earlier in this chapter.

Phospholipids are the major class of lipid in membranes. The enzymes responsible for the synthesis of phospholipids reside in the cytoplasmic surface of the cisternae of the ER. As phospholipids are synthesized at that site, they probably self-assemble into thermodynamically stable bimolecular layers, thereby expanding the membrane and perhaps promoting the detachment of so-called **lipid vesicles** from it. It has been proposed that these vesicles travel to other sites, donating their lipids to other membranes; however, little is known about this matter. As indicated above, cytosolic proteins that take up phospholipids from one membrane and release them to another (ie, **phospholipid exchange proteins**) have been demonstrated; they probably play a role in contributing to the specific lipid composition of various membranes.

It should be noted that the **lipid compositions** of the ER, Golgi, and plasma membrane differ, the latter two membranes containing **higher amounts of cholesterol, sphingomyelin, and glycosphingolipids,** and **less phosphoglycerides** than does the ER. Sphingolipids pack more densely in membranes than do phosphoglycerides. These differences affect the structures and functions of membranes. For example, the **thickness of the bilayer** of the GA and PM is greater than that of the ER, which affects which particular transmembrane proteins are found in these organelles. Also, **lipid rafts** (see Chapter 40) are believed to be formed in the GA.

Lipids & Proteins Undergo Turnover at Different Rates in Different Membranes

It has been shown that the half-lives of the lipids of the ER membranes of rat liver are generally shorter than those of its

Preproalbumin $\xrightarrow{\text{Signal peptidase}}$ Signal peptide + Proalbumin $\xrightarrow{\text{Furin}}$ Hexapeptide + Albumin

FIGURE 46–12 **Cleavage of preproalbumin to proalbumin and of the latter to albumin.** Furin cleaves proalbumin at the C-terminal end of a basic dipeptide (ArgArg).

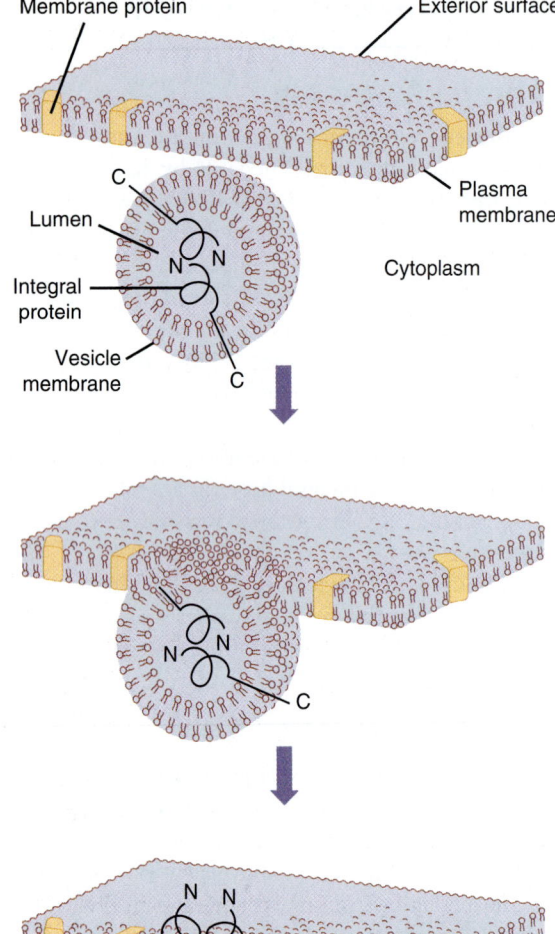

FIGURE 46–13 **Fusion of a vesicle with the plasma membrane preserves the orientation of any integral proteins embedded in the vesicle bilayer.** Initially, the amino terminal of the protein faces the lumen, or inner cavity, of such a vesicle. After fusion, the amino terminal is on the exterior surface of the plasma membrane. That the orientation of the protein has not been reversed can be perceived by noting that the other end of the molecule, the carboxyl terminal, is always immersed in the cytoplasm. The lumen of a vesicle and the outside of the cell are topologically equivalent. (Redrawn and modified, with permission, from Lodish HF, Rothman JE: The assembly of cell membranes. Sci Am [Jan] 1979;240:43.)

proteins, so that the **turnover rates of lipids and proteins are independent.** Indeed, different lipids have been found to have different half-lives. Furthermore, the half-lives of the proteins of these membranes vary quite widely, some exhibiting short (hours) and others long (days) half-lives. Thus, individual lipids and proteins of the ER membranes appear to be inserted into it relatively independently; this is the case for many other membranes.

The biogenesis of membranes is thus a complex process about which much remains to be learned. One indication of

TABLE 46–11 Some Major Features of Membrane Assembly

- Lipids and proteins are inserted independently into membranes.
- Individual membrane lipids and proteins turn over independently and at different rates.
- Topogenic sequences [eg, signal (amino terminal or internal) and stop-transfer] are important in determining the insertion and disposition of proteins in membranes.
- Membrane proteins inside transport vesicles bud off the endoplasmic reticulum on their way to the Golgi; final sorting of many membrane proteins occurs in the trans-Golgi network.
- Specific sorting sequences guide proteins to particular organelles such as lysosomes, peroxisomes, and mitochondria.

the complexity involved is to consider the number of **post-translational modifications** that membrane proteins may be subjected to prior to attaining their mature state. These include disulfide formation, proteolysis, assembly into multimers, glycosylation, addition of a glycophosphatidylinositol (GPI) anchor, sulfation on tyrosine or carbohydrate moieties, phosphorylation, acylation, and prenylation—a list that is not complete. Nevertheless, significant progress has been made; **Table 46–11** summarizes some of the major features of membrane assembly that have emerged to date.

Various Disorders Result from Mutations in Genes Encoding Proteins Involved in Intracellular Transport

Some disorders reflecting abnormal **peroxisomal** function and abnormalities of protein synthesis in the **ER** and of the synthesis of **lysosomal proteins** have been listed earlier in this chapter (see Tables 46–3 and 46–8, respectively). Many other mutations affecting folding of proteins and their intracellular transport to various organelles have been reported, but are not discussed here [eg, Alzheimer disease (see Chapter 57)] and Huntington disease. The elucidation of the causes of these various **conformational disorders** has contributed significantly to our understanding of **molecular pathology.** The term **"diseases of proteostasis deficiency"** has also been applied to diseases due to misfolding of proteins. Proteostasis is a composite word derived from protein homeostasis. Normal proteostasis is due to a balance of many factors, such as synthesis, folding, trafficking, aggregation, and normal degradation. If any one of these is disturbed (eg, by mutation, aging, cell stress or injury), a variety of disorders can occur, depending on the particular proteins involved.

Apart from the possibility of **gene therapy,** it is hoped that attempts to restore at least a degree of normal folding to misfolded proteins by **administration to affected individuals of small molecules** that interact specifically with such proteins will be of therapeutic benefit. This is a very active area of research.

SUMMARY

- Many proteins are targeted to their destinations by signal sequences. A major sorting decision is made when proteins are partitioned between cytosolic and membrane-bound polyribosomes by virtue of the absence or presence of an N-terminal signal peptide.

- Pathways of protein import into mitochondria, nuclei, peroxisomes, and the endoplasmic reticulum are described.

- Numerous proteins synthesized on membrane-bound polyribosomes proceed to the Golgi apparatus and the plasma membrane in transport vesicles.

- Many glycosylation reactions occur in compartments of the Golgi, and proteins are further sorted in the *trans*-Golgi network.

- The role of chaperone proteins in the folding of proteins is presented and the UPR is described.

- ERAD is briefly described and the key role of ubiquitin in protein degradation is shown.

- A model describing budding and attachment of transport vesicles to a target membrane is summarized.

- Certain proteins (eg, precursors of albumin and insulin) are subjected to proteolysis while inside transport vesicles, producing the mature proteins.

- Small GTPases (eg, Ran, Rab) and GEFs play key roles in many aspects of intracellular trafficking.

- The complex process of membrane assembly is discussed briefly. Asymmetry of both lipids and proteins is maintained during membrane assembly.

- Many disorders have been shown to be due to mutations in genes or to other factors that affect the folding of various proteins. These conditions have been referred to as conformational diseases, or alternatively as diseases of proteostatic deficiency. Apart from gene therapy, the development of small molecules that interact with misfolded proteins and help restore at least some of their function is an important area of research.

REFERENCES

Alberts B, Johnson A, Lewis J, et al: *Molecular Biology of the Cell*, 5th ed. Garland Science, 2008. (An excellent textbook of cell biology, with comprehensive coverage of trafficking and sorting).

Alder NN, Johnson AE: Cotranslational membrane protein biogenesis at the endoplasmic reticulum. J Biol Chem 2004;279:22787.

Bonifacino JS, Glick BS: The mechanisms of vesicle budding and fusion. Cell 2004;116:153.

Cooper GM, Hausman RE: *The Cell: A Molecular Approach*. Sinauer Associates, Inc. 2009. (An excellent textbook of cell biology, with comprehensive coverage of trafficking and sorting).

Lai E, Teodoro T, Volchuk A: Endoplasmic reticulum stress: signaling the unfolded protein response. Physiology 2007;22:193.

Lodish H, Berk A, Krieger M, et al: *Molecular Cell Biology*, 6th ed. WH Freeman & Co., 2008. (An excellent textbook of cell biology, with comprehensive coverage of trafficking and sorting).

Neupert W, Herrmann JM: Translocation of proteins into mitiochondria. Annu Rev Biochem 2007;76:723.

Platta HW, Erdmann R: The peroxisomal protein import machinery. FEBS Lett 2007;581;2811.

Pollard TD, Earnshaw WC: *Cell Biology*, 2nd ed. WB Saunders, 2008. (An excellent textbook of cell biology, with comprehensive coverage of trafficking and sorting).

Powers ET, Morimoto RI, Dillin A, et al: Biological and chemical approaches to diseases of proteostasis deficiency. Annu Rev Biochem 2009;78:959.

Romisch K: Endoplasmic-reticulum–associated degradation. Annu Rev Cell Dev Biol 2005;21:435.

Stewart M: Molecular mechanisms of the nuclear protein import cycle. Nature Rev Mol Cell Biol 2007;8:195.

Glycoproteins

Robert K. Murray, MD, PhD

O B J E C T I V E S

After studying this chapter,
you should be able to:

- Have a general appreciation of the importance of glycobiology and glycomics, and in particular of glycoproteins, in health and disease.
- Know the principal sugars found in glycoproteins.
- Be aware of the several major classes of glycoproteins (*N*-linked, *O*-linked, and GPI-linked).
- Understand the major features of the pathways of biosynthesis and degradation of *O*- and *N*-linked glycoproteins.
- Understand the importance of advanced glycation end-products in causing tissue damage in diabetes mellitus.
- Be able to indicate the involvement of glycoproteins in inflammation and in a host of conditions including I-cell disease, congenital disorders of glycation, paroxysmal nocturnal hemoglobinuria and cancer.
- Be familiar with the concept that many microorganisms, such as influenza virus, attach to cell surfaces via sugar chains.

BIOMEDICAL IMPORTANCE

Glycobiology is the study of the roles of sugars in health and disease. The **glycome** is the entire complement of sugars, whether free or present in more complex molecules, of an organism. **Glycomics,** an analogous term to genomics and proteomics, is the comprehensive study of glycomes, including genetic, physiologic, pathologic, and other aspects.

One major class of molecules included in the glycome is **glycoproteins.** These are proteins that contain oligosaccharide chains (glycans) covalently attached to their polypeptide backbones. It has been estimated that approximately 50% of eukaryotic proteins have sugars attached, so that **glycosylation** (enzymic attachment of sugars) is the most frequent posttranslational modification of proteins. Nonenzymic attachment of sugars to proteins can also occur, and is referred to as **glycation.** This process can have serious pathologic consequences (eg, in poorly controlled diabetes mellitus). Glycoproteins are one class of **glycoconjugate** or **complex carbohydrate**—equivalent terms used to denote molecules containing one or more carbohydrate chains covalently linked to protein (to form glycoproteins or proteoglycans) or lipid (to form glycolipids). (**Proteoglycans** are discussed in Chapter 48 and **glycolipids** in Chapter 15.) Almost all the **plasma proteins** of humans—with the notable exception of albumin—are glycoproteins. Many **proteins of cellular membranes** (Chapter 40) contain substantial amounts of carbohydrate. A number of the **blood group substances** are glycoproteins, whereas others are glycosphingolipids. Certain **hormones** (eg, chorionic gonadotropin) are glycoproteins. A major problem in cancer is **metastasis,** the phenomenon whereby cancer cells leave their tissue of origin (eg, the breast), migrate through the bloodstream to some distant site in the body (eg, the brain), and grow there in an unregulated manner, with catastrophic results for the affected individual. Many cancer researchers think that alterations in the structures of glycoproteins and other glycoconjugates on the surfaces of cancer cells are important in the phenomenon of metastasis.

GLYCOPROTEINS OCCUR WIDELY & PERFORM NUMEROUS FUNCTIONS

Glycoproteins occur in most organisms, from bacteria to humans. Many viruses also contain glycoproteins, some of which have been much investigated, in part because they often play key roles in viral attachment to cells (eg, HIV-1 and influenza A virus). Numerous proteins with diverse functions

TABLE 47–1 Some Functions Served by Glycoproteins

Function	Glycoproteins
Structural molecule	Collagens
Lubricant and protective agent	Mucins
Transport molecule	Transferrin, ceruloplasmin
Immunologic molecule	Immunoglobulins, histocompatibility antigens
Hormone	Chorionic gonadotropin, thyroid-stimulating hormone (TSH)
Enzyme	Various, for example, alkaline phosphatase
Cell attachment-recognition site	Various proteins involved in cell–cell (eg, sperm–oocyte), virus–cell, bacterium–cell, and hormone–cell interactions
Antifreeze	Certain plasma proteins of cold-water fish
Interact with specific carbohydrates	Lectins, selectins (cell adhesion lectins), antibodies
Receptor	Various proteins involved in hormone and drug action
Affect folding of certain proteins	Calnexin, calreticulin
Regulation of development	Notch and its analogs, key proteins in development
Hemostasis (and thrombosis)	Specific glycoproteins on the surface membranes of platelets

TABLE 47–2 Some Functions of the Oligosaccharide Chains of Glycoproteins

- Modulate physicochemical properties, for example, solubility, viscosity, charge, conformation, denaturation, and binding sites for various molecules, bacteria viruses, and some parasites.
- Protect against proteolysis, from inside and outside of cell.
- Affect proteolytic processing of precursor proteins to smaller products.
- Are involved in biologic activity, for example, of human chorionic gonadotropin (hCG).
- Affect insertion into membranes, intracellular migration, sorting and secretion.
- Affect embryonic development and differentiation.
- May affect sites of metastases selected by cancer cells.

Source: Adapted, with permission, from Schachter H: Biosynthetic controls that determine the branching and heterogeneity of protein-bound oligosaccharides. Biochem Cell Biol 1986;64:163.

are glycoproteins (**Table 47–1**); their carbohydrate content ranges from 1% to over 85% by weight.

Many studies have been conducted in an attempt to define the precise roles oligosaccharide chains play in the functions of glycoproteins. **Table 47–2** summarizes results from such studies. Some of the functions listed are firmly established; others are still under investigation.

OLIGOSACCHARIDE CHAINS ENCODE BIOLOGIC INFORMATION

An enormous number of glycosidic linkages can be generated between sugars. For example, three different hexoses may be linked to each other to form over 1000 different trisaccharides. The conformations of the sugars in oligosaccharide chains vary depending on their linkages and proximity to other molecules with which the oligosaccharides may interact. It is now established that certain oligosaccharide chains encode **biologic information** and that this depends upon their constituent sugars, their sequences, and their linkages. For instance, mannose 6-phosphate residues target newly synthesized lysosomal enzymes to that organelle (see later). The biologic information that sugars contain is expressed via interactions between

specific sugars, either free or in glycoconjugates, and proteins (such as lectins; see below) or other molecules. These interactions lead to changes of the cellular activity. Thus, deciphering the so-called **sugar code of life** (one of the principal aims of glycomics) entails elucidating all of the interactions that sugars and sugar-containing molecules participate in, and also the results of these interactions on cellular behavior. This will not be an easy task, considering the diversity of glycans found in cells.

TECHNIQUES ARE AVAILABLE FOR DETECTION, PURIFICATION, STRUCTURAL ANALYSIS & SYNTHESIS OF GLYCOPROTEINS

A variety of methods used in the detection, purification, and structural analysis of glycoproteins are listed in **Table 47–3**. The conventional methods used to purify proteins and enzymes are also applicable to the purification of glycoproteins. Once a glycoprotein has been purified, the use of **mass spectrometry** and **high-resolution NMR spectroscopy** can often identify the structures of its glycan chains. Analysis of glycoproteins can be complicated by the fact that they often exist as **glycoforms;** these are proteins with identical amino acid sequences but somewhat different oligosaccharide compositions. Although linkage details are not stressed in this chapter, it is critical to appreciate that the precise natures of the linkages between the sugars of glycoproteins are of fundamental importance in determining the structures and functions of these molecules.

Impressive advances are also being made in **synthetic chemistry,** allowing synthesis of complex glycans that can be tested for the biologic and pharmacologic activity. In addition, methods have been developed that use simple organisms, such as yeasts, to secrete human glycoproteins of therapeutic value (eg, erythropoietin) into their surrounding medium.

TABLE 47–3 Some Important Methods Used to Study Glycoproteins

Method	Use
Periodic acid–Schiff reagent	Detects glycoproteins as pink bands after electrophoretic separation.
Incubation of cultured cells with a radioactive sugar	Leads to detection of glycoproteins as radioactive bands after electrophoretic separation.
Treatment with appropriate endo- or exoglycosidase or phospholipases	Resultant shifts in electrophoretic migration help distinguish among proteins with N-glycan, O-glycan, or GPI linkages and also between high mannose and complex N-glycans.
Sepharose-lectin column chromatography	To purify glycoproteins or glycopeptides that bind the particular lectin used.
Compositional analysis following acid hydrolysis	Identifies sugars that the glycoprotein contains and their stoichiometry.
Mass spectrometry	Provides information on molecular mass, composition, sequence, and sometimes branching of a glycan chain.
NMR spectroscopy	To identify specific sugars, their sequence, linkages, and the anomeric nature of glycosidic linkages.
Methylation (linkage) analysis	To determine linkages between sugars.
Amino acid or cDNA sequencing	Determination of amino acid sequence.

EIGHT SUGARS PREDOMINATE IN HUMAN GLYCOPROTEINS

About 200 monosaccharides are found in nature; however, only eight are commonly found in the oligosaccharide chains of glycoproteins (**Table 47–4**). Most of these sugars were described in Chapter 14. N-acetylneuraminic acid (NeuAc) is usually found at the termini of oligosaccharide chains, attached to subterminal galactose (Gal) or N-acetylgalactosamine (Gal-NAc) residues. The other sugars listed are generally found in more internal positions. **Sulfate** is often found in glycoproteins, usually attached to Gal, GalNAc, or GlcNAc.

NUCLEOTIDE SUGARS ACT AS SUGAR DONORS IN MANY BIOSYNTHETIC REACTIONS

It is important to understand that in most biosynthetic reactions, it is not the free sugar or phosphorylated sugar that is involved in such reactions, but rather the corresponding **nucleotide sugar.** The first nucleotide sugar to be reported was uridine diphosphate glucose (UDP-Glc); its structure is shown in Figure 19–2. The common nucleotide sugars involved in the biosynthesis of glycoproteins are listed in Table 47–4; the reasons some contain UDP and others guanosine diphosphate (GDP) or cytidine monophosphate (CMP) are not clear. Many of the glycosylation reactions are

TABLE 47–4 The Principal Sugars Found in Human Glycoproteins[1]

Sugar	Type	Abbreviation	Nucleotide Sugar	Comments
Galactose	Hexose	Gal	UDP-Gal	Often found subterminal to NeuAc in N-linked glycoproteins. Also, found in the core trisaccharide of proteoglycans.
Glucose	Hexose	Glc	UDP-Glc	Present during the biosynthesis of N-linked glycoproteins but not usually present in mature glycoproteins. Present in some clotting factors.
Mannose	Hexose	Man	GDP-Man	Common sugar in N-linked glycoproteins.
N-Acetylneuraminic acid	Sialic acid (nine C atoms)	NeuAc	CMP-NeuAc	Often the terminal sugar in both N- and O-linked glycoproteins. Other types of sialic acid are also found, but NeuAc is the major species found in humans. Acetyl groups may also occur as O-acetyl species as well as N-acetyl.
Fucose	Deoxyhexose	Fuc	GDP-Fuc	May be external in both N- and O-linked glycoproteins or internal, linked to the GlcNAc residue attached to Asn in N-linked species. Can also occur internally attached to the OH of Ser (eg, in t-PA and certain clotting factors).
N-Acetylgalactosamine	Aminohexose	GalNAc	UDP-GalNAc	Present in both N- and O-linked glycoproteins.
N-Acetylglucosamine	Aminohexose	GlcNAc	UDP-GlcNAc	The sugar attached to the polypeptide chain via Asn in N-linked glycoproteins; also found at other sites in the oligosaccharides of these proteins. Many nuclear proteins have GlcNAc attached to the OH of Ser or Thr as a single sugar.
Xylose	Pentose	Xyl	UDP-Xyl	Xyl is attached to the OH of Ser in many proteoglycans. Xyl in turn is attached to two Gal residues, forming a link trisaccharide. Xyl is also found in t-PA and certain clotting factors.

[1]Structures of glycoproteins are illustrated in Chapter 14.

involved in the biosynthesis of glycoproteins utilize these compounds (see below). The **anhydro nature** of the linkage between the phosphate group and the sugars is of the high energy, high-group-transfer-potential type (Chapter 11). The sugars of these compounds are thus **"activated"** and can be transferred to suitable acceptors provided appropriate transferases are available.

Most nucleotide sugars are formed in the cytosol, generally from reactions involving the corresponding nucleoside triphosphate. CMP-sialic acids are formed in the nucleus. Formation of uridine diphosphate galactose (UDP-Gal) requires the following two reactions in mammalian tissues.

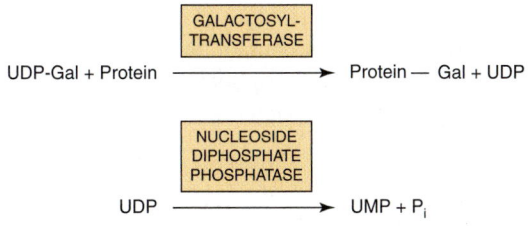

UTP + Glucose 1-phosphate ⇌ **UDP-Glc PYROPHOS-PHORYLASE** ⇌ UDP-Glc + Pyrophosphate

UDP-Glc ⇌ **UDP-Glc EPIMERASE** ⇌ UDP-Gal

Because many glycosylation reactions occur within the lumen of the Golgi apparatus, **carrier systems** (permeases, transporters) are necessary to transport nucleotide sugars across the Golgi membrane. Systems transporting UDP-Gal, GDP-Man, and CMP-NeuAc into the cisternae of the Golgi apparatus have been described. They are **antiport** systems; that is, the influx of one molecule of nucleotide sugar is balanced by the efflux of one molecule of the corresponding nucleotide (eg, UMP, GMP, or CMP) formed from the nucleotide sugars. This mechanism ensures an adequate concentration of each nucleotide sugar inside the Golgi apparatus. UMP is formed from UDP-Gal in the above process as follows.

UDP-Gal + Protein → **GALACTOSYL-TRANSFERASE** → Protein — Gal + UDP

UDP → **NUCLEOSIDE DIPHOSPHATE PHOSPHATASE** → UMP + P_i

EXO- & ENDOGLYCOSIDASES FACILITATE STUDY OF GLYCOPROTEINS

A number of **glycosidases** of defined specificity have proved useful in examining structural and functional aspects of glycoproteins (**Table 47–5**). These enzymes act at either external (exoglycosidases) or internal (endoglycosidases) positions of oligosaccharide chains. Examples of exoglycosidases are **neuraminidases** and **galactosidases;** their sequential use removes terminal NeuAc and subterminal Gal residues from

TABLE 47–5 Some Glycosidases Used to Study the Structure and Function of Glycoproteins[1]

Enzymes	Type
Neuraminidases	Exoglycosidase
Galactosidases	Exo- or endoglycosidase
Endoglycosidase F	Endoglycosidase
Endoglycosidase H	Endoglycosidase

[1]The enzymes are available from a variety of sources and are often specific for certain types of glycosidic linkages and also for their anomeric natures. The sites of action of endoglycosidases F and H are shown in Figure 47–5. F acts on both high-mannose and complex oligosaccharides, whereas H acts on the former.

most glycoproteins. **Endoglycosidases F** and **H** are examples of the latter class; these enzymes cleave the oligosaccharide chains at specific GlcNAc residues close to the polypeptide backbone (ie, at internal sites; Figure 47–5) and are thus useful in releasing large oligosaccharide chains for structural analyses. A glycoprotein can be treated with one or more of the above glycosidases to analyze the effects on its biologic behavior of removal of specific sugars.

THE MAMMALIAN ASIALOGLYCOPROTEIN RECEPTOR IS INVOLVED IN CLEARANCE OF CERTAIN GLYCOPROTEINS FROM PLASMA BY HEPATOCYTES

Experiments performed by Ashwell and his colleagues in the early 1970s played an important role in focusing attention on the functional significance of the oligosaccharide chains of glycoproteins. They treated rabbit ceruloplasmin (a plasma protein; see Chapter 50) with neuraminidase in vitro. This procedure exposed subterminal Gal residues that were normally masked by terminal NeuAc residues. Neuraminidase-treated radioactive ceruloplasmin was found to disappear rapidly from the circulation, in contrast to the slow clearance of the untreated protein. Very significantly, when the Gal residues exposed to treatment with neuraminidase were removed by treatment with a galactosidase, the clearance rate of the protein returned to normal. Further studies demonstrated that liver cells contain a **mammalian asialoglycoprotein receptor** that recognizes the Gal moiety of many desialylated plasma proteins and leads to their endocytosis. This work indicated that an individual sugar, such as Gal, could play an important role in governing at least one of the biologic properties (ie, time of residence in the circulation) of certain glycoproteins. This greatly strengthened the concept that oligosaccharide chains could contain biologic information.

LECTINS CAN BE USED TO PURIFY GLYCOPROTEINS & TO PROBE THEIR FUNCTIONS

Lectins are **carbohydrate-binding proteins** that agglutinate cells or precipitate glycoconjugates; a number of lectins are themselves glycoproteins. Immunoglobulins that react with sugars are not considered lectins. Lectins contain at least two sugar-binding sites; proteins with a single sugar-binding site will not agglutinate cells or precipitate glycoconjugates. The specificity of a lectin is usually defined by the sugars that are best at inhibiting its ability to cause agglutination or precipitation. Enzymes, toxins, and transport proteins can be classified as lectins if they bind carbohydrate. Lectins were first discovered in plants and microbes, but many lectins of animal origin are now known. The mammalian asialoglycoprotein receptor described above is an important example of an animal lectin. Some important lectins are listed in **Table 47–6**. Much current research is centered on the roles of various animal lectins in the mechanisms of action of glycoproteins, some of which are discussed below (eg, with regard to the selectins).

Numerous lectins have been purified and are commercially available; three plant lectins that have been widely used experimentally are listed in **Table 47–7**. Among many uses, lectins have been employed to purify specific glycoproteins, as tools for probing the glycoprotein profiles of cell surfaces, and as reagents for generating mutant cells deficient in certain enzymes involved in the biosynthesis of oligosaccharide chains.

TABLE 47–6 Some Important Lectins

Lectins	Examples or Comments
Legume lectins	Concanavalin A, pea lectin.
Wheat germ agglutinin	Widely used in studies of surfaces of normal cells and cancer cells.
Ricin	Cytotoxic glycoprotein derived from seeds of the castor plant.
Bacterial toxins	Heat-labile enterotoxin of *E coli* and cholera toxin
Influenza virus hemagglutinin	Responsible for host-cell attachment and membrane fusion.
C-type lectins	Characterized by a Ca^{2+}-dependent carbohydrate recognition domain (CRD); includes the mammalian asialoglycoprotein receptor, the selectins, and the mannose-binding protein.
S-type lectins	β-Galactoside-binding animal lectins with roles in cell–cell and cell–matrix interactions.
P-type lectins	Mannose 6-P receptor.
I-type lectins	Members of the immunoglobulin super-family, for example, sialoadhesin mediating adhesion of macrophages to various cells.

TABLE 47–7 Three Plant Lectins and the Sugars with Which They Interact[1]

Lectin	Abbreviation	Sugars
Concanavalin A	ConA	Man and Glc
Soybean lectin		Gal and GalNAc
Wheat germ agglutinin	WGA	Glc and NeuAc

[1]In most cases, lectins show specificity for the anomeric nature of the glycosidic linkage (α or β); this is not indicated in the table.

THERE ARE THREE MAJOR CLASSES OF GLYCOPROTEINS

Based on the nature of the linkage between their polypeptide chains and their oligosaccharide chains, glycoproteins can be divided into three major classes (**Figure 47–1**). (1) Those containing an **O-glycosidic linkage** (ie, O-linked), involving the hydroxyl side chain of serine or threonine and a sugar such as N-acetylgalactosamine (GalNAc-Ser[Thr]); (2) those containing an **N-glycosidic linkage** (ie, N-linked), involving the amide nitrogen of asparagine and N-acetylglucosamine (GlcNAc-Asn); and (3) those linked to the carboxyl terminal amino acid of a protein via a phosphoryl-ethanolamine moiety joined to an oligosaccharide (glycan), which in turn is linked via glucosamine to phosphatidylinositol (PI). This latter class is referred to as **glycosylphosphatidylinositol-anchored** (**GPI-anchored,** or **GPI-linked**) glycoproteins. Members of this class, among other functions, are involved in directing certain glycoproteins to the apical or basolateral areas of the plasma membrane (PM) of some polarized epithelial cells (see Chapter 40 and below). Other minor classes of glycoproteins also exist.

The number of oligosaccharide chains attached to one protein can vary from one to 30 or more, with the sugar chains ranging from one or two residues in length to much larger structures. Many proteins contain more than one type of sugar chain; for instance, **glycophorin,** an important red cell membrane glycoprotein (Chapter 52), contains both O- and N-linked oligosaccharides.

GLYCOPROTEINS CONTAIN SEVERAL TYPES OF O-GLYCOSIDIC LINKAGES

At least four subclasses of O-glycosidic linkages are found in human glycoproteins. (1) The **GalNAc-Ser(Thr)** linkage shown in Figure 47–1 is the predominant linkage. Two typical oligosaccharide chains found in members of this subclass are shown in **Figure 47–2**. Usually a Gal or a NeuAc residue is attached to the GalNAc, but many variations in the sugar compositions and lengths of such oligosaccharide chains are found. This type of linkage is found in **mucins** (see below). (2) **Proteoglycans** contain a **Gal-Gal-Xyl-Ser** trisaccharide

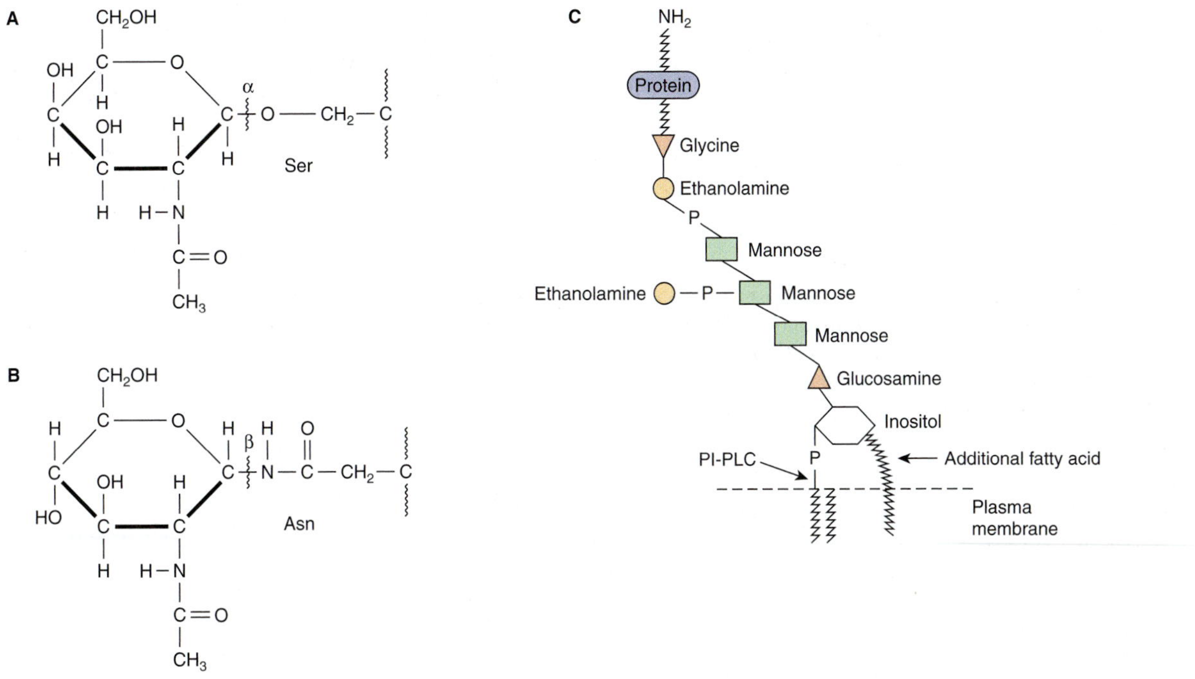

FIGURE 47–1 Depictions of (A) an *O*-linkage (*N*-acetylgalactosamine to serine), (B) an *N*-linkage (*N*-acetylglucosamine to asparagine), and (C) a glycosylphosphatidylinositol (GPI) linkage. The GPI structure shown is that linking acetylcholinesterase to the plasma membrane of the human red blood cell. The carboxyl terminal amino acid is glycine joined in amide linkage via its COOH group to the NH₂ group of phosphorylethanolamine, which in turn is joined to a mannose residue. The core glycan contains three mannose and one glucosamine residues. The glucosamine is linked to inositol, which is attached to the phosphatidic acid. The site of action of PI-phospholipase C (PI-PLC) is indicated. The structure of the core glycan is shown in the text. This particular GPI contains an extra fatty acid attached to inositol and also an extra phosphorylethanolamine moiety attached to the middle of the three mannose residues. Variations found among different GPI structures include the identity of the carboxyl terminal amino acid, the molecules attached to the mannose residues, and the precise nature of the lipid moiety.

(the so-called link trisaccharide). (3) **Collagens** contain a **Gal-Hydroxylysine (Hyl)** linkage. (Subclasses [2] and [3] are discussed further in Chapter 48.) (4) Many **nuclear proteins** (eg, certain transcription factors) and **cytosolic proteins** contain side chains consisting of a single GlcNAc attached to a serine or threonine residue (**GlcNAc-Ser[Thr]**).

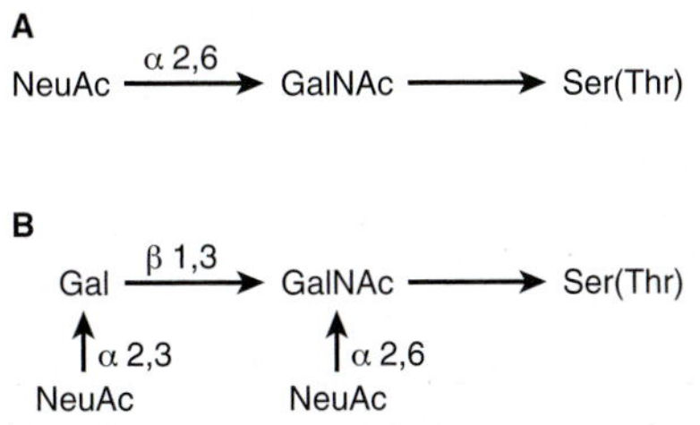

FIGURE 47–2 Structures of two *O*-linked oligosaccharides found in (A) submaxillary mucins and (B) fetuin and in the sialoglycoprotein of the membrane of human red blood cells. (Modified and reproduced, with permission, from Lennarz WJ: *The Biochemistry of Glycoproteins and Proteoglycans.* Plenum Press, 1980. Reproduced with kind permission from Springer Science and Business Media.)

Mucins Have a High Content of *O*-Linked Oligosaccharides & Exhibit Repeating Amino Acid Sequences

Mucins are glycoproteins with two major characteristics: (1) a high content of **O-linked oligosaccharides** (the carbohydrate content of mucins is generally more than 50%); and (2) the presence of **variable numbers of tandem repeats (VNTRs)** of peptide sequence in the centre of their polypeptide backbones, to which the *O*-glycan chains are attached in clusters (**Figure 47–3**). These sequences are rich in serine, threonine, and proline. Although *O*-glycans predominate, mucins often contain a number of *N*-glycan chains. Both **secretory** and **membrane-bound** mucins occur. The former are found in the mucus present in the secretions of the gastrointestinal, respiratory, and reproductive tracts. **Mucus** consists of about 94% water and 5% mucins, with the remainder being a mixture of various cell molecules, electrolytes, and remnants of cells. Secretory mucins generally have an oligomeric structure and thus often have a very high molecular mass. The oligomers are composed of monomers linked by disulfide bonds. Mucus exhibits a high **viscosity** and often forms a **gel.** These qualities are functions of its content of mucins. The high content

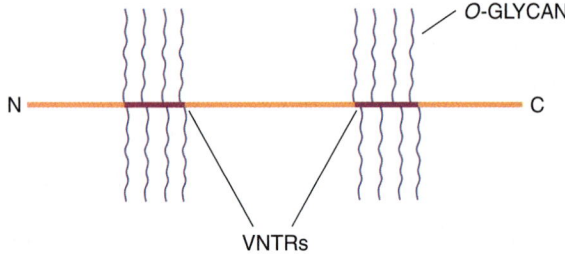

FIGURE 47–3 **Much simplified schematic of a mucin.**
O-glycans (blue) are shown attached to two of many VNTR regions (red). *N*-glycans may also be present. Mucins generally contain cysteines (not shown) near their N and C termini, which are involved in polymerization via disulfide bridges. Other domains (D) near their N-termini are also involved in polymerization. Membrane-bound mucins contain transmembrane and cytosolic domains, in addition to larger extracellular domains containing *O*-glycans.

of *O*-glycans confers an extended structure on mucins. This is in part explained by steric interactions between their GalNAc moieties and adjacent amino acids, resulting in a chain-stiffening effect so that the conformations of mucins often become those of rigid rods. Intermolecular noncovalent interactions between various sugars on neighboring glycan chains contribute to gel formation. The high content of **NeuAc** and **sulfate** residues found in many mucins confers a negative charge on them. With regard to their functions, mucins help **lubricate** and form a **protective physical barrier** on epithelial surfaces. Membrane-bound mucins participate in various **cell–cell interactions** (eg, involving selectins; see below). The density of oligosaccharide chains makes it difficult for **proteases** to approach their polypeptide backbones, so that mucins are often resistant to their action. Mucins also tend to "mask" certain surface antigens. Many cancer cells form excessive amounts of mucins; perhaps the mucins may mask certain surface antigens on such cells and thus protect the cells from immune surveillance. Mucins also carry cancer-specific peptide and carbohydrate epitopes (an epitope is a site on an antigen recognized by an antibody, also called an antigenic determinant). Some of these epitopes have been used to stimulate an immune response against cancer cells.

The **genes** encoding the polypeptide backbones of a number of mucins derived from various tissues (eg, pancreas, small intestine, trachea and bronchi, stomach, and salivary glands) have been cloned and sequenced. These studies have revealed new information about the polypeptide backbones of mucins (size of tandem repeats, potential sites of *N*-glycosylation, etc) and ultimately should reveal aspects of their genetic control. Some important properties of mucins are summarized in **Table 47–8**.

The Biosynthesis of *O*-Linked Glycoproteins Uses Nucleotide Sugars

The polypeptide chains of *O*-linked and other glycoproteins are encoded by mRNA species; because most glycoproteins are

TABLE 47–8 **Some Properties of Mucins**

- Found in secretions of the gastrointestinal, respiratory, and reproductive tracts and also in membranes of various cells.
- Exhibit high content of *O*-glycan chains, usually containing NeuAc.
- Contain repeating amino acid sequences rich in serine, threonine, and proline.
- Extended structure contributes to their high viscoelasticity.
- Form protective physical barrier on epithelial surfaces, are involved in cell–cell interactions, and may contain or mask certain surface antigens.

membrane-bound or secreted, they are generally translated on membrane-bound polyribosomes (Chapter 37). Hundreds of different oligosaccharide chains of the *O*-glycosidic-type exist. These glycoproteins are built up by the **stepwise donation of sugars from nucleotide sugars,** such as UDP-GalNAc, UDP-Gal, and CMP-NeuAc. The enzymes catalyzing this type of reaction are membrane-bound **glycoprotein glycosyltransferases.** Generally, synthesis of one specific type of linkage requires the activity of a correspondingly specific transferase. The factors that determine which specific serine and threonine residues are glycosylated have not been identified but are probably found in the peptide structure surrounding the glycosylation site. The enzymes assembling *O*-linked chains are located in the Golgi apparatus, sequentially arranged in an assembly line with terminal reactions occurring in the *trans*-Golgi compartments.

The major features of the biosynthesis of *O*-linked glycoproteins are summarized in **Table 47–9**.

N-LINKED GLYCOPROTEINS CONTAIN AN Asn-GlcNAc LINKAGE

N-Linked glycoproteins are distinguished by the presence of the Asn–GlcNAc linkage (Figure 47–1). It is the major class of glycoproteins and has been much studied, since the most

TABLE 47–9 **Summary of Main Features of *O*-Glycosylation**

- Involves a battery of membrane-bound glycoprotein glycosyltransferases acting in a stepwise manner; each transferase is generally specific for a particular type of linkage.
- The enzymes involved are located in various subcompartments of the Golgi apparatus.
- Each glycosylation reaction involves the appropriate nucleotide sugar.
- Dolichol-P-P-oligosaccharide is not involved, nor are glycosidases; and the reactions are not inhibited by tunicamycin.
- *O*-Glycosylation occurs posttranslationally at certain Ser and Thr residues.

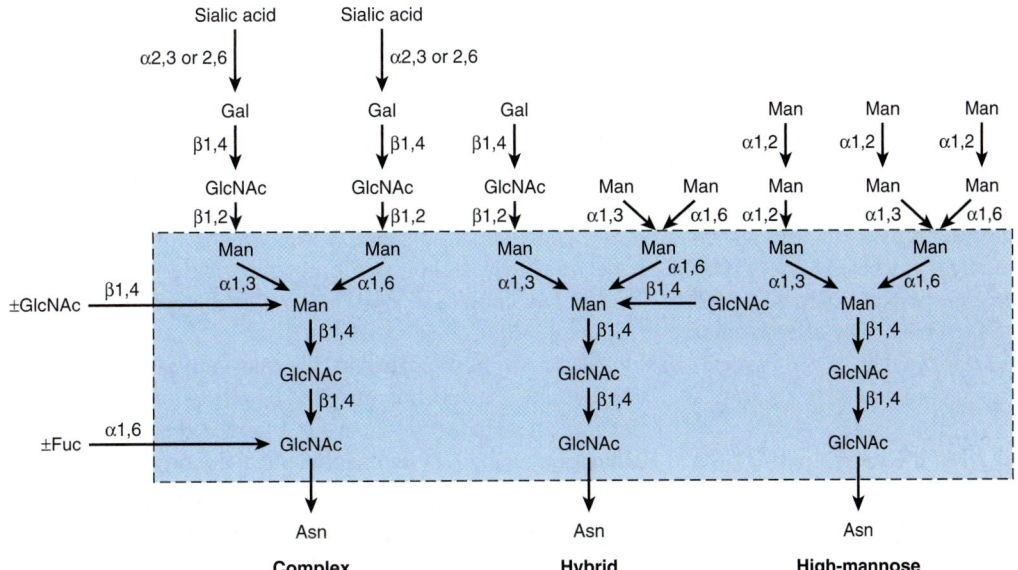

FIGURE 47–4
Structures of the major types of asparagine-linked oligosaccharides. The boxed area encloses the pentasaccharide core common to all N-linked glycoproteins. (Reproduced, with permission, from Kornfeld R, Kornfeld S: Assembly of asparagine-linked oligosaccharides. Annu Rev Biochem 1985;54:631. Copyright © 1985 by Annual Reviews. Reprinted with permission.)

readily accessible glycoproteins (eg, plasma proteins) mainly belong to this group. It includes both **membrane-bound** and **circulating** glycoproteins. The principal difference between this and the previous class, apart from the nature of the amino acid to which the oligosaccharide chain is attached (Asn vs Ser or Thr), concerns their biosynthesis.

Complex, Hybrid & High-Mannose Are the Three Major Classes of N-Linked Oligosaccharides

There are three major classes of N-linked oligosaccharides: **complex, hybrid,** and **high-mannose** (**Figure 47–4**). Each type shares a common pentasaccharide, $Man_3GlcNAc_2$—shown within the boxed area in Figure 47–4 and depicted also in **Figure 47–5**—but they differ in their outer branches. The presence of the **common pentasaccharide** is explained by the fact that all three classes share an initial common mechanism of biosynthesis. Glycoproteins of the complex type generally contain terminal NeuAc residues and underlying Gal and GlcNAc residues, the latter often constituting the disaccharide N-acetyllactosamine. Repeating **N-acetyllactosamine units**—[Galβ1–3/4GlcNAcβ1–3]$_n$ (poly-N-acetyllactosaminoglycans)—

are often found on N-linked glycan chains. I/i blood group substances belong to this class. The majority of complex-type oligosaccharides contain two, three, or four outer branches (Figure 47–4), but structures containing five branches have also been described. The oligosaccharide branches are often referred to as **antennae,** so that bi-, tri-, tetra-, and penta-antennary structures may all be found. A bewildering number of chains of the complex type exist, and that indicated in Figure 47–4 is only one of many. Other complex chains may terminate in Gal or Fuc. High-mannose oligosaccharides typically have two to six additional Man residues linked to the pentasaccharide core. Hybrid molecules contain features of both of the two other classes.

The Biosynthesis of N-Linked Glycoproteins Involves Dolichol-P-P-Oligosaccharide

Leloir and his colleagues described the occurrence of a **dolichol-pyrophosphate-oligosaccharide (Dol-P-P-oligosaccharide),** which subsequent research showed to play a key role in the biosynthesis of N-linked glycoproteins. The oligosaccharide chain of this compound generally has the structure R-Glc-NAc$_2$Man$_9$Glc$_3$ (R = Dol-P-P). The sugars of this compound are first assembled on the Dol-P-P backbone, and the oligosaccharide chain is then transferred en bloc to suitable Asn residues of acceptor apoglycoproteins during their synthesis on membrane-bound polyribosomes. All N-glycans have a common pentasaccharide core structure (Figure 47–5).

To form **high-mannose** chains, only the Glc residues plus certain of the peripheral Man residues are removed. To form an oligosaccharide chain of the **complex type,** the Glc residues and four of the Man residues are removed by glycosidases in the endoplasmic reticulum and Golgi. The sugars characteristic of complex chains (GlcNAc, Gal, and NeuAc) are added

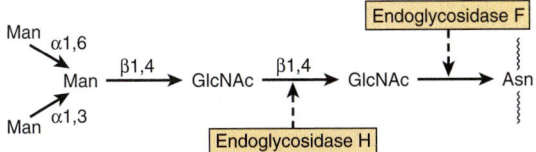

FIGURE 47–5 Schematic of the pentasaccharide core common to all N-linked glycoproteins and to which various outer chains of oligosaccharides may be attached. The sites of action of endoglycosidases F and H are also indicated.

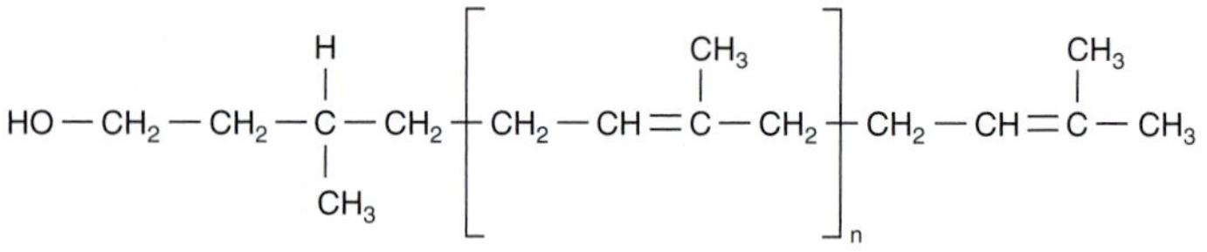

FIGURE 47–6   **The structure of dolichol.** The phosphate in dolichol phosphate is attached to the primary alcohol group at the left-hand end of the molecule. The group within the brackets is an isoprene unit ($n = 17$–20 isoprenoid units).

$$HO-CH_2-CH_2-\underset{\underset{CH_3}{|}}{\overset{\overset{H}{|}}{C}}-CH_2\Bigg[CH_2-CH=\underset{}{\overset{\overset{CH_3}{|}}{C}}-CH_2\Bigg]_n CH_2-CH=\overset{\overset{CH_3}{|}}{C}-CH_3$$

by the action of individual glycosyltransferases located in the Golgi apparatus. The phenomenon whereby the glycan chains of *N*-linked glycoproteins are first partially degraded and then in some cases rebuilt is referred to as **oligosaccharide processing.** **Hybrid chains** are formed by partial processing, forming complex chains on one arm and Man structures on the other arm.

Thus, the initial steps involved in the biosynthesis of the *N*-linked glycoproteins differ markedly from those involved in the biosynthesis of the *O*-linked glycoproteins. The former involves Dol-P-P-oligosaccharide; the latter, as described earlier, does not.

The process of *N*-glycosylation can be broken down into two stages: (1) assembly of Dol-P-P-oligosaccharide and transfer of the oligosaccharide; and (2) processing of the oligosaccharide chain.

Assembly & Transfer of Dolichol-P-P-Oligosaccharide

Polyisoprenol compounds exist in both bacteria and eukaryotic cells. They participate in the synthesis of bacterial polysaccharides and in the biosynthesis of *N*-linked glycoproteins and GPI anchors. The polyisoprenol used in eukaryotic tissues

is **dolichol,** which is, next to rubber, the longest naturally occurring hydrocarbon made up of a single repeating unit. Dolichol is composed of 17–20 repeating isoprenoid units (**Figure 47–6**).

Before it participates in the biosynthesis of Dol-P-P-oligosaccharide, dolichol must first be phosphorylated to form dolichol phosphate (Dol-P) in a reaction catalyzed by **dolichol kinase** and using ATP as the phosphate donor.

Dolichol-P-P-GlcNAc (Dol-P-P-GlcNAc) is the key lipid that acts as an acceptor for other sugars in the assembly of Dol-P-P-oligosaccharide. It is synthesized in the membranes of the endoplasmic reticulum from Dol-P and UDP-GlcNAc in the following reaction, catalyzed by GlcNAc-P transferase.

$$\text{Dol-P + UDP-GlcNAc} \rightarrow \text{Dol-P-P-GlcNAc + UMP}$$

The above reaction, which is the first step in the assembly of Dol-P-P-oligosaccharide, and the other later reactions are summarized in **Figure 47–7**. The essential features of the subsequent steps in the assembly of Dol-P-P-oligosaccharide are as follows:

1. A second GlcNAc residue is added to the first, again using UDP-GlcNAc as the donor.
2. Five Man residues are added, using GDP-mannose as the donor.

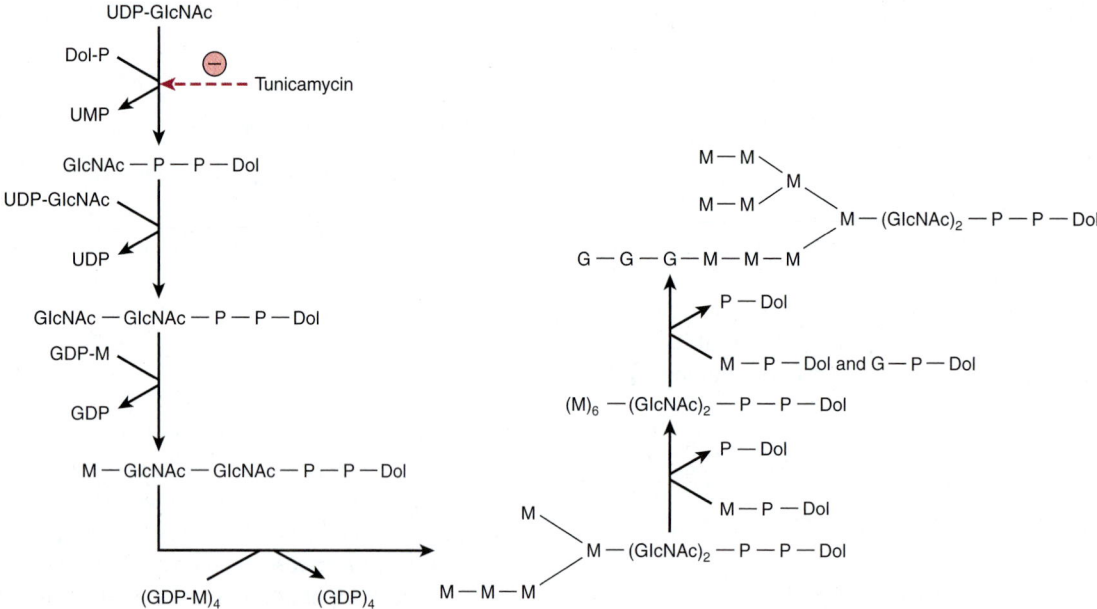

FIGURE 47–7   **Pathway of biosynthesis of dolichol-P-P-oligosaccharide.** The specific linkages formed are indicated in Figure 47–8. Note that the first five internal mannose residues are donated by GDP-mannose, whereas the more external mannose residues and the glucose residues are donated by dolichol-P-mannose and dolichol-P-glucose. (UDP, uridine diphosphate; Dol, dolichol; P, phosphate; UMP, uridine monophosphate; GDP, guanosine diphosphate.)

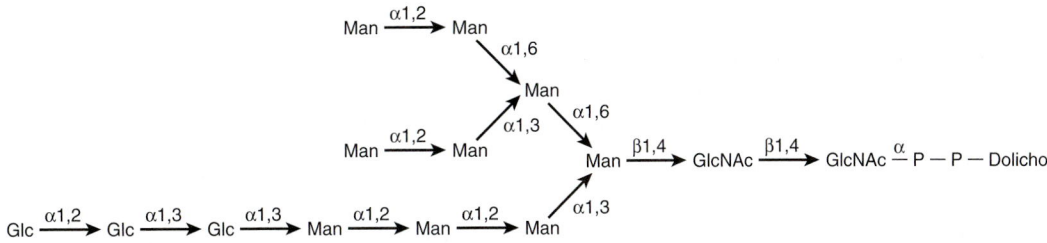

FIGURE 47–8 **Structure of dolichol-P-P-oligosaccharide.** (With permission from Li E, et al: Structure of the lipid-linked oligosaccharide precursor of the complex-type oligosaccharides of the vesicular stomatitis virus G protein. J Biol Chem 1978;253:7762.)

3. Four additional Man residues are next added, using Dol-P-Man as the donor. Dol-P-Man is formed by the following reaction.

$$\text{Dol-P} + \text{GDP-Man} \rightarrow \text{Dol-P-Man} + \text{GDP}$$

4. Finally, the three peripheral glucose residues are donated by Dol-P-Glc, which is formed in a reaction analogous to that just presented except that Dol-P and UDP-Glc are the substrates.

It should be noted that the first seven sugars (two GlcNAc and five Man residues) are donated by nucleotide sugars, whereas the last seven sugars (four Man and three Glc residues) added are donated by dolichol-sugars. The net result is assembly of the compound illustrated in **Figure 47–8** and referred to in shorthand as Dol-P-P-GlcNAc$_2$Man$_9$Glc$_3$.

The oligosaccharide linked to dolichol-P-P is transferred en bloc to form an *N*-glycosidic bond with one or more specific Asn residues of an acceptor protein emerging from the luminal surface of the membrane of the endoplasmic reticulum. The reaction is catalyzed by **oligosaccharide: protein transferase,** a membrane-associated enzyme complex. The transferase will recognize and transfer any substrate with the general structure Dol-P-P-(GlcNAc)$_2$-R, but it has a strong preference for the Dol-P-P-GlcNAc$_2$Man$_9$Glc$_3$ structure. Glycosylation occurs at the Asn residue of an Asn-XSer/Thr tripeptide sequence, where X is any amino acid except proline, aspartic acid, or glutamic acid. A tripeptide site contained within a β turn is favored. Only about one-third of the Asn residues that are potential acceptor sites are actually glycosylated, suggesting that factors other than the tripeptide are also important.

The acceptor proteins are of both the secretory and integral membrane class. Cytosolic proteins are rarely glycosylated. The transfer reaction and subsequent processes in the glycosylation of *N*-linked glycoproteins, along with their subcellular locations, are depicted in **Figure 47–9.** The other product of the oligosaccharide: protein transferase reaction is dolichol-P-P, which is subsequently converted to dolichol-P by a phosphatase. The dolichol-P can serve again as an acceptor for the synthesis of another molecule of Dol-P-P-oligosaccharide.

Processing of the Oligosaccharide Chain

1. Early Phase: the various reactions involved are indicated in Figure 47–9. The oligosaccharide: protein transferase catalyzes reaction 1 (see above). Reactions 2 and 3 involve the removal of the terminal Glc residue by glucosidase I and of the next two Glc residues by glucosidase II, respectively. In the case of **high-mannose** glycoproteins, the process may stop here, or up to four Man residues may also be removed. However, to form **complex** chains, additional steps are necessary, as follows. Four external Man residues are removed in reactions 4 and 5 by at least two different mannosidases. In reaction 6, a GlcNAc residue is added to the Man residue of the Man 1–3 arm by GlcNAc transferase I. The action of this latter enzyme permits the occurrence of reaction 7, a reaction catalyzed by yet another mannosidase (Golgi α-mannosidase II) and which results in a reduction of the Man residues to the core number of three (Figure 47–5).

An important additional pathway is indicated in reactions I and II of Figure 47–9. This involves enzymes destined for **lysosomes.** Such enzymes are targeted to the lysosomes by a specific chemical marker. In reaction I, a residue of GlcNAc-1-P is added to carbon 6 of one or more specific Man residues of these enzymes. The reaction is catalyzed by a GlcNAc phosphotransferase, which uses UDPGlcNAc as the donor and generates UMP as the other product:

UDP-GlcNAc + Man — Protein $\xrightarrow{\text{GlcNAc PHOSPHO-TRANSFERASE}}$

GlcNAc-1-P-6-Man — Protein + UMP

In reaction II, the GlcNAc is removed by the action of a phosphodiesterase, leaving the Man residues phosphorylated in the sixth position:

GlcNAc-1-P-6-Man — Protein $\xrightarrow{\text{PHOSPHO-DIESTERASE}}$

P-6-Man — Protein + GlcNAc

Man 6-P receptors, located in the Golgi apparatus, bind the Man 6-P residues of these enzymes and direct them to the lysosomes. Fibroblasts from patients with **I-cell disease** (see below) are severely deficient in the activity of the GlcNAc phosphotransferase.

2. Late Phase: to assemble a typical complex oligosaccharide chain, additional sugars must be added to the structure formed in reaction 7. Hence, in reaction 8, a second GlcNAc is added to the peripheral Man residue of the other arm of the bi-antennary structure shown in Figure 47–9; the enzyme catalyzing this step is GlcNAc transferase II. Reactions 9, 10,

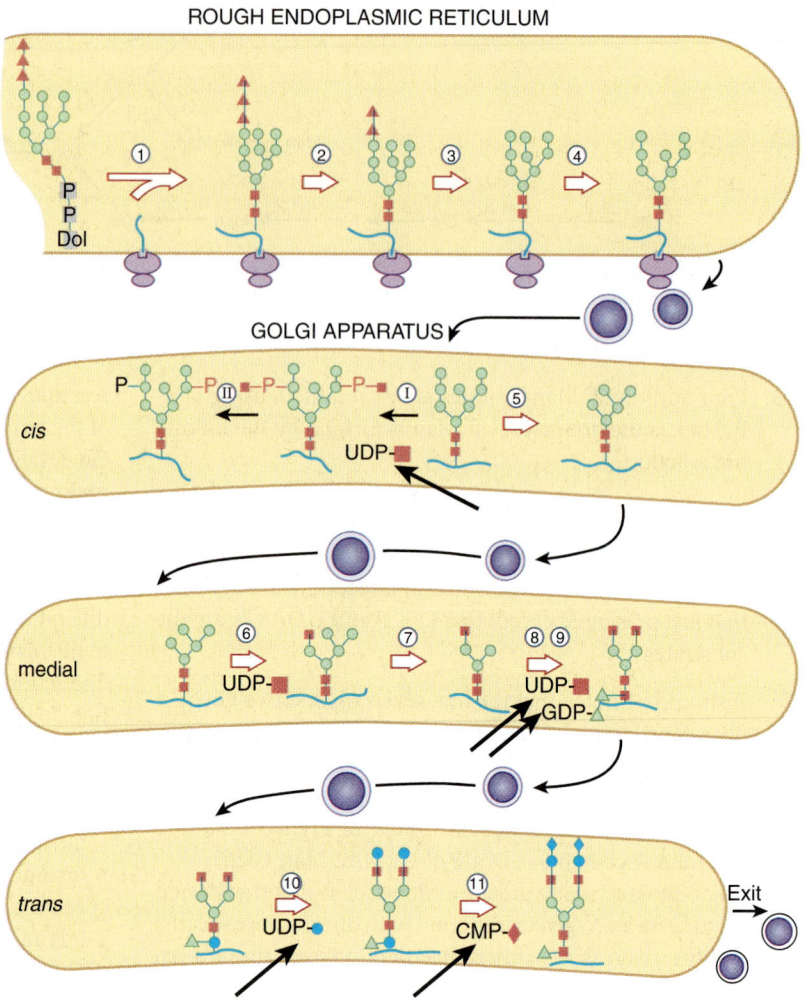

FIGURE 47–9 **Schematic pathway of oligosaccharide processing.**
The reactions are catalyzed by the following enzymes: ① oligosaccharide: protein transferase; ② α-glucosidase I; ③ α-glucosidase II; ④ endoplasmic reticulum α 1,2-mannosidase; I *N*-acetylglucosaminylphosphotransferase; II *N*-acetylglucosamine-1-phosphodiester α-*N*-acetylglucosaminidase; ⑤ Golgi apparatus α-mannosidase I; ⑥ *N*-acetylglucosaminyltransferase I; ⑦ Golgi apparatus α-mannosidase II; ⑧ *N*-acetylglucosaminyltransferase II; ⑨ fucosyltransferase; ⑩ galactosyltransferase; ⑪ sialyltransferase. The thick arrows indicate various nucleotide sugars involved in the overall scheme. (Solid square, *N*-acetylglucosamine; open circle, mannose; solid triangle, glucose; open triangle, fucose; solid circle, galactose; solid diamond, sialic acid.) (Reproduced, with permission, from Kornfeld R, Kornfeld S: Assembly of asparagine-linked oligosaccharides. Annu Rev Biochem 1985;54:631. Copyright © 1985 by Annual Reviews. Reprinted with permission.)

and 11 involve the addition of Fuc, Gal, and NeuAc residues at the sites indicated, in reactions catalyzed by fucosyl, galactosyl, and sialyl transferases, respectively. The assembly of poly-*N*-acetyllactosamine chains requires additional GlcNAc transferases.

The Endoplasmic Reticulum & Golgi Apparatus Are the Major Sites of Glycosylation

As indicated in Figure 47–9, the endoplasmic reticulum and the Golgi apparatus are the major sites involved in glycosylation processes. The assembly of Dol-P-P-oligosaccharide occurs on both the cytoplasmic and luminal surfaces of the ER membranes. Addition of the oligosaccharide to protein occurs in the rough endoplasmic reticulum during or after translation. Removal of the Glc and some of the peripheral Man residues also occurs in the endoplasmic reticulum. The Golgi apparatus is composed of *cis*, medial, and *trans* cisternae; these can be separated by appropriate centrifugation procedures. Vesicles containing glycoproteins bud off in the endoplasmic reticulum and are transported to the *cis*-Golgi. Various studies have shown that the enzymes

involved in glycoprotein processing show differential locations in the cisternae of the Golgi. As indicated in Figure 47–9, Golgi α-mannosidase I (catalyzing reaction 5) is located mainly in the *cis*-Golgi, whereas GlcNAc transferase I (catalyzing reaction 6) appears to be located in the medial Golgi, and the fucosyl, galactosyl, and sialyl transferases (catalyzing reactions 9, 10, and 11) are located mainly in the *trans*-Golgi. The major features of the biosynthesis of *N*-linked glycoproteins are summarized in **Table 47–10** and should be contrasted with those previously listed (Table 47–9) for *O*-linked glycoproteins.

Some Glycan Intermediates Formed During *N*-Glycosylation Have Specific Functions

The following are a number of specific functions of *N*-glycan chains that have been established or are under investigation: (1) The involvement of the **mannose 6-P signal** in targeting of certain lysosomal enzymes is clear (see above and discussion of I-cell disease, in the following). (2) It is likely that the large *N*-glycan chains present on newly synthesized glycoproteins may assist in keeping these proteins in a **soluble**

TABLE 47–10 Summary of Main Features of *N*-Glycosylation

- The oligosaccharide Glc$_3$Man$_9$(GlcNAc)$_2$ is transferred from dolichol-P-P-oligosaccharide in a reaction catalyzed by oligosaccharide:protein transferase, which is inhibited by tunicamycin.

- Transfer occurs to specific Asn residues in the sequence AsnX-Ser/Thr, where X is any residue except Pro, Asp, or Glu.

- Transfer can occur cotranslationally in the endoplasmic reticulum.

- The protein-bound oligosaccharide is then partially processed by glucosidases and mannosidases; if no additional sugars are added, this results in a high-mannose chain.

- If processing occurs down to the core heptasaccharide (Man$_5$[GlcNAc]$_2$), complex chains are synthesized by the addition of GlcNAc, removal of two Man, and the stepwise addition of individual sugars in reactions catalyzed by specific transferases (eg, GlcNAc, Gal, NeuAc transferases) that employ appropriate nucleotide sugars.

chaperone (Chapter 46) and lectin. Binding to calnexin prevents a glycoprotein from aggregating. It has been found that calnexin will bind specifically to a number of glycoproteins (eg, the influenza virus hemagglutinin [HA]) that possess the **monoglycosylated core structure.** This species is the product of reaction 2 shown in Figure 47–9, but from which the terminal glucose residue has been removed, leaving only the innermost glucose attached. Calnexin and the bound glycoprotein form a complex with **ERp57,** a homolog of protein disulfide isomerase (PDI), which catalyzes disulfide bond interchange, facilitating proper folding. The bound glycoprotein is released from its complex with calnexin-ERp57 when the sole remaining glucose is hydrolyzed by glucosidase II and **leaves the ER if properly folded.** If **not properly folded,** an ER **glucosyltransferase** recognizes this and **reglucosylates** the glycoprotein, which **rebinds** to the calnexin-Erp57 complex. If now properly folded, the glycoprotein is again deglucosylated and leaves the ER. If not capable of proper folding, it is **translocated out of the ER** into the cytoplasm, where it is **degraded** (compare Figure 46–8). This so-called **calnexin cycle** is illustrated in **Figure 47–10.** In this way, calnexin retains certain partly folded (or misfolded) glycoproteins and releases them when further folding has occurred. The glucosyltransferase, by **sensing** the folding of the glycoprotein and

state inside the lumen of the endoplasmic reticulum. (3) One species of *N*-glycan chains has been shown to play a role in the folding and retention of certain glycoproteins in the lumen of the endoplasmic reticulum. **Calnexin** is a protein present in the endoplasmic reticulum membrane that acts as a

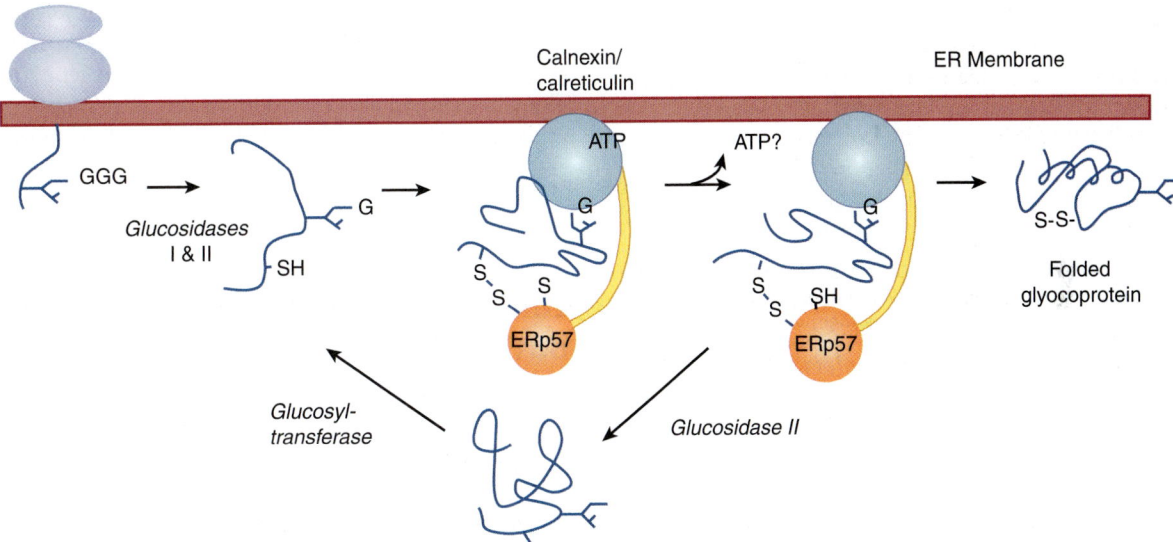

FIGURE 47–10 Model of the calnexin cycle. As a nascent (growing) polypeptide chain enters the ER, certain Asn residues are glycosylated by addition of Glc$_3$Man$_9$GlcNAc$_2$ (see the text). The outermost two molecules of glucose are removed via the actions of glucosidases I and II. This exposes the innermost molecule of glucose, which is recognized by the lectin sites of calnexin and calreticulin. In their ATP-bound state, calnexin and calreticulin bind to the monoglucosylated oligosaccharide (via their lectin sites) as well as to hydrophobic segments of the unfolded glycoprotein (via their polypeptide binding or chaperone sites). Glycoprotein dissociation involves the action of glucosidase II to remove the terminal glucose and also a change in affinity of the polypeptide binding site. After dissociation, if folding does not occur rapidly, the glycoprotein is reglucosylated by an ER glucosyltransferase, which acts only on non-native protein conformers (conformer = a protein in one of several possible conformations). The reglucosylated glycoprotein can then rebind to the ATP form of calnexin/calreticulin. Thus, both the glucosyltransferase and calnexin/calreticulin act as folding sensors. This cycle of binding and release has three functions: it prevents glycoprotein aggregation; it retains non-native conformers in the ER until a native structure is acquired (quality control); and binding to calnexin/calreticulin brings ERp57 into proximity with the non-native glycoprotein. ERp57 catalyzes disulfide bond formation and isomerization within the glycoprotein substrate, assisting it to assume its native conformation. If the glycoprotein is not capable of proper folding, it is translocated out of the ER into the cytoplasm for proteosomal degradation (compare Figure 46–8). Calreticulin, a soluble ER protein, plays a similar role to calnexin. (G, glucose.) (Figure and legend generously supplied by Dr D B Williams, and modified slightly with his permission.)

only reglucosylating misfolded proteins, is a key component of the cycle. The calnexin cycle is an important component of the **quality control systems** operating in the lumen of the ER. The soluble ER protein **calreticulin** performs a similar function.

Several Factors Regulate the Glycosylation of Glycoproteins

It is evident that glycosylation of glycoproteins is a complex process involving a large number of enzymes. It has been estimated that some 1% of the human genome may be involved with glycosylation events. Another index of its complexity is that more than ten distinct GlcNAc transferases involved in glycoprotein biosynthesis have been reported, and others are theoretically possible. Multiple species of the other glycosyltransferases (eg, sialyltransferases) also exist. Controlling factors of the first stage of *N*-linked glycoprotein biosynthesis (ie, **oligosaccharide assembly and transfer**) include (1) the presence of suitable acceptor sites in proteins, (2) the tissue level of Dol-P, and (3) the activity of the oligosaccharide: protein transferase.

Some factors known to be involved in the regulation of **oligosaccharide processing** are listed in **Table 47–11**. Two of the points listed merit further comment. First, **species variations**

TABLE 47–11   Some Factors Affecting the Activities of Glycoprotein Processing Enzymes

Factor	Comment
Cell type	Different cell types contain different profiles of processing enzymes.
Previous enzyme	Certain glycosyltransferases act only on an oligosaccharide chain if it has already been acted upon by another processing enzyme.[1]
Development	The cellular profile of processing enzymes may change during development if their genes are turned on or off.
Intracellular location	For instance, if an enzyme is destined for insertion into the membrane of the ER (eg, HMG-CoA reductase), it may never encounter Golgi-located processing enzymes.
Protein conformation	Differences in conformation of different proteins may facilitate or hinder access of processing enzymes to identical oligosaccharide chains.
Species	Same cells (eg, fibroblasts) from different species may exhibit different patterns of processing enzymes.
Cancer	Cancer cells may exhibit processing enzymes different from those of corresponding normal cells.

[1]For example, prior action of GlcNAc transferase I is necessary for the action of Golgi α-mannosidase II.

among processing enzymes have assumed importance in relation to the production of glycoproteins of therapeutic use by means of recombinant DNA technology. For instance, **recombinant erythropoietin** (epoetin alfa; EPO) is sometimes administered to patients with certain types of chronic anemia in order to stimulate erythropoiesis. The half-life of EPO in plasma is influenced by the nature of its glycosylation pattern, with certain patterns being associated with a short half-life, appreciably limiting its period of therapeutic effectiveness. It is thus important to harvest EPO from host cells that confer a pattern of glycosylation consistent with a normal half-life in plasma. Second, there is great interest in analysis of the activities of glycoprotein-processing enzymes in various types of **cancer cells.** These cells have often been found to synthesize different oligosaccharide chains (eg, they often exhibit greater branching) from those made in control cells. This could be due to cancer cells containing different patterns of glycosyltransferases from those exhibited by corresponding normal cells, due to specific gene activation or repression. The differences in oligosaccharide chains could affect adhesive interactions between cancer cells and their normal parent tissue cells, contributing to metastasis. If a correlation could be found between the activity of particular processing enzymes and the **metastatic properties** of cancer cells, this could be important as it might permit synthesis of drugs to inhibit these enzymes and, secondarily, metastasis.

The genes encoding many glycosyltransferases have already been **cloned,** and others are under study. Cloning has revealed new information on both protein and gene structures. The latter should also cast light on the mechanisms involved in their **transcriptional control,** and **gene knockout studies** are being used to evaluate the biologic importance of various glycosyltransferases.

Tunicamycin Inhibits *N*- But Not *O*-Glycosylation

A number of compounds are known to inhibit various reactions involved in glycoprotein processing. **Tunicamycin, deoxynojirimycin,** and **swainsonine** are three such agents. The reactions they inhibit are indicated in **Table 47–12**. These agents can be used experimentally to inhibit various stages of glycoprotein biosynthesis and to study the effects of specific alterations upon the process. For instance, if cells are grown in the presence of tunicamycin, no glycosylation of their normally *N*-linked glycoproteins will occur. In certain cases, lack of glycosylation has been shown to increase the susceptibility of these proteins to proteolysis. Inhibition of glycosylation does not appear to have a consistent effect upon the secretion of glycoproteins that are normally secreted. The inhibitors of glycoprotein processing listed in Table 47–12 do not affect the biosynthesis of *O*-linked glycoproteins. The extension of *O*-linked chains can be prevented by GalNAc-benzyl. This compound competes with natural glycoprotein substrates and thus prevents chain growth beyond GalNAc.

TABLE 47–12 Three Inhibitors of Enzymes Involved in the *N*-Glycosylation of Glycoproteins and Their Sites of Action

Inhibitor	Site of Action
Tunicamycin	Inhibits GlcNAc-P transferase, the enzyme catalyzing addition of GlcNAc to dolichol-P, the first step in the biosynthesis of oligosaccharide-P-P-dolichol.
Deoxynojirimycin	Inhibitor of glucosidases I and II.
Swainsonine	Inhibitor of mannosidase II.

TABLE 47–13 Some GPI-Linked Proteins

- Acetylcholinesterase (red cell membrane)
- Alkaline phosphatase (intestinal, placental)
- Decay-accelerating factor (red cell membrane)
- 5′-Nucleotidase (T lymphocytes, other cells)
- Thy-1 antigen (brain, T lymphocytes)
- Variable surface glycoprotein (*Trypanosoma brucei*)

SOME PROTEINS ARE ANCHORED TO THE PLASMA MEMBRANE BY GLYCOPHOSPHATIDYL-INOSITOL STRUCTURES

GPI-linked glycoproteins comprise the third major class of glycoprotein. The GPI structure (sometimes called a "sticky foot") involved in linkage of the enzyme acetylcholinesterase (ACh esterase) to the plasma membrane of the red blood cell is shown in Figure 47–1. GPI-linked proteins are anchored to the outer leaflet of the plasma membrane by the fatty acids of phosphatidylinositol (PI). The PI is linked via a GlcN moiety to a glycan chain that contains various sugars (eg, Man, GlcN). In turn, the oligosaccharide chain is linked via phosphorylethanolamine in an amide linkage to the carboxyl terminal amino acid of the attached protein. The core of most GPI structures contains one molecule of phosphorylethanolamine, three Man residues, one molecule of GlcN, and one molecule of phosphatidylinositol, as follows:

Ethanolamine-phospho →6Manα1 →
2Manα1 →6Manα1 →GlcNα1 →
6 — *myo* – inositol –1– phospholipid

Additional constituents are found in many GPI structures; for example, that shown in Figure 47–1 contains an extra phosphorylethanolamine attached to the middle of the three Man moieties of the glycan and an extra fatty acid attached to GlcN. The functional significance of these variations among structures is not understood. This type of linkage was first detected by the use of bacterial PI-specific phospholipase C (PI-PLC), which was found to release certain proteins from the plasma membrane of cells by splitting the bond indicated in Figure 47–1. Examples of some proteins that are anchored by this type of linkage are given in **Table 47–13**. At least three possible functions of this type of linkage have been suggested. (1) The GPI anchor may allow greatly enhanced **mobility** of a protein in the plasma membrane compared with that observed for a protein that contains transmembrane sequences. This is perhaps not surprising, as the GPI anchor is attached only to the outer leaflet of the lipid bilayer, so that it is freer to diffuse than a protein anchored via both leaflets of the bilayer. Increased mobility may be important in facilitating rapid responses to appropriate stimuli. (2) Some GPI anchors may connect with **signal transduction** pathways. (3) It has been shown that GPI structures can **target** certain proteins to apical domains and also basolateral domains of the plasma membrane of certain polarized epithelial cells. The biosynthesis of GPI anchors is complex and begins in the endoplasmic reticulum. The GPI anchor is assembled independently by a series of enzyme-catalyzed reactions and then transferred to the carboxyl terminal end of its acceptor protein, accompanied by cleavage of the preexisting carboxyl terminal hydrophobic peptide from that protein. This process is sometimes called **glypiation**. An acquired defect in an early stage of the biosynthesis of the GPI structure has been implicated in the causation of **paroxysmal nocturnal hemoglobinuria** (see below).

ADVANCED GLYCATION END-PRODUCTS (AGEs) ARE THOUGHT TO BE IMPORTANT IN THE CAUSATION OF TISSUE DAMAGE IN DIABETES MELLITUS

Glycation refers to nonenzymic attachment of sugars (mainly glucose) to amino groups of proteins and also to other molecules (eg, DNA, lipids). Glycation is distinguished from **glycosylation** because the latter involves enzyme-catalyzed attachment of sugars. When glucose attaches to a protein, intermediate products formed include **Schiff bases.** These can further be rearranged by the **Amadori rearrangement** to **ketoamines** (see **Figure 47–11**). The overall series of reactions is known as the **Maillard reaction.** These reactions are involved in the **browning** of certain foodstuffs that occurs on storage or processing (eg, heating). The end-products of glycation reactions are termed **advanced glycation end-products (AGEs).**

The major medical interest in AGEs has been in relation to them causing **tissue damage in diabetes mellitus,** in which

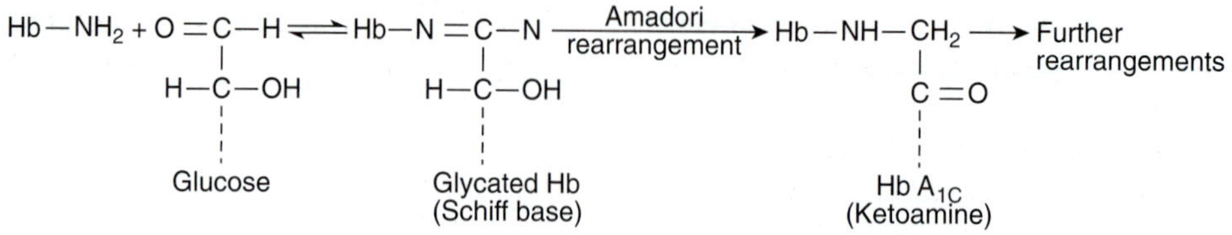

$$\text{Hb}-\text{NH}_2 + \text{O}=\overset{\overset{\displaystyle |}{\text{C}}-\text{H}}{\underset{\underset{\displaystyle \text{Glucose}}{\text{H}-\text{C}-\text{OH}}}{|}} \rightleftharpoons \overset{\overset{\displaystyle |}{\text{Hb}-\text{N}=\text{C}-\text{N}}}{\underset{\underset{\displaystyle \substack{\text{Glycated Hb}\\(\text{Schiff base})}}{\text{H}-\text{C}-\text{OH}}}{|}} \xrightarrow[\text{rearrangement}]{\text{Amadori}} \overset{\overset{\displaystyle |}{\text{Hb}-\text{NH}-\text{CH}_2}}{\underset{\underset{\displaystyle \substack{\text{Hb A}_{1C}\\(\text{Ketoamine})}}{\text{C}=\text{O}}}{|}} \longrightarrow \substack{\text{Further}\\\text{rearrangements}}$$

FIGURE 47–11 **Formation of AGEs from glucose.** Glucose is shown interacting with the amino group of hemoglobin (Hb) forming a Schiff base. This is subject to the Amadori rearrangement, forming a ketoamine. Further rearrangements can occur, leading to other AGEs.

the level of blood glucose is often consistently elevated, promoting increased glycation. At constant time intervals, the extent of glycation is more or less proportional to the blood glucose level. It has also been suggested that AGEs are involved in other processes, such as **aging.**

Glycation of collagen and other proteins in the ECM alters their properties (eg, increasing the **cross-linking of collagen**). Cross-linking can lead to accumulation of various plasma proteins in the walls of blood vessels; in particular, accumulation of **LDL** can contribute to **atherogenesis.** AGEs appear to be involved in both **microvascular** and **macrovascular** damage in diabetes mellitus (**Figure 47–12**). Also endothelial cells and macrophages have AGE receptors on their surfaces. Uptake of glycated proteins by these receptors can activate the transcription factor **NF-kB** (see Chapter 50), generating a variety of **cytokines** and **proinflammatory molecules.** It is thus believed that AGEs are one significant contributor to some of the pathologic finding found in diabe. Nonenzymic attachment of glucose to **hemoglobin A** present in red blood cells (ie, formation of **HbA$_{1c}$**) occurs in normal individuals and is increased in patients with diabetes mellitus whose blood sugar levels are elevated. As discussed in Chapter 6, measurement of HbA$_{1c}$ has become a very important part of the **management of patients with diabetes mellitus.**

GLYCOPROTEINS ARE INVOLVED IN MANY BIOLOGIC PROCESSES & IN MANY DISEASES

As listed in Table 47–1, glycoproteins have many different functions; some have already been addressed in this chapter and others are described elsewhere in this text (eg, transport molecules, immunologic molecules, and hormones). Here, their involvement in two specific processes—**fertilization** and **inflammation**—will be briefly described. In addition, the **bases of a number of diseases** that are due to abnormalities in the synthesis and degradation of glycoproteins will be summarized.

Glycoproteins Are Important in Fertilization

To reach the plasma membrane of an oocyte, a sperm has to traverse the **zona pellucida (ZP),** a thick, transparent, noncellular envelope that surrounds the oocyte. The zona pellucida contains three glycoproteins of interest, ZP1–3. Of particular note is ZP3, an *O*-linked glycoprotein that functions as a receptor for the sperm. A protein on the sperm surface, possibly galactosyl transferase, interacts specifically with oligosaccharide chains of ZP3; in at least certain species (eg, the mouse),

FIGURE 47–12 **Some consequences of the formation of AGEs.** Hyperglycemia (eg, occurring in poorly controlled diabetes) leads to the formation of AGEs. These can occur in proteins of the ECM or plasma. In the ECM, they can cause increased cross-linking of collagen, which can trap proteins such as LDL (contributing to atherogenesis) and damage basement membranes in the kidneys and other sites. Thickening of basement membranes can also occur by binding of glycated proteins to them. AGEs can attach to AGE receptors on cells, activating NFkB (see Chapter 50), which has several consequences (as shown). Damage to renal basement membranes, thickening of these membranes in capillaries and endothelial dysfunction are found in ongoing uncontrolled diabetes mellitus.

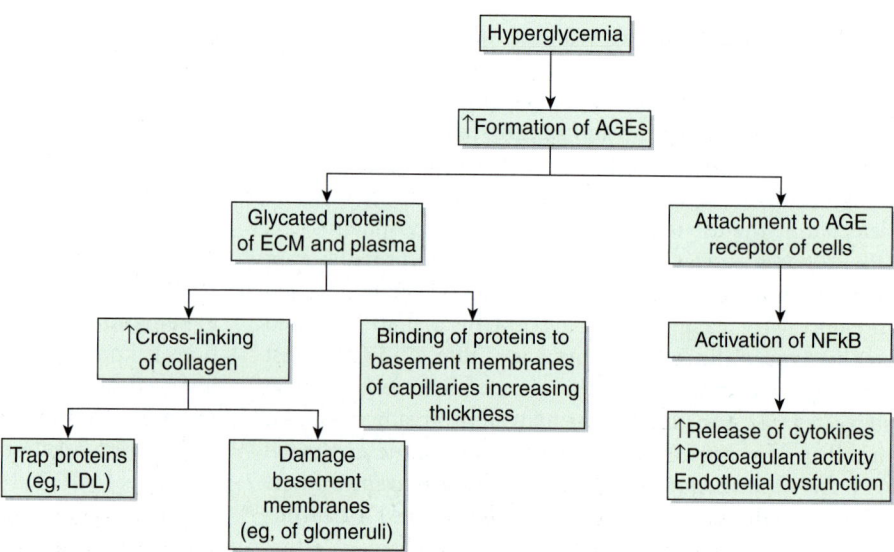

this interaction, by transmembrane signaling, induces the **acrosomal reaction,** in which enzymes such as proteases and hyaluronidase and other contents of the acrosome of the sperm are released. Liberation of these enzymes helps the sperm to pass through the zona pellucida and reach the plasma membrane (PM) of the oocyte. In hamsters, it has been shown that another glycoprotein, PH-30, is important in both the binding of the PM of the sperm to the PM of the oocyte, and also in the subsequent fusion of the two membranes. These interactions enable the sperm to enter and thus fertilize the oocyte. It may be possible to **inhibit fertilization** by developing drugs or antibodies that interfere with the normal functions of ZP3 and PH-30 and which would thus act as contraceptive agents.

Selectins Play Key Roles in Inflammation & in Lymphocyte Homing

Leukocytes play important roles in many inflammatory and immunologic phenomena. The first steps in many of these phenomena are interactions between circulating leukocytes and **endothelial cells** prior to passage of the former out of the circulation. Work done to identify specific molecules on the surfaces of the cells involved in such interactions has revealed that leukocytes and endothelial cells contain on their surfaces specific lectins, called **selectins,** that participate in their intercellular adhesion. Features of the three major classes of selectins are summarized in **Table 47–14.** Selectins are single-chain Ca^{2+}-binding transmembrane proteins that contain a number of domains (**Figure 47–13**). Their amino terminal ends contain the lectin domain, which is involved in binding to specific carbohydrate ligands.

The adhesion of neutrophils to endothelial cells of postcapillary venules can be considered to occur in four stages, as shown in **Figure 47–14.** The initial baseline stage is succeeded by **slowing or rolling** of the neutrophils, mediated by selectins. Interactions between L-selectin on the neutrophil surface and CD34 and GlyCAM-1 or other glycoproteins on the endothelial surface are involved. These particular interactions are initially short-lived, and the overall binding is of relatively low affinity, permitting rolling. However, during this stage, **activation** of the neutrophils by various chemical mediators (discussed below) occurs, resulting in a change of shape of the neutrophils and firm adhesion of these cells to the endothelium. An additional set of **adhesion molecules** is involved in firm adhesion, namely, LFA-1 and Mac-1 on the neutrophils and ICAM-1 and ICAM-2 on endothelial cells. LFA-1 and Mac-1 are CD11/CD18 integrins (see Chapter 52 for a discussion of integrins), whereas ICAM-1 and ICAM-2 are members of the immunoglobulin superfamily. The fourth stage is **transmigration** of the neutrophils across the endothelial wall. For this to occur, the neutrophils insert pseudopods into the junctions between endothelial cells, squeeze through these junctions, cross the basement membrane, and then are free to migrate in the extravascular space. Platelet-endothelial cell adhesion molecule-1 (PECAM-1) has been found to be localized at the junctions of endothelial cells and thus may have a

TABLE 47–14 Some Molecules Involved in Leukocyte–Endothelial Cell Interactions

Molecule	Cell	Ligands
Selectins		
L-selectin	PMN, lymphs	CD34, Gly-CAM-1[1], sialylLewis[x], and others
P-selectin	EC, platelets	P-selectin glycoprotein ligand-1 (PSGL-1), sialyl-Lewis[x], and others
E-selectin	EC	Sialyl-Lewis[x] and others
Integrins		
LFA-1	PMN, lymphs	ICAM-1, ICAM-2
(CD11a/CD18)		
Mac-1	PMN	ICAM-1 and others
(CD11b/CD18)		
Immunoglobulin Superfamily		
ICAM-1	Lymphs, EC	LFA-1, Mac-1
ICAM-2	Lymphs, EC	LFA-1
PECAM-1	EC, PMN, lymphs	Various platelets

Source: Modified, with permission, from Albelda SM, Smith CW, Ward PA: Adhesion molecules and inflammatory injury. FASEB J 1994;8:504.
Abbreviations: CD, cluster of differentiation; EC, endothelial cell; ICAM, intercellular adhesion molecule; LFA-1, lymphocyte function-associated antigen-1; lymphs, lymphocytes; PECAM-1, platelet endothelial cell adhesion cell molecule-1, PMN, polymorphonuclear leukocytes.
[1]These are ligands for lymphocyte L-selectin; the ligands for neutrophil L-selectin have not apparently been identified.

role in transmigration. A variety of biomolecules have been found to be involved in **activation** of neutrophil and endothelial cells, including tumor necrosis factor, various interleukins, platelet activating factor (PAF), leukotriene B$_4$, and certain complement fragments. These compounds stimulate various signaling pathways, resulting in changes in cell shape and function, and some are also chemotactic. One important functional change is **recruitment of selectins** to the cell surface, as

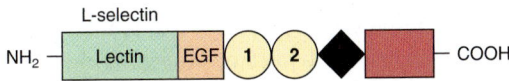

FIGURE 47–13 Schematic of the structure of human L-selectin. The extracellular portion contains an amino terminal domain homologous to C-type lectins and an adjacent epidermal growth factor-like domain. These are followed by a variable number of complement regulatory-like modules (numbered circles) and a trans-membrane sequence (black diamond). A short cytoplasmic sequence (red rectangle) is at the carboxyl terminal. The structures of P- and E-selectin are similar to that shown except that they contain more complement-regulatory modules. The numbers of amino acids in L-, P-, and E- selectins, as deduced from the cDNA sequences, are 385, 789, and 589, respectively. (Reproduced, with permission, from Bevilacqua MP, Nelson RM: Selectins. J Clin Invest 1993;91:370.)

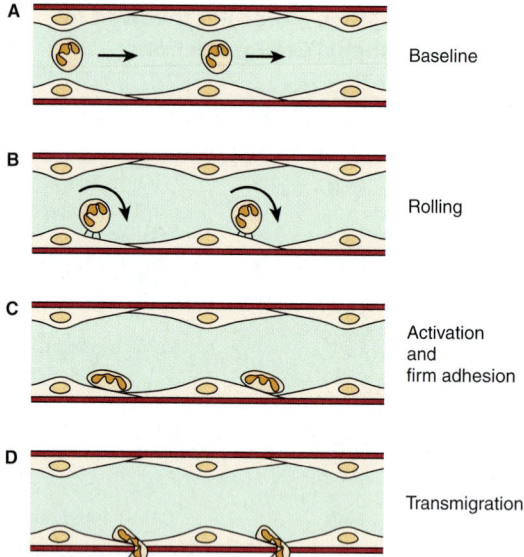

FIGURE 47–14 Schematic of neutrophil–endothelial cell interactions. **(A)** Baseline conditions: Neutrophils do not adhere to the vessel wall. **(B)** The first event is the slowing or rolling of the neutrophils within the vessel (venule) mediated by selectins. **(C)** Activation occurs, resulting in neutrophils firmly adhering to the surfaces of endothelial cells and also assuming a flattened shape. This requires interaction of activated CD18 integrins on neutrophils with ICAM-1 on the endothelium. **(D)** The neutrophils then migrate through the junctions of endothelial cells into the interstitial tissue; this requires involvement of PECAM-1. Chemotaxis is also involved in this latter stage. (Reproduced, with permission, from Albelda SM, Smith CW, Ward PA: Adhesion molecules and inflammatory injury. FASEB J 1994;8;504.)

in some cases selectins are stored in granules (eg, in endothelial cells and platelets).

The precise chemical nature of some of the ligands involved in selectin–ligand interactions has been determined. All three selectins bind **sialylated and fucosylated oligosaccharides,** and in particular all three bind **sialyl-Lewis^x (Figure 47–15),** a structure present on both glycoproteins and glycolipids. Whether this compound is the actual ligand involved in vivo is not established. Sulfated molecules, such as the sulfatides (Chapter 15), may be ligands in certain instances. This basic knowledge is being used in attempts to synthesize compounds that block selectin–ligand interactions and thus may inhibit the inflammatory response. Approaches include administration of specific monoclonal antibodies or of chemically synthesized analogs of sialyl-Lewis^x, both of which bind selectins. **Cancer cells** often exhibit sialyl-Lewis^x and other selectin ligands on their surfaces. It is thought that these ligands play a role in the invasion and metastasis of cancer cells.

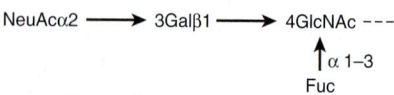

FIGURE 47–15 Schematic representation of the structure of sialyl-Lewis^x.

TABLE 47–15 Some Diseases Due to or Involving Abnormalities in the Biosynthesis of Glycoproteins

Disease	Abnormality
Cancer	Increased branching of cell surface glycans or presentation of selectin ligands may be important in metastasis.
Congenital disorders of glycosylation[1]	See Table 47–16.
HEMPAS[2] (OMIM 224100)	Abnormalities in certain enzymes (eg, mannosidase II and others) involved in the biosynthesis of N-glycans, particularly affecting the red blood cell membrane.
Leukocyte adhesion deficiency, type II (OMIM 266265)	Probably mutations affecting a Golgi-located GDP-fucose transporter, resulting in defective fucosylation.
Paroxysmal nocturnal hemoglobinuria (PNH) (OMIM 311770)	Acquired defect in biosynthesis of the GPI[3] structures of decay accelerating factor (DAF) and CD59.
I-cell disease (OMIM 252500)	Deficiency of GlcNAc phosphotransferase, resulting in abnormal targeting of certain lysosomal enzymes.

[1]The OMIM number for congenital disorder of glycosylation type Ia is 212065.
[2]Hereditary erythroblastic multinuclearity with a positive acidified serum lysis test (congenital dyserythropoietic anemia type II). This is a relatively mild form of anemia. It reflects at least in part the presence in the red cell membranes of various glycoproteins with abnormal N-glycan chains, which contribute to the susceptibility to lysis.
[3]Glycosylphosphatidylinositol.

Abnormalities in the Synthesis of Glycoproteins Underlie Certain Diseases

Table 47–15 lists a number of conditions in which abnormalities in the synthesis of glycoproteins are of importance. As mentioned above, many **cancer cells** exhibit different profiles of oligosaccharide chains on their surfaces, some of which may contribute to metastasis.

The **congenital disorders of glycosylation (CDG)** are a group of disorders of considerable current interest. The major features of these conditions are summarized in **Table 47–16.**

Leukocyte adhesion deficiency (LAD) II is a rare condition probably due to mutations affecting the activity of a Golgi-located GDP-fucose transporter. It can be considered a congenital disorder of glycosylation. The absence of fucosylated ligands for selectins leads to a marked decrease in neutrophil rolling. Subjects suffer life-threatening, recurrent bacterial infections, and also psychomotor and mental retardation. The condition appears to respond to oral fucose.

Hereditary erythroblastic multinuclearity with a positive acidified lysis test—congenital dyserythropoietic anemia type II—is another disorder in which abnormalities in the processing of N-glycans are thought to be involved. Some cases have been claimed to be due to defects in alpha–mannosidase II.

PNH is an acquired mild anemia characterized by the presence of hemoglobin in urine due to hemolysis of red cells,

TABLE 47–16 Major Features of the Congenital Disorders of Glycosylation

- Autosomal recessive disorders.
- Multisystem disorders that have probably not been recognized in the past.
- Generally affect the central nervous system, resulting in psychomotor retardation and other features.
- Type I disorders are due to mutations in genes encoding enzymes (eg, phosphomannomutase-2 [PMM-2], causing CDG Ia) involved in the synthesis of dolichol-P-P-oligosaccharide.
- Type II disorders are due to mutations in genes encoding enzymes (eg, GlcNAc transferase-2, causing CDG IIa) involved in the processing of N-glycan chains.
- At least 15 distinct disorders have been recognized.
- Isoelectric focusing of transferrin is a useful biochemical test for assisting in the diagnosis of these conditions; truncation of the oligosaccharide chains of this protein alters its isoelectric focusing pattern.
- Oral mannose has proved of benefit in the treatment of CDG Ia.

Abbreviation: CDG, congenital disorder of glycosylation.

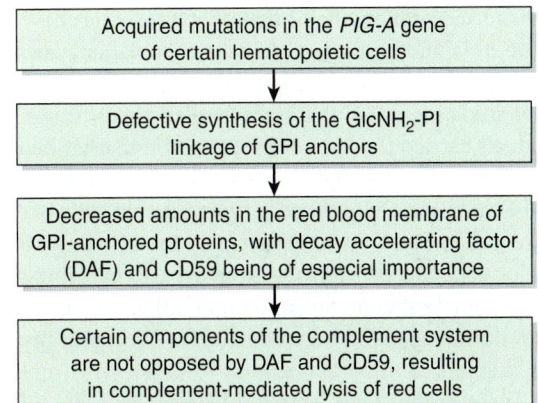

FIGURE 47–16 Scheme of causation of paroxysmal nocturnal hemoglobinuria (OMIM 311770).

particularly during sleep. This latter phenomenon may reflect a slight drop in plasma pH during sleep, which increases susceptibility to lysis by the complement system (Chapter 50). The basic defect in PNH is the acquisition of somatic mutations in the *PIG-A* (for PI glycan class A) gene of certain hematopoietic cells. The product of this gene appears to be the enzyme that links glucosamine to PI in the GPI structure (Figure 47–1). Thus, proteins that are anchored by a GPI linkage are deficient in the red cell membrane. Two proteins are of particular interest: **decay accelerating factor** and another protein designated **CD59.** They normally interact with certain components of the complement system (Chapter 50) to prevent the hemolytic actions of the latter. However, when they are deficient, the complement system can act on the red cell membrane to cause hemolysis. A monoclonal antibody to C5, a terminal component of the complement system, has proven useful in the management of PNH by inhibiting the complement cascade. PNH can be diagnosed relatively simply, as the red cells are much more sensitive to hemolysis in normal serum acidified to pH 6.2 (Ham's test); the complement system is activated under these conditions, but normal cells are not affected. **Figure 47–16** summarizes the etiology of PNH.

Study of the **congenital muscular dystrophies (CMDs)** has revealed that certain of them (eg, the Walker–Warburg syndrome, muscle-eye-brain disease, Fukuyama CMD) are the result of defects in the synthesis of glycans in the protein α-dystroglycan (α-DG). This protein protrudes from the surface membrane of muscle cells and interacts with laminin-2 (merosin) in the basal lamina (see Figure 49–11). If the glycans of α-DG are not correctly formed (as a result of mutations in genes encoding certain glycosyltransferases), this results in defective interaction of α-DG with laminin, which in turn leads to the development of a CMD.

Rheumatoid arthritis is associated with an alteration in the glycosylation of circulating immunoglobulin G (IgG) molecules (Chapter 50), such that they lack galactose in their Fc regions and terminate in GlcNAc. **Mannose-binding protein** (MBP, not to be confused with the mannose 6-P receptor), a C-lectin synthesized by liver cells and secreted into the circulation, binds mannose, GlcNAc, and certain other sugars. It can thus bind agalactosyl IgG molecules, which subsequently activate the complement system (see Chapter 50), contributing to chronic inflammation in the synovial membranes of joints.

MBP can also bind the above sugars when they are present on the surfaces of certain bacteria, fungi, and viruses, preparing these pathogens for opsonization or for destruction by the complement system. This is an example of **innate immunity,** not involving immunoglobulins or T lymphocytes. Deficiency of this protein in young infants as a result of mutation renders them very susceptible to **recurrent infections.**

I-Cell Disease Results from Faulty Targeting of Lysosomal Enzymes

As indicated above, Man 6-P serves as a chemical marker to target certain lysosomal enzymes to that organelle. Analysis of cultured fibroblasts derived from patients with I-cell (inclusion cell) disease played a large part in revealing the above role of Man 6-P. I-cell disease is an uncommon condition characterized by severe progressive psychomotor retardation and a variety of physical signs, with death often occurring in the first decade. Cultured cells from patients with I-cell disease were found to lack almost all of the normal lysosomal enzymes; the lysosomes thus accumulate many different types of undegraded molecules, forming inclusion bodies. Samples of plasma from patients with the disease were observed to contain very high activities of lysosomal enzymes; this suggested that the enzymes were being synthesized but were failing to reach their proper intracellular destination and were instead being secreted. Cultured cells from patients with the disease were noted to take up exogenously added lysosomal enzymes obtained from normal subjects, indicating that the cells contained a normal receptor on their surfaces for endocytic uptake of lysosomal enzymes. In addition, this finding suggested that lysosomal enzymes from patients with I-cell disease **might lack a recognition marker.** Further studies revealed that lysosomal enzymes from normal individuals carried the Man 6-P recognition marker described above, which interacted with a specific intracellular protein, the Man 6-P receptor. Cultured cells from patients with I-cell disease were then found to be **deficient** in the activity of the *cis*-Golgi-located **GlcNAc phosphotransferase,** explaining how their lysosomal enzymes failed to acquire the Man 6-P marker. It is now known that there are **two Man 6-P receptor proteins,** one of high (275 kDa) and one of low (46 kDa) molecular mass. These proteins are **lectins, recognizing Man 6-P.** The former is cation-independent and also binds IGF-II (hence it is named the Man 6-P–IGFII receptor), whereas the latter is cation-dependent in some species and does not bind IGF-II. It appears that both receptors function in the intracellular sorting of lysosomal enzymes into clathrin-coated vesicles, which occurs in the *trans*-Golgi subsequent to synthesis of Man 6-P in the *cis*-Golgi. These vesicles then leave the Golgi and fuse with a prelysosomal compartment. The **low pH** in this compartment causes the lysosomal enzymes to **dissociate** from their receptors and subsequently enter into lysosomes. The receptors are recycled and reused. Only the smaller receptor functions in the endocytosis of **extracellular** lysosomal enzymes, which is a minor pathway for lysosomal location. Not all cells employ the Man 6-P receptor to target their lysosomal enzymes (eg, hepatocytes use a different but undefined pathway); furthermore, not all lysosomal enzymes are targeted by this mechanism. Thus, biochemical investigations of I-cell disease not only led to elucidation of its basis, but also contributed significantly to knowledge of how newly synthesized proteins are targeted to specific organelles, in this case the lysosome. **Figure 47–17** summarizes the causation of I-cell disease.

Pseudo-Hurler polydystrophy is another genetic disease closely related to I-cell disease. It is a **milder condition,** and patients may survive to adulthood. Studies have revealed that

the **GlcNAc phosphotransferase** involved in I-cell disease has several domains, including a catalytic domain and a domain that specifically recognizes and interacts with lysosomal enzymes. It has been proposed that the defect in pseudo-Hurler polydystrophy lies in the latter domain, and the retention of some catalytic activity results in a milder condition.

Genetic Deficiencies of Glycoprotein Lysosomal Hydrolases Cause Diseases Such as α-Mannosidosis

Glycoproteins, like most other biomolecules, undergo both synthesis and degradation (ie, turnover). Degradation of the oligosaccharide chains of glycoproteins involves a battery of lysosomal hydrolases, including α-neuraminidase, β-galactosidase, β-hexosaminidase, α- and β-mannosidases, α-*N*-acetylgalactosaminidase, α-fucosidase, endo-β-*N*-acetyl-glucosaminidase, and aspartylglucosaminidase. The sites of action of the last two enzymes are indicated in the legend to Figure 47–5. Genetically determined defects of the activities of these enzymes can occur, resulting in abnormal degradation of glycoproteins. The accumulation in tissues of such degraded glycoproteins can lead to various diseases. Among the best recognized of these diseases are mannosidosis, fucosidosis, sialidosis, aspartylglycosaminuria, and Schindler disease, due respectively to deficiencies of α-mannosidase, α-fucosidase, α-neuraminidase, aspartylglucosaminidase, and α-*N*-acetyl-galactosaminidase. These diseases, which are relatively uncommon, have a variety of manifestations; some of their major features are listed in **Table 47–17.** The fact that patients affected

TABLE 47–17 Major Features of Some Diseases[1] Due to Deficiencies of Glycoprotein Hydrolases[2]

• Usually exhibit mental retardation or other neurologic abnormalities, and in some disorders coarse features or visceromegaly (or both).
• Variations in severity from mild to rapidly progressive.
• Autosomal recessive inheritance.
• May show ethnic distribution (eg, aspartylglycosaminuria is common in Finland).
• Vacuolization of cells observed by microscopy in some disorders.
• Presence of abnormal degradation products (eg, oligosaccharides that accumulate because of the enzyme deficiency) in urine, detectable by TLC and characterizable by GLC-MS.
• Definitive diagnosis made by assay of appropriate enzyme, often using leukocytes.
• Possibility of prenatal diagnosis by appropriate enzyme assays.
• No definitive treatment at present.

[1]α-Mannosidosis, β-mannosidosis, fucosidosis, sialidosis, aspartylglycosaminuria, and Schindler disease.
[2]OMIM numbers: α-mannosidosis, 248500; β-mannosidosis, 248510; fucosidosis, 230000; sialidosis, 256550; aspartylglycosaminuria, 208400; Schindler disease, 609241.

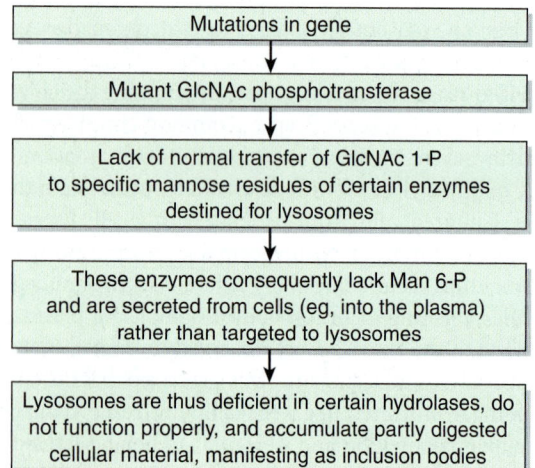

FIGURE 47–17 Summary of the causation of I-cell disease (OMIM 252500).

by these disorders all show signs referable to the **central nervous system** reflects the importance of glycoproteins in the development and normal function of that system.

THE GLYCANS OF GLYCOCONJUGATES ARE INVOLVED IN THE BINDING OF VIRUSES, BACTERIA & CERTAIN PARASITES TO HUMAN CELLS

A principal feature of glycans, and one that explains many of their biologic actions, is that they **bind** specifically to a variety of molecules such as proteins or other glycans. One reflection of this is their ability to bind certain viruses, many bacteria and some parasites.

Influenza virus A binds to cell surface glycoprotein receptor molecules containing NeuAc via a protein named **hemagglutinin (H)**. It also possesses a **neuraminidase (N)** that plays a key role in allowing elution of newly synthesized progeny from infected cells. If this process is inhibited, spread of the viruses is markedly diminished. Inhibitors of this enzyme (eg, zanamivir, oseltamivir) are now available for use in treating patients with influenza. Influenza viruses are classified according to the type of hemagglutinin and neuraminidase that they possess. There are at least 16 types of hemagglutinin and nine types of neuraminidase. Thus, **avian influenza virus** is classified as H5N1. There is great interest in how this virus attaches to human cells, in view of the possibility of a pandemic occurring. It has been found that the virus preferentially attaches to glycans terminated by the disaccharide **galactose → α 2,3-NeuAc** (**Figure 47–18**). However, the predominant disaccharide terminating glycans in cells of

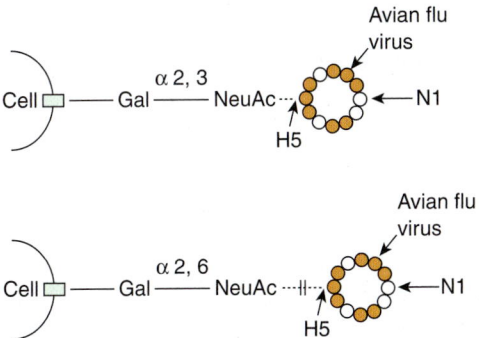

FIGURE 47–18 **Schematic representation of binding of the avian influenza virus (H5N1) to a respiratory epithelial cell.** The viral hemagglutin (HA) mediates its entry to cells by binding to a glycan on the cell surface that is terminated by the disaccharide galactose → α 2,3-NeuAc. It will not bind to a glycan terminated by galactose → α 2,6-NeuAc, which is the type predominantly found in the human respiratory tract. If the viral HA alters via mutation to be able to bind to the latter disaccharide, this could greatly increase its pathogenicity for humans. (H5, hemagglutinin type 5; N1, neuraminidase type 1.)

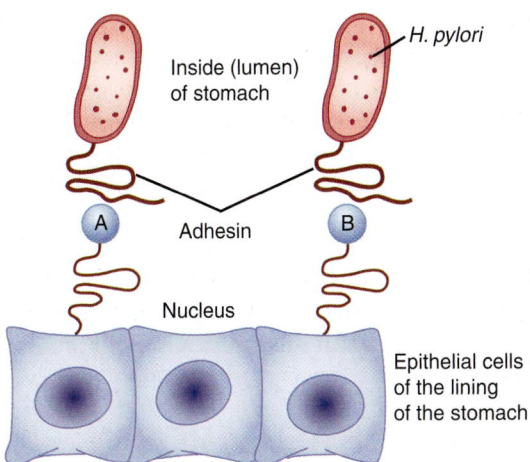

FIGURE 47–19 **Attachment of *Helicobacter pylori* to epithelial cells of the stomach.** Adhesin, a protein present in the tail of *H pylori*, interacts with two different glycans (structures shown in the figure) present in glycoproteins on the surface of gastric epithelial cells. This provides an attachment site for the bacterium. Subsequently it liberates molecules, such as ammonia, that contribute to initiating peptic ulceration. (A) NeuAcα2,3Galβ1,4—Protein (Neuraminylgalactose); (B) Fucα1,2Galβ1,3GlcNAc—Protein (Lewis^B substance).

the human respiratory tract is **galactose → α 2,6-NeuAc.** If a change in the structure of the viral hemagglutinin (due to mutation) occurs that allows it to bind to the latter disaccharide, this could greatly increase the potential infectivity of the virus, possibly resulting in very serious consequences.

Human immunodeficiency virus type 1 (HIV-1), thought by most to be the cause of AIDS, attaches to cells via one of its surface glycoproteins (gp120) and uses another surface glycoprotein (gp 41) to fuse with the host cell membrane. **Antibodies** to gp 120 develop during infection by HIV-1, and there has been interest in using the protein as a vaccine. One major problem with this approach is that the structure of gp 120 can change relatively rapidly due to mutations, allowing the virus to escape from the neutralizing activity of antibodies directed against it.

Helicobacter pylori is believed to be the major cause of **peptic ulcers.** Studies have shown that this bacterium binds to at least two different glycans present on the surfaces of epithelial cells in the stomach (see **Figure 47–19**). This allows it to establish a stable attachment site to the stomach lining, and subsequent secretion of ammonia and other molecules by the bacterium are believed to initiate ulceration.

Similarly, many **bacteria that cause diarrhea** are also known to attach to surface cells of the intestine via glycans present in glycoproteins or glycolipids.

The basic cause of **cystic fibrosis** (CF) is mutations in the gene encoding CFTR (see Chapters 40 and 57). A major problem in this disease is recurring lung infections by bacteria such as *Pseudomonas aeruginosa*. In CF, a relative dehydration of respiratory secretions occurs secondary to changes in electrolyte composition in the airway as a result of mutations in CFTR. Bacteria such as *P aeruginosa* attach to the sugar chains of mucins and find the dehydrated environment in the bronchioles a favorable location in which to multiply.

The attachment of *Plasmodium falciparum*—one of the types of plasmodia causing **malaria**—to human cells is mediated by a GPI present on the surface of the parasite.

Various researchers are **analyzing the surfaces of viruses, bacteria, parasites** and **human cells** to determine which molecules are involved in attachment. It is important to define the precise nature of the interactions between invading organisms and host cells, as this will hopefully lead to the development of drugs or other agents that will specifically inhibit attachment.

THE PACE OF RESEARCH IN GLYCOMICS IS ACCELERATING

Research on glycoconjugates in the past has been hampered by the lack of availability of suitable technics to determine the structures of glycans. However, appropriate analytical technics are now available (some of which are listed in Table 47–3), as are powerful new genetic technics (eg, knockouts and knockdowns using RNAi molecules). It is certain that research in glycomics will not only provide a wealth of structural information on glyconconjugates, helping to disclose "the sugar code of life," but will also uncover many new important biologic interactions that are sugar-dependent and will provide targets for drug and other therapies.

SUMMARY

- Glycoproteins are widely distributed proteins—with diverse functions—that contain one or more covalently linked carbohydrate chains.

- The carbohydrate components of a glycoprotein range from 1% to more than 85% of its weight and may be simple or very complex in the structure. Eight sugars are mainly found in the sugar chains of human glycoproteins: xylose, fucose, galactose, glucose, mannose, *N*-acetylgalactosamine, *N*-acetylglucosamine and *N*-acetylneuraminic acid.

- At least certain of the oligosaccharide chains of glycoproteins encode biologic information; they are also important to glycoproteins in modulating their solubility and viscosity, in protecting them against proteolysis, and in their biologic actions.

- The structures of oligosaccharide chains can be elucidated by gas–liquid chromatography, mass spectrometry, and high-resolution NMR spectrometry.

- Glycosidases hydrolyze specific linkages in oligosaccharides and are used to explore both the structures and functions of glycoproteins.

- Lectins are carbohydrate-binding proteins involved in cell adhesion and many other biologic processes.

- The major classes of glycoproteins are *O*-linked (involving an OH of serine or threonine), *N*-linked (involving the *N* of the amide group of asparagine), and GPI-linked.

- Mucins are a class of *O*-linked glycoproteins that are distributed on the surfaces of epithelial cells of the respiratory, gastrointestinal, and reproductive tracts.

- The endoplasmic reticulum and Golgi apparatus play a major role in glycosylation reactions involved in the biosynthesis of glycoproteins.

- The oligosaccharide chains of *O*-linked glycoproteins are synthesized by the stepwise addition of sugars donated by nucleotide sugars in reactions catalyzed by individual specific glycoprotein glycosyltransferases.

- In contrast, the synthesis of *N*-linked glycoproteins involves a specific dolichol-P-P-oligosaccharide and various glycotransferases and glycosidases. Depending on the enzymes and precursor proteins in a tissue, it can synthesize complex, hybrid, or high-mannose types of *N*-linked oligosacccharides.

- Glycoproteins are implicated in many biologic processes. For instance, they have been found to play key roles in fertilization and inflammation.

- A number of diseases involving abnormalities in the synthesis and degradation of glycoproteins have been recognized. Glycoproteins are also involved in many other diseases, including influenza, AIDS, rheumatoid arthritis, cystic fibrosis and peptic ulcer.

- Developments in the new field of glycomics are likely to provide much new information on the roles of sugars in health and disease and also indicate targets for drug and other types of therapies.

REFERENCES

Chandrasekeran A, Srinivasan A, Raman R, et al: Glycan topology determines human adaptation of avian H5N1 virus hemagglutinin. Nat Biotechnology 2008;26:107.

Freeze HH: Congenital disorders of glycosylation: CDG-I, CDG-II, and beyond. Curr Mol Med 2007;7:389.

Kiessling LL, Splain RA: Chemical approaches to glycobiology. Annu Rev Biochem. 2010;79:619.

Kornfeld R, Kornfeld S: Assembly of asparagine-linked oligosaccharides. Annu Rev Biochem 1985;54:631.

Ohtsubo K, Marth JD: Glycosylation in cellular mechanisms of health and disease. Cell 2006;126:855.

Pilobelli KT, Mahal LK: Deciphering the glycocode: the complexity and analytical challenge of glycomics. Curr Opin Chem Biol 2007;11:300.

Ramasamy R, Yan SF, Herold K, et al: Receptor for advanced glycation end products: fundamental roles in the inflammatory response: winding the way to the pathogenesis of endothelial dysfunction and atherosclerosis. Ann NY Acad Sci 2008;1126:7.

Scriver CR, Beaudet AI, Sly WS, et al (editors): *The Metabolic and Molecular Bases of Inherited Disease.* 8th ed. McGraw-Hill, 2001. (Various chapters in this text and its updated online version [see References for Chapter 1] give in-depth coverage of topics such as I-cell disease and disorders of glycoprotein degradation.)

Taylor ME, Drickamer K: *Introduction to Glycobiology.* Oxford University Press, 2003.

Varki A, Cummings RD, Esko JD, et al: *Essentials of Glycobiology.* 2nd ed. Cold Spring Harbor Laboratory Press, 2008.

Von Itzstein M, Plebanski M, Cooke RM, Coppel RL: Hot, sweaty and sticky: the glycobiology of *Plasmodium falciparum.* Trends Parasitol 2008;24:210.

Werz DB, Seeberger PH: Carbohydrates are the next frontier in pharmaceutical research. Chemistry 2005;11:3194.

The Extracellular Matrix

Robert K. Murray, MD, PhD & Frederick W. Keeley, PhD

OBJECTIVES

After studying this chapter, you should be able to:

- Appreciate the importance of the extracellular matrix (ECM) and its components in health and disease;
- Describe the structural and functional properties of collagen and elastin, the major proteins of the ECM;
- Indicate the major features of fibrillin, fibronectin, and laminin, other important proteins of the ECM;
- Describe the properties and general features of the synthesis and degradation of glycosaminoglycans and proteoglycans, and their contributions to the ECM;
- Give a brief account of the major biochemical features of bone and cartilage.

BIOMEDICAL IMPORTANCE

Most mammalian cells are located in tissues where they are surrounded by a complex **ECM** often referred to as **"connective tissue."** The ECM contains three major classes of biomolecules: (1) **structural proteins,** for example, collagen, elastin, and fibrillin-1; (2) certain **specialized proteins** such as fibronectin and laminin; and (3) **proteoglycans,** whose chemical natures are described below. The ECM has been found to be involved in many normal and pathologic processes—for example, it plays important roles in development, in inflammatory states, and in the spread of cancer cells. Involvement of certain components of the ECM has been documented in both **rheumatoid arthritis** and **osteoarthritis.** Several diseases (eg, osteogenesis imperfecta and a number of types of the Ehlers–Danlos syndrome) are due to genetic disturbances of the synthesis of collagen. Specific components of proteoglycans (the glycosaminoglycans; GAGs) are affected in the group of genetic disorders known as the **mucopolysaccharidoses.** Changes occur in the ECM during the **aging process.** This chapter describes the basic biochemistry of the three major classes of biomolecules found in the ECM and illustrates their biomedical significance. Major biochemical features of two specialized forms of ECM—bone and cartilage—and of a number of diseases involving them are also briefly considered.

COLLAGEN TYPE I IS COMPOSED OF A TRIPLE HELIX STRUCTURE & FORMS FIBRILS

Collagen, the major component of most connective tissues, constitutes approximately 25% of the protein of mammals. It provides an extracellular framework for all metazoan animals and exists in virtually every animal tissue. At least 28 distinct types of collagen made up of over 30 distinct polypeptide chains (each encoded by a separate gene) have been identified in human tissues. Although several of these are present only in small proportions, they may play important roles in determining the physical properties of specific tissues. In addition, a number of proteins (eg, the C1q component of the complement system, pulmonary surfactant proteins SPA and SPD) that are not classified as collagens have collagen-like domains in their structures; these proteins are sometimes referred to as "noncollagen collagens."

Table 48–1 summarizes information on many of the types of collagens found in human tissues; the nomenclature used to designate types of collagen and their genes is described in the footnote.

In **Table 48–2**, the types of collagen listed in Table 48–1 are subdivided into a number of classes based primarily on the structures they form. In this chapter, we shall be primarily

TABLE 48–1   **Types of Collagen and Their Genes[1]**

Type	Genes	Tissue
I	COL1A1, COL1A2	Most connective tissues, including bone
II	COL2A1	Cartilage, vitreous humor
III	COL3A1	Extensible connective tissues such as skin, lung, and the vascular system
IV	COL4A1–COL4A6	Basement membranes
V	COL5A1–COL5A3	Minor component in tissues containing collagen I
VI	COL6A1–COL6A3	Most connective tissues
VII	COL7A1	Anchoring fibrils
VIII	COL8A1–COL8A2	Endothelium, other tissues
IX	COL9A1–COL9A3	Tissues containing collagen II
X	COL10A1	Hypertrophic cartilage
XI	COL11A1, COL11A2, COL2A1	Tissues containing collagen II
XII	COL12A1	Tissues containing collagen I
XIII	COL13A1	Many tissues
XIV	COL14A1	Tissues containing collagen I
XV	COL15A1	Many tissues
XVI	COL16A1	Many tissues
XVII	COL17A1	Skin hemidesmosomes
XVIII	COL18A1	Many tissues (eg, liver, kidney)
XIX	COL19A1	Rhabdomyosarcoma cells

Source: Adapted slightly from Prockop DJ, Kivirrikko KI: Collagens: molecular biology, diseases, and potentials for therapy. Annu Rev Biochem 1995;64:403. Copyright © 1995 by Annual Reviews, www.annualreviews.org. Reprinted with permission.

[1]The types of collagen are designated by Roman numerals. Constituent procollagen chains, called proα chains, are numbered using Arabic numerals, followed by the collagen type in parentheses. For instance, type I procollagen is assembled from two proα1(I) and one proα2(I) chains. It is thus a heterotrimer, whereas type 2 procollagen is assembled from three proα1(II) chains and is thus a homotrimer. The collagen genes are named according to the collagen type, written in Arabic numerals for the gene symbol, followed by an A and the number of the proα chain that they encode. Thus, the COL1A1 and COL1A2 genes encode the α1 and α2 chains of type I collagen, respectively. At least 28 types of collagen have now been recognized.

concerned with the fibril-forming **collagens I and II,** the major collagens of skin and bone and of cartilage, respectively. However, mention will be made of some of the other collagens.

COLLAGEN IS THE MOST ABUNDANT PROTEIN IN THE ANIMAL WORLD

All collagen types have a **triple helical structure.** In some collagens, the entire molecule is triple helical, whereas in others the triple helix may involve only a fraction of the structure. Mature collagen type I, containing approximately 1000 amino acids, belongs to the former type; in it, each polypeptide subunit or alpha chain is twisted into a left-handed polyproline helix of three residues per turn (**Figure 48–1**). Three of these alpha chains are then wound into a **right-handed superhelix,**

TABLE 48–2   **Classification of Collagens, Based Primarily on the Structures That They Form**

Class	Type
Fibril-forming	I, II, III, V, and XI
Network-like	IV, VIII, X
FACITs[1]	IX, XII, XIV, XVI, XIX
Beaded filaments	VI
Anchoring fibrils	VII
Transmembrane domain	XIII, XVII
Others	XV, XVIII

Source: Based on Prockop DJ, Kivirrikko KI: Collagens: molecular biology, diseases, and potentials for therapy. Annu Rev Biochem 1995;64:403. Copyright © 1995 by Annual Reviews. Reprinted with permission.

[1]FACITs = fibril-associated collagens with interrupted triple helices. Additional collagens to these listed above have been recognized.

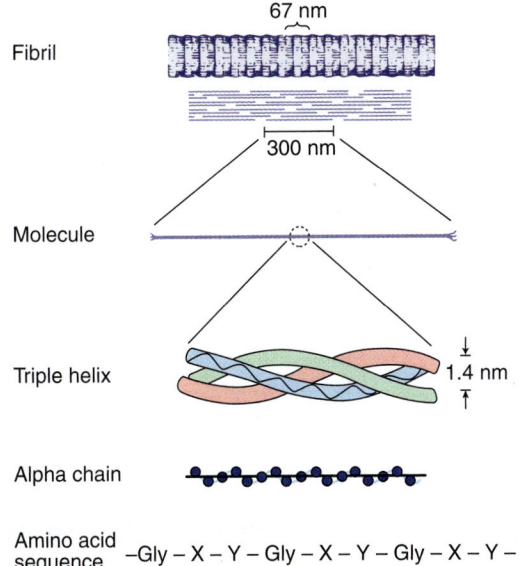

Fibril

67 nm

300 nm

Molecule

Triple helix

1.4 nm

Alpha chain

Amino acid sequence   –Gly – X – Y – Gly – X – Y – Gly – X – Y –

FIGURE 48–1 **Molecular features of the collagen structure from primary sequence up to the fibril.** Each individual polypeptide chain is twisted into a left-handed helix of three residues (Gly-X-Y) per turn, and all of these chains are then wound into a right-handed superhelix. (Slightly modified and reproduced, with permission, from Eyre DR: Collagen (1980), "Molecular diversity in the body's protein scaffold". Science 207:1315. Reprinted with permission from AAAS.)

forming a rodlike molecule 1.4 nm in diameter and about 300 nm long. A striking characteristic of collagen is the occurrence of **glycine** residues at every third position of the triple helical portion of the alpha chain. This is necessary because glycine is the only amino acid small enough to be accommodated in the limited space available down the central core of the triple helix. This **repeating structure**, represented as (Gly-X-Y)$_n$, is an absolute requirement for the formation of the triple helix. While X and Y can be any other amino acids, about 100 of the X positions are proline and about 100 of the Y positions are hydroxyproline. Proline and hydroxyproline confer **rigidity** on the collagen molecule. **Hydroxyproline** is formed by the posttranslational hydroxylation of peptide-bound proline residues catalyzed by the enzyme **prolyl hydroxylase**, whose cofactors are **ascorbic acid** (vitamin C) and α-ketoglutarate. Lysines in the Y position may also be posttranslationally modified to hydroxylysine through the action of **lysyl hydroxylase**, an enzyme with similar cofactors. Some of these hydroxylysines may be further modified by the addition of galactose or galactosyl-glucose through an **O-glycosidic linkage (see Chapter 47)**, a glycosylation site that is unique to collagen.

Collagen types that form long rod-like fibers in tissues are assembled by lateral association of these triple helical units into a **"quarter staggered" alignment** such that each is displaced longitudinally from its neighbor by slightly less than one-quarter of its length (Figure 48–1, upper part). This arrangement is responsible for the banded appearance of these fibers in connective tissues. Collagen fibers are further

stabilized by the formation of **covalent cross-links,** both within and between the triple helical units. These cross-links form through the action of **lysyl oxidase,** a copper-dependent enzyme that oxidatively deaminates the ε-amino groups of certain lysine and hydroxylysine residues, yielding reactive aldehydes. Such aldehydes can form aldol condensation products with other lysine- or hydroxylysine-derived aldehydes or form Schiff bases with the ε-amino groups of unoxidized lysines or hydroxylysines. These reactions, after further chemical rearrangements, result in the stable covalent cross-links that are important for the tensile strength of the fibers. Histidine may also be involved in certain cross-links.

Several collagen types do not form fibrils in tissues (Table 48–2). They are characterized by interruptions of the triple helix with stretches of protein lacking Gly-X-Y repeat sequences. These non-Gly-X-Y sequences result in areas of globular structure interspersed in the triple helical structure.

Type IV collagen, the best-characterized example of a collagen with discontinuous triple helices, is an important component of **basement membranes,** where it forms a mesh-like network.

Collagen Undergoes Extensive Posttranslational Modifications

Newly synthesized collagen undergoes extensive **posttranslational modification** before becoming part of a mature extracellular collagen fiber (**Table 48–3**). Like most secreted proteins, collagen is synthesized on ribosomes in a precursor form, **preprocollagen,** which contains a leader or signal sequence that directs the polypeptide chain into the lumen of the endoplasmic reticulum. As it enters the endoplasmic reticulum, this leader sequence is enzymatically removed. **Hydroxylation** of proline and lysine residues and **glycosylation** of

TABLE 48–3 Order and Location of Processing of the Fibrillar Collagen Precursor

Intracellular
1. Cleavage of signal peptide
2. Hydroxylation of prolyl residues and some lysyl residues; glycosylation of some hydroxylysyl residues
3. Formation of intrachain and interchain S–S bonds in extension peptides
4. Formation of triple helix
Extracellular
1. Cleavage of amino and carboxyl terminal propeptides
2. Assembly of collagen fibers in quarter-staggered alignment
3. Oxidative deamination of ε-amino groups of lysyl and hydroxylysyl residues to aldehydes
4. Formation of intra- and interchain cross-links via Schiff bases and aldol condensation products

hydroxylysines in the **procollagen** molecule also take place at this site. The procollagen molecule contains polypeptide extensions (**extension peptides**) of 20–35 kDa at both its amino and carboxyl terminal ends, neither of which is present in mature collagen. Both extension peptides contain cysteine residues. While the amino terminal propeptide forms only intrachain disulfide bonds, the carboxyl terminal propeptides form both intrachain and interchain disulfide bonds. Formation of these disulfide bonds assists in the **registration** of the three collagen molecules to form the triple helix, winding from the carboxyl terminal end. After formation of the triple helix, no further hydroxylation of proline or lysine or glycosylation of hydroxylysines can take place. **Self-assembly** is a cardinal principle in the biosynthesis of collagen.

Following **secretion** from the cell by way of the Golgi apparatus, extracellular enzymes called **procollagen aminoproteinase** and **procollagen carboxyproteinase** remove the extension peptides at the amino and carboxyl terminal ends, respectively. Cleavage of these propeptides may occur within crypts or folds in the cell membrane. Once the propeptides are removed, the triple helical collagen molecules, containing approximately 1000 amino acids per chain, **spontaneously assemble** into collagen fibers. These are further stabilized by the formation of **inter- and intrachain cross-links** through the action of lysyl oxidase, as described previously.

The same cells that secrete collagen also secrete **fibronectin,** a large glycoprotein present on cell surfaces, in the extracellular matrix, and in blood (see below). Fibronectin binds to aggregating precollagen fibers and alters the kinetics of fiber formation in the pericellular matrix. Associated with fibronectin and procollagen in this matrix are the **proteoglycans** heparan sulfate and chondroitin sulfate (see below). In fact, **type IX collagen,** a minor collagen type from cartilage, contains an attached glycosaminoglycan chain. Such interactions may serve to regulate the formation of collagen fibers and to determine their orientation in tissues.

Once formed, collagen is relatively **metabolically stable.** However, its breakdown is increased during starvation and various inflammatory states. Excessive production of collagen occurs in a number of conditions, for example, hepatic cirrhosis.

A Number of Genetic Diseases Result from Abnormalities in the Synthesis of Collagen

About 30 genes encode the collagens, and their pathway of biosynthesis is complex, involving at least eight enzyme-catalyzed posttranslational steps. Thus, it is not surprising that a number of diseases (**Table 48–4**) are due to **mutations in collagen genes** or in **genes encoding some of the enzymes** involved in these post-translational modifications. The diseases affecting bone (eg, osteogenesis imperfecta) and cartilage (eg, the chondrodysplasias) will be discussed later in this chapter.

The **Ehlers–Danlos syndrome** comprises a group of inherited disorders whose principal clinical features are

TABLE 48–4 Diseases Caused by Mutations in Collagen Genes or by Deficiencies in the Activities of Posttranslational Enzymes Involved in the Biosynthesis of Collagen

Gene or Enzyme	Disease[1]
COL1A1, COL1A2	Osteogenesis imperfecta, type 1[2] (OMIM 166200)
	Osteoporosis[3] (OMIM 166710)
	Ehlers-Danlos syndrome type VII autosomal dominant (OMIM 130060)
COL2A1	Severe chondrodysplasias Osteoarthritis[3] (OMIM 165720)
COL3A1	Ehlers-Danlos syndrome type IV (OMIM 130050)
COL4A3–COL4A6	Alport syndrome (including both autosomal and X-linked forms) (OMIM 104200)
COL7A1	Epidermolysis bullosa, dystrophic (OMIM 131750)
COL10A1	Schmid metaphysial chondrodysplasia (OMIM 156500)
Lysyl hydroxylase	Ehlers-Danlos syndrome type VI (OMIM 225400)
Procollagen N-proteinase	Ehlers-Danlos syndrome type VII autosomal recessive (OMIM 225410)
Lysyl hydroxylase	Menkes disease[4] (OMIM 309400)

Source: Adapted from Prockop DJ, Kivirrikko KI: Collagens: molecular biology, diseases, and potentials for therapy. Annu Rev Biochem 1995;64:403. Copyright © 1995 by Annual Reviews. Reprinted with permission.
[1]Genetic linkage to collagen genes has been shown for a few other conditions not listed here.
[2]At least four types of osteogenesis imperfecta are recognized; the great majority of mutations in all types are in the COL1A1 and COL1A2 genes.
[3]At present applies to only a relatively small number of such patients.
[4]Secondary to a deficiency of copper (Chapter 50).

hyperextensibility of the skin, abnormal tissue fragility, and increased joint mobility. The clinical picture is variable, reflecting underlying extensive genetic heterogeneity. At least 10 types have been recognized, most but not all of which reflect a variety of lesions in the synthesis of collagen. **Type IV** is the most serious because of its tendency for spontaneous rupture of arteries or the bowel, reflecting abnormalities in type III collagen. Patients with **type VI,** due to a deficiency of lysyl hydroxylase, exhibit marked joint hypermobility and a tendency to ocular rupture. A deficiency of procollagen N-proteinase, causing formation of abnormal thin, irregular collagen fibrils, results in **type VIIC,** manifested by marked joint hypermobility and soft skin.

The **Alport syndrome** is the designation applied to a number of genetic disorders (both X-linked and autosomal) affecting the structure of **type IV** collagen fibers, the major collagen found in the basement membranes of the renal glomeruli (see discussion of laminin, in the following). Mutations in several genes encoding type IV collagen fibers have been demonstrated. The presenting sign is hematuria, and patients may eventually develop end-stage renal disease. Electron microscopy reveals characteristic abnormalities of the structure of the basement membrane and lamina densa.

In **epidermolysis bullosa,** the skin breaks and blisters as a result of minor trauma. The dystrophic form is due to mutations in *COL7A1,* affecting the structure of **type VII** collagen. This collagen forms delicate fibrils that anchor the basal lamina to collagen fibrils in the dermis. These anchoring fibrils have been shown to be markedly reduced in this form of the disease, probably resulting in the blistering. Epidermolysis bullosa simplex, another variant, is due to mutations in keratin 5 (Chapter 49).

Scurvy affects the structure of collagen. However, it is due to **a deficiency of ascorbic acid** (Chapter 44), and is not a genetic disease. Its major signs are bleeding gums, subcutaneous hemorrhages, and poor wound healing. These signs reflect impaired synthesis of collagen due to **deficiencies of prolyl and lysyl hydroxylases,** both of which require ascorbic acid as a cofactor.

In **Menkes disease** deficiency of copper results in defective cross-linking of collagen and elastin by the copper-dependent enzyme lysyl oxidase. (Menkes disease is discussed in Chapter 50.)

ELASTIN CONFERS EXTENSIBILITY & RECOIL ON LUNG, BLOOD VESSELS & LIGAMENTS

Elastin is a connective tissue protein that is responsible for properties of extensibility and elastic recoil in tissues. Although not as widespread as collagen, elastin is present in large amounts, particularly in tissues that require these physical properties, for example, lung, large arterial blood vessels, and some elastic ligaments. Smaller quantities of elastin are also found in skin, ear cartilage, and several other tissues. In contrast to collagen, there appears to be only one genetic type of elastin, although variants arise by alternative splicing (Chapter 36) of the hnRNA for elastin. Elastin is synthesized as a soluble monomer of ~70 kDa called **tropoelastin.** Some of the prolines of tropoelastin are hydroxylated to **hydroxyproline** by prolyl hydroxylase, though hydroxylysine and glycosylated hydroxylysine are not present. Unlike collagen, tropoelastin is not synthesized in a pro-form with extension peptides. Furthermore, elastin does not contain repeat Gly-X-Y sequences, triple helical structure, or carbohydrate moieties.

After secretion from the cell, certain lysyl residues of tropoelastin are oxidatively deaminated to aldehydes by **lysyl oxidase,** the same enzyme involved in this process in collagen. However, the major cross-links formed in elastin are the **desmosines,** which result from the condensation of three of these lysine-derived aldehydes with an unmodified lysine to form a tetrafunctional cross-link unique to elastin. Once cross-linked in its mature, extracellular form, elastin is highly insoluble and **extremely stable** and has a very low-turnover

TABLE 48–5 Major Differences Between Collagen and Elastin

Collagen	Elastin
1. Many different genetic types	One genetic type
2. Triple helix	No triple helix; random coil conformations permitting stretching
3. (Gly-X-Y)$_n$ repeating structure	No (Gly-X-Y)$_n$ repeating structure
4. Presence of hydroxylysine	No hydroxylysine
5. Carbohydrate-containing	No carbohydrate
6. Intramolecular aldol cross-links	Intramolecular desmosine cross-links
7. Presence of extension peptides during biosynthesis	No extension peptides present during biosynthesis

rate. Elastin exhibits a variety of random coil conformations that permit the protein to stretch and subsequently recoil during the performance of its physiologic functions.

Table 48–5 summarizes the main differences between collagen and elastin.

Deletions in the elastin gene (located at 7q11.23) have been found in approximately 90% of subjects with the **Williams–Beuren syndrome** (OMIM 194050), a developmental disorder affecting connective tissue and the central nervous system. The mutations, by affecting synthesis of elastin, probably play a causative role in the **supravalvular aortic stenosis** often found in this condition. Fragmentation or, alternatively, a decrease of elastin is found in conditions such as pulmonary emphysema, cutis laxa, and aging of the skin.

MARFAN SYNDROME IS DUE TO MUTATIONS IN THE GENE FOR FIBRILLIN-1, A PROTEIN PRESENT IN MICROFIBRILS

The **Marfan syndrome** is a relatively prevalent inherited disease affecting connective tissue; it is inherited as an autosomal dominant trait. It affects the **eyes** (eg, causing dislocation of the lens, known as ectopia lentis), the **skeletal system** (most patients are tall and exhibit long digits [arachnodactyly] and hyperextensibility of the joints), and the **cardiovascular system** (eg, causing weakness of the aortic media, leading to dilation of the ascending aorta). Abraham Lincoln may have had this condition. Most cases are caused by mutations in the gene (on chromosome 15) for **fibrillin-1;** missense mutations have been detected in several patients with the Marfan syndrome. This results in abnormal fibrillin and/or lower amounts being deposited in the ECM. There is evidence that the cytokine **TGF-β** normally binds to fibrillin-1, and if this binding is decreased (due to lower amounts of fibrillin-1), this can lead to an excess of the cytokine. The excess of TGF-β may contribute to the pathology (eg, in the aorta and aortic valve) found in the syndrome. This finding may lead to the development of therapies for the condition using drugs that antagonize TGF-β (eg, Losartan).

Fibrillin-1 is a large glycoprotein (about 350 kDa) that is a structural component of microfibrils, 10- to 12-nm fibers found in many tissues. It is secreted (subsequent to a proteolytic cleavage) into the ECM by fibroblasts and becomes incorporated into the insoluble **microfibrils,** which appear to provide a **scaffold** for deposition of elastin. Of special relevance to the Marfan syndrome, fibrillin-1 is found in the zonular fibers of the **lens,** in the **periosteum,** and associated with elastin fibers in the **aorta** (and elsewhere); these locations respectively explain the ectopia lentis, arachnodactyly, and cardiovascular problems found in the syndrome. Other proteins (eg, emelin and two microfibril-associated proteins) are also present in microfibrils. It appears likely that their abnormalities may cause other connective tissue disorders. A gene for another fibrillin—**fibrillin-2**—exists on chromosome 5; mutations in this gene are linked to causation of **congenital contractural arachnodactyly** (OMIM 121050), but not to the Marfan syndrome. Fibrillin-2 may be important in deposition of microfibrils early in development. The probable sequence of events leading to Marfan syndrome is summarized in **Figure 48–2.**

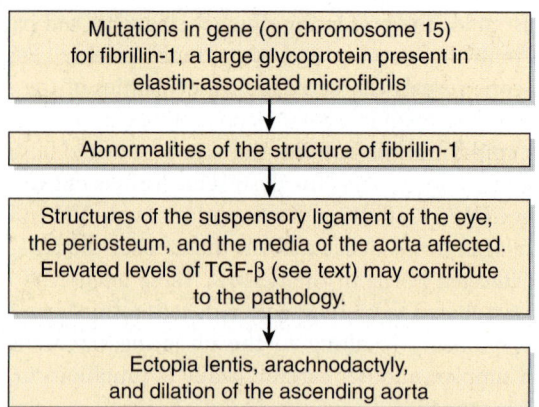

FIGURE 48–2 Probable sequence of events in the causation of the major signs exhibited by patients with the Marfan syndrome (OMIM 154700).

FIBRONECTIN IS AN IMPORTANT GLYCOPROTEIN INVOLVED IN CELL ADHESION & MIGRATION

Fibronectin is a major glycoprotein of the extracellular matrix, also found in a soluble form in plasma. It consists of two identical subunits, each of about 230 kDa, joined by two disulfide bridges near their carboxyl terminals. The gene encoding fibronectin is very large, containing some 50 exons; the RNA produced by its transcription is subject to considerable alternative splicing, and as many as 20 different mRNAs have been detected in various tissues. Fibronectin contains three types of repeating motifs (I, II, and III), which are organized into functional **domains** (at least seven); functions of these domains include binding **heparin** (see below) and fibrin, collagen, DNA, and cell surfaces (**Figure 48–3**). The amino acid sequence of the fibronectin receptor of fibroblasts has been derived, and the protein is a member

of the transmembrane integrin class of proteins (Chapter 51). The **integrins** are heterodimers, containing various types of α and β polypeptide chains. Fibronectin contains an Arg-Gly-Asp (RGD) sequence that binds to the receptor. The **RGD sequence** is shared by a number of other proteins present in the ECM that bind to integrins present in cell surfaces. Synthetic peptides containing the RGD sequence inhibit the binding of fibronectin to cell surfaces. **Figure 48–4** illustrates the interaction of collagen, fibronectin, and laminin, all major proteins of the ECM, with a typical cell (eg, fibroblast) present in the matrix.

The fibronectin receptor interacts indirectly with **actin** microfilaments (Chapter 49) present in the cytosol (**Figure 48–5**). A number of proteins, collectively known as **attachment proteins,** are involved; these include talin, vinculin, an actin-filament capping protein, and α-actinin. Talin interacts with the receptor and vinculin, whereas the latter two interact with actin. The interaction of fibronectin with its receptor provides one route whereby the **exterior of the cell can communicate with the interior** and thus affect cell behavior. Via the interaction with its cell receptor, fibronectin plays an important role in the **adhesion** of cells to the ECM. It is also involved in **cell migration** by providing a binding site for cells and thus helping them to steer their way through the ECM. The amount of fibronectin around many **transformed cells** is sharply reduced, partly explaining their faulty interaction with the ECM.

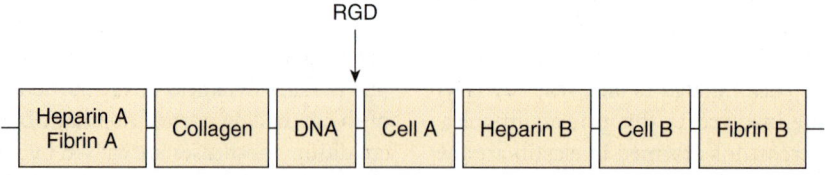

FIGURE 48–3 **Schematic representation of fibronectin.** Seven functional domains of fibronectin are represented; two different types of domain for heparin, cell binding, and fibrin are shown. The domains are composed of various combinations of three structural motifs (I, II, and III), not depicted in the figure. Also not shown is the fact that fibronectin is a dimer joined by disulfide bridges near the carboxyl terminals of the monomers. The approximate location of the RGD sequence of fibronectin, which interacts with a variety of fibronectin integrin receptors on cell surfaces, is indicated by the arrow. (Redrawn after Yamada KM: Adhesive recognition sequences. J Biol Chem 1991;266:12809.)

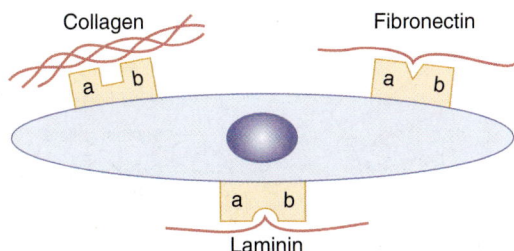

FIGURE 48–4 Schematic representation of a cell interacting through various integrin receptors with collagen, fibronectin, and laminin present in the ECM. (Specific subunits are not indicated.) (Redrawn after Yamada KM: Adhesive recognition sequences. J Biol Chem 1991;266:12809.)

LAMININ IS A MAJOR PROTEIN COMPONENT OF RENAL GLOMERULAR & OTHER BASAL LAMINAS

Basal laminas are specialized areas of the ECM that surround epithelial and some other cells (eg, muscle cells); here we discuss only the laminas found in the **renal glomerulus.** In that structure, the basal lamina is contributed by two separate sheets of cells (one endothelial and one epithelial), each disposed on opposite sides of the lamina; these three layers make up the **glomerular membrane.** The primary components of the basal lamina are three proteins—laminin, entactin, and type IV collagen—and the GAG **heparin** or **heparan sulfate.** These components are synthesized by the underlying cells. Type IV collagen was described above.

Laminin (a glycoprotein of about 850 kDa and 70 nm length) consists of three distinct elongated polypeptide chains (α, β and γ chains) linked together to form a complex, elongated shape (see Figure 49–11, **in which laminin is called merosin**). There are a number of genetic variants of laminin, details of which will not be presented here. Laminin has potential binding sites for type IV collagen, heparin, and integrins on cell surfaces. The collagen interacts with laminin (rather than directly with the cell surface), which in turn interacts with integrins or other laminin receptor proteins, thus anchoring the lamina to the cells. **Entactin,** also known as "nidogen," is a glycoprotein containing an RGD sequence; it binds to laminin and is a major cell attachment factor. The relatively thick basal lamina of the renal glomerulus has an important role in **glomerular filtration,** regulating the passage of large molecules (most plasma proteins) across the glomerulus into the renal tubule. The glomerular membrane allows small molecules, such as **inulin** (5.2 kDa), to pass through as easily as water. On the other hand, only a small amount of the protein **albumin** (69 kDa), the major plasma protein, passes through the normal glomerulus. This is explained by two sets of facts. (1) The **pores** in the glomerular membrane are large enough to allow molecules up to about 8 nm to pass through. (2) Albumin is smaller than this pore size, but it is prevented from passing through easily by the **negative charges** of heparan sulfate and

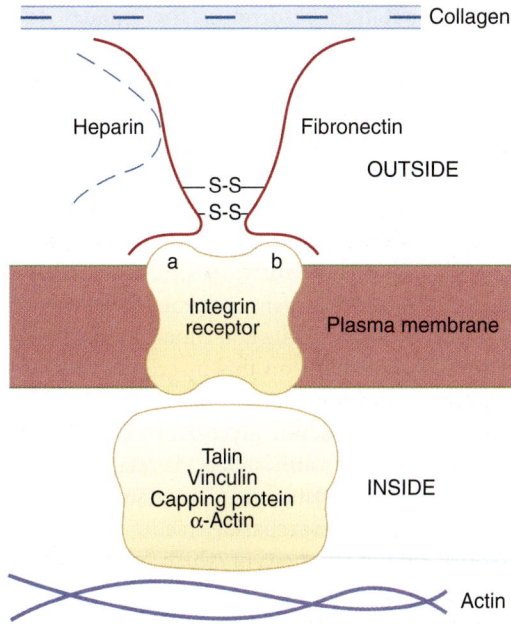

FIGURE 48–5 Schematic representation of fibronectin interacting with an integrin fibronectin receptor situated in the exterior of the plasma membrane of a cell of the ECM and of various attachment proteins interacting indirectly or directly with an actin microfilament in the cytosol. For simplicity, the attachment proteins are represented as a complex.

of certain sialic acid-containing glycoproteins present in the lamina. These negative charges repel albumin and most plasma proteins, which are negatively charged at the pH of blood. The normal structure of the glomerulus may be severely damaged in certain types of **glomerulonephritis** (eg, caused by antibodies directed against various components of the glomerular membrane). This alters the pores and the amounts and dispositions of the negatively charged macromolecules referred to above, and relatively massive amounts of albumin (and of certain other plasma proteins) can pass through into the urine, resulting in severe **albuminuria.**

PROTEOGLYCANS & GLYCOSAMINOGLYCANS

The Glycosaminoglycans Found in Proteoglycans Are Built Up of Repeating Disaccharides

Proteoglycans are proteins that contain covalently linked glycosaminoglycans. At least 30 have been characterized and given names such as syndecan, betaglycan, serglycin, perlecan, aggrecan, versican, decorin, biglycan, and fibromodulin. They vary in tissue distribution, nature of the core protein, attached glycosaminoglycans, and function. The proteins bound covalently to glycosaminoglycans are called **"core proteins"**; they have proved difficult to isolate and characterize, but the use of recombinant DNA technology is beginning to yield important information about their structures. The amount of **carbohydrate** in a proteoglycan is usually much greater than that found in a glycoprotein and may comprise up to 95% of its weight. **Figures 48–6** and **48–7** show the general structure of one particular proteoglycan, **aggrecan,** the major type found in cartilage. It is very large (about 2×10^3 kDa), with its overall structure resembling that of a bottlebrush. It contains a long strand of hyaluronic acid (one type of GAG) to which link proteins are attached **noncovalently.** In turn, these latter interact noncovalently with core protein molecules from which chains of other GAGs (keratan sulfate and chondroitin sulfate in this case) project. More details on this macromolecule are given when cartilage is discussed below.

There are at least seven **glycosaminoglycans (GAGs):** hyaluronic acid, chondroitin sulfate, keratan sulfates I and II, heparin, heparan sulfate, and dermatan sulfate. A GAG is an unbranched polysaccharide made up of repeating disaccharides, one component of which is always an **amino sugar** (hence, the name GAG), either D-glucosamine or D-galactosamine. The other component of the repeating disaccharide (except in the case of keratan sulfate) is a **uronic acid,** either L-glucuronic acid (GlcUA) or its 5′-epimer, L-iduronic acid (IdUA). With the exception of hyaluronic acid, all

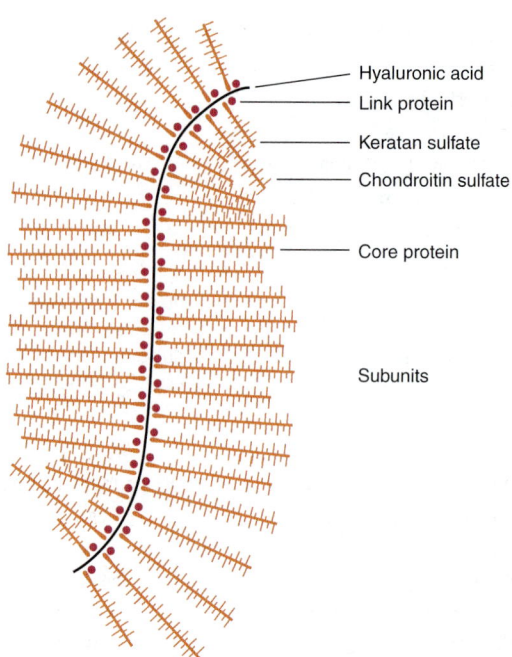

FIGURE 48–7   **Schematic representation of the proteoglycan aggrecan.** (Reproduced, with permission, from Lennarz WJ: *The Biochemistry of Glycoproteins and Proteoglycans.* Plenum Press, 1980. Reproduced with kind permission from Springer Science and Business Media.)

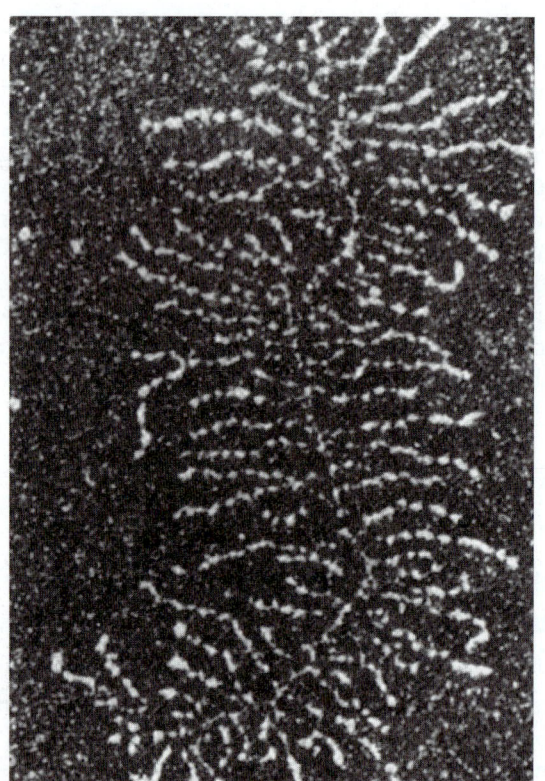

FIGURE 48–6   **Darkfield electron micrograph of a proteoglycan aggregate in which the proteoglycan subunits and filamentous backbone are particularly well extended.** (Reproduced, with permission, from Rosenberg L, Hellman W, Kleinschmidt AK: Electron microscopic studies of proteoglycan aggregates from bovine articular cartilage. J Biol Chem 1975;250:1877.)

the GAGs contain **sulfate groups,** either as *O*-esters or as *N*-sulfate (in heparin and heparan sulfate). Hyaluronic acid affords another exception because there is no clear evidence that it is attached covalently to protein, as the definition of a proteoglycan given above specifies. Both GAGs and proteoglycans have proved difficult to work with, partly because of their complexity. However, they are major components of the ECM; they have a number of important biologic roles; and they are involved in a number of disease processes—so that interest in them is increasing rapidly.

Biosynthesis of Glycosaminoglycans Involves Attachment to Core Proteins, Chain Elongation & Chain Termination

Attachment to Core Proteins

The linkage between GAGs and their core proteins is generally one of three types.

1. An *O*-**glycosidic bond** between **xylose** (Xyl) and **Ser,** a bond that is unique to proteoglycans. This linkage is formed by transfer of a Xyl residue to Ser from UDP-xylose. Two residues of Gal are then added to the Xyl residue, forming a **link trisaccharide,** Gal-Gal-Xyl-Ser. Further chain growth of the GAG occurs on the terminal Gal.

2. An *O*-**glycosidic bond** forms between **GalNAc** (*N*-acetylgalactosamine) and **Ser (Thr)** (Figure 47–1A), present in keratan sulfate II. This bond is formed by donation to Ser

(or Thr) of a GalNAc residue, employing UDP-GalNAc as its donor.

3. An **N-glycosylamine bond** between **GlcNAc** (*N*-acetyl-glucosamine) and the amide nitrogen of **Asn,** as found in *N*-linked glycoproteins (Figure 47–1B). Its synthesis is believed to involve dolichol-P-Poligosaccharide.

The synthesis of the core proteins occurs in the **endoplasmic reticulum,** and formation of at least some of the above linkages also occurs there. Most of the later steps in the biosynthesis of GAG chains and their subsequent modifications occur in the **Golgi apparatus.**

Chain Elongation

Appropriate **nucleotide sugars** and highly specific Golgi-located **glycosyltransferases** are employed to synthesize the oligosaccharide chains of GAGs. The **"one enzyme, one linkage"** relationship appears to hold here, as in the case of certain types of linkages found in glycoproteins. The enzyme systems involved in chain elongation are capable of high-fidelity reproduction of complex GAGs.

Chain Termination

This appears to result from (1) **sulfation,** particularly at certain positions of the sugars, and (2) the **progression** of the growing GAG chain away from the membrane site where catalysis occurs.

Further Modifications

After formation of the GAG chain, **numerous chemical modifications** occur, such as the introduction of sulfate groups onto GalNAc and other moieties and the epimerization of GlcUA to IdUA residues. The enzymes catalyzing sulfation are designated **sulfotransferases** and use **3'-phosphoadenosine-5'-phosphosulfate** [PAPS; active sulfate] (see Figure 32–11) as the sulfate donor. These Golgi-located enzymes are highly specific, and distinct enzymes catalyze sulfation at different positions (eg, carbons 2, 3, 4, and 6) on the acceptor sugars. An **epimerase** catalyzes conversions of glucuronyl to iduronyl residues.

Various Glycosaminoglycans Exhibit Differences in Structure & Have Characteristic Distributions

The seven GAGs named above **differ** from each other in a number of the following properties: amino sugar composition, uronic acid composition, linkages between these components, chain length of the disaccharides, the presence or absence of sulfate groups and their positions of attachment to the constituent sugars, the nature of the core proteins to which they are attached, the nature of the linkage to core protein, their tissue and subcellular distribution, and their biologic functions.

The structures (**Figure 48–8**) and the distributions of each of the GAGs will now be briefly discussed. The major features of the seven GAGs are summarized in **Table 48–6.**

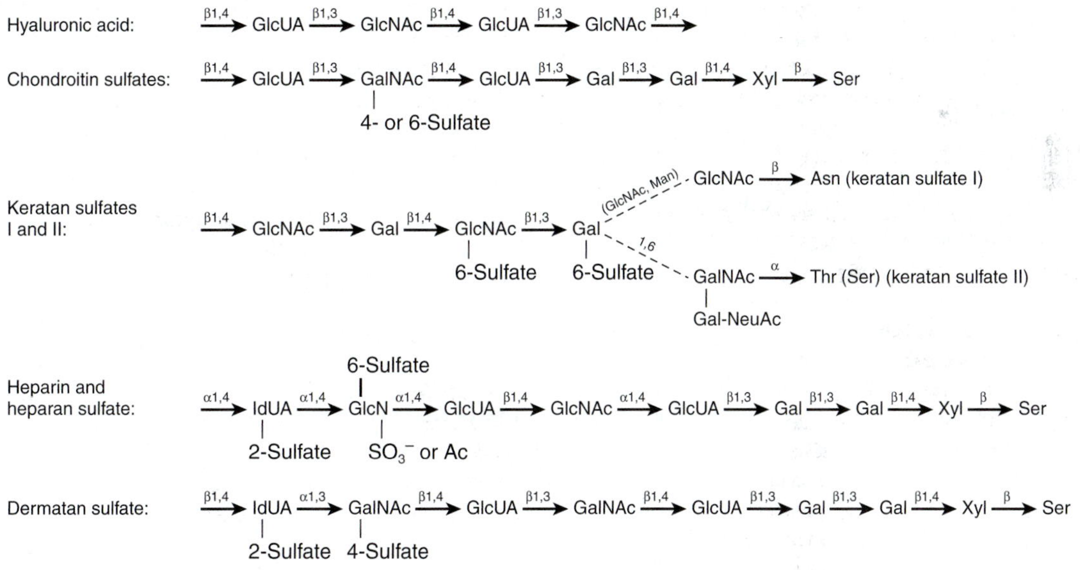

FIGURE 48–8 **Summary of structures of glycosaminoglycans and their attachments to core proteins.** (Ac, acetyl; Asn, L-asparagine; Gal, D-galactose; GalN, D-galactosamine; GlcN, D-glucosamine; GlcUA, D-glucuronic acid; IdUA, L-iduronic acid; Man, D-mannose; NeuAc, *N*-acetylneuraminic acid; Ser, L-serine; Thr, L-threonine; Xyl, D-xylose.) The summary structures are qualitative representations only and do not reflect, for example, the uronic acid composition of hybrid glycosaminoglycans such as heparin and dermatan sulfate, which contain both L-iduronic and D-glucuronic acid. Neither should it be assumed that the indicated substituents are always present, for example, whereas most iduronic acid residues in heparin carry a 2'-sulfate group, a much smaller proportion of these residues are sulfated in dermatan sulfate. The presence of link trisaccharides (Gal-GalXyl) in the chondroitin sulfates, heparin, and heparin, and dermatan sulfates is shown. (Slightly modified and reproduced, with permission, from Lennarz WJ: *The Biochemistry of Glycoproteins and Proteoglycans.* Plenum Press, 1980. Reproduced with kind permission from Springer Science and Business Media.)

TABLE 48–6 Major Properties of the Glycosaminoglycans

GAG	Sugars	Sulfate[1]	Linkage of Protein	Location
HA	GlcNAc, GlcUA	Nil	No firm evidence	Synovial fluid, vitreous humor, loose connective tissue
CS	GalNAc, GlcUA	GalNAc	Xyl-Ser; associated with HA via link proteins	Cartilage, bone, cornea
KS I	GlcNAc, Gal	GlcNAc	GlcNAc-Asn	Cornea
KS II	GlcNAc, Gal	Same as KS I	GalNAc-Thr	Loose connective tissue
Heparin	GlcN, IdUA	GlcN GlcN IdUA	Ser	Mast cells
Heparan sulfate	GlcN, GlcUA	GlcN	Xyl-Ser	Skin fibroblasts, aortic wall
Dermatan sulfate	GalNAc, IdUA, (GlcUA)	GalNAc IdUa	Xyl-Ser	Wide distribution

[1]The sulfate is attached to various positions of the sugars indicated (see Figure 48–8).
Note that all of the GAGs (except the keratan sulfates) contain a uronic acid (glucuronic or iduronic acid).

Hyaluronic Acid

Hyaluronic acid consists of an unbranched chain of repeating disaccharide units containing GlcUA and GlcNAc. The hyaluronic acid is present in bacteria and is widely distributed among various animals and tissues, including synovial fluid, the vitreous body of the eye, cartilage, and loose connective tissues.

Chondroitin Sulfates (Chondroitin 4-Sulfate & Chondroitin 6-Sulfate)

Proteoglycans linked to **chondroitin sulfate** by the Xyl-Ser oglycosidic bond are prominent components of **cartilage** (see below). The repeating disaccharide is similar to that found in hyaluronic acid, containing GlcUA but with **GalNAc** replacing GlcNAc. The GalNAc is substituted with **sulfate** at either its 4′ or its 6′ position, with approximately one sulfate being present per disaccharide unit.

Keratan Sulfates I & II

As shown in Figure 48–8, the keratan sulfates consist of repeating **Gal-GlcNAc** disaccharide units containing **sulfate** attached to the 6′ position of GlcNAc or occasionally of Gal. Type I is abundant in **cornea,** and type II is found along with chondroitin sulfate attached to hyaluronic acid in **loose connective tissue.** Types I and II have different attachments to protein (Figure 48–8).

Heparin

The repeating disaccharide contains **glucosamine** (GlcN) and either of the two uronic acids (**Figure 48–9**). Most of the amino groups of the GlcN residues are **N-sulfated,** but a few are acetylated. The GlcN also carries a sulfate attached to carbon 6.

Approximately 90% of the uronic acid residues are **IdUA.** Initially, all of the uronic acids are GlcUA, but a 5′-epimerase converts approximately 90% of the GlcUA residues to IdUA after the polysaccharide chain is formed. The protein molecule of the heparin proteoglycan is unique, consisting exclusively of serine and glycine residues. Approximately two-thirds of the serine residues contain GAG chains, usually of 5–15 kDa but occasionally much larger. Heparin is found in the granules of **mast cells** and also in liver, lung, and skin.

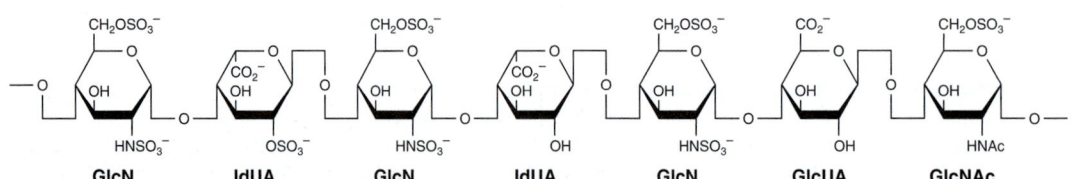

FIGURE 48–9 **Structure of heparin.** The polymer section illustrates structural features typical of heparin; however, the sequence of variously substituted repeating disaccharide units has been arbitrarily selected. In addition, non-O-sulfated or 3-O-sulfated glucosamine residues may also occur. (Modified, redrawn, and reproduced, with permission, from Lindahl U et al: Structure and biosynthesis of heparin-like polysaccharides. Fed Proc 1977;36:19.)

Heparan Sulfate

This molecule is present on many **cell surfaces** as a proteoglycan and is extracellular. It contains **GlcN** with fewer *N*-sulfates than heparin, and, unlike heparin, its predominant uronic acid is **GlcUA.**

Dermatan Sulfate

This substance is widely distributed in animal tissues. Its structure is similar to that of chondroitin sulfate, except that in place of a GlcUA in β-1,3 linkage to GalNAc it contains an **IdUA** in an α-1,3 linkage to **GalNAc.** Formation of the IdUA occurs, as in heparin and heparan sulfate, by 5′-epimerization of GlcUA. Because this is regulated by the degree of sulfation and because sulfation is incomplete, dermatan sulfate contains **both** IdUA-GalNAc and GlcUA-GalNAc disaccharides.

Deficiencies of Enzymes That Degrade Glycosaminoglycans Result in Mucopolysaccharidoses

Both **exo-** and **endoglycosidases** degrade GAGs. Like most other biomolecules, GAGs are subject to **turnover,** being both synthesized and degraded. In adult tissues, GAGs generally exhibit relatively **slow** turnover, their half-lives being days to weeks.

Understanding of the degradative pathways for GAGs, as in the case of glycoproteins (Chapter 47) and glycosphingolipids (Chapter 24), has been greatly aided by elucidation of the specific enzyme deficiencies that occur in certain **inborn errors of metabolism.** When GAGs are involved, these inborn errors are called **mucopolysaccharidoses** (**Table 48–7**).

Degradation of GAGs is carried out by a battery of **lysosomal hydrolases.** These include certain **endoglycosidases,**

TABLE 48–7 Biochemical Defects and Diagnostic Tests in Mucopolysaccharidoses (MPS) and Mucolipidoses (ML)

Name	Alternative Designation[1,2]	Enzymatic Defect	Urinary Metabolites
Mucopolysaccharidoses			
Hurler (OMIM 607014), Scheie (OMIM 607016), Hurler-Scheie (OMIM 607015)	MPS I	α-L-Iduronidase	Dermatan sulfate, heparan sulfate
Hunter (OMIM 309900)	MPS II	Iduronate sulfatase	Dermatan sulfate, heparan sulfate
Sanfilippo A (OMIM 252900)	MPS IIIA	Heparan sulfate *N*-sulfatase (sulfamidase)	Heparan sulfate
Sanfilippo B (OMIM 252920)	MPS IIIB	α-*N*-Acetylglucosaminidase	Heparan sulfate
Sanfilippo C (OMIM 252930)	MPS IIIC	α-Glucosaminide N-acetyltransferase	Heparan sulfate
Sanfilippo D (OMIM 252940)	MPS IIID	*N*-Acetylglucosamine 6-sulfatase	Heparan sulfate
Morquio A (OMIM 253000)	MPS IVA	Galactosamine 6-sulfatase	Keratan sulfate, chondroitin 6-sulfate
Morquio B (OMIM 253010)	MPS IVB	β-Galactosidase	Keratan sulfate
Maroteaux-Lamy (OMIM 253200)	MPS VI	*N*-Acetylgalactosamine 4-sulfatase (arylsulfatase B)	Dermatan sulfate
Sly (OMIM 253220)	MPS VII	β-Glucuronidase	Dermatan sulfate, heparan sulfate, chondroitin 4-sulfate, chondroitin 6-sulfate
Mucolipidoses			
Sialidosis (OMIM 256550)	ML I	Sialidase (neuraminidase)	Glycoprotein fragments
I-cell disease (OMIM 252500)	ML II	N-acetylglucosamine-1-phosphotransferase (acid hydrolases thus lack phosphomannosyl residues)	Glycoprotein fragments
Pseudo-Hurler polydystrophy (OMIM 252600)	ML III	As for ML II but deficiency is incomplete	Glycoprotein fragments

Source: Modified and reproduced, with permission, from DiNatale P, Neufeld EF: The biochemical diagnosis of mucopolysaccharidoses, mucolipidoses and related disorders. In: *Perspectives in Inherited Metabolic Diseases,* vol 2. Barr B et al (editors). Editiones Ermes (Milan), 1979.

[1]Fibroblasts, leukocytes, tissues, amniotic fluid cells, or serum can be used for the assay of many of the above enzymes. Patients with these disorders exhibit a variety of clinical findings that may include cloudy corneas, mental retardation, stiff joints, cardiac abnormalities, hepatosplenomegaly, and short stature, depending on the specific disease and its severity.

[2]The term MPS V is no longer used. The existence of MPS VIII (suspected glucosamine 6-sulfatase deficiency: OMIM 253230) has not been confirmed. At least one case of hyaluronidase deficiency (MPS IX; OMIM 601492) has been reported.

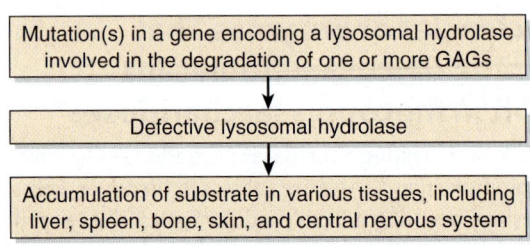

FIGURE 48–10 Simplified scheme of causation of a mucopolysaccharidosis, such as the Hurler syndrome (OMIM 607014), in which the affected enzyme is α-ʟ-iduronidase. Marked accumulation of the GAGs in the tissues mentioned in the figure could cause hepatomegaly, splenomegaly, disturbances of growth, coarse facies, and mental retardation, respectively.

various **exoglycosidases,** and **sulfatases,** generally acting in sequence to degrade the various GAGs. A number of them are indicated in Table 48–7.

The **mucopolysaccharidoses** share a common mechanism of causation, as illustrated in **Figure 48–10.** They are usually inherited in an **autosomal recessive** manner, with **Hurler** and **Hunter syndromes** being perhaps the most widely studied. None is common. **General features** of these conditions are summarized in **Table 48–8** and **laboratory tests** of use in their diagnosis are summarized in **Table 48–9.** In some cases, **a family history** of a mucopolysaccharidosis is obtained.

The term **"mucolipidosis"** was introduced to denote diseases that combined features common to both mucopolysaccharidoses and sphingolipidoses (Chapter 24). Three mucolipidoses are listed in Table 48–7. In **sialidosis** (mucolipidosis I, ML-I), various oligosaccharides derived from glycoproteins and certain gangliosides can accumulate in tissues. **I-cell disease** (ML-II) and **pseudo-Hurler polydystrophy** (MLIII) are described in Chapter 47. The term "mucolipidosis" is retained because it is still in relatively widespread clinical usage, but it is not appropriate for these two latter diseases since the mechanism of their causation involves **mislocation** of certain lysosomal enzymes. Genetic defects of the

TABLE 48–8 Summary of the Major Features of the Mucopolysaccharidoses

They exhibit a chronic progressive course.
They affect a number of organ systems (ie, they are multisystem disorders).
Many patients exhibit organomegaly (eg, hepato- and splenomegaly may be present).
Patients often exhibit dysostosis multiplex (characterized by severe abnormalities in the development of cartilage and bone, and also mental retardation).
Patients often exhibit abnormal facies (facial appearance).
Other signs sometimes found are abnormalities of hearing, of vision, of the cardiovascular system, and of mental development.

TABLE 48–9 Some Laboratory Tests Used in the Diagnosis of a Mucopolysaccharidosis

Urinalysis for presence of increased amounts of GAGs.
Assays of suspected enzymes in white blood cells, fibroblasts, or possibly serum.
Tissue biopsy with subsequent analysis of GAGs by electrophoresis.
Use of specific gene tests.
Prenatal diagnosis can now be performed in at least certain cases using amniotic fluid cells or chorionic villus biopsy.

catabolism of the oligosaccharide chains of glycoproteins (eg, mannosidosis, fucosidosis) are also described in Chapter 47. Most of these defects are characterized by increased excretion of various fragments of glycoproteins in the urine, which accumulate because of the metabolic block, as in the case of the mucolipidoses.

Hyaluronidase is one important enzyme involved in the catabolism of both hyaluronic acid and chondroitin sulfate. It is a widely distributed endoglycosidase that cleaves hexosaminidic linkages. From hyaluronic acid, the enzyme will generate a tetrasaccharide with the structure $(GlcUA\beta\text{-}1,3\text{-}GlcNAc\text{-}\beta\text{-}1,4)_2$, which can be degraded further by a β-glucuronidase and β-N-acetylhexosaminidase. Surprisingly, only one case of an apparent genetic deficiency of this enzyme appears to have been reported (OMIM 601492).

Proteoglycans Have Numerous Functions

As indicated above, **proteoglycans** are remarkably complex molecules and are found in **every tissue** of the body, mainly in the ECM or "ground substance." There they are associated with each other and also with the other major structural components of the matrix, collagen and elastin, in quite specific manners. Some proteoglycans bind to collagen and others to elastin. These interactions are important in determining the structural organization of the matrix. Some proteoglycans (eg, decorin) can also **bind growth factors** such as TGF-β, modulating their effects on cells. In addition, some of them interact with certain **adhesive proteins** such as fibronectin and laminin (see above), also found in the matrix. The GAGs present in the proteoglycans are **polyanions** and hence bind polycations and cations such as Na^+ and K^+. This latter ability attracts water by osmotic pressure into the extracellular matrix and contributes to its turgor. GAGs also **gel** at relatively low concentrations. Because of the long extended nature of the polysaccharide chains of GAGs and their ability to gel, the proteoglycans can act as **sieves,** restricting the passage of large macromolecules into the ECM but allowing relatively free diffusion of small molecules. Again, because of their extended structures and the huge macromolecular aggregates they often form, they occupy a **large volume** of the matrix relative to proteins.

Some Functions of Specific GAGs & Proteoglycans

Hyaluronic acid is especially high in concentration in embryonic tissues and is thought to play an important role in permitting **cell migration** during morphogenesis and wound repair. Its ability to attract water into the extracellular matrix and thereby "loosen it up" may be important in this regard. The high concentrations of hyaluronic acid and chondroitin sulfates present in **cartilage** contribute to its compressibility (see below).

Chondroitin sulfates are located at sites of calcification in endochondral **bone** and are also found in **cartilage.** They are also located inside certain **neurons** and may provide an endoskeletal structure, helping to maintain their shape.

Both **keratan sulfate I** and **dermatan sulfate** are present in the **cornea.** They lie between collagen fibrils and play a critical role in corneal transparency. Changes in proteoglycan composition found in corneal scars disappear when the cornea heals. The presence of dermatan sulfate in the **sclera** may also play a role in maintaining the overall shape of the eye. Keratan sulfate I is also present in **cartilage.**

Heparin is an important **anticoagulant.** It binds with factors IX and XI, but its most important interaction is with **plasma antithrombin** (discussed in Chapter 51). Heparin can also bind specifically to **lipoprotein lipase** present in capillary walls, causing a release of this enzyme into the circulation.

Certain proteoglycans (eg, **heparan sulfate**) are associated with the plasma membrane of cells, with their core proteins actually spanning that membrane. In this, they may act as **receptors** and may also participate in the mediation of the **cell growth** and **cell–cell communication.** The attachment of cells to their substratum in culture is mediated at least in part by the heparan sulfate. This proteoglycan is also found in the **basement membrane of the kidney** along with type IV collagen and laminin (see above), where it plays a major role in determining the charge selectivity of glomerular filtration.

Proteoglycans are also found in **intracellular locations** such as the nucleus; their function in this organelle has not been elucidated. They are present in some storage or secretory granules, such as the chromaffin granules of the adrenal medulla. It has been postulated that they play a role in release of the contents of such granules. The various functions of GAGs are summarized in **Table 48–10.**

Associations with Major Diseases & with Aging

Hyaluronic acid may be important in permitting **tumor cells to migrate** through the ECM. Tumor cells can induce fibroblasts to synthesize greatly increased amounts of this GAG, thereby perhaps facilitating their own spread. Some tumor cells have less heparan sulfate at their surfaces, and this may play a role in the **lack of adhesiveness** that these cells display.

The intima of the **arterial wall** contains hyaluronic acid and chondroitin sulfate, dermatan sulfate, and heparan sulfate proteoglycans. Of these proteoglycans, dermatan sulfate binds plasma low-density lipoproteins. In addition, dermatan sulfate

TABLE 48–10 Some Functions of Glycosaminoglycans and Proteoglycans

- Act as structural components of the ECM
- Have specific interactions with collagen, elastin, fibronectin, laminin, and other proteins such as growth factors
- As polyanions, bind polycations and cations
- Contribute to the characteristic turgor of various tissues
- Act as sieves in the ECM
- Facilitate cell migration (HA)
- Have role in compressibility of cartilage in weight-bearing (HA, CS)
- Play role in corneal transparency (KS I and DS)
- Have structural role in sclera (DS)
- Act as anticoagulant (heparin)
- Are components of plasma membranes, where they may act as receptors and participate in cell adhesion and cell-cell interactions (eg, HS)
- Determine charge selectiveness of renal glomerulus (HS)
- Are components of synaptic and other vesicles (eg, HS)

Abbreviations: CS, chondroitin sulfate; DS, dermatan sulfate; ECM, extracellular matrix; HA, hyaluronic acid; HS, heparan sulfate; KS I, keratan sulfate I.

appears to be the major GAG synthesized by arterial smooth muscle cells. Because it is these cells that proliferate in **atherosclerotic lesions** in arteries, dermatan sulfate may play an important role in development of the atherosclerotic plaque.

In various types of **arthritis,** proteoglycans may act as **autoantigens,** thus contributing to the pathologic features of these conditions. The amount of chondroitin sulfate in cartilage diminishes with age, whereas the amounts of keratan sulfate and hyaluronic acid increase. These changes may contribute to the development of **osteoarthritis,** as may increased activity of the enzyme aggrecanase, which acts to degrade aggrecan. Changes in the amounts of certain GAGs in the skin are also observed with **aging** and help to account for the characteristic changes noted in this organ in the elderly.

An exciting new phase in proteoglycan research is opening up with the findings that mutations that affect individual proteoglycans or the enzymes needed for their synthesis alter the regulation of **specific signaling pathways** in *Drosophila* and *Caenorhabditis elegans*, thus affecting **development;** it already seems likely that similar effects exist in mice and humans.

BONE IS A MINERALIZED CONNECTIVE TISSUE

Bone contains both **organic** and **inorganic** material. The **organic** matter is mainly **protein.** The principal proteins of bone are listed in **Table 48–11; type I collagen** is the major protein, comprising 90–95% of the organic material. Type V

TABLE 48–11   The Principal Proteins Found in Bone[1]

Proteins	Comments
Collagens	
Collagen type I	Approximately 90% of total bone protein. Composed of two α1(I) and one α2(I) chains.
Collagen type V	Minor component.
Noncollagen proteins	
Plasma proteins	Mixture of various plasma proteins.
Proteoglycans[2] CS-PG I (biglycan)	Contains two GAG chains; found in other tissues.
CS-PG II (decorin)	Contains one GAG chain; found in other tissues.
CS-PG III	Bone-specific.
Bone SPARC[3] protein (osteonectin)	Not bone-specific.
Osteocalcin (bone Gla protein)	Contains γ-carboxyglutamate (Gla) residues that bind to hydroxyapatite. Bone-specific.
Osteopontin	Not bone-specific. Glycosylated and phosphorylated.
Bone sialoprotein	Bone-specific. Heavily glycosylated, and sulfated on tyrosine.
Bone morphogenetic proteins (BMPs)	A family (eight or more) of secreted proteins with a variety of actions on bone; many induce ectopic bone growth.
Osteoprotegerin	Inhibits osteoclastogenesis

[1]Various functions have been ascribed to the noncollagen proteins, including roles in mineralization; however, most of them are still speculative. It is considered unlikely that the noncollagen proteins that are not bone-specific play a key role in mineralization. A number of other proteins are also present in bone, including a tyrosine-rich acidic matrix protein (TRAMP), some growth factors (eg, TGFβ), and enzymes involved in collagen synthesis (eg, lysyl oxidase).
[2]CS-PG, chondroitin sulfate–proteoglycan; these are similar to the dermatan sulfate PGs (DS-PGs) of cartilage (TABLE 48–13).
[3]SPARC, secreted protein acidic and rich in cysteine.

collagen is also present in small amounts, as are a number of noncollagen proteins, some of which are relatively specific to bone. The **inorganic** or mineral component is mainly crystalline **hydroxyapatite**—$Ca_{10}(PO_4)_6(OH)_2$—along with sodium, magnesium, carbonate, and fluoride; approximately 99% of the body's calcium is contained in bone (Chapter 44). Hydroxyapatite confers on bone the strength and resilience required by its physiologic roles.

Bone is a **dynamic structure** that undergoes continuing cycles of remodeling, consisting of resorption followed by deposition of new bone tissue. This remodeling permits bone to adapt to both physical (eg, increases in weight bearing) and hormonal signals.

The major cell types involved in bone resorption and deposition are **osteoclasts** and **osteoblasts** (**Figure 48–11**). The former are associated with resorption and the latter with deposition of bone. Osteocytes are descended from osteoblasts; they also appear to be involved in maintenance of bone matrix but will not be discussed further here.

Osteoclasts are multinucleated cells derived from pluripotent hematopoietic stem cells. Osteoclasts possess an apical membrane domain, exhibiting a ruffled border that plays a key role in bone resorption (**Figure 48–12**). A proton-translocating **ATPase** expels protons across the ruffled border into the resorption area, which is the microenvironment of low pH shown in the figure. This lowers the local pH to 4.0 or less, thus increasing the solubility of hydroxyapatite and allowing demineralization to occur. Lysosomal acid proteases are released that digest the now accessible matrix proteins. **Osteoblasts**—mononuclear cells derived from pluripotent mesenchymal precursors—synthesize most of the proteins found in bone (Table 48–11) as well as various growth factors and cytokines. They are responsible for the deposition of the new bone matrix (osteoid) and its subsequent mineralization. Osteoblasts **control mineralization** by regulating the passage of calcium and

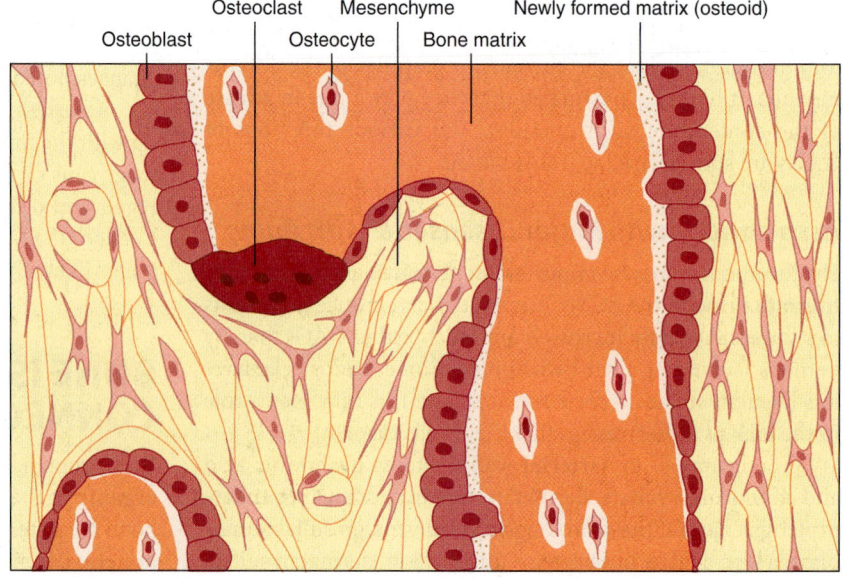

FIGURE 48–11   Schematic illustration of the major cells present in the membranous bone. Osteoblasts (lighter color) are synthesizing type I collagen, which forms a matrix that traps cells. As this occurs, osteoblasts gradually differentiate to become osteocytes. (Reproduced, with permission, from Junqueira LC, Carneiro J: *Basic Histology: Text & Atlas,* 10th ed. McGraw-Hill, 2003.)

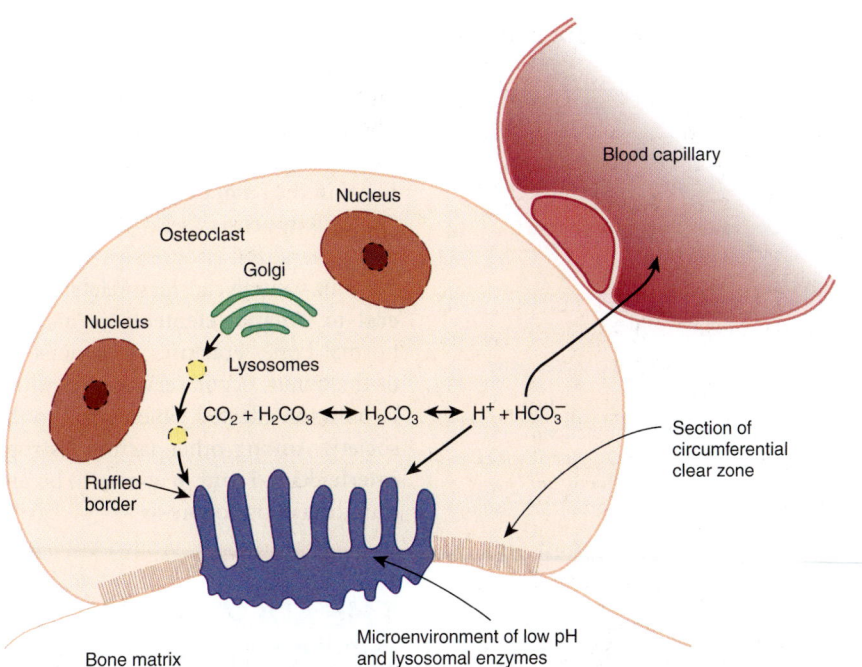

FIGURE 48–12 **Schematic illustration of some aspects of the role of the osteoclast in bone resorption.** Lysosomal enzymes and hydrogen ions are released into the confined microenvironment created by the attachment between the bone matrix and the peripheral clear zone of the osteoclast. The acidification of this confined space facilitates the dissolution of calcium phosphate from bone and is the optimal pH for the activity of lysosomal hydrolases. The bone matrix is thus removed, and the products of bone resorption are taken up into the cytoplasm of the osteoclast, probably digested further, and transferred into capillaries. The chemical equation shown in the figure refers to the action of carbonic anhydrase II, described in the text. (Reproduced, with permission, from Junqueira LC, Carneiro J: *Basic Histology: Text & Atlas,* 10th ed. McGraw-Hill, 2003.)

phosphate ions across their surface membranes. The latter contain **alkaline phosphatase,** which is used to generate phosphate ions from organic phosphates. The mechanisms involved in mineralization are not fully understood, but several factors have been implicated. Alkaline phosphatase contributes to mineralization, but in itself is not sufficient. Small vesicles (matrix vesicles) containing calcium and phosphate have been described at sites of mineralization, but their role is not clear. **Type I collagen** appears to be necessary, with mineralization being first evident in the gaps between successive molecules. Recent interest has focused on **acidic phosphoproteins,** such as bone sialoprotein, acting as sites of nucleation. These proteins contain motifs (eg, poly-Asp and poly-Glu stretches) that bind calcium and may provide an initial scaffold for mineralization. Some macromolecules, such as certain proteoglycans and glycoproteins, can also act as **inhibitors** of nucleation.

It is estimated that approximately 4% of compact bone is **renewed annually** in the typical healthy adult, whereas approximately 20% of trabecular bone is replaced.

Many factors are involved in the **regulation of bone metabolism,** only a few of which will be mentioned here (see Case no. 15 on osteoporosis, Chapter 57). Some **stimulate osteoblasts** (eg, parathyroid hormone and 1,25-dihydroxycholecalciferol) and others **inhibit** them (eg, corticosteroids). Parathyroid hormone and 1,25-dihydroxycholecalciferol also stimulate osteoclasts, whereas calcitonin and estrogens inhibit them.

BONE IS AFFECTED BY MANY METABOLIC & GENETIC DISORDERS

A number of the more important examples of metabolic and genetic disorders that affect bone are listed in **Table 48–12**.

Osteogenesis imperfecta (brittle bones) is characterized by abnormal fragility of bones. The scleras are often abnormally thin and translucent and may appear blue owing to a deficiency of connective tissue. **Four types** of this condition (mild, extensive, severe, and variable) have been recognized, of which the extensive type occurring in the newborn is the most ominous. Affected infants may be born with multiple fractures and not survive. Over 90% of patients with osteogenesis imperfecta have mutations in the *COL1A1* and *COL1A2* genes, encoding proα1(I) and proα2(I) chains, respectively. Over 100 mutations in these two genes have been documented and include partial gene deletions and duplications. Other mutations affect RNA splicing, and the most frequent type results in the **replacement of glycine** by another bulkier amino acid, affecting formation of the triple helix. In general, these mutations result in decreased expression of collagen or in structurally abnormal pro chains that assemble into **abnormal fibrils,** weakening the overall structure of bone. When one abnormal chain is present, it may interact with two normal chains, but

TABLE 48–12 Some Metabolic and Genetic Diseases Affecting Bone and Cartilage

Disease	Comments
Dwarfism	Often due to a deficiency of growth hormone, but has many other causes.
Rickets	Due to a deficiency of vitamin D during childhood.
Osteomalacia	Due to a deficiency of vitamin D during adulthood.
Hyperparathyroidism	Excess parathormone causes bone resorption.
Osteogenesis imperfecta (eg, OMIM 166200)	Due to a variety of mutations in the COL1A1 and COL1A2 genes affecting the synthesis and structure of type I collagen.
Osteoporosis (OMIM 166710)	Commonly postmenopausal or in other cases is more gradual and related to age; a small number of cases are due to mutations in the COL1A1 and COL1A2 genes and possibly in the vitamin D receptor gene
Osteoarthritis	A small number of cases are due to mutations in the COL1A genes
Several chondrodysplasias	Due to mutations in COL2A1 genes
Pfeiffer syndrome[1] (OMIM 101600)	Mutations in the gene encoding fibroblast growth receptor 1 (FGFR1)
Jackson-Weiss (OMIM 123150) and Crouzon (OMIM 123500) syndromes[1]	Mutations in the gene encoding FGFR2
Achondroplasia (OMIM 100800) and thanatophoric dysplasia[2] (OMIM 187600)	Mutations in the gene encoding FGFR3

[1]The Pfeiffer, Jackson-Weiss, and Crouzon syndromes are craniosynostosis syndromes; craniosynostosis is a term signifying premature fusion of sutures in the skull.
[2]Thanatophoric (Gk *thanatos* "death" + *phoros* "bearing") dysplasia is the most common neonatal lethal skeletal dysplasia, displaying features similar to those of homozygous achondroplasia.

folding may be prevented, resulting in enzymatic degradation of all of the chains. This is called **"procollagen suicide"** and is an example of a dominant negative mutation, a result often seen when a protein consists of multiple different subunits.

Osteopetrosis (marble bone disease), characterized by **increased bone density,** is due to inability to resorb bone. One form occurs along with renal tubular acidosis and cerebral calcification. It is due to mutations in the gene (located on chromosome 8q22) encoding **carbonic anhydrase II** (CA II), one of four isozymes of carbonic anhydrase present in human tissues. The reaction catalyzed by carbonic anhydrase is

$$CO_2 + HO_2 \leftrightarrow HO_2CO_3 \leftrightarrow HO^+ + HCO_3^-$$

In osteoclasts involved in bone resorption, CA II apparently provides protons to neutralize the OH$^-$ ions left inside the cell

when H$^+$ ions are pumped across their ruffled borders (see above). Thus, **if CA II is deficient in activity** in osteoclasts, normal bone resorption does not occur, and osteopetrosis results. The mechanism of the cerebral calcification is not clear, whereas the renal tubular acidosis reflects deficient activity of CA II in the renal tubules.

Osteoporosis (see Case History no. 15 in Chapter 57) is a generalized progressive reduction in bone tissue mass per unit volume causing skeletal weakness. The ratio of **mineral** to **organic elements** is unchanged in the remaining normal bone. Fractures of various bones, such as the head of the femur, occur very easily and represent a huge burden to both the affected patients and to the health care budget of society. Among other factors, **estrogens** and the cytokines **interleukins-1 and -6** appear to be intimately involved in the causation of osteoporosis.

THE MAJOR COMPONENTS OF CARTILAGE ARE TYPE II COLLAGEN & CERTAIN PROTEOGLYCANS

The **principal proteins** of hyaline cartilage (the major type of cartilage) are listed in **Table 48–13**. **Type II collagen** is the principal protein (**Figure 48–13**), and a number of other

TABLE 48–13 The Principal Proteins Found in Cartilage

Proteins	Comments
Collagen proteins	
Collagen type II	90–98% of total articular cartilage collagen. Composed of three α 1(II) chains.
Collagens V, VI, IX, X, XI	Type IX cross-links to type II collagen. Type XI may help control diameter of type II fibrils.
Non-collagen proteins	
Proteoglycans	The major proteoglycan of cartilage.
Aggrecan	Found in some types of cartilage.
Large non-aggregating proteoglycan	
DS-PG I (biglycan)[1]	Similar to CS-PG I of bone.
DS-PG II (decorin)	Similar to CS-PG II of bone.
Chondronectin	May play role in binding type II collagen to surface of cartilage.
Anchorin C II	May bind type II collagen to surface of chondrocyte.

[1]The core proteins of DS-PG I and DS-PG II are homologous to those of CS-PG I and CS-PG II found in bone (TABLE 48–11). A possible explanation is that osteoblasts lack the epimerase required to convert glucuronic acid to iduronic acid, the latter of which is found in dermatan sulfate.

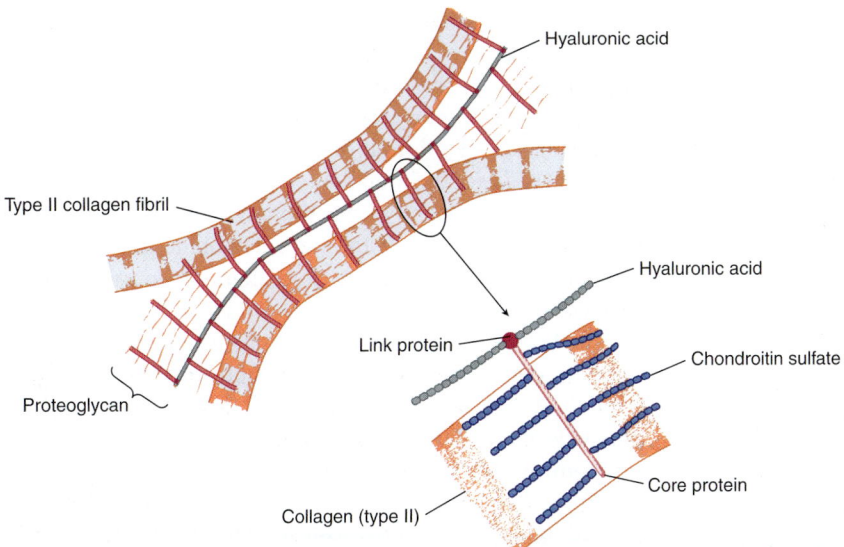

FIGURE 48–13 **Schematic representation of the molecular organization in the cartilage matrix.** Link proteins noncovalently bind the core protein (lighter color) of proteoglycans to the linear hyaluronic acid molecules (darker color). The chondroitin sulfate side chains of the proteoglycan electrostatically bind to the collagen fibrils, forming a cross-linked matrix. The oval outlines the area enlarged in the lower part of the figure. (Reproduced, with permission, from Junqueira LC, Carneiro J: *Basic Histology: Text & Atlas,* 10th ed. McGraw-Hill, 2003.)

minor types of collagen are also present. In addition to these components, elastic cartilage contains elastin and fibroelastic cartilage contains type I collagen. Cartilage contains a number of **proteoglycans,** which play an important role in its compressibility. **Aggrecan** (about 2×10^3 kDa) is the major proteoglycan. As shown in **Figure 48–14,** it has a very complex structure, containing several GAGs (hyaluronic acid, chondroitin sulfate, and keratan sulfate) and both link and core proteins. The core protein contains three domains: A, B, and C. The hyaluronic acid binds noncovalently to domain A of

the core protein as well as to the link protein, which stabilizes the hyaluronate–core protein interactions. The keratan sulfate chains are located in domain B, whereas the chondroitin sulfate chains are located in domain C; both of these types of GAGs are bound covalently to the core protein. The core protein also contains both *O-* and *N-*linked oligosaccharide chains.

The other proteoglycans found in cartilage have simpler structures than aggrecan.

Chondronectin is involved in the attachment of type II collagen to chondrocytes.

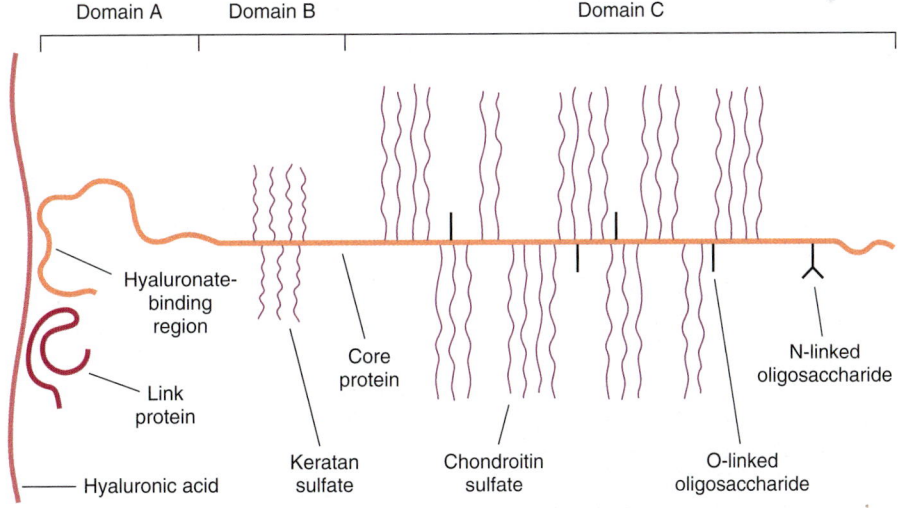

FIGURE 48–14 **Schematic diagram of the aggrecan from the bovine nasal cartilage.** A strand of hyaluronic acid is shown on the left. The core protein (about 210 kDa) has three major domains. Domain A, at its amino terminal end, interacts with approximately five repeating disaccharides in hyaluronate. The link protein interacts with both hyaluronate and domain A, stabilizing their interactions. Approximately 30 keratan sulfate chains are attached, via GalNAc-Ser linkages, to domain B. Domain C contains about 100 chondroitin sulfate chains attached via Gal-Gal-Xyl-Ser linkages and about 40 *O-*linked oligosaccharide chains. One or more *N-*linked glycan chains are also found near the carboxyl terminal of the core protein. (Moran LA, et al, *Biochemistry,* 2nd edition, © 1994, p. 9–43. Adapted by permission of Pearson Education, Inc., Upper Saddle River, NJ.)

Cartilage is an avascular tissue and obtains most of its nutrients from synovial fluid. It exhibits slow but continuous **turnover.** Various **proteases** (eg, collagenases and stromalysin) synthesized by chondrocytes can **degrade collagen** and the other proteins found in cartilage. Interleukin-1 (IL-1) and tumor necrosis factor α (TNFα) appear to stimulate the production of such proteases, whereas transforming growth factor β (TGFβ) and insulin-like growth factor 1 (IGF-I) generally exert an anabolic influence on the cartilage.

THE MOLECULAR BASES OF THE CHONDRODYSPLASIAS INCLUDE MUTATIONS IN GENES ENCODING TYPE II COLLAGEN & FIBROBLAST GROWTH FACTOR RECEPTORS

Chondrodysplasias are a mixed group of hereditary disorders affecting cartilage. They are manifested by shortlimbed dwarfism and numerous skeletal deformities. A number of them are due to a variety of mutations in the *COL2A1* gene, leading to abnormal forms of type II collagen. One example is the **Stickler syndrome,** manifested by degeneration of the joint cartilage and of the vitreous body of the eye.

The best known of the chondrodysplasias is **achondroplasia,** the most common cause of **short-limbed dwarfism.** Affected individuals have short limbs, normal trunk size, macrocephaly, and a variety of other skeletal abnormalities. The condition is often inherited as an autosomal dominant trait, but many cases are due to new mutations. The molecular basis of achondroplasia is outlined in **Figure 48–15.** Achondroplasia is not a collagen disorder but is due to mutations in the gene encoding **fibroblast growth factor receptor 3 (FGFR3).** Fibroblast growth factors are a family of at least nine proteins that affect the growth and differentiation of cells of mesenchymal and neuroectodermal origin. Their **receptors** are transmembrane proteins and form a subgroup of the family of receptor tyrosine kinases. FGFR3 is one member of this subgroup of four and mediates the actions of FGF3 on the cartilage. In almost all cases of achondroplasia that have been investigated, the mutations were found to involve nucleotide 1138 and resulted in substitution of arginine for glycine (residue number 380) in the transmembrane domain of the protein, rendering it inactive. No such mutation was found in unaffected individuals.

Rather amazingly, other mutations in the same gene can result in **hypochondroplasia, thanatophoric dysplasia** (types I and II) and the **SADDAN phenotype** (severe achondroplasia with developmental delay and acanthosis nigricans [the latter is a brown to black hyperpigmentation of the skin]).

As indicated in Table 48–12, **other skeletal dysplasias** (including certain craniosynostosis syndromes) are also due to mutations in genes encoding FGF receptors. Another type of skeletal dysplasia (diastrophic dysplasia) has been found to be due to mutation in a sulfate transporter. Thus, thanks to recombinant DNA technology, a new era in understanding of skeletal dysplasias has begun.

SUMMARY

- The major components of the ECM are the structural proteins collagen, elastin, and fibrillin-1, a number of specialized proteins (eg, fibronectin and laminin), and various proteoglycans.

- Collagen is the most abundant protein in the animal kingdom; approximately 28 types have been isolated. All collagens contain greater or lesser stretches of triple helix and the repeating structure $(Gly-X-Y)_n$.

- The biosynthesis of collagen is complex, featuring many post-translational events, including hydroxylation of proline and lysine.

- Diseases associated with impaired synthesis of collagen include scurvy, osteogenesis imperfecta, Ehlers–Danlos syndrome (many types), and Menkes disease.

- Elastin confers extensibility and elastic recoil on tissues. Elastin lacks hydroxylysine, Gly-X-Y sequences, triple helical structure, and sugars but contains desmosine and isodesmosine cross-links not found in collagen.

- Fibrillin-1 is located in microfibrils. Mutations in the gene encoding fibrillin-1 cause Marfan syndrome. The cytokine TGF-β appears to contribute to the cardiovascular pathology.

- The glycosaminoglycans (GAGs) are made up of repeating disaccharides containing a uronic acid (glucuronic or iduronic) or hexose (galactose) and a hexosamine (galactosamine or glucosamine). Sulfate is also frequently present.

- The major GAGs are hyaluronic acid, chondroitin 4- and 6-sulfates, keratan sulfates I and II, heparin, heparan sulfate, and dermatan sulfate.

- The GAGs are synthesized by the sequential actions of a battery of specific enzymes (glycosyltransferases, epimerases, sulfotransferases, etc) and are degraded by the sequential action of lysosomal hydrolases. Genetic deficiencies of the latter result in mucopolysaccharidoses (eg, the Hurler syndrome).

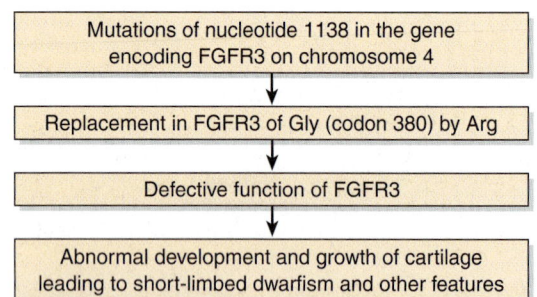

Mutations of nucleotide 1138 in the gene encoding FGFR3 on chromosome 4
Replacement in FGFR3 of Gly (codon 380) by Arg
Defective function of FGFR3
Abnormal development and growth of cartilage leading to short-limbed dwarfism and other features

FIGURE 48–15 Simplified scheme of the causation of achondroplasia (OMIM 100800). In most cases studied so far, the mutation has been a G to A transition at nucleotide 1138. In a few cases, the mutation was a G to C transversion at the same nucleotide. This particular nucleotide is a real "hot spot" for mutation. Both mutations result in replacement of a Gly residue by an Arg residue in the transmembrane segment of the receptor. A few cases involving replacement of Gly by Cys at codon 375 have also been reported.

- GAGs occur in tissues bound to various proteins (linker proteins and core proteins), constituting proteoglycans. These structures are often of very high molecular weight and serve many functions in tissues.

- Many components of the ECM bind to proteins of the cell surface named integrins; this constitutes one pathway by which the exteriors of cells can communicate with their interiors.

- Bone and cartilage are specialized forms of the ECM. Collagen I and hydroxyapatite are the major constituents of bone. Collagen II and certain proteoglycans are major constituents of cartilage.

- The molecular causes of a number of heritable diseases of bone (eg, osteogenesis imperfecta) and of cartilage (eg, the chondrodystrophies) are being revealed by the application of recombinant DNA technology.

REFERENCES

Baldridge D, Shchelochkov O, Kelley B, Lee B: Signaling pathways in human skeletal dysplasias. Annu Rev Genomics Human Genet 2010;11:189.

Couchman JR: Transmembrane signaling proteoglycans. Annu Rev Cell Develop Biol 2010;26:89.

Fauci AS, Braunwald E, Kasper DL, et al: *Harrison's Principles of Internal Medicine,* 17th ed. McGrawHill, 2008. (Chapter 357, *Heritable Disorders of Connective Tissue;* Chapter 355, *Lysosomal Storage Diseases;* Chapter 326, *Osteoarthritis;* Chapter 346, *Bone and Mineral Metabolism in Health and Disease;* Chapter 349, *Paget Disease and Other Dysplasias of Bone*).

Karsenty G, Kronenberg HM, Settembre C: Genetic control of bone formation. Ann Rev Cell Develop Biol 2009;25:629.

Khosla S, Westendorf JJ, Oursler MJ: Building bone to reverse osteoporosis and repair fractures. J Clin Invest 2008; 118:421.

Neufeld EF: From serendipity to therapy. Annu Rev Biochem 2011;80. (Describes pioneering work on the causes and treatment of mucopolysaccharidoses).

Rowe RG, Weiss SJ: Navigating ECM barriers at the invasive front: the cancer cell–stroma interface. Annu Rev Cell Develop Biol 2009;25:567.

Scriver CR, Beaudet AL, Valle D, et al (editors): *The Metabolic and Molecular Bases of Inherited Disease,* 8th ed. McGraw-Hill, 2001. (This comprehensive four-volume text and the updated online version [see Chapter 1] contain chapters on disorders of collagen biosynthesis and structure, Marfan syndrome, the mucopolysaccharidoses, achondroplasia, Alport syndrome, and craniosynostosis syndromes.)

Seeman E, Delams PD: Bone quality—the material and structural basis of bone strength and fragility. N Engl J Med 2006;354:2250.

Shoulders MD, Raines RT: Collagen structure and stability. Ann Rev Biochem 2009;78:929.

Muscle & the Cytoskeleton

Robert K. Murray, MD, PhD

OBJECTIVES

*After studying this chapter,
you should be able to:*

- Understand the general biochemical features of skeletal, cardiac, and smooth muscle contraction.
- Know the biologic effects of nitric oxide (NO).
- Indicate the different metabolic fuels required for a sprint and for the marathon.
- Know the general structures and functions of the major components of the cytoskeleton, namely microfilaments, microtubules, and intermediate filaments.
- Understand the bases of malignant hyperthermia Duchenne and Becker muscular dystrophies, inherited cardiomyopathies, the Hutchinson–Gilford syndrome (progeria), and several skin diseases due to abnormal keratins.

BIOMEDICAL IMPORTANCE

Proteins play an important role in **movement** at both the organ (eg, skeletal muscle, heart, and gut) and cellular levels. In this chapter, the roles of specific proteins and certain other key molecules (eg, Ca^{2+}) in **muscular contraction** are described. A brief coverage of **cytoskeletal proteins** is also presented.

Knowledge of the molecular bases of a number of conditions that affect muscle has advanced greatly in recent years. Understanding of the molecular basis of **Duchenne-type muscular dystrophy** was greatly enhanced when it was found that it was due to mutations in the gene encoding dystrophin (see case history no. 7 in Chapter 57). Significant progress has also been made in understanding the molecular basis of **malignant hyperthermia,** a serious complication for some patients undergoing certain types of anesthesia. **Heart failure** is a very common medical condition, with a variety of causes; its rational therapy requires understanding of the biochemistry of heart muscle. One group of conditions that cause heart failure are the **cardiomyopathies,** some of which are genetically determined. NO has been found to be a major regulator of smooth muscle tone. Many widely used **vasodilators**—such as nitroglycerin, used in the treatment of angina pectoris—act by increasing the formation of NO. Muscle, partly because of its mass, plays major roles in the **overall metabolism** of the body.

MUSCLE TRANSDUCES CHEMICAL ENERGY INTO MECHANICAL ENERGY

Muscle is the major biochemical **transducer** (machine) that converts potential (chemical) energy into kinetic (mechanical) energy. Muscle, the largest single tissue in the human body, makes up somewhat less than 25% of body mass at birth, more than 40% in the young adult, and somewhat less than 30% in the aged adult. We shall discuss aspects of the three types of muscles found in vertebrates: **skeletal, cardiac,** and **smooth.** Both skeletal and cardiac muscles appear **striated** upon microscopic observation; smooth muscle is **nonstriated.** Although skeletal muscle is under voluntary nervous control, the control of both cardiac and smooth muscle is involuntary.

Sarcoplasm of Muscle Cells Contains ATP, Phosphocreatine, & Glycolytic Enzymes

Striated muscle is composed of multinucleated muscle fiber cells surrounded by an electrically excitable plasma membrane, the **sarcolemma.** An individual muscle fiber cell, which may extend the entire length of the muscle, contains a bundle of many **myofibrils** arranged in parallel, embedded

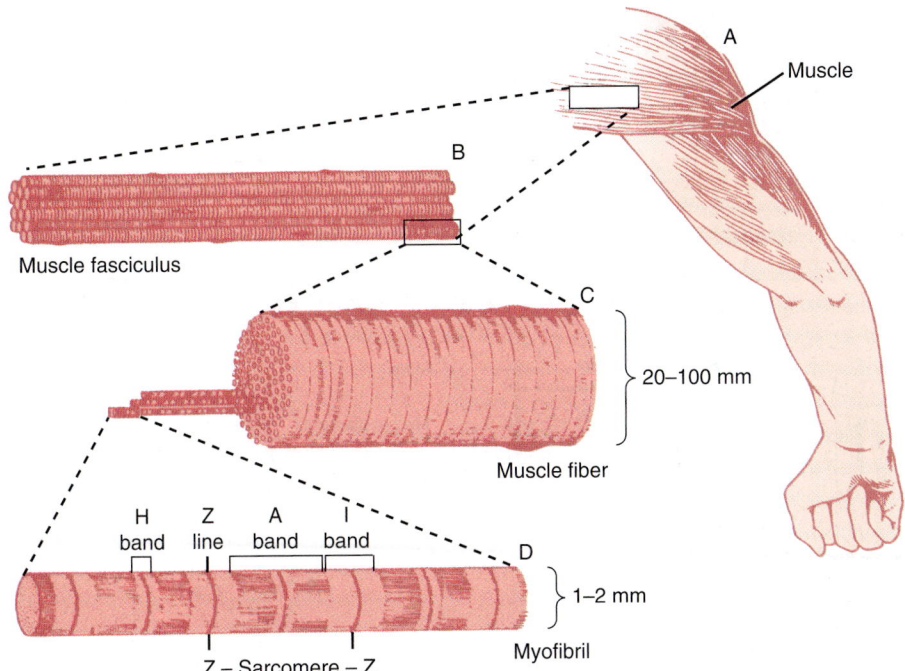

A — Muscle

Muscle fasciculus

B

C

20–100 mm

Muscle fiber

H band   Z line   A band   I band   D

1–2 mm

Myofibril

Z – Sarcomere – Z

FIGURE 49–1 **The structure of voluntary muscle.** The sarcomere is the region between the Z lines. (Drawing by Sylvia Colard Keene. Reproduced, with permission, from Bloom W, Fawcett DW: *A Textbook of Histology,* 10th ed. Saunders, 1975.)

in intracellular fluid termed **sarcoplasm.** Within this fluid is contained glycogen, the high-energy compounds ATP and phosphocreatine, and the enzymes of glycolysis.

Sarcomere Is the Functional Unit of Muscle

An overall view of voluntary muscle at several levels of organization is presented in **Figure 49–1**.

When the **myofibril** is examined by electron microscopy, alternating dark and light bands (anisotropic bands, meaning birefringent in polarized light, and isotropic bands, meaning not altered by polarized light) can be observed. These bands are thus referred to as **A and I bands,** respectively. The central region of the A band (the H band) appears less dense than the rest of the band. The I band is bisected by a very dense and narrow **Z line** (**Figure 49–2**).

The **sarcomere** is defined as the region between two Z lines (Figures 49–1 and 49–2) and is repeated along the axis of a fibril at distances of 1500–2300 nm depending upon the state of contraction.

The **striated** appearance of voluntary and cardiac muscle in light microscopic studies results from their high degree of organization, in which most muscle fiber cells are aligned so that their sarcomeres are in parallel register (Figure 49–1).

Thick Filaments Contain Myosin; Thin Filaments Contain Actin, Tropomyosin & Troponin

When **myofibrils** are examined by electron microscopy, it appears that each one is constructed of two types of

longitudinal filaments. One type, the **thick filament,** confined to the A band, contains chiefly the protein myosin. These filaments are about 16 nm in diameter and arranged in the cross-section as a hexagonal array (Figure 49–2, center; right-hand cross-section).

The **thin filament** (about 7 nm in diameter) lies in the I band and extends into the A band but not into its H zone (Figure 49–2). Thin filaments contain the proteins actin, tropomyosin, and troponin (**Figure 49–3**). In the A band, the thin filaments are arranged around the thick (myosin) filament as a secondary hexagonal array. Each thin filament lies symmetrically between three thick filaments (Figure 49–2, center, mid cross section), and each thick filament is surrounded symmetrically by six thin filaments.

The thick and thin filaments interact via **cross bridges** that emerge at intervals of 14 nm along the thick filaments. As depicted in Figure 49–2, the cross bridges (drawn as arrowheads at each end of the myosin filaments, but not shown extending fully across to the thin filaments) have opposite polarities at the two ends of the thick filaments. The two poles of the thick filaments are separated by a 150-nm segment (the M band, not labeled in the figure) that is free of projections.

The Sliding Filament Cross-Bridge Model Is the Foundation on Which Current Thinking About Muscle Contraction Is Built

This model was proposed independently in the 1950s by Henry Huxley and Andrew Huxley and their colleagues. It was largely based on careful morphologic observations on

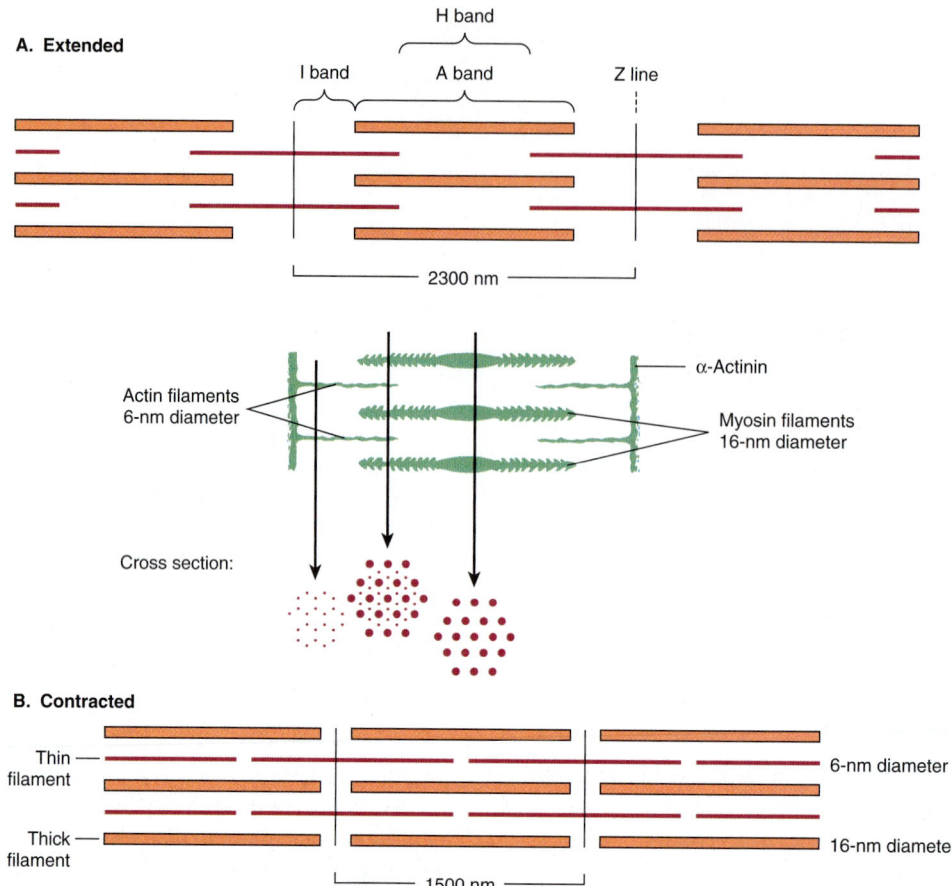

FIGURE 49–2   Arrangement of filaments in striated muscle. (**A**) Extended. The positions of the I, A, and H bands in the extended state are shown. The thin filaments partly overlap the ends of the thick filaments, and the thin filaments are shown anchored in the Z lines (often called Z disks). In the lower part of Figure 49–2(A), "arrowheads," pointing in opposite directions, are shown emanating from the myosin (thick) filaments. Four actin (thin) filaments are shown attached to two Z lines via α-actinin. The central region of the three myosin filaments, free of arrowheads, is called the M band (not labeled). Cross sections through the M bands, through an area where myosin and actin filaments overlap and through an area in which solely actin filaments are present, are shown. (**B**) Contracted. The actin filaments are seen to have slipped along the sides of the myosin fibers toward each other. The lengths of the thick filaments (indicated by the A bands) and the thin filaments (distance between Z lines and the adjacent edges of the H bands) have not changed. However, the lengths of the sarcomeres have been reduced (from 2300 to 1500 nm), and the lengths of the H and I bands are also reduced because of the overlap between the thick and thin filaments. These morphologic observations provided part of the basis for the sliding filament model of muscle contraction.

resting, extended, and contracting muscle. Basically, when muscle contracts, there is no change in the lengths of the thick and thin filaments, but the H zones and the I bands shorten (see legend to Figure 49–2). Thus, the arrays of inter-digitating filaments must **slide past one another** during con-traction. **Cross-bridges** that link thick and thin filaments at certain stages in the contraction cycle generate and sustain the tension. The tension developed during muscle contrac-tion is proportionate to the filament overlap and to the num-ber of cross bridges. Each cross-bridge head is connected to the thick filament via a flexible fibrous segment that can bend outward from the thick filament. This flexible segment facili-tates contact of the head with the thin filament when neces-sary but is also sufficiently pliant to be accommodated in the interfilament spacing.

ACTIN & MYOSIN ARE THE MAJOR PROTEINS OF MUSCLE

The mass of a muscle is made up of 75% water and more than 20% protein. The two major proteins are actin and myosin.

Monomeric **G-actin** (43 kDa; G, globular) makes up 25% of muscle protein by weight. At physiologic ionic strength and in the presence of Mg^{2+}, G-actin polymerizes noncovalently to form an insoluble double helical filament called F-actin (Figure 49–3). The **F-actin** fiber is 6–7 nm thick and has a pitch or repeating structure every 35.5 nm.

Myosins constitute a family of proteins, with at least 12 classes having been identified in the human genome. The myosin discussed in this chapter is **myosin-II,** and when myosin is referred to in this text, it is this species that is

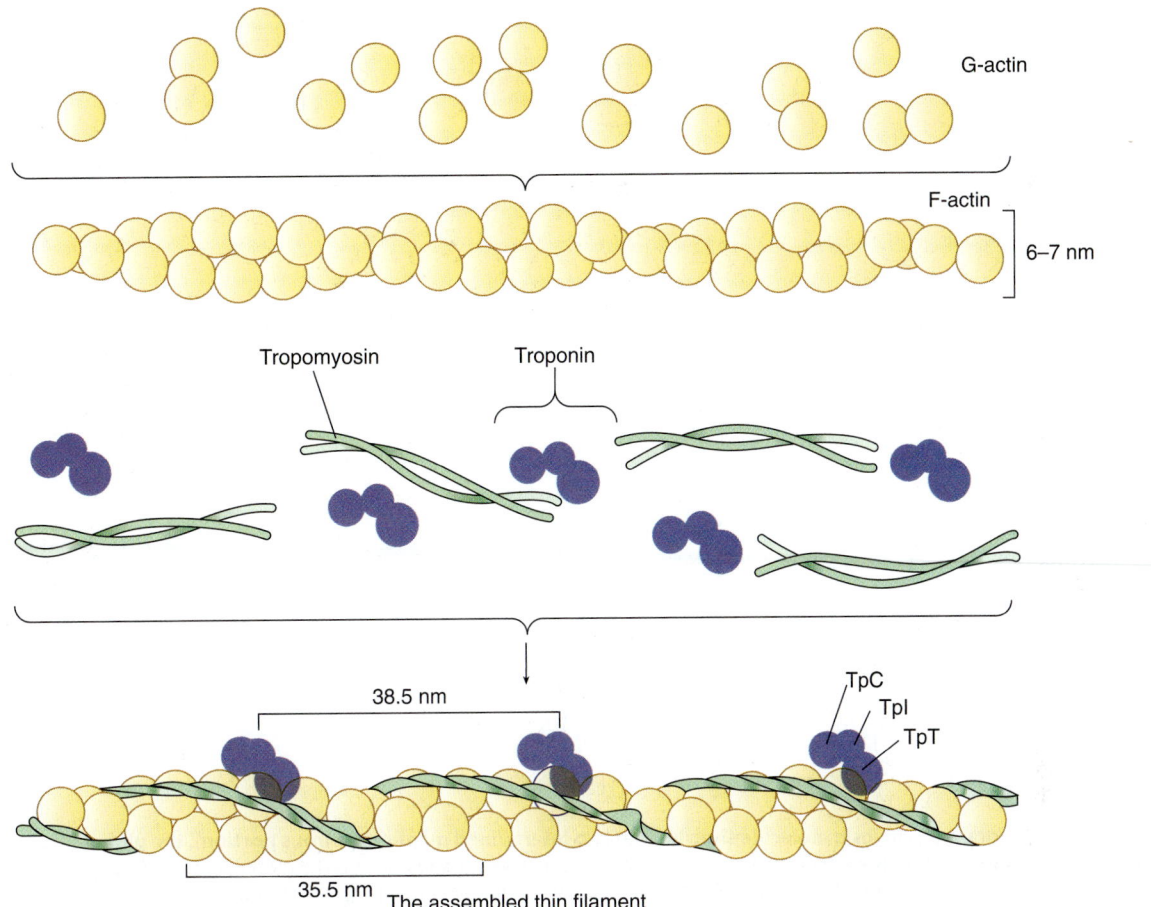

FIGURE 49–3 **Schematic representation of the thin filament, showing the spatial configuration of its three major protein components: actin, myosin, and tropomyosin.** The upper panel shows individual molecules of G-actin. The middle panel shows actin monomers assembled into F-actin. Individual molecules of tropomyosin (two strands wound around one another) and of troponin (made up of its three subunits) are also shown. The lower panel shows the assembled thin filament, consisting of F-actin, tropomyosin, and the three subunits of troponin (TpC, TpI, and TpT).

meant unless otherwise indicated. Myosin-I is a monomeric species that binds to cell membranes. It may serve as a linkage between microfilaments and the cell membrane in certain locations.

Myosin contributes 55% of muscle protein by weight and forms the **thick filaments.** It is an asymmetric hexamer with a molecular mass of approximately 460 kDa. Myosin has a fibrous tail consisting of two intertwined helices. Each helix has a globular head portion attached at one end (**Figure 49–4**). The hexamer consists of one pair of **heavy (H) chains** each of approximately 200 kDA molecular mass, and two pairs of **light (L) chains** each with a molecular mass of approximately 20 kDa. The L chains differ, one being called the **essential** light chain and the other the **regulatory** light chain. Skeletal muscle myosin binds actin to form **actomyosin** (actin–myosin), and its intrinsic ATPase activity is markedly enhanced in this complex. Isoforms of myosin exist whose amounts can vary in different anatomic, physiologic, and pathologic situations.

The structures of actin and of the head of myosin have been determined by X-ray crystallography; these studies have confirmed a number of earlier findings concerning their structures and have also given rise to much new information.

Limited Digestion of Myosin with Proteases Has Helped to Elucidate Its Structure & Function

When myosin is digested with **trypsin,** two myosin fragments (meromyosins) are generated. **Light meromyosin** (LMM) consists of aggregated, insoluble α-helical fibers from the tail of myosin (Figure 49–4). LMM exhibits no ATPase activity and does not bind to F-actin.

Heavy meromyosin (HMM; molecular mass about 340 kDa) is a soluble protein that has both a fibrous portion and a globular portion (Figure 49–4). It exhibits **ATPase** activity and binds to F-actin. Digestion of HMM with **papain** generates two subfragments, S-1 and S-2. The S-2 fragment is fibrous in character, has no ATPase activity, and does not bind to F-actin.

S-1 (molecular mass approximately 115 kDa) does exhibit ATPase activity, binds L chains, and in the absence of ATP will

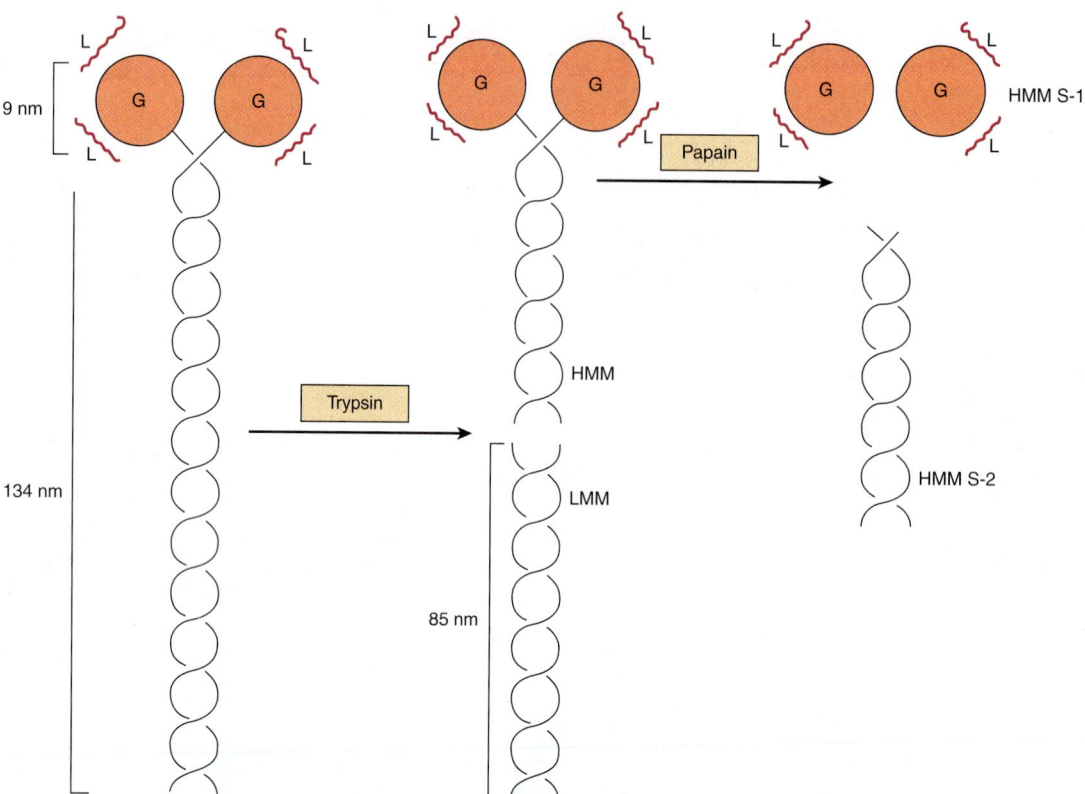

FIGURE 49–4 **Diagram of a myosin molecule showing the two intertwined α-helices (fibrous portion), the globular region or head (G), the light chains (L), and the effects of proteolytic cleavage by trypsin and papain.** The globular region (myosin head) contains an actin-binding site and an L chain-binding site and also attaches to the remainder of the myosin molecule.

bind to and decorate actin with "arrowheads" (**Figure 49–5**). Both S-1 and HMM exhibit **ATPase** activity, which is accelerated 100- to 200-fold by complexing with F-actin. As discussed below, F-actin greatly enhances the rate at which myosin ATPase releases its products, ADP and P_i. Thus, although **F-actin** does not affect the hydrolysis step per se, its ability

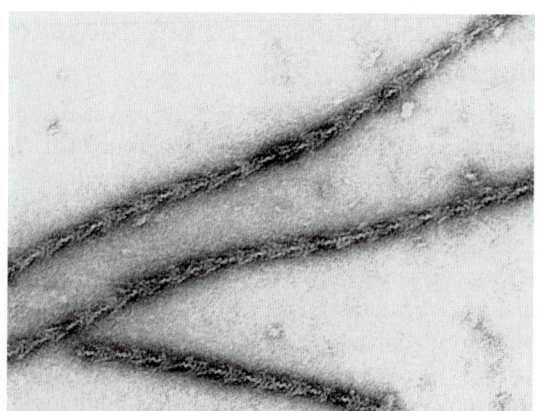

FIGURE 49–5 **The decoration of actin filaments with the S-1 fragments of myosin to form "arrowheads."** (Courtesy of JA Spudich.)

to **promote release** of the products produced by the ATPase activity greatly accelerates the overall rate of catalysis.

CHANGES IN THE CONFORMATION OF THE HEAD OF MYOSIN DRIVE MUSCLE CONTRACTION

How can hydrolysis of ATP produce macroscopic movement? Muscle contraction essentially consists of the cyclic **attachment and detachment** of the S-1 head of myosin to the F-actin filaments. This process can also be referred to as the making and breaking of cross-bridges. The attachment of actin to myosin is followed by **conformational changes** that are of particular importance in the S-1 head and are dependent upon which nucleotide is present (ADP or ATP). These changes result in the **power stroke**, which drives movement of actin filaments past myosin filaments. The energy for the power stroke is ultimately supplied by **ATP,** which is hydrolyzed to ADP and P_i. However, the power stroke itself occurs as a result of **conformational changes** in the myosin head when **ADP** leaves it.

The major biochemical events occurring during one cycle of muscle contraction and relaxation can be

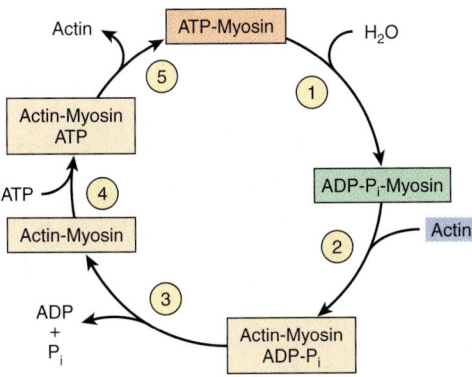

FIGURE 49–6 The hydrolysis of ATP drives the cyclic association and dissociation of actin and myosin in five reactions described in the text. (Modified, with permission, from Stryer L: *Biochemistry*, 2nd ed. Freeman, 1981. Copyright © 1981 by W. H. Freeman and Company.)

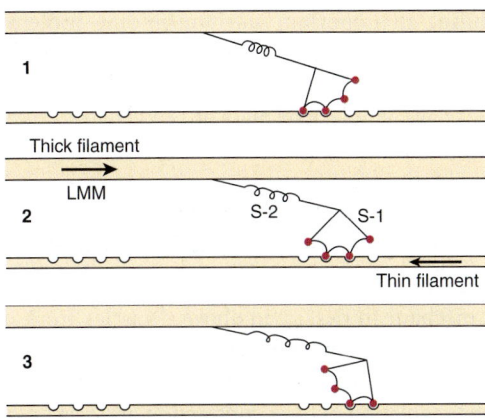

FIGURE 49–7 Representation of the active cross bridges between thick and thin filaments. This diagram was adapted by AF Huxley from HE Huxley: the mechanism of muscular contraction. Science 1969;164:1356. The latter proposed that the force involved in muscular contraction originates in a tendency for the myosin head (S-1) to rotate relative to the thin filament and is transmitted to the thick filament by the S-2 portion of the myosin molecule acting as an inextensible link. Flexible points at each end of S-2 permit S-1 to rotate and allow for variations in the separation between filaments. This figure is based on HE Huxley's proposal, but also incorporates elastic (the coils in the S-2 portion) and stepwise-shortening elements (depicted here as four sites of interaction between the S-1 portion and the thin filament). (See Huxley AF, Simmons RM: Proposed mechanism of force generation in striated muscle. Nature [Lond] 1971;233:533.) The strengths of binding of the attached sites are higher in position 2 than in position 1 and higher in position 3 than position 2. The myosin head can be detached from position 3 with the utilization of a molecule of ATP; this is the predominant process during shortening. The myosin head is seen to vary in its position from about 90° to about 45°, as indicated in the text. (S-1, myosin head; S-2, portion of the myosin molecule; LMM) (see legend to Figure 49–4). (Reproduced from Huxley AF: Muscular contraction. J Physiol 1974;243:1. By kind permission of the author and the *Journal of Physiology*.)

represented in the five steps shown in **Figure 49–6** as follows:

1. In the **relaxation phase** of muscle contraction, the S-1 head of myosin hydrolyzes ATP to ADP and P_i, but these products remain bound. The resultant ADP-P_i-myosin complex has been energized and is in a so-called high-energy conformation.

2. When **contraction** of muscle is stimulated (via events involving Ca^{2+}, troponin, tropomyosin, and actin, which are described below), actin becomes accessible and the S-1 head of myosin finds it, binds it, and forms the actin-myosin-ADP-P_i complex indicated.

3. Formation of this complex **promotes the release of P_i**, which initiates the power stroke. This is followed by release of ADP and is accompanied by a large conformational change in the head of myosin in relation to its tail (**Figure 49–7**), pulling actin about 10 nm toward the center of the sarcomere. This is the **power stroke**. The myosin is now in a so-called low-energy state, indicated as actin–myosin.

4. Another molecule of ATP binds to the S-1 head, forming an actin-myosin-ATP complex.

5. Myosin-ATP has a low affinity for actin, and **actin is thus released.** This last step is a key component of relaxation and is dependent upon the binding of ATP to the actin–myosin complex.

Another cycle then commences with the hydrolysis of ATP (step 1 of Figure 49–6), re-forming the high-energy conformation.

Thus, hydrolysis of ATP is used to drive the cycle, with the actual power stroke being the conformational change in the S-1 head that occurs upon the release of ADP. The **hinge regions** of myosin (referred to as flexible points at each end of S-2 in the legend to Figure 49–7) permit the large range of movement of S-1 and also allow S-1 to find actin filaments.

If intracellular **levels of ATP drop** (eg, after death), ATP is not available to bind the S-1 head (step 4 above), **actin does not dissociate,** and relaxation (step 5) does not occur. This is the explanation for **rigor mortis,** the stiffening of the body that occurs after death.

Calculations have indicated that the **efficiency** of contraction is about 50%; that of the internal combustion engine is less than 20%.

Tropomyosin & the Troponin Complex Present in Thin Filaments Perform Key Functions in Striated Muscle

In striated muscle, there are two other proteins that are minor in terms of their mass but important in terms of their function. **Tropomyosin** is a fibrous molecule that consists of two chains, alpha and beta, that attach to F-actin in the groove between its filaments (Figure 49–3). Tropomyosin is present in all muscular and muscle-like structures. The **troponin complex** is unique to striated muscle and consists of three polypeptides. **Troponin T** (TpT) binds to tropomyosin as well as to the other two troponin components. **Troponin I** (TpI) inhibits the F-actin–myosin interaction and also binds to the other components of troponin. **Troponin C** (TpC) is a calcium-binding polypeptide that is structurally and functionally analogous to

calmodulin, an important calcium-binding protein widely distributed in nature. Four molecules of calcium ion are bound per molecule of troponin C or calmodulin, and both molecules have a molecular mass of 17 kDa.

Ca²⁺ Plays a Central Role in Regulation of Muscle Contraction

The contraction of muscles from all sources occurs by the general mechanism described above. Muscles from different organisms and from different cells and tissues within the same organism may have different molecular mechanisms responsible for the regulation of their contraction and relaxation. In all systems, **Ca²⁺** plays a key regulatory role. There are two general mechanisms of **regulation** of muscle contraction: **actin-based** and **myosin-based.** The former operates in skeletal and cardiac muscles, the latter in smooth muscle.

Actin-Based Regulation Occurs in Striated Muscle

Actin-based regulation of muscle occurs in vertebrate skeletal and cardiac muscles, both striated. In the general mechanism described above (Figure 49–6), the only potentially limiting factor in the cycle of muscle contraction might be ATP. The skeletal muscle system is **inhibited** at rest; this inhibition is relieved to activate contraction. The inhibitor of striated muscle is the **troponin system,** which is bound to tropomyosin and F-actin in the thin filament (Figure 49–3). In striated muscle, there is no control of contraction unless the tropomyosin–troponin systems are present along with the actin and myosin filaments. As described above, **tropomyosin** lies along the groove of F-actin, and the three components of **troponin**—TpT, TpI, and TpC—are bound to the F-actin–tropomyosin complex. TpI prevents binding of the myosin head to its F-actin attachment site either by altering the conformation of F-actin via the tropomyosin molecules or by simply rolling tropomyosin into a position that directly blocks the sites on F-actin to which the myosin heads attach. Either way prevents activation of the myosin ATPase that is mediated by binding of the myosin head to F-actin. Hence, the TpI system blocks the contraction cycle at step 2 of Figure 49–6. This accounts for the **inhibited state** of relaxed striated muscle.

The Sarcoplasmic Reticulum Regulates Intracellular Levels of Ca²⁺ in Skeletal Muscle

In the sarcoplasm of resting muscle, the concentration of Ca²⁺ is 10^{-8} to 10^{-7} mol/L. The resting state is achieved because Ca²⁺ is pumped into the sarcoplasmic reticulum (SR) through the action of an active transport system, called the Ca²⁺ ATPase (**Figure 49–8**), initiating relaxation. The SR is a network of fine membranous sacs. Inside the SR, Ca²⁺ is bound to a specific Ca²⁺-binding protein-designated **calsequestrin.** The

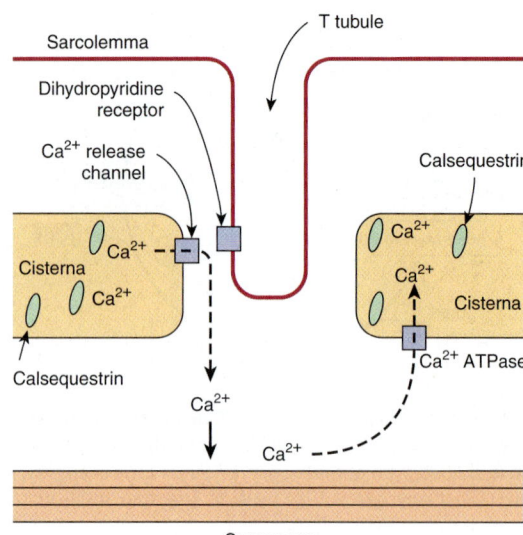

FIGURE 49–8 Diagram of the relationships among the sarcolemma (plasma membrane), a T tubule, and two cisternae of the SR of skeletal muscle (not to scale). The T tubule extends inward from the sarcolemma. A wave of depolarization, initiated by a nerve impulse, is transmitted from the sarcolemma down the T tubule. It is then conveyed to the Ca²⁺ release channel (RYR), perhaps by interaction between it and the dihydropyridine receptor (slow Ca²⁺ voltage channel), which are shown in close proximity. Release of Ca²⁺ from the Ca²⁺ release channel into the cytosol initiates contraction. Subsequently, Ca²⁺ is pumped back into the cisternae of the SR by the Ca²⁺ ATPase (Ca²⁺ pump) and stored there, in part bound to calsequestrin.

sarcomere is surrounded by an excitable membrane (the T-tubule system) composed of transverse (T) channels closely associated with the SR.

When the sarcolemma is excited by a **nerve impulse,** the signal is transmitted into the T-tubule system and a **Ca²⁺ release channel** in the nearby SR opens, releasing Ca²⁺ from the SR into the sarcoplasm. The concentration of Ca²⁺ in the sarcoplasm rises rapidly to 10^{-5} mol/L. The Ca²⁺-binding sites on TpC in the thin filament are quickly occupied by Ca²⁺. The TpC-4Ca²⁺ interacts with TpI and TpT to alter their interaction with tropomyosin. Accordingly, tropomyosin moves out of the way or alters the conformation of F-actin so that the myosin head–ADP-P_i (Figure 49–6) can interact with F-actin to start the contraction cycle.

The Ca²⁺ release channel is also known as the **ryanodine receptor** (RYR). There are two isoforms of this receptor, RYR1 and RYR2, the former being present in skeletal muscle and the latter in heart muscle and brain. **Ryanodine** is a plant alkaloid that binds to RYR1 and RYR2 specifically and modulates their activities. The Ca²⁺ release channel is a homotetramer made up of four subunits of kDa 565. It has transmembrane sequences at its carboxyl terminal, and these probably form the Ca²⁺ channel. The remainder of the protein protrudes into the cytosol, bridging the gap between the SR and the transverse tubular membrane. The channel is ligand-gated, Ca²⁺ and ATP working synergistically in vitro, although how it operates in vivo is not

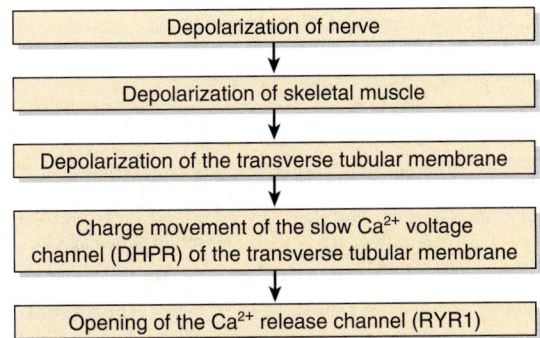

FIGURE 49–9 Possible chain of events leading to opening of the Ca²⁺ release channel. As indicated in the text, the Ca²⁺ voltage channel and the Ca²⁺ release channel have been shown to interact with each other in vitro via specific regions in their polypeptide chains. (DHPR, dihydropyridine receptor; RYR1, RYR 1.)

TABLE 49–1 Sequence of Events in Contraction and Relaxation of Skeletal Muscle

Steps in contraction

1. Discharge of motor neuron.
2. Release of transmitter (acetylcholine) at motor endplate.
3. Binding of acetylcholine to nicotinic acetylcholine receptors.
4. Increased Na⁺ and K⁺ conductance in endplate membrane.
5. Generation of endplate potential.
6. Generation of action potential in muscle fibers.
7. Inward spread of depolarization along T tubules.
8. Release of Ca²⁺ from terminal cisterns of sarcoplasmic reticulum and diffusion to thick and thin filaments.
9. Binding of Ca²⁺ to troponin C, uncovering myosin binding sites of actin.
10. Formation of cross-linkages between actin and myosin and sliding of thin on thick filaments, producing shortening.

Steps in relaxation

1. Ca²⁺ pumped back into sarcoplasmic reticulum.
2. Release of Ca²⁺ from troponin.
3. Cessation of interaction between actin and myosin.

Source: Reproduced, with permission, from Ganong WF: *Review of Medical Physiology,* 21st ed. McGraw-Hill, 2003.

clear. A possible sequence of events leading to opening of the channel is shown in **Figure 49–9**. The channel lies very close to the **dihydropyridine receptor** (DHPR), a voltage-dependent calcium channel of the transverse tubule system (Figure 49–8). Experiments in vitro employing an affinity column chromatography approach have indicated that a 37-amino-acid stretch in RYR1 interacts with one specific loop of DHPR.

Relaxation occurs when sarcoplasmic Ca^{2+} falls below 10^{-7} mol/L owing to its resequestration into the SR by Ca^{2+} ATPase. TpC-4Ca^{2+} thus loses its Ca^{2+}. Consequently, **troponin**, via interaction with tropomyosin, **inhibits** further myosin head and F-actin interaction, and in the presence of ATP the myosin head detaches from the F-actin.

Thus, Ca^{2+} controls skeletal muscle contraction and relaxation by an allosteric mechanism mediated by TpC, TpI, TpT, tropomyosin, and F-actin.

A **decrease** in the concentration of **ATP** in the sarcoplasm (eg, by excessive usage during the cycle of contraction–relaxation or by diminished formation, such as might occur in ischemia) has two major effects: (1) The **Ca²⁺ ATPase** (Ca²⁺ pump) in the SR ceases to maintain the low concentration of Ca^{2+} in the sarcoplasm. Thus, the interaction of the myosin heads with F-actin is promoted. (2) The ATP-dependent **detachment of myosin heads** from F-actin cannot occur, and rigidity (contracture) sets in. The condition of **rigor mortis,** following death, is an extension of these events.

Muscle contraction is a delicate dynamic balance of the attachment and detachment of myosin heads to F-actin, subject to fine regulation via the nervous system.

Table 49–1 summarizes the overall events in contraction and relaxation of skeletal muscle.

Mutations in the Gene Encoding the Ca²⁺ Release Channel Are One Cause of Human Malignant Hyperthermia

Some genetically predisposed patients experience a severe reaction, designated **malignant hyperthermia (MH),** on exposure to certain anesthetics (eg, halothane) and depolarizing skeletal muscle relaxants (eg, succinylcholine). The reaction consists primarily of rigidity of skeletal muscles, hypermetabolism, and high fever. A **high cytosolic concentration of Ca²⁺** in skeletal muscle is a major factor in its causation. Unless malignant hyperthermia is recognized and treated immediately, patients may die acutely of ventricular fibrillation or survive to succumb subsequently from other serious complications. Appropriate treatment is to stop the anesthetic and administer the drug **dantrolene** intravenously. Dantrolene is a skeletal muscle relaxant that acts to inhibit release of Ca²⁺ from the SR into the cytosol, thus preventing the increase of cytosolic Ca²⁺ found in MH.

MH also occurs in **swine.** Susceptible animals homozygous for MH respond to stress with a fatal reaction (**porcine stress syndrome**) similar to that exhibited by humans. If the reaction occurs prior to slaughter, it affects the quality of the pork adversely, resulting in an inferior product. Both events can result in considerable economic losses for the swine industry.

The finding of a high level of cytosolic Ca²⁺ in muscle in MH suggested that the condition might be caused by abnormalities of the Ca²⁺ ATPase or of the **Ca²⁺ release channel.** No abnormalities were detected in the former, but sequencing of cDNAs for the latter protein proved insightful, particularly in swine. All cDNAs from **swine** with MH so far examined have shown a substitution of T for C1843, resulting in the

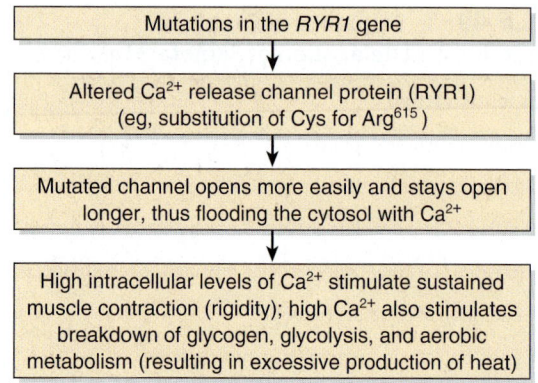

FIGURE 49–10 **Simplified scheme of the causation of malignant hyperthermia (OMIM 145600).** Many different point mutations have been detected in the *RYR1* gene, some of which are associated with central core disease (OMIM 117000). It is estimated that at least 50% of families with members who have malignant hyperthermia are linked to the *RYR1* gene. Some individuals with mutations in the gene encoding DHPR have also been detected; it is possible that mutations in other genes for proteins involved in certain aspects of muscle metabolism will also be found.

substitution of Cys for Arg[615] in the Ca^{2+} release channel. The mutation affects the function of the channel in that it opens more easily and remains open longer; the net result is massive release of Ca^{2+} into the cytosol, ultimately causing sustained muscle contraction.

The picture is more complex in **humans,** since MH exhibits **genetic heterogeneity.** Members of a number of families who suffer from malignant hyperthermia have not shown genetic linkage to the *RYR1* gene. Some humans susceptible to MH have been found to exhibit the same mutation found in swine, and others have a variety of point mutations at different loci in the *RYR1* gene. Certain families with MH have been found to have mutations affecting the **DHPR.** It is possible that mutations affecting other muscle proteins, such as **calsequestrin-1,** a SR Ca^{2+}-binding protein that modulates RyR1 function may also cause MH. **Figure 49–10** summarizes the probable chain of events in malignant hyperthermia. The major promise of these findings is that, once additional mutations are detected, it will be possible to **screen,** using suitable DNA probes, for individuals at risk of developing MH during anesthesia. Current screening tests (eg, the in vitro caffeine-halothane test) are relatively unreliable. Affected individuals could then be given **alternative anesthetics,** which would not endanger their lives. It should also be possible, if desired, to eliminate MH from swine populations using suitable breeding practices.

Another condition due to mutations in the *RYR1* gene is **central core disease.** This is a rare myopathy presenting in infancy with hypotonia and proximal muscle weakness. Electron microscopy reveals an absence of mitochondria in the center of many type I (see below) muscle fibers. Damage to mitochondria induced by high intracellular levels of Ca^{2+} secondary to abnormal functioning of *RYR1* appears to be responsible for the morphologic findings.

TABLE 49–2 Some Other Important Proteins of Muscle

Protein	Location	Comment or Function
Titin	Reaches from the Z line to the M line	Largest protein in body. Role in relaxation of muscle.
Nebulin	From Z line along length of actin filaments	May regulate assembly and length of actin filaments.
α-Actinin	Anchors actin to Z lines	Stabilizes actin filaments.
Desmin	Lies alongside actin filaments	Attaches to plasma membrane (plasmalemma).
Dystrophin	Attached to plasmalemma	Deficient in Duchenne muscular dystrophy. Mutations of its gene can also cause dilated cardiomyopathy.
Calcineurin	Cytosol	A calmodulin-regulated protein phosphatase. May play important roles in cardiac hypertrophy and in regulating amounts of slow and fast twitch muscles.
Myosin-binding protein C	Arranged transversely in sarcomere A-bands	Binds myosin and titin. Plays a role in maintaining the structural integrity of the sarcomere.

MUTATIONS IN THE GENE ENCODING DYSTROPHIN CAUSE DUCHENNE MUSCULAR DYSTROPHY

A number of **additional proteins** play various roles in the structure and function of muscle. They include titin (the largest protein known), nebulin, α-actinin, desmin, dystrophin, and calcineurin. Some properties of these proteins are summarized in **Table 49–2.**

Dystrophin is of special interest. As discussed in case no. 9 of Chapter 57, mutations in the gene encoding this protein have been shown to be the cause of **Duchenne muscular dystrophy** and the milder **Becker muscular dystrophy.** They are also implicated in some cases of **dilated cardiomyopathy** (see below). As shown in **Figure 49–11,** dystrophin forms a part of a large complex of proteins that attach to or interact with the plasmalemma. Dystrophin links the actin cytoskeleton to the ECM and appears to be needed for assembly of the synaptic junction. Impairment of these processes by formation of defective dystrophin is thought to be critical in the causation of Duchenne muscular dystrophy. Mutations in the genes encoding some of the components of the **sarcoglycan complex** shown in Figure 49–11 are responsible for **limb-girdle** and certain **other congenital forms** of muscular dystrophy.

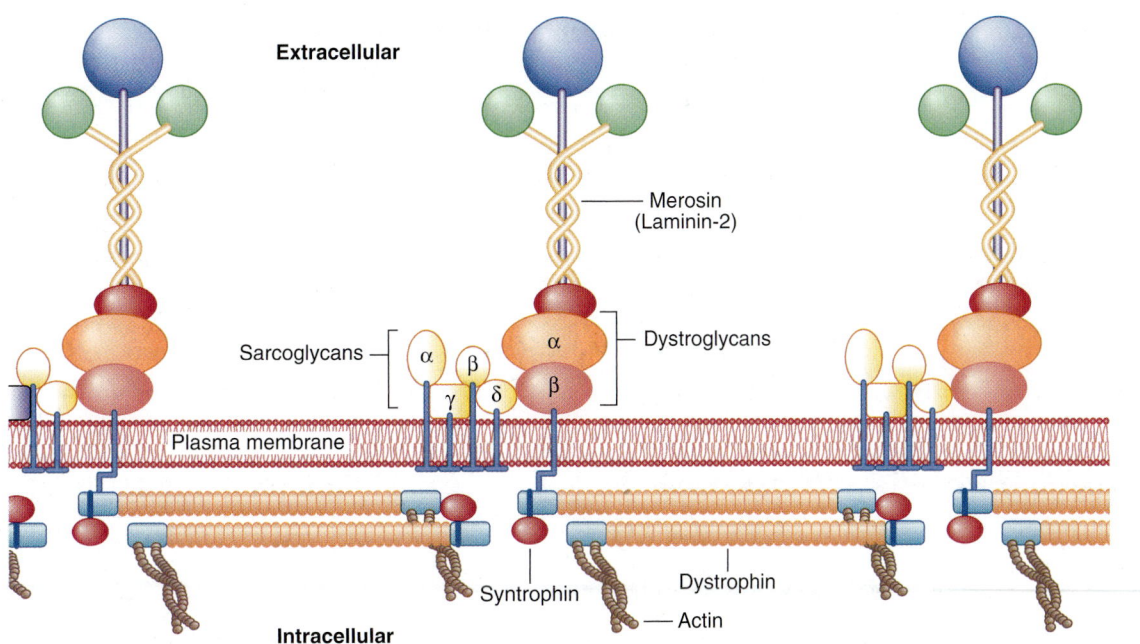

FIGURE 49–11 **Organization of dystrophin and other proteins in relation to the plasma membrane of muscle cells.** Dystrophin is part of a large oligomeric complex associated with several other protein complexes. The dystroglycan complex consists of α-dystroglycan, which associates with the basal lamina protein merosin (also named laminin-2, see Chapter 48), and α-dystroglycan, which binds α-dystroglycan and dystrophin. Syntrophin binds to the carboxyl terminal of dystrophin. The sarcoglycan complex consists of four transmembrane proteins: α-, β-, γ -, and δ-sarcoglycan. The function of the sarcoglycan complex and the nature of the interactions within the complex and between it and the other complexes are not clear. The sarcoglycan complex is formed only in striated muscle, and its subunits preferentially associate with each other, suggesting that the complex may function as a single unit. Mutations in the gene encoding dystrophin cause Duchenne and Becker muscular dystrophy. Mutations in the genes encoding the various sarcoglycans have been shown to be responsible for limb-girdle dystrophies (eg, OMIM 604286) and mutations in genes encoding other muscle proteins cause other types of muscular dystrophy. Mutations in genes encoding certain glycosyltransferases involved in the synthesis of the glycan chains of α-dystroglycan are responsible for certain congenital muscular dystrophies (see Chapter 47). (Reproduced, with permission, from Duggan DJ et al: Mutations in the sarcoglycan genes in patients with myopathy. N Engl J Med 1997;336:618. Copyright © 1997 Massachusetts Medical Society. All rights reserved.)

Mutations in genes encoding **several glycosyltransferases** involved in the synthesis of the sugar chains of **α-dystroglycan** have been found to be the cause of certain types of **congenital muscular dystrophy** (see Chapter 47).

CARDIAC MUSCLE RESEMBLES SKELETAL MUSCLE IN MANY RESPECTS

The general picture of muscle contraction in the heart resembles that of skeletal muscle. Cardiac muscle, like skeletal muscle, is **striated** and uses the actin–myosin–tropomyosintroponin system described above. Unlike skeletal muscle, cardiac muscle exhibits **intrinsic rhythmicity,** and individual myocytes communicate with each other because of its syncytial nature. The **T-tubular system** is more developed in cardiac muscle, whereas the **SR** is less extensive and consequently the intracellular supply of Ca^{2+} for contraction is less. Cardiac muscle thus relies on **extracellular** Ca^{2+} for contraction; if isolated cardiac muscle is deprived of Ca^{2+}, it ceases to beat within approximately 1 min, whereas skeletal muscle can continue to contract with-

out an extracellular source of Ca^{2+} for a longer period. **Cyclic AMP** plays a more prominent role in cardiac than in skeletal muscle. It modulates intracellular levels of Ca^{2+} through the activation of protein kinases; these enzymes phosphorylate various transport proteins in the sarcolemma and SR and also in the troponin–tropomyosin regulatory complex, affecting intracellular levels of Ca^{2+} or responses to it. There is a rough correlation between the phosphorylation of TpI and the increased contraction of cardiac muscle induced by catecholamines. This may account for the **inotropic effects** (increased contractility) of β-adrenergic compounds on the heart. Some differences among skeletal, cardiac, and smooth muscle are summarized in **Table 49–3.**

Ca²⁺ Enters Myocytes Via Ca²⁺ Channels & Leaves Via the Na⁺–Ca²⁺ Exchanger & the Ca²⁺ ATPase

As stated, **extracellular** Ca^{2+} plays an important role in contraction of cardiac muscle but not in skeletal muscle. This means that Ca^{2+} both enters and leaves myocytes in a regulated manner. We shall briefly consider three transmembrane proteins that play roles in this process.

TABLE 49–3 Some Differences Among Skeletal, Cardiac, and Smooth Muscle

Skeletal Muscle	Cardiac Muscle	Smooth Muscle
1. Striated	1. Striated	1. Nonstriated
2. No syncytium	2. Syncytial	2. Syncytial
3. Small T tubules	3. Large T tubules	3. Generally rudimentary T tubules.
4. Sarcoplasmic reticulum well developed and Ca^{2+} pump acts rapidly.	4. Sarcoplasmic reticulum present and Ca^{2+} pump acts relatively rapidly.	4. Sarcoplasmic reticulum often rudimentary and Ca^{2+} pump acts slowly.
5. Plasmalemma lacks many hormone receptors.	5. Plasmalemma contains a variety of receptors (eg, α- and β-adrenergic).	5. Plasmalemma contains a variety of receptors (eg, α- and β-adrenergic).
6. Nerve impulse initiates contraction.	6. Has intrinsic rhythmicity.	6. Contraction initiated by nerve impulses, hormones, etc.
7. Extracellular fluid Ca^{2+} not important for contraction.	7. Extracellular fluid Ca^{2+} important for contraction.	7. Extracellular fluid Ca^{2+} important for contraction.
8. Troponin system present.	8. Troponin system present.	8. Lacks troponin system; uses regulatory head of myosin.
9. Caldesmon not involved.	9. Caldesmon not involved.	9. Caldesmon is important regulatory protein.
10. Very rapid cycling of the cross-bridges.	10. Relatively rapid cycling of the cross-bridges.	10. Slow cycling of the cross-bridges permits slow, prolonged contraction and less utilization of ATP.

Ca^{2+} Channels

Ca^{2+} enters myocytes via these channels, which allow entry only of Ca^{2+} ions. The major portal of entry is the L-type (long-duration current, large conductance) or **slow Ca^{2+}** channel, which is voltage-gated, opening during depolarization induced by spread of the cardiac action potential and closing when the action potential declines. These channels are equivalent to the dihydropyridine receptors of skeletal muscle (Figure 49–8). Slow Ca^{2+} channels are **regulated** by cAMP-dependent protein kinases (stimulatory) and cGMP-protein kinases (inhibitory) and are blocked by so-called calcium channel blockers (eg, verapamil). **Fast** (or T, transient) Ca^{2+} channels are also present in the plasmalemma, though in much lower numbers; they probably contribute to the early phase of increase of myoplasmic Ca^{2+}.

The resultant increase of Ca^{2+} in the myoplasm acts on the Ca^{2+} release channel of the SR to open it. This is called **Ca^{2+}-induced Ca^{2+} release** (CICR). It is estimated that approximately 10% of the Ca^{2+} involved in contraction enters the cytosol from the extracellular fluid and 90% from the SR. However, the former 10% is important, as the rate of increase of Ca^{2+} in the myoplasm is important, and entry via the Ca^{2+} channels contributes appreciably to this.

Ca^{2+}–Na$^+$ Exchanger

This is the principal route of **exit** of Ca^{2+} from myocytes. In resting myocytes, it helps to maintain a low level of free intracellular Ca^{2+} by exchanging one Ca^{2+} for three Na$^+$. The energy for the uphill movement of Ca^{2+} out of the cell comes from the downhill movement of Na$^+$ into the cell from the plasma. This exchange contributes to relaxation, but may run in the reverse direction during excitation. Because of the Ca^{2+}–Na$^+$ exchanger, anything that causes intracellular Na$^+$ (Na^+_i) to rise will secondarily cause Ca$^{2+}_i$ to rise, causing more forceful contraction. This is referred to as a **positive inotropic effect**. One example is when the drug **digitalis** is used to treat heart failure. Digitalis inhibits the sarcolemmal Na$^+$–K$^+$-ATPase, diminishing exit of Na$^+$ and thus increasing Na^+_i. This in turn causes Ca^{2+} to increase, via the Ca^{2+}–Na$^+$ exchanger. The increased Ca$^{2+}_i$ results in increased force of cardiac contraction (see **Figure 49–12**), of benefit in heart failure.

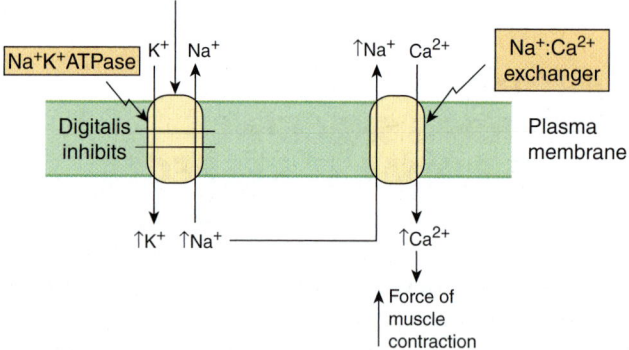

FIGURE 49–12 **Scheme of how the drug digitalis (used in the treatment of certain cases of heart failure) increases cardiac contraction.** Digitalis inhibits the Na$^+$K$^+$ ATPase (see Chapter 40). This results in less Na$^+$ being pumped out of the cardiac myocyte and leads to an increase of the intracellular concentration of Na$^+$. In turn, this stimulates the Na$^+$-Ca^{2+} exchanger so that more Na$^+$ is exchanged outward, and more Ca^{2+} enters the myocyte. The resulting increased intracellular concentration of Ca^{2+} increases the force of muscular contraction.

TABLE 49–4 Major Types of Ion Channels Found in Cells

Type	Comment
External ligand gated	Open in response to a specific extracellular molecule, for example, acetylcholine.
Internal ligand gated	Open or close in response to a specific intracellular molecule, for example, a cyclic nucleotide.
Voltage gated	Open in response to a change in membrane potential, for example, Na^+, K^+, and Ca^{2+} channels in heart.
Mechanically gated	Open in response to change in mechanical pressure.

Ca²⁺ ATPase

This Ca^{2+} pump, situated in the sarcolemma, also contributes to Ca^{2+} exit but is believed to play a relatively minor role as compared with the Ca^{2+}–Na^+ exchanger.

It should be noted that there are a variety of **ion channels** (Chapter 40) in most cells, for Na^+, K^+, Ca^{2+}, etc. Many of them have been cloned in recent years and their dispositions in their respective membranes worked out (number of times each one crosses its membrane, location of the actual ion transport site in the protein, etc). They can be classified as indicated in **Table 49–4.** Cardiac muscle is rich in ion channels, and they are also important in skeletal muscle. Mutations in genes encoding ion channels have been shown to be responsible for a number of relatively rare conditions affecting muscle. These and other diseases due to mutations of ion channels have been termed **channelopathies;** some are listed in **Table 49–5.**

Inherited Cardiomyopathies Are Due to Disorders of Cardiac Energy Metabolism or to Abnormal Myocardial Proteins

An **inherited cardiomyopathy** is any structural or functional abnormality of the ventricular myocardium due to an inherited cause. There are nonheritable types of cardiomyopathy, but these will not be described here. As shown in **Table 49–6,** the causes of inherited cardiomyopathies fall into two broad classes: (1) disorders of **cardiac energy metabolism,** mainly reflecting mutations in genes encoding enzymes or proteins involved in fatty acid oxidation (a major source of energy for the myocardium) and oxidative phosphorylation; (2) mutations in genes encoding proteins involved in or **affecting myocardial contraction,** such as myosin, tropomyosin, the troponins, and cardiac myosin-binding protein C. Mutations in the genes encoding these latter proteins cause familial hypertrophic cardiomyopathy, which will now be discussed.

TABLE 49–5 Some Disorders (Channelopathies) Due to Mutations in Genes Encoding Polypeptide Constituents of Ion Channels

Disorder[1]	Ion Channel and Major Organs Involved
Central core disease (OMIM 117000)	Ca^{2+} release channel (RYR1) Skeletal muscle
Hyperkalemic periodic paralysis (OMIM 170500)	Sodium channel Skeletal muscle
Hypokalemic periodic paralysis (OMIM 170400)	Slow Ca^{2+} voltage channel (DHPR) Skeletal muscle
Malignant hyperthermia (OMIM 145600)	Ca^{2+} release channel (RYR1) Skeletal muscle
Myotonia congenita (OMIM 160800)	Chloride channel Skeletal muscle

Source: Data in part from Ackerman NJ, Clapham DE: Ion channels—basic science and clinical disease. N Engl J Med 1997;336:1575.
[1]Other channelopathies include the long QT syndrome (OMIM 192500); pseudoaldosteronism (Liddle syndrome, OMIM 177200); persistent hyperinsulinemic hypoglycemia of infancy (OMIM 601820); hereditary X-linked recessive type II nephrolithiasis of infancy (Dent syndrome, OMIM 300009); and generalized myotonia, recessive (Becker disease, OMIM 255700). The term "myotonia" signifies any condition in which muscles do not relax after contraction.

Mutations in the Cardiac β-Myosin Heavy Chain Gene Are One Cause of Familial Hypertrophic Cardiomyopathy

Familial hypertrophic cardiomyopathy is one of the most frequent hereditary cardiac diseases. Patients exhibit hypertrophy—often massive—of one or both ventricles, starting early in life, and not related to any extrinsic cause such as hypertension. Most cases are transmitted in an autosomal dominant manner; the rest are sporadic. Until recently, its cause was

TABLE 49–6 Biochemical Causes of Inherited Cardiomyopathies[1]

Cause	Proteins or Process Affected
Inborn errors of fatty acid oxidation	Carnitine entry into cells and mitochondria
	Certain enzymes of fatty acid oxidation
Disorders of mitochondrial oxidative phosphorylation	Proteins encoded by mitochondrial genes
	Proteins encoded by nuclear genes
Abnormalities of myocardial contractile and structural proteins	β-Myosin heavy chains, troponin, tropomyosin, dystrophin

Source: Based on Kelly DP, Strauss AW: Inherited cardiomyopathies. N Engl J Med 1994;330:913.
[1]Mutations (eg, point mutations, or in some cases deletions) in the genes (nuclear or mitochondrial) encoding various proteins, enzymes, or tRNA molecules are the fundamental causes of the inherited cardiomyopathies. Some conditions are mild, whereas others are severe and may be part of a syndrome affecting other tissues.

obscure. However, this situation changed when studies of one affected family showed that a **missense mutation** (ie, substitution of one amino acid by another) in the **β-myosin heavy chain gene** was responsible for the condition. Subsequent studies have shown a number of missense mutations in this gene, all coding for highly conserved residues. Some individuals have shown other mutations, such as formation of an α/β-myosin heavy chain hybrid gene. Patients with familial hypertrophic cardiomyopathy can show great variation in clinical picture. This in part reflects **genetic heterogeneity;** that is, mutation in a number of **other genes** (eg, those encoding cardiac actin, tropomyosin, cardiac troponins I and T, essential and regulatory myosin light chains, cardiac myosin-binding protein C, titin, and mitochondrial tRNA-glycine and tRNA-isoleucine) may also cause familial hypertrophic cardiomyopathy. In addition, mutations at different sites in the gene for β-myosin heavy chain may affect the function of the protein to a greater or lesser extent. The missense mutations are clustered in the head and head-rod regions of the myosin heavy chain. One hypothesis is that the mutant polypeptides ("poison polypeptides") cause formation of abnormal myofibrils, eventually resulting in compensatory hypertrophy. Some mutations alter the **charge** of the amino acid (eg, substitution of arginine for glutamine), presumably affecting the **conformation** of the protein more markedly and thus affecting its function. Patients with these mutations have a significantly shorter life expectancy than patients in whom the mutation produced no alteration in charge. Thus, definition of the precise mutations involved in the genesis of FHC may prove to be of the important prognostic value; it can be accomplished by appropriate use of the polymerase chain reaction on genomic DNA obtained from one sample of blood lymphocytes. **Figure 49–13** is a simplified scheme of the events causing familial hypertrophic cardiomyopathy.

Another type of cardiomyopathy is termed **dilated cardiomyopathy.** Mutations in the genes encoding dystrophin, muscle LIM protein (so-called because it was found to contain a cysteine-rich domain originally detected in three proteins: Lin-II, Isl-1, and Mec-3), the cyclic response-element binding protein (CREB), desmin, and lamin have been implicated in the causation of this condition. The first two proteins help organize the contractile apparatus of cardiac muscle cells, and CREB is involved in the regulation of a number of genes in these cells. Current research is not only elucidating the molecular causes of the cardiomyopathies but is also disclosing mutations that cause **cardiac developmental disorders** (eg, septal defects) and **arrhythmias** (eg, due to mutations affecting ion channels).

Ca²⁺ Also Regulates Contraction of Smooth Muscle

While all muscles contain actin, myosin, and tropomyosin, only vertebrate **striated** muscles contain the **troponin system.** Thus, the mechanisms that regulate contraction must differ in various contractile systems.

Smooth muscles have molecular structures similar to those in striated muscle, but the sarcomeres are not aligned so as to generate the striated appearance. Smooth muscles contain α-actinin and tropomyosin molecules, as do skeletal muscles. They **do not have the troponin system,** and the light chains of smooth muscle myosin molecules differ from those of striated muscle myosin. Regulation of smooth muscle contraction is **myosin-based,** unlike striated muscle, which is actin-based. However, like striated muscle, smooth muscle contraction is **regulated by Ca²⁺.**

Phosphorylation of Myosin Light Chains Initiates Contraction of Smooth Muscle

When smooth muscle myosin is bound to F-actin in the absence of other muscle proteins such as tropomyosin, there is **no detectable ATPase activity.** This absence of activity is quite unlike the situation described for striated muscle myosin and F-actin, which has abundant ATPase activity. Smooth muscle myosin contains **light chains** that prevent the binding of the myosin head to F-actin; they **must be phosphorylated** before they allow F-actin to activate myosin ATPase. The ATPase activity then attained hydrolyzes ATP about 10-fold more slowly than the corresponding activity in skeletal muscle. The phosphate on the myosin light chains may form a chelate with the Ca²⁺ bound to the tropomyosin-TpC-actin complex, leading to an increased rate of formation of cross bridges between the myosin heads and actin. The phosphorylation of light chains **initiates** the attachment–detachment contraction cycle of smooth muscle.

Myosin Light Chain Kinase Is Activated by Calmodulin-4Ca²⁺ & Then Phosphorylates the Light Chains

Smooth muscle sarcoplasm contains a **myosin light chain kinase** that is calcium dependent. The Ca²⁺ activation of myosin light chain kinase requires binding of **calmodulin-4Ca²⁺** to its kinase subunit (**Figure 49–14**). The calmodulin-4Ca²⁺-activated light chain kinase phosphorylates the light chains,

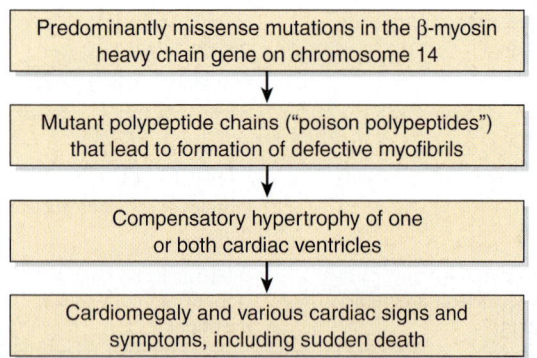

FIGURE 49–13 Simplified scheme of the causation of familial hypertrophic cardiomyopathy (OMIM 192600) due to mutations in the gene encoding β-myosin heavy chain. Mutations in genes encoding other proteins (see text) can also cause this condition.

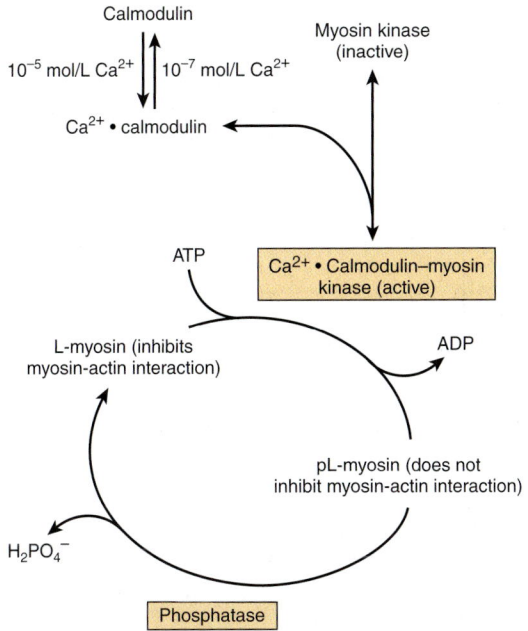

FIGURE 49–14 **Regulation of smooth muscle contraction by Ca²⁺.** The pL-myosin is the phosphorylated light chain of myosin and L-myosin is the dephosphorylated light chain. (Adapted, with permission, from Adelstein RS, Eisenberg R: Regulation and kinetics of actin–myosin ATP interaction. Annu Rev Biochem 1980;49:921. Copyright © 1980 by Annual Reviews, www.annualreviews.org.)

which then ceases to inhibit the myosin–F-actin interaction. The contraction cycle then begins.

Another non-Ca²⁺-dependent pathway exists in smooth muscle for initiating contraction. This involves **Rho kinase,** which is activated by a variety of stimuli (not shown in Figure 49–14). This enzyme phosphorylates myosin light chain phosphatase, inhibiting it, and thus increasing the phosphorylation of the light chain. Rho kinase also directly phosphorylates the light chain of myosin. Both of these actions increase the contraction of smooth muscle.

Smooth Muscle Relaxes When the Concentration of Ca²⁺ Falls Below 10⁻⁷ Molar

Relaxation of smooth muscle occurs when sarcoplasmic Ca²⁺ falls below 10⁻⁷ mol/L. The **Ca²⁺** dissociates from calmodulin, which in turn dissociates from the myosin light chain kinase, inactivating the kinase. No new phosphates are attached to the p-light chain, and **light chain protein phosphatase,** which is continually active and calcium independent, removes the existing phosphates from the light chains. Dephosphorylated myosin p-light chain then inhibits the binding of myosin heads to F-actin and the ATPase activity. The myosin head detaches from the F-actin in the presence of ATP, but it cannot reattach because of the presence of dephosphorylated p-light chain; hence, **relaxation** occurs.

Table 49–7 summarizes and compares the regulation of actin–myosin interactions (activation of myosin ATPase) in striated and smooth muscles.

The myosin light chain kinase is not directly affected or activated by **cAMP.** However, cAMP-activated protein kinase can phosphorylate the myosin light chain kinase (not the light chains themselves). The phosphorylated myosin light chain kinase exhibits a significantly lower affinity for calmodulin-Ca²⁺ and thus is less sensitive to activation. Accordingly, an **increase in cAMP dampens the contraction response** of smooth muscle to a given elevation of sarcoplasmic Ca²⁺. This molecular mechanism can explain the relaxing effect of β-adrenergic stimulation on smooth muscle.

Another protein that appears to play a Ca²⁺-dependent role in the regulation of smooth muscle contraction is **caldesmon** (87 kDa). This protein is ubiquitous in smooth muscle

TABLE 49–7 Actin-Myosin Interactions in Striated and Smooth Muscle

	Striated Muscle	Smooth Muscle (and Nonmuscle Cells)
Proteins of muscle filaments	Actin Myosin Tropomyosin Troponin (TpI, TpT, TpC)	Actin Myosin[1] Tropomyosin
Spontaneous interaction of F-actin and myosin alone (spontaneous activation of myosin ATPase by F-actin)	Yes	No
Inhibitor of F-actin–myosin interaction (inhibitor of F-actin-dependent activation of ATPase)	Troponin system (TpI)	Unphosphorylated myosin light chain
Contraction activated by	Ca²⁺	Ca²⁺
Direct effect of Ca²⁺	4Ca²⁺ bind to TpC	4Ca²⁺ bind to calmodulin
Effect of protein-bound Ca²⁺	TpC·4Ca²⁺ antagonizes TpI inhibition of F-actin–myosin interaction (allows F-actin activation of ATPase)	Calmodulin·4Ca²⁺ activates myosin light chain kinase that phosphorylates myosin p-light chain. The phosphorylated p-light chain no longer inhibits F-actin–myosin interaction (allows F-actin activation of ATPase)

[1]Light chains of myosin are different in striated and smooth muscles.

and is also found in nonmuscle tissue. At low concentrations of Ca^{2+}, it binds to tropomyosin and actin. This **prevents interaction of actin with myosin,** keeping muscle in a relaxed state. At higher concentrations of Ca^{2+}, Ca^{2+}-calmodulin binds caldesmon, **releasing it from actin.** The latter is then free to bind to myosin, and contraction can occur. Caldesmon is also subject to phosphorylation–dephosphorylation; when phosphorylated, it cannot bind actin, again freeing the latter to interact with myosin. Caldesmon may also participate in organizing the structure of the contractile apparatus in smooth muscle. Many of its effects have been demonstrated in vitro, and its physiologic significance is still under investigation.

As noted in Table 49–3, slow cycling of the cross-bridges permits slow prolonged contraction of smooth muscle (eg, in viscera and blood vessels) with less utilization of ATP compared with striated muscle. The ability of smooth muscle to maintain force at reduced velocities of contraction is referred to as the **latch state;** this is an important feature of smooth muscle, and its precise molecular bases are under study.

Nitric Oxide (NO) Relaxes the Smooth Muscle of Blood Vessels & Also Has Many Other Important Biologic Functions

Acetylcholine is a vasodilator that acts by causing relaxation of the smooth muscle of blood vessels. However, it does not act directly on smooth muscle. A key observation was that if **endothelial cells** were stripped away from underlying smooth muscle cells, acetylcholine no longer exerted its vasodilator effect. This finding indicated that vasodilators such as acetylcholine initially interact with the endothelial cells of small blood vessels via receptors. The receptors are coupled to the phosphoinositide cycle, leading to the intracellular release of Ca^{2+} through the action of inositol trisphosphate. In turn, the elevation of Ca^{2+} leads to the liberation of **endothelium-derived relaxing factor (EDRF),** which diffuses into the adjacent smooth muscle. There, it reacts with the heme moiety of a soluble guanylyl cyclase, resulting in activation of the latter, with a consequent elevation of intracellular levels of **cGMP (Figure 49–15).** This in turn stimulates the activities of certain cGMP-dependent protein kinases, which probably phosphorylate specific muscle proteins, causing relaxation; however, the details are still being clarified. The important coronary artery vasodilator **nitroglycerin,** widely used to relieve angina pectoris, acts to increase intracellular release of EDRF and thus of cGMP.

Quite unexpectedly, EDRF was found to be the gas **NO.** NO is formed by the action of the enzyme NO synthase, which is cytosolic. The endothelial and neuronal forms of NO synthase are activated by Ca^{2+} (**Table 49–8**). The substrate is **arginine,** and the products are citrulline and NO:

$$\text{Arginine} \xrightarrow{\text{NO SYNTHASE}} \text{Citrulline + NO}$$

NO synthase catalyzes a five-electron oxidation of an amidine nitrogen of arginine. L-hydroxyarginine is an

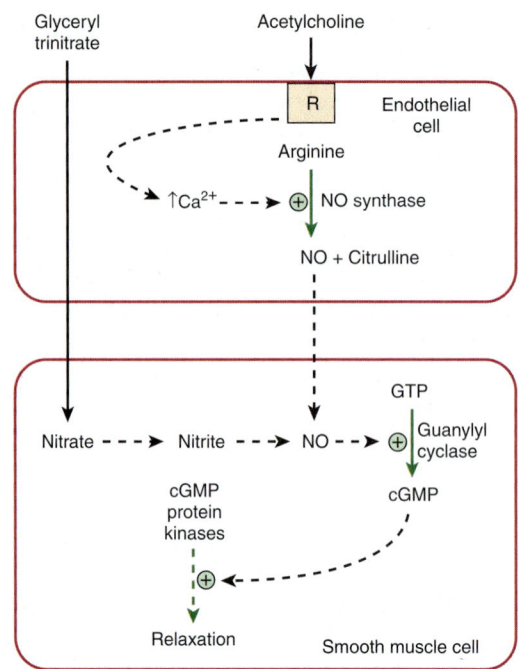

FIGURE 49–15 **Diagram showing formation in an endothelial cell of nitric oxide (NO) from arginine in a reaction catalyzed by NO synthase.** Interaction of an agonist (eg, acetylcholine) with a receptor (R) probably leads to intracellular release of Ca^{2+} via inositol trisphosphate generated by the phosphoinositide pathway, resulting in activation of NO synthase. The NO subsequently diffuses into adjacent smooth muscle, where it leads to activation of guanylyl cyclase, formation of cGMP, stimulation of cGMP protein kinases, and subsequent relaxation. The vasodilator nitroglycerin is shown entering the smooth muscle cell, where its metabolism also leads to formation of NO.

intermediate that remains tightly bound to the enzyme. NO synthase is a very complex enzyme, employing five redox cofactors: NADPH, FAD, FMN, heme, and tetrahydrobiopterin. NO can also be formed from **nitrite,** derived from vasodilators such as glyceryl trinitrate during their metabolism. NO has a very short half-life (approximately 3–4 seconds) in tissues because it reacts with oxygen and superoxide. The product of the reaction with superoxide is **peroxynitrite** ($ONOO^{-}$), which decomposes to form the highly reactive OH· radical. NO is inhibited by hemoglobin and other heme proteins, which bind it tightly. **Chemical inhibitors of NO synthase** are now available that can markedly decrease formation of NO. Administration of such inhibitors to animals and humans leads to vasoconstriction and a marked elevation of blood pressure, indicating that NO is of major importance in the maintenance of blood pressure in vivo. Another important cardiovascular effect is that by increasing synthesis of cGMP, it acts as an **inhibitor of platelet aggregation** (Chapter 51).

Since the discovery of the role of NO as a vasodilator, there has been intense experimental interest in this molecule. It has turned out to have a variety of physiologic roles, involving virtually every tissue of the body (**Table 49–9**). Three major isoforms of NO synthase have been identified, each

TABLE 49–8 Summary of the Nomenclature of the NO Synthases and of the Effects of Knockout of Their Genes in Mice

Subtype	Name[1]	Comments	Result of Gene Knockout in Mice[2]
1	nNOS	Activity depends on elevated Ca^{2+}; first identified in neurons; calmodulin-activated	Pyloric stenosis, resistant to vascular stroke, aggressive sexual behavior (males)
2	iNOS[3]	Independent of elevated Ca^{2+}; prominent in macrophages	More susceptible to certain types of infection
3	eNOS	Activity depends on elevated Ca^{2+}; first identified in endothelial cells	Elevated mean blood pressure

Source: Adapted from Snyder SH: NO. Nature 1995;377:196.
[1]e, endothelial; i, inducible; n, neuronal.
[2]Gene knockouts were performed by homologous recombination in mice. The enzymes are characterized as neuronal, inducible (macrophage), and endothelial because these were the sites in which they were first identified. However, all three enzymes have been found in other sites, and the neuronal enzyme is also inducible. Each gene has been cloned, and its chromosomal location in humans has been determined.
[3]iNOS is Ca^{2+}-independent but binds calmodulin very tightly.

of which has been cloned, and the chromosomal locations of their genes in humans have been determined. Gene knockout experiments have been performed on each of the three isoforms and have helped establish some of the postulated functions of NO.

To summarize, research in the past decade has shown that NO plays an important role in many physiologic and pathologic processes.

SEVERAL MECHANISMS REPLENISH STORES OF ATP IN MUSCLE

The **ATP** required as the constant energy source for the contraction–relaxation cycle of muscle can be generated (1) by glycolysis, using blood glucose or muscle glycogen, (2) by oxidative phosphorylation, (3) from creatine phosphate, and (4) from two molecules of ADP in a reaction catalyzed by adenylyl kinase (**Figure 49–16**). The amount of ATP in skeletal muscle is only sufficient to provide energy for contraction for a few seconds, so that ATP must be constantly renewed from one or more of the above sources, depending upon metabolic conditions. As discussed below, there are at least **two distinct types of fibers** in skeletal muscle, one predominantly active in **aerobic** conditions and the other in **anaerobic** conditions; not unexpectedly, they use each of the above sources of energy to different extents.

Skeletal Muscle Contains Large Supplies of Glycogen

The sarcoplasm of skeletal muscle contains large stores of **glycogen**, located in granules close to the I bands. The release of glucose from glycogen is dependent on a specific muscle **glycogen phosphorylase** (Chapter 19), which can be activated by Ca^{2+}, epinephrine, and AMP. To generate glucose 6-phosphate for glycolysis in skeletal muscle, glycogen phosphorylase b must be activated to phosphorylase a via phosphorylation by phosphorylase b kinase (Chapter 19). Ca^{2+} promotes the activation of phosphorylase b kinase, also by phosphorylation. Thus, Ca^{2+} both initiates muscle contraction and activates a pathway to provide necessary energy. The hormone **epinephrine** also activates glycogenolysis in muscle. **AMP,** produced by breakdown of ADP during muscular exercise, can also activate phosphorylase b without causing phosphorylation. Muscle glycogen phosphorylase b is inactive in **McArdle disease,** one of the glycogen storage diseases (Chapter 19).

TABLE 49–9 Some Physiologic Functions and Pathologic Involvements of Nitric Oxide (NO)

- Vasodilator, important in regulation of blood pressure.
- Involved in penile erection; sildenafil citrate (Viagra) affects this process by inhibiting a cGMP phosphodiesterase.
- Neurotransmitter in the brain and the peripheral autonomic nervous system.
- Role in long-term potentiation.
- Role in neurotoxicity.
- Low level of NO involved in causation of pylorospasm in infantile hypertrophic pyloric stenosis.
- May have role in relaxation of skeletal muscle.
- May constitute part of a primitive immune system.
- Inhibits adhesion, activation, and aggregation of platelets.

Under Aerobic Conditions, Muscle Generates ATP Mainly by Oxidative Phosphorylation

Synthesis of ATP via **oxidative phosphorylation** requires a supply of oxygen. Muscles that have a high demand for oxygen as a result of sustained contraction (eg, to maintain posture) store it attached to the heme moiety of **myoglobin.** Because of the heme moiety, muscles containing myoglobin are red, whereas muscles with little or no myoglobin are white.

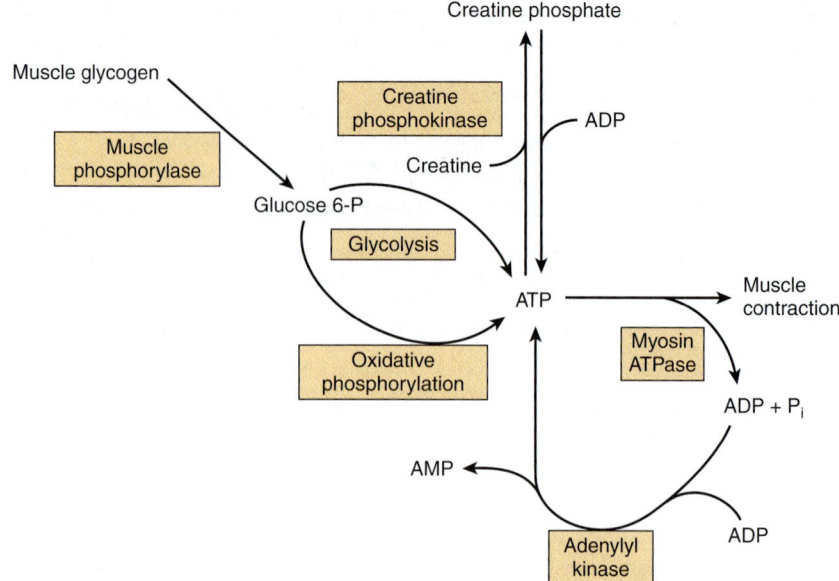

FIGURE 49–16 The multiple sources of ATP in muscle.

Glucose, derived from the blood glucose or from endogenous glycogen, and **fatty acids** derived from the triacylglycerols of adipose tissue are the principal substrates used for aerobic metabolism in muscle.

Creatine Phosphate Constitutes a Major Energy Reserve in Muscle

Creatine phosphate prevents the rapid depletion of ATP by providing a readily available high-energy phosphate that can be used to regenerate ATP from ADP. Creatine phosphate is formed from ATP and creatine (Figure 49–16) at times when the muscle is relaxed and demands for ATP are not so great. The enzyme catalyzing the phosphorylation of creatine is **creatine kinase** (CK), a muscle-specific enzyme with clinical utility in the detection of acute or chronic diseases of muscle.

SKELETAL MUSCLE CONTAINS SLOW (RED) & FAST (WHITE) TWITCH FIBERS

Different types of fibers have been detected in skeletal muscle. One classification subdivides them into type I (slow twitch), type IIA (fast twitch-oxidative), and type IIB (fast twitch-glycolytic). For the sake of simplicity, we shall consider only two types: type I (slow twitch, oxidative) and type II (fast twitch, glycolytic) (**Table 49–10**). The **type I** fibers are red because they contain myoglobin and mitochondria; their metabolism is aerobic, and they maintain relatively sustained contractions. The **type II** fibers, lacking myoglobin and containing few mitochondria, are white: they derive their energy from anaerobic glycolysis and exhibit relatively short durations of contraction. The **proportion** of these two types of fibers varies among the muscles of the body, depending on the

function (eg, whether or not a muscle is involved in sustained contraction, such as maintaining posture). The proportion also varies with **training;** for example, the number of type I fibers in certain leg muscles increases in athletes training for marathons, whereas the number of type II fibers increases in sprinters.

A Sprinter Uses Creatine Phosphate & Anaerobic Glycolysis to Make ATP, Whereas a Marathon Runner Uses Oxidative Phosphorylation

In view of the two types of fibers in skeletal muscle and of the various energy sources described above, it is of interest to compare their involvement in a sprint (eg, 100 meters) and in the marathon (42.2 km; just over 26 miles) (**Table 49–11**).

The major sources of energy in the **100-m sprint** are **creatine phosphate** (first 4–5 sec) and then **anaerobic glycolysis**, using muscle glycogen as the source of glucose. The two main sites of metabolic control are at **glycogen phosphorylase**

TABLE 49–10 Characteristics of Type I and Type II Fibers of Skeletal Muscle

	Type I Slow Twitch	Type II Fast Twitch
Myosin ATPase	Low	High
Energy utilization	Low	High
Mitochondria	Many	Few
Color	Red	White
Myoglobin	Yes	No
Contraction rate	Slow	Fast
Duration	Prolonged	Short

TABLE 49–11 Types of Muscle Fibers and Major Fuel Sources Used by a Sprinter and by a Marathon Runner

Sprinter (100 m)	Marathon Runner
Type II (glycolytic) fibers are used predominantly	Type I (oxidative) fibers are used predominantly
Creatine phosphate is the major energy source during the first 4–5 s	ATP is the major energy source throughout
Glucose derived from muscle glycogen and metabolized by anaerobic glycolysis is the major fuel source	Blood glucose and free fatty acids are the major fuel sources
Muscle glycogen is rapidly depleted	Muscle glycogen is slowly depleted

and at **PFK-1.** The former is activated by Ca^{2+} (released from the SR during contraction), epinephrine, and AMP. PFK-1 is activated by AMP, P_i, and NH_3. Attesting to the efficiency of these processes, the flux through glycolysis can increase as much as 1000-fold during a sprint.

In contrast, in the **marathon, aerobic metabolism** is the principal source of ATP. The major fuel sources are **blood glucose** and **free fatty acids,** largely derived from the breakdown of triacylglycerols in adipose tissue, stimulated by epinephrine. Hepatic glycogen is degraded to maintain the level of blood glucose. Muscle glycogen is also a fuel source, but it is degraded much more gradually than in a sprint. It has been calculated that the amount of glucose in the blood, glycogen in the liver, glycogen in muscle, and triacylglycerol in adipose tissue is sufficient to supply muscle with energy during a marathon for 4, 18, 70, and approximately 4000 min, respectively. However, the rate of oxidation of fatty acids by muscle is slower than that of glucose, so that oxidation of glucose and of fatty acids is a major source of energy in the marathon.

A number of procedures have been used by athletes to counteract muscle fatigue and inadequate strength. These include **carbohydrate loading, soda (sodium bicarbonate) loading, blood doping** (administration of red blood cells), and ingestion of **creatine** and **androstenedione.** Their rationales and efficacies will not be discussed here.

SKELETAL MUSCLE CONSTITUTES THE MAJOR RESERVE OF PROTEIN IN THE BODY

In humans, **skeletal muscle protein** is the major nonfat source of stored energy. This explains very large losses of muscle mass, particularly in adults, resulting from prolonged caloric undernutrition.

The study of **tissue protein breakdown** in vivo is difficult, because amino acids released during intracellular breakdown of proteins can be extensively reutilized for protein synthesis within the cell, or the amino acids may be transported to other organs where they enter anabolic pathways. However, actin and myosin are methylated by a posttranslational reaction, forming **3-methylhistidine.** During intracellular breakdown of actin and myosin, 3-methylhistidine is released and excreted into the urine. The urinary output of the methylated amino acid provides a reliable index of the rate of myofibrillar protein breakdown in the musculature of human subjects.

Various features of muscle metabolism, most of which are dealt with in other chapters of this text, are summarized in **Table 49–12.**

TABLE 49–12 Summary of Major Features of the Biochemistry of Skeletal Muscle Related to Its Metabolism[1]

- Skeletal muscle functions under both aerobic (resting) and anaerobic (eg, sprinting) conditions, so both aerobic and anaerobic glycolysis operate, depending on conditions.
- Skeletal muscle contains myoglobin as a reservoir of oxygen.
- Skeletal muscle contains different types of fibers primarily suited to anaerobic (fast twitch fibers) or aerobic (slow twitch fibers) conditions.
- Actin, myosin, tropomyosin, troponin complex (TpT, TpI, and TpC), ATP, and Ca^{2+} are key constituents in relation to contraction.
- The Ca^{2+} ATPase, the Ca^{2+} release channel, and calsequestrin are proteins involved in various aspects of Ca^{2+} metabolism in muscle.
- Insulin acts on skeletal muscle to increase uptake of glucose.
- In the fed state, most glucose is used to synthesize glycogen, which acts as a store of glucose for use in exercise; "preloading" with glucose is used by some long-distance athletes to build up stores of glycogen.
- Epinephrine stimulates glycogenolysis in skeletal muscle, whereas glucagon does not because of the absence of its receptors.
- Skeletal muscle cannot contribute directly to blood glucose because it does not contain glucose-6-phosphatase.
- Lactate produced by anaerobic metabolism in skeletal muscle passes to liver, which uses it to synthesize glucose, which can then return to muscle (the Cori cycle).
- Skeletal muscle contains phosphocreatine, which acts as an energy store for short-term (seconds) demands.
- Free fatty acids in plasma are a major source of energy, particularly under marathon conditions and in prolonged starvation.
- Skeletal muscle can utilize ketone bodies during starvation.
- Skeletal muscle is the principal site of metabolism of branched chain amino acids, which are used as an energy source.
- Proteolysis of muscle during starvation supplies amino acids for gluconeogenesis.
- Major amino acids emanating from muscle are alanine (destined mainly for gluconeogenesis in liver and forming part of the glucose-alanine cycle) and glutamine (destined mainly for the gut and kidneys).

[1]This Table brings together material from various chapters in this book.

TABLE 49–13 Some Properties of Microfilaments and Microtubules

	Microfilaments	Microtubules
Protein(s)	Actin	α-and β-tubulins
Diameter	8–9 nm	25 nm
Functions	Structural, motility	Structural, motility, polarity

Note: Some properties of intermediate filaments are described in Table 49–14.

THE CYTOSKELETON PERFORMS MULTIPLE CELLULAR FUNCTIONS

Non-muscle cells perform mechanical work, including self-propulsion, morphogenesis, cleavage, endocytosis, exocytosis, intracellular transport, and changing cell shape. These cellular functions are carried out by an extensive intracellular network of filamentous structures constituting the **cytoskeleton.** The cell cytoplasm is not a sac of fluid, as once thought. Essentially all eukaryotic cells contain three types of filamentous structures: **actin filaments** (also known as microfilaments), **microtubules,** and **intermediate filaments.** Each type of filament can be distinguished biochemically and by the electron microscope.

Some properties of these three structures are summarized in **Tables 49–13** and **49–14.**

Non-muscle Cells Contain Actin That Forms Microfilaments

G-actin is present in most if not all cells of the body. With appropriate concentrations of magnesium and potassium chloride, it spontaneously polymerizes to form double helical **F-actin** filaments like those seen in muscle. There are at least two types of actin in non-muscle cells: β-actin and γ-actin. Both types can coexist in the same cell and probably even copolymerize in the same filament. In the cytoplasm, **F-actin** forms **microfilaments** of 7–9.5 nm that frequently exist as bundles of a tangled-appearing meshwork. These bundles are prominent just underlying the plasma membrane of many cells and are there referred to as **stress fibers.** The stress fibers disappear as cell motility increases or upon malignant transformation of cells by chemicals or oncogenic viruses.

Although not organized as in muscle, actin filaments in non-muscle cells interact with **myosin** to cause cellular movements.

Microtubules Contain α- & β-Tubulins

Microtubules, an integral component of the cellular cytoskeleton, consist of cytoplasmic tubes 25 nm in diameter and often of extreme length (see **Figure 49–17**). Microtubules are necessary for the formation and function of the **mitotic spindle** and thus are present in all eukaryotic cells. They are also involved in the intracellular movement of endocytic and

TABLE 49–14 Classes of Intermediate Filaments of Eukaryotic Cells and Their Distributions

Proteins	Molecular Mass (kDa)	Distributions
Lamins		
A, B, and C	65–75	Nuclear lamina
Keratins		
Type I (acidic)	40–60	Epithelial cells, hair, nails
Type II (basic)	50–70	As for type I (acidic)
Vimentin-like		
Vimentin	54	Various mesenchymal cells
Desmin	53	Muscle
Glial fibrillary acid protein	50	Glial cells
Peripherin	66	Neurons
Neurofilaments		
Low (L), medium (M), and high (H)[1]	60–130	Neurons

Note: Intermediate filaments have an approximate diameter of 10 nm and have various functions. For example, keratins are distributed widely in epithelial cells and adhere via adapter proteins to desmosomes and hemidesmosomes. Lamins provide support for the nuclear membrane.
[1]Refers to their molecular masses.

exocytic **vesicles** and form the major structural components of **cilia** and **flagella.** Microtubules are a major component of **axons** and **dendrites,** in which they maintain structure and participate in the axoplasmic flow of material along these neuronal processes.

Microtubules are cylinders of 13 longitudinally arranged protofilaments, each consisting of dimers of **α-tubulin** and **β-tubulin,** closely related proteins of approximately 50 kDa molecular mass. The tubulin dimers assemble into protofilaments and subsequently into sheets and then cylinders. A microtubule-organizing center, located around a pair of centrioles, nucleates the growth of new microtubules. A third species of tubulin, **γ-tubulin,** appears to play an important role in this assembly. **GTP** is required for assembly. A variety of proteins are associated with microtubules (**microtubule-associated proteins [MAPs],** one of which is **tau**) and play important roles in microtubule assembly and stabilization. Microtubules are in a state of dynamic instability, constantly assembling and disassembling. They exhibit **polarity** (plus and minus ends); this is important in their growth from centrioles and in their ability to direct intracellular movement. For instance, in axonal transport, the protein **kinesin,** with a myosin-like ATP-ase activity, uses hydrolysis of ATP to move vesicles down the axon toward the positive end of the microtubular formation. Flow of materials in the opposite direction, toward the negative end, is powered by **cytosolic dynein,** another protein with ATPase

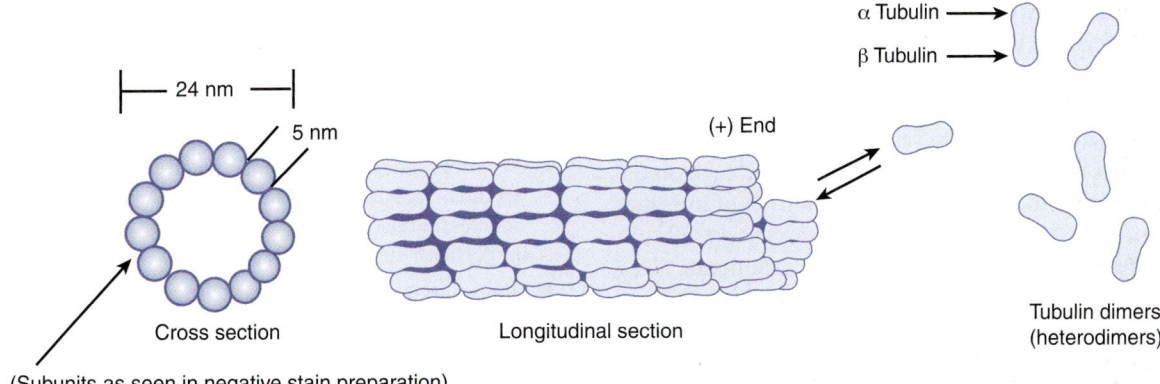

FIGURE 49–17 **Schematic representation of microtubules.** The upper left-hand corner shows a drawing of microtubules as seen in the electron microscope following fixation with tannic acid in glutaraldehyde. The unstained tubulin subunits are delineated by the dense tannic acid. Cross sections of tubules reveal a ring of 13 subunits of dimers arranged in a spiral. Changes in microtubule length are due to the addition or loss of individual tubulin subunits. Characteristic arrangements of microtubules (not shown here) are found in centrioles, basal bodies, cilia, and flagellae. (Reproduced, with permission, from Junqueirai LC, Carneiro J, Kelley RO: *Basic Histology,* 7th ed. Appleton & Lange, 1992.)

activity. Similarly, **axonemal dyneins** power ciliary and flagellar movement. Another protein, **dynamin,** uses GTP and is involved in endocytosis. Kinesins, dyneins, dynamin, and myosins are referred to as **molecular motors.**

An absence of dynein in cilia and flagella results in immotile cilia and flagella, leading to male sterility, situs inversus and chronic respiratory infection, a condition known as **Kartagener syndrome** (OMIM 244400). Mutations in genes affecting the synthesis of dynein have been detected in individuals with this syndrome.

Certain **drugs** bind to microtubules and thus interfere with their assembly or disassembly. These include **colchicine** (used for treatment of acute gouty arthritis), **vinblastine** (a vinca alkaloid used for treating certain types of cancer), **paclitaxel** (Taxol) (effective against ovarian cancer), and **griseofulvin** (an antifungal agent).

Intermediate Filaments Differ from Microfilaments & Microtubules

An intracellular fibrous system exists of filaments with an axial periodicity of 21 nm and a diameter of 8–10 nm that is intermediate between that of microfilaments (6 nm) and microtubules (23 nm). At least four classes of **intermediate filaments** are found, as indicated in Table 49–14.

They are all elongated, fibrous molecules, with a central rod domain, an amino terminal head, and a carboxyl terminal tail. They form a structure like a rope, and the mature filaments are composed of tetramers packed together in a helical manner. They are important structural components of cells, and most are **relatively stable** components of the cytoskeleton, not undergoing rapid assembly and disassembly and not disappearing during mitosis, as do actin and many microtubular filaments.

An important exception to this is provided by the **lamins,** which, subsequent to phosphorylation, disassemble at mitosis and reappear when it terminates. **Lamins** form a meshwork in apposition to the inner nuclear membrane.

Mutations in the gene encoding **lamin A** and lamin C cause the Hutchinson–Gilford progeria syndrome **(progeria)** [OMIM 176670], characterized by the appearance of **accelerated aging** and other features. A farnesylated form (see Figure 26–2 for the structure of farnesyl) of prelamin A accumulates in the condition, because the site of normal proteolytic action to cleave off the farnesylated portion of lamin A is altered by mutation. Lamin A is an important component of the structural scaffolding that maintains the integrity of the nucleus of a cell. It appears that the accumulation of the farnesylated prelamin A makes nuclei unstable, altering their shape, and somehow this predisposes to the development of signs of premature aging. Experiments in mice have indicated that administration of a farnesyltransferase inhibitor may ameliorate the development of misshapen nuclei. Children affected by this condition often die in their teens of atherosclerosis. A brief scheme of the causation of progeria is shown in **Figure 49–18**.

Keratins form a large family, with about 30 members being distinguished. As indicated in Table 49–14, two major types of keratins are found; all individual keratins are **heterodimers** made up of one member of each class.

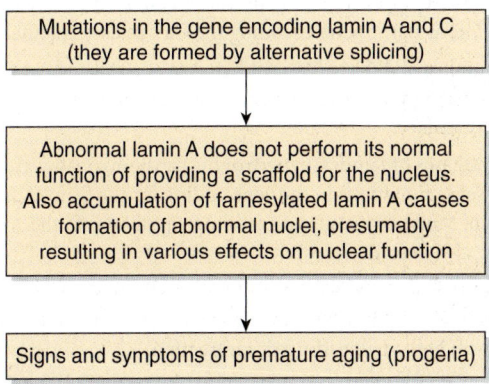

FIGURE 49–18 **Scheme of the causation of progeria (Hutchinson–Gilford syndrome, OMIM 176670).**

Vimentins are widely distributed in mesodermal cells, and desmin, glial fibrillary acidic protein, and peripherin are related to them. All members of the vimentin-like family can copolymerize with each other.

Intermediate filaments are very prominent in nerve cells; neurofilaments are classified as low, medium, and high on the basis of their molecular masses. The **distribution of intermediate filaments** in normal and abnormal (eg, cancer) cells can be studied by the use of immunofluorescent techniques, using antibodies of appropriate specificities. These antibodies to specific intermediate filaments can also be of use to pathologists in helping to decide the origin of certain dedifferentiated malignant tumors. These tumors may still retain the type of intermediate filaments found in their cell of origin.

A number of **skin diseases,** mainly characterized by blistering, have been found to be due to mutations in genes encoding **various keratins.** Two of these disorders are epidermolysis bullosa simplex (OMIM 131800) and epidermolytic palmoplantar keratoderma (OMIM 144200). The **blistering** found in these disorders probably reflects a diminished capacity of various layers of the skin to resist mechanical stresses due to abnormalities in the keratin structure.

SUMMARY

- The myofibrils of skeletal muscle contain thick and thin filaments. The thick filaments contain myosin. The thin filaments contain actin, tropomyosin, and the troponin complex (troponins T, I, and C).

- The sliding filament cross-bridge model is the foundation of current thinking about muscle contraction. The basis of this model is that the interdigitating filaments slide past one another during contraction and cross bridges between myosin and actin generate and sustain the tension.

- The hydrolysis of ATP is used to drive movement of the filaments. ATP binds to myosin heads and is hydrolyzed to ADP and P_i by the ATPase activity of the actomyosin complex.

- Ca^{2+} plays a key role in the initiation of muscle contraction by binding to troponin C. In skeletal muscle, the SR regulates distribution of Ca^{2+} to the sarcomeres, whereas inflow of Ca^{2+} via Ca^{2+} channels in the sarcolemma is of major importance in cardiac and smooth muscle.

- Many cases of malignant hyperthermia in humans are due to mutations in the gene encoding the Ca^{2+} release channel.

- A number of differences exist between skeletal and cardiac muscle; in particular, the latter contains a variety of receptors on its surface.

- Some cases of familial hypertrophic cardiomyopathy are due to missense mutations in the gene coding for the β-myosin heavy chain. Mutations in genes encoding a number of other proteins have also been detected.

- Smooth muscle, unlike skeletal and cardiac muscle, does not contain the troponin system; instead, phosphorylation of myosin light chains initiates contraction.

- NO is a regulator of vascular smooth muscle; blockage of its formation from arginine causes an acute elevation of blood pressure, indicating that regulation of blood pressure is one of its many functions.

- Duchenne-type muscular dystrophy is due to mutations in the gene, located on the X chromosome, encoding the protein dystrophin.

- Two major types of muscle fibers are found in humans: white (anaerobic) and red (aerobic). The former are particularly used in sprints and the latter in prolonged aerobic exercise. During a sprint, muscle uses creatine phosphate and glycolysis as energy sources; in the marathon, oxidation of fatty acids is of major importance during the later phases.

- Non-muscle cells perform various types of mechanical work carried out by the structures constituting the cytoskeleton. These structures include actin filaments (microfilaments), microtubules (composed primarily of α-tubulin and β-tubulin), and intermediate filaments. The latter include lamins, keratins, vimentin-like proteins, and neurofilaments. Mutations in the gene encoding lamin A cause progeria, a condition characterized by the appearance of premature aging. Mutations in genes for certain keratins cause a number of skin diseases.

REFERENCES

Alberts B, Johnson A, Lewis J, et al: *Molecular Biology of the Cell*, 5th ed. Garland Science, 2008. (Contains excellent coverage of muscle and the cytoskeleton).

Barrett KE, Barman SM, Boitano S, Brooks HL: *Ganong's Review of Medical Physiology*, 23rd ed. McGraw-Hill Lange, 2010. (Contains excellent coverage of skeletal, cardiac and smooth muscle structure and function).

Brosnan JT, Brosnan ME: Creatine: endogenous metabolite, dietary and therapeutic supplement. Annu Rev Nutr 2007;27:241.

Cooper GM, Hausman RE: *The Cell: A Molecular Approach*, 5th ed. Sinauer Associates Inc., 2009. (Contains excellent coverage of muscle and the cytoskeleton).

Lodish H, Berk A, Kaiser CA, et al: *Molecular Cell Biology*, 6th ed. WH Freeman & Co, 2008. (Contains excellent coverage of muscle and the cytoskeleton).

Murad F: Nitric oxide and cyclic GMP in cell signalling and drug development. N Engl J Med 2006;355:2003.

Murphy RT, Starling RC: Genetics and cardiomyopathy: where are we now? Cleve Clin J Med 2005;72:465.

Neubauer S: The failing heart—an engine out of fuel. N Engl J Med 2007;356:1140.

Pollard TD, Earnshaw WC: *Cell Biology,* 2nd ed. Saunders, 2008. (Contains excellent coverage of muscle and the cytoskeleton).

Sanders KM: Regulation of smooth muscle excitation and contraction. Neurogastroenterol Motil 2008;20 Suppl 1:39.

Scriver CR, Beaudet AL, Valle D, et al (editors): *The Metabolic and Molecular Bases of Inherited Disease,* 8th ed. McGraw-Hill, 2001. (This comprehensive four-volume text and its updated online edition [see Chapter 1] contain chapters on malignant hyperthermia, channelopathies, hypertrophic cardiomyopathy, the muscular dystrophies, and disorders of intermediate filaments.)

Sweeney HL, Houdusse A: Structural and functional insights into the myosin motor mechanism. Annu Rev Biophys 2010;39:539.

Taimen P, Pfleghaar K, Shimi T, et al: A progeria mutation reveals functions for lamin A in nuclear assembly, architecture, and chromosome organization. Proc Natl Acad Sci USA 2009;106(49):20788.

Plasma Proteins & Immunoglobulins

Robert K. Murray, MD, PhD, Molly Jacob, MB BS, MD, PhD, & Joe Varghese, MB BS, MD

OBJECTIVES

After studying this chapter, you should be able to:

- List the major functions of blood.
- Explain the functions of the major plasma proteins, including albumin, haptoglobin, transferrin, ceruloplasmin, α_1-antitrypsin, and α_2-macroglobulin.
- Describe how iron homeostasis is maintained and how it is affected in certain disorders.
- Describe the general structures and functions of the five classes of immunoglobulins and the uses of monoclonal antibodies.
- Appreciate that the complement system is involved in a number of important biological processes.
- Indicate the causes of Wilson disease, Menkes disease, the lung and liver diseases associated with α_1-antitrypsin deficiency, amyloidosis, multiple myeloma, and agammaglobulinemia.

BIOMEDICAL IMPORTANCE

The fundamental role of blood in the maintenance of **homeostasis** (see Chapter 51) and the ease with which blood can be obtained have meant that the study of its constituents has been of central importance in the development of biochemistry and clinical biochemistry. The basic properties of a number of **plasma proteins,** including the **immunoglobulins** (antibodies), are described in this chapter. Changes in the amounts of various plasma proteins and immunoglobulins occur in many diseases and can be monitored by electrophoresis or other suitable procedures. As indicated in an earlier chapter, alterations of the activities of certain **enzymes** found in plasma are of diagnostic use in a number of pathologic conditions. Plasma proteins involved in blood coagulation are discussed in Chapter 51.

THE BLOOD HAS MANY FUNCTIONS

The functions of blood—except for specific cellular ones such as oxygen transport and cell-mediated immunologic defense—are carried out by plasma and its constituents (**Table 50–1**).

Plasma consists of water, electrolytes, metabolites, nutrients, proteins, and hormones. The water and electrolyte composition of plasma is practically the same as that of all extracellular fluids. Laboratory determinations of levels of Na^+, K^+, Ca^{2+}, Mg^{2+}, Cl^-, HCO_3^-, $PaCO_2$, and of blood pH are important in the management of many patients.

PLASMA CONTAINS A COMPLEX MIXTURE OF PROTEINS

The concentration of total protein in human plasma is approximately 7.0–7.5 g/dL and comprises the major part of the solids of the plasma. The proteins of the plasma are actually a complex mixture that includes not only simple proteins but also conjugated proteins such as **glycoproteins** and various types of **lipoproteins.** Use of proteomic techniques is allowing the isolation and characterization of previously unknown plasma proteins, some present in very small amounts (eg, detected in hemodialysis fluid and in the plasma of patients with cancer), thus expanding **the plasma proteome.** Thousands of **antibodies** are present in human plasma, although the amount of any one antibody is usually quite low under normal circumstances. The relative dimensions and molecular masses of some of the most important plasma proteins are shown in **Figure 50–1**.

The **separation** of individual proteins from a complex mixture is frequently accomplished by the use of solvents or electrolytes (or both) to remove different protein fractions in accordance with their solubility characteristics. This is the basis

TABLE 50–1   Major Functions of Blood

1. **Respiration**—transport of oxygen from the lungs to the tissues and of CO_2 from the tissues to the lungs
2. **Nutrition**—transport of absorbed food materials
3. **Excretion**—transport of metabolic waste to the kidneys, lungs, skin, and intestines for removal
4. Maintenance of the normal **acid–base balance** in the body
5. Regulation of **water balance** through the effects of blood on the exchange of water between the circulating fluid and the tissue fluid
6. Regulation of **body temperature** by the distribution of body heat
7. **Defense** against infection by the white blood cells and circulating antibodies
8. Transport of **hormones** and regulation of metabolism
9. Transport of **metabolites**
10. **Coagulation**

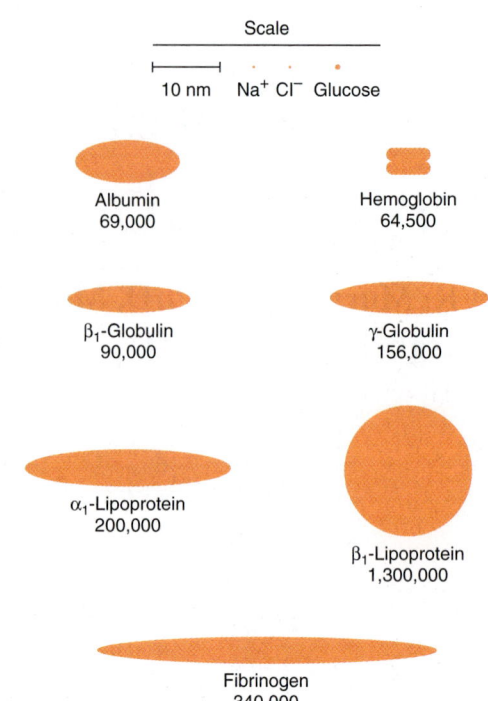

FIGURE 50–1   Relative dimensions and approximate molecular masses of protein molecules in the blood (Oncley).

of the so-called salting-out methods, which find some usage in the determination of protein fractions in the clinical laboratory. Thus, one can separate the proteins of the plasma into three major groups—**fibrinogen, albumin,** and **globulins**—by the use of varying concentrations of sodium or ammonium sulfate.

The most common method of analyzing plasma proteins is by **electrophoresis.** There are many types of electrophoresis, each using a different supporting medium. In clinical laboratories, **cellulose acetate** is widely used as a supporting medium. Its use permits resolution, after staining, of plasma proteins into five bands, designated albumin, α_1, α_2, β, and γ fractions,

respectively (**Figure 50–2**). The stained strip of cellulose acetate (or other supporting medium) is called an electrophoretogram. The amounts of these five bands can be conveniently quantified by use of densitometric scanning machines. Characteristic changes in the amounts of one or more of these five bands are found in many diseases.

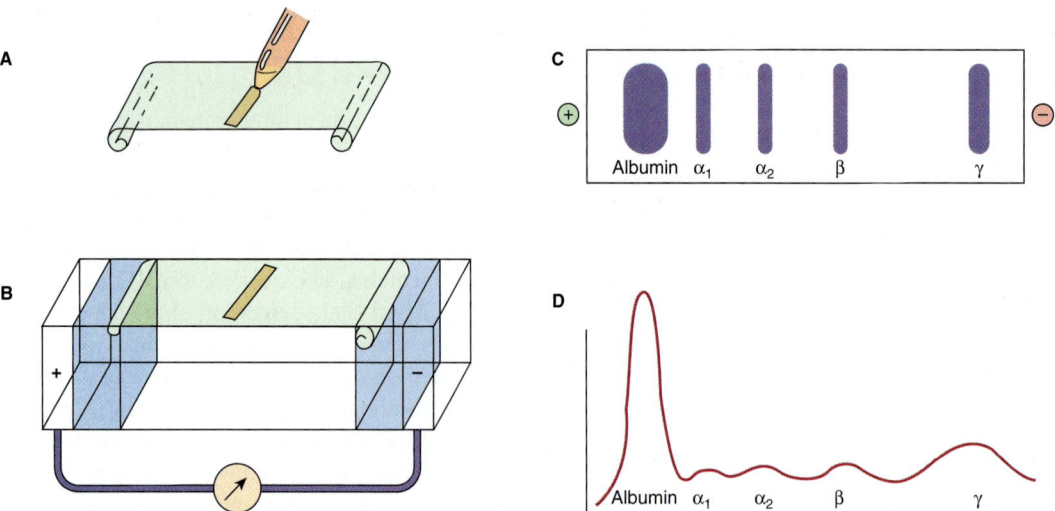

FIGURE 50–2   Technique of cellulose acetate zone electrophoresis. **(A)** A small amount of serum or other fluid is applied to a cellulose acetate strip. **(B)** Electrophoresis of sample in electrolyte buffer is performed. **(C)** Separated protein bands are visualized in characteristic positions after being stained. **(D)** Densitometer scanning from cellulose acetate strip converts bands to characteristic peaks of albumin, α_1-globulin, β_2-globulin, β-globulin, and γ-globulin. (Reproduced, with permission, from Parslow TG et al (editors): *Medical Immunology,* 10th ed. McGraw-Hill, 2001.)

The Concentration of Protein in Plasma Is Important in Determining the Distribution of Fluid Between Blood & Tissues

In arterioles, the **hydrostatic pressure** is about 37 mmHg, with an interstitial (tissue) pressure of 1 mmHg opposing it. The **osmotic pressure** (oncotic pressure) exerted by the plasma proteins is approximately 25 mmHg. Thus, a net outward force of about 11 mmHg drives fluid out into the interstitial spaces. In venules, the hydrostatic pressure is about 17 mmHg, with the oncotic and interstitial pressures as described above; thus, a net force of about 9 mmHg attracts water back into the circulation. The above pressures are often referred to as the **Starling forces.** If the concentration of plasma proteins is markedly diminished (eg, due to severe protein malnutrition), fluid is not attracted back into the intravascular compartment and accumulates in the extravascular tissue spaces, a condition known as **edema.** Edema has many causes; protein deficiency is one of them.

Plasma Proteins Have Been Studied Extensively

Because of the relative ease with which they can be obtained, plasma proteins have been studied extensively in both humans and animals. Considerable information is available about the biosynthesis, turnover, structure, and functions of the major plasma proteins. Alterations of their amounts and of their metabolism in many disease states have also been investigated. In recent years, many of the genes for plasma proteins have been cloned and their structures determined.

The preparation of **antibodies** specific for the individual plasma proteins has greatly facilitated their study, allowing the precipitation and isolation of pure proteins from the complex mixture present in tissues or plasma. In addition, the use of **isotopes** has made possible the determination of their pathways of biosynthesis and of their turnover rates in plasma.

The following generalizations have emerged from studies of plasma proteins.

Most Plasma Proteins Are Synthesized in the Liver

This has been established by experiments at the whole animal level (eg, hepatectomy) and by use of the isolated perfused liver preparation, of liver slices, of liver homogenates, and of in vitro translation systems using preparations of mRNA extracted from liver. However, the γ-globulins are synthesized in plasma cells and certain plasma proteins are synthesized in other sites, such as endothelial cells.

Plasma Proteins Are Generally Synthesized on Membrane-Bound Polyribosomes

They then traverse the major secretory route in the cell (rough endoplasmic membrane → smooth endoplasmic membrane → Golgi apparatus → secretory vesicles) prior to entering the plasma. Thus, most plasma proteins are synthesized as **preproteins** and initially contain amino terminal signal peptides (Chapter 46). They are usually subjected to various posttranslational modifications (proteolysis, glycosylation, phosphorylation, etc) as they travel through the cell. Transit times through the hepatocyte from the site of synthesis to the plasma vary from 30 min to several hours or more for individual proteins.

Most Plasma Proteins Are Glycoproteins

Accordingly, they generally contain either N- or O-linked oligosaccharide chains, or both (Chapter 47). Albumin is the major exception; it does not contain sugar residues. The oligosaccharide chains have various functions (Table 47-2). Removal of terminal sialic acid residues from certain plasma proteins (eg, ceruloplasmin) by exposure to neuraminidase can markedly shorten their half-lives in plasma (Chapter 47).

Many Plasma Proteins Exhibit Polymorphism

A **polymorphism** is a Mendelian or monogenic trait that exists in the population in at least two phenotypes, neither of which is rare (ie, neither of which occurs with frequency of <1–2%). The ABO blood group substances (Chapter 52) are the best known examples of human polymorphisms. Human plasma proteins that exhibit polymorphism include α_1-antitrypsin, haptoglobin, transferrin, ceruloplasmin, and immunoglobulins. The polymorphic forms of these proteins can be distinguished by different procedures (eg, various types of electrophoresis or isoelectric focusing), in which each form may show a characteristic migration. Analyses of these human polymorphisms have proved to be of genetic, anthropologic, and clinical interest.

Each Plasma Protein Has a Characteristic Half-Life in the Circulation

The **half-life** of a plasma protein can be determined by labeling the isolated pure protein with ^{131}I or Cr^{51} under mild, nondenaturing conditions. The labeled protein is freed of unbound free isotope and its specific activity (disintegrations per minute per milligram of protein) determined. A known amount of the radioactive protein is then injected into a normal adult subject, and samples of blood are taken at various time intervals for determinations of radioactivity. The values for radioactivity are plotted against time, and the half-life of the protein (the time for the radioactivity to decline from its peak value to one-half of its peak value) can be calculated from the resulting graph, discounting the times for the injected protein to equilibrate (mix) in the blood and in the extravascular spaces. The half-lives obtained for albumin and haptoglobin in normal healthy adults are approximately 20 and 5 days, respectively. In certain diseases, the half-life of a protein may be markedly altered. For instance, in some gastrointestinal diseases such as regional ileitis (Crohn's disease), considerable amounts of plasma proteins, including albumin, may be lost into the bowel through the inflamed intestinal mucosa. Patients with this condition have a **protein-losing gastroenteropathy,** and the half-life of injected iodinated albumin in these subjects may be reduced to as little as 1 day.

The Levels of Certain Proteins in Plasma Increase During Acute Inflammatory States or Secondary to Certain Types of Tissue Damage

These proteins are called **"acute-phase proteins"** (or reactants) and include **C-reactive protein** (CRP, so named because it reacts with the C polysaccharide of pneumococci), α1-antitrypsin, haptoglobin, α₁-acid glycoprotein, and **fibrinogen.** The elevations of the levels of these proteins vary from as little as 50% to as much as 1000-fold in the case of CRP. Their levels are also usually elevated during chronic inflammatory states and in patients with cancer. These proteins are believed to play a role in the body's response to inflammation. For example, C-reactive protein can stimulate the classic complement pathway (see below), and α₁-antitrypsin can neutralize certain proteases released during the acute inflammatory state. CRP is used as a marker of tissue injury, infection, and inflammation, and there is considerable interest in its use as a predictor of certain types of cardiovascular conditions secondary to atherosclerosis. The cytokine (= a protein made by cells that affects the behavior of other cells) interleukin-1 (IL-1), a polypeptide released from mononuclear phagocytic cells, is the principal—but not the sole—stimulator of the synthesis of the majority of acute phase reactants by hepatocytes. Additional molecules such as IL-6 are involved, and they as well as IL-1 appear to work at the level of gene transcription.

 Nuclear factor kappa-B (NFkB) is a transcription factor that has been involved in the stimulation of synthesis of certain of the acute phase proteins. This important factor is also involved in the expression of many cytokines, chemokines, growth factors, and cell adhesion molecules implicated in immunologic phenomena. Normally, it exists in an inactive form in the cytosol but is activated and translocated to the nucleus via the action of a number of molecules (eg, interleukin-1) produced in processes such as inflammation, infection, and radiation injury.

 Table 50–2 summarizes the functions of many of the plasma proteins. The remainder of the material in this chapter presents basic information regarding selected plasma proteins: albumin, haptoglobin, transferrin, ceruloplasmin, α₁-antitrypsin, α₂-macroglobulin, the immunoglobulins, and the complement system. The lipoproteins are discussed in Chapter 25. New information is constantly forthcoming on plasma proteins and their variants (including those discussed here), as the techniques of proteomics, particularly sensitive new methods of determining proteins sequences by mass spectrometry (see Chapter 4), are applied to their study. A number of laboratories are participating in efforts to determine the complete **human plasma protein proteome.** It is believed that this will shed further light on **genetic variations** in humans and also provide many new **biomarkers** to aid in the diagnosis of many diseases. (A biomarker has been defined as a

characteristic that is objectively measured and evaluated as an indicator of normal biologic processes, pathogenic processes, or pharmacologic responses to a therapeutic intervention.)

Albumin Is the Major Protein In Human Plasma

Albumin (69 kDa) is the major protein of human plasma (3.4–4.7 g/dL) and makes up approximately 60% of the total plasma protein. About 40% of albumin is present in the plasma, and the other 60% is present in the extracellular space. The liver produces about 12 g of albumin per day, representing about 25% of total hepatic protein synthesis and half its secreted protein. Albumin is initially synthesized as a **preproprotein.** Its **signal peptide** is

TABLE 50–2 Some Functions of Plasma Proteins

Function	Plasma Proteins
Antiproteases	Antichymotrypsin α₁-Antitrypsin (α₁-antiproteinase) α₂-Macroglobulin Antithrombin
Blood clotting	Various coagulation factors, fibrinogen
Enzymes	Function in blood, for example, coagulation factors, cholinesterase Leakage from cells or tissues, for example, aminotransferases
Hormones	Erythropoietin[1]
Immune defense	Immunoglobulins, complement proteins, and β₂-microglobulin
Involvement in inflammatory responses	Acute phase response proteins (eg, C-reactive protein, α₁-acid glycoprotein [orosomucoid])
Oncofetal	α₁-Fetoprotein (AFP)
Transport or binding proteins	Albumin (various ligands, including bilirubin, free fatty acids, ions [Ca²⁺], metals [eg, Cu²⁺, Zn²⁺], metheme, steroids, other hormones, and a variety of drugs) Ceruloplasmin (contains Cu²⁺; albumin probably more important in physiologic transport of Cu²⁺) Corticosteroid-binding globulin (transcortin) (binds cortisol) Haptoglobin (binds extracorpuscular hemoglobin) Lipoproteins (chylomicrons, VLDL, LDL, HDL) Hemopexin (binds heme) Retinol-binding protein (binds retinol) Sex-hormone-binding globulin (binds testosterone, estradiol) Thyroid-binding globulin (binds T₄, T₃) Transferrin (transport iron) Transthyretin (formerly prealbumin; binds T₄ and forms a complex with retinol-binding protein)

[1]Various other protein hormones circulate in the blood but are not usually designated as plasma proteins. Similarly, ferritin is also found in plasma in small amounts, but it too is not usually characterized as a plasma protein.

removed as it passes into the cisternae of the rough endoplasmic reticulum, and a **hexapeptide** at the resulting amino terminal is subsequently cleaved off farther along the secretory pathway (see Figure 46–12). The synthesis of albumin is depressed in a variety of diseases, particularly those of the liver. The plasma of patients with **liver disease** often shows a decrease in the ratio of albumin to globulins (decreased albumin–globulin ratio). The synthesis of albumin decreases relatively early in conditions of protein malnutrition, such as kwashiorkor.

Mature human albumin consists of one polypeptide chain of 585 amino acids and contains 17 disulfide bonds. By the use of proteases, albumin can be subdivided into three **domains,** which have different functions. Albumin has an ellipsoidal shape, which means that it does not increase the viscosity of the plasma as much as an elongated molecule such as fibrinogen does. Because of its relatively low molecular mass (about 69 kDa) and high concentration, albumin is thought to be responsible for 75–80% of the **osmotic pressure** of human plasma. Electrophoretic studies have shown that the plasma of certain humans lacks albumin. These subjects are said to exhibit **analbuminemia.** One cause of this condition is a mutation that affects splicing. Subjects with analbuminemia show only moderate edema, despite the fact that albumin is the major determinant of plasma osmotic pressure. It is thought that the amounts of the other plasma proteins increase and compensate for the lack of albumin.

Another important function of albumin is its ability to **bind various ligands.** These include free fatty acids (FFA), calcium, certain steroid hormones, bilirubin, and some of the plasma tryptophan. In addition, albumin appears to play an important role in transport of copper in the human body (see below). A variety of drugs, including sulfonamides, penicillin G, dicumarol, and aspirin, are bound to albumin; this finding has important pharmacologic implications.

Preparations of human albumin have been widely used in the treatment of hemorrhagic shock and of burns. However, some recent studies question the value of this therapy.

Haptoglobin Binds Extracorpuscular Hemoglobin, Preventing Free Hemoglobin from Entering the Kidney

Haptoglobin (Hp) is a plasma glycoprotein that binds extracorpuscular hemoglobin (Hb) in a tight noncovalent complex (Hb–Hp). The amount of haptoglobin in human plasma ranges from 40 to 180 mg of hemoglobin-binding capacity per deciliter. Approximately 10% of the hemoglobin that is degraded each day is released into the circulation and is thus extracorpuscular. The other 90% is present in old, damaged red blood cells, which are degraded by cells of the histiocytic system. The molecular mass of hemoglobin is approximately 65 kDa, whereas the molecular mass of the simplest polymorphic form of haptoglobin (Hp 1-1) found in humans is approximately 90 kDa. Thus, the Hb–Hp complex has a molecular mass of approximately 155 kDa. Free hemoglobin passes through the glomerulus of the kidney, enters the tubules, and tends to

A. Hb $\rightarrow$ Kidney $\rightarrow$ Excreted in urine or precipitates in
(MW 65,000)   tubules; iron is lost to body

B. Hb   +   Hp $\rightarrow$ Hb : Hp complex $\leftrightarrow$ Kidney
(MW 65,000) (MW 90,000) (MW 155,000)
Catabolized by liver cells;
iron is conserved and reused

FIGURE 50–3 **Different fates of free hemoglobin and of the hemoglobin–haptoglobin complex.**

precipitate therein (as can happen after a massive incompatible blood transfusion, when the capacity of haptoglobin to bind hemoglobin is grossly exceeded) (**Figure 50–3**). However, the Hb–Hp complex is too large to pass through the glomerulus. The function of Hp thus appears to prevent loss of free hemoglobin into the kidney. This conserves the valuable iron present in hemoglobin, which would otherwise be lost to the body.

Human haptoglobin exists in **three polymorphic forms,** known as Hp 1-1, Hp 2-1, and Hp 2-2. Hp 1-1 migrates in starch gel electrophoresis as a single band, whereas Hp 2-1 and Hp 2-2 exhibit much more complex band patterns. Two genes, designated Hp^1 and Hp^2, direct these three phenotypes, with Hp 2-1 being the heterozygous phenotype. It has been suggested that the haptoglobin polymorphism may be associated with the prevalence of many inflammatory diseases.

The levels of haptoglobin in human plasma vary and are of some diagnostic use. Low levels of haptoglobin are found in patients with **hemolytic anemias.** This is explained by the fact that whereas the half-life of haptoglobin is approximately 5 days, the half-life of the Hb–Hp complex is about 90 min, the complex being rapidly removed from plasma by hepatocytes. Thus, when haptoglobin is bound to hemoglobin, it is cleared from the plasma about 80 times faster than normally. Accordingly, the level of haptoglobin falls rapidly in situations where hemoglobin is constantly being released from red blood cells, such as occurs in hemolytic anemias. Haptoglobin is an acute phase protein, and its plasma level is elevated in a variety of inflammatory states.

Haptoglobin-related protein is another protein found in human plasma. It bears a high degree of homology to haptoglobin and it appears to bind hemoglobin. Its level is elevated in some patients with cancers, although the significance of this is not understood.

Certain other plasma proteins **bind heme** but not hemoglobin. **Hemopexin** is a β_1-globulin that binds free heme. **Albumin** will bind some metheme (ferric heme) to form methemalbumin, which then transfers the metheme to hemopexin.

IRON IS A VERY IMPORTANT BODY CONSTITUENT AND IS ZEALOUSLY CONSERVED

Transferrin is an important plasma protein involved in iron transport. Before discussing it and a number of other proteins involved in iron homeostasis, we shall describe certain aspects of iron and its metabolism.

TABLE 50–3 Distribution of Iron in a 70-kg Adult Male[1]

Transferrin	3–4 mg
Hemoglobin in red blood cells	2500 mg
In myoglobin and various enzymes	300 mg
In stores (ferritin)	1000 mg
Absorption	1 mg/d
Losses	1 mg/d

[1]In an adult female of similar weight, the amount in stores would generally be less (100–400 mg) and the losses would be greater (1.5–2 mg/d).

Iron is important in the human body because of its occurrence in many hemoproteins such as hemoglobin, myoglobin, and the cytochromes (including the cytochrome P450 group of enzymes). A normal 70 kg adult male has 3–4 g of iron in his body, the distribution of which is shown in **Table 50–3**. Under normal circumstances, the body guards its content of iron zealously. A healthy adult male loses only about 1 mg/day, which is replaced by absorption from the intestine. Adult females (premenopausal) require about 1.5 mg/day and are more prone to develop states of iron deficiency because of blood loss during menstruation.

Absorption of Dietary Iron Occurs in the Duodenum

Iron is ingested in the diet either as non-heme or heme iron. Absorption of iron by enterocytes of the proximal duodenum is a highly regulated process (**Figure 50–4**). **Inorganic dietary iron** in the ferric state (Fe^{3+}) is reduced to its ferrous form (Fe^{2+}) by a brush border membrane-bound ferrireductase, **duodenal cytochrome b (Dcytb)**. Vitamin C, gastric acid, and a number of other reducing agents present in food may also favor reduction of ferric to ferrous iron. The transfer of iron across the apical membrane of the enterocytes is accomplished via the **divalent metal transporter 1 (DMT1 or SLC11A2)**. DMT1 is relatively non-specific and may also be involved in the transport of other divalent cations, such as Mn^{2+}, Co^{2+}, Zn^{2+}, Cu^{2+}, and Pb^{2+}. Once inside the enterocytes, iron can either be stored as **ferritin** or transferred across the basolateral membrane into the circulation by the iron exporter protein, **ferroportin or iron-regulated protein 1 (IREG1 or SLC40A1)**. This process occurs in conjunction with **hephaestin**, a copper-containing ferroxidase homologous to ceruloplasmin, which oxidizes Fe^{2+} to Fe^{3+}. Iron is transported in plasma in the Fe^{3+} form by the transport protein, **transferrin**. Excess iron that is stored in the enterocytes as ferritin is lost when the enterocytes are sloughed off into the gut lumen.

Heme iron in the diet is taken up by enterocytes by mechanisms independent of those involved in the uptake of dietary inorganic iron. Heme in enterocytes is broken down by heme oxygenase (HO, see Chapter 31) to release iron, which is either stored as ferritin or transported into circulation by ferroportin.

Transferrin Shuttles Iron to Sites Where It Is Needed

Free iron is extremely toxic because of its ability to catalyze the formation of harmful oxygen free radicals, via the Fenton

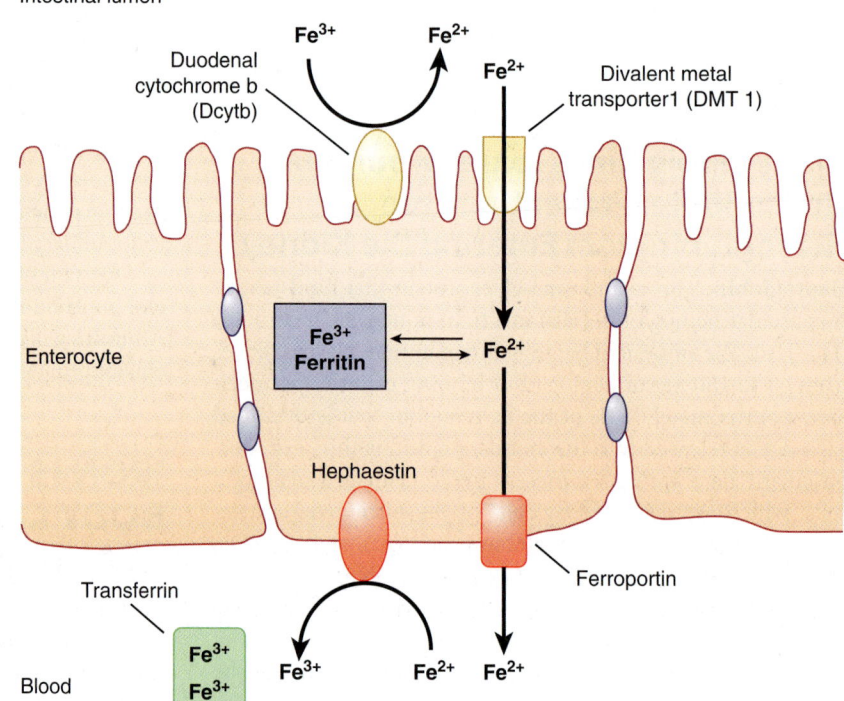

FIGURE 50–4 Non-heme iron transport in enterocytes. Ferric iron is reduced to the ferrous form by a luminal ferrireductase, duodenal cytochrome b (Dcytb). Ferrous iron is transported into the enterocyte via divalent metal transporter-1 (DMT-1). Within the enterocyte, iron is either stored as ferritin, or transported out of the cell, across the baslolateral membrane, by ferroportin (Fp). Ferrous iron is oxidized to its ferric form by the ferroxidase, hephaestin. Ferric iron is then bound by transferrin in blood which transports it to various sites in the body. (Based on Andrews NC: Forging a field: the golden age of iron biology. Blood 2008;112(2);219.)

$$Fe^{2+} + H_2O_2 \longrightarrow Fe^{3+} + OH^{\bullet} + OH^{-}$$

FIGURE 50–5 **The Fenton reaction.** Free iron is extremely toxic as it can catalyze the formation of hydroxyl radical (OH$^{\bullet}$) from hydrogen peroxide (see also Chapter 52). The hydroxyl radical is a transient but highly reactive species and can oxidize cellular macromolecules resulting in tissue damage.

reaction (**Figure 50–5**). In biological systems, iron is always bound to proteins in order to limit the generation of toxic radicals. In the plasma, iron is tightly bound to the plasma protein, **transferrin (Tf),** which plays a central role in transporting iron around the body to sites where it is needed. It is a β_1-globulin with a molecular mass of approximately 76 kDa. It is a glycoprotein and is synthesized in the liver. It transports iron in circulation to sites where iron is required (eg, the bone marrow). It has two high-affinity binding sites for Fe^{3+} iron. Transferrin, when bound to two atoms of iron, is called **holotransferrin (Tf-Fe).** The concentration of Tf in plasma is approximately 300 mg/dL. This amount of transferrin can bind a total of approximately 300 μg of iron (per deciliter of plasma). This represents the **total iron-binding capacity (TIBC)** of plasma. Normally, transferrin is about **30% saturated** with iron. Transferrin saturation decreases to less than 16% during severe iron deficiency and may increase to more than 45% in iron overload conditions.

Glycosylation of transferrin is impaired in **congenital disorders of glycosylation** (Chapter 47) and **chronic alcoholism,** resulting in increased circulating levels of **carbohydrate-deficient transferrin (CDT).** CDT can be measured by isoelectric focussing (IEF) and is used as a marker of chronic alcoholism.

The Transferrin Cycle Helps in Cellular Uptake of Iron

Transferrin receptor 1 (TfR1) is present on the surface of almost all cells, especially erythroid precursors in the bone marrow. Transferrin binds to these receptors and is internalized by receptor-mediated endocytosis (similar to LDL receptors described in Chapter 25). The acidic pH inside the late endosome causes iron to dissociate from transferrin (Tf). The dissociated iron leaves the endosome via DMT1 to enter the cytoplasm. Unlike the protein component of LDL, apoTf (Tf without iron bound to it) is not degraded within the endosome. Instead, it remains associated with its receptor and returns to the plasma membrane. It can then dissociate from its receptor, re-enter plasma, and pick up more iron to be delivered to cells. This is called the **transferrin cycle (Figure 50–6).**

Transferrin receptor 2 (TfR2) is expressed primarily on the surface of hepatocytes and also in the crypt cells of the small intestine. It has a low affinity for Tf-Fe and does not appear to be involved in iron uptake by cells. It plays a role in sensing body iron stores in association with other proteins, as discussed later.

Iron in Senescent Erythrocytes Is Recycled by Macrophages

Normally erythrocytes have a lifespan of approximately 120 days. Senescent or damaged erythrocytes are phagocytosed by macrophages of the reticuloendothelial system (RES) present in the spleen and liver. Around 200 billion erythrocytes (in about 40 mL of blood) are catabolized every day in this way. Within the macrophage, heme derived from hemoglobin is broken down by **heme oxygenase,** converting it to biliverdin. Carbon monoxide and iron are released as by-products. Iron released from heme is exported from the phagocytic vesicle in the macrophage by **NRAMP 1** (natural resistance-associated macrophage protein 1), a transporter homologous to DMT1. It is subsequently transported into the circulation via ferroportin in the plasma membrane of the macrophage (**Figure 50–7**). Therefore, ferroportin plays a central role, not only in iron absorption in the intestine, but also in iron release from macrophages. **Ceruloplasmin** (see below) is a copper-containing plasma protein synthesized by liver. It has ferrioxidase activity. Ceruloplasmin is required for the oxidation of Fe^{2+} to Fe^{3+}. Fe^{3+} is then bound to transferrin in blood. The iron released from macrophages in this way (about 25 mg per day) is recycled and forms the major source of iron for the body. In comparison, intestinal iron absorption contributes only 1–2 mg of the body's daily iron needs.

Ferritin Stores Iron in Cells

Under normal circumstances, **ferritin** stores excess iron in various tissues and constitutes approximately 1 g of the total body iron content. Ferritin has a molecular mass of approximately 440 kDa. It is composed of 24 subunits, which surround 3000–4500 ferric atoms. The subunits may be of the H (heavy) or the L (light) type. The H-subunit possesses ferroxidase activity which is required for iron-loading of ferritin. The function of the L subunit is not clearly known but is proposed to play a role in ferritin nucleation and stability. Normally, there is a small amount of ferritin in human plasma (50–200 μg/dL) proportionate to the total stores of iron in the body. Plasma ferritin levels are, thus, considered to be an **indicator of body iron stores.** However, it is not known whether ferritin in plasma is derived from damaged cells or actively secreted by cells.

Hemosiderin is an ill-defined molecule and appears to be a partly degraded form of ferritin that contains iron. It can be detected in tissues, under conditions of iron overload (**hemosiderosis**), by histological stains (eg, Prussian blue).

Intracellular Iron Homeostasis Is Tightly Regulated

Syntheses of TfR1 and ferritin are reciprocally linked to intracellular iron content. When iron levels are high, ferritin is synthesized to store iron and, since no further uptake of iron is required, synthesis of TfR1 is inhibited. Conversely, when iron

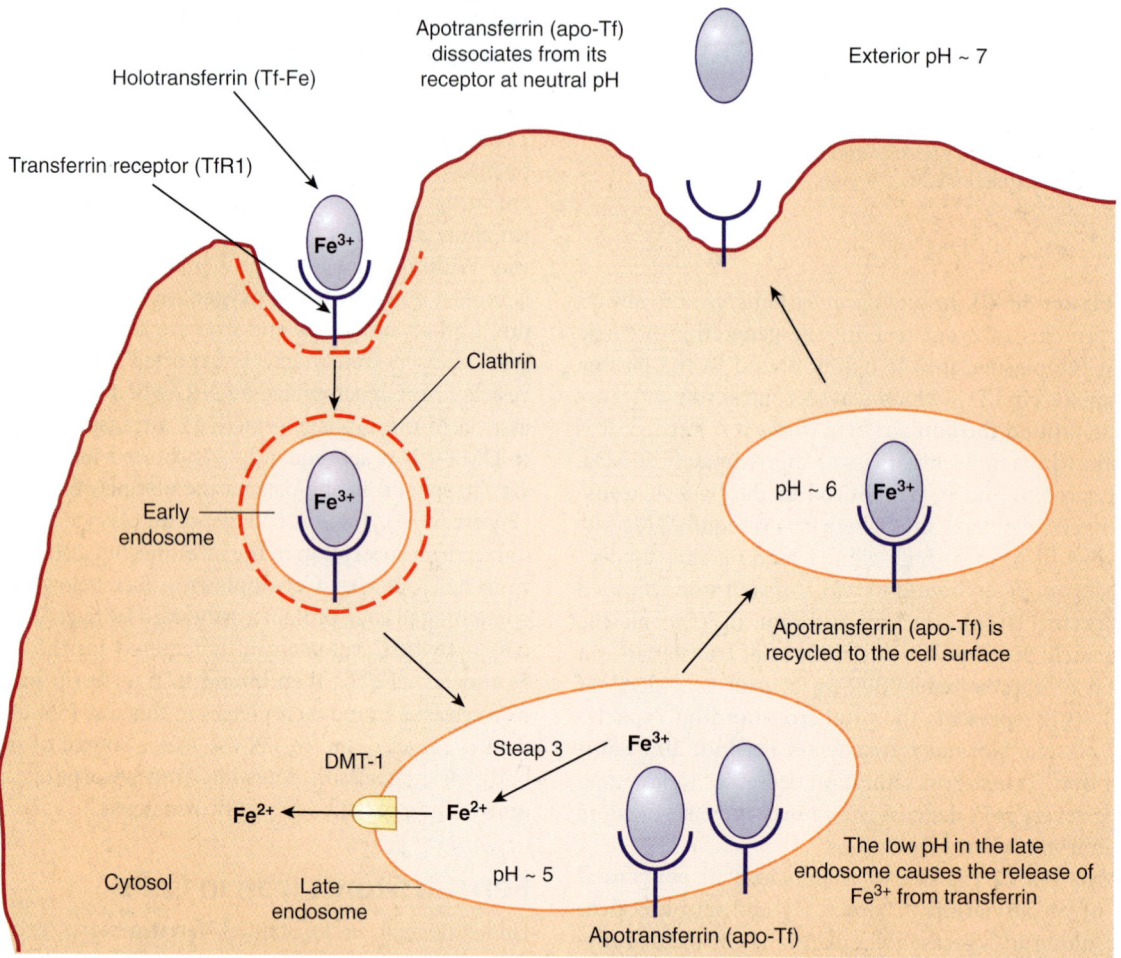

FIGURE 50–6 **The transferrin cycle.** Holotransferrin (Tf-Fe) binds to transferrin receptor 1 (TfR1) present in clathrin-coated pits on the cell surface. The TfR1–Tf–Fe complex is endocytosed and endocytic vesicles fuse to form early endosomes. The early endosomes mature to late endosomes, which have an acidic pH inside. The low pH causes release of iron from its binding sites on transferrin. Apotransferrin (apo-Tf) remains bound to TfR1. Ferric iron is converted to its ferrous form by the ferrireductase, Steap 3. Ferrous iron is then transported into the cytosol via DMT1. The TfR1-apo-Tf complex is recycled back to the cell surface. At the cell surface, apo-Tf is released from TfR1. TfR1 then binds to new Tf-Fe. This completes the transferrin cycle. (Based on Figure 17–48 in Lodish H et al: *Molecular Cell Biology*, 6th ed. WH Freeman, 2008).

levels are low, ferritin is not synthesized while TfR1 is, in order to promote uptake of iron from transferrin in blood.

The mechanisms involved in the regulation of syntheses of ferritin and TfR1 have been elucidated (**Figure 50–8**). This is brought about through regulation of the stability of the mRNAs for ferritin and TfR1. The ferritin and TfR1 mRNAs contain **iron response elements (IREs),** which form hairpin loops at their 5′ and 3′ untranslated regions (UTRs), respectively. IREs are bound by **iron regulatory proteins (IRPs).** The IRPs are sensitive to intracellular iron levels and are induced by low levels. They bind IREs only when intracellular iron levels are low. Binding of IRP to the IRE at the 3′ UTR of TfR1 mRNA stabilizes the TfR1 mRNA, thus increasing TfR1 synthesis and expression on the cell surface. On the other hand, binding of IRP to the IRE at the 5′ UTR of ferritin mRNA blocks translation of ferritin. Similarly, in the absence of IRP binding to IRE (which happens in the presence of high levels of iron), translation of ferritin mRNA is facilitated and TfR1 mRNA

is rapidly degraded. The net result is that, when intracellular iron levels are high, ferritin is synthesized but TfR1 is not, and when intracellular iron levels are low, TfR1 is synthesized and ferritin is not. This is a classical example of control of expression of proteins at the translational level.

Hepcidin Is the Chief Regulator of Systemic Iron Homeostasis

Hepcidin is a protein that is known to play a central role in iron homeostasis in the body. It is synthesized by the liver as an 84-amino-acid precursor protein (prohepcidin). Prohepcidin is cleaved to generate bioactive hepcidin, which is a 25-amino-acid peptide. **Hepcidin binds to the cellular iron exporter, ferroportin, and triggers its internalization and degradation.** Thus, as shown in **Figure 50–9**, hepcidin decreases iron absorption in the intestine (producing a "mucosal block") and also prevents recycling of iron from macrophages. These effects

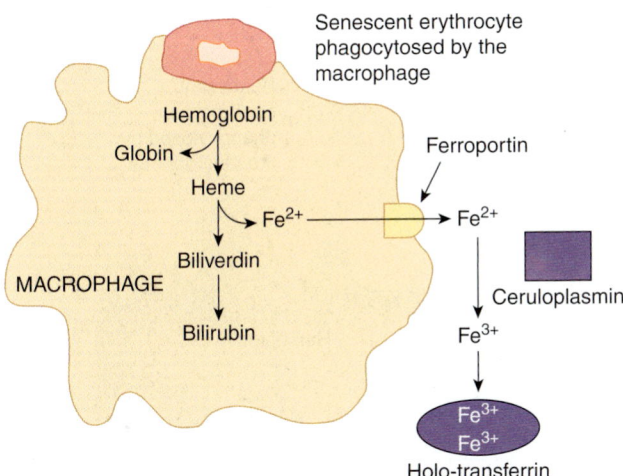

FIGURE 50–7 **Recycling of iron in macrophages.** Senescent erythrocytes are phagocytosed by macrophages. Hemoglobin is degraded and iron is released from heme by the action of the enzyme heme oxygenase. Iron, in the ferrous form, is then transported out of the macrophage via ferroportin (Fp). In the plasma, it is oxidized to the ferric form by ceruloplasmin before binding to transferrin (Tf). Iron circulates in blood tightly bound to Tf.

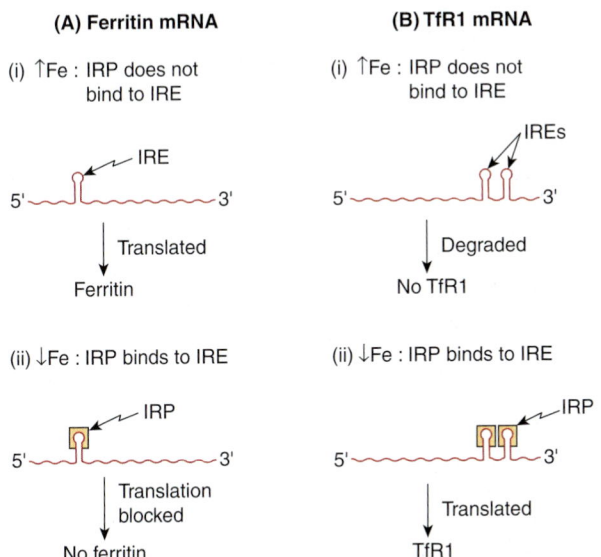

FIGURE 50–8 **Schematic representation of the reciprocal relationship between synthesis of ferritin and the transferrin receptor (TfR1).** The mRNA for ferritin is represented on the left, and that for TfR1 on the right of the diagram. At high concentrations of iron, the iron bound to the IRP prevents that protein from binding the IREs on either type of mRNA. The mRNA for ferritin is able to be translated under these circumstances, and ferritin is synthesized. On the other hand, when the IRP is not able to bind to the IRE on the mRNA for TfR1, that mRNA is degraded. In contrast, at low concentrations of iron, the IRP is able to bind to the IREs on both types of mRNA. In the case of the ferritin mRNA, this prevents it from being translated. Hence ferritin is not synthesized. In the case of the mRNA for TfR1, binding of the IRP prevents that mRNA from being degraded, it is translated, and TfR1 is synthesized. IRP, iron regulatory protein; IRE, iron response element.

result in a reduction in circulating iron levels (hypoferremia). In addition, it reduces placental iron transfer (not shown in Figure 50–9). When plasma iron levels are high, hepatic synthesis of hepcidin increases, thus reducing iron absorption and macrophage iron recycling. The opposite occurs when plasma iron levels are low.

The expression of hepcidin in the liver is a highly regulated process, many aspects of which are still under investigation. Hepcidin expression is regulated by systemic iron availability, erythropoiesis, inflammation, and hypoxia, among other signals. Studies on patients with hereditary hemochromatosis (described below and Case 10 in Chapter 57) have yielded a large amount of information regarding hepcidin regulation.

Hepatocytes have an "iron-sensing complex" on their surface. There are several proteins that constitute this complex (see **Figure 50–10**). These include the **HFE protein** (which is most commonly mutated in hereditary hemochromatosis), **TfR1**, **TfR2**, and **hemojuvelin (HJV)**. The HFE protein is a major histocompatibility (MHC) class 1-like molecule that is expressed on the cell surface, bound to **β_2-microglobulin** (a component of class I MHC molecules, not shown in Figure 50–10) and TfR1. HFE binds to TfR1 at a site that overlaps with its binding site for holotransferrin (Tf-Fe). Tf-Fe, therefore, competes with HFE for binding to TfR1. When displaced from TfR1 by Tf-Fe (which happens when Tf-Fe levels are high), HFE binds to TfR2, which is also expressed on the surface of hepatocytes. The HFE-TfR2 complex is further stabilized by binding to Tf-Fe. This complex triggers an intracellular signaling cascade which ultimately results in upregulation of expression of the hepcidin gene (*HAMP*). The critical roles of HFE, HJV, and TfR2 in hepcidin regulation are demonstrated by the fact that mutations in their genes (like mutations in *HAMP*) are all characterized by low circulating levels of hepcidin and iron overload.

Bone morphogenetic proteins (BMPs), especially BMP6, play an important role in regulating basal hepcidin expression. BMPs act by mechanisms that are distinct from HFE, but considerable cross-talk exists between these pathways. BMP levels are regulated by iron by mechanisms that have not yet been fully elucidated. BMP binds to its cell-surface receptors (BMPR). This binding is facilitated by HJV, which acts as a BMP co-receptor. The activation of the BMPR-HJV complex causes phosphorylation of SMAD (intracellular signaling proteins) (Figure 50–10), which subsequently results in transcriptional activation of hepcidin.

Hepcidin levels are also regulated by **erythropoietic signals.** For example, it is known that hepcidin levels are downregulated in β-thalassemia major, which is characterized by ineffective erythropoiesis and iron overload. Recently, **growth differentiation factor 15 (GDF15)** and **twisted gastrulation 1 (TWSG1)** have been shown to be critical molecules secreted by erythroblasts. These **downregulate** hepatic hepcidin expression in β-thalassemia.

Inflammation is known to induce hepcidin expression. **Interleukin–6 (IL-6),** an inflammatory cytokine, signals through the JAK-STAT (Janus Kinase—Signal Transducer and Activator of Transcription) pathway to mediate this effect

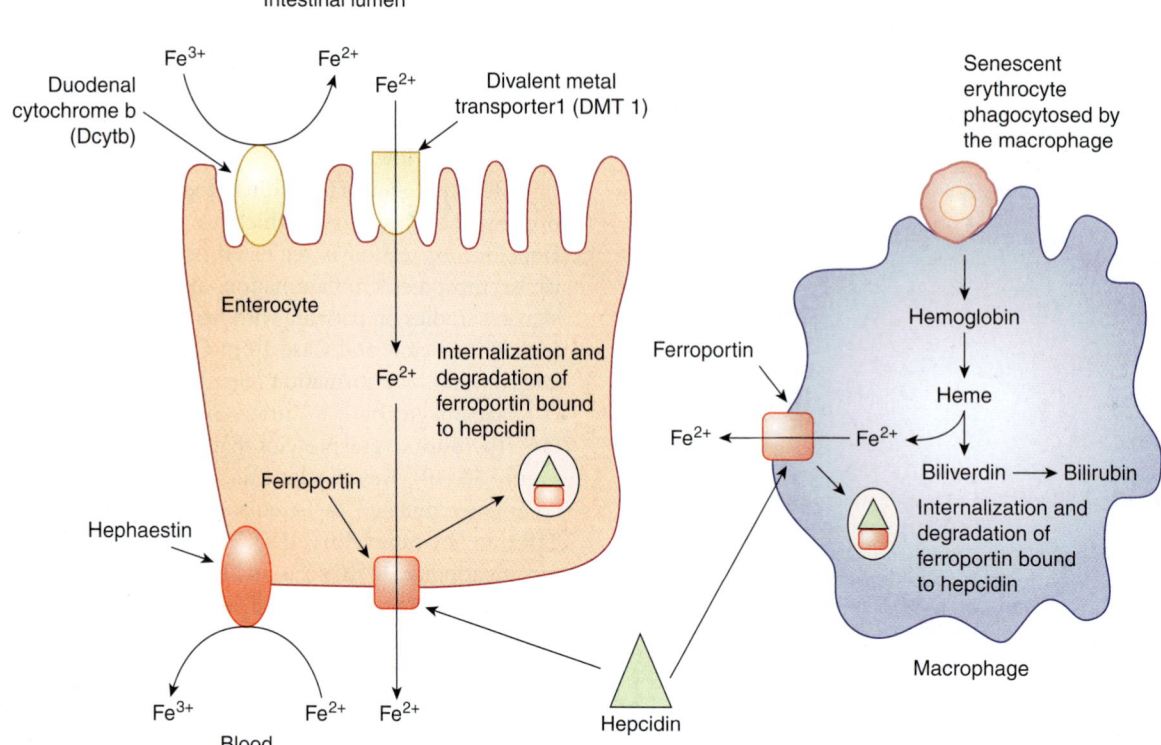

FIGURE 50–9 **Role of hepcidin in systemic iron regulation.** Hepcidin binds to and triggers the internalization and degradation of ferroportin expressed on the surface of enterocytes and macrophages. This decreases iron absorption from the intestine and inhibits iron release from macrophages, leading to hypoferremia. (Based on Andrews NC: Forging a field: the golden age of iron biology. Blood 2008;112(2): 219.)

(Figure 50–10). Anemia that is associated with chronic inflammation (**anemia of inflammation or AI**) is probably due to inflammation-mediated upregulation of hepcidin. AI manifests as a microcytic, hypochromic anemia that is refractory to iron supplementation. In addition to the above-mentioned factors, **hypoxia** is also known to induce hepcidin and this effect is mediated by stabilization of the hypoxia-inducible factors 1 and 2 (HIF-1 and HIF-2).

Iron Deficiency Is Highly Prevalent

Iron deficiency is extremely common in many parts of the world, especially in developing countries. There are **several causes** for iron deficiency. Decreased intake of iron in the diet and malabsorption are the most common causes in developing countries. Premenopausal women are more prone to developing iron deficiency since they have higher daily requirements of iron, owing to loss of blood during menstruation. Chronic blood losses, due to gastrointestinal bleeding or excessive menstruation, are also common causes for iron deficiency.

The **development** of iron deficiency anemia progresses through three stages—negative iron balance, iron-deficient erythropoiesis, and finally, iron-deficiency anemia. **Negative iron balance** is the initial stage where intestinal iron absorption is insufficient to meet the body's demands. This results in the body's iron stores being mobilized to meet requirements. This

leads to progressive depletion of the iron stores. At this stage, all laboratory tests are normal, except for a low serum ferritin, which, as described above, is a marker of body iron stores. If this situation persists, iron stores are depleted and serum ferritin levels fall below 15 μg/dL. At this stage, the transferrin levels in blood increase resulting in an increase in the **TIBC**. The **transferrin saturation,** however, decreases and may fall below 20%. Hemoglobin synthesis is impaired. This is the stage of **iron-deficient erythropoiesis.** If iron deficiency is not corrected at this stage, hemoglobin levels in blood gradually begin to fall, resulting in the final stage of **iron-deficiency anemia.** This stage is characterized by a **hypochromic, microcytic blood picture.** Patients typically present with fatigue, pallor, and reduced exercise capacity. **Table 50–4** gives a summary of changes in common laboratory tests at various stages in the development of iron deficiency anemia.

Red-cell protoporphyrin levels are increased in iron deficiency. The presence of protoporphyrin in erythrocytes reflects impaired ferrochelatase-catalyzed incorporation of iron into the protoporphyrin IX ring. **Serum transferrin receptor protein** or **soluble transferrin receptor (sTfR)** is also considered to be a useful marker of iron deficiency. Under normal conditions, TfR1 is highly expressed on the surface of erythroid cells and a certain proportion is released into the circulation by proteolytic cleavage. This is referred to as sTfR. Increased serum sTfR levels reflect increased expression of TfR1 on the erythroid cell surface during iron

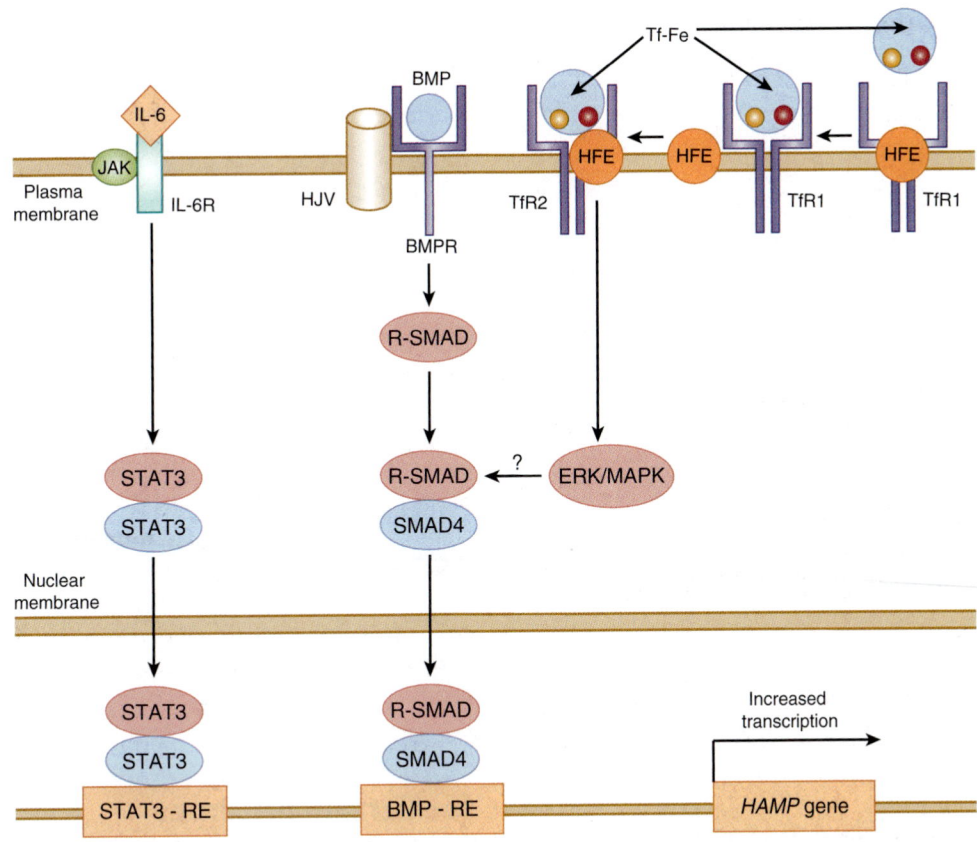

FIGURE 50–10 **Regulation of hepcidin gene expression.** Tf-Fe (holotransferrin) competes with HFE for binding to TFR1. High levels of Tf-Fe displace HFE from its binding site on TfR1. Displaced HFE binds to TfR2 along with Tf-Fe to signal via the ERK/MAPK pathway to induce hepcidin. BMP binds to its receptor BMPR and HJV (co-receptor) to activate R-SMAD. R-SMAD dimerizes with SMAD4, translocates to the nucleus where it binds to the BMP-RE, resulting in transcriptional activation of hepcidin as shown. IL-6, which is a biomarker of inflammation, binds to its cell-surface receptor and activates the JAK-STAT pathway. STAT3 translocates to the nucleus where it binds to its response element (STAT-RE) on the hepcidin gene to induce it. BMP-RE, BMP response element; BMP, bone morphogenetic protein; BMPR, bone morphogenetic protein receptor; ERK-MAPK, extracellular signal-regulated kinase/mitogen-activated protein kinase; *HAMP*, gene encoding hepcidin antimicrobial peptide (hepcidin); HJV, hemojuvelin; IL-6, interleukin 6; IL-6R, interleukin 6 receptor; JAK, Janus-associated kinase; SMAD, Sma and MAD (Mothers Against Decapentaplegic)-related protein; STAT, signal transduction and activator of transcription; STAT3-RE, STAT 3 response element; TfR1, transferrin receptor 1; TfR2, transferrin receptor 2. (Redrawn from Hentz MW, Muckenthaler MU, Gali B et al: Two to tango: regulation of mammalian iron metabolism. Cell 2010;142:24.)

deficiency. Estimation of the serum sTfR level is especially useful in distinguishing between anemia due to iron deficiency and that due to chronic inflammation (AI, described above). It is elevated in the former and remains within the reference range in the later condition. Serum ferritin is not useful in this situation because, ferritin, being an acute phase protein, increases during inflammation, even in the presence of anemia.

TABLE 50–4 **Changes in Various Laboratory Tests Used to Assess Iron-Deficiency Anemia**

Parameter	Normal	Negative Iron Balance	Iron-Deficient Erythropoiesis	Iron-Deficiency Anemia
Serum ferritin (µg/dL)	50–200	Decreased <20	Decreased <15	Decreased <15
Total iron binding capacity (TIBC) (µg/dL)	300–360	Slightly increased >360	Increased >380	Increased >400
Serum iron (µg/dL)	50–150	Normal	Decreased <50	Decreased <30
Transferrin saturation (%)	30–50	Normal	Decreased <20	Decreased <10
RBC protoporphyrin (µg/dL)	30–50	Normal	Increased	Increased
Soluble transferrin receptor (µg/L)	4–9	Increased	Increased	Increased
RBC morphology	Normal	Normal	Normal	Microcytic Hypochromic

Modified, with permission, from Figure 98–2, page 630, Harrison's Principles of Internal Medicine, 17th ed. Fauci AS et al (editors). McGraw-Hill, 2008.

TABLE 50–5   Iron Overload Conditions

Hereditary Hemochromatosis
• HFE- related hemochromatosis (type 1)
• Non- HFE-related hemochromatosis
○ Juvenile hemochromatosis (type 2)
▪ Hepcidin mutation (type 2A)
▪ Hemojuvelin mutation (type 2B)
○ Transferrin receptor 2 mutation (type 3)
○ Ferroportin mutation (type 4)
Secondary Hemochromatosis
• Anemia characterized by ineffective erythropoiesis (eg, thalassemia major)
• Repeated blood transfusions
• Parenteral iron therapy
• Dietary iron overload (Bantu siderosis)
Miscellaneous Conditions Associated with Iron Overload
• Alcoholic liver disease
• Nonalcoholic steatohepatitis
• Hepatitis C infection

Hereditary Hemochromatosis Is Characterized by Iron Overload

Iron overload conditions, also called hemochromatosis, are characterized by increased intestinal iron absorption resulting in increased total body iron stores. The term **hemosiderosis** is used to signify the presence of stainable iron in tissues, which is often a characteristic feature of iron overload conditions.

Iron overload may be hereditary or secondary (**Table 50–5**). **Hereditary hemochromatosis** most often is caused by mutations in the *HFE* gene. Rarer forms of this disease may arise from mutations in the hepcidin (*HAMP*), TfR2, HJV, and ferroportin genes. **Secondary iron overload** is usually associated with ineffective erythropoiesis, as seen in the thalassemia syndromes. Repeated blood transfusions can also result in progressive iron overload. Additional details of the causes, pathogenesis, and clinical manifestations of hereditary hemochromatosis are discussed under Case 10, Chapter 57.

Ceruloplasmin Binds Copper, & Low Levels of This Plasma Protein Are Associated With Wilson Disease

Ceruloplasmin (about 160 kDa) is an α_2-globulin. It has a blue color because of its high copper content and carries 90% of the copper present in plasma. Each molecule of ceruloplasmin binds six atoms of copper very tightly, so that the copper is not readily exchangeable. **Albumin** carries the other ~10% of the plasma copper, but binds the metal less tightly than does ceruloplasmin. Albumin thus donates its copper to tissues more readily than ceruloplasmin and appears to be more important

TABLE 50–6   Some Important Enzymes That Contain Copper

• Amine oxidase
• Copper-dependent superoxide dismutase
• Cytochrome oxidase
• Tyrosinase

than ceruloplasmin in copper transport in the human body. Ceruloplasmin exhibits a copper-dependent **oxidase** activity, but its physiologic significance has not been clarified apart from possible involvement in the oxidation of Fe^{2+} in transferrin to Fe^{3+} (see above). The amount of ceruloplasmin in plasma is decreased in liver disease. In particular, low levels of ceruloplasmin are found in **Wilson disease** (hepatolenticular degeneration), a disease due to abnormal metabolism of copper. In order to clarify the description of Wilson disease, we shall first consider **the metabolism of copper** in the human body and then **Menkes disease,** another condition involving abnormal copper metabolism.

Copper Is a Cofactor for Certain Enzymes

Copper is an essential trace element. It is required in the diet because it is the metal cofactor for a variety of enzymes (see **Table 50–6**). Copper plays important roles in cellular respiration (cytochrome *c* oxidase), iron homeostasis (ceruloplasmin), melanin formation (tyrosinase), neurotransmitter production (various enzymes), synthesis of connective tissue (lysyl oxidase), and protection against oxidants (eg, superoxide dismutase). It accepts and donates electrons and is involved in reactions involving dismutation, hydroxylation, and oxygenation. However, **excess copper** can cause problems because it can oxidize proteins and lipids, bind to nucleic acids, and enhance the production of free radicals. It is thus important to have mechanisms that will maintain the amount of copper in the body within normal limits. The body of the normal adult contains about 100 mg of copper, located mostly in bone, liver, kidney, and muscle. The daily intake of copper is about 2–4 mg, with about 50% being absorbed in the stomach and upper small intestine and the remainder excreted in the feces. Copper is carried to the liver **bound to albumin,** taken up by liver cells, and part of it is excreted in the bile. Copper also leaves the liver attached to **ceruloplasmin,** which is synthesized in that organ.

The Tissue Levels of Copper & of Certain Other Metals Are Regulated in Part by Metallothioneins

Metallothioneins are a group of small proteins (about 6.5 kDa), found in the cytosol of cells, particularly of liver, kidney, and intestine. They have a high content of cysteine and can **bind copper, zinc, cadmium,** and **mercury.** The SH groups of cysteine are involved in binding the metals. Acute intake (eg, by injection) of copper and of certain other metals increases the amount (induction) of these proteins in tissues, as does administration of certain hormones or cytokines. These proteins may function to store the above metals in a nontoxic form and are involved in their overall metabolism in the body. Sequestration of copper also diminishes the amount of this metal available to generate free radicals.

Menkes Disease Is Due to Mutations in the Gene Encoding a Copper-Binding P-Type ATPase

Menkes disease ("kinky" or "steely" hair disease) is a disorder of copper metabolism. It is X-linked, affects only male infants, involves the nervous system, connective tissue, and vasculature, and is usually fatal in infancy. Early diagnosis is important, because injections of copper may be effective if the condition is treated promptly. In 1993, it was reported that the basis of Menkes disease was mutations in the gene (the *ATP7A* gene) for a **copper-binding P-type ATPase** (the ATP7A protein). Interestingly, the enzyme showed structural similarity to certain metal-binding proteins in microorganisms. This ATPase is thought to be responsible for directing the efflux of copper from cells. When altered by mutation, copper is not mobilized normally from the intestine, in which it accumulates, as it does in a variety of other cells and tissues, from which it cannot exit. Despite the accumulation of copper, the activities of many copper-dependent enzymes are decreased, perhaps because of a defect of its incorporation into the apoenzymes. Normal liver expresses very little of the ATPase, which explains the absence of hepatic involvement in Menkes disease. This work led to the suggestion that liver might contain a different copper-binding ATPase, which could be involved in the causation of Wilson disease. As described below, this turned out to be the case.

Wilson Disease Is Also Due to Mutations in a Gene Encoding a Copper-Binding P-Type ATPase

Wilson disease is a genetic disease in which copper fails to be excreted in the bile and accumulates in liver, brain, kidney, and red blood cells. It can be regarded as an inability to maintain a near-zero copper balance, resulting in **copper toxicosis.** The increase of copper in liver cells appears to inhibit the coupling of copper to apoceruloplasmin and leads to low levels of ceruloplasmin in plasma. As the amount of copper accumulates, patients may develop a hemolytic anemia, chronic liver disease (cirrhosis and hepatitis), and a neurologic syndrome owing to accumulation of copper in the basal ganglia and other centers. A frequent clinical finding is the **Kayser–Fleischer ring.** This is a green or golden pigment ring around the cornea due to deposition of copper in Descemet's membrane. The major laboratory tests of copper metabolism are listed in **Table 50–7.** If Wilson disease is suspected, a **liver biopsy** should be performed; a value for liver copper of over 250 μg/g dry weight

TABLE 50–7 Major Laboratory Tests Used in the Investigation of Diseases of Copper Metabolism

Test	Normal Adult Range
Serum copper	10–22 μmol/L
Ceruloplasmin	200–600 mg/L
Urinary copper	<1 μmol/24 h
Liver copper	20–50 μg/g dry weight

Source: Based on Gaw A et al: *Clinical Biochemistry.* Churchill Livingstone, 1995. Copyright © 1995 Elsevier Ltd. Reprinted with permission from Elsevier.

along with a plasma level of ceruloplasmin of under 20 mg/dL is diagnostic.

The cause of Wilson disease was also revealed in 1993, when it was reported that a variety of mutations in a gene encoding a **copper-binding P-type ATPase** (ATP7B protein) were responsible. The gene (*ATP7B*) is estimated to encode a protein of 1411 amino acids, which is highly homologous to the product of the gene affected in Menkes disease. In a manner not yet fully explained, a nonfunctional ATPase causes defective excretion of copper into the bile, a reduction of incorporation of copper into apoceruloplasmin, and the accumulation of copper in liver and subsequently in other organs such as brain.

Treatment for Wilson disease consists of a diet low in copper along with lifelong administration of **penicillamine,** which chelates copper, is excreted in the urine, and thus depletes the body of the excess of this mineral.

Another condition involving ceruloplasmin is **aceruloplasminemia.** In this genetic disorder, levels of ceruloplasmin are very low and consequently its ferroxidase activity is markedly deficient. This leads to failure of release of iron from cells, and iron accumulates in certain brain cells, hepatocytes, and pancreatic islet cells. Affected individuals show **severe neurologic signs** and have diabetes mellitus. Use of a chelating agent or administration of plasma or ceruloplasmin concentrate may be beneficial.

Genetic Deficiency of α₁-Antiproteinase (α₁-Antitrypsin) Is Associated With Emphysema & One Type Of Liver Disease

α_1-Antiproteinase (about 52 kDa) was formerly called α_1-antitrypsin, and this name is retained here. It is a single-chain protein of 394 amino acids, contains three oligosaccharide chains, and is the major component (>90%) of the α_1 fraction of human plasma. It is synthesized by hepatocytes and macrophages and is the principal **serine protease inhibitor** (**serpin,** or **Pi**) of human plasma. It inhibits trypsin, elastase, and certain other proteases by forming complexes with them. At least 75 **polymorphic forms** occur, many of which can be separated by electrophoresis. The major genotype is MM, and

A. Active elastase + α_1–AT → Inactive elastase: α_1–AT complex → No proteolysis of lung → No tissue damage

B. Active elastase + ↓ or no α_1–AT → Active elastase → Proteolysis of lung → Tissue damage

FIGURE 50–11   Scheme illustrating (A) normal inactivation of elastase by α_1-antitrypsin and (B) situation in which the amount of α_1-antitrypsin is substantially reduced, resulting in proteolysis by elastase and leading to tissue damage.

its phenotypic product is PiM. There are two areas of clinical interest concerning α_1-antitrypsin. A deficiency of this protein has a role in certain cases (approximately 5%) of **emphysema.** This occurs mainly in subjects with the **ZZ genotype,** who synthesize PiZ, and also in PiSZ heterozygotes, both of whom secrete considerably less protein than PiMM individuals. Considerably less of this protein is secreted as compared with PiM. When the amount of α_1-antitrypsin is deficient and polymorphonuclear white blood cells increase in the lung (eg, during pneumonia), the affected individual lacks a countercheck to proteolytic damage of the lung by proteases such as elastase (**Figure 50–11**). It is of considerable interest that a particular **methionine** (residue 358) of α-antitrypsin is involved in its binding to proteases. **Smoking** oxidizes this methionine to methionine sulfoxide and thus inactivates it. As a result, affected molecules of α_1-antitrypsin no longer neutralize proteases. This is particularly devastating in patients (eg, PiZZ phenotype) who already have low levels of α_1-antitrypsin. The further diminution in α_1-antitrypsin brought about by smoking results in increased proteolytic destruction of lung tissue, accelerating the development of emphysema. **Intravenous administration of α_1-antitrypsin** (augmentation therapy) has been used as an adjunct in the treatment of patients with emphysema due to α_1-antitrypsin deficiency. Attempts are being made, using the techniques of protein engineering, to replace methionine 358 by another residue that would not be subject to oxidation. The resulting "mutant" α_1-antitrypsin would thus afford protection against proteases for a much longer period of time than would native α_1-antitrypsin. Attempts are also being made to develop **gene therapy** for this condition. One approach is to use a modified adenovirus (a pathogen of the respiratory tract) into which the gene for α_1-antitrypsin has been inserted. The virus would then be introduced into the respiratory tract (eg, by an aerosol). The hope is that pulmonary epithelial cells would express the gene and secrete α_1-antitrypsin locally. Experiments in animals have indicated the feasibility of this approach.

Deficiency of α_1-antitrypsin is also implicated in one type of **liver disease** (α_1-antitrypsin deficiency liver disease). In this condition, molecules of the **ZZ phenotype** accumulate and aggregate in the cisternae of the endoplasmic reticulum of hepatocytes. Aggregation is due to formation of **polymers** of mutant α_1-antitrypsin, the polymers forming via a strong interaction between a specific loop in one molecule and a prominent β-pleated sheet in another (loop-sheet polymerization). By mechanisms that are not understood, **hepatitis** results with consequent **cirrhosis** (accumulation of massive amounts of collagen, resulting in fibrosis). It is possible that

administration of a synthetic peptide resembling the loop sequence could inhibit loop-sheet polymerization. Diseases such as α_1-antitrypsin deficiency, in which cellular pathology is primarily caused by the presence of aggregates of aberrant forms of individual proteins, have been named **conformational diseases** (see also Chapter 46). Most appear to be due to the formation of β-sheets by conformationally unstable proteins, which in turn leads to formation of aggregates. Other members of this group of conditions include Alzheimer disease, Parkinson disease, and Huntington disease.

At present, severe α_1-antitrypsin deficiency liver disease can be successfully treated by **liver transplantation.** In the future, introduction of the gene for normal α_1-antitrypsin into hepatocytes may become possible, but this would not stop production of the PiZ protein. **Figure 50–12** is a scheme of the causation of this disease.

α_2-Macroglobulin Neutralizes Many Proteases & Targets Certain Cytokines to Tissues

α_2-**Macroglobulin** is a large plasma glycoprotein (720 kDa) made up of four identical subunits of 180 kDa. It comprises 8–10% of the total plasma protein in humans. Approximately 10% of the **zinc** in plasma is transported by α_2-macroglobulin, the remainder being transported by albumin. The protein is synthesized by a variety of cell types, including monocytes, hepatocytes, and astrocytes. It is the major member of a group

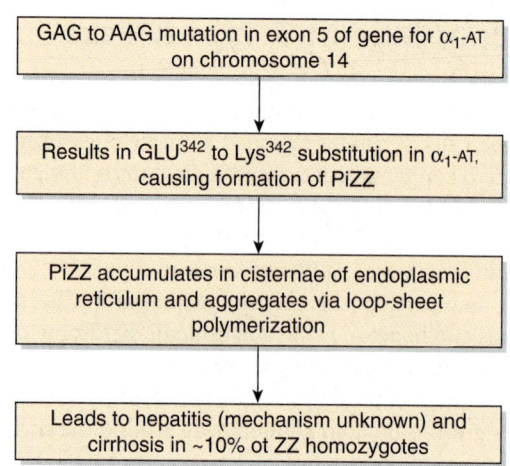

FIGURE 50–12   Scheme of causation of α_1-antitrypsin-deficiency liver disease. The mutation shown causes formation of PiZZ (OMIM 107400). (α_1-AT, α_1-antitrypsin.)

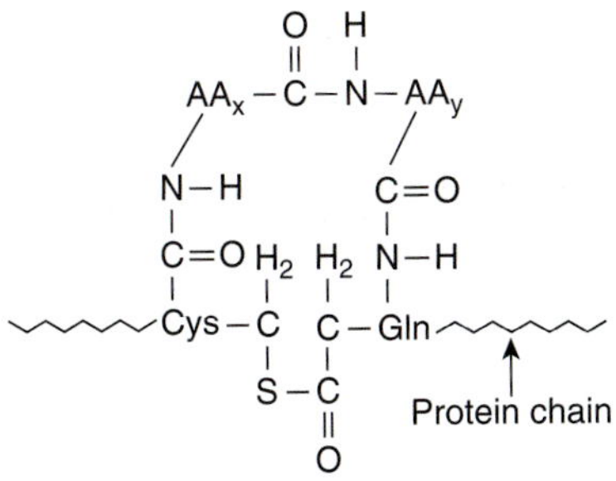

FIGURE 50–13 **An internal cyclic thiol ester bond, as found in α$_2$-macroglobulin.** AA$_x$ and AA$_y$ are neighboring amino acids to cysteine and glutamine.

TABLE 50–8 A Classification of Amyloidosis

Type	Protein Implicated
Primary	Principally light chains of immunoglobulins
Secondary	Serum amyloid A (SAA)
Familial	Transthyretin; also rarely apolipoprotein A-1, cystatin C, fibrinogen, gelsolin, lysozyme
Alzheimer disease	Amyloid β peptide (see Chapter 57, case no. 2)
Dialysis-related	β$_2$-microglobulin

Note: Proteins other than these listed have also been implicated in amyloidosis.

of plasma proteins that include complement proteins C3 and C4. These proteins contain a unique **internal cyclic thiol ester bond** (formed between a cysteine and a glutamine residue, see **Figure 50–13**) and for this reason have been designated as the **thiol ester plasma protein family.** This bond is highly reactive and is involved in some of the biologic actions of α$_2$-macroglobulin.

α$_2$-Macroglobulin binds many proteinases and is thus an important **panproteinase inhibitor.** The α$_2$-macroglobulin-proteinase complexes are rapidly cleared from the plasma by a receptor located on many cell types. In addition, α$_2$-macroglobulin binds many **cytokines** (platelet-derived growth factor, transforming growth factor-β, etc) and appears to be involved in targeting them toward particular tissues or cells. Once taken up by cells, the cytokines can dissociate from α$_2$-macroglobulin and subsequently exert a variety of effects on cell growth and function. The binding of proteinases and cytokines by α$_2$-macroglobulin involves different mechanisms that will not be considered here.

Amyloidosis Occurs by the Deposition of Proteins or Protein Fragments in Various Tissues

Amyloidosis is the accumulation of various insoluble fibrillar proteins between the cells of tissues to an extent that affects function. The accumulation is generally due to either **increased production** of certain proteins or **accumulation of mutated forms** of other proteins (see below). One or more organs or tissues may be affected, and the clinical picture depends on the sites and extent of deposition of amyloid fibrils. The fibrils generally represent proteolytic fragments of various plasma proteins and possess a **β-pleated sheet structure.** The term "amyloidosis" is a misnomer, as it was originally thought that the fibrils were starch-like in nature.

Amyloidosis is now generally classified as AX, where A represents amyloidosis and X the protein in the fibrils. However, this system will not be used here. A simple **classification** of amyloidosis is shown in **Table 50–8.** **Primary** amyloidosis is usually due to a monoclonal plasma cell disorder in which the protein that accumulates is a fragment of a **light chain** (see

below) of an immunoglobulin. **Secondary** amyloidosis usually occurs secondary to chronic infections or cancer and is due to accumulation of degradation products of **serum amyloid A** (SAA). Increased synthesis of SAA occurs in chronic inflammatory states due to elevated levels of certain inflammatory cytokines that stimulate the liver to produce more of this protein. **Familial** amyloidosis results from accumulation of mutated forms of certain plasma proteins, particularly **transthyretin** (see Table 50–2). Over 80 mutated forms of this protein have been documented. Other plasma proteins can also accumulate in other rare types of familial amyloidosis. In patients undergoing long-term chronic dialysis, the plasma protein **β$_2$-microglobulin** can accumulate, because it is retained in the plasma by dialysis membranes. Accumulation of an amyloid-type protein is believed to be a crucial factor in the causation of Alzheimer disease (see Case 2, Chapter 57). In all, at least 20 different proteins have been implicated in the different types of amyloidosis. The precise factors that determine the deposition of proteolytic fragments in tissues await elucidation. Amyloid fibrils generally have a **P component** associated with them, which is derived from **serum amyloid P component,** a plasma protein closely related to C-reactive protein. Tissue sections containing amyloid fibrils interact with **Congo red dye** and display striking green birefringence when viewed by polarizing microscopy. Deposition of amyloid occurs in patients with a variety of disorders; **treatment of the underlying disorder** should be provided if possible.

In general, experimental approaches to the treatment of amyloidosis can be considered under three headings: (1) preventing production of the precursor protein; (2) stabilizing the structures of precursor proteins so that they do not convert to β-pleated sheet structures; and (3) destabilizing amyloid fibrils so that they reconvert to their normal conformations. For instance, regarding the third approach, several **small ligands** bind avidly to amyloid fibrils. For example, **iodinated anthracycline** binds specifically and with high affinity to all natural amyloid fibrils and promotes their disaggregation in vitro. Another similar approach has been the development of the drug **eprodisate.** Amyloid fibrils bind to glycosaminoglycans (see Chapter 48) in tissues. Eprodisate binds to the GAGs, and thus disrupts the binding of the fibrils to these molecules. It is hoped that molecules affecting any of the three processes just mentioned may prove useful in the treatment of amyloidosis.

PLASMA IMMUNOGLOBULINS PLAY A MAJOR ROLE IN THE BODY'S DEFENSE MECHANISMS

The immune system of the body consists of three major components: **B lymphocytes, T lymphocytes** and **the innate immune system.** The B lymphocytes are mainly derived from bone marrow cells in higher animals and from the bursa of Fabricius in birds. The T lymphocytes are of thymic origin. The **B cells** are responsible for the synthesis of circulating, humoral antibodies, also known as **immunoglobulins.** The **T cells** are involved in a variety of important **cell-mediated immunologic processes** such as graft rejection, hypersensitivity reactions, and defense against malignant cells and many viruses. The **innate immune system** defends against infection in a nonspecific manner and unlike B and T cells is **not adaptive.** It contains a variety of cells such as phagocytes, neutrophils, natural killer cells, and others. Case number 1 in Chapter 57 describes one condition in which there is a genetic deficiency of T cells due to mutation in the gene encoding adenosine deaminase. There are a variety of other conditions in which **various components of the immune system are deficient due to mutations.** Most of these are characterized by **recurrent infections,** which must be treated vigorously by, for example,

administration of immunoglobulins (if these are deficient) and appropriate antibiotics.

This section considers only the plasma immunoglobulins, which are synthesized mainly in **plasma cells.** These are specialized cells of B-cell lineage that synthesize and secrete immunoglobulins into the plasma in response to exposure to a variety of **antigens.**

All Immunoglobulins Contain a Minimum of Two Light & Two Heavy Chains

Immunoglobulins contain a minimum of two identical light (L) chains (23 kDa) and two identical heavy (H) chains (53–75 kDa), held together as a tetramer (L_2H_2) by disulfide bonds. The structure of IgG is shown in **Figure 50–14**; it is **Y-shaped**, with binding of antigen occurring at both tips of the Y. Each chain can be divided conceptually into specific domains, or regions, that have structural and functional significance. The half of the light (L) chain toward the carboxyl terminal is referred to as the **constant region (C_L),** while the amino terminal half is the **variable region** of the light chain (V_L). Approximately one-quarter of the heavy (H) chain at the amino terminals is referred to as its **variable region** (V_H), and the other three-quarters of the heavy chain are referred to as the **constant regions (C_H1, C_H2, C_H3)** of that H chain. The portion of the immunoglobulin molecule that binds the specific antigen is formed by the amino terminal portions (variable regions) of

FIGURE 50–14   **Structure of IgG.** The molecule consists of two light (L) chains and two heavy (H) chains. Each light chain consists of a variable (V_L) and a constant (C_L) region. Each heavy chain consists of a variable region (V_H) and a constant region that is divided into three domains (C_H1, C_H2, and C_H3). The C_H2 domain contains the complement-binding site and the C_H3 domain contains a site that attaches to receptors on neutrophils and macrophages. The antigen-binding site is formed by the hypervariable regions of both the light and heavy chains, which are located in the variable regions of these chains (see Figure 50–10). The light and heavy chains are linked by disulfide bonds, and the heavy chains are also linked to each other by disulfide bonds. (Reproduced, with permission, from Parslow TG et al (editors): *Medical Immunology,* 10th ed. McGraw-Hill, 2001.)

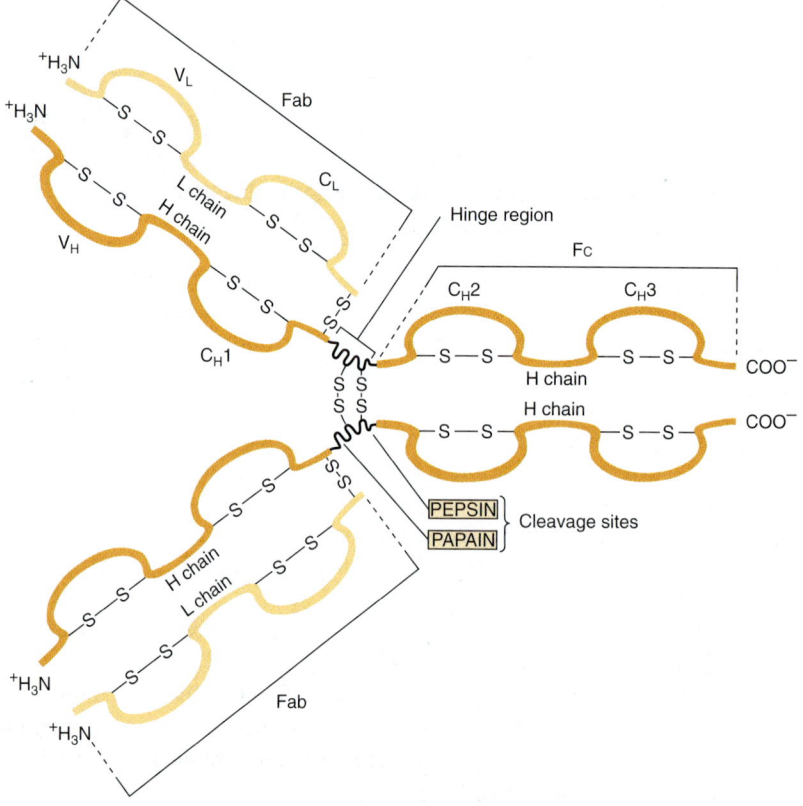
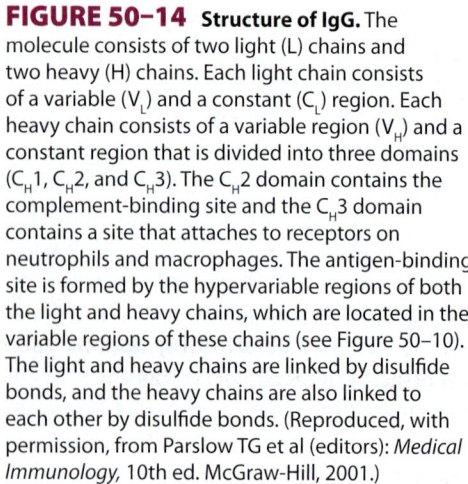

TABLE 50–9 Properties of Human Immunoglobulins

Property	IgG	IgA	IgM	IgD	IgE
Percentage of total immunoglobulin in serum (approximate)	75	15	9	0.2	0.004
Serum concentration (mg/dL) (approximate)	1000	200	120	3	0.05
Sedimentation coefficient	7S	7S or 11S[1]	19S	7S	8S
Molecular weight (× 1000)	150	170 or 400[1]	900	180	190
Structure	Monomer	Monomer or dimer	Monomer or pentamer	Monomer	Monomer
H-chain symbol	γ	α	μ	δ	ε
Complement fixation	+	—	+	—	—
Transplacental passage	+	—	—	?	—
Mediation of allergic responses	—	—	—	—	+
Found in secretions	—	+	—	—	—
Opsonization	+	—	—[2]	—	—
Antigen receptor on B cell	—	—	+	?	—
Polymeric form contains J chain	—	+	+	—	—

Source: Reproduced, with permission, from Levinson W, Jawetz E: *Medical Microbiology and Immunology,* 7th ed. McGraw-Hill, 2002.
[1]The 11S form is found in secretions (eg, saliva, milk, and tears) and fluids of the respiratory, intestinal, and genital tracts.
[2]IgM opsonizes indirectly by activating complement. This produces C3b, which is an opsonin.

both the H and L chains—that is, the V_H and V_L domains. The domains of the protein chains consist of two sheets of antiparallel distinct stretches of amino acids that bind antigen.

As depicted in Figure 50–14 digestion of an immunoglobulin by the enzyme **papain** produces two antigen-binding fragments (**Fab**) and one crystallizable fragment (**Fc**), which is responsible for functions of immunoglobulins other than direct binding of antigens. Since there are two Fab regions, IgG molecules bind two molecules of antigen and are termed **divalent.** The site on the antigen to which an antibody binds is termed an **antigenic determinant,** or **epitope.** The area in which papain cleaves the immunoglobulin molecule—that is, the region between the C_H1 and C_H2 domains—is referred to as the "**hinge region.**" The hinge region confers **flexibility** and allows both Fab arms to move independently, thus helping them to bind to antigenic sites that may be variable distances apart (eg, on bacterial surfaces). Fc and hinge regions differ in the different classes of antibodies, but the overall model of antibody structure for each class is similar to that shown in Figure 50–14 for IgG.

All Light Chains Are Either Kappa or Lambda in Type

There are two general types of light chains, **kappa (κ)** and **lambda (λ),** which can be distinguished on the basis of structural differences in their C_L regions. A given immunoglobulin molecule always contains two κ or two λ light chains—never a mixture of κ and λ. In humans, the κ chains are more frequent than λ chains in immunoglobulin molecules.

The Five Types of Heavy Chain Determine Immunoglobulin Class

Five classes of H chain have been found in humans (**Table 50–9**), distinguished by differences in their C_H regions. They are designated λ, α, μ δ, and ε. The μ and ε chains each have four C_H domains rather than the usual three. The type of H chain determines the class of immunoglobulin and thus its effector function. There are thus five immunoglobulin classes: **IgG, IgA, IgM, IgD,** and **IgE.** The biologic functions of these five classes are summarized in **Table 50–10.**

No Two Variable Regions Are Identical

The **variable regions** of immunoglobulin molecules consist of the V_L and V_H domains and are quite heterogeneous. In fact, no two variable regions from different humans have been found to have identical amino acid sequences. However, amino acid analyses have shown that the variable regions comprise **relatively invariable regions** and other **hypervariable regions** (**Figure 50–15**). L chains have three hypervariable regions (in V_L) and H chains have four (in V_H). These **hypervariable regions** comprise the **antigen-binding site** (located at the tips of the Y shown in Figure 50–14) and dictate the amazing specificity of antibodies. For this reason, hypervariable regions are also termed **complementarity-determining regions (CDRs).** About 5 to 10 amino acids in each hypervariable region (CDR) contribute to the antigen-binding site. CDRs are located on small loops of the variable domains, the surrounding polypeptide regions between the hypervariable regions being termed

TABLE 50–10   **Major Functions of Immunoglobulins**

Immunoglobulin	Major Functions
IgG	Main antibody in the secondary response. Opsonizes bacteria, making them easier to phagocytose. Fixes complement, which enhances bacterial killing. Neutralizes bacterial toxins and viruses. Crosses the placenta.
IgA	Secretory IgA prevents attachment of bacteria and viruses to mucous membranes. Does not fix complement.
IgM	Produced in the primary response to an antigen. Fixes complement. Does not cross the placenta. Antigen receptor on the surface of B cells.
IgD	Found on the surfaces of B cells where it acts as a receptor for antigen.
IgE	Mediates immediate hypersensitivity by causing release of mediators from mast cells and basophils upon exposure to antigen (allergen). Defends against worm infections by causing release of enzymes from eosinophils. Does not fix complement. Main host defense against helminthic infections.

Source: Reproduced, with permission, from Levinson W, Jawetz E: *Medical Microbiology and Immunology,* 7th ed. McGraw-Hill, 2002.

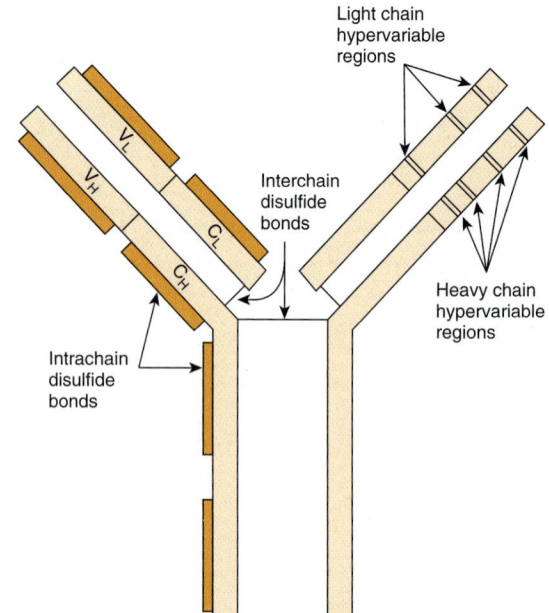

FIGURE 50–15 **Schematic model of an IgG molecule showing approximate positions of the hypervariable regions in heavy and light chains.** The antigen-binding site is formed by these hypervariable regions. The hypervariable regions are also called complementarity-determining regions (CDRs). (Modified and reproduced, with permission, from Parslow TG et al (editors): *Medical Immunology,* 10th ed. McGraw-Hill, 2001.)

framework regions. CDRs from both V_H and V_L domains, brought together by folding of the polypeptide chains in which they are contained, form a single hypervariable surface comprising the **antigen-binding site.** Various combinations of H and L chain CDRs can give rise to many antibodies of different specificities, a feature that contributes to the tremendous diversity of antibody molecules and is termed **combinatorial diversity.** Large antigens interact with all of the CDRs of an antibody, whereas small ligands may interact with only one or a few CDRs that form a pocket or groove in the antibody molecule. The essence of antigen-antibody interactions is **mutual complementarity** between the surfaces of CDRs and epitopes. The interactions between antibodies and antigens involve **noncovalent forces and bonds** (electrostatic and van der Waals forces and hydrogen and hydrophobic bonds).

The Constant Regions Determine Class-Specific Effector Functions

The **constant regions** of the immunoglobulin molecules, particularly the C_H2 and C_H3 (and C_H4 of IgM and IgE), which constitute the Fc fragment, are responsible for the **class-specific effector functions** of the different immunoglobulin molecules (Table 50–9, bottom part), eg, complement fixation or transplacental passage.

Some immunoglobulins such as immune IgG exist only in the basic tetrameric structure, while others such as IgA and IgM can exist as higher order polymers of two, three (IgA), or five (IgM) tetrameric units (**Figure 50–16**).

The L chains and H chains are synthesized as separate molecules and are subsequently assembled within the B cell or plasma cell into mature immunoglobulin molecules, all of which are **glycoproteins.**

Both Light & Heavy Chains Are Products of Multiple Genes

Each immunoglobulin **light chain** is the product of at least three separate structural genes: a **variable region** (V_L) gene, a **joining region** (*J*) gene (bearing no relationship to the J chain of IgA or IgM), and a **constant region** (C_L) gene. Each **heavy chain** is the product of at least **four** different genes: a **variable region** (V_H) gene, a **diversity region** (*D*) gene, a **joining region** (*J*) gene, and a **constant region** (C_H) gene. Thus, the "one gene, one protein" concept is not valid. The molecular mechanisms responsible for the generation of the single immunoglobulin chains from multiple structural genes are discussed in Chapters 35 and 38.

Antibody Diversity Depends on Gene Rearrangements

Each person is capable of generating antibodies directed against perhaps 1 million different antigens. The generation of such immense **antibody diversity** depends upon a number of factors including the existence of multiple gene segments (V, C, J, and D segments), their recombinations (see Chapters 35

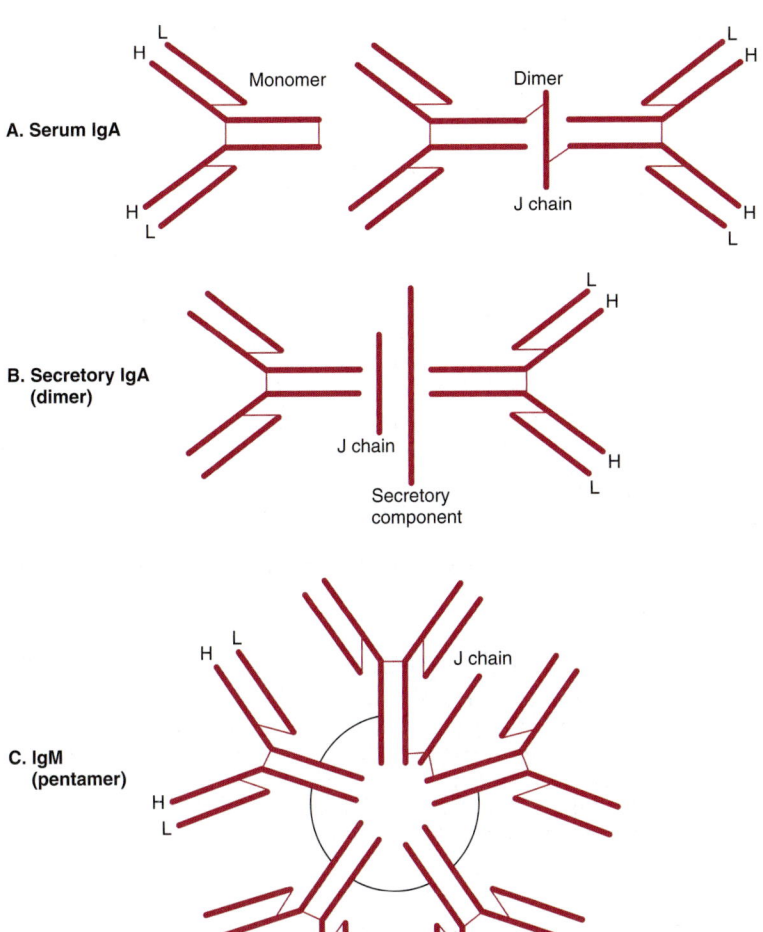

A. Serum IgA

Monomer   Dimer

J chain

B. Secretory IgA
(dimer)

J chain

Secretory
component

C. IgM
(pentamer)

J chain

FIGURE 50–16 **Schematic representation of serum IgA, secretory IgA, and IgM.** Both IgA and IgM have a J chain, but only secretory IgA has a secretory component. Polypeptide chains are represented by thick lines; disulfide bonds linking different polypeptide chains are represented by thin lines. (Reproduced, with permission, from Parslow TG et al (editors): *Medical Immunology,* 10th ed. McGraw-Hill, 2001.)

and 38), the combinations of different L and H chains, a high frequency of somatic mutations in immunoglobulin genes, and **junctional diversity.** The latter reflects the addition or deletion of a random number of nucleotides when certain gene segments are joined together and introduces an additional degree of diversity. Thus, the above factors ensure that **a vast number of antibodies** can be synthesized from several hundred gene segments.

Class (Isotype) Switching Occurs During Immune Responses

In most humoral immune responses, antibodies with identical specificity but of different classes are generated in a specific chronologic order in response to the immunogen (immunizing antigen). For instance, antibodies of the IgM class normally precede molecules of the IgG class. The switch from one class to another is designated **"class or isotype switching,"** and its molecular basis has been investigated extensively. A single type of immunoglobulin light chain can combine with an antigen-specific μ chain to generate a specific IgM molecule. Subsequently, the same antigen-specific light chain combines with a γ chain with an identical V_H region to generate an IgG molecule with antigen specificity identical to that of the original IgM molecule. The same light chain can also combine with an α heavy chain, again containing the identical V_H region, to form an IgA molecule with identical antigen specificity. These three classes (IgM, IgG, and IgA) of immunoglobulin molecules against the same antigen have **identical variable domains** of both their light (V_L) chains and heavy (V_H) chains and are said to share an **idiotype.** (Idiotypes are the antigenic determinants formed by the specific amino acids in the hypervariable regions.) The **different classes** of these three immunoglobulins (called **isotypes**) are thus determined by their **different C_H** regions, which are combined with the same antigen-specific V_H regions.

Both Over- & Underproduction of Immunoglobulins May Result in Disease States

Disorders of immunoglobulins include **increased production** of specific classes of immunoglobulins or even specific immunoglobulin molecules, the latter by clonal tumors of plasma cells called myelomas. **Multiple myeloma** is a neoplastic condition; electrophoresis of serum or urine will usually reveal a large increase of one particular immunoglobulin or one particular light chain (the latter termed a Bence-Jones protein). **Decreased production** may be restricted to a single class of immunoglobulin molecules (eg, IgA or IgG) or may involve underproduction of all classes of immunoglobulins (IgA, IgD, IgE, IgG, and IgM). A severe reduction in synthesis of an immunoglobulin class due to a genetic abnormality can result in a serious immunodeficiency disease—for example, **agammaglobulinemia,** in which production of IgG is markedly affected—because of impairment of the body's defense against microorganisms.

Hybridomas Provide Long-Term Sources of Highly Useful Monoclonal Antibodies

When an antigen is injected into an animal, the resulting antibodies are **polyclonal,** being synthesized by a mixture of B cells. Polyclonal antibodies are directed against a number of different sites (epitopes or determinants) on the antigen and thus are **not monospecific.** However, by means of a method developed by Kohler and Milstein, almost limitless amounts of a single monoclonal antibody specific for one epitope can be obtained.

The method involves **cell fusion,** and the resulting permanent cell line is called a **hybridoma.** Typically, B cells are obtained from the spleen of a mouse (or other suitable animal) previously injected with an antigen or mixture of antigens (eg, foreign cells). The B cells are mixed with mouse **myeloma cells** and exposed to polyethylene glycol, which causes cell fusion. A summary of the principles involved in generating hybridoma cells is given in **Figure 50–17**. Under the conditions used, only the hybridoma cells multiply in cell culture. This involves plating the hybrid cells into hypoxanthine-aminopterin-thymidine (HAT)-containing medium at a concentration such that each dish contains approximately one cell. Thus, a **clone** of hybridoma cells multiplies in each dish. The culture medium is harvested and screened for antibodies that react with the original antigen or antigens. If the immunogen is a mixture of many antigens (eg, a cell membrane preparation), an individual culture dish will contain a clone of hybridoma cells synthesizing a monoclonal antibody to one specific antigenic determinant of the mixture. By harvesting the media from many culture dishes, a battery of monoclonal antibodies can be obtained, many of which are specific for individual components of the immunogenic mixture. The hybridoma cells can be frozen and stored and subsequently thawed when more of the antibody is required; this ensures its long-term supply. The hybridoma cells can also be grown in the abdomen of mice, providing relatively large supplies of antibodies. Attempts to produce **human** monoclonal antibodies are underway.

Because of their **specificity,** monoclonal antibodies have become extremely **useful reagents** in many areas of biology and medicine. For example, they can be used to measure the amounts of many individual proteins (eg, plasma proteins),

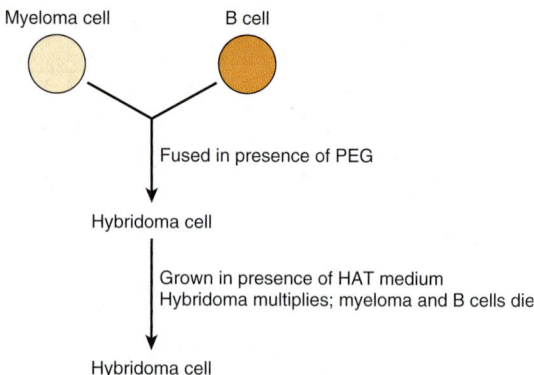

FIGURE 50–17 **Scheme of production of a hybridoma cell.** The myeloma cells are immortalized, do not produce antibody, and are HGPRT⁻ (rendering the salvage pathway of purine synthesis [see Chapter 33] inactive). The B cells are not immortalized, each produces a specific antibody, and they are HGPRT⁺. Polyethylene glycol (PEG) stimulates cell fusion. The resulting hybridoma cells are immortalized (via the parental myeloma cells), produce antibody, and are HGPRT⁺ (both latter properties gained from the parental B cells). The B cells will die in the medium because they are not immortalized. In the presence of HAT, the myeloma cells will also die since the aminopterin in HAT suppresses purine synthesis by the de novo pathway by inhibiting reutilization of tetrahydrofolate (see Chapter 33). However, the hybridoma cells will survive, grow (because they are HGPRT⁺), and—if cloned—produce monoclonal antibody. (HAT, hypoxanthine, aminopterin, and thymidine; HGPRT, hypoxanthineguanine phosphoribosyl transferase.)

to determine the nature of infectious agents (eg, types of bacteria), and to subclassify both normal (eg, lymphocytes) and tumor cells (eg, leukemic cells). In addition, they are being used to direct therapeutic agents to tumor cells and also to accelerate removal of drugs from the circulation when they reach toxic levels (eg, digoxin).

For **therapeutic use in humans,** monoclonal antibodies made in mice can be **humanized.** This can be achieved by attaching the CDRs (the sites that bind antigens) onto appropriate sites in a human immunoglobulin molecule. This produces an antibody that is very similar to a human antibody, thus markedly **lessening immunogenicity** and the chances of an anaphylactic reaction.

The Complement System Comprises Some 20 Plasma Proteins & Is Involved in Cell Lysis, Inflammation, & Other Processes

Plasma contains approximately 20 proteins that are members of **the complement system.** This system was discovered when it was observed that the addition of fresh serum containing antibodies directed to a bacterium caused its **lysis.** Unlike antibodies, the factor was **labile** when heated at 56°C. Subsequent work has resolved the proteins of the system and how they function; most have been cloned and sequenced. The complement system is involved in the ability to **lyse** various cells, but also in aspects of **inflammation** (eg, chemotaxis and phagocytosis) and in the **clearance of antigen–antibody complexes** from the circulation. Deficiencies of various components of the system due to mutations cause **complement deficiency disorders.** The details of this system are relatively complex, and a textbook of immunology should be consulted. The basic concept is that the **normally inactive proteins** of the system, when triggered by various stimuli, become **activated by proteolysis** and interact in a specific sequence with one or more of the other proteins of the system. **The overall result of activation of the pathway is cell lysis** and generation of **peptide or polypeptide fragments** that are involved in aspects of inflammation. The complement system resembles blood coagulation (Chapter 51) in that it involves both **conversion of inactive precursors to active products by proteases** and **a cascade with amplification.**

SUMMARY

- Plasma contains many proteins with a variety of functions. Most are synthesized in the liver and are glycosylated.

- Albumin, which is not glycosylated, is the major protein and is the principal determinant of intravascular osmotic pressure; it also binds many ligands, such as drugs and bilirubin.

- Haptoglobin binds extracorpuscular hemoglobin, prevents its loss into the kidney and urine, and hence preserves its iron for reutilization.

- Transferrin binds iron, transporting it to sites where it is required. Ferritin provides an intracellular store of iron. The regulation of body iron levels involves a battery of proteins, some of which—such as ferroportin and hepcidin—have only been discovered relatively recently. Iron deficiency anemia is a very prevalent disorder. Hereditary hemochromatosis is a genetic disease involving excess absorption of iron; it is discussed in Chapter 57 (Case 10). A number of laboratory tests are available for assessing the status of iron (eg, excess or deficiency) in the human body, and many different proteins are involved in different aspects of its metabolism.

- Ceruloplasmin contains substantial amounts of copper, but albumin appears to be more important with regard to its transport. Both Wilson disease and Menkes disease, which reflect abnormalities of copper metabolism, have been found to

be due to mutations in genes encoding copper-binding P-type ATPases.

- α_1-Antitrypsin is the major serine protease inhibitor of plasma, in particular, inhibiting the elastase of neutrophils. Genetic deficiency of this protein is a cause of emphysema and can also lead to liver disease.

- α_2-Macroglobulin is a major plasma protein that neutralizes many proteases and targets certain cytokines to specific organs.

- Immunoglobulins play a key role in the defense mechanisms of the body, as do proteins of the complement system. Some of the principal features of these proteins are briefly described.

REFERENCES

Andrew NC: Forging a field: the golden age of iron biology. Blood 2008;112:219.

Burtis CA, Ashwood EA, Bruns DE (editors): *Tietz Textbook of Clinical Chemistry and Molecular Diagnostics*, 4th ed. Elsevier Saunders, 2006. (Chapters 20, 26, and 31 give extensive coverage of plasma proteins, complement proteins, immunoglobulins, C-reactive protein, hemoglobin, iron, and bilirubin.)

Craig WY, Ledue TB, Ritchie RF: *Plasma Proteins: Clinical Utility and Interpretation.* Foundation for Blood Research, 2008.

Fauci AS, Branwald E, Kasper DL, et al: *Harrison's Principles of Internal Medicine,* 17th ed. McGraw-Hill, 2008 (Chapters 58, 98, and 308 contain coverage of anemia and polycythemia, iron deficiency and other hypoproliferative anemias, and an introduction to the immune system).

Ganz T: Iron homeostasis: fitting the puzzle pieces together. Cell Metab 2008;7:288.

Hentz MW, Muckenthaler MU, Gali B, et al: Two to tango: regulation of mammalian iron metabolism. Cell 2010;142:24.

Lab Tests Online: http://www.labtestsonline.org/ (A comprehensive web site provided by the American Association of Clinical Chemists that provides information on the measurement and significance of the various plasma proteins discussed in this Chapter, and also on most other lab tests.)

Levinson W: *Review of Medical Microbiology and Immunology,* 11th ed. Appleton & Lange, 2010. (Good description of the basics of Immunology).

Murphy KM, Travers P, Walport M: *Janeway's Immunobiology,* 7th ed. Garland Science Publishing, 2007.

Schaller H, Gerber S, Kaempfer U, et al: *Human Blood Plasma Proteins: Structure and Function.* Wiley, 2008.

Hemostasis & Thrombosis

Peter L. Gross, MD, MSc, FRCP(C), Robert K. Murray, MD, PhD,
& Margaret L. Rand, PhD

- Understand the significance of hemostasis and thrombosis in health and disease.
- Outline the pathways of coagulation that result in the formation of fibrin.
- Identify the vitamin K-dependent coagulation factors.
- Provide examples of genetic disorders that lead to bleeding.
- Describe the process of fibrinolysis.
- Outline the steps leading to platelet aggregation.
- Identify the antiplatelet drugs and their mode of inhibition of platelet aggregation.

BIOMEDICAL IMPORTANCE

Basic aspects of the proteins of the blood coagulation system and of fibrinolysis are described in this chapter. Some fundamental aspects of platelet biology are also presented. Hemorrhagic and thrombotic states can cause serious medical emergencies, and thromboses in the coronary and cerebral arteries are major causes of death in many parts of the world. Rational management of these conditions requires a clear understanding of the bases of blood coagulation, fibrinolysis, and platelet activation.

HEMOSTASIS & THROMBOSIS HAVE THREE COMMON PHASES

Hemostasis is the cessation of bleeding from a cut or severed vessel, whereas **thrombosis** occurs when the endothelium lining blood vessels is damaged or removed (eg, upon rupture of an atherosclerotic plaque). These processes involve blood vessels, platelet aggregation, and plasma proteins that cause formation or dissolution of platelet aggregates and fibrin.

In hemostasis, there is initial vasoconstriction of the injured vessel, causing diminished blood flow distal to the injury. Then, hemostasis and thrombosis share **three phases:**

1. Formation of a loose and temporary **platelet aggregate** at the site of injury. Platelets bind to collagen at the site of vessel wall injury, and form thromboxane A_2 and release ADP, which activate other platelets flowing by the vicinity

of the injury. (The mechanism of platelet activation is described below.) Thrombin, formed during coagulation at the same site, causes further platelet activation. Upon activation, platelets change shape and, in the presence of fibrinogen and/or von Willebrand factor, aggregate to form the hemostatic plug (in hemostasis) or thrombus (in thrombosis).

2. Formation of a **fibrin mesh** that binds to the platelet aggregate, forming a more stable hemostatic plug or thrombus.

3. Partial or complete **dissolution** of the hemostatic plug or thrombus by plasmin.

There Are Three Types of Thrombi

Three types of thrombi or clots are distinguished. All three contain **fibrin** in various proportions.

1. The **white thrombus** is composed of platelets and fibrin and is relatively poor in erythrocytes. It forms at the site of an injury or abnormal vessel wall, particularly in areas where blood flow is rapid (arteries).

2. The **red thrombus** consists primarily of red cells and fibrin. It morphologically resembles the clot formed in a test tube and may form in vivo in areas of retarded blood flow or stasis (eg, veins) with or without vascular injury, or it may form at a site of injury or in an abnormal vessel in conjunction with an initiating platelet plug.

3. A third type is **fibrin deposits** in very small blood vessels or capillaries.

We shall first describe the coagulation pathway leading to the formation of fibrin. Then, we shall briefly describe some aspects of the involvement of platelets and blood vessel walls in the overall process. This separation of clotting factors and platelets is artificial since both play intimate and often mutually interdependent roles in hemostasis and thrombosis, but it facilitates description of the overall processes involved.

Both Extrinsic & Intrinsic Pathways Result in the Formation of Fibrin

Two pathways lead to **fibrin clot** formation: the **extrinsic** and the **intrinsic** pathways. These pathways are not independent, as previously thought. However, this artificial distinction is retained in the following text to facilitate their description.

Initiation of fibrin clot formation in response to **tissue injury** is carried out by the **extrinsic pathway**. The **intrinsic pathway** is activated by negatively charged surfaces **in vitro**, eg, glass. Both pathways lead to activation of **prothrombin to thrombin** and the thrombin-catalyzed cleavage of **fibrinogen** to form the **fibrin** clot. The pathways are complex and involve many different proteins (**Figures 51–1** and **51–2; Table 51–1**). In general, as shown in **Table 51–2**, these proteins can be classified into **five types**: (1) zymogens of serine-dependent proteases, which become activated during the process of coagulation; (2) cofactors; (3) fibrinogen; (4) a transglutaminase, which stabilizes the fibrin clot; and (5) regulatory and other proteins.

The Extrinsic Pathway Leads to Activation of Factor X

The **extrinsic pathway** involves tissue factor, factors VII and X, and Ca^{2+} and results in the production of factor Xa (by convention, activated clotting factors are referred to by use of the suffix a). It is initiated at the site of **tissue injury** with the exposure of **tissue factor** (Figure 51–1), located in the subendothelium and on activated monocytes. Tissue factor interacts with and activates **factor VII** (53 kDa, a zymogen containing vitamin K–dependent γ -carboxyglutamate [Gla] residues; see Chapter 44), synthesized in the liver. It should be noted that in the Gla-containing zymogens (factors II, VII, IX, and X), the Gla residues in the amino terminal regions of the molecules serve as high-affinity binding sites for Ca^{2+}. Tissue factor acts as a cofactor for **factor VIIa,** enhancing its enzymatic activity to activate **factor X** (56 kDa). The reaction by which **factor X** is activated requires the assembly of components, termed the **extrinsic tenase complex,** on a cell membrane surface exposing the procoagulant phospholipid phosphatidylserine; these components are Ca^{2+}, tissue factor, factor VIIa, and factor X. Factor VIIa cleaves an Arg–Ile bond in factor X to produce the two-chain serine protease, **factor Xa.** Tissue factor and factor VIIa also activate factor IX in the intrinsic pathway. Indeed, **the formation of complexes between tissue factor and factor VIIa is now considered to be the key process involved in initiation of blood coagulation in vivo.**

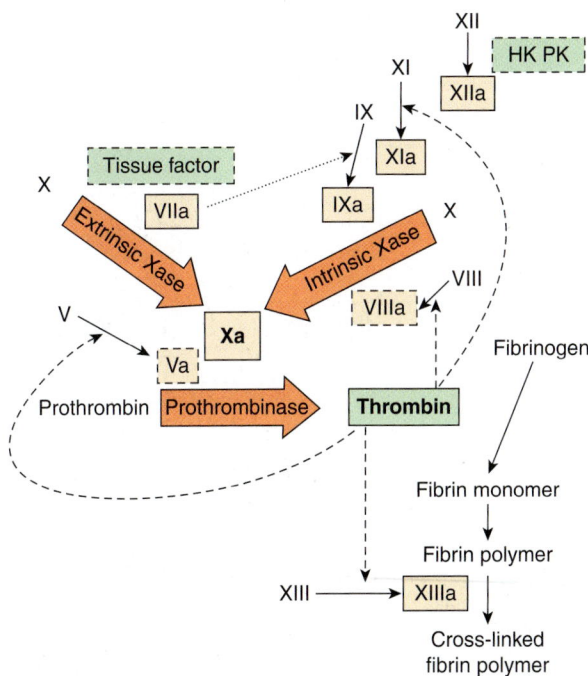

FIGURE 51–1 **The pathways of blood coagulation, with the extrinsic pathway indicated at the top left and the intrinsic pathway at the top right.** The pathways converge in the formation of factor Xa and culminate in the formation of cross-linked fibrin. Complexes of tissue factor and factor VIIa activate not only factor X (extrinsic Xase [tenase]) but also factor IX in the intrinsic pathway (dotted arrow). In addition, thrombin feedback activates at the sites indicated (dashed arrows) and also activates factor VII to factor VIIa (not shown). The three predominant complexes, extrinsic Xase, intrinsic Xase, and prothrombinase, are indicated in the arrows; the reactions require anionic procoagulant phospholipid membrane and calcium. Activated proteases are in solid-outlined boxes; active cofactors are in dash-outlined boxes and inactive factors are not in boxes. (HK, high-molecular-weight kininogen; PK, prekallikrein.)

Tissue factor pathway inhibitor (TFPI) is a major physiologic inhibitor of coagulation. It is a protein that circulates in the blood associated with lipoproteins. TFPI directly inhibits factor Xa by binding to the enzyme near its active site. This factor Xa–TFPI complex then inhibits the factor VIIa–tissue factor complex.

The Intrinsic Pathway Also Leads to Activation of Factor X

The activation of **factor Xa** is the major site where the intrinsic and extrinsic pathways converge (Figure 51–1). The **intrinsic pathway** (Figure 51–1) involves factors XII, XI, IX, VIII, and X as well as prekallikrein, high-molecular-weight (HMW) kininogen, Ca^{2+}, and phospholipid. It results in the production of **factor Xa** that is cleaved by the tenase complex, with factor IXa as the serine protease and factor VIIIa as the cofactor, of the intrinsic pathway. Activation of **factor X** provides an important **link** between the intrinsic and extrinsic pathways.

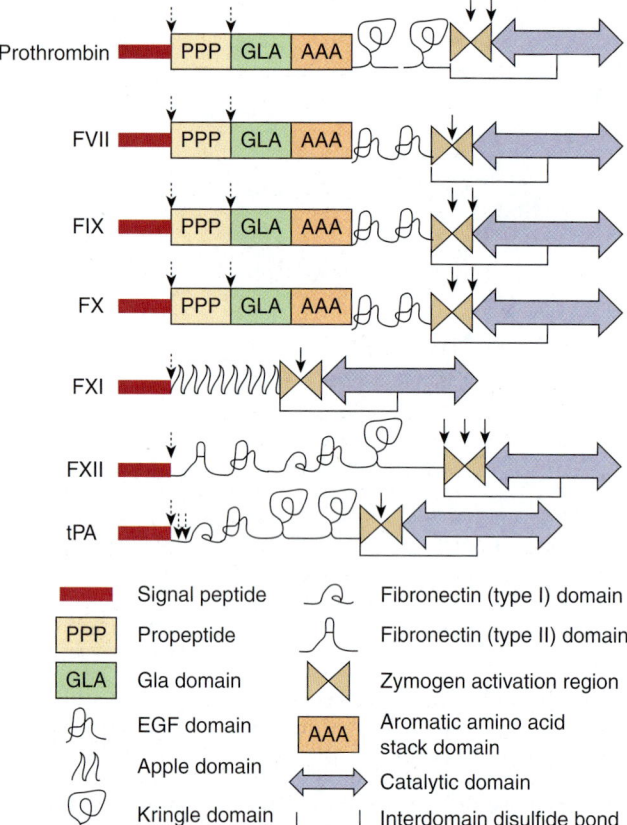

FIGURE 51–2 **The structural domains of selected proteins involved in coagulation and fibrinolysis.** The domains are as identified at the bottom of the figure and include signal peptide, propeptide, Gla (γ-carboxyglutamate) domain, epidermal growth factor (EGF) domain, apple domain, kringle domain, fibronectin (types I and II) domain, the zymogen activation region, aromatic amino acid stack, and the catalytic domain. Interdomain disulfide bonds are indicated, but numerous intradomain disulfide bonds are not. Sites of proteolytic cleavage in synthesis or activation are indicated by arrows (dashed and solid, respectively). FVII, factor VII; FIX, factor IX; FX, factor X; FXI, factor XI; FXII, factor XII; tPA, tissue plasminogen activator. (Adapted, with permission, from Furie B, Furie BC: The molecular basis of blood coagulation. Cell 1988;53:505.)

The **intrinsic pathway** can be initiated with the "contact phase" in which prekallikrein, HMW kininogen, factor XII, and factor XI are exposed to a negatively charged activating surface. Kaolin can be used for in vitro tests as an initiator of the intrinsic pathway. When the components of the contact phase assemble on the activating surface, factor XII is activated to **factor XIIa** upon proteolysis by kallikrein. This factor XIIa, generated by kallikrein, attacks prekallikrein to generate more kallikrein, setting up a reciprocal activation. Factor XIIa, once formed, activates **factor XI** to **XIa** and also releases **bradykinin** (a nonapeptide with potent vasodilator action) from HMW kininogen.

Factor XIa in the presence of Ca^{2+} activates factor IX (55 kDa, a Gla-containing zymogen), to the serine protease, **factor IXa.** This, in turn, also cleaves an Arg–Ile bond in factor X to produce **factor Xa.** This latter reaction requires the assembly

TABLE 51–1 Numerical System for Nomenclature of Blood Clotting Factors

Factor	Common Name	
I	Fibrinogen	These factors are usually referred to by their common names
II	Prothrombin	
III	Tissue factor	
IV	Ca^{2+}	Ca^{2+} is usually not referred to as a coagulation factor
V	Proaccelerin, labile factor, accelerator (Ac-) globulin	
VII[1]	Proconvertin, serum prothrombin conversion accelerator (SPCA), cothromboplastin	
VIII	Antihemophilic factor A, antihemophilic globulin (AHG)	
IX	Antihemophilic factor B, Christmas factor, plasma thromboplastin component (PTC)	
X	Stuart-Prower factor	
XI	Plasma thromboplastin antecedent (PTA)	
XII	Hageman factor	
XIII	Fibrin stabilizing factor (FSF), fibrinoligase	

Note: The numbers indicate the order in which the factors have been discovered and bear no relationship to the order in which they act.
[1]There is no factor VI.

of components, called **the intrinsic tenase complex,** on a membrane surface: Ca^{2+} and factor VIIIa, as well as factors IXa and X.

Factor VIII (330 kDa), a circulating glycoprotein, is not a protease precursor but a cofactor that serves as a receptor for factors IXa and X on the platelet surface. Factor VIII is activated by minute quantities of thrombin to form **factor VIIIa,** which is in turn inactivated upon further cleavage by thrombin.

The role of the **initial steps of the intrinsic pathway** in initiating coagulation has been called into question because patients with a hereditary deficiency of factor XII, prekallikrein or HMW kininogen do not exhibit bleeding problems. Similarly, patients with a deficiency of factor XI may not have bleeding problems. The intrinsic pathway largely serves to **amplify factor Xa** and ultimately **thrombin formation,** through feedback mechanisms (see below). The intrinsic pathway may also be important in **fibrinolysis** (see below) since kallikrein, factor XIIa, and factor XIa can cleave plasminogen and kallikrein can activate single-chain urokinase.

Factor Xa Leads to Activation of Prothrombin to Thrombin

Factor Xa, produced by either the extrinsic or the intrinsic pathway, activates **prothrombin** (factor II) to **thrombin** (factor IIa) (Figure 51–1).

TABLE 51–2 The Functions of the Proteins Involved in Blood Coagulation

Zymogens of Serine Proteases	
Factor XII	Binds to negatively charged surface, eg, kaolin, glass; activated by high-molecular-weight kininogen and kallikrein
Factor XI	Activated by factor XIIa
Factor IX	Activated by factor XIa and factor VIIa
Factor VII	Activated by factor VIIa, factor Xa, and thrombin
Factor X	Activated on the surface of activated platelets by tenase complex (Ca²⁺, factors VIIIa and IXa) and by factor VIIa in the presence of tissue factor and Ca²⁺
Factor II	Activated on the surface of activated platelets by prothrombinase complex (Ca²⁺, factors Va and Xa) [Factors II, VII, IX, and X are Gla-containing zymogens] (Gla = γ-carboxyglutamate)
Cofactors	
Factor VIII	Activated by thrombin; factor VIIIa is a cofactor in the activation of factor X by factor IXa
Factor V	Activated by thrombin; factor Va is a cofactor in the activation of prothrombin by factor Xa
Tissue factor (factor III)	A glycoprotein located in the subendothelium and expressed on activated monocytes to act as a cofactor for factor VIIa
Fibrinogen	
Factor I	Cleaved by thrombin to form fibrin clot
Thiol-Dependent Transglutaminase	
Factor XIII	Activated by thrombin; stabilizes fibrin clot by covalent cross-linking
Regulatory and Other Proteins	
Protein C	Activated to activated protein C (APC) by thrombin bound to thrombomodulin; then degrades factors VIIIa and Va
Protein S	Acts as a cofactor of protein C; both proteins contain Gla (γ-carboxyglutamate) residues
Thrombomodulin	Protein on the surface of endothelial cells; binds thrombin, which then activates protein C

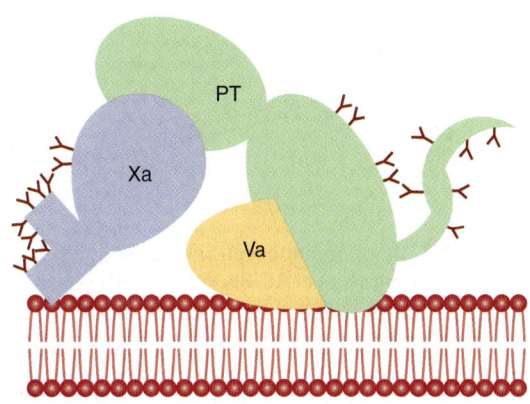

FIGURE 51–3 Diagrammatic representation (not to scale) of the binding of factors Va, Xa, and prothrombin (PT) to the plasma membrane of the activated platelet. A central theme in blood coagulation is the assembly of protein complexes on membrane surfaces. Gamma-carboxyglutamate residues (indicated by Y) on vitamin K-dependent proteins bind calcium and contribute to the exposure of membrane binding sites on these proteins. (Adapted, with permission, from Furie B, Furie BC: The molecular basis of blood coagulation. Cell 1988;53:505.)

functions as a cofactor in a manner similar to that of factor VIII in the tenase complex. When activated to **factor Va** by traces of thrombin, it binds specifically to the platelet membrane (**Figure 51–3**) and forms a complex with factor Xa and prothrombin. It is subsequently inactivated by activated protein C (see below), thereby providing a means of limiting the activation of prothrombin to thrombin. **Prothrombin** (72 kDa; Figure 51–3) is a single-chain glycoprotein synthesized in the liver. The amino terminal region of prothrombin (Figure 51–2) contains 10 Gla residues, and the serine-dependent active protease site is in the catalytic domain close to the carboxyl terminal region of the molecule. Upon binding to the complex of factors Va and Xa on the platelet membrane (Figure 51–3), prothrombin is cleaved by factor Xa at two sites to generate the active, two-chain thrombin molecule, which is then released from the platelet surface.

Conversion of Fibrinogen to Fibrin Is Catalyzed by Thrombin

Thrombin, produced by the prothrombinase complex, in addition to having a potent stimulatory effect on platelets (see below), **converts fibrinogen to fibrin** (Figure 51–1). **Fibrinogen** (factor I, 340 kDa; see Figures 51–1 and 51–4; Tables 51–1 and 51–2) is a soluble plasma glycoprotein that consists of three nonidentical pairs of polypeptide chains (Aα, Bβ,γ)₂ covalently linked by disulfide bonds. The Bβ and γ chains contain asparagine-linked complex oligosaccharides. All three chains are synthesized in the liver; the three genes are on the same chromosome, and their expression is coordinately regulated in humans. The amino terminal regions of the six chains are held in close proximity by a number of disulfide bonds, while the carboxyl terminal regions are spread apart, giving rise to a highly asymmetric, elongated molecule (**Figure 51–4**). The

The activation of prothrombin, like that of factor X, occurs on a membrane surface and requires the assembly of a **prothrombinase complex,** consisting of Ca²⁺, factor Va, factor Xa, and prothrombin. The assembly of the prothrombinase and tenase complexes takes place on the membrane surface of platelets activated to expose the acidic (anionic) phospholipid **phosphatidylserine,** which is normally on the internal side of the plasma membrane of resting, nonactivated platelets.

Factor V (330 kDa), a glycoprotein with homology to factor VIII and ceruloplasmin, is synthesized in the liver, spleen, and kidney and is found in platelets as well as in plasma. It

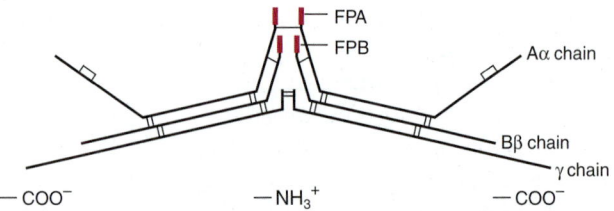

FIGURE 51–4 **Diagrammatic representation (not to scale) of fibrinogen showing pairs of Aα, Bβ, and γ chains linked by disulfide bonds.** (FPA, fibrinopeptide A; FPB, fibrinopeptide B.)

Aα and Bβ portions of the A and B chains, designated **fibrinopeptide A (FPA) and fibrinopeptide B (FPB),** respectively, at the amino terminal ends of the chains, bear excess negative charges as a result of the presence of aspartate and glutamate residues, as well as an unusual tyrosine O-sulfate in FPB. These negative charges contribute to the solubility of fibrinogen in plasma and also serve to prevent aggregation by causing electrostatic repulsion between fibrinogen molecules.

Thrombin (34 kDa), a serine protease formed by the prothrombinase complex, hydrolyzes the four Arg–Gly bonds between the fibrinopeptides and the α and β portions of the Aα and Bβ chains of fibrinogen (**Figure 51–5A**). The release of the fibrinopeptides by thrombin generates **fibrin monomer,** which has the subunit structure $(α, β, γ)_2$. Since FPA and FPB contain only 16 and 14 residues, respectively, the fibrin molecule retains 98% of the residues present in fibrinogen. The removal of the fibrinopeptides exposes binding sites that allow the molecules of fibrin monomers to aggregate spontaneously in a regularly staggered array, forming an insoluble fibrin clot. This initial fibrin clot is rather weak, held together only by the noncovalent association of fibrin monomers.

In addition to converting fibrinogen to fibrin, thrombin also converts **factor XIII to factor XIIIa.** The latter is a highly specific **transglutaminase** that covalently cross-links

fibrin molecules by forming peptide bonds between the amide groups of glutamine and the ε-amino groups of lysine residues (**Figure 51–5B**), yielding a more stable fibrin clot with increased resistance to proteolysis. This fibrin mesh serves to stabilize the hemostatic plug or thrombus.

Levels of Circulating Thrombin Are Carefully Controlled

Once active thrombin is formed in the course of hemostasis or thrombosis, its concentration must be carefully controlled to prevent further fibrin formation or platelet activation. This is achieved in **two ways.** Thrombin circulates as its inactive precursor, prothrombin, which is activated as a result of a cascade of enzymatic reactions, each converting an inactive zymogen to an active enzyme and leading finally to the conversion of prothrombin to thrombin (Figure 51–1). At each point in the cascade, **feedback mechanisms** produce a delicate balance of activation and inhibition. The concentration of factor XII in plasma is approximately 30 μg/mL, while that of fibrinogen is 3 mg/mL, with intermediate clotting factors increasing in concentration as one proceeds down the cascade, showing that the clotting cascade provides **amplification.** The second means of controlling thrombin activity is **the inactivation of any thrombin** formed by **circulating inhibitors,** the most important of which is antithrombin (see below).

The Activity of Antithrombin, an Inhibitor of Thrombin, Is Increased by Heparin

Four naturally occurring **thrombin inhibitors** exist in normal plasma. The most important is **antithrombin,** which contributes approximately 75% of the antithrombin activity. Antithrombin can also inhibit the activities of factors IXa, Xa, XIa,

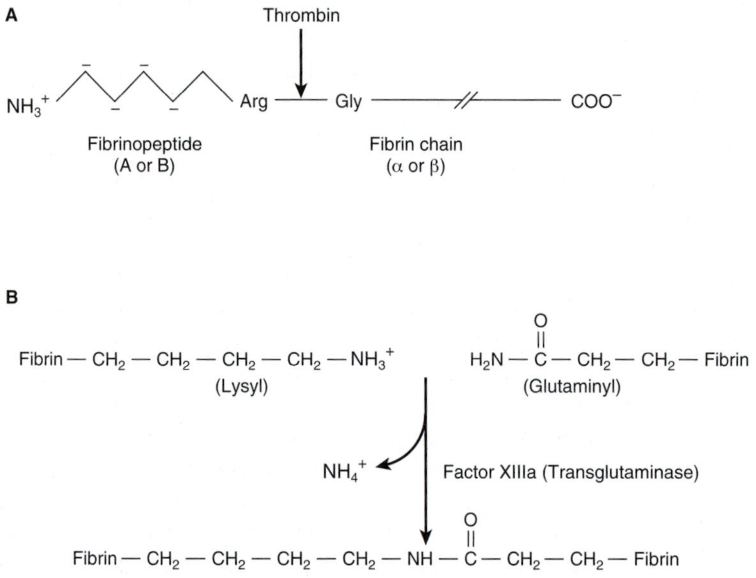

A

$$NH_3^+ \quad \text{———Arg} \xrightarrow{\text{Thrombin}} \text{Gly} \text{——} // \text{——} COO^-$$

Fibrinopeptide (A or B)   Fibrin chain (α or β)

B

Fibrin — CH_2 — CH_2 — CH_2 — CH_2 — NH_3^+ H_2N — $\overset{O}{\overset{||}{C}}$ — CH_2 — CH_2 — Fibrin
(Lysyl)   (Glutaminyl)

$NH_4^+ \longleftarrow$ Factor XIIIa (Transglutaminase)

Fibrin — CH_2 — CH_2 — CH_2 — CH_2 — NH — $\overset{O}{\overset{||}{C}}$ — CH_2 — CH_2 — Fibrin

FIGURE 51–5 **Formation of a fibrin clot.**
(A) Thrombin-induced cleavage of Arg-Gly bonds of the Aα and Bβ chains of fibrinogen to produce fibrinopeptides (left-hand side) and the α and β chains of fibrin monomer (right-hand side). **(B)** Cross-linking of fibrin molecules by activated factor XIII (factor XIIIa).

XIIa, and VIIa complexed with tissue factor. α_2-**Macroglobulin** contributes most of the remainder of the antithrombin activity, with **heparin cofactor II** and α_1-**antitrypsin** acting as minor inhibitors under physiologic conditions.

The endogenous activity of antithrombin is greatly potentiated by the presence of sulfated glycosaminoglycans (heparans) (Chapter 48). These bind to a specific cationic site of antithrombin, inducing a conformational change and promoting its binding to thrombin as well as to its other substrates. This is the basis for the use of **heparin,** a derivatized heparan, in clinical medicine to inhibit coagulation. The anticoagulant effects of heparin can be antagonized by strongly cationic polypeptides such as **protamine,** which bind strongly to heparin, thus inhibiting its binding to antithrombin.

Low-molecular-weight heparins (LMWHs), derived from enzymatic or chemical cleavage of unfractionated heparin, are finding increasing clinical use. They can be administered subcutaneously at home, have greater bioavailability than unfractionated heparin, and do not need frequent laboratory monitoring.

Individuals with **inherited deficiencies of antithrombin** are prone to develop venous thrombosis, providing evidence that antithrombin has a physiologic function and that the coagulation system in humans is normally in a dynamic state.

Thrombin is involved in an additional regulatory mechanism that operates in coagulation. It combines with **thrombomodulin,** a glycoprotein present on the surfaces of endothelial cells. The complex activates **protein C** on the **endothelial protein C receptor.** In combination with **protein S,** activated protein C (APC) degrades factors Va and VIIIa, limiting their actions in coagulation. A genetic deficiency of either protein C or protein S can cause venous thrombosis. Furthermore, patients with **factor V Leiden** (which has a glutamine residue in place of an arginine at position 506) have an increased risk of venous thrombotic disease because factor V Leiden is resistant to inactivation by APC. This condition is termed APC resistance.

Coumarin Anticoagulants Inhibit the Vitamin K–Dependent Carboxylation of Factors II, VII, IX, & X

The **coumarin drugs** (eg, warfarin), which are used as anticoagulants, inhibit the vitamin K-dependent carboxylation of Glu to Gla residues (see Chapter 44) in the amino terminal regions of factors II, VII, IX, and X and also proteins C and S. These proteins, all of which are synthesized in the liver, are dependent on the Ca^{2+}-binding properties of the Gla residues for their normal function in the coagulation pathways. The coumarins act by **inhibiting the reduction of the quinone derivatives of vitamin K to the active hydroquinone forms** (Chapter 44). Thus, the administration of vitamin K will bypass the coumarin-induced inhibition and allow the posttranslational modification of carboxylation to occur. **Reversal** of coumarin inhibition by vitamin K requires 12–24 h, whereas

reversal of the anticoagulant effects of heparin by protamine is almost instantaneous.

Heparin and **warfarin** are widely used in the treatment of thrombotic and thromboembolic conditions, such as deep vein thrombosis and pulmonary embolism. Heparin is administered first, because of its prompt onset of action, whereas warfarin takes several days to reach full effect. Their effects are closely monitored by use of appropriate tests of coagulation (see below) because of the risk of producing hemorrhage.

New oral inhibitors of thrombin (dabigatran) or of factor Xa (rivaroxaban and others) are also used in the treatment of thrombotic conditions.

There Are Several Hereditary Bleeding Disorders, Including Hemophilia A

Inherited deficiencies of the clotting system that result in bleeding are found in humans. The most common is deficiency of factor VIII, causing **hemophilia A,** an X chromosome-linked disease. **Hemophilia B, also X chromosome-linked,** is due to a deficiency of factor IX and has recently been identified as the form of hemophilia that played a major role in the history of the royal families of Europe; its clinical features are almost identical to those of hemophilia A, but the conditions can be separated on the basis of specific assays that distinguish between the two factors.

The **gene for human factor VIII** has been cloned and is one of the largest so far studied, measuring 186 kb in length and containing 26 exons. A variety of mutations in the factor VIII and IX genes have been detected leading to diminished activities of the factor VIII and IX proteins; these include partial gene deletions and point and missense mutations. **Prenatal diagnosis** by DNA analysis after chorionic villus sampling is now possible.

In the past, treatment for patients with hemophilia A and B consisted of administration of **cryoprecipitates** (enriched in factor VIII) prepared from individual donors or lyophilized factor VIII or IX **concentrates** prepared from very large plasma pools. It is now possible to prepare factors VIII and IX by **recombinant DNA technology.** Such preparations are free of contaminating viruses (eg, hepatitis A, B, C, or HIV-1) found in human plasma, but are expensive; their use may increase if cost of production decreases.

The most common hereditary bleeding disorder is **von Willebrand disease,** with a prevalence of up to 1% of the population. It results from a deficiency or defect in **von Willebrand factor,** a large multimeric glycoprotein that is secreted by endothelial cells and platelets into the plasma, where it stabilizes factor VIII. von Willebrand factor also promotes platelet adhesion at sites of vessel wall injury (see below).

Fibrin Clots Are Dissolved by Plasmin

As stated above, the coagulation system is normally in a state of dynamic equilibrium in which fibrin clots are constantly being laid down and dissolved. This latter process is termed

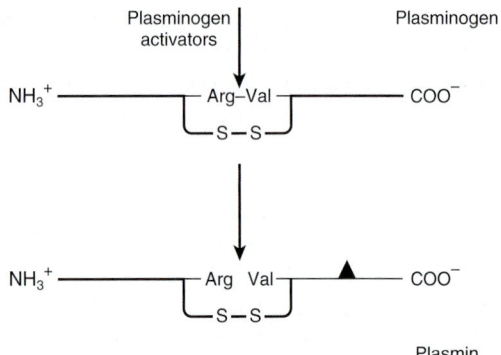

FIGURE 51–6 **Activation of plasminogen.** The same Arg-Val bond is cleaved by all plasminogen activators to give the two-chain plasmin molecule. The solid triangle indicates the serine residue of the active site. The two chains of plasmin are held together by a disulfide bridge.

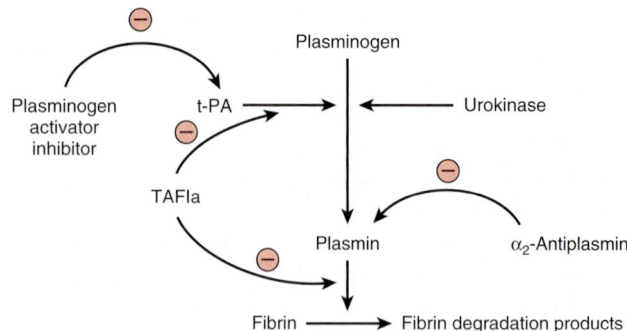

FIGURE 51–7 **Initiation of fibrinolysis by the activation of plasmin.** Scheme of sites of action of tissue plasminogen activator (t-PA), urokinase, plasminogen activator inhibitor, α_2-antiplasmin, and thrombin-activatable fibrinolysis inhibitor (TAFIa) (the last three proteins exert inhibitory actions).

fibrinolysis. **Plasmin,** the serine protease mainly responsible for degrading fibrin and fibrinogen, circulates in the form of its inactive zymogen, **plasminogen** (90 kDa), and any small amounts of plasmin that are formed in the fluid phase under physiologic conditions are rapidly inactivated by the fast-acting plasmin inhibitor, α_2-antiplasmin. Plasminogen binds to fibrin and thus becomes incorporated in clots as they are produced; since plasmin that is formed when bound to fibrin is protected from α_2-antiplasmin, it remains active. **Activators of plasminogen** of various types are found in most body tissues, and all cleave the same Arg–Val bond in plasminogen to produce the two-chain serine protease, plasmin (**Figure 51–6**). The **specificity of plasmin for fibrin** is another mechanism to regulate fibrinolysis. Via one of its kringle domains, plasmin (ogen) specifically binds lysine residues on fibrin and so is increasingly incorporated into the fibrin mesh as it cleaves it. (Kringle domains [Figure 51–2] are common protein motifs of about 100-amino-acid residues in length, that have a characteristic covalent structure defined by a pattern of three disulfide bonds.) Thus, the carboxypeptidase **TAFIa (activated thrombin activatable fibrinolysis inhibitor)** (**Figure 51–7**), which removes terminal lysines from fibrin, is able to inhibit fibrinolysis. Thrombin activates TAFI to TAFIa, thereby inhibiting fibrinolysis during clot formation.

Tissue plasminogen activator (t-PA) (Figures 51–2 and 51–7) is a serine protease that is released into the circulation from vascular endothelium under conditions of injury or stress and is catalytically inactive unless bound to fibrin. Upon binding to fibrin, t-PA cleaves plasminogen within the clot to generate plasmin, which in turn digests the fibrin to form soluble degradation products and thus dissolves the clot. Neither plasmin nor the plasminogen activator can remain bound to these degradation products, and so they are released into the fluid phase, where they are inactivated by their natural inhibitors. Prourokinase is the precursor of a second activator of plasminogen, **urokinase.** Originally isolated from urine, it is now known to be synthesized by cell types such as monocytes

and macrophages, fibroblasts, and epithelial cells. Its main action is probably in the degradation of extracellular matrix. Figure 51–7 indicates the sites of action of five proteins that influence the formation and action of plasmin.

Recombinant t-PA & Streptokinase Are Used as Clot Busters

Alteplase, t-PA produced by recombinant DNA technology, is used therapeutically as a fibrinolytic agent, as is **streptokinase.** However, the latter is less selective than t-PA, activating plasminogen in the fluid phase (where it can degrade circulating fibrinogen) as well as plasminogen that is bound to a fibrin clot. The amount of plasmin produced by therapeutic doses of streptokinase may exceed the capacity of the circulating α_2-antiplasmin, causing fibrinogen as well as fibrin to be degraded and resulting in the bleeding often encountered during fibrinolytic therapy. Because of its relative **selectivity** for degrading fibrin, recombinant t-PA has been widely used to restore the patency of coronary arteries following thrombosis. If administered early enough, before irreversible damage of heart muscle occurs (about 6 h after onset of thrombosis), t-PA can significantly reduce the mortality rate from myocardial damage following coronary thrombosis. Streptokinase has also been widely used in the treatment of coronary thrombosis, but has the disadvantage of being antigenic.

t-PA has also been used in the treatment of ischemic stroke, peripheral arterial occlusion, and pulmonary embolism.

There are a number of disorders, including cancer and sepsis, in which **the concentrations of plasminogen activators increase.** In addition, the **antiplasmin activities** contributed by α_1-antitrypsin and α_2-antiplasmin may be impaired in diseases such as cirrhosis. Since certain bacterial products, such as streptokinase, are capable of activating plasminogen, they may be responsible for the diffuse hemorrhage sometimes observed in patients with disseminated bacterial infections.

Platelet Aggregation Requires Outside-In and Inside-Out Transmembrane Signaling

Platelets normally circulate in an unstimulated disk-shaped form. During hemostasis or thrombosis, they become **activated** and help form hemostatic plugs or thrombi. Three major steps are involved: (1) adhesion to exposed collagen in blood vessels, (2) release (exocytosis) of the contents of their storage granules, and (3) aggregation.

Platelets adhere to collagen via specific receptors on the platelet surface, including the glycoprotein complexes GPIa–IIa (α2β1 integrin; Chapter 52) and GPIb–IX–V, and GPVI. The binding of GPIb–IX–V to collagen is mediated via von Willebrand factor; this interaction is especially important in platelet adherence to the subendothelium under conditions of high shear stress that occur in small vessels and partially stenosed arteries.

Platelets adherent to collagen change shape and spread out on the subendothelium. They release the contents of their **storage granules** (the dense granules and the alpha granules); secretion is also stimulated by thrombin.

Thrombin, formed from the coagulation cascade, is the most potent activator of platelets and initiates activation by interacting with its receptors PAR (protease-activated receptor)-1, PAR-4, and GPIb–IX–V on the platelet plasma membrane (**Figure 51–8A**). The further events leading to platelet activation upon binding to PAR-1 and PAR-4 are examples of outside-in **transmembrane signaling,** in which a chemical messenger outside the cell generates effector molecules inside the cell. In this instance, thrombin acts as the external chemical messenger (stimulus or agonist). The interaction of thrombin with its G protein-coupled receptors PAR-1 and PAR-4 stimulates the activity of an intracellular **phospholipase Cβ.** This enzyme hydrolyzes the membrane phospholipid **phosphatidylinositol 4,5-bisphosphate** (PIP$_2$, a polyphosphoinositide) to form the two internal effector molecules, 1,2-diacylglycerol and 1,4,5-inositol trisphosphate.

Hydrolysis of PIP$_2$ is also involved in the action of many hormones and drugs. **Diacylglycerol** stimulates **protein kinase C,** which phosphorylates the protein **pleckstrin** (47 kDa). This results in aggregation and release of the contents of the storage granules. ADP released from dense granules can also activate platelets via its specific G protein-coupled receptors (Figure 51–8A), resulting in aggregation of additional platelets. IP$_3$ causes release of Ca^{2+} into the cytosol mainly from the dense tubular system (or residual smooth endoplasmic reticulum from the megakaryocyte), which then interacts with calmodulin and myosin light chain kinase, leading to phosphorylation of the light chains of myosin. These chains then interact with actin, causing changes of platelet shape.

Collagen-induced activation of a platelet **cytosolic phospholipase A$_2$** by increased levels of intracellular Ca^{2+} results in liberation of arachidonic acid from platelet membrane phospholipids, leading to the formation of **thromboxane A$_2$** (Chapter 23). Thromboxane A$_2$, in turn, by binding to its G protein-coupled TP receptor, can further activate phospholipase C, promoting platelet aggregation (Figure 51–8A).

Activated platelets, besides forming a platelet aggregate, accelerate the **activation of factor X and prothrombin** by exposing the anionic phospholipid phosphatidylserine on their membrane surface (Figure 51–1).

All of the **aggregating agents,** including thrombin, collagen, ADP, and others such as platelet-activating factor, via an inside-out signaling pathway, modify the platelet surface **glycoprotein complex GPIIb-IIIa** (αIIbβ3; Chapter 52) so that the receptor has a higher affinity for **fibrinogen** or **von Willebrand factor** (**Figure 51–8B**). Molecules of divalent fibrinogen or multivalent von Willebrand factor then link adjacent activated platelets to each other, forming a platelet aggregate. von Willebrand factor-mediated platelet aggregation occurs under conditions of high shear stress. Some agents, including epinephrine, serotonin, and vasopressin, exert synergistic effects with other aggregating agents.

Endothelial Cells Synthesize Prostacyclin & Other Compounds That Affect Clotting & Thrombosis

The **endothelial cells** in the walls of blood vessels make important contributions to the overall regulation of hemostasis and thrombosis. As described in Chapter 23, these cells synthesize the prostanoid **prostacyclin** (PGI$_2$), a potent inhibitor of platelet aggregation. Prostacyclin acts by stimulating the activity of adenylyl cyclase in the surface membranes of platelets via its G protein-coupled receptor (Figure 51–8A). The resulting increase of intraplatelet **cAMP** opposes the increase in the level of intracellular Ca^{2+} produced by IP$_3$ and thus inhibits platelet activation. This is in contrast with the effect of the prostanoid thromboxane A$_2$ that is formed by activated platelets, which is that of promoting aggregation. Endothelial cells play other roles in the regulation of thrombosis. For instance, these cells possess an **ADPase,** which hydrolyzes ADP, and thus opposes its aggregating effect on platelets. In addition, these cells appear to synthesize **heparan sulfate,** an anticoagulant, and they also synthesize **plasminogen activators,** which may help dissolve thrombi. **Table 51–3** lists some molecules produced by endothelial cells that affect thrombosis and fibrinolysis. **Nitric oxide** (endothelium-derived relaxing factor) is discussed in Chapter 49.

Analysis of the mechanisms of **uptake of atherogenic lipoproteins,** such as LDL, by endothelial, smooth muscle, and monocytic cells of arteries, along with detailed studies of how these lipoproteins damage such cells is a key area of study in elucidating the mechanisms of **atherosclerosis** (Chapter 26).

Aspirin Is One of Several Effective Antiplatelet Drugs

Certain **drugs** (antiplatelet drugs) inhibit platelet responses. The most commonly used antiplatelet drug is **aspirin** (acetylsalicylic acid), which irreversibly acetylates and thus inhibits the platelet

A.

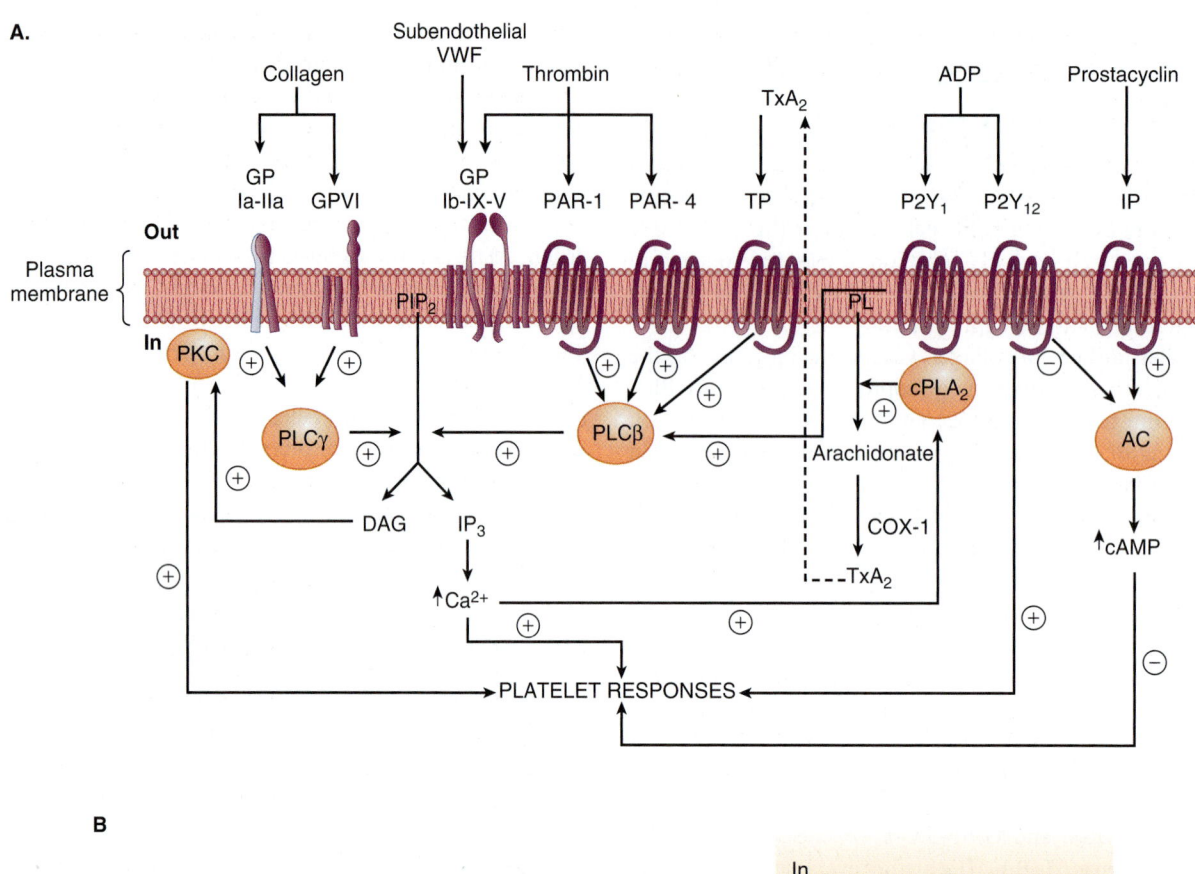

B

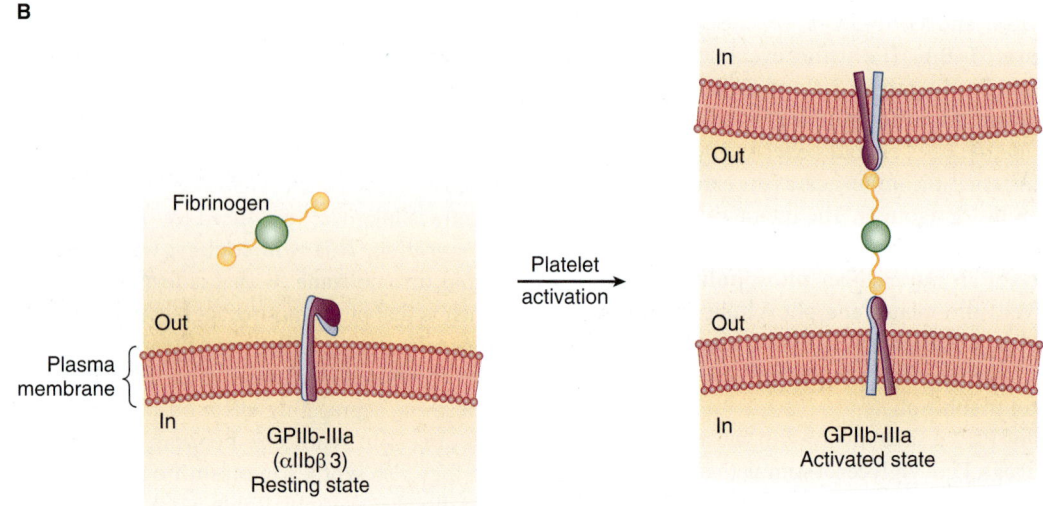

FIGURE 51–8   **(A) Diagrammatic representation of platelet activation by collagen, thrombin, thromboxane A₂ and ADP, and inhibition by prostacyclin.** The external environment, the plasma membrane, and the inside of a platelet are depicted from top to bottom. Platelet responses include, depending on the agonist, change of platelet shape, release of the contents of the storage granules, and aggregation. (AC, adenylyl cyclase; cAMP, cyclic AMP; COX-1, cyclooxgenase-1; cPLA₂, cytosolic phospholipase A₂; DAG, 1,2-diacylglycerol; GP, glycoprotein; IP, prostacyclin receptor; IP₃, inositol 1,4,5-trisphosphate; P2Y₁, P2Y₁₂, purinoceptors; PAR, protease activated receptor; PIP₂, phosphatidylinositol 4,5-bisphosphate; PKC, protein kinase C; PL, phospholipid; PLCβ, phospholipase Cβ; PLCγ, phospholipase Cγ; TP, thromboxane A₂ receptor; TxA₂, thromboxane A₂; VWF, von Willebrand factor.) The G proteins that are involved are not shown. **(B) Diagrammatic representation of platelet aggregation mediated by fibrinogen binding to activated GPIIb-IIIa molecules on adjacent platelets.** Signaling events initiated by all aggregating agents transform GPIIb-IIIa from its resting state to an activated form that can bind divalent fibrinogen or multivalent von Willebrand factor at the high shear that occurs in small vessels.

TABLE 51–3 Molecules Synthesized by Endothelial Cells That Play a Role in the Regulation of Thrombosis and Fibrinolysis

Molecule	Action
ADPase (CD39, an ectoenzyme)	Degrades ADP (an aggregating agent of platelets) to AMP + Pi
Nitric oxide (NO)	Inhibits platelet adhesion and aggregation by elevating levels of cGMP
Prostacyclin (PGI$_2$, a prostaglandin)	Inhibits platelet aggregation by increasing levels of cAMP
Thrombomodulin (a glycoprotein)	Binds thrombin, which then cleaves protein C to yield activated protein C; this in combination with protein S degrades factors Va and VIIIa, limiting their actions
Endothelial protein C receptor (EPCR, a glycoprotein)	Facilitates protein C activation by the thrombin-thrombomodulin complex
Tissue plasminogen activator (t-PA, a protease)	Activates plasminogen to plasmin, which digests fibrin; the action of t-PA is opposed by plasminogen activator inhibitor-1 (PAI-1)

Source: Adapted, with permission, from Wu KK: Endothelial cells in hemostasis, thrombosis and inflammation. Hosp Pract (Off Ed) 1992;27:145.

cyclooxygenase system (COX-1) involved in formation of thromboxane A$_2$ (Chapter 15), a potent aggregator of platelets and also a vasoconstrictor. Platelets are very sensitive to aspirin; as little as 30 mg/d (one regular aspirin tablet contains 325 mg) effectively eliminates the synthesis of thromboxane A$_2$. Aspirin also inhibits production of prostacyclin (PGI$_2$, which opposes platelet aggregation and is a vasodilator) by endothelial cells, but unlike platelets, these cells regenerate cyclooxygenase within a few hours. Thus, the overall balance between thromboxane A$_2$ and prostacyclin can be shifted in favor of the latter, opposing platelet aggregation. Indications for treatment with aspirin include management of acute coronary syndromes (angina, myocardial infarction), acute stroke syndromes (transient ischemic attacks, acute stroke), severe carotid artery stenosis, and primary prevention of these and other atherothrombotic diseases.

Other antiplatelet drugs include clopidogrel, a specific inhibitor of the P2Y$_{12}$ receptor for ADP, and antagonists of ligand binding to GPIIb–IIIa (eg, abciximab) that interfere with fibrinogen and von Willebrand factor binding and thus platelet aggregation.

Laboratory Tests Measure Coagulation, Thrombolysis, & Platelet Aggregation

A number of **laboratory tests** are available to measure the phases of hemostasis described above. The tests include platelet count, bleeding time/closure time, platelet aggregation, activated partial thromboplastin time (aPTT or PTT), prothrombin time (PT), thrombin time (TT), concentration of fibrinogen, fibrin clot stability, and measurement of fibrin degradation products. The **platelet count** quantitates the number of platelets. The **skin bleeding time** is an overall test of platelet and vessel wall function, while **the closure time** measured using the platelet function analyzer PFA-100 is an in vitro test of platelet-related hemostasis. **Platelet aggregation** measures responses to specific aggregating agents. **aPTT** is a measure of the intrinsic pathway and **PT** of the extrinsic pathway, with aPTT being used to monitor heparin therapy and PT, to measure the effectiveness of oral anticoagulants such as warfarin. The reader is referred to a textbook of hematology for a discussion of these tests.

SUMMARY

- Hemostasis and thrombosis are complex processes involving coagulation factors, platelets, and blood vessels.

- Many coagulation factors are zymogens of serine proteases, becoming activated, then inactivated during the overall process.

- Both extrinsic and intrinsic pathways of coagulation exist, the former initiated in vivo by tissue factor. The pathways converge at factor Xa, ultimately resulting in thrombin-catalyzed conversion of fibrinogen to fibrin, which is strengthened by covalent cross-linking, catalyzed by factor XIIIa.

- Genetic disorders that lead to bleeding occur; the principal disorders involve factor VIII (hemophilia A), factor IX (hemophilia B), and von Willebrand factor (von Willebrand disease).

- Antithrombin is an important natural inhibitor of coagulation; genetic deficiency of this protein can result in thrombosis.

- For their activity, factors II, VII, IX, and X and proteins C and S require vitamin K-dependent γ-carboxylation of certain glutamate residues, a process that is inhibited by the anticoagulant warfarin.

- Fibrin is dissolved by plasmin. Plasmin exists as an inactive precursor, plasminogen, which can be activated by tissue plasminogen activator (t-PA). Both t-PA and streptokinase are widely used to treat early thrombosis in the coronary arteries.

- Thrombin and other agents cause platelet aggregation, which involves a variety of biochemical and morphologic events. Stimulation of phospholipase C and the polyphosphoinositide pathway is a key event in platelet activation, but other processes are also involved.

- Aspirin is an important antiplatelet drug that acts by inhibiting production of thromboxane A$_2$.

REFERENCES

Colman RW, Marder VJ, Clowes AW, et al (editors): *Hemostasis and Thrombosis: Basic Principles and Clinical Practice*, 5th ed. Lippincott Williams & Wilkins, 2006.

Fauci AS, Braunwald E, Kasper DL, et al: *Harrison's Principles of Internal Medicine*, 17th ed. McGraw-Hill, 2008.

Hoffman R, Benz Jr EJ, Shattil SJ, et al (editors): *Hematology: Basic Principles and Practice*, 4th ed. Elsevier Churchill Livingston, 2005.

Israels SJ (editor): *Mechanisms in Hematology*, 4th ed. Core Health Sciences Inc, 2011. (This text has many excellent illustrations of basic mechanisms in hematology.)

Michelson AD (editor): *Platelets*, 2nd ed. Elsevier, 2007.

Red & White Blood Cells

Robert K. Murray, MD, PhD

OBJECTIVES

After studying this chapter, you should be able to:

- Understand the concept of stem cells and their importance.
- Summarize the causes of the major disorders affecting red blood cells.
- Discuss the general structure of the red blood cell membrane.
- Know the biochemical bases of the ABO blood group substances.
- Indicate the major biochemical features of neutrophils and understand the basis of chronic granulomatous disease.
- Appreciate the importance of integrins in health and disease.

BIOMEDICAL IMPORTANCE

Blood cells have been studied intensively because they are obtained easily, because of their functional importance, and because of their involvement in many disease processes. The structure and function of **hemoglobin**, the **porphyrias**, **jaundice**, and aspects of **iron metabolism** are discussed in previous chapters. **Table 52–1** summarizes the causes of a number of important diseases affecting red blood cells; some are discussed in this chapter, and the remainder are discussed elsewhere in this text. **Anemia** is a very prevalent condition with many causes. The **discovery of the causes of certain types of anemias** (eg, of pernicious anemia [a form of B_{12} deficient anemia] and of sickle cell anemia) has been an area where the reciprocal relationship between medicine and biochemistry referred to in Chapter 1 has been extremely beneficial. The World Health Organization (WHO) defines **anemia** as a hemoglobin level of <130 g/L in men and <120 g/L in females. There are many causes of anemia; only the most prevalent or biochemically relevant are mentioned here. A simplified classification of the causes of anemia is given in **Table 52–2**. It has been estimated that some 300,000 children are born each year with a severe inherited disorder of hemoglobin, the majority in low- or middle-income countries. Because infant mortality is decreasing, many of these children will survive to present a global health problem. Certain of the **blood group systems,** present on the membranes of erythrocytes and other blood cells, are of extreme importance in relation to blood transfusion and tissue transplantation. Every organ in the body can be affected by **inflammation;** neutrophils play a central role in acute inflammation, and other white blood cells, such as lymphocytes, play important roles in chronic inflammation.

Leukemias, defined as malignant neoplasms of blood-forming tissues, can affect precursor cells of any of the major classes of white blood cells; common types are acute and chronic myelocytic leukemia, affecting precursors of the neutrophils; and acute and chronic lymphocytic leukemias. Knowledge of the molecular mechanisms involved in the causation of the leukemias is increasing rapidly, but is not discussed in any detail in this text. Combination chemotherapy, using combinations of various chemotherapeutic agents, all of which act at one or more biochemical loci, has been remarkably effective in the treatment of certain of these types of leukemias. Understanding the role of red and white cells in health and disease requires a knowledge of certain fundamental aspects of their biochemistry.

ALL BLOOD CELLS DERIVE FROM HEMATOPOIETIC STEM CELLS

Figure 52–1 summarizes the derivation of the various types of blood cells from **hematopoietic stem cells.** The first solid evidence for the existence of stem cells, and in particular hematopoietic stem cells, was reported from studies done in mice by Ernest McCulloch and James Till in 1963. In recent years, interest in stem cells has grown enormously, and they are now of interest to almost every area of medicine and the health sciences. A stem cell is a cell with a unique capacity to produce unaltered daughter cells (ie, **self-renewal**) and to generate specialized cell types (**potency**). Stem cells may be **totipotent** (capable of producing all the cells in an organism), **pluripotent** (able to differentiate into cells of any of the three germ layers), **multipotent** (produce only cells of a closely

TABLE 52–1 Summary of the Causes of Some Important Disorders Affecting Red Blood Cells

Disorder	Sole or Major Cause
Iron deficiency anemia	Inadequate intake or excessive loss of iron
Methemoglobinemia	Intake of excess oxidants (various chemicals and drugs)
	Genetic deficiency in the NADH-dependent methemoglobin reductase system (OMIM 250800)
	Inheritance of HbM (OMIM 141900)
Sickle cell anemia (OMIM 603903)	Sequence of codon 6 of the β chain changed from GAG in the normal gene to GTG in the sickle cell gene, resulting in substitution of valine for glutamic acid
α-Thalassemias (OMIM 141800)	Mutations in the α-globin genes, mainly unequal crossing-over and large deletions and less commonly nonsense and frameshift mutations
β-Thalassemias (OMIM 141900)	A very wide variety of mutations in the β-globin gene, including deletions, nonsense and frameshift mutations, and others affecting every aspect of its structure (eg, splice sites, promoter mutants)
Megaloblastic anemias	
Deficiency of vitamin B_{12}	Decreased absorption of B_{12}, often due to a deficiency of intrinsic factor, normally secreted by gastric parietal cells
Deficiency of folic acid	Decreased intake, defective absorption, or increased demand (eg, in pregnancy) for folate
Hereditary spherocytosis[1] (OMIM 182900)	Deficiencies in the amount or in the structure of α or β spectrin, ankyrin, band 3 or band 4.1
Glucose-6-phosphate dehydrogenase (G6PD) deficiency[1] (OMIM 305900)	A variety of mutations in the gene (X-linked) for G6PD, mostly single-point mutations
Pyruvate kinase (PK) deficiency[1] (OMIM 266200)	A variety of mutations in the gene for the R (red cell) isozyme of PK
Paroxysmal nocturnal hemoglobinuria[1] (OMIM 311770)	Mutations in the PIG-A gene, affecting synthesis of GPI-anchored proteins

[1]The last four disorders cause hemolytic anemias, as do a number of the other disorders listed. Most of the above conditions are discussed in other chapters of this text. OMIM numbers apply only to disorders with a genetic basis.

TABLE 52–2 A Brief Classification of the Causes of Anemia

A. Blood loss: Acute, chronic

B. Deficiencies causing defects of erythropoiesis (eg, of iron, folate, vitamin B_{12}, and other factors)

C. Hemolysis:

    i. Due to extrinsic factors: eg, various antibodies, hemolysins, snake venoms, etc

    ii. Due to intrinsic factors:

        Mutations in genes encoding red cell membrane proteins (eg, hereditary spherocytosis and hereditary elliptocytosis)

        Enzymopathies of the red cells (eg, glucose 6-phosphate dehydrogenase, pyruvate kinase, and others)

        Hemoglobinopathies (particularly HbS) and thalassemias, parasitic infections (eg, plasmodia in malaria)

Note: Figure 52–3 also indicates the causes of hemolytic anemias. In anemia, the red cells may be **larger** than normal (macrocytes, as in folate and B12 deficiencies, of **normal size** (normocytes, as in acute blood loss or bone marrow failure) or **smaller** than normal (microcytes, as in iron deficiency anemia). They may also **stain** more intensely than usual (hyperchromic), normally (normochromic) or paler than usual (hypochromic). These differences in intensity of staining qualitatively reflect higher, normal, or lower contents of hemoglobin.

other white blood cells at the stage of the common myeloid progenitors. Each pathway is **regulated** by various factors (eg, stem cell factor, thrombopoietin, various interleukins, erythropoietin, etc), and key specific **transcription factors** (not indicated in the figure) are also involved at the stages indicated.

Stem cell factor is a cytokine that plays an important role in the proliferation of hematopoietic stem cells and some of their progeny. **Thrombopoietin** is a glycoprotein that is important in regulating the production of platelets by the bone marrow. **Interleukins** are cytokines produced by leukocytes; they regulate various aspects of hematopoiesis and of the immune system.

THE RED BLOOD CELL IS SIMPLE IN TERMS OF ITS STRUCTURE & FUNCTION

The **major functions of the red blood cell** are relatively simple, consisting of delivering oxygen to the tissues and of helping in the disposal of carbon dioxide and protons formed by tissue metabolism. Thus, it has a much simpler structure than most human cells, being essentially composed of a membrane surrounding a solution of hemoglobin (this protein forms about 95% of the intracellular protein of the red cell). There are no intracellular organelles, such as mitochondria, lysosomes, or Golgi apparatus. Human red blood cells, like most red cells of animals, are nonnucleated. However, the red cell is not metabolically inert. **ATP** is synthesized from **glycolysis** and is important in processes that help the red blood cell maintain its biconcave shape

related family) or **unipotent** (produce only one type of cell). Stem cells are also classified as **embryonic** and **adult;** the latter are more limited in their capabilities to differentiate than the former, although genetic approaches to overcoming this restriction are becoming available.

As shown in Figure 52–1, **red blood cells** and **platelets** share a common pathway of differentiation until the stage of megakaryocyte erythroid progenitors. Cells of **lymphoid origin** branch off at the stage of multipotent progenitors, and

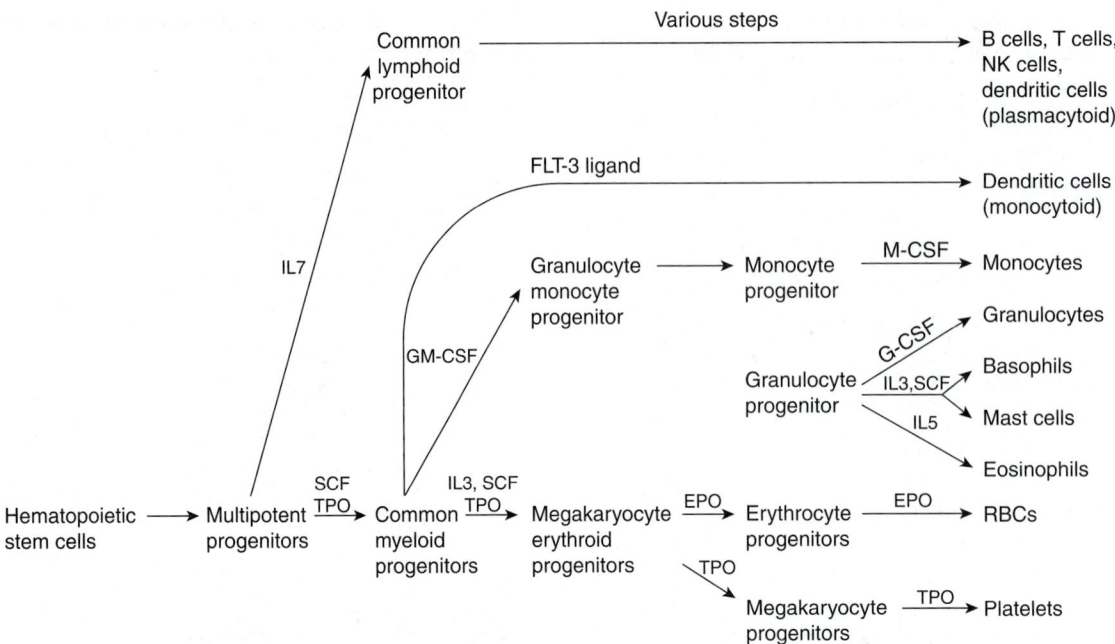

FIGURE 52–1 **Simplified scheme of differentiation of red blood cells and other blood cells from the hematopoietic stem cell.** Sites of action of interleukins (IL-7, IL-3, and IL-5), stem cell factor (SCF), thrombopoietin (TPO), FLT-3 ligand (a growth factor), granulocyte-macrophage colony-stimulating factor (GM-CSF), erythropoietin (EPO), monocyte colony-stimulating factor (M-CSF), and granulocyte colony-stimulating factor (G-CSF) are shown. Sites of action of important transcription factors are not shown. Various steps in the development of lymphoid cells (top part of the figure) have been omitted and abbreviated to one step. (Modified, with permission, from Scadden DT, Longo DL in Fauci AS et al (editors), *Harrison's Principles of Internal Medicine,* 17th ed. McGraw-Hill, 2008. Chapter 68)

and also in the regulation of the **transport of ions** (eg, by the Na⁺–K⁺-ATPase and the anion exchange protein [see below]) and of **water** in and out of the cell. The biconcave shape increases the surface-to-volume ratio of the red blood cell, thus facilitating gas exchange. The red cell contains cytoskeletal components (see below) that play an important role in determining its shape.

About Two Million Red Blood Cells Enter the Circulation per Second

The **lifespan** of the normal red blood cell is 120 days; this means that slightly less than 1% of the population of red cells (~200 billion cells) is replaced daily (or ~2 million per second). The new red cells that appear in the circulation still contain ribosomes and elements of the endoplasmic reticulum. The RNA of the ribosomes can be detected by suitable stains (such as cresyl blue), and cells containing it are termed reticulocytes; they normally number about 1% of the total red blood cell count. The lifespan of the red blood cell can be dramatically shortened in a variety of **hemolytic anemias.** The number of reticulocytes is markedly increased in these conditions, as the bone marrow attempts to compensate for rapid breakdown of red blood cells by increasing the amount of new, young red cells in the circulation.

Erythropoietin Regulates Production of Red Blood Cells

Human **erythropoietin (EPO)** is a glycoprotein of 166 amino acids (molecular mass about 34 kDa). Its amount in plasma can be measured by radioimmunoassay. It is the major regulator of human erythropoiesis (Figure 52–1). As shown in the figure, earlier stages in the development of red blood cells involve stem cell factor, thrombopoietin, and interleukin-3. EPO is synthesized mainly by the kidney and is released in response to hypoxia into the bloodstream, in which it travels to the bone marrow. There it interacts with progenitors of red blood cells via **a specific receptor.** The receptor is a transmembrane protein consisting of two different subunits and a number of domains. It is not a tyrosine kinase, but it stimulates the activities of specific members of this class of enzymes involved in downstream signal transduction.

The availability of **a cDNA for EPO** has made it possible to produce substantial amounts of this hormone for analysis and for therapeutic purposes; previously the isolation of erythropoietin from human urine yielded very small amounts of the protein. The major use of **recombinant EPO** has been in the treatment of a small number of **anemic states,** such as that due to renal failure. As described in Chapter 47, attempts have been made to prolong the half-life of EPO (thus lengthening its activity) in the circulation by altering the nature of its sugar chains.

MANY GROWTH FACTORS REGULATE PRODUCTION OF WHITE BLOOD CELLS

A large number of **hematopoietic growth factors** have been identified in recent years in addition to erythropoietin. This area of study adds to knowledge about the differentiation of blood cells, provides factors that may be useful in treatment, and also has implications for understanding of the abnormal growth of blood cells (eg, the leukemias). Like erythropoietin, most of the growth factors isolated have been glycoproteins, are very active in vivo, and in vitro interact with their target cells via specific cell surface receptors, and ultimately (via intracellular signals) affect gene expression, thereby promoting differentiation. Many have been cloned, permitting their production in relatively large amounts. Two of particular interest are **granulocyte-** and **granulocyte-macrophage colony-stimulating factors** (G-CSF and GM-CSF, respectively). As indicated in Figure 52–1, G-CSF is relatively specific, inducing mainly granulocytes, whereas GMCSF induces a wider variety of white blood cells. When the production of neutrophils is severely depressed, this condition is referred to as **neutropenia.** It is particularly likely to occur in patients treated with certain chemotherapeutic regimens and after bone marrow transplantation. These patients are liable to develop overwhelming infections. G-CSF has been administered to such patients to boost production of neutrophils.

THE RED BLOOD CELL HAS A UNIQUE & RELATIVELY SIMPLE METABOLISM

Various aspects of the **metabolism of the red cell,** many of which are discussed in other chapters of this text, are summarized in **Table 52–3.**

The Red Blood Cell Has a Glucose Transporter in Its Membrane

The entry rate of glucose into red blood cells is far greater than would be calculated for simple diffusion. Rather, it is an example of **facilitated diffusion** (Chapter 40). The specific protein involved in this process is called the **glucose transporter** (GLUT1) or glucose permease. Some of its properties are summarized in **Table 52–4.** The process of entry of glucose into red blood cells is of major importance because it is the major fuel supply for these cells. About 12 different but related glucose transporters have been isolated from various human tissues; unlike the red cell transporter, some of these are insulin-dependent (eg, in muscle and adipose tissue). There is considerable interest in the latter types of transporter because defects in their recruitment from intracellular sites to the surface of skeletal muscle cells may help explain the **insulin resistance** displayed by patients with type 2 diabetes mellitus.

TABLE 52–3 Summary of Important Aspects of the Metabolism of the Red Blood Cell

- The RBC is highly dependent upon glucose as its energy source; its membrane contains high-affinity glucose transporters.
- Glycolysis, producing lactate, is the site of production of ATP.
- Because there are no mitochondria in RBCs, there is no production of ATP by oxidative phosphorylation.
- The RBC has a variety of transporters that maintain ionic and water balance.
- Production of 2,3-bisphosphoglycerate, by reactions closely associated with glycolysis, is important in regulating the ability of Hb to transport oxygen.
- The pentose phosphate pathway is operative in the RBC (it metabolizes about 5–10% of the total flux of glucose) and produces NADPH; hemolytic anemia due to a deficiency of the activity of glucose-6-phosphate dehydrogenase is common.
- Reduced glutathione (GSH) is important in the metabolism of the RBC, in part to counteract the action of potentially toxic peroxides; the RBC can synthesize GSH and requires NADPH to return oxidized glutathione (G-S-S-G) to the reduced state.
- The iron of Hb must be maintained in the ferrous state; ferric iron is reduced to the ferrous state by the action of an NADH-dependent methemoglobin reductase system involving cytochrome b_5 reductase and cytochrome b_5.
- Synthesis of glycogen, fatty acids, protein, and nucleic acids does not occur in the RBC; however, some lipids (eg, cholesterol) in the red cell membrane can exchange with corresponding plasma lipids.
- The RBC contains certain enzymes of nucleotide metabolism (eg, adenosine deaminase, pyrimidine nucleotidase, and adenylyl kinase); deficiencies of these enzymes are involved in some cases of hemolytic anemia.
- When RBCs reach the end of their lifespan, the globin is degraded to amino acids (which are reutilized in the body), the iron is released from heme and also reutilized, and the tetrapyrrole component of heme is converted to bilirubin, which is mainly excreted into the bowel via the bile.

Reticulocytes Are Active in Protein Synthesis

The mature red blood cell cannot synthesize protein. **Reticulocytes** are active in protein synthesis. Once reticulocytes enter the circulation, they lose their intracellular organelles (ribosomes, mitochondria, etc) within about 24 h, becoming young red blood cells and concomitantly losing their ability to synthesize protein. **Extracts of rabbit reticulocytes** (obtained by injecting rabbits with a chemical—phenylhydrazine—that causes a severe hemolytic anemia, so that the red cells are almost completely replaced by reticulocytes) are widely used as an in vitro system for synthesizing proteins. Endogenous mRNAs present in these reticulocytes are destroyed by use of a nuclease, whose activity can be inhibited by the addition of Ca^{2+}. The system is then programmed by adding purified mRNAs or whole-cell extracts of mRNAs, and radioactive

TABLE 52–4 Some Properties of the Glucose Transporter of the Membrane of the Red Blood Cell (GLUT1)

- It accounts for ~2% of the protein of the membrane of the RBC.
- It exhibits specificity for glucose and related D-hexoses (L-hexoses are not transported).
- The transporter functions at ~75% of its V_{max} at the physiologic concentration of blood glucose, is saturable, and can be inhibited by certain analogs of glucose.
- Some 12 similar but distinct glucose transporters have been detected to date in mammalian tissues, of which the red cell transporter is one.
- It is not dependent upon insulin, unlike the corresponding carrier in muscle and adipose tissue.
- Its complete amino acid sequence (492 amino acids) has been determined.
- It transports glucose when inserted into artificial liposomes.
- It is estimated to contain 12 transmembrane helical segments.
- It functions by generating a gated pore in the membrane to permit passage of glucose; the pore is conformationally dependent on the presence of glucose and can oscillate rapidly (~900 times/s).

proteins are synthesized in the presence of ^{35}S-labeled L-methionine or other radiolabeled amino acids. The radioactive proteins synthesized are separated by sodium dodecyl sulfate polyacrylamide gel electrophoresis (SDS-PAGE) and detected by radioautography.

With regard to **protein synthesis,** it is of interest to note that certain disorders due to genetic abnormalities cause impairment of ribosome structure and function and have been named **ribosomopathies.** These include some cases of **Diamond–Blackfan anemia,** in which mutations in a ribosomal RNA processing gene (RPS19) result in red cell hypoplasia. The **5q-syndrome** presents with a similar clinical picture and is due to an insufficiency of ribosomal protein RPS 14.

Superoxide Dismutase, Catalase, & Glutathione Protect Blood Cells from Oxidative Stress & Damage

Several powerful **oxidants** are produced during the course of metabolism, in both blood cells and most other cells of the body. These include superoxide ($O_2^{-\cdot}$, hydrogen peroxide (H_2O_2), peroxyl radicals (ROO$^\cdot$), and hydroxyl radicals (OH$^\cdot$) and are referred to as **reactive oxygen species (ROS). Free radicals** are atoms or groups of atoms that have an unpaired electron (see Chapters 15 & 45). OH$^\cdot$ is a particularly reactive molecule and can react with proteins, nucleic acids, lipids, and other molecules to alter their structure and produce tissue damage. The reactions listed in **Table 52–5** play an important role in forming these oxidants and in disposing of them; each of these reactions will now be considered in turn.

Superoxide is formed (reaction 1) in the red blood cell by the auto-oxidation of hemoglobin to methemoglobin (approximately 3% of hemoglobin in human red blood cells has been calculated to auto-oxidize per day); in other tissues, it is formed by the action of enzymes such as cytochrome P450 reductase and xanthine oxidase. When stimulated by contact with bacteria, **neutrophils** exhibit a **respiratory burst** (see below) and produce superoxide in a reaction catalyzed by NADPH oxidase (reaction 2). Superoxide spontaneously dismutates to form H_2O_2 and O_2; however, the rate of the same reaction is speeded up tremendously by the action of the enzyme **superoxide dismutase** (reaction 3). **Hydrogen peroxide** is subject to a number of fates. The enzyme **catalase,** present in many types of cells, converts it to H_2O and O_2 (reaction 4). Neutrophils possess a unique enzyme, **myeloperoxidase,** which uses H_2O_2 and halides to produce hypohalous acids (reaction 5); this subject is discussed further below. The selenium-containing enzyme **glutathione peroxidase** (Chapter 21) will also act on reduced glutathione (GSH) and H_2O_2 to produce oxidized glutathione (GSSG) and H_2O (reaction 6); this enzyme can also use other peroxides as substrates. OH$^\cdot$ and OH$^-$ can be formed from H_2O_2 in a nonenzymatic reaction catalyzed by Fe^{2+} (the **Fenton reaction,**

TABLE 52–5 Reactions of Importance in Relation to Oxidative Stress in Blood Cells and Various Tissues

1. Production of superoxide (by-product of various reactions)	$O_2 + e^- \rightarrow O_2^{-\cdot}$
2. NADPH oxidase	$2\,O_2 + NADPH \rightarrow 2\,O_2^{-\cdot} + NADP + H^+$
3. Superoxide dismutase	$O_2^{-\cdot} + O_2^{-\cdot} + 2\,H^+ \rightarrow H_2O_2 + O_2$
4. Catalase	$H_2O_2 \rightarrow 2\,H_2O + O_2$
5. Myeloperoxidase	$H_2O_2 + X^- + H^+ \rightarrow HOX + H_2O\ (X^- = Cl^-, Br^-, SCN^-)$
6. Glutathione peroxidase (Se-dependent)	$2\,GSH + R\text{-}O\text{-}OH \rightarrow GSSG + H_2O + ROH$
7. Fenton reaction	$Fe^{2+} + H_2O_2 \rightarrow Fe^{3+} + OH^\cdot + OH^-$
8. Iron-catalyzed Haber–Weiss reaction	$O_2^{-\cdot} + H_2O_2 \rightarrow O_2 + OH^\cdot + OH^-$
9. Glucose-6-phosphate dehydrogenase (G6PD)	$G6P + NADP \rightarrow 6\ Phosphogluconate + NADPH + H^+$
10. Glutathione reductase	$G\text{-}S\text{-}S\text{-}G + NADPH + H^+ \rightarrow 2\,GSH + NADP$

reaction 7). O_2^- and H_2O_2 are the substrates in the iron-catalyzed **Haber–Weiss reaction** (reaction 8), which also produces $OH^{\cdot}$ and OH^-. Superoxide can release iron ions from ferritin. Thus, production of $OH^{\cdot}$ may be one of the mechanisms involved in tissue injury due to iron overload in hemochromatosis (see Case no. 10 Chapter 57).

Chemical compounds and reactions capable of generating potential toxic oxygen species can be referred to as **pro-oxidants.** On the other hand, compounds and reactions disposing of these species, scavenging them, suppressing their formation, or opposing their actions are **antioxidants** and include compounds such as NADPH, GSH, ascorbic acid, and vitamin E. In a normal cell, there is an appropriate pro-oxidant:antioxidant balance. However, this balance can be shifted toward the pro-oxidants when production of oxygen species is increased greatly (eg, following ingestion of certain chemicals or drugs) or when levels of antioxidants are diminished (eg, by inactivation of enzymes involved in disposal of oxygen species and by conditions that cause low levels of the antioxidants mentioned above). This state is called "**oxidative stress**" (see Chapter 45) and can result in serious cell damage if the stress is massive or prolonged.

ROS are now thought to play an important role in many types of **cellular injury** (eg, resulting from administration of various toxic chemicals or from ischemia), some of which can result in cell death. Indirect evidence supporting a role for these species in generating cell injury is provided if administration of an enzyme such as superoxide dismutase or catalase is found to protect against cell injury in the situation under study.

Deficiency of Glucose-6-Phosphate Dehydrogenase Is Frequent in Certain Areas & Is an Important Cause of Hemolytic Anemia

NADPH, produced in the reaction catalyzed by the X-linked glucose-6-phosphate dehydrogenase (Table 52–5, reaction 9) in the **pentose phosphate pathway** (Chapter 21), plays a key role in supplying reducing equivalents in the red cell and in other cells such as the hepatocyte. Since the pentose phosphate pathway is virtually its sole means of producing NADPH, the red blood cell is very sensitive to oxidative damage if the function of this pathway is impaired (eg, by enzyme deficiency). One function of NADPH is to reduce GSSG to GSH, a reaction catalyzed by glutathione reductase (reaction 10).

Deficiency of the activity of **glucose-6-phosphate dehydrogenase,** owing to mutation, is extremely frequent in some regions of the world (eg, tropical Africa, the Mediterranean, certain parts of Asia, and in North America among blacks). It is the most common of all **enzymopathies** (diseases caused by abnormalities of enzymes), and some 140 genetic variants of the enzyme have been distinguished; at least 400 million people are estimated to have a variant gene. It is thought that an abnormal form of this enzyme confers resistance to malaria. The disorder resulting from deficiency of glucose-6-phosphate dehydrogenase is **hemolytic anemia.** When an abnormal form of an enzyme causes pathology it is referred to as an **enzymopathy.** Consumption of **broad beans** (*Vicia faba*) by individuals deficient in activity of the enzyme can precipitate an acute attack of hemolytic anemia because they contain potential oxidants. In addition, a number of drugs (eg, the antimalarial drug **primaquine** [the condition caused by intake of primaquine is called **primaquine-sensitive hemolytic anemia**] and **sulfonamides**) and chemicals (eg, naphthalene) precipitate an attack, because their intake leads to generation of H_2O_2 or O_2^-. Normally, H_2O_2 is disposed of by **catalase** and **glutathione peroxidase** (Table 52–5, reactions 4 and 6), the latter causing increased production of GSSG. GSH is regenerated from GSSG by the action of the enzyme **glutathione reductase,** which depends on the availability of NADPH (reaction 10). The red blood cells of individuals who are deficient in the activity of glucose-6-phosphate dehydrogenase cannot generate sufficient NADPH to regenerate GSH from GSSG, which in turn impairs their ability to dispose of H_2O_2 and of oxygen radicals. These compounds can cause oxidation of critical SH groups in proteins and possibly peroxidation of lipids in the membrane of the red cell, causing lysis of the red cell membrane. Some of the SH groups of hemoglobin become oxidized, and the protein precipitates inside the red blood cell, forming **Heinz bodies,** which stain purple with cresyl violet. The presence of Heinz bodies indicates that red blood cells have been subjected to oxidative stress. **Figure 52–2** summarizes the possible chain of events in hemolytic anemia due to deficiency of glucose-6-phosphate dehydrogenase.

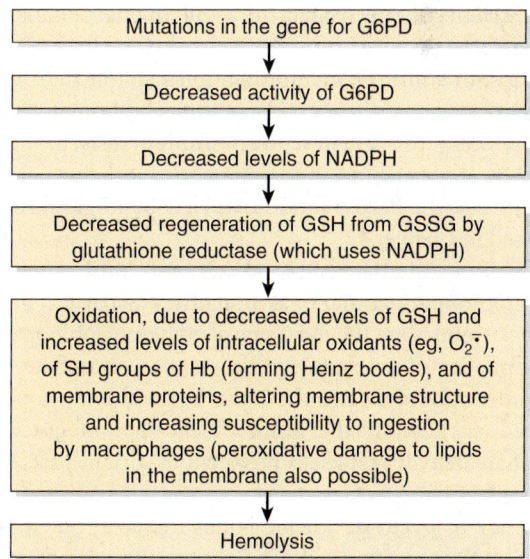

FIGURE 52–2   Summary of probable events causing hemolytic anemia due to deficiency of the activity of glucose-6-phosphate dehydrogenase (G6PD) (OMIM 305900).

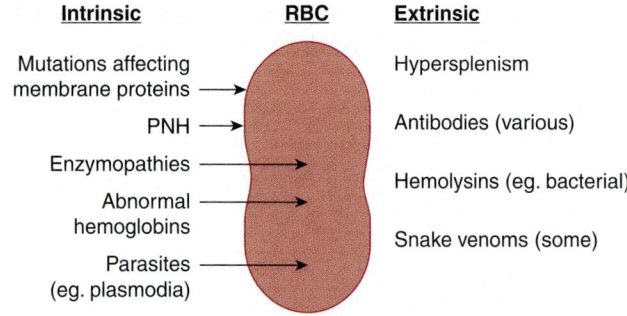

FIGURE 52–3 **Schematic diagram of some causes of hemolytic anemias.** Extrinsic causes are causes outside the red cell; they include hypersplenism, various antibodies, certain bacterial hemolysins and some snake venoms. Causes intrinsic to the red cells include mutations affecting the structures of membrane proteins (eg, in hereditary spherocytosis and hereditary elliptocytosis), PNH (paroxysmal nocturnal hemoglobinuria, see Chapter 47), enzymopathies, abnormal hemoglobins, and certain parasites (eg, plasmodia causing malaria).

TABLE 52–6 Laboratory Investigations that Assist in the Diagnosis of Hemolytic Anemia

General tests and findings
Increased nonconjugated (indirect) bilirubin
Shortened red cell survival time as measured by injection of autologous ^{51}Cr-labeled red cells
Reticulocytosis
Low level of plasma haptoglobin
Specific tests and findings
Hb electrophoresis (eg, HbS)
Red cell enzymes (eg, G6PD or pyruvate kinase deficiency)
Osmotic fragility (eg, hereditary spherocytosis)
Coombs test[1]
Cold agglutinin

[1]The direct Coombs test detects the presence of antibodies on red cells, whereas the indirect test detects the presence of circulating antibodies to antigens present on red cells.

Hemolytic Anemias Are Caused by Abnormalities Outside, Within or Inside the Red Cell Membrane

Various causes of hemolytic anemias are summarized in **Figure 52–3**. Causes **outside the membrane** (ie, extrinsic) include **hypersplenism,** a condition in which the spleen is enlarged from a variety of causes and red blood cells become sequestered in it. **Various antibodies** (eg, transfusion reactions and anti-Rh antibodies, the presence in plasma of warm, and cold antibodies that lyse red blood cells) also fall in this class, as do **hemolysins** released by various infectious agents, such as certain bacteria (eg, certain strains of *Escherichia coli* and clostridia). Some snakes release **venoms** that act to lyse the red cell membrane (eg, via the action of phospholipases or proteinases).

Causes **within the membrane** (intrinsic) include abnormalities of proteins. The most important conditions are **hereditary spherocytosis** and **hereditary elliptocytosis,** principally caused by abnormalities in the amount or structure of spectrin (see below). **Paroxysmal nocturnal hemoglobinuria** is discussed in Chapter 47.

Causes **inside the red blood cell** (also intrinsic) include **hemoglobinopathies** and **enzymopathies.** Sickle cell anemia and thalassemias are the most prevalent hemoglobinopathies. Abnormalities of enzymes in the pentose phosphate pathway and in glycolysis are the most frequent enzymopathies involved, particularly the former. Deficiency of **glucose-6-phosphate dehydrogenase** is prevalent in certain parts of the world and is a frequent cause of hemolytic anemia (see above). Deficiency of **pyruvate kinase** is not frequent, but it is the second commonest enzyme deficiency resulting in hemolytic anemia; the mechanism appears to be due to impairment of glycolysis, resulting in decreased formation of ATP, affecting various aspects of membrane integrity. **Parasitic** infections (eg, the plasmodia causing malaria) are also important causes of hemolytic anemias in certain geographic areas.

Laboratory investigations that aid in the diagnosis of hemolytic anemia are listed in **Table 52–6**.

Methemoglobin Is Useless in Transporting Oxygen

The ferrous iron of hemoglobin is susceptible to oxidation by superoxide and other oxidizing agents, forming **methemoglobin,** which cannot transport oxygen. Only a very small amount of methemoglobin is present in normal blood, as the red blood cell possesses an effective system (the NADH-cytochrome b_5 methemoglobin reductase system) for reducing heme Fe^{3+} back to the Fe^{2+} state. This system consists of **NADH** (generated by glycolysis), a flavoprotein named **cytochrome b_5** reductase (also known as methemoglobin reductase), and **cytochrome b_5.** The Fe^{3+} of methemoglobin is reduced back to the Fe^{2+} state by the action of reduced cytochrome b_5:

$$Hb-Fe^{3+}+Cyt\,b_{5red}\rightarrow Hb-Fe^{2+}+Cyt\,b_{5ox}$$

Reduced cytochrome b_5 is then regenerated by the action of cytochrome b_5 reductase:

$$Cyt\,b_{5ox}+NADH\rightarrow Cyt\,b_{5red}+NAD$$

Methemoglobinemia Is Inherited or Acquired

Methemoglobinemia can be classified as either **inherited** or **acquired** by ingestion of certain drugs and chemicals. Neither type occurs frequently, but physicians must be aware of them.

The inherited form is usually due to deficient activity of **cytochrome** b_5 **reductase**, but mutations can also affect the activity of cytochrome b_5. Certain **abnormal hemoglobins** (eg, HbM) are also rare causes of methemoglobinemia. In HbM, mutation changes the amino acid residue to which heme is attached, thus altering its affinity for oxygen and favoring its oxidation. Ingestion of **certain drugs** (eg, sulfonamides) or **chemicals** (eg, aniline) can cause acquired methemoglobinemia. Cyanosis (bluish discoloration of the skin and mucous membranes due to increased amounts of deoxygenated hemoglobin in arterial blood, or in this case due to increased amounts of methemoglobin) is usually the presenting sign in both types and is evident when 10% of hemoglobin is in the "met" form. Diagnosis is made by spectroscopic analysis of blood, which reveals the characteristic absorption spectrum of methemoglobin. Additionally, a sample of blood containing methemoglobin cannot be fully reoxygenated by flushing oxygen through it, whereas normal deoxygenated blood can. Electrophoresis can be used to confirm the presence of an abnormal hemoglobin. Ingestion of **methylene blue** or **ascorbic acid** (both reducing agents) is used to treat mild methemoglobinemia due to enzyme deficiency. Acute massive methemoglobinemia (due to ingestion of chemicals) should be treated by intravenous injection of methylene blue.

MORE IS KNOWN ABOUT THE MEMBRANE OF THE HUMAN RED BLOOD CELL THAN ABOUT THE SURFACE MEMBRANE OF ANY OTHER HUMAN CELL

A variety of biochemical approaches have been used to study the membrane of the red blood cell. These include analysis of membrane proteins by SDS-PAGE, the use of specific enzymes (proteinases, glycosidases, and others) to determine the location of proteins and glycoproteins in the membrane, and various techniques to study both the lipid composition and disposition of individual lipids. Morphologic (eg, electron microscopy and freeze-fracture electron microscopy) and other techniques (eg, use of antibodies to specific components) have also been widely used. When red blood cells are lysed under specific conditions, their membranes will reseal in their original orientation to form **ghosts** (right-side-out ghosts). By altering the conditions, ghosts can also be made to reseal with their cytosolic aspect exposed on the exterior (inside-out ghosts). Both types of ghosts have been useful in analyzing the disposition of specific proteins and lipids in the membrane. In recent years, cDNAs for many proteins of this membrane have become available, permitting the deduction of their amino sequences and domains. All in all, more is known about the membrane of the red blood cell than about any other membrane of human cells (**Table 52–7**).

TABLE 52–7 Summary of Biochemical Information about the Membrane of the Human Red Blood Cell

- The membrane is a bilayer composed of ~50% lipid and 50% protein.
- The major lipid classes are phospholipids and cholesterol; the major phospholipids are phosphatidylcholine (PC), phosphatidylethanolamine (PE), and phosphatidylserine (PS) along with sphingomyelin (Sph).
- The choline-containing phospholipids, PC and Sph, predominate in the outer leaflet and the amino-containing phospholipids (PE and PS) in the inner leaflet.
- Glycosphingolipids (GSLs) (neutral GSLs, gangliosides, and complex species, including the ABO blood group substances) constitute about 5–10% of the total lipid.
- Analysis by SDS-PAGE shows that the membrane contains ~10 major proteins and >100 minor species.
- The major proteins (which include spectrin, ankyrin, the anion exchange protein, actin, and band 4.1) have been studied intensively, and the principal features of their disposition (eg, integral or peripheral), structure, and function have been established.
- Many of the proteins are glycoproteins (eg, the glycophorins) containing O- or N-linked (or both) oligosaccharide chains located on the external surface of the membrane.

Analysis by SDS-PAGE Resolves the Proteins of the Membrane of the Red Blood Cell

When the membranes of red blood cells are analyzed by **SDS-PAGE,** about 10 major proteins are resolved (**Figure 52–4**), several of which have been shown to be **glycoproteins.** Their migration on SDS-PAGE was used to name these proteins, with the slowest migrating (and hence highest molecular mass) being designated band 1 or **spectrin.** All these major proteins have been isolated, most of them have been identified, and considerable insight has been obtained about their functions (**Table 52–8**). Many of their amino acid sequences have also been established. In addition, it has been determined which are integral or peripheral membrane proteins, which are situated on the external surface, which are on the cytosolic surface, and which span the membrane (**Figure 52–5**). Many minor components can also be detected in the red cell membrane by use of sensitive staining methods or two-dimensional gel electrophoresis. One of these is the glucose transporter described above.

The Major Integral Proteins of the Red Blood Cell Membrane Are the Anion Exchange Protein & the Glycophorins

The **anion exchange protein (band 3)** is a transmembrane glycoprotein, with its carboxyl terminal end on the external surface of the membrane and its amino terminal end on the cytoplasmic surface. It is an example of a **multipass** membrane

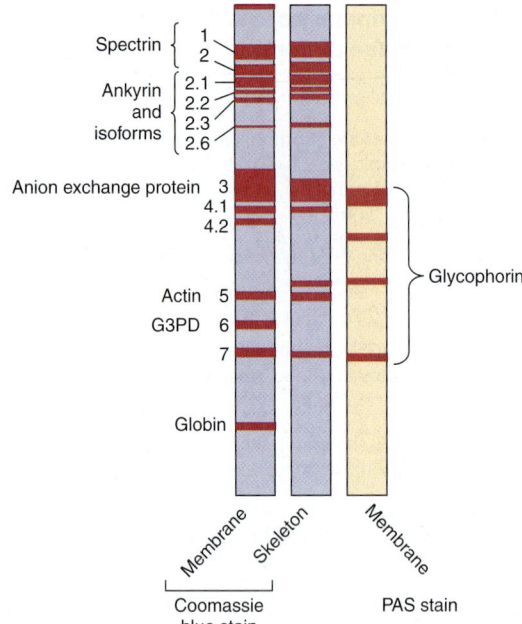

FIGURE 52–4 **Diagrammatic representation of the major proteins of the membrane of the human red blood cell separated by SDS-PAGE.** The bands detected by staining with Coomassie blue are shown in the two left-hand channels, and the glycoproteins detected by staining with periodic acid-Schiff (PAS) reagent are shown in the right-hand channel. The direction of electrophoretic migration is from top to bottom. (Reproduced, with permission, from Beck WS, Tepper RI: Hemolytic anemias III: membrane disorders. In: *Hematology*, 5th ed. Beck WS (editor). The MIT Press, 1991.)

TABLE 52–8 **Principal Proteins of the Red Cell Membrane**

Band Number[1]	Protein	Integral (I) or Peripheral (P)	Approximate Molecular Mass (kDa)
1	Spectrin (α)	P	240
2	Spectrin (β)	P	220
2.1	Ankyrin	P	210
2.2	Ankyrin	P	195
2.3	Ankyrin	P	175
2.6	Ankyrin	P	145
3	Anion exchange protein	I	100
4.1	Unnamed	P	80
5	Actin	P	43
6	Glyceraldehyde-3-phosphate dehydrogenase	P	35
7	Tropomyosin	P	29
8	Unnamed	P	23
	Glycophorins A, B, and C	I	31, 23, and 28

Source: Adapted from Lux DE, Becker PS: Disorders of the red cell membrane skeleton: hereditary spherocytosis and hereditary elliptocytosis. in: *The Metabolic Basis of Inherited Disease,* 6th ed. Scriver CR et al (editors). McGraw-Hill, 1989. Chapter 95.
[1]The band number refers to the position of migration on SDS-PAGE (see Figure 52–4). The glycophorins are detected by staining with the periodic acid-Schiff reagent. A number of other components (eg, 4.2 and 4.9) are not listed. Native spectrin is $\alpha_2\beta_2$.

protein, extending across the bilayer approximately 14 times. It probably exists as a dimer in the membrane, in which it forms a tunnel, permitting the exchange of chloride for bicarbonate. Carbon dioxide, formed in the tissues, enters the red cell as bicarbonate, which is exchanged for chloride in the lungs, where carbon dioxide is exhaled. The amino terminal end binds many proteins, including hemoglobin, proteins 4.1 and 4.2, ankyrin, and several glycolytic enzymes. Purified band 3 has been added to lipid vesicles in vitro and has been shown to perform its transport functions in this reconstituted system.

Glycophorins A, B, and C are also transmembrane glycoproteins but of the **single-pass** type, extending across the membrane only once. A is the major glycophorin, is made up of 131 amino acids, and is heavily glycosylated (about 60% of its mass). Its amino terminal end, which contains 16 oligosaccharide chains (15 of which are O-glycans), extrudes out from the surface of the red blood cell. Approximately 90% of the sialic acid of the red cell membrane is located in this protein. Its transmembrane segment (23 amino acids) is α-helical. The carboxyl terminal end extends into the cytosol and binds to protein 4.1, which in turn binds to spectrin. **Polymorphism** of this protein is the basis of the MN blood group system (see below). Glycophorin A contains binding sites for influenza virus and for *Plasmodium falciparum,* the cause of one form of malaria. Intriguingly, the function of red blood cells of individuals who lack glycophorin A does not appear to be affected.

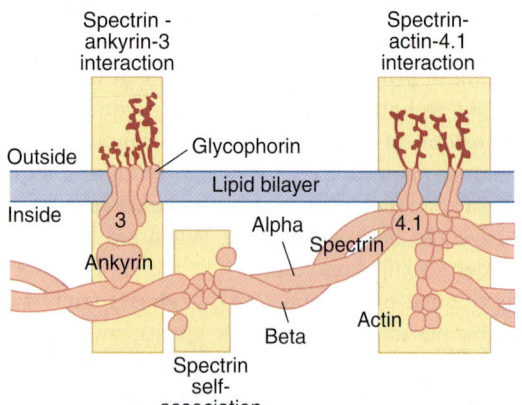

FIGURE 52–5 **Diagrammatic representation of the interaction of cytoskeletal proteins with each other and with certain integral proteins of the membrane of the red blood cell.** (Reproduced, with permission, from Beck WS, Tepper RI: Hemolytic anemias III: membrane disorders. In: *Hematology*, 5th ed. Beck WS (editor). The MIT Press, 1991.)

Spectrin, Ankyrin, & Other Peripheral Membrane Proteins Help Determine the Shape & Flexibility of the Red Blood Cell

The red blood cell must be able to squeeze through some tight spots in the microcirculation during its numerous passages around the body; the sinusoids of the spleen are of special importance in this regard. For the red cell to be easily and reversibly **deformable,** its membrane must be both fluid and flexible; it should also preserve its biconcave shape since this facilitates gas exchange. Membrane **lipids** help determine membrane fluidity. Attached to the inner aspect of the membrane of the red blood cell are a number of **peripheral cytoskeletal proteins** (Table 52–8) that play important roles in respect to preserving shape and flexibility; these will now be described.

Spectrin is the major protein of the cytoskeleton. It is composed of two polypeptides: spectrin 1 (α chain) and spectrin 2 (β chain). These chains, measuring approximately 100 nm in length, are aligned in an antiparallel manner and are loosely intertwined, forming a dimer. Both chains are made up of segments of 106 amino acids that appear to fold into triple-stranded α-helical coils joined by nonhelical segments. One dimer interacts with another, forming a head-to-head tetramer. The overall shape confers **flexibility** on the protein and in turn on the membrane of the red blood cell. At least four **binding sites** can be defined in spectrin: (1) for self-association, (2) for ankyrin (bands 2.1, etc), (3) for actin (band 5), and (4) for protein 4.1.

Ankyrin is a pyramid-shaped protein that **binds spectrin.** In turn, ankyrin binds tightly to band 3, securing attachment of spectrin to the membrane. Ankyrin is sensitive to proteolysis, accounting for the appearance of bands 2.2, 2.3, and 2.6, all of which are derived from band 2.1.

Actin (band 5) exists in red blood cells as short, double-helical filaments of F-actin. The tail end of spectrin dimers binds to actin. Actin also binds to protein 4.1.

Protein 4.1, a globular protein, binds tightly to the tail end of spectrin, near the actin-binding site of the latter, and thus is part of a protein 4.1-spectrin-actin ternary complex. Protein 4.1 also binds to the integral proteins, glycophorins A and C, thereby attaching the ternary complex to the membrane. In addition, protein 4.1 may interact with certain membrane phospholipids, thus connecting the lipid bilayer to the cytoskeleton.

Certain other proteins (4.9, adducin, and tropomyosin) also participate in **cytoskeletal assembly.**

Abnormalities in the Amount or Structure of Spectrin Cause Hereditary Spherocytosis & Elliptocytosis

Hereditary spherocytosis is a genetic disease, transmitted as an autosomal dominant, that affects about 1:5000 North Americans. It is characterized by the presence of spherocytes (spherical red blood cells, with a low surface-to-volume ratio) in the peripheral blood, by a **hemolytic anemia** (see Figure 52–3), and by splenomegaly. The spherocytes are not as deformable as are

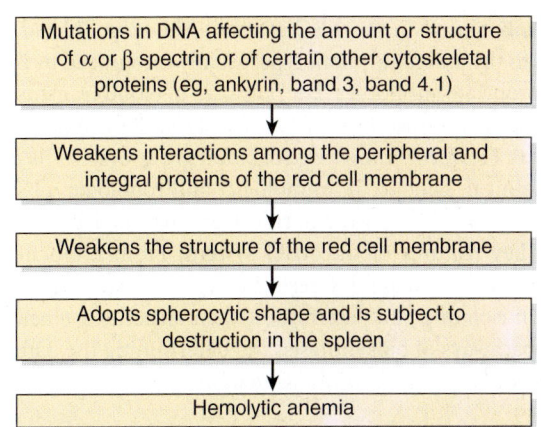

FIGURE 52–6   **Summary of the causation of hereditary spherocytosis (OMIM 182900).** Approximately 50% of cases are due to abnormalities in ankyrin and 25% to abnormalities in spectrin.

normal red blood cells, and they are subject to destruction in the spleen, thus greatly shortening their life in the circulation. Hereditary spherocytosis is **curable by splenectomy** because the spherocytes can persist in the circulation if the spleen is absent.

The spherocytes are much more susceptible to osmotic lysis than are normal red blood cells. This is assessed in the **osmotic fragility test,** in which red blood cells are exposed in vitro to decreasing concentrations of NaCl. The physiologic concentration of NaCl is 0.85 g/dL. When exposed to a concentration of NaCl of 0.5 g/dL, very few normal red blood cells are hemolyzed, whereas approximately 50% of spherocytes would lyse under these conditions. The explanation is that the spherocyte, being almost circular, has little potential extra volume to accommodate additional water and thus lyses readily when exposed to a slightly lower osmotic pressure than is normal.

One cause of hereditary spherocytosis (**Figure 52–6**) is a deficiency in the amount of **spectrin** or abnormalities of its structure, so that it no longer tightly binds the other proteins with which it normally interacts. This weakens the membrane and leads to the spherocytic shape. Abnormalities of **ankyrin** and of **bands 3, 4.1,** and **4.2** are involved in other cases.

Hereditary elliptocytosis is a genetic disorder that is similar to hereditary spherocytosis except that affected red blood cells assume an elliptic, disk-like shape, recognizable by microscopy. It is also due to abnormalities in **spectrin;** some cases reflect abnormalities of band **4.1** or of **glycophorin** C.

THE BIOCHEMICAL BASES OF THE ABO BLOOD GROUP SYSTEM HAVE BEEN ESTABLISHED

Approximately 30 human **blood group systems** have been recognized, the best known of which are the **ABO, Rh (Rhesus),** and **MN** systems. The term **"blood group"** applies to a defined system of red blood cell antigens (blood group substances) controlled by a genetic locus having a variable number of alleles (eg, A, B, and O in the ABO system). The term **"blood type"** refers to the antigenic phenotype, usually recognized by the use of appropriate antibodies. For purposes of blood transfusion, it is particularly important to know the basics of the **ABO** and **Rh** systems. However, knowledge of blood group systems is also of biochemical, genetic, immunologic, anthropologic, obstetric, pathologic, and forensic interest. Here, we shall discuss only some key features of the **ABO system.** From a biochemical viewpoint, the major interests in the ABO substances have been in isolating and determining their structures, elucidating their pathways of biosynthesis, and determining the natures of the products of the A, B, and O genes.

The ABO System Is of Crucial Importance in Blood Transfusion

This system was first discovered by Landsteiner in 1900 when investigating the basis of compatible and incompatible transfusions in humans. The membranes of the red blood cells of most individuals contain one blood group substance of type A, type B, type AB, or type O. Individuals of **type A** have anti-B antibodies in their plasma and will thus agglutinate type B or type AB blood. Individuals of **type B** have anti-A antibodies and will agglutinate type A or type AB blood. **Type AB** blood has neither anti-A nor anti-B antibodies and has been designated the **universal recipient. Type O** blood has neither A nor B substances and has been designated the **universal donor.** The explanation of these findings is related to the fact that the body does not usually produce antibodies to its own constituents. Thus, individuals of type A do not produce antibodies to their own blood group substance, A, but do possess antibodies to the foreign blood group substance, B, possibly because similar structures are present in micro-organisms to which the body is exposed early in life. Since individuals of type O have neither A nor B substances, they possess antibodies to both these foreign substances. The above description has been simplified considerably; eg, there are two subgroups of type A: A_1 and A_2.

The genes responsible for production of the ABO substances are present on the long arm of chromosome 9. There are **three alleles,** two of which are codominant (A and B) and the third (O) recessive; these ultimately determine the four phenotypic products: the A, B, AB, and O substances.

The ABO Substances Are Glycosphingolipids & Glycoproteins Sharing Common Oligosaccharide Chains

The **ABO substances** are complex oligosaccharides present in most cells of the body and in certain secretions. On membranes of red blood cells, the oligosaccharides that determine the specific natures of the ABO substances appear to be mostly present in **glycosphingolipids,** whereas in secretions the same oligosaccharides are present in **glycoproteins.** Their presence in secretions is determined by a gene designated *Se* (for **secretor**), which codes for a specific **fucosyl (Fuc) transferase** in secretory organs, such as the exocrine glands, but which is not active in red blood cells. Individuals of *SeSe* or *Sese* genotypes secrete A or B antigens (or both), whereas individuals of the *sese* genotype do not secrete A or B substances, but their red blood cells can express the A and B antigens.

H Substance Is the Biosynthetic Precursor of Both the A & B Substances

The ABO substances have been isolated and their structures determined; simplified versions, showing only their nonreducing ends, are presented in **Figure 52–7**. It is important to first appreciate the structure of the **H substance** since it is the precursor of both the A and B substances and is the blood group substance found in persons of type O. H substance itself is formed by the action of a **fucosyltransferase,** which catalyzes the addition of the terminal fucose in $\alpha 1 \rightarrow 2$ linkage onto the terminal Gal residue of its precursor:

$$GDP\text{-}Fuc + Gal\text{-}\beta\text{-}R \rightarrow Fuc\text{-}\alpha 1,2\text{-}Gal\text{-}\beta\text{-}R + GDP$$
$$\text{Precursor} \qquad\qquad \text{H substance}$$

The H locus codes for this fucosyltransferase. The *h* allele of the H locus codes for an inactive fucosyltransferase; therefore, individuals of the *hh* genotype cannot generate H substance, the precursor of the A and B antigens. Thus, individuals of the *hh* genotype will have red blood cells of type O, even though they may possess the enzymes necessary to make the A or B substances (see below). They are referred to as being Bombay phenotype (O_h).

FIGURE 52–7 **Diagrammatic representation of the structures of the H, A, and B blood group substances.** R represents a long complex oligosaccharide chain, joined either to ceramide where the substances are glycosphingolipids, or to the polypeptide backbone of a protein via a serine or threonine residue where the substances are glycoproteins. Note that the blood group substances are biantennary; ie, they have two arms, formed at a branch point (not indicated) between the GlcNAc—R, and only one arm of the branch is shown. Thus, the H, A, and B substances each contain two of their respective short oligosaccharide chains shown above. The AB substance contains one type A chain and one type B chain.

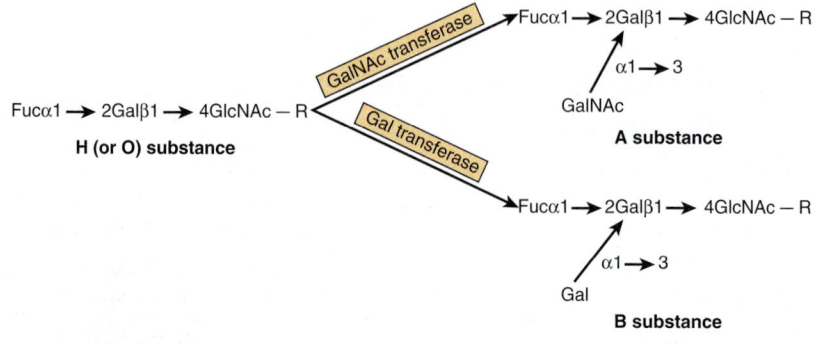

The *A* Gene Encodes a GalNAc Transferase, the *B* Gene a Gal Transferase, & the *O* Gene an Inactive Product

In comparison with blood group H substance (Figure 52–7), **A substance** contains an additional GalNAc and **B substance** an additional Gal, linked as indicated. Anti-A antibodies are directed to the additional GalNAc residue found in the A substance, and anti-B antibodies are directed toward the additional Gal residue found in the B substance. Thus, GalNAc is the **immunodominant sugar** (ie, the one determining the specificity of the antibody formed) of blood group A substance, whereas Gal is the immunodominant sugar of the B substance. In view of the structural findings, it is not surprising that A substance can be synthesized in vitro from O substance in a reaction catalyzed by a GalNAc transferase, employing UDP-GalNAc as the sugar donor. Similarly, blood group B can be synthesized from O substance by the action of a Gal transferase, employing UDP-Gal. It is crucial to appreciate that the product of the *A* gene is the **GalNAc transferase** that adds the terminal GalNAc to the O substance. Similarly, the product of the *B* gene is the **Gal transferase** adding the Gal residue to the O substance. Individuals of **type AB** possess both enzymes and thus have two oligosaccharide chains (Figure 52–6), one terminated by a GalNAc and the other by a Gal. Individuals of type O apparently synthesize an inactive protein, detectable by immunologic means; thus, H substance is their ABO blood group substance.

In 1990, a study using cloning and sequencing technology described the nature of the differences between the glycosyltransferase products of the *A*, *B*, and *O* genes. A difference of four nucleotides is apparently responsible for the distinct specificities of the A and B glycosyltransferases. On the other hand, the *O* allele has a single base-pair mutation, causing a **frameshift mutation** resulting in a protein-lacking transferase activity.

NEUTROPHILS HAVE AN ACTIVE METABOLISM & CONTAIN SEVERAL UNIQUE ENZYMES & PROTEINS

The major biochemical features of **neutrophils** are summarized in **Table 52–9**. Prominent features are active aerobic glycolysis, active pentose phosphate pathway, moderately active oxidative phosphorylation (because mitochondria are relatively sparse), and a high content of lysosomal enzymes. Many of the enzymes listed in Table 52–5 are also of importance in the oxidative metabolism of neutrophils (see below). **Table 52–10** summarizes

TABLE 52–9 Summary of Major Biochemical Features of Neutrophils

- Active glycolysis
- Active pentose phosphate pathway
- Moderate oxidative phosphorylation
- Rich in lysosomes and their degradative enzymes
- Contain certain unique enzymes (eg, myeloperoxidase and NADPH oxidase) and proteins
- Contain CD 11/CD18 integrins in plasma membrane

TABLE 52–10 Some Important Enzymes and Proteins of Neutrophils[1]

Enzyme or Protein	Reaction Catalyzed or Function	Comment
Myeloperoxidase (MPO)	$H_2O_2 + X^-$ (halide) $+ H^+ \rightarrow HOX + H_2O$ (where $X^- = Cl^-$, HOX = hypochlorous acid)	Responsible for the green color of pus Genetic deficiency can cause recurrent infections
NADPH oxidase	$2O_2 + NADPH \rightarrow 2O_2^- + NADP + H^+$	Key component of the respiratory burst Deficient in chronic granulomatous disease
Lysozyme	Hydrolyzes link between *N*-acetylmuramic acid and *N*-acetyl-D-glucosamine found in certain bacterial cell walls	Abundant in macrophages
Defensins	Basic antibiotic peptides of 20–33 amino acids	Apparently kill bacteria by causing membrane damage
Lactoferrin	Iron-binding protein	May inhibit growth of certain bacteria by binding iron and may be involved in regulation of proliferation of myeloid cells
CD11a/CD18, CD11b/CD18, CD11c/CD18[2]	Adhesion molecules (members of the integrin family)	Deficient in leukocyte adhesion deficiency type I (OMIM 116920)
Receptors for Fc fragments of IgGs	Bind Fc fragments of IgG molecules	Target antigen-antibody complexes to myeloid and lymphoid cells, eliciting phagocytosis and other responses

[1]The expression of many of these molecules has been studied during various stages of differentiation of normal neutrophils and also of corresponding leukemic cells employing molecular biology techniques (eg, measurements of their specific mRNAs). For the majority, cDNAs have been isolated and sequenced, amino acid sequences deduced, genes have been localized to specific chromosomal locations, and exons and intron sequences have been defined. Some important proteinases of neutrophils are listed in Table 52–13.

[2]CD = cluster of differentiation. This refers to a uniform system of nomenclature that has been adopted to name surface markers of leukocytes. A specific surface protein (marker) that identifies a particular lineage or differentiation stage of leukocytes and that is recognized by a group of monoclonal antibodies is called a member of a cluster of differentiation. The system is particularly helpful in categorizing subclasses of lymphocytes. Many CD antigens are involved in cell–cell interactions, adhesion, and transmembrane signaling.

TABLE 52–11 Sources of Biomolecules with Vasoactive Properties Involved in Acute Inflammation

Mast Cells and Basophils	Platelets	Neutrophils	Plasma Proteins
Histamine	Serotonin	Platelet-activating factor (PAF)	C3a, C4a, and C5a from the complement system
		Eicosanoids (various prostaglandins and leukotrienes)	Bradykinin and fibrin degradation products from the coagulation system

the functions of some proteins that are relatively unique to neutrophils.

Neutrophils Are Key Players in the Body's Defense Against Bacterial Infection

Neutrophils are motile phagocytic cells of the innate immune system that play a key role in acute inflammation. When bacteria enter tissues, a number of phenomena result that are collectively known as the "acute inflammatory response." They include (1) the increase of vascular permeability, (2) entry of activated neutrophils into the tissues, (3) activation of platelets, and (4) spontaneous subsidence (resolution) if the invading micro-organisms have been dealt with successfully.

A variety of molecules are released from cells and plasma proteins during acute inflammation whose net overall effect is to increase vascular permeability, resulting in tissue edema (**Table 52–11**).

In acute inflammation, neutrophils are recruited from the bloodstream into the tissues to help eliminate the foreign invaders. The neutrophils are attracted into the tissues by **chemotactic factors,** including complement fragment C5a, small peptides derived from bacteria (eg, *N*-formyl-methionyl-leucyl-phenylalanine), and a number of leukotrienes. To reach the tissues, circulating neutrophils must pass through the capillaries. To achieve this, they marginate along the vessel walls and then adhere to endothelial (lining) cells of the capillaries.

Integrins Mediate Adhesion of Neutrophils to Endothelial Cells

Adhesion of neutrophils to endothelial cells employs specific adhesive proteins (**integrins**) located on their surface and also specific receptor proteins in the endothelial cells. (See also the discussion of **selectins** in Chapter 47.)

The **integrins** are a superfamily of surface proteins present on a wide variety of cells. They are involved in the **adhesion** of cells to other cells or to specific components of the extracellular matrix. They are **heterodimers,** containing an α and a β subunit linked noncovalently. The subunits contain extracellular, transmembrane, and intracellular segments. The **extracellular segments** bind to a variety of ligands such as specific proteins of the extracellular matrix and of the surfaces of other cells. These ligands often contain ArgGly-Asp (R-G-D) sequences. The **intracellular domains** bind to various proteins of the cytoskeleton, such as actin and vinculin. The integrins are proteins that **link the outsides of cells to their insides,** thereby helping to integrate responses of cells (eg, movement and phagocytosis) to changes in the environment.

Three **subfamilies** of integrins were recognized initially. Members of each subfamily were distinguished by containing a common β subunit, but they differed in their subunits. However, more than three β subunits have now been identified, and the classification of integrins has become rather complex. Some integrins of specific interest with regard to neutrophils are listed in **Table 52–12.**

TABLE 52–12 Examples of Integrins That Are Important in the Function of Neutrophils, of Other White Blood Cells, and of Platelets[1]

Integrin	Cell	Subunit	Ligand	Function
VLA-1 (CD49a)	WBCs, others	α1β1	Collagen, laminin	Cell-ECM adhesion
VLA-5 (CD49e)	WBCs, others	α5β1	Fibronectin	Cell-ECM adhesion
VLA-6 (CD49f)	WBCs, others	α6β1	Laminin	Cell-ECM adhesion
LFA-1 (CD11a)	WBCs	αLβ2	ICAM-1	Adhesion of WBCs
Glycoprotein IIb/IIIa	Platelets	αIIbβ3	ICAM-2	
Fibrinogen, fibronectin, von Willebrand factor	Platelet adhesion and aggregation			

[1]LFA-1, lymphocyte function-associated antigen 1; VLA, very late antigen; CD, cluster of differentiation; ICAM, intercellular adhesion molecule; ECM, extracellular matrix. A deficiency of LFA-1 and related integrins is found in type I leukocyte adhesion deficiency (OMIM 116920). A deficiency of platelet glycoprotein IIb/IIIa complex is found in Glanzmann thrombasthenia (OMIM 273800), a condition characterized by a history of bleeding, a normal platelet count, and abnormal clot retraction. These findings illustrate how fundamental knowledge of cell surface adhesion proteins is shedding light on the causation of a number of diseases.

A deficiency of the β_2 subunit (also designated CD18) of LFA-1 and of two related integrins found in neutrophils and macrophages, Mac-1 (CD11b/CD18) and p150,95 (CD11c/CD18), causes **type 1 leukocyte adhesion deficiency,** a disease characterized by recurrent bacterial and fungal infections. Among various results of this deficiency, the adhesion of affected white blood cells to endothelial cells is diminished, and lower numbers of neutrophils thus enter the tissues to combat infection.

Once having passed through the walls of small blood vessels, the neutrophils migrate toward the highest concentrations of the chemotactic factors, encounter the invading bacteria, and attempt to attack and destroy them. The neutrophils must be **activated** in order to turn on many of the metabolic processes involved in phagocytosis and killing of bacteria.

Activation of Neutrophils Is Similar to Activation of Platelets & Involves Hydrolysis of Phosphatidylinositol Bisphosphate

The mechanisms involved in platelet activation are discussed in Chapter 51 (see Figure 51–8). The process involves interaction of the stimulus (eg, thrombin) with a receptor, activation of G proteins, stimulation of phospholipase C, and liberation from phosphatidylinositol bisphosphate of inositol triphosphate and diacylglycerol. These two second messengers result in an elevation of intracellular Ca^{2+} and activation of protein kinase C. In addition, activation of phospholipase A_2 produces arachidonic acid that can be converted to a variety of biologically active eicosanoids.

The process of **activation of neutrophils** is essentially similar. They are activated, via specific receptors, by interaction with bacteria, binding of chemotactic factors, or antibody–antigen complexes. The resultant **rise in intracellular Ca^{2+}** affects many processes in neutrophils, such as assembly of microtubules and the actin–myosin system. These processes are respectively involved in secretion of contents of granules and in motility, which enables neutrophils to seek out the invaders. The activated neutrophils are now ready to destroy the invaders by mechanisms that include production of active derivatives of oxygen.

The Respiratory Burst of Phagocytic Cells Involves NADPH Oxidase & Helps Kill Bacteria

When neutrophils and other phagocytic cells engulf bacteria, they exhibit a rapid increase in oxygen consumption known as **the respiratory burst.** This phenomenon reflects the rapid utilization of oxygen (following a lag of 15–60 s) and production from it of large amounts of **reactive derivatives,** such as O_2^-, H_2O_2, OH', and OCl^- (hypochlorite ion). Some of these products are potent microbicidal agents.

The **electron transport chain system** responsible for the respiratory burst (named NADPH oxidase) is composed of several components. One is **cytochrome b_{558},** located in the plasma membrane; it is a heterodimer, containing two polypeptides of 91 kDa and 22 kDa. When the system is activated (see below), two cytoplasmic polypeptides of 47 kDa and 67 kDa are recruited to the plasma membrane and, together with cytochrome b_{558}, form **the NADPH oxidase** responsible for the respiratory burst. The reaction catalyzed by NADPH oxidase, involving formation of superoxide anion, is shown in Table 52–5 (reaction 2). This system catalyzes the one-electron reduction of oxygen to superoxide anion. The NADPH is generated mainly by the pentose phosphate cycle, whose activity increases markedly during phagocytosis.

The above reaction is followed by the spontaneous production (by spontaneous dismutation) of **hydrogen peroxide** from two molecules of superoxide:

$$O_2^- + O_2^- + 2H^+ \rightarrow H_2O_2 + O_2$$

The **superoxide ion** is discharged to the outside of the cell or into phagolysosomes, where it encounters ingested bacteria. Killing of bacteria within phagolysosomes appears to depend on the combined action of elevated pH, superoxide ion, or further oxygen derivatives (H_2O_2, OH', and HOCl [hypochlorous acid; see below]) and on the action of certain bactericidal peptides (defensins) and other proteins (eg, cathepsin G and certain cationic proteins) present in phagocytic cells. Any superoxide that enters the cytosol of the phagocytic cell is converted to H_2O_2 by the action of **superoxide dismutase,** which catalyzes the same reaction as the spontaneous dismutation shown above. In turn, H_2O_2 is used by myeloperoxidase (see below) or disposed of by the action of glutathione peroxidase or catalase.

NADPH oxidase is inactive in resting phagocytic cells and is **activated** upon contact with various ligands (complement fragment C5a, chemotactic peptides, etc) with receptors in the plasma membrane. The events resulting in activation of the oxidase system have been much studied and are similar to those described above for the process of activation of neutrophils. They involve **G proteins,** activation of **phospholipase C,** and generation of **inositol 1,4,5-triphosphate** (IP_3). The last mediates a transient increase in the level of cytosolic Ca^{2+}, which is essential for induction of the respiratory burst. **Diacylglycerol** is also generated and induces the translocation of protein kinase C into the plasma membrane from the cytosol, where it catalyzes the **phosphorylation** of various proteins, some of which are components of the oxidase system. A second pathway of activation not involving Ca^{2+} also operates.

Mutations in the Genes for Components of the NADPH Oxidase System Cause Chronic Granulomatous Disease

The importance of the **NADPH oxidase system** was clearly shown when it was observed that the respiratory burst was defective in **chronic granulomatous disease,** a relatively uncommon condition characterized by recurrent infections and widespread

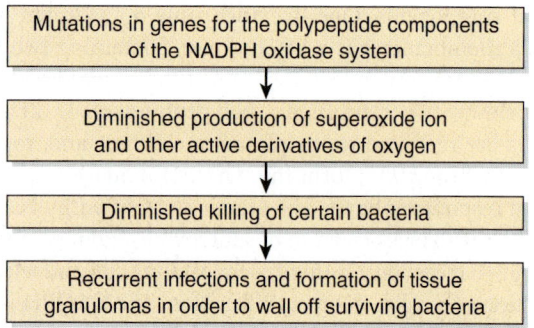

FIGURE 52–8 Simplified scheme of the sequence of events involved in the causation of chronic granulomatous disease (OMIM 306400). Mutations in any of the genes for the four polypeptides involved (two are components of cytochrome b_{558} and two are derived from the cytoplasm) can cause the disease. The polypeptide of 91 kDa is encoded by a gene in the X chromosome; approximately 60% of cases of chronic granulomatous disease are X-linked, with the remainder being inherited in an autosomal recessive fashion.

TABLE 52–13 Proteinases of Neutrophils and Antiproteinases of Plasma and Tissues[1]

Proteinases	Antiproteinases
Elastase	α_1-Antiproteinase (α_1-antitrypsin)
Collagenase	α_2-Macroglobulin
Gelatinase	Secretory leukoproteinase inhibitor
Cathepsin G	α_1-Antichymotrypsin
Plasminogen activator	Plasminogen activator inhibitor–1
	Tissue inhibitor of metalloproteinase

[1]The Table lists some of the important proteinases of neutrophils and some of the proteins that can inhibit their actions. Most of the proteinases listed exist inside neutrophils as precursors. The proteinases listed can digest many proteins of the extracellular matrix, causing tissue damage. The overall balance of proteinase:antiproteinase action can be altered by activating the precursors of the proteinases, or by inactivating the antiproteinases. The latter can be caused by proteolytic degradation or chemical modification, eg, Met-358 of α_1-antiproteinase inhibitor is oxidized by cigarette smoke.

granulomas (chronic inflammatory lesions) in the skin, lungs, and lymph nodes. The granulomas form as attempts to wall off bacteria that have not been killed, owing to genetic deficiencies in the NADPH oxidase system. The disorder is due to mutations in the genes encoding the four polypeptides that constitute the NADPH oxidase system. Some patients have responded to treatment with gamma interferon, which may increase transcription of the 91-kDa component if it is affected. Attempts are being made to develop gene therapy for this condition. The probable sequence of events involved in the causation of chronic granulomatous disease is shown in **Figure 52–8**.

Neutrophils Contain Myeloperoxidase, Which Catalyzes the Production of Chlorinated Oxidants

The enzyme **myeloperoxidase,** present in large amounts in neutrophil granules and responsible for the green color of pus, can act on H_2O_2 to produce hypohalous acids.

$$H_2O_2 + X^- + H^+ \xrightarrow{\text{MYELOPEROXIDASE}} HOX + H_2O$$

$(X^- = Cl^-, Br^-, I^- \text{ or } SCN^-; HOCl = \text{hypochlorous acid})$

The H_2O_2 used as substrate is generated by the NADPH oxidase system. Cl^- is the halide usually employed since it is present in relatively high concentration in plasma and body fluids. **HOCl,** the active ingredient of household liquid bleach, is a powerful oxidant and is highly microbicidal. When applied to normal tissues, its potential for causing damage is diminished because it reacts with primary or secondary amines present in neutrophils and tissues to produce various nitrogen-chlorine derivatives; these **chloramines** are also oxidants, though less powerful than HOCl, and act as microbicidal agents (eg, in sterilizing wounds) without causing tissue damage.

The Proteinases of Neutrophils Can Cause Serious Tissue Damage if Their Actions Are Not Checked

Neutrophils contain a number of proteinases (**Table 52–13**) that can hydrolyze elastin, various types of collagens, and other proteins present in the extracellular matrix. Such enzymatic action, if allowed to proceed unopposed, can result in serious damage to tissues. Most of these proteinases are **lysosomal enzymes** and exist mainly as inactive precursors in normal neutrophils. Small amounts of these enzymes are released into normal tissues, with the amounts increasing markedly during inflammation. The activities of elastase and other proteinases are normally kept in check by a number of **antiproteinases** (also listed in Table 52–13) present in plasma and the extracellular fluid. Each of them can combine—usually forming a noncovalent complex—with one or more specific proteinases and thus cause inhibition. In Chapter 50, it was shown that a genetic deficiency of α_1-**antiproteinase inhibitor** (α_1-antitrypsin) permits elastase to act unopposed and digest pulmonary tissue, thereby participating in the causation of emphysema. α_2-**Macroglobulin** is a plasma protein that plays an important role in the body's defense against excessive action of proteases; it combines with and thus neutralizes the activities of a number of important proteases (Chapter 50).

When increased amounts of chlorinated oxidants are formed during inflammation, they affect the proteinase: antiproteinase equilibrium, tilting it in favor of the former. For instance, some of the proteinases listed in Table 52–13 are **activated** by HOCl, whereas certain of the antiproteinases are **inactivated** by this compound. In addition, the tissue inhibitor of metalloproteinases and α_1-antichymotrypsin can be hydrolyzed by activated elastase, and α_1-antiproteinase inhibitor can be hydrolyzed by activated collagenase and gelatinase. In most circumstances, **an appropriate balance** of proteinases and antiproteinases is achieved. However, in certain instances, such as in the lung when α_1-antiproteinase inhibitor is deficient or when large amounts of neutrophils accumulate in tissues because of inadequate drainage, considerable **tissue damage** can result from the unopposed action of proteinases.

RECOMBINANT DNA TECHNOLOGY HAS HAD A PROFOUND IMPACT ON HEMATOLOGY

Recombinant DNA technology has had a major impact on many aspects of hematology. The bases of the **thalassemias** and of many **disorders of coagulation** (Chapter 51) have been greatly clarified by investigations using cloning and sequencing. The study of oncogenes and chromosomal translocations has advanced understanding of the **leukemias.** As discussed above, cloning techniques have made available therapeutic amounts of **erythropoietin** and **other growth factors.** Deficiency of **adenosine deaminase,** which affects lymphocytes particularly, is the first disease to be treated by gene therapy (see Case no. 1, Chapter 57). Like many other areas of biology and medicine, hematology has been and will continue to be revolutionized by this technology. **Systems biology** is another approach that is beginning to be applied to normal and abnormal hematopoiesis. It depends largely on mathematical, engineering, and computational tools to further understand complex biological processes, such as hematopoiesis. Advances in genomics and proteomics will also be critical for its future development.

SUMMARY

- Some types of anemias are very prevalent conditions. Major causes of anemia include blood loss, deficiencies of iron, folate and vitamin B_{12}, and various factors causing hemolysis.

- The red blood cell is simple in terms of its structure and function, consisting principally of a concentrated solution of hemoglobin surrounded by a membrane.

- The production of red cells is regulated by erythropoietin, whereas other growth factors (eg, granulocyte and granulocyte-macrophage colony-stimulating factors) regulate the production of white blood cells.

- The red cell contains a battery of cytosolic enzymes, such as superoxide dismutase, catalase, and glutathione peroxidase, to dispose of powerful oxidants (ROS) generated during its metabolism.

- Genetically determined deficiency of the activity of glucose-6-phosphate dehydrogenase, which produces NADPH, is an important cause of hemolytic anemia.

- Methemoglobin is unable to transport oxygen; both genetic and acquired causes of methemoglobinemia are recognized.

- Considerable information has accumulated concerning the proteins and lipids of the red cell membrane. A number of cytoskeletal proteins, such as spectrin, ankyrin, and actin, interact with specific integral membrane proteins to help regulate the shape and flexibility of the membrane.

- Deficiency of spectrin results in hereditary spherocytosis and hereditary elliptocytosis, both causes of hemolytic anemia.

- The ABO blood group substances in the red cell membrane are complex glycosphingolipids; the immunodominant sugar of A substance is *N*-acetyl-galactosamine, whereas that of the B substance is galactose. O substance does not contain either of these two sugar residues in the particular linkages found in the A and B substances.

- Neutrophils play a major role in the body's defense mechanisms. Integrins on their surface membranes determine specific interactions with various cell and tissue components.

- Leukocytes are activated on exposure to bacteria and other stimuli; NADPH oxidase plays a key role in the process of activation (the respiratory burst). Mutations in this enzyme and associated proteins cause chronic granulomatous disease.

- The proteinases of neutrophils can digest many tissue proteins; normally, this is kept in check by a battery of antiproteinases. However, this defense mechanism can be overcome in certain circumstances, resulting in extensive tissue damage.

- The application of recombinant DNA technology is revolutionizing the field of hematology.

REFERENCES

Fauci AS, Braunwald E, Kasper DL, et al (editors): *Harrison's Principles of Internal Medicine,* 17th ed. McGraw-Hill, 2008. (Chapters 58, 61, & 98–108 deal with various blood disorders. Chapters 66–68 deal with various aspects of hematopoietic and other stem cells).

Hofmann R, Benz Jr EJ, Shattal SJ, et al (editors): *Hematology: Basic Principles and Practice,* 4th ed. Elsevier Churchill Livingston, 2005.

Imlay JA: Cellular defenses against superoxide and hydrogen peroxide. Annu Rev Biochem 2008;77:755.

Israels SJ (editor): *Mechanisms in Hematology,* 4th 3rd ed. Core Health Sciences Inc, 2011.

Naria A, Ebert BL: Ribosomopathies: human disorders of ribosome dysfunction. Blood 2010;115:3196.

Orkin SH, Higgs DR: Sickle cell disease at 100 years. Science 2010;329:291.

Scriver CR, Beaudet AL, Valle D, et al (editors): *The Molecular Bases of Inherited Disease,* 8th ed. McGraw-Hill, 2001. (This text is now available online and updated as *The Online Metabolic & Molecular Bases of Inherited Disease* at www.ommbid.com Subscription is required, although access may be available via university and hospital libraries and other sources). A number of the chapters concern topics described in this chapter.

van den Berg JM, van Koppen E, Ahlin A, et al: Chronic granulomatous disease: the European experience. PLoS ONE 2009;4:e5234.

Weatherall DJ: The inherited diseases of hemoglobin are an emerging global health problem. Blood 2010;115:4331.

Whichard ZL, Sarkar CA, Kimmel M, Corey SJ: Hematopoiesis and its disorders: a systems biology approach. Blood 2010;115:2339.

Yonekawa K, Harlan JM: Targeting leukocyte integrins in human diseases. J Leukoc Biol 2005;77:129.

53

Metabolism of Xenobiotics

Robert K. Murray, MD, PhD

OBJECTIVES

After studying this chapter, you should be able to:

- Discuss how drugs and other xenobiotics are metabolized in the body.
- Describe the two general phases of xenobiotic metabolism, the first involving mainly hydroxylation reactions catalyzed by cytochrome P450 species and the second conjugation reactions catalyzed by various enzymes.
- Indicate the metabolic importance of glutathione.
- Appreciate that xenobiotics can cause pharmacologic, toxic, immunologic, and carcinogenic effects.
- Comprehend how knowledge of pharmacogenomics should help to personalize drug use.

BIOMEDICAL IMPORTANCE

Increasingly, humans are subjected to exposure to various foreign chemicals (**xenobiotics**)—drugs, food additives, pollutants, etc. The situation is well summarized in the following quotation from Rachel Carson: "As crude a weapon as the cave man's club, the chemical barrage has been hurled against the fabric of life." **Understanding how xenobiotics are handled at the cellular level** is important in learning how to cope with the chemical onslaught, and thus helping to **preserve the environment.** For example, building on such information, attempts are being made to modify microorganisms by introducing genes that encode various enzymes involved in metabolizing specific xenobiotics to harmless products. These modified organisms will then be used to help dispose of various pollutants that contaminate the planet.

Knowledge of the metabolism of xenobiotics is basic to a rational understanding of pharmacology and therapeutics, pharmacy, toxicology, management of cancer, and drug addiction. All these areas involve administration of, or exposure to, xenobiotics.

HUMANS ENCOUNTER THOUSANDS OF XENOBIOTICS THAT MUST BE METABOLIZED BEFORE BEING EXCRETED

A **xenobiotic** (Gk *xenos* "stranger") is a compound that is foreign to the body. The principal classes of xenobiotics of medical relevance are **drugs, chemical carcinogens,** and **various compounds** that have found their way into our environment by one route or another, such as polychlorinated biphenyls (PCBs) and certain insecticides. More than 200,000 manufactured environmental chemicals exist. Most of these compounds are subject to metabolism (chemical alteration) in the human body, with the liver being the main organ involved; occasionally, a xenobiotic may be excreted unchanged. **At least 30 different types of enzymes** catalyze reactions involved in xenobiotic metabolism; however, this chapter will only cover a selected group of them.

It is convenient to consider the metabolism of xenobiotics in two phases. In **phase 1,** the major reaction involved is **hydroxylation,** catalyzed mainly by members of a class of enzymes referred to as **monooxygenases** or **cytochrome P450s.** Hydroxylation may terminate the action of a drug, though this is not always the case. In addition to hydroxylation, these enzymes catalyze **a wide range of reactions,** including those involving deamination, dehalogenation, desulfuration, epoxidation, peroxygenation, and reduction. Reactions involving hydrolysis (eg, catalyzed by esterases) and certain other non-P450-catalyzed reactions also occur in phase 1.

In **phase 2,** the hydroxylated or other compounds produced in phase 1 are converted by specific enzymes to **various polar metabolites** by **conjugation** with glucuronic acid, sulfate, acetate, glutathione, or certain amino acids, or by **methylation.**

The overall purpose of the two phases of metabolism of xenobiotics is to increase their **water solubility (polarity)** and thus **excretion** from the body. Very hydrophobic xenobiotics would persist in adipose tissue almost indefinitely if they were not converted to more polar forms. In certain cases, phase 1 metabolic reactions convert xenobiotics from **inactive** to **biologically active** compounds. In these instances, the original xenobiotics are referred to as **"prodrugs"** or **"procarcinogens."** In other cases, additional phase 1 reactions (eg, further hydroxylation reactions) convert the active compounds to less active or inactive forms prior to conjugation. In yet other cases, it is the conjugation reactions themselves that convert the active products of phase 1 reactions to less active or inactive species, which are subsequently excreted in the urine or bile. In a very few cases, conjugation may actually increase the biologic activity of a xenobiotic.

The term **"detoxification"** is sometimes used for many of the reactions involved in the metabolism of xenobiotics. However, the term is not always appropriate because, as mentioned above, in some cases the reactions to which xenobiotics are subject actually **increase** their biologic activity and toxicity.

ISOFORMS OF CYTOCHROME P450 HYDROXYLATE A MYRIAD OF XENOBIOTICS IN PHASE 1 OF THEIR METABOLISM

Hydroxylation is the chief reaction involved in phase 1. The responsible enzymes are called **monooxygenases** or **cytochrome P450s.** It is estimated that there are some 57 cytochrome P450 genes present in humans. The reaction catalyzed by a monooxygenase (cytochrome P450) is as follows:

$$RH + O_2 + NADPH + H^+ \rightarrow R\text{-}OH + H_2O + NADP$$

RH above can represent a very wide variety of xenobiotics, including drugs, carcinogens, pesticides, petroleum products, and pollutants (such as a mixture of PCBs). In addition, **endogenous compounds,** such as certain steroids, eicosanoids, fatty acids, and retinoids, are also substrates. The substrates are generally **lipophilic** and are rendered more **hydrophilic** by hydroxylation.

Cytochrome P450 is considered the **most versatile biocatalyst** known. The actual reaction mechanism is complex and has been briefly described previously (Figure 12–6). It has been shown by the use of $^{18}O_2$ that one atom of oxygen enters R–OH and one atom enters water. This dual fate of the oxygen accounts for the former naming of monooxygenases as **"mixed-function oxidases."** The reaction catalyzed by cytochrome P450 can also be represented as follows:

$$\text{Reduced cytochrome P450} \quad \text{Oxidized cytochrome P450}$$
$$RH + O_2 \rightarrow R\text{-}OH + H_2O$$

Cytochrome P450 is so named because the enzyme was discovered when it was noted that preparations of **microsomes** that had been chemically reduced and then exposed to carbon monoxide **exhibited a distinct peak at 450 nm.** Microsomes contain fragments of the endoplasmic reticulum, where much of the P450 content of cells is located (see below). Among reasons that this enzyme is important is the fact that **approximately 50% of the common drugs humans ingest are metabolized by isoforms of cytochrome P450;** these enzymes also act on various carcinogens and pollutants. The major cytochrome P450s in drug metabolism are members of the CYP1, CYP2, and CYP3 families (see below).

Isoforms of Cytochrome P450 Make Up a Superfamily of Heme-Containing Enzymes

The following are important points concerning cytochrome P450s.

1. Because of **the large number of isoforms** (about 150) that have been discovered, it became important to have a **systematic nomenclature** for isoforms of P450 and for their genes. This is now available and in wide use and is based on structural homology. The abbreviated root symbol CYP denotes a cytochrome P450. This is followed by an Arabic number designating the **family;** cytochrome P450s are included in the same family if they exhibit 40% or more amino acid sequence identity. The Arabic number is followed by a capital letter indicating the **subfamily,** if two or more members exist; P450s are in the same subfamily if they exhibit greater than 55% sequence identity. The **individual** P450s are then arbitrarily assigned Arabic numerals. Thus, CYP1A1 denotes a cytochrome P450 that is a member of family 1 and subfamily A and is the first individual member of that subfamily. The nomenclature for the **genes** encoding cytochrome P450s is identical to that described above except that italics are used; thus, the gene encoding CYP1A1 is *CYP1A1.*

2. Like **hemoglobin,** they are hemoproteins.

3. They are widely distributed across species, **including bacteria.**

4. They are present in highest amount in **liver cells** and enterocytes but are probably present in all tissues. In liver and most other tissues, they are present mainly in the **membranes of the smooth endoplasmic reticulum,** which constitute part of the **microsomal fraction** when tissue is subjected to subcellular fractionation. In hepatic microsomes, cytochrome P450s can comprise as much as 20% of the total protein. P450s are found in most tissues, though often in low amounts compared with liver. In the **adrenal,** they are found in **mitochondria** as well as in the endoplasmic reticulum; the various hydroxylases present in that organ play an important role in cholesterol and steroid biosynthesis. The mitochondrial cytochrome P450 system differs from the microsomal system in that it uses an NADPH-linked flavoprotein, **adrenodoxin reductase,** and a nonheme iron–sulfur protein, **adrenodoxin.** In addition, the specific P450 isoforms involved in steroid biosynthesis are generally much more restricted in their substrate specificity.

5. At least six different species of cytochrome P450 are present in the endoplasmic reticulum of human liver, each with wide and somewhat overlapping **substrate specificities** and acting on both xenobiotics and endogenous compounds. The genes for many isoforms of P450 (from both humans and animals such as the rat) have been isolated and studied in detail in recent years. The combination of there being **a number of different types** and **each having a relatively wide substrate specificity** explains why the cytochrome P450 family can metabolize thousands of different chemicals.

6. NADPH, not NADH, is involved in the reaction mechanism of cytochrome P450. The enzyme that uses NADPH to yield the reduced cytochrome P450, shown at the left-hand side of the above equation, is called **NADPH-cytochrome P450 reductase.** Electrons are transferred from NADPH to NADPH-cytochrome P450 reductase and then to cytochrome P450. This leads to the **reductive activation of molecular oxygen,** and one atom of oxygen is subsequently inserted into the substrate. **Cytochrome b_5,** another hemoprotein found in the membranes of the smooth endoplasmic reticulum (Chapter 12), may be involved as an electron donor in some cases.

7. Lipids are also components of the cytochrome P450 system. The preferred lipid is **phosphatidylcholine,** which is the major lipid found in membranes of the endoplasmic reticulum.

8. Most isoforms of cytochrome P450 are **inducible.** For instance, the administration of phenobarbital or of many other drugs causes hypertrophy of the smooth endoplasmic reticulum and a three- to fourfold increase in the amount of cytochrome P450 within 4–5 days. The mechanism of induction has been studied extensively and in most cases involves **increased transcription of mRNA** for cytochrome P450. However, certain cases of induction involve **stabilization of mRNA, enzyme stabilization,** or **other mechanisms** (eg, an effect on translation).

Induction of cytochrome P450 has important clinical implications since it is a biochemical mechanism of **drug interaction.** A drug interaction has occurred when the effects of one drug are altered by prior, concurrent, or later administration of another. As an illustration, consider the situation when a patient is taking the anticoagulant **warfarin** to prevent blood clotting. This drug is metabolized by **CYP2C9.** Concomitantly, the patient is started on **phenobarbital** (an inducer of this P450) to treat a certain type of epilepsy, but the dose of warfarin is not changed. After 5 days or so, the level of CYP2C9 in the patient's liver will be elevated three- to fourfold. This in turn means that warfarin will be **metabolized much more quickly than before,** and its **dosage will have become inadequate.** Therefore, the **dose must be increased** if warfarin is to be therapeutically effective. To pursue this example further, **a problem** could arise later on **if the phenobarbital is discontinued** but the increased dosage of warfarin stays the same. The patient will be at risk of bleeding since the high dose of warfarin will be even more active than before, because the level of CYP2C9 will decline once phenobarbital has been stopped.

Another example of enzyme induction involves **CYP2E1,** which is induced by consumption of **ethanol.** This is a matter for concern, because this P450 metabolizes certain widely used solvents and also components found in tobacco smoke, many of which are established **carcinogens.** Thus, if the activity of CYP2E1 is elevated by induction, this may increase the risk of carcinogenicity developing from exposure to such compounds.

9. Certain isoforms of cytochrome P450 **(eg, CYP1A1)** are particularly involved in the metabolism of polycyclic aromatic hydrocarbons (PAHs) and related molecules; for this reason they were formerly called **aromatic hydrocarbon hydroxylases (AHHs).** This enzyme is important in the metabolism of PAHs and in carcinogenesis produced by these agents. For example, in the lung it may be involved in the conversion of inactive PAHs (procarcinogens), inhaled by smoking, to active carcinogens by hydroxylation reactions. Smokers have higher levels of this enzyme in some of their cells and tissues than do nonsmokers. Some reports have indicated that the activity of this enzyme may be elevated (induced) in **the placenta** of a woman who smokes, thus potentially altering the quantities of metabolites of PAHs (some of which could be harmful) to which the fetus is exposed.

10. Certain cytochrome P450s exist in **polymorphic forms** (genetic isoforms), some of which exhibit low catalytic activity. These observations are one important explanation for the variations in drug responses noted among many patients. One P450 exhibiting polymorphism is **CYP2D6,** which is involved in the metabolism of **debrisoquin** (an antihypertensive drug; see Table 53–2) and **sparteine** (an antiarrhythmic and oxytocic drug). Certain polymorphisms of CYP2D6 cause poor metabolism of these and a variety of other drugs so that they can accumulate in the body, resulting in untoward consequences. Another interesting polymorphism is that of **CYP2A6,** which is involved in the metabolism of **nicotine** to conitine. Three *CYP2A6* alleles have been identified: a wild type and two null or inactive alleles. It has been reported that individuals with the null alleles, who have impaired metabolism of nicotine, are apparently protected against becoming tobacco-dependent smokers (Table 53–2). These individuals smoke less, presumably because their blood and brain concentrations of nicotine remain elevated longer than those of individuals with the wild-type allele. It has been speculated that inhibiting CYP2A6 may be a novel way to help prevent and to treat smoking.

Table 53–1 summarizes some principal features of cytochrome P450s.

CONJUGATION REACTIONS PREPARE XENOBIOTICS FOR EXCRETION IN PHASE 2 OF THEIR METABOLISM

In phase 1 reactions, xenobiotics are generally converted to more polar, hydroxylated derivatives. In phase 2 reactions, these derivatives are conjugated with molecules such as glucuronic acid, sulfate, or glutathione. This renders them even more water-soluble, and they are eventually excreted in the urine or bile.

TABLE 53–1 Some Properties of Human Cytochrome P450s

- Involved in phase I of the metabolism of innumerable xenobiotics, including perhaps 50% of the drugs administered to humans; they may increase, decrease or not affect the activities of various drugs.

- Involved in the metabolism of many endogenous compounds (eg, steroids).

- All are hemoproteins.

- Often exhibit broad substrate specificity, thus acting on many compounds; consequently, different P450s may catalyze formation of the same product.

- Extremely versatile catalysts, perhaps catalyzing about 60 types of reactions.

- However, basically they catalyze reactions involving introduction of one atom of oxygen into the substrate and one into water.

- Their hydroxylated products are more water-soluble than their generally lipophilic substrates, facilitating excretion.

- Liver contains highest amounts, but found in most if not all tissues, including small intestine, brain, and lung.

- Located in the smooth endoplasmic reticulum or in mitochondria (steroidogenic hormones).

- In some cases, their products are mutagenic or carcinogenic.

- Many have a molecular mass of about 55 kDa.

- Many are inducible, resulting in one cause of drug interactions.

- Many are inhibited by various drugs or their metabolic products, providing another cause of drug interactions.

- Some exhibit genetic polymorphisms, which can result in atypical drug metabolism.

- Their activities may be altered in diseased tissues (eg, cirrhosis), affecting drug metabolism.

- Genotyping the P450 profile of patients (eg, to detect polymorphisms) may in the future permit individualization of drug therapy.

Five Types of Phase 2 Reactions Are Described Here

Glucuronidation

The glucuronidation of bilirubin is discussed in Chapter 31; the reactions whereby xenobiotics are glucuronidated are essentially similar. UDP-glucuronic acid is the glucuronyl donor, and a variety of glucuronosyltransferases, present in both the endoplasmic reticulum and cytosol, are the catalysts. Molecules such as 2-acetylaminofluorene (a carcinogen), aniline, benzoic acid, meprobamate (a tranquilizer), phenol, and many steroids are excreted as glucuronides. The glucuronide may be attached to oxygen, nitrogen, or sulfur groups of the substrates. Glucuronidation is probably the most frequent conjugation reaction.

Sulfation

Some alcohols, arylamines, and phenols are sulfated. The **sulfate donor** in these and other biologic sulfation reactions

(eg, sulfation of steroids, glycosaminoglycans, glycolipids, and glycoproteins) is **adenosine 3′-phosphate-5′-phosphosulfate (PAPS)** (Chapter 24); this compound is called "active sulfate."

Conjugation With Glutathione

Glutathione (γ-glutamylcysteinylglycine) is a **tripeptide** consisting of glutamic acid, cysteine, and glycine (Figure 3–3). Glutathione is commonly abbreviated GSH (because of the sulfhydryl group of its cysteine, which is the business part of the molecule). A number of potentially toxic electrophilic xenobiotics (such as certain carcinogens) are conjugated to the nucleophilic GSH in reactions that can be represented as follows:

$$R + GSH \rightarrow R\text{–}S\text{–}G$$

where R = an electrophilic xenobiotic. The enzymes catalyzing these reactions are called **glutathione *S*-transferases** and are present in high amounts in liver cytosol and in lower amounts in other tissues. A variety of glutathione *S*-transferases are present in human tissue. They exhibit different substrate specificities and can be separated by electrophoretic and other techniques. If the potentially toxic xenobiotics were not conjugated to GSH, they would be free to combine covalently with DNA, RNA, or cell protein and could thus lead to serious cell damage. GSH is therefore an important **defense mechanism** against certain toxic compounds, such as some drugs and carcinogens. If the levels of GSH in a tissue such as liver are **lowered** (as can be achieved by the administration to rats of certain compounds that react with GSH), then that tissue can be shown to be more susceptible to injury by various chemicals that would normally be conjugated to GSH. Glutathione conjugates are subjected to **further metabolism** before excretion. The glutamyl and glycinyl groups belonging to glutathione are removed by specific enzymes, and an acetyl group (donated by acetyl-CoA) is added to the amino group of the remaining cysteinyl moiety. The resulting compound is a **mercapturic acid,** a conjugate of L-acetylcysteine, which is then excreted in the urine.

Glutathione has other important functions in human cells apart from its role in xenobiotic metabolism.

1. It participates in the decomposition of potentially toxic **hydrogen peroxide** in the reaction catalyzed by glutathione peroxidase (Chapter 21).

2. It is an important **intracellular reductant and antioxidant,** helping to maintain essential SH groups of enzymes in their reduced state. This role is discussed in Chapter 21, and its involvement in the hemolytic anemia caused by deficiency of glucose-6-phosphate dehydrogenase is discussed in Chapters 21 and 52.

3. A metabolic cycle involving GSH as a carrier has been implicated in the **transport of certain amino acids** across membranes in the kidney. The first reaction of the cycle is shown below.

$$\text{Amino acid} + \text{GSH} \rightarrow \text{γ-Glutamyl amino acid} + \text{Cysteinylglycine}$$

This reaction helps transfer certain amino acids across the plasma membrane, the amino acid being subsequently hydrolyzed from its complex with GSH and the GSH being resynthesized from cysteinylglycine. The enzyme catalyzing the above reaction is **γ-glutamyltransferase (GGT).** It is present in the plasma membrane of renal tubular cells and bile ductule cells, and in the endoplasmic reticulum of hepatocytes. The enzyme has diagnostic value because it is released into the blood from hepatic cells in **various hepatobiliary diseases.**

Other Reactions

The two most important other reactions are acetylation and methylation.

1. **Acetylation**—Acetylation is represented by

$$X + Acetyl\text{-}CoA \rightarrow Acetyl\text{-}X + CoA$$

where X represents a xenobiotic. As for other acetylation reactions, **acetyl-CoA** (active acetate) is the acetyl donor. These reactions are catalyzed by **acetyltransferases** present in the cytosol of various tissues, particularly liver. The drug **isoniazid,** used in the treatment of tuberculosis, is subject to acetylation. **Polymorphic types** of acetyltransferases exist, resulting in individuals who are classified as **slow or fast acetylators,** and influence the rate of clearance of drugs such as isoniazid from blood. Slow acetylators are more subject to certain toxic effects of isoniazid because the drug persists longer in these individuals.

2. **Methylation**—A few xenobiotics are subject to methylation by methyltransferases, employing *S*-adenosylmethionine (Figure 29–18) as the methyl donor.

THE ACTIVITIES OF XENOBIOTIC-METABOLIZING ENZYMES ARE AFFECTED BY AGE, SEX, & OTHER FACTORS

Various factors affect the activities of the enzymes metabolizing xenobiotics. The activities of these enzymes may differ substantially among **species.** Thus, for example, the possible **toxicity** or **carcinogenicity** of xenobiotics cannot be extrapolated freely from one species to another. There are significant differences in enzyme activities among individuals, many of which appear to be due to **genetic factors.** The activities of some of these enzymes vary according to **age** and **sex.**

Intake of various xenobiotics such as phenobarbital, PCBs, or certain hydrocarbons can cause **enzyme induction.** It is thus important to know whether or not an individual has been exposed to these inducing agents in evaluating biochemical responses to xenobiotics. (It is always important when taking a clinical history to ask whether the patient has been taking any drugs or other therapeutic preparations). Metabolites of certain xenobiotics can **inhibit** or **stimulate** the activities of xenobiotic-metabolizing enzymes. Again, this can affect the doses of certain drugs that are administered to patients. Various **diseases** (eg, cirrhosis of the liver) can affect the activities of drug-metabolizing enzymes, sometimes necessitating adjustment of dosages of various drugs for patients with these disorders.

RESPONSES TO XENOBIOTICS INCLUDE PHARMACOLOGIC, TOXIC, IMMUNOLOGIC, & CARCINOGENIC EFFECTS

Xenobiotics are metabolized in the body by the reactions described above. When the xenobiotic is a drug, phase 1 reactions may produce its active form or may diminish or terminate its action if it is pharmacologically active in the body without prior metabolism. The diverse effects produced by drugs comprise the area of study of pharmacology; here it is important to appreciate that drugs act primarily through biochemical mechanisms. **Table 53–2** summarizes four important reactions to drugs that reflect **genetically determined differences** in enzyme and protein structure among individuals—part of the field of study known as **pharmacogenetics.** This area of science has been defined as **the study of the contribution of genetic factors to variation in drug response and toxicity.**

Polymorphisms that affect drug metabolism can occur in any of the **enzymes** involved in drug metabolism (including cytochrome P450s), in **transporters** and in **receptors.**

Certain xenobiotics are very toxic even at low levels (eg, cyanide). On the other hand, there are few xenobiotics, including drugs, that do not exert some toxic effects if sufficient amounts are administered. The **toxic effects of xenobiotics** cover a wide spectrum, but the major effects can be considered under three general headings (**Figure 53–1**).

TABLE 53–2 **Some Important Drug Reactions Due to Mutant or Polymorphic Forms of Enzymes or Proteins[1]**

Enzyme or Protein Affected	Reaction or Consequence
Glucose-6-phosphate dehydrogenase (G6PD) [mutations] (OMIM 305900)	Hemolytic anemia following ingestion of drugs such as primaquine
Ca^{2+} release channel (ryanodine receptor) in the sarcoplasmic reticulum [mutations] (OMIM 180901)	Malignant hyperthermia (OMIM 145600) following administration of certain anesthetics (eg, halothane)
CYP2D6 [polymorphisms] (OMIM 124030)	Slow metabolism of certain drugs (eg, debrisoquin), resulting in their accumulation
CYP2A6 [polymorphisms] (OMIM 122720)	Impaired metabolism of nicotine, resulting in protection against becoming a tobacco-dependent smoker

[1]G6PD deficiency is discussed in Chapters 21 and 52 and malignant hyperthermia in Chapter 49. At least one gene other than that encoding the ryanodine receptor is involved in certain cases of malignant hypertension. Many other examples of drug reactions based on polymorphism or mutation are available.

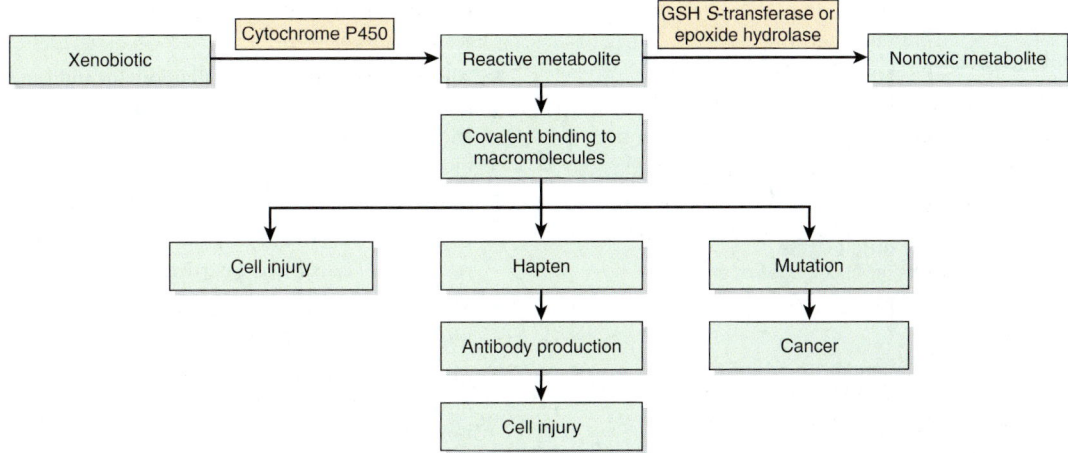

FIGURE 53–1 **Simplified scheme showing how metabolism of a xenobiotic can result in cell injury, immunologic damage, or cancer.** In this instance, the conversion of the xenobiotic to a reactive metabolite is catalyzed by a cytochrome P450, and the conversion of the reactive metabolite (eg, an epoxide) to a nontoxic metabolite is catalyzed either by a GSH S-transferase or by epoxide hydrolase.

The first is **cell injury** (cytotoxicity), which can be severe enough to result in cell death. There are many mechanisms by which xenobiotics injure cells. The one considered here is **covalent binding to cell macromolecules** of reactive species of xenobiotics produced by metabolism. These macromolecular targets include **DNA, RNA,** and **protein.** If the macromolecule to which the reactive xenobiotic binds is essential for short-term cell survival, for example, a protein or enzyme involved in some critical cellular function such as oxidative phosphorylation or regulation of the permeability of the plasma membrane, then severe effects on cellular function could become evident quite rapidly.

Second, the reactive species of a xenobiotic may bind to a protein, altering its **antigenicity.** The xenobiotic is said to act as a **hapten,** that is, a small molecule that by itself does not stimulate antibody synthesis but will combine with antibody once formed. The resulting **antibodies** can then damage the cell by several immunologic mechanisms that grossly perturb normal cellular biochemical processes.

Third, reactions of activated species of chemical carcinogens with **DNA** are thought to be of great importance in **chemical carcinogenesis.** Some chemicals (eg, benzo[α] pyrene) require activation by monooxygenases in the endoplasmic reticulum to become carcinogenic (they are thus called **indirect carcinogens**). The activities of the monooxygenases and of other xenobiotic-metabolizing enzymes present in the endoplasmic reticulum thus help to determine whether such compounds become carcinogenic or are "detoxified." Other chemicals (eg, various alkylating agents) can react directly (direct carcinogens) with DNA without undergoing intracellular chemical activation.

The enzyme **epoxide hydrolase** is of interest because it can exert a protective effect against certain carcinogens. The products of the action of certain monooxygenases on some procarcinogen substrates are **epoxides.** Epoxides are highly reactive and mutagenic or carcinogenic or both. Epoxide hydrolase—like cytochrome P450, also present in the membranes of the endoplasmic reticulum—acts on these compounds, converting them into much less reactive dihydrodiols. The reaction catalyzed by epoxide hydrolase can be represented as follows.

$$-\overset{|}{\underset{\underset{O}{\diagdown\diagup}}{C}}-\overset{|}{\underset{}{C}}- + H_2O \longrightarrow -\overset{|}{\underset{HO}{C}}-\overset{|}{\underset{OH}{C}}-$$

Epoxide   **Dihydrodiol**

PHARMACOGENOMICS WILL DRIVE THE DEVELOPMENT OF NEW & SAFER DRUGS

As indicated above, **pharmacogenetics is the study of the contribution of genetic factors to variation in drug response and toxicity.** As a result of the progress made in sequencing the human genome, a new field of study—**pharmacogenomics**—has developed recently. It has been defined as **the use of genomic information and technologies to optimize the discovery and development of drug targets and drugs.** It builds on pharmacogenetics, but covers a wider sphere of activity. Information from genomics, proteomics, bioinformatics, and other disciplines such as biochemistry and toxicology will be integrated to make possible the synthesis of newer and safer drugs. As the sequences of all our genes and their encoded proteins are determined, this will reveal many new **targets for drug actions.** It will also reveal **polymorphisms** (this term is briefly discussed in Chapter 50) of enzymes and proteins **related to drug metabolism, action, and toxicity.** Microarrays capable of detecting them will be constructed, permitting **screening of individuals** for potentially harmful polymorphisms prior to the start of

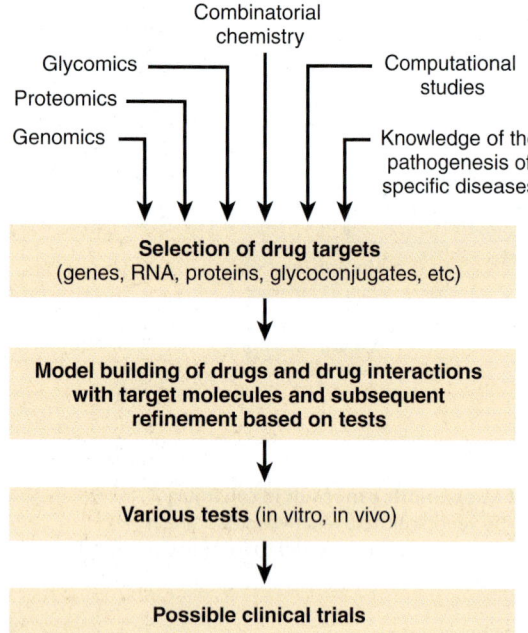

FIGURE 53–2 **Simplified scheme of some approaches to the development of new drugs.**

drug therapy. Already **gene chips** are available for analyzing certain P450 genotypes (eg, for CYP2D6, whose gene product is involved in the metabolism of many antidepressants, antipsychotics, β-blockers, and some chemotherapeutic agents). **Figure 53–2** summarizes some approaches to developing new drugs. Major thrusts of new drug development are to **enhance treatment** and to **provide safer, personalized drugs,** taking into account polymorphisms, and other **genetic** and **environmental** factors. It has been estimated that some 100,000 deaths from adverse drug reactions occur each year in the United States alone. It is hoped that new information provided by studies in the various areas indicated in Figure 53–2 and in other areas will translate into successful therapies and also **eventually** into a new era of personalized therapeutics. However, much work remains to be done before this is achievable.

SUMMARY

- Xenobiotics are chemical compounds foreign to the body, such as drugs, food additives, and environmental pollutants; more than 200,000 have been identified.

- Xenobiotics are metabolized in two phases. The major reaction of phase 1 is hydroxylation catalyzed by a variety of monooxygenases, also known as the cytochrome P450s. In phase 2, the hydroxylated species are conjugated with a variety of hydrophilic compounds such as glucuronic acid, sulfate, or glutathione. The combined operation of these two phases renders lipophilic compounds into water-soluble compounds that can be eliminated from the body.

- Cytochrome P450s catalyze reactions that introduce one atom of oxygen derived from molecular oxygen into the substrate, yielding a hydroxylated product. NADPH and NADPH

cytochrome P450 reductase are involved in the complex reaction mechanism.

- All cytochrome P450s are hemoproteins and generally have a wide substrate specificity, acting on many exogenous and endogenous substrates. They represent the most versatile biocatalyst known.

- Some 57 cytochrome P450 genes are found in human tissue.

- Cytochrome P450s are generally located in the endoplasmic reticulum of cells and are particularly enriched in liver.

- Many cytochrome P450s are inducible. This has important implications in phenomena such as drug interaction.

- Mitochondrial cytochrome P450s also exist and are involved in cholesterol and steroid biosynthesis. They use a nonheme iron-containing sulfur protein, adrenodoxin, not required by microsomal isoforms.

- Cytochrome P450s, because of their catalytic activities, play major roles in the reactions of cells to chemical compounds and in chemical carcinogenesis.

- Phase 2 reactions are catalyzed by enzymes such as glucuronosyltransferases, sulfotransferases, and glutathione S-transferases, using UDP-glucuronic acid, PAPS (active sulfate), and glutathione, respectively, as donors.

- Glutathione not only plays an important role in phase 2 reactions but is also an intracellular reducing agent and is involved in the transport of certain amino acids into cells.

- Xenobiotics can produce a variety of biologic effects, including pharmacologic responses, toxicity, immunologic reactions, and cancer.

- Catalyzed by the progress made in sequencing the human genome, the new field of pharmacogenomics offers the promise of being able to make available a host of new rationally designed, safer drugs.

REFERENCES

Caskey CT: Using genetic diagnosis to determine individual therapeutic utility. Annu Rev Med 2010;61:1.

Human Cytochrome P450 (CYP) Allele Nomenclature Committee. http://www.imm.ki.se/CYPalleles/

Ingelman-Sundberg M: Pharmacogenomic biomarkers for prediction of severe adverse drug reactions. N Engl J Med 2008;358:637.

Kalant H, Grant DM, Mitchell J (editors): *Principles of Medical Pharmacology,* 7th ed. Saunders Elsevier, 2007. (Chapters 4 [Drug Biotransformation by Riddick DS] and 10 [Pharmacogenetics and Pharmacogenomics by Grant DM and Kalow W] are particularly relevant to this Chapter).

Kamali F, Wynne H: Pharamacogenetics of Warfarin. Annu Rev Med 2010;61:63.

Katzung BG, Masters SB, Trevor AJ (editors): *Basic & Clinical Pharmacology,* 11th ed. McGraw-Hill, 2009.

Lee C, Morton CC: Structural genomic variation and personalized medicine. N Engl J Med 2008;358:740.

Pharmacogenomics. Human Genome Project Information. http://www.ornl.gov/sci/techresources/Human_Genome/medicine.pharma.shtml

Shurin SB, Nabel EG: Pharmacogenomics—ready for prime time? N Engl J Med 2008:358:1061.

The Biochemistry of Aging

Peter J. Kennelly, PhD

- Describe the essential features of wear and tear theories of aging.
- List at least four common environmental factors known to damage biological macromolecules such as proteins and DNA.
- Describe why nucleotide bases are especially vulnerable to damage.
- Describe the most physiologically important difference between mitochondrial and nuclear genomes.
- Describe the oxidative theory of aging and name the primary sources of reactive oxygen species (ROS) in humans.
- Describe three mechanisms by which cells prevent or repair damage inflicted by ROS.
- Describe the basic tenets of metabolic theories of aging.
- Describe the mechanism of the telomere "countdown clock."
- Describe our current understanding of the genetic contribution to aging.
- Explain the evolutionary implications of a genetically encoded lifespan.

BIOMEDICAL IMPORTANCE

Consider the various stages in the lifespan of *Homo sapiens*. Infancy and childhood are characterized by continual growth in height and body mass. Basic motor and intellectual skills develop: walking, language, etc. Infancy and childhood also represent a period of vulnerability wherein a youngster is dependent upon adults for water, food, shelter, protection, and instruction. Adolescence witnesses a final burst of growth in the body's skeletal framework. More importantly, a series of dramatic developmental changes occur—an accumulation of muscle mass, loss of residual "baby fat," maturation of the gonads and brain tissue, and the emergence of secondary sex characteristics—that transform the dependent child into a strong, independent, and reproductively capable adult. Adulthood, the longest stage, is a period devoid of dramatic physical growth or developmental change. With the notable exception of pregnancy in females, it is not unusual for adults to maintain the same body weight, overall appearance, and general level of activity for two or three decades.

Barring fatal illness or injury, the onset of the final stage of life, old age, is signaled by a resurgence of physical and physiological change. Hair begins to noticeably thin, turning white or gray as it loses its pigmentation. Skin loses its suppleness and accumulates blemishes. Individuals appear to shrink as both muscle and bone mass are progressively lost. Attention span and recall decline. Eventually, inevitably, life itself comes to an end as one or more essential bodily functions ceases to operate.

Understanding the underlying causes and instigating triggers of aging and the changes that accompany it is of great biomedical importance. Hutchison–Gilford, Werner's, and Down syndrome are three human genetic diseases whose pathologies include an acceleration of many of the physiological events associated with aging. Slowing or stopping some of the degenerative processes that cause or accompany aging can render the last stage of life much more vital, productive, and fulfilling. Co-opting the factors responsible for triggering cell death may enable physicians to selectively destroy harmful tissues and cells such as tumors, polyps, and cysts without collateral damage to healthy tissues.

TABLE 54–1   Average Life Expectancy by Decade, USA

| Sample Period | Average Life Expectancy (Years) | |
	From Birth	If Survived to Age 5
1900–1902	49.24	59.98
1909–1911	51.49	61.21
1919–1921	56.40	62.99
1929–1931	59.20	64.29
1939–1941	63.62	67.49
1949–1951	68.07	70.54
1959–1961	69.89	72.04
1969–1971	70.75	72.43
1979–1981	73.88	75.00
1989–1991	75.37	76.22
1999–2001	76.83	77.47

Adapted from Table 12 of the *National Vital Statistics Reports* (2008) 57, vol. 1.

LIFESPAN VERSUS LONGEVITY

From Paleolithic times through Greece's Golden Age to Medieval times the average life expectancy for a newborn baby remained relatively constant, oscillating within the range of 25–35 years. Beginning with the Renaissance, this number has gradually increased, so that by the beginning of the twentieth century the average life expectancy of persons born in developing countries reached the mid-40s. Today, 100 years later, the current world average is 67 years, and that for developed nations is approaching 80. This has led to speculation in the popular press about how long this trend might be expected to continue. Can future generations expect to live past the century mark? Is it possible that human beings possess the potential, barring accidents and with proper care and maintenance, to live indefinitely?

Unfortunately, this extrapolation is unlikely to be realized because it is based on a misunderstanding of the term **life expectancy.** Life expectancy is calculated by averaging over all births. Hence, it is dramatically influenced by infant mortality rates. While the life expectancy of a Roman child was 25 years, if one calculated the expected lifespan only for those persons who survived infancy, which we will refer to as **longevity,** the average nearly doubled to 48. When one factors out the dramatic decline in infant mortality rates that has taken place over the past century and a half, the apparent doubling in the human lifespan largely, *but not entirely,* disappears. As can be seen in **Table 54–1**, the predicted longevity of a 5-year-old child in the United States has increased from 70.5 in 1950 to 77.5 years in 2000. Is there some sort of upper limit to the lifespan of a properly nourished, well-maintained human being? Perhaps not.

AGING & MORTALITY: NONSPECIFIC OR PROGRAMMED PROCESSES?

Are aging and death nondeterminant or **stochastic** processes in which living creatures inevitably reach a tipping point where they succumb to a lifetime's accumulation of damage from disease, injury, and simple wear and tear? While the human body has a certain capacity to repair and replace at the molecular and cellular levels, this capacity is variable and finite. No matter how much attention is devoted to care and maintenance, like an automobile or some other sophisticated mechanical device, sooner or later some key component of our bodies will wear out. An alternative school of thought posits that aging and death are genetically programmed processes analogous to puberty, which have evolved through a process of natural selection.

Aging and death are, in all likelihood, multifactorial processes to which numerous factors, some nondeterminant and others programmed, make important contributions. While much work remains to be done before the precise makeup of this mechanistic mosaic can be determined, a large range of potential contributors have been identified. Several of the more prominent of these are presented in the sections that follow.

WEAR & TEAR THEORIES OF AGING

Many theories regarding aging and mortality hypothesize that the human body eventually succumbs to the accumulation of damage over time due to a variety of environmental factors that are reactive with organic biomolecules. These theories note that while repair and turnover mechanisms exist to restore or replace many classes of damaged molecules, these mechanisms are not absolutely perfect. Hence, some damage leaks through—damage that will inevitably accumulate over time, particularly in long-lived cell populations that experience little, if any, turnover (**Table 54–2**). Ironically, many of the agents that are most damaging to proteins, DNA, and other biomolecules are also essential for terrestrial life: water, oxygen, and sunlight.

Hydrolytic Reactions Can Damage Proteins and Nucleotides

Water is a relatively weak nucleophile. However, because of its ubiquitous presence and high concentration (>55 M, see Chapter 2), even this weak nucleophile will react with susceptible targets inside the cell. In proteins, hydrolysis of peptide bonds leads to cleavage of the polypeptide chain. The amide

TABLE 54–2 Time Required for All of The Average Cells of this Type to be Replaced

Tissue or Cell Type	Turnover
Intestinal epithelium	34 hrs[1]
Epidermis	39 days[2]
Leukocyte	<1 yr[3]
Adipocytes	9.8 yrs[3]
Intercostal skeletal muscle	15.2 yrs[3]
Cardiomyocytes	≥100 yrs[3]

Data from:
[1]Potten CS, Kellett M, Rew DA, et al: Proliferation in human gastrointestinal epithelium using bromodeoxyuridine in vivo. Gut 1992; 33:524.
[2]Weinstein GD, McCullough JL, Ross P: Cell proliferation in normal epidermis. J Invest Dermatol 1984; 82:623.
[3]Spalding KL, Arner E, Westermark PO, et al: Dynamics of fat cell turnover in humans. Nature 2008; 453:783.

bonds most frequently targeted by water are those found on the side chains of the amino acids asparagine and glutamine, presumably because they are more exposed, on average, to solvent than the amide bonds in the protein's backbone. Hydrolysis leads to the replacement of the neutral amide group with an acidic carboxylic acid group, forming aspartate and glutamate, respectively

(**Figure 54–1, parts A and B**). This change leads to the introduction of both a negative charge and of a potential proton donor or acceptor to the affected region of the protein. As the protein population within a living organism is subject to continual turnover, in most cases the chemically modified protein will be degraded and replaced by a newly synthesized protein.

Of perhaps greater potential biological consequence are the reactions of the nucleotide bases in DNA with water. The amino groups projecting from the heterocyclic aromatic rings of the nucleotide bases cytosine, adenine, and guanine are each susceptible to hydrolytic attack in which the amino group is replaced by a carbonyl to form uracil, hypoxanthine, and xanthine, respectively (**Figure 54–1, part C**). If the affected base is located in the cell's DNA, the net result is a mutation that, if left unrepaired, can potentially perturb gene expression or produce a dysfunctional gene product. The bond between the nucleotide base and the deoxyribose moiety in DNA is also vulnerable to hydrolysis. In this instance the base is completely eliminated, leaving a gap in the sequence (**Figure 54–1, part D**) which, if left unrepaired, can lead to either a substitution or a frame-shift mutation (see Chapter 37).

Many other bonds within biological macromolecules also possess the potential to be cleaved by random chemical hydrolysis. Included in this list are the ester bonds that bind fatty acids to their cognate glycerolipids, the glucosidic bonds

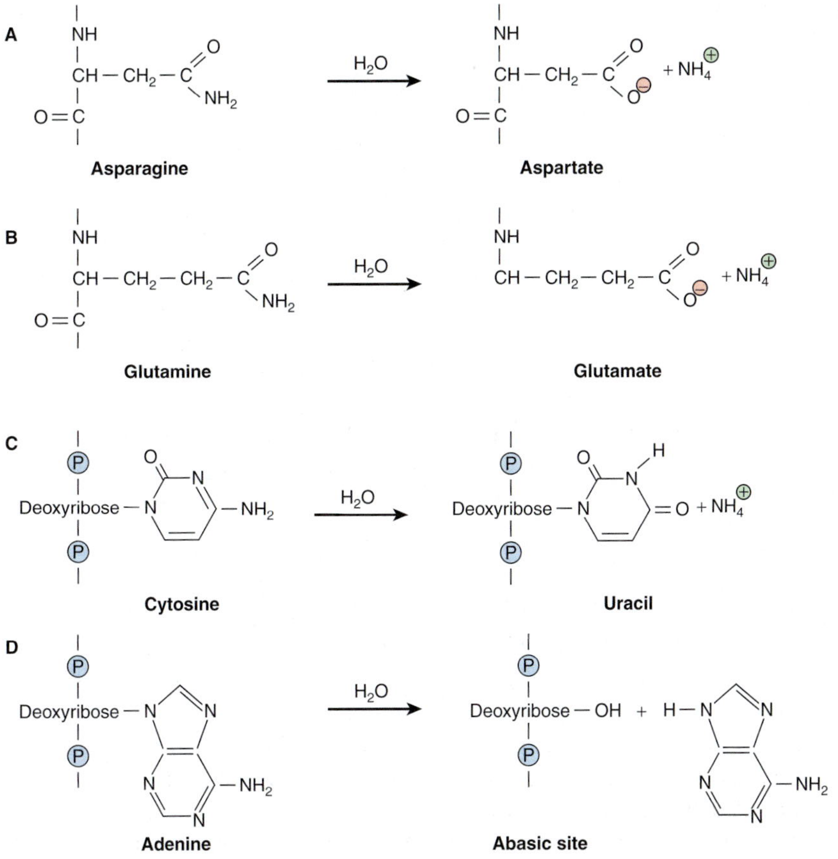

FIGURE 54–1 Examples of hydrolytic damage to biological macromolecules. Shown are a few of the ways in which water can react with and chemically alter proteins and DNA: **(A)** Net substitution of aspartic acid via hydrolytic deamidation of the neutral side chain of asparagine. **(B)** Net substitution of glutamic acid via hydrolytic deamidation of the neutral side chain of glutamine. **(C)** Net mutation of cytosine to uracil by water. **(D)** Formation of an abasic site in DNA via hydrolytic cleavage of a ribose-base bond.

that link the monosaccharide units of carbohydrates, and the phosphodiester bonds that hold polynucleotides together and link the head groups of phospholipids to their diacylglycerol partners. However, these reactions appear to take place too infrequently (phosphodiester hydrolysis) or to generate insufficiently perturbing products to manifest significant biological consequences.

Respiration Generates Reactive Oxygen Species

Numerous biological processes require enzyme-catalyzed oxidation of organic molecules by molecular oxygen (O_2). These processes include the hydroxylation of proline and lysine side chains in collagen (Chapter 5), the detoxification of xenobiotics by cytochrome P450 (Chapter 53), the degradation of purine nucleotides to uric acid (Chapter 33), the reoxidation of the prosthetic groups in the flavin-containing enzymes that catalyze oxidative decarboxylation (eg, the pyruvate dehydrogenase complex, Chapter 18) and other redox reactions (eg, amino acid oxidases, Chapter 28), and the generation of the chemiosmotic gradient in mitochondria by the electron transport chain (Chapter 13). Redox enzymes frequently employ prosthetic groups such as flavin nucleotides, iron–sulfur centers, or heme-bound metal ions (Chapters 12 and 13) to assist in the difficult task of generating and stabilizing the highly reactive free radical and oxyanion intermediates formed during catalysis. The electron transport chain employs specialized carriers such as ubiquinone and cytochromes to safely transport single, unpaired electrons among and within its various multiprotein complexes.

Occasionally, these highly reactive intermediates escape into the cell in the form of ROS such as superoxide and hydrogen peroxide (**Figure 54–2, part A**). By virtue of its structural and functional complexity and extremely high level of electron flux, "leakage" from the electron transport chain constitutes by far and away the major source of ROS in most mammalian cells. In addition, many mammalian cells synthesize and release the second messenger nitric oxide (NO·), which contains an unpaired electron, to promote vasodilation and muscle relaxation in the cardiovascular system (Chapter 49).

Reactive Oxygen Species Are Chemically Prolific

The extremely high reactivity of ROS makes them extremely dangerous. ROS can react with and chemically alter virtually any organic compound, including proteins, nucleic acids, and lipids. In some cases, reaction leads to the cleavage of covalent bonds. They also display a strong tendency to form **adducts,** the products of the direct addition of two (or more) compounds, with nucleotide bases, polyunsaturated fatty acids, and other biological compounds possessing multiple double bonds (**Figure 54–3**). Adducts formed with nucleotide bases can be especially dangerous because of their

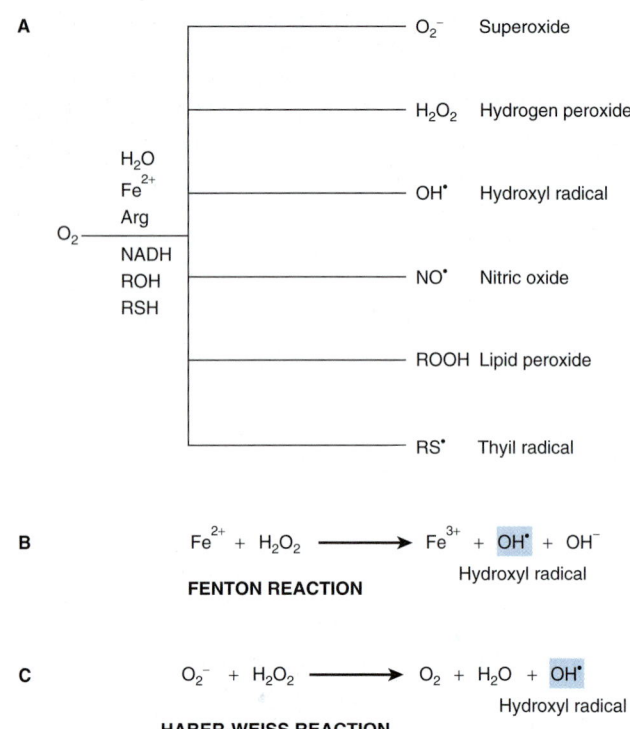

FIGURE 54–2 **Reactive oxygen species (ROS) are toxic by-products of life in an aerobic environment. (A)** Many types of ROS are encountered in living cells. **(B)** Generation of hydroxyl radical via the Fenton reaction. **(C)** Generation of hydroxyl radical by the Haber–Weiss reaction.

potential, if left uncorrected, to cause misreads that introduce mutations into DNA.

The ease with which oxygen evokes the chemical changes that turn household butter rancid is a testament to the reactivity of unsaturated fats, those containing one or more carbon–carbon double bond (Chapter 23) with ROS. Lipid peroxidation can lead to the formation of cross-linked lipid–lipid and lipid–protein adducts and a loss of membrane fluidity and integrity. Loss of membrane integrity, in turn, can—in the case of the mitochondria—undermine the efficiency with which the electron transport chain converts reducing equivalents to ATP, leading to greater production of deleterious ROS. Loss of membrane integrity can also trigger **apoptosis,** the programmed death of a cell.

Chain Reactions Multiply the Destructiveness of ROS

The destructiveness inherent in the high reactivity of many of these ROS, particularly free radicals, is exacerbated by their capacity to participate in chain reactions in which the product of the reaction between the free radical and some biomolecule is a damaged biomolecule and another species containing a highly reactive unpaired electron. The chain will terminate when a free radical is able to acquire another lone electron to form a relatively innocuous electron pair without generating a

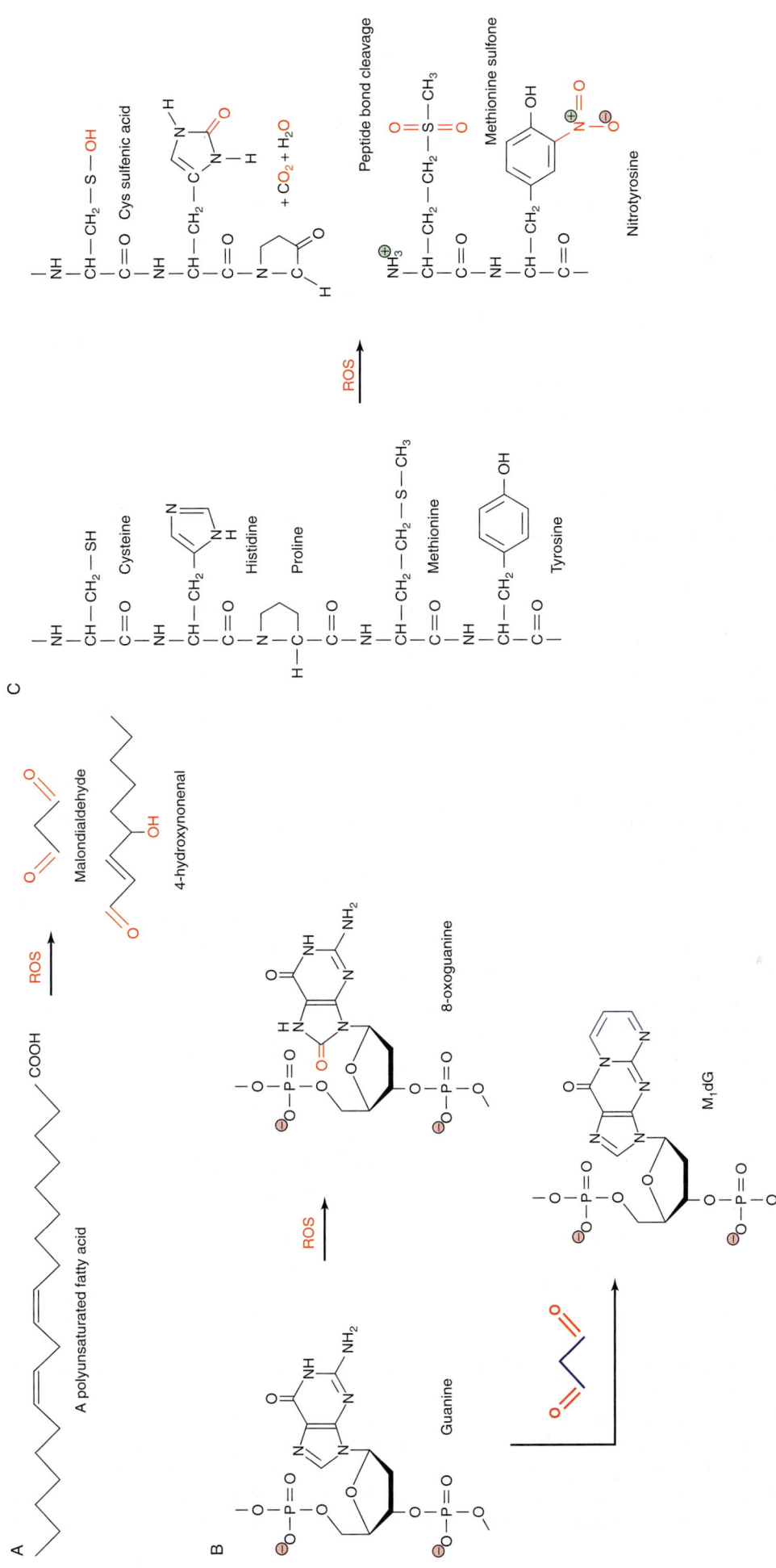

FIGURE 54–3 **ROS react directly and indirectly with a wide range of biological molecules. (A)** Peroxidation of unsaturated lipids generates reactive products such as malondialdehyde and 4-hydroxynonenal. **(B)** Guanine can be directly oxidized by ROS to produce 8-oxoguanine or form an adduct, M₁dG, with the ROS product malondialdehyde. **(C)** Common reactions of proteins with ROS, including oxidation of amino acid side chains and cleavage of peptide bonds. Oxygen atoms derived from ROS are marked in red. Carbon atoms derived from malondialdehyde in M₁dG are colored blue. The complete chemical name for M₁dG is 3-(2-Deoxy-ᴅ-erythro-pentofuranosyl)pyrimido(1,2-α)purin-10(3H)-one.

new unpaired electron as a by-product. Such is the case when one free radical encounters another. The two "odd" electrons combine to form a pair. Alternatively, the ROS may be eliminated by one of the cell's suite of dedicated antioxidant enzymes (Chapters 12 and 52).

The reactivity, and hence destructiveness, of individual ROS varies. Hydrogen peroxide, for example, is less reactive than superoxide, which in turn is less reactive than hydroxyl radical (OH$^\cdot$). Unfortunately, two pathways exist in living organisms by which highly toxic hydroxyl radical can be generated from less destructive ROS. If ferric iron is present, for example, the Fenton reaction can transform hydrogen peroxide into hydroxyl radicals (Figure 54–2, part B). The ferrous (+3) iron, in turn, can be reduced back to the ferric (+2) state by other hydrogen peroxide molecules, permitting the iron to act catalytically to produce additional hydroxyl radicals. Hydroxyl radical can also be generated when superoxide and hydrogen peroxide disproportionate, a process called the Haber–Weiss reaction (Figure 54–2, part C).

Free Radicals and the Mitochondrial Theory of Aging

In 1956, Denham Harmon proposed the so-called free radical theory of aging. It had been reported that the toxicity of hyperbaric oxygen treatment and radiation could be explained by a factor common to both, the generation of ROS. This report dovetailed nicely with Harmon's own observation that lifespan was inversely related to metabolic rate and, by extrapolation, respiration. He therefore postulated that the cumulative damage was caused by the continual and inescapable production of ROS.

In more recent years, the proponents of the free radical theory of aging have focused attention on the mitochondria. Not only is the mitochondria host to the major source of ROS in the cell, the electron transport chain, but oxidative damage to the components of this pathway could lead to increased leakage of hydrogen peroxide, superoxide, etc, into the cytoplasm. Damage to the mitochondria would be likely to adversely affect the efficiency with which it performs its most important function, the synthesis of ATP. A significant slowing in the rate of ATP synthesis could readily lead to the types of wholesale declines in physiological function that occur in aging.

A second contributor to the proposed self-perpetuating cycle of mitochondrial redox damage is the fact that several components of the electron transport chain are encoded by the mitochondrion's indigenous genome. The mitochondrial genome is a much reduced, vestigial remnant of the genome of the ancient bacterium that was the precursor of the current organelle. Through a process called **endosymbiosis,** primitive eukaryotes became dependent upon surrounding bacteria to provide certain materials, and vice versa. Eventually, the smaller bacterium was absorbed by and lived within the interior of its eukaryotic host. Over time most, but not all, of the genes contained in the bacterial genome were either eliminated as superfluous to the needs of the new fusion organism

TABLE 54–3 Genes Encoded by the Genome of Human Mitochondria

rRNA	12S, 16S rRNA
tRNA	22 tRNAs (2 for Leu and Ser)
Subunits of NADH-ubiquinone oxidoreductase (Complex I, >40 total)	ND 1-6, ND 4L
Subunits of ubiquinol-cytochrome c oxidoreductase (Complex III, 11 total)	Cytochrome b
Subunits of cytochrome oxidase (Complex IV, 13 total)	COX I, COX II, COX III
Subunits of the F_1, F_0 ATPase (ATP synthase, 12 total)	ATPase 6, ATPase 8

or were transferred into the host cell's nuclear DNA. At present, the genome of the human mitochondrion encodes a small and a large ribosomal RNA, 22 tRNAs, and polypeptide subunits for complexes I, III, and IV of the electron transport chain as well as the F_1, F_0 ATPase (**Table 54–3**). The mitochondrial genome lacks the surveillance and repair mechanisms that help maintain the integrity of nuclear DNA. Hence, mutations induced by adducts or reaction with ROS, and any functional defects resulting from these mutations, become a permanent feature of each individual mitochondrion's genome, which will continue to accumulate mutations with time.

While the mitochondrial hypothesis is no longer viewed as providing a unifying explanation for all of the changes that are associated with human aging and its comorbidities, it likely is an important contributor. Powerful circumstantial evidence for this is provided by the central role played by this organelle in the sensor-response pathways that trigger apoptosis.

Mitochondria Are Key Participants in Apoptosis

Apoptosis imbues higher organisms with the ability to selectively eliminate cells that are rendered superfluous by developmental changes, such as those that take place on a continual basis during embryogenesis, or which have been damaged beyond repair. During developmental tissue remodeling, the apoptotic cell death program is triggered by receptor-mediated signals. In the case of damaged cells, any one of several interior indicators may serve as trigger: ROS, viral dsRNA, DNA damage, and heat shock. These signals trigger the opening of the permeability transition pore complex embedded in the mitochondrial outer membrane, through which molecules of the small ($\approx$12.5 kDa) electron carrier protein cytochrome c escapes into the cytoplasm. Here, cytochrome c serves as the core protein for nucleating a multiprotein complex, called the apoptosome, that initiates a cascade of proteolytic activation events targeting the proenzyme forms of a series of cysteine proteases known as caspases. The terminal caspases, numbers 3 and 7, break down structural proteins in the cytoplasm and chromatin proteins in the nucleus; events that lead to the death of the affected cell and its elimination by phagocytosis. Needless to say, the presence of an intrinsic, receptor-mediated cell death pathway offers the hope that we can eliminate harmful cells, such as cancer, by learning how to selectively activate their apoptotic pathway.

Ultraviolet Radiation Can Be Extremely Damaging

The term **ultraviolet (UV) radiation** refers to those wavelengths of light that lie immediately beyond the blue or short wavelength end of the visible spectrum. While the human eye cannot detect these particular wavelengths of light, they are strongly absorbed by organic compounds possessing aromatic rings or multiple, conjugated double bonds such as the nucleotide bases of DNA and RNA; the aromatic side chains of the amino acids phenylalanine, tyrosine, and tryptophan; polyunsaturated fatty acids; heme groups; and cofactors and coenzymes such as flavins, cyanocobalamine, etc. Absorption of this short wavelength, high-energy light can cause the rupture of covalent bonds in proteins, DNA, and RNA; the formation of thymine dimers in DNA (**Figure 54–4**); cross-linking of proteins; and the generation of free radicals including ROS. While UV radiation does not penetrate beyond the first few layers of skin cells, the high efficiency of absorption leads to the rapid accumulation of damage to the limited population of skin cells that are impacted. Because the nucleotide bases of DNA and RNA are particularly effective at absorbing UV radiation, it is highly mutagenic. Prolonged exposure to intense sunlight can lead to the accumulation of multiple DNA lesions that can overwhelm a cell's intrinsic repair capacity. It is thus relatively common for persons whose work or lifestyle involves prolonged exposure to sunlight to manifest aberrant skin tissue, in the form of both moles and cancerous myelomas. Many of the latter can proliferate and spread with great rapidity, necessitating careful surveillance and rapid medical intervention.

Protein Glycation Often Leads to the Formation of Damaging Cross-links

When amino groups such as those found on the side chain of lysine or some of the nucleotide bases are exposed to a reducing sugar such as glucose, a reversible adduct is slowly generated through the formation of a Schiff's base between the aldehyde or ketone group of the sugar and the amine. Over time, the glycated protein undergoes a series of rearrangements to form **Amadori** products, which contain a conjugated carbon–carbon double bond that can react with the amino group on a neighboring protein (**Figure 54–5**). The net result is the formation of a covalent crosslink between two proteins or other biological macromolecules that can, in turn, undergo further glycation and crosslink to yet another macromolecule. These crosslinked aggregates are sometimes called Advanced Glycation Endproducts or AGEs.

The physiological impact of protein glycation can be especially pronounced when long-lived proteins such as collagen

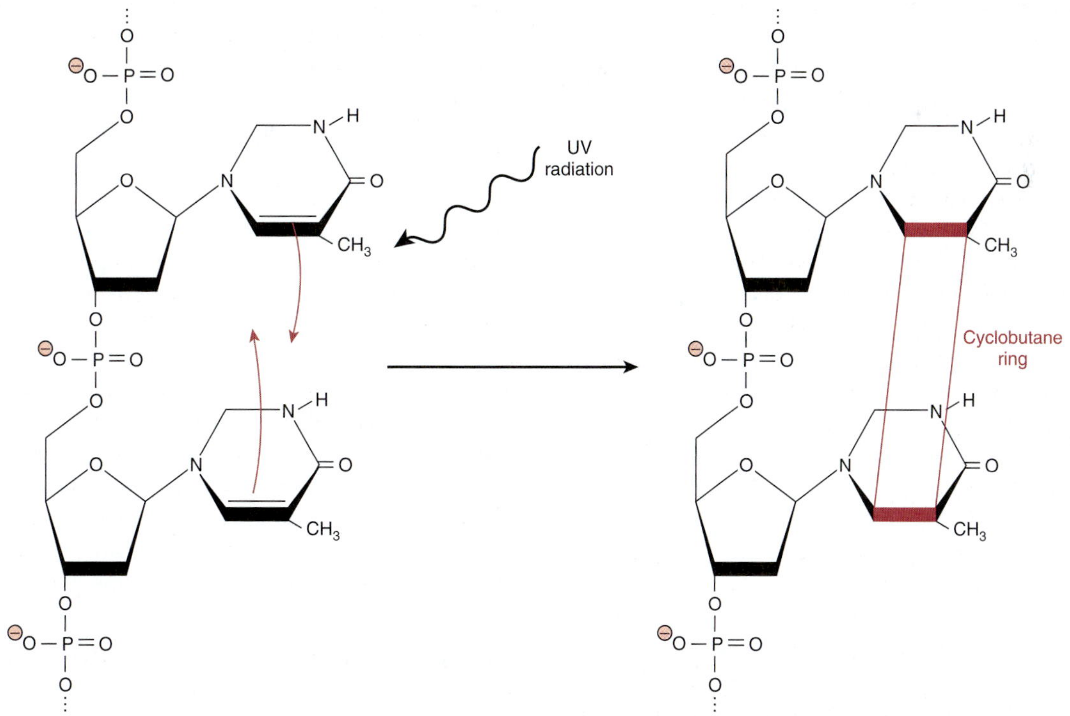

FIGURE 54–4 **Formation of a thymine dimer following excitation by UV light.** When consecutive thymine bases are stacked one above the other in a DNA double helix, absorption of UV light can lead to the formation of a cyclobutane ring (red, not to scale) covalently linking the two bases together to form a thymine dimer.

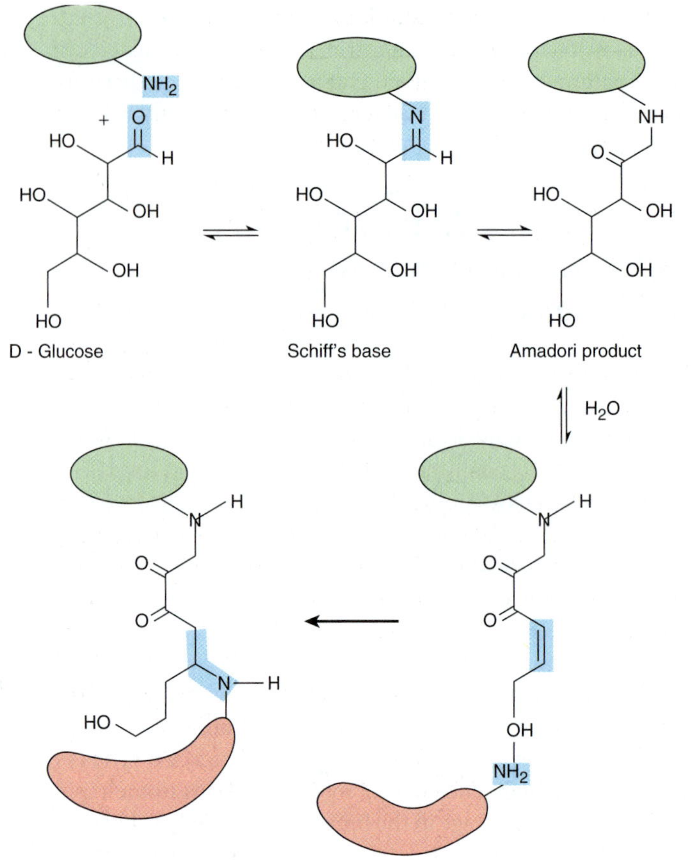

FIGURE 54–5 **Protein glycation can lead to the formation of protein–protein crosslinks.** Shown are the sequence of reactions that generate the Amadori product on the surface of the protein marked in green, and the subsequent formation of a protein–protein crosslink via an amino group on the surface of a second, red, protein.

or β-crystallins are involved. Their persistence provides the opportunity for multiple glycation and crosslinking events to occur. The progressive crosslinking of the collagen network in vascular endothelial cells leads to the progressive loss of elasticity and thickening of the basement membrane in blood vessels, promoting plaque formation. The overall result is a progressive increase in the heart's workload. In the eye, the accumulation of aggregated proteins compromises the opacity of the lens and eventually manifests itself in the form of cataracts. Impairment of glucose homeostasis renders diabetics particularly susceptible to the formation of advanced glycation end products. In fact, the glycation of hemoglobin and serum albumin are used as biomarkers for the diagnosis of diabetes and the assessment of its treatment.

MOLECULAR REPAIR MECHANISMS COMBAT WEAR & TEAR

Enzymatic and Chemical Mechanisms Intercept Damaging ROS

A corollary to the wear and tear theory of aging is that longevity reflects the effectiveness and robustness of the molecular prevention, repair, and replacement mechanisms in a given

species and the individuals within it. Enzymes such as superoxide dismutase and catalase protect the cell by converting superoxide and hydrogen peroxide, respectively, to less reactive products, thereby preventing potential molecular damage before it occurs (Chapter 52). For example, fruit flies that have been genetically altered to express elevated levels of superoxide dismutase exhibit significantly extended life spans.

In the cytoplasm, the cysteine-containing tripeptide glutathione acts as a chemical redox protectant by reacting directly with ROS to generate less reactive compounds such as water. Oxidized glutathione, which consists of two tripeptides linked by an S–S bond, is then enzymatically reduced to maintain the pool of protectant (Chapter 52). Glutathione can also react directly with cysteine sulfenic acids and disulfides on proteins to restore them to their reduced state, and form adducts with toxic xenobiotics (Chapter 53). Other biomolecules such as ascorbic acid and vitamin E also possess antioxidant properties, which accounts for the fact that many "popular" diets target foods rich in these compounds in an effort to buttress the body's ability to neutralize ROS and slow aging.

The Integrity of DNA Is Maintained by Proofreading and Repair Mechanisms

In addition to the prophylactic measures mentioned above, living organisms possess a limited capacity to replace or repair damaged macromolecules. The majority of this capacity is directed toward maintaining the integrity of the nuclear (but not the mitochondrial) genome, which is to be expected given DNA's unique information storage function, the vulnerability of heterocyclic aromatic nucleotide bases to chemical assault and UV radiation, and the fact that—by contrast to almost every other macromolecule—each cell contains only a single unique copy of each chromosome. A **somatic cell** is one that forms part of the body of an organism. Maintaining the integrity of the genome begins at replication, where careful proofreading is performed to insure that the new genome formed in the process of somatic cell division faithfully replicates the template that directed it synthesis. In addition, most living organisms possess an impressive cadre of enzymes whose role is to inspect and correct aberrations that either escaped proofreading or were subsequently generated through the action of water (double strand breaks, loss of a nucleotide base, and deamidation of cytosine), UV radiation (thymine dimers and strand breaks), or exposure to chemical modifiers (adduct formation). This multilayered system is composed of mismatch repair enzymes, nucleotide excision repair enzymes, and base excision repair enzymes as well as the Ku system for repairing double-strand breaks in the phosphodiester backbone (Chapter 35). As a last resort, cells harboring damaging mutations are subject to removal by apoptosis.

Nevertheless, despite the many precautions taken to insure fidelity during replication and to repair subsequent damage listed above, some mutations inevitably slip through. Indeed, some leakage in the surveillance and repair system is necessary in order to generate the genetic variability that drives evolution. The **somatic mutation theory of aging** proposes that it also serves a second purpose as a driver of the aging process. Simply put, the accumulation of mutant cells over time must inevitably lead to compromised biological function that manifests itself, at least in part, as the physical changes we associate with aging.

Some Types of Protein Damage Can Be Repaired

In contrast to DNA, a cell's capacity to repair damage to other biomolecules is relatively limited. For the most part, cells appear to rely on routine turnover, wherein the global population of a given biomolecule is degraded and replaced by new synthesis on a continuing, or constitutive, basis (Chapter 9), to remove aberrant lipids, carbohydrates, and proteins. Some proteins, particularly the fibrous proteins that contribute to the structural integrity of tendons, ligaments, bones, matrix, etc, undergo little if any turnover. These long-lived proteins tend to accumulate damage over many years, contributing to the loss of elasticity in vascular tissues and joints, loss of lens opacity, etc. The most prominent mechanisms for the repair of damaged proteins target the sulfur atoms contained in the side chains of cysteine and methionine, and the isoaspartyl groups formed by the shift of the peptide bond to side-chain carboxyl group.

The side-chain sulfhydryl group of cysteine frequently plays important catalytic, regulatory, and structural roles in proteins that are dependent upon its oxidation state. However,

both its sulfhydryl group and the sulfur ether of methionine are extremely vulnerable to oxidation (Figure 54–3, part C). As is the case for many other oxidized biomolecules, the tripeptide glutathione can react directly with cysteine-disulfides, cysteine sulfenic acids, and methinonine sulfoxide to regenerate cysteine and methionine, respectively. In addition, disulfide reductases and methionine sulfoxide reductases provide an enzyme-catalyzed reduction mechanism using NADPH as electron donor. Unfortunately, the reduction potential of glutathione and NADPH is only sufficient only to reduce the lowest oxidation states of these sulfur atoms: cysteine disulfides or sulfenic acids and methionine sulfoxide. Cysteine sulfinic acid, cysteine sulfonic acid, and methinoine sulfone are refractory to reduction under physiologic conditions.

Aspartic acid residues possess the precise geometry needed to enable the side-chain carboxyl group to react with the amino group within the peptide bond formed with its alpha carboxyl group. The resulting cyclic diamide can then reopen to form either the original peptide bond or an isoaspartyl residue in which the side-chain carboxyl now forms part of the protein's peptide backbone (**Figure 54–6**). Methylation of the alpha carboxyl group provides a leaving group that promotes the reformation of the cyclic diamide, which can then reopen to form the normal peptide bond linkage (Figure 54–6).

Aggregated Proteins Are Highly Refractory to Degradation or Repair

Modifications to a protein's composition or conformation that cause it to adhere to other protein molecules can lead to the formation of toxic aggregates, called **amyloid.** Such aggregates form the hallmark of several neurodegenerative diseases, including Parkinson's, Alzheimer's, Huntington's disease, spinocerebellar ataxias, and the transmissible spongiform encephalopathies. The toxic effects of these insoluble aggregates are exacerbated by their persistence, as in this state most are generally refractory to the catalytic action of the proteases normally responsible for their turnover.

AGING AS A PREPROGRAMMED PROCESS

While molecular wear and tear undoubtedly contribute to aging, several observations suggest a role for programmed, deterministic mechanisms as well. For example, rather than "rusting" gradually, many of the physical manifestations of aging—liver spots, grayness, trembling hands, memory lapses—generally surface late in adulthood and progress at a rapid pace, as if the molecular maintenance mechanisms responsible for their repair and replacement had suddenly received a command to cease operation. Female menopause provides an unambiguous example of an age-associated physiological change that is genetically programmed and hormonally controlled. The paragraphs below describe several current

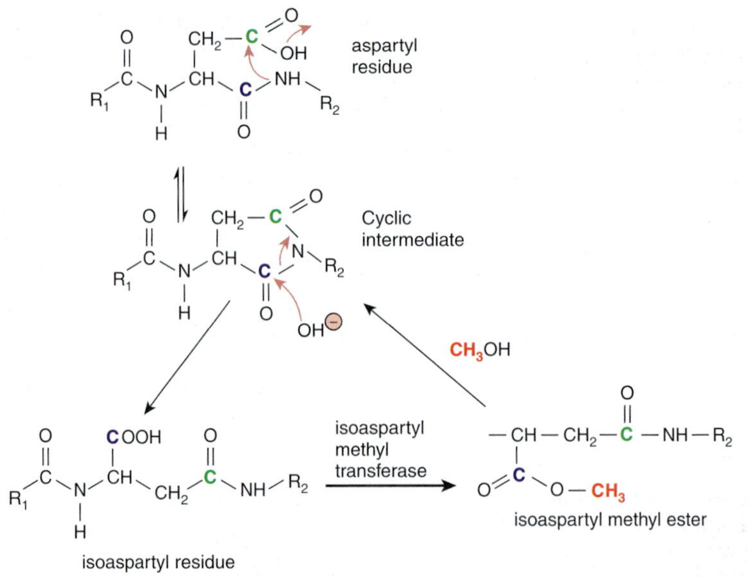

FIGURE 54–6 **Formation of an isoaspartyl linkage in a polypeptide backbone and its repair via the intervention of isoaspartyl methyltransferase.** Shown is the sequence of chemical and enzyme-catalyzed reactions that lead to formation of an isoaspartyl linkage and restoration of a normal peptide linkage. The carbons corresponding to the alpha and side-chain carboxylic acid groups in aspartic acid are colored blue and green, respectively. Red arrows denote the routes of nucleophilic attack during the cyclization and hydrolysis reactions. The methyl group added by isoaspartyl methyltransferase is colored pink.

Metabolic Theories of Aging: "The Brighter the Candle, the Quicker it Burns"

One of the many variants of the famous quote attributed to the ancient Chinese philosopher Lao Tzu summarizes the salient features of **metabolic theories of aging.** Its origins can be traced to the observation that the larger members of the animal kingdom tend to live longer than the smaller ones (**Table 54–4**). Reasoning that the causal basis for this correlation lay in something connected with size, rather than size itself, many scientists focused their attention on the organ most closely associated with life and vitality—the heart. In general, the resting heart rate of small animals such as hummingbirds, 250 beats per minute, tends to be higher than those of large animals such as whales, 10–30 beats per minute. Estimates of the cumulative number of times each vertebrate animal's heart beat over the course of a lifetime exhibited an amazing convergence on 1.0×10^9 beats: one billion.

The so-called **heartbeat hypothesis** posited that every living creature is capable of performing only a finite number of heartbeats and/or breaths. A more nuanced variation of this basic idea, variously referred to as the **metabolic** or **rate of living hypothesis,** was put forward by Raymond Pearl in the 1920s. Pearl proposed that an individual's lifespan was reciprocally linked to their basal metabolic rate. In other words, those who "burned the candle at both ends," so to speak, burnt out sooner. A new round of calculations revealed that, while animals differ markedly in size, longevity, and heart rate; over their lifetime each expends a similar amount of total metabolic energy *per unit body mass*, 7×10^5 J/g. While intuitively appealing, a mechanistic link between lifespan and body size and energy

expenditure or metabolic rate has proven elusive. Adherents of the mitochondrial theory of aging suggest that what is being "counted" is not heartbeats or energy, but the ROS that are the by-product of respiration. Over time the continued generation of energy and related consumption of O_2 consumed leads to the accumulation of ROS-induced damage to DNA, proteins, and lipids until, eventually, a tipping point is reached. The rationale behind calorically restricted diets as a means to prolong life is based on the rationale that burning fewer calories will lead to a concomitant reduction in the production of damaging ROS.

Telomeres: A Molecular Countdown Clock?

A second school of thought holds that the putative countdown clock that controls aging and lifespan does not sense heartbeats, energy, or ROS. Rather, it uses **telomeres** to track the number of times each somatic cell divides.

Telomeres are composed of long strings of GT-rich hexanucleotide repeats that cap the ends of eukaryotic chromosomes. Unlike the closed circular DNA of bacterial genomes, the genomic DNA of eukaryotes is linear. If left unprotected, the exposed ends of these linear polynucleotides would be available to participate in potentially carcinogenic genetic recombination events. A second function of telomeres is to provide some disposable DNA to accommodate the wastage that occurs when linear DNA molecules are replicated.

This wastage is a consequence of the fact that all DNA polymerases work unidirectionally, 3′ to 5′ (Chapter 35). While in closed circular DNA this is not a problem, when trying to replicate the 5′ ends of a linear double stranded DNA via discontinuous 3′ to 5′ synthesis and ligation of small **Okazaki** fragments, there is simply not enough room at the end to accommodate the small RNA primer, polymerase, etc. Synthesis of the 5′ end of each strand will generally fall 100 bp or more short. Each time a cell divides, its genome will be

TABLE 54-4 Life Span Versus Body Mass for Several Mammals

Species	Approximate Mass (kg)	Mean Expectation of Life at Maturity (years)
White-footed mouse	0.02	0.28
Deer mouse	0.02	0.43
Bank vole	0.025	0.48
Eastern chipmunk	0.1	1.63
American pika	0.13	2.33
Golden mantled grd. squirrel	0.155	2.12
Red squirrel	0.189	2.45
Belding's ground squirrel	0.25	1.78
Uinta ground squirrel	0.35	1.72
Eastern gray squirrel	0.6	2.17
Arctic ground squirrel	0.7	1.71
Eastern cottontail	1.25	1.48
Striped skunk	2.25	1.90
American badger	7.15	2.33
North American river otter	7.2	3.79
Bobcat	7.5	2.48
North American beaver	18	1.52
Impala	44	4.80
Bighorn sheep	55	5.48
Wild boar	85	1.91
Warthog	87	2.82
Nilgiri tahr	100	4.71
Blue wildebeest	165	4.79
Red deer stag	175	4.90
Waterbuck	200	5.87
Burchell's zebra	270	7.95
African buffalo	490	4.82
Hippopotamus	2390	16.40
African elephant	4000	19.10

Adapted from Millar JS, Zammuto: Life histories of mammals: An analysis of life tables. Ecology 1983; 64:631.

shortened further (**Figure 54-7**). The telomeres provide an innocuous source of DNA whose decreasing length is of little consequence to the cell. However, once the supply of telomere DNA is exhausted, roughly 100 cell divisions for humans, all mitosis ceases and the somatic cell enters a state of **replicative senescence.** As more and more cells within the body enter senescence, it gradually loses the capacity to replace lost or damaged cells.

Organisms are able to produce progeny that contain full-length telomeres thanks to the intervention of the enzyme **telomerase.** Telomerase is a ribonucleoprotein that is expressed in stem cells and most cancer cells, but not in somatic cells. Using an RNA template, telomerase adds GT-rich hexanucleotide repeat sequences ranging from a few hundred (yeast) to several thousand (humans) nucleotides in length to the ends of linear DNA molecules to restore their telomeres to full length. When somatic cells are genetically engineered in the laboratory to express telomerase, they continue to divide in culture long after an unaltered control cell line stops dividing. The ability to prevent replicative senescence using an enzyme that maintains telomeres at full-length represents the most compelling evidence of the operation of a telomere clock.

Kenyon Used a Model Organism to Discover the First Aging Genes

Many advances in biomedical science are the product of research that uses a variety of so-called model organisms as their test subject. The fruit fly, *Drosophila melanogaster,* has yielded a rich harvest of information concerning the genes that guide cellular differentiation and organ development. Baker's yeast and the African clawed frog *Xenopus laevis,* have served as the workhorses for dissecting the complex signal transduction circuitry that orchestrates the cell division cycle. A variety of cultured mammalian cell lines serve as surrogates for adipocytes, kidney cells, tumors, dendrites, etc. While at first glance it would appear that many of these model systems share little in common with humans, each possesses unique attributes that render them convenient vehicles for addressing certain problems and exploring specific systems.

Caenorhabditis elegans is a worm that has served as an important subject for the study of developmental biology. *C. elegans* is transparent and grows rapidly, attributes which facilitated tracing the entire developmental program for all 959 cells found in the mature adult to the fertilized egg. In the early 1990s, Cynthia Kenyon and colleagues observed that worms carrying mutations of the gene encoding an insulin receptor-like molecule, *daf-2,* lived 70% longer than their wild-type counterparts. Equally important, the mutant worms behaved in a manner resembling that of young wild-type *C. elegans* for much of this period. This is an important distinction. To qualify as an "aging gene," its manipulation must accomplish more than merely prolonging old age by delaying the point at which life ceases. It must impact the schedule of changes associated with aging.

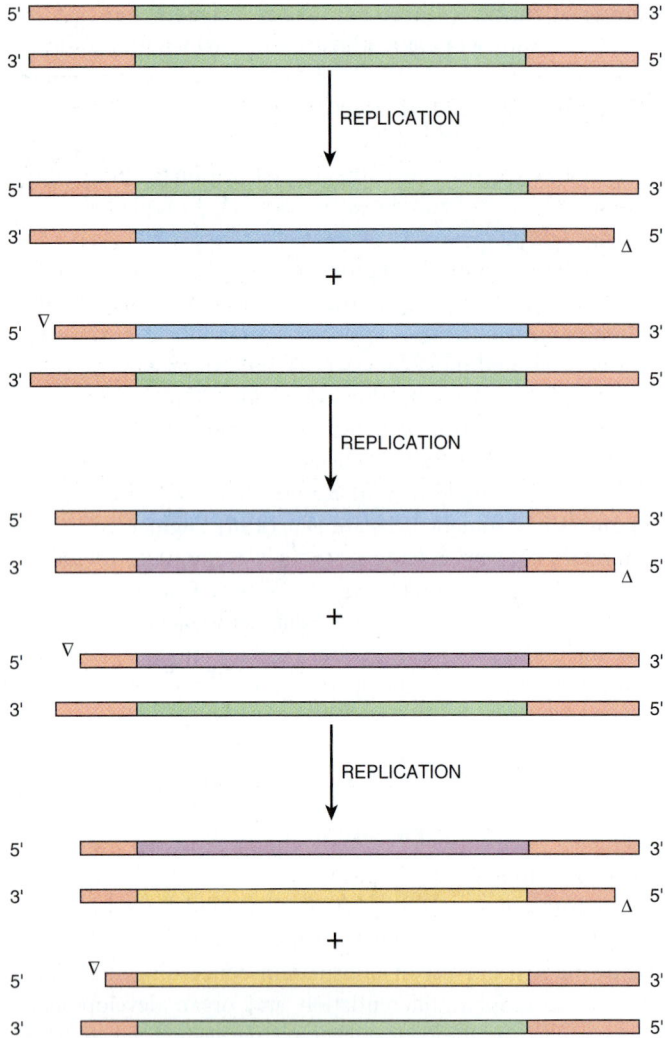

FIGURE 54–7   The telomeres at the ends of eukaryotic chromosomes progressively shorten with each cycle of replication. Shown is a schematic diagram of the linear DNA of a eukaryotic chromosome (green) containing telomeres at each end (red). During the first replication, new DNA strands are synthesized (green) using the original chromosome as template. For simplicity, the next two replication cycles (purple, yellow) show the fate of only the lower of the two nucleotide products from the preceding replicative cycle. Open arrowheads denote the site of incomplete strand synthesis. The model assumes that the single strand overhangs at the ends of each chromosome are trimmed at the completion of each cycle of cell division. Note the progressive shortening of the telomere repeats.

Investigation of further aging genes indicate they code for either one of a small set of transcription factors that include PHA-4 or DAF-16 that presumably control expression of aging critical genes, or signaling proteins such as DAF-2 that probably activate PHA-4, DAF-16, etc, in response to specific environmental signals. Much remains to be learned about the extent to which aging is controlled by genetic programming events, and how these gene products interact with the many other factors that influence vitality and longevity including nutrition, environmental stress, etc.

WHY WOULD EVOLUTION SELECT FOR LIMITED LIFESPANS?

The idea that animals would have evolved mechanisms designed specifically to limit their lifespan would appear, at first glance, to be highly counterintuitive. If the driving force behind evolution is the selection for traits that enhance fitness and survival, shouldn't this translate into an ever-increasing life expectancy? While maximizing lifespan may represent a desirable trait from the point of view of the individual, it does not necessarily follow that this applies to a population or species as a whole. A genetically programmed limit on lifespan could benefit the group by eliminating the drain on available resources imposed by members no longer actively involved in the production, development, and training of offspring. Indeed, the current three generation lifespan can be rationalized as providing time (a) for newborns to develop into reproductively active young adults, (b) for these young adults to protect and nurture their offspring, and (c) to serve as a source of guidance and assistance to young adults facing the challenges of childbirth and childrearing.

SUMMARY

- Aging and longevity are controlled via the complex and largely cryptic interplay between random and deterministic factors that include genetic programming, environmental stresses, lifestyle, cellular countdown clocks, and molecular repair processes.

- Wear and tear theories of aging hypothesize that the changes associated with old age and death itself reflect the accumulation of damage over time.

- The ubiquitous and life-essential environmental elements water, oxygen, and light possess an intrinsic capacity to damage biological macromolecules.

- ROS such as hydroxyl radical and superoxide are particularly problematic as they are highly reactive, often participate in chain reactions that multiply their impact, and are continually generated as a by product of the complex network of redox reactions taking place in the electron transport chain.

- The reactivity of their unsaturated ring systems and ability to absorb UV light render the nucleotide bases of DNA particularly vulnerable to UV or chemical damage.

- Mutations resulting from errors caused by missing or chemically modified nucleotide bases can be particularly harmful, as they may result in oncogenic transformation or render a cell vulnerable to further damage.

- Mitochondria occupy a central place in many theories of aging and death. This prominence can be attributed to several factors. Mitochondria are the site of the electron transport chain, by far the largest source of ROS in the cell.

- The efficient production of ATP is essential to cell vitality. Mitochondria play a central role in apoptosis, programmed cell death. Mitochondria lack the capacity to repair damage to their DNA.

- In eukaryotic cells, long repeating sequences called telomeres cap the ends of their linear chromosomes. These telomeres progressively shorten each time a somatic cell divides. When a somatic cell's telomeres become too short, it enters replicative senescence. Thus, telomeres are hypothesized to serve as a countdown clock for somatic cells.

■ Animal lifespan may be genetically programmed. Mutation of the *daf-2* gene in *Caenorhabditis elegans* yielded worms whose lifespan was 70% longer than wild type.

■ Evolutionary selection of a limited lifespan optimizes the vitality of the species rather than that of its individual members.

REFERENCES

Aguzzi A, O'Connor T: Protein aggregation diseases: pathogenicity and therapeutic perspectives. Nat Drug Discov 2010;9:237.

Anderson S, Bankier AT, Barrell BG, et al: Sequence and organization of the human mitochondrial genome. Nature 1981; 290:457.

Arias E, Curtin LR, Wei R, et al: U.S. decennial life tables for 1999–2001, United States life tables. Natl Vital Stat Rep 2008;57:1.

Berlett BS, Stadtman ER: Protein oxidation in aging, disease, and oxidative stress. J Biol Chem 1997;272:20313.

Clarke S: Aging as war between chemical and biochemical processes: Protein methylation and the recognition of age-damaged proteins for repair. Ageing Res Rev 2003;2:263.

Eisenberg DTA: An evolutionary overview of human telomere biology: the thrifty telomere hypothesis and notes on potential adaptive paternal effects. Am J Hum Biol 2011; 23:149.

Kenyon CJ: The genetics of aging. Nature 2010;464:504.

Knight JA: The biochemistry of aging. Adv Clin Chem 2000;35:1.

Speakman JR: Body size, energy metabolism and lifespan. J Exp Biol 2005;208:1717.

Ulrich P, Cerami A: Protein glycation, diabetes, and aging. Recent Prog Horm Res 2001;56:1.

Cancer: An Overview

Robert K. Murray, MD, PhD, Molly Jacob, MB BS, MD, PhD,
& Joe Varghese, MB BS, MD

OBJECTIVES

*After studying this chapter,
you should be able to:*

- Present an overview of important aspects of the biochemical and genetic features of cancer cells.
- Describe important properties of oncogenes and tumor suppressor genes.
- Briefly describe the concepts of genomic instability, aneuploidy, and angiogenesis in tumors.
- Discuss the use of tumor markers for following responses to treatments and to detect recurrences.
- Appreciate that recent understanding of the biology of cancer has led to the development of various new therapies.

BIOMEDICAL IMPORTANCE

Cancers constitute the **second most common cause of death,** after cardiovascular disease, in the USA and many other countries. Approximately 6–7 million people around the world die from cancer each year, and this figure is projected to increase. Humans of all ages develop cancer, and a wide variety of organs are affected. Worldwide, the main types of cancer accounting for mortality are those involving the lung, stomach, colon, rectum, liver and breast. Other types of cancers that lead to death include cervical, esophageal, and prostate cancers. Skin cancers are very common, but apart from melanomas, are generally not as aggressive as those mentioned above. The incidence of many cancers increases with **age.** Hence, as people live longer many more will develop the disease. Hereditary factors play a role in some types of tumors. Apart from great individual suffering caused by the disease, the economic burden to society is immense.

SOME GENERAL COMMENTS ON NEOPLASMS

A neoplasm refers to any abnormal new growth of tissue. It may be benign or malignant in nature. The term "cancer" is usually associated with malignant tumors. Tumors can arise in any organ in the body and result in different clinical features, depending on the location of the growth.

Cancer cells are characterized by certain key properties: (1) they proliferate rapidly and display diminished growth control, (2) they display loss of contact inhibition in vitro, and (3) they **invade** local tissues and **spread** (metastasize) to other parts of the body. These properties are characteristic of cells of malignant tumors. It is the last property that is generally responsible for the deaths of patients who have cancer. Cells of benign tumors also show diminished control of growth, but do not invade local tissue or spread to other parts of the body. Other important properties of cancer cells are as follows: (1) they are **self-sufficient in growth signals,** (2) **they are insensitive to anti-growth signals,** (3) they stimulate local **angiogenesis,** and (4) they are often able to **evade apoptosis.** These points are summarized in **Figure 55–1.**

Figure 55–2 shows a number of other important properties associated with cancer cells. These various points will be discussed below.

The **central issues** in cancer are to elucidate the biochemical and genetic mechanisms that underlie the uncontrolled growth of cancer cells, their ability to invade and metastasize and to develop successful treatments that destroy cancer cells, while causing minimal damage to normal cells. Considerable progress has been made in understanding the basic nature of cancer cells, with a central finding being that cancer is a disease due to abnormalities in key genes. However, many aspects of the behavior of cancer cells, in particular their ability to spread, have yet to be fully explained. In addition, despite improvements in treatment of certain types of cancers, therapies are still often unsuccessful. The study of cancer (oncology) is a huge area, so this short chapter can only introduce the reader to some key concepts.

A **Glossary** at the end of this chapter summarizes the meanings of many of the terms used.

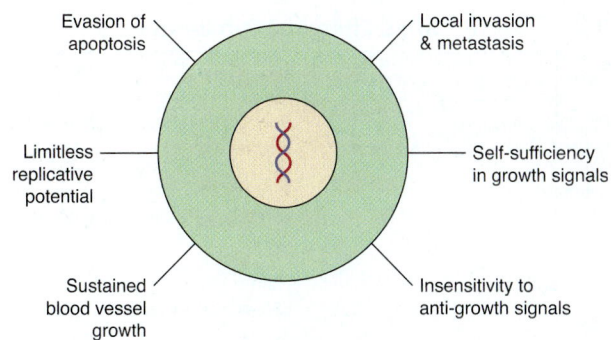

FIGURE 55-1 **Six major features of cancer cells.** Other important properties of cancer cells are shown in Figure 55-2. (With permission, After Hanahan D, Weinberg RA. The Hallmarks of Cancer. Cell 2000; 100:57).

FUNDAMENTAL FEATURES OF CARCINOGENESIS

Nonlethal genetic damage is the initiating event in carcinogenesis. There are principally four classes of genes, which when affected by such damage, can result in the development of a tumor. These are **proto-oncogenes, tumor suppressor genes, genes involved in DNA repair,** and **those that are involved in apoptosis.** Cancer is of **clonal origin,** with a single abnormal cell multiplying to become a mass of cells forming a tumor. Carcinogenesis is thus **a multistep process,** with multiple genetic alterations occurring in cells, transforming normal cells into malignant ones. Hence, a tumor often takes several years to develop.

CAUSES OF GENETIC DAMAGE

Genetic damage can be due to acquired or inherited mutations. The former occur due to exposure to environmental carcinogens while the latter are hereditary. Such hereditary abnormalities result in a number of **familial conditions** that predispose to hereditary cancer. These mutations are found in specific genes (eg, tumor suppressor genes) present in the germ cells and are discussed later.

Spontaneous mutations, some of which may predispose to cancer, occur at a frequency of approximately 10^{-7} to 10^{-6} per cell per generation. This rate will increase in tissues subject to a high rate of proliferation, increasing the generation of cancer cells from affected parent cells. **Oxidative stress** (see Chapter 45), by producing increased numbers of reactive oxygen species, may be a factor in increasing the mutation rate.

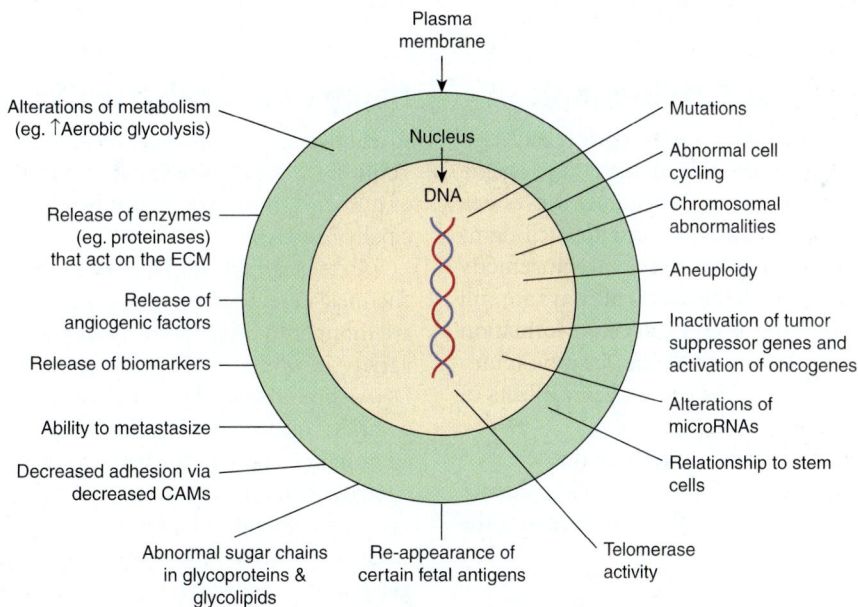

FIGURE 55-2 **Some biochemical and genetic changes occurring in human cancer cells.** Many changes, in addition to those indicated in Figure 55-1, are observed in cancer cells, only some of which are shown here. The roles of mutations in activating oncogenes and inactivating tumor suppressor genes are discussed in the text. Abnormalities of cell cycling and of chromosome structure, including aneuploidy, are common. Alterations of microRNA molecules that regulate gene activities have been reported, and the relationship of stem cells to cancer cells is a very active area of research. Telomerase activity is often detectable in cancer cells. Tumors sometimes synthesize certain fetal antigens, which may be measurable in the blood. Changes in plasma membrane constituents (eg, alteration of the sugar chains of various glycoproteins—some of which are cell adhesion molecules—and glycolipids) have been detected in many studies, and may be of importance in relation to decreased cell adhesion and metastasis. Various molecules can pass out of cancer cells and can be detected in the blood as tumor biomarkers. Angiogenic factors and various proteinases are also released by some tumors. Many changes in metabolism have been observed; for example, cancer cells often exhibit a high rate of aerobic glycolysis. (CAM, cell adhesion molecule; ECM, extracellular matrix.)

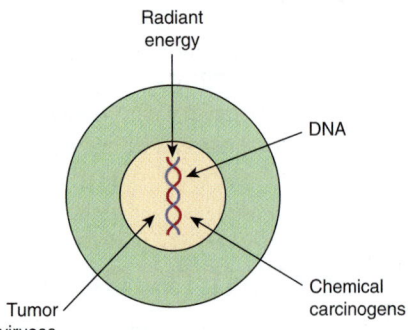

FIGURE 55–3 Radiant energy, chemical carcinogens and certain viruses can cause cancer.

RADIANT ENERGY, CHEMICALS, AND CERTAIN VIRUSES ARE THE MAJOR KNOWN CAUSES OF CANCER

In general, there are three classes of carcinogens, exposure to which result in tumor formation. These are **radiant energy, chemicals,** and **certain oncogenic viruses** (see **Figure 55–3**). The first two cause mutations in DNA, and the third class generally acts by introducing novel genes into normal cells.

We shall only describe briefly how radiant energy, chemicals, and oncogenic viruses cause cancer.

Radiant Energy can be Carcinogenic

Ultraviolet rays, x-rays, and **γ-rays** are mutagenic and carcinogenic. Extensive studies have shown that these agents can damage DNA in a number of ways, including the lesions listed in **Table 55–1**. Mutations in DNA, due to such damage, are thought to be the basic mechanism of carcinogenicity caused by radiant energy although the exact pathways are still under investigation. X-rays and γ-rays can cause formation of reactive oxygen species (ROS), which can also be mutagenic and probably contribute to the carcinogenic effects of radiant energy.

Exposure to ultraviolet radiation is common due to exposure to sunlight, which is its main source. Ample evidence exists to show that such radiation is linked to cancers of the

TABLE 55–1   Some Types of DNA Damage Caused by Radiant Energy

- Formation of pyrimidine dimers.
- Formation of apurinic or apyrimidinic sites by elimination of corresponding bases.
- Formation of single- or double-strand breaks or cross-linking of DNA strands.

TABLE 55–2   Some Chemical Carcinogens

Class	Compound
Polycyclic aromatic hydrocarbons	Benzo[a]pyrene, dimethylbenz-anthracene
Aromatic amines	2-Acetylaminofluorene, N-methyl-4-aminoazobenzene (MAB)
Nitrosamines	Dimethylnitrosamine, diethylnitrosamine
Various drugs	Alkylating agents (eg, cyclophosphamide), diethylstilbestrol
Naturally occurring compounds	Dactinomycin, aflatoxin B_1

Note: As listed above, some drugs used as chemotherapeutic agents (eg cyclophosphamide) can be carcinogenic. Diethylstilbestrol was formerly taken by women as an estrogenic agent; if they were pregnant, some of their daughters developed vaginal cancer.

skin. The risk of developing a skin cancer due to ultraviolet radiation increases with increasing frequency and intensity of exposure and decreasing melanin content of skin.

DNA damage produced by environmental agents is usually removed by DNA repair mechanisms. Individuals who have an inherited inability to repair DNA, as is seen in xeroderma pigmentosa (see Chapter 57) and ataxia telangiectasia, have increased risk of developing a malignancy.

Many Chemicals are Carcinogenic

A wide variety of chemical compounds are carcinogenic (see **Table 55–2** and **Figure 55–4**). It is estimated that perhaps **80%** of human cancers are caused by environmental factors, principally chemicals.

Extensive studies have been performed in the field of chemical carcinogenesis. Overall, most chemical carcinogens are thought to **interact covalently with DNA,** forming a wide variety of **adducts.** Depending on the extent of damage to DNA and its repair by DNA repair systems (see Chapter 35), a variety of mutations in DNA can result from exposure of an animal or human to chemical carcinogens, some of which contribute to the development of cancer.

Some chemicals **interact directly** with DNA (eg, methchlorethamine and β-propiolactone), but others (**procarcinogens**) require conversion by enzyme action to become **ultimate carcinogens** (**Figure 55–5**). Most ultimate carcinogens are **electrophiles** (molecules deficient in electrons) and readily attack nucleophilic (electron-rich) groups in DNA. Conversion of chemicals to **ultimate carcinogens** is principally due to the actions of various species of cytochrome P450 located in the endoplasmic reticulum (ER) (see Chapter 53). This fact is used in the Ames assay (see below), in which an aliquot of post-mitochondrial supernatant (containing ER) is added to the assay system as a source of cytochrome P450 enzymes.

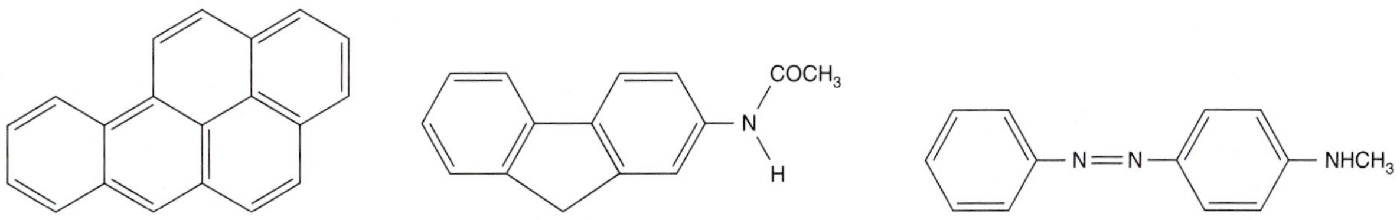

Benzo[a]pyrene   **2-Acetylaminofluorene**   ***N*-methyl-4-aminoazobenzene**

FIGURE 55–4 Structures of three experimentally widely used chemical carcinogens.

Chemical carcinogenesis comprises two stages—**initiation** and **promotion.** Initiation is the stage where exposure to a chemical causes irreversible DNA damage and is a necessary initial event for a cell to become cancerous. Promotion comprises the stage at which an initiated cell begins to grow and proliferate. The cumulative effect of these stages is a neoplasm.

Chemical carcinogens can be identified by screening for their mutagenicity. A simple way to do this is by using the **Ames assay** (**Figure 55–6**). This relatively simple test, which detects mutations in *Salmonella typhimurium* caused by chemicals, has proven very valuable for screening purposes. A refinement of the Ames test is to add an aliquot of endoplasmic reticulum (ER) to the assay, to make it possible to indentify procarcinogens. Very few, if any, compounds that have tested negative in the Ames test have been shown to cause tumors in animals. However, animal testing is required to show unambiguously that a chemical is carcinogenic.

It should be noted that compounds that alter epigenetic factors (eg, stilbestrol), thus perhaps leading to cancer, would not test positive in the Ames test, as they are not mutagenic.

Approximately 15% of Human Cancers may be Caused by Viruses

The study of **tumor viruses** has contributed very significantly to the understanding of cancer. For example, discovery of both oncogenes and tumor suppressor genes (see below) emerged from studies of oncogenic viruses. Both DNA and RNA viruses have been identified as being able to cause cancers

in humans (**Table 55–3**). The details of how each of these viruses causes cancer will not be described here. In general, the genetic material of viruses is incorporated into the genome of the host cell. In the case of RNA viruses, this would occur after reverse transcription of the viral RNA to viral DNA. Such integration of viral DNA (called the provirus) with the host DNA results in various events such as **deregulation of the cell cycle, inhibition of apoptosis,** and **abnormalities of cell signaling pathways.** All these events are discussed later in this chapter. The **DNA viruses** often act by down-regulating the tumor suppressor genes *P53* and *RB* (see below). RNA viruses often carry oncogenes in their genomes; how oncogenes act to cause malignancy is discussed below. It has been estimated that about 15% of human tumors may be caused by viruses.

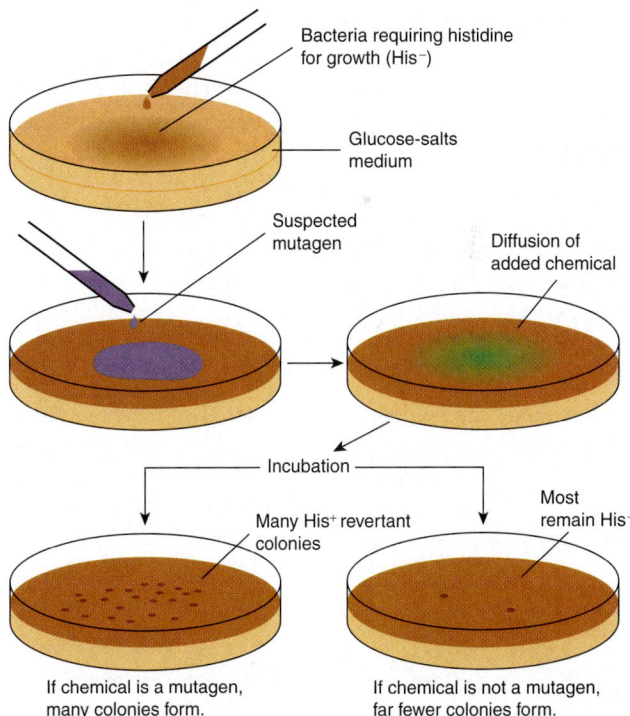

FIGURE 55–6 The Ames assay to screen for mutagens. The chemical tested will increase the frequency of reversion of His⁻ to His⁺ cells if it is a mutagen and, therefore, a potential carcinogen. A control plate (not shown) contains the liquid in which the suspected mutagen is dissolved. Reproduced, with permission, from Nester EW et al: *Microbiology: A Human Perspective.* 5th ed. McGraw-Hill, 2007.

A. Direct carcinogen → DNA

B. Pro-carcinogen → Enzyme → Ultimate carcinogen → DNA

FIGURE 55–5 **(A) Direct and (B) indirect carcinogens.** Direct carcinogens can interact with DNA without prior enzyme activation. Indirect carcinogens are activated by an enzyme (eg a cytochrome P450 species) to the ultimate carcinogen and then interact with DNA.

TABLE 55–3 Some Viruses That Cause or Are Associated With Human Cancers

Virus	Genome	Cancer
Epstein-Barr virus	DNA	Burkitt's lymphoma, nasopharyngeal cancer, B cell lymphoma
Hepatitis B	DNA	Hepatocellular carcinoma
Hepatitis C	RNA	Hepatocellular carcinoma
Human herpesvirus type I	DNA	Kaposi's sarcoma
Human papilloma viruses (certain types)	DNA	Cancer of the cervix
Human T-cell leukemia virus type 1	RNA	Adult T-cell leukemia

Note: It has been estimated that virus-linked human cancers are responsible for ~15% of total cancer incidence.

TABLE 55–4 Mechanisms of Activating Oncogenes

Mechanism	Explanation
Mutation	A classic example is point mutation of the *RAS* oncogene. This results in the gene product, a small GTPase, having less activity in tumors and in resultant stimulation of the activity of adenylyl cyclase.
Promoter insertion	Insertion of a viral promoter region near a gene activates it
Enhancer insertion	Insertion of a viral enhancer region near a gene activates it
Chromosomal translocation	The basis is that a piece of one chromosome is split off and joined to another. Classic examples are these involved in Burkitt's lymphoma (see Figure 55–8) and in the Philadelphia chromosome (see the Glossary).
Gene amplification[1]	Abnormal multiplication of a gene occurs, resulting in many copies. This can occur with oncogenes and also genes involved in tumor drug resistance.

[1]Gene amplification may be recognized as homogenously stained regions on chromosomes, or as double minute chromosomes.

ONCOGENES AND TUMOR SUPPRESSOR GENES PLAY KEY ROLES IN CAUSING CANCER

Over the past 30 years or so, major advances have been made in understanding how cancer cells develop and grow. Two key findings were the discoveries of **oncogenes** and **tumor suppressor genes**. These discoveries pointed to specific mechanisms by which cell growth and division could be disturbed, resulting in abnormal growth. The overall effects of oncogenes and loss of activity of tumor suppressor genes are summarized in **Figure 55–7**.

Oncogenes are Derived from Proto-oncogenes and Encode a Wide Variety of Proteins that Affect Cell Growth and Cell Death

An **oncogene** can be defined as an altered gene whose product acts in a dominant manner to accelerate cell growth or

Proto-oncogenes   → Oncogenes

                                     → ↑Growth → → → Cancer
                                            rate

Tumor suppressor   → Inactivated
genes

FIGURE 55–7 Oncogenes and loss of activity of tumor suppressor genes drive cell growth towards cancer. Oncogenes encode various proteins that can drive the growth of cancer cells. Oncogenes are derived from proto-oncogenes. Tumor suppressor genes encode proteins that normally suppress cell growth, but which are inactivated when altered by mutations. MicroRNA molecules (not shown here) are also affected by mutations, and this can affect their normal regulatory functions. In addition, epigenetic changes (also not shown) affect gene expression, and hence growth of cancer cells.

cell division. It is derived by "activation" of normal cellular **proto-oncogenes** (which encode growth stimulating proteins). The mechanisms involved in such activation are listed in **Table 55–4**.

The Table lists an example of a **point mutation** occurring in the *RAS* oncogene, which encodes a small GTPase. Loss of the activity of this G protein (see Chapter 42) results in chronic stimulation of the activity of adenylyl cyclase, leading to cell proliferation. Another way an oncogene can be activated is via **insertion of a promoter** (see **Figure 55–8(A)**), in which integration of a retroviral provirus (ie, a DNA copy of the RNA genome of a tumor virus such as Rous sarcoma virus, made by reverse transcriptase) activates *MYC*, a neighboring host gene. Overproduction of the protein encoded by *MYC* (a transcription factor) stimulates cell proliferation. As the legend to Figure 55–8(A) indicates, a similar type of effect results from **enhancer insertion. Chromosomal translocations** are found quite frequently in cancer cells, with about a hundred different examples having been documented. The translocation found in cases of Burkitt's lymphoma is illustrated in **Figure 55–8(B)**. The overall effect of this translocation is also to activate *MYC*, resulting in cell proliferation. Yet another mechanism of oncogene activation is via **gene amplification,** which occurs quite commonly in various cancers. In this case, multiple copies of an oncogene are formed resulting in increased production of a growth-promoting protein.

Once oncogenes are activated, **how do their protein products act to promote development of cancer? Figure 55–9** shows certain of the ways in which they operate. Some affect cell signaling pathways (eg, the product of an oncogene may act as growth factor, a growth factor receptor, a G-protein or as a downstream signaling molecule). Others act to alter

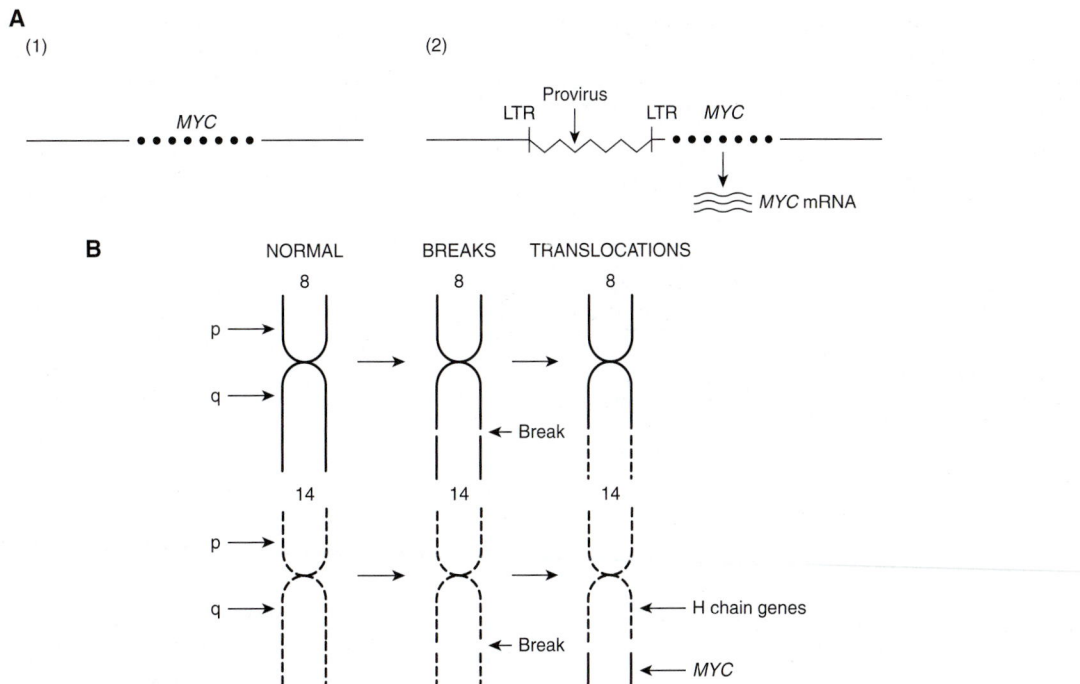

FIGURE 55–8 **(A) Schematic representation of how promoter insertion may activate a proto-oncogene.** (1). Normal chicken chromosome showing an inactive *MYC* gene. (2) An avian leukemia virus has integrated in the chromosome in its proviral form (a DNA copy of its RNA genome) adjacent to the *MYC* gene. Its right-hand long terminal repeat (LTR), containing a strong promoter (see Chapter 36), lies just upstream of the *MYC* gene and activates that gene, resulting in transcription of *MYC* mRNA. For simplicity, only one strand of DNA is depicted and other details have been omitted. Enhancer insertion acts similarly, except that the site of integration may be downstream or considerably upstream, and it cannot act as a promoter. Instead, a specific proviral sequence acts as an enhancer element (see Chapter 36), leading to activation of the *MYC* gene and its transcription. **(B) Schematic representation of the reciprocal translocation involved in Burkitt's lymphoma.** The chromosomes involved are 8 and 14. A segment from the end of the q arm of chromosome 8 breaks off and moves to chromosome 14. The reverse process moves a small segment from the q arm of chromosome 14 to chromosome 8. The *MYC* gene is contained in the small piece of chromosome 8 that was transferred to chromosome 14; it is thus placed next to genes transcribing the heavy chains of immunoglobulin molecules, and itself becomes activated. Many other translocations have been identified, with perhaps the best known being that involved in formation of the Philadelphia chromosome (see the Glossary).

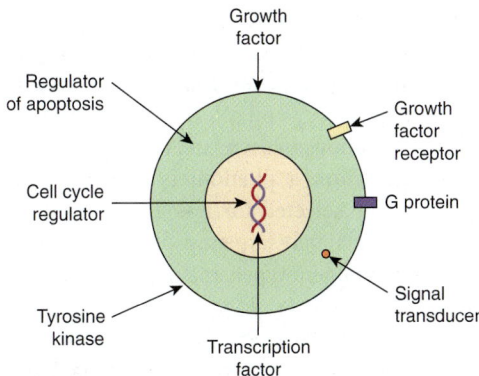

FIGURE 55–9 **Some ways in which proteins encoded by oncogenes work.** The Figure shows examples of various proteins encoded by oncogenes. The proteins are listed below with the corresponding oncogene given in parentheses along with its OMIM number. A growth factor, fibroblast growth factor 3 (*INT2*,164950); a growth factor receptor, epidermal growth factor receptor [EGFR] (*HER1*, 131550); a G protein (*H-RAS-1*, 190020); a signal transducer (*BRAF*, 164757); a transcription factor (*MYC*,190080); a tyrosine kinase and involved in cell-cell adhesion (*SRC*, 190090); a cell cycle regulator (*PRAD*, 168461); a regulator of apoptosis (*BCL2*, 151430).

transcription or to deregulate the cell cycle. Yet others can affect cell–cell interactions or the process of apoptosis. These mechanisms help to explain many of the major features of cancer cells shown in Figure 55–1, such as their limitless replicative potential, their signaling defects, their ability to invade and spread, and their evasion of apoptosis.

Certain **tumor viruses** (eg, retroviruses) **contain oncogenes.** It was the study of such tumor viruses (eg, Rous sarcoma virus [RSV], a retrovirus) that first revealed the presence of oncogenes. Further study showed that viral oncogenes were derived from cellular proto-oncogenes that the tumor viruses had picked up during their passage through host cells.

Tumor Suppressor Genes Act to Inhibit Cell Growth and Cell Division

A **tumor suppressor gene** produces a protein product that normally suppresses cell growth or cell division. When such a gene is altered by mutation, the inhibitory effect of its product is lost or diminished, leading to increased cell growth or cell division. As first suggested by AG Knudson, based on studies

TABLE 55–5   Some Differences Between Oncogenes and Tumor Suppressor Genes

Oncogenes	Tumor Suppressor Genes
Mutation in one of the two alleles is sufficient	Both alleles must be affected
Gain of function of a protein that signals cell division	Loss of function of a protein
Mutation arises in somatic cells, not inherited	Mutation present in germ cell (can be inherited), or in somatic cell
Some tissue preference	Often strong tissue preference (eg effect of *RB* gene in the retina)

Data from Levine AJ: The p53 tumor suppressor gene. N Engl J Med 1992;326:1350.

TABLE 55–6   Some Properties of a Few Important Oncogenes and Tumor Suppressor Genes

Name	Properties
MYC	An oncogene (OMIM 190080) that encodes a DNA-binding factor that can alter transcription. Involved in cell growth, cell cycle progression, and DNA replication. Mutated in a variety of tumors.
P53	A tumor suppressor gene (OMIM 191170) that responds to various cellular stresses. It induces cell cycle arrest, apoptosis, senescence, DNA repair and is involved in some aspects of regulation of cellular metabolism. It has been named "the guardian of the genome." Mutated in some 50% of human tumors. The nomenclature P53 refers to the approximate molecular mass of the protein encoded by *P53*, as calculated from SDS-PAGE.
RAS	A family of oncogenes encoding small GTPases. They were initially identified as being the transforming genes of certain murine sarcoma viruses. Important members of the family are K-RAS (Kirsten), H-RAS (Harvey) (OMIM 190020), and N-RAS (neuroblastoma). Persistent activation of these genes due to mutations contributes to the development of a variety of cancers.
RB	A tumor suppressor gene (OMIM 180200) encoding the RB protein. RB regulates the cell cycle by binding the elongation factor E2F. It represses the transcription of various genes involved in the S phase of the cycle. Mutation of the *RB* gene is the cause of retinoblastoma, but it is also involved in the genesis of certain other tumors.

of the inheritance of retinoblastomas, both copies of a tumor suppressor gene must be affected for it to lose its inhibitory effects on growth.

A useful distinction has been made between **gatekeeper** and **caretaker** functions of **tumor** suppressor genes. The former **control cell proliferation,** and include mainly genes that act to **regulate the cell cycle and apoptosis.** The latter are concerned with **preserving the integrity of the genome,** and include genes whose products are involved in recognizing and **correcting DNA damage** and maintaining **chromosomal integrity** during cell division.

Many oncogenes and tumor suppressor genes have now been identified. Only a few are mentioned here. Some differences between oncogenes and tumor suppressor genes are listed in **Table 55–5.**

Table 55–6 lists some of the properties of two of the most studied oncogenes (*MYC* and *RAS*), and two of the most studied tumor suppressor genes (*P53* and *RB*).

Studies of the Development of Colorectal Cancers Have Illuminated the Involvements of Specific Oncogenes and Tumor Suppressor Genes

Many types of tumors have been analyzed for genetic changes. One of the most informative areas in this respect has been analyses of the **development of colorectal cancers** by Vogelstein and colleagues. Their work, and that of others, has shown the involvement of various oncogenes and tumor suppressor genes in human cancer. (Case 4 in Chapter 57 describes the history of a patient with colorectal cancer). These workers analyzed various oncogenes, tumor suppressor genes, and certain other relevant genes in samples of **normal colonic epithelium,** of **dysplastic epithelium** (a preneoplastic condition, characterized by abnormal development of epithelium), of various stages of **adenomatous polyps,** and of

adenocarcinomas. Some of their major findings are summarized in **Figure 55–10.** It can be seen that certain genes were found to be mutated at relatively specific stages of the total sequence shown. Functions of the various genes identified are listed in **Table 55–7.** The **overall sequence** of changes can vary somewhat from that shown, and other genes may also be involved. Similar studies have been performed on a number of **other human tumors** revealing somewhat different patterns of activation of oncogenes and mutations of tumor suppressor genes. Further mutations in these and other genes are involved in **tumor progression,** a phenomenon whereby **clones** of tumor cells become selected for fast growth rate and ability to spread. Thus, a relatively large tumor may contain a variety of cells with different genotypes, making successful treatment more difficult.

Several other inferences can be made from these results and those from other similar studies. The first of these is that cancer is truly **a genetic disease,** but in a somewhat different sense from the normal meaning of the phrase, insofar as many of the gene alterations are due to somatic mutations. Secondly, carcinogenesis is, as mentioned above, a **multistep process.** It is estimated that in most cases a minimum of five to six genes must be mutated for cancer to occur. Thirdly, additional subsequent mutations are thought to confer selective advantages on **clones** of cells, some of which acquire the ability to metastasize successfully (see below). Fourthly, many

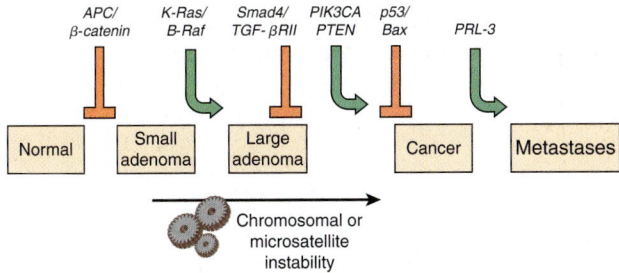

FIGURE 55–10 Multistep genetic changes associated with the development of colorectal cancers. Mutations in the *APC* gene initiate the formation of adenomas. One sequence of mutations in an oncogene and in various tumor suppressor genes that can result in further progression to large adenomas and cancer is indicated. Patients with familial adenomatous polyposis (OMIM 175100) inherit mutations in the *APC* gene and develop numerous dysplastic aberrant crypt foci (ACF), some of which progress as they acquire the other mutations indicated in the Figure. The tumors from patients with hereditary nonpolyposis colon cancer (OMIM 120435) go through a similar though not identical series of mutations; mutations in the mismatch repair system (see Chapter 35) speed up this process. *K-RAS* is an oncogene, and the other specific genes indicated are tumor suppressor genes. The chromosomal locations of the various genes shown here are known. The sequence of events shown here is not invariable in the development of all colorectal cancers. A variety of other genetic alterations have been described in a small fraction of advanced colorectal cancers. These may be responsible for the heterogeneity of biological and clinical properties observed among different cases. Instability of chromosomes and microsatellites (see Chapter 35) occurs in many tumors, and likely involves mutations in a considerable number of genes. (Reproduced, with permission, from Bunz F, Kinzler KW, Vogelstein B: Colorectal Tumors, Fig. 48-2, The Online Metabolic & Molecular Bases of Inherited Disease, www.ommbid.com)

TABLE 55–7 Some Genes Associated with Colorectal Carcinogenesis

Gene[1]	Action of Encoded Protein
APC (OMIM 611731)	Antagonizes WNT[2] signaling; if mutated, WNT signaling is enhanced, stimulating cell growth
β-CATENIN (OMIM 116806)	Encodes β-catenin, a protein present in adherens junctions, which are important in the integrity of epithelial tissues
K-RAS (OMIM 601599)	Involved in tyrosine kinase signaling
BRAF (OMIM 164757)	A serine/threonine kinase
SMAD4 (OMIM 600993)	Affects signaling by transforming growth factor-beta (TGF-β)
TGF-βRII	Acts as a receptor for TGF-β[3]
PI3KCA (OMIM 171834)	Acts as a catalytic subunit of phosphatidylinositol 3-kinase
PTEN (OMIM 601728)	A protein tyrosine phosphatase with an area of homology to tensin, a protein that interacts with actin filaments at focal adhesions
P53 (OMIM 191170)	The product, P53, is induced in response to DNA damage and is also a transcription factor for many genes involved in cell division (see Table 55–10)
BAX (OMIM 600040)	Acts to induce cell death (apoptosis)
PRL3 (OMIM 606449)	A protein-tyrosine phosphatase

Abbreviations: *APC,* adenomatous polyposis coli gene; *BAX,* encodes BCL2-associated X protein (BCL2 is a repressor of apoptosis); *BRAF,* the human homolog of an avian proto-oncogene; *K-RAS,* Kirsten–Ras-associated gene; *PI3KCA,* encodes the catalytic subunit of phosphatidylinositol 3-kinase; *PRL3,* encodes a protein-tyrosine phosphatase with homology to PRL1, another protein-tyrosine phosphatase found in regenerating liver; *PTEN,* encodes a protein-tyrosine phosphatase and tensin homolog; *P53,* encodes a polypeptide of molecular mass 53,000; *SMAD4,* the homolog of a gene found in Drosophila.

Note: The various genes listed are either oncogenes, tumor suppressor genes or genes whose products are closely associated with the products of these two types of genes. The cumulative effects of mutations in the genes listed are to drive colonic epithelial cells to proliferate and eventually become cancerous. They achieve this mainly via effects on various signaling pathways affecting cellular proliferation. Other genes and proteins not listed here are also involved. This table and Figure 55–10 vividly show the importance of a detailed knowledge of cell signaling for understanding the genesis of cancer.

[1]*K-RAS* and *BRAF* are oncogenes; the other genes listed are either tumor suppressor genes or genes whose products are associated with the actions of the products of tumor suppressor genes.

[2]The WNT family of secreted glycoproteins is involved in a variety of developmental processes. Tensin is a protein that interacts with actin filaments at focal adhesions.

[3]TGF-β is a polypeptide (a growth factor) that regulates proliferation and differentiation in many cell types.

of the genes implicated in colorectal carcinogenesis and other types of cancers are involved in **cell signaling events,** showing the central role that alterations in signaling play in the development of cancer.

GROWTH FACTORS & ABNORMALITIES OF THEIR RECEPTORS AND SIGNALING PATHWAYS PLAY MAJOR ROLES IN CANCER DEVELOPMENT

There Are Many Growth Factors

A large variety of polypeptide growth factors that work on human tissues and cells have been identified. Some are listed in **Table 55–8**. Here we focus mostly on their relationship with cancer.

Growth factors can act in an **endocrine, paracrine,** or **autocrine** manner and affect a wide variety of cells to produce **a mitogenic response.** As described earlier (Chapter 52), they play an important role in the differentiation of hematopoietic cells.

Growth inhibitory factors also exist. For example, transforming factor beta (TGF-β) exerts inhibitory effects on the growth of certain cells. Thus, chronic exposure to increased amounts of a growth factor or to decreased amounts of a growth inhibitory factor can alter the balance of cellular growth.

TABLE 55–8   Some Polypeptide Growth Factors

Growth Factor	Function
Epidermal growth factor (EGF)	Stimulates growth of many epidermal and epithelial cells
Erythropoietin (EPO)	Regulates development of early erythropoietic cells
Fibroblast growth factors (FGFs)	Promote proliferation of many different cells
Interleukins	Interleukins exert a variety of effects on cells of the immune system
Nerve growth factor (NGF)	Trophic effect on certain neurons
Platelet-derived growth factor (PDGF)	Stimulates growth of mesenchymal and glial cells
Transforming growth factor-alpha (TGFα)	Similar to EGF
Transforming growth factor-beta (TGFβ)	Exerts both stimulatory and inhibitory effects on certain cells

Many other growth factors have been identified. Growth factors may be made by a variety of cells, or may have mainly one source. Many different interleukins have now been isolated; along with the interferons and some other proteins/polypeptides, they are referred to as cytokines.

Growth Factors Work Via Specific Receptors and Transmembrane Signaling to Affect the Activities of Specific Genes

Growth factors produce their effects by interacting with **specific receptors** on cell surfaces, initiating **various signaling events** (Chapter 42). Many receptors for growth factors have been cloned. They generally have short membrane-spanning segments and external and cytoplasmic domains. A number (eg, those for epidermal growth factor [EGF], insulin and platelet-derived growth factor [PDGF]) have **tyrosine kinase** activities. The kinase activity, located in the cytoplasmic domains, causes autophosphorylation of the receptor protein and also phosphorylates certain other proteins.

Consideration of how **PDGF** acts illustrates how one particular growth factor brings about its effects. Interaction of PDGF with its receptor stimulates the activity of phospholipase C. This acts to split phosphatidylinositol bisphosphate (PIP_2) into inositol trisphosphate (IP_3) and diacylglycerol (DAG) (see Figure 42–6). Increased IP_3 stimulates the release of intracellular Ca_2^+ and DAG increases the activity of protein kinase C (PKC). Hydrolysis of DAG may release arachidonic acid, which can stimulate production of prostaglandins and leukotrienes, each of which has various biologic effects. Exposure of target cells to PDGF can result in rapid (minutes to 1–2 h) activation of certain cellular proto-oncogenes (eg, *MYC* and *FOS*), which participate in stimulation of mitosis via effects on the cell cycle (see below). The bottom line is that growth factors interact with specific receptors, which

TABLE 55–9   Measures That Might Prevent Approximately 50% of Cancers if Introduced on a Population-Wide Basis

- Reduce tobacco use
- Increase physical activity
- Control weight
- Improve diet
- Limit alcohol
- Use safer sex practices
- Routine cancer screening tests
- Avoid excess exposure to the sun

Date from Stein CJ, Colditz GA: Modifiable risk factors for cancer. Brit J Cancer 90:299 (2004).

stimulate specific signaling pathways to increase or decrease the activities of various genes that affect cell division.

MANY CANCERS CAN BE PREVENTED BY MODIFYING RISK FACTORS

Modifiable risk factors have been linked to a wide variety of cancers. It has been estimated that over half of all cancers in developed countries could be prevented if the measures summarized in **Table 55–9** were introduced on a population-wide basis. **Smoking** is still a major cause of cancer across the globe. It cannot be overemphasized that **prevention** and **early detection of cancer** are most critical if the disease is to be beaten.

ABNORMALITIES OF THE CELL CYCLE ARE UBIQUITOUS IN CANCER CELLS

Knowledge of **the cell cycle** is necessary for understanding many of the mechanisms involved in the development of cancer. It is also of importance because many anti-cancer drugs act only against cells that are dividing, or in a certain phase of the cycle.

Basic aspects of the cell cycle were described in Chapter 35. As shown in Figure 35–20, the cycle has four phases: G_1, S, G_2, and M. If cells are not cycling, they are said to be in G_0 phase and are termed quiescent. Cells can be recruited into the cycle from G_0 by various influences (eg, certain growth factors). Generation time is the time needed for a cell in G_0 to enter the cycle and give rise to two daughter cells. The cells of a cancer usually have a shorter generation time than normal cells, and there are less of them in G_0 phase.

The roles of various **cyclins, cyclin-dependent kinases** (CDKs), and a number of **other important molecules** that affect the cell cycle (eg, the genes *RB and P53*) were also described in Chapter 35. The points in the cycle at which some of these molecules act are indicated in Figure 35–21 and Table 35–7.

Because a major property of cancer cells is **uncontrolled growth,** many aspects of their **cell cycle** have been studied in considerable depth. Only a few results can be mentioned here. A variety of mutations affecting **cyclins** and **CDKs** have been reported. Many products of **proto-oncogenes** and **tumor suppressor genes** play important roles in regulating the normal cycle. A wide variety of mutations have been found in these types of genes, including *RAS, MYC, RB, P53* (which are among the most studied, see below) and many others.

For example, as discussed in Chapter 35, the protein product of the **RB gene** is a cell cycle regulator. It acts via binding to the transcription factor E2F, blocking progression of the cell from G1 to S phase. Loss of the RB protein due to mutations thus removes this element of control of the cell cycle.

When damage to DNA occurs (by radiation or chemicals), the **P53 protein** increases in amount and activates transcription of genes that delay transit through the cycle. If the damage is too severe to repair, P53 activates genes that cause apoptosis (see below). If P53 is absent or inactive due to mutation, apoptosis does not occur and cells with damaged DNA persist, perhaps becoming progenitors of cancer cells.

GENOMIC INSTABILITY AND ANEUPLOIDY ARE IMPORTANT CHARACTERISTICS OF CANCER CELLS

As referred to above and also later in this Chapter, cancer cells have many mutations. One possible explanation for their **genomic instability** is that they have **a mutator phenotype.** This was originally postulated by Loeb and colleagues to be due to cancer cells having acquired mutations in genes involved in **DNA replication** and **DNA repair,** thus allowing mutations to accumulate. The concept was later expanded to include mutations that affect **chromosomal segregation, DNA damage surveillance,** and processes such as **apoptosis.**

The term genomic instability is frequently used to refer to two abnormalities shown by many cancer cells, **microsatellite** and **chromosomal instability (CI). Microsatellite instability** was described briefly in Chapter 35. It involves expansion or contraction of microsatellites, usually due to abnormalities of mismatch repair or to replication slippage. **CI** occurs more often than microsatellite instability, and the two are often mutually exclusive. It refers to gain or loss of chromosomes caused by abnormalities of chromosomal segregation during mitosis.

Another area of interest regarding CI is **copy number variation** (CNV) (see the Glossary). Associations of various CNVs with many cancers have been identified, and their precise roles in cancer are under investigation.

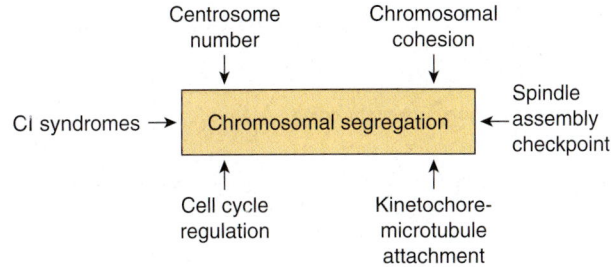

FIGURE 55–11 **Some factors involved in chromosomal segregation which are relevant to understanding chromosomal instability (CI) and aneuploidy.** CI syndromes include Bloom syndrome (OMIM 210900) and others. (Based on Thompson SL et al: Mechanisms of chromosomal instability. Curr Biol 2010;20(6):R285).

An important aspect of CI is **aneuploidy,** a very common feature of solid tumors. Aneuploidy exists when the chromosomal number of a cell is not a multiple of the haploid number. The degree of aneuploidy often correlates with **a poor prognosis.** This has suggested that abnormalities of chromosomal segregation may contribute to tumor progression by increasing genetic diversity. Some scientists believe that aneuploidy is a fundamental aspect of cancer.

Much research is proceeding on determining the basis of CI and aneuploidy. As shown in **Figure 55–11**, a number of different processes are involved in normal chromosomal segregation. Each of the processes shown is complex, involving various organelles and also many individual proteins. A textbook of Cell Biology should be consulted for details of these important processes. Studies are in progress to compare these processes in normal and tumor cells, and to determine which of the differences detected may be contributors to CI and aneuploidy. One hope of this line of research is that it might be possible to **develop drugs** that diminish or even prevent CI and aneuploidy.

MANY CANCER CELLS DISPLAY ELEVATED LEVELS OF TELOMERASE ACTIVITY

There has been considerable interest in the involvement of telomeres (see Chapter 35) in a number of diseases and also in aging. With respect to cancer, when tumor cells divide rapidly their telomeres often shorten. Such telomeres (usually detected in leukocytes because of ease of obtaining them) have been implicated as a risk factor for many, but not all, solid tumors, (eg, breast cancer). **Short telomeres** appear to be of predictive value regarding the progression of chronic inflammatory diseases (such as ulcerative colitis and Barrett's esophagus) to cancer. Abnormalities of telomere structure and function can contribute to CI (see above). The activity of **telomerase,** the main enzyme involved in synthesizing telomeres, is frequently elevated in cancer cells, providing one mechanism for over-

TABLE 55–10 **Some Hereditary Cancer Conditions**

Condition	Gene	Major Function	Major Clinical Feature
Adenomatous polyposis of the colon (OMIM 175100)	*APC*	See Table 55–7	Development of many early-onset adenomatous polyps, which are immediate precursors of colorectal cancers
Breast cancer 1, early onset (OMIM 113705)	*BRCA1*	DNA repair	About 5% of women in N America with breast cancer carry mutations in this gene or in *BRCA2*. Also substantially increases risk of ovarian cancer
Breast cancer 2, early onset (OMIM 600185)	*BRCA2*	DNA repair	As stated above for *BRCA1*. Mutations in this gene also increase the risk of ovarian cancer, but to a lesser extent
Hereditary nonpolyposis cancer, type I (OMIM 120435)	*MSH2*	DNA mismatch repair	Early onset of colorectal cancers
Li–Fraumeni syndrome (OMIM 151623)	*P53*	See Table 55–6	A rare syndrome involving cancers at different sites, developing at an early age
Neurofibromatosis, type 1 (OMIM 162200)	*NF1*	Encodes neurofibromin	Varies from a few café au lait spots to development of thousands of neurofibromas
Retinoblastoma (MIM 180200)	*RB1*	See Table 55–6	[1]Hereditary or sporadic retinoblastoma

Many other hereditary cancer conditions have also been identified.

[1]In hereditary retinoblastoma, one allele is mutated in the germ line, requiring only one subsequent mutation for a tumor to form. In sporadic retinoblastoma, neither allele is mutated at birth, so that subsequent mutations in both alleles are required.

coming telomere shortening. Selective inhibitors of telomerase have been considered as possible drugs for treating cancer, but have not as yet been translated into successful clinical use.

A NUMBER OF CANCERS HAVE A HEREDITARY PREDISPOSITION

It has been known for many years that certain cancers have a hereditary basis. It is estimated that about 5% of cancers fall into this category. The discovery of oncogenes and tumor suppressor genes has allowed investigations of the basis of this phenomenon. Many hereditary types of cancer have now been recognized; only a few of these are listed in **Table 55–10**. In a number of cases, where a hereditary syndrome is suspected, appropriate genetic screening of families has allowed early interventions to be made. For example, some young women who have inherited either a mutated *BRCA1* or *BRCA2* gene have opted for prophylactic mastectomies to prevent cancer of the breast occurring in later life.

WHOLE GENOME SEQUENCING OF TUMOR CELLS IS PROVIDING NEW INSIGHTS REGARDING CANCER

Since the completion of the Human Genome Project, the technology of large-scale sequencing and the analysis and interpretation of sequence data have advanced considerably. Sequencing has become much cheaper. This has led to studies, the aim of which is to **completely sequence the genomes of a large number of different types of tumors.** In this way, a comprehensive **catalog** of the types and numbers of mutations found in various types of cancers will eventually become available. It is hoped that such studies will lead to results that will also impact **diagnostic testing** and point to useful, new **prognostic biomarkers.** It should also reveal the involvement of **many new genes** in carcinogenesis, some of which may provide leads to new therapeutic approaches. To date, **about 350 genes involved in cancer have been identified;** it has been predicted that as many as 2000 may exist. It is of particular interest to identify mutations in genes that cause and accelerate cancers; these are known as **driver** mutations, whereas other mutations are called **passenger** mutations.

Studies on a number of types of cancer have already afforded interesting results. For example, in a study of DNA from a person suffering from **acute myelocytic leukemia,** about 750 somatic mutations were detected. Only 12 of these were in genes encoding proteins or regulatory RNAs; mutations in such genes are most likely to exert serious effects. (Coding areas of the genome only occupy about 2% of genomic DNA.) Mutations in other genomic sites may or may not have consequences. Whole genome sequencing on seven human primary prostate cancers has revealed the presence of surprisingly many genomic rearrangements.

A fascinating example of the information that can be provided by genome sequencing of cancers is provided by the results of a recent study of **pancreatic carcinomas.** These are among the most deadly cancers. However, an issue that had not been resolved is whether their lethality is due to their **aggressiveness** (ability to metastasize and grow), or to **late**

diagnosis. In the study under consideration, the genomes of seven primary pancreatic cancers were sequenced, as were the genomes of metastases from them obtained at autopsy. Approximately 61 known cancer-related mutations were detected in each metastasis. Using a "molecular clock" technique borrowed from evolutionary biology, it was calculated how long it took the metastases to accumulate these mutations. Prior knowledge of the overall sequence of mutations made this possible. It was estimated that it took **just over 10 years** from the time of the initiating mutation for non-metastatic primary tumors to develop in the pancreas. **Another 5 years** was required for such tumors to gain metastatic potential. Thereafter, **about 2 years** elapsed before the tumors metastasized and death resulted. Thus, it was suggested that the evolution of many pancreatic cancers is **a relatively slow process,** and that pancreatic cancers are not highly aggressive. The problem is that they are difficult to diagnose. The hope is that methods such as **detection of mutations in pancreatic cancer cells present in stool specimens,** development of **new blood biomarkers** for pancreatic cancers, and perhaps **new imaging techniques** will allow **early diagnosis,** always a critical factor in the management of cancer. Within a relatively few years, **complete sequence data** on the genomes of **many important types of tumors** should be available, and, hopefully, will considerably advance understanding of the basic biology of cancer and provide information relevant to better drug design and less toxic treatment.

Distinct from gene sequencing studies, there have been many reports of involvements of **microRNAs** in cancer. Because of their importance in many aspects of gene control, the study of these molecules is already forming an important area of cancer research.

Several hundred types of cancers have been identified by classical histopathology. Thus, **cancer is not one single disease.** The work on genome sequencing of cancers, revealing the large number of mutations in cancer cells, suggests that there may well be not just hundreds of different cancers, but that **each cancer may be genetically different.**

CANCER CELLS HAVE ABNORMALITIES OF APOPTOSIS THAT PROLONG THEIR LIVES

Apoptosis is a genetically regulated program that, when activated, causes cell death. The main players are proteolytic enzymes named **caspases,** which normally exist as inactive **procaspases.** The name caspase reflects that they are cysteine proteases that split peptide bonds on the C-terminal end of aspartate residues. About 15 human caspases are known, although not all participate in apoptosis. When those involved in apoptosis are activated (mainly 2, 3, 6, 7, 8, 9 and 10), they participate in a **cascade** of events (compare with the coagulation cascade, Chapter 51) that ultimately kills cells by digesting various proteins and other molecules. The **upstream**

caspases (eg, 2, 8, and 10) at the beginning of the cascade are often called **initiators,** and those downstream at the end of the pathway (eg, 3, 6, and 7) are called **effectors** or **executioners.** **Caspase-activated DNase** (CAD) fragments DNA, producing a characteristic laddering pattern detected by electrophoresis. **Microscopic features** of apoptosis include condensation of chromatin, changes of nuclear shape and membrane blebbing. The dead cells are rapidly disposed of by phagocytic activity, avoiding an inflammatory reaction.

Apoptosis differs from **necrosis,** a pathologic form of cell death that is not genetically programmed. Necrosis occurs on exposure to external agents, such as certain chemicals and extreme heat (eg, burns). Various hydrolytic enzymes (proteases, phospholipases, nucleases, etc) are involved in necrosis. Release of cell contents from dying cells can cause local inflammation, unlike apoptosis.

The overall process of apoptosis is **complex** and it is a matter of life and death (excuse the pun!) that it is **tightly regulated.** It includes proteins that act as receptors, adapters, procaspases and caspases, and pro- and antiapoptotic factors. There are **extrinsic** and i**ntrinsic pathways,** with **mitochondria** being important participants in the intrinsic pathway.

Figure 55–12 shows a much simplified diagram of some of the key events in apoptosis. Two major pathways are involved, the death receptor (extrinsic) pathway and the mitochondrial (intrinsic) pathway.

Major features of **the death receptor pathway** are shown on the left-hand side of the Figure. **External signals** initiating apoptosis include the cytokine tumor necrosis factor-α (TNF-α) and FAS ligand. A number of death receptors have been identified. They are transmembrane proteins and some interact with **adapter proteins** (such as FADD). These complexes in turn interact with **procaspase-8,** resulting in its conversion to **caspase-8** (an initiator). **Caspase-3** (an effector) is activated via a series of further reactions. It digests important structural proteins such as lamin (this is associated with nuclear condensation), various cytoskeletal proteins, and enzymes involved in DNA repair, causing cell death.

Regulation of this pathway occurs at several levels. **FLIP** inhibits the conversion of procaspase-8 to its active form. **Inhibitors of apoptosis** (IAPs) inhibit the conversion of procaspase-3 to its active form. These effects can be overcome by the protein **SMAC,** which is released from mitochondria.

The **mitochondrial pathway** can be initiated by exposure to reactive oxygen species, DNA damage and other stimuli. This results in pores forming in the outer mitochondrial membrane, through which **cytochrome c** escapes into the cytoplasm. In the cytoplasm, cytochrome c interacts with **APAF-1, procaspase-9** and **ATP** to form a multi-protein complex known as an **apoptosome.** As a result of this interaction, **procaspase-9** is converted to its active form and, in turn, acts on **procaspase-3** to produce **caspase-3.**

Regarding **regulation,** activation of the *P53* gene upregulates transcription of *BAX*. **BAX** is proapoptotic, in that it causes loss of mitochondrial membrane potential, helping

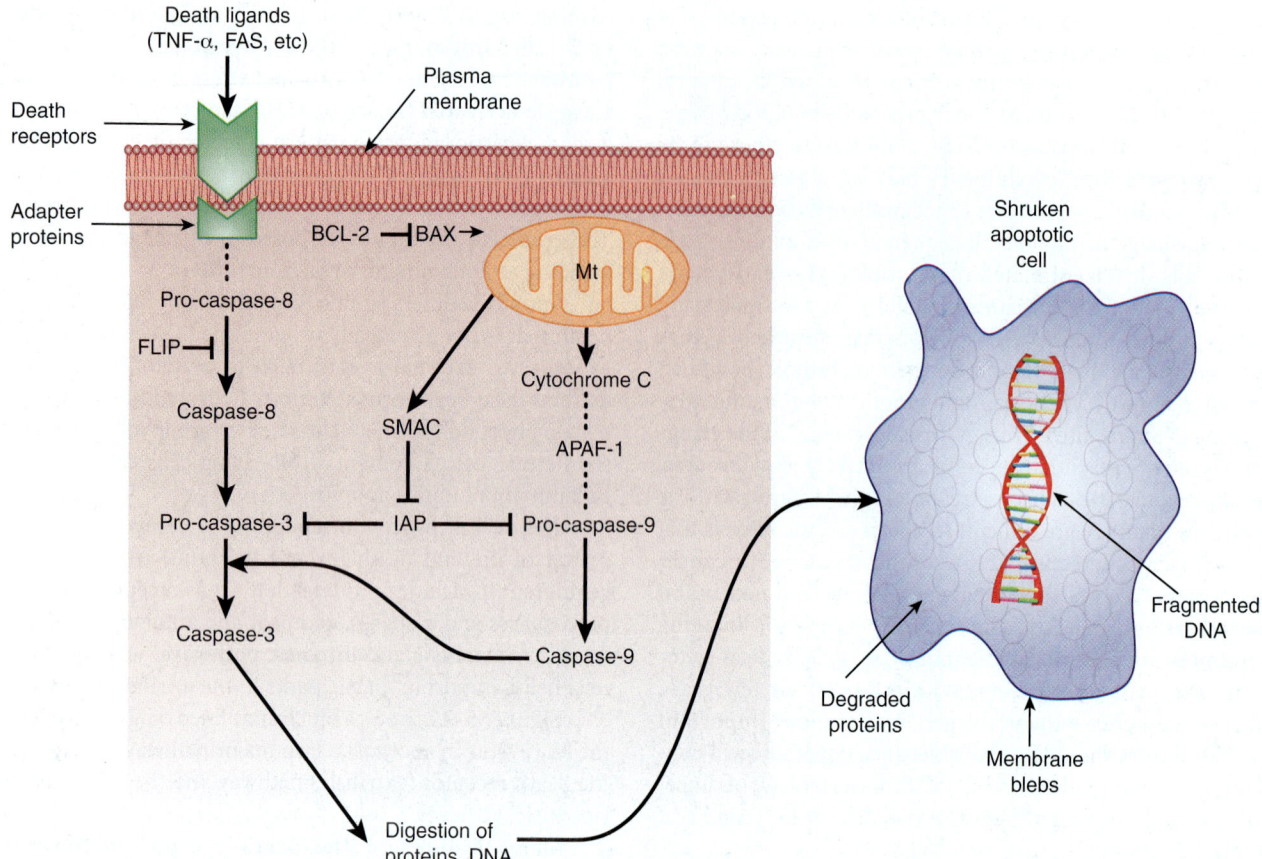

FIGURE 55–12 **Scheme of apoptosis, much simplified.** The left-hand side represents major events in the extrinsic pathway. Death signals include TNF-α and FAS (present on the surface of lymphocytes and some other cells). The signals (ligands) interact with specific death receptors (there are a number of them). The activated receptor then interacts with an adapter protein (FADD is one of a number of them), and then forms a complex with pro-caspase 8. (The complex is indicated by the ⋯ between the receptor and pro-caspase-8 in the Figure). Through a series of further steps, active caspase-3 is formed, which is a major effector (executioner) of cell damage. Regulation of the extrinsic pathway can occur due to the inhibitory effect of FLIP on the conversion of pro-caspase-8 to caspase-8, and also the inhibitory effect of IAP on pro-caspase-3. The right-hand side represents major events in the intrinsic (mt) pathway. Various cell stresses affect the permeability of the mt outer membrane, resulting in efflux of cytochrome c into the cytoplasm. This forms a multi-protein complex with APAF-1 and pro-caspase-9, called an apoptosome. Via these interactions, pro-caspase -9 is converted to caspase-9. This, in turn, can act on pro-caspase-3 to convert it to its active form. Regulation of the intrinsic pathway can occur at the level of BAX, which facilitates increasing mt permeability permitting efflux of cytochrome c, and is thus pro-apoptotic. BCL-2 opposes this effect of BAX and is thus anti-apoptotic. IAP also inhibits pro-caspase-9, and this effect of IAP can be overcome by SMAC. (APAF-1, apoptopic protease activating factor-1; BAX, BCL-2 associated X protein; BCL-2, B-cell CLL/lymphoma 2 (CLL represents chronic lymphatic leukemia); FADD, FAS-associated via death domain; FAS, FAS antigen; FLICE, FADD-like ICE; FLIP, FLICE inhibitory protein; IAP, inhibitor of apoptosis proteins; ICE, interleukin1-β convertase; SMAC, second mitochondria-derived activator of caspase.) ⊣ signifies opposes the action of.

initiate the mitochondrial apoptotic pathway. On the other hand, **BCL-2** inhibits the loss of the membrane potential, and is thus antiapoptotic. **IAPs** inhibit conversion of procaspase 9 to caspase-9; **SMAC** can overcome this.

Note that the death pathway uses **caspase-8** as an initiator, whereas the mitochondrial pathway uses **caspase-9.** These two pathways can interact. In addition, there are also other pathways of apoptosis not discussed here.

Cancer Cells Evade Apoptosis

Cancer cells have developed **ways of evading apoptosis,** and thus of continuing to grow and divide. In general, these result from mutations that cause loss of function of proteins that are proapoptotic, or from overexpression of anti-apoptotic genes. One such example concerns loss of function of the *P53 gene,* perhaps the most commonly mutated gene in cancers. Resultant loss of up-regulation of proapoptotic *BAX* (see above) shifts the balance in favor of antiapoptotic proteins. **Overexpression** of many anti-apoptotic genes is a frequent finding in cancers. The resulting evasion of apoptosis favors the continuing growth of cancers. Attempts are being made to develop **drugs** or **other compounds** that will specifically turn on apoptosis in cancer cells, terminating their lifespans.

As indicated above, apoptosis is a **complex, multiregulated pathway** with numerous participants, all of which are not mentioned here in this simplified account. It is also involved

TABLE 55–11 Summary of Some Important Points Regarding Apoptosis

- It involves a genetically programmed series of events and differs from necrosis.

- The entire series of reactions is a cascade, similar to blood coagulation.

- It is characterized by cell shrinking, membrane blebbing, absence of inflammation, and a distinct pattern (laddering) of degradation of DNA.

- Many caspases (proteinases) are involved; some are initiators and others effectors (executioners).

- There are both external and internal (mt) pathways.

- FAS and other receptors are involved in the death receptor (external) pathway of apoptosis.

- Cellular stress and other factors activate the mt pathway; release of cytochrome c into the cytoplasm is an important event in this pathway.

- Apoptosis is regulated by a balance between inhibitors (antiapoptotic) and activators (proapoptotic).

- Cancer cells have acquired mutations that enable them to evade apoptosis, thus prolonging their existence.

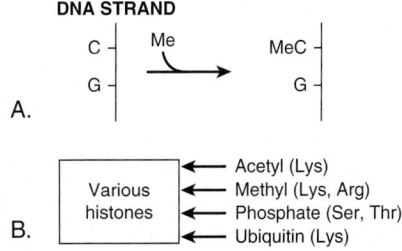

FIGURE 55–13 Some factors involved in epigenetics. **(A)** Methylation of cytosine to form 5′methylcytosine. The cytosine is usually located next to a guanine residue, forming a CpG island. Methylation of cytosine by a methyltransferase is associated with silencing of the activities of certain genes. **(B)** Post-translational modifications of various histones. Specific residues in specific histones are modified by various enzymes, changing the conformations and activities of the modified histones. For example, acetylation of N-terminal lysines in certain histones is associated with opening up of chromatin and with increased transcription of certain genes. **(C)** Chromatin remodelling: see Figure 36–10.

in various developmental and physiological processes. It may seem paradoxical, but regulated cell death is as important in maintaining health as is formation of new cells. In addition to cancer, apoptosis is implicated in other diseases, including certain autoimmune and chronic neurological disorders, such as Alzheimer disease and Parkinson disease, where **excessive cell death** (rather than excessive growth) is a feature.

Table 55–11 summarizes some of the principal features of apoptosis.

EPIGENETIC MECHANISMS ARE INVOLVED IN CANCER

There is growing evidence that epigenetic mechanisms (see Chapter 36) are involved in the causation of cancer. **Methylation of specific cytosine bases** in genes is implicated in turning off the activities of certain genes. Changes from normal in methylation/demethylation of cytosine residues in specific genes have been detected in cancer cells. **Post-translational modifications of histones,** such as acetylation, methylation, phosphorylation, and ubiquitination, also affect gene expression. Changes in **acetylation of histones H3 and H4,** affecting gene transcription, have been found in cancer cells. Mutations affecting the structures of protein complexes (eg, the SW1/SNF complexes) involved in **chromatin remodeling** can also affect gene transcription. Indeed, several of the components of the SW1/SNF complexes may act as tumor suppressor genes. Some of these points about epigenetics are summarized in **Figure 55–13**.

A matter of particular interest regarding epigenetic changes is that many of them are potentially **reversible.** In this regard, **5′azadeoxycytidine** is an inhibitor of methyltransferases, and **suberoylanilide hydroxamic acid** (SAHA, Vorinostat) acts to deacetylate histones. Both of these agents have been used to treat certain types of leukemias and lymphomas.

The increasing use of **screening techniques** for studying epigenetic changes (eg, analysis of the methylome [the sum total of methylation modifications in the genome]) in more types of cancers is likely to add considerably to knowledge in this area.

THERE IS MUCH INTEREST IN THE ROLE OF STEM CELLS IN CANCER

Stem cells were discussed briefly in Chapter 52. Many scientists are currently investigating the role of stem cells in cancer. Cancer stem cells are believed to harbor mutations that, either by themselves or in collaboration with further mutations, make these cells cancerous. They may be detected by the use of specific surface markers, or other techniques. It appears that **surrounding tissues** (eg, components of the extracellular matrix) can significantly influence the behavior of these cells. An important concept driving some of the research in this area is the belief that one of the reasons that cancer chemotherapy is often not successful is that **a pool of cancer stem cells exists** that is not susceptible to conventional chemotherapy. Reasons for this include the facts that many stem cells are relatively dormant, have active DNA repair systems, express drug transporters that can expel anticancer drugs, and are often resistant to apoptosis.

Evidence is accumulating that cancer stem cells do indeed play key roles in many types of neoplasia. If so, development of therapies with high specificity for killing these cells will prove of extreme value.

TUMORS OFTEN STIMULATE ANGIOGENESIS

Tumor cells need an adequate blood supply to provide nutrients for their survival. Both tumor cells and cells in tissues surrounding tumors have been found to **secrete factors** that stimulate the growth of new blood vessels (ie, **angiogenesis**). There has been much interest in tumor angiogenesis, partly because if it could be specifically inhibited, this could provide a selective method of killing tumor cells.

The growth of blood vessels supplying tumor cells can be stimulated by **hypoxia** and other factors. Hypoxia causes elevated levels of **hypoxia-inducible factor-1** (HIF-1), which in turn increases levels of **vascular endothelial growth factor** (VEGF), a major stimulant of angiogenesis. Some **five types of VEGF** have been identified (A-E), with most interest focusing on VEGF A. They interact with specific tyrosine kinase receptors on endothelial and lymphatic cells. These receptors, via signaling pathways, cause up-regulation of the NF-κB pathway (see Chapter 50), resulting in proliferation of endothelial cells and formation of new blood vessels. Blood vessels supplying tumors are not normal; their structure is often disorganized and they are much leakier than normal blood vessels. Molecules other than VEGFs, such as angipoietin, beta-fibroblast growth factor, TGF-α, and placental growth factor, also stimulate angiogenesis. Certain other molecules also **inhibit** blood vessel growth (eg, angiogenin and endostatin).

Monoclonal antibodies to VEGF A have been developed (eg, bevacizumab or Avastatin) and have been used in the treatment of certain types of cancer (eg, colon and breast). They bind to VEGF and prevent it acting. They were found to increase the overall patient survival, but most patients eventually relapsed. It is felt that they are best used in combination with other anti-cancer therapies. Monoclonal antibodies to other growth factors that stimulate angiogenesis are also being developed and tried out therapeutically, as are small molecule inhibitors of angiogenesis. Inhibitors of angiogenesis are useful in **other conditions,** such as "wet" age-related macular degeneration and diabetic retinopathy, in which proliferation of blood vessels is a feature.

METASTASIS IS THE MOST SERIOUS ASPECT OF CANCER

It has been estimated that about **85% of the mortality** associated with cancer results from metastasis. Spread of cancer is usually via lymphatics or blood vessels. Metastasis is a complex process, and its molecular bases are yet to be elucidated.

Figure 55–14 is a simplified scheme of metastasis. The earliest event is **detachment** of tumor cells from the primary

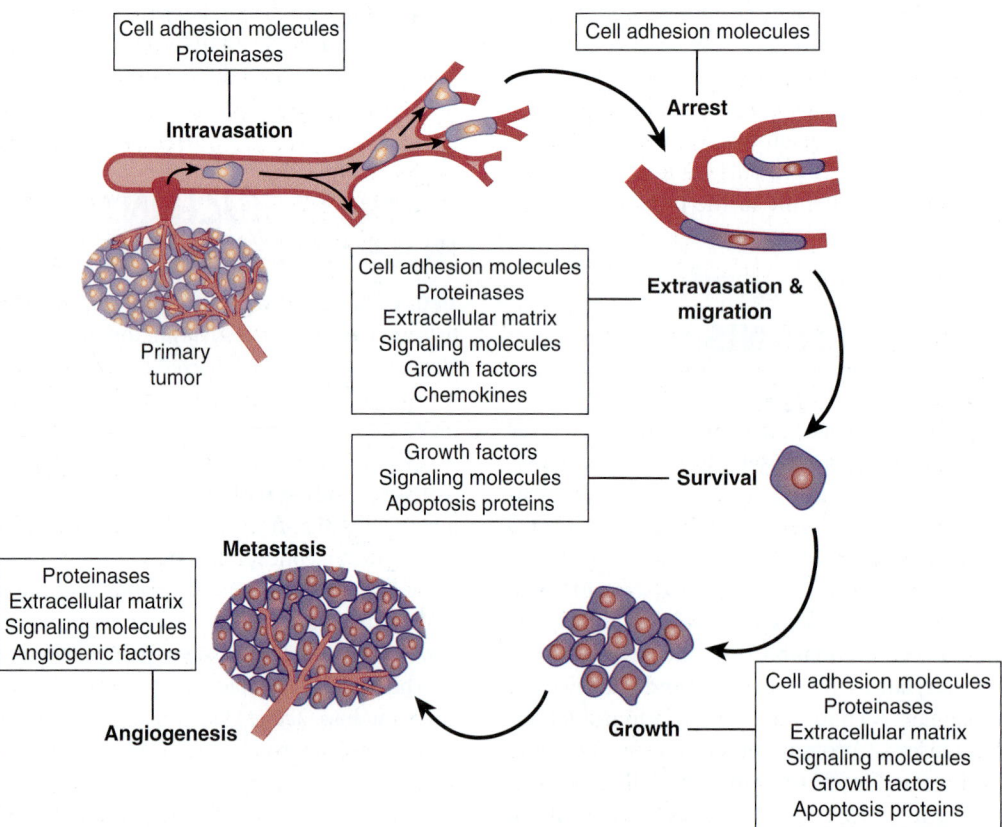

FIGURE 55–14 Simplified scheme of metastasis. Schematic representation of the sequence of steps in metastasis, indicating some of the factors believed to be involved. From Tannock IF et al: *The Basic Science of Oncology.* 4th ed. McGraw-Hill, 2005.

TABLE 55–12 **Some Important Cell Adhesion Molecules (CAMs)**

• Cadherins
• Immunoglobulin (Ig) superfamily (Ig CAMs)
• Integrins
• Selectins

CAMs may be homophilic or heterophilic. Homophilic CAMs interact with identical molecules on neighboring cells, whereas heterophilic CAMs interact with different molecules. Cadherins are homophilic, selectins and integrins are heterophilic and Ig CAMs may be either. Integrins are discussed briefly in Chapter 52, and selectins in Chapter 47.

tumor. The cells can then gain access to the circulation (or lymphatics), a process termed **intravasation.** Once in the circulation, they tend to **arrest** in the nearest small capillary bed. In that site, they **extravasate** and **migrate** through the neighboring ECM, before finding a site to settle. Thereafter, if they **survive** host defense mechanisms, they **grow** at variable rates. To ensure growth, they need an **adequate blood supply,** as discussed above. Some aspects of these events are now discussed in more detail.

Many studies **have shown changes from normal of molecules** on the surfaces of cancer cells. These changes may permit **decreased cell adhesion** and allow individual cancer cells to detach from the parent cancer. Molecules on cell surfaces involved in cell adhesion are called **cell adhesion molecules** (CAMs) (see **Table 55–12**). Decreases of the amounts of **E-cadherin,** a molecule of major importance in the adhesion of many normal cells, may help to account for the decreased adhesiveness of many cancer cells. Many studies have shown changes in the **oligosaccharide chains of cell surface glycoproteins,** due to altered activities of various glycosyltransferases (see Chapter 47). One important change is an increase of the activity of **GlcNAc transferase V.** This enzyme catalyzes transfer of GlcNAc to a growing oligosaccharide chain, forming a $\beta 1\text{->}6$ linkage and allowing further growth of the chain. It has been proposed that such elongated chains participate in an **altered glycan lattice** at the cell surface. This may cause structural re-organization of receptors and other molecules, perhaps predisposing to the spread of cancer cells.

An important property of many cancer cells is that they can release various **proteinases** into the ECM. Of the four major classes of proteinases (serine, cysteine, aspartate and metallo-), particular interest in cancer has focused on the metalloproteinases (MPs), which constitute a very large family of metal-dependent (usually zinc) enzymes. A number of studies have shown increased activity in tumors of **MPs** such as MP-2 and MP-9 (also known as gelatinases). These enzymes are capable of degrading proteins in the basement membrane and in the ECM, such as collagen and others, facilitating the spread of tumor cells. Inhibitors of these enzymes have been developed, but so far these have not had any clinical success.

A factor that allows increased movement of cancer cells is **epithelial mesenchymal transition.** This is a change of cell morphology and function from epithelial to mesenchymal type, perhaps induced by growth factors. The mesenchymal type has more actin filaments, permitting increased movement, an essential property of cells that metastasize.

The **ECM** plays an important role in metastasis. There is evidence of **communication** by signaling mechanisms between cancer cells and cells of the ECM. The **types of cells** in the ECM can also affect metastasis. As mentioned above, **proteinases** that degrade proteins in the ECM can facilitate spread of cancer cells. In addition, the ECM contains various **growth factors** that can influence tumor behavior.

On their travels, tumor cells are exposed to various cells of the **immune system** (such as T cells, NK cells, and macrophages), and must be able to survive exposure to them. Some of these surveillance cells secrete various **chemokines,** small proteins that can attract various cells such as leukocytes, sometimes causing an inflammatory response to tumor cells.

It has been estimated that only about 1:10,000 cells may have the **genetic capacity** to successfully colonize. Certain tumor cells show a predilection to **metastasize to specific organs** (eg, prostate cells to bone); specific cell surface molecules may be involved in this tropism.

Various studies have shown that **certain genes enhance metastasis,** whereas others act as **metastasis suppressor genes.** How exactly these genes work is the subject of ongoing studies. However, their existence raises the possibility that increasing the expression of a tumor suppressor gene by a drug or other method might be of therapeutic benefit.

Table 55–13 summarizes some important points regarding metastasis.

TABLE 55–13 **Some Important Points Regarding Metastasis**

• An epithelial-mesenchymal cell transition is often found in cancers, allowing increased movement of potentially metastatic cells.
• Metastasis is relatively inefficient (only about 1:10,000 tumor cells may have the genetic potential to colonize).
• Metastatic cells must evade various cells of the immune system to survive.
• Changes in cell surface molecules (eg CAMs and others) are involved.
• Increased proteinase activity (eg of MP-2 and MP-9) facilitates invasion.
• The existence of metastasis enhancer and suppressor genes has been shown.
• Some cancer cells metastasize preferentially to specific organs.
• Metastasis gene signatures may be detected by gene microarray analysis; they can be of prognostic value.

Abbreviations: CAM, cell adhesion molecule; MP, metalloproteinase.

CANCER CELLS EXHIBIT A HIGH RATE OF AEROBIC GLYCOLYSIS

Many aspects of the **metabolism** of cancer cells (eg, of carbohydrates, lipids, amino acids, and nucleic acids) have been and are being studied. One current approach is analysis of blood and urine by **mass spectrometry** looking for alterations in metabolite profile that can **help detect cancer at an early stage.** Here we shall describe only a few of the studies performed on the metabolism of glucose.

Glucose and the amino acid **glutamine** are two of the most abundant metabolites in plasma, and together they account for much of the carbon and nitrogen metabolism in human cells. In the 1920s, the German biochemist Otto Warburg and his colleagues made a seminal discovery regarding the biochemistry of cancer cells. They found that cancer cells take up large amounts of **glucose** and metabolize it to lactic acid, even in the presence of oxygen (the so-called **Warburg effect**). The reasons for this **high rate of aerobic glycolysis** are still under investigation. Warburg postulated that it could be due to a **defect in the respiratory chain,** so that tumor cells compensated for this by producing more ATP via glycolysis. Subsequent work has produced a number of other explanations, including that tumor cells have **less mitochondria** than normal cells and that they contain a mitochondrial-bound **isozyme of hexokinase (HK-2)** that is not subject to feedback control, allowing increased uptake of glucose. Recent work has focused on the finding that tumor cells exhibit **a different isozyme of pyruvate kinase (PK)** (see Chapter 18). Normal cells contain PK-1 and tumor cells contain PK-2; they are produced by alternative splicing of the same gene product. For complex reasons that are still to be fully elucidated, this change of isozyme profile is associated with, or leads to **less production of ATP** (see **Figure 55–15**). It also is thought to allow increased use of metabolites supplied by glycolysis for **building up the biomass** (proteins, lipids, nucleic acids, etc) required for proliferation of cancer cells. This offers an explanation for the presumed selective advantage conferred on tumor cells by having a high rate of glycolysis.

Despite angiogenesis, many solid tumors have localized areas of **poor blood supply,** and thus show high rates of **anaerobic glycolysis.** This leads to excessive production of lactic acid and local **acidosis.** It has been postulated that local production of acid may allow tumor cells to invade more easily. The **low oxygen tension** in areas of tumors with poor blood supply stimulates the formation of hypoxia-inducible factor-1 (HIF-1). This transcription factor, whose activity is turned on by low oxygen tension, up-regulates—among other actions—the activities of at least eight genes controlling synthesis of glycolytic enzymes.

The **pH** and **oxygen tension** in tumors are important factors affecting the actions of anti-cancer drugs and other treatments. For example, the anti-cancer efficacy of radiation treatment of cancers is significantly lower in hypoxic conditions. Chemicals have been developed to **inhibit glycolysis** in tumor cells, and perhaps selectively kill them (see **Table 55–14**). Although found to have variable effectiveness in preclinical studies, so far none of them have attained much

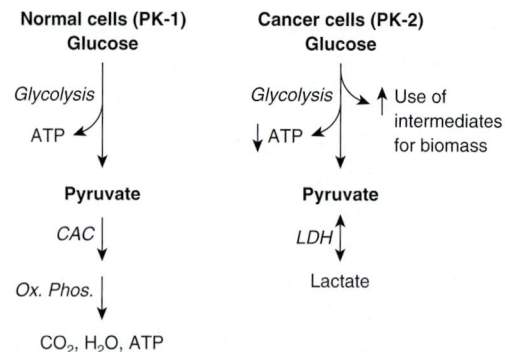

FIGURE 55–15 **Pyruvate kinase isozymes and glycolysis in normal and in cancer cells.** In normal cells, the major source of ATP is oxidative phosphorylation. Some ATP is obtained from glycolysis. The major pyruvate kinase (PK) isozyme in normal cells is PK-1. In cancer cells, aerobic glycolysis is prominent, lactic acid is produced via the action of lactate dehydrogenase (LDH) and production of ATP from oxidative phosphorylation is diminished (not shown in the Figure). In cancer cells, PK-2 is the major PK isozyme. For complex reasons not as yet fully understood, this change of isozyme profile in cancer cells is associated with decreased net production of ATP from glycolysis, but increased use of metabolites to build up biomass. (CAC, citric acid cycle; OX PHOS, oxidative phosphorylation.)

clinical use. They include **3-bromopyruvate** (an inhibitor of HK-2) and **2-deoxy-D-glucose** (an inhibitor of HK-1). Another compound, **dichloroacetate** (DCA), inhibits the activity of pyruvate dehydrogenase kinase, and thus stimulates the activity of pyruvate dehydrogenase (see Chapter 18), diverting substrate from glycolysis into the citric acid cycle. So far, none of these has attained much clinical use.

MITOCHONDRIA SHOW ALTERATIONS IN CANCER CELLS

Mitochondria are involved in various aspects of cancer. The possibility that a **diminished number** of mitochondria are involved in the Warburg effect in cancer cells was mentioned above. Mitochondria generate **reactive oxygen species** (ROS), which

TABLE 55–14 **Some Compounds that Inhibit Glycolysis and Have Been Found to Display Variable Anticancer Activity**

Compound	Enzyme Inhibited
3-Bromopyruvate	Hexokinase II
2-Deoxy-D-glucose	Hexokinase I
Dichloroacetate	Pyruvate dehydrogenase kinase
Iodoacetate	Glyceraldehyde-phosphate dehydrogenase

The rationale for development of these agents is that glycolysis is usually much more active in tumor cells, so that inhibiting it may damage them more than normal cells. Inhibition of PDH kinase results in stimulation of PDH, diverting pyruvate away from glycolysis.
Abbreviation: PDH, pyruvate dehydrogenase.

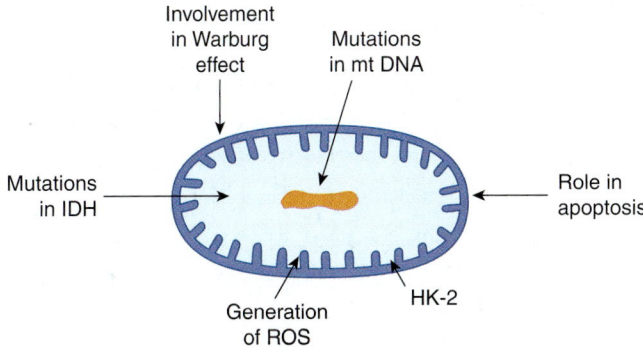

FIGURE 55–16 Some involvements of mitochondria in cancer. Mitochondria are involved in a number of ways in cancer, not all of which are shown here. Possible explanations for the Warburg effect are discussed in the text. Mutations in mtDNA are found in cancer cells, although their significance is not fully understood. Mitochondria play an important role in apoptosis, with release of cytochrome c into the cytoplasm being an important event. Many cancer cells have an isozyme of hexokinase (HK-2), that plays a role in the increased uptake of glucose by tumor cells. Reactive oxygen species (ROS) are generated in mitochondria, and may be important in causing mutations. Recent studies have shown that certain tumor cells have an abnormal isozyme of isocitrate dehydrogenase (IDH), the significance of which is under study.

TABLE 55–15 Some Useful Tumor Biomarkers Measurable in Blood

Tumor Biomarker	Associated Cancer
Alpha-fetoprotein (AFP)	Hepatocellular carcinoma, germ cell tumor
Calcitonin (CT)	Thyroid (medullary carcinoma)
Carcinoembryonic antigen (CEA)	Colon, lung, breast, pancreas, ovary
Human chorionic gonadotropin (hCG)	Trophoblastic disease, germ cell tumor
Monoclonal immunoglobulin	Myeloma
Prostate-specific antigen (PSA)	Prostate

Most of these tumor biomarkers are also elevated in the blood of patients with noncancerous diseases. For example, CEA is elevated in a variety of non-cancerous gastrointestinal disorders, and PSA is elevated in prostatitis and benign prostatic hyperplasia. This is why interpretation of elevated results of tumor markers must be made with caution and why their main uses are to follow effectiveness of treatments and also recurrences. A number of other quite widely used tumor biomarkers are also available.

can damage DNA, and they are also involved in **apoptosis** (see above). Mutations affecting **mitochondrial (mt) DNA** and thus various mt proteins have also been detected although their precise roles in the biology of cancer cells have not been established. The possible involvement of mitochondrial-bound **hexokinase-2** in the Warburg effect was mentioned above. Another mitochondrial enzyme that has attracted considerable attention recently is **isocitrate dehydrogenase** (IDH) (see Chapter 18). Mutations of this enzyme have been detected in gliomas and acute myeloid leukemia. These cause an elevation of 2-hydroxyglutarate, which may have effects on the tumor cells. **Figure 55–16** summarizes some of these points regarding mitochondria.

TUMOR BIOMARKERS CAN BE MEASURED IN BLOOD SAMPLES

Biochemical tests (see Chapter 56) are often helpful in the management of patients with cancer (eg, some patients with advanced cancers may have elevated levels of plasma calcium, which can cause serious problems if not attended to). Many cancers are associated with the abnormal production of enzymes, proteins, and hormones that can be measured in plasma or serum. These molecules are known as **tumor biomarkers**. Some of them are listed in **Table 55–15**.

However, significant elevations of some of the biomarkers listed in Table 55–15 also occur in a variety of **non-cancerous conditions**. For example, elevations of the level of **prostate-specific antigen** (PSA), a glycoprotein synthesized by prostate cells, occur not only in patients with cancer of the prostate, but also in patients with prostatitis and benign prostatic hyperpla-

sia (BPH). Similarly, elevations of **carcinoembryonic antigen** (CEA) are found not only in patients with various types of cancer, but also in heavy smokers and people with ulcerative colitis and cirrhosis. The fact that elevations of tumor biomarkers are usually not specific for cancer has meant that measurements of most of them are not used primarily for diagnosis of cancer. Their main uses have been in following the **effectiveness of treatments** and in **detecting early recurrence.** The use of CEA in the management of a patient with colorectal cancer is discussed briefly in Chapter 57, Case 4. As with other laboratory tests (Chapter 56), **the entire clinical picture** must be considered when interpreting the results of measurements of tumor biomarkers.

It is hoped that ongoing research on **proteomics** will provide new tumor biomarkers of **increased sensitivity and specificity,** and those capable of alerting one to the presence of **cancers at an early stage** of their development.

Use of **microarray analysis** of cancer cells has revealed certain useful biomarkers. It has also been of use in subclassifying tumors (eg, breast) and has provided useful information regarding prognosis and prediction of response to therapy. This is a rapidly expanding area of laboratory analysis.

KNOWLEDGE OF MECHANISMS INVOLVED IN CARCINOGENESIS HAS LED TO THE DEVELOPMENT OF NEW THERAPIES

One of the great hopes of cancer research is that revealing the fundamental mechanisms involved in cancer will **lead to new and better therapy.** This has already occurred to a certain extent, and it is hoped that ongoing developments will accelerate this process.

TABLE 55–16 Some Anticancer Agents That are Based on Recent Advances in Knowledge of Cancer Biology

Class	Example	Used To Treat
Inhibitors of signal transduction	Imatinib, an inhibitor of tyrosine kinase	CML
Monoclonal antibodies	Trastuzumab, a Mab to the HER2/Neu receptor	Late stage breast cancer
Anti-angiogenesis agents	Bevacizumab, a Mab to VEGF A	Colon and breast cancers
Anti-hormonal agents	Tamoxifen, antagonist of the estrogen receptor	Breast cancer
Affect differentiation	All-trans-retinoic acid, targets the retinoic leukemia acid receptor on promyelocytic leukemia cells causing them to differentiate	Promyelocytic leukemia
Affect epigenetic changes	5′ Azadeoxycytidine, inhibits DNA methyltransferases	Certain leukemias
	SAHA inhibits histone deacetylases	Cutaneous T-cell lymphoma

In some cases, the above agents may have been replaced by other more effective agents. Also, certain of the above are used to treat conditions other than these listed.

Abbreviation: CML, chronic myelocytic leukemia; Mab, monoclonal antibody; SAHA, suberoylanilide hydroxamic acid (Vorinostat); VEGF A, vascular endothelial growth factor A.

Classical chemotherapeutic drugs include alkylating agents, platinum complexes, antimetabolites, spindle poisons, and others. These will not be discussed here.

Among the classes of drugs developed more recently are **inhibitors of signal transduction** (including tyrosine kinase inhibitors), **monoclonal antibodies** directed to various target molecules, **inhibitors of hormone receptors, drugs that affect differentiation, anti-angiogenesis agents,** and **biologic response modifiers.** Examples of each of these are listed in **Table 55–16.** A few comments will now be made concerning some of these drugs.

The finding of widespread defects in signaling mechanisms in cancer cells, and in particular the detection of mutations in **tyrosine kinases,** has led to the development of inhibitors of these enzymes. The most dramatic success has probably been the introduction of imatinib for the treatment of chronic myelocytic leukemia (CML). Imatinib is an orally administered drug that inhibits the tyrosine kinase formed due to the *ABL-BCR* chromosomal translocation involved in the genesis of CML. This drug has produced complete remissions in many patients. It can be combined with other drugs. **Other tyrosine kinase inhibitors** have also been developed. Two of these are Erlotinib and Gefinib, which **inhibit the epidermal growth factor receptor (EGFR).** This molecule is over-expressed in certain lung (eg, non-small cell cancers) and breast cancers, resulting in aberrant signaling. It is important to appreciate that the design of these anti-cancer drugs and of other drugs requires **detailed structural knowledge** (as provided by X-ray crystallography, NMR studies, model building, etc) of the molecules being targeted. Another class of drugs that have proven useful are **monoclonal antibodies** to various molecules on the surfaces of neoplastic cells. A few of these are listed in Table 55–16.

Other approaches to treating cancer that are being used or being developed, but are not listed in Table 55–16, include various types of **gene therapy** (including siRNAs, Chapter 34), **immunotherapy** (see below), **oncolytic viruses** (viruses that preferentially invade tumor cells and kill them), **inhibitors**

of the **progesterone receptor, aromatase inhibitors** (see Chapter 41) (for some breast and ovarian cancers), **telomerase inhibitors,** applications of **nanotechnology** (eg, nanoshells and other nanoparticles), **phototherapy** (see Chapter 31), and drugs that will **selectively target cancer stem cells.**

It is important to appreciate that all anticancer drugs have **side effects,** sometimes severe, and that **resistance** to many of them can develop after variable time periods. The biochemistry of **how cancer cells develop resistance to drugs** is an important area of research that will not be described here. The overall thrust of drug development for cancer therapy is to use new information emerging from basic studies of cancer biology to develop safer and more effective agents. Understanding **genetic differences** in handling drugs (see Chapter 53) may also help to **personalize anti-cancer drug treatments.**

Figure 55–17 summarizes some of the **targets** for drug therapy and some emerging **therapies** that have developed from relatively recent studies of basic aspects of cancer.

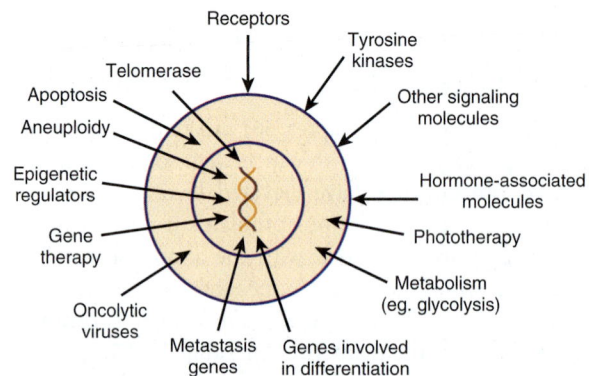

FIGURE 55–17 Some targets for anti-cancer drugs and some emerging therapies, both of which have developed from relatively recent research. Not shown in the Figure are anti-angiogenic agents, applications of nanotechnology, therapies directed against cancer stem cells and immunologic approaches. Most of the targets and therapies indicated are discussed briefly in the text.

THERE ARE MANY IMMUNOLOGIC ASPECTS OF CANCER

Tumor immunology is a large subject. Only a few comments will be made here concerning it. It seems probable that the decline in immune responsiveness that accompanies **aging** plays a role in the increased incidence of cancer in older people. One long-standing hope has been that immunologic approaches, because of their **specificity,** might be able to kill off cancer cells selectively. There are many ongoing trials investigating this possibility. These involve the use of various **antibodies, vaccines,** and the **use of various types of T cells** that generally have been manipulated in one way or another to increase their ability to kill neoplastic cells.

Chronic inflammation involves aspects of immune function. There is evidence that it can **predispose to cancer** (eg, the incidence of colorectal cancer is much higher than normal in individuals who have had long-standing ulcerative colitis). Some inflammatory cells produce relatively large amounts of **reactive oxygen species,** which can cause damage to DNA, and perhaps contribute to oncogenesis. It has also been reported that **low doses of aspirin** may lower the risk of development of colorectal cancer, perhaps via its anti-inflammatory action.

SUMMARY

- Cancer is due to mutations in genes controlling cell multiplication, cell death (apoptosis) and cell–cell interactions (eg, cell adhesion). Other important aspects of cancer are widespread defects in cell signaling pathways, stimulation of angiogenesis, and aneuploidy.

- The great majority of cancers are due to mutations affecting somatic cells. However, many hereditary cancer conditions have been identified.

- Major classes of genes involved in cancer are oncogenes and tumor suppressor genes. Mutations affecting genes directing the synthesis and expression of microRNAs are also important.

- Epigenetic changes are increasingly being recognized in cancer (and in other diseases); one reason for interest in them is that they may be reversible changes.

- The biology of metastasis is being explored intensively; the discovery of metastasis enhancer and suppressor genes, among other findings, may lead to new therapies.

- The process of apoptosis was described briefly. Cancer cells acquire mutations that permit them to evade apoptosis, thus prolonging their existence.

- Whole genome sequencing of cancers is helping to reveal the important mutations present in many types of cancer and is yielding new information on the evolution of cancer cells.

- Cancer cells show various alterations of metabolism. One major finding that has attracted much attention is the high rate of aerobic glycolysis exhibited by many cells. Possible explanations for this phenomenon are described. Mitochondria show some alterations in many cancer cells.

- Overall, the development of cancer is a multi-step process involving genetic changes that confer selective advantages on clones of cells, some of which eventually acquire the ability to metastasize successfully. Because of the diversity of mutations, it is possible that no two tumors have identical genomes.

- Tumor markers may help in the early diagnosis of cancer. They are of particular use in following the response of cancer to treatment and in detecting recurrences.

- Advances in understanding the molecular biology of cancer cells have led to the introduction of a number of new therapies, with others in the pipeline.

REFERENCES

Berger M, Lawrence, MS, Demechelis F, et al: The genomic complexity of primary human prostate cancer. Nature 2011; 470(7333):214.

Bielas JH, Loeb KR, Rubin BP, et al: Human cancers express a mutator phenotype. Proc Natl Acad Sci USA 2006;103(48):18238.

Couzin-Frankel J: Immune therapy steps up the attack. Science 2010;330:440.

Croce CM: Oncogenes and cancer. N Engl J Med 2008;358(5):502. (One of a series of 12 articles on the molecular origins of cancer published in this journal between Jan 31, 2008 and Dec 17, 2009).

Dick JE: Stem cell concepts renew cancer research. Blood 2008;112:4793.

Fauci AS, Braunwald E, Kasper DL, et al (editors): *Harrison's Principles of Internal Medicine.* 17th ed. McGraw-Hill, 2008. (A number of Chapters cover various aspects of cancer, but Chapters 77–81 and 103–106 are particularly relevant to this Chapter).

Green DR: *Means to an End: Apoptosis and Other Cell Death Mechanisms.* Cold Spring Harbor Press, 2010.

Kaiser J: Epigenetic drugs take on cancer. Science 2010;330:576. (Also see other articles on Epigenetics in the same issue of Science),

Lodish H, Berk A, Kaiser CA, et al: *Molecular Cell Biology.* 6th ed. WH Freeman & Co, 2008. (Contains a comprehensive chapter on cancer).

Mukerjee S: *The Emperor of All Maladies: A Biography of Cancer.* Scribner, 2010.

Nussbaum RL, McInnes, RR, Willard HF: *Thompson and Thompson Genetics in Medicine.* 7th ed. Saunders Elsevier, 2007.

Tannock IF, Hill RP, Bristow RG, Harrington L (editors): *The Basic Science of Oncology.* 4th ed. McGraw-Hill, 2005.

Thompson SL, Bakhoum SF, Compton DA: Mechanisms of chromosomal instability. Curr Biol 2010; 20(6): R285.

Weinberg R: *The Biology of Cancer,* Garland Science. 2006.

Yachida S, Jones S, Bozic I, et al: Distant metastasis occurs late during the genetic evolution of pancreatic cancer. Nature 2010 (Oct 28);467(7319):1114.

Journals dedicated to cancer research include the British Journal of Cancer, Cancer Research, Journal of the National Cancer Institute and Nature Reviews Cancer.

Useful Web Sites:

American Cancer Society. http://www.cancer.org

National Cancer Institute, U.S. National Institute of Health. http://www.cancer.gov

The Cancer Genome Atlas. http://cancergenome.nih.gov

GLOSSARY

Adenomatous polyp: A benign tumor of epithelial origin which has the potential to become a carcinoma. Adenomas are often polypoid. A polyp is a growth that protrudes from a mucous membrane; most are benign, but some polyps can become malignant.

Ames assay: An assay system devised by Dr. Bruce Ames that uses specially designed *Salmonella typhimurium* to detect mutagens. Most carcinogens are mutagens, but if mutagenicity of a chemical is detected, ideally it should be tested for carcinogenicity by animal testing.

Aneuploidy: Refers to any condition in which the chromosome number of a cell is not an exact multiple of the basic haploid number. Aneuploidy is found in many tumor cells and may play a fundamental role in the development of cancer.

Angiogenesis: The formation of new blood vessels. Angiogenesis is often active around tumor cells, ensuring that they obtain an adequate blood supply. A number of growth factors are secreted by tumor and surrounding cells (eg, vascular endothelial growth factor, VEGF) and are involved in this process.

Apoptosis: Cell death due to activation of a genetic program that causes fragmentation of cellular DNA and other changes. Caspases play a central role in the process. Many positive and negative regulators affect it. The protein p53 induces apoptosis as a response to cell DNA damage. Most cancer cells exhibit abnormalities of apoptosis, due to various mutations that help to ensure their prolonged survival.

Biologic response modifiers: Molecules produced by the body or in the laboratory that when administered to patients alter the body's response to infection, inflammation and other processes. Examples include monoclonal antibodies, cytokines, interleukins, interferons, and growth factors.

Bloom syndrome: One of the CI syndromes. Due to mutations in a DNA helicase. Subjects are sensitive to DNA damage and may develop various tumors.

Burkitt's lymphoma: This is a B cell lymphoma, endemic in parts of Africa, where it mainly affects the jaw and facial bones. It is also found elsewhere. A reciprocal translocation involving the *C-MYC* gene on chromosome 8 and the immunoglobulin heavy chain gene on chromosome 14 is characteristic.

Cancer: A malignant growth of cells.

Cancer stem cell: A cell within a tumor that has the capacity to self-renew and to give rise to the heterogeneous lineages of cancer cells found in the tumor.

Carcinogen: Any agent (eg, a chemical or radiation) capable of causing cells to become cancerous.

Carcinoma: A malignant growth of epithelial origin. A cancer of glandular origin or showing glandular features is usually designated as an adenocarcinoma.

Caspases: Proteolytic enzymes that play a central role in apoptosis, but are also involved in other processes. Some 15 are present in humans. Caspases hydrolyse peptide bonds just C-terminal to aspartate residues.

Cell cycle: The various events pertaining to cell division that occur as a cell goes from one mitosis to another.

Centriole: An array of microtubules that is paired and found in the center of a centrosome. (Also see **Centrosome.**)

Centromere: The constricted region of a mitotic chromosome where chromatids are joined together. It is in close proximity to the kinetochore. Abnormalities of centromeres may contribute to CI. (Also see **Kinetochore.**)

Centrosome: A centrally located organelle that is the primary microtubule-organizing center of a cell. It acts as the spindle pole during cell division.

Chromatid: A single chromosome.

Chromatin remodeling: This involves conformational changes of nucleosomes brought about by the actions of multiprotein complexes (such as the SW1/SNF complex). These changes alter gene transcription (turning it on or off, depending on specific conditions). The complexes contain domains homologous to ATP-dependent helicase; these are involved in the changes of conformation. Mutations affecting proteins of the complexes, such as may be found in cancer cells, can affect gene expression. (See also **Epigenetics.**)

Chromosomal instability (CI): The rate of gain or loss of whole chromosomes or segments of them caused by abnormalities of chromosome segregation during mitosis. (See also **Genome instability** and **Microsatellite instability.**) There are a number of disorders that are named CI syndromes because they are associated with chromosomal abnormalities. One such is Bloom syndrome, in which a high frequency of sister chromatid exchanges is observed. An increased incidence of various cancers is found in these conditions.

Chromosomal passenger complex: A complex of proteins that plays a key role in regulating mitosis. At the centromere, it directs alignment of the chromosome and participates in spindle assembly. Its proteins include aurora B kinase and survivin. Mutations affecting its proteins may contribute to CI and aneuploidy.

Chromosomal translocation: When part of one chromosome becomes fused to another, often causing activation of a gene at the site. The Philadelphia chromosome (see below) is one of many examples of a chromosomal translocation involved in the causation of cancer.

Clone: All the cells of a clone are derived from one parent cell.

Copy number variations (CNVs): Variations (because of duplications or deletions) among individuals as to the number of copies they have of particular genes. CNVs are being increasingly recognized for various genes, and some may be associated with various diseases, including certain types of cancer.

Driver mutation: A mutation in a gene that either helps cause cancer or accelerates it. Mutations found in tumors that do not cause cancer or its progression are called passenger mutations.

Epigenetic: Refers to changes of gene expression without change of the sequence of bases in DNA. Factors causing epigenetic changes include methylation of bases in DNA, posttranslational modifications of histones and chromatin remodeling.

FAS receptor: A receptor that initiates apoptosis when it binds its ligand, FAS. FAS is a protein present on the surfaces of activated natural killer cells, cytotoxic T lymphocytes and other sources.

Genome instability: This refers to a number of alterations of the genome, of which the two principal ones are CI and microsatellite instability. In general, it reflects the fact that the genomes of cancer cells are more susceptible to mutations than are normal cells, in part due to impairment of DNA repair systems.

Growth factors: A variety of polypeptides secreted by many normal and tumor cells. These molecules act via autocrine (affects the cells that produce the growth factor), paracrine (affects neighboring cells) or endocrine (travels in the blood to affect distant cells) modes. They stimulate proliferation of target cells via interactions with specific receptors. They also have many other biologic properties.

Hypoxia-inducible factors: A family of transcription factors (at least three) important in directing cellular responses to varying levels of oxygen. Each factor is made up of a different oxygen-regulated α-subunit and a common constitutive β-subunit. At physiological levels of oxygen, the α-subunit undergoes rapid degradation, initiated by prolyl hydroxylases. HIFs have various functions; eg HIF-1 up-regulates various genes encoding enzymes of glycolysis, and also the expression of vascular endothelial growth factor (VEGF).

Kinetochore: A structure that forms on each mitotic chromosome adjacent to the centromere. Mutations affecting the structures of its component proteins could contribute to causing CI. (See also **Centromere.**)

Leukemias: A variety of malignant diseases in which various white cells (eg, myeloblasts, lymphoblasts, etc) proliferate in an unrestrained manner. Leukemias may be acute or chronic.

Loss of heterozygosity (LOH): This occurs when there is loss of the normal allele (often encoding a tumor suppressor gene) from a pair of heterozygous chromosomes, allowing the results of the defective allele to be manifest clinically.

Lymphoma: A group of neoplasms arising in the reticuloendothelial and lymphatic systems. Major members of the group are Hodgkin and non-Hodgkin lymphomas.

Malignant cells: They are cancer cells—cells with the ability to grow in an unrestrained manner, to invade, and to spread (metastasize) to other parts of the body.

Metastasis: The ability of cancer cells to spread to distant parts of the body and grow there.

Microarray analysis: Analysis using a wafer, often made of silicon, on which thousands of different nucleic acids have been spotted. Samples of cDNA or genomic DNA are applied and allowed to hybridize (binding between complementary sequences). The interactions can be quantitated to assess gene expression or other parameters. Various commercial microarrays ("gene chips") are available. This technique is being increasingly applied to patients with cancers to help in tumor classification, and in predictions of prognosis and response to drug therapy.

Microsatellite instability: Expansion or contraction of short tandem repeats (microsatellites) due to replication slippage, abnormalities of mismatch repair or of homologous recombination. For **Microsatellites,** see Chapter 35.

Nanotechnology: The development and application of devices which are only a few nanometers in size. (10^{-9} m = 1 nm). Some are being applied to cancer therapy. For example, **nanoshells** (very small spherical particles with a silica core and a gold covering) tuned to near-infrared light have been administered to mice with tumors, in which the nanoshells accumulate. The tumors were subsequently subjected to near infrared laser light. This heated the tumors selectively, killed them, and there was no sign of recurrence on follow-up. (Morton JG et al: Nanoshells for photothermal cancer therapy. Methods Mol Biol 2010;624:101. June 25, 2010).

Necrosis: Cell death induced by chemicals or tissue injury. Various hydrolytic enzymes are released and digest cellular molecules. It is not a genetically programmed process, as is apoptosis. Affected cells usually burst and release their contents, causing local inflammation.

Neoplasm: Any new growth of tissue, benign or malignant

Oncogene: A mutated proto-oncogene whose protein product is involved in the transformation of a normal cell to a cancer cell.

Oncology: The area of medical science that concerns itself with all aspects of cancer (causes, diagnosis, treatment, etc).

Philadelphia chromosome: A chromosome formed by a reciprocal translocation between chromosomes 9 and 22. It is the cause of chronic myeloid leukemia (CML). To form the abnormal chromosome, part of the *BCR* (breakpoint cluster region) gene of chromosome 22 fuses with part of the *ABL* gene (encodes a tyrosine kinase) of chromosome 9, directing the synthesis of a chimeric protein that has unregulated tyrosine kinase activity and drives cell proliferation. The activity of this kinase is inhibited by the drug Imatinib (Gleevec), which has been successfully used to treat CML. (See also **Chromosomal translocation.**)

Procarcinogen: A chemical that becomes a carcinogen when altered by metabolism.

Proto-oncogene: A normal cellular gene, which when mutated can give rise to a product that stimulates the growth of cells, contributing to the development of cancer.

Replication slippage: A process in which, because of misalignment of DNA strands where repeat sequences occur, DNA polymerase pauses and dissociates, resulting in deletions or insertions of repeat sequences.

Retinoblastoma: A rare tumor of the retina. Mutation of the *RB* tumor suppressor gene plays a key role in its development. Patients with hereditary retinoblastomas have inherited one mutated copy of the RB gene, and need only one further mutation to develop the tumor. Patients with sporadic retinoblastomas are born with two normal copies, and require two mutations to inactivate the gene.

Rous sarcoma virus (RSV): An RNA tumor virus that causes sarcomas in chickens. It was discovered in 1911 by Peyton Rous. It is a retrovirus, using reverse transcriptase in its replication; the DNA copy of its genome subsequently integrates into the host cell genome. It has been widely used in studies of cancer, and its use has led to many important findings.

Sarcoma: A malignant tumor of mesenchymal origin (eg, from cells of the extracellular matrix or other sources).

Telomeres: Structures at the ends of chromosome that contain multiple repeats of specific hexanucleotide DNA sequences. The telomeres of normal cells shorten on repeated cell division, which may result in cell death. The enzyme telomerase replicates telomeres and is often expressed in cancer cells, helping them to evade cell death. Telomerase is usually not detected in normal somatic cells.

Translocation: The displacement of one part of a chromosome to a different chromosome or to a different part of the same chromosome. Classic examples are the translocation found in Burkitt's lymphoma (see above) and the translocation between chromosomes 9 and 22, which causes the appearance of the Philadelphia chromosome found in chronic myelogenous leukemia. Translocations have been found in many cancer cells.

Transformation: The process by which normal cells in tissue culture become changed to abnormal cells (eg, by oncogenic viruses or chemicals), some of which may be malignant.

Tumor: Any new growth of tissue, but usually refers to a benign or malignant neoplasm.

Tumor suppressor gene: A gene whose protein product normally restrains cell growth, but when its activity is lost or reduced by mutation contributes to the development of a cancer cell.

Clinical Biochemistry

Joe Varghese, MB BS, MD, Molly Jacob, MB BS, MD, PhD,
& Robert K. Murray, MD, PhD

OBJECTIVES

After studying this chapter, you should be able to:

- Briefly describe the importance of biochemical tests in clinical medicine.
- Discuss major considerations to be kept in mind when ordering and interpreting results of laboratory tests.
- Name and have a general understanding of the principal tests used to assess liver, kidney, thyroid, and adrenal function.

IMPORTANCE OF LABORATORY TESTS IN CLINICAL MEDICINE

Laboratory (lab) tests play an important role in helping physicians and other healthcare workers make diagnoses and other clinical judgments. In this chapter, only biochemical tests (and not hematological, microbiological, immunological, or other types of tests) are discussed, with a few exceptions. In addition, only a brief overview of this subject is presented; medical students will receive much further coverage as they progress through the years of their medical education.

Lab tests constitute only one part of the diagnostic process in clinical medicine. In fact, it has been stated that an experienced physician may arrive at a relatively accurate diagnosis in ~80% of cases, based solely on a thorough history and physical examination; some may, however, doubt the validity of this statement. However, there is no doubt that, nowadays, biochemical and other lab tests are almost always an important part of the overall diagnostic process. The use of biochemical investigations and lab tests is, however, not confined to diagnoses of diseases. **Table 56–1** summarizes some of the different uses of biochemical tests, with examples for each.

Two important questions that one should answer prior to ordering any lab investigation are: "What useful information will I get by ordering this test?" and "Will it help the patient?" The concepts discussed in this chapter will contribute to answering these questions and help one make wise use of lab tests in the management of patients.

CAUSES OF ABNORMALITIES IN LEVELS OF ANALYTES MEASURED IN THE LABORATORY

A myriad of conditions and disorders can lead to abnormalities of levels of various molecules (analytes) measured in clinical labs. Some of these are listed in **Table 56–2**.

It is clear from this Table that conditions that cause abnormal levels of analytes are diverse. For instance, when tissue injury occurs, it results in damage to cell membranes and an increase in the permeability of the plasma membrane of affected cells. This leads to **leakage of intracellular molecules into the blood** (eg, leakage of troponins into blood following a myocardial infarction), causing their blood levels to increase. In other cases, the **synthesis of certain molecules is increased or decreased** (eg, C-reactive protein [CRP] in inflammatory states or specific hormones in certain endocrine disorders). Kidney and liver failure lead to the accumulation of a number of molecules (eg, creatinine and ammonia respectively) in the blood, due to **an inability of the organ concerned to excrete or metabolize the analyte** concerned.

VALIDITY OF LABORATORY RESULTS

Good diagnostic labs are subject to inspection and regulatory procedures on a regular basis. These assess the validity of their results and ensure **quality control** of their reports.

TABLE 56–1 Major Uses of Biochemical Tests with Selected Examples for Each

Uses	Selected Examples
Confirm diagnosis of specific diseases	Use of plasma cardiac troponin I (cTI) levels in early diagnosis of myocardial infarction.
Suggest rational treatment of disease.	An elevated low-density lipoprotein (LDL) cholesterol level is an indication for therapy with cholesterol-lowering drugs (such as statins) in persons at risk for cardiovascular diseases.
Are used as screening tests for the early diagnosis of certain diseases.	Measurement of thyroid stimulating hormone (TSH) levels in neonates helps in diagnosis of congenital hypothyroidism.
Assist in monitoring the progress of certain diseases	Serum alanine transaminase (ALT) activity is used to monitor the progress of viral hepatitis
Help in assessment of the response of diseases to therapy	In patients with hypo- or hyperthyroidism, measurement of TSH levels helps in monitoring response of patients to treatment.
Reveal the fundamental causes and mechanisms of disease	Demonstrate the nature of the genetic defect in cystic fibrosis.

TABLE 56–2 Common Causes for Abnormalities in Analytes Measured in Clinical Labs

Condition	Selected Examples
Certain physiological conditions	High hCG levels in pregnancy, high blood lactate levels following strenuous exercise.
Changes in body fluid volume	Hypernatremia (high serum sodium levels) may accompany dehydration due to excessive sweating or vomiting.
Changes in pH balance	Serum bicarbonate levels are low in metabolic acidosis (eg, diabetic ketoacidosis) and high in metabolic alkalosis (eg, severe vomiting)
Changes in endocrine function	Serum TSH is low in primary hyperthyroidism and high in primary hypothyroidism
Changes in nutritional status	Serum albumin is low in protein-energy malnutrition (PEM)
Cell injury or death	Serum creatine kinase–MB (CK-MB) and cardiac-specific troponins are elevated following cardiac injury. Serum pancreatic amylase is elevated in acute pancreatitis.
Acute or chronic inflammatory processes (including infections)	C-reactive protein (CRP) is elevated in inflammatory conditions.
Genetic diseases	Plasma phenylalanine levels are elevated in phenylketonuria
Organ failure	Serum creatinine and urea levels are elevated in renal failure
Trauma	Serum myoglobin levels may be elevated following muscle injury/trauma.
Cancer	Various tumor markers are elevated in specific cancers, eg, alpha fetoprotein in hepatocellular carcinoma.
Drugs	Cancer chemotherapeutics increase serum uric acid levels.
Poisons	Organophosphorus poisons decrease butyrylcholinesterase activity in blood.
Others	Stress can increase serum cortisol and catecholamine levels

Such measures will ensure that the value of the concentration, activity, or amount of a substance in a specimen reported from a clinical lab represents the best value obtainable with the method, reagents and instruments used and technical personnel involved in obtaining and processing the specimen. In addition, it is important for medical personnel to possess basic knowledge about the validity of lab results and their interpretation. It is also good practice to visit labs and to discuss, with appropriate lab personnel, questions, and problems that may arise with regard to values of lab results.

Accuracy is the degree of agreement of an estimated value of an analyte with the "true" value of the analyte in the sample. **Precision** denotes the reproducibility of an analysis and is the ability of the method used to consistently produce the same value when an analyte in a sample is repeatedly measured. It is expressed as the variation seen when these repeated measurements of the analyte are done. Precision is not absolute, but is subject to variation inherent in the complexity of the method used, the stability of reagents, the accuracy of the primary standard used, the sophistication of the equipment used for the assay and the skill of the technical personnel involved. Each lab should maintain data on precision (reproducibility) that can be expressed statistically in terms of the standard deviation (SD) from the mean value obtained by repeated analyses of the same sample.

For example, the precision in determination of cholesterol in serum in a good lab may be the mean value of repeated estimations with a SD of 5 mg/dL. The 95% confidence limits for this assay are ± 2 SD or ±10 mg/dL. This means that any value reported is "accurate" within a range of 20 mg/dL. Therefore, a reported value of 200 mg/dL signifies that, in 95% of cases, the true value lies between 190 and 210 mg/dL. Similarly, for the determination of serum potassium levels in a specimen, a SD of ± 0.1 mmol/L indicates that the 95% confidence limit of ± 2SD for this assay is ± 0.2 mmol/L. Thus, a potassium value of 5.5 mmol/L indicates that, in 95% of cases, the true value lies in the range 5.3–5.7 mmol/L. Values of 5.3 and 5.7 mmol/L may be obtained on repeated analysis of the sample and will still be within the limits of precision of the test.

ASSESSMENT OF VALIDITY OF A LAB TEST

The clinical value of a lab test depends on its **specificity, sensitivity,** and the **prevalence** of the disease in the population tested.

Sensitivity is the **percentage of positive results in patients with the disease.** Ideally, a test should have 100% sensitivity. However, this is seldom attained. For example, the carcinoembryonic antigen (CEA) test has a lower than ideal sensitivity; only 72% of those with carcinoma of the colon test positive when the disease is extensive, and only 20% test positive when they have early disease. Lab tests often have lower sensitivity in the early stages of many diseases, in contrast to their higher sensitivity in well-established disease. In biochemical analysis, sensitivity refers to the ability of the method to detect small changes in the levels of the analyte. The lowest concentration of the analyte that can reliably be detected is called the **limit of detection.** Usually, a highly sensitive test will have a very low limit of detection.

Specificity refers to the **percentage of negative results among people who do not have the disease.** Ideally, tests should be 100% specific. The CEA test for carcinoma of the colon has a variable specificity; about 3% of non-smoking individuals give a false-positive result (97% specificity), whereas 20% of smokers give a false-positive result (80% specificity). In biochemical analysis, specificity may also indicate if substances or factors other than the one being measured influence the assay in any way (positive or negative **interference**).

The **predictive value of a positive test** (positive predictive value) defines the percentage of positive results that are true positives. Similarly, the **predictive value of a negative test** (negative predictive value) defines the percentage of negative results that are true negatives. This is related to the **prevalence** of the disease. For example, in a group of patients in a urology ward, the prevalence of renal disease is higher than in the general population. In this group, the level of serum creatinine will have a higher predictive value than in the general popula-tion. Formulae for calculating sensitivity, specificity, and predictive values of a diagnostic test are shown in **Table 56–3.**

An **ideal** diagnostic test is one that has 100% sensitivity and 100% specificity. However, this is not true for most, if not all, tests available nowadays. Before ordering a test, it is important to attempt to determine whether the sensitivity, specificity, and predictive value of the test are adequate to provide useful information. The result obtained should **influence diagnosis, prognosis, or therapy** or lead to a **better understanding of the disease process** and **benefit the patient.**

VARIABLES THAT AFFECT VALUES OF ANALYSIS

Apart from age and sex, many **other factors (called pre-analytical variables)** may affect values of analytes and influence their normal ranges. These include **race, environment, posture (supine vs. sitting), diurnal** and other **cyclic variations, pregnancy, fasting** or **postprandial state, foods eaten, drugs,** and **level of exercise.** For example, it is important to take a blood sample for analysis of triglycerides after fasting for 12 h. It is also always important to ask a patient if he/she is on any medications.

Values of analytes may also be influenced by the **method of collection** of the specimen. Inaccurate collection of urine over a 24-h period, hemolysis of a blood sample, addition of an inappropriate anticoagulant, and contaminated glassware or other apparatus are other examples of pre-analytical errors that may occur.

Errors may also be associated with the analysis of samples. **Random errors** are those errors that are not easily identified and are commonly associated with manual assays. Automation of analysis can significantly lower random errors. On the other hand, **systematic errors** are errors inherently associated with the method of analysis and result in inaccurate results. These can often be identified and corrected if quality control procedures are adequately followed.

TABLE 56–3 Two-By-Two Table Illustrating Concepts of Sensitivity, Specificity and Predictive Values

		Does the Patient have the Disease?	
		Yes	**No**
What is the result of the test?	**Positive**	True positive (a)	False positive (b)
	Negative	False negative (c)	True negative (d)

Sensitivity	=	$\dfrac{\text{True positive (a)} \times 100}{\text{Number of patients who have the disease (a + c)}}$
Specificity	=	$\dfrac{\text{True negative (d)} \times 100}{\text{Number of patients who do not have the disease (b + d)}}$
Positive predictive value	=	$\dfrac{\text{True positive (a)} \times 100}{\text{Number of patients who have a positive test (a + b)}}$
Negative predictive value	=	$\dfrac{\text{True negative (d)} \times 100}{\text{Number of patients who have a negative test (c + d)}}$

INTERPRETATION OF LAB TESTS

Normal values are generally considered to be those that fall within 2 standard deviations (SD) (± 2 SD) of the mean value for a healthy population. This span of values constitutes a **reference range,** which is constructed from results for an analyte obtained from a particular population (say, healthy male adults, 20–50 years of age). Other reference ranges are constructed and used for the same analyte for other healthy populations, such as adult females, neonates, infants, adolescents, and elderly subjects. These ranges usually encompass 95% of the selected population. **Normal** or **reference ranges** vary with the method employed, the analytical instrument used, and conditions of collection and preservation of specimens. The normal ranges established by individual labs should be clearly expressed to ensure proper interpretation of results of lab tests.

Interpretation of lab results must always be related to **the condition of the patient.** A low value may be the result of a deficit, or of dilution of the substance measured (eg, low serum sodium). Deviation from normal may be associated with a specific disease, or with some drug consumed by the subject. For example, elevated levels of uric acid may occur in patients with gout or be due to treatment with certain diuretics or with anticancer drugs.

The **role of a clinician** in assessing the probability of disease in the individuals tested cannot be overemphasized. There needs to be a reasonable certainty about the presence or absence of a disease before a surrogate marker for the disease is checked for. This will ensure most optimal interpretation of test results. Whenever an **unusual or unexpected result** is obtained, one may wish to consult a clinical biochemist before initiating any treatment based on the result, to ensure that no pre-analytical error has occurred. If no such error is detected, the test should be repeated to rule out an analytical mistake.

THE REALITY OF LABORATORY TESTS—A PERSPECTIVE ON THEIR USE

One must always keep in mind that **laboratory tests serve as surrogate markers for tissue pathology.** They do not provide definitive evidence of such pathology and hence should not be used as the sole means by which a diagnosis is made on a patient. Information obtained from laboratory tests needs to be combined with a clinical history and data from other investigations to arrive at a diagnostic decision. The results of many diagnostic tests used in clinical practice are often categorized as positive or negative. This makes for mathematical and clinical convenience; however, such dichotomous categorizations may not represent clinical reality, as states of disease often lie on a spectrum, and such rigid demarcations can lead to errors in diagnosis.

AUTOMATION OF LABORATORY TESTS

Most modern clinical laboratories use a high degree of **automation.** Automated analyzers improve efficiency and reduce random errors that are invariably associated with manual methods. The pre-analytical phase of laboratory testing (eg, sample processing and transport) may also be automated, thus reducing the lag time between collection of the sample and analysis.

ORGAN FUNCTION TESTS

Tests that provide information on the functioning of particular organs are often grouped together as organ function tests and are sometimes ordered together by a clinician. Commonly done organ function tests are briefly discussed below.

Liver Function Tests (LFTs)

Liver function tests (LFTs) are a group of tests that help in diagnosis, monitoring therapy, and assessing prognosis of liver disease. Major tests in this category are listed in **Table 56–4.** Each test assesses a specific aspect of liver function. Increases in **serum bilirubin** levels occur due to many causes and result in jaundice. Low levels of **total serum protein** and **albumin** are seen in chronic hepatic disorders, such as cirrhosis. **Prothrombin time** (PT) may be prolonged in acute disorders of the liver due to impaired synthesis of coagulation factors.

Activities of **serum alanine aminotransferase** (ALT) and **aspartate aminotransferase** (AST) are significantly elevated several days before the onset of jaundice in acute viral hepatitis. ALT is considered to be more specific for the liver than AST, as the latter may be elevated in cases of cardiac or skeletal muscle injury while the former is not. **Serum alkaline phosphatase** (ALP) activity is elevated in obstructive jaundice. A high activity of serum ALP may also be seen in bone diseases.

The liver is also the primary site of detoxification of ammonia (in the urea cycle). Elevation of **blood ammonia** levels is an important sign of liver failure and plays an important role in the pathogenesis of hepatic encephalopathy in patients with liver cirrhosis and portal hypertension. Blood ammonia levels are also elevated in disorders of the urea cycle.

The **albumin:globulin ratio** (A:G ratio) often provides useful clinical information. The normal ratio varies from 1.2:1 to 1.6:1. A reversal of the A:G ratio may be seen in conditions where the albumin levels are low (hypoalbuminemia) or where globulins are abnormally high, eg, multiple myeloma. Reversal of the A:G ratio is often the first investigation that raises suspicion of multiple myeloma.

Renal Function Tests

The major renal function tests are listed in **Table 56–5.** A complete **urine analysis (urinalysis)** provides valuable information on renal function. It includes assessment of the physical and chemical characteristics of urine. The physical characteristics

TABLE 56–4 Important Liver Function Tests

Test	Aspect of Liver Function Assessed	Major Utility
1. Serum bilirubin levels (total and conjugated)	Indicator of the ability of the liver to conjugate and excrete bilirubin (conjugation and excretory function)	Aids in the differential diagnosis of jaundice (see Table 31–3)
2. Total serum protein and albumin	Measure of the biosynthetic function of the liver, as the liver is the primary site of synthesis of most plasma proteins.	Indicator of severity of chronic liver disease
3. Prothrombin time	Measure of the biosynthetic function of the liver, as several coagulation factors are synthesized in the liver.	Indicator of severity of acute liver disease
4. Serum enzymes:		
a. Aspartate transaminase (AST)	Serves as marker of injury to hepatocytes that contain AST in abundance.	Activities of serum AST and ALT are early indicators of liver damage. They also help in monitoring response to treatment.
b. Alanine transaminase (ALT)	Serves as marker of injury to hepatocytes that contain ALT in abundance.	
c. Alkaline phosphatase (ALP)	Serves as marker of biliary obstruction.	Aids in diagnosis of obstruction of the biliary tract.
5. Blood ammonia	Indicator of the ability of the liver to detoxify ammonia.	Levels are elevated in cirrhosis of liver with portal hypertension and in disorders of the urea cycle.

to be assessed include **urine volume** (this requires a timed urine sample, usually 24 h), **odor, color, appearance** (clear or turbid), **specific gravity,** and **pH. Protein, glucose, blood, ketone bodies, bile salts,** and **bile pigments** are abnormal constituents of urine, that appear in different disease conditions (**Table 56–6**). Most of these parameters can now be estimated semiquantitatively at the bedside using **disposable "dipstick" strips.** Dipsticks are plastic strips on which specific chemicals are impregnated. When the portion of the strip that contains the chemicals is dipped into a sample of urine, they react with specific constituents of urine to produce a color change that

TABLE 56–5 Major Renal Function Tests

1. Urine analysis
   a. Physical characteristics—assessment of volume, color, odor, appearance, specific gravity, and pH.
   b. Chemical characteristics—checking for the presence of protein, reducing sugar, ketone bodies, blood, bile salts, and bile pigments.
   c. Microscopy—checking for the presence of WBCs, RBCs, and casts.
2. Serum markers of renal function
   a. Serum creatinine
   b. Serum urea (or blood urea nitrogen [BUN])
3. Estimation of glomerular filtration rate (GFR)
   a. Creatinine clearance
   b. Inulin clearance
4. Tests of renal tubular function
   a. Water deprivation test
   b. Urine acidification test

is proportional to the concentration of that substance in the sample of urine.

Serum urea and creatinine are markers of renal function (see Table 56–5). Both these substances are primarily excreted in the urine. Deterioration of renal function is, therefore, associated with increases in the serum levels of these substances. **Creatinine** is considered a better marker of renal function than **urea** because its blood level is not significantly affected by non-renal factors, thus making it a specific indicator of renal function. A number of "pre-renal" factors (dietary protein intake, renal perfusion, etc) and "post-renal" factors significantly increase blood urea levels.

Normally, the **total amount of protein** excreted in urine over 24 h is less than 150 mg (and less than 30 mg of albumin) and is not detectable by routine tests. The presence of protein in urine in excess of this is referred to as **proteinuria.** Proteinuria is an important sign of renal disease. The most common cause of proteinuria is loss of integrity of the glomerular basement membrane (glomerular proteinuria), as seen in nephrotic syndrome and diabetic nephropathy. As listed in Table 56–6, other causes of proteinuria also occur (overflow, tubular and postrenal). The major protein found in glomerular proteinuria is **albumin,** which is the hallmark of this condition. **Microalbuminuria** is defined as the presence of 30–300 mg of albumin in a 24-h urine collection. It is considered to be an early and independent predictor of renal damage and cardiovascular mortality in diabetes mellitus.

Even though **serum creatinine** is considered a specific marker of renal function, a significant increase in its blood level is seen only after a ~50% decline in the glomerular filtration rate (GFR) has occurred. It is therefore a test of poor sensitivity. Measurement of **creatinine clearance,** on the other hand, which gives an estimate of the GFR, helps in early detection of renal

TABLE 56–6 Some Abnormal Constituents of Urine

Constituent	Clinical Significance	Examples of Conditions in which Present
Protein	**Glomerular proteinuria** refers to the presence of albumin in urine due to a breach in the integrity of the glomerular basement membrane.	Nephrotic syndrome, acute glomerulonephritis, diabetic nephopathy, etc
	Overflow proteinuria is due to the presence of abnormally high levels of low molecular weight proteins in the plasma that are filtered by the glomerulus and thus appear in the urine.	Multiple myeloma (light chains of immunoglobulins appear in urine, resulting in Bence–Jones proteinuria)
	Tubular proteinuria refers to the presence of low molecular weight proteins (like β_2-microglobulin) in urine, due to impaired reabsorption of these proteins by the proximal tubule.	Fanconi's syndrome, nephrotoxicity due to aminoglycoside antibiotics, heavy metals, etc
	Postrenal proteinuria refers to the presence of proteins in urine derived from the urinary tract.	Urinary tract infection (UTI) resulting in inflammatory exudates in urine
Glucose	**Hyperglycemic glucosuria:** Presence of glucose in urine is usually seen when plasma glucose rises above the renal threshold of ~180 mg/dL.	Uncontrolled diabetes mellitus
	Renal glucosuria: Presence of glucose in urine due to impaired reabsorption of glucose in the proximal tubule	Fanconi's syndrome and inherited defects in the sodium glucose transporter 2 (SGLT2)
Ketone bodies	Detectable levels in urine (ketonuria) are seen in conditions characterized by increased ketogenesis.	Diabetic ketoacidosis and starvation ketoacidosis.
Blood	**Hematuria** refers to the presence of red blood cells in urine, due to bleeding into the urinary tract.	Renal stones or urinary tract infections
	Hemoglobinuria refers to the presence of hemoglobin in urine, which occurs due to intravascular hemolysis.	Incompatible blood transfusions, malaria etc
Bile salts and bile pigments	Presence of these in urine is associated with obstruction of the biliary tract	Gall stone or carcinoma of the head of pancreas obstructing the common bile duct

failure. **Clearance** refers to the volume of plasma from which a particular substance is completely cleared by the kidney in unit time (usually a minute). It is calculated by the formula

$$\text{Clearance (mL/min)} = \frac{U \times V}{P}$$

where U = concentration of the measured analyte in a timed sample of urine (usually 24 h), P = plasma concentration of the analyte, and V = volume of urine produced per minute (calculated by dividing the value for the volume of urine collected over 24 h by 1440 [24×60]).

Inulin clearance is considered the gold standard method for measuring GFR, as it satisfies all the criteria essential for a substance to be used in clearance tests (**Table 56–7**). However, **creatinine clearance** is widely used due to the **ease of estimation of creatinine** (by Jaffe's method) and the fact that it is an **endogenous substance** (as opposed to inulin, which is exogenous in origin and has to be infused intravenously at a constant rate).

An important **drawback** associated with the use of clearance tests to estimate GFR is the need for an accurately timed urine sample. However, this problem can be overcome by employing formulae, which can be used to calculate an estimated value for GFR (EGFR), using serum creatinine values alone by correcting for age, sex, and body weight. One such formula is the **Cockcroft–Gault formula,** shown below.

$$\text{Estimated GFR (mL/min)} = \frac{(140 - \text{age}) \times \text{weight(kg)} \times 0.85 \text{ (if female)}}{72 \times \text{serum creatinine (mg/dL)}}$$

Thyroid Function Tests

The thyroid gland secretes the thyroid hormones—thyroxine or tetraiodothyronine (T_4) and triiodothyronine (T_3). Clinical conditions associated with increased or decreased synthesis of thyroid hormones (hyperthyroidism and hypothyroidism respectively) occur commonly. A clinical diagnosis of a thyroid disorder is confirmed with the help of thyroid function tests. The main thyroid function tests commonly done in clinical practice are shown in **Table 56–8**.

Measurement of **thyroid stimulating hormone (TSH)** is often the first test done in the assessment of thyroid function (also see Case 11, Chapter 57). The serum concentration of TSH is high in primary hypothyroidism and low or undetectable in primary hyperthyroidism. Measurement of free

TABLE 56–7 Characteristics of an Ideal Substance to be Used for Clearance Tests

1. It should have a fairly constant blood level.

2. It should be excreted from the body only in the urine.

3. It should be freely filtered at the glomerulus.

4. It should neither be reabsorbed nor secreted by the renal tubules[1].

[1]Creatinine satisfies all these criteria except for the fact that it is secreted by the renal tubules to a small, but variable, extent. Creatinine clearance, therefore, overestimates the GFR. This becomes particularly significant when GFR is estimated in late stages of chronic renal failure when the GFR is expected to be very low.

TABLE 56-8 Major Thyroid Function Tests

1. Serum levels of thyroid stimulating hormone (TSH)

2. Serum levels of free thyroxin (T_4) and triiodothyronine (T_3)

3. Tests for autoimmune thyroid diseases include tests for anti-TSH receptor, antithyroglobulin, antimicrosomal, and antithyroperoxidase antibodies

TABLE 56-10 Commonly Used Adrenal Function Tests

Basal hormone levels:
- Serum cortisol levels (8 A.M. and midnight)
- Urinary cortisol levels (in a 24-h urine collection)
- Serum ACTH levels (8 A.M.)

Suppression tests (to confirm adrenal hyperfunction):
- Dexamethasone suppression test

Stimulation tests (to confirm adrenal hypofunction):
- Synacthen (synthetic ACTH) stimulation test

thyroxine levels will help establish the diagnosis in most cases where an abnormal TSH value is obtained (**Table 56–9**). This strategy has proven to be cost-effective and clinically efficient in the diagnosis of thyroid disorders.

Serum levels of **total thyroxine** are seldom measured nowadays, as reliable assays to measure **free thyroxine** are now available commercially. Levels of total thyroxine are affected by changes in levels of thyroid-binding globulin (TBG), in the absence of thyroid disease. Additional tests such as measurement of **thyroid auto-antibodies** can be done to diagnose specific diseases related to the thyroid. For example, Graves' disease is commonly associated with the presence of **anti-TSH receptor antibodies,** while autoimmune thyroiditis (Hashimoto's thyroiditis) is associated with the presence of **antithyroid peroxidase (antimicrosomal) antibodies.**

Adrenal Function Tests

A clinical diagnosis of adrenal hyperfunction (Cushing's syndrome) or hypofunction (Addison's disease) is confirmed by adrenal function tests. The commonly used tests for this purpose are listed in **Table 56–10**.

Secretion of **cortisol** from the adrenal gland shows a regular diurnal variation. Serum cortisol levels are highest during the early hours of morning and least at midnight. Loss of diurnal variation is one of the earliest signs of adrenal hyperfunction. Estimations of serum cortisol in blood samples drawn at 8 A.M. and at midnight are, therefore, useful as screening tests. A diagnosis of adrenal hyperfunction is confirmed by demonstration of failure of suppression of 8 A.M. cortisol levels following administration of 1 mg dexamethasone (a potent synthetic glucocorticoid) at midnight (**dexamethasone suppression test**). Measurement of **adrenocorticotropic hormone** (ACTH) can help differentiate hypercortisolism due

excessive ACTH production (ACTH-dependent) from those where ACTH production is normal or suppressed (ACTH-independent). Failure to increase serum cortisol levels following a single dose of synacthen (a synthetic analogue of ACTH) is diagnostic of adrenal hypofunction (**Synacthen stimulation test**). Additional biochemical tests and imaging techniques (CT or MRI scan) may be required to diagnose the exact cause of adrenal hyper- or hypo-function.

Table 56–11 lists the principal biochemical laboratory tests that are performed in many hospitals.

There are numerous other biochemical tests that have been developed, but many of them are only available in large regional hospital laboratories

SUMMARY

- Any laboratory investigation that is ordered should provide useful information for the diagnosis, prognosis, and management of the patient and should be directly beneficial to the patient.

- A good lab test should be accurate and precise. Accuracy refers to the degree of agreement with the "true" value. Precision refers to the reproducibility of the analysis.

- While interpreting results of a test, one should be aware of the sensitivity, specificity and predictive value of the test. Sensitivity refers to the percentage of positive results in patients with the disease. Specificity is the percentage of negative results among people who do not have the disease. Positive predictive value refers to the percentage of positive results that are true positives. Diagnostic tests should be highly sensitive and specific.

- Several preanalytical variables can significantly affect the results of measurement of biochemical analytes. These factors need to be kept in mind while ordering and interpreting the results of a test.

- A high degree of automation is employed in most clinical laboratories for the analysis of routine tests.

- Tests that provide information on the functioning of particular organs are often grouped together as organ function tests

- Creatinine clearance provides useful information on the glomerular filtration rate and is therefore an important renal function test.

TABLE 56-9 Lab Diagnoses of Thyroid Disorders

		TSH Level	
		High	**Low**
Free thyroxine level	**High**	Secondary hyperthyroidism[1]	Primary hyperthyroidism
	Low	Primary hypothyroidism	Secondary hypothyroidism[1]

[1]Secondary hyper- and hypothyroidism are much rarer than primary hyper- and hypothyroidism.

TABLE 56–11 Reference Values for Selected Biochemical Laboratory Tests[1]

Blood Analyte	SI Units[2]	Conventional Units
1. Anion gap	7–16 mmol/L	7–16 meq/L
2. Arterial blood gas analysis (ABG)		
a. pH	7.35–7.45	7.35–7.45
b. Bicarbonate	22–30 mmol/L	22–30 meq/L
c. pCO_2	4.3–6.0 kPa	32–45 mm Hg
d. pO_2	9.6–13.8 kPa	72–104 mm Hg
3. Electrolytes and other ions		
a. Sodium	136–146 mmol/L	136–146 meq/L
b. Potassium	3.5–5.0 mmol/L	3.5–5.0 meq/L
c. Chloride	102–109 mmol/L	102–109 meq/L
d. Calcium (total)	2.2–2.6 mmol/L	8.7–10.2 mg/100 mL
e. Phosphorus (inorganic)	0.81–1.4 mmol/L	2.5–4.3 mg/100 mL
f. Magnesium	0.62–0.95 mmol/L	1.5–2.3 mg/100 mL
4. Glucose		
a. Fasting		
i. Normal	4.2–6.1 mmol/L	75–110 mg/100 mL
ii. Impaired glucose tolerance	6.2–6.9 mmol/L	111–125 mg/dL
iii. Diabetes mellitus	> 7.0 mmol/L	> 125 mg/dL
b. 2-h postprandial	3.9–6.7 mmol/L	70–120 mg/100 mL
5. Glycated hemoglobin (HbA_{1c})	0.04–0.06 Hb fraction	4.0–6.0%
6. Iron homeostasis parameters		
a. Ferritin		
i. Male	29–248 µg/L	29–248 ng/mL
ii. Female	10–150 µg/L	10–150 ng/mL
b. Iron	7–25 µmol/L	41–141 µg/dL
c. Iron binding capacity	45–73 µmol/L	251–406 µg/dL
d. Transferrin saturation	0.16–0.35	16–35%
e. Transferrin	2.0–4.0 g/L	200–400 mg/dL
7. Kidney function tests:		
a. Creatinine		
i. Male	53–106 µmol/L	0.6–1.2 mg/dL
ii. Female	44–80 µmol/L	0.5–0.9 mg/dL
b. Urea	5.4–14.3 mmol/L	15–40 mg/100 mL
c. Blood urea nitrogen (BUN)	2.5–7.1 mmol/L	7–20 mg/100 mL
d. Creatinine clearance	1.5–2.2 mL/s	91–130 mL/min
8. Lipid profile:		
a. Total cholesterol		
i. Desirable	< 5.17 mmol/L	<200 mg/dL
ii. Borderline high	5.17–6.18 mmol/L	200–239 mg/dL
iii. High	> 6.21 mmol/L	> 240 mg/dL
b. LDL cholesterol		
i. Optimal	< 2.59 mmol/L	<100 mg/dL
ii. Above optimal/near optimal	2.59–3.34 mmol/L	100–129 mg/dL
iii. Borderline high	3.36–4.11 mmol/L	130–159 mg/dL
iv. High	4.14–4.89 mmol/L	160–189 mg/dL
v. Very high	> 4.91 mmol/L	> 190 mg/dL
c. HDL cholesterol		
i. Low	< 1.03 mmol/L	< 40 mg/dL
ii. High	> 1.55 mmol/L	60 mg/dL
d. Triglycerides (fasting)	0.34–2.26 mmol/dL	30–200 mg/dL

(continued)

TABLE 56–11 Reference Values for Selected Biochemical Laboratory Tests[1] *(continued)*

Blood Analyte	SI Units[2]	Conventional Units
9. Liver function tests:		
a. Total Protein	67–86 g/L	6.7–8.6 g/dL
b. Albumin	40–50 g/L	4.0–5.0 g/dL
c. Bilirubin		
i. Total	5.1–22 µmol/L	0.3–1.3 mg/dL
ii. Direct	1.7–6.8 µmol/L	0.1–0.4 mg/dL
iii. Indirect	3.4–15.2 µmol/L	0.2–0.9 mg/dL
d. Alanine transaminase (ALT/SGPT)	0.12 – 0.70 µkat/L	7–41 U/L
e. Aspartate transaminase (AST/SGPT)	0.2–0.65 µkat/L	12–38 U/L
f. Alkaline phosphatase (ALP)	0.56–1.63 µkat/L	33–96 U/L
g. Gamma Glutamyl Transferase (γGT/GGT)	0.15–0.99 µkat/L	9–58 U/L
10. Osmolality	275–295 mOsmol/kg	275–295 mOsmol/kg
11. Thyroid profile:		
a. Thyroid stimulating hormone (TSH)	0.34–4.25 mIU/L	0.34–4.25 µIU/mL
b. Thyroxine		
i. Free	10.3–21.9 pmol/L	0.8–1.7 ng/dL
ii. Total	70–151 nmol/L	5.4–11.7 µg/dL
12. Troponin T	0–0.1 µg/L	0–0.1 ng/mL
13. Uric acid		
i Male	0.18–0.41 µmol/L	3.1–7.0 mg/100 dL
ii Female	0.15–0.33 µmol/L	2.5–5.6 mg/dL
Cerebrospinal fluid (CSF)[3]		
1. Glucose	2.22–3.89 mmol/L	40–70 mg/dL
2. Protein (lumbar)	0.15–0.5 g/L	15–50 mg/dL
3. Red cells	0	0
4. Leukocytes (WBCs)	0–5 mononuclear cells/µL	0–5 mononuclear cells/mm[3]
Urine		
1. Acidity, titratable	20–40 mmol/d	20–40 meq/d
2. Creatinine	8.8–14 mmol/d	1.0–1.6 g/d
3. Albumin		
i. Normal	0.0–0.03 g/d	0–30 mg/d
ii. Microalbuminuria	0.03–0.30 g/d	30–300 mg/d
iii. Clinical albuminuria	> 0.3 g/d	> 300 mg/d
4. pH	5.0–9.0	5.0–9.0
5. Total protein	< 0.15 g/d	< 150 mg/d

[1]A variety of factors such as the population studied, the duration and means of transport of specimen, laboratory methods and instrumentation, etc., can influence the reference values. Therefore, the reference or "normal" vaues given in this Table may not be appropriate for all laboratories. Whenever possible, reference values provided by the laboratory performing the test should be utilized in the interpretation of laboratory data.

[2]It is recommended that the system of international units (SI) be used in the reporting of all laboratory values. However, many laboratories and doctors prefer the familiar "conventional units" and continue to use them for reporting and interpreting lab results. Therefore, both systems are provided in this Table.

[3]Since CSF concentrations are equilibrium values, measurement of the same parameters in blood plasma obtained at the same time are recommended.

(The results listed here are from Harrison's Principles of Internal Medicine, Fauci AS et al [editors], Appendix: Laboratory Values of Clinical Importance, by Kratz A et al, 17th ed. McGraw-Hill, 2008, with permission.)

- Measurement of TSH, using an accurate and sensitive immunoassay, is often the first test done in the assessment of thyroid function. High levels are seen in primary hypothyroidism and low levels in primary hyperthyroidism.

REFERENCES

Beckett G, Walker S, Rae P, Ashby P: *Clinical Biochemistry.* 8th ed. Wiley-Blackwell, 2010.

Bishop ML, Fody EP, Schoeff LE: *Clinical Chemistry Techniques, Principles, Correlations.* 6th ed. Wolters Kluwer, Lippincott Williams & Wilkins, 2010.

Burtis CA, Ashwood ER, Bruns DE (editors): *Tietz Textbook of Clinical Chemistry and Molecular Diagnostics.* 4th ed. Elsevier Saunders, 2006.

Gaw A, Murphy MJ, Cowan RA, et al: *Clinical Biochemistry.* 4th ed. Churchill Livingstone, 2008.

Kratz A, Pesce MA, Fink DJ: Appendix: Laboratory Values of Clinical Importance. *Harrison's Principles of Internal Medicine.* 17th ed. Fauci AS et al (editors). McGraw-Hill, 2008.

Krieg AF, Gambino R, Galen RS: Why are clinical laboratory tests performed? When are they valid? JAMA 1975;233:76.

Lab Tests Online: www.labtestsonline.org (A comprehensive web site provided by the American Association of Clinical Chemists that provides accurate information on most if not all the lab tests mentioned in this text).

Laposaka M: *Laboratory Medicine.* McGraw-Hill Lange, 2010.

MedlinePlus: http://www.nlm.nih.gov/medlineplus/encyclopedia. html (The A.D.A.M. Medical Encyclopedia includes over 4000 articles about diseases, lab tests and other matters)

Biochemical Case Histories

57

Robert K. Murray, MD, PhD & Peter L. Gross, MD, MSc, FRCP(C)

O B J E C T I V E S

After studying this chapter, you should be able to:

- Appreciate the importance of a sound knowledge of Biochemistry and Genetics in understanding and managing many clinical conditions.
- Understand the general features and some aspects of management of the following conditions: adenosine deaminase deficiency; Alzheimer disease; cholera; colorectal cancer; cystic fibrosis; diabetic ketoacidosis; Duchenne muscular dystrophy; acute ethanol intoxication; acute gout; hereditary hemochromatosis; primary hypothyroidism; kwashiorkor and protein-energy malnutrition; myocardial infarction; obesity; primary osteoporosis; xeroderma pigmentosum.

INTRODUCTION

In this final chapter, 16 case histories are presented and discussed. They illustrate the importance of knowledge of Biochemistry for **the understanding of disease.** Of course, as has been shown throughout the text, Biochemistry is also crucial for the **understanding of health and wellness.**

Most of the diseases discussed here are **prevalent,** or relatively prevalent, in a global sense. (Prevalence is the proportion of persons in a given population that has a particular disease at a point or interval of time.) However, two (xeroderma pigmentosum and severe combined immunodeficiency disease due to deficiency of adenosine deaminase [ADA]) are relatively rare. They are included because they illustrate two crucial biologic facts: the importance of **DNA repair** and of the **immune system** as protective mechanisms. In addition, ADA deficiency is the first disease for which **gene therapy** was performed in humans.

The **reference values** for laboratory tests cited in the cases below may differ from these listed by laboratories with which the reader may be familiar. This is because reference values from different laboratories vary somewhat, in part due to different methodologies. In this chapter, laboratory results are generally given as **SI units** (Systeme International d'Unites). Chapter 56 presents some of the basic principles concerned with the ordering and interpretation of laboratory tests. **Table 56–11**

lists the majority of the normal values for the lab tests referred to in this chapter, and gives both SI and "conventional" values (as widely used in the United States).

The **doses of drugs** administered in the treatment of the cases described here are not generally given; the reader should check these out on her/his own.

CASE 1: ADENOSINE DEAMINASE (ADA) DEFICIENCY CAUSING SEVERE COMBINED IMMUNODEFICIENCY DISEASE (SCID)

Causation

Genetic (due to mutations in the gene encoding ADA). Deficiency of ADA affects the immune system.

History and Physical Examination

A young girl aged 11 months was brought by her parents to a children's hospital. She had had a number of attacks of pneumonia and thrush (oral infection usually due to *Candida albicans*) since birth. The major findings of a thorough workup were very low levels of circulating lymphocytes (ie, severe lymphopenia) and low levels of circulating immunoglobulins. The attending pediatrician suspected SCID.

Laboratory Findings

Analysis of a sample of red blood cells revealed a very low activity of ADA and also a very high level (about 50 times normal) of dATP. This confirmed the diagnosis of SCID due to deficiency of ADA, the enzyme that converts adenosine to inosine (Chapter 33):

$$Adenosine \rightarrow Inosine + NH_4^+$$

Treatment

Appropriate **antibiotic** treatment was commenced and the child was given periodic injections of **immune globulin.** In addition, she was started on weekly intramuscular injections of **bovine ADA conjugated to polyethylene glycol.** Bovine ADA is relatively nonimmunogenic, and conjugation to polyethylene glycol prolongs its half-life. It has been shown to be beneficial in the treatment of ADA deficiency. Her parents were informed that **bone marrow transplantation** was the most appropriate therapy, but the treatment was declined. In view of reports of successes with **ADA gene therapy,** this treatment was offered, and the parents consented. The treatment was approved by the hospital Ethics Committee. Lymphocytes and mononuclear cells were isolated from her blood using a gradient of Ficoll (a neutral highly branched polysaccharide). They were then grown in the presence of interleukin-2 (to stimulate cell division) and infected with a modified retrovirus containing inserts encoding ADA and also a gene (the NeoR gene) encoding an enzyme governing neomycin resistance, which was used to show that gene transfer had been achieved. An alternative at the present time would be to use **bone marrow stem cells** (which have been reported to give good increases of both B and T cells), but these were not available at the time of treatment. The autologous gene-treated cells were then injected intravenously. The child received similar injections once a month over the next year, and in addition continued to receive polyethylene glycol-conjugated ADA. Measurement of the activity of ADA revealed a sustained elevation (about 20% of normal) of the enzyme in circulating lymphocytes after 6 months of treatment; analyses using the PCR technique with NeoR probes revealed that approximately the same percentage of lymphocytes contained inserted genetic material.

Discussion

Deficiency of the activity of ADA is inherited as an **autosomal recessive** condition. It accounts for approximately 15% of cases of SCID; other causes involve mutations in a variety of genes affecting the function of cells of the immune system. Most of the mutations in the gene for ADA so far detected have been base substitutions, though deletions have also been detected. These mutations result in diminished activity or stability of ADA. The block of activity of ADA results in accumulation of adenosine, which in turn results in accumulation of deoxyadenosine and **dATP.** Elevated levels of the latter are **toxic,** particularly to **T lymphocytes,** which normally exhibit a high activity of ADA. Thus, lymphocytes are injured or killed,

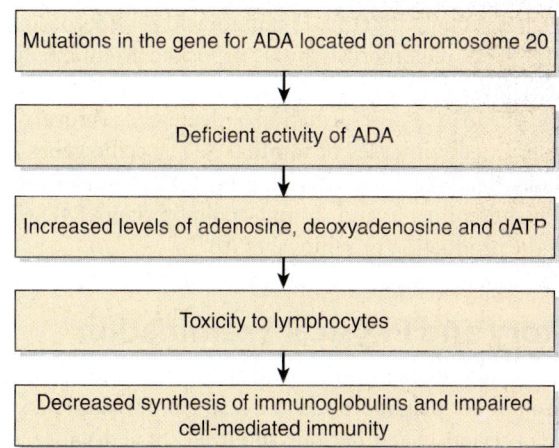

FIGURE 57–1 Summary of probable events in the causation of SCID due to ADA deficiency (OMIM 102700).

resulting in **impairment of both cell and humoral immunity,** because impairment of T-cell function can secondarily affect B-cell function.

Adenosine deaminase deficiency has become quite prominent because it is the first disease to be treated by somatic cell gene therapy. Several patients have been treated by protocols similar to that described above, which is an example of **ex vivo** (the lymphocytes and mononuclear cells were removed from the body prior to insertion of the ADA gene) gene therapy. One reason for selecting ADA deficiency as a suitable condition for somatic gene therapy was that cells that express the gene for ADA would have **a selective advantage for growth** over uncorrected cells. Gene therapy is discussed briefly in Chapter 39. Important points concerning gene therapy include that the **level of expression** of the affected protein should be sufficient to sustain normal function, that ideally the inserted gene should show **normal regulation,** and that **no significant undesirable side effects** should occur (eg, cancer due to insertional mutagenesis). In relation to the last point, the gene that is being delivered may insert into a gene that is essential for normal cell growth, and if this occurs that may result in the cell becoming cancerous (an example of **insertional mutagenesis**), as has been noted in some instances of gene therapy.

A simplified scheme of the events involved in the causation of ADA deficiency is given in **Figure 57–1.** Safe and effective treatment of ADA deficiency by gene therapy has been reported recently (see References).

CASE 2: ALZHEIMER DISEASE (AD)

Before studying this case, the reader is advised to consult the material in Chapter 5 on AD.

Causation

Deposition of **amyloid β peptide** ($A\beta_{42}$) in certain parts of the brain is believed by many neuroscientists to be one major cause of AD. It is thought that this 42-amino-acid peptide, occurring as beta sheets, oligomerizes and is deposited around neurons; the oligomers may be toxic to neurones. Deposition of $A\beta_{42}$ may be due to excessive formation or decreased removal of the peptide. In certain cases of **familial** AD, specific genes have been identified (eg, these encoding amyloid precursor protein [APP], presenilins 1 and 2, and apolipoprotein E4), which affect the production or removal of $A\beta_{42}$.

History and Physical Examination

A 72-year-old woman who lived on her own was found wandering around her neighborhood at 2 A.M. Her husband had died 3 years previously and her one son lived some distance away. The lady was confused and taken to the hospital. The son was notified and came immediately to see his mother. She was not able at the time of admission to give a clear history. The son volunteered that she had been diagnosed by a neurologist as having early AD, but had refused to go into a nursing home. She had home help during the day, and had not previously wandered out of her home. Sometimes a lady friend visited and spent the night at her house. In fact, she had appeared relatively normal prior to the present situation, as her son spoke to her on the phone every day of the week. However, her short-term memory had become worse in recent months, and he had become concerned about her. She was on a medication (donepezil) for AD. Otherwise she had no significant medical history. She was kept in the hospital for a couple of days, during which time her family doctor and the neurologist were consulted.

Treatment

There is **no specific treatment** for AD at the present time. Donepezil and several other drugs that are used in the management of AD **inhibit the activity of cholinesterase,** an enzyme that hydrolyzes acetylcholine (ACh) to acetate and choline. They are used because some studies had shown lower than normal levels of ACh in specimens of brains from subjects who had died of AD. They appear to produce a modest improvement of brain function and memory in some patients. Memantine, a drug that **antagonizes *N*-methyl-D-aspartate receptors,** may cause some slowing of the progression of AD. Symptoms such as depression, agitation, anxiety, and insomnia can be treated by appropriate drugs and antipsychotics may be required if psychosis intervenes. A possible preventive role of **omega-3 fatty acids** is under study. However, overall, there is as yet **no effective therapy** for AD. This patient was kept on donepezil and admitted to a nursing home that specialized in caring for patients with AD. Apart from high-quality basic care, the nursing home offered a variety of programs, including music therapy and exercise programs.

Discussion

AD is a progressive neurodegenerative condition in which **decline of general cognitive function** occurs, usually accompanied by **affective** and **behavioral disturbances.** At least 2 million people in the United States suffer from AD, and its prevalence is likely to increase as more people live longer. Some cases have a **familial (genetic)** basis, but the great majority (~90%) appears to be **sporadic.** AD is the most common cause of **dementia,** which can be defined as a progressive decline in intellectual functions, due to an organic cause, that interferes substantially with an individual's activities. AD places a tremendous burden on families and on the health care system, as, sooner or later, most patients cannot look after themselves. The usual age at onset is over 65 years, but the disease can have an early onset (eg, in the 40s), particularly in cases where there is a genetic predisposition (see below). Survival ranges from 2 to 20 years. It is estimated that about 40% of people over 85 years of age have variable degrees of AD. **Loss of short-term memory** is often the first sign. AD usually **progresses** inexorably, and many patients are eventually completely incapacitated.

The **diagnosis** is usually one of exclusion. A complete neurologic exam is necessary and also a recognized mental status exam. Other forms of dementia (Lewy body, vascular, etc) must be excluded, as must other organic and psychiatric problems; various lab tests may thus be indicated to do this. In certain cases an MRI or CT scan may be indicated; these will usually reveal variable degrees of cortical atrophy and enlargement of ventricles if AD is present. Considerable research is underway to **develop laboratory tests** (eg, on blood or cerebrospinal fluid) that will assist in making the unequivocal diagnosis of AD.

The basic **pathologic picture** is of a degenerative process characterized by the death and consequent loss of cells in certain areas of the brain (eg, the cortex, hippocampus and certain other sites). **Apoptosis** (a programmed type of cell death in which various mechanisms, particularly the activities of proteolytic enzyme known as caspases, are activated within a cell leading to rapid cell death, see Chapter 55) may be involved in the cell death occurring in AD. At the microscopic level, **neuritic plaques** containing aggregated amyloid β peptide (Aβ, a peptide of 42 amino acids, occurring in beta sheets) surrounded by nerve cells containing **neurofibrillary tangles** (paired helical filaments formed from a hyperphosphorylated form of the microtubule associated protein, **tau**) are hallmarks. Deposits of $A\beta_2$ are frequent in small blood vessels.

Intensive research is under way to determine the cause of AD. Particular interest has focused on the presence of **$A\beta_{42}$,** the major constituent of the plaques found in AD. The term "amyloid" refers to a group of diverse extracellular protein deposits found in many different diseases (see Chapter 50). Amyloid proteins usually stain blue with iodine, like starch, which accounts for the name (amylo denotes starch). The **amyloid cascade hypothesis** proposes that deposition of $A\beta_{42}$ is the cause of the pathologic changes observed in the brains of the victims of AD and that other changes, such as neurofibrillary tangles and vascular alterations, are secondary. $A\beta_{42}$ is derived from a larger precursor protein named **amyloid precursor protein (APP),** whose gene is located on

Step 1: Cleavage by either α- or β-secretase

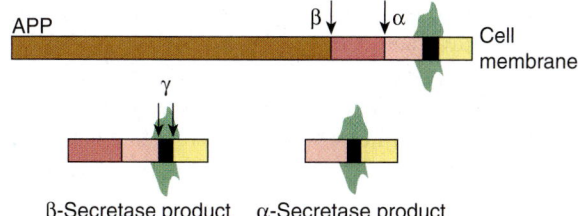

Step 2: Cleavage by γ-secretase

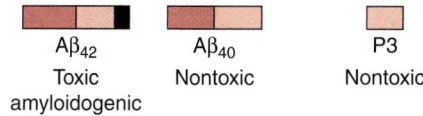

FIGURE 57–2 **Simplified scheme of the formation of Aβ$_{42}$.** Amyloid precursor protein (APP) is digested by β-, α-, and γ-secretases. A key initial step (**Step 1**) is the digestion by either β-secretase or α-secretase, producing smaller nontoxic products. Cleavage of the β-secretase product by γ-secretase (**Step 2**) results in either the toxic Aβ$_{42}$ (containing 42 amino acids) or the nontoxic Aβ$_{40}$ peptide. Cleavage of the α-secretase product by γ-secretase produces the nontoxic P3 peptide. Excess production of Aβ$_{42}$ is a key initiator of cellular damage in Alzheimer disease (AD). Among research efforts on AD have been attempts to develop therapies to reduce accumulation of Aβ$_{42}$ by inhibiting β- or γ-secretases, promoting α-secretase activity or clearing Aβ$_{42}$ by use specific antibodies. (Reproduced, with permission, from Fauci AS et al [editors,] *Harrison's Principles of Internal Medicine,* 17th ed, McGraw-Hill, 2008, p. 2542.)

TABLE 57–1 **Some Genes Involved in Familial Types of Alzheimer Disease (AD)[1]**

Gene	Type of AD	Chromosome	Protein Product
APP	AD1, Familial (OMIM 104300)	21	APP
APOE4	AD2, Late onset (OMIM 104310)	19	ApoE4
PS1	AD3, Early onset (OMIM 104311)	14	Presenilin 1
PS2	AD4, Familial (OMIM 606889)	1	Presenilin 2

[1]In general, the products of these genes act by increasing the production of Aβ$_{42}$ (APP, PS1 and PS2) or by decreasing its clearance (APOE4). Presenilins 1 and 2 may be involved in the action of γ-secretase.
Abbreviations: APOE4, apolipoprotein E4; APP, amyloid precursor protein; OMIM, Online Mendelian Inheritance in Man catalog number; PS, presenilin.

A second part of the amyloid cascade hypothesis is that Aβ or Aβ-containing fragments are directly or indirectly **neurotoxic.** There is evidence that exposure of neurons to Aβ can increase their intracellular concentrations of **Ca^{2+}.** Some **protein kinases,** including that involved in phosphorylation of **tau,** are regulated by levels of Ca^{2+}. Thus, the increase of Ca^{2+} may lead to hyperphosphorylation of tau and formation of the paired helical filaments present in the neurofibrillary tangles. Interference with **synaptic function** is also probable, perhaps secondary to neuronal damage.

Further research may reveal unexpected developments that alter the validity of the amyloid cascade theory as presented above.

Work on AD has shown the probable importance of an **abnormally folded peptide** in the causation of this important brain disease. It is hoped that further research on AD may result in drugs that will prevent, arrest, or even reverse AD. For example, it may be possible to **develop small molecules** that prevent formation or deposition of Aβ$_{42}$, prevent its aggregation, or accelerate its clearance. In addition, it is possible that specific antibodies to Aβ$_{42}$ or tau could prevent their putative toxic actions.

AD is one of the so-called **conformational diseases** (Chapters 46 and 50), in which abnormally folded proteins play a central role in the causation of a disease. Other examples of these diseases are cystic fibrosis (this chapter), alpha$_1$-antitrypsin disease (Chapter 50), and the prion diseases (Chapter 5).

The study of **various neurodegenerative diseases** is providing dramatic evidence of **the importance of protein structure and function** in their causation. For example, abnormal forms of the protein huntingtin play an important role in Huntington disease, abnormalities of α-synuclein play a role in some cases of Parkinson disease, and prions have been found to play key roles in the causation of bovine spongiform encephalopathy (BSE) and Creutzfeldt–Jacob disease. The application of genomic and proteomic techniques is also beginning to

chromosome 21 close to the area affected in Down syndrome (trisomy 21). Individuals with Down syndrome who survive to age 50 often suffer from AD.

As shown in **Figure 57–2,** APP is a transmembrane protein that can be cleaved by proteases known as secretases. In step 1, APP is cleaved by either β-secretase or α-secretase to produce small nontoxic products. Then in step 2, cleavage of the β-secretase product by γ-secretase results in either the toxic Aβ$_{42}$ (containing 42 amino acids) or the nontoxic Aβ$_{40}$ peptide. Cleavage of the α-secretase product by γ-secretase produces the nontoxic P3 peptide. When split off from its parent protein, Aβ$_{42}$ forms an insoluble extracellular deposit. **Aggregation** of Aβ$_{42}$, produced by its **oligomerization** and formation of **beta sheets,** is thought by some to be a key event in causing AD. Recent studies have suggested that **impairment of clearance** of Aβ$_{42}$ may be an important part of the problem in AD.

Mutations in certain genes have been found in some patients with AD (familial AD). These mutations often predispose to early onset AD. One of these genes is that encoding APP. **Table 57–1** summarizes some aspects of the principal genes so far discovered. In general, the effects of the products of these genes are to enhance deposition of amyloid or to diminish its removal. Precise dissection of their mechanisms of action is underway.

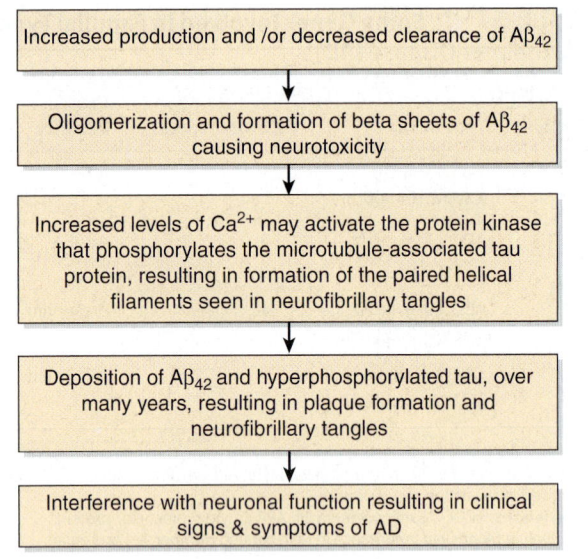

Increased production and /or decreased clearance of $A\beta_{42}$

↓

Oligomerization and formation of beta sheets of $A\beta_{42}$ causing neurotoxicity

↓

Increased levels of Ca^{2+} may activate the protein kinase that phosphorylates the microtubule-associated tau protein, resulting in formation of the paired helical filaments seen in neurofibrillary tangles

↓

Deposition of $A\beta_{42}$ and hyperphosphorylated tau, over many years, resulting in plaque formation and neurofibrillary tangles

↓

Interference with neuronal function resulting in clinical signs & symptoms of AD

FIGURE 57–3 **A tentative scheme of the possible sequence of events in at least certain cases of AD.**

throw light on the **causation of major psychiatric disorders,** such as bipolar disease and schizophrenia. The importance of genetic and biochemical approaches in understanding disease processes has never been more clear. A simplified scheme of the causation of AD is shown in **Figure 57–3**.

CASE 3: CHOLERA

Causation

Infection by *Vibrio cholerae.*

History and Physical Examination

A 21-year-old female medical student working in a developing country suddenly began to pass profuse watery stools almost continuously. She soon started to vomit, her general condition declined abruptly, and she was rushed to the local village hospital. On admission, she was cyanotic, skin turgor was poor, blood pressure was 70/50 mmHg (normal 120/80 mmHg), and her pulse was rapid and weak. The doctor on duty diagnosed cholera, took a stool sample, and started treatment immediately.

Treatment

Treatment consisted of **intravenous** administration of a solution made up in the hospital containing 5 g NaCl, 4 g $NaHCO_3$, and 1 g KCl per liter of pyrogen-free distilled water. This solution was initially given rapidly (100 mL per h) until her blood pressure and pulse rate were normalized. She was also given the antibiotic **doxycycline.** On the second day, she was able to take the **oral rehydration solution** recommended by the

World Health Organization (WHO) for the treatment of cholera, consisting of 20 g of glucose, 3.5 g of NaCl, trisodium citrate, and dihydrate 2.9 g (or 2.5 g $NaHCO_3$), and 1.5 g KCl per liter of drinking water.

She took amounts moderately exceeding the volume of her daily stools. **Solid food** was reinstituted on the fourth day after admission. She continued to recover rapidly and was discharged 7 days after admission.

Discussion

Cholera is an important infectious disease endemic in certain Asian countries and other parts of the world. Fecal contamination of water and food is the principal method of transmission. It is due to *Vibrio cholerae,* a bacterium that secretes a protein **enterotoxin.** The toxin is actually encoded by a bacteriophage (CTXφ) resident in *V cholerae.* The enterotoxin is made up of **one A subunit** (composed of one A1 and one A2 peptide joined by a disulfide link) and **five B subunits** and has a molecular mass of approximately 84 kDa. In the small intestine, the toxin **attaches** by means of the **B subunits** binding to the **ganglioside GM1** (Figure 15–17) present in the plasma membrane of mucosal cells (**Figure 57–4**). The A subunit then dissociates, and the **A1 peptide** passes across to the inner aspect of the plasma membrane. It catalyzes the **ADP-ribosylation** (using **NAD^+** as donor) of the GTP-binding regulatory component (G_s) of adenylate cyclase, upregulating the activity of this enzyme. Thus, **adenylyl cyclase** becomes chronically activated (Chapter 42). This results in an elevation of **cAMP,** which activates **protein kinase A** (PKA). This in turn via phosphorylation of CFTR and of a Na^+–H^+ exchanger leads to **inhibition of absorption of Na^+** and **enhancement of secretion of Cl^-.** Thus, massive amounts of NaCl accumulate inside the lumen of the intestine, attracting water by osmosis and contributing to the liquid stools characteristic of cholera.

Cholera toxin may also affect **other molecules** involved in intestinal secretion (eg, prostaglandins and nerve histamine receptors). The histologic structure of the small intestine remains remarkably unaffected, despite the loss of large amounts of Na^+, Cl^- water, HCO_3, and K^+. It is the loss of these constituents that results in the marked fluid loss (dehydration), low blood volume, acidosis, and K^+ depletion found in serious cases of cholera, and which can prove fatal unless appropriate replacement therapy (as described above) is begun immediately. A person suffering from cholera can lose as much as 1 L of fluid per hour.

The recognition and easy availability of appropriate replacement fluids, such as **oral rehydration solution,** has led to tremendous improvement in the treatment of cholera. It should be emphasized that **glucose** is an essential component of oral rehydration solution (Chapter 40). Cholera toxin inhibits absorption of Na^+ by intestinal cells, but it does not inhibit glucose-facilitated Na^+ transport into these cells, so glucose will be absorbed and used to supply energy.

Figure 57–5 summarizes the mechanisms involved in the causation of the diarrhea of cholera.

(b) The A component of the toxin causes ADP-ribosylation of a G protein that controls activation of adenylyl cyclase, locking the G protein in the "active" mode.

(a) The B component of the toxin attaches to specific receptors on cell membrane; A component penetrates membrane.

Plasma membrane of intestinal cell

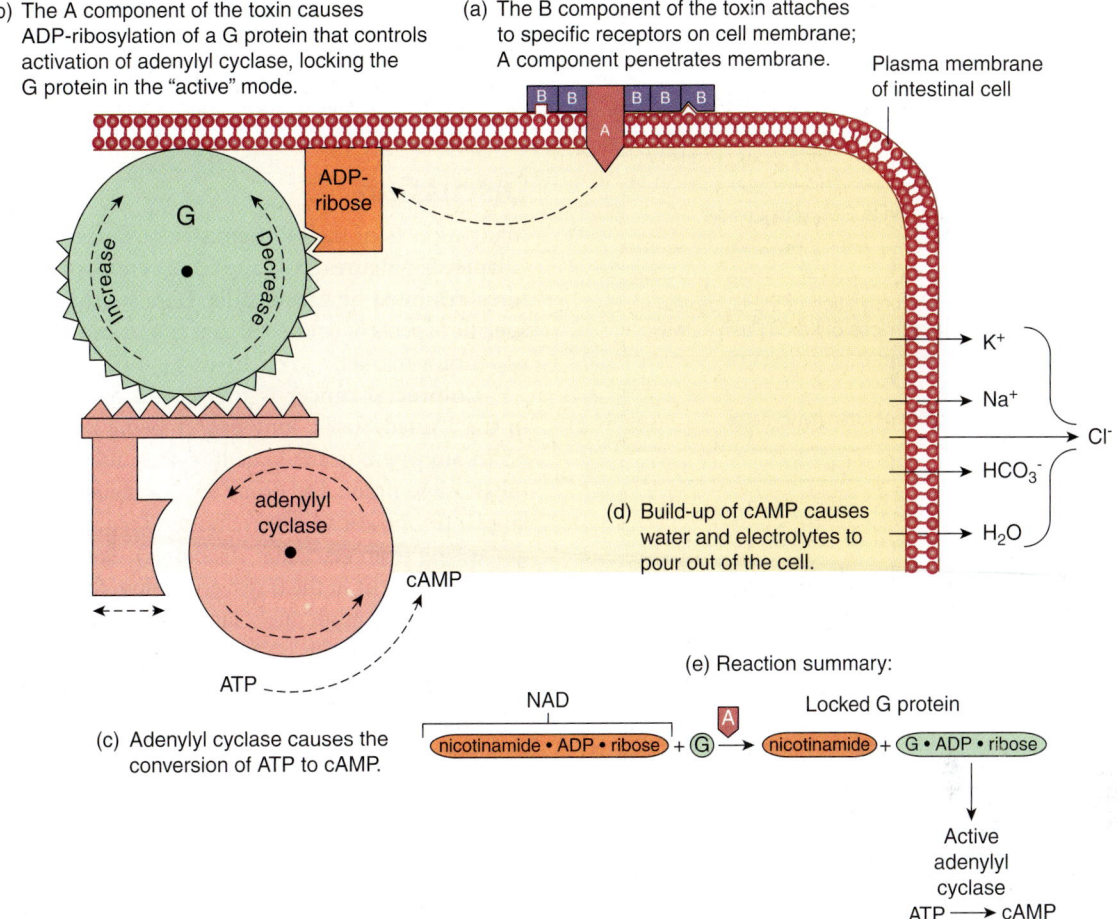

G

Increase   Decrease

ADP-ribose

K⁺

Na⁺

Cl⁻

HCO₃⁻

(d) Build-up of cAMP causes water and electrolytes to pour out of the cell.

H₂O

adenylyl cyclase

cAMP

ATP

(c) Adenylyl cyclase causes the conversion of ATP to cAMP.

(e) Reaction summary:

NAD   A   Locked G protein

nicotinamide • ADP • ribose + G → nicotinamide + G • ADP • ribose

Active adenylyl cyclase

ATP ⟶ cAMP

FIGURE 57–4 Diagrammatic representation of the mechanism of action of cholera toxin (CT). The toxin attaches to the plasma membrane via interaction of its B subunits with the ganglioside GM₁. The A subunit crosses the membrane and catalyzes the addition of the ADP-ribose component of NAD⁺ to the G protein involved in stimulating adenylyl cyclase (NAD⁺ → ADP-ribose + nicotinamide). The addition of ADP-ribose to the G protein locks adenylyl cyclase in its active conformation, increasing the intracellular level of cAMP. This leads to phosphorylation of several membrane transporters, in turn resulting in accumulation of the ions shown and of water in the intestinal lumen, thus producing often massive diarrhea. (Reproduced, with permission, from Nester EW et al, Microbiology: A Human Perspective, 5th ed. McGraw Hill, 2007.)

CASE 4: COLORECTAL CANCER

Causation

Environmental and **genetic.** Most if not all cancers are thought to originate by the occurrence of mutations in key genes regulating cell growth (oncogenes and tumor suppressor genes, see Chapter 55) The mutations may be inherited, but much more often various environmental influences (eg chemicals, radiation, and some viruses) are involved.

History, Physical Examination, and Results of Tests

A 62-year-old male consulted his family physician. He had noted that he had passed some fresh bright red blood in his stools several times in the previous 3 months, which he

attributed to hemorrhoids. Over the previous 12 months his appetite had decreased and he had lost over 10 lbs. He had always been in good health until the past year, and was not on any medications. He had no other complaints.

In view of the history of rectal bleeding, weight loss, and the patient's age, his physician suspected that he might have colorectal cancer and requested the patient to submit three consecutive daily specimens of feces for the **fecal occult blood test.** Shortly thereafter the physician received a report indicating that the results were positive. He also ordered a **complete blood count** and estimations of levels of serum iron, iron-binding capacity, and ferritin. The results showed a microcytic anemia (see Chapter 52), often found in patients with colorectal cancer because of bleeding from the tumor. A rectal examination was negative. No abnormalities were noted in chest x-rays.

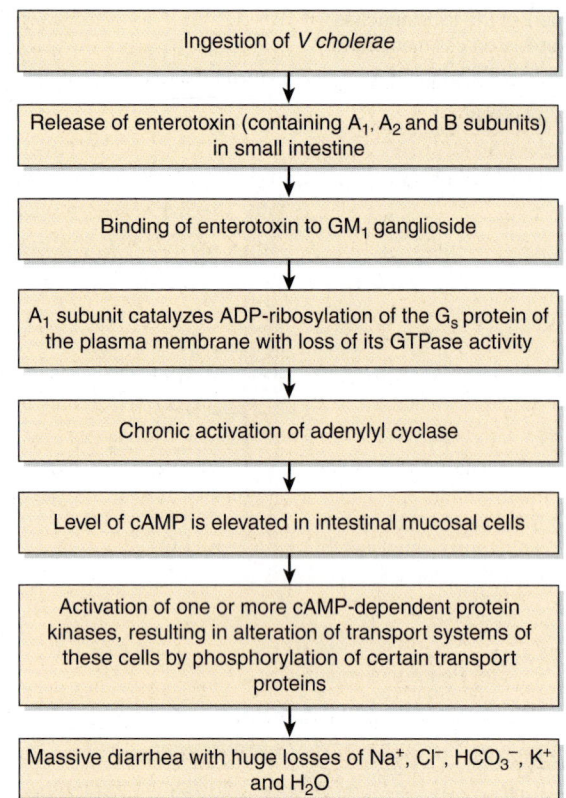

FIGURE 57–5 Summary of mechanisms involved in the causation of the diarrhea of cholera.

The physician arranged a consult 4 days later with a gastroenterologist. **Colonoscopy** was performed 1 week later. This revealed the presence of a moderately large tumor (approximately 5 × 6 cm) in the middle of the transverse colon. Measurement of **carcinoembryonic antigen** (CEA), a biomarker for colorectal cancer (see below for further comments and also Chapter 7), was ordered. It was elevated (20 µg/L, normal 0–3 µg/L). **Surgery** was scheduled 2 weeks later, when the tumor was resected and end-to-end anastomosis performed. The regional lymph nodes were also excised and submitted along with the tumor specimen to the pathology lab. No local invasion by the tumor was noted, and no tumor was visible elsewhere in the abdomen, including the liver. The subsequent pathology report described the tumor as a relatively well-differentiated adenocarcinoma, invading the muscular mucosa. No tumor cells were noted in the lymph glands; no distant metastases were noted at the time of surgery. The TNM stage was T1N0M0 (cancer limited to the mucosa and submucosa, with an approximate 5-year survival rate of >90% [T = tumor, N = nodes, M = metastases]). (The interested reader should check the staging of tumors of the large intestine in a textbook of pathology.) In view of these findings, no chemotherapy or radiation therapy was considered necessary. Determination of CEA several weeks after surgery showed it had declined to normal levels. The patient was advised to return for follow-up at regular intervals, when, among other tests, samples of blood for measurements of CEA were taken; they remained at normal levels. A follow-up colonoscopy was performed 3 years after the operation; no tumor was seen in the colon. The patient was alive and well at 5 years after operation.

Discussion

Many aspects of the biochemistry of cancer are discussed in Chapter 55. Figures 55–1 and 55–2 summarize important features exhibited by cancer cells. Here we shall focus on a few specific aspects of colorectal cancer and also on the use of CEA as a tumor marker.

Colorectal cancer is the second most common cancer in the United States, lung cancer being number one. It can occur anywhere in the large intestine, although the rectum is the most common site. Some 95% of malignant tumors in the large intestine are adenocarcinomas (cancers of epithelial origin arising from glandular structures). About 10% of colorectal cancers occur in the transverse colon. In this case, although the tumor was moderately large, no extension from the primary site of the tumor occurred, no local nodes were involved, and no distant metastases had occurred. This was fortunate for the patient, as it meant there was an excellent prognosis and also he did not have to be subjected to chemotherapy or radiotherapy.

Most colorectal adenocarcinomas originate from **adenomatous polyps**. A polyp is a growth, usually benign, protruding from a mucous membrane. There are a variety of types. The one of interest here is the adenomatous polyp. Most tumors of the colon arise from such polyps although the majority of such polyps do not progress to cancer.

There are a number of well-defined **genetic syndromes** that predispose to colorectal cancer. The most common is **hereditary nonpolyposis colorectal cancer** (HNPCC), in which mutations in various genes involved in DNA mismatch repair are involved (see Chapter 35). Another relatively rare condition is **familial adenomatous polyposis** (adenomatous polyposis coli, APC) in which hundreds or thousands of polyps appear in the colon and rectum. The *APC* **gene** is located on chromosome 5q21 and many mutations have been described. Overall, it has been estimated that approximately 20% of colorectal cancers have a genetic basis.

Various **environmental factors** have been proposed as being involved in the causation of colorectal cancer. These include diets rich in **saturated fat,** high in **calories,** low in **calcium,** and low in **fiber.** How exactly each of these proposed factors operates is the subject of ongoing research. For example, dietary fat appears to enhance the production of cholesterol and bile acids by the liver. When bile acids are excreted into the bowel, bacterial enzymes may act on them to convert them to secondary bile acids, which are thought to be tumor promoters. A **tumor promoter** is a molecule that along with an **initiator** (ie a molecule that causes a mutation in DNA) leads a cell to become cancerous. **Inflammatory bowel disease** (eg, ulcerative colitis) is another predisposing factor to colorectal cancer.

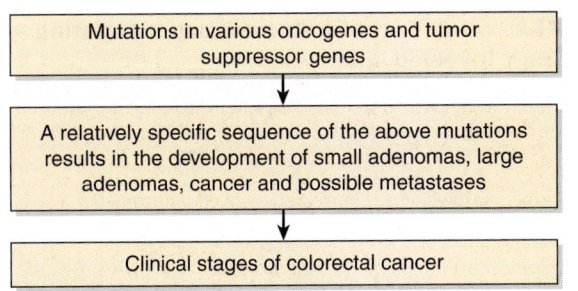

FIGURE 57–6 Simplified scheme of the multistep causation of colorectal cancer.

The pioneering work of Vogelstein and colleagues on the sequence of changes occurring in **tumor suppressor genes and oncogenes** in dysplastic epithelium, adenomatous polyps and adenocarcinomas of the colon was described in Chapter 55 (see Figure 55–10 and Table 55–9). The reader is advised to read the appropriate section in that Chapter.

The use of tumor markers in the management of cancer was discussed in Chapter 55. CEA was one of the tumor markers described (see Table 55–15). It is a glycoprotein present in the plasma membranes of many cells. It is released into the plasma in a number of conditions, in which it can be measured by a radioimmunoassay. Levels of CEA are increased in serum in colorectal cancer, but also in other alimentary and nonalimentary cancers, and in certain noncancerous conditions.

Less than 50% of individuals with localized colorectal cancer have elevated levels of CEA, so it is **not a useful screening test** for this condition. Its main use is to **monitor the effects of treatment** and to detect **recurrence** of cancers such as colorectal cancer, by following changes in its level. In the present case, the level of CEA remained low for 5 years after surgery, suggesting no recurrence of cancer had occurred. A goal (probably unattainable) of research in this area would be to develop highly specific biomarkers for very early colorectal cancer and for other very early tumors that would be positive in 100% of cases and negative in 100% of normal individuals!

Figure 57–6 summarizes some major factors leading to the development of colorectal cancer.

CASE 5: CYSTIC FIBROSIS (CF)

Causation

Genetic (mutations in the gene encoding the cystic fibrosis transmembrane regulator [CFTR] protein).

History and Physical Examination

A 1-year-old girl, an only child of Caucasian background, was brought to the clinic at the Hospital for Sick Children by her mother. She had been feverish for the past 24 hours and was coughing frequently. The mother stated that her daughter had experienced three attacks of "bronchitis" since birth, each of which had been treated with antibiotics by their family physician. The mother had also noted that her daughter had been passing somewhat bulky, foul-smelling stools for the past several months and was not gaining weight as expected. In view of the history of pulmonary and gastrointestinal problems, the attending physician suspected that the patient might have CF although no family history of this condition was elicited.

Laboratory Findings

Chest x-rays showed signs consistent with bronchopneumonia. Culture of sputum revealed predominantly *Pseudomonas aeruginosa*. Fecal fat was increased. A quantitative pilocarpine iontophoresis sweat test was performed, and the sweat Cl⁻ was 70 mmol/L (>60 mmol/L is abnormal); the test was repeated a week later with similar results.

Treatment

The child was given an appropriate **antibiotic** and referred to the **cystic fibrosis clinic** for further care. A **comprehensive program** was instituted to look after all aspects of her health, including psychosocial considerations. She was started on a **pancreatic enzyme preparation** (given with each meal) and placed on a **high-calorie diet** supplemented with **multivitamins** and **vitamin E. Postural drainage** was begun for the thick pulmonary secretions. Subsequent infections were treated promptly with appropriate antibiotics and with an aerosolized recombinant preparation of **human DNase** that digests the DNA of microorganisms present in the respiratory tract. At age 6 years, she had grown normally, had been relatively free of infection for a year, was attending school, and making satisfactory progress. Serious chronic cases of CF in which the lungs are severely compromised are candidates for **lung transplants,** although the efficacy of this treatment has been challenged recently.

Research on **gene therapy** for CF is under examination (eg, using recombinant viruses encoding the CFTR protein). Another line of research is investigating whether **small molecules** can be found for clinical use that help abnormally folded CFTR molecules re-fold into at least partially active molecules.

Discussion

CF is a **prevalent** and usually serious genetic disease among whites in North America. It affects approximately 1:2500 individuals and is inherited as an **autosomal recessive** disease; about 1 person in 25 is a carrier. It is a disease of the **exocrine glands,** with the **respiratory and gastrointestinal tracts** being most affected. A diagnostic hallmark is the presence of **high amounts of NaCl in sweat,** reflecting an underlying abnormality in exocrine gland function (see below). **Pilocarpine iontophoresis** has generally been used to allow collection of sufficient amounts of sweat for analysis. Iontophoresis is a process by which drugs are introduced into the body (in this case the skin) via use of an electrical current. Its use is diminishing as the availability of specific genetic probes increases.

The **classic presentation** of CF is that of a young child with a history of recurrent pulmonary infection and signs of exocrine insufficiency (eg, fatty, bulky stools due to a lack of pancreatic lipase), as in the present case. However, the disease is **clinically heterogeneous,** which at least partly reflects heterogeneity at the molecular level (see below). Approximately 15% of patients may have sufficient pancreatic function to be classified as "pancreatic sufficient."

For reasons related to abnormalities in Cl⁻ and Na⁺ transport (see below), the **pancreatic ducts** and the ducts of certain other exocrine glands become filled with **viscous mucus,** which leads to their **obstruction.** This mucus is also present in the **bronchioles,** leading to their obstruction; this favors the growth of certain bacteria (eg, *Staphylococcus aureus* and *Pseudomonas aeruginosa*) that cause **recurrent bronchopulmonary infections,** eventually seriously compromising lung function. In turn, the pulmonary disease can lead to right ventricular hypertrophy and possible heart failure. Patients usually die of **pulmonary infection** or **heart failure.** In recent years, more patients have been living into their 30s and later, as the condition is now diagnosed earlier and appropriate comprehensive therapy started. Sometimes, problems due to lack of pancreatic secretions can be present **at birth,** the infants presenting with intestinal obstruction due to very thick meconium **(meconium ileus).** Other patients, less severely affected, may not be diagnosed until they are in their teens or later. CF also affects the **genital tract** and most males and many females are **infertile.**

In 1989, results of a collaborative program between Canadian and American scientists revealed the nature of the genetic lesion in the majority of sufferers from CF. The gene involved in CF was the first to be cloned solely on its position determined by **linkage analysis (positional cloning)** and constituted an enormous amount of painstaking labor and a tremendous triumph for **"reverse genetics".** By reverse genetics is meant that **the gene was isolated based on its map location,** and not with the availability of chromosomal rearrangements or deletions (in contrast to, for example, the isolation of the gene involved in Duchenne muscular dystrophy). The success of this Herculean effort showed that, at least in theory, the molecular basis of any genetic disease could be revealed by similar approaches. More recent advances (eg, outcomes of the Human Genome Project) have further facilitated identification of "disease genes."

Table 57–2 summarizes the major strategies used in detecting the gene involved in the causation of CF. The protein product of the gene encodes an integral membrane protein of approximately 170 kDa. It was named **CFTR** and was found to be a **cAMP-responsive chloride transporter,** helping to explain the high chloride content of sweat found in patients with CF.

Some of the characteristics of the **gene** and of the **CFTR** protein are listed in **Table 57–3.** The major mutation initially located in the gene was deletion of three bases encoding phenylalanine residue 508 (ΔF508); in North America,

TABLE 57–2 Some Strategies Used in Isolating the Gene Involved in CF

- From study of a large number of families with CF, assignment of the gene to chromosome 7 was made by demonstrating linkage to several RFLPs on that chromosome.
- Further narrowing to a smaller region of chromosome 7 was accomplished by use of additional RFLPs.
- Chromosome jumping and chromosome walking were used to isolate clones.
- The affected region was sequenced by looking for mutations in DNA that were not present in DNA from normal individuals, for exons expressed as mRNAs in tissues affected by CF (eg, pancreas and lung), for sequences conserved across species, and for an open reading frame (indicating an expressed protein).

approximately 70% of CF carriers have this mutation. Subsequent work has revealed well **over 1000 different mutations** in the gene. A variety of different types of mutations have been found, including small deletions, insertions, and missense and nonsense mutations. Because of the importance of early diagnosis, in certain countries, **all newborn infants** are now undergoing genetic screening for CF. Elsewhere, techniques to detect deletion of ΔF508 and a number of the other most frequent mutations are being used to **confirm the diagnosis** of CF, to **detect carriers** and in **prenatal diagnosis.**

The **CFTR protein** (**Figure 57–7**) consists of two similar halves, each containing six transmembrane regions and a nucleotide (ATP)-binding fold (NBF). The two halves of the molecule are joined by a regulatory domain. F508 is located in NBF1. The protein shows similarities in structure to certain other proteins that use ATP to transport molecules across cells membranes (eg, P-glycoprotein, involved in resistance to certain cancer chemotherapeutic agents).

Normally, CFTR is synthesized on bound polyribosomes and exported to the plasma membrane, where it functions.

TABLE 57–3 Some Characteristics of the Gene for the CFTR Protein and of the Protein Itself

- About 250,000-bp gene on chromosome 7.
- 25 exons.
- mRNA of 6129 bp.
- Transmembrane protein of 1480 amino acids.
- CFTR contains two NBFs and one regulatory domain.
- Commonest mutation in CF is deletion of ΔF508 present in the first NBF.
- CFTR is a cAMP-responsive chloride transporter.
- Shows homology to other proteins that use ATP to affect transport across membranes (eg, P-glycoprotein).

Abbreviations: CFTR, cystic fibrosis transmembrane regulatory protein; F, phenylalanine; NBF, nucleotide-binding fold.

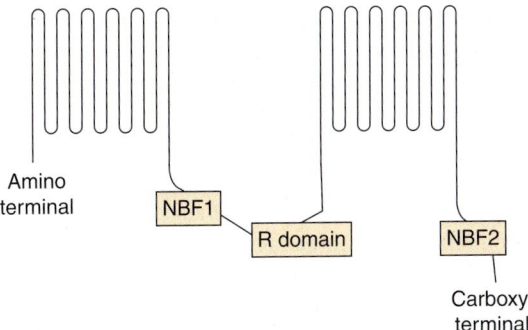

FIGURE 57–7 **Diagram of the structure of the CFTR protein (not to scale).** The protein contains 12 transmembrane segments, two nucleotide-binding folds or domains (NBF1 and NBF2), and one regulatory (R) domain. NBF1 and NBF2 bind ATP and couple its hydrolysis to transport of Cl^-; they are called ATP-binding cassettes (ABCs), a feature found in many membrane transporters. Phe 508, the major locus of mutations in CF, is located in NBF1.

Mutations can affect CFTR is a number of ways, summarized briefly in **Table 57–4**. Many mutations affect the **folding** of the protein, markedly reducing its function; this classifies CF as a **conformational disease** or a disease due to a deficiency in proteostasis (see Chapter 46 and the discussion of Alzheimer disease in this chapter). Mutations affect many other proteins in a similar manner to those summarized in Table 54–4, affecting their synthesis, processing, or function.

Figure 57–8 summarizes some of the mechanisms involved in the causation of CF.

CASE 6: DIABETIC KETOACIDOSIS (DKA)

Causation

Endocrine (due to deficiency of insulin).

History and Physical Examination

A 14-year-old girl was admitted to a children's hospital in coma. Her mother stated that the girl had been in good health until approximately 2 weeks previously, when she developed a sore throat and moderate fever. She subsequently lost her

TABLE 57–4 **Some Mechanisms by Which Mutations May Affect the CFTR[1] Protein**

I	Reduce or abolish its synthesis
II	Block its intracellular processing
III	Alters its regulation of chloride flux
IV	Alters conductance of chloride channel

[1]Many of the above mechanisms involve abnormalities of folding of the CFTR protein, and thus CF can be classified as a conformational disease or a disease due to a deficiency of proteostasis (see Chapter 46). Until gene therapy for CF is developed, various scientists are trying to find small molecules that interact with abnormally folded CFTR and partially restore its function.

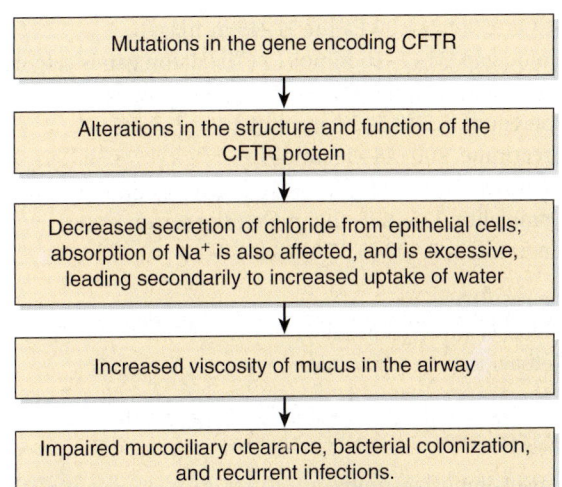

FIGURE 57–8 **Summary of possible mechanisms involved in cells in the airways of individuals with cystic fibrosis (OMIM 219700) who have pulmonary pathology.** In individuals of Caucasian origin, 70% of the mutations occur at one locus, resulting in deletion of ΔF508 from the CTR protein. However, over 1000 mutations have been identified in the *CFTR* gene. Basically, the CFTR protein acts normally as a cAMP-regulated transporter involved in secretion of Cl^-, but in addition normally inhibits absorption of Na^+ by a Na^+ channel. The viscosity of the mucus in the pancreatic ductules is also increased, leading to their obstruction. The details of how abnormalities of CFTR affect ion transport in the pancreas are somewhat different than in the lung.

appetite and generally did not feel well. Several days before admission, she began to complain of undue thirst and also started to get up several times during the night to urinate. Her family doctor was out of town and her mother felt reluctant to contact another physician. However, on the day of admission the girl had started to vomit, had become drowsy and difficult to arouse, and accordingly had been brought to the emergency department. On examination she was dehydrated, her skin was cold, she was breathing in a deep sighing manner (Kussmaul respiration), and her breath had a fruity odor. Her blood pressure was 90/60 and her pulse rate 115/min. She could not be aroused. A diagnosis of type 1 diabetes mellitus (formerly called insulin-dependent) with resulting ketoacidosis and coma (DKA) was made by the intern on duty.

Laboratory Findings

The admitting diagnosis was confirmed by the laboratory findings, kindly supplied by Dr. ML Halperin:

Plasma or serum results (normal levels in parentheses, SI Units):

- Glucose, 50 (4.2–6.1 mmol/L)
- Ketoacids ++++ (trace)
- Bicarbonate, 6 (22–30 mmol/L)
- Urea nitrogen, 15 (2.5–7.1 mmol/L)
- Arterial blood pH, 7.07 (7.35–7.45)
- Na^+, 136 (136–146 mmol/L)
- Cl^-, 100 (102–109 mmol/L)

- pCO$_2$ 2.7 (4.3–6.0 kPa [or 32–45 mmHg])
- Anion gap, 31 (7–16 mmol/L) (The anion gap is calculated from plasma Na$^+$ – [Cl$^-$ + HCO$_3^-$].)
- Potassium, 5.5 (3.5–5.0 mmol/L)
- Creatinine, 200 (44–80 μmol/L)
- Albumin 50 (41–53 g/L)
- Osmolality, 325 (275–295 mOsm/kg serum water)
- Hematocrit, 0.500 (0.354–0.444)

Urine results:
- Glucose, ++++ (normal –)
- Ketoacids, ++++ (normal –)

Treatment

The most important initial measures in treatment of diabetes ketoacidosis are intravenous administration of **insulin** and **saline** solution. This patient was given intravenous insulin (10 units/h) added to 0.9% NaCl. Glucose was withheld until the level of plasma glucose fell below 15 mM. Insulin and glucose facilitate entry of K$^+$ into cells. KCl was also administered cautiously, with plasma K$^+$ levels monitored every hour initially. **Continual monitoring of K$^+$ levels** is extremely important in the management of diabetic ketoacidosis because inadequate management of K$^+$ balance is the main cause of death. Bicarbonate is not needed routinely, but may be required if acidosis is very severe.

Discussion

The precise **cause** of type 1 (insulin-dependent) diabetes mellitus has not been elucidated, and is under intense investigation. Genetic, environmental and immunologic factors have all been implicated. A very tentative scheme of the chains of events is the following. Patients with this type of diabetes have a **genetic susceptibility** (a large number of genes, including histocompatibility genes located on chromosome 6, have been implicated), which

may predispose to a **viral infection** (eg, by coxsackie or rubella viruses). The infection and consequent inflammatory reaction may alter the antigenicity of the surface of the pancreatic B cells and set up an autoimmune reaction involving both cytotoxic antibodies and T lymphocytes. This leads eventually to widespread destruction of beta cells, resulting in type I diabetes mellitus. Perhaps the **sore throat** this patient had several weeks before admission reflected the initiating viral infection.

The marked **hyperglycemia, glucosuria, ketonemia,** and **ketonuria** confirmed the diagnosis of DKA. The **low pH** indicated a severe acidosis due to the greatly increased production of acetoacetic acid and β-hydroxybutyric acid. The low levels of **bicarbonate** and **pCO$_2$** confirmed the presence of a **metabolic acidosis** with partial respiratory compensation (the hyperventilation). Calculation of the **anion gap** is useful in a number of metabolic situations. In this case it is elevated because of the presence of excess ketoacids in the blood. There are a number of other causes of elevation of the anion gap, including lactic acidosis and intoxication by methanol, ethylene glycol, and salicylates.

The elevated values of **urea** and **creatinine** indicated some renal impairment (due to diminished renal perfusion because of low blood volume secondary to dehydration), dehydration, and increased degradation of protein. A high plasma level of **potassium** is often found in DKA owing to a lowered uptake of potassium by cells in the absence of insulin. Thus, the clinical picture in DKA reflects the abnormalities in the metabolism of carbohydrate, lipid, and protein that occur when plasma levels of insulin are sharply reduced. The increased **osmolality** of plasma due to hyperglycemia also contributes to the development of coma in diabetic ketoacidosis. It should be apparent that the rational treatment of a patient with DKA depends on thorough familiarity with the actions of insulin.

A general scheme of the events occurring in DKA is given in **Figure 57–9**.

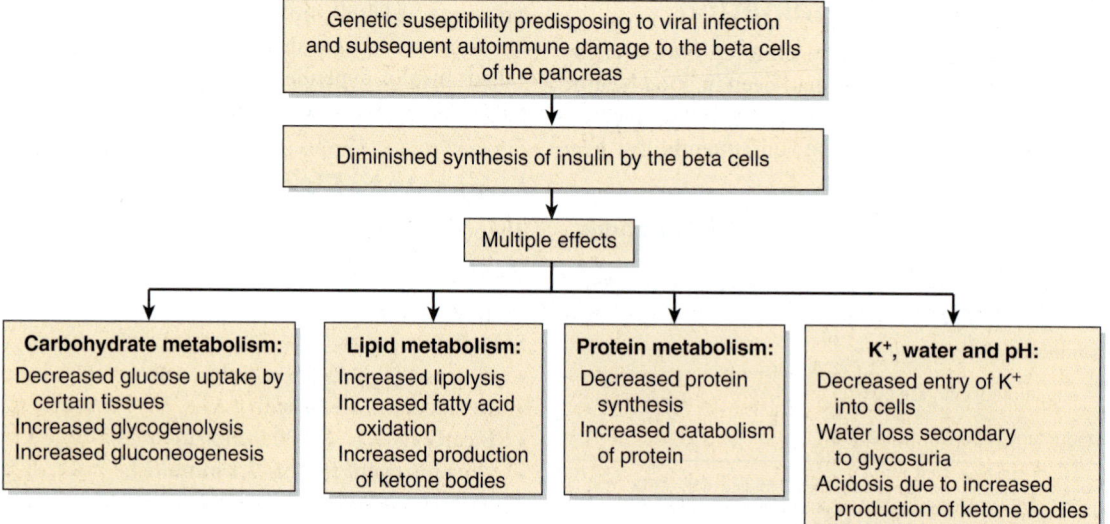

FIGURE 57–9 Summary of some mechanisms involved in the causation of the ketoacidosis of type 1 diabetes mellitus (OMIM 222100).

CASE 7: DUCHENNE MUSCULAR DYSTROPHY (DMD)

Causation

Genetic (mutations in the gene encoding the protein dystrophin).

History and Physical Examination

A 4-year-old boy was brought to a children's hospital clinic. His mother was concerned because she had noticed that her son was walking awkwardly, fell over frequently, and had difficulty climbing stairs. There were no siblings, but the mother had a brother who died at age 19 of muscular dystrophy. The pediatrician on call noted muscular weakness in both the pelvic and shoulder girdle. Modest enlargement of the calf muscles was also noted. Because of the muscle weakness and its distribution, the pediatrician made a provisional diagnosis of DMD.

Laboratory and Other Findings

The activity in serum of **creatine kinase** (CK) was markedly increased. It was decided to proceed directly to **mutation analysis** using a sample of the patient's lymphocytes. This showed a large deletion in the gene for dystrophin, confirming the diagnosis of DMD. This saved the patient from being tested by electromyography and also from having a muscle biopsy performed; these tests, along with Western blotting for detection of dystrophin, were routinely performed prior to the availability of mutation analysis, and still may be performed in certain circumstances.

Discussion

The family history, the typical distribution of muscular weakness, the elevation of the activity in serum of CK, and the results of mutation analysis confirmed the provisional diagnosis of DMD. This is a severe **X-linked** degenerative disease of muscle. It has a prevalence of approximately 1:3500 live male births. It affects **young boys,** who first show loss of strength in their proximal muscles, leading to a waddling gait, difficulty in standing up, and eventually very severe weakness. Death generally occurs from respiratory insufficiency.

The **cause of DMD** was revealed in 1986–1987. Various studies led to localization of the defect to the middle of the short arm of the X chromosome and to subsequent identification of a segment of DNA that was **deleted** in patients with DMD. Using the corresponding nondeleted segment from normal individuals, a cDNA was isolated derived by reverse transcription from a transcript (mRNA) of 14 kb that was expressed in fetal and adult skeletal muscle. This was cloned and the protein product was identified as **dystrophin,** a 400-kDa red-shaped protein of 3685 amino acids. Dystrophin was absent or markedly reduced in electrophoretograms of extracts of muscle from patients with DMD and from mice with an X-linked muscular

dystrophy. Antibodies against dystrophin were used to study its localization in muscle; it is associated with the **sarcolemma** (plasma membrane) of normal muscle and was absent or markedly deficient in patients with DMD. A less severe reduction in the amount of dystrophin, or a reduction in its size, is the cause of **Becker muscular dystrophy,** a milder type of muscular dystrophy. Whereas the gene **deletions** and **duplications** found in DMD tend to cause frame-shift mutations, the same types of mutations in Becker MD are generally in-frame, and thus synthesis of dystrophin is not as affected in the latter.

Dystrophin appears to have four domains, two of which are similar to domains present in α-actinin (another muscle protein) and one to a domain in spectrin, a protein of the cytoskeleton of the red blood cell. As shown in Figure 49–11, dystrophin interacts with actin, syntrophin, and β-dystroglycan. Its **function** is not clear, but it may play a role in transmembrane signaling and in stabilizing the cytoskeleton and sarcolemma.

Deficiency of dystrophin may affect the integrity of the sarcolemma, resulting in increased osmotic fragility of dystrophic muscle or permitting excessive influx of Ca^{2+}. The **gene** coding for dystrophin is the largest human gene recognized to date (~2500 kb, 79 exons), which helps explain the observation that approximately one third of cases of DMD are new mutations. Attempts are being made to **produce dystrophin** by recombinant DNA technology and perhaps eventually administer it to patients. The availability of **probes** for dystrophin facilitates **prenatal diagnosis** of DMD by chorionic villus sampling or amniocentesis. The demonstration of lack of dystrophin as a cause of DMD has been one of the major accomplishments of the application of molecular biology to the study of human diseases.

There are many **different types** of muscular dystrophy. The molecular causes of many of them have been elucidated. Not surprisingly, perhaps, they have been found to be due to a variety of mutations in genes that encode specific muscle proteins, such as those shown in Figure 49–11 and others not shown. The various muscular dystrophies can be classified on the basis of their clinical features (eg, affecting the limb girdle, etc), or increasingly on the basis of the genes or proteins affected by the causative mutations. Dystrophin also occurs in heart muscle and the brain. Its occurrence in the former can result in **cardiomyopathy.** The absence of dystrophin in **brain** results in an IQ of less than 75 being observed in some 25% of boys with DMD.

Treatment

At present, no specific therapy for DMD exists. Treatment in this case was thus essentially **symptomatic.** He was enrolled in a special **muscular dystrophy clinic,** started on **prednisone** (which can slow down the progress of DMD for a few years) and encouraged to undertake mild **exercise.** A **physiotherapist** was available when needed. **Portable respirators** have proven very useful when breathing is affected. The mother was advised to seek genetic **counseling** regarding future

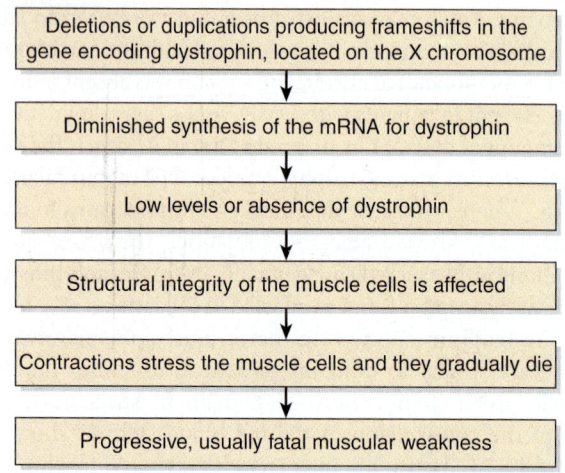

FIGURE 57–10 Summary of mechanisms involved in the causation of Duchenne muscular dystrophy (OMIM 310200).

pregnancies. At different times, a variety of therapeutic measures intended to benefit patients with DMD have been used. These include the use of myoblast transfer (to supply dystrophin), antisense oligonucleotides (to skip mutated dystrophin gene exons), CoE Q10 (to possibly increase muscle strength), creatine monohydrate (to perhaps help build up muscle mass), pentoxyfylline (anti-inflammatory), and gentamicin (may ignore premature stop codons in the dystrophin gene). None appears to have been very successful. A small trial using dystrophin gene transfer was scheduled to start in the United States in January 2008.

Figure 57–10 summarizes the mechanisms involved in the causation of DMD.

CASE 8: ETHANOL INTOXICATION, ACUTE

Causation

Chemical (due to excess intake of ethanol).

History and Physical Examination

A 52-year-old man was admitted to the emergency department in a coma. Apparently, he had become increasingly depressed after the death of his wife 1 month previously. Before her death he had been a moderate drinker, but his consumption of alcohol had increased markedly over the last few weeks. He had also been eating poorly. His married daughter had dropped round to see him on Sunday morning and had found him unconscious on the living room couch. Two empty bottles of rye whiskey were found on the living room table. On examination, he could not be roused, his breathing was deep and noisy, alcohol could be smelled on his breath, and his temperature was 35.5°C (normal; 36.3–37.1°C). The diagnosis on admission was coma due to excessive intake of alcohol.

Laboratory Findings

The pertinent laboratory results were alcohol 500 mg/dL, glucose 2.7 mmol/L (normal: 4.2–6.1), lactate 8.0 mmol/L (normal: 0.5–1.6), and blood pH of 7.21 (normal: 7.35–7.45).

These results were consistent with the admitting diagnosis, accompanied by a metabolic acidosis.

Treatment

He was started on intravenous normal saline, and then, because of the very high level of blood alcohol and the coma, it was decided to start **hemodialysis** immediately. This directly eliminates the toxic ethanol from the body but is only required in very serious cases of ethanol toxicity. In this case, the level of blood alcohol fell rapidly and the patient regained consciousness later the same day. Intravenous **glucose** (5%) was administered after dialysis was stopped to counteract the hypoglycemia that this patient exhibited. The patient made a good recovery and was referred for **psychiatric counseling.**

Discussion

Excessive consumption of alcohol is a major health problem in most societies. The present case deals with the acute, toxic effects of a very large intake of ethanol. A related problem, which is not discussed here but which has many biochemical aspects, is the development of **liver cirrhosis** in individuals who maintain a high intake of ethanol (eg, 80 g of absolute ethanol daily for more than 10 years).

From a biochemical viewpoint, the major question concerning the present case is how does ethanol produce its diverse acute effects, including coma, lacticacidosis, and hypoglycemia? The clinical viewpoint is how best to treat this condition.

The **metabolism of ethanol** was described in Chapter 25; it occurs mainly in the liver and involves two routes. The first and major route uses **alcohol dehydrogenase** and **acetaldehyde dehydrogenase,** converting ethanol via acetaldehyde to acetate (Chapter 25), which is then converted to acetyl-CoA. NADH + H⁺ are produced in both of these reactions. The intracellular **ratio of NADH/NAD⁺** can thus be increased appreciably by ingestion of large amounts of ethanol. In turn, this can affect the Keq of a number of important metabolic reactions that use these two cofactors. High levels of NADH favor **formation of lactate** from pyruvate, accounting for the lactic acidosis. This diminishes the concentration of pyruvate (required for the **pyruvate carboxylase** reaction, Chapter 17) and thus **inhibits gluconeogenesis.** In severe cases, when liver glycogen is depleted and no longer available for glycogenolysis, **hypoglycemia** results. The second route involves a **microsomal cytochrome P450** (microsomal ethanol oxidizing system), also producing acetaldehyde (Chapter 25). **Acetaldehyde** is a highly reactive molecule and can form adducts with proteins, nucleic acids, and other molecules. It appears likely that its ability to react with various molecules is involved in the causation of the toxic effects of ethanol.

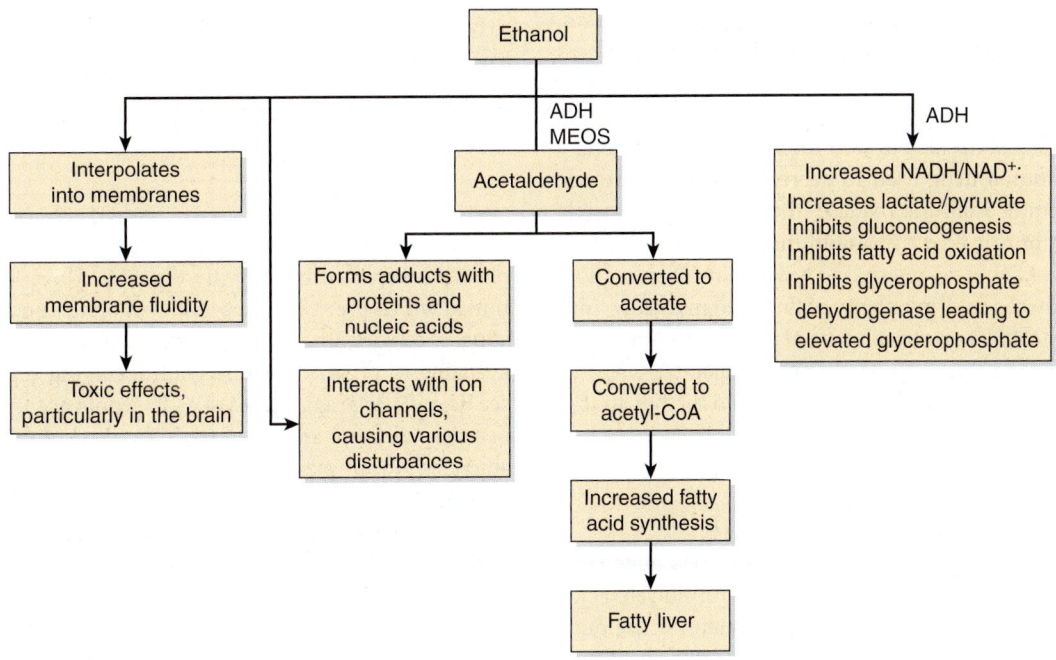

FIGURE 57–11 **Summary of some mechanisms involved in acute ethanol toxicity.** (ADH, alcohol dehydrogenase; MEOS, microsomal ethanol oxidizing system, involving cytochrome P450 species CYP2E1.)

Apart from the metabolic studies just mentioned, early studies suggested that ethanol produced many of its acute intoxicating effects by **disordering lipids** in the cell membranes of **neurons.** More recent studies have focused on **proteins,** particularly **receptors** and **ion channels** (see Chapters 40 and 41), which play important roles in the normal function of neurons and other cells. A variety of studies have revealed that ethanol **inhibits,** or in some cases **activates** various receptors and ion channels. These include GABA (γ-amino-butyric acid), nicotinic acetylcholine, glutamate, and NMDA (N-methyl-D-aspartate) receptors, and various K^+ and Ca^{2+} channels. Alterations of the activities of receptors can affect intracellular signaling pathways, contributing to the effects of ethanol. One way in which ethanol may disturb the functions of these proteins is by **replacing water molecules** at various critical sites. Further research should help account for the acute effects of ethanol on mental status. **Figure 57–11** summarizes some of the major mechanisms involved in the causation of toxicity by ethanol.

CASE 9: GOUT, ACUTE

Before studying this case, the reader is advised to consult the material in Chapter 33 dealing with uric acid.

Causation

Deposition of crystals of monosodium urate (MSU) in one or more joints and various tissues. The great majority of cases (~90%) are associated with **decreased renal excretion of MSU,** but in certain cases, **increased production of MSU** is involved, and increased **dietary intake of purines** may play a role.

History and Physical Examination

A moderately obese 64-year-old male appeared at the emergency department complaining of severe pain of 12-hour duration in his left big toe. He stated that he regularly had at least two to three drinks of Scotch whisky every evening after work. He had no other significant medical history. On examination, his left big toe was found to be red and markedly swollen around the metacarpophalangeal joint, and exquisitely sensitive. There was no evidence of arthritis elsewhere. Because of the history and location of the affected joint, the intern on duty suspected that the patient was having an attack of acute gout. She ordered a number of lab tests, including a white cell count, determination of serum uric acid, and X-ray examination of the affected joint. The serum level of uric acid, was 0.61 mmol/L (normal, 0.18–0.41 mmol/L in males); the white cell count was at the high end of normal. The x-ray findings were nonspecific; no indication of chronic arthritis was evident. Under local anesthesia, arthrocentesis was performed on the affected joint, and a small amount of synovial fluid withdrawn and sent to the laboratory for detection of cells and crystals. Typical needle-shaped crystals of MSU showing negative birefringence were detected in the synovial fluid, as were neutrophils.

Treatment

The patient was started on a suitable dose of a **nonsteroidal anti-inflammatory drug** (NSAID) to relieve the acute inflammation and pain. He was referred to a rheumatologist for ongoing treatment; she continued administration of the NSAID.

Several months later he had another acute attack of joint pain, this time in his right knee. His plasma level of uric acid was 0.57 mmol/L. **Daily excretion of uric acid** was measured and found to be ~9.0 mmol (~1500 mg) (normal 3.6–5.4 mmol/d). The rheumatologist decided to start him on long-term therapy with **allopurinol,** a drug used to decrease formation of uric acid by inhibiting xanthine oxidase, the enzyme responsible for formation of uric acid from xanthine (see Chapter 33). In addition, the patient was referred to a dietitian regarding losing weight, advised to drink plenty of fluids, to markedly limit his intake of alcohol, to restrict intake of purine-rich foods (eg, anchovies and red meat), and started on a regular exercise program. Until the present, the patient has had no further attacks of acute arthritis.

Discussion

Uric acid is formed from purine nucleosides (eg, adenosine and guanosine) produced by the breakdown of nucleic acids and other molecules (eg, ATP), and in humans is **the end-product of purine catabolism.** The daily synthesis rate is estimated to be 1.8 mmol (~300 mg), with a total body pool of approximately 7.2 mmol (1200 mg in adult males, and about one-half that in females). In individuals with gout, the total body pool can be as large as 180 mmol (30,000 mg).

The enzyme involved in formation of uric acid is **xanthine oxidase** (see Chapter 33). Humans do not possess the peroxisomal enzyme uricase (urate oxidase), which is involved in the degradation of uric acid to allantoin. Uricase is now available as a medication to help lower levels of uric acid. Approximately 70% of uric acid is excreted by the kidneys and the remaining by the gut. Uric acid exhibits **antioxidant properties** (see Chapter 45); the possible significance of this is under investigation.

Gout is a type of **arthritis,** acute or chronic, due to deposition of crystals of MSU usually in relatively avascular areas, such as cartilage and tissues around joints, and also where body temperature is lower (eg, the ears, distal ends of the limbs). When crystals of MSU are deposited in synovial fluid, they elicit **an inflammatory reaction.** In acute gout, this consists mainly of neutrophils. The inflammatory reaction causes the characteristic signs and symptoms of heat, pain, swelling, and redness experienced in acute gout. It is generally important to ascertain that the **characteristic crystals** of MSU are actually present in the synovial fluid of an affected joint, as other crystals (eg, calcium pyrophosphate) may cause signs and symptoms similar to gout. One joint is usually affected initially (ie, monoarticular arthritis), often the metacarpophalangeal joint of the big toe, as in this case. One factor helping to account for this is that the temperatures of the joints of the lower extremities are lower than elsewhere in the body.

MSU has a **solubility** value in plasma of ~0.42 mmol/L at 37°C. It is much more soluble than uric acid, which is the major ionic species below pH 5.75 (the pKa value for the dissociation of uric acid to urate). This difference is particularly important in **urine,** where **calculi** of uric acid may form at

FIGURE 57–12 **One common sequence of events leading to chronic gout.**

acidic values of pH. When the above concentration is exceeded, plasma is **supersaturated** with respect to MSU. The concentration at which precipitation of MSU in tissues and joints occurs appears to vary, for unknown reasons. Before treatments were available to prevent chronic gout occurring (eg, allopurinol, see below), large aggregates of MSU would accumulate in various tissues; these are called **tophi,** and can attain a considerable size. Tophi may still occur if gout is not diagnosed and treated early.

Gout is usually preceded and accompanied by **hyperuricemia** (plasma uric acid level >0.41 mmol/L). The sequence shown in **Figure 57–12** is often involved. Chronic gout can be prevented if appropriate treatment is instituted following an attack of acute gout. Hyperuricemia is much commoner in **males,** although its incidence in females increases after menopause. It should be noted that approximately 30% of people experiencing an attack of gout may have normal levels of MSU in the plasma. **Hyperuricemia** is caused by decreased renal excretion, increased production, or increased intake of uric acid. **Decreased renal excretion** is involved in the great majority of cases of gout, and genetic factors are likely involved. Many kidney diseases affect renal excretion, as does acidosis caused by various metabolic conditions. A variety of **drugs** (eg, certain diuretics and also salicylates) interfere with excretion of uric acid. Handling of MSU by the kidneys is complex, including phases of glomerular filtration, reabsorption, secretion, and further reabsoption in various parts of the renal tubule. The precise contributions of alterations of these phases to the causation of hyperuricemia have not been clearly defined as yet. **Increased production** can occur due to certain enzyme abnormalities (eg, deficiency of hypoxanthine-guanine phosphoribosyl transferase [HGPRT] and overactivity of PRPP synthetase) although these are uncommon (see Chapter 33). In Lesch–Nyhan syndrome, mutations of the gene encoding HGPRT are involved (see Chapter 33), and gout can be a feature. **Death of cancer cells** caused by chemotherapy leads to degradation of their nucleic acids, and thus to increased formation of purines. **Increased intake** can occur via ingestion of purine-rich foods, such as sweetbreads and certain meats, although this is not thought to be a major contributor to elevation of serum uric acid.

The role of **intake of alcohol** in precipitating gout has long been recognized. Ingestion of ethanol can lead to formation of **lactic acid,** which inhibits secretion of uric acid. In addition, ethanol appears to **promote breakdown of ATP,** leading to increased production of purines from which uric acid is formed. Also, the solubility of MSU is markedly diminished as the **pH** in tissues drops, a situation favored by increased production of lactic acid.

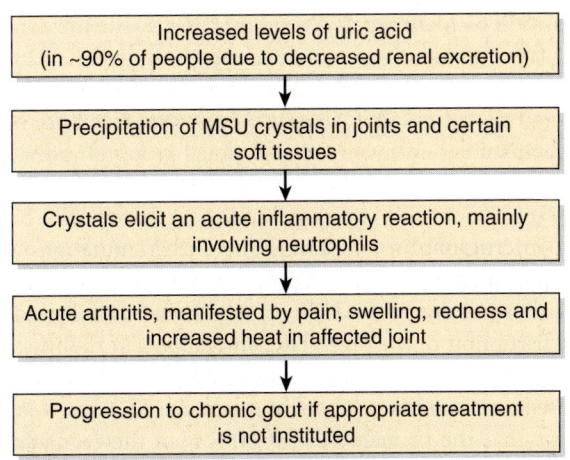

FIGURE 57-13 Simplified scheme of some of the events involved in the causation of gout.

Renal stones (urate calculi) develop frequently in patients with chronic gout; the risks of these are lessened by therapy with allopurinol.

For **treatment** of the acute inflammation and pain of acute gout, an **NSAID** is generally used. **Colchicine** is also effective in blocking inflammation caused by MSU crystals. It is known to bind to free tubulin causing the depolymerization of microtubules (Chapter 49); this may prevent the movement of neutrophils into an area containing crystals of MSU. However, it may cause nausea and vomiting. A **corticosteroid** or **ACTH** may also be employed for their anti-inflammatory effects. For long-term management, intended to prevent or reverse any complications that may have arisen, **allopurinol** is used to chronically inhibit the production of uric acid from xanthine. **Uricase** may also be used in some cases.

Uricosuric drugs (which increase the rate of excretion of uric acid) may be used instead of allopurinol if under-excretion of urate is involved; these include probenecid, sulfinpyrazone, and benzbromarone. Any **accompanying conditions** should be addressed (eg, obesity, hypertension, hypertriglyceridemia, alcoholism, renal disease). If gout is treated early, it is compatible with a normal lifespan.

A simplified scheme of the causation of acute gout is shown in **Figure 57-13.**

CASE 10: HEREDITARY HEMOCHROMATOSIS

Before studying this case, the reader is advised to consult the material in Chapter 50 on iron and its metabolism.

Causation

Genetic (due to mutations in the *HFE* gene or certain other genes whose protein products affect the metabolism of iron).

History and Physical Examination

A 50-year-old man visited his family doctor complaining of fatigue, low libido, and moderate generalized joint pains of approximately 1-year duration. The joint pains were mostly in the fingers, wrists, hips, knees, and ankles. His parents, both dead, were born in Scotland but emigrated to Canada in early adulthood. The patient had no siblings and did not smoke or drink. He occasionally took acetaminophen for his joint pains, but otherwise was not receiving any medication. An uncle had died of liver cancer about 10 years previously. In addition to stiffness and slight swelling over some joints, the physician noted grayish skin pigmentation, most prominent in exposed parts, and for that reason referred the patients to an internist, who also noted that the liver edge was firm and palpable just below the costal margin. The internist suspected hereditary hemochromatosis and ordered appropriate laboratory tests as well as x-rays of the hands, hips, knees, and ankles.

Laboratory Findings

Normal reference values in parentheses:

- Hb, 120 g/L (133–162 g/L, males)
- RBC, 4.6×10^{12}/L (4.30–5.60×10^{12}/L, males)
- Glucose (fasting), 5 mmol/L (4.2–6.1 mmol/L)
- Alanine aminotransferase [ALT], 1.8 μkat/L or 105 units/L (0.12–0.70 μkat/L or 7–41 units/L)
- Plasma iron, 50 μmol/L (7–25 μmol/L)
- Total iron-binding capacity, 55 μmol/L (45–73 μmol/L)
- Transferrin saturation with iron 82% (16–35%)
- Serum ferritin, 3200 μg/L, (29–248 μg/L, males)

X-rays of the joints showed loss of articular cartilage, narrowing of joint spaces, and diffuse demineralization.

In view of the above findings, it was decided to perform a **liver biopsy.** Histologic examination revealed moderate periportal fibrosis. Hemosiderin (aggregates of ferritin micelles) was visible as golden brown granules in both parenchymal and bile duct epithelial cells; with Prussian blue staining, iron was markedly visible in these cells. **Quantitative measurement of iron** in the biopsy material revealed a marked elevation of iron (8100 μg/g dry weight; normal: 300–1400 μg). The laboratory and other findings were consistent with a diagnosis of hereditary hemochromatosis.

Treatment

The treatment of hereditary hemochromatosis is relatively simple, consisting of withdrawing blood from the patient until the excess iron in the body is brought down to near normal values. This is accomplished initially by weekly **phlebotomy** of approximately 500 mL of blood (250 mg of iron). The efficacy of treatment is assessed by monthly monitoring of serum ferritin and transferrin saturation. Once these parameters have reached satisfactory levels, phlebotomy is required only once every 3 months.

Chelation therapy (eg, with deferoxamine) is rarely required and is much more expensive, as well as less effective than phlebotomy. Foods high in iron should be avoided.

Complications such as hepatic cirrhosis, diabetes mellitus, infertility, and cardiac problems must be treated appropriately. **Screening** of family members is recommended in order to give them early treatment if necessary. If the condition is recognized early (eg, before hepatic cirrhosis develops), the prognosis is excellent. **Hepatocellular carcinoma** may develop in patients with cirrhosis. The **arthropathy** of hemochromatosis does not respond well to treatment and may continue relentlessly despite normalization of body iron parameters.

Discussion

This Section of Case 10 was written by Drs Joe Varghese and Moly Jacob.

The key feature of hemochromatosis is **an increase in total body iron** sufficient to cause tissue damage. Total body iron ranges between 2.5 and 3.5 g in normal adults; in hemochromatosis, it usually exceeds 15 g. It is critical to diagnose the condition early to prevent complications such as cirrhosis and liver cancer. Hepatomegaly, skin pigmentation, diabetes mellitus, heart disease, arthropathy, and hypogonadism are usually found in full-blown cases. However, it is common to find patients with only one or two of the above-mentioned features. Therefore, to achieve an early diagnosis, a high index of suspicion is necessary. **Elevated levels** of **transferrin saturation** and **serum ferritin** are the most useful tests for early diagnosis.

In hereditary hemochromatosis, the **absorption of iron** from the small intestine is greatly increased. There are no physiological mechanisms to remove excess iron from the body. Free iron is **toxic** because of its ability to generate free radicals (Chapter 45). The accumulated iron damages organs such as the liver, pancreatic islets, and the heart. **Melanin** and iron accumulate in the skin, accounting for the slate-gray color often seen. The precise cause of melanin accumulation is not clear. The frequent co-existence of diabetes mellitus (due to pancreatic islet damage) and the skin pigmentation led to use of the term **bronze diabetes** for this condition.

Hereditary hemochromatosis is an **autosomal recessive** disorder. It is common in Europe (carrier frequency of approximately 1:10), particularly in Ireland and Scotland. Emigrants from these countries have contributed to dissemination of the affected gene around the world. Since 1976, it has been known that there is **an association** between HLA antigens and hereditary hemochromatosis. In 1996, Feder and colleagues isolated a gene, now known as *HFE*, located on chromosome 6 (6p21.3) in close proximity to the major histocompatibility complex (MHC) genes. The encoded product was found to be related to the MHC class 1 antigens. *HFE* has been found to exhibit **three missense mutations** in individuals with hereditary hemochromatosis. The most frequent mutation is one that changes a cysteinyl residue at position 282 to a tyrosyl residue (C282Y), resulting in conformational changes in the structure of the protein. The other mutation involves a change of a histidyl residue

at position 63 to an aspartyl residue (H63D). The incidence of these two mutations varies in different ethnic groups. C282Y is less frequent in Italians than in Northern Europe. A third less frequent mutation (S65C) has also been found, but has not as yet been studied in much detail. A small group of individuals are compound heterozygotes (C282Y/H63D).

HFE is normally expressed on the surface of cells, bound to β_2-**microglobulin** and **TfR1**. The C282Y mutation inhibits the formation of this complex, resulting in decreased cell surface expression of the HFE protein. It was initially thought that disruption of the HFE-TfR1 interaction in the intestinal crypt cell was responsible for increased duodenal iron absorption seen in hereditary hemochromatosis. However, it is now known that the primary role of HFE is in the regulation of hepatic hepcidin expression (described in Chapter 50). Mutations in the HFE gene are associated with **low circulating hepcidin levels.** Hepcidin binds to and triggers the internalization and degradation of ferroportin, which is the exporter of iron from enterocytes, macrophages, and other cells. Low levels of hepcidin therefore **increase ferroportin expression** on the basolateral membrane of enterocytes. Mutations in **genes other than HFE** (hepcidin, TfR2, HJV) have also been implicated in hereditary hemochromatosis, though these are comparatively rare (see Table 50–5). The proteins encoded by these genes play important roles in hepatic hepcidin expression. **A low plasma hepcidin level** is a characteristic finding in these conditions. A tentative scheme of the main events in causation of hereditary hemochromatosis is given in **Figure 57–14**.

Penetrance of hereditary hemochromatosis is low, and is lower in women than in men. Penetrance is the fraction of individuals with a genotype known to cause a disease that has signs and symptoms of the disease. Genetic testing has been

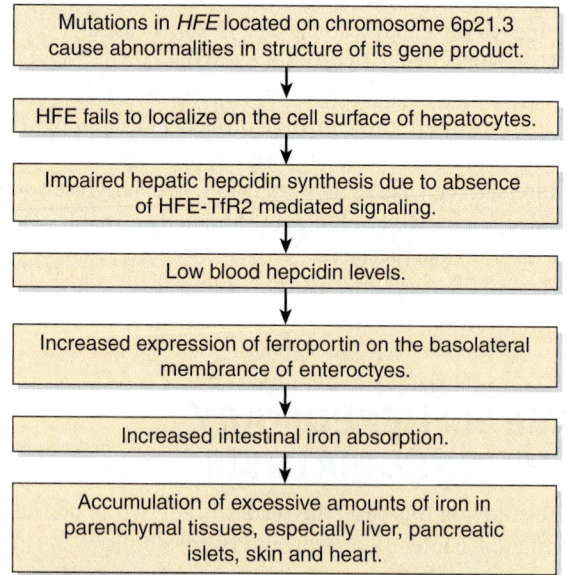

FIGURE 57–14 Tentative scheme of the main events in causation of hereditary hemochromatosi (OMIM 235200). The two principal mutations seen in HFE are CY282Y and H63D. Mutations in genes other than *HFE* also cause hemochromatosis.

evaluated but is not presently recommended, except in family members of patients with hereditary hemochromatosis. **Tests for HFE mutations** in individuals with elevated serum iron concentrations may be useful.

CASE 11: HYPOTHYROIDISM, PRIMARY

Before studying this case, the reader is advised to consult the material in Chapter 41 dealing with the thyroid gland.

Causation

Primary hypothyroidism is a state due to **deficiency of the thyroid hormones,** usually due to impaired function, damage to or surgical removal of the thyroid gland. Specific causes are discussed below.

History, Physical Examination, and Laboratory Tests

A 57-year-old woman visited her family physician complaining of chronic fatigue and sluggishness for a number of years; this was her first visit to her doctor for 5 years. On questioning, a history of constipation and feeling cold (cold intolerance) was elicited. She had two adult children; her last menstrual period had occurred some seven years previously. A sister had pernicious anemia and a maternal aunt had had "a thyroid problem."

On examination, the patient was moderately obese. She answered questions slowly, with little change of expression, her voice sounded coarse, and her tongue appeared moderately swollen. Some puffiness around her cheeks was also evident. Palpation of the neck revealed that her thyroid gland was rather firm in consistency, and modestly enlarged. Her blood pressure was mildly elevated and her deep tendon reflexes were delayed. Some clinical findings in a case of hypothyroidism are summarized in **Figure 57–15.** On the basis of the history and clinical examination, her doctor suspected that the lady had hypothyroidism. She ordered various tests to investigate this possibility; the following results were these of relevance:

- Thyroid stimulating hormone (TSH): 20 mIU/L (normal range, 0.34–4.25 mIU/L)
- Thyroxine (T_4), free: 4.0 pmol/L (normal 10.3–21.9 pmol/L)
- Anti-thyroperoxidase (anti-TPO) antibodies: ++++ (normal trace)
- Cholesterol, total: 6.20 mmol/L (normal <5.17 mmol/L)
- Chest x-ray: Revealed a small pericardial effusion.
- ECG: Revealed bradycardia and low voltage complexes, but no evidence of ischemia or arrhythmias.
- Hemoglobin and RBC count: Results consistent with a mild normocytic anemia.

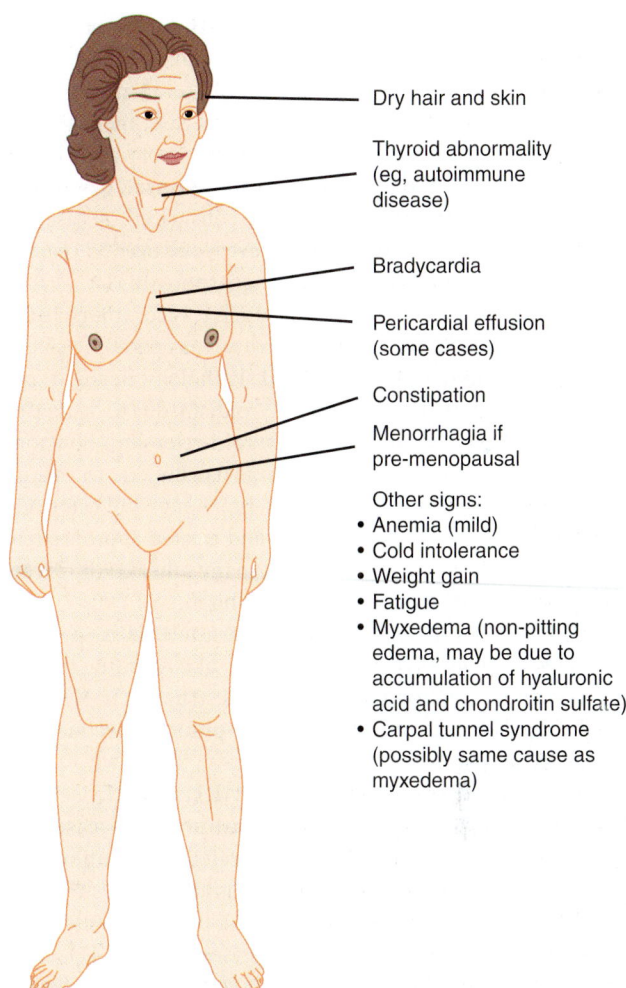

Dry hair and skin

Thyroid abnormality (eg, autoimmune disease)

Bradycardia

Pericardial effusion (some cases)

Constipation

Menorrhagia if pre-menopausal

Other signs:
- Anemia (mild)
- Cold intolerance
- Weight gain
- Fatigue
- Myxedema (non-pitting edema, may be due to accumulation of hyaluronic acid and chondroitin sulfate)
- Carpal tunnel syndrome (possibly same cause as myxedema)

FIGURE 57–15 **Some of the major signs of hypothyroidism.**

Treatment

The clinical history, physical examination, and lab results were all consistent with primary hypothyroidism. Accordingly, the patient was started on a low dose of **thyroxine** (T_4). It is important to begin therapy with a small dose of T_4, as larger doses can precipitate serious cardiac events, due to the changes in metabolism caused by administration of the hormone. The dosage of T_4 is gradually increased at 6–8 week intervals, up to a maximum of approximately 125 µg. Assessment of progress is made by assays of TSH, which should eventually decline to normal range and be sustained there. Regular assessments are important. Once started, T_4 therapy is generally **continued for life.**

Discussion

Primary hypothyroidism is a relatively **prevalent** condition, and is probably the most common endocrine problem (excluding diabetes mellitus) seen in clinical practice. The most frequent cause worldwide is **deficient intake of iodine.** In North America (as in the present case) and other developed countries, a major cause is **Hashimoto disease,** an autoimmune condition affecting the thyroid. **Other causes** include ^{131}I ablation of the thyroid, surgical resection of the thyroid, and the use of drugs for treating hyperthyroidism. It is more common in **females** than in males. In this case, the diagnosis was relatively easy because of the classical history and clinical findings. However, it often has an **insidious onset,** developing gradually over years, and may not be considered. A case can be made for routine assay of TSH in everyone over 35 years of age, but there is as yet no consensus regarding this.

Secondary hypothyroidism is much less common, and is due to decreased secretion of TSH by various pathologic conditions affecting the pituitary gland. Pathology in the **hypothalamus** can cause **tertiary** hypothyroidism, due to decreased secretion of the hypothalamic thyroid-releasing hormone (TRH). **Congenital** hypothyroidism is usually due to various blocks in manufacture of the thyroid hormones and can result in **cretinism** if not diagnosed early and treated appropriately. All newborns in North America and many other countries are routinely screened for levels of TSH at birth.

Detection of increased levels of serum **TSH** is the most useful test for hypothyroidism. As levels of circulating thyroid hormones (T_4, T_3) drop due to thyroid destruction in Hashimoto disease, feedback inhibition on the pituitary declines and levels of TSH rise.

The presence of **elevated TSH** and **decreased T_4** is highly indicative of hypothyroidism. In **Hashimoto disease,** the thyroid gland becomes heavily infiltrated by **lymphocytes** and other inflammatory cells, which gradually destroy and replace much of the gland resulting in a progressive decrease of secretion of the thyroid hormones (see Chapter 41), causing the hypothyroid state. The lymphocytes include activated $CD4^+$ T cells specific for various thyroid antigens. **Various autoantibodies** can be detected in serum from patients with Hashimoto disease; among these often measured at present are **anti-thyroperoxidase (anti-TPO) antibodies,** which serve as markers of Hashimoto disease. Very often there is a **family history** of the disease or of other autoimmune conditions, indicating a **genetic** contribution. In the present case, a sister of the patient's mother had a family history of "thyroid disease" and a sister of the patient had pernicious anemia, another autoimmune condition.

It is important to consider hypothyroidism as a diagnosis, as **early treatment** can make a huge difference to a patient's quality of life.

A simplified scheme of the causation of hypothyroidism is given in **Figure 57–16**.

CASE 12: KWASHIORKOR, ONE TYPE OF PROTEIN-ENERGY MALNUTRITION (PEM)

Before studying this case, the reader is advised to consult the material in Chapters 43 and 44 on nutrition. Chapter 43 contains information on energy balance, kwashiorkor, marasmus, and essential amino acids.

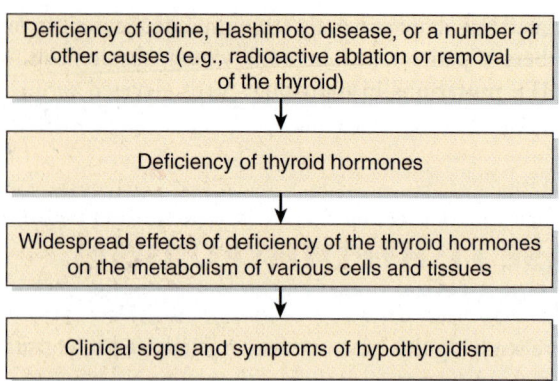

FIGURE 57–16 **Simplified scheme of the causation of primary hypothyroidism.**

Causation

Inadequate food intake, leading to deficiency of energy and also protein is the cause of both marasmus and kwashiorkor. Deficiency of antioxidant nutrients and the stress of infection precipitate kwashiorkor in undernourished children.

History and Physical Examination

A 2-year-old African girl was brought by her mother to the outpatient department of the local hospital. The mother had four children, the youngest of whom was 3 months of age and still being breast-fed. The father had broken his leg in an accident during the previous year and had not been able to work since. Family income was thus low, and they were not able to buy milk and meat on a regular basis. Their main subsistence food was a starchy gruel, high in carbohydrate and low in protein, and even that had been in short supply recently. The mother stated that the daughter had been eating poorly for the past several months, had intermittent diarrhea for that time, had recently developed a cough and fever, and had become very irritable, weak, and apathetic.

On examination, she was found to be both underweight for her height and small for her age. Her temperature was 40.5°C. Her midarm circumference was a little below normal. Her skin was flaky and her hair dry, brittle, and easily pluckable. The abdomen was distended and the liver moderately enlarged. Peripheral edema was evident. Rales were heard over the lower lobes of both lungs.

The doctor on duty made the diagnosis of kwashiorkor, diarrhea, pneumonia, and possible bacteremia.

Laboratory Findings

Samples of blood were taken for analysis. Results were subsequently reported as hemoglobin 6.0 g/dL (normal for a 2 year old: 11–14 g/dL), total serum protein 4.4 g/dL (normal 6.0–8.0 g/dL), and albumin 2.2 g/dL (normal 3.5–5.5 g/dL). Stool and blood samples were taken for culture; a gram-negative anaerobe was later reported in both. The neutrophil count was elevated (consistent with a bacterial pneumonia) and her lymphocyte count was markedly depressed. Chest x-rays revealed mottled opacities in the lower lobes of both lungs, consistent with bilateral acute bronchopneumonia.

Treatment

In many cases it is better not to treat children with mild to moderate kwashiorkor in the **hospital,** because this only increases the chance of infection. However, in view of the fever, weakness, drowsiness, and severe edema, this patient was admitted. She was immediately started on an appropriate **antibiotic** and **intravenous saline-dextrose** infusion. Sadly, her condition worsened and she died approximately 12 h after admission. Autopsy findings were compatible with kwashiorkor and also revealed severe fatty liver and bilateral bronchopneumonia.

Discussion

PEM is the most common nutritional disorder in many parts of the world. As many as one billion people suffer from various degrees of severity of PEM. It is due to inadequate dietary intake of protein and energy; kwashiorkor is commonly precipitated by the oxidative stress of an infection. PEM is almost always accompanied by deficiencies of **other nutrients** (eg, vitamins, minerals, etc). Children and the elderly are particularly susceptible, but it may occur at any age.

PEM may be defined as **primary** (due directly to deficiency of protein and energy intake), or **secondary** (due to increased nutrient needs, decreased absorption of nutrients, or increased loss of nutrients). This was a case of primary kwashiorkor.

Many of the features of primary PEM represent **adaptations** to the dietary deficiencies of energy and protein. For example, physical activity decreases in the face of deficient intake of nutrients. Stores of glycogen in muscle and liver are only capable of supplying energy for a short time (a day or two), so fat stores are mobilized to produce energy.

Eventually, when these are exhausted, **protein is catabolized** (mainly in muscles) to supply amino acids and energy. Thus, patients with PEM exhibit little activity, have diminished or no body stores of fat, and show muscle wasting, depending on the severity of the condition.

PEM has been classified as **edematous** (kwashiorkor) or **nonedematous** (marasmus). The precise cause of the edema in kwashiorkor is still under study. **Hypoalbuminemia** (due to deficient supply of amino acids to synthesize the protein) is likely one contributing factor (see Chapter 50), although this is not settled. Increased vascular permeability due to oxidative stress secondary to infection may also be important. Deficiency of the amino acid **methionine,** a precursor of cysteine, may also contribute. **Cysteine** is one of the three amino acids present in **glutathione,** the body's major antioxidant. If the tissues levels of glutathione decline, this could result in free radicals damaging various molecules and tissues (see Chapter 45)

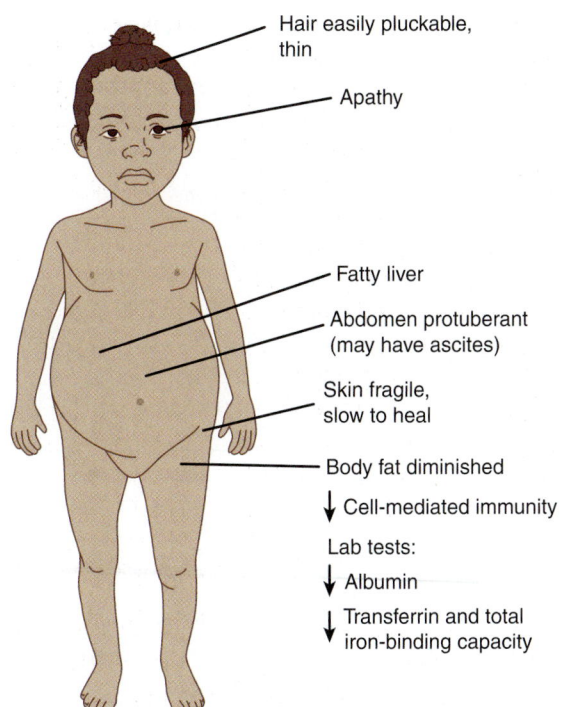

FIGURE 57–17 Some of the major signs of kwashiorkor.

Labels in figure:
- Hair easily pluckable, thin
- Apathy
- Fatty liver
- Abdomen protuberant (may have ascites)
- Skin fragile, slow to heal
- Body fat diminished
- ↓ Cell-mediated immunity
- Lab tests:
- ↓ Albumin
- ↓ Transferrin and total iron-binding capacity

and perhaps damaging cell membranes, increasing their permeability.

Marasmus is the predictable outcome of severe energy deficiency, and the affected child is less than 60% of expected weight for age. In **kwashiorkor** the child is 60–80% of expected weight for age, and is edematous. If the child is less than 60% of weight for age and edematous, this is **marasmic kwashiorkor,** the most severe and serious type of PEM. This patient exhibited primarily signs of kwashiorkor. The **hallmarks of kwashiorkor** are hypoalbuminemia, fragile skin (eg, poor wound healing, ulcers), easy hair pluckability, and edema (see **Figure 57–17**). Kwashiorkor is the word used by members of the Ga tribe in Ghana to describe "the sickness the older child gets when the next child is born." It follows weaning from breast milk and exposure to a diet low in protein and high in carbohydrate. **Fatty liver** is often found in kwashiorkor because of depressed synthesis of apoliproteins in the liver, resulting in accumulation of triglycerides. The poor skin and hair seen in kwashiorkor are mainly due to protein deficiency. **Hypoalbuminemia** is a common feature. While deficiency of protein can cause hypoalbuminemia, chronic inflammation can also contribute by suppressing synthesis of albumin. Total iron-binding capacity and levels of transferrin are also depressed.

Hormones may be important in the generation of PEM. The exposure to a relatively high intake of carbohydrate is thought by some to keep levels of insulin high and levels of epinephrine and cortisol low in kwashiorkor, as opposed to marasmus. The combination of low insulin and high cortisol greatly favors **catabolism of muscle;** thus, **muscle wasting** is greater in marasmus than in kwashiorkor. In addition,

TABLE 57–5 Some Differences between Kwashiorkor and Marasmus

	Kwashiorkor	Marasmus
Edema	Present	Absent
Onset	Rapid (eg, weeks), often associated with stress such as infections	Gradual (months to years)
Hypolbuminemia	Present and may be severe	Mild if present
Muscle wasting	Absent or mild	May be very severe
Body fat	Diminished	Absent

Note: Patients with marasmic kwashiorkor exhibit variable combinations of the above features.

because of the lower levels of epinephrine, fat is not mobilized to the same extent in kwashiorkor. The **immune system** is impaired in PEM, particularly T-cell function. Individuals are thus very susceptible to **infections** (eg, causing diarrhea), and infections worsen the situation by placing a higher metabolic demand on the body (eg, through fever). Some differences between kwashiorkor and marasmus are summarized in **Table 57–5**.

PEM is entirely **preventable** by a well-balanced diet containing adequate amounts of the major macronutrients, micronutrients, vitamins, and minerals.

Figure 57–18 summarizes some of the mechanisms involved in kwashiorkor.

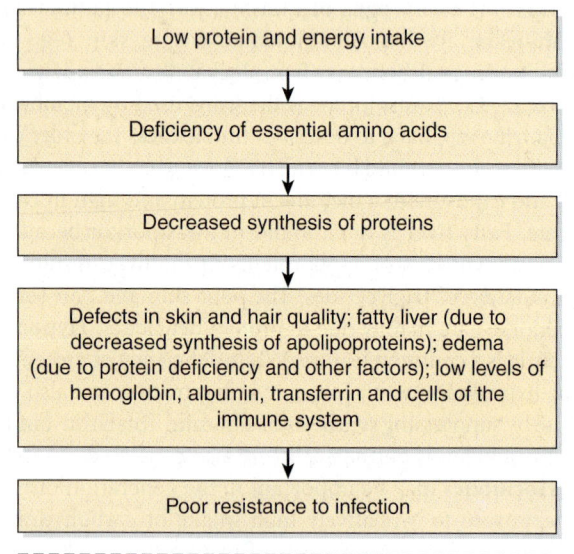

In comparison, a child with severe marasmus would show marked loss of muscle mass.

FIGURE 57–18 Summary of some of the factors involved in the causation of kwashiorkor.

CASE 13: MYOCARDIAL INFARCTION (MI)

Etiology

Lack of oxygen and various metabolites (due to blockage of blood flow in a coronary artery to an area of myocardium). **Genetic** and other factors predispose to this situation.

History and Physical Examination

A 46-year-old businessman was admitted to the emergency department of his local hospital complaining of severe retrosternal pain of 1-h duration. He had previously been admitted to the hospital once for treatment of a small MI, but despite this he continued to smoke heavily. He had been advised to take a mainly vegetarian diet, restrict his salt intake, join an exercise program, and prescribed an HMG-CoA reductase inhibitor (a statin) (his total cholesterol and LDL cholesterol had been elevated and HDL depressed), and a combination of a thiazide diuretic and an ACE (angiotensin-converting enzyme) inhibitor for moderate hypertension. He was also taking one aspirin (81 mg) per day. His blood pressure was 150/90 mmHg (before this incident it had been 140/80 mmHg; it was likely elevated because of stress), pulse was 60/min, and he was sweating profusely. There was no evidence of cardiac failure. His father had died at age 52 from a "heart attack," and one of his two siblings had had an MI at age 49. Because of the admitting diagnosis of probable MI, he was given morphine to relieve his pain and apprehension and also for its coronary dilator effect, and immediately transferred to a cardiac care unit, where continuous electrocardiographic monitoring was started at once.

Laboratory Findings

The initial ECG showed S-T segment elevation and other changes in certain leads, indicative of an acute anterior transmural left ventricular infarction. Blood was taken initially and at regular intervals thereafter for measurement of troponin T; on admission the level was within normal limits, but had risen eightfold by 6 h after admission.

Some proteins and enzymes that have been used in the diagnosis of a myocardial infarction are indicated in **Figure 57–19**. The use of enzymes and proteins in the diagnosis of MI and other diseases is discussed in Chapter 7. Levels of total cholesterol and the LDL/HDL cholesterol ratio were within normal limits (<5.17 mmol/L and 4:1, respectively), and triacylglycerols were 1.50 mmol/L (normal <2.26 mmol/L).

Treatment

The attending cardiologist, after reviewing all aspects of the case, decided to administer **tissue plasminogen activator (t-PA)** (see Chapter 51) intravenously because of the diagnosis of anterior transmural MI. Some 1.5 h had elapsed since the onset of symptoms. Chest pain began to disappear after 12 h, and the patient felt increasingly comfortable. He was discharged

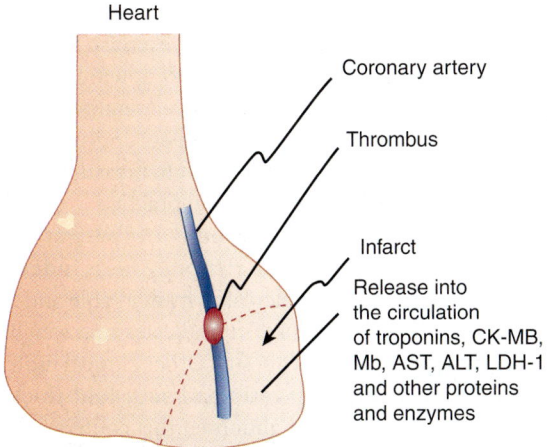

Heart

Coronary artery

Thrombus

Infarct

Release into the circulation of troponins, CK-MB, Mb, AST, ALT, LDH-1 and other proteins and enzymes

FIGURE 57–19 **Diagrammatic representation of a coronary artery thrombus resulting in release of various proteins and enzymes into the circulation from an area of myocardial infarction (MI).** Various proteins and enzymes are released from infarcted tissue at different rates, and they exhibit different half-lives in the circulation. At the present time, troponins are widely used to assist in the diagnosis of MI, but the other enzymes shown are still used to variable extents, and were more widely used in the past. (Mb, myoglobin; CK-MB, the MB isozyme of creatine kinase; AST, aspartate aminotransferase; ALT, alanine aminotransferase; LDH-1, the heart muscle isozyme of lactate dehydrogenase.)

from hospital 7 days later under the care of his family doctor. He was instructed to continue his medication, to attend a cardiac rehabilitation program, and to stop smoking.

Discussion

An MI is generally caused by an **occlusive thrombus** lying in close proximity to an unstable **atherosclerotic plaque** that has often ruptured recently. The rupture of the plaque helps generate the thrombus. Usually the diagnosis can be made from the clinical history, the electrocardiographic results, and serial measurements of a cardiac biomarker, such as **troponin T.** Measurement of this protein has replaced that of CK-MB in many hospitals (see Chapter 7).

Major aims of treatment are to prevent death from cardiac arrhythmias by administration of appropriate drugs and to limit the size of the infarction. In this case, the decision was made to limit the size of the infarct by **administration of t-PA**, which can dissolve or limit the growth of the thrombus (see Chapter 51) if given up to 12 h of onset of symptoms, although preferably sooner. An alternative would have been **percutaneous coronary intervention** (PCI), consisting of percutaneous transluminal coronary angioplasty (PTCA), with or without insertion of a stent.

Atherosclerosis, coronary thrombosis, and **MI** are described here very briefly; a textbook of pathology should be consulted for detailed descriptions. Atherosclerosis consists of patchy plaques in the intima of medium and large arteries. **Endothelial dysfunction** is believed to play an important role in the genesis of atherosclerosis. Platelets and fibrin can

deposit on the luminal aspect of an artery and monoclonally derived smooth muscle cells present in the medial layer of artery grow into the intimal lesion, attracted by growth factors released by macrophages and platelets (eg, platelet-derived growth factor). Plasma lipoproteins, glycosaminoglycans, collagen, and calcium subsequently accumulate in a lesion called a **fatty streak.** The presence of **oxidized LDL** in atherosclerotic lesions appears to be particularly important, in that it encourages recruitment of macrophages (inflammatory cells) and stimulates the release of growth factors. **Inflammation** is thus thought to be a key factor in atherosclerosis, as reflected by accumulation of macrophages and lymphocytes. The elevated plasma level of **C-reactive protein** (CRP) (Chapter 50) is also a reflection of chronic inflammation.

As the above processes progress, the fatty streak evolves into an **intimal plaque** Inflammation and hemorrhage into the plaque can occur, leading to rupture of its surface and exposure of its underlying constituents to the blood. Platelets will adhere to the exposed collagen, and a **thrombus** is initiated (Chapter 51).

Risk factors for atherosclerosis include age, family history, male sex, high levels of LDL and low levels of HDL, hypertension, diabetes mellitus, and smoking. This patient had elevated levels of total cholesterol and LDL and depressed levels of HDL before starting dietary and drug treatment.

If the thrombus in a coronary artery occludes approximately 90% of the vessel wall, blood flow through the affected vessel may cease (**total ischemia**) and the oxygen supply of the affected area of mycocardium will be rapidly compromised. The normal metabolism of the myocardium is **aerobic,** with most of its ATP being derived from oxidative phosphorylation. The anoxia secondary to total ischemia results in a switch to **anaerobic glycolysis,** which generates only about one-tenth of the ATP produced by oxidative phosphorylation. Not only does this switch in metabolism occur, but the flow of substrates into the myocardium via the blood and the removal of metabolic products from it are also greatly reduced. This accumulation of intracellular metabolites increases the intracellular oncotic pressure, resulting in cell swelling, affecting the permeability of the plasma membrane. Thus, the affected myocardium exhibits depletion of ATP, accumulation of lactic acid, development of severe acidosis, and marked reduction of contractile force. The **precise metabolic changes that commit a cell to dying** are being investigated in many laboratories; this is a very important area of research. Changes under study include depletion of ATP, activation of intracellular phospholipases (resulting in damage to cellular membranes), activation of proteases, and accumulation of intracellular Ca^{2+}.

Some of the mechanisms involved in the causation of an acute MI are summarized in **Figure 57–20.**

CASE 14: OBESITY

Before studying this case history the reader is advised to consult the material on triacylglycerols and adipose tissue in Chapter 25 and the material on nutrition in Chapters 43 and 44.

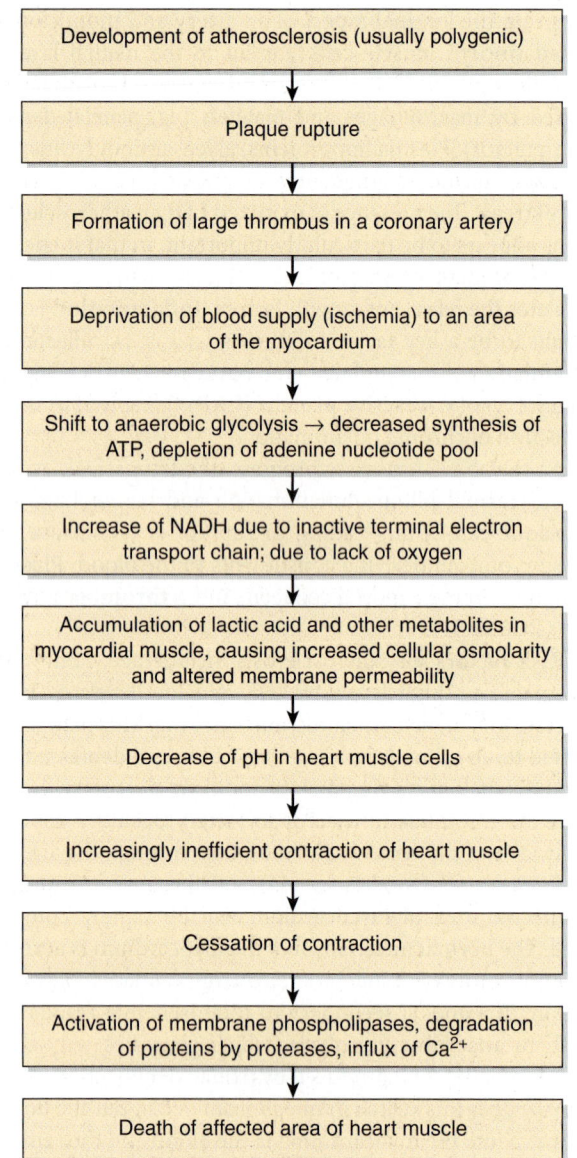

FIGURE 57–20 **Summary of mechanisms involved in the causation of an acute myocardial infarction.** The arrows do not in all cases imply a strict causal relationship.

Causation

Many factors contribute to obesity (genetic, environmental, cultural, etc). However, the central issue is that energy intake exceeds energy expenditure, resulting in the storage of triacylglycerols in adipose tissue.

History, Physical Examination, and Laboratory Findings

A 30-year-old woman visited her family doctor complaining of being markedly overweight. She was married, but had no children. She also complained that her periods were irregular. She gave no significant past history and was not on any drugs. She stated that she had always tended to being overweight, and

that her mother and two sisters were also overweight. She had a sedentary job in an office and did not exercise regularly. In addition, she said that she had "a healthy appetite," and both she and her husband (who was also overweight) often consumed a variety of fast foods. Many obese individuals deny that they eat excessively, and it is difficult to measure food consumption precisely in ordinary medical practice.

On examination, she was obviously overweight (200 lb, 91 kg) for her size (5'4", 163 cm). Her **body mass index** (BMI = weight kg/height m^2) was calculated from a table and found to be ~34. A BMI of 25–29.9 kg/m^2 indicates overweight and a BMI of >30 kg/m^2 indicates obesity. Other clinical indicators of obesity include **waist–hip ratio** and **skin fold thickness.** If fat accumulates around the abdomen this confers an **apple shape,** whereas fat accumulating around the buttocks confers a **pear shape.** The former may be more serious, as abdominal (visceral) fat upon lipolysis may release fatty acids into the portal vein, leading to fat deposition in the liver and muscles. **More precise tools** for measuring obesity are also available (eg, bioelectrical impedance analysis and underwater weighing). Her blood pressure was 140/95 and her total cholesterol was 6.1 mmol/L (borderline high). Urinalysis was negative. Her fasting blood glucose was 6.6 mmol/L (borderline high).

Treatment

Her physician advised her that she was obese, but not morbidly obese (BMI >40). The laboratory results indicated that her levels of cholesterol and blood glucose were borderline high, as was her blood pressure. He told her that further elevation of these values would predispose her to **heart disease, diabetes mellitus, hypertension,** and **the metabolic syndrome.** The latter is characterized by excess abdominal fat, high blood glucose (resistance to insulin), abnormal blood lipids (increased LDL and decreased HDL), and high blood pressure. He also pointed out a number of **other complications** to which obesity predisposed her (eg, reproductive problems such as irregular periods, gallbladder disease, deep vein thrombosis, sleep apnea, etc). He indicated to the patient that treatment of obesity was not easy, and that she would have to undergo a **permanent change in lifestyle** if she wished therapy to be successful and weight loss to be maintained. The **major approaches to treatment** of obesity were summarized: (1) diet, (2) exercise, (3) behavioral therapy, (4) drugs (eg, to suppress appetite centrally, or acting as an inhibitor of lipase activity in the gut, thus reducing absorption of fatty acids), and (5) surgery (eg, laparoscopic adjustable gastric banding and other procedures) in very severe cases.

He stated that in her case he believed satisfactory progress could be made, at least initially, through use of the first two lines of therapy. He told the patient that he seldom recommended drugs for treatment of obesity, and that surgery was generally restricted to people with morbid obesity (BMI >40) who had not responded to other approaches. He also encouraged her to get her husband involved because of his weight problem, and because it would be mutually beneficial if he also were on a weight reduction program.

TABLE 57–6 Summary of Dietary Advice Regarding Weight Loss

Acquire information regarding general nutrition and calories and study food labels.
Start a low calorie diet (about 1200 calories/day), containing appropriate amounts of carbohydrates, proteins and fats.
Eat more frequent, smaller meals.
Reduce intake of simple sugars and refined carbohydrates, and increase intake of complex carbohydrates (grains, etc) and low glycemic index foods.
Reduce intake of red and processed meat.
Reduce intake of saturated fats, and increase intake of monounsaturated and polyunsaturated fats and of omega-3 fatty acids.
Reduce salt intake.
Increase consumption of fresh fruits, vegetables, legumes, nuts, and low-fat dairy foods.
Markedly reduce intake of fast foods and calorie-rich soft drinks.
Avoid any of the fad diets.
Drink good quality water.
Take vitamin and mineral supplements.
Consult a dietitian for further details on diet and nutrition.
Join an organization that specializes in advising people on weight loss and runs a daily exercise program.

Regarding **diet,** her doctor (who had a special interest in implementing healthful nutrition in his practice) discussed general features of a suitable diet and also the advantages of **regular daily exercise.** Specific dietary recommendations that he made are listed in **Table 57–6.**

She subsequently joined **a weight loss organization** and found that it stressed changes in **behavior** (eg, developing sensible eating habits, rewards for good results, group counseling) and also provided support and encouragement from the other members. The patient was very keen to lose weight, and over the next year she closely followed the various advices given. She lost 34 lbs over that period of time. She felt much better and also stated that her periods had regularized and she was hoping to get pregnant. In addition, her blood pressure and her levels of total cholesterol and blood glucose were down to normal values. She was absolutely determined to continue on the weight loss program, as was her husband. Her sole regret was that she had not lost even more weight. Many patients under treatment for obesity eventually regain lost weight, for a variety of reasons.

Discussion

Obesity is a very prevalent condition, which is on the increase. Almost **one third of adults** in the Unites States are obese, and, alarmingly, more and more **children** are obese. Some talk of an epidemic of obesity in Western society. Dissecting all the factors contributing to this is complex, but **increased energy intake** and **decreased energy expenditure** play important roles. The

increased intake of fast foods and high calorie soft drinks and TV watching by "couch potatoes" are important contributors.

While the dangers of obesity have been widely circulated, it is of interest that some obese individuals feel that they are being unfairly targeted by health professionals and that the dangers of obesity have been overemphasized. It can be debated that it is better to be somewhat obese and fit, than to be of normal weight and unfit. In addition, certain individuals claim that they enjoy being obese.

While obesity itself is relatively easy to recognize and define (eg, we can somewhat arbitrarily use a BMI >30), pinning down the specific factors that contribute to individual cases is not that easy.

In the present case, **a number of contributory factors** are apparent. For example, she led a sedentary lifestyle, she frequently ate fast foods, she did not exercise regularly, etc. However, what was her precise caloric intake? What was her precise energy expenditure? Were the various mechanisms controlling appetite (see **Figure 57–21**) working properly? What role if any did genetic factors play in her obesity (she gave a family history of obesity)? These factors are not easy to quantify for a physician in her/his office.

Because of its medical importance, much research is going on presently in the field of obesity. This spans basic research on the adipocyte and signaling mechanisms to epidemiologic studies on various populations and their susceptibility to obesity. Here we shall only touch briefly on three areas: (1) regulation of appetite and food intake; (2) some genetic aspects; and (3) some aspects of energy expenditure.

Figure 57–21 summarizes a body of knowledge concerning the **regulation of appetite.** The **hypothalamus** plays a key role

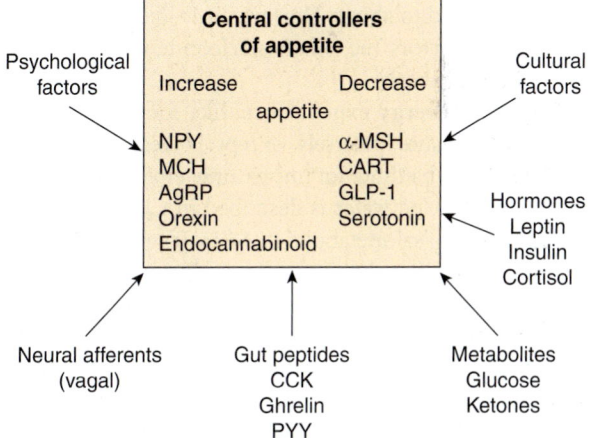

FIGURE 57–21 Factors that regulate appetite through effects on central neural circuits. Factors that increase or decrease appetite are listed. (NPY, neuropeptide Y; MCH, melanin-concentrating hormone; Ag RP, Agouti-related peptide; MSH, melanocyte-stimulating hormone; CART, cocaine- and amphetamine-related transcript; GLP-1, glucagon-related peptide-1, CCK, cholecystokinin; PYY, peptide YY.) The reader is advised to consult a textbook of Physiology for descriptions of the actions of these various molecules. (Reproduced, with permission, from Harrison's *Principles of Internal Medicine,* 17th ed, Fauci AS et al [editors], McGraw-Hill, 2008.)

in the central regulation of appetite. Factors that both increase and suppress appetite are shown. Psychologic, neural, and cultural factors all play a role. A variety of peptides, affecting specific areas of the hypothalamus, are indicated. In addition, levels of circulating metabolites (eg, glucose) and of hormones affect the hypothalamic centers. Regarding hormones, for example, individuals with Cushing syndrome (high levels of cortisol or related hormones) are obese and display a characteristic distribution of body fat. Particular interest has focused on **leptin,** a polypeptide released from adipocytes that acts mainly on the hypothalamus. Elevated levels of leptin decrease food intake and also increase energy expenditure. Leptin was discovered through studies of genetically obese mice (ob/ob). In humans, leptin is the product of the *OB* gene. The levels of leptin are elevated in most obese humans, suggesting that in some manner they may be resistant to its action.

Genetic influences on obesity have been recognized. Identical twins tend to have similar body weights. Occasional cases of mutations in the genes encoding leptin and the leptin receptor have been identified. Cases of obese individuals having mutations in the gene encoding proopiomelanortin (POMC), which is processed to form α-melanocyte hormone (α-MSH), a powerful appetite suppressant, have been described. In addition, mutations in the gene encoding the type-4 receptor for α-MSH and in the genes encoding two proteolytic enzymes involved in the conversion of POMC to α-MSH have been reported.

Children with the **Prader–Willi syndrome** (due to deletion of a part of chromosome 15; affected individuals overeat, among other signs) and with the **Laurence–Moon–Biedl syndrome** (an autosomal recessive genetic disorder) are characteristically obese. It should be stressed that the great majority of obese individuals do not apparently have mutations in the genes referred to above. However, it is likely that multiple subtle genetic factors (eg, single nucleotide polymorphisms, SNPs) influence obesity.

Regarding **energy expenditure,** like food intake, this is difficult to measure precisely except in research facilities. A sophisticated method for measuring energy expenditure using doubly labeled water is described in Chapter 44, as are the concepts of basal metabolic rate (BMR) and other factors involved in daily energy expenditure. It appears that most obese individuals have higher energy expenditure than people of normal weight, because their lean body mass is increased. Another important issue is whether obese individuals actually eat more than non-obese individuals. It appears likely that they do although many deny this. In the present case, it seems likely from her history that the patient did overeat.

An issue that has been the subject of debate is the possible role of **variations in diet-induced thermogenesis** (see Chapter 25) in contributing to obesity. Brown adipose tissue contains a mitochondrial protein known as **thermogenin** (uncoupling protein–1) that dissipates energy as heat. While brown adipose tissue is not a prominent component of adults (unlike newborns), it does occur. Also it appears to be diminished in amount in at least certain obese individuals, which

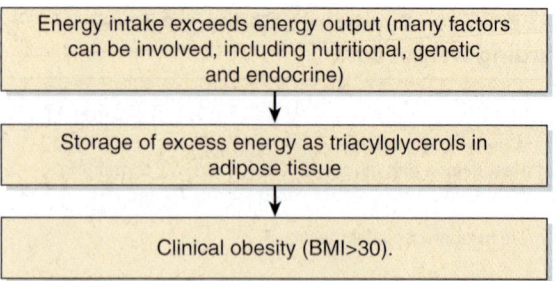

FIGURE 57–22 Simplified scheme of the causation of obesity.

could mean that they dissipate less energy as heat than non-obese individuals and have more available for other purposes. **Two other uncoupling proteins** are known to exist in human tissues, although their overall contribution to diet-induced thermogenesis is not clearly established. Thus, many factors can contribute to obesity and it is not easy to assess their contribution in most cases seen in clinical practice. At the present time, it is reasonable to treat obesity primarily by reducing intake of food, eating a healthful diet, increasing physical activity and offering appropriate support and encouragement. One way in which ethanol may disturb the functions of these proteins is by **replacing water molecules** at various critical sites.

Figure 57–22 summarizes some major factors involved in the causation of obesity.

CASE 15: OSTEOPOROSIS, PRIMARY (POSTMENOPAUSAL)

Before studying this case history the reader is advised to peruse the material in Chapter 48 on bone and the material in Chapter 44 on vitamin D.

Causation

Osteoporosis is loss of bone mass with preservation of the normal ratio of organic matrix (mostly proteins) to mineral. A **variety of factors** (endocrine, nutritional, lack of physical activity, etc) contribute to its development. In the postmenopausal type of osteoporosis dealt with here, **estrogen deficiency** is the major factor.

History, Physical Examination, and Investigations

A 64-year-old woman presented at emergency department, having tripped in her garden and apparently fallen rather gently onto her right forearm. Nevertheless, she suspected that she had fractured a bone in her arm, because of the pain and swelling just above her right wrist joint. X-rays showed a fracture of the distal end of the radius, with displacement. The radius also showed moderate reduction of radiodensity, suggestive of osteoporosis. The fracture was reduced, an appropriate cast was applied and she was asked to report to her family physician 2 weeks later. The emergency department physician gave her a note to give to her own physician, which mentioned that, because of the mildness of her fall, the resultant fracture, and the reduced radiodensity of her radius, he suspected that she might have osteoporosis. The patient attended her family doctor 2 weeks later. She had four adult children. Her last menstrual period had been some 5 years previously, and she had only attended her physician very irregularly over the years for the occasional minor ailment. She was not on any medication, did not take any vitamin or mineral supplements, and had never taken hormonal treatment for menopause. She ate very few fruits and vegetables, and overall she consumed a high carbohydrate diet along with liberal amounts of fried foods. She smoked one pack of cigarettes a day and also had several drinks of vodka each evening. In addition, she rarely exercised. She complained of chronic lower backache, but otherwise gave no significant history. In view of the fracture and the suggestion of osteoporosis, her family doctor ordered **dual-energy x-ray absorptiometry (DEXA)** of the lumbar spine and hip areas to assess bone density. He also ordered x-rays of the lower spine because of the history of lower back pain. In addition, he ordered determinations of serum Ca, P, alkaline phosphatase, 25-hydroxyvitamin D, parathyroid hormone, and complete urinalysis (including 24-h calcium) to check for any other bone disease (eg, due to vitamin D deficiency, hyperparathyroidism, or multiple myeloma). The results of her DEXA scan showed a marked reduction (over 3 standard deviations; over 2.5 is diagnostic for osteoporosis) from the average value in 25-year-old women of her race, compatible with severe osteoporosis. The X-rays of her lower spine showed decreased radiodensity, but no fractures. Results of her blood and urine tests were within normal limits, suggesting that she had no other serious bone disorder.

When bone resorption occurs, there is increased production of **collagen cross-link products.** These include *N*-telopeptide, deoxypyridinoline and *C*-telopeptide. Also, in some cases of osteoporosis, **increased bone formation** occurs; bone alkaline phosphatase and osteocalcin are markers of this. These various markers are not in themselves diagnostic of osteoporosis, but can be measured in serum or urine as **indicators of response to therapy.** For example, a 30% decrease of these markers would suggest successful therapy. They were not measured in this case, and in fact are not measured in many clinical laboratories.

Treatment

The history, DEXA scan, and the results of the other tests were consistent with a diagnosis of relatively severe osteoporosis. She was advised to start on an **exercise program** immediately at a gym, initially under the supervision of a personal trainer. She was also referred to **a dietitian** to change her dietary habits; recommendations included eating regular, daily portions of fresh fruits and vegetables, and consuming a more balanced diet (eg, reducing the starchy and fried foods). It was recommended that she **stop smoking** and cut down on her consumption of **alcohol,** as both of these may contribute to

osteoporosis. She was also started on appropriate doses of **calcium citrate** and **vitamin D.** In addition to the above, her physician recommended that she start on a **bisphosphonate** (eg, alendronate or risedronate), giving detailed instructions as to how to take the drug. These two drugs, N-containing bisphosphonates, are taken up by osteoclasts and inhibit formation of farnesyl diphosphate. This in turn inhibits **prenylation of certain proteins** (see Figure 26–2) and their attachment to the plasma membrane, negatively affecting osteoclast activity and thus inhibiting bone resorption. **Other drugs** that may be used in the treatment of osteoporosis in selected cases include salmon calcitonin, estrogen, selective estrogen modulators (eg, raloxifene), and parathyroid hormone. Estrogen was formerly widely prescribed early in menopause to reduce osteoporosis, and appeared to be relatively effective. However, the Women's Health Initiative Study indicated that the risks of estrogen therapy outweighed the benefits in this situation.

The patient was followed up over the next few years. She lost considerable weight and maintained her exercise program. In general, she felt much more healthy than she had done in the previous 20 years. Her fracture healed satisfactorily, no further loss of bone mass occurred, but normal bone mass was not restored. She was advised on how to take precautions to avoid falls and it was also recommended that she wear hip pads.

Discussion

Osteoporosis can be defined as **reduction of bone mass or density.** Often it is first detected after a fracture, as the loss of bone tissue predisposes to fractures. The World Health Organization (WHO) has suggested that osteoporosis exists when **bone density falls 2.5 standard deviations or more below the mean for young healthy adults of the same race and gender.** In the United States, some 8 million women and 2 million men have osteoporosis, and many others are at risk of developing it.

Two terms related to osteoporosis are osteopenia and osteomalacia. **Osteopenia** is decreased bone mass, embracing both osteoporosis and osteomalacia. In **osteoporosis,** bone mass decreases due to decreased bone formation and increased resorption, but a normal ratio of bone mineral (hydroxyapatite) to bone matrix (mostly collagen type 1) is preserved. **Osteomalacia** is also an example of osteopenia, but in it there is decreased mineralization; its most common cause is deficiency of vitamin D.

Primary osteoporosis can be divided into three types: **idiopathic** (uncommon, and occurs in children and young adults), **postmenopausal,** and **involutional** (elderly) types. This case is an example of **postmenopausal osteoporosis,** and decline in estrogen levels is a major factor in its causation. This type can also occur in males due to a decline of serum testosterone, which acts to increase osteoclastic activity. **Involutional** osteoporosis occurs in older individuals and is due to the decline of the number of osteoblasts with age. Postmenopausal and involutional osteoporosis may coexist.

FIGURE 57-23 **Simplified scheme of various factors involved in the causation of osteoporosis.** (OC, osteoclastic, OB, osteoblastic, IGF-1 and IGF-2, insulin-like growth factors 1 and 2; TGF-β, transforming growth factor β; TNF-α, tumor necrosis factor α.) The hormones listed have a variety of effects on bone. IGF-1 and IGF-2 have anabolic effects on bone, but may also stimulate turnover of bone. RANK-ligand is a cytokine that is involved in communication between osteoblasts and osteoclasts; when it interacts with osteoclasts, it increases their activity. The other cytokines are made by osteoblasts. Their synthesis is increased or decreased by estrogen deficiency, with the overall effect of extending the life span of osteoclasts (by inhibiting apoptosis).

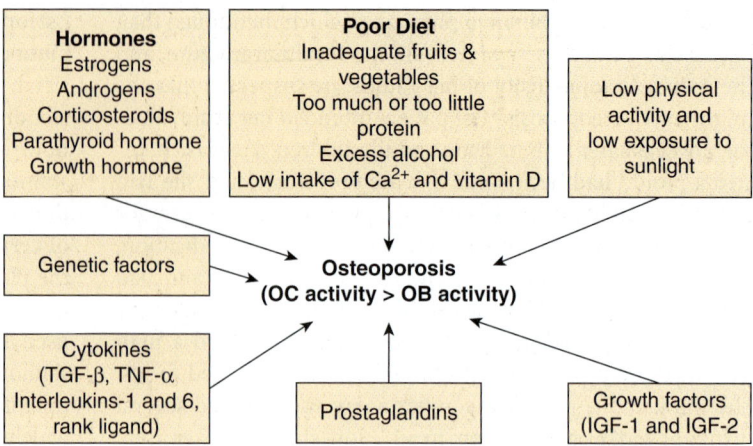

Secondary osteoporosis is relatively uncommon, and may be due to a variety of conditions (eg, chronic renal disease, various drugs [especially corticosteroids], a number of endocrine disorders, malabsorption syndrome, etc).

Many factors are involved in the causation of osteoporosis (see **Figure 57-23**). An in-depth understanding of them requires knowledge of bone **modeling,** of a variety of **cytokines,** of the actions of **a number of hormones,** and of various **nutritional** and **genetic factors,** among other considerations. (Cytokines are a heterogeneous group of proteins released by various cells and which have autocrine or paracrine effects.) Here we only indicate the complexity of a complete understanding of osteoporosis by briefly mentioning the main players. The central point is that osteoporosis occurs when **bone resorption** (osteoclastic [OC]) activity) **exceeds bone formation** (osteoblastic [OB] activity). In regards to post-menopausal osteoporosis in particular, one important consideration is that **estrogen loss** appears to increase the secretion of a number of cytokines that lead to recruitment of osteoclasts. Also, estrogen loss diminishes the secretion of certain other cytokines that promote osteoblastic activity. The overall effect is thus an imbalance between osteoclasts and osteoblasts, in favor of the former.

Figure 57-24 is a simplified schematic representation of the causation of osteoporosis.

Alterations of the levels of various factors (hormones, cytokines, nutritional) result in an increase of osteoclastic activity over osteoblastic activity. In post-menopausal osteoporosis, loss of estrogen is critical.

↓

Resorption of bone with loss of matrix (mainly collagen type I, but also other proteins such as osteocalcin) with preservation of the ratio of matrix: mineral (hydroxyapatite).

↓

Fragility of bones, often resulting in fractures (hip fractures are among the most serious).

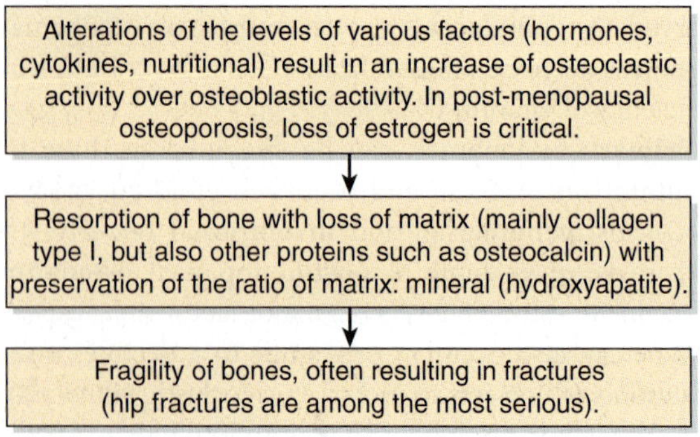

FIGURE 57-24 **Simplified scheme of some major factors in the causation of osteoporosis.**

CASE 16: XERODERMA PIGMENTOSUM (XP) (PIGMENTED DRY SKIN)

The reader should consult the material in Chapter 35 on DNA repair and on XP before reading this case.

Causation

Genetic (a mutation in a gene directing synthesis of a DNA repair enzyme in the nucleotide excision repair pathway) and **physical** (exposure to ultraviolet [UV] irradiation) (**Figure 57-25**).

History and Physical Examination

An 8-year-old boy, an only child, presented at a dermatology clinic with a skin tumor on his right cheek. He had always

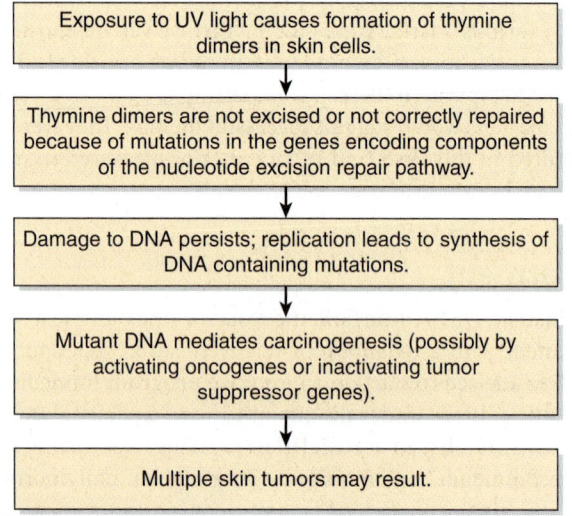

FIGURE 57-25 **Summary of mechanisms involved in the causation of xeroderma pigmentosum (OMIM 278730 and other entries).**

avoided exposure to sunlight because it made his skin blister. His skin had scattered areas of hyperpigmentation and other areas where it looked mildly atrophied. There was no family history of a similar disorder. Because of the presence of a skin tumor at such a young age, the history of avoidance of sunlight, and the other milder skin lesions, the dermatologist made a provisional diagnosis of XP.

Laboratory Findings

Histologic examination of the excised tumor showed that it was a **squamous cell carcinoma** (a common type of tumor skin cancer in older people, but very unusual in a boy of this age). A small piece of skin was removed for **preparing fibroblasts** to be grown in tissue culture. A research laboratory in the hospital specialized in radiation biology and was set up to measure the amount of **thymine dimers** (see below) formed following exposure to UV light. The patient's fibroblasts and control fibroblasts were exposed to UV light, and cell samples were taken at 8-h intervals for a total of 32 hours postirradiation. Extracts of DNA were prepared and the numbers of dimers remaining at each time point indicated were determined. Whereas only 24% of the dimers formed persisted in DNA extracted from the normal cells at 32 h, approximately 95% were found in the extract from the patient's cells at 32 h. This showed that the UV-induced lesions had not been repaired, and thus confirmed the diagnosis of XP.

Discussion

XP is a rare **autosomal recessive** condition in which the mechanisms for repair of DNA subsequent to damage by UV irradiation are defective. This arises because of mutations in the genes encoding components of the **nucleotide excision pathway** of DNA repair (nucleotide excision repair, NER; see Chapter 35). The major damage inflicted on DNA by UV irradiation is the formation of thymine (pyrimidine) dimers (also known as cyclobutane pyrimidine dimers), where covalent bonds are formed between carbon atoms 5 and 5 and 6 and 6 of adjacent intrachain thymine residues, resulting in the dimers. Other types of damage can also occur.

NER has two subpathways: global genome repair and transcription-coupled repair. The former scans the whole genome and removes damaged areas rapidly. The second is linked to transcription, operates slowly, and removes damage from the transcribed strand of DNA.

Detailed analyses of NER involved in the removal of thymine dimers have been performed. The pathway is highly conserved across species, indicating its importance. In general, the pathway involves four major steps, all involving various enzymes: (1) **recognition** of damaged DNA; (2) **excision** of damaged region; (3) **filling in** of the gap by DNA polymerase; and (4) **ligation** of the filled in area. In *E coli,* an endonucleolytic cleavage, catalyzed by a specific **endonuclease** (also called an excinuclease), occurs on either side of the damage, releasing a 12- to 13-base-pair oligonucleotide. The polymerization step involves **DNA polymerase** and the final step is sealing by **DNA ligase.**

The NER pathway operates in **humans,** and its details are still being elucidated. It appears to be generally similar to the pathway in *E coli.* The most noticeable difference is that a much larger oligonucleotide (about 30 bases) is excised in humans. The products of at least seven genes (these encoding XPA through XPG) have been implicated in NER in humans, and all have been cloned. Mutations in any one of these genes cause XP. In the boy whose situation is discussed here, the specific gene involved was not determined.

Because the responsible genes have been cloned, **prenatal diagnosis** of XP is now possible using appropriate probes. The NER pathway is also involved in processes other than DNA repair, such as recombination, replication, and transcription. The involvement of the seven genes mentioned above in DNA repair was originally shown in the following manner. It was observed that when cultured cells from individuals with XP were cocultured with cells from other individuals under conditions in which **cell fusion** occurred, the defect in DNA repair could sometimes be corrected. This indicated that one set of cells was providing a normal gene product to the other, thus correcting the defect. In this manner, at least **seven complementation groups,** corresponding to the seven genes and their products mentioned above, have been recognized.

If UV damage is not repaired, **mutations in DNA** will result, chromosomal abnormalities may occur and **cancer** may ensue. Patients with XP often suffer from a variety of skin cancers from an early age.

As described in Chapter 35, there are **other pathways** of DNA repair operative in humans. They are all important in preserving the integrity of DNA, and abnormalities of some of them have been associated with cancer (eg, mismatch repair).

The parents of the boy in the present case were told that he would have to be **watched closely** throughout life for the development of new skin cancers. In addition, he was advised to **avoid sunlight** and to use an appropriate **sunscreen ointment.** Although XP is a rare condition, the existence of a variety of mechanisms for repairing DNA following exposure to different types of irradiation and to chemical damage is of **great protective importance.** Without their existence, life on this planet would be fraught with even more danger that it presently is! For instance, it has been estimated that patients with XP have a 1000-fold greater chance of developing skin cancer than do normal individuals.

Figure 57–25 summarizes the mechanisms involved in the causation of XP.

EPILOG

Remarkable progress has been made in biochemistry and in related fields such as genetics and cell biology. Many of the discoveries in these disciplines have had great impact on medicine and related life sciences. Crucial insights into the natures of many diseases have emerged from studies of their molecular

TABLE 57–7   Some Major Challenges Facing Medicine and Allied Health Sciences[1]

Topic	Challenge
Aging	Understanding its molecular bases and perhaps modifying some of its effects.
Various types of arthritis and osteoporosis	Understanding their molecular bases (eg, further study of the roles of the ECM in their causation) and improving present therapies.
Cancers	Understanding their molecular bases (eg, additional studies on oncogenes, tumor suppressor genes and mechanism of metastasis), developing better biomarkers for earlier diagnosis, and improving present therapies.
Cardiovascular diseases (eg, myocardial infarctions and strokes)	Understanding their molecular bases (eg, increased knowledge of atherosclerosis and thrombosis) and improving present therapies.
Chronic neurodegenerative diseases (eg, Alzheimer disease)	Understanding their molecular bases (eg, additional insights into the roles of various proteins in their causation) and improving present therapies.
Diabetes mellitus	Obtaining further insights as to its causes and effects (eg, obtaining a complete picture of all aspects of insulin action and of mechanisms of tissue damage such as glycation) and improving therapy.
Environmental medicine	Requires a concerted effort among scientists and health workers to prevent further degradation of the environment and forestall potentially serious health effects.
Genetic diseases	Establishing their molecular bases and developing gene therapy and other treatments (eg, use of small molecules to help affected proteins fold properly).
Infections, including AIDS and tropical diseases	Understanding their molecular bases, preventing their spread (eg, by increased knowledge of the biochemistry of microorganisms and of mechanisms of their attachment to cells) and improving present therapies.
Nutrition	Improving the standard world-wide and combatting problems such as protein-energy-malnutrition and obesity.
Poverty	Encouragement of a global attack on poverty, which is a root cause of many physical and mental disorders.
Psychiatric diseases	Understanding their molecular bases (eg, determining which genes are involved in diseases such as schizophrenia and bipolar disorders) and improving present therapies.
Wellness	Educating populations regarding its maintenance (cellular health) and instituting measures (eg, nutrition, exercise, vaccines, mental health, avoidance of toxins) to help prevent illnesses.

[1]Biochemical and various related approaches (genetic, cell biologic, immunologic, pathologic, pharmacologic, etc) will be critical in addressing many of these challenges, as will public health approaches.

bases. New drugs and other treatments are constantly being developed based on such discoveries. However, it is obvious that there are still many major challenges facing medical science. **Table 57–7** summarizes some of them. The authors of this text and other biochemists believe that the application of biochemical, genetic, and allied approaches to these problems and to others not listed will pay rich dividends from which people all across the globe will benefit. Hopefully some of the readers of this text will contribute to such endeavors.

REFERENCES

Aiuti A, et al: Gene therapy for immunodeficiency due to adenosine deaminase deficiency. N Engl J Med 2009;360:447.

Beers MH, Porter RS, Jones TV, et al (editors): *The Merck Manual of Diagnosis and Therapy.* 18th ed. Merck Research Laboratories, 2006. (This book is available free online at http://merck.com/mmpe/index.html and contains coverage of many of the conditions discussed here).

Brunicardi FC, Anderson D, Billiar T, et al: *Schwartz's Principles of Surgery.* 9th ed. McGraw-Hill, 2009. (Chapter 29 contains a discussion of colorectal cancer).

Brunton LL, Blumenthal D, Buxton I, et al: *Goodman and Gilman's The Pharmacological Basis of Therapeutics.* 11th ed. McGraw-Hill, 2005 (Chapter 22 contains a discussion of acute ethanol intoxication).

Burtis CA, Ashwood ER, Bruns DE (editors): *Tietz Textbook of Clinical Chemistry and Molecular Diagnostics.* 4th ed. Elsevier Saunders, 2006. (Chapters 23 [Tumor Markers] and 49 [Mineral and Bone Metabolism] are of particular interest to the contents of this Chapter.)

Fauci AS, Braunwald E, Kasper DL, et al (editors): *Harrison's Principles of Internal Medicine.* 17th ed. McGraw-Hill, 2008. (Contains comprehensive descriptions of many of the conditions described here).

Freedman DH: How to Fix the Obesity Crisis. Sci American 2011; 304 (No. 2):40.

Kohn DB, Candotti F: Gene therapy fulfilling its promise. N Engl J Med 2009;360:518. (Discusses aspects of the successful treatment of ADA deficiency.)

Neogi T: Gout. N Engl J Med 2011;364 (No. 5):443.

Riordan JR: CFTR function and prospects for therapy. Annu Rev Biochem 2008;77:70.

Scriver CR, Beaudet AL, Valle D, et al (editors): *The Metabolic and Molecular Bases of odf Inherited Disease.* 8th ed. McGraw-Hill, 2001. (The online version of this text contains comprehensive up-dated descriptions of many of the conditions discussed in this Chapter).

Shils ME, Shile M, Ross AC, et al (editors): *Modern Nutrition in Health and Disease.* 10th ed. Lippincott Williams & Wilkins, 2006. (Contains comprehensive coverage of nutritional topics, including PEM [Chapter 57] and obesity [Chapters 63 & 64]).

Exam Questions

1. The glycemic index of a food is a way of assessing how rapidly the carbohydrate in that food is digested and absorbed
   Which of the following is the best definition of glycemic index?
   A. The decrease in the blood concentration of glucagon after consuming the food compared with an equivalent amount of white bread.
   B. The increase in the blood concentration of glucose after consuming the food.
   C. The increase in the blood concentration of glucose after consuming the food compared with an equivalent amount of white bread.
   D. The increase in the blood concentration of insulin after consuming the food.
   E. The increase in the blood concentration of insulin after consuming the food compared with an equivalent amount of white bread.

2. Which of the following will have the lowest glycemic index?
   A. A baked apple
   B. A baked potato
   C. An uncooked apple
   D. An uncooked potato
   E. Apple juice

3. Which of the following will have the highest glycemic index?
   A. A baked apple
   B. A baked potato
   C. An uncooked apple
   D. An uncooked potato
   E. Apple juice

4. Which of the following best describes the digestion and absorption of dietary triacylglycerol?
   A. Absorption of partially hydrolysed lipids in lipid micelles that are then incorporated directly into chylomicrons.
   B. Absorption of triacylglycerol in lipid micelles that is then incorporated into chylomicrons.
   C. Complete hydrolysis to free fatty acids and glycerol in the intestinal lumen.
   D. Hydrolysis to free fatty acids and monoacylglycerol in the intestinal lumen followed by complete hydrolysis of monoacylglycerol in mucosal cells, and re-esterification to triacylglycerol matching the pattern of dietary triacylglycerol.
   E. Hydrolysis to free fatty acids and monoacylglycerol in the intestinal lumen followed by partial hydrolysis of monoacylglycerol in mucosal cells and re-esterification to triacylglycerol with a different pattern of fatty acids from that in the dietary triacylglycerol.

5. Plant sterols and stanols inhibit the absorption of cholesterol from the gastro-intestinal tract. Which of the following best describes how they act?
   A. They are incorporated into chylomicrons in place of cholesterol.
   B. They compete with cholesterol for esterification in the intestinal lumen, so that less cholesterol is esterified.
   C. They compete with cholesterol for esterification in the mucosal cell, and unesterified cholesterol is actively transported out of the cell into the intestinal lumen.
   D. They compete with cholesterol for esterification in the mucosal cell, and unesterified cholesterol is not incorporated into chylomicrons.
   E. They displace cholesterol from lipid micelles, so that it is not available for absorption.

6. Which one of following statements about energy metabolism is CORRECT?
   A. Adipose tissue does not contribute to basal metabolic rate (BMR).
   B. Physical activity level (PAL) is the sum of physical activity ratios for different activities throughout the day, multiplied by the time spent in each activity, expressed as a multiple of BMR.
   C. Physical activity ratio (PAR) is the energy cost of physical activity throughout the day.
   D. Resting metabolic rate (RMR) is the energy expenditure of the body when asleep.
   E. The energy cost of physical activity can be determined by measuring respiratory quotient (RQ) during the activity.

7. Which one of following statements about nitrogen balance is CORRECT?
   A. If the intake of protein is greater than requirements, there will always be positive nitrogen balance.
   B. In nitrogen equilibrium the excretion of nitrogenous metabolites is greater than the dietary intake of nitrogenous compounds.
   C. In positive nitrogen balance the excretion of nitrogenous metabolites is less than the dietary intake of nitrogenous compounds.
   D. Nitrogen balance is the ratio of intake of nitrogenous compounds/output of nitrogenous metabolites from the body.
   E. Positive nitrogen balance means that there is a net loss of protein from the body.

8. Which of the following vitamins provides the cofactor for reduction reactions in fatty acid synthesis?
   A. Folate
   B. Niacin
   C. Riboflavin
   D. Thiamin
   E. Vitamin B_6

9. Which of the following vitamins provides the cofactor for transamination of amino acids?
   A. Folate
   B. Niacin
   C. Riboflavin
   D. Thiamin
   E. Vitamin B_6

10. Which of the following vitamins provides the cofactor for transfer of one-carbon units?
    A. Folate
    B. Niacin
    C. Riboflavin
    D. Thiamin
    E. Vitamin B_6

11. Which of the following vitamins is essential for fatty acid synthesis?
    A. Biotin
    B. Folate
    C. Vitamin B_6
    D. Vitamin B_{12}
    E. Vitamin C

12. Which one of these vitamins is involved in calcium homeostasis?
    A. Vitamin B_{12}
    B. Vitamin B_6
    C. Vitamin D
    D. Vitamin E
    E. Vitamin K

13. Which one of these vitamins is involved in blood clotting?
    A. Vitamin B_6
    B. Vitamin B_{12}
    C. Vitamin D
    D. Vitamin E
    E. Vitamin K

14. Deficiency of which one of these vitamins may lead to megaloblastic anemia?
    A. Vitamin B_6
    B. Vitamin B_{12}
    C. Vitamin D
    D. Vitamin E
    E. Vitamin K

15. Which one of these vitamins may mask the anemia of vitamin B_{12} deficiency?
    A. Biotin
    B. Folate
    C. Riboflavin
    D. Thiamin
    E. Vitamin B_6

16. Deficiency of which one of these vitamins may lead to hemolytic anemia?
    A. Vitamin B_6
    B. Vitamin B_{12}
    C. Vitamin D
    D. Vitamin E
    E. Vitamin K

17. Deficiency of which one of these vitamins is a major cause of blindness worldwide?
    A. Vitamin A
    B. Vitamin B_{12}
    C. Vitamin B_6
    D. Vitamin D
    E. Vitamin K

18. Which one of the following is NOT a source of oxygen radicals?
    A. Action of superoxide dismutase
    B. Activation of macrophages
    C. Non-enzymic reactions of transition metal ions
    D. Reaction of β-carotene with oxygen
    E. Ultraviolet radiation

19. Which one of the following is NOT the result of oxygen radical action?
    A. Activation of macrophages
    B. Modification of bases in DNA
    C. Oxidation of amino acids in apoproteins of LDL
    D. Peroxidation of unsaturated fatty acids in membranes
    E. Strand breaks in DNA

20. Epidemiological evidence and laboratory studies suggest that antioxidant nutrients such as vitamins C and E and β-carotene are protective against atherosclerosis and some cancers. However, intervention trials with antioxidant supplements have given disappointing results, and in many cases there have been more deaths from coronary heart disease and cancer in the intervention group than in those receiving placebo. Which of the following best explains this paradox?
    A. Antioxidants are lipid-soluble and therefore cannot act in the cytosol or extracellular fluid.
    B. Antioxidants form stable radicals that penetrate deeper into tissues, causing more damage.
    C. Antioxidants form stable radicals that quench the radical chain reaction.
    D. The doses of antioxidant have generally been too low for any beneficial effects to be seen.
    E. The intervention trials have generally been of too short a duration for any beneficial effects to be seen.

21. Select the one FALSE statement:
    A. Glycosylation occurs solely in the Golgi apparatus.
    B. There are no structural differences between free and bound ribosomes.
    C. Proteins destined for the plasma membrane and for secretion are generally synthesized on membrane-bound polyribosomes.
    D. The proton-motive force across the inner mitochondrial membrane is derived from the electric potential and the pH gradient.
    E. Certain pancreatic proteins destined for export are carried in secretory vesicles.

22. Select the one FALSE statement:
    A. The great majority of mitochondrial proteins are encoded by the nuclear genome.
    B. Ran proteins, like ARF and Ras proteins, are monomeric GTPases.
    C. One cause of Refsum disease is mutations in genes encoding peroxisomal proteins.
    D. Peroxisomal proteins are synthesized on cytosolic polyribosomes.
    E. Import of proteins into mitochondria involves proteins known as importins.

23. Select the one FALSE statement:
    A. N-terminal signal peptides directing nascent proteins to the ER membrane contain a hydrophobic sequence.
    B. Post-translational translocation of proteins to the ER does not occur in mammalian species.
    C. The SRP contains one RNA species.
    D. N-glycosylation is catalyzed by oligosaccharide:protein transferase.
    E. Type I membrane proteins have their N-termini facing the lumen of the ER.

24. Select the one FALSE statement:
    A. Chaperones often exhibit ATPase activity.
    B. Protein disulfide isomerase and peptidyl prolyl isomerase are enzymes involved in helping proteins fold properly.
    C. Ubiquitin is a small protein involved in protein degradation by lysosomes.
    D. Mitochondria contain chaperones.
    E. Retrotranslocation across the ER membrane is involved in helping dispose of misfolded proteins.

25. Select the one FALSE statement:
    A. Rab is a small GTPase involved in vesicle targeting.
    B. COPII vesicles are involved in anterograde transport of cargo from the ER to the ERGIC or Golgi apparatus.
    C. Brefeldin A prevents GTP binding to ARF, and thus inhibits formation of COP I vesicles.
    D. Botulinum toxin B acts by cleaving synaptobrevin, inhibiting release of acetylcholine at the neuromuscular junction.
    E. Furin converts preproalbumin to proalbumin.

26. All of the following are glycoproteins EXCEPT:
    A. Collagen
    B. TSH
    C. Albumin
    D. IgG
    E. Transferrin

27. All of the following sugars are found in glycoproteins EXCEPT:
    A. Fructose
    B. Fucose
    C. Galactose
    D. Mannose
    E. Xylose

28. Select the one FALSE statement:
    A. Mucins contain predominantly O-linked glycans.
    B. O-linked sugar chains are built up by the stepwise donation of sugars from nucleotide-sugars.
    C. Dolichol-pyrophosphate–oligosaccharide donates all of the sugars found in mature N-linked glycoproteins.
    D. N-acetylneuraminic acid is commonly found at the termini of N-linked sugar chains.
    E. Calnexin retains partly folded or misfolded proteins in the ER until proper folding has occurred.

29. Select the one FALSE statement:
    A. Glucose can attach to proteins via a non-enzymic reaction to form a Schiff base.
    B. Abnormal glycation end-products are thought to play a role in the tissue damage that occurs in diabetes mellitus.
    C. Glycation of hemoglobin only occurs in individuals with diabetes mellitus.
    D. Abnormal glypiation is involved in the causation of paroxysmal nocturnal hemoglobinuria.
    E. The attachment of *P falciparum* to some human cells is mediated by a GPI structure on the surface of the parasite.

30. Select the one FALSE statement:
    A. The initial attachment of neutrophils to endothelial cells of small blood vessels in acute inflammation involves interactions between L-selectin and endothelial cell glycoproteins.
    B. I-cell disease is due to mutations in the gene encoding a GalNAc phosphotransferase.
    C. The binding of avian influenza virus to human cells involves an interaction between hemagglutinin and cell surface N-acetylneuraminic acid.
    D. HIV-1 attaches to human cells via a glycoprotein (gp120) present on its surface.
    E. *H pylori* attaches to human cells via an interaction between adhesin and cell surface glycans.

31. Select the one FALSE statement:
    A. Collagen has a triple helical structure, forming a right-hand superhelix.
    B. Proline and hydroxyproline confer rigidity on collagen.
    C. Collagen contains one or more O-glycosidic linkages.
    D. Collagen lacks cross-links.
    E. Deficiency of vitamin C impairs the action of prolyl and lysyl hydroxylases.

32. Select the one FALSE statement:
    A. Elastin contains hydroxyproline, but not hydroxylysine.
    B. Elastin contains cross-links formed by desmosines.
    C. No genetic diseases due to abnormalities of elastin have as yet been identified.
    D. Unlike collagen, there is only one gene encoding elastin.
    E. Elastin does not contain any sugar molecules.

33. Select the one FALSE statement:
    A. Marfan syndrome is due to mutations in the gene encoding fibrillin-1, a major constituent of microfibrils.
    B. All cases of Ehlers–Danlos syndrome are due to mutations affecting the genes encoding the various types of collagen.
    C. Laminin is found in renal glomeruli along with entactin, type IV collagen, and heparin or heparan sulfate.
    D. Mutations affecting type IV collagen can cause serious renal disease.
    E. Mutations in the collagen *1A1* gene can cause osteogenesis imperfecta.

34. Select the one FALSE statement:
    A. Most but not all GAGs contain an amino sugar and a uronic acid.
    B. All GAGs are sulfated.
    C. GAGs are built up the actions of glycosyltransferases using sugars donated by nucleotide-sugars.
    D. Glucuronic acid can be converted to iduronic acid by an epimerase.
    E. The proteoglycan aggrecan contains hyaluronic acid, keratan sulfate, and chondroitin sulfate.

35. A male infant is failing to thrive and, on examination, is noted to have hepatomegaly and splenomegaly, among other findings. Urinalysis reveals the presence of both dermatan sulfate and heparan sulfate. You suspect the patient has Hurler syndrome. From the following list, select the enzyme that you would wish to have assayed to support your diagnosis:

    A. β-Glucuronidase.
    B. β-Galactosidase.
    C. α-L-Iduronidase
    D. α-N-Acetylglucosaminidase
    E. Neuraminidase

36. You see a child in clinic who is well below average height. You note that the child has short limbs, normal trunk size, macrocephaly, and a variety of other skeletal abnormalities. You suspect that the child has achrondoplasia. Select from the following list the test that would best confirm your diagnosis:

    A. Measurement of growth hormone
    B. Assays for enzymes involved in the metabolism of GAGs
    C. Tests for urinary mucopolysaccharides
    D. Gene tests for abnormalities of the fibroblast growth factor receptor 3
    E. Gene tests for abnormalities of growth hormone

37. Regarding muscle proteins, select the one FALSE statement:

    A. Actin is a major constituent of thin filaments.
    B. F-actin can polymerize under physiological condition to G-actin.
    C. The head region of myosin-II, the major constituent of thick filaments, has ATPase activity.
    D. Tropomyosin is an important constituent of the thick filament.
    E. F-actin greatly promotes the release of ADP and P_i from myosin ATPase.

38. Regarding the process of muscle contraction, select the one FALSE statement:

    A. Binding of Ca^{2+} to troponin C uncovers the myosin binding sites of actin, permitting actin and myosin to interact.
    B. Release of P_i from the actin-myosin-ADP-P_i complex initiates the power stroke.
    C. Release of ADP from the actin-myosin-ADP complex is accompanied by a large conformational change in the head of myosin in relation to its tail.
    D. Myosin-ATP has a high affinity for actin.
    E. If levels of ATP are low, rigor mortis can ensue because of failure of release of actin from the actin-myosin complex.

39. During anesthesia using halothane you notice that your patient's temperature is rising rapidly and you suspect malignant hyperthermia (MH). Select the one FALSE statement:

    A. MH can be due to mutations affecting the Ca^{2+} release channel (RYR).
    B. MH can be due to mutations affecting the Na^+K^+-ATPase.
    C. MH can also be due to mutations affecting the dihydropyridine receptor, a voltage-gated slow K type Ca^{2+} channel.
    D. A high intracellular concentration of Ca^{2+}, causing rigidity of muscles, is found in MH.
    E. Appropriate treatment of MH is i.v. administration of dantrolene to inhibit release of Ca from the SR into the cytosol.

40. Concerning different types of muscle, select the one FALSE statement:

    A. Skeletal muscle lacks caldesmon.
    B. Cardiac muscle lacks the troponin system.
    C. The concentration of extracellular Ca^{2+} is important for the contraction of cardiac and smooth muscle.
    D. Smooth muscle exhibits slow cycling of cross-bridges, allowing prolonged contraction.
    E. Skeletal, cardiac, and smooth muscles all contain the actin-myosin system.

41. Select the one FALSE statement regarding nitric oxide (NO):

    A. NO acts as a vasodilator.
    B. NO can be formed from one particular amino acid via the action of NO synthase.
    C. NO activates adenylate cyclase and the resulting cAMP inhibits the action of certain protein kinases, causing muscle relaxation.
    D. NO can inhibit platelet aggregation.
    E. NO has a very short half-life in tissues and can lead to the generation of OH· radicals.

42. Select the one FALSE statement:

    A. Type I muscle fibers contain myoglobin and mitochondria; their metabolism is predominantly aerobic.
    B. Type II muscle fibers derive their energy predominantly from anaerobic glycolysis.
    C. The amounts of type I and type II fibers can be altered by training.
    D. In a marathon, blood glucose and free fatty acids are major energy sources.
    E. In a 100 m sprint, anaerobic glycolysis is the sole energy source.

43. Select the one FALSE statement:

    A. Microfilaments are composed of actin and myosin.
    B. Microtubules contain α and β tubulins; the drugs colchicine and vinblastine bind to microtubules and inhibit their assembly.
    C. Intermediate filaments include lamins and keratins.
    D. Mutations affecting keratins are one cause of blistering.
    E. Mutations in the gene encoding lamin A and lamin C cause progeria (accelerated aging).

44. Select the one FALSE statement:

    A. Many but not all plasma proteins are synthesized by hepatocytes.
    B. Haptoglobin is an acute phase glycoprotein that binds hemoglobin in the plasma and prevents it from entering the kidneys.
    C. Many plasma proteins, such as haptoglobin, transferrin, and α1-antitrypsin, exhibit polymorphisms.
    D. C-reactive protein (CRP) is a biomarker for many inflammatory states.
    E. NFkB is an acute phase protein whose plasma level is elevated in acute inflammation.

45. Select the one FALSE statement:

    A. Albumin is synthesized as a preproprotein.
    B. Albumin is the major plasma protein by mass; it is not an acute phase protein.
    C. Subjects with analbuminemia display severe edema.

D. Albumin binds many ligands, such as bilirubin, free fatty acids, copper, and certain drugs.

E. The plasma half-life of albumin can be markedly shortened in severe gastroenteropathies.

46. Select the one FALSE statement:

A. Transferrin shuttles ferric iron around the circulation.

B. Iron is absorbed in the duodenum in the ferrous state; a duodenal cytochrome b reduces ferric iron to the ferrous state.

C. Inside enterocytes, hephaestin oxidizes ferrous to ferric iron and the latter is then transferred to the plasma by the action of ferroportin.

D. Transferrin binds to transferrin receptor 1 (TfR1), is taken up by receptor-mediated endocytosis, and is then degraded inside endosomes.

E. Except in inflammation, levels of plasma ferritin are generally an indicator of body iron stores.

47. Select the one FALSE statement:

A. When intracellular levels of iron are high, ferritin is not synthesized but TfR1 is.

B. Hepcidin decreases iron absorption in the intestine by binding ferroportin and triggering its degradation.

C. The HFE protein can influence iron metabolism by up-regulating expression of hepcidin.

D. Levels of bone morphogenetic protein 6 can also affect regulation of hepcidin expression.

E. Ceruloplasmin, a copper-binding plasma protein, plays a role in iron metabolism by oxidizing ferrous iron to the ferric state.

48. You see a 50-year-old woman in clinic who is pale and tired. You suspect she has iron deficiency anemia. You order various lab tests, as shown below. Which result is NOT consistent with your provisional diagnosis?

A. Low levels of plasma ferritin

B. Decreased saturation of transferrin

C. Decreased level of hemoglobin

D. Decreased level of red cell protoporphyrin

E. Increase of serum soluble TfR

49. You see a 57-year-old man in clinic who exhibits a green pigment ring around his cornea and also some signs of neurologic impairment. You suspect he has Wilson disease. Select the one FALSE statement regarding this condition:

A. Copper accumulates in the liver and brain.

B. It is caused by mutations in the copper-binding P-type ATPase that is involved in Menkes disease.

C. There is an increase of copper in Descemet's membrane.

D. Levels of ceruloplasmin are generally low.

E. The condition responds to treatment with penicillamine, which chelates copper and removes it from the body in the urine.

50. Regarding amyloidosis, select the one FALSE statement:

A. It can be caused by a defect in amylase.

B. It can be caused by deposition of light chain fragments of immunoglobulins.

C. It can be caused by accumulation of degradation products of serum amyloid A.

D. It can be caused by accumulation of mutated plasma proteins such as transthyretin.

E. It can be caused by accumulation of β_2-microglobulin.

51. Select the one FALSE statement:

A. Most but not all immunoglobulins are glycoproteins.

B. All immunoglobulins contain a minimum of two light and two heavy chains.

C. The type of heavy chain that an immunoglobulin contains determines its class.

D. The hypervariable regions of immunoglobulins comprise the antigen-binding sites and dictate the specificity of immunoglobulins.

E. The constant regions of immunoglobulins determine class-specific effector functions such as complement fixation and transplacental passage.

52. Select the one FALSE statement:

A. Junctional diversity reflects the addition or deletion of a random number of nucleotides when certain gene segments of antibodies are joined together.

B. In response to an immunogen, IgM molecules normally precede the appearance of IgG molecules in plasma; this is known as class switching.

C. Bence-Jones proteins are heavy chains of immunoglobulins that are overproduced in multiple myeloma.

D. Humanization of monoclonal antibodies involves attaching the complementarity-determining regions onto appropriate sites in a human immunoglobulin molecule, thus lessening immunogenicity.

E. The complement system consists of approximately 20 proteins and is involved in cell lysis, Inflammation, and clearance of antigen–antibody complexes from the circulation.

53. Which one of the following statements regarding the blood coagulation pathways is NOT CORRECT?

A. The components of the extrinsic Xase (tenase) complex are factor VIIa, tissue factor, Ca^{2+}, and factor X.

B. The components of the intrinsic Xase (tenase) complex are factor IXa, factor VIIIa, Ca^{2+}, and factor X.

C. The components of the prothrombinase complex are factor Xa, factor Va, Ca^{2+}, and factor II (prothrombin).

D. The extrinsic and intrinsic Xase complexes and prothrombinase complex require anionic procoagulant phosphatidylserine on LDL (low density lipoprotein) for their assembly.

E. Fibrin formed by cleavage of fibrinogen by thrombin is covalently cross-linked by the action of factor XIIIa, which itself is formed by the action of thrombin on factor XIII.

54. On which one of the following coagulation factors does a patient taking warfarin for his thrombotic disorder have decreased Gla (γ-carboxyglutamate) residues?

A. Tissue factor

B. Factor XI

C. Factor V

D. Factor II (prothrombin)

E. Fibrinogen

55. A 65-year-old male suffers a myocardial infarction and is given tissue plasminogen activator within 6 hours of onset of the thrombosis to achieve which one of the following?

    A. Prevent activation of the extrinsic pathway of coagulation
    B. Inhibit thrombin
    C. Enhance degradation of factors VIIIa and Va
    D. Enhance fibrinolysis
    E. Inhibit platelet aggregation

56. Which one of the following statements regarding platelet activation in hemostasis and thrombosis is NOT CORRECT?

    A. Platelets adhere directly to subendothelial collagen via GPIa-IIa and GPVI, while binding of GPIb-IX-V is mediated via von Willebrand factor.
    B. The aggregating agent thromboxane A_2 is formed from arachidonic acid liberated from platelet membrane phospholipids by the action of phospholipase A_2.
    C. The aggregating agent ADP is released from the dense granules of activated platelets.
    D. The aggregating agent thrombin activates intracellular phospholipase $C\beta$, which forms the internal effector molecules 1,2-diacylglycerol and 1,4,5-inositol trisphosphate from the membrane phospholipid phosphatidylinositol 4,5-bisphosphate.
    E. The ADP receptors, the thromboxane A_2 receptor, the thrombin PAR-1 and PAR-4 receptors, and the fibrinogen GPIIb-IIIa receptor are all examples of G protein-coupled receptors.

57. A 15-year-old female presented at clinic with bruises on her lower extremities. Of the following, which is *least likely* to explain the bleeding signs exhibited by this individual?

    A. Hemophilia A
    B. von Willebrand disease
    C. A low platelet count
    D. Aspirin ingestion
    E. A platelet disorder with absence of storage granules

58. Select the one FALSE statement:

    A. Alpha-thalassemias are due to mutations affecting the alpha chains of hemoglobin.
    B. Deficiency of either folic acid or vitamin B12 causes a megaloblastic anemia.
    C. Hereditary spherocytosis is due to mutations affecting certain proteins of the red cell membrane.
    D. Paroxysmal nocturnal hemoglobinuria (PNH) is due to mutations affecting synthesis of GPI-anchored proteins in the red cell membrane.
    E. Mutations in the pyruvate kinase gene also cause PNH.

59. Select the one FALSE statement:

    A. The red blood cell is highly dependent on glucose for its metabolism.
    B. The red blood cell has a glucose transporter (GLUT1) that is estimated to contain 12 transmembrane helical segments.
    C. The red blood cell has an active pentose phosphate shunt that generates NADPH.

D. The red blood cell has an active citric acid cycle.
E. The red cell contains certain enzymes involved in nucleotide metabolism, deficiencies of which can cause hemolytic anemia.

60. Regarding the anemia due to deficiency of G-6-P dehydrogenase, select the one FALSE statement:

    A. It occurs extremely frequently in various parts of the world due to mutations in the gene encoding the enzyme.
    B. It is a hemolytic anemia, and depending on severity, levels of conjugated bilirubin are often elevated as are levels of haptoglobin (Hp).
    C. Mutant forms of the enzyme do not produce NADPH at normal levels.
    D. Levels of reduced glutathione (GSH) are low in affected red cells because NADPH is needed to regenerate GSH from oxidized glutathione (GSSG).
    E. Broad beans, because of their content of potential oxidants, may precipitate an attack, as may certain drugs such as primaquine (an anti-malarial).

61. All of the following proteins are present in the red cell membrane EXCEPT:

    A. Pyruvate kinase
    B. Spectrin
    C. Ankyrin
    D. Glycophorin
    E. Anion exchange protein

62. Regarding the ABO blood group substances, select the one FALSE statement:

    A. They can be either glycosphingolipids or glycoproteins, depending on location.
    B. Individuals with blood group AB have antibodies to both A and B blood group substances.
    C. Blood group H substance is formed from its precursor by the action of a fucosyltransferase.
    D. Blood group B is formed from its precursor by the action of a galactosyltransferase.
    E. Individuals of blood group O lack the galactosyltransferase.

63. Regarding various white blood cells, select the one FALSE statement:

    A. Neutrophils possess integrins that are involved in their adhesion to endothelial cells.
    B. A deficiency of a subunit common to several integrins affects the ability of neutrophils to bind to endothelial cells, and results in one type of leukocyte adhesion deficiency disease.
    C. Lysozyme, which is abundant in macrophages, hydrolyzes the linkage between N-acetyl-neuraminic acid and N-acetyl-D-glucosamine found in certain bacterial cell walls, causing lysis.
    D. Lactoferrin is a protein synthesized by neutrophils that binds iron, which may inhibit the growth of certain bacteria.
    E. Myeloperoxidase can produce hypochlorous acid; this enzyme is responsible for the green color of pus.

64. Select the one FALSE statement:
    A. NADPH oxidase is inactive in resting phagocytic cells.
    B. The active form of NADPH oxidase contains cytochrome P450 and at least two other polypeptides.
    C. The oxidase can be activated when phagocytic cells contact various ligands.
    D. Activation of the oxidase leads to the production of superoxide from molecular oxygen, a phenomenon known as the respiratory burst.
    E. Patients with mutations in any of the polypeptides present in the active enzyme do not generate sufficient superoxide to kill invading bacteria and fungi, and are susceptible to recurrent infections (chronic granulomatous disease).

65. Select the one FALSE statement:
    A. Cytochrome P450 is a hemoprotein present in high concentration in the ER of liver that plays a key role in drug and xenobiotic metabolism.
    B. The main reaction catalyzed by the enzyme is hydroxylation of a large number of different substrates.
    C. It uses NADPH and requires NADPH-cytochrome P450 reductase for activity.
    D. The reaction catalyzed by cytochrome P450 produces stoichiometric amounts of superoxide.
    E. The enzyme is inducible by various drugs, a fact that may require changing the dose of other drugs that are being taken simultaneously.

66. Phase 2 reactions include all of the following EXCEPT:
    A. Hydroxylation
    B. Glucuronidation
    C. Sulfation
    D. Methylation
    E. Acetylation

67. Select the one FALSE statement:
    A. Glutathione (GSH) is a dipeptide derived from glutamic acid and cysteine.
    B. Reduced GSH can be conjugated to a number of toxic electrophilic molecules, thus lessening their toxicity.
    C. Metabolism of glutathione conjugates can lead to the production and subsequent urinary excretion of mercapturic acids.
    D. GSH helps to maintain the SH groups of certain proteins in the reduced state.
    E. GSH participates in the transport of certain amino acids across the plasma membrane of cells in a reaction catalyzed by γ-glutamyltransferase.

68. Which of the following is NOT a feature of the mitochondrial hypothesis of aging?
    A. Reactive oxygen species are generated as a by-product by the electron transport chain.
    B. Mitochondria lack the capacity to repair damaged DNA.
    C. Many of the complexes in the electron transport chain are constructed from a mixture of subunits encoded by the nuclear genome and by the mitochondrial genome.
    D. Damaged mitochondria form protease-resistant aggregates.
    E. Damaged mitochondria can trigger apoptosis (programmed cell death).

69. Which of the following is NOT a component of the cell's suite of damage repair and prevention agents?
    A. Superoxide dismutase
    B. Caspase 7
    C. Glutathione
    D. Isoaspartyl methyltransferase
    E. Catalase

70. The cellular component that is most vulnerable, in terms of both susceptibility and potential consequences, is:
    A. Phospholipids of membranes
    B. Thymine dimers
    C. Cysteine side-chains
    D. The electron transport chain
    E. DNA

71. Which of the following is an element of the metabolic theory of aging?
    A. Large animals generally live longer because, statistically speaking, their larger chromosomes can adsorb more damage before suffering a mutation.
    B. Calorically restricted diets tend to be life-extending because metabolic activity must decrease when the availability of nutrients is limited.
    C. The heart is the most essential organ for life.
    D. Damage by reactive oxygen species is multiplied by their tendency to participate in chain reactions.
    E. Heartbeats are measured through the progressive shortening of telomeres.

72. Regarding chemical carcinogenesis, select the one FALSE statement:
    A. Approximately 80% of human cancers may be due to environmental factors.
    B. In general, chemical carcinogens interact non-covalently with DNA.
    C. Some chemicals are converted to carcinogens by enzymes, usually cytochrome P450 species.
    D. Most ultimate carcinogens are electrophiles and attack nucleophilic groups in DNA.
    E. The Ames assay is a useful test for screening chemicals for mutagenicity; however, animal testing is required to show that a chemical is carcinogenic.

73. Regarding viral carcinogenesis, select the one FALSE statement:
    A. Approximately 15% of human cancers may be caused by viruses.
    B. Only RNA viruses are known to be carcinogens.
    C. RNA viruses causing or associated with tumors include hepatitis C virus.
    D. Retroviruses possess reverse transcriptase, which copies RNA to DNA.
    E. Tumor viruses act by deregulating the cell cycle, inhibiting apoptosis and interfering with normal cell signaling processes.

74. Regarding oncogenes and tumor suppressor genes, select the one FALSE statement:

    A. Both copies of a tumor suppressor gene must be mutated for its product to lose its activity.

    B. Mutation of an oncogene occurs in somatic cells and is not inherited.

    C. The product of an oncogene shows a gain of function that signals cell division.

    D. *RB* and *P53* are tumor suppressor genes; *MYC* and *RAS* are oncogenes.

    E. Mutation of one tumor suppressor gene or one oncogene is thought to be sufficient to cause cancer.

75. Regarding growth factors, select the one FALSE statement:

    A. They include a large number of polypeptides, most of which stimulate cell growth.

    B. Growth factors can act in an endocrine, paracrine, or autocrine manner.

    C. Certain growth factors, such as TGF-β, can act in a growth inhibitory manner.

    D. Some receptors for growth factors have tyrosine kinase activity; mutations of these receptors occur in cancer cells.

    E. PDGF stimulates phospholipase A$_2$, which hydrolyzes PIP$_2$ to form DAG and IP$_3$, both of which are second messengers.

76. Regarding the cell cycle, select the one FALSE statement:

    A. The cell cycle has 4 phases (G$_1$, S, G$_2$ and M).

    B. Cancer cells usually have a shorter generation time than normal cells and there are less of them in G$_0$ phase.

    C. A variety of mutations in cyclins and CDKs have been reported in cancer cells.

    D. RB is a cell cycle regulator; it binds to transcription factor E2F, thus allowing progression of the cell from G$_1$ to S phase.

    E. When damage to DNA occurs, P53 increases in amount and activates transcription of genes that delay transit through the cycle.

77. Regarding chromosomes and genomic instability, select the one FALSE statement:

    A. Cancer cells may have a mutator phenotype, which means that they have mutations in genes that affect DNA replication and repair, chromosomal segregation, DNA damage surveillance, and apoptosis.

    B. Chromosomal instability refers to gain or loss of chromosomes caused by abnormalities of chromosomal segregation during mitosis.

    C. Microsatellite instability involves expansion or contraction of microsatellites due to abnormalities of nucleotide excision repair.

    D. Aneuploidy (when the chromosomal number of a cell is not a multiple of the haploid number) is a common feature of tumor cells.

    E. Abnormalities of chromosome cohesion and of kinetochore-microtubule attachment may contribute to chromosomal instability and aneuploidy.

78. Select the one FALSE statement:

    A. The activity of telomerase is frequently elevated in cancer cells.

    B. A number of cancers have a hereditary predisposition, including Li-Fraumeni syndrome and retinoblastoma.

    C. The products of *BRCA1* and *BRCA2* (responsible for hereditary breast cancer types I and II) appear to be involved in DNA repair.

    D. Tumor cells usually exhibit a high rate of anaerobic glycolysis; this may be at least partly explained by the presence in many tumor cells of the PK-2 isozyme, which is associated with lesser production of ATP and possibly increased use of metabolites to build up biomass.

    E. Dichloroacetate, a compound found to display some anti-cancer activity, inhibits pyruvate carboxylase, and thus diverts pyruvate away from glycolysis.

79. Select the one FALSE statement:

    A. Whole-genome sequencing is revealing important new information about the numbers and types of mutations in cancer cells.

    B. Abnormalities of epigenetic mechanisms, such as demethylation of cytosine residues, abnormal modification of histones, and aberrant chromatin remodeling are being increasingly detected in cancer cells.

    C. Persistence of cancer stem cells (which are often relatively dormant and have active DNA repair systems) may help to explain some of the shortcomings of chemotherapy.

    D. Angiogenin is a potent stimulator of angiogenesis.

    E. Chronic inflammation, possibly via increased production of reactive oxygen species, predisposes to development of certain types of cancer.

80. Regarding apoptosis, select the one FALSE statement:

    A. Apoptosis can be initiated by the interaction of certain ligands with specific receptors on cell surface.

    B. Cell stress and other factors activate the mitochondrial pathway of apoptosis; release of cytochrome P450 into the cytoplasm is an important event in this pathway.

    C. A distinct pattern of fragments of DNA is found in apoptotic cells; it is caused by caspase-activated DNase.

    D. Caspase 3 digests cell proteins such as lamin, certain cytoskeletal proteins, and various enzymes, leading to cell death.

    E. Cancer cells have acquired various mutations that allow them to evade apoptosis, prolonging their existence.

81. Select the one FALSE statement:

    A. Proteins involved in cell adhesion include cadherins, integrins, and selectins.

    B. Decreased amounts of E-cadherin on the surfaces of cancer cells may help account for the decreased adhesiveness shown by tumor cells.

    C. Increased activity of GlcNAc transferase V in cancer cells may lead to an altered glycan lattice at the cell surface, perhaps predisposing to their spread.

    D. Cancer cells secrete metalloproteinases that degrade proteins in the ECM and facilitate their spread.

    E. All tumor cells have the genetic capacity to colonize.

82. Of the following, the *best test* of renal function is:
    A. Measurement of blood urea
    B. Measurement of urinary protein
    C. Creatinine clearance
    D. Measurement of urinary ammonia
    E. Measurement of urinary volume

83. A test that assesses the *synthetic* function of liver is:
    A. Serum albumin levels
    B. Serum conjugated (direct) bilirubin levels
    C. Serum alanine transaminase (ALT)
    D. Serum alkaline phosphatase (ALP)
    E. Blood ammonia

84. All of the following are characteristics of a substance whose clearance is indicative of the glomerular filtration rate (GFR) EXCEPT:
    A. It should have stable blood levels.
    B. It should be freely filtered at the glomerulus.
    C. It should be completely reabsorbed by the renal tubule.
    D. It should not be secreted by the renal tubule.
    E. It should not be metabolized in the body.

85. The first test that should be done in the assessment of thyroid function is the measurement of:
    A. Total thyroxine
    B. Thyroid stimulating hormone (TSH)
    C. Free thyroxine and triiodothyronine
    D. Thyrotropin releasing hormone (TRH)
    E. Thyroid binding globulin (TBG)

86. In relation to adenosine deaminase (ADA) deficiency, select the one FALSE statement:
    A. Deficiency of ADA accounts for the majority of cases of severe combined immunodeficiency disease (SCID).
    B. Increased levels of dATP resulting from deficiency of ADA are toxic to T lymphocytes.
    C. Conjugation of ADA to polyethylene glycol (PEG) prolongs the life in the circulation of the enzyme.
    D. Integration of a gene via gene therapy can, in some cases, cause cancer by insertional mutagenesis.
    E. Criteria to be satisfied prior to administration of gene therapy include that the gene administered should show sufficient levels of expression, it should be regulated, and there should be no significant side effects.

87. In relation to Alzheimer disease (AD), select the one FALSE statement:
    A. Only about 10% of cases of AD appear to have a genetic basis.
    B. Deposition of amyloid β-peptide $(A\beta_{42})$ followed by its aggregation secondary to oligomerization and formation of β-sheets is thought by some to play a central role in the causation of AD.
    C. Aβ42 is derived from amyloid precursor protein (APP), a transmembrane protein; APP is a substrate for several proteases (secretases).

D. Genes involved in AD include *APOE4*, which appears to increase deposition of $A\beta_{42}$.
    E. Another important feature of AD is deposition of a phosphorylated form of tau, a microtubule-associated protein, which forms neuritic tangles.

88. In relation to cholera, select the one FALSE statement:
    A. Cholera is due to infection by the bacterium *V cholera*, which secretes an enterotoxin.
    B. Cholera toxin has A and B subunits; the B subunits interact with the ganglioside GM1 present on the plasma membrane of intestinal cells.
    C. The A subunit uses NAD to ribosylate the Gs subunit of adenylate cyclase, up-regulating it and increasing formation of cAMP, which in turn activates protein kinase A (PKA).
    D. PKA phosphorylates various target proteins, resulting in massive loss of NaCl into the gut.
    E. Therapy for treatment of cholera includes immediate replacement of lost fluid and administration of an appropriate antibiotic; thereafter, taking oral rehydration solution (containing glucose, sodium chloride, sodium citrate, and sodium bicarbonate) has proven very effective.

89. Regarding colorectal cancer, select the one FALSE statement:
    A. Colorectal cancers can arise from adenomatous polyps.
    B. Specific oncogenes and tumor suppressor genes have been shown to be involved in its causation.
    C. Environmental factors – such as a diet high in saturated fat and low in fiber – have also been proposed to be involved in its causation.
    D. CEA is a glycoprotein that is released into the plasma from the surface membranes of certain cells.
    E. CEA is an excellent test, with high sensitivity and high specificity, for detecting the presence of colorectal cancer.

90. Regarding cystic fibrosis (CF), select the one FALSE statement:
    A. CF is a genetic disease due to mutations in the *CFTR* gene.
    B. The *CFTR* gene was discovered because of its association with a characteristic massive deletion located on chromosome 7.
    C. Over 1,000 different mutations have been reported in the gene since its discovery.
    D. The CF gene encodes a cAMP-responsive chloride transporter; abnormalities of the transporter lead to decreased secretion of chloride from epithelial cells and high chloride content in sweat.
    E. Viscous mucus can obstruct pancreatic ducts, leading to a deficiency of various digestive enzymes in the gut causing malnutrition, and its presence in the respiratory tract favors the growth of nasty bacteria such as *Pseudomonas aeruginosa*.

91. Regarding diabetic ketoacidosis (DKA), select the one FALSE statement:

    A. Over-production of ketone bodies in DKA is caused by breakdown of fats due to lack of insulin; it results in acidosis.

    B. The ketone bodies are β-hydroxybutyric acid and acetoacetic acid.

    C. The anion gap (plasma $Na^+ - [Cl^- + HCO_3^-]$) is elevated in DKA, but also in other conditions such as lactic acidosis and intoxication by salicylates.

    D. Plasma levels of K^+ are often elevated in DKA because of lack of insulin.

    E. Appropriate initial treatment of DKA is administration of insulin and i.v. saline; glucose and KCl are added later.

92. Regarding Duchenne muscular dystrophy (DMD), select the one FALSE statement:

    A. Creatine kinase MB is an important enzyme to measure in the diagnosis of DMD.

    B. DMD is an X-linked degenerative disease of muscle.

    C. The protein affected in both DMD and Becker muscular dystrophy is dystrophin.

    D. Dystrophin is a very large protein associated with the sarcolemma of muscle.

    E. Mutations affecting muscle proteins encoded by various genes have been shown to be the causes of a variety of other types of muscular dystrophy.

93. Regarding ethanol intoxication, select the one FALSE statement:

    A. Increased production of NADH, via the alcohol dehydrogenase (ADH) reaction, favors the formation of lactate from pyruvate.

    B. The resulting diminution of the level of pyruvate required for the pyruvate carboxylase reaction inhibits gluconeogenesis and may promote hypoglycemia.

    C. Acetaldehyde, produced by the ADH reaction, is a highly reactive molecule, and may be responsible for some of the toxic effects of ethanol.

    D. Ethanol may interpolate into membranes and also interacts with ion channels, affecting their functions.

    E. Ethanol is metabolized exclusively by alcohol dehydrogenase.

94. Regarding gout, select the one FALSE statement:

    A. Gout is due to accumulation of uric acid in one or more joints and other tissues, causing acute or chronic inflammation.

    B. Uric acid is the end-product of purine and pyrimidine metabolism.

    C. Uric acid is produced from xanthine via the action of xanthine oxidase; humans lack the enzyme uricase, so we cannot produce allantoin.

    D. Decreased excretion of uric acid is the major cause of gout.

    E. Acute attacks of gout are treated using anti-inflammatory drugs or colchicine; allopurinol, which inhibits xanthine oxidase, is a useful drug for the longer-term treatment of gout.

95. Regarding hereditary hemochromatosis, select the one FALSE statement:

    A. The hallmark of hereditary hemochromatosis is an increase of total body iron, sufficient to cause tissue damage.

    B. Free iron is toxic because it can generate free radicals via the Fenton reaction.

    C. Elevated levels of transferrin saturation and of serum ferritin are the most useful tests for early diagnosis.

    D. The most common cause of the condition is mutations in the gene encoding the protein HFE; the primary role of HFE is regulation of the level of hepcidin.

    E. Chelation therapy to remove excess iron is the preferred treatment for this condition.

96. Regarding hypothyroidism, select the one FALSE statement:

    A. Causes of primary hypothyroidism include deficient intake of iodine and Hashimoto disease (an autoimmune condition).

    B. Symptoms include chronic fatigue, sluggishness, constipation, and cold intolerance.

    C. Congenital hypothyroidism can and should be detected by routine screening of levels of TSH at birth.

    D. Elevated levels of TSH and of T_4 are highly indicative of hypothyroidism.

    E. Treatment of primary hypothyroidism consists of judicious administration of thyroxine (T_4), generally for life.

97. Regarding protein-energy malnutrition (PEM), select the one FALSE statement:

    A. Oxidative stress, by affecting vascular permeability, may contribute to the clinical picture of kwashiorkor.

    B. Signs of kwashiorkor include thin hair, apathy, fatty liver, protuberant abdomen, fragile skin, and diminished body fat.

    C. Edema is usually a feature of kwashiorkor; contributing factors to its development may be hypoalbuminemia and deficient dietary intake of methionine.

    D. Low levels of insulin and cortisol contribute to the muscle wasting seen in marasmus.

    E. PEM is entirely preventable by a well-balanced diet.

98. Regarding myocardial infarction (MI), select the one FALSE statement:

    A. The major cause of an MI is an occlusive thrombus occurring in close proximity to an atherosclerotic plaque which may have ruptured recently.

    B. Serial measurements of CK-MB is the best lab test in helping to confirm the diagnosis of MI.

    C. The presence of oxidized LDL in a plaque encourages recruitment of inflammatory cells, which are thought to be important contributors to atherosclerosis.

    D. Contributors to cell death in an MI include depletion of ATP, activation of various degradative enzymes, and accumulation of intracellular Ca^{2+}.

    E. One treatment for MI is administration of t-PA; this enzyme can help dissolve the thrombus and prevent further cardiac damage when given as soon as possible.

99. Regarding obesity, select the one FALSE statement:
    A. A body mass index (BMI) of over 30 is indicative of obesity.
    B. Obesity predisposes to various conditions including the metabolic syndrome; this includes excess abdominal fat, high blood glucose, increased LDL and decreased HDL, and high blood pressure.
    C. Neuropeptide Y and melanocyte stimulating hormone (α-MSH) increase appetite.
    D. Elevated levels of leptin, a polypeptide released by adipocytes, decrease food intake and also increase energy expenditure.
    E. Brown adipose tissue contains a mitochondrial protein, thermogenin, that dissipates energy as heat; differences in amounts of this protein and perhaps of other uncoupling proteins may play a role in the causation of obesity.

100. Regarding osteoporosis, select the one FALSE statement:
    A. Osteoporosis is a reduction of bone mass or density.
    B. In osteoporosis, a normal ratio of bone mineral (hydroxyapatite) to bone matrix (mostly collagen type I) is preserved.
    C. In osteomalacia, as caused by deficiency of vitamin C, decreased mineralization is present.
    D. Decline in levels of estrogen appears to increase the secretion of a number of cytokines that lead to the recruitment of osteoclasts, stimulating increased resorption.
    E. Levels of serum Ca, P, alkaline phosphatase, 25-hydroxyvitamin D, and parathyroid hormone may all be normal in osteoporosis.

101. Regarding xeroderma pigmentosum (XP), select the one FALSE statement:
    A. This condition is due to mutations in any one of at least 7 genes involved in the mismatch DNA repair pathway.
    B. UV irradiation can cause the formation of thymine dimers, in which covalent bonds are formed between adjacent intrachain thymine residues.
    C. The formation of thymine dimers can be measured in fibroblasts taken from a patient.
    D. Patients with XP develop cancers at an early age because of the defect in DNA repair; UV irradiation may activate oncogenes or inactivate tumor suppressor genes.
    E. Patients with XP should be followed up closely and advised to avoid exposure to sunlight and to use appropriate sunscreen ointments.

REFERENCES

MacDonald RG, Chaney WG: USMLE Road Map:Biochemistry. McGraw-Hill Lange, 2007.

Toy EC, Seifert WE, Strobel HW, Harms KP: Case Files Biochemistry. McGraw-Hill Lange, 2008.

Appendix

Selected Worldwide Web Sites

The following is a list of web sites that readers may find useful. The sites have been visited at various times by one or more of the authors. Most are located in the United States, but many provide extensive links to international sites and to databases (eg, for protein and nucleic acid sequences) and online journals. RKM would be grateful if readers who find other useful sites would notify him of their URLs by e-mail (rmurray6745@rogers.com) so that they may be considered for inclusion in future editions of this text.

Readers should note that URLs may change or cease to exist.

Access to the Biomedical Literature

High Wire Press: http://highwire.stanford.edu/
>   (Extensive lists of various classes of journals—biology, medicine, etc.—and offers also the most extensive list of journals with free online access.)

National Library of Medicine: http://www.nlm.nih.gov/
>   (Free access to Medline via PubMed.)

General Resource Sites

Access Medicine from McGraw-Hill: http://accessmedicine.com/features.aspx
>   A subscription site containing online basic science and medical texts and many other resources.

The Biology Project (from the University of Arizona): http://www.biology.arizona.edu/default.html
>   (Contains excellent biochemical coverage of enzymes, membranes, etc.)

Harvard University Department of Molecular & Cellular Biology Links: http://mcb.harvard.edu/BioLinks.html
>   (Contains many useful links.)

Lister Hill National Center for Biomedical Communications: http://www.lhncbc.nlm.nih.gov/
>   (Includes Profiles in Science, Genetics Home Reference, information on clinical trials, and a comprehensive guide to newborn screening codes).

Medical Encyclopedia in MedlinePlus: http://www.nlm.nih.gov/medlineplus/encyclopedia.html
>   (Includes several thousand articles about diseases. Also tests, medical photographs, and illustrations.)

MITOPENCOURSEWARE: http://ocw.mit.edu/courses/biology
>   (Various Biology Courses given at MIT are accessible via this web site.

The Official Web Site of the Nobel Prize: http://nobelprize.org/nobel_prizes/ medicine/laureates/
>   (The site contains the Nobel Lectures of Nobel Prize winners and thus provides a rich source of biomedical information.

Sites on Specific Topics

American Heart Association: http://www.heart.org/
>   (Valuable information on nutrition, on the role of various biomolecules—eg, cholesterol, lipoproteins—in heart disease, and on the major cardiovascular diseases.)

Cancer Genome Anatomy Project (CGAP): http://www.cgap.nci.nih.gov/
>   (An interdisciplinary program that helped generate information and technical tools to assist in deciphering the molecular anatomy of the cancer cell.)

Carbohydrate Chemistry and Glycobiology: A Web Tour: http://sciencemag.org/site/feature/data/carbohydrates.xhtml
>   (Contains links to organic chemistry, carbohydrate chemistry, and glycobiology.)

European Bioinformatics Institute: http://www.ebi.ac.uk/
>   (Contains databases on genomes, nucleotide sequences, protein sequences, etc.)

GeneCards: http://www.genecards.org/
>   (A database of human genes, their products, and their involvements in disease; from the Weizmann Institute of Science.)

GeneTests: http://www.ncbi.nlm.nih.gov/sites/GeneTests
(A medical genetics information resource for physicians and others with comprehensive articles [under GeneReviews] on many genetic diseases.)

Genetics Home Reference: http://ghr.nlm.nih.gov/
(A guide to understanding genetic conditions, primarily for lay people, containing information on more than 550 health conditions and more than 800 genes).

Howard Hughes Medical Institute: http://www.hhmi.org/
(An excellent site for following current biomedical research. Contains a comprehensive Research News Archive.)

Human Gene Mutation Database: http://www.hgmd/org
(An extensive tabulation of mutations in human genes from the Institute of Medical Genetics in Cardiff, Wales.)

Human Genome Project Information: http://www.genomics.energy.gov
(From the U.S. Department of Energy; also contains general information on genomics and on microbial genomes.)

J. Craig Venter Institute: http://www.jcvi.org
(Contains information on synthetic biology, sequences of bacterial genomes, and other information).

Karolinska Institute: Diseases, Disorders and Related Topics http://www.mic.stacken.kth.se/Diseases/
(A comprehensive site containing links to information on most classes of disease).

Lipids Online: http://lipidsonline.org/
(A resource from Baylor College of Medicine containing educational resources on atherosclerosis).

MITOMAP: http://www.mitomap.org/
(A human mitochondrial genome database.)

National Cancer Institute: http://www.cancer.gov/
(Contains information on various types of cancer and also on current research on cancer)

National Center for Biotechnology Information: http://ncbi.nlm.nih.gov/
(Provides access to biomedical and genomic information)

National Human Genome Research Institute: http://www.genome.gov/
(Extensive information about the Human Genome Project and subsequent work.)

National Institutes of Health: http://www.nih.gov/
(Includes links to the separate Institutes and Centers that constitute NIH, covering a wide range of biomedical research.)

Office of Rare Diseases: http://rarediseases.info.nih.gov
(Access to information on most rare diseases, including current research, and a Program on Undiagnosed Diseases)

OMIM (Online Mendelian Inheritance in Man): http://www.ncbi.nlm.nih.gov/omim
(An extremely comprehensive resource on human genetic diseases, initiated by Dr. Victor A. McKusick, considered by many to be the Father of modern human genetics.)

RCSB Protein Data Bank: http://www.pdb.org
(A repository for the processing and distribution of three-dimensional biologic macromolecular structure data.)

Society for Endocrinology: http://www.endocrinology.org/
(The site aims to advance education and research in endocrinology for the public benefit.)

Society for Neuroscience: http://www.sfn.org
(Contains useful information on a variety of topics in neuroscience.)

The Broad Institute: http://www.broad.mit.edu/
(The Broad Institute is a research collaboration of MIT, Harvard and its affiliated hospitals—created to bring the power of genomics to medicine.)

The Endocrine Society: http://www.endo-society.org/
(The Society is devoted to research on hormones and to the clinical practice of endocrinology.)

The Protein Kinase Resource: http://pkr.genomics.purdue.edu/
(Information on the protein kinase family of enzymes.)

The UCSD-Nature Signaling Gateway: http://www.signalinggateway.org/
(A comprehensive resource for anyone interested in signal transduction.)

The Wellcome Trust Sanger Institute: http://www.sanger.ac.uk/
(A genome research center whose purpose is to increase knowledge of genomes, particularly through large-scale sequencing and analysis.)

Biochemical Journals and Reviews

The following is a partial list of biochemistry journals and review series and of some biomedical journals that contain biochemical articles. Biochemistry and biology journals now usually have Web sites, often with useful links, and some journals are fully accessible without charge. The reader can obtain the URLs for the following by using a search engine.

- *Annual Reviews of Biochemistry, Cell and Developmental Biology, Genetics, Genomics and Human Genetics*
- *Archives of Biochemistry and Biophysics (Arch. Biochem. Biophys.)*
- *Biochemical and Biophysical Research Communications (Biochem. Biophys. Res. Commun.)*
- *Biochemical Journal (Biochem J.)*
- *Biochemistry (Biochemistry)*
- *Biochemistry (Moscow) (Biochemistry [Mosc])*

- *Biochemistry and Cell Biology (Biochem. Cell Biol.)*
- *Biochimica et Biophysica Acta (Biochim. Biophys. Acta)*
- *Biochimie (Biochimie)*
- *European Journal of Biochemistry (Eur. J. Biochem.)*
- *Indian Journal of Biochemistry and Biophysics (Indian J. Biochem. Biophys.)*
- *Journal of Biochemistry (Tokyo) (J. Biochem. [Tokyo])*
- *Journal of Biological Chemistry (J. Biol. Chem.)*
- *Journal of Clinical Investigation (J. Clin. Invest.)*
- *Journal of Lipid Research (J. Lipid Res.)*

- *Nature (Nature)*
- *Nature Genetics (Nat. Genet.)*
- *Proceedings of the National Academy of Sciences USA (Proc. Natl. Acad. Sci. USA)*
- *Public Library of Science (PLoS) Journals; PLOS One and the journals on Biology, Genetics, Medicine, and other subjects are available free online (e.g. http://www.plosone.org)*
- *Science (Science)*
- *Trends in Biochemical Sciences (Trends Biochem. Sci.)*

The Answer Bank

Section I

1. **D.**	2. **A.**	3. **D.**	4. **C.**
5. **A.**	6. **E.**	7. **B.**	8. **C.**
9. **A.**	10. **D.**	11. **E.**	12. **C.**
13. **B.**	14. **E.**	15. **D.**	16. **D.**
17. **E.**	18. **B.**	19. **B.**	20. **C.**
21. **D.**	22. **B.**	23. **B.**	24. **C.**

Section II

1. **E.** 2. **A.** 3. **E.**
4. **D.** Thiamin diphosphate is a coenzyme of pyruvate dehydrogenase.
5. **E.** Xylulose, which is excreted in essential pentosuria, is a reducing compound and will therefore give a positive result with the alkaline copper reagent.
6. **D.** After 24 hours his liver and muscle glycogen reserves will be more or less completely depleted, and his plasma glucose will have fallen to about 3–4 mmol/L. There is no glycogen in the blood-stream. In response to low insulin and high glucagon, he will liberate free fatty acids from adipose tissue as a source of metabolic fuel for tissues that can metabolize fatty acids, so sparing glucose for brain and red blood cells.
7. **C.** As he becomes progressively more starved, his liver will synthesize ketone bodies as an additional fuel for muscle, which cannot meet all of its energy needs from fatty acid metabolism. This spares glucose for the brain and red blood cells.
8. **D.** 9. **C.** 10. **C.** 11. **E.**
12. **D.** 13. **D.**
14. **A.** Chylomicrons are synthesized in the intestinal mucosa, containing mainly triacylglycerol from dietary lipids, and peripheral tissues take up fatty acids by the action of extracellular lipoprotein lipase. Chylomicron remnants are cleared by the liver.
15. **E.** VLDL is secreted by the liver, containing both newly synthesized triacylglycerol and triacylglycerol from chylomicron remnants, and peripheral tissues take up fatty acids by the action of extracellular lipoprotein lipase.
16. **D.** Intermediate density lipoprotein results from the removal of triacylglycerol from very low-density lipoprotein by peripheral tissues. It then takes up cholesterol and proteins from high-density lipoprotein to become low-density lipoprotein, which is normally cleared by the liver.
17. **A.** Chylomicrons are synthesized in the intestinal mucosa, containing mainly triacylglycerol from dietary lipids, and peripheral tissues take up fatty acids by the action of extracellular lipoprotein lipase. Chylomicron remnants are cleared by the liver. Ketone bodies and non-esterified fatty acids are elevated in fasting, not after a meal.
18. **E.** Chylomicrons are mainly cleared by peripheral tissues within about 2 hours after a meal, and the chylomicron remnants are cleared by the liver. The residual triacylglycerol, plus triacylglycerol newly synthesized in the liver are secreted in very low-density lipoprotein as a source of fuel for peripheral tissues. Ketone bodies and non-esterified fatty acids are elevated in fasting, not after a meal.
19. **D.**
20. **C.**
21. **C.** Statins inhibit the activity of 3-hydroxy-3-methylglutaryl-CoA reductase, the enzyme that converts 3-hydroxy-3-methylglutaryl-CoA to mevalonate in the cholesterol biosynthesis pathway.

22. **A.**	23. **B.**	24. **E.**	25. **D.**
26. **C.**	27. **B.**	28. **D.**	29. **C.**

Section III

1. **B.** The insertion of seleneocysteine into a peptide occurs *during* translation, and is directed by a specific tRNA, tRNA^Sec.
2. **D.** Phenylalanine hydroxylase does not convert tyrosine to phenylalanine.
3. **E.** Histamine.
4. **B.** Pyridoxal-dependent transamination is the first reaction in degradation of all the common amino acids except threonine, lysine, proline, and hydroxyproline.
5. **A.** Alanine.
6. **A.** The carbon skeleton of alanine contributes the most to hepatic gluconeogenesis.
7. **B.** ATP and ubiquitin participate in the degradation of nonmembrane-associated proteins and proteins with *short* half lives.
8. **C.** Clinical signs of metabolic disorders of the urea cycle include respiratory *alkalosis*, not acidosis.
9. **E.** *Cytosolic* fumarase and *cytosolic* malate dehydrogenase convert fumarase to oxaloacetate following a *cytosolic* reaction of the urea cycle.
10. **B.** *Serine* provides the thioethanol moiety of coenzyme A.
11. **C.** Decarboxylation of *glutamate* forms GABA.
12. **E** is not a hemoprotein. In cases of hemolytic anemia albumin can bind some metheme, but unlike the other proteins listed, albumin is not a hemoprotein.
13. **B.** Acute intermittent porphyria is due to mutations in the gene for uroporphyrin I synthase.
14. **A.** Bilirubin is a *linear* tetrapyrrole.
15. **D.** The severe jaundice, upper abdominal pain, and weight loss plus the lab results indicating an obstructive type of jaundice are consistent with cancer of the pancreas.

Section IV

1. **D.** β,γ-Methylene and β,γ-imino purine and pyrimidine triphosphates do not readily release the terminal phosphate by hydrolysis or by phosphoryl group transfer.
2. **D.**
3. **E.** Pseudouridine is excreted unchanged in human urine. Its presence there is not indicative of pathology.
4. **A.** Metabolic disorders are infrequently associated with defects in pyrimidine catabolism, which forms water-soluble products.

5. **B.**	6. **D.**	7. **B.**	8. **C.**
9. **C.**	10. **D.**	11. **E.**	12. **B.**
13. **D.**	14. **D.**	15. **E.**	16. **A.**
17. **C.**	18. **B.**	19. **D.**	20. **B.**
21. **C.**	22. **A.**	23. **C.**	24. **A.**
25. **E.**	26. **B.**	27. **A.**	28. **E.**
29. **C.**	30. **A.**	31. **A.**	32. **C.**
33. **D.**	34. **E.**	35. **C.**	36. **B.**
37. **C.**	38. **E.**	39. **D.**	40. **D.**
41. **B.**	42. **A.**	43. **A.**	44. **E.**
45. **C.**	46. **A.**	47. **C.**	48. **D.**
49. **C.**	50. **B.**	51. **E.**	52. **C.**
53. **D.**	54. **A.**	55. **E.**	56. **A.**
57. **E.**	58. **C.**	59. **A.**	60. **D.**
61. **D.**	62. **E.**	63. **A.**	64. **C.**
65. **C.**	66. **E.**	67. **D.**	

Section V

1. **B.** Glycolipids are located on the outer leaflet.
2. **A.** Alpha-helices are major constituents of membrane proteins.
3. **E.** Insulin also increases glucose uptake in muscle.
4. **A.** Its action maintains the high intracellular concentration of potassium compared with sodium.

5. **D.**	6. **B.**	7. **C.**	8. **B.**
9. **D.**	10. **A.**	11. **E.**	12. **B.**
13. **D.**	14. **E.**	15. **B.**	16. **C.**
17. **A.**	18. **C.**	19. **A.**	20. **B.**
21. **D.**	22. **A.**		

Section VI

1. **C.**	2. **D.**	3. **E.**	4. **E.**
5. **C.**	6. **B.**	7. **C.**	8. **B.**
9. **E.**	10. **A.**	11. **A.**	12. **C.**
13. **E.**	14. **B.**	15. **B.**	16. **D.**

17. **A.**
18. **A.** Superoxide dismutase serves to remove the superoxide radical.
19. **A.** Activation of macrophages leads to production of oxygen radicals.
20. **B.**
21. **A.** Glycosylation can occur at other sites, for example in the ER.
22. **E.** Importins are involved in import of proteins into the nucleus.
23. **B.** Some mammalian proteins are translocated post-translationally.
24. **C.** Ubiquitin is involved in protein degradation by proteasomes.
25. **E.** Furin converts proalbumin to albumin.
26. **C.**
27. **A.**
28. **C.** Some of the original sugars donated by the dolichol compound are subsequently removed and replaced by other sugars.
29. **C.** Glycation of hemoglobin occurs in normal individuals, but usually at a significantly lower level.
30. **B.** I-cell disease is due to mutations in a gene encoding a GlcNAc phosphotransferase.
31. **D.** Collagen contains cross-links.
32. **C.** Deletions in the elastin gene are responsible for many cases of Williams-Beuren syndrome.
33. **B.** Some cases of Ehlers-Danlos syndrome are not due to mutations affecting the genes encoding the various types of collagen.
34. **B.** Hyaluronic acid is not sulfated.
35. **C.** Hurler syndrome is due to a deficiency of α-L-iduronidase.
36. **D.** Achrondoplasia is due to mutations in the FGFR3 gene.
37. **D.** Tropomyosin is a constituent of the thin filament.
38. **D.** Myosin-ATP has a low affinity for actin, promoting release of actin from actin-myosin.
39. **B.** There is no evidence that the Na$^+$ K$^+$-ATPase is involved in MH.
40. **B.** Cardiac muscle contains the troponin system, and smooth muscle lacks it.
41. **C.** NO activates a guanylate cyclase, not adenylate cyclase.
42. **E.** Creatine phosphate is an important source of energy during the first few seconds.
43. **A.** Microfilaments do not contain myosin.
44. **E.** NFkB is a transcription factor, not a plasma protein.
45. **C.** Subjects with analbuminemia show only moderate edema; an increase in the plasma concentration of other proteins appears to compensate for the deficiency of albumin.
46. **D.** Transferrin is not degraded inside endosomes, but is instead re-used.
47. **A.** When intracellular levels of iron are high, the synthesis of ferritin is increased and that of TfR1 decreased.
48. **D.** Increased levels of red cell protoporphyrin are usually found in iron deficiency anemia. Note that tests D and E are not usually a standard part of the work-up of a patient with iron deficiency anemia.
49. **B.** A different copper-binding ATPase is involved in the causation of Menkes disease.
50. **A.** Amylase has nothing to do with amyloidosis.
51. **A.** All immunoglobulins are glycoproteins.
52. **C.** Bence-Jones proteins are light chains.
53. **D.**
54. **D.** Of the listed proteins, only factor II is a vitamin K-dependent coagulation factor.
55. **D.**
56. **E.** GPIIb-IIIa (integrin αIIbβ3) is not a G protein-coupled receptor.
57. **A.** Hemophilia A, being an X chromosome-linked disease, is very unlikely to occur in a female.
58. **E.** Mutations in the pyruvate kinase gene cause a hemolytic anemia, but not PNH.
59. **D.** The red blood cell lacks mitochondria and therefore does not possess the citric acid cycle.
60. **B.** In a hemolytic anemia levels of non-conjugated bilirubin are often elevated and levels of Hp decrease because of the release of hemoglobin into the plasma.

61. **A.** Pyruvate kinase is present in the cytosol.
62. **B.** Individuals of blood group AB have neither anti-A nor anti-B antibodies.
63. **C.** Lysozyme hydrolyzes the linkage between N-acetylmuramic acid and N-acetyl-D-glucosamine.
64. **B.** The enzyme contains cytochrome b$_{558}$, not cytochrome P450.
65. **D.** Superoxide is not a product of the reaction catalyzed by cytochrome P450.
66. **A.** Hydroxylation is a phase 1 reaction.
67. **A.** GSH is a tripeptide containing glutamic acid, cysteine, and glycine.
68. **D.**   69. **B.**   70. **E.**   71. **B.**
72. **B.** Most chemical carcinogens interact covalently with DNA.
73. **B.** Certain DNA viruses are also known to be carcinogens.
74. **E.** Mutations in approximately 5-6 of these genes are thought to be necessary for carcinogenesis.
75. **E.** PDGF stimulates phospholipase C, not phospholipase A.
76. **D.** Binding of RB to E3F blocks progression of the cell from G1 to S phase.
77. **C.** Microsatellite instability is caused by abnormalities of mismatch repair.
78. **E.** Dichloroacetate inhibits pyruvate dehydrogenase kinase.
79. **D.** Angiogenin is an inhibitor of angiogenesis.
80. **B.** Cytochrome c is released from mitochondria.
81. **E.** Only about 1:10, 000 cancer cells may have the capacity to colonize.
82. **C.** Creatinine clearance is an estimate of the glomerular filtration rate (GFR). Therefore, measurement of creatinine clearance can help detect renal failure in its early stages. Measurement of blood urea and creatinine are not sensitive markers of renal function. Proteinuria is an important sign of kidney disease, but urinary protein may also be elevated in other pathological conditions and in physiological conditions such as pregnancy, prolonged standing, exposure to severe cold, strenuous exercise, etc. Urinary ammonia levels depend on a number of factors such as the acid-base status of the body and liver function. Urine volume depends on the amount of water consumed and loss of water due to sweating, etc.
83. **A.** The liver synthesizes most of the plasma proteins, including albumin. A low level of serum albumin is characteristically seen in chronic liver failure. Hypoalbuminemia may also be caused by loss of albumin in urine (e.g., nephrotic syndrome) or malnutrition (e.g., kwashiorkor). Measurement of total and conjugated bilirubin gives information on the ability of the liver to conjugate and excrete bilirubin. Alanine transaminase (ALT) is a marker of hepatocyte injury, and alkaline phosphatase (ALP) is a marker of biliary obstruction. Blood ammonia levels indicate the ability of the liver to detoxify ammonia.

84. **C.** An ideal substance, clearance of which is representative of the GFR, should have stable concentrations in the blood, be filtered freely at the glomerulus, and should neither be reabsorbed nor be secreted by the renal tubule. A substance that is metabolized by the body will not have stable blood levels and therefore is not suitable to measure the GFR.
85. **B.** TSH levels are usually greatly increased in primary hypothyroidism and suppressed or undetectable in primary hyperthyroidism. Since the large majority of cases of hypo- and hyperthyroidism are due to disorders primarily related to the thyroid gland, measurement of TSH has proven to be a cost-effective and clinically efficient strategy in the diagnosis of thyroid disorders. In addition, highly sensitive assays for TSH are now available commercially. Total thyroxine levels can be affected by changes in levels of thyroid-binding globulin, even in the euthyroid state. Free thyroxine and triiodothyronine measurements are technically difficult and expensive and are therefore not preferred as a first test in the assessment of thyroid function.
86. **A.** ADA deficiency only accounts for about 15% of cases of SCID.
87. **D.** *APOE4* appears to decrease clearance of Aβ_{42}.
88. **C.** The A subunit ADP-ribosylates the Gs subunit.
89. **E.** CEA is not an excellent test for detection of colorectal cancer because its sensitivity and specificity are relatively low.
90. **B.** The CF gene was discovered using "reverse genetics", not because of a massive deletion on chromosome 7.
91. **B.** Acetone is also a ketone body.
92. **A.** CK-MM is the isozyme of creatine kinase that should be measured if DMD is suspected. Often measurement of total CK activity is used.
93. **E.** Ethanol is also metabolized by a microsomal cytochrome P450.
94. **B.** Uric acid is not produced from the metabolism of pyrimidines.
95. **E.** The preferred treatment is phlebotomy.
96. **D.** Levels of T$_4$ are decreased in primary hypothyroidism.
97. **D.** High levels of cortisol favor muscle wasting.
98. **B.** Use of measurements of troponin T has replaced the use of CK-MB, partly because it is a more sensitive biochemical marker of damage to cardiac muscle.
99. **C.** α-MSH decreases appetite.
100. **C.** Osteomalacia is due to deficiency of vitamin D, not vitamin C.
101. **A.** XP involves mutations in the nucleotide excision DNA repair pathway.

Index

Note: Page numbers followed by *f* indicate figures; and page numbers followed by *t* indicate tables.